SLOW DYNAMICS IN COMPLEX SYSTEMS

Related Titles from AIP Conference Proceedings

627   Computing Anticipatory Systems: CASYS 2001 - Fifth International Conference
Edited by D. M. Dubois, September 2002, 0-7354-0081-4

577   Density Functional Theory and Its Application to Materials
Edited by V. Van Doren, C. Van Alsenoy, and P. Geerlings, July 2001, 0-7354-0016-4

519   Statistical Physics: 3rd Tohwa University International Conference
Edited by Michio Tokuyama and H. Eugene Stanley, June 2000, 1-56396-940-8

517   Computing Anticipatory Systems: CASYS'99—Third International Conference
Edited by Daniel M. Dubois, June 2000, 1-56396-933-5

503   Two Dimensional Correlation Spectroscopy
Edited by Y. Ozaki and I. Noda, March 2000, 1-56396-916-5

489   Physics of Glasses: Structure and Dynamics
Edited by Philippe Jund and Rémi Jullien, October 1999, 1-56396-903-3

469   Slow Dynamics in Complex Systems: Eighth Tohwa University International Symposium
Edited by Michio Tokuyama and Irwin Oppenheim, April 1999, 1-56396-811-8

465   Computing Anticipatory Systems (CASYS-98): Second International Conference
Edited by Daniel M. Dubois, March 1999, 1-56396-863-0

437   Computing Anticipatory Systems (CASYS-97)
Edited by Daniel M. Dubois, July 1998, 1-56396-827-4

To learn more about these titles, or the AIP Conference Proceedings Series, please visit the webpage **http://proceedings.aip.org**

SLOW DYNAMICS IN COMPLEX SYSTEMS

3rd International Symposium on
Slow Dynamics in Complex Systems

Sendai, Japan   3–8 November 2003

EDITORS
Michio Tokuyama
Institute of Fluid Science, Tohoku University
Sendai, Japan

Irwin Oppenheim
Massachusetts Institute of Technology
Cambridge, Massachusetts

SPONSORING ORGANIZATIONS
Institute of Fluid Science, Tohoku University
Ministry of Education, Culture, Sports, Science and
 Technology of Japan

Melville, New York, 2004
AIP CONFERENCE PROCEEDINGS ■ VOLUME 708

Editors:

Michio Tokuyama
Theoretical Fluid Dynamics Laboratory
Complex Flow Division
Institute of Fluid Science
Tohoku University
Sendai 980-8577
JAPAN

E-mail: tokuyama@ifs.tohoku.ac.jp

Irwin Oppenheim
Department of Chemistry
Massachusetts Institute of Technology
77 Massachusetts Avenue
Cambridge, MA 02139
USA

E-mail: irwin@mit.edu

The article on pp. 515-522 was authored by U.S. Government employees and is not covered by the below mentioned copyright.

Authorization to photocopy items for internal or personal use, beyond the free copying permitted under the 1978 U.S. Copyright Law (see statement below), is granted by the American Institute of Physics for users registered with the Copyright Clearance Center (CCC) Transactional Reporting Service, provided that the base fee of $22.00 per copy is paid directly to CCC, 222 Rosewood Drive, Danvers, MA 01923. For those organizations that have been granted a photocopy license by CCC, a separate system of payment has been arranged. The fee code for users of the Transactional Reporting Service is: 0-7354-0183-7/04/$22.00.

© 2004 American Institute of Physics

Individual readers of this volume and nonprofit libraries, acting for them, are permitted to make fair use of the material in it, such as copying an article for use in teaching or research. Permission is granted to quote from this volume in scientific work with the customary acknowledgment of the source. To reprint a figure, table, or other excerpt requires the consent of one of the original authors and notification to AIP. Republication or systematic or multiple reproduction of any material in this volume is permitted only under license from AIP. Address inquiries to Office of Rights and Permissions, Suite 1NO1, 2 Huntington Quadrangle, Melville, NY 11747-4502; phone: 516-576-2268; fax: 516-576-2450; e-mail: rights@aip.org.

L.C. Catalog Card No. 2004104516
ISBN 0-7354-0183-7
ISSN 0094-243X

Printed in the United States of America

CONTENTS

Preface xvii
Organization of Symposium xix
Invited Speakers and Discussion Leaders xx
Papers Presented But Not Included in These Proceedings xxi
Supporting Agencies xxii
Group Photograph xxv

I. COMPLEX FLUIDS

Colloidal Suspensions Driven by External Fields 3
      H. Löwen, C. N. Likos, R. Blaak, S. Auer, V. Froltsov, J. Dzubiella, A. Wysocki, and H. M. Harreis

Universal Features of Collective Interactions in Hard-Sphere Systems at Higher
Volume Fractions 8
      M. Tokuyama, Y. Terada, H. Yamazaki, and I. Oppenheim

Neutron and Light Scattering Studies of the Liquid-to-Glass and Glass-to-Glass Transitions in
a Copolymer Micellar System 16
      S.-H. Chen, W.-R. Chen, and F. Mallamace

Glassy Dynamics in Gelling Systems: From Chemical Gels to Colloidal Glasses 28
      E. Del Gado, A. Fierro, L. de Arcangelis, and A. Coniglio

Mode Coupling Theories for Jamming and Gelation 33
      M. E. Cates

Glass Transition in a Two-Dimensional System of Magnetic Colloids 40
      H. König, K. Zahn, and G. Maret

General Nonlinear 2-Fluid Hydrodynamics of Complex Fluids and Soft Matter 46
      H. Pleiner and J. L. Harden

Recovery of Polymer Blends after Melt Elongation: Analysis of a Model for Small and Large
Capillary Numbers 52
      U. A. Handge

Visco-Elastic Relaxation in Novel Mosaic Phase of Non-Symmetric Chiral
Twin Liquid Crystals 56
      J. Yamamoto, I. Nishiyama, and H. Yokoyama

Aging and Shear Rejuvenation of Soft Glassy Materials 60
      D. Bonn, H. Tanaka, P. Coussot, and J. Meunier

Scattering Probes of Complex Fluids and Solids 64
      R. Vavrin, A. Stradner, J. Kohlbrecher, F. Scheffold, and P. Schurtenberger

Temporal Heterogeneity of the Slow Dynamics of a Colloidal Paste 68
      P. Ballesta, C. Ligoure, and L. Cipelletti

Dynamics around the Sol-Gel Transition in Thermoreversible Atactic Polystyrene Gels 72
      A. Matic, J. Mattsson, and R. Bergman

Local Particle Rearrangements in a Two-Dimensional Binary Colloidal Glass Former 76
      H. König

What Is the Excluded Volume for an Excluded Volume Chain? 80
      M. W. Dreischor and C. P. Lowe

Flow Instabilities in Complex Fluids: Nonlinear Rheology and Slow Relaxations 84
      A. Aradian and M. E. Cates

Molecular Dynamics Simulation of Dendrimers: Structural Formation and Internal
Charge Distribution 88
      T. Terao and T. Nakayama

Droplet Density Dependences of the Static and Dynamic Structures in a Ternary
Microemulsion System 92
      M. Nagao, H. Seto, Y. Kawabata, and M. Shibayama

Anomalous Change in Lamellar Spacing by Shear Flow in Nonionic Surfactant/Water System 94
      T. Kato, K. Minewaki, K. Miyazaki, Y. Kawabata, and M. Imai

Texture Observed in a Simple Shear Flow and after Cessation of the Flow for Liquid Crystalline Polymers 96
    N. Mori, N. Moriguchi, T. Yamamoto, and K. Yasuda

Dynamics of Nano-Sized Colloidal Particles in a Lyotropic Lamellar Phase 98
    Y. Kimura and D. Mizuno

Segregation of Binary Granular Mixtures in a Horizontal Rotating Cylinder 100
    H. Takano, M. Furuta, T. Ishikawa, and T. Tanizawa

Micelle Formation in a Highly Charged Polyelectrolyte Solution 102
    T. Fujima and H. Frusawa

Liquid-Liquid Extraction in Microfluidic System Using Dispersed Liquid Droplet 104
    M. Kumemura and T. Korenaga

Morphology Transition from Sphere to Rod by Confining the Polymer Chains in a Dilute Microemulsion System 106
    K. Nakaya, M. Imai, S. Komura, and N. Urakami

Inter-Lamellar Interaction Mediated by Amphiphilic Triblock Copolymer 108
    T. Masui, M. Imai, and K. Nakaya

Molecular Assembly under a Focused Laser 110
    M. Ichikawa, Y. Matsuzawa, Y. Koyama, and K. Yoshikawa

Surface Undulation Appearing by Continuous Temperature Elevation of Supercooled Liquids on Metal Substrates 112
    H. Nakayama, K. Ishikawa, H. Umeyama, and K. Ishii

Fluctuation Modes Prior to Lamellar-Double Gyroid Transition of a Nonionic Surfactant/Water System 114
    M. Imai, K. Nakaya, and T. Kawakatsu

Structural Changes of an Immiscible Polymer Blend under Shear Flows and Electric Fields 116
    H. Orihara and T. Shibuya

Neutron Spin Echo Study on Slow Dynamics of Lipid Bilayers in the $DPPC/D_2O/CaCl_2$ System 118
    T. Takeda, N. L. Yamada, Y. Kawabata, H. Seto, and M. Nagao

Temperature- and Pressure-Dependences of a Bending Modulus of Surfactant Monolayers in a Ternary Microemulsion Composed of $AOT/D_2O/Decane$ 120
    H. Seto, Y. Kawabata, M. Nagao, and T. Takeda

Adsorption of Magnetic Nanoparticles onto Polyacrylamide Chains in Dilute Polymer Solutions and Ferrogel Networks 122
    D. El kharrat, O. Sandre, P. Licinio, and R. Perzynski

Slow Dynamics Approaching the Glass Transition in Repulsive Magnetic Fluids 124
    G. Mériguet, E. Dubois, V. Dupuis, and R. Perzynski

The Dynamics of Granular Matter 126
    B. M. Schulz, M. Schulz, and S. Herminghaus

Conformational Transitions of a Semiflexible Polymer in a Liquid Crystalline Phase 128
    A. Matsuyama

The Unbinding Transition of Mixed Fluid Membranes 130
    S. Komura and D. Andelman

Contributions of Steady Heat Conduction to the Rate of Chemical Reaction 132
    H.-D. Kim and H. Hayakawa

Theory for the Switching of Nematic Director Deformations 134
    H. Matsuda, T. Koda, and S. Ikeda

Dynamics of Phase Separation in Confined Two-Component Fluid Membranes 136
    H. Wada

Membrane-Membrane Interaction and Free Energy of Multilayer Membrane System 138
    T. Yamamoto and Y. Iwamoto

Dynamic Critical Phenomena of Polymer Solutions 140
    A. Furukawa

Stable and Time-Dependent Vortex-Patterns in 2D Schrödinger Equations and a Consideration of Biological Memories 142
    T. Kobayashi

String Tension and Surface Tension of Elastic Membranes with Fluidity 144
    H. Koibuchi, N. Kusano, A. Nidaira, and K. Suzuki

Configuration of Nematic Liquid Crystals around Particles under External Fields 146
    J. Fukuda, H. Stark, M. Yoneya, and H. Yokoyama
Brownian Dynamics Simulation for Suspension of Oblate Spheroidal Particles in Simple
Shear Flow 148
    T. Yamamoto, T. Suga, and N. Mori
Multifractal Analysis for the Dynamical Heterogeneity in Strongly Correlated
Many-Body Systems. 150
    O. Narikiyo and W. Sakikawa
Molecular Dynamics Simulation of Mixtures of Hard Rod-Like Molecules 152
    T. Koda and S. Ikeda
Initial Deformation Process in Granular Matter.. 154
    S. Kitsunezaki
The Simulation of the Inelastic Impact.. 156
    H. Kuninaka and H. Hayakawa
Molecular Dynamics Study on a Self-organized Shock Wave in an Inelastic Hard Disk System 158
    M. Isobe
Dynamics of Orientational Phase Ordering Coupled to Elastic Degrees of Freedom 160
    N. Uchida
The Instantaneous-Normal-Mode Analysis on the Potential Energy Landscapes of the
Lennard-Jones 2n-n Fluids... 162
    T.-M. Wu and S. L. Chang
Brownian Dynamics Simulation of Highly Charged Colloidal Suspensions 164
    Y. Terada and M. Tokuyama
Computer Simulations of Two Kinds of Polydisperse Hard-Sphere Systems; Atomic Systems
and Colloidal Suspensions.. 166
    T. Shimura, H. Yamazaki, M. Tokuyama, and Y. Terada
Brownian Dynamics Study of Dilute Suspensions of Magnetic Particles in a Static Field 168
    J. Akiyama, M. Tokuyama, and Y. Terada
Orientation Orders of Small Anisotropic Molecules in Confinement............................... 170
    X. Zhou, H. Chen, and M. Iwamoto
Particle Modeling of Plasma Confinement by Multipolar Magnetic Fields.......................... 172
    H. Takekida and K. Nanbu
Computer Simulation of Nano Metallic-Particle Synthesis in an Advanced RF Inductively
Coupled Plasma 174
    M. Shigeta and H. Nishiyama

II. POLYMER DYNAMICS

Entanglement Concept Revisited... 179
    K. Kremer and S. Leon
Slow Dynamics and Mechanical Properties of Polymer Glasses: Ageing Properties 183
    J. Rault
Intermediate Length Scale Dynamics in Polymer Melts 189
    D. Richter, M. Monkenbusch, A. Arbe, and J. Colmenero
Glassy Dynamics of Polymer Thin Films 197
    T. Kanaya, T. Miyazaki, R. Inoue, H. Yamano, K. Nishida, I. Tsukushi, and K. Shibata
Slow Diffusive Motions in a Monolayer of Tetracosane Molecules Adsorbed on Graphite............ 201
    H. Taub, F. Y. Hansen, L. Criswell, D. Fuhrmann, K. W. Herwig, A. Diama, H. Mo, R. M. Dimeo,
    D. A. Neumann, and U. G. Volkmann
Power Laws in the Dynamics of Polymer Solutions 205
    T. Uematsu, C. Svanberg, M. Nydén, and P. Jacobsson
Direct Measurement of Polymer Segment Orientation and Distortion in Shear: Semi-Dilute
Solution Behavior of a Conjugated System 209
    E. K. Hill, Y. Wei, and D. E. Dunstan
Polymer Film Dynamics with Coherent X-Ray Scattering....................................... 213
    H. Kim, A. Rühm, L. B. Lurio, J. K. Basu, J. Lal, S. G. J. Mochrie, and S. K. Sinha

Rearrangement Dynamics in Lamellar Forming Block Copolymers under an Electric Field 217
    K. S. Lyakhova, A. V. Zvelindovsky, and G. J. A. Sevink

Thermoreversible Gelation Driven by Coil-to-Helix Transition of Polymers 221
    F. Tanaka and Y. Tamura

Analysis of Shear-Thickening in Physical Gel by Transient Network Theory 225
    T. Indei and T. Arimitsu

Polyelectrolyte Adsorption on Charged Substrate 229
    C.-H. Cheng and P.-Y. Lai

Analysis of the Center of Mass-, Rotational- and Intramolecular Diffusive Motions in a Monolayer Film of Intermediate-Length Alkane Molecules Adsorbed on a Solid Surface 233
    F. Y. Hansen and H. Taub

Nanostructure Formation in Polymer Thin Films: Dissipative Particle Dynamics Simulation Using a Space-Time Coarse-Grained Model 237
    C. C. Liew and M. Mikami

Simulating the Long-Time Viscoelastic Response of Long Polymers 241
    C. P. Lowe, A. F. Bakker, and M. W. Dreischor

Dynamics in Self-Assembling Polymer Films 245
    K. Fukunaga and T. Hashimoto

Dynamics of Granular Chain 247
    K. To

Two-Dimensional Fluorescence Spectroscopy: Medium Dependence and Thermal Effect of Inhomogeneity in Glass 249
    T. Muromoto, Y. Mori, Y. Nagasawa, H. Miyasaka, and T. Okada

Dynamics of Thin Films of cis-Polyisoprene 251
    K. Fukao

Electric Birefringence of Amorphous Polymers around the Glass Transition Temperature 253
    T. Inoue

Melting Behavior of Polystyrene Surface Studied by X-Ray Reflectivity 255
    A. Kitahara, S. Doi, and I. Thakahashi

Simulations of Gaussian and Excluded-Volume Chains in Curved Slits 257
    Y. Y. Suzuki, T. Dotera, and M. Hirabayashi

Hydrodynamic Coarsening and Pattern Selection in Two-Dimensional Quenched Diblock Copolymers 259
    Y. Yokojima and Y. Shiwa

Primitive Chain Network Model for Entangled Polymer Blends 261
    Y. Masubuchi, J. Takimoto, M. Doi, G. Ianniruberto, F. Greco, and G. Marrucci

Diffusion of Particle in Hyaluronan Solution, a Brownian Dynamics Simulation 263
    M. Takasu and J. Tomita

Dynamical Self-Consistent Field Theory for Inhomogeneous Polymer Systems 265
    T. Kawakatsu

Voronoi Space Division of a Polymer 267
    N. Tokita, M. Hirabayashi, C. Azuma, and T. Dotera

A Possibility of Controlling Polymer Entanglement Structure under the Flow Using the Computational Simulations 269
    A. Kuroda and K. Koyama

A Short Analysis of Two Novel Methods for Modeling Polymers 271
    E. A. Koopman and C. P. Lowe

Modelling the Flow of Polymer Solutions in Confined Geometries Using a Dissipative Gas Solvent Model 273
    A. Berkenbos and C. P. Lowe

Three Regimes of a Steady-State Rimming Flow of the Liquid Polymers 275
    S. Fomin and T. Hashida

Chain Dynamics on Polymer Crystallization from a Stretched Amorphous State: A Molecular Dynamics Simulation 277
    A. Koyama, T. Yamamoto, K. Fukao, and Y. Miyamoto

Relaxation and Self-Diffusion of a Polymer Chain in a Melt 279
    K. Hagita and H. Takano

Dynamics of Knotted Polymers 281
    P.-Y. Lai

III. BIOLOGICAL SYSTEMS

Charge Inversion of a Macroion in Electrolyte Solvent: A Rotating Rod with Polyelectrolyte Counterions 285
    M. Tanaka

Dynamics of Protein Extraction and Extension by Force Spectroscopy and Molecular Dynamics Simulation 291
    A. Ikai, R. Afrin, R. Hertadi, and S. Ohta

A Brownian Ratchet Model of Actin Polymerization Motor by Using Extended Scaled Particle Theory 294
    M. Irisa

Modelling Polymersomes: A Prototype for Complex Cellular Structures 298
    G. J. A. Sevink and J. G. E. M. Fraaije

Hierarchical Regularity in Multi-Basin Dynamics on Protein Landscapes 302
    Y. Matsunaga, K. S. Kostov, and T. Komatsuzaki

Low-Order Chaos in Sympathetic Nerve Activity Causes $1/f$ Fluctuation of Heartbeat Intervals 306
    M. Osaka, H. Kumagai, K. Sakata, T. Onami, K. H. Chon, M. A. Watanabe, and T. Saruta

Molecular Dynamics of 8-Oxoguanine Lesioned B-DNA Molecule—Structure and Energy Analysis 310
    M. Pinak, P. O'Neill, H. Fujimoto, and T. Nemoto

Electrostatic Effects in Phase Transitions of Biomembranes between Cubic Phases and Lamellar Liquid-Crystalline (L_α) Phase 314
    S. M. Masum, S. J. Li, Y. Tamba, Y. Yamashita, and M. Yamazaki

Membrane Fusion of Giant Liposomes of Neutral Phospholipid Membranes Induced by La^{3+} and Gd^{3+} 316
    T. Tanaka, R. Sano, A. Yamagami, and M. Yamazaki

Ultrafast Dynamics in Low Temperature Saccharide Glasses: A Photon Echo Study 318
    Y. Nagasawa, Y. Nakagawa, Y. Mori, T. Muromoto, and T. Okada

Mechanical Response of Single Filamin A (ABP-280) Molecules and Its Role in the Actin/Filamin A Gel 320
    R. Sano, S. Furuike, T. Ito, K. Ohashi, and M. Yamazaki

Oscillatory Reaction of Enzyme Caused by Gradual Entry of Substrate 322
    T. Hideshima

Reentrant Collapsing Transition of Single DNA Molecules: Elastic Response Depending on Spermidine Concentration 324
    Y. Murayama and M. Sano

Study of Fluidity of Lipid Membranes Using Single Molecule Microscopy 326
    M. Hibino and M. Kobayashi

The DNA Adsorption by the Charged Cholesterol Monolayer at the Air-Liquid Interface 328
    T.-L. Lin, Y. Hu, J.-C. Wu, C.-P. Yang, U.-S. Jeng, and M.-C. Shih

Entropy Driven Unidirectional Motion of Brownian Particle Inside a Three-Dimensional Tube: Entropy Ratchet 330
    S.-Y. Sheu and D.-Y. Yang

Free Energy Calculations of the Stacked and Unstacked States for DNA Dimers by Replica-Exchange Umbrella Sampling 332
    K. Murata, Y. Sugita, and Y. Okamoto

Phenomenological Models of Raft Structure 334
    H. Shirotori, S. Komura, T. Kato, and P. D. Olmsted

The Principle for Artificial Molecular Machine and Noise: Slow Stochastic Dynamics in Complex Systems 336
    H. Matsuura, M. Nakano, and T. Nemoto

A Lattice Model of the Protein Diffusion in Membranes 338
    S. Kinouchi, K. Tamura, S. Komura, T. Kato, and Y. Y. Suzuki

Necklace on a Strongly Charged Polyelectrolyte...........340
    T. Sakaue

Protein Folding Dynamics: Ergodic Behavior in Principal Component Space...........342
    Y. Matsunaga and T. Komatsuzaki

A Coarse-Graining of Energy Landscape of Proteins—Structural Stability of the Most Stable States...........344
    K. Hoshino, Y. Matsunaga, M. Miller, D. J. Wales, and T. Komatsuzaki

Charge Conductivity in Peptides: Dynamic Simulations of a Bifunctional Model Supporting Experimental Data...........346
    S.-Y. Sheu, D.-Y. Yang, H. L. Selzle, and E. W. Schlag

Irreversibility on the Structural Transition under Strain in a Single Semi-Flexible Polymer...........348
    N. Yoshinaga and K. Yoshikawa

Generalized-Ensemble Monte Carlo Algorithms for Simulations of Proteins...........350
    A. Mitsutake, Y. Sugita, and Y. Okamoto

Molecular Dynamics Study of the Lipid Bilayers: Effects of the Chain Branching on the Structure and Dynamics...........352
    W. Shinoda, M. Mikami, T. Baba, and M. Hato

Computer Simulation of Bacterial Colony Formation with Multiplying Rods Producing a Chemotactic Factor...........354
    Y. Ueno, R. Morikawa, and M. Hayashi

Application of the Tsallis Statistics to the Molecular Dynamics Simulation...........356
    I. Fukuda and H. Nakamura

Slow Protein Dynamics to be Detected in Inelastic Neutron Scattering Spectra Studied by Molecular Simulation...........358
    Y. Joti, A. Kitao, and N. Go

Theoretical Study on Local Backbone Dynamics of Cupredoxin...........360
    H. Saito, A. Sugiyama, T. Yoshimoto, H. Nagao, T. Sakurai, and K. Nishikawa

Theoretical Study on Vibrational Dynamics of Cupredoxin...........362
    A. Sugiyama, H. Saito, T. Yoshimoto, H. Nagao, T. Sakurai, and K. Nishikawa

Deformation of Crosslinked Semiflexible Polymer Networks...........364
    D. A. Head, A. J. Levine, and F. C. MacKintosh

IV. COMPLEX SYSTEMS

Pioneering Work of THz Wave and Its Application for Molecular Sciences...........369
    J.-i. Nishizawa

Molecular Dynamics Compared to Hydrodynamics for Rayleigh-Taylor Instability...........376
    K. Kadau, T. C. Germann, N. G. Hadjiconstantinou, G. Dimonte, P. S. Lomdahl, B. L. Holian, and B. J. Alder

Dynamics of Liquid-Vapor Phase Transition under High Frequency Vibrations...........379
    D. A. Beysens

The Flow and Adsorption of DNA Polymers near Surfaces...........386
    R. G. Larson, L. Li, M. Chopra, and M. A. Burns

Experimental Study of Self Organized Criticality on a Three Dimensional Pile of Rice...........390
    C. M. Aegerter, K. A. Lörincz, M. S. Welling, and R. J. Wijngaarden

Sorption Dynamics of Cr(VI) on Used Black Tea Leaves...........394
    M. A. Hossain, M. Kumita, and S. Mori

Kikuchi-Kossel Diffraction Line Analysis on Crystallization in Salt-Free Aqueous Colloidal Suspensions...........398
    I. S. Sogami, T. Shinohara, M. Okuno, T. Kurokawa, M. M. Arishiro, T. Itoh, M. Tanigawa, and T. Yoshiyama

Phase Separation of Polymer Mixtures Driven by Temporally and Spatially Periodic Forcing...........402
    S. Komatsu, S. Nishigami, S. Yoshida, T. Norisuye, and Q. Tran-Cong-Miyata

Statistical Mechanics of Phase Unwrapping Problem by the Q-Ising Model...........406
    Y. Saika and H. Nishimori

Feedback Coupling and Chemical Reactions...........410
    S. Trimper and K. Zabrocki

Phase Transitions of Nematic Rubbers 414
    K. Okumura and P.-G. de Gennes

Driven Motion of Extended Defects Wetted by a New Phase 418
    A. L. Korzhenevskii, R. Bausch, and R. Schmitz

Effects of Temperature Chaos on Rejuvenation and Memory in Migdal-Kadanoff Spin Glasses 422
    M. Sasaki and O. C. Martin

Searching for Backbones—An Efficient Parallel Algorithm for Finding Groundstates in Spin Glass Models 426
    J. J. Schneider

Convective Flow Driven by Chemical Reaction 430
    H. Kitahata, N. Magome, and K. Yoshikawa

Crack Patterns in Drying Process Show Memories Contained Inside Granular Networks 432
    A. Nakahara and Y. Matsuo

Diffusing-Wave Spectroscopy of Gelling Dairy Systems 434
    Y. Hemar, P. Hebraud, R. Sarcia, and D. N. Pinder

Dynamic Structure of Liquid Se, Te and Se-Te Mixtures by Neutron Scattering Measurements 436
    A. Chiba, M. Yao, Y. Ohmasa, J. Taylor, and S. M. Bennington

Slow Dynamics Induced by the Metal-Nonmetal Transition in Liquids 438
    H. Kajikawa, M. Yao, H. Kohno, and K. Kobayashi

Experimental Study of One-Dimensional Spinodal Decomposition in Liquid Crystals 440
    T. Nagaya and J.-M. Gilli

Sonoluminescing Bubbles from a Diffusing Glycerol Droplet 442
    S. Hayashi and N. Harba

Shear-Induced Structure and Velocity Fluctuations in Particulate Suspensions Probed by Ultrasonic Correlation Spectroscopy and Rheology 444
    A. L. Strybulevych, D. M. Leary, and J. H. Page

Generalization of the Ohta-Kawasaki Theory for Microphase Separation of Block Copolymer Melts 446
    T. Uneyama, Y. Masubuchi, J. Takimoto, and M. Doi

Dynamics in a Bistable-Element-Network with Delayed Coupling and Local Noise 448
    D. Huber and L. Tsimring

Photo-Induced Wave Propagation in Langmuir Monolayers 450
    T. Okuzono, Y. Tabe, and H. Yokoyama

Molecular Dynamics Simulation of Liquid Crystalline Polymer Networks and Flexible Polymer Network in Liquid Crystal Solution 452
    A. Zarembo, A. Darinskii, I. Neelov, N. Balabaev, and F. Sundholm

Irreversible Sequential Adsorption of Line Segments with Diffusional Relaxation on a One-Dimensional Lattice 454
    K. E. Lee, H. S. Park, and J. W. Lee

Complex History-Dependent Freeze-Slip Transition in the Population of Phase Oscillators 456
    A. Awazu

One Idea of Portfolio Risk Control Focusing on States of Correlation 458
    N. Nishiyama

Dynamics in a Complex-Fracture-Subterranean-System with Application to HDR Geothermal Reservoirs 460
    K. Yoshida, S. Fomin, Z. Jing, and T. Hashida

The Stability of the Critical Scaling Against the Time-Dependent Perturbation 462
    He. Park and Hy. Park

Internal Motion of Confined Molecules in Fullerene 464
    Y. Shigeta

Granular Flow Simulation by Granular Element Method 466
    Y. Kishino

Molecular Hydrodynamics: From Kubo to Smoluchowski Example of a Montmorillonite Clay 468
    J.-F. Dufrêche, N. Malikova, V. Marry, F. Grün, M. Jardat, E. Dubois, and P. Turq

V. GLASS TRANSITION

Vibrational Dynamics and Thermodynamics, Ideal Glass Transitions and Folding Transitions, in Liquids and Biopolymers 473
    C. A. Angell, L.-M. Wang, S. Mossa, Y. Yue, and J. R. D. Copley

Dynamic Heterogeneities in Liquid Water 483
    N. Giovambattista, S. V. Buldyrev, F. W. Starr, and H. E. Stanley

Self-Diffusion and Spatially Heterogeneous Dynamics in Supercooled Liquids near T_g 491
    S. F. Swallen, O. Urakawa, M. Mapes, and M. D. Ediger

Crystallisation and Local Order in Glass-Formimg Binary Mixtures 496
    J. R. Fernández and P. Harrowell

The Structure of Energy Landscape and the Non-Arrhenius Behaviour of Supercooled Liquids 503
    M. Schulz

Computer Simulation of the Glass Transition in Thin Films 509
    K. Binder, F. Varnik, J. Baschnagel, P. Scheidler, and W. Kob

Johari-Goldstein or Primitive Relaxation: Terminator of Caged Dynamics and Precursor of α-Relaxation 515
    K. L. Ngai

Johari-Goldstein Relaxations During Physical Aging of Propylene Glycol Oligomers under High Pressure 523
    C. M. Roland and R. Casalini

Non-Debye Dielectric Response and non-Arrhenius Kinetics in Complex Systems at Mesoscale 527
    Y. Feldman, A. Puzenko, Y. Ryabov, and A. Gutina

Dynamic Crossover in Complex Systems: From a "Simple" Liquid to a Protein 533
    A. Sokolov and V. Novikov

Two-Order-Parameter Model of Liquid: Water-Like Thermodynamic Anomaly, Liquid-Liquid Transition, and Liquid-Glass Transition 541
    H. Tanaka

Bulk Nonequilibrium Alloys by Stabilization of Supercooled Liquid: Fabrication and Functional Properties 547
    A. Inoue and A. Takeuchi

Logarithmic Decay in a Two-Component Model 559
    M. Sperl

The Boson Peak and the Phonons in Glasses 565
    S. Ciliberti, T. S. Grigera, V. Martín-Mayor, G. Parisi, and P. Verrocchio

Structural Relaxations in Silica Glass 571
    A. J. Ikushima, K. Saito, and H. Kakiuchida

Dynamics of a Rod in a Homogeneous/Inhomogeneous Frozen Disordered Medium: Correlation Functions and Non-Gaussian Effects 576
    A. J. Moreno and W. Kob

Microscopic Dynamics in Non-Simple Liquid Metals 583
    S. Hosokawa and W.-C. Pilgrim

Volume Effects on the Molecular Rearrangements in Vicinity of Glass Transition 587
    M. Paluch, K. Grzybowska, A. Grzybowski, and C. M. Roland

Glassy Crystals: Above and Below Tg 591
    M. Descamps and J. F. Willart

Neutron Scattering and Dielectric Study on the Structural and Dynamical Peculiar Properties of Poly(Vinyl Chloride) 594
    A. Arbe, A. Moral, A. Alegría, J. Colmenero, W. Pyckhout-Hintzen, D. Richter, B. Farago, and B. Frick

Effect of Polymer-Substrate Interactions on the Glass Transition of Polymer Thin Films 598
    O. K. C. Tsui, T. P. Russell, and C. J. Hawker

Universal Reference Temperature for Melt Viscosity Temperature Relationship 601
    N. Okui

The Role of Configurational Entropy in Chemical Vitrification 604
    S. Corezzi

A Crossover Region in the Glassy Freezing of Deuteron Dipole Glass 608
    Y.-S. Choi and J.-J. Kim

Confined and Bulk Dynamics of a Simple Glass-Former 611
    C. Svanberg, R. Bergman, P. Jacobsson, and L. Börjesson

Different Routes to an Understanding of the Excess Wing in the Dielectric Loss of Glass-Formers 615
    R. Bergman, J. Mattsson, and C. Svanberg

Observation on Surface Change of Fragile Glass: Temperature–Time Dependence Studied by X-Ray Reflectivity 619
    H. Kikkawa, A. Kitahara, and I. Takahashi

Competition between Crystallization and Glass-Transition Processes in Binary Amorphous Molecular Systems 623
    K. Ishii, M. Murai, M. Yamamoto, M. Takei, and H. Nakayama

Structural Relaxation and Low-Energy Excitation in Amorphous Ice and Related Glasses 627
    O. Yamamuro

A Unified Theory of the Liquid-Glass Transition 631
    T. Kitamura

Supercooled Liquids under Shear: A Mode-Coupling Theory Approach 635
    K. Miyazaki, R. Yamamoto, and D. R. Reichman

Fluctuation-Dissipation Relations in Ageing and Driven Non-Mean Field Glass Models 639
    S. M. Fielding and P. Sollich

Stochastic Approach to Glass Transition 643
    T. Ishii

Molecular Dynamics Simulations and Neutron Spin Echo Experiments of Difluorotetrachloroethane Glassy Crystal 647
    F. Affouard, E. Cochin, R. Decressain, M. Descamps, and W. Haeussler

Single Particle Jumps in a Glass: A Computer Simulation 651
    K. Vollmayr-Lee

Self-Atomic Motions in Glass-Forming Polymers: Neutron Scattering and Molecular Dynamics Simulations Results 655
    J. Colmenero, F. Alvarez, A. Narros, A. Arbe, M. Monkenbusch, D. Richter, and B. Farago

Broadband Dielectric Study on Alpha- and Beta-Process for Poly(Ethylene Glycol)-Water Mixtures 659
    N. Shinyashiki, S. Sudo, M. Shimomura, and S. Yagihara

Mechanical Relaxations in Metallic Glasses at Higher Temperatures 661
    Y. Hiki, T. Aida, and S. Takeuchi

Concentration Dependence of the α- and β-Processes for Alcohol-Water Mixtures 663
    S. Sudo, M. Shimomura, S. Tsubotani, N. Shinyashiki, and S. Yagihara

Low-Temperature Specific Heat and Brillouin Scattering Measurements on Hydrogen-Bonded Glasses 665
    M. A. Ramos, C. Talón, R. J. Jiménez-Riobóo, and S. Vieira

Ultra-Slow Dielectric Relaxation Process in Polyols 667
    Y. Yomogida, A. Minoguchi, and R. Nozaki

Terahertz Time Domain Spectroscopy of Boson Peak 669
    S. Kojima, H. Kitahara, S. Nishizawa, and M. Wada Takeda

Slow Dynamics and Dielectric Relaxation in Water/Glycerol Mixtures 671
    Y. Hayashi, Y. E. Ryabov, A. Gutina, and Y. Feldman

Relaxation Dynamics in Supercooled Liquids Studied by Time-Resolved Spectroscopy 673
    M. Kobayashi, Y. Tsujimi, and T. Yagi

Thermal Properties of Supercooled Water Confined within Silica Gel Pores 675
    S. Maruyama, K. Wakabayashi, and M. Oguni

Intramolecular Rotational Diffusion Crossover in Supercooled Liquids 677
    N. Yonekura

History Memorized on Glass Transition 679
    Y. Miyamoto, H. Yamao, and K. Sekimoto

Dielectric Study on Poly(Vinyl Pyrrolidone)-Alcohol Mixtures 681
    D. Imoto, S. Sudo, N. Shinyashiki, and S. Yagihara

Broadband Dielectric Study on Dynamics of Poly(Vinyl Pyrrolidone)-Poly (Ethylene Glycol) Blend 683
    S. Tsubotani, S. Sudo, H. Nakamura, N. Shinyashiki, S. Yagihara, and R. J. Sengwa

Glass Transition in the Stable Crystalline State of 3-Chlorothiophene 685
    H. Fujimori, A. Todoroki, T. Asaji, and M. Oguni

Origin of the Exceptional Behaviors of Lower Alcohols in the Supercooled Liquid State 687
    Y. Hiejima and M. Yao

Study of Dielectric Relaxations in Glucose-Water Mixtures 689
    J. Y. Oh, J.-A. Seo, H. K. Kim, and Y.-H. Hwang

Ultra-Slow Dynamics in Glass-Forming Polybutadiene 691
    R. Inoue, N. Takahashi, K. Nishida, and T. Kanaya

High-Resolution Brillouin Scattering Study of Intermediate Glass-Forming Materials 693
    Y. Ike and S. Kojima

SAXS Analysis of Rapidly Solidified $Al_{92}V_3Fe_3Zr_2$ Ribbon 695
    T. Kamiyama, H. M. Kimura, and A. Inoue

Study of Glass Transition Temperatures in Sugar Mixtures by DSC 697
    J.-A. Seo, S. J. Kim, J. Oh, Y. S. Yang, H. K. Kim, and Y.-H. Hwang

NMR Hole-Burning Experiments on Superionic Conductor Glasses 699
    J. Kawamura, N. Kuwata, and T. Hattori

Ground-State Memories Survive Strenuous Thermal Fluctuations: Dynamics of Dimerized Spin Chains 701
    I. Sawada

Exact Non-Equilibrium Fluctuation Dissipation Relations for Multi-Spin Observables in the Glauber-Ising Spin Chain 703
    P. Mayer and P. Sollich

Duality Symmetry, the Disorder Parameter, and the Glass Transition 705
    I. Kanazawa

Glass Transition of Hard Sphere Systems—Molecular Dynamics and Density Functional Approaches 707
    K. Kim and T. Munakata

Transition from Annealed to Quenched Dynamics 709
    F. Tagawa and T. Odagaki

How Reproducible is the Structure of Dynamic Heterogeneity in Glass Forming Liquids? 711
    A. Widmer-Cooper and P. Harrowell

Computer Simulations of a Model Glass with the Internal Structures 713
    T. Muranaka

Simulation of the Effect of Interstitials to Shear Modulus in Aluminum and an Ionic Crystal 715
    O. Lizhi and D.-M. Zhu

Supercooled Liquids under Shear: Computational Approach 717
    R. Yamamoto, K. Miyazaki, and D. R. Reichman

VI. OTHER RELATED TOPICS

Atomic Dynamics of a Bulk and Hyperquenched Metallic Glass 721
    J.-B. Suck

New Aspect of the Spontaneous Formation of a Bilayer Lipid Membrane 724
    H. Fujiwara, M. Fujihara, T. Koyama, and T. Ishiwata

Nonperturbative Anharmonic Phenomena in Crystal Lattice Dynamics 727
    M. I. Katsnelson and A. V. Trefilov

Dynamics and Its Stability of Boltzmann-Machine Learning Algorithm for Gray Scale Image Restoration 731
    J.-i. Inoue and K. Tanaka

Simple Models of Unusual Elastic Properties 735
    K. W. Wojciechowski

Temperature Dependence of the Ultrafast Solvation of a Dye Molecule in Alcohol 739
    H. Murakami and M. Tanaka

Dynamical Susceptibility Close to a Critical Point in $Sr_3Ru_2O_7$ 741
    R. A. Borzi, S. A. Grigera, and A. P. Mackenzie

Deformations of Adhering Elastic Tubes 743
    K. Tamura, S. Komura, and T. Kato

Exact Renormalization Group for the Brazovskii Model..745
    Y. Shiwa

A Concept of Effectively Global Search in Optimization by Local Search Heuristics................747
    M. Hasegawa

Behaviors of Thermodynamic Quantities of a Noise-Driven Nonlinear Oscillator....................749
    M. Akimoto and A. Suzuki

Thermal Contact in Quantum Systems..751
    A. Iwaya and A. Suzuki

Theoretical Study on Photoexcited States of Strongly Correlated Electron Systems..................753
    K. Sugiyama, T. Yamaguchi, I. Sakamoto, T. Yoshimoto, H. Nagao, and K. Nishikawa

Theoretical Study on the Effect of Solvent and Intermolecular Fluctuations in Proton Transfer Reactions: General Theory..755
    N. Kato, T. Ida, and K. Endo

Bottleneck in Energy Relaxation and Its Self-Organization..757
    H. Morita and K. Kaneko

Exact Relations for Quantum Many-Body Correlation Functions......................................759
    T. Toyoda, H. Koizumi, K. Ito, and K. Takiuchi

New Canonical Transformations to Eliminate External Fields in Quantum Many-Body Problems........761
    M. Fujita, D. Anma, T. Fukuda, H. Koizumi, and T. Toyoda

Quantum Many-Body Virial Theorem and Matsubara Green's Function..................................763
    D. Anma, T. Fukuda, M. Fujita, K. Takiuchi, and T. Toyoda

Quasi-Particle Lifetime of Quantum Coulomb Systems...765
    T. Fukuda, M. Fujita, D. Anma, K. Ito, and T. Toyoda

Behavior of Observables in Nonequilibrium State..767
    H. Majima and A. Suzuki

Resonant Charge-Exchange Processes in Nonideal Plasmas...769
    M.-Y. Song and Y.-D. Jung

Light Transmittance of Solid Polymeric Film Including Spherulites................................771
    T. Taniguchi, N. Kobayashi, M. Doi, M. Sugimoto, and K. Koyama

Localization in Disordered Two-Chain System with Long-Range Correlation..........................773
    H. Yamada

Molecular Dynamics Simulations of "The Cooperativity Blockage Effect" in Alkali Metasilicate....775
    J. Habasaki, K. L. Ngai, and Y. Hiwatari

Kinetic Monte Carlo Simulation of Via Filling..777
    Y. Kaneko, Y. Hiwatari, K. Ohara, and T. Murakami

Simulation Study on Strain-Mediated Coarsening of Quantum Dots...................................779
    Y. Enomoto, H. Itamoto, and M. Okada

Simulation Study on Slow Dynamics in Magnetic Fluids...781
    M. Okada and Y. Enomoto

Vortex Dynamics in Superconducting Films with Twin Boundary Networks.............................783
    H. Itamoto and Y. Enomoto

Numerical Simulation for Collisions of a Rigid Disk on Fluid Surface.............................785
    S. Nagahiro and Y. Hayakawa

Laser Control of Non-Stationary Proton State in Hydrogen-Bonded System...........................787
    T. Yamaguchi, I. Sakamoto, Y. Ohta, H. Nagao, and K. Nishikawa

Simulation of Vortex Creep in Type-II Superconductors..789
    R. Kato and Y. Enomoto

Proton Dynamics Simulation of p-Chloro and p-Bromobenzyl Alcohol Crystals....................791
    T. Ida, D. Matsumoto, M. Hamada, M. Mizuno, K. Endo, and M. Hashimoto

Control of Cis-Trans Isomerization by Stimulated Raman Adiabatic Passage Method..................793
    T. Yamaguchi, Y. Ohta, K. Sugiyama, H. Nagao, and K. Nishikawa

Quantum Algorithm in Quantum Network Systems...795
    I. Sakamoto, T. Yamaguchi, H. Nagao, and K. Nishikawa

Quantum Molecular Dynamics Simulation of Guest Molecules in Gas Hydrate..........................797
    D. Matsumoto, T. Ida, N. Kato, M. Mizuno, and K. Endo

Dendritic Side Branching Structure of CML Model 799
  M. Ohtaki, H. Honjo, and H. Sakaguchi

Program 801
Participants 811
Author Index 823

PREFACE

This volume contains the papers presented at the 3rd International Symposium on "Slow Dynamics in Complex Systems", which was hosted by the Institute of Fluid Science, Tohoku University, Japan, and was held mainly at the LaLaLa Hall and the Institute of Fluid Science from 3 to 8 November 2003. There were 130 participants from 26 foreign countries and 253 from Japan. There were 30 invited talks, 76 oral presentations, and 205 poster presentations. The success of the Symposium was guaranteed by the high level of the presentations and also by the enthusiastic participation of 383 persons.

An exciting aspect of the Symposium was the presence of four Nobel Prize Winners who presented hour long talks and participated actively in the discussions. S. Chu's topic was "Watching Molecular Systems Work, One at a Time"; I. Giaever's was "Electrical Impedance Analysis of Mammalian Cells"; A. J. Heeger's was " Ultrafast Photoinduced Electron Transfer: "Superquenching" as a Route to Biosensors using Luminescent Conjugated Polymers", and R. B. Laughlin's was "Configurational Memory of RNA Polymerase in Transcription Regulation".

The Symposium covered five general topics all of which were interrelated and shared the property of slow dynamics. The topics were complex fluids, polymer dynamics, biological systems, complex systems, and glass transitions. The speakers were experimentalists, theorists, and computer scientists and it was impressive how closely these disparate groups worked together.

The systems studied covered almost all states of matter including solids, liquids, complex solutions, polymers, and suspensions. Significant progress was made on a variety of topics. Among these were experimental and theoretical studies of colloidal systems; experiments on glass to glass transitions in micellar systems; theoretical studies of polyelectrolytes and polymer melts and networks; theoretical and computer studies of hydrodynamics in suspensions and Rayleigh-Taylor and Rayleigh-Couette instabilities; theoretical and experimental studies of the glass transition; computer simulations of the glass transition in thin films; vibrational motions in glass forming liquids and glasses; the effects of shear on supercooled liquids; engineering and experimental studies of metallic glasses; mode-coupling studies of complex glass formation; and Lorentz gas studies of the translational and rotational motion of a rigid rod.

The discussions between five different groups were extremely interesting and effective. We hope that this volume reflects the atmosphere of this Symposium. The success of this Symposium was also due to the efforts of the discussion leaders, P. Harrowell, K. L. Ngai, P. Pincus, and D. A. Weitz, who helped the participants review and extend their present understanding of slow dynamics, giving valuable comments and even suggesting some possible future directions of our research.

While organizing this Symposium, we were fortunate to receive abundant help from many people and organizations. On behalf of the organizing committee, we would like to thank them for their support and encouragement. The Symposium was also aided in one form or another by the supporting agencies listed on page xxii in this volume, which are gratefully acknowledged. The organizers would also like to thank members of the international advisory committee whose names are listed on page xix for their encouragement and pertinent advice.

Last but not least, we are grateful to Ms. Megumi Kusano, Mr. Tsutomu Shimura, Ms. Yayoi Terada, and Ms. Takako Ueno, and all others who have contributed to organizing and running the Symposium.

Michio Tokuyama
Irwin Oppenheim

ORGANIZATION OF SYMPOSIUM

ORGANIZING COMMITTEE:

M. Tokuyama (Chairperson)	Tohoku University
M. Doi	Nagoya University
T. Hashimoto	Kyoto University
Y. Hiwatari	Kanazawa University
A. Ikushima	Toyota Technological Institute
A. Onuki	Kyoto University
M. Shibayama	University of Tokyo
Y. Shiwa	Kyoto Institute of Technology
H. Tanaka	University of Tokyo
T. Toyoda	Tokai University

INTERNATIONAL ADVISORY COMMITTEE:

C. A. Angell	Arizona State University
K. Binder	Universität Mainz
P. G. de Gennes	College de France
K. L. Ngai	Naval Research Laboratory
I. Oppenheim	M.I.T.
P. Pincus	U.C.S.B.
H. E. Stanley	Boston University
D. A. Weitz	Harvard University

INSTITUTE OF FLUID SCIENCE STEERING COMMITTEE:

J. Tani (Chairperson)	T. Hayase	K. Hayashi
T. Ikohagi	O. Inoue	K. Kamijo
H. Kobayashi	Y. Kohama	S. Maruyama
K. Nanbu	T. Niioka	H. Nishiyama
S. Obayashi	S. Samukawa	A. Sasoh
T. Takagi	K. Takayama	M. Tokuyama

LOCAL COMMITTEE:

Y. Enomoto	Nagoya Institute of Technology
Y.-H. Hwang	Pusan National University
Y. Kaneko	Kyoto University
T. Muranaka	Aichi Institute of Technology
I. Sawada	Ishikawa National College of Technology
T. Shimura	Tohoku University
A. Suzuki	Tokyo University of Science
Y. Terada	Tohoku University

SECRETARIAT:

T. Ueno	Tohoku University
M. Kusano	Tohoku University

INVITED SPEAKERS:

Complex Fluids
    M. E. Cates   University of Edinburgh
    S.-H. Chen   M.I.T.
    H. Löwen   Heinrich-Heine-Universität Düsseldorf
    G. Maret   University of Konstanz
    D. A. Weitz   Harvard University

Polymer Dynamics
    A. J. Heeger   U.C.S.B.
    K. Kremer   Max Planck Institute for Polymer Research
    P. Pincus   U.C.S.B.
    D. Richter   Forschungszentrum Jülich
    M. Rubinstein   University of North Carolina

Biological Systems
    S. Chu   Stanford University
    I. Giaever   Rensselaer Polytechnic Institute and Applied Biophysics, Inc.

Complex Systems
    B. J. Alder   Lawrence Livermore National Laboratory
    D. Beysens   Commissariat à l'Energie Atomique
    R. Larson   University of Michigan
    J. Nishizawa   Iwate Prefectural University/Semiconductor Research Institute
    R. B. Laughlin   Stanford University

Glass Transition
    C. A. Angell   Arizona State University
    K. Binder   Johannes Gutenberg Universität Mainz
    M. D. Ediger   University of Wisconsin-Madison
    Y. Feldman   The Hebrew University of Jerusalem
    P. Harrowell   University of Sydney
    A. Inoue   Tohoku University
    W. Kob   Université de Montpellier II
    K. L. Ngai   Naval Research Laboratory
    S. Sastry   Jawaharlal Nehru Centre for Advanced Scientific Research
    A. Sokolov   The University of Akron
    M. Sperl   Technische Universität München
    H. E. Stanley   Boston University
    G. Tarjus   Université Pierre et Marie Curie

DISCUSSION LEADERS:

    P. Harrowell   University of Sydney
    K. L. Ngai   Naval Research Laboratory
    P. Pincus   U.C.S.B.
    D. A. Weitz   Harvard University

PAPERS PRESENTED
BUT NOT INCLUDED IN THESE PROCEEDINGS

Watching Molecular Systems Work, One at a Time
Steven Chu
Physics Department, Stanford University

Electrical Impedance Analysis of Mammalian Cells
Guo Chen, Charlie R. Keese, and <u>Ivar Giaever</u>
Rensselaer Polytechnic Institute and Applied BioPhysics, Inc. Troy

Ultrafast Photoinduced Electron Transfer: "Superquenching" as a Route to Biosensors Using Luminescent Conjugated Polymers
Brent S. Gaylord, <u>Alan J. Heeger</u>, and Guillermo C. Bazan
Center for Polymers and Organic Solids, UCSB

Configurational Memory of RNA Polymerase in Transcription Regulation
R. B. Laughlin
Department of Physics, Stanford University

Polyelectrolyte Animals and Slow Modes
<u>P. Pincus</u> and M. Henle
Department of Physics and Materials Research Laboratory, UCSB

"Gelling" Transition of Hydrophobic Polyelectrolytes
<u>Michael Rubinstein</u>[1] and Andrey V. Dobrynin[2]
[1] Department of Chemistry, University of North Carolina
[2] Institute of Materials Science and Department of Physics, University of Connecticut

Dynamics and the Glass Transition in Liquids
Srikanth Sastry
Jawaharlal Nehru Centre for Advanced Scientific Research, Jakkur Campus

The Glass Transition of Liquids: A Theoretical Approach in Terms of Frustration
G. Tarjus
Laboratoire de Physique Theorique des Liquides, Université Pierre et Marie Curie

Jamming Phase Diagram for Colloidal Particles
V. Prasad,[1] V. Trappe,[2] J. Conrad,[1] A. D. Dinsmore,[3] P. N. Segre[4], and <u>D. A. Weitz</u>[1]
[1] Department of Physics and DEAS, Harvard University
[2] Department of Physics, University of Fribourg
[3] Department of Physics, University of Massachusetts
[4] Biotechnology Science Group, SD46, Marshall Space Flight Center

SUPPORTING AGENCIES:

This symposium is partially supported by

    Fuji Xerox Co., Ltd.

    Inoue Foundation for Science

    Institute of Physics Publishing

    Izumi Science and Technology Foundation

    KEIHIN CORPORATION

    MBK MICROTEK INC.

    Ministry of Education, Culture, Sports, Science and Technology of Japan

    NEC Corporation

    Nippon Sheet Glass Foundation for Materials Science and Engineering

    Science Foundation for Fluid Mechanics and Instruments

    Sendai Tourism & Convention Bureau

    SGI Japan, Ltd.

    The Asahi Glass Foundation

    The Kao Foundation For Arts And Sciences

    Tohoku University

    TOHOKU UNIVERSITY FOUNDATION

    TOSHIBA CORPORATION Digital Media Network Company

In cooperation with

    City of SENDAI

    Combustion Society of Japan

    FM Izumi

    FM Jonpa 78.8MHz

    Higashi Nippon Broadcasting Co., LTD.

    Japan Broadcasting Corporation Sendai Station NHK Sendai

    Japan Society of Applied Electromagnetics and Mechanics

    Japan Society of Fluid Mechanics

    Japanese Liquid Crystal Society

    Kahoku Shimpo Publishing Co.

    MIYAGI PREFECTURAL BOARD OF EDUCATION

    Miyagi Prefecture

    Miyagi Television Broadcasting Company Ltd.

Nippon Telegraph and Telephone East Corporation

Science News

Sendai Board of Education

Sendai Television Broadcasting Corp.

Society of Automotive Engineers of Japan Inc.

The 77 Bank, Ltd.

The Asahi Shimbun

The Biophysical Society of Japan

The Chemical Society of Japan

The Heat Transfer Society of Japan

The Japan Fluid Power System Society

The Japan Society for Aeronautical and Space Sciences

The Japan Society of Applied Physics

The Japan Society of Mechanical Engineers

The Japanese Society for Multiphase Flow

The Mainichi Newspapers

THE NIKKAN KOGYO SHIMBUN, LTD.

The Physical Society of Japan

The Sendai Chamber of Commerce and Industry

The Society of Polymer Science, Japan

The Society of Rheology, Japan

The Visualization Society of Japan

The Yomiuri Shimbun

Tohoku Broadcast Co., LTD.

Tohoku Electric Power

The 3rd International Symposium on Slow Dynamics in Complex Systems
November 3-8, 2003, Sendai, Japan

I. COMPLEX FLUIDS

Colloidal suspensions driven by external fields

H. Löwen *, C. N. Likos *, R. Blaak *, S. Auer †, V. Froltsov *, J. Dzubiella †,
A. Wysocki * and H. M. Harreis *

*Institut für Theoretische Physik II, Heinrich-Heine-Universität Düsseldorf, Universitätsstraße 1, D-40225 Düsseldorf, Germany
†Department of Chemistry, University of Cambridge, Lensfield Road, Cambridge CB2 1EW, United Kingdom

Abstract. Colloidal suspensions have been proven to play a pivotal role of model systems in order to understand the principles of equilibrium phase transitions such as freezing and fluid-fluid demixing. One of the main reasons for that is that real-space studies are possible thanks to the mesoscopic length scale of the particle size. The same model character of colloidal suspensions holds in non-equilibrium situations as e.g. represented by an external driving field (such as shear, gravity, an electric and/or magnetic field). In this paper some current examples of non-equilibrium transitions are reviewed where recent progress has been made by theory and computer simulation. In particular, we discuss the competition between phase separation and lane formation in driven colloidal mixtures, crystal nucleation in charged suspensions under shear and chain formation of two-dimensional superparamagnetic suspensions induced by an external magnetic field.

INTRODUCTION

Suspensions of mesoscopic colloidal particles are excellent realizations of classical statistical models since their interactions are tunable. One further advantage of colloidal suspensions lies in the fact that the particle configurations can be watched in real-space, e.g. by using confocal microscopy, which enables a direct comparison between experiments and theory. While in the past two decades most of the investigations of colloidal dispersions were done in the bulk either under equilibrium conditions or regarding the kinetic glass transition in order to explore the thermodynamics, structure and bulk phase behaviour, more recent studies exploit the fascinating possibility to expose colloids to external driving fields [1, 2] and to study thus non-equilibrium dynamics in a controlled way. One of the most intriguing possibilities is to fix and move the colloidal particles by using optical tweezers. In non-equilibrium situations, however, the *dynamics* of the colloids will enter explicitly. Hence a theoretical description is more difficult as long as the long-ranged hydrodynamic interactions induced by the solvent flow will play a significant role.

In this paper we review some progress in the area of colloidal suspensions driven by external fields. In particular, three examples are discussed in detail, all of which have to do with certain aspects of slow dynamics in such complex fluids. It is known that binary mixtures of colloidal suspensions when driven by a constant external field (such as gravity or an electric field) can exhibit formation of particles lanes provided the driving forces acting on the two different particles species differ. These lanes can be intuitively understood by watching pedestrian motion in pedestrian zones [3] and are also mesoscopic analogs to the so-called two-stream instability in plasmas [4]. Here we study the competition between lane formation in a fluid-fluid phase-separating mixture and study the effect of anisotropic coarsening which is a slow dynamical process. The second topic concerns the presence of a shear field. It is known that typically a colloidal solid is molten by shear. But if the shear rate is reduced such shear-molten fluids can recrystallize into a solid. The question is how crystal nucleation rates are affected by shear. Since nucleation is a rare event, this intimately has to do with slow dynamics. Finally we study the chain formation in anisotropically interacting magnetic colloidal spheres exposed to an external magnetic field. If the attraction between the particles is strong enough they form chains and the chains form aggregates. The dynamics towards the agregates is very slow and the question is whether the aggregates finally crystallize into a lattice [5] or whether a liquid-chain phase is stable [6, 7, 8].

COMPETITION BETWEEN LANE FORMATION AND PHASE SEPARATION IN DRIVEN COLLOIDAL MIXTURES

Spinodal decomposition of a phase-separating binary fluid mixture is a well-studied dynamical coarsening

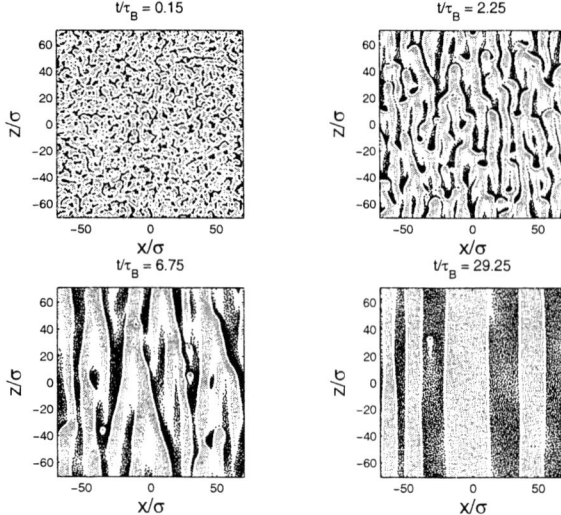

FIGURE 1. Simulation snapshots for a) $t = 0.15$, b) $t = 2.25\tau_B$, c) $t = 6.75\tau_B$, d) $t = 29.25\tau_B$. Simulation parameters are: $\kappa\sigma = 2.34$, $U_0 = 15$, $\frac{F\sigma}{k_B T} = 120$, $\rho\sigma^2 = 0.4$, $N_1 = N_2 = 4000$, $\Delta = 1.6$

process which slows down significantly until complete separation into two macroscopic portions of two fluids is reached [9]. Typically the structure grows with a power-law in time. When combined with another non-equilibrium situation, new pattern structures and growth laws do arise. One example is phase separation under shear [10], another situation occurs if the mixtures is driven by a constant external force which acts differently on the two different particle species. An experimental realization of such a situation is a sedimenting and phase-separating colloid-polymer mixture [11]. In the absence of phase separation, i.e. for a stable mixed fluid, it has been shown by computer simulation [12, 13, 14] and by theory [15] that - upon a critical force difference - the mixture spontaneously forms lanes containing only particles moving alike. The direction of the lanes is along the driving force direction.

Here we study the competition between phase separation and lane formation. We consider a two-dimensional model system interacting via a set of Yukawa pair potentials

The asymmetric binary colloidal mixture comprises $N_1 + N_2$ Brownian colloidal particles in an area S [12]. N_1 particles are of type 1, the other N_2 are of type 2 with partial number densities $\rho_1 = N_1/S$ and $\rho_2 = N_2/S$. In the following we set $\rho_1 = \rho_2 = \rho$. The colloidal suspension is held at fixed temperature T via the bath of microscopic solvent particles. Two colloidal particles are interacting via effective Yukawa potentials as follows:

$$\frac{V_{ij}(r)}{k_B T} = U_0 \sigma_{ij} \frac{\exp(-\kappa(r - \sigma_{ij}))}{r}, \quad (1)$$

where $(ij) = (11),(12),(22)$. Here r is the center-to-center separation, U_0 is the interaction strength measured in terms of the thermal energy $k_B T$ and κ is the inverse screening length. The set of diameters, σ_{ij}, is nonadditive and given by

$$\sigma_{11} = \sigma_{22} = \sigma \quad (2)$$
$$\sigma_{12} = \sigma(1 + \Delta) \quad (3)$$

where Δ is the dimensionless nonadditive parameter.

The dynamics of the colloids is completely overdamped Brownian motion. The friction constant is $\xi = 3\pi\eta\sigma$ with η denoting the shear viscosity of the solvent. The constant external force acting on the ith particle of species j, $\vec{F}_i^{(j)}$, has the same amplitude but an opposite direction for the both constituents of the binary mixture. It is $\vec{F}_i^{(1)} = F\vec{e}_y$ and $\vec{F}_i^{(2)} = -F\vec{e}_y$ where $\vec{e}_y$ is a unit vector along the y-direction of the system.

The stochastic Langevin equations for the colloidal trajectories $\vec{r}_i^{(j)}(t)$ ($j = 1,2$) (with $i = 1,...,N_1$ for $j = 1$ and $i = 1,...,N_2$ for $j = 2$) read as

$$\xi \frac{d\vec{r}_i^{(j)}}{dt} = -\vec{\nabla}_{\vec{r}_i^{(j)}} [\sum_{k=1}^{N_{j'}} V_{jj'}(|\vec{r}_i^{(j)} - \vec{r}_k^{(j')}|) + \sum_{k=1, k \neq i}^{N_j} V_{jj}(|\vec{r}_i^{(j)} - \vec{r}_k^{(j)}|)] + \vec{F}_i^{(j)} + \vec{K}_i^{(j)}(t), \quad (4)$$

where j' is the complementary index to j ($j' = 1$ if $j = 2$ and $j' = 2$ if $j = 1$). The right-hand-side includes all forces acting onto the colloidal particles, namely the force resulting from inter-particle interactions, the external constant force, and the random forces $\vec{K}_i^{(j)}$ describing the collisions of the solvent molecules with the ith colloidal particle of species j. The latter are Gaussian random numbers with zero mean, $\overline{\vec{K}_i^{(j)}} = 0$, and variance

$$\overline{(\vec{K}_i^{(k)})_\alpha(t)(\vec{K}_j^{(n)})_\beta(t')} = 2k_B T \xi \delta_{\alpha\beta} \delta_{ij} \delta_{kn} \delta(t - t'). \quad (5)$$

The subscripts α and β stand for the two Cartesian components. Note that within this simple Langevin picture, hydrodynamic interactions are ignored.

We solve the Langevin equations of motion by Brownian dynamics simulations [16, 17, 18] using a finite time-step and the technique of Ermak [19, 20]. We use a square cell of length ℓ with periodic boundary conditions. The typical size of the time-step Δt was $0.0002\tau_B$, where $\tau_B = \xi\sigma^2/k_B T$ is a suitable Brownian timescale. We simulated typically 2×10^4 time steps which corresponds to a simulation time of $4\tau_B$.

A set of different snapshots are presented in Figure 1 for different times. The starting configuration at $t = 0$ (see Fig. 1.a) is a completely mixed configuration as equilibration for $\Delta = 0$ and $F = 0$. One clearly sees an anisotropic coarsening due to the external drive. Two extreme limit can be understood in more detail: first, if the driving field is much smaller than the fluid-fluid equilibrium line tension γ, then the traditional isotropic phase separation will dominate at small times. At an interface the external field will then lead to a Rayleigh-Taylor instability leading to finger fomation inside the phase separated region. This is presumbaly what has been seen in experiments of sedimenting colloid-polymer mixtures [11] and was checked for pure interfacial situations [21]. On the other hand, if the driving force is much larger than γ, the system directly relaxes into the laning state.

CRYSTAL NUCLEATION UNDER SHEAR

In the last years, remarkable progress has been made to calculate the free energy barrier for crystal nucleation via smart simulation methods using the umbrella sampling technique. In three spatial dimensions, results for the homogeneous crystal nucleation rate and the structure of the critical nucleus were obtained for Lennard-Jones sytems [22], hard spheres [23] and Yukawa particles [24]. Under linear shear flow of a given shear-rate $\dot{\gamma}$, the nucleation rate is expected to change drastically since usually a crystal is getting less stable with respect to a fluid phase (shear-thinning or shear-induced melting).

In a recent work [25], Brownian dynamics computer simulations of charged colloids as modelled by a Yukawa interaction without hydrodynamic interactions have been performed to address this problem. The pair potential reads as (see Eqn. (1))

$$\frac{V(r)}{k_B T} = U_0 \sigma \frac{\exp(-\kappa(r-\sigma))}{r}. \quad (6)$$

The negative logarithm of the probability to find a solid-like cluster containing n solid-like particle normalized to unity for $n = 1$ is plotted versus n in Figure 2. It is tempting to interpret this data as a free energy even in the non-equilibrium steady-state situation setting the barrier height and the critical nucleus size. We have tested our Brownian dynamics data in the zero-shear limit against Monte-Carlo data and find good agreement, see again Fig. 2. For increasing shear rates $\dot{\gamma}$, the barrier and the cluster size do increase. Further simulations will explore the structure of the critical nucleus and will compare to data to classical nucleation theory.

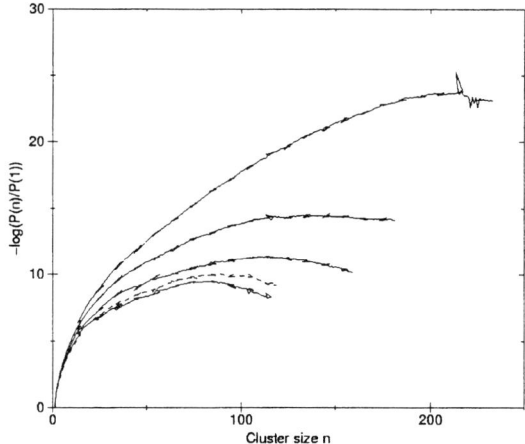

FIGURE 2. The negative logarithm of the probability $P(n)$ of finding a cluster which contains n solid particles, normalized to unity for $n = 1$. The solid curves correspond to Brownian dynamics simulations at different shear rates, the dashed line is the result for Monte Carlo simulations without shear. Simulations were performed for $\kappa \sigma = 5$, $U_0 \sigma = 20$, pressure is $\frac{P\sigma^3}{k_B T} = 30$ and from top to bottom shear rates $\dot{\gamma} \tau_B = 0.04, 0.02, 0.01, 0$.

TWO-DIMENSIONAL MAGNETIC COLLOIDAL SUSPENSIONS IN AN EXTERNAL MAGNETIC FIELD

Systems of colloidal particles at a liquid–gas interface controlled by magnetic interactions are valuable realizations of two dimensional model systems to study the properties of their phase transitions and response to external fields. Here we consider two-dimensional macroscopic assemblies of paramagnetic particles, each carrying a magnetic moment $\mathbf{m}_i$, under the influence of an arbitrary external magnetic field $\mathbf{B}$.

The physical setup is schematically depicted in Fig. 3. The total potential energy of the system reads as:

$$V_{tot} = \sum_{i<j} \left[u_0(r_{ij}) + u_{\text{dd}}(\mathbf{r}_{ij}, \mathbf{m}_i, \mathbf{m}_j) \right] - \sum_i \mathbf{B} \cdot \mathbf{m}_i, \quad (7)$$

where $u_0(r)$ is a truncated and shifted Lenard-Jones potential which reads as

$$u_0(r) = \begin{cases} 4\varepsilon\left(\left(\frac{\sigma_0}{r}\right)^{12} - \left(\frac{\sigma_0}{r}\right)^6\right) + \varepsilon & \text{for } r \leq \sqrt[6]{2}\sigma_0 \\ 0 & \text{else.} \end{cases} \quad (8)$$

with length and energy scale σ_0 and ε and the dipole-dipole interaction is:

$$u_{\text{dd}}(\mathbf{r}, \mathbf{m}_i, \mathbf{m}_j) = -\frac{1}{r^3}\left[3(\hat{\mathbf{r}} \cdot \mathbf{m}_i)(\hat{\mathbf{r}} \cdot \mathbf{m}_j) - \mathbf{m}_i \cdot \mathbf{m}_j\right]. \quad (9)$$

Of particular interest are so-called *super-paramagnetic* particles [26] for which the magnetic moment completely aligns with the external field, if the latter is strong

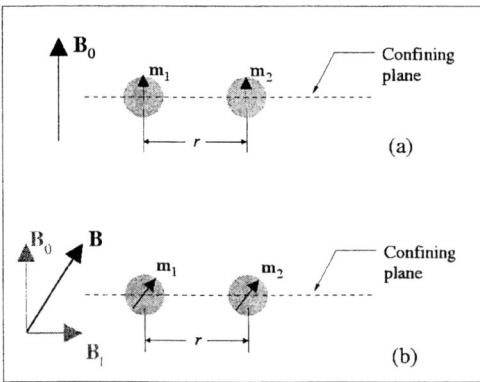

FIGURE 3. Sketch of two super-paramagnetic particles confined on a plane, under the influence of an external magnetic field **B**. In (a), the field is perpendicular to the plane and only the the isotropic part, $\sim r^{-3}$ of the interaction u_{dd} survives. In (b), there is an in-plane field component, rendering the interaction anisotropic, $u = u(\mathbf{r})$ in this case.

enough. In this case, we have $\mathbf{m}_i = \chi\mathbf{B}$, the last term in Eq. (7) becomes an irrelevant constant $N\chi B^2$ and the magnetic field plays the role of tuning the repulsions between the particles through its influence on the magnitude and orientation of the $\mathbf{m}_i$'s see Eq. (9) above.

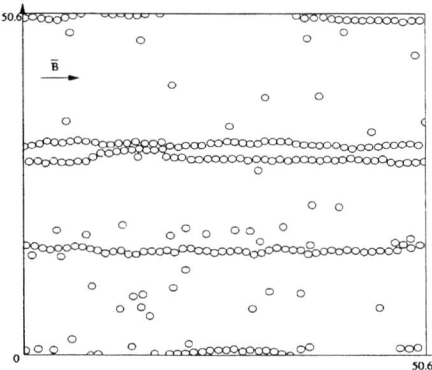

FIGURE 4. Chain formation in two dimensional dipolar colloids as obtained via a Molecular dynamics simulation. The starting configuration was homogeneous. The total simulation time was $35000t^*$ with $t^* = t\sqrt{\varepsilon}/\sigma/\sqrt{m}$ where m denotes the mass of the particles.

Novel experimental methods have shed new light on the statics and dynamics of two-dimensional systems of interacting magnetic colloids [26, 27, 28]. In the following we shall concentrate of the chain formation of two-dimensional super-paramagnetic colloids with an in-plane external magetic field where other preliminary computer simulations have already been published [29]. Snapshots of molecular dynamics simulations of $N = 256$ magnetic particles in an square box are presented in Figure 4. The parameters are number density $n^* = n\sigma_0^2 = 0.1$, the angle between the magnetic field and the plane is zero. The relative strength of the dipol interaction with respect to the thermal energy k_BT is $\lambda = m^2/\sigma^3 kT = 8$.

Starting from a homogeneous disordered configuration, one clearly sees the formation of chain-like configurations. The results are obtained via molecular dynamics. Although this is not the proper dynamics of the colloids, it provides nevertheless qualitative insight into the dynamics of chain formation.

CONCLUSIONS

In conclusion, we have briefly described three different examples of slow dynamics in colloidal suspensions driven by an external field: i) phase separation kinetics under an external driving field, ii) crystal nucleation under shear, and iii) chain formation in an external magnetic field. They all demonstrate that the formation of new complex structures such as phase-separating patterns and critical nuclei which initiate crystal birth are formed on time-scales which are considerably slower than the typical time scale characterizing single particle diffusion. The self-assembly of many particles leads to interesting structures on intermediate transient time-scales.

We think that colloidal suspensions in non-equilibrium will be valuable model systems to study further fundamental questions of slow dynamics. In particular, pattern formation in non-equilibrium and glass and gel formation in external fields are key areas in which progress can be expected in the near future.

ACKNOWLEDGMENTS

I thank D. Frenkel for helpful comments. Financial support from the DFG (Sonderforschungsbereich TR6) is gratefully acknowledged.

REFERENCES

1. Löwen, H., *J. Phys.: Condensed Matter* **13**, R415-R432 (2001).
2. Sullivan, M., Zhao, K., Harrison, C., Austin, R. H., Megens, M., Hollingsworth, A., Russel, W. B., Cheng, Z. D., Mason, T., and Chaikin, P. M., *J. Phys.: Condensed Matter* **15**, S11-S18 (2003).
3. Helbing, D., *Rev. Mod. Phys.* **73**, 1067-1142 (2001).
4. Anderson, D., Fedele, R., and Lisak, M., *American Journal of Physics* **69**, 1262-1266 (2001).
5. Yethiraj, A., and van Blaaderen, A., *Nature* **421**, 513-517 (2003).
6. Osipov, M. A., Teixeira, P. I. C., and Telo da Gama, M. M., *Phys. Rev. E* **54**, 2597-2609 (1996).

7. Groh, B., and Dietrich, S., *Phys. Rev. E* **63**, 021203 (1-11) (2001).
8. Weis, J. J., Tavares, J. M., and Telo da Gama, M. M., *J. Phys.: Condensed Matter* **14**, 9171-9186 (2002).
9. Onuki, A., *Phase Transition Dynamics*, Cambridge University Press, Cambridge, 2002.
10. Bray, A. J., *Philos. Trans. Roy. Soc. A* **361**, 781-792 (2003).
11. Aarts, D. G. A. L., van der Wiel, J. H., and Lekkerkerker, H. N. W., *J. Phys.: Condensed Matter* **15**, S245-S250 (2003).
12. Dzubiella, J., Hoffmann, G. P., and Löwen, H., *Phys. Rev. E* **65**, 021402 (1-6) (2002).
13. Dzubiella, J., and Löwen, H., *J. Phys.: Condensed Matter* **14**, 9383-9395 (2002).
14. Löwen, H., and Dzubiella, J., *Faraday Discussion* **123**, 99-105 (2003).
15. Chakrabarti, J., Dzubiella, J., and Löwen, H., *Europhys. Letters* **61**, 415-422 (2003).
16. Hoffmann, G. P., and Löwen, H., *Phys. Rev. E* **60**, 3009-3014 (1999).
17. Hoffmann, G. P., and Löwen, H., *J. Phys.: Condens. Matter* **12**, 7359-7370 (2000).
18. Löwen, H., Hansen, J. P., and Roux, J. N., *Phys. Rev. A* **44**, 1169-1181 (1991).
19. Allen, M. P., and Tildesley, D. J., *Computer Simulations of Liquids*, Clarendon Press, Oxford, 1989.
20. Ermak, D. L., *J. Chem. Phys* **62**, 4189-4196 (1975).
21. Wysocki, A., and Löwen, H., to be published.
22. ten Wolde, P. R., Montero-Ruiz, M., and Frenkel, D., *Phys. Rev. Lett.* **75**, 2714-2717 (1995).
23. Auer, S., and Frenkel, D., *Nature* **409**, 1020-1023 (2001).
24. Auer, S., and Frenkel, D., *J. Phys.: Condensed Matter* **14**, 7667-7680 (2002).
25. Auer, S., Blaak, R., Frenkel, D., and Löwen, H., to be published.
26. Zahn, K., Lenke, R., and Maret, G., *Phys. Rev. Lett.* **82**, 2721-2724 (1999).
27. Zahn, K., Méndez-Alcaraz, J. M., and Maret, G., *Phys. Rev. Lett.* **79**, 175-178 (1997).
28. Zahn, K., and Maret, G., *Phys. Rev. Lett.* **85**, 3656-3659 (2000).
29. Satoh, A., Chantrell, R. W., Kamiyama, S.-I., and Coverdale, G. N., *Journal of Colloid and Interface Science* **178**, 620-627 (1996).

Universal Features of Collective Interactions in Hard-Sphere Systems at Higher Volume Fractions

M. Tokuyama*, Y. Terada*, H. Yamazaki[†] and I. Oppenheim**

Institute of Fluid Science, Tohoku University, Sendai 980-8577, Japan
[†]*Fuji Photo Film Co. Ltd., Nishiazabu 2-26-30, Tokyo 106-8620, Japan*
**Department of Chemistry, Massachusetts Institute of Technology, Cambridge, MA 02139, USA*

Abstract. In order to investigate the universal features of collective behavior due to the many-body interactions, we perform two types of computer simulations on hard-sphere systems, a Brownian-dynamics simulation on polydisperse suspensions of hard spheres, where the hydrodymamic interactions between particles are neglected, and a molecular-dynamics simulation on atomic systems of hard spheres. Thus, we show that the long-time self-diffusion coefficient in atomic systems has the same form as that derived theoretically by Tokuyama and Oppenheim (TO) for the monodisperse suspension by taking into account the many-body hydrodynamic interactions, except that the singular point is now replaced by a new one. We also show that the difference between two coefficients in both systems can be well explained by the short-time self-diffusion coefficient derived theoretically for a wide range of volume fractions.

INTRODUCTION

We consider N different hard spheres with radius a_i and mass m_i ($i = 1,.....N$) in a cubic box of length L at a constant temperature T. We assume that the distribution of radii obeys a Gaussian distribution with the standard deviation σ divided by the average radius a and the mass m_i is proportional to a_i^3. Then, the particle volume fraction is given by $\phi_{eq} = (4\pi a^3 n_{eq}/3)(1 + 3\sigma^2)$, where $n_{eq} = N/L^3$. In the present paper, we discuss three different systems. The first is a suspension of monodisperse hard spheres with both hydrodynamic and direct interactions between particles. This was theoretically analyzed by Tokuyama and Oppenheim (TO) [1, 2]. The second is suspensions of monodisperse and polydisperse hard spheres without hydrodynamic interactions. The third is atomic systems of monodisperse and polydisperse hard spheres. The last two systems are investigated by computer simulations. Thus, we show that the long-time collective behavior is universal for different systems.

BASIC EQUATIONS FOR SUSPENSIONS OF HARD SPHERES

We first summarize and discuss the basic equations for a monodisperse suspension of hard spheres with $a_i = a$ and $m_i = m$, where the spheres are suspended in an equilibrium fluid with a viscosity η_0 and $\sigma = 0$. The present system has four characteristic lengths and times [1, 2]; the molecular radius r_m, the microscopic time t_m, the average moving distance of a particle $r_B (= a(t_B/t_D)^{1/2})$, the Brownian relaxation time $t_B (= m/\zeta_0)$, the screening length $\ell_H (= (6\pi a n_{eq})^{-1/2})$, within which the hydrodynamic interactions between particles become important, the screening time $t_H (= \rho a^2/\eta_0 \phi)$, in which the hydrodynamic interactions become important, the particle radius a, and the structural-relaxation time $t_D (= a^2/D_0)$, which is a time required for a particle to diffuse over a distance a, where ρ is the fluid mass density, $D_0 (= k_B T/\zeta_0)$ the single-particle diffusion coefficient, and $\zeta_0 (= 6\pi \eta_0 a)$ the friction coefficient. In this paper we deal with concentrated suspensions in which the following inequalities hold: $r_m \ll r_B \ll \ell_H \leq a$ and $t_m \ll t_B \ll t_H \ll t_D$. Depending on the space-time scales of interest, therefore, there exist two characteristic stages. The first is a kinetic stage [K], where the space-time cutoffs (r_c, t_c), which are the minimum wavelength and time of the dynamic process of interest, are set as $r_m \ll r_c \leq \ell_H$ and $t_m \ll t_c \leq t_H$. The second is a suspension-hydrodynamic stage [SH], where $\ell_H \leq a \ll r_c$ and $t_H \ll t_c \leq t_D$.

We first review the Langevin equations in a kinetic stage [K]. Let $\mathbf{X}_i(t)$ and $\mathbf{u}_i(t)$ denote the position and the velocity of the ith particle at time t, respectively. Then, those are described by the Langevin equations discussed elsewhere [1, 2]

$$\frac{d}{dt}\mathbf{X}_i(t) = \mathbf{u}_i(t). \qquad (1)$$

$$m\frac{d}{dt}\boldsymbol{u}_i(t) = -\sum_{j=1}^{N}\boldsymbol{\zeta}(\boldsymbol{X}_{ij}(t))\cdot\boldsymbol{u}_j(t)$$
$$+\sum_{j(\ne i)}\boldsymbol{F}(\boldsymbol{X}_{ij}(t)) + \boldsymbol{R}(\boldsymbol{X}_i(t),t), \quad (2)$$

where $\boldsymbol{F}(\boldsymbol{X}_{ij}(t))$ is the force between particles i and j, and $\boldsymbol{X}_{ij} = \boldsymbol{X}_i - \boldsymbol{X}_j$. Here the random force $\boldsymbol{R}(\boldsymbol{X}_i(t),t)$ obeys a Gaussian, Markov process with zero mean and satisfies

$$<\boldsymbol{R}(\boldsymbol{X}_i,t)\boldsymbol{R}(\boldsymbol{X}_j,t')> = 2k_BT\boldsymbol{\zeta}(\boldsymbol{X}_{ij})\delta(t-t'), \quad (3)$$

where the brackets $<\cdots>$ indicate an equilibrium ensemble average. The friction tensors satisfy

$$\boldsymbol{\zeta}(\boldsymbol{X}_{ij}) = \zeta_0\boldsymbol{1}\delta_{ij} - \sum_{k=1}^{N}\boldsymbol{\zeta}(\boldsymbol{X}_{ik})\cdot\boldsymbol{g}(\boldsymbol{X}_{kj})(1-\delta_{kj}), \quad (4)$$

where the tensors $\boldsymbol{g}(\boldsymbol{X}_{ij})$ represent the solvent-mediated hydrodynamic interactions between particles and lead to corrections to the friction coefficient ζ_0. The explicit forms of $\boldsymbol{g}(\boldsymbol{X}_{ij})$ are given, to order $(a/X_{ij})^3$, by

$$\boldsymbol{g}(\boldsymbol{X}_{ij}) = \frac{3}{4}\frac{a}{X_{ij}}(\boldsymbol{1}+\boldsymbol{x}_{ij}\boldsymbol{x}_{ij})$$
$$+ \frac{1}{2}\left(\frac{a}{X_{ij}}\right)^3(\boldsymbol{1}-3\boldsymbol{x}_{ij}\boldsymbol{x}_{ij}) + O(a/X_{ij})^4, \quad (5)$$

respectively, where $\boldsymbol{x}_{ij} = \boldsymbol{X}_{ij}/X_{ij}$ and $X_{ij} = |\boldsymbol{X}_{ij}|$. The first term of Eq. (5) represents the Oseen tensor and the second the dipole tensor. Here the hydrodynamic interactions $\boldsymbol{g}(\boldsymbol{r})$ between particles can be classified into two types, depending on the range of interactions; the long-range hydrodynamic interactions between particles over a distance of order ℓ_H, which lead to divergent integrals, and the short-range hydrodynamic interactions between particles over a distance of order a. On the other hand, the force $\boldsymbol{F}(\boldsymbol{r})$ in Eq. (2) represents the direct (collision) interactions between particles over a distance of order a. Equations (1) and (2) are the basic equations to discuss the colloidal suspension of concentrated hard spheres in stage [K]. Because of the long-range interactions, however, it is beyond our capacity to deal with Eq. (2) analytically. Hence one must further reduce them to obtain a more macroscopic equation.

We now review a formal derivation of a nonlinear stochastic diffusion equation for the number density in a suspension-hydrodynamic stage [SH]. As discussed in the previous paper [2], the relevant variable to describe the dynamics of a colloidal suspension in the stage [SH] is given by the number density

$$N(\boldsymbol{r},t) = \sum_{i=1}^{N}\Delta(\boldsymbol{X}_i(t)-\boldsymbol{r}), \quad (6)$$

where $\Delta(\boldsymbol{r})(= L^{-3}\sum_{|\boldsymbol{k}|\le 1/r_c}\exp(-i\boldsymbol{k}\cdot\boldsymbol{r}))$ indicates the coarse-grained δ function. Taking a time derivative of $N(\boldsymbol{r},t)$ then leads to

$$\frac{\partial}{\partial t}N(\boldsymbol{r},t) = -\frac{1}{m}\nabla\cdot\boldsymbol{P}(\boldsymbol{r},t), \quad (7)$$

where the momentum density is given by $\boldsymbol{P}(\boldsymbol{r},t) = \sum_{i=1}^{N}m\boldsymbol{u}_i(t)\Delta(\boldsymbol{X}_i(t)-\boldsymbol{r})$. As discussed elsewhere [2], in order to find a closed equation for $N(\boldsymbol{r},t)$, one further needs another equations for $\boldsymbol{P}(\boldsymbol{r},t)$ and the energy density given by $\boldsymbol{E}(\boldsymbol{r},t) = (1/2)m\sum_{i=1}^{N}\boldsymbol{u}_i(t)\boldsymbol{u}_i(t)\Delta(\boldsymbol{X}_i(t)-\boldsymbol{r})$. Taking a time derivative of those densities, using Eqs. (1) and (2), and employing a projection operator method [3] to eliminate the irrelevant processes related to the kinetic processes, such as the terms $N(\boldsymbol{r},t)\boldsymbol{R}(\boldsymbol{r},t)$ and $\boldsymbol{P}(\boldsymbol{r},t)\boldsymbol{R}(\boldsymbol{r},t)$, in an appropriate manner, one can obtain, up to lowest order in ∇,

$$\frac{\partial}{\partial t}\boldsymbol{P}(\boldsymbol{r},t) = -\frac{1}{t_B}\left[\boldsymbol{P}(\boldsymbol{r},t) + 2t_B\nabla\cdot\boldsymbol{E}(\boldsymbol{r},t)\right.$$
$$-N(\boldsymbol{r},t)\int d\boldsymbol{r}'\left\{\int d\boldsymbol{r}''\frac{\boldsymbol{\zeta}(\boldsymbol{r}-\boldsymbol{r}')}{\zeta_0}\cdot\boldsymbol{g}(\boldsymbol{r}'-\boldsymbol{r}'')\boldsymbol{P}(\boldsymbol{r}'')\right.$$
$$\left.\left.+\frac{m}{k_BT}D_0\boldsymbol{F}(\boldsymbol{r}-\boldsymbol{r}')\right\}N(\boldsymbol{r}',t) + \boldsymbol{f}^p(\boldsymbol{r},t)\right], \quad (8)$$

$$\frac{\partial}{\partial t}\boldsymbol{E}(\boldsymbol{r},t) = -\frac{1}{t_B}\left[2\boldsymbol{E}(\boldsymbol{r},t) - k_BTN(\boldsymbol{r},t) + \boldsymbol{f}^e(\boldsymbol{r},t)\right]. \quad (9)$$

Here the new random forces $\boldsymbol{f}^\alpha(\boldsymbol{r},t)(\alpha = p,e)$ satisfies

$$<\boldsymbol{f}^\alpha(\boldsymbol{r},t)\cdot\boldsymbol{f}^\beta(\boldsymbol{r}',t')> = 2D_0\delta(t-t')$$
$$\times \int\frac{d\boldsymbol{r}'}{V}\frac{\boldsymbol{\zeta}(\boldsymbol{r}-\boldsymbol{r}')}{\zeta_0}\boldsymbol{A}^\alpha(\boldsymbol{r},t)\cdot\boldsymbol{A}^\beta(\boldsymbol{r}',t), \quad (10)$$

where $\boldsymbol{A}^p(\boldsymbol{r},t) = N(\boldsymbol{r},t)\boldsymbol{1}$ and $\boldsymbol{A}^e(\boldsymbol{r},t) = \boldsymbol{P}(\boldsymbol{r},t)$. We note here that Eqs. (8) and (9) contains two types of interactions, the hydrodynamic interactions through $\boldsymbol{g}(\boldsymbol{r})$ and the direct interactions through $\boldsymbol{F}(\boldsymbol{r})$.

In order to solve Eqs. (8) and (9) for $N(\boldsymbol{r},t)$ self-consistently, one may first use two types of expansions, the expansion in the spatial gradient $\nabla(\propto a/r_c)$ and the expansion in the slowness parameter $\partial/\partial t(\propto t_B/t_c)$ [2, 4]. In fact, we find that on the time scale of order t_D, $\partial\boldsymbol{P}(\boldsymbol{r},t)/\partial t = \partial\boldsymbol{E}(\boldsymbol{r},t)/\partial t \simeq \boldsymbol{0}$. Then, one may further apply the projector method again for a reduced equation for $N(\boldsymbol{r},t)$ to eliminate the irrelevant processes. After a long calculation, one can finally obtain a nonlinear stochastic equation for the local volume fraction given by $\Phi(\boldsymbol{r},t)(= 4\pi a^3 N(\boldsymbol{r},t)/3)$, up to order ∇^2, [4]

$$\frac{\partial}{\partial t}\Phi(\boldsymbol{r},t) = \nabla\cdot\left[D_S^l(\Phi(\boldsymbol{r},t))\nabla\Phi(\boldsymbol{r},t)\right] + \boldsymbol{\xi}(\boldsymbol{r},t), \quad (11)$$

where $\xi(r.t)$ denotes the Gaussian, Markov random force (see Ref. [4] for details). Here the long-time self diffusion coefficient $D_S^L(\Phi)$ is given by [1, 2]

$$D_S^L(\Phi) = D_S^S(\Phi) \frac{1 - \frac{9}{32}\Phi}{1 + \left(\frac{\Phi D_S^S}{\phi_g^{TO} D_0}\right)\left(1 - \frac{\Phi}{\phi_g^{TO}}\right)^{-\gamma}}, \quad (12)$$

where $\gamma = 2$ here. Here D_S^S is the short-time self-diffusion coefficient given by [1]

$$D_S^S(\Phi) = D_0/[1 + h(\Phi)] \quad (13)$$

with the non-memory (short-time) hydrodynamic effect

$$h(\Phi) = \frac{2A^2}{1-A} - \frac{B}{1+2B} - \frac{AB(2+B)}{(1+B)(1-A+B)}. \quad (14)$$

where $A = \sqrt{9\Phi/8}$ and $B = 11\Phi/16$. The factor $(9\Phi/32)$ in Eq. (12) represents the coupling effect between the direct interactions and the hydrodynamic interactions, while it reduces to 2Φ for binary collisions without the hydrodynamic interactions. Here ϕ_g^{TO} is the theoretical glass transition volume fraction given by

$$\phi_g^{TO} = (4/3)^3/(7\ln 3 - 8\ln 2 + 2) \approx 0.57184\cdots. \quad (15)$$

We note here that the singular term in the denominator of Eq. (12) results from the many-body, long-range hydrodynamic interactions through the Oseen tensor. Hence the diffusion coefficient $D_S^L(\Phi(r.t))$ becomes smaller and smaller near ϕ_g^{TO} as time goes on since $D_S^L(\phi_g^{TO}) = 0$. This is the so-called dynamic anomaly.

Equation (11) is a starting equation to describe the dynamics of diffusion processes in colloidal suspensions at higher volume fractions. It enables us to describe not only the dynamics of spatial heterogeneities but also the dynamics of density fluctuations near the glass transition. It can be solved in two cases, a nonequilibrium case and an equilibrium case, separately. This is discussed next.

Let decompose $\Phi(r.t)$ into an average part $\phi(r.t)$ and a fluctuating part $\delta\phi(r.t)$; $\Phi(r.t) = \phi(r.t) + \delta\phi(r.t)$. Here $\overline{\delta\phi(r.t)} = 0$, where the bar denotes the average over an appropriate initial ensemble. Then, $\phi(r.t) \neq \phi_{eq}$ for a nonequilibrium case and $\phi(r.t) = \phi_{eq}$ for an equilibrium case. This decomposition is essential since the relative magnitude of the density fluctuation to the average density is small even near the colloidal glass transition, $|\delta\phi(r.t)/\phi(r.t)| \ll 1$. This is because the glass transition is not a critical phenomenon since there exist no correlation length diverging even at the glass transition point.

In a nonequilibrium case where $\phi(r.t) \neq \phi_{eq}$, one can decompose Eq. (11) into the nonlinear deterministic diffusion equation for $\phi(r.t)$

$$\frac{\partial}{\partial t}\phi(r.t) = \nabla \cdot \left[D_S^L(\phi(r.t))\nabla\phi(r.t)\right], \quad (16)$$

and the linear stochastic diffusion equation for $\delta\phi(r,t)$

$$\frac{\partial}{\partial t}\delta\phi(r.t) = \nabla^2\left[D_S^L(\phi(r.t))\delta\phi(r.t)\right] + \xi(r.t). \quad (17)$$

The dynamics of spatial heterogeneities is described by the solution of Eq. (16). On the other hand, the density fluctuations are described by Eq. (17) and are observed through the self-intermediate scattering function given by $F_s(k.t) = \int dr \exp(ik \cdot r) < \delta\phi(r.t)\delta\phi(0.0) > /(4\pi a^3/3)^2$.

In order to solve Eq. (16) numerically, one must fix the values of two parameters, ϕ_{eq} and z_0, as the initial conditions. Here z_0 measures how the system is spatially nonuniform initially and is given by

$$z_0 = 1 - \frac{1}{L^3}\int dr \left|1 - \frac{\phi(r,t=0)}{\phi_{eq}}\right|. \quad (18)$$

where $z_0 = 1.0$ in an equilibrium state. Starting from nonequilibrium random configurations, the smoothing process of $\phi(r.t)$ starts to occur, leading to the equilibrium volume fraction ϕ_{eq} for long times of order $t_L(= a^2/D_S^L(\phi_{eq}))$. As discussed in Ref. [5], there exist four characteristic time stages near the glass transition. The first is the early stage where $t \ll t_\gamma(= a^2/D_S^S(\phi_{eq}))$. The spatial configurations are random and are described by $\phi(r.t) = \exp[tD_S^S\nabla^2]\phi(r.0)$. The density fluctuations are described by Eq. (17) and obeys the exponential decay $F_S(k.t) = \exp[-k^2 D_S^S t]$. After this stage, the finite-sized, glassy domains with $\phi(r.t) \geq \phi_g^{TO}$ are formed for $\phi_\beta \leq \phi_{eq} < \phi_g^{TO}$, where ϕ_β indicates the crossover volume fraction [5]. The smoothing process of $\phi(r.t)$ is then slowing down due to those domains. On the time scale of order t_β, therefore, those glassy domains affect the dynamics of the density fluctuations. This is the so-called β-relaxation stage where $t_\gamma \ll t \leq t_\alpha$, where t_α denotes the α-relaxation time. Depending on the time scale, the scattering function $F_S(k.t)$ obeys two types of power-law decays. In the fast β-relaxation stage where $t_\gamma \ll t \leq t_\beta$, small aggregates of glassy domains are formed and influence the dynamics of the density fluctuations, leading to the critical decay

$$F_s(k.t) = f_0 - f_1(t/t_\beta)^{b'}. \quad (19)$$

where $b'(0 < b' < 1)$ is an exponent to be dertermined, t_β the β-relaxation time, and f_i a positive constant. This decay continues up to the time scale of order t_β. In the slow β-relaxation stage ($t_\beta \leq t \leq t_\alpha$), the aggregates grow to the larger clusters and lead to the power-law decay of von Schweidler type

$$F_s(k.t) = f_2 - f_3(t/t_\beta)^b. \quad (20)$$

where $b(0 < b < 1)$ is an exponent to be dertermined. After this stage, the glassy domains further grow to larger

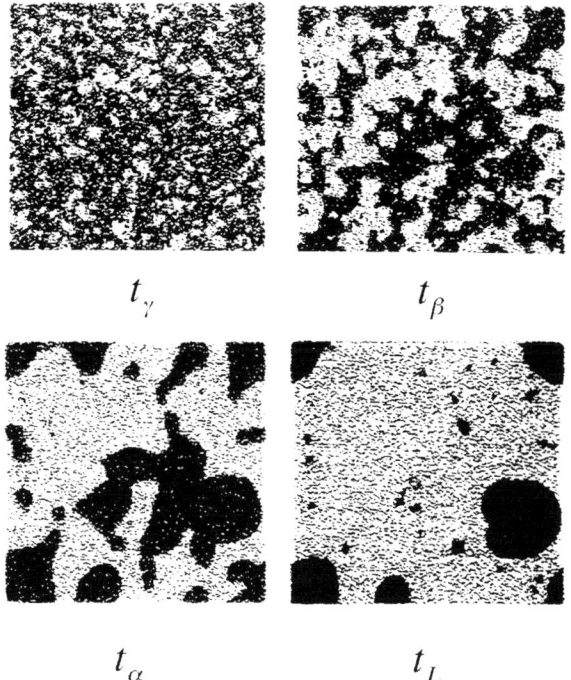

FIGURE 1. Snapshots of nonequilibrium configurations $\phi(\mathbf{r},t)$ in a $x-y$ plane for times $t_\gamma/t_D \simeq 4, t_\beta/t_D \simeq 800, t_\alpha/t_D \simeq 2 \times 10^4$, and $t_L/t_D \simeq 4 \times 10^5$ at $\phi_{eq} = 0.566$ and $z_0 = 0.8$. The glassy regions with $\phi(\mathbf{r},t) \geq \phi_g^{TO}$ are colored black, while the liquid regions are colored white.

TABLE 1. Time exponents for $\phi_{eq} = 0.566$ at $k = 1.3a$.

	b'	b	β	η
nonequilibrium ($z_0 = 0.8$)	0.37	0.72	0.59	3.41
equilibrium ($z_0 = 1.0$)	—	0.97	0.95	2.1

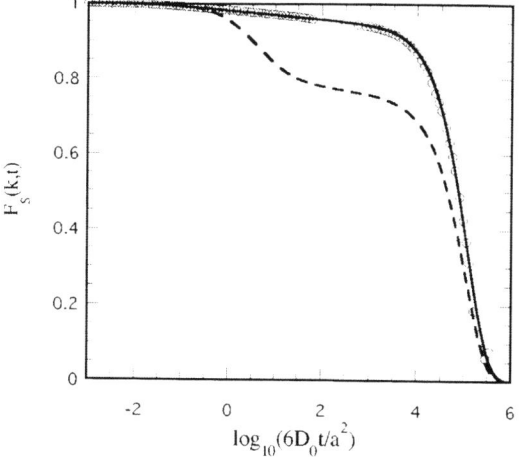

FIGURE 2. $F_S(k,t)$ versus time for $\phi_{eq} = 0.566$ at $k = 1.3a$. The dashed line indicates the nonequilibrium solution of Eq. (17) and the solid line the equilibrium solution given by Eq. (26). The symbols indicate the experimental data from Ref. [6]

clusters and affect the dynamics of the density fluctuations, leading to the stretched exponential decay of Kohlrausch-Williams-Watts type

$$F_s(k,t) \sim f_4 \exp[-(t/t_\alpha)^\beta], \qquad (21)$$

where $\beta (0 < \beta < 1)$ is a stretched exponent to be dertermined. This is the so-called α-relaxation stage ($t_\alpha \leq t \leq t_L$). Here we note that the α-relaxation time t_α satisfies $t_\alpha \propto |1 - \phi_{eq}/\phi_g^{TO}|^{-\eta}$, where $\eta = \gamma/\beta$. After this stage, the glassy domains disappear very slowly and the system gradually reaches the equilibrium state on the time scale of order t_L. Then, the spatial configurations become random and the density fluctuations obey the exponential decay $F_S(k,t) = \exp[-k^2 D_S^L t]$. The time exponents are listed in Table 1. In Fig. 1, the snapshot of the long-lived spatial heterogeneities is shown. The long-lived glassy clusters are shown to exist near ϕ_g^{TO} [5]. Figure 2 shows the numerical solution of Eq. (17).

In an equilibrium case where $\phi(\mathbf{r},t) = \phi_{eq}$, one can expand Eq. (11) in powers of $|\delta\phi/\phi_{eq}|$ near ϕ_g^{TO} and obtain the nonlinear stochastic diffusion equation for $\delta\phi(\mathbf{r},t)$, up to order $|\delta\phi/\phi_{eq}|^3$, [4]

$$\frac{\partial}{\partial t}\delta\phi(\mathbf{r},t) = \nabla^2 \left[D_S^L(\phi_{eq}) \delta\phi(\mathbf{r},t) \right.$$
$$+ D_S^S(\phi_{eq}) \left\{ -\omega(\phi_{eq})\delta\phi(\mathbf{r},t)^2 \right.$$
$$\left. \left. + \kappa(\phi_{eq})\delta\phi(\mathbf{r},t)^3 \right\} \right] + \xi(\mathbf{r},t). \quad (22)$$

where the random force $\xi(\mathbf{r},t)$ satisfies

$$<\xi(\mathbf{r},t)\xi(\mathbf{r}',t')> = 2\delta(t-t')\phi_{eq}D_S^L(\phi_{eq})$$
$$\times \nabla \cdot \nabla' \left[\frac{4\pi}{3}a^3\delta(t-t') + \phi_{eq}\{g(|\mathbf{r}-\mathbf{r}'|) - 1\} \right]. \quad (23)$$

Here $g(r)$ represents the radial distribution function and the coefficients ω and κ are known smooth functions of ϕ_{eq}. Since $g(r)$ is not known, Eq. (22) must be calculated self-consistently.

In contrast to the nonequilibrium case, the long-lived spatial heterogeneities are caused by nonlinear fluctuations. Although they lead to a distinct two-step relaxation, only the glassy domains related to the α-relaxation process seem to survive because the small domains related to the β-relaxation process are easily destroyed by

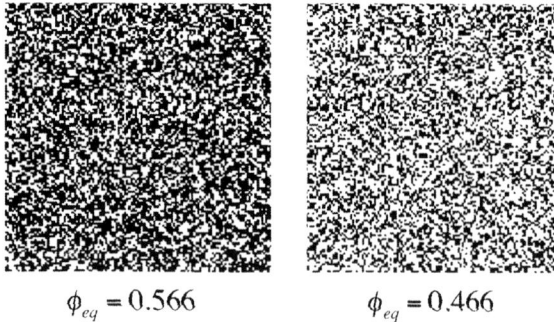

$\phi_{eq} = 0.566$ $\phi_{eq} = 0.466$

FIGURE 3. Snapshots of the density fluctuations, projected onto a $x - y$ plane, for different volume fractions. The glassy domains are clored black and the liquid domains are white.

fluctuations. In a nonequilibrium case, the clusters are stable for the fluctuations. This difference is clearly seen in Table 1.

Because the radial distribution function $g(r)$ is not known, it is rather difficult to calculate Eq.(22) numerically. In Fig. 3 the snapshots of the density fluctuations are shown for the special case where $g(r)$ is assumed to be one. Near the glass transition, the nonlinear fluctuations play an important role, forming the clusters of glassy domains, while away from the glass transition only the linear fluctuations exist, causing random configurations. In suspensions, the self-intermediate scattering function $F_S(k,t)$ can be written as $F_S(k,t) = \exp[-k^2 M_2(t)/6]$ [7], where $M_2(t)$ denotes the mean-square displacement given by

$$M_2(t) = \frac{1}{N}\sum_{i=1}^{N} <[\boldsymbol{X}_i(t) - \boldsymbol{X}_i(0)]^2>. \quad (24)$$

Instead of solving Eq. (22), therefore, one can derive a mean-field nonlinear equation for $M_2(t)$ [8]

$$\frac{d}{dt}M_2(t) = 6D_S^L(\phi_{eq}) + 6[D_S^S(\phi_{eq}) - D_S^L(\phi_{eq})]e^{-\lambda M_2(t)}. \quad (25)$$

where $\lambda(\phi_{eq})$ is a free parameter to be determined from the fitting with experimental data. Equation (25) can be easily solved to give

$$M_2(t) = \frac{1}{\lambda}\ln\left[1 + \frac{D_S^S}{D_S^L}\left\{e^{t/t_\beta} - 1\right\}\right], \quad (26)$$

where $t_\beta = 1/(6\lambda D_S^L)$. For short times $t \ll t_0 (= 1/(6\lambda D_S^S))$, Eq. 26 reduces to $M_2(t) \simeq 6D_S^S t$, leading to the exponential decay $F_S(k,t) = \exp[-k^2 D_S^S t]$. In the fast β-relaxation stage where $t_0 \leq t \leq t_\beta$, the scattering function obeys the logarithmic decay

$$F_s(k,t) = g_0 - g_1 \ln(D_0 t/a^2), \quad (27)$$

where $g_0 = (1+t_D/t_0)^{-a_k}$ and $g_1 = a_k g_0/(1+t_0/t_D)$. Here $a_k = k^2/(6\lambda)$. In the slow β-relaxation stage where $t_\beta \leq t \leq t_\alpha$, it obeys the power-law decay of von Schweidler type

$$F_s(k,t) = (1+t/t_0)^{-a_k} - g_2(t/t_\beta)^b. \quad (28)$$

where g_2 is a positive constant. In the α-relaxation stage where $t_\alpha \leq t \leq t_L$, it obeys the stretched exponential decay of Kohlrausch-Williams-Watts type

$$F_s(k,t) = (1+t/t_0)^{-a_k}\exp[-(t/t_\alpha)^\beta]. \quad (29)$$

The time exponents are listed in Table 1. After this stage, the scattering function obeys the exponential decay $F_S(k,t) = \exp[-k^2 D_S^L t]$, where $M_2(t) \simeq 6D_S^L t$.

Both in a nonequilibrium case and in an equilibrium case, the finite-sized, long-lived clusters are considered to be the origin of well-known two-step relaxation. Once the diffusion coefficient $D_S^L(\Phi)$ is given in the form of Eq. (12), Eq.(11) is shown to describe the characteristic features observed by experiments near the glass transition very well. Although the diffusion coefficient given by Eq. (12) mainly results from the many-body, long-range hydrodynamic interactions, that singular form is considered to be rather universal even for different many-body interactions. We next discuss this.

COMPUTER SIMULATIONS ON HARD-SPHERE SYSTEMS

In this section we investigate how the form of the self-diffusion coefficient given by Eq. (12) is universal even for different interactions. As discussed in the previous papers [9, 10], we consider two kinds of hard-sphere systems, the suspension of hard spheres and the atomic system of hard spheres, and perform two types of computer simulations, a Brownian-dynamics (BD) simulation on polydisperse suspensions and a molecular-dynamics (MD) simulation on atomic systems. In the following, we thus test whether the following form of the diffusion coefficient holds or not:

$$\frac{D_S^{L(p)}(\phi_{eq})}{d_p} = \frac{\mu(\phi_{eq})}{D_0}\frac{1-\nu(\phi_{eq})}{1+(\frac{D_S^S}{D_0})(\frac{\phi_{eq}}{\phi_c})\left(1-\frac{\phi_{eq}}{\phi_c(\sigma)}\right)^{-\gamma}}, \quad (30)$$

where $\mu(\phi_{eq})$ is a function of ϕ_{eq}, and D_S^S is given by Eq. (13). Here γ is an exponent to be determined and $\phi_c(\sigma)$ the singular point to be determined. For the suspensions of neutral colloids, TO theory finds that $\mu = D_S^S$, $\nu = 9\phi_{eq}/32$, and $\phi_c = \phi_g^{TO}$. The coefficient d_p comes from the fact that the position vector and the time for the suspension ($p = S$) and those for the atomic system ($p =$

A) are scaled by different parameters from each other. In fact, we have $d_p = D_0$ in the suspension, while $d_p = d_0 (= a v_0)$ in the atomic system, where $v_0 = \sqrt{d k_B T / m}$ denotes the average velocity of an atom. Here we note that in both simulations $\nu = 0$ since there is no coupling between the hydrodynamic interactions and collisions.

We first discuss the suspensions. For simplicity, we neglect the hydrodynamic interactions between particles because it is not easy to treat the Oseen tensor numerically. Hence the system contains mainly two kinds of interactions: the direct interaction between particles and the interaction between a particle and solvent particles. In a kinetic stage of the polydisperse suspensions, the velocity $\boldsymbol{u}_i(t)$ of ith sphere then obeys the Langevin equation

$$m \frac{d}{dt} \boldsymbol{u}_i(t) = -\zeta_i \boldsymbol{u}_i(t) + \sum_{j(\neq i)} \boldsymbol{F}(\boldsymbol{X}_{ij}(t)) + \boldsymbol{R}_i(t). \quad (31)$$

where $\zeta_i = 6\pi \eta a_i$ and the Gaussian random force $\boldsymbol{R}_i(t)$ satisfies

$$< \boldsymbol{R}_i(t) \boldsymbol{R}_j(t') > = 2 k_B T \zeta_i \delta_{ij} \delta(t-t') \mathbf{1}. \quad (32)$$

Since $d\boldsymbol{u}_i(t)/dt \simeq 0$ on a time scale of order t_D, use of Eq.(31) leads to

$$\frac{d}{dt} \boldsymbol{X}_i(t) = \frac{1}{\zeta_i} \sum_{j \neq i} \boldsymbol{F}(\boldsymbol{X}_{ij}(t)) + \boldsymbol{\xi}_i(t). \quad (33)$$

Here the reduced random force $\boldsymbol{\xi}_i(t) (= \boldsymbol{R}_i(t)/\zeta_i)$ satisfies

$$< \boldsymbol{\xi}_i(t) \boldsymbol{\xi}_j(t') > = 2 D_{0i} \delta(t-t') \delta_{i,j} \mathbf{1}. \quad (34)$$

where $D_{0i} = k_B T / \zeta_i$. In order to calculate Eq. (33) numerically, we first scale the position vector $\boldsymbol{X}_i$ with radius a, time with t_D, radius a_i with a, and D_S^L with D_0. Then, we employ the forward Euler difference scheme to integrate Eq. (33) with time step 10^{-3} under periodic boundary and appropriate initial conditions together with the momentum and the energy conservation laws, where elastic binary collisions between particles are assumed and the total number of particles N is chosen to be 10976. The simulation results are compared with Eq.(26).

Since the above system of suspensions lacks one of important interactions, that is, hydrodynamic interactions, we still need to investigate the system which contains complete mechanisms. It is an atomic system of hard spheres, where only the direct interactions between particles play an important role. In the atomic systems, the velocity $\boldsymbol{u}_i(t)$ of ith particle obeys the Newton equation

$$m_i \frac{d}{dt} \boldsymbol{u}_i(t) = \sum_{j \neq i} \boldsymbol{F}(\boldsymbol{X}_{ij}(t)). \quad (35)$$

Before we solve Eq.(35) numerically under the same initial and boundary conditions as those in BD, we also scale the position vector $\boldsymbol{X}_i$ with radius a, the velocity $\boldsymbol{u}_i(t)$ with v_0, time t with $t_0 (= a/v_0)$, mass m_i with m, and D_S^L with d_0. Similar to Eq. (25), $M_2(t)$ is described by the mean-field equation [9]

$$\frac{d}{dt} M_2(t) = 2d D_S^L + 2d \left[\frac{v_0^2}{d} t - D_S^L \right] e^{-\lambda M_2(t)}. \quad (36)$$

which is easily solve to give

$$M_2(t) = \frac{1}{\lambda} \ln \left[1 + 2 \left(\frac{t_\beta}{t_A} \right)^2 \{ e^{t/t_\beta} - (1 + t/t_\beta) \} \right]. \quad (37)$$

where $t_A = 1/(v_0 \lambda^{1/2})$ denotes the short time for an atom to move over a distance of order $\lambda^{-1/2}$. The simulation results are compared with Eq.(37).

Both simulations, MD and BD, are done for two cases separately, a monodisperse case where spheres are all identical, that is, $a_i = a$, $m_i = m$, and $\sigma = 0$, and a polydisperse case, where $\sigma = 0.06$ here. In both simulations, we start from two kinds of non-equilibrium initial states. One is a disordered initial state [D] which shows a random configuration obtained by using the Jodrey and Tory's algorithm [11]. The other is an ordered initial state [O] which shows a face-centered-cubic configuration. Then, we wait for a long time enough to reach a final state in which the mean-square displacement grows linearly in time. We then use this final state as an initial state and repeat the simulations again, until the whole time behavior of the mean-square displacement coincides with a previous one. Depending on the values of the volume fractions, there are three phase regions: a fluid region for $0 < \phi < \phi_f(\sigma)$, a metastable region for $\phi_f(\sigma) \leq \phi < \phi_m(\sigma)$, and a crystal region for $\phi_m(\sigma) \leq \phi$, where $\phi_f(\sigma)$ and $\phi_m(\sigma)$ are the so-called freezing and melting volume fractions, respectively. From our simulations, we thus find $\phi_f(0) \simeq 0.51$, $\phi_f(0.06) \simeq 0.53$, $\phi_m(0) \simeq 0.54$, and $\phi_m(0.06) \simeq 0.57$.

In Figs. 4 and 5, the mean-square displacement $M_2(t)$ is shown in an equilibrium liquid state. Both simulation results are shown to be in good agreement with the mean-field solutions given by Eqs. (26) and (37). We note here that the mean-square displacement satisfies the following asymptotic solutions: $M_2(t) \simeq 6 D_0 t$ for $t \leq t_D$ and $6 D_S^L t$ for $t_D \ll t$ in suspensions, and $M_2(t) \simeq (v_0 t)^2$ for $t \leq t_0$ and $6 D_S^L t$ for $t_0 \ll t$ in atomic systems. The short-time behavior in both systems are different from each other since in suspensions it is governed by the short-time diffusion process, while in atomic systems it is governed by the ballistic motion. Thus, we find the long-time self-diffusion coefficients $D_S^{L(p)}(\phi)/d_p$ for different volume fractions, where $d_p = d_0$ for atomic systems and $d_p = D_0$ for suspensions. In Fig. 6, we plot the long-time self-diffusion coefficient $D_S^{L(p)}(\phi_{eq})/d_p$ versus ϕ_{eq}. Then, we

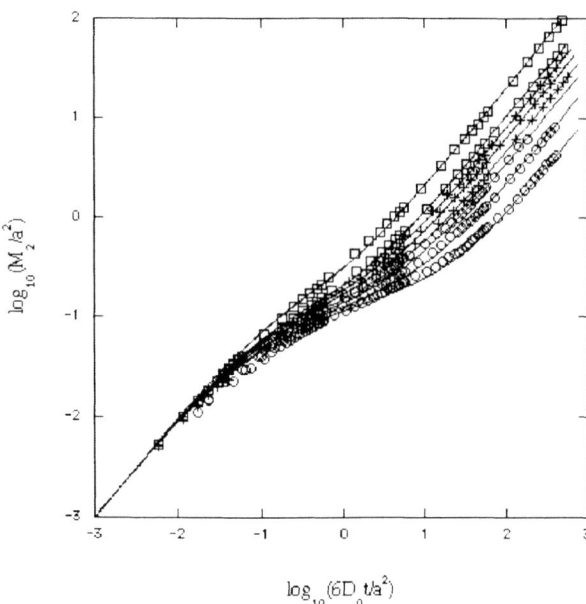

FIGURE 4. A log-log plot of the mean-square displacement $M_2(t)$ for suspensions versus time for different volume fractions (from left to right) 0.45, 0.50, 0.51, 0.52, 0.53, 0.54, 0.55, 0.56. The solid lines indicate the mean-field results given by Eq. (26). The symbols indicate the BD results: the open squares are for a monodisperse equilibrium fluid state, the crosses are for a monodisperse metastable fluid state, and the open circles are for a polydisperse metastable fluid state.

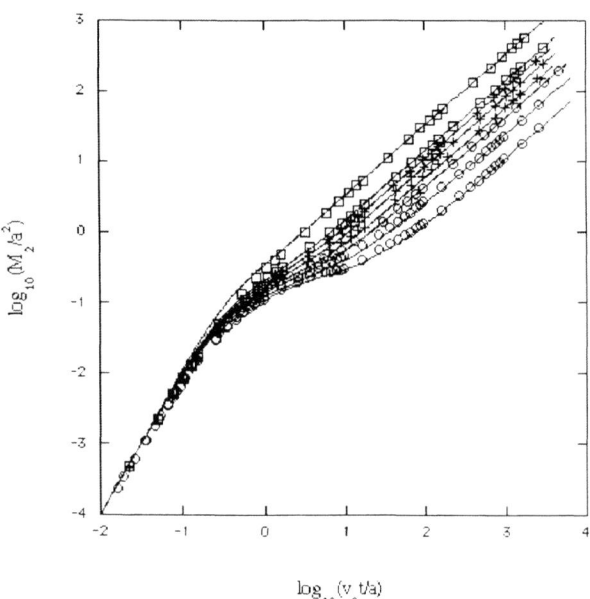

FIGURE 5. A log-log plot of the mean-square displacement $M_2(t)$ for atomic systems versus time. The solid lines indicate the mean-field results given by Eq. (36). The details are the same as in Fig. 4.

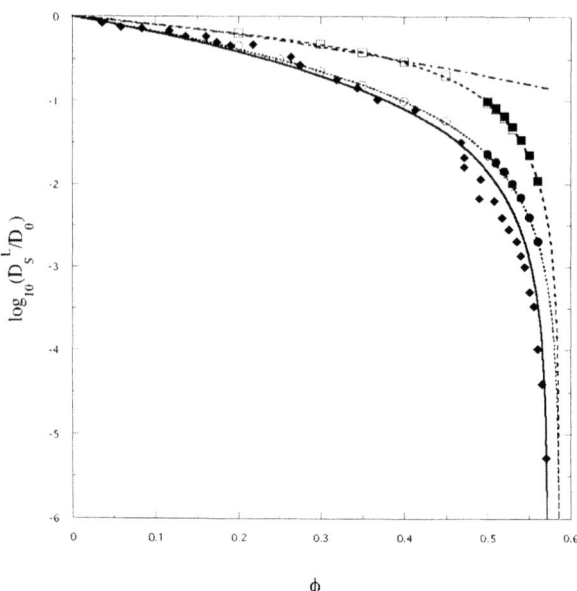

FIGURE 6. A log plot of the long-time self-diffusion coefficient $D_S^{L(p)}(\phi)/d_p$ versus ϕ. The open symbols indicate the simulation results for a monodisperse case; the squares for $D_S^{L(C)}/D_0$ and the circles for $D_S^{L(A)}/d_0$. The filled symbols indicate those for a polydisperse case. The solid line is the theoretical results given by Eqs. (12). The dotted and dashed lines indicate the diffusion coefficients $D_S^{L(A)}/d_0$ and $D_S^{L(C)}/D_0$ given by Eq. (30), respectively. The filled diamonds indicate the experimental data from Ref. [6].

TABLE 2. Coefficients and exponents.

	γ	μ	ν	ϕ_c
BD ($\sigma = 0$)	2.0	D_0	0	0.586
BD ($\sigma = 0.06$)	2.0	D_0	0	0.586
MD ($\sigma = 0$)	2.0	D_S^S	0	0.586
MD ($\sigma = 0.06$)	2.0	D_S^S	0	0.586
TO ($\sigma = 0$)	2.0	D_S^S	$\frac{9}{32}\phi_{eq}$	ϕ_g^{TO}

show that the simulation results are described by Eq. (30) very well if the coefficients μ, ν, the exponent γ, and the singular point ϕ_c are set as in Table 2. Here we note that the singular point $\phi_c(\sigma)$ is not a sensitive function of σ. In fact, Doliwa and Heuer [12] have also found by a Monte-Carlo simulation that $\phi_c(0.1) \simeq 0.587$. In the simulations $\mu = D_0$ comes from the fact that the systems do not contain the hydrodynamic interactions, while in the atomic systems $\mu = D_S^S$ comes from the fact that the systems contain the full mechanisms. In order to check whether this is correct or not, we also plot the short-time self-diffusion coefficient $D_S^S(\phi)$ versus ϕ in Fig. 7. For comparison, the experimental data are also shown. The simulation results are shown to agree with

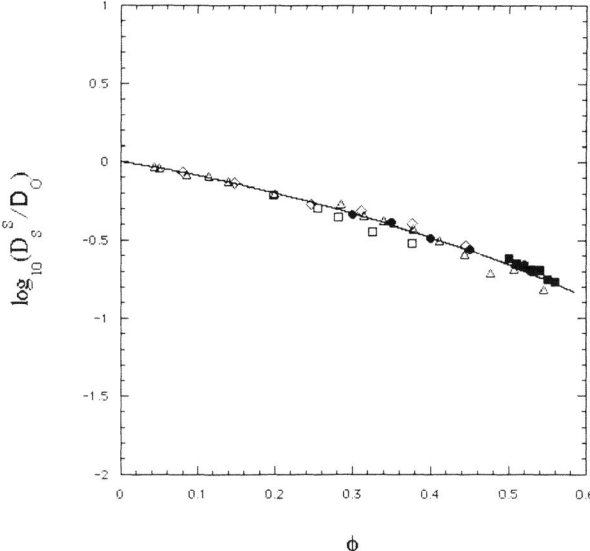

FIGURE 7. A log plot of the short-time self-diffusion coefficient $D_S^S(\phi)$ versus ϕ. The solid line indicates the theoretical short-time self-diffusion coefficient $D_S^S(\phi)$ given by Eq. (8). The filled symbols indicate the simulation results given by $[D_S^{L(A)}/d_0]/[D_S^{L(C)}/D_0]$; the circles for a monodisperse case and squares for a polydisperse case. Shown are also the experimental data from Ref. [13] (squares), Ref. [14] (triangles), and Ref. [15] (diamonds).

the theoretical and experimental results very well within errors for a wide range of volume fractions. Hence the following relation holds:

$$\frac{D_S^{L(A)}(\phi_{eq})}{d_0} = \frac{D_S^S(\phi_{eq})}{D_0}\frac{D_S^{L(C)}(\phi_{eq})}{D_0}. \quad (38)$$

CONCLUSION

In this paper, we have tested whether the long-time collective behavior is universal or not by performing two types of computer simulations on hard-sphere systems, BD on colloidal suspensions and MD on atomic systems. Thus, we have shown in both systems that the many-body collision interactions lead to the same singular behavior of the long-time self-diffusion coefficient as that obtained by the many-body hydrodynamic interactions, except that the singular points are modified.

Next, we point out that two types of hydrodynamic interactions are necessary to recover the theoretical result given by Eq.(12). One is the short-time (uncorrelated) hydrodynamic inetractions, leading to D_S^S. Then, $D_S^{L(C)}$ reduces to $D_S^{L(A)}$. The other is the long-time (correlated) hydrodynamic interactions. This effect must shift the singular point $\phi_c \simeq 0.586$ to ϕ_g^{TO}. Thus, $D_S^{L(A)}$ may finally reduce to D_S^L. Finally, the diffusion coefficient similar to Eq.(31) also holds for magnetic colloids, where the dipole interactions play an important role. This will be discussed elsewhere.

ACKNOWLEDGMENTS

This work was partially supported by Grants-in-aid for Science Research with No. 14540348 from Ministry of Education, Culture, Sports, Science and Technology of Japan. Numrical computations for this work were performed on the ORIGIN 2000 machine at the Institute of Fluid Science, Tohoku University.

REFERENCES

1. M. Tokuyama and I. Oppenheim, Phys. Rev. E **50**, R16-R19 (1994).
2. M. Tokuyama and I. Oppenheim, Physica A **216**, 85-119 (1995).
3. M. Tokuyama, Physica A **102**, 399-430 (1980); **109**, 128-160 (1981).
4. M. Tokuyama, Physica A **294**, 23-43 (2001).
5. M. Tokuyama, Y. Enomoto, and I. Oppenheim, Physica A **270**, 380-402 (1999).
6. W. van Megen, T. C. Mortensen, S. R. Williams, J. Müler, Phys. Rev. E **58**, 6073-6085 (1998).
7. P. N. Segrè and P. N. Pusey, Phys. Rev. Lett. **77**, 771-774 (1996).
8. M. Tokuyama, Physica A **289**, 57-85 (2001).
9. M. Tokuyama, H. Yamazaki, and Y. Terada, Phy. Rev. E **67**, 062403 (2003).
10. M. Tokuyama, H. Yamazaki, and Y. Terada, Physica A **328**, 367-379 (2003).
11. W. S. Jodrey and E. M. Tory, Phys. Rev. A **32**, 2347-2351 (1985).
12. B. Doliwa and A. Heuer, Phys. Rev. E **61**, 6898-6908 (2000).
13. P.N. Pusey and W. van Megen, Nature (London) **320**, 340-342 (1986).
14. R. H. Ottewill and N. S. J. Williams, Nature **325**, 232-234 (1987).
15. A. van Veluwen and H. N. W. Lekkerkerker, Phys. Rev. A **38**, 3758-3763 (1988).

Neutron and Light Scattering Studies of the Liquid-to-Glass and Glass-to-Glass Transitions in a Copolymer Micellar System

Sow-Hsin Chen*, Wei-Ren Chen* and Francesco Mallamace[†]

*Department of Nuclear Engineering, Massachusetts Institute of Technology, Cambridge MA 02139-4307, USA.
[†]Dipartimento di Fisica e Istituto Nazionale per la Fisica della Materia, Universita' di Messina, Messina, Italy.

Abstract. Recent mode coupling theory (MCT) calculations for a hard-sphere system with a short-range attraction show that one may observe a new type of structurally arrested state originating from clustering effect, called the "attractive glass", as a result of the attractive interaction. This is in addition to the well-known glass-forming mechanism due to the cage effect in the hard sphere system, called the repulsive glass. The calculations also indicate that, if the range of attraction is sufficiently short compared to the diameter of the hard sphere, within a certain interval of the volume fraction and the effective temperature, the two glass-forming mechanisms can compete with each other. For example, by varying, the effective temperature at appropriate volume fractions, one may observe respectively, the glass-to-liquid-to-glass re-entrance or the glass-to-glass transitions. Here we present experimental evidence for both transitions, obtained from small-angle neutron scattering (SANS) and photon correlation spectroscopy (PCS) measurements taken from dense $L64$ copolymer micellar solutions in heavy water. We show, by varying the temperature in the predicted volume fraction range triggers a sharp transition between the two types of glass. In particular, according to MCT, there is an end point (called A_3 singularity) of this glass-to-glass transition line, beyond which the long-time dynamics of the two glasses become identical. Our findings confirm this theoretical prediction. Surprisingly, although the Debye-Waller factors (DWF), the long-time limit of the coherent intermediate scattering functions, of these two glasses obtained from PCS measurements indeed become identical at the predicted volume fraction, they exhibit distinctly different intermediate time relaxation. Furthermore, our SANS results on the local structure obtained from volume fractions beyond the end point are characterized by the the same features as the repulsive glass obtained before the end point. A complete phase diagram giving the boundaries of the structural arrest transitions for $L64$ micellar system is given.

INTRODUCTION

Gelation phenomena in colloids, such as in protein gels or in gel-like emulsion-polymer mixtures, is a common experience in our daily life [1]. The gel state of a soft matter system is a nonergodic state or a "structurally arrested" state, in which the local particle configuration is deviated from the thermodynamical equilibrium state, formed due to the short-range effective attractive interaction between particles. Although the structurally arrested state is an ubiquitous state of matter in our environment, traditionally this state of matter is rather ill-characterized physically: It is amorphous but classified neither as a gas, a liquid nor as a crystal. Recent progress [2] in the mode coupling theory (MCT) calculations opens up a possibility to gain a deeper physical insight into the detailed slow dynamical behavior of the system involved in these structurally arrested states.

The structural arrest transition which causes the colloidal system to transform to an amorphous solid state is regarded as a result of a kinetic glass transition (KGT) predicted by the MCT bifurcation equation. The KGT has been studied extensively both experimentally and theoretically for a class of systems which can be modelled as a hard sphere system in the past two decades [3, 4, 5].

At low volume fractions, the behavior of the hard sphere system is fluid-like. As the volume fraction increases, the test-particle time correlation function ($self - ISF$) and the deneity-density correlattion function (ISF) of the particles exhibit a two-stage relaxation process. The initial decay of the $self - ISF$ corresponds to rattling of a typical particle confined within a transient cage formed by its neighbors, followed by a slow decay, resulting from relaxation of the cage itself and the escape of the trapped particle by re-arranging its nearest neighbor configuration. The latter process leads to a possibility of particle diffusion through coupling to the structural relaxation process. If a system can be manipulated so as to avoid crystallization at the volume fraction of 0.495, for example by artificially creating a polidispersity in sizes of a few percent, at a critical volume fraction ϕ_c, which was predicted to be 0.516 (but determined to be 0.58 experimentally), the KGT can be ob-

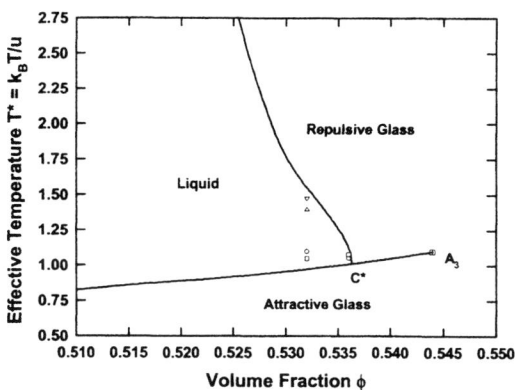

FIGURE 1. Theoretical phase diagram predicted by MCT calculations using a short-range attractive square well potential with $\varepsilon = 0.03$. The calculations predict the attractive glass-to-liquid-to-repulsive glass re-entrant transition, the attractive glass-to-repulsive glass transition and the end point of the glass-glass transition line. Symbols (open circles, squares and triangles) represent the effective temperature T^*, determined by fitting the experimental SANS data taken in the liquid state with the method explained in the text, for different volume fractions.

served [6]. At the KGT, both the particle diffusion and the long-time density fluctuations freeze and the system undergoes an ergodic-to-nonergodic transition. Although this characteristic feature has been confirmed by computer molecular-dynamics simulations [7] and laboratory experiments [6, 8] involving several different glass-formers, there have been some anomalous dynamical observations, which cannot be interpreted in terms of the theory based on the hard sphere potential alone [9].

A more complete picture of the KGT is obtained by modeling the inter-particle potential more accurately. Recent MCT calculations [10, 11, 12] show that if a system is characterized by a hard core plus an additional short-range attractive interaction, for example, by an adhesive hard sphere system (AHS), a different structural arrest scenario emerges. Theoretically, the phase behavior of the AHS is characterized by an effective temperature $T^* = k_B T/u$, the volume fraction of the particles ϕ and the fractional attractive well width $\varepsilon = \Delta/R$, where $-u$ is the depth of the attractive square well, Δ the width of the well and R the diameter of the particle. In this case, for a given ε, aside from the volume fraction, a second external control parameter, the effective temperature T^*, is introduced into the description of the phase behavior of the system and the loss of ergodicity can take place either by increasing the volume fraction or by changing the effective temperature.

Figure 1 gives part of the predicted phase diagram based on the AHS with $\varepsilon = 0.03$. In this case, it is possible for the system to undergo a re-entrant (glass-to-liquid-to-glass) transition (for example at $\phi = 0.532$ in Fig. 1) by varying the effective temperature. At high effective temperatures and at sufficiently high volume fractions, the system evolves into the well-known structurally arrested (glassy) state, called a "repulsive glass", as a result of the cage effect, a manifestation of the excluded volume effect due to the existence of the hard core. However, at relatively low effective temperatures, an "attractive glass" can form in which motion of the typical particle is constrained in stead by the cluster formation with neighboring particles. Moreover, in an AHS, aside from the hard-core diameter, the range of the attractive well, should come into play (parameter ε). The MCT calculations show that, with sufficiently small ε, variation of the two control parameters, T^* and ϕ, allows the transition between these two distinct forms of glass. Thus there is a branch of the KGT line across which transitions between the attractive glass and the repulsive glass are predicted. Of particular interest is the occurrence of an A_3 singularity (see Fig. 1) at which point the glass-to-glass transition line terminates. MCT suggests that the long-time dynamics of the two distinct structurally arrested states become identical at and beyond this point.

Several ongoing experimental investigations have confirmed some of the theoretical predictions, such as the re-entrant glass-to-liquid-to-glass transition phenomenon [13, 14] and the logarithmic relaxation of the glassy dynamics in the liquid states in the vicinity of the A_3 singularity [15]. They are considered to be the signatures of the glassy dynamics in the two-length-scale system. Yet, except for some recent experimental reports on a dense micellar system [16, 17], a detailed investigation of the glass-to-glass transition is just emerging, and the physical insight into the slow dynamics in the two glassy states need to be obtained.

The study of KGT by neutron scattering method is traditionally dominated by the use of the inelastic neutron scattering (INS) and the spin-echo spectroscopy techniques. However, by analyzing an extensive set of small-angle neutron scattering (SANS) intensity distribution obtain from L64 micellar solutions, which exposes all the characteristics of the AHS from our previous works [16, 17], we demonstrated that SANS is also a suitable tool to investigate the boundary of the KGT: By analyzing SANS intensity distributions, the difference in the structure factor before and after the KGT can be visualized vividly. Supplemental to the photon correlation spectroscopy (PCS), it provides rapid measurements of the peak of the structure factor over a wide temperature and volume fraction ranges, which is more difficult to do with INS or spin echo spectroscopy. By this novel method, we successfully confirmed the predicted re-entrant phenomenon, the glass-to-glass transition and most importantly, the existence of the A_3 singularity, in quantitative agreement with predictions of MCT. Furthermore, we are able, using SANS technique, to mapped out the complete structural arrest phase boundaries for the *L*64 micellar system.

EXPERIMENT

The micellar system we study is a triblock copolymer *L*64, one member of Pluronic family, used extensively in industrial applications. After necessary purification procedure [18] to remove hydrophobic impurities, the polymer is dissolved in deuterated water (D_2O). Pluronic is made of polyethylene oxide (*PEO*) and polypropylene oxide (*PPO*). The chemical formula of *L*64 is $(PEO)_{13}(PPO)_{30}(PEO)_{13}$, having a molecular weight of 2990 Dalton. At low temperatures, both *PEO* and *PPO* are hydrophilic, so that *L*64 chains readily dissolve in water, and the polymers exist as unimers. As the temperature increases there is a decrease in the probability of hydrogen-bond formation between water and polymer molecules, *PPO* tends to become less hydrophilic faster than *PEO*. This creates an inbalance of hyrophilicity between the end-blocks and the middle-block of the polymer molecule and the copolymer molecules acquire surfactant properties in the aqueous environment and self-assemble to form micelles. Thus the micellar formation is initiated at a well-defined Critical Micellar Temperature-Concentration (CMT-CMC) line. As the temperature further increases, water becomes progressively a poor solvent to both *PPO* and *PEO* chains, and the effective micelle-micelle interaction becomes attractive. The evidence for the increased short-range micellar attraction as a function of temperature comes from the existence of a lower consolute critical point at $C = 5.0$ *wt%* and $T = 330.9\ K$ [19] and a percolation line detected by a jump of zero-shear viscosity of more than two orders of magnitude [20]. We have explored the dynamically arrested states and their structure in this micellar system, as a function of temperature at high volume fractions.

SANS measurements were performed at *NG*7, 40 *m* SANS spectrometer in the NIST Center for Neutron Research, and at SAND station at the Intense Pulsed Neutron Source (IPNS) in Argonne National Laboratory. At *NG*7, incident monochromatic neutrons of wave length $\lambda = 5$ Å with $\Delta\lambda/\lambda = 10\%$ was used. Sample to detector distance was fixed at 6 *m*, covering the magnitude of wave vector transfer (k) range 0.008 Å^{-1} to 0.3 Å^{-1}. At IPNS, a pulse of white neutrons was selected with an effective wave length range from 1.5 Å to 14 Å. In *SAND* all these neutrons are utilized by encoding their individual time-of-flight and their scattering angles determined by their detected position at a *2D* area detector. The *2D* area detector has an active area of $40 \times 40\ cm^2$ and the sample to detector distance is 2 *m*. This configuration allows a reliable k-range covering from 0.004 Å^{-1} to 0.6 Å^{-1}. k-resolution functions of both of these SANS spectrometers are Gaussian and well characterized. It is essential that we apply these resolution broadening to the theoretical cross-section when fitting the intensity data. Sample liquid was contained in a flat quartz cell with 1 *mm* path length.

The Photon Correlation Spectroscopy (PCS) measurements were made at a scattering angle $\theta = 90°$, using a continuous wave solid state laser (Verdi-Coherent) operating at 50 *mW* ($\lambda = 5120$ Å) and an optical scattering cell of a diameter 1 *cm* in a refractive index matching bath. The intensity data were also corrected for turbidity and multiple scattering effects. PCS data have been taken using a digital correlator with a logarithmic sampling time scale which allows an accurate description of the intermediate Scattering Function (*ISF*), from 1 μ sec up to 100 *secs*. Following the method used for colloidal hard spheres [8] we measure the correlation function

$$g^{(2)}(k,\tau) = \langle\langle I(k,0)I(k,\tau)\rangle\rangle / \langle\langle I(k,0)\rangle\rangle^2 \quad (1)$$

where $I(k,\tau)$ is the intensity of light scattered at wave vector k and at a delay time τ. The first bracket denotes the time average, and the second bracket denotes the positional average over different parts of the sample. Since the system shows a structural arrest transition, and therefore a nonergodic behavior, particular care has been taken in averaging over many different positions in the sample for the measurement of the long time part of the time correlation function. For each sample and each temperature we have performed more than 200 positional average measurements, observing a large scattering area corresponding to three or more independent Fourier components and changing the position of the sample in order to observe different scattering volumes. The ISF $g^{(1)}(k,\tau)$ can be obtained from equation (1) in a straightforward way using the Siegert relation.

SANS DATA ANALYSIS

The absolute intensity (in unit of cm^{-1}) of small-angle neutron scattering from a system of mono-dispersed micelles can be expressed by the following formula:

$$I(k) = cN\left[\sum_i b_i - \rho_w v_p\right]^2 \overline{P}(k)S(k) \quad (2)$$

where c is the concentration of polymer (number of polymers/cm^3), N the aggregation number of polymers in a micelle, $\sum_i b_i$ sum of coherent scattering lengths of atoms comprising a polymer molecule, ρ_w the scattering length density of D_2O, v_p the molecular volume of the polymer. $\overline{P}(k)$ is the normalized intra-particle structure factor calculated by the modified cap-and-gown model [16] and $S(k)$ the inter-micellar structure factor of a spherical particle system interacting via an attractive square well potential with a repulsive hard core. $S(k)$ is calculated analytically by solving the Ornstein-Zernike (OZ) equation in Percus-Yevick (PY) approximation, for this square well potential, to the first order in a series of small ε expansion [22]. The detailed analytical expression of $S(k)$ is given in the following section[16]. It is important to note that the $\overline{P}(k)$ in the modified cap-and-gown model is a function the aggregation number N only and $S(k)$ is the function of the aggregation number N, the volume fraction of micelles ϕ, the fractional well width parameter ε and the effective temperature T^* Thus an absolute SANS intensity distribution in the liquid state can be calculated by the above equation (2) unambiguously with the four parameters: N, ϕ, ε and T^*.

The analytical expression for the inter-micellar structure factor $S(k)$:

A square well potential with a repulsive core of diameter R' and the real particle diameter R is used to model a hard sphere with an adhesive surface layer. The pairwise interaction potential is defined as

$$V(r) = \begin{array}{ll} +\infty & for \ 0 < r < R' \\ -u & for \ R' < r < R \\ 0 & for \ r > R \end{array} \quad (3)$$

We define here $\varepsilon = \frac{R-R'}{R}$ as the fractional well width parameter and $T^* = k_B T/u$ as the effective temperature. Then the inter-micellar structure factor is a function of four pameters: the real particle diameter R, the volume fraction ϕ, the fractional well width parameter ε and the effective temperature T^*. It should be noted that real particle diameter R is tied uniquely to the aggregation number N and the volume fraction ϕ through a relation $\phi = \frac{c\pi R^3}{6N}$.

The Ornstein-Zernike (OZ) equation in Percus-Yevick (PY) approximation for this square well potential can be solved analytically to the first order in ε expansion [22]. The result of the structure factor is given as follows:

$$\frac{1}{S(Q)} - 1 = 24\phi \left[\alpha f_2(Q) + \beta f_3(Q) + \frac{1}{2}\phi \alpha f_5(Q)\right] + $$
$$4\phi^2 \lambda^2 \varepsilon^2 \left[f_2(\varepsilon Q) - \frac{1}{2}f_3(\varepsilon Q)\right] + $$
$$2\phi^2 \lambda^2 \left[f_1(Q) - \varepsilon^2 f_1(\varepsilon Q)\right] - $$
$$\frac{2\phi\lambda}{\varepsilon}\left[f_1(Q) - (1-\varepsilon)^2 f_1((1-\varepsilon)Q)\right] - $$
$$24\phi \left[f_2(Q) - (1-\varepsilon)^3 f_2((1-\varepsilon)Q)\right] \quad (4)$$

where
$$\begin{aligned} Q &= kR \quad (5)\\ \phi &= \frac{c\pi R^3}{6N} \\ \alpha &= \frac{(1+2\phi-\mu)^2}{(1-\phi)^4} \\ \beta &= -\frac{3\phi(2+\phi)^2 - 2\mu(1+7\phi+\phi^2) + \mu^2(2+\phi)}{2(1-\phi)^4} \\ \mu &= \lambda\phi(1-\phi) \\ \lambda &= \frac{6(\Delta - \sqrt{\Delta^2 - \Gamma})}{\phi} \\ \Delta &= \tau + \frac{\phi}{(1-\phi)} = \frac{1}{12\varepsilon}\exp(-\frac{u}{k_B T}) + \frac{\phi}{(1-\phi)} \\ \Gamma &= \frac{\phi(1+\phi/2)}{3(1-\phi)^2} \\ f_1(x) &= \frac{1-\cos x}{x^2} \\ f_2(x) &= \frac{\sin x - x\cos x}{x^3} \\ f_3(x) &= \frac{2x\sin x - (x^2 - 2)\cos x - 2}{x^4} \\ f_5(x) &= \frac{(4x^3 - 24x)\sin x - (x^4 - 12x^2 + 24)\cos x + 24}{x^6} \end{aligned}$$

This formula is accurate in the k region around the first diffraction peak for $\varepsilon < 0.05$ [21]. It should be noted that the real volume of a micelle is significantly larger than the volume of N dry polymer chains. It consists of the volume of the N dry polymer chains plus the volumes of the associated hydration water per polymer chain in the corona region. The real volume of the micelle, rather than the polymer volume, determines the thermodynamic quantities such as the osmotic compressibility and the scattering intensities. It further determines the dynamical properties, such as diffusivity and viscosity, of the polymeric micellar solution[22]. Obtaining the real volume of a micelle in a self consistent way is not a trivial matter. It requires the knowledge of the detailed microstructure such as hydration number per polymer chain. This information of the associated solvent molecules is not available from the theoretical consideration but can be extracted accurately from the analysis of the SANS intensity distribution through the volume fraction ϕ and the aggregation number N.

To illustrate the model fitting, a SANS intensity distribution obtained from a 48.5 *wt.%* micellar solution at 333 *K.* in an absolute scale is shown in the upper panel of Fig. 2 as an example. Symbols give the experimental data and the dash line gives the model fitting taking into account the resolution correction. The same data plotted in a log-log scale is given in the inset. It can be seen that

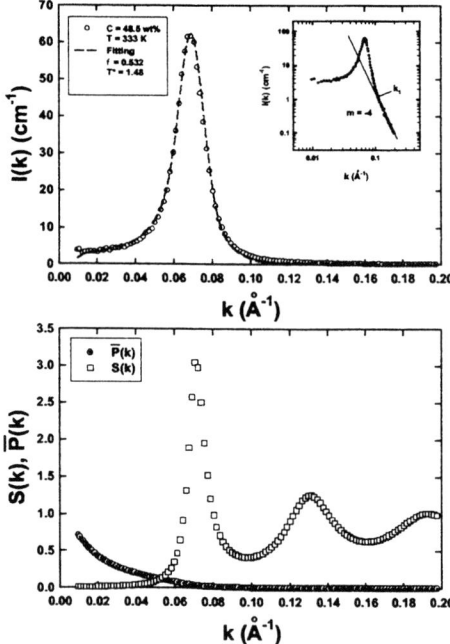

FIGURE 2. The upper panel shows a typical SANS intensity distribution in an absolute scale and its model fit, taking into account the effect of resolution function. Symbols represent the experimental data and the dash line the theory convoluted with the resolution function. The inset gives the same data but plotted in a log-log scale. It can be seen that the Porod's law is satisfied at large k, as expected in a two-phase system with a sharp interface. The lower panel gives the normalized intra-particle structure factor $\bar{P}(k)$ (circles) and the inter-particle structure factor $S(k)$ (squares) used to fit the data in the upper panel. The $S(k)$ is calculated by solving the OZ equation with a square-well inter-micellar potential and the $\bar{P}(k)$ is calculated using the modified cap-and-gown model [16] for the polymer segmental distribution function in a micelle. The observed SANS data is the product of these two functions and therefore it is clear that the interaction peak in the SANS data is primarily due to the first diffraction peak in the inter-micellar structure factor.

for sufficiently large k, the SANS absolute intensity decreases as k^{-4}, in agreement with the Porod's law as expected in a two-phase system with a sharp interface. The lower panel gives the normalized intra-particle structure factor $\bar{P}(k)$ (circles) and the inter-particle structure factor $S(k)$ (squares) for this case. The observed SANS data is proportional to the product of these two functions. The fact that the first diffraction peak of $S(k)$ occurs at a relatively smooth tail part of $\bar{P}(k)$ implies that the interaction peak in the SANS intensity distribution is primarily reflecting the width of the first diffraction peak of $S(k)$.

SCALING PLOT OF SANS INTENSITY DISTRIBUTION

At high enough polymer concentration, SANS intensity distribution from the $L64/D_2O$ micellar system generally consists of a single, sharp interaction peak. Since there is a single peak in the observed SANS intensity distribution, we can assume that the system is characterized by a single length scale $\Lambda = \frac{1}{k_{max}}$ where k_{max} is the peak position of the intensity distribution. It is well known that the absolute intensity in a two-phase system (the micelles and the solvent) is given by a 3D Fourier transform of the Debye correlation function $\Gamma(r)$. The Debye correlation function is a two-point correlation function of the local scattering length density in the system. In this case it must be of the form $\Gamma(\frac{r}{\Lambda})$ since Λ is the unique length scale in the system. Therefore

$$I(k) = <\eta^2> \int_0^\infty dr 4\pi r^2 j_0(kr) \Gamma\left(\frac{r}{\Lambda}\right) \quad (6)$$

where $<\eta^2>$ is the so-called invariant of the intensity distribution.

By making a transformation of variables

$$x = \frac{r}{\Lambda} = k_{max} r$$
$$y = \frac{k}{k_{max}}$$

it is straight forward to show that

$$\frac{k_{max}^3 I(k)}{<\eta^2>} = \int_0^\infty dx 4\pi x^2 j_0(xy) \Gamma(x) \quad (7)$$

From the above equation, it can be seen that the scaled intensity $\frac{k_{max}^3 I(k)}{<\eta^2>}$ is a unique function of the scaled magnitude of the scattering vector, $y = \frac{k}{k_{max}}$. Therefore, if we plot the scaled intensity distributions at different temperatures as a function of y, they should collapse into one single master curve in a single-phase amorphous state.

More specifically, take Fig. 2 as an example, the physical meaning of the scale intensity $\frac{k_{max}^3 I(k)}{<\eta^2>}$ can be interpreted as follows:

The invariant $<\eta^2>$ is mathematically defined as

$$<\eta^2> = \frac{1}{2\pi^2} \int_0^\infty k^2 I(k) dk \quad (8)$$

From Fig. 2 one can see that the minimum wave vector signaling the onset of the Porod's law is at $k_1 \approx 0.09 \text{ Å}^{-1}$. According to the inset of Fig.2, Porod's law $I(k) = Ak^{-4}$ is valid, when $k_1 = 0.09 \text{ Å}^{-1}$. $I(k_1) = 1 \text{ cm}^{-1} = 10^{-8} \text{ Å}^{-1}$, thus $A = 6.5 \times 10^{-13} \text{ Å}^{-5}$. Therefore, the integral in

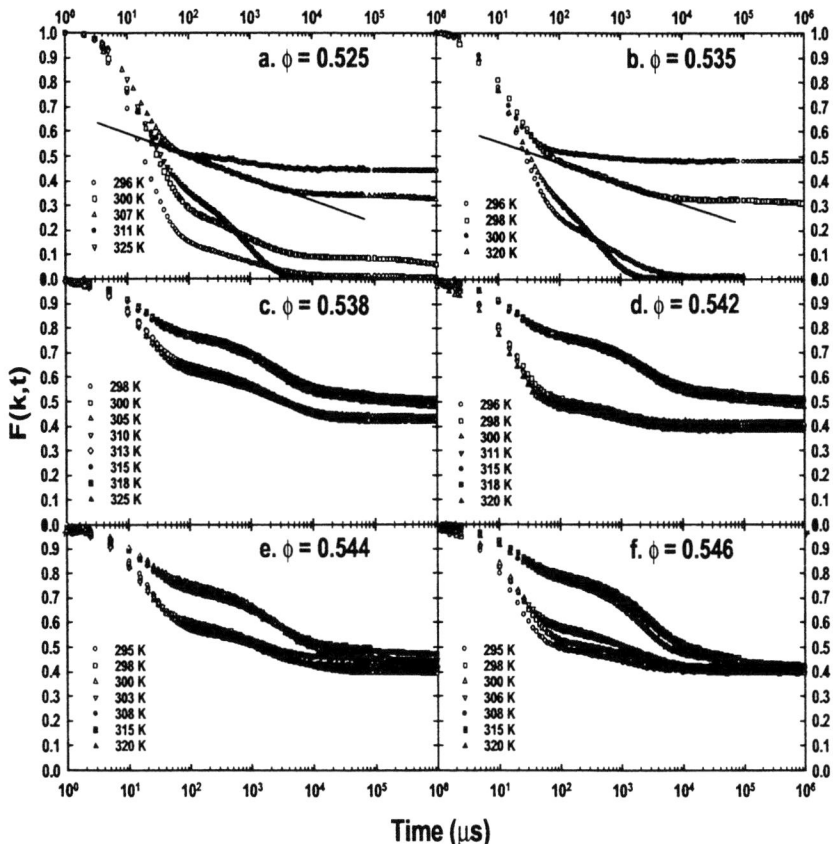

FIGURE 3. Fig. 3a & 3b show ISFs measured at $\phi = 0.525$ and $\phi = 0.535$ respectively, where the liquid to attractive glass transition is found, as a function of temperature. In the liquid state, the long-time limit of ISF (DWF), f_q, is zero, while in the attractive glass state f_q is about 0.5. The structural arrest transition is thus characterized by a discontinuous change of f_q, called a bifurcation transition. The observed occurrence of a region of the logarithmic time dependence, preceding the plateau region for the system in the ergodic state just before the transition, are highlighted by a straight line in the linear-log plots respectively. Fig. 3c & 3d show ISFs measured at $\phi = 0.538$ and $\phi = 0.542$. According to MCT, in this volume fraction range, there is a possibility to observe the glass-to-glass transition by varying the effective temperature T^*. Since the depth of the potential well $-u$ is temperature dependent, and increases on heating, the effective temperature T^* actually decreases as temperature rises, making the transition from the repulsive glass to the attractive glass possible. By comparing the long-time limit of the ISFs, the two different types of the glasses can be identified by the respective DWFs, $f_q^A \sim 0.5$ and $f_q^R \sim 0.4$. The reason for observing two different values of DWFs can be interpreted as the different degrees of localization of the density fluctuation for the two types of glasses. Fig. 3e shows the ISFs measured at $\phi = 0.544$, where according to MCT, the long-time limit of ISF's of the two glasses become identical. In our measurement, the long-time limit of ISFs of the attractive glass gradually decreases from 0.5, and get closer to 0.4, the long-time limit of ISFs of the repulsive glass. Although the DWFs of these two types of glasses are almost identical, it is essential to recognize that there is a significant difference between the dynamics of their intermediate time relaxations. Fig. 3f shows the ISFs measured at $\phi = 0.546$. At this volume fraction, the long-time limit of ISFs of the attractive glass approaches closely to 0.4, which is the long-time limit of ISFs of the repulsive glass. It indicates that beyond this volume fraction, the two glasses merge into a single repulsive glass. Judging from Fig. 3e and Fig. 3f, the experimentally determined volume fraction for A_3 singularity lie somewhere between $\phi = 0.544$ and $\phi = 0.546$.

$\langle\eta^2\rangle$ can be divided into two parts and the scale intensity $\frac{k_{max}^3 I(k)}{\langle\eta^2\rangle}$ can be rewritten as

$$\begin{aligned}\frac{k_{max}^3 I(k)}{\langle\eta^2\rangle} &= \frac{k_{max}^3 I(k)}{\frac{1}{2\pi^2}\int_0^\infty k^2 I(k)dk} \\ &= \frac{k_{max}^3 I(k)}{\frac{1}{2\pi^2}\left[\int_0^{k_1} k^2 I(k)dk + \int_{k_1}^\infty k^2 I(k)dk\right]} \\ &= \frac{k_{max}^3 I(k)}{\frac{1}{2\pi^2}\left[\int_0^{k_1} k^2 I(k)dk + \int_{k_1}^\infty k^2 \frac{6.5\times 10^{-13}}{k^4} dk\right]} \\ &= \frac{k_{max}^3 I(k)}{\frac{1}{2\pi^2}\int_0^{k_1} k^2 I(k)dk + 3.3\times 10^{-13}} \quad (9)\end{aligned}$$

Furthermore, $\langle\eta^2\rangle$ can be evaluated numerically and result is $4.7\times 10^{-12} \text{ Å}^{-4}$. Since the contribution from the Porod's part is only $3.3\times 10^{-13} \text{ Å}^{-4}$, which is about 7.0 % of $\langle\eta^2\rangle$, $\frac{k_{max}^3 I(k)}{\langle\eta^2\rangle}$ can thus be approximated as

$$\frac{k_{max}^3 I(k)}{\langle\eta^2\rangle} \approx \frac{k_{max}^3 I(k)}{\frac{1}{2\pi^2}\int_0^{k_1} k^2 I(k)dk} \quad (10)$$

As we see from Fig. 2, $\overline{P}(k)$ varies slowly within the k range of the first diffraction peak of $S(k)$, and is negligibly small beyond k_1, the scale intensity $\frac{k_{max}^3 I(k)}{\langle\eta^2\rangle}$ can thus be well approximated by substituting equation (2) into equation (10) and canceling the common smooth factor $\overline{P}(k)$ from the numerator and the denominator of the equation to obtain the result:

$$\frac{k_{max}^3 I(k)}{\langle\eta^2\rangle} \approx \frac{k_{max}^3 S(k)}{\frac{1}{2\pi^2}\int_0^{k_1} k^2 S(k)dk}, for k \leq k_1 \quad (11)$$

Therefore, without losing generality, the scaled intensity is proportional to the inter-particle structure factor $S(k)$ in the region of its first diffraction peak. Examples and discussion of the scaling plots are given in the next section.

RESULTS AND DISCUSSION

Intermediate Scattering Function (ISF) Measured by Photon Correlation Spectroscopy (PCS)

To test the phase behavior predicted by MCT calculations for an AHS system, ISFs were measured for a set of volume fractions within the interval $0.525 < \phi < 0.546$, where the re-entrant phenomenon of the glass-to-liquid-to-glass transition, glass-to-glass transition and the A_3 singular point were predicted for the case of $\varepsilon = 0.03$, as shown in Fig. 1. Fig. 3a & 3b show ISFs obtained by PCS for two volume fractions, $\phi = 0.525$ and $\phi = 0.535$ respectively, at different temperatures. As can be seen, on increasing temperature, the system, starting from a liquid state at a lower temperature, where the ISF decays to zero at long time, approaches a KGT, characterized by a diverging α relaxation time, and thus the ISF tends to a finite plateau at long time ($F(k, t\to\infty) = f(k) > 0$). Just before such an ergodic-to-nonergodic transition taking place, the ISF measured in an ergodic state exhibits a logarithmic relaxation (indicated by a straight line fit) at an intermediate time followed by a power-law decay before the final α relaxation sets in, in agreement with the MCT prediction. The non-ergodic state ($T = 300$ K, for $\phi = 0.535$), as indicated by the ISF having a finite plateau at long time, represents the attractive glass for which the Debye-Waller factor (DWF, the height of the plateau) is about 0.5. As discussed above, an important prediction of the MCT calculation for an AHS is a suggestion that there exists an attractive branch of KGT line near the cusp singularity separating two different glass phases [23, 10] Motivated by this prediction, we extended the study of the micellar system to higher temperatures. Starting from a glass state, upon increasing the temperature further, the measured ISF reveals a surprising re-entry from the glass to the liquid state, as seen by the ISF again decaying to zero at higher temperatures.

The ISFs measure at $\phi = 0.538$ and $\phi = 0.540$, where the glass-to-glass transition is predicted, are given in Fig. 3c & 3d. In the two-length-scale AHS, aside from the volume fraction ϕ, the glass transition can be triggered by varying the effective temperature T^* as well. Due to the fact that the depth of the potential well $-u$ is temperature dependent and increases on heating to a certain extent, the effective temperature T^* actually decreases as temperature rises for certain interval of temperatures. Variation of temperature thus makes the transition between two distinct types of glass possible. Because the long-time limit of the ISF (DWF) reflects the degree of localization of the density fluctuation, these two different types of glass, if they have different degree of local disorder, can be identified by their different values of DWF. From Fig. 3c & 3d it is obvious that all the ISFs can be grouped into two distinct sets of curves having two different values of DWF – one at $f(k) = 0.5$ (attractive glass) and the other at $f(k) = 0.4$ (repulsive glass). The two panels, 3c and 3d, confirm the predicted glass-to-glass transition.

Fig. 3e shows the ISFs measured at $\phi = 0.544$, which, according to MCT, is the end point of the predicted glass-to-glass transition line and therefore the long-time limit of ISF of the two glasses should become identical at this point. Our measured ISFs verify this prediction, showing two nearly identical values of DWF, for the two glassy states, with an average value of $f(k) \approx 0.45$. Although

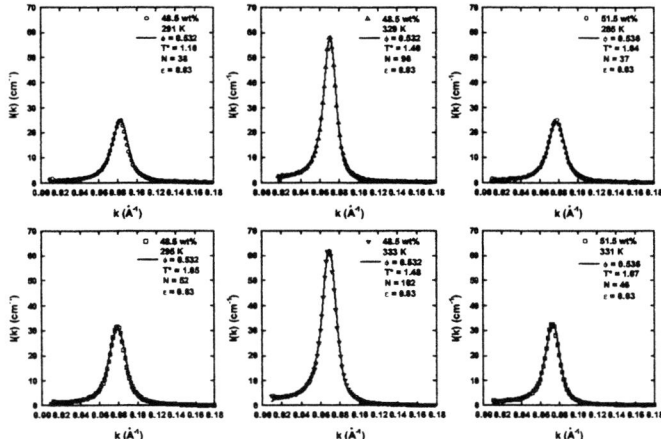

FIGURE 4. The theoretical fits to SANS data taken at $NG7$ SANS spectrometer at the NIST Center for Neutron Research, for Pluronic $L64$ micellar solutions at various concentrations and temperatures. Symbols are experimental data and lines are the fits. The fits in absolute intensity scale take into account the effects of the resolution and the incoherent scattering backgrounds. The scattering intensities increase at the higher temperature liquid phase as compared to the lower temperature liquid phase and the positions of the peak shifts toward smaller k due to the enhanced self-association (larger aggregation number) as the consequence of increased hydrophobicity of the polymer segments at higher temperatures. It is important to note that from these fits, we obtain unique values of four parameters: the volume fraction ϕ, the fractional well width ε, the effective temperature $T^* = k_B T/u$ and the aggregation number of the micelle N.

the DWFs of these two types of glasses are identical, it is essential to recognize that there is a significant difference between the dynamics of their intermediate time relaxations (the β relaxation region).

The ISFs measured at $\phi = 0.546$, which is a volume fraction beyond the A_3 point, is shown in Fig. 3f. This figure shows a similar features as in Fig. 3e, with a merged DWF of $f(k) \approx 0.4$. It strongly hints that beyond the volume fraction $\phi = 0.544$, the system exists in a repulsive glass state. It is interesting to note here a critical point-like characteristics of the A_3 point.

Fitting Results of SANS Data

Small-angle neutron scattering experiments were performed on Pluronic $L64/D_2O$ micellar solutions at different concentrations and temperatures. Part of the experimental results as well as their theoretical fits are given in Fig. 4. Symbols are experimental data and lines are the fits. The fits are in an absolute intensity scale taking into account the effects of the instrument resolution and the incoherent background. It can be seen that for samples at both volume fractions, $\phi = 0.532$ and $\phi = 0.536$, the peak intensity is generally higher for the sample at the higher temperature liquid phase. Furthermore, the position of the peak shifts toward smaller k as temperature increases. This is due to the enhanced self-association of the micelle (larger aggregation number) as the consequence of the increased hydrophobicity of the polymer segments at higher temperatures. It is important to note here that in spite of the micelle growth as temperature increases, SANS data analyses show that the volume fraction of the micelles remains constant, for a given weight fraction of $L64$, at all temperatures studied.

Scaling Plots of SANS Data

SANS intensity distributions and their scaling plots for a sample at $51.5wt\%$, or $\phi = 0.536$, at a series of temperatures ranging from $291\,K$ to $340\,K$ are given in Fig. 5. A temperature dependent degree of disorder characterizes the system. As the temperature rises, the system experiences a liquid-to-glass-to-liquid transition. However, in addition to all the similarities, when the temperature increases to $340\,K$, the system is driven into another glassy state peaked at $k = 0.082\,\text{Å}^{-1}$. From the scaling plots given in the bottom two panels, one can tell the differences between these two disordered glassy states. While the narrowest peak ($333K$ to $343\,K$) is resolution limited, the slightly broader peak ($290\,K$ to $328\,K$) is also nearly resolution limited, but lower in the scaled intensity. Since the difference in local structure of different states is reflected in their scaling plots (See discussion in section IV), we conclude that the degree of disorder are different for these two amorphous states. It can be interpreted that, by varying the temperature, the system shows

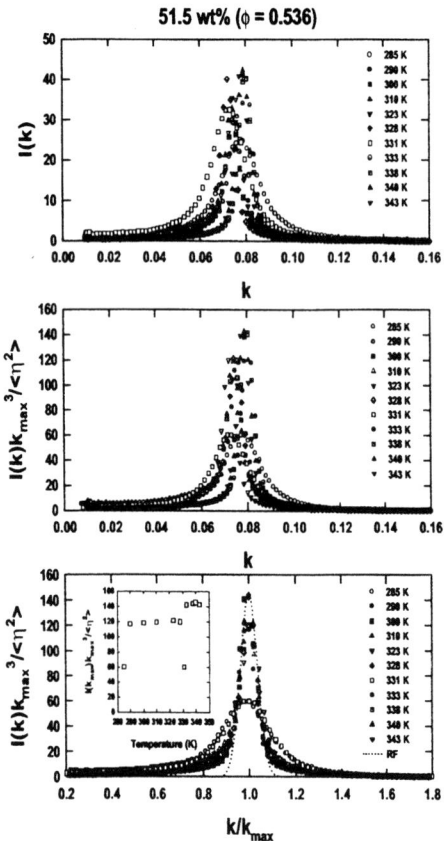
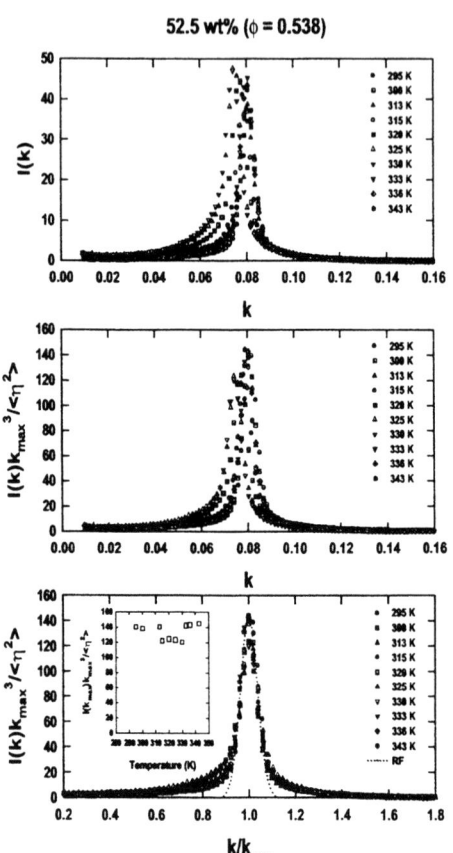

FIGURE 5. SANS intensity distributions and the scaling plots at $\phi = 0.536$, at a temperatures range spanning from 285 K to 343 K. The system is characterized by a temperature-dependent degree of disorder. As temperature increases, a liquid-to-attractive galss-to-liquid-to-repulsive glass transition is observed: The top panel gives SANS intensity distributions as a function of temperature. The broad peaks (from 291 K to 295 K, 329 K to 335 K) represent the liquid states and the narrower ones (from 298 K to 332 K) represent the attractive glass states. When the temperature increases to 333 K, the system transits to the repulsive glass state. The other two panels show the scaling plots of the SANS intensity distributions. Both of the narrow peaks are resolution-limited but the slight difference in scaling height is due to the fact that the two non-ergodic states have different degree of disorder. Fig. 5 gives a convincing evidence of the predicted re-entrant glass-to-liquid-to-glass transition. The re-entrant transition can be seen more clearly from the inset.

FIGURE 6. SANS intensity distributions and the scaling plots at $\phi = 0.538$, at a temperatures range spanning from 295 K to 343 K. From the SANS intensity distributions given in the top panel, it can be seen that the much broader peaks (liquid state) disappear. Variation of temperature triggers the transition between the two amorphous solid states with different degrees of disorder. The variation of the peak heights of the scaling plots as a function of temperature given in the inset shows a re-entrant repulsive glass-to-attractive glass-to-repulsive glass transition.

a liquid-to-glass-to-liquid–to-glass transition. In order to make this point more clearly, the peak heights of the scaling plots as a function of temperature are given in the insets of the bottom panels of Fig. 5. The transition temperatures between different amorphous states can be visualized clearly from them. These two figures give a firm evidence of the re-entrant glass-to-liquid-to-glass transition which is in good agreement with the prediction of MCT for an AHS system. We would like to mention here that the KGT boundaries determined by SANS agree with the results obtained from the latest specific heat measurements [26].

According to the MCT calculations for an AHS with sufficiently short-range attraction, a glass-to-glass transition is predicted [10]. Although the transitions between different amorphous glassy states are not uncommon in pure substance such as H_2O, Si, Ge et al [24], yet there is no detail investigation of the glass-to-glass transition in a micellar system so far except for some recent reports [16, 17]. Combing with the PCS (see Fig. 3c & 3d), a concrete evidence of such a transition in a micellar system was revealed by SANS. From SANS intensity distributions, taken at $\phi = 0.538$, given in Fig. 6, one can see that the much broader peaks (liquid state) disappear and variation of temperature triggers the transition between

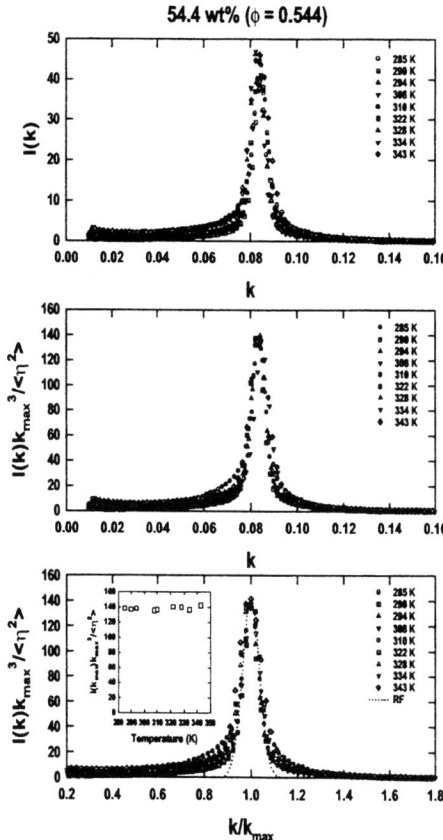

FIGURE 7. SANS intensity distributions and the scaling plots of SANS intensities at $\phi = 0.544$, which is predicted to be the volume fraction where the A_3 singularity point is located, at temperatures range spanning from 285 K to 343 K. From the top panel, it can be seen that all the peaks of the SANS intensity distribution are located at the same k (0.0836 Å^{-1}). Furthermore, from the scaling plots given in the bottom two panels, all the scaling plots collapse into one single master curve. It suggests that the two different types of glass become identical in their local structure at this volume fraction. The inset given in the bottom panel shows that all the scaling peaks have the identical height (about 140), indicating the two glasses indeed have the same degree of local order.

the two amorphous solid states with different degrees of disorder. By increasing temperature, the variation of the peak heights of the scaling plots given in the insect shows a re-entrant repulsive glass-to-attractive glass-to-repulsive glass transition.

Perhaps the most important prediction of the AHS system is the existence of the end point of the glass-to-glass transition line, the so-called A_3 singularity. Comparing with the critical point of the equilibrium states, certain degree of similarity between them can be found. Therefore, it is intriguing to speculate the extent to which one can draw the analogy between the A_3 singularity and the ordinary equilibrium critical point. In the case of

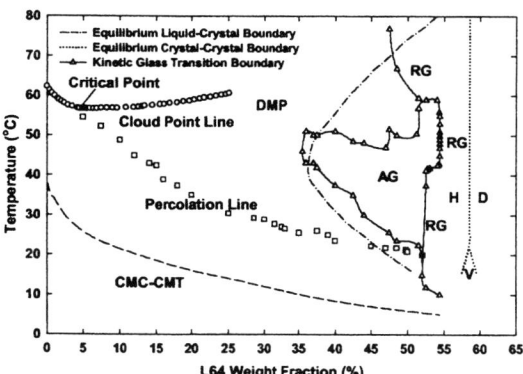

FIGURE 8. The phase diagram (in the temperature-$L64$ weight fraction plane) of Pluronic $L64$ in the $D2O$ solution. The phase diagram contains the critical point (black square), the CMC-CMT line (dash line), the cloud point line (circles), the percolation line (squares), the equilibrium liquid-to-crystal phase boundary (dash-dot line), the equilibrium crystal-to-crystal phase boundary (dot line) and the re-entrant kinetic glass transition (KGT) boundaries (triangles) which are determined by SANS experiments.

$\varepsilon = 0.03$, the volume fraction of the A_3 point is predicted to be $\phi(A_3) = 0.544$. The SANS intensity distribution and its associated scaling plot are shown in Fig. 7. As one can see, all the scaling intensity curves collapse into one single master curve, indicating the fact that first diffraction peak of the structure factor for all states are identical. It suggests that the local structures of the two glasses are identical. The insect given in the bottom panel shows all the scaling peaks have the identical height (about 140), indicating the two glasses indeed have same degree of local order.

Fig. 8 is the experimental phase diagram of $L64/D_2O$ system in the L64 weight fraction range $0 < c < 65$ wt%. Besides the cloud point curve, its associated critical point and the percolation points, which we have discussed before [20, 19, 18], we have added the newly found KGT boundaries (triangles) at high concentrations. The dash-dot line depicts the equilibrium phase boundary between the disordered micellar phase and the ordered liquid crystalline (hexagonal) phase given by Zhang et al [25]. Clearly, the amorphous states in the region of phase space under discussion are metastable states of the system, such as in supercooled liquids. In our system the crystallization would not happen unless the system is disturbed by an applied shear or placed very near a surface[19]. Thus it is clear that KGT is observed in a metastable state of the micellar solution which will last as long as we observe (days).

The phase diagram shown in Fig. 9 summarizes the essential results of the extensive SANS and PCS data

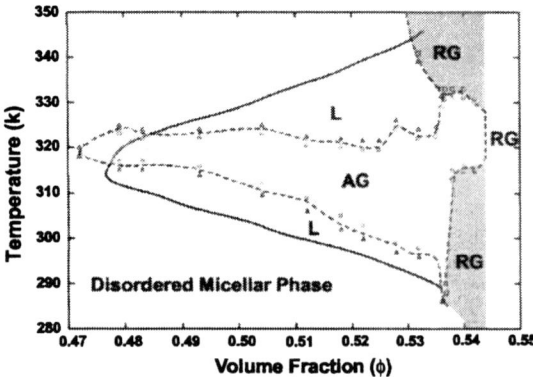

FIGURE 9. The experimental phase diagram of $L64/D_2O$ micellar system. The solid line represents the equilibrium phase boundary of the disorder micellar liquid states and the hexagonal liquid crystalline states [25]. The dash line gives the kinetic glass transition boundary which is determined by SANS and PCS. The Symbols represent the phase points where parts of the experimental data were taken. The triangles represents liquid state (L), the circles the attractive glass (AG), the squares the repulsive glass (RG). This figure gives several important information about this system: Within the region where the true ground state is the hexagonal liquid crystalline state, only the metastable attractive glass is observed. The repulsive glass is found in the region where the volume fraction is larger than 0.536. It is interesting to see that there is a pocket of the attractive glass imbedded in-between two separate repulsive glass regions, spanning the volume fraction range between 0.536 and 0.544. Furthermore, it is important to note that the two different glasses become identical in their local structures and the long-time dynamics when the volume fraction exceed $\phi(A_3) = 0.544$.

analysis. It contains the known equilibrium, liquid-to-hexagonal crystalline phase boundary (solid line) [25], the experimentally determined KGT lines (dash lines) and the phase points where parts of the experimental data are taken (symbols). This figure gives several important information about this system: First, only the metastable attractive glass is observed within the region where the true ground state is the hexagonal liquid crystalline phase. Next, the repulsive glass only exists in the region where the volume fraction is larger than 0.536. It is interesting to see that there is a pocket of the attractive glass imbedded in-between two separate repulsive glass regions spanning the volume fraction range between 0.536 and 0.544, where the re-entrant glass-to-glass transition is observed. Furthermore, from Fig. 3 and Fig. 7, it is important to note that the two different glasses become identical in their local structures and the long-time dynamics when the volume fraction exceed $\phi(A_3) = 0.544$. Furthermore, judging from the DWF and the peak height of the scaled intensity, the merged identical glassy state is the repulsive glass.

CONCLUSIONS

In summary, we use PCS to verify the fact that $L64/D_2O$ micellar system follows the overall structural arrest transition phase behavior predicted by the MCT, calculated using a square well potential with a short-range attraction relative to the micellar size. In particular, we show experimentally, the existence of a glass-to-glass transition line which starts at point C^*, where the two glass phases and the liquid phase co-exist, and ends at point A_3, where the two glass phases merge [17]. Our SANS experiment further shows that while the local structures of the attractive and the repulsive glasses are in general different, they become identical at the predicted volume fraction of the A_3 singularity, independent of temperature. However, our PCS results indicate that the relaxation of the two glasses are different in the intermediate time region even at the A_3 point. The central result of this article is the use of the SANS method to pinpoint the exact volume fractions where the C^* point and the A_3 singularity are located [10] and furthermore, to verify that for volume fractions beyond the A_3 point, the system remains in the repulsive glass phase. We are able to map out the whole structural arrest transition boundaries in the $L64/D_2O$ micellar system using SANS method alone. From SANS experiments, we observe a significant difference in the local structure when crossing the KGT boundaries at various temperatures for all the volume fractions studied. It is our conjecture that our SANS data are reflecting the ageing effect of the sample in the time scale of our measurements.

ACKNOWLEDGMENT

The research at MIT is supported by a grant from Materials Science Division of US DOE, DE-FG02-90ER45429. The research in Messina is supported by INFM-PRA98 and MURST-PRIN2000. We are indebted to Dr. Emiliano Fratini of Department of Chemistry, University of Florence, for purifying $L64$ samples. We are grateful to NIST CNR and IPNS (ANL) respectively for granting the beam time of $NG7$ SANS and of SAND station and to Dr. Charles J. Glinka and Dr. Papanan Thiyagarajan for technical assistances.

All correspondences regarding this paper should be addressed to Sow-Hsin Chen(sowhsin@mit.edu)

REFERENCES

1. D. H. Everett, *Basic Principles of Colloid Science*, (Royal Society of Chemistry, Letchworth, U. K. 1988).
2. K. Dawson, Curr. Opin. Coll. Int. Sci. **7**, 218 (2002).

3. W. Götze, in *Liquids, Freezing and the Glass Transition*, edited by J. P. Hansen, D. Levesque and J. Zinn-Justin (North Holland, Amsterdam, Netherlands 1991), p.287.
4. W. Götze and L. Sjögren, Rep. Prog. Phys. **55**, 241 (1992).
5. W. Götze, J. Phys.: Condens. Matter **11**, A1 (1999).
6. P. N. Pusey and W. van Megan, Phys. Rev. Lett. **59**, 2083 (1987).
7. M. Fuchs, Transport Theor. Stat. Phys. **24**, 855-880 (1995).
8. W. van Megan and S. M. Underwood, Phys. Rev. E **49**, 4206 (1994).
9. E. Bartsch, M. Antonietti, W. Schupp, H. Sillescu, J. Chem. Phys. **97**, 3950 (1992).
10. K. Dawson, G. Foffi, M. Fuchs, W. Götze, F. Sciortino, M. Sperl, P. Tartaglia, Th. Voigtmann and E. Zaccarelli, Phys. Rev. E **63**, 011401 (2001).
11. J. Bergenholtz, M. Fuchs, Phys. Rev. E **59**, 5706 (1999).
12. J. Bergenholtz, M. Fuchs, J. Phys.: Condens. Matter **11**, 10171 (1999).
13. T. Eckert and E. Bartsch, Phys. Rev. Lett. **89**, 125701 (2002).
14. K. N. Pham, A. M. Puertas, J. Bergenholtz, S. U. Egelhaaf, A. Moussaid, P. N. Pusey, A. B. Schofield, M. E. Cates, M. Fuchs and W. C. K. Poon, Science **296**, 104 (2002).
15. F. Mallamace, P. Gambadauro, N. Micali, P. Tartaglia, C. Liao and S. H. Chen, Phys. Rev. Lett. **84**, 5431 (2000).
16. Wei-Ren Chen, Sow-Hsin Chen and Francesco Mallamace, Phys. Rev. E **66**, 021403 (2002).
17. Sow-Hsin Chen, Wei-Ren Chen and Francesco Mallamace, Science **300**, 619 (2003).
18. S. H. Chen, C. Liao, E. Fratini, P. Baglioni, and F. Mallamace, Coll. & Surfaces A **183-185**, 95 (2001).
19. C. Liao, S. M. Choi, F. Mallamace and S. H. Chen, J. Appl. Crystallogr. **33**, 677 (2000).
20. L. Lobry, N. Micali, F. Mallamace, C. Liao and S. H. Chen, Phys. Rev E **60**, 7076 (1999).
21. G. Foffi, E. Zaccarelli, F. Sciortino, P. Tartaglia and K. A. Dawson, J. Stat. Phys. **10**, 363 (2000)
22. Y. C. Liu, S. H. Chen and J. S. Huang, Phys. Rev. E **54**, 1698 (1996).
23. L. Fabbian, W. Götze, F. Sciortino, P. Tartaglia and F. Thiery, Phys. Rev. E **59**, R1347 (1999).
24. E. G. Ponyatovsky, O. I. Barkalov, Mater. Sci. Rep. **8**, 147 (1992).
25. K. Z. Zhang, B. Lindman and L. Coppola, Langmuir **11**, 538 (1995).
26. F. Mallamace, C. Ferrari, A. Mazzaglia, P. Salvetti, E. Tombari and S. H. Chen, to be published.

Glassy dynamics in gelling systems: From chemical gels to colloidal glasses

E. Del Gado*, A. Fierro†, L. de Arcangelis** and A. Coniglio†

*Laboratoire des Verres, Université Montpellier II, France, and INFM, Napoli, Italy
†Dipartimento di Scienze Fisiche,Università di Napoli, and INFM, Italy
**Dipartimento di Ingegneria dell'Informazione,Seconda Università di Napoli, and INFM, Italy

Abstract. The study of our minimal statistical mechanics model for gelling systems, by means of numerical simulations, suggests a unifying picture for gelation phenomena, connecting classical gelation and recent results on colloidal systems. By varying the model parameters the slow dynamics present a crossover from the classical polymer gelation to dynamics more typical of colloidal systems, with a glassy regime that is interpreted in terms of effective clusters.

INTRODUCTION

In colloidal systems a strong short range attraction produces a diffusion limited cluster-cluster aggregation process with a gel formation (colloidal gelation) at low density as a permanent spanning structure is formed [1, 2, 3]. The viscoelastic response typically displays power law behavior, and the relaxation process becomes critically slow. This is the typical viscoelastic behaviour also observed in the formation of polymer gels, where the gelation is instead due to the chemical bonding and the gel structure is different (usually in polymer gels one has a random percolation structure) [4, 5, 6, 7, 8]. A weaker attraction in colloids is not able to produce a permanent gel but the formation of stable or metastable structures is still detected [9, 10, 11] together with a slowing down in the dynamics. At high densities the short range attraction is able to enhance the caging effect typical of the glassy regime, and to produce a glass-like kinetic arrest at density values lower than the hard-sphere case, and depending on the strength of the attraction [12, 13]. Therefore, for different values of the attraction strength and of the density, colloidal systems should eventually cross over from a gel-like to a glass-like behaviour. How this takes place and which is the role of the formation of stable or metastable structures is still not clear.

In order to investigate this problem we have introduced a model for gelling systems and studied the dynamic behavior by means of numerical simulations [14, 15]. Here we present the results of extensive Monte-Carlo simulations on cubic lattices. In the following sections the model and the numerical simulations are described, and the relaxation properties are studied by means of time autocorrelation functions.

MODEL AND NUMERICAL SIMULATIONS

The system we study is a solution of tetrafunctional monomers with excluded volume interactions. Each monomer occupies a lattice elementary cell and, to take into account the excluded volume interaction, two occupied cells cannot have common sites. At $t = 0$ we fix the fraction ϕ of present monomers respect to the maximum number allowed on the lattice, and randomly quench bonds between them. The four possible bonds per monomers, randomly selected, are formed with probability p_b along lattice directions between monomers that are nearest neighbours and next nearest neighbours. According to bond-fluctuation dynamics [16] the monomers diffuse on the lattice via random local movements and the bond length may vary but not be larger than l_0. The value of l_0 is determined by the self-avoiding walk condition and on the cubic lattice is $l_0 = \sqrt{10}$ in lattice spacing units.

We have first considered this model in the case of permanents bonds, i.e. once formed at $t = 0$ the bonds cannot break, nor new bonds can be formed during the dynamic evolution of the system. This corresponds to the case of chemical gelation, that can be tipically obtained by irradiating the monomeric solution.

As in colloids the aggregation is due to a short range attraction and in general the monomers are not permanently bonded, we have introduced in our model a finite bond lifetime τ_b and study the effect on the dynamics. The features of this model with finite τ_b can be realized in a microscopic model: a solution of monomers interacting via an attraction of strength $-E$ and excluded volume repulsion. Due to monomers diffusion the aggregation process eventually takes place. The finite bond lifetime τ_b corresponds to an attractive interaction of strength $-E$ that does not produce permanent bonding between monomers, and $\tau_b \sim e^{E/KT}$. Due to the finite τ_b, in the simulations during the monomer diffusion the bonds between monomers are broken with a frequency $1/\tau_b$. Between monomers separated by a distance less than l_0 a bond is formed with a frequency f_b. For each value of τ_b we fix f_b so that the fraction of present bonds is always the same [14, 15]. The finite bond lifetime τ_b obviously introduces a correlation in the bond formation during the simulation and may eventually lead to a phase separation between a low density and a high density phase: For the values of τ_b and f_b here considered there is no evidence of phase separation.

We let the monomers diffuse to reach the stationary state and then study the system for different values of the monomers concentration. We have considered $p_b = 1$, for which the system presents a percolation transition at $\phi_c = 0.718 \pm 0.005$. Varying the monomer concentration we have studied the autocorrelation function of density fluctuation $f_{\vec{q}}(t)$ given by

$$f_{\vec{q}}(t) = \frac{<\rho_{\vec{q}}(t+t')\rho_{-\vec{q}}(t')>}{<|\rho_{\vec{q}}(t')|>^2} \quad (1)$$

where $\rho_{\vec{q}}(t) = \sum_{i=1}^{N} e^{-i\vec{q}\cdot\vec{r}_i(t)}$, $\vec{r}_i(t)$ is the position of the $i-th$ monomer at time t, N is the number of monomers and the average $\langle ... \rangle$ is performed over the time t'. Due to the periodic boundary conditions the values of the wave vector $\vec{q}$ on the cubic lattice are $\vec{q} = \frac{2\pi}{L}(n_x, n_y, n_z)$ with $n_x, n_y, n_z = 1...L/2$ integer values. We also study the mean square displacement of the particles $\langle \vec{r}^2(t) \rangle = \frac{1}{N}\sum_{i=1}^{N}\langle (\vec{r}_i(t+t') - \vec{r}_i(t'))^2 \rangle$. These quantities have been calculated on a cubic lattice of size $L = 16$. The data have been averaged over ~ 10 up to 10^5 time intervals and over different initial configurations of the sample.

RELAXATION PROPERTIES

In Fig.1 we present these time autocorrelation functions as function of the time calculated on a cubic lattice of size $L = 16$. The data have been averaged over ~ 10 up to 10^5 time intervals and over 20 different initial configurations of the sample. As the monomer concentration ϕ approaches the percolation threshold ϕ_c, $f_{\vec{q}}(t)$ displays a long time decay well fitted by a stretched exponential law $\sim e^{-(t/\tau)^\beta}$ with a $\beta \sim 0.30 \pm 0.05$. At the percolation threshold the onset of a power law decay is observed as it is shown by the double logarithmic plot of Fig.1 with an exponent c [5, 6, 7, 8]. As the monomer concentration is increased above the percolation threshold

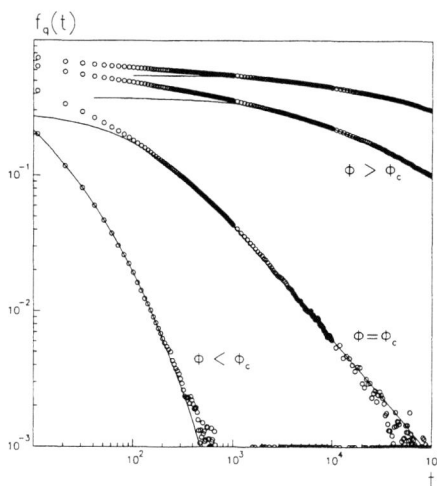

FIGURE 1. Double logarithmic plot of the autocorrelation functions $f_{\vec{q}}(t)$ as function of the time for $q \sim 1.36$ and $\phi = 0.6, 0.718, 0.8, 0.85$. For $\phi < \phi_c$ the long time decay is well fitted by a function (full line) $\sim e^{-(t/\tau)^\beta}$ with $\beta \sim 0.3$. At the percolation threshold and in the gel phase in the long time decay the data are well fitted by a function $\sim (1 + \frac{t}{t'})^{-c}$.

in the gel phase, the long time power law decay of the relaxation functions can be fitted with a decreasing exponent c, varying from $c \sim 1$. at ϕ_c to $c \sim 0.2$ well above ϕ_c, where a nearly logarithmic decay appears. This behaviour well agrees with the one observed in gelling systems investigated in the experiments of refs. [5, 6, 7, 8]. It is interesting to notice that this kind of decay with a stretched exponential and a power law reminds the relaxation behaviour found in spin-glasses [17].

In Fig.2 we present the time autocorrelation functions $f_{\vec{q}}(t)$ as function of the time calculated in the model with finite bond lifetime ($\tau_b = 400$) on a cubic lattice of size $L = 16$. The data have been averaged over ~ 10 up to 10^5 time intervals and over 20 different initial configurations of the sample. With a finite bond lifetime, close to ϕ_c $f_{\vec{q}}(t)$ is well fitted by a stretched exponential decay, but no onset of a power law decay is observed. For high monomer concentrations it exhibits a two-step decay, that closely resembles the one observed in supercooled liquids. We fit these curves using the mode-coupling β-correlator [18], corresponding to a short time power law $\sim f + \left(\frac{t}{\tau_s}\right)^{-a}$ and a long time von Schweidler law $\sim f - \left(\frac{t}{\tau_l}\right)^b$, giving the exponents $a \sim 0.33 \pm 0.01$ and $b \sim 0.65 \pm 0.01$ (the full lines in Fig.2). At long times the different curves obtained for different ϕ collapse into a unique master curve by opportunely rescaling the time via a factor $\tau(\phi)$. The master curve is well fitted by a stretched exponential decay with

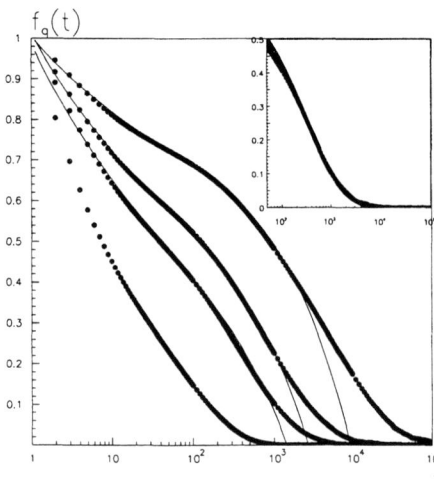

FIGURE 2. The $f_{\vec{q}}(t)$ as function of the time for $q \sim 1.36$ calculated on a cubic lattice of size $L = 16$ for $\tau_b = 400 MCstep/particle$. (from left to right $\phi = 0.7, 0.8, 0.85, 0.9$). The full lines correspond to the fit with the mode-coupling β-correlator. In the inset, by rescaling the time via a factor $\tau(\phi)$, the data for $\phi = 0.8, 0.85, 0.89$ collapse onto a unique master curve.

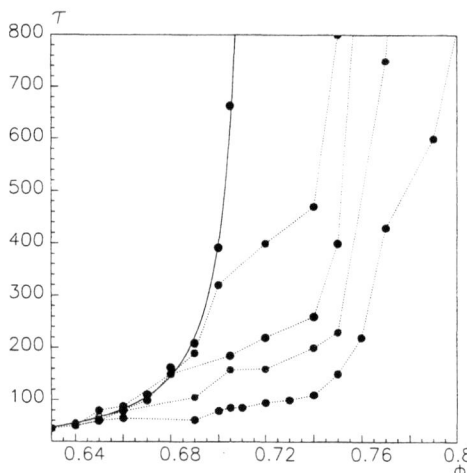

FIGURE 3. The average relaxation time as function of the density; from left to right: the data for the permanent bonds case diverge at the percolation threshold with a power law (the full line); the other data refer to finite $\tau_b = 3000, 1000, 400, 100 MCstep/particle$ decreasing from left to right (the dotted lines are a guide to the eye)

$\beta \sim 0.50 \pm 0.06$. The characteristic time $\tau(\phi)$ diverges at a value $\phi_g \sim 0.96 \pm 0.05$ with the exponent $\gamma \sim 2.3 \pm 0.1$. This value well agrees with the mode-coupling prediction $\gamma = 1/2a + 1/2b$ [18]. A glassy pattern, characterized by a plateau, is also found in the mean square displacement $\langle \overline{r^2}(t) \rangle$ (Fig.4), with the diffusion coefficient going to zero as ϕ approaches ϕ_g. Analogous results are found for different τ_b, ranging from 100 to $3000 MCS/p$. In Fig.3 the relaxation time τ, calculated as $f_{\vec{q}}(\tau) \sim 0.1$, is plotted as function of the monomer concentration ϕ for different τ_b. For comparison we have also shown the behaviour of τ in the case of permanent bonds, which displays a power law divergence at the percolation threshold ϕ_c. We notice that for finite bond lifetime τ_b the relaxation time increases following the permanent bonds case (chemical gelation), up to some value ϕ^* and then deviates from it. The longer the bond lifetime the higher ϕ^* is. In the high monomer concentration region, well above the percolation threshold, the relaxation time in the finite bond lifetime case again displays a steep increase and a power law divergence at some higher value. This truncated critical behaviour followed by a glassy-like transition has been actually detected in some colloidal systems in the viscosity behaviour [19, 20].

These results can be explained by considering that only clusters whose diffusion relaxation time is smaller than τ_b will behave as in the case of permanent bonds. Larger clusters will not persist and their full size will not be relevant in the dynamics: the finite bond lifetime induces an effective cluster size distribution with a cut-off, which keeps the macroscopic viscosity finite [21]. As the concentration increases the final growth of the relaxation time is due to the crowding of the particles. In Fig.4 we directly compare the time autocorrelation functions and the particle mean square displacement obtained at the same concentrations in the finite bond lifetime case and in the case with permanent bonds. We observe that for short time (of the order of τ_b) the relaxation process coincides. Analogous results are found for all the values of τ_b considered in this study [15]. This suggests that on time scales smaller than τ_b the relaxation process must be on the whole the same as in the case of permanent bonds, where permanent clusters are present in the system. That is, for non permanent bonds, this first relaxation is due to relaxation processes that take place over length of the order of the effective cluster radius scales. Over the considered time scale, the system is kinetically arrested in the permanent bond case, once that the percolating cluster is formed. The system with finite τ_b may still relax instead, as the size of the effective clusters is small compared to the sample size. This gives an interpretation in terms of the effective clusters for the two step glassy behavior of the relaxation functions. The first relaxation should be due to the motion of a cluster within the cage formed by the other clusters whereas the second relaxation is due to the cage opening. For high temperature $\tau_b \to 0$, the clusters reduce to single monomers. When τ_b is large enough (strong attraction) the cluster effect will dominate and the slow dynamics will exhibit features more closely re-

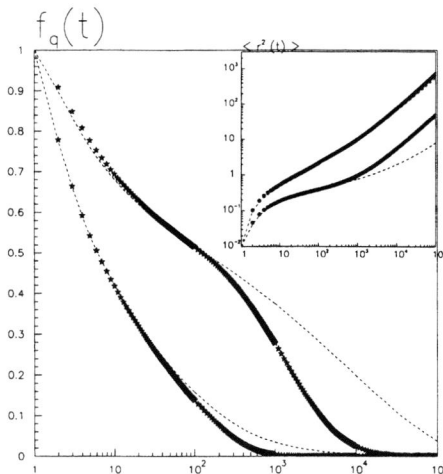

FIGURE 4. The $f_{\vec{q}}(t)$ as function of the time for $\phi = 0.718$ and $\phi = 0.8$ in the case of $\tau_b = 1000 MCstep/particle$ and at the same densities in the case of permanent bonds (dashed lines). In the inset $\langle \vec{r}^2(t) \rangle$ for the same values of ϕ for $\tau_b = 1000 MCS/particle$ and in the case of permanent bonds (dashed line).

lated to chemical gelation (Fig.1). The only difference is that in the limit $\tau_b \to \infty$ we expect that the spanning cluster will have the structure of the cluster-cluster irreversible aggregation model instead of random percolation [14, 15].

CONCLUSION

Our study shows that in a gelation phenomenon with permanent bond formation (chemical gelation or irreversible cluster aggregation) the divergence of the relaxation time is due to the formation of a macroscopic critical cluster and the autocorrelation function exhibits a one step decay related to the relaxation of such spanning cluster. That is, in this case the percolating cluster of bonds is the stress-bearing network which produces the divergence of the relaxation time.

In the case of non permanent bonds the relaxation time increases due the formation of effective clusters. The size of these effective clusters due to the finite lifetime does not diverge and the relaxation time exhibits a pseudo divergence corresponding to a state which we call pseudo gel or soft gel. As the density ϕ increases the clusters will get crowded until a glass transition of clusters is reached. In the $\tau_b - \phi$ plane therefore we have two lines: a pseudo gel line and a cluster glass transition line. We expect that as τ_b diverges (low temperature) the two lines will end up at zero density in the cluster-cluster aggregation point. However in the low density region in general we expect

that the line will interfere with the phase separation curve which needs to be treated with much more care. This problem is under investigation.

The effective clusters do not coincide with pairwise bonded particles: A cluster can be identified in a statistical sense as group of monomers which keeps its identity (i.e. the bonds unbroken) when diffusing a distance equal to its diameter. In terms of such clusters we give an interpretation of the two step relaxation (which is well fitted by the Mode-Coupling theory preditions) exhibited by the time autocorrelation function at high monomer concentrations. The first relaxation should be due to the motion of a cluster within the cage formed by the other clusters whereas the second relaxation is due to the cage opening resulting in a structural relaxation. Therefore the effective clusters play the role of single molecules in an ordinary super-cooled liquid or in a colloidal hard sphere system. This picture supports the jamming of clusters that has been suggested on the basis of experimental observations on colloidal gelation in ref. [11].

In conclusion, these results suggest a unifying approach for chemical gelation, colloidal gelation and colloidal glass transition. In chemical gelation and colloidal gelation the cluster formation should produce the slow dynamics. In colloidal systems for weak attraction and high concentration the system crosses over from colloidal gelation to colloidal glass due to the jamming of effective clusters.

ACKNOWLEDGMENTS

This research has been supported by Marie Curie Fellowship of the European Community Programme FP5 under contract number HPMF-CI2002-01945. It has been partially supported by MIUR-FIRB-2002 and by the INFM Parallel Computing Initiative.

REFERENCES

1. Dinsmore, A., and Weitz, D., *J. Phys. : Condens. Matter*, **14**, 7581–7597 (2002).
2. Meakin, P., *Phys. Rev. Lett*, **51**, 1119–1122 (1983).
3. Kolb, M., Botet, R., and Jullien, R.; *Phys. Rev. Lett*, **51**, 1123–1126 (1983).
4. Adam, M., Lairez, D., Karpasas, M., and Gottlieb, M., *Macromolecules*, **30**, 5920–5929 (1997).
5. Martin, J., Wilcoxon, J., and Odinek, J., *Phys. Rev. A*, **43**, 858–872 (1991).
6. Ikkai, F., and Shibayama, M., *Phys. Rev. Lett*, **82**, 4946–4949 (1999).
7. Ren, S., and Sorensen, C., *Phys. Rev. Lett*, **70**, 1727–1730 (1993).
8. Lang., P., and Burchard, W., *Macromolecules*, **24**, 814–815 (1991).

9. Trappe, V., Prasad, V., Cipelletti, L., Segre, P., and Weitz, D., *Nature*, **411**, 772–775 (2001).
10. Trappe, V., and Weitz, D., *Phys. Rev. Lett*, **85**, 449–452 (2000).
11. Segré, P. N., Prasad, V., Schofield, A. B., and Weitz, D. A., *Phys. Rev. Lett*, **86**, 6042–6045 (2001).
12. Gang, H., Krall, A. H., Cummins, H. Z., and Weitz, D. A., *Phys. Rev. E*, **59**, 715–721 (1999).
13. Fabbian, L., Götze, W., Sciortino, F., Tartaglia, P., and Thiery, F., *Phys. Rev. E*, **59**, R1347–R1350 (1999).
14. DelGado, E., Fierro, A., de Arcangelis, L., and Coniglio, A., *Europhys. Lett.*, **63**, 1–7 (2003).
15. DelGado, E., Fierro, A., de Arcangelis, L., and Coniglio, A., *submitted to Phys. Rev. E* (2003).
16. Deutsch, H. P., and Binder, K., *J. Chem. Phys.*, **94**, 2294–2304 (1991).
17. Ogielski, A. T., *Phys. Rev. B*, **32**, 7384–7398 (1985).
18. Göetze, W., "," in *Liquid, Freezing and Glass Transition*, edited by D. L. J.P. Hansen and P. Zinn-Justin, Elsevier, 1991, pp. 287–504.
19. Mallamace, F., Chen, S., Liu, Y., Lobry, L., and Micali, N., *Physica A*, **266**, 123–135 (1999).
20. Laflèche, F., Durand, D., and Nicolai, T., *Macromolecules*, **36**, 1331–1340 (2002).
21. Coniglio, A., *J. Phys. : Condensed Matter*, **13**, 9039–9053 (2001).

Mode Coupling Theories for Jamming and Gelation

M. E. Cates

School of Physics, University of Edinburgh, JCMB Kings Buildings, Edinburgh EH9 3JZ, United Kingdom

Abstract. The mode coupling theory (MCT) for the glass transition is in reasonable accord with observations on attractive colloids at high densities. This paper presents a brief review of the MCT. This is followed by a more detailed discussion of two recent extensions of it. These address (i) nonlinear rheology and jamming in colloids under shear and (ii) weak gelation of colloids at low densities.

INTRODUCTION

Hard-sphere colloids with short-range attractions can undergo several types of arrest. At high densities they show two distinct glass transitions (repulsion-driven and attraction-driven), with a re-entrant dependence on attraction strength [1]. This scenario was first predicted by mode coupling theory (MCT) [2, 3, 4], and depends on both the attraction range ξ and well-depth ε. MCT is remarkably successful, at least for large volume fractions ϕ ($\gtrsim 0.4$).

At lower volume fractions, however, there is no comparable theoretical framework. This is a serious deficit: 'weak gelation', in which bonding is strong but not so strong as to be irreversible, can lead to nonergodic soft solids, of nonzero static elastic modulus, at volume fractions of just a few percent [5]. It might be argued that a finite modulus requires a percolating network of bonds whose lifetime exceeds that of the experiment. However this is simplistic: as shown by the case of repulsive glasses, a finite modulus can arise with no bonding at all. We argue here (following [6]) that the rigidity of weak gels arises not from bond percolation but from ergodicity breaking, just as it does in glass formation. This suggests that an extension of the MCT approach towards the regime weak gelation could be fruitful. Recent progress in that direction is discussed in the second half of this paper.

This is preceded by a discussion of another area in which recent extensions of MCT may play an important role: the nonlinear rheology of colloids. This has two main aspects. One is shear thinning: for example, a suspension that is in the nonergodic (glass) phase has a yield stress, defined as a nonzero steady-state stress in the limit where a steady flow rate tends smoothly to zero. The second aspect is shear thickening, in which a fluid phase appears to arrest, or at least become more sluggish, as a result of an applied stress. Our discussion of these topics is chiefly geared to hard sphere colloids, though the tools developed are more general.

First, though, let us summarize briefly the broader issue of arrest in colloidal suspensions, and how MCT (in is standard, unextended form) handles this.

Arrest in colloidal fluids

Colloidal fluids can be studied relatively easily by light scattering [7, 8]. This measures the dynamic structure factor $S(q,t_1-t_2) = \langle \rho(\mathbf{q},t_1)\rho(-\mathbf{q},t_2)\rangle/N$ and also the static one, $S(q) = S(q,0)$. Here $\rho(\mathbf{r},t) = \sum_i \delta(\mathbf{r}_i(t) - \mathbf{r}) - N/V$; this is the real space particle density (with the mean value subtracted), and $\rho(\mathbf{q},t)$ is its Fourier transform. For particles of radius a with short-range repulsions, $S(q)$ exhibits a peak at a value q^* with $q^*a = \mathcal{O}(1)$. The dynamic structure factor $S(q,t)$, at any q, decays monotonically from $S(q)$ with increasing t. In an ergodic phase, $S(q,t)$ decays to zero eventually: all particles can move, and the density fluctuations have a finite correlation time. In an arrested phase, which is nonergodic, this is not true. Instead the limit $S(q,\infty)/S(q) = f(q)$ defines the *nonergodicity parameter*. The presence of nonzero $f(q)$ signifies frozen-in density fluctuations. Although $f(q)$ is strongly wavevector dependent, it is common to quote only $f(q^*)$ [9]. The above formulas assume time-translation invariance; nonergodic systems can violate this (showing aging phenomena) in which case $S(q,t_1-t_2)$ as defined above must be written $S(q,t_1,t_2)$ with two time arguments.

In many colloidal materials the effective interparticle interaction $u(r)$ comprises a hard sphere repulsion, operative at separation $2a$, perhaps combined with an attraction at larger distance. (For simplicity one can imagine a square well potential of depth ε and range $\xi = a\delta$, with $\delta < 1$ typically.) Colloidal fluids of this type are found to undergo transitions into two different broad classes of nonergodic phase. One is the colloidal glass, in which arrest is caused by the trapping of each particle in a cage of neighbours. This occurs even for $\varepsilon = 0$ (i.e. hard spheres) at volume fractions above about $\phi \equiv 4\pi a^3 N/3V \simeq 0.58$. The nonergodicity parameter for the colloidal glass obeys $f(q^*) \simeq 0.8$. The second arrested state is called the attractive glass (which seems to be connected to the colloidal gel at lower densities). Unlike the repulsive glass, the arrest here is driven by attractive interactions, resulting in a bonded, network-type structure. Such gels can be unambiguously found, for short range attractions, whenever $\varepsilon \gtrsim 5 - 10$. (Here $\varepsilon \equiv \epsilon/k_B T$.) Hence it is not necessary that the local bonds are individually irreversible (this happens, effectively, at $\varepsilon \gtrsim 15 - 20$); and when they are not, the arrest is a collective, not just a local, phenomenon.

Mode Coupling Theory (MCT)

We do not review MCT in detail here. One widely used form of the theory [10] is based on projection methods. However, in a stripped-down version (see e.g. [11, 12]) the resulting equations can be viewed as a fairly standard one-loop selfconsistent approximation to a dynamical theory for the particle density field.

We take $\beta = 1$, bare particle diffusivities $D_0 = 1$, and start from the overdamped Langevin equations $\dot{\mathbf{r}}_i = \mathbf{F}_i + \mathbf{f}_i$ for independent particles of unit diffusivity subjected to external forces $\mathbf{F}_i$ and random forces $\mathbf{f}_i$. One proceeds by a standard route to a Smoluchowski equation $\dot{\Psi} = \Omega \Psi$ for the N-particle distribution function Ψ, with evolution operator $\Omega = \sum_i \nabla_i \cdot (\nabla_i - \mathbf{F}_i)$. Now take the forces $\mathbf{F}_i$ to originate (via $\mathbf{F}_i = -\nabla_i H$) from an interaction Hamiltonian

$$H = -\frac{1}{2} \int d^3 \mathbf{r} d^3 \mathbf{r}' \rho(\mathbf{r}) \rho(\mathbf{r}') c(|\mathbf{r} - \mathbf{r}'|) \quad (1)$$

where $Nc(q) = V[1 - S(q)^{-1}]$. This is a harmonic expansion in density fluctuations; $c(q)$ is the direct correlation function, and this form ensures that $S(q)$ is recovered in equilibrium. We neglect solvent-mediated dynamic forces: in principle these couplings mean that the noise in the Langevin equation should be correlated between particles, in contrast to the independent white noise assumed here. In addition we neglect anharmonic terms in H; to regain the correct higher order density correlators (beyond the two point correlator $S(q)$) in equilibrium, these terms would have to be put back.

From the Smoluchowski equation (or the corresponding nonlinear Langevin equation for the density $\rho(\mathbf{r})$[11, 13]), one can derive a hierarchy of equations of motion for correlators such as $S(q,t)$, more conveniently expressed via $\Phi(q,t) \equiv S(q,t)/S(q)$. Factoring arbitrarily the four-point correlators that arise in this hierarchy into products of two Φ's, one obtains a closed equation of motion for the two point correlator

$$\dot{\Phi}(q,t) + \Gamma(q) \left[\Phi(q,t) + \int_0^t m(q,t-t') \dot{\Phi}(q,t') dt' \right] = 0 \quad (2)$$

where $\Gamma(q) = q^2/S(q)$ is an initial decay rate, and the memory function obeys

$$m(\mathbf{q},t) = \sum_\mathbf{k} V_{\mathbf{q},\mathbf{k}} \Phi(\mathbf{k},t) \Phi(\mathbf{k}-\mathbf{q},t) \quad (3)$$

with the vertex

$$V_{\mathbf{q},\mathbf{k}} = \frac{N}{2V^2 q^4} S(q) S(k) S(|\mathbf{k}-\mathbf{q}|) g(\mathbf{q},\mathbf{k}) \quad (4)$$

$$g(\mathbf{q},\mathbf{k}) \equiv [\mathbf{q}.\mathbf{k} c(k) + \mathbf{q}.(\mathbf{k}-\mathbf{q}) c(|\mathbf{k}-\mathbf{q}|)]^2 \quad (5)$$

Equations 2-5 are slightly simpler than the ones used in molecular glasses because they neglect inertial terms; they completely define the MCT as usually applied in colloidal systems [10].

These MCT equations exhibit a bifurcation that corresponds to a sudden arrest transition, upon smooth variation of either the density ϕ or any interaction parameters that control $c(q)$ (equivalently, $S(q)$). Here the nonergodicity parameters $f(q)$, suddenly jump from zero to nonzero values. Near this (on the ergodic side), $\Phi(q,t)$ develops interesting behaviour. Viewed as a function of time, it decays onto a plateau of height $\tilde{f}(q)$, stays there for a long time, and then finally decays again at very late times. The two decays are called β and α respectively. Upon crossing the bifurcation, their relaxation times diverge smoothly with the parameters; upon crossing the locus of this divergence, $\tilde{f}(q) \to f(q) \equiv S(q,\infty)$, so that $f(q)$ jumps discontinuously from zero to a finite value.

COLLOID RHEOLOGY

MCT for shear thinning

In Ref.[14], a theory is propounded, along MCT lines, of colloidal suspensions under flow. The work was intended mainly to address the case of repulsion-driven glasses, and to study the effect of imposed shear flow either on a glass, or on a fluid phase very near the glass transition. In either case, simplifications might be expected because the bare diffusion time $\tau_0 = a^2/D_0$ is small compared to the 'renormalized' one $\tau = a^2/D$, which in fact diverges (enslaved to the α relaxation time) as the glass transition is approached. If the imposed steady shear rate is $\dot\gamma$, then for $\dot\gamma\tau_0 \ll 1 \leq \dot\gamma\tau$, one can hope that the details of the local dynamics are inessential and that universal features related to glass formation should dominate. Note, however, that by continuing to use a quadratic H (Eq.1), Ref.[14] assumes that, even under shear, the system remains 'close to equilibrium' in the sense that the density fluctuations that build up remain small enough for a harmonic approximation to be useful. This may well be inadequate for hard spheres, but a systematic means of improvement upon it is not yet available. A related but distinct approach, similarly inspired by MCT but with a different set of approximations, was presented in Ref.[15].

The basic route followed in Ref.[14] is quite similar to that laid out above for standard MCT, modulo the fact that an imposed shear flow is now present. A key simplification is to neglect velocity fluctuations so that the imposed shear flow is locally identical to the macroscopic one; this cannot be completely correct, but allows progress to be made. For related earlier work see Refs.[16, 17].

The calculations of Ref.[14] (see also [18, 19]) give several interesting results. First, any nonzero shear rate, however small, restores ergodicity for all wavevectors (including ones which are transverse to the flow and do not undergo direct advection). This is important, since it is the absence of ergodicity that normally prevents MCT-like theories being used inside the glass phase, at $T < T_g$ or $\phi > \phi_g$. Here we may use the theory in that region, so long as the shear rate is finite.

In the liquid phase ($\phi < \phi_g$) the resulting flow curve $\sigma(\dot\gamma)$ shows shear thinning at $\dot\gamma\tau \gtrsim 1$, which is when the shearing becomes significant on the timescale of the slow relaxations. This is basically as expected. Less obviously, throughout the glass, one finds that the limit $\sigma(\dot\gamma \to 0+) \equiv \sigma_Y$ is nonzero. This quantity is called the yield stress and represents the minimum stress that needs to be applied before the system will respond with a steady-state flow. (For lower stresses, various forms of creep are possible, but the flow rate vanishes in steady state.)

The prediction of a yield stress in colloidal glasses is significant, because glasses, operationally speaking, are normally defined by the divergence of the viscosity. However, it is quite possible for the viscosity to diverge without there being a yield stress, for example in 'power law fluids' where $\sigma(\dot\gamma) \sim \dot\gamma^p$ with $0 < p < 1$ [20]. This does not happen in the calculation of Ref.[14], where the yield stress jumps discontinuously from zero to a nonzero value, σ_Y^c, at ϕ_g. The existence of a yield stress seems to be in line with most experimental data on the flow of colloidal glasses, although one must warn that experimentalists' definitions of what a yield stress is, do vary across the literature [21]. Ours is defined as the limiting stress achieved in a sequence of experiments at ever decreasing $\dot\gamma$, ensuring that *a steady state is reached* for each shear rate before moving onto the next one. The latter requirement may not be practically achievable since the equilibration time could diverge smoothly at small $\dot\gamma$: one would expect to have to wait at least for times t such that $\dot\gamma t \gtrsim 1$. But unless the flow curve has unexpected structure (absent in this approach) at small shear rates, the required extrapolation should be safe.

Schematic MCT models

It has long been known that the key mathematical structure behind Eqs. 2–5 can be captured by low-dimensional schematic models in which the full **q** dependence is suppressed [22, 10]. In other words, one chooses a single mode, with a representative wavevector around the peak of the static structure factor, and writes mode coupling equations for this mode treated by itself. At a phenomenological level, one can capture the physics similarly even with shearing present (despite the more complicated vectorial structure that in reality this implies). Specifically one can define [14] the $F_{12}^{\dot\gamma}$ model — the sheared extension of a well known static model, F_{12} — via

$$\dot\Phi(t) + \Gamma\left[\Phi(t) + \int_0^t m(t-t')\dot\Phi(t')dt'\right] = 0 \quad (6)$$

with memory function (now schematically incorporating shear)

$$m(t) = [v_1\Phi(t) + v_2\Phi^2(t)]/(1+\dot\gamma^2 t^2) \quad (7)$$

The vertex parameters $v_{1,2}$ are smooth functions of the volume fraction ϕ and of the interactions. To calculate flow curves, etc., one also needs a recipe for computing stresses: here we take the first moment of the correlator to fix the time scale for stress relaxation (which is, in suitable units, simply the viscosity):

$$\eta = \int_0^\infty \Phi(t)dt \quad (8)$$

This simplest of schematic models gives very similar results to a much more sophisticated (but still schematic) approximation of the full equations [14], with $\sigma - \sigma_Y \sim \dot\gamma^{0.16}$ and $\sigma_Y - \sigma_Y^c \sim (\phi - \phi_g)^{1/2}$. Such predictions can be compared with experiment [18] and suggest that the more advanced schematic models are at least semi-quantitative. The essential physics captured by these models is the loss of structural memory caused by shear: by forcing particles to flow past each other, the correlations encoded in local cages are disrupted and this leads to loss of memory. This memory loss applies even to modes with wavevector in the vorticity direction which do not feel any direct effect of the shearing [14].

Shear thickening and jamming

The calculations described above predict generic shear thinning behaviour: advection kills fluctuations, reducing the α relaxation time, which causes the system to flow more easily at higher stresses. However, in some colloidal systems, the reverse occurs. This is shear thickening, and gives a flow curve $\sigma(\dot\gamma)$ with upward curvature. In extreme cases, an essentially vertical portion of the curve is reported [23, 24]. One interpretation of the latter scenario (called 'discontinuous shear thickening') is that the underlying flow curve is actually S-shaped. Since any part of the curve with negative slope is mechanically unstable (a small increase in the local shear rate would cause an acceleration with positive feedback), this allows a hysteresis cycle in which, at least according to the simplest models, discontinuous vertical jumps on the curve bypass the unstable section (see Figure 1).

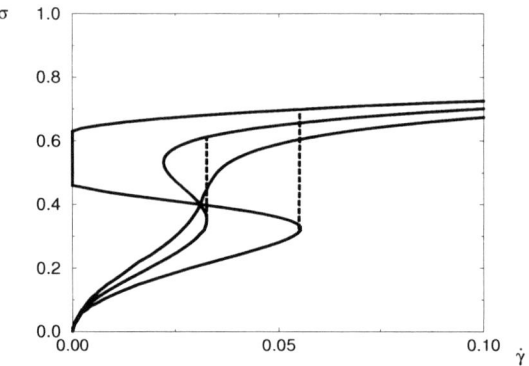

FIGURE 1. Three possible flow curves for a shear thickening material. The monotonic curve corresponds to continuous shear thickening. The remaining two curves are S-shaped; one expects, on increasing the shear rate, the stress to jump from the lower to upper branch at (or before) the vertical dashed line shown in each case. One curve shows the full jamming scenario: the existence of an interval of stress, here between 0.45 and 0.63, within which the flow rate is zero, even in a system ergodic at rest. (Stress and strain rate units are arbitrary.)

If this viewpoint is adopted, there seems to be nothing to prevent the upper, re-entrant part of the curve from extending right back to the vertical axis (see Figure 1) in which case there is zero steady-state flow within a certain interval of stress. The system has both an upper and a lower yield stress delimiting this region. (If it is nonergodic at rest, it could also have a regular yield stress on the lower part of the curve near the origin – we ignore this here.) This case has been called 'full jamming' [25]. Although mostly a theoretical speculation, at least one experimental report of this kind of behaviour has appeared in the literature recently [26].

The above discussion suggests that shear thickening and full jamming might be viewed as a stress-induced glass transition of some sort [27]. If so, it is natural to ask whether this idea can be accommodated within an MCT-like approach. Since the analysis of Ref. [14] gives only shear thinning, this is far from obvious. In particular, a stress-induced glass transition would require the vertex V to 'see' the stress; this might require one to go beyond harmonic order in the density, that is, it might require improvement to Eq.1. Indeed, since it is thought that jamming arises by the growth of chainlike arrangements of strong local compressive contacts [27], it is very reasonable to assume that correlators beyond second order in density should enter.

In Ref.[28], Holmes et al develop a schematic model along the lines of Eqs.6–8 to address shear thickening (with, for simplicity, $v_2 = 0$). This is the $F_1^{\dot\gamma,\sigma}$ model

$$\dot\Phi(t) + \Gamma\left[\Phi(t) + \int_0^t m(t-t')\dot\Phi(t')dt'\right] = 0 \quad (9)$$

with memory function

$$m(t) = [v_0 + \alpha\sigma]\exp[-\dot\gamma t]\Phi(t) \quad (10)$$

and viscosity $\eta = \sigma/\dot\gamma$ obeying Eq.8. The memory function now schematically incorporates both the loss of memory by shearing and a stress-induced shift of the glass transition. (Without stress or shear, the latter occurs at $v_0 = 4$.) The choice of an exponential strain rate dependence is purely for algebraic convenience, whereas the form in Eq.7 is closer to the one found in the full **q**-dependent vertex under shear [14]. The choice of a linear dependence of the vertex on stress (rather than the quadratic one that would arise in a Taylor expansion about the quiescent state) can be viewed as a linearization about a finite stress chosen to lie close to the full jamming region: this, rather than the behaviour at very small stresses, is the interesting region of the model. In any case, the qualitative scenarios that emerge from Eqs.8–10 are relatively robust to the precise details of the model [28].

This model results in a 'full jamming' scenario (qualitatively similar to Fig.1) as part of a wider range of rheological behaviour. Note that if, as seems likely, α depends on the details of interparticle interactions, then the evolution between scenarios does too. This makes sense since one would certainly expect hard particles to be more 'jammable' than soft ones.

A very interesting aspect of shear thickening fluids is their tendency to exhibit unstable shear rates under conditions of constant stress or vice versa. To study this requires a time-resolved description of the glass transition, which is not currently available with MCT (since that is essentially a steady-state theory). However, progress can be made with other types of model [29, 30] and this is explored in a separate paper elsewhere in these proceedings [31].

WEAK GELATION

MCT takes its structural input from equilibrium liquid state theory; it cannot address states of arrest where this structure is strongly perturbed. This matters relatively little at $\phi \gtrsim 0.4$, where each particle interacts with many others, and not much room is left for structural development upon a quench. But more severe consequences must be expected at low volume fractions where strongly nonuniform, ramified gels arise. Here the pathway to complete nonergodicity (starting from a homogenized fluid sample, say) must involve a nontrivial episode of structure formation, akin to irreversible cluster aggregation (ICA). Such kinetics certainly dominates for irreversible ("strong") gelation ($\varepsilon^{-1} = 0$), where particles aggregate on contact into clusters, with various kinetic universality classes [32]. Relative simplicity is restored at low ϕ thanks to the invariance of this aggregation process under coarse graining in the (ordered) limit δ, $\varepsilon^{-1} \to 0$ and $\phi \to 0$ (recall we define $\delta \equiv \xi/a$ and $\varepsilon \equiv \varepsilon/k_B T$). This scaling limit is controlled by an ICA 'fixed point', where details of the short-ranged attraction are irrelevant. The resulting fractal clusters grow indefinitely only if $\phi = 0$; for $\phi > 0$ they eventually form a percolating gel of locally ICA-like structure [33].

In Ref.[34], Kroy et al explore how ICA might connect to weak gelation through a unified (albeit speculative) MCT-based scenario of colloidal arrest. They consider the changes in the system parameters $(\phi, \delta, \varepsilon^{-1})$, upon coarse graining in the vicinity of the ICA fixed point. This leads to a schematic description of the suspension in terms of an effective theory for a dense liquid of 'renormalized particles' or coarse-grained clusters. Applying MCT to this theory, a new condition for arrest *of the clusters* is found [34], and the arrested state identified with the weak gel phase. If this condition is not met, the authors predict instead a fluid of clusters (or 'cluster phase' [35]) that is ergodic at large scales.

Double ergodicity breaking

Within this scenario, weak colloidal gelation emerges as a double ergodicity breaking: once when the original suspension becomes kinetically unstable against aggregation, and a second time when the fluid of clusters arrests. It is supposed for simplicity that the criterion for the initial instability is given by bare MCT. We assume $1 \ll \varepsilon < \infty$: bonds can break with a small but finite probability. Initially this has almost no effect on the aggregation, which flows towards the ICA fixed point. But as clusters grow, the chances of fragmentation increase. The ICA fixed point is unstable and the system eventually flows away from it.

The flow takes place in a parameter space comprising effective system variables $(\tilde{\phi}, \tilde{\delta}, \tilde{\varepsilon}^{-1})$. These coarse-grained parameters describe the density and interaction of aggregated clusters. Kroy et al write:

$$\tilde{\phi} = \phi\, N^{3/d_c - 1} \tag{11}$$

$$\tilde{\delta} = \delta\, N^{-\nu} \tag{12}$$

$$\tilde{\varepsilon} = \varepsilon - \chi \ln N - f(N) \tag{13}$$

and call the combination of MCT with Eqs.(11–13) 'cluster mode–coupling theory' (CMCT). N is the number of particles in a typical cluster, which is a parametric label along flow trajectories, each of which is indexed by values of $(\phi, \delta, \varepsilon^{-1})$ in the initial state. In Eqs.(11,12), evolution of the effective volume fraction of clusters (viewed as quasi–spherical, smooth objects) involves a scaling index d_c, and that of the effective relative range δ of their attraction another positive exponent ν.

Somewhat more subtle is the evolution of the bond energy ε in Eq.(13). This accounts for bond breaking through a renormalization of the cluster–cluster attraction ε by a breaking entropy, which is written as the logarithm of a breaking degeneracy $N^\chi \leq N$. This acknowledges that a ramified cluster falls apart when any of the N^χ bonds along its backbone is broken (with probability $\sim e^{-\varepsilon}$). For completeness, Eq.(13) also includes a term $f(N)$ arising from interactions. For purely short range attractions this term should vanish, but it allows us to address also the case of a weak long range repulsion, such as has recently been shown to arise from poorly screened dissociated surface charges even for supposedly neutral colloids in organic solvents [36]. Depending on the strength and range of the repulsion, the resulting $f(N)$ may dominate the logarithmic term in Eq.(13).

Initially, all three exponents introduced in the CMCT equations are properties of the ICA fixed point (typically one expects $d_c \approx 2$, with $\nu \approx \chi \approx 1/d_c$, in three dimensions [32]). However, they will start evolving once the flow moves away from the fixed point, as the internal structure of clusters evolves. An explicit treatment of this drift is not attempted in Ref.[34] where d_c, ν, χ are treated as 'negotiable constants'.

Results

The resulting CMCT flow within the renormalized parameter space $(\tilde{\phi}, \tilde{\delta}, \tilde{\varepsilon}^{-1})$ is discussed fully in Ref.[34]. The flow is bounded by the MCT transition surface in this space, which divides ergodic from nonergodic regions. Suppose we start within a nonergodic region, at some low density ϕ. As a result of the initial, short-length-scale ergodicity breaking, this system will start to aggregate. But due to the parameter renormalizations

Eqs.(11–13), its trajectory up increasing N can cause it to cross the MCT transition surface. At this point, the system ceases to aggregate and one has a semi-ergodic phase of finite clusters: the cluster fluid. If a series of such cluster fluids is now created by increasing ϕ at fixed ε (or vice versa), it is found that the last of these to enter an ergodic phase does so just beside a certain cusp on the MCT surface [34]; this cusp lies at the junction between repulsive stabilized glasses and those stabilized by attraction [2, 3, 4].

A system whose parameter flow under renormalization encounters the MCT surface at this cusp is called the 'marginal gel'. In Ref.[34] it is argued that this may have some special kinetic signatures (such as logarithmic relaxations) as a result of proximity to certain higher order singularities that reside quite nearby [2, 3, 4]. For the marginal gel, which is the first fully nonergodic phase formed in a sequence of initial states of increasing ϕ (or ε), the renormalization trajectory just meets the cusp. Aggregation now ceases, not because the aggregating system has entered a fluid phase at large length scales (as arose in the cluster fluid) but because the renormalized parameters are now those of a *repulsive* glass.

This state is fully nonergodic at large scales, and thus has finite modulus — as do all the non-marginal cases arising at larger ϕ (or ε). The resulting new "cluster glass" phase is tentatively identified with the weak gel. Note that the condition for gelation is not that of percolation: weak gelation arises in systems having reversible bonding so that the mere existence of a connected network of bonds at any instance is not relevant. What is relevant is the breakdown of ergodicity at large length scales; it is this that CMCT attempts to predict. A typical phase diagram predicted by this approach is given in Ref.[34]; this has a strong qualitative resemblence to an experimental one published by Segrè *et al*, [35], who were the first to report the existence of cluster phases. Their observations of a large exponent (5.5 ± 1 at $\phi \approx 0.1$) for the divergence of the terminal time, and of slow, possibly logarithmic, structural relaxation close to the gel line at low ϕ, are likewise both suggestive that the CMCT scenario, though speculative, is valid to some degree.

Complicating factors

CMCT clearly offers nontrivial predictions for cluster phases, weak gels and relaxation anomalies. But it neglects two important pieces of physics which may complicate the experimental situation considerably. Firstly, it allows clusters to dissociate by bond breaking, but neglected any internal restructuring within the clusters. By allowing these to coarsen into globules, this could cause both the cluster fluid and weak gel phases to have finite lifetime. This issue is less important in the presence of a weak, long-range repulsion (e.g. [36]) causing positive increasing $f(N)$. This, like bond-breaking, drives a flow towards smaller ε, but can do so without giving rise to internal restructuring of the clusters. In favorable cases the $f(N)$ term could dominate the bond-breaking term at large ε, and allow the CMCT scenario to establish on a timescale during which local reconstruction remains negligible.

The second omission from the CMCT theory as summarized above is the interplay of cluster formation with liquid-gas type phase separation. This is explored further in Ref.[34]; it leads to some shifts of the kinetic boundaries between equilibrium fluid, cluster fluid phase, and weak gel phase. But – at least for the parameters relevant to the experiments of Segrè *et al* [35] – these shifts are all relatively modest [34].

CONCLUSION

In the first part of the paper, I reviewed recent progress on shear thinning and thickening in colloids. These can be addressed, at least in steady state, using extensions of MCT to allow for advection of fluctuations and, in the thickening case, to allow within a schematic model for a posited direct dependence of the MCT vertex on stress. The latter involves a large element of speculation but may offer a useful pointer to where a more complete theory will one day be gained.

In the second part of the paper, I discussed a new framework to describe weak gelation based on mode-coupling theory applied to clusters (CMCT). This certainly also contains speculative elements but gives nontrivial predictions, several of which compare well to existing experimental data. Much depends on issues of time scale separation, but even if the resulting weak gels are not true examples of broken ergodicity and are only temporary structures, CMCT may offer valuable insights into their behavior.

In summary, MCT has had remarkable successes in the colloidal domain, and offers new ways of looking at some important problems such as nonlinear colloidal rheology and weak gelation. This paper has not addressed the ongoing debate about whether MCT does indeed offer a good starting point for continuing theoretical work on colloidal arrest. The author's views on this heated topic are summarised in Ref.[13].

ACKNOWLEDGMENTS

I thank M. Fuchs, C. Holmes, K. Kroy and W. Poon for their valued collaboration on work described here.

REFERENCES

1. Pham, K. N., Puertas, A. M., Bergenholtz, J., Egelhaaf, S. U., Moussaid, A., Pusey, P. N., Schofield, A. B., Cates, M. E., Fuchs, M. and Poon, W. C. K., *Science* **296**, 104–106 (2002). Poon, W. C. K., Pham, K. N., Egelhaaf, S. U. and Pusey, P. N., *J. Phys. Cond. Mat.* **15**, S269–S275 (2003).
2. Bergenholtz, J., and Fuchs, M., *Phys. Rev. E* **59**, 5706–5715 (1999).
3. Dawson, K., Foffi, G., Fuchs, M., Goetze, W., Sciortino, F., Sperl, M., Tartaglia, P., Voigtmann, T., Zaccarelli, E., *Phys. Rev. E* **63**, 011401 (2001).
4. Fabbian, L., Goetze, W., Sciortino, F., Tartaglia, P., and Thiery, F., *Phys. Rev. E* **59**, R1347–R1350 (1999).
5. Poon, W. C. K., and Haw, M. D. *Adv. Colloid Interface Sci.* **73**, 71–126 (1997).
6. Bergenholtz, J., Fuchs, M., and Voigtmann, T., *J. Phys. Cond. Mat.* **12**, 6575–6583 (2000).
7. Cates, M. E., and Evans, M. R., Eds. *Soft and Fragile Matter: Nonequilibrium Dynamics, Metastability and Flow*, IOP Publishing, Bristol (2000).
8. Pine, D. J., "Light Scattering and Rheology of Complex Fluids far from Equilibrium" in [7], pp.9–47.
9. Kob, W., " Supercooled Liquids and Glasses" in [7], pp. 259–284.
10. Goetze, W., and Sjoegren, L., *Rep. Prog. Phys.* **55**, 241–348 (1992).
11. Ramaswamy, S., "Self-Diffusion of Colloids at Freezing', in *Theoretical Challenges in the Dynamics of Complex Fluids*, McLeish, T. C. B., Ed., pp.7-20, Kluwer, Dordrecht 1987.
12. Kawasaki, K., and Kim, B., *J. Phys. Cond. Mat.* **14**, 1627–1636 (2002).
13. Cates, M. E., cond-mat/0211066, to appear in Ann. Henri Poincaré.
14. Fuchs, M., and Cates M. E., *Phys. Rev. Lett.* **89** 248303 (2002); *Faraday Discussion* **123** 267–286 (2002).
15. Miyazaki, K., and Reichman, D. R., *Phys. Rev. E* **66**, 050501(R) (2002).
16. Indrani, A. V., and Ramaswamy, S., *Phys. Rev. E* **52**, 6492–6496 (1995).
17. Cates, M. E., and Milner, S. T., *Phys. Rev. Lett.* **62**, 1856–1859 (1989).
18. Fuchs, M., and Cates, M. E., *J. Phys. Cond. Mat.* **15**, S401–S406 (2003).
19. Fuchs, M., and Cates, M. E., in preparation.
20. Fielding, S. M., Sollich, P., and Cates, M. E., *J. Rheol.* **44**, 323–369 (2000); Sollich, P., Lequeux, F., Hebraud, P. and Cates, M. E., *Phys. Rev. Lett.* **78**, 2020–2023 (1997). Sollich, P., *Phys. Rev. E* **58** 738–759 (1998).
21. Barnes, H. A., Hutton, J. F., and Walters, K., *An Introduction to Rheology*, Elsevier, Amsterdam 1989.
22. Goetze, W., *Z. Phys. B* **60**, 195–211 (1985).
23. Laun, H. M., *J. Non-Newtonian Fluid Mech.* **54**, 87–108 (1994).
24. Bender, J. and Wagner, N. J., *J. Rheol.* **40**, 899–916 (1996).
25. Head, D. A., Ajdari, A., and Cates, M. E., *Phys. Rev. E* **64**, 061509 (2001).
26. Bertrand, E., Bibette, J., and Schmitt, V., *Phys. Rev. E* **66**, 06040(R) (2002).
27. Cates, M. E., Wittmer, J. P., Bouchaud, J.-P. and Claudin, P., *Phys. Rev. Lett.* **81**, 1841–1845 (1998). Liu, A. J., and Nagel, S. R., *Nature* **396**, 21–21 (1998). Ball, R. C. and Melrose, J. R., *Adv. Colloid Interface Sci.* **59**, 19–30 (1995).
28. Holmes, C., Fuchs, M., and Cates, M. E., *Europhys. Lett.* **63**, 240–246 (2003).
29. Derec, C., Ajdari, A., and Lequeux, F., *Eur. Phys. J. E* **4**, 355–361 (2001).
30. Cates, M. E., Head, D. A., and Ajdari, A., *Phys. Rev. E* **66**, 025202(R) (2002).
31. Aradian, A., and Cates, M. E., cond-mat/0310660.
32. Vicsek, T. *Fractal Growth Phenomena*, World Scientific, Singapore 1992. 2nd ed.
33. Hasmy, A. and Jullien, R., *Phys. Rev. E* **53**, 1789–1794 (1996).
34. Kroy, K., Cates, M. E., and Poon, W. C. K., in preparation; cond-mat/0310566
35. Segrè, P. N., Prasad, V., Schofield, A. B. and Weitz, D., *Phys. Rev. Lett.* **86**, 6042–6045 (2001).
36. Yethiraj, A. and van Blaaderen, A. *Nature* **421**, 513–517 (2003).

Glass transition in a two-dimensional system of magnetic colloids

H. König[*], K. Zahn[†] and G. Maret[*]

[*]*University of Konstanz, Department of Physics, 78457 Konstanz, Germany*
[†]*present address: Ascom Systec Ltd., 5504 Mägenwil, Switzerland*

Abstract. We describe experiments on binary mixtures of superparamagnetic colloidal particles confined by gravity to a flat horizontal water-air interface. The colloids repel each other because of their magnetic dipole moments induced by a vertical external magnetic field B. By tuning B, the effective temperature of the system can be adjusted over several orders of magnitude. Particle coordinates are monitored by video-microscopy over more than five decades in time. Measured radial pair-distribution functions $g(r)$ and mean-square displacements illustrate that this system is an ideal model of a two-dimensional (2D) glass former. We find that the effects of small amounts of aggregated particles only weakly affect the averaged structure and dynamics. Locally, a small number of elementary structural elements are observed each characterized by a special triangular shape. These triangles arrange in dense mostly space-filling arrays and account for the essential features of $g(r)$. The long-time α-relaxation is related to drifts of arrays as well as erosion due to single particle and collective hopping events.

Glasses are solids without structural long-range order and with dynamics frozen-in at least at time scales of typical experiments [1]. One of the most striking and yet unexplained phenomenon in glass physics is the widely universal slowing down of structural relaxations when approaching the glass transition from the liquid side. Physical quantities as different as viscosity, frequency dependent dielectric constants or density correlations obtained from inelastic neutron scattering behave very similar in many different atomic, molecular or macro-molecular systems [2]. Recent experiments on colloidal glasses [3, 4] as well as computer simulations [5, 6, 7] suggest that essential properties of glass formers can be understood by studying simple model systems. Both techniques provide time dependent particle coordinates opening a new way to investigate the microscopic structural properties of glasses and their influence on local dynamic relaxations near the glass transition. While simulations have dealt with three dimensional and two dimensional systems, we are not aware of any experiment on true two-dimensional colloidal glasses.

In this paper, we describe an *experimental* model system which shows all features of a glass former in 2D [8]. Binary mixtures of superparamagnetic colloidal particles are confined by gravity to a horizontal water-air interface in a flat hanging drop geometry. The interparticle magnetic dipole interaction is controlled by a magnetic field applied vertically in order to adjust the effective system temperature. The interaction strength is precisely calibrated. Time-dependent particle positions are determined by video-microscopy and partial radial pair-distribution functions $g_{ij}(r)$ and mean-square displacements $\langle \Delta r^2(t) \rangle = \langle (\vec{r}_i(0) - \vec{r}_i(t))^2 \rangle$ were calculated. The shapes of these functions do not significantly depend on small amounts of aggregates and compare surprisingly well with those of 3D glass formers [9, 10]. Locally, a small number of elementary structural elements of hexagonal, squared body-centered, and 10-fold symmetry are observed. They arrange in dense mostly space-filling arrays and account for the essential features of $g_{ij}(r)$ [11, 12].

A. Experimental set-up: The commercial [13] monodisperse colloidal PMMA particles used are porous and doped with magnetite nanocrystals making them superparamagnetic. The big (b) (Dynabeads M-450, uncoated) and small (s) (dried Dynabeads M-280 uncoated) particles have diameters $d_b = 4.7 \mu m$ and $d_s = 2.8 \mu m$, and densities $\rho_b \approx 1.5 \cdot 10^3 kg/m^3$ and $\rho_s \approx 1.3 \cdot 10^3 kg/m^3$, respectively. Their magnetic susceptibilities were determined by comparison $g(r)$ measured in the liquid state with Brownian-dynamics simulations [14]. We thus obtained low field magnetic susceptibilities $\chi_b = 6.2 \cdot 10^{-11} Am^2/T$ and $\chi_s = 6.6 \cdot 10^{-12} Am^2/T$. χ_b was found constant within the range of B-values used (0 to $\approx 4mT$), while χ_s lowers due to magnetic saturation above $3mT$ reaching about 80% of the low-B-value at $4mT$. These values agree with measurements of magnetophoretic mobilities of individual particles and direct SQUID-magnetometry [15]. In order to prevent aggregation, the big Dynabeads were stabilized by sodium dodecyl sulfonate (SDS) while the small Dynabeads were found to be stable in pure water.

The colloidal suspension was filled into the cylindrical holes of a glass cell as sketched in Fig. 1. The larger hole was 8*mm* in diameter and 1*mm* in height. The cell was mounted upside down in the sample holder which was temperature stabilized by a large copper block. The liquid drop was suspended by interfacial tension, that is by pinning the water-air interface at the edge of the cylindrical cell due to hydrophobic treatment of the outer cell surface. The colloids sediment on the interface and form a monolayer with average density dependent on the amount of colloids present in the initial bulk suspension. The curvature of the drop was adjusted by tuning the water volume through a computer controlled nano-syringe. In order to minimize convection due to filling, the volume of the large cell was adjusted by adding/removing water in the small satellite cell connected to the main cell by a thin capillary. The interface was kept flat by observation of the sharpness of particle images everywhere in the focal plane of a 40x microscope lens. Thus a flatness of $< 1\mu m$ at a cell diameter of 8*mm* was obtained. Additionally, the horizontal orientation of the liquid-air interface was controlled through adjustment of the optical table to $\pm 10^{-4}$ *rad*.

The system can be considered almost ideally two-dimensional: First, the thermal gravitational length of the big and small colloids is about 8*nm* and 62*nm*, respectively, much smaller than both particle diameters. Secondly, the gravitational force of the colloids is essentially unable to bulge the surface against surface tension. Finally, vertical particle displacements due to thermal capillary waves are less than $\sim 1 nm$ and, thus, can be neglected [16]. In order to reduce collective particle drifts induced by convective flow of water, thermal gradients and disturbances of the water-air interface were suppressed by surrounding the entire experimental setup by a polystyrene box and air conditioning of the laboratory. Well equilibrated samples could be conserved for several weeks.

A homogeneous magnetic field, B, was applied perpendicular to the water-air interface by coils inducing vertical magnetic moments $\vec{M} = \chi \cdot \vec{B}$ in the colloidal particles. The magnetic field generated by the induced magnetic moments themselves is only of the order of 1% of B (at maximum concentration) and therefore can be neglected. Hence, two induced magnetic moments interact by a repulsive dipole potential proportional to $\chi^2 B^2/r^3$, where r is the center-to-center distance between two colloids. Other contributions to the interaction potential can be neglected [14].

The interaction strength is characterized by an interaction parameter, Γ, defined by the ratio of the magnetic interaction potential $E_{magn}(B)$ to the thermal energy $k_B T$. Γ corresponds to an inverse system temperature, T_{sys} tun-

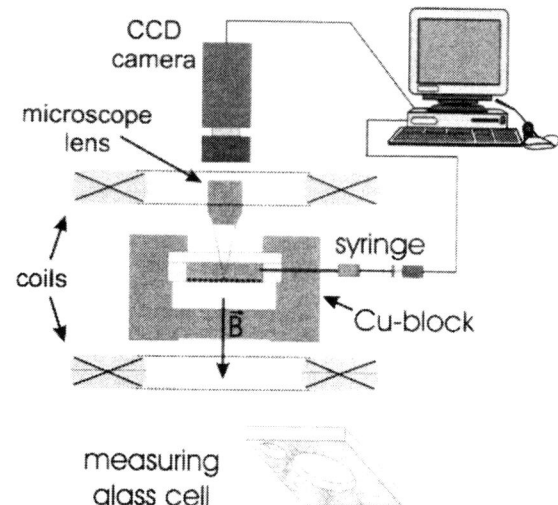

FIGURE 1. Sketch of the experimental set-up (see text for details).

able by B. With the particle area density, ρ, corresponding to the averaged particle distance $\bar{r} = \rho^{-1/2}$, and the number ratio ξ of the small to all particles the Γ-value of the binary 2D colloidal mixture can be defined as:

$$\Gamma = \frac{E_{magn}}{k_B T} = \frac{\mu_0}{4\pi} \frac{B^2 (\rho \pi)^{3/2}}{k_B T} \left(\xi \cdot \chi_s + (1-\xi) \cdot \chi_b \right)^2. \quad (1)$$

ρ and T remain constant during an experiment, thus T_{sys} decreases for increasing Γ.

Typically 1,000 particles out of approximately 100,000 in the cell were observed by video-microscopy, at locations far away from the cell edge, in order to avoid effects of the cell boundaries. Time-dependent positions of particles were determined in real-time by digitization on a frame grabber card and image processing on a PC. Every $\approx 0.3s$ a complete set of uniquely labelled particle coordinates were determined depending on the number of observed particles and saved for evaluation later on: Each video frame was converted into a binary image using a brightness cut-off value such that colloids appeared as black discs on white background. The center of mass of a disc's area gives the particle position which was calculated with an accuracy better than $1\mu m$. The disc's area indicated whether a given particle was a small or a big one.

For purposes of more efficient data storage, and in order to display the mean-square displacements equidistant on a logarithmic time-scale over many decades, a multiple-τ-algorithm was used. In this algorithm the sampling time Δt is doubled after every 1,000 sampling intervals. The initial maximum sampling rate is used for

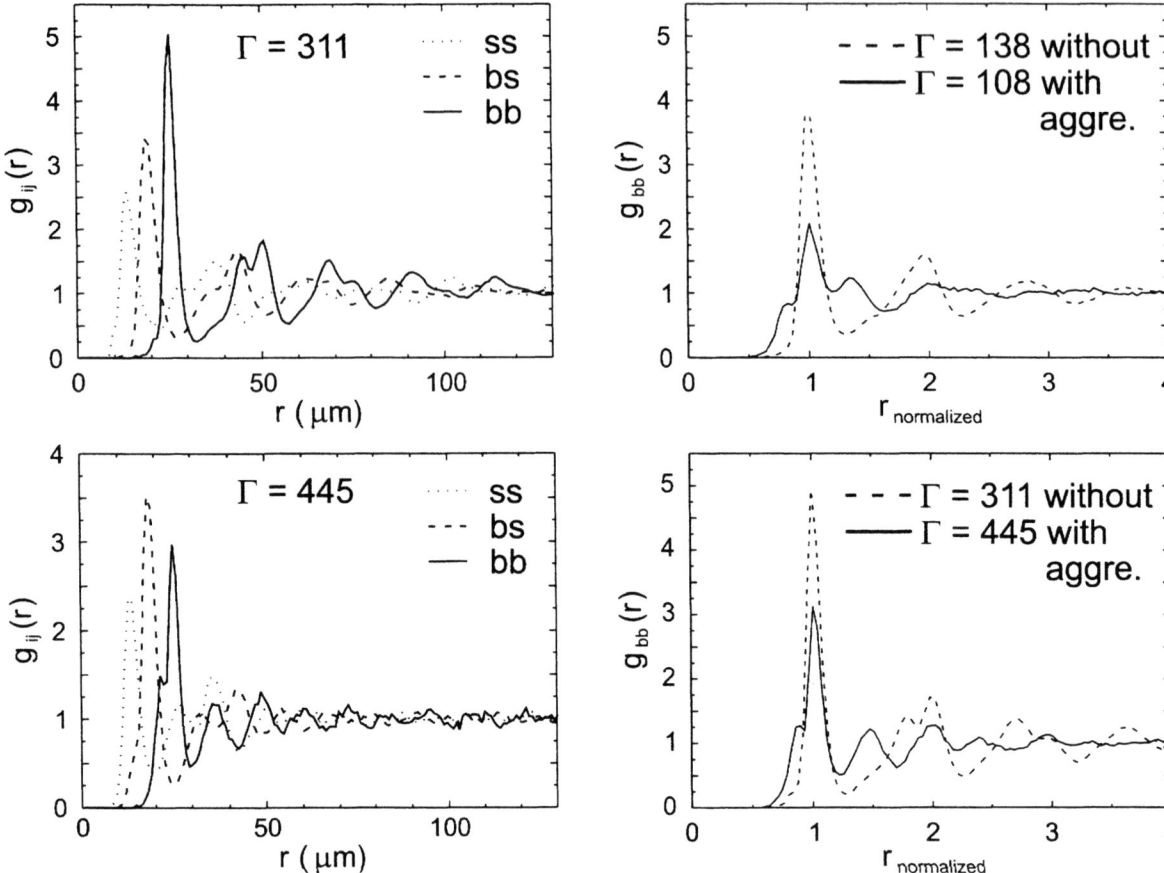

FIGURE 2. Partial radial pair-distribution functions $g_{ij}(r)$, $g_{bs}(r)$ and $g_{ss}(r)$ for a sample with a low number of aggregates ($\Gamma = 311$, $\xi = 0.29$, $\rho = 1.78 \cdot 10^{-3} \mu m^{-2}$) and for a sample with a high number of aggregates ($\Gamma = 445$, $\xi = 0.57$, $\rho = 2.08 \cdot 10^{-3} \mu m^{-2}$).

FIGURE 3. Comparison of big/big radial pair-distribution functions $g_{bb}(r)$ for low (a) and for high (b) Γ-values. Dashed curves correspond to a sample with few (<10) aggregates, continuous lines refer to a sample with (≈ 50) explored aggregates. r-values are normalized to the average interparticle distance r_{bb} obtained from the positions of the biggest peak of $g_{bb}(r)$.

data processing in order to enable a reliable identification of particles between consecutive time steps.

Some uncertainties in the trajectories originate either from drift effects caused by convective flow of water or from aggregates of a few colloids (mostly two small or one small and one big particle) having different magnetic susceptibilities and different diffusion coefficients than isolated particles. Although the drift was typically smaller than $1\mu m/h$, we minimized this effect by subtracting the average displacement of all particles from the displacement of each particle. The overall stability of the system itself was sufficiently good to obtain reliable data up to more than $10^5 s$.

B. Structure: Partial pair-distribution functions $g_{ij}(r)$ for the different pairs ij of big and small colloids were obtained by spatial averaging over typically about 100 independent particle configurations. Sufficient ensemble averaging was obtained by additional time averaging. In Fig. 2 $g_{ij}(r)$ for the three pairs (bb), (bs) and (ss) are compared for two different samples. The highest peaks correspond to the most frequent interparticle distances which are rather sharp. A series of additional peaks at higher distances are observed for all three partial $g_{ij}(r)$ indicating substantial local order of the binary mixture. With increasing Γ the peaks of $g_{ij}(r)$ become sharper and narrower and more features appear at higher distances, in particular a splitting of the second maximum. These observations are well-known "finger prints" of amorphous structures of glass formers.

Fig. 3 shows $g_{bb}(r)$ for two binary suspensions, one containing very few aggregates within the picture explored, while the other contains about 5 times more aggregates. For the sample containing more aggregates additional peaks in $g_{bb}(r)$, in particular at distances shorter than the average distance between big particles are seen.

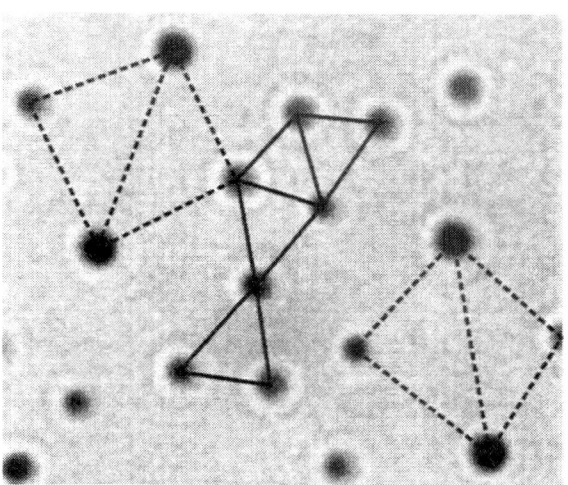

FIGURE 4. Hexagonal arrangements of small particles (solid lines) and rectangular triangles of one small and two big colloids (dashed lines).

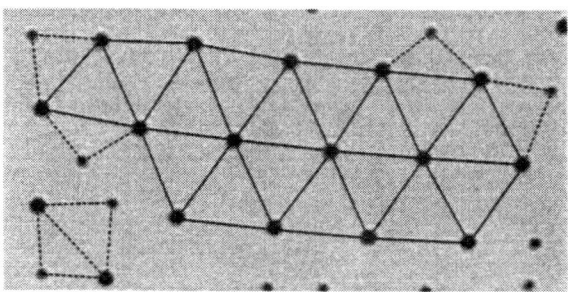

FIGURE 5. Hexagonal arrangements of big particles (solid lines) and rectangular triangles of one small and two big colloids (dashed lines).

Visual inspection of the pictures reveal that these shorter distances mostly originate from aggregates of two or three small particles (sticking together at mutual surface contact) which are processed by the particle recognition algorithm as one big particle. This observation indicates that $g(r)$ is rather sensitive to aggregates of this type. Note that, in contrast, the relaxation dynamics discussed below turns out rather insensitive to these aggregates.

The observed features in $g_{ij}(r)$ are related to highly ordered local arrangements of particles [12] which can be directly seen by inspection of local configurations. A few typical examples are given in the series of pictures Fig. 4 to Fig. 8 which were taken on different spots of the same sample for $\Gamma = 411$. Such structures are disturbed by thermal particle fluctuations. Because of the purely repulsive dipole interaction it seems natural to expect local hexagonal packings in cases were only one species of particles happens to be around, Fig. 4

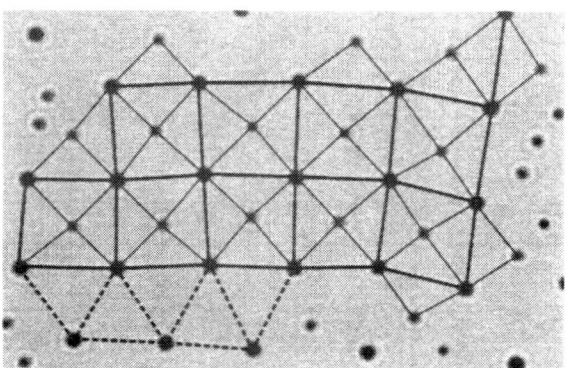

FIGURE 6. Quadratic arrangements of equal numbers of big and small particles (solid lines) and neighboring hexagonal triangles of only big colloids (dashed lines).

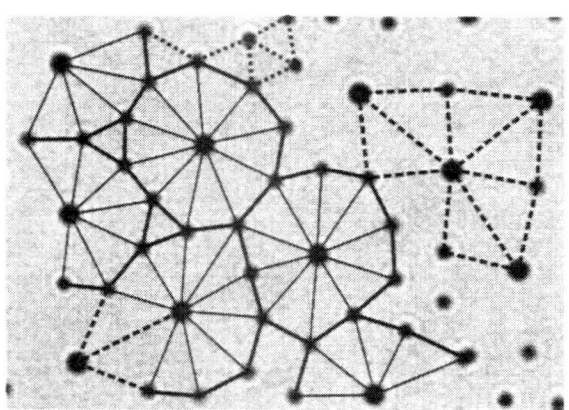

FIGURE 7. Arrangements of big particles surrounded by many small particles (solid lines). A configuration frequently observed consists of 10 small particles forming a ring-like structure around a big particle. Additionally, rectangular triangles of one small and two big colloids (dashed lines) and hexagonal triangles of only small particles (dotted lines) are depicted.

and Fig. 5. At locations of equal numbers of small and big particles a square body-centered lattice decorated like a checkerboard pattern is observed. Packings at high local densities of small particles are shown in Fig. 7 and Fig. 8. Because of the much smaller repulsion between two small particles as compared to the cases small/big and big/big, small particles get to substantially shorter mutual distances. This seems to stabilize lozenge-type configurations consisting of two big particles at the small angles of the lozenges and two small particles at the big angles, respectively. Often, 10 small particles are observed in a ring-like configuration around one big particle. Then, the small/big internal angle of the lozenge is close to 36/144 degrees, respectively. This type of lozenge is also the smaller one of the two tiles required

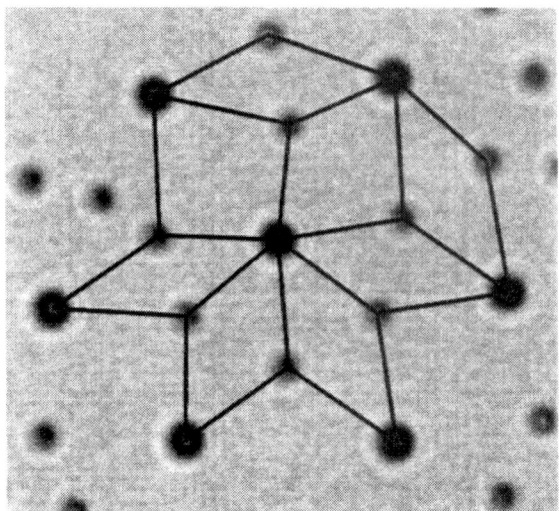 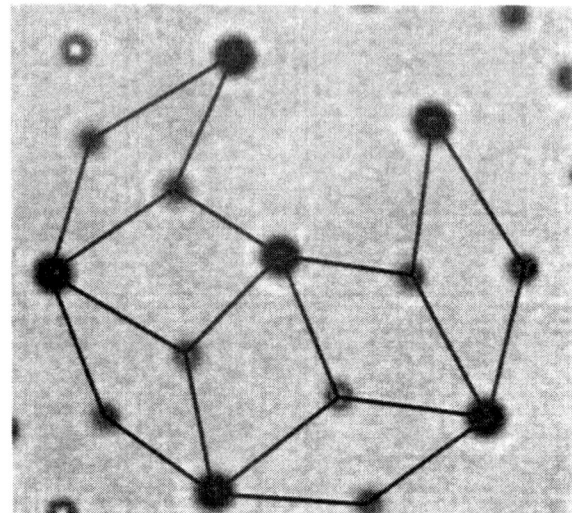

FIGURE 8. Local configurations revealing some lozenges close to Penrose tiles with small angles of 36 or 72 degrees at the big colloid ($\Gamma = 411$).

to obtain a quasi-crystal with Penrose tiling. At some locations the two Penrose tiles seem to appear, see Fig. 8. Similar observations are reported in [17].

Different local-ordered elementary particle arrangements coexist in direct neighborhood. These observations suggest that the amorphous structure of our 2D system is built up of different local structures which depend on local concentration and on local composition. The highly ordered local structures essentially account for all peaks observed in the partial radial pair distribution functions [12]. In addition, they give rise to significant features in the 3-point and 4-point correlation functions [11, 12].

C. Dynamics: The overall dynamics of our system was studied in real space by mean-square displacements of individual particles, e.g. the time-dependent particle deviations from their initial positions: $\langle \Delta r^2(t) \rangle = \langle (\vec{r}_i(0) - \vec{r}_i(t))^2 \rangle$. The mean-square displacements show a short-time behavior which is independent of Γ. In the time-range up to about $2s$, the particles obey a diffusive motion with a diffusion constant somewhat smaller than the free Stokes diffusion due to the hydrodynamic interaction with the liquid-air interface [14]. For longer time-scales and increasing Γ, the slope of $\langle \Delta r^2(t) \rangle$ diminishes, and a increasingly pronounced plateau appears. As in atomic or molecular glasses this plateau may be attributed to a so-called cage-effect. A particle stays in the cage of its neighbors until structural α-relaxations allow it to escape. The long-time α-relaxations cause an additional increase of the mean-square displacements at long times. An example is shown in Fig. 9.

It can be seen that short-time diffusion, plateau region and long-time diffusion are essentially insensitive to the

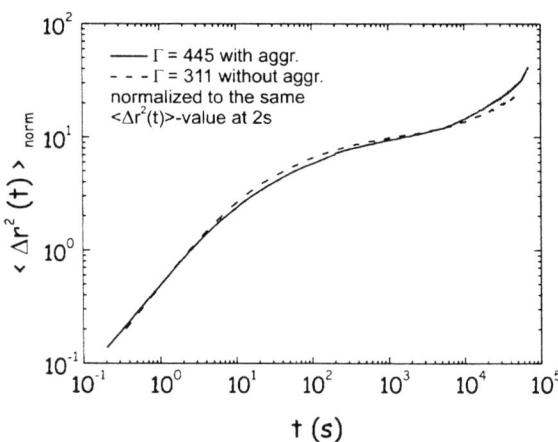

FIGURE 9. Mean square displacements of the two samples shown in Fig. 2. Because the two samples possessed different particle densities, e.g. different particle distances, the $\langle \Delta r^2(t) \rangle$ of the sample with $\Gamma = 445$ was brought into line with the short-time behavior of the other sample at $t = 2s$.

presence of some aggregates. This argues that the details of the local structure do not really matter for the relaxation dynamics and may suggest that other more polydisperse systems may relax in a qualitatively similar way.

Fig. 10 shows particle trajectories of a sample at rather high Γ. These data clearly reveal a strongly heterogeneous relaxation behavior: There are regions were particle motions are essentially restricted to diffusive motion around a given position, both for small and big particles. Some of these areas are seen to obey slow convective

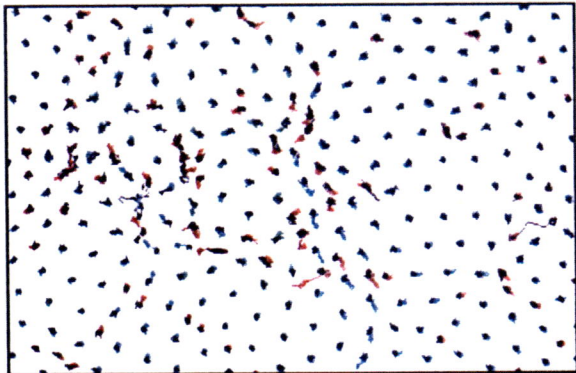

FIGURE 10. Trajectory plot of a part of the sample with $\Gamma = 311$ of Fig. 2 monitored over a time interval of $34450s$ for time steps between two data files of $\sim 252s$. The color of the trajectory becomes darker as time goes on. Green: big particles, red: small particles.

drifts without much changes of the relative particle positions. Finally some particles have excursions much larger than the average diffusive displacement, sometimes even larger than the average interparticle distance. These particles have clearly left their cages, sometimes as isolated particles, sometimes as groups of two or more particles in a string-like motion. Thus the relaxation dynamics of this system is clearly quite heterogeneous. Slow collective drifts and single or multiple hopping events out of cages coexist. In a forthcoming study [12] a correlation between the locally ordered viz. disordered structures with the collective viz. single hopping dynamics will be discussed.

D. Conclusions: 2D binary colloidal suspensions with knob-tunable magnetic interparticle interactions are ideal systems to study structural and dynamic processes near the glass transition. The local structure remains liquid-like with more pronounced short-range order for decreasing T_{sys}. Correspondingly, the particle dynamics slows down, as seen by an increased cage-effect. Regular elementary structures (such as triangular, square or rhombic particle arrangements) are clearly observed, presumably much easier to detect than equivalent clusters in 3D.

ACKNOWLEDGMENTS

This work was supported by the Deutsche Forschungsgemeinschaft, in the frame of SFB 513, project B6.

REFERENCES

1. Ediger, M. D., Angell, C. A., and Nagel, S. R., *J. Chem. Phys.* **100**, 13200-13212 (1996).
2. Vilgis, T. A., "Models for Transport and Relaxation in Glass Forming and Complex Fluids: Universality?", in *Disorder Effects on Relaxational Proceses, Glasses, Polymers, Proteins*, edited by R. Richert, and A. Blumen, Springer-Verlag, Berlin, 1994, pp. 153-191.
3. Weeks, E. W., Crocker, J. C., Levitt, A. C., Schofield, A., and Weitz, D. A., *Science* **287**, 627-631 (2000).
4. Gasser, U., Schofield, A., and Weitz, D. A., *J. Phys.: Condens. Matter* **15**, 375-380 (2003).
5. Perera, D. N., and Harrowell, P., *Phys. Rev. E* **59**, 5721-5743 (1999).
6. Binder, K., Baschnagel, J., Kob, W., and Paul, W., "Simulation of models for the glass transition: Is there progress?", in *Bridging the time scales: molecular simulations for the next decade*, edited by Nielaba, P., Mareschal, M., and Ciccotti, G., Springer-Verlag, Berlin, 2002, pp. 199-228.
7. Donati, C., Glotzer, S. C., Poole, P. H., Kob, W., and Plimpton, S. J., *Phys. Rev. E* **60**, 3107-3119 (1999).
8. Similar binary 2D mixtures were used earlier to investigate hydrodynamic interactions in the liquid state far away from the glass transition: Kollmann, M., Hund, R., Rinn, B., Nägele, G., Zahn, K., König, H., Maret, G., and Klein, R., *Europhys. Lett.*, 919-925 (2002).
9. Kondo, T., Tsumuraya, K., and Watanabe, M. S., *J. Chem. Phys.* **93**, 5182-5186 (1990).
10. Miyagawa, H., Hiwatari, Y., Bernu, B., and Hansen, J. P., *J. Chem. Phys.* **88**, 3879-3886 (1988).
11. König, H., *Local particle rearrangements in a two-dimensional binary colloidal glass former*, in *the same issue*.
12. König, H., to be published.
13. DYNAL PARTICLES AS, *http://www.dynalbiotech.com*.
14. Zahn, K., Mèndez-Alcaraz, J. M., and Maret, G., *Phys. Rev. Lett.* **79**, 175-178 (1997).
15. Zahn, K., unpublished.
16. Wille, A., Zahn, K., Valmont, F., and Maret, G., *Europhys. Lett.* **57**, 219-225 (2000).
17. Wen, W., Zhang, L., and Shen, P., *Phys. Rev. Lett.* **85**, 5464-5467 (2000).

General Nonlinear 2-Fluid Hydrodynamics of Complex Fluids and Soft Matter

H. Pleiner* and J.L. Harden[†]

Max Planck Institute for Polymer Research, 55021 Mainz, Germany
[†]*Department of Chemical and Biomolecular Engineering, Johns Hopkins University, Baltimore, MD 21218, USA*

Abstract. We discuss general 2-fluid hydrodynamic equations for complex fluids, where one kind is a simple Newtonian fluid, while the other is polymeric/elastomeric, thus being applicable to polymer solutions and swollen elastomers. The procedure can easily be generalized to other complex fluid solutions. Special emphasis is laid on such nonlinearities that originate from the 2-fluid description, like the transport part of the total time derivatives. It is shown that the proper velocities, with which the hydrodynamic quantities are convected, cannot be chosen at will, since there are subtle relations among them. Within allowed combinations the convective velocities are generally material dependent. The so-called stress division problem, i.e. how the elastic stresses are distributed between the two fluids, is shown to depend partially on the choice of the convected velocities, but is otherwise also material dependent. A set of reasonably simplified equations is given as well as a linearized version of an effective concentration dynamics that may be used for comparison with experiments.

INTRODUCTION

The thermodynamic and hydrodynamic properties of multi-component complex fluids are determined by the microscopic degrees of freedom of their constituents and the coupling between these degrees of freedom. Such systems can exhibit rather rich phase behavior and dynamics, especially when one or more components is a structured or macromolecular fluid [1]. Due in part to the coupling of internal degrees of freedom, these systems can also exhibit novel flow-induced structural evolution phenomena, including shear-induced phase transformations and flow alignment of constituents on microscopic to mesoscopic length scales. Such structural evolution in turn leads to nonlinear rheological behavior, such as stress overshoots in response to imposed rates of strain, plasticity, and thixotropy.

Due to the overwhelming complexity of the microscopic description of these systems, such a detailed description is often not well suited for analysis of the macroscopic dynamical behavior. Instead, explicit macroscopic models have been developed for this purpose. Some such models have been obtained by a suitable coarse-graining procedure starting from a microscopic theory. Others are purely phenomenological models constrained only by conservation laws, symmetry considerations and thermodynamics. The so-called "two-fluid" models for binary systems of distinct components or phases are useful examples of such a macroscopic approach [2]. In the two-fluid description, each component or phase is treated as a continuum described by local thermodynamic variables (e.g. temperature, density, and relevant order parameters), and dynamical quantities (e.g. velocity or momentum). In general, these variables for the constituents are coupled. For instance, the effective friction between components in a binary fluid mixture leads to a drag force in the macroscopic description that is proportional to the local velocity difference.

Two-fluid models have been employed in many different physical contexts. The two-fluid approach is a key element of many traditional models for multi-phase flow of bubbly liquids, fluid suspensions of particulates, and binary mixtures of simple fluids [3]. Other examples in condensed matter physics include two-fluid models for superfluid helium [4], dynamics of plasmas [5], transport in superconductors [6], viscoelasticity of concentrated fluid emulsions [7], flow-induced ordering of wormlike micelle solutions [8], flow of colloidal suspensions [9]. Two-fluid models have been used extensively to model a wide range of dynamical phenomena in polymer solutions and binary blends, including the hydrodynamic modes of quiescent polymer solutions [10, 11], kinetics of polymer dissolution [12], hydrodynamics and rheology of polymer solutions and blends [13]-[19], and polymer migration and phase separation under flow [20]-[27].

These examples share certain general features. In each, two distinct species or coexisting phases (gas and liquid, normal fluid and superfluid, polymer and solvent, mesogens and solvent etc.) with mass densities ρ_1 and ρ_2, which are conserved individually in the absence of chemical reactions, move with distinct velocities $\vec{v}_1$ and $\vec{v}_2$, respectively. Due to (usually strong) internal friction, the momenta of the constituent species, $\rho_1\vec{v}_1$ and $\rho_2\vec{v}_2$, are not conserved individually. Of course, total momentum is conserved. In most cases of fluid mixtures the friction is so strong that the velocity difference $\vec{v}_1 - \vec{v}_2$ is nonzero for very short times only, i.e. it is a very rapidly relaxing quantity that is not included in the hydrodynamic description for binary mixtures. However, there are systems and situations, where the relaxation of the relative momenta is slow enough to have a significant influence even on the hydrodynamic time scale. Then a two-fluid description is appropriate and useful.

In this communication we focus on a general nonlinear two-fluid description of complex fluids, where one species is a viscous Newtonian fluid and the other a polymer or elastomer. Emphasis is placed on the rigorous derivation of the dynamic equations within the framework of hydrodynamics as contrasted to ad-hoc treatments. The resulting equations are rather general and complicated. They can and have to be simplified for special applications or systems by appropriate and well-defined approximations. One of the advantages of starting from the general theory is the possibility to identify and characterize the approximations made. The hydrodynamic method, described in some detail in [28]-[30], is quite general and rigorous, being based on symmetries, conservation laws, and thermodynamics.

The linear hydrodynamic description of a single component elastic media has been given in [31], generalized to the nonlinear domain in [32, 33], where in addition the necessary steps are described when the elastic degree of freedom is not constant (as in elastomers), but decaying in time (as in viscoelastic media). In the following sections, we provide a detailed analysis of two-fluid models for isotropic polymers and elastomers (e.g. entangled polymer solutions and gels) in a simple viscous solvent. We close with a discussion of our general results and their possible implications for experiments.

THERMODYNAMICS

The hydrodynamics of fluid mixtures as described above is governed by conservation laws (individual masses, total momentum and total energy), balance equations for the liquid crystalline degrees of freedom, for the transient elasticity of polymers and for the relaxation of relative momentum. There are different ways of writing the appropriate equations. One popular choice is to use equations for individual mass densities and individual momentum densities, another to use the mass density and one concentration variable and the total momentum density and the relative velocity difference. Since they both have their advantages and disadvantages we will present both ways of description, and show how they are connected. We will use an isotropic elastomer (or viscoelastic media) as the second, complex fluid.

The starting point of any macroscopic description is the total energy E of the system as a function of all the relevant variables. Since the energy is a first order Eulerian form of the extensive quantities, we can write

$$E = \varepsilon V = \int \varepsilon dV \qquad (1)$$
$$= E(M_1, M_2, V, \vec{G}_1, \vec{G}_2, S, M_2 U_{ij})$$

The masses, M_1, M_2 and momenta $\vec{G}_1, \vec{G}_2$ of species 1 and 2 are related to the appropriate (volume) densities by $\rho_1 = M_1/V$, $\rho_2 = M_2/V$, $\vec{g}_1 = \vec{G}_1/V = \rho_1\vec{v}_1$, $\vec{g}_2 = \vec{G}_2/V = \rho_2\vec{v}_2$, thus defining the two velocities $\vec{v}_{1,2}$ whose components we write as $v_i^{(1,2)}$. The entropy density is $\sigma = S/V$. The elastic degree of freedom of species 2 is described by the Eulerian strain tensor U_{ij} [32], which is symmetric and often related to a displacement vector $\vec{u}$ by $2U_{ij} = \nabla_i u_j + \nabla_j u_i + (\nabla_k u_i)(\nabla_k u_j)$. Introducing thermodynamic derivatives (partial derivatives where all other variables are kept fixed) we define temperature T, thermodynamic pressure p, chemical potentials μ_1, μ_2 and velocities $\vec{v}_1, \vec{v}_2$ of the two fluids, as well as the elastic stress Φ_{ij} conjugate to the elastic strain

$$\begin{aligned} T &= \frac{\partial E}{\partial S} = \frac{\partial \varepsilon}{\partial \sigma}, & \mu_1 &= \frac{\partial E}{\partial M_1} = \frac{\partial \varepsilon}{\partial \rho_1}, \\ \mu_2 &= \frac{\partial E}{\partial M_2} = \frac{\partial \varepsilon}{\partial \rho_2} & p &= -\frac{\partial E}{\partial V}, \\ \vec{v}_1 &= \frac{\partial E}{\partial \vec{G}_1} = \frac{\partial \varepsilon}{\partial \vec{g}_1}, & \vec{v}_2 &= \frac{\partial E}{\partial \vec{G}_2} = \frac{\partial \varepsilon}{\partial \vec{g}_2} \\ \Phi'_{ij} &= \frac{\partial E}{\partial (M_2 U_{ij})} = \frac{\partial \varepsilon}{\partial (\rho_2 U_{ij})} \equiv \rho_2^{-1} \Phi_{ij} \end{aligned} \qquad (2)$$

Expanding eq.(1) into first order differentials, the condition $dV = 0$ leads to an expression for the pressure

$$p = -\varepsilon + T\sigma + \rho_1\mu_1 + \rho_2\bar{\mu}_2 + \vec{v}_1 \cdot \vec{g}_1 + \vec{v}_2 \cdot \vec{g}_2 \qquad (3)$$

where we have introduced the effective chemical potential of the elastomer $\bar{\mu}_2 = \mu_2 + \rho^{-1}\Phi_{ij}U_{ij}$. In addition, the differentials are related by the Gibbs relation

$$\begin{aligned} d\varepsilon &= Td\sigma + \mu_1 d\rho_1 + \bar{\mu}_2 d\rho_2 + \vec{v}_1 \cdot d\vec{g}_1 + \vec{v}_2 \cdot d\vec{g}_2 \\ &\quad + \Phi_{ij} dU_{ij} \end{aligned} \qquad (4)$$

From eqs.(3, 4) the expression for the differential pressure results (Gibbs-Duhem relation) that is useful in switching from pressure to chemical potentials or vice versa

$$dp = \sigma dT + \rho_1 d\mu_1 + \rho_2 d\bar{\mu}_2 + \vec{g}_1 \cdot d\vec{v}_1 + \vec{g}_2 \cdot d\vec{v}_2 \\ - \Phi_{ij} dU_{ij} \quad (5)$$

A second set of equations is obtained by switching to the total density, $\rho = \rho_1 + \rho_2$, and the total momentum, $\vec{g} = \vec{g}_1 + \vec{g}_2 = \rho_1 \vec{v}_1 + \rho_2 \vec{v}_2$, which are the sums of the original quantities and which are both conserved quantities. The two-fluid nature has then to be represented by additional variables. A natural choice seems to be the use of the density and momentum differences. However the latter choice is problematic, since it necessarily implies the conjugate quantities also to be the (arithmetic) sums and differences of the original conjugate quantities. Thus, the conjugate to $\vec{g}$ would be $\vec{v}_1 + \vec{v}_2$, which does not reflect correctly the possible one-fluid limits $\rho_1 \to 0$ or $\rho_2 \to 0$. The physically acceptable conjugate to the total momentum is the mean velocity $\vec{v}$ defined by $\rho^{-1}\vec{g}$. Insisting on $\vec{v}$, the mean velocity, to be the conjugate of the total momentum $\vec{g}$, the choice of the remaining variable describing the different velocities is severely limited. Compatibility with (4) allows as variable only the velocity difference $\vec{w} \equiv \vec{v}_1 - \vec{v}_2$ (with $\vec{m} \equiv \rho^{-1}\rho_1\rho_2\vec{w}$ as conjugate quantity) or more generally $\alpha\vec{w}$ as variable with $\alpha^{-1}\rho_1\rho_2\rho^{-1}\vec{w}$ as conjugate, where α can be freely choosen. There is no a-priori advantage for any of the choices and we will stick to $\alpha = 1$.[1] From $\vec{w} = \vec{g}_1/\rho_1 - \vec{g}_2/\rho_2$ one gets

$$\vec{v}_1 = \rho^{-1}\vec{g} + (1-\phi)\vec{w}, \qquad \vec{v}_2 = \rho^{-1}\vec{g} - \phi\vec{w} \quad (6)$$

The representation of the two different densities is less problematic. A convenient choice for that variable is the concentration, $\phi = \rho_1/\rho$, with $\rho_2/\rho = 1 - \phi$. If the expansion coefficients of the two fluids are the same, ϕ can be interpreted as the volume fraction as well. Instead of ϕ one could have used, e.g. the density difference $\rho_1 - \rho_2$ (or any other linear combination of ρ_1 and ρ_2 different from ρ) as variable without much change.

After some trivial algebra eqs.(3-5) can be written in the new variables as

$$p = -\varepsilon + T\sigma + \rho\mu + \rho^{-1}\vec{g}^2 \quad (7)$$

$$d\varepsilon = T d\sigma + \Pi' d\phi + \mu d\rho + \vec{v} \cdot d\vec{g} + \vec{m} \cdot d\vec{w} \\ + \Phi_{ij} dU_{ij} \quad (8)$$

$$dp = \sigma dT + \rho d\mu + \vec{g} \cdot d\vec{v} - \vec{m} \cdot d\vec{w} - \Pi' d\phi \\ - \Phi_{ij} dU_{ij} \quad (9)$$

where we have introduced the relative pressure Π' and the total chemical potential μ

$$\Pi' = \rho(\mu_1 - \bar{\mu}_2) + \vec{w} \cdot \vec{g} + \rho\vec{w}^2(1-2\phi) \equiv \rho\Pi$$
$$\mu = \mu_1\phi + \bar{\mu}_2(1-\phi) + \vec{w}^2\phi(1-\phi) \quad (10)$$

or vice versa

$$\mu_1 = \mu + \rho^{-1}\rho_2(\Pi - \vec{w}\cdot\vec{v}_1)$$
$$\mu_2 = \mu - \rho^{-1}\rho_1(\Pi + \vec{w}\cdot\vec{v}_2) \quad (11)$$

with the mean velocity $\vec{v}$ and the weighted relative momentum $\vec{m}$ given by

$$\vec{v} = \phi\vec{v}_1 + (1-\phi)\vec{v}_2 = \rho^{-1}(\vec{g}_1 + \vec{g}_2)$$
$$\vec{m} = \rho(1-\phi)\phi\vec{w} = (\rho_2\vec{g}_1 - \rho_1\vec{g}_2)\rho^{-1} \quad (12)$$

The Gibbs relations connects variables that show different rotational behavior. Energy, entropy, the densities and the concentration are scalar quantities that do not change under (rigid) rotations, i.e. $d\varepsilon = d\sigma = d\rho = d\rho_1 = d\rho_2 = d\phi = 0$. The vectors and tensors are transformed according to $dv_i^{(1,2)} = \Omega_{ij}v_j^{(1,2)}, dg_i = \Omega_{ij}g_j, dw_i = \Omega_{ij}w_j, dU_{ij} = \Omega_{jk}U_{ik} + \Omega_{ik}U_{kj}$, where $\Omega_{ij} = -\Omega_{ji}$ is any constant antisymmetric matrix. The rotational invariance of the Gibbs relation (4,8) then leads to the relation

$$U_{ik}\Phi_{kj} = U_{jk}\Phi_{ki} \quad (13)$$

which has to be fulfilled by the conjugate quantities. There are no contributions from the momenta and velocities, since $\vec{g} \parallel \vec{v}, \vec{w} \parallel \vec{m}$, and $\vec{g}_{1,2} \parallel \vec{v}_{1,2}$. Relation (13) is useful for reformulating the total stress tensor (see below), in particular to symmetrize it explicitly.

Having set up the thermodynamics of the relevant variables we are now in a position to establish the structure of the dynamic equations.

SIMPLIFIED ELASTOMERIC TWO-FLUID EQUATIONS

Recently we have derived the most general and complete set of 2-fluid equations [34]. These equations are for most purposes unnecessarily complicated and can be simplified using reasonable assumptions. Starting from the correct general equations such assumptions, clearly spelled out, lead to controlled approximations and to a set of 2-fluid equations, whose limitations and implicit assumptions are clear and well defined in contrast to most ad-hoc approaches.

Here we want to display explicitly 2-fluid hydrodynamics under the following assumptions:
a) convection with natural velocities (for $U_{ij}, \vec{g}_2, \rho_2$ and

---

[1] The choice $\alpha = \rho_1\rho_2\rho^{-1}$ would just interchange the roles of $\vec{w}$ and $\vec{m}$ as variable and conjugate.

$\vec{g}_1, \rho_1$ this is $\vec{v}_2$ and $\vec{v}_1$, respectively);
b) the linearized elastic force acts on the elastomeric fluid (index 2) only;
c) global incompressibility, $\delta\rho = 0$ (i.e. $\delta\rho_1 = -\delta\rho_2$); and
d) linearizing the phenomenological dissipative currents, but keeping quadratic nonlinearities otherwise. Then the following set of equations is obtained:
The incompressibility condition (in 3 equivalent versions),

$$0 = \text{div}\,\vec{v} \quad (14)$$
$$0 = \vec{w}\cdot\vec{\nabla}\rho_1 + \rho_1\text{div}\,\vec{v}_1 + \rho_2\text{div}\,\vec{v}_2 \quad (15)$$
$$0 = \vec{w}\cdot\vec{\nabla}\phi + \phi\,\text{div}(1-\phi)\vec{w} - (1-\phi)\,\text{div}\phi\vec{w} \quad (16)$$

the concentration dynamics (in 3 equivalent versions),

$$\dot{\phi} + \nabla_i(\phi v_i + \phi(1-\phi)w_i) - d_{ij}\nabla_i\nabla_j(\mu_1 - \bar{\mu}_2)$$
$$-\phi(1-\phi)d_{ij}^{(T)}\nabla_j\nabla_i T = 0 \quad (17)$$

$$\dot{\rho}_1 + \vec{v}_1\cdot\vec{\nabla}\rho_1 + \rho_1\text{div}\vec{v}_1 - \rho\,d_{ij}\nabla_i\nabla_j(\mu_1 - \bar{\mu}_2)$$
$$-\frac{\rho_1\rho_2}{\rho}d_{ij}^{(T)}\nabla_i\nabla_j T = 0 \quad (18)$$

$$\dot{\rho}_2 + \vec{v}_2\cdot\vec{\nabla}\rho_2 + \rho_2\text{div}\vec{v}_2 + \rho\,d_{ij}\nabla_i\nabla_j(\mu_1 - \bar{\mu}_2)$$
$$+\frac{\rho_1\rho_2}{\rho}d_{ij}^{(T)}\nabla_i\nabla_j T = 0 \quad (19)$$

the entropy dynamics (heat conduction equation),

$$\dot{\sigma} + v_i\nabla_i\sigma + \frac{\beta}{\rho}\nabla_i(\rho_1\rho_2 w_i) - \kappa_{ij}\nabla_i\nabla_j T$$
$$-\frac{\rho_1\rho_2}{\rho}d_{ij}^{(T)}\nabla_i\nabla_j(\mu_1 - \bar{\mu}_2) = 0 \quad (20)$$

and the elasticity dynamics

$$\dot{U}_{ij} + v_k^{(2)}\nabla_k U_{ij} - \frac{1}{2}(\nabla_j v_i^{(2)} + \nabla_i v_j^{(2)})$$
$$-\frac{\rho_1}{2}(w_i\nabla_j + w_j\nabla_i)\ln\frac{\rho_2}{\rho} + U_{ki}\nabla_j v_k^{(2)} + U_{kj}\nabla_i v_k^{(2)}$$
$$+\zeta_l\delta_{ij}\Phi_{kk} + \zeta_{tr}(\Phi_{ij} - \frac{1}{3}\delta_{ij}\Phi_{kk}) - \xi_1\delta_{ij}\Delta\Phi_{kk}$$
$$-\xi_2\Delta\Phi_{ij} - \xi_3(\nabla_i\nabla_j\Phi_{kk} + \delta_{ij}\nabla_k\nabla_l\Phi_{kl})$$
$$-\xi_4(\nabla_i\nabla_k\Phi_{jk} + \nabla_j\nabla_k\Phi_{ik}) = 0 \quad (21)$$

Since elasticity is assumed to be related to fluid 2, the linear and quadratic couplings to flow are related to velocity $\vec{v}_2$ only. The quadratic couplings are of the "lower convected" type, well-known for the single fluid description [32] - [35] and are of great importance in the viscoelastic case. In addition, there are nonlinear couplings to the concentration variable (the cubic one has been suppressed), which are not possible in a 1-fluid description. The ζ-terms describe relaxation of elastic strains and are absent in a permanent network, where only the strain diffusion terms ($\sim \xi_{1,2,3,4}$) are present. Despite the global incompressibility assumption $\text{div}\,\vec{v} = 0$, the trace of the elastic tensor, U_{kk}, does not vanish even in linear order, since neither $\text{div}\vec{v}_1 = 0$, nor $\text{div}\vec{v}_2 = 0$, nor $\text{div}\vec{w} = 0$, generally.

For the momentum balance of the two different species we get

$$\rho_1\dot{v}_i^{(1)} + \rho_1 v_j^{(1)}\nabla_j v_i^{(1)} + \frac{\rho_1}{\rho}\nabla_i(p + \frac{1}{2}\rho_2 w_j(v_j^1 + v_j^2))$$
$$+\frac{\rho_1\rho_2}{\rho}\nabla_i(\mu_1 - \bar{\mu}_2) + \frac{\rho_1}{\rho}\Phi_{kj}\nabla_i U_{kj} - \rho_1\Phi_{ij}\nabla_j\ln\frac{\rho_2}{\rho}$$
$$+\frac{\rho_1\rho_2}{\rho}\beta\nabla_i T + \xi\rho_1\rho_2 w_i - v_{ijkl}^{(1)}\nabla_j\nabla_l v_k^{(1)}$$
$$- v_{ijkl}^{(12)}\nabla_j\nabla_l v_k^{(2)} = 0 \quad (22)$$

$$\rho_2\dot{v}_i^{(2)} + \rho_2 v_j^{(2)}\nabla_j v_i^{(2)} + \frac{\rho_2}{\rho}\nabla_i(p - \frac{1}{2}\rho_1 w_j(v_j^1 + v_j^2))$$
$$-\frac{\rho_1\rho_2}{\rho}\nabla_i(\mu_1 - \bar{\mu}_2) - \frac{\rho_1}{\rho}\Phi_{kj}\nabla_i U_{kj} + \rho_1\Phi_{ij}\nabla_j\ln\frac{\rho_2}{\rho}$$
$$-\frac{\rho_1\rho_2}{\rho}\beta\nabla_i T - \nabla_j\Phi_{ij} + \nabla_j(\Phi_{jk}U_{ik} + \Phi_{ik}U_{jk})$$
$$-\xi\rho_1\rho_2 w_i - v_{ijkl}^{(2)}\nabla_j\nabla_l v_k^{(2)} - v_{ijkl}^{(12)}\nabla_j\nabla_l v_k^{(1)} = 0 \quad (23)$$

The ξ-terms describe the coupling of the the two momenta due to the difference of the two velocities (friction). Note that although we made the approximation that the linear elastic stress does only act on fluid 2, there are inevitably nonlinear contributions to fluid 1, too. There is also a (nonlinear) coupling of fluid 1 to the concentration, if elastic distortions are present.

In order to facilitate actual calculations we also give eqs.(22,23) as dynamic equations for the total momentum and for the relative velocity

$$\rho\dot{v}_i + \nabla_i p + \nabla_j\left(\rho v_i v_j + \frac{\rho_1\rho_2}{\rho}w_i w_j\right) + 2\nabla_j(\Phi_{jk}U_{ik})$$
$$-\nabla_j\Phi_{ij} - v_{ijkl}\nabla_j\nabla_l v_k - \frac{\rho_1\rho_2}{\rho}v_{ijkl}^{(c)}\nabla_j\nabla_l w_k = 0 \quad (24)$$

$$\dot{w}_i + \left(v_j + \frac{\rho_2 - \rho_1}{\rho}w_j\right)\nabla_j w_i + \rho\xi w_i$$
$$+\nabla_i\left(\mu_1 - \bar{\mu}_2 + \vec{v}\cdot\vec{w} + \frac{\rho_2-\rho_1}{2\rho}\vec{w}^2\right) + \nabla_j\frac{1}{\rho_2}\Phi_{ij}$$
$$+\frac{1}{\rho_2}\Phi_{kj}\nabla_i U_{kj} - \frac{2}{\rho_2}\nabla_j(\Phi_{kj}U_{ik})$$
$$-\frac{\rho_1\rho_2}{\rho}v_{ijkl}^{(m)}\nabla_l\nabla_j w_k - v_{ijkl}^{(c)}\nabla_j\nabla_l v_k = 0 \quad (25)$$

From (24) angular momentum conservation is obvious, since the total stress tensor (defined by $\rho\dot{v}_i + \nabla_j\sigma_{ij} = 0$ in the incompressible limit) is symmetric due to the relation (13), if the viscosities are of the usual form $v_{ijkl} = $

$\nu_\perp (\delta_{ik}\delta_{jl} + \delta_{jk}\delta_{il} - \frac{2}{3}\delta_{ij}\delta_{kl}) + \nu_b \delta_{ij}\delta_{kl}$. Note that the viscosities introduced in (22,23), $\nu_{ijkl}^{(1,2,12)}$, are different from ν_{ijkl} and $\nu_{ijkl}^{(c,m)}$ used in (24,25).

In order to conserve the global incompressibility condition for all times, i.e. div$\vec{v} = 0$, the pressure has to fulfill the relation

$$\begin{aligned}\Delta p &= -\nabla_i \nabla_j (\rho_1 v_i^{(1)} v_j^{(1)} + \rho_2 v_i^{(2)} v_j^{(2)}) + \nabla_i \nabla_j \Phi_{ij} \\ &\quad - \nabla_i \nabla_j (\Phi_{kj} U_{ik} + \Phi_{ik} U_{jk}) + \nu_{ijkl} \nabla_i \nabla_j \nabla_l v_k \\ &\quad + \rho_1 \rho_2 \rho^{-1} \nu_{ijkl}^{(c)} \nabla_i \nabla_j \nabla_l w_k \end{aligned} \quad (26)$$

In contrast to 1-fluid descriptions for simple fluids, where the incompressibility condition leads to a considerable mathematical simplification, this is no longer the case for a 2-fluid description due the complicated form of (26), even if incompressibility is a very good approximation in physical terms. In particular, Δp is not only connected to elastic compressions (U_{kk}), but also to shear deformations, even in linear order.

The static relations between the conjugate quantities to the variables close the system of equations

$$\delta T = TC_V^{-1}\delta\sigma + \alpha_\phi^{-1}\delta\phi + \alpha_3^{-1} U_{kk} \quad (27)$$

$$\begin{aligned}\Phi_{ij} &= c_{tr}(U_{ij} - \frac{1}{3}\delta_{ij} U_{kk}) + c_l U_{kk} \\ &\quad + \alpha_3^{-1}\delta\sigma + \rho^{-1}\kappa_u^{-1}\delta\phi \end{aligned} \quad (28)$$

$$\begin{aligned}\delta(\mu_1 - \bar\mu_2) &= \rho^{-1}\kappa_\phi^{-1}\delta\phi + \rho^{-1}\alpha_\phi^{-1}\delta\sigma \\ &\quad + \kappa_u^{-1} U_{kk}\end{aligned} \quad (29)$$

with $\delta\phi = \rho^{-1}\delta\rho_1 = -\rho^{-1}\delta\rho_2$. Note that $\delta\mu$ is not needed, but follows from δp via eq. (5) or (9).

DISCUSSION

Within the general framework of hydrodynamics and thermodynamics we have set up a consistent nonlinear 2-fluid description of complex fluids, in particular for polymer solutions or swollen elastomers. Such a general theory [34] determines the frame for any ad-hoc model, which has to be a special case of the general one. The comparison with the general theory also reveals implicit and explicit assumptions, approximations and possible generalizations of a given model. A "simple" or "natural" choice in a given model may not be mandatory, but rather imply a presumption.

Quite generally we find that neither the velocity, with which a certain variable is convected, nor the stress division between the different fluids can be determined by general principles, but is rather system or material dependent. On the other hand, there are certain restrictions and interrelations among the convective velocities and other physical effects that limit the possible choices. For the two densities ρ_1, ρ_2 e.g., the "natural" choice for the convection velocities (taken in the preceding section) seems to be their native velocities $\vec{v}_1$ and $\vec{v}_2$, respectively. This implies that the total density is convected with the mean velocity $\vec{v}$ (as required by mass transport, but manifest only, if the incompressibility assumption is lifted), while the concentration ϕ is convected with $(1/\rho)(\rho_2 \vec{v}_1 + \rho_1 \vec{v}_2)$. Another obvious ("simple") choice would be the mean velocity as convection velocity for both, the total density as well as the concentration implying that also ρ_1 and ρ_2 are convected with $\vec{v}$. However, the actual convection velocity depends on the value of a material dependent (reactive) flow parameter (call it γ) and is not restricted to the two choices mentioned above ($\gamma = 0$ and $= -1$ for the "natural" and the "simple" one).

In the case of visco-elastic and elastic media, which are described by a dynamic equation for the (Eulerian) strain tensor U_{ij}, there are two velocities involved. One is the usual convection velocity ($v_k \nabla_k U_{ij}$) and the other one occurs in the "lower convected" part ($U_{kj}\nabla_i v_k + U_{ki}\nabla_j v_k$). There is no fundamental reason for the two to be equal, nor to be one of the obvious choices ($\vec{v}$ or $\vec{v}_2$).

For the evolution equations of the momenta special care has to be taken to get a description, which is compatible with general laws. The currents and quasi-currents that enter the description in terms of either the total momentum and the velocity difference or in terms of the two individual momenta are not the same comparing eqs.(22, 23) with (24, 25). In particular, the momenta $\vec{g}_{1,2}$ are convected with $\vec{v}_{1,2}$ implying that the total momentum $\vec{g}$ and the relative velocity $\vec{w}$ are convected with $\vec{v}$ and $(1/\rho)(\rho_2\vec{v}_1 + \rho_1\vec{v}_2)$, respectively. Instead of this "natural" choice there are other possibilities governed by some phenomenological parameters (independent from γ), e.g. the "simple" choice that all 4 quantities are convected with $\vec{v}$. Even the convection of the entropy can be tuned by choosing a parameter $\beta \equiv \beta_0 + \beta_{00}\sigma$ where $\beta_{00} = 1/\rho_1, = 0, = -1/\rho_2$ leads to the convective velocity to be $\vec{v}_1, \vec{v}, \vec{v}_2$, respectively.

In the preceding section it was assumed that (linear) elastic stress is carried only by fluid 2. Generally, however, the distribution of the elastic stress among the two fluids is governed by a phenomenological coefficient (call it $\lambda^{(U)}$). For, respectively, $2\lambda^{(U)} = 1/\rho_2, = 0,$ or $= -1/\rho_1$, the elastic stress is carried only by fluid 2, is equally distributed between 1 and 2, or carried only by fluid 1; but $\lambda^{(U)}$ can have any value in between.

A prominent feature of the 2-fluid description is the coupling of the concentration dynamics to the velocity difference. Linearizing and Fourier transforming the dynamic equations, thus eliminating $\vec{w}$ from e.g. the concentration dynamics and neglecting fourth order gradient terms we get

$$i\omega\phi - d^{eff}\Delta\Pi - \frac{\rho_1\rho_2}{\rho^2}d^{(T)eff}\Delta T - 2\lambda^{(\phi)}\nabla_i\nabla_j\Phi_{ij} = 0 \tag{30}$$

with frequency dependent effective diffusion and thermo-diffusion coefficients

$$d^{eff} = d + \frac{\rho_1\rho_2}{\rho^2}\frac{(\gamma+1)^2}{\rho\xi+i\omega} \tag{31}$$

$$d^{(T)eff} = d^{(T)} + \frac{\beta(\gamma+1)}{\rho\xi+i\omega} \tag{32}$$

and the dynamic coupling to the elastic degree of freedom by

$$\lambda^{(\phi)} = \frac{\rho_1\rho_2}{\rho}\lambda^{(U)}\frac{1+\gamma}{\rho\xi+i\omega} \tag{33}$$

where the dispersion step around $\omega \approx \rho\xi$ is due to the friction ($\sim \xi$) between the two fluid momenta. Again these possible additions to the concentration dynamics, however, depend on the choices for the convection velocities (i.e. on γ and β) as well as on the way how the elastic stress has been divided among the two fluids (on $\lambda^{(U)}$). E.g. if the stress is equally distributed among the 2 fluids and if both densities are convected with $\vec{v}$, there are no frequency dependent additions to the concentration dynamics at all.

Recently, 2-fluid descriptions of diffusion in polymeric systems have been given [36, 37] based on the GENERIC approach making use of Poisson brackets. A detailed comparison with these formulations is beyond the scope of this manuscript and will be discussed elsewhere.

ACKNOWLEDGMENTS

This research was supported in part by the National Science Foundation under Grant No. PHY99-07949.

REFERENCES

1. Larson R.G., *The Structure and Rheology of Complex Fluids*, Oxford University Press (1999).
2. Reichl L.E., *A Modern Course in Statistical Physics*, 2nd Ed., Wiley (1997).
3. Drew D.A. and Passman S.L., *Theory of multi-component fluids*, in *Applied Mathematical Sciences* Vol. **135**, Springer Verlag (1998).
4. Hohenberg P.C. and Martin P.C., *Annals Phys.* **34** 291-359 (1965).
5. Sugiyama L.E. and Park W., *Phys. Plasmas* **7**, 4644-58 (2000).
6. Lee D.K.K. and Lee P.A., *Physica B* **261**, 481-2 (1999).
7. Hebraud P., Lequeux F., and Palierne J.F., *Langmuir* **16**, 8296-9 (2000).
8. Kadoma I.A. and van Egmond J.W., *Phys. Rev. Lett.* **80**, 5679-82 (1998).
9. Lhuillier D., *J. Non-Newtonian Fluid Mech.* **96**, 19-30 (2001).
10. Brochard F. and deGennes P.G., *Macromolecules* **10**, 1157-61 (1977).
11. Harden J.L., Pleiner H., and Pincus P.A., *J. Chem. Phys.* **94**, 5208-21 (1991).
12. Brochard F. and deGennes P.G., *PhysicoChemical Hydrodynamics* **4**, 313-22 (1983).
13. Doi M., in *Dynamics and Patterns in Complex Fluids. New Aspects of the Physics-Chemistry Interface*, eds. Onuki A. and Kawasaki K., Springer Proceedings in Physics, Vol. **52**, Springer Verlag, Berlin, (1990).
14. Onuki A., *Phys. Rev. Lett.* **62**, 2472-5 (1989).
15. Milner S.T., *Phys. Rev. Lett.* **66**, 1477-80 (1991).
16. Milner S.T., *Phys. Rev. E* **48**, 3674-91 (1993).
17. Helfand E. and Fredrickson G.H., *Phys. Rev. Lett.* **62**, 2468-71 (1989).
18. Ji H. and Helfand E., *Macromolecules* **28**, 3869-80 (1995).
19. Saito S., Takenaka A., Toyoda N., and Hashimoto T., *Macromolecules* **34**, 6461-73 (2001).
20. Doi M. and Onuki A., *J. Phys. (France)* **112**, 1631-56 (1992).
21. Onuki A., Yamamoto R., and Taniguchi T., *J. Phys. II (France)* **7**, 295-304 (1997).
22. Okuzono T., *Phys. Rev. E* **56**, 4416-26 (1997).
23. Sun T., Balazs A.C., and Jasnow D., *Phys. Rev. E* **59**, 603-11 (1999).
24. Araki T. and Tanaka H., *Macromolecules* **34**, 1953-63 (2001).
25. Ianniruberto G. and Marrucci G., *J. Non-Newtonian Fluid Mech.* **54**, 231-9 (1994).
26. Ianniruberto G., Greco F., and Marrucci G., *Ind. Eng. Chem. Res.* **33**, 2404-11 (1994).
27. Apostolakis M.V., Mavrantzas V.G., and Beris A.N., *J. Non-Newtonian Fluid Mech.* **102**, 409-45 (2002).
28. Brand H.R. and Pleiner H., Hydrodynamics and Electrohydrodynamics of Liquid Crystals, in *Pattern Formation in Liquid Crystals*, eds. Buka A. and Kramer L., Springer New York, p. 15-67 (1995).
29. Lubensky T.C. and Chaikin P.M., *Principles of Condensed Matter Physics*, Cambridge University Press (1995).
30. Forster D., *Hydrodynamic Fluctuations, Broken Symmetry and Correlation Functions*, Benjamin, Reading Mass. (1975).
31. Martin P.C., Parodi O., and Pershan P.S., *Phys. Rev. A* **6**, 2401-24 (1972).
32. Temmen H., Pleiner H., Liu M., and Brand H.R., *Phys. Rev. Lett.* **84**, 3228-31 (2000); **86**, 745 (2001).
33. Pleiner H., Liu M., and Brand H.R., *Rheol. Acta* **39**, 560-5 (2000).
34. Pleiner H. and Harden J.L., General 2-Fluid Hydrodynamics of Complex Fluids and Soft Matter, in *Nonlinear Problems of Continuum Mechanics, Special issue of Notices of Universities. South of Russia. Natural sciences*, p.46 - 61 (2003).
35. Grmela M., *Phys. Lett. A* **296**, 97-104 (2002).
36. Elafif A., Grmela M., and Lebon G., *J. Non-Newtonian Fluid Mech.* **86**, 253-75 (1999).
37. El Afif A. and Grmela M., *J. Rheol.* **46**, 591-628 (2002).

Recovery of polymer blends after melt elongation: Analysis of a model for small and large capillary numbers

U.A. Handge

Institute of Polymers, Department of Materials, ETH Zürich, ML J 16, CH-8092 Zürich, Switzerland

Abstract. Applying an effective medium approximation, we theoretically investigate the recovery of binary blends of immiscible polymers after melt elongation. In our model, we consider effective values for the Hencky strain rates of the disperse and the matrix phase. We derive temporal evolution equations which allow calculation of the transient recovered stretch. Numerical solutions of this set of equations are presented and discussed. Our analysis reveals that the capillary number strongly influences the recovery process. By comparing the predictions of our model with experiments, we show that our model captures the basic features of the experimental data, i.e. the time scale of the recovery process and the equilibrium value of the recovered stretch.

INTRODUCTION

The formation of an emulsion of two immiscible Newtonian fluids results in a viscoelastic fluid with a complex morphology. The viscoelasticity of the emulsion is caused by the interfacial tension at the interphase between the two phases: Since the interfacial tension is associated with an elastic energy, the combination of the viscosities of the two components of the emulsion with the interfacial tension leads to viscoelastic effects. Blends of two immiscible polymers also form two phase viscoelastic systems. In addition to the interfacial tension, the viscoelasticity of the pure blend components significantly contributes to the viscoelasticity of the polymer blend. The rheology of emulsions and polymer blends has attracted much attention since many decades, see Refs. [1-3] for an overview. Generally, for blends of highly viscous polymers the stresses that are caused by the viscoelasticity of the blend components are much larger than the interfacial stress. This was explicitly shown for polystyrene/poly(methyl methacrylate) blends in Ref. [4]. Consequently, large macroscopic effects are only observable when the externally applied stresses are negligible or comparable to the interfacial stress. Important examples are shear oscillations at low frequencies and recovery experiments after shear deformation and melt elongation. In this work, we focus on the later situation and theoretically study the recovery of blends of immiscible polymers after melt elongation.

In simple elongation, the sample is extended uniaxially with a constant Hencky strain rate $\dot{\varepsilon}_0$. Then the macroscopic Hencky strain ε_s increases linearly with time t and $\varepsilon_s = \dot{\varepsilon}_0 t$ holds. For a single drop of a Newtonian fluid that is embedded in an infinite matrix of a Newtonian fluid, the ratio of the viscous stress of the matrix to the interfacial stress can be estimated by the capillary number Ca. In simple elongation, the capillary number is defined by [5]

$$\text{Ca} = \eta_m \dot{\varepsilon}_0 r_0 / \alpha. \quad (1)$$

In Eq. (1) η_m is the viscosity of the matrix, r_0 the radius of the spherical drop and α the interfacial tension. In recovery experiments, the sample is cut immediately after the macroscopic extension of the sample has been terminated. Then the sample recovers freely. The retraction of the sample is caused by the recovery of the macromolecules, the surface tension induced recovery portion and the contribution of the interfacial tension to the recovery for blends of immiscible polymers [6]. In this work, we analyze the recovery portion which is driven by the interfacial tension.

THE MODEL

In this study, we focus on droplet morphologies where the minor phase consists of spherical drops that are spatially uniformly distributed in the continuous matrix. The volume fraction of the minor phase is denoted by Φ. In order to simplify the analysis, we assume that the radii of the spherical drops are all equal and are denoted by r_0. In addition, we assume that the pure blend components are Newtonian fluids. Since the time scale of the interfacial tension induced recovery generally is larger than the relaxation times of the polymers, this assumption is admissible for many polymer blends. In the following, we

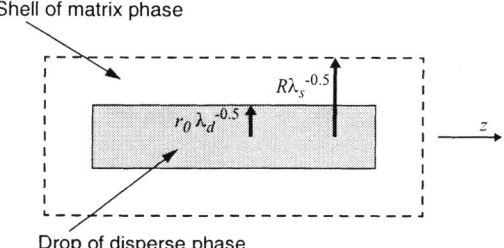

FIGURE 1. Schematic illustration of a highly elongated drop of the disperse phase that is surrounded by a cylindrical shell of matrix material (dashed line).

denote the shear viscosity of the disperse phase by η_d and the shear viscosity of the matrix phase by η_m.

During elongation, the polymer blend is stretched with Hencky strain rate $\dot{\varepsilon}_0$. The stretch ratio of the sample is denoted by $\lambda_s(t) = L(t)/L_0$, where L_0 is the initial length of the sample and $L(t)$ is the length of the sample at time t. The Hencky strain ε_s of the sample is related to the stretch ratio by $\lambda_s = \exp(\varepsilon_s)$. Because of the macroscopic deformation of the sample, the drops of the minor phase are also elongated and are deformed into an ellipsoidal shape. The stretch ratio λ_d of the drops is defined by $\lambda_d = a/r_0$. Here a is the length of the semiaxis of the elongated drop which is parallel to the loading direction. For a single drop that is embedded in an infinite matrix, Delaby et al. calculated the deformation of the drop for small and large capillary numbers [7]:

$$\lambda_d(t) = \begin{cases} \exp\{2F(p)\text{Ca}[1-\exp(-t/\tau_{\text{rel}})]\} & \text{for Ca} \ll 1 \\ 1 + 5[\lambda_s(t)-1]/(2p+3) & \text{for Ca} \gg 1. \end{cases} \quad (2)$$

Here we have $F(p) = (19p+16)/(16p+16)$, the viscosity ratio $p = \eta_d/\eta_m$ and $\tau_{\text{rel}} = \eta_m r_0(2p+3)(19p+16)/[40\alpha(p+1)]$. Equation (2) reveals that the stretch ratio λ_d of the drops is influenced by the capillary number. For Ca $\ll$ 1 the stretch ratio λ_d of the drops depends on Ca, whereas for Ca $\gg$ 1 the stretch ratio λ_d is independent of Ca. In this study, we assume that Eq. (2) is a rough approximation for the true deformation of the disperse phase. The extension of the sample is terminated when the maximum Hencky strain $\varepsilon_{\max}$ is attained at time $t = t_{\max} = \varepsilon_{\max}/\dot{\varepsilon}_0$. Then the sample is cut at one boundary and begins to recover freely. The transient retraction of the sample is usually quantified by the recovered stretch λ_r. It is defined by

$$\lambda_r(t') = L_A/L(t'), \quad (3)$$

where L_A is the length of the sample at cutting time $t = t_{\max}$ and $L(t')$ denotes the length of the sample at recovery time $t' = t - t_{\max}$.

In this work, we consider effective values for the Hencky strain rates $\dot{\varepsilon}_d$ of the disperse phase and $\dot{\varepsilon}_m$ of the matrix. These effective strain rates linearly superpose to the total strain rate $\dot{\varepsilon}_s$ of the sample. This yields

$$\dot{\varepsilon}_s = \Phi\dot{\varepsilon}_d + (1-\Phi)\dot{\varepsilon}_m. \quad (4)$$

Because of the macroscopic extension of the sample, the drops of the disperse phase are elongated and the interfacial area increases. Thus elastic energy is stored in the sample up to the termination of elongation. In subsequent recovery, the elastic energy is released which causes the retraction of the sample. In this study, we assume that the interfacial energy E_α is completely dissipated by the viscous flow of the two phases. Thus

$$dE_\alpha/dt = -3V_0[\Phi\eta_d\dot{\varepsilon}_d^2 + (1-\Phi)\eta_m\dot{\varepsilon}_m^2] \quad (5)$$

holds where V_0 denotes the volume of the sample. The factor 3 in Eq. (5) results from the elongational viscosity which was derived by Trouton [8] and equals $3\eta_d$ for the disperse phase and $3\eta_m$ for the matrix material.

The interfacial area A_d of an elongated drop with stretch ratio λ_d was calculated by Gramespacher and Meissner [6]. Modeling the ellipsoidal drops by cylinders with radius r_0, they derived the approximation $A_d = 2\pi r_0^2(2\lambda_d^{1/2} + \lambda_d^{-1} - 1)$ which we use in the following. The number N of all drops of the disperse phase is $N = 3\Phi V_0/(4\pi r_0^3)$. Since the total interfacial area is given by $A_\alpha = NA_d$, we have

$$A_\alpha = 3\Phi V_0(2\lambda_d^{1/2} + \lambda_d^{-1} - 1)/(2r_0). \quad (6)$$

The interfacial energy E_α equals $E_\alpha = \alpha A_\alpha$. Inserting $E_\alpha = \alpha A_\alpha$ into Eq. (5) yields

$$[\Phi/(2\tau_m)]\left(\lambda_d^{-1} - \lambda_d^{1/2}\right)\dot{\varepsilon}_d = p\Phi\dot{\varepsilon}_d^2 + (1-\Phi)\dot{\varepsilon}_m^2 \quad (7)$$

with $\tau_m = \eta_m r_0/\alpha$. We now consider the strain rates of the drops and the matrix during elongation and recovery. The z axis is parallel to the loading direction. Initially at time $t = 0$ each drop is surrounded by an approximately cylindrical shell with outer radius R, see Fig. 1. We define the volume V_{sh} of each shell such that the ratio of the shell volume to the volume V_d of a drop is given by $V_{sh}/V_d = (1-\Phi)/\Phi$. Therefore the sample can be completely split up into the drops and the shells which each of them surround one drop. In Ref. [4] it was shown that in each cross section of the sample the area fraction Φ_A which is occupied by the drops equals on an average the volume fraction Φ of the drops: $\Phi_A = \Phi$. Consequently, at $t = 0$ the outer radius of the cylindrical shell and the initial radius of the drop are related by $\pi r_0^2 = \pi\Phi R^2$. Hence we have $r_0 = \sqrt{\Phi}R$. During elongation and subsequent recovery, the drop and the sample

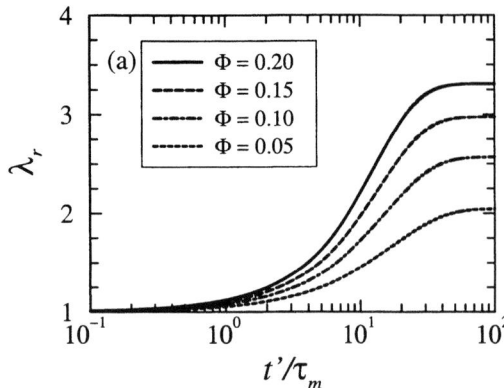
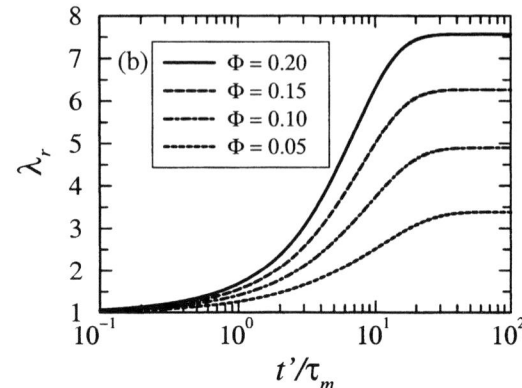

FIGURE 2. Transient recovered stretch λ_r versus recovery time t'. The viscosity ratio is $p = 1$ and the stretch ratio λ_{max} of the sample is $\lambda_{max} = \exp(3) = 20.1$. The results are obtained by solving Eqs. (8) and (9) numerically. The Ca values are (a) Ca $\ll$ 1 with Ca = 0.5 and $t_{max}/\tau_m = 10$ and (b) Ca $\gg$ 1.

are deformed. Then the semiaxes of the drop perpendicular to the loading direction are given by $r_0/\sqrt{\lambda_d}$. The outer radius of the shell is $R/\sqrt{\lambda_s}$. Note that the deformation of the outer radius of the sample and of the shell are equal, since the sample can be completely split up into all shells which each of them surround one droplet. During recovery, the drops contract in the direction parallel to the z direction and expand perpendicular to this direction. This leads to a radial velocity $v_r(r)$ in the matrix which is calculated in the following. The velocity gradient perpendicular to the z direction inside a drop is $-\dot{\varepsilon}_d/2$. The matrix does not contribute to the radial velocity $v_r(r)$. Since the ratio of the drop radius to the outer radius of the shell is $\sqrt{\Phi\lambda_s/\lambda_d}$ we have $v_r(r) = -\dot{\varepsilon}_d r\sqrt{\Phi\lambda_s/\lambda_d}/2$, see also Ref. [4]. Inserting this expression and $v_\varphi = 0$ into the equation of continuity for incompressible fluids $\partial(rv_r)/\partial r + \partial v_\varphi/\partial \varphi + r\partial v_z/\partial z = 0$ we find $\partial v_z/\partial z = \dot{\varepsilon}_d\sqrt{\Phi\lambda_s/\lambda_d}$ for the velocity gradient in the z direction. In simple elongation, the Hencky strain rate equals the gradient of the velocity and thus $\dot{\varepsilon}_m = \partial v_z/\partial z = \dot{\varepsilon}_d\sqrt{\Phi\lambda_s/\lambda_d}$ holds. Inserting $\dot{\varepsilon}_m = \dot{\varepsilon}_d\sqrt{\Phi\lambda_s/\lambda_d}$ into Eq. (7), we obtain an ordinary differential equation for λ_d:

$$d\lambda_d/dt' = \left(1 - \lambda_d^{3/2}\right)/[2\tau_m(p + (1-\Phi)\lambda_s/\lambda_d)]. \quad (8)$$

Combining the mixing rule $\dot{\varepsilon}_s = \Phi\dot{\varepsilon}_d + (1-\Phi)\dot{\varepsilon}_m$ with $\dot{\varepsilon}_m = \dot{\varepsilon}_d\sqrt{\Phi\lambda_s/\lambda_d}$ and Eq. (8) leads to

$$d\lambda_s/dt' = \left[\Phi + (1-\Phi)\sqrt{\Phi\lambda_s/\lambda_d}\right]\lambda_s\dot{\lambda}_d/\lambda_d. \quad (9)$$

In Eq. (9) $\dot{\lambda}_d = d\lambda_d/dt'$ holds. At $t' = 0$ we have $\lambda_s(t'=0) = \lambda_{max} = \exp(\varepsilon_{max})$. The stretch ratio λ_d of the drops at $t' = 0$ equals to their stretch ratio at cutting time $t = t_{max}$ and thus is given by Eq. (2):

$$\lambda_d(t'=0) = \quad (10)$$

$$\begin{cases} \exp\{2F(p)\text{Ca}[1 - \exp(-t_{max}/\tau_{rel})]\} & \text{for Ca} \ll 1 \\ 1 + 5(\lambda_{max} - 1)/(2p + 3) & \text{for Ca} \gg 1. \end{cases}$$

RESULTS

In this study, we solved Eqs. (8) and (9) numerically for various sets of parameters using a Runge-Kutta algorithm. The results are presented in Fig. 2 for the parameters $p = 1$ and $\lambda_{max} = \exp(3) = 20.1$ and two values of the capillary number. In Fig. 2(a) we consider the case Ca $\ll$ 1 with Ca = 0.5 and $t_{max}/\tau_m = 10$. The data of Fig. 2(b) are obtained for a large capillary number Ca $\gg$ 1. Figure 2 reveals that the transient recovered stretch increases with time and attains an equilibrium value. The steady state value is attained when the drops of the minor phase have been retracted to spheres. The equilibrium value of λ_r increases with Φ, since for larger Φ values more drops contribute to the retraction of the sample. In Fig. 2 the equilibrium value of λ_r depends roughly linearly on Φ. Comparing the transient recovered stretch for small and large capillary numbers, one finds that λ_r increases more rapidly for a larger Ca value and attains a larger equilibrium value for Ca $\gg$ 1 than for Ca $\ll$ 1.

Figure 3 depicts the numerical solutions of Eqs. (8) and (9) for $\Phi = 0.15$ and $\lambda_{max} = \exp(4) = 54.6$. The ratio of the zero shear viscosities is $p = 0.1, 1.0$ and 10.0, respectively. In Fig. 3(a) we have Ca = 0.3 and $t_{max}/\tau_m = 10$. For these parameters, the equilibrium value of λ_r attains the largest value for $p = 1.0$. In Fig. 3(b) we consider the case of a large capillary number. Here the curve for $p = 0.1$ attains most rapidly its equilibrium value which is larger than the equilibrium values for $p = 1.0$ and $p = 10.0$. The recovered stretch for the blend with $p = 10.0$ increases most slowly and attains the smallest equilibrium value of the blends in Fig. 3(b).

Gramespacher and Meissner experimentally studied the recovery of polystyrene (PS)/poly(methyl methacry-

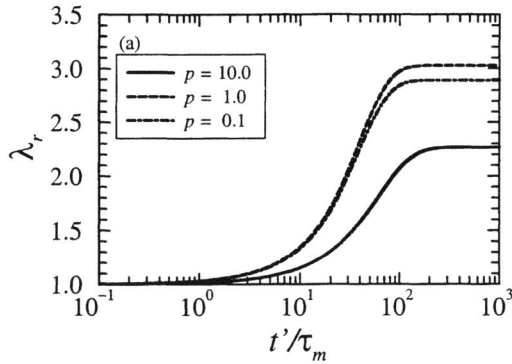 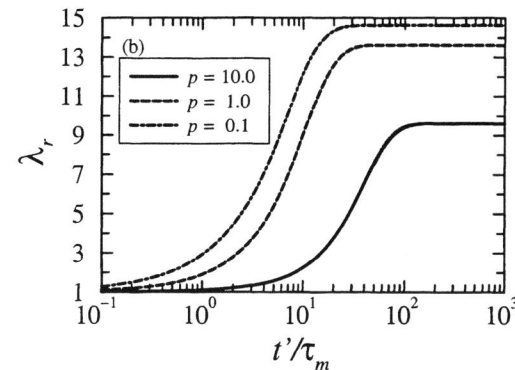

FIGURE 3. Transient recovered stretch $\lambda_r(t')$ for $\Phi = 0.15$ and $\lambda_{max} = \exp(4) = 54.6$. The capillary number is (a) Ca = 0.3 and $t_{max}/\tau_m = 10$ and (b) Ca $\gg$ 1.

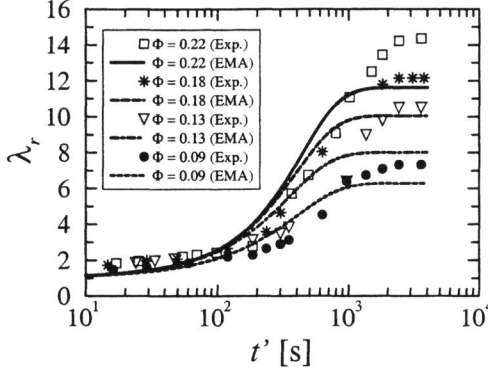

FIGURE 4. Transient recovered stretch λ_r of PS/PMMA blends versus recovery time t' for $\dot{\varepsilon}_0 = 0.1\,\text{s}^{-1}$ and $\lambda_{max} = \exp(3.5) = 33.1$. The experimental data are taken from Ref. [6]. The lines are the results of our effective medium approximation (EMA).

late) (PMMA) blends after melt elongation [6]. Their results are depicted in Fig. 4 for the parameters $\dot{\varepsilon}_0 = 0.1\,\text{s}^{-1}$ and $\lambda_{max} = \exp(3.5) = 33.1$. The Hencky strain rate $\dot{\varepsilon}_0 = 0.1\,\text{s}^{-1}$ corresponds to Ca $\gg$ 1 for the PS/PMMA system of Ref. [6]. This can be verified by inserting the zero shear viscosity of the PMMA matrix $\eta_m = 1.22 \cdot 10^5\,\text{Pas}$, the interfacial tension $\alpha = 2 \cdot 10^{-3}\,\text{Nm}^{-1}$ and the average droplet radius $r_0 = 0.5\,\mu\text{m}$ for the blend with $\Phi = 0.09$ into Eq. (1) which yields Ca = 3.1. In Fig. 4 we also plot the results of our effective medium approximation for large Ca. The numerical solutions of Eqs. (8) and (9) agree qualitatively well with the experimental data of Ref. [6].

CONCLUSIONS

In this study, we theoretically investigated the time-dependent recovery of blends of immiscible polymers after melt elongation. In our work, we consider effective values for the Hencky strain rates of the disperse and the matrix phase. Applying an effective medium approximation, we derive two temporal evolution equations for the stretch ratio of the disperse phase and the blend. Our analysis explicitly takes into account the value of the capillary number. Numerical solutions of the time evolution equations show that the capillary number strongly influences the recovery process. We also compare the results of our model with experimental data. The experimental values for the time scale of the recovery and the equilibrium value of the recovered stretch agree qualitatively well with the results of our model. Consequently, the effective medium approach which we apply here is a powerful tool in order to model the recovery of two phase polymer blends.

ACKNOWLEDGMENTS

The discussions with Professor Chr. Friedrich, Professor J. Meissner and Professor H.C. Öttinger are gratefully acknowledged.

REFERENCES

1. Rallison, J.M., *Ann. Rev. Fluid Mech.*, **16**, 45-66 (1984).
2. Stone, H.A., *Ann. Rev. Fluid Mech.*, **26**, 65-102 (1994).
3. Tucker III, C.L., and Moldenaers, P., *Ann. Rev. Fluid Mech.*, **34**, 177-210 (2002).
4. Handge, U.A., *J. Rheol.*, **47**, 969-978 (2003).
5. Acrivos, A., and Lo, T.S., *J. Fluid Mech.*, **86**, 641-672 (1978).
6. Gramespacher, H., and Meissner, J., *J. Rheol.*, **41**, 27-44 (1997).
7. Delaby, I., Ernst, B., and Muller, R., *Rheol. Acta*, **34**, 525-533 (1995).
8. Trouton, F.T., *Proc. R. Soc. London Ser. A*, **77**, 426-440 (1906).

Visco-elastic Relaxation in Novel Mosaic Phase of Non-Symmetric Chiral Twin Liquid Crystals

Jun Yamamoto[*], Isa Nishiyama[*] and Hiroshi Yokoyama[#*]

[*]Japan Science and Technology Agency, Yokoyama Nano-structured Liquid Crystal Project, 5-9-9 Tokodai, Tsukuba, Ibaraki 300-263 Japan
[#]Nanotechnology Research Institute, National Institute of Advanced Industrial Science and Technology, 1-1-4 Umezono, Tsukuba, Ibaraki 305-8568 Japan

Abstract. We have found the finite elasticity and characteristic visco-elastic relaxation in novel mosaic phases of the non-symmetric chiral twin liquid crystal. X-ray scattering experiment shows the mosaic phase possesses smectic layer structure, but its intensity rather weak in comparison with smectic phase, and appears in the higher temperature and optical purity region than TGB phase. Thus, defects of the smectic layers are three dimensionally dispersed in the mosaic phase, and interacted with each other by the spatial deformation of the layers. Correlation of defects produces the macroscopic mobility and the unique visco-elastic properties. These relaxation phenomena can be expected to relate to the intrinsic feature of the collective hydrodynamic modes in the defect lattices systems.

INTRODUCTION

Chirality tends to introduce the helical twist into the liquid crystal system, and competes with the flat homogenous layer structure in the smectic phase or uniform orienational order in the nematic phase [1]. Various types of modulated phases are produced by the competition between chirality and the uniformity of the liquid crystals. Cholesteric blue phase is well-known system of this kind of modulated phases, where the twist of the nematic director is regularly ordered in 3-dimensional space, but the position of molecules are completely random as usual isotropic liquid phases [1]. In other words, "defects" of the director form the super lattice in visible light wave-length scale. TGB phase is also predicted from the theoretical view point [2] and already confirmed experimentally as chirality modulated smectic systems [3]. Recently, chirality modulated 3-dimensional smectic liquid crystals, such as the smectic blue phases [4] and the smectic Q phase [5] are reported.

In these defect lattice system, background liquid crystals order play role of the "field" of the apparent inter-defect interaction originated from the long-range elastic distortion of the liquid crystalline order by defect themselves. Thus, the correlation of defects is expected to produce the macroscopic mobility and the unique visco-elastic properties, such as solid like elastic response and sound modes of the collective vibration of the defects. In this paper, we report the unique visco-elastic properties of the novel mosaic phase [6] which is one of chirality modulated 3-dimensional smectic liquid crystals.

MOSAIC PHASE

Non-symmetric chiral twin molecule consists of two equivalent chiral liquid crystal molecules connected by the non-symmetric spacer unit. Novel mosaic phase appears in narrow temperature region between isotropic and ferro smectic phase [6]. Figure 1a shows the texture of the mosaic phase of B7R. In case of shorter spacer B5R, the mosaic texture is more clearly observed as shown in Fig. 1b. Mosaic texture appears both in cooling and heating process, but the size and shape of mosaic is strongly dependent on the process. Mosaic phase has macroscopic anisotropy, but it is so weak to identify in the thin sample cell.

Figure 2 shows the temperature dependence of the low angle X-ray scattering. Layer structure appears below 131°C where the peak intensity abruptly increases. Again peak intensity more increases at 127°C, where the macroscopic anisotropy appears as indicated in the inset photograph of X-ray sample capillary (Fig.3a). Under polarizing microscope observation, the phase sequence of the B7R can be determined. In the intermediate mosaic phase, there is no macroscopic change in the sample capillary (Fig.3b); nevertheless the microscopic layer structure already exists. On the other hand, layer repeat distance continuously shrinks with decreasing in temperature. In the three tilted smectic phases, the shrinkage of the layer repeat distance can be explained by the temperature dependence of the molecular tilt angle to the smectic layers. Thus, it is characteristic to the mosaic phase that there exists finite microscopic layer structure, but its intensity is weaker and repeat distance is longer than lower temperature classical tilted smectic phases.

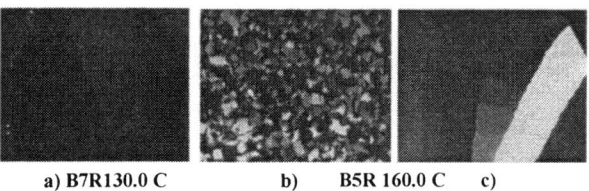

FIGURE 1. a) Photograph of the mosaic phase of B7R in the visco-elastic sample cell by polarizing microscope. b) and c) mosaic phase of B5R. Domain size of the mosaic texture is dependent on the cooling rate.

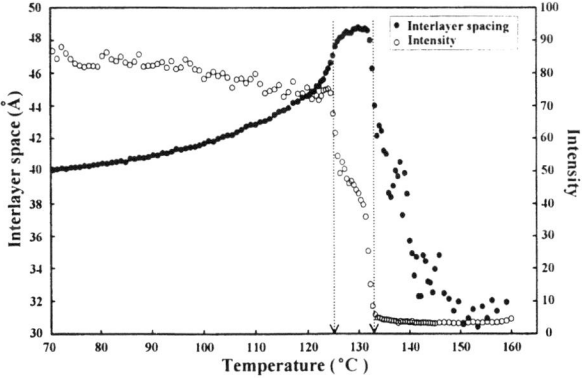

FIGURE 2. Temperature dependence of the small angle X-ray scattering of B7R. Closed circles indicate the layer repeat distance, open circles the intensity of the X-ray scattering.

FIGURE 3. Photographs of X-ray sample capillary under crossed Nichol simultaneously obtained with X-ray scattering. In mosaic phase (129.0°C), weakly scattering region can be identified different from the isotropic phase (134.0°C). Below 124°C, strong macroscopic anisotropy appears and sample becomes evidently turbid.

VISCOELASTIC PROPERTIES

In this paper, we focus our attention to the visco-elastic properties of the mosaic phase. Liquid crystal sample is sandwiched between two glass plates in the measurement cell. Complex mechanical transfer function $Z(\omega)$ at low frequency can be observed by applying the tiny longitudinal strain produced by the PZT transducer [7]. Figure 4 and 5 show frequency dependencies of the real (elastic) part and imaginary (viscous) part of $Z(\omega)$ respectively. On the cooling process, the frequency independent real part of $Z(\omega)$ appears at the isotropic to mosaic phase transition temperature(T_{IM}). Constant real part is strongly dependent on the temperature in the region $T_{IM}-T<1.5°C$, as shown by open symbol data in Fig. 3. Imaginary part of $Z(\omega)$ is almost proportional to the frequency in this temperature region. So the viscosity, which is the proportionality coefficient of the imaginary part of $Z(\omega)$ against frequency, is also abruptly increase with decrease in the temperature.

In lower temperature region $1.5°C<T_{IM}-T<4.5°C$ (mosaic-ferro smectic phase transition), real part of the $Z(\omega)$ has large frequency dependent part in addition to the frequency independent part. Since all close symbol data can extrapolate to the same value at the low frequency limit as can be seen in the figure 3, so that the constant part is almost independent on the temperature. These values are almost similar to the layer compression modulus of conventional smectic phases. Imaginary part of $Z(\omega)$ also deviates from the linear dependence to the frequency in this temperature region. Frequency dependent real part is almost proportional to the ω^2, and its amplitude is almost equivalent to the deviation from the linear frequency dependence in the imaginary part. Thus, we can conclude that large frequency dependence both in the real and imaginary parts are originated by the visco-elastic relaxation. Unfortunately the relaxation frequency is higher than the limit of our experiment for pure B7R, but roughly estimated as 10kHz-100kHz. It is obvious that the relaxation frequency is slowing down towards the low temperature mosaic to ferro smectic phase transition.

Figure 6 and 7 show the temperature dependence of the real and imaginary part of $Z(\omega)$ respectively. Two separated temperature regions are more evident as shown in vertical line in Fig. 6. Visco-elastic relaxation exists in lower temperature region, whereas real part has no frequency dependence in higher temperature region. It is much obvious that the low frequency limit of the real part of Z(ω) is completely independent of the temperature as indicated by dotted line in Fig. 6.

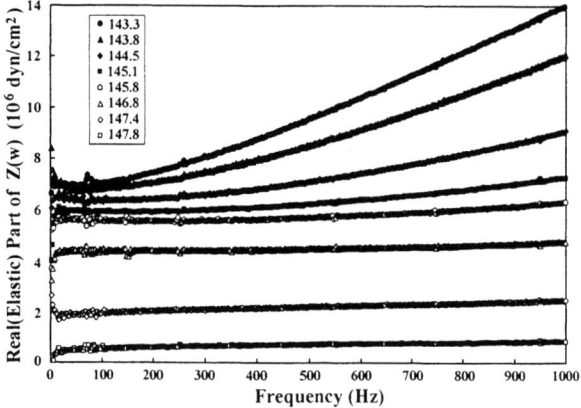

FIGURE 4. Frequency dependence of the real (elastic) part of Z(ω). Each line represents the temperature dependence as indicated in the figure.

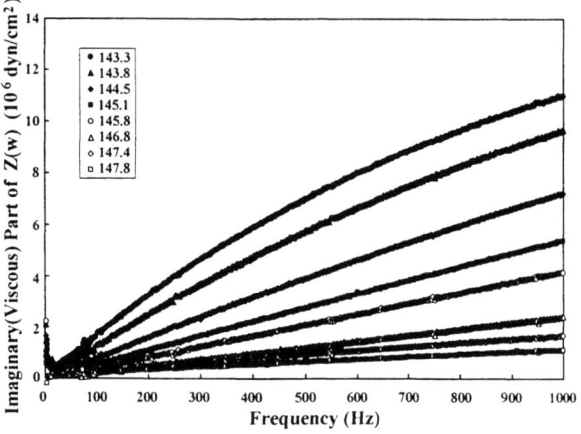

FIGURE 5. Frequency dependence of the imaginary (viscous) part of Z(ω). Each line represents the temperature dependence as indicated in the figure.

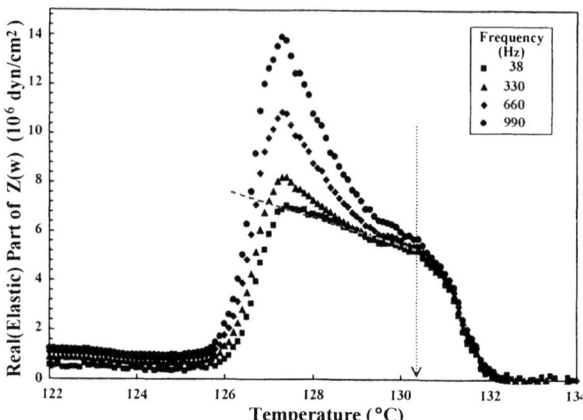

FIGURE 6. Temperature dependence of the real (elastic) part of Z(ω). Each line represents the frequency dependence as indicated in the figure.

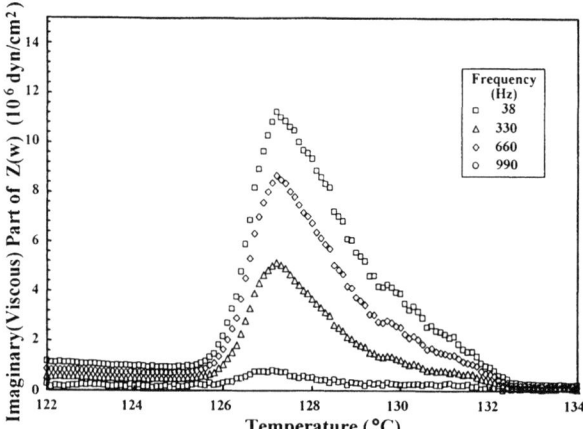

FIGURE 7. This is the Style for Figure Captions. Center this text if it doesn't run for more than one line.

ORIGIN OF THE VISCOELASTIC RELAXATION AND MICROSCOPIC STRUCTURE IN MOSAIC PHASE

It should be noted that above characteristics of the visco-elastic behavior in the mosaic phase is qualitatively the same both in cooling and heating process, and also independent of the cooling or heating rate. On the other hand macroscopic shape and size of the mosaic texture in the measurement cell, which is simultaneously observed by the polarizing microscope, is strongly dependent on the process. Thus, we can conclude that the visco-elastic relaxation is not originated from the macroscopic mosaic domain structure, or dynamically produced structural defects. These are suggest that the relaxation phenomena is intrinsic to the internal structure of the mosaic phase.

Existence of the macroscopic elasticity at low frequency directly confirmed that system has complete long-range order. In the cubic phase, smectic layers are inter-connected in 3-dimensions, and the macroscopic elasticity also appears. However, the cubic phase does not show the relaxation phenomena as reported in this paper, and macroscopically looks isotropic. For this reason, the mosaic phase is different from the cubic phase. If the optical purity decreases by mixing the opposite handedness B7S, TGB phase appear in between the mosaic and ferro smectic phase. So, the mosaic phase expected to be more disordered phase compare to the TGB phase, nevertheless the there exist smectic layer as already be confirmed in X-ray experiment. Thus, it can be concluded that the macroscopic elasticity is supported by the 3-dimensional distribution of the smectic layers, and the relaxation phenomena is originated from the collective modes of the 3-dimensionally distributed regular TGB defects.

REFERENCES

1. Kitzwerow, H-S., *Chriality of the Liquid Crystals*, edited by Kitzerow, H-S., and Bahr, C, Springer-Verlag, New York, 2001, pp. 296-354

2. Renn, S., and Lubensky, T.C., *Phys. Rev. A.* **38**, 2132-2147 (1988).

3. Goodby, J.W., Waugh, M.A., Stein, S.M., Chin, E., Pindak, R., and Patel, J.S., *Nature* **337**, 449-452 (1989).

4. Li, M-H., Laux, V., Nguyen, H-T., Sigaud, G., Barois P., and Iseart, N. *Liq Cryst.* **23**, 389-408 (1997).

5. Levelut, A.M., Hallouin, E., Bennemann, D., Heppke, G., and Lötzsch, D., *J. Phys. II*, **71**, 981-989 (1997).

6. Nishiyama, I., Yamamoto, J., Goodby, J.W., and Yokoyama, H., *J. Mater. Chem.*, **12**, 1709-1716 (2002).

7. Yamamoto, J., and Okano, K., *Jpn. J. Appl. Phys.*, .**31**, 754-763 (1991).

Aging and shear rejuvenation of soft glassy materials

Daniel Bonn[1,2], Hajime Tanaka[3], Philippe Coussot[4], and Jacques Meunier[1]

[1] Laboratoire de Physique Statistique, Ecole Normale Supérieure, 24, rue Lhomond, 75231, Paris Cedex 05, France
[2] van der Waals-Zeeman Institute, University of Amsterdam, Valckenierstraat 65, 1018XE Amsterdam, the Netherlands
[3] Institute of Industrial Science, University Tokyo, Minato-ku, Tokyo 153-8505, Japan
[4] Laboratoire des Matériaux et des Structures de Génie Civil, 2 Allée Kepler, 77420 Champs-sur-Marne, France

Abstract. Structured fluids (concentrated suspensions, emulsions, gels···.) typically exhibit an apparent yield stress. We show here that for a number of these fluids, a unique yield stress cannot be defined. Instead, when solicited above a critical stress, typical yield stress fluids (gels, clay suspensions) and soft glassy materials (colloidal glasses) start flowing abruptly and subsequently accelerate. We demonstrate that the competition between the spontaneous restructuration (aging) and the destruction of the internal structure ('shear rejuvenation') lead to a bifurcation in rheological behavior. For a stress smaller than a (time-dependent) critical value, the viscosity increases in time and the material eventually stops flowing. For a slightly larger stresses the viscosity decreases continuously in time and the flow accelerates. Thus the viscosity jumps discontinuously to infinity at the critical stress.

INTRODUCTION

A large number of structured fluids such as clays, foams and gels are believed to exhibit a yield stress: if they are solicited below a certain critical (yield) stress, they do not flow, but respond elastically to the deformation. An ideal yield stress fluid can be defined as a fluid for which the stress (the shear force per unit area) goes to a finite nonzero value, the yield stress, if the shear rate (the velocity gradient) goes to zero. As a consequence, the steady-state viscosity (the ratio of shear stress and shear rate) diverges in a continuous fashion when the yield stress is approached from above [1-3].

In practice, however, it is often difficult to experimentally determine a yield stress for a given fluid: very different answers can be obtained depending on the experimental protocol [1-4]. We show here that one of the reasons for the difficulty of determining a yield stress is possibly the rearrangements the flow may cause in the internal structure of the fluids. Indeed, for typical yield stress fluids such as clay suspensions, emulsions and foams, the yield stress is mostly due to the microstructure of the fluid that resists large rearrangements: the system is jammed [5] and stops or starts flowing abruptly. When submitted to flow, this microstructure is partly destroyed, which is generally observed in rheological tests as a viscosity that slowly decreases in time: the system is said to be thixotropic [4]. In addition, for most of these systems at rest the microstructure reforms or evolves spontaneously: the system is said to age. If the microstructure re-establishes at rest, experimentally, one observes an increase of an apparent yield stress with time. The mechanical behavior of these systems consequently results from the competition between aging and progressive 'rejuvenation' (destruction of the microstructure) by the shear flow.

Due to the effect of flow on the microstructure, for different 'typical' yield stress fluids, it is therefore impossible to unambiguously define a yield stress. Our experiments show that typical yield stress fluids (gels, clay suspensions) and soft glassy materials (colloidal glasses) do NOT have such a well-defined yield stress. Instead, they exhibit a critical stress at which the viscosity jumps discontinuously to infinity.

We will focus here on the behavior of the colloidal glass of Laponite, synthetic clay consisting of discoid colloidal particles with a diameter of 25 nm and a height of 1 nm. We use this system as it allows determining both the microscopic dynamics and the rheological behavior in detail. Solutions of Laponite are strongly visco-elastic, even at very low particle concentrations. The formation of a gel, evidenced by the existence of a fractal network, has been invoked in explaining the visco-elasticity. We previously studied the structure and viscosity of Laponite using static light scattering. [6]. Contrary to previous observations, we find no evidence of a fractal-like organization of the colloidal particles. Therefore, it has now been firmly established that it is in fact a colloidal glass stabilized by electrostatic repulsions [6], without any structuration at large length scales, making its microstructure fundamentally different from that of gels or foams.

Glasses are a non-equilibrium form of matter and are, maybe for that reason, still ill-understood. The usual way of looking at the glass transition is given by the so-called mode-coupling theory [6]. In this theory, the glass transition is a strong ergodic to non-ergodic transition. In real systems, however, the 'transition' always appears rounded. The rounding of the transition is due to the appearance of a 'slow mode' in the system. The aging is then the nonequilibrium evolution of a system quenched into a glassy state. Understanding the aging processes in a glassy system is crucial for the description of glassy dynamics.

A recent careful inspection of the mode-coupling equations [6] presents the first detailed description of the aging process. The evolution of the system is described in terms of the correlation and response functions of the system. The key result of the theory is that the diffusion may be looked upon as a cage-diffusion process. The particles reside in dynamic cages formed by their neighbours, and escape from these cages after a characteristic time that depends on the aging time t_w.

EXPERIMENTAL

The experiments are performed by dissolving the particles at a concentration of 3.5 wt% in ultrapure water (corresponding to a volume fraction $\phi=0.014$) with NaOH to obtain a pH=10, which fixes the ionic strength at 10^{-4} M. We dissolve the Laponite under vigorous stirring during 15 min. and subsequently pass the solution through a Millipore Millex AA (0.8 μm) filter, to obtain a reproducible initial liquid state. In order to allow for the Diffusive Wave Spectroscopy measurements of the dynamics, we add a small amount of 0.5 μm latex particles to the samples after filtration. We subsequently let the sample age.

RESULTS AND DISCUSSION

A direct measurement of the viscosity during the aging shows that the viscosity increases in time. These measurements are performed on a Reologica Stress-Tech rheometer in a Couette geometry with a gap of 1 mm. In order not to disturb the system by a continuous shear, the measurements are performed using small oscillations at a frequency of 0.1 Hz at an imposed stress of 0.5 Pa. Such oscillation measurements yield the dynamic moduli G' and G'', the storage and loss modulus, respectively. The complex viscosity η^* can be calculated from these quantities by using $\eta^* = (G'2 + G''2)^{1/2}/\omega$. In Fig.1 we plot the complex viscosity, as a function of aging time. It can be observed that the viscosity changes by four orders of magnitude over a time that is on the order of an hour.

The key structural parameter that characterizes the aging is the slowing down of the diffusion of the colloidal particles [6]. The diffusion process can be characterized by looking at the correlation functions of particle positions for different aging times. This we do by Diffusing Wave Spectroscopy [7]. By our method of preparation, we have in fact performed a quench into the liquid (sol) phase. We can subsequently follow the dynamics of glass formation in time, using Diffusive Wave Spectroscopy to measure the diffusion of the colloidal particles. This is depicted in Fig.2, where the measured correlation functions are shown as a function of the age of the samples. The characteristic diffusion time extracted from these measurements are shown in Fig.1. The results can indeed be described in terms of a cage-diffusion process [6]. For short aging times, the dominant dynamical process is the escape of the particles from dynamic 'cages' formed by neighboring particles. However, at long times the cages stiffen; the particles cannot escape anymore, and the diffusion becomes much slower, concomitant with the increase in viscosity of the sample (Fig.1).

If now, after a certain aging time, the sample is submitted to a continuous shear, both the viscosity and the diffusion time decrease. Therefore, this shear melting reverses the aging process, and consequently allows to 'rejuvenate' the glass. The incipient flow therefore destructures the material, entailing a viscosity decrease. If the measurements are performed under constant stress conditions, the shear rate will consequently increase, and the destructuration will be stronger. This will in turn accelerate the flow again and so on: an avalanche behavior results.

Combined with the spontaneous restructuration at rest that leads to a viscosity increase, we show that a bifurcation in the rheological behaviour occurs: for a given load, the fluid either stops flowing altogether, or fluidizes, leading to rapid flows. Creep tests were carried out with a Reologica Stress-Tech rheometer. We used two geometries: parallel plates and a vane geometry [8]. For the latter, the outer cylinder was covered with sandpaper. For the parallel plate geometry, both surfaces were covered with sandpaper. For the different geometries the ratio of both the free and the sheared surface to sheared volume differ; therefore if important perturbing effects (such as edge effects, evaporation, wall slip, etc) would yield different apparent viscosities. Good quantitative agreement was found between the two different geometries proving the validity of the data.

For each test the material was presheared and left at rest for a given fixed time to obtain a reproducible initial state. Starting from this state, we observe for the three

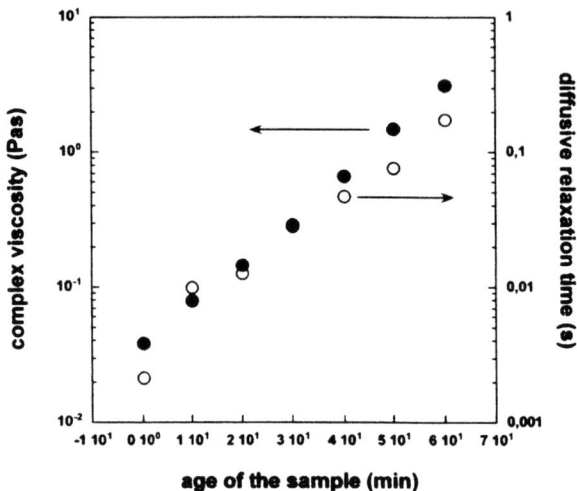

FIGURE 1. Complex viscosity and diffusive relaxation time (see Fig.2) as a function of the age of the sample. Both behave very similarly and show a roughly exponential increase with the age of the sample.

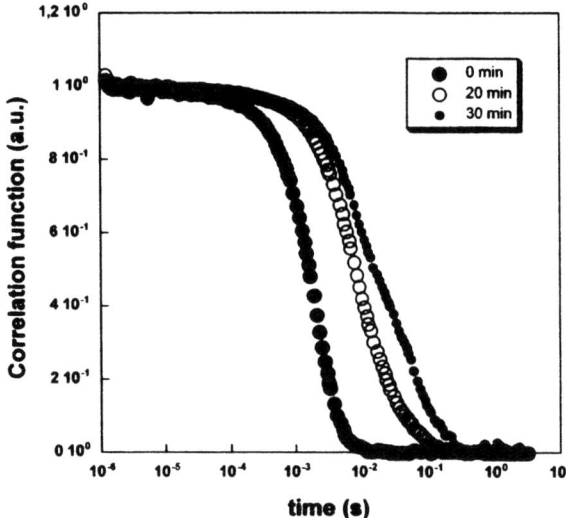

FIGURE 2. Aging of the Laponite sample: correlation functions measured using Diffusive Wave Spectroscopy at different ages of the sample. The decay of the correlation function is a measure for the time a particle needs to 'forget' its initial position (at time zero). Therefore, as the sample ages, the diffusive relaxation time increases. The aging time are in minutes.

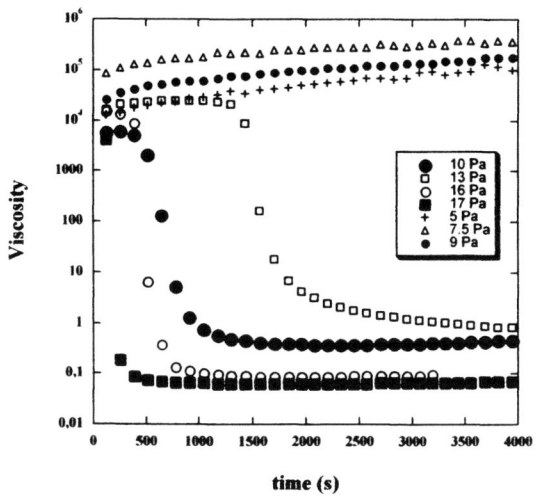

FIGURE 3. Bifurcation in the rheological behaviour: viscosity as a function of time for the colloidal glass: 3% Laponite RD in water at pH=10.

different materials that above a critical stress, the viscosity decreases to reach a low steady state value after a long time. For a stress smaller than this critical value the viscosity increases indefinitely. Therefore, around the critical stress σ_c the flow either stops completely or evolves through the avalanche effect discussed above towards a rapid flow. For all these materials, the competition between aging and shear rejuvenation thus shows up as a bifurcation in the viscosity (Fig.3). The important observation is that for stresses slightly smaller than σ_c, the viscosity increases in time, until the flow is halted altogether: the steady-state viscosity is infinite. On the other hand, for a stress only slightly above σ_c, the viscosity reaches a (low) steady state value η_0. This implies that at the critical stress, the steady-state viscosity jumps discontinuously from infinity to a finite and low value at σ_c. This is in striking contrast with what an ideal yield stress fluid would do. In addition, due to the aging of the systems, this critical stress is not an intrinsic property, but depends on the (shear) history of the sample. All these observations show that yield stress is ill-defined. That the phenomenon is general for 'typical' yield stress fluids, follows from rheometrical tests on a weakly flocculated Bentonite clay suspension, a polymer gel and the colloidal glass (see [9] for more details). Therefore, yield stress cannot be considered separately from thixotropy:. we show here for a number of 'typical' yield stress fluids that both are strongly interconnected and lead to a bifurcation in the rheological behavior. This highly non-linear behaviour should be taken into account for the formulation and handling of paints, inks, cement, muds, etc.

ACKNOWLEDGMENTS

LPS de l'ENS is UMR 8550 of the CNRS associated with the universities Paris 6 and Paris 7.

REFERENCES

1. Bird, R. B., Dai, G. C., and Yarusso, B. Y., *Rev. Chem. Eng.* **1**, 1-70 (1982).
2. Nguyen, Q. D., and Boger, D. V., *Ann. Rev. Fluid Mech.* **24**, 47-88 (1992).
3. Barnes, H. A., *J. Non-Newt. Fluid Mech.* **81**, 133-178 (1999).
4. Sollich, P., Lequeux F., Hebraud, P., and Cates, M. E., *Phys. Rev. Lett.* **78**, 2020-2023 (1997).
5. Coussot, P. and Ancey, C., *Rheophysics of pastes and suspensions* (EDP Sciences, Paris, 1999).
6. Bonn, D., Tanaka, H., Kellay, H., Wegdam, G., Meunier, J., *Langmuir* **15**, 7534-7536 (1999); Bonn, D., Tanaka, H., Kellay, H., Wegdam, G., and Meunier, J., *Europhys. Lett.* **45**, 52-57 (1998); Abou, B., Bonn, D., and Meunier, J., *Phys. Rev. E* **64**, 021510 (2001); for the mode-coupling theory, see e.g. Bouchaud, J. -P., Cugliandolo, L., Kurchan, J., and Mezard, M., *Physica A* **226**, 243-273 (1996).
7. Knaebel, A., Bellour, M., Munch, J. P., Viasnoff, V., Lequeux, F., and Harden, J. M., *Europhys. Lett.* **52**, 73-79 (2000).
8. Alderman, N. J., Meeten, G. H., and Sherwood, J. D., *J. Non-Newt. Fluid Mech.* **39**, 291-310 (1991).
9. Coussot, P., Nguyen, Q. D., Huynh, H. T., and Bonn, D., *Phys. Rev. Lett.* **88**, 175501 (2002) ; Bonn, D., Tanase, S., Abou, B., Tanaka, H., and Meunier, J., *Phys. Rev. Lett.* **89**, 015701 (2002); Coussot, P. Nguyen, Q. D., Huynh, H. T., and Bonn, D., *J. Rheol.* **46**, 573-589 (2002).

Scattering Probes of Complex Fluids and Solids

R. Vavrin, A. Stradner, J. Kohlbrecher, F. Scheffold[1] and P. Schurtenberger

Physics Department, University of Fribourg, CH-1700 Fribourg, Switzerland
Paul Scherrer Institute, CH-5232 Villingen, Switzerland

Abstract. We have studied the dynamical and structural properties of destabilized nanoparticle suspensions by a combination of small-angle neutron scattering (SANS) and diffusing wave spectroscopy (DWS). SANS and DWS provide structural and dynamic information about opaque samples online during the gelation process. With our new experimental setup we access a broad range of length and time scales perfectly suited for the (non-invasive) investigation of dense nano- and mesostructured complex fluids and solids.

INTRODUCTION

Many soft materials are slowly relaxing complex fluids or even viscoelastic solids. This means that on the time scale accessible to experiment these systems do not fully relax and in dynamic light scattering (DLS) experiments they appear nonergodic. Ergodicity however is an important condition for the applicability of DLS. Only recently it became possible to overcome these limitations and to access internal dynamic properties of arrested media [1, 2]. In our work particular emphasis is given to dense complex fluids, such as colloidal suspensions and gels. For these systems strong multiple scattering of light further complicates the situation. To characterize the sample properties over the full range of interest we employ a rather unique set of static and dynamic light and small angle neutron scattering (SANS) experiments [3]. In moderately turbid systems we use multiple scattering suppression schemes, such as 3D-dynamic light scattering [4]. These techniques provide exactly the same kind of information as conventional DLS even in the presence of significant multiple scattering. For even more turbid samples diffusing wave spectroscopy (DWS) is perfectly suited to study the internal dynamics on nanometer length scales by measuring the intensity fluctuations of the diffusively transmitted light. In this paper we focus on the problem of the liquid to solid transition in suspensions of colloidal nanoparticles. Such gels are formed in the presence of attractive interactions between particles. Individual aggregation clusters grow until a space filling network is reached. But even when the gel seems to have reached a final state the gel dynamic and elastic properties continue to mature and age [5-7].

Particle gels have been for a long time (and still are) subject of intense research due to their importance both in fundamental and applied science. Applications of gels and sol-gel processing cover such different areas as ceramics processing, cosmetics and consumer products, food technology, to name only a few.

EXPERIMENTS

Samples

The sample has been prepared from a solution of polystyrene spheres (Interfacial Dynamics Corporation, nominal diameter 19nm, polydispersity 16.3%), particle volume fraction 3.8%, in a buoyancy-matching mixture of water and heavy water. It is kept in a stoppered quartz cell, thickness L=2mm (Figure 1). To effectively compensate the electrostatic repulsion we use an enzyme-catalyzed internal chemical reaction (the urease catalyzed hydrolysis of urea) to increase in-situ the solvent ionic strength

---

[1] Corresponding author: Frank.Scheffold@unifr.ch

without any local gradient [8]. In the stable suspension the urea represents 4.5% of the solvent volume leading to a final ionic strength of ca. 0.16 M. We add urease at a temperature T=3-5°C where its activity is sufficiently reduced. Afterwards the sample is kept at 25°C during the whole gelation process.

FIGURE 1 Nanoparticle suspension before (a) and after gelation (b).

Small angle neutron scattering

The structural properties of dense colloidal suspensions were studied using time resolved small-angle neutron scattering. Details of the instrument and data treatment are given in Refs. [4] (see also Figure 2). In order to increase the experimental resolution, we have used a set of focussing neutron lenses [9].

The measured intensity I(q) as a function of the scattering vector q contains contributions from the particle structure [SANS particle form factor F(q)] as well as from interparticle interaction effects if present [structure factor S(q)], $I(q) \propto F(q)S(q)$ for monodisperse spherical particles. To determine F(q) experimentally we have measured I(q) using a dilute suspension (0.1%, data not shown). Fitting the form factor of a sphere to the SANS data we find a radius of 12.4 nm in good agreement with the intensity weighted radius when taking into account the polydispersity (adjusted value 20%). We note that at the chosen instrument configuration (sample detector distance 20.3m, collimation length 18m, neutron wavelength λ=1.27nm) the smearing induced by the instrument is weak and has been neglected for the analysis of the gel structure.

Photon transport

In the limit of very strong multiple scattering photons travel along random paths with a characteristic step length l^*. The inverse transport mean free path, or optical density, $1/l^*$ is directly related to the material microstructure [10]. We now take advantage of the fact that light and neutron scattering from nanoparticle aggregates is quite similar. Both cases can be treated in the limit of weak scattering using the 1st Born approximation (called Rayleigh-Gans-Debye approximation in optics [4, 11]). Thus the structural information extracted from SANS can be used directly to determine the optical density of the gel [4].

$$l^*_{susp}/l^*_{gel} = 4k_0^4 \bigg/ \int_0^{2k_0} \left[I_{gel}(q)/I_{susp}(q) \right] q^3 dq \quad (1)$$

This expression reduces to one for a suspension of non- interacting particles $l^*_{gel} = l^*_{susp}$. The corresponding l^*_{susp} can be determined from scattering theory as described in ref. [10] (in our case $l^*_{susp} = 15$ mm).

Diffusing Wave Spectroscopy

If the optical properties of the sample are known we can access the local dynamic properties with DWS. Laser light transmitted through the sample fluctuates due to internal motion. While in conventional light scattering experiments the sample has to be almost transparent (and hence often highly diluted), diffusing wave spectroscopy (DWS) extends "conventional" dynamic light scattering (DLS) to media with strong multiple scattering [10, 12]. To analyze the light scattering signal we use the following expression for the field auto correlation function (from ref.[4]):

$$g_1(t) = \frac{(L/l^* + 4/3)\sqrt{k_0^2 \delta^2 (1 - e^{-(t/\tau)^p})}}{\sinh\left[(L/l^* + 4/3)\sqrt{k_0^2 \delta^2 (1 - e^{-(t/\tau)^p})}\right]} \quad (2)$$

The local gel dynamic properties are analyzed in the frame of the fractal gel model of Krall and Weitz[5]: Gel sub-clusters fluctuate with a maximum excursion δ^2 and a characteristic time scale τ. This time scale is directly linked to the gel elastic modulus by the following approximate expression

$$G_0 = 6\pi\eta/\tau \quad (3)$$

The dynamics of sub-clusters is determined by a superposition of many gel modes leading to a non-diffusive motion characterized by the exponent $p < 1$. Initially derived for mature gels we also tentatively apply this expression during the sol-gel transition [6].

Setup

We built an experimental environment at the Swiss neutron source (Villingen) for combined SANS and DWS. Besides the combined use, both experiments follow traditional schemes, e.g. those described in ref. [1, 13] We use a 35mW HeNe Laser operating at 632.8nm. Two-cell DWS (TCDWS) has been used (as described in ref.[14]) in order to be able to study solid-like samples [1, 14].

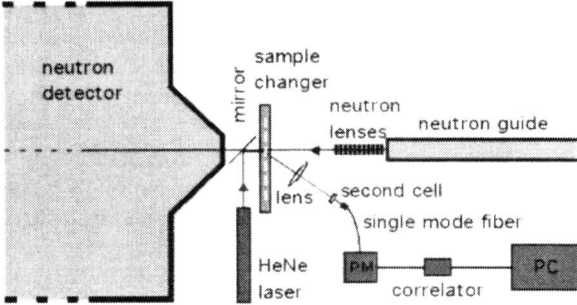

FIGURE 2. Experimental setup for simultaneous SANS and DWS measurements.

RESULTS AND DISCUSSION

We observe a continuous increase of the neutron scattering intensity over several hours. Optically the sample transforms from almost transparent to strongly opaque (Figure 1). The experimental data has been modelled over the full range using the well known Fisher Burford structure factor for gelation clusters [15]

$$S(q) = \left(1 + \left[2/3d_f\right] q^2 R_c^2\right)^{-d_f/2} \quad (4)$$

The cluster radius R_c obtained from the fit increases from 80nm (2h) to approx. 450nm after 10h (data not shown). For the apparent fractal dimension we find values increasing from 1.6 to roughly 2.4 (a more detailed analysis will be presented elsewhere).

From the neutron scattering data (extrapolated to $q = 10^{-4} nm^{-1}$) we numerically obtain the sample optical density via Eq.(1): L/l^* increases from an initial value of 0.13 to more than 4.

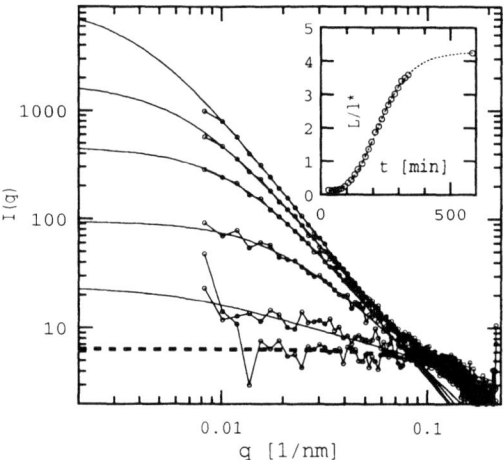

FIGURE 3 Time resolved SANS (measurement time ca. 15min) for t= 43, 93, 158, 221, 284, 585 min. Dashed line: Single particle form factor. Symbols: SANS scattering intensity during destabilization and gelation. Lines: Fisher-Burford fits. Inset: Optical density L/l^* from Eq.(1). The dotted line is a guide to the eye only.

The knowledge of l^* enables us to follow the local dynamic changes in situ without any adjustable parameter. Though in principle DWS is best used for $L/l^*>5$ we think that our analysis still provides valuable information already for $L/l^*>2$; in particular since the relevant dynamic properties change orders of magnitude while the DWS error is less than 30% for $L/l^*>2$ (Note that in our case scattering is strongly forward peaked, which improves the DWS accuracy) [16].

Figure 4 shows a selection of measured autocorrelation functions (measurement time 15 min) during the aggregation and gelation process. The corresponding fits with Eq.(3) are shown as solid lines [with δ,τ,p as adjustable parameters, p is decreasing from ca. 0.7 to 0.5 over the range of data shown]. From the fitted parameters we can deduce approximate values of the gel modulus (inset). It is clearly seen that while the gel structure remains almost unchanged after 300min the gel modulus is still increasing significantly.

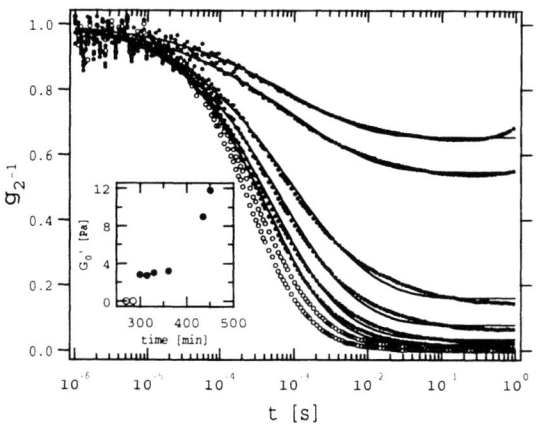

FIGURE 4 Diffusing Wave Spectroscopy. Intensity autocorrelation function before (open symbols) and after gelation (solid symbols). Solid lines: fit with Eq.(2). Inset: Gel elastic modulus derived from the DWS data using Eq.(4).

Further experiments are currently underway to study in more detail the kinetics and concentration dependence of this gelation process. Here we have shown that the simultaneous use of SANS and DWS provides a powerful tool to study this class of systems.

ACKNOWLEDGMENTS

We gratefully acknowledge financial support from the Swiss national science foundation. SANS experiments were performed at the Swiss Spallation Neutron Source, Paul Scherrer Institute, Villigen (CH).

REFERENCES

1. Scheffold, F. and Schurtenberger, P., *Soft Materials* **1**, 139-165 (2003).
2. Pusey, P.N. and van Megen, W., *Physica A* **157**, 705-741 (1989)
3. Romer, S., et al., *J. Appl. Cryst.* **36**, 1-6 (2003)
4. Rojas, L.F., et al., *Faraday Disc.* **123**, 385-400 (2003)
5. Krall, A.H. and Weitz, D.A., *Phys. Rev. Lett.* **80**, 778-781 (1998).
6. Romer, S., Scheffold, F. and Schurtenberger, P., *Phys. Rev. Lett* **85**, 4980-4983 (2000).
7. Cipelletti, L., et al., *J. Phys.: Condens. Matter* **15**, 257-262 (2003).
8. Wyss, H.M., et al., *J. Colloid Interface Sci.* **241**, 89-97 (2001).
9. Eskildsen, M.R., et al., *Nature* **391**, 563-566 (1998).
10. Fraden, S. and Maret, G., *Phys. Rev. Lett.* **65**, 512-515 (1990).
11. v.d. Hulst, H.C., *Light scattering by small particles*, Dover Publications, New York, 1981
12. Pine, D.J., et al., *Phys. Rev. Lett.* **60**, 1134-1137 (1988).
13. Rojas-Ochoa, L.F., et al., *Phys. Rev. E* **65**, art. no. 051403 (2002).
14. Scheffold, F., et al., *Phys. Rev. E* **63**, art. no. 061404 (2001).
15. Fisher, M.E. and F.J. Burford, *Phys. Rev.* **156**, 583-622 (1967).
16. Kaplan, P.D., et al., *Appl. Opt.* **32**, 3828-3836 (1993).

Temporal heterogeneity of the slow dynamics of a colloidal paste

Pierre Ballesta[*], Christian Ligoure[*] and Luca Cipelletti[*][†]

[*]GDPC UMR 5581 Université Montpellier 2 and CNRS, Montpellier, France
[†]email: lucacip@gdpc.univ-montp2.fr

Abstract. We investigate the slow dynamics of a soft glass, a concentrated suspension of polydisperse colloidal particles, by using multispeckle Diffusing Wave Spectroscopy (DWS). Two distinct regimes are observed: for small sample age, t_w, the dynamics smoothly slows down, as revealed by a nearly linear increase of the characteristic relaxation time, τ_s, measured by DWS. At longer ages, the dynamics is quasi stationary, but τ_s exhibit anomalously large fluctuations in time. The time scale of the dynamical fluctuations is found to be slightly shorter than the average relaxation time. The variance of the intensity correlation function is maximum for time delays comparable to the average relaxation time, in striking analogy with recent simulations of glass-forming liquids.

INTRODUCTION

In the past years, experimental and theoretical work has increasingly pointed to the fundamental role of dynamical heterogeneity in the slow dynamics of glassy systems [1, 2]. While investigations on hard condensed matter systems, such as structural or spin glasses, are relatively abundant, until recently much less attention has been devoted to soft matter systems, an exception being the confocal microscopy study of dynamical heterogeneity in hard sphere colloidal glasses [3, 4]. Soft glasses are of great interest, both in view of the numerous industrial applications (e.g. in the food, personal care, and paint industries) and because they can serve as model systems, since the relevant time and length scale are relatively easily accessible and because the interactions between particles can be controlled to a great extent. Therefore, glassy –or *jammed* [5]– soft materials provide a unique opportunity to better understand dynamical heterogeneity and to address issues such as the life time and the spatial extent of dynamical fluctuations.

MATERIALS AND METHODS

In this paper we present an experimental investigation of temporal heterogeneity in the slow dynamics of a concentrated colloidal suspension. The colloids are xenospheres [6], of typical size 20 μm and large polydispersity, due to the presence of fragmented spheres, as revealed by scanning electron microscopy. They are formed by permanently aggregated primary particles, latexes of poly-vinylchlorure covered by a surfactant, whose size is of the order of 0.2 μm. The xenospheres are suspended in an organic solvent, dioctylphthalate, at a mass fraction of 40%, corresponding to a volume fraction $\phi \approx 66.4\%$ (note that, due to the large polidispersity, the maximum packing fraction is estimated to be as high as 75.2%). Similar colloidal pastes are used as a precursor in the industrial production of PVC foams. After loading the sample in a cell, we vigorously shake it (using a vortex mixer) to erase any memory of the shear imposed during the cell filling and to re-initialize the dynamics [7]. We define the age t_w of the sample as the time elapsed since mixing.

The dynamics of the suspension is probed by Diffusing Wave Spectroscopy (DWS) [8], a non-invasive light scattering technique for highly turbid samples. We use a combination of the multispeckle technique [9] and the Time Resolved Correlation method [10] to measure the slow relaxation mode of the paste and to characterize its temporal heterogeneity. In our experiment, a laser beam of width 1 mm impinges onto a 2 mm thick cell containing the sample. The cell is thermostated at 23 ± 0.05 °C. Due to the high turbidity of the sample, the incoming photons are scattered a large number of times before exiting the cell. The speckle pattern due to the interference of the scattered photons is recorded by a CCD camera. Any motion in the sample determines a change in the speckle pattern. Accordingly, the dynamics of the sample is measured by quantifying the temporal fluctuations of the speckle pattern. Following the TRC scheme, we measure the degree of correlation $c_I(t_w, \tau)$ between speckle patterns at time t_w and $t_w + \tau$:

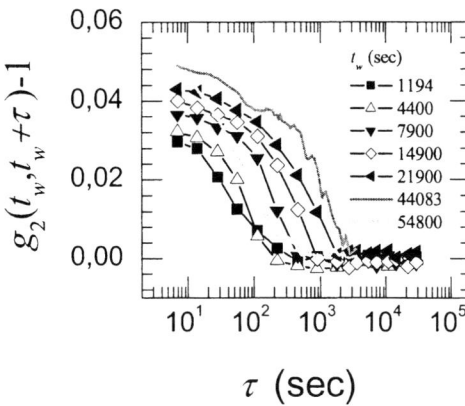

FIGURE 1. Final decay of the intensity autocorrelation function measured by multispeckle DWS for a concentrated colloidal paste. The curves are labelled by sample age. For $t_w < 28000$ sec the dynamics steadily slows down (symbols). At larger t_w, large fluctuations of the characteristic time are observed and $g_2 - 1$ is much more noisy (lines).

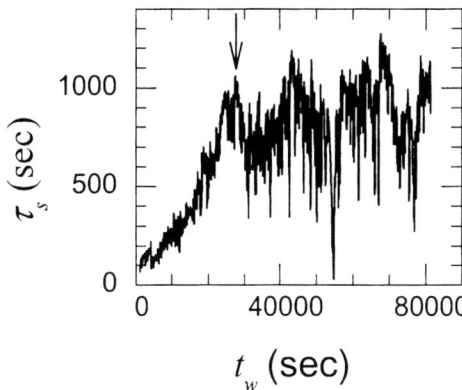

FIGURE 2. Age dependence of the characteristic time τ_s of the slow dynamics of a concentrated colloidal paste. The arrow indicates the cross-over between the initial regime, where $\tau_s \propto t^{1.1 \pm 0.1}$, and the quasi-stationary regime, where large fluctuations of τ_s are observed

$c_I(t_w, \tau) = \frac{<I_p(t_w)I_p(t_w+\tau)>_p}{<I_p(t_w)>_p<I_p(t_w+\tau)>_p} - 1$, where $I_p(t)$ is the time-dependent intensity at pixel p and $<...>_p$ denotes the average over the CCD pixels. Note that, contrary to previous experiments, c_I is averaged only over pixels, and not over time or over several runs. This allows us to probe the dynamics with a time resolution equal to the CCD frame rate (in this experiment, a frame is grabbed every 7 sec) and to fully characterize its temporal heterogeneity. In the following, we will first describe the behavior of the two-time intensity autocorrelation function $g_2(t_w, t_w + \tau)$, which we obtain by averaging $c_I(t_w, \tau)$ over a short time window of 70 sec, centered around t_w. We will then focus on the temporal heterogeneities of the dynamics by analyzing the fluctuations of $c_I(t_w, \tau)$ at a fixed lag τ.

RESULTS AND DISCUSSION

Figure 1 shows $g_2(t_w, t_w + \tau)$ as a function of τ for several times t_w after initializing the sample. Because of the limited speed of the CCD camera, only the final decay of $g_2(t_w, t_w + \tau)$ is accessible. However, a fast relaxation mode must also be present, which is responsible for the reduced value of the correlation function at the smallest lag accessible to the CCD (in our setup, the upper limit of $g_2 - 1 > 0.3$). Initially, the amplitude of the final relaxation mode increases, thus suggesting that the fast motion of the particles is increasingly constrained, and its characteristic time appears to grow steadily with sample age (symbols in Fig. 1). This aging behavior is typical of glassy materials and has been observed in several colloidal systems [11]. For $t_w > 28000$ sec, a totally different behavior is observed (lines in Fig. 1). Strikingly, the characteristic time of the final decay of $g_2(t_w, t_w + \tau)$ exhibits large fluctuations: the dynamics appears to slow down or accelerate in a random fashion. Moreover, the shape of the correlation functions becomes quite erratic: sudden drops and oscillations are often observed. (see for example the curve for $t_w = 54800$ sec).

To better investigate this surprising behavior, we fit $g_2(t_w, t_w + \tau) - 1$ by a stretched exponential function $a \exp[-(\tau/\tau_s)^p]$. This functional form is found to fit well most of the data and to provide a reasonable estimate of the relaxation time even when $g_2(t_w, t_w + \tau)$ strongly deviates from a smooth behavior. We plot the age evolution of the characteristic time τ_s obtained from the fits in Fig. 2. The smooth growth of τ_s for $t_w < 28000$ sec confirms the existence of an initial aging regime, as suggested by the evolution of the correlation functions shown in Fig. 1. In this regime, the growth of τ_s is nearly linear: a fit to a power law in the range 3000 sec $< t_w <$ 28000 sec (not shown) yields an exponent of 1.1 ± 0.1. A similar linear growth of the characteristic relaxation time has been observed in many glassy systems, both in hard and soft condensed matter and is often referred to as simple aging.

As can be seen in Fig. 2, after the initial simple aging regime the slow dynamics becomes almost stationary, yet extremely heterogeneous in time. A linear fit to the data shows that in the quasi-stationary regime τ_s increases –on average– by less than 20% over 14 hours, to be compared to a 32-fold increase over 8 hours in the initial aging regime. Surprisingly, the TRC method reveals that the relaxation time exhibits extremely large fluctuations, much greater than those in the initial regime. In-

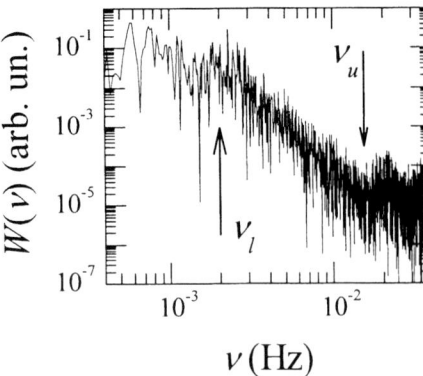
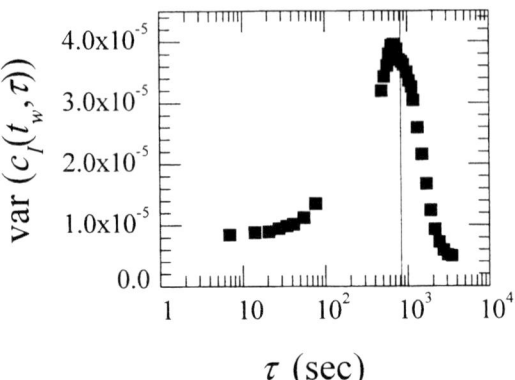

FIGURE 3. Power spectrum of the fluctuations of the characteristic time of the slow dynamics in the quasi-stationary regime. The two arrows indicate the lower and upper cutoff frequency, corresponding to 500 and 70 sec, respectively.

FIGURE 4. Variance of the degree of correlation c_I as a function of the time delay between CCD images, in the quasi-stationary regime. The vertical line indicates the average relaxation time of $g_2 - 1$ over the same time interval.

deed, the peak-to-peak amplitude of the variation of τ_s is comparable to its mean value ($<\tau_s>_{t_w}= 830$ sec) and occasionally τ_s drops by more than a factor of 10.

Dynamical heterogeneities have been observed in the past in a variety of glassy systems, and are generally associated to spatial variations of the dynamics, with regions of "fast-moving" molecules or particles coexisting with regions of slower particles. For hard condensed matter systems, the size of these regions is estimated to be of the order of a few nm [1], that is of the order of 10 molecules. Confocal microscopy experiments on supercooled colloidal fluids have shown that the most mobile particles form clusters whose maximum size is also of the order of 10 particles [4]. It is therefore surprising that dynamical heterogeneities are observed in our experiment, where the dynamics is averaged over the whole scattering volume, which contains more than $3\,10^5$ particles. Indeed, if the size of the regions over which the dynamics is correlated was the same as in previous experiments, more than 300 such regions would be contained in the scattering volume at any time, thus reducing drastically the temporal fluctuations of the dynamics measured by DWS. Therefore, our experiments suggest that the size of dynamically correlated regions in highly concentrated colloidal pastes may be larger.

In order to quantify the time scale of the fluctuations of the dynamics in the quasi-stationary regime, we calculate the power spectrum of $\tau_s(t_w)$, $W(v)$. The results are shown in Fig. 3: at high frequency ($v > v_u \approx 0.014$ Hz, corresponding to a time scale of 70 sec) $W(v)$ is flat. This upper cutoff frequency is most likely related to the slight smoothing of the c_I data (average over a time window of 70 sec), which we perform when calculating $g_2 - 1$, as explained above. Between $v_l \approx 0.002$ Hz and v_u, the power spectrum rapidly increases with decreasing frequency, while below v_l $W(v)$ is again flat, as one would expect for random, uncorrelated fluctuations. Therefore, the frequency dependence of the power spectrum indicates that the fluctuations of the dynamics occur mainly on time scales up to $v_l^{-1} = 500$ sec. Similar results are obtained when calculating the time autocorrelation function of τ_s (not shown): a decay is observed, whose characteristic time is of the order of 280 sec. Thus, the time scale of the fluctuations of the dynamics is of the order of a few hundreds seconds, close to but slightly smaller than the time-averaged relaxation time, $<\tau_s>_{t_w}= 830$ sec.

The behavior of the characteristic time τ_s as a function of t_w provides a convenient means to describe dynamical heterogeneity. Further information, however, can be obtained by analyzing the fluctuations of the degree of correlation $c_I(t_w, \tau)$ for all time delays τ. Figure 4 shows the variance of $c_I(t_w, \tau)$, σ^2, as a function of τ, calculated from the data in the quasi-stationary regime. We recall that in an usual light scattering experiment $g_2(\tau) - 1$ is obtained by averaging $c_I(t_w, \tau)$ over time. Therefore, σ^2 represents the "spread" around the mean value of the intensity correlation function at any given time delay τ. For 77 sec $< \tau <$ 490 sec the variance is artifactually enhanced due to quasi-periodic fluctuations of the cell temperature: these data are not shown in Fig. 4. These fluctuations, whose amplitude is of the order of 0.05°C, slightly change the solvent refractive index and thus the phase of the scattered photons, resulting in a periodic loss of correlation. However, they do not affect the particle dynamics, as we checked in control experiments where the sample temperature was varied in a controlled, periodic way. The variance of c_I is maximum for a delay of 700 sec, very close to the time-averaged relaxation time, indicated by the vertical line in Fig. 4. This behavior is very different from that observed for the dynamics of a diluted suspension of colloids undergoing Brownian motion, for which no dynamical heterogeneities are expected. Indeed, we find that for Brownian particles $\sigma^2(\tau)$ is constant for all delays.

The fact that the maximum of the variance is very close to $<\tau_s>_{t_w}$, together with the observation that the typical life time of the fluctuations of the dynamics is again of the order of $<\tau_s>_{t_w}$, strongly suggest that the only relevant time scale is that of the relaxation of $g_2 - 1$, and that large fluctuations of τ are an intrinsic feature of the slow relaxation mode. Indeed, a preliminary analysis of the data, where we assume $g_2(t_w, t_w + \tau) - 1 = a * \exp[-(\tau/\tau_s(t_w))^{p(t_w)}]$ at all t_w, shows that, up to second order in the fluctuations $\delta \tau_s$ and δp of $\tau_s(t_w)$ and $p(t_w)$, the variance of c_I is peaked for $\tau \approx <\tau_s>_{t_w}$. This analysis will be presented in a forthcoming paper.

Remarkable analogies exist between the experiments presented here and recent simulations of glass-forming liquids [12]. Lačević et al. introduce a time-dependent 'order parameter' $Q(t)$ that compares the system configuration at two times separated by t and then calculate the (normalized) variance of $Q(t)$, $\chi_4(t)$. $\chi_4(t)$ is found to be peaked around the characteristic time of the final relaxation of $Q(t)$, much as, in our experiments, σ^2 is peaked around $<\tau_s>_{t_w}$. Indeed, the precise form of $Q(t)$ is irrelevant for the main results of their theory: we can thus identify our $c_I(t_w, \tau)$ with $Q(t = \tau)$, since $c_I(t_w, \tau)$ compares two speckle patterns, and thus two system configurations, separated by τ. Similarly, $\sigma^2(\tau)$ can be identified with the generalized susceptibility $\chi_4(t = \tau)$. Therefore, TRC provides a novel means to access experimentally $\chi_4(t)$. The similarities between vastly different systems hint at the generality of the behavior of dynamical fluctuations in glassy systems. Further insight will be gained by a systematic exploration of the behavior of the fluctuations when approaching the fluid-solid transition. For our system, contrary to molecular glasses, this transition is not achieved by decreasing the temperature, but rather by increasing the particle volume fraction. Preliminary results hint at a non-monotonic variation of σ^2 with ϕ.

ACKNOWLEDGMENTS

It is a pleasure to acknowledge E. Pitard, L. Berthier, L. Ramos, V. Trappe, and H. Bissig for many illuminating discussions, and N. Peron for the scanning electron microscope images. We thank M. Cloître for introducing us to the PVC colloidal pastes and P. Bergougnon (Atofina) for providing us with the sample. This work was partially supported by CNRS and Région Languedoc-Roussillon (projet "Dynamiques lentes et vieillissement de matériaux désordonnés") and CNES (grant no. 03/CNES/4800000123). L. C. thanks the French Ministère des Affaires Etrangères for supporting his participation to the conference "Slow Dynamics in Complex System".

REFERENCES

1. Ediger, M. D., *Annu. Rev. Phys. Chem.*, **51**, 99–128 (2000).
2. Glotzer, S. C., *Journal of Non-Crystalline Solids*, **274**, 342–355 (2000).
3. Kegel, W. K., and van Blaaderen, A., *Science*, **287**, 290–293 (2000).
4. Weeks, E., Crocker, J., Levitt, A., Schofield, A., and Weitz, D., *Science*, **287**, 627–630 (2000).
5. Liu, A., and Nagel, S., *Nature*, **396**, 21 (1998).
6. Herk, H. H., Bikoles, N. M., Overgerger, C. G., and Menzes, G., "Vinyl chloride polymerization," in *Encyclopedy of polymer science and engineering*, edited by H. E. Mark, Wiley-Interscience, New York, 2003, vol. 17, 3rd edn.
7. Viasnoff, V., and Lequeux, F., *Phys. Rev. Lett.*, **89**, 065701 (2002).
8. Weitz, D. A., and Pine, D. J., "Diffusing-wave spectroscopy," in *Dynamic Light scattering*, edited by W. Brown, Clarendon Press, Oxford, 1993, pp. 652–720.
9. Viasnoff, V., Lequeux, F., and Pine, D. J., *Rev. Sci. Instrum.*, **73**, 2336–2344 (2002).
10. Cipelletti, L., Bissig, H., Trappe, V., Ballesta, P., and Mazoyer, S., *J. Phys.: Condens. Matter*, **15**, S257–S262 (2003).
11. Cipelletti, L., and Ramos, L., *Curr. Opin. Colloid Interface Sci.*, **7**, 228–234 (2002).
12. Lačević, N., Starr, F. W., Schroder, T. B., Norikov, V. N., and Glotzer, S. C., *Phys. Rev. E*, **66**, 030101 (2002).

Dynamics around the sol-gel transition in thermoreversible atactic polystyrene gels

A. Matic, J. Mattsson[†] and R. Bergman

Department of Applied Physics, Chalmers University of Technology, SE-412 96 Göteborg, Sweden
[†]Department of Physics and DEAS, Harvard University, Cambridge, MA 02138, USA

Abstract. We present a dielectric relaxation spectroscopy study of the dynamics in the thermoreversible polymer gel system atactic polystyrene/toluene. We observe three distinct relaxational process, the α- and β-relaxations related to solvent dynamics and a process related to the dynamics of the polymer network. While both the α-relaxation and the network processes are slowed down with increasing polymer content, the β-relaxation remains practically unchanged. The behaviour of the relaxational process related to the polymer network, shows no distinct change at gelation, suggesting that there are no major changes taking place in the polymer network on the length scale of the observed relaxation. Furthermore, it is possible to construct a master curve of the temperature dependence of the relaxation time of the matrix relaxation by a simple temperature scaling.

INTRODUCTION

Thermoreversible physical gels form as the result of molecular association, where crosslinks are formed by reversible physical interactions. This is in contrast to the permanent chemical bonds of chemical gels. A large variety of gel systems exist and in general a gel can be formed from synthetic as well as biological macromolecules and the exact nature of the cross-links in physical gels varies between different types of systems. Despite the fact that the physical gels incorporate a wide range of systems, often of very different chemical nature, they display several striking similarities in physical properties and the existence of universal dynamical and/or structural features have been proposed [1, 2, 3, 4].

Recently, a number of intriguing similarities have been found between the dynamical behaviour of gel- and glass-forming materials [3, 4]. For instance, both the glass and gel transitions are reversible, they show typical kinetic features and occur as a result of the physical arrest of either molecules or molecular structures. The obvious difference between the systems is one of length and correspondingly time-scales, with the ones for gels being 10-100 times longer than those for glasses. Recent computer simulations on physically associating systems [5] have further strengthened these arguments. In that work a striking resemblance was found between the chain dynamics and diffusion at the sol-gel transition and the dynamics of supercooled liquids.

The archetypal thermoreversible physical gel system consists of a polymer and a solvent. The gelation ability and gelation temperature in such systems is strongly dependent on the type of solvent used, the molecular weight of the polymer and the polymer concentration [1]. Depending on the particular combination of polymer and solvent, the characteristics and interactions in the systems vary. The cross-links in these systems can be induced by e. g. crystallization, chain overlap or formation of polymer-solvent complexes.

In this work we have investigated the thermoreversible system atactic-polystyrene (aPS) dissolved in toluene. For gel systems based on aPS it has been reported that gels can be formed over a relatively large concentration range [6]. However, despite a large number of studies focusing both on structural and dynamical aspects the gelation mechanism in this system is still unclear [1, 6, 7]. We report experimental data from dielectric relaxation spectroscopy (DRS) where we follow the dynamics over a large temperature and frequency range. The aPS/toluene system is particularly suitable for DRS experiments since the polymer and the solvent has similar dielectric strengths and thus both the solvent and polymer dynamics can be monitored simultaneously in the experiment.

EXPERIMENTAL

The gel samples were prepared from atactic polystyrene of M_w=513 000 and M_w/M_n=1.12 (Polymer Source Inc.) and toluene 99.6% (Aldrich Inc.). The polymer was dissolved in the appropriate amount of solvent and allowed to equilibrate at room temperature for at least one week before the experiments were performed. Samples in the concentration range 10-30 %(w/w) were prepared.

The dielectric relaxation spectroscopy (DRS) experiments were performed on a broad band (10^{-2} – 10^7 Hz) and high resolution Novocontrol Alpha dielectric spectrometer. The gel samples were placed between gold plated electrodes with 20 mm diameter. The complex dielectric permittivity, $\varepsilon^*(f) = \varepsilon'(f) - i\varepsilon''(f)$, was recorded isothermally every second or third degree over the temperature range 100-300 K with a temperature stability better than 0.1 K.

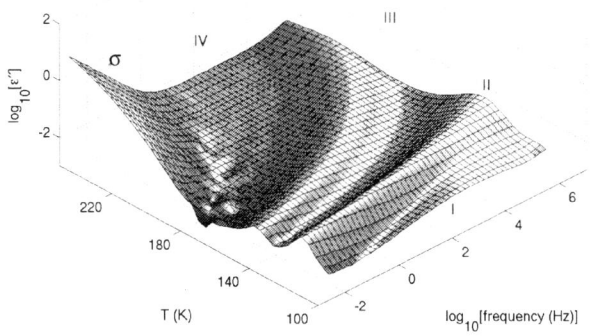

FIGURE 1. Dielectric loss data for a 20% w/w aPS/toluene solution. Four relaxational processes can be distinguished.

RESULTS AND DISCUSSION

In the DRS experiment we can distinguish four different contributions to the dielectric loss, $\varepsilon''(f)$, marked as I-IV in figure 1. At high temperatures and low frequencies a DC-conductivity, σ, contribution is also present. The processes I-III are well defined and a quantitative analysis can be performed. The fourth contribution, IV, is very broad and partly covered by the conductivity contribution, preventing a detailed analysis of this relaxation.

The two fastest relaxational processes (I and II) can be assigned to the β- and α-relaxations of the solvent respectively. The third process, III, is markedly slower than the α-relaxation and can be tentatively assigned to dynamics of the polymer network. This assignment is supported by the dependence of these relaxations on the polymer concentration. With increasing polymer content the α-relaxation (II) decreases in strength and slows down, resulting in a more pronounced β-process (I), see figure 2. One can also note a broadening of the α process with increasing polymer concentration. This result is in agreement with what has been previously reported

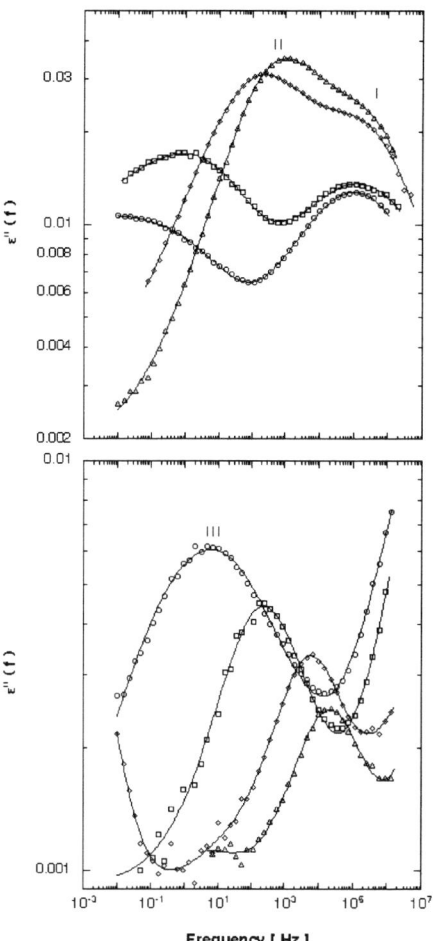

FIGURE 2. Dielectric loss spectra for the concentrations 10 (triangles), 14 (diamonds), 23 (squares) and 29 (circles) %w/w. Upper panel shows the region of the α and β relaxations corresponding to the solvent dynamics at T=128 K. Lower panel shows the region of the relaxation related to the polymer network at T $\approx$ 177 K. Solid lines show the fit to the data as described in the text.

for glass formers subject to geometrical confinement [8, 9]. Furthermore, it is also evident from figure 2 that the strength of relaxation III increases with increasing polymer concentration supporting the assignment of this mode to polymer network dynamics. Accompanying the increasing strength is a slowing down of the dynamics at higher polymer concentrations, marked by the shift of the maximum of the relaxation loss peak to lower frequencies in figure 2.

In order to parametrize the temperature dependence of the three relaxational processes the data was analyzed using the symmetric Cole-Cole equation [10] for the β (I) and the network (III) relaxation. The α-relaxation (II) was described by a recently proposed empirical equation that parametrizes the observed asymmetric broadening [10]. The relaxation times obtained from the detailed analysis are displayed in figure 3, where data for pure toluene is also included for comparison [11]. A comparison with the data for pure toluene corroborates the assignment of the relaxational processes I and II to the β- and α-relaxations of the solvent. The faster β process is largely unaffected by the presence of the polymer, whereas the α-relaxation slows down with increasing polymer content, as discussed above. In the dynamic regime where the α- and the β-relaxation merge the present analysis is not valid and a more elaborate analysis, based on e.g. the Ansatz by Williams, has to be performed [12].

For the third relaxation process (III), the results from the analysis show a clear slowing down of the dynamics with increasing polymer concentration and also reveal a non-Arrhenius behavior. This process is, in contrast to the α- and β-relaxations, present in the region of the gel transition (T_{gel} 170 for 14%(w/w) and 230 K for 29%(w/w) [6]). It is of interest to note that there is no dramatic change in the behavior at the reported gelation temperatures and that the process is still present for temperatures above the transition. Thus, assuming that this relaxation is related to network dynamics, there is no major change of the polymer network at the gel transition, at least not on the length scale associated with the relaxation. One can then speculate that the gel transition is not a result of changes in the structural arrangements of the network and that the network develops already above the gelation point [5].

To further investigate the nature of the network related relaxation process we analyze the shape of the temperature dependence of the relaxation time. This can be performed e.g. by trying to construct a master curve for the relaxation time by temperature scaling. In figure 4 we have constructed such a master curve by scaling the temperature for each concentration by an arbitrary temperature T. In order to relate the behavior to the sol-gel transition we initially set T (23%) to the reported gelation temperature [6], T_{gel}=211 K. T for the other concentrations was freely adjusted to obtain the best possible matching of the curves. This simple temperature scaling results in a perfect master curve, see figure 4. Furthermore, the obtained scaling temperatures, T, fall close to the reported gelation temperatures (for the 10% concentration unfortunately no data exist). A more detailed study, including more concentrations, is needed to determine if a relation exists between T and T_{gel} or if the current observations are a mere coincidence. Such a study is currently underway.

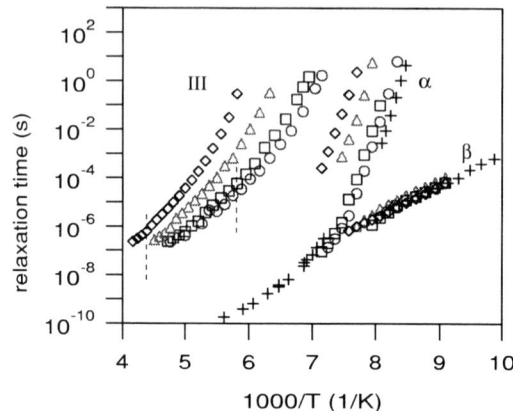

FIGURE 3. Relaxation times for the three relaxation processes as a function of inverse temperature. The symbols represent: circles - 10%, squares - 14%, triangles - 23%, diamonds- 29% and + toluene data from ref. [11]. The vertical dashed lines indicat reported gelation temperatures for the 14% and 29% concentration respectively [6].

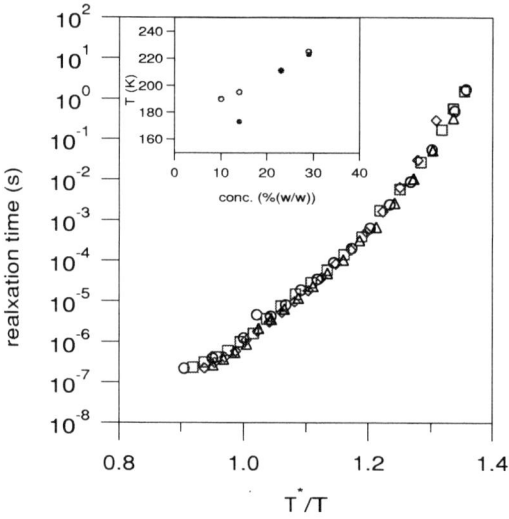

FIGURE 4. Temperature scaling of the polymer network related relaxation process (III) as described in the text. The symbols represent: circles - 10%, squares - 14%, triangles - 23%, diamonds-29%. Inset shows the scaling parameter T (open circles) and the reported gelation temperatures [6] (full circles) as a function of concentration.

CONCLUSIONS

In this work we have presented results on the dynamics in the thermoreversible aPS/toluene gel-forming system. The DRS experiment clearly show the presence of three relaxational processes. The two fastest processes can be assigned to the α- and β-relaxations of the solvent, whereas the third process is associated with the polymer network. With increasing polymer concentration the α-relaxation is reduced in strength, slowed down and broadened whereas as the β-relaxation is unaffected. The relaxation related to the polymer network increases in strength and becomes slower with polymer concentration. The temperature dependence of this relaxation is found to be non-Arrhenius and there is no marked change in the behavior around the gelation temperature. In fact, we show that a master curve can be obtained for this relaxation by a simple temperature scaling.

ACKNOWLEDGMENTS

Financial support from the Swedish Research Council (AM and RB) and the Wenner-Gren and Hans Werthén foundations (JM) is gratefully acknowledged.

REFERENCES

1. Nijenhuis, K. T., *Thermoreversible Networks. Viscoelastic Properties and Structure of Gels in Advances in Polymer Science*, Springer-Verlag, Berlin, 1997.
2. de Gennes, P.-G., *Scaling Concepts in Polymer Physics*, Cornell University Press, New York, 1979.
3. Ren, S. Z., and Sorensen, C. M., *Phys. Rev. Lett.*, **70**, 1727–1730 (1993).
4. Ikkai, F., and Shibayama, M., *Phys. Rev. Lett.*, **82**, 4946–4949 (1999).
5. Kumar, S., and Douglas, J. F., *Phys. Rev. Lett.*, **87**, 188301 (2001).
6. Tan, H.-M., Moet, A., Hiltner, A., and Baer, E., *Macromelecules*, **16**, 28–34 (1983).
7. Boyer, R. F., Baer, E., and Hiltner, A., *Macromelecules*, **18**, 427–434 (1985).
8. Svanberg, C., Bergman, R., and Jacobsson, P., *Phys. Rev. B*, **66**, 54304 (2002).
9. Bergman, R., Mattsson, J., Svanberg, C., Schwartz, G., and Swenson, J., *Europhys. Lett.* (in press 2003).
10. Bergman, R., *J. Appl. Phys.*, **88**, 1356–1365 (2000).
11. Kudlik, A., Benkhof, S., Blochowicz, T., Tschirwitz, C., and Rössler, E., *J. Mol. Struct.*, **479**, 201–218 (1999).
12. Williams, G., *Adv. Polymer Sci.*, **33**, 60–92 (1979).

Local particle rearrangements in a two-dimensional binary colloidal glass former

Hans König

Department of Physics, University of Konstanz, Universitätsstraße 10, D-78457 Konstanz, Germany

Abstract. In a two-dimensional (2D) glass former composed of two kinds of different sized and repulsively interacting colloidal particles, the time-dependent particle positions were observed by video-microscopy. Analyzing the local particle arrangements by 3-point correlation functions, we find four different local density-optimized configurations of nearest neighboring particles, which we call elementary triangles (ET), one for each three-particle combination of small and big particles. These four ET form a random tiling in the 2D monolayer, which is not space filling. Therefore, a heterogeneous local particle packing does not have long-range order and shows structural frustrations. Furthermore, an analysis of structural relaxations, using triangles of nearest neighboring particles (TNNP) in the monolayer, suggests that hopping processes are the reason for the rearrangements of the particles. In the outlook, we propose a concept of local density-optimized crystallite-clusters to describe the glass transition as a percolation of stable local density-optimized triangles.

In supercooled liquids the particle dynamics of mean-square displacements can be classified in three time zones. For the short-time dynamics the particles diffuse like free ones. In the intermediate time region the particles are captured by their surrounded neighbors for a while, like in a cage, until in the long-time zone structural relaxations allow the particles to escape their vicinity. For decreasing temperatures, such cages become stronger and remain visible longer, as the system solidifies at the glass transition temperature due to the cage-effect becoming the stable state. Considering the short-range order in supercooled liquids by pair-distribution functions (PDF), the local packing becomes better structured for decreasing temperatures while at the glass transition temperature no discontinuous changes occur. This is why a glassy state can be considered as a supercooled melt, forming a solid with liquid-like structure and with frozen-in particle dynamics [1], [2], [3].

Up to now no theory is able to describe the properties of glass formers or to figure out the interconnection between the microscopic particle dynamics and the local particle configurations. But, undoubtedly, the cage-effect and structural relaxations are clearly related to the local microscopic structure. Therefore, in our opinion, the glass transition can only be understood if the interplay between the microscopic dynamics and the local particle configurations in the amorphous structure is recognized.

In this work, novel microscopic structural analyses are introduced which try to combine the local particle arrangements with the dynamics of the local structure. Therefore, in a binary 2D colloidal glass former of known time-dependent particle positions, the structure of triangles of nearest neighboring particles and their time-dependent changes have been analyzed simultaneously.

EXPERIMENT

Time-dependent particle positions of a 2D binary colloidal suspension of repulsively interacting paramagnetic PMMA colloids were directly observed by video-microscopy. The particles lay on a completely flat adjusted water-air interface of hanging droplet geometry, confined due to gravity. In the monolayer, the big (b) and small (s) colloids interacted with their induced magnetic moments, tunable by an external magnetic field B_{ext}.

During the measurements, the particle area density, ρ, the ratio, $\xi = n_s/(n_s+n_b)$, of the number n_s of s particles to the number of all colloids $n_s + n_b$, the room temperature, T, as well as the magnetic susceptibilities χ_b and χ_s of the b and s colloids remain constant. Therefore, the strength of the particle repulsion can be controlled via B_{ext}. Because of the experimental conditions, other in-plane particle interaction potentials can be neglected.

The 2D system is characterized by a dimensionless interaction parameter, $\Gamma(B_{ext})$, which is the magnetic energy E_{magn} divided by the thermal energy $k_B T$. Thus, Γ corresponds to an inverse effective system temperature, T_{sys}, tuneable by B_{ext}. The full form of Γ is given by:

$$\Gamma = \frac{E_{magn}}{k_B T} = \frac{\mu_0 \sqrt{\pi}}{4} \frac{B_{ext}^2 \rho^{3/2}}{k_B T} (\xi \chi_s + (1-\xi)\chi_b)^2. \quad (1)$$

More details of the experiment are presented in this issue [4]. There it is also shown that the investigated 2D system behaves like a glass former. Additionally, we have to point out that Zahn and Maret have used the same 2D setup (but using monodisperse colloidal suspensions) as model system for studying the 2D melting scenario of Kosterlitz and Thouless [5].

MICROSCOPIC STRUCTURE

Up to now, microscopic amorphous 2D structures were mainly investigated by the Voronoi construction [1], which divides the area into cells around each particle, or by bond-order parameter methods [6], which characterize the orientational structure by an order parameter. In this work, a novel microscopic structure analysis is introduced: We investigate the structural and dynamical properties of our colloidal model system by analyzing triangles of nearest neighboring particles (TNNP) which are defined as the smallest area units of the local microscopic particle arrangements in the binary 2D sample. Such analysis is not common since we no longer consider properties of single particles or pair-distances, but triangles.

In the first step, we are interested in the local density-optimized packing of the particles investigating a highly supercooled 2D binary monolayer, i.e. of well-defined short-range order. Therefore we use 3-point correlation functions (3-PCF) and define the distances of the two fixed s or b particles in a narrow interval around the first peak of their corresponding partial PDF. This determination definitely guarantees pairs of nearest neighbors. All such pairs in the monolayer are transformed on the x-axis of a new coordinate system. Additionally, in the 3-PCF all s and b third particles are plotted on top of each other lying in a given region, which is aligned to the position of the two fixed particles in the monolayer (see Fig. 1).

In the 3-PCF, around both fixed particles there is first a zone with no particles followed by a ring of s (blue) and then of b colloids (red) with high particle probabilities (Fig. 1). The ring-radii correspond to the first maximum positions of the PDF of the corresponding pair-combination. At the point of intersection for the two rings of the same kind of colloids, an accumulation point is seen where particles are simultaneously nearest neighbors of both fixed particles. Hence, the local density-optimized packing of each of the four different 3-particle combinations of b and s colloids (bbb, bbs, bss, sss) can be read off from the isosceles triangles in Fig. 1. Thus, for each of the four 3-particle combinations of b and s particles exactly one local density-optimized triangular structure can be found. These triangles we call elementary triangles (ET). For the bbb and sss 3-particle combi-

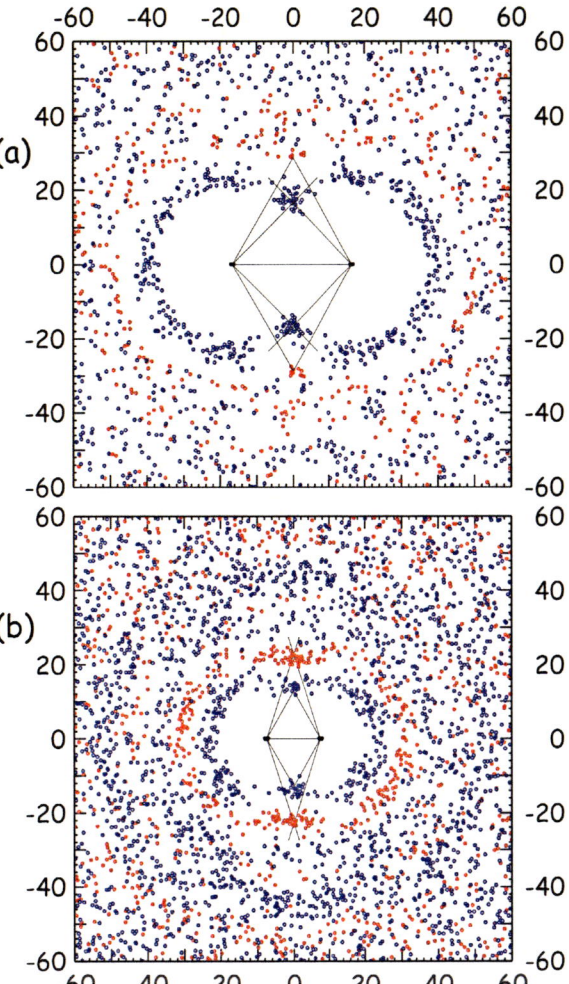

FIGURE 1. 3-point correlation functions for (a) two b or (b) two s colloids fixed at the first maximum of the corresponding partial PDF, sitting symmetrically around the origin on the x-axis. The units are μm. (a) The two fixed b colloids have distances in the interval $[32\mu m, 34\mu m]$. (b) The two fixed s colloids have distances in the interval $[14\mu m, 16\mu m]$. The elementary triangles are indicated by solid lines. Black: The two fixed particles; Red: b colloids; Blue: s colloids. $\Gamma = 411$, $\xi = 0.7$, $\rho = 2.34 \cdot 10^{-3} \mu m^{-2}$, $B_{ext} = 7mT$.

nations the ET are equilateral or hexagonal, the bbs ET are rectangular and the bss ET have a $36°$ angle near the b particle. Nevertheless, the particles undergo thermal fluctuations while in reality the TNNP of the monolayer deviate from the shape of the idealized ET, seen in Fig. 1. However, if the shape of a TNNP is comparable with that of the corresponding ET we call it ET-like TNNP.

Since in Fig. 1 neither orientational order nor any long-range order is seen, the 2D sample is amorphous although ET are present. By simple arguments the apparent contradiction of global disorder and local order can be solved as follows.

 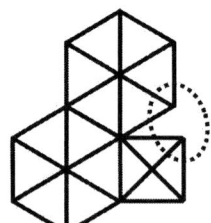 

FIGURE 2. Structural frustration in near-zone for the tiling of ET. In the examples shown the tiling produce forbidden *ss* and *bb* pair-distances because they are shorter than the first maximum distances of the corresponding partial PDF.

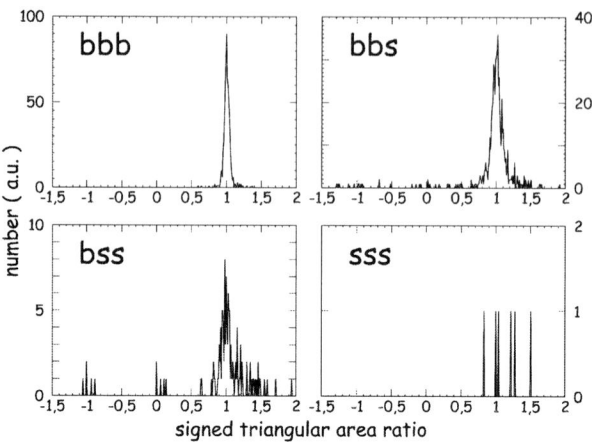

FIGURE 3. Triangle-area ratio for $\Delta t = 3600s$, separated in the different 3-particle combinations. $\Gamma = 142$, $\xi = 0.28$, $\rho = 2.62 \cdot 10^{-3} \mu m^{-2}$, $B_{ext} = 1.9 mT$.

The reason for the amorphous structure is that the microscopic particle distribution of the *b* and *s* colloids locally varies in the 2D sample. Thus, the different ET-like TNNP lie side by side in the monolayer and tend to form clusters without structural mismatch. We call such clusters of the same ET 'crystallite clusters' (CC) and of different ET 'multi crystallite cluster' (MCC). However, the mixed tiling of the different ET-like TNNP cannot cover the monolayer without structural frustrations of the local structure (Fig. 2). At that places the repulsive particle interaction bents the shape of the TNNP in the monolayer. Therefore, structural frustrations are responsible for broadened peaks of the short-range order in the partial PDF. Also, the increasing number of possible mixed tilings of ET for increasing interparticle distances lead to the loss of long-range order.

It can be shown [7] that for decreasing system temperature the accumulation points in 3-PCF become less sharp and the TNNP gets more ET-like. This means that the microscopic particle configurations become more ordered as well as more densely packed, which simultaneously corresponds to a slow-down of the particle dynamics displacements and to more pronounced heterogeneous relaxations.

HETEROGENEOUS STRUCTURAL DYNAMICS

In the 2D binary colloidal suspension the TNNP can directly be distinguished in the monolayer by construction rules which are reported in detail in [7], [8]. For the following it is important to note that investigations of the structural dynamics, using the angles, the area or the edges of the TNNP, are independent of the translation or the rotation of the TNNP in the monolayer.

Here, we introduce a new function, called triangular area ratio (TAR), in order to investigate a special structural relaxation process in 2D, the hopping process. We store the particles of each of the N TNNP in the monolayer on the condition that the initial area $F_i(0)$ of each TNNP with the number i in the first data file ($t = 0s$), which is calculated by vector product between the particle bond vectors, is defined positively. After the time Δt the area $F_i(\Delta t)$ is calculated once again. Dividing the final through the initial area of each TNNP and plotting this ratio $F_i(\Delta t)/F_i(0)$ in a histogram, one obtains the TAR for the four different 3-particle combinations, shown in Fig. 3.

The peaks in Fig. 3 can be interpreted as follows. The peak around 1 represents the TNNP, which have not significantly changed their area during Δt. If the peak around 1 is sharp for a special 3-particle combination, the structure has remained stable. However the peak at −1 for *bbs* and *bss* TNNP represents triangles with the same area, but with opposite direction of the particle arrangement. This peak indicates triangles which we call hopping triangles. There, during the time interval Δt, the hopping particle of such a TNNP has passed through between two neighboring particles, and thus it has changed its neighborhood. Since the final hopping triangle conserves the 3-particle combination, its area as well as its structure are expected to be the same as before. That is why the hopping particle would also give a second peak in the van Hove-function [2], [9] which one uses for the characterization of hopping processes. Because not only the hopping triangle but also the TNNP, which are built up by the hopping particle, typically have a TAR-value unequal to 1 in Δt. We call such TNNP 'hopping connected triangles'. From them there are signals in TAR around zero for aligned particles and somewhere in the positive range for TNNP of the same direction of the particle arrangement. More details can be found in [7] and will be published elsewhere.

Analyzing the TAR of Fig. 3, the following conclusions can be drawn about the structural dynamics of the different 3-particle combination in Δt. The *bbb* TNNP are the most stable structures. Hopping processes are found for *bbs* and *bss* TNNP. The very low number of *sss* TNNP show no hopping process, but their structure fluctuates strongly. For the sake of completeness it should be remarked that the time period Δt beyond which hopping processes can be observed corresponds to the time range where the cage-effect in the 2D sample starts to crumble [7], [8]. Furthermore, it is also possible to analyze the hopping process directly in the monolayer.

Since hopping processes need less densely packed regions, the reverse statement is allowed that particles which are surrounded by only local density-optimized ET, undergo a 'perfect cage-effect'. Therefore there is, as we say, a 'blocking effect' for microscopic structural dynamics which is why also in fast regions of a monolayer, perfectly captured particles can only follow their neighbors.

CONCLUSIONS

In this paper, the statical structure and the dynamics of the microscopic particle configurations of a binary 2D colloidal suspension are analyzed by means of 3-PCF and of TNNP. Several new conclusions are drawn: First, the amorphous structure of the investigated 2D glass formers can be described by four different idealized ET and structural frustration. Secondly, hopping processes in the 2D sample are responsible for the structural relaxations of the TNNP. The special feature of the TNNP analyses by their areas or their angles is that the translation or rotation of the triangles themselves in the monolayer are ignored. This is why only the microscopic structural relaxations of the triangles are considered.

OUTLOOK

In the free volume theory the idea of 'solid-like' and 'liquid-like' regions and of a percolation, in order to explain the glass transition, has been mentioned more than 40 years ago [10], [11], [12]. Against this background, our new description of the glass transition should shortly be introduced here, which we call concept of local density-optimized crystallite clusters (CLDOCC). Those considerations base on structural and time-dependent analysis of TNNP in the monolayer and allow for the first time to distinguish the dynamics of 2D structures microscopically. Therefore, not single particles but TNNP are in the center of interest.

CLDOCC treats the glass transition as a percolation of the different ET-like TNNP to one big MCC. There, the solidification occurs because different ET-like TNNP stabilize each other by the blocking effect of local density-optimized packed particles. In between the connected crystallite clusters less densely packed, spatially separated, and structurally frustrated regions with frozen-in structural relaxations are found. That is why relaxation processes cannot happen in a glassy state any more.

In another paper of this issue [4], where the same 2D binary colloidal suspension is considered, additional aspects of the local and dynamical investigations of the microscopic particle configurations are presented which confirm the idea of CLDOCC. Nevertheless, further analysis is necessary.

In future the predictions of the CLDOCC has to be checked in other experiments. Since the considerations of the microscopic structural dynamics do not involve restriction on the dimension or the mixture of different kinds of particles these ideas can easily be treated also for samples in 3D, of more component systems or with other particle interactions. In 3D, however, the ET have to be transformed into elementary tetrahedrons representing the local density-optimized packing of four particles.

ACKNOWLEDGMENTS

The author thanks G. Maret, K. Zahn, R. Haussmann, and U. Gasser for discussions and useful advices and the Deutsche Forschungsgemeinschaft for financial support.

REFERENCES

1. Kondo, T., Tsumuraya, K., and Watanabe, M. S., *J. Chem. Phys.* **93**, 5182-5186 (1990).
2. Miyagawa, H., Hiwatari, Y., Bernu, B., and Hansen, J. P., *J. Chem. Phys.* **88**, 3879-3886 (1988).
3. Ediger, M. D., Angell, C. A., and Nagel, S. R., *J. Chem. Phys.* **100**, 13200-13212 (1996).
4. König, H., Zahn, K., and Maret, G., *Glass transition in a 2D system of magnetic colloids*, in *the same issue*.
5. Zahn, K., and Maret, G., *Phys. Rev. Lett.* **85**, 3656-3659 (2000).
6. Steinhardt, P. J., Nelson, D. R., and M. Ronchetti, *Phys. Rev. B* **28**, 784-805 (1983).
7. König, H., *Mikroskopische Prozesse am Glasübergang einer binären paramagnetischen 2D Kolloidsuspension: Lokal-dichteoptimierte Kristallit-Cluster*, PhD Thesis, University of Konstanz, 2002, pp.1-228, http://www.ub.uni-konstanz.de/kops/volltexte/2003/998.
8. König, H., *to be published*.
9. Roux, J. N., Barrat, J. L., and Hansen, J. P., *J. Phys.: Condens. Matter 1* **89**, 7171-7186 (1989).
10. Cohen, M. H., and Turnbull, D., *J. Chem. Phys.* **31**, 1164-1169 (1959).
11. Turnbull, D., and Cohen, M. H., *J. Chem. Phys.* **52**, 3038-3041 (1970).
12. Cohen, M. H., and Grest, G. S., *Phys. Rev. B* **20**, 1077-1098 (1979).

What is the excluded volume for an excluded volume chain?

M.W. Dreischor, C.P. Lowe

*Department of Chemical Engineering,
University of Amsterdam,
Nieuwe Achtergracht 166, 1018 WV, Amsterdam,
The Netherlands*

Abstract. Solvated polymers are generally expanded due to the effective interaction between the monomers. In Flory's lattice theory for polymers this "excluded volume" is the same for a polymer solvated by its own monomers and a polymer in the absence of solvent. The magnitude of the effect is determined by a parameter υ, that is directly related to the lattice spacing. Computer simulations using a simple off-lattice model (dissipative particle dynamics) show that this is not true. Only an off lattice analogue of Flory's theory that takes into account the compressibility of the solvent quantitatively predicts the degree of expansion as a function of solvent density. We find that for the solvated model chains the distribution of mass around the centre of mass is to a good approximation Gaussian. This means that the diffusion coefficient can be calculated by simply treating the chain as an ideal chain with a Kuhn length that is modified to take into account the expansion.

INTRODUCTION

The dynamics of long polymer chains are slow by atomic standards. The true long time dynamic behaviour can be reached only on time-scales extending to minutes. For atomic systems this is picoseconds. To access long times in a simulation we must simplify our representation of a polymer dramatically and sacrifice an atomic level of detail. One way to do this is to take a theoretical model for the real long polymer and simulate that instead. The simplest theoretical model that gives plausible dynamics is the ideal chain [1]. Here the polymer is modelled as a set of non-interacting beads connected by harmonic springs. This potential is related to the Kuhn length of the polymer (the root mean square separation between monomers).

$$U = k_B T \sum_{i=1}^{N-1} \frac{3}{2b^2} (\vec{r}_i - \vec{r}_{i+1})^2 \quad (1)$$

Because the ideal chain model neglects interactions between monomers (and solvent), it only strictly applies for a polymer in the absence of solvent or in a theta solvent. That is, a solvent in which the effective interactions between the monomers, mediated by the solvent, are zero. In general this will not be the case. We want to mesoscopically model the dynamics of solvated polymer chains. The diffusion of a polymer is sensitive to its size so to do realistic dynamics we need to understand what drives the effect and to what extent a real polymer will be expanded (or collapsed).

In this article we consider an off lattice model, Dissipative Particle Dynamics (DPD), that aims to simulate polymers at the mesoscopic level. It has been used to look at the scaling behaviour of solvated polymers [2] but, with the exception of Groot and Warren [3], little attempt has been made to compare the results quantitatively to well known polymer physics theories. Classic theories of the excluded volume effect are based on lattice models and the mapping from from a lattice (with its well defined but artificial length scale, the lattice spacing) to off lattice systems is somewhat ambiguous. If this mapping cannot be made, we run into the problem, is this a limitation of the simulation method or the theory? To illustrate this point we compare the excluded volume effect of a polymer chain solvated in an ideal gas solvent with the excluded volume effect of a polymer chain solvated in its own monomers, using DPD merely as the simulation tool. The excluded volume in the case of an ideal solvent is directly related to the second Virial coefficient, as was previously shown by Fisher [4]. Since we can easily calculate the second Virial for the DPD model, we can accurately predict the excluded volume effect based on essentially an off lattice analogue of Flory's Self Avoiding Walk theory. In the case of a polymer chain solvated by its own monomers, the magnitude of the excluded volume effect is greatly reduced when compared to the ideal solvent, something you would not expect based on Flory's theory. Furthermore it appears that in the limit of incompressibility the excluded volume is determined by the number density of the solvent and not the second Virial coefficient. We describe a simple off-lattice theory that accounts for this.

FLORY'S THEORY

The volume of an ideal polymer chain V_0 composed of N non-interacting monomers scales as $V_0 \propto b^3 N^{3/2}$. Here b represents the average separation between two connected monomers in the chain, also known as the Kuhn-length. This implies that with an increasing number of monomers in the chain the actual monomer density decreases, since $\rho = N/V_0 \propto 1/(b^3 \sqrt{N})$. For a long time this led people to the believe that even if the monomers have some excluded volume v, it would not affect the size of the polymer chain if the chain was long enough. Experiments on polymers however showed that there still was a significant size effect even for very long chains. Flory [6] was the first to suggest a theory that explained this "excluded volume effect". Considering a lattice self-avoiding random walk he derived an approximate expression for the expansion of a polymer chain as a function of the number of monomers:

$$\alpha^5 - \alpha^3 = v_L \frac{N^2}{V_0} \qquad (2)$$

where the expansion parameter $\alpha = (V/V_0)^{1/3}$, V being the actual volume of the expanded chain and v_L is the volume of one lattice site. If N is very large this leads to a scaling of the length $l = V^{1/3}$ of $l \propto N^{3/5}$, as opposed to $l \propto N^{1/2}$ for the ideal chain. This scaling is very close to the best current estimates for the self-avoiding random walk [7]. However, the theory still has some limitations. The expression involves the lattice volume. A lattice volume does not have an unambiguous interpretations in terms of a real system. This problem can be solved by using an off-lattice analogue to Flory's approach, in which case $v = 2B_2$, where B_2 is the second Virial coefficient of the monomers [4]. The second problem in Flory's original analysis was that it did not take into account interactions between the polymer and the surrounding solvent. These interactions were later incorporated by introducing contact energies between the two species and weighting the polymer configurations accordingly [1]. By using Flory-Huggins equation of state for lattice polymers, assuming a completely filled lattice, one finds

$$v = v_L(1 - 2\chi) \qquad (3)$$

where χ is an interaction parameter related to the contact energies between the different sites. This interaction parameter is always inversely proportional the the temperature. At low temperatures it will be positive leading to a negative value for v and thus a collapsed polymer chain. At high temperatures it is positive and the polymer chain will be expanded. At a temperature T_F (the Flory temperature), $\chi = 1/2$ and $v = 0$. At this temperature the polymer chain will have the same size as an ideal chain. If the solvent and the monomers are the same species (the 'symmetric' solvent), in other words they have the same contact energies, $\chi = 0$ and $v = v_L$ at all temperatures and the polymer chain behaves as if there is no interaction with the solvent.

DISSIPATIVE PARTICLE DYNAMICS

Flory's theory has two main limitations. It is based on a lattice model and real polymers do not exist on a lattice. Secondly it assumes that the lattice is fully occupied. In a real system this corresponds to assuming incompressibility. For reasons of computational expediency, most computer simulations are also carried out with lattice models so suffer similar drawbacks. A relatively new computational method, Dissipative Particle Dynamics [2] (DPD), is a particle based simulation method without these drawbacks. It is off-lattice and has finite compressibility. Because it also conserves momentum and is Galilean invariant it can also reproduce plausible (hydrodynamic) behaviour. The particles in DPD interact through three simple forces. A random force, a dissipative force and a conservative force. The combination of the random and dissipative forces acts as a thermostat whereas the conservative force induces the non-ideal behaviour of a DPD fluid. A DPD fluid without these interactions would just be an ideal gas. In most cases a very simple interaction potential is used

$$u_{ij} = \frac{a_{ij}}{2}\left(1 - \frac{r_{ij}}{r_c}\right)^2 \qquad (4)$$

where a_{ij} is an interaction parameter and r_c is the cut-off radius of the potential This "soft" potential is computationally convenient. but introduces a weakness in the method. It begs the question, what does this represent in the real world? Groot and Warren suggested that in a mixture the a_{ij} parameters can be mapped onto the χ parameters of Flory-Huggins theory. This gives the potential used in the model some theoretical basis. With this interpretation, DPD could be viewed as a dynamic off-lattice compressible analogue of the lattice models commonly used to model solvated polymers. It therefore makes sense to use it to study diffusion of solvated polymers.

It is possible to simulate polymer chains simply by introducing spring potentials between N consecutive particles. Here we use the harmonic potential (equation 1). This means that we can specify the Kuhn length of the equivalent ideal chain, the parameter b appearing in Flory's result. Here we restrict ourselves to the symmetric case (solvent and monomers are the same). Without ambiguity this corresponds to $\chi = 1/2$ in terms of Flory-Huggins theory. To compare directly with Flory's theory we still need to identify what volume in our off-lattice model corresponds to the lattice volume appearing in equation 3. The off-lattice version of Flory's model neglects the solvent and gives $v = 2B_2$, where B_2 is the second Virial coefficient. However, according to lattice models the excluded volume parameter should be the same in the case of a symmetric solvent. The second Virial coefficient can be evaluated analytically for the DPD potential. According to lattice theories, for the symmetric case we consider here the solvent has no effect so the excluded volume parameter should be independent of the solvent density.

RESULTS

We calculated the radius of gyration of the model DPD polymer as a function of the solvent density for various short polymer chains. The diffusion coefficient of this "excluded volume" chain was also calculated. We compared the values to hydrodynamic result for an ideal chain

$$\frac{D}{D_0} = \frac{1}{N} + \frac{a}{N^2} \left\langle \sum_i \sum_{j \neq i} \frac{1}{|(\vec{r}_i - \vec{r}_j)|} \right\rangle \quad (5)$$

calculating the equilibrium average using a Monte Carlo method. Here a is the hydrodynamic radius of a monomer. For chains consisting of $N > 16$ beads we could find no difference, to an accuracy of a few percent, between the value we calculated and the value for an ideal chain if we used an effective Kuhn length. This effective Kuhn length is simply the Kuhn length that an ideal chain with the same number of beads would need in order to give the same radius of gyration as the excluded volume chain. The reason for this is that the distribution of mass around the centre of mass in the polymer is to a very good approximation Gaussian (see figure 1). The equilibrium average in equation 5 can itself be written in terms of this distribution [8] so a simple re-scaling suffices. Note that in contrast the distribution of the end to end distance is very non-Gaussian but this is not directly related to the diffusion coefficient. So the dynamics of an excluded volume chain, at least within the context of the DPD model, are relatively straightforward to understand, so long as we know the degree of expansion of the chain. How well does Flory's theory describe this? In figure 2 we have plotted the excluded volume parameter v as a function of the number of beads in the chain, for differing solvent densities. The ideal solvent result is equivalent to the limit of vanishing solvent density. To

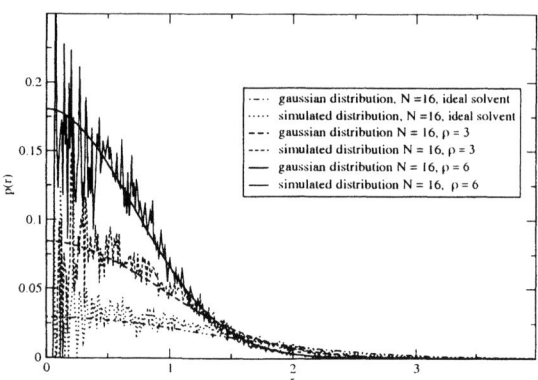

FIGURE 1. Distribution of the monomers of a 16 bead chain around the center of mass. The different data are for different solvent densities, "ideal" being the zero density limit. Densities are in units where the range of the potential r_c is unity.

use equation 2 we also need to specify the equivalent ideal chain volume V_0. This we take as the volume of a uniform spherical distribution of the monomers with the same radius of gyration as the polymer. Flory's result is also shown in the figure (equation 2) and denoted "ideal solvent theory". According to lattice models it should hold for all solvent densities. Clearly it does not. Before discussing this though, we should point out that it does extremely well for quantitatively predicting the ideal solvent case. Normally the result is taken as a scaling argument with little attention given to the actual excluded volume parameter. However, with increasing N it appears to be an increasingly good estimate. As real polymers have many more monomers than we have here, it should hold to a good approximation in practice.

Where the theory fails is in that the excluded volume parameter shows a pronounced decrease with increasing solvent density.

AN OFF-LATTICE MODEL

Clearly, the results above show that the mapping of the excluded volume parameter from lattice to continuum theories is non-trivial. The latter is of course more important because real systems do not exist on a lattice. With this in mind we re-examined the thermodynamic model from which one can derive equation 2 from thermodynamic arguments. The details of this will be described elsewhere, here we give the result and a brief summary of how it is arrived at.

The model assumes that our polymer solvent system consists of two phases. A mixed monomer-solvent phase with a spherical volume V surrounded by a pure solvent phase. The amount of solvent in the monomer solvent phase we determine from the condition that the chemical

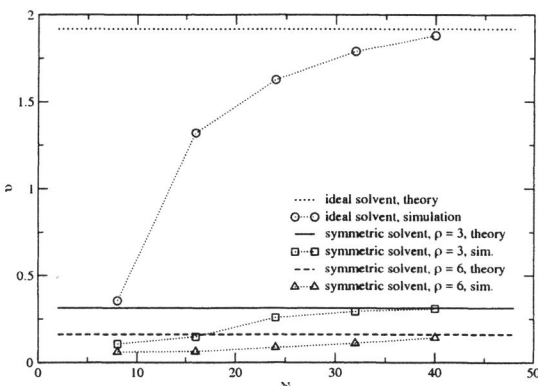

FIGURE 2. The excluded volume parameter as a function of the number of beads in the chain N at different solvent densities, "ideal" being the zero density limit. Lengths are measured in units of the interaction radius r_c. The theoretical result is a numerical solution to (equation 8)

potential of the solvent inside the mixed phase equals that of the surrounding pure solvent. We then assume that we can write the total free energy of the system as a summation of the chemical potentials of the two species:

$$G(\alpha) = \mu_p + n_s \mu_s + \Delta G_\alpha \quad (6)$$

where μ_p is the chemical potential of the polymer ($\mu_p = N\mu_m$) and ΔG_α is the penalty for either expanding or contracting the polymer chain due to the spring connectivity. According to Flory this penalty is given by:

$$\frac{\Delta G_\alpha}{k_B T} = \frac{3}{2}(\alpha^2 - 1) - 3\ln(\alpha) \quad (7)$$

where k_B is Boltzmann's constant. We then minimize the total free energy with respect to α to get the following expression for α:

$$\alpha^5 - \alpha^3 = \frac{(1-\kappa)}{\rho_0} \frac{N^2}{V_0} \quad (8)$$

where ρ_0 is the number density of the solvent and $\kappa^{-1} = 1/k_B T \delta P/\delta \rho$ is the dimensionless compressibility of the solvent. In other words $\upsilon = (1-\kappa)/\rho_0$. In the limit of zero density this reduces to Flory's result. However we now have a non-trivial dependence of υ on the density. This is the theoretical result plotted in figure 2 and clearly it quantitatively describes the reduction in the excluded volume parameter with increasing solvent density that we observe.

DISCUSSION

We used dissipative particle dynamics as an off lattice model to study the dynamics of model polymers solvated in their own monomers. So long as the number of beads exceeds a modest 16 the diffusion coefficient is given by the ideal chain result with an effective Kuhn length that reflects the expansion of the chain. This is because the distribution of mass around the center of mass is Gaussian, just as it is for an ideal chain. This led us to consider how one might predict the degree of expansion. The off-lattice analogue of Flory's result led to a surprisingly good quantitative estimate of the degree of expansion of the chain in the absence of solvent. However, the degree of expansion of the chain clearly depended on the solvent density. According to lattice models it should not. An off-lattice compressible theory, however, successfully explained the density dependence. Note that in the incompressible limit ($\kappa = 0$) equation 8 implies that $\upsilon = 1/\rho_0$. That is, the specific volume of the solvent is the excluded volume parameter. Contrary to Flory's theory this result shows that there is a difference between a polymer solvated in a noninteracting solvent, where $\upsilon = 2B_2$ and an incompressible symmetric solvent where the excluded volume is determined by the specific volume of the solvent. The DPD model is in fact somewhat artificial in that because there is no hard core repulsion the specific volume of the solvent can be arbitrarily small. In a real system it cannot exceed the actual volume of the particles. The density dependence is therefore more dramatic. This makes it a sensitive test for the theory. The theory itself applies for any symmetric system for which the equation of state is known. It will be interesting to see what it predicts for a more realistic equation of state.

REFERENCES

1. *"The theory of polymer dynamics"* M. Doi and S.F Edwards, Clarendon, Oxford (1986).
2. N. A. Spenley, Europhys. Lett. **49**m 534 (2000).
3. R.D. Groot and P.B. Warren, J. Chem. Phys. **107**, 4423 (1997).
4. M.E. Fisher, J. Phys. Soc. Jpn **26** (Suppl.), 44 (1969).
5. M. Doi, *"Introduction to polymer physics"*, Clarendon Press (1996).
6. P.J. Flory, J. Chem. Phys. **17**, 303 (1949).
7. B. Li, N. Madras and A.D. Sokal, J. Stat. Phys. **80**, 661 (1995).
8. B. Dunweg, D. Reith, M. Steinhauser and K. Kremer, J. Chem. Phys. **117**, 914 (2002).

Flow instabilities in complex fluids: Nonlinear rheology and slow relaxations

A. Aradian[*] and M. E. Cates[*]

[*]School of Physics, University of Edinburgh, JCMB Kings Buildings, Edinburgh EH9 3JZ, United Kingdom
E-mails: A.Aradian@ed.ac.uk, M.E.Cates@ed.ac.uk

Abstract. We here present two simplified models aimed at describing the long-term, irregular behaviours observed in the rheological response of certain complex fluids, such as periodic oscillations or chaotic-like variations. Both models exploit the idea of having a (non-linear) rheological equation, controlling the temporal evolution of the stress, where one of the participating variables (a "structural" variable) is subject to a distinct dynamics with a different relaxation time. The coupling between the two dynamics is a source of instability.

INTRODUCTION

Complex fluids are known to exhibit a wide range of unconventional behaviour when forced to flow, due to the intricate couplings between their structure at the mesoscopic scale and the imposed flow. In this work, we investigate theoretically certain situations where, under steady external drive, the long-term behaviour of the fluids is intrinsically unstable: the fluids never reach a steady state, and rather respond in a unsteady way (as shown by time measurements of e.g., the shear stress or the shear rate). Such situations include the appearance of *sustained temporal oscillations* in surfactant solutions [1, 2] as well as polymer solutions [3]. In other cases, some apparently totally erratic temporal responses have been found, in dense colloidal suspensions [4] and several surfactant systems [5, 6]. In the latter, there are some indications [5, 6] that the obtained signal is in fact the result of a *deterministic chaotic dynamics*. Such a chaotic behaviour at virtually zero Reynolds number (and thus no inertia) must stem from a nonlinearity within the constitutive properties of the material, and has thus been dubbed "rheological chaos", or "rheochaos".

General principles

These unsteady behaviours raise numerous questions from the theoretical viewpoint. Various hypotheses can be made as to what mechanisms give rise to such temporal instabilities (oscillations or rheochaos). The path that we are currently exploring relies on two main physical features shared by many complex fluids:

(i) An underlying tendency to form shear-banded flows.

(ii) A dependence of the present state of the fluid on past history, due to the presence of slow-relaxing structural modes.

The appearance of shear-bands in shear flows is indeed one of the most frequent mechanical instability observed in complex fluids. Slow-relaxing "structural modes", on the other hand, are directly linked to the commonly observed memory effects – i.e., long-term metastability – in complex fluids, where the shear history of a sample can be remembered for a surprisingly long time. One way to interpret this is that the previous flow has disturbed in some way one or several quantities characterizing the *structure* of the fluid (for instance, the micellar length in a worm-like micelle system, or the local density in colloidal systems, etc.), and that these structural variables then only relax over long periods, because they involve, e.g., collective motions, or are controlled by a slow, independent physico-chemical process.

In this article, we give a short overview of our ongoing work on this subject, presenting two related models which include the above-mentioned physical ingredients, as well as some preliminary results.

A SHEAR-THICKENING MODEL WITH MEMORY

The first model is a toy rheological equation which describes a shear-thickening fluid with memory, within a purely scalar approach where only the shear component σ of the stress tensor is considered. Imagining that the fluid is sheared in a Couette cell, we take z as the vorticity direction (the axial direction), and will consider spatial variations in that direction only. In the proposed model, the evolution with time t of the shear stress $\sigma(z,t)$ is then given by (in units where the elastic modulus is one)

$$\dot\sigma(z,t) = \dot\gamma - R(\sigma) - \lambda \int_{-\infty}^{t} M(t-t')\,\sigma(z,t')\,dt' + \kappa \nabla^2 \sigma(z,t) \quad (1)$$

with $\dot\sigma \equiv \partial\sigma/\partial t$. The shear rate, $\dot\gamma$, is supposed *uniform* in the z-direction (this is related to a low-Reynolds assumption). The positive term $R(\sigma) = a\sigma - b\sigma^2 + c\sigma^3$ is a non-linear, instantaneous relaxation term, with a, b, c positive constants chosen so that $R(\sigma)$ has a decreasing portion. This creates a tendency for the fluid to form "vorticity" shear bands (stacked in the z-direction). The integral term over past states of the stress represents the delayed relaxation of slow structural modes, with $\lambda > 0$ a parameter (homogeneous to a jump rate) and M a decaying memory function; we will here specialise to an exponential memory, $M(t) = \tau_s^{-1}\exp(-t/\tau_s)$, and τ_s will be the typical relaxation time of the slow structures. Finally, the inclusion of a non-local term, in the form of stress diffusion with diffusivity κ, is required to describe interfaces between different bands in inhomogeneous flows.

The original version of this model, proposed by Cates et al. [7], was purely temporal (no space variable z, $\kappa = 0$) and already proved capable of long-term unsteady responses: when $\dot\gamma$ is externally fixed in a certain range, the stress displays sustained (space-homogeneous) oscillations akin to those of the van der Pol oscillator[1]. With the spatially-resolved version presented here, we are able to study the *spatio-temporal* dynamics of such unstable behaviour.

Qualitative features of the model

The qualitative features of the model are best explained by rewriting the integro-differential equation (1) as an equivalent differential system[2]:

$$\dot\sigma = \dot\gamma - R(\sigma) - \lambda m + \kappa \nabla^2 \sigma \quad,\quad \dot m = -\frac{m-\sigma}{\tau_s} \quad (2)$$

The memory integral within eq. (1) now appears as an auxiliary variable $m(z,t)$, which follows its own dynamics and at each instant tries to relax towards the current value of the stress $\sigma(z,t)$.

The equilibrium flow curve $\sigma(\dot\gamma)$ for the model is obtained when $\dot\sigma = \dot m = 0$. Flowing states on this curve are spatially uniform, and on each point of the curve, the memory m has had time to relax to the equilibrium

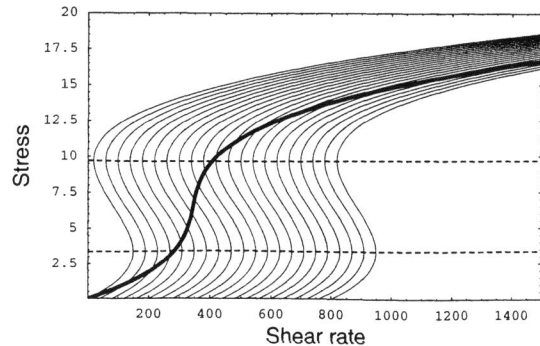

FIGURE 1. Long-term flow curve (thick line) and underlying short-term curves (obtained, from left to right, with fixed values of the memory $m=0$ to $m=20$, in steps of one). The stress range between the dotted lines is unstable. (Parameters: $H=1$, $\lambda = 40$, $\tau_s = 100$, $a = 100$, $b = 20$, $c = 1.02$, $\kappa = 0.01$.)

value of the stress ($m = \sigma$). However, when τ_s is large (slow memory relaxation), this equilibrium flow curve may only be observed on long enough timescales. On timescales much shorter than τ_s, the fluid will behave as if the memory m were static: one has therefore also to consider a set of "instantaneous" flow curves which correspond to the relaxation of the stress (i.e., $\dot\sigma = 0$) *at fixed m*. As the memory slowly evolves, the fluid will then accordingly "jump" from one instantaneous flow curve to the other.

It is then easy to understand qualitatively why the fluid has an unstable behaviour in a certain range of stress. In Figure 1, the long-term, equilibrium flow curve is drawn together with the set of instantaneous curves: the sole inspection of the equilibrium curve, which has no decreasing portion (for the choice of parameter values considered), could let one think that all states are stable; but one observes that there is a region where the long-term curve is in fact crossed by *decreasing parts* of short-term curves. Consequently, at these intersection points, the fluid has an instantaneous tendency to destabilize, thereby precluding the establishment of the equilibrium state [7].

Numerical study of the model

We now briefly present some of the results that have been found so far on the spatio-temporal dynamics of the model.

The model was studied numerically through a spectral Galerkin truncation [8], where $\sigma(z,t)$ and $m(z,t)$ are decomposed as a finite sum of spatial Fourier modes, of the form $\sigma_n(t)\cos(q_n z)$ with wave-vector $q_n = n\pi/H$ for the stress (H is the total height of our imaginary Couette device), plus a similar set of memory modes $m_n(t)$. (We usually take the first ten modes into account in our nu-

---

[1] Regimes of temporal chaos were also found when the relaxation term R was furthermore allowed to become slightly delayed. However, the physical meaning of this modification and its relevance remain uncertain.

[2] This transformation is possible with an exponential memory, but not in the general case.

merics.) The system of equations (2) then reduces to a set of coupled ordinary differential equations governing the temporal evolution of $\sigma_n(t)$ and $m_n(t)$.

As is the case for conventional shear bands, two different protocols can be followed in the numerical study of the instability (details will appear elsewhere [9]): either working at fixed shear rate $\dot{\gamma}$ (i.e., with a shear-controlled Couette cell), or working at fixed torque, or equivalently at fixed average stress $\langle\sigma\rangle$ (stress-controlled Couette). These two protocols lead to rather different results, as we shall now see.

Working at fixed $\dot{\gamma}$

We here work with a fixed, externally imposed value of $\dot{\gamma}$, and compute the values of the different modes $\sigma_n(t)$ and $m_n(t)$, from which we reconstruct $\sigma(z,t)$ and $m(z,t)$. For numerical solutions carried within the unstable window, our results show that, in the vast majority of cases, the spatio-temporal dynamics of the instability is in fact essentially purely temporal: regardless of their initial magnitude, all the modes rapidly decay and disappear, except for the spatially uniform mode $\sigma_0(t)$. This mode then oscillates alone, with a fixed period and a regular shape, as in the temporal version of ref. [7].

In a few, small regions of the parameter space explored, it has been possible to obtain a slightly richer dynamics, where some of the lower modes survive and undergo small-scale periodic oscillations, alongside a large uniform mode oscillation which still dominates.

Working at fixed $\langle\sigma\rangle$

We now work at imposed values of the spatial average of the stress $\langle\sigma\rangle$; in terms of Fourier modes, this corresponds to fixing the value of the uniform Fourier mode $\sigma_0(t)=$ const., which will thus not be able to oscillate at all. A very rich spatio-temporal behaviour then arises, as the higher spatial modes will now, by necessity, be involved in the instability.

Different unstable features may appear [9]; In Figure 2-*a*, we present a most striking one, called "flip-flop shear-bands": a shear-banded profile appears, with a low shear band and a high shear band separated by an interface; but these bands are unstable, and periodically flip, the higher band becoming the lower and vice-versa. The "flipping time" is very short as compared to the latency time between two flips. Figure 2-*b* shows the corresponding temporal evolution of the total stress $\sigma(z,t)$.

Much more irregular-looking time variations can also be obtained for different parameter choices, as Fig. 2-*c* shows for the stress. We emphasize however that these patterns remain time-periodic; so far no chaotic response has been found in the present model.

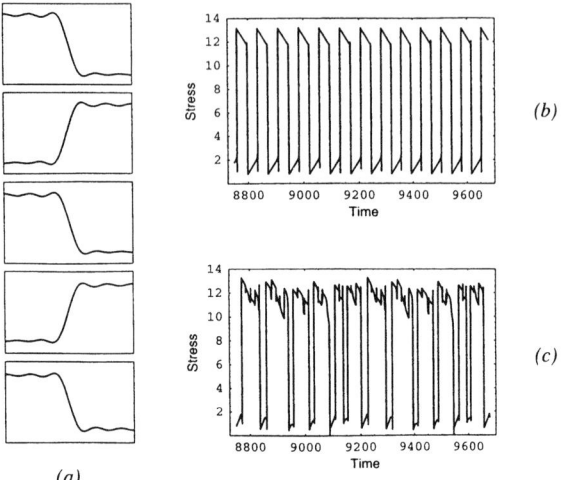

FIGURE 2. (*a*) "Flip-flop" shear-bands for $\langle\sigma\rangle = 7.0$: From top to bottom, successive snapshots of the stress $\sigma(z,t)$ vs. position z, at times $t = 8800, 8840, 8880, 8920, 8960$. (*b*) Time series of the stress at position $z = 0$ for the same value of $\langle\sigma\rangle$. (*c*) Time series of the stress at position $z = 0$ for $\langle\sigma\rangle = 9.0$. The time window corresponds to two periods of the signal. (Parameters same as Fig. 1.)

A MICELLAR MODEL WITH "FLUIDITY"

We would now like to introduce another fluid model, based on the same general ideas as previously, but more specifically oriented towards solutions of wormlike micelles, and polymers. The shear stress Σ in the fluid is locally the sum of a polymer or micellar part, σ, and a Newtonian part $\eta\dot{\gamma}$ corresponding to the solvent:

$$\Sigma = \sigma + \eta\dot{\gamma} \qquad (3)$$

The model's equations are as follows:

$$\dot{\sigma} = -\frac{\sigma}{\tau} + \frac{G(\tau)}{\tau}g(\dot{\gamma}\tau) + \kappa\nabla^2\sigma \qquad (4)$$

$$\dot{\tau} = -\frac{\tau - \tau_{\text{eq}}(\dot{\gamma})}{\tau_s} \qquad (5)$$

The stress evolution equation (4) has a classical form, and describes the relaxation (with a timescale τ) of the "polymer" stress towards a value which, at a given $\dot{\gamma}$, is controlled by $G(\tau)g(\dot{\gamma}\tau)$. $G(\tau)$ is the elastic modulus and generally depends on the value of τ (see below). The function g is hump-shaped, hence conferring on the fluid a tendency to shear-thinning, and shear-banding, with bands in the velocity gradient direction (i.e., the radial direction of the Couette cell).

Similarly to the model with memory, one of the variables involved in the stress equation is subject to a distinct dynamics: this is now the Maxwell time τ of the fluid, which, as stated by eq. (5), relaxes towards a shear-rate dependent equilibrium value $\tau_{eq}(\dot\gamma)$ with a characteristic time τ_s. We note that having a dynamical Maxwell time is indeed very similar to the "fluidity" model introduced by Derec et al. in the context of paste flow (see [10]).

Here again, τ is a "structural" variable in the sense that it reflect variations in the local structure of the fluid: for semi-dilute solutions of wormlike micelles, it will typically relate to local variations of the mean chain length of the micelles. In this interpretation, the chain length distribution, and consequently the mean chain length, may change only through the action of some micelle-micelle chemical reactions, which will have their own timescale τ_s – this time will thus control the relaxation of τ, as in eq. (5). On the other hand, the dependence of the equilibrium value τ_{eq} on the shear rate can be interpreted as a displacement of the chemical equilibrium by the mechanical shearing (for example, decreasing τ_{eq}, by helping chain scission, or, increasing τ_{eq}, by helping polymerisation). For more dilute solutions (non-entangled), one may rather interpret the variations of τ_{eq} as related to the strong shear-thickening transition usually observed upon varying $\dot\gamma$: the structural changes occurring at the transition (gelation) will then strongly alter τ_{eq}.

In accord with the model with memory of the previous section, we have focussed on slow relaxations of the structural variable: $\tau_s \gg 1$. Then one can again construct a "long-term", master flow curve where all variables are equilibrated ($\dot\sigma = \dot\tau = 0$), and an underlying set of "short-term" curves where τ is fixed (quasi-static) and the stress is equilibrated with respect to that value of τ. As seen in the previous model, an increasing portion of the "long-term" curve is here also made unstable when crossed by decreasing portions of underlying curves.

Two variants of the model can be studied, depending on whether the elastic modulus G is affected or not by changes in the structure of the fluid, that is, changes in τ. Variant 1 corresponds to $G(\tau) = G_0$ being constant, which would be suitable for a semi-dilute, entangled solution of micelles, where (reasonable) changes in the mean micellar length leave the modulus unchanged. In variant 2, G varies with τ, as would e.g. be the case in the shear-thickening transition of dilute micelles, where the significant structural changes affecting τ will affect G as well. Work on both variants of the model is currently in progress, but results so far are promising. Figure 3 shows an example of results in variant 2.

To conclude, we would like to mention extremely interesting results by S. Fielding and P. Olmsted [11]: working independently on essentially the same model as eqs. (4)-(5), they have been able, within variant 1, to obtain chaotic-like signals in a regime where *both* the underlying curves and the master flow curve have decreasing portions, *and* where the structural relaxation occurs on timescales comparable to the stress relaxation ($\tau_s \simeq 1$).

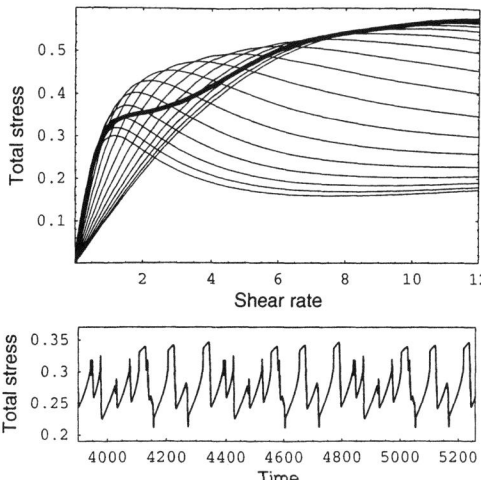

FIGURE 3. Upper figure: Long-term flow curve (thick line) and underlying short-term curves in variant 2. Lower figure: very irregular (though periodic) time series of the total stress $\Sigma(t)$ for $\langle\dot\gamma\rangle = 4.0$.

Many questions are still open in the study of the type of models which have been described in this article. Are the existence of distinct "structural" relaxations really the origin of the instabilities observed in experiments? Is the regime just described the only one displaying chaotic-like behaviour, and relatedly, what is required for regular or irregular-looking periodic motion to destabilize into chaos?

ACKNOWLEDGMENTS

The authors would like to thank S. Fielding and P. Olmsted for sharing their work with them prior to publication. AA is funded by an EPSRC Post-doctoral Fellowship in Theoretical Physics (GR/R95098).

REFERENCES

1. Wunenburger, A.-S., Colin, A., Leng, J., Arnéodo, A. and Roux, D., *Phys. Rev. Lett.* **86**, 1374–1377 (2001).
2. Courbin, L., Panizza, P., and Salmon, J.-B., to be published.
3. Hilliou, L., and Vlassopoulos, D., *Ind. Eng. Chem. Res.* **41**, 6246–6255 (2002).
4. Laun, H. M., *J. Non-Newt. Fluid Mech.* **54**, 87–108 (1994).
5. Bandyopadhyay, R., Basappa, G., and Sood, A. K., *Phys. Rev. Lett.* **84**, 2022–2025 (2000). Bandyopadhyay, R., and Sood, A. K., *Europhys. Lett.* **56** 447–453 (2001).
6. Salmon, J.-B., Colin, A., and Roux, D., *Phys. Rev. E* **66**, 031505-1–031505-13 (2002).
7. Cates, M. E., Head, D. A., and Ajdari, A., *Phys. Rev. E* **66**, 025202-1–025202-4 (2002).
8. Boyd, J. P., *Chebyshev and Fourier Spectral Methods*, Dover Publications, New York, 2000.
9. Aradian, A., and Cates, M. E., in preparation.
10. Derec, C., Ajdari, A., and Lequeux, F., *Eur. Phys. J. E* **4**, 355–361 (2001).
11. Fielding, S. M., and Olmsted, P. D., cond-mat/0310658.

Molecular Dynamics Simulation of Dendrimers: Structural Formation and Internal Charge Distribution

Takamichi Terao and Tsuneyoshi Nakayama

Department of Applied Physics, Hokkaido University, Sapporo 060-8628, Japan

Abstract. We study the structural formation of dendrimers by stochastic molecular dynamics simulations. The density profile under different pH condition is clarified numerically. We find that the structural change of dendrimers strongly depends on the solvent quality. We also investigate the effective interaction between nanosized dendrimer molecules. It is demonstrated that the effective interaction between like-charged dendrimers becomes attractive when the electrostatic coupling is strong.

INTRODUCTION

Many kinds of dendrimers, with different initiator cores and branches, have been synthesized in this decade, for which intensive studies have been performed on their practical use as a new functional material in biochemical and medical applications[1, 2]. Dendrimers have numerous potential applications such as utilizing their inner space (*i.e.,* nanosized molecular encapsulation) in drug-delivery systems, and as contrast agents for visualizing blood vessels and bloodstreams in magnetic resonance imaging(MRI).

It is quite significant to understand the conformation of dendrimers in solution for their application as nanocapsules. Dendrimers are formed by a step-by-step iterative reaction starting from a core. The resulting tree-like molecules have a well-defined number of end groups and narrow molecular weight distributions. A key problem is their spatial structures and the location of the end groups. Small-angle neutron scattering (SANS) and small-angle X-ray scattering (SAXS) experiments have been performed in order to elucidate the spatial structure of dissolved dendrimers[3, 4, 5]. SANS is the optimal tool for probing the location of end groups, because their contrast can be greatly enhanced using the deuterium labeling technique. These results indicate that the size variation of dendrimers is still a curious problem.

We present the results of stochastic molecular dynamics simulations in order to reveal the nano-sized structural formation of charged dendrimers as well as neutral dendrimers in solution. We calculate the density distribution function of dendrimers under different pH conditions. We also study the effective interaction between charged dendrimers as a function of distance between two molecules. The effective interaction between charged dendrimers becomes attractive when the electrostatic coupling is strong.

MODELS

We study the conformation of dendrimers in solution using a stochastic molecular dynamics method. The solvent is treated as a continuum, which acts as a heat bath for the molecules and produces a viscous drag when each segment moves. The equation of motion for i-th segment with mass m is thus given by

$$m\frac{d^2}{dt^2}\mathbf{r}_i(t) = -\nabla U_i - m\Gamma \frac{d}{dt}\mathbf{r}_i(t) + \mathbf{W}_i(t) , \quad (1)$$

where Γ, $\mathbf{r}_i(t)$, and $\mathbf{W}_i(t)$ are the friction coefficient that couples the monomers to the heat bath, the positional vector of i-th segment, and the random force of the heat bath acting on each segment, respectively. $\mathbf{W}_i(t)$ is a Gaussian white noise given by

$$\langle \mathbf{W}_i(t) \cdot \mathbf{W}_j(t') \rangle = 6mk_B T \Gamma \delta_{ij} \delta(t-t') . \quad (2)$$

where k_B is the Boltzmann constant and T is the temperature. The potential U_i consists of three terms such as

$$U_i = \sum_j U_{LJ}(r_{ij}) + U_{FENE} + U_c , \quad (3)$$

where r_{ij} is the distance between monomers i and j. Here $U_{LJ}(r)$ is a Lennard-Jones potential between any two monomers written by

$$U_{LJ}(r_{ij}) = \begin{cases} \phi(r_{ij}) - \phi(r_c) & \text{for } r_{ij} \leq r_c \\ 0 & \text{for } r_{ij} > r_c \end{cases} . \quad (4)$$

Here $\phi(r_{ij})$ is defined to be

$$\phi(r_{ij}) \equiv 4\varepsilon \left[(\sigma/r_{ij})^{12} - (\sigma/r_{ij})^6 \right] , \quad (5)$$

where ε, σ and r_c are the unit of energy and the diameter of particles and the cutoff radius, respectively. For an infinitely good solvent, the cutoff distance r_c is taken as $r_c = 2^{1/6}\sigma$, such that the potential is purely repulsive. To introduce the effect of solvent quality, we also extended the range of the interaction to be $r_c = 2.5\sigma$. U_{FENE} denotes the bonding interaction between neighboring segments given by

$$U_{\text{FENE}}(r_{ij}) = \begin{cases} -0.5KR_0^2 \ln\left[1 - (r_{ij}/R_0)^2\right] & \text{for } r_{ij} \leq R_0 \\ 0 & \text{for } r_{ij} > R_0 \end{cases} \quad (6)$$

where K is the bonding constant and R_0 is the maximum extension of the bond. In the following simulation, we use the values $R_0 = 1.5\sigma$ and $K = 30.0\varepsilon/\sigma^2$, where these parameters prevent bond crossings.

The implementation of the long-range Coulomb interaction requires special care to calculate the Coulomb sum. We impose periodic boundary conditions in x-, y-, and z-directions, respectively, and the indices n_x, n_y, and n_z run over the periodic images of the systems. The long-ranged nature of the Coulomb interaction was numerically treated via the efficient method proposed by Lekner[6, 7, 8, 9].

Our computer simulation was performed within a setup which is schematically shown in Fig. 1. We consider single dendrimer molecule in a cubic box of length L with periodic boundary condition in all three dimension. Solid circles, open circles, and gray circles in Fig. 1 denote positively-charged monomers, neutral monomers, and negatively-charged counterions, respectively. In the following, we consider two different models to treat the effect of pH conditions on polyelectrolyte dendrimers such as poly(amidoamine) (PAMAM) dendrimers. Under a high pH condition, no amines are assumed to be protonated. Under a low pH condition all the amines are protonated (shown as black circles in Fig. 1). The initial configuration of dendrimer molecules is built as follows: The segment of each chain is attached by self-avoiding random walks with a distance of $\sim \sigma$ in a cubic box of system size L. A dendrimer with G generation is built by adding $M_b - 1$ chains (of length M_n) to each of the free ends of a dendrimer with $G-1$ generation, where M_b and M_n are the branching factors and the length per monomer between each branching point, respectively. The same procedure is continued in order to build dendrimers of desired generation numbers.

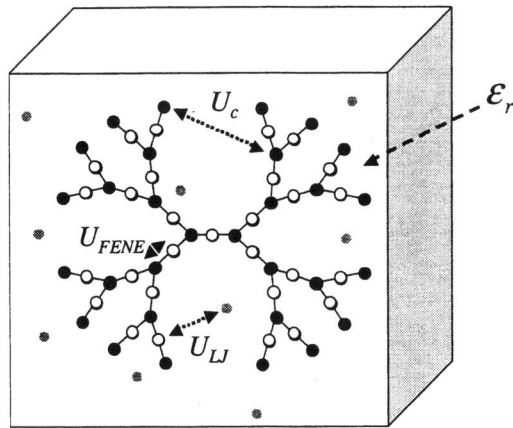

FIGURE 1. Schematic picture of the system. Solid circles, open circles, and gray circles denote charged monomers, neutral monomers, and counterions, respectively.

NUMERICAL RESULTS

The monomer density profiles

At first, we calculate the density profiles of dendrimers in an aqueous solution. The temperature T and the friction coefficient Γ are taken to be $k_B T/\varepsilon = 1.2$ and $\Gamma = 0.5\tau^{-1}$, respectively, and a time step $\delta t = 0.001\tau$ or $\delta t = 0.002\tau$ is selected, where τ is the unit of time defined as $\tau \equiv \sigma(m/\varepsilon)^{1/2}$. In the following calculations, the values of M_b and M_n are set to be $M_b = 3$ and $M_n = 4$, respectively. In actual calculations, we carry out the Monte Carlo(MC) simulation with short MC steps($\sim 10^4$ MCS) for structural relaxation, prior to molecular dynamics(MD) simulations with large MD steps ($\geq 10^6$ steps). In the MD simulation, physical quantities are calculated after a sufficient number of MD steps are discarded to reach an equilibrium. These MC and MD codes are parallelized efficiently using the message-passing interface (MPI). In Fig. 2, open squares and open circles denote the distribution function $n(r)$ of neutral dendrimer and charged dendrimers, respectively. Fig. 2(a) and Fig. 2(b) show the calculated results with cutoff radius in Eq. (4), $r_c = 2^{1/6}\sigma$ and $r_c = 2.5\sigma$, to evaluate the effect of solvent quality. In Fig. 2(a), the difference on the density profiles between these two results is small, indicating that the structure of dendrimer molecule in these systems is less sensitive to the effect of solvent quality. In Fig. 2(b), on the other hand, the change of the density distribution function $n(r)$ becomes more drastic. These results indicate that the gyration radius R_g, which can be calculated from $n(r)$, also changes in Fig. 2(b).

For smaller generation number of dendrimers (or shorter length between nodes), free volume inside dendrimer molecule becomes small, and excluded volume

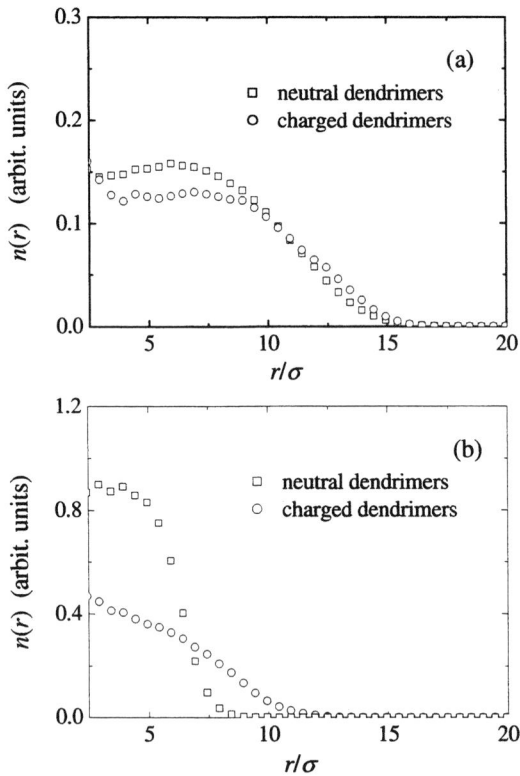

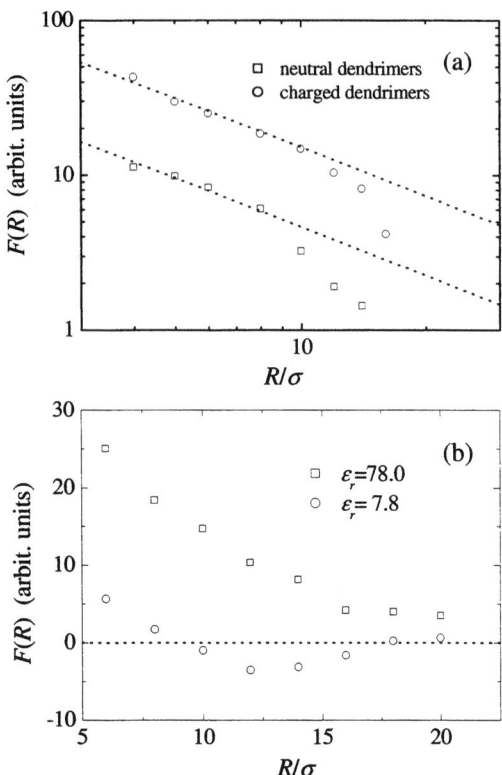

FIGURE 2. The profiles of monomer density distribution $n(r)$. (a) in the case of $r_c = 2^{1/6}\sigma$ (Eq. 4). (b) $r_c = 2.5\sigma$.

FIGURE 3. Effective force between dendrimers $F(R)$. (a) under different pH conditions. (b) under different strength of electrostatic couplings.

effect becomes dominant. In this case, it is assumed that the gyration radius of dendrimers slightly depends on pH condition in solution. Conversely, for larger generation number of dendrimers (or larger length between nodes), electrostatic interaction becomes dominant to determine the conformation of molecules, which cause that the structure of dendrimers strongly depends on pH condition in solution. Our results have pointed out that not only pH condition but also the solvent quality will strongly influence on the gyration radius of dendrimers in an aqueous solution.

Effective force between dendrimers

Secondly, we clarify the effective interaction between dendrimer molecules in an aqueous solution. Recently, neutron scattering experiments have been performed on dendrimer solutions. To understand the scattering properties in concentrated systems, it is necessary to know about the effective interaction between dendrimers. In the following computer simulations, we consider three different systems such as (i) electrically neutral dendrimers, (ii) charged dendrimers in the medium with relative dielectric constant $\varepsilon_r = 78.0$, and (iii) charged dendrimers with different dielectric constant $\varepsilon_r^l = \varepsilon_r/10$.

We study a pair of dendrimer molecules which are confined in a cubic box of length L. A pair of dendrimers are placed symmetrically along the body diagonal of the cube such that the center of the cube coincides with the center of mass of the two molecules. Their positions are denoted by $\mathbf{R}_1$ and $\mathbf{R}_2$ ($R \equiv |\mathbf{R}_1 - \mathbf{R}_2|$). Furthermore the box contains N_c counterions carrying an opposite charge $+e$ when two dendrimer molecules are charged. Using molecular dynamics simulations, the effective force $\mathbf{F}_i$ acting on each dendrimer molecule i ($i = 1, 2$) is calculated. Since by symmetry, $\mathbf{F}_2 = -\mathbf{F}_1$, we projected $\mathbf{F}_i$ onto the body diagonal defining

$$F(R) \equiv \frac{\mathbf{F}_1 - \mathbf{F}_2}{2} \cdot \frac{(\mathbf{R}_1 - \mathbf{R}_2)}{|\mathbf{R}_1 - \mathbf{R}_2|}. \tag{7}$$

In Eq. (7), the effective interaction is repulsive when

$F(R) > 0$, and vice versa. Results for the effective force $F(R)$ is shown in Fig. 3(a). Open squares and open circles denote the results for neutral dendrimers and charged dendrimers ($\varepsilon_r = 78.0$), respectively. In Fig. 3(a), the value of $2R_g$ is shown by vertical arrow. Both results show that the effective force $F(R)$ decays as a power law such as

$$F(R) \sim \frac{1}{R^\gamma} \quad \text{for} \quad R \ll 2R_g, \quad (8)$$

with the exponent $\gamma = 1.0 \pm 0.1$. This implies that effective interaction between dendrimers $V(R)$ obeys $V(R) \propto -\log(R/\sigma)$ for $R \ll 2R_g$, and it decays more rapidly for $R > 2R_g$.

We also show the calculated results on the effective force with strong electrostatic couplings. In Fig. 3(b), the effect of strong electrostatic couplings is treated by smaller dielectric constant of the medium. Open squares and open circles denote the results for $\varepsilon_r = 78.0$ and $\varepsilon_r' = \varepsilon/10$, respectively. We find that the effective interaction between like-charged dendrimers can be attractive when the electrostatic couplings is strong.

CONCLUSIONS

In conclusion, we have studied the structural formation of dendrimers by stochastic molecular dynamics simulations. Recently, neutron scattering experiments have been performed on dendrimer solutions, and it is an important problem to clarify the condition which a dendrimer molecule obeys a structural change. We have calculated the density profile of dendrimers under different pH conditions. We have also evaluated the effect of solvent quality by controlling the cutoff radius in Eq. (4). For an infinitely good solvent ($r_c = 2^{1/6}\sigma$), the density profile is less sensitive to pH condition in an aqueous solution. For a different solvent condition ($r_c = 2.5\sigma$), on the other hand, the change of the density distribution function $n(r)$ becomes more drastic. The latter result indicates that the gyration radius R_g of dendrimers also changes as a function of pH condition.

We have also clarified the effective interaction between dendrimers in an aqueous solution. It is necessary to understand the effective interaction between dendrimers to understand the scattering properties in concentrated systems. We have considered three different systems such as electrically neutral dendrimers, charged dendrimers, and charged dendrimers in the medium with smaller dielectric constant. We have studied a pair of dendrimer molecules which are confined in a cubic box. Using molecular dynamics simulations, the effective force acting on each dendrimer molecule has been calculated. For neutral dendrimers and charged dendrimers with relative dielectric constant $\varepsilon_r = 78.0$, the effective force $F(R)$ between molecules becomes repulsive and decays as a power law such as $F(R) \sim 1/R^\gamma$ ($\gamma = 1.0 \pm 0.1$) for $R \ll 2R_g$. This implies that effective interaction between dendrimers $V(R)$ obeys $V(R) \propto -\log(R/\sigma)$ for $R \ll 2R_g$, and dendrimers can be regarded as a colloidal system with very *soft* interaction in a solution. We have also clarified the effective force between charged dendrimers with strong electrostatic couplings by setting smaller dielectric constant of the medium. We have demonstrated that the effective interaction between like-charged dendrimers can be attractive when the electrostatic couplings is strong.

Our result is reminiscent of the like-charged attraction between DNA in an aqueous solution. It has been pointed out that strongly-charged polyelectrolytes show various characteristic behaviors when multivalent salts are added. In such a system, a charged macroion binds many counterions on its surface, and the fluctuation-induced attraction is induced by strong Coulomb coupling. This resemblance gives a deep insight into the physical description of the Coulomb screening effect on strongly-charged soft matters, and motivates the development of novel, potential applications such as physical encapsulation of guest molecules, specific drug targeting, and pH-sensitive controlled release.

ACKNOWLEDGMENTS

This work was supported in part by a Grant-in-Aid from the Japan Ministry of Education, Science, and Culture for Scientific Research. The authors thank the Supercomputer Center, Institute of Solid State Physics, University of Tokyo for the use of the facilities.

REFERENCES

1. Fischer, M., and Vögtle, F., *Angew. Chem. Int. Ed.*, **38**, 884–905 (1999).
2. Narayanan, V. V., and Newkome, G. R., *Top. Curr. Chem.*, **197**, 19–77 (1998).
3. Rosenfeldt, S., Dingenouts, N., Ballauff, M., Werner, N., Vögtle, F., and Lindner, P., *Macromolecules*, **35**, 8098–8105 (2002).
4. Rietveld, I. B., Bouwman, W. G., Baars, M. W. P. L., and Heenan, R. K., *Macromolecules*, **34**, 8380–8383 (2001).
5. Topp, A., Bauer, B. J., Klimash, J. W., Spindler, R., Tomalia, D. A., and Amis, E. J., *Macromolecules*, **32**, 7226–7231 (1999).
6. Lekner, J., *Physica A*, **176**, 485–498 (1991).
7. Terao, T., and Nakayama, T., *Phys. Rev. E*, **63**, 041401(1)–041401(6) (2001).
8. Terao, T., and Nakayama, T., *Phys. Rev. E*, **65**, 021405(1)–021405(5) (2002).
9. Terao, T., *Phys. Rev. E*, **66**, 046707(1)–046707(5) (2002).

Droplet density dependences of the static and dynamic structures in a ternary microemulsion system

M. Nagao*, H. Seto†, Y. Kawabata** and M. Shibayama*

Institute for Solid State Physics, The University of Tokyo, Tokai 319-1106, Japan
†Department of Physics, Kyoto University, Kyoto 606-8502, Japan
**Faculty of Science, Tokyo Metropolitan University, Hachioji 192-0397, Japan

Abstract. A neutron spin echo (NSE) experiment was performed in order to clarify the droplet density dependence of the dynamic structures in a ternary microemulsion system consisting of AOT, water and decane. At the droplet concentration, ϕ, of 0.05, the dynamic mode due to the peanuts-like deformation of water droplets and the translational diffusion of them were analyzed, and the estimated decay time for the droplet deformation was about 10 ns. At $\phi = 0.3$, another dynamical mode due to the packing properties of water droplet is expected to be observed. Assuming that the deformation motion of the water droplets was independent of ϕ, the dynamical mode due to the structure factor was estimated. The order of the decay time was one order of magnitude larger than that of the droplet deformation, and was smaller than that of the translational diffusion.

INTRODUCTION

A ternary microemulsion, consisting of AOT (dioctyl sulfossucinate sodium salt), water, and oil has been intensively investigated by scattering techniques. Thus, this system is suitable to extend the structural parameters depending on the external field, such as the concentration of ingredients, temperature, pressure and so on.

A water-in-oil droplet structure is known to form under a wide concentration region with a fixed water to AOT ratio at the vicinity of room temperature [1]. A droplet density, ϕ, dependences of the static and the dynamic structures were investigated by Sheu et al. [2] by means of small-angle neutron scattering (SANS), quasi-elastic light scattering (QELS), and neutron spin echo (NSE). They obtained that the photon correlation function, $C(t)$, obtained by QELS was represented by a stretched exponential form, with the stretched exponent, β, depending on ϕ. It was unity for low ϕ, while decreased with increasing ϕ below $\phi \sim 0.6$. It was concluded that the decrease of β was due to the dynamic correlations between the spherical water droplets. This means that the dynamical mode due to the inter-droplet structure affects the experimental data, and it is difficult to understand such information independently from the experimental data, so far.

Quite recently, we succeeded in separation of the form factor, $F(q)$ (indicating the intra-structure), and the structure factor, $S(q)$ (the inter-structure), from the SANS data by using the relative form factor method [3]. Using the procedure, the ϕ dependence of $F(q)$ without the influence of $S(q)$ was deduced. The radius of droplets kept almost constant, while the polydispersity decreased with increasing ϕ below $\phi \sim 0.6$. In this contribution, we tried to separate the dynamic structure between the intra- and the inter-droplet fluctuations from the NSE data.

EXPERIMENTAL

We performed a NSE experiment at the ISSP-NSE of JRR-3M, JAERI, Tokai. The measured samples were $\phi = 0.05$ and 0.3 for the film contrast. In the measured spatial and temporal domains ($0.02 \leq q \leq 0.10 \text{Å}^{-1}$ and $0.14 \leq t \leq 14.7$ ns), it is known that the deformation motion of water droplets as well as the translational diffusion of the droplets are expected to be obtained by means of NSE from the dilute droplet microemulsion.

The translational diffusion coefficient, D_{tr}, was measured by means of dynamic light scattering (DLS). A DLS/SLS-5000, ALV, Langen, Germany, with a 22mW He-Ne laser (Uniphase, USA) was used for the measurement at the scattering angle of 90°. The observed $C(t)$ showed the decay curve with $\beta = 1$, and D_{tr} was estimated to be $1.97 \times 10^{-11} \text{m}^2/\text{s}$. Both the NSE and DLS measurements were done at the temperature of 25°C.

RESULTS AND DISCUSSION

The obtained intermediate scattering function by NSE, $I(q,t)/I(q,0)$, from $\phi = 0.05$ was fitted by using a model describing the deformation motion of the spherical shell under the concept of the Helfrich's bending Hamiltonian [4]. Farago and Gradzielski [5] proposed the following equation,

$$I(q,t) = (4\pi)^2 [F_s(q) + F_{sc}(q) + F_d(q,t)] e^{-D_{tr}q^2 t}. \quad (1)$$

Where, $F_s(q)$ describes the scattering from a sphere in the absence of fluctuations, $F_{sc}(q)$ gives a time independent correction for the fluctuations, and $F_d(q,t)$ contains the correlators of the mode amplitudes. The exponential factor takes into account the translational diffusion. In the case when the contribution of F_s to the scattering intensity is small, the term of F_d mainly contributes to $I(q,t)$. Therefore, we fitted $I(q,t)/I(q,0)$ for $\phi = 0.05$ at $q = 0.05 \text{Å}^{-1}$, where the scattering intensity shows a fringe due to the shape of the form factor of the sphere, and got a decay time of about 10 ns.

On the other hand, the dynamical mode due to the inter-droplet structure will affect $I(q,t)$ for $\phi = 0.3$. In order to deduce such a contribution, we propose the following analysis procedure. We assume that the double exponential decay function is observed as follows,

$$\frac{I(q,t)}{I(q,0)} = a\exp(-t/\tau_d) + (1-a)\exp(-t/\tau_p), \quad (2)$$

where, τ_d and τ_p indicate the characteristic decay times of each dynamical mode. When it is assumed that the intra- and inter-droplet structure fluctuations are independent each other, the decay mode due to the form factor and the structure factor are assigned as follows,

$$F(q,t)/F(q,0) = \exp(-t/\tau_d), \quad (3)$$
$$S(q,t)/S(q,0) = \exp(-t/\tau_p), \quad (4)$$

and

$$\frac{I(q,t)}{I(q,0)} = a\frac{F(q,t)}{F(q,0)} + (1-a)\frac{S(q,t)}{S(q,0)}, \quad (5)$$

where, $F(q,t)/F(q,0)$ and $S(q,t)/S(q,0)$ indicate the intermediate scattering function due to the form factor and the structure factor, respectively.

We assume that the deformation motion of the shape fluctuations is independent of ϕ, and the translational diffusion is changed to the collective motion of the packing properties of the water droplets with increasing ϕ. In this case, $F(q,t)/F(q,0)$ at $\phi = 0.3$ is written as,

$$\frac{F(q,t)}{F(q,0)} = \left[\frac{I(q,t)}{I(q,0)}\right]_{\phi=0.05} \exp(D_{tr}q^2 t), \quad (6)$$

where, $F(q,t)/F(q,0)$ expressed only the deformation motion of the spherical shell and the translational diffusion was neglected by the exponential term.

When the value of a is obtained, $S(q,t)/S(q,0)$ is evaluated from the NSE data by using eq. (5). Figure 1

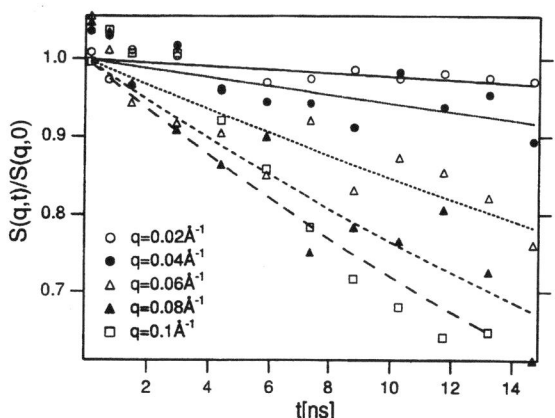

FIGURE 1. The q dependence of the evaluated decay function due to the structure factor, namely, $S(q,t)/S(q,0)$. Lines are fitting results according to eq. (4). The order of τ_p is 10 to 100 ns depending on q.

shows the evaluated $S(q,t)/S(q,0)$ with assuming $a = 0.5$, and eq. (4) is applied. The order of the characteristic decay time, $\tau_p \sim 100$ ns, is one order of magnitude larger than that of τ_d. This indicates that it is possible to assume that the dynamical mode due to the intra- and inter-droplet structure are independent.

In this result, the characteristic decay time due to the inter-droplet structure fluctuations is slower than that from the intra-droplet structure fluctuations and is faster than that from the translational diffusion at $\phi = 0.05$ especially at low q region.

ACKNOWLEDGMENTS

One of the authors (M.N.) was supported by the Grant-in-Aid for Encouragement of Young Scientists (B) (No. 13740234 and 15740260) from the Japanese Ministry of Education, Culture, Sports, Science and Technology. The NSE experiment was done under approval of the Neutron Scattering Program Advisory Committee (Proposal No. 02.207 and 03.210.f).

REFERENCES

1. Kotlarchyk M., Chen S.H., Huang J.S., and Kim M.W., *Phys. Rev. A* **29**, 2054-2069 (1983).
2. Sheu E.Y., Chen S.H., Huang J.S., and Sung J.C., *Phys. Rev. A* **39**, 5867-5876 (1989).
3. Nagao M., Seto H., Shibayama M., and Yamada N.L., *J. Appl. Cryst.* **36**, 602-606 (2003).
4. Milner S.T., and Safran S.A., *Phys. Rev. A* **36**, 4371-4379 (1987).
5. Farago B., and Gradzielski M., *J. Chem. Phys.* **114**, 10105-10122 (2001).

Anomalous Change in Lamellar Spacing by Shear Flow in Nonionic Surfactant/Water System

T. Kato[1], K. Minewaki[1], K. Miyazaki[1], Y. Kawabata[1], and M. Imai[2]

[1] *Department of Chemistry, Tokyo Metropolitan University, Minamiohsawa, Hachioji, Tokyo 192-0397, Japan*
[2] *Department of Physics, Ochanomizu University, Bunkyo, Tokyo 112-0012, Japan*

Abstract. Small-angle neutron scattering (SANS) is measured on a lamellar phase in $C_{16}E_7$ (hepta(oxyethylene glycol)-n-hexadecylether) /water system (40 - 55 wt% of $C_{16}E_7$) at 70 °C under shear flow. As the shear rate increases from 0.3 to 1 s^{-1}, the repeat distance (d) is reduced suddenly and discontinuously. With the further increase in the shear rate, d increases slightly after taking a minimum (d^*). These results are obtained for all the principal orientations of lamellae. While d at rest increases from 6.5 nm to 8.5 nm with decreasing concentration from 55 to 40 wt%, d^* is almost independent of concentration and nearly equal to the thickness of bilayers (~ 5 nm) obtained from the line shape analysis of small angle X-ray scattering at rest. These results suggest segregation into surfactant-rich and water-rich regions.

INTRODUCTION

In recent 10 years, much attention has been paid to the effects of shear flow on the structure of surfactant self-assembly with the development of the apparatus which enable us to observe the structures under shear in real or reciprocal space. Among them, the lamellar phase has become of interest and has been studied by using microscopy, NMR, and small angle scattering of light, X-ray, and neutron. [1, 2]. In the present study, we have measured small-angle neutron scattering (SANS) on the lamellar phase in nonionic surfactant $C_{16}H_{33}(OC_2H_4)_7OH$ ($C_{16}E_7$) /D_2O system at 70°C paying attention to the behaviors at relatively low shear rates, $10^{-3} \sim 10$ s^{-1}. It has been found that the repeat distance is decreased exceedingly and discontinuously by shear flow, suggesting segregation into concentrated lamellar and water-rich regions. Preliminary results at 55 wt% have been reported already [3, 4].

EXPERIMENTAL

$C_{16}E_7$ was purchased from Nikko Chemicals, Inc. in crystalline form (> 98 %) and used without further purification. Deuterium oxide purchased from ISOTEC, Inc. (99.9%) was used after being degassed by bubbling of nitrogen to avoid oxidation of the ethylene oxide group of surfactants.

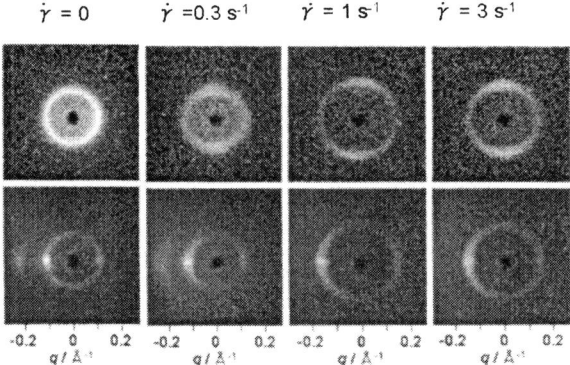

FIGURE 1. Examples of observed 2-D diffraction patterns for radial (q_ω vs. q_v, top) and tangential (q_ω vs. $q_{\nabla v}$, bottom) configurations for different shear rates at 48 wt% and 70°C (q_ω, q_v, and $q_{\nabla v}$ are absolute values of the scattering vectors along the directions vorticity, flow, and velocity gradient, respectively).

Measurements of SANS were carried out at the instrument SANS-U of Institute for Solid State Physics of University of Tokyo in JRR-3M at Tokai with a Couette shear cell [5].

RESULTS AND DISCUSSION

Figure 1 shows examples of the observed 2-D SANS patterns for different shear rates. These data were reduced to 1-D data in the vorticity (q_ω), flow (q_v), and velocity gradient ($q_{\nabla v}$) directions by integrating the scattering intensity over a segment of width $\Delta\phi = \pm 10°$ where ϕ is the azimuthal angle.

In the range $0 - 0.3$ s^{-1}, the peak height of first reflection decreases with increasing shear rate while the repeat distance (d) does not change very much. At 1 s^{-1}, d is reduced suddenly. During time resolved SANS measurements at 1 s^{-1}, we observed coexistence of two peaks, suggesting discontinuous reduction of d. With the further increase in the shear rate, d increases slightly after taking a minimum (referred to as d^* hereafter). Such a specific shear rate dependence of d has been obtained for three principal orientations of lamellae, i.e., perpendicular, parallel, and transverse orientations, with the layer normal along the vorticity, velocity gradient, and flow directions, respectively.

In Fig. 2, the repeat distances at rest (d_0) and d^* are plotted against the surfactant concentration, together with d_0 and the total thickness of bilayers (δ_0, the sum of the thicknesses of the hydrophobic and hydrophilic layers) at rest obtained from analyses of peak position and line shape of small angle X-ray scattering (SAXS) reported before [6]. This figure demonstrates that d^* depends on concentration only slightly in spite of the variation of d_0 values. From our preliminary measurements on a 40 wt% sample, it has been found that d^* is again closed to δ_0. In this case, the change of d corresponds to about 40% of d_0. Such a large amount of reduction has not been reported so far.

Figure 2 demonstrates also that d^* is nearly equal to δ_0. Although the thickness of bilayers may decrease slightly by shear flow [7], these results suggest that the water layer between the surfactant bilayers is excluded and that the system segregates into concentrated lamellar and water-rich regions.

In the range $0 - 0.3$ s^{-1}, as mentioned before, the peak heights decrease with increasing shear rate for all the principal directions before jumping of d. It has been found that the reduction in the peak height originates from the increase in the line width. This suggests that the lamellar domains are substantially shrunk by shear flow, which leads to the reduction of the "patch" or "collision" length which dominates lamellar spacing.

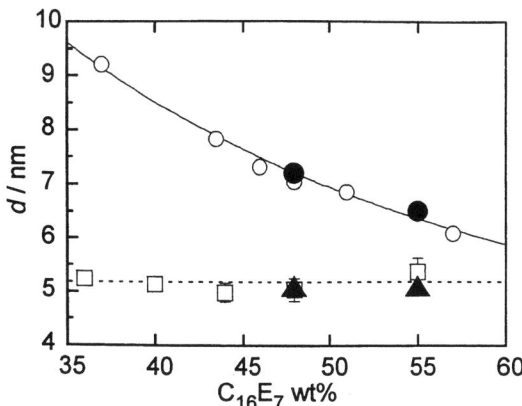

FIGURE 1. Concentration dependence of the repeat distance (d) at 70°C at rest (filled circles) and d^* (filled triangles). Open circles and squares indicate repeat distance and total thickness of bilayers at rest obtained from SAXS [6].

ACKNOWLEDGMENTS

We thank Prof. Yoshiaki Takahashi of Kyusyu University and Dr. Michihiro Nagao of Institute for Solid State Physics of University of Tokyo for assistance with shear-SANS measurements.

REFERENCES

1. Diat, O., Roux, D., and Nallet, F., *J. Phys II France*, **3**, 1427-1452 (1993).

2. See the references in Marlow, S. W., and Olmsted, P. D., *Eur. Phys. J. E*, **8**, 485-497 (2002).

3. Minewaki, K., Kato, T., Yoshida, H., and Imai, M., *J. Thermal Analysis Calorimetry*, **57**, 753-758 (1999).

4. Minewaki, K., Kato, and Imai, M., *Studies in Surface Science and Catalysis*, **132**, 185-188 (2001).

5. Takahashi, Y., Noda, M., Naruse, M., Kanaya, T., Watanabe, H., Kato, T., Imai, M., and Matsushita, Y., *J. Soc. Rheol. Jpn.*, **28**, 187-191 (2000).

6. Minewaki, K., Kato, T., Yoshida, H., and Imai, M., and Ito, K., *Langmuir*, **17**, 1864-1871 (2001).

7. Imai, M., Nakaya, K., and Kato, T., *Eur. Phys. J. E*, **5**, 391-402 (2001).

Texture Observed in a Simple Shear Flow and after Cessation of the Flow for Liquid Crystalline Polymers

N.Mori*, N.Moriguchi*, T.Yamamoto*, and K.Yasuda*

*Department of Mechanophysics Engineering, Osaka University
2-1 Yamadaoka, Suita, Osaka 565-0871, Japan*

Abstract. The texture development of liquid crystalline solution of hydroxypropylcellulose (HPC-L) after cessation of shear flow was studied. The wide stripe texture (WST) was observed after elimination of the band texture. The WST was dependent on shear rate, sample thickness and temperature.

INTRODUCTION

One of the famous textures observed in liquid crystalline polymers is the banded texture perpendicular to the shear direction appearing after the cessation of shearing. The behavior of the band formation has been investigated by many researchers [1-2].

In shear flows liquid crystalline polymers exhibit complex behaviors of molecular orientation such as tumbling, which may induce defects. In the transient shear flow of mono-domain nematic LCPs, the texture develops as a function of shearing strain at low shear rates and a striated texture parallel to the flow direction is formed [3-4]. The texture evolution at low shear rates was explained by considering the balance between the elastic forces and the viscous forces in tumbling nematic LCPs [4]. In the present paper, we will present optical observation of the texture evolution for liquid crystalline polymer solution of HPC during a shear flow and after cessation of the shear flow. Especially, we will focus on the texture seen in a long-time relaxation process after cessation of the shearing.

EXPERIMENTAL

Shear flows were imposed to a sample using a Cambridge Shear System (CSS-450) having parallel plate geometry, as shown in Fig 1. Observations of texture developments during shear and after cessation of the shear were carried out with a polarizing microscope (OLYMPUS BX51) under crossed polarizers with the polarizer in the flow direction and a quarter-wave plate. The CSS-450 system allows us to control a temperature of the sample in the measurement; the present experiments were carried out at several temperatures to examine the temperature dependence. Before imposing specified shear rate to the sample, we applied preshearing for 2 minutes at $\dot{\gamma} = 0.1$ s^{-1} and then stopped it for 2 minutes in every experiment to make the sample relax.

A liquid crystalline polymer solution used was 50 wt% aqueous solution of hydroxypropylcellulose (HPC-L, supplied by Nippon Soda)

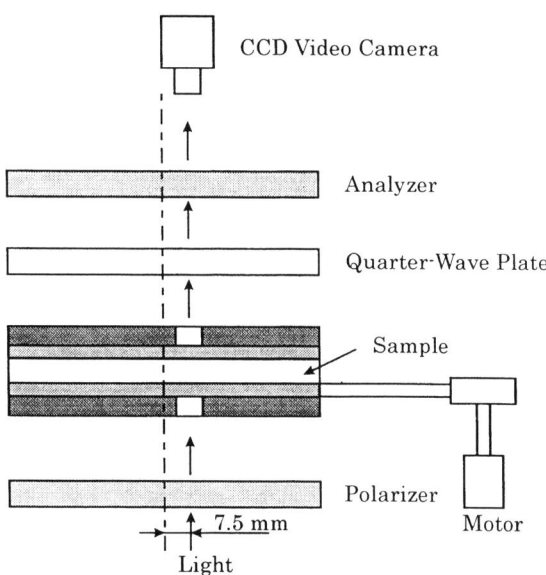

Figure 1. Experimeatal apparatus

EXPERIMENTAL RESULT

A time series of pictures in Fig.1 demonstrates the development of textures after imposition of the shear flow and after its cessation at $\dot{\gamma} = 5.0$ s^{-1} and $T = 30$ °C for the sample thickness $h = 500$ μm. The dimension of an image area in Fig.2 is 446 μm × 336 μm. After imposition of the shear flow, thin striated textures develop as seen in Figs.2(b)-(d). Just after cessation of the shear flow, the striated texture become vivid in Fig.2 (e) and then the banded textures perpendicular to the flow direction appear and gradually hide the striated texture (Figs.2(f)-(g)). The banded textures disappear before long and becomes featureless (Fig.2(e)). Wide stripe textures (WST) parallel to the flow direction, however, appear again in Fig.2 (f). This texture persists for a long time. As far as the authors know, there is no study on the WST.

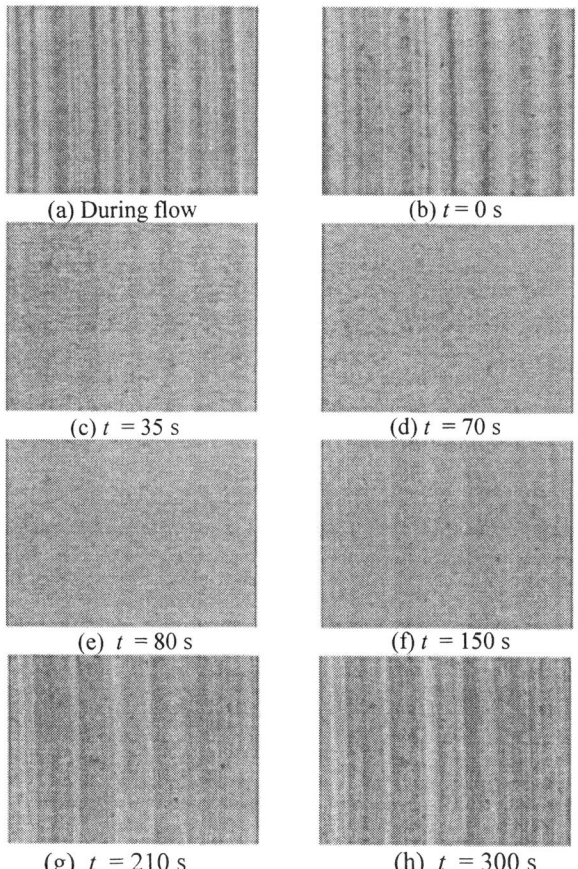

FIGURE 2. Development of textures during steady shear flow and after cessation of the shear flow ($h = 500$ μm, $T = 30$ °C, $\dot{\gamma} = 3$ s^{-1}). The flow direction is from bottom to top.

Shear rate significantly affects the WST. Figure3 shows the images when the texture sufficiently relaxed for various shear rates. At low shear rates as $\dot{\gamma} = 0.5$ s^{-1} (Fig3(a)), the WST is not formed. On the other hand, the stripe is divided into lumps as the shear rate increases, and finally small lumps are scattered at $\dot{\gamma} = 50$ s^{-1} (Fig3(f)).

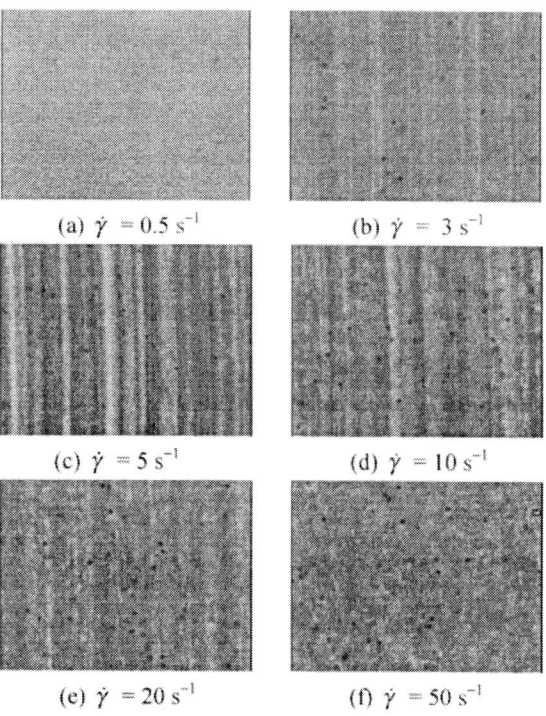

FIGURE 3. The relaxed texture on $h = 500$μm, $T = 30$ °C

We also found the dependence of the WST on a gap between two parallel plates. At $\dot{\gamma} = 3$ s^{-1}, the WST is clearly observed for $h = 300$ μm and $h = 500$ μm, but the image is featureless for $h = 100$ μm. The similar tendency was obtained at other shear rates. Furthermore, liquid temperature and shear rate affected the induction time from the cessation of the shearing to the emergence of the WST.

CONCLUSION

The wide stripe texture was found in a long-time relaxation process after cessation of shearing. The WST seems to be closely related to the striated texture during the shearing.

REFERENCES

1. Christopher,V. and Wendy, S. P., *Polymer*, **36**, 1731-1741 (1995).
2. Harrison, P., and Navard, P., *J. Rheol. Acta*, **38**, 569-593 (1999).
3. Larson, R. G., and Mead, D. W., *Liq. Cryst.*, **12**, 751-768 (1992).
4. Yan, N. X. et al., *Macromolecules*, **27**, 2784-2788 (1994).

Dynamics of Nano-sized Colloidal Particles in a Lyotropic Lamellar Phase

Yasuyuki Kimura* and Daisuke Mizuno*

*Department of Applied Physics, School of Engineering, University of Tokyo,
7-3-1 Hongo, Bunkyo-ku, Tokyo, 113-8656 Japan*

Abstract. Transport of nano-sized colloidal particles in a dilute lyotropic lamellar phase of a nonionic surfactant has been studied by AC electrophoretic light scattering. The frequency dispersion of complex electrophoretic mobility shows two relaxation processes at about 1kHz (HF relaxation) and a few Hz (LF relaxation). These relaxations are originated from the hindrance of diffusion of particles in characteristic local structures of lamellar phase. The HF relaxation is found to relate the local deformation of membranes induced by particles. The LF one relates the confinement of a particle within persistence length of lamellar orientation.

INTRODUCTION

One can find hierarchical structures with various shapes and sizes in biological systems. Among these structures, lipid bilayer is one of the basic and common elements as a cell membrane. It is important to study the transport property of nano-sized colloidal particles such as protein molecules and charged polymers in bilayer matrices to understand the functions and dynamics of living systems. One of typical structures made up of bilayers is lamellar structure, but local transport property in lyotropic lamellar phase has not been studied much. In this study, we have studied the transport of nano-sized colloidal particles in the lyotropic lamellar phase whose period is larger than the size of particles.

Recently, the local mechanical properties of soft matter has been studied by analyzing the motion of micrometer-sized probe particles embedded in soft matter [1]. In these studies, the diffraction limit and strong scattering from matrices prevents us from extracting information on the motion of probe particles with much smaller size. To overcome these difficulties, we have measured the motion of charged particles under a sinusoidal electric field by newly developed AC electrophoretic light scattering [2].

In this study, we dispersed polystyrene latex particles (47nm in diameter) in the lyotropic lamellar phase (period d: 60-200nm) made up of a nonionic surfactant, n-dodecyl pentaethyleneglycol monododecylether ($C_{12}E_5$), hexanol and water. The surfactant bilayer in our system has small bending elasticity about $k_B T$ at room temperature and its lamellar structure is stabilized by steric repulsion between undulating membranes.

AC ELECTROPHORETIC MOBILITY

We find two relaxation processes in the frequency dispersion of $\mu^*(\omega)$ at about 1kHz (HF relaxation: f_H) and a few Hz (LF relaxation: f_L) as shown in Figure 1 [3]. At these frequencies, the mobility decreases with decreasing frequency. This indicates that there is some potential barrier for particles to travel in longer distance. We can roughly estimate its size from $f_{H(L)}$, $\mu_{I(II)}$ and mobility in aqueous solution. The estimated size of potential for respective relaxation is as large as $\xi = d/2$ for HF relaxation and a persistence length of lamellar structure for LF relaxation. Recently, we have found a Maxwell-Wagner relaxation originated from the structure of lamellar at over 100kHz by dielectric measurement in the same system [4]. Therefore, the electric field applied to colloidal particles in the regions I-III is parallel to membranes. We can discuss the origin of these relaxations by studying the dependence of $\mu^*(\omega)$ on the lamellar period d.

DISCUSSIONS

1. Effective drag coefficient in lamellar phase

The particles between membranes induce the distortion field whose size is as large as the period of lamellar structure d around them [5]. In the region I, the particles relatively freely diffuse in this distortion field. But the confinement of particles between membrane walls increases the effective drag coefficient for particles and the mobility decreases with decreasing d. The dependence of effective drag coefficient on d is much stronger but is similar to that predicted for a spherical particle between

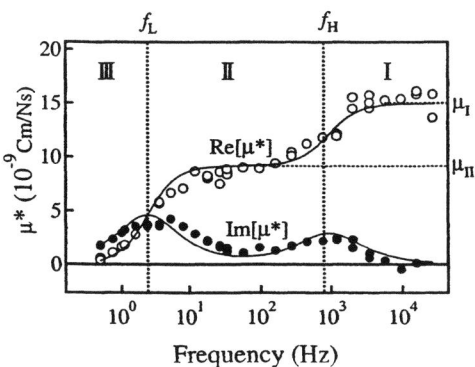

FIGURE 1. Frequency dispersion of electrophoretic mobility $\mu^*(\omega)$ of colloid particles in lamellar phase of $d=63$nm.

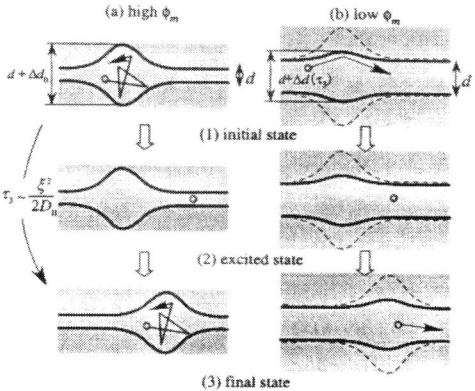

FIGURE 2. Diffusion process of colloidal particles between soft membranes. A colloidal particle which fluctuates in a distortion field (1) hops out (2), then creates a new distortion field around it (3). At high ϕ_m, distortion field grows completely (a), while at low ϕ_m, the particle moves too quick to form distortion field fully (b).

parallel walls [6]. Therefore, we can roughly discuss the drag coefficient in a nanometer-sized structure by continuum hydrodynamics.

2. HF relaxation

In the region II, the particles have to drag the distortion field around them to travel in longer distance. The excitation energy necessary to escape the distortion field is approximately given by $k_B T \ln(1 + \Delta d_0/d)$, where Δd_0 is the excess distortion induced by osmotic pressure of a particle. The relative amplitude μ_{II}/μ_{III} can be estimated as $\mu_{II}/\mu_{III} \sim d/(d + \Delta d_0)$. Since the static value of Δd_0 is estimated as $0.4d$ [5], μ_{II}/μ_{III} will be about 2/3 independent of the concentration of membrane ϕ_m. The observed ratio μ_{II}/μ_{III} decreases with increasing ϕ_m from about 1 to 0.7. To explain this contradiction, we have to consider the dynamic process for creation of distortion field. By calculating the step response of distortion field from hydrodynamic equation of lyotropic lamellar phase [7], the characteristic time for formation of the distortion field τ_R as $\tau_R \sim 3\eta d^3/k_B T$ (η: solvent viscosity). Since the distortion only grows when a colloidal particle stays inside it, the amplitude of the distortion field becomes a function of time τ_3 which is necessary for a particle to jump out one site. Since τ_3 can be estimated as $\tau_3 \sim \xi^2/2D_{II}$ (D: diffusion constant), τ_3 is proportional to ϕ_m^{-2}. On the other hand, the time τ_R is proportional to ϕ_m^{-3}. Therefore, as illustrated in Figure 2, at high ϕ_m where $\tau_3 > \tau_R$, a distortion field can fully grow around a particle. On the contrary, at low ϕ_m where $\tau_3 > \tau_R$, a particle will escape the distortion field before distortion field grows to its equilibrium one.

3. LF relaxation

In the region III, the mobility $\mu^*(\omega)$ decreases to almost zero. This indicates that almost all particles are trapped within the domain as large as 500nm. The existence of trapping site is confirmed by the direct observation of fluorescent particles dispersed in lamellar phase. Since lamellar structure in this study is not macroscopically oriented, there must be defects in a scale larger than d. In fact, the vesicle-like structure or folded lamellar structure surrounded by perforated lamellar has been frequently observed by freeze fracture electron microscopy [8]. The LF relaxation relates such multi-lamellar vesicle structure. The trapping site is composed of multilamellar vesicle where colloidal particles cannot move across but colloidal particles can move inside and outside of the vesicle-like region, which is composed of perforated lamellar. Even in the region III, particles might diffuse in longer length scale if the reorganization of lamellar structure or renewal of trapping path occurs.

REFERENCES

1. MacKintosh, F. C., and Schmidt, C. F., *Curr. Opin. Colloid Interface Sci.*, **4**, 300–307 (1999).
2. Mizuno, D., Kimura, Y., and Hayakawa, R., *Langmuir*, **16**, 9547–9554 (2000).
3. Mizuno, D., Kimura, Y., and Hayakawa, R., *Phys. Rev. Lett.*, **87**, 88104-1–4 (2001).
4. Mizuno, D., Nishino, T., Kimura, Y., and Hayakawa, R., *Phys. Rev. E*, **67**, 061505-1–4 (2002).
5. Sens, P., Turner, M. S., and Pincus, P., *Phys. Rev. E*, **55**, 4394–4405 (1997).
6. Happel, J., and Brenner, H., *Low Reynolds Number Hydrodynamics*, Kluwer, Dordrecht, 1991.
7. Brochard, F., and de Gennes, P. D., *Pramana, Suppl.*, **1**, 1–21 (1975).
8. Hoffmann, H., Munkert, U., Thunig, C., and Valiente, M., *Colloid Polymer Sci.*, **163**, 217–228 (1994).

Segregation of Binary Granular Mixtures in a Horizontal Rotating Cylinder

H.Takano, M.Furuta, T.Ishikawa and T.Tanizawa

Kochi National College of Technology, Nankoku, Kochi, Japan

Abstract. The segregation phenomena of binary granular mixtures in a horizontal rotating cylinder are investigated. For the sake of simplicity, the glass cylinder which is a half-filled with binary mixture of glass particles is adopted. The mixture consists of the same volumes of small size (diameter:0.2mm) and large size (diameter:0.6mm) glass particles. The rotation speed dependence of the long time scale segregation phenomena is studied. In a low rotation speed region, the initial narrow bands, which consist of small particles, gradually combined with each other and finally form one large band. On the other hand, in a high rotation speed region the initial narrow bands never combine with each other. The segregation phenomena in our case is thought to be caused by the drift motion of small particles along a inner axial core.

We propose a new type model for numerical simulation which is able to take account of both effects of surface flow and inner drift of small particles. This model can reproduce the consecutive segregation phenomena obatained in our experiments.

INTRODUCTION

A segregation phenomenon of granular mixture in a horizontal rotating cylinder was first observed by Y.Oyama [1] in 1939. Since then to 1990, the segregation problem of this kind was studied by only a few groups [2]. For the last decade, Oyama's segregation problem has been paid considerable attention by many authors [3, 4, 5, 6, 7, 8]. It is widely believed that there are three processes in Oyama's segregation phenomenon. First step, a radial segregation occurs quite rapidly. The smaller particles form an inner core. Second step, an axial segregation occurs. More than ten bands of small particles appear along the axial line of the cylinder. Final step, the small particle bands combine with each other and finally they form one large band. Moreover, it was reported that there are two kinds of axial segregations. One kind of segregation [6] is that the small particle bands are connected with each other through the axial core formed by small particles. The other [7] is that the small particle bands are independent with each other. These results imply that the two types of axial segregation may be changed with each other by the experimental conditions, like as size and sort of particles, rotation speed and so on.

EXPERIMENT

In this paper we study the rotation speed dependence of the long time scale segregation phenomena in a horizontal rotating cylinder. For the sake of simplicity, a

FIGURE 1. The rotation speed dependence of final states is shown. The rotation speeds are 7rpm, 11rpm, 16rpm, 20rpm, 25rpm and 32rpm in order of height, respectively.

glass cylinder and glass particles are used in our experiment. The outline of our apparatus is as follows. The diameter and the length of the glass cylinder are 30 mm and 460 mm, respectively. The rotation speed can be changed from 7 rpm to 32 rpm. The sizes of glass particles are 0.214 ± 0.036 mm and 0.605 ± 0.105 mm, respectively. The density of glass particles is 1.5 g/cm3. The glass cylinder is a half-filled with binary mixture which consists of 50% small glass particles and 50% large glass particles. The rotation speed dependence of the final states is shown in Figure 1. In the case of 7 rpm, an unstable narrow small particle band appears and disappears near the center of the cylinder. The most of small particles is thought to form a center core (The final state is a radial segregation.). In the middle speed region of 11 rpm, 16 rpm, and 20 rpm, the small particles form one large band after several days through the states of radial and axial segregations. It is noticed that the ratio of the small particle band length against the cylinder length increases according to the increase of rotation speed, i.e. 38% (11 rpm), 42% (16 rpm), 49% (20 rpm). This fact indicates that the axial core part of small particles existing in the large particle band will disappear in the high-speed region more than 20 rpm. In the high-speed region of 25 rpm and 32 rpm, the small particles form more than ten narrow bands after the radial segregation. These narrow bands never combine with each other during several days. These narrow bands are considered to be independent mutually.

NUMERICAL MODEL

Though the simulation model based on a surface flow was proposed by T.Yanagita [9] in 1999, our experimental results require a model which takes account of both mechanisms of surface flow and inner drift of small particles. So, we propose a new type of model for numerical simulation. For the sake of simplicity, a hexagonal prism constructed with piling triangular plane lattices is used instead of the cylinder. Each lattice point can be occupied by small or large particles. The initial state is that the half of lattice points is randomly occupied by particles. The hexagonal prism system rotates step by step and it turns full circle by 8×6 steps (here the number 8 means a side length of hexagon). The surface particles randomly fall down along the surface, if the angle of surface exceeds the repose angles of each sort of particles. Moreover, small particles in large particle rich region can drift along the axial direction if the sites on the route are occupied by large particles. The results of our numerical simulation are shown in figure 2. We can observe that three initial bands form and then the left hand side band absorbs other bands. It is noticed that the drift motion of small particles is the primary cause in the band formation and that the surface flow plays only a supporting role in our model.

CONCLUSION AND DISCUSSION

We investigate the rotation speed dependence of segregation phenomena experimentally. The followings are concluded from our experiments.

1. In the low rotation speed region, the initial several

FIGURE 2. The rotation number dependence in our simulation is shown. The number of rotation is 500turns, 1000turns, 1500turns, 2000turns and 2500turns in order of height, respectively. The side length and the height of hexagonal prism are 8units and 300units, respectively. Here, one unit means a lattice constant.

bands combine into one band within two or three days.

2. In the high rotation speed region, the initial bands are held as it is and they don't combine with each other.

The segregation phenomena in our case can be thought to be caused by the small particle drift motion. So, we propose the numerical model which takes account of small particle drift motion. The formation of initial bands and the combination among them can be reproduced with our numerical model. The reappearance of the rotation speed dependence is the problem left in the future.

REFERENCES

1. Oyama, Y., *Bull. Inst. Phys. Chem. Res. Jpn. Rep.* **18**, 600-639(1939).
2. Donald, M.B., and Roseman, B., *Br. Chem. Eng.* **7**, 749-753 and 823-827(1962).
3. Gupta, D., Khakhar, D.V., and Bhatia, S.K., *Chem. Eng. Sci.* **46**, 1513-1517(1993).
4. Nakagawa, M., *Chem. Eng. Sci.* **49**, 2540-2544(1994).
5. Zik, O., Levine, D., Lipson, S., Shtrikman, S., and Stavans, J., *Phys. Rev. Lett.* **73**, 644-647(1994).
6. Hill, K.M., Caprihan, A., and Kakalios, J., *Phys. Rev. Lett.* **78**, 50-53(1997).
7. Nakagawa, M., Altobelli, S.A., Caprihan, A., and Fukushima, E., *Chem. Eng. Sci.* **52**, 4423-4428(1997).
8. Dury, C.M., Ristow, G.H., Moss, J.L., and Nakagawa, M., *Phys. Rev. E* **57**, 4491-4497(1998).
9. Yanagita, T., *Phys. Rev. Lett.* **82**, 3488-3491(1999).

Micelle Formation in a Highly Charged Polyelectrolyte Solution

Takuya Fujima* and Hiroshi Frusawa[†]

*Molecular Spectroscopy Laboratory, The Institute of Physical and Chemical Research (RIKEN),
2-1 Hirosawa, Wako-city, Saitama 351-0198, Japan.
[†]Soft Matter Laboratory, Kochi University of Technology, Tosa-Yamada, Kochi 782-8502, Japan.

Abstract. A highly charged polyelectrolyte, sodium polystyrenesulfonate, was investigated in various quality of solvent by using dynamic light scattering and viscosity measurements. As the solvent quality was lowered, the relaxation dynamics by the scattering measurements exhibited a drastic change from two relaxations to single one. Furthermore, an obvious reduction of zero-shear viscosity coincided with the transition. These results indicate that the polyelectrolyte solution system in the solvent region has formed mono-disperse micelles by strong segregation. The micelle size was estimated to be about 20 nm by Einstein-Stokes relation. This micelle formation of polyelectrolyte solution system has been achieved experimentally for the first time despite theoretical predicts so far.

Theoretical models [1, 2] predict the phase separation of polyelectrolyte solutions in microscopical order. However, experimental evidence for this phenomenon has rarely been reported. While microphase separations weakly segregated have been observed for weakly charged polyelectrolyte solutions recently [3], there have been no reports on the formation of strongly segregated micelles, which we would like to explore here.

We have used, other than previous works, a highly charged polyelectrolyte, sodium polystyrenesulfonate (NaPSS). Also, solvents adopted are not aqueous but organic: mixture of a good solvent, ethylene glycol (EG), and a poor solvent, cyclohexanol (CHN). Because these two solvents are compatible despite their opposite quality for NaPSS, we can control the quality of solvent continuously between good and poor solvent, which is an intrinsic parameter for this work.

NaPSS which has a molecular weight of 500,000 was purchased from Scientific Polymer and purified through following procedures: first, we dissolved NaPSS powder in ultrapure water and kept it with a surplus amount of amphoteric ion-exchange resin, MB-1 from ORGANO, for four days to eliminate impurity salt and replace the counter ion of NaPSS, Na^+, with H^+. Next, the accurate quantity of the polyelectrolytes was determined by neutralizing titration using NaOH aqueous solution, which also changed the HPSS back to NaPSS. Then we obtained pure and dry NaPSS by freeze-dry method after filtration with 0.25 μm-pore-size membrane for dust removal. Both of two solvent components, EG and CHN, were purchased from Nacalai Tesque and used after the same filtration as above. The solute concentration in all samples was fixed at 20 mM in terms of monomer concentration (C_m). The volume fraction of CHN in mixed solvents (ϕ) varied between 20% and 75%. Incidentally, samples of higher ϕ were not available for measurement because of precipitation. All experiments were carried out under a constant temperature, 30 °C.

We mainly performed dynamic light scattering (DLS) measurements in a scattering angle range between 30° and 150° by an ALV system, which revealed obvious change in dynamics property of the system. When we lowered the quality of solvent by the change of fraction of solvent (increasing ϕ), a transition on its dynamics occurred, from two relaxation system to single relaxation one as shown in Fig. 1 especially about results measured at a scattering angle of 90°.

The two relaxations in good solvent region (low ϕ) are well known for this type of systems. The slower mode and the faster one are correspondent to heterogeneous fluctuation of the polymer chains and a coupling motion between counter ions and the polymer chains, respectively. The heterogeneity of the system is also detected by SANS as an upturn of scattering intensity with unspecified scattering wave number [4].

The origin of single relaxation in the poor solvent region is now attributed to strongly segregated micelles, which is also indicated by the scattering wave number (q) dependence of the relaxation time (τ). τ^{-1} is proportional to q^2 in the single relaxation region in Fig. 1, which indicates that the single relaxation is from self-diffusion process of spheres or micelles. In the good solvent re-

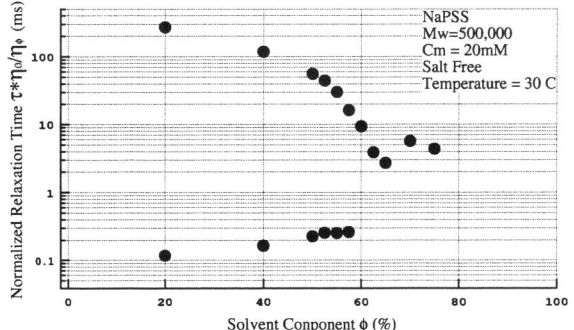

FIGURE 1. A solvent quality dependence of the relaxation times by dynamic light scattering experiments at a scattering angle of 90°. The influence from the variation of solvent viscosity at each solvent fraction on relaxation times has been normalized by the ratio of each solvent viscosity and that at $\phi = 0$.

gion, however, the exponents about the slower mode are around 3. Additionally, the size of the formed micelles is about 20 nm at the region of just after the transition, estimated by Einstein-Stokes relation.

Furthermore, results of visco-elasticity measurements supports this micelle formation. We investigated the solvent quality dependence of zero-shear viscosity by DYNALYSER VAR-CF system of REOLOGICA using a cone-plate type rotator with a diameter of 40 mm and an angle of 1°. The results show, in Fig. 2, clear decrease as the solvent quality lowered although the viscosity of solvent slightly increases during this solvent-quality change. This viscosity reduction indicates an unraveling of the polymer-chain entanglement which is well known to be caused by "coil"-formation chains in good solvent. This unravelment is a strong segregation itself leading to the micelle formation.

Another thing to be adverted here is that this zero-shear viscosity behavior is obviously opposite to that in the weakly charged polyelectrolyte system [3] in which the zero-shear viscosity exhibits a steep increase around the transition from well dissolved to weakly segregated (networking) region.

Additionally, this system kept clear transparency for human eyes even after the transition of micelle formation, though it turned misty in the region of adequately poor solvent. This microscopic phase separation or micelle is stable for several years. This stability originates in the electrostatic repulsion among the micelles. Therefore, slight addition of salt (NaCl) caused steep precipitation of the micelles.

Furthermore, added-salt-concentration dependence of this system was also investigated also at the same polymer concentration (C_m=20 mM) and a fixed solvent component (ϕ=55 %). Adding salt (NaCl) drove samples in the good solvent region to the micelle forming transition around an added salt concentration (C_s) of 1.5 mM similar to the solvent-quality dependence mentioned above. Further added salt after the micelle formation caused augmentation of the micelle size and precipitation at last (C_s: around and above 6 mM).

ACKNOWLEDGMENTS

We are grateful to T. Koyama, T. Araki and H. Tanaka (Univ. Tokyo) for their great help on viscosity measurements. We also thank K. Ito (Univ. Tokyo) and M. Takagi (TOPPAN) for useful discussions.

REFERENCES

1. V. Yu. Borue and I. Ya. Erukhimovich, *Macromolecules* **21**, 3240-3249 (1988).
2. J. L. Barrat and J. F. Joanny, *Adv. Chem. Phys.* **94**, 1-66 (1996).
3. O. Braun, F. Boue, and F. Candau, *Eur. Phys. J. E* **7**, 141-151 (2002), and references therein.
4. M. Sedlák, *Langmuir* **15**, 4045-4051 (1999).

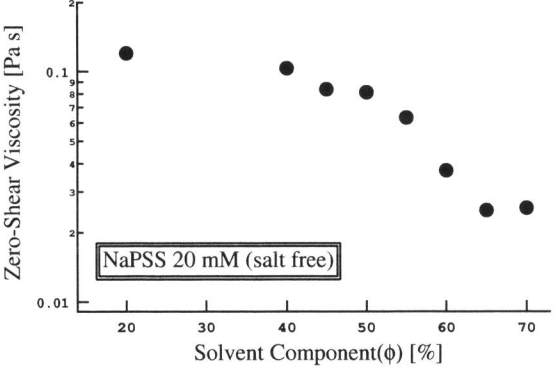

FIGURE 2. Solvent quality dependence of the zero-shear viscosity. The data have been normalized by solvent viscosity at each solvent fraction.

Liquid-Liquid Extraction in Microfluidic System Using Dispersed Liquid Droplet

Momoko Kumemura and Takashi Korenaga

Department of Chemistry, Tokyo Metropolitan University, 192-0397, JAPAN

Abstract. The formation of liquid droplets, which were generated at the confluence of two-microchannels, was nvestigated in the microchip. Tributyl phosphate (TBP)-droplets were formed precisely and continuously at 90-200 μm-diameter at the range of 5-55 mm/s in the main-channel and 47-370 mm/s in the other-channel and a variation in the formation of TBP-droplet was observed in a microchannel. And, the extraction of Aluminum-DHAB chelate (DHAB=2,2'-dihydroxyazobenzen) was studied by the laser-induced fluorescence. The fluorescence intensity of Al^{3+}-DHAB chelate in the droplet was gradually increased with the distance from the confluence of two-channel. The extraction using droplet in the microchannel has finished for 18 second.

INTRODUCTION

In the conventional liquid-liquid extraction with a separating funnel, the formation of small droplets with the large specific area is important for the efficient extraction. However, it is need that routine work as shaking for a long time. The micrototal analysis system (μ-TAS) [1] or lab-on-a-chip [2], is remarkably progressing in recent years. One of the advantages of utilizing microchip for analysis is a short diffusion length for mass transfer. The molecular transportation time (diffusion time) is given by equation (1).

$$T = L^2 D^{-1} \qquad (1)$$

Where T, L and D are the molecular transportation time, diffusion distance (microchannel width) and diffusion coefficient, respectively [3]. Therefore, transportation time of molecule is short in two liquid phases in microchannel.

The aim of this study is demonstration of the liquid-liquid extraction in a microchip as a tool of more useful and quickly extraction.

EXPERIMENTAL

Microchip

Four microchips were purchased from Institute of microchemical technology (Kanagawa, JAPAN), which were fabricated by photolithography and a wet chemical-etching method. The size of a main microchannel is 600 μm (wide)-50 μm (deep). The main microchannel was confluent with the other microchannel with different size. The sizes of other microchannel were (a) 300 μm-50 μm; (b) 100 μm -38μm; (c) 67 μm-20 μm; and (d) 39 μm-5 μm, respectively. Droplets were formed at the confluent of two microchannels in microchip, by flowing water as continuous phase in the main microchannel and tributyl phosphate (TBP) as dispersed phase in the other microchannel.

RESULT AND DUSSCUSSION

Droplet-formation in the microchannel

In microchip (a), a slug was observed at 55 mm/s-continuous phase velocity, 1 mm/s-disperse phase velocity. A slug gradually changing the formation, the slug became to a droplet at 6mm from the confluent point. In other microchips, these phenomena were not observed.

The relationship between the velocity of continuous phase flow and the droplet diameter in microchip (d) was shown in FIGURE 1. Droplet diameter was decreased with an increase in the velocity of continuous phase, and it reached 200 μm at 55mm/s (47mm/s-dispersed phase velocity). From using the other microchips, the results were shown as follows: microchip (a); droplet diameter were 153-185 μm at 45-78 mm/s-continuous phase velocity, at 1 mm/s-dispersed phase velocity: microchip (b); droplet diameter were 105-168 μm at 10-64 mm/s-continuous phase velocity, at 5 mm/s-dispersed phase velocity: microchip (c); droplet diameter were 93-150 μm at 5-52 mm/s-continuous phase velocity, at 12-49 mm/s-dispersed phase velocity. It was suggested that the droplet-diameter depend on the channel size and the flow velocity in both microchannels.

The relationship between the velocity of dispersed phase flow and the droplet diameter in microchip (d) was shown in FIGURE 2 The droplet-diameter was increased in proportion to the velocity of dispersed phase flow at 43-373 mm/s.

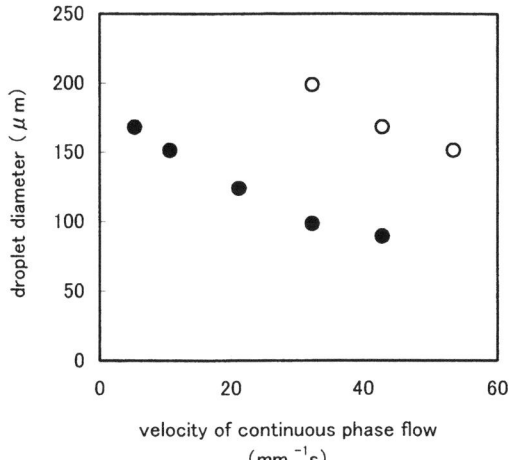

FIGURE 1. Effect of velocity of continuous phase flow on droplet size. Velocity of dispersed phase flow; ○: 90 mm/s, ●: 370 mm/s

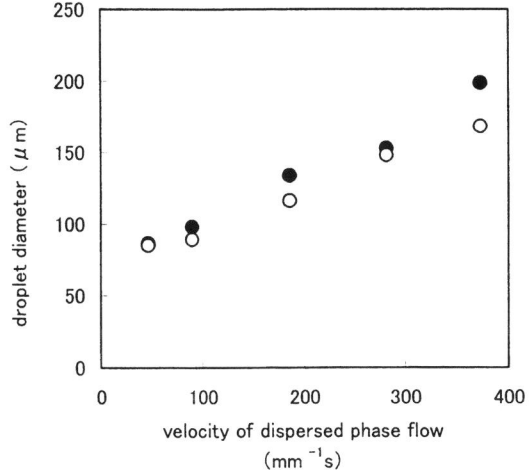

FIGURE 2. Effect of velocity of dispersed phase flow on droplet size. Velocity of continuous phase flow; ●: 32 mm/s, ○: 42 mm/s

Effect of viscosity

Relationship between a droplet size and the viscosity of solvent was investigated. It was suggested that a droplet size was varied from the viscosity of solvent in the dispersed phase. Chloroform, 1-butanol, TBP was flowed as dispersed phase, water was used as continuous phase. In the result, this droplet-diameter increased with a decreasing in the viscosity of solvent, as it was expected.

Liquid-liquid extraction using dispersed droplet

Liquid-liquid extraction in the microchip was estimated as amount of Al^{3+}- 2,2'-dihydroxyazobenzen (DHAB) chelate. Aqueous Al^{3+}-DHAB chelate solution (pH=6.3) and TBP were delivered by syringe pumps (IC3200; kd Scientific, USA), respectively. In the microchannel, TBP-droplets were formed continuously in the aqueous Al^{3+}-DHAB chelate solution, and then Al^{3+}-DHAB chelate was extracted with the TBP-droplets. Al^{3+}-DHAB chelate fluoresces in the TBP-droplets, but does not in the aqueous phase. The excitation light from the Ar^+ laser (488 nm wavelength, 2101-40MLA; uniphase, USA) was introduced to the microscope. And the fluorescence of the Al^{3+}- DHAB chelate in a TBP-droplet was detected by the polychrometer-multichannel photodetector system (PMA-11; Hamamatsu Photonics, Hamamatsu, JAPAN).

The fluorescence intensity of Al^{3+}-DHAB chelate in the droplet was gradually increased and retained constant at 200 mm with the distance from the confluence of two-channels. Extraction time was estimated, it was about 18 s, and it was about 1/10 for extraction time by a conventional extraction with separating funnel. In this experiment, it has found that the extraction in the microchannel can be achieved quickly with droplets.

REFERENCES

[1] Manz, A., Granber, N., and Widmer, H. M., *Sens. Actuators*, **B1**, 244-248 (1990).
[2] Terry, S. C., Jerman, J. H., and Angell, J. B., *IEEE Trans. Electron Devices*, **ED-26**, 1880-1886 (1979).
[3] Lide, D. R., *CRC Handbook of Chemistry and Physics* 78 th edn., edited by D. R. Lide, CRC Press, Boca Raton, Florida, 1997, pp.5-93.

Morphology Transition from Sphere to Rod by Confining the Polymer Chains in a Dilute Microemulsion System

Kaori Nakaya[*], Masayuki Imai[*], Shigeyuki Komura[**], and Naohito Urakami[***]

[*]Ochanomizu University, 2-1-1 Otsuka, Bunkyo-ku, Tokyo, 112-0012, Japan
[**]Tokyo Metropolitan University, 1-1 Minami-Osawa, Hachioji-shi, Tokyo, 192-0397, Japan
[***]Yamaguchi University, 1667-1 Yoshida, Yamaguchi-shi, Yamaguchi, 753-8512, Japan

Abstract. In this study we investigated the morphology transition of microemulsion droplet induced by polymer confinement using a small angle neutron scattering (SANS) technique. By confining the polymer chain strongly the scattering profiles showed the following changes; 1) a characteristic scattering peak corresponding to the size of droplet shifts to the higher q side, and 2) the scattering intensity in the low q region increases considerably. These changes of the scattering profiles can be described by a rod (or sphero-cylinder) model. Thus, the strong confinement of polymer chains in droplets induces the morphological transition from spherical to rod-like droplet.

INTRODUCTION

Microemulsion is a unique system in which water and oil coexist stably through a surfactant monolayer. It shows a variety of morphologies in membrane structure, such as globular (droplet), layered (lamellar) or network state (bicontinuous) depending on compositions or external fields. Recently an increasing interest is directed toward membrane-polymer system [1]. In the complex system of microemulsion and polymer, it is anticipated that it appears various phenomena based on a balance of energy and entropy, because each component has large internal degree of freedom. In the previous paper [2], we studied the static behavior of the membrane structure of a dilute microemulsion droplet confining a single polymer chain weakly. In this case the balance of entropy loss due to the polymer confinement and membrane elastic energy governs the droplet size. The purpose of this study is to clarify the morphology of the membrane confining polymer chains very strongly.

EXPERIMENT

We used water-in-oil microemulsion droplets, which consisted of AOT, isooctane, and water. As a confined polymer it was adopted the water-soluble polymer, gelatin with the radius of gyration R_g=81Å. The radius of the droplet (R) was controlled by changing the water to surfactant ratio (ω_0) and fixed at ω_0=41.1 in this study. The volume fraction of the dispersed phase that consists of water and AOT (and polymer) was fixed to 0.07. The static structure of the membrane was followed by the small angle neutron scattering (SANS) technique. The SANS measurements were performed using ISSP-SANS-U (Tokai) and KENS-SWAN (Tsukuba).

RESULTS AND DISCUSSION

Figure 1 shows the SANS scattering profiles for water-in-oil microemulsion droplet with and without polymer in the film contrast condition. By confining the polymer chains strongly, the peak observed in the droplet without the polymers shifted to the higher q side, and the scattering intensity increased in low q region. The scattering profiles of droplet are described by form factor $P(q)$ and structure factor $S(q)$. We fitted the experimental profiles with the spherical shell model [3] as $P(q)$ and Percus-Yevic model as $S(q)$. The scattering profile of droplet without the polymers was fitted very well, which gives R=64Å. Thus in this study we confined the polymer chain inside of a microemulsion droplet strongly. On the other hand, the scattering profile for microemulsion droplet + polymer system could not described by the spherical model well. Then we adopted the rod (or sphero-cylinder)

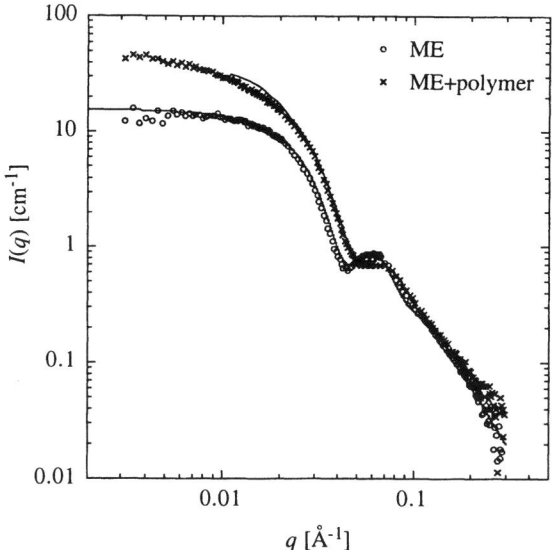

FIGURE 1. SANS profiles for the system of AOT-isooctane-water microemulsion droplet of ω_0=41.1 without (opened circle) and with (cross) gelatin at 30°C. The solid lines are curves fitted with the sphere and the rod model.

model; $P(q)$ is approximated by the rotational ellipsoid and $S(q)$ for the rod system was obtained from the Monte Carlo simulation. The scattering curve with polymer could be described well with the rod model. This result suggests that the morphological transition of membranes from sphere to rod is induced by strong confinement of the polymer chains. It was found that the radii of the minor axis and the major axis are about 45 and 360 Å, respectively. Using these parameters the number of the polymer chains was calculated to about 0.9 chains per rod-like droplet.

In order to elucidate this morphological change, we estimate the total free energy of each structure with three terms, the elastic energy of the membrane, the conformation entropy loss by the confinement of polymers, and the interaction between segments (excluded volume effect)

$$F_{tot} = F_{mem} + F_{conf} + F_{ex} \qquad (1)$$

The elastic energy of a membrane is given by the Helfrich expression [4]

$$F_{mem} = \int \left[2\kappa(H - H_0)^2 + \bar{\kappa}K \right] dS \qquad (2)$$

where κ is the bending modulus, $\bar{\kappa}$ is the saddle-splay modulus, H is the mean curvature, and K is the Gaussian curvature. The free energy of the polymer confinement can be estimated from the entropic confinement contribution [5] as

$$F_{conf} \sim T\left(\frac{R_F}{D}\right)^2 \qquad (3)$$

where T is temperature and the free energy of the interaction between segments can be described as

$$F_{ex} = v\rho^2 V \qquad (4)$$

where v is the excluded volume parameter, ρ is the density of segments, and V is volume of one droplet. On the basis of this concept, the total free energy of each structure was calculated. As the result, we found that the interaction between segments F_{ex} becomes dominant in the system, and stabilizes the rod shape compared with the spherical shape by confining polymer chains strongly.

ACKNOWLEDGMENTS

This work was supported by Grant-in-Aid for Exploratory Research (No.13740252, 15740261) from the Ministry of Education, Sciences, and Culture of Japan. In addition, this work was done under the approval of the Neutron Scattering Program Advisory Committee, Japan. We are grateful to T. Kawakatsu and S.A. Safran for helpful discussions.

REFERENCES

1. For example: (a) Lal, J., and Auvray, L., *J.Phys.II (France)* **4**, 2119-2125 (1994), (b) Bellocq, A. M., *Langmuir* **14**, 3730-3739 (1998).

2. Nakaya, K., and Imai, M., *J. Phys. Soc. of Japan* **70** Suppl. A, 338-340 (2001).

3. Grazielski, M., Langevin, D., and Farago, B., *Phys.Rev.E* **53**, 3900-3919 (1996).

4. Helfrich, W., *Z.Naturforsch* **28c**, 693-703 (1973).

5. de Genne, P. G., *Scaling Concepts in Polymer Physics*, Cornell University Press, New York, 1979, pp.18-19.

Inter-Lamellar Interaction Mediated by Amphiphilic Triblock Copolymer

Tomomi Masui, Masayuki Imai and Kaori Nakaya

Department of Physics, Ochanomizu University, 2-1-1 Otsuka, Bunkyo-ku, Tokyo 112-8610, Japan

Abstract. We incorporated amphiphilic triblock copolymer into a lyotropic lamellar system and studied the effects of the hydrophilic chain length on the elastic nature of the lamellar membranes. The triblock copolymer consists of a hydrophobic chain (PPO) bounded by two identical hydrophilic chains (PEO). The hydrophobic chain anchors in the lamellar membranes and the hydrophilic chain decorates the membrane. We estimated the layer compression modulus by combining a small angle x-ray scattering (SAXS) and a neutron spin echo (NSE) technique. The increase of the hydrophilic chain length changes the bending rigidity of membrane and enhances the repulsive forces between membranes. We discuss the results in terms of the entropic undulation interaction of lamellar membranes and comformational entropy of the hydrophilic chains.

INTRODUCTION

The lyotropic lamellar phase consists of surfactant membranes periodically stacked in space, separated by solvent layers. In the lamellar phase, adjacent membranes interact by various forces (van der Waals, hydration, electrostatic and undulation interactions). The previous experimental studies have shown that the addition of non-adsorbed polymers in the lamellar phase induces attractive force between membranes [1], while the addition of the end grafted polymers induced repulsive force [2]. In this study we investigate effects of the hydrophilic chain length of the grafted polymer on the elastic nature of the membranes.

EXPERIMENTAL

The lamellar phase used in this study is composed of membranes of the nonionic surfactant $C_{12}E_5$ (Nikko Chemical) and water. The grafted polymer is amphiphilic triblock copolymer with trade name pluronics (Asahi Denka and BASF). It consists of two identical polyoxyethylene (PEO) blocks symmetrically bounded to a central hydrophobic polyoxypropylene (PPO) blocks. The PPO chains anchor in the membrane and the PEO chains decorate the membrane. We change the degree of polymerization of PEO (N_{PEO}= 3, 9, 13, 20, 75), while the degree of polymerization PPO (N_{PPO}=13) keeps constant. The surfactant/polymer mole ratio was fixed at 1.00/0.02. The SAXS measurements were performed using Rigaku NANO-Viewer and the dynamics experiments were carried out using a NSE spectroscopy at JRR-3M reactor (JAERI).

RESULTS AND DISCUSSION

We first estimated the effects of the hydrophilic chain length on bending rigidity of membrane κ by means of NSE measurement. Figure 1 shows the NSE relaxation curves for the lamellar phase of surfactant/water/amphiphilic triblock copolymer systems as a function of N_{PEO}. The relaxation rate decreases with N_{PEO}. We analyzed the relaxation profiles using Zilman and Granek model [3]. The bending rigidity of membrane κ increases with N_{PEO}. This means that hydrophilic chains decorating membranes enable to stiffen the membrane.

In order to examine the interactions between membranes, we estimated the layer compression modulus B from SAXS profiles, because it is related to

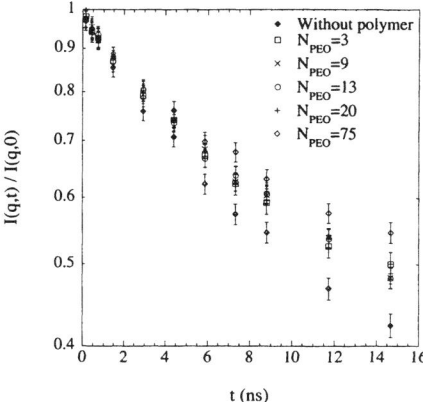

FIGURE 1. NSE relaxation profiles for lamellar phase of surfactant/water/amphiphilic triblock copolymer systems.

the free energy of interaction between adjacent membranes V

$$B = d \frac{\partial^2 V}{\partial \overline{d}^2}. \qquad (1)$$

In Figure 2, we show the change of SAXS profiles of the lamellar phase of surfactant/water/amphiphilic triblock copolymer systems. Here we kept the surfactant weight fraction at $\rho_s = 0.30$. The increase of N_{PEO} brings the sharpening of the Bragg peak and the emergence of high order harmonics. This means hydrophilic chain length induces additional repulsive interaction between membranes. We evaluated B following procedure. First we extract the Caille parameter η from SAXS profiles using Nallet *et al.* model [4]. The η is defined in terms of the smectic elastic constants by

$$\eta = \frac{q_0^2 k_B T}{8\pi \sqrt{K\overline{B}}}, \qquad (2)$$

where K is the bulk bending modulus (related to the bending rigidity κ and the smectic periodicity d according to $K=\kappa/d$) and q_0 is the position of the first Bragg singularity, $q_0 = 2\pi/d$. Then we can obtained B using η and κ. The layer compression modulus B also increases with N_{PEO} as shown in Figure 3.

We now discuss the possible reason for the observed increasing of B. Here we consider two additional repulsive interactions mediated by amphiphilic triblock copolymer. One is renormalizing the Helfrich's interaction between membranes due to the increase of the effective bilayer thickness [2]. And the other is entropic repulsive force given rise to the reduction in the number of available conformations of the polymer chains [5]. In Figure 3, we compared the measured layer compression modulus B_{exp} and the calculated values B_{theo} as a function of N_{PEO}. The calculated values qualitatively described the experimentally obtained variations of B.

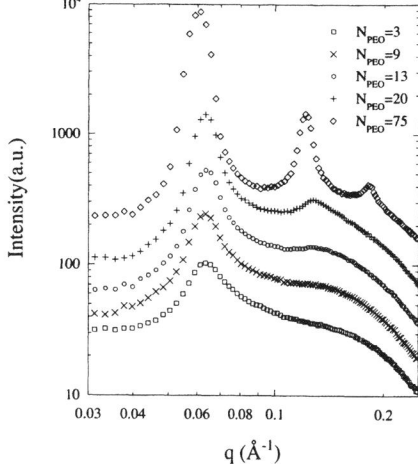

FIGURE 2. SAXS profile for the lamellar phase of surfactant/water/amphiphilic triblock copolymer systems.

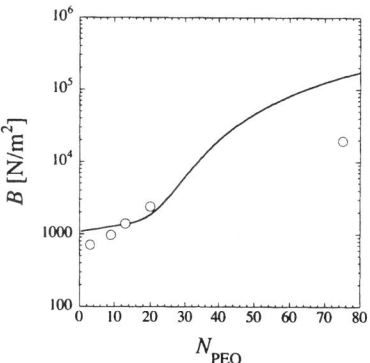

FIGURE 3. Experimentally evaluated B_{exp} and the caluculated B_{theo} (solid line) as a function of N_{PEO}.

REFERENCES

1. Ligoure, C., et al., *J. Phys. II.* **7**, 473-491 (1997).

2. Castro-Roman, F., Porte, G., and Ligoure, C., *Phys. Rev. Lett.* **82**, 109-112 (1999).

3. Zilman, A. G., and Granek, R., *Phys Rev Lett.* **77**, 4788-4791 (1996).

4. Nallet, F., et al., *J. Phys. II.* **3**, 487-502 (1993).

5. Dolan, A., and Edwards, F., *Proc. Royal Soc. Lond. A.* **337**, 509-516 (1974).

Molecular assembly under a focused laser

Masatoshi Ichikawa*, Yukiko Matsuzawa[†], Yoshiyuki Koyama** and Kenichi Yoshikawa*

*Department of Physics, Graduate School of Science, Kyoto University and CREST, Kyoto 606-8502, Japan
[†]Department of Ecological Engineering, Toyohashi University of Technology, Toyohashi, Aichi 441-8580, Japan
**Department of Home Economics, Otsuma Women's University, 12 Sanban-cho, Chiyoda-ku, Tokyo 102-8357, Japan

Abstract. We report a novel method to assemble colloidal particles to an ordered structure. Individual DNA molecules folded by a sticky condensing agent can be prefabricated into a linear micrometer-sized pearling chain under trapping field with a focused IR laser. Using this method, the macromolecules are assembled in a sequential order, and then a complex structure is fabricated by arraying them on a solid substrate. This approach may be useful for the further development of micro-manufacture using macromolecules as nano-elements.

A focused laser, or optical tweezers, trap a dielectric particle on the laser focus in aqueous solution [1]. Continuous trapping forms a cluster of the particles on the focus during the laser irradiation. We have found that sticky particles assemble linearly under the trapping field [2]. In the present paper, we report the process of linear assembling and discuss the experimental results with the aid of simulation.

EXPERIMENTS

A naked DNA thermally fluctuates in water solution with random coil shape, and is condensed to be a globular or a folded DNA particle by adding various reagents [3, 4]. Under dilute condition of DNA, each DNA molecules are collapsed by condensing agent, and behave as colloidal particles. A single DNA chain with about 166 kbp folded by multivalent-cations, polyethylene glycol derivatives with both cholesteryl- and amino-pendant groups (Chol-PEG-A) or either amino-pendant groups (PEG-A), are prepared as a sticky particle and as a non-sticky particle, respectively. The behavior of folded DNA particles correspond to the colloidal particles of about 50 to 100 nm in radii [5].

In the experiments as in Fig 1, we observed assembling phenomena of DNA particles complex with Chol-PEG-A. The process seems as follows: The sticky particles, under dilute dispersion, are falling to the beam focus along the light line, since the optical trapping potential is proportional to the light intensity. When another floating particle has been caught in the focus, the falling particle is forced to align just before a collision and adhesion to the other one. Repetition of collisions and coalitions around the focus generates a linear structure as shown in Fig. 1(a). On the other hand, DNA particles collapsed by PEG-A also gather on the focus, but do not form a linear structure.

From the experimental observations, it is expected that the light scattering, or light pressure, performs as a directional pushing force and plays a roll in making the straight bars. Convection induced by light irradiation and thermal effect of laser heating may have additional effect to induce the uni-directional motion. Next, we discuss the effects of the pushing force for linear assembling.

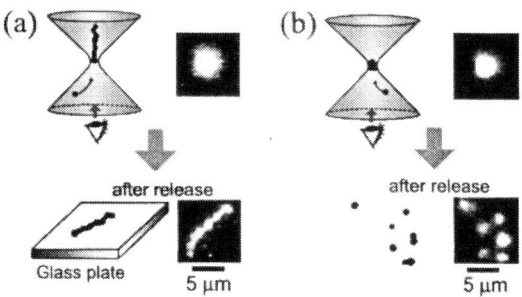

FIGURE 1. The experiments on the assembling process of DNA particles with different properties. (Reference [2]) (a) The folded DNA particles with Chol-PEG-A assemble to micrometer-sized linear bar under laser trapping field. Laser power for optical trapping is about 500 mW. (b) The DNA particles complex with PEG-A are once collected by the optical tweezers. Since its non-sticky or weak repulsive nature of folded DNA, each particle re-disperses after laser off.

DISCUSSIONS

To make clear effective components for linear assembling, we have confirmed them through Brownian dynamics simulations. Optical tweezers are roughly composed of two forces, attractive trapping force and one way pushing force. First one is called optical gradient force, which comes from induced dipole interaction and attracts a dielectric object with the direction to higher light intensity. The latter, which derives from scattering force, etc., pushes particles along the light line with the direction following the laser beam.

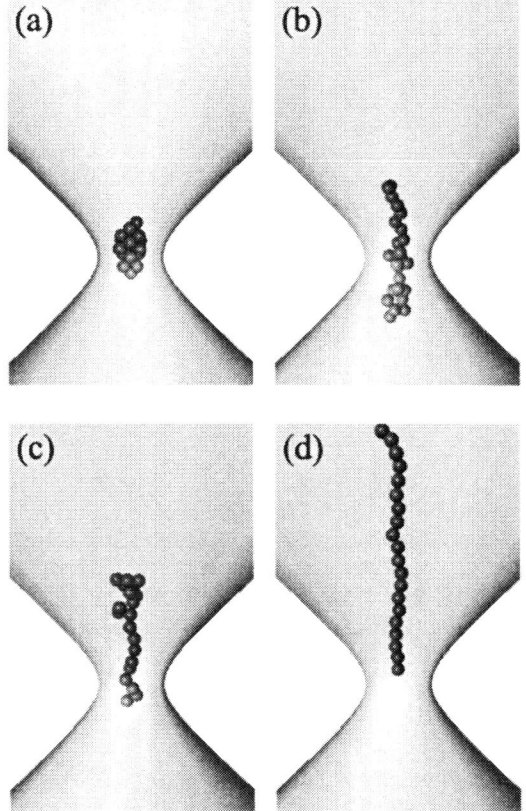

FIGURE 2. Results of the simulations. Optical cones in the snap shots show converged Gausian beam with 1 micrometer in beam waist diameter. The laser beams with 1 W power are incident from bottom side of the panels. The particle is 100 nm in radius and is 1.60 in refractive index. (a) An aggregate of non-sticky particles is generated. (b) An assembly of sticky particles formed under optical gradient force without uni-directional pushing force is shown. (c) A pushing force corresponding to the Rayleigh scattering is applied to the condition of (b). (d) A pushing force three times larger than the Rayleigh scattering is added to the condition of (b).

Figure 2(a) shows an assembly of no-sticky particles with an only excluded volume interaction under the optical trapping. The conditions and the product on the focus correspond to the Fig. 1(b) experiment. The cluster forms dense packing in the small area by means of the strong collective force. The second panel, the assembly shown as in Fig. 2(b) is generated under gradient force only. The situation is similar to an experiment with opposed dual beams system. The product is taken with certain linear shape than random aggregation, for example, diffusion limited aggregation patterns. Since the experiments show more straight assemblies, it indicates that scattering force could not be ignored for the assembling process, and had a kind of effect to make the linear bars. The more straight assemblies have been achieved in simulations with existence of pushing forces as denoted in Figs. 2(c) and (d). The laser trapping including the effects of Rayleigh scattering and other pushing forces makes straight assemblies being similar to the experiments (as in Figs. 1(a) and 2(d)). The simulations give suggestions that pushing forces, including not only light scattering but also absorbing and convection etc., are effective for making a linear assembly in the experiment.

Finally, we briefly explain the process of linear assembling from simulative and experimental observations: Particles are falling to the focus along the laser light line because a trapping potential is proportional to the light intensity. Thus the particles around the focus are arranged their collisions to be made linear bar with taking them along the beam line. A linear object in optical tweezers tends to stand before the light from the focus. In addition the one way pushing forces enhance a probability of one side approach of particles. Therefore the collision point between the particle and the bar is fixed near the focus, i.e. the edge of the bar on the focus. The linear assembly grows with the direction before the beam.

In conclusion, optical trapping potential with light pressure generates a linear assembly from sticky DNA particles. We also have demonstrated selective or sequential assembling by use of the method, and have made a complex structure from the assembled bars on the glass plate through the optical and mechanical manipulation [2, 6]. Such a novel method is expected to be applicable to assemble colloidal particles to an ordered structure.

REFERENCES

1. Ashkin, A., Dziedzic, J. M., Bjorkholm, J. E., and Chu, S., *Opt. Lett.*, **11**, 288–290 (1986).
2. Ichikawa, M., Matsuzawa, Y., Koyama, Y., and Yoshikawa, K., *Langmuir*, **19**, 5444–5447 (2003).
3. Bloomfield, V. A., *Biopolymers*, **31**, 1471–1481 (1991).
4. Minagawa, K., Matsuzawa, Y., Yoshikawa, K., Khokhlov, A. R., and Doi, M., *Biopolymers*, **34**, 555–558 (1994).
5. Yoshikawa, K., Yoshikawa, Y., Koyama, Y., and Kanbe, T., *J. Am. Chem. Soc.*, **119**, 6473–6477 (1997).
6. Matsuzawa, Y., Hirano, K., Mizuno, A., Ichikawa, M., and Yoshikawa, K., *Appl. Phys. Lett.*, **81**, 3494–3496 (2002).

Surface Undulation Appearing by Continuous Temperature Elevation of Supercooled Liquids on Metal Substrates

Hideyuki Nakayama, Kenji Ishikawa, Hirobumi Umeyama, and Kikujiro Ishii

Department of Chemistry, Gakushuin University, 1-5-1 Mejiro, Toshimaku, Tokyo, 171-8588 Japan

Abstract. Curious light scattering occurs during continuous elevation of temperature of supercooled liquid films on metal substrates. The phenomenon is observed for ethylbenzene and some related compounds when the temperature of the sample prepared by vapor deposition on cold substrate was raised to the point a little above the glass-transition temperature but still below the temperature where spontaneous crystallization takes place. The relaxation time of particular Fourier components of structural inhomogeneity that gives the light scattering is well related to the viscosity of the sample. Surface undulation of viscous liquid is suggested to occur being caused by the difference between the thermal-expansion coefficients of the supercooled liquid and metal substrate.

We have been studying the crystallization and glass-transition phenomena of amorphous molecular films prepared by vacuum deposition of organic vapor on cold metal substrates [1,2]. In the case of ethylbenzene (EB) and several related compounds, the samples undergo first the glass transition, and thus are in the supercooled liquid state until they undergo spontaneous crystallization finally by further temperature elevation. For monitoring the thickness and quality of the deposited samples, we record during the experiment the intensity of laser light that is reflected from the sample. This light is actually the superposition of the lights reflected at the top surface of the sample and at the interface between the sample and substrate, but we call this light the transmission light for convenience.

We have previously found that the samples of EB, propylbenzene, and isopropylbenzene show curious dip in the transmission-light intensity when the sample temperature is continuously elevated in their super-cooled-liquid states [1,2]. Interestingly, the intensity is improved for a while before the sample undergoes the final crystallization. We have studied X-ray diffraction and Raman scattering to see the structural change in the liquid in the temperature range of the dip, but no indication of the change was observed.

Recently, we found that a serious elastic light scattering occurs at the sample, diminishing the amount of the light entering into the detector of the transmission light. We report in this paper the results on EB samples, and discuss the cause of the light scattering in question. Undulation that appears at the surface of viscous liquids is suggested to be the cause.

The apparatus was essentially the same as that described in the previous paper [1]. Sample vapor was introduced into vacuum chamber and deposited on a gold-plated copper substrate at 78 K. The film thickness was made to be about 10 μm by monitoring the interference fringe of the transmission laser light [1,3]. We used a He-Ne laser and an Ar$^+$ ion laser, and employed several kinds of optical configuration for monitoring the light scattering. Each configuration will be described where the corresponding experimental result is quoted.

Figure 1 shows the intensity changes of the transmission light (A) and scattered light (B) observed for an EB sample. The light wavelength was 515 nm, and the incidence angle was 75 degree from the normal of the sample. The scattered light was collected by a lens placed in the direction of the sample normal. The sample was in the supercooled-liquid state at 120 K that is 3 K higher than the glass-transition temperature T_g. Plot A shows the dip of the transmission light in question around 124 K. The decrease around 130 K is due to the crystallization of the supercooled liquid. For the scattered light (plot B), it should be noted that a peak was observed in a temperature range remarkably narrower than that of the dip of plot A. Such a light scattering may be given by a Fourier component (periodical structure) of some inhomogeneity appearing in the sample, and is worthy to be called diffraction.

To confirm the above inference, we performed another experiment in which we used two lasers with different wavelengths, and recorded the normally scattered lights by alternately opening the shutters for lasers. Typical results are plotted in Fig. 2 clearly

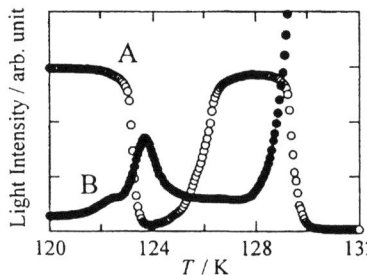

Fig. 1 Changes of transmission light (A) and scattered light (B) observed during the temperature elevation. See the text for the optical setup.

indicating that the change of the scattered intensity is different for the lights with different wavelengths. Thus different structural components are considered to have caused the scattering of lights of different wavelengths.

To explore the nature of the structural inhomogeneity causing the light scattering, we tried to monitor the decay of the scattering intensity at constant temperatures. The optical setup for the measurement was the same as the preceding experiment, but in this case, the sample temperature was first raised to an aimed annealing temperature and kept there during the further intensity monitoring. Typical results are displayed in Fig. 3. The abscissa of the figure is the time after the start of the experiment, and the change of scattered intensities is plotted along the ordinate as well as the sample temperature. The temperature was raised continuously to about 122.8 K, and was kept there afterwards. The intensity change of the scattered lights turned from increase to decrease when the temperature elevation was stopped. We estimated the relaxation time τ of the structural inhomogeneity from the inclination of the logarithmic plot of the scattering intensity. The results in Fig. 3 give 56 and 91 s respectively for the lights with wavelengths of 477 and 633 nm.

From similar experiments at different annealing temperatures, we obtained the temperature dependence of τ. The results plotted in Fig. 4 indicate that τ increases steeply as the annealing temperature is lowered. It is well known that viscosity of supercooled liquid increase steeply if the temperature is lowered toward T_g. We thus compared the temperature dependence of τ with that of viscosity.

Figure 5 shows the correlation between τ for samples at several annealing temperatures and viscosity η of EB liquid estimated for the corresponding temperatures. η in the low-temperature region was estimated from the literature data [4] by assuming the empirical Vogel-Fulcher equation. Although there may be a large ambiguity in the absolute magnitude of η, the results in Fig. 5 suggest that the inhomogeneity causing the light scattering in question relaxes being accompanied with viscous flow of the material. We consider that the surface undulation of the film sample may appear in the course of the thermal expansion of viscous liquid on the metal substrate. In this process, the stress due the difference in the thermal-expansion coefficients of the organic liquid and metal substrate, and also some inhomogeneity originally existed in the supercooled liquid, may play important roles.

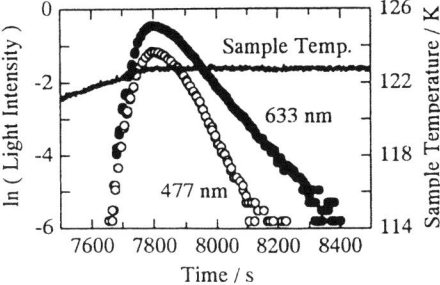

Fig. 3 Changes of scattering intensity at different wavelengths during the annealing. Temperature elevation was stopped at about 7800 s after the start of the experiment. Light intensity is indicated on the left in logarithmic scale, and the temperature is indicated on the right.

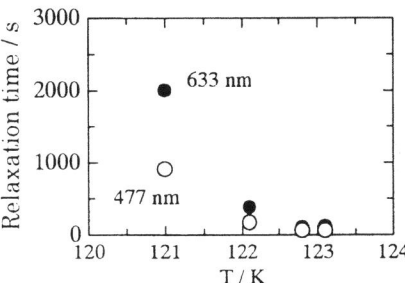

Fig. 4 Decay times of light scattering at 477 and 633 nm during the annealing of EB samples at different temperatures.

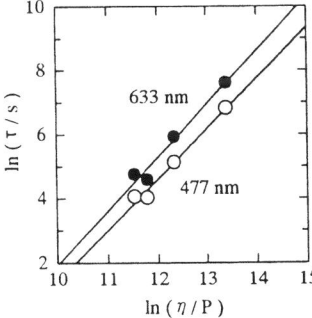

Fig. 5 Correlation between the decay time τ of scattering from EB samples at different annealing temperatures with the viscosity η at corresponding temperatures.

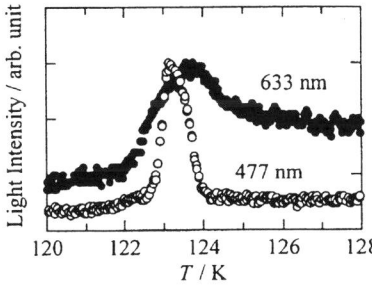

Fig. 2 Changes of scattered intensity of lights with different wavelengths during the temperature elevation. The optical configuration around the sample was the same as for the results in Fig. 1.

REFERENCES

1. K. Ishii, H. Nakayama, T. Okamura, M. Yamamoto, and T. Hosokawa, *J. Phys. Chem.*, **107** (2003) 876-881, and the references cited therein.
2. K. Ishii, T. Okamura, N. Ishikawa, and H. Nakayama, *Chem. Lett.* **30** (2001) 52-53.
3. K. Ishii, M. Yoshida, K. Suzuki, H. Sakurai, T. Shimayama, and H. Nakayama, *Bull. Chem. Soc. Jpn.*, **74** (2001) 435-440.
4. A. J. Barlow, J. Lamb, and A. J. Matheson, *Proc. Roy. Soc. A***292** (1966) 322-342.

Fluctuation Modes Prior to Lamellar-Double Gyroid Transition of a Nonionic Surfactant/Water System

M. Imai*, K. Nakaya* and T. Kawakatsu**

*Department of Physics, Ochanomizu University, Bunkyo, Tokyo 112-8610, Japan
**Department of Physics, Tohoku University, Aoba, Sendai 980-8578, Japan

Abstract. The nature of fluctuation modes of lamellar structure in a nonionic surfactant/water system has been investigated using a small angle x-ray scattering and a neutron spin echo technique. Approaching temperature from lamellar phase to double-gyroid phase, a diffuse scattering peak appears in the small angle scattering profile prior to the transition. This diffuse scattering is originated from the least stable fluctuation modes of lamellar structure predicted by Qi and Wang. The intermediate scattering profiles of the lamellar phase can be described by an undulation fluctuation mode and a least stable fluctuation mode of the lamellar structure.

INTRODUCTION

One of the most fascinating properties of surfactant/water systems and block copolymers is their ability to form a variety of ordered mesophases, such as hexagonally packed cylinder (C), lamellar (L), body-centered cubic (BCC) micelle and bicontinuous double-gyroid (DG) structures. It is interesting to note that the DG phase is an almost unique three dimensional periodic constant mean curvature (CMC) surface observed in surfactant/water systems and diblock copolymers, in spite that there are many other CMC minimal surfaces, such as double-diamond and double-P surfaces. From the phase transition kinetics point of view, characteristic intermediate structures are observed on the kinetic pathway from L phase to DG phase experimentally [1,2]. In the case of surfactant/water systems, the L to DG transition proceeds through a characteristic rhombohedral (R) network structure with $R_{\bar{3}c}$ symmetry, which is a subgroup of $I_{a\bar{3}d}$ symmetry for the DG structure. The intermediate structure is metastable and transform to the most stable DG phase spontaneously. Thus the R phase plays an important role to form the DG phase from the L phase.

Recently, we found that a modulation fluctuation layer (MFL) structure appears prior to the L to R transition in nonionic surfactant/water systems. The MFL structure gives a diffuse scattering peak at the first bragg peak position of the R phase. Thus the MFL may assist the formation of the R structure. In this study we investigate structure of the MFL using a small angle x-ray scattering (SAXS) and a neutron spin echo (NSE) technique and examine the MFL in terms of the least stable fluctuation modes (LSFMs) of the lamellar structure base on the Qi and Wang framework [3].

EXPERIMENT

In this study we used $C_{16}E_6$ as a surfactant, because the phase behavior is well investigated and the MFL structure of $C_{16}E_6$/water system gives fairly strong scattering intensity. A 55 wt% D_2O solution of $C_{16}E_6$ sample is used for the SAXS and the NSE experiments. The SAXS measurements were performed using BL-15A instrument at the photon factory (PF) in the high energy accelerator research organization (KEK) and Rigaku NANO-Viewer with confocal mirror. The dynamics experiment were carried out using a NSE spectrometer at C2-2 port of JRR-3M at Japan Atomic Energy Research Institute, Tokai.

RESULTS AND DISCUSSIONS

The 55 wt% $C_{16}E_6$ sample showed the L to R phase transition at 34 °C and in the temperature range between 42 and 34°C we observed the MFL structure

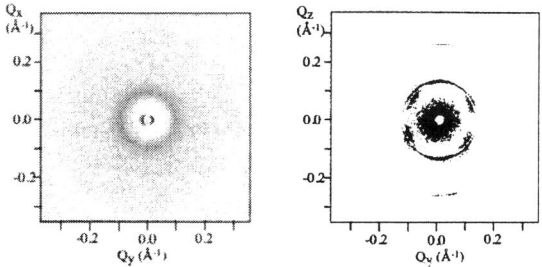

FIGURE 1. Two-dimensional scattering patterns (a) $S_\perp(Q_x, Q_y)$ and (b) $S_\parallel(Q_z, Q_x)$ for MFL structure.

in the SAXS patterns. Here we show scattering patterns of the MFL structure using the highly oriented sample in Figure 1. The $S_\perp(Q_x, Q_y)$ pattern where the x-ray beam irradiates perpendicular to the lamellar plane, shows a diffuse ring having radius of $\sim \sqrt{3}/2\, Q^*$ (Q^*: peak position of first lamellar peak). Here x and y axes are in the lamellar plane and z axis is perpendicular to the lamellar plane. On the other hand $S_\parallel(Q_y, Q_z)$ pattern where the x-ray beam irradiates parallel to the lamellar plane shows four diffuse spots at $Q_z \sim \pm Q^*/2$ and $Q_y \sim \pm \sqrt{3}/2\, Q^*$. In $S_\parallel(Q_y, Q_z)$ pattern we subtracted the Bragg peak component.

Qi and Wang [3] showed that the LSFMs of L structure lie on two rings at $Q_z = \pm Q^*/2$ and $|Q| = Q^*$. Thus the scattering from the LSFMs has a ring with $|Q_x| = |Q_y| = \sqrt{3}/2\, Q^*$ in (Q_x, Q_y) plane and four spots at $Q_z = \pm Q^*/2$ and $Q_y = \pm \sqrt{3}/2\, Q^*$ in (Q_y, Q_z) plane. Our experimental results demonstrate that the MFL structure observed in a nonionic surfactant/water system corresponds to the LSFMs of L structure predicted by Qi and Wang. The LSFMs appear at 47 °C and increase their scattering intensity until ~ 36 °C, but the increase of the diffuse scattering intensity is suppressed in the vicinity of T_{LR} and then the MFL structure transformed to the R structure.

In order to make clear the dynamical nature of the MFL structure, we measured intermediate scattering functions of MFL structures using the NSE technique. From the SAXS experiments, at 51 °C the lamellar fluctuations are governed by the undulations, whereas at 36°C the lamellar fluctuations consist of the undulation fluctuations and modulation fluctuations (LSFMs). Then we assumed that the $S(Q,t)$ profile consists of two components, the undulation fluctuation mode and the modulation fluctuation mode and the obtained intermediate scattering functions were fitted by a double-exponential function

$$S(Q,t) = a(Q)\exp(-\omega_1(Q)t) + (1-a(Q))\exp(-\omega_2(Q)t) \quad (1)$$

where ω_1 and ω_2 are relaxation rates. Figure 2 shows the Q dependence of the relaxation rates at 51 °C (UFL) and 36 °C (MFL). At 51°C, the ω_1 has a

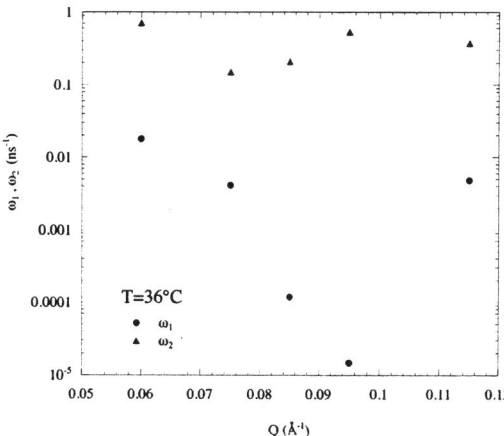

FIGURE 2. Q dependence of the relaxation rates w_1 and w_2 at 36 °C (MFL structure).

minimum at $Q \sim 0.09$ Å^{-1} corresponding to the first lamellar peak position, whereas the ω_2 is almost independent of Q. Thus the dynamical nature at 51°C is governed by the ω_1 mode. The relaxation rate of ω_1 at first lamellar peak position ($Q = 0.095$ Å^{-1}) is quite small ($\sim 10^{-5}$ ns^{-1}), indicating the scattering process at the Bragg peak position is purely elastic. Contrariwise at the tail of first lamellar peak, the ω_1 increases to ~ 0.01 ns^{-1}, indicating that a quasi-elastic process is involved at the Q positions. Taking into account that the profile of the first lamellar peak is well described by the scattering function for undulationally fluctuating lamellar (Caillé model), we attribute the ω_1 mode to the "out-of-plane" undulation fluctuations of lamellar structure. The MFL phase at 36°C, the undulation mode ω_1 shows the similar profile at 51 °C, whereas the ω_2 mode has a minimum at $Q \sim 0.075$ Å^{-1} corresponding to the peak position of the diffuse scattering. The temperature dependence of the relaxation rate ω_2 at $Q = 0.075$ Å^{-1} agrees well with the temperature dependence of the diffuse scattering intensity at $Q = 0.075$ Å^{-1} observed in the SAXS measurements. Thus the behavior of ω_2 mode is governed by the the LSFMs of L structure and the characteristic LSFMs have the relaxation time of about 8 ns.

REFERENCES

1. Hajduk, D.A., Ho, R.-M., Hillmyer, M.A., Bates F.S., and Almdal, K. *J. Phys. Chem.* B, **102**, 1356-1363(1998).
2. Imai, M., Saeki, A., Teramoto, T., *et al.*, J. Chem. Phys. **115**, 10525-10531 (2001)
3. Qi, S., and Wang, Z.-G., Macromolecules **30**, 4491-4497 (1997).

Structural Changes of an Immiscible Polymer Blend under Shear Flows and Electric Fields

Hiroshi Orihara[1] and Tetsunori Shibuya[2]

[1] *Center for Integrated Research in Science and Engineering, Nagoya University, Nagoya 464-8603, Japan*
[2] *Department of Applied Physics, Graduate School of Engineering, Nagoya University, Nagoya 464-8603, Japan*

Abstract. By using a new system that combines a confocal scanning laser microscope (CSLM) and a rheometer, we observed the structural change and measured the shear stress simultaneously in an immiscible polymer blend. The blend used is composed of a liquid crystalline plolymer (LCP) and a petroline mineral oil (PMO). When subjected to an external electric field, we observed a change from an LCP-droplet-dispersed structure to a network structure, corresponding to an increase of the shear stress.

INTRODUCTION

Several polymer blends have been found to show a large electrorheological (ER) effect[1]; when they are subjected to an electric field the viscosity increases. They are usually composed of two mutually immiscible polymers. We mainly used two kinds of liquid crystalline polymers (LCPs) and a polydimethylsiloxane (DMS), where the LCPs are isotropic at temperatures at which the ER effect appears. These blends are classified into two types[2,3]. In both types there are small droplets dispersed in the matrix; in one type (Type I) the droplets consist of LCP, while in the other (Type II) they consist of DMS. So far it has been found that the ER effect is related to the morphological changes of the blends under an electric field and shear[2-5].

Quite recently, we have constructed a new system that combines a confocal scanning laser microscope (CSLM) and a rheometer. By using it, we found that a network of LCP was formed in Type I blend by applying both a shear flow and an electric fields[6]. But the observation was made after removing them. It was difficult to observe the structure under a shear flow and an electric field, since the refractive indices of LCP and DMS are different and so the blend is not so optically transparent. In order to overcome the difficulty we used a petroline mineral oil (PMO), the refractive index of which is closer to that of LCP than DMS.

EXPERIMENT

The blend used in the present experiment was a mixture of an LCP (80 Pa s at 20 °C) and a PMO (0.03 Pa s at 25 °C) with the ratio of 1:3 in weight. The dielectric constant of the LCP is about 10, while that of the PMO are about 3, and the conductivity of the LCP is about 10^{-7} $\Omega^{-1}m^{-1}$, while that of the PMO is less than $10^{-14} \Omega^{-1}m^{-1}$. The sample was sandwiched between a bottom glass plate with an ITO-coated electrode and the rotating metal disk of a rheometer (M10 and RS20, Haake). The diameter of the disk was 20 mm. Since we used the parallel-plate rotational viscometer, the shear rate depends on its position and so we define the shear rate as the one at the periphery of the upper disk. The shear stress at the edge of the top plate was calculated from the torque by assuming that the fluids are Newtonian. The observation was made through the bottom glass plate with a confocal scanning laser microscope (Fluoview FV300, Olympus). To observe the structure clearly and to distinguish LCP and PMO, a small amount of fluorescent dye, IANBD amide (Molecular Probes), was doped to the LCP before mixing the LCP and the PMO. It was confirmed that the dye did not dissolve in the PMO. The wavelength of excitation was 488 nm. Electric fields were applied to the blend by a synthesizer (Model 1940, NF Electric Instruments) and a high voltage amplifier (Model 609C-6, Trek).

RESULTS AND DISCUSSIONS

The transient response of the blend was observed when subjected to an ac electric field of 2 kV$_{p-p}$/mm and 10 Hz under a constant shear rate of 40 s^{-1}. Figures 1 and 2 show the structural change and the time dependence of the shear stress, respectively. In Fig. 2 the fluid flows from right to left. Before applying the field (Fig. 2(a)), bright LCP droplets are seen. Although the droplets are elliptic and the major axes are not along the flow direction, this was caused by the artificial effect that the scanning rate was slow compared with the flow speed.

Immediately after applying the field, the shear stress suddenly increases as shown in Fig. 1. Corresponding to this increase the droplets are elongated along the flow direction and linked together to form stripe pattern (Fig. 2(b)). This fast increase in shear stress is called the fast mode and its origin is conjectured as follows. There are two kinds of torques exerted on the elongated droplets; one is the hydrodynamic torque due to the shear flow and the other is the electrical torque, which keeps the elongated droplet along the field. In the steady state the former must balance the latter, leading to the tilt of the elongated droplet from the direction of the field and the increase of apparent shear stress. At the rate stage of the first mode, on the other hand, taking into account that the viscosity of LCP is much larger than that of DMS, the LCP bridges between the top and bottom plates may increase the apparent viscosity.

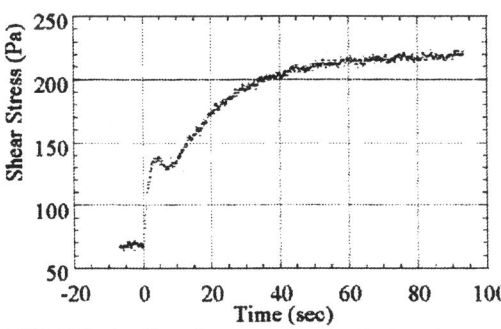

FIGURE 1. Transient response of the shear stress after applying an ac electric field at 0 sec.

Following the first mode, the shear stress gradually increases as shown in Fig. 1. This gradual increase is called the second mode. Corresponding to this increase in shear stress, the LCP stripes begin to link together (Fig. 2(c)) and finally change into a tree-dimensionally linked network (Fig. 2(d)). This indicates that the second mode is caused by the increase in the degree of LCP linkage. This network formation under shear flow and electric field is a significant feature of the present immiscible blend. The viscosity increase is due to the linkage of LCP with higher viscosity and the interface tension between LCP and PMO.

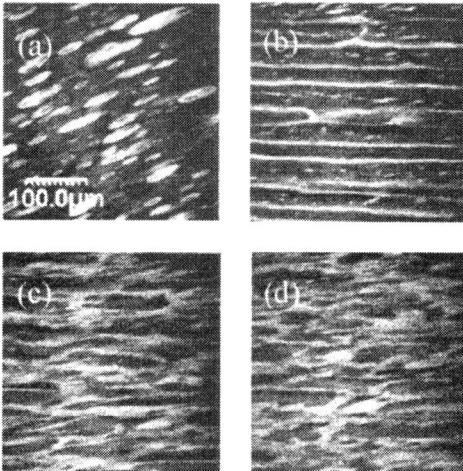

FIGURE 2. CLSM images (a) before applying the field, (b) at 3 sec, (c) at 34 sec and (d) at 74 sec.

ACKNOWLEDGMENTS

We would like to thank Dr. Inoue of Asahi Chemical Industry for supplying the LCP. This work was partly supported by a Grand-in Aid for Scientific Research (B) (Grant No. 13555046) from JSPS and a Grand-in-Aid for Scientific Research on Priority Areas (Grant No. 14045236) from MEXT

REFERENCES

1. Inoue, A., and Maniwa, S., *J. Appl. Polym. Sci.* **55**, 113-118 (1995).

2. Kimura, H., Aikawa, K., Masubuchi, Y., Takimoto, J., Koyama, K., and Minagawa, K., *Rheol. Acta* **37**, 54-60 (1998).

3. Orihara, H., Doi, M., and Ishibashi, Y., *Int. J. Mod. Phys.* B**13**, 1949-1955 (1999).

4. Tajiri, K., Ohta, K., Nagaya, T., Orihara, H., Ishibashi, Y., Doi, and M., Inoue, A., *J. Rheol.* **41**, 335-341 (1997).

5. Kimura, H., Aikawa, K., Masubuchi, Y., Takimoto, J., Koyama, K., and Uemura, T., *J. Non-Newtonian Fluid Mechanics* **76**, 199-211(1998).

6. Orihara, H., Ikeyama, Y., Ujiie, S., and Inoue, A., *J. Rheol.* **47**, 1299-1310 (2003).

Neutron Spin Echo Study on Slow Dynamics of Lipid Bilayers in the DPPC/D$_2$O/CaCl$_2$ System

T. Takeda[a], N. L. Yamada[b], Y. Kawabata[c], H. Seto[d] and M. Nagao[e]

[a]Faculty of Integrated Arts and Sciences, Hiroshima University, Higashi-Hiroshima 739-8521, Japan
[b]Graduate School of Bio-Sphere Science, Hiroshima University, Higashi-Hiroshima 739-8521, Japan
[c]Graduate School of Science, Tokyo Metropolitan University, Hachioji 192-0397, Japan
[d]Graduate School of Science, Kyoto University, Kyoto 606-8502, Japan
[e]NSL, ISSP, University of Tokyo, Tokai, Naka, Ibaraki 319-1106, Japan

Abstract. In order to study slow dynamics of lipid bilayers, neutron spin echo (NSE) experiments were carried out on the dilute lamellar phase in the DPPC/D$_2$O/CaCl$_2$ system. From the NSE results, we estimated the bending modulus κ of the bilayer using the theory presented by Zilman and Granek (Phys. Rev. Letters **77** (1996) 4788). The estimated values of κ decrease monotonically with increasing temperature and with increasing the lamellar repeat distance d_l. They depend strongly on d_l though d_l is longer than 500Å. We discuss validity of the theory for the analysis of the NSE results.

INTRODUCTION

In the dipalmitoylphosphatidylcholine(DPPC) /water system, there are various multilamellar phases, in which lipid bilayers and water are stacked alternately. The lamellar repeat distance d_l varies greatly with an addition of salt in the DPPC/water system [1]. In order to study dynamics in undulation of lipid bilayers, NSE experiments were carried out on the dilute lamellar liquid crystalline phase in the DPPC/D$_2$O/CaCl$_2$ system with d_l longer than 500 Å in order to avoid the effect on the single membrane dynamics from neighbouring sheets of membranes.

EXPERIMENTAL

For sample preparation, DPPC was dispersed in D$_2$O with 7mM CaCl$_2$. Small angle X-ray scattering (SAXS) experiments were carried out using BL15A at PF in KEK and BL40B2 at SPring8 and small angle neutron scattering (SANS) experiments using SANS-U at C1-2 port of JRR-3M in JAERI(Tokai). The NSE experiments were performed using ISSP-NSE at C2-2 port of JRR-3M [2].

RESULTS AND DISCUSSION

In SANS experiments, we observed the peaks corresponding to d_l. On the other hand, we observed the scattering corresponding to the correlation between the polar head parts of a lipid bilayer instead of the lamellar peaks for the samples with d_l larger than 280Å in SAXS experiments. The disappearance of diffraction peaks in SAXS reflects the X-ray form factor of a lipid bilayer for the lamellar peaks which approaches to zero because the undulation of the bilayer increases with increasing d_l. The values of d_l obtained from the SANS experiments are roughly in inverse proportion to the DPPC concentration c in the dilute lamellar region as shown in Fig. 1.

The scattering vector Q and time t dependent intermediate functions $I(Q,t)$ obtained from the NSE experiments were well fitted to the following equation,

$$I(Q,t) = I(Q,0)\exp[-(\Gamma t)^{2/3}] \quad . \quad (1)$$

The relaxation rates Γ obtained from the fitting to Eq.(1)

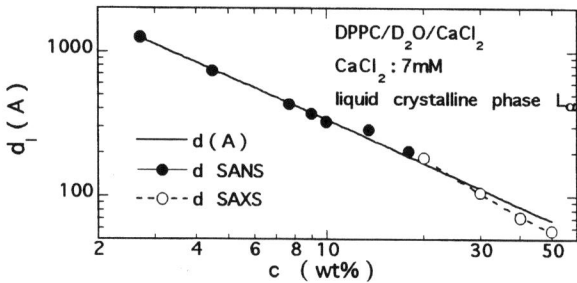

Fig. 1. DPPC concentration c dependence of the lamellar repeat distance d_l in DPPC/D$_2$O/CaCl$_2$ system with 7mM CaCl$_2$. The solid line indicates c^{-1} proportionality.

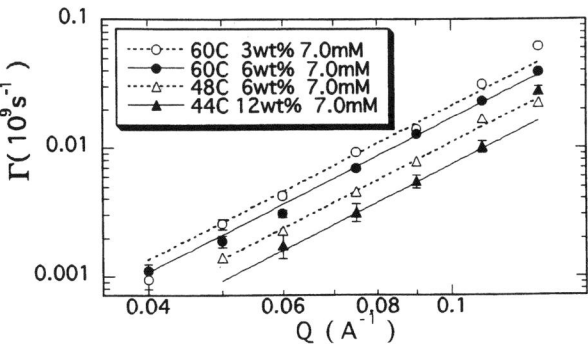

Fig. 2. Q dependence of the relaxation rate Γ obtained using the fitting to Eq.(1). The lines are fitting curves to Eq.(2) at Q lower than 0.13 Å^{-1} and look like indistinguishable from fitting one to aQ^3 where a is constant

increased as Q^3 over the range of Q from 0.05 Å^{-1} to 0.12 Å^{-1} as shown in Fig. 2. So with the case of the non-ionic surfactant n-dodecyl pentaoxylethylene glycol ether ($C_{12}E_5$)/n-octane/ D_2O system [3], these NSE results supported the theory presented by Zilman and Granek (ZG) [4]. They predicted a stretched exponential relaxation of $I(Q,t)$ for large Q in sponge and lamellar phases as follows Eq. (1) where the relaxation rate Γ is given by

$$\Gamma = 0.025\gamma(k_BT/\kappa)^{1/2}(k_BT/\eta)Q^3 \quad . \quad (2)$$

Here, κ is the bending modulus of the membrane and η the viscosity of the surrounding medium. For $k_BT \ll \kappa$, γ is given by

$$\gamma = 1 - 3\ln(Q\xi)k_BT/4\pi\kappa \quad , \quad (3)$$

where ξ is a typical size of the mesoscopic structure.

The rates Γ obtained from the fitting to Eq.(1) were also well fitted to Eq.(2) with Eq.(3) with $\xi = d_l$ as shown in Fig. 2. The values of κ estimated using Eq.(2) from

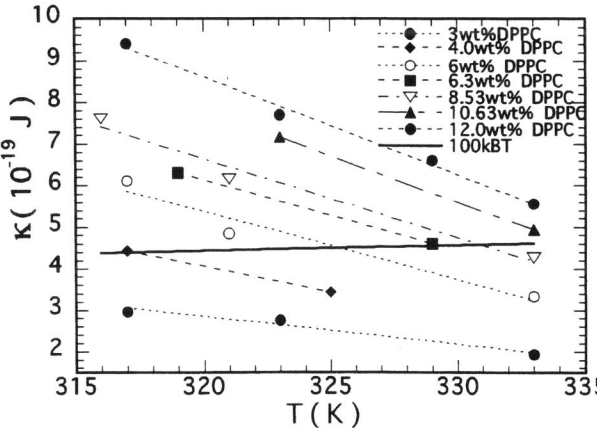

Fig. 3. Temperature dependence of the estimated bending modulus κ from NSE experiments. The solid line indicates $100k_BT$.

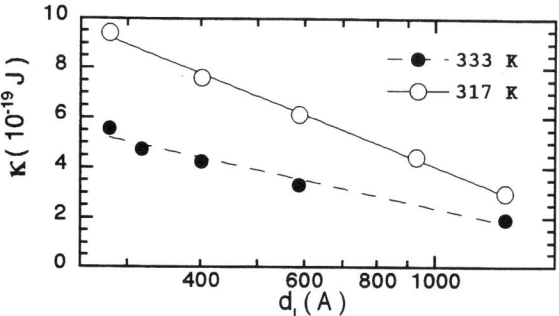

Fig. 4. Dependence of the estimated κ on the lamellar repeat distance d_l obtained from NSE experiments.

the results of the NSE experiments decrease monotonically with increasing temperature as shown in Fig. 3 and with increasing d_l as shown in Fig. 4. The values from the sample with 3wt% DPPC are nearly the same as that obtained using the method of electric-field-induced bending deformation of the cylindrical tubes [5]. The estimated values of κ depend strongly on d_l though d_l is longer than 500Å as shown in Fig. 4. Since κ of the lipid bilayer is not considered to depend so strongly on d_l, the above analysis using ZG theory should be improved. The deviation at Q higher than 0.13 Å^{-1} as shown in Fig. 2 suggests that the effect of the thickness of the membrane which is neglected in the ZG model plays a significant role in dynamics at higher Q.

Since $I(Q,t)$ and $\Gamma(Q,\kappa)$ behave in the manner predicted by ZG, their theory describes well membrane undulations in these complex fluids involving and is useful to analyze NSE data though the analysis using ZG is open to further improvement.

ACKNOWLEDGEMENTS

These experiments at JRR-3 were done under the approval of the Neutron Scattering Program Advisory Committee. One of the authers (T. T.) was financially supported by the Grant-in-Aid for Scientific Research (No. 09640466, No. 13440119) from the Japanese Ministry of Education, Culture, Sports, Science and Technology.

REFERENCES

1. Takeda, T., Ueno, S., Kobayashi, H., Komura, S., Seto, H. and Toyoshima, Y., *Physica B* **213&214** 763-765 (1995).
2. Takeda, T., Seto, H., Kawabata, Y. et al, *J. Phys. Chem. Solids* **60** 1599-1601 (1999).
3. Takeda, T., Kawabata, Y., Seto, H., Komura, S. and Nagao, M., *J. Phys. Soc. Jpn.* **70** Suppl. A 323-325 (2001).
4. Zilman, A. G. and Granek, R., *Phys. Rev. Letters* **77** 4788-4791 (1996).
5. Mishima, M., Nakamae, S., Ohshima, H. and Kondo, T., *Chem. Phys. Lipids* **110** 27-33(2001).

Temperature- and Pressure-dependences of a Bending Modulus of Surfactant Monolayers in a Ternary Microemulsion Composed of AOT / D$_2$O / decane

H. Seto[*], Y. Kawabata[†], M. Nagao[**] and T. Takeda[‡]

[*]*Department of Physics, Kyoto University, Kyoto 606-8502, Japan*
[†]*Department of Chemistry, Tokyo Metropolitan University, Hachioji, 192-0397, Japan*
[**]*Institute for Solid State Physics, The University of Tokyo, Tokai 319-1106, Japan*
[‡]*Faculty of Integrated Arts and Sciences, Hiroshima University, Higashihiroshima 739-8521, Japan*

Abstract. A bending modulus of surfactant monolayers in a microemulsion system composed of AOT, D$_2$O and deuterated decane were investigated by means of neutron spin echo. In this system, a water-in-oil droplet structure at ambient temperature and pressure decomposes into two-phases both with increasing temperature and pressure. However, the bending modulus slightly decreased with increasing temperature while it increased with increasing pressure. This behavior was explained in terms of a microscopic model proposed by Würger (Phys. Rev. Lett. **85** (2000) 337). It was confirmed that an increase of an interaction between hydrocarbon tails of AOT molecules with increasing pressure could be an origin of the pressure-induced phase transition.

INTRODUCTION

In these decades much interest has been focused on the structural formation of microemulsion systems. It has been recognized that the elastic properties of surfactant monolayers play a key role to control the intricate phase behavior. Helfrich introduced the concepts of a spontaneous curvature, R_s^{-1}, a bending modulus κ and a saddle-splay modulus $\bar{\kappa}$ and a bending energy of surfactant layers to be, [1]

$$H = \int \left[\frac{\kappa}{2} \left(\frac{1}{R_1} + \frac{1}{R_2} - \frac{2}{R_s} \right)^2 + \bar{\kappa} \frac{1}{R_1 R_2} \right] dS, \quad (1)$$

where R_1^{-1} and R_2^{-1} are the two principal curvatures. Some of these elastic parameters have been well investigated by means of neutron spin echo (NSE), because only surfactant layers could be observed by neutrons in case that deuterated oil and water were used. [2, 3]

A ternary mixture of AOT (Aerosol-OT), water and n-decane is the most popular system to investigate the structural formation of microemulsion systems. At ambient temperature and pressure, a one-phase water-in-oil droplet structure is conformed, and it decomposes into a coexisting phase with a droplet-rich phase and a droplet-poor both with increasing temperature and pressure. This evidence indicated that the effects of temperature and pressure are the same on the phase behavior. [4] However, the microscopic origins of the structural changes with varying temperature and pressure should not be the same. [5, 6] In order to clarify the origins of the temperature- and the pressure-induced phase transitions, a dynamical behavior of AOT monolayers was investigated by means of neutron spin echo (NSE).

EXPERIMENTAL RESULTS

5.4 vol. % of AOT, 4.6 vol. % of D$_2$O, and 90 vol. % of d-decane was mixed. The R_0 was evaluated by small-angle neutron scattering (SANS); it decreased from 33 Å to 28 Å with increasing temperature between 23 and 80°C, while it remained to be 32 Å with increasing pressure up to 66.5 MPa. The polydispersity index p, the phase separation temperature and pressure were also evaluated by SANS.

NSE experiments were performed at ISSP-NSE spectrometer at JRR-3M, JAERI, Tokai. Temperature-run experiments at ambient pressure were done between 10 and 65 °C with a water-cooling / heating bath system controlled within 0.1 °C, and pressure-run experiments at room temperature were done at $P = 0.1, 20, 40,$ and 60 MPa with a non-magnetic high-pressure cell made of stainless steel. [7]

The NSE data were analyzed in terms of the model proposed by Milner and Safran. [2, 8] From the analysis, the decay rate of the peanuts-like deformation $l = 2$

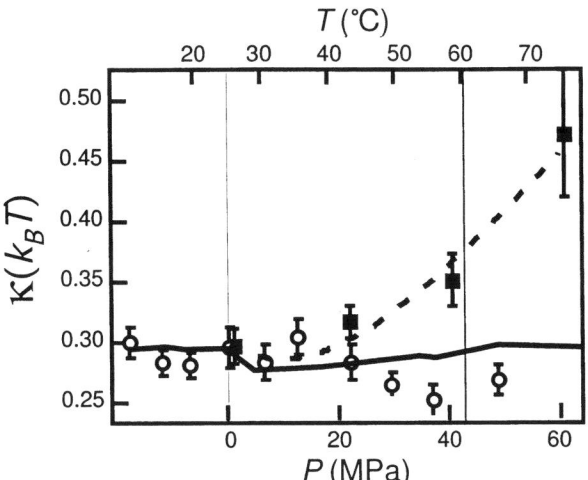

FIGURE 1. Temperature- and pressure-variation of the bending modulus. The vertical axis indicates the bending elasticities. The thin vertical lines indicate ambient temperature and pressure (left), and the phase separation temperature and pressure (right), respectively. The open circles are the temperature-dependence and the full square the pressure-dependence. The lines are the calculated values following the Würger's model.

mode, Γ_2, was obtained. The bending modulus of the AOT monolayers could be estimated from the relation,

$$\kappa = \frac{1}{48}\left(\frac{k_B T}{\pi p^2} + \Gamma_2 R_0^3 \frac{23\eta' + 32\eta}{3}\right), \quad (2)$$

where η and η' the viscosities of inside and outside of a droplet, respectively. [9]

In Figure 1, temperature- and pressure-dependences of the bending modulus κ were shown. With increasing temperature, κ slightly decreased, on the other hand, it increased with increasing pressure. This means that the AOT monolayers become floppy with increasing temperature while stiff with increasing pressure. This tendency is the same as a result obtained by Nagao et al. for the same system at dense droplet region. [6] In their data analysis, another model describing a single membrane fluctuation [10] was used. The consistency of these independent experiments and analyses assured a model independence of the result.

DISCUSSION AND CONCLUSION

One possible microscopic model to explain the bending modulus κ is the one proposed by Würger. [11] They omitted a contribution of the molecular conformation to the bending elasticities, and retain the configurational average in the surfactant layer only. It depends on the inter-molecular pair potential v and the area per molecule a_H, which change with the inter-molecular distances thus with curvature. They simplified the inter-molecular potential by introducing a model of dangling rods that consists of N rigid molecules with a Lennard-Jones potential between hydrophobic tails with the potential minimum existing at $r = \sigma = 4.85$ Å and the potential depth being $v(\sigma) = -\varepsilon = -15$ meV. The temperature dependence of the bending modulus was well reproduced as shown by the full line in Figure 1 with the temperature dependence of a_H estimated from the SANS experiment.

In order to explain the pressure-dependence of κ, we assumed that the interaction between hydrophobic tails of AOT molecules depends linearly on pressure; $\sigma = 4.85 + xP, \varepsilon = 15 + yP$. The dashed line in Figure 1 indicates the fit result of the pressure dependence of the observed bending modulus with this assumption. The fit parameters were obtained to be $x = -0.029$ Å/MPa and $y = 0.037$ meV/MPa. This result was consistent with the fact that the attractive interaction between hydrophilic tails of surfactant molecules increases with increasing pressure. [5]

From these experiments and analysis, the temperature- and pressure-dependences of the bending modulus of AOT monolayers was investigated by means of neutron spin echo. It was clearly shown that the monolayers becomes floppy with increasing temperature, while rigid with increasing pressure. These dependencies were interpreted with a microscopic model describing the bending elasticities by an interaction between hydrophobic tails of surfactant molecules. These results confirmed that the microscopic origins of the temperature- and pressure-induced structural change of the microemulsion system composed from AOT, water and decane.

REFERENCES

1. Helfrich W., *Z. Naturforsch.* **28c**, 693 - 703 (1973).
2. Huang J. S., Milner S. T., Farago B. and Richter D., *Phys. Rev. Lett.* **59**, 2600 - 2603 (1987).
3. Hellweg T. and Langevin D., *Physica A* **264**, 370 - 387 (2000).
4. Nagao M. and Seto H., *Phys. Rev. E* **59**, 3169 - 3176 (1999).
5. Seto H., Okuhara D., Kawabata Y., Takeda T., Nagao M., Suzuki J, Kamikubo H. and Amemiya Y., *J. Chem. Phys.* **112**, 10608 - 10614 (2000).
6. Nagao M., Seto H., Takeda T. and Kawabata Y., *J. Chem. Phys.* **115**, 10036 - 10044 (2001).
7. Kawabata Y., Seto H., Nagao M. and Takeda T., *J. Neut. Res.* **10**, 131 - 136 (2002).
8. Milner S. T. and Safran S. A., *Phys. Rev. A* **36**, 4371 - 4379 (1987).
9. Seki. K and Komura S., *Physica A* **219**, 253 (1995).
10. Zilman A. G. and Granek R., *Phys. Rev. Lett.* **77**, 4788 - 4791 (1996).
11. Würger A., *Phys. Rev. Lett.* **85**, 337 - 340 (2000).

Adsorption of Magnetic Nanoparticles onto Polyacrylamide Chains in Dilute Polymer Solutions and Ferrogel Networks

Delphine El kharrat*, Olivier Sandre*, Pedro Licinio*[†], and Régine Perzynski[†]

*Laboratoire Liquides Ioniques et Interfaces Chargées UMR7612 Centre National de la Recherche Scientifique
[†]Laboratoire Milieux Désordonnés et Hétérogènes UMR7603 Centre National de la Recherche Scientifique
/ Université Pierre et Marie Curie – 4, place Jussieu *case 63 / [†]case 78, 75252 Paris cedex 05, France

Abstract. We study iron oxide nanoparticles stabilized by citrate ligands interacting with long linear poly(acrylamide) chains in the regime where the diluted chains are decorated by many smaller magnetic nanoparticles. The strength of adsorption of the particles onto polymer increases when the unbound citrate concentration decreases, as evidenced by the faster translational dynamics of the particle–polymer complexes and the slower rotational dynamics of the nanoparticles.

INTRODUCTION

A novel class of nanocomposite materials called "ferrogels" with possible applications as magnetic actuators can be obtained by embedding magnetic nanoparticles in a soft polymer matrix highly swollen by water [1]. In a previous work, we have studied such magnetic hydrogels made of the combination of a poly(acryamide) hydrogel cross-linked by N,N'-methylene-bis-acrylamide and an aqueous "citrated ferrofluid" [2]. The latter is a suspension of iron oxide nanoparticles in water at pH≈7 stabilized by surface charges provided by citrate ligands in equilibrium with unbound tri-sodium citrate electrolyte. We found that the swelling degree Q of these ferrogels (defined as the amount of absorbed water per dry polymer weight) is lower for ferrogels compared to undoped hydrogels. In addition, Q presents a minimum value as a function of salinity at low citrate salt concentration. To understand this specific role of the citrate ligands on the polymer–particle interactions, we study here a somehow simpler system consisting of linear poly(acrylamide) chains having a calculated radius of gyration in water around 100nm mixed with the same magnetic nanoparticles in different citrate buffers. The resulting mixtures are fluid and can be studied by different dynamical methods: macroscopic viscosimetry, dynamic light scattering and oscillatory magneto-birefringence in cross-fields, the latter probing the micro-rheology in the local environment of the nanoparticles [3].

MATERALS

We used a commercial poly(acrylamide) (PAM) homopolymer (Polysciences) known as to strongly interact with metal oxide colloids. This very long linear PAM (5–6x10^6g/mol) is commonly used as a flocculent. As for the magnetic nanoparticles, they came from ionic ferrofluids prepared by alkaline co-precipitation of $FeCl_2$ and $FeCl_3$ followed by complete oxidation using $Fe(NO_3)_3$ leading to positively charged γ-Fe_2O_3 (maghemite) nanoparticles in HNO_3 (pH=1.2). A size-sorting process yielded a fraction of narrower distribution of diameters, described by a Log-normal law of parameters d_0=6.6nm and σ=0.21. Finally the iron oxide surface was coated by tri-sodium citrate ligands leading to a stable dispersion of negatively charged nanoparticles at pH=7.2.

After complete dissolution in tri-sodium citrate of the polymer at a concentration 0.6 g/L lower than c* (which is estimated about 2.4 g/L), the citrated ferrofluid was introduced in the solutions and allowed to equilibrate overnight. Two values of the final salt concentration [Na_3Cit] were examined, respectively a low concentration (8mM) and a larger one (50mM). The volume ratio of the nanoparticles relatively to the polymer was kept at a constant value (=2) all along the dilutions required for the viscosimetry experiment.

TABLE 1. Intrinsic viscosity [η] (g^{-1}.mL) measured by capillary viscosimetry for PAM (5–6x10^6) solutions mixed with 2 volume equivalents of citrated ferrofluid.

[sodium citrate] =	50mM	8mM
Polymer only	535	595
Polymer + particles	495	435

RESULTS

Capillary viscosimetry shows that the conformation of chains with and without nanoparticles does not vary a lot at high citrate concentration (8% decrease at 50mM citrate), whereas the presence of nanoparticles decreases the chain swelling significantly when the citrate concentration is lower (27% decrease at 8mM). The same effect of lowering the citrate concentration on the shrinking of the polymer chains by the nanoparticles is also observed by dynamical light scattering, the hydrodynamic diameter of the polymer-ferrofluid complexes varying from $d_H=107$nm at 50mM of unbound citrate down to $d_H=78$nm at 8mM.

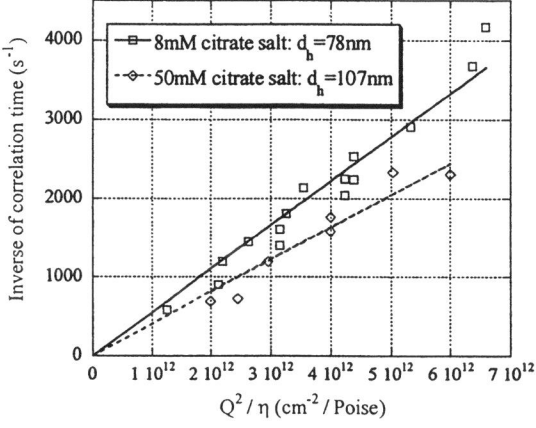

FIGURE 1. Dynamic light scattering: the hydrodynamic diameters of the polymer–particles complexes at the two citrate salt concentrations are measured from the slope of the main relaxation rate (from CONTIN fit of the correlogram) vs. the square of wave vector divided by viscosity.

With cross-fields magneto-birefrengence we get the frequency response of nanoparticles rotation in their local environment. In the presence of polymer, the visco-elastic behavior is evidenced by the phase shift at low frequency, increasing when [Na$_3$Cit] decreases. Therefore we can conclude that the nanoparticles strongly interact with the long linear poly(acrylamide) chains and that this coupling becomes stronger at lower unbound citrate salt concentration in equlibrium with the citrate ligands. An adsorption of the particles onto the chains is thus very probable, for it explains the specific effect of the citrate concentration by a competition between the citrate ligands and the polar amide groups of the polymer to access the surface of iron oxide. In the case of complete adsorption, we calculate that on average 10^2 particles can adsorb on a single chain. Having proved that the polymer–particles complexes behave as microgels swelling and shrinking in a similar way as cross-linked ferrogels, we showed by AFM that they also exhibit a necklace morphology.

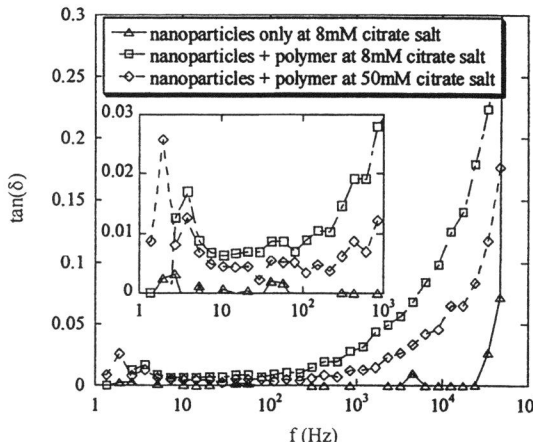

FIGURE 2. Cross-fields dynamical birefringence: the non zero phase shift angle between the magnetically induced birefringence and the ac H field is a sign of visco-elasticity.

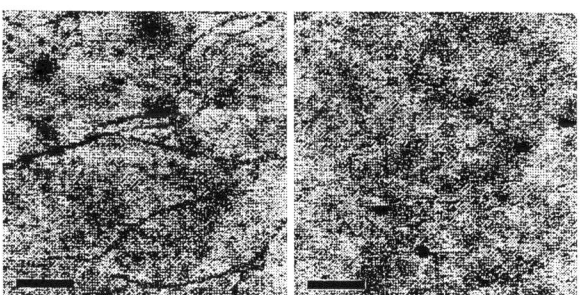

FIGURE 3. AFM pictures of PAM solutions mixed with 2 vol. of citrated ferrofluid deposited on mica at 8mM (left) and 50mM citrate (right, scale bars=500nm). Necklaces of spheres about 80nm in diameter linked to rigid segments up to 1μm long are thought to be particles-shielded polymers.

REFERENCES

1. Zrínyi, M., Barsi, L., and Büki, A., *J. Chem. Phys.* **104**, 8750-8756 (1996).

2. Galicia, A. Sandre, O. Cousin, F. Guemghar, D. Ménager, C., and Cabuil, V., *J. Phys. Cond. Mat.* **15**, S1379-S1402 (2003).

3. Hasmonay, E. Dubois, E. Neveu, S. Bacri, J.-C., and Perzynski, R., *Eur. Phys. J.* B **21**, 19-29 (2001).

Slow dynamics approaching the glass transition in repulsive magnetic fluids

G. Mériguet*, E. Dubois*, V. Dupuis[†] and R. Perzynski[†]

*Laboratoire des Liquides Ioniques et Interfaces Chargées LI2C, UMR CNRS 7612,
case 51, 4 place Jussieu, 75252 Paris cedex 05, France*
[†]*Laboratoire des Milieux Désordonnés et Hétérogènes LMDH, UMR CNRS 7603
case 78, 4 place Jussieu, 75252 Paris cedex 05, France*

Abstract. We study the dynamics of concentrated ionic magnetic colloidal dispersions, which are constituted of $\gamma-Fe_2O_3$ nanoparticles dispersed in water, and stabilized with electrostatic interparticle repulsion, using magneto-optical birefringence measurements. By gradually increasing the volume fraction Φ of the particles at constant ionic strength in the repulsive region of the phase diagram, we observe a dramatic increase of the characteristic time associated with the rotation of the particles that we induce by applying a field pulse. This increase is reminiscent of the divergence of the relaxation time observed at the approach of a glass transition and confirms the existence of a glassy phase in these magnetic colloids.

Ionic magnetic colloidal dispersions are constituted of nanometric magnetic particles dispersed in water [1, 2]. Widely used for technical applications, they are also of fundamental interest as versatile systems of interacting hard spheres in which the interparticle interactions can be tuned by varying external parameters.

The colloidal dispersions in which we are interested here are constituted of nanocrystals of maghemite γ-Fe_2O_3 (diameter around 8 nm) dispersed in water, and stabilized by electrostatic repulsion [2] (ionic strengths $I = 0.003\,M$ and $I = 0.03\,M$). We have previously shown that the interparticle interactions can be continuously tuned by varying the osmotic pressure π through the ionic strength I [3]. These parameters, as well as the volume fraction ϕ control the phase behavior and the resulting phase diagram is presented in fig. 1. At high π, only fluid and solid phases are obtained while, at low ϕ, gas, liquid and solid phases are obtained, a unique behavior for an electrostatically stabilized colloid.

We focus here on the transition toward the solid phases, in the region of high π above the critical point (repulsive fluid-solid transition). The amorphous structure factors evidenced in the solids by Small Angle Neutron Scattering strongly suggest they are glasses [5]. Macroscopically, I being constant, the viscosity increases with the volume fraction until obtaining samples that do not flow anymore, and that we call solids. However, between fluids and solids, a domain of samples flowing over long periods of times (weeks or months) exists.

The dynamical properties of the nanoparticles in these

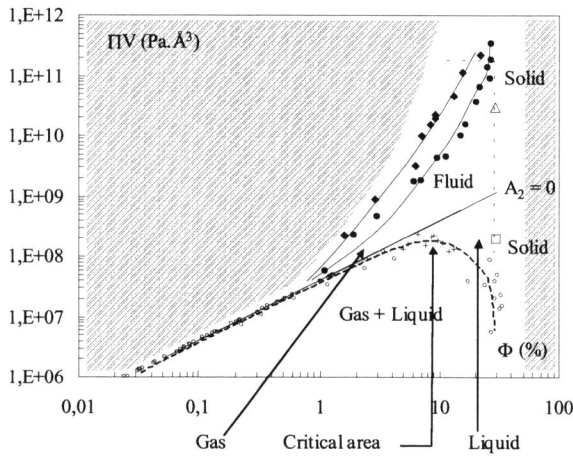

FIGURE 1. Phase diagram ΠV versus Φ from ref. [4]. The open dots correspond to the coexistence lines of the gas-liquid transition and the dashed line is a guide for the eye showing the location of the critical point. The straight line corresponds to ΠV for $A_2 = 0$ and separates the repulsive (top) region of the phase diagram from the attractive (bottom). The dotted line is an evaluation of the frontier between solid and fluid phases. Filled diamonds and filled circles correspond to the two sets of samples (ionic strengths $I = 0.003\,M$ and $I = 0.03\,M$ respectively) studied in this paper.

dispersions are studied using several techniques specific to these magnetic dispersions. (i) Each particle bears a dipole moment μ ($\sim 10^4 \mu_B$), which aligns along an applied magnetic field **H**. The measurement of the magnetization **M** versus **H** gives the ability of orientation of the dipole moments. (ii) Each particle also bears a uniaxial optical anisotropy linked to μ through the magnetic energy of anisotropy of the nanocrystal ($\sim 5\ 10^{-21}$ J) [6].

In a fluid dispersion, without field, the crystal axes are oriented at random and no optical birefringence is observed. The orientation of the particles via their magnetic moment under an applied field induces a magnetobirefringence Δn in the solution due to the mechanical orientation of the nanoparticle axes. For dilute samples (independent nanoparticles) the dynamical optical response after a low field pulse is a quasi-exponential relaxation of Δn. Its analysis allows determining the characteristic time of rotational diffusion τ of the nanoparticles in the fluid dispersion ($\sim 4\ \mu s$). The relaxation of μ (and **M**), being driven by a Néel process, is here much faster ($\sim 10^{-9}$ s).

While *increasing ϕ towards the glass transition*, the time τ increases drastically. The relaxation becomes non exponential: much slower relaxation times appear. The signal is here fitted with a pure exponential relaxation (τ_1) and a stretched exponential relaxation (τ_2, exponent β ranging between 0.5 and 0.25). The time τ shifts from microseconds for dilute dispersions to hours for concentrated dispersions (fig. 1). *There is a variation of 9 orders of magnitude of the rotational dynamic in these systems!*

$$I(t) = I_1 \exp\left(-\frac{t}{\tau_1}\right) + I_2 \exp\left(-(\frac{t}{\tau_2})^\beta\right) \quad (1)$$

This slowing down is comparable in magnitude to what is observed with more standard molecular systems, when the glass transition is approached. It corresponds to the ϕ domain (above $\phi=15\%$) where a divergence of the macroscopic viscosity is observed. It could be related here to the growing of correlation domains inside the colloid. However it is associated with a decrease of the maximum signal $\Delta n/\phi$ above $\phi = 15\%$ while on the contrary, the normalized magnetization M/ϕ remains constant. It means that mechanical rotation of the nanocrystals is hindered while the dipole μ rotates. In the solid, $\Delta n/\phi$ is very small and the measured relaxation time is then back to microseconds. Still few particles seem able to rotate individually.

Preliminary results with cobalt ferrite nanoparticles show similar trends. However, because their magnetic energy of anisotropy is much higher ($\sim 2\ 10^{-19}$ J), the dipole moment μ is then strongly linked to the nanocrystal structure. Thus magnetic relaxation can also exhibit slow dynamics.

In this paper, we have investigated by dynamic magneto-optical birefringence measurements how the rotational dynamics of γ-Fe_2O_3 nanoparticles dispersed in water changes as we increase the volume fraction of the particles. We observe a dramatic increase of the rotational characteristic time and the appearance of a stretched exponential relaxation that reminds of what is observed at the approach of a conventional glass transition. In the near future, we plan to extend this study to other types of nanoparticles such as the cobalt ferrites mentioned previously, in order to probe the effect of the dipolar interactions on the observed glass transition. We also want to investigate the fluid-solid transition in the attractive region of the phase diagram which may be of a different nature from the one we have evidenced here as can be inferred from the strongly different structure factors observed in SANS [5]. At last, we want to compare the freezing of the rotational degrees of freedom of the particles observed here with the sol-gel transition observed when the aggregation of the particles is strongly enhanced by reducing the electrostatic repulsions that stabilize the dispersion [7].

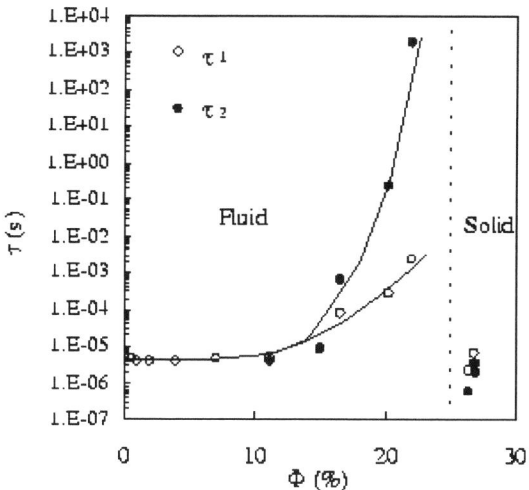

FIGURE 2. Plot of relaxation time τ versus volume fraction for maghemite nanoparticles in water.

REFERENCES

1. Rosensweig, R., *Ferrohydrodynamics*, Cambridge University Press, Cambridge, 1985.
2. Berkovsky, B., *Magnetic Fluids and Applications Handbook*, Begell House, New York, 1996.
3. Cousin, F., Dubois, E., and Cabuil, V., *Phys. Rev. E*, **68**, 021405-1–9 (2003).
4. Dubois, E., Perzynski, R., Boué, F., and Cabuil, V., *Langmuir*, **13**, 5617–5625 (2000).
5. Cousin, F., Dubois, E., Cabuil, V., Boué, F., and Perzynski, R., *Brazilian J. Phys.*, **31**, 350–355 (2001).
6. Gazeau, F., Bacri, J.-C., Gendron, F., Perzynski, R., Raikher, Y. L., Stepanov, V., and Dubois, E., *J. Magn. Magn. Mat.*, **186**, 175–187 (1998).
7. Hasmonay, E., Bee, A., Bacri, J.-C., and Perzynski, R., *J. Phys. Chem.*, **103**, 6421–6428 (1999).

The dynamics of granular matter

B.M. Schulz[*], M. Schulz[†] and S. Herminghaus[**]

[*]FB Physik, Martin-Luther-Universität Halle-Wittenberg, 06099 Halle (Saale), Germany
[†]Abteilung für Theoretische Physik, Universität Ulm, 89069 Ulm, Germany
[**]Abteilung Angewandte Physik, Universität Ulm, 89069 Ulm, Germany

Abstract. We present a numerical study of a shear-induced solid-fluid transition in wet granular matter in order to show the self-organized critical behavior close to the transition point. The continuous time simulation is based on a simple model considering both the cohesive forces induced by the adsorbed liquid amount and the repulsive forces due to the hard core interaction of the granules. Dissipation is assumed to be entirely due to the hysteretic character of the cohesive forces. The aim of our analysis concerns the crossover from a solid like behavior to a mobile ergodic state under the influence of an external force field F exceeding the critical force F_c. Diffusion coefficients, dissipation and kinetic order parameters can be expressed as characteristic scaling laws.

Granular matter, like sand, soil, or gravel, has been studied closely by scientists and engineers for over hundred years. However, since these materials are inherently far off thermal equilibrium, attempts to treat granular matter with the methods of statistical physics are comparably young [1]. By far most of the work has been concerned with dry granular matter although the much more complicated case of wet granular material is also the much more important one. Any finite humidity leads to a thin layer of water on virtually all surfaces. Dramatic effects like soil liquefaction, with devastating land slides as a possible consequence, are believed to be due to the interplay of the liquid with the grain piling [2].

As a handy illustration, let us consider a sand castle. It is clear that its stability is due to its moisture content, since from dry sand no stable shape can be created. Similarly, stability is lost as well if the sand is immersed in water. It is thus obviously the presence of liquid/vapor interfaces which provides the mechanical stability. In fact, gravitational shear induces no flow at all, since the sand castle will not yield at any perceivable rate to gravity. This state of the material may be called solid. If, however, it is subject to a critical shear force, it starts 'flowing', i.e., yielding to the applied force by changing its shape. This state will be called fluid. If the moisture content is small but finite, there is a liquid bridge between any two grains touching each other, which causes an attractive force upon the grains. An external force will entail the extension, and eventual rupture, of some of the bridges. In the present study, we assume the grains to be completely frictionless, such that all of the dissipation is due to the rupture of liquid bridges.

In order to obtain a sufficiently reasonable description of wet granular matter we consider a dense system of $N \sim 10^4$ spherical beads in a cubic box with cyclic boundary conditions [3]. The radii of the beads are chosen at random within a moderate range $\delta R/R \sim 0.1$. The characteristic feature of our system is a hysteretic force modeling the liquid bridges between the beads [3]. This force is set to zero as long as the center of mass distance, δx_{ab}, of two approaching particles of radius R_a and R_b is larger than $R_a + R_b$. As soon as $\delta x_{ab} = R_a + R_b$, a liquid bridge is formed, and the interaction force $F_{ab} = F_0(\Phi(\xi) - 1)$ is switched on, with $\xi = \delta x_{ab}/(R_a + R_b)$. The δ-like function $\Phi(\xi)$ models the mutual hard core repulsion of the beads while the constant attractive force is an idealized representation which neglects the curvature of the liquid layers, the roughness of the granules and all effects related to the conservation of the liquid volume. However, the main effect is well represented by this simple assumption. If the distance δx_{ab} exceeds for the first time after the collision the critical value $R_a + R_b + R_{crit}$, the liquid bridge snaps, and the interaction is reset to $F_{ab} = 0$. The hysteresis spanning the range between $R_a + R_b$ and $R_a + R_b + R_{crit}$ is the only source of dissipation in our model. In our simulations, we have set $R_{crit}/R \sim 0.2$. Shear is applied to the system by means of a space dependent external force field $\mathbf{F}(\mathbf{x})$, which acts upon each particle individually. A cosine profile, $\mathbf{F}(\mathbf{x}) = \mathbf{e}_x F cos(2\pi z/L)$, is chosen. Obviously, the system is characterised by the relative force scale F/F_0. Hence, we may set $F_0 = 1$ for the following investigations.

On the basis of this model, we have performed standard simulations with variable values of the applied shear, F, and found three regimes. At small values on

F, we observe a solid like behavior, where no substantial displacement of beads is observed. After a short relaxation regime, an arbitrary initial configuration reaches a frozen state. At large amplitudes F, the system exhibits fluid like behavior with a constant rate of bridge formation and rupture. The long time behavior of each individual particle can be well described by an anisotropic Brownian diffusion due to the symmetry breaking with respect to the external shear field. The most interesting regime is the crossover from the solid like regime to the fluid regime which occurs in the vicinity of a critical force. As F_c is approached from below, the system develops a propensity to forming avalanches. This can be seen in the total actual velocity obtained from all beads of the system, where periods of high mobility alternate with long intervals where the particles are largely at rest.

For a quantitative analysis of this effect, we define an order parameter characterising the strength of the observed cascades in the mobility. Each cascade corresponds to a significant jump of the total kinetic energy over a relatively short time interval $\Delta t = t_{m+1} - t_m$. It should be remarked that these jumps are not real discontinuities. They become strong but regular changes of the kinetic energy for $\Delta t \to 0$ because of the underlying Newtonian equations of motion. We stress the weight of the cascades by defining the kinetic order parameter

$$\chi^{(n)} = \left\langle \sum_{t_m=0}^{t_{max}} |E(t_{m+1}) - E(t_m)|^n \right\rangle \quad (1)$$

where n is sufficiently large. As demonstrated in the figure, the data are in agreement with a power law, $\chi^{(n)} \sim |F - F_c|^{-\gamma_n}$, over more that 3 decades in $F - F_c$. The critical exponent, γ_n, of the divergence depends, in principle, upon n. However, it acquires a rather stable value of $\gamma_n/n = 0.24$ in a wide range $n = 4...12$. Larger values of n favor some few cascades and deteriorate the statistics, while the influence of small energy fluctuations increases rapidly for $n < 4$. Cascades can be also observed above the treshold, but here they are confined to the initial stage of the simulation. The frequency of their occurrence then decreases rapidly, such that a more or less continuous flow of the granular medium is rapidly established. Vice versa, below the critical threshold the frequency of cascades decreases due to the gradual solidification of the system. The diffusion coefficients characterizing the long time behavior of a single bead may be also expressed by a power law $D \sim |F - F_c|^\beta$, with $\beta = 0.62 \pm 0.05$. Within our accuracy, there is no difference in the scaling behavior for the three directions of motion (x, y, and z). Another critical behavior can be obtained for the time averaged number Γ of ruptured liquid bonds per unit time. We obtain the scaling behavior $\Gamma \sim |F - F_c|^{\bar\beta}$ with $\bar\beta = 0.73 \pm 0.06$. All these results indicate that the solid-fluid transition of our model system,

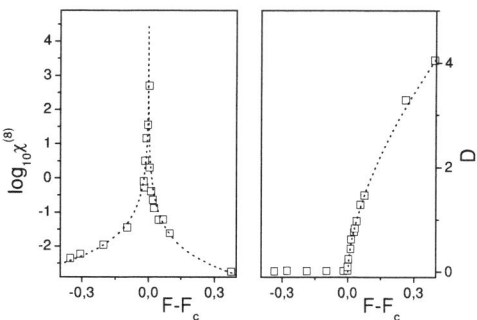

FIGURE 1. Kinetic order parameter as a function of $F - F_c$ for $n = 8$. The cascade like motion dominates close to the critical threshold and pushes the order parameter. The full lines are fits corresponding to $\chi^{(8)} \sim |F - F_c|^{\gamma_8}$ with $\gamma_n = 1.94 \pm 0.08$ (left). Diffusion coefficients in the direction of the force field orientation. The dotted line corresponds to a scaling law in the vicinity of the critical force, F_c (right).

and potentially of wet granular matter in general, can be interpreted as a dynamical critical phenomenon.

In conclusion, we state that a dynamic solid fluid transition, which is much reminiscent of the typical mechanical behavior of wet sand or soil, can be observed in a very simple model which neglects internal friction, and has a hysteretic interaction force as the only source of dissipation. The transition can be characterised by rather well defined critical exponents in all relevant quantities investigated. It is clear that further investigations are required to characterize the universality class of this fascinating phenomenon. In particular, it must be studied whether the critical exponents are universal, or depend upon the form of the force law, or other characteristic parameters, such as the rupture distance R_{crit}. This will be left to further work.

ACKNOWLEDGMENTS

This work was supported by the Deutsche Forschungsgemeinschaft (SFB 569 and Priority program 1052).

REFERENCES

1. Edwards, S. F., and Oakescott, R. B. S., *Physica A*, **157**, 1080 – 1090 (1989).
2. Hong, S. W., Kim, M. M., Yang, G. S., Lee, S. R., Chung, S. S., Ihm, C. C., Kim, H. T., and Park, J. B., *Soil mechanics and geotechnical engineering - Eleventh Asian regional conference, Proceedings, Seoul, Korea, 16-20 August 1999*, Balkema, Rotterdam, 1999, pp. 1–213.
3. Schulz, B. M., Schulz, M., and Herminghaus, S., *Physical Review E*, **67**, 052301 (2003).

Conformational transitions of a semiflexible polymer in a liquid crystalline phase

Akihiko Matsuyama

Department of Biochemical Engineering and Science, Faculty of Computer Science and System Engineering, Kyusyu Institute of Technology, Kawazu 680-4, Iizuka, Fukuoka, 820-8502, Japan

Abstract. Conformations of a single semiflexible polymer chain dissolved in a low molecular weight liquid crystalline solvent (nematogen) are examined by using a mean field theory. We takes into account a stiffness and partial orientational ordering of the polymer. As a result of an anisotropic coupling between the polymer and nematogen, we predict a discontinuous (or continuous) phase transition from a condensed-rodlike conformation to a swollen-one of the polymer chain, depending on the stiffness of the polymer. We also study a single polymer chain confined in a smectic phase.

INTRODUCTION

Mixtures of a flexible polymer and a nematogen show a macroscopic phase separation between an isotropic and a nematic phase below the nematic-isotropic transition (NIT) temperature of the pure nematogen[1]. Flexible polymers present a weak anisotropy in a nematic phase[2]. In contrast, liquid crystalline polymers, or stiffer polymers, have good miscibility with nematogens due to the strong anisotropic coupling between the polymer and the nematogen. Anisotropy of the conformation for liquid crystalline polymers has been experimentally[3, 4] and theoretically[5, 6, 7] studied in melt and in dilute nematic solutions. It is now important to consider the conformation of a polymer chain with various degrees of stiffness dissolved in nematogens. Recently we presented a mean field theory to describe partial orientational ordering (induced rigidity) of semiflexible polymers dissolved in nematogens and showed various phase diagrams for the mixtures[8].

In this paper we theoretically study the conformation of a semiflexible polymer dissolved in nematic solvents by combining the previous model[8] with an elastic free energy of the chain. We show a discontinuous (or continuous) conformational transition between two different nematic states, depending on the stiffness of the polymer. We also study a single polymer chain confined in a smectic phase.

Consider a single linear polymer chain dissolved in nematogens. In order to take into account the stiffness of the polymer, we here assume that two neighboring bonds on the polymer chain have either bent (gauche state) or straightened (trans state) conformations. Hereafter we refer the segments in straightened bonds as "rigid" seg-

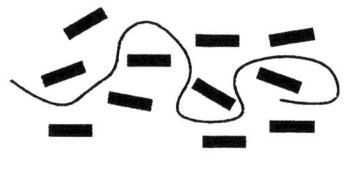

in nematic phase

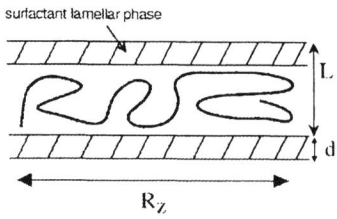

in smectic phase

FIGURE 1. Polymer in a liquid crystalline phase

ments.

Let $V = R^3$ be the volume of the region occupied by a polymer, n be the number of segments on the polymer chain and n_l be the axial ratio of the nematogen. The volume fraction of the polymer in the volume V is given by $\phi = a^3 n/V$, where a^3 is the volume of an unit segment. To derive an equilibrium conformation of the polymer, we consider thermodynamics of our systems. The free energy density of our system can be given by

$$f = f_{el} + f_{bent} + f_{mix} + f_{nem}. \quad (1)$$

The first term shows the elastic free energy due to the deformation of the polymer chain. The second term shows the free energy change needed to straighten bent bonds

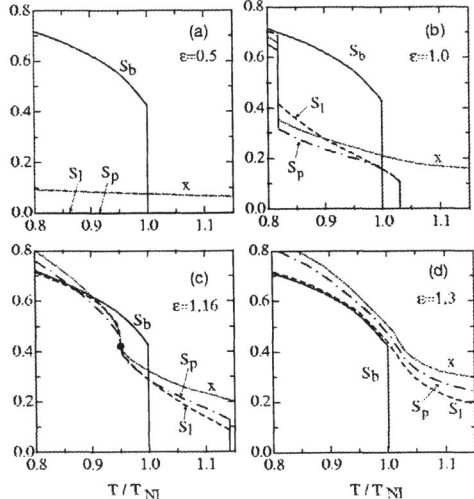

FIGURE 2. Orientational order parameters

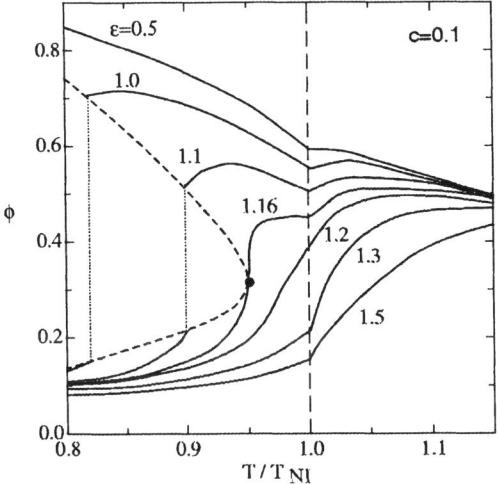

FIGURE 3. Equilibrium volume fraction of the polymer

on the polymer. The third term in Eq. (1) is the free energy of the isotropic mixing for the polymer and nematogens. The last term in Eq. (1) shows the free energy for the nematic ordering. The details refer to Ref. [9].

PHASE TRANSITION

Figure 2 shows the orientational order parameters and the fraction x of the rigid segments on the polymer plotted against the reduced temperature T/T_{NI}, where T_{NI} is the NIT temperature of the pure nematogen outside the polymer. The value of stiffness parameter ε of the polymer is changed from (a) to (d). The larger values of ε correspond to the stiffer chains. The solid curve refers to the order parameter S_b of the nematogen outside the polymer and the dash-dotted line shows the order parameter S_p of the polymer. The short-dashed line shows the order parameter S_l of the nematogen inside the polymer and the dotted line corresponds to the fraction x of the rigid segments. For $\varepsilon = 0.5$, or a flexible polymer, the polymer is in an isotropic state for all temperatures and there is no anisotropic coupling between the polymer and nematogen. When $\varepsilon = 1.0$, we find two phase transition temperatures: one is the temperature T_{NI}^p at high temperatures where the NIT of the polymer takes place and the other is the T_{NN} at low temperatures where the first-order phase transition between two different nematic states takes place. At the nematic state of the high temperature side ($T_{NN} < T < T_{NI}^p$), the fraction x of rigid segments is small.

Figure 3 shows the equilibrium volume fraction ϕ (swelling curve) of the polymer plotted against the reduced temperature for various values of ε. When $\varepsilon = 0.5$, as decreasing temperature, the polymer is continuously condensed (or the volume fraction of the polymer is increased). The swelling curve has as kink at T_{NI}. For stiffer polymers, we find two different types in the rodlike conformation of the polymer chain: one is the swollen-rodlike conformation at $T < T_{NN}$ and the other is the condensed-rodlike conformation at $T_{NN} < T < T_{NI}^p$. Above the critical stiffness, the polymer is continuously swollen with decreasing temperature.

In conclusion, we have predicted two different rodlike conformations of a polymer chain in a nematic solvent. The phase transition from a condensed-rodlike conformation to a swollen-one is strongly affected by the stiffness of a polymer and the anisotropic interaction between polymer segments.

REFERENCES

1. Kronberg B., Bassignana I. and Patterson D., *J. Phys. Chem.* **82**, 1714-1719 (1978).
2. Dubault A, Ober R., Veyssie M., and Cabane B., *J. Phys.* (France) **46**, 1227-1232 (1985).
3. D'Allest J. F., Maissa P., ten Bosch A., Sixou P., Blumstein A., Blumstein R., Teixeira J. and Noirez L., *Phys. Rev. Lett.* **61**, 2562-2565 (1988).
4. Volino F., Gauthier M. M., Giroud-Godquin A. M. and Blumstein R. B., *Macromolecules* **18**, 2620-2670 (1985).
5. Warner M., Gunn J. M. F. and Baumgartner A., *J. Phys. A* **18**, 3007-3019 (1985).
6. Carri G. A. and Muthukumar M., *J. Chem. Phys.* **109**, 11117-11128 (1998).
7. Olmsted P. D. and Milner S. T., *Macromolecules* **27**, 6648-6600 (1994).
8. Matsuyama A. and Kato T., *Phys. Rev. E* **59**, 763-770 (1999).
9. Matsuyama A., *Phys. Rev. E* **67**, 042701-1-4 (2003).

The unbinding transition of mixed fluid membranes

S. Komura* and D. Andelman[†]

Department of Chemistry, Tokyo Metropolitan University, Tokyo 192-0397, Japan
[†]*School of Physics and Astronomy, Raymond and Beverly Sackler Faculty of Exact Sciences, Tel Aviv University, Ramat Aviv 69978, Tel Aviv, Israel*

Abstract. A phenomenological model for the unbinding transition of multicomponent fluid membranes is proposed, where the unbinding transition is described using a theory analogous to Flory-Huggins theory for polymers. The coupling between the lateral phase separation of inclusion molecules and the membrane-substrate potential provides a rich phase behavior. Our model describes the first-order nature of the unbinding transition in multicomponent membranes as was observed in a recent experiment. In particular, we predict different scenarios of phase coexistence between bound and unbound membrane states.

INTRODUCTION

Adhesion of membranes and vesicles is responsible for cell-cell adhesion which plays an important role in all multicellular organisms. In general, bio-adhesion is governed by the interplay of various generic and specific interactions. Specific interactions act between complementary pairs of proteins such as ligand and receptor, or antibody and antigen. The problem of adhesion of multicomponent membranes is intimately related to that of domain formation. In a recent experiment by Marx et al. [1], the role of long-range repulsions due to thermal fluctuations (Helfrich repulsion) of adhering membranes has been addressed. Analyzing the probability distribution of the membrane-substrate spacing for various multicomponent membranes, a phase separation between two distinct lipopolymer-poor and lipopolymer-rich states having two different spacings from the substrate was suggested.

In this paper, we propose a simple phenomenological model for a multicomponent (mixed) fluid membrane which can undergo simultaneously a lateral phase separation and an unbinding transition. The model is motivated by the experiment [1] and relies on the coupling between the inclusion concentration and the membrane-substrate spacing. The lateral phase separation of the inclusion affects the second virial coefficient of the unbinding transition which is taken into account in analogy to the Flory-Huggins theory for polymers. Our model exhibits various types of phase coexistence, including a phase separation between bound and unbound states as well as between two unbound ones. The former phase coexistence indicates the first-order nature of the unbinding transition, as was anticipated in the experiment.

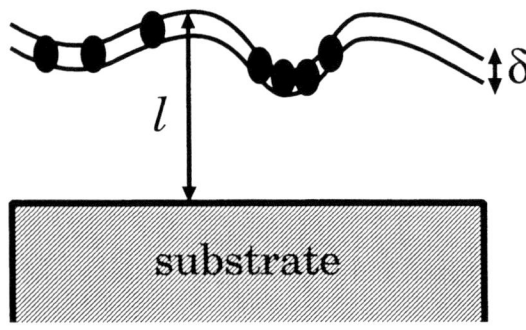

FIGURE 1. A mixed fluid membrane adhering to a substrate. Black filled ovals indicate inclusions such as proteins or lipopolymers. The height of the upper membrane leaflet from the substrate is ℓ, whereas the thickness of the bilayer membrane is δ.

MODEL

Fluid membranes in a lamellar stack or close to a substrate experience steric repulsion arising from their reduced undulation entropy due to the confinement effect. The corresponding interaction energy per unit area has been given by Helfrich as

$$v_s(\ell) = \frac{b(k_B T)^2}{\kappa(\ell - \delta)^2}, \quad (1)$$

where k_B is the Boltzmann constant, T the temperature, κ the bending rigidity of the membrane having thickness δ, and ℓ the average height of the upper membrane lipid leaflet from the substrate (see Fig. 1). While a simple superposition of the Helfrich repulsion, Eq. (1), and other direct interactions gives an incorrect (first-order) description of the unbinding transition, a simple theory

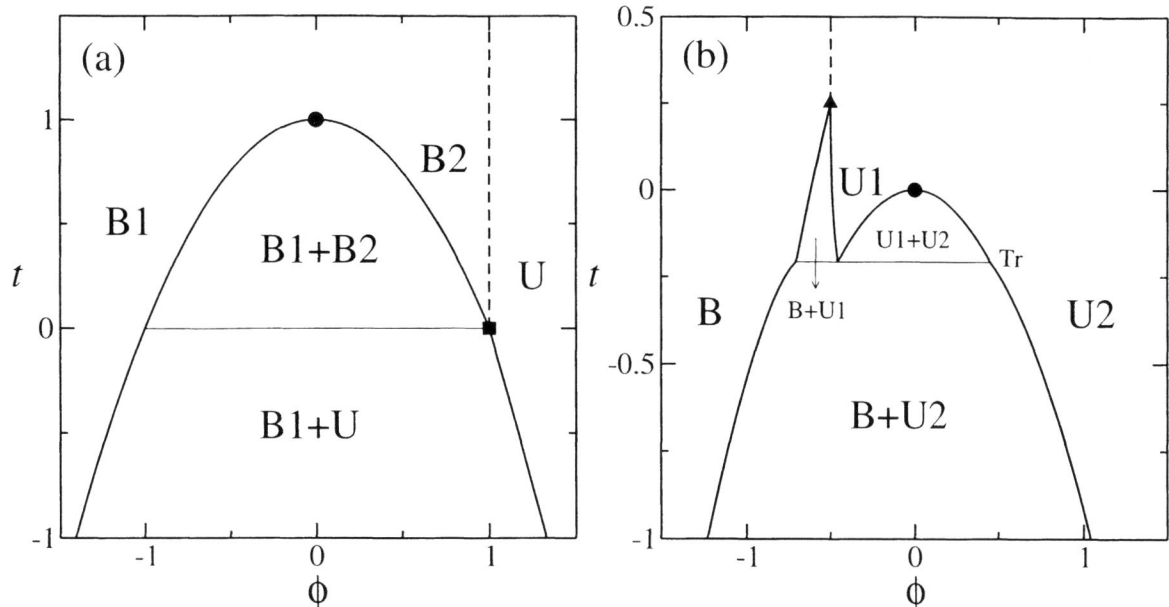

FIGURE 2. The phase diagrams for (a) $\chi = 1$ and (b) $\chi = -0.5$ as a function of ϕ and t when $\gamma = 1$. The continuous line is a first-order line, whereas the dashed line is a second-order one. The critical point, the critical end-point, and the tricritical point are indicated by a circle, square, and triangle, respectively. The bound and the unbound phases are denoted by B and U, respectively.

for the unbinding transition in a bulk of lamellar phase was considered by Milner and Roux [2]. Following the spirit of Flory-Huggins theory for polymers, the Helfrich estimate of the entropy is taken into account accurately, whereas the other interactions are approximately incorporated via a second virial term. Then the free energy *per unit area* of a single membrane can be expressed as

$$f(w) = -k_B T \delta \bar{\chi} w + \frac{b(k_B T)^2}{\kappa \delta^2} w^2, \quad (2)$$

where $w = \delta/\ell \geq 0$ cannot be negative, $\bar{\chi}$ is the second virial coefficient. Minimization of $f(w)$ with respect to w shows that this free energy has a continuous second-order transition at $\bar{\chi} = 0$ between a bound state ($w > 0$) for $\bar{\chi} > 0$ and an unbound state ($w = 0$) for $\bar{\chi} < 0$.

We now consider a two-component membrane adhering to a substrate as in Fig. 1. The overall membrane state is characterized by its average distance ℓ from the substrate. The internal degree of freedom, on the other hand, corresponds to the membrane composition. The interaction between two inclusions is attractive leading to a condensation transition, and the concentration with respect to the critical concentration is defined as ϕ. The proposed free energy *per unit area* of a single mixed membrane undergoing the unbinding transition is

$$f(\phi, w) = -\mu\phi + \frac{1}{2}t\phi^2 + \frac{1}{4}\phi^4 - \chi w + \frac{1}{2}w^2 + \gamma\phi w, \quad (3)$$

with the constraint $w \geq 0$. Here all energy terms have been scaled by $2b(k_B T)^2/\kappa \delta^2$, and are now dimensionless. The first three terms in Eq. (3) depend only on ϕ, and correspond to the Landau free energy of a two-component membrane undergoing a lateral inclusion-lipid phase separation, μ is the chemical potential conjugate to ϕ, and $t \sim (T - T_c)$ the reduced temperature. The next two terms depend only on w, and represent the unbinding transition of a single membrane as have been described by Eq. (2). In the above, χ is the scaled virial coefficient. The last term is the lowest order coupling term between ϕ and w with a dimensionless coupling coefficient $\gamma > 0$. The physical meaning of this bilinear term is as follows. When the mixed membrane is quenched into the two-phase region, an inclusion-poor phase ($\phi < 0$) coexists with an inclusion-rich phase ($\phi > 0$). This can lead to different direct interactions and hence different second virial coefficients χ for each of the domains. We model this situation by considering an effective virial term in Eq. (3) as $-\chi_{\text{eff}} w = -(\chi - \gamma\phi)w$, which leads to the coupling term ϕw. A typical example of the obtained phase diagrams for $\chi = 1$ and -0.5 are given in Fig. 2 when $\gamma = 1$. Our theory describes the coexistence between the two unbound states.

REFERENCES

1. Marx, S., Schilling, J., Sackmann, E., and Bruinsma, R., *Phys. Rev. Lett.* **88**, 1381021-1381024 (2002).
2. Milner, S.T., and Roux, D., *J. Phys. I France* **2**, 1741-1754 (1992).

Contributions of Steady Heat Conduction to the Rate of Chemical Reaction

Kim Hyeon-Deuk* and Hisao Hayakawa[†]

Graduate School of Human and Environmental Studies, Kyoto University, Kyoto 606-8501, JAPAN
[†]*Department of Physics, Kyoto University, Kyoto 606-8501, JAPAN*

Abstract. We have derived the effect of steady heat flux on the rate of chemical reaction based on the line-of-centers model using the explicit velocity distribution function of the steady-state Boltzmann equation for hard-sphere molecules to second order. We have found that the second-order velocity distribution function plays an essential role for the calculation of it. This indicates the significance of the second-order coefficients in the solution of the steady-state Boltzmann equation as terms which reflect the local nonequilibrium effect.

CHEMICALLY REACTING GAS

In the early stage of a chemical reaction between monatomic molecules:

$$A + A \rightarrow \text{products}, \quad (1)$$

the rate of chemical reaction is not affected by the existence of products. From the viewpoint of kinetic collision theory[1], the rate of chemical reaction (1) can be described as

$$R = \int d\mathbf{v} \int d\mathbf{v}_1 \int d\mathbf{k} \int f f_1 g \sigma(g), \quad (2)$$

where $\mathbf{v}$ and $\mathbf{v}_1$ are the velocities of the molecules, $g = |\mathbf{v} - \mathbf{v}_1|$ their relative speed, $\mathbf{k}$ the solid angle, $f = f(\mathbf{r}, \mathbf{v})$ and $f_1 = f(\mathbf{r}, \mathbf{v}_1)$ are the distributions of $\mathbf{v}$ and $\mathbf{v}_1$ at $\mathbf{r}$, respectively.

The line-of-centers model proposed by Present has been accepted as a standard model to describe the chemical reaction in gases.[1] It assumes the chemical cross-section as

$$\sigma(g) = \begin{cases} 0 & g < \sqrt{\frac{4E^*}{m}} \\ \frac{d^2}{4}\left(1 - \frac{4E^*}{mg^2}\right) & g \geq \sqrt{\frac{4E^*}{m}} \end{cases}, \quad (3)$$

with m mass of the molecules and E^* the threshold energy of the chemical reaction. d is regarded as a distance between centers of monatomic molecules at contact.

NONEQUILIBRIUM EFFECT ON THE RATE OF CHEMICAL REACTION

In order to calculate the rate of chemical reaction (2), we expand the velocity distribution function f to second order as

$$f = f^{(0)} + f^{(1)} + f^{(2)} = f^{(0)}(1 + \phi^{(1)} + \phi^{(2)}), \quad (4)$$

around the local Maxwellian, $f^{(0)} = n(m/2\pi\kappa T)^{3/2} \exp[-mv^2/2\kappa T]$, with n the density of molecules, κ the Boltzmann constant and T the temperature defined from the kinetic energy. Substitution of eq.(4) into eq.(2) leads to

$$R = R^{(0)} + R^{(1)} + R^{(2)}, \quad (5)$$

up to second order. The zeroth-order term of R, the rate of chemical reaction of the equilibrium theory, becomes $R^{(0)} = \int d\mathbf{v} \int d\mathbf{v}_1 \int d\mathbf{k} \int f^{(0)} f_1^{(0)} g \sigma(g) = 4n^2\sigma^2 \left(\frac{\pi\kappa T}{m}\right)^{\frac{1}{2}} e^{-\frac{E^*}{\kappa T}}$. The first-order term of R, i.e. $R^{(1)}$, does not appear because $\phi^{(1)}$ is an odd functions of $\mathbf{c}$. The second-order term of R, i.e. $R^{(2)}$, is divided into

$$R^{(2,A)} = \int d\mathbf{v} \int d\mathbf{v}_1 \int d\mathbf{k} \int f^{(0)} f_1^{(0)} \phi^{(1)} \phi_1^{(1)} g\sigma(g), \quad (6)$$

and

$$R^{(2,B)} = \int d\mathbf{v} \int d\mathbf{v}_1 \int d\mathbf{k} \int f^{(0)} f_1^{(0)} [\phi^{(2)} + \phi_1^{(2)}] g\sigma(g). \quad (7)$$

Since the integrations (6) and (7) have the cutoff from eq.(3), the explicit forms of $\phi^{(1)}$ and $\phi^{(2)}$ of the steady-state Boltzmann equation for hard-sphere molecules are required to calculate $R^{(2,A)}$ and $R^{(2,B)}$, respectively.

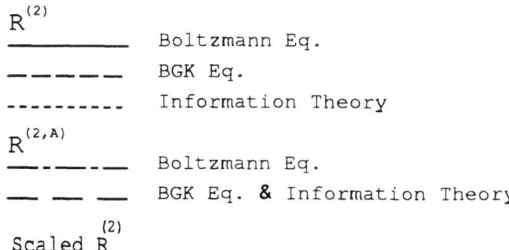

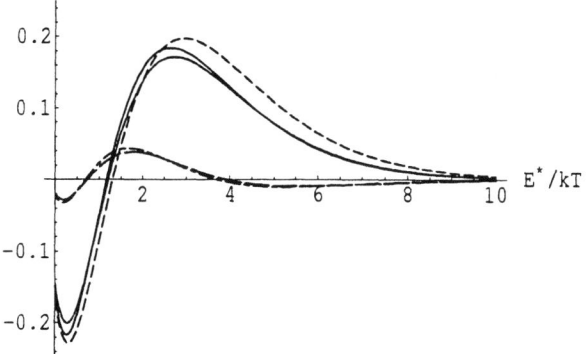

FIGURE 1. Both of $R^{(2)}$ and $R^{(2,A)}$ are scaled by $\pi^{1/2}d^2m^{1/2}J_x^2/\kappa^{5/2}T^{5/2}$. Here J_x means a steady heat flux.

Although Burnett had determined the second-order pressure tensor for the Boltzmann equation, he had not derived the explicit second-order velocity distribution function of the Boltzmann equation.[2, 3] Therefore, none has succeeded to obtain the correct reaction rate of Present's model except for Fort and Cukrowski who adopted information theory[4] as the nonequilibrium velocity distribution function to second order.[5]

We have recently derived the explicit velocity distribution function of the steady-state Boltzmann equation for hard-core molecules to second order in density and the temperature gradient.[3] This enables us to calculate the effect of steady heat flux on the rate of chemical reaction based on the line-of-centers model.[6]

RESULTS AND DISCUSSION

In this proceeding, we show only the graphical results of $R^{(2)}$ compared with those of $R^{(2,A)}$ in Fig.1.

We have found that $R^{(2,B)}$ plays an essential role for the evaluation of $R^{(2)}$, and that there are no qualitative differences in $R^{(2)}$ of the steady-state Boltzmann equation, the steady-state Bhatnagar-Gross-Krook(BGK) equation and information theory. It should be mentioned that, however, we have found qualitative differences among these theories in pressure tensor and the kinetic temperature as Table 1 shows.[3] We have also found that the steady-state BGK equation belongs to the same universality class as Maxwell molecules, and that information theory is inconsistent with the steady-state Boltzmann equation.[7]

The nonequilibrium effect on the rate of chemical reaction will substantiate significance of the second-order coefficients in the solution of the steady-state Boltzmann equation, although their importance has been demonstrated only for descriptions of shock wave profiles and sound propagation phenomena. This indicates the significance of the second-order coefficients as terms which reflect the local nonequilibrium effect.

We also propose a *thermometer* of a monatomic dilute gas system under a steady heat flux.[6] We mean that we can measure the temperature T around a heat bath at T_0 in the nonequilibrium steady-state system indirectly with the aid of the nonequilibrium effect on the rate of chemical reaction. The nonequilibrium effect in the early stage of chemical reaction around the heat bath can be measured experimentally. Thus, one can compare the experimental result with the theoretical result by setting $T = T_0$. The difference between the former and the latter will indicate that the temperature T around the heat bath is not identical with T_0, but $T = T_0 + \Delta$ where Δ depends upon the steady heat flux in general.

TABLE 1. Numerical constants for the pressure tensor $P_{ij} = n\kappa T[\delta_{ij} + \lambda_P^{ij}\frac{mJ_x^2}{n^2\kappa^3 T^3}]$ and the each component of the kinetic temperature $T_i = T[1 + \lambda_{T_i}\frac{mJ_x^2}{n^2\kappa^3 T^3}]$. Note that the off-diagonal components of λ_P^{ij} are zero, and that $\lambda_P^{yy} = \lambda_P^{zz} = -\lambda_P^{xx}/2$ and $\lambda_{T_y} = \lambda_{T_z} = -\lambda_{T_x}/2$.

	λ_P^{xx}	λ_{T_x}
Hard-Core	-4.600×10^{-2}	-2.300×10^{-2}
Maxwell	0	0
BGK eq.	0	0
Information theory	$\frac{12}{25}$	$\frac{6}{25}$

REFERENCES

1. R. D. Present, Kinetic Theory of Gases, (Mcgraw-Hill, New York, 1958).
2. D. Burnett, Proc. Lond. Math. Soc. **40**, 382-435 (1935).
3. Kim. H.-D. and H. Hayakawa, J. Phys. Soc. Jpn. **72**, 1904-1916 (2003).
4. D. Jou, J. Casas-Vázquez and G. Lebon: Extended Irreversible Thermodynamics, (Springer, Berlin, 2001).
5. J. Fort and A. S. Cukrowski, Chem. Phys. **222**, 59-69 (1997).
6. Kim. H.-D. and H. Hayakawa, Chem. Phys. Lett. **372**, 314-319 (2003).
7. Kim. H.-D. and H. Hayakawa, J. Phys. Soc. Jpn. **72**, 2473-2476 (2003).

Theory for The Switching of Nematic Director Deformations

Hiromitsu Matsuda*, Tomonori Koda* and Susumu Ikeda*

Faculty of Engineering, Yamagata University, 4-3-16 Jonan, Yonezawa 992-8510, JAPAN

Abstract. Director of nematic liquid crystal fluctuate with relaxation time which is infinity at long wave length. We calculated effect of fluctuation on transmitted light along layered nematic medium. We found fluctuation of retardation and distance transmitted is expressed by a type of Langevin equation.

INTRODUCTION

Nematic liquid crystal has fluctuation whose relaxation time is infinity at long wave length. Nematic liquid crystal looks muddy white in the bulk system and microscope image of nematic liquid crystal is always fluctuating. It is important for application of liquid crystal to consider how the fluctuation affects retardation of transmitted light. In this study, we calculated the effect of fluctuation using equation of light transmission[1] and formula of Frank elastic free energy.

MODEL AND THEORY

As Fig.1 shows, we set z-axis along the normal of layered nematic medium. We assumed director $\mathbf{n}$ is function of z. $\mathbf{E}(z)$ in Fig.1 is electric field of light at z. We denote polar angle and azimuthal angle of director, respectively, as θ and ϕ. Retardation rotates electric field along z. Change in electric field for distance Δz is described as

$$\begin{pmatrix} E_x(z+\Delta z) \\ E_y(z+\Delta z) \end{pmatrix} = \mathbf{ABA}^* \begin{pmatrix} E_x(z) \\ E_y(z) \end{pmatrix}. \quad (1)$$

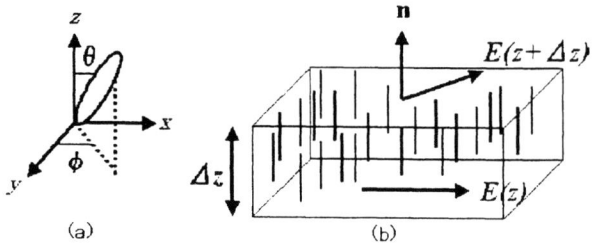

FIGURE 1. Orientational model (a) orientation of liquid crystal (b) transmitted light in the cell

where $E_x(z)$ and $E_y(z)$ are x- and y-component of $\mathbf{E}(z)$, respectively. And

$$\mathbf{A} = \begin{pmatrix} \cos\phi & -\sin\phi \\ \sin\phi & \cos\phi \end{pmatrix}, \quad (2)$$

$$\mathbf{B} = \begin{pmatrix} e^{2\pi i \frac{(n_e+\delta n_e(z))\Delta z}{\lambda}} & 0 \\ 0 & e^{2\pi i \frac{n_0 \Delta z}{\lambda}} \end{pmatrix}. \quad (3)$$

and $\mathbf{A}^*$ is the transpose of $\mathbf{A}$. In eq.(3), λ is wave length of the incident light in the vacuum. $n_e + \delta n_e(z)$ and n_0 are extraordinary and ordinary index. They are expressed as

$$n_e + \delta n_e(z) = \frac{n_\parallel n_\perp}{\sqrt{C}}, \quad (4)$$

$$n_0 = n_\perp. \quad (5)$$

where

$$C = n_\parallel^2 \cos^2(\theta + \delta\theta(z)) + n_\perp^2 \sin^2(\theta + \delta\theta(z)). \quad (6)$$

In eqs.(4)-(6) $n_\parallel$ and $n_\perp$ are refractive index parallel and perpendicular to director, respectively. n_0 is constant.

We assume that electric field is described as

$$\mathbf{E}(z) \equiv E_0 e^{iU(z)}. \quad (7)$$

In this case, we have

$$dU(z) = \frac{2\pi(n_e+\delta n_e(z))}{\lambda} dz. \quad (8)$$

The $U(z)$ is related to retardation $R = \lambda U(z)/2\pi$. Equation(8) indicates that $U(z)$ and distance transmitted is expressed by a type of Langevin equation. To solve eq.(8), $\delta n_e(z)$ should be calculated from Frank elastic free energy. Frank elastic free energy is

$$f_d = \frac{1}{2}\left[K_1(\nabla \cdot \mathbf{n})^2 + K_2(\mathbf{n} \cdot \nabla \times \mathbf{n})^2 + K_3(\mathbf{n} \times \nabla \times \mathbf{n})^2\right]. \quad (9)$$

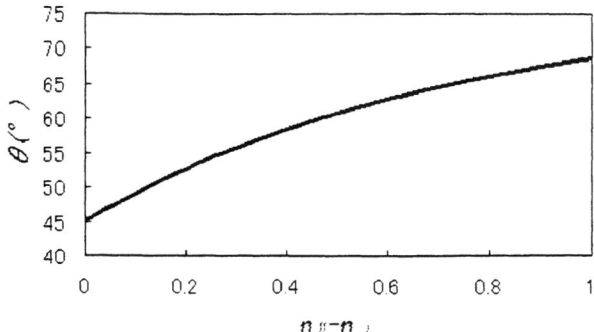

FIGURE 2. Polar angle maximized fluctuation ($n_\perp = 1.0$)

where K_1, K_2 and K_3 are Frank elastic constant.

Fluctuation of director and extraordinary index are related in Fourier space as

$$\begin{aligned}
<|\delta\theta(\mathbf{q})|^2> &= <|\delta n(\mathbf{q})|^2> \\
&= \frac{kT}{VKq^2}, \quad (10) \\
<|\delta n_e(\mathbf{q})|^2> &= \frac{n_\parallel^2 n_\perp^2 (n_\parallel - n_\perp)^2 \sin^2\theta \cos^2\theta}{C^3} \\
&\quad \times <|\delta\theta(\mathbf{q})|^2> \\
&= \frac{kT}{VKq^2} \frac{n_\parallel^2 n_\perp^2 (n_\parallel - n_\perp)^2 \sin^2\theta \cos^2\theta}{C^3}.
\end{aligned}$$
(11)

where V is volume, K is Frank elastic constant and q is wave number. For eq.(11), we used the one-constant approximation: $K = K_1 = K_2 = K_3$.

Using eqs.(8) and (11), we have

$$<U(z)> = \frac{2\pi n_e z}{\lambda}. \quad (12)$$

$$<U(z)^2> = \left(\frac{2\pi n_e z}{\lambda}\right)^2 + \frac{kT}{2K} \frac{n_\parallel^2 n_\perp^2 (n_\parallel - n_\perp)^2 \sin^2\theta \cos^2\theta}{C^3} \frac{z}{\lambda^2}. \quad (13)$$

where $<\cdots>$ indicates thermal average.

Equations(12) and (13) give

$$\frac{\sqrt{<U(z)^2> - <U(z)>^2}}{<U(z)>} \propto \frac{1}{\sqrt{z}}. \quad (14)$$

Retardation of nematic liquid crystal is proportional to $1/\sqrt{z}$. Fig.2 shows the dependence of the polar angle which maximize $<U(z)^2>$ on birefringence. Effect of fluctuation is minimized in the case of homeotropic and

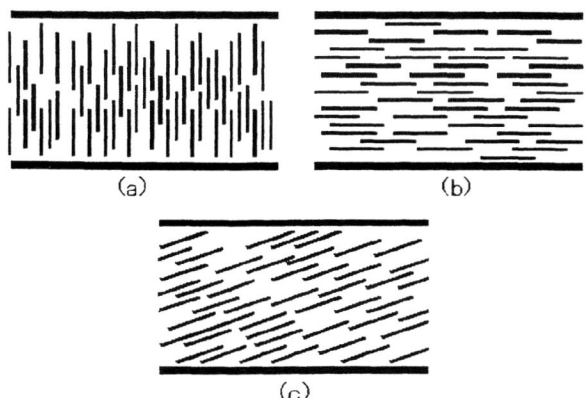

FIGURE 3. Nematic structure in the cell(a)homeotropic structure. (b)homogeneous structure. (c)tilted structure

homogeneous alignment(Fig.3(a)(b)). The polar angle that maximize effect of fluctuation is increasing function of birefringence like Fig.3(c).

CONCLUDING REMARKS

In this paper we examined relation of light and distance transmitted. The relation is expressed as a type of Langevin equation. Retardation of nematic liquid crystal is proportional to $1/\sqrt{z}$. Effect of fluctuation is minimized in the case of homeotropic and homogeneous alignment. The polar angle that maximize fluctuation is increasing function of birefringence.

ACKNOWLEDGMENTS

This work was supported by a Grant-in-Aid from the Ministry of Education, Culture, Sports, Science and Technology of Japan, No. 15607002.

REFERENCES

1. Ondris-Crawford, R., Boyko, E. P., Wagner, B. G., Erdmann, J. H., Zumer, S., and Doane, J. W., *J. Appl. Phys.* **69**, 6380-6386(1991).

Dynamics of phase separation in confined two-component fluid membranes

Hirofumi Wada[†]

Department of Physics, University of Tokyo, Hongo, Tokyo, 113-0033, Japan
Department of Mathematical and Life Sciences, Hiroshima University, Higashi Hiroshima, 739-8526, Japan

Abstract. We investigate the dynamics of phase separation in two-component fluid membranes confined between parallel and rigid bounding plates. Through numerical simulation, we find a considerably slow coarsening of the order parameter field within the time region we studied. In particular, a microphase separation is obtained as an equilibrium phase for sufficiently large values of the elastic coupling, which is in agreement with the previous mean-field analysis.

INTRODUCTION

Fluid membranes composed of several kinds of surfactants or phospholipids can be seen in a variety of systems from biological cell membranes to synthetic vesicles. One of the most fascinating phenomena can be found when the internal degrees of freedom of a membrane cause dramatic changes of its shape stability [1]. Although few direct experimental evidences are available, it has been pointed out in a number of theoretical and numerical studies that a coupling of a local curvature to a local composition of amphiphilies plays the crucial role in the formation of various distinct shapes of membranes [2, 3, 4, 5].

Another example of interest is provided by the entropically driven steric repulsion between fluctuating membranes, the so-called Helfrich interaction [6]. It is expected that a combination of the composition-curvature coupling and the long-ranged Helfrich repulsion will lead to novel in-plane phase separation dynamics. Two-component fluid membrane confined between parallel and rigid bounding plates can provide a particularly simple model for such problems [7]. Here we are concerned with the nonlinear dynamics of that system [8].

MODEL

Within the flat membrane approximation, a set of kinetic equations for the membrane displacement $h(\mathbf{r},t)$ and the appropriately rescaled composition difference (of, say external additive) $\phi(\mathbf{r},t)$ are given by

$$\frac{\partial \phi}{\partial t} = L_\phi \nabla^2 \left[-M\nabla^2 \phi + \frac{\partial f_0}{\partial \phi} + \frac{\lambda^2}{\kappa}\phi + \lambda \nabla^2 h \right], \quad (1)$$

$$\frac{\partial h}{\partial t} = -L_h \left[\kappa \nabla^4 h + \frac{\partial f_s}{\partial h} + \lambda \nabla^2 \phi \right], \quad (2)$$

where M is the positive constant, $f_0(\phi)$ the areal Ginzburg-Landau free energy density, λ the coupling constant (taken to be positive in our treatment), κ the bending rigidity, f_s the energy density of the Helfrich repulsion, L_h and L_ϕ the kinetic coefficients, respectively. Hydrodynamic effects are entirely neglected here. The equilibrium aspects of this problem have been analyzed by the mean-field theory [7]. To investigate the nonlinear dynamics, we numerically solve Eqs. (1) and (2) on a 128×128 square lattice by changing the elastic coupling λ.

RESULTS

Fig. 1 shows the typical time evolutions of the order parameter for the critical quench case $\langle \phi \rangle = 0$. The corresponding snapshot of the membrane shape at $t = 10^5$ is also displayed in Fig. 3. We find that the pattern does not change significantly after about time steps $t \sim 50000$, which implies that both ϕ and h fields have their stable bicontinuous structures as their equilibrium phases. Such modulated patterns obtained here are consistent with the predictions from the previous mean-field analysis.

To see the kinetics of ordering more quantitatively, we monitor the inverse of the time dependent characteristic length scale defined by $\langle q(t) \rangle = \sum_{|\mathbf{q}|>0} |\mathbf{q}|^{-1} S(\mathbf{q},t) / \sum_{|\mathbf{q}|>0} |\mathbf{q}|^{-2} S(\mathbf{q},t)$, where $S(\mathbf{q},t) = \langle \phi_\mathbf{q}(t)\phi_{-\mathbf{q}}(t) \rangle$ is the structure factor. In Fig. 4, the calculated data of $\langle q(t) \rangle$ is plotted as a function of t for $\langle \phi \rangle = 0$. We also include the result for $\lambda = 0$, which exhibits $t^{-1/3}$ evolution law. It is remarkable

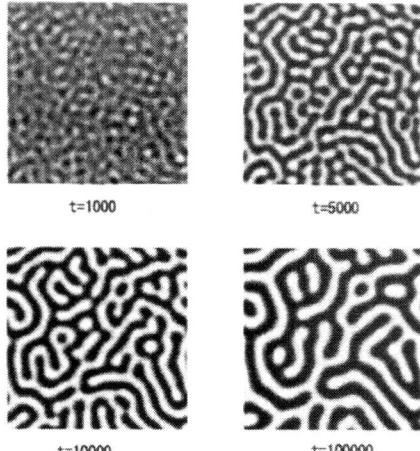

FIGURE 1. Time evolution of ϕ for critical quench case $\langle\phi\rangle = 0$. The white region represents the additive-rich phase $\phi > 0$.

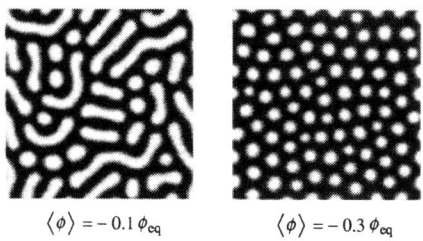

FIGURE 2. Domain patterns obtained for off-critical quench cases $\langle\phi\rangle = -0.1\phi_{eq}$ (left) and $\langle\phi\rangle = -0.3\phi_{eq}$ (right) at time steps $t = 10^5$.

that considerably slow, but algebraic domain growth is found for $\lambda \sim 0.50$ (at least within the time region we simulated). The effective growth exponents defined as $z = -d\langle\ln\langle q(t)\rangle\rangle/d(\ln t)$ are estimated by the least-square fits to the data in $5 \times 10^4 < t \leq 2 \times 10^5$, which give $z = 0.24$ for $\lambda = 0.50$ and $z = 0.15$ for $\lambda = 0.55$, for example. Notice that significant slowing down of

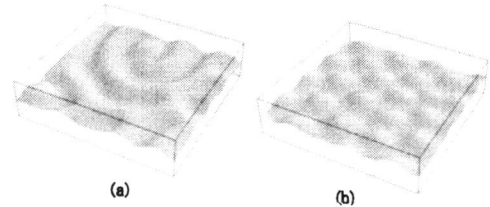

FIGURE 3. Snapshot of the membrane shape h on a 64×64 portion of the lattice for (a) critical quench case and for (b) off-critical quench case $\langle\phi\rangle = -0.3\phi_{eq}$ at time step $t = 10^5$. The bounding box depicted in the Figure represents the confining geometry in the exact scale. Note that there is some exaggeration along the vertical direction.

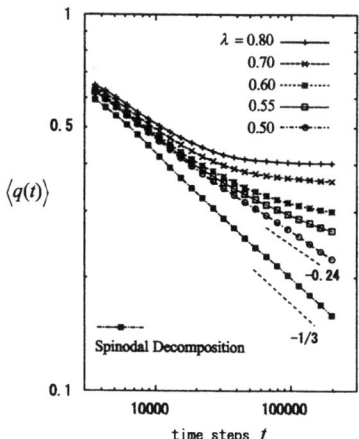

FIGURE 4. Time evolutions of the inverse characteristic length scale $\langle q(t)\rangle$ of the order parameter for various values of λ. The data for $\lambda = 0$ (usual spinodal decomposition) and $\lambda = 0.50$ represent the average over three 128×128 system with different initial conditions, while the others are obtained from one system, respectively.

coarsening kinetics starts earlier as the coupling strength increases.

In the early stage of phase separation, the elastic coupling does not affect the coarsening kinetics significantly, because the membrane shape can follow the composition change immediately [5]. However, as the phase separation proceeds, the shape change cannot keep up with the coarsening of the order parameter, since the large deformation of the membrane shape is energetically disfavored by the Helfrich repulsion. Therefore, once the elastic energy becomes dominant over the interfacial energy, the coarsening begins to slow down drastically.

ACKNOWLEDGMENTS

The author is grateful to Professor Takao Ohta for a number of illuminating discussions.

REFERENCES

1. Leibler, S., *J. Phys. (Paris)* **47**, 507-516 (1986).
2. Andelman, D., Kawakatsu, T., and Kawasaki, K., *Europhys. Lett.* **19**, 57-62 (1992).
3. Taniguchi, T., *Phys. Rev. Lett.* **76**, 4444-4447 (1996).
4. Sunil Kumar, P.B., Gompper, G., and Lipowsky, R., *Phys. Rev. Lett.* **86**, 3911-3914 (2001).
5. Uchida, N., *Phys. Rev. E* **66**, 040902(R) 1-4 (2002).
6. Helfrich, W., *Z. Naturforsch.* 33A, 305-315 (1978).
7. MacKintosh, F.C., *Phys. Rev. E* **50**, 2891-2897 (1994).
8. Wada, H., to appear in *J. Phys. Soc. Jpn.*

Membrane-Membrane Interaction and Free Energy of Multilayer Membrane System

Takao Yamamoto* and Yukitoshi Iwamoto[†]

*Department of Physics, Faculty of Engineering, Gunma University, Kiryu, Gunma, 376-8515, Japan
[†]Department of Biological and Chemical Engineering, Faculty of Engineering, Gunma University, Kiryu, Gunma, 376-8515, Japan

Abstract. The effect of the membrane-membrane interactions in multilayer membrane systems on the free energy is investigated. An approximation by which the free energy increment Δf induced by the interactions is expressed in terms of the membrane-membrane distance distribution function is introduced. The analytic expression of the distance distribution function is derived on the basis of the Monte Carlo calculation. It is found that Δf behaves as $\Delta f \sim \rho^4$ for the small membrane density ρ.

INTRODUCTION

Statistical-mechanical properties of fluid bilayer lipid membrane are expressed in terms of the curvature elastic energy[1-3]. For an isolated membrane sheet, the Hamiltonian is given by

$$H_{\text{one}}(\{u\}) = \int dxdy \frac{1}{2} K |\nabla^2 u(x,y)|^2, \quad (1)$$

where the x-y plane is chosen to be parallel to the mean membrane plane, $z = u(x,y)$ is the shape of the membrane, K is the rigidity of the membrane and $\nabla = (\partial/\partial x, \partial/\partial y)$.

Let us consider the n-layer membrane system embedded in an $L \times L \times L$ space. Denoting the shape of the j-th ($j = 1, 2, \cdots, n$) membrane of the system by $z = u_j(x,y)$ and the membrane-membrane interaction potential by $V(z)$, we have the Hamiltonian of the multilayer system as

$$H(\{u_j(x,y)\}_{j,(x,y)}) = H_0 + H_{\text{I}} \quad (2)$$

$$H_0 = \sum_{j=1}^{n} H_{\text{one}}(\{u_j(x,y)\}) \quad (3)$$

$$H_{\text{I}} = \sum_{i<j} \int dxdy V(u_j(x,y) - u_i(x,y)). \quad (4)$$

The partition function and the free energy per unit volume $f(\rho)$ ($\rho = n/L$ being the density of the membrane) are respectively given by

$$Z = \int_{u_j < u_{j+1}} \prod_j \prod_{(x,y)} du_j(x,y) e^{-\beta H}, \quad (5)$$

$$f(\rho) = -\frac{k_B T}{L^3} \ln Z, \quad (6)$$

where $u_j < u_{j+1}$ denotes the non-crossing nature of the membrane and $\beta = 1/(k_B T)$ (k_B being the Boltzmann constant and T being the temperature). The partition function is rewritten as

$$Z = Z_0 < \exp[-\beta H_{\text{I}}] >_0, \quad (7)$$

where Z_0 and $< \cdots >_0$ are respectively the partition function and the thermal average for the system with $V = 0$. The free energy per volume is given by

$$f(\rho) = -k_B T \frac{1}{L^3} \ln Z = f_0(\rho) + \Delta f(\rho) \quad (8)$$

$$f_0(\rho) = -k_B T \frac{1}{L^3} \ln Z_0 \quad (9)$$

$$\Delta f(\rho) = -k_B T \frac{1}{L^3} \ln < \exp[-\beta H_{\text{I}}] >_0, \quad (10)$$

where $f_0(\rho)$ is the free energy for the $V = 0$ system (the "pure" non-crossing membrane system). For the "pure" non-crossing membrane system, the free energy behaves as[4,5]

$$f_0(\rho) \equiv f(\rho)|_{V=0} = \gamma \rho + B_0 \frac{(k_B T)^2}{K} \rho^3 + \mathcal{O}(\rho^4), \quad (11)$$

where γ is the free energy per project area of the isolated membrane and B_0 is a numerical constant.

APPROXIMATION METHOD

We pay attention to the deviation of the free energy $\Delta f(\rho)$ from f_0 induced by the membrane-membrane interaction V. We adopt the following approximation;

$$< \exp[-\beta H_{\text{I}}] >_0 = < \prod_{i<j} \prod_{(x,y)}$$

$$\times e^{-\beta V(u_i(x,y)-u_j(x,y))} >_0$$
$$\simeq \prod_i \prod_{(x,y)} < e^{-\beta V(u_i(x,y)-u_{i+1}(x,y))} >_0$$
$$= \left[\int_0^\infty ds\, e^{-\beta V(s)} P_0(s) \right]^{nL^2}, \quad (12)$$

where, $P_0(s)$ is the membrane-membrane distance distribution function defined by

$$P_0(s) = < \delta(u_{j+1}(x,y) - u_j(x,y) - s) >_{V=0}, \quad (13)$$

Therefore, using $P_0(s)$, we have an approximate expression:

$$\Delta f(\rho) \simeq -k_B T \rho \ln[\int_0^\infty ds \exp(-\beta V(s)) P_0(s)]. \quad (14)$$

MONTE CARLO ANALYSIS

To obtain $P_0(s)$, we performed a Monte-Carlo (MC) calculation based on the solid-on-solid (SOS) membrane model[6]. In the SOS model, the shape of the j-th membrane ($j = 1, 2, \cdots, n$) is expressed by a set of the discrete variables $z = \tilde{z} = \tilde{u}_j(\tilde{x}, \tilde{y})$, where $\tilde{x}, \tilde{y}$ and $\tilde{u}_j$ have an integer. The n membranes sit on the $\tilde{L} \times \tilde{L} \times \tilde{M}$ simple cubic lattice. The energy for the j-th membrane is given by

$$E_j = \tfrac{1}{2} J \sum_{\tilde{x}=1}^{\tilde{L}} \sum_{\tilde{y}=1}^{\tilde{L}} [(\tilde{u}_j(\tilde{x}+1, \tilde{y}) + \tilde{u}_j(\tilde{x}-1, \tilde{y})$$
$$- 2\tilde{u}_j(\tilde{x}, \tilde{y}))$$
$$+ (\tilde{u}_j(\tilde{x}, \tilde{y}+1) + \tilde{u}_j(\tilde{x}, \tilde{y}-1) - 2\tilde{u}_j(\tilde{x}, \tilde{y}))]^2, \quad (15)$$

where J is the "microscopic" rigidity. The temperature was chosen as $T = 5J/(2k_B)$. It was verified that the long-range behaviors of the SOS model and of the original continuum model are same. For equilibration, 5×10^6 Monte Carlo steps (MCS) were required. Averages were taken over 1×10^7 MCS. The number of the membrane n and the size $\tilde{L}$ were respectively chosen as $n = 20 \sim 80$ and $\tilde{L} = 150 \sim 200$. For the five membrane densities $\tilde{\rho} = 1/10, 1/12, 1/14, 1/16$, and $1/20$, the discrete version $\tilde{P}_0(\tilde{s}) = < \delta_{\tilde{u}_{j+1} - \tilde{u}_j, \tilde{s}} >_{MC}$ for $P_0(s)$ was calculated, where $< \cdots >_{MC}$ stands for the MC thermal average. In Figure 1, the result for the system with $\tilde{\rho} = 1/20$ is shown. The plot lies on the curve expressed by $\ln \tilde{P}_0(\tilde{s}) = -8.30 + 2.18 \ln \tilde{s} - 0.0030 \tilde{s}^2$ very well. From the results for the five membrane densities, we obtained the relation

$$\tilde{P}_0(\tilde{s}) = \tilde{C}_0 \tilde{\rho}^3 \tilde{s}^2 \exp(-\tilde{C}_1 \tilde{\rho}^2 \tilde{s}^2), \quad (16)$$

where $\tilde{C}_0$ and $\tilde{C}_1$ are constants. Hence, for the original continuum model, the membrane-membrane distance distribution function is expected to be expressed as

$$P_0(s) = C_0 \rho^3 s^2 \exp(-C_1 \rho^2 s^2), \quad (17)$$

where C_0 and C_1 are constants.

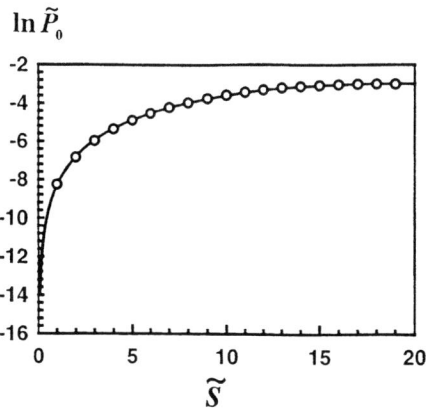

FIGURE 1. Plot of the relationship between $\ln \tilde{P}_0(\tilde{s})$ and the distance $\tilde{s}$ for the system with $\tilde{\rho} = 1/20$

DISCUSSIONS

Using Eqs. (14) and (17), we can calculate the change of the free energy $\Delta f(\rho)$ induced by the membrane-membrane interactions. For the short-range interactions and the power law interactions $V(r) = A/r^\alpha$ with $\alpha \geq 3$ (A being a constant), we obtain

$$\Delta f(\rho) = k\rho^4 + \mathscr{O}(\rho^5), \quad (18)$$

where k is a positive constant for the repulsive membrane-membrane interactions and a negative constant for the attractive interactions. For the repulsive interaction systems, we can propose an inequality to evaluate the exponent of the lowest power term in $\Delta f(\rho)$ exactly. For the attractive interaction systems, the phase separation phenomena are analyzed on the basis of Eqs.(11) and (18). There are three typical regions in the phase diagram. In the high temperature region, a uniform phase appears. In the middle temperature region, a dense and a dilute phase coexist. In the low temperature region, all the membranes bunch; the phase composed of only the "solvent" and a dense phase coexist.

This work was partially supported by the Grant-in-Aid for Science Research from Ministry of Education, Science, Sports and Culture (Grant # 14540370).

REFERENCES

1. Helfrich, W., *Z.Naturforsh* **28c**, 693-703 (1973).
2. Brochard, F. and Lennon, F., *J. Phys. (Paris)* **36**, 1035-1047 (1975).
3. Brochard, F., de Gennes, P.G. and Pfeuty, P., *J. Phys. (Paris)* **37**, 1099-1104 (1976).
4. Helfrich, W., *Z. Naturforsh* **33a**, 305-315 (1978).
5. Bachmann, M., Kleinert, H. and Pelster, A., *Phys. Rev. E* **63**, 051709-1-051709-10 (2001).
6. Yamamoto, T. and Kawashima, Y., *J. Phys. Soc. Jpn.* **69**, 2715-2718 (2000).

Dynamic Critical Phenomena of Polymer Solutions

Akira Furukawa

Department of Physics, Graduateschool of Science, Kyoto University, Kyoto 606-8502

Abstract. The kinetic coefficients of critical polymer solutions are calculated by the mode coupling theory. We predict that the critical divergence of viscosity should be suppressed with increasing the molecular weight. The diffusion constant and the dynamic structure factor are also calculated. The present results explicitly show that the critical dynamics of polymer solutions should be affected by an extra spatio-temporal scale intrinsic to polymer solutions, and are consistent with the experiment of Tanaka, *et al.*

INTRODUCTION

Dynamic critical phenomena of classical fluids have been successfully investigated by the mode coupling theory [1] and the dynamic renormalization group theory [2]. It has been believed that the critical dynamics of complex fluids such as polymer solutions also belongs to the model H universality. Contrary to this conventional understanding, some experimental works have questioned that the dynamic critical phenomena of polymer solutions can be really categorized into the model H universality.

Recently H. Tanaka and his co-workers [3] have reported experimental work of dynamic critical phenomena of polymer solutions. Their accurate measurements for critical anomaly of viscosity show that the dynamic exponent y_c significantly decreases as the molecular weight M_w increases. That is, in polymer solutions with large molecular weight the critical divergence of viscosity is suppressed. They also measured the dynamic structure factor by the light scattering experiment. It has been found that in the case of large molecular weight the dynamic structure factor cannot be expressed by the Kawasaki scaling function even in the vicinity of the critical point. On the basis of such experimental facts, they concluded that the critical dynamics of polymer solutions exhibits a non-universal nature and cannot be classified into the model H universality in a *practical sense* [3].

Motivated by the experiments reported by Tanaka, *et al* [3], we have investigated the dynamic critical phenomena of polymer solutions by the mode coupling theory [4]. In our analysis we have used the two-fluid model [5, 6] as the basic equations describing the dynamics of polymer solutions. Near the critical point in semi-dilute polymer solutions, the dynamics of the critical fluctuations is strongly influenced by the non-linear hydrodynamic interaction arising from the streaming type mode coupling. As a result, the kinetic coefficients, observed near the critical point, are renormalized by the non-linear hydrodynamic interaction. In our analysis such a non-linear hydrodynamic interaction is successfully taken into account by the mode coupling theory. We then show that the viscoelasticity affects the critical dynamics of polymer solutions, resulting in a suppression of the divergence of viscosity. The detailed calculation of this work and other many results have already been presented elsewhere [4, 7].

MODE COUPLING APPROACH

Based on the mode coupling theory [1], we can derive a set of self-consistent equations for the self energies, taking the non-linear hydrodynamic interaction into account. Then, renormalizing the non-linear hydrodynamic interaction terms, the transport coefficients can be calculated (see Ref. [4] for details).

Renormalized Diffusion Constant

From our mode coupling analysis [4], in the vicinity of the critical point, the long-wavelength expression of the diffusion constant D is obtained as

$$D = \frac{k_B T}{6\pi\xi(\eta_p + \eta_0)}\left(1 + \frac{\eta_p}{\eta_0 A}\right), \quad (1)$$

where ξ is the thermal correlation length, η_0 and η_p are the solvent and the polymeric viscosity, respectively. The parameter A is given by

$$A = \frac{\xi}{\xi_{ve}}\sqrt{1 + \frac{\eta_p}{\eta_0}}. \quad (2)$$

where ξ_{ve} is a viscoelastic length or a magic length [5, 6], which characterizes the cooperative length scale intrinsic to entangled polymer solutions. For polymer solutions the two-fluid model gives the viscoelstic length as $\xi_{ve} \cong (\eta_p/\eta_0)^{\frac{1}{2}} \xi_b$ [6], where ξ_b is the blob length. In Eq. (1), the first term comes from the usual Kawasaki function, but an additional term $\eta_p/\eta_0 A$ arises from the viscoelsticity. In the case of $\eta_p \gg \eta_0$ and $\xi \cong \xi_{ve}$, where the viscoelasticity is relatively strong, the second term dominates the first term.

It must be worthwhile to mention the following: We have also calculated the dynamic structure factor with the renormalized relaxation rates by the non-linear hydrodynamic interaction. The dynamic scaling is violated and the self energy cannot be expressed by the Kawasaki scaling function. This is evident in Fig. 1(a) and 1(d) of Ref. [4], where the viscoelasticity is found to be relatively strong.

Renormalized Shear Viscosity

As is well known, the shear viscosity exhibits a weak divergence at the critical point. From our mode coupling analysis [4], the anomalous part of the shear viscosity $\Delta \eta$ is given in the following form,

$$\Delta \eta \cong \frac{8\bar{\eta}}{15\pi^2} \left[\frac{3\pi}{40} \frac{1}{1+\frac{\eta_p}{\eta_0 A}} + \ln\left(\frac{\xi+\tau}{\xi_b+\tau}\right) + \frac{\eta_0}{\bar{\eta}} \ln(\xi_b \Lambda_0) \right], \quad (3)$$

where $\tau = 8\eta_p \xi / 3\pi \eta_0 A$, $\bar{\eta} = \eta_0 + \eta_p$ is the bare shear viscosity, and Λ_0 is the microscopic cut-off wave number.

For $\eta_p \gg \eta_0$ and $\xi_{ve} \cong \xi$, it is readily shown that the resultant renormalized shear viscosity $\eta = \bar{\eta} + \Delta \eta$ does not exhibit critical anomaly.

$$\frac{\eta}{\bar{\eta}} \cong 1. \quad (4)$$

That is, the critical divergence of the shear viscosity is suppressed as the molecular weight increases.

For $\eta_p \cong \eta_0$ and $\xi \gg \xi_{ve}$, where the viscoelasticity is relatively weak, we obtain the anomalous shear viscosity as follows:

$$\frac{\eta}{\bar{\eta}} \cong 1 + \frac{8}{15\pi^2} \ln\left[\frac{\xi \Lambda_0}{(\xi_b \Lambda_0)^{\frac{\eta_p}{\bar{\eta}}}} \right]. \quad (5)$$

Because of the small coefficient of the logarithmic term, it may well be exponentiated with the small exponent $x_\eta = 8/15\pi^2 \cong 0.054$ as $\eta/\bar{\eta} \cong [\xi \Lambda_0/(\xi_b \Lambda_0)^{\eta_p/\bar{\eta}}]^{x_\eta}$. In the low molecular weight limit ($\eta_p = 0$) the expressions reduce to those of the simple classical fluids, with $y_c = x_\eta v \cong 0.034$ [2, 8].

The above results are due to the existence of an extra length-scale ξ_{ve} intrinsic to the viscoelasticity. As is well known [1, 2], in the case of classical fluids the behavior of the critical anomaly of the shear viscosity does not depend on the detail of the material. That is, the behavior of the critical anomaly does not depend on the material parameter, and is universal (model H). Contrary to this, as shown in the present analysis, in the case of polymer solutions the critical divergence of the shear viscosity strongly depend on the molecular weight. Now, a few comments must be made on those points. When the correlation length ξ is very large compared with the other length scale, it is shown that our results certainly exhibit the universal behavior. However, as noted by Tanaka[3], it is difficult to experimentally access such a temperature region for polymer solutions whose polymer has a very high molecular weight. In this sense we can say that the model H universality does not hold, at least in a *practical sense* [3].

ACKNOWLEDGMENTS

The author wishes to thank Prof. Akira Onuki for useful comment.

REFERENCES

1. K. Kawasaki: Ann. Phys. **61**, 1-56 (1970).
2. P.C. Hohenberg and B.I. Halperin: Rev. Mod. Phys. **49**, 435-479 (1976).
3. H. Tanaka, Y. Nakanishi and N. Takubo: Phy. Rev. E **65**, 021802 (2002).
4. A. Furukawa: J. Phys. Soc. Jpn **72**, 1436-1445 (2003).
5. F. Brochard and P.G. de Gennes: Macromolecules **10**, 1157-1161 (1977).
6. M. Doi and A. Onuki: J. Phys.(Paris) II **2**, 1631-1656 (1992).
7. A. Furukawa: J. Phys. Soc. Jpn **71**, 209-212 (2003).
8. T. Ohta: J. Phys. C **10**, 791-793 (1977).

Stable and Time-Dependent Vortex-Patterns in 2D Schrödinger Equations and A Consideration of Biological Memories

Tsunehiro Kobayashi

Department of General Education, Tsukuba College of Technology, Ibaraki 305-0005, Japan

Abstract. Various vortex patterns are investigated in terms of the quantum flows. A possibility of biological memories is also proposed by using the vortex patterns.

ZERO-ENERGY FLOWS

Problems of nested vortices appear in the wide range of present-day physics such as solid state physics, non-neutral plasma, cosmology and biosciences. In those problems some kind of quantum flows are very important roles. Recently it has been found that quantum flows have the exactly zero-energy eigenvalue and infinitely degenerate in two-dimensional (2D) Schrödinger equations with central potentials $V_a(\rho) = -a^2 g_a \rho^{2(a-1)}$ ($a \neq 0$ and $\rho = \sqrt{x^2 + y^2}$) [1-4]. As far as the zero-energy eigenstates (ψ_0) are concerned, the Schrödinger equations for all a can be reduced to the following equation in terms of the conformal transformations $\zeta_a = z^a$ with $z = x + iy$;

$$[-\frac{\hbar^2}{2m}\Delta_a - g_a] \psi_0(u_a, v_a) = 0, \quad (1)$$

where $\Delta_a = \partial^2/\partial u_a^2 + \partial^2/\partial v_a^2$, using $\zeta_a = u_a + iv_a$ [2,3]. It is transparent that the zero-energy states are written by the same plane-wave solutions for all a in the ζ_a space and then they represent the stationary flows in the ζ_a plane. Furthermore we see that the zero-energy states degenerate infinitely. For $g_a' > 0$, putting the function $f_n^\pm(u_a; v_a)e^{\pm i k_a u_a}$ with $k_a = \sqrt{2mg_a}/\hbar$ into (1), where $f_n^\pm(u_a; v_a)$ are polynomials of degree n ($n = 0, 1, 2, \cdots$), we obtain the equation for the polynomials as $[\Delta_a \pm 2ik_a\frac{\partial}{\partial u_a}]f_n^\pm(u_a; v_a) = 0$. A few examples of $f^\pm$ are given by $f_0^\pm = 1$, $f_1^\pm = 4k_a v_a$, $f_2^\pm = 4(4k_a^2 v_a^2 + 1 \pm 4ik_a u_a)$. The general forms are obtained by using the solutions of the 2D parabolic potential barrier (2D PPB) [2,3].

Note that the zero-energy eigenfunctions are not normalizable. This fact means that the probability density ($\rho(t, x, y)$) and the probability current ($j(t, x, y)$) lose the meanings. We, however, see that the velocity defined by
$$v(x, y; t) = j(x, y; t)/\rho(x, y; t),$$
can have the well-defined meaning because of the cancellation of the normalization factors. It is known that vortices appear at the nodal points of the wave function. Remembering the infinite degeneracy, we can construct wave functions having the nodal points at arbitrary positions in terms of the superposition of the zero-energy solutions [2-4]. In this paper we study stable and unstable vortex patterns by using the eigenfunctions of the conjugate spaces of Gel'fand triplets in the 2D PPB, where all the eigenfunctions are solved [1]. We also show that such vortex patterns are very suitable objects to describe dynamics inside living beings [6].

VOTEX PATTERNS IN 2D PPB

Let us investigate vortex patterns in the 2D PPB [5].

A. Stable vortex patterns: A few examples for the superposition with a few vortices are presented here.

(1) A pattern with three vortices at the origin and the two points of $(\pm c/\beta, \pm c/\beta)$ in Fig. 1a is given by

$$\Phi_3^{+-}(x, y) = [4\beta^2 xy(\beta^2 xy - c^2) + 2i\beta^2(x^2 - y^2)]e^{i\beta^2(x^2-y^2)/2}. \quad (2)$$

(2) A pattern with four vortices at $(\pm c/\beta, \pm c/\beta)$ in Fig. 1b is given by

$$\Phi_4^{+-}(x, y) = [(\beta^2 xy - c^2)(\beta^2 xy + c^2) + i\beta^2(x^2 - y^2)/2]e^{i\beta^2(x^2-y^2)/2}. \quad (3)$$

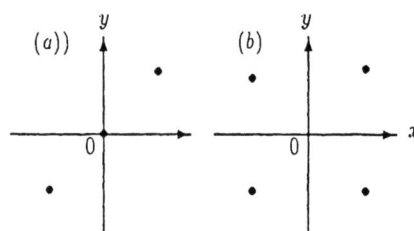

FIG. 1. Stable patterns with three vortices (a) and four vortices (b). • denotes a vortex.

B. Moving vortex patterns: Let us study time-dependent vortex patterns, where vortices can move, and sometimes be created and annihilated. Such patterns are obtained by the superposition of stable and unstable flows. In the following considerations the unstable flows are put in the stable flows at $t = 0$.

(1) A pair annihilation pattern in Fig. 2a is written by

$$\Phi^{+-}_{1pair}(x,y,t) = [2\beta x(\beta^3 xy^2 - \theta(t)c^3 e^{-\gamma t}) + i\beta^2(x^2 - y^2)]e^{i\beta^2(x^2-y^2)/2}, \quad (4)$$

where $\theta(t)$ is taken as $\theta(t) = 0$ for $t < 0$ and $= 1$ for $t \geq 0$. It has two nodal points at $(ce^{-\gamma t/3}/\beta, \pm ce^{-\gamma t/3}/\beta)$ for $t \geq 0$. The vortices are created at $t = 0$ and move toward the origin as $t \to \infty$ and annihilate there.

(2) A two-pairs annihilation pattern is given by

$$\Phi^{+-}_{2pairs}(x,y,t) = [4\beta^2 x^2(\beta^2 y^2 - \theta(t)c^2 e^{-2\gamma t}) - 2i(\theta(t)c^2(1 - e^{-2\gamma t}) - \beta^2(x^2 - y^2))]e^{i\beta^2(x^2-y^2)/2} \quad (5)$$

It has four nodal points at $(\pm c/\beta, \pm ce^{-\gamma t}/\beta)$. In Fig. 2b two pairs are created at $t = 0$ and annihilate as $t \to \infty$.

We stress that one can make much more interesting vortex patterns in terms of simple superposition [5].

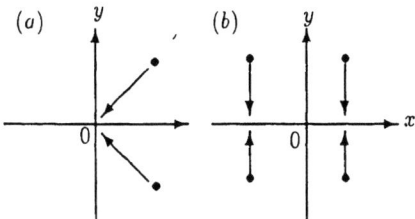

FIG. 2. A pair (a) and two pairs (b) annihilation patterns.

A BIOLOGICAL MEMORIES

We see that standing waves of the zero-energy flows can be confined in quadrangles in the ζ_a plane, which are $4a$ gons in the xy plane. Let us study a vortex formation in a simple example where a standing wave $\psi_{cc} = \cos k_u u \cdot \cos k_v v$ is confined in a quadrangle, that is, the half of the wave length of k_u is confined in the u direction, whereas the $3/2$ wave lengths of k_v is confined in the v direction. Let us put a different standing wave, for instance, the standing wave with the wave number vector $\vec{q} = (q_u, q_v)$, which is described by the wave function $\psi_{sc} = i \sin q_u u_a \cdot \cos q_v v_a$, into the quadrangle prepared. Note that in the process where the new flow is added we can use the free motions of the direction perpendicular to the ζ_a plane. We see that the ranges of u_a and v_a in the quadrangle are given by $|u_a| < u_B \equiv \pi/2k_u$ and $|v_a| < v_B \equiv 3\pi/2k_v$. The total wave function is written by $\Psi = \psi_{cc} + C\psi_{cc}$, where C is a complex number. In order that the inserted wave ψ_{sc} is stable in the quadrangle, the wave function must vanish on the boundaries of the quadrangle, that is, on the lines fulfilling $|u_a| = u_B$ or $|v_a| = v_B$. This constraint gives us the relations $q_u = 2(L_u + 1)k_u$ and $q_v = 2(L_v + 1)k_v/3$, where L_u and L_v are zero or positive integers. The correspondences between the vortex patterns and the wave numbers q_u are seen in Fig. 3.

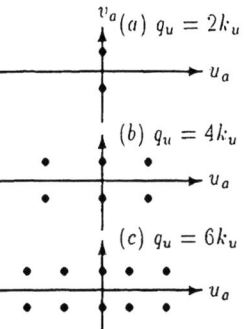

FIG. 3. Vortex patterns in the quadrangle.

In this vortex-formation process *we find out the selection rule for the wave number vector of the inserted wave $\vec{q}$ such that the ratios of q_u/k_u and q_v/k_v must be rational numbers.* We can read off the number L_u (or L_v) from the observation of the vortex patterns. This fact means that the system analyzing the wave number vector of the inserted wave, namely, Fourier analyzer in terms of quantum vortices can be constructed by using many quadrangles having different sizes and shapes. We can also use the system as the memories [6]. Namely the set of various vortex patterns awaked in the system by a stimulus can be considered as the memory corresponding to the stimulus.

This system has the following very interesting properties to describe the system in living beings: It has (i) the absolute energy saving mechanism, (ii) the infinite variety of the vortex patterns corresponding to our memories, (iii) the perfect stability from any disturbances and (iv) the huge flexibility arising from the infinite degeneracy.

REFERENCES

1. T. Kobayashi and T. Shimbori, *J. Phys. A* **33** 7637-7652 (2000).
2. T. Kobayashi, *Physica A* **303** 469-480 (2002).
3. T. Kobayashi and T. Shimbori, *Phys. Rev. A* **65** 042108 1-10 (2002).
4. T. Kobayashi, *Int. J. Theor. Phys.*, **42** 2265-2284 (2003).
5. T. Kobayashi, preprints, quant-ph/0302148 (2003).
6. T. Kobayashi, preprint, physics/0307031 (2003).

String Tension and Surface Tension of Elastic Membranes with Fluidity

Hiroshi Koibuchi*, Nobuyuki Kusano*, Atsusi Nidaira* and Komei Suzuki*

*Ibaraki College of Technology, Nakane 866, Hitachinaka, Ibaraki 312-8508, Japan

Abstract. The string tension and the surface tension of a model of elastic and phantom membranes are calculated by the grand canonical Monte Carlo simulation. It is shown by scaling arguments that both of them vanish at the critical point of the phase transition.

INTRODUCTION

Models of membranes have played an important role in understanding physical system of two dimensional surfaces [1]. The string tension τ and the surface tension σ of a model of elastic and phantom membranes with fluidity were studied by Ambjorn et.al [2] about 10 years ago.

They used the canonical ensemble to extract information about the scaling properties of these quantities. By using the canonical MC technique, they found that the phase transition of the model are characterized by a vanishing string tension τ at the critical point and a vanishing surface tension σ at the critical point [3].

The same model was recently reported [4, 5] to undergo a second order phase transition of shape fluctuation. The canonical MC technique was also used in [4].

However, both τ and σ are naturally defined in the grand-canonical ensemble. Hence, we study the scaling properties of τ and σ by using a grand-canonical MC technique, and show some preliminary results of their scaling properties in this presentation.

MODEL AND MC TECHNIQUE

The partition function is defined by

$$Z(b,\mu;V_d) = \sum_N \sum_T \int \prod_{i=1}^N dX_i \exp[-S(X,T)+\mu N],$$

$$S(X,T) = S_1 + bS_2 \qquad (1)$$

where $\sum_N, \sum_T$ are the sum over N and the sum over triangulations T respectively, and surfaces are allowed to self-intersect. The coefficients b and μ are the bending rigidity and the chemical potential respectively. The V_d in $Z(b,\mu;V_d)$ denotes the d-dimensional boundary of volume V_d.

In order to calculate τ, we use triangulated spherical surfaces with two fixed vertices separated by $V_1 = L$. Then, we expect

$$Z(b,\mu;V_1) \sim \exp(-\tau L) \qquad (2)$$

in the limit $L \to \infty$. We show in Fig. 1(a) a schematic drawing of the surface, and in Fig. 1(b) the corresponding string of length $V_1 = L$, whose energy is given by τL.

FIGURE 1. (a) A spherical surface with two fixed vertices separated by $V_1 = L$ for the calculation of τ, (b) the corresponding string of the length $V_1 = L$.

To calculate σ, we use surfaces with boundaries which correspond to 2-dimensional boundary of volume $V_2 = A$. Then, we expect

$$Z(b,\mu;V_2) \sim \exp(-\sigma A) \qquad (3)$$

in the limit $A \to \infty$. Figure 2 shows that a square surface of $V_2 = A$ is continuously deformed to a cylindrical surface with boundaries, where the periodic boundary condition is assumed.

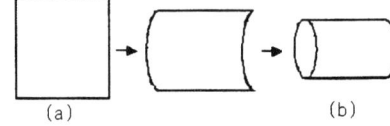

FIGURE 2. (a) A square of area $V_2 = A$, (b) a cylindrical surface with boundaries for the calculation of σ.

By using the scale invariance of the partition function, we have

$$\tau(b,L) = \frac{2\langle S_1 \rangle - 3\langle N \rangle}{L}, \quad (4)$$

$$\sigma(b,A) = \frac{2\langle S_1 \rangle - 3\langle N \rangle}{2A}. \quad (5)$$

It is expected that τ and σ scale according to

$$\tau \sim \left(\frac{L}{\langle N \rangle}\right)^{\nu_\tau}, \quad \sigma \sim \left(\frac{A}{\langle N \rangle}\right)^{\nu_\sigma} \quad (6)$$

for $b \to b_c$ if the model undergoes a second order phase transition at finite critical bending rigidity b_c.

The dynamical variables of the model are X_i and T. The total number of vertices N can also be viewed as a dynamical variable of the model, because it is summed over in the partition function. The MC updates of the variables are identical with those described in [5].

RESULTS

The MC results of τ are shown in Fig. 3. τ is calculated on surfaces with boundary points separated by length $L = 10, 14, 20$. Figure 3(a) shows the specific heat C_{S_2} vs. the reduced bending rigidity

$$\lambda = \frac{b}{b_c} - 1, \quad (7)$$

where C_{S_2} is defined by a fluctuation of S_2 such that

$$C_{S_2} = \frac{b^2}{N}\left(\langle S_2^2 \rangle - \langle S_2 \rangle^2\right). \quad (8)$$

Figure 3(b) shows τ vs. $L/\langle N \rangle$. We see in Fig.3(b) the expected scaling property of τ given by (6) at $\lambda = 0$; $\tau(\lambda = 0) \to 0$ in the limit $N \to \infty$.

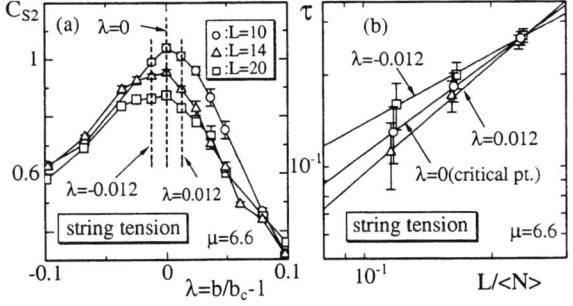

FIGURE 3. (a) C_{S_2} vs. λ and (b) τ vs. $L/\langle N \rangle$. The chemical potential $\mu = 6.6$ corresponds to $N \simeq 850$.

C_{S_2} are computed with the cylindrical surfaces of projected area $A = 400, 625$ and $A = 900$ with fixed boundaries. Figure 4(a) shows C_{S_2} vs. λ, and Fig.4(b) shows σ vs. $A/\langle N \rangle$. We see in Fig.4(b) the expected scaling property of σ given by (6) at $\lambda = 0$; $\sigma(\lambda = 0) \to 0$ in the limit $N \to \infty$.

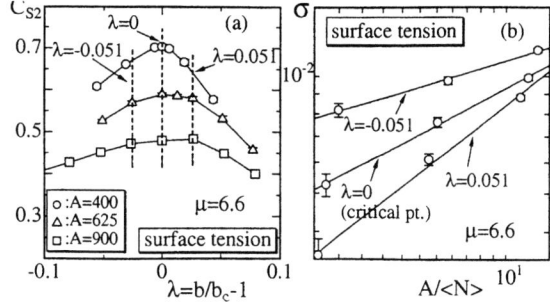

FIGURE 4. (a) C_{S_2} vs. λ and (b) σ vs. $A/\langle N \rangle$. The chemical potential $\mu = 6.6$ corresponds to $N \simeq 800$.

The snapshots of surfaces used for the calculation of τ and σ are shown in Figs. 5(a) and 5(b).

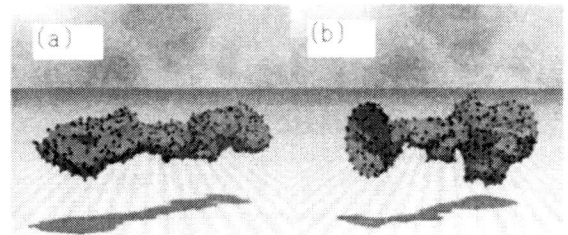

FIGURE 5. Snapshots of (a) spherical surface of $L = 20$ at $b = 1.75$, and (b) cylindrical surface of $A = 400$ at $b = 1.68$. $N = 800 \sim 850$ in (a) and (b).

SUMMARY

We have studied the scaling property of the string tension τ and the surface tension σ of a model of elastic and phantom membranes by using the grand canonical MC technique on dynamically triangulated surfaces. We find that both τ and σ vanish at the critical point. These results are consistent with those obtained by the canonical MC in [2].

REFERENCES

1. David F., in *Two Dimensional Quantum Gravity and Random Surfaces*. edited by Gross D.J., Piran T., and Weinberg S., World Scientific, Singapore, 1989, Vol.8, pp.81-124.
2. Ambjorn J., Irback A., Jurkiewicz J. and Peterson B., *Nucl.Phys.* B **393**, 571-600 (1993).
3. Wheater J.F., *J. Phys. A:Math.Gen.* **27**, 3323-3353 (1994).
4. Koibuchi H., *Phys. Lett. A* **300**, 586-590 (2002).
5. Koibuchi H. et.al, *Phys. Lett. A* **319**, 44-52 (2003).

Configuration of Nematic Liquid Crystals around Particles under External Fields

Jun-ichi Fukuda*, Holger Stark†, Makoto Yoneya* and Hiroshi Yokoyama**

*Yokoyama Nano-structured Liquid Crystal Project, 5-9-9 Tokodai, Tsukuba 300-2635, Japan
†Universität Konstanz, Fachbereich Physik, D-78457 Konstanz, Germany
**Nanotechnology Research Institute, AIST, 1-1-4 Umezono, Tsukuba 305-8568, Japan

Abstract. We present the results of our numerical attempts to simulate the configuration of a nematic liquid crystal around a spherical particle. We focus on the effect of an external field, such as a magnetic field or a flow field, on the director configuration and the topological defects accompanied by the particles. The use of adaptive mesh refinement together with a tensor order parameter for the description of the orientational order makes it feasible to tract the dynamics of a nematic liquid crystal without any special treatment of the topological defects.

INTRODUCTION

Liquid crystal colloid dispersions have been attracting much interests in technology as well as in fundamental science as a novel class of composite materials[1, 2]. The surfaces of particles immersed in a liquid crystal induce elastic distortions of the positional or orientational order of liquid crystal molecules because the particle surface tends to align molecules along some direction with respect to it (surface anchoring). The orientational profile of liquid crystals around the particles depends sensitively on the surface properties, size and shape of the particles. In the case of strong anchoring, topological defects are formed close to the particles. Those topological defects include a point-like defect referred to as a hyperbolic hedgehog, and a Saturn ring that surrounds the particle as the name implies. Topological defects have long been an important subject of condensed matter physics and the formation of topological defects due to foreign inclusions provides one of the interesting and important problems.

However, numerical approaches to the topological defects in a liquid crystal colloid dispersion are highly challenging, because it possesses two largely different characteristic lengths; the size of the particles ($\sim \mu$m) and that of topological defect cores (~ 10 nm). Simultaneous numerical treatment of two different length scales is highly difficult, because when numerical resolution is low, the shorter characteristic length scale cannot be described, while when numerical resolution is high enough, huge numerical resources will be necessary. To overcome these difficulties, we have devised an adaptive mesh refinement (AMR) scheme, in which fine grids are generated only around the defect core region where fine resolution will be necessary. Moreover the use of a tensor order parameter Q_{ij} to describe the orientational order of a nematic liquid crystal eliminates the difficulty of the director (unit vector) description where topological defects must be treated as singularities. We have already shown[3, 4, 5] that our numerical scheme can be applied to the investigation of dynamical behavior as well as fine equilibrium structures of topological defects. In this paper we will mainly aim at investigating how the orientation profile around particles is influenced by external perturbations such as magnetic or electric field, and imposed hydrodynamic flow.

RESULTS

In Fig. 1, we present one of the typical examples of our simulation results showing a transition from a hedgehog configuration to a Saturn ring around a spherical particle imposing strong normal anchoring after applying a magnetic field H parallel to the horizontal z-axis, the axis of rotational symmetry. The details of our simulations presented here are found in Refs. [4, 5]. Such a transition was recently observed in a beautiful experiment of Loudet and Poulin[6] and so far as we know our simulation is the first one that reproduces their experiment successfully. We emphasize that owing to the use of the AMR scheme, "numerical pinning" is absent, which enables us to investigate the dynamics of topological defects.

Fig. 2 shows the time evolution of the orientation profile around a fixed particle after the application of a hy-

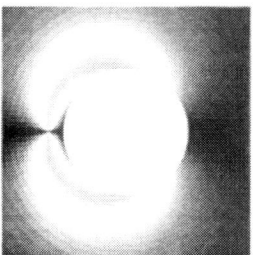  

FIGURE 1. Time evolution of the orientation profile under the magnetic field parallel to the (horizontal) z-axis. The gray-scale plots of Q_{zz}^2 are shown. The leftmost figure is the initial condition, the equilibrium hedgehog profile under no external field and the time intervals between adjacent two figures are equal.

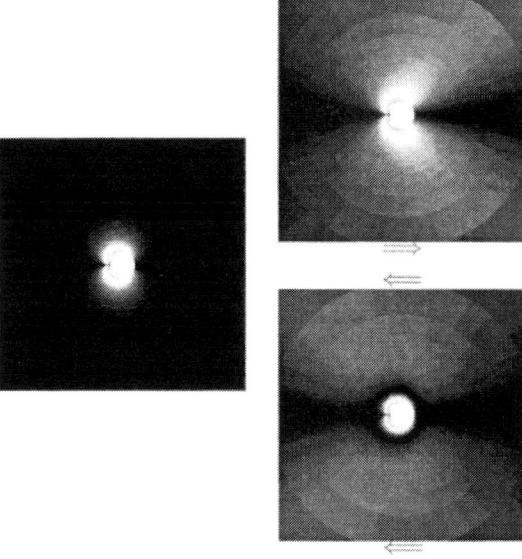

FIGURE 2. Time evolution of the orientation profile (gray-scale plots of Q_{zz}^2) under a hydrodynamic flow, which direction at infinity is shown by arrows. The left figure is the initial condition, the equilibrium hedgehog profile under no fluid flow.

drodynamic flow which direction is along the horizontal axis at infinity. The details of the numerical calculation are given in Ref. [5] and we note here that we numerically solve the set of hydrodynamic equations for the order parameter Q_{ij} and the fluid velocity $\mathbf{v}$ given by Olmsted and Goldbart[7]. The Ericksen number, which characterizes the ratio of the viscous force of a fluid to the elastic force of a liquid crystal and whose precise definition is given in Ref. [5], is set to $|Er| = 1$. To our knowledge, this is the first numerical study focusing on the dynamics of a nematic liquid crystal around a particle under a uniform flow imposed. There have been only few numerical studies[8, 9] that discussed the effect of hydrodynamic flow on the configuration of a nematic liquid crystal around a particle, all of which showed only the stationary profile of a liquid crystal long after the application of the flow. In contrast to the case of a magnetic field, where the equation of motion for Q_{ij} is invariant under the transformation $\mathbf{H} \leftrightarrow -\mathbf{H}$, the flows with different directions yield different orientation profiles as is evident in Fig. 2, because the hydrodynamic equations are not invariant under the transformation $\mathbf{v} \leftrightarrow -\mathbf{v}$.

These results indicate that we can expect a rich variety of responses of a nematic liquid crystal around a particle by applying various types of external perturbations and that there are therefore many ways of controlling the profile of the orientation and the topological defects of a nematic liquid crystal around a particle.

REFERENCES

1. Poulin, P., *et al.*, *Science* **275**, 1770-1773 (1997).
2. Stark, H., *Phys. Rep.* **351**, 387-474 (2001) and references therein.
3. Fukuda, J., and Yokoyama, H., *Eur. Phys. J. E* **4**, 389-396 (2001); Fukuda, J., Yoneya, M., and Yokoyama, H., *Phys. Rev. E* **65**, 041709(1-4) (2002).
4. Fukuda, J., Yoneya, M., and Yokoyama, H., *Mol. Cryst. Liq. Cryst.* (to be published).
5. Fukuda, J., et al., *J. Phys.: Condens. Matter* (submitted).
6. Loudet, J.C., and Poulin P., *Phys. Rev. Lett.* **87**, 165503(1-4) (2001).
7. Olmsted, P.D., and Goldbart, P., *Phys. Rev. A* **41** 4578-4581 (1990); *ibid.* **46**, 4966-4993 (1992).
8. Chono, S., and Tsuji, T., *Mol. Cryst. Liq. Cryst.* **309**, 217-236 (1998).
9. Stark, H., and Ventzki, D., *Europhys. Lett.* **57**, 60-66 (2002); Stark, H., Ventzki, D., and Reichert, M., *J. Phys.: Condens. Matter* **15**, S191-S196 (2003).

Brownian Dynamics Simulation for Suspension of Oblate Spheroidal Particles in Simple Shear Flow

Takehiro Yamamoto*, Takanori Suga* and Noriyasu Mori*

Department of Mechanophysics Engineering, Graduate School of Engineering, Osaka University, 2–1, Yamadaoka, Suita, Osaka 565–0871, Japan

Abstract. Brownian dynamics simulations have been carried out for suspensions of oblate spheroidal particles interacting via the Gay-Berne potential. The system changed from isotropic phase to nematic one with increasing the particle concentration. In addition, the behavior of particles in simple shear flows was simulated; the shear was imposed on the systems in nematic phase at rest. The systems exhibited various motions of the director depending on the shear rate, e.g. continuous rotations of director at low shear rates and flow aligning at high shear rates.

INTRODUCTION

It has been known that flow induces the change in structure of suspensions and the relation between the structural change and particle motion in suspensions is an interesting subject. The Brownian dynamics (BD) simulation is a useful technique to investigate the structural change at a particulate level. We used a suspension of oblate spheroidal particles as a model of suspension of disc-like particles, and carried out the BD simulation to analyze the orientational behavior of particles during the system changes its structure.

NUMERICAL SCHEME

We investigated both the phase transition in an equilibrium state and the orientation behavior of suspension of oblate spheroidal particles under simple shear flows. We used the system of the suspension of the particles interacting via the Gay-Berne (GB) potential[1]. The equations of the GB potential are not indicated here but are available in a previous paper[2]. We used the same notation as that used in the previous paper. In the present simulation, we applied $\varepsilon_e/\varepsilon_s=5$ and $(\mu, \nu)=(1, 0.625)$ to a system of $r_a=\sigma_e/\sigma_s=0.2$, and $\varepsilon_e/\varepsilon_s=5$ and $(\mu, \nu)=(1, 1)$ to $r_a=0.3$: The ratio of σ_e to σ_s corresponds to the aspect ratio of the particle. These values were determined to make the well depths of the potential for the two systems almost the same.

The BD simulation was performed by solving translational and rotational equations of motion with respect to the center of gravity of the particles[2]. All the simulations were carried out for the number of particles $N=256$ in a cubic box using the Lees-Edwards periodic boundary condition[3]. The dimensionless time step $\Delta t^*(=\Delta t/\tau_r)$ used ranges from 10^{-4} to 10^{-3}, where $\tau_r = 3\pi\eta_s\sigma_0/(4k_BT)$. During the simulation, we fixed the dimensionless temperature: $T^*(=Tk_B/\varepsilon_0)=1$. The simple shear is imposed upon the system that is in nematic phase at equilibrium.

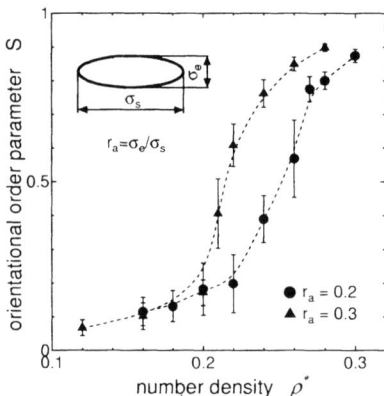

FIGURE 1. Dependence of the orientational order parameter S on the number density ρ^* for the systems with $r_a=0.2$ and 0.3. The error bars indicate the standard deviation of fluctuation in S.

RESULTS AND DISCUSSIONS

Figure 1 shows the dependence of an order parameter S on the dimensionless number density of the particles $\rho^*(=\rho\sigma_0^{-3})$ for the systems with $r_a=0.2$ and 0.3. The order parameter is defined by $S=\sum_{i=1}^{N}(3\cos^2\beta_i-1)/(2N)$, where β_i is an angle between a vector indicating the orientation of the ith particle and the director, a unit vector indicating the mean direction of the particles. S takes a value between 0 and 1, which means a random state and a perfect alignment, respectively. The both systems change from an isotropic phase to a nematic phase with increasing ρ^* as shown in Fig.1.

case of r_a=0.3, qualitatively the same results were obtained.

Figure 3 shows the shear rate dependence of the average orientational order parameter. In both cases of r_a, the rotation and the aligning are observed at low shear rates and at high shear rates, respectively. In addition, S increases consistently with $\dot{\gamma}^*$ in the aligning state, while it has an inflection point in the rotating state. The behavior of system changes from the rotation to the aligning at higher $\dot{\gamma}^*$ for the system of more plate-like particles, i.e. r_a=0.2.

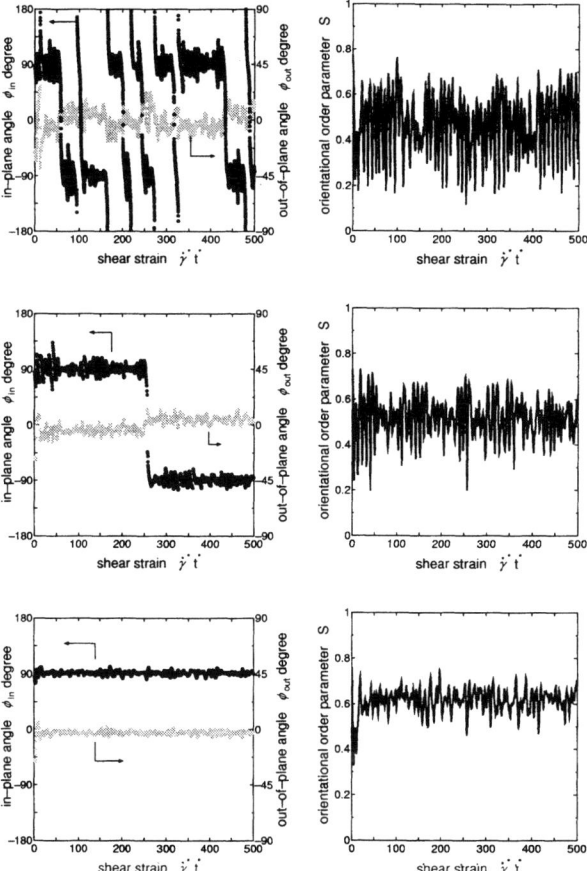

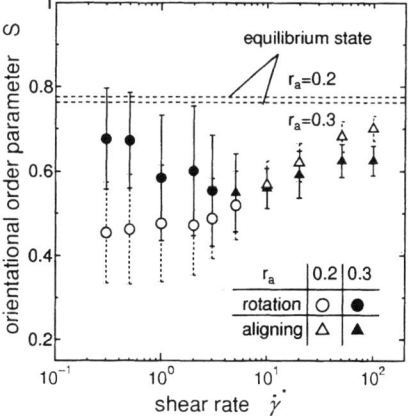

FIGURE 3. Shear rate dependence of the average orientational order parameter S for the systems with r_a=0.2 and 0.3. The circles and triangles indicate that particles are rotating and aligning, respectively. The error bars indicate the standard deviation of fluctuation in S.

FIGURE 2. Changes in the orientation angles ϕ_{in} and ϕ_{out} and the orientational order parameter S with time at $\dot{\gamma}^*$=2 (top), 5 (center), and 20 (bottom) for the system with r_a=0.2.

We next analyzed the orientational behavior of particles under simple shear flows at various shear rates. We chose two systems with (ρ^*, r_a)=(0.27, 0.2) and (0.24, 0.3); both the systems are in nematic phase at rest. Figure 2 shows temporal change in orientation angles, ϕ_{in} and ϕ_{out}, of the system with r_a=0.2 after the imposition of the shear flow at the dimensionless shear rates $\dot{\gamma}^*(=\dot{\gamma}/\tau_r)$ of 2, 5, and 20. The director is expressed by both the in-plane angle ϕ_{in} and the out-of-plane angle ϕ_{out}: ϕ_{in} is the angle between the projection of the director onto the vorticity plane and the flow direction; ϕ_{out} the angle between the projection and the director. The director continuously rotates out of the shear plane at low shear rates: This result indicates that the director shows a kayaking motion. At high shear rates, the flow aligning is predicted. In the

CONCLUSION

The BD simulation for the suspensions of oblate spheroidal particles predicted the phase transition depending on the number density. The phase changed at higher density for the system with more plate-like particles. Furthermore, the simulation under simple shear flows showed that the particles continuously rotated at low shear rates and aligned at high shear rates. The threshold shear rate also is higher for more plate-like particles.

REFERENCES

1. Gay, J.G. and Berne, B.J., *J. Chem. Phys.* **74**, 3316–3319 (1981).
2. Mori, N., Fujioka, H., Semura, R., and Nakamura, K., *Rheol. Acta* **42**, 102–109 (2003).
3. Lees, A.W. and Edwards, S.F., *J. Phys. C* **5**, 1921–1928 (1972).

Multifractal Analysis for the Dynamical Heterogeneity in Strongly Correlated Many-Body Systems

O. Narikiyo and W. Sakikawa

Department of Physics, Kyushu University, Fukuoka 810-8560, Japan

Abstract. By calculating the non-equilibrium parameter of the probability distribution function and the singularity spectrum of multifractal we have quantified the dynamical heterogeneity in strongly correlated many-body systems.

INTRODUCTION

We have numerically studied the dynamical heterogeneity in strongly correlated many-body systems, for example, the critical spin state[1], the supercooled liquid near the glass transition[2] and the turbulence[3]. These systems are scale-invariant and multifractal. In this paper we briefly review how we quantify the dynamical heterogeneity and discuss in detail the turbulence simulation data recently obtained.

In order to quantify the dynamical heterogeneity we have two measures. One is the non-equilibrium parameter q of the Rényi-Tsallis distribution function. The other is the singularity spectrum $f(\alpha)$. These two measures are closely related and important ingredients of the multifractal analysis.

In the numerical study of the critical spin state[1] we have found that the q-parameter represents the degree of non-equilibrium. We have actually quantified the deviation of the q-parameter from the equilibrium value according to the spatio-temporal scale of the observation.

In the numerical study of the the supercooled liquid near the glass transition[2] we have found that the broken-bond distribution which reflects the nature of the cooperatively rearranging region is well described by the singularity spectrum $f(\alpha)$ of multifractal. The width of the spectrum becomes broader as approaching the glass transition. Such a broadening is similar to that observed in the numerical study of the Anderson-localization transition.

TURBULENCE SIMULATION

In the numerical study of the turbulence[3] we have adopted the lattice Boltzmann method. Our numerical turbulence on 200^3 cubic lattice points is sustained by a random forcing and the Reynolds number is about 500.

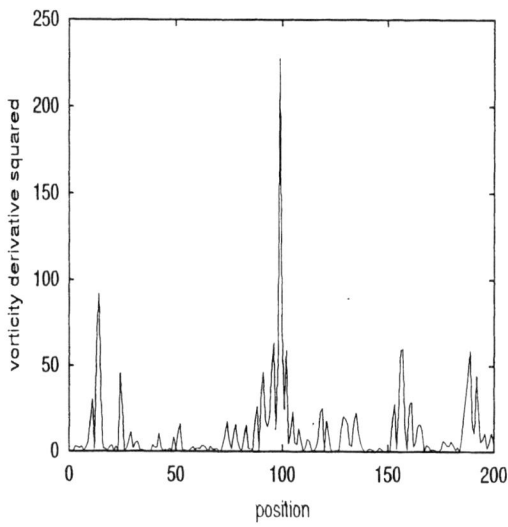

FIGURE 1. The intermittent distribution of the strength of the vortices.

mean-field analysis

Turbulence is one of the typical phenomena with multiscale motions. Each phenomena of a scale strongly couples with all the other scales of turbulent motion. In order to analyze such a system a scale-dependent entropy, the so-called ε-entropy $h(\varepsilon)$, works well. For example, the time series of the velocity field in turbulence leads to a non-trivial scaling relation, $h(\varepsilon) \propto \varepsilon^{-3}$, expected from Kolmogorov's scaling. Here ε is the scale of the observation. The existing experimental data are consistent with this scaling. By our numerical experiment simulating the Navier-Stokes equation we have shown the consistency of the scaling. In contrast to the evaluation of the energy spectrum, which is usually employed for testing Kolmogorov's scaling and determined by the two-point correlation function, Kolmogorov's scaling is easily observed in the ε-entropy, since it is determined by the mean of the exit time from the observation window and fluctuations are averaged out. Such a unifractal scaling is a mean-field description and fluctuations can be taken into account in a multifractal analysis as shown in the following. In the above mentioned mean-field description the velocity difference in temporal and spatial directions have the same fractal scaling exponent. Thus we have confirmed Kolmogorov's scaling and Taylor's hypothesis at the same time.

fluctuation analysis

In order to analyze the fluctuations we use two measures, the non-equilibrium parameter q of the Rényi-Tsallis distribution function and the singularity spectrum $f(\alpha)$.

The probability distribution function (PDF) for the velocity or vorticity field is non-Gaussian. The non-Gaussianity is quantified by the q-parameter. The strong non-Gaussianity is observed at small spatio-temporal scale comparable to that for the coherent vortex in the PDF for the velocity or vorticity difference between two space-time points. By using the wavelet denoising we have clarified that the non-Gaussianity or dynamical heterogeneity results from the existence of the coherent vortex which is a strongly-correlated non-equilibrium region.

The strength of the vortices is shown in Fig. 1 to be intermittent. The spatial distribution of the coherent vortices is quantified by the singularity spectrum of multifractal $f(\alpha)$ as shown in Fig. 2.

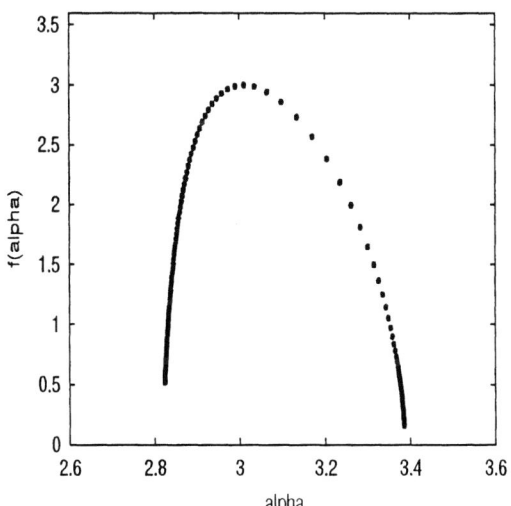

FIGURE 2. The singularity spectrum $f(\alpha)$ of the spatial distribution of the coherent vortices.

The remainder of this subsection is devoted to some speculations. As seen in our previous study[2] the width of the singularity spectrum depends on the degree of intermittency so that we expect some Reynolds-number dependence of the spectrum. It has been discussed by many authors and should be claryfied in future systematic study. In our present study the density of the vortex is dilute. Thus the singularity spectrum in Fig. 2 describes the spatial distribution of relatively free vortices. As the Reynolds number is increased the density increases. In this case the interaction among vortices becomes important and the correlation length of the vorticity fluctuation becomes large. While we can observe the vortex only as an individual elementary excitation in our numerical experiment, some collective excitation is expected to dominate at higher Reynolds number. In the limit of divergently large Reynolds number the correlation length becomes divergently large so that we can expect full scale-invariance. We can find a resemblance to the case of the scaling theory in polymers where an ideal scaling relation is realized for dense solutions where polymers are strongly entangled. In the limit of high Reynolds number each boxes counting the coherent vortex for calculating $f(\alpha)$ will be filled by almost equal number of vortices so that intermittency will disappear and unifractal Kolmogorov's scaling will prevail.

summary for turbulence simulation

Numerical simulation data, in real space and time, for a forced turbulence on the basis of the lattice Boltzmann method have been analyzed by unifractal and multifractal schemes.

Our new findings are summarized into two points. First in the unifractal analysis using the exit-time statistics we have verified Kolmogorov's scaling and Taylor's hypothesis at the same time. Second in the analysis using the Rényi-Tsallis PDF and the wavelet denoising we have clarified that the coherent vortices sustain the power-law velocity correlation in the non-equilibrium state.

Finally in the multifractal analysis it is clarified that the intermittent distribution of the coherent vortices in space-time is described as a multifractal.

REFERENCES

1. Sakikawa, W., and Narikiyo, O., *J. Phys. Soc. Jpn.*, **71**, 1200-1201 (2002).
2. Sakikawa, W., and Narikiyo, O., *J. Phys. Soc. Jpn.*, **72**, 450-451 (2003).
3. Sakikawa, W., and Narikiyo, O., cond-mat/0307604.

Molecular dynamics simulation of mixtures of hard rod-like molecules

Tomonori Koda* and Susumu Ikeda*

*Faculty of Engineering, Yamagata University
4-3-16 Jonan, Yonezawa 992-8510, Japan*

Abstract. We performed molecular dynamics simulation of binary system of 432 hard spheres and 1008 parallel hard spherocylinders. We calculated mean square displacements of molecules. Smectic layer structure was monitored through structure factor along molecular long axis. Results indicate that the smectic layer formation decreases diffusion of molecules along the layer normal.

INTRODUCTION

Theories and computer simulations have shown that, without attractive intermolecular force, hard rod-like molecules form liquid crystals. Onsager presented theoretical evidence that hard rods form the nematic liquid crystal[1]. Theories have shown that hard rods also form the smectic liquid crystal[2, 3, 4, 5].

It has been indicated that appearing liquid crystal structures depend on molecular shape. The spherocylinder is a cylinder each end of which is capped with a hemisphere. It is known that mono-disperse spherocylinders show both the nematic and the smectic phase depending on density and aspect ratio[6, 7, 8, 9].

In this study, we consider a binary mixture of hard spherocylinders and hard spheres. We performed molecular dynamics (MD) simulation and discuss dynamical behavior of the present rod-sphere mixtures.

MODEL AND SIMULATION

Figure 1 shows a spherocylinder and a sphere. The length of the spherocylinder is denoted as L and the diameter as D. In the present case, the diameter of the sphere is equal to the diameter of the spherocylinder. We used $L/D = 5$ for the present study. We consider a system that is composed of 432 spheres and 1008 spherocylinders. All the spherocylinders are parallel aligned to z-direction; we neglect orientation fluctuation of spherocylinders.

Before starting MD simulation, we ran isobaric Monte Carlo (MC) simulation in the manner described by Stroobants, Lekkerkerker, and Frenkel[6]. The system size fluctuates under the isobaric MC simulation. After the MC simulation for equilibration of lattice constants, we fixed the system size for MD simulation.

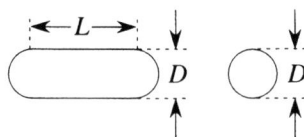

FIGURE 1. Spherocylinder and sphere.

MD of the present model was driven by elastic collisions between sphere and sphere, between sphere and spherocylinder, and between spherocylinder and spherocylinder. To ensure the parallel alignment of the model, we considered processes that do not change angular momentum of cylinders.

We define volume fraction η of the system as

$$\eta = \frac{N_c v_c + N_s v_s}{V} \quad (1)$$

where V is volume of the system, N_c number of spherocylinders, N_s number of spheres, v_c volume of the cylinder, and v_s volume of the sphere.

Denoting masses of the spherocylinder and the sphere, respectively, by m_c and m_s, we used

$$m_c/m_s = v_c/v_s = 1 + (3L/2), \quad (2)$$

for the present simulation.

We calculated mean square displacements that are defined as

$$M_{xy}(t) = <[y(t) - y(0)]^2 + [x(t) - x(0)]^2>, \quad (3)$$
$$M_z(t) = <[z(t) - z(0)]^2>, \quad (4)$$

where x, y and z is, respectively, x-, y-, z-component of position of molecule, t time, and $<\bullet>$ means average of quantity •. Because the spherocylinders are aligned to z-direction, M_{xy} corresponds to the mean square displacement perpendicular to molecular long axis, while M_z corresponds to the mean square displacement along molecular long axis.

RESULTS

Figure 2 shows mean square displacements of spherocylinders and spheres at $\eta = 0.213$. Results of Fig. 2 indicate that mean square displacements are proportional to time in the present time region of samplings. We fitted the data by

$$M_\alpha = c_\alpha t + b_\alpha, \quad (5)$$

where c_α and b_α are fitting parameters, and α is to be replaced by xy or z.

For evaluation of formation of smectic layer structure, we monitored longitudinal structure factor of spherocylinders:

$$s(k) = \frac{1}{N_c} \left\langle \left| \sum_{j=1}^{N_c} \exp[-ikz_j] \right|^2 \right\rangle, \quad (6)$$

where z_j is z-component of position of j-th molecule, and we numbered molecules as the j-th molecule is spherocylinder when $1 \leq j \leq N_c$.

Figure 3 shows volume fraction dependence of fitting coefficient of eq. (5) and the first maximum of the structure factor of eq. (6).

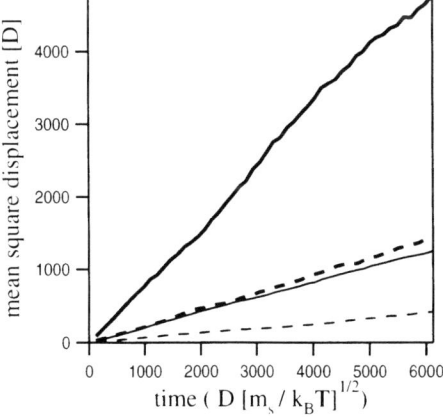

FIGURE 2. Mean square displacements at $\eta = 0.213$. Thin solid line is M_{xy} of spherocylinders. Thick solid line is M_z of spherocylinders. Thin dashed line is M_{xy} of spheres. Thick dashed line is M_z of spheres.

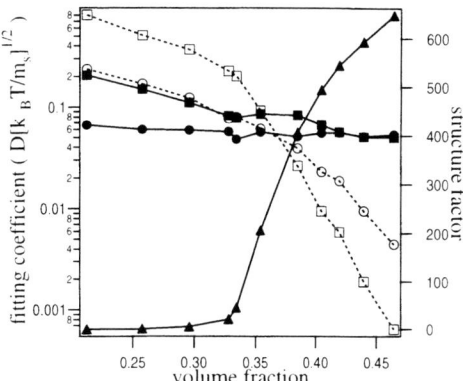

FIGURE 3. Fitting coefficients and structure factor depending on volume fraction. Open squares, c_z of spherocylinders; open circles, c_z of spheres; solid squares, c_{xy} of spherocylinders; solid circles, c_{xy} of spheres; solid triangles, the first maximum in the longitudinal structure factor.

CONCLUSION

We performed MD simulation of the binary system of 432 spheres and 1008 aligned spherocylinders. The first maximum of the longitudinal structure factor of Fig. 3 shows that the smectic layer formation starts about $\eta = 0.35$. Though the quantitative dynamical feature of the present results depends on eq. (2) which defines a relation between masses of sphere and spherocylinder, present results indicates following conclusions. The smectic layer formation disturbs diffusion of spherocylinders and spheres along layer normal. It does not give so much effect to diffusion along the direction perpendicular to the layer normal.

REFERENCES

1. Onsager, L., *Ann. N. Y. Acad. Sci.*, **51**, 627–659 (1949).
2. Hoshino, M., Nakano, H., and Kimura, H., *J. Phys. Soc. Jpn.*, **46**, 1709–1715 (1979).
3. Hoshino, M., Nakano, H., and Kimura, H., *J. Phys. Soc. Jpn.*, **47**, 740–745 (1979).
4. Mulder, B., *Phys. Rev. A*, **35**, 3095–3101 (1987).
5. Kimura, H., and Tsuchiya, M., *J. Phys. Soc. Jpn.*, **59**, 3563–3570 (1990).
6. Stroobants, A., Lekkerkerker, H. N. W., and Frenkel, D., *Phys. Rev. A*, **36**, 2929–2945 (1987).
7. Veerman, J. A. C., and Frenkel, D., *Phys. Rev. A*, **43**, 4334–4343 (1991).
8. McGrother, S. C., Williamson, D. C., and Jackson, G., *J. Chem. Phys.*, **104**, 6755–6771 (1996).
9. Bolhuis, P., and Frenkel, D., *J. Chem. Phys.*, **106**, 666–687 (1997).

Initial Deformation Process in Granular Matter

So Kitsunezaki

*Department of Physics, Graduate School of Human Culture,
Nara Women's University, Nara 630-8506.*

Abstract. Applying quasi-static deformation to a dry granular system induces a series of faults. We numerically investigate idealized systems with regular initial arrangements of particles and find that microscopic shear zones develop through a fingering-like instability before the creation of faults. Considering infinitesimal deformations of the system, we find that this instability is caused by interactions among slips at contact points between particles, which are communicated through the displacement field.

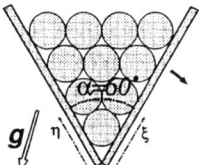

FIGURE 1. The initial arrangement of particles.

INTRODUCTION

It is well known from experimental studies that a series of faults, or shear bands, is created when a granular system is subject to deformation. Such behavior has been observed in many types of systems, from sand to nearly monodisperse glass beads [1, 2]. Static and cohesionless granular materials are regarded as systems that are fractured everywhere, with frictional interactions acting among constituent particles. Although the shear bands that develop in granular systems are silmilar to those in plastic materials, these two cases are qualitatively different when considered microscopically [3, 4, 5].

We study monodisperse sysems of particles with regular initial arrangements and investigate the stress that arises as a result of quasi-static deformation.

DEVELOPMENT OF MICROSCOPIC SHEAR ZONES

We first numerically simulated 2-dimensional granular systems in a V-shaped container using the discrete elements method (DEM). We prepared triangular initial arrangements in which there were no tangential forces acting on any particles, as depicted in the schematic picture of Fig.1. We introduced the reference coordinates (ξ, η) along the walls of the initial arrangements. We applied shear strain to the system by opening one of the walls of its container quasi-staticaly. As the wall is opened widely, a serise of faults similar to those observed in experiments also appear in this system [6].

In our simulations, two linear elastic forces with spring constants k_n and k_t act in the normal and tangential directions on each contact point among particles. Particles are assumed to slip when the tangential elastic force increases over the maximal frictional force with coefficient μ.

As the wall is opened, the arrangements of particles do not change until the creation of a fault, although left and right contacts are lost and the numbers of contact points become 4 for most of the particles. We focus on the change of the stress in this stage. Figure 2 displays the grey-scale image of the internal frictional angles in the system whose V container is tilted by $15°$ from the direction of gravity and the wall is opened only by $0.3°$. We observe that the heterogeneity of the stress develops to several microscopic shear zones in the system. These shear zones appear from the free surface and compete among one another to increase in length. This process is similar to that seen in system with a fingering-like instability. As the wall is opened, they also migrate to the negative ξ direction in Fig.2 and some of them develop into macroscopic faults.

LINEARIZED MODEL

We consider the infinitesimal deformation to investigate the instability of stress. Here we ignore weak contacts to right and left particles in the initial states and regard the

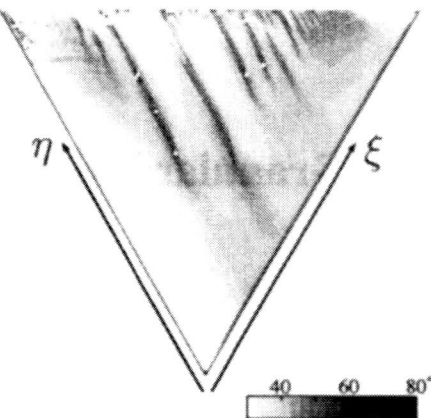

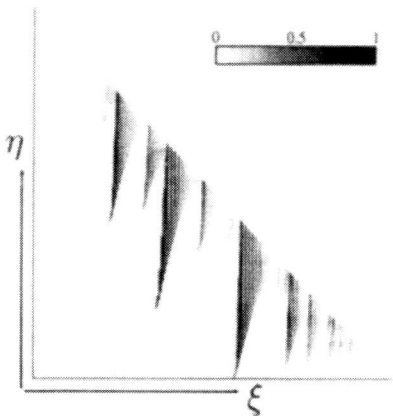

FIGURE 2. Internal frictional angles in a DEM simulation using the parameters $k_n = 10^4$, $k_t = 500$ and $\mu = 0.5$. The height of the system is 200 particle radii.

FIGURE 3. A snapshot of the slip displacements $\Psi_{\eta ij}$ in a simulation of the linearized model. $k_t = 0.1$, $\mu = 0.5$ and $\alpha = 60°$ are used where k_t is rescaled as described in the text.

arrangement as a rhombic lattice with an opening angle α. There are 2 contact points per particle which are in the ξ and η directions from the center of each particle, and the neighbors do not change while the lattice is deformed slightly.

We set the mass and the diameter of a particle and k_n to be 1 by appropriately rescaling. $\mathbf{u}_{ij}$ and ϕ_{ij} indicate the displacement and the rotation angle of the particle placed at $(\xi, \eta) = (i, j)$, and we refer the ξ and η components of $\mathbf{u}_{ij}$ as U_{ij} and V_{ij}. Considering the linear springs on ξ- and η-contact points of every particles, we describe the energy of the system as the following equations.

$$E = \sum_{ij} \left\{ \mathbf{g} \cdot \mathbf{u}_{ij} + \frac{1}{2}(N_{\xi ij}^2 + N_{\eta ij}^2 + k_t T_{\xi ij}^2 + k_t T_{\eta ij}^2) \right\}, \quad (1)$$

$$T_{\xi ij} \equiv \frac{\partial_i V_{ij} - \cos\alpha \partial_i U_{ij}}{\sin\alpha} - \frac{\phi_{ij} + \phi_{i+1\,j}}{2} - \Psi_{\xi ij}, \quad (2)$$

$$T_{\eta ij} \equiv -\frac{\partial_j U_{ij} - \cos\alpha \partial_j V_{ij}}{\sin\alpha} - \frac{\phi_{ij+1} + \phi_{ij}}{2} - \Psi_{\eta ij}, \quad (3)$$

$$N_{\xi ij} \equiv \partial_i U_{ij} \quad \text{and} \quad N_{\eta ij} \equiv \partial_j V_{ij}, \quad (4)$$

where we define $\partial_i A_{ij} \equiv A_{i+1\,j} - A_{ij}$, $\partial_j A_{ij} \equiv A_{i\,j+1} - A_{ij}$.

We introduced the slip displacements $\Psi_{\xi ij}$ and $\Psi_{\eta ij}$ in Eqs.(2) and (3) so that the tangential elastic forces $k_t T_{\xi ij}$ and $k_t T_{\eta ij}$ are relaxed by slip. The equilibrium state of $(U_{ij}, V_{ij}, \phi_{ij})$ is given by minimizing E for the fixed slip displacements. Assuming that the Coulomb conditions

$$k_t |T_{\xi ij}| \leq -\mu N_{\xi ij} \quad \text{and} \quad k_t |T_{\eta ij}| \leq -\mu N_{\eta ij} \quad (5)$$

are always satisfied for every contacts in the equilibrium states, we obtain the update rules for $\Psi_{\xi ij}$ and $\Psi_{\eta ij}$.

The numerical simulations of these equations give results corresponding to the instability of stress observed in the DEM simulations. Figure 3 shows the grey-scale image of $\Psi_{\eta ij}$, where ξ and η are displayed as orthogonal coordinates. The system is placed in an opening V-shaped container as described first, and the number of particles on the ξ or η axis is 200.

The behavior of these equations is controlled by two parameters: the ratio of the tangential spring constant to the normal spring constant, k_t, and the frictional coefficient μ. The history of deformation is memorized in the slip displacements of contact points.

Considering the limit that the tangential interactions are small, Eqs. (1)-(5) can be reduced to simpler equations by a perturbative approach. For the case that shear zones appear along one direction as shown in Fig.3, they are reduced to the equations which do not contain $\Psi_{\xi ij}$ and U_{ij}. Hence the instability of stress and the development of microscopic shear zones are essentially thought to result from the interactions between the displacement V_{ij} and the slip $\Psi_{\eta ij}$.

This study is supported by a Grant-in-Aid from the Japan Science and Technology Corporation.

REFERENCES

1. Terada, T., and Miyabe, N., *Bull. Eartq. Res. Inst.* **4**, 33-56 (1928), **6**, 109-126 (1929), **7**, 65-93 (1929).
2. Hubbert, M.K., *Bull. Geol. Soc. Am.* **62**, 355-372 (1951).
3. Williams, J.R., and Rege, N., *Powder Technol.* **90**, 187-194 (1997).
4. Roux, J.-N., and Combe, G., *C. R. Phys.* **3**, 131-140 (2002).
5. Kuhn, M.R., *Granular Matter* **4**, 155-166 (2003).
6. Kitsunezaki, S., and Kurumatani, A., "Stripe patterns induced by slow deformation of a container," in *Powder and Grain 2001*, edited by Y. Kishino, Balkema, Lisse, 2001, pp. 309-312.

The Simulation of the Inelastic Impact

Hiroto Kuninaka[*] and Hisao Hayakawa[†]

[*]*Graduate School of Human and Environmental Studies, Kyoto University, Sakyo-ku, Kyoto, Japan, 606-8501*
[†]*Department of Physics, Yoshida-south campus, Kyoto University, Sakyo-ku, Kyoto, Japan, 606-8501*

Abstract. The coefficient of normal restitution (COR) in an oblique impact is theoretically studied. Using a two-dimensional lattice models for an elastic disk and an elastic wall, we investigate the dependency of COR on an incident angle and demonstrate that COR can exceed one and have a peak against an incident angle in our simulation. Finally, we explain these phenomena based upon the phenomenological theory of elasticity.

INTRODUCTION

The coefficient of restitution (COR) e is introduced to determine the post-collisional velocity in the normal collision of two materials and defined by $\mathbf{u}' \cdot \mathbf{n} = -e\mathbf{u} \cdot \mathbf{n}$, where $\mathbf{u}$ and $\mathbf{u}'$ are respectively the velocity of the contact point of two colliding materials before and after the impact, and $\mathbf{n}$ is the normal unit vector of the tangential plane of them. In the oblique impact, it has become clear that COR depends on the incident angle as well as on the impact velocity. Louge and Adams[1] recently reported that COR increases as a linear function of the tangent of the incident angle in the oblique impact of a hard aluminum oxide sphere on a thick elasto-plastic plate. They also suggested that COR can exceed 1 in most grazing impacts. In this proceeding, we carry out the two-dimensional simulation of the oblique impact and investigate the dependency of COR on the incident angle based upon the theory of elasticity.

MODEL

Our numerical model is a two-dimensional model which is composed of an elastic disk and an elastic wall[2]. Each of them is composed of randomly placed mass particles connected by nonlinear springs each other. Numbers of mass particles are 400 for the disk and 2000 for the wall. The width and the height of the wall are $8R$ and $2R$ respectively where R is the radius of the disk. Both sides and bottom of the wall are fixed. Spring potential is described as $V(x) = \frac{1}{2}k_a x^2 + \frac{1}{4}k_b x^4$, where x is a stretch from natural length. The values of k_a are $1.0 \times mc^2/R^2$ for the disk and $k_a = 1.0 \times 10^{-2} mc^2/R^2$ for the wall, where c is the one-dimensional sound of velocity. And we adopt $k_b = k_a \times 10^{-3}/R^2$ for each of them. Poisson's

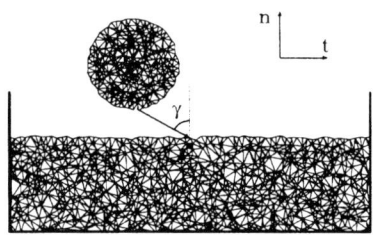

FIGURE 1. The elastic disk and wall consisted of random lattice.

ratio v can be evaluated from the strains of the band of random lattice in vertical and horizontal directions to the applied force. We obtain Poisson's ratio v of the model as $v = (7.50 \pm 0.11) \times 10^{-2}$.

In our simulation, we define the incident angle γ by the angle between the normal vector of the wall and the initial velocity vector of the disk(Fig.(1)). We fix the initial colliding velocity of the disk as $|\mathbf{v}(0)| = 0.1c$ to control the normal and tangential components of the initial colliding velocity as $v_t(0) = |\mathbf{v}(0)| \sin\gamma$ and $v_n(0) = |\mathbf{v}(0)| \cos\gamma$, respectively. From the normal components of the contact point velocities before and after the collision, we calculate COR for each γ. We use the fourth order symplectic numerical method for the numerical scheme of integration with the time step $\Delta t = 10^{-3} R/c$.

RESULTS AND DISCUSSION

Figure 2 is the COR against the tangent of the incident angle $\Psi_1 = \tan\gamma$ in our simulation. The cross points are the average and the error bars are the standard deviation of 100 samples for each incident angle. This result shows that the COR increases as Ψ_1 increases to exceed 1, and has a peak around $\Psi_1 = 5.0$. This behavior is contrast to that in the experiment by Louge and Adams[1].

To explain this result, we consider the correction of COR by the local deformation of the wall. We assume that the normal unit vector $\mathbf{n}$ to the surface of the wall rotates toward the incoming disk by an angle α to become $\mathbf{n}^\alpha$. If we define $e = -(\mathbf{u}' \cdot \mathbf{n})/(\mathbf{u} \cdot \mathbf{n})$ and $e^\alpha = -(\mathbf{u}' \cdot \mathbf{n}^\alpha)/(\mathbf{u} \cdot \mathbf{n}^\alpha)$, the relation between e and e^α becomes

$$e = (e^\alpha + \Psi_2^\alpha \tan\alpha)/(1 - \Psi_1^\alpha \tan\alpha), \quad (1)$$

where $\Psi_1^\alpha = (\Psi_1 - \tan\alpha)/(1 + \Psi_1 \tan\alpha)$ and $\Psi_2^\alpha = (\Psi_1 - \tan\alpha)/(1 + \Psi_1 \tan\alpha) - 3(1 + e^\alpha)(\mu + \tan\alpha)/(1 - \mu \tan\alpha)$. As for Ψ_2^α, we use the phenomenological theory for the oblique impact by Walton and Braun[3]. The correction angle α can be estimated by the theory of elasticity. If we express the contact area by a parabora, $\tan\alpha$ equals to $|x_c - x_a|(1-2\theta)/R(2-2\theta)$ with $\theta = (1/\pi)\arctan(1-2\nu)/(\mu(2-2\nu))$, where μ is the coefficient of friction and x_c and x_a are the x coordinates of both ends of the contact area. μ can be calculated from the simulation data through the definition $\mu = |J_t|/|J_n|$, where J_n and J_t are the normal and tangential components of the impulse. The relation between μ and Ψ_1 are cross points in Fig. 3. The solid curve in Fig. 2 is Eq. (1) with $e^\alpha = 0.95$ which is COR in the normal impact. Our numerical results can be reproduced by our phenomenological theory.

The relation between μ and Ψ_1 can be explained as follows. We assume that jags are uniformly placed on the surface of the wall with the density ρ per unit length and the tangential velocity of the disk is decreased by η when the disk interacts with the one jag. The tangential and normal impulses can be calculated by calculating the number of jags the disk interacts during collision time[4] as $J_t = -m\eta\rho|\mathbf{v}(0)|\sin\gamma\pi(R/c)\sqrt{\ln(4c/|\mathbf{v}(0)|\cos\gamma)}$ and $J_n = -m(e+1)|\mathbf{v}(0)|\cos\gamma$. We also assume that the tangential impulse decreases by $J_t' = -m\zeta|\mathbf{v}(0)|\sin\gamma$ which is proportional to the initial tangential velocity with the proportionality constant ζ. Thus, μ can be calculated by the ratio of $|J_t - J_t'|$ to J_n as

$$\mu = \left| \zeta\tan\gamma - \eta\rho\tan\gamma\pi\frac{R}{c}\sqrt{\ln\left(\frac{4c}{|\mathbf{v}(0)|\cos\gamma}\right)} \right| / (e+1). \quad (2)$$

The solid curve in Fig.3 is Eq.(2) with $\zeta = 0.317$ and $\eta\rho = 0.0416c/R$. Our numerical results can be well reproduced by our phenomenological theory of the coefficient of friction.

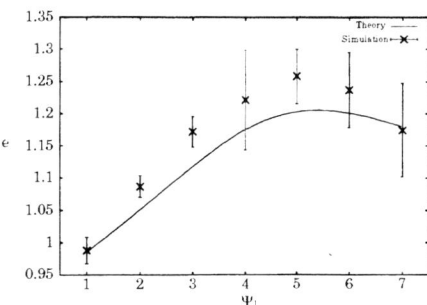

FIGURE 2. Numerical and theoretical results of the relation between Ψ_1 and COR.

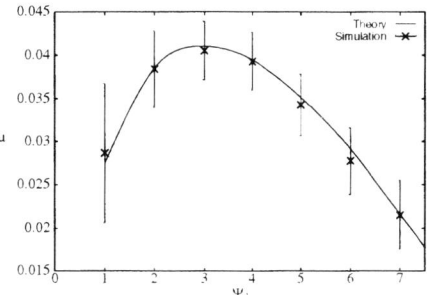

FIGURE 3. Numerical and theoretical results of the relation between Ψ_1 and μ.

SUMMARY

In the present study, we have shown that COR can exceed 1 also in our two-dimensional simulation and depends on μ in the oblique impact. Our results can be explained by our simple phenomenological theory.

ACKNOWLEDGMENTS

This study is partially supported by the Grant-in-Aid of Ministry of Education, Science and Culture, Japan (Grant No. 15540393).

REFERENCES

1. Louge. M. Y., and Adams. M. E., *Phys. Rev. E.* **65**, 021303-1–021303-6 (2002).
2. Kuninaka, H., and Hayakawa, H., *J. Phys. Soc. Jpn.*, **72**, 1655–1663 (2003).
3. Walton, O. R., and Braun, R. L., *J. Rheol.*, **30**, 949–980 (1986).
4. Hayakawa, H., and Kuninaka, H., *Chem. Eng. Sci.*, **57**, 239–252 (2002).

Molecular Dynamics Study on a Self-organized Shock Wave in an Inelastic Hard Disk System

Masaharu Isobe

Graduate School of Engineering, Nagoya Institute of Technology, Nagoya, 466-8555, Japan

Abstract. We discovered new macroscopic phenomena, self-organized shock wave, through the study of the well-defined simple microscopic model by performing an large scale event-driven molecular dynamics simulation. These phenomena are excited by collective motion of collisions between an inelastic hard disk and the shock front propagate much larger the particle diffusion.

INTRODUCTION

The event-driven molecular dynamics simulation in hard disk system was worked out for the first time by Alder and Wainwright[1], in which they discovered fluid-solid phase transition at critical packing fraction ($v_c \sim 0.70$). Recently, a freely evolving inelastic hard sphere system is studied as an ideal microscopic model of the granular system[2], in which the particles lose energy through the collision. The dynamics of granular materials becomes one of the important topics in the studies of nonlinear, dissipative, and nonequilibrium statistical physics[3]. In this stage, there is one interesting question what happens in the macroscopic system composed of the inelastic hard spheres in which they " obtain " small energy through the collision? Since both a sufficient large system and a long-time relaxation are needed, there are few studies on such a cooperative motion of density fluctuation in the microscopic molecular level[4].

In this paper, we provide one of an ideal example to investigate relationship between the microscopic small dissipation and the macroscopic patterns in non-equilibrium and show that the system evolves to an ordered state, in which the stable propagating density wave patterns are formed, after a long time, starting from equilibrium in the two-dimensional hard disk system.

MODEL & NUMERICAL SETTING

In the study of dissipative structure composed of distinct elements, it is necessary to describe fluctuations suitably in the microscopic level, especially far from equilibrium. To analyze the shocks, macroscopic Navier-Stokes equation might be not appropriate in which it is based on local equilibrium assumption. In the length of mean free path, mesoscopic kinetic theory (Boltzmann equation) is often used to study the behavior of statistical properties on the distribution of particles. However, it was applied especially in equilibrium system and there is no proof that it works in non-equilibrium system such as shocks or not. To understand the correct behavior of such system, the particle-model with molecular dynamics method, which can describe microscopic fluctuation directly, might be the most relevant.

Our system consists of more than 0.25 million hard disks (up to a few million) placed in a square box with periodic boundaries, the unit of length being the diameter of a disk. The freely evolving inelastic hard disk system is completely described by three parameters: the number of particle N, the packing fraction v and the restitution coefficient r. When two disks, i and j, with respective velocities $\mathbf{v}_i$ and $\mathbf{v}_j$ collide, the velocities after the collision, $\mathbf{v}'_i$ and $\mathbf{v}'_j$, are given by

$$\mathbf{v}'_i = \mathbf{v}_i - \frac{1}{2}(1+r)[\mathbf{n} \cdot (\mathbf{v}_i - \mathbf{v}_j)]\mathbf{n} \quad (1)$$

$$\mathbf{v}'_j = \mathbf{v}_j + \frac{1}{2}(1+r)[\mathbf{n} \cdot (\mathbf{v}_i - \mathbf{v}_j)]\mathbf{n}. \quad (2)$$

where $\mathbf{n}$ is the unit vector parallel to the relative position of the two colliding particles in contact. Initially, the system is prepared as the equilibrium state by the long enough preliminary run with the restitution coefficient $r = 1$, in which the density is uniform and the disk velocities are Maxwell-Boltzmann distribution. The system evolves by collision until 9000 collisions per particle, which is called event-driven; the author developed a special algorithm based on event-driven molecular dynamics[5]. Since the total energy is monotonically increasing in freely evolving an inelastic hard disk system, a steady state can be realized by scaling the velocity of the entire particle to the total energy remaining constant. The great advantage of this procedure is that the trajectory of particle does not change compared with the non-scaled case.

RESULTS

We give in Fig. 1 typical snapshots for self-organized patterns. The round shape patterns grow spontaneously from the initial equilibrium state after a long transient and propagating as the density wave when the restitution coefficient is set slightly more than the unity.

We calculated that the time evolution of density fluctuation in case of the particle number and the packing fraction are fixed at $(N, v) = (512^2, 0.25)$, respectively. We found that the density wave is formed earlier and more clearly when the restitution coefficient r becomes far from equilibrium. Even after shock front are organized, the mean square displacement of particle within the unit of scaled time remains relatively small value (0.04 in $r = 1.0137$), but the shock front velocity is very large which is calculated by the about estimation from series of snapshots (2.0 in $r = 1.0137$). Therefore, this organization of shock front wave is collective mode and does not relate to particle diffusion.

We investigated that the onset time for forming shock wave with respect to the restitution coefficient when the packing fraction is 0.25. We found that the onset time diverges in the limit of the restitution coefficient being the unity (i.e. equilibrium system). On the other hand, when the restitution coefficient becomes large, the onset time becomes faster. Though the accurate estimation for the onset time is difficult when the particle number is small and the restitution coefficient is near the unity, it agrees with the universal curve by changing the number of particles. When the restitution coefficient becomes much larger than the unity, the inelastic collapse, which is well-known phenomenon in the granular inelastic hard disk system, again appears even in two-dimensional system, in which a few particles have almost all kinetic energy. In the small system less than 10,000 particles, the deviation of density fluctuation from the equilibrium cannot be distinguished. That might be the main reason why this phenomenon has not been found until now. What is most surprising is the dependence of packing fraction. When the packing fraction is set above that of Alder transition (v_c), the shock wave have never appeared. Since the sound velocity might become high in such a solid state, we consider, the shock wave cannot be organized.

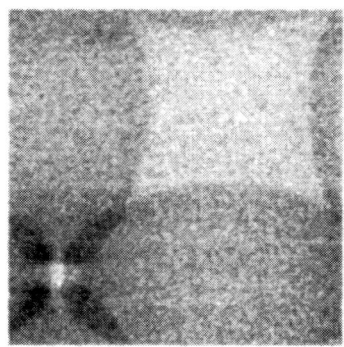

FIGURE 1. The typical snapshot for self-organized shock wave is shown. The round shape patterns grow spontaneously from the initial equilibrium state after many collisions. The system has 0.25 million particles. The packing fraction is 0.25 and the restitution of coefficient is 1.0141, respectively. The density field after 2400 collisions per particle is shown

DISCUSSION & CONCLUSION

We show the fascinating density wave which appears after a long transient time when the restitution coefficient is set slightly more than the unity in the large-scale ideal inelastic hard disk system. These results were obtained by only solving a simple Newton's equation system for inelastic hard disks. We call this collective behavior "Self-organized Shock Wave (SSW)". What is the essential interaction between microscopic elements and the macroscopic ordered patterns? Theoretical investigation on the scenario for organization of shocks is still under way. Moreover, the comparison with the experimental situation such as chemical reaction system, combustion, and laser oscillations is, of course, very important.

ACKNOWLEDGMENTS

This work was supported by the Ministry of Education, Science, Sports and Culture, Grant-in-Aid for Scientific Research (C), 15560042, 2003.

REFERENCES

1. Alder, B.J., and Wainright, T.E., *J. Chem. Phys.* **27**, 1208-1209 (1957).
2. Goldhirsh, I., and Zanetti, G., *Phys. Rev. Lett.* **70**, 1619-1622 (1993).
3. Jaeger, H.M., Nagel, S.R., and Behringer, R.P., *Rev. Mod. Phys.* **68**, 1259-1273 (1996).
4. Kawakatsu, T., and Ueda, A., *J. Phys. Soc. Jpn.* **56**, 847-850 (1987).
5. Isobe, M, *Int. J. Mod. Phys. C* **10**, 1281-1293 (1999).

Dynamics of Orientational Phase Ordering Coupled to Elastic Degrees of Freedom

Nariya Uchida

Department of Physics, Tohoku University, Sendai 980-8578, Japan

Abstract. We take Ginzburg-Landau approaches to the dynamics of orientational phase ordering coupled to elastic degrees of freedom. In nematic fluid membranes, subdiffusively slow shape relaxation limits the defect coarsening kinetics. In lamellar binary gels, coarsening is frozen by random stress that acts via a coupling between strain and layer orientation.

INTRODUCTION

Dynamics of orientational phase ordering such as those in nematic liquid crystals and vector spin models possess certain universal characters, as was clarified in last decade. In some complex fluids, however, elastic degrees of freedom coupled to the orientational order give rise to non-universal features. Here we give two examples that demonstrate the roles of such couplings in the coarsening dynamics and patterns of topological defects.

NEMATIC MEMBRANES

First we consider model fluid membranes that carry in-plane long-range orientational order. It has recently been observed that the so-called gemini surfactants with chiral counter-ions form ribbon- or helix-like sheets of stacked bilayers [1]. They have been modeled in terms of the coupling between membrane curvature and in-plane nematic order. For simplicity, here we assume a non-chiral membrane with an almost-flat configuration expressed in the Monge gauge as $z = h(x,y)$. The curvature tensor is approximated as $H_{ij} = \partial_i \partial_j h$ $(i, j = x, y)$ under the assumption $|\nabla h| \ll 1$. We consider the isotropic-nematic transition described by the order parameter $Q_{ij} = S(n_i n_j - \frac{1}{2}\delta_{ij})$. We write the free energy as the sum of the Landau, Frank, bending, and coupling contributions as

$$F = \int d^2r \left[f_L(Q_{ij}) + \frac{M}{2}(\partial_i Q_{jk})^2 + \frac{\kappa}{2}H_{ii}^2 + \alpha Q_{ij} H_{ij} \right].$$

where $f_L(Q_{ij}) = \frac{A}{2}Q_{ij}^2 + \frac{C}{4}Q_{ij}^4$. We numerically solved the kinetic equations

$$\frac{\partial Q_{ij}}{\partial t} = -\Gamma_Q \left(\frac{\delta F}{\delta Q_{ij}} \right)_s$$

where s denotes the symmetric traceless part, and

$$\frac{\partial h}{\partial t} = -\Gamma_h \frac{\delta F}{\delta h} = -\Gamma_h \left(\nabla^2 \nabla^2 h + \alpha \partial_i \partial_j Q_{ij} \right).$$

In Fig. 1 we show the orientational and curvature corre-

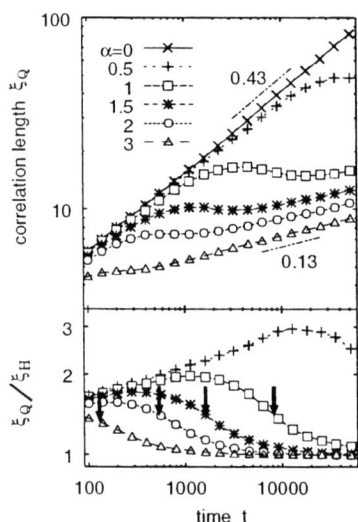

FIGURE 1. Coarsening dynamics of nematic membranes. Top: Orientational correlation length ξ_Q versus time. Bottom: The ratio between orientational and curvature correlation lengths. Arrows indicate the time at which the curvature free energy becomes twice the Frank free energy.

lation lengths ξ_Q and ξ_H versus time. The coarsening is slowed down as the curvature free energy becomes non-negligible compared to the Frank one, and converges to a power law with the exponent 0.12 ± 0.02, which is much smaller than the exponent 0.43 for the XY model. We can interpret the slowing down as follows. In the initial stage, the order parameter and curvature fields coarsen in a different manner, since the former is driven by orientational diffusion while the membrane height obeys the subdiffisive equation given above. Their coupling becomes important when ξ_Q exceeds $\xi_c = \sqrt{\kappa M}/\alpha$, after which curvature relaxation limits defect coarsening. The former obeys $\xi_H \propto t^{1/4}$ according to a simple dimensional analysis. The deviation of the numerical value from $1/4$ may be partly explained as logarithmic corrections in 2D. Due to the logarithmic dependence of the Frank free energy on defect separation, the apparent coarsening exponent for the XY model is smaller than the scaling value $1/2$. In addition, the effective curvature free energy (after integrating out h assuming local equilibrium) diverges logarithmically. We also incorporated the long-range hydrodynamic interaction due to solvent flow. It slightly accelerates the coarsening in consistency with a dimensional argument.

LAMELLAR BINARY GELS

The dynamics of isotropic-lamellar transition in its late stage can also be regarded as an orientational ordering process. For block copolymer melts, the correlation length ξ of layer normal is known to grow as $\xi \sim t^{1/4}$ [4]. Here we consider microphase separation in binary gels, where crosslinking of the two polymer species in the isotropic phase causes a long-range attraction between them in the lamellar phase. We extend de Gennes's classical model [5] to inhomogeneous systems by incorporating strain-orientation coupling and quenched random stress. Segregation of the two components induces strains that are parallel to the layer normal on average and modulated by the random disorder. This can be formulated in terms of the concentration gradient tensor,

$$Q_{ij} = (\partial_i \psi)(\partial_j \psi) - \frac{1}{d}(\nabla^2 \psi)\delta_{ij}.$$

Under the assumption of incompressibility $\nabla \cdot u = 0$, we write the optimal value of the elastic strain tensor $U_{ij} = \frac{1}{2}\left(\partial_i u_j + \partial_j u_i\right)$ as

$$U_{ij}^0(r) = \alpha Q_{ij}(r) + R_{ij}(r).$$

Here, α is the anisotropy parameter and $R_{ij}(r)$ is the (dimensionless) random stress which is spatially uncorrelated and is parametrized by its strength $R = \sqrt{R_{ij}^2}$. We decompose the free energy into two parts as $F = F_\psi + F_{el}$ where F_ψ is the free energy of an orientationally homogeneous system and F_{el} is the elastic contribution due to heterogeneity. Explicitly, they read

$$F_\psi = \int d^d r \left[f_L(\psi) - \frac{K}{2}\psi \frac{1}{\nabla^2}\psi + \frac{L}{2}(\nabla\psi)^2 + \frac{M}{4}(\nabla\psi)^4 \right],$$

where $f_L(\psi) = \frac{A}{2}\psi^2 + \frac{C}{4}\psi^4$ and

$$F_{el} = \mu \int d^d r \left(U_{ij} - U_{ij}^{(0)} \right)^2,$$

where μ is the shear modulus. After integrating out the strain field assuming local mechanical equilibrium, we have an effective elastic free energy $\tilde{F}_{el} = \tilde{F}_{el}[\psi(r)]$. We numerically solved the kinetic equation

$$\frac{\partial \psi}{\partial t} = \Gamma \nabla^2 \frac{\delta}{\delta \psi}\left(F_\psi + \tilde{F}_{el} \right)$$

in 2D, and found that the elastic coupling slightly reduces the growth exponents for S_{max} and ξ, from the non-elastic exponents $1/5$ and $1/4$ [4], respectively. We may argue, as in the membrane case, that the deviation is due to a logarithmic anomaly in the effective elastic interaction. When the randomness is turned on, the correlation length converges to a finite value. The dependence of the equilibrium correlation length on the disorder strength is weaker than a power law, which is in contrast with the exponential dependence found in the random field XY model and disordered nematic elastomers [6]. The details of our results will be presented elsewhere.

SUMMARY

In this paper, we have shown that both dynamic and static properties of an elastic medium can affect orientational ordering dynamics through energetic couplings. It might be interesting to seek analaougous effects in collective reorientation dynamics of anisotropic elements coupled to elastic media, such as proteins attached to membranes [7] and cells on extra-cellular matrices [8].

ACKNOWLEDGMENTS

This work is supported by Grant-in-Aid for Scientific Research from Ministry of Education, Culture, Sports, Science and Technology of Japan.

REFERENCES

1. Oda, R. et al., *Nature* **399**, 566-569 (1999).
2. Blundell R. E. and Bray, A. J., *Phys. Rev. E* **49**, 4925-4937 (1994).
3. Uchida, N., *Phys. Rev. E* **66**, 040902 (2002).
4. Christensen, J. J. and Bray, A. J., *Phys. Rev. E* **58**, 5364-5370 (1998).
5. de Gennes, P. G., *J. Physique* **40**, L69-L72 (1979).
6. Uchida, N. *Phys. Rev. E* **62**, 5119-5136 (2000).
7. Kim K. S. et al., *Phys. Rev. E* **61**, 4281-4285 (2000).
8. Bischofs I. B. et al., preprint cond-mat/0309427.

The instantaneous-normal-mode analysis on the potential energy landscapes of the Lennard-Jones 2n-n fluids

Ten-Ming Wu and S. L. Chang

Institute of Physics, National Chiao-Tung University, Hsin-Chu, Taiwan

Abstract. We have examined the instantaneous-normal-mode spectra of the Lennard-Jones 2n-n fluids with an aim to characterize the potential energy landscape (PEL) of a simple system. Under the isothermal-isochoric ensemble, the fluid density and the value of n are two independent parameters to control the PEL. We perform two ways of investigation by changing the value of n at a fixed fluid density and temperature and varying the fluid density for several different n values.

INTRODUCTION

Energy landscape paradigm has been the frontier of the theoretical approaches in our understanding of complex systems in physics and biology [1]. However, due to the complexity of a surface in a space of much higher dimensions, a properly conceptual description even for the potential energy surface (PES) of a simple system, N structureless particles interacting via a pairwise additive potential, is still a matter impossible. For a simple system, the topography of the PES [2] is generally determined by both the repulsive and the attractive interactions between particles. For the repulsive part, which mainly determines the system structures at high densities, its effects on the PES have been investigated through the truncated Lennard-Jones (LJ) fluids, in which particles interact only with the repulsive part of the LJ potential [3, 4]. On the other hand, the attractive interactions play an important role on many physical systems, like C_{60} [5] and the colloid-polymer mixtures [6]. In this proceedings, we study the roles of the interaction range in the PES of a simple system.

In our analysis for PES, we use the instantaneous normal mode (INM) formalism [7, 8], which was originally developed for describing the short-time dynamics of liquids and has been one approach to explore the potential energy landscapes of disordered systems. The INM formalism provides the information in regard to the local-curvature distribution of the PES of a system at the simulated thermodynamic state. For a configuration, the eigenvalues of the Hessain matrix are the local curvatures of the PES along the corresponding eigenvectors, which are referred as the INMs; the square roots of the eigenvalues give the INM frequencies. After an ensemble average for configurations taken at different snapshots during the evolution of the system, the INM spectrum is obtained.

For fluid systems, since the analyzed configurations are not exactly at the local minima of the PES, the INM spectrum has two branches for either real or imaginary INM frequencies, corresponding to the positive or the negative local curvatures of the PES, respectively. So far, various models and physical systems have been investigated via the INM formalism.

PAIR POTENTIAL AND SIMULATIONS

Consider systems of particles with the same mass and interacting pairwisely with the LJ 2n-n potential

$$\phi(r) = 4\varepsilon\left[\left(\frac{\sigma}{r}\right)^{2n} - \left(\frac{\sigma}{r}\right)^{n}\right], \quad (1)$$

where ε is the well depth of the potential, and σ the particle diameter. At fixed ε and σ, n is a parameter for tuning the range of the potential: the larger the value of n, the shorter the range. The potential with $n = 6$ has a long interaction range. For $n = 12$, the potential has a medium range, and is very similar to the one describing the C_{60} system. For $n = 18$, the potential becomes short-ranged and is close to the hard-sphere attractive Yukawa potential [9], which is used to describe the interaction between colloids mixed in a non-absorbing polymer.

As our previous simulations [10], we performed the isothermal-isochoric MD simulations so that the density and the value of n are two independent parameters to control the PES. Two series of simulations were carried out: one for the LJ 2n-n fluids with $n = 6$, 12 and 18 at a liquid-like thermodynamic state with the same temperature ($T^* = 1.0$) and density ($\rho^* = 0.8$), and the other for the fluids with the three n values at a supercritical temperature ($T^* = 1.4$) but different densities (with ρ^* from

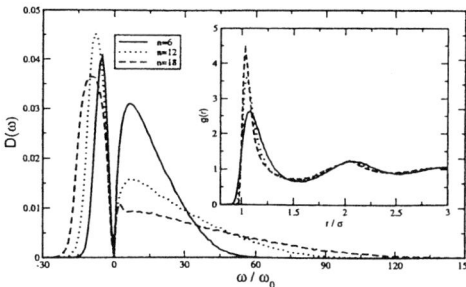

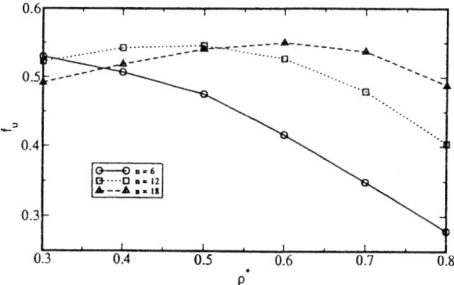

FIGURE 1. INM DOS $D(\omega)$ of the LJ 2n-n fluids at a liquid-like state with $\rho^* = 0.8$ and $T^* = 1.0$ for $n = 6$ (solid line), 12 (dotted line) and 18 (dashed line). The inset shows the radial distribution functions of the three fluids.

FIGURE 2. Density dependence of the fractions of imaginary INMs in the LJ 2n-n fluids at $T^* = 1.4$. The solid, dotted, and dashed lines are for $n = 6$, 12, and 18, respectively. All symbols are the simulation results.

0.3 to 0.8). Here, the LJ reduced units are used. Therefore, the results of the first series simulations manifest the effects mainly due to the range of the potential, and those of the second series exhibit the density dependence of the PES characters of the fluids with different interaction ranges.

RESULTS AND CONCLUSIONS

The INM spectra and the radial distribution functions, $g(r)$, of the three LJ 2n-n fluids in our first series simulation are presented in Fig. 1. Apparently, the first peak of $g(r)$ becomes sharper with increasing n, but the second maximum and beyond are almost the same for the three fluids. Since the fluids are at the same density and temperature, this comparison indicates that as the pair potential is shortened in range, the cohesiveness within the first shell of each particle becomes stronger, and makes the first peak of $g(r)$ sharper. As switching to the INM spectrum, both the real-frequency and the imaginary-frequency spectra are broadened with increasing n; especially, the real-frequency spectrum, which is plotted along the positive axis in Fig. 1, has a much extended tail. The physical reason for this long tail has been attributed to the stiffness of the repulsive part of the pair potential in a previous study [3], and our present results are consistent with this explanation. In addition to the long tail, the shape of the real-frequency spectrum of $n = 18$ is somewhat different in quality from those of $n = 6$ and 12, with the appearance of a small peak at the low-frequency region. The fraction f_u of the imaginary-frequency INMs, a quantity related to the diffusivity, increases with n from 0.26 for $n = 6$ to 0.39 for $n = 12$ and 0.47 for $n = 18$.

For the results from our second series simulations, the INM spectra will be presented elsewhere. Here, we only show in Fig. 2 the density dependence of f_u for $n = 6$, 12 and 18 in the density range of our calculations. The figure indicates that f_u decreases in the high-density region for the three values of n. This is consistent with the fact that, no matter what the pair potential is, the self-diffusion constant of a simple fluid decreases with the increase in density. Within the density range in our present consideration, f_u for $n = 6$ increaes monotonically with the decrease in density; however, f_u has a maximum around $\rho^* = 0.6$ for $n = 18$, and between $\rho^* = 0.4$ and $\rho^* = 0.5$ for $n = 12$. Thus, for the LJ 2n-n fluids, as the range of the potential becomes shorter, the density at the maximum f_u raises.

In conclusions, for the LJ 2n-n fluids, at the liquid-like states, f_u increases by reducing the range of the pair potential; however, the situation is expected to be reversed at the gas-like densities. Thus, the density dependence of f_u shows a maximum at some density, which increases with the n value.

ACKNOWLEDGMENTS

T. M. Wu thanks the National Science Council of Taiwan, R. O. C. for financial support under Grand No. NSC 92-2112-M009020.

REFERENCES

1. Franeufelder, H., *et al.*, *Physica D* **107**, (1997).
2. Stillinger, F.H., *Science* **267**, 1935-1939 (1995).
3. Wu, T.M., Ma, W.J., and Tsay, S.F., *Physica A* **254**, 257-271 (1998).
4. Ma, W.J. and Wu, T.M., *Physica A* **281**, 393-403 (2000).
5. Hagen, M.H., Meijer, E.J., Mooij, G.C.A.M., Frenkel, D., and Lekkerkerker, H.N.W., *Nature* **365**, 425-426 (1993).
6. Poon, W.C.K., *J. Phys. Condens. Matter* **14**, R859-R880 (2002), and references therein.
7. Stratt, R.M., *Acc. Chem. Res.* **28**, 201-207 (1995).
8. Keyes, T., *J. Phys. Chem.* **101**, 2921-2930 (1997).
9. Hasegawa, M., *J. Chem. Phys.* **108**, 208-217 (1998).
10. Wu, T. M., and Chang, S. L., *Phys. Rev. E* **59**, 2993-3000 (1999).

Brownian Dynamics Simulation of Highly Charged Colloidal Suspensions

Yayoi Terada* and Michio Tokuyama*

Institute of Fluid Science, Tohoku University, Sendai 980-8577, Japan

Abstract. We perform the Brownian-dynamics simulations on dilute suspensions of highly charged colloids with Tokuyama attractive potential. We then show that there exist two kinds of droplet phases in addition to a gas phase, a liquid-droplet phase and a crystal-droplet phase. The detailed structures of those droplets are analyzed by the calculating the radial distribution function.

Recently, we have shown that there exist three phases in dilute highly charged colloidal suspensions, a gas phase, a liquid-droplet phase, and a crystal-droplet phase [1]. In this paper, we discuss the detailed structures of those phases.

We perform the Brownian-dynamics simulations on the charged colloidal suspensions by employing an effective attractive force between colloidal particles proposed by Tokuyama [2]. We consider a simple model system, which consists of N highly charged colloidal particles with bare charge Ze and radius a and N_c counter ions with charge $-qe$ and radius a_c in an equilibrium solvent with a dielectric constant ε and temperature T, where $Z \gg q$ and $a \gg a_c$. Here the electrical neutrality is satisfied as $NZ = N_c q$. The volume fraction of the colloidal particles ϕ is given by $\phi = 4\pi a^3 N/3V$, where the total volume of the system is given by V. Then, the particle motion is described by the Langevin-like equation discussed elsewhere [2]

$$\frac{d}{dt}\mathbf{r}_i(t) = \frac{D_0}{k_B T}\sum_{j\neq i}^{N}\mathbf{F}_T(\mathbf{r}_{ij}(t)) + \boldsymbol{\xi}_i(t), \quad (1)$$

where $\boldsymbol{\xi}_i(t)$ is a Gaussian, Markov random velocity with zero mean, and satisfies

$$<\boldsymbol{\xi}_i(t)\boldsymbol{\xi}_j(t')> = 2D_0\delta(t-t')\delta_{i,j}\mathbf{1}. \quad (2)$$

Here D_0 is a diffusion coefficient of a single particle and $\mathbf{F}_T(\mathbf{r}_{ij})$ is the Tokuyama force between particles i and j and is given by [2]

$$\mathbf{F}_T(\mathbf{r}_{ij}) = -\nabla U(\mathbf{r}_{ij})$$
$$= k_B T(Zql_B)^2 \left[z^2 e^{-r_{ij}/\lambda_m} - e^{-r_{ij}/\lambda}\right]\frac{\mathbf{r}_{ij}}{r_{ij}^4}, \quad (3)$$

where $U(\mathbf{r}_{ij})$ is a potential, $l_B (= e^2/\varepsilon k_B T)$ the Bjerrum length, $\mathbf{r}_i$ the position vector of particle i, $r_{ij} = |\mathbf{r}_{ij}|$,

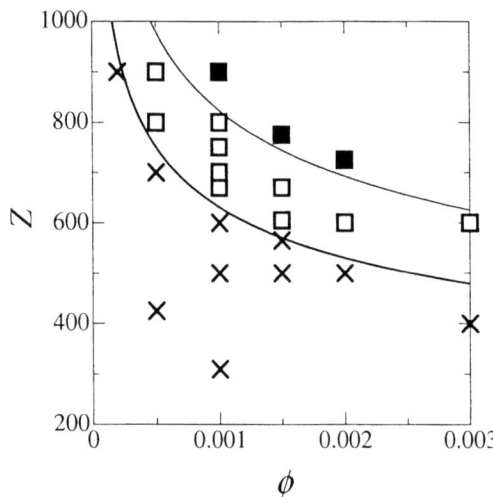

FIGURE 1. Phase diagram in parameter space ϕ and Z at $q = 1$. The symbols indicate each phase; a gas phase (×), a liquid-droplet phase (□), and a crystal-droplet phase (■). The lower and upper transition lines are drawn at $U_{min}/k_B T \simeq -1.0$ and -3.0, respectively.

$z = Z/q$, and $\mathbf{r}_{ij} = \mathbf{r}_i - \mathbf{r}_j$. λ $(= \sqrt{a/3\phi(r)Zql_B})$ is a Debye screening length, $\lambda_m = z^{1/2}\lambda$, and $\phi(r)$ the local density. In the following, we simply assume that $\phi(r)$ is constant to be ϕ. As one of the parameters, Z, ϕ, and q, increases with the other two parameters being fixed, $\mathbf{F}_T$ becomes a more attractive interaction. We are only interested in highly charged dilute suspensions with $\phi \ll 1$. Hence, we can safely neglect the hydrodynamic interactions between the colloidal particles. We simulate Eq. (1) in a cubic simulation cell, where N is chosen to be 3000 here.

In Fig.1, the phase diagram is shown in the parameter space ϕ and Z at $q = 1$. There exist a gas phase, a

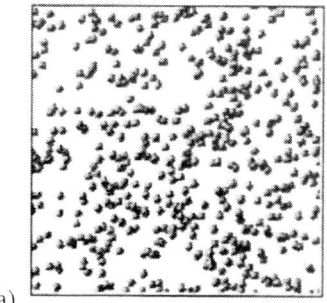 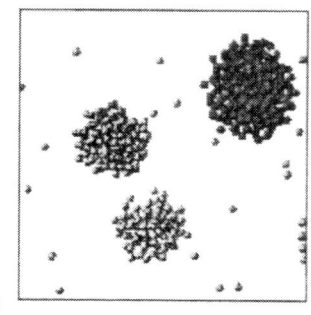 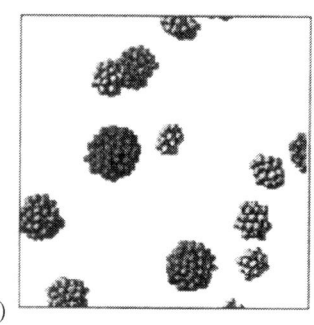

FIGURE 2. The typical snapshots of particle configurations for different values of Z at $q = 2$ and $\phi = 0.001$; (a) $Z = 400$ on a gas phase, (b) $Z = 500$ on a liquid-droplet phase, and (c) $Z = 700$ on a crystal-droplet phase.

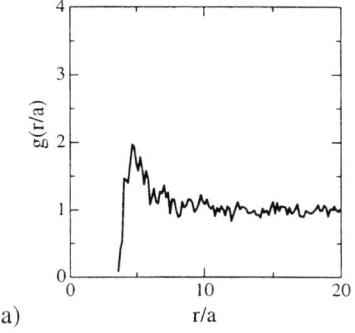 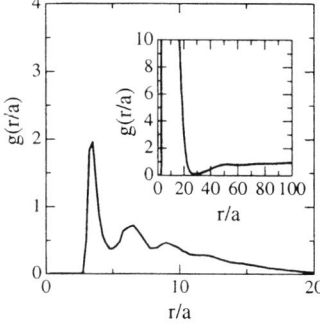 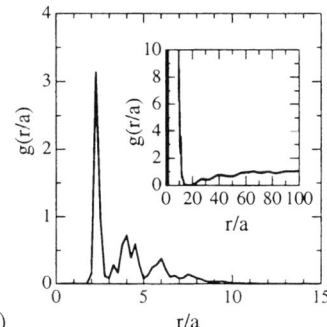

FIGURE 3. The radial distribution functions $g(r)$ versus distance r for different values of Z at $q = 2$ and $\phi = 0.001$; (a) $Z = 400$, (b) $Z = 500$, and (c) $Z = 700$. $g(r)$ in (a) is calculated for the whole system. $g(r)$ in (b) and (c) are calculated for the largest droplet, where those in the insets are calculated for the whole system.

liquid-droplet phase, and a crystal-droplet phase. The simulation results show that each transition line between phases is determined only by the value of the potential minimum U_{min}, which is given by

$$\phi = \left(\frac{-4U_{min}}{3\Gamma^3 k_B T}\right) \frac{\ln z}{z^{-u}(1-\sqrt{z})^2 + 2f(z)\ln z} \quad (4)$$

with $f(z) = G(u \ln z) - z^3 G(u\sqrt{z} \ln z)$ analytically, where $u = 2/(\sqrt{z} - 1)$ and $G(x) = \int_x^\infty y^{-1} e^{-y} dy$. A gas phase exists for $U_{min}/k_B T > -1.0$, a liquid-droplet phase for $-1.0 \geq U_{min}/k_B T > -3.0$, and a crystal-droplet phase for $-3.0 \geq U_{min}/k_B T$. The phase diagrams are also obtained for values larger than $q = 2$, where the transition lines shift to lower values of Z.

Figure 2 (a), (b), and (c) show typical snapshots on each phase after a long time. The radial distribution functions $g(r)$ of each snapshot on Fig. 2 are also shown in Fig. 3. The configuration of the particles in Fig. 2 (a) is similar to those of the theoretical results in a gas phase, where $g_t(r) = \exp[-U(r)/k_B T]$. In Fig.3 (b), the simulation results does not agree with $g_t(r)$ no longer and shows the liquid-like order in the droplets. On the other hand, the configurations of colloids in droplets in Fig.2 (c) are in local order and the peak positions of $g(r)$ agree with those of face-centered cubic crystal within the fourth nearest peaks. We note that the droplets in both droplet phases are dispersed randomly, because $g(r)$ in the insets of Fig.3 (b) and (c) show disorder-state outside the droplets.

ACKNOWLEDGMENTS

This work was partially supported by Grants-in-aid for Science Research with No. 14540348 from Ministry of Education, Culture, Sports, Science and Technology of Japan. The calculations were performed using the ORIGIN 2000 in the Institute of Fluid Science, Tohoku University.

REFERENCES

1. Terada, Y., and Tokuyama, M., *Physica A* (in press).
2. Tokuyama, M., Phys. Rev. E **59**, R2550-R2553 (1999);Phys. Rev. E **58**, R2729-R2732(1998).

Computer Simulations of Two Kinds of Polydisperse Hard-Sphere Systems; Atomic Systems and Colloidal Suspensions

Tsutomu Shimura[*], Hiroyuki Yamazaki[†], Michio Tokuyama[*] and Yayoi Terada[*]

[*]*Institute of Fluid Science, Tohoku University, Sendai 980-8577, Japan*
[†]*Fuji Photo Film Co., Ltd, Nishiazabu 2-26-30, Minato-ku, Tokyo 106-8620, Japan*

Abstract. We perform two kinds of computer simulations on polydisperse hard-sphere systems; a molecular-dynamics simulation on atomic systems and a Brownian-dynamics simulation on colloidal suspensions. By the analyses of the mean square displacement and the radial distribution function, the simulation results suggest that the long-time behavior of colloidal suspensions is exactly the same as that of atomic systems. It is also shown that there exist three phase regions, a liquid phase region, a metastable phase region, and a crystal phase region, where the freezing and melting points in polydisperse case are shifted to the values higher than in monodisperse case.

INTRODUCTION

A considerable number of studies have been made on the hard-sphere systems because of their simplicity. The polydisperse hard-sphere systems have been simulated to compare with experiments on hard-sphere colloids with subtle polydispersity [1,2]. Recently Tokuyama et al have performed two kinds of computer simulations on polydisperse hard-sphere systems; a molecular-dynamics simulation on atomic systems and a Brownian-dynamics simulation on colloidal suspensions to investigate the many-body collision interactions [3]. In this paper, we discuss three kinds of phases, depending on the values of the volume fraction ϕ in both systems. Our simulations have been performed for longer times with larger number of particles than those in previous work [2].

SIMULATION MODEL AND RESULTS

The atomic particle obeys Newton equation with the force $\boldsymbol{F}_{ij}$ between particle i and particle j. On the other hand, the colloidal particles dispersed in solvent obey Langevin equations. On a time scale of order t_D, the position $\boldsymbol{X}_i(t)$ of the i th particle is described by

$$\frac{d}{dt}\boldsymbol{X}_i(t) = \frac{1}{\gamma_i}\sum_{j\neq i}\boldsymbol{F}_{ij}(t) + \boldsymbol{R}_i(t), \quad (1)$$

$$<\boldsymbol{R}_i(t)\boldsymbol{R}_j(t')> = 2D_{0i}\delta_{ij}\delta(t-t')\mathbf{1}, \quad (2)$$

where $\gamma_i(=6\pi\eta a_i)$ represents the friction coefficient, a_i the particle radius, $D_{0i}(=k_BT/\gamma_i)$ the single-particle diffusion coefficient, $\boldsymbol{R}_i(t)$ the Gaussian random velocity with zero mean, and η the viscosity of the solvent. Here $t_D(=a^2/D_0)$ indicates structural relaxation time, where $D_0(=k_BT/(6\pi\eta a))$ denotes the mean single particle diffusion coefficient, k_B Boltzmann constant, T temperature, and a mean particle radius. The distribution of radii obeys a Gaussian distribution with standard deviation σ divided by a. For simplicity, we have neglected the hydrodynamic interactions between particles. We assume the elastic binary collisions between particles. Our simulations have been performed in the cubic cell which consists of N polydisperse hard spheres with periodic boundary condition, where $N = 10976$ and $\sigma = 0.06$ here. Our simulations start from two nonequilibrium configurations; (I) a disordered configuration [4] and (II) an ordered face centered cubic crystal configuration.

The mean square displacement is given by

$$M_2(t) = \frac{1}{N}\sum_{i=1}^{N} <[\boldsymbol{X}_i(t)-\boldsymbol{X}_i(0)]^2>. \quad (3)$$

Figure 1 shows a log-log plot of $M_2(t)$ versus time t for different volume fractions. For short time, $M_2(t) \propto t^2$ for atomic systems, while $M_2(t) \propto t$ for colloidal suspensions. On the other hand, the long-time behavior of colloidal suspensions is the same as that of atomic systems in Fig. 1. That is, $M_2(t) \propto t$ in liquid phase and $M_2(t) = const.$ in solid phase in both systems. In order to investigate the detailed structures, it is convenient to introduce the radial distribution function by

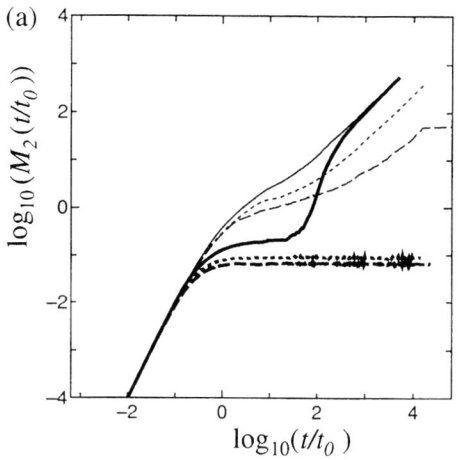
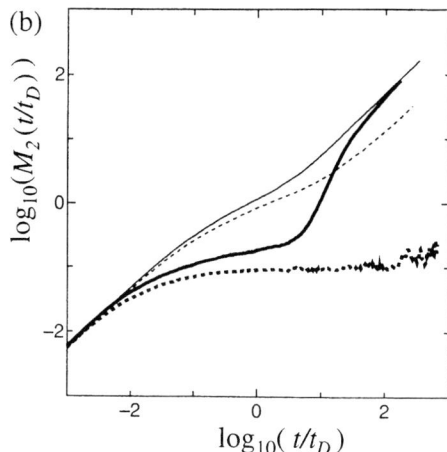

FIGURE 1. A log-log plot of the mean square displacement $M_2(t)$. (a) the atomic systems and (b) the colloidal suspensions with $\sigma = 0.06$ polydispersity. Here $t_0 = a/v_0$, where v_0 is the mean velocity of particles. Solid lines, dotted lines, and broken lines stand for the simulation results at $\phi = 0.51$, 0.55, and 0.57, respectively. The thick line indicates the results for the ordered initial configuration, and the thin line for the disordered initial configuration.

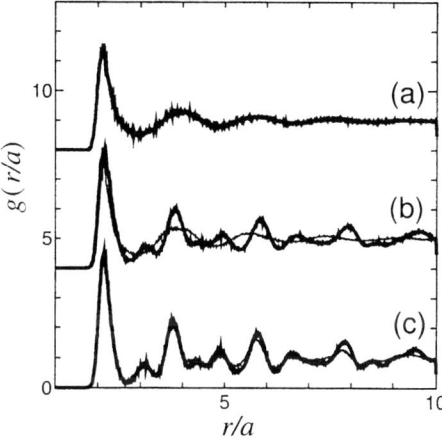

FIGURE 2. The radial distribution function $g(r)$ of atomic systems with $\sigma = 0.06$ polydispersity for (a) $\phi = 0.51$, (b) $\phi = 0.55$, and (c) $\phi = 0.57$. The details are the same as in Fig. 1.

$$g(r) = \frac{V}{N^2} \left\langle \sum_{i}^{N} \sum_{j(\neq i)}^{N} \delta(r - r_{ij}) \right\rangle. \quad (4)$$

In Fig. 2, $g(r)$ of atomic systems is shown for different volume fractions in an equilibrium state, which are similar to those in colloidal suspensions. For $\phi = 0.51$, $g(r)$ shows liquid like order in Fig. 2 (a), and $M_2(t)$ grows linearly in time for longer times. For $\phi = 0.57$, $g(r)$ shows face-centered cubic crystal order in Fig. 2 (c) and $M_2(t)$ becomes constant for longer times. For those volume fractions the long time behaviors starting from different initial configurations become identical. On the other hand, for $\phi = 0.55$, the long time behavior strongly depends on the initial configurations. In case (I) the system remains in a liquid state within our simulation time, while in case (II) it remains in a crystal state.

The simulation results suggest that there exist three phase regions, a liquid phase region for $\phi < \phi_f(\sigma)$, a metastable phase region for $\phi_f(\sigma) \leq \phi < \phi_m(\sigma)$, and a crystal phase region for $\phi_m(\sigma) \leq \phi$, where $\phi_f(\sigma)$ denotes the freezing volume fraction and $\phi_m(\sigma)$ the melting volume fraction. Our simulations show that $\phi_f(\sigma = 0.06) \simeq 0.53$, $\phi_m(\sigma = 0.06) \simeq 0.57$, $\phi_f(\sigma = 0) \simeq 0.51$, and $\phi_m(\sigma = 0) \simeq 0.54$. Thus the freezing and melting points in polydisperse case are shifted to the values higher than in monodisperse case.

ACKNOWLEDGMENTS

This work was partially supported by Grants-in-aid for Science Research with No. 14540348 from Ministry of Education, Culture, Sports, Science and Technology of Japan. The calculations were performed using the ORIGIN 2000 in the Institute of Fluid Science, Tohoku University.

REFERENCES

1. Phan, S.-E., Russel, W. B., Zhu, J., and Chailin, P. M., *J. Chem. Phys.* **108**, 9789-9795 (1998).
2. Sear, R. P., *J. Chem. Phys.* **113**, 4732-4739 (2000).
3. Tokuyama, M., Yamazaki, H., and Terada, Y., *Physica A* **328**, 367-379 (2003); *Phys. Rev.* E**67**, 062403 (2003).
4. Jodrey, W. S., Tory, E. M., *Phys. Rev.* A**32**, 2347-2351 (1985).

Brownian Dynamics Study of Dilute Suspensions of Magnetic Particles in a Static Field

J.Akiyama[*], M.Tokuyama[†] and Y.Terada[†]

[*]Graduated School of Engineering,Tohoku University, Sendai 980-8579, Japan
[†]Institute of Fluid Science,Tohoku University, Sendai 980-8579, Japan

Abstract. Brownian-dynamics simulations without the hydrodynamic interactions between particles are performed to investigate two-dimensional dilute suspensions of magnetic colloidal particles under a static magnetic field. From the detailed analyses of the numerical results, we obtain the phase diagram for the magnitude of magnetic potential J and the area fraction σ. The results also suggests that the phase transition line is determined only by the value of the parameter $\Gamma(=J\sigma^{\frac{3}{2}})$, where $\Gamma \simeq 68$ at the melting point.

INTRODUCTION

Recently, a lot of attentions have been paid for suspensions of magnetic colloidal particles, because of their interesting physical behavior and their useful applications as MR fluids. Under a magnetic field, particles interacts each other by anisotropic magnetic dipole potential. In a three-dimensional system, the particles form chained clusters. Hwang et al have experimentally studied the long-time self-diffusion process of formed chained clusters and found the glasslike transition [1]. In a two-dimensional system, Zahn et al have found the melting transition experimentally [2, 3].

In this paper, we perform Brownian-dynamics simulations to investigate the static and dynamic properties of dilute two-dimensional suspensions of magnetic particles.

SIMULATION MODEL

We consider the two-dimensional monodisperse suspension of magnetic particles, where the magnetic field is applied in a manner similar to the experiments by Zahn et al [2, 3]. We assume that the particles interact only by magnetic potential and do not collide directly each other because of dilute suspension. On a time scale of diffusion $t_D(=a^2/D_0)$, where a is the radius of the particle and D_0 the self diffusion coefficient of the particle, the position $\boldsymbol{r}_i$ of particle i obeys the Langevin like equation in a dimensionless form.

$$\frac{d\boldsymbol{r}_i}{dt} = -J\boldsymbol{\nabla}_i \sum_{i(\neq j)} U_{ij} + \boldsymbol{R}_i, \quad (1)$$

where $\boldsymbol{R}_i$ is the reduced random velocity and satisfies the fluctuation-dissipation theorem

$$<\boldsymbol{R}_i(t)\boldsymbol{R}_j(s)> = 2\delta_{ij}\delta(t-s)\mathbf{1}. \quad (2)$$

The reduced magnetic potential $U_{ij} = 1/r_{ij}^3$ is given as a function of r_{ij}, where $r_{ij} = |\boldsymbol{r}_i - \boldsymbol{r}_j|$. Here the ratio J of magnetic potential to thermal energy k_BT is given by

$$J = \frac{\mu_0|\boldsymbol{M}|^2}{4\pi a^3 k_B T}, \quad (3)$$

where μ_0 is the magnetic permeability. Because of the dilute suspension, all of the magnetic dipole moments on the particles are identical to be $\boldsymbol{M}$. Here $\boldsymbol{M}$ is given by $\boldsymbol{M} = \frac{4}{3}\pi\chi a^3 \boldsymbol{H}$, where χ is the susceptibility of the particles and $\boldsymbol{H}$ the magnetic field. The hydrodynamic interactions between particles are neglected because of dilute suspensions. In our simulation, the total number N of particles is chosen to be 2024 and the periodic boundary condition is employed. The length of each direction (x or y axis) of the system is determined to satisfy the hexagonal lattice condition in a crystal phase.

RESULTS

It is shown that there exist two phases, a liquid phase and a crystal phase. The phase diagram in parameter space the area fraction σ and J is shown in Fig. 1. The melting line shown in the figure is determined by

$$J = \Gamma\sigma^{-\frac{3}{2}} \quad \text{at} \quad \Gamma = \Gamma_m, \quad (4)$$

where $\Gamma_m \simeq 68$.

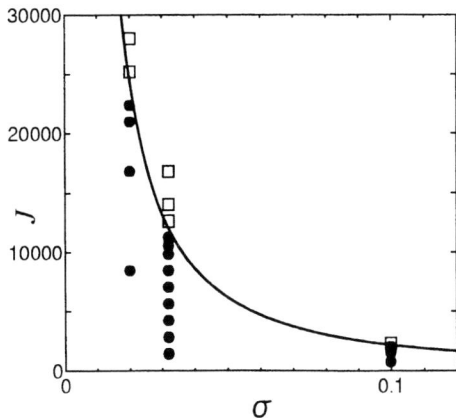

FIGURE 1. Phase diagram: Circles denote liquid phase and squares crystal phase. The solid line indicates the melting line $J = \Gamma \sigma^{-\frac{3}{2}}$ at $\Gamma = \Gamma_m$.

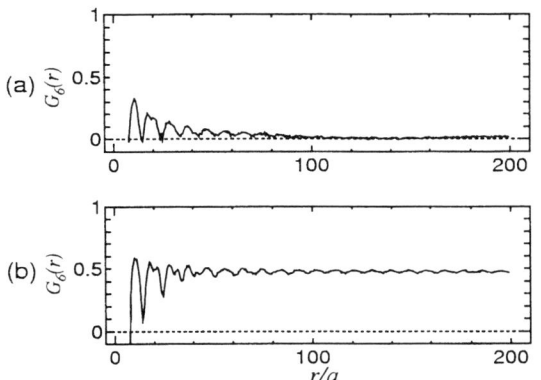

FIGURE 2. Bond orientation function versus distance: (a): The result at $\Gamma = 64$ ($J = 11200, \sigma = 0.032$). (b): The result at $\Gamma = 72$ ($J = 12600, \sigma = 0.032$).

Since the angular correlation is the most typical feature of static properties in a two-dimensional system, it is convenient to introduce bond orientation function by

$$G_6(r) = \frac{<\Psi_6(r)\Psi_6(r')>}{<\rho(r)\rho(r')>} \quad (5)$$

with

$$\Psi_6(r) = \sum_{i=1}^{N} \delta(r - r_i) \frac{1}{n_i} \sum_{j=1}^{n_i} e^{6i\theta_{ij}}, \quad (6)$$

where $\rho(r) = \sum_{i=1}^{N} \delta(r - r_i)$. Here n_i is the number of neighboring particles of particle i within the distance of the first peak of the radial distribution function at each area fraction σ. θ_{ij} represents the angle respect to a fixed reference axis. Figure 2(a) shows that $G_6(r)$ decays exponentially to zero with increasing the distance at $\Gamma = 64$. Since this reveals the short-range angular order, the system is considered to be in a liquid phase. On the other hand, in Fig. 2(b), $G_6(r)$ converges to a non-zero finite value at $\Gamma = 72$. This long-range order suggests a crystal phase. From those data, there occur a melting transition by increasing Γ, where the melting point Γ_m is around 68. And we also mention here that $G_6(r)$ is the same as that for different J and σ if the value of Γ is fixed.

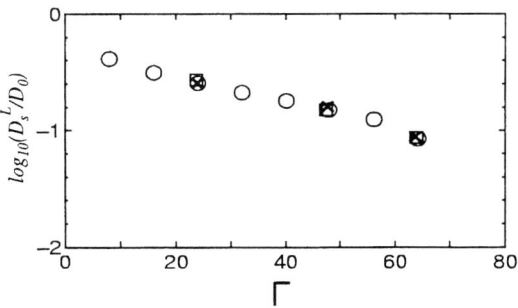

FIGURE 3. Γ dependence of D_s^L: Circles denote D_s^L at $\sigma = 0.032$, squares at $\sigma = 0.02$, and crosses at $\sigma = 0.1$.

In Fig. 3, the Γ dependence of the long-time self-diffusion coefficient D_s^L is shown for different values of σ. D_s^L is calculated from mean-square displacement given by $M_2(t) = \frac{1}{N} \sum_{i=1}^{N} < |r_i(t) - r_i(0)|^2 >$. In fact, it reduces for $t/t_D \gg 1$ to

$$M_2(t) \simeq 4D_s^L t. \quad (7)$$

D_s^L decreases as Γ increases until the phase transition occurs. Here we note that D_s^L is a function of Γ only.

We have shown from detailed analyses of the simulation results that Γ is an only control parameter to describe the long-time behavior of dilute suspensions of magnetic colloids since the hydrodynamic interactions and the direct interactions can be neglected.

ACKNOWLEDGMENTS

This work was partially supported by Grants-in-aid for Science Research with No. 14540348 from Ministry of Education, Culture, Sports, Science and Technology of Japan. The calculations were performed using the ORIGIN 2000 in the Institute of Fluid Science, Tohoku University.

REFERENCES

1. Hwang, Y.H., and Wu, X-l., *Phys. Rev. Lett.* **74**, 2284-2288(1995).
2. Zahn, K., Lenke, R., and Maret, G., *Phys. Rev. Lett.* **82**, 2721-2724(1999).
3. Zahn, K., and Maret, G., *Phys. Rev. Lett.* **85** 3656-3659(2000).

Orientation Orders of Small Anisotropic Molecules in Confinement

Xin Zhou*, Hu Chen, and M. Iwamoto*

*Department of Physical Electronics, Tokyo Institute of Technology, O-okayama 2-12-1, Meguro-ku, Tokyo 152-8552, Japan
Department of Civil Engineering, National University of Singapore, 10 Kent Ridge Crescent, Singapore 119260

Abstract. We have studied orientation-ordered transitions of small anisotropic molecules in confinement based on standard constant-pressure Monte Carlo molecular simulation. These molecules are modeled by the hard Gaussian overlap (HGO) model with a small elongation parameter. Two different confining geometries (slit pores and mixtures of HGO) are studied, which confining surfaces are composed of two hard parallel plane and the molecular surfaces of some large HGO molecules, respectively. In both cases, there is no attractive interaction between the confining surfaces and the small molecules which favors the orientation alignment. We found, although the small HGO molecules cannot form stable orientation-ordered phases in bulk due to their too small molecular elongation, a liquid-crystal (LC) phase form in the nanometre-scale confinement. It means that the required molecular elongation for forming LC phases will decrease in confinement. Our obtained result implies that small anisotropic molecules might show liquid crystal behavior in confinement.

INTRODUCTION

The behavior of liquid crystals (LCs) have been widely studied based on molecular simulations, in which some anisotropic molecular models are employed to model interactions between LC molecules. It is generally confirmed that LC phases can form only when the anisotropic parameters of these models are greater than some critical values. On the other hand, confinement in nanometer scale is expected to induce surface phase transitions not observed in bulk systems in fluids[1]. Whether do the confinement change the required molecular condition for forming LC phases? In another words, are there some LC-like phases near confining surfaces for the molecules with small anisotropic parameters? The answer to the above question is important not only for the sake of properties of confined fluids, but implies the possibility of existing small molecule liquid crystals.

METHODS AND RESULTS

Many different models are used in the literature on the simulations of LCs. Among them, the hard Gaussian overlap (HGO) model is simplest and widely used. The HGO molecules with a larger elongation parameter (k) can form LC phase in bulk, but the molecules with smaller k cannot form stable orientation-ordered phases in bulk[2,3]. The model is thought as an approximation for the excluded volume of hard ellipsoids of revolution.

In this paper, based on standard constant-pressure Monte Carlo molecular simulations, we study LC phases of the small HGO molecules confined in two different confined geometries, (1) two parallel hard

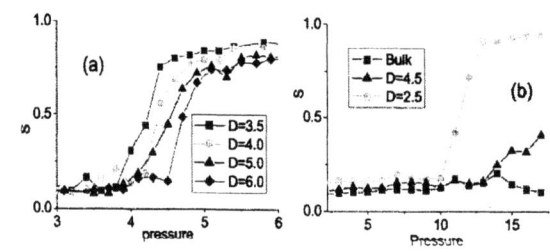

FIGURE 1. Orientation order parameter S versus pressure P of HGO in slit pores with different D and in bulk, where the number of particles N = 256. (a) k=3; (b) k=2.2.

walls (slit pores); (2) networks composed of large HGO molecules (mixtures of small and large HGO molecules). The orientational order parameter (S) of LC phases and the equation of state is calculated in the simulations. In mixture case, we also separately calculate the order parameter of the large, small HGO molecules and all molecules (S_L, S_S and S_a).

First, we study the HGO molecules with k=3.0 confined in slit pores with the interwall separation D. For the molecules, there is an isotropic-nematic (IN) phase transition at P≈5 in bulk[2]. We find the IN transition point shift to lower pressure direction as D decreases (see Fig.1 (a)). The shift indicates that the nematic phase forms more easily in confinement than in bulk. Thus, we suspect the required smallest molecular elongation (k) for forming orientation-ordered phase in confinement may be smaller than that in bulk. We simulate the molecules with k=2.2, the results are shown in Fig. 1 (b). The system is compressed up to a typical high density of fluids (fluids are thought as freezing at higher density[4]), in bulk, there is not phase transition. But for the confined system in a slit pore with D=2.5, an obvious isotropic-anisotropic phase transition is found. More researches show the anisotropic phase is a smectic

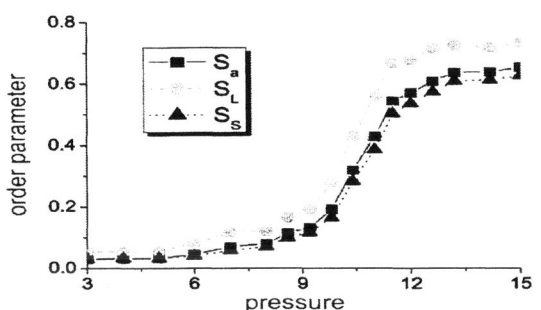

FIGURE 2. Orientation order parameters of large, small and all molecules versus pressure P in a HGO binary mixture, where N=1372, x=0.25, k_S=2.2 and k_L=3.0.

phase rather than a nematic phase. It implies the molecular-scale confinement decrease the required molecular elongation k to form a LC phase.

In mixtures of large and small HGO molecules, the surfaces of the large molecules can be thought as forming complex pores to confine the small molecules. The complex pores are also expected to induce surface phase transition. If the large molecules align, the effect may be global. We simulate the mixtures with low concentration (x) of large molecules. The small molecules are selected so that they cannot form LC phase alone. In Fig. 2, we show the results of a mixture with N=1372 molecules. Here, x = 0.25, the elongations of large and small molecules are 3.0 and 2.2, respectively. From the figure, there is a transition of the order parameter S at P≈10, and S_S is approaching to the value of S_L. The result clearly shows an IN phase transition in whole mixture including large and small HGO molecules. Since the concentration of large molecules is low, we expect the mixture show some dynamic characteristics of small-molecular LCs. The dependences of the effects on the k_L, k_S and x in the mixtures have also been studied.

CONCLUSIONS

In our simulations, the found surface-induced phase transitions completely ascribe to entropy of the system. A suitable surface attractive interaction is expected to strengthen the effects. As next works, it is necessary to study the global phase diagrams of small anisotropic molecules with attractive interaction in confinement. The molecular elongation of usual LC materials is large than 4, but our results indicate that small modeling molecules which k is about 2 may form LC phases in confined geometries. The confinement-induced LC phases present here may be useful. For example, we can consider small anisotropic molecules with dispersion of large molecules (usual LC molecules or polymers). In these mixtures, the viscosity should be smaller than pure large-molecular systems, they may be LCs with shorter response time in external field. It is useful to study the confined effects based on more reliable molecular models.

ACKNOWLEDGMENTS

X. Zhou is financially supported by the Grants-in-Aid for Scientific Research of JSPS; H. Chen is supported by the Singapore Millennium Scholarship. X. Z. would like to acknowledge Douglas Cleaver and Fred Barmes.

REFERENCES

1. Gelb, L. D., et al., *Rep. Prog. Physics* **62**, 1573-1659 (1999).

2. Miguel, E. D., and Rio, E. M. D., *J. Chem. Physics* **118**, 1852-1858 (2003).

3. Cleaver, D. J., et al., *Phys. Rev. E* **54**, 559-567 (1996).

4. Rigby, M., *Mol. Physics*, **68**, 687-697 (1989).

Particle Modeling of Plasma Confinement by Multipolar Magnetic Fields

Hideto TAKEKIDA* and Kenichi NANBU*

*Institute of Fluid Science, Tohoku University,
1-1, Katahira 2-chome, Aoba-ku, Sendai 980-8577*

Abstract. Multipolar magnetic fields are widely used to enhance plasma density. The effect of plasma confinement by multipolar magnetic fields is studied numerically using a self-consistent particle modeling of discharges in a cylinder.

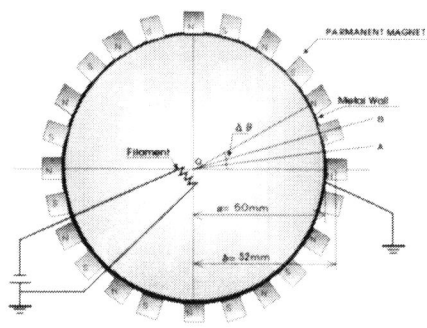

FIGURE 1. Schematic of apparatus

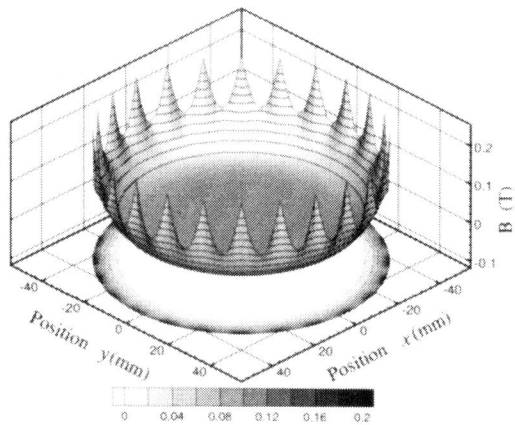

FIGURE 2. Magnetic field in the apparatus

INTRODUCTION

Multipolar plasma chambers have widely been used for research on basic and applied researches on low temperature plasmas. The multipolar confinement realizes enhanced plasma density, homogeneous plasmas of a large volume, and quiescent plasmas. However, detailed theoretical studies on the multipolar confinement are few.

In this work, we examine the effect of plasma confinement using a self-consistent particle modeling of discharges in a cylinder sustained by electron emission from a heated filament at the central axis of the cylinder. The goal is to present design criteria of multipolar magnetic fields. Use of the present method makes it possible to treat three roles of plasma confinement all together; (1)magnetic mirror effect near the cusp of two poles, (2)supressed electron diffusion in the direction normal to magnetic field line, and (3)electron repelling in the sheath on the chamber wall[1].

OUTLINE OF SIMULATION PROCEDURE

The schematic of a reactor with multipolar magnetic field is shown in Fig.2. The reactor has the straight magnets of N-pole and S-pole altenately around the wall. The cylinder has an infinite length. We simulate discharge sustained by a hot cathod at the center. The chamber wall is grounded. Numerical analysis of the magnetic field is carried out by using the idea of potential flow; the magnetic field B is regarded as velocity field, N-pole being line source and S-pole line sink. The field B is given by $B = \nabla \phi$, where the potential $\phi(x,y)$ is $\phi = \sum_i \phi_i$. Here ϕ_i denotes ith magnet. The summation is over all N- and S- poles. The total is $n(=24)$. The function $\phi_i(x,y)$ is given by

$$\phi_i(x,y) = \frac{q_i}{4\pi} \int_{-\infty}^{\infty} \frac{dz_i}{\sqrt{(x-x_i)^2 + (y-y_i)^2 + (z-z_i)^2}} \quad (1)$$

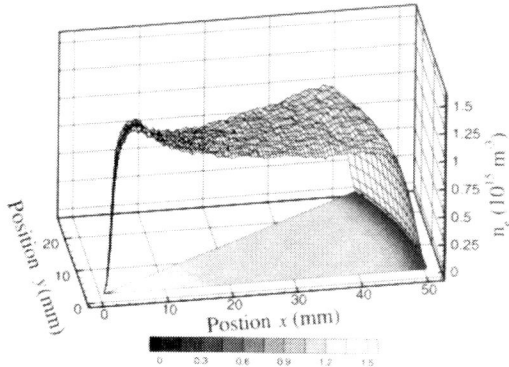

FIGURE 3. Electron number density (B=0T)

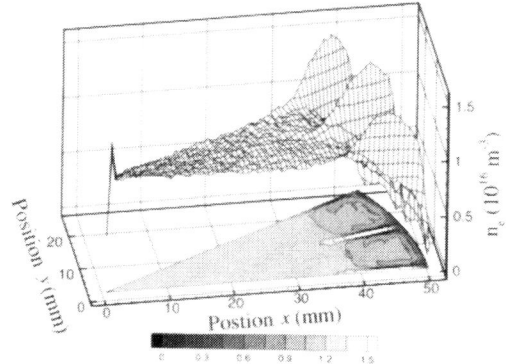

FIGURE 4. Electron number density (B=0.1T)

where (x_i, y_i, z_i) is the position of the element of an infinitesimally small magnet with length dz_i. The components of B is given by

$$B_x = \sum_{i=1}^{n} \frac{\partial \phi}{\partial x} = \sum_{i=1}^{n} \frac{q_i}{2\pi} \frac{x - x_i}{(x - x_i)^2 + (y - y_i)^2} \quad (2)$$

$$B_y = \sum_{i=1}^{n} \frac{\partial \phi}{\partial y} = \sum_{i=1}^{n} \frac{q_i}{2\pi} \frac{y - y_i}{(x - x_i)^2 + (y - y_i)^2} \quad (3)$$

Figure 2 shows the magnetic field in the apparatus for $n=24$. We set $q_i = 2.28 \times 10^{-3}$ T·m. This choice gives $|B_\theta| = 0.1$T at $r = 50$mm and $\varphi = \Delta\theta/4$. The magnetic field is concentrated close to the wall. We see that there is almost no magnetic field in the inner region.

We use PIC/MC method to simulate the discharge. We first divide the computational domain into small cells and put simulated particles. Then we trace the motion of each particles using the modified Verlet method and hence find the distribution of charge density ρ from all positions of the simulated particles. Next, we obtain a renewed electric field E from the charge density. The renewed field is used in solving the equation of particles. The three N,S,N-poles make one period and periodical boundary condition is used. We use the fast-fourier-transform(FFT) for calculation of the electric field. Computational domain is a fan-shaped one with angle $\Delta\theta$.

Gas is argon. As for collisions of charged particles, we consider e^--Ar and Ar$^+$-Ar collisions using the Monte Carlo method. The collision judgement and the determination of collisional event were done using Nanbu's method[2].

RESULTS AND DISCUSSION

Calculation was done for the gas pressure p_g of 1mTorr, gas temperature T_g of 300K. We consider the magnetic field of 0T or 0.1T.

Figures 3 and 4 shows the electron number density for magnetic field of 0T and 0.1T. We see that the electron

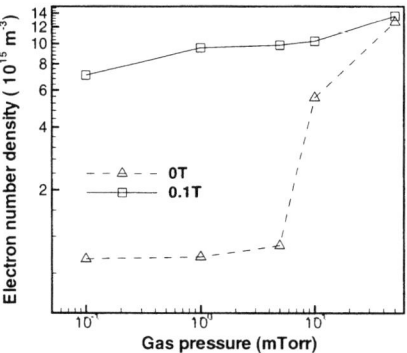

FIGURE 5. Electron number density for various gas pressure

number density for $B=0.1$T is 9 times higher than that for $B=0$T. In addtion, when magnetic field is added, there appear peaks near the wall. This is due to a back and forth movement of electrons trapped by the multipolar magnetic fields.

Figure 5 shows elctron number density for various gas pressure. The multipolar magnetic field has a great effect on the plasma density when the pressure is lower than 5mTorr.

CONCLUSION

Effect of a multipolar magnetic field on plasma density is clarified using the PIC/MC simulation.

REFERENCES

1. Moisan, M. and Pelletier, J., *Microwave Excited Plasma*, Elsevier, Amsterdam, 1992, pp.249-384.
2. Nanbu, K., *Jpn. J. Appl. Phys.* **33**, 4752-4753 (1994).

Computer Simulation of Nano Metallic-Particle Synthesis in an Advanced RF Inductively Coupled Plasma

Masaya Shigeta and Hideya Nishiyama

Institute of Fluid Science, Tohoku University, 2-1-1 Katahira, Aoba-ku, Sendai 980-8577, JAPAN

Abstract. In the present study, the thermofluid field and the electromagnetic field of an advanced RF inductively coupled plasma, which is expected to be utilized for nano metallic-particle synthesis, are investigated. The influence of the operating conditions on the mechanism of the nano metallic-particle synthesis and the finally obtained diameter is clarified by computer simulation.

INTRODUCTION

An radio frequency inductively coupled plasma (RF-ICP) has advantages of large volume, clean, high energy and chemical reactivity. It has, therefore, been extensively used in the reactive plasma spraying, decomposition of hazardous matters and synthesis of nano-particles, for example [1]. For higher efficiency and convenience in these processes, an advanced RF-ICP combined with DC plasma jet is introduced.

Particularly in the synthesis of nano metallic-particles, the steep temperature gradient at the tail of the advanced RF-ICP leads to the synthesis of ultrafine powders with high purity. Nano metallic-particles are produced through evaporation caused by high energy density of DC and RF plasma and also recondensation of larger particles injected into the advanced RF-ICP. Nowadays, it is considerably important to control the diameters of nano metallic-particles for changeable functions and the extensive applications [2-4].

In the present study, computer simulation is conducted to clarify the influence of the operating conditions on the mechanism of the nano metallic-particle synthesis and the finally obtained diameter.

GOVERNING EQUATIONS

Plasma Flow

The governing equation of the induction electromagnetic fields is expressed by using the azimuthal component of vector potential A_θ [5]

$$\frac{\partial^2 A_\theta}{\partial z^2} + \frac{1}{r}\frac{\partial}{\partial r}(r\frac{\partial A_\theta}{\partial r}) - \frac{A_\theta}{r^2} = i\mu_0 \sigma_c \omega A_\theta \quad (1)$$

where μ_0, σ_c and ω are the permeability in vacuum, electrical conductivity and angular frequency. i is equal to $\sqrt{-1}$. The governing equations of continuity, momentum, energy, and plasma species per unit volume are summarized in the following general form in cylindrical coordinates:

$$\frac{\partial}{\partial z}(\psi u \phi) + \frac{1}{r}\frac{\partial}{\partial r}(r\psi v \phi) = \frac{\partial}{\partial z}\left(\Gamma\frac{\partial \phi}{\partial z}\right) + \frac{1}{r}\frac{\partial}{\partial r}\left(r\Gamma\frac{\partial \phi}{\partial r}\right) + S_C \quad (2)$$

where ϕ corresponds to physical variables such as u, v, w, h, and n. ψ and Γ correspond to density and diffusion coefficient, respectively. S_C is each source term.

Nano Particle Synthesis

Supersaturated vapor creates nuclei by homogeneous nucleation. Homogeneous nucleation rate is given as [6]

$$J = \frac{\beta_{ij} n_s^2 S}{12}\sqrt{\frac{\Theta}{2\pi}}\exp\left(\Theta - \frac{4\Theta^3}{27(\ln S)^2}\right) \quad (3)$$

where β_{ij}, n_s, S and Θ are the collision frequency function, number density in the saturated condition, supersaturation ratio and normalized surface tension, respectively. In the condition of the high concentration and the low supersaturation, heterogeneous condensation occurs on the surface of the particles. The net molecular flux from the vapor to the condensed phase considering all the range of Knudsen number Kn is expressed by [7]

$$F = 2\pi D d_p (n_1 - n_s)\left\{\frac{1+Kn}{1+1.7Kn+1.333Kn^2}\right\} \quad (4)$$

where D, d_p and n_1 are the diffusion coefficient, particle diameter and number density of monomer.

RESULTS

Plasma flow is produced by injecting argon gas from the top nozzles and applying RF electromagnetic field. Table 1 shows the operating conditions of the conventional and advanced RF-ICPs. Q_1, Q_2, Q_3, f,

TABLE 1. Operating Conditions

	Q_1	Q_2	Q_3	f	P_{DC}	P_{RF}
Conventional	5.0	8.0	30.0	6.0	0.0	16.0
Advanced	22.7	8.0	30.0	6.0	4.0	12.0

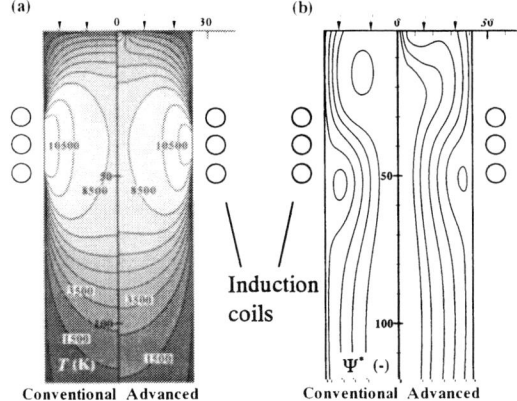

FIGURE 1. Thermofluid fields of conventional and advanced RF-ICPs; (a) Plasma temperature and (b) Normalized stream function

P_{RF} and P_{DC} correspond to the injected flow rate for Nozzle 1 - 3 (Sl min^{-1}), coil frequency (MHz), input power of RF and DC (kW), respectively.

Figure 1 shows the thermofluid fields of conventional and advanced RF-ICPs. In Fig. 1 (a), high temperature regions exist near the induction coils due to the Joule heating. The conventional RF-ICP has a larger high temperature zone near the wall due to the higher applied coil power, while the advanced RF-ICP has relatively longer high temperature zone from the inlet to the downstream region owing to the DC plasma jet. In Fig. 1 (b), a conventional RF-ICP has a recirculating zone in the upstream region of the induction coils since the flow is pinched by Lorentz force caused by the induced electromagnetic fields. On the other hand, the recirculating zone in the upstream region of the induction coils does not exist in an advanced RF-ICP since the momentum of DC plasma jet overcomes the pinch effect.

Figure 2 shows the titanium vapor consumption rate evolution caused by homogeneous nucleation and heterogeneous condensation. The number of the nuclei produced by nucleation increases drastically and then the vapor condenses on the nuclei surface. Nucleation and condensation in the advanced RF-ICP start later than those in the conventional one due to the higher temperature in the downstream region as shown in figure 1 (a).

Figure 3 shows the number density of synthesized nano titanium-particles. The particles produced in an advanced RF-ICP have the larger number density and smaller diameter than in a conventional RF-ICP

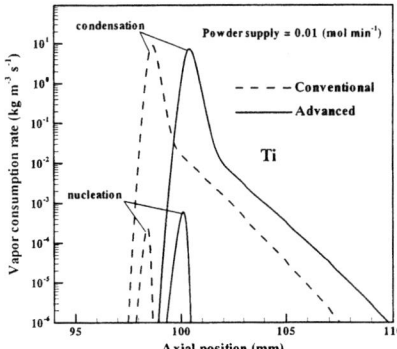

FIGURE 2. Titanium vapor consumption rate evolution

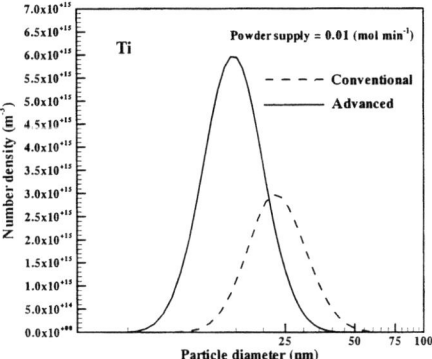

FIGURE 3. Number density of nano titanium-particles

owing to the higher quenching rate by high velocity of DC plasma jet. The more particles are produced, the smaller the quantity of vapor consumption per one nucleus becomes.

ACKNOWLEDGEMENTS

This work was partly supported by a grant-in-aid for Scientific Research (B) from the Japan Society for Promotion Science (2003). This numerical investigation was conducted under the research project (No. E01JUN03) at the Advanced Fluid Information Research Center of the Institute of Fluid Science, Tohoku University, Japan.

REFERENCES

1. Japan Society of Mechanical Engineers, *Functional Fluids and Intelligent Fluids*, Corona Pub. Corp., Japan, 2000, pp. 2-10.
2. Siegel, R. W., *Mater. Sci. Eng.* **A168**, 189-197 (1993).
3. Girshick, S. L., et al, *J. Aerosol Sci.* **24**-3 367-382 (1993).
4. Desilets, M., et al, *J. Phys. D: Appl. Phys.* **30**, 1951-1960 (1997).
5. Mostaghimi, J., and Boulos, M. I., *Plasma Chem. Plasma Processing* **9**, 25-44 (1989).
6. Girshick, S. L., Chiu, C. P., and McMurry, P. H., *Aerosol Science and Technology* **13**, 465-477 (1990).
7. Friedlander, S. K., *Smoke, Dust and Haze*, John Wiley & Sons, 1977, pp. 241-242.

II. POLYMER DYNAMICS

Entanglement Concept Revisited

K. Kremer and S. Leon

Max Planck Institute for Polymer Research, 55021 Mainz, Germany

Abstract. Different methods to determine the entanglement molecular weight M_e or chain length N_e are discussed and compared as a function of e.g. chain stiffness. In addition to simple model systems, polycarbonate as a specific example is considered as well.

INTRODUCTION

Polymeric melts and networks are prototype complex systems with a characteristically slow dynamics. In the simplest case such melts can be envisaged as a melt of many interpenetrating coils of more or less flexible threads. The threads are a highly idealized model of long chain molecules of many repeat units. Those repeat units can be simple identical synthetic units (e.g. polyethylene $(CH_2)_N$) or rather complicated units, especially for polymers of biological origin. Their number in a single polymer can range from below 100 up to thousands and millions. The dynamics of such systems is strongly affected and in most cases dominated by the fact, that these chains cannot pass through but only along each other. For melts this leads to the extraordinary long relaxation times of the order $O(N^{3.4})$ and correspondingly huge viscosities and a viscoelastic plateau modulus regime, which can last for very long times [1]. Within the regime of the plateau modulus polymer melts and networks behave approximately the same. The onset of the plateau regime is characterized by a specific length scale and a corresponding chain length N_e, the entanglement chain length. In this contribution the dependency of N_e on different aspects of the chain chemistry as well as experimental and theoretical approaches to determine it will be shortly discussed.

REPTATION CONCEPT AND SIMULATION TEST

Since the first papers, which treated the elasticity of long chain polymer networks on the basis of a confinement to tube like spatial objects by S. F. Edwards in 1967 [2] and the subsequent extension of de Gennes to the reptation motion picture in 1971 [3] a detailed qualitative and quantitative test of this concept was and is in the focus of many theoretical and experimental studies. When the chains are short, namely $N < N_e$, the dynamics can be well understood by the Rouse model. There the chain motion is treated as the Brownian motion of a single random walk chain in a viscous background. All the complicated many chain interactions as well as the fact that chains cannot simply cross through each other are represented by a friction coefficient ζ. The longest relaxation time $\tau_R \propto N^2$, the chain diffusion constant $D = kT/\zeta N$ and the viscosity is $\eta \propto N$. While this ansatz works very well for melts of short chains, it brakes down when the chains exceed a chemistry dependent characteristic molecular weight M_e or number of beads N_e respectively. The relaxation slows down dramatically, $D \propto N^{-2}$ to $N^{-2.4}$, $\eta \propto N^{3.4}$ and the longest relaxation time $\tau_d \propto N^{3.4}$. The reason for this is, that the fact that the chains cannot simply pass through each other can no longer be neglected. Actually this effect becomes dominant and eventually leads to the confinement of the motion along the coarse grained backbone of the chains, which is also the backbone of the reptation tube.

While in the beginning the tube or entanglement concept itself was highly disputed there now is a wide spread consensus that it qualitatively captures the essential motion and relaxation mechanisms [4]. This was due to a number of very different studies [5][6], where computer simulations were the first to observe a reptation model typical slowing down in the bead motion of a polymer melt on the scale of the tube diameter d_T directly. Combined with neutron spin echo experiments, which came shortly thereafter growing evidence was collected. By now, also NMR experiments directly followed the mean square displacements of beads and reproduced rather accurately the typical power laws in time for the bead motion as expected by the reptation model and are in agreement to earlier simulation studies. All this together gives convincing evidence for the general concept. This however was only the starting point for many additional investigations, which aim at a solid quantitative understanding of material properties. It is clear that from a theoretical or simulation point of view it is impossible to make quantitative predictions for all available materials. It is also however a clear goal to understand the influence of structural, chemical properties on the melt and network properties. This is essential for an improved computationally aided materials development. While such a focus is more heading towards (technical) applications, there are still many other problems to be solved on the basis of the reptation concept. One reason for this is that the reptation theory from its very beginning is a single chain concept, where all the consequences of the complicated many chain interactions on the dynamics result in the tube confinement and a subsequent Rouse motion of the chain along the backbone of the tube. Data analysis of different experimental methods thus involves different approximations and thus it is not surprising that the results are coinciding qualitatively but not necessarily quantitatively. Therefore, research in the last few years centered around questions like

- Asymptotic power law for diffusion D(N) and viscosity $\eta(N)$
- How and why do different measurements of N_e give rather different results?
- What is an entanglement, or is this a well posed questions at all?

Within the reptation concept the longest relaxation time of the chains $\tau_\alpha \propto N^3$. This in turn leads to a viscosity of $\eta \propto N^3$ and a chain diffusion constant $D \propto N^{-2}$. For many years experiments on the viscosity of long chains polymer melts found $\eta \propto N^{3.4}$ and consequently $\tau_\alpha \propto N^{3.4}$. On the other hand the amplitude of the viscosity turned out to be somewhat smaller than expected from experiments on the plateau modulus. This indicated that one probably still was not really in the asymptotic regime. Meanwhile, dedicated experiments as well as further theoretical studies gave clear evidence for the asymptotic N^3 power law, however, only for chain lengths much beyond any technically relevant values [7]. In parallel significantly improved experiments on the chain diffusion constant seem to indicate a stronger slowing down of the motion, $D \propto N^{-x}, x \approx 2.4$, than expected from the asymptotic theory [8]. So far, it is not entirely clear whether this can be attributed to a similar crossover effect as for the viscosity or whether so called correlation hole effects come into play in addition. The correlation hole is a consequence of the self density of the individual chains $\rho_{self} = N/\langle R^2(N)\rangle^{3/2}$, which is (relatively) the deeper shorter tube chain is or the shorter the piece of the chain is, one is looking at [3]. One consequence is that even in the Rouse regime, where the mean square displacement of the center of mass of a chain always

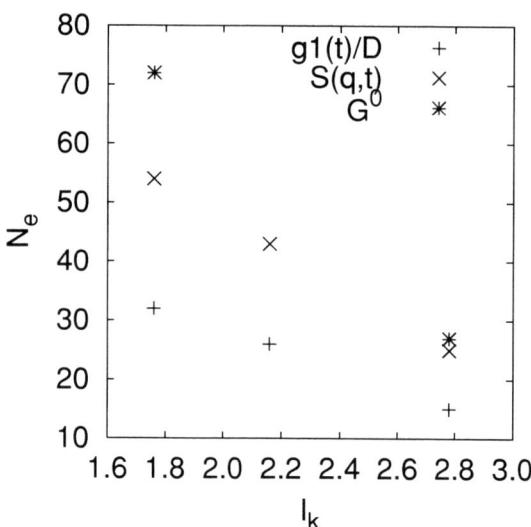

FIGURE 1. Values of the entanglement number of beads N_e for melts of bead spring polymer chains with variable intrinsic stiffness resulting the Kuhn lengths indicated. All melts are at a number density of the beads of $\varsigma = 0.85$. Chain lengths vary between $N = 350$ and $N = 700$.

should be diffusive, $g_3(t) = \langle (\vec{r}_{CM}(t) - \vec{r}_{CM}(0)^2) \rangle = 6Dt$, $\vec{r}_{CM}$ being the center of mass of a chain and D the chain diffusion constant, one typically observes $g_3 \propto t^{0.8}$ for $g_3(t) < \langle R_G^2 \rangle$, the radius of gyration. To our knowledge there is no systematic study of this effect on the overall relaxation time of the chains. It should also be noted that a conformational relaxation, as e. g. defined by the decay of the end to end vector auto correlation function $\langle \vec{R}(0) \vec{R}(t) \rangle / \langle R^2 \rangle$, can occur in principle without any global diffusion of the chain. First studies to investigate such questions are underway. However the general applicability of the reptation picture is not questioned anymore [4]. The second group of problems, namely that different experimental techniques might lead to rather different values for N_e, the entanglement molecular weight, is of concern in two ways. In principle this is not of a problem if one measurement would be sufficient to properly predict the outcome of the other experiments. Then it would be clear that the same physical quantity is measured. Unfortunately, there is no established relation between the value of N_e coming out of the diffusion constant $D(N)$, the intermediate single chain scattering function $S(q,t)$ and the plateau modulus G_N^0. A typical example is given in Fig. 1, where for simple bead spring chain melts of different intrinsic stiffness N_e values from three different methods are given.

The data seem to show no indication that the ratio between the different values remains the same. This not only holds for simulations of highly idealized polymers,

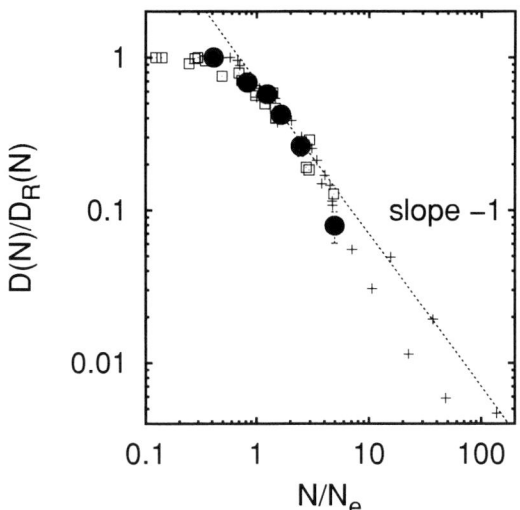

FIGURE 2. Normalized diffusion constants vs N/N_e for experiments on PE; Monte Carlo simulations of the bond fluctuation model and molecular dynamics simulations of simple bead spring chains as well as BPA-PC [10].

but also for actual chemical systems. For instance, polycarbonate (BPA-PC) is known to have an extraordinary small entanglement molecular weight of only about $M_e \approx 5...7 M_{Mon}$, depending on system, temperature, etc. Of course, the monomers are rather extended objects, but even keeping this in mid, $N_e \approx 6$ is extremely small compared to e. g. polyethylene ($N_e \approx 100 - 120$) or polystyrene ($N_e \approx 160$). This value experimentally is determined from the plateau modulus via the relation $G^\circ = \frac{4}{5} \frac{\rho kT}{N_e}$. Recently, Leon et al. [10] in a simulation of a coarse grained model of BPA-PC at $T = 570K$, the typical process temperature, also found this value from the modulus in simulations. In contrast to this a very careful analysis of the mean square displacements, where a time-temperature superposition had to be applied to adjust for the chain length dependence of the bead friction constant, revealed from this a value of $N_e \approx 12$. Though this last value is somewhat preliminary, the error bars are clearly smaller than the observed differences in N_e determined from the modulus and the diffusion constants. To get N_e out of the diffusion constant, N_e had to be fitted in a way that $D(N)/D_{Rouse}(N)$ with D_{Rouse} the hypothetical Rouse diffusion constant of chains of length N, plotted vs. N/N_e follows the same universal curve that includes a number of different simulations as well as experiment, cf. Figure 2.

Currently, a complete study on BPA-PC, where all typical experimentally employed measures of N_e are evaluated is underway.

The above comparison shows another complication in the discussion of what the entanglement molecular weight is and how it should be determined. For the simple bead spring chains, where the only variation came from an increasing intrinsic stiffness, we found that N_e determined from the modulus always was larger than the one from the scattering function and the one obtained directly by the bead diffusion. This order for PC, which can be (very roughly) viewed as a chain of banana like beads joint at pivot junctions (carbonate group), is reversed. Thus measuring N_e from one quantity generally cannot be used to infer the result of a different experiment. This is why simulations are very important in this context, as they are currently the only case, where "all measurements" can be made on exactly the same system and under well defined conditions. Currently we also employed a topological analysis, which determines the primitive path of the chains [11]. This provides a basis for a link between chain conformation and packing and the modulus. As it turns out, the construction of the primitive path simultaneously for all chains in the system, leads to a value of N_e, which quantitatively coincides with the results from the elastic modulus. In addition they fit excellently to the empirical finding that the modulus is proportional to p^{-3}, p being the so called packing length [10]. For BPA-PC this analysis yields $N_e \approx 6$ and agrees to the value obtained by a direct numerical "elongation experiment".

A condition for all of this of course are well equilibrated conformations. While the general quality of both experimental and simulation data is steadily increasing, the situation remains unsatisfactory. To perform simulations on polymer melts, one is looking for ways to generate independent equilibrated starting conformations, without running the systems for the full physical relaxation time by a standard MD or Monte Carlo procedure. While it is beyond the scope of this short overview to discuss this in detail, we just want to mention that there are several recent developments, which address this problem, both for all atom or unified atom models as well as for more idealized coarse grained models [9]. Basis for this is however a reference system of relatively short chains, which has to be equilibrated the very conservative way. Once this is at hand more advanced approaches can be employed.

Acknowledgement One of us (KK) acknowledges a longstanding and very fruitful collaboration of the entanglement problem with G. S. Grest and R. Everaers. SL acknowledges the support by the Bundesministerium für Bildung und Forschung, BMBF, grant No. 03N6015.

REFERENCES

1. M. Doi and S. F. Edwards, *Theory of Polymer Dynamics*, Clarendon, Oxford (1986)
2. S. F. Edwards, *Proc. Phys. Soc.*, **91**, 9 (1967)
3. P. G. de Gennes, *J. Chem. Phys*, **55**, 572 (1971)
4. T. C. B McLeish, *Adv. in Phys.*, **5**, 1389 (2002)
5. K. Kremer, G. S. Grest and I. Carmesin, *Phys. Rev. Lett.*, **61**, 566 (1989)
6. K. Kremer and G. S. Grest, *J. Chem. Phys.*, **92**, 5057 (1990)
7. R. Colby, L. J. Fetters and W. W. Graessley *Macromolecules*, **20**, 2226 (1987)
8. T. P. Lodge, *Phys. Rev. Lett.*, **83**, 3218 (1999)
9. R. Auhl, R. Everaers, G. S. Grest, K. Kremer and S. J. Plimpton, *J. Chem. Phys.*, **(2003)**,
10. S. Leon, K. Kremer, *to be published* **(2004)**
11. R. Everaers et al., *preprint* **(2003)**

Slow Dynamics and Mechanical Properties of Polymer Glasses : Ageing Properties

Jacques Rault

Physique des Solides, Université Paris-Sud, Orsay, France

Abstract. The kinetics of ageing, in particular the memory effects, (volume, enthalpy, creep and yield stress) are explained in term of the Vogel law (α motion) and Arrhenius law (β motion).

INTRODUCTION

Since the pioneer work of Kovacs[1] and Struik[2] a great amount of works have appeared on the kinetic of aging and rejuvenation (deaging) of glass forming materials. References can be found in the review papers[3-9]. When a glass is annealed at a temperature below Tg (measured at 10^{-2} Hz) the volume V and enthalpy H relaxations follow the well known sigmoidal curves with the aging time t_a, as shown in fig.1. Two characteristic times τ_i and τ_f can be defined. The incubation time τ_i has the same activation energy E_β as the individual β motion and τ_f follows the Vogel law of the cooperative α motion[10,11]. This is illustrated for PS in fig.1. In the aging time domain $\tau_i < t_a < \tau_f$ the relaxation laws are :

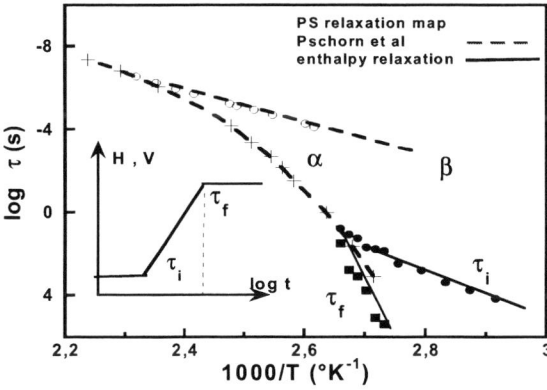

FIGURE 1. Relaxation map of PS. Below Tg, the relaxation times determined by enthalpy relaxation and dilatometry measurements verify the Vogel law.

$$\Delta V = k_d \log(t_a/\tau_i) \quad ; \quad k_d = (C_2/C_1)\Delta\alpha \quad (1)$$

$$\Delta H = k_H \log(t_a/\tau_i) \quad ; \quad k_H = (C_2/C_1)\Delta C_p \quad (2)$$

$\Delta\alpha$ and ΔC_p are the expansion coefficient and capacity jumps at Tg. C_1 and C_2 are the WLF constants, τ_i is the induction time (dependant on T). These laws[11] which are experimentally verified are easily found if one states that the aging time t_a corresponds to the α relaxation time (given by the Vogel law); in this law the experimental temperature T is replaced by the fictive temperature T', extrapolated temperature of the liquid which would have the same value $V(t_a)$ and $H(t_a)$. The aim of this paper is to propose simple laws concerning the mechanical properties of aged polymer glass. One distinguishes simple aging and complex aging; (memory effects).

SIMPLE AGEING

Yield Stress σ_Y.

Aging at one temperature T > Tg produces an increase of the Tg, a densification of the material, an increase of the relaxation enthalpy H and of the yield stress σ_Y. These effects are illustrated in fig.2. In polymers the important points are the following[12]:

a) The difference ΔH, $\Delta\sigma$ and ΔTg between the aged and virgin samples follow the same kinetics, $\sim \log t_a$.

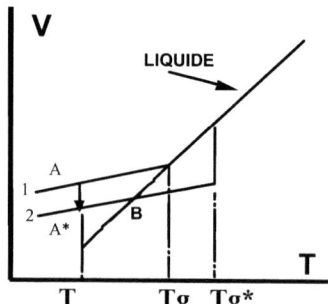 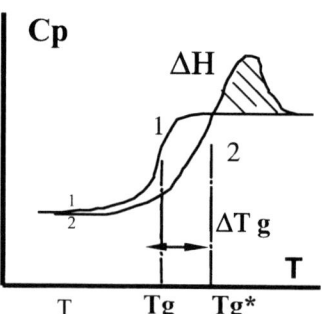 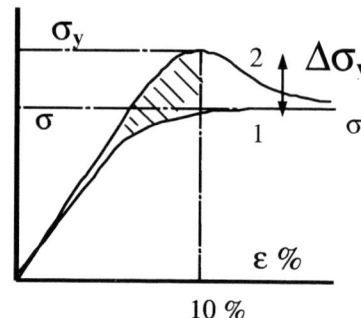

FIGURE 2 Effect of aging on the specific volume (a) heat capacity (b) stress-strain curve Aging at T induces densification at point A* the equivalent temperature is T_B and the glass temperatures of virgin samples (1) and aged samples (2) are Tg and Tg*, measured at the same heating rate.

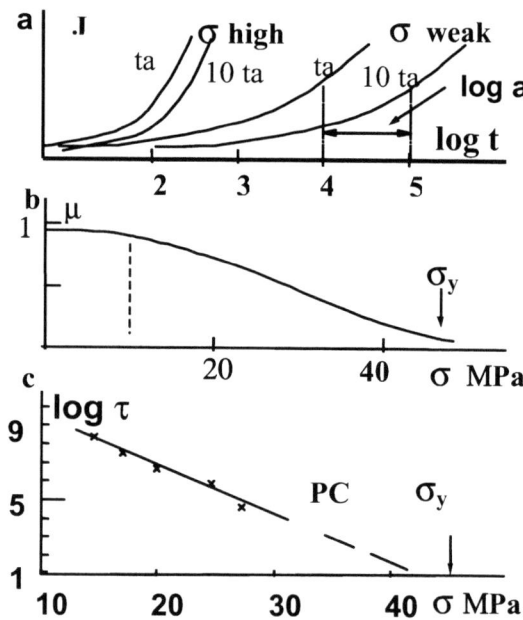

b) After the yield stress, the new material has the same properties H and σ_y as the virgin one, the sample has been rejuvenated.

c) In all the polymers the yield strain ε_Y is of the order of 0.1 and the cohesive energy E_c is proportional to the glass temperature (measured at constant heating rate). The mechanical energy $\varepsilon_Y \Delta\sigma_Y$ and the enthalpy relaxation are equal, this lead to the relation

$$\Delta H = 0.1\ \Delta\sigma_y = E_c\ \Delta Tg/Tg \qquad (3)$$

which is independent of the aging temperature and aging time. The yield strain ε_Y in polymer is about 0.1 and Ec is the cohesive energy. The data of Cook et al[13] on DGEBA epoxy verify this linear variation between ΔH and $\Delta\sigma_y$.

d) The yield stress (at constant strain rate and constant aging time) varies linearly with the temperature $\sigma_Y = \sigma_0\ (1 - T/Tg)$. The slope σ_0/Tg does not vary conspicuously with the nature of the polymer (0.5 MPa K^{-1}) then one assumes σ_0 independent of the aging. The Tg* of aged sample obeys the relation :

$$Tg^* = Tg\ (1 + k_T \log t_a/t_0)) \qquad (4)$$

for short time t_0 the glass temperature is Tg, then σ_Y* value of aged sample at first order is :

$$\sigma_Y^* = \sigma_0\ (1 - T/Tg^*) \approx \sigma_Y + \sigma_0\ (T/Tg)k_T \log t_a/t_0 \quad (5)$$

This relation has been verified experimentally by several authors. Typically the yield stress and the Tg measured by DSC, increase respectively of 3.5 MPa and 3 K per decade of time. From the results of Ricco et al[14] and Chow[15] one verifies the relation between modulus and σ_Y : $\Delta E/E = \Delta\sigma_Y / \sigma_Y \sim \log t_a$

FIGURE 3 a) Compliance J of a glass under a small and a high stress aged at two different times ta and 10 ta. b) aging factor µ as function of the stress c) Relaxation τ deduced from the creep curves and rel.6 .Rejuvenation occurs gradually, for the yield stress σ_y there is no aging effect and the relaxation time is of the order of the time of measurement. (a and b after Struik[2]).

Creep Compliance J.

In fig.3 one sketches the variation of the creep compliance of amorphous solids as reported in the Struik book[2]. The creep compliance can be put on the following form for t< ta :

$$J(t) = J_V^0\ \exp\left(\frac{t}{\tau}\right)^n \approx J_V^0\left(\frac{t}{\tau}\right)^n \qquad (6)$$

which is the Andrade form at short time. Or on the proposed form

$$J(t) = J_C^o \left[1 - \exp-\left(\frac{t}{\tau^*}\right)^n\right] + J_V^o \quad (7)$$

where J_V^o and J_C^o are the compliance of the glass and rubber states (unrelaxed and relaxed states). Rel.7 has more physical grounds, it indicates that the solid (glass below Tg) and the liquid (above Tg) behave in a similar manner. For the two relations, one verifies that the Kohlrausch exponent n is 1 and 0 for respectively the bifurcation T* and Vogel T_0 temperatures.

The only difference between solid (T<Tg) and liquid (T>Tg) is the presence of observable aging effects. The two forms for t << ta fit with the same accuracy the experimental creep curves for T<Tg. Obviously the relaxation times τ and τ^* are different $(\tau^*/\tau)^n = J_C^0/J_V^0 \sim 10^3$, but vary in the same way with the stress and the aging time ta. In fig.3c one gives the variation of τ (rel.6) with the applied stress deduced from the data of Read[12,16] on PC. One draws these important conclusions :

a) The relaxation time deduced from the creep curves varies linearly with the stress. For the yield stress the relaxation time is of the order of the time of measurement 10 s.

b) The Struik factor μ can be written:

$$\mu = d \log a / d \log ta = 1 - \sigma / \sigma_Y \quad (8)$$

log a is the translation factor for superposing the two creep curves of samples aged at different times. Here one does not discuss the small translation parallel to the ordinate which is in fact necessary. This indicates that rejuvenation occurs gradually and disappears completely at the yield stress.

These rejuvenation (de-aging) effects can be observed on fig.4 deduced from fig.7 of McKenna[5]. Creep of epoxy glass aged at Tg- 9°C have been measured for 4 different applied stress. The slope d log a / d log ta decreases linearly with σ and extrapolates to zero at the yield stress about 40 MPa.

Equilibrium Time

The creep results reported in fig.4 show two important properties:

a) The aging time ta* to reach the equilibrium (no more aging effect) is independent on the stress

b) The value ta* is about the relaxation time of the α motions given by the Vogel law (fig.1).

This effect of saturation at ta> ta* on yield stress has been rarely studied. The approach of the equilibrium of the relaxation enthalpy of polymer glasses during aging has been extensively studied. As reported above (rel.3) Tg, ΔH, σ_Y and ΔV follows the same kinetics. In fig.5 one compare the $\Delta H(ta)$ curves of PS isotropic and drawn samples ($\lambda_X = \lambda_Y = 2$) aged at Tg - 5°C

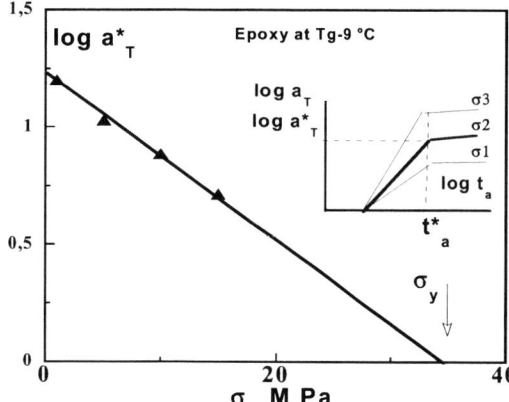

FIGURE 4 Creep translation factor log a versus the applies stress during creep at room temperature. For long aging time ta > ta* at Tg-9 °C equilibrium value log ta* are obtained. When σ exceeds the yield stress σ_Y no aging effect is observed. (from McKenna data[5])

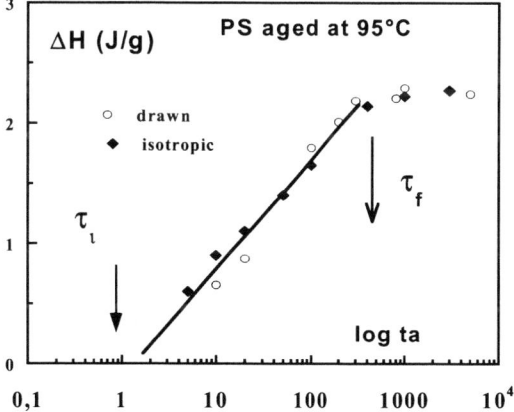

FIGURE 5 Relaxation enthalpy of drawn and isotropic PS annealed at Tg-5 °C , τ_f is the Vogel time, τ_i the incubation time

During aging one has noted that dimensional relaxation of the drawn sample occurs. For all the temperatures one finds that the kinetics of drawn and isotropic materials are exactly the same, ie same equilibrium time which follows the Vogel law. This effect and that reported by McKenna (fig.4) are puzzling.

Finaly equilibrium times has also been observed by yielding stress measurements. Chow[15] reported important variations of t_a^* in PS with the supercooling Tg-T in agreement with the WLF (Vogel) law.

COMPLEX AGEING : MEMORY EFFECT

Here one analyses the properties of glasses aged at temperature T_1 during the time t_1 and then aged at a higher temperature T_2 during the time ta. The other situation $T_1 > T_2$ does not lead to memory effect. These memory effects in PS has been studied by various authors[1,2,6,7,11]. In fig.6a one gives as an example[11] the variation of the relaxation enthalpy ΔH of PS with the annealing time ta at the final temperature (T_2=Tg-5°C),

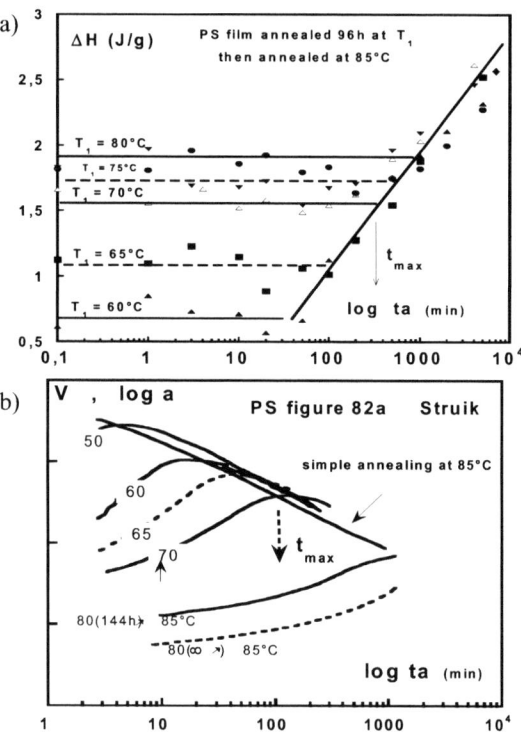

FIGURE 6 Enthalpy relaxation ΔH (a) and volume V and creep factor log a (b) as function of the annealing time ta at 85°C. The PS samples has previously been annealed at different temperature $T_1<T_2$ for different times. In (b) the samples have been annealed first during 96 hours at different temperature T_1.

In fig.6b one recall the Struik results of volume relaxation and creep factor, log a, using the same thermal treatment. From these results and those reported in ref.11 one concludes :

a) the extrapolated time ta at the final temperature is obviously the Vogel time τ_f (fig.1) , it does not depend on the previous thermal treatment.

b) Volume and creep factor log a pass through a maximum at the annealing time t_{max} . This phenomena called memory effect has been observed on PS by Adachi et al.[6] and Hozumi et al[7] . One must note that Adachi et al measured the rate enthalpy change in PS during the thermal treatments , they found a maximum in agreement with the volume measurement . Here one measure the difference in enthalpy between the two materials, aged and non aged.

c) The relaxation enthalpy at short time is constant and then increases when the enthalpy ΔH intersect the linear curve $\Delta H(\log ta)$ of the virgin material annealed at the final temperature T_2 (simple ageing). One has found that this time called equivalent time t_{equ}. or induction time[11] is equal to the time t_{max} corresponding to the maximum of the log a(ta) and V(ta) curves.

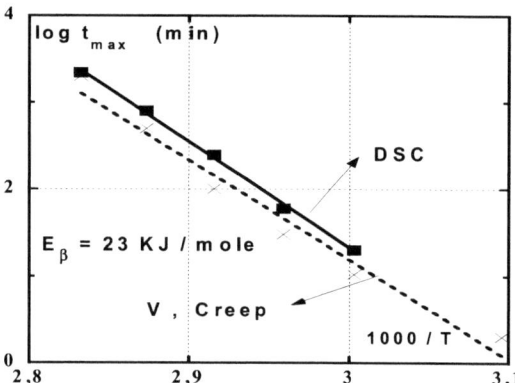

FIGURE 7 Arrhenius plot of the time t_{max} of PS measured by DSC (from ref.11), dilatometry and creep (from ref.2). Complex ageing corresponding to the experiments of fig.6 $T_2 = 85°C$, T_1 is varying.

d) The equivalent time t_{max} is not dependant on the initial temperature T_1, but only on the starting enthalpy ΔH_{start} acquired at T_1. Rel.2 gives the relation between t_{max} and the incubation time τ_i (function of T_2) is deduced from rel.2 :

$$\Delta H_{start} = 0.5 \log (t_{max} / \tau_i) \;\; ; \; (Jg^{-1}) \qquad (9)$$

e) Adachi and Kotaka[6] made similar experiments on PS, but T_2 and T_1 were constant , only the aging time ta at T_1 was varying. They observed a linear law $\log t_{max} \sim \log t_a$, this a direct consequence of rel.2,9.

f) The time t_{max} deduced from enthalpy relaxation verifies the Arrhenius law as it is found for the incubation time τ_i. As observed in fig. 6a, t_{max} can be considered as a new incubation time at T_2 of the glass previously aged at T_1 As shown in fig.7 the various times t_{max} measured by volume creep and DSC are equal, and varying linearly with $1/T_1$, the activation energy corresponds to that of the β process (for PS, E_β ~10 to 20 KJ/mole for most polymers)

CONCLUSION

A careful analyze of the creep and dilatometry data of Struik and our data on relaxation enthalpy shows that the kinetics of aging are analogous in case of simple annealings. The concept of fictive temperature T' first introduced by Tool[17] explains the observed properties. After ageing the equivalent (fictive) temperature is related to the change of enthapy and volume relaxation by $\Delta H = \Delta C_p (T_g - T')$, $\Delta H = V_0 \Delta\alpha (T_g - T')$, then one postulates that the Kohlrausch exponent has the form

$$n = (T' - T_0) / (T^* - T^0) , \quad (10)$$

In the liquid state, $T > T_g$ (at equilibrium $T = T'$) this law is verified[10], the bifurcation temperature T^* where the α and β process merge (fig.1) is of the order of $T_g + 100$ °C. For the (supercooled) liquid below T_g, the relaxation time (equilibrium time) is the ageing time ta:

$$\log t_a / t_g = - C_1 (T' - T_g) / (T' - T_g + C_2) \quad (11)$$

the WLF constants are $C_1 = 16$ and $C_2 = 50$ °C, then one finds the Kohlrausch exponent at first order:

$$n = n_g - k_n \log (t_a / t_g) \; ; \; k_n \sim C_2/(C_1(T^* - T_0)) \quad (12)$$

for PS the value found $k_n = 0.022$ is near the experimental value (0.03) found by Plazek et al[8].

In conclusion the variation of n with the ageing is very weak, then one can assume at first order that the creep curves after ageing are superimposable by the translation log a as assumed by most of the authors. It is important to recall that the principle of T-t superposition is not verified in the liquid (T>Tg) if one wants to compare the experimental results at two very different temperatures. Near T* the absorption peak J'' has the Debye width 1.14 decades (n ~1), near Tg the width w is about 3 decades ($n_g = 1/w \sim 0.3$), there is obviously no possible superposition of the curves J''(log t) and J'(log t). Superposition is observed in the solid state T<Tg because the fictive temperatures of two samples aged differently are never very different, a few degrees below Tg.

In the model of author derived from the Ngai model[18] the relaxation time in the liquid state (equilibrium) is

$$\tau = \tau_0 (\tau_\beta / \tau_0)^{1/n} \quad (13)$$

which verifies the conditions $\tau = \tau_\beta$ for n=1 at T* and $\tau = \infty$ for n=0 at the Vogel temperature T_0. This relation with rel.10 leads to the well known Vogel (WLF) law without adjustable parameter. The above equation shows that τ and n are coupled parameters, as pointed out by Ngai. Taking into account the aging effect (T' ≠ T) rel.12 and 13 give at first order :

$$\log \frac{\tau}{\tau_g} \approx \left(\frac{k_n}{n_g^2} \log(\frac{\tau_\beta}{\tau_0}) \right) \log (\frac{t_a}{t_g}) + C \quad (14)$$

A Tg $\tau_g \sim t_g \sim 10$ s is the reference time. The Struik factor is then :

$$\mu = \frac{k_n}{n_g^2} \log \frac{\tau_\beta}{\tau_0} \approx 1 \quad (15)$$

Typicaly for PS, $\tau_\beta = 10^{-5}$ s at Tg and $\tau_0 = 10^{-10}$ s, the Kohlrausch coefficient at Tg is n ~ 0.4, and k_n ~ 0.022. μ is of the order of unity (0.7). It is important to remark that μ scale like log τ_β. Therefore if the stress during creep is high, the process is dependant on the activation volume V_a the new activation energy of the β process is reduced to $E_\beta - V_a \sigma = E_\beta(1 - V_a \sigma / E_\beta)$. The Struik factor μ(σ) is then

$$\mu(\sigma) = \mu(\sigma=0) (1 - \sigma / \sigma_0) \; ; \; \sigma_0 = E_\beta / V \quad (16)$$

which is the experimental rel.8

In conclusion during simple aging (annealing at one temperature) the different aging properties of the glass (V, H, σ_Y, n, μ) are ruled by the Vogel (WLF) law. The different laws concerning the mechanical properties do not involve any adjustable parameter. In the future it would be necessary to study the aging properties of drawn samples, some recent experiments have shown that the time for obtaining the dimensional relaxation is about the Vogel time τ_f, this important property should be verified.

In two step ageing (at two temperatures $T_1<T_2$) the kinetics of enthalpy relaxation present some differences with that of V and J (more exactly log a) kinetics. For any properties (H,V,log a) one defines the same equivalent time t_{max}. For long annealing at T_2 the kinetics is not different from the simple annealing experiment at T_2; the sample has lost its thermal history. For short annealing at T_2 various authors have shown that V and log a follow the same kinetics, here one has shown that t_{max} can be considered as a new incubation time. The incubation times τ_i and t_{max} are several orders of magnitude higher than the relaxation time of the β process but have the same activation energy. Comparison between simple and complex aging are sketched in fig.8.

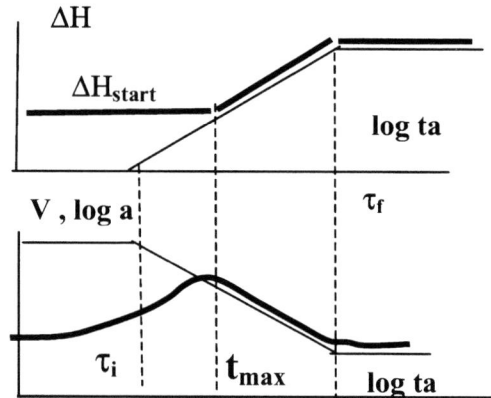

FIGURE 8 Ageing properties of glass at a temperature T_2: comparison between simple annealing at T_2 (thin lines) and two steps annealings at T_1 and T_2 (heavy lines). The new incubation time t_{max} in complex annageing depends only on the starting enthalpy not on T_1. If T_2 increases, the incubation time τ_i decreases, all the oblique curves are shifted towards short times and the differences ΔH and ΔV at the plateau decrease (see fig.7 of ref.11).

The main problem to solve in the future is the understanding of the relaxation process in the short aging time domains, ta < τ_i and ta < t_{max} for simple and complex aging. Also it would be interesting to compare the aging properties of fragile and strong glasses in the isotropic and oriented states.

REFERENCES

1 Kovacs, A. J., *J. Polym. Sci.* **30**, 131-147 (1958*). Fortschr. Hochpolym. Forsch. (Adv. Polym. Sci.)* **3**, 394-507 (1964)

2 Struik L.C., *Physical Aging in Amorphous Polymers and Others Materials*, Elsevier, Amsterdam, 1978

3 McKenna, G. B., Simon, S-L. Time Dependent Volume and Enthalpy Responses in Polymers in *Time Dependent and Nonlinear Effects in Polymers and Composites*, edited by R. A. Schapery and C. T. Sun, *ASTM Stand. Techn. Publ*, West Conshohocken, 2000 pp.18-46

4 Simon, S-L; Plazek, D. J., Sobieski, J. W., McGregor E.T. *J. Polym. Science-Part (Physics.)* **35 (6)**, 929-936 (1997)

5 McKenna, G. B. *J. Non-Cryst. Solids* **172-174**, 756 –764 (1994)

6 Adachi, K., Kotaka, T. *Polym. J.* **14**, 959-970 (1982)

7 Hozumi, S. Wakabayashi, T. Sugihara, K. *Polym. J.* **1 (6)**, 632-638 (1970)

8 Plazek, D. J., Ngai, K. I., Rendell, R. W. *Polym. Eng. Sci.* **24 (14)**, 1111 (1984)

9 Crissman, J. M., McKenna, G. B. *J. Polym. Sci. Phys.* B **28**, 1463-1473 (1990)

10 Rault J,. *Non Cryst. Solids* **271**, 177-217 (2000)

11 Rault, J., *J. Phys. Cond. Matter.* **15**, S1193-S1213 (2003)

12 Rault J., *Les Polymères Solides : Amorphes, Elastomères, Semi-cristallins,* Cepadues, Toulouse 2002

13. Cook, W.D. Mehrabi M.,. Edward G.H. *Polymer* **40**, 1209-1218 (1998)

14 Ricco,T. Smith, T. L., *Polymer* **26**, 1979-1984 (1985)

15 Chow,T. S., *Polymer* **34**, 541-545 (1993)

16 Read, B. E., *J. Non-Cryst. Solids* **131-133**, 408-419 (1991)

17 Tool, A. Q., *J. Am. Ceram. Soc.* **29**, 240-253 (1946)

18 Ngai K. L., *Comments Solid State Phys.* **9 (4)**, 127-140 (1979) and *J. Phys. Cond. Matter.* **15 (11)**, S1107-S1125 (2003)

Intermediate Length Scale Dynamics in Polymer Melts

D. Richter[a], M. Monkenbusch[a], A. Arbe[b], J. Colmenero[b]

[a] *Institut für Festkörperforschung, Forschungszentrum Jülich, D-52425 Jülich, Germany*
[b] *Departamento de Fisica de Materiales, Universidad del Pais Vasco, and Unidad de Fisica de Materiales (CSIC-UPV/EHU), Apartado 1072, E-20080 San Sebastian, Spain*

Abstract. We report on neutron spin echo investigations of the intermediate scale dynamics of polyisobutylene (PIB) studying both the self and the collective motions. The momentum transfer dependencies of the self correlation times are found to follow a $Q^{-2/\beta}$ law in agreement with the picture of Gaussian dynamics. In the full Q range of observation their temperature dependence is weaker than that of the rheological shift factor. Studying the single chain dynamic structure factor, deviations from the universal Rouse dynamics become evident at intermediate length scales. In solutions they can be directly related to intrachain viscosity effects arising from the intrachain rotational potential. Invoking a similar mechanism in the melt leads to a consistent description of the self and the single chain correlation functions suggesting that the weaker temperature dependence relates to configurational dynamics. The same is true for the stress relaxation times as seen in sound wave damping. The collective times show both temperature dependencies. At the structure factor peak they follow the temperature dependence of the viscosity but below the peak one finds the stress relaxation behaviour.

INTRODUCTION

On length scales, where the influence of local potentials have ceased and topological effects are not yet important, the dynamics of polymer chains in the melt are governed by chain connectivity and the entropy driven Rouse dynamics prevails [1]. This entropy driven dynamics displays a number of universal features such as the sublinear time dependence of the mean square displacement $<\Delta r(t)^2> \approx t^{0.5}$ or the scaling with a universal variable combining time and space coordinates [2]. On short scales on the other hand, to a large extend the specific molecular chemistry of a given monomer determines the local packing and thereby the dynamical features. The α-relaxation and the glass transition relate to these local structures. This is also true for the secondary relaxation which e.g. determines the ductility of a material. The materials properties depend on the universal and the local properties and it is one of the great challenges of polymer science to bridge the scales between the universal and the specific. Experiments at intermediate scales connecting both regimes are of great importance to learn more about this connection [3]. Quasielastic neutron scattering offers unique opportunities to access these intermediate scales from different point of views. Depending on the deuteration or hydrogenation of the polymers, it is possible to observe the single chain dynamics – this is realized, if a fraction of protonated chains is observed in a deuterated matrix - or the collective dynamics - it is seen if a fully deuterated material is studied - or finally the single particle motion – the so called self-motion – it is observed, if a protonated sample is investigated. Each of these different correlation functions reveal different aspects of the chain motions and altogether promise deeper insight into the chain dynamics in this crucial regime.

In this manuscript we will discuss a series a neutron spin echo experiments on polyisobutylene, where all three different dynamic correlation functions have been studied. We will present experimental results on the self-correlation function [3], revealing the self-motion of the chain protons, we will then display results on the single chain dynamics [4,5], where the chain relaxation of a single chain in the environment of the other chains in the melts is studied. Finally, data on the collective dynamics are presented, where in particular the regime of density fluctuations at low Q in front of the first maximum of the structure factor is emphasized [3].

EXPERIMENTAL

The collective dynamics as well as the structure factor were studied on fully deuterated polyisobutylene (d-PIB), while the self-motion was investigated on a protonated material (h-PIB). The h-PIB was purchased from the American Polymer Standard Cooperation (Mentor, OH). The synthesis of the d-PIB including the sources and the purification of the materials has been described elsewhere [4]. The h-PIB sample had a molecular weight of $M_w = 24.900$ with a polydispersity of $\frac{M_w}{M_n} = 1.23$. The molecular weight of the d-PIB sample amounted to $M_w = 72.000$ with a polydispersity of $\frac{M_w}{M_n} = 1.05$.

In the low Q regime the absolute intensity of the scattered neutrons was studied at the small angle scattering instrument D22 at the ILL. The static structure factor was investigated at the triple axis instrument SV4 at the FRJ-2 reactor in Jülich. The dynamic experiments were performed at the neutron spin echo instruments IN11C and IN15, both at the ILL in Grenoble.

THE SELF-CORRELATION FUNCTION

Fig. 1 displays a set of NSE spectra taken at a neutron wavelength $\lambda = 6$Å from the h-PIB material at $T = 390$K for a number of different Q values. For the smallest $Q = 0.48$Å^{-1} also data taken at $\lambda = 10$Å are shown. In going from larger to smaller Q the relaxation process is continuously slowed down indicating an important Q-dispersion. The data were evaluated in terms of a Kohlrausch-Williams-Watts (KWW-function) given by the general expression

$$S_{self}(Q,t) = \exp\left(-\frac{\langle u^2 \rangle Q^2}{3}\right) \exp\left(-\left(\frac{t}{\tau_{self}(Q)}\right)^\beta\right) \quad (1)$$

where the first exponential function is the Debye-Waller factor with $\langle u^2 \rangle$, the average mean square displacement of the proton motions faster than the relaxation processes within window of the spin echo machine ($\langle u^2 \rangle = 0.375$Å^2), the stretching exponent β was fixed to $\beta = 0.55$, a value that was obtained in earlier measurements for the relaxation at the structure factor maximum [6] and $\tau_{self}(Q)$ the Q dependent correlation time for the self-motion.

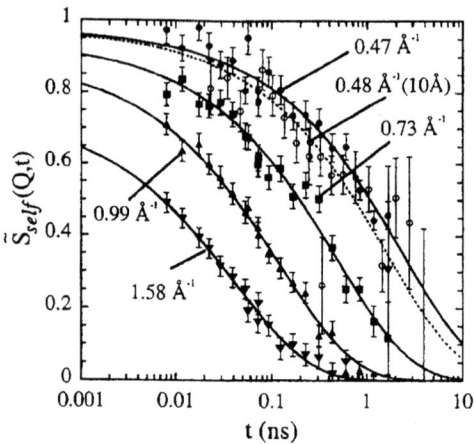

FIGURE 1. Time evolution of the self-correlation function measured on a fully protonated PIB sample by means of IN11C. The symbols (full: incoming wavelength $\lambda = 6$Å; empty: $\lambda = 10$Å) correspond to the different Q values indicated. Lines are the resulting KWW fit curves (Eq.[1]) (solid: $\lambda = 6$Å; dotted: $\lambda = 10$Å).

Fig. 2 displays the resulting characteristic times $\tau_{self}(Q)$ as the function of momentum transfer Q and temperature. As a guide to the eye the solid lines mark the $Q^{-2/\beta}$ power law expected for dynamics in Gaussian approximation [7,8]. In the Q-regime $Q \leq 1$Å^{-1} the Q-dependence of all data sets is well represented by such an asymptote. At larger Q's a tendency towards a crossover leading to a weaker power law for $\tau_{self}(Q)$ is apparent. The temperature dependence is well described by a common shift factor $a(T)_{self}$ with an apparent activation energy $E_{self} = 0.43$eV. Unexpectedly, the temperature dependence is weaker than that of the viscosity relaxation, which in our temperature regime exhibits the significantly higher apparent activation energy of $E_{sl} = 0.67$eV [6]. In the lower part of Fig. 2, the such shifted data are presented and we realize that the weaker activation energy holds for the full Q range.

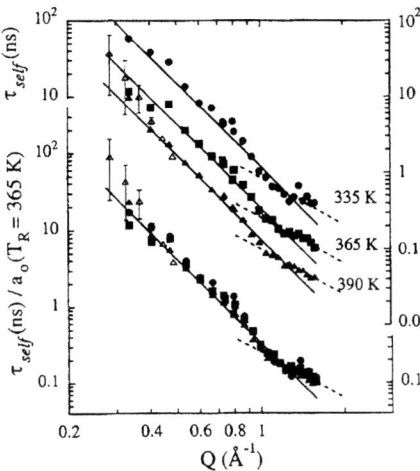

FIGURE 2. Upper part: Momentum transfer dependence of the characteristic time of the KWW functions describing the self-correlation function at 335K (circles), 365K (squares), and 390K (triangles). Lower part: Scaling representation: 335K and 390K data have been shifted to the reference temperature 365K applying a shift factor corresponding to an activation energy of 0.43eV. Full symbols correspond to results from measurements performed with incoming wavelength $\lambda = 6\text{Å}$ and empty symbols to $\lambda = 10\text{Å}$. Solid (dotted) lines through the points represent $Q^{-2/0.55}$ power laws.

In Gaussian approximation the self-correlation function assumes the form

$$S_{self}(Q,t) = \exp\left[-\frac{Q^2}{6}\langle r^2(t)\rangle\right] \quad (2)$$

where $\langle r^2(t)\rangle$ is the time dependent mean square displacement of the scattering centers – in our case protons. For an homogeneous ensemble, where during the observation time each proton exhibits the same average fate the combination of Eq.[1] and [2] leads to the power law dependence of the characteristic time $\tau_{self} = a(T)_{self} \cdot Q^{-2/\beta}$ seen in Fig. 2.

From Equ's. [1] and [2] also the time dependent proton mean square displacement may be directly obtained.

$$\langle r^2(t)\rangle = \frac{6}{Q^2}\left(\frac{t}{a_{self}(T)Q^{-2/\beta}}\right)^{\beta} = 6\left(\frac{t}{a_{self}(T)}\right)^{\beta} \quad (3)$$

For example at 390K within an observation time of 5.5ns the proton mean square displacement amounts to 43.5Å^2.

The relation between shape and dispersion of the self-correlation function $\tau(Q) \sim Q^{-2/\beta}$ qualifies it as Gaussian and suggests an interpretation of the observed motion as sublinear diffusion with $\langle r(t)^2\rangle \approx t^{\beta}$. The Gaussianity of the self-correlation function in the Q regime $Q \leq 1\text{Å}^{-1}$ by now has been shown for a nine different polymer melts and seems to be a general phenomenon [9].

SINGLE CHAIN DYNAMICS

Fig. 3b displays the single chain dynamic structure factor from a PIB melt at $T = 470$K. The solid lines give the results of a fit with the Rouse model describing the entropy driven universal chain dynamics [4]. From the figure it is obvious, that for small Q's $Q \leq 0.15\text{Å}^{-1}$ this model is well adapted to describe the observed relaxation processes. In going to larger Q's systematic deviations towards a slower relaxation are evident. Obviously, in the Q-regime beyond 0.15Å^{-1} local chain properties begin to play a role rendering the Rouse model inapplicable. The question arises now, what are the leading effects, which cause these deviations from the universal Rouse dynamics. Theoretical ideas in two directions have been brought forward. The first favours chain rigidity effects [10], the second dissipation by intrachain potentials related to jumps across rotational barriers [11].

In order to address this problem, we have compared the dynamics of polyisobutylene with that of polydimethylsiloxane. Both polymers exhibit a similar local chain stiffness which is described by the characteristic ratio C_{∞} which for PIB comes out to 6.7 and for PDMS to 6.2, but very different rotational potentials $(E_{rot}(\text{PDMS}) \approx 0.1\text{kcal/mol}, E_{rot}(\text{PIB}) \approx 3\text{kcal/mol})$. The chain length were taken such that both chains had about the same radius of gyration ($R_g = 19\text{Å}$ (PIB); and $R_g = 21\text{Å}$ (PDMS)) in order to have similar diffusional properties.

Fig. 3a presents the outcome of an experiment on the PDMS melt at $T = 373$K. The solid lines again display a fit at the Rouse model. It is obvious the dynamics of PDMS is well described by the Rouse model over the full accessible Q regime. From this observation an immediate conclusion arises. Since both chains have the same local rigidity, the reason for the deviations of the PIB dynamics from the Rouse model cannot relate to the chain stiffness but rather should connect to the very different rotational potentials of both polymers.

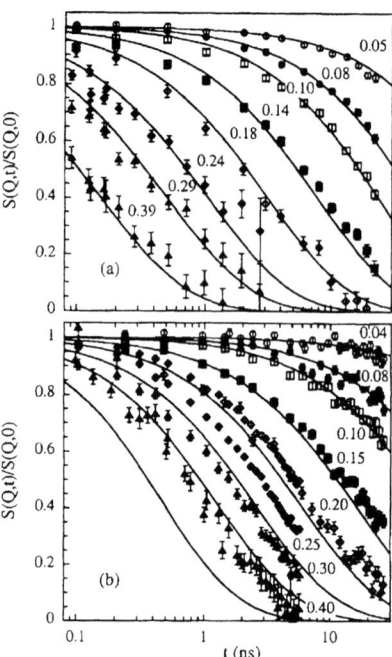

FIGURE 3. Chain dynamic structure factor of (a) PDMS at 373K and (b) PIB at 417K measured in the melt. Each symbol corresponds to the same or very close values of Q for both polymers, which are indicated in the figures. Solid lines show the Rouse prediction.

In order to describe the effect of such a rotational transition, Allegra and Ganazzoli [11] have developed a model assuming local dissipative relaxation processes, that limit the entropy relaxation in polymer melts towards shorter length scales. Such a process could be e.g. composed by combined jumped processes over rotational barriers and are behind an internal viscosity force arising from local departures from configurational equilibrium. In the model the dissipative process is described by a Debye process with a single relaxation time τ_0. This local mode couples to the Rouse modes, hybridizes with them and distorts the Rouse spectrum. This distorted Rouse spectrum then causes deviations from the simple Rouse model at larger momentum transfers Q.

In the polymer melt single rotational jump processes are impossible and a clear connection between the characteristic time τ_0 and the rotational barrier is impossible. Therefore, we performed experiments in PDMS and PIB solutions, in order to investigate the effect of the rotational transition [6]. Fig. 4 displays experimental results on PDMS and PIB in toluene. At low Q for both polymers the same dynamic structure factor is measured – since both coils have the same size, their translational diffusion coefficients are equal. The data taken over a temperature range $251 < T < 378K$ were fitted with the Allegra internal viscosity model revealing the characteristic time τ_0 as a function of temperature. A very good fit is achieved for all temperatures. The results for τ_0 are displayed in

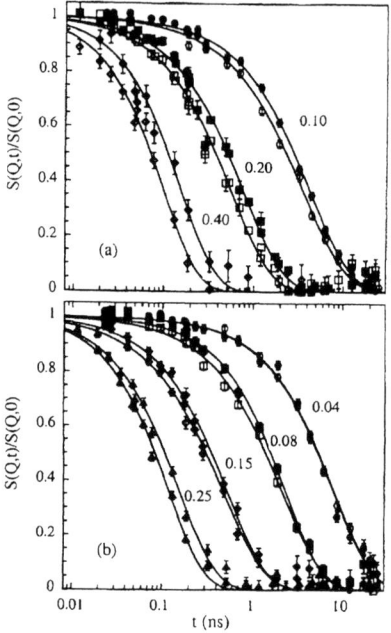

FIGURE 4. Chain dynamic structure factor of PDMS (empty symbols) and PIB (full symbols) in toluene solution at 300K (a) and 378K (b). The corresponding Q values are indicated. Lines through the points are guides to eye.

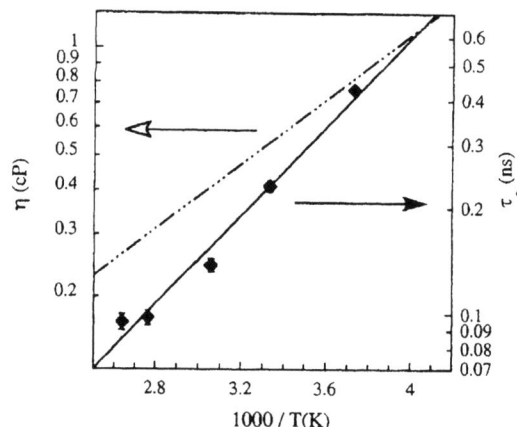

FIGURE 5. T dependence of the solvent viscosity (dashed line) and the characteristic time τ_0 deduced for the conformational transitions in PIB (♦). The solid line through the points corresponds to the fit to an Arrhenius law.

Fig. 5. $\tau_0(T)$ follows in activated behaviour with an activation energy $E = 3.1 \pm 0.3$ Kcal/mol. The pre-exponential factor of about 1ps lies well in the microscopic range. The found activation energy in very good agreement with the expected value for the

C-C rotational barrier in this polymer proving the general correctness of the Allegra approach.

The experimental spectra on the PIB melts are also well described by this model. However, the activation energy of the characteristic time now results in $E_0 = 0.43 \pm 0.1$ eV [4].

This value is very close to the activation energy observed for the self-motion and about three times higher than the rotational barrier. Dwelling on this agreement, we now may calculate the prediction of the Allegra model for the self-time using the results from the single chain dynamics measurements.

Fig. 6 compares the extracted average relaxation times on the basis of the Allegra model with those from the incoherent measurements at 390K. In the Q range of applicability nearly quantitative agreement is achieved.

Thus, we have established a relationship between the observed self-correlation function and the intrachain viscosity effect present in the single chain structure factor. With Eq.[3] we may also evaluate the mean square proton displacement related to this, the Rouse relaxation limiting process in PIB. Inserting τ_0 in Eq.[3] we obtain $\langle r^2(\tau_0)\rangle = 27.7 \text{Å}^2$ 84% of the interchain distance of $d_{chain} = 6.3$ Å.

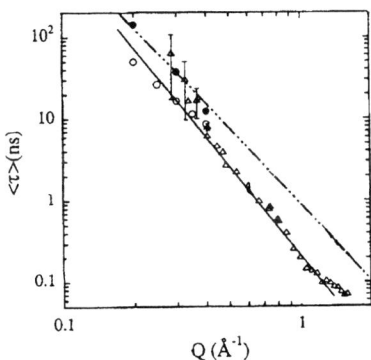

FIGURE 6. Momentum transfer dependence of the average times at 390K corresponding to: the self-correlation function ($\triangle$) and the single chain structure factor ($\bullet$). The deduced values for the self-correlation obtained from the description of the single chain dynamic structure factor in terms of the model proposed by Allegra and Ganazzoli are also shown ($\circ$). The solid line represents at $Q^{-2/0.55}$ power law, and the dashed-dotted the extrapolation of the single chain structure factor times to higher Q-values.

Summarizing, the measurements on the single chain dynamic structure factor have shown, that the Rouse dynamics is limited by intrachain friction, where the torsional potentials dissipate the energy. In polymer solution there is a direct relation to the intrachain potential, while in melts the dissipation process appears to happen with an higher effective barrier indicating correlated configurational jumps. The observed activation energy agrees with that of the self-motion. Applying the Allegra model and calculating the self-correlation function leads to a consistent description of both, the self and single chain dynamics. Thus, we may conclude that apparently the configurational dynamics dominates the self-correlation function in the Q and time range of observation.

COLLECTIVE DYNAMICS

These experiments are performed on fully deuterated PIB-melts revealing the dynamics associated with the short range order around the structure factor maximum and the density fluctuations at lower Q [3]. Fig. 7 displays the static structure factor as obtained from a measurement with the thermal triple axis spectrometer SV4 at the Jülich reactor FRJ-2. The crosses display the results of a polarisation analysis performed at the spin echo instrument IN11C at the ILL. This measurement allowed to determine the level of incoherent scattering at the triple axis instrument (~ 7900 counts). With the intensity scaling in place the coherent and incoherent NSE data in Fig. 7 are placed "up to scale". Finally, the $S(Q)$ scale is adjusted such as to place the "1" as the line around which the oscillations of the second peak take place. $S(Q)$ is characterized by a strong first peak at $Q_{max} = 1 \text{Å}^{-1}$ relating to an average interchain distance of $d_{chain} = 6.3$ Å. Other than the first peak the second broad peak is little affected by temperature and must relate to intrachain correlations.

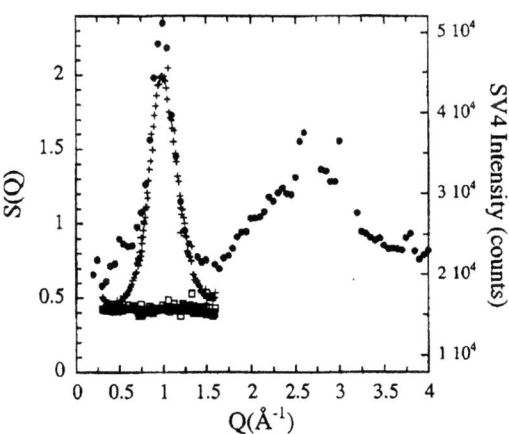

FIGURE 7. Differential scattering cross section measured for PIB by means of the triple axis spectrometer SV4 ($\bullet$). The corresponding scale (right) is chosen such that the incoherent contribution coincides with the origin for the static structure factor scale (left). NSE spectrometer IN11C

results on the static structure factor are shown for comparison (+) as well as the incoherent contribution measured also by IN11C (□).

We now concentrate on the low Q-regime, where primary scattering due to density fluctuations takes place. The sequence of a low intensity plateau followed by a strong peak makes multiple scattering (MS) a serious problem in the low Q-regime of glasses and liquids. This type of scattering is avoided if the incoming neutron wavevector $k_i = 2\pi/\lambda$ is smaller than $Q_{max}/2$. On the other hand it is strongest, if the amorphous halo is seen under 90° scattering angle. Fig. 8 displays a series of small angle scattering experiments performed at the small angle neutron scattering (SANS) instrument D22. By means of polarisation analysis studies at the NSE instrument the data were corrected for the incoherent contribution.

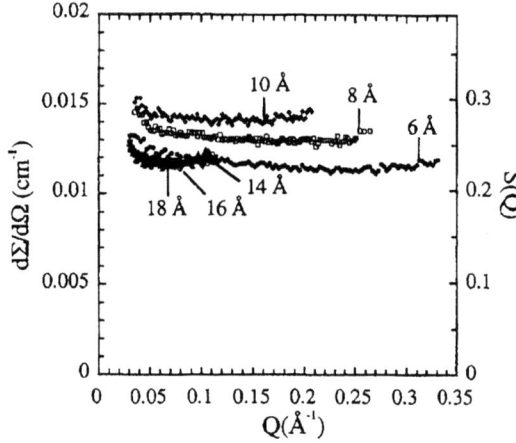

FIGURE 8. Differential scattering cross section measured for PIB by means of the SANS instrument D22. The different symbols correspond to the different incoming wavelengths indicated. From the scale on the right the corresponding values of $S(Q)$ can be read.

The observed wavelength dependent SANS intensities are highest at $\lambda = 10$Å, where $S(Q_{max})$ appears under 90°. Going away from this condition reduces the MS contributions. For wavelengths above 12Å $k_i < Q_{max}/2$ is fulfilled and the MS-contribution should be minimal. There an absolute cross section $\frac{d\Sigma}{d\Omega} = 0.0115 cm^{-1}$ is found. The data also show that at $\lambda = 6$Å, where most of the dynamic experiments were performed, MS was not important. By multiplying with the average cross section per average atom volume the absolute cross section may be converted into $S(Q \to 0) = 0.22 \pm 0.02$.

Dynamic results on the collective dynamics were obtained in a Q regime 0.2Å$^{-1} \leq Q \leq 1.5$Å^{-1} covering well the region of the first structure factor peak and that of the density fluctuations at lower Q. Using the results on the self-correlation function the obtained spectra were corrected for the incoherent contribution which amounted to about 25% at low Q. A fit with a KWW function according to Eq.[1], where the amplitude was taken as a separate fit parameter $A_{pair}(Q)$, led to the Q-dependent relaxation times depicted in Fig. 9.

For all three temperatures we observe the same general behaviour with the collective relaxation times displaying a low Q-plateau like behaviour with some tendency to increase at the lowest Q. The low Q behaviour is followed by an increase of the relaxation times to a maximum value that shifts from around $Q = 0.8$Å^{-1} at 390K to $Q = 1$Å^{-1} at $T = 335$K. Towards higher Q a strong decrease of the relaxation times is observed. The relative peak height of $\tau_{pair}(Q)$ around the structure factor maximum increases with decreasing temperature indicating different temperature dependencies in different Q-regimes.

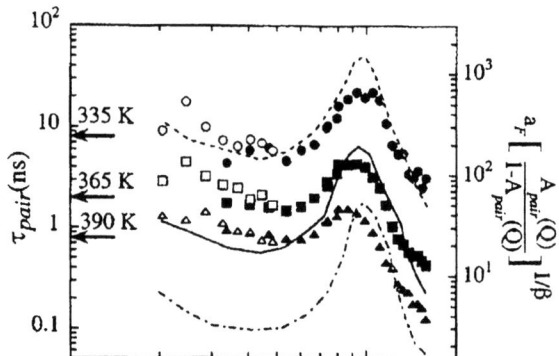

FIGURE 9. Momentum transfer dependence of the characteristic time of the KWW functions describing the dynamic structure factor at 335K (circles), 365K (squares) and 390K (triangles). The lines display the Q-dependence of the characteristic times predicted by MCT (Eq.[5]) affected by the rheological shift factor for 335K (dashed), 365K (solid) and 390K (dashed dotted). The arrows show interpolated mechanical susceptibility relaxation times at the temperatures indicated.

In order to display this observation more quantitatively in Fig. 10 we present the associated activation energies as a function of Q. Horizontal lines indicate the effective activation energy of the viscosity relaxation (E_α), of the self-motion (E_{self}) and the β-process (E_β).

The experimental Q-dependent activation energies reach E_{self} in the low Q-regime, while E_α is reached at Q_{max} and beyond. Thus, the structural relaxation, which is observed at the first structure factor peak coincides with the activation energy for macroscopic flow, while the density fluctuations appear to relax with the weaker temperature dependence found for the self-motions and the conformational dynamics.

The coherent scattering in the low Q-regime is due to long range density fluctuations in disordered matter. With the atom number density ρ and the isothermal compressibility κ_T, in thermodynamic equilibrium at constant pressure, the structure factor in the limit $Q \to 0$ becomes

$$S(Q \to 0) = \frac{k_B T}{m v_{l_0}^2} - \frac{k_B T}{m v_l^2} + \frac{k_B T^2 \alpha^2}{m C_p} \quad (4)$$

Thereby m is the average atomic mass, v_{l_0} the relaxed low frequency longitudinal velocity of sound, α the thermal expansion coefficient, C_p the heat capacity per mass unit and v_l the unrelaxed velocity of sound. Eq.[4] accounts already for the fact that because of kinetic constraints for cold neutrons the Brillouin lines are not excited.

Marvin et al. have studied ultrasonic attenuation in PIB [12] yielding v_{l_0} = 1418m/s and v_l = 2653m/s. At room temperature we have α = 7.7 x 10^{-4}K^{-1} and C_p = 1680 J/kg/K. With these values Eq.[4] yields 0.183. 80% of that value comes from the adiabatic compressibility and only 20% from entropy fluctuations. The experimental structure factor in the low Q regime $S(Q \to 0) \cong 0.22$ at room temperatures compares well with this theoretical prediction. Also the result for the partition in adiabatic compressibility and entropy fluctuations is found in the data.

(E_α), for the dielectric β relaxation (E_β), and for the self motion (E_0).

Fig. 11 displays the Q-dependent of the collective fluctuations as obtained T = 390K. At low Q this amplitude amounts to about 0.8 reflecting the fact, that the adiabatic compressibility decays with a stress relaxation time (ns-time scale), while the entropy fluctuations are expected to decay with the thermal conductivity time well in the picosecond regime outside of the experimental window of the spin echo instrument.

The major and unexpected feature of the collective dynamics is the observation of a broad weakly Q dependent plateau of the collective or pair relaxation times τ_{pair} at Q values smaller than those of the structure factor peak. Wavelength dependence studies below and above the MS threshold have confirmed that the phenomenon is in the genuine.

In the low Q-limit the coherent scattering relates to the imaginary part of the susceptibility $\chi_{11} = 1/C_{11}$, where C_{11} is the elastic modulus of the longitudinal sound waves. The elastic modulus relaxes with the stress relaxation time τ_M. This time may be accessed from longitudinal sound wave damping experiments in the megahertz region (298K) [12] and at higher temperatures from the damping of the light scattering Brillouin lines (473K) [13]. Taking into account the relation between modulus and susceptibility a temperature dependence of the τ_M may be deduced with E_A = 0.48eV. In Fig. 9 the arrows gives the interpolated mechanical susceptibility relaxation times. As may be seen they nearly quantitatively agree with the pair relaxation times in the low Q regime. Obviously the low Q collective relaxation data follow the temperature dependence of the stress relaxation, while the structural relaxation as observed at the structure factor maximum clearly displays a stronger temperature dependence.

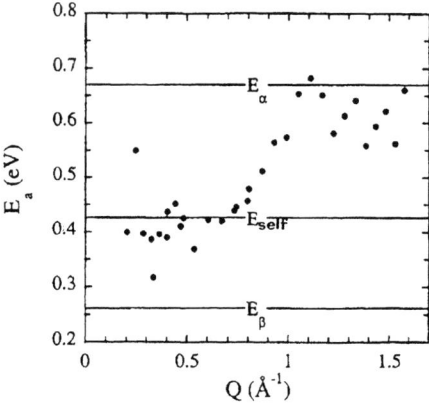

FIGURE 10. Momentum transfer dependence of the activation energy observed for the collective dynamics. Horizontal lines show the values obtained from the rheological shift factor in the temperature range investigated

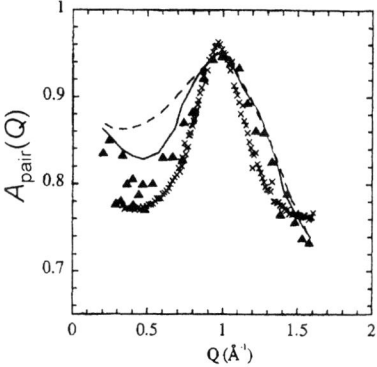

FIGURE 11. Momentum transfer dependence of the amplitude of the KWW functions describing the dynamic

structure factor. For 390K all the values obtained are shown (▲), while the dashed and solid line represent the smoothed behaviour for 335K and 365K, respectively. The static structure factor as obtained with IN11C is shown in arbitrary units for comparison (×).

For the Q dependence of the characteristic time of the structural relaxation the mode coupling theory (MCT) predicts

$$\tau_{pair} \propto \left[\frac{f_Q}{h_Q}\right]^{2/b} \quad (5)$$

where f_Q and h_Q are the relaxation strength of the α and the fast β relaxations respectively and $b \approx \beta$ is the von Schweidler exponent. Using the coherent scattering amplitudes (see Fig. 11) we have $f_Q = A_{pair}(Q)$ and $h_Q = 1 - A_{pair}(Q)$. MCT predicts universality of the time scales. This prediction is incompatible with the observation of the two different temperature dependencies for the structural and the stress relaxation. Nevertheless, we would like to explore the predicted Q dependence. To do so we take the shift factor of the structural relaxation and calculate following Eq.[5] the Q and T dependencies predicted by MCT. They are presented in Fig. 9 by the lines. The agreement is good for the lowest temperature (335K), where the modulation in the peak region as well as the modulation in the plateau are fairly well reproduced by MCT. However, the thermal evolution obtained from experiments and the theoretical prediction are clearly different.

SUMMARY AND OUTLOOK

We have studied the intermediate scale dynamics of PIB investigating all three correlation functions accessible for neutron scattering revealing the self motion the collective motion and the relaxation behaviour of the single chain. The main outcome of this investigation is that the self-motion, the density fluctuations at low Q as well as the intrachain dissipation follow a significantly weaker temperature dependence (0.45eV), than the structural relaxation or the viscous flow ($E_{st} = 0.67$eV). This implies that the stress relaxation displays a weaker temperature dependence than the structural relaxation – a very astonishing result. Relating to the results on the self-correlation function and the intrachain dissipation, it is suggestive to conclude that the stress relaxation relates to conformational motion, while the structural relaxation appears further translational dynamics.

With the self-motion measured over a wide Q range, the meansquared proton displacement has been obtained explicitly for each temperature. Given the different temperature dependencies of the self-correlation function and the structural relaxation, we find that the mean square proton displacement measured at the structural relaxation time increases from 17Å^2 at 390K to 30Å^2 at 335K. Thus, the average meansquared proton displacement needed to relax the structure significantly increases with decreasing temperature.

The results leave us with a number of challenges, which need to be answered in the future:
(i) Are the different mechanisms for stress relaxation and structural relaxation in polymer melts?
(ii) What is the connection between the sublinear diffusion regime seen in the self-motion and the Rouse regime?
(iii) Is the paradigm of one single temperature scale of MCT generally invalid in polymer melts?
(iv) Do the secondary relaxations also couple to the density fluctuations? And if so, how do they affect the collective dynamics at low Q?
(v) Finally, does the increase of the mean square displacement at the structural time signify a length scale, which increases with decreasing temperature?

REFERENCES

1. Rouse, P.E., Jr., J. Chem. Phys. **21**, 1272 (1953).
2. Doi, M., Edwards, S.F., *The theory of polymer dynamics*, Oxford Science Publications, Oxford University Press (1986).
3. Farago, B., Arbe, A., Colmenero, J., Faust, R., Buchenau, U., Richter, D., Phys. Rev. E **65**, 051803 (2002).
4. Richter, D., Monkenbusch, M., Allgaier, J., Arbe, A., Colmenero, J., Farago, B., Cheol Bae, Y., Faust, R. J. Chem. Phys. **111**, 6107-6120 (1999).
5. Arbe, A., Monkenbusch, M., Stellbrink, J., Richter, D., Farago, B., Almdahl, K., Faust, R., Macromolecules **34**, 1281-1290 (2001).
6. Arbe, A., Colmenero, J., Frick, B., Monkenbusch, M., Richter, D., Macromolecules **31**, 4926-4934 (1998).
7. Colmenero, J., Alegría, A., Arbe, A., Frick, B., Phys. Rev. Lett. **68**, 478-481 (1992).
8. Arbe, A., Colmenero, J., Monkenbusch, M., Richter, D., Phys. Rev. Lett. **81**, 590-593 (1998).
9. Colmenero, J., Arbe, A., Alegria, A., Monkenbusch, M., Richter, D., J. Phys. Condensed Matter **11**, A363-A370 (1999).
10. Harnau, L., Winkler, R.G., Reinecker, P., J. Chem. Phys. **102**, 7750-7757 (1995); **104**, 6355-6368 (1996).
11. Allegra, G., Ganazzoli, F., Macromolecules **14**, 1110-1119 (1981).
12. Marvin, R.S., Aldrich, R., Sack, H.S., J. Appl. Phys. **25**, 25 (1954)
13. Patterson, G.D., J. Polym. Science, Polym. Phys. Ed. **15**, 455 (1977)
14. e.g. Götze, W., Sjörgen, L, Rep. Prog. Phys. **55**, 241 (1992)

Glassy Dynamics of Polymer Thin Films

T. Kanaya*, T. Miyazaki*, R. Inoue,*, H. Yamano*, K. Nishida*, I. Tsukushi[†] and K. Shibata**

*Institute for Chemical Research, Kyoto University, Uji, Kyoto fu 611-0011
[†]Chiba Institute of Technology, Narashino, Chiba-ken 275-0023
**Japan Atomic Energy Research Institute, Tokai, Ibaraki-ken 319-1106

Abstract. We have investigated thermal expansion behavior of polystyrene thin films using X-ray reflectivity and dynamics in the glassy state using inelastic neutron scattering. It was found that the thermal expansivity decreases with the film thickness below twice of the radius of gyration. This has been assigned to the chain deformation in the confinement, leading to an idea of stress-induced hardening. This idea was confirmed by the thickness dependence of the mean square displacement evaluated in the inelastic neutron scattering experiments.

INTRODUCTION

Properties of polymer thin films and/or polymer surfaces are very different from those of the bulk and related to many phenomena such as adhesion, wetting, surface friction, and hence are important from viewpoints of not only science but also industrial applications. Glass transition of thin films and/or surfaces is also one of the most interesting phenomena because many properties such as mechanical and thermal ones drastically change at the glass transition temperature T_g. Aiming to elucidate the special nature of glass transition of thin films and/or surfaces, many studies have been performed using many techniques such as ellipsometry, X-ray and neutron reflectometry, positron annihilation, dielectric relaxation, Brillouin light scattering and atomic force microscopy [1]. One of the objectives of these studies is to elucidate the characteristic length scale responsible for glass transition, which increases as the temperature is lowered towards T_g [2]. In confinement systems, the correlation length is truncated by the dimension of the restrictive geometry when it reaches the system size, giving information about the characteristic length scale without knowing any physical nature of the correlations. However, many experiments revealed that pure finite size effects are hardly extracted from these experiments for thin polymer films supported on a solid substrate because surface and interface effects on T_g are not negligible [3]. Regarding the thickness dependence of thermal expansivity in glassy and molten states there are still contradictory experimental results, depending on experimental techniques and thermal history of thin films [4]. In previous papers [5, 6], we have investigated annealing effects on thickness of deuterated polystyrene thin films and found that the thermal expansivity is smaller than that in bulk even for well-annealed thin films. In this paper, we have investigated the thermal expansion behavior of polystyrene (PS) thin films using X-ray reflectivity (XR) as well as dynamics of PS thin films using an inelastic neutron scattering technique. One of the purposes of this study is to elucidate the cause of the small thermal expansivity in very thin films.

EXPERIMENTAL

In this study, we used polystyrene (PS) with molecular weight $M_w = 3.0 \times 10^5$ for both XR measurements and inelastic neutron scattering measurements. The molecular weight distribution of PS is $M_w/M_n = 1.06$, where M_w and M_n are the weight average and the number average of molecular weight, respectively. Polystyrene thin films for XR measurements were prepared on cleaned Si (111) wafers by spin-coating the toluene solution. Thickness of polymer film was controlled by varying the polymer concentration in the solution. For inelastic neutron scattering measurements PS thin films were prepared on glass plates and floated on water surface. The film on water was collected on Al foil 15 μm thick. 299 films on Al foil were stacked and placed into a cylindrical Al cell 14 mm in diameter.

XR measurements were performed using a home-built X-ray reflectometer which was based on a conventional powder diffractometer. The sample was maintained in a chamber with beryllium windows under in vacuum during the measurements. Inelastic neutron scattering measurements were performed on an inverted geometry time-of-flight (TOF) spectrometer LAM-40 at KEK, Tsukuba. The energy resolution and the range of length of scattering vector Q of LAM-40 are 0.2 meV and 0.2 to 2.6 Å^{-1}, respectively.

RESULTS AND DISCUSSION

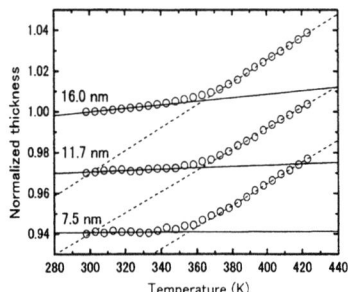

FIGURE 1. Temperature dependence of thickness of PS thin films with initial thickness d_0=16.0, 11.7 and 7.5 nm.

Fig. 1 shows a typical example of the temperature dependence of thickness of PS thin films with various initial values. The thicknesses were normalized to that at 298 K for each sample. For all the samples, discontinuous change of thermal expansivity was clearly observed, showing a definite glass transition temperature T_g. The expansivities in the glassy and molten states, which are shown by dashed and solid lines in the figure, respectively, are also definitely determined in the measurements. In this paper, we will focus on the thermal expansivity of PS thin films in the glassy state. Thermal expansivity below the glass transition temperature T_g was evaluated from the slope of the straight line in Fig. 1 and plotted as a function of the thickness in Fig. 2. The expasivity in the thickness range above 30 nm is $\sim 1.3 \times 10^{-4}$, which is close to the expected value (1.1×10^{-4} K^{-1}) from the bulk assuming that thin films are constrained along the substrate [7]. On the other hand, as the thickness decreases it begins to decrease at $\sim$ 30 nm, and becomes almost zero below $\sim$10 nm.

It is known that unrelaxed structure due to lack of annealing makes apparent small or negative expansivity for thin films [5, 6]. However, the present result is not the case because the samples were annealed at 423 K (50 K above T_g) long enough (38 h) before the measurements. As mentioned above, the expansivity begins to decrease at $\sim$ 30 nm, which is very close to twice of the radius of gyration of a chain $2R_g$ (= 30 nm). This implies that the decrease in the expansivity is due to the confinement in the thin films. When polymer chains are confined in films thinner than $2R_g$, they are deformed and the intra-

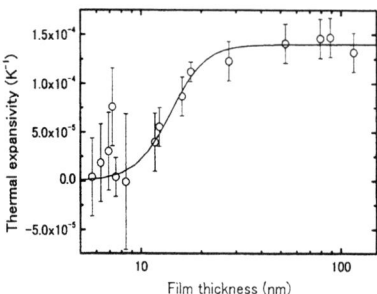

FIGURE 2. Thickness dependence of thermal expansivity of PS thin films in glassy state.

chain potentials between non-bonded atoms are stressed and hardened (stress-induced hardening). Taking into account anharmonic contributions, we assume that the potential is described as $U(u) = fu^2 - gu^3 - hu^4$, where u is the displacement from its equilibrium position, f is the harmonic force constant, and g and h are positive constants representing the anharmonic contributions. Using this potential, the mean value of the displacement is given by

$$<u> = \frac{3g}{4f^2} k_B T \quad (1)$$

where k_B is the Boltzmann constant. This is the origin of the thermal expansivity due to the anharmonic contribution. If the stress-induced hardening due to the confinement occurs, the force constant f must become larger, leading to the smaller thermal expansivity. Therefore, the thermal expansivity in the glassy state decreases with the thickness below $\sim 2R_g$. This idea of stress-induced hardening can be confirmed by measureing the force constant f or related quantities. For this purpose we performed inelastic neutron scattering experiments on PS thin films 40 and 100 nm thick.

Fig. 3 shows the observed dynamic scattering laws $S(Q,\omega)$ of PS thin film 100 nm thick at 11, 80 150 and 230 K. A very broad peak or shoulder is observed at $\sim$ 1.5 meV, which is the so-called boson peak. The peak position is not different from that of bulk [8]. The inelastic scattering spectra were scaled using the Bose-Einstein population factor and the result is shown in Fig. 4 in an energy region of 0.5 to 4 meV. As reported for bulk PS [8], the dynamic scattering law can be well scaled above $\sim$ 1.5 meV while excess inelastic scattering intensity is observed below $\sim$ 1.5 meV. This anharmonic excess scattering is well known as the fast process. Although it is considered that the boson peak and the fast process are essential to understand the nature of glasses as well as the glass transition, they are out of scope of this paper. These problems will be discussed in a separated paper.

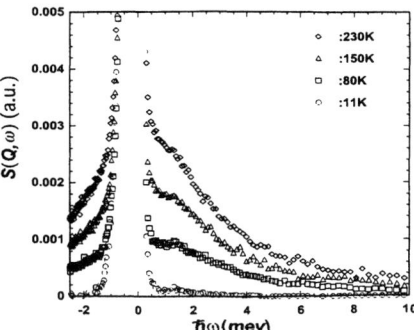

FIGURE 3. Dynamic scattering law $S(Q,\omega)$ of PS thin film 100 nm thick at 11, 80, 150 and 230K.

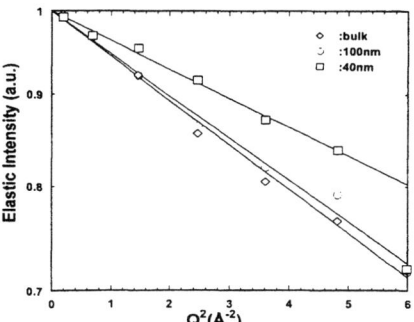

FIGURE 5. Elastic scattering intensity of PS thin films versus Q^2 at 230K.

Here in this paper, we would like to concentrate on the mean square displacement $<u^2>$.

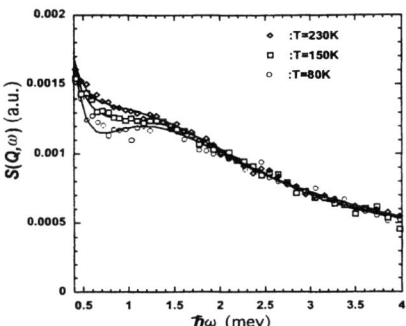

FIGURE 4. Bose-scaled dynamic scattering law of PS thin films 100 nm thick.

As well known, Q dependence of incoherent elastic scattering intensity $I_{el}(Q)$ is described by

$$I_{el}(Q) \sim [\exp(-<u^2>_{el} Q^2)]. \quad (2)$$

Logarithm of the elastic scattering intensity $I_{el}(Q)$ is plotted against Q^2 for bulk PS, PS thin films 100 and 40 nm thick in Fig. 5 where $I_{el}(Q)$ was divided by the corresponding $I_{el}(Q)$ at 11 K to remove the effect of coherent scattering. From the straight lines in the figure we have calculated $<u^2>_{el}$. Exactly speaking, this $<u^2>_{el}$ corresponds to $<u^2>_{el,T} - <u^2>_{el,11K}$ where $<u^2>_{el,T}$ and $<u^2>_{el,11K}$ are the mean square displacement at T and 11 K, respectively, because $I_{el}(Q)$ was divided by that at 11 K.

Thus evaluated mean square displacement $<u^2>_{el}$ is shown for PS bulk, PS thin films 100 and 40 nm thick in Fig. 6 as a function of temperature. The mean square displacement of thin film 40 nm thick is the smallest among three and increases with thickness at 150 and 230K while at 80 K they are almost identical within the experimental error. The mean square displacement is related to the harmonic force constant f through the equation (3)

$$f \sim \frac{k_B T}{<u^2>}. \quad (3)$$

The present result that $<u^2>$ decreases with deacreasing thickness (see Fig. 6) means that the force constant increases with decreasing the film thickness. This supports the idea of stress-induced hardening due to the confinement of polymer chains in the thin film.

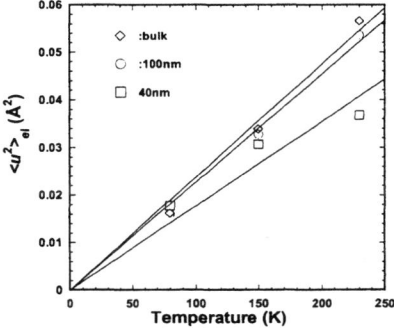

FIGURE 6. Temperature dependence of mean square displacement $<u^2>_{el}$ evaluated from elastic scattering.

We can also evaluate the mean square displacement from the Q dependence of the incoherent inelastic scattering intensity. In case of harmonic oscillation, it is given by

$$I_{ine}(Q) \sim Q^2[\exp(-<u^2>_{ine} Q^2)]. \quad (4)$$

The observed inelastic scattering intensity integrated in an energy region between 1 and 4 meV, corresponding to the boson peak region, is plotted versus Q for PS thin film 100 nm thick in Fig. 7 and eq. (4) was fitted to the observed data. In the fitting, we included the contribution from the multiple scattering. The results of fits are indicated by solid curves.

The evaluated mean square displacement $<u^2>_{ine}$ is shown in Fig. 8 as a function of temperature. It is surprising that the mean square displacement evaluated from

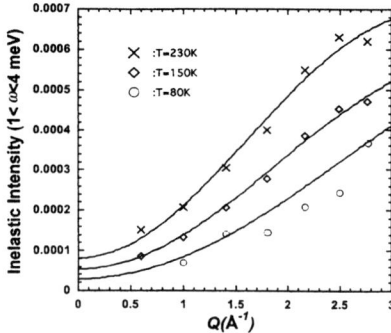

FIGURE 7. Inelastic scattering intensity of PS thin films 100nm thick versus Q.

the inelastic scattering intensity increases with decreasing the film thickness on contrary to the results on the elastic scattering. How can we understand?

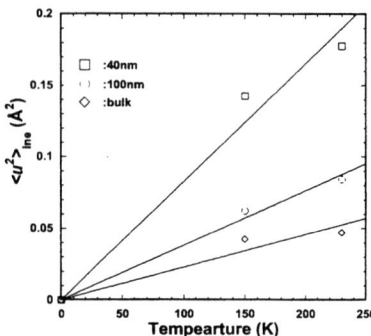

FIGURE 8. ITemperature dependence of mean square displacement $<u^2>_{ine}$ evaluated from inelastic scattering.

Our tentative interpretation is as follows. The mean square displacement $<u^2>_{el}$ evaluated from the elastic intensity is an average over all motions outside the energy resolution. On the other hand, $<u^2>_{ine}$ from the inelastic intensity corresponds to the Debye-Waller factor of the boson peak mode. As reported by Buchenau et al. [9] the inelastic mean square displacement $<u^2>_{ine}$ due to the boson peak mode is larger than the elastic (average) one $<u^2>_{el}$. Therefore, if we assume that relative number of the boson peak modes increases with decreasing the film thickness, we can understand the present result. It is obvious that the relative ratio of the surface region increases with decreasing the thickness, and if the boson peak modes exist in the surface region more than in the bulk region, they relatively increases with decreasing the film thickness. Therefore, the mean square displacement evaluated from the boson peak mode increases with decreasing the thickness. This scenario qualitatively explains the present results. Finally we have to mention why the boson peak mode are in the surface region more than in bulk region. As elucidated in the annealing experiments of the boson peak in poly (methyl methacrylate) (PMMA) [10] as well as the experiment of the boson peak of vapor-deposited propylene glass [11], the boson peak intensity (or number of the boson peak modes) decreases with increasing density of glass. It is easily expected that the surface region is less dense than the bulk region, and hence relative number of the boson peak modes increases with decreasing the thickness.

CONCLUSIONS

We have shown that the thermal expansivity in the glassy state of PS thin films decreases with the film thickness below about twice of the radius of gyration $2R_g$. This has been assigned to the deformation of polymer chains in the confinement, predicting the increase in the force constant (stress-induced hardening). This idea was confirmed by the fact that the average mean square displacement evaluated from the elastic scattering intensity decreases with the film thickness. Another interesting finding in the inelastic scattering experiments is the mean square displacement from the boson peak increases with decreasing the thickness. This has been explained by an idea that the boson peak modes are in the surface region more than in the bulk region.

REFERENCES

1. Karim, A., and Kumar, S., *Polymer Surfaces, Interfaces and Thin Films*, World Scientific, Singapore, 2000.
2. Donth, E., *Relaxation and thermodynamics of polymers: glass transition*, Akademie-Verlag, Berlin, 1992.
3. Keddie, J. L., Jones, R. A., and Cory, R. A., *Faraday Discuss*, **98**, 219–230 (1994).
4. Forrest, J. A., and Jones, R. A. L., *Polymer Surfaces, Interfaces and Thin Films*, World Scientific, Singapore, 2000, pp. 251–294.
5. Kanaya, T., Miyazaki, T., Watanabe, H., Nishida, K., Yamano, H., Tasaki, S., and Bucknall, D. B., *Polymer*, **44**, 3769–3773 (2002).
6. Miyazaki, T., Nishida, K., and Kanaya, T., *Phys. Rev. E* (2003), in press.
7. Wallace, W. E., v. Zanten, J. H., and Wu, W., *Phys. Rev. E*, **52**, R3329–R3332 (1995).
8. Kanaya, T., Kawaguchi, T., and Kaji, K., *J. Chem. Phys.*, **104**, 3841–3850 (1996).
9. Linder, K., Frick, B., and Buchenau, U., *Physica A*, **201**, 112–114 (1993).
10. Kanaya, T., Miyakawa, M., Kawaguchi, T., and Kaj, K., "Workshop on Non Equilibrium Phenomena in Supercooled Fluids, Glasses and Amorphous Materials," World Scientific, Singapore, 1995, pp. 301–302.
11. Yamamuro, O., Tsukushi, I., Matsuo, T., Takeda, K., Kanaya, T., and Kaji, K., *J. Chem. Phys.*, **106**, 2997–3002 (1997).

Slow Diffusive Motions in a Monolayer of Tetracosane Molecules Adsorbed on Graphite

H. Taub*, F. Y. Hansen[†], L. Criswell*, D. Fuhrmann*[¶], K. W. Herwig[‡], A. Diama*, H. Mo*, R. M. Dimeo[§], D. A. Neumann[§], and U. G. Volkmann[††]

*Department of Physics and Astronomy, University of Missouri-Columbia, Columbia, Missouri 65211, USA
[†]Department of Chemistry, Technical University of Denmark, 207 DTU, DK-2800 Lyngby, Denmark
[¶]Infineon Technologies, Memory Products, Balanstr. 73, D-81541 Munich, Germany
[‡]Spallation Neutron Source, Oak Ridge National Laboratory, 701 Scarboro Rd., Oak Ridge, TN 37830, USA
[§]Center for Neutron Research, National Institute of Standards and Technology, Gaithersburg, Maryland 20899-8562, USA
[††]Facultad de Física, Pontificia Universidad Católica de Chile, Santiago 22, Chile

Abstract. Monolayers of intermediate-length alkane molecules such as tetracosane (n-$C_{24}H_{50}$ or C24) serve as prototypes for studying the interfacial dynamics of more complex polymers, including bilayer lipid membranes. Using high-resolution quasielastic neutron scattering (QNS) and exfoliated graphite substrates, we have investigated the relatively slow diffusive motion in C24 monolayers on an energy/time scale of ~1–36 μeV (~0.1–4 ns). Upon heating, we first observe QNS in the crystalline phase at ~160 K. From the crystalline-to-smectic phase transition at ~215 K to a temperature of ~230 K, we observe the QNS energy width to be dispersionless, consistent with molecular dynamics simulations showing rotational motion of the molecules about their long axis. At 260 K, the QNS energy width begins to increase with wave vector transfer, suggesting onset of nonuniaxial rotational motion and bounded translational motion. We continue to observe QNS up to the monolayer melting temperature at ~340 K where our simulations indicate that the only motion slow enough to be visible within our energy window results from the creation of *gauche* defects in the molecules.

INTRODUCTION

Over the past three decades nuclear magnetic resonance (NMR) has been used to investigate the conformational dynamics of lipid molecules in membranes [1]. In particular, measurements of the nuclear quadrupole splitting at selectively deuterated methylene groups along the alkyl chains have yielded the distribution of *gauche* defects along the chains in the membrane's fluid phase [2].

Recently, we have shown that the melting transition in a monolayer of intermediate-length alkane molecules (n-C_nH_{2n+2}), such as tetracosane (n = 24) and dotriacontane (n = 32), adsorbed on a graphite basal-plane surface provides an interesting analog of the gel-to-fluid transition in bilayer lipid membranes [3]. That is, our molecular dynamics (MD) simulations provide evidence of a monolayer melting transition in which intramolecular and translational order are lost simultaneously [3]. Figure 1 shows a top view of the tetracosane (C24) molecules in the MD simulation cell above and below the melting transition at ~340 K. Below the transition, we see translational order manifested by a lamellar structure. We refer to this structure as the "smectic" phase, since both the simulations and quasielastic neutron scattering measurements (see below) indicate some translational diffusive motion within the lamellae. Above the melting transition, the individual molecules have transformed to a more globular shape, i.e. have undergone a "chain-melting," and the monolayer has lost the translational order characterized by its lamellar structure. As in the case of the lipid membrane, the simultaneous "chain" melting and lattice melting of the alkane monolayer result from an abrupt increase in the number of *gauche* defects within the central region of each molecule.

Here we describe how high-resolution quasielastic neutron scattering (QNS) can be used to investigate the diffusive motion in a C24 monolayer occurring on a time scale of ~0.1–4 ns. We present evidence that

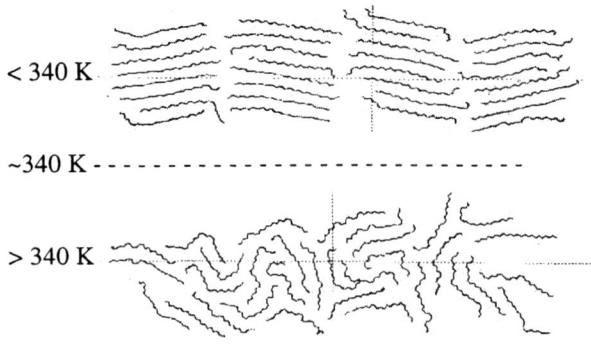

FIGURE 1. Top view showing configuration of C24 molecules in MD simulation cell in the "smectic" phase (upper panel) and the fluid phase (lower panel).

just below the monolayer melting temperature the principal contribution to the observed QNS is from the diffusive motions associated with conformational changes in the C24 molecules.

EXPERIMENT

The QNS experiments were performed on the new High Flux Backscattering Spectrometer (HFBS) at the Center for Neutron Research, National Institute of Standards and Technology [4]. The HFBS has a neutron energy resolution of ~1.0 μeV and a dynamic range of 0–36 μeV. Spectra were collected in a neutron wave vector transfer (Q) range from 0.6 Å^{-1} to 1.7 Å^{-1}. The large solid angle subtended by the silicon analyzing crystals used to select the final neutron energy makes the HFBS particularly effective for inherently small samples such as adsorbed monolayers.

The samples were prepared by depositing C24 from its vapor phase onto an exfoliated graphite substrate (Grafoil) in an evacuated cell at a temperature of ~200 °C. The cells had an annular shape (to reduce multiple scattering) with a height of 30 mm and an inner and outer diameter of 16 mm and 22 mm, respectively. Preparation details have been described elsewhere [5]. Neutron diffraction measurements at the University of Missouri Research Reactor on the same samples used for the QNS experiments were used to verify the film structure. Consistent with earlier studies [6], room temperature measurements confirmed the presence of the low-density "smectic" monolayer C24 phase shown in Fig. 1 (upper panel) and the higher-density crystalline phase, depending on coverage.

RESULTS AND DISCUSSION

We began our measurements at low temperature where we anticipated the molecular diffusive motion would be simplest to describe. Elastic neutron diffraction measurements on deuterated samples show that, upon cooling, the C24 monolayer undergoes a transition at ~215 K from the "smectic" to a commensurate phase in which the monolayer contracts in a direction parallel to the lamellae boundaries [7]. The nearest-neighbor molecular separation in this direction becomes 4.26 Å = √3 a_g where a_g = 2.46 Å is the graphite basal-plane lattice constant compared to ~4.6 Å in the "smectic" phase. This higher-density crystalline phase is similar in structure to that found at room temperature when a C24 monolayer is grown from a heptane (n-C$_7$H$_{16}$) solution or when a partial layer of either heptane or C24 is adsorbed above the C24 monolayer [6].

In Fig. 2, we show quasielastic spectra taken on the HFBS at two C24 coverages, θ = 1.0 and 1.15 layers, where unity coverage is defined as a complete monolayer of the "smectic" phase. At all temperatures in Fig. 2, both samples are in the higher-density crystalline phase described above. Due to the compression of the first layer when a partial second layer is added, we estimate that at θ = 1.15 there is about 7% of a second layer present. The solid line in the spectra of Fig. 2 represents the HFBS' energy resolution function. It is actually the quasielastic spectrum observed from the sample at 100 K where all diffusive motion within the film is frozen out.

For both coverages, we begin to observe QNS having an energy width greater than that of the HFBS' resolution function at a temperature of ~160 K. MD simulations indicate [8] that this results from motion in the crystalline phase involving a uniaxial rotation of the C24 molecules about their long axis aligned parallel to the graphite basal-plane surface (the molecules are nearly in the all-*trans* configuration).

Above ~215 K at both coverages, we begin to observe QNS with sufficient intensity to determine the Q dependence of its energy width. At these temperatures, the 1.0-layer sample has transformed to the "smectic" phase, while the 1.15-layer sample remains crystalline. We have fit the quasielastic intensity at each Q to the sum of a Lorentzian and a delta-function convoluted with the instrumental resolution function. As shown in Fig. 3, the half-width at half-maximum (HWHM) of the Lorentzian component is essentially dispersionless at temperatures of 215 K and 230 K for both coverages. Our MD simulations indicate that in the "smectic" phase the

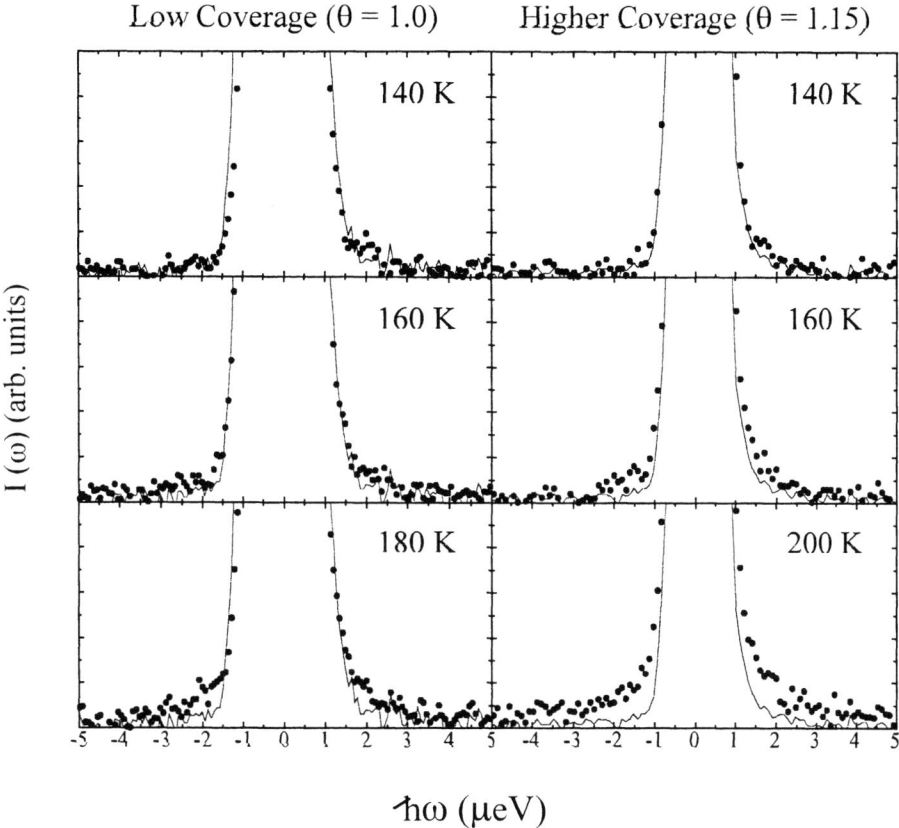

FIGURE 2. Quasielastic scattered neutron intensity measured on the HFBS from crystalline C24 monolayers at two different coverages as the temperature is increased. All QNS detected in the range 0.6–1.6 Å^{-1} has been summed.

diffusive motion at 215 K and 230 K is still dominated by a uniaxial rotation of the molecules about their long axis aligned parallel to the graphite surface. Such a model is consistent with the dispersionless HWHM of the Lorentzian component of the QNS. Note that at 215 K and 230 K the 1.15-layer sample (right panel) is still in the crystalline phase, and its HWHM's are somewhat smaller than for the 1.0-layer sample.

We see in Fig. 3 that at a temperature of 260 K the energy width of the QNS of the 1.0-layer sample more than triples, and the HWHM also begins to increase with Q. Our simulations suggest that this Q-dependence results primarily from nonuniaxial rotational motion and, to a lesser degree, translational diffusion.

It is interesting to compare the HWHM's for the higher-coverage sample at 260 K. They have increased very little above their values at 230 K, and they remain dispersionless. In fact, the magnitudes of the HWHM for the 1.15-layer sample at 260 K are about the same as for the 1.0-layer sample at 230 K. These results suggest that both the nonuniaxial rotational motion and translational diffusion are inhibited in the 1.15-layer sample, which remains in the crystalline phase.

At a temperature of 330 K, our measurements support the prediction of the MD simulations [8] that QNS should be observable due to the creation and annihilation of *gauche* defects in the C24 molecules. In Fig. 4, we have analyzed the Q dependence of the HFBS spectra of the 1.0-layer sample at temperatures of 285 K and 330 K in the same way as for lower temperatures in Fig. 3: (a) the Lorentzian HWHM characterizing the energy width of the observed QNS; and (b) the relative intensity of the Lorentzian component in the spectra. Although the intensity becomes very weak at 330 K, it is clear that there is still QNS within the dynamic range of the HFBS. Comparing with Fig. 3, we see that at each Q the energy width of the QNS continues to increase with temperature above 260 K. There is also a large increase in the slope of the HWHM vs. Q between 260 K and 285 K. For comparison, we have also plotted in Fig. 4 (dashed line) the HWHM of a Lorentzian fitted

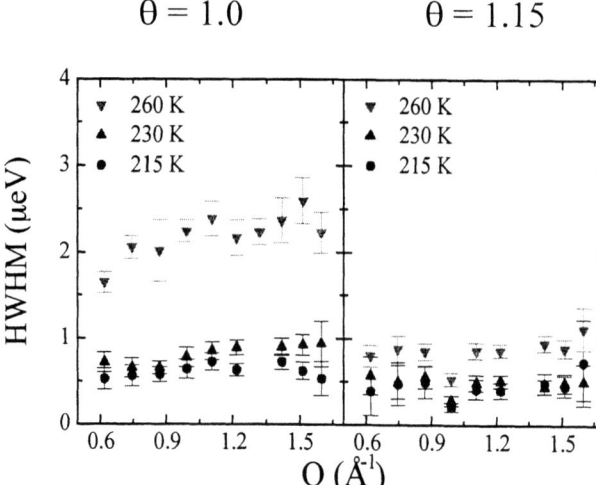

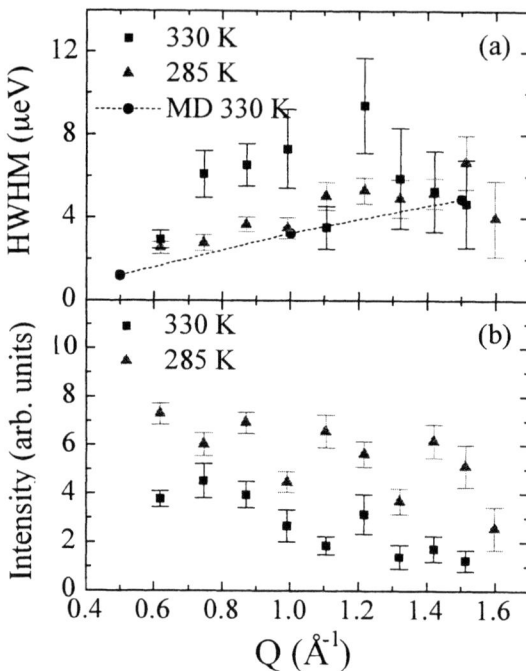

FIGURE 3. Results of fitting the quasielastic spectra at temperatures ≤ 260 K for the low-coverage sample (left panel) and higher coverage sample (right panel). The energy width (HWHM) of the Lorentzian component is plotted vs. wave vector transfer Q.

to the incoherent inelastic structure factor calculated from the MD simulation at 330 K after removing the center-of-mass motion from the C24 molecules. These calculated HWHM's characterize the contribution to the QNS from both rotational and intramolecular diffusive motions. We conclude that both their magnitude and Q-dependence are in qualitative agreement with experiment.

In summary, our QNS measurements and MD simulations indicate two principal temperature ranges of interest. The first of these is below 230 K where the diffusive motion primarily involves rotation about the long axis of the molecule aligned parallel to the surface. The second temperature range of interest is above 285 K where we identify the observed QNS with motions involving conformational changes of the molecules as predicted by the simulations.

FIGURE 4. Results of fitting the quasielastic spectra at higher temperatures for the 1.0-layer sample: (a) the HWHM and (b) the intensity of the Lorentzian component vs. Q. In (a), the dashed line indicates values calculated from the MD simulations at 330 K from Ref. 8.

ACKNOWLEDGMENTS

This work was supported by the National Science Foundation (NSF) under Grant Nos. DMR-9802476 and DMR-0109057, by the Chilean government under FONDECYT Grant No. 1010548, and by the U.S. Department of Energy through grant DE-FG02-01ER45912. The neutron scattering facilities used in this work are supported in part by the NSF under Agreement No. DMR-0086210.

REFERENCES

1. See, e.g., Pastor, R.W., Venable, R.M., and Feller, S.E., *Acc. Chem. Res.* **35**, 438-446 (2002).

2. Douliez, J.-P., Léonard, A., and Dufourc, E.J., *Biophys. Jour.* **68**, 1727-1739 (1995).

3. Hansen, F.Y. et al., *Phys. Rev. Lett.* **83**, 2362-2365 (1999).

4. Meyer, A. et al., *Rev. Sci. Instrum.* **74**, 2759-2777 (2003).

5. Herwig, K.W., Matthies, B., and Taub, H., in *Neutron Scattering in Materials Science*, edited by D.A. Neumann, T.P. Russell, and B.J. Wuensch, Materials Research Society Symposium Proceedings 376, Pittsburgh, Pennsylvania, 1994, pp. 757-762.

6. Herwig, K.W., Matthies, B., and Taub, H., *Phys. Rev. Lett.* **75**, 3154-3157 (1995).

7. Fuhrmann, D. et al., *Surf. Sci.* **482-485**, 77-82 (2001); Matthies, B. et al., unpublished.

8. Hansen, F.Y., and Taub, H., this proceedings, 2003.

Power Laws in the Dynamics of Polymer Solutions

T. Uematsu*, C. Svanberg*, M. Nydén† and P. Jacobsson*

*Department of Applied Physics, Chalmers University of Technology, SE-412 96 Göteborg, Sweden
†Department of Applied Surface Chemistry, Chalmers University of Technology, SE-412 96 Göteborg, Sweden

Abstract. The dynamics of polymer solutions in the semidilute to concentrated regimes has been examined, using solutions of poly(methyl methacrylate) in propylene carbonate over the wide concentration range from $\phi \approx 0.001$ to $\phi \approx 0.6$. Here ϕ is the volume fraction of polymer in the solutions. Four power law regimes of the dynamical correlation length, $\xi_h \sim \phi^{-\alpha}$ are observed when the local viscosity is taken into account. Crossovers from $\alpha = 0.75$, via $\alpha = 0.5$, to $\alpha = 1$ corresponding to transitions from a good solvent, via a marginal solvent, to a theta solvent behavior have been predicted due to the gradual reduction of the segmental flexibility between the entanglement points with increasing ϕ. For the first time a clear experimental validation of these transitions is found experimentally presented. The fourth regime ($\alpha \approx 2$) is observed in the highly concentrated region where the static correlation length reaches the persistence length of the polymer. This dynamical regime was not included in the original theoretical treatment but can be explained within the same framework.

INTRODUCTION

Polymer solution dynamics in the semidilute regime has attracted considerable scientific attention over the past three decades since the seminal theoretical work by de Gennes[1]. According to the theory, a simple power law behavior of the distance between interchain contacts with the different chains, or the dynamical correlation length for the hydrodynamic interaction, ξ_h, is predicted:

$$\xi_h \sim \phi^{-\alpha}. \qquad (1)$$

Here ϕ is the volume fraction of the polymer, and α is 0.75, 0.5 and 1 in a good, a marginal and a theta solvent, respectively. These simple power laws have been experimentally confirmed for several systems in the semidilute regime (typically $\phi \leq 0.1$)[2].

When the solution is further concentrated so that ξ_h approaches a scale, where the molecular structure governs the chain conformation, the power law prediction becomes inapplicable since the prediction only applies in the long chain limit. In this concentration range, instead the exponent α is predicted to become dependent of ϕ[3, 4, 5]. According to this theory, which is hereafter referred as the "polymer fraction dependence theory", transitions from $\alpha = 0.75$, over $\alpha = 0.5$, to $\alpha = 1$ are expected with increasing ϕ even in a good solvent. These transitions can be explained as a consequence of the gradual reduction of the segmental flexibility between the entanglement points with increasing the concentration, as depicted in Figure 1. A good solvent behavior ($\alpha = 0.75$) is observed only when intrachain two-body excluded-volume interactions dominate so that the segments can be swollen. Thus, the segments should be sufficiently flexible for the intrachain interactions. When the concentration is increased the segments between the entanglement points become so rigid that the intrachain interactions are prevented. Then a marginal solvent behavior ($\alpha = 0.5$) enters because the segments within the entanglement points become ideal sequences. When the solution is further concentrated, even interchain two-body excluded-volume interactions vanish. Then a theta solvent behavior ($\alpha = 1$) is observed since the segments even beyond the entanglement points become ideal sequences. Finally when ξ_h reaches the persistence length of the polymer, the theory predicts no further concentration dependence ($\alpha = 0$) of ξ_h since the segments between the entanglement points were thought to behave like a rigid rod. However, the static correlation lengths, ξ_s, which are expected to be identical with ξ_h[1, 3, 4], obey a strong power law ($\alpha \approx 2$) in various polymer solutions[6, 7, 8]. To explain this contradiction, we recently proposed a flexible rod model[6] yielding the exponent $\alpha = 2$, in which the motions of the monomers are only correlated between the entanglement points, as shown in Figure 1(IV).

Experimentally, the validity of the polymer fraction dependence theory has been examined with Photon Correlation Spectroscopy, PCS, by probing the collective diffusion coefficient, D_c via the Stokes-Einstein relation[1], i.e.,

$$\xi_h = \frac{k_B T}{6\pi \eta_{\text{local}} D_c}. \qquad (2)$$

Here k_B, T, and η_{local} are the Bolzmann constant, the

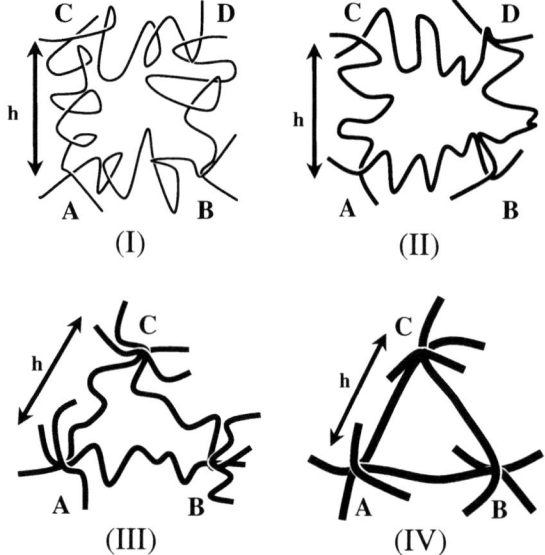

FIGURE 1. Qualitative pictures of different cooperative dynamical regimes in semidilute and concentrated solutions: A, B, C, and D are contact points between different chains. (I), (II), (III), and (IV) indicate a good solvent, a marginal solvent, a theta solvent, and a flexible rod behavior regime, respectively. The thickness of the solid lines represents the scale of the interchain contacts in each regime. In all the regimes, the motions of the monomers are only correlated within the interchain contacts, between which ξ_h is recognized as a characteristic distance.

absolute temperature and the local viscosity of the solutions, respectively. However, the results up to now are experimentally controversial. For example, according to Brown et al. [2], no crossovers in ξ_h can be observed and, instead a common power law ($\alpha \approx 0.7$) describes the concentration dependence of ξ_h up to $\phi \approx 0.3$ for the well-studied polystyrene solutions. This finding is puzzling since, according to the theory, a crossover to a theta solvent behavior should have occurred before ξ_s reaches the persistence length at this concentration. Furthermore, numerous studies of these solutions [9] show that the collective diffusion constant, D_c, strongly decreases at $\phi > 0.5$, thus, causing a sharp *increase* in ξ_h according to equation 2. This is clearly in conflict with both the $\alpha = 2$ and $\alpha = 0$ scenarios expected at high ϕ.

In previous attempts to derive ξ_h via equation 2, a major problem is the assumption that η_{local} is equal to the pure solvent viscosity. However, η_{local} is expected to increase dramatically with concentration since the solvent self-diffusion constants, D_{sol}, in various solutions decrease by more than a factor of 3 at $\phi = 0.3$[10]. Therefore the bulk viscosity of the solvent is no more the relevant factor for the cooperative dynamics in the intermediate and concentrated concentration. We therefore have evaluated ξ_h taking into account η_{local} calculated from a rewritten Stokes-Einstein relation[6]:

$$\eta_{\text{local}} = \frac{k_B T}{6\pi D_{sol} R_{sol}}. \quad (3)$$

Here R_{sol} is the hydrodynamic radius of the solvent molecule. D_{sol} has been obtained using pulsed field gradient NMR, pfg-NMR. Note that equation 3 neglects the influence of topological elasticity due to the entanglements since the cooperative dynamics is dominantly driven by the isothermal osmotic modulus[9].

In the present paper the concentration dependence of ξ_h in solutions of poly(methyl methacylate), PMMA, in propylene carbonate, PC, from the semidilute to concentrated regimes is investigated. Using PCS and pfg-NMR we have obtained ξ_h from D_c and D_{sol} via equation 2 and 3, and confirmed the validity of the polymer fraction dependence theory after taking η_{local} into account.

EXPERIMENTAL SECTION

The polymer solutions were prepared by mixing PC (Aldrich, HPLC grade) with two standard monodisperse high molecular weight PMMA samples (Fluka, $M_w = 2\,480\,000$, $M_w/M_n = 1.16$ and $M_w = 850\,000$, $M_w/M_n = 1.05$), and a conventional high molecular weight PMMA sample (Aldrich Chemicals, $M_w = 996\,000$, $M_w/M_n = 6.97$). Conventional PMMA was used for the intermediate and concentrated regime, where the polydispersity do not to have any influence on the cooperative dynamics. The samples preparation procedure follows the description given in ref. [6] and the polymer concentrations ranged from 0 to roughly 60 vol% PMMA.

In order to determine ξ_h via equation 2, D_c at 300 K was measured using PCS. For the investigation of η_{local} via equation 3, D_{sol} measurements were performed at 300 K using pfg-NMR. The details of the experimental set-ups are given in ref. [6].

RESULTS

The PCS measurements yield an auto-correlation function with three relaxation processes (see ref. [6]). The fast process, which is generally observed as a simple exponential decay, is attributed to the so-called collective diffusion process according to a number of studies of polymer solutions, see ref. [11] and references therein. The other two processes are related to the segmental mobility of the polymer matrix, and the reptation process of the polymer or the heterogeneity of the system, respectively[6]. From the fast process relaxation time, τ_{fast}, D_c is obtained in the short scattering vector limit

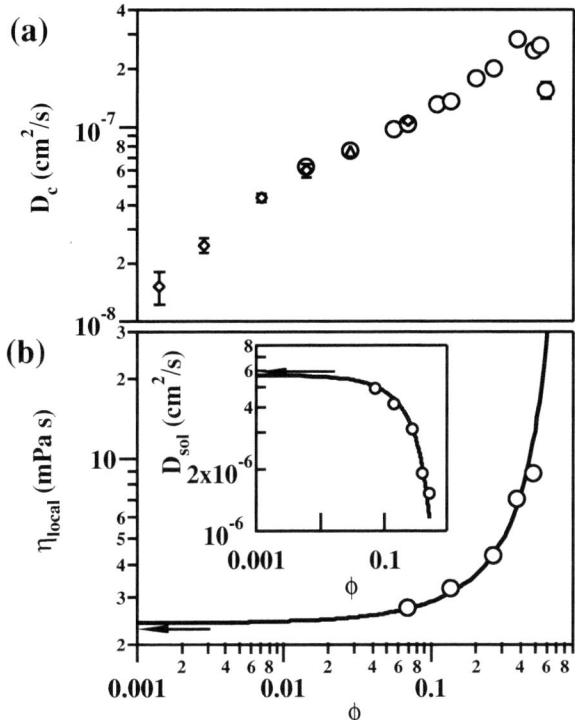

FIGURE 2. (a) Polymer volume fraction, ϕ, dependence of D_c at 300 K: The circles, triangles, and diamonds represent the data for the solutions with conventional PMMA of $M_w = 996\,000$, standard monodisperse PMMA of $M_w = 850\,000$ and $M_w = 2\,480\,000$, respectively. (b) The ϕ dependence of η_{local} at 300 K is shown in the main panel. In the inset the corresponding data for D_{sol} is presented. The circles and lines represent the experimental data for the conventional PMMA, and a free-volume theory based model proposed by Vrentas[10], respectively. The arrows in the main figure and in the inset are data at $\phi = 0$.

as:

$$D_c = \left.\frac{1}{\tau_{\text{fast}} q^2 (1-\phi)}\right|_{q \to 0}. \quad (4)$$

Here q is the scattering vector and $(1-\phi)$ corrects for the solvent backflow[1]. A double logarithmic plot of D_c as a function of ϕ at 300 K is shown in Figure 2(a). D_c increases monotonously with increasing ϕ until the highly concentrated regime ($\phi \approx 0.4$), and then substantially drops. The behavior is thus very similar to the one reported for well-studied polystyrene solutions[9].

The inset in Figure 2(b) shows that D_{sol} decreases significantly from the pure solvent regime in the same manner as found for other polymer solutions[10]. Evidentially, D_{sol} can be well described by a free-volume theory based model proposed by Vrentas[10]. From the bulk viscosity of the pure PC, $\eta_s = 2.25$ mPa s at 300 K[6], the hydrodynamic radius of PC molecules, $R_{sol} = 0.16$ nm at 300 K was obtained, which is roughly identical with the rotation radius of a PC molecule[12]. Assuming that R_{sol} does not vary with the concentration[13], η_{local} at 300 K was obtained via equation 3. Figure 2(b) clearly demonstrates that η_{local} significantly increases with the concentration, and that the η_{local} factor hence plays a crucial role in the determination of ξ_h.

DISCUSSION

Figure 3(a) shows ξ_h obtained via equation 2, using bulk viscosity data ($\eta_s = 2.25$ mPa s at 300 K[6]) as a function of ϕ. Up to $\phi \approx 0.5$, a common power law ($\alpha = 0.47$), which is close to a marginal solvent behavior ($\alpha = 0.5$), can qualitatively describe the data. However, the physical interpretation is difficult because already at $\phi = 0.3$ ξ_s is roughly equal to the persistence length of PMMA ($\approx$ 1 nm)[14, 15]. According to the polymer fraction dependence theory, before ξ_s reaches the persistence length, a theta solvent behavior ($\alpha = 1$) should have been observed. Furthermore, in the highly concentrated regime ($\phi > 0.5$), ξ_h increases significantly, which is seriously in contradiction with both the theory and the experimental results of ξ_s.

In Figure 3(b), ξ_h obtained via equation 2, using local viscosity data, is shown. Over the whole investigated concentration range ξ_h is monotonously decreasing with the concentration. A closer inspection yields three crossovers, i.e. four power law regimes, in agreement with the polymer fraction dependence theory. In order to examine the power law exponents and the crossover concentrations, the following relation is employed to describe our data:

$$\frac{\xi_h}{\xi_0} = \begin{cases} \phi^{-\alpha_1} & \phi \leq \phi_{12} \\ \phi_{12}^{\alpha_2-\alpha_1} \phi^{-\alpha_2} & \phi_{12} < \phi \leq \phi_{23} \\ \phi_{12}^{\alpha_2-\alpha_1} \phi_{23}^{\alpha_3-\alpha_2} \phi^{-\alpha_3} & \phi_{23} < \phi \leq \phi_{34} \\ \phi_{12}^{\alpha_2-\alpha_1} \phi_{23}^{\alpha_3-\alpha_2} \phi_{34}^{\alpha_4-\alpha_3} \phi^{-\alpha_4} & \phi_{34} < \phi \end{cases}. \quad (5)$$

Here ξ_0 is the amplitude, α_i is the power law exponent of the regime i, ϕ_{ij} is the crossover polymer fraction from regime i to regime j. This relation is found to yield not only an excellent description of our data but also α_i and ϕ_{ij} with relatively small standard deviations as shown in Figure 3(b). The obtained exponents are roughly equal to 0.75, 0.5, 1, and 2 with increasing concentration. This is a result in excellent agreement with the polymer fraction dependence theory and the extension to the flexible rod model. Furthermore, according to the theory, the crossover concentrations are expected to occur at:

$$\tilde{\phi}_{12} = \frac{3}{4\pi} \frac{(1-2\chi)}{n^3}, \quad \phi_{23}^\dagger \approx (1-2\chi). \quad (6)$$

Here χ is the Flory-Huggins interaction parameter, and n, which is proportional to the number of bonds in a Kuhn

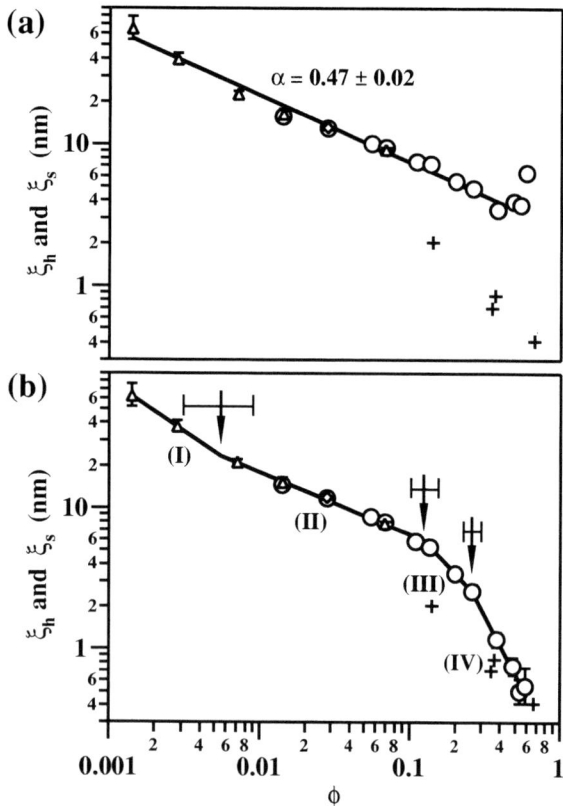

FIGURE 3. (a) Polymer volume fraction, ϕ, dependence of ξ_h at 300 K obtained using bulk viscosity data. The solid line is a least square fit with a single power law, i.e. $\xi_h \sim \phi^{-\alpha}$. (b) The ϕ dependence of ξ_h at 300 K obtained using local viscosity data. The solid line represents a least square fit line using equation 5. The obtained exponents are 0.7 ± 0.1, 0.44 ± 0.03, 1.1 ± 0.3 and 2.0 ± 0.1, corresponding to the (I), (II), (III) and (IV) regimes. The arrows indicate the crossover concentrations with error bars. For comparison, literature data for ξ_s (the crosses) at room temperature are also included in both figures[15]. The data symbols in both figures are used in the same in Figure 2.

segment, is given by $C_\infty/6$, where C_∞ is the characteristic ratio[3]. In our system, $n = 1.36$ and $\chi \approx 0.42$[6], which yields $\tilde{\phi}_{12} \approx 0.01$ and $\phi_{23}^\dagger \approx 0.1$. This is in nice agreement with the values $\phi_{12} = 0.006 \pm 0.003$ and $\phi_{23} = 0.12 \pm 0.02$ that we obtained from the curve-fit. The third crossover at ϕ_{34} is predicted to occur when ξ_h and ξ_s reach the persistence length (≈ 1 nm)[14]. Therefore, ϕ_{34} is also consistent with the prediction since the third crossover occurs at $\phi \approx 0.3$, where both the correlation lengths are roughly equal to the persistence length.

CONCLUSIONS

We have examined the concentration dependence of ξ_h in solutions of poly(methyl methacrylate) in propylene carbonate ranging from the semidilute to the highly concentrated regime at 300 K. Without taking account of η_{local}, we observed a single power law without any crossover, in contradiction to the theoretical expectations. However, after taking into account η_{local}, sharp crossovers between a good solvent, a marginal solvent, a theta solvent, and a flexible rod behavior are observed. This is in excellent agreement with the polymer fraction dependence theory proposed two decades ago.

ACKNOWLEDGMENTS

Financial support from the Swedish Foundation for Strategic Research and the Swedish Research Council is gratefully acknowledged.

REFERENCES

1. de Gennes, P. G., *Macromolecules*, **9**, 594–598 (1976).
2. Brown, W., and Nicolai, T., *Colloid. Polym. Sci.*, **268**, 977–990 (1990).
3. Schaefer, D. W., *Polymer*, **25**, 387–394 (1984).
4. Schaefer, D. W., and Han, C. C."Quasielastic Light Scattering from Dilute and Semidilute Polymer Solutions," in *Dynamic Light Scattering: Applications of Photon Correlation Spectroscopy*, edited by R. Pecora, Plenum Press, New York, 1985, pp. 181–243.
5. Farnoux, B., Boué, F., Cotton, J. P., Daoud, M., Jannink, G., Nierlich, M., and de Gennes, P. G., *J. Phys. (Paris)*, **39**, 77–86 (1978).
6. Uematsu, T., Svanberg, C., Nydén, M., and Jacobsson, P., *Phys. Rev. E*, **68**, (in press in November, 2003).
7. Brown, W., Mortensen, K., and Floudas, G., *Macromolecules*, **25**, 6904–6908 (1992).
8. Horkay, F., Hecht, A. -M., Stanley, H. B., and Geissler, E., *Eur. Polym. J.*, **30**, 215–219 (1994).
9. Brown, W., Johnsen, R. M., Konak, C., and Dvoranek, L., *J. Chem. Phys.*, **95**, 8568–8577 (1991).
10. Waggoner, R. A., Blum, F. D., and MacElory, J. M. D., *Macromolecules*, **26**, 6841–6848 (1993).
11. Wang, C. H., *Progr. Colloid Polym. Sci.*, **91**, 138–141 (1993).
12. Andersson, D., Svanberg, C., Swenson, J., Howells, W. S., Börjesson, L., *Physica B*, **301**, 44–48 (2001).
13. Nicolai, T., and Brown, W., *Macromolecules*, **29**, 1698–1704 (1996).
14. Tricot, M., *Macromolecules*, **19**, 1268–1270 (1986).
15. Svanberg, C., Pyckhout-Hintzen, W., and Börjesson, L., (unpublished): ξ_s data for the solutions of fully deuterated PMMA ($M_w = 1\,587\,000$) with PC, obtained by Small Angle Neutron Scattering experiments.

Direct measurement of polymer segment orientation and distortion in shear: Semi-dilute solution behavior of a conjugated system

Elisabeth K. Hill, Yalin Wei & Dave E. Dunstan

Department of Chemical & Biomolecular Engineering
The University of Melbourne, 3010
Australia.
Author for correspondence (e-mail: *davided@unimelb.edu.au*)

Abstract. Direct measurement of the shear induced backbone segment orientation and deformation of the polymeric chromophore polydiacetylene 4-butoxycarbonylmethylurethane (4BCMU) in semi-dilute solution has been performed using an extended dichroism technique. At low shear rates the random coil, visco-elastic polymer shows orientation of the segments in the flow direction. At high shear rates the segments orient perpendicular to the flow direction with a reduction in the average conjugation length. The results show that coil shrinkage occurs for this visco-elastic polymer at high shear rates. We postulate that the initial alignment of the prolate random coils is unstable at high shear rates when the coils "ball up" in the flow.

INTRODUCTION

Understanding the molecular basis of polymer visco-elasticity is the Holy Grail of polymer physics [1]. While a significant number of theoretical works have modeled this complex and fascinating behavior, significant discrepancies between theory and experiment still exist [2,3,4]. Of particular importance is the inability of current models to predict shear thinning in simple flow [4]. The polymer is generally modeled as beads on springs to encapsulate the hydrodynamic friction and elastic deformation [5,6]. Recent Brownian dynamics simulations by Petera and Muthukumar have shown that the size of the chains shrinks for very high shear rates when hydrodynamic interactions are considered for chains in dilute solution [7].

A number of methods have been employed in order to gain an experimental understanding of complex fluid flow [1,8,9].

4BCMU solutions exhibit dramatic and reversible solvatochromism and thermochromism associated with backbone conformational changes [10]. These properties arise from the highly conjugated backbone which absorbs light in a manner such that the distribution of conjugated segment lengths correlates with the absorption spectrum [11-14]. The transition dipole of the absorption is parallel to each segment length [15,16]. Polarized light is therefore selectively absorbed by the subset of segments oriented in the polarization direction. Consequently, any flow induced change in the orientation of the polymer backbone conjugation/segment lengths results in an absorbance magnitude change as observed in a classical dichroism measurement [8]. The measured spectral shift then corresponds to a change in the distribution of the conjugation/segment lengths. This experimental system is therefore an extended dichroism measurement which enables both orientation and distortion of the polymer backbone segments to be determined in shear.

EXPERIMENTAL SECTION

4BCMU used in these experiments was prepared by the method of Patel, Chance and Witt [17] to produce a polymer with an average molecular weight of 800,000 and a polydispersity of 1.8 as determined using Dawn Wyatt/Waters GPC light scattering apparatus. The critical overlap concentration is 0.007 g/ml and intrinsic viscosity is 300 ml/g in chloroform as measured with capillary viscometry (C*[η] = 2.1). The Mark-Houwink exponent of 0.8 indicates that 4BCMU is a random coil in CHCl₃ [10]. Lim & Heeger have used light scattering to show that the persistence length of 4BCMU in chloroform is of order 2-3 nm consistent with values calculated from quantum mechanical arguments [10,15]. A contour length of 800 nm is calculated from the monomer size and molecular weight. The ratio of the contour length to persistence length of order 250 indicates that the chains exist as a random coil in this good solvent [10,11,15]. Allegra et al. have also used light scattering to interpret a persistence length of the 4BCMU of order 20 nm in CHCl₃. This yields a ratio of contour length to persistence length of 40 which may still be considered a random coil [13]. A number of studies using quasi-elastic light scattering, small angle neutron scattering and high-pressure optical absorption have also shown that the polydiacetylene chain is a random coil in chloroform solvent [14,15,16].

Optical rheometry measurements were undertaken in a custom-built Couette cell comprising a pair of co-axial quartz cylinders which were placed in the beam of a Cary 3E spectrophotometer [18,19]. A Glan-Taylor polarizer (Harrick) was included in the beam line in order to select light polarized either parallel or perpendicular to the shear direction as defined by a flow line. (See FIG. 3 of Ref. [20]). The gap to radius ratio in the Couette cell was maintained such that the shear was uniform across the gap (gap = 0.025 mm, outer radius = 4.925 mm, giving a cylinder radius ratio of 0.995). An angular velocity range of 0 to 6.28 rad.s⁻¹ accurate to 1% was used. All measurements were performed in laminar flow as confirmed by direct visual observation of 0.3 μm tracer particles in the polymer solutions. Laminar flow was observed for shear rates significantly higher than those reported in this work in accord with theory [51]. The absence of thermal effects that could markedly affect the spectra was demonstrated by reproducing the data three times with fresh solution, using short shear periods and long equilibration intervals between measurements. No systematic trends with time indicative of shear-induced temperature increases were observed. The different shear rates were measured by random variation with reversibility observed.

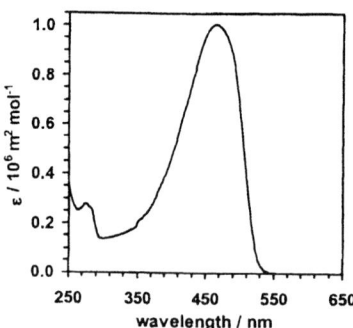

FIG 1. Absorption spectrum for concentrated 4BCMU in chloroform (11 mg/ml) recorded with unpolarized light and no applied shear.

The zero shear absorption spectrum for this polymer is provided in FIG. 1. The absorption maximum occurs at 468 nm and is characteristic of 4BCMU in its random coil form. The effect of small spectral shifts on the extinction coefficient values was modeled with simple quadratic curve fitting procedures. It was confirmed that, because the spectrum exhibits a broad flat peak, small changes in the maximum absorption wavelength (<8 nm) do not introduce significant errors to the measured absorbance changes.

The rheological properties of 4BCMU have been previously measured in the semi-dilute range, and exhibit visco-elastic behavior typical of polymers in semi-dilute solution [1,19].

RESULTS AND DISCUSSION

Rheooptic measurements of semi-dilute solutions of 4BCMU in CHCl₃ show complex orientational behavior as a function of shear rate as shown in FIGs. 2a. and 2b.

The data show an initial increase (positive change) in the extinction coefficient parallel to the shear direction, with a complementary decrease in the perpendicular direction (negative change). This is consistent with an increase in the proportion of segments aligned parallel to the shear direction. At shear rates greater than 500 s⁻¹ the trend is reversed. A decrease is observed in the parallel extinction coefficient, whereas the perpendicular extinction efficient increases relative to the zero shear case. This is consistent with an increase in the proportion of polymer segments aligned perpendicular to the shear direction. This data uniquely shows that the ensemble of polymer segments orient parallel to the shear direction at low shear rates and align perpendicular to the shear at high shear rates. It should be noted that these shear-induced changes in extinction coefficient represent up to 8 % of the measured quiescent values. The low shear data is consistent with prior birefringence studies on other polymer systems that have reported polymer alignment in the flow direction [21]. Taylor et al. have shown that no birefringence is observed for semi-dilute solutions of the yellow form of 4BCMU in THF in extensional shear [21].

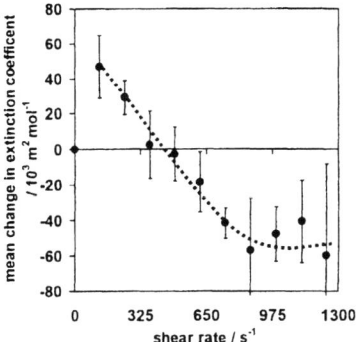

FIG 2a. The measured change in extinction coefficient for concentrated 4BCMU in chloroform versus shear rate with 0° polarized light (parallel to the shear direction) at λ = 468 nm (λ_{max} in zero shear). The data points were determined as mean values from repeated measurements of three concentrations in the semi-dilute range (8, 11 and 13 mg/ml). All samples yield similar data within the error bars shown. The curve drawn through the points serves merely to guide the eye.

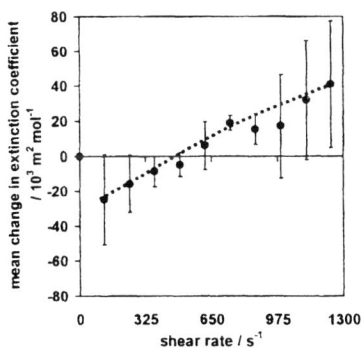

FIG 2b. The change in extinction coefficient for concentrated 4BCMU in chloroform versus shear rate with 90° polarized light (perpendicular to the shear direction). All other conditions were as described above.

Interestingly, orientational behavior similar to that observed in the present work at high shear rates has previously been reported in one rheooptic birefringence study of high molecular weight polystyrene in dioxane in simple shear flow [22].

Further to this measured complex orientational behavior, the 4BCMU polymer coil exhibits a small shear-dependent blue shift with increasing shear rate. A spectral shift of this nature corresponds to a change in the distribution of conjugation lengths to shorter values. We conclude that this is a shear-induced effect, since thermal effects have been eliminated from these experiments and shear induced polymer aggregation would lead to a red shift in the spectra [12]. FIG. 3 shows the normalized difference spectra for 4BCMU at a number of shear rates. All spectra were normalized to a peak value of 1 at λ_{max} and the normalized zero shear spectrum was subtracted from each one. This method largely removes changes in extinction coefficient due to orientation and the shear induced spectral shift is revealed. It should be noted that a change in the conjugation length by one monomer unit corresponds to a 7 nm shift in the absorbance wavelength [11]. The data shows that there is an increase in the proportion of shorter conjugation lengths within the polymer ensemble, and a simultaneous reduction in the relative number of longer conjugation lengths in shear.

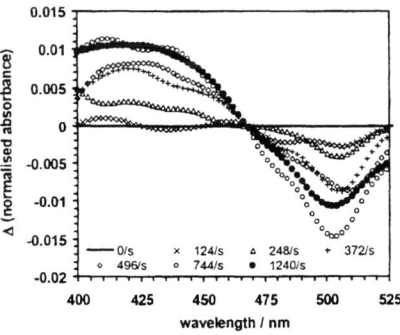

FIG 3. Normalized absorbance difference spectra for concentrated 4BCMU in chloroform (11 mg/ml), measured at a range of shear rates. The zero shear spectrum was normalized and subtracted from each normalized spectrum collected in shear. Unpolarized light was used for this experiment.

The Rouse/Zimm-type theories assume fixed segment lengths, with non-local coil deformation arising at length scales above the segment size [2,3,23,24]. The current experiment senses local, segmental effects and shows that these systematically reduce in size in shear. We hypothesize that this demonstrates that a reduction in the global polymer size is occurring in flow at high shear. As the segments orient perpendicular to the shear direction the average segment length decreases at high shear rates. The coil will therefore be reduced in size at the higher shear rates and not stretched as much of the theory would suggest [2,3] These results confirm the recent Brownian dynamics simulations of bead-rod chains in shear flow by Petera and Muthukumar [7].

In summary, the experimental findings of this work demonstrate that, in semi-dilute solution, 4BCMU polymer segments align with the shear direction at low shear rates. At elevated shear rates (above 500 s^{-1}) the proportion of segments aligned perpendicular to the shear direction increases. A measurable distortion, namely a reduction in the average conjugation lengths of the polymer, is observed for this visco-elastic polymer solution at high shear rates. We conclude that a reduction in the ensemble segment lengths occurs at high shear consistent with a reduction in the size of the polymer confirming the recent Brownian dynamics simulations [7].

ACKNOWLEDGEMENTS

YW would like to gratefully acknowledge the University of Melbourne for the award of a postgraduate scholarship.

REFERENCES

1. Larson, R.G., *The Structure and Rheology of Complex Fluids*, Oxford University Press, New York, 1999.
2. Doi, M. and Edwards, S.F., *The Theory of Polymer Dynamics*, Oxford University Press, Oxford, 1986.
3. de Gennes, P.G., *Scaling Concepts in Polymer Physics*, Cornell University Press, Ithaca, 1991.
4. Mead, D.W., Larson, R.G. and Doi, M., *Macromolecules* **31**, 7895-7914 (1998).
5. Flory, P.J., *Principles of Polymer Chemistry*, Cornell University Press, London, 1953.
6. Bird, R. B., Curtis, C. F., Armstrong, R. C. and Hassager, O., *Dynamics of Polymeric Liquids: Kinetic Theory*, Wiley Interscience, New York, 1987.
7. Petera, D. and Muthukumar, M., *J. Chem. Phys.* **111**, 7614-7623 (1999).
8. Fuller, G. G., *Optical Rheometry of Complex Fluids*, Oxford University Press, New York, 1995.
9. *Flow Induced Structure in Polymers*, edited by Nakatani, A.I. and Dadmun, M.D., ACS Symposium Series 597, American Chemical Society, Washington DC, 1995.
10. Lim, K. C. and Heeger, A. J., *J. Chem. Phys.* **82**, 522-528 (1985).
11. Lim, K. C., Fincher, C. R. and Heeger, A. J., *Phys. Rev. Lett.* **50**, 1934-1938 (1983).
12. Peiffer, D. G., Chung, T. C., Schulz, D. N., Agarawal, P. K., Garner, R. T. and Kim, M. W., *J. Chem. Phys.* **85**, 4712-4728 (1986).
13. Allegra, G., Bruckner, S., Schmidt, M. and Wegner, G., *Macromolecules* **19**, 399-412 (1986).
14. Chu, B. and Xu, R. L., *Acc. Chem. Res.* **24**, 384-387 (1991).
15. Patel, G. N. and Miller, G. G., *J. Macromol. Sci. Phys.* **B20**, 111-114 (1981).
16. Wolfe, D. and Baker, G. L., *Polymer Preprints* **35**, 275-277 (1994).
17. Patel, G. N., Chance, R. R. and Witt, J. D., *J. Chem. Phys.* **70**, 4387-4394 (1979).
18. Gason, S. J., Dunstan, D. E., Smith, T. A., Chan, D. Y. C., White, L. R. and Boger D. V., *J. Phys. Chem. B* **101**, 7732-7737 (1997).
19. Gason, S. J., *Ph. D. Thesis*. University of Melbourne, Melbourne, 2001.
20. Larson, R. G., Shaqfeh, E. S. G. and Muller, S. J., *J. Fluid. Mech.* **218**, 573-581 (1990).
21. Taylor, M. A., Batchelder, D. N. and Odell, J. A., *Polymer Commun.* **29**, 253- 255 (1988).
22. Frisman, E. V. and Mao, S., *Vysokomol. Soedin.* **6**, 34-38 (1964).
23. Rouse, P. E., *J. Chem. Phys.* **21**, 1272-1281 (1953).
24. Zimm, B. H., *J. Chem. Phys.* **24**, 269-278 (1956).

Polymer Film Dynamics with Coherent X-Ray Scattering

Hyunjung Kim[1], A. Rühm[2], L. B. Lurio[3], J. K. Basu[4], J. Lal[5], S. G. J. Mochrie[6], and S. K. Sinha[7]

[1] Department of Physics, Sogang University, Seoul 121-742, Korea
[2] Max-Planck-Institut für Metallforschung, Stuttgart, Germany
[3] Department of Physics, Northern Illinois University, DeKalb, IL 60115, USA
[4] Materials Research Laboratory, University of Illinois, Urbana-Champaign, IL 61801, USA
[5] Intense Pulsed Neutron Source, Argonne National Laboratory, Argonne, IL 60439, USA
[6] Departments of Physics and Applied Physics, Yale University, New Haven, CT 06520, USA
[7] Department of Physics, University of California San Diego, La Jolla, CA 92093 and LANSCE, Los Alamos National Laboratory, Los Alamos, NM 87545, USA

Abstract. Surface x-ray photon correlation spectroscopy (SXPCS) is applied for probing the dynamics of surface height fluctuations as a function of lateral length scale. We present the first experimental verification of the theoretical predictions for the thickness, wave vector, and temperature dependence of the capillary wave relaxation times for the supported polymer films above the glass transition temperature. Measurements were performed on polystyrene (PS) films of thicknesses varying from 84 to 333 nm.

INTRODUCTION

The transition from a molten state to a glass is one of the least-well understood phenomena in physics. To study this transition, many investigators have turned to polymers for experimental and theoretical insight [1]. An especially important effect that occurs in polymer films is an unusual depression in the glass transition temperature compared with the bulk material [2-7]. Among the proposed explanations for this effect is the notion that a low-viscosity surface layer remains at temperatures below the glass transition temperature, even as the rest of the film enters the glass state. According to this view, the surface effect is minimal but starts to dominate the polymer behavior as the film thickness decreases.

The slow modes of surface in viscoelastic liquid films have been calculated to be strongly overdamped capillary waves where the relaxation times are determined by the viscosity, the surface tension, the film thickness, and the wavelength of the capillary waves [8,9]. However, it is unclear to what extent these theories are still valid in situations where the film thickness approaches the typical length scale of the polymer chains, i.e., the radius of gyration (R_g), and how satisfactorily fluctuations over certain lateral length scales are described by these theories. In this paper, we present a new method of x-ray photon correlation spectroscopy (XPCS) for probing the dynamics of surface height fluctuations as a function of lateral length scale. This emerging technique applies the principles of dynamic light scattering in the x-ray regime. The short wavelength and slow time scales characteristic of XPCS extend the phase space accessible to scattering studies beyond some restrictions by light and neutron. In surface x-ray photon correlation spectroscopy (SXPCS), highly coherent x-rays using reflection geometry impinges on the polymer surface at grazing incidence.

METHODS AND MATERIALS

Our films were prepared by dissolving polystyrene (PS) of M_W=123 000 g/mol (M_W/M_n=1.08) in toluene and then spin casting onto optically flat silicon substrates. These samples were then annealed in vacuum for 12 h at 150°C to ensure complete solvent removal. The thicknesses of the PS films investigated by XPCS were 84, 170, 177, 312.5, and 333 nm. The XPCS experiments were performed at beam line 8-ID at the Advanced Photon Source (APS). A Germanium (111) channel-cut double-bounce monochromator located 64.4 m from the source and an adjustable slit

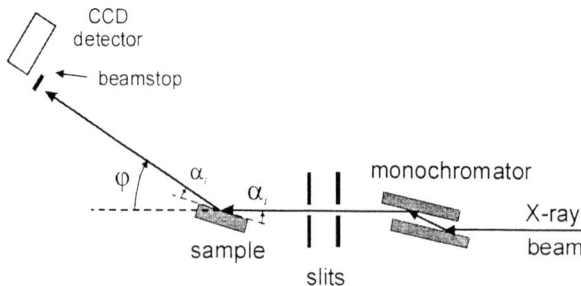

FIGURE 1. The schematic diagram of the experimental setup for SXPCS.

system 40 cm upstream of the sample are used to prepare a coherent monochromatic x-ray beam with a wavelength of 0.16 nm and a cross section of 20×20 μm^2. In this configuration, the charge-coupled device (CCD) pixel size of 22.5×22.5 μm^2 corresponds to the theoretical speckle size. The experimental geometry is illustrated schematically in Fig. 1. By arranging for the x-ray incidence angle (0.14°) to lie below the critical angle for total external reflection (0.16°) we were able to restrict the x-ray penetration into the film to a depth of 9 nm, far less than any of the film thicknesses studied here. Thus, scattering from the film-substrate interface is negligible, and only fluctuations of the polymer/vacuum interface are probed. In-plane wave vectors up to 10^{-2} nm^{-1}, where van der Waals interactions are negligible [10], could be accessed. The off-specular diffuse scattering [11] from the polymer surface was recorded with a direct-illumination CCD camera, located 3545 mm downstream of the sample.

The polymer surface is partially coherently illuminated, giving rise to a speckled scattering pattern which varies in time as the surface modes undergo random thermal fluctuations. The normalized intensity-intensity time autocorrelation function, g_2, then yields the sample's surface dynamics. The typical flux was 8×10^8 photons/sec. To avoid x-ray sample damage, the x-ray exposure of any position on the sample was limited to about 10 minutes after which time the sample was shifted to illuminate a fresh area.

RESULTS AND DISCUSSION

The normalized intensity-intensity time autocorrelation function is

$$g_2(\mathbf{q},t) = \frac{\langle I(\mathbf{q},t')I(\mathbf{q},t+t')\rangle}{\langle I(\mathbf{q},t')\rangle^2} \quad (1)$$

where $I(\mathbf{q},t')$ is the scattering intensity at wave vector transfer $\mathbf{q}$ at time t'. In Eq. (1), the angular brackets refer to averages over time t' and t denotes the delay time [12]. In Fig. 2, experimental correlation functions acquired from the 84nm-thick film at 160°C are shown as symbols for four different in-plane wave vectors $q_\parallel$. The lines show single-exponential fits, i.e., $g_2 = 1 + \beta \exp(-2t/\tau)$, where β is the speckle contrast and $\tau = \tau(q_\parallel)$ is the relaxation time for equilibrium surface height fluctuations. The $q_\parallel$ dependence of the best-fit relaxation times is displayed in Fig. 3(a) at three different temperatures for the 177 nm-thick film. The time constants for 170°C are fastest, those for 160°C are slower, and those for 150°C are slowest. In each case, however, the $q_\parallel$ dependence appears to be similar, with larger length-scale asperities relaxing more slowly than smaller ones. Shown in Fig. 3(b) are the relaxation times at 160°C for three films of different thickness. The time constants decrease monotonically with increasing thickness. The fact that the surface dynamics depend on film thickness demonstrates that, although our measurements are

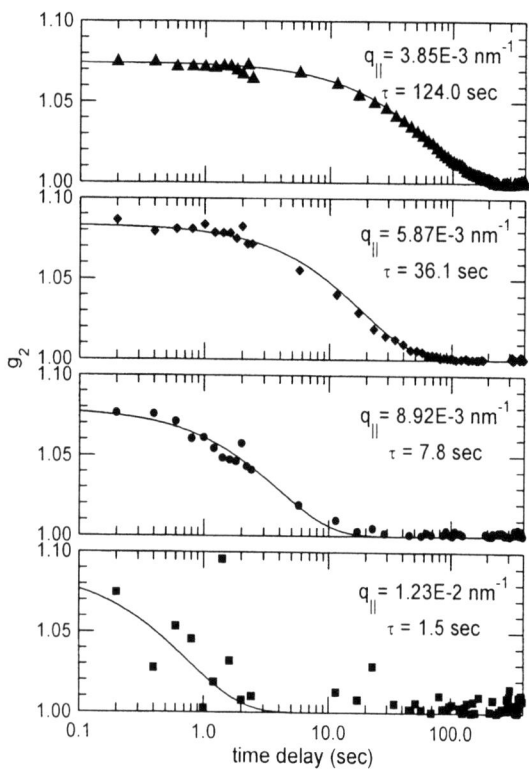

FIGURE 2. Autocorrelations obtained at four different in-plane wave vectors measured on a sample thickness of 84 nm at 160 °C (shown as symbols), compared with single-exponential fits (shown as lines). The time constant τ at each $q_\parallel$ is also displayed.

sensitive only to surface motions, those motions in turn depend on molecular movements throughout the film. From the theory [9] of the dynamics of capillary waves on viscous liquid films, we have earlier deduced [13] the expression for the relaxation time τ for capillary waves as functions of the viscosity, η, the surface tension, γ, the thickness, h, and the in-plane wave vector, $q_\parallel$, in the overdamped regime. From that expression, τ/h should be solely a function of $q_\parallel h$ and directly proportional to the ratio η/γ as shown in Eq. (2).

$$\frac{\tau}{h} \approx \frac{2\eta}{\gamma} \frac{\left[\cosh^2(q_\parallel h) + (q_\parallel h)^2\right]}{(q_\parallel h)\left[\sinh(q_\parallel h)\cosh(q_\parallel h) - q_\parallel h\right]} \quad (2)$$

To test the predicted scaling behavior, we have plotted the quantity τ/h versus $q_\parallel h$ for different film thicknesses at 150, 160, and 170°C in Fig. 4. At each temperature, data from different samples collapse to form a single curve, confirming the anticipated scaling with film thickness.

Experiment and theory [Eq. (2)] show excellent agreement with a single fit parameter, the ratio η/γ. Knowing the surface tension (γ) at each temperature [14], we may obtain the film viscosity (η), because the fits to τ/h versus $q_\parallel h$ determine η/γ. The viscosity obtained from these fits is plotted versus temperature in the inset of Fig. 4 together with the corresponding bulk viscosity interpolated from Ref. [15] for a molecular weight of 123 000. Evidently, bulk and film viscosities agree within the accuracy of our measurements. However, the temperature dependence of the film viscosity might be weaker than that of the bulk viscosity.

In addition, it was possible to place limits on the extent to which viscosity inhomogeneities in the film were present. A Navier-Stokes model was used to calculate relaxation times for a film with two layers having different viscosities but the same density and no interfacial tension. This model did not fit the observed data as well, however [13]. By using this method, we were able to rule out a surface layer thicker than 10 nm having one-tenth the bulk viscosity.

In summary, using SXPCS we measured the relaxation times of overdamped capillary waves for thin polystyrene films of molecular weight 123 000 at various temperatures above T_g, and we verified scaling relations for τ/h vs. $q_\parallel h$ as predicted by the theory of such capillary waves. We used these results and an analysis of the static off-specular scattering to obtain viscosities and surface tensions for these films at various temperatures which are in good agreement

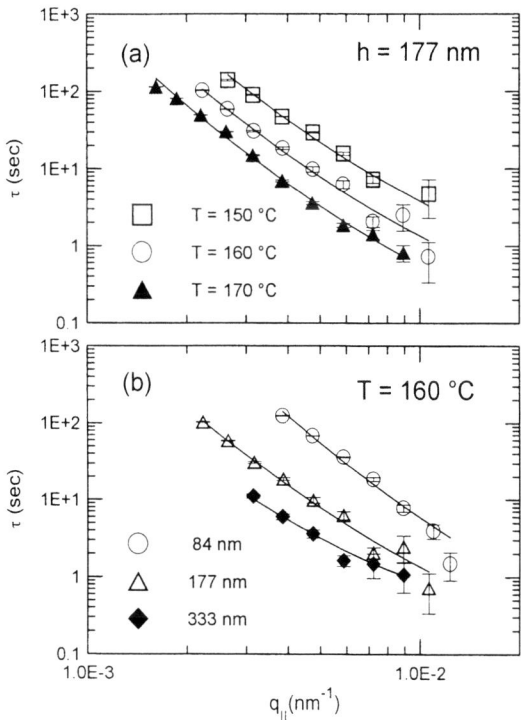

FIGURE 3. (a) Measured time constant τ vs. in-plane wave vector $q_\parallel$ for the 177 nm-thick film at 150°C (squares), 160°C (circles), and 170°C (triangles). (b) τ vs. $q_\parallel$ at 160 °C for films of thickness 84 nm (circles), 177 nm (triangles), and 333 nm (diamonds).

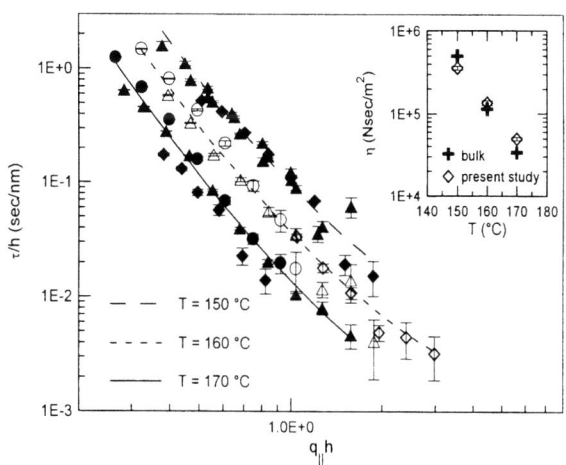

FIGURE 4. τ/h vs. $q_\parallel h$ for film thicknesses 84 nm (circles), 170, 177 nm (triangles), and 312.5, 318, 333 nm (diamonds). Lines represent least-squares fit based on Eq. (2).

with bulk values interpolated to the above molecular weight. The investigations to a wider range of M_W and thinner films are in progress in order to test what is the dominating length scale controlling the surface dynamics.

ACKNOWLEDGMENTS

H. K. acknowledges the supports from the contribution of Advanced Backbone IT Technology Development Project (IMT2000-B3-2) of the Ministry of Information and Communication and the Sogang University Research Grants in 2003. Work at MIT and Yale was supported by the NSF (DMR 0071755). Work was also partly supported by NSF (DMR-0209542). 8-ID is supported by the DOE Facilities Initiative Program DE-FG02-96ER45593 and NSERC. The APS is supported by the U.S. DOE, Office of Basic Science, under W-31-109-ENG-38.

REFERENCES

1. Forrest, J. A., and Jones, R. A. L., "The Glass Transition and Relaxation Dynamics in Thin Polymer Films," in *Polymer Surfaces, Interfaces and Thin Films*, edited by Karim, A., and Kumar, S., World Scientific, Singapore, 2000, pp. 251-294.

2. Keddie, J. L., Jones, R. A. L., and Cory, R. A., *Europhys. Lett.*, **27**, 59-64 (1994).

3. DeMaggio, G. B., et al., *Phys. Rev. Lett.* **78**, 1524-1527 (1997).

4. Kawana, S and Jones, R. A. L., *Phys. Rev. E* **63**, 021501 (2001).

5. Mayes, A. M., *Macromolecules* **27**, 3114-3115 (1994).

6. de Gennes, P. G., *Eur. Phys. J. E* **2**, 201-205 (2000).

7. McCoy, J. D. and Curro, J. G., *J. Chem. Phys.* **116**, 9154-9157 (2002).

8. Harden, J. L., Pleiner, H., and Pincus, P. A., *J. Chem. Phys.* **94**, 5208-5221 (1991).

9. Jäckle, J., *J. Phys. Condens. Matter* **10**, 7121-7131 (1998).

10. Seemann, R., Herminghaus, S. and Jacobs, K., *Phys. Rev. Lett.* **86**, 5534-5537 (2001).

11. Sinha, S. K., et al., *Phys. Rev. B* **38**, 2297-2311 (1988): Braslau, A. et al., *Phys. Rev. Lett.* **54**, 114-117 (1985): Sanyal, M. K. et al., *Phys. Rev. Lett.* **66**, 628-631 (1991).

12. Lumma, D., et al., *Rev. Sci. Instrum.* **71**, 3274-3289 (2000).

13. Kim, H., et.al., *Phys. Rev. Lett.* **90**, 068302 (2003).

14. Wu, S., "Surface and Interfacial Tensions of Polymers, Oligomers, Plasticizers, and Organic Pigments," in *Polymer Handbook* edited by Brandrup, J., Immergut, E. H., and Grulke, E. A., Wiley, New York, 1999, 4th ed., p. VI-540.

15. Plazek, D. J. and O'Rourke, V. M., *J. Polym. Sci. A2* **9**, 209-243 (1971).

Rearrangement dynamics in lamellar forming block copolymers under an electric field

K.S. Lyakhova*, A.V. Zvelindovsky* and G.J.A. Sevink*

Leiden University, LIC, P.O. Box 9502, 2300 RA Leiden, The Netherlands

Abstract. By means of dynamic self-consistent field simulation we study the phase behavior of thin block copolymer films under an electric field. We concentrate on the rearrangement dynamics in lamellar forming systems. We present a dynamic picture of defect formation and annihilation.

INTRODUCTION

Block copolymers are fascinating materials capable of forming a large variety of soft nanostructures depending on the chemical composition of blocks and the architecture of relatively simple molecules [1]. A challenging task for the application of block copolymers in nanomanufactured devices is to form perfect arrays of nanostructures. One possible way to do so is by the application of an external field to a block copolymer sample, which is very often a thin film. Examples of such fields commonly used in experimental practice are shear flow [2], electric field [3] and temperature gradients [4]. All these fields affect block copolymers in a different fashion. A systematic study of the phase behavior under each of these fields is therefore required.

Recently we have shown that the phase behavior of cylinder forming block copolymers in thin films is extremely complex even without external fields [5]. The experimental situation is found to be well explained by means of dynamic self-consistent field theory (known as dynamic density functional theory [6]). Two parameters are found to control the phase behavior in a thin films, namely the film thickness and an effective interaction of blocks with the film surfaces. In the present work we describe the phase behavior of symmetric diblock copolymer system under the influence of an externally applied electric field. Some preliminary results have been recently reported in [7, 8, 9, 10].

Most of the previous theoretical analysis is based on the calculation of an extra contribution to the free energy due to electrostatics for different orientations of structures, and then comparing the energies of the static patterns [11, 12, 13, 14]. As to the dynamical behavior, Onuki and Fukuda examined the dynamics of undulation instability of lamellae induced by the electric field [15].

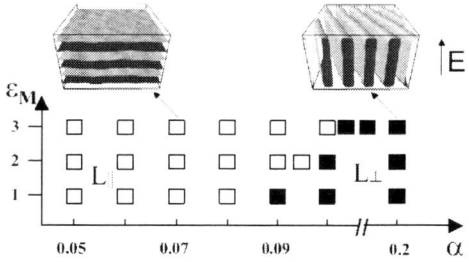

FIGURE 1. Phase diagram of A_4B_4-copolymer film. The surfaces are at the top and the bottom of the box. The electric field is applied in the vertical direction. Only the isodensity surfaces of A-component with mean value 0.5 are shown.

To answer a question *how* a phase transition proceeds in time one needs to develop a different approach and to examine a dynamic equation of pattern evolution. The knowledge of *dynamical pathways* of phase transitions is essential in understanding fundamentals of block copolymer physics. One of the justifying examples is the fact that the metastable states, which block copolymer system visits during its evolution, can be very long-living. Often the system can be trapped in those states so long, that they require specially designed experiments to distinguish them from an equilibrium phase [16].

METHOD

The phase separation can be monitored by the scalar order parameter $\psi(\mathbf{r},t)$, which is the normalized deviation of the density of a polymer component from its average value. In the case of an incompressible diblock copolymer melt the system is described by only a single order parameter. Although a diblock copolymer solution requires an extra order parameter for the solvent, we use

FIGURE 2. Dynamics of orientational phase transition from Fig. 1: $\varepsilon_M = 1$ and $\tilde{\alpha} = 0.1$ (a); $\varepsilon_M = 3$ and $\tilde{\alpha} = 0.12$ (b). From the top to the bottom: 2100, 2300, 2400, 2600, 3400, 8000 timespteps.

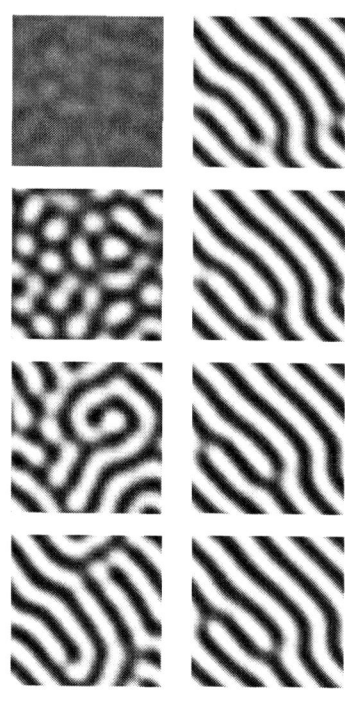

FIGURE 3. The dynamics of the phase transition for the system from the Fig. 2b shown via 2D slices at $z =$ parallel to the xy plane. From the top to the bottom and from the left to the right: 2100, 2200, 2300, 2600, 2800, 8000 timesteps.

a simplified model with only one order parameter in the present study. As we have shown recently, such description is well justified and gives an excellent agreement with experiments in the case of a nonselective or almost nonselective solvent [5]. The time evolution of the order parameter in the simplest case follows a diffusion type equation [17]

$$\dot{\psi} = M\nabla^2 \mu + \eta \qquad (1)$$

with the constant mobility M, the thermal noise η, and a proper choice of the boundary conditions [18]. The chemical potential in the presence of an electric field **E** has the form [19]

$$\mu = \mu^0 - \frac{E^2}{8\pi}\frac{\partial \varepsilon}{\partial \psi} \qquad (2)$$

where μ^0 is the chemical potential in the absence of the electric field, and ε is the dielectric constant of the polymeric material, which can be approximated as

$$\varepsilon \approx \varepsilon_0 + \varepsilon_1 \psi$$

for small ψ. The electric field inside the material **E** deviates from the applied electric field $\mathbf{E_0} = (0,0,E_0)$ satisfying the Maxwell equation

$$\mathrm{div}\varepsilon\mathbf{E} = \mathbf{0} \qquad (3)$$

where

$$\mathbf{E} = \mathbf{E_0} - \nabla\varphi.$$

Keeping only leading terms, one can rewrite [1] in the form

$$\dot{\psi} = M\nabla^2 \mu^0 + \alpha\nabla_z^2 \psi + \eta \qquad (4)$$

$$\alpha \equiv ME_0^2 \frac{\varepsilon_1^2}{4\pi\varepsilon_0} \qquad (5)$$

The chemical potential without the electrostatic contribution μ^0 is calculated using self-consistent field theory for the ideal Gaussian chains with the mean field interactions between copolymer blocks A and B, described by a parameter ε_{AB} [18].

The model system we study in the following is an A_4B_4 melt in three dimensions with mean field interactions $\varepsilon_{AB} = 7$ J/mol. For the simulations, the electric field strength is parameterized by

$$\tilde{\alpha} \equiv \frac{\alpha}{kTMv}. \qquad (6)$$

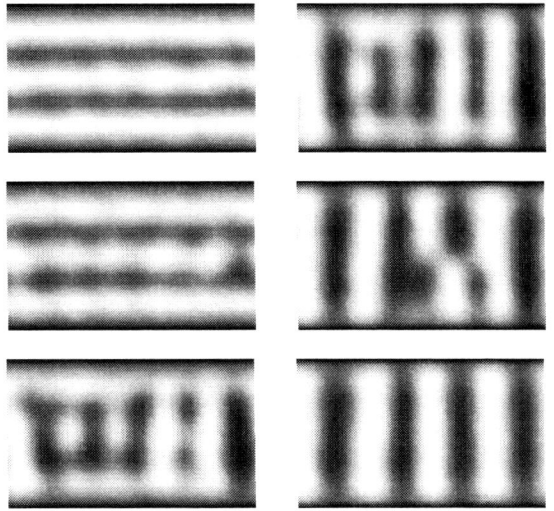

FIGURE 4. The dynamics of the phase transition for the system from the Fig. 2b shown via 2D slices at $x = 16$ parallel to the zy plane. From the top to the bottom and from the left to the right: 2100, 2200, 2300, 2400, 3000, 8000 timesteps.

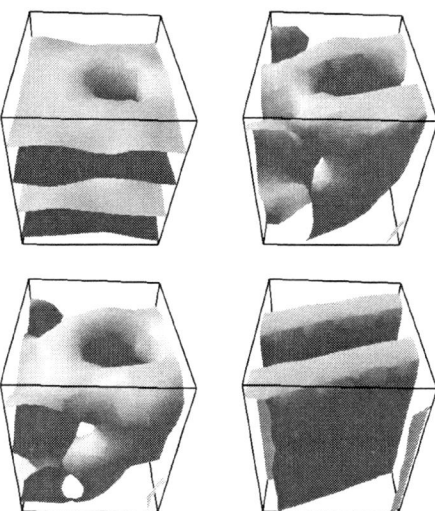

FIGURE 5. Individual defect dynamics of the phase transition for the system from the Fig2a (crop of a large box). From the top to the bottom and from the left to the right: 2200, 2300, 2400, 8000 timesteps.

We consider the thin film confined between two solid interfaces. All simulations were performed in a box of a size $32 \times 32 \times 20$. The two substrates are located at $z = 1$ and $z = 20$, and span the box in the x and y direction completely. The rigid wall boundary condition are used at the substrates while in x and y direction the periodical boundary conditions are applied. The interaction of A and B blocks with the substrates is described by parameter $\varepsilon_M = \varepsilon_{AM} - \varepsilon_{BM}$, which shows preferential attraction of one of the blocks to the surface in comparison with another. We are interested in a generic qualitative picture of the transitions, therefore, all analysis is done in dimensionless units.

RESULTS AND DISCUSSION

We have chosen a system that forms three alternating pairs of parallel lamella in the absence of electric field. Competition between preferential interaction of one of the polymer blocks with the film surface and electrostatic forces results into the phase diagram shown in figure 1. For all points of the diagram first the phase separation was reached without electric field (dimensionless time $\tau = 2000$). Then the electric field was applied and all the simulations were carried till $\tau = 8000$. When the electric field increases above some critical value all lamella swap their orientation. The stronger the interaction with the surface, the larger the critical value of electric field is. The coexistence region of these two phases is a line: there were no mixed morphologies observed as stable structures for this film thickness.

In figure 2 we show the dynamics of the orientational transition from the lamella parallel to the surface ($L_\parallel$) to the lamella perpendicular to the surface ($L_\perp$) for two values of surface interaction ε_M: a) $\varepsilon_M = 1$ and b) $\varepsilon_M = 3$. In both cases the initial stage is $L_\parallel$. After the electric field is applied (case a)) first undulations appear. The undulation lead to a mixed structure ($\tau = 2400$) which quickly changes to a $L_\perp$ structure ($\tau = 2600$) with some defects. These defects disappear quite quickly and the defectless $L_\perp$ is formed throughout the slit. The situation in case b) is somewhat different. The initial stages are the same, but the defect in the $L_\perp$ observed at $\tau = 2600$ is persistent even after a long time of simulation. Only a small defect rearrangement is visible between $\tau = 2600$ and $\tau = 8000$ while in case a) all defects already disappeared. In figure 3 we show the process of transition of the system from figure 2b in detail by considering two-dimensional (2D) slices though the slit in the direction parallel to the solid surface: the xy plane. First we see that the value of the density of A-component is almost homogeneous, corresponding to parallel lamellae. After the electric field is switched on, we see formation of holes in the lamella, corresponding to lamellar undulations. The regions of perpendicular structures that are formed when time progresses, rearrange themselves quickly into parallel stripes. It is remarkable that the whole process of orientational phase transition as seen in the 2D slices is very similar to a phase separation of a 2D lamellar system following a temperature quench. First the system is homogeneous, then fluctuations lead to the formation of spherical micelles, followed by their merge to elongated micelles, and finally to a formation of lamellar stripes. This suggest that the paths through the energy landscapes of these two phenomena are very similar.

In figure 4 we observe 2D slices of the same process but in the perpendicular yz direction. The process in this plane is totally different from the described above. The system is phase separated in all images, which reflects the phase separation in 3D box. The only observed process is deformation and rearrangement of domains. Remarkably, the reorientation starts in the middle of the box, whilst the lamellae next to the walls survive longer in their orientation parallel to the wall.

Figure 5 shows a detailed insight in the topological nature of undulations, appearance and time evolution of defects. They start as holes between neighboring lamellae with a well established overshooting rim around. While the hole extends, the rim falls apart in various nuclii for the new lamellae in the perpendicular direction.

CONCLUSION

We gave a detailed description of rearrangement dynamics of lamellar forming system under applied electric field. Dynamic picture of defect formation and annihilation is presented. The extension of our of dynamic self-consistent field simulation for cylinder and sphere forming block copolymers under electric field is in progress.

ACKNOWLEDGMENTS

We thank professor T. Kawakatsu for stimulating discussions during his stay in Leiden.

REFERENCES

1. Hamley, I. W., *The physics of block copolymers*, Oxford Univ. Press, Oxford, 1998.
2. Chen, Z.-R., Kornfield, J. A., Smith, S. D., Grothaus, J. T., and Satkowski, M. M., *Science*, **227**, 1248–1253 (1997).
3. Böker, A., Elbs, H., Hänsel, H., Knoll, A., Ludwigs, S., Zettl, H., Urban, V., Abetz, V., Müller, A. H. E., and Krausch, G., *Phys. Rev. Lett.*, **89**, 135502 (2002).
4. Hashimoto, T., Bodycomb, J., Funaki, Y., and Kimishima, K., *Macromolecules*, **32**, 952–954 (1999).
5. Knoll, A., Horvat, A., Lyakhova, K. S., Krausch, G., Sevink, G. J. A., Zvelindovsky, A. V., and Magerle, R., *Phys. Rev. Lett.*, **89**, 035501 (2002).
6. Fraaije, J. G. E. M., *J. Chem. Phys.*, **99**, 9202–9212 (1993).
7. Zvelindovsky, A. V., and Sevink, G. J. A., *Phys. Rev. Lett.*, **90**, 049601 (2003).
8. Böker, A., Elbs, H., Hänsel, H., Knoll, A., Ludwigs, S., Zettl, H., Zvelindovsky, A. V., Sevink, G. J. A., Urban, V., Abets, V., and Müller, A. H. E., *Macromolecules*, **36**, 8078–8087 (2003).
9. Kyrylyuk, A. V., Zvelindovsky, A. V., Sevink, G. J. A., and Fraaije, J. G. E. M., *Macromolecules*, **35**, 1473–1476 (2002).
10. Kyrylyuk, A. V., Sevink, G. J. A., Zvelindovsky, A. V., and Fraaije, J. G. E. M., *Macromol. Theory Simul.*, **12**, 508–511 (2003).
11. Amundson, K., Helfand, E., Quan, X. N., Hudson, S. D., and Smith, S. D., *Macromolecules*, **27**, 6559–6570 (1994).
12. Tsori, Y., and Andelman, D., *Macromolecules*, **35**, 5161–5170 (2002).
13. Böker, A., Knoll, A., Elbs, H., Abetz, V., Müller, A. H. E., and Krausch, G., *Macromolecules*, **35**, 1319–1325 (2002).
14. Gurovich, E., *Phys. Rev. Lett.*, **74**, 482–485 (1995).
15. Onuki, A., and Fukuda, J., *Macromolecules*, **28**, 8788–8795 (1995).
16. Park, C., Simmons, S., Fetters, L. J., Hsiao, B., Yeh, F., and Thomas, E. L., *Polymer*, **41**, 2971–2977 (2000).
17. Onuki, A., *Phase Transition Dynamics*, Cambridge Univ. Press, Cambridge, 2002.
18. Sevink, G. J. A., Zvelindovsky, A. V., van Vlimmeren, B. A. C., Maurits, N. M., and Fraaije, J. G. E. M., *J. Phys. Chem.*, **110**, 2250–2256 (1999).
19. Landau, L. D., and Lifshitz, E. M., *Electrodynamics of Continuous Media*, Pergammon, Oxford, 1960.

Thermoreversible Gelation driven by Coil-to-Helix Transition of Polymers

Fumihiko Tanaka and Yunoshin Tamura

*Department of Polymer Chemistry, Graduate School of Engineering,
Kyoto University, Katsura, Nishikyo-ku, Kyoto 615-8510, Japan*

Abstract. This paper theoretically studies thermoreversible gelation driven by aggregation of helices formed on the polymer chains. Two fundamentally different cases of (i) multiple association of single helices, and (ii) association by multiple helices with multiplicity k (such as double helices ($k = 2$), triple helices ($k = 3$), etc.) are treated on the basis of different equations. The helix length distribution on a polymer chain (or assemble of chains for multiple helices) is derived as a function of polymer concentration and temperature. Theretical calculation of the total helix content in the solution is compared with experimental data of optical rotation in ι-carrageenan solutions at different polymer concentrations. It is shown that at low temperature there is a sharp transition from network to bundle state (pair, triplet, etc.). To confirm such a network/pairing transition, we carried out Monte Carlo simulation of polymer solution in which hydrogen-bonded zipper-like crosslinks are formed.

INTRODUCTION

The formation and materials properties of biopolymer gels and networks have been the subject of a great deal of work by many researchers. In biopolymer gelation, activation of the particular functional groups on a polymer chain accompanied by a proper three dimensional conformation change is a necessary prerequisite for the interchain cross-linking. For instance, water-soluble natural polymers such as gelatin, polysaccharides (agar, alginate, carrageenan, gellan, etc.) change their conformation from the random coil state to a partially helical state, and then the helical parts aggregate to form extended network junctions. In our recent study[1], we developed statistical-mechanical theory for the study of gelation strongly coupled to polymer conformational transitions. In this study, we focus our interest to thermoreversible gelation driven by coil-to-helix transition, and consider the case where helix formation is faster than their association. The helices formed after the solution is quenched from the high-temperature uniform liquid state quickly reach almost equilibrium length, and then, by further adjustment during the lapse of time, associate with each other into junctions of the true equilibrium length that minimizes the free energy of the network as a whole. If the solution is slowly annealed (at the rate slower than helix association), on the contrary, the equilibrium network structure at a given temperature is not the one formed at the earlier stage, so that polymers try to adjust their helix length and spatial distribution of junction zones to reach the true equilibrium from the *wrong initial conditions*. The topological constraints introduced in the network in the preceding stage is so strong that the network continues to change but can never reach equilibrium.

We consider model polymer solutions in which polymers in random-coil conformation (reference conformation) at a given temperature and concentration first form partial helices after being cooled, and then helices aggregate into multiple junctions (Figure 1).

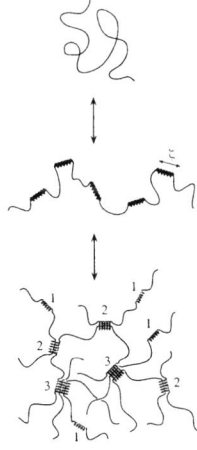

FIGURE 1. Thermoreversible gelation in helix-forming polymer solutions. In the case of multiple association of single helices, polymers form partial helices of variable length ζ upon cooling, and helices aggregate into network junctions with multiplicity indicated by the figures.

The multiplicity of a junction is defined by the number of helices combined to it. Therefore it is 1 for unassociated helices, 2 for pairwise junctions and 3 for triple junction etc.

THEORETICAL METHOD

Consider a polymer chain carrying the total number n of statistical units. In order for helices to be generated on this chain, helix sequences must be selected from the finite length n. Let j_ζ be the number of helices with length $\zeta = 1, 2, 3, \cdots, n$ (counted in terms of the statistical units). The number of possible ways to select these sequences is first counted by combinatorial consideration. We then regard these helices as functional groups and express the free energy of the reacting solution as a functional of the helix distribution function[3]. Our theoretical framework of associating polymer solutions[1, 2] has been applied to find the free energy. Finally, we minimize this free energy by changing the distribution function, and find the most probable one. It obeys a power law of the parameter $t = 1 - \nu/(1 - \theta)$ as

$$j_\zeta/n = (1 - \theta)\eta_\zeta u(z_\zeta) t^{\zeta+1}, \quad (1)$$

where η_ζ is the statistical weight of a helix of length ζ measured relative to the random coil, θ is the helix content per chain, ν the average number of helices on a chain divided by the total number n of the repeat units.

The function $u(z)$ is the junction function for giving a correct statistical weight to each multiplicity[2], and is given by

$$u(z) = \sum_{k \geq 1} \gamma_k z^{k-1} \quad (2)$$

with constant coefficients γ_k. For instance, it is

$$u(z) = 1 + z^{k-1} \quad (3)$$

for k-ple association of single helices, and

$$u(z) = z^{k-1} \quad (4)$$

for association by k-ple helices. (The first term for $k = 1$ doesn't exist in the latter because there is no isolated helix.) The papameter z_ζ is related to the polymer concentration by the relation

$$z_\zeta = (1 - \theta)\phi\lambda_\zeta \eta_\zeta t^{\zeta+1}. \quad (5)$$

The sol/gel transition point is found on the temperature-concentration plane. It is found that, in the case of k-ple helices, the condition $n\theta/\bar{\zeta} = k/(k-1)$ is fulfilled at the gel point, $\bar{\zeta}$ the average helix length. Hence, the independent measurement of θ and $\bar{\zeta}$ gives the multilicity k.

RESULTS

For double helices, we employ Zimm-Bragg form $\eta_\zeta = \sigma\lambda(T)^\zeta$ for the weight, where σ is the helix initiation factor and λ is the association constant per one hydrogen bond[3]. Figure 2 shows helix content, number of helices and average helix length as functions of the temperature for polymers carrying $n = 100$ repeat units. The temperature is measured in terms of $\ln \lambda \sim |\epsilon_A|/k_B T$. The polymer volume fraction is changed from curve to curve. The coil-to-helix transition takes place at around $\ln \lambda = 0$, and slightly shifts to higher temperature with the polymer concentration. The helix initiation parameter is fixed at $\sigma_2 = 1.0$ by assuming the simplest case where there is no restriction for a pair of chains to start winding. At high temperatures, helix content increases with polymer concentration, but at low temperatures it decreases because θ is defined by the total number of repeat units in helices on a single chain. To obtain the total helix content in the solution, the number of polymer chains must be multiplied to θ. This total content of helices in the solution is an increasing function of the polymer concentration at all temperature regions. It is expected to be proportional to the rotation angle of the polarization plane in optical measurements.

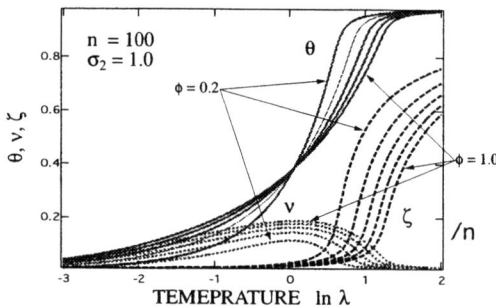

FIGURE 2. The helix content θ, number of helices ν, mean helix length $\bar{\zeta}$ per chain shown as functions of the temperature. Temperature is measure in terms of $\ln \lambda = const + |\epsilon_A|/k_B T$ by using the association constant λ per monomer. Polymer volume fraction is changed from figure to figure. The total DP of a polymer is fixed at $n = 100$. The helix initiation factor is fixed at $\sigma_2 = 1.0$. With increase in the polymer concentration, the helix number increases, while the mean helix length decreases.

Theretical calculation of the total helix content $\theta \times \phi$ in the solution is compared with experimental data of optical rotation in degraded ι-carrageenan ($n = 50$) solutions at different polymer concentrations (Figure 3). At low temperature, there is a sharp transition from network to bundle state (pair, triplet, etc.). This network-to-bundle transition becomes a real phase transition in the limit of infinite chain length.

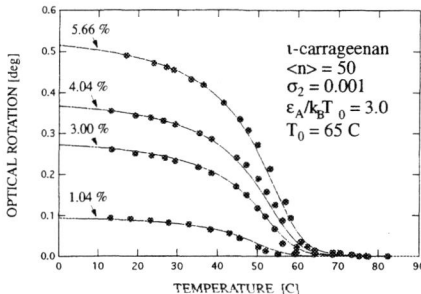

FIGURE 3. Comparison of the experimentally measured optical rotation angle (circles) and theoretically calculated total content of helices (lines) in a solution. Experimental data were obtained from degraded ι-carrageenan solution with 0.1 mol salt. Polymer concentration is varied from curve to curve. The proportionality constant between theoretical helix content and optical rotation is found from the data at the highest concentration of 5.66% measured.

MONTE CARLO SIMULATION OF ZIPPER GELS

In order to study formation of double helices and confirm network/pair transition, we carried out Monte Carlo simulation. We use bead-spring model chain with bending energy and excluded-volume interaction energy. Beads are assumed to form pairwise hydrogen bonds of energy $-\epsilon$. To simplify the problem, helices are replaced by zipper-like sequential hydrogen bonds. To assure zipper formation, we introduce additional stabilization energy $-\Delta\epsilon$ between the two neighboring hydrogen bonds (see Figure 4). When a new bond is created in the position next to the already formed bond, the total energy gain becomes $\epsilon + \Delta\epsilon$. The relation between these two energy parameters and Zimm-Bragg's parameters used in the theoretical study of coil/helix transition of biopolymers[5, 6] is given by $\sigma = \exp(-\Delta\epsilon/k_BT)$ for the helix initiation factor, and $\lambda(T) = \exp[(\epsilon + \Delta\epsilon)/k_BT]$ for the statistical weight of a monomer in the helix state. The total energy due to hydrogen bonds is therefore given by

$$H_{\mathrm{HB}} = -\epsilon \sum_{\zeta=1}^{n} \zeta J_\zeta - \Delta\epsilon \sum_{\zeta=1}^{n} (\zeta - 1) J_\zeta, \quad (6)$$

where J_ζ is the total number of zipper-like crosslinks with length ζ in the system. The length distribution function of zippers *on a single chain* is given by $j_\zeta \equiv 2 < J_\zeta > /N$, where $< J_\zeta >$ is the canonical average of the number of zippers.

Figure 5 shows (a) finite-size scaling for finding the gel concentration, and (b) phase diagram derived by the MC simulation. The gel point is defined here by the percolation point where the largest connected cluster spans the entire simulation box. As seen from Fig. 5, the average junction length reaches 0.5 at the gel point for a

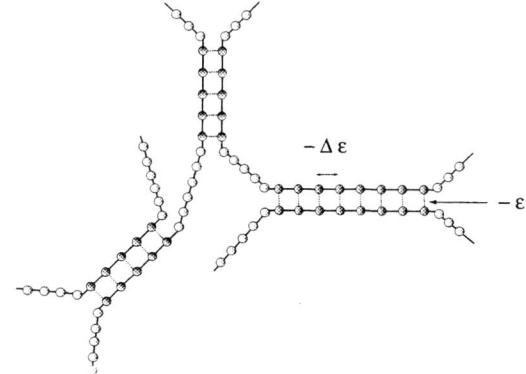

FIGURE 4. Two kinds of hydrogen-bonding energy. In addition to the usual energy gain ϵ on binding a pair of beads, we assume stabilization energy gain $\Delta\epsilon$ between the neigboring hydrogen bonds.

temperature range $\epsilon/k_BT \simeq 1 - 3$, so that thermal dissociation of junctions is highly improbable. The rheological gel point is therefore expected to be identical to the percolation point in this system. To study the effect of the system size, we calculate the percolation probability Π by changing the number N of chains as $N = 30, 60, 90, 120$ at a given temperature. It is calculated by examining whether or not the end parts of the largest cluster are cyclically connected through the boundaries of the simulation box *in all three directions*. The results are plotted against the polymer volume fraction. From such curves, the concentration interval Δ where Π lies from 0.2 to 0.8 is found for each number N, and the midpoint value ϕ^* in the region Δ is regarded as the percolation point found for a given value of N. Then, they are plotted against Δ in Figure 5(a). We extrapolate the result into $\Delta \to 0$, and find the gel point in the $N \to \infty$ limit.

In the phase diagram, large symbols show sol/gel transition line for given values of N. In our theoretical work[3], we find that the gel point for multiple association with multiplicity k is given by the condition $\nu = k/(k-1)$, where ν is the average number of zippers per chain. Since $k = 2$ for pairwise association, the gel point is found by the condition $\nu = 2$ for the present system. In the phase diagram, small black circles connected by the dotted line show the observed contour given by this condition for $N = 90$. We can see very good agreement.

At high temperature and low concentration, we have solutions of separate chains. Across the upper branch of the sol/gel transition line, networks take TypeI form in which random coils are cross-linked by short zippers. At low temperature in the gel region, networks are long zippers crosslinked by short random coils (TypeII network). Since zippers are practically rigid rods, TypeII is a network of rigid rods connected by free joints. They show

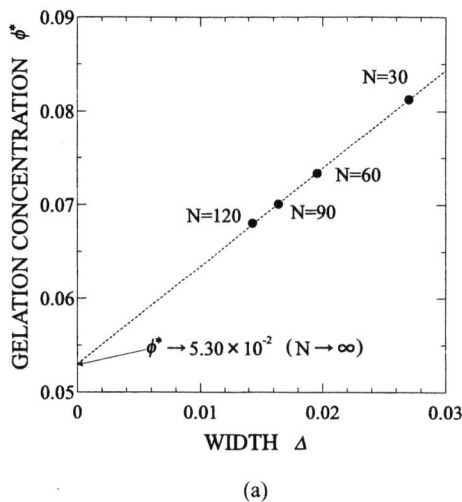
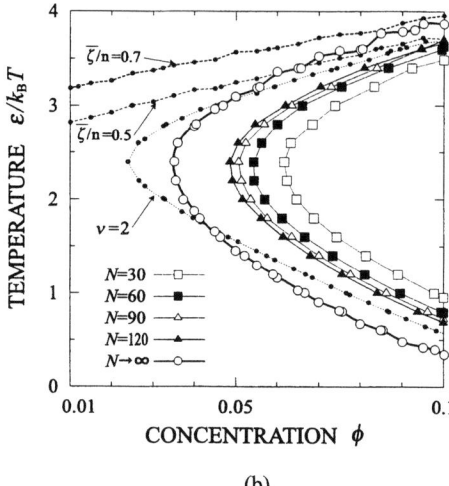

FIGURE 5. (a) Finite-size scaling for finding the gel concentration at $\epsilon/k_B T = 1.4$. Polymer volume fraction ϕ^* (black circles) at the percolation point is plotted against the width Δ of the region obtained from the condition $0.2 \leq \Pi \leq 0.8$ for $N = 30, 60, 90, 120$. The midpoint in the region Δ is regarded as the percolation point. (b) Phase diagrams on the temperature-concentration plane. Sol/gel transition points (large symbols) for $N = 30, 60, 90, 120, \infty$ are drawn on the temperature-concentration plane, together with the contours with constant zipper length (small black circles with broken lines), and constant zipper number ($\nu = 2$) for $N = 90$ (small black circles with dotted line).

highly nonlinear elastic properties.

At lower temperatures, zippers grow so long that polymers cannot form network structure. Most of chains are in paired state. They are sparsely connected to each other by short random coils. We call this pairing phase. Broken lines show the contours at constant zipper length for $N = 90$. Along these lines, the average zipper length reaches 50%, and 70% of the total chain length. Transition from network to pairing phase (*network/pair transition*) becomes a real thermodynamic transition in the limit of infinite chain length.

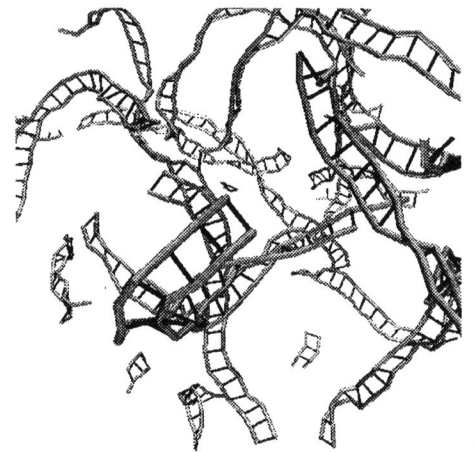

FIGURE 6. Typical snapshot of MC simulation at low temperature $\epsilon/k_B T = 4.0$ in the pairing phase where most of chains are paired by hydrogen-bonded zippers.

Figure 6 shows some Monte Carlo snapshots at different temperatures along the fixed polymer volume fraction $\phi = 0.05$. At the highest temperature $\epsilon/k_B T = 0.2$, we have a uniform molecularly dispersed solution. Chains are separated from each other. As temperature goes down, short zippers start to form. We find TypeI network at this temperature ($\epsilon/k_B T = 2.2$). Random coils are cross-linked by short zippers. As the solution is further cooled, zippers grow, and the network turns into TypeII where long zippers are cross-linked by short random coils. At the lowest temperature $\epsilon/k_B T = 4.0$ outside the gel region, most chains are paired, and we can see long ladders, most of them are separated from each other.

REFERENCES

1. Tanaka, F. *Macromolecules*, **33**, 4249-4263 (2000).
2. Tanaka, F. ; Stockmayer, W.H. *Macromolecules*, **27**, 3943-3954 (1994).
3. Tanaka, F. *Macromolecules* **36**, 5392-5405 (2003).
4. Reid, D.S.; Bryce, T.A.; Clark, A.H.; Rees, D.A. *Faraday Disc. Chem. Soc.* **57**, 230-237 (1974).
5. Poland, D.; Scheraga, H.A. *Theory of Helix-Coil Transitions in Biopolymers*; Academic Press (1970).
6. Zimm, B.H. ; Bragg, J.K. *J. Chem. Phys.* **31**, 526-535(1959).

Analysis of Shear-Thickening in Physical Gel by Transient Network Theory

Tsutomu Indei* and Toshihico Arimitsu[†]

*Department of Polymer Chemistry, Graduate School of Engineering, Kyoto University, Kyoto 615-8510, Japan
[†]Institute of Physics, University of Tsukuba, Ibaraki 305-8571, Japan

Abstract. The shear-thickening phenomenon observed in HEUR aqueous solutions by Jenkins, Sileibi and El-Aasser is analyzed on the basis of transient network theory. Effects of looped chains which disconnect their one end from the junction due to the collisions with other chains under shear flow, are taken into account. This effect consequently enhances the number of elastically effective chains (active chains), leading to the gradual increase in the shear viscosity at relatively lower shear rate. It is also shown that the nonlinear force sustained by well-stretched active chains under the flow induces the strong and sharp enhancement of the shear viscosity at moderate shear rate. These results indicate that both effects are indispensable to understand the shear-thickening.

INTRODUCTION

It is known that some sorts of physical gels show an anomalous behavior called shear-thickening, i.e., their viscosity increase with increasing the shear rate. These physical gels are mostly formed by associating polymers carrying functional groups at their both ends, such as HEUR [1, 2, 3, 4, 5, 6, 7]. HEUR is a poly(ethylene oxide) carrying alkyl groups at its both ends. They form a physically cross-linked network in water due to temporary association among their hydrophobic end-groups. Another example is telechelic ionomers in apolar solvent [8, 9]. Although several theories have been proposed trying to explain this peculiar phenomenon [10, 11, 12, 13], its mechanism is still unclear.

We analyze shear-thickening on the basis of transient network theory [14, 15] which describes polymer networks made up of linear chains carrying associative groups at their both ends (recently, transient network theory was extended so that trifunctional polymer chains with different species of associative groups can be treated [16]). Chains in the network are assumed to be classified into three categories according to the way of connecting their associative groups to junctions. The three categories are 1) active chain: its both ends are incorporated into distinct junctions, 2) dangling chain: one of its end is a member of a junction and 3) loop: its both ends are incorporated into the same junction (Fig. 1). For simplicity, free chains floating in the solvent are not taken into account. Loops have been observed in HEUR aqueous solutions [3], where they form flowerlike micelles as schematically shown in Fig. 1. No topological entanglements are considered here.

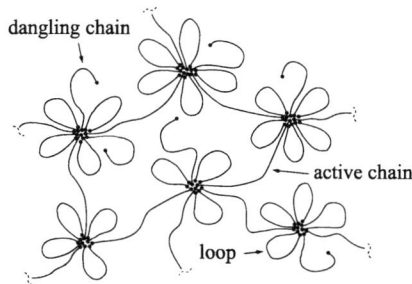

FIGURE 1. Schematic representation of a network formed by linear chains carrying associative groups at their both ends.

TRANSIENT NETWORK THEORY

Basic Equation

Let $\phi^i(\mathbf{r},t)d\mathbf{r}$ be the number of i-chains per unit volume having end-to-end vector $\mathbf{r} \sim \mathbf{r}+d\mathbf{r}$ at time t. The superscript i represents the type of chains, i.e., $i=a$ for active chains, $i=d$ for dangling chains and $i=l$ for loops. The number $v^i(t)$ of i-chains (per unit volume) is then given by

$$v^i(t) = \int d\mathbf{r}\, \phi^i(\mathbf{r},t). \quad (1)$$

Since the total number n of chain conserves, we have a relation $\sum_i v^i(t) = n$. An i-chain alters its type as time goes on by detaching its associative group from the junction or by connecting its free group with the junction nearby. Taking it into account, we have the equation for number-conservation of *i*-chains

$$\frac{\partial}{\partial t}\phi^i(\mathbf{r},t) + \frac{\partial}{\partial \mathbf{r}} \cdot \left(\dot{\mathbf{r}}^i(\mathbf{r},t)\phi^i(\mathbf{r},t) \right)$$
$$= -\sum_{j(\neq i)} W_{ji}(\mathbf{r})\phi^i(\mathbf{r},t) + \sum_{j(\neq i)} W_{ij}(\mathbf{r})\phi^j(\mathbf{r},t). \quad (2)$$

where $\dot{\mathbf{r}}^i(\mathbf{r},t)$ is the rate of deformation of *i*-chains under the macroscopic flow, and $W_{ij}(\mathbf{r})$ is the transition probability during unit time from *j*-chains to *i*-chains. In the following, we put $\beta \equiv W_{da}$, $p \equiv W_{ad}$, $v \equiv W_{ld}$ and $u \equiv W_{dl}$. The first term in the right-hand side of (2) represents the number of *i*-chains which become other types of chains in unit time, whereas the second term stands for the number of $j(\neq i)$-chains which convert to *i*-chains. That is, the first and the second terms represent annihilation and creation of *i*-chains, respectively. Note that each type of chains are characterized by means of $\dot{\mathbf{r}}^i(\mathbf{r},t)$ and $W_{ij}(\mathbf{r})$.

Rate of Deformation Vector

Active Chain. We adopt the affine deformation assumption for active chains. This assumption states that $\dot{\mathbf{r}}^a(\mathbf{r},t) = \hat{\kappa}(t)\mathbf{r}$, where $\hat{\kappa}(t)$ is the velocity gradient tensor representing the deformation added to the system. In the case of steady shear flow with the shear rate $\dot{\gamma}$, $\kappa_{12} = \dot{\gamma}$ and the other components are zero.

Dangling Chain. The shear-thickening occurs around the shear rate $\dot{\gamma} \simeq 10 \sim 10^3 \text{sec}^{-1}$ as shown in Fig. 3, while the Rouse relaxation time τ_R is estimated as $10^{-5} \sim 10^{-4}$sec (see Table 1). That is, even under the flow, dangling chains are essentially in an equilibrium state without tension. The rate of deformation vector for dangling chains is then effectively given by $\dot{\mathbf{r}}^d(\mathbf{r},t) = 0$. We assume that these dangling chains are Gaussian whose end-to-end vector distribution is given by $\psi(\mathbf{r}) = (3/2\pi Na^2)^{3/2} \exp(-3r^2/2Na^2)$, where N is the number of segments and a is the characteristic length of a segment. The chain distribution function for the dangling chain is written as $\phi^d(\mathbf{r},t) = v^d(t)\psi(\mathbf{r})$.

Loop. The end-to-end distance of a loop is *always* zero. Hence, $\dot{\mathbf{r}}^l(\mathbf{r}=0,t) = 0$.

Transition Rates

From Active to Dangling Chains (β). The detaching rate β of one end of active chain from the junction is supposed to be an increasing function with respect to its end-to-end length $r = |\mathbf{r}|$ since, the longer it becomes under the flow, the easier its one end detaches itself from the junction due to the enhanced tension between its two ends. We assume $\beta(r)$ increases abruptly around a certain end-to-end length r^* and put approximately

$$\beta(r) = \begin{cases} \beta_0 & (r \leq r^*) \\ \infty & (r > r^*) \end{cases}, \quad (3)$$

where β_0 is constant independent of r.

From Dangling to Active Chains (p). We regard the connection rate p of a free end of a dangling chain with the junction as a constant. The connecting process does not seem to be strongly influenced by the end-to-end length of the dangling chain [15].

From Dangling Chains to Loops (v). The connecting process of a free end of a dangling chain to the junction with which its other end has already stuck takes place only when its end-to-end length is 0. Therefore, we put $v(\mathbf{r}) = v_0 \delta(\mathbf{r})V$, where $V = (2\pi Na^2/3)^{3/2}$. $v_0 = \int d\mathbf{r}\, v(\mathbf{r})\psi(\mathbf{r})$ is a mean transition rate from a dangling chain to a loop.

From Loops to Dangling Chains (u). We assume that the open process of a loop is influenced by the collision with other chains, and put

$$u(\dot{\gamma}) = \beta_0 + v_{in}(\dot{\gamma})\beta_0, \quad (4)$$

where $v_{in}(\dot{\gamma})$ is the number of segments colliding with the loop during its lifetime $1/u(\dot{\gamma})$. The first term in the right-hand side of (4) represents the breakage rate at $\dot{\gamma} = 0$. This term is necessary since even at equilibrium state, a loop has possibility to disconnect its one end from the junction due to its thermal fluctuation. Under the shear flow, on the contrary, the second term appears as a result of collisions. Segments, which hit a backbone of a loop, stimulate immediately its ends through the backbone and enhances the possibility to open the loop. We take this effect into account by assuming that the detaching rate becomes $v_{in}(\dot{\gamma})$ times larger than the one at equilibrium state (the second term in (4)).

We can estimate $v_{in}(\dot{\gamma})$ as follows. The radius of the region occupied by a loop where the collision with other chains works effectively, is given by αs_y, where $s_y = Na^2/36$ is the *y*-component of the radius of gyration of the loop, and α is an adjustable parameter which should be determined from experiments. From α, we know how much of the domain occupied by the loop is responsible for the collisions leading to its open process. By supposing that segments are distributed uniformly in space, the number $\tilde{v}_{in}(y)dy$ of segments which enter into a region $y \sim y+dy$ $(y>0)$ and $-z(y) \sim z(y)$ in this domain

TABLE 1. Properties of HEUR in water. M_n: number-average molecular weight [1, 17], n_{rep}: number of EO units ($M_0 = 44$ is the molecular weight of an EO unit), l: contour length ($l_0 = 3.6$Å is the length of an EO unit), $\sqrt{\langle r^2 \rangle_0}$: observed root-mean-square of the end-to-end length [17], N: number of segments, a: length of each segment, Φ: volume fraction for $c = 1$wt% solution (N_A is the Avogadro constant) and τ_R: Rouse relaxation time. $\eta_s = 0.89 \times 10^{-3}$ Pa·sec is the viscosity of pure water at $T = 298$K, 1atm, and k_B is the Boltzmann constant.

M_n	n_{rep} $= M_n/M_0$	l (Å) $= l_0 n_{rep}$	$\sqrt{\langle r^2 \rangle_0}$ (Å)	N $= l^2/\langle r^2 \rangle_0$	a (Å) $= l/N$	Φ $= 10^4 c N_A N a^3 / M_n$	τ_R (sec) $= 2\eta_s N^2 a^3 / (\pi k_B T)$
34,200	777	2,800	186	227	12.3	7.4×10^{-2}	1.3×10^{-5}
51,000	1,160	4,180	231	327	12.8	8.1×10^{-2}	3.1×10^{-5}
67,600	1,540	5,540	265	437	12.7	8.0×10^{-2}	5.4×10^{-5}
84,300	1,920	6,910	301	527	13.1	8.5×10^{-2}	8.6×10^{-5}

in unit time is estimated as $\tilde{v}_{in}(y) dy = nN \cdot 2z(y) dy \cdot |v_x(y)| = 2nN\sqrt{(\alpha s_y)^2 - y^2}\,|\dot{\gamma}| y dy$. Integrating it from 0 to αs_y and multiplying by 2 to take account of the region $y < 0$, we have the number of segments entering into this domain in unit time: $2 \int_0^{\alpha s_y} dy\,\tilde{v}_{in}(y) = \frac{\alpha^3}{162} n a^3 N^{5/2} |\dot{\gamma}|$. By definition, $v_{in}(\dot{\gamma})$ is given by dividing it by $u(\dot{\gamma})$. Finally, solving (4) with respect to $u(\dot{\gamma})$, we obtain

$$u(\dot{\gamma})/\beta_0 = \frac{1 + \sqrt{1 + \frac{2}{81}\alpha^3 \Phi N^{3/2} |\dot{\gamma}|/\beta_0}}{2}, \quad (5)$$

where $\Phi \equiv Nna^3$ is the volume fraction of chains.

ANALYSIS OF EXPERIMENTS

Now we analyze the steady shear viscosity observed for HEUR aqueous solutions [1], with the help of the formula $\eta(\dot{\gamma}) = \int d\mathbf{r}\,(xy/r)\,f(r)\,\phi^a(\mathbf{r})/\dot{\gamma}$, where $f(r)$ is the tension between chain ends given by the random-flight model: $f(r) = L^{-1}(r/Na) k_B T/a$. Since τ_R is regarded as 0, dangling chains and loops do not contribute to the viscosity of the system.

We can estimate the values of N, a and Φ from the experimental conditions [1, 17] as listed in Table 1. We can also estimate the cut-off length r^* as follows. An end of an active chain dissociates from the junction when it climbs over the potential barrier W of the association. Supposing that the effective range of the hydrophobic interaction is roughly a, then r^* is given by the condition $W \simeq f(r^*)a$. Assuming that $r^* \lesssim l$, we have $W/k_B T \simeq (1 - r^*/l)^{-1}$ leading to $r^*/l \simeq 1 - k_B T/W$. According to Annable et al. [2], $W = 28 k_B T$ ($T = 298$K) for HEUR ($M_w \simeq 35,000$). In this case, r^* is estimated as $r^* \simeq 0.96l$. We will adopt the value $0.96l$ for r^* in the following.

Although there are still four parameters in our model (β_0, p, v_0 and α), we can eliminate one of three β_0, p, v_0 by making use of the zero-shear viscosity obtained from the experiments [1] through a relation $\eta_0 = n/(1 + (1 + v_0/\beta_0)\beta_0/p) \cdot k_B T/\beta_0$. Hereafter, we adopt β_0, v_0/p (and α) as adjustable parameters.

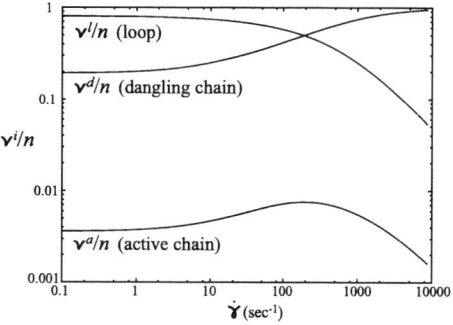

FIGURE 2. The number of active chains, dangling chains and loops plotted against the shear rate for $\alpha = 0.45$, $\beta_0 = 8.7$sec^{-1} and $v_0/p = 223$ ($N = 227$, $r^*/l = 0.96$).

Shear Viscosity

The number of each type of chains are shown in Fig. 2 for $\alpha = 0.45$, $\beta_0 = 8.7$sec^{-1} and $v_0/p = 223$ ($N = 227$, $r^* = 0.96l$). With increasing the shear rate, the number v^l of loops decreases and the number v^d of dangling chains increases, since the transition probability u from a loop to a dangling chain increases with shear rate. As a result, the number v^a of active chains also increases with the same rate as the dangling chains (p does not depend on $\dot{\gamma}$). For higher shear rates, the population of active chains whose end-to-end length reaches r^* becomes large. Since such active chains detach from the network, the total number v^a of active chains decreases leading to the shear-thinning (see below).

The steady shear viscosity for optimal values of α, β_0 and v_0/p for each molecular weight listed in Table 1 is shown in Fig. 3. We see that the calculated viscosity coincides well with experimental data for the broad shear rate (except for $M_n = 84,300$ at higher shear rate). The gradual increase of the viscosity at lower shear rates is attributed to the increasing number of active chains with shear rate in this region (Fig. 2). In addition, stretched active chains along the flow cause a strong nonlinear

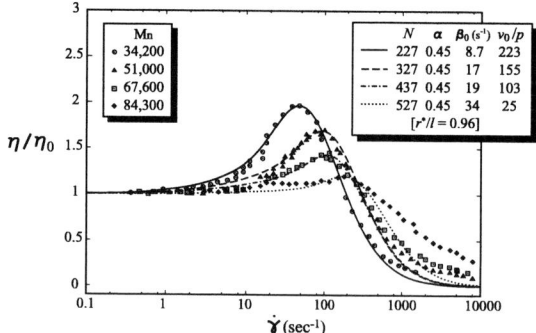

FIGURE 3. The steady shear viscosity plotted against the shear rate. The lines represent theoretical results obtained for each M_n listed in Table 1. The dots represent the experimental results observed for $T = 298K$ and $c = 1$wt% [1]. The zero-shear viscosity η_0 is 3.0Poise for $M_n = 34,200$, 0.94Poise for $M_n = 51,000$, 0.53Poise for $M_n = 67,600$ and 0.26Poise for $M_n = 84,300$ [1].

force. It adds the sharp enhancement in the viscosity at the shear rate where the end-to-end length of the active chain reaches r^*. Note that $\alpha = 0.45$ does not depend on N. It indicates that the region with about half of the radius of gyration of the loop contributes to the open process of the loop regardless of the number of segments.

We see that $p \propto N^{-1.2}$ and $v_0 \propto N^{-2.4}$ from the data for $N = 227, 327$ and 437. The dependence of p on N is roughly interpreted as follows. Since the polymer concentration is fixed, the number of hydrophobic end-groups in the network decreases as the molecular weight increases, thereby diminishing the number of junctions. It reduces the chance for a free end of a dangling chain to catch a junction nearby. Assuming that the number of junctions is proportional to the number of end-groups, we obtain $p \propto N^{-1}$ for fixed concentration. The dependence of v_0 on N is also reasonable. The probability for a self-avoiding chain to form a loop is approximately proportional to N^{-2} [18]. Since a flower-like micelle contains several number of loops [4], the exponent is expected to be less than -2 due to the excluded volume effects of other loops.

CONCLUSION

Steady shear viscosity of HEUR aqueous solutions observed by Jenkins *et al.* [1] was analyzed on the basis of transient network theory [14, 15]. Three types of chains were assumed to exist in the transient network, i.e., active chains, dangling chains and loops. We took into account the open process of a loop caused by the collision with other chains under the shear flow. This process produces the weak enhancement of the number of active chains, leading to the weak and wide increase in the shear viscosity. The idea of increase in the number of active chains around the shear-thickening region has been suggested from observations that the plateau modulus in this region is larger than that in the Newtonian region [5]. It was also shown that the nonlinear force of active chains described by the inverse Langevin function induces the strong and sharp enhancement of the steady shear viscosity. The effect of the nonlinear force has been discussed by other researchers in different approaches both theoretically [10, 12] and experimentally [19]. The success in the analysis of the shear viscosity in HEUR aqueous solutions reveals that both the increase in the number of active chains and the nonlinear force of active chains are indispensable to understand the shear-thickening behavior.

REFERENCES

1. Jenkins, R. D., Silebi, C. A., and El-Aasser, M. S., *ACS Symp. Ser.*, **462**, 222–233 (1991).
2. Annable, T., Buscall, R., Ettelaie, R., and Whittlestone, D., *J. Rheol.*, **37**, 695–726 (1993).
3. Yekta, A., Duhamel, J., Adiwidjaja, H. Brochard, P., and Winnik, M. A., *Langmuir*, **9**, 881–883 (1993).
4. Yekta, A., Xu, B., Duhamel, J., Adiwidjaja, H., and Winnik, M., *Macromolecules*, **28**, 956–966 (1995).
5. Tam, K. C., Jenkins, R. D., Winnik, M. A., and Bassett, D. R., *Macromolecules*, **31**, 4149–4159 (1998).
6. Le Meins, J. F., and Tassin, J. F., *Macromolecules*, **34**, 2641–2647 (2001).
7. Ma, S. X., and Cooper, S. L., *Macromolecules*, **34**, 3294–3301 (2001).
8. Maus, C., Fayt, R., Jérôme, R., and Ph., T., *Polymer*, **36**, 2083–2088 (1995).
9. Bhargava, S., and Cooper, S. L., *Macromolecules*, **31**, 508–514 (1998).
10. Vrahopoulou, E. P., and McHugh, A. J., *J. Rheol.*, **31**, 371–384 (1987).
11. Wang, S. Q., *Macromolecules*, **25**, 7003–7010 (1992).
12. Marrucci, G., Bhargava, S., and Cooper, S. L., *Macromolecules*, **26**, 6483–6488 (1993).
13. Vaccaro, A., and Marrucci, G., *J. Non-Newtonian Fluid Mech.*, **92**, 261–273 (2000).
14. Tanaka, F., and Edwards, S. F., *Macromolecules*, **25**, 1516–1523 (1992).
15. Tanaka, F., and Edwards, S. F., *J. Non-Newtonian Fluid Mech.*, **43**, 247–271, 273–288, 289–309 (1992).
16. Indei, T., and Tanaka, F. (2004), to be published in *J. Rheol.*.
17. Jenkins, R. D., Bassett, D. R., Silebi, C. A., and El-Aasser, M. S., *J. Appl. Polym. Sci.*, **58**, 209–230 (1995).
18. de Gennes, P. G., *Scaling Cocepts in Polymer Physics*, Cornell University Press, Ithaca, 1979, pp. 38–43.
19. Séréro, Y., Jacobsen, V., and Berret, J.-F., *Macromolecules*, **33**, 1841–1847 (2000).

Polyelectrolyte Adsorption on Charged Substrate

Chi-Ho Cheng* and Pik-Yin Lai*

*Department of Physics and Center for Complex Systems, National Central University, Taiwan, Republic of China.

Abstract. The behavior of a polyelectrolyte adsorbed on a charged substrate of high-dielectric constant is studied by both Monte-Carlo simulation and analytical methods. It is found that in a low enough ionic strength medium, the adsorption transition is first-order where the substrate surface charge still keeps repulsive. The monomer density at the adsorbed surface is identified as the order parameter. It follows a linear relation with substrate surface charge density because of the electrostatic boundary condition at the charged surface. During the transition, the adsorption layer thickness remains finite. A new scaling law for the layer thickness is derived and verified by simulation.

INTRODUCTION

The problem of polymer adsorption on an attractive surface has drawn considerable interest due to its relation to surface effects in critical phenomena and practical importance in technology and biology. It is well established that the adsorption transition is continuous if its attraction on the surface is short-ranged [1]. On the other hand, long-ranged electrostatic interactions in polyelectrolyte systems pose many challenging theoretical problems. Recently the macroion adsorption on an electrostatically attractive interface [2, 3] and the associated charge inversion phenomena of adsorbed polyelectrolytes [4, 5] acquire lots of attention.

Previous analytical approaches impose the continuity of monomer density across the charged surface, the surface monomer density is then set to zero [6, 7]. The polyelectrolyte adsorption problem becomes the competition between the electrostatic force from charged surface and the configurational entropy of the polyelectrolyte itself. The transition has been shown to be continuous. However, we found that the above treatment is not quite adequate since the electrostatic boundary condition at the charged surface is not faithfully respected. Furthermore, the electrostatic boundary condition may induce different physics in the presence of a substrate under the charged surface. In this paper, we study the adsorption of a polyelectrolyte on a high-dielectric substrate at low ionic strength (e.g. aqueous solution with a metal substrate) in which image charge interaction is attractive. The adsorption transition occurs where the surface charges are repulsive instead of the attractive case that were usually studied. The problem is tackled by Monte-Carlo simulations and analytical methods taking full account of appropriate boundary conditions. It is found that the order of adsorption transition, the physical mechanism, and the scaling behavior are all different from those of the attractive surfaces.

MODEL

A polyelectrolyte carrying positive charges is immersed in a medium ($z > 0$) of dielectric constant ε. At $z = 0$ there is an impenetrable surface of uniformly surface charge density σ. Below that ($z < 0$), it is a substrate of dielectric constant ε'. Denote the charge on a polymer segment ds by $q_0 ds$, the Hamiltonian is written as

$$\mathcal{H} = \frac{3k_B T}{2l_0^2}\int_0^N ds\left(\frac{\partial \vec{r}(s)}{\partial s}\right)^2 + \frac{1}{2}\int_0^N ds\int_0^N ds'\{$$
$$\Gamma\frac{e^{-\kappa|\vec{r}(s)-\vec{r}(s')|}}{|\vec{r}(s)-\vec{r}(s')|} - \Gamma'(2-\delta_{s,s'})\frac{e^{-\kappa|\vec{r}(s)-\vec{r}'(s')|}}{|\vec{r}(s)-\vec{r}'(s')|}\}$$
$$+h\int_0^N ds\,\kappa^{-1}e^{-\kappa\vec{r}(s)\cdot\hat{z}}$$
$$+\omega\int_0^N ds\int_0^N ds'\delta(\vec{r}(s)-\vec{r}(s')) \quad (1)$$

where s is the variable to parametrize the chain, l_0 the bare persistence length, and κ^{-1} the Debye screening length. $\vec{r}(s) = (x(s), y(s), z(s))$, $\vec{r}'(s') = (x(s'), y(s'), -z(s'))$ are the positions of the monomers and their electrostatic images, respectively. $\Gamma = q_0^2/\varepsilon$, $\Gamma' = \Gamma(\varepsilon' - \varepsilon)/(\varepsilon' + \varepsilon)$, and $h = 4\pi q_0 \sigma/(\varepsilon' + \varepsilon)$ are the coupling parameters governing the strengths of Coulomb interactions among the monomers themselves, between the polymer and its image, and between the polymer and the charged surface, respectively. In good solvent regime, $\omega > 0$ in the last term.

MONTE-CARLO RESULTS

The continuum model is discretized to perform Monte-Carlo simulations. $\vec{r}(s)$ is replaced by a chain of beads $\vec{r}_i$ ($i = 1, \ldots, N$) with hard-core excluded volume of finite radius a. Polymer lengths up to $N = 120$ are employed. The units of length and energy are $2a$ and $q_0^2/2\varepsilon a$, respectively. The ratios ε'/ε are studied from 2 to 12.5 (aqueous solution with a metallic substrate). Runs up to 10^9 MC steps are performed.

The adsorption behavior can be characterized by normalized monomer density $\rho(z)$. $\rho_a \equiv \rho(z=a)$, the probability density of monomers adsorbed on the substrate, is chosen as the order parameter to describe the adsorption transition. ρ_a as a function of the surface charge density σ for various $\varepsilon'/\varepsilon > 1$ is shown in Fig.1. It is seen that ρ_a vanishes abruptly when σ increases up to its threshold value σ_t indicating a first-order transition. At low enough ionic strength that κ^{-1} is much larger than absorption layer thickness, $\sigma_t > 0$ for $\varepsilon'/\varepsilon > 1$. The discontinuous drop of ρ_a across the transition decreases to zero as $\varepsilon'/\varepsilon \to 1$. We have also verified that the sharp jump in energy (latent heat) across the transition is proportional to N as expected for a first-order transition. Same results are obtained for larger κ^{-1}.

Furthermore, ρ_a is linear in σ with the slope depending on ε'/ε. Such a linear relation between ρ_a and σ can be understood from the electrostatic boundary conditions that the system has to satisfy. The electric potential $\phi(z)$ in the neighborhood of $z=0$ boundary obeys

$$-\frac{\partial \phi}{\partial z}\bigg|_{z=0^+} + \frac{\partial \phi}{\partial z}\bigg|_{z=0^-} = -\frac{4\pi}{\varepsilon}\left(\frac{2\sigma}{\varepsilon'/\varepsilon + 1} + \sigma_p\right) \quad (2)$$

where σ_p is the polarization surface charge density induced by the polyelectrolyte only. Notice that σ_p depends only on ε'/ε. It is independent of σ in the adsorbed regime near the transition. The reason is based on the electric blob picture which will be explained in next section. If one treats the polyelectrolyte as a marcomolecule with a well-defined surface, its surface charge density at $z = a$ should be proportional to the monomer density ρ_a. It also applies to the electric field in the $z < 0$ region. One have

$$K\frac{\partial \phi}{\partial z}\bigg|_{z=a^-} = -\frac{4\pi}{\varepsilon}\rho_a, \quad (3)$$

$$K\frac{\partial \phi}{\partial z}\bigg|_{z=0^-} = -\frac{4\pi}{\varepsilon}\frac{2\varepsilon'}{\varepsilon'+\varepsilon}\rho_a \quad (4)$$

where $K > 0$ is the proportionality constant. Applying the electric field continuity from $z = 0^+$ to $z = a^-$, and

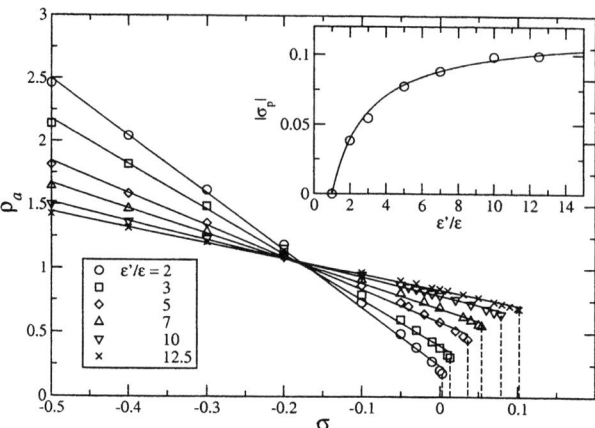

FIGURE 1. Monte-Carlo results for the normalized monomer density at the surface, ρ_a, as a function of surface charge density σ (in unit of $q_0/4a^2$) for different ε'/ε at $\kappa^{-1} = 25$. The fitted straight lines are terminated at their adsorption transition points. The vertical dashed lines are drawn as guides to the eyes. Inset: The polarization surface charge density due to the charged polymer, $|\sigma_p|$, as a function of ε'/ε. σ_p saturates as $\varepsilon'/\varepsilon \to \infty$. The sign of σ_p is opposite to q_0 and is thus negative. The solid curve is fitted from Eq.(6) with $\sigma_{\text{poly}} = 0.118$.

using Eqs.(2)-(4), one get the linear behavior

$$\rho_a = -\frac{2K}{\varepsilon'/\varepsilon - 1}\left(\sigma + \frac{\varepsilon'/\varepsilon + 1}{2}\sigma_p\right). \quad (5)$$

Notice that K and σ_p depends only on ε'/ε. As shown in Fig.1, the slope K decreases monotonically with ε'/ε. Lines with different values of ε'/ε intersect at a common point implies the naive intuition that a more polarizable substrate gives a stronger adsorption is not always true.

Without substrate surface charge, set $\sigma = 0$ in Eq.(5), one get

$$\sigma_p = -\frac{\rho_a|_{\sigma=0}}{K}\frac{\varepsilon'/\varepsilon - 1}{\varepsilon'/\varepsilon + 1}. \quad (6)$$

The dependence on ε'/ε can be considered as a flat surface of surface charge density $\sigma_{\text{poly}} \equiv \rho_a|_{\sigma=0}/K$ located above the substrate. K are obtained by linear fittings in Fig.1, and hence σ_p as a function of ε'/ε is plotted in the inset. σ_p increases from zero to a saturated value as $\varepsilon'/\varepsilon \to \infty$. Presumably, Eq.(6) suggests that it is possible to experimentally detect the sign and magnitude of the charge of the polyelectrolyte by measuring the substrate polarization.

ANALYTICAL RESULTS

At low ionic strength, a polyelectrolyte can be treated as electric blobs arranged longitudinally. The rod-like polyelectrolyte tends to lie down on the substrate so as to lower its energy. It is thus rigid (rod-like) in x-y plane, but still flexible in the z-direction. In the adsorbed regime, decreasing σ (chain is attracted more by the surface) would cause the rearrangement of the electric blobs such that the fluctuations in the z-direction is reduced (and hence the layer thickness decreases), but the fluctuations in the other two directions and the size of the blobs are basically unchanged. The effective in-plane surface charge distribution of the polyelectrolyte, and hence σ_p, is not affected by σ. The excluded volume effect is ignored because it takes almost no effect in z-direction. The effect from self-electrostatic interaction can be absorbed by renormalising the bare persistence length l_0 to l.

Because the monomer would feel the strongest attraction from its direct image around the adsorption regime, the Γ'-term in Eq.(1) is approximated by the interaction of every monomer and its corresponding image only. The residual attraction from the images of other monomers could be absorbed into the coupling parameter Γ' by renormalising q_0 to q. The partition function is reduced to

$$Z = \int \mathscr{D}[\vec{r}(s)] \exp[\int_0^N ds \{ -\frac{3}{2l^2}\left(\frac{\partial \vec{r}(s)}{\partial s}\right)^2 + \frac{\beta \Gamma'}{4}\frac{e^{-2\kappa \vec{r}(s)\cdot \hat{z}}}{\vec{r}(s)\cdot \hat{z}} - \beta h \kappa^{-1} e^{-\kappa \vec{r}(s)\cdot \hat{z}} \}]. \quad (7)$$

$\vec{r}(s)$ is transformed to $\rho(\vec{r}) \equiv \frac{1}{N}\int_0^N ds \delta(\vec{r}-\vec{r}(s))$ by introducing an auxiliary field. Ground state dominance in large-N limit is then applied. By variational principle, one obtains the Edwards-Schrödinger equation,

$$\left(-\frac{l^2}{6}\frac{d^2}{dz^2} - \frac{\beta \Gamma'}{4}\frac{e^{-2\kappa z}}{z} + \beta h \kappa^{-1} e^{-\kappa z}\right)\psi(z) = \varepsilon_0 \psi(z) \quad (8)$$

where ε_0 acts as a Lagrange multiplier to enforce the constraint of the ground state wavefunction normalization. The monomer density is given by $\rho(z) = |\psi(z)|^2$. Eq.(8) also describes a quantum particle at its ground state moving under a combined potential of a 1d screened Coulomb attraction and an almost linear potential. However, the boundary condition expressed by Eq.(5) is different from that of the hard-wall one $\psi|_s = 0$ usually employed for the quantum particle. Instead $\psi|_s = \sqrt{\rho_a} \neq 0$ for the present problem implies that the steric force felt by the polyelectrolyte from the charged surface should be modified [8]. Setting $\psi|_s = 0$ [6, 7, 9] in previous studies is not completely correct.

During the adsorption, the rod-like polyelectrolyte tends to lie down on the charged surface. The thickness of the adsorption layer is of the same order of the gyration radius in z-direction. At low ionic strength corresponding to Debye length much larger than layer thickness, the polyelectrolyte can only feel the potential barrier formed by $V(z)$ as shown in Fig2. Analytically, the original potential $V(z)$ in Eq.(8) is replaced by

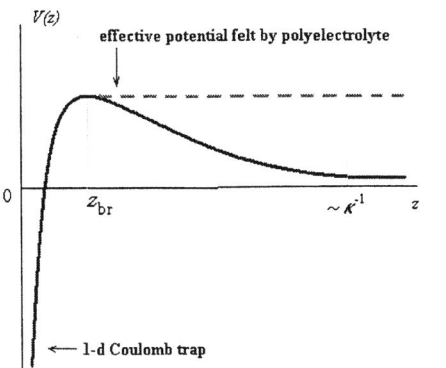

FIGURE 2. Effective Potential felt by the polyelectrolyte

$$V_{\text{mod}}(z) = \begin{cases} +\infty, & z < a \\ V(z), & a \leq z < z_{\text{br}} \\ V(z_{\text{br}}), & z \geq z_{\text{br}} \end{cases} \quad (9)$$

where z_{br} is chosen such that $V'(z_{\text{br}}) = 0$, and $V(z_{\text{br}})$ is the barrier height. In the limit of $\sigma = \kappa = 0$, the analytic solution exists in the form

$$\psi(z) = W_{\lambda,1/2}(\frac{3\beta \Gamma'}{2l^2 \lambda}z) \; ; \; \varepsilon_0 = -\frac{3\beta^2 \Gamma'^2}{32l^2}\frac{1}{\lambda^2} \quad (10)$$

where $W_{\lambda,1/2}$ is the Whittaker's notation of the confluent hypergeometric function [10], and λ is the least value satisfying the boundary condition. Bound state exists for arbitrary $\varepsilon'/\varepsilon > 1$. It implies $\sigma_t > 0$ at low enough ionic strength.

No exact solution exists for $\sigma, \kappa > 0$ in general but one can analyze it around the transition. In the region of interest, the Coulomb term dominates over the almost linear term. One can approximate the binding energy by Eq.(10). The polyelectrolyte undergoes a de-sorption transition when the binding energy is equal to barrier height $V(z_{\text{br}})$. After some algebra, it shows $\sigma_t \sim (\varepsilon'/\varepsilon - 1)$ for $\varepsilon'/\varepsilon \gg 1$ and $\sigma_t \sim (\varepsilon'/\varepsilon - 1)^3$ for $\varepsilon'/\varepsilon \sim 1$. This analytic result is consistent with simulation as shown in Fig.3a.

An approximate solution for density profile $\rho(z)$ for $\sigma > 0$ can be studied by the variational wavefunction

$$\psi(z) = \sqrt{\rho_a}(1 + \mu \alpha(z-a))e^{-\frac{1}{2}\alpha(z-a)} \quad (11)$$

where α^{-1} is the decay length. μ is positive because the wavefunction is restricted to be nodeless. α and μ are not independent but related via the wavefunction normalization condition. The inverse decay length is calculated to

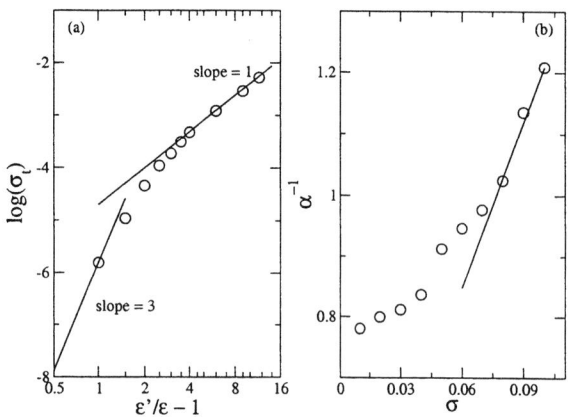

FIGURE 3. (a) Simulation results for the surface charge density at the transition, σ_t, as a function of ε'/ε in logarithmic scale at $\kappa^{-1} = 25$. The straight lines indicate slopes of 1 and 3 as suggested in the text. (b) Simulation results for the decay length α^{-1} (which is proportional to the layer thickness) as a function of σ (in unit of $q_0/4a^2$) for $\varepsilon'/\varepsilon = 12.5$, $\kappa^{-1} = 25$. The straight line is a linear fit to points near the transition. $\sigma_t = 0.102$ in this case. α^{-1} is obtained from exponential fitting to the tail of corresponding density profile.

be

$$\alpha = \frac{3\beta\Gamma'}{2l^2} + \rho_a \quad (12)$$

where the leading term is independent of σ. Near the transition, the decay length α^{-1} and hence the adsorption layer thickness $D \sim \alpha^{-1}$ remains finite. Combining Eqs.(5) and (12), one get the scaling behavior

$$D_t - D \sim (\alpha_t^{-1} - \alpha^{-1}) \sim (\sigma_t - \sigma) \quad (13)$$

where D_t is the threshold layer thickness. α^{-1} as a function of σ from simulations is shown in Fig.3b.

It would be beneficial to compare with the case that a polyelectrolyte is adsorbed onto an attractive charged substrate of low-dielectric constant (e.g. DNA in aqueous solution onto a charged lipid membrane). Its asymptotic solution for $z \to +\infty$ to Eq.(8) reproduces the usual scaling $D \sim \alpha^{-1} \sim |\sigma|^{-\frac{1}{3}}$ [4]. The adsorption onto a low-dielectric substrate is continuous. The order parameter ρ_a vanishes continuously across the transition, and the layer thickness swells to infinity as the polyelectrolyte is desorbed [11].

DISCUSSION

A strongly charged polyelectrolyte immersed in a salt solution will attract oppositely charged ions to condense until its effective charge density reaches the Manning threshold [12]. This means that one can just renormalize q_0 in our system to $2ea/l_B$ if q_0 is larger than $2ea/l_B$ (l_B is the Bjerrum length). Similarly, the strongly charged surface of bare charge density larger than $\kappa/(\pi l_B)$ is just renormalized back to $\kappa/(\pi l_B)$ [13].

Although we focus on the adsorption of a single polyelectrolyte, our results may provide a starting point to study the charge inversion and multi-layer adsorption. [14]. At low ionic strength, polyelectrolytes are adsorbed in a multi-layer structure because of strong Coulomb repulsion. Each layer is composed of the parallel 1d Wigner crystal [15]. From our physical picture of a single polyelectrolyte adsorption, the upper bound for the thickness of the multi-layer structure is $z_{br} \sim \sigma^{-1/2}(\varepsilon'/\varepsilon - 1)^{1/2}$. This suggests one can easily control just one single layer adsorbed onto high-dielectric substrate by tuning the surface charge density. Rigorous treatment based on this physical picture will be elaborated elsewhere.

REFERENCES

1. de Gennes, P.G., *Scaling Concepts in Polymer Physics*, Cornell Univ., New York, 1979, pp.252-254.
2. Gelbart, W.M., Bruinsma, R.F., Pincus, P.A., and Parsegian, V.A., *Physics Today* **53**, 38-44, (2000).
3. Grosberg, A.Y., Nguyen, T.T., and Shklovskii, B.I., *Rev. Mod. Phys.* **74**, 329-345 (2002).
4. Joanny, J.F., *Eur. Phys. J.* **B9**, 117-122 (1999).
5. Nguyen, T.T., Grosberg, A.Y., and Shklovskii, B.I., *Phys. Rev. Lett.* **85**, 1568-1571 (2000).
6. Wiegel, F.W., *Introduction to Path-Integral Methods in Physics and Polymer Science*, World Sci., Singapore, 1986, pp.39-40.
7. Muthukumar, M., *J. Chem. Phys.* **86**, 7230-7235 (1987).
8. Cheng, C.H. and Lai, P.Y. (unpublished).
9. Netz, R.R. and Andelman, D., *Phys. Rep.* **380**, 1-95 (2003).
10. Loudon, R., *Am. J. Phys.* **27**, 649 (1959).
11. Cheng, C.H. and Lai, P.Y. (unpublished).
12. Manning, G.S., *J. Chem. Phys.* **51**, 924-933 (1969).
13. Bocquet, L., Trizac, E., and Aubouy, M., *J. Chem. Phys.* **117**, 8138-8152 (2002).
14. Dobrynin, A.V., Deshkovski, A., and Rubinstein, M., *Macromolecules* **34**, 3421-3436 (2001).
15. Shklovskii, B.I., *Phys. Rev. Lett.* **82**, 3268-3271 (1999).

Analysis of the center of mass-, rotational- and intramolecular diffusive motions in a monolayer film of intermediate–length alkane molecules adsorbed on a solid surface

F. Y. Hansen* and H. Taub[†]

*Department of Chemistry, Technical University of Denmark, 207-DTU, DK-2800 Lyngby, Denmark.
[†]Department of Physics and Astronomy, University of Missouri–Columbia, Columbia, Missouri 65211, USA.

Abstract. The intramolecular diffusive motions associated with the creation and annihilation of *gauche* defects in intermediate–length alkane molecules adsorbed on a solid surface are relatively slow and on a time scale of nanoseconds. This has been shown by molecular dynamics (MD) simulations and by high–energy–resolution quasielastic neutron scattering (QNS) on monolayers of tetracosane (n-$C_{24}H_{50}$) adsorbed on graphite basal–plane surfaces. QNS with the high energy resolution makes it now possible to probe diffusive motions on the same time scale as in nuclear magnetic resonance (NMR) experiments and, in addition, to obtain information about length scales and character of motions through the wavevector dependence of the QNS.

INTRODUCTION

Films of organic molecules on solid surfaces are of great importance in material science and occur in many technological applications such as in adhesion and coatings [1]. Of particular interest are films of n–alkanes (C_nH_{2n+2}) and branched alkanes because they are the principal constituents of commercial lubricants [2], and they are also prototypes of more complex polymers.

In the area of biological physics, lipid molecules play an important role as building blocks of the membranes that form the cell walls. The lipid molecules consist of a polar head group to which are attached two alkane chains with typically 14-20 carbon atoms [3]. The phase transition and diffusive motion of the lipid molecules in the membranes are the result of a very delicate balance between the head group and alkane chain properties. By studying the simpler alkane subsystem on a microscopic level, useful information for the understanding of phase transitions and molecular diffusion in the much more complicated bilayer lipid membranes may be obtained.

Previously, we have conducted extensive molecular dynamics (MD) simulations of the structure, melting transition, and diffusive motion in monolayer films of tetracosane (n=24) and dotriacontane (n=32) molecules adsorbed on the basal–planes of graphite [4]. For our neutron scattering experiments, we have chosen a large–surface–area substrate of exfoliated graphite. At low temperature (< 215 K), the adsorbed molecules form a 2D crystalline structure which is transformed at 215 K into a smectic crystalline phase that melts at about 340–350 K. The simulated structures and melting transitions have been compared with the results of neutron diffraction measurements, while the diffusive motions inferred from the simulations have been used to interpret the results of QNS experiments.

The low–energy conformation of an alkane molecule is one where all dihedral bonds (the C-C bonds along the carbon backbone of the molecule) are in the *trans*–conformation. The potential energy function for rotation around the dihedral bonds also has a local minimum at approximately 120 degrees from the *trans* conformation, and this conformation is usually referred to as a *gauche* conformation or a *gauche* defect. One of the predictions of the MD simulations [4] was that the lattice melting of the ordered low–temperature structure was driven by a chain melting. That is, well below the melting temperature (340 K for tetracosane (C24) and 350 K for dotriacontane (C32) monolayers), *gauche* defects are confined to the ends of the otherwise all–*trans* carbon backbone of the molecules. However, beginning at about 10 K below the melting temperature, *gauche* defects penetrate into the central region of the molecules. When their number reaches a certain level, a simultaneous chain melting and lattice melting occurs. This type of melting has also been observed in the gel-to-fluid transition of bilayer lipid membranes by NMR [3, 4].

Our MD simulations have shown that the diffusive

motions associated with the creation and annihilation of *gauche* defects are relatively slow and on the time scale of nanoseconds. We have now been able to corroborate the MD predictions by using QNS measurements taken on the High Flux Backscattering Spectrometer (HFBS) at the Center for Neutron Research, National Institute of Standards and Technology (NIST) [5]. The instrument has a very high energy resolution of ~1 μeV allowing us to probe motions on the nanosecond time scale as well as a sufficient throughput to detect scattering from a molecular monolayer adsorbed on an exfoliated graphite substrate. This is the same time scale that can be probed by NMR; but, in addition, the wave vector dependence of the QNS gives information about length scales in the systems and may be used in the analysis of the types of motions involved.

In this paper, we describe a method for projecting out the translational, rotational, and intramolecular diffusive motions of a molecule. This allows us to determine the signature of the various kinds of molecular diffusive motion in a QNS experiment and compare the results directly with experiments. The time scales of these motions are often quite different and, by choice of a spectrometer with the proper energy resolution, it may be possible to study each of these motions separately. Details of the experiments at the new HFBS at NIST are given in the paper by Taub *et al.* [6].

TIME CORRELATION FUNCTIONS IN NEUTRON SCATTERING

In a neutron scattering experiment, we distinguish between coherent and incoherent scattering where the intensity of the former is proportional to the coherent scattering function $S_{coh}(\mathbf{Q},\omega)$ and the latter to the incoherent scattering function $S_{inc}(\mathbf{Q},\omega)$ [7]. $\mathbf{Q}$ is the wavevector transfer and $\hbar\omega$ the energy change of the scattered neutrons. For a fluid and a powder, the scattering functions only depend on the magnitude $|\mathbf{Q}|$ of the wavevector which will be the case here. The scattering functions reflect the dynamics of the system and may be calculated in an MD simulation. For $\hbar\omega = 0$, we have elastic scattering. Of particular interest here are the scattering functions for $\hbar\omega \to 0$, usually referred to as QNS, and the motions associated with that scattering are referred to as *diffusive*, since they are not oscillatory in this limit. Examples are translational and rotational diffusive motions.

All experiments are done on protonated alkane molecules. Since the hydrogen atom has a very large incoherent scattering cross section compared to the coherent cross sections for both hydrogen and carbon atoms, the scattering will be strongly dominated by the incoherent scattering. We shall therefore only be interested in the incoherent scattering function $S_{inc}(Q,\omega)$. It is the time Fourier transform of the intermediate self-scattering function $F_s(Q,t)$

$$S_{inc}(Q,\omega) = \frac{1}{2\pi}\int_{-\infty}^{\infty} dt \exp(-i\omega t) F_s(Q,t) \quad (1)$$

that is given by the expression

$$F_s(\mathbf{Q},t) = \frac{1}{N}\sum_{j=1}^{N} <\exp(i\mathbf{Q}\cdot(\mathbf{r}_j(t)-\mathbf{r}_j(0)))>. \quad (2)$$

This function is readily calculable from a time series of atomic positions $\mathbf{r}_j(t)$ generated in a MD simulation. The sum extends over all N atoms in the system, and the bracket $<\cdots>$ indicates an ensemble average. The simulations have been done by canonical sampling as described in an earlier publication [4].

Let us rewrite the intermediate scattering function in the following way

$$\begin{aligned} F_s(Q,t) &= F_s(Q,\infty) + F_s'(Q,t) \quad \text{with} \\ F_s'(Q,t) &\to 0 \quad \text{for} \quad t \to \infty. \end{aligned} \quad (3)$$

Then the scattering function will be

$$S_{inc}(Q,\omega) = 2\pi F_s(Q,\infty)\delta(\omega) + S_{inc}'(Q,\omega). \quad (4)$$

This shows that the incoherent scattering function, and hence the incoherent scattering intensity, may consist of two contributions. The first term on the right hand side gives an elastic contribution when $F_s(Q,\infty)$ is different than zero, that is, when the atoms do not move far away from their original positions as in rotational or vibrational motion. When the atoms move far away, there is no elastic contribution, since $F_s(Q,\infty)$ will be zero. The second term in the expression gives the frequency dependent part of the scattering function and reflects the kind of diffusive motion involved.

ANALYSIS OF THE MOLECULAR MOTIONS

Any motion of the atoms in a molecule may be described as a combination of a center of mass motion, a rotational motion, and an intramolecular motion. The time scales of these motions are often different, and it will be useful to have a method that allows us to study the various types of motion in an MD simulation. Of particular interest here, the intramolecular motion.

Let us call the position vector of atom i in molecule m at time t $\mathbf{r}_{im}(t)$. Then the center of mass position vector is given by the standard expression [8]

$$\mathbf{R}_m(t) = \frac{\sum_{i=1}^{M} m_i \mathbf{r}_{im}(t)}{\sum_{i=1}^{M} m_i}. \quad (5)$$

The rotational motion of a molecule may be determined in the following way. At time t_1, let the three principal axes of rotation be $\mathbf{e}_1(t_1)$, $\mathbf{e}_2(t_1)$ and $\mathbf{e}_3(t_1)$, such that

$$[\mathbf{e}_1(t_1), \mathbf{e}_2(t_1), \mathbf{e}_3(t_1)] = [\mathbf{i}, \mathbf{j}, \mathbf{k}] [E(t_1)]. \quad (6)$$

$\mathbf{i}, \mathbf{j}, \mathbf{k}$ are the basis vectors for the laboratory–fixed cartesian coordinate system. The elements of the *i*th column of the (3x3) matrix $E(t_1)$ are the cartesian coordinates of the *i*th principal axis $\mathbf{e}_i(t_1)$; they are determined by a diagonalization of the inertia tensor for the molecule at time t_1 [8]. We may invert Eq. (6) and find

$$[\mathbf{e}_1(t_1), \mathbf{e}_2(t_1), \mathbf{e}_3(t_1)][E(t_1)]^{-1} = [\mathbf{i}, \mathbf{j}, \mathbf{k}]. \quad (7)$$

At some later time t_2, the three principal axes of rotation are $\mathbf{e}_1(t_2)$, $\mathbf{e}_2(t_2)$ and $\mathbf{e}_3(t_2)$, that is

$$[\mathbf{e}_1(t_2), \mathbf{e}_2(t_2), \mathbf{e}_3(t_2)] = [\mathbf{i}, \mathbf{j}, \mathbf{k}][E(t_2)]. \quad (8)$$

Before we can project out the rigid rotational displacement of a given atom in going from time t_1 to time t_2, we need to determine the atomic position vectors with respect to the center of mass coordinate $\mathbf{R}_m(t_1)$ of molecule m. Let us call this "relative" atomic position vector $\mathbf{s}_{im}(t_1)$, then

$$\mathbf{s}_{im}(t_1) = \mathbf{r}_{im}(t_1) - \mathbf{R}_m(t_1). \quad (9)$$

The "relative" position vector $\mathbf{s}_{im}(t_1)$ may be written in terms of the cartesian components arranged in the (3x1) column vector $[s_{im}(t_1)]$ like

$$\mathbf{s}_{im}(t_1) = [\mathbf{i}, \mathbf{j}, \mathbf{k}][s_{im}(t_1)] = [\mathbf{i}, \mathbf{j}, \mathbf{k}][r_{im}(t_1) - R_m(t_1)] \quad (10)$$

and from Eq. (7), in the basis of $\mathbf{e}_1(t_1), \mathbf{e}_2(t_1), \mathbf{e}_3(t_1)$

$$\mathbf{s}_{im}(t_1) = [\mathbf{e}_1(t_1), \mathbf{e}_2(t_1), \mathbf{e}_3(t_1)][E(t_1)]^{-1}[s_{im}(t_1)]. \quad (11)$$

If the displacement of the *i*th atom in molecule m from time t_1 to time t_2 is given by a rigid rotation of the molecule, as defined by the rotation of the principal axes of rotation, then the coordinates of atom i in the $\mathbf{e}_1(t_2), \mathbf{e}_2(t_2), \mathbf{e}_3(t_2)$ system should be the same as in the $\mathbf{e}_1(t_1), \mathbf{e}_2(t_1), \mathbf{e}_3(t_1)$ system, that is

$$\begin{aligned}\mathbf{s}^r_{im}(t_2) &= [\mathbf{e}_1(t_2), \mathbf{e}_2(t_2), \mathbf{e}_3(t_2)][E(t_1)]^{-1}[s_{im}(t_1)] \\ &= [\mathbf{i}, \mathbf{j}, \mathbf{k}][E(t_2)][E(t_1)]^{-1}[s_{im}(t_1)] \\ &\equiv [\mathbf{i}, \mathbf{j}, \mathbf{k}] Rot_m(t_2, t_1)[s_{im}(t_1)]. \end{aligned} \quad (12)$$

We have used Eq. (8), and $\mathbf{s}^r_{im}(t_2)$ is the position vector of atom i in molecule m relative to the center of mass position at time t_2, if the atom has performed a rigid rotational motion from time t_1 to time t_2. $Rot_m(t_2,t_1)$ is a (3x3) matrix operator given by

$$Rot_m(t_2, t_1) = [E(t_2)][E(t_1)]^{-1} \quad (13)$$

which may be used to determine the "relative" position vector of atom i, when it performs a rigid rotation from time t_1 to time t_2 around the center of mass.

The MD simulation generates the atomic positions at time t_2. From those we may determine the center of mass position at time t_2 using Eq. (5), and the "relative" atomic positions $\mathbf{s}_{im}(t_2)$ at that time are given by

$$\begin{aligned}\mathbf{s}_{im}(t_2) &= \mathbf{r}_{im}(t_2) - \mathbf{R}_m(t_2) \\ &= \mathbf{s}^r_{im}(t_2) + \mathbf{s}^{int}_{im}(t_2).\end{aligned} \quad (14)$$

Since $\mathbf{s}_{im}(t_2)$ is known from the MD simulation and $\mathbf{s}^r_{im}(t_2)$ may be determined from Eq. (12), the intramolecular displacement of atom i in molecule m from t_1 to t_2, $\mathbf{s}^{int}_{im}(t_2)$, is given by this equation.

RESULTS AND DISCUSSION

The energy resolution and dynamical range in the MD simulation were chosen to match those in the experiment. The energy resolution is given as $1/\tau_{max}$ (s^{-1}) where τ_{max} (s) is the length of the time series in the MD simulation, and the dynamic range $1/(2d\tau)$ (s^{-1}) is determined by the time $d\tau$ (s) between two consecutive recordings of the atomic positions in the time series.

It is clear from the definition in Eq. (1) that the intermediate scattering function $F_s(Q,t)$ is equal to unity at $t = 0$. As discussed previously, it decays either to a non–zero constant level or to zero at long times, depending on the character of the motion. As an example, let us look at the intermediate scattering function in Fig. 1 for the rotational and intramolecular motions of C24 molecules in a monolayer. We note that the intermediate scattering functions level off at a non–zero level at long times. This implies that there also will be elastic scattering from these modes. In Fig. 1(a) we note that the intermediate scattering function for both modes has dropped directly to the long–time level at the first time step after the initial time for both modes. This means that the modes are too fast on the time scale chosen in the simulations for us to follow the decay. In order to do so we need to choose a smaller $d\tau$. The functions may be approximated by a delta function $\delta(t)$ at time zero and by a constant at any other time. Since the scattering function $S_{inc}(Q,\omega)$ is the Fourier transform of the intermediate scattering function, Eq.(1), the delta function will produce a constant intensity, independent of ω, and the constant time–independent part an elastic scattering at $\omega = 0$.

It was found that the translational mode also was too fast at about 260 K with the given time resolution, so the scattered intensity will consist of only an elastic part and a featureless constant quasielastic scattering. Since the modes will be faster with increasing temperature, we would not expect to find any other quasielastic scatter-

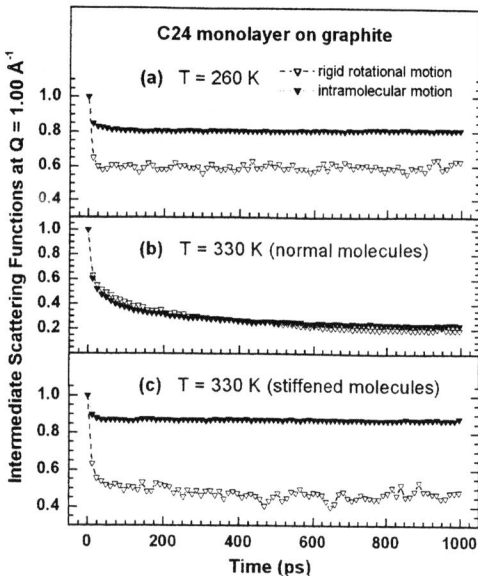

FIGURE 1. Intermediate scattering functions for the rotational- and intramolecular motions of C24 molecules.

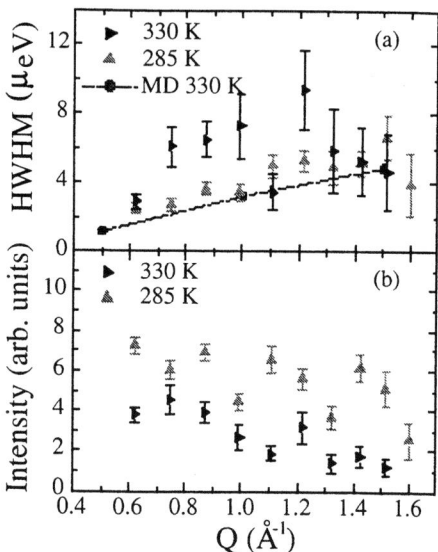

FIGURE 2. Half-width at half-maximum (HWHM) of the scattering function and scattered intensity for a monolayer of C24 molecules.

ing within the dynamic range of the HFBS unless some new slow modes were activated. This happened at about 330 K as can be concluded from Fig. 1(b). Contrary to what we saw in Fig. 1(a), it is now possible to follow the decay of both the rotational and intramolecular modes for several hundred picoseconds before they reach a constant non–zero plateau. Our MD simulations of the melting transition [4] had shown that *gauche* defects started to penetrate into the central region of the molecules at about 330 K, 10 K below the melting temperature. The appearance of the new slow modes was therefore interpreted as originating from the motion associated with the creation and annihilation of the defects. The slow–down of the rotational motions also is explained by the change in shape of the molecules, when *gauche* defects are introduced. This leads to a larger moment of inertia for the rotation around the long axis that gives the primary contribution to the scattering from the rotational diffusive motion. To substantiate this conclusion, we stiffened the molecules around the dihedral torsion angles to prevent any formation of *gauche* defects and found that the slow modes disappeared as is evident from Fig. 1(c).

The half–width at half–maximum (HWHM) of the scattering function $S_{inc}(Q, \omega)$ for both experiments and simulation has been plotted in Fig. 2(a) as a function of Q. We note that the experimental results at 285 K have no counterpart in the simulation, because all modes were found to be outside the dynamical range of the HFBS at that temperature. The reason is that in the united atom model used in the simulations the moment of inertia for rotation about the long molecular axis is about a factor of two too small. This makes the rotational motion faster than in the experiments.

At 330 K, on the other hand, we find qualitative agreement of the HWHM of the QNS between experiments and the simulations and interpret this as an experimental verification of the melting mechanism proposed in Ref. [4]. In addition, our simulations show a Q–dependence of the QNS intensity similar to that measured [see Fig. 2(b)]. We conclude that our results demonstrate the potential of QNS for investigating slow modes on a nanosecond time scale in samples as small as a molecular monolayer.

REFERENCES

1. Xia, T.K. *et al.*, *Phys. Rev. Lett.* **69**, 1967-1970, (1992).
2. Xia, T.K. *et al.*, *Science* **261**, 1310-1312, (1993).
3. Small, D.M. *The Physical Chemistry of Lipids*, Plenum, New York, 1986, pp. 1-19.
4. Hansen, F.Y. *et al.*, *Phys. Rev. Lett.* **83**, 2362-2365, (1999).
5. Meyer, A. *et al.*, *Rev. Sci. Instrum.* **74**, 2759-2777, (2003).
6. Taub, H. *et al.*, this proceedings, 2003.
7. Marshall, W. and Lovesey, S.W., *Theory of Thermal Neutron Scattering*, Oxford University Press 1971, Oxford, pp. 38-45, 370-383.
8. Goldstein, H. *Classical Mechanics*, Addison–Wesley Publishing Company, Reading 1980, pp. 188-203.

Nanostructure Formation in Polymer Thin Films: Dissipative Particle Dynamics Simulation Using a Space-Time Coarse-Grained Model

Chee Chin Liew[‡] and Masuhiro Mikami

Research Institute for Computational Sciences (RICS) and Research Consortium for Synthetic Nano-Function Materials Project (SYNAF), AIST, 1-1-1 Umezono, Tsukuba 305-8568, Japan.

Abstract. We have devised a soft-attractive-and-repulsive pair-potential model, which has a hardness parameter (β) and a smooth cut-off, to adopt coarse-graining procedures for soft-matter. A model with β–3.0 analytically represents a time-averaged $U_{eff}(r)$ over a long time-span, and gives excellent energy conservation at a very long time-step. In this work, we extended our model to represent copolymer systems, and combined it with dissipative particle dynamics (DPD) method. We have performed DPD simulations to study the nano-pattern formation in symmetric block copolymer film, and found that the simulated surface structures resembled those patterns observed in experiment of copolymers thin films on a solid substrate.

INTRODUCTION

Well-controlled surface nanostructures of polymers on solid substrate are potentially suitable for a large number of technological applications. The length-scale and slow dynamics of such system, however, is beyond the range of atomistic molecular simulations (MD). Dissipative particle dynamics (DPD) method [1-3] seems to be promising for simulations of such meso-scale systems. DPD is based on simulations of particles with "soft-repulsive" and dissipative-and-random forces. The dissipative-and-random forces enhanced collisions among particles, hence effectively stretched the characteristic time-scale of the simulated system, and ensured that DPD is simulating a Hamiltonian system in canonical ensemble [2]. The soft-particle represent a group or segment of molecules, and allow much larger time-step than MD simulation [3]. A fundamental problem of DPD, however, is that there is no straightforward way to derive the model for a realistic molecular system. Namely, the use of the "soft-repulsive" model in DPD results in a loss of connections to the phase behavior of the underlying molecular systems. Practically, it is impossible for simulation of soft-matter at free surface.

We have developed a series of particle models that are consistent in phase behavior at various space-time coarse-grained levels [4]. Here, we extended our model for DPD simulations of surface nano-pattern in polymer thin films. Using a simple model for block copolymer, we found that the simulated structure resembled those patterns observed in experiment of copolymers thin films on a solid substrate.

COARSE-GRAINED MODEL

We have divided the essential coarse-graining procedures into (i) Grouping, (ii) Packing, and (iii) Time-averaging, see [4], as illustrated in Figure 1. Grouping (Gp) can be considered as an extension of united-atom to a larger scale for monomer or segment in different coarse-grain level for a large molecule. Packing (Pk) is meant to represent the effective potential ($U_{eff}(r)$) for a group of solvents or small molecules that are loosely gathering. This is, in fact, the original concept for the "fluid particles" in DPD; however, we required that they are consistent in phase behavior. Time-averaging (Ta) is meant to consider the

---

[‡] Corresponding author. Present address: Polymer Physics, BASF-AG, GKP/M-G200, D-67056 Ludwigshafen, Germany.

$U_{eff}(r)$ between two groups of particles over a finite time-span. Forrest and Suter [5] have performed MD simulations of polymer melts using a LJ(12-6) based united-atom chains, and they found that the $U_{eff}(r)$ pre-averaged over a long time-span are much softer and shallower than the original potential.

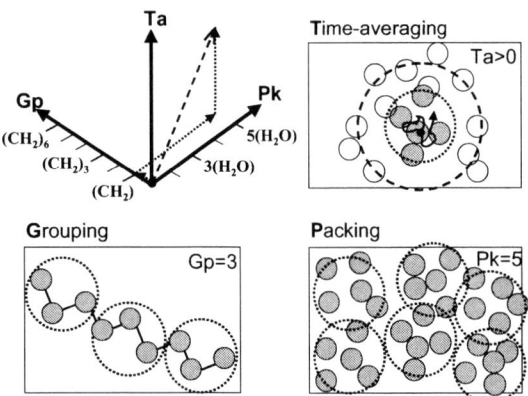

FIGURE 1. Schematic illustration of three essential coarse-graining procedures for soft matter.

To adopt all these procedures, we have devised a potential model, which has an adjustable hardness parameter (β) and a smooth cut-off (r_{cut}), in our previous work [4]. It is a combination of a Morse-like function with a damping function, $w(r_{ij})$, to give a smooth cut-off and limit the interaction range,

$$U(r_{ij}) = \varepsilon \left\{ \left(1 - \exp\left[-\beta\left(\frac{r_{ij}}{r_{min}} - 1\right)\right]\right)^2 - 1 \right\} w(r_{ij}) \quad (1)$$

where $\mathbf{r}_{ij} = \mathbf{r}_i - \mathbf{r}_j$, $r_{ij} = |\mathbf{r}_{ij}|$, and ε determines the depth of the potential minimum which is located at r_{min}. The damping function is written as

$$w(r) = \begin{cases} 1.0 & \text{for} \quad r_{ij} < r_{min} \\ \left[1 - \left(\frac{r_{ij} - r_{min}}{r_{cut} - r_{min}}\right)^n\right]^n & \text{for} \quad r_{min} \leq r_{ij} \leq r_{cut} \\ 0.0 & \text{for} \quad r_{ij} > r_{cut} \end{cases} \quad (2)$$

where n is a small integer. We found that $n=2\sim4$ gives smooth functions at the cut-off region and does not strongly alter the features of the Morse force functions. We are using $n=2$ in this work. Note that β is a dimensionless parameter, and when $\beta=6$, the function is very similar to the LJ(12-6) model. The distance at $U(r_{ij})=0$, denoted by r_0, is defined as the reduced unit of length; $r_0 = r_{min}(1-\ln 2/\beta)$. We take ε as the reduced unit of energy, so for a particle with mass, m, the time-scale is $t_{sc} = r_0(m/\varepsilon)^{1/2}$ and the reduced time is $t^* = t/t_{sc}$. We have shown that for models of $\varepsilon=1$, $r_0=1$ and $r_{cut}=2.6r_0$, the liquid-vapor coexistence curves of the model with $\beta=5.0\sim3.0$ are in good agreement with LJ(12-6) model [4].

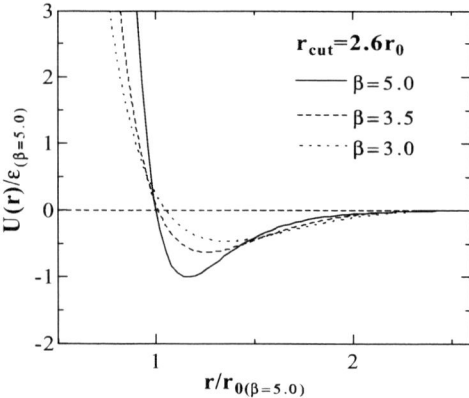

FIGURE 2. Potential models of $\beta=5.0$ to 3.0 that give similar liquid-vapor coexistence curves as that of LJ(12-6).

The functions of models with ε and r_0 being adjusted to give the similar critical point as that of $\beta=5.0$, shown in Figure 2, strongly resembled those time-coarse-grained $U_{eff}(r)$ of Forrest and Suter [5]. This is not enough to justify, but it indicates that a model with $\beta<5.0$ could analytically represent time-coarse-grained $U_{eff}(r)$ at longer time-spans. A clear difference is that the tabulated $U_{eff}(r)$ of [5] were obtained by time-averaging at a specific temperature and density, while our potential models reproduce the phase behavior at a wider temperature range.

Considering a model of $\beta=3.0$ as an analytical function of $U_{eff}(r)$ over a long time-span for a model with $\beta=5.0$, is in accord with the fact that t_{sc} of $\beta=3.0$ is larger than that of $\beta=5.0$, since ε of $\beta=3.0$ is about a half of that of $\beta=5.0$. Furthermore, since the model of $\beta=3.0$ has a very smooth-and-soft function, it gives excellent energy-conservation in NVE-MD run using a very long time-step ($\Delta t^*=0.02$) in comparison to that of LJ model ($\Delta t^*=0.004$).

COMBINATION WITH DPD METHOD

The new coarse-grained model can be combined with any particle simulation method, such as MD, hybrid Monte Carlo, or Brownian Dynamics. As we have shown in our work [4], combining the model with dissipative-and-random forces of DPD will enable us to reproduce the dynamic of the particles for simulations of soft-matter. The dissipative-and-random forces of DPD are given by

$$F^D = -\gamma w^D(r_{ij})(\mathbf{v}_{ij} \cdot \mathbf{r}_{ij})\mathbf{r}_{ij}/r_{ij}^2 \quad (3)$$

$$F^R = \sigma w^R(r_{ij})\xi_{ij}\mathbf{r}_{ij}/r_{ij} \quad (4)$$

where $\mathbf{v}_{ij} = \mathbf{v}_i - \mathbf{v}_j$; ξ_{ij} is an independent random number with Gaussian statistic for each pair of particle; $w^D(r_{ij})$ and $w^R(r_{ij})$ are weight functions that are related to each other, and the friction and random amplitudes, namely γ and σ, are related to $k_B T$:

$$w^D(r_{ij}) = [w^R(r_{ij})]^2 \; ; \; \sigma^2 = 2\gamma k_B T. \quad (5)$$

For our new model we defined

$$w^D(r_{ij}) = \begin{cases} (1-r_{ij}/r_d)^2 & \text{for } r_{ij} < r_d \\ 0 & \text{for } r_{ij} \geq r_d \end{cases} \quad (6)$$

where the forces are vanishing at r_d, a range in between r_{min} and r_{cut}, and we have shown that r_d can be used as a parameter to adjust the dynamics properties of a coarse-grained model particle [4].

MODEL FOR BLOCK POLYMER

For polymer systems, the soft-particles are threaded together in linear chains using the harmonic bond stretching: $U^{bond} = K^{bond}(r-r_{eq})^2$, where r is the bond length, and $r_{eq} = r_0$, $K^{bond}/\varepsilon = 200$ were employed. As space is limited, detailed modeling of polymers will be presented in another paper.

For comparison, the polymer model can be related to the Flory-Huggins χ-parameter. For polymer mixture of two components, A and B, that do not favor contact, the χ-parameter can be obtained from simulations of polymer segregation, as have been done in [3] using the following equation:

$$\chi N = \frac{\ln[(1-\phi_A)/\phi_A]}{1-2\phi_A} \quad (7)$$

where ϕ_A is the volume fractions of the A component, and N is the polymer length. Mixtures of homo-polymers with $N=N_A=N_B$ is modeled by using particle model of $\beta=3.0$ with $\varepsilon=1.0$, $r_0=1.0$ for both segments of A and B. In this work, however, we are limiting the non-bonded particle interactions to a shorter range by $r_{cut} = 2.0 r_0$, and we defined the A-B interaction as

$$\varepsilon_{AB} = (1-\Delta)\sqrt{\varepsilon_{AA}\varepsilon_{BB}} \quad (8)$$

where Δ is a parameter to adjust the affinity of A-B segments.

Figure 3 shows the results for polymers at particle density $\rho^* = \rho/r_0^{-3} = 0.8$ and temperature at $T^* = k_B T/\varepsilon = 2.0$. The phase separation or order-disorder transition (ODT) of symmetric block copolymer in bulk system occurs when $\chi N > 10.5$. For our model with $\Delta=0.2$ and $N=10$, we estimated $\chi N=9$, as in Figure 3.

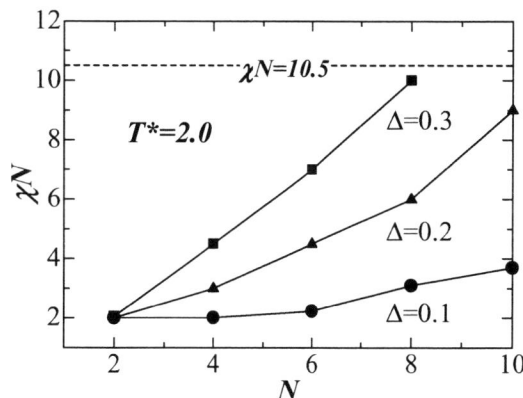

FIGURE 3. χN as a function of N for mixture of homopolymers of $N=N_A=N_B$ with various A-B affinity Δ-parameters.

NANO-PATTERN OF BLOCK COPOLYMER

We have performed DPD simulations of nano-pattern in symmetric block copolymer film of $N=10$ (or A_5-B_5) and $N=20$ (or A_{10}-B_{10}) with A-B affinity parameter of $\Delta=0.2$ on a solid substrate. Samples of copolymer thin firms were generated at temperature $T^*=1.0$ to 1.5. From the results of the χN-parameters in bulk systems (Figure 3), we can presume that the simulated systems are all at the regime of $\chi N > 10.5$ or at low temperature where $T < T_{ODT}$.

The size of the simulation box is $100 \times 100 \times 30 r_0^3$ and a layer of substrate particles (=A) is fixed at the bottom of the simulation box using harmonic bonding ($K^{bond}/\varepsilon=10$). The CPU time requirement for a DPD simulation for $t^*=1 \times 10^3$ using $\Delta t^*=0.02$ for a system of 45000 particles as shown in figure 4, was about 6 hours on a Linux-PC with a P4-1.8GHz processor. We found the simulated surface patterns, such as that of "bi-continuous" pattern in Figure 4, resembled those pattern observed in experimental work of symmetric PS-b-PMMA block copolymer thin film on a SiOx substrate at the temperature regime T< T_{ODT} [6].

In conclusion, we have developed a simple model based on the "soft-attractive-and-repulsive" model for particle dynamics simulation of copolymers. We have shown that DPD simulation using the new model allows us to study nanostructure formation in block copolymer thin films on free surface.

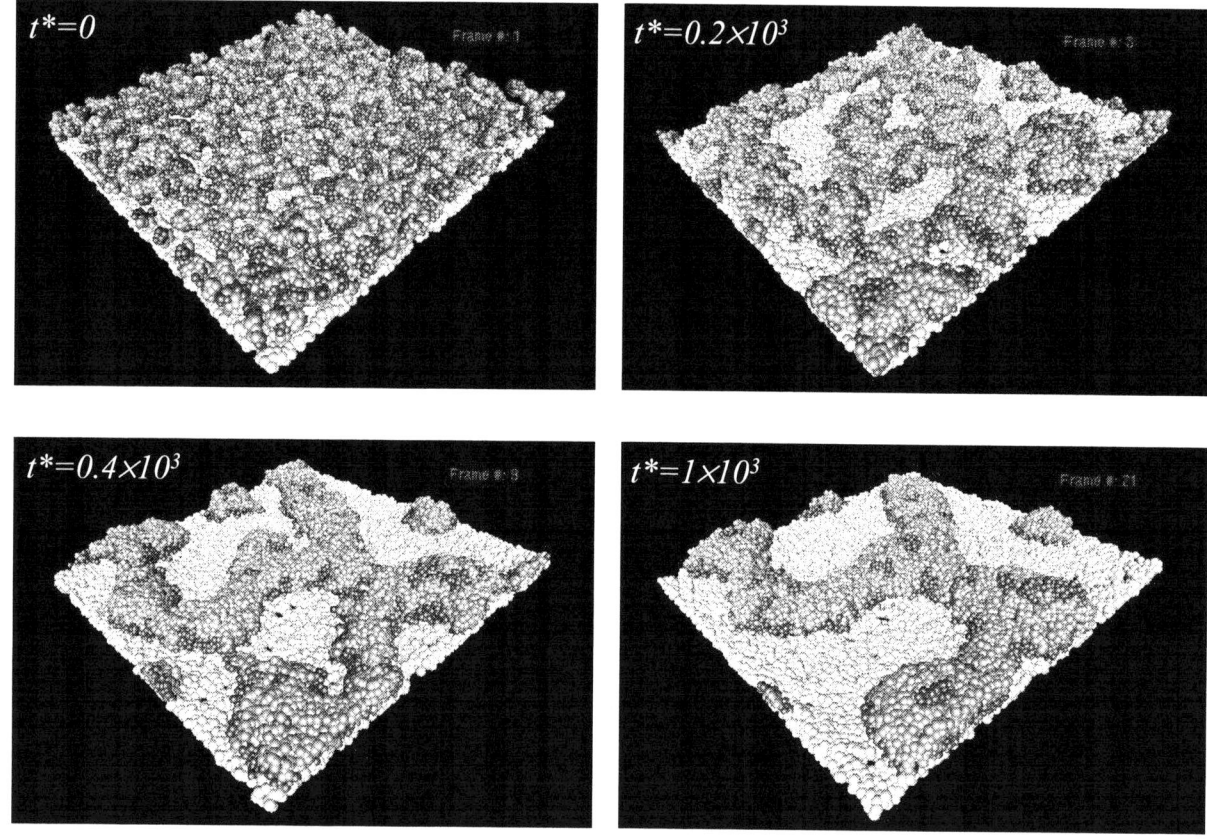

FIGURE 4. Structure formation of block copolymers (A_{10}-B_{10}) on substrate (A) at T^*=1.25.

ACKNOWLEDGMENTS

This work was partly supported by NEDO under the Nanotechnology Materials Program.

REFERENCES

1. Hoogerbrugge, P. J., and Koelman, J. M. V. A., *Europhys. Letters* **19**, 155-160 (1992).

2. Espanol, P., and Warren, P. B., *Europhys. Letters* **30**, 191-196 (1995).

3. Groot, R. D. and Warren, P. B., *J. Chem. Phys.* **107**, 4423-4435 (1997).

4. Liew, C. C., and Mikami, M., *Chem. Phys. Letters* **368**, 346-351 (2003).

5. Forrest, B. M., and Suter, U. W., *J. Chem. Phys.* **102**, 7256-7266 (1995).

6. Green, P. F., and Limary, R., *Advances in Colloid. and Interface Sciences* **94**, 53-81 (2001).

Simulating the long-time viscoelastic response of long polymers

C.P. Lowe[*], A.F. Bakker[†] and M.W. Dreischor[*]

[*]Department of Chemical Engineering,
University of Amsterdam,
Nieuwe Achtergracht 166, 1018 WV Amsterdam,
The Netherlands
[†]Faculty of Applied Science,
Delft University of Technology,
Lorentzweg 1, 2628 CJ Delft,
The Netherlands

Abstract. Simulating the long time dynamics of even a simple model polymer with a realistic number of monomers is impractical. We therefore use a renormalization procedure to represent an original polymer with fewer monomers. By showing that the results become independent of the number of beads in the chain, we can infer the stress-stress correlation function for a long polymer in both the short and long time regimes. To do so we simulate a chain of at most 32 monomers. Our value for the polymer contribution to the viscosity in the long polymer limit is consistent with theory. However, it comes toward the lower end of the range of theoretical predictions.

INTRODUCTION

Because polymers are very large molecules their dynamics are, by atomic standards, slow. Thus, if one is interested in studying numerically the long time dynamics, a computational approach which resolves atomistic detail is impractical. The system we consider here is a single ideal chain. That is, the polymer is modelled as a set of connected particles with a root mean square separation between adjacent beads b (the Kuhn length). This model is a minimal representation of a real (flexible) polymer but predicts reasonably well the dynamic properties of dilute polymer solutions [1]. Despite its simplicity, it is still not possible to simulate even an ideal chain with the typical number of monomers commonly encountered in reality ($N > 10^6$). Consequently, it is necessary to adopt a more mesoscopic approach whereby a long polymer is drastically simplified ("renormalized") so that it can be simulated on long time scales. However, we need to be able to argue that in some meaningful sense we still capture the dynamics of the real polymer. The dynamics of a polymer (in dilute solution) are largely determined by the hydrodynamic interactions between the monomers making up the polymer [2]. For example, the centre of mass diffusion coefficient D in terms of the monomer diffusion coefficient D_0 is

$$\frac{D}{D_0} = \frac{1}{N} + \left(\frac{8a}{3b}\sqrt{\frac{6}{\pi}}\right)\frac{1}{\sqrt{N}} \quad (1)$$

where a is the hydrodynamic radius of a monomer. This is related to D_0 by $a = k_B T/6\pi\eta D_0$, where η is the solvent viscosity, T the temperature and k_B Boltzmann's constant. The second term on the right hand side is the contribution from the the hydrodynamic interactions between the polymer beads. Equation 1 shows that for a long polymer (large N) the hydrodynamic term dominates. Because the hydrodynamic radius is less than the Kuhn length, for a short polymer the hydrodynamic term is proportionally smaller and will not typically dominate. So the hydrodynamics of long polymers is not the same as short ones. The idea we pursue here is the following. Suppose we have a long ideal polymer chain of N^* monomers. We now try to model it by reducing the number of monomers to a value $N(< N^*)$. To keep a relation to the original polymer we keep the diffusion coefficient and radius of gyration of the new "renormalized" polymer the same as for the original. We can do this by dimensional analysis. Assuming that a relevant length is the radius of gyration of the polymer r_g and that the diffusion coefficient D defines the relevant time scale, we simply measure lengths in units of r_g and time in units of r_g^2/D. Now we see how the dynamic behaviour of the renormalized chain depends on N. If we can show that in reduced units (characteristic of the long polymer) the results are independent of N then they apply equally well for the original long polymer. Of course there is no guarantee this will work. If a property depends on the original polymer in all its detail, no coarse graining will work. Here we examine how well we can do for perhaps the most important dynamic property of a polymer solution, its viscoelastic response. We begin by describing the model that we use to include thermal fluctuations and mimic long polymer hydrodynamics.

DESCRIPTION OF THE MODEL

From a computational point of view the main difficulty is that both thermal fluctuations and hydrodynamics must be included. The approach we choose is to use a simple particle model for the solvent - a dissipative ideal gas. There are no static interactions between solvent particles so the solvent has an ideal gas equation of state. The method has been used widely as a test case for dynamic particle models [3] but little used as a practical tool. The model polymer consists of beads, with adjacent beads connected by an harmonic potential of the form

$$U = k_B T \sum_{i=1}^{N-1} \frac{3}{2b^2} (\vec{r}_i - \vec{r}_{i+1})^2 \qquad (2)$$

This allows us to specify the Kuhn length. As the solvent is ideal it will not influence the static properties of the chain. It remains exactly an ideal chain. The solvent's role here is simply to propagate the hydrodynamic interactions between polymer beads.

The ideal gas solvent is dissipative because total energy is not conserved. It is maintained at a constant temperature by a Lowe-Andersen thermostat [3]. Dynamically, the thermostat makes a contribution to the viscosity so it is possible to satisfy the condition for liquid like ($D_s << v$) rather than gas like ($D_s \sim v$) dynamics. Here D_s is the solvent diffusion coefficient. The equilibrium distribution is that of the canonical (NVT) ensemble. The Lowe-Andersen thermostat conserves momentum and is Galilean invariant. These are important pre-requisites for reproducing the correct hydrodynamics. In this respect it is similar to dissipative particle dynamics (DPD) [4]. It has the advantage that a simple algorithm suffices to update the equations of motion and still satisfy detailed balance. In practice this means that static properties of the system (temperature, Kuhn length etc.) are correct. As the dynamic properties also depend on the static properties, this is an important pre-requisite. The procedure simply consists of using a velocity Verlet algorithm to integrate the normal (conservative) equations of motion over a time-step Δt. Pairs of particles within a distance r_c (an interaction radius for the thermostat) of each other are identified. With a probability $\Gamma \Delta t$, the particles undergo "bath" collisions, in that their relative velocity along the line of centres is re-drawn from a Maxwellian. The individual particle velocities are then updated such that momentum is conserved. The bath collisions take on the role of the dissipative force in DPD. While simpler and more efficient than DPD when viewed purely as a thermostat, this method may seem a little more crude in that it does not involve a distance dependent weight function for this dissipative interaction. However, in DPD this function is somewhat arbitrary and, as we will see, this does not seem to be a problem. The beads making up the model polymer have exactly the same dissipative interaction with surrounding solvent but do not interact with each other (the dynamic interaction between beads comes only through the springs and solvent). This means that we can, without ambiguity, identify the monomer diffusion coefficient as being the diffusion coefficient of the solvent particles ($D_s = D_0$).

Despite its simplicity, the model still has a number of parameters we need to specify. The dissipative ideal gas itself is characterized by just two parameters. Firstly, a typical interparticle separation $\lambda = (1/\rho)^{1/3}$, where ρ is the number density. Secondly, a parameter $\Lambda = \sqrt{k_B T / \Gamma^2 r_c^2 m}$, where m is the solvent particle mass, characterizing the ratio of the time it takes particles to displace to the mean bath collision time. These parameters are fixed by the requirement for realistic relative time-scales. There are three relevant time scales. First we have a sonic time τ_s (as with all particle model solvents, the solvent is compressible). This is the time it takes sound to travel a characteristic distance l i.e. $\tau_s = l/c_s$ where c_s is the speed of sound. Second we have a viscous time $\tau_v = l^2/v$, where v is the kinematic viscosity of the solvent. This is the time it takes transverse momentum to diffuse a distance l. Finally we have a diffusive time $\tau_D = l^2/D$, where D is the polymer diffusion coefficient. This is the time it takes a particle to diffuse a distance l. For a real long polymer we have $\tau_s < \tau_v << \tau_D$. The diffusive time is the longest and this defines the "long" time-scale. Ensuring that these time-scale take realistic relative values gives us very little freedom, values of $\Lambda = 0.03$ and $r_c^3 = 6/\pi \rho$ are about optimal.

With the polymer present we have an additional parameter a/b. This parameter is set to 0.25. The reason for this is that for this value the polymer diffusion coefficient scales as $1/\sqrt{N}$ to a very good approximation for $N > 4$. The reason for this is that there is a small N correction term in equation 1 that scales as $1/N$ and for this value of a/b it almost exactly cancels the non-hydrodynamic term [5]. The reason for doing this is that, in terms of polymer theory, it imposes dynamic scaling [6] even for short chains. As we are using scaling arguments to try to work out long polymer dynamics from short polymer dynamics this makes sense. It is our attempt to correct for the problem outlined above, that the dynamics of short and long polymers generally differ. Using this value for a/b requires that a typical polymer bead separation must be of the order of a typical solvent particle separation. Despite this, we have shown that the hydrodynamic interactions between beads are still modelled accurately.

RESULTS

The simulations were carried out using 10^4 solvent particles and chains consisting of $N = 4, 8, 16$ and 32 beads. The Lowe-Anderson thermostat has a maximum timestep ($\Gamma \Delta t = 1$) and we used half this value. The polymer temperature and Kuhn length calculated from the simulations were, to the accuracy we calculated them ($> 0.1\%$), the same as the set values. From the data, we calculated the polymer contribution to the scaled stress stress correlation function $\Sigma^*(t)$

$$\Sigma^*(t) = \frac{1}{k_B T V \phi_p \eta_s} <\sigma(0)\sigma(t)> \quad (3)$$

where η_s is the solvent viscosity, $\phi_p (= 4\pi r_g^3/3)$ is the "volume fraction" occupied by the polymer and $\sigma(t) (= f_{ijx} r_{ijy})$ is the instantaneous polymer contribution to the shear stress. Here f_{ijx} is the x component of the force on particle i due to particle j and r_{ijy} the y component of the vector joining i and j. By integrating $\Sigma^*(t)$ we can calculate the polymer contribution to the total viscosity of the system η

$$\frac{\eta - \eta_s}{\eta_s \phi_p} = \int_0^\infty \Sigma^*(t) dt \quad (4)$$

Taking the Laplace transform of equation 3 gives the frequency dependent viscosity, although we do not discuss this here. At sufficiently long times, from the renormalization argument we expect that the only relevant timescale will be the characteristic time it takes a polymer to diffuse a distance the order of its own size, τ_p. This being the case, we expect that the function $\tau_p \Sigma^*(t/\tau_p)$ will become independent of N for sufficiently large N. In figure 1 we have plotted the data in this form and indeed for sufficiently long times the data scale onto each other even for $N = 4$. With increasing N the curves superimpose down to shorter times. Where we find data independent of N we have effectively solved the long-polymer case because we can substitute any value of N, however large, and work back to actual values. Therefore, even with such a small number of beads we can calculate the long polymer stress-stress correlation function from some time onward. The shorter the times we need to resolve the more beads we need. It should also be noted that even for a fairly modest polymer, for example polyethylene with 10^6 monomers, $\tau_p \sim 10^{-2}$. By atomic standards this is a very long time indeed.

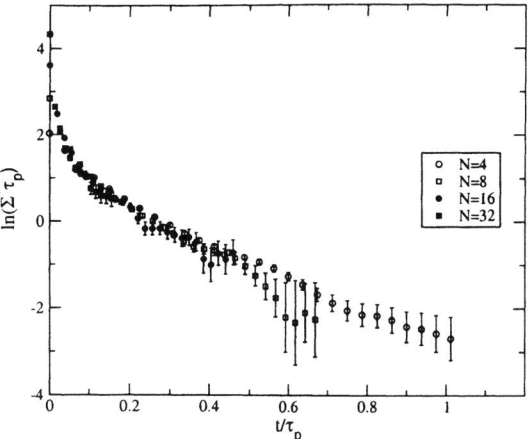

FIGURE 1. Dimensionless stress-stress correlation function Σ^* as a function of the time relative to the positional diffusion time t/τ_p

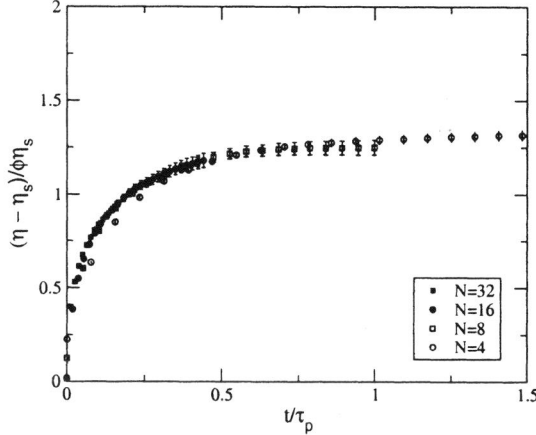

FIGURE 2. Time dependent polymer contribution to the viscosity $\eta(t)$ as a function of the time relative to the positional diffusion time t/τ_p

By integrating the dimensionless stress stress correlation function we can calculate the polymer contribution to the viscosity. The time integrals are plotted in figure 2. For the $N = 16$ and 32 cases there is no statistically significant difference indicating that they track the long polymer result. For the $N < 16$ results we have only used data from times where the stress-stress correlation function does not differ from the $N = 32$ and 16 data. The contribution to the integral from shorter times is taken from the latter. The viscosity asymptotes to a value $\frac{\eta - \eta_s}{\eta_s \phi_p} = 1.35$. The pre-average Zimm model predicts this quantity to be 1.49 [1]. More sophisticated hydrodynamic theories give values between 1.29 and 1.65 [7], our value is therefore consistent with the lower theoretical estimates.

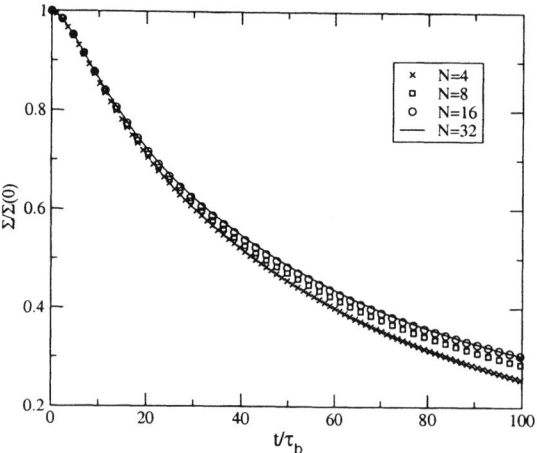

FIGURE 3. Normalized stress-stress correlation function $\Sigma(t)/\Sigma(0)$ as a function of the time relative to the time to diffuse a Kuhn length time t/τ_b

So far we have considered the relevant quantities that we need to preserve in our renormalized model are the centre of mass diffusion coefficient and radius of gyration. As we showed above, this works for the long times we are interested in and it is the long time behaviour that determines the viscosity. What about short times? At short enough times we saw that this would not work. However, at short times one might expect that the relevant time-scale determining the decay of the stress will be the time it takes a bead to diffuse the order of a Kuhn length with the monomer diffusion coefficient, $\tau_b = b^2/D_0$. We take the monomer diffusion coefficient because the beads will not yet have had time to "feel" each others presence. We carry out the analysis using these variables, plotting the normalized function $\Sigma^*(t)/\Sigma^*(0)$, $\Sigma^*(0)$ is an equilibrium average, independent of N. The results are shown in figure 3. From $t = 0$, up to some value which increases with increasing N, the curves superimpose. Again, where this is the case we have the true answer for any ideal polymer, independent of N. Therefore, as well as determining the long time limit of the stress-stress correlation function of a long chain we can determine the short time behaviour. For the polyethylene example quoted above $\tau_b \sim 10^{-11} s$ so the range plotted in the figure would correspond to $10^{-9} s$. Short as far as the polymer is concerned but still long by atomic standards.

DISCUSSION

We have shown that it is possible to infer the viscoelastic response of a long polymer chain by simulating a relatively small chain. This was possible for both the short and long time regimes. Our methodology relies on a renormalization argument, whereby the original polymer is simplified into a smaller polymer whilst keeping a length and time scale characteristic of the original long polymer the same. In this we have been successful in what we set out to do. For intermediate times we still do not know the asymptotic result, but the range of this intermediate regime decreases as we simulate polymers with increasing N. It should be noted that this drastic simplification is essential if polymers are going to be simulated on long time-scales. The longest polymer we simulated $N = 32$ was simulated for 10^7 time steps and even then the statistics are not good enough to determine the asymptotic viscosity. By using the data from the short chains we could correct for this. The value we obtained for the asymptotic long polymer contribution to the viscosity was within the bounds of theoretical estimates but differed by a statistically significant margin from the pre-averaged result. We could use the renormalized result to determine the short time form of the stress-stress correlation function as well. This cannot be predicted from the Zimm model because the scaling involves the time to diffuse a Kuhn length. The theory takes the continuous limit of the chain, for which this time-scale is zero. It is connected with the discreteness of the chain that the Zimm model does not resolve.

The methodology we used for the calculations is straightforward. It accurately includes thermal fluctuations and hydrodynamics. Because it involves an explicit solvent it might not be optimal for this problem. At least, not if one is prepared to neglect the time dependence of the hydrodynamic interactions. However, the computational cost scales linearly with the number of polymers and it is straightforward to include solid boundaries. For many polymer systems and or flow in complex geometries, both things that we are interested in, these advantages come to the fore.

REFERENCES

1. Doi, M., and Edwards, S.F., *"The theory of polymer dynamics"*, Clarendon, Oxford, 1986.
2. Zimm, B.H., *J. Chem. Phys* **24**, 269-278 (1956).
3. Lowe, C.P., *Europhys. Lett.* **47**, 145-151 (1999).
4. Spenley, N.A., *Europhys. Lett.* **49**, 534-540 (2000).
5. Dunweg, B., Reith, D., Steinhauser, M., and Kremer, K., *J. Chem. Phys.* **117**, 914-924 (2002).
6. de Gennes, P.G., *"Scaling concepts in polymer physics"*, Cornell University Press, Ithaca, 1979.
7. Yamakawa, H., *"Modern theory of polymer solutions"*, Harper & Row, New York, 1971.

Dynamics in Self-Assembling Polymer Films

Kenji Fukunaga*, and Takeji Hashimoto[¶]

*Polymer Laboratory, UBE Industries, Ltd. 8-1 Goi-minamikaigan, Ichihara, Chiba 290-0045, Japan
[¶]Department of Polymer Chemistry, Graduate School of Engineering, Kyoto University, Kyoto 606-01, Japan

Abstract. Thin films of lamella-forming block copolymers are annealed by the solvent-vapor treatment and the ordering process of them is investigated. The influence of the boundary surface strongly adsorbing the block component and the influence of the unconfined free surface are discussed.

INTRODUCTION

Block copolymers comprised of two or more polymer chains covalently bonded at their chain ends are microphase-separated into novel morphologies with nano-periodicity and various symmetries. In the case of thin films, the microphase-separation is influenced by the surfaces. Lamella-forming block copolymer thin films are suitable to explore the phenomena. Generally the surface induce orientation of the lamellae parallel to the film surface (parallel lamellae)[1]. However, a substrate that strongly adsorbs the one component of the block copolymer develops the special structure, laterally microphase-separated structure, on top of it [2]. This special structure may influence the ordering of the lamellae stacked on top of it. The lamellar film frequently develops a significantly stepped surface. When the steps move in order to minimize the surface free energy, they may influence the ordering of the lamellae too. In this study, the influences of the substrate and the free surface on the ordering of the thin films of the lamella-forming block copolymers are explored by using the solvent-vapor treatment of the system.

EXPERIMENTAL SECTION

The lamella-forming block copolymers used in this work are shown in Table 1. Thin films were deposited onto either native oxide silicon (SiO_x) or polyimide (PIM) substrate by dip coating from a some 1wt% solution in tetrahydrofuran (THF). Both substrates preferentially adsorb P2VP chains. For equilibration the samples were exposed to saturated THF-vapor in a closed chamber. After the vapor exposure, the sample was promptly dried in atmosphere. The samples were investigated by scanning force microscopy (SFM) and transmission electron microscopy (TEM). Details of the treatment are described elsewhere [3].

RESULTS AND DISCUSSION

The as-prepared thin films of SVT1 formed already multi-layered lamellar structure on SiO_x substrate as shown in Figure 1a. The thin film structure did not differ on the different substrate (PIM) which adsorbs P2VP chain more weakly. However, in the case of SVT of 1.7 times larger molecular weight (SVTh), the as-prepared films on PIM substrate exhibited distorted non-equilibrium structure without long-range order. After the THF-vapor treatment, this sample developed three-phase coexisting lamellar micro-domains parallel to the film plane (Figure 2a). However, in the case of SVTh, the difference in the substrate caused striking difference in the lamellar structure first formed after the vapor treatment. On SiO_x (Figure 2b), the spacing of the lamellae formed after 1min of the vapor treatment was considerably smaller than that observed on PIM. The PVP block is more strongly adsorbed by SiO_x surface than PIM surface. Due to the strong interaction, PVP chains adsorbed to SiO_x take a conformation more expanded parallel to the surface than those on PIM [3].

TABLE 1. Materials used in this study

sample	Total molecular weight, Mw 10³g/mol	Volume fraction		
		PS	PVP	PBMA
PS-*b*-PVP-*b*-PBMA (SVTl)	172	0.20	0.34	0.46
PS-*b*-PVP-*b*-PBMA (SVTh)	293	0.18	0.39	0.43
PS-*b*-PVP (SV)	233	0.44	0.56	

PS = polystyrene, PVP = poly(2-vinylpyridine), PBMA = poly(*tert*-butylmethacrylate)

The difference in the conformations of PVP adsorbed to the substrate causes different configurations of the SVT chains in the first layer on the substrate. In the first layer, PS and PBMA domains ordered in parallel to the surface are more spaced on SiO_x than on PIM. This laterally expanded first layer morphology develops the compressed lamellae on SiO_x. The first layer structure influences several layers stacked on top of it, as shown in Figure 2. Although the lamella spacing of SVT2 films on PIM did not change after the longer vapor treatment, the compressed lamellae observed on SiO_x showed thickening to the stable thickness, which did not much differ from that observed on PIM, by the vapor treatment.

The local thickness of the thin films of the lamella-forming block copolymers is quantized by the lamella spacing, i.e., the many terraces of the well-defined thickness develop. In the case of the solvent-vapor treatment, the terraces coarsen with a treatment time. The SVT thin films significantly dewetted the substrate and thickened; the step move due to the film thickening was observed to occur over much larger distances than the heat treatment process. After the long vapor-treatment, some particular defects (string-like defects) in the parallel lamellar structure were observed as shown in Figure 3a. The defects seem to originate from the moving steps. In order to explore the formation of the string-like defects, we did the same sort of experiments on the SV diblock copolymer system. String-like defects were observed as well in this system (Figure 3b). The defects seem to be developed on the step as well as the SVT system. Due to the step move, the defects seem to be left on the lower terrace.

In conclusion, the special structure of the first layer that is induced by the strong interaction between the component of the block copolymer and the substrate was shown to influence the ordering in the layers stacked on top of it. The phenomena are especially observed in the slow ordering of the triblock terpolymer of high molecular weight. Furthermore, we have shown that the unconfined free surface of the lamellar block copolymer thin films possibly develops the defects when the steps move during the annealing processes.

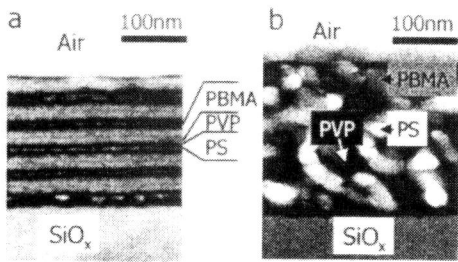

FIGURE 1. TEM micrographs of the as-prepared film of SVTl (a) and SVTh (b) on SiO_x substrate.

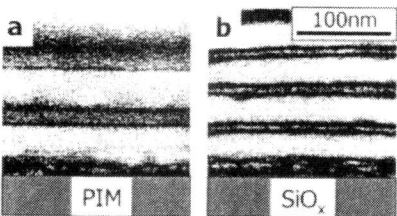

FIGURE 2. TEM micrographs of the SVTh film on PIM (a) and on SiO_x (b) after the THF-vapor treatment for 1min.

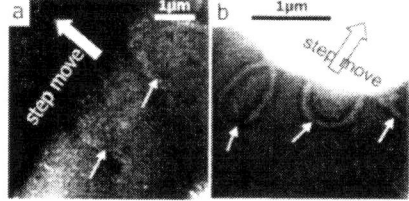

Figure 3. Plan-view TEM image of the SVTh lamellar film on SiO_x substrate after the THF-vapor treatment for 1week (a). The higher terrace appears as a darker region in the left-hand part of part a. Part b shows an SFM height image of the SV lamellar film on SiO_x substrate after the THF-vapor treatment for 3days. The higher terrace appears as a brighter region in part b. In both images, string-like defects in the lamellar film are indicated by the white arrows.

REFERENCES

1. Russell, T. P., et al., *Macromolecules* **22**, 4600-4606 (1989).

2. Fukunaga, K., et al., *Macromolecules* **33**, 947-953 (2000).

3. Fukunaga, K., et al., *Macromolecules* **36**, 2852-2861 (2003)

Dynamics of Granular Chain

Kiwing To

Institute of Physics, Academia Sinica, Nankang, Taipei, Taiwan 11529 R.O.C.

Abstract. We report experimental results of a granular chain on a vertically vibrating platform under gravity. One end of the granular chain was confined to a small region while the other end performed random motion. The conformation of the chain was captured by digital camera and the end-to-end distance R_n was measured. The power spectrum of R_n was found to decay in a power law with an exponent of 0.7.

INTRODUCTION

Linear granular chains have been used as macroscopic analogy to linear polymer molecules. These had been studied on knot dynamics in granular chains [1], conformation of granulr chain in solution [2], and entropic tightening phenomenon [3] as in topologically-constrained polymer [4]. In this paper, we report our study on the dynamical behavior of a vibrating granular chain with one end of the chain being confined to a small region while the other end moved under the influence of the chain and the vibrating floor. We measured the temporal fluctuation of the end-to-end distance R_n and found that the power spectrum of R_n decay in a power law with an exponent of 0.7.

EXPERIMENTAL

Figure 1 show the schematic diagram of the experimental setup. Granular chain made of 137 metallic balls of 1.4 mm diameter was placed on a vertically vibrating table of 30 cm diameter. A transparent plexi-glass plate was placed 3 cm above the table and one end of the chain was attached to the center of the plexi- glass using a thread. A digital camera was installed above the plexi-glass to capture the conformation of the chain with a rate of one picture per second. The free end of the chain was colored red so that it could be detected easily by imaging analysis programs.

We fixed the vibrating frequency, f at 20 Hz and performed the experiment with vibrating amplitude, a set to 1 mm, 2.5 mm and 3 mm, respectively. Hence, the acceleration of the vibrating table were $\Gamma = a(2\pi f)^2 = 16 m/s^2, 40 m/s^2$, and $47 m/s^2$, respectively. Images (see Figure 2) were taken every second and a total of 40,000 images were recorded for each acceleration setting. From

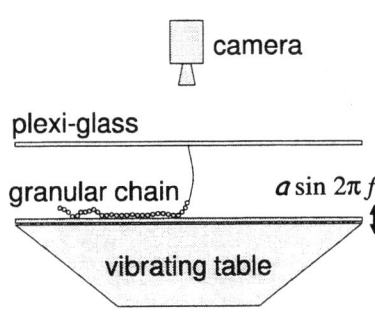

FIGURE 1. Schematic diagram of the experimental setup.

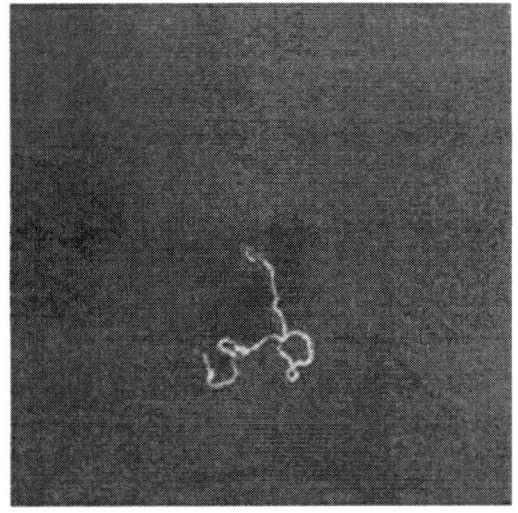

FIGURE 2. A typical image captured in the experiment. The actual dimension of the image is $192 \times 192 mm^2$.

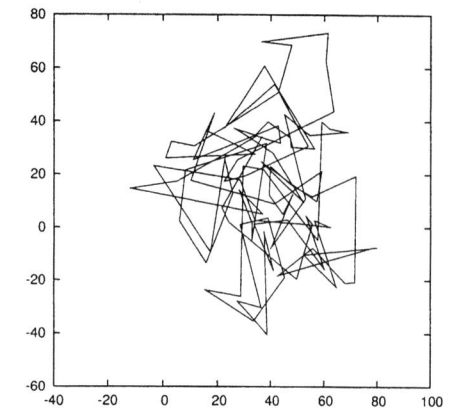

FIGURE 3. Trajectory (100 steps) of the free end for $\Gamma = 47m/s^2$. The spatial unit is in *mm*.

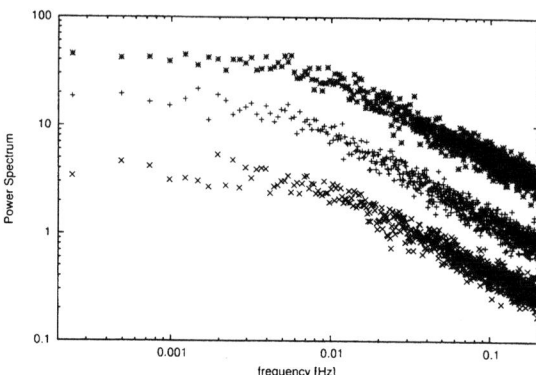

FIGURE 5. Power spectra of R_n for $\Gamma = 16m/s^2$ (lower, ×), $40m/s^2$ (middle, +) and $47m/s^2$ (upper, *).

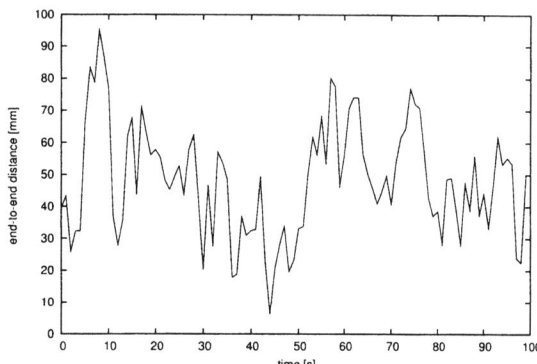

FIGURE 4. Temporal record of R_n for the data in Figure 3.

these images, the locations of the free end were recorded, time variation of R_n were measured and the power spectra of R_n were obtained by a Fast Fourier Transform algorithm.

RESULTS AND DISCUSSION

Figure 3 shows the trajectory of the free end $\Gamma = 47m/s^2$ and Figure 4 is the temporal record of R_n for the trajectory. These figures suggest that the free end was trapped within a certain region instead of undergoing random diffusion. This is reasonable because the maximum extension of the chain is 20 cm. However, the mean location of the free end position is not at the origin where the fixed end was confined. This was probably because of the vibrating table was not perfectly horizontal. Such behavior was also observed in the experiments of other accelerations. With the temporal records of R_n, we calculated the power spectra for each acceleration setting. Figure 5 shows the result obtained.

Note that the spectra for $\Gamma = 40m/s^2$ and $47m/s^2$ were displaced upward for better viewing the data. Otherwise, all the data would collapse onto one curve. Hence, the spectra are practically the same. This implies that changing the vibrating amplitude by a factor of three did not change very much the dynamical behavior of the granular chain. On the other hand, the spectrum show a power law decay in frequency with an exponent of about 0.7. More detailed studies of this system is currently undertaking.

ACKNOWLEDGMENTS

This work was supported by the National Science Council of R.O.C. under grant number: NSC92-2112-M-001-028.

REFERENCES

1. Ben-Naim E., Daya Z.A., Vorobieff P., and Ecke R.E., *Phys. Rev. Lett.* **8**, 1414-1417 (2001).
2. Prenitis J.J. and Sisan D.R., *Phys. Rev.* E**65**, 031306 (2002).
3. Hastings M.B., Daya Z.A., Ben-Naim E., and Ecke R.E., *Phys. Rev.* E**66**, 025102(R) (2002).
4. Shaw S.Y. and Wang J.C., *Science* **260**, 533-536 (1993).

Two-Dimensional Fluorescence Spectroscopy: Medium Dependence and Thermal Effect of Inhomogeneity in Glass

Takayuki Muromoto, Yoshio Mori, Yutaka Nagasawa,
Hiroshi Miyasaka, Tadashi Okada

*Department of Materials Engineering Science, Graduate School of Engineering Science,
Osaka University, Osaka 560-8531, Japan*

Abstract. Two-dimensional fluorescence spectroscopy (2DFS) utilizing Nile blue 690 dye (NB) was applied to obtain information of inhomogeneity in liquid and polymer glass systems at temperatures of 298 K and 10 K. The emission peak shift and the bandwidth of the diagonal and anti-diagonal cross section indicated increase of inhomogeneity in polymer compared to the liquid solution, especially at low temperature. This result indicates the possibility of quantitative analysis of the line-broadening mechanism in condensed phase utilizing 2DFS.

INTRODUCTION

Two-dimensional fluorescence spectroscopy (2DFS) was originally applied to the elucidation of the dynamics of solvent relaxation by Görlach et al [1]. They pointed out 2DFS became elongated along the diagonal direction with decreasing temperature, of which results were interpreted as due to the glass transition of the medium leading to increase in the inhomogeneity. Although Litvinyuk performed detailed analysis based on the computer simulation [2], the clear-cut insight into the inhomogeneity appearing on 2DFS is still an open question. In the present study, we have analyzed the results of 2DFS by using quantitative approach based on the 2D hole-burning model originally introduced by Tokmakoff [3].

EXPERIMENT

Nile blue 690 dye (NB) was doped into poly(methylmethacrylate) (PMMA) glass film and methanol. The fluorescence spectra were measured by HITACHI F850E fluorescence spectrophotometer. The 2DFS is composed of a number of 1-D fluorescence spectra recorded at various excitation wavelengths and fitted by four Gaussian functions. Peak shifts were obtained from these fitted 1-D spectra.

RESULTS AND DISCUSSION

The 2DFS of NB in methanol solution and in PMMA matrix are shown in Figure 1. In the methanol solution (Fig. 1 (a)), the emission and excitation peak shifts were simple straight lines. The counter lines of the 2DFS in methanol solution showed circular shapes. FWHM of the diagonal cross section (inhomogeneous origin) was 673cm^{-1} and the anti-diagonal half width half maximum (HWHM) (homogeneous origin) was 368cm^{-1} (Fig. 2). The ratio of anti-diagonal HWHM to the half of the diagonal FWHM was almost 1:1 (circular). These observations indicate "apparent" homogeneity attained by the rapid motions of methanol molecules around the solute NB. In other words, all of the solvent relaxation process occurred in a shorter time scale than the excited state lifetime.

In PMMA at 298 K (Fig 1 (b)), the peak shifts were complicated curves depending on the excitation frequency. The counter lines of the 2DFS in PMMA showed ellipsoidal shapes elongated along the diagonal line. The diagonal FWHM in PMMA (893cm^{-1}) was larger than that in methanol (673cm^{-1}) (Fig. 2 (a)). The anti-diagonal HWHM in PMMA (315cm^{-1}) was smaller than that in methanol (368cm^{-1}) (Fig. 2 (b)). The ratio of anti-diagonal HWHM to the half of the diagonal FWHM was 2:3 (ellipsoidal), indicating larger inhomogeneity. Some of the relaxation dynamics in glass takes longer than the excited state lifetime of NB.

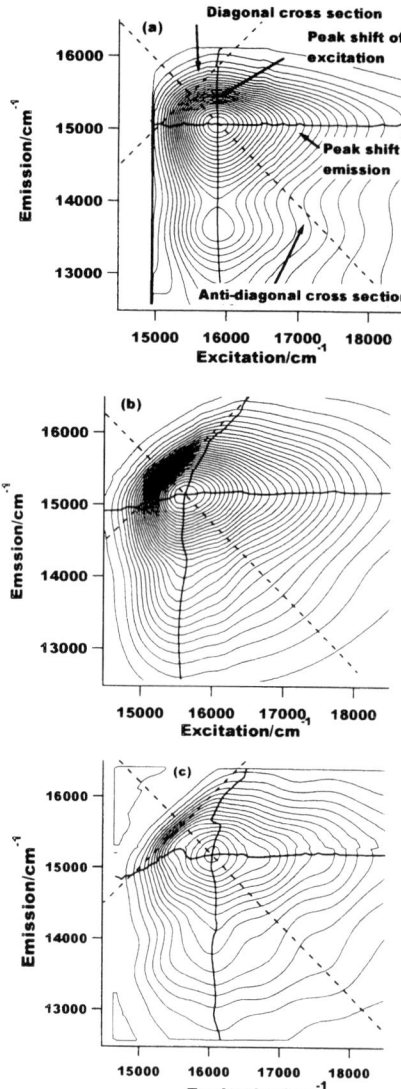

FIGURE 1. 2DFS of Nile blue in methanol (a), in PMMA at 298K (b) and at 10K (c).

In PMMA at 10K (Fig. 1 (c)), the peak shift curves and the form of 2DFS were much complicated than the ones observed at 298 K. This indicates that the thermal fluctuation of the system has decreased and the vibrational structure has become more apparent. However, both diagonal FWHM (925cm^{-1}) and anti-diagonal HWHM (419cm^{-1}) were larger than those at 298 K (Fig. 2 (b)). Increase of the diagonal FWHM indicates increase of inhomogeneity or decrease of the thermal fluctuation, although, increase of anti-diagonal HWHM indicates increase of homogeneity which contradicts with other observations. In the anti-diagonal cross section of NB emission in PMMA at 10 K shown in Figure 2 (b), one can notice a manifestation of a shoulder on the lower frequency side of the peak. For the analysis of this spectrum, more detailed information on 0-0 transition frequency and vibrational structure is necessary.

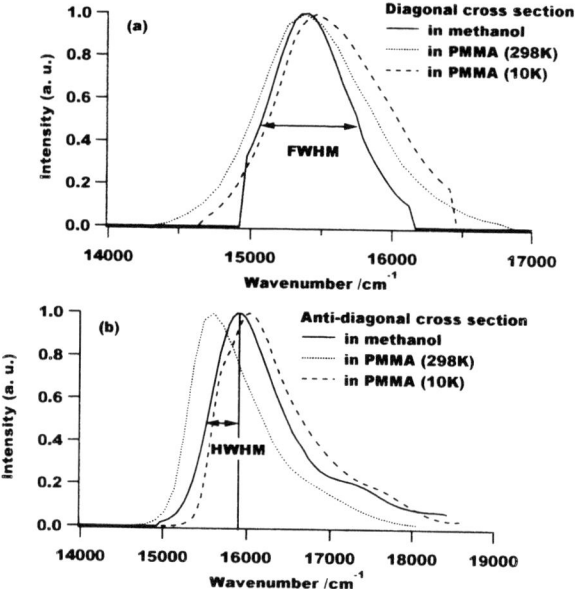

FIGURE 2. Diagonal cross sections (a) and anti-diagonal cross sections (b) for 2DFS in various systems.

As a conclusion, we have compared the 2DFS in methanol solution and in PMMA glass, and quantitatively showed that inhomogeneity increased in PMMA. However, for the analysis of the temperature dependence, more detailed information was necessary. In the future, we are planning to combine 2DFS analysis with nonlinear spectroscopy such as photon echo measurements.

REFERENCES

1. Görlach, E., Gygax, H., Lubini, P., Wild, P., *Chemical Physics* **194** 185-193 (1995).

2. Litvinyuk, I. V., *J. Phys. Chem. A* **101**, 813-816 (1997).

3. Tokmakoff, A., *J. Phys. Chem. A*, **104**, 4247-4255 (2000).

Dynamics of Thin Films of cis-Polyisoprene

Koji Fukao

Department of Polymer Science, Kyoto Institute of Technology, Matsugasaki, Sakyo-ku, Kyoto 606-8585, Japan

Abstract. Motion of entire polymer chains, namely, the normal mode, has been investigated for thin films of cis-polyisoprene. The dielectric loss spectrum of the normal mode is much more sensitive to the decrease in film thickness than that due to the α-process. The broadening of dielectric loss spectra of the normal mode is observed for film thickness below about 150 nm, while the position of the loss peak does not change in the thickness range down to about 50 nm. Anomalous increase in dielectric loss between the α-process and the normal mode was observed, which is consistent with the recent report on the existence of an additional relaxation process.

INTRODUCTION

From the viewpoint of polymer dynamics, the dynamics of polymer chains in confined geometry such as thin films is a fascinating topics [1]. For this study, dielectric relaxation spectroscopy (DRS) is a powerful tool for investigating dynamics near glass transitions, because this technique can cover over a very wide frequency range. This method probes a range of mobilities, from a local motion, a segmental motion to motion of entire polymer chains. In this paper, dynamics of the normal mode in thin films of polyisoprene (PIP) is investigated by DRS.

EXPERIMENTS

The molecular weight, the distribution of molecular weights and the radius of gyration of polymers used in this study are listed in Table 1 for cis-polyisoprene (PIP). Thin films were made by spin-coat method onto Al-deposited glass substrate from the cyclohexane solution. Dielectric relaxation spectroscopy (DRS) was performed for thin films prepared in the above way. An LCR meter (HP4284A) was used for DRS in the frequency range from 20Hz to 1MHz. The dielectric measurements during the heating and cooling processes with a rate of 0.5 - 1.0 K/min were performed repeatedly several times. [2, 3].

RESULTS AND DISCUSSION

In the case of PIP, the motion of entire polymer chains, the normal mode, has been investigated in addition to the α-relaxation process. The dielectric loss peak due to the normal mode is much more sensitive to the decrease in film thickness d than that due to the α-process. The broadening of dielectric loss peak due to the normal mode is observed for film thickness below about 150 nm, while the position of the loss peak does not change in the thickness range down to about 50 nm. The existence of an additional relaxation process between the α-process and the normal mode, which has already been reported by Kremer's group, was confirmed [4, 5].

Figure 1 shows a typical result of master curves obtained from dielectric loss spectra at various temperatures for PIP with $M_n=1.0\times10^4$. In case of $d=439$nm, there are two loss peaks. One is due to the α-process, which is associated with the segmental motion of polymer chains, the other is due to the normal mode, which is associated with the entire motion of polymer chains and is called reptation. As the thickness decreases, it is found that additional relaxation process (Hereafter, this new process is called N'-mode.) appears between the normal mode and the α-process at the expense of the relaxation strength of the normal mode. By using Havriliak-Negami (HN) equations, the observed data can well be repro-

TABLE 1. The number averaged molecular weight M_n, the ratio of the weight averaged molecular weight M_w to M_n, radius of gyration of cis-isoprene R_g

M_n	M_w/M_n	R_g (nm)
2.56×10^3	1.08	1.7
1.0×10^4	1.04	3.3
1.7×10^4	1.04	4.3
3.0×10^4	1.04	5.8

duced as shown in the curves in Fig.1.

Figure 2 shows the thickness dependence of dielectric relaxation strength $\Delta\varepsilon$ of the normal mode and the N'-mode. The sum of dielectric relaxation strengths of both processes is also given in Fig.2. As thickness decreases, $\Delta\varepsilon$ of the normal mode ($\Delta\varepsilon_N$) decreases and $\Delta\varepsilon$ of the N'-mode ($\Delta\varepsilon_{N'}$) increases. Furthermore, the sum of the dielectric relaxation strengths of both processes remains almost constant regardless of thickness for thin films of PIP with M_n larger than 10^4. This result suggests that the fraction of the regions related to the normal mode decreases with decreasing film thickness, while that to the N'-mode increases at the expense of the region related to the normal mode. It should be noted here that the N'-mode is observed only for M_n larger than 10^4. Below this molecular weight, no N'-mode is observed within the experimental accuracy and $\Delta\varepsilon$ of the normal mode and the α-process decreases with decreasing thickness. For PIP, the critical molecular weight of entanglement is almost equal to 1×10^4. Therefore, the above result of the dielectric relaxation may be associated with the existence of entanglement.

Judging from the thickness dependence of $\Delta\varepsilon_N$ and $\Delta\varepsilon_{N'}$, it is reasonable that the N'-mode originates from the normal mode. Because the peak position of the N'-mode is located in the higher frequency side of the normal mode, the effective chain length of polymer chains associated with the N'-mode is shorter than that of the original chain length. It can be expected that the regions including chain segments of shorter effective length exist near the surface or the interface between the polymers and the substrate [4, 6].

ACKNOWLEDGMENTS

The work was supported by a Grant-in-Aid for Scientific Research (No.14540377) from Japan Society for the Promotion of Science.

REFERENCES

1. "Proceedings of International Workshop on Dynamics in Confinement", *J. Phys. IV France* **10**(2000).
2. Fukao, K. and Miyamoto, Y., *Europhys. Lett.* **46**, 649-654 (1999).
3. Fukao, K. and Miyamoto, Y., *Phys. Rev.* **E61**, 1743-1754 (2000).
4. Serghei, A. and Kremer, K., *Phys. Rev. Lett.*, submitted.
5. Fukao, K., *Euro. Phys. J. E*, in press.
6. Petychakis, L., Floudas, G., and Fleischer, G., *Europhys. Lett.*, **40**, 685-690 (1997).

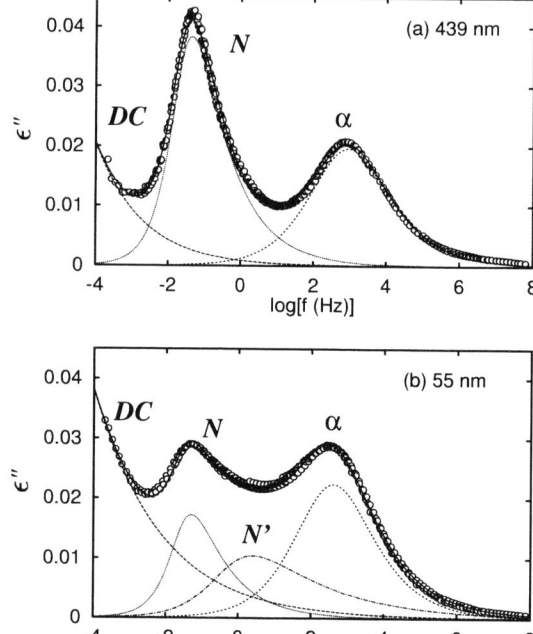

FIGURE 1. Master curves for dielectric loss as a function of frequency for thin films of PIP with $M_n=1.0\times 10^4$ for various film thickness: (a) d=439nm, (b) d=55nm. The reference temperature T_r is 229K. The original data are measured over the temperature range from 205 K to 308 K. The solid curves are calculated by using the best fitting parameters. The four components (the α-process, the normal mode, the N'-mode, and dc conductivity) are shown by different styles of lines.

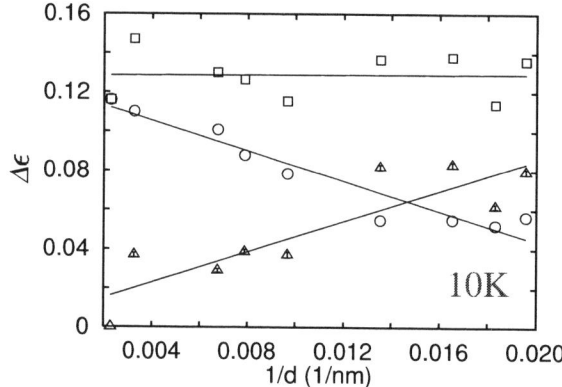

FIGURE 2. Thickness dependence of dielectric strength of the normal mode in thin films of cis-isoprene with $M_n=1.0\times 10^4$. The symbol $\circ$ corresponds to $\Delta\varepsilon$ of the normal mode, $\triangle$, to $\Delta\varepsilon$ of the N'-mode, and square, to the sum of dielectric relaxation strength of the normal mode and the N'-mode.

Electric Birefringence Of Amorphous Polymers Around The Glass Transition Temperature

T. Inoue

*Institute for Chemical Research, Kyoto University
Uji, Kyoto 611-0011, JAPAN*

Abstract. Frequency dependence of dynamic electric birefringence (Kerr effect) using sinusoidal electric fields was measured for polystyrene (M_w = 1050) around the glass transition zone. A weak dispersion due to the induced dipole moments was observed. The imaginary part of the complex Kerr coefficient was found to be proportional to the imaginary part of the rubbery component of the complex shear modulus. This result is in accord with a theoretical prediction for Rouse-Zimm chain by Stockmayer and Baur.

INTRODUCTION

Polymeric materials become anisotropic and birefringent when they are subjected to external fields such as strain fields, electric fields, and magnetic fields.[1] For the case of polymer melts, the strain-induced birefringence can be related to the stress through the stress-optical rule, SOR. Validity of the SOR indicates that the molecular origin of the stress and birefringence of rubbery materials is the orientation of the main chain. On the other hand, in the glassy state, the SOR does not hold. We have found that the strain-induced birefringence can be related with the stress through the modified stress-optical rule, MSOR.[2] The MSOR says that not only the rubbery stress but also the glassy stress contribute to the stress and the birefringence in the glassy state. Thus, the MSOR is effective to understand the polymer dynamics in the glassy zone.

Electric birefringence (Kerr effect) has been used to investigate polymer dynamics. However, measurements on bulk polymers, particularly on polymers having only the induced dipoles are quite limited. In this study, we report the dynamic electric birefringence data of polystyrene around the glass transition zone. Our interest is how the electric birefringence can be related with the stress. We will show that the dispersion of the Kerr coefficient can be described with the reorientation process of the main chain as the rubbery component of the stress is.

EXPERIMENTAL

The complex Kerr coefficient, $K^*(\omega) = \Delta n/E^2$ for a polystyrene (A1000) with M_W = 1,050 was measured with a home-made apparatus over the frequency region of 0.1Hz - 50kHz. The operating electric field was typically 100kV/m. The measurements were performed at isothermal condition at several temperatures (20 - 60°C). For each measurement, temperature was slowly decreased from a higher temperature and kept at the desired temperatures for a long time (typically one day) to minimize the residual birefringence. All measurements were performed above the glass transition temperature (20°C).

Measurements of the complex shear modulus $G^*(\omega) = G'(\omega) + iG''(\omega)$ and strain-induced birefringence was reported elsewhere.[3] The separation of the modulus into the two components was performed with the MSOR.[2,3]

$$G^*(\omega) = G_R^*(\omega) + G_G^*(\omega) \quad (1)$$

$$G^*(\omega) = G_R^*(\omega) + G_G^*(\omega) \quad (2)$$

Here, O^* is the complex ratio of birefringence to the strain. C_i (i=R, G) is associating stress-optical coefficient for component i. According to the molecular interpretation of MSOR, the R component is originated by main chain orientation.[6]

RESULTS AND DISCUSSION

Figure 1 shows frequency dependence of the real and imaginary parts of $K^*(\omega)$, $K'(\omega)$ and $K''(\omega)$, for A1000. DC component of the Kerr coefficient, $K_{DC}(\omega)$, is also included. Here, we used the method of reduced variables: K_{DC}, K', and K'' data at different temperatures are shifted along the abscissa. Temperature dependence of the shift factor, a_T, was comparable to the factor for the viscoelastic quantities, $G^*(\omega)$. A weak dispersion is observed at frequencies of $1 < \log(\omega/s^{-1}) < 3$. It is known that contribution of the permanent dipole to the electric birefringence is negligible small for the case of PS. Consistently, the dispersion of K_{DC} is not observed. Also shown in Figure. 1 is dielectric loss of the same polystyrene, ε''. Dispersion of ε'' is observed at higher frequencies compared with K''.

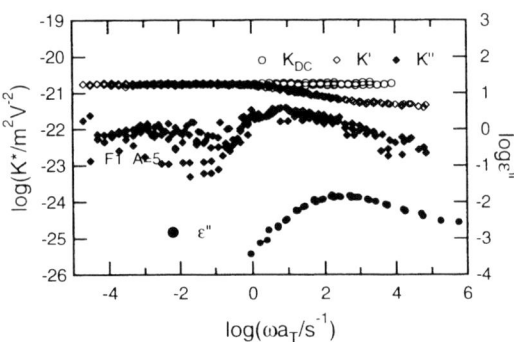

FIGURE 1. Frequency dependence of complex Kerr coefficient and imaginary part of dielectric constant for PS (Mw=1050).

Electric birefringence of the beads-springs model was calculated by Stockmayer and Baur.[5] According to their theory, for polymers having only the induced dipoles, a simple relationship holds well between the imaginary parts of the complex Kerr coefficient and the shear modulus.

$$K''(2\omega) \propto G''_{BS}(\omega). \quad (3)$$

Here $G''_{BS}(\omega)$ represents imaginary part of shear modulus for the beads-springs models. In order to examine Equation (3), $K''(2\omega)$ and $G_R''(\omega)$ are compared in Figure. 2. We may conclude that Equation (3) holds well at frequencies $-1 < \log(\omega/s^{-1}) < 3$, if we regard $G_{BS}''(\omega) = G_R''(\omega)$. Similar results were obtained for PSs having higher molecular weights (F1: M_w=10,500 and A5000: M_w=5000). Thus, we conclude that $K'''(2\omega)$ is proportional to $G_R''(\omega)$ for polymers having only induced dipoles.

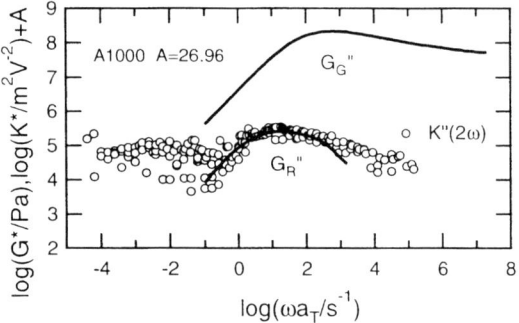

FIGURE 2. Comparison of the imaginary parts of Kerr coefficient and shear modulus.

A comment may be needed on the molecular weight range where Equation (3) holds well. The Rouse segment size for PS is estimated to be 850 in a previous study.[1] For the case of F1 (M_w=10,500), G_R^* can be described with the Rouse model with 12 modes, and therefore validity of Equation (3) may be reasonable. On the other hand, for the case of A1000, the whole chain size is comparable to one segment size. For such a low molecular weight polymer, validity of the beads-spring models is still open to question. However, it may be worthwhile to note that for the case of rode-like molecules having induced dipoles, Equation (3) can be derived theoretically.

REFERENCES

1. Riande, E., and Saiz, E., *Dipole Moment and Birefringence of polymers*, Prentice Hall, 1992, part 2.

2. Inoue, T., Okamoto, H., and Osaki, O., *Macromolecules* **24**, 5670-5675 (1991).

3. Inoue, T., Onogi, T., and Osaki, K., *J. Polym. Sci. Polym. Phys.* **38**, 954-964 (2000).

4. Inoue, T., Matsui, H., and Osaki, K., *Rheol. Acta* **36**, 239-244 (1997).

5. Stockmayer W. H., Baur, M. E., *J. Am. Chem. Soc.* **86**, 3485-3489(1964).

Melting Behavior of Polystyrene Surface Studied by X-ray Reflectivity

A. Kitahara [a], S. Doi [b], and I. Thakahashi [a]

[a] *Advanced Research Center of Science, School of Science and Engineering, Kwansei Gakuin University (ARCS-KGU), Sanda 669-1337, Japan*
[b] *Fujitsu Laboratories Ltd., Morinosato-Wakamiya, Atsugi 243-0197, Japan*

Abstract. We investigate thermal behavior of polystyrene (PS) surfaces by X-ray reflectivity. A very thick, atomically flat PS samples are employed for quantitative determination of surface morphology in the vicinity of glass transition temperature (T_g). From the specular reflectivities, mobility of PS molecules in the surface region is enhanced at temperatures above 340K, which is 30K lower than T_g.

INTRODUCTION

Polymer surfaces play an important role in our daily life. For lubrication, biomaterials, coatings and adhesion, behavior of polymer molecules in surface region is known to be crucial. To apply polymeric materials more intelligently, it is necessary to understand the physical properties of polymer molecules near the surfaces. Therefore, there have been many theoretical works and experiments on polymer surfaces [1-4]. However, most of them have been focused on thin polymer films. Thus, they seem to be affected by the substrates and film thickness comparable to radius of gyration R_g which may be referred to as "Confined Geometry". In order to avoid the effects of the confined geometry and elucidate the nature of polymer molecules in surface region, we prepare atomically flat polystyrene (PS) surfaces with macroscopic thickness and perform X-ray reflectivity (XR) measurements around the glass transition temperature T_g.

EXPERIMENT

PS surfaces were prepared by the following procedure: (1) a lump of melted atactic PS ($M_w=310 \times 10^3$, $R_g=170$ Å, $T_g=373$ K) without any solvent was dropped onto a polished silicon wafer, the surface of which has a native oxide layer of 18 Å thickness and root mean square (rms) surface roughness σ of 6.4 Å; (2) annealed up to 460 K for 6 hours and cooled down to room temperature within 2 hours in low vacuum (60 torr); (3) the PS of which dimension was $20 \times 20 \times 2$ mm^3 was removed from the silicon matrix (Fig. 1). We measured the PS surface which had stuck to the silicon wafer.

XR was measured at the beamline 4C of the Photon Factory where the wavelength λ was tuned to 1.54 Å. The PS samples were mounted on a heater in air; the temperature was controlled with a precision of ±0.1 K. A heating and cooling rate was 2 K/min. Specular XR was measured by transverse scans for investigation of surface morphology. Prior to the XR data collection, the sample temperature was kept for 1.5 hours for relaxation of PS molecules at each temperature. The XR measurements took almost 2 hours for each temperature.

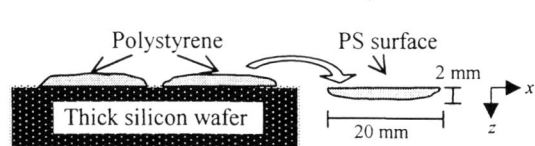

FIGURE 1. Making polystyrene surfaces with macroscopic thickness. Polystyrene is annealed on thick silicon wafer. Thus, the surface is free from thermal concave.

RESULTS AND DISCUSSION

Figure 2 shows the observed XR $R_{OBS}(q_z)$, where q_z means the momentum transfer of incident and reflected X-rays in direction normal to the PS surface: The magnitude of q_z is $4\pi\sin\theta/\lambda$, where 2θ is the scattering angle. $R_{OBS}(q_z)$ at 300 K (open circles) and that at 345 K (triangles) reveal the Fresnel-type reflectivity $R_F(q_z)$, indicating that the surface of the present PS is atomically smooth. However, they show a slight deviation from $R_F(q_z)$ which affords the surface morphology of non-ideally deformed surfaces. Therefore, we analyzed the $R_{OBS}(q_z)$ by using kinematical theory. In the kinematical limit the specular XR is expressed as,

$$R_{OBS}(q_z) = R_F(q_z)\left|\frac{1}{\rho_{bulk}}\int\frac{d\rho(z)}{dz}\exp(iq_z z)dz\right|^2 \quad (1)$$

where ρ_{BULK} denotes the average electron density of the matrix and $\rho(z)$ is laterally-averaged electron density profile as a function of z (depth from the average surface that is defined as z=0) [5]. By using this expression, $\rho(z)$ is obtained by standard fitting procedures. The best-fitted XRs are expressed by the solid curves in Fig. 2; the agreement between observation and calculation is fairly good. The fitted $\rho(z)$ at 300 K and 345 K are also shown in the inset.

The value of σ indicating the surface width in the inset of Fig. 2 is 5-6 Å, which is almost the same as that of the silicon wafer. Thus, we can conclude that our procedure can yield the atomically flat PS surfaces, which afford the quantitative study on the PS surfaces at various temperatures. The fitted $\rho(z)$ at 300 K clearly indicates a low density (about 85% of bulk density) region about 100 Å thick at the surface. It claims that some PS molecules remain on the silicon wafer, after the removal of the PS sample from the silicon wafer (procedure (3)). If so, there should be some free volume randomly distributed on the surface where the PS molecules were pulled out by the silicon wafer. Since many particles comparable to R_g on the PS removed silicon wafer are confirmed by AFM measurements, such a speculation seems to be quite plausible for us.

In the heating from 300 K to 345 K, ROBS.(qz) showed slight variation as indicated in Fig. 2, although profile of XR in the transverse scans showed a complex variation in this temperature range, which will be published elsewhere. The fitted $\rho(z)$ at 345 K shows interesting features: (i) Density of the surface region becomes higher; (ii) The range of the surface region becomes smaller; (iii) however the rms surface

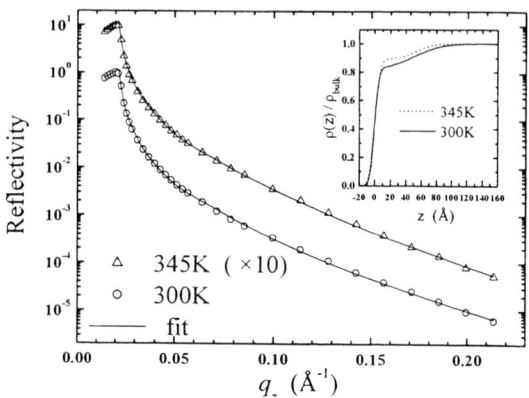

FIGURE 2. Specular X-ray reflectivity $R(q_z)$ at 300K (circles), and at 345K (triangles). For clarity $R(q_z)$ at 435K is shifted vertically. The solid curves are calculated by using the fitted parameters. The inset displays the best-fitted density profiles at the surface. Both of them are normalized by the electron density in bulk.

mobility in the surface region is greatly enhanced at 345 K. If we recall this temperature is about 30 K lower than T_g, we have to recognize that such a variation on the PS surface with macroscopic thickness is quite unusual. Consequently, we succeeded to make the unexpectedly active, but atomically flat polymer surface free from the constraint of confined geometry. It surely gives us a key to understand the nature of polymer surface near the glass transition temperature. Measurements of specular XR at higher temperatures close to T_g, and measurements of off-specular XR below and above T_g are now in progress, which would give us fruitful information on surface morphology in lateral directions. Proposal number of Photon Factory is 2002G043.

REFERENCES

1. Keddie, J. L., Jones, R. A. L., and Cory, R. A., *Faraday Discuss* **98**, 219-230 (1994).

2. Tanaka, K., Kajiyama, T., Takahara, A., and Tasaki, S., *Macromolecules* **35**, 4702-4706 (2002).

3. Dalnoki-Veress, K., Forrest, J. A., Murray, C., Gigault, C., and Dutcher, J. R., *Phys. Rev. E* **63**, 31801-1-31801-12 (2001).

4. Saulnier, F., Raphaël, E., and de Gennes, P.-G., *Phys. Rev. E* **66**, 61607-1-61607-10 (2002).

5. Tolan, M., *X-Ray Scattering from Soft-Matter Thin Films*, Berlin: Springer, 1999, pp. 75-80.

Simulations of Gaussian and Excluded-Volume Chains in Curved Slits

Yasuo Y. Suzuki*, Tomonari Dotera† and Megumi Hirabayashi**

*Faculty of Engineering, Takushoku University, Hachioji, Tokyo 193-0985, Japan
†Department of Polymer Chemistry, Kyoto University, Kyoto Daigaku Katsura 1, Kyoto, Japan
**Saitama Study Center, the University of the Air, 682-2 Nishiki-cho, Saitama, Japan

Abstract. We propose polymer models for Monte Carlo simulation and apply them to a polymer chain confined in a thin box which has both curved and flat sides, and show that either a Gaussian or an excluded-volume chain spends more time in the curved region than in the flat region. The ratio of the probability of finding a chain in the curved region and in flat region increases exponentially with increasing chain length up to a certain length defined by the size of box.

INTRODUCTION

Yaman et al. [1] showed that an ideal (Gaussian) polymer confined between cylindrical shells has a lower free energy than one confined between two flat surfaces. The difference is

$$\Delta F \sim -kTN\frac{l^2}{24a^2}, \quad (1)$$

where N, l, and a corresponds to the number of links, the size of the link, and the radius of the inner surface of the shell, respectively. It does not depend on the slit width d. They discussed that polymer chains confined in bilayer membranes reduce the effective bending rigidity of the membranes and might induce spontaneous curvature in the system, e.g. leading to transitions from lamellar to bicontinuous phases. [2]

Their studies of the confinement effects are limited to ideal polymers described by the continous random walk. It is assumed in their analysis that $l \ll d$, $d \ll a$, and $d \ll R_g$ where R_g denotes a typical size (gyration radius) of the chain in free space. Their result is, therefore, valid asymptotically at large N in a small curvature slit. The ideal chain can act as an unperturbed model of a perturbation calculation for the "real" chain in a good solvent characterized by excluded-volume interactions. It is, however, difficult for confined polymers and few publications have addressed the excluded-volume effects. [3]

MODELS AND METHOD

Our polymer models are off-lattice and consist of $N+1$ beads and N bonds; they are relaxed versions of the random flight model with variable bond lengths. (1) Con-

FIGURE 1. Race-track slit: the width of the slit is d.

nectivity: $|r_i - r_{i+1}| \leq 1.0$, where $r_i = (r_{ix}, r_{iy}, r_{iz})$ stands for the position of the i-th bead. (2) Excluded-volume: $|r_i - r_j| \geq 2r_e$, for all i, j, where r_e is the radius of beads. (3) Geometric constraint: the position of beads are in a confining geometry, here a race-track box (Fig. 1). Note that we impose the last condition only on beads, but not on bonds. Thus, we even allow a bond that cuts through the curved wall as long as two beads attached to the bond satisfy the geometric constraint. We consider two models: Model-G which requires (1) and (3), and Model-E which requires (1)–(3). They correspond to the Gaussian chain and the excluded-volume chain respectively.

Our Monte Carlo (MC) algorithm is the following. i) Select a bead at random. ii) Generate a trial jump: $r_i \to r_i + \Delta$. Each component of the jump vector Δ_μ ($\mu = 1, 2, \cdots, n$) is independently determined by the Gaussian distribution with zero mean value and a standard deviation of 0.3 for Model-G and 0.15 for Model-E. The Gaussian distribution is generated by the Box-Muller method. [4] iii) Check if the new position satisfies above mentioned conditions: (1) and (3) for Model-G, (1)–(3) for Model-E; if it does, accept the jump; if not, abandon the jump. iv) Return to i). For a simulation with $N+1$

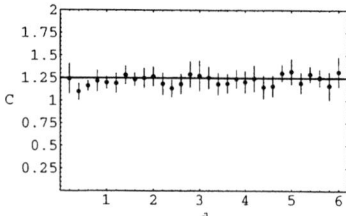

FIGURE 2. Variation of slit width in Model-I with $a = 4.75$. The line indicates $C = 1.249$ which is predicted by the theory.

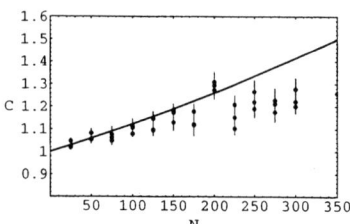

FIGURE 3. Variation of the chain length in Model-I with $d = 0.5, 0.7, 1.0$. The line is drawn according to the theoretical curve with parameters $a = 4.75$, $\langle l^2 \rangle = 0.6$: $C = \exp(N/902.5)$.

beads, $N + 1$ cycles comprise one MC step.

For Model-E, we use the same race-track geometry as used for Model-G. Note that the width of the race-track $d = 0.5$ is defined based on the center of the beads. When the width is defined based on the surface of the beads, it is $d + 2r_e = 1.0$ and the radius of the inside wall is $a - r_e = 4.5$ where $r_e = 0.25$ is the radius of the beads.

RESULTS AND DISCUSSIONS

For each model and each polymer length, we carried out 5 runs of simulation with 4×10^8 MC steps after an initial randomized stage (10^6 MC steps). We have calculated the ratio of probability C finding beads of polymer between in a curved region and in a flat region as a function of chain length N with the formula:

$$C = e^{-\Delta F/kT} = \exp\left(\frac{Nl^2}{24a^2}\right). \quad (2)$$

In Model-G, our simulation results (Fig. 2) does not depend on d. The chain length dependency fits the analytical formula (Fig. 3) as far as the chain lengths are short. In Model-E, our results (Fig. 4) are also well described by the same formula for short chains. However, they show a deviation when the chain gets longer.

Snap-shots show that shapes of confined chains are elongated compared with the chain in free space. The gyration radius of confined chain with $N + 1 = 125 \sim 175$ (for Model-G, $d = 0.5 \sim 1.0$) and with $N + 1 = 75$ (for

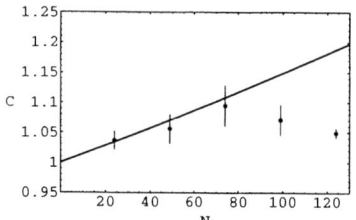

FIGURE 4. Results of Model-E with $d = 0.5$ and $a = 4.75$. The line is drawn according to the theoretical curve for a Gaussian chain with parameters $a = 4.5$, $\langle l^2 \rangle = 0.664$: $C = \exp(N/720.9)$.

Model-E, $d = 0.5$) is comparable to the section length $L \sim \pi a$. When the size of chains exceed a single section, the probability does not follow Eq. 2. For longer chains, a single curved section cannot accommodate the whole chain. Therefore, deviation from the theoretical curve is consistent with our analysis. The longer Gaussian chains can be squeezed into the curved section, however, the excluded-volume chains can not. The probability ratio C become constant beyond the threshold length for Model-G and reach a maximum at the threshold for Model-E.

We present new polymer models for Monte Carlo simulation for efficient changes of the polymer conformation in confining geometries by the variable bond lengths and the variable jump distances. We have investigated the curvature effect on a polymer confined in a thin box. Our results agree quantitatively with the prediction by the analytical theory up to a certain polymer length determined by the confining geometry. Beyond the length, the effect stays constant for the Gaussian chain and it decreases for the excluded-volume chain. The effect is screened by the excluded-volume interactions for longer chain.

Due to restriction on the scale of confining geometry, our simulations only cover the linear region of C as a function of chain length ($N \ll a^2/l^2$). However, our results may still be useful under present circumstances where there is no experimental data for either ideal or excluded-volume chains and there is no analytical theory for excluded-volume chains.

REFERENCES

1. Yaman K., Pincus P., Solis F., and Witten T.A., *Macromolecules*, **30**, 1173-1178(1997).
2. Dotera T., and Suzuki Y.Y., *Phys. Rev. E*, **62**, 5318-5323(2000).
3. Eisenriegler E., Hanke A., and Dietrich S., *Phys. Rev. E*, **54**, 1134-1152(1996).
4. Press W.H., Teukolsky S.A., Vetterling W.T., and Flannery B.P., *Numerical Recipes in C*, Cambridge University Press, Cambridge, 1988.

Hydrodynamic Coarsening and Pattern Selection in Two-dimensional Quenched Diblock Copolymers

Y. Yokojima[*†] and Y. Shiwa[*]

[*]Division of Materials Science, Kyoto Institute of Technology, Matsugasaki, Sakyo-ku, Kyoto 606-8585, Japan
[†]Vehicle Control Engineering Div., Vehicle Development Headquaters, Toyota Techno Service Corp., Shimoichiba, Toyota, Aichi 471-0875, Japan

Abstract. We investigate the hydrodynamic coarsening of microphase separation in two-dimensional diblock copolymers using a model set of equations which incorporates the coupling of the flow velocity field to the order parameter. Cell-dynamical-system simulations are carried out to study how the hydrodynamic flow controls the coarsening dynamics of lamellar structures. We extract the growth exponent for the characteristic length scale by measuring the orientational correlation function. Moreover, a global wave number attained after a long time as a function of the strength of hydrodynamic coupling is determined analytically by means of a random phase approximation. The result is compared with the simulation results.

The theoretical model that describes the dynamics of microphase separation consists of the following set of equations for the order parameter $\psi(r,t)$ and the velocity field $v(r,t)$, slowly varying with horizontal coordinates $r=(x,y)$ [1].

$$\partial_t \psi = \nabla^2(-\tau\psi + \psi^3 - \nabla^2\psi) - B\psi - (v \cdot \nabla)\psi, \quad (1)$$

$$(c^2 - \nabla^2)\nabla^2\zeta = g\hat{z} \cdot \left[\nabla(\nabla^2 + B\nabla^{-2})\psi \times \nabla\psi\right]. \quad (2)$$

Here $\hat{z}$ is the unit vector along the z axis, and the velocity v is defined in terms of the vertical vorticity potential, ζ, with $v=(\partial_y\zeta, -\partial_x\zeta)$. The positive constants τ, B, c^2 and g are phenomenological parameters, the latter two determining the strength of the hydrodynamic coupling of ψ's.

In order to simulate Eqs. (1) and (2), we employ the cell-dynamical-system (CDS) method on a square lattice. Explicitly, we solve the following CDS model [2, 1].

$$\psi(n, t+1) - \psi(n,t) = -\langle\langle \mathcal{J}(n,t)\rangle\rangle + \mathcal{J}(n,t) - B\psi(n,t) - v(n,t) \cdot [\nabla]_d \psi(n,t), \quad (3)$$

$$(c^2 - [\nabla^2]_d)[\nabla^2]_d \zeta = g\hat{z} \cdot \{[\nabla]_d(D(\langle\langle\psi\rangle\rangle - \psi) + B[\nabla^{-2}]_d \psi) \times [\nabla]_d \psi\}. \quad (4)$$

Here $\psi(n,t)$ is the order parameter in the nth cell at time t, and $\mathcal{J} \equiv A\tanh\psi + D(\langle\langle\psi\rangle\rangle - \psi) - \psi$ is the effective chemical potential. The double angular brackets denote an isotropized average of a neighborhood of cells; $\langle\langle X\rangle\rangle = (1/6)\sum_{nn} X + (1/12)\sum_{nnn} X$, $nn(nnn)$ representing nearest (next-nearest) neighbor cells. The $[\mathcal{O}]_d$ denotes the discrete version of the enclosed operator $\mathcal{O}$; the discrete gradient was center difference evaluated and for the Laplacian we used the identification: $[\nabla^2]_d \mathcal{O} = 3(\langle\langle\mathcal{O}\rangle\rangle - \mathcal{O})$. The operator $[\nabla^{-2}]_d$ is the inverse of the discrete Laplacian $[\nabla^2]_d$, and is computed using fast Fourier transform techniques.

We have simulated Eqs. (3) and (4) on a system of size 1024×1024 with periodic boundary conditions. The initial conditions were a random distribution of ψ of amplitude 0.1. The parameters used were $A=1.22$, $D=0.45$, $c^2=2$ and $B=0.02$, and we will present results for $g=0$ and $g=5$ to assess the importance of the hydrodynamic interactions. To quantify the temporal change in orientation of the lamellae, we have computed a correlation function of the local orientation field, $\theta(r,t)$, of the lamellar patterns:

$$C_2(R,t) \equiv \langle \exp\{2i[\theta(r+R,t) - \theta(r,t)]\}\rangle \quad (5)$$

averaging over the spatial coordinate r and R for fixed $R \equiv |R|$. The local orientation θ is defined as the angle in the direction normal to the lamellar axis. We can then extract the orientational correlation length, $\xi(t)$, as the value of r at which $C_2(r,t)$ reaches the value of $1/2$. The result is demonstrated in Fig. 1(left). The notable feature is that the $\xi(t)$ grows as a power law $\xi \sim t^\gamma$ in the late stages of microphase separation with $\gamma=0.21$ and 0.30 for $g=0$ and 5, respectively.

To understand the underlying cause of the difference, we show in Fig. 1(right) the temporal change in the strength of the hydrodynamic flux ($\mathcal{F}_h$) and the dissipative or diffusion flux ($\mathcal{F}_d$). They are defined, respectively,

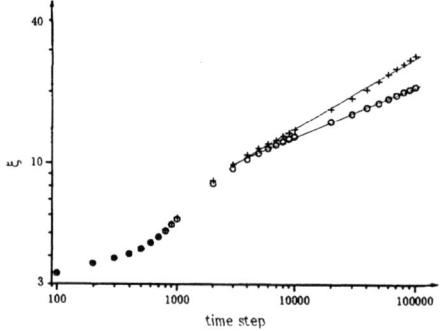 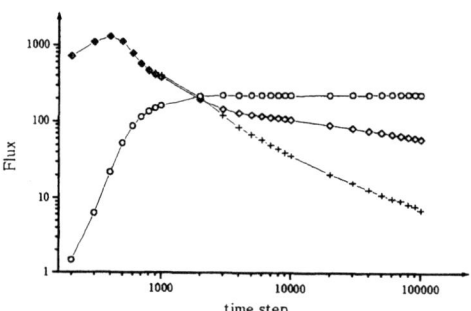

FIGURE 1. Time evolution of the orientational correlation length ξ for $g = 0$ (circles) and $g = 5$ (crosses). Shown on the right is the plot of the hydrodynamic flux $\mathcal{F}_h$ (circles) and the diffusion flux $\mathcal{F}_d$ (crosses for $g = 0$ and diamonds for $g = 5$).

by

$$\mathcal{F}_h \equiv \langle |\psi v| \rangle, \quad \mathcal{F}_d \equiv \langle |-\nabla \delta H/\delta \psi| \rangle. \quad (6)$$

where $H \equiv H\{\psi\}$ is the free energy functional pertinent to Eq. (1) in the absence of the velocity field. Comparing the relative importance of the two transport mechanisms, we can deduce that the hydrodynamic flow is responsible for the increased orientational order in lamellar patterns. In fact, we found that, once the hydrodynamic flux becomes dominant, the orientational characteristic length ξ starts growing faster than the other length scale which is probed by the structure factor, $S(q,t) = \langle \tilde{\psi}(q,t)\tilde{\psi}^*(q,t) \rangle$, $\tilde{\psi}(q,t)$ being the Fourier transform of ψ.

Finally we consider how the presence of the hydrodynamic coupling affects the preferred wave number of the final stationary lamellar states. We identify the asymptotic peak position, q_e, of $S(q,t)$ with the selected wave number of the system. Equation fo motion for $S(q,t)$ is obtained from Eqs. (1) and (2). In order to solve the equation, we assume that the four-point correlation is determined entirely by the pair correlation function, and that two modes of different wavevectors are poorly correlated. The equation then becomes soluble for small g. The result is

$$q_e \simeq q_e^{(0)} + g q_e^{(1)}, \quad (7)$$

$$q_e^{(1)} = \frac{1}{64c^2\sqrt{\epsilon}} \frac{(q_e^{(0)2} - q_0^2)^2}{q_e^{(0)3}}$$

$$\times \left[c^2 - q_0^2 + \frac{q_e^{(0)2}(q_0^2 - c^2) - (q_0^2 + c^2)^2}{\sqrt{(q_e^{(0)2} + q_0^2 + c^2)^2 - 4q_e^{(0)2}q_0^2}} \right], \quad (8)$$

with $q_e^{(0)} = B^{1/4}$ and $q_0^4 = q_e^{(0)4} - \epsilon$, where $\epsilon(>0)$ is determined via

$$(\epsilon - B)\left(\sqrt{\epsilon} + \frac{3}{8}\right)^2 + \frac{\tau^2}{4}\epsilon = 0. \quad (9)$$

Since it is readily proved that $q_e^{(1)}$ is negative definite, we see that there is a trend to smaller wave number at late times with added flow. As shown in Fig. 2, this analytical result (7) agrees semiquantitatively with the simulation data.

In closing we remark that in Ref. [3] the CDS simulation has been performed to study the coarsening of hexagonal patterns in microphase separation. Interestingly, in this case, it is found that the hydrodynamic flow is ineffective in enhancing the domain growth.

REFERENCES

1. Shiwa, Y., *Phys. Rev. E* **61**, 2924-2928 (2000).
2. Oono, Y., and Shiwa, Y., *Mod. Phys. Lett. B* **1**, 49-55 (1987).
3. Yokojima, Y., and Shiwa, Y., *Phys. Rev. E* **65**, 056308(11 pages) (2002).

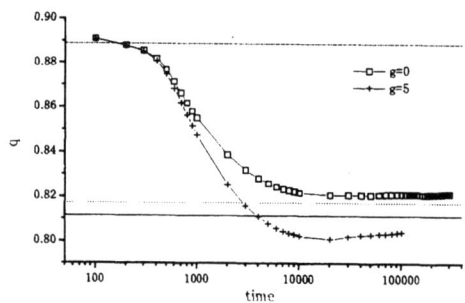

FIGURE 2. Peak position of the structure factor $S(q,t)$ versus time. The top dash-dotted line represents the wavenumber of the linearly most unstable mode, while the other two lines correspond to the theoretical prediction for the selected mode without ($g = 0$, dotted line) and with the hydrodynamic interactions ($g = 5$, continuous line).

Primitive Chain Network Model for Entangled Polymer Blends

Yuichi Masubuchi*, Jun-ichi Takimoto*, Masao Doi*, Giovanni Ianniruberto[†], Francesco Greco** and Giuseppe Marrucci[†]

*Department of Computational Science and Engineering, Nagoya University, Japan
[†]Dipartimento di Ingegneria Chimica, Università "Federico II", Napoli, Italy
**Istituto CNR per i Materiali Compositi e Biomedici, Napoli, Italy

Abstract. The primitive-chain network model for Brownian simulation of entangled polymers is extended so as to include the case of polymer blends. Predictions for the phase diagram and for the linear viscoelastic response appear consistent with earlier works.

INTRODUCTION

Although the melt rheology of polymer mixtures has industrial significance just as that of homopolymers, calculation methods for mixtures are not well developed. Widely used bead-spring simulations are useful if the molecules are short enough or if only the local dynamics is relevant, e.g., close to the glass transition. Conversely, they become too much time consuming for well entangled melts, and especially for polymer blends. Another option to investigate blend systems are the density functional theories, which however, by definition, can hardly account for entanglement effects, and have difficulties in calculating the rheological response of the system. In this study, we extend the primitive chain network model [1, 2] to polymer mixtures. In that model, entangled molecules are coarse-grained at the level of the distance between consecutive entanglements, and their motion obeys Langevin equations. The coarse graining adopted allows simulation of slow dynamics such as the growth of phase domains [3].

MODEL

In the primitive chain network model, entanglements are represented as sliplinks connecting two chains, and subchains between consecutive entanglements are replaced with linear springs. Each entanglement node obeys the following Langevin equation:

$$(\zeta_\alpha + \zeta_\beta)(\dot{R} - \kappa \cdot R) =$$

$$3kT \left[\frac{n_{0\alpha}}{a_\alpha^2}\left(\frac{r_{i+1}}{n_{i+1}} - \frac{r_i}{n_i}\right) + \frac{n_{0\beta}}{a_\beta^2}\left(\frac{r_{j+1}}{n_{j+1}} - \frac{r_j}{n_j}\right) \right]$$
$$- \nabla(n_{0\alpha}\mu_\alpha + n_{0\beta}\mu_\beta) + f \qquad (1)$$

Here R is node position, and κ is the velocity gradient; r and n are current values of the end-to-end vector and monomer number of a subchain, respectively; a, n_0, ζ are equilibrium values of the subchain size, monomer number, and friction coefficient, respectively. Finally, f is a random force, and μ is the monomer chemical potential. Equation (1) is written for an entanglement between an α and a β chain.

The chemical potential is obtained from the free energy

$$F = F_{mix} + F_{vol} \qquad (2)$$

where F_{mix} is the Flory-Huggins term for binary polymer blends [4], and F_{vol} is a phenomenological free energy that enforces the incompressibility constraint:

$$\frac{F_{mix}}{kT} = \frac{\phi_\alpha}{n_{0\alpha}Z_\alpha}\ln\phi_\alpha + \frac{\phi_\beta}{n_{0\beta}Z_\beta}\ln\phi_\beta + \chi\phi_\alpha\phi_\beta \qquad (3)$$

$$\frac{F_{vol}}{kT} = \left(\frac{N_\alpha + N_\beta}{N_0} - 1\right)^2 \qquad (4)$$

In Equation (4), ϕ is the monomer volume fraction, χ is the Flory interaction parameter, and Z is the number of subchains per chain. In Equation (4), N is the local number density of monomers for each species, and N_0 the average monomer density in the sample.

Monomer transport through a sliplink is described by

$$\zeta_\alpha \frac{\dot{n}_i}{\rho} = \frac{3kT n_{0\alpha}}{a_\alpha^2}\left(\frac{r_{i+1}}{n_{i+1}} - \frac{r_i}{n_i}\right) - \nabla n_{0\alpha}\mu_\alpha + f \qquad (5)$$

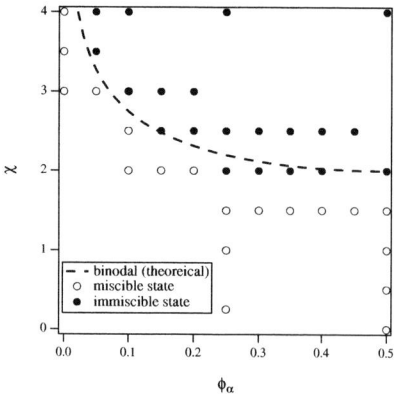

FIGURE 1. Phase diagram.

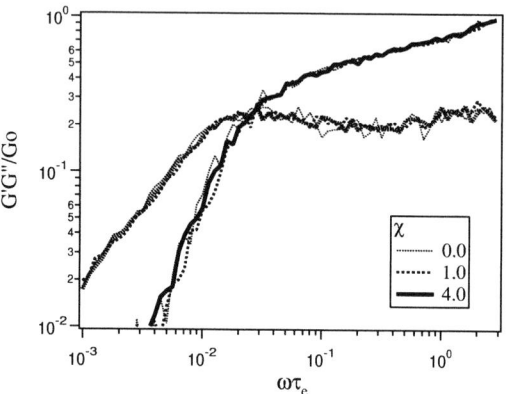

FIGURE 2. Linear viscoelastic response for polymer blends with various χ parameter.

where ρ is the local monomer linear density along the chain, and f is a 1D Brownian force.

Binary polymer blends were investigated with all parameters equal for the two polymers, and $Z = 10$ for both. Units for the calculation were: a for length, kT for energy, $\tau_e = \zeta a^2/6kT$ for time.

RESULTS AND DISCUSSION

Figure 1 shows the phase diagram of the system. The curve is the theoretical bimodal, calculated from the free energy of the system as given by Equation (4). The dots are simulation results, which indicate that the tube model agrees reasonably well with the equilibrium theory. It is noteworthy that the phase separation kinetics predicted by the model [5] is also consistent with the conventional picture for phase separation kinetics [3].

Figure 2 shows the linear viscoelastic response of the equimolar system with various χ parameters. It has been experimentally reported that, if the chains in the pure melt have the same relaxation spectrum, the blend (either miscible or immiscible) maintains the same spectrum up to the reptation time (see e.g. [6]). The simulation results are consistent with this observation. Note that the long time behavior corresponding to the kinetics of the interface does not appear in our results because of computational limitations of system size and calculation time.

CONCLUSIONS

The primitive chain network model which can calculate entangled polymer dynamics was extended to polymer blends by introducing (through a chemical potential) inter-polymer interactions described by a χ parameter.

A typical phase diagram obtained by the simulation is consistent with the theoretical prediction based on free energy. The linear viscoelastic response with various χ parameter is consistent with experiments where the χ parameter does not influence the chain dynamics up to the reptation time, for both the miscible and the immiscible case. From these results the method appears reliable. Further studies on quantitative comparison with experiments as well as further extensions to block co-polymers are in progress.

ACKNOWLEDGMENTS

This study was supported within the Industrial Technology Research Grant Program for 2003 by the New Energy and Industrial Technology Development Organization (NEDO) of Japan (No. 01A31001) and by Grants in Aid for Scientific Research [KAKENHI] 2003 from the Ministry of Education, Culture, Sports, Science and Technology of Japan.

REFERENCES

1. Masubuchi, Y., Takimoto, J.-I., Koyama, K., Ianniruberto, G., Marrucci, G., and Greco. F., *J. Chem. Phys.*, **115** 4387-4394 (2001).
2. Masubuchi, Y., Ianniruberto, G., Marrucci, G., and Greco. F., *J. Chem. Phys.*, **119** 6925-6930 (2003).
3. Onuki, A., *Phase Transition Dynamics*, Cambridge University Press, 2002.
4. Flory, P. J., *Principles of polymer chemistry*, Cornell University Press, Ithaca, 1953.
5. Masubuchi, Y., Ianniruberto, G., Marrucci, G., and Greco. F., *Model. Sim. Mat. Sci. Eng.*, in print.
6. Aoki, Y., and Tanaka, T., *Macromolecules*, **32** 8560-8565 (1999).

Diffusion of Particle in Hyaluronan Solution, a Brownian Dynamics Simulation

Masako Takasu* and Jungo Tomita

Department of Computational Science, Kanazawa University, Kakuma, Kanazawa 920-1192 Japan
* takasu@cphys.s.kanazawa-u.ac.jp

Abstract. Diffusion of a particle in hyaluronan solution is investigated using Brownian dynamics simulation. The slowing down of diffusion is observed, in accordance with the experimental results. The temperature dependence of the diffusion is calculated, and a turnover is obtained when the temperature is increased.

INTRODUCTION

Extracellular matrix is found in connective tissues outside the cells, and is a network of macromolecules. The matrix is important for the survival and various functions of cells. Masuda et al[1] measured the diffusion in hyaluronan solution with pulsed field gradient NMR and photochemical quenching. We performed simulation to investigate the diffusion of particles in extracellular matrix such as hyaluronan.

METHOD

Brownian dynamics method (see for example, [2]) gives the equation of motion for the diffusing particle and the hyaluronan polymers, described by,

$$m \frac{d^2 \mathbf{r}_i}{dt^2} = -\frac{\partial U}{\partial \mathbf{r}_i} - \zeta \frac{d \mathbf{r}_i}{dt} + \mathbf{g}_i(t). \quad (1)$$

Here, $\mathbf{g}_i(t)$ is the random force caused by water molecules in the system.

$$< \mathbf{g}_i(t) > = 0, \quad < \mathbf{g}_i(t) \mathbf{g}_j(t') > = 6 k_B T \zeta \delta_{ij} \delta(t - t') \quad (2)$$

Repulsive force is assumed for the interaction between polymers and between a particle and polymers.

$$U_{\text{rep}}(r) = \varepsilon \left(\frac{\sigma}{r} \right)^{12} \quad (3)$$

The monomers are connected by spring potential.

$$U_{\text{bond}}(r) = \frac{1}{2} k_{\text{bond}} (r - r_0)^2 \quad (4)$$

The polymers also have angle potential.

$$U_{\text{bend}}(\theta_{\alpha\beta\gamma}) = \frac{1}{2} k_{\text{bend}} (\cos \theta_{\alpha\beta\gamma} - \cos \theta_0)^2 \quad (5)$$

RESULTS

A snapshot of the simulation is shown in Fig. 1.

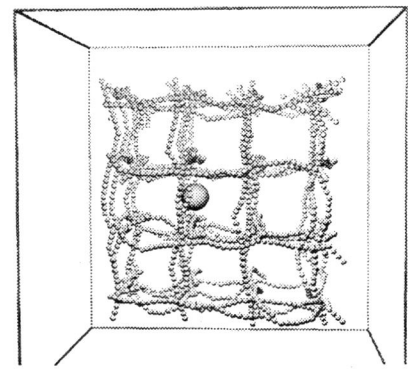

FIGURE 1. A snapshot of our simulation of a particle diffusing in extracellular matrix

We investigate the effect of fluctuations of chains and the dynamics of both the particle and the polymer chains. The diffusion constant is obtained by the mean square displacement of the particle.

$$< |r(t) - r(t_0)|^2 > = 2dDt \quad (6)$$

As the time scale of the observation becomes larger, the diffusion becomes slower, in accordance with the experimental results of Masuda et al[1].

We also found that the temperature dependence of diffusion shows a turnover, as shown in Fig.2. The decrease of mobility of the particle at high temperatures is due

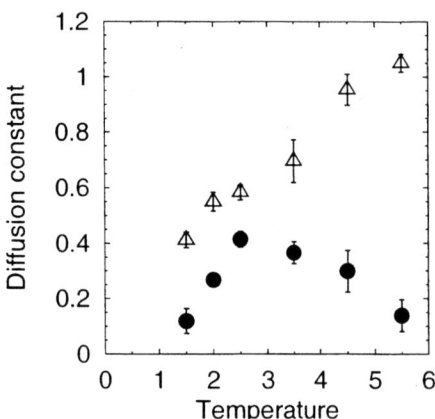

FIGURE 2. Diffusion constant as a function of temperature. The circles denote long-time diffusion. The triangles denote short-time diffusion, with one tenth the time scale of long-time diffusion.

to the movements of hyaluronan chains. This turnover is seen only when the particle size is optimal. If the particle is too small, the diffusion is only slightly affected by the polymer chains and the turnover is not observed. On the other hand, if the particle is too large, the particle is trapped in a cell, and the diffusion constant does not show large dependence on the temperature.

We also investigated the positions and movements of the polymer chains as the particle diffuses. When the particle moves to the next cell, the square 'window' of polymers showed outward movements.

DISCUSSION

Our calculations have shown the slowdown of diffusion for the longer time-scale in accordance with the experimental results. The temperature dependence shows a turnover. The experimental situation of Masuda et al. corresponds to the lower temperature of our simulation. The changing of temperature is not feasible in biological situation, but may apply to some experimental cases in laboratory.

For simplicity, our model is the case of regular crosslinking between the polymer chains. In experimental situation the linkers are placed randomly. The case of random and time-dependent linkers should be investigated in the future.

Diffusion of larger molecule such as collagen is currently investigated.

ACKNOWLEDGMENTS

The authors would like to thank the useful discussion with Dr. K. Ushida and Dr. A. Masuda. This work is partially founded by a Grant-in-Aid from the Ministry of Education, Culture and Sports, Science and Technology.

REFERENCES

1. Masuda, A., Ushida, K., Koshino, H., Yamashita, K., and Kluge, T., *J. Am. Chem. Soc.* **123**, 11468-11471 (2001).
2. Noguchi, H., and Takasu, M., *J. Chem. Phys.* **115**, 9547-9551 (2001).

Dynamical Self-Consistent Field Theory for Inhomogeneous Polymer Systems

Toshihiro Kawakatsu

Department of Physics, Tohoku University, Sendai 980-8578, Japan

Abstract. We propose a dynamical extension of the self-consistent field theory for inhomogeneous polymer systems by combining path integral calculation of the chain conformation and the reptation theory. Stress-strain relation of sheared two polymer brushes is studied using this technique, which shows an important effect of the non-equilibrium chain conformation under flow field.

INTRODUCTION

Self-consistent field (SCF) theory is now widely used in studying mesoscopic structures of inhomogeneous polymer systems. This theory has successfully been applied to equilibrium phenomena such as the microphase separation of block copolymers, polymer brushes, and so on [1,2].

The main assumption of the SCF theory is that the chain conformation is in equilibrium under the influence of the mean field that accounts for the effects of the other chains. However, in order to study non-equilibrium systems such as a polymer mixture undergoing phase separation process or a concentrated polymer solution subjected to an external deformation, this equilibrium assumption should be released.

Recently, several groups proposed dynamical extensions of the SCF theory and performed computer simulations on phase separation dynamics of block copolymer melts, thin polymer films, and so on [3-6]. In most of these dynamical theories, a conventional Fick's law of linear diffusion of the segments driven by the gradient of the chemical potential is assumed. It is obvious, however, that such an assumption cannot be justified in strongly non-equilibrium systems where the chains are largely deformed. In the present study, we propose an SCF theory where the dynamical processes associated with strongly deformed polymer chains are taken into account, and evaluate its usefulness and efficiency.

COMBINATION OF SCF THEORY AND REPTATION DYNAMICS

In the usual SCF theory, the conformational entropy of the polymer chain is evaluated using the path integral formalism assuming that the chain conformation obeys the canonical ensemble. In the case of a sheared polymer melt, however, the assumption of such local equilibrium becomes unreliable due to the large deformation of the chain. To overcome this difficulty, we extend the path integral formalism by introducing a tensor that specifies the probability distribution of the bond vectors, and combined it with the reptation dynamics [7]. In this formalism, the polymer path integral $Q(s,r;s',r')$, i.e. the statistical weight of a subchain between the s-th and s'-th segments whose ends are found at r and r', is given by

$$\frac{\partial}{\partial s}Q(s,r;s',r') = \frac{1}{2}\nabla\nabla:\{A(s,r)Q(s,r;s',r')\} \\ - \nabla\cdot\{\overline{u}(s,r)Q(s,r;s',r')\} \\ - \frac{1}{k_B T}V(r)Q(s,r;s',r') \quad (1)$$

where $\overline{u}(s,r)$ and $A(s,r)$ are the average and

covariance of the distribution of the s-th bond vector, and $V(r)$ is the self-consistent field. The temporal evolutions of $\overline{u}(s,r)$ and $A(s,r)$ are determined by the reptation theory and the Stokes equation for the flow field [7].

We applied our proposed method to a sheared two parallel plates on which melt brushes are grafted [7]. Figure 1 shows typical time evolution of the shear stress and a typical chain conformation in the deformed state. Due to the disentanglement between the two brushes, the shear stress and the density profile of the segments change considerably. Such a behavior is consistent with a microscopic Monte Carlo simulation using a many chain system [8]. In the present model, the history of the imposed flow or imposed deformation is memorized in the form of a deformation of the chain conformation in the same way as the realistic dense polymer systems. Thus, the model is expected to reproduce the non-linear rheological behavior under time-dependent external deformations such as the double step shear process.

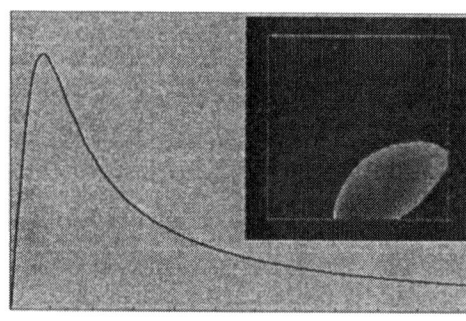

FIGURE 1. Time evolution of the shear stress and the single chain conformation at $t = 1.0$.

TREATMENT OF FLOW FIELD

In the above-mentioned model, we assumed that the flow field is a constant shear flow, and neglected the induced flow effect for simplicity. The correct treatment of the flow field requires us to solve the complex viscoelastic flow equation coupled with the SCF equation. Such a simulation technique for solving the viscoelastic flow equation has been developed [9]. As a first step, we solved a flow equation for a model viscoelastic fluid (so-called Johnson-Segalman model) instead of the full SCF model. The model successfully reproduced the formation of the shear banding in binary mixture. By combining this technique with the dynamical SCF model, we could predict the characters of the complex polymer flow on the mesoscopic scales.

CONCLUSION

As a conclusion, we have proposed a new dynamic SCF theory to study rheological properties of the dense polymer systems in the nonlinear viscoelastic regime. By applying it to sheared polymer brushes, we demonstrated its usefulness. This technique could also be applied to many non-equilibrium phenomena in dense polymer systems, such as adhesion, phase separation, and so on.

ACKNOWLEDGMENTS

This work is supported by the Scientific Research Fund of the Ministry of Education, Culture, Sports, Science and Technology, Japan.

REFERENCES

1. Fleer, G. J., Cohen Stuart, M. A., Schetjens, J. M. H. M., Cosgrove, T., and Vincent, B., *Polymers at Interfaces*, Chapman & Hall, London, 1993.
2. Matsen, M. W., and Schick, M., *Phys. Rev. Lett.* **72**, 2660-2663, (1994).
3. Fraaije, J. G. E. M., *J. Chem. Phys.* **99**, 9202-9212 (1993).
4. Sevink, G. J. A., Zvelindovsky, A. V., van Vlimmeren, B. A. C., Maurits, N. M., and Fraaije, J. G. E. M., *J. Chem. Phys.* **110**, 2250-2256 (1999).
5. Morita, H., Kawakatsu, T. and Doi, M., *Macromolecules* **34**, 8777-8783 (2001).
6. Morita, H., Kawakatsu, T., Doi, M., Yamaguchi, D., Takenaka, M., and Hashimoto, T., *Macromolecules* **35**, 7473-7480 (2002).
7. Shima, T., Kuni, H., Okabe, Y., Doi, M., Yuan, X. F., and Kawakatsu, T., *Macromolecules*, in press.
8. Neelov, I.M., Borisov, O.V., Binder, K.; *J. Chem. Phys.* **108**, 6973-6988 (1998).
9. Jupp, L., Kawakatsu, T., and Yuan, X. F., *J. Chem. Phys.* **119**, 6361-6372 (2003).

Voronoi Space Division of a Polymer

Nakako Tokita*, Megumi Hirabayashi*, Chiaki Azuma† and Tomonari Dotera**

*Saitama Study Center, the University of the Air, Saitama 331-0851, Japan
†Setagaya Study Center, the University of the Air, Tokyo 154-0002, Japan
**Department of Polymer Chemistry, Kyoto University, Kyoto Daigaku Katsura, Kyoto 615-8510, Japan

Abstract. In terms of Voronoi division we study the local geometry of a grafted polymer having 52 ends in united-atom molecular dynamics simulations. The volume of a Voronoi polyhedron for a chain end is larger than that for an internal or junction atom, and that it is the most sensitive to temperature. Chain ends dominantly localize at the surface of the globule: While the ratio of surface atoms is only 24% of all atoms, the ratio of ends at the surface is 91% out of all ends. The shape of Voronoi polyhedra for internal atoms is prolate even in the bulk. We find that two specific faces play a significant role in the Voronoi space division of covalently bonding polymers: Two bonding faces occupy 38% of the total surface area of a Voronoi polyhedron and determine the prolate shape.

Voronoi analysis has been applied to study the distribution of atoms in simple liquids and non-crystalline metallic solids. Recently, the analysis has been employed to study local geometry of a protein [1] and a glass-forming melt [2]. When it is applied to polymers, a fundamental question arises concerning the volume and the shape of Voronoi polyhedra: Are there any specific features of polymers in contrast to simple liquids or metallic glasses? Does chain connectivity have influence upon the volume and the shape of Voronoi polyhedra? In this paper, we show that the existence of bonds affects Voronoi space division.

Here, united-atom molecular dynamics simulations of a single 500-mer polyethylene linked by 50 hexyl groups ($-C_6H_{13}$), i.e. a grafted polymer with 52 ends, are carried out. In the molecule there are three kinds of topologically different atoms with respect to the number of bonds: end atoms with one bond, internal atoms with two bonds, and junction atoms with three bonds. We find that the volume of a polyhedron strongly depends on the number of bonds: Polyhedra associated with ends have the largest volume [Fig.1] with the largest local thermal expansion [Fig.2].

We assign atoms whose inverse volume is smaller than 0.02 Å^{-3} to surface atoms. Then we find that the chain ends dominantly localize at the surface of the globule. In fact, from 50 K to 450 K surface atoms are 24% of all atoms, while ends at the surface are 91% of all ends. Furthermore, not only end atoms appear on the surface, but also ends pull up some connected atoms. About 60% of the next to ends are surface atoms [Fig.3]. As expected, the closer to ends, the higher the probability of surface atoms is, although there is a zigzag move due

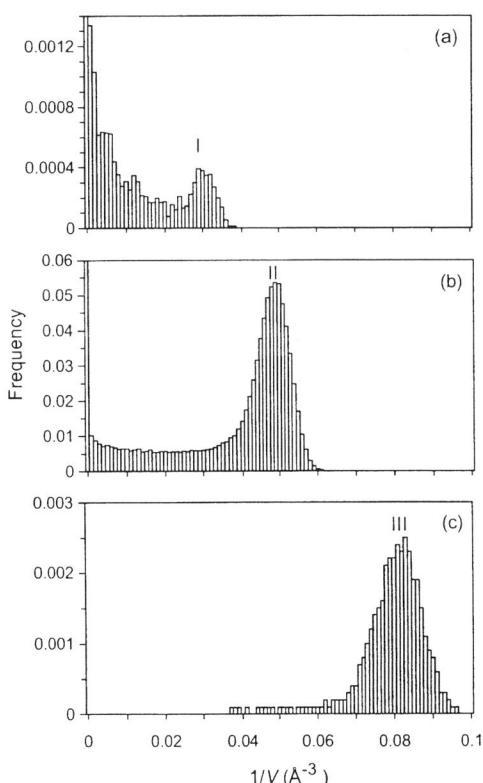

FIGURE 1. Histograms of the frequency distribution of the inverse polyhedron volume $1/V$ at 300 K for three types of atoms, averaged over ten independent runs. (a) End atoms. (b) Internal atoms. (c) Junction atoms. The volume for (a) is the largest and for (c) the smallest. The frequencies at $1/V = 0$ in (a) and (b) are 0.028 and 0.074, respectively.

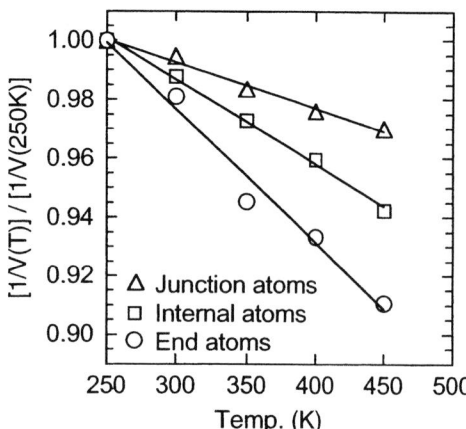

FIGURE 2. Mean value of $1/V$ for bulk atoms divided by that of 250 K versus temperature. Slopes correspond to the local thermal expansion coefficients: $\kappa = -[1/(1/V)][\Delta(1/V)/\Delta T]$. The steepest slope for end atoms implies the largest expansion.

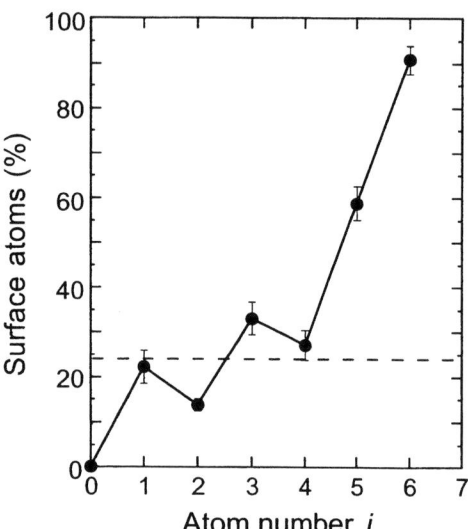

FIGURE 3. Probability of surface atoms at 300K versus the number from a junction atom in a side chain: $i = 0$ (junction), $i = 6$ (end). The dash line indicates the probability of surface atoms for all atoms, 23.9%. Error bars come from ten independent molecular dynamics starting structures.

to covalent bonding.

The shape of a Voronoi polyhedron containing an internal atom is prolate even in the bulk, and near the surface it becomes more prolate [Fig.4], which can be shown by evaluating shape factors $g_1 \sim g_2 < g_3$ defined by $g_n = L_n^2/(L_1^2 + L_2^2 + L_3^2)$, where L_1^2, L_2^2, and L_3^2 are eigen values of the gyration tensor calculated by using coordinates of vertices of a polyhedron.

We propose the concept of *bonding faces* defined as the faces that bisect covalent bonds. For typical internal

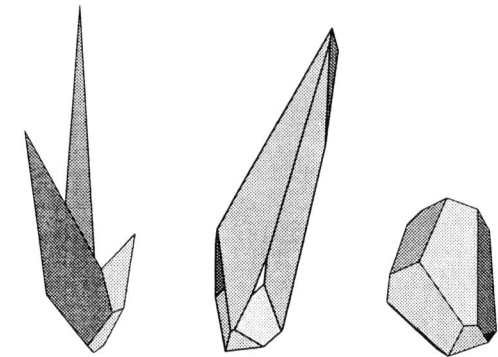

FIGURE 4. Voronoi polyhedra for internal atoms. From left, an open polyhedron ($1/V=0$) and a polyhedron ($1/V = 0.013$ Å^{-3}) at the surface, and a polyhedron ($1/V = 0.053$ Å^{-3}) inside the globule.

atoms, two bonding faces out of 15.3 faces (average) occupy 38.2% of the total surface area of a polyhedron, and the principal direction corresponding to the largest shape factor for the polyhedron is almost parallel to the edge between two bonding faces. Therefore, we conclude that the two bonding faces determine the prolate shape and play significant roles in the Voronoi space division of covalently bonded chains.

In summary, the volume and shape of Voronoi polyhedra strongly depend on the existence of bonds. Soyer et al. [1] pointed out a close relation to random packings of hard spheres and Starr et al. [2] found a universal feature of Voronoi volume distribution functions for dense liquids. The former analayzed an averaged structure (barycenter model) and the latter simulated the bead-spring model, both of which lost chemical character. In contrast to these works, we emphasize the importance of the number of bonds and bonding faces, which represent the characteristic feature of chain molecules.

ACKNOWLEDGMENTS

C.A. was supported by the special research grant of the University of the Air. N.T. and T.D. are grateful to Prof. Susumu Fujiwara for helpful discussion and Norishige Matsuzaki and Kotaro Nomura for technical assistance.

REFERENCES

1. Soyer, A., Chomilier, J., Mornon, J.-P., Jullien, R., and Sadoc, J.-F., *Phys. Rev. Lett.* **85**, 3532 (2000).
2. Starr, F. W., Sastry, S., Douglas, J. F., and Glotzer, C., *Phys. Rev. Lett.* **89**, 125501-1 (2002).

A Possibility of Controlling Polymer Entanglement Structure under the Flow using the Computational Simulations

A. Kuroda* and K. Koyama[†]

*Venture Business Laboratory, Yamagata University, 4-3-16 Jonan, Yonezawa, Yamagata 992-8510, Japan
[†]Department of Polymer Science and Engineering, Yamagata University, 4-3-16 Jonan, Yonezawa, Yamagata 992-8510, Japan

Abstract. Recently there have been a number of enquiries, and requests from the textile industry for the development of high-strength fibers. Increasing the degree of crystallinity by chain entanglement control is a method to manufacture high strength fibers. The objective of the research project was characterization and evaluation of chain entanglement behavior of polymeric materials through theoretical, and computational approaches to achieve the final target of an optimum design for the melt-spinning system.

The first part of the paper analyses the results of experiments on chain entanglement behavior using OCTA simulation system (Dual Slip-Link Model Simulator). The second part of the paper deals with the theoretical analysis of the chain entanglement behavior using Mead-Larson-Doi theory. The possibility of polymer chain entanglement control is also discussed.

INTRODUCTION

Recently there are many requests about the development of high-strength fibers from the textile industry and so on. For this we have to control the chain entanglement in the polymeric materials under the flow for making high crystalline materials. This research is aimed to characterize and evaluate the chain entanglement behavior using our simulator system by the theoretical and computational approach; and our final target is to set the melt-spinning system design.

SIMULATION

Simulation Result

Firstly, we calculated the chain entanglement behavior using OCTA simulation system which is the Dual Slip-Link model simulator[1] based on Mead-Larson-Doi theory[2] developed by JCII project.

Fig.1 and Fig.2 shows the chain entanglement behavior under the accelerating flow, we can see the number of entanglement decreases with external shear rate growth. In general, it can be explained that, the entanglement relaxation time is enough slow compared with the inverse of shear rate, the entanglement formation cannot catch up and disentanglement. In this figure, we can also observe the gap of entanglement at zero shear rates for large poly-

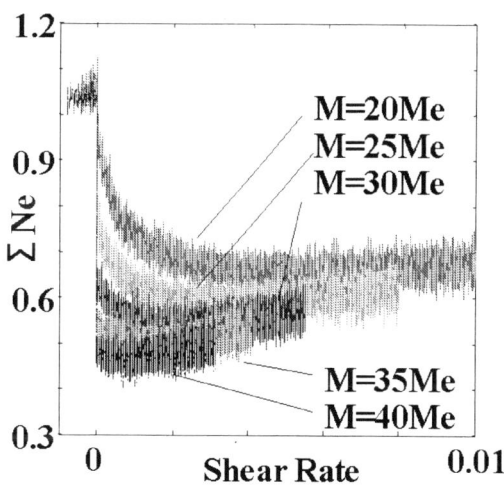

FIGURE 1. The number of entanglement vs shear rate $[\times \tau_e^{-1}]$.

mer. This gap appears at about 10xMe and grows up with the molecular weight.

Theoretical Analysis

Next, we analyze this behavior using Mead-Larson-Doi theory, which was written by relaxation type partial differential equations such as

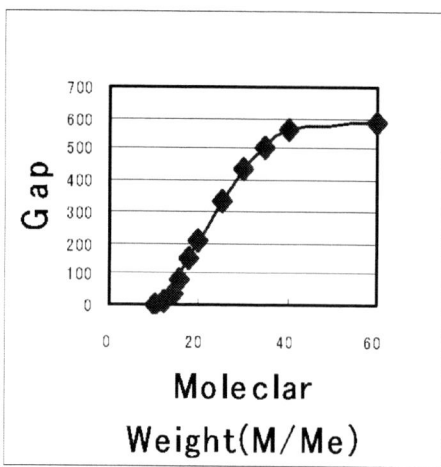

FIGURE 2. The number entanglement gap vs molecular weight.

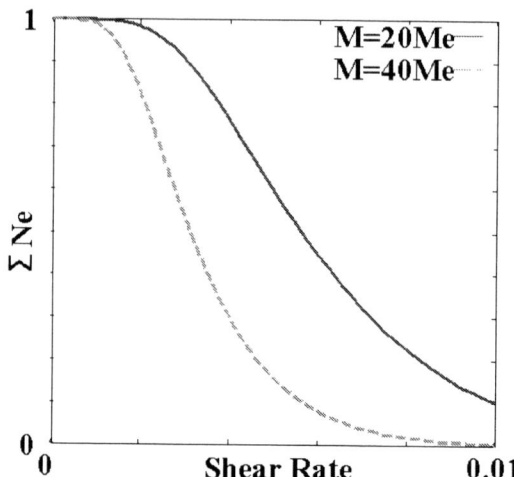

FIGURE 3. The analysis of the number of entanglement vs shear rate using Mead-Larson-Doi type relaxation eq.(1).

$$\frac{\partial P(t,t')}{\partial t} = [-\frac{1}{\lambda^2 \tau_d} - f(\lambda)\kappa : \mathbf{S} - \frac{\dot{\lambda}}{\lambda}]P(t,t')$$

$$\mathbf{S} = \int_{-\infty}^{t} dt'(\frac{\partial}{\partial t'}P(t-t'))\mathbf{Q}(\mathbf{E}(t,t'))$$

$$\dot{\lambda} = \lambda \kappa : \mathbf{S} - \frac{1}{\tau_s}(\lambda - 1) - \frac{1}{2}(\kappa : \mathbf{S} - \frac{\dot{\lambda}}{\lambda})(\lambda - 1) \quad (1)$$

$$\sigma = 5G_N^0 \lambda^2 \mathbf{S}$$

$$\mathbf{Q} = \langle \frac{\mathbf{E} \cdot \mathbf{u}' \mathbf{E} \cdot \mathbf{u}'}{|\mathbf{E} \cdot \mathbf{u}'|^2} \rangle.$$

Here λ is normalized polymer length, and $\mathbf{S}$ is orientation tensor described by $\mathbf{Q}$ tensor and survival probability $\mathbf{P}$. To know the number of entanglement, we need to calculate the distance of two neighbor links. This length has same behavior as polymer length except total length relaxation (Rause relaxation). Then we are able to write the segment length l as

$$\dot{l} = l\kappa : \mathbf{S} \quad -\frac{1}{2}(\kappa : \mathbf{S} - \frac{\dot{l}}{l})(l-1). \quad (2)$$

Then we estimated the number of entanglement as λ/l. Fig.3 shows the number of entanglement against shear rate using Mead-Larson-Doi type relaxation equations. We can understand the decreasing behavior against shear rate, but we do not observed the gap behavior. This gap appears not due to the phase transition, but the cause of non-equilibrium effect. If we calculate above the simulation under the enough relaxation time, this gap may not appear. But the flow in the polymer processing machine does not wait to relax, and then this gap effect has enough meanings for example as a melt fracture effect.

CONCLUSION

Polymeric materials have some nonlinear effects such as memory effect, and we can observe some strange effects for example the behavior under the double step shear flow. Our analysis mainly indicated the number of entanglement was proportional to flow. These strange effects were observed by rheological stress measurement, and eq.(1) shows stress tensor descried by $\mathbf{S}$ tensor and polymer length. We consider that the origin of these strange phenomena is mainly not polymer length, but $\mathbf{S}$ tensor effect. The behavior of the polymer length is simple, however our non-equilibrium gap phenomena may be useful for industrial technology.

ACKNOWLEDGMENTS

This research is supported by the national project, High-Strength Fiber Project, which has been entrusted to the (JCII) Japan Chemical Innovation Institute by the NEDO (New Energy and Industrial Technology Development Organization) under METI's Program for the Scientific Technology Development for Industries that Create New Industries.

REFERENCES

1. Takimoto, J.-I., Tasaki, H., and Doi, M., *Polym. Preprints Jpn.* **49**, 2512 (2000).
2. Mead, D. W., Larson, R. G., and Doi, M., *Macromolecules* **31**, 7895-7914 (1998).

A short analysis of two novel methods for modeling polymers

E.A. Koopman* and C.P. Lowe*

Nieuwe achtergracht 166, 1018WV Amsterdam, The Netherlands

Abstract. We will compare two different new computational techniques for modeling polymer behavior. The description of the polymers is identical in both methods but the solvent around the polymers is different. The first method is dissipative particle dynamics. This describes the solvent as lumps of fluid interacting through a soft potential and a thermostat. The other method is a lattice Boltzmann method. This is a mesoscopic method to solve the Navier-Stokes equations. We find that both resolve the hydrodynamic interactions between beads surprisingly well even for small Kuhn lengths.

INTRODUCTION

Polymer dynamics modeling on a mesoscopic scale is a relatively new field of research. On a macroscopic scale it is often impossible to capture polymer behavior in equations. Microscopic descriptions on a molecular level on the other hand are often computationally too complex. One solution to this is to describe the polymers and the solvent on a length scale which is larger than the typical size of an atom. It is possible to get correct polymer behavior in an ideal gas solvent when the polymer is described as a set of beads interconnected by a simple spring potential. Such a system has the correct thermodynamic behavior but it lacks hydrodynamics.

Dissipative particle dynamics (DPD) was introduced as a method to model both the thermodynamics and the hydrodynamics of a fluid [1]. It consists of a number of particles interacting through a soft potential. A number of possible thermostats can be included in the model to get the correct thermodynamics. The thermostat we used is the Lowe-Andersen thermostat [2]. This thermostat works by thermalizing the relative velocities of particle pairs. We include polymers by just connecting certain fluid particles with a spring. If the potential between polymer and solvent particles is identical we effectively model a polymer solluted in it's own monomers.

The other method we used is an adaption of the method proposed by Ahlrichs and Dünweg [3]. They use a lattice Boltzmann fluid to model the solvent. Lattice Boltzmann (LB) is traditionally used to model macroscopic fluid dynamics [4], but Ladd introduced a way to include thermal fluctuations in the model [5]. Ahlrichs and Dünweg combined this with realistic model polymers. But combining a molecular description of the polymers with a mesoscopic representation is somewhat odd so we use "effective" polymers as used in DPD.

METHODS

Dissipative particle dynamics

DPD was recently introduced as a new way to model fluid properties on a mesoscopic length scale [1]. The fluid is supposedly represented as small lumps which interact through a soft repulsive potential which typically looks like $u(r) = 1 - (\frac{r}{r_c})^2$ within a certain cutoff radius r_c. The model usually also includes a thermostat. For this there are a number of options. Originally the thermostatting was done in an indirect way by applying a random force and a dissipative force to particle pairs. This type of thermostat has some drawbacks. One of them is that it does not even guarantee the system has the right temperature. A thermostat which gives the correct thermal behavior is the Lowe-Andersen thermostat. This works by calculating the relative velocities between particle pairs which are close and applying a new relative velocity taken from a Maxwell Boltzmann distribution.

One major drawback of this method is one inherent to most particle based methods. Since it has a large number of pair interactions the computational time needed to do a simulations scales with the number of particles squared. A few computational tricks can be applied to speed up the simulations, neighbor lists for example, but massive simulations typically take days or even weeks.

Polymers can be modeled using DPD by including harmonic spring potentials between particle pairs. Without the repulsive potential the method reproduces the theoretical results for an ideal chain in an ideal gas solvent. When including the repulsive potential a number of other non ideal effects such as excluded volume can be observed.

Lattice Boltzmann

Lattice Boltzmann is a lattice description of a fluid (an excellent overview is found in [4]). Each grid node has a fluid density, velocity and stress tensor. Every time step the grid is updated by letting bits of fluid stream to neighboring grid points where those bits collide to give a new density, velocity and stress tensor. Originally this was only used to solve the Navier-Stokes equations. In 1994 Ladd proposed a modification of the model to include thermal fluctuations. This is done by applying small fluctuations to the stress tensor every time step. Ahlrichs and Dunweg suggested a way to combine this method with molecular dynamics to model polymer behavior. Since lattice Boltzmann can resolve the time evolution of the solvent much faster then particle methods such as molecular dynamics or DPD this should give a large computational advantage.

The way they combine the particle description of the polymer with the is by a friction force. The force applied by the fluid on the particles is given by $\vec{f} = -\zeta(v_{polymer} - v_{fluid})$ with ζ a representing the amount of friction. A similar force is applied by the particles on the fluid. Ahlrichs and Dünweg then show that this system exhibits both thermodynamic and hydrodynamic behavior.

For testing the static properties we used a simpler version of the polymers. Instead of using the Lennard-Jones potential and a non harmonic spring force we just use a harmonic spring. Since this type of polymer representation works in DPD we also expect it to work with lattice Boltzmann.

RESULTS AND DISCUSSION

Since the lattice Boltzmann method is a new method we focused first on its static properties. Static properties such as the Kuhn length and temperature all agreed with theoretical predictions. One difference we found with the method Ahlrichs and Dünweg used is that our method does not allow for temperatures as high as they used. Since we can not directly relate the numerical temperature to a real life temperature we do not know if this is a real problem. One possible cause is that they use a trick to correct negative fluid densities. We simply abort our simulations in that case.

The effect of hydrodynamics on the system was measured by doing a simulation where the polymer is dragged through the fluid with a very small force. This force must give a velocity which is much smaller than the thermal fluctuations, consistent with linear response theory. We chose a Kuhn length to be 4 times the effective hydrodynamic radius of a single bead. With this ratio the behavior of a short chain scales like that of a long chain. It also ensures we are away from the free-draining regime where hydrodynamic interactions are negligible. For DPD this requires that the Kuhn length be the order of a solved particle separation and for LB that it is less than a lattice spacing. Can these methods resolve the hydrodynamic interactions on this very short length scale?

For the DPD system we calculate the friction coefficient for a model polymer in a fluctuating solvent. For LB we have so far only considered the purely dissipative case for friction perpendicular to a rigid link of length b. The results are shown in in table 1.

TABLE 1. Results for the quantity $\frac{\zeta}{N\zeta_0}$.

N	DPD measured	DPD theory	LB measured	LB theory
2	0.74	0.71	0.863	0.842
4	0.54	0.53		
8	0.38	0.37		
16	0.27	0.27		
32	0.17	0.17		

CONCLUSIONS

Both methods do a surprisingly good job of describing hydrodynamic interactions on the very short length scales needed for a practical simulation. We add the proviso that as of yet we have not found such good agreement for the fluctuating LB case.

ACKNOWLEDGMENTS

This research was funded by the Dutch Polymer Institute.

REFERENCES

1. Groot, R., and Warren, P., *J. Chem. Phys*, **107**(11), 4423–4435 (1997).
2. Lowe, C., *Europhys. Lett.*, **47**, 145–151 (1999).
3. Ahlrichs, P., and Dunweg, B., *J. Chem. Phys.*, **111**(17), 8225–8239 (1999).
4. Chen, C., and Doolen, G., *Ann.Rev.Fluid Mech.*, **30**, 329–364 (1998).
5. Ladd, A., *J. Fluid Mech.*, **271**, 285–339 (1994).

Modelling the flow of polymer solutions in confined geometries using a dissipative gas solvent model

A. Berkenbos* and C.P. Lowe[†]

*Department of Chemical Engineering, University of Amsterdam
Nieuwe Achtergracht 166, 1018 WV Amsterdam, The Netherlands
Email: Berkenbos@science.uva.nl
[†]Email: Lowe@science.uva.nl

Abstract. Particle methods are promising techniques for simulating mesoscopic systems where both hydrodynamics and fluctuations are important. An example is the flow of polymers in confined geometries. Studying this phenomenon requires solid/fluid boundaries. We describe a method for incorporating them in the context of a particle model. Our method accurately reproduces the flow profile right up to a planar boundary with no apparent artifacts. We studied a single model polymer in a fluid flowing between two plates and observed an increase of the polymer velocity relative to the mean flow velocity. This is because the size of the polymer effective excludes it from low velocity regions near the plates.

INTRODUCTION

At low Reynold numbers the velocity profile of a pressure driven fluid flow trough a tube is parabolic. Fluid particles in the centre of the tube have the highest velocity, while the velocity of the fluid particles at the wall is zero (the "stick" boundary condition). For larger particles, polymers for example, it is more difficult for the centre of mass to get close to the wall. Therefore a polymer in the solvent is less influenced by the low velocities near the wall and the mean velocity of the polymer will on average be higher than the mean velocity of the solvent. This is the principle behind a separation technique called hydrodynamic chromatography. It is also a general property of polymers carried by a flow through a porous medium. To model this behaviour we need to model the hydrodynamic interactions between polymer segments. As polymers are mesoscopic, Brownian motion cannot be neglected either. Using a simple particle based polymer/solvent model we can include both these effects in a computationally tractable manner. This will capture the effect of both the flow distorting the polymer and polymer perturbing the flow.

Several particle based models for simulating fluctuating hydrodynamics are available[1, 2]. It has however proved problematic to introduce solid boundaries with these methods, notably with respect to the behavior of the particles near the wall. Most models have problems obtaining zero velocity precisely at the specified wall position, or worse, they introduce artificial density differences close to the wall[3, 4, 5]. This is a pity because one advantage of this type of model over lattice models is that the particle positions are continuous in space. One would imagine that it should be possible to simulate accurately a boundary located at an arbitrary point in space.

Here we describe a method for introducing solid boundaries that we have tested for flat plates. We have used the model to calculate the difference between the velocity of a single polymer and a fluid flowing between two plates. The methodology, however, generalizes to more complex geometries and the many polymer case.

DESCRIPTION OF THE MODEL

As a solvent model we use a simple ideal gas maintained at a given temperature by a Lowe-Andersen Thermostat[2]. This guarantees momentum conservation, Galillean invariance and isotropy. It also enables one to work at a high (fluid like) value for the ratio between the kinematic viscosity and the diffusion coefficient (the Schmidt number) while being computationally simpler than DPD[6]. The polymers are represented as gaussian chains of, in this case, sixteen beads connected by an harmonic potential. This assumes theta conditions for a real polymer system. The root mean square separation between adjacent beads is equal to the average solvent interparticle distance. This gives long polymer scaling for the centre of mass polymer diffusion coefficient for all N[7].

In common with other workers[1, 8] we introduce a layer of "boundary" particles outside the wall. This

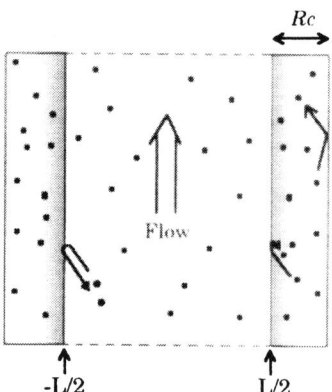

FIGURE 1. Simple representation of the system. The arrows represent the two types of collisions involves in our model. The darker area is the extra layer of wall particles.

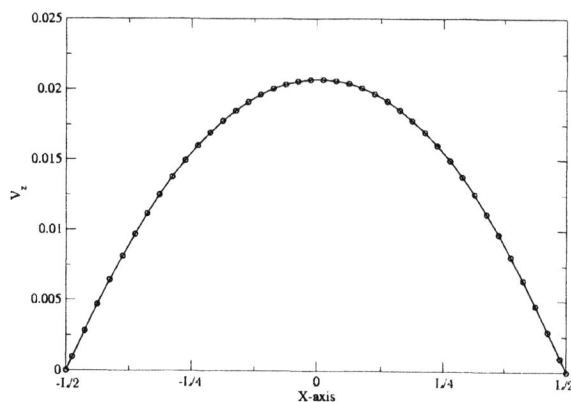

FIGURE 2. Average velocity profile along the X-axis of the system (the walls are at x = 0.5 and at x = - 0.5) over 10 runs of 10^5 steps, the error bars are smaller than the symbols.

is necessary because otherwise the bath collision frequency drops as particles move within r_c of the boundary, decreasing the viscosity and inducing severe boundary layer effects. A stick boundary collision is enforced at the boundary for both particles in the system and boundary particles by a simple bounce back rule: at the moment of impact the velocity is reversed. This makes the average of the pre- and post-collisional velocity zero parallel and perpendicular to the wall. The boundary particles are contained by means of a second wall where we impose a slip boundary condition. This is accomplished by only reversing the component of the velocity perpendicular to the wall (see Figure 1) . We keep the total force zero by giving all the wall particles an additional force opposite to the flow direction. As a result of this the average force on the wall is zero and that ensures that the derivative of the steady state velocity field will be continuous on both sides of the wall.

RESULTS AND DISCUSSION

Figure 2 shows that the method reproduces the analytic parabolic flow profile, going to zero at precisely the point where we locate the walls. We also noted that the density of the model solvent was constant across the entire system. Introducing the model polymer, we kept the flow rate low enough so that the structure of the polymer was not significantly perturbed. The ratio of the plate separation (L) to the radius of gyration (r_g) of the polymer was relatively high ($L \sim 13 r_g$). Nonetheless we could still observe that the average velocity of the polymer $\bar{v}_p$ was indeed higher than the average velocity of the solvent $\bar{v}_s$. We found the ratio $\bar{v}_p/\bar{v}_s = 1.2 \pm 0.02$. This is consistent with a simple estimate based on assuming that the polymer is uniformly distributed within the range $-(L/2 - 1.6 r_g) < x < +(L/2 - 1.6 r_p)$ and that the flow field is unperturbed by the polymer. Integrating the velocity profile between these limits gives an average velocity similar to $\bar{v}_p$. This effective radius is close to the value $1.5 r_g$ that one gets from modelling the polymer as a uniform spherical distribution of mass with the same inertial moment as the polymer.

We conclude that our new method for simulating boundaries in a particle system generates both a uniform solvent density profile and the parabolic velocity profile with zero velocity at the wall. Introducing a model polymer into the flow, the average velocity is higher than the average solvent velocity by an amount consistent with a simple model calculation. Having developed the methodology our aim is to extend this work to more complex problems involving polydisperse polymer solutions and more complex geometries.

REFERENCES

1. Hoogerbrugge, P., and Koelman, J., *Europhys. Lett.*, **19**, 155–160 (1992).
2. Lowe, C., *Europhys. Lett.*, **47**, 145–151 (1999).
3. Kong, Y., Manke, C., Madden, W., and Schlijper, A., *Int. J. Thermophys.*, **15**, 1093–1101 (1994).
4. Jones, J., Lal, M., Ruddock, J., and Spenley, N., *Faraday Discuss.*, **112**, 129–142 (1999).
5. Revenga, M., Zuñiga, I., Español, P., and Pagonabaragga, I., *Int. J. Mod. Phys. C*, **9**, 1319–1328 (1998).
6. Nikunen, P., Karttunen, M., and Vattulainen, I., *Comp. Phys. Comm.*, **153**, 407–423 (2003).
7. Dreischor, M., and Lowe, C., *Lecture Notes in Physics (in press)* (2003).
8. Willemsen, S., Hoefsloot, H., and Iedema, P., *Int. J. Mod. Phys. C*, **11**, 881–890 (2000).

Three Regimes of a Steady-State Rimming Flow of the Liquid Polymers

Sergei Fomin and Toshiyuki Hashida

Fracture Research Institute, School of Engineering, Tohoku University, Sendai, Japan

Abstract. Rimming flow on the inner surface of a horizontal rotating cylinder is investigated. Simple lubrication theory is applied since the Reynolds number is small and liquid film thin. Since the Deborah number is very small the flow is viscometric. A general constitutive law for this kind of flow requires only a single function relating shear stress and shear rate that corresponds to a generalized Newtonian liquid. For this case the run-off condition for rimming flow is derived. Provided the run-off condition is satisfied, the existence of a steady-state solution is proved. In the bounds implied by this condition film thickness admits a continuous solution, which corresponds to subcritical and critical flow regimes. In the supercritical case when the mass of liquid polymer exceeds a certain value or the speed of rotation is less than an indicated limit, a discontinuous solution is possible and a hydraulic jump may occur. As an example, the location and height of the hydraulic jump is determined numerically for Ellis model.

The problem of rotational flow on the inner and/or on the outer wall of a hollow horizontal cylinder has been of interest for many years due to its wide range of applications in industry. A schematic sketch of the process is presented in the upper-left corner of Fig. 1(c). Rimming flow at high rotation rates, as a limiting case when the motion of the liquid is a small perturbation from a rigid-body motion, was analyzed in [1]. However, as it was demonstrated later [2-4], already at relatively low angular velocities the flow can settle into a steady two-dimensional flow. Based on the previous studies three flow regimes can be identified. When the mass flux through the cross-section of the liquid layer on the wall of the cylinder q is below its maximal supportable value q_{max} then fluid film thickness is a continues and smooth function of an angular coordinate. This regime is called subcritical. The critical steady state regime occurs when the mass flux q is equal to its critical value q_{max}. In addition to these two regimes, a third regime, which was named supercritical, is also possible for $q=q_{max}$. This regime is associated with the existence of a steady-state puddle on the rising wall of the cylinder, which occurs as a result of an excessive mass of the liquid loaded on the wall or a too low rotational rate. In [4] it was shown that for the Newtonian model the straightforward leading order steady state lubrication theory could work well to study hydraulic jumps locations and heights in the supercritical regime. Although the aforementioned investigations highlight the main characteristics of the rimming flow, due respect to the effect of non-Newtonian properties was not given.

Our main concern is rotational moulding of highly viscous polymers, which exhibit Newtonian behavior at low shear rates with transition to power-law-shear-thinning at moderate shear rates. The angular velocity of the cylinder, Ω, is relatively low, the liquid film is thin, the effect of the centripetal force is negligible small (in contrast to the case studied in [1]) and rimming flow is mainly dominated by the interaction of the gravity and viscous forces. In our recent paper related to non-Newtonian fluids [5] only subcritical and critical flow regimes were analyzed. The latter was done numerically on the basis of Carreau-Yasuda constitutive model. In the present studies we extend our previous estimates to the supercritical flow regime. We assume that the ratio $\delta = h_0/r_0$ of the liquid layer characteristic thickness, h_0, to the radius of the cylinder, r_0, is small and, hence, the simple lubrication theory can be applied. Using a scale analysis, a theoretical description for a steady-state non-Newtonian flow is obtained. The main non-dimensional complexes that define the process are derived and their typical numerical values are computed, e.g. it was found that

$\delta = (\rho g r_0 / \mu \Omega)^{1/2} \approx 0.02$, where ρ and μ are fluid density and typical viscosity, respectively. Since Deborah number is small (10^{-2}-10^{-1}), the flow is assumed to be viscometric. A general constitutive law for this kind of flow requires only a single function relating shear stress τ and shear rate $\dot{\gamma}$ that corresponds to a generalized Newtonian model. Ignoring the terms of $O(\delta^2)$, the non-dimensional constitutive equation for this model reduces to $\tau = \mu(|\dot{\gamma}|)\dot{\gamma}$, where $\mu(|\dot{\gamma}|)$ is generalized viscosity, $\dot{\gamma} = \partial v_\theta / \partial R$, R and v_θ are the non-dimensional radial coordinate and azimuthal fluid velocity, respectively. Hence, the non-dimensional volume flux q can be readily computed from equation $q = \int_0^h v_\theta dR$, which after integration by parts and some trivial manipulations, reduces to

$$q = h - \text{sgn}(\cos\theta) h^2 \int_0^1 y G[h|\cos\theta|y] dy, \quad (1)$$

where h is the unknown non-dimensional thickness of the liquid layer and function G as the inverse of function $\mu(\dot{\gamma})\dot{\gamma}$. It is physically evident that function G, should be analytic and an increasing function of its argument. Obviously, solution $h = h(\theta)$ of equation (1) is periodic, even, and symmetric respective to $\theta=0$. Hence, the azimuthal distribution of h can be represented by its variation on the interval $[0, \pi]$. Differentiating equation (1) with respect to h and equalizing this derivative to 0 leads to equation $1 - h \text{sgn}(\cos\theta) G[h|\cos\theta|] = 0$. Once solution of the latter equation $h = h_*(\theta)$ is found, the maximal supportable mass flux q_{max} can be readily computed from equation (1) setting in it $h = h_*$ and $\theta = 0$.

Qualitative analysis of equation (1) for the subcritical regime ($q<q_{max}$) shows that: (A) at the interval $[\pi/2,\pi]$ equation has the unique solution $h = h(\theta)$ which satisfy an inequality $0 < h < q$; (B) on the interval $[0,\pi/2]$ equation (1) has two solutions $h = h_1(\theta)$ and $h = h_2(\theta)$, which satisfy inequalities $q < h_1 < h_*$ and $h_2 > h_*$, and $h_2 \to \infty$ for $\theta \to \pi/2$. Apparently, solution $h = h_2(\theta)$ for this regime has no physical meaning and, hence, should be omitted.

Quantitative analysis of equation (1) for the critical flow regime ($q=q_{max}$) leads to the following conclusions: (A) for this regime equation (1) reduces to an identity at $h = h_*$ and $\theta=0$. Hence, in the point $\theta=0$ functions $h_1(\theta)$ and $h_2(\theta)$ intersect, i.e. $h_1(0) = h_2(0) = h_*(0)$; (B) $h_1(\theta)$ decreases on $[0,\pi/2]$. At $\theta = 0$ it reaches maximal value and for $\theta \to \pi/2 - 0$, $h_1 \to q$. In contrast, $h_2(\theta)$ exhibits monotonous growth on $[0,\pi/2]$ and $h_2 \to \infty$ for $\theta \to \pi/2 - 0$. It can be readily shown that the unique solution $h = h(\theta)$ defined on $[\pi/2, \pi]$ is a smooth continuation of function $h_1(\theta)$ in the domain $[\pi/2, \pi]$.

A supercritical regime, which also may take place for $q=q_{max}$, is associated with existence of hydraulic jump on the rising wall of the cylinder, which occurs as a result of an excessive mass of the liquid, M, loaded on the wall, or in mathematical terms, when condition $(M/\rho r_0^2)\delta^{1/2} > H_c$ is satisfied, where H_c is non-dimensional mean fluid layer thickness in the critical regimes. The above analysis is illustrated by the numerical computations for Ellis model, $\dot{\gamma} = -\tau[1 + (W|\tau|)^{1/n-1}]$, presented in Figs. 1(a,b,c). In this model parameter $W = \Omega/\dot{\gamma}_t\delta$, where $\dot{\gamma}_t$ is a transitional value of shear rate, below which fluids exhibit Newtonian and above shear-thinning behavior.

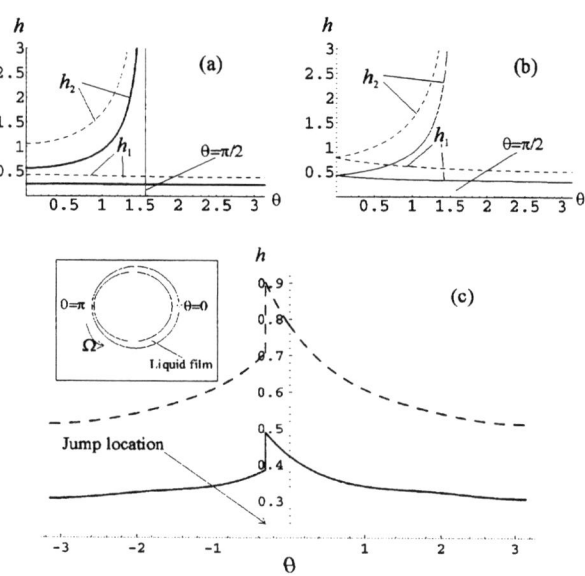

FIGURE 1. Thickness of the liquid layer (Ellis model at $n=1/3$) along the inner wall of the cylinder for different flow regimes: (a) - subcritical; (b) - critical; (c) - supercritical. Solid lines correspond to $W=5$ and dashed lines to $W=1$.

REFERENCES

1. Ruschak, K.J., and Scriven, L.E., *J. Fluid Mech.* **76**, 113-125 (1976).
2. Moffatt, H.K., *J. de Mecanique* **16**, 651-673 (1977).
3. Melo, F., *Phys. Rev. E* **48**, 2704-2712 (1993).
4. O'Brien, S.B.G., and Gath, E.G., *Phys. Fluids* **10**, 1040-1042 (1998).
5. Fomin, S., Watterson, J., Raghunathan, S., and Harkin-Jones, E., *Phys. Fluids* **14** 3350-3353 (2002).

Chain dynamics on polymer crystallization from a stretched amorphous state: a molecular dynamics simulation

Akira Koyama[a], Takashi Yamamoto[b], Koji Fukao[c], and Yoshihisa Miyamoto[a]

[a]*Graduate School of Human and Environmental Studies, Kyoto University, Kyoto 606-8501 Japan*
[b]*Faculty of Science, Yamaguchi University, Yamaguchi 753-8512 Japan*
[c]*Department of Polymer Science and Engineering, Kyoto Institute of Technology, Kyoto 606-8585 Japan*

Abstract. We have performed the molecular dynamics (MD) simulations of polymer crystallization from a stretched amorphous state, employing a linear poly(ethylene) molecular model. During the simulation, the crystal domain emerged and grew in the amorphous state, whereas the shape of the whole polymer chain remained effectively unchanged. The MD simulation suggests the ordering initiates from the segments with a spatial scale smaller than the stem length, and from many places in a chain. In this study, the change in the molecular motion of subchain during crystallization and its dependence on the size of the subchain are examined.

INTRODUCTION

The molecular dynamics (MD) simulation has recently come to be recognized as a very promising tool to understand crystallization in systems of chain molecules, since they can directly provide us with molecular trajectories that are experimentally hard to access. In very recent literatures we can notice a surge of interest in reproducing chain folded crystallization from the melt and solution both during the primary nucleation and the growth [1-5], We have also reported simulation results of polymer crystallization from a stretched amorphous state [5], where the time development of various static order parameters were discussed, whereas changes in dynamic properties during the crystallization remained not to be analyzed in detail. In this presentation we report the results of our recent simulation of polymer crystallization from the bulk oriented amorphous state, and discuss the changes in the molecular motion of subchain during the crystallization.

SIMULATION

In the present simulation we employed the single linear poly(ethylene)(PE) model consisting of 5000 united atoms. The simulations were performed under usual periodic boundary condition. First we prepared the well-relaxed isotropic melt at 600 K, and quench them to glassy state at 100 K. Next we generated the oriented amorphous state by drawing the isotropic glassy sample uniaxially up to 400 % at 100 K. Crystallization from the oriented amorphous state was then simulated for 30-60 ns, under constant temperatures (280-380 K) and a pressure (1 atm).

RESULTS

The structural changes in the present simulation were analyzed by use of the real-space images and its Fourier transforms (structure functions). During the

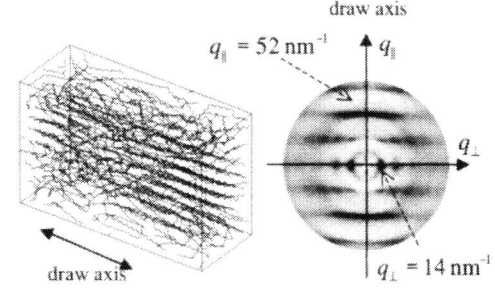

FIGURE 1. Snapshot (left side) and structure function averaged around the draw axis (right side) of the system at the final state of 30 ns (330 K).

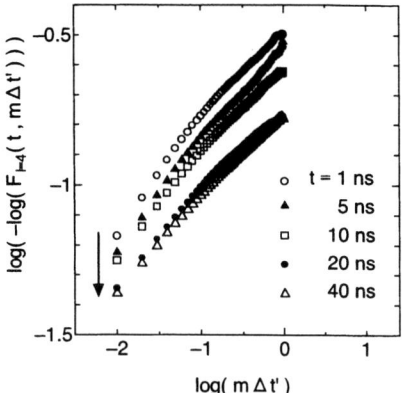

FIGURE 2. Changes in the $F_{l=4}(t, m\Delta t')$ vs. $m\Delta t'$ profiles during the crystallization at 320 K.

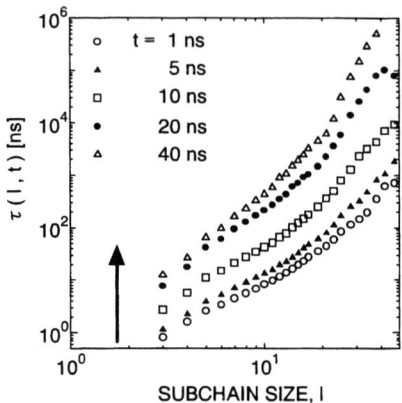

FIGURE 3. Changes in the $\tau(l, t)$ v.s. l profiles during the crystallization at 320 K.

simulations, the crystallization proceeds and the highly ordered crystal structures were successfully obtained at the temperatures between 300 and 360 K. The structure function shows an ordered crystalline pattern (Fig. 1) which indicates that the monomer heights between neighboring stems are arranged in the crystal.

Now our concern is the molecular motion during the crystallization. The rearrangement motion of the subchains made up of l monomers (united atoms) is estimated by a time dependent autocorrelation function of orientation unit vectors, which is defined by

$$F_l(t, m\Delta t') \equiv 2 \langle | \mathbf{u}_{l,i}(t + m\Delta t') \cdot \mathbf{u}_{l,i}(t) | \rangle - 1,$$

where $\mathbf{u}_{l,i}(t)$ is the i-th orientation unit vector at a time t, m is an integer, $\Delta t'$ (=0.01 ns) is the time interval, and $\langle ... \rangle$ denotes the ensemble average. The orientation unit vector of each subchain is defined as the unit vector along the principal axis of radius of gyration tensor of the subchain, which gives the maximum Eigen value.

The plots of $F_{l=4}(t, m\Delta t')$ vs. $m\Delta t'$ at 320 K are shown in Fig. 2. As the crystallization proceeds, $F_{l=4}(t, m\Delta t')$ decreases at a given time $m\Delta t'$ and the slope also decreases, which corresponds to the power of the decay time in the exponent when the stretched exponential decay is assumed.

The $F_l(t, m\Delta t')$ profiles obtained at the each subchain size l and simulation time t were fitted to the stretched exponential function defined as

$$F_{FIT,l}(t, m\Delta t') \equiv \exp(-(m\Delta t'/\tau(l,t))^{\beta(l,t)}),$$

where $\tau(l, t)$ and $\beta(l, t)$ are the characteristic time and the exponent, respectively.

Figure 3 shows the change in $\tau(l, t)$ during the crystallization at 320 K. Even at the initial oriented amorphous state of $t=1$ ns, $\tau(l>10, t)$ are beyond the characteristic time of crystallization which is about 10 ns at the temperature, as reported previously. It suggests that the molecular rearrangement did not occur frequently in the range $l>10$ during the crystallization. Then, the value of $\tau(l, t)$ becomes larger in the whole l range as the crystallization proceeds. The detailed comparison between the time developments of the dynamic parameters and the order parameters during the crystallization is under progress.

In this study, we examine the subchain motion during an MD simulation of polymer crystallization. It is found that the molecular motion at the subchain size $l<10$ concerns more effectively to the crystallization than that at $l>10$

REFERENCES

1. Meyer, H., Müller-Plathe, F., *J. Chem. Phys.* **115**, 7807-7810 (2001).

2. Yamamoto, T., *J. Chem. Phys.* **115**, 8675-8680 (2001).

3. Koyama, A., Yamamoto, T., Fukao, K., Miyamoto, Y., *Phys. Rev.* **E 65**, 050801 (2002).

4. Lavine, M. S., Waheed, N., Rutledge, G. C., *Polymer* **44**, 1771-1779 (2003).

5. Koyama, A., Yamamoto, T., Fukao, K., Miyamoto, Y., *J. Macromol. Sci. Part B–Physics* **B42**, 821-831 (2003).

Relaxation and Self-Diffusion of a Polymer Chain in a Melt

Katsumi Hagita* and Hiroshi Takano*

*Department of Physics, Faculty of Science and Technology, Keio University,
Yokohama 223-8522, Japan*

Abstract. Relaxation and self-diffusion of a polymer chain in a melt are discussed on the basis of the results of our recent Monte Carlo simulations of the bond fluctuation model, where only the excluded volume interaction is considered. Polymer chains are located on an $L \times L \times L$ simple cubic lattice under periodic boundary conditions. Each chain consists of N segments, each of which occupies $2 \times 2 \times 2$ unit cells. The results for $N = 32, 48, 64, 96, 128, 192, 256, 384$ and 512 at the volume fraction $\phi \simeq 0.5$ are examined, where $L = 128$ for $N \leq 256$ and $L = 192$ for $N \geq 384$. The longest relaxation time τ is estimated by solving generalized eigenvalue problems for the equilibrium time correlation matrices of the positions of segments of a polymer chain. The self-diffusion constant D is estimated from the mean square displacements of the center of mass of a single polymer chain at the times larger than τ. From the data for $N = 256, 384$ and 512, the apparent exponents x_r and x_d, which describe the power law dependences of τ and D on N as $\tau \propto N^{x_r}$ and $D \propto N^{-x_d}$, are estimated to be $x_r \simeq 3.5$ and $x_d \simeq 2.4$, respectively. For $N = 192, 256, 384$ and 512, $D\tau/\langle R_e^2 \rangle$ appears to be a constant, where $\langle R_e^2 \rangle$ denotes the mean square end-to-end distance of a polymer chain.

INTRODUCTION

Recently, slow dynamics of a polymer chain in a melt has been studied by large-scale simulations,[1, 2, 3, 4] in order to examine the predictions of the reptation theory. [5, 6] According to the reptation theory, the longest relaxation time τ and the self-diffusion constant D of the center of mass of a polymer chain of N segments in concentrated polymer systems behave as $\tau \propto N^3$ and $D \propto N^{-2}$, respectively, for sufficiently large values of N. The theoretical result $\tau \propto N^3$ has been found to disagree with the typical experimental result $\tau \propto N^{3.4}$.[6] The contour length fluctuation[6] is considered to explain this disagreement. Although the exponent for D observed in the experiments has been believed to agree with that predicted by the theory, the recent experiment[7] reported $D \propto N^{-2.4}$ for hydrogenated polybutadiene.

In this article, we discuss the behaviors of the longest relaxation time τ and the self-diffusion constant D obtained by our recent Monte Carlo simulations.[1, 2]. In these simulations, the bond fluctuation model has been used,[8] where only the excluded volume interaction is taken into account. Polymer chains are located on an $L \times L \times L$ simple cubic lattice under periodic boundary conditions. Each chain consists of N segments, where each segment occupies $2 \times 2 \times 2$ unit cells. The N-dependences of τ and D are examined for $N = 32, 48, 64, 96, 128, 192, 256, 384$ and 512 at the volume fraction $\phi \simeq 0.5$, where $L = 128$ for $N \leq 256$ and $L = 192$ for $N \geq 384$.

METHOD

Left-eigenfunctions and eigenvalues of the time-evolution operator of the master equation of the system can be regarded as relaxation modes and rates, respectively. [9, 10] The eigenvalue problem can be approximately solved as a variational problem [9, 10] by choosing a trial function for the pth slowest relaxation mode as $\mathbf{X}_p = \sum_{i=1}^{N'} f_{p,i}^{(n)} \bar{\mathbf{R}}_i^{(n)}(t_0/2; Q)$ with $N' = N/n$.[1, 11] Here, $A(t; Q)$ represents the expected value of quantity A after period t starting from configuration Q of the system. The quantity $\bar{\mathbf{R}}_i^{(n)}$ is given by $\bar{\mathbf{R}}_i^{(n)} = \frac{1}{n}\sum_{k=1}^{n} \mathbf{R}_{(i-1)n+k}$ with $\mathbf{R}_j = \mathbf{r}_j - \mathbf{r}_c$ and $\mathbf{r}_c = \frac{1}{N}\sum_{i=1}^{N} \mathbf{r}_i$, where $\mathbf{r}_i$ denotes the position of the ith segment of a polymer chain. For this trial function, the variational problem is reduced to the generalized eigenvalue problem $\sum_{j=1}^{N'} C_{i,j}^{(n)}(t_0+t_1) f_{p,j}^{(n)} = e^{-\lambda_p t_1} \sum_{j=1}^{N'} C_{i,j}^{(n)}(t_0) f_{p,j}^{(n)}$ for $C_{i,j}^{(n)}(t) = \frac{1}{3}\langle \bar{\mathbf{R}}_i^{(n)}(t) \cdot \bar{\mathbf{R}}_j^{(n)}(0) \rangle$ with $\sum_{i=1}^{N'}\sum_{j=1}^{N'} f_{p,i}^{(n)} C_{i,j}^{(n)}(t_0) f_{q,j}^{(n)} = \delta_{p,q}$.[9, 10, 11] From the generalized eigenvalue problem, the p-th relaxation mode $\mathbf{X}_p$, which is described by $f_{p,i}^{(n)}$, and the corresponding relaxation rate λ_p are determined. The longest relaxation time τ is given as the inverse of the slowest relaxation rate λ_1 estimated by this method.

The self-diffusion constant D is estimated from the elapsed time t dependence of the mean square displacement $\langle [\mathbf{r}_c(t) - \mathbf{r}_c(0)]^2 \rangle$ of the center of mass $\mathbf{r}_c$ of a polymer chain by fitting the data points at times longer than τ to $\langle [\mathbf{r}_c(t) - \mathbf{r}_c(0)]^2 \rangle = 6Dt + \text{constant}$.[2]

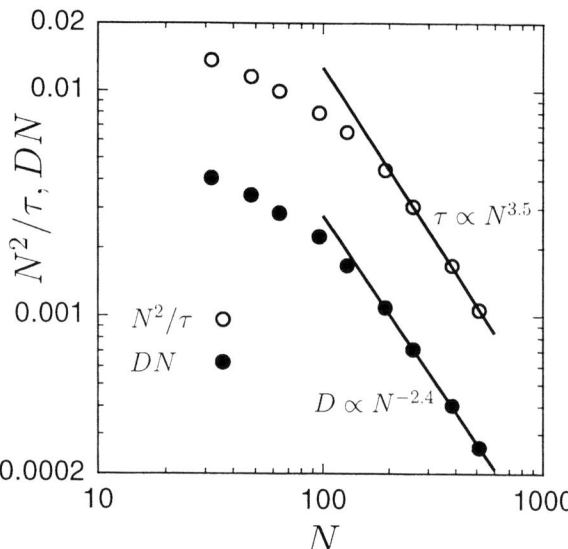

FIGURE 1. Log-log plot of N^2/τ versus N[1, 2] and DN versus N.[2] Solid lines represent the results of the fit of the data for $N = 256, 384$ and 512.[1, 2]

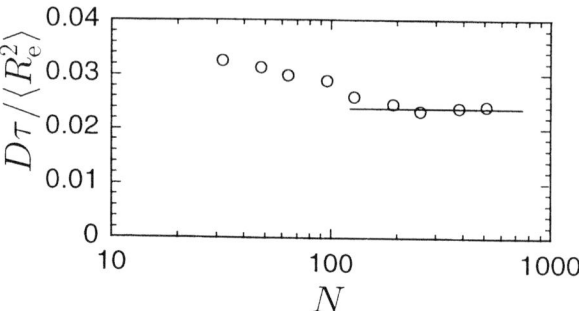

FIGURE 2. Semilog plot of $D\tau/\langle R_e^2 \rangle$ versus N. Solid line represents the average of the values of $D\tau/\langle R_e^2 \rangle$ for $N = 192, 256, 384$ and 512.[2]

RESULTS

Figure 1 shows a log-log plot of N^2/τ versus N[1] and DN versus N.[2] It can be seen that the apparent exponents x_r and x_d, which describe the power law dependences of τ and D on N as $\tau \propto N^{x_r}$ and $D \propto N^{-x_d}$, respectively, increase as N increases. From the data for $N = 256, 384$ and 512, the apparent exponents are estimated to be $x_r \simeq 3.5$ and $x_d \simeq 2.4$, which agree with the experimental results $\tau \propto N^{3.4}$[5, 6] and $D \propto N^{-2.4}$.[7]

The reptation theory predicts that the ratio $D\tau/\langle R_e^2 \rangle$ is independent of N for sufficiently large N, where $\langle R_e^2 \rangle$ denotes the mean square end-to-end distance of a polymer chain. The apparent exponent 2ν of the power law dependence $\langle R_e^2 \rangle \propto N^{2\nu}$ has been estimated to be $2\nu \simeq 1.0$ for $N = 256, 384$ and 512.[1, 2] Thus, it seems that the relation $-x_d + x_r - 2\nu \simeq 0$ seems to hold for $N = 256, 384$ and 512. This means that $D\tau/\langle R_e^2 \rangle$ is independent of N even in the range of N where the theoretical asymptotic behaviors of τ and D are not seen.

According to the reptation theory, the ratio $D\tau/\langle R_e^2 \rangle$ is a constant $1/(3\pi^2) \simeq 0.034$, which contains no adjustable parameter.[6] Figure 2 shows the N-dependence of $D\tau/\langle R_e^2 \rangle$.[2] It has been found that $D\tau/\langle R_e^2 \rangle$ seems to converge to a constant value around 0.024 for $N \geq 192$.[2] If this value can be regarded as the large N limit of $D\tau/\langle R_e^2 \rangle$, it reasonably agrees with the value predicted by the reptation theory.

The reptation theory predicts that $D\tau/\langle R_e^2 \rangle$ is independent of the volume fraction ϕ of monomers.[6] The study of ϕ dependence of D, τ and $\langle R_e^2 \rangle$ is in progress.

ACKNOWLEDGMENTS

The authors are grateful to Professor K. Binder, Professor M. Doi, Professor S. F. Edwards, Professor P. G. de Gennes, Professor E. M. Terentjev, Dr. G. S. Grest and Dr. W. Paul for their comments. The authors thank Research Center for Computational Science of Okazaki National Research Institutes for the use of the Fujitsu VPP 5000 and Hokkaido University Computing Center for the use of the Hewlett-Packard V2500.

REFERENCES

1. Hagita, K. and Takano, H.: *J. Phys. Soc. Jpn.* **71**, 673-676 (2002).
2. Hagita, K. and Takano, H.: *J. Phys. Soc. Jpn.* **72**, 1824-1827 (2003).
3. Kreer, T., Baschnagel, J., Müller, M. and Binder, K.: *Macromolecules* **34**, 1105-1117 (2001).
4. Dünweg, B., Grest, G. S. and Kremer, K.: *Numerical Methods for Polymeric Systems*, ed. S. Whittington, IMA Volumes in Mathematics and its Applications **102**, Springer-Verlag, New York, 1998, pp. 159-196.
5. De Gennes, P. G.: *Scaling Concepts in Polymer Physics*, Cornell University Press, Ithaca, 1984.
6. Doi, M. and Edwards, S. F.: *The Theory of Polymer Dynamics*, Oxford University Press, Oxford, 1986.
7. Tao, H., Lodge, T. P. and Meerwall, E. D.: *Macromolecules* **33**, 1747-1758 (2000).
8. Carmesin, I. and Kremer, K.: *Macromolecules* **21**, 2819-2823 (1988).
9. Koseki, S., Hirao, H. and Takano, H.: *J. Phys. Soc. Jpn.* **66**, 1631-1637 (1997).
10. Takano, H. and Miyashita, S.: *J. Phys. Soc. Jpn.* **64**, 3688-3698 (1995).
11. Hagita, K., Ishizuka, D. and Takano, H.: *J. Phys. Soc. Jpn.* **70**, 2897-2902 (2001).

Dynamics of Knotted Polymers

Pik-Yin Lai

Department of Physics and Center for Complex Systems, National Central University, Chung-Li, Taiwan 320.

Abstract. Most people should have the frustrated experience of trying to untie an entangled mess of a knotted string. It seems that the more crossing the string has, it would take more time and patience for one to untie it. And certainly random and/or violent movements of the string segments would make thing even worse in most cases. To our surprise, our recent studies[1, 2]on the non-equilibrium relaxation of a closed knotted polymer cut at some point and relaxed by Brownian motion to a linear chain, indicates that it is not always true that knots with more essential crossings will, on average, take more time to untie. The detail topology of the knot type and the topological interactions are important in the untying relaxation dynamics.

DYNAMICS OF UNTYING A KNOT

The simplest topological invariant quantity to describe the complexity of a closed knot is the number of essential crossings C, i.e. the minimum number of crossings in any planar projection of the knot. Topological interactions manifest most prominently not only in this macroscopic string knot, but also dominate the dynamics of long polymer molecules. A particular topological state of a knot will remain until the knot is cut. Flexible polymer knots with strict topological constraint of no segment crossing are studied in our simulations. A well-equilibrated polymer knot is cut at a randomly chosen segment and allow to relax to it new equilibrium state, which is the indistinguishable from the free linear chain. In the course of the relaxation, the topological information of the original knot is lost and the characteristic relaxation time measures its rate. The average relaxation time is the typical time scale needed to untie the cut knot by random Brownian motion. Some knot is found to have a longer relaxation time than other knots having more crossings. Remarkably, when all the knots are arranged into homologous groups, the relaxation times, τ, increase monotonically and linearly with C for all the groups we studied. Fig. 1 shows the relaxation times extracted from the time-dependent radius of gyration of a cut knot as classified into homologous groups. Our observation indicates that conventional labelling of knots can further be parametrized naturally into groups in a way that has a direct physical meaning in terms of the topological interactions in a knot. By cutting the knot and releasing the strong topological constraints, the chain releases some sort of "topological free energy" and relaxes to a free linear chain and such the classification into knot groups emerges naturally. The linear stepwise increase of the relaxation time with the essential crossing suggests

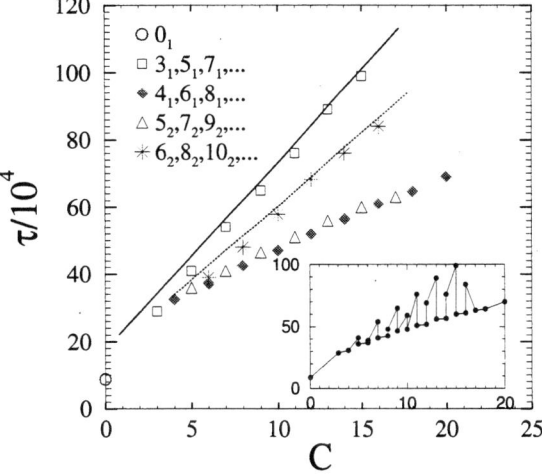

FIGURE 1. Monte Carlo simulation data for the non-equilibrium relaxation time of the radius of gyration of the knot vs. C. Each knotted polymer consists of 180 monomers. Inset: Irregular variation of τ with C without classification. Uncertainties are about the size of the symbols.

that the topological energy spectrum has equal spacing of for knots within a group.

DIFFUSION AT EQUILIBRIUM

On the other hand, the equilibrium relaxation dynamics[3] of the uncut prime knots do not display the classification into group as in the case of the non-equilibrium relaxation of cut knots. In this case, the time auto-correlation functions for the radius of gyration of the non-trivial knots can be fitted by a sum of two

TABLE 1. Table for the different knot groups in this paper with the empirical formulae for the mean writhe number and scaled non-equilibrium relaxation times. τ_0 is the relaxation time of the trivial knot.

Knot Group	$\langle Wr \rangle$	τ/τ_0
$(2,C)$ torus knots $(3_1, 5_1, 7_1, ...)$	$\frac{10}{7}C - \frac{6}{7}$	$\frac{2}{3}C + \frac{4}{3}$
Even Twist Knots $(4_1, 6_1, 8_1, ...)$	$\frac{4}{7}C - \frac{16}{7}$	$\frac{4}{15}C + \frac{8}{3}$
Odd Twist Knots $(5_2, 7_2, 9_2, ...)$	$\frac{4}{7}C + \frac{12}{7}$	$\frac{4}{15}C + \frac{8}{3}$
$(6_2, 8_2, 10_2, ...)$	$\frac{10}{7}C - \frac{40}{7}$	$\frac{1}{2}C + \frac{4}{3}$

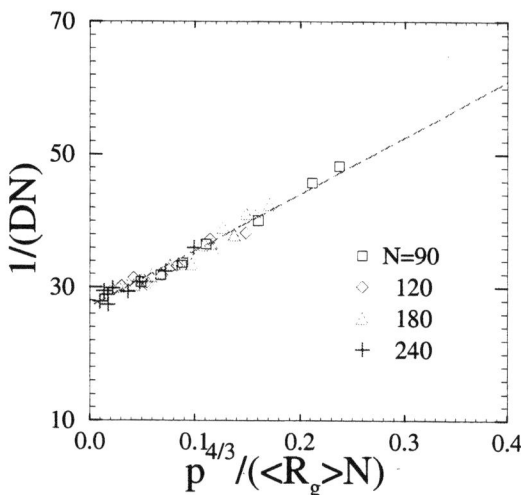

FIGURE 2. A plot of $1/(DN)$ versus $p^{4/3}/(\langle R_g \rangle N)$ for various types of knots of different lengths. The dashed curve denotes the linear behavior given in Eq. (1).

exponential decays of long and short characteristic relaxation times. These two relaxation times decrease with the number of essential crossings C. The faster relaxation follows the Rouse behavior and scales as $N^{1+2\nu}$ and its dependence on C is consistent with the scaling analysis using the blob picture. The mean-square displacement of the center of mass of the knots obeys the free diffusion behavior compatible with the Rouse dynamics. The diffusion coefficients of the knots, $D \propto 1/N$ for large N, but D decreases for knots with increasing C. The relaxation motion of monomers in a knot can be pictured as confined inside an imaginary inflated knotted tube. Grosberg et al.[4] introduced a new topological invariant p defined as the aspect ratio of the length(L) to the diameter(d) of a knotted polymer at its maximum inflated state, $p = L/d$. The diffusion coefficient of the knot can be calculated according to the Einstein relation: $D = k_B T/\mu_t$ where k_B is the Boltzmann constant and μ_t is the total friction coefficient. In Rouse model, the friction coefficient is proportional to the number monomers N in the macromolecule, i. e. $N\xi$, where ξ is the monomer-solvent friction coefficient. For a linear polymer chain, the monomers tend to avoid each other in good solvents and the probability of two monomers in direct close contact is small. However, for knotted polymers, monomers are forced to be in close contact because of the topological constraint. During the relaxation process, monomers will slide onto each other, and extra friction results. The collision probability among the monomers is greatly increased as the number of crossings increases. Using an analog to electric resistance to estimate this internal friction[1], the monomer-monomer friction coefficient is assumed to be proportional to the ratio of length to cross-section area of the maximally inflated knot. Thus the total friction coefficient can be expressed as $\mu_t = N\xi + L\zeta$, where ζ represents the monomer-monomer friction coefficient and $\zeta \simeq \zeta_o/d^2$ for some characteristic monomer-monomer friction ζ_o. Following the idea on the construction of the maximally inflated tube, one has $L \sim R_g p^{2/3}$ and $d \sim R_g p^{-1/3} \sim N^\nu p^{-3/5}$. Then $\mu_t = N\xi + \zeta_o N^{-\nu} p^{8/5}$ and hence

$$\frac{k_B T}{D} = N\xi + \frac{p^{4/3}}{R_g}\zeta_o. \quad (1)$$

Values of p was calculated independently and Fig. 2 plots $1/(DN)$ against $p^{4/3}/(\langle R_g \rangle N)$ for various prime knots with different lengths, the data collapse rather well and fall onto a straight line verifying the scaling result in Eq. (1). One recovers the standard Rouse behavior of $D \sim 1/N$ in Eq. (1) for knots with $p << N$ (i.e. for knots with smaller values of C).

ACKNOWLEDGMENTS

This research is supported by National Council of Science of Taiwan. The author thank Prof. Y.J. Sheng and H.K. Tsao for their collaborations in part of this work.

REFERENCES

1. Sheng, Y.-J., Lai, P.-Y. and Tsao, H.-K., *Phys. Rev. E* **58**, R1222-R1225 (1998); Physica A **281**, 381-392 (2000).
2. Lai, P.-Y., Sheng, Y.-J. and Tsao, H.-K., *Phys. Rev. Lett.* **87**, 1755031-1755034 (2001).
3. Lai, P.-Y., *Phys. Rev. E* **66**, 0210851-0210858 (2002).
4. Grosberg, A. Yu., Feigel, A. and Rabin, Y., *Phys. Rev. E* **54**, 6618-6622 (1996).

III. BIOLOGICAL SYSTEMS

Charge Inversion of a Macroion in Electrolyte Solvent: A Rotating Rod with Polyelectrolyte Counterions

Motohiko Tanaka

National Institute for Fusion Science, Toki 509-5292, Japan

Abstract. Electrophoresis study of charge inversion by molecular dynamics simulation is first reviewed. The cases of spherical and infinite-rod macroions with isolated (spherical) counterions are presented. Then, the charge inversion of a rotating rod macroion of the finite length is examined in the presence of polyelectrolyte counterions. Discrete surface charges (and/or non-smooth surface) are necessary for charge inversion of the rotating rod, which tends to be oriented along the applied electric field due to induced dipole moment.

INTRODUCTION

Charge inversion is the phenomenon principally in electrolyte solvent where each macroion attracts large number of counterions so that the net charge of the aggregates becomes reversed in sign [1, 2, 3, 4, 5, 6, 7, 8, 9, 10, 11, 12]. The charge inversion phenomenon takes place when both of (i) strong electrostatic interactions and (ii) asymmetry in radii and/or valences between counterions and coions, are satisfied [9, 10]. Such conditions are satisfied in high-salt water solutions at room temperature. The phenomenon is not only an interesting physiochemical process in strongly correlated systems, but also has biological applications to the gene therapy as the means of delivering DNAs to living cells [13, 14, 15, 16].

In this article, we first review our electrophoresis study of charge inversion by molecular dynamics simulations [17, 18]. Then, we describe new charge inversion results for a rotating rod of finite length in the presence of polymer (polyelectrolyte) counterions.

Simulation method and parameters are summarized below. The simulation system contains one macroion, many counterions, coions and neutral particles, the last of which represent the solvent. The units of length, charge and mass are, a, e, and m, respectively. Our choice of the temperature corresponds to $a \sim 1.4$Å in water and $m \sim 40$ a.m.u. A macroion with radius $R_0 = 5a$ and negative charge Q_0 is surrounded by the N^+ number of counterions of a positive charge $Z^+ e$ and the N^- coions of a negative charge $-Z^- e$. The surface charge density of the macroion is between $\sigma_{sp} = 0.048 e/a^2$ (0.39 C/m^2) and $0.26 e/a^2$ (2.1 C/m^2). A rod-shaped macroion is also used whose surface charge density is equal to that of DNA, $\sigma_{rod} \approx 0.02 e/a^2$ (0.17 C/m^2). The system is maintained in overall charge neutrality, $Q_0 + N^+ Z^+ e - N^- Z^- e = 0$. The radii of counterions and coions are a^+ and a^-, respectively, with the counterion radius being fixed at $a^+ = a$, and the radius of neutral particles is $a/2$. Approximately one neutral particle is distributed in every volume element $(2.1a)^3 \approx (3\text{Å})^3$ inside the simulation domain, excluding the locations already occupied by ions. These particles are placed in a cubic box of size $L = 32a$, with periodic boundary conditions in all three directions.

The Newton equations of motion are solved for each particle with the Coulombic and Lennard-Jones potential forces under a uniform applied electric field E ($E > 0$),

$$m_i \frac{d\mathbf{v}_i}{dt} = -\nabla \Phi_i(\mathbf{r}_i) + q_i \mathbf{E} \qquad (1)$$
$$\frac{d\mathbf{r}_i}{dt} = \mathbf{v}_i$$

The potential Φ_i is the sum of the Coulombic potential

$$\phi_C = \sum_j q_i q_j / \varepsilon r_{ij} \qquad (2)$$

and the repulsive Lennard-Jones potential which takes care of the volume exclusion effect among particles,

$$\phi_{LJ} = 4\varepsilon_{LJ}[(A/r_{ij})^{12} - (A/r_{ij})^6] \qquad (3)$$

for $r_{ij} = |\mathbf{r}_i - \mathbf{r}_j| \leq 2^{1/6} A$, and otherwise $\phi_{LJ} = -\varepsilon_{LJ}$ to exclude the attraction part. Here $\mathbf{r}_i$ is the position vector of the i-th particle, and A is the sum of the radii of two interacting particles.

To account for the periodic boundary conditions for the inverse-square (Coulombic) forces, the Ewald sum needs to be taken [19]. Numerically, this is accomplished by the particle-particle-particle-mesh (PPPM) method

[20, 21]. To treat the rigid-body rotation of the rod macroion in the final section, the Euler equation with the polar coordinates is solved through the quaternion method [22].

We relate ε_{LJ} with the temperature by $\varepsilon_{LJ} = k_B T$, and choose $k_B T = e^2/5\varepsilon a$ (we assume spatially homogeneous dielectric constant ε). The Bjerrum length is thus $\lambda_B = e^2/\varepsilon k_B T = 5a$, which is 7Å in water.

The applied electric field brings in no momentum into the simulation system as it is neutral. However, it does work on the charged particles and heats the solvent through collisions. Thus, it is necessary to drain this Joule heat, and a heat bath is adopted; the macroion is located at the center of the bath at every moment. Although the heat bath generally suppresses long-range interactions, it has no side effects for the present simulation of neutral electrolyte solvent, because the velocity gradients around the macroion are electrostatically screened at short distances [23, 24, 17].

CHARGE INVERSION STUDY BY ELECTROPHORESIS

Charge inversion is the phenomenon that occurs principally without an applied electric field E. Even if the electric field is applied, charge inversion is not affected if the field is weak $E \ll Q_0/\varepsilon R_0^2 \sim 10^6$V/cm. Traditionally, the charge inversion phenomenon has been quantified by two independent criteria, (i) the integrated charge obtained from the radial distribution functions of ions [3, 8, 9, 10, 11], and (ii) the electrophoretic mobility of the macroion [17, 18]. It was found that two criteria agree well except around the threshold for charge inversion [18].

Figure 1 is a bird's-eye view of (a) all the ions and (b) ions in the vicinity of the macroion. Counterions are shown in light gray and coions in dark gray (nearly 4000 solvent neutral particles are not shown) [17]. The macroion has charge $Q_0 = -30e$ and radius $R_0 = 3a$, and counterions are trivalent. The macroion is predominantly covered by the counterions, and coions condense to the topside of the surface counterions because of repulsion from the macroion. Similarly to the case without the electric field [10], the radially integrated charge has a sharp positive peak at a distance about a from the macroion surface. This peak is due to the positive counterions being adsorbed right on the macroion surface. The peak integrated charge for this case is $Q_{peak} \approx 1.6|Q_0|$.

Figure 2 shows the drift speed V_{drift} of the macroion at steady state as a function of the applied electric field E [17]. When the onset conditions are satisfied, the drift speed becomes a linearly increasing function of the (weak) electric field. This means constant electrophoretic mobility $\mu = V_{drift}/E$, and hence, constant net inverted charge $Q^* \sim \nu\mu$, over a wide range of the parameter space (ν the solvent friction). For a very strong electric field comparable to that produced by the bare macroion charge $E \sim Q/\varepsilon R^2$, counterions and coions can no longer stably attach to the macroion. The mobility decreases with the electric field, and flips back to non-reversed [17].

There is a threshold of surface charge density of the macroion σ for charge inversion to take place [17, 18]. Typically, it is $\sigma \sim 0.05e/a^2$ (0.4C/m^2) for a spherical macroion. The corresponding correlation energy of surface counterions is

$$Z^2 e^2 / 2\varepsilon R_{WS} \sim 5k_B T, \quad (4)$$

where the Wigner-Seitz cell radius $R_{WS} = (Ze/\pi\sigma)^{1/2}$ is half the spacing between the surface counterions. The reversed mobility increases either with ionic strength and/or valence of counterions for small ionic strength and/or valence.

The asymmetry of radii and/or valences between counterions and coions causes or enhances charge inversion. Figure 3 shows the dependence of the reversed mobility against the ratio of coion to counterion radii a^-/a^+. (The normalization is $\mu_0 = v_0/(|Q_0|/\varepsilon R_0^2) \approx 21(\mu\text{m/sec})/(\text{V/cm})$ with v_0 the thermal speed of neutral particles.) With the ratio of the radii, the mobility increases linearly up to $a^-/a^+ \leq 1.5$. This is because coions with larger size geometrically compete with each other to condense on the surface counterions, which is consistent with Monte Carlo simulations of charge inversion of the finite-size coions [25] and the condensation of the $Z:1$ ions with the size asymmetry [26].

Large asymmetry of valences also causes charge inversion. In Fig.4, (reversed) mobility is plotted against the interaction energy of counterions and coions $Z^-Z^+ e^2/2\varepsilon\gamma a$. Here, counterions are either divalent, trivalent or tetravalent, and coion valence ranges from unity to five; the asymmetry becomes larger on the left-hand side of the figure [18]. Interestingly, mobilities for different valences, Z^+ and Z^-, collapse to a master curve whose index is the interaction energy. Here, the γ factor stands for screening by coion condensation onto the surface counterions, which is a general feature of charge inversion, as seen in Fig.1.

At small ionic strength, the monovalent salt enhances reversed mobility of a strongly charged macroion, as shown with filled squares in Fig.5 [18]. Monovalent salt ions fill the vacancies on the macroion not occupied by multivalent counterions due to repulsion among them. The electrostatic energy of the whole system is thereby globally minimized. It is quite remarkable that, even for the case where excess counterions are not present, $Q_0 + eZ^+ N^+ = 0$, charge inversion is invoked by the ad-

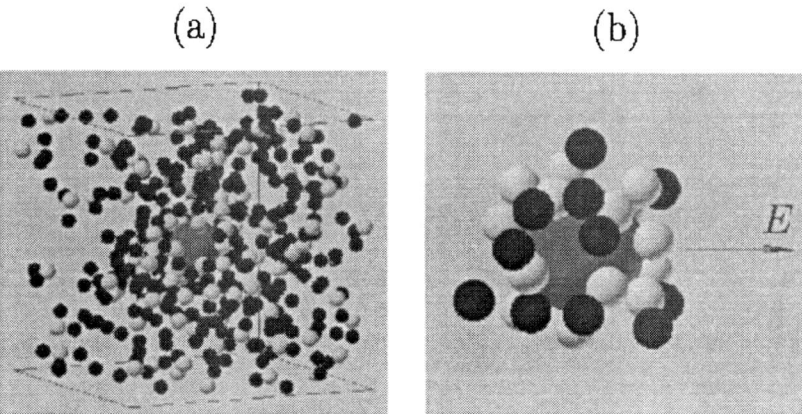

FIGURE 1. Bird's-eye view of (a) all the ions in the simulation domain and (b) the screening ion atmosphere within $3a$ from the macroion surface. A macroion with charge $Q_0 = -30e$ and radius $R_0 = 3a$ is a large sphere located in the middle; trivalent counterions and monovalent coions are shown by light and dark gray spheres, respectively. The arrow to the right shows the direction of the electric field (x-axis), with $E = 0.3\varepsilon/ea$.

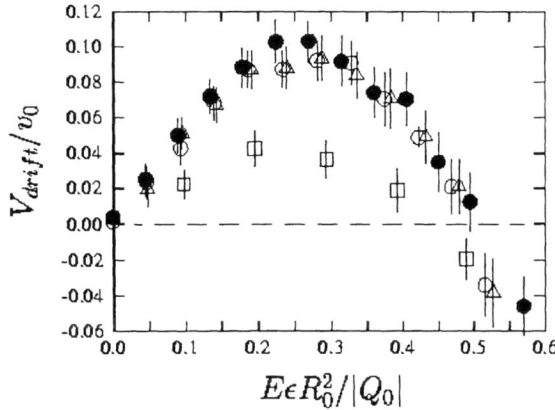

FIGURE 2. Dependence of the macroion drift speed V_{drift} (v_0: the thermal speed of solvent particles) on the electric field E for a macroion of various radii and charges. Counterions are trivalent and the Bjerrum length is $\lambda_B = e^2/\varepsilon k_B T = 5a$ where a is the radius of counterions.

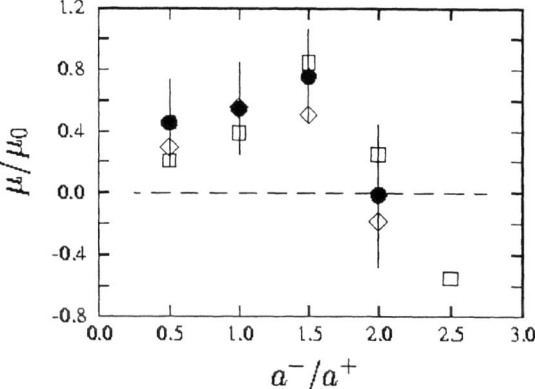

FIGURE 3. The electrophoretic mobility of the macroion μ against the ratio of coion and counterion radii a^-/a^+. Counterions are divalent (diamonds), trivalent (circles), and pentavalent (squares) with radius $a^+ = a$. The macroion radius is $R_0 = 5a$ and surface charge density $\sigma_{sp} \sim 0.26e/a^2$ (2.1 C/m²). The Bjerrum length is $\lambda_B = 5a$.

dition of monovalent salt when the macroion is strongly charged (open squares). Coexistence of multivalent and monovalent counterions should be energetically more favorable to achieve charge inversion. On the other hand, either at large ionic strength or for a weakly charged macroion (filled square), monovalent salt simply screens and suppresses charge inversion.

An infinite rod macroion shows similar but somewhat different dependences on the monovalent salt. Two different settings are compared: (i) a cylindrical macroion with polyelectrolyte counterions and (ii) a spherical macroion with isolated counterions, for the same surface charge densities. (The rod moves in the (x,z) space with its axis oriented perpendicularly to the applied electric field, which should give rise to maximum mobility compared to a rotating rod.) At zero salt, they both have the same mobility, while for finite ionic strength the former is more persistent to added monovalent salt than the latter [18]. Furthermore, it was demonstrated that the mobility of an elongated macroion can be reversed or enhanced by mechanical twining of polyelectrolyte counterions around the rod axis [27]. These findings support the advantage of using polyelectrolyte counterions such as spermidine and spermine in charge inversion of biological matters including DNA.

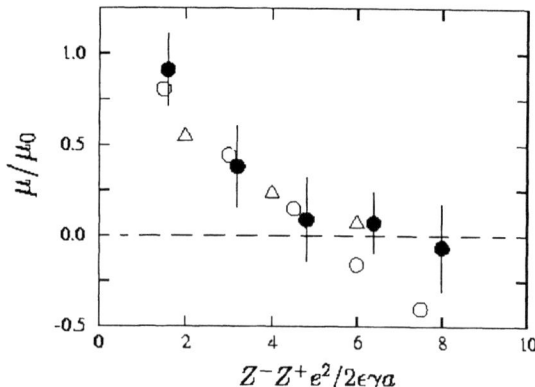

FIGURE 4. The macroion mobility is shown against the interaction energy of counterions and coions $Z^+Z^-e^2/2\epsilon\gamma a$, for the fixed counterion valences, $Z^+ = 2$ (triangles), $Z^+ = 3$ (open circles), and $Z^+ = 4$ (solid circles). The γ factor is 1.0, 2.0 and 2.5 for $Z^+ = 2$, 3 and 4, respectively. Coions and counterions are of the same radius, $a^- = a^+ = a$, and the temperature is $e^2/\epsilon a k_B T = 5$.

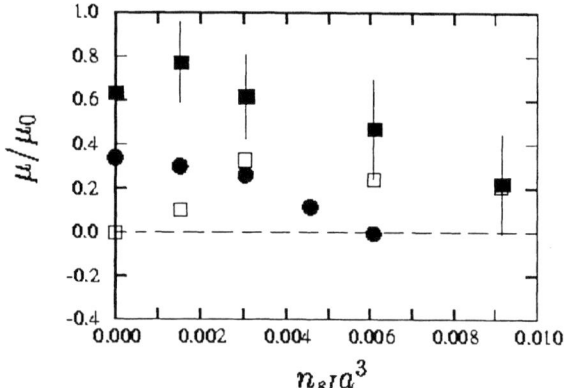

FIGURE 5. The electrophoretic mobility of the spherical macroion is shown against ionic strength of monovalent salt n_{sl} (1 Mol/l salt corresponds to $0.0017/a^3$). The surface charge density of the macroion is $\sigma_{sp} = Q_0/4\pi R_0^2 \sim 0.26 e/a^2$, with excess Z-ions (filled squares), and without excess Z-ions (open squares). Also, the mobility for the macroion $\sigma_{sp} \sim 0.080 e/a^2$ with excess Z-ions (filled circles) is shown. Here, $\mu_0 \approx 21 (\mu m/sec)/(V/cm)$.

ROTATING ROD OF FINITE LENGTH

The charge inversion of an infinite rod macroion was studied by molecular dynamics simulations [28, 18]. The effect of finite length was examined for a static (perpendicular) rod previously [18]. It was shown that the rod length ℓ does not affect the mobility if $\ell/R \geq 3$ (R: the rod radius). In the below, a macroion is a finite-length rod ($\ell = 16a$, $R = 4a$) that can rotate as a rigid body. The macroion charge $Q_{rod} = -14e$ is provided by localized discrete unit charges aligned on the double helices, which are shown with white spheres in Fig.6. The average surface charge density is chosen equal to that of the DNA $\sigma_{DNA} \sim 0.023 e/a^2$ (0.19 C/m^2), and the initial rod axis is placed perpendicularly to the electric field ($\Theta = 90°$). The number of coions is nearly fixed $N^- = 31$, which determines that of counterions by charge neutrality. Coion radius is larger than that of counterions, $a^- = 1.5 a^+$. The applied electric field is weak compared to the self electric field produced by the rod macroion, $E_{rod} \sim Q/2\pi R\ell$, so that the electrophoresis occurs well in the linear regime of Fig.1.

When the counterions are made of polyelectrolyte, where each chain consists of three trivalent monomers $3e - 3e - 3e$, the peak integrated charge far exceeds bare macroion charge in the top panel of Fig.8(a), $Q_{peak}/|Q_0| \sim 2.5$, implying strong adsorption of polyelectrolyte counterions to the macroion, and therefore charge inversion. This is verified in the radial distribution functions (RDF) of Fig.7(a), which depicts the charge densities of counterions (open bars) and coions (shaded bars), and that of the integrated charge (inset panel). Also in the dynamical criterion, the mobility of the macroion is reversed on time average, as seen for the drift speed in the bottom panel of Fig.8(a).

The angle Θ between the rod axis and the electric field in the middle panel of Fig.8(a) shows that the rod axis tends to be oriented parallel to the electric field, similarly with the case of a very strong electric field [29]. This arises from minimization of the electrostatic energy for the macroion with the induced dipole moment due to adsorbed counterions and coions. Namely, more positive (negative) ions tend to stay at the positive (negative)-x side of the rod which rotates the rod if it is not aligned parallel to the applied electric field. The dipole moment along the rod axis is measured. For the instantaneous value, large fluctuations preclude the detection of the dipole moment. However, the time integration clearly shows the buildup of the constant and negative dipole moment pointing along the rod axis, whose time history is closely analogous to that of the rod angle in the middle panel of Fig.8(a).

We note in passing that, for the rotating rod, the surface charge of the macroion needs to be localized and discrete to pin down the counterions. Otherwise for the uniform surface charge and the smooth cylinder, the counterions and coions slip on the macroion surface and easily desorbed from the tips of the rod. This strongly suggests the importance of the actual non-smooth surface of the macroion such as the DNA, which is characterized by side groups branches and specific *ion traps*. Indeed, the inclusion of the attractive Lennard-Jones potential $\epsilon_{LJ} = k_B T$ in the molecular dynamics simulation makes the mobility reversed for the rod macroion with DNA's surface charge density [27].

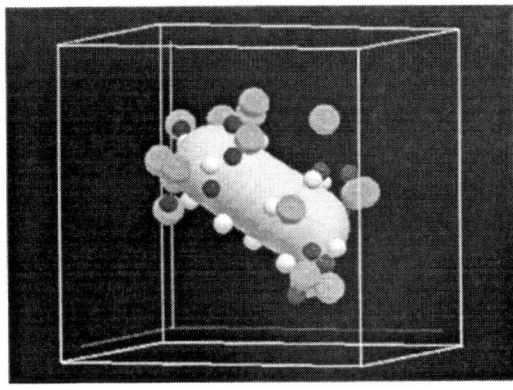 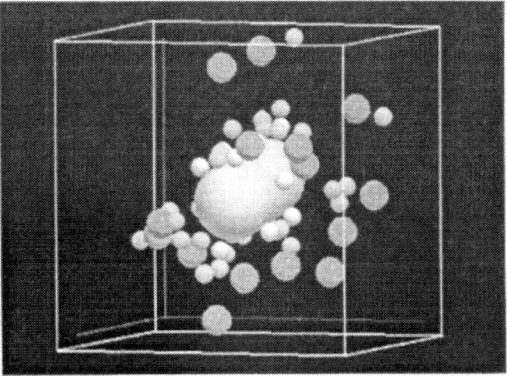

FIGURE 6. The bird's-eye view of the rotating rod macroion of the finite axis-length corresponding to the runs in Fig.8(a) (left) and Fig.8(b) (right). Discrete macroion surface charges (unit charge e) are shown with white spheres, counterions with dark gray spheres and coions with large spheres. The applied electric field points rightward along the box edge.

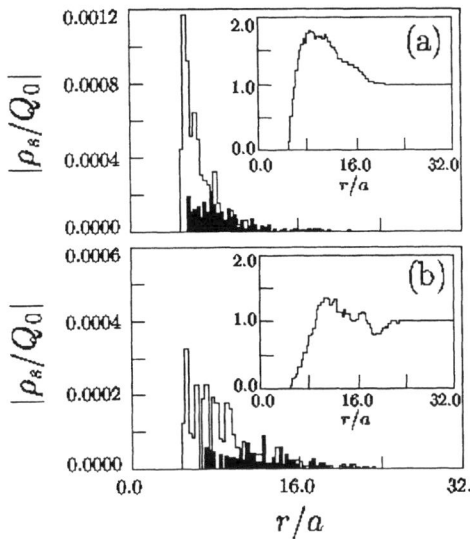

FIGURE 7. Radial distribution functions of charge density of counterions (open bars) and coions (shaded bars), and the integrated charge (inset panel) for the rod macroions in Fig.8. The finite-length rod macroion is put in the solution of polyelectrolyte counterions, each chain of which consists of three (a) trivalent, and (b) monovalent monomers.

On the other hand, for the polyelectrolyte counterions consisting of all unit charges $e-e-e$ with other conditions fixed, the mobility is not reversed. The adsorption of counterions to the very surface of the macroion is not significant either, as seen in Fig.6(b), which is just around the charge neutralization level $Q_{peak}/|Q_0| \sim 1$ (see Fig.8(b)). In this case, the rod does not rotate but stays around the initial angle. Although the RDF of the integrated charge in Fig.7(b) (inset panel) is peaked above neutrality, it is associated with an overshoot to below unity after the primary peak. The macroion with this type of the RDF in *charge inversion* usually shows non-reversed mobility, including the case of spherical macroions [18].

ACKNOWLEDGMENTS

It is a great pleasure of the author to thank Prof. A.Yu.Grosberg for a series of close collaborations on molecular dynamics study of charge inversion. This work was supported by the Grant-in-Aid No.15035218 (2003) from the Ministry of Education, Science and Culture of Japan. The computation was performed mainly with the supercomputers of the University of Minnesota Supercomputing Institute.

REFERENCES

1. Grosberg, A.Yu., Nguyen, T., and Shklovskii, B., *Reviews Modern Phys.*, **74**, 329–345 (2002).
2. Quesada-Perez, M., Gonzalez-Tovar, E., Martin-Molina, A., Lozada-Cassou, M., and Hidalgo-Alvarez, R., *ChemPhysChem.*, **4**, 234–248 (2003).
3. Gonzales-Tovar, E., Lozada-Cassou, M., and Henderson, D.J., *J. Chem.Phys.*, **83**, 361–372 (1985).
4. Elimelech, M., and O'Melia, C.R., *Colloids Surface*, **44**, 165–178 (1990).
5. Bastos, D., and De Las Nieves, F.S., *Colloid Poly. Sci.*, **271**, 860–867 (1993).
6. Walker, H.W., and Grant, S.B., *Colloids Surfaces*, **A119**, 229–239 (1996).
7. Netz, R.R. and Joanny, J.F., *Macromolecules*, **32**, 9013–9026 (1999).
8. Nguyen, T., Grosberg, A.Yu., and Shklovskii, B.I., *Phys.Rev.Lett.*, **85**, 1568–1571 (2000).
9. Messina, R., Holm, C., and Kremer, K., *Phys.Rev. Lett.*, **85**, 872–875 (2000).
10. Tanaka, M., and Grosberg, A.Yu., *J.Chem.Phys.*, **115**, 567–574 (2001).

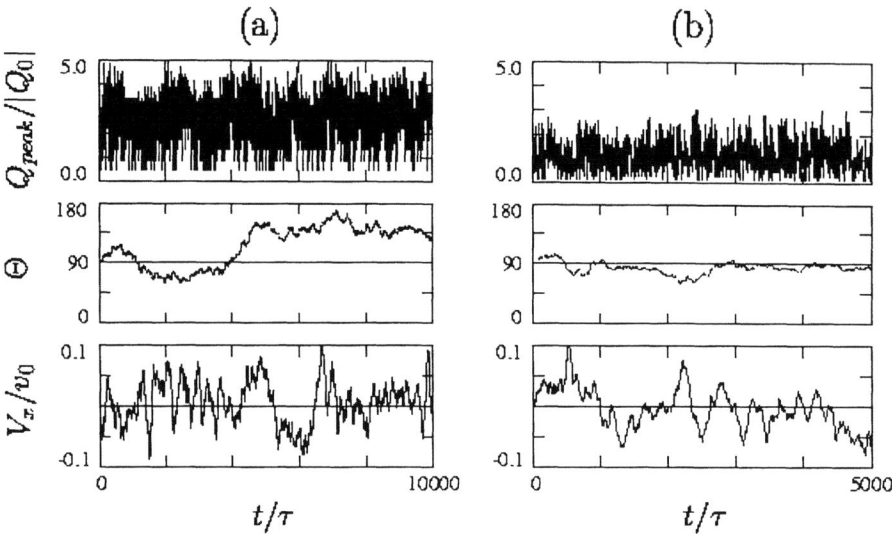

FIGURE 8. Time histories of the integrated peak charge Q_{peak}, the angle between the rod axis and the electric field Θ, and the drift speed of the macroion V_x, from top to bottom. The macroion is the finite-length rotating rod shown in Fig.6, the counterions are polyelectrolyte, each of which is a chain of three (a) *trivalent* and (b) *monovalent* counterions, and the coions are spheres of large radius. The surface charge density of the macroion is chosen the same as that of the DNA, which is provided by localized discrete unit charges aligning along double helices on the rod. The electric field is weak, (a) $E = 0.02\varepsilon/ea$ and (b) $E = 0.005\varepsilon/ea$. The time normalization is $\tau \approx 1$ ps.

11. Lozada-Cassou, M., and Gonzales-Tovar, E., *J. Colloid Interf.Sci.* **239**, 285–295 (2001).
12. Levin, Y., *Rep.Prog.Phys.*, **65**, 1577–1632 (2002).
13. Sukhishvili, S.A., Obolskii, O.L., Astafieva, L.V., Kabanov, A.V., and Yaroslavov, A.A., *Vysokomol.Soed.*, **35**, 1895–1899 (1993) [*Polymer Sci.*, **A35**, 1895–1899 (1993).]
14. Kabanov, A.V., and Kabanov, V.A., *Bioconj.Chem.*, **6**, 7–20 (1995).
15. Gelbart, W.R., Bruinsma, R., Pincus, P., and Parsegian, A., *Physics Today*, **53**, 38-44 (2000).
16. Ewert, K., Ahmad, A., Evans, H.M., Schmidt, H.W., and Safinya, C.R., *J.Med.Chem.*, **45**, 5023–5029 (2002).
17. Tanaka, M., and Grosberg, A.Yu., *Euro.Phys.J.*, **E7**, 371–379 (2002).
18. Tanaka, M., *Phys.Rev.E*, **68**, 061501 (2003).
19. Ewald, P.P., *Ann.Physik*, **64**, 253–287 (1921).
20. Eastwood, J.W., and Hockney, R.W., *J.Comput.Phys.*, **16**, 342–359 (1974).
21. Deserno, M., and Holm, C., *J.Chem.Phys.*, **109**, 7678–7693 (1998).
22. Evans, D.J., *Mol.Phys.*, **34**, 317–325 (1977).
23. Long, D., Viovy, J.-L., and Ajdari, A., *Phys.Rev.Lett.*, **76**, 3858–3861 (1996).
24. Viovy, J.-L., *Rev.Mod.Phys.*, **72**, 813–872 (2000).
25. Greberg, H., and Kjellander, R., *J.Chem.Phys.*, **108**, 2940–2953 (1998).
26. Panagiotopoulos, A.Z., and Fisher, M.E., *Phys.Rev.Lett.*, **88**, 045701 (2002).
27. Tanaka, M., *J.Phys.: Condensed Matters* (submitted) (2003); *cond-mat* /0311009 (2003).
28. Deserno, M., Jimenez-Angeles, F., Holm, C., and Lozada-Cassou, M., *J.Phys.Chem.*, **B105**, 10983–10991 (2001).
29. Netz, R.R., *Phys.Rev.Lett.* **90**, 128104 (2003).

Dynamics of Protein Extraction and Extension by Force Spectroscopy and Molecular Dynamics Simulation

Atsushi Ikai, Rehana Afrin, Rukman Hertadi, and Satoko Ohta

Laboratory of Biodynamics, Graduate School of Biodynamics and Biotechnology, Tokyo Institute of Technology, Nagatsuta, Midoriku, Yokohama, Japan

Abstract. Dynamics of protein stretching and protein extraction depends on the force loading rate but the molecular dynamics simulations in pico to nano second range seem to be able to reproduce some of the main feature of experimental results done at 9 orders of magnitude slower force loading rate. This is due to relatively large values of Δx, the activation distance, of the structure to be broken. Consequently, major brittle structures in a protein molecule observed in molecular dynamics simulation remain brittle in actual experiment done at much slower loading rate. Extraction of protein molecules from lipid bilayer is an extreme case where the force loading rate dependence of the force is, in a relative scale, almost negligible for a wide range of loading rate difference.

INTRODUCTION

The atomic force microscope (AFM) allowed us to record the force vs. extension relationship of single protein molecules during their forced unfolding from the globular native conformation to a perfectly extended one in liquid medium. The first such experiment was performed by Mitsui et al. [1] which was closely followed by many attempts to stretch various protein molecules [2,3,4,5,6,7]. This paper summarizes our most recent work on mechanical stretching of globular protein molecules and application of such knowledge to the extraction of membrane proteins using covalent crosslinkers. Our recent work is concerned with mechanical extension of carbonic anhydrase and OspA and revealed the presence of one or more brittle structures within globular protein molecules. Preliminary results of molecular dynamics simulations supported such experimental observations.

METHODS AND MATERIALS

Proteins: Bovine carbonic anhydrase II and the lyme disease related protein OspA were derivatized so that both proteins had single cysteine residues at their N- and C-termini in addition to N-terminal his-tag sequences to facilitate their purification on a Ni-chelating chromatography. Cysteine residues were used to covalently immobilize proteins between the substrate and the AFM tip for protein stretching.

AFM experiments: A Nanoscope IV multimode AFM (Veeco Japan, Tokyo) was used to stretch protein molecules which were sandwiched between the AFM tip and the substrate both of which were silanized and chemically modified to facilitate chemical crosslinking them with protein molecules having –SH groups. Details have been given previously [4].

RESULTS AND DISCUSSION

Figure 1 shows the relationship between the tensile force and the extension of the proteins called OspA as reported in [7]. One of the characteristic features of the curves is the presence of two force peaks. Such force peaks in the force-extension curve represent breakdown phenomena of brittle structures at the threshold magnitude of tensile force corresponding to the maximum of each force peak. The tensile force curve from the start of extension up to the first force peak represents forced extension of non-brittle substructure of the proteins provided that the proteins retained their native conformation before tensile deformation was started. The straight line after each force peak corresponds to the jump of the cantilever as a response to sudden unfolding of a brittle substructure, and the subsequent monotonously increasing curve up to the next force peak is the mechanical response of unfolded chain. Thus the force curve in Fig. 1 revealed the presence of substructures of different mechanical properties, one continuously stretchable and the other is less easily stretchable and breaks as a unit with a threshold force. The latter type of substructure is in another word brittle in nature.

Although much has been discussed about the importance of softness or local softness of a protein molecule in its functional expression, detection of substructures with distinct mechanical softness has been first reported in this and previous ones on bovine carbonic anhydrase II and calmodulin, both from our laboratory. Mechanical unfolding work on titin and other tandemly repeated structure of many globular units has shown the brittle nature of constituent subunits. Since the mechanical properties are, unlike thermodynamic properties, time dependent measurables, it is interesting to ask what is the structural basis for the co-presence of brittle and non-brittle substructures in a single protein molecule. It is intuitively straightforward to relate a mechanically brittle structure to a thermodynamically cooperative one but a brittle response may arise from two mechanisms, one a truly cooperative, crystal like substructure and the other due to the presence of highly localized but strong interactions such as disulfide bonds acting as a "gate keeper" for the rest of the structure. In some cases, orientation of anti-parallel β-sheet has been proposed to be a deciding factor of brittleness in mechanical unfolding [8].

Since the thermodynamic structural transitions of the proteins mentioned above as studied in solution are all highly cooperative, the observed clear demarcation of brittle and non-brittle substructures is intriguing. Possibility that the proteins were half unfolded before stretching is unlikely because, at least in the case of bovine carbonic anhydrase II, it was shown that the protein retained its inhibitor binding capacity in AFM experiments. In the case of OspA, cyclic repetition of approach and extension showed reformation of brittle and non-brittle structures (Fig. 1). The rest of the discussion will be based on the assumption that the proteins retained their native conformations, or at least the substantial part of them, before the force curve measurement.

Experimentally, much attention has been paid to brittle breakdowns of globular units in repeating substructures of titin and other tandem proteins and in the case of titin, the presence of two polypeptide strands in anti-parallel β-sheet configuration that are pulled in opposite directions in a shearing manner has been identified more or less as the "gate keeper" for the breakdown of the rest of the structure. Besides randomly coiled polypeptides after the breakdown of brittle structures, non-brittle breakdown has been observed for non-native form of bovine carbonic anhydrase II having almost indistinguishable CD spectrum from the native form (type II in ref.[5]) and purely α-helical polyglutamic acid [9] and poly-alanine based polypeptide [Afrin and Ikai, to be published] has been observed. A common feature for all of these is a lack of tertiary structure. breakdown has been observed for non-native form of bovine carbonic anhydrase II.

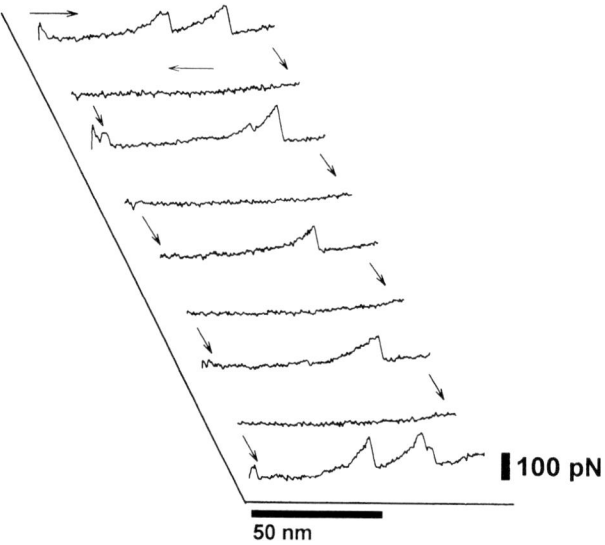

FIGURE 1. Force extension curves of OspA during four cycles of extension and retraction of the same polypeptide chain (reprinted from [7] ©2003 with permission from Elsevier).

Molecular dynamics simulation of bovine carbonic anhydrase II stretching is now almost completed in our laboratory and the result indicates that the experimentally observed major brittle fracture corresponds to a breakdown of the central β-sheet core which has been known to be a thermodynamically highly stable structure. In the case of OspA, previous solution studies suggested that there are two almost independently folded globular regions in a molecule separated by a flat sheet of central β-sheet array. These examples suggest that mechanically brittle substructures correspond to local substructures having extra thermodynamic stabilities within a globular molecule. Other factors such as shearing separation of anti-parallel β-sheet may also play an important role, but pure shearing proposed as a model in computer simulation may be rare in actual cases, because, unless the two polypeptide strands are strongly fixed in their strictly parallel positions assisted by a tight local folding, diagonally applied tensile force at the opposing ends of the two strands should reorient them so that there is always a component of force acting to separate them laterally. Lateral separation of strands is expected to require much less force than longitudinal, *i.e.*, shearing separation. In any case, we think involvement of local tertiary folding of the polypeptide chain is an important factor for the appearance of force peaks in the force extension curve.

In our molecular dynamics simulation of bovine carbonic anhydrase II stretching, we observed appearance of several force. Experimentally, Alam et al. reported the appearance of at most a single force peak [5] but in our preliminary work in a mildly denaturing condition of 2 M guanidinium chloride, a known denaturant of proteins, we observed force curves represented by a curve with at least two force peaks. We think that most global features of experimental force curves can be reproduced by molecular dynamics simulations despite a large, up to 10^9 times, difference in time scale. This is because peak force depends linearly on the logarithm of the force loading rate and the slope is proportional to $k_B T/\Delta x$ where k_B, T, and Δx are: the Boltzmann constant, temperature in K and the activation distance of the bond to be ruptured [10]. If we adopt a typical value of Δx for mechanical unfolding of proteins, the slope will be as small as 20-30 pN per one order change in the force loading rate. For a 10^9 times difference in the loading rate, the estimated change in the force would be at the most 300 pN. This is a large change but if the force to rupture brittle structure of a protein is in several hundreds to 1000 pNs, we should expect that similar rupture events might be observed in experiment and in molecular dynamics simulations.

When the force spectroscopy was applied to extract membrane proteins from a live cell surface, we observed that the force required for uprooting membrane proteins distributed from 200-2000 pN with the most frequent values in the range of 400-600 pN [11] (Fig.2). The dependency of the extraction force on the force loading rate was small verifying that the observed events corresponded to protein extraction from lipid bilayer membrane of several nm in thickness as anticipated by Bell [10]. Although extraction of intrinsic membrane proteins from a live cell surface involves not only separation of the proteins from lipid bilayer but also detachment of them from all possible associations with intracellular components such as cytoskeletons, the observed force values were in the right range expected for protein extraction. In this case the expected slope of the force vs. force loading rate line would be as small as 2-3 pN per one order difference in the loading rate and at most 30 pN difference in nine order difference. In general, dynamic events involving deformation of proteins and biological structures have small dependence on the force loading rate and the force measured in ordinary experimental setups applies to biological time scale and even to that of molecular dynamics simulations.

A more precisely designed experiment using

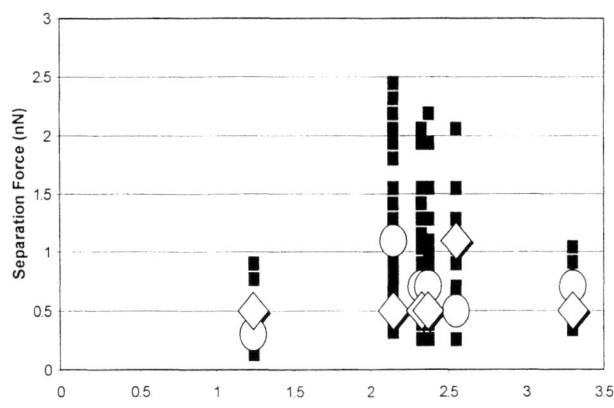

FIGURE 2. Dependence of separation force (extraction force of membrane proteins) on the force loading rate (reproduced from ref. [11] with permission).

artificial lipid vesicles with embedded proteins shouldyield conclusive force values for protein extraction from a lipid bilayer membrane.

REFERENCES

1. Mitsui, K., Hara, M. and Ikai, A., *FEBS Letters*, **385** 29-33 (1996).
2. Zlatanova, J., Lindsay, S.M., Leuba, S.H. *Prog. Biophys. Mol. Biol.* **74**, 37-61 (2000).
3. Rief, M., Pascual, J., Saraste, M. and Gaub, H.E. *J. Mol. Biol.* **286**, 553-561 (1999).
4. Wang, T. and Ikai, A. *Jpn. J. Appl. Phys.* **38**, 3912-3917 (1999).
5. Alam, M.T., Yamada, T., Carlsson, U. and Ikai, A. *FEBS Lett.* **519**, 35-40 (2002).
6. Hertadi, R. and Ikai, A. *Protein Sci.* **11**, 1532-1538 (2002).
7. Hertadi, R. Gruswitz F, Silver L, Koide A, Koide S, Arakawa H, and Ikai A. *J. Mol. Biol.* **333**, 933-1002 (2003)
8. Rohs, R., Etchebest, C., and Lavery, R., *Biophys. J.* **76** 2760–2768 (1999).
9. Idiris, A., Alam, M.T., and Ikai, A., *Engineering* **13** 763-770 (2000)
10. Bell, G.I. *Science* **200,** 618-627 (1978).
11. Afrin, R., Arakawa, H., Osada, T., and Ikai, A., *Cell Biochem. Biophys.* **39** 101-117 (2003).

A Brownian Ratchet Model of Actin Polymerization Motor by using Extended Scaled Particle Theory

Masayuki Irisa

Faculty of Computer Science and Systems Engineering, Kyushu Institute of Technology, Iizuka, Fukuoka, 820-8502 JAPAN

Abstract. One of the types of movements, e.g. protoplasmic movement, observed in cells are caused by the polymerization of the actin molecules. Recently, attempts to reveal the mechanism of force generation by using a Brownian ratchet model were reported. The load, e.g. the cell membrane, is pushed by the actin polymerization occurred at the head (barbed end) of the filament close to the load. Relative position between the head of the actin filament and the load are fluctuated as the Brownian motion. We focused on the association of the actin monomer at the head of the actin filament. To simplify the problem, association of the two actin monomers are modeled as the hard fused spheres and studied theoretically by using the extended scaled particle theory (XSPT) which we have developed. The mean force between two molecules is attractive as the result of the excluded volume effect.

Our results by using XSPT show the probability that the force of the actin filament against the load is determined not only by the fluctuation of position of the head of the actin filament but also by the fluctuation of position of the surrounding free actin monomers moving around the head of the filament. This suggests that thermal motion of the surrounding free actin monomers is also the origins of the force of the actin polymerization motor through the excluded volume effect.

INTRODUCTION

One of the types of movements, e.g. protoplasmic movement, observed in cells are caused by the polymerization of the actin molecules [1] [2]. In addition, an artificial system of liposomes including actin molecules, ATP, and salt solution is able to work as the actin polymerization motor [3]. Recently, attempts to reveal the mechanism of force generation by using a Brownian ratchet model were reported [4] [5]. Experimental results support the fact that the released chemical energy from hydrolysis of nucleotides, ATP, is not directly related with the generation of the mechanical force. Chemical energy is used to orient the direction of the force [1]. The load, e.g. the cell membrane, is pushed by the actin polymerization occurred at the head of the filament close to the load. Relative position between the head of the actin filament and the load are fluctuated as the Brownian motion. We focused on the association of the actin monomer at the head of the actin filament. It is well known that the actin filament exists only in the salt solution. Hydrophobic interaction, especially core repulsion, is considered for the actin association in this study.

Generally, the intermolecular interaction between macromolecules affects the association. One of the interactions between macromolecules is the attraction through the excluded volume effect. Asakura-Oosawa proposed the simple equation (AO theory) to estimate the attraction [6]. Their theory is well agreed with the experimental results in spite of the usage of rough assumptions [7, 8, 9]. In the AO theory, the intermolecular force is derived from the geometric consideration of the excluded volume of the macromolecule and the microscopic osmotic pressure at the molecular size. On the other hand, we have recently extended the widely used solution theory, scaled particle theory (SPT) [10], to the arbitrary shaped solute molecule [11][12]. The extended SPT (XSPT) has been used in the calculation of the solvation free energy of the protein molecule [13].

The XSPT can also be used to calculate the excluded volume effect of a suspended macromolecule against the surrounding free macromolecules moving around. The excluded volume effect of the head of the actin filament by the surrounding free actin monomers in solution is calculated in this study. In the calculation of the excluded volume effect, core repulsion is treated as the direct interaction between the molecules following the manner in the studies of "macromolecular crowding", "depletion force", or "attraction through the excluded volume effect". As the result of excluded volume effect, the mean force between two molecules is attractive.

The Brownian ratchet model of actin polymerization proposed by Oster [4][5] is that the actin monomer becomes to be the pawl by binding to the head of the actin filament as the stochastic process. The head of the filament close to the load is bending to "bite" the actin monomer by the thermal fluctuation.

METHOD

Solution Theory

In the case of the interaction of two macromolecules close to each other in the macromolecular solution, the excluded volume of the two macromolecules can be regarded as one cavity where the center of the surrounding macromolecule cannot enter. Then the attractive force through the excluded volume effect can be interpreted as the minimization process of the work required to make the cavity in the macromolecular solution. Usually in the XSPT, the contribution from the repulsive force between the solute and the solvent is calculated as the work to make the cavity in the solvent. When the XSPT is applied to the macromolecular solution, it is assumed that the degrees of freedom of the solvent molecules are neglected by following "primitive model" in the traditional polymer/colloid theory. The basis of the XSPT is the theoretical derivation of the microscopic pressure on the solute molecule by using the statistical mechanics.

We focused on the association of the actin monomer at the head of the actin filament. To simplify the problem, association of the two actin monomers are modeled as the hard fused spheres and studied theoretically by using XSPT.

Extended Scaled Particle Theory

The solvation free energy of the fused hard spheres in the dilute hard sphere solution, g_C, is [11][12]

$$g_C = A + B + \frac{1}{2}C + PV_c(1) \qquad (1)$$

$$A = -k_B T \ln(1 - \rho V_c(0)), \quad B = \frac{k_B T}{1-\rho V_c(0)} \rho (\frac{\partial V_c}{\partial \lambda})_{\lambda=0},$$
$$C = \frac{k_B T}{1-\rho V_c(0)} \rho (\frac{\partial^2 V_c}{\partial \lambda^2})_{\lambda=0} + \frac{k_B T}{(1-\rho V_c(0))^2} \rho^2 \left((\frac{\partial V_c}{\partial \rho})_{\lambda=0}\right)^2.$$

where $V_c(\lambda)$ is the excluded volume of the λ-fold scaled solute, P is the pressure of the system, k_B is the Boltzmann constant, T is temperature, ρ is the number density of the solvent. The value of $V_c(0)$ is equal to the volume of the solvent molecule. The Equation (1) uses charging formula in the statistical mechanics by scaling the solute molecule which result the changes of the shape of the corresponding cavity in the hard sphere solvent.

The potential of mean force

The XSPT can be used as the more precise version of the AO theory when the size ratio between the focused macromolecule and the surrounding macromolecule in the solution is nearly equal. We found that the XSPT has a term which converges to the AO theory when the suspended two macromolecules are extremely larger than the other macromolecules.

This system can be modeled by following the traditional "primitive model" in the polymer/colloid theory. The solution of the system is composed of two hard fused spheres as two molecules and hard spheres as surrounding molecules. The positions and orientations of two molecules are suspended and the surrounding hard spheres in the liquid state are moving around. In this work, the interaction through the excluded volume effect between the macromolecules are calculated by using XSPT. The work required to suspend two macromolecules, W, is

$$W = A + B + \frac{1}{2}C + PV_c(1), \quad \eta = \frac{4}{3}\pi r_v^3 \rho \qquad (2)$$

$$\frac{P}{\rho k_B T} = \frac{1 + \eta + \eta^2}{(1-\eta)^3} \qquad (3)$$

$$V_c(1) = \frac{1}{6}(\frac{\partial^3 V_c}{\partial \lambda^3})_{\lambda=0} \qquad (4)$$

where P is the osmotic pressure and calculated by using the equation of state derived by SPT for the hard sphere liquid [10], r_v is the radius of the surrounding macromolecule moving around, A, B, and C are the same expression in Equation (1). But the ρ and $V_c(\lambda)$ are not the same as those in Equation (1). The ρ in Equation (2) is the number density of the surrounding macromolecules. The $V_c(\lambda)$ in Equation (1) is the excluded volume of the solute molecule against the solvent molecule in the dilute solution, but the $V_c(\lambda)$ in Equation (2) is the excluded volume of the "two" macromolecules against the surrounding macromolecules. The term C in Equation (2) is roughly proportional to the surface area of the cavity made by two macromolecules. The term, $PV_c(1)$, in Equation (2) converges to the AO theory when the focused two macromolecules are extremely larger than the other macromolecules. The Equation (4) is the modification of the interpretation of the term $PV_c(1)$ in order to derive the radial distribution function of hard spheres by using XSPT. We do not go into detail of this modification in this paper. When the direct intermolecular attraction like Lennard-Jones type interaction decreases the osmotic pressure, the term C in Equation (2) becomes to be important. Effect of the terms is roughly proportional to the accessible surface area.

Configuration of two actin molecules

The arrangement of the actin filament can be viewed as a one-stranded left-handed helix. Two configurations of the two suspended actin monomers close to each other chosen from the model structure [14] of the actin filament which is constructed by using the experimental results of XSAS and computer modeling (Fig. 1). One configuration, Lorenz 2, is the location of molecules along the spiral path of the monomers in the filament. The other configuration, Lorenz 2' is the location of molecules along the axis of the filament. In this work, the radii of the atoms in the actin monomer have the same value, 0.19 nm. The shape of the excluded volume is the suspended fused spheres whose sphere radius has the value of atom radius plus molecule radius moving around.

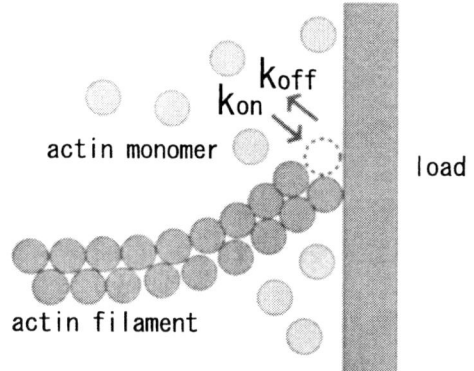

FIGURE 2. Brownian ratchet model, k_{on} and k_{off} are the kinetic rate constants of binding and dissociation.

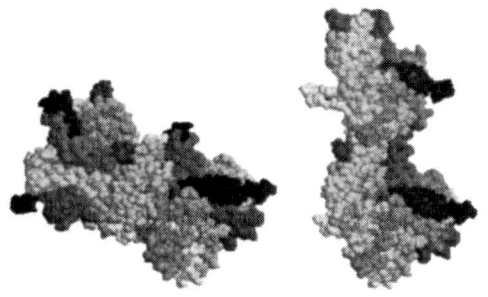

FIGURE 1. Two types of configurations, Lorenz 2 (left) and Lorenz 2'(right), of two suspended actin monomers.

Brownian Ratchet Model

Oster et al. proposed the Brownian ratchet model for the force generation by the actin polymerization motor [4][5](Fig. 2). The load is pushed by the head of the actin filament where the free actin monomer binds. Bending occurred at the actin filament by the thermal fluctuation makes the gap between the head of the actin filament and the load. The gap enables the monomer actin molecules to come in and bind to the filament head. The directly binded ATP is used to dissociate the monomer from the tail (pointed end) of the filament. The chemical energy of ATP is used to increase the concentration of the free actin monomers around the head of the actin filament. The rate of the binding of the actin monomer is slower than the rate of the bending of the filament.

RESULTS

Excluded volume of suspended two actin molecules against actin monomers moving around is calculated analytically by using the technique of alpha shape in the field of computational geometry [15]. In XSPT, the excluded volume and derivatives of excluded volumes by the scaling parameter are used to derive the work required to make a excluded volume of suspended two actin monomers against actin monomers moving around. Results of the work to suspend the two actin molecules are listed in Table 1. The number density of the free actin monomers is determined from the packing fraction η. The potential of mean force of the attraction through the excluded volume effect is listed in Table 2.

TABLE 1. Work ($\frac{W}{k_B T}$) required to make a excluded volume of suspended two actins against actin monomers moving around.

η	Lorenz 2	Lorenz 2'
0.01	0.173048	0.185606
0.05	0.948918	1.016490
0.1	2.145541	2.293700
0.2	5.640930	5.996338
0.3	11.650483	12.282182

TABLE 2. Potential of mean force ($\frac{U}{k_B T}$) of two actin monomers through excluded volume effect.

η	Lorenz 2	Lorenz 2'
0.01	-0.047782	-0.035225
0.05	-0.250906	-0.183335
0.1	-0.535240	-0.387081
0.2	-1.233016	-0.877608
0.3	-2.172497	-1.540798

The absolute values of the potential of the attraction between the actin molecules are small at $\eta = 0.01$ because the size of the focused and surrounding actin molecules is nearly equal. The effect of the terms except for $PV_c(1)$ in Equation (2) is negligible at low number density of the surrounding macromolecules. On the other hand, when η is large, the contribution of $PV_c(1)$ in Equation (2) is comparable to the contributions from other terms in Equation (2). These results show that the associated two actin monomers in Lorenz 2 is more stabilized than those in Lorenz 2'.

DISCUSSION

Our results show that the configuration of Lorenz 2 is more stable than that of Lorenz 2'. It is considered by the experiments that the actin dimer configuration resembles the configuration of Lorenz 2. And the trimer of the actin molecules is known as the core of the actin polymerization from the experiments. The potential of mean force of attraction between two actin molecules through the excluded volume effect is decreased depending on the concentration of the free actin monomers. Free actin monomers stabilize the two actin monomers close to each other in the configuration of Lorenz 2.

One of the key of the Brownian ratchet model in the studies of force generation by the actin polymerization is estimation of the rate of the insertion of the actin monomers into the actin filament head close to the load. In this study, we focused on the excluded volume effect on the binding of the actin monomers.

The excluded volume effect affect not only the binding of the actin monomers to the head of the actin filament but also the pushing of the actin filament against the load. The load and the head of the filament close to the load suffer the excluded volume effect by the surrounding free actin molecules. Therefore the excluded volume effect require the change of the Brownian ratchet model proposed by Oster on two points; the kinetic rate constant of binding to the head of the filament and the elastic constant of the filament. The attractive mean force through the excluded volume effect increases both the rate of the polymerization of the actin filament and the elasticity of the filament close to the load. These effect is enhanced by the concentration of the free actin monomers.

Our results by using XSPT show the probability that the force of the actin filament against the load is determined not only by the fluctuation of position of the head of the actin filament but also by the fluctuation of position of the surrounding free actin monomers moving around the head of the filament. This suggests that thermal motion of the surrounding actin monomers is also the origins of the force of the actin polymerization motor through the excluded volume effect. This work is the first step to understand the force generation of the actin polymerization motor by using the Brownian ratchet model.

ACKNOWLEDGMENTS

This work was supported, in part, from the KAKENHI (14540473) and the Research for the Future Program (JSPS-RFTF98P01101) of the Japan Society for the Promotion of Science, and from the Ministry of Education, Science, Sports and Culture in Japan.

REFERENCES

1. Theriot, J. A., *Traffic 2000* **1**, 19–28 (2000).
2. Mogilner, A., and Oster, G., *Current Biology* **13**, R721–R733 (2003).
3. Miyake, H., and Hotani, H., *Proc. Natl. Acad. Sci. USA* **89**, 11547–11551 (1992).
4. Mogilner, A., and Oster, G., *Biophysical Journal* **71**, 3030–3045 (1996).
5. Oster, G., *Nature* **417**, 25–25 (2002).
6. Asakura, S., and Oosawa, F., *J. Poly. Sci.* **33**, 183–192 (1958).
7. Gast, E. P., Hall, C. K., and Bussel, W. B., *J. Colloid Interface Sci.* **96**, 251–267 (1983).
8. Gast, A. P., Russel, W. B., and Hall, C. K., *J. Colloid Interface Sci.* **109**, 161–171 (1986).
9. Patel, P. D., and Russel, W. B., *J. Colloid Interface Sci.* **131**, 201–210 (1989).
10. Reiss, H., Fish, H. L., and Lebowitz, J. L., *J. Chem. Phys.* **31**, 369–380 (1959).
11. Irisa, M., Nagayama, K., and Hirata, F., *Chem. Phys. Letters* **207**, 430–435 (1993).
12. Irisa, M., Takahashi, T., Nagayama, K., and Hirata, F., *Molecular Physics* **85**, 1227–1238 (1995).
13. Irisa, M., Takahashi, T., Hirata, F., and Yanagida, T., *J. Mol. Liquids* **65/66**, 381–384 (1995).
14. Lorenz, M., Popp, D., and Holmes, K. C., *J. Mol. Biol.* **234**, 826–836 (1993).
15. Edelsbrunner, H., Facello, M., Fu, P., and Liang, J., "Measuring Proteins and Voids in Proteins," in *Proceedings of the 28th Annual Hawaii International Conference on System Science*, edited by IEEE Computer Society, IEEE Computer Society Press, Los Alamitos, 1995, pp. 256–264.

Modelling polymersomes: a prototype for complex cellular structures

G.J.A. Sevink* and J.G.E.M. Fraaije*

Leiden Institute of Chemistry, PO Box 9502, 2300 RA Leiden, The Netherlands

Abstract. Self-organisation of small amphiphilic molecules is a key technique in many applications of modern nano-technology. Thin polymer films have been extensively studied theoretically and experimentally because of their rich phase behaviour and use as templates in lithographic processes, optical devices and surfaces with molecular recognition capabilities. Encapsulating polymeric vesicles or polymersomes can be applied in very diverse applications ranging from drug delivery, templates for heterogeneous catalysts, aerosols and personal care products. Moreover, there is some understanding that polymersomes with internal structures can serve as a scaffold for the understanding of complex biological structures, such as the mitochondrium and other (sub-)cellular stuctures. The experimental technique of making polymersomes is relatively new, and the kinetics of their formation delicate, and often not well understood. As a result, the internal and external structures of experimental polymersomes are very diverse, and often highly depend on the method of preparation. Here we report the results of field-theoretic computer simulations of remarkable structures in dispersed droplets of a polymer surfactant. The preparation method is that of quenching a homogenous droplet of polymer surfactant in an aqueous bath. In the discussion part we shortly discuss the road ahead: the use of our method as a tool for understanding complex biological systems.

INTRODUCTION

Controlled structuring of materials on all length scales is required to tune their specific chemical and physical properties. In this respect, the complex molecular organisation and different microdomain structuring of block copolymer melts has been the focus of intense efforts over the last two decades [1]. Adding solvent to amphiphilic block copolymer melts produces an even richer zoo of marvellous structures on a supra-molecular scale of 10 to 100 nm [2, 3]. Recently, vesicles with a typical size of 10 μm made of a diblock copolymer membrane, so-called polymersomes, could be obtained in aqueous solution [4]. Polymersomes are attractive alternatives for lipid vesicles, due to their stability under varying external conditions and the fact that the building blocks (the block copolymer molecules) can be tuned with respect to desired properties. Polymersomes and polymer surfactant assemblies have many applications in soft nanotechnology, in drug delivery, templates for heterogeneous catalysts, aerosols and personal care products. Patterned nanostructures are the key in all these applications, and a better understanding of their formation is of paramount interest.

Similar as in studies of lipid aggregates, there are several experimental methods for generating polymersomes, for example by dispersing insoluble polymer surfactant through sonification, or through destabilizing a suitable surface film. As mentioned, isolated closed block copolymer membrane structures are found in various systems [4, 5]. In solutions of crew-cut amphiphilic polymers Eisenberg and co-workers have determined a entire wonderland of structures [2]. Some exhibit onion phases, others are bicontinuous. The recently discovered high-genus giant superstructures ($\gg 1$ micron) [6] have walls formed either by a double bilayer, or a tubular network with hexagonal symmetry. There are a few difficult questions in the formation of polymersomes, one of which is whether such systems are in local or global equilibrium. To what extent is the kinetic pathway responsible for the (final) complex morphologies? Eisenberg prepares his systems by a very slow quench from good solvent, with demonstrated exchange between different aggregates through fission and fusion phenomena [5]. Only recently it was acknowledged [7, 8] that the formation of such structures is a complex interplay between kinetic and thermodynamic factors.

In this paper, we trespass some of the kinetic factors by considering a quench of isolated droplets in an aqueous environment. Through self-consistent-field simulations of initially homogeneous diblock copolymer droplets we discovered a wonderland of structures depending on only two variables: the block ratio f, the ratio between the hydrophilic and hydrophobic blocks, and the initial radius of the droplet. This work can be seen as an extension of previous work on block copolymers confined in thin films [9]. However, in this case, the confinement is soft: the droplet is able to adjust its shape, in response

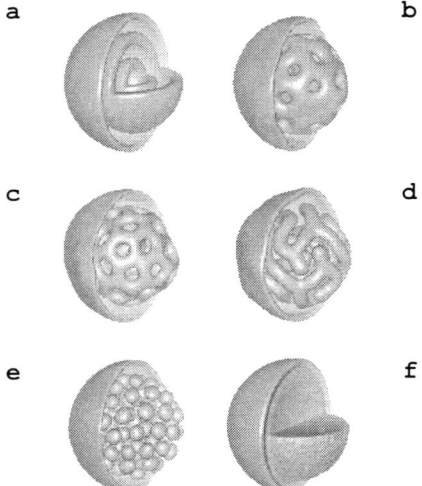

FIGURE 1. Morphologies of $A_{N-M}B_M$ polymer surfactant nanodroplets (isosurfaces partly removed for visualisation). Solvophobic A concentration field for different block ratios $f = M/N$. $0.35(a), 0.30(b), 0.25(c), 0.20(d), 0.15(e), 0.10(f)$. (Reprinted with permission from Fraaije et al, accepted as a communication by *Macromolecules*. Copyright 2003, American Chemical Society)

to the microdomain structures that are formed inside the droplet. The remarkable bicontinuous structures (Figures 1-2) can be viewed as analogons to confinement induced stable perforated lamallar structures [9, 10] in cylinder-forming triblock copolymer systems. Although quenching of monodisperse homogeneous droplets is an experimental challenge, the proposed preparation method itself is robust and produces in parallel: the self-organisation of the block copolymer inside the droplet is fully autonomous.

MODEL

The simulation parameters are for diblock polymer surfactant $A_{N-M}B_M$ with total length $N = 20$ and fraction $f = M/N$, in weakly selective solvent and mild segregation, $\chi_{AS} = 1.7$, $\chi_{AB}N = 40$, $\chi_{AS} - \chi_{BS} = 0.3$, so that A is slightly more solvophobic and B slightly more solvophilic. These are essentially the parameters we verified before by comparison with experimental microphase diagrams of concentrated Poly Propylene Oxide – Poly Ethylene Oxide aqueous solutions in ambient conditions [11], in which case each bead or statistical unit corresponds to 3 – 4 monomers. One should realize that in the mean-field model the above mentioned parameters are the only relevant parameters. Any polymer surfactant solution that can be represented by the same parameters will behave in exactly the same way.

The free energy model for a necklace of beads in a mean field environment is given by [11, 12]:

$$F[\rho] = -kT \ln \frac{\Phi_P^{n_P}\Phi_W^{n_W}}{n_P! n_W!} - \sum_I \int_V d\mathbf{r} U_I(\mathbf{r})\rho_I(\mathbf{r}) + F_{MF}[\rho].$$
(1)

Here Φ is the intramolecular partition function for a mixture of Gaussian chains and solvent, U_I is the external potential conjugate to the particle density ρ_I, V the system volume, n_p (n_W) is the number of polymer (water) molecules and $F_{MF}[\rho]$ is the mean-field contribution due to the non-ideal interactions [11, 12]. In the calculations, we first generate the initially homogenous droplet of polymer surfactant in an aqueous bath, by setting $\rho_A(\mathbf{r}) = 1 - f$, $\rho_B(\mathbf{r}) = f$ and $\rho_S = 0$ for $\mathbf{r} \leq R^0$; $\rho_A(\mathbf{r}) = 0$, $\rho_B(\mathbf{r}) = 0$ and $\rho_S = 1$ for $\mathbf{r} > R^0$, where the midpoint of the initial sphere $\mathbf{r}$ is translated to the centre of the simulation boxes. Then we use the bijective relation (3) to find, by means of an iterative steepest descent method, the correct external potential fields U_I, starting from initial values $U_I(\mathbf{r}) = 0$ in order not to bias the search. Once the corresponding fields U_I are generated, the free energy is minimized. The minimization of the free energy is by dynamic iteration, adapted from the external potential dynamics algorithm for collective Rouse dynamics that our group proposed before [12]. The dynamic equations are

$$\frac{\partial \rho_I}{\partial \tau} = \nabla^2[\rho_I - f_I[\mu_{MF,I}]]$$
(2)

where $f_I[X]$ is the density functional of the polymer (or solvent) molecule

$$f_I[X](\mathbf{r}) = \mathcal{N}\sum_s \theta_{Is} \int_{V^N} d\{\mathbf{R}\}\delta(\mathbf{r}-\mathbf{R}_s)e^{\frac{-1}{kT}(H_{id}+\Sigma_{s'}X_{s'}(\mathbf{R}_{s'}))}$$
(3)

where $\mathcal{N}$ is a normalisation constant, θ_{Is} is 1 if bead s is of type I, and 0 otherwise, the integration is over all coordinates of the chain, H_{id} is the intramolecular Gaussian chain Hamiltonian, and X is the field variable. Notice that in the proposed new density dynamics algorithm, the mean field chemical potential is used as the field variable X, so that now $X_I = \mu_{MF,I}[\rho]$, whereas we used $X_I = U_I$ in the previous algorithms [11, 12]. It can be shown that the density dynamics equations have the same kinetic coefficients as collective Rouse dynamics, which is proper for the spinodal-like microphase separation internal to the droplets. In the notation of [12], the External Potential Dynamics algorithm is obtained by left-commutation of mobility and gradient operators $\nabla \cdot P_{IJ}\nabla\mu_J \to P_{IJ}\nabla^2\mu_J$. The present Density Dynamics algorithm is obtained by right-commutation $\nabla \cdot P_{IJ}\nabla\mu_J \to \nabla^2 P_{IJ}\mu_J$, which is valid provided $\mu_J << kT$. As a technical note, the algorithm is fully explicit in the concentration variables, and thereby avoids calculation of the external potentials.

The free energy model is that for an nVT ensemble, and accordingly we do not calculate the global equilibrium in an open system, but rather a local equilibrium morphology of an isolated droplet. The situation is analogous to that of the classic study of the shape of an isolated lipid vesicle, when interactions between vesicles are less relevant. With the selected values of the Flory-Huggins parameters, and N and M, the polymers are all insoluble, hence the deformations of the droplets are at constant mass of polymer.

RESULTS

The simulations proceed by a sudden quench of a homogenous droplet in a solvent bath. Following the quench, the droplet takes up or releases additional solvent locally and globally, depending on the particular morphology being formed. The polymer concentration outside the droplet is zero, and this remains so during the morphology adaptation. The simulations are stopped when the order parameters do not change any more: the free energy is then steady in a minimum, and the pattern is a solution to the self-consistent field condition $\rho_I = f_I$. We have found it advantageous to add a small uncorrelated white noise field to the mean-field chemical potentials. The uncorrelated noise does not obey the fluctuation-dissipation theorem, but it helps small barrier crossings in the free energy landscape. The droplets (initial radius R^0 cell units, each cell has the size of the statistical unit) are placed in the centre of the box with sufficient space between the droplet surface and the boundaries of the computational box, thereby avoiding artefacts resulting from the periodic boundary conditions. In all cases the droplets develop an outer fuzzy layer of the solvophilic B block, but since the confinement of the polymers is soft, the droplet surface is not necessarily spherical. The surface topography is that of small valleys, ridges and bumps, reflecting the underlying morphology, very much like earths topography is an image of deeper events. The internal structures, shown in Figure 1, depend on the size ratio $f = M/N$ similar to bulk block copolymer systems. More symmetric polymers $f = 0.35$ form into an onion structure of alternating A and B layers (Figure 1a); slightly less symmetric polymers $f = 0.30 - 0.25$ form a bicontinuous phase (Figure 1b-c), then at $f = 0.20$ a cylindrical phase (Figure 1d) and an inverted micellar phase (Figure 1e) at $f = 0.15$. Too asymmetric polymers $f = 0.1$ do not form any internal structure (Figure 1f). In equilibrium the droplets contain an appreciable amount of solvent (ca. 15%), distributed inhomogeneously over the solvophobic A and solvophilic B-rich layers.

In the case of $f = 0.25$, Figure 2, the layers are perforated with pores, and the entire structure is bicontinuous. The droplet strikingly resembling a buckyball, with in each inner layer a mixed pore pattern of pentagons, hexagons and septagons. In bulk solution or melt systems, perforated lamellae consist ideally of a perfectly hexagonal array of pores. In the curved nanodroplets, by rule of geometry, a perfect array of hexagons is impossible to form, and the perforation is mixed. In bulk, a mixture of 85% of this particular polymer surfactant and 15% solvent forms a gyroid bicontinuous structure (data not shown).

From Figure 3, where the radial positions of internal maxima of A concentration are plotted as a function of initial droplet size R^0, it can be observed that, with increasing R^0, the position of the A shells shifts linearly. When an additional layer can be formed, the inner layers move outward. Since the domain size is constant, the pore density in each layer is constant too, and the number of pores in each shell is determined from its radius. There is no intrinsic magic '60' number, as in C_{60}, associated with additional stability of the nanodroplets. The resemblance with a carbon buckyball is purely coincidental, based on a geometrical rule for packing pores in a spherical shell.

DISCUSSION

The previous section deals with micro-structuring in isolated droplets of surfactant and focusses on the thermodynamic driving forces. The question how the formation of polymersomes takes place in general has not been addressed. Can our method be used to consider both thermodynamics and kinetics simultaneously? Yes, and no. Our method describes pattern formation by minimising the free energy, following the pathway determined by the systems natural diffusion dynamics. The method is very flexible and can access huge systems due to the parallel software; many molecular species can be easily included and different processing conditions (shear, confinements, (local) reactions) can be added. However, the dynamic factors considered in our method are limited: for instance entanglement and hydrodynamic contributions are not included. Still, we have evidence that in many cases the small-scale dynamics is less affected by these missing factors than one would think, and that the diffusive dynamics descibes the dynamics of phase transitions quite well, including many transient states. At the same time we work hard to also consider the missing factors. So, where do we stand? A first step to modelling apparent mitochondrical structures can be seen in figure 4. The system is a short symmetric diblock copolymer in

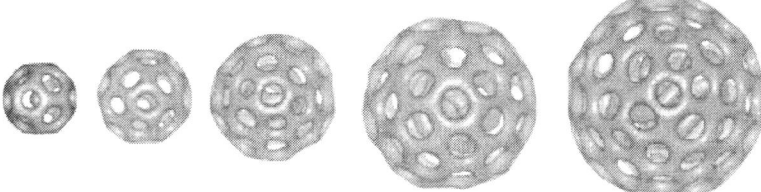

FIGURE 2. Morphologies of $A_{N-M}B_M$ polymer surfactant nanodroplet ($f = M/N = 0.25$) for different initial radii R^0. From left to right: $R^0 = 20, 23, 26, 30, 33$ (in units of polymer bead size). Notice that in this viewing mode the outer fuzzy shell is not visible. (Reprinted with permission from Fraaije et al, accepted as a communication by *Macromolecules*. Copyright 2003, American Chemical Society)

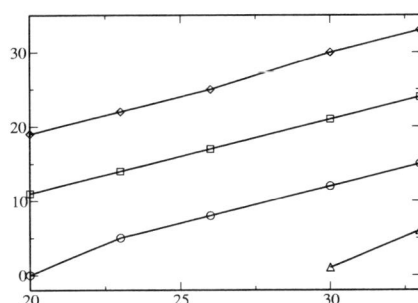

FIGURE 4. Final stage of structure formation of a A_4B_4 melt inside a single aggregate (only A isosurface is shown) after 5×10^4 dimensionless simulation steps. The diblock copolymer was initially homogeneously distributed inside a spherical droplet.

FIGURE 3. Shell position versus initial radius for different layers. (Reprinted with permission from Fraaije et al, accepted as a communication by *Macromolecules*. Copyright 2003, American Chemical Society)

bad solvent; the initial configuration is an isolated spherical droplet as before. Under the influence of the structure formation inside, the droplet elongates and forms self-shaking hands after Escher in an attempt to generate lamellae perpendicular to the polymer-water interface. We can easily add membrane forming amphiphiles and metaboles, in the form of local reactive networks. The conclusion must be that we think the tools are already there; all we lack is a data-to-model step. By trial and error, and some creative thinking, we hope to facilitate this step in the near future.

ACKNOWLEDGMENTS

The supercomputer resources were provided by the High-Performance Computing Centre, University of Groningen. We thank Andrei Zvelindovsky (Leiden University, The Netherlands) and Adi Eisenberg (McGill University, Canada) for stimulating discussions.

REFERENCES

1. Bates, F. S., and Fredrickson, G. H., *Physics Today*, **52**, 32–38 (1999).
2. Cameron, N. S., Corbierre, M. K., and Eisenberg, A., *Can. J. Chem.*, **77**, 1311–1326 (1999).
3. Alexandridis, P., and Spontak, R. J., *Curr. Opin. Colloid & Interface Sci.*, **4**, 130–139 (1999).
4. Discher, B. M., Won, Y. Y., Ege, D. S., Lee, J. C. M., Bates, F. S., Discher, D. E., and Hammer, D. A., *Science*, **284**, 1143–1146 (1999).
5. Discher, D. E., and Eisenberg, A., *Science*, **297**, 967–973 (2002).
6. Haluska, C. K., Gozdz, W. T., Dobereiner, H. G., Forster, S., and Gompper, G., *Phys. Rev. Lett.*, **89**, 238302 (2002).
7. Lasic, D. D., *Science*, **351**, 613 (1991).
8. Leng, J., Egelhaaf, S. U., and Cates, M. E., *Europhys. Lett.*, **59**, 311–317 (2002).
9. Knoll, A., Horvat, A., Lyakhova, K. S., Krausch, G., Sevink, G. J. A., Zvelindovsky, A. V., and Magerle, R., *Phys. Rev. Lett.*, **89**, 035501 (2002).
10. Huinink, H. P., van Dijk, M. A., Brokken-Zijp, J. C. M., and Sevink, G. J. A., *Macromolecules*, **34**, 5325–5330 (2001).
11. van Vlimmeren, B. A. C., Maurits, N. M., Zvelindovsky, A. V., Sevink, G. J. A., and Fraaije, J. G. E. M., *Macromolecules*, **32**, 646–656 (1999).
12. Maurits, N. M., and Fraaije, J. G. E. M., *J. Phys. Chem.*, **107**, 5879–5889 (1997).

Hierarchical Regularity in Multi-Basin Dynamics on Protein Landscapes

Yasuhiro Matsunaga*, Konstatin S. Kostov† and Tamiki Komatsuzaki*

Department of Earth and Planetary Sciences, Faculty of Science, Kobe University
†*Department of Chemistry, University of Chicago*

Abstract. We analyze time series of potential energy fluctuations and principal components at several temperatures for two kinds of off-lattice 46-bead models that have two distinctive energy landscapes. The less-frustrated "funnel" energy landscape brings about stronger nonstationary behavior of the potential energy fluctuations at the folding temperature than the other, rather frustrated energy landscape at the collapse temperature. By combining principal component analysis with an embedding nonlinear time-series analysis, it is shown that the fast fluctuations with small amplitudes of 70-80% of the principal components cause the time series to become almost "random" in only 100 simulation steps. However, the stochastic feature of the principal components tends to be suppressed through a wide range of degrees of freedom at the transition temperature.

The questions, "what kinds of mechanisms carry a protein into an unique native state", and "what is the best reaction coordinate to describe the dynamics of protein folding" have been among the most intriguing subjects over the past decades. Protein folding may be well interpreted as a normal Brownian process of a few collective coordinates on a thermodynamic potential such as the "funnel" landscape at least for small proteins[1]. The diffusive nature may, however, depend on the choice of the viewpoint from which one sees the dynamical events[2, 3]. The fraction of native contacts Q is often taken as a reaction coordinate or global order parameter. However, it is not self-evident[4] that Q is always appropriate to represent the progress of folding, and many different sets of contacts may yield the same value. There exists the nontrivial question if motions along this coordinate correspond to such a slow step that dynamical contributions of all the other degrees of freedom are averaged out, resulting in an effective single dominant free energy barrier of folding. Many non-Brownian, dynamical transitions have been reported in simulation studies on phase transitions of small clusters, the Lennard-Jones liquid, liquid water, and proteins[5]. For instance, García and Hummer[5] showed how non-Brownian, hierachical strange kinetics emerge in multi-basin dynamics in cytochrome c in aqueous solution at 300-550 K for at least 1.5 ns. For two-basin dynamics, we clarified[2, 3] that, irrespective of the system, there exists a dynamical hierarchy in the region of saddles, in which even at higher energies above the threshold where mode-mode mixing wipes out most invariants of motion, the reacting system climbs through a deterministic path along which it persists its dynamical memory buried in a stochastic sea.

The observed kinetics of protein folding is a consequence of averaging over an ensemble of many activated barrier crossings with multiple time scales. The direct observation of dynamical behavior of a single molecule, so far buried in an ensemble average, should provide us with a new magnifying glass, which enables us to "see" the dynamical structure of complex systems. The purpose of the present article is to scrutinize how time series of *scalar* quantities can shed light on the complexity of protein dynamics by means of several nonlinear time series analyses. The analyses we use here have been originally developed in the field of nonlinear science to discriminate chaotic and stochastic dynamics, and are, in principle, applicable to any time series of any observable.

THEORY

What can we learn or deduce from an (observed) scalar time series about the multivariate state (or phase) space buried in the observations? The so-called embedding theorem[6] attributed to Takens provides us with a clue to the answer of such a question. Suppose that we have a nonlinear dynamical system, i.e., first-order d-dimensional ordinary differential equations,

$$\frac{d\mathbf{x}(t)}{dt} = \mathbf{F}(\mathbf{x}(t)), \quad (1)$$

where $\mathbf{x}(t)$ and $\mathbf{F}$ are the d-dimensional vector representations of the state space variables $(x_1(t), x_2(t), \ldots, x_d(t))$ and the one-to-one map $(F_1, F_2, \ldots, F_d)$, respectively. It is supposed, that all degrees of freedom (dof) more or less influence one another through explicit or implicit couplings, and the d-dimensional manifold is compact and smooth in the state space $\mathbf{x}$. Let there exist a scalar quantity $s(t)$ with an infinitesimal precision and suppose that $s(t)$ is derived by a smooth transformation h from $\mathbf{x}(t)$, i.e., $s(t) = h(\mathbf{x}(t))$. The embedding theorem states that, in principle, from the knowledge of the infinite time series $s(t)$ an equivalent state space $\mathbf{y}(t)$ can be reconstructed preserving the differential properties of the state space of the original multivariate variables $\mathbf{x}(t)$. A time delay coordinate system $\mathbf{y}_d(n)$,

$$\mathbf{y}_d(n) = (s(n), s(n+\tau), \ldots, s(n+(d-1)\tau)) \quad (2)$$

has been often used to reconstruct the state space, where the d-dimensional state space variables $\mathbf{y}_d(n)$ are represented in a discrete form with a time interval τ without loss of generality. How can one find the *unknown* dimension d from the time series of $s(n)$? If there *actually* exists a dynamical system behind the observation of $s(t)$, any orbit $\mathbf{y}$ should never cross with itself in the state space because of the uniqueness of the solution. If a smaller dimension to reconstruct the state space is used, the orbit $\mathbf{y}$ will have self-intersections and can not be "unfolded" due to the insufficient size of the chosen dimensionality. In other words, if a minimum dimension to unfold the orbit in the time delay coordinate system is reached, this implies that a state space $\mathbf{y}$ equivalent to the original $\mathbf{x}$ can be reconstructed in such a sense that it has the same differential properties of the original manifolds. The embedding theorem holds irrespective of the choice of the delay time τ, but, in practice, the observed time series are always contaminated by noise, computer round-off errors, or a finite resolution of observations, and are sampled up to a certain *finite* time. In this regard, the dimension, referred as the minimum embedding dimension d_L hereinafter, reached for the reconstructed space $\mathbf{y}$ is not necessarily equal to the *global* dimension d of the original state space of $\mathbf{x}$. Rather, it must provide us with different *local* topographies of the reconstructed state space landscape, resulting in a different dimensionality ($d_L \leq d$).

To choose an appropriate time delay τ for an observable $s(t)$, we chose a prescription[6] based on the concept of average mutual information in information theory. All realistic *finite* time series of interest are contaminated by noise and round-off errors, and moreover both stochastic and high-dimensional chaotic time series may result in a certain *high* dimension d for its *finiteness* of the time series. To discriminate stochastic time series from high-dimensional chaos, we chose Cao's algorithm[7] and found the minimum embedding dimension d_L for each principal components $s(t)$ with the delay time τ.

MODEL AND CALCULATIONS

As an illustrative vehicle, we apply these analyses to time series at a wide range of temperatures of a coarse-grained, off-lattice "3 color, 46-bead" protein model [8]. The model is composed of hydrophilic (L), hydrophobic (B), and neutral (N) beads, interacting with the following potential:

$$\begin{aligned}V = & \sum_i^{bonds} K_r (r_i - r_0^i)^2 + \sum_i^{angles} K_\theta (\theta_i - \theta_0^i)^2 \\ & + \sum_i^{dihedral} [A(1+\cos\Phi_i) + B(1+\cos 3\Phi_i)] \\ & + \sum_{i<j-3}^{non-bonded\ pairs} 4\varepsilon S_1 \left[\left(\frac{\sigma}{R_{ij}}\right)^{12} - S_2 \left(\frac{\sigma}{R_{ij}}\right)^6\right],\end{aligned}$$

where the van der Waals (vdW) interactions are used to mimic the hydrophilic, hydrophobic, and neutral characters of the beads. The sequence, $B_9N_3(LB)_4N_3B_9N_3(LB)_5L$, folds into a lowest energy β-barrel structure with four strands. This model, referred to as BLN hereinafter, exhibits not an ideal "funnel-like," but a rather frustrated potential energy topography. We examine both the original BLN model, and a less-frustrated Gō -like BLN model[9] in which only 47 native contact pairs of hydrophobic beads possess attractive vdW interactions, while the interactions between all the other pairs are repulsive, responsible for excluded volume (the potential parameters were reported in [8]). For the constant-temperature MD simulation we used Berendsen's method[10], which can control the temperature well with minimal local disturbance to the system. The trajectory calculations are performed at a wide range of temperatures ranging from 0.2 (where the system is at the native state) to 5.0 (at the denatured state)[ε/k_B]. At each temperature the trajectory data are collected at every 100 steps for up to 10^7 steps (a time step is $0.0025\, t^* (= \sigma\sqrt{M/\varepsilon})$). Here we analyze how the distinct protein landscapes yield the different dynamical behaviours by the projections of the bead's Cartesian coordinates into the principal component space.

RESULTS AND DISCUSSIONS

Figure 1 shows the so-called Allan variance of the time series of the potential energies of these models at several temperatures. The Allan variance $\sigma_A^2(N)$ is defined by

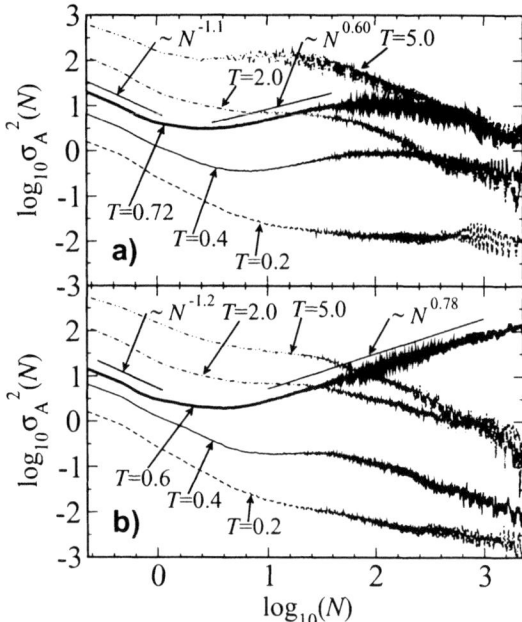

FIGURE 1. The Allan variance of potential energy fluctuations a) BLN model, b) Gō -like BLN model. The transition temperature are 0.72 for BLN model and 0.6 for Gō -like BLN model, respectively.

$$\sigma_A^2(N) = \frac{1}{2}\left\langle \left(\frac{1}{N}\sum_{i=1}^N s(i) - \frac{1}{N}\sum_{i=1}^N s(i+N)\right)^2 \right\rangle, \quad (3)$$

and measures the degree of nonstationarity of a given time series $s(t)$: if $s(t)$ is stationary, the following scaling relation should be satisfied obeying the law of large numbers,

$$\sigma_A^2(N) \sim \mathcal{O}(N^{-\gamma}) \quad (\gamma: \text{a positive constant}). \quad (4)$$

If $s(t)$ is nonstationary, $\gamma < 0$. These figures demonstrate that, for both the BLN and Gō -like models, significant nonstationary features emerge about the collapse and folding temperatures in $V(t)$, which diminish departing from these transition temperatures. The emergence of the nonstationary feature is more pronounced in the Gō -like model than in the BLN model. In both models at these transition temperatures $\sigma_A^2(N)$ obey the simple scaling relation $\sim N^{-\gamma}$ for $0 \le t \le 10^{0.2}t^*(= 1.6t^* \simeq 10^3$ simulation steps) after which the index γ changes from positive (stationary regime: $\gamma \simeq -1.1$(BLN), $\gamma \simeq -1.2$(Gō -like)) to negative (nonstationary regime: $\gamma \simeq 0.60$(BLN), $\gamma \simeq 0.78$(Gō -like)). The nonstationary regime turns to another stationary regime around $100t^*(\simeq 10^5$ steps) for the BLN model, while it even continues for more than 10^5 steps for the Gō -like model. It implies that at short time scales, proteins move about only within an individual rugged basin, and exhibit a *local* stationary behavior

due to the fast (chaotic) fluctuations. The longer the time is, the more the protein crosses between large basins, bringing about slow large fluctuations at the transition temperature. The longer persistence of nonstationarity of the Gō -like model at T_f compared with the original BLN model indicates that a less-frustrated "funnel" energy landscape leads to longer memory persistence of the process in the large. Such a hierachical nature of stationarity of the potential fluctuations in time and temperature was also observed in their power spectra[8].

Let us now examine how the principal components (PCs) shed light on the multi-dimensional dynamics of proteins. How long does memory persist in the principal components $\mathbf{Q}(t)$ and what are the characteristic time scales inherent to them? Figure 2 shows the first minimum time t_{min} of each average mutual information $I_i(t)$ of each individual principal component $Q_i(t)$ for both the BLN and Gō -like models at $T = 0.2 - 5.0$. The average mutual information I_{AB} between measurements A and B is defined by

$$I_{AB} = \sum_{a,b} P_{AB}(a,b) \log_2\left[\frac{P_{AB}(a,b)}{P_A(a)P_B(b)}\right], \quad (5)$$

where $P_{AB}(a,b)$ is the joint probability density for measurements A and B to observe a and b, respectively; $P_A(a)$ and $P_B(b)$ are the individual probability densities for the measurements A and B, respectively. The average mutual information I_{AB}, an average of $I_{AB}(a,b)$ over all measurements $\{a\}$ and $\{b\}$ implies the degree of mutual correlation of the two measurements A and B. The average mutual information between $Q_i(t_0)$ and $Q_i(t_0+t)$, $I_i(t)$, is defined by

$$I_i(t) = \sum P(Q_i(t_0), Q_i(t_0+t))$$
$$\times \log_2\left[\frac{P(Q_i(t_0), Q_i(t_0+t))}{P(Q_i(t_0))P(Q_i(t_0+t))}\right]. \quad (6)$$

We took the time $t_{min}(\equiv \tau)$ of the first minimum of $I_i(t)$, as a characteristic time scale inherent to the time series. In general, one can expect that the longer t_{min}, the longer the memory persistence in the signal $Q_i(t)$. Except for the PCs with very small variances, whose indexes are greater than ~ 100 for the BLN and ~ 80 for the Gō -like models, t_{min} of almost of all the other PCs exhibit a longer increase at the transition temperatures, $T = 0.72$ (BLN) and $T = 0.6$ (Gō -like), than those at the other temperatures.

Now, let us look deeply into the question of what the dimensionality of the state space is, buried in the complexity of the time series of the protein dynamics. Figure 3 shows the embedding dimensions of the principal component time series for the original BLN model at a wide range of temperatures. The more the variance of the PC fluctuation decreases, i.e., the more the index i

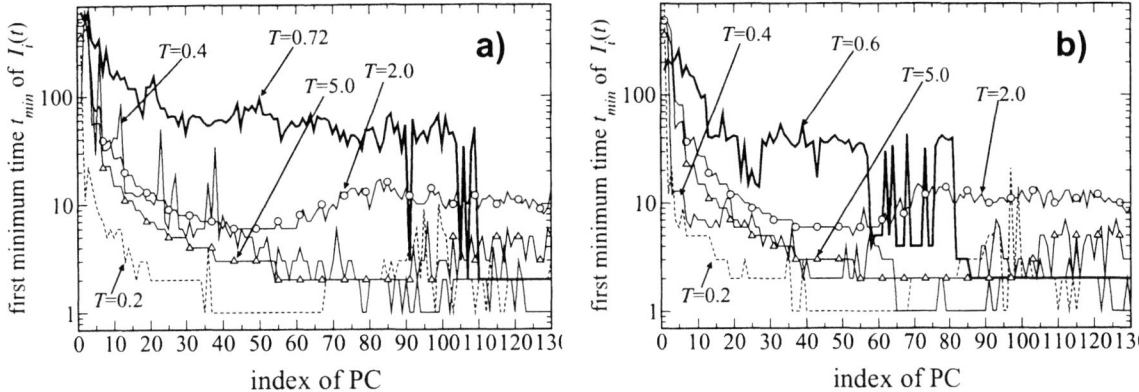

FIGURE 2. The first minimum time τ of the average mutual information of all principal components. a) BLN model. b) Gō-like BLN model.

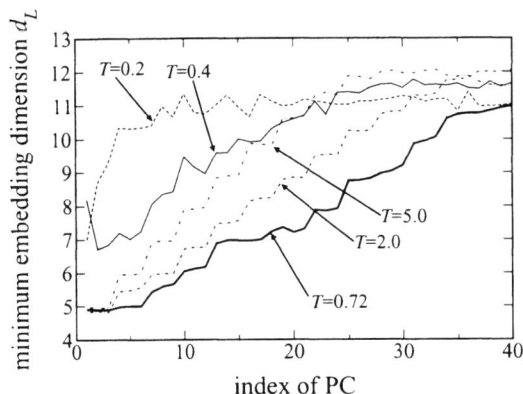

FIGURE 3. The minimum embedding dimension d_L estimated by the principal components at $T = 0.2, 0.72, 5.0$ for BLN model.

of the Q_i increases, the more the dimensionality tends to increase (the time series Q_i for $i > 30$ are regarded as approximately "random" and their finite dimensionality should arise from the finiteness of the sampling length). An observation from the embedding analysis that is perhaps even more striking appears in the behavior at the collapse temperature: the larger the variance of the PC, the lower dimensionality is necessary to reconstruct the state space, and higher and lower from the collapse temperature the dimensionality of the state space increases.

"What is the implication of the dimension d_L evaluated by scalar *finite* time series of each principal component of the *same* system?" In principle, the time series of the principal components analyzed here should not be random, because all the time series originally arise from the deterministic equation of motion. In this report, at each temperature for both the models, we analyzed the time series collected at every one hundred simulation steps for up to 10^7 steps, i.e., we could access the time series $\mathbf{Q}(t)$ not with an infinitesimal resolution but with only a finite resolution. The fast chaotic fluctuations of the principal components $Q_i(t)$ ($i > 30$) with small variances make the $\mathbf{Q}(t)$ "lose" memory of the process in such an short duration, resulting in a "random" stochastic time series. One can not extract the geometric information of the underlying dynamical structure from any observable composed of such "lost" degrees of freedom.

This work was supported by the Japan Society for the Promotion of Science, Grant-in-Aid for Research on Priority Areas "Genome Information Science" and "Optical Control of Strong Laser Field" of the Ministry of Education, Science, Sports and Culture of Japan.

REFERENCES

1. Takada, S., *Proc. Natl. Acad. Sci. USA*, **96**, 11698-11700 (1999).
2. Komatsuzaki, T., and Berry, R.S., *Proc. Nat. Acad. Sci. USA* **78**, 7666-7671 (2001).
3. Komatsuzaki, T., and Berry, R.S., *Adv. Chem. Phys.* **123**, 79-152 (2002).
4. Karplus, M., *J. Phys. Chem. B* **104**, 11-27 (2000).
5. García, A.E., and Hummer, G., *Proteins* **36**, 175-191 (1999).
6. Abarbanel, H. D-I., *Analysis of Observed Chaotic Data*; Springer-Verlag; New York, 1995.
7. Cao, L., *Physica D* **110**, 43-50 (1997).
8. Matsunaga, Y., Kostov, K.S., and Komatsuzaki, T., *J. Phys. Chem. A*, **106**, 10898-10907 (2002).
9. Nymeyer, H., García, A.E., and Onuchic, J.N., *Proc. Natl. Acad. Sci. USA* **95**, 5921-5928 (1998).
10. Berendsen, H.J.C., Postma, J.P.M., van Gunsteren, W.F., DiNola, A., and Haak, J.R., *J. Chem. Phys.*, **81**, 3684-3690 (1984).

Low-order chaos in sympathetic nerve activity causes *1/f* fluctuation of heartbeat intervals

Motohisa Osaka[*], Hiroo Kumagai[†], Katsufumi Sakata[†], Toshiko Onami[†], Ki H. Chon[¶], Mari A. Watanabe[‡], and Takao Saruta[†]

[*]*Department of Life Information Sciences, Institute of Gerontology, Nippon Medical School, Kawasaki 211-8533, Japan*
[†]*Department of Internal Medicine, Keio University School of Medicine, Tokyo 160-8582, Japan*
[¶]*Department of Electrical Engineering, City College of New York, New York, NY 10031, USA*
[‡]*Institute of Biomedical and Life Sciences, Glasgow University, Glasgow, G12 8QQ, UK*

Abstract. The mechanism of *1/f* scaling of heartbeat intervals remains unknown. We recorded heartbeat intervals, sympathetic nerve activity, and blood pressure in conscious rats with normal or high blood pressure. Using nonlinear analyses, we demonstrate that the dynamics of this system of 3 variables is low-order chaos, and that sympathetic nerve activity leads to heartbeat interval and blood pressure changes. It is suggested that *1/f* scaling of heartbeat intervals results from the low-order chaos of these variables and that impaired regulation of blood pressure by sympathetic nerve activity is likely to cause experimentally observable steeper scaling of heartbeat intervals in hypertensive (high blood pressure) rats.

INTRODUCTION

The power spectra of heartbeat intervals from healthy individuals exhibit a scale invariant *1/f* pattern in the low-frequency range ($f < 0.1$ Hz) [1-3]. Recent studies show that loss of this *1/f* slope [4] and loss of heartbeat-interval multifractality [5] are closely correlated to the prognosis and severity of heart disease, but the mechanism underlying the heartbeat-interval power law scaling remains unknown [6,7]. Scale invariance is commonly associated with chaotic dynamics, so nonlinearity in the dynamics of the system regulating heartbeat intervals would be a prime suspect. Physiologically, heartbeat intervals are determined by sympathetic nerve activity and blood pressure in a complex interaction that involves the brainstem and feedback loops, but the details of the interaction are unknown for low-frequency oscillations. One study has suggested that the low-frequency component of the heartbeat-interval power spectra is independent of sympathetic nerve activity [8]. However, we recently showed that low-frequency blood pressure oscillations arise from sympathetic nerve activity, and that the elevated levels of sympathetic nerve activity in spontaneously hypertensive rats reduced the nonlinear correlation between sympathetic nerve activity and blood pressure that exists in normotensive (normal blood pressure) rats [9]. We therefore hypothesize that sympathetic nerve activity could also be responsible for the *1/f* slope in heartbeat intervals and examine this hypothesis, using normotensive Wistar-Kyoto rats and spontaneously hypertensive rats [10]. The latter have higher sympathetic nerve activity, and are widely acknowledged to be an appropriate model of essential hypertension in man.

EXPERIMENTAL METHODS

The following is a brief description of our experimental methods. We used telemetry to record electrocardiograms [9]. The transmitter was implanted in the peritoneal cavity 1 to 2 days before the experiment. The arterial pressure signal from a transducer attached to the left femoral artery catheter

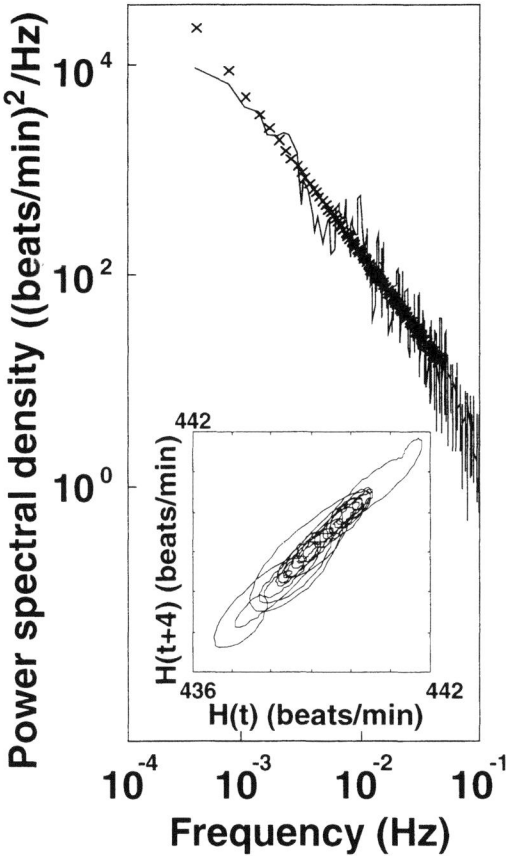

FIGURE 1. Low-frequency portion of heart rate oscillation power spectrum from a normotensive rat. The crosses denote the regression line fitted over the range of 0.0005 to 0.02 Hz to the experimental data. Inset: Phase portrait.

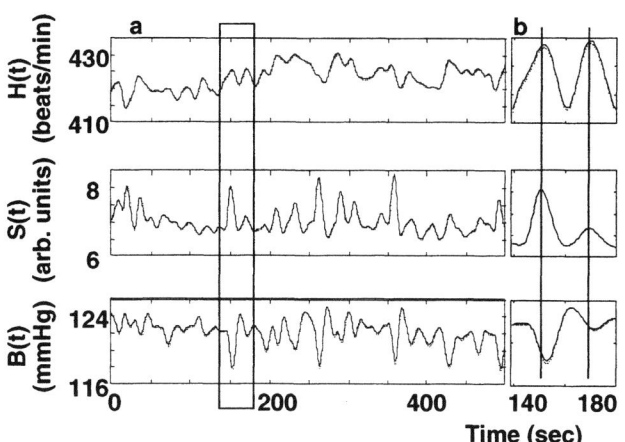

FIGURE 2. a, Filtered recording from the rat in Figure 1 (solid line) and fit of a nonlinear autoregressive model (dotted line). The two lines are so close as to be indistinguishable. $S(t)$ and $H(t)$ share the same polarity and approximate timing of their extrema. Panel **b** shows the box in panel **a** on an expanded time scale. $S(t)$ peaks clearly precede $H(t)$ peaks and $B(t)$ nadirs.

was amplified. Multifiber recordings of renal sympathetic nerve activity [11] were made from electrodes placed on the left renal nerve fascicle. Neural recording electrodes were connected to a high-impedance probe, which was connected to a differential amplifier with a band-pass filter of 50 to 1000 Hz. The filtered neurogram was integrated by a resistance-capacitance circuit (time constant = 20 ms). Electrocardiogram, blood pressure, and renal sympathetic nerve activity were simultaneously recorded for over 100 minutes in conscious, unrestrained rats, both normotensive and spontaneously hypertensive (7 each). The signals were digitized with an A/D converter and sampled at 2 kHz. A smoothed instantaneous heart rate time series was constructed from heartbeat intervals between the R waves of the electrocardiogram using an algorithm proposed by Berger et al. [12]. The time series of heart rate, blood pressure, and renal sympathetic nerve activity were splined and sampled at 64 Hz so that values of the entire constructed time series were made to occur simultaneously. To ensure accurate preservation of the low-frequency (< 0.1 Hz) signals, we used a Butterworth filter, which lost no more than 0.5 dB in the passband < 0.1 Hz and had at least 15 dB of attenuation in the stopband > 0.15 Hz. We visually confirmed the absence of aliasing errors due to digitization and filtering.

MATHEMATICAL METHODS AND RESULTS

Figure 1 shows a typical heart-rate power spectrum from a normotensive rat plotted with log-log axes. Regression analysis was performed between 0.0005 and 0.02 Hz. In a similar manner to the power spectra in healthy humans [1, 3], the slope of the line fitted to the data is approximately −1, indicating a $1/f$ relationship between frequency and power. The mean and standard error of the slope from normotensive rats was −1.35 ± 0.11 (n = 7). In contrast, the corresponding slope in spontaneously hypertensive rats was significantly greater, at -1.95 ± 0.12 (n = 7) (p < 0.05).

Several methods have been proposed to detect chaotic behavior in biological systems which tend to be noisy [13-15]. We used a recently developed algorithm [16], because it was the only method thought to be sensitive in cases where stochastic and deterministic components are both involved. This method first fits a nonlinear autoregressive model to a time series, followed by an estimation of the characteristic exponents of the model over the observed probability distribution of states for the system. More specifically, the model for a system output y can be written as:

$$y_n = F[y_{n-1}, \ldots, y_{n-k}] + k\varepsilon_n,$$

where F is a nonlinear (polynomial) function corresponding to the deterministic part, k is a constant, and ε_n are independent, identically distributed Gaussian random variables. The $k\varepsilon_n$ term corresponds to the stochastic part. We resampled the filtered heart rate, renal sympathetic nerve activity, and blood pressure data at 2 Hz ($H(t)$, $S(t)$ and $B(t)$) and applied the above algorithm for data lengths of 500 s. The solid lines in Fig. 2a show a 500 s segment of $H(t)$, $S(t)$ and $B(t)$ recorded from a normotensive rat, the dotted lines, the result of the fit. The fit is remarkably accurate, considering that, for example, only 11 coefficients are needed to define 1000 points (500 s, 2 Hz) for the $S(t)$ trace in Fig. 2a (equation given in Appendix). It must be noted, however, that the coefficients found for a particular data segment do not describe the coefficients for an adjacent data segment because the stochastic component brings about the sensitive dependence on initial conditions in this chaotic system (see below), resulting in rapid diversion from predicted values.

From the models, we obtained the characteristic exponents of the system by rewriting the model as a j-dimensional system, where j is the value of the largest delay in the model. The maximum dimension (delay) for any of the rats was low, at 5. The largest Lyapunov exponents of $H(t)$, $S(t)$ and $B(t)$ were calculated for 5 random segments in each animal, and were always found to be positive, thereby indicating chaotic dynamics.

To find further evidence of low-order chaos, we constructed phase portraits in 2-dimensional state space $(S(t), S(t+T))$. Fig. 3α shows that some of the cycles in the phase portrait seem to have approximately the same period. Similarly, Fig. 2b shows cycles with a ~20 s period (0.05 Hz), and Fig. 1, cycles of 0.01 Hz, 0.005 Hz and 0.002 Hz as well.

We constructed stroboscopic plots of $(S(t), S(t+T))$ for an incident wave of 0.005 Hz. In Fig. 3α (a-l) we can see stretching, folding, and compression, processes peculiar to low-order chaos. However, these processes were observed for 40 minutes at most, because the circulatory system was not being forced by a single oscillator at 0.005 Hz. Similarly, Fig. 3β (incident wave of 0.002 Hz) shows evidence of determinism for $S(t)$-$H(t)$ dynamics. The correlation dimensions of such phase portraits of $S(t)$ were also found to be low (2.35 ± 0.10 in normotensive rats and 2.33 ± 0.10 in spontaneously hypertensive rats), another sign of chaotic dynamics. In summary, evidence for chaotic dynamics in the 3-variable system consisted of continuous broad band power spectra, positive Lyapunov exponents, trajectories typical of low-dimensional chaos, and phase portraits with a low number of correlation dimensions.

As Figure 2 shows, we confirmed quantitatively that the $S(t)$ peaks precede the $H(t)$ peaks and $B(t)$ nadirs, using the mutual information method [9, 17]. We interpret this result as an indication that slow oscillations of heart rate and blood pressure are produced by slow oscillations of sympathetic nerve activity.

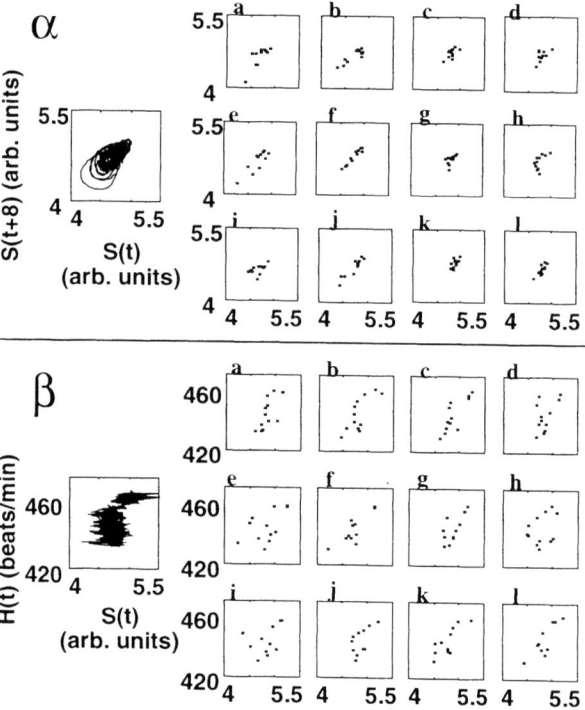

FIGURE 3. Phase portraits and stroboscopic plots.

The usual physiological interaction between blood pressure, heart rate and sympathetic nerve activity is known as the baroreflex, wherein an increase (decrease) in blood pressure is compensated for by decreases (increases) in heart rate and vascular resistance, mediated by sympathetic nerve activity emanating from a reflex center in the brainstem. In other words, blood pressure drives sympathetic nerve activity. However, Fig. 2a & b show that $S(t)$ peaks precede peaks of $H(t)$ and nadirs of $B(t)$, which is consistent with a recent report that sympathetic nerve activity precedes blood pressure in conscious rats [9]. Thus, correlation in the low-frequency band (< 0.1 Hz) is actually baroreflex-independent. This bolsters the view that sympathetic nerve activity may play a causative role in hypertension. This hypothesis is supported by the observation that some of the nonlinear components of blood pressure regulation remain after baroreceptor denervation (neural incapability of monitoring blood pressure) [18]. The sympathetic nerve activity was significantly higher in our spontaneously hypertensive rats than in the normotensive rats (16.1 ± 6.2 versus 7.2 ± 2.5 arbitrary units; $P < 0.01$), despite their higher blood pressure (156 ± 17 versus 110 ± 7 mmHg; $P < 0.001$). It seemed obvious that the interaction between sympathetic nerve activity and blood pressure might be impaired in spontaneously hypertensive rats, and that this might also cause the $1/f$ slope to be steeper.

CONCLUSIONS

Spontaneously hypertensive rats, an animal model for human hypertension, have steep scaling of the low-frequency portion of their heartbeat-interval power spectra. Application of a nonlinear autoregressive algorithm shows that, hypertensive or not, low-frequency heartbeat-interval power spectra characteristics are determined by the low-dimensional chaotic dynamics of a system of 3 variables, heartbeat intervals, blood pressure and sympathetic nerve activity, that is different from interactions between the variables at higher frequencies (baroreflex). It is suggested that decreased sensitivity of blood pressure to sympathetic nerve activity is the cause of the steep scaling seen in the hypertensive animals.

APPENDIX

For example, the model equation for renal sympathetic nerve activity in Fig. 2a is as follows:

$S(t) = 1.8306\ S(t-1) + 0.2496\ S(t-2) - 1.5032\ S(t-3)$
$- 0.0781\ S(t-4) + 0.5007\ S(t-5)$
$- 0.1178\ S(t-1)S(t-1) + 0.3873\ S(t-1)S(t-2)$
$+ 0.011\ S(t-1)S(t-4) - 0.4232\ S(t-2)S(t-2)$
$+ 0.2139\ S(t-3)S(t-3) - 0.0711\ S(t-3)S(t-5)$.

ACKNOWLEDGMENTS

We thank Sunao Murashige and "Creation and sustenance of Diversity — to explore a new theory of diversity of complex systems" held at the International Institute of Advanced Studies at Kyoto for helpful discussions (2001).

REFERENCES

1. Kobayashi, M., and Musha, T., *IEEE Trans. Biomed. Eng.* **29**, 456-457 (1982).
2. Peng, C. K., et al., *Phys. Rev. Lett.* **70**, 1343-1346 (1993).
3. Saul, J.P., et al., in *Computers in Cardiology* edited by IEEE Computer Society Press, Silver Spring, 1987, pp. 419-422.
4. Bigger, J. T., et al., *Circulation* **93**, 2142-2151 (1996).
5. Ivanov, P. C., et al., *Nature* **399**, 461-465 (1999).
6. Goldberger, A. L., et al., *Biophys. J.* **48**, 525-528 (1985).
7. Turcott, R. G., and Teich, M., *Ann. Biomed. Eng.* **24**, 269-293 (1996).
8. Koh, J., et al., *J. Physiol. (Lond)* **474**, 483-495 (1994).
9. Sakata, K., et al., *Circulation* **106**, 620-625 (2002).
10. Yamori, Y., *Handbook of Hypertension,* Amsterdam: Elsevier Science B. V., 1994, Vol. 16, pp. 346-350.
11. Kumagai, H., et al., *Circ. Res.* **67**, 1309-1322 (1990).
12. Berger, R. D., et al., *IEEE Trans. Biomed. Eng.* **33**, 900-904 (1986).
13. Glass, L., *Nature* **410**, 277-284 (2001).
14. Sugihara, G., and May, R. M., *Nature* **344**, 734-741 (1990).
15. Pei, X., and Moss, F., *Nature* **379**, 618-621 (1996).
16. Chon, K., et al., *Physica D* **99**, 471-486 (1997).
17. Osaka, M., et al., *Phys. Rev. E* **67**, 041915-1-4 (2003).
18. Wagner, C. D., et al., *Am. J. Physiol.* **269**, H1760-H1766 (1995).

Molecular Dynamics of 8-oxoguanine Lesioned B-DNA Molecule - Structure and Energy Analysis

M. Pinak[1], P. O'Neill[2], H. Fujimoto[1] and T. Nemoto[3]

[1] Japan Atomic Energy Research Institute, Tokai, Japan, [2] Radiation and Genome Stability Unit, MRC, Harwell, Didcot, UK, [3] Research Organization for Information Science and Technology, Tokai, Japan

Abstract. The molecular dynamics (MD) simulation of DNA mutagenic oxidative lesion - 7,8-dihydro-8-oxoguanine (8-oxoG), complexed with the repair enzyme - human oxoguanine glycosylase 1 (hOGG1) was performed for 1 nanosecond (ns) in order to describe the dynamical process of DNA-enzyme complex formation. After 900 picoseconds of MD the lesioned DNA and enzyme formed a complex that lasted until the end of the simulation at 1 ns. The amino group of arginine 324 was located close to the phosphodiester bond of nucleotide with 8-oxoG enabling chemical reactions between amino acid and lesion. Phosphodiester bond at C5' of 8-oxoG was displaced to the position close to the amino group of arginine 324. In the background simulation of the identical molecular system with the native DNA, neither the complex nor the water mediated hydrogen bond network were observed. The electrostatic energy is supposed to be significant factor causing the disruption of DNA base stacking in DNA duplex and may also to serve as a signal toward the repair enzyme informing on the presence of the lesion.

INTRODUCTION

Several metabolic pathways in living organisms generate reactive oxidative radicals and these radicals can attack DNA molecule to yield both DNA base damage and strand breakage. In many cases the produced specific DNA damage may either block replication and transcription or generate mutation by miscoding during replication. To counter the threat posed by the genotoxic lesion on DNA, cells express enzymes that function solely to recognize and repair structural aberrations in their genomes [1, 2, 3]. It is known that sequence specific DNA binding by repair and regulatory enzymes occurs as a result of multistage hydrogen bonding and van der Waals interactions between the DNA recognition amino acid chains of enzyme and nucleotide base sites of DNA. However, the underlying mechanisms by which repair enzymes recognize aberrant sites on DNA are still subject of a debate [4, 5].

Among the oxidative DNA lesions, the 7,8-dihydro-8-oxoguanine (8-oxoG), formed by oxidation of a guanine base in DNA, is considered to be one of the major endogenous mutagens contributing broadly to spontaneous cell transformation. Its frequent mispairing with adenine during replication increases the number of G-C $\rightarrow$ T-A transversion mutations. This mutation is among the most common somatic mutations detected in human cancers [6, 7].

The 8-oxoG is recognized and repaired with specific repair enzyme human oxoguanine glycosylase 1 (hOGG1) that is located in a region on the short arm of chromosome 3 (3p26). Loss of one hOGG1 allele increases the mutagenic burden imposed by guanine oxidation. The biochemical activity of this enzyme was extensively studied revealing that the enzyme recognizes 8-oxoG:C base pairs, catalyzes expulsion of the aberrant base and cleaves the DNA backbone [8, 9].

The present molecular dynamics (MD) computational study of the 8-oxoG lesioned DNA complexed with human repair enzyme oxoguanine glycosylase hOGG1 is aiming to describe dynamical mechanisms by which repair enzyme recognizes oxidatively damaged guanine on DNA and prevents its malignant transformation. The description of the dynamical process of forming DNA-enzyme complex, in addition to the existing crystal structures, may help to determine factors that are crucial for successful recognition of the oxidative DNA lesion.

MATERIALS AND METHODS

The classical MD simulation of the 8-oxoG lesioned DNA and repair enzyme hOGG1 was performed with the program AMBER 5.0 [10]. The module NUCGEN of AMBER 5.0 software package was used to prepare the native sequence of the 15 base pairs B-DNA duplex (DNA 15-mer). This module generated Cartesian coordinates of double helical B-DNA. The DNA sequence - d(GCGTCCAGGTCTACC)$_2$ was the same as in crystal structure of the complex [11]. The standard force field parameters included in the parm96.dat force field [12] were modified to satisfy the molecular parameters of the 8-oxoG.

The 8-oxoG was constructed as the 7,8-dihydro-8-oxoguanine by oxidizing of C8 of guanine. During this oxidation the double bond C8-N7 was transformed into the single one and N7 was hybridized (using graphical molecular software INSIGHT II [13], (Fig. 1). The potential energy of the constructed molecule was minimized in two two-stage energy minimization processes at 30 K. In this process the parameters of the double bond C8-N7 were transformed to a single bond and stabilized at the length of 1.52 Å. Atomic charges of the 8-oxoG were taken as those calculated by Poltev et al [14]. The 8-oxoG replaced the native guanine at the position 8 of the DNA 15-mer.

RESULTS AND DISCUSSION

Analysis of movement of DNA and enzyme

The DNA molecule and the enzyme were inserted into the center of simulated systems and solavetd by water molecules. The major consideration in the selection of their initial positions was to minimize initial mutual van der Waals interactions between the DNA and the enzyme.

Minimization was intended to eliminate an interaction prior the beginning of the MD simulation that would favor formation or certain configuration of resulting complex. The initial position of the enzyme with respect to the DNA was determined using the molecular graphic software INSIGHT II in such a manner that the enzyme faced the 8-oxoG lesion and that there was a minimal overlap of van der Waals surfaces of the DNA molecule and the enzyme (distance between the closest atoms of enzyme and DNA was ~5 Å). The constructed structure was partially optimized by energy minimization, with bonds of both solute molecules kept constrained in order to relieve bad contacts between them.

During 1 ns of MD simulation the enzyme in system with lesioned DNA approached DNA and formed a close contacts between several atoms. From initial positions, the enzyme and DNA during MD simulation distanced from each other, then approached around 400 ps, distanced again and finally formed a very tight contact after 900 ps of MD (closest heavy atoms around 2 Å). The formed bi-molecular complex lasted stable afterwards.

Fig. 2 shows detail view on each contact region with overlapping of van der Waals surfaces between respective parts of sugar-phosphate backbone of the DNA and the amino acids of the enzyme. These van der Waals contacts between the DNA and the enzyme were established after 900 ps of MD simulation. In the case of control simulation with non-lesioned DNA, the enzyme and DNA remained well separated and no van der Waals contacts were formed.

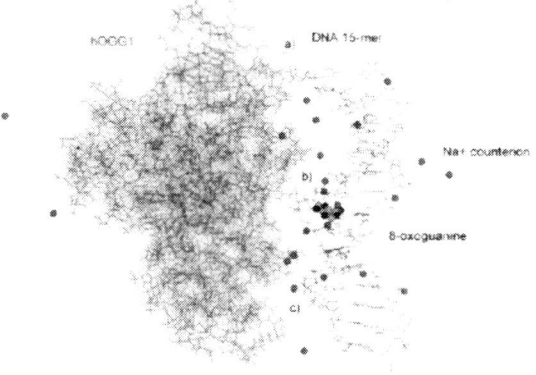

FIGURE 2. Molecular configuration of the lesioned DNA and the enzyme at 1 ns of MD simulation. The enzyme and the DNA formed a close van der Waals contacts between several atoms in three regions (encircled): a) arginine 277 - guanine G5, b) arginine 324 - guanine, 8-oxoG, adenine, and c) serine 54, glutamine 53 - adenine, cytosine.

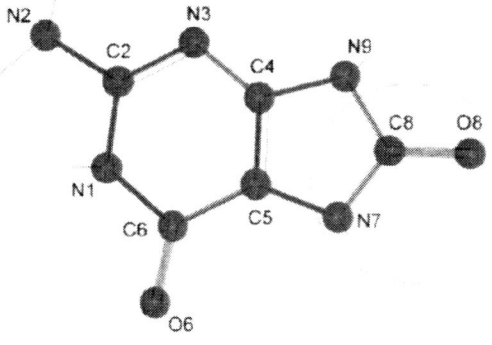

FIGURE 1. Molecule of the 8-oxoG. O8 is added at C8 transforming the double bond C8-N7 of native guanine into single one. N7 atom is hybridized.

The direct contact between amino acids and 8-oxoG base has not been established and instead, the aggregation of both molecules was caused by an extensive van der Waals interaction between enzyme and DNA strand at the 8-oxoG containing strand and favorable electrostatic interaction between amino acids and DNA backbone enhanced by the presence of lesion. This feature is common for most DNA-binding proteins which contact backbone phosphates. There were three regions forming van der Waals contacts:
- arginine 277 - guanine G5;
- arginine 324 – guanine, 8-oxoG , adenine;
- serine 54, glutamine 53 - adenine, cytosine.

Van der Waals Contacts and Electrostatic Interaction Between DNA and Enzyme

Interaction between van der Waals surfaces affects the binding affinity directly and also deters the exclusion of water and counterions from the interface. This way it may serve as a major driving force for the enzyme-DNA binding. The van der Waals contacts between the DNA and the enzyme were established after 900 ps of MD simulation.

Since the aggregation of both molecules was not observed in system with native DNA molecule (control simulation), it is estimated that in the case of lesioned DNA the electrostatic interaction caused by the presence of 8-oxoG lesion contributed to the recognition. This electrostatic interaction enhanced the existing van der Waals interaction between both molecules.

The amino group of arginine 324 is located close to C5' atom of 8-oxoG, forming 2 hydrogen bonds between O1P (8-oxoG) and N atoms (arginine 324), (Fig. 3). The amino group of the arginine is very basic due to its positive charge ^+H_2N. This positive charge establishes an attractive force to the negative charge of DNA phosphate group. In addition to the observed van der Waals interaction, the electrostatic attractive interaction between the amino group and negative charge of the DNA phosphate group contributes to the stability of the DNA-enzyme complex. The amino group of arginine 324 is also located close to the phosphodiester bond of the nucleotide with 8-oxoG that is supposed to be sequentially cleaved during the repair process.

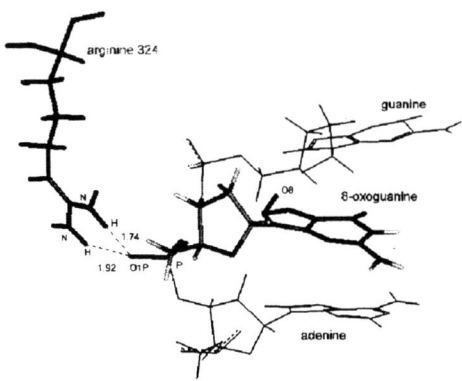

FIGURE 3. Detail view on the contact site between arginine 324 and nucleotide with 8-oxoG. The amino group of arginine 324 is located close to C5' atom of 8-oxoG, forming 2 hydrogen bonds between O1P (8-oxoG) and N atoms (arginine 324). The attractive electrostatic interaction between ^+H_2N and DNA phosphate group contributes to the stability of the complex.

CONCLUSIONS

The MD simulation of the 8-oxoguanine lesioned DNA segment (15 base pairs) in complex with specific repair enzyme hOGG1 is being discussed in this paper. The extensive van der Waals interaction between enzyme and DNA strand at the 8-oxoG containing strand caused the aggregation of both molecules and recognition was enhanced by the electrostatic interaction between amino acids and DNA backbone at the lesion site. The lesioned DNA and the enzyme made van der Waals molecular contacts at three regions after 900 ps of MD simulation and maintained them until the end of simulation at 1 ns. One of the regions with overlapping van der Waals surfaces encloses the phosphodiester bond of nucleotide with 8-oxoG that is supposed to be sequentially cleaved during the repair. The same MD simulation was performed also for an identical molecular system with the native DNA molecule instead of lesioned one. In this system, the native DNA did not form any van der Waals contacts with the enzyme and number of water molecules located close to both solute molecules was significantly lower. The lack of van der Waals contacts and non-existence of water mediated hydrogen bonds network between the native DNA and the enzyme suggest that the DNA-enzyme complex is formed only for the lesioned DNA molecule. In the formation of complex, the structural complementarity and existence of an enzyme pocket that enables insertion of damaged base are among important factors. The entire process of approaching both molecules before establishing a surface contacts can be explained by reported favoring electrostatic and van der Waals interactions. The achieved results shall also serve as a template for studies of other factors contributing to proper recognition of the oxidative base lesion and for the formation of complex.

ACKNOWLEDGMENTS

The authors wish to acknowledge staff of Center for Computational Science And Engineering of JAERI and of the Research Organization for Information Science and Technology for the installation, maintenance and adjustment of the AMBER 5.0 code on supercomputers VPP5000 and SR8000. The valuable support from the all members of the Radiation Risk Analysis Laboratory, JAERI Tokai Research Establishment is also highly acknowledged.

REFERENCES

1. Harrison, S. C.; Aggarwal, A. K. *Annu. Rev. Biochem.* **59**, 933-969 (1990).
2. Ham, J.; Thompson, A.; Needham, M.; Webb; P. and Parker; M. *Nucleic Acid Res.* **16(12)**, 5263-5276 (1988).
3. Beato, M. *Cell* **56**, 335-344 (1989).
4. Matthews, B. W. *Nature* **335**, 294-295 (1988).
5. Harris, L. F.; Sulliwan, M. R.; Hickok, D. F. *Proc. Natl. Acad. Sci. USA* **90**, 5534-5538 (1993)
6. Hollestein, M.; Shomer, B.; Greenblatt, M.; Soussi, T.; Hovig, E.; Montesano, R.; Harris, C.C. *Nucl. Acid Res.* **24**, 141-146 (1996).
7. Hazra, T. K.; Muller, J. G.; Manuel, R. C.; Burrows, C. J.; Lloyd; R. S.; Mitra, S. *Nucl Acid Res.* **29 (9)** 1967-1974 (2001).
8. Chevillard, S.; Radicella, J. P.; Levalois, C.; Lebeau, J.; Poupon, M. F.; Oudard, S.; Dutrillaux, B.; Boiteux, S. *Oncogene* **16**, 3083-3086 (1998)
9. Dherin, C.; Radicella, J. P.; Dizdaroglu, M; Boiteus, S. *Nucl. Acid Res.* **27**, 4001-4007 (1999).
10. Case, D. A.; Pearlman, D. A.; Caldwell, J. W.; Cheathman III, T. E.; Ross, W. S.; Simmerling, C. L.; Darden, T. A.; Merz, K. M.; Stanton, R. V.; Cheng, A. L.; Vincent, J. J.; Crowley, M.; Ferguson, D. M.; Radmer, R. J.; Seibel, G. L.; Weiner, P. K.; Kollman, P. A. *AMBER 5.0*, 1997, University of California San Francisco.
11. Bruner, S. D.; Norman D. P.; Verdine, G. L. *Nature* **403**, 859-866 (2000).
12. Wilkinson, A.; Weiner, P.; Van Gunsteren, W. in Wilfred, F., Van Gunsteren, W., Weiner, P. (eds.) Computer Simulation of Biomolecular Systems: Theoretical and Experimental Applications, Vol. 1, 1997.
13. InsightII 97.0, Molecular Simulations, Inc., San Diego, CA, USA, 1977.
14. Poltev, V. I.; Smirnov, S. L.; Issarafutdinova, O. V.; Lavery, R. *J Biomol Struct Dyn.* **11(2)**, 293-301 (1993).

Electrostatic Effects in Phase Transitions of Biomembranes between Cubic Phases and Lamellar Liquid-Crystalline (L_α) phase

Shah Md. Masum,[a] Shu Jie Li,[a] Yukihiro Tamba,[a] Yuko Yamashita,[b] Masahito Yamazaki [a,b]

[a] *Materials Science, Graduate School of Science and Engineering, Shizuoka University, 836 Oya, Shizuoka, Japan.*
[b] *Dept. Physics, Fac. Science, Shizuoka University, 836 Oya, Shizuoka 422-8529, Japan*

Abstract. Elucidation of the mechanisms of transitions between cubic phase and liquid-crystalline (L_α) phase, and between different IPMS cubic phases, are essential for understanding of dynamics of biomembranes and topological transformation of lipid membranes. Recently, we found that electrostatic interactions due to surface charges of lipid membranes induce transition between cubic phase and L_α phase, and between different IPMS cubic phases. As electrostatic interactions increase, the most stable phase of a monoolein (MO) membrane changes: $Q^{224} \Rightarrow Q^{229} \Rightarrow L_\alpha$. We also found that a de novo designed peptide partitioning into electrically neutral lipid membrane changed the phase stability of the MO membranes. As peptide-1 concentration increased, the most stable phase of a MO membrane changes: $Q^{224} \Rightarrow Q^{229} \Rightarrow L_\alpha$. In both cases, the increase in the electrostatic repulsive interaction greatly reduced the absolute value of spontaneous curvature of the MO monolayer membrane. We also investigated factors such as poly (L-lysine) and osmotic stress to control structure and phase stability of DOPA/MO membranes. Based on these results, we discuss the mechanism of the effect of electrostatic interactions on the stability of cubic phase.

INTRODUCTION

The biological and physicochemical aspects of cubic phases of lipid membranes have attracted much attention. One family of cubic phases, which includes Q^{224} phase (Schwartz's D surface), Q^{229} phase (P surface) and Q^{230} phase (G surface), has an infinite periodic minimal surface (IPMS) consisting of bicontinuous regions of water and hydrocarbon. In these cubic phase membranes, the minimal surface (defined to have zero mean curvature and negative Gaussian curvature at all points) is located at the bilayer midplane. Transmission electron microscopy has revealed regular 3-D structures of biomembranes similar to cubic phases in various cells. Cubic phase membranes and also transitions from cubic phases to the L_α phase play important roles in biomembrane dynamics such as membrane fusion, control of membrane protein functions, and intracellular structures of membranes and their structural changes. Elucidation of mechanisms of transitions between cubic phase and L_α phase, and between different cubic phases, is essential for understanding of biomembrane dynamics and topological transformation of lipid membranes. However, there has been only limited research on phase transitions and stability of cubic phases; only the effects of water content and temperature on their stability have been investigated.

EFFECT OF NEGATIVELY-CHARGED LIPIDS ON STABILITY OF CUBIC PHASES OF BIOMEMBRANES

Recently, we have systematically investigated the effects of electrostatic interactions due to surface charges on the structure and stability of cubic-phase membranes, and have found that electrostatic interactions due to surface charges of the membrane induce transitions between cubic phase and L_α phase, and between different IPMS cubic phases. For example, monoolein (MO) membranes are in the Q^{224} phase in excess water over a wide range of temperature. We doped a negatively-charged lipid (such as dioleoylphosphatidic acid [DOPA] and oleic acid [OA]) in the MO membrane and investigated structures of DOPA/MO membranes and OA/MO membranes using small-angle X-ray diffraction (SAXS) [1, 2]. As electrostatic interactions increase (i.e. surface charge density increases or salt concentration in bulk phase decreases), the most stable phase of a MO membrane changes: $Q^{224} \Rightarrow Q^{229} \Rightarrow L_\alpha$. The increase in DOPA concentration in the DOPA/MO membrane reduced the absolute value of spontaneous curvature, $|H_0|$, of the monolayer DOPA/MO membrane. $|H_0|$ of DOPA/MO monolayers increased with an increase in NaCl concentration. On the basis of these results, the stability of cubic-phase membranes can be explained by the spontaneous curvature of the monolayer membrane and a curvature elastic energy of the membrane.

Next, we investigated factors to control structure and phase stability of DOPA/MO membranes. 10%-DOPA/90%-MO membranes were in the Q^{229} phase, and at $\geq$ 16 mM poly (L-lysine), 10%-DOPA/90%-MO membranes were in the Q^{230} phase. Secondly, we investigated effects of osmotic stress on phase stability of DOPA/MO membranes, using poly (ethylene glycol) with molecular weight 7,500 (PEG-6K). For 10%-DOPA/90%-MO membranes, with an increase in PEG-6K concentration, i.e., with an increase in osmotic stress, the most stable phase changed as follows; Q^{229} (P) $\Rightarrow Q^{224}$ (D) $\Rightarrow Q^{230}$ (G). Based on these results, we discuss the mechanism of effects of two factors (poly (L-lysine) and osmotic stress) on structure and phase stability of DOPA/MO membranes.

EFFECT OF CHARGED PEPTIDES PARTITIONING INTO MEMBRANE INTERFACE ON CUBIC PHASES

We have recently succeeded in designing of the positively-charged peptide (e.g. peptide-1 <WLFLLKKK>) which can be partitioned in a lipid membrane interface composed of electrically neutral lipids such as phosphatidylcholine [3]. This peptide-1 also changed the phase stability of the MO membranes. As peptide-1 concentration increased, the most stable phase of a MO membrane changes: $Q^{224} \Rightarrow Q^{229} \Rightarrow L_\alpha$ [4]. As NaCl concentration increased, these phase transitions were inhibited. These results indicate that peptide-1 was partitioned in the membrane interface of the MO membrane, and that electrostatic interactions due to peptide-1 in the membrane interface make the Q^{229} phase more stable than the Q^{224} phase, and that, with larger electrostatic interactions, the L_α phase is more stable than these cubic phases. Increased peptide-1 concentration reduced $|H_o|$ of the MO/peptide-1 monolayer membrane, which is a main factor in the cubic-to-L_α phase transition [4].

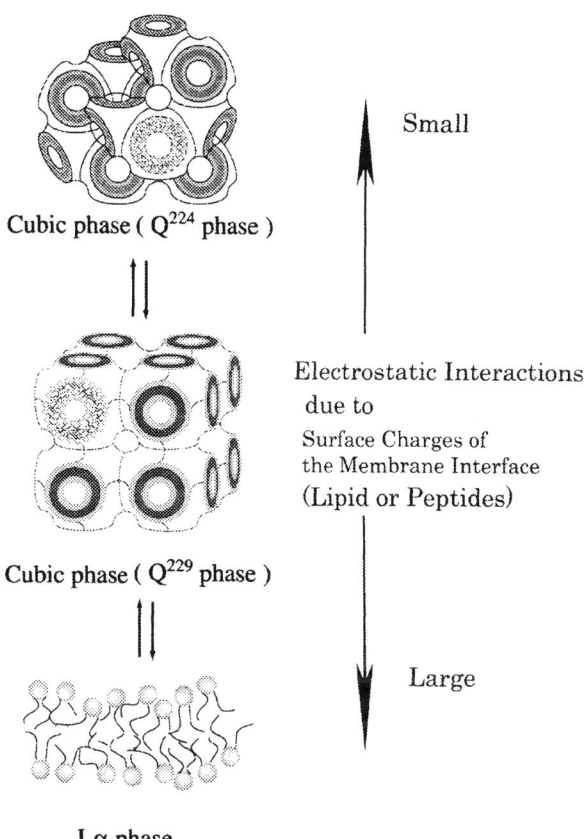

FIGURE 1. A scheme of the effect of electrostatic interactions on stability of cubic phases and L_α phase.

REFERENCES

1. Aota-Nakano, Y., Li, S.J., and Yamazaki, M. *Biochim Biophys Acta*, **1461**, 96-102 (1999)

2. Li, S.J., Yamashita, Y., and Yamazaki, M. *Biophys. J.* **81**, 983-993 (2001)

3. Yamashita, Y., Masum, S.M., Tanaka, T. and Yamazaki, M. *Langmuir*, **18**, 9638-9641 (2002)

4. Masum, S.M., Li, S.J., Tamba, Y., Yamashita, Y., Tanaka, T., and Yamazaki, M. *Langmuir*, **19**, 4745-4753 (2003)

Membrane Fusion of Giant Liposomes of Neutral Phospholipid Membranes Induced by La^{3+} and Gd^{3+}

Tomoki Tanaka,[a] Ryoko Sano,[b] Ayumi Yamagami,[b] and Masahito Yamazaki [a,b]

[a] *Materials Science, Graduate School of Science and Engineering, Shizuoka University, 836 Oya, Shizuoka,*
[b] *Dept. Physics, Fac. Science, Shizuoka University, 836 Oya, Shizuoka 422-8529, Japan*

Abstract. Membrane fusions of vesicles of biomembranes and also cell fusion play various important roles in cells. We have found that 100 μM~1 mM La^{3+} induced a membrane fusion of giant liposomes of DOPC (dioleoyl-PC, C18:1) and DPOPE (dipalmitoleoyl-PE, C16:1) mixture. The mechanism of this fusion can be explained by the increase in the lateral compression pressure in the phospholipid membranes induced by La^{3+}.

INTRODUCTION

Membrane fusions of vesicles of biomembranes and also cell fusion play various important roles in cells. Although many researches have been done on the membrane fusions, their mechanisms are still unclear. Membrane fusion should be considered as 2 steps; one is the strong association of vesicles, and the other is the instability of membrane structures in the associated vesicles. For example, in the membrane fusions induced by SNARE proteins, the mechanism for the association between vesicles and cell membranes has been revealed, but the mechanism of their membrane fusion is not still clear.

EFFECT OF La^{3+} AND Gd^{3+} ON PHOSPHOLIPID MEMBRANES

Recently, we investigated effects of La^{3+} and Gd^{3+} on phospholipids membranes such as phosphatidylcholine (PC) and phosphatidylethanolamine (PE), which don't have a net charge. The addition of 10~100 μM La^{3+} (or Gd^{3+}) through a 10-μm diameter micropipette near the dioleoyl-PC (DOPC)-giant liposome triggered several kinds of shape changes; such as the discocyte via stomatocyte to inside budded shape transformation, and the two-spheres connected by a neck to prolate transformation. The chain-melting phase transition temperature and the L$_{\beta'}$ to P$_{\beta'}$ phase transition temperature of DPPC-MLV increased with an increase in La^{3+} concentration, indicating that the lateral compression pressure of the membrane increases with an increase in La^{3+} concentration. The shape changes of the giant liposomes induced by these lanthanides can be explained reasonably by the decrease in the area difference between two monolayers on the basis of the bilayer-couple model [1]. Moreover, we have found that in the PE membrane La^{3+} can stabilize the H$_{II}$ phase rather than the L$_\alpha$ phase, which can be explained by the La^{3+}-induced increase in the lateral compression pressure of the membrane at the local sites [2].

La^{3+}-INDUCED MEMBRANE FUSION OF DPOPE/DOPC-GIANT LIPOSOMES

We have a hypothesis that the lateral tension of lipid membranes plays an important role in the membrane fusion. We investigated effect of La^{3+} on giant liposomes of DOPC and DPOPE (dipalmitoleoyl-PE, C16:1) mixture (DPOPE/DOPC-giant liposome). Fig.1B shows a shape change of two spherical 30mol% DPOPE/70mol%DOPC-giant liposome in 2%(w/v) PEG6K aqueous solution induced by addition of 100 μM La^{3+} through a 10-μm diameter micropipette near the giant liposomes. During the addition of La^{3+}, two giant liposomes associated each other (Fig. 1B-(1)) and then, the contact area between these giant liposome gradually increase (Fig.1B-(2)(3)). Further addition of La^{3+} induced a membrane fusion between

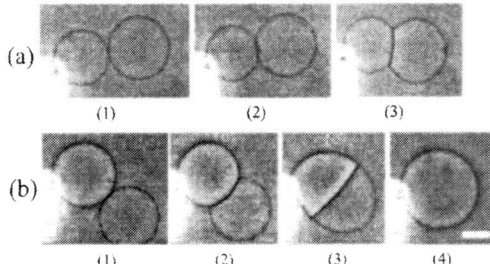

FIGURE 1. (a) Association of DOPC giant liposome induced by 100 μM La^{3+}. (b) Membrane fusion of 30% DPOPE/70%DOPC-giant liposome induced by 100 μM La^{3+}.

these giant liposomes to produce a larger spherical giant liposome (Fig.1B-(4)). 1 mM La^{3+} also induced membrane fusions between these vesicles, but 10 μM La^{3+} did not induce membrane fusion and membrane association. In contrast, when we added 100 μM~1 mM La^{3+} near two DOPC giant liposomes, the association between two liposomes occurred, but a membrane fusion did not occur (Fig. 1A). Efficiency of the membrane fusion depends on concentration of DPOPE in a DPOPE/DOPC membrane; at 10 mol% DPOPE, 50% of associated giant liposomes were fused, and at ≥30 mol%, 100% of associated giant liposomes were fused. During the membrane fusion between two DPOPE/DOPC giant liposomes, total volume was kept, i.e., summation of volumes of two vesicles (V_1+V_2) was equal to that of fused vesicle (V_3). However, total surface area (A_t) decreased on the membrane fusion, i.e., summation of surface areas of two vesicles (A_1+A_2) was less than that of fused vesicle (A_3).

To elucidate a mechanism of the La^{3+}-induced membrane fusion, we followed a process of membrane fusion of two 30%DPOPE/70%DOPC-giant liposomes induced by addition of 100 μM La^{3+}. At first, two giant liposomes were associated each other strongly and a partition membrane between two giant liposomes, which is composed of two bilayers of each giant liposome, was formed. Then, a partition membrane was suddenly broken at one end, and then the area of the partition membrane rapidly decreased. Finally it became a small structure which is difficult to be defined it or it was curled to form a large spherical vesicle inside the giant liposome. The area of the partition was almost equal to the decrease in A_t at the membrane fusion. The mechanism of this fusion can be explained by the increase in the lateral compression pressure in the membranes induced by La^{3+}.

REFERENCES

1. Tanaka, T., Tamba, Y, Masum, S.M., Yamashita, Y., and Yamazaki, M. *Biochim. Biophys. Acta*, **1564**, 173-182 (2002)

2. Tanaka, T., Li, S.J., Kinoshita, K., and Yamazaki, M. *Biochim. Biophys. Acta*, **1515**, 189-201 (2001)

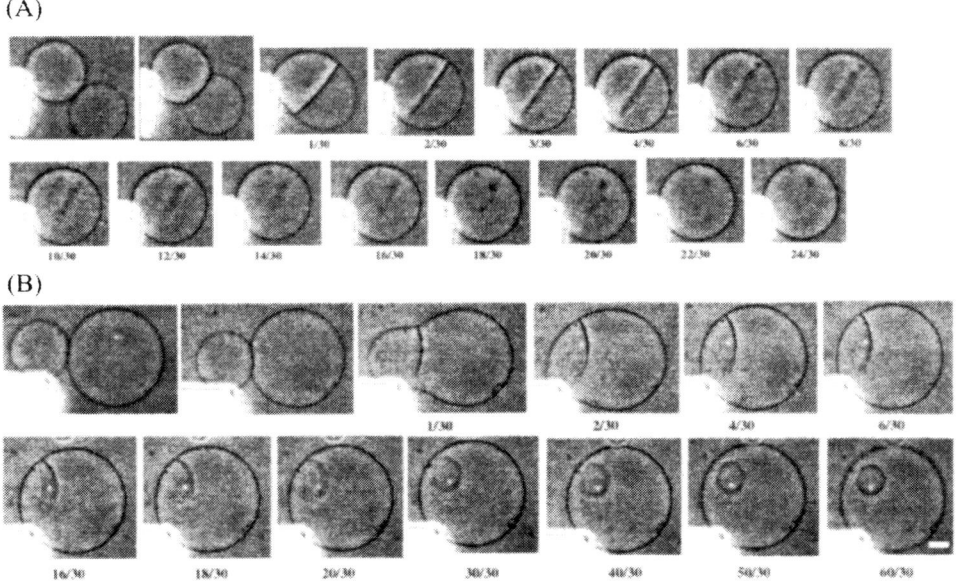

FIGURE 2. Phase contrast microscope images of intermediate structures of membrane fusion of 30%DPOPE/70%DOPC-giant liposome induced by addition of 100 μM La^{3+}. Digits under the photographs are the time (in sec). The bar is 10 μm.

Ultrafast Dynamics in Low Temperature Saccharide Glasses: A Photon Echo Study

Yutaka Nagasawa, Yukako Nakagawa, Yoshio Mori, Takayuki Muromoto, and Tadashi Okada

Graduate School of Engineering Science and Research Center for Materials in Extreme Conditions, Osaka University, Osaka 560-8531 Japan

Abstract. Saccharides are used as protectant by many organisms such as insects and amphibians. The glass transition of the saccharides is considered to be the key factor in the protection of the biological tissue against freezing and dehydration. The molecular dynamics of saccharide glasses were studied by photon echo spectroscopy and it revealed that electronic dephasing time is much longer in saccharide glasses compared to artificial polymer glass, polyvinylalcohol (PVA), at temperature of 10 K. Critically damped oscillation which can be assigned to the phonon mode of the saccharide glass was also observed.

INTRODUCTION

Saccharides are known to play an important role in "cryptobiosis" (hidden life), which is the amazing ability of some microorganisms, insects, reptiles, amphibians, and plants to survive hazardous environmental conditions, such as dehydration and freezing. For example, microorganism called tardigrade can revive from years of severe dehydration (anhydrobiosis).[1] Gray tree frog living in Canada can revive from days of frozen state (cryobiosis).[2] The glass transition and water substitution by the saccharides are considered to be the important factor in protecting proteins and membranes from these serious life threatening events. It is known that trehalose and glucose are responsible for the anhydrobiosis of tardigrade and for the cryobiosis of the gray tree frog, respectively. Trehalose is a disaccharide of glucose, which recently became the center of great attention as a cryoprotectant.

To understand the dynamical aspects of the glassy state of these saccharides in a molecular level, femtosecond three-pulse photon echo (3PPE) measurement was carried out. 3PPE is a nonlinear coherent spectroscopy that monitors the fluctuation of the molecules through the dephasing of an electronic coherence.[3] In this report, we have compared the 3PPE signal from a dye doped saccharide glasses and that from a hydrogen-bonding artificial polymer glass. The results indicated that the electronic dephasing was significantly slower in saccharide glasses.

EXPERIMENT

Ultrashort laser pulse centered at 635 nm with pulse duration of ~26 fs (fwhm) was used as a light source (the second harmonic generation of cavity-dumped chromium: forsterite laser). The laser beam was split into three beams with similar intensity (~300 pJ) and was used for the 3PPE measurement. For each measurement, the time delay between the first and the second pulse was scanned while the delay between the second and the third pulse was set at a certain value. A photodiode and a lock-in amplifier were utilized for the signal detection. Glass films of trehalose, glucose, and polyvinylalcohol (PVA) were produced with an organic dye, oxazine 4 (OX4), doped as a probe molecule. The optical densities of the samples were set close to 1.0 at the absorption maximum. The sample thickness was about 100 μm for PVA and 200-300 μm for the saccharides. 3PPE measurements were carried out with the samples set in a closed-cycled helium gas cryostat.

RESULTS AND DISCUSSION

Figure 1 (a) shows the 3PPE signal at 240 K. At this temperature, all of the echo signals decayed very rapidly, indicating a strong thermal fluctuation and ultrafast electronic dephasing. When the temperature was decreased to 10 K (Figure 1 (b)), clear difference was observed between the signals. All of the signals had a very rapid decay that occurred within 100 fs and there was a much slower decaying component for the saccharide glasses. The signal completely decayed after 200 fs for PVA. Coherent oscillations caused by the intramolecular vibrations of OX4 were clearly observed in the signals from saccharide glasses. The oscillations were mainly caused by the ring breathing mode of OX4 at ~580 cm^{-1}. The 3PPE signal at 10 K indicates slower electronic dephasing and much restricted molecular fluctuation for saccharide glasses.

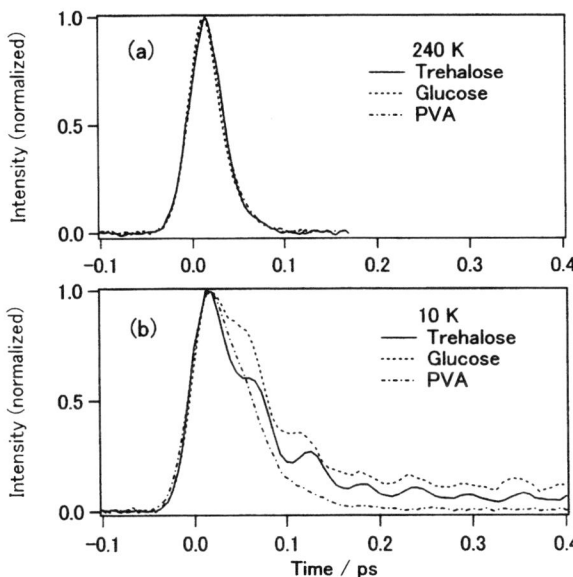

FIGURE 1. 3PPE signals from OX4 doped in saccharides and PVA glasses. Temperatures are (a) 240 K and (b) 10 K. The echo intensity was measured as a function of the delay time between the first and the second laser pulse with the delay between the second and the third pulse fixed to 387 fs.

The signal decay showed an Arrenius type of temperature dependence and the decay at 10 K was the fastest for PVA and the slowest for glucose glass. It can be considered that the intermolecular fluctuation (phonon mode) responsible for the electronic dephasing has the highest frequency for glucose and the lowest for PVA. The modes with higher frequencies will be deactivated more rapidly when the temperature is reduced, hence resulting in slower dephasing of the electronic coherence. However, from the longer scan of the echo signal shown in Figure 2, it can be seen that there is a recurrence in the signal, indicating a critically damped oscillation caused by the phonon mode. Because the maximum of the recurrence occurred near 0.9 ps in both trehalose and glucose, the frequency of the phonon mode seems to be similar in both glasses. A different reason has to be attributed to the difference observed for the echo signals of the saccharide glasses, probably the larger inhomogeneity in glucose. It seems that PVA has the largest free volume where the molecules can move quite freely and gave rise to low-frequency phonon modes. On the other hand, saccharides have less free volume because of the smaller molecular size enabling the molecules to be tightly packed. Moreover, these samples should contain large amount of water molecules, however for saccharides, the motion of the water molecules seems to be also restricted. Photon echo studies of other artificial organic polymer glasses utilizing similar dye molecules and at similar temperatures never revealed such a slow electronic dephasing as the ones in saccharides.[3, 4] The slow echo decays we have observed indicates the significance of the saccharides as cryoprotectants.

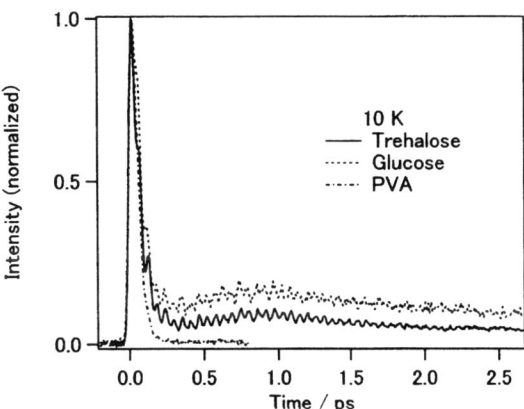

FIGURE 2. Longer scan of the 3PPE signal from saccharide and PVA glasses at 10 K. The delay between the second and the third pulse was fixed to 387 fs.

REFERENCES

1. Crowe, J. H., and Cooper Jr., A. F., *Scientific American* **225**, Dec., 30-36 (1971).

2. Storey, K. B. and Storey, J. M., *Scientific American* **263**, Dec., 62-67 (1990).

3. Nagasawa, Y., Seike, K., Muromoto, T., Okada, T., *J. Phys. Chem. A* **107**, 2431-2441 (2003).

4. Bardeen, C. J., Gerullo, G., Shank, C. V., *Chem. Phys. Lett.* **280**, 127-133 (1997).

Mechanical Response of Single Filamin A (ABP-280) Molecules and Its Role in the Actin/Filamin A Gel

Ryoko Sano,[a] Shou Furuike,[b] Tadanao Ito,[c] Kazuyo Ohashi,[d] and Masahito Yamazaki[a,b]

[a] Dept. Physics, Fac. Science, Shizuoka University, 836 Oya, Shizuoka 422-8529, Japan
[b] Materials Science, Graduate School of Science and Engineering, Shizuoka University, 836 Oya, Shizuoka, Japan
[c] Dept. Biophysics, Graduate School of Science, Kyoto University, Kyoto, 606-8502, Japan
[d] Dept. Biology, Fac. Science, Chiba University, 1-33, Yayoi-cho, Inage-ku, Chiba, 263-8522, Japan

Abstract. Actin/filamin A gel plays important roles in mechanical response of cells. We found a force (50 to 220 pN)-induced unfolding of single filamin A molecules using AFM, and have proposed a hypothesis on the role of single filamin A in the novel property of viscoelasticity of actin/filamin A gel. We also investigated structure and its dynamics of actin/filamin A gel formed in a giant liposome using fluorescence microscopy.

INTRODUCTION

Filamin A (ABP-280) is one of the actin filament cross-linking protein, which can produce isotropic cross-linked 3-D orthogonal networks with actin filaments *in vivo* and *in vitro*, which is called actin/filamin A (actin/ABP-280) gel or actin/filamin A network (Fig.1). It is present at the leading edge and cortex of many kinds of nonmuscle cells. Filamin A can also link the actin cytoskeleton to the plasma membrane via its association with membrane proteins. Filamin A and actin/filamin A gel have been considered to play important roles in the maintenance of membrane stability, mechanical response of cells, and cell locomotion [1]. Human filamin A (hsFLNa) is a dimeric protein with equivalent 280-kDa subunits that associate with each other at their C-terminal domains. It has two N-terminal actin-binding domains per dimer, and can thus cross-link actin filaments. Most of filamin A is a semiflexible rod composed of 24 tandem repeats, and its amino acid sequence predicts stretches of anti-parallel β-sheets in an immunoglobulin (Ig) fold.

MECHANICAL UNFOLDING OF SINGLE FILAMIN A MOLECULES

To elucidate the mechanical properties of the actin/filamin A gel and the complex of membrane protein–filaminA–actin cytoskeleton, we stretched single hsFLNa molecules in aqueous solution, and measured their force–extension relationship at room temperature [2]. In the force-extension curves, we often observed a large attractive force between the tip and the gold surface to a large extension, sometimes more than 500 nm. This indicates that the filamin A molecule adsorbed on both the tip and the gold surface was stretched as the distance between the tip and the gold surface increased (i.e., extension). In most cases, we observed a periodic increase and decrease in force

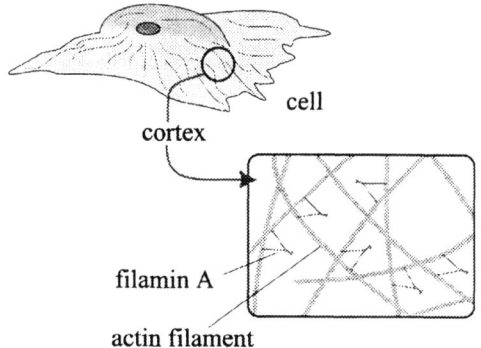

FIGURE 1. A schematic picture of actin/filamin A gel.

during the extension; i.e., so-called sawtooth pattern of the force-extension curve (Fig. 2). The abrupt decrease in force in the sawtooth patterns corresponds to the force-induced unfolding of an individual Ig-fold domain of filamin A. At a pulling speed of 0.37 μm/s, the unfolding force ranged from 50 to 220 pN, indicating that filamin A has various kinds of Ig-fold domains with different mechanical properties, which is due to the wide variation in values of activation energy and the width of activation barrier of 24 Ig-fold domains of the filamin A at the unfolding transition [2].

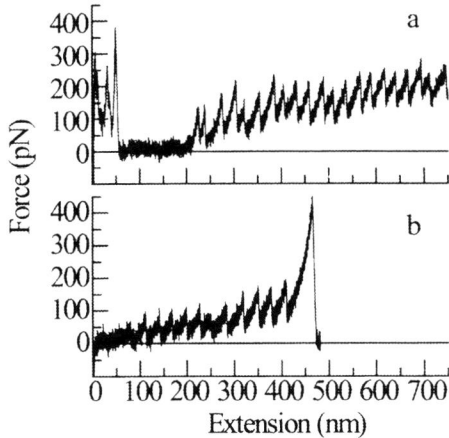

FIGURE 2. Two force-extension curves of single hsFLNa molecules in aqueous buffer. Pulling speed was 0.37 μm/s

This unfolding is reversible, i.e., the refolding of the unfolded chain of the filamin A occurs when the external force is removed. As a result of this reversible unfolding of Ig-fold domains, filamin A molecule can be stretched to several times the length of its native state. On the basis of this new feature of filamin A as the "large-extensible linker", we have proposed our hypothesis for the mechanical role of single filamin A molecules in actin/filamin A gel and in the actin cytoskeletons in cells [1].

VISCOELASTIC PROPERTY OF ACTIN/FILAMIN A GEL

The actin/filamin A gel shows the following characteristic properties *in vitro*. At low shear strains ($\gamma < 20\%$), it behaves as a linearly deforming gel; at moderate shear strains ($20\% < \gamma < 100\%$), it behaves as a non-linearly deforming gel with deformability that increases with increasing shear strain; at high shear strains ($\gamma > 100\%$), it behaves as a non-Newtonian fluid. In addition, its rheological properties show reversibility; after the removal of a large shear strain, the rheological properties return to their original values.

ROLE OF FILAMIN A IN MECHANICAL RESPONSE OF GEL

To investigate unfolding and refolding of filamin A in the actin/filamin A gel using fluorescence microscope, hsFLNa was labeled with a fluorescent dye (Oregon green 488 maleimide). At native state of hsFLNa, the fluorescence intensity was low due to the fluorescence self-quenching, and in the presence of 5M guanidine hydrochloride it increased to the 3.5-fold of that at the native state because unfolding of hsFLNa reduced the self-quenching. To investigate effect of shear stress to the actin/ filamin A gel using fluorescence microscope, we prepared the gel (actin/filaminA =100/1) in a giant liposome in F-buffer (10 mM imidazole buffer (pH 7.5), 100 mM KCl, 2 mM $MgCl_2$, 0.2 mM $CaCl_2$, 0.2 mM ATP) by a new method developed by us [3]. At first, we observed fluorescent-labeled F-actin in the gel using Texas Red-X phalloidin (Fig. 3). We could not observe any undulation motion and translational motion of F-actin, although in the absence of filamin A, large undulation and translation of F-actin were observed. In the conference, we will present new data on actin/filamin A gel using the above methods, and discuss a role of hsFLNa in the viscoelasticity of actin/filamin A gel.

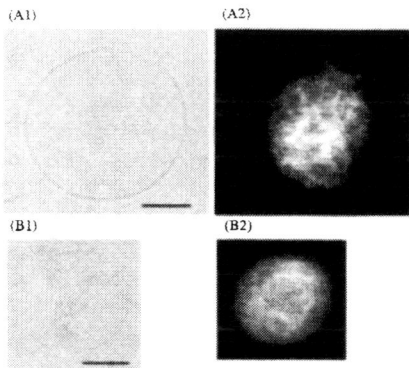

FIGURE 3. (A1)(B1) Phase-contrast microscopy images and (A2)(B2)fluorescence microscopy images of actin/filamin A gel in giant liposome. Bar is 10 μm.

REFERENCES

1. Yamazaki, M., Furuike, S., and Ito, T. *J. Muscle Research and Cell Motility*, **23**, 525-534 (2002)

2. Furuike, S., Ito, T., and Yamazaki, M. *FEBS Lett.* **498**, 72-75 (2001)

3. Yamashita, Y., Oka, M., Tanaka, T., and Yamazaki, M. *Biochim. Biophys. Acta*, **1561**, 129-134 (2002)

Oscillatory Reaction of Enzyme Caused by Gradual Entry of Substrate

Taketoshi Hideshima

Faculty of Science, Chiba University
Yayoi-cho, Inage-ku, Chiba, Japan, 263-8522

Abstract. We have been investigating the oscillatory reaction of enzyme by using semipermeable membrane, by which gradual entry of substrate into enzyme solution is obtained. In the present paper, we investigated oscillatory reaction for enzyme solution wrapped by liposome.

INTRODUCTION

Oscillatory reactions of enzyme are well known in those of allosteric enzymes and peroxidase. We found oscillatory reaction caused by different mechanism from these enzymes. Gradual entry of substrate causes oscillation. For gradual entry of substrate, we used semipermeable membrane or oil-water system. Using these method, we could oscillatory reactions in many enzymes.

Here we present the oscillatory reaction in the presence of liposome for further comprehension of living system.

METHODS

As shown in Figure 1, dialysis membrane was sandwiched between two cells. In one cell, enzyme solution was put, in the other cell, substrate.

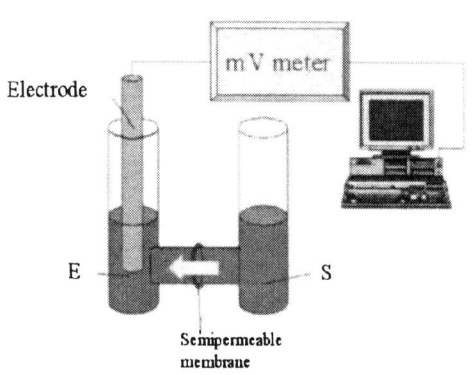

FIGURE 1. Apparatus for measuring oscillatory reaction

The enzyme can not migrate via the semipermeable membrane, while the substrate, which is small molecule, can permeate through it. The oscillatory reaction induced by gradual entry of substrate into the enzyme solution was measured by ORP electrode or selective electrode such as oxygen electrode.

OSCILLATORY REACTION OF CATALASE

Using this method as shown in Figure.1, we could observe oscillatory reactions in many enzymes. In the case of catalase, the period of oscillation depends on the concentration of hydrogen peroxide which is substrate, but is not independent of enzyme concentration, pH and temperature. Figure 2. shows the scheme of reaction.

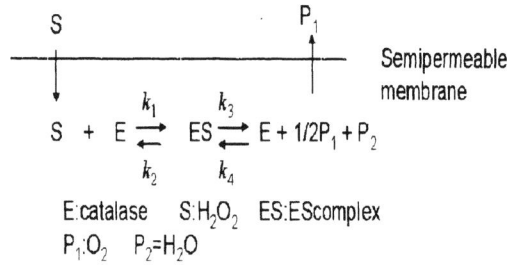

FIGURE 2. Scheme of oscillatory reaction by catalase

Rate equations of each species are as follows:

$$\frac{d[S]}{dt} = k_S([S]_e - [S]) - k_1[E][S] + k_2[ES] \quad (1)$$

$$\frac{d[ES]}{dt} = k_1[E][S] - (k_2 + k_3)[ES] + k_4[E][P_1]^{\frac{1}{2}}[P]^2 \quad (2)$$

$$\frac{d[P_1]}{dt} = k_3[ES] - k_4[E][P_1]^{\frac{1}{2}}[P]^2 - k_P[P_1] \quad (3)$$

, where $[S]_e$ denotes the concentration of substrate at equilibrium. The rate constants of reaction, permeation rate of substrate and elution rate of product are important factors for causing the oscillatory reaction. Increase in initial concentration of substrate leads to the increase of permeation rate. Consequently, it was found that the increase in permeation rate decreases the period of oscillation.

OSCILLATORY REACTION IN THE PRESENCE OF LIPOSOME

In the present paper, oscillatory reaction in the presence of liposome of phospholipid was investigated. As a lipid, dimyristoylphosphatidylcholine was used. Solution including liposome was prepared as follows: At first, toluene including phospholipid (1mM/l) was layered on aqueous solution of enzyme. After 1 day, only enzyme solution was used for experiment. Experimental results are shown in Figure 3.

a)
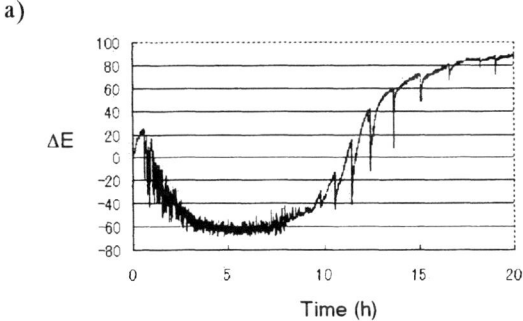

b)
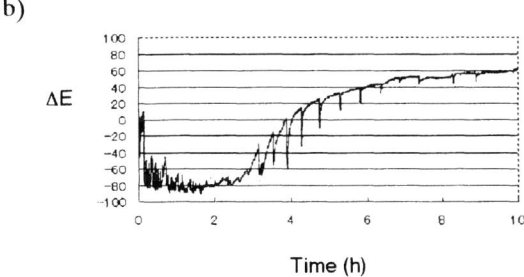

FIGURE 3. Time course of potential in the presence of catalase wrapped with lposome:[catlase]=0.1 mg/ml, [H_2O_2]=1.28 mM(a), and 0.64mM(b).

Downward oscillation of long period was observed after irregular oscillation with short period. The increase in substrate concentration decreased the time of irregular oscillation. The rise in potential denotes the increase in H_2O_2 concentration, while the fall, the increase in O_2 concentration. These results are quietly different from those in the absence of liposome. On the other hand, when enzyme was dissolved in the sonicated solution of lipid, the mode of oscillation was similar to that in the absence of liposome as shown in Fig.4(b).

a)
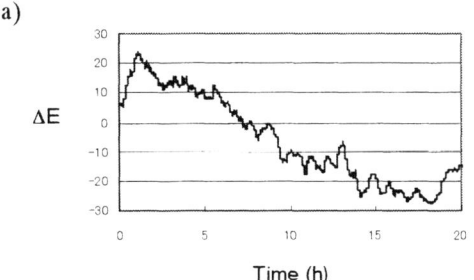

b)
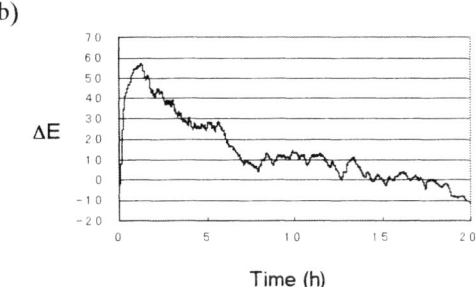

FIGURE 4. Time course of potential in the absence of catalase wrapped with lposome:[catlase]=0.1 mg/ml, [H_2O_2]=1.28 mM. (a) without liposome, (b) with liposome.

Enzymes used in Fig.3 seem to be wrapped by liposome. Substrate permeates through not only semipermeable membrane but also the membrane of lipids.

REFERENCES

1. Hideshima, T, *Biophys. Chem.*, **38**, 265-268 (1990)

2. Hideshima, T, *Biophys. Chem.*, **39**, 171-175 (1990)

3. Hideshima, T and Inoue, T, *Biophys. Chem.*, **63**, 81-86 (1991)

Reentrant Collapsing Transition of Single DNA Molecules: Elastic Response Depending on Spermidine Concentration

Yoshihiro Murayama and Masaki Sano

Department of Physics, University of Tokyo, Tokyo 113-0033, Japan

Abstract. The elastic response of single DNA molecules at various concentrations of the trivalent cation, spermidine, was measured. We found that the force-extension (f-x) curves in collapsed state showed either plateaus or stick-release patterns depending on the spermidine concentration. The periodic stick-release response with a certain characteristic length indicates abrupt release of toroidal structure. At high concentration of spermidine, we observed the reelongation of collapsed DNA, implying a reentrant transition at the single molecule level.

Like-charge attractions have been widely observed in biopolymer systems such as DNA, actin, and microtubule. For DNA in aqueous solution containing monovalent ions, chains repel each other, however, multivalent ions induce the DNA molecules to collapse or condense [1]. DNA condensation may contribute to the packaging of DNA in viruses and cells, gene therapy, and the conformational changes of chromosomal DNA. Although substantial knowledge about DNA condensation have been accumulated, we still do not understand the mechanical properties of the collapsed DNA. Recent single molecule techniques enable us to observe and manipulate single molecules and yield detailed information which could be helpful to reveal the mechanism of like-charge attraction in polyelectrolyte systems [2]. The DNA condensation occurs when the concentration of multivalent cations N exceeds a critical value N_c where approximately 90% of the DNA's charge is neutralized. Recent studies have found that DNA molecules dissolves for N above a critical concentration N_d and the condensation occurs within the range $N_c < N < N_d$ [3] where the short range attraction overcomes Coulomb repulsion due to the net charge of DNA [4].

We measured the elastic response of single DNA molecules at various spermidine^{3+} (SPD) concentrations [5] and found that the force-extension $(f - x)$ curves show wormlike chain (WLC) behavior for $N < N_c$, force plateaus for $N \sim N_c$, and stick-release patterns for $N_c < N$. Moreover, the WLC behavior recurs for still larger N, which implies a reentrant transition at the single molecule level.

We used a dual-trap optical tweezers to stretch individual double-stranded DNA molecules of 15.7 kilobasepairs (kbp), and the contour length 5.3 μm tethered between two protein-coated polystyrene beads [5]. We ob-

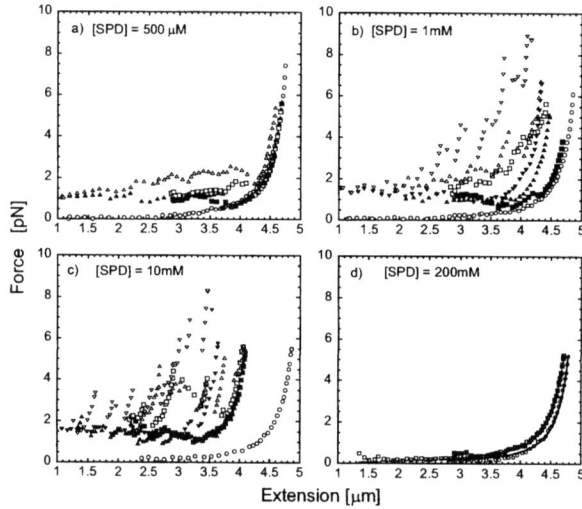

FIGURE 1. F-x curves at various SPD concentrations. $N = 500$ μM (a), 1 mM (b), 10 mM (c). ○: $N = 0$ mM (before solution exchange), ■: first relaxation after solution exchange, □: first stretching, ▲: second relaxation, △: second stretching, ▼: third relaxation, ▽: third stretching (data not shown for the third stretching and relaxation for $N = 500$ μM). Each data point is the average value of 20 measurements completed within 10 s. After solution exchange, DNA molecules were initially relaxed except for $N = 10$ mM. (d) The elastic response of a reelongated DNA molecule at $N = 200$ mM. ○: before solution exchange. ■: first stretching after solution exchange, □: first relaxation after solution exchange. The solid and dotted lines describe the WLC with $P = 39$ nm, $L = 5.2$ μm and with $P = 25$ nm, $L = 5.2$ μm, respectively.

tained $f - x$ curves before and after solution exchange for concentrations of SPD from 200 μM to 200 mM.

Figure 1 shows the $f - x$ curves before and after solution exchange for various N. In the absence of SPD, the elastic response obeys the WLC model [6], which gives the force f versus the extension x in terms of the contour length L and the persistence length P. In the WLC regime, $f - x$ curves showed no hysteresis in stretching or relaxation. In the presence of SPD, the elastic response depended on N. We observed WLC behavior at $N = 200$ μM, however, for 500 μM $\leq N \leq$ 100 mM, intramolecular collapse occurred and the elastic response differed from WLC behavior. At $N = 500$ μM, both stretching and relaxation had a force plateau of 1–2 pN [Fig. 1(a)]. At $N = 1$ mM, the plateau during both stretch and release persisted to an extension of 2 μm [Fig. 1(b)]. At higher concentrations of SPD, force plateaus of 0.5–2 pN appeared only during relaxation. Stretching showed a different elastic response, stick-release, for $N > 1$ mM. The force first gradually increased with increasing extension then abruptly decreased periodically during stretching [Fig. 1(c)]. We repeatedly observed stick-release during stretching and force plateaus during relaxation. Figure 1(d) shows that at $N = 200$ mM, the elastic response and WLC behavior resumed. Thus, the condensation of a single DNA molecule is a reentrant transition. Nguyen et al. showed [4] that the net charge of DNA is fully neutralized at $N = N_0$ where DNA collapse is most favorable. At concentrations N_c and N_d, the short range attraction energy ε equals the Coulomb repulsion energy for negatively and positively charged DNA molecules, respectively. Thus, at $N = 200$ mM [Fig. 1(d)], the DNA molecule is negatively charged before solution exchange and positively charged after, although its WLC elastic response is similar. Indeed, we observed the force increased at first and then decreased during solution exchange for 200 mM SPD.

Figure 1 shows that stick-release patterns have a characteristic length. The histogram of the distance between neighboring peaks characterizes the periodicity of the stick-release pattern. The histogram peaks at 0.3±0.05 μm (data not shown), indicating that collapsed state has a structure with a characteristic length of 0.3 μm. A peak in the histogram implies that the collapsed structures frequently unravel with a length of 0.3 μm, which is close to the circumference of the toroidal structure revealed by electron microscopy in condensed DNA [7]. Unraveling of the toroid in well defined quanta would explain the stick-release periodicity.

Since the force is solely entropic in the uncollapsed state, the excess work for the collapsed state, $\Delta W = \int_0^{x^*} (f_c - f_u) dx$, gives the energy of attraction between molecule segments, where f_c and f_u are the collapsed and uncollapsed forces respectively, and x_c is the extension at which the force responses coincide. Figure 2 shows the calculated energy at various N. The energy is maximal between 1 and 10 mM and the maximum value

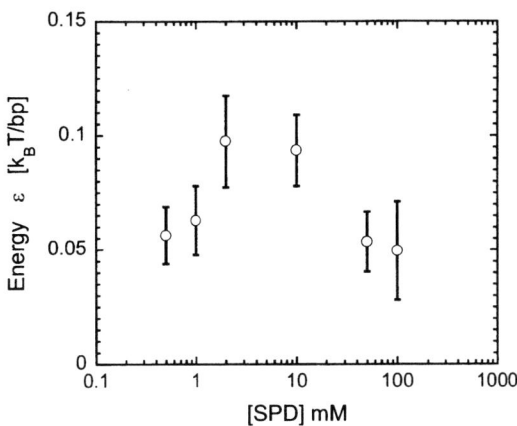

FIGURE 2. Dependence of the energy ε on the SPD concentration

in our experiments was $0.10\, k_B T$/bp at $N = 2$ mM, where k_B is Boltzmann's constant and T is the absolute temperature. Nguyen et al.'s theory estimates N_0 and ε using experimental N_c and N_d. For our experiment, $N_0 = 11 \pm 3$ mM and $\varepsilon = 0.09 \pm 0.02 k_B T$/bp. The experimental and theoretical energies agree, and N_0 is almost the N where the energy peaks.

ACKNOWLEDGMENTS

We thank K. S. Sogawa and H. Higuchi for helpful advice on sample preparation and experimental setup, I. Rouzina for helpful comments. This work was supported by Grant in Aid of the Ministry of Education, Sports, Science, and Culture, No. 14340119, and the Mitsubishi Science Foundation.

REFERENCES

1. Bloomfield, V. A., *Biopolymers*, **44**, 269–282 (1997).
2. Gelbart, W. M., Bruinsma, R. F., Pincus, P. A., and Parsegian, V. A., *Phys. Today*, **53**, 38–44 (2000).
3. Pelta, J., Livolant, F., and Sikorav, J.-L., *J. Biol. Chem.*, **271**, 5656–5662 (1996).
4. Nguyen, T. T., Rouzina, I., and Shklovskii, B. I., *J. Chem. Phys.*, **112**, 2562–2568 (2000).
5. Murayama, Y., Sakamaki, Y., and Sano, M., *Phys. Rev. Lett.*, **90**, 018102(1–4) (2003).
6. Marko, J. F., and Siggia, E. D., *Macromolecules*, **28**, 8759–8770 (1995).
7. Arscott, P. G., Li, A.-Z., and Bloomfield, V. A., *Biopolymers*, **30**, 619–630 (1990).

Study of Fluidity of Lipid Membranes Using Single Molecule Microscopy

M. Hibino and M. Kobayashi

Department of Applied Chemistry, Muroran Institute of Technology, 27-1 Mizumoto-cho, Muroran 050-8585, Japan

Abstract. The motion of single dye molecules in supported phospholipid membranes can be visualized by using total internal reflection fluorescence (TIRF) microscopy. For demonstration of the potentials of observing individual molecules we imaged a fluorescence labeled molecule in membranes made from dipalmitoyl phosphocholine (DPPC). The position of individual lipid molecules in gel phase was immobile within an accuracy of 20 nm. Trajectories of individual molecules in fluid membranes made from palmitoyloleoyl phosphocholine (POPC) showed that lipid motion occurs random with the mean square displacements described by a lateral diffusion constant 1.1-1.2 x 10^{-8} cm^2/s.

INTRODUCTION

The lateral mobility of lipids and proteins in cell membranes has been found to be a major principle controlling the structure and function of biomembranes, such as large-scale aggregation and diffusion-controlled biomolecular reactions [1,2]. Thus the mobility has been extensively studied experimentally and theoretically. Recently developed ultrasensitive optical microscopy has made possible the detection and characterization of single fluorescence labeled molecules. Moreover we have been imaging single biomolecules in the mobile state with a spatial resolution of nanometer [3,4]. The ability to observe individual molecules has given us new insights into biological system. We report here visualization of the diffusional path of individual molecules in a supported phospholipid membrane by using lipids carrying one fluorophore.

EXPERIMENTAL

The supported phospholipid membranes were deposited on glass substrates by the Langmuir-Blodgett technique. There are no fluorescence molecules in a first lipid monolayer. A second monolayer contained small amounts of fluorescence labeled lipid (octadecyl rhodamine; R-18 or tetramethylrhodaminethiocarbamoyl dihexadecanoyl phosphoethanolamine; TRITC DHPE). This cell was mounted on the microscope stage. We used an inverted fluorescence microscopy equipped with a x100 objective (numerical aperture = 1.4). For excitation, the 532nm line of a YAG laser running in TEM$_{00}$ mode was coupled into the epiport of the microscope. The laser beam was incident on a microscope cover glass and was totally reflected at the glass-to-solution interface (see Fig.1). For observation of rapid movement of the single fluorescence labeled lipid molecule, an ICCD camera was used.

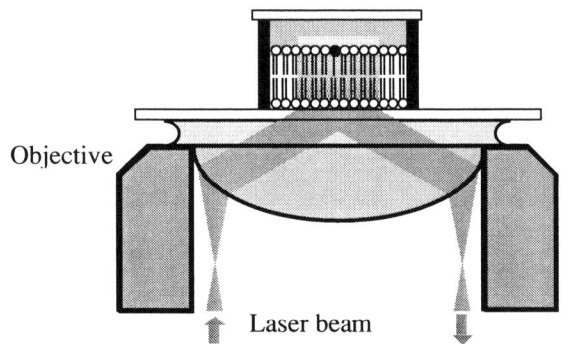

FIGURE 1. Schematic drawing showing the principle of single lipid molecule imaging in a phospholipid membrane with a low-background total internal reflection fluorescence microscope. The laser beam was incident on a microscope cover glass and was totally reflected. An ICCD camera was used for observation of rapid movement of the single fluorescence labeled lipid molecule at the video rate.

RESULTS AND DISCUSSION

We have been able to directly observe single lipid molecules in the membrane which had been labeled with fluorescent dye molecules (see Fig. 2). In Fig.3, we showed a trajectory of a fluorescence labeled molecule in membranes made from dipalmitoyl phosphocholine (DPPC) for demonstration of the potentials of observing individual molecules. A position of the molecule in gel phase was immobile within an accuracy of 20 nm. This positional accuracy allowed us to visualize a trajectory of individual fluorophore movements for time intervals of 33 ms. In Fig.4, the mobility of individual molecules was observed in fluid membranes made from palmitoyloleoyl phosphocholine (POPC). Trajectories showed that lipid motion occurs random with the mean square displacements described by a same lateral diffusion constant 1.1-1.2×10^{-8} cm^2/s for R-18 and TRITC DHPE (see Fig.5). In comparison, the mobility of lipids in membranes containing 30% cholesterol yielded an increase in the mean diffusion constant, 1.2×10^{-8} cm^2/s.

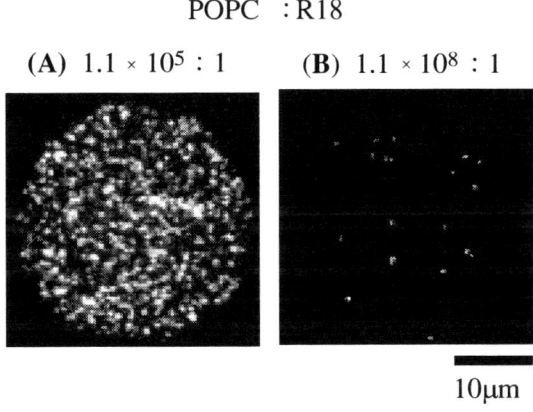

FIGURE 2. Fluorescence images for two surface densities of labeled lipids.

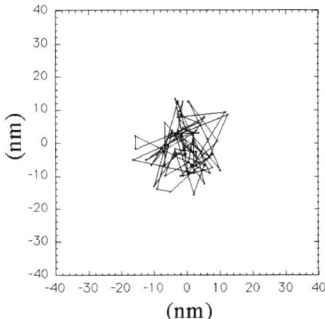

FIGURE 3. Two-dimensional trajectory of an individual lipid molecule in a DPPC membrane.

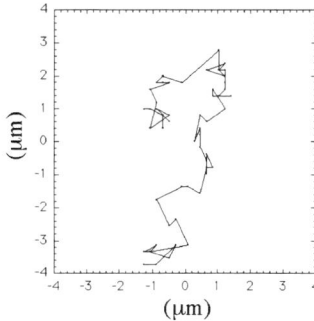

FIGURE 4. Two-dimensional trajectory of an individual lipid molecule in a POPC membrane.

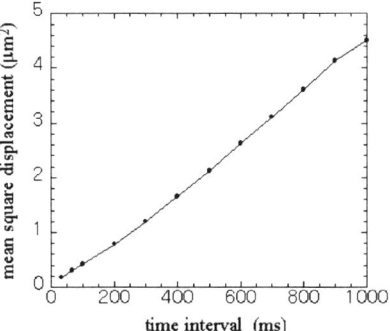

FIGURE 5. Mean square displacement versus time interval plot for trajectories. For each fluorescence labeled molecule in POPC bilayer, the tracking program computed the mean square displacement (MSD) for every time interval by using the formula:

$$\text{MSD}(t) = \langle (x_{i+n}-x_i)^2 + (y_{i+n}-y_i)^2 \rangle$$

Where (x_{i+n}, y_{i+n}) describes the particle position following a time interval t, given by $n \times$ video frame time after starting at position (x_i, y_i). i ranges from 1 to N-n, where N is the total number of molecule positions recorded, and n takes on values 1, 2, 3…N-1. For a two dimensional random walk,

$$\text{MSD} = \langle r^2 \rangle = 4Dt,$$

Where D is the lateral diffusion coefficient and t is the time interval. By using this equation, D was computed from the linear portion of the MSD versus time plot.

REFERENCES

1. Jacobson, K., and Dietrich, C., *Trends Cell Biol.* **9**, 87-91 (1999).
2. Simons, K., and Ikonen E., *Nature* **387**, 569-572 (1997).
3. Schmidt, Th, Schütz, G. J., Baumgartner, W., Gruber, H. J., and Schindler H., *Proc. Natl. Acad. Sci. USA* **93**, 2926-2929 (1996).
4. Sonnleitner, A., Schütz, G. J., and Schmidt, Th., *Biophys. J.* **77**, 2638-2642 (1999).

The DNA Adsorption by the Charged Cholesterol Monolayer at the Air-liquid Interface

Tsang-Lang Lin *, Yuan Hu *, Jui-Ching Wu *, Chun-Pang Yang *, U-Ser Jeng **, M.-C. Shih***

*Department of Engineering and System Science, National Tsinghua University, Hsinchu, Taiwan 300
** National Synchrotron Radiation Research Center, Hsinchu 300, Taiwan
*** Department of Physics, National Chung-Hsin University, Tai-Chung 402, Taiwan

Abstract. The adsorption of DNA by the 3-,-[N-(N',N'-dimethyl amino ethane) carbamoyl] cholesterol (DC-Chol) monolayer at the air-liquid interface was studied by using the Langmuir-Blodgett film balance. With the presence of 1 μM DNA in the subphase, the surface pressure increases right at the beginning of the compression. The liquid expanded phase of the DC-Chol disappears due to the adsorption of DNA. The AFM image of the prepared DC-Chol/DNA film has tree-branch-like fractal structure with a height of 2 nm that correspond to the diameter of DNA.

INTRODUCTION

The DNA/liposome complex was widely studied as they can be used in gene therapy. Cationic lipids were often used in forming the liposomes. The negatively charged DNA can be adsorbed by the cationic lipids due to the electrostatic interactions. F. Tang et al. compared the transfection activity and toxicity of cholesteryl hemidithiodiglycolyl tris (aminoethyl) amine (CHDTAEA) with a dithiodiglycolyl linker, its nondisulfide analogue cholesteryl hemisuccinyl tris(aminoethyl)-amine (CHSTAEA) and 3-,-[N-(N',N'-dimethyl amino ethane) carbamoyl] cholesterol (DC-Chol) [1]. The results showed that both CHDTAEA and CHSTAEA have higher transfection activity than DC-Chol at almost all mixing ratios with DOPE. The study by N. J. Zuidam and Y. Barenholz showed that only about 50% of the DC-Chol in DC-Chol/DOPE (1/1) liposomes was charged at pH 7.4 as compared to the nearly 100% charged DOTAP and DOTAP/DOPE (1/1) cases at the same pH [2]. Compared with DC-Chol, the DOTAP has higher CAC (7×10^{-5} M), and the dilution to below its CAC could induce liposome instability. K. Kago et al. investigated the lipid-DNA complex at the air-water interface by in-situ X-ray reflectivity method [3]. It was found that the DNA might not closely tight to the lipid monolayer. It is fundamentally important to understand how the DNA is adsorbed to the free surface of the lipid monolayer (such as the lipid monolayer at the air-water interface) or bilayer (such as the unilamellar versicle).

Here we will report our studies of the DNA interaction with the mixed DC-Chol monolayer by measuring the surface pressure versus molecular volume isotherms with a Langmuir-Blodgett film balance. The adsorption pattern of DNA will be controlled by the morphology of cholesterol domains, for instance disk-like, in the mixed lipid monolayers with neutral or zwitterionic lipids (no net charge). Patterned DNA absorption may also be useful in preparing delicate biosensors. On the other hand, the absorption of DNA may also induce structural changes to the morphology of the mixed lipid monolayers at the air-water interface, for instances, a re-orientation of the lipid head group conformation, as a consequence, a reduction of the tilt angle of the lipid chains.

RESULTS AND DISCUSSIONS

The DNA was obtained from Sigma, D6898, type XIV with 700 bp. DC-Chol ($C_{32}H_{57}N_2O_2Cl$) was also obtained from Sigma with a 95% purity. As shown in Fig. 1 of the surface pressure-area (π-A) isotherm results, that DNA (1 μM) in water solution adsorbs to the surface DC-Chol monolayer and it affects the isotherm significantly. The earlier rising surface pressure of the DC-Chol monolayer on DNA solution, compared to that on pure water, can be attributed to the additional charge interactions and physical contact interactions between the DNA that adsorbed underneath the surface monolayer.

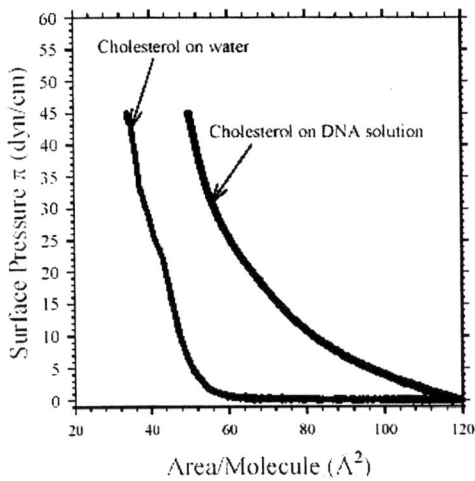

Figure 1. Surface pressure-area isotherms for DC-Chol on pure water and DNA water solution, respectively. The DNA used has 700 bp.

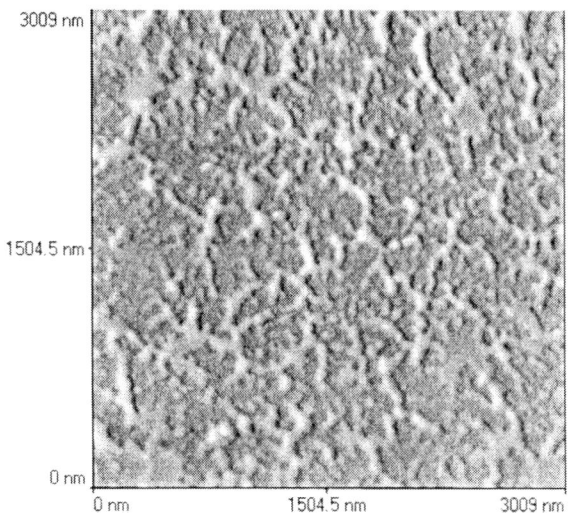

FIGURE 2. AFM images for the LB film of DC-Chol/DNA film prepared at a surface pressure of 38 dyne/cm.

Fig. 2 shows the AFM (atomic force microscopy) image of the prepared DC-Chol/DNA film on mica substrate. As shown in Fig. 2, the thin tree-branch-like pattern, of a typical height around 2 nm than the neighboring plateau area. The 2 nm height corresponds to the cross section of the DNA. Such a structure would not appear for the pure cholesterol LB film. From the AFM image for the LB film preparing from the DNA adsorbed DC-Chol monolayer, it is evident that DNA is absorbed to the positively charged DC-Chol monolayer. However, the DNA adsorption coverage is much less than 100% of the surface area. Several possible effects may contribute to this. For example, not all the DC-Chol molecules are charged at the current pH condition [2]. Only part of the DC-Chol molecules which are charged have the strong affinity to adsorb the DNA. The already adsorbed DNA may also hinder the additional DNA adsorption at its neighboring area due to steric and charge repulsion. For very high surface coverage of DNA absorption, the DNA will have to be aligned to their neighbors with the unfavorable decrease of entropy.

We also studied the absorption kinetics with different waiting times of 30, 60 and 120 min before starting the isotherm measurements. The three measured isotherms are almost identical. This indicates that the adsorption saturation of DNA by the DC-Chol can be reached less than 30 min and a typical waiting time of 30 min is enough.

ACKNOWLEDGEMENTS

This research is supported by the National Science Council, grant NSC9102113-M007-019.

REFERENCES

1. Tang, F., and Hughes J. A., *Bioconjugate Chem.* **10**, 791-796 (1999).
2. Zuidam, N. J., and Barenholz Y., *Biochimica et Biophysica Acta* **1329**, 211-222 (1997).
3. Kago, K., Matsuoka, H., Yoshitome, R., Yamaoka, H., Ijiro, K., and Shimomura, M., *Langmuir* **15**, 5193-5196 (1999).

Entropy Driven Unidirectional Motion of Brownian Particle Inside A Three-Dimensional Tube: Entropy Ratchet

Sheh-Yi Sheu* and Dah-Yen Yang[†]

*Department of Life Science, National Yang-Ming University, Taipei, Taiwan
[†]Institute of Atomic and Molecular Science, Academia Sinica, Taipei, Taiwan

Abstract. We show the existence of unidirectional flow in gated ion channel flow. The driving force is the asymmetric entropy barrier.

Recently, the unidirectional motion of a Brownian particle inside a tube (naro-tube or microtube) in biomolecular systems has attracted considerable general interest. In general, channels open and close stochastically, allowing ionic current to flow from one side of a membrane to the other side, i.e., permeation.

In order to study the unidirectional Brownian motion inside a tube, a model ideally contains a 3-D tubular structure. Our blob model of a 3-D tube contains a periodic asymmetric angular entropy potential surface (AEPS) with fluctuating barrier heights that are governed by gating behavior. In the present work, we show that a nonzero net flow exists inside our 3-D tube model when the periodic set of blobs has a different blob size and different gate modulation frequency. This proves that an entropy-driven mechanism can exist in an ion channel.

Let us consider a spherical cavity with the ligand initially located at its center (see Fig. 1). On the surface of this cavity, there are two punctures that acts as gates whose modulation are regulated by protein dynamics. The ligand diffuses inside the cavity with diffusion constant D_0 and is bounced back from the cavity surface until it reaches the gate parts. Note that the ligand can only escape through the gate parts. Instead, the rest part of the cavity surface acts as a reflecting wall, assuming that the gate can adsorb the ligand with reaction rate constant k_0 when it is opened.

The escape process can then be explained as the Brownian particle initially situated at the cavity center and subsequently moves downward along the r axis to the blob surface. This process is driven by the radial entropy potential (see Fig. 2).

We adopt the modulated radiation boundary condition at $t > 0$, as $\theta_i(t)$ is the modulated opening angle of gate i. Furthermore, the gates when they are opened can absorb the Brownian particle with the reaction rate constants $k_{i,0} (i = 1, 2$ for gates).

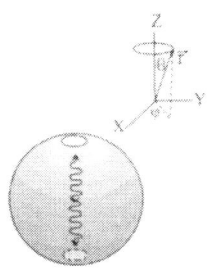

FIGURE 1. Schematic representation of a two-gate cavity. The modulation of the gates is controlled by the polar angles $\theta_1(t)$ and $\pi - \theta_2(t)$, which are in the ranges of $0 \leq \theta_i(t) \leq \theta_{i,max}$ where $i = 1, 2$. The major escape way of the Brownian particle is controlled by the gate size and its modulation. Hence the selectivity concept is introduced.

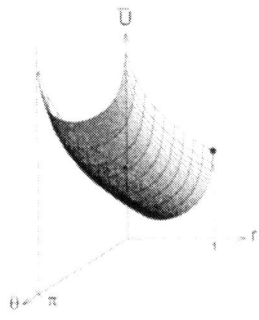

FIGURE 2. Two-dimensional entropy potential surface for ligand motion inside a two-gate spherical cavity. These two spots decide the barrier height for particle to climb. Due to different spot or gate size, the barrier heights are different. Hence the entropy potential is asymmetric.

Since the gate is regarded as sink in diffusion-controlled reaction systems, two nearby gates on the same blob surface exhibit a strong competition effect.

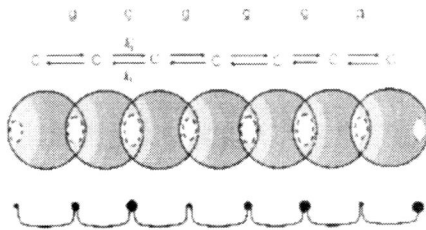

FIGURE 3. Blob model of a 3-D tube. Two blobs are connected through a junction gate. Three blobs are collected in one unit and the tube consists of infinite units. Moreover, in each unit, the gates have different size and modulation frequency. Below the tube, we portray the connected AEPS and the radial entropy potential surface is perpendicular to the AEPS. Two nearby angular entropy potential surfaces are connected through a junction spot.

When both of the maximum gate-opening angles $\theta_{1,max}$ and $\theta_{2,max}$ are small and different in size, we arrive at an upper bound of the total escape rate as

$$k_{0,tot}^{escape} \leq k_{1,0}^{escape} + k_{2,0}^{escape} \quad (1)$$

Here $k_{i,0}^{escape}$ is the one-gate escape rate for gate i. The equal sign in Eq. (1) is satisfied when these two gates are static, or these two gates have the same size and also the same modulation frequency.

We resolve the competition effect via the quantity

$$\Delta = k_{0,tot}^{escape} - k_{1,0}^{escape} - k_{2,0}^{escape}$$
$$= 0, \text{ two gates with the same gate size}$$
$$\text{and gate modulation frequency}$$
$$\neq 0, \text{ others} \quad (2)$$

The most probable situation is $\Delta < 0$, i.e., the destructive competition or slowing downs the total escape process. We obtain the individual escape rates in the two-gate system as $k_1 = k_{1,0}^{escape} + \frac{k_{2,0}^{escape}}{k_{1,0}^{escape}+k_{2,0}^{escape}}\Delta$ and $k_2 = k_{2,0}^{escape} + \frac{k_{1,0}^{escape}}{k_{1,0}^{escape}+k_{2,0}^{escape}}\Delta$. With the above equations we find that $k_1 + k_2 = k_{0,tot}^{escape}$. In this context, when $k_1 \neq k_2$, the escape pathway for the Brownian particle is selected by these two gates due to the gate modulation that is determined by protein fluctuation.

We now can examine the transition rates between two connected blobs that share the same junction gate (see Fig. 3). We define $c_j = blob_j$ and g_j being the gate between c_j and c_{j+1}. The transition rates at the junction gate g_j are $k_{c_j}^+$ and $k_{c_{j+1}}^-$. Here the superscript $+(-)$ denotes the transition from $c_j(c_{j+1})$ to $c_{j+1}(c_j)$. If the gate on the 3-D tube are not static and their maximum gate-opening angles satisfy the inequality $\theta_{g_{j-1},max} < \theta_{g_j,max} < \theta_{g_{j+1},max}$ or the size of c_j is not equal to the size of c_{j+1}, then at that junction gate g_j the escape rate $k_{c_j}^+$ is not equal to the escape rate $k_{c_{j+1}}^-$. This means that at the junction gate between two connected blobs the forward $k_{c_j}^+$ and backward $k_{c_{j+1}}^-$ transition rates are different. Similarly, different blob size may generate asymmetric forward and backward transition rates at the same gate. Moreover, inside each blob, due to the competition effect, $k_{c_j}^+$ is not equal to $k_{c_j}^-$. Thus, these are the onset of the nonzero net flow.

It is verified that the net flows passing through every gate, due to the periodic structure, are all the same and are given by

$$J = k_{c_{i+1}}^+ k_{c_{i+2}}^+ k_{c_{i+3}}^+ - k_{c_{i+1}}^- k_{c_{i+2}}^- k_{c_{i+3}}^- \quad (3)$$

Since $k_{c_{i+1}}^+ k_{c_{i+2}}^+ k_{c_{i+3}}^+$ is not equal to $k_{c_{i+1}}^- k_{c_{i+2}}^- k_{c_{i+3}}^-$, the net flow at any gate is nonzero and hence the current flows unidirectional. It is clear in this study that the molecular motor is entropy driven and does not require the application of any external fluctuation forces. Moreover, we would like to stress that there is an additional concrete explanation for the unidirectional flow. By using the existing entropy potential surface for two-gate blob in Fig. 2, the periodical asymmetric AEPS of our blob model of the 3-D tube is constructed in Fig. 3. The gate fluctuation includes the fluctuating barrier height in the AEPS. The escape pathway of the Brownian particle is selected by the fluctuating activation energy. Therefore, a periodic asymmetric AEPS induces a nonzero net flow inside the 3-D tube. Our APES differs from the well-know molecular motor model suggested by Magnasco. The periodic asymmetric potential in his model is of enthalpy origin and requires the application of external fluctuation forces. However, the potential in our 3-D tube model is of entropy origin and does not require any applied external fluctuation forces. Note in particular that the gate fluctuation or the fluctuation barrier height in our model is regulated by protein dynamics.

In summary, the escape process for a two-gate blob is entropy driven and the Brownian particle escape pathway is selected by these two gates due to the gate regulation. Based on the two-gate blob system, we have constructed a 3-D tube model that consists of a sequence of connected two-gate blobs. Our result suggests that the net flow in the 3-D tube is nonzero.

REFERENCES

1. Sheu, S. Y., and Yang, D. Y., *J. Chem. Phys.* **112**, 408-415(2000).
2. Sheu, S. Y., and Yang, D. Y., *J. Chem. Phys.* **114**, 3325-3329(2001).

Free energy calculations of the stacked and unstacked states for DNA dimers by replica-exchange umbrella sampling

Katsumi Murata[1], Yuji Sugita[2], and Yuko Okamoto[1,3]

[1] Department of Functional Molecular Science
The Graduate University for Advanced Studies, Okazaki, Aichi 444-8585, Japan
[2] Department of Structural Biology, Institute of Molecular and Cellular Biosciences
University of Tokyo, Yayoi, Bunkyo-ku, Tokyo 113-0032, Japan
[3] Department of Theoretical Studies Institute for Molecular Science, Okazaki, Aichi 444-8585, Japan

Abstract. We report the free energy difference between the stacked state and unstacked state for DNA dimers. The free energy difference was generated from replica-exchange umbrella sampling molecular dynamics simulations which allows the simulation to sample much wider conformational space and, therefore, to give more accurate free energy than by the conventional umbrella sampling (US). From the free energy difference, we observed good stacking for all 16 possible DNA dimers and sequence-dependent stacking stability. This sequence dependence of the stacking free energy is in agreement with the experimental results.

INTRODUCTION

The three-dimensional structure of DNA is stabilized mainly by base stacking interactions. A number of experimental studies have determined the stacking free energies of nucleic acid bases, nucleosides, nucleotides in aqueous solution have been determined [1-4]. These values follow the stability order purine-purine > purine-pyrimidine > pyrimidine-purine > pyrimidine-pyrimidine. The potential of mean force (PMF) calculations have also been used to investigate the free energy of the stacking process of all 16 possible DNA dimers using the conventional umbrella sampling (US) [5]. In this study, good stacking was observed for almost all dimers, but very poor stacking was also observed for some dimers. The poor stacking is presumably caused because the simulation got trapped in a few of a huge number of local-minimum-energy states. This multiple-minima problem can be overcome by the *generalized-ensemble algorithms* (for reviews, see Refs. [6]). In a generalized-ensemble simulation, each state is weighted by a non-Boltzmann probability weight factor so that a random walk in potential energy space may be realized. The random walk allows the simulation to escape from any energy barrier and to sample much wider configurational space than by conventional methods. In the present study, we employ one of the recently developed generalized-ensemble algorithms, replica-exchange umbrella sampling (REUS) [7]. We have performed the PMF calculations of all 16 possible DNA dimers in aqueous solution by REUS and compared the results with those by the conventional US.

RESULTS AND DISCUSSION

We performed the potential of mean force (PMF) calculations using the replica-exchange umbrella sampling (REUS). All 16 possible dimers (namely, we have in the 5' → 3' directions dApdA, dApdC, dApdG, dApdT, dCpdA, dCpdC, dCpdG, dCpdT, dGpdA, dGpdC, dGpdG, dGpdT, dTpdA, dTpdC, dTpdG, dTpdT) were considered. Although good stacking was observed for all DNA dimers from the PMF profiles obtained by REUS, we have to evaluate the stacking stability more quantitatively. Therefore the free energy difference between stacked and unstacked states was calculated by

TABLE 1.	The free energy difference ΔW (kcal/mol) between stacked and unstacked states.							
	dimer	ΔW	dimer	ΔW	dimer	ΔW	dimer	ΔW
purine-purine	dApdA	-5.1	dApdG	-4.6	dGpdA	-6.5	dGpdG	-4.8
Purine-pyrimidine	dApdT	-3.5	dApdC	-2.5	dGpdT	-5.4	dGpdC	-3.8
pyrimidine-purine	dTpdA	-5.6	dTpdG	-5.2	dCpdA	-4.3	dCpdG	-5.4
pyrimidine-pyrimidine	dTpdT	-3.6	dTpdC	-3.5	dCpdT	-3.3	dCpdC	-1.9

$$\Delta W = \int_{4.0}^{6.0} W(R)dR - \int_{8.0}^{10.0} W(R)dR. \quad (1)$$

where $W(R)$ is the potential of mean force along the reaction coordinate R. The values of ΔW for all DNA dimers are listed in Table 1, where the trapezoidal rule was used for the integral in Eq. (17) with the bin size of 0.1 Å.

From the results in Table 1, we have shown that all DNA dimers follow the stability order purine-purine > purineu-pyrimidine or pyrimidine-purine > pyrimidine-pyrimidine, which is in agreement with the experiments [1-4].

ACKNOWLEDGMENTS

The calculations were carried out on the computers at the Research Center for Computational Science, Okazaki National Research Institutes, and this work was supported, in part, by NAREGI Nanoscience Project, Ministry of Education, Culture, Sports, Science and Technology, Japan.

REFERENCES

1. Ts'o, P. O. P., Melvin, I.S., Olson, A.C., *J. Am. Chem. Soc.* **85** 1289-1296 (1962).

2. Solie, T. N., and Schellman, J. A., *J. Mol. Biol.* **33** 61-77 (1968).

3. Nakano, N. I., and Igarashi, S.J., *Biochemistry* **9** 577-583 (1970).

4. Mitchell, P. R., and Sigel, H., *Eur. J. Biochem.* **88** 149-154 (1978).

5. Norberg, J., and Nilson, L., *Biophys. J.* **69** 2277−2285(1995).

6. Mitsutake, A., Sugita, Y., Okamoto, Y., *Biopoymers (Pept. Sci.)* **60** 96-123 (2001).

7. Sugita, Y., Kitao, A., Okamoto, Y., *J. Chem. Phys.* **113** 6042-6051 (2000).

Phenomenological models of raft structure

H. Shirotori*, S. Komura*, T. Kato* and P. D. Olmsted[†]

Department of Chemistry, Tokyo Metropolitan University, Tokyo 192-0397, Japan
[†]*Department of Physics and Astronomy, University of Leeds, Leed LS2 9JT, UK*

Abstract. We propose two phenomenological models describing the phase behavior of lipid-lipid systems and lipid-cholesterol systems in order to understand the "rafts" in cell membranes. In our models, the coupling between the lateral phase separation and the internal degree of freedom of a lipid membrane is considered. The calculated phase diagrams are in semiquantitative agreement with the experimental phase diagrams.

INTRODUCTION

Cell membranes are mainly composed of lipids, cholesterols (Chol), and proteins. Recent studies have shown that dynamical domains called "rafts" exist in the cell membranes in which the above components are inhomogeneously distributed [1]. These domains mainly consist of Chol and lipids with saturated hydrocarbon chains. The raft structure is related to the signaling and the material transfer in cells. For example, it is known that proteins are selectively included or excluded from the domains. The formation of raft structure has attracted both physical and biological interests.

It is considered that at least saturated lipid, unsaturated lipid and Chol are necessary to form rafts in cell membranes. In this paper, we focus on the saturated-unsaturated lipid systems and the lipid-Chol systems as a starting point to understand the rafts. We propose two simple phenomenological models describing the phase behavior of these systems. Within the mean-field treatment, we can produce the phase diagrams which are in semiquantitative agreement with the experimental ones.

MODELS

Saturated-Unsaturated Lipid Systems. It is kown that lipid bilayers exhibit first-order phase transition between the liquid crystalline phase and the gel phase. We call these two phases the disordered phase and the ordered phase, respectively. As the contributions to the total free energy of a bilayer membrane, we consider the mixing free energy of lipids ($f_1^{\ell\ell}$), and the stretching free energy of lipid chains ($f_2^{\ell\ell}$). The total free energy is approximated as the sum of these two energies: $f^{\ell\ell} = f_1^{\ell\ell} + f_2^{\ell\ell}$.

We use a lattice model to express the mixing free energy. The lattice size of saturated lipid and unsaturated lipid are assumed to be the same. Let us define the number of saturated lipid and unsaturated lipid by N_s and N_u, respectively, and the molar fraction of saturated lipids by $x = N_s/(N_s + N_u)$. Using the Bragg-Williams theory, the mixing free energy per lattice site is given by

$$f_1^{\ell\ell} = k_B T [x \log x + (1-x) \log(1-x)] + \frac{J}{2} x(1-x), \quad (1)$$

where k_B is the Boltzmann constant, T is the temperature, and $J > 0$ is the attractive interaction parameter.

Next we consider the stretching free energy of the lipid chains. To express the first-order phase transition, we use a Landau free energy [2]. The order parameter is defined by $\psi \equiv (\delta - \delta_0)/\delta_0$, where δ is the bilayer thickness, and δ_0 is that of the disordered phase. The stretching free energy of a lipid chain is then expressed by

$$f_2^{\ell\ell} = \frac{1}{2} a_2' [T - T^*(x)] \psi^2 + \frac{1}{3} a_3 \psi^3 + \frac{1}{4} a_4 \psi^4, \quad (2)$$

where we define as $a_2' > 0$, and require $a_3 < 0$ and $a_4 > 0$ to ensure the thicker ordered phase. The reference temperature $T^*(x)$ corresponds to the critical temperature if the cubic term is absent. Here we assume that $T^*(x)$ has a linear dependence on x: $T^*(x) = x T_s^* + (1-x) T_u^*$, where T_s^* and T_u^* are the reference temperatures of the single component bilayer composed of saturated and unsaturated lipids, respectively.

To obtain the phase diagram, we minimize $f^{\ell\ell}$ with respect to ψ first. Figure 1 (a) shows one of the typical phase diagrams of the saturated-unsaturated lipid systems. This type of the phase diagram was experimentally observed for aqueous dispersions of a DEPC-DPPC binary system.

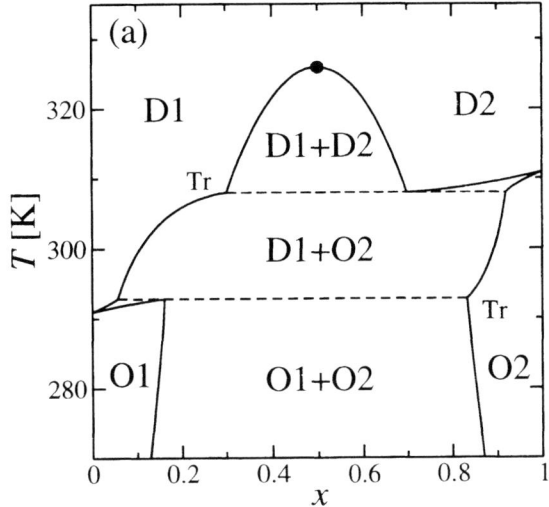 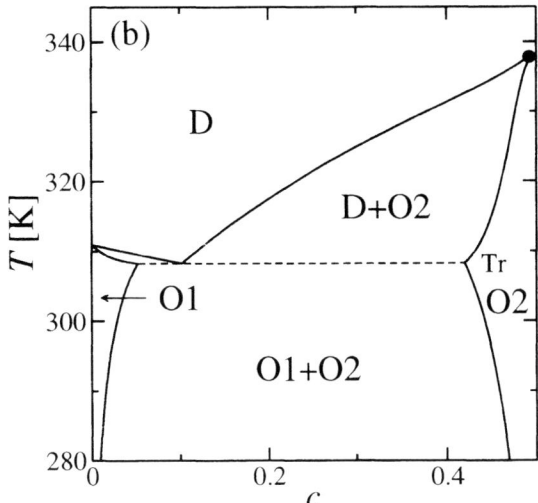

FIGURE 1. The numerically calculated phase diagrams for (a) the saturated-unsaturated lipid system, and (b) the lipid-Chol system. The critical point is indicated by a filled circle, and the triple point by Tr. The ordered and the disordered phase are denoted by O and D, respectively. The parameters are $J = 1.8 \times 10^{-20}$J, $a'_2 = 2.4 \times 10^{-21}$J, $a_3 = -1.1 \times 10^{-18}$J, $a_4 = 2.2 \times 10^{-18}$, $T_s^* = 260$K, $T_u^* = 240$K, $\Gamma_1 = 4.0 \times 10^{-20}$J, $\Gamma_2 = 1.7 \times 10^{-19}$J, $T^* = 260$K.

Lipid-Cholesterol system. For the lipid-Chol systems, it is known that phase separation occurs. At low Chol concentration, the high temperature disordered phase and the low temperature ordered phase are stable. At high Chol concentration, the so-called *liquid-ordered* (*lo*) phase which is peculiar to this system exists. In order to explain the phase behavior of the lipid-Chol systems, the "dual effect" of the Chol has been suggested. Due to the rigid molecular structure of the cholesterol, the hydrocarbon chain in the ordered phase becomes disordered, whereas it becomes ordered in the disordered phase. The phase diagram of the lipid-Chol system was reported for a mixture of DPPC and Chol [3]. The two-phase coexisting state between the disordered phase and the *lo*-phase is the basic structure of the rafts in cell membranes.

We consider three contributions to the total free energy of a bilayer. These are the mixing free energy between lipid and Chol (f_1^{lc}), the stretching free energy of lipid chains (f_2^{lc}), and the coupling energy (f_3^{lc}). The total free energy is approximated as the sum of these three energies: $f^{lc} = f_1^{lc} + f_2^{lc} + f_3^{lc}$.

Since the two-phase coexistence region is limited for $c < 0.5$ in the experiment [3], we assume that all the cholesterols form dimers with lipids. Following the Flory-Huggins theory, the mixing free energy becomes

$$f_1^{lc} = k_B T [c \log 2c + (1 - 2c) \log(1 - 2c)], \quad (3)$$

where c is the molar concentration of Chol and is restricted to $c < 0.5$.

The stretching free energy of lipid chains can be expressed as before:

$$f_2^{lc} = \frac{1}{2} a'_2 [T - T^*] \psi^2 + \frac{1}{3} a_3 \psi^3 + \frac{1}{4} a_4 \psi^4, \quad (4)$$

where T^* is a constant.

The simplest but still meaningful coupling energy to account for the above mentioned dual effect of Chol can be phenomenologically expressed by

$$f_3^{lc} = \frac{1}{2} \Gamma_1 c \psi - \frac{1}{2} \Gamma_2 c^2 \psi, \quad (5)$$

where $\Gamma_1 > 0$ and $\Gamma_2 > 0$ are the constants. We assume that the presence of Chol decreases the chains ordering at low Chol concentration and increases the chains ordering at high Chol concentration. Since both of the coupling terms are linear in ψ, they act as an external field on the order parameter ψ. The second term in Eq. (5) has a tendency to eliminate the first-order transition, and it overwhelms the first term at high Chol concentration. We consider that this effect explains the existence of the *lo*-phase. Figure 1 (b) shows one of the typical phase diagrams of the lipid-Chol systems. The O2-phase corresponds to the *lo*-phase.

REFERENCES

1. Simons, K., and Ikonen, E., *Nature* **387**, 569-572 (1997).
2. Goldstein, R. E., and Leibler, S., *Phys. Rev. A* **40**, 1025-1035 (1989).
3. Vist, M. R., and Davis, J. H., *Biochemistry* **29**, 451-464 (1990).

The Principle for Artificial Molecular Machine and Noise: Slow Stochastic Dynamics in Complex Systems

[1)]H. Matsuura, [2)]M. Nakano, [3)]T. Nemoto

1) National Graduate Institute for Policy Studies, 2)University of Occupational and Environmental Health, 3) Tokyo metropolitan Collage

Abstract. A novel Nano-Molecular Machine Model (DS-NMM) was proposed in order to further the understanding of the movement of the actin-myosin system in muscle. DS-NMM is comprised of numerous inclined rods extending from a central body in a manner similar to myosin; furthermore, its movement is forward in one direction upon independent vibration of these rods. DS-NMM can convert thermal noise to unidirectional motion employing stochastic resonance and inclined rods in random open fields. Concrete estimates were obtained of the physical characteristics of DS-NMM as a conceptual model of the actin-myosin system. When DS-NMM displays radius of 5 nm and mass of 10^{-21} kg, which are characteristic of a myosin molecule, the estimated frequency of stochastic resonance closely approaches phonon frequency, 10^{11} rad/s. The amplitude of stochastic resonance is 2×10^{-11} m. Moreover, physical work of this machine is approximately 10^{-18} joules, which is nearly equal to the energy output generated upon hydrolysis of one ATP molecule (10^{-10} N/head). We concluded that the actin-myosin system in muscle derives its movement from the random motion of water molecules

INTRODUCTION

In 1994, we proposed a model, referred to as the stochastic inclined rod model (SIRM), which was designed to estimate the input and output energies. In 1998 and 1999, we clearly demonstrated on the basis of the numerical calculations that our model, SIRM, was a type of nanomachine utilizing stochastic resonance as expected.

The SIRM operates based on a vastly different principle relative to these aforementioned models. The principle on which the SIRM model operates involves stochastic (random thermal noise) motion of water molecules, or thermal white noise, around the actin-myosin. In contrast, the former models assume that the random motion of the water molecules disturbs the motion of the actin-myosin system.

Very recently, Yanagida measured the motion of the myosin head on an actin filament using fluorescence microscopy, which revealed the walking behavior of myosin heads. The new experimental facts support the SIRM model in principle. Yanagida proposed an interpretation based on the stochastic process, stochastic dynamics and the actin-myosin system of Brownian motion.

Therefore, the objective of this paper is to show directly that our novel improved SIRM model (DS-NMM) agrees with the recent experimental data of Yanagida et al. Moreover, we attempt to offer an explanation regarding the role of ATP hydrolysis and a mechanism governing bipedal locomotion of bicephalic myosin. In addition, we intend to present evidence in support of the sliding mechanism by application of the DS-NMM for natural actin-myosin

STRUCTURE OF DS-SIRM

We presume that our model (SIRM or DS-NMM) apparently exists in an open dissipative system as well as in natural muscle; moreover, the sliding or kinetic energy of this system is supplied by an external heat or energy source in the form of ATP hydrolysis. The New Stochastic Inclined Rods Model, which incorporates double springs (DS-NMM), initially proposed in the present paper[1]-[3], possesses complex and real structure in comparison to our previous SIRM. However, it is more likely to mimic a natural actin-myosin system, thus providing a more accurate explanation regarding the role of ATP hydrolysis and the elementary process involved in the sliding mechanism In DS-NMM[4], the actin-myosin system is comprised of nine parts as shown in Figure 1: the myosin bundle (thick filaments), the soft myosin rod (myosin tail), the neck (2nd flexible hinge region: 2nd soft spring), posterior element of the myosin head (2nd domain of the myosin head), ATP binding site, 1st flexible hinge region (1st hard spring), anterior element of the myosin head (interacting with actin filaments, 1st domain of the myosin head), an actin filament (F-actin composed of G-actin) and the Z-membrane, which provides elasticity for the actin filament. An anterior portion of the myosin head interacts with an actin filament through a generalized intermolecular potential U_a. This potential indicates summation of a repulsive force and an attractive force from the G-actin to the 1st domain of the myosin head as Lennard-Jones potential commonly adopted in molecular dynamics. The 1st domain and the 2nd domain of the myosin head are subject to another intermolecular force by the 1st hinge region, the 1st spring, the potential of which is expressed approximately as vibration of the 1st spring in a manner similar to a harmonic oscillator. This potential represents the force that attempts to maintain the length of the 1st hinge region (natural length of the 1st spring) as a common spring; therefore, this potential consistently works to bring the myosin head to its proper position. ATP binding at an ATP binding site functions as a wedge bending the 1st spring, which induces a change in conformation between the 1st domain and the 2nd domain as exhibited in X-ray analysis. Moreover, this ATP binding affects the spring of the 2nd hinge region, thus altering the physical relationship between the 2nd domain and the myosin tail (rod), i.e., allosteric effects as in the case of an enzyme.

FORMALISM OF DS-SIRM

The equations of motion at the center of gravity (x, y) of the 1st domain are

$$m_1 \ddot{x} = -\frac{\partial}{\partial x}(U_a + U_{1S}) + F_{1x}(t) - \alpha_1 \dot{x} \quad (1)$$

$$m_1 \ddot{y} = -\frac{\partial}{\partial y}(U_a + U_{1S}) + F_{1y}(t) - \beta_1 \dot{y} \quad (2)$$

In our DS-NMM, the equation of myosin bundle is (X, Y) of the myosin bundle, a similar equation:

$$m_B \ddot{X} = -\frac{\partial}{\partial X} U_{2S} + F_X(t) - \eta \dot{X} \quad (3)$$

Here, $F_{1x}(t), F_{1y}(t), F_{2x}(t), F_{2y}(t)$ and $F_X(t)$ are fluctuations in the thermal noise. We assume that the potential of G-actin exhibits a spherical shape (the shape of G-actin is spherical as shown in Figure 1):

$$U_a = \sum_{n,j} U_a^0 \left(\sqrt{(x-x_j)^2 + (y-y_j)^2} - R \right)^{-n+1} \quad (4)$$

In the case of ATP binding with ATPase, the 1st spring is inclined relative to the y-axis:

$$U_{1S-ATP} = A \exp\left(-\sqrt{(x-x_2)^2 - (y-y_2)^2} + L\right) \\ + A \exp\left(\sqrt{(x-x_2) + (y-y_2)^2} - L\right) + \frac{1}{2} KL^2 (\theta - \theta_0)^2 \quad (5)$$

Thus, the solution for myosin bundle is expressed as

$$X = Const. + \frac{KA}{\eta} t + \left\{ \frac{KA \sin \gamma t}{\eta \gamma} - \frac{KA(\eta M \cos \gamma t + M^2 \gamma \sin \gamma t)}{\eta(\eta^2 + \gamma^2 M^2)} \right\} \quad (6)$$
$$+ c \exp(-\rho t) \cdot \cos(\omega t + \vartheta)$$
$$+ (thermal\ noise\ terns, rapidly\ damping\ irregular\ motion)$$

This equation shows that DS-SIRM, accepted an energy of 1-ATP hydrolysis, moves in one direction by the mechanism of the hitting motion and SR. With using numerical method for Eq.(1)-Eq.(6), we can calculate the Signal-Noise Ratio(SNR, Fig.1) and can obtain the displacement of myosin head(Fig.2).

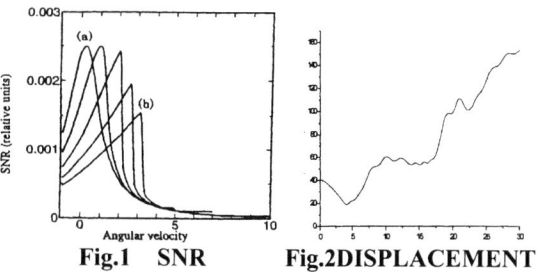

Fig.1   SNR   **Fig.2 DISPLACEMENT**

The Fig.1 shows that that the graph of the SN ratio consistently exhibits one peak, which indicates the existence of the resonance with noise. The straight line in Fig.2 displays the position of the myosin head; this trace displays irregularities and randomness akin to Brownian particles, which exhibit translational motion as a whole. The SR is a kind of the quantum the coherent state as shown in our paper[4].

REFERENCES

1. Matsuura, H., Nakano, M,. *Jr. of Biomed. Fuzzy Sys. Association*, **3(1)**, 23-31(1997).

2. Matsuura, H., Nakano, H., *Jr. of Biomed. Fuzzy Sys. Association*, **4(1)**, 47-52,(1998).

3. Bao-quan, A., Liu,W., Liu,L., H. Matsuura, Nakano, M., *INFORMATION*, **6(2)**, 187-196 (2003).

4. Matsuura, H., Nakano, M., *Jr. of Biomed. Fuzzy Sys. Association*, **8(1)**, 1-13(2002).

A lattice model of the protein diffusion in membranes

S. Kinouchi*, K. Tamura*, S. Komura*, T. Kato* and Y. Y. Suzuki[†]

*Faculty of Science, Tokyo Metropolitan University, Tokyo 192-0397, Japan
[†]Faculty of Engineering, Takushoku University, Tokyo 193-0985, Japan

Abstract. Diffusion of a protein in a biological membrane is studied by Monte Carlo computer simulations. The membrane is modeled as a two-dimensional lattice in which a protein and lipids diffuse under the action of Brownian motion. We calculate the diffusion coefficient of the protein as the concentration of lipids and the protein size are changed. These results are compared with our analytical calculation.

INTRODUCTION

Biological membranes consist of various lipid molecules and protein molecules, and their fundamental structure is a lipid bilayer including proteins. Since the "fluid mosaic model" was proposed by Singer and Nicolson in 1972, lipids in membranes have been considered that they are distributed uniformly, and can move almost freely. Later, dynamical domains which are organized by sphingolipids and cholesterol were observed in biological membranes. These domains are called "rafts".

Recently, new techniques of microscopy have made it possible to observe the motion of individual proteins or small clusters of lipids on the cell surface [1]. In these experiments, proteins or lipids are labeled with a highly fluorescent label or with colloidal gold microspheres. The shape of the trajectories implies various biologically important processes, such as binding to immobile species, free diffusion, hindered diffusion, directed transport, and trapping of particles in bounded microdomains.

In this paper, by using a lattice-model of a biological membranes, we investigate both numerically and analytically how the diffusion of a protein depends on its size and the concentration of lipids. The present study is the extension of Ref. [2] in which the concentration of lipids was fixed.

MODEL AND RESULTS

The key assumption embodied in the stochastic model is that the force exerted by the medium on the diffusing molecule consists of a stationary, rapidly fluctuating random force due to molecular bombardments which are independent of the velocity of the diffusing molecule. The drag force proportional to the velocity is supposed to

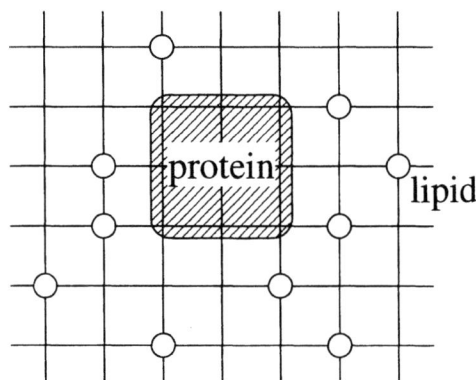

FIGURE 1. A two-dimensional square lattice model of a biomembrane. The square represents a protein, and white circles are lipids. The size of a protein is denoted by the number of sites occupied by it ($M^2 = 3 \times 3$ in this case).

be smeared out, because the water molecules immediately conduct away the momentum from the diffusing molecule.

We consider a two-dimensional square lattice model as shown in Fig. 1. A protein is represented by the square which covers $M \times M$ lattice sites, and lipids are represented by the small particles which occupy a single site. We have performed Monte Carlo (MC) computer simulations for the stochastic model with a 128×128 square lattice. Periodic boundary conditions are used in the simulations. The lipids, whose concentration is denoted by c, are initially distributed on the lattice so that they do not overlap with the $M \times M$ square (protein) or with each other.

In the computer simulation, a lattice site is randomly chosen initially. When there is a particle on the selected site, the particle is moved randomly into one of the four nearest neighbor lattice sites. The movement is executed

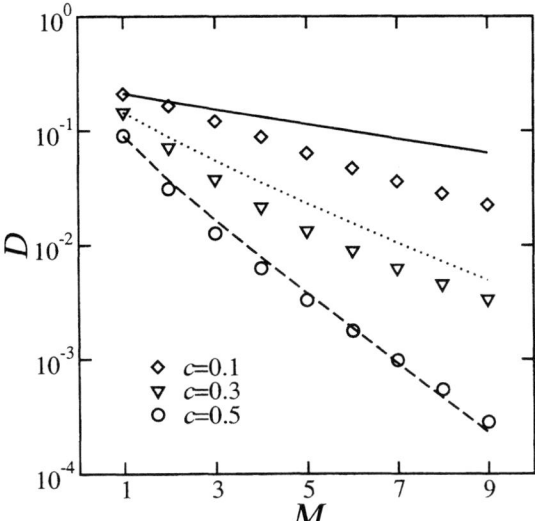

FIGURE 2. The protein diffusion coefficient D as a function of its size M. The symbols are MC results. The solid, dotted, and dashed curves represent Eq. (3) for $c = 0.1$, 0.3, and 0.5, respectively.

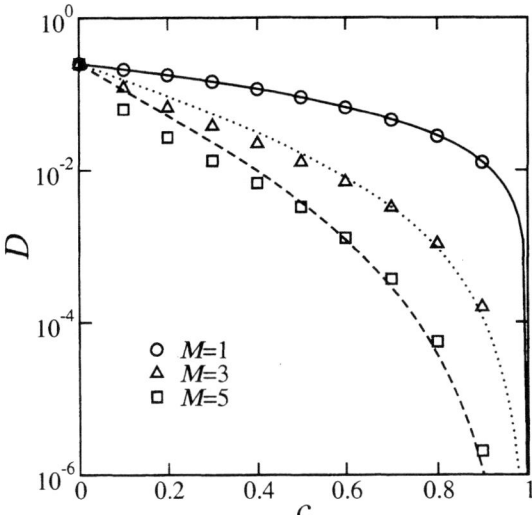

FIGURE 3. The protein diffusion coefficient D as a function of the lipid concentration c. The symbols are MC results. The solid, dotted, and dashed curves represent Eq. (3) for $M = 1$, 3, and 5, respectively.

if and only if it does not lead to any overlapping either among particles or between the square and the particles. A set of attempts which, on average, covers the entire lattice defines one Monte Carlo step (1 MCS).

The diffusion coefficient of a k MCS interval is calculated by

$$D(k) = \frac{1}{4}\frac{\langle r^2(k)\rangle}{k}, \quad (1)$$

where $r(k)$ is the absolute value of the displacement vector during an interval of k MCS's, and the average $\langle \cdots \rangle$ is taken for the displacement data over one simulation.

In the long-time limit ($k \to \infty$), $D(k)$ converges to a fixed value. Figure 2 shows the calculated diffusion coefficient D as a function of the protein size M, whereas Fig. 3 gives that as a function of the lipid concentration c. Note that the case of $M = 1$ is identical with self-diffusion of lipids. We see in Fig. 2 that the size dependence of the diffusion coefficient becomes more pronounced for larger c. On the other hand, the lipid concentration dependence of the diffusion coefficient is stronger for larger M as shown in Fig. 3.

As an analytical approach, we extend the Nakazato-Kitahara theory [3]. The number of sites is denoted by N, and the number of sites occupied by lipids is denoted by N_0. The average concentration of lipids is then defined as $c = N_0/(N - M^2)$. After some calculations, the diffusion coefficient is written in the form

$$D(M,c,k) = (1-c)^M f(M,c,k) D_0, \quad (2)$$

where D_0 is the diffusion coefficient in the case of $c = 0$, and $f(M,c,k)$ is called as the "correlation factor". The analytical approximate solution of D is obtained as

$$D(M,c,k) = \frac{1}{4}(1-c)^M \left(1 - \frac{A}{B}\right). \quad (3)$$

where

$$A = 2\alpha(k) \sum_{i=1}^{M} \binom{M}{i} c^i (1-c)^{M-i}, \quad (4)$$

$$B = 1 + (1-c)^M - \alpha(k)\left[1 + (1-c)^M - 2\sum_{i=1}^{M} \binom{M}{i} c^i (1-c)^{M-i}\right]. \quad (5)$$

Here, $\alpha(k)$ is the lattice Green's function. In the long-time limit ($k \to \infty$), $\alpha(k)$ approaches to 0.363. Equation (3) is plotted in Figs. 2 and 3 by various lines. When $M = 1$, the analytical result is in complete agreement with the MC simulation result as seen in Fig. 3. In Fig. 3, the size dependence is well described by Eq. (3) for larger c.

REFERENCES

1. Anderson, C. M., Georgiou, G. N., Morrison, I. E. G., Stevenson, G. V W., and Cherry, R. J., *J. Cell. Sci.* **101**, 415-425 (1992).
2. Suzuki, Y. Y., and Izuyama, T., *J. Phys. Soc. Jpn.* **58**, 1104-1119 (1989).
3. Nakazato, K., and Kitahara, K., *Prog. Theor. Phys.* **64**, 2261-2264 (1980).

Necklace on a Strongly Charged Polyelectrolyte

Takahiro Sakaue

Department of Physics, Kyoto University, Japan

Abstract. Motivated by recent experimental observations of single DNA molecules in the intra-chain segregated state, we study the collapse transition of a single polyelectrolyte chain by the addition of condensing guest molecules. In this proceeding, we specifically concentrate on the appearance of "rings-on-a string" structures, which are observed using atomic force microscopy in the experiment with gemini surfactant as a condensing agent. We show that owing to the chain stiffness and electrostatic interaction, a semiflexible polyelectrolyte takes "rings-on-a-string" structures within a certain range of guest molecules concentration.

INTRODUCTION

A DNA molecule is a highly charged polymer, and assumes a disordered swollen coil in usual aqueous solutions. Although DNA molecules are complicated heteropolymers with hierarchical structure, many of the interesting behaviors can be captured by the coarse-grained homopolymer model with uniform rigidity and/or charge density. Among them, the drastic conformational change induced by the addition of so-called condensing agents has been attracting much attention both from physical and biological points of view. The conformatioal transition of a polymer chain from a swollen coil state into a compact state, *i.e.*, coil-globule transition, has been intensively studied during the past few decades[1]. Although the theory is well developed for the ideal case of simple homo-polymer, there still remain some gaps between our understanding and the transition behavior of DNA molecules. The possible picks for these gaps may be the effect of chain stiffness and/or electrostatics, which leads to the variation in the transition behavior.

One of the most interesting examples is the appearance of the intra-chain segregated, or, partially collapsed state of the single DNA molecules. It has been shown that this phenomenon is induced by the various kinds of condensing agents. Recent atomic force microscopy observation[2] has revealed the fine structure of the intra-chain segregated DNA molecule. As is schematically shown in Fig. 1, the collapsed parts assume the form of the ordered torus, which are interconnected by the disordered coil parts. Motivated by this experimental finding, we consider the collapse transition of a single model polyelectrolyte (PE) by the addition of guest molecules, and propose a possible scenario for the appearance of "rings-on-a-string" structure.

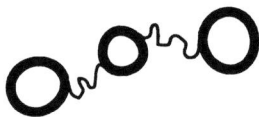

FIGURE 1. A schematic picture of "rings-on-a-string" structure observed by atomic force microscopy. In this experiment, a di-cationic surfactant (gemini surfactant) is used as a condensing agent.

THEORETICAL MODEL

Situation and Assumptions

We consider a single semiflexible PE in the large, but, finite volume. The system contains a moderate concentration of 1:1 salt, thus, the electrostatic interactions are screened on the order of PE segment diameter. The control parameter is the concentration of oppositely charged guest molecules with some collapsing ability. Each PE segment takes two distinct states; either densely packed torus or coil. Accordingly, guest molecules as a condensing agent are considered to be inside the torus, or, condensed on the coil, or, free in the outer solution. The information on how PE segments and guest molecules are packed inside the torus and the mechanism on the formation of the complex are important issues, but still unclear, especially for the complicated guest molecules. In the case of surfactant guest molecules, it is expected that the hydrophobic interaction may play some role to stabilize the torus, and that torus may not be electrostatically neutral. In the present proceeding, by assuming the net effective charge inside the torus, we construct our theoretical model, and ask the possible mechanism responsible for the rings-on-a-string structure.

Free energy

Let us briefly describe our model. We consider the free energy of the single PE with n_t tori interconnected by $n_c (= n_t, \text{ or } n_t \pm 1)$ coils.

$$F(n_t, n_c, \{N_{t,i}\}, \{N_{c,j}\}) = \sum_{i=1}^{n_t} F_t(N_{t,i}) + \sum_{j=1}^{n_c} F_c(N_{c,j}) + F_{out},$$

where $F_t(N_{t,i})$, $F_c(N_{c,j})$ and F_{out} are the free energy of i th torus with $N_{t,i}$ segments, j th coil with $N_{c,j}$ segments and outer solution, respectively. The free energy of the torus is decomposed as a sum of the volume, surface, bending, electrostatic energy and configurational entropy of condensed guest molecules. The free energy of the coil consists of the electrostatic energy and configurational entropy of guest molecules condensed onto the coil and the conformational freedom of the PE coil. Finally, the free energy of the outer solution mostly arises from the configurational entropy of free guest molecules.

The torus is characterized by two radii of curvature; the average radius R, and thickness r of the torus. They are determined by the balance of the bending, surface and electrostatic energy of the torus. It can be shown that for a relatively highly charged chain, the competition between latter two factors leads to the finite thickness almost independent of the chain length. The distribution of guest molecules is determined by the binding equilibrium among three states (note that, in our model, the fraction condensed into the torus is a fixed given parameter).

RESULTS

We have constructed the model free energy for the single PE composed of n_t tori interconnected by n_c coils with the segment distribution $\{N_{t,i}, N_{c,j}\}$. Now it is possible to calculate the partition function with n_t tori, $Z(n_t)$, by taking summation over the all possible segment distribution for n_t tori with the constraint that the total segment number is constant. Then, the thermodynamic properties of the model system are readily discussed.

As an example of the result, Fig. (2) shows the dependence of the probability to find the structure with n_t tori $P(n_t)$ on the guest molecules concentration c_{guest}. At first, a PE coil is collapsed to the intra-chain segregated structure with one torus. With the increase of c_{guest}, the intra-chain segregated structures with two and three tori begin to appear.

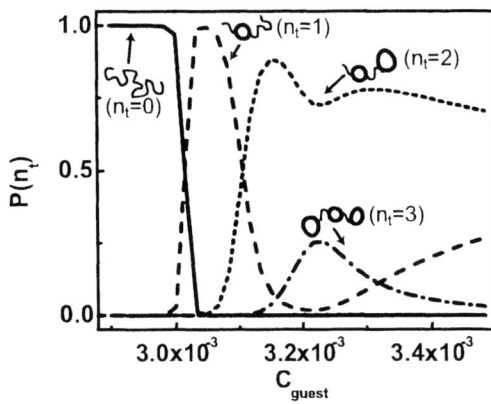

FIGURE 2. Dependences of $P(n_t)$ on c_{guest}. The used parameters are the followings; chain length $= 1.4 \times 10^4$, Kuhn length $= 20$, chain charge density $= 0.7$, Bjerrum length $= 1$, cohesive energy density, surface tension and the degree of surviving charge of torus$= -0.4, 0.15$ and 0.2, respectively (units of length and energy are the segment diameter and thermal energy, respectively). The system volume is roughly the same as that of the PE chain in coil state.

DISCUSSIONS AND SUMMARY

We have proposed a simple model for the novel PE structure, multiple tori interconnected by strings in the coil state. The reason for the appearance of the intra-chain segregated structure with one torus or multiple tori is ascribed to the unique property of the torus made from a charged chain. The transition manner is expected to depend on the system properties, such as PE chain length, linear charge density and system volume, On the other hand, such structures are energetically too unstable, and never expected in the case of the neutral chain.

ACKNOWLEDGMENTS

I am grateful to N. Miyazawa (Tokyo Medical and Dental Univ.) and K. Yoshikawa (Kyoto Univ.) for very useful discussions on the rings-on-a-string structure of a single DNA molecule.

REFERENCES

1. Grosberg, A. Y., and Khokhlov, A. R., *Statistical Physics of Macromolecules*, American Institute of Physics, New York, 1994.
2. Miyazawa, N., Sakaue, T., Mayama, H., Zana, R., and Yoshikawa, K., in preparation.

Protein folding dynamics: ergodic behavior in principal component space

Yasuhiro Matsunaga* and Tamiki Komatsuzaki*

Department of Earth and Planetary Sciences, Faculty of Science, Kobe University

Abstract. Using a novel method for determining the time scale for the self-averaging of properties of many body systems, the effective ergodic behavior of a coarse-grained protein model was investigated in terms of both the chemical characters and the principal components. The ergodic behavior was examined in detail with a focus on the time scale needed to obtain effective ergodicity.

In present, it is believed that small single-domain protein folding can be regarded as a normal Brownian process of a few collective coordinates on a thermal potential such as the "funnel" landscape. The diffusive feature may, however, depend on the choice of the viewpoint from which one sees the dynamical events. We have investigated non-stationarity and non-Markovianity of a coarse-grained protein model in terms of principal components [1]. In this presentation, the ergodic behavior was examined in detail with a focus on the time scale needed to obtain ergodicity using the ergodic measure introduced by D. Thirumalai et al [2].

The fluctuation metric of the ergodic measure [2, 3] is defined by

$$\Omega(t) = \sum_{j}^{N}[f_j(t) - \bar{f}(t)]^2, \quad (1)$$

where $f_j(t)$ is the time average for atom (or mode) j of property F of the system,

$$f_j(t) = \frac{1}{t}\int_0^t ds F_j(s) \quad (2)$$

and $\bar{f}$ is the average over M atoms (or modes) of property F at time t,

$$\bar{f}(t) = \frac{1}{M}\sum_{j}^{M} f_j(t). \quad (3)$$

In this study, we take F_j to be the kinetic energy of the jth atom (or mode). The criterion for ergodicity requires $\Omega(t)$ to decay to zero with time.

Principal component analysis is often used to capture *essence* of complex dynamics in proteins. Let $\mathbf{X}(t)$ be the coordinates of a protein at time t. The variance-covariance matrix $\mathbf{R}$ is defined by

$$\mathbf{R} = \langle(\mathbf{X}(t) - \langle\mathbf{X}(t)\rangle)(\mathbf{X}(t) - \langle\mathbf{X}(t)\rangle)^\mathrm{T}\rangle, \quad (4)$$

where $\langle\cdot\rangle$ denotes the time average and $^\mathrm{T}$ denotes transposition. Then we can define a set of principal components $\mathbf{Q}$ by using the eigenvectors $\mathbf{U}$ that diagonalize $\mathbf{R}$:

$$\mathbf{RU} = \mathbf{U}\mathbf{r} \quad (\mathbf{U}^\mathrm{T}\mathbf{U} = \mathbf{I}) \quad (5)$$

The eigenvalue r_i, the ith element of the diagonal matrix $\mathbf{r}$, represents the variance of the ith collective coordinate

$$Q_i = \sum_{j=1} u_{ji} X_j. \quad (6)$$

A protein model we chose is a coarse-grained, off-lattice, 3-color, 46-bead protein model [5]. The model is composed of hydrophobic (B), hydrophilic (L), and neutral (N) beads for which the van der Waals interactions are used to mimic chemical character of each bead. $B_9N_3(LB)_4N_3B_9N_3(LB)_5L$ folds into a lowest-energy β-barrel structure with four strands (Fig. 1). The topograpy of the potential energy landscape fot this model has been well surveyed in terms of its disconnectivity graph [4]. The ergodic measure was calculated along each principal component and for each bead at several temperatures (temperatures were controlled by Nosé-Hoover chain algorithm).

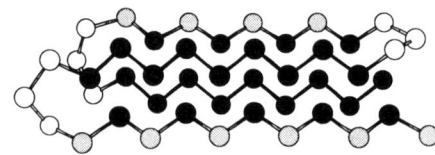

FIGURE 1. Side and end views of the global minimum of the 46-beads model. Hydrophobic, hydrophilic, and neutral beads are shaded dark gray, light gray and white, respectively.

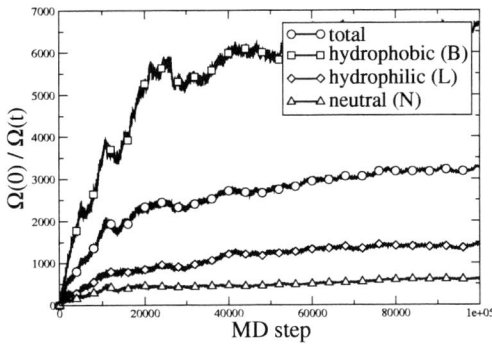

FIGURE 2. $\Omega(0)/\Omega(t)$ as a function of time.

In Fig. 2, several forms of $\Omega(0)/\Omega(t)$ for the 46-beads model system are plotted. In addition, the plots for the the hydrophobic beads, the hydrophilic, and the neutral alone are displayed. At low temperature, the hydrophobic group become ergodic more rapidly rather than other residue groups. This is because that the hydrophobic group has stronger interaction each other rather than other groups. The result indicates that the environment of the hydrophobic residue is not unlike that of another residue.

For example, some $\Omega(0)/\Omega(t)$ are shown at low temperature where the system resides near the most stable configuration. Here, $\Omega(0)/\Omega(t)$ are elucidated with $N = M = 46$ (total) and N = number of B, L and N beads(B,L,N) and $M = 46$. This indicates significant inhomogeneous nature of ergodic behavior depending on the kinds of beads. The most weakly interacting neutral beads linking two β strands are slowest to be thermalized, but the hydrophobic beads are much more faster to be thermalized in the "core" of protein than the others. We will discuss how such an ergodic behaivor changes as temperature increases, reflecting the dynamical behaviors, in terms of both the beads and the principal components.

REFERENCES

1. Matsunaga, Y., Kostov, K.S., and Komatsuzaki, T., *J. Phys. Chem. A* **106**, 10898-10907 (2002).
2. Thirumalai, D., Mountain, R.D., and Kirkpatrick, T.R., *Phys. Rev. A* **39**, 3563-3574 (1989).
3. Sagnella, D.E., Straub, J.E., and Thirumalai, D., *J. Chem. Phys.* **113**, 7702-7711 (2000).
4. Wales, D.J., Doye, J.P.K., Miller, M.A., Mortenson, P.N., and Walsh, T.R., *Adv. Chem. Phys.* **115**, 1-114 (2000).
5. Honeycutt, J.D., and Thirumalai, D., *Biopolymers* **32**, 695-709 (1992).

A Coarse-Graining of Energy Landscape of Proteins
- structural stability of the most stable states -

Kyoko Hoshino[*], Yasuhiro Matsunaga[*], Mark Miller[†], David J. Wales[**] and Tamiki Komatsuzaki[*]

[*]*Nonlinear Science Laboratory, Department of Earth and Planetary Science, Faculty of Science, Kobe University, Nada, Kobe 657-8501 Japan*
[†]*FOM Institute, Netherlands*
[**]*University of Cambridge, UK*

Abstract. The minimal sets of degrees of freedom to persist the multidimensional topographical feature are investigated for the funnel, and frustrated energy landscapes of proteins. The robustness of local energy topographies near the global energy minimum structures against a perturbation is also discussed.

The questions, "how degrees of freedom can be reduced as few as possible, to persist the topographical feature of the funnel, and frustrated energy landscapes of proteins?" and "how is the energy landscapes topography persisted or destroyed near the most stable structure when it is under perturbations?" are among the outstanding open questions in protein folding. Our vehicles to attack these problems are 46-beads model [1] (called BLN), an off-lattice model of bead sequences that folds into a b-barrel native structure, which is 46 monomers long with 3-letter codes (See Fig.1) and Gō-like BLN model, which has an energy bias toward global energy minimum [2].

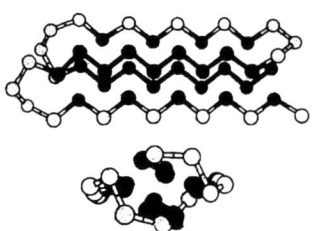

FIGURE 1. The side and end views of the global minimum of the BLN model. Hydrophobic, hydrophilic, and neutral beads are represented by shaded dark gray, white and light gray, respectively. [3]

By eigen-vector following search algorithm [3], are sampled as thoroughly as possible the saddle and minimum energy structures of 46-beads model and the Gō-like model (1000 2000 stationary points), for which the principal component (PC) analysis is employed. The 99.9% of the total variance ($\sigma_{total} \equiv \sum_{i=1}^{138} \sigma_i^2$) for each model is reproduced by 40-50 principal components. We chose the n (46-92 in our case) largest eigenvalues' PCs from 138 degrees of freedom by which the original Cartesian coordinates were elucidated approximately: the inverse transformation matrix is multiplied to them with putting zeros into the residual PCs. How the two distinct energy landscape topographies can be approximated in terms of a reduced set of PCs? In Fig.2, the difference (ΔV) between the approximated potential energies with n=46 and 92 and the exact those are plotted for both of models. It is found that, the lower the potential energy, the less the deviation, and the deviation in Gō-like BLN model is lesser than that in BLN model. Here the bonding, and non-bonding energy terms contribute mostly to the energy deviation due to truncating 46 small variance's PCs, and the bending, and the torsional energies don't contribute so much. By introducing a new topological index $N_i(\varepsilon)$, i.e., the number of stationary points within the multidimensional sphere whose radius from the ith point is ε, which is taken to be one tenth of the square of largest eigenvalue of PCs, and the local topological feature of the energy landscape is estimated near each stationary point. As a preliminary result, it is found that the more the value of $N_i(\varepsilon)$, the better the energy landscape can be reproduced in terms of a reduced set of degrees of freedom. The degrees of freedom in the funnel energy landscapes can be reduced lesser than those in the frustrated energy landscapes. It is because, as shown in Fig.3, the stationary points with large $N_i(\varepsilon)$ are more densely distributed near the most stable structure in Gō-like model than those in BLN model. To grasp the concernment between $N_i(\varepsilon)$ the connectivity of the stationary points and topographical feature of the energy landscapes, is shown the disconnectivity graph [4] for the BLN model in Fig.4 (BLN), based on sample of 500 minima and 636 transition states, and for Gō-like model in (Gō-like BLN), based on a sample of 500 minima and 805 transition states. It is clear that the pathway from nearby low energy structure is shorter with larger $N_i(\varepsilon)$ (4-5) in Gō-like model, compared with frustrated BLN model whose $N_i(\varepsilon)$ is smaller (1-2). As a conjecture, one may expect that the funnel-energy landscape more tend to persist the local energy topography near the native structure against some mutations or changes in environment, if one can regard the lack of some noisy PCs as a perturbation to the system.

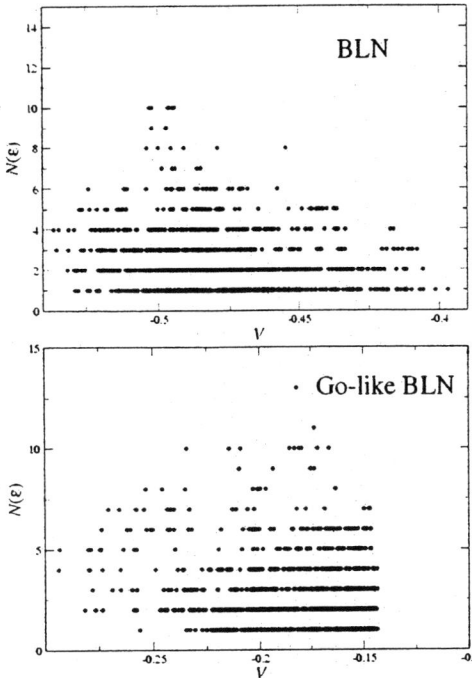

FIGURE 3. The relationship between V and $N_i(\varepsilon)$. $N_i(\varepsilon) = 1$ means there is no stationary points besides the center's point. e was taken to be 1/10 of the square of largest eigenvalue of PCs.

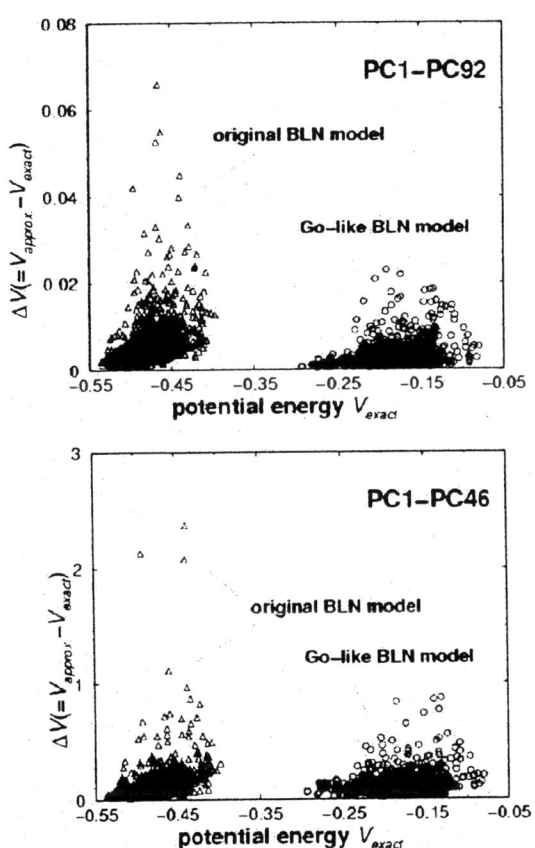

FIGURE 2. The difference (ΔV) between the approximate potential energy and the original potential energy.

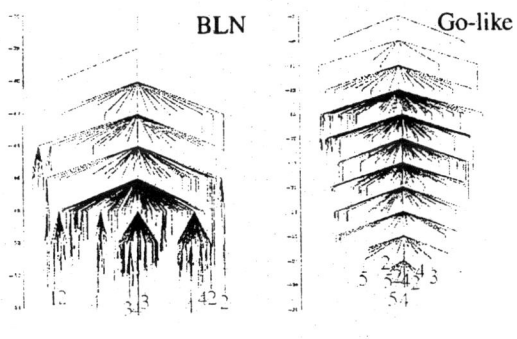

FIGURE 4. Disconnectivity graph [4] of BLN (left) and Gō-like BLN (right) models with $N_i(\varepsilon)$. The number in the figures denote the values $N_i(\varepsilon)$ of the ith stationary point.

REFERENCES

1. Honeycutt, J.D., and Thirumalai, D., *Biopolymers* **32**, 695-709 (1992).
2. Nymeyer, H., García, A.E., and Onuchic, J.N., *Proc. Natl. Acad. Sci. U.S.A.* **95**, 5921-5928 (1998).
3. Miller, M.A., and Wales, D.J., *J. Chem. Phys.* **111**, 6610-6616 (1999).
4. Becker, O.M., and Karplus, M., *J. Chem. Phys.* **106**, 1495-1517 (1997).

Charge Conductivity In Peptides: Dynamic Simulations Of A Bifunctional Model Supporting Experimental Data

Sheh-Yi Sheu*, Dah-Yen Yang†, H. L. Selzle** and E. W. Schlag**

*Department of Life Science, National Yang-Ming University, Taipei, Taiwan
†Institute of Atomic and Molecular Science, Academia Sinica, Taipei, Taiwan
**Institut fur Physikalische und Theoretische Chemie, Technische Universitat Munchen, D-85747 Garching, Lichtenbergstrasse 4, Germany

Abstract. We proposed a bifunctional model for charge conducting along polypeptide chain. By computing the efficiency, a β-value, our results agree with experimental value.

Charge conductivity in biomolecules has become a general topic of substantial current interest. This phenomenon is associated with the fascinating issue to what extent these systems can be classified as molecular wires and thus with the question of the proper mechanism for signal transduction in biomolecules such as proteins.

In our bifunctional model, the charge is transported along a polypeptide chain. At each C_α-atom, the torsional angles ψ and φ are constrained in the Ramachandran plot (see Figure 1(a)). The $\overrightarrow{C_\alpha N}$ and $\overrightarrow{C_\alpha C}$ of the C_α-atom form a hinge. The charge is transported from the C-side of the C_α-atom to the N-side. Before the charge is transported, it waits in the C-side until the O-O atom between two connected amino acids collide. The rotational motion of the ψ and φ angles is similar to a Brownian particle moving inside a 2D box with a static gate where the O-O atom collide with each other and charge starts to transfer(see Figure 1(b)). In our MD simulation, we provide charge energy E to the atoms attached to the ψ axis, i.e., N-, O-, and H-atoms. We now define the efficiency as

$$\text{efficiency} = \frac{successful\ configurations}{total\ configurations} \quad (1)$$

The interesting new sidelight on these MD calculations is that we find that not all initial starting configurations of the calculation lead to a firing state, but in fact a fraction of initial states are dissipated, leading to an efficiency in the process of less than unity. Such an inefficiency in charge transport is experimentally well-known and leads to β-values in the efficiency Y of charge transfer with distance

$$Y = Ae^{-\beta R} \quad (2)$$

with typical values in condensed media of β near $1.0 \text{\AA}^{-1}$. Here A is the prefactor.

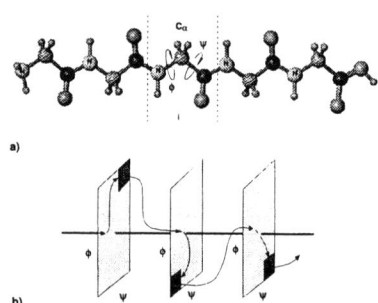

FIGURE 1. Charge transport in a polypeptide. (a) Polypeptide chain. Charge is first created on the donor and then hop through the amino acid chain until reaching the acceptor. (b) On each amino acid, the motion of the rotors is mapped into a Ramachandran plot. Here a simple two-dimensional (2D) area in phase space represents the hinge, i.e., the junction of two amino acids. The exit or gate part (orange) is the charge ratchet position. After the motion of rotors reaches the gate part, charge jumps to the next amino acid. The iteration of the previous procedure makes the charge hops to the final site.

In Figure 2, we excite the C-side of the C_α-hinge at the Val_{10} site of a gas phase polypeptide Mb_{20} with excitation energy $2667\ K$. The efficiency is estimated for the MD simulation time from beginning to $1500\ fs$. Its value is close to unity even at the fourth site away from the local heating site. At the fourth site Leu_{14}, the efficiency decreases abruptly. When we check the first passage time distribution curve in Figure 2A, the first peak (a) is extremely high. This peak is situated at 1 fs right after the C-side of the C_α-hinge is excited. This confirms the ballistic motion of the carbonyl group that moves directly toward the carbonyl group on the N-side of the C_α-hinge. The second peak (b) that occurs at 3

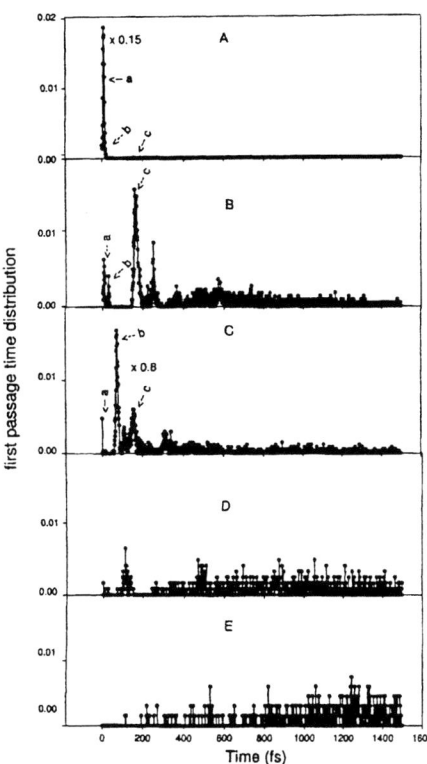

FIGURE 2. First passage time distribution vs time. In this simulation the excitation energy is ca. 2667 K, and the local heating site is at Val_{10}., while the background temperature is 300 K. At the initial stage, there are three peaks generated in figure (A). Peak a denotes the rotational motion of the C-side carbonyl group directly hits the N-side carbonyl group starting from the native structure of Mb_{20}. Peak b corresponds to a short wandering in the initial direction of the C-side carbonyl group. In peak c shows a random motion of the C-side carbonyl group inside the BS box. Here the distribution in the peak c part is similar to a thermal distribution. Hence the Figures (B)-(E) express the propagation of the peaks to the right. This peak propagation shows an energy transfer along the polypeptide chain. The efficiency drops at Leu_{14} (or E). Note that the polypeptide sequence is (N-terminal) Glu_1-Asp_2-Leu_3-$Lysn_4$-$Lysn_5$-Hsd_6-Gly_7-Val_8-Thr_9-Val_{10}-Leu_{11}-Thr_{12}-Ala_{13}-Leu_{14}-Gly_{15}-Ala_{16}-Ile_{17}-Leu_{18}-$Lysn_{19}$-$Lysn_{20}$(C-terminal).

fs is due to the initial random distribution of the "trans" configurations. The third peak (c) situated at 310 fs is due to the random motion of the carbonyl group and the rotational energy is dissipated to vibration modes or thermal motion. This certainly demonstrates that at the local heating site the energy is mainly kept in the rotational degrees of freedom (peak a) and a very small ratio of the energy is dissipated (peak c).

At the site Leu_{14} near the local heating site, the efficiency is unity and the first peak (a) in Figure 2B is lowered. However, the second peak (b) and the third peak (c) are shifted to the right. These peaks are also broadened due to the random motion. In particular, peak (c) is broadened and is similar to Gaussian distribution. This ensures that peak (c) corresponds to a stochastically motion inside the BS box. In the Figure 2C,D, the peaks are broadened and shifted to the right. At the fourth site, in Figure 2E, the efficiency changes abruptly and peak (c) almost disappears.

The same polypeptide Mb_{20} was embedded in a TIP3 water ball with 611 H_2Os. We estimate the successful run with efficiency equal to 0.008, i.e. the β-value =1.32 Å^{-1}, a value in astonishing agreement with experiment. The charge transport along each individual chain inside the β-sheet is seen to have the same efficiency as it has in the α-helix. But the sum of the efficiency of the β-sheet is the geometric sum of each individual chain. Hence, for the example Azurin, the β-sheet contains about three chains. Its efficiency is expected to be about three times higher than that of each individual α-helix chain. We thus have an efficiency of the β-sheet ca. 0.0244; i.e. β-valuet=1.0 Å^{-1}. Therefore, for the theoretical predicted collision distance for firing, the calculated efficiency here predicts a β-value that corresponds closely to that of known experiments.

To summarize, local excitation in our bifunctional model based on two rotors motions around a C_α-hinge and the charge transfer occurring in an O-O atoms collision can be mapped into an escape process inside a subregion with a gate in the bounds of the Ramachandran plot or the BS box. Before the charge is transferred, it waits in the C-side of C_α-hinge. Since it carries charge energy, the corresponding ψ angle is locally excited. By using MD simulation, we introduce a local excitation procedure, which in MD generates the efficiency of successful collision between O-O atoms. The good agreement with the β-value between the simulation result and experimental data suggests the unique result that real protein charge transport depends heavily on protein dynamics. Furthermore, the β-sheet here is seen to be superior in charge transfer to the α-helix just as a result of parallel path and not intrinsically.

REFERENCES

1. Schlag, E. W., Sheu, S. Y., Yang, D. Y., Selzle, H. L., and Lin, S. H., *Proc. Natl. Acad. Sci. USA* **97**, 1068-1072(2000).
2. Schlag, E. W., Sheu, S. Y., Yang, D. Y., Selzle, H. L., and Lin, S. H., *J. Phys. Chem.* **B104**, 7790-7794(2000).
3. Sheu, S. Y., Schlag, E. W., Yang, D. Y., Selzle, H. L., *J. Phys. Chem.* **A105**, 6353-6361(2001).
4. Sheu, S. Y., Yang, D. Y., Selzle, H. L. and Schlag, E. W., *J. Phys. Chem.* **A106**, 9390-9396(2002).
5. Schlag, E. W., Sheu, S. Y., Yang, D. Y., Selzle, H. L., and Lin, S. H., "Signal transport in peptides", *(J. Phys. Chem. submitted)*.

Irreversibility on the Structural Transition under Strain in a Single Semi-flexible Polymer

Natsuhiko Yoshinaga and Kenichi Yoshikawa

Department of Physics, Graduate School of Science, Kyoto University & CREST

Abstract. We present a result of molecular dynamics simulation on mechanical unfolding of a semi-flexible polymer in comparison with a flexible polymer. The result shows that the force response changes remarkably depending on the initial morphology of the folded compact state. We also obtain the marked hysteresis in the loading-unloading cycle of a semi-flexible polymer even under slow loading velocity, whereas a flexible polymer does not exhibit hysteresis.

INTRODUCTION

Recent development in the technique of single molecule observation and manipulation enables us to examine directly the statistical properties of individual small systems.[1] These experiments provide us the deeper understanding of the structure of complex systems such as proteins and, moreover, of the microscopic energetics used in the molecular motors.

Among such single molecule manipulation, polymer pulling experiments with laser tweezers or a atomic force microscopy have been widely studied for the last decade. [2, 3] The advantages of these measurements is that we can obtain force and displacement not only as one of the observable variables which microscopic information are reduced to, but also as set of intensive and extensive variables conjugated each other. Thus we can evaluate the work that is done to the system by integrating the force with the displacement.

On the other hand, most biopolymers including DNA and many of proteins, have finite stiffness. Nevertheless the properties of a single semi-flexible polymer are not completely understood. It has been clarified in the recent experiment and simulation that a single semi-flexible polymer exhibits discrete transition by lowering the solvent quality[4]. Folded semi-flexible polymer exhibit a ordered structure, such as a toroid and a rod. This implies that a semi-flexible polymer exhibit multi-stability on its folded state.

It is, therefore, interesting to investigate the unfolding-folding transition in a single semi-flexible polymer under the strain. We perform molecular dynamics simulation of a flexible and semi-flexible polymer in the loading or unloading process.

MODEL

To examine the unfolding kinetics, we use Langevin dynamics simulation of beads-spring model, using following potential

$$V_{\text{beads}} = \sum_i \frac{k_a}{2}(|\mathbf{r}_{i+1} - \mathbf{r}_i| - a)^2 \quad (1)$$

$$V_{\text{bend}} = \sum_i \frac{k_\theta}{2}(1 - \cos\theta_i)^2 \quad (2)$$

$$V_{\text{LJ}} = 4\varepsilon \sum_{i,j}\left(\left(\frac{a}{|\mathbf{r}_i - \mathbf{r}_j|}\right)^{12} - \left(\frac{a}{|\mathbf{r}_i - \mathbf{r}_j|}\right)^6\right), \quad (3)$$

where $\mathbf{r}_i$ is the coordinate of the ith monomer and θ_i is the angle between the vector $\mathbf{r}_{i+1} - \mathbf{r}_i$ and $\mathbf{r}_{i-1} - \mathbf{r}_i$. The monomer size a is used as a unit length and $k_B T$ as a unit energy. The monomer-monmer interaction is included by Lenneard-Jones Potential controlled by ε. The bending elasticity is $k_\theta = 100$ for the semi-flexible polymer and $k_\theta = 5$ for the flexible polymer. To characterize the stiffness of a polymer chain, persistence length l_p is convenient. In our case, $l_p \sim 13.5$ when $k_\theta = 100$ and $l_p \sim 3.5$ when $k_\theta = 5$.

We consider the $N = 200$ homopolymer, which has enough length to form ordered structure (toroid, rod *etc.*) in a semi-flexible polymer.[5] Among such ordered structure, toroids are the energetically most favorable in our parameter set.

The structure of a polymer is characterized by the averaged local monomer density ρ. It is defined as,

$$\rho = \langle \sum_j H(r_c^2 - |\mathbf{r}_i - \mathbf{r}_j|^2) \rangle, \quad (4)$$

where $H(x)$ is Heaviside function and the average is performed over the whole monomers. In this work, we set $r_c = 3.0$ and $\rho_c = 10.0$.

We fix one end and move the other end at uniform velocity v, so that the end-to-end distance is z the force f applied to the Nth monomer is monitored during the operation. Typical snapshots are shown in Fig. 1.

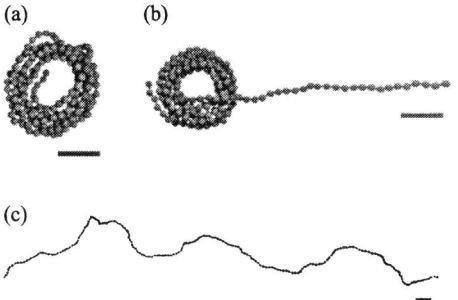

FIGURE 1. Typical snapshots of a semi-flexible polymer in the pulling process. (a) initial toroidal state. (b) phase segregated state at intermediate stage. (c) coiled state. The scale bar shows $5a$, where a is the diameter of a monomer.

RESULTS AND DISCUSSIONS

Figure 2 shows the averaged local monomer density and the force of a flexible and semi-flexible polymer in the cycle. As we can see, a semi-flexible polymer behaves in rather complicated manner than flexible polymer as represented in the large hysteresis.

The force in a flexible polymer increase in small z, and maintain constant value, then increase again. Correspondingly, the structure of a flexible polymer changes from spherical globular shape to ellipsoidal one and to the phase segregation of a coil and globule, then to coil. This behavior can be observed both in the loading and unloading process, and thus a flexible polymer does not exhibit hysteresis in the cycle. Note that this result of force response is consistent with the previous theory.[6]

In the loading process of a semi-flexible polymer, the toroidal semi-flexible polymer takes the phase segregated state of a coil and a toroid. The monomer density, in this state, decreases in a stepwise manner, correspondingly the force exhibits stick and release (saw tooth) pattern, which has been observed in the experiment for DNA.[3] At $z \sim 140$, the polymer makes a transition to a coiled state. In the unloading process, on the other hand, the phase separated state of a rod and a coil is appeared in the initial step, then the polymer become single phase of a rod. Final structures in the unloading process are mainly rods, together with some probability of toroids.

This remarkable deviation of the force response in a semi-flexible polymer from that in a flexible polymer reflects the variety of ordered structures. This fact shows that we need to consider time evolution of the additional variable expressing the folded order structure to characterize the kinetics of a semi-flexible polymer. A flexible polymer, on the other hand, can be characterized only by its end-to-end distance z. Thus we expect that stiffness, reflecting the existence of various floded structures, plays essential role on non-trivial behavior in the force response of DNAs and proteins.

We should note a time scale in our model in relation to the actual experiments. The folding transition time is the order of $10^5 - 10^6$ steps, which is corresponded to about 0.1 - 1.0 sec in the experiment of DNA.[7] Thus our loading and unloading speed $v = 0.0001$ corresponds to the realistic slow value.

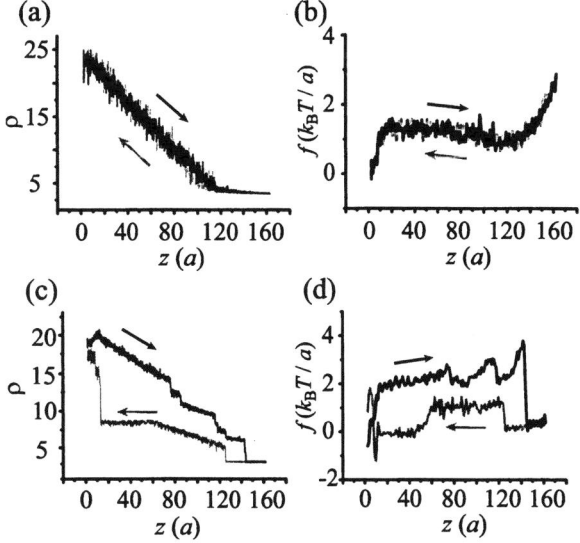

FIGURE 2. The averaged local monomer density ρ (a,c) and the force f (b,d) of a flexible (a,b) and semi-flexible (c,d) polymer in the loading (the arrow facing right) or unloading (the arrow facing left) process. We show the loading and unloading process in black and gray respectively.

REFERENCES

1. Yoshinaga, N., Yoshikawa, K., and Kidoaki, S., *J. Chem. Phys.*, **116**, 9926–9929 (2002).
2. Rief, M., Gautel, M., Oesterhelt, F., Frenandez, J. M., and Gaub, E., *Science*, **276**, 1109–1112 (1997).
3. Murayama, Y., and Sano, M., *J. Phys. Soc. Jpn.*, **70**, 345–348 (2001).
4. Takahashi, M., Yoshikawa, K., Vasilevskaya, V. V., and Khokhlov, A. R., *J. Chem. Phys. B*, **101**, 9396–9401 (1997).
5. Noguchi, H., Saito, S., Kidoaki, S., and Yoshikawa, K., *Chem. Phys. Lett.*, **261**, 527–533 (1996).
6. Halperin, A., and Zhulina, B., *Europhys. Lett.*, **15**, 417–421 (1991).
7. Yoshikawa, K., and Matsuzawa, Y., *J. Am. Chem. Soc.*, **118**, 929–930 (1996).

Generalized-Ensemble Monte Carlo Algorithms for Simulations of Proteins

Ayori Mitsutake*, Yuji Sugita† and Yuko Okamoto**

*Department of Physics, Faculty of Science and Technology, Keio University, Yokohama 223-8522, Japan
†Institute of Molecular and Cellular Biosciences, University of Tokyo, Bunkyo-ku, Tokyo 113-0032, Japan
**Department of Theoretical Studies, Institute for Molecular Science, Okazaki, Aichi 444-8585, Japan

Abstract. We have developed new generalized-ensemble algorithms for Monte Carlo version. Some of new generalized-ensemble algorithms are the replica-exchange multicanonical algorithm, multicanonical replica-exchange method, and the replica-exchange simulated tempering. In this paper, the comparisons of performances of these algorithms are made, taking an example of a 17-residue helical peptide.

In complex systems such as spin glasses and biopolymers, it is very difficult to obtain accurate canonical distributions at low temperatures by conventional Monte Carlo (MC) and molecular dynamics (MD) simulation methods. This is because simulations at low temperatures tend to get trapped in one of huge number of local-minimum-energy states. One way to overcome this multiple-minima problem is to perform a simulation in a *generalized ensemble* where each state is weighted by a non-Boltzmann probability weight factor so that a random walk in potential energy space may be realized (for reviews, see Refs. [1, 2]). The random walk allows the simulation to escape from any energy barrier and to sample much wider phase space than by conventional methods. Monitoring the energy in a single simulation run, one can obtain not only the global-minimum-energy state but also canonical ensemble averages as a function of temperature by the single-histogram and/or multiple-histogram reweighting techniques.

Two of the most well-known generalized-ensemble algorithms are perhaps *multicanonical algorithm* (MUCA) and *simulated tempering* (ST). A simulation in MUCA performs a free 1D random walk in potential energy space. A simulation in ST performs a free 1D random walk in temperature space, which in turn induces a random walk in potential energy space and allows the simulation to escape from states of energy local minima again.

The generalized-ensemble algorithms are powerful, but in the above two methods the probability weight factors are not *a priori* known and have to be determined by iterations of short trial simulations. This process can be non-trivial and very tedious for complex systems with many local-minimum energy states.

The *replica-exchange method* (REM) alleviates this difficulty. In this method, a number of non-interacting copies of the original system (or replicas) at different temperatures are simulated independently and simultaneously by the conventional MC or MD methods. Every few steps, pairs of replicas are exchanged with a specified transition probability. The weight factor is just the product of Boltzmann factors, and so it is essentially known.

However, REM also has a computational difficulty: As the number of degrees of freedom of the system increases, the required number of replicas also increases, and so does the required computation time. This is why we want to combine the merits of MUCA and ST and those of REM so that we can determine the weight factors for MUCA and ST with ease and save the computation time greatly.

The combined methods of MUCA and REM were developed for molecular dynamics [3] and for Monte Carlo versions [4, 5]. The methods are referred to as the *replica-exchange multicanonical algorithm* (REMUCA) and the *multicanonical replica-exchange algorithm* (MUCAREM) [3, 4, 5]. In REMUCA, a short replica-exchange simulation is performed, and the multicanonical weight factor is determined by the multiple-histogram reweighting techniques. In MUCAREM the production run is a REM simulation with a few replicas not in the canonical ensemble but in the multicanonical ensemble.

Moreover, we presented a combined method of ST and REM [6]. In the new method, which we refer to as the *replica-exchange simulated tempering* (REST), a short replica-exchange simulation is performed, and the simulated tempering weight factor is determined by the multiple-histogram reweighting techniques as in REMUCA.

We recently applied these algorithms to a system of a 17-residue fragment of ribonuclease T1. Its amino acid sequence is SSDVSTAQIAAYKLHED. In this paper, the comparisons of performances of these algorithms are made, taking a system of a 17-residue helical peptide. These algorithms for Monte Carlo version were explained in detail in Refs. [4, 5, 6]. We only show a few of results of REM, REMUCA, MUCAREM, and REST simulations.

REMUCA and REST consist of two simulations: a short REM simulation and a subsequent production run of MUCA and ST simulations, respectively. A short REM simulation which determines the weight factors of a production run of MUCA and ST is referred to as REM1. In MUCAREM we used 4 replicas with multicanonical ensembles. The same density of states that was obtained from REM1 was used. A production run of the original REM simulation was also performed for comparison of performance with REMUCA, MUCAREM, and REST. It is referred to as REM2. In REM1, REM2, and REST 14 replicas were used with 14 different temperatures, ranging from 200 K to 700 K. The temperatures were distributed exponentially between 200 K and 700 K (see Ref. [5]). The number of MC sweeps for REM1 was 60,000 for each replica. The total number of MC sweeps for the four production runs (REM2, MUCA, MUCAREM, and REST) was all set equal (i.e., 3,000,000 MC sweeps). In REM1 and REM2 a replica exchange was tried every 25 MC sweeps. In REST a temperature exchange was tried every 100 MC sweeps. The REMUCA and MUCAREM simulations were explained in detail in Refs. [4, 5]. The details of the REST simulation will be given elsewhere [7].

We observed random walks in the potential energy space in these simulations (data not shown). We can obtain proper weight factors of MUCA and ST simulations. To check the validity of the canonical-ensemble expectation values calculated by the new algorithms, we compare the average potential energy as a function of temperature in Fig. 1. The results of the conventional canonical MC simulations at four temperatures ($T = 200, 294, 542,$ and 700 K) are also plotted. All the results from the four generalized-ensemble production runs (REM2, REMUCA, MUCAREM, REST) are more or less similar, while those from the regular canonical simulations strongly deviate at lower temperatures (200 K and 294 K). At 200 K, the energy in the usual canonical MC simulation fluctuates around -180 kcal/mol, while that in the new simulations at 200K fluctuates around lower energy (-200 kcal/mol). Typical snapshots from these canonical MC simulations and REM2 at a few of temperatures are shown in Fig. 1. At 200 K, an ideal helix structure (global-minimum-energy state) is dominant for REM2, whereas partially unfolded helical structures corresponding to local-minimum-energy states are obtained for the

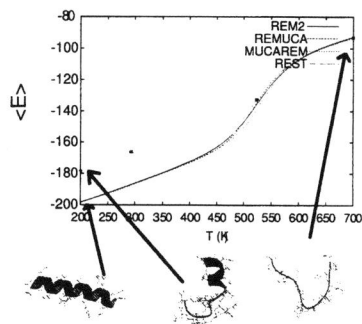

FIGURE 1. Averages total energy as a function of temperature in REM2, REMUCA, MUCAREM, and REST

canonical simulation. These results suggest that the new simulations avoid getting trapped in states of energy local minima and find the lowest-energy state.

To study the efficiency of the new algorithms, we calculated the number of tunneling processes of the potential energies during REM2, MUCA, MUCAREM and REST. One tunneling event is defined by a trajectory that goes from $E_{14}=-93$ kcal/mol to $E_1=-199$ kcal/mol and back. Here, we define E_1 and E_{14} by the canonical expectation values of the total potential energy at temperatures 200 K and 700 K, respectively. We consider that the more tunneling events we observe during a fixed number of MC sweeps, the more efficient the method is as a generalized-ensemble algorithm. In REM2 and MUCAREM, we added the number of tunneling events of each replica to get the total number of tunneling process. Note that the total number of MC sweeps is set equal (i.e., 3,000,000 MC sweeps). The total number of tunneling events of REM2, MUCA, MUCAREM, and REST were 5, 12, 17, and 7, respectively. The number of tunneling process of REM2 is the lowest among the three. These results suggest that the new algorithms are more efficient than the regular REM.

REFERENCES

1. Hansmann, U.H.E., and Okamoto, Y., *Annual Reviews of Computational Physics VI*, edited by D. Stauffer, World Scientific, Singapore, 1999, pp. 129-157.
2. Mitsutake, A., Sugita, Y., and Okamoto, Y., *Biopolymers (Peptide Science)* **60**, 96-123 (2001).
3. Sugita, Y., and Okamoto, Y., *Chem. Phys. Lett.* **314**, 141-151 (1999).
4. Mitsutake, A., Sugita, Y., and Okamoto, Y., *J. Chem. Phys.* **118**, 6664-6675 (2003).
5. Mitsutake, A., Sugita, Y., and Okamoto, Y., *J. Chem. Phys.* **118**, 6676-6688 (2003).
6. Mitsutake, A., and Okamoto, Y., *Chem. Phys. Lett.* **332**, 131-138 (2000).
7. Mitsutake, A., and Okamoto, Y., in preparation.

Molecular dynamics study of the Lipid Bilayers: Effects of the Chain Branching on the Structure and Dynamics

W. Shinoda[1,*], M. Mikami[1], T. Baba[2], and M. Hato[2]

[1]Research Institute for Computational Sciences, AIST, Tsukuba Central 2, Umezono 1-1-1, Tsukuba 305-8568, Japan, [2]Nanotechnology Research Institute, AIST, Tsukuba Central 5, Higashi 1-1-1, Tsukuba 305-8565, Japan

Abstract. We studied effects of lipid chain branching on structural and dynamical properties of the lipid bilayers by a comparative molecular dynamics simulation of dipalmitoyl phosphatidylcholine (DPPC) and diphytanoyl phosphatidylcholine (DPhPC) bilayers. *Trans-gauche* isomerization rate at the dihedrals along the hydrophobic main chain was significantly reduced by chain branching. The slower conformational motion of branched chains lead to slower wobbling of the chain and slower rotational and translational motions of the lipid molecules, compared with straight-chained counterpart. In contrast, headgroup motion was slightly enhanced by the chain branching. The slower dynamics of the branched hydrophobic chains accounts for the high structural bilayer stability and low solute permeability of the branched DPhPC bilayer.

INTRODUCTION

Lipids bearing highly branched hydrophobic chains form a bilayer, which is believed to be stable with low permeability to proton and other ionic or nonionic solutes. Some of the present authors have synthesized a novel, branch-chained glycolipid to develop a stable bilayer matrix for membrane proteins, and the bilayer was actually shown to have low proton permeability.[1] It has not been well understood, however, how the branched chain does affect the bilayer properties on the molecular level. To understand the effect of the chain branching on the physical properties of the bilayer, a series of molecular dynamics (MD) simulations has been undertaken for the two systems: straight-chained DPPC and branch-chained DPhPC.

METHODS

Both DPPC and DPhPC bilayers consisted of 72 lipid and 2,088 water molecules. The lipid molecules were modeled by the CHARMM PARM27 force field with the modified TIP3P water. Firstly, 2.5ns-NPT-MD simulation has been carried out for each system at the same thermodynamic condition; $P=0.1$MPa and $T=323$K. In order to study permeation rate of water and small neutral solutes, free energy profiles of several molecules (H_2O, NH_3, O_2, CO, NO, and CO_2) across the lipid bilayers were calculated by the Widom insertion method with the cavity-based sampling.[2] Furthermore, to investigate long-time dynamics of lipid molecules, another 10ns-NVE-MD simulation has been undertaken for each bilayer system.

RESULTS AND DISCUSSION

The cross-sectional areas of lipid calculated using the last 1.5ns NPT-MD trajectories were 62.0Å^2 in DPPC and 76.8Å^2 in DPhPC membranes, both of which were in good agreement with those obtained by experimental measurements.[3] The detailed analysis of the lipid chain conformation revealed that the branched chain have a high *gauche* probability at the dihedrals in the vicinity of *tert*-carbons, and as a result, the chain selectively bend at the branched segments. Furthermore, a chain caught between the two chains of the neighboring lipid in the same leaflet of the bilayer was frequently observed in DPhPC bilayer. Higher structural correlation among adjacent branch chains gave rise to slow chain dynamics. The *trans-gauche* isomerization rate of the dihedrals in the main chain was measured as a relaxation time of the auto-correlation function of the defined state function. Indeed, the branched chain was shown to have longer relaxation time than the straight counterpart; for the middle part of the chains, the relaxation time of the correlation function of DPhPC (200~300ps) was approximately five times longer than that of DPPC (~50ps). The slower conformational motion of branched chains leads to slower wobbling motion of the chain and slower rotational and translational motions of the lipid molecules, compared with straight-chained counterpart. In contrast, headgroup motions such as rotation of the PN vector were slightly enhanced by the chain branching. (TABLE 1)

TABLE 1. Dynamical properties of a lipid molecule.

	DPPC	DPhPC
D^a [cm^2/s]	3.0×10^{-7}	1.8×10^{-7}
τ_{chain}^b [ns]	1.66	3.32
τ_{PN}^b [ps]	530	450

[a] Lateral diffusion coefficient calculated from mean square displacement of the lipid mass center. [b] Relaxation times of rotational autocorrelation function ($P2$) of the chain vector and PN (phosphorus-to-nitrogen) vector, respectively.

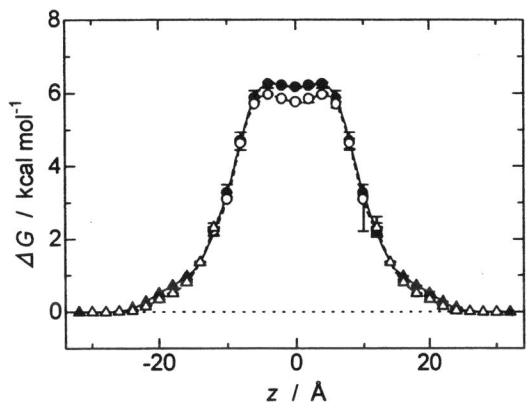

Fig. 1: Free energy profiles of water across lipid bilayers. Closed and open symbols denote DPPC and DPhPC, respectively. Circles and triangles are the profile estimated by cavity based insertion and probability ratio method, respectively.

In order to see the effect of the chain branching on the permeability of lipid bilayers, free energy profiles

ACKNOWLEDGMENT

This work is partly supported by NAREGI Nanoscience Project, Ministry of Education, Culture, Sports, Science and Technology, Japan.

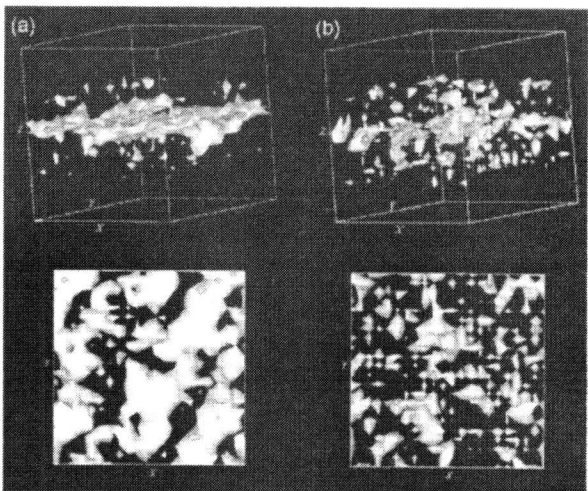

Fig. 2: Cavity distribution in the lipid bilayer averaged over 1.5ns MD trajectories. (a) DPPC and (b) DPhPC. The isosurface shows the probability of cavity at a level of 0.15. The bottoms are the view from the bilayer normal, z.

of small neutral molecules across the membranes have been calculated by the Widom insertion method with the cavity-based sampling. The calculated free energy profiles revealed that there was little change of free energy barriers between DPPC and DPhPC bilayers. For example, free energy profiles of water across the bilayers were plotted in Fig. 1. The result implied that the variation of permeation rate across these two lipid bilayers (generally lower in the branched membrane) was mainly explained by the changes of mobility of the penetrants inside the membranes.

Cavity distribution analysis over the bilayer systems clearly showed that the large cavity formation was relatively prevented in the branched DPhPC bilayer (Fig. 2). The difference suggested that the mobility of the penetrants in the bilayers was reduced as a result of the lipid-chain branching. This was confirmed by the calculation of the diffusion coefficient of water molecule inside the bilayers. Thus, the slow dynamics of the branched chain accounts for not only the high structural stability but also low water permeability of branched DPhPC bilayer.

REFERENCES

1. Baba, T., Minamikawa, H., Hato, M., and Handa, T., *Biophys. J.* **81**, 3377-3386 (2001).
2. Jedlovszky, P., Mezei, M., *J. Am. Chem. Soc.* **122**, 5125-5131(2000).
3. Nagle, J. F., Zhang, R., Tristram-Nagle, S., Sun, W., Petrache, H. I., and Suter, R. M., *Biophys. J.* **70**, 1419-1431 (1996); We, Y., He, K., Ludtke, S. J., and Huang, H. W., *Biophys. J.* **68**, 2361-2369 (1995).

Computer Simulation of Bacterial Colony Formation with Multiplying Rods Producing a Chemotactic Factor

Yoshihiro Ueno*, Ryota Morikawa and Masaki Hayashi

*School of Life Science, Tokyo University of Pharmacy and Life Science,
1432-1 Horinouchi, Hachioji, Tokyo 192-0392, Japan*

Abstract. A multiplying-rod model for individual bacteria is useful to study the bacterial colony microscopically. Here we present an additional model which specifically includes the lubricant secreted by the bacteria. The model is analyzed by Monte Carlo method and a branched or dense colony is reproduced, which is observed in experiments.

INTRODUCTION

Bacteria exhibit various colony patterns according to the substrate softness and nutrient concentration [1]. They often cooperate under adverse conditions by secreting the lubricating fluid [2]. Particularly colony pattern of bacteria species B. subtilis has been vigorously studied from both experimental and theoretical viewpoints [3, 4, 5, 6, 7, 8]. From the experimental studies, a morphological phase diagram of colonies of B. subtilis is determined by varying both the concentration of nutrient and the substrate softness. The phase diagram is composed of five patterns, DLA-like, Eden, DBM-like, concentric ring and homogeneous disk-like [4].

In order to study the bacterial colony microscopically we had proposed multiplying-rod model treating bacterial individuals [9]. Here we present an additional model which includes the lubricating factor secreted by the bacteria. We study this model by means of Monte Carlo simulation. By varying the substrate softness and nutrient concentration, growth of the colony in various environments is investigated. In the result of the simulation, several patterns observed in experiments have been globally reproduced.

MODEL

We consider a hard-rod system in two dimensions. The rods move ahead and rotate continuously on the surface of agar plates [10]. When rods move on the surface, friction occurs between the bacteria and the agar surface. The mimic bacteria secrete lubricant and decrease friction [2]. The rod ingests the nutrient and grows until the bacteria divide into two new individuals when the length of the rod doubles. Therefore the number of rods increases with the simulation step and will form a bacterial colony.

There is a direct repulsive interaction between rods. The potential energy between the rods i and j is represented as

$$u(r_{i,j}) = \begin{cases} (2a/r_{i,j})^{12} & (r_{i,j} > 2a) \\ \infty & (r_{i,j} \leq 2a) \end{cases} \quad (1)$$

where r_{ij} is the distance between central segments of rods i and j. Since $u(r_{ij})$ includes a softcore term $(2a/r_{ij})^{12}$, the rods can push and shove each other [6]. As a result, more active rods push other rods. This biological repulsion corresponds to a 'short-range repulsive chemotaxis' [2].

The movement of the rod is classified as passive or active [10]. A passive movement is caused by a fluctuation of surrounding mediums and thus it depends on the temperature T and the viscosity of the mediums. From the Stokes' law, the mean displacement $\sqrt{\langle \delta_P^2 \rangle} = \sqrt{2\Delta\tau k_B T / f_{xy}}$ and the mean rotational angle $\sqrt{\langle \theta_P^2 \rangle} = \sqrt{2\Delta\tau k_B T / f_r}$ of the rod for a unit time $\Delta\tau$ are adopted [10]. Here f_{xy} is the viscous drag coefficient of the rod moving at random and f_r is the rotational frictional drag coefficient of the minor axis. For example, f_{xy} is given $f_{xy} = 3\pi\eta l_b / \ln(l_b/a)$. k_B is Boltzmann's constant. η is the coefficient of viscosity of the surrounding mediums, which depends on not only the concentration of agar but also on the lubricant such as a surfactant secreted by the bacteria. l_b is length of the rod. a is a unit length of the simulation. An increase in the amount of lubricant decreases the friction between the bacteria and the agar surface. We define the coefficient of viscosity $\eta = \eta_0 q^{-\frac{\lambda}{\lambda_M}}$

where η_0 is the initial coefficient of viscosity, λ is the amount of lubricant secreted by bacteria and λ_M is the maximal amount each lattice has.

On the other hand, a mimic bacteria moves actively due to either 'run' or 'tumble' mode. The active movement is dependent on bacterial activity instead of thermal fluctuation $k_B T$ and proportional to the reciprocal of η in the same way as the passive movement [10]. The alternation of two modes is determined by the recent memory for an amount of nutrients near the bacteria [11]. When the amount of nutrients near the bacteria is more than the memory, the bacteria will switch its movement from 'tumble' to 'run' and vice versa. In the result, the bacteria automatically moves towards a nutrient-rich area.

The quantities of the nutrient and lubricant in the thin agar plate are represented by a mesoscopic value $\phi_{i,j}$ and $\lambda_{i,j}$ at the intersecting point on a triangular lattice. In the case of nutrient diffusion, each $\phi_{i,j}$ varies due to the their diffusion under the constraint of a Ginzburg-Landau expanded free energy $V = \alpha \sum_{i,j} \phi_{i,j}^2$. The parameter α corresponds to the diffusion coefficient. In addition to the potential energy V, the ingestion by bacteria also affects the diffusion of the nutrient.

SIMULATION AND RESULTS

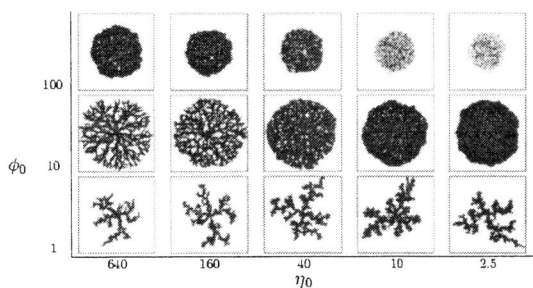

FIGURE 1. Patterns of bacterial colonies for various values of initial quantities of the nutrient ϕ_0 and coefficients of viscosity η_0

We prepare a two-dimensional square space $2048a \times 2048a$ which roughly corresponds to a 1mm × 1mm dimensional agar plate and initially place a few rods on the center of the space. We also set a value of the nutrient homogeneously in the agar plate as $\phi_{i,j} = \phi_0$.

Simulations are run up to $1 \sim 50 \times 10^4$ Monte Carlo steps (MCS) at various coefficients of viscosity $\eta_0 = 1.25 \sim 640$ and initial quantities of the nutrient $\phi_0 = 1 \sim 100$. Parameters chosen are $k_B T = 1$, $\Delta \tau = 1$, $q = 10$ and $\alpha = 1$. In the result of the simulation, we obtain typical colony patterns shown in Fig. 1. At a nutrient-poor and solid agar medium, the bacterial colony shows the DLA-like pattern. Using a box count method, the fractal dimension of 1.655 is obtained in a good linear regression line. Eden-like pattern appears in the region of the nutrient-rich and solid agar medium shown. In the region of the solid agar medium, movement of rods are restricted and growth of the colony is controlled by an extension and a division of the rods. On the other hand, rods actively move at semi-soft and soft agar medium and form DBM-like or homogeneous disk-like colonys.

SUMMARY

In order to study the morphology of bacterial colonies microscopically, we performed Monte Carlo simulations of a multiplying-rod model for various values of the viscosity of the surrounding mediums and the nutrient concentration. We obtained the same colony patterns that are found in the experimental studies except for a concentric ring-like pattern [4].

ACKNOWLEDGMENTS

The authors wish to thank Dr. Y. Yamazaki and M. Kasahara for their helpful advices.

REFERENCES

1. Singleton, P. and Sainbury D., *Introduction to Bacteria* Wiley, New York, 1981.
2. Kozlovsky, Y., Cohen, I., Golding, I. and Ben-Jacob, E., *Phys. Rev. E*, **59**, 7025-7034 (1999).
3. Fujikawa, H. and Matsushita, M., *J. Phys. Soc. Jpn.*, **58**, 3875-3878 (1989).
4. Ohgiwari, M., Matsushita M. and Matsuyama, T., *J. Phys. Soc. Jpn.*, **61**, 816-822 (1992).
5. Wakita, J., Komatsu, K., Nakahara, A., Matsuyama, T. and Matsushita, M., *J. Phys. Soc. Jpn.*, **63**, 1205-1211 (1994).
6. Wakita, J., Shimada, H., Itoh, H., Matsuyama, T. and Matsushita, M., *J. Phys. Soc. Jpn.*, **70**, 911-919 (2001).
7. Kawasaki, K., Mochizuki, A., Matsushita, M., Umeda, T. and Shigesada, N., *J. theor. Biol.*, **188**, 177-185 (1997).
8. Matsushita, M., Wakita, J., Itoh, H., Ràfols, I., Matsuyama, T., Sakaguchi, H. and Mimura, M., *Physica A* **249**, 517-524 (1998).
9. Morikawa, R., Kasahara, M., Ueno, Y. and Hayashi, M., *Forma*, **18**, 59-65 (2003).
10. Berg, H. C., *Random Walks in Biology*, expanded Ed., Princeton University Press, NJ, (1992).
11. Macnab, R. M. and Koshland, D. E., Jr., *Proc. Natl. Acad. Sci. USA*, **69**, 2509-2512 (1972).

Application of the Tsallis Statistics to the Molecular Dynamics Simulation

Ikuo Fukuda* and Haruki Nakamura*[†]

*National Institute of Advanced Industrial Science and Technology, 2-41-6 Koto-ku, Tokyo 135-0064, Japan
[†]Institute for Protein Research, Osaka University, 3-2 Yamadaoka, Suita, Osaka 565-0871, Japan

Abstract. We have developed a deterministic algorithm that produces the Tsallis distribution in continuous systems. Using this, efficient sampling of the states is performed to construct the Boltzmann-Gibbs distributions.

Tsallis statistics based on the Tsallis entropy, characterized by a nonextensive property, is considered as one parameter extension of the Boltzmann-Gibbs (BG) statistics, by the 'Tsallis index' q [1]. The Tsallis distribution can be obtained by extremization of the Tsallis entropy, and the density we use is

$$\rho_{\text{Tsallis}}(x,p) = [1-(1-q)\beta E(x,p)]^{q/(1-q)}, \quad (1)$$

where $E(x,p)$ is an energy of an objective system represented by coordinates $x \equiv (x_1,\ldots,x_n)$ and momenta $p \equiv (p_1,\ldots,p_n)$, and β is related to the temperature of the system. The Tsallis index q measures the difference from the BG statistics, e.g., shown as

$$\lim_{q \to 1} \rho_{\text{Tsallis}}(x,p) = \exp[-\beta E(x,p)]. \quad (2)$$

This new statistics has been applied to various areas and the efficiencies have been reported [1]

Our purpose is to obtain a deterministic dynamics that realizes the Tsallis distribution represented by (x,p) with the index $q \geq 1$. We expect that it can be used for effective sampling of states, particularly for complex molecule systems obeying the BG distribution. The Tsallis distribution with $q > 1$ shows more slow decreasing with E than that for the BG distribution. Thus, high energy states can be realized more frequently than those in the BG distribution, and the traps into energy wells may be effectively avoided. We show an ODE, Tsallis dynamics (TD), which can produce the Tsallis distribution on the basis of the Nosé-Hoover (NH) method [2]. Realization of the distribution is examined in numerical simulations for some fundamental interaction models and peptide systems. Reconstruction of the BG distributions from this Tsallis distributed system is shown, and the efficiency is compared with those of other methods.

TD for the use in molecular dynamics, for a system defined by energy $E(x,p) \equiv \frac{1}{2}\sum_{i=1}^n p_i^2/m_i + U(x)$ with mass parameter m_i and potential function U, is represented by

$$\begin{aligned}
\dot{x}_i &= p_i/(m_i/g(x,p)), \quad i=1,\ldots,n, \\
\dot{p}_i &= -g(x,p)D_i U(x) - \tau(\zeta)p_i, \quad i=1,\ldots,n, \quad (3) \\
\dot{\zeta} &= \sum_{j=1}^n p_j^2/(m_j/g(x,p)) - nk_B T',
\end{aligned}$$

where $g(x,p) \equiv q/[1-(1-q)\beta E(x,p)]$, τ is a suitably defined function of extended variable ζ, and $T' \equiv 1/k_B \beta > 0$ is the "renormalized temperature" [1]. Equation (3) can realize the Tsallis distribution under an ergodic assumption and some related conditions [3]. This equation seems to be the NH equation with deformed mass $m_i/g(x,p)$ and deformed force $-g(x,p)D_i U(x)$, caused by the index q. If $q=1$, then $g(x,p)=1$ and thus this equation equals to the NH equation [with generalized friction $\tau(\zeta)$], which generates the BG distribution. This situation corresponds to Eq. (2). The BG distributions at arbitrary temperatures can be obtained from Eq. (3), using the "reweighting" technique [4].

Compared with BG dynamics (BGD) that directly produce the BG distribution, TD has an advantage in multi-well problems and exhibits statistically stable results. Figure 1 shows time dependence of the total error of the marginal distribution density (MDD) in the BG distribution for x for 1-dimensional double-well (1DW), $U(x) = 10k_B T(x^2-1)^2$. Here we showed the results obtained by the TD with $q=3$ and the Nosé-Hoover chain (NHC) [5] BGD with thermostat masses $Q_1 = Q_2 = 0.05$. Oscillating behavior with large amplitude for the error in the NHC was caused from the small number of the transitions between the two potential wells [which were $3-6$ in Fig. 1 (b)]. In contrast, many transitions occurred in the TD [about $700-800$ in Fig. 1 (a)], and no considerable dependence for the error on the initial value was found. This difference comes from the jumping ability

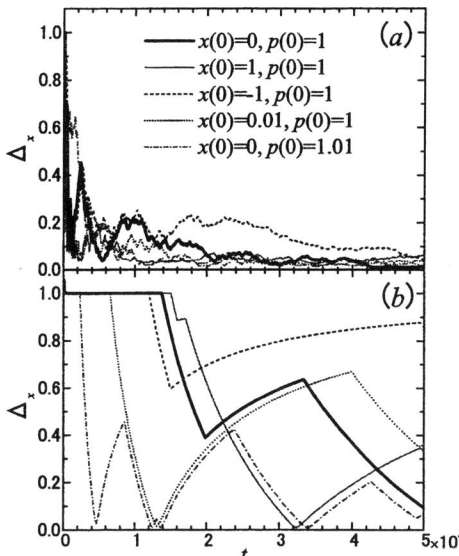

FIGURE 1. Total error of the MDD vs. time, by (a) TD and (b) NHC, for the 1DW is shown for indicated initial values.

between these two dynamics of TD and BGD, based on the Tsallis distribution and the BG distribution, respectively.

Our protocol was applied to the peptide system of Ac-Ala-Ala-NMe *in vacuo*. Figure 2 shows the potential energy distribution obtained by the TD with $\alpha = 50$ [$q \equiv 1 + 1/(3N/2 - 1 + \alpha)$; N is the number of atoms], $E_0 = 20$ kcal/mol [for shifting U so that Eq. (1) is well-defined], and $T' = 50$ K; and that by the conventional multicanonical molecular dynamics (MCMD) [6]. Characteristic feature for TD is the long tail in the distribution, but the reweighing results for the BG distribution between these methods agreed as in Fig. 2. This shows the validity of the current method for the sampling in the U space. We changed each parameter individually from the above values, and show the distributions along with the distributions obtained by BGD in Fig. 3. With increasing q, the increase was observed for: a minimum of U obtained in the simulation; its maximum; and the absolute value of their difference. Similar behavior was observed for the other parameters. These results show the ability of wide-region covering in the U space and extrapolability for the above parameters, in the TD.

ACKNOWLEDGMENTS

The New Energy and Industrial Technology Development Organization is acknowledged for financial support.

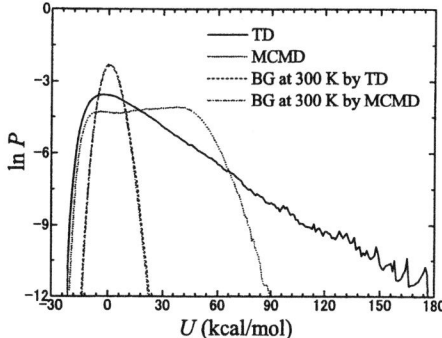

FIGURE 2. Potential energy distribution for the Ac-Ala-Ala-NMe system. Simulation results by TD and MCMD and their reweighing BG distributions at 300 K are shown.

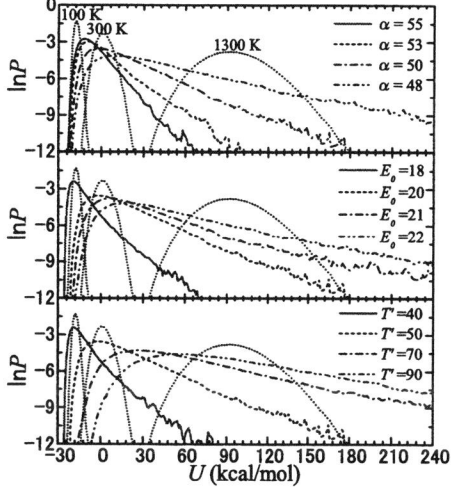

FIGURE 3. Potential energy distribution for the Ac-Ala-Ala-NMe system, by TD using the indicated parameter value. The dotted lines show distributions by BGD at 100, 300, 1300 K.

REFERENCES

1. Tsallis, C., *J. Stat. Phys.* **52**, 479-487 (1988); *Braz. J. Phys.* **29**, 1-35 (1999); Tsallis, C., Mendes, R. S., and Plastino, A. R., *Physica A* **261**, 534-554 (1998).
2. Nosé, S., *J. Chem. Phys.* **81**, 511-519 (1984); Hoover, W. G., *Phys. Rev. A* **31**, 1695-1697 (1985).
3. Fukuda I., and Nakamura, H., *Phys. Rev. E* **65**, 026105 (2002).
4. Ferrenberg, A. M., and Swendsen, R. H., *Phys. Rev. Lett.* **61**, 2635-2638 (1988).
5. Martyna, G. J., Klein, M. L., and Tuckerman, M., *J. Chem. Phys.* **97**, 2635-2643 (1992).
6. Nakajima, N., Nakamura, H., and Kidera, A., *J. Phys. Chem. B* **101**, 817-824 (1997).

Slow Protein Dynamics to be Detected in Inelastic Neutron Scattering Spectra Studied by Molecular Simulation

Yasumasa Joti[*], Akio Kitao[†], Nobuhiro Go[*]

[*] *Neutron Science Research Center, Japan Atomic Energy Research Institute, Kyoto 619-0215, Japan*
[†] *Institute of Molecular and Cellular Biosciences, University of Tokyo, Tokyo 113-0032, Japan*

Abstract. The dynamic structure factors were calculated by using the results of biomolecular simulations at the room and cryogenic temperatures. Three types of simulation, normal mode analysis, molecular dynamics in vacuum, and molecular dynamics in water were applied to HEW Lysozyme. At the room temperature, the shapes of the three dynamic structure factors are in good agreement in the high frequency regions (> 60 cm^{-1}), but considerably different in the low frequency regions (< 60 cm^{-1}). At the cryogenic temperature, the so-called boson peak (~ 30 cm^{-1}) is observed only in the results of molecular dynamics in water. The slow dynamics of protein, found in these low frequency regions, are likely to play important roles to protein function.

INTRODUCTION

Inelastic and quasielastic neutron scattering give the information on the dynamics of biological macromolecules. On the other hand, molecular simulations are also essential tool for studying the protein dynamics, for their ability to simulate the detailed motion of all atoms, which cannot be observed directly by experiments. The combination of computer simulations with neutron scattering experiments allows us to characterize a wide range of dynamical phenomena in condensed-phase biomolecular systems.

The fundamental quantity measured by inelastic neutron scattering experiment is the dynamic structure factor. In this work, the dynamic structure factors were calculated by using the results of biomolecular simulations at the room and cryogenic temperatures. Three types of simulation, normal mode analysis, molecular dynamics in vacuum, and molecular dynamics in water of HEW Lysozyme were carried out. These calculations were done using the force-field parameters of AMBER parm99 [1]. From comparison of these calculated dynamic structure factors in the frequency range lower than 80 cm^{-1}, we discuss protein slow dynamics, which are likely to be related to protein function.

THE CALCULATED DYNAMIC STRUCTURE FACTORS

The shapes of three dynamic structure factors are in good agreement in the high frequency regions (> 60 cm^{-1}), but considerably different in the low frequency regions (< 60 cm^{-1}) at the room temperature, 300 K (Figure 1a). In normal mode analysis, the conformational energy surface of protein is assumed to be multi-dimensional parabola. However, real energy surface is not parabolic. It is known that there are many conformational substates in the protein energy surface, and that jumping among different conformational substates is crucial for protein functions. The effects of solvent on the collective motion of protein are also thought to be important for protein functions. Our results show that such anharmonicity and solvent effects on the protein dynamics can be detected in the low frequency regions (< 60 cm^{-1}), by inelastic neutron scattering experiments.

To elucidate the effects of solvent on the protein dynamics, we introduce the friction coefficient as a solvent effect [3,4]. The effect of frinction originated from solvent environment can be included in the calculation of the dynamic structure factor by assuming each harmonic mode to act as an independent Langevin oscillator. Here we use the friction coefficient of 10 cm^{-1} for all modes, which is estimated based on the calculation in the references [3,4,5]. At the room temperature the shapes of two dynamic structure factors, one by molecular dynamics in water and the other by Langevin mode, are in good agreement in the frequency regions higher than 8 cm^{-1}. The difference in the frequency regions lower than 8 cm^{-1} is originated from the anharmonicity of protein energy surface.

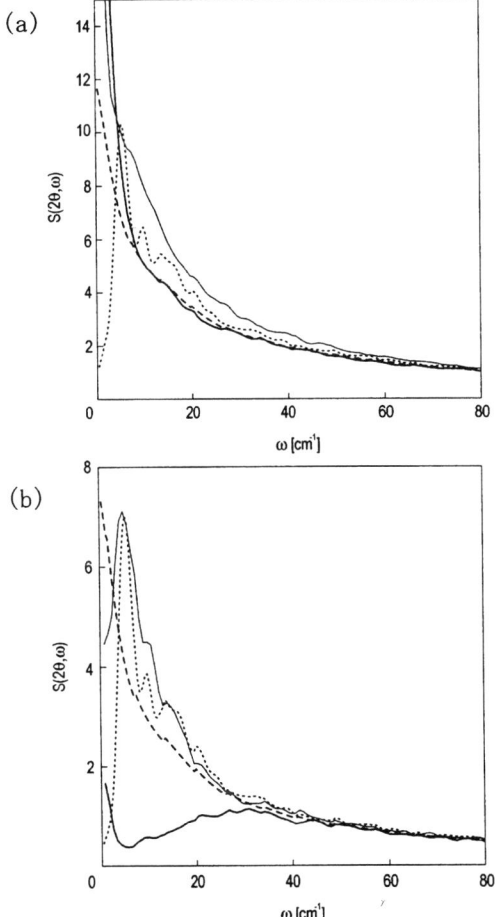

FIGURE 1. Dynamic structure factors, $S(2\theta,\omega)$, as a function of angular frequency, ω, at a fixed angle $2\theta=135°$, calculated by using the results of molecular simulations at (a) 300 K and (b) 100 K, molecular dynamics in water (*thick solid line*), molecular dynamics in vacuum (*thin solid line*), normal mode analysis (*dotted line*), and Langevin mode analysis (*broken line*). Incident neutron wave vector is 1.5 [Å^{-1}].

However, the calculated results at cryogenic temperature, 100 K, are rather different from those at room temperature, 300 K. The so-called boson peak, a broad peak at low frequency regions (15 cm^{-1} to 50 cm^{-1}), is observed only in the result of molecular dynamics in water (Figure 1b). Peaks are observed at around 5 cm^{-1} in the cases of both normal mode and molecular dynamics in vacuum. A single peak is seen at 0 cm^{-1} in the result by Langevin mode.

From detailed analysis of simulation results, we conclude the origin of the protein boson peak as follows. The effects of solvent on the protein dynamics can be classified into two features; one is the friction effect, as introduced in Langevin mode. Another is the change of the potential energy surface in solution [6]. Because of the rearrangement of intermolecular atom packing topology between protein and water molecules, fine structures appear in the conformational energy surface along the direction of collective motions of protein. At the room temperature, these fine structures have no effect on the magnitude of protein fluctuation, because energy barriers are relatively low and jumping among different conformational substates take place frequently. However, at the cryogenic temperature, the protein dynamics are confined in one minimum of these fine structures. This is the reason why protein boson peak can be observed only in the analysis of molecular dynamics in water.

ACKNOWLEDGMENTS

This work was supported by Grant-in-Aid for Young Scientists (B) to YJ and for Scientific Research on Priority Areas (C) "Genome Information Science" to AK from the Ministry of Education, Culture, Sports, Science and Technology of Japan.

REFERENCES

1. Wang, P., Cieplak, P., and Kollman, P.A., *J. Comput. Chem.* **21**, 1049-1074 (2000).

2. Rasmussen B. F., Stock, A. M., Ringe, D., and Petsko, G. A., *Nature* **357**, 423-424 (1992).

3. Kitao, A., Hirata, F., and Go, N., *Chem. Phys. lett.* **158**, 447-472 (1991).

4. Hayward, S., Kitao, A., Hirata, F., and Go, N., *J. Mol. Biol.* **234**, 1207-1217 (1993).

5. Kitao, A., Hayward, S., and Go, N., *Proteins* **33**, 496-517 (1998).

6. Kitao, A., Hirata, F., and Go, N., *J. Phys. Chem.* **97**, 10231-10235 (1993).

Theoretical Study on Local Backbone Dynamics of Cupredoxin

H. Saito[a], A. Sugiyama[b], T. Yoshimoto[b], H. Nagao[b], T. Sakurai[c], K. Nishikawa[b]

[a]*Department of Social Work, Faculty of Social Work, Kinjo University, Matto, Ishikawa 924-8511, Japan*
[b]*Department of Computational Science, Faculty of Science, Kanazawa University, Kanazawa 920-1192, Japan*
[c]*Department of Chemistry, Faculty of Science, Kanazawa University, Kanazawa 920-1192, Japan*

Abstract. The backbone dynamics of azurin has been observed by 3.0 ns molecular dynamics simulation at 300K. The local correlation function and the spectral density are derived form the torsion and bending motion of NH bond existed in peptide plane of protein. We have presented the square order parameter S^2 which is comparable value with NMR experiment.

INTRODUCTION

The mononuclear blue copper proteins[1] (cupredoxins) are ubiquitous metallproteins found in plants and bacteria. In these proteins, it is generally accepted that a more accurate understanding of the dynamic properties of proteins would be of significance in explaining motional contributions to protein stability, folding, reactivity at active site, and so on. Additionally, it is well known that the fast dynamics of the protein (nano- to picosecond time scale) contributes to temperature factors in X-ray structures and to nuclear magnetic resonance (NMR) spin relaxation[2].

In this study, we investigate the local backbone dynamics of azurin by molecular dynamics simulations, and extract the backbone fluctuation and the squared order parameter S^2 at each residue. From the simulated data, we show the backbone dynamics of azurin with precise stability and flexibility of the structure. A typical local correlation function and spectral density are also shown.

COMPUTATONAL METHODS

Initial coordinates of azurin are from the X-ray crystal structure of Pseudomonas aeruginosa oxidized (Cu^{2+}) azurin at 0.193 nm resolution (4AZU entry of Brookhaven Protein Data Bank). In particular, the entry coordinates have been derived from the second segment (denoted by B) of the tetramer composing the crystal unit. The MD trajectories of azurin have been generated by an integration step of 0.002 ps using the AMBER 7 program package with the TIP3P water model[3]. A cut-off radius of 0.8 nm for the non-bonded interactions and 1.0 nm for the Coulomb interactions has been used. The system was energetically equilibrated using NVT ensemble simulation for 1.5 ns at 300K and then a NVT-MD simulation (300K) for analysis has been performed for 1.5 ns. In all simulation time, the SHAKE constraint algorithm has been used at H bond to fix the vibrational motion.

The local correlation function $C(t)$ describing the internal motion of the NH bond vectors is defined as

$$C_\mu(t) = \langle P_2(\mu(t') \cdot \mu(t'+t)) \rangle = \{3\langle[\mu(t') \cdot \mu(t'+t)]^2\rangle - 1\}/2 \quad (1)$$

where $\mu(t)$ is the orientation of the interatomic unit vector at time t, and P_2 is the second-rank-Legendre polynomial. The spectral density is obtained by the Fourier cosine transform of eq. (1):

$$J(\omega) = 2\int_0^\infty C(t)\cos(\omega t)dt \quad (2)$$

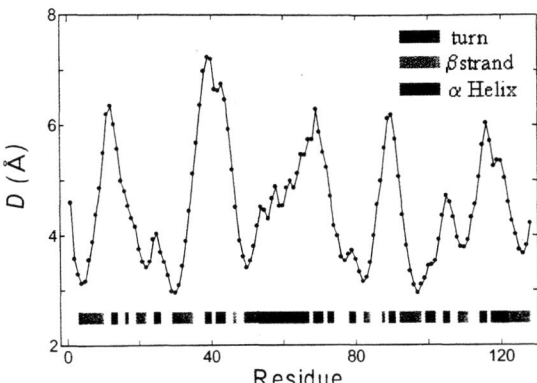

FIGURE 1. The displacements of backbone N atom at each residue.

Assuming that the simulation time has been sufficiently extended to reach the long-term limit of the correlation function, the values of order parameters, that measures the angular restriction of the motion of the vector $\mu(t)$, can be calculated as the following time average[4]:

$$S^2 = \frac{1}{2}\left\{\frac{3}{n^2}\sum_{j=0}^{n-1}\sum_{i=0}^{n-1}[\mu(i\Delta t)\mu(j\Delta t)]^2\right\} - \frac{1}{2}. \quad (3)$$

Here, if reorientation of $\mu(t)$ is completely restricted, then $S^2 = 1$, and if $\mu(t)$ is completely free, then $S^2 = 0$. This order parameter can be readily evaluated from the simulation trajectories, and compared to experimental data.

RESULTS AND DISCUSSION

In order to see the isotropic fluctuation of backbone structure, we have calculated the displacements of the backbone N atoms with respect to the average structure. The results are shown in Figure 1. We can see that large fluctuation is observes in the turn and α-Helix structure regions. The largest fluctuation point is observed at residue 39 belonging to the turn structure. On the other hands, β-strand shows obvious stability than another structure.

To characterize the motional process in detail, the spectral density was calculated by Fourier transform of the local correlation function. As an example, the simulated local correlation function and the spectral density for the residue 39 are shown in Fig.2. In this figure, it is shown that two peaks indicate the torsional motions to the peptide plane appear in 650-800cm^{-1}.

The squared order parameter S^2 obtained form NH bond in peptide plane has relevance for the torsion and bending motion in backbone. Therefore, we have calculated the squared order parameters S^2 for NH bond at each residue in peptide plane. The results are shown in Figure 3. In this figure, S^2 values of residues 36, 40, 75 and 115 are not presented, because these residues are prorin amino acids which are not included NH bond. Figure 3 shows that the values S^2 of residue 74 and 116 belonging to turn structure are obviously small. Therefore, we can find that the sufficient flexibility in term of torsion and bending motion exists at these residues in azurin protein.

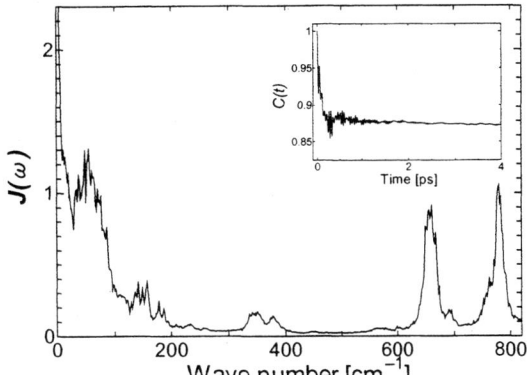

FIGURE 2. Spectral density and local correlation function at residue 120

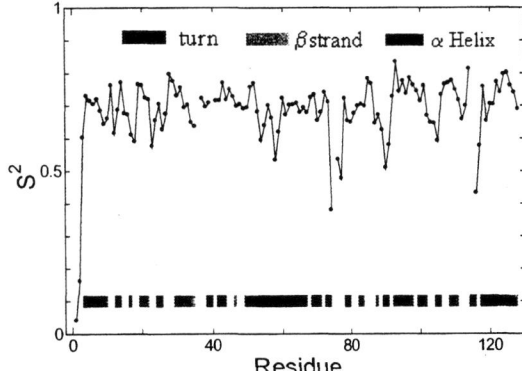

FIGURE 3. Simulated squared order parameters at each residue.

REFERENCES

1. Gray, H.B., Solomon, E.I., *Copper Proteins*, John Wiley, New York, 1981, **3**, pp.1-39.
2. Fushman, D., Cowburn, D., *Structure, Motion, Interaction and Expression of Biological Macromolecules* edited by R. Sarma, M. Sarma, Adenine Press: Albany, New York, 1998, pp. 63-77.
3. Pearlman, D.A., Case. D.A., Caldwell, J.W., Ross, W.S., Cheatham, III, T.E., DeBolt, S., Ferguson, D., Seibel, G., Kollman, P., *Comput. Phys. Commun.* **91**, 1-44 (1995).
4. Levitt, M., *J. Mol. Biol.* **168**, 621-657 (1983).

Theoretical Study on Vibrational Dynamics of Cupredoxin

A. Sugiyama*, H. Saito†, T. Yoshimoto*, H. Nagao*, T. Sakurai** and K. Nishikawa*

*Department of Computational Science, Faculty of Science, Kanazawa Univeristy,
Kakuma, Kanazawa 920-1192, Japan
†Department of Social Work, Faculty of social Work, Kinji University, Matto, Ishikawa 924-8511, Japan
**Department of Chemistry, Faculty of Science, Kanazawa Univeristy, Kakuma, Kanazawa 920-1192, Japan

Abstract. Azurin is a relatively small metalloprotein of the mononuclear blue copper protein found in plants and bacteria. In this study, we investigate the vibrational motion of the active site of azurin by molecular dynamics (MD) calculations and estimatethe Resonance Raman(RR) spectrum of azurin from the analysis of the time correlation function, and compare the experimentally measured RR spectrum with one obtained from MD calculations.

INTRODUCTION

Azurin[1] is a relatively small metalloprotein of the mononuclear blue copper protein found in plants and bacteria. The metalloproteins have attracted a great deal of interest because of the functionality such as electron transfer dynamics coupled with ready accessibility of Cu(I) and Cu(II) oxidation states.

These unique properites and the highly different reactivities as compared to smaller inorganic analogues are assumed to be enforced by the protein biopolymer leading to imposition of peculiar geometric and electronic configurations at the active site[2]. In azurin, the active site consists of a cupper center atom and five residues (Histidine46, Cystein112, Histidine117, Methionine121, Glycine45) as shown in Figure 1. In particular, it is well known that Cu-S(Cys112) interactions are evident in the High-resolution resonance Raman (RR) spectra whose spectra appear in the region between $350 cm^{-1}$ and $450 cm^{-1}$. Additionally, it is reported that RR spectrum appears at about $400 cm^{-1}$ at room tempreture[3]. The RR spectra has been reported for bacterially expressed azurin from *Pseudomonas aeruginosa* (WT) [4].

In this study, the vibrations of Cu-X (X= S(Cys112), S(Met121), N(His46), N(His117), O(Gly45)) in oxdized and reduced azurins is investigated by molecular dynamics (MD) simulation, and compares computational results with experimental ones.

THEORY AND METHOD

We have estimated the charge distribution of the optimized geometry configuration of oxdized and

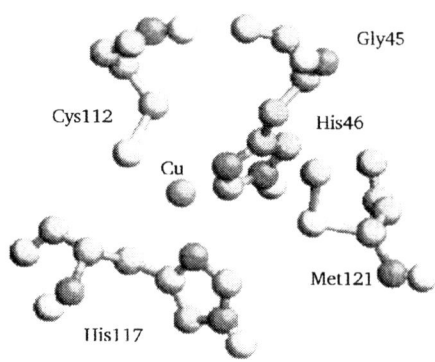

FIGURE 1. Active site of azurin

reduced azurins by semiempitical molecular orbital method(AM1) by MOPAC program package[5]. In the oxdized azurin, the total charge at the active site is 1.0, and reduced one is 0. The initial coordinates of azurin is taken from Protein Date Bank named 4AZU. It is X-ray crystal structure of Pseudomonas aeruginosa azurin at 0.193 nm resolution. The MD simulation curried out under the NVT ensemble condition(300K). The MD trajectories of azurin have been generated by an integration step of 0.002 ps using the AMBER 7 program package with the TIP3P water model[6]. The SHAKE algorithm for all bonds with hydrogen atom is used in the simulation. Cut-off radii are 8 Å and 10 Å for the forces of the van der Waals and for the calculation of Ewald sum, respectively. We have used the value of the harmonic force constant $K(kcal/mol/Å^2)$ and the bond distances at near acitve site of azurin given in Ref[7].

RESULTS AND DISCUSSION

We have calculated the correlation functions of the cupper-residue distance in order to obtain the RR spectra. Figures 2 and 3 show the RR spectra obtained from the Fourier transform of the correlation function. We can find the frequence peak at near 500cm^{-1}, which corresponds to the vibrational peak of Cu-S(Cys112) in oxdized azurin. On the other hand, the peak of reduced azurin shifts lower than that of oxdized azurin(450cm^{-1}).

From figure 2, we find the frequence peak at near 320cm^{-1} with second high intensity which corresponds to the vibration of Cu-N(His117). Other peaks of Cu-N(His46) is also at near 300cm^{-1} and 270cm^{-1}. It may be difficult to experimetaly observe these frequence peaks of Cu-N bonds. On the other hand, those of reduced azurin are found in the region from about 290cm^{-1} to 200cm^{-1}. We also observe the frequence peaks of Cu-S(Met121) and Cu-O(Gly45) at lower frequence of other peaks.

The average bond distances for Cu-X bond are listed in Table 1. We found that the average distance of oxdized azurin is shorter than one of reduced type from Table 1. The frequency peak of Cu-S(Cys112) obtained from this MD simulation is larger than that of experiment. In this different, generally, the RR spectrum obtained from experiments ovserves the frequence of vibration with electric excited states of the materials. In this study, we have used charge distribution of the electric ground state. To compare experimental results more accurately, it is better that we use charge distribution of the excited states. We expect that the peak of Cu-S(Cys112) become smaller than that of this calculation, because the harmonic force constant is smaller in the excited state. These results will be presented elsewhere[8]. For the more accurate simulation, it is necessary to estimate parameters in the MD simulation for the electric state by using more high level molecular orbital methods.

In summary, we have investigated the RR spectra of oxdized and redcuced azurins by MD simulation. The present results are qualitatively good agreement with experimental results.

REFERENCES

1. Gray.B.H, and Solomon.I.E, *Copper Proteins*, John Wiley, New York, 1981, vol. 3, pp. 1–39.
2. Nakamura.A., Ueyama.N., and Yamaguchi.K., *Organometallic conjugation*, Kodansha-Springer, Tokyo, 2002.
3. Dave.B.C., and Dave.Bakul.C.;Germanas, *J.A.Chem.Soc.*, **115**, 12175–12179 (1993).
4. Czernuszewicz.S.R., Dave.C.B., and Fermanas.P.J., *Spectroscopic Methods in Bionorganic Chemistry*, Eds.E.I.Solomon and K.O.Hodgson, 1998, chap. 12, p. 220.
5. Stewart.P.J.J., and Fjitsu.Limited., *MOPAC2002*, Tokyo, Japan, 2001.
6. Pearlman, Case.D.A., and Caldwell.D.A., *Comput. Phys. Commun*, **91**, 1–41 (1995).
7. *Density Functional Theory Applied to Coppe Proteins*, John Wiley, New York, 2002, chap. 6, pp. 113–139, URL http://www.ub.rug.nl/eldoc/dis/science/m.seart.
8. Sugiyama.A., Saito.H., Yoshimoto.T., Nagao.H., Sakurai.T., and Nishikawa.K., *to be published* (2003).

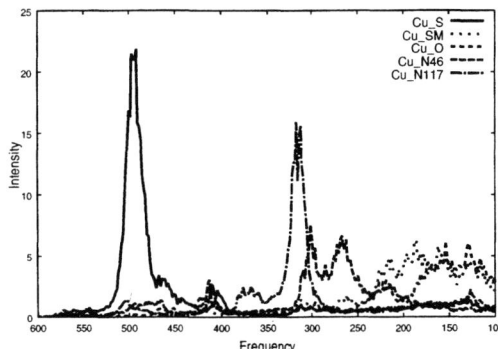

FIGURE 2. Simulated RR spectra for each Cu-residue bonds (oxidized azurin)

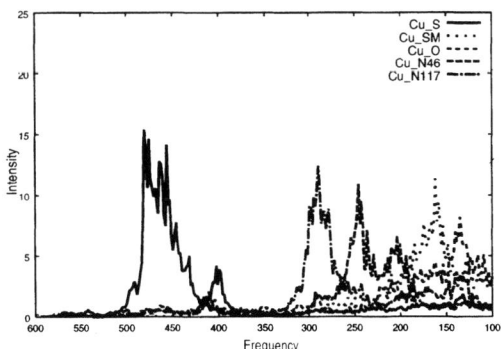

FIGURE 3. Simulated RR spectra for each Cu-residue bonds (reduced azurin)

TABLE 1. Average Cupper - Residues distances. comparison oxidized state with reduced one (Å)

Residue	Oxidized state	Rediced state
S(Cys112)	2.271 Å	2.281 Å
S(Met121)	3.245	3.230
O (Gly45)	2.897	2.903
N (His46)	2.069	2.078
N (His117)	1.971	1.981

Deformation of crosslinked semiflexible polymer networks

D.A. Head*†, A.J. Levine**† and F.C. MacKintosh*†

*Division of Physics and Astronomy, Vrije Universiteit, 1081 HV Amsterdam, The Netherlands.
†The Kavli Institute for Theoretical Physics, University of California, Santa Barbara CA 93106, USA
**Department of Physics, University of Massachusetts, Amherst MA 01060, USA.

Abstract. Motivated by the cellular cytoskeleton, we investigate the mechanical properties of networks of crosslinked semiflexible polymers in two dimensions and at zero temperature. A novel simulation method is constructed which allows us to probe the linear response of such networks to an imposed macroscopic shear. Above a rigidity percolation transition, we observe two regimes: an *affine* regime in which the bulk of the elastic energy is due to filament stretching, and a *non–affine* regime which is dominated by bending modes. We quantify the crossover between these regimes and re–express it in terms of more natural parameters for experiments.

INTRODUCTION

The mechanical, motile and structural properties of cells are defined in part by an assemblage of filamentous proteins known as the *cytoskeleton* [1]. These filaments are somewhat thick, varying in diameter from around 5—9nm for actin to 25nm for microtubules, and therefore their *persistence length* (defined as the contour length over which the filament orientation is preserved under thermal fluctuations) tends to be large, of the order of microns and comparable to typical filament lengths [2]. It is therefore to be expected that dense systems of these so–called *semiflexible* polymers will behave in a fundamentally different manner to flexible polymers, such as polyethylene, whose persistence length can usually be treated as zero. In contrast to flexible polymers [3], the properties of condensed semiflexible polymer systems have only recently begun to be investigated and many basic issues have yet to be answered.

Associated with the cytoskeleton are a range of smaller proteins that perform a wide variety of tasks, such as controlling the rate at which the filament polymerises or severing it along its length [1]. Of interest here are *crosslinking* proteins that can attach to separate filaments, allowing the formation of a crosslinked network that at sufficiently high densities may become a solid–like *gel*. These crosslinks may be passive, such as filamin dimers, or active, such as myosin–II minifilaments that can induce a relative force impulse between two actin filaments in the presence of energy–providing ATP molecules. Active crosslinks present a uniquely biological problem that may require the invention of novel physical concepts, such as the proposed enhanced temperature along the filament contour [4]. However, before addressing such advanced issues, it is perhaps sensible to first understand the mechanical properties of networks formed by passive crosslinking.

A basic question when considering the response of a crosslinked semiflexible polymer network to an applied shear is this: is the displacement field at small length scales *affine* i.e. a scaled–down version of the macroscopic strain, or in some manner *non–affine*? This is clearly of crucial importance when relating the macroscopic shear modulus to the properties of individual filaments; indeed, different assumptions regarding affinity produce disparate predictions for the rheology of entangled actin solutions [5]. Visualisation experiments on sheared networks are difficult to perform and have not yet been attempted, to our knowledge. Instead, we have devised a numerical scheme in which the filaments are explicitly represented, thus allowing for direct measurement of the degree of affinity of the deformed network.

MODEL

Filaments are assumed to deform in response to an applied force by a combination of two forms: *bending* modes in which displacements are transverse to the filament contour, and *stretching* modes along the contour. These are described by the Hamiltonian $\mathcal{H}$, which can be expressed per unit length δs of filament as

$$\frac{\delta \mathcal{H}}{\delta s} = \frac{\mu}{2}\left(\frac{\delta l}{\delta s}\right)^2 + \frac{\kappa}{2}\left(\frac{\delta \theta}{\delta s}\right)^2, \quad (1)$$

where δl is the longitudinal extension (or compression), and $\delta \theta$ gives the local deflection angle. This second term derives from the wormlike chain model [5]. The two coefficients μ and κ, which together define a length scale $l_b = \sqrt{\kappa/\mu}$, are here treated as parameters of the system.

Although localised regions of the cytoskeleton may take a variety of geometric forms, such as parallel bundles of stress fibres [1], we are interested here in generic properties and thus consider random networks that are isotropic and homogeneous when viewed on sufficiently large length scales. Filaments of monodisperse length L are deposited with random orientation and position onto a rectangular, two–dimensional shear cell. Every intersection between filaments is taken to be an inextensible, freely–rotating crosslink. The mean distance between crosslinks (as measured along a filament) is then l_c which, together with L and l_b, define the relevant length scales for our problem.

Once the random network has been constructed, the system Hamiltonian $\mathcal{H}$ is constructed by integrating (1) along all filament lengths. Note that we have linearised the local behaviour, so a linear macroscopic response is assured. Since we are interested in the mechanical, zero–temperature behaviour, we minimise $\mathcal{H}$ with respect to node positions using the conjugate gradient method. The shear modulus G and any other required quantities can then easily be extracted. The procedure is then repeated until reliable estimates of these quantities are found.

RESULTS

Our core finding is summarised in Fig. 1. For low network concentrations of monomer and/or low filament lengths, the elastic moduli vanish and the system is in a liquid–like state, at least with regards linear response. The moduli become non–vanishing at a *rigidity percolation* transition given by the solid line in the diagram, which corresponds to a constant value of $L/l_c \approx 5.71$, or an average of around 4.71 crosslinks per filament. This transition appears to be continuous, with an associated divergent susceptibility and power–law decay of fluctuation correlations.

Away from the transition, two regimes are observed. First, there is a non–affine regime in which the predominant contribution to the total elastic energy comes from the bending of filaments. As either L or c increases, we pass through a somewhat broad crossover region into the second, affine regime where the filaments primarily stretch. The dashed line between the two regimes can be written as $L \sim \lambda$, where λ is some combination of l_c and l_b; for the range of densities $13 < L/l_c < 47$ (which includes biologically relevant networks), we have found $\lambda = \sqrt[3]{l_c^4/l_b}$ to give a satisfactory fit.

A full description of our results, including confirmation of the degree of affinity and speculation about the

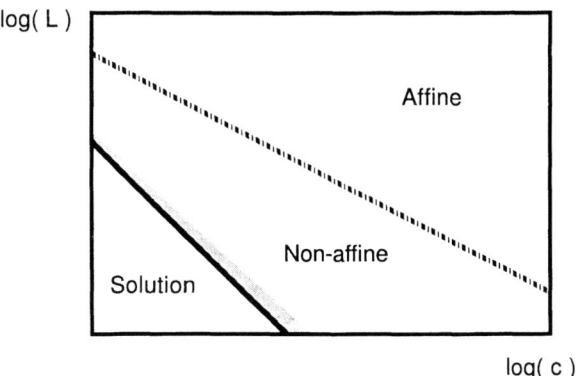

FIGURE 1. Schematic diagram of the different regimes observed in the simulations, shown here for the concentration of polymerised protein $c \sim l_c^{-1}$ against the molecular weight of a filament $\propto L$. The crossover between the affine and non–affine regimes is broad, and moves upwards for increasing filament flexibility. The grey area denotes the scaling regime of the sol-gel transition, in which network deformation is also non–affine.

effects of temperature, can be found in [6]. Related work can be found in [7].

ACKNOWLEDGEMENTS

DAH was partly funded by a European Community Marie Curie Fellowship. This work is supported in part by the National Science Foundation under Grant Nos. DMR98-70785 and PHY99-07949.

REFERENCES

1. Alberts, A., Bray, D., Lewis, J., Raff, M., Roberts, K., and Watson, J.D., *Molecular Biology of the Cell*, Garland, New York, 1994.
2. Frey, E., Kroy, K., and Wilhelm, J., in *Dynamical Systems in Physics and Biology*, EDP Sciences–Springer, Berlin, 1998.
3. Doi, M., and Edwards, S.F., *The Theory of Polymer Dynamics*, Clarendon, Oxford, 1986.
4. Liverpool, T.B., Maggs, A.C., and Ajdari, A., *Phys. Rev. Lett.* **86**, 4171–4174 (2001).
5. MacKintosh, F.C., Käs, J., and Janmey, P.A., *Phys. Rev. Lett.* **75**, 4425–4428 (1995); Kroy, K., and Frey, E., *Phys. Rev. Lett.* **77**, 306–309 (1996).
6. Head, D.A., Levine, A.J., and MacKintosh, F.C., *Phys. Rev. Lett.* **91**, 108102 (2003); Head, D.A., MacKintosh, F.C., and Levine, A.J., *Phys. Rev. E* **68**, 025101(R) (2003); Head, D.A., Levine, A.J., and MacKintosh, F.C., *Phys. Rev. E* (accepted for publication).
7. Wilhelm, J., and Frey, E., *Phys. Rev. Lett.* **91**, 108103 (2003).

IV. COMPLEX SYSTEMS

Pioneering Work of THz Wave and Its Application for Molecular Sciences

Jun-ichi Nishizawa

Semiconductor Research Institute, Kawauchi, Aoba-ku, Sendai 980-0862, Japan
Photodynamics Research Center, RIKEN, 519-1399, Aoba, Aramaki, Aoba-ku, Sendai 980-0845, Japan

Abstract. THz wave has been developed by the application of resonance in chemical bonding inducing phonon phenomena. Especially, success of the realization of sweep oscillator enhances the measurement of the spectrum of organic chemical substances even biochemical substances. The result can be applied in many directions in the field of food, bio, medical and drugs.

INTRODUCTION

Atoms in the molecule or in the crystal are connected each other by the van der Waals force and by the binding force induced from covalent linkage or ionic linkage, in covalent binding conditions and ionic binding condition, respectively. It looks like binded by spring, and each atom with the mass M seems to be bound by a spring with the constant S, then the resonant frequency can be represent to be proportional to $\sqrt{S/M}$.

The resonant frequency water vapor is well known to be 2.45 GHz and in the more complicated polymer seems to have much higher frequency in the range of THz, because of the lighter weight and tighter coupling.

In this case, the atom corresponds to a group of atoms and the coupling force between two group of atoms is large therefore oscillating frequency became very high.

Every combinations of composing atoms into two groups coupled each other and as the result, one polymer seems to have many resonance phenomena and these are usually in the THz wave frequency region. However, very few were measured at the age of 1960, using Fourier transform infrared (FTIR).

Reason was that there has not been any method to generate THz wave, and another method to generate THz has been used to be only by the use of non-linear dielectrics from GHz oscillations only.

In 1964, the author published the fore scope of next research just after the success of semiconductor laser in four establishment; GE, RCA, IBM and Lincoln Research Laboratory corresponding to the military contract in United States, after the patent application by the author in 1957 about the possibility of semiconductor injection laser diode, but there could not obtain any financial support.

The first of forecast after the laser study was the development of glass fiber for optical communication and the second was the development of defect free GaAs crystals and the third was the generation of THz wave by the use of resonant oscillation of molecules and crystals, et. al.

And the top two was succeeded by the authors group at the beginning based on the idea of focusing by the method to settle core structure at the center and now glass fiber had laid under the oceans and through the continent and serving for the human society in many way already, and the very high quality GaAs had been produced by Sumitomo and others over more than half of the total world production leading ultra-bright LED by Stanley Electric Co., Ltd. and also very high reliability laser diode and others, by the application of optimum vapor pressure of arsenics from the top of the melt through the process of higher

Table 1 Chronological Table for Terahertz Generation

Year	Author(s)	Description
1910	D. Hondros & P. Debye	Dielectric waveguide
1936	Seki & H. Seimiya	Light transmission via dielectric rod
1954	M. T. Weiss & E. M. Gyorgy	Dielectric waveguide
1957	J. Nishizawa	Proposal of laser, in particular, semiconductor laser
1958	A. L. Shawlow & C. H. Townes	Proposal of gas laser
1960	T. H. Maiman	Realization of Ruby laser
1961	A. Javan	Realization of gas laser
1962	M. I. Naithan, R. N. Hall, T. M. Quist, et al.	Realization of semiconductor laser
1963 April	J. Nishizawa	Proposal of terahertz wave generation via molecular and lattice vibrations (Denshikagaku p.17, p.30)
1963 May	R. Loudon	Proposal of terahertz wave generation via molecular and lattice vibrations
1964	J. Nishizawa & I. Sasaki	Proposal of optical fiber communication and focusing optical fiber (Diminution of Leakage Loss)
1965	J. Nishizawa	Proposal of terahertz wave generation via molecular and lattice vibrations together with tunneling (Denshigijutsu)
1966	K. C. Kao	Estimation of low absorption loss optical fiber
1969	R. H. Pantell	Observation of frequency shift via lattice vibrations
1971	J. Nishizawa	Proposal of generation of terahertz wave as a result of mixing of original wave with the wave induced by Raman effect
1973	P. P. Solokin, J. J. Wynne & J. R. Lankard	Four wave parametric effect in alkaline metals
1973	J. Nishizawa	Proposal of Ideal Static Induction Transistor (Ballistic SIT) as terahertz operating devices
1979	J. Nishizawa & K. Suto	Realization of semiconductor Raman laser with lattice vibration
1979	J. Nishizawa	Realization of TUNNETT diode oscillating at 0.33 THz
1983	K. Suto & J. Nishizawa	Difference frequency wave generation via semiconductor Raman laser
1997	K. Suto & J. Nishizawa	Semiconductor waveguide Raman amplifier
1999	J. Nishizawa, P. Plotka & T. Kurabayashi	Realization of Ballistic SIT (scattering free SIT)
2000	K. Kawase & H. Ito	THz wave parametric generation by injection seeding
2000	J. Nishizawa	Proposal of application of THz wave to diagnosis and medical treatment of cancer

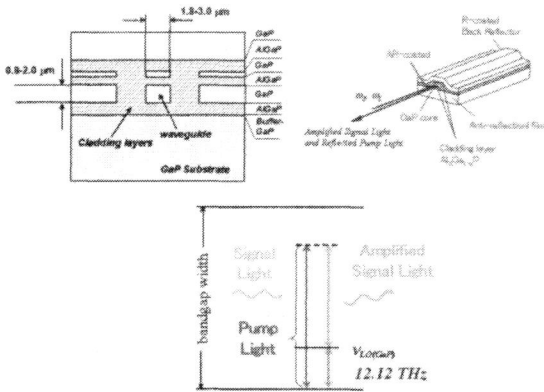

Fig. 1 Cross sectional view of Raman amplifier waveguide and its operation and energy relations.

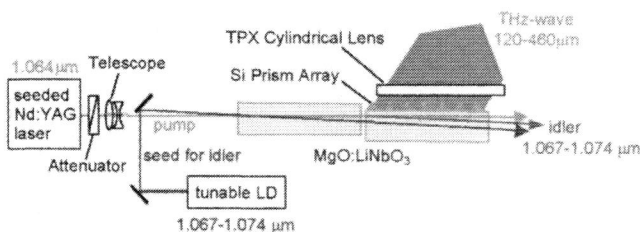

Fig. 2 Injection-Seeded THz-wave Parametric Generator (IS-TPG).

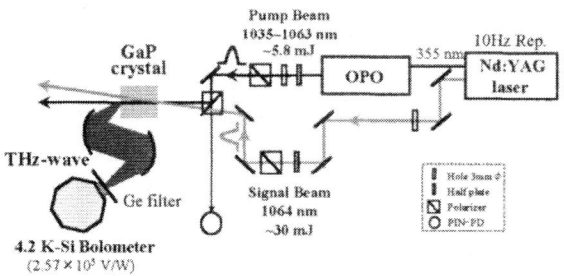

Fig. 3 Schematics of the experimental set-up used for THz-wave generation in GaP crystals.

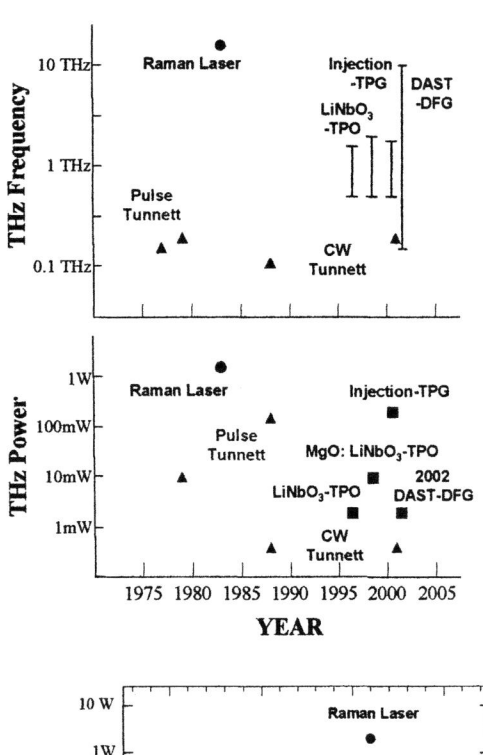

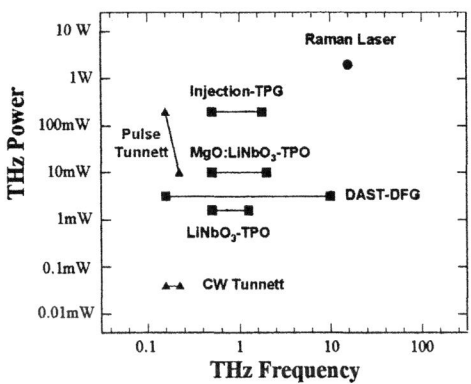

Fig. 4 Progress of out put power and oscillating frequency per year.

temperature to compensate the evaporation of arsenics from the top.

About the third items, authors group continued over 40 years with the support by Japanese electronic manufacturers and by Japanese Ministry of communications and Japanese Science and Technology Agency. And in 1979, author succeeded to realize the Raman Laser based on the YAG laser and 1983, with its mixed back realized 12 THz by GaP high quality crystals with Prof. K. Suto, we can say that this is the first generation of coherent wave in THz region. [1] Experimental set up is shown in Fig. 1. Mixed back of the out put power of Raman Laser with the original pumping power induces Raman frequency of 12 THz.

Afterwards, RIKEN asked me to settle SENDAI branch as the director in 1990 and named as Photodynamics Research Center following my idea to study photo effect mainly THz region, and when 2nd period started in 1998, again I was asked to work as the 3rd director. I invited Prof. H. Ito to contribute to realize a terahertz wave oscillation using dielectrics like lithium niobate and Dr. K. Kawase joined to his group earnestly worked and succeeded to realize sweep frequency oscillator between 1~4.8 THz by the use of phonon polaritons in LiNbO3 optical parametric oscillation [2] and Prof. K. Suto's group succeeded to realize 0.3~7.5 THz by GaP. [3]

Outline of their generation equipment are shown in Figs. 2 and 3, respectively. When, Dr. K. Kawase reported me the success of generation of THz wave in the range of 1 THz, the author suggested to measure the absorption in some materials and he measured the absorption spectrum of water vapor [4], it was already enough to analyze the internal atomic or atom cluster vibration in the polymer.

Now in our laboratory, two groups are running with their peculiar equipment; Raman in GaP and parametric oscillations in LiNbO3 et. al. Fig. 4 shows the progress of generation of THz waves.

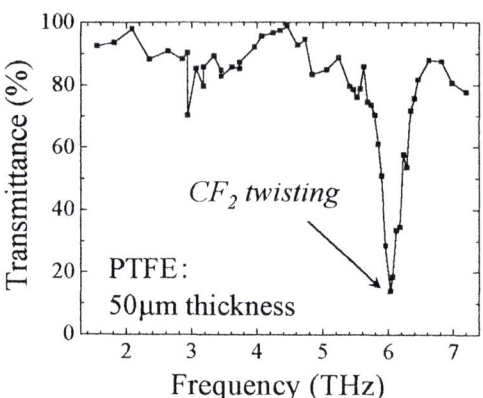

Fig. 5 Teflon (PTFE) terahertz absorption spectrum measured by GaP tunable terahertz source.

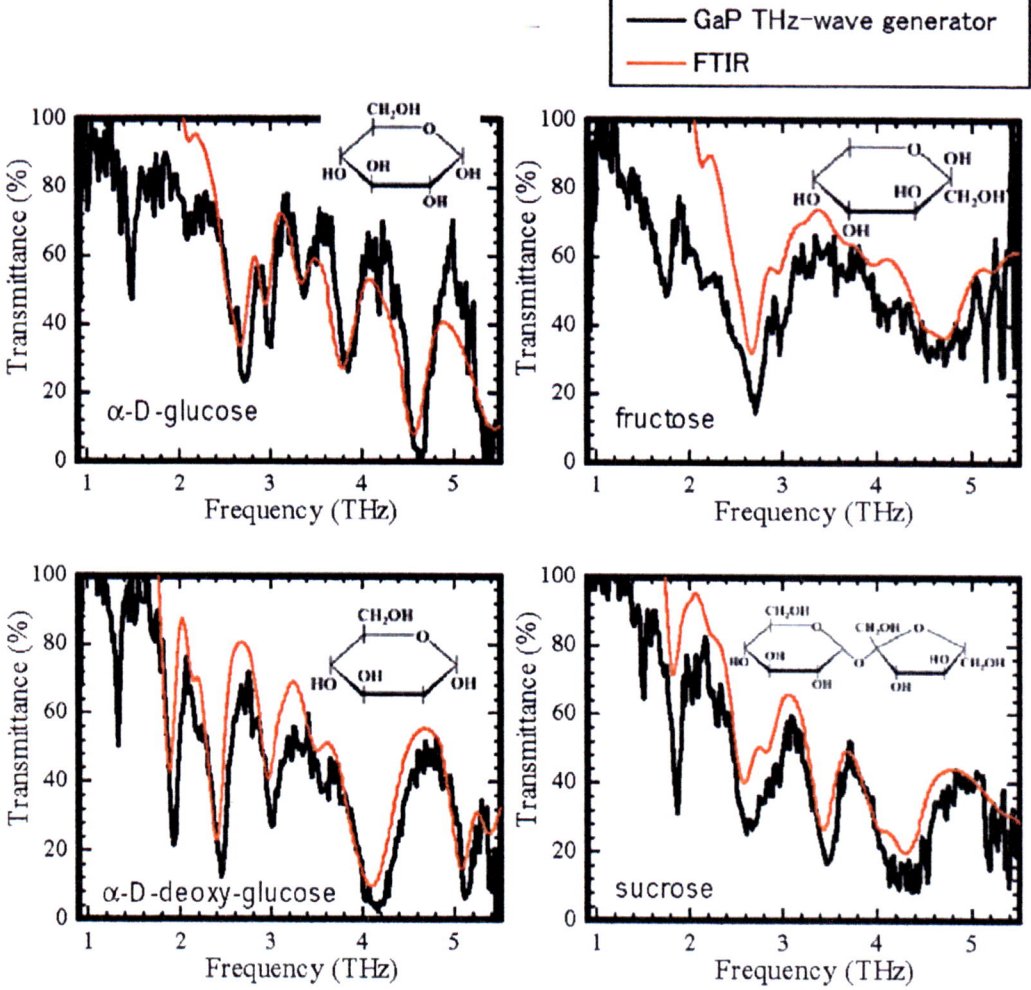

Fig. 6 Transmittance of Sugars in THz band.

MEASURED RESULTS OF ABSORPTION SPECTRUM

Fig. 5 shows the result of measurement of Teflon using the sweepable frequency generator of GaP Raman effect and it shows very beautiful absorption caused from trusting of CF_2 clearly. [5]

In this manner, spectrum can be used for the analysis of chemical bonding in many substances.

Fig. 6 shows the result of measurement of some sorts of sugar pressed pellet with glucose and the ratio of the transmitted THz light through the pellet to that through the pellet of pressed pure polyethylene powder are shown in Fig. 6 and the value seems more reasonable than that by FTIR, in many point as an example the result by THz are scarcely overwhelm the limit of 100 % introducing to that by FTIR.

It is clear enough that the results suggest the very valuable result to analyze the lattice vibrations [6] in these polymers. Also, it was already applied for the sample composed of the same material but different crystalline type or different molecular structure like o-, m– and p-Methyl benzoic acid and others. These seem to be effective for the analysis of molecular structure of many molecules like DNA, SARS and others.

Once, it was mentioned that the water adsorbed film will interference the measurement of the data from the base material.

However, coherent anti-Stokes Raman spectroscopy gave enough data. Pump light source excite the anti-Stokes Raman phenomena when it was illuminated by the light source just with the wavelength a little bit longer simultaneously to protect the Stokes Raman radiation, then the pump light source induces the anti-Stokes Raman radiation and many data have been measured already PPLN, Albumin, Lysozyme and Riboflavin and others, even under the water [6].

It seems to be already clear that some sorts of structural defects in polymer can be detected and measured using THz wave reflections and transmittance. But the sensitivity has not yet clear, which seems to be correlated with the resistivity oppose to burning out.

PHONON GENERATION AND ABSORPTION

As is easing expected that the phonon in Raman effect or other effect concerned with lattice vibrations,

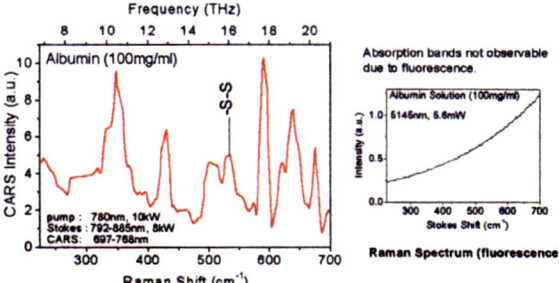

- CARS spectrum of protein molecules (albumin)
- THz-frequency vibrations measured for the first time.
- characteristic vibrational modes in the THz region.

Fig. 7(a) Low-Frequency CARS Spectrum of Albumin in Water.

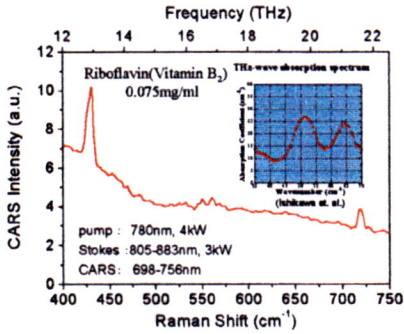

- Low frequency CARS spectrum of Riboflavin.
- → vibrational bands observed at 430 and 720 cm⁻¹ for the first time.
- Measurement of dilute solutions with high sensitivity.

Fig. 7(b) CARS Spectrum of Riboflavin in Water.

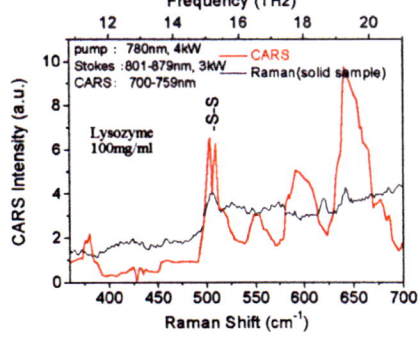

- CARS spectrum of protein molecules (lysozyme) for the first time.
- characteristic vibrational modes in the THz region.
- more sensitive than Raman spectroscopy.

Fig. 7(c) CARS Spectrum of Lysozyme in Water.

generations and/or absorption of phonons should be considered. In the case of Coherent Anti-Stokes Raman Spectroscopy (CARS), generated photons by Stokes Raman influences anti-Stokes Raman effect.

Fig. 7 shows an example of the result, measured by H. Ito group in the case of Albumin in water, and shows the results can be obtained in spite of the covered water.

In these cases, mechanical structure of the surface seems to influence the reflection of phonon wave. Therefore, even standing wave can be realized which influences the phenomena remarkably. In some cases out put THz intensity or directionality is expected to be control. This kind of study can be represented as phonon engineering.

FUTURE PROGRESS

Fig. 8 shows an example of recent application; map of transmitted(b) or reflected(c) THz wave can be easily observed with a rod shifting gradually along the x-axis as a lens. The result can be obtained, as shown in (b) by T. Sasaki, T. Yamada and K. Suto in my group. And also (c) shows the progressed display of cancered liver which was also taken in our research group. The lower half, which was cancered, loses the reflection with the increase of frequency from 0.63 THz to 1.5 THz and top half where not yet cancered keeps reflections even with higher frequency. With the more precise spectroscopic measuring experiment, the diagnosis can be expected more precise and accurate in future.

And also, the diagnosis from the external camera picture through opened window next to the diseased part by surgical operation before the cut-off of liver can be expected and even the ambitious experimental study from out side of body is expected.

Very large power illumination is expected to enhance the lattice vibration specially limited and to destroy the specified bonds, which can be expected to induce new method for medical treatment like improved hot cure. Also, simultaneous drug injection with THz irradiation seems to enhance by the cutting or activating special bond, and it in creases the effect of medicals, as a result of very accurate activation of chemical reactions by THz wave.

Finally, fundamental measurement of spectrum of a large number of organic chemical substances seems to be necessitated and these will help quantitatively the understandings of chemical bonding and reactions.

These results seems also can be extended toward the virus and bacteria to fix the organic chemical structure and to search the corresponding chemical drugs for the treatment after diagnosis based on the measurement of spectrum of THz reflections and transmission.

Also, as a result of the fact that the molecular structure of virus or bacteria became clear, it is expected to be delighted the search of the most suitable medical drug for each virus or bacteria. Before that, as an example, the structural difference between original and mutated virus or bacteria as shown Fig. 6.

And, corresponding each structures, which seems to be most effective, most suitable medical drug can be choose based on the molecular structures. Therefore, these can be called as molecular theory for pharmacy, or molecular pharmacology.

These are only some examples to be thought realizable in very new future.

This work has been supported and continued by Semiconductor Research Institute and also by Ministry of Education, Culture, Sports, Science and Technology and by Japan Society for the Promotion of Science. The author should like to represent hearty thanks to many corresponding ladies and gentlemen and many co-workers.

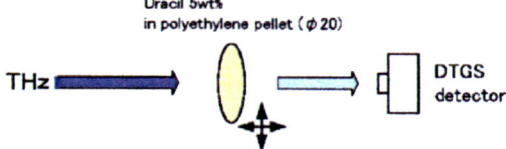

Fig. 8(a) Set-up for terahertz imaging.

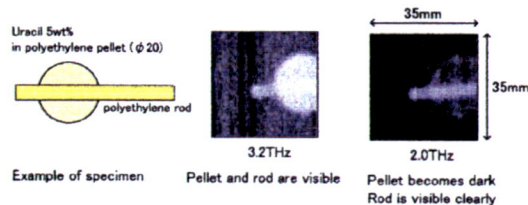

Fig. 8(b) Pellet contains 5% Uracil and it absorbs 3.2 THz wave.

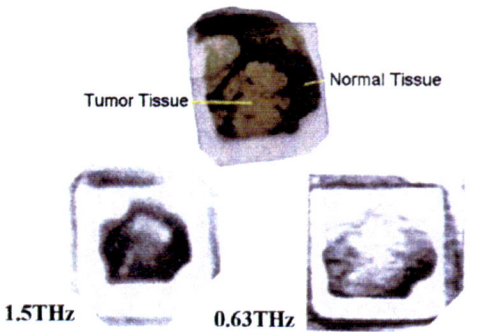

Fig. 8(c) Partly cancered liver (visibly light photo upper) changes the THz reflection with higher frequency.

REFERENCES

1) Suto K. and Nishizawa J., *IEEE J. Quantum Electron.* **QE-19**, 1251-1254 (1983)

2) Kawase K., Minamide H., Imai K., Shikata J. and Ito H., *Appl. Phys. Letters* **80**, 195-197 (2002)

3) Tanabe T., Suto K., Nishizawa J., Saito K. and Kimura T., *J. Phys. D: Appl. Phys.* **36**, 953-957 (2003)

4) Kawase K., Shikata J. and Ito H., *J. Phys. D: Appl. Phys.* **34**, R1-14 (2001)

5) Tanabe T., Suto K., Sasaki T., Nishizawa J., Saito K. and Kimura T., *Conf. Digest, The 22th Int. Conf. on Infrared and Millimeter Waves*, p. 197 (Sept. 29-Oct. 2, 2003, Otsu, Japan)

6) Shikata J., Nakazawa M., Matsumoto T. and Ito H., *Technical Digest, 11th Int. Conf. on Terahertz Electronics*, p. 44 (Sept. 24-26, 2003, Sendai, Japan)

Molecular Dynamics Compared to Hydrodynamics for Rayleigh-Taylor Instability

K. Kadau*, T.C. Germann*, N.G. Hadjiconstantinou[†], G. Dimonte*, P.S. Lomdahl*, B.L. Holian* and B.J. Alder**

*Los Alamos National Laboratory, PO Box 1663 Los Alamos, NM 87545, USA
[†]Massachusetts Institute of Technology, 77 massachusetts avenue cambridge, ma 02139-4307, USA
**Lawrence Livermore National Laboratory, 7000 East Ave., Livermore, CA 94550-9234, USA

Abstract. A massively parallel molecular-dynamics (MD) simulation of some 100 million particles is compared to theoretical Navier-Stokes (NS) predictions, continuum NS simulations, and experimental observations of the Rayleigh-Taylor fluid instability. Smaller MD simulations were performed to verify the initial exponential rise of a single bubble, as predicted analytically for both miscible and immiscible fluids. The penetration of spikes of heavy fluid into light, along with bubbles of light rising into heavy, grows as the square of time at long times—with experiment and both atomistic and continuum computer simulation in 20% agreement. However, part of the differences may be due to dependence upon initial conditions, whose influence on the growth of individual modes was studied for two different cases: an initial interface formed by a single sinusoidal perturbation vs. one perturbed only by thermal fluctuations.

INTRODUCTION

The question to be addressed in this talk is whether it is possible by means of the latest generation of supercomputers to solve the Liouville equation to investigate quantitatively complex hydrodynamic flows. The Liouville equation represents nothing more than the simultaneous Newtonian equations of motion of classically behaving particles or large scale MD. Such supercomputers have been able to follow 10 to the 9 particles for 10 to the -8 seconds and hence the systems are at most, in 3 dimensions, 10 to the 3 particles wide or at most 10 to the 3 nanometers, hence the term nanohydrodynamics. More massively parallel computers in the foreseeable future might increase the number of particles that might be followed by 2 orders of magnitude, but the time that can be followed is unlikely to exceed the 10 to the -8 limit, because these calculations are logically sequential [1]. Favoring this attempt is the observation some years ago of the long-time tail, where hydrodynamics was found to be quantitatively applicable for low velocity flow over a sphere at 10 nanometers and 10 to the -12 seconds [2].

Solving the Liouville equation has some great advantages over solving the usual continuum NS equations, which after all are derived from it under some approximations. The principal approximations are the neglect of physical non-linearities and fluctuations. The particle method also allows a molecular mechanism to be investigated, for example, for the onset of instabilities, and for imposing more realistic boundary conditions, as in the example studies here, a surface roughness generated only by natural fluctuations. Numerically, the advantages are also great, namely it is a gridless method that is unconditionally stable. The disadvantage that it is computer intensive is mitigated by the fact that in a complex flow the number of grid points required to spatially resolve the flow is comparable to the number of particles required [3]. Since to advance a grid point in a NS calculation is only somewhat faster than advancing a particle in a MD calculation, it turns out that typical Rayleigh-Taylor(RT) calculations consume a comparable amount of computer time.

The one unavoidable disadvantage of MD is that the external fields that have to be applied, because the systems are so small, are unrealistically large. From non-equilibrium MD studies of the transport coefficients, where the stress that had to be applied to overcome the fluctuations to obtain the viscosity, for example, was 6 orders of magnitude larger than in a typical experiment, it was nevertheless found that reasonable viscosities could be obtained [4]. However there is a breaking point where the stresses are so large that the physics of the process changes and shear thinning is observed. In the RT problem huge gravity had to be applied to reach nanometer and nanosecond scales, and the hope is that the acceleration is a scalable variable. One can, of course, make a limited check on that by varying the acceleration by an order of magnitude, but the ultimate check is by comparison to experiment or with the NS equations, where scalability is assumed.

THE RAYLEIGH-TAYLOR PROBLEM

As an initial demonstration the well studied RT instability was chosen using the Scalable Parallel Short-range MD (SPaSM) code with pair-wise additive Lennard-Jones interactions [5]. The particles in each of the fluids interact with the same potential, except that one is more massive than the other. Particles in the two different fluids either interact also with the same potential, for the miscible case, where the surface tension is zero, or with only the pure repulsive part of the Lennard-Jones potential in the immiscible case. The heavy fluid is initially placed on top of the light fluid under the influence of a gravitational field, g. The density ratio is characterized by the Atwood number, A. The initial conditions were set up such that the system was as close as possible, given the equation of state, to a mechanically stable compressed fluid under the influence of the gravitational field. The interface roughness was then only due to fluctuations. Most runs employed a thin slab geometry containing 3 million particles ;small enough and thus short enough runs(about 10 hours) to investigate a wide range of parameters space. However, a few runs were also performed in a full three dimensional geometry containing up to 100 million particles.

RESULTS

To demonstrate the quantitative validity of the particle approach, the early time growth of an imposed single sinusoidal mode perturbation of the interface was compared to analytical results of linear stability based on continuum hydrodynamics [6]. The amplitude of the perturbation grows exponentially, where the growth rate depends on the surface tension and the viscosities of the 2 fluids, and is a function of the imposed wave number. This dependence on wave number reaches a maximum, which represents the fastest growing or most unstable mode. It is important for quantitative purposes to solve the linear stability equations numerically rather than depend on inaccurate analytical approximations. The MD calculations require an initial mode with a significant amplitude and an even larger amplitude was present before reliable data could be taken. Thus, deviations of about 10% of the growth rate at the maximum from theory can be qualitatively explained by the non-linearity resulting from saturation effects due to the large amplitude, as judged from the solutions of the continuum equations.

The growth rate from an interface perturbed only by fluctuations was found to be dominated quickly by the wave number corresponding to the maximum, which then determines the initial number of bubbles and spikes. For example, the 3 dimensional system should have 3.5 spikes along the edge of the simulation cell, while actually 4 were observed at early times. In some cases the deviations from the theoretical predictions of the most unstable mode were less accurate due to the presence of multiple modes and the finite edge length.

At later times the spikes and bubbles develop mushrooms at their tips that eventually interact and merge in a process that can be described as turbulent. The penetration distance, h, of the spikes and bubbles is characterized at long times by the only relevant distance scale applicable [7], namely gt^2 multiplied by the coefficient α, so that $h = \alpha g t^2$, where h is determined at the position where the fluid concentration has changed by 95%. The long time growth of the bubbles and spikes is thus determined from the slopes α of the evolution of the boundaries. Somewhat surprisingly, there was no statistically significant difference between the average value of alpha as determined from the large number of simulations done with the slab geometry at A=0.867 and a single fully 3 dimensional system.

The growth of the bubbles and spikes, as given by the value of alpha, were found to agree with experiment [8], as well as Youngs' theoretical model [9]. This remarkable agreement with experiment at late times is a significant validation of nanohydrodynamics. Most simulations were performed with a large value of g, however, an order of magnitude smaller value did not indicate any difference in that alpha value within statistical fluctuations. At the longest time of penetration of the spikes in the simulations, the depth reached 60% of the experimental value, measured in terms of the most unstable wavelength. However, the Reynolds number in the simulation reached only 1500, compared to experimental values up to 100 000. NS simulations reach Reynolds numbers of 5500. Different continuum hydrodynamics calculation show a range of alpha values for bubbles between 0.03 and 0.08 [7] compared to an experimental range of 0.05 to 0.08. The relatively large variation of alpha in the hydrodynamic calculations reflect numerical difficulties as well as different initial conditions. For spikes the value of alpha is not independent of A as for bubbles, but tend toward the free fall limit at A=1 of 0.5. In this limit the spikes are not hindered by any friction, whereas for bubbles the dynamics is determined by a balance between friction and buoyancy.

Another advantage of the MD simulations is that it is possible to quantitatively determine how the alpha value depends on initial conditions, such as the surface tension and imposed disturbances. Thus it was found that the growth of the spikes in the miscible case is reduced by about 20% compared to the immiscible case at high Atwood number. In the miscible case the interface gets blurred by diffusion, which can be interpreted as a reduction of the effective Atwood number that leads to a reduction in the growth of the spikes but not bubbles, since their alpha value is nearly independent of A. It was found that an initial single mode perturbation of the interface enhances the mixing process at short times, however, at distances large compared to the initial amplitude, the effect is washed out and the alpha corresponds to a value from a fluctuation perturbed interface.

DISCUSSION

It is, in any case, questionable whether the alpha value alone characterizes the long time behavior. For example, from experiment as well as from simulations, oscillations in the bubble growth about the limiting slope are observed. These might have to do with the compressibility of the fluid, often ignored in continuum calculations, coupled with viscous effects that have a major influence on bubble dynamics. Of even more significance is the observation of spikes in a slab that was 5 times longer than so far reported. The purpose of that run was to reach longer times, since in previous runs the spikes reached the bottom of the simulation cell. This run showed an abruptly changed behavior from $\alpha g t^2$ observed in the first half of the run. This could be traced to a new physical mechanism, namely the mushrooms at the tips of the spikes breaking off and forming droplets. This phenomenon could be an artifact due to an insufficient number of particles in the simulations or the large g value, but it could also be real, since thinning of the spikes as they advance is experimentally observed. A detailed comparison with continuum calculations is in progress and should tell whether the effect is real, but it is already clear that these simulations can throw much light on complex flows.

REFERENCES

1. Kadau, K., Germann, T.C., and Lomdahl, P.S., *Int. J. Mod. Phys. C* **15** in press (2004).
2. Alder, B.J., and Wainwright, T.E., *Phys. Rev. A* **1**, 18 (1970).
3. Cook, A.W., and Dimotakis, P.E., *J. Fluid Mech.* **443**, 69 (2001).
4. Holian, B.L., and Evans, D.J., *J. Chem. Phys.* **78**, 5147 (1983).
5. Beazley, D.M., and Lomdahl, P.S., *Computers in Physics* **11**, 230 (1997).
6. Chandrasekhar, S., *Hydrodynamic and Hydromagnetic Stability*, Oxford University Press, Oxford, 1961, Chapt.X.
7. Glimm, J., Grove, J.W., Li, X.L., Oh, W., and Sharp, D.H., *J. Comp. Phys.* **169**, 652 (2001).
8. Dimonte, G., *Phys. Plasma* **6**, 2009 (1999).
9. Youngs, D.L., *Phys. Fluids A* **3**, 1312 (1990).

DYNAMICS OF LIQUID-VAPOR PHASE TRANSITION UNDER HIGH FREQUENCY VIBRATIONS

Daniel A. Beysens[*]

ESEME, Service des Basses Températures, Département de Recherche Fondamentale sur la Matière Condensée, Commissariat à l'Energie Atomique, 17, rue des Martyrs, 38054 Grenoble Cedex 9, France

[*]*Postal address : Institut de Chimie de la Matière Condensée de Bordeaux, 87, avenue du Docteur A. Schweitzer, 33608 Pessac Cedex, France*

Abstract. Liquid-vapor phase transition of hydrogen near its critical point ($T_c = 33$ K) is investigated under high frequency vibrations (amplitude range [53 -500] µm, frequency range [5 - 50 s^{-1}]). Gravity effects are compensated in a high magnetic field gradient as provided by a 10 T superconducting coil. The experiments are performed in the temperature range within [0.17 – 1.1] mK from T_c, at critical and off-critical density. The pattern shows up as interconnected gas-liquid domains or bubbles. When the domain size becomes larger than the viscous boundary layer, growth is accelerated and the domains eventually elongate in the direction perpendicular to the vibration (interconnected pattern) or align in periodic planes in the same direction perpendicular to vibration (bubble pattern). We explain the experimental findings by the presence of inertial velocity gradients between the vapor and liquid domains, which favor coalescence and speedup domain growth.

BACKGROUND

The investigation of the mechanisms that triggers gas-liquid phase separation – a very out-of-equilibrium process - is facilitated by performing studies near the critical point [1]. Here one benefits of the universal, scaling laws and the critical slowing down of the dynamics. As gravity effects are very strong during phase transition where the liquid, denser phase and the lighter, vapor phase are strongly convected, any investigation of the phase separation process needs the suppression of such gravity effects [2-3]. This suppression can be performed under microgravity conditions as provided by either free fall (sounding rockets) or in satellites where the centrifugational force exactly compensates the weight. In this work, we have used a strong magnetic field gradient to compensate the gravity forces by magnetic forces [4].

Let us describe a typical phase separation experiments. The sample is quenched from an initial state (temperature T_i) where it is homogeneous to another state (temperature T_f) below the coexistence curve where it is no longer thermodynamically stable. It has been found [5-6] that phase separation in gas-liquid systems (and liquid mixtures) obey two universal scaling laws of growth that is monitored by the volume fraction

$$\phi = \frac{v_i}{v_g + v_l} \quad (1)$$

of the minority phase i (= g or l; $v_{g,l}$ denotes the volume of gas g and liquid l). Volume fraction is a direct measurement of the average distance L_m between the growing domains; with D the domain diameter,

$$L_m = D(\gamma/\phi)^{1/3} . \quad (2)$$

The factor γ depends on the space arrangement of the drops. A very dense face centered cubic gives $\gamma \approx 0.74$. We can also consider as a lower bound the random close packing arrangement where $\gamma \approx 0.64$. The value of γ is not very sensitive to the particular space arrangement and the median value $\gamma = 0.69$ appears to be a reasonable choice.

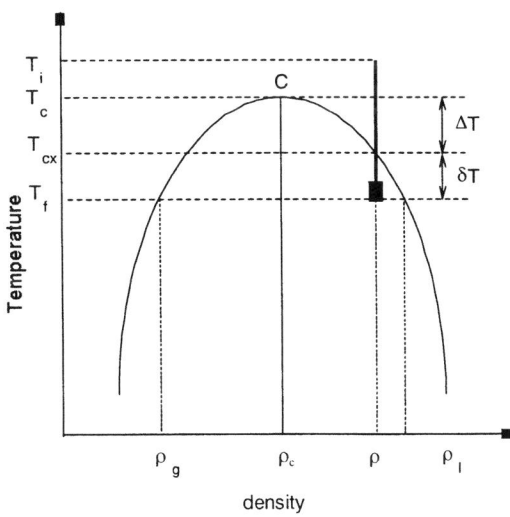

FIGURE 1. Schematic phase diagram for simple fluids

The equilibrium volume fraction ϕ of the minority phase is related to T_c, the coexistence temperature T_{cx} and the quench depth $\delta T = T_{cx} - T_f$ by (Fig. 1):

$$\phi = \frac{\rho_l - \rho}{\rho_l - \rho_g} = \frac{1}{2}\left[1 - \left(1 + \frac{\delta T}{\Delta T}\right)\right]^{-\beta} \quad (3)$$

Here ρ_g, ρ_l, stands for gas and liquid density, respectively and $\beta = 0.325$ a universal exponent. In the critical region, the density difference $\Delta\rho = \rho_l - \rho_g$ of the two phases writes as [7]

$$\Delta\rho = \rho_l - \rho_g = 2\rho_c B \tau^\beta. \quad (4)$$

Here $\tau = (T-T_c)/T_c$, ρ_c is the critical density and B is a critical amplitude (see table 1). T_c is the critical temperature. $\Delta T = T_c - T_{cx}$ is calculated from Eq. 4.

When $\phi < 30\%$, growth is driven by droplet coalescences that are induced by Brownian motion and droplet-droplet coalescence [6]. The typical evolution is

$$L_m = e\left(\frac{k_B T}{6\pi\eta}\right)^{1/3} t^{1/3}. \quad (5)$$

Here L_m is the typical distance between the domains, t is time, k_B is the Boltzmann constant and e is a constant ($e \approx 2\pi/0.9$ [6]); T stands for temperature, in K; η is the shear viscosity, which remains nearly constant in the vicinity of the critical point. (For numerical values, see Table 1). Note that the growth law is independent of the temperature difference $T-T_c$, a remark that will be of importance below.

When $\phi > 30\%$ [5-6], the flow resulting from a coalescence event is able to induce another coalescence, and so on... This process leads to an interconnected pattern with pseudo - wavelength L_m whose growth at late time is limited by the flow resulting from the coalescences. In the viscous limit, one gets [8],

$$L_m = b(\sigma/\eta) t, \quad (6)$$

$b \approx 0.03$ being a universal constant, independent of the fluid properties [5-6, 8]. σ is the liquid-vapor interfacial tension which behaves with the reduced temperature as

$$\sigma = \sigma_0 \tau^{2\nu}. \quad (7)$$

Here σ_0 is a system dependent, "critical" amplitude and $\nu = 0.63$ is the universal exponent of the correlation-length ξ (see Eq. 8 below).

The t and $t^{1/3}$ growth can be scaled by natural length and time scales. The correlation length ξ of the order parameter fluctuations (density for gas-liquids, concentration for liquid mixtures), is the natural lengthscale of the process and obeys the power law with the reduced temperature

$$\xi = \xi_0 \tau^{-\nu}. \quad (8)$$

Here ξ_0 is a system dependent, "critical" amplitude and $\nu = 0.63$ is the universal exponent. The decay time t_ξ of the critical fluctuations is the natural time scale [1]

$$t_\xi = 6\pi\eta\xi^3/(k_B T). \quad (9)$$

We consider all critical amplitudes below T_c. When expressed in units of the correlation length $K_m^* = 2\pi\xi/L_m$ and decay time $t^* = t/t_\xi$, the data for all liquid mixtures and gas-liquids obtained in gravity-free experiment reasonably fit on the same masters curves, supporting universality (Fig. 2) [5]. There exists some deviation, which can be accounted with the particular thermalization process of simple fluids, the so-called "piston effect", [9]. All the experiments show that $K_m^* \approx 1$ when $t^* \approx 1$, which means that nucleation proceeds from fluctuations of size the correlation length, in agreement with the concept of «generalized nucleation» [10].

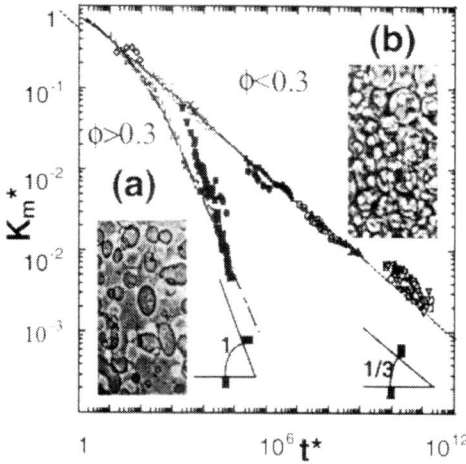

FIGURE 2. Growth laws K_m^* vs t^* in fluids (SF_6, CO_2, data points) and liquid mixtures (partially deuterated cyclohexane and methanol, letters and squares) when gravity effects are absent. (See Ref. [5])

Note that thermalization in such compressible fluids is very fast due to the very particular "Piston Effect" of the thermal boundary layer which contracts and depressurize adiabatically the sample [1].

In the following, we investigate the effect of "high frequency" linear vibrations on such growing pattern in the simple fluid H_2. By high frequency we mean frequency f larger than the typical inverse evolution times, e.g. $tf > 1$. Note the difficulty of working near the low T_c (33 K) of H_2. With the same techniques of analysis, the approach of the critical point has to be closer by a factor of 10 when compared to fluids where T_c is around room temperature.

EXPERIMENTAL

The large diamagnetic susceptibility of H_2 enables gravity forces to be compensated by the magnetic field (B) gradient such as $BgradB \approx 491$ $T^2 m^{-1}$ [4]. This value is provided near the end of a superconductive coil. Details can be found in [4]. The cavity containing the fluid is cylindrical (3 mm in diameter) with its axis horizontal, perpendicular to the coil axis (Figs. 3-4). It is made of sapphire and is closed by two parallel sapphire windows distant of 3 mm and sealed by indium rings. The gravity forces are ideally compensated at a unique point; however, in such a small sample the residual magnetic forces, which are directed towards the coil and are proportional to the distance to it, remains lower than $\approx 1.5 \cdot 10^{-2} g_o$ (here g_o is the earth acceleration value). The use of sapphire prevents heating by eddy currents when the sample is under vibrations. A parallel beam illuminates the sample, and the sample is observed with a CCD camera. The cell is filled with a capillary that can be sealed with an ice floe. It is hold in a thermostat with 0.3 mK accuracy temperature regulation. Temperature can be varied between 9 K and 40 K, pressure up to 2 MPa. The temperature quenches are very fast (of order a few ms) as sapphire exhibit an increase in heat conductivity around 30 K, leading to a thermal diffusivity of order of 5 $m^2 s^{-1}$. In addition, the sample is placed on a vibration device with amplitude [0 - 0.5 mm] and frequency [0 - 50 s^{-1}]. The experiment is computer-monitored.

The cell is filled at critical density as checked by the position of the meniscus that remains in the middle of the vessel without magnetic compensation within 1 mK from the critical temperature. When H_2 is cooled in the cell, the n-H_2 - p-H_2 equilibrium is shifted and the percentage of p-H_2 increases from 25 % at room temperature to 96 % at 30 K [4]. The useful data are listed in Table 1. The slow conversion of n-H_2 to p-H_2 is followed in time by determining at regular time interval the temperature at which the meniscus disappears (critical temperature of the ortho-para Hydrogen mixture). We find that the critical temperature decays exponentially with a time constant of order 50 hours. When the critical temperature has reached a constant value within ± 1 mK, equilibrium is supposed to be attained and the experiment starts.

The cell is vibrated by means of an electric stepper motor equipped with an adjustable eccentricity cam. The motor is fixed at the top of the cryostat. The alternative motion is transmitted to the cell by means of a thin tube 4.5x5 mm in diameter that induces an angular vibration with the rotation axis at $D_0 = 48$ mm below the center of the sample. The latter is thus submitted to a nearly linear vibration (axis X) directed perpendicularly to the axis of the coil. In Fig. 3 is given the schematic diagram of the assembly and in Fig. 4 a photo of the cell. As the vibration amplitude remains small, the vertical motion does not exceed $a^2/D_0 \approx 5$ μm and can be neglected. In addition, the centrifuge force $a^2 \omega^2 / D_o$ remains much smaller than the translational acceleration force $a\omega^2$, in the ratio a/D_o.

The typical wavelength of the pattern is analyzed by measuring in several directions the spatial frequency of the domains directly on the pictures. The resolution is of order 20 μm, the uncertainty is of order 10%. We analyze in the following experiments performed with vibrations of frequency in the range [5 - 50 s^{-1}], amplitude range [53 -500] μm, temperature range [0.17 -1.1] mK.

FIGURE 4. Photo of the cell and its vibration device.

RESULTS AND ANALYSIS

The first difficulty in such experiments is the temperature detection. When the vibration is set, magnetic-induced currents in the temperature sensor and its connecting wires prevents accurate temperature measurements and temperature regulation. Locating the sensor on the axis of rotation can minimize these effects, but they cannot be completely avoided. We therefore adopted the following procedure. First, the temperature was regulated without vibrations at about 1 mK above T_c, where the fluid is homogeneous. The average regulating power (≈ 600 mW) is measured.

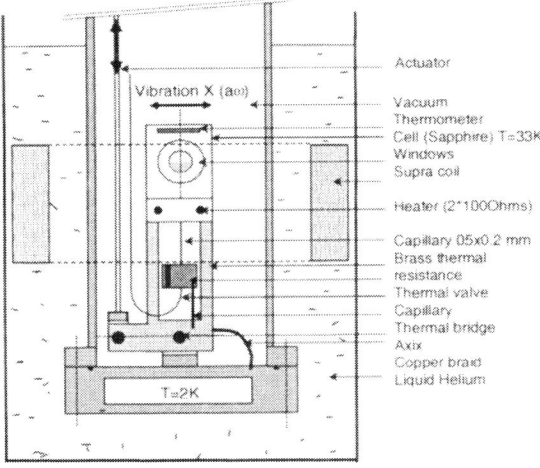

FIGURE 3. The sample in the magnetic device and its stirring mechanism.

Then the vibration (a, ω) is set; simultaneously the temperature regulation mode is removed and a DC power is sent instead in the sample heater. This power is slightly less than that above, because the vibration sends some power in the sample through eddy currents in the small metallic parts (screw, indium rings). For $a = 0.3$ mm and $f = 20$ s^{-1}, the needed power is 585 mW. We then decrease the power with small values (typically 0.5 mW). The temperature decreases according to an exponential evolution, with typical time a few ms. With several trials, it is possible to observe phase separation with different growth rates, corresponding to different τ. The phase separation process is always very fast, lasting between 0.5 s to 15s (Figs. 5-6, 8-9). The validity of the procedure has been verified by checking without vibrations the growth laws (see insert in Fig. 6).

TABLE 1. p-H$_2$ data. (a): from Ref. [11] and Refs. therein. (b) From an extrapolation at T_c of the tables [12]. (c) from the amplitude relationship $\xi^2 = R\,k_B T_c/\sigma$, with $R = 0.1$ the universal amplitude ratio [11].

ρ_c [a]	T_c [a]	η [b]	ξ_0 [c] $T<T_c$	σ_0 [a]	B [a]
(Kg.m^{-3})	(K)	(Pa.s)	(m)	(N.m^{-1})	
31.4	33	2.7 10^{-6}	8.9 10^{-11}	5.7 10^{-3}	1.61

Phase transition at $\phi > 0.3$

A typical evolution pattern is shown in Fig. 5. The pattern remains generally isotropic during the experimental time (Figs. 5 a-b), except for the highest values of frequency and amplitude ($a \geq 0.4$ mm, $f \geq 40$ s^{-1}). Here, at late times, the domains grow more rapidly perpendicularly to the vibration direction and become anisotropic (Fig. 5c, case noted "c").

When analyzed (Fig.6), the pattern exhibits at early time a linear growth similar to the evolution without vibration, case "a". When the domain wavelength L_m exceeds a given size L_o, another (linear) growth law is observed (Fig. 6), case "b". Due to experimental limitations in resolution and cell dimension, this cross-over is visible only for a vibration velocity $a\omega > 10^4$ µm s^{-1}.

A difficulty in these experiments is concerned with the fact that temperature cannot be measured. One notes, however, that without vibrations, the domain evolution is well known. We therefore use Eq. 6 to infer the τ value:

$$\tau = (\eta/b\sigma_0)]^{1/2\nu}\,U^{1/2\nu} \sim U^{0.79}\,. \qquad (10)$$

In other words, $U^{0.79}$ will be used as a substitute to temperature.

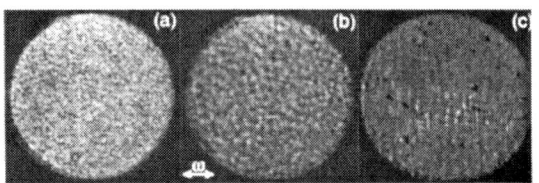

FIGURE 5. Phase transition in H$_2$ under vibration. (a – c): gas – liquid interconnected pattern after a quench of about 0.6 mK below T_c. (a) vibration of 0.3 mm amplitude and 20.3 s^{-1} frequency, time 0.6 s after quench; (b) vibration of same amplitude and frequency, 1.7 s after quench; (c) vibration of 0.53 mm amplitude and 40 s^{-1} frequency, 3.6 s after quench.

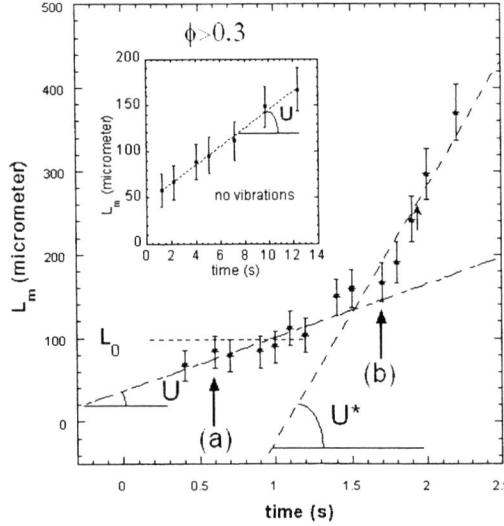

FIGURE 6. Typical evolution of the pseudo-wavelength L_m between interconnected phase separating domains (vibration $a = 0.3$ mm, $f = 20.3$ s^{-1}). The evolution for $L_m < L_0$ corresponds to $T_c - T = 1.06$ mK. The insert shows typical evolution without vibrations. U and U^* are growth velocities. The arrows correspond to Fig. 5 pictures.

The boundary layer theory of liquids under vibration [13] shows that the motion of the two phases are strongly coupled by viscosity as long as the boundary layer thickness

$$\delta = (2\mu/\omega)^{1/2} \qquad (11)$$

remains larger than L_m. Here $\mu = \eta/\rho$ is the kinematic viscosity. In this case, the two phases move coherently and the growth is not affected by vibrations. However, as soon as $L_m > \delta$, the motion of the two-phases are locally decoupled so that they do not exhibit the same velocity anymore due to inertial effects. Then the cross-over length between the two growth laws should correspond to the viscous boundary layer, $L_o \sim \delta$. In Fig. 7 is plotted the crossover length L_0 with respect to ω in a log-log plot. A fit to $L_0 = l_0\omega^{-x}$ gives $x = 0.53 \pm 0.06$, a value in good accord with Eq. 11. When $x = 1/2$ is imposed in the fit, the amplitude is $l_0 = 1660 \pm 50$ µm s$^{1/2}$. This value has to be compared with

$(2\mu)^{1/2} = 420$ μm s$^{1/2}$. The factor 4 difference between the amplitudes is not meaningful at this stage of the evaluation of the cross-over and the region $L_m < L_0 \sim \delta$ can then be assessed with confidence to be the region where vibrations do not affect the growth. The slope U during case (a) will therefore be used together with Eqs. 6-10 to infer the reduced temperature τ.

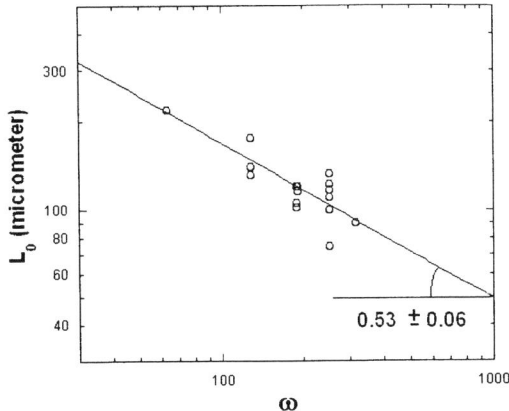

FIGURE 7. Variation of the crossover length L_0 (see Fig.6) with respect to the angular frequency ω (log - log plot). The line is the best fit to a power law.

Analysis

The acceleration of the growth in the region $L_m > \delta$ is linked to the presence of inertial velocity difference ΔV in the gas and liquid domains, in relation with the gas and liquid density difference $\Delta\rho$. Note that, according to Eqs. 4 and 10, the temperature behavior of $\Delta\rho$ can be written in function of U as:

$$\Delta\rho \sim \tau^\beta \sim U^{\beta/2\nu} = U^{0.26}. \quad (12)$$

The velocity difference ΔV between the gas and liquid phases can be written as

$$\Delta V = a\omega(\Delta\rho/\rho_c). \quad (13)$$

This velocity difference can act on the growing domains in different ways.

(1) The simplest effect comes from dimensional analysis considerations. The viscous capillary flow should exhibit a larger growth velocity U^*. This increase is expected to be proportional to the velocity ΔV:

$$\Delta U_G \equiv U^* - U = f\Delta V = fa\omega(\Delta\rho/\rho_c), \quad (14)$$

with f a numerical factor of order unity. This relation implies hydrodynamic non linearities and does not provide any details on the mechanisms at the origin of the growth speeding up. It corresponds to stage (b) and stage (c) where domains are isotropic.

(2) Another mechanism is connected with the effect of a periodic shear flow between the domains of amplitude S. The velocity difference is of order ΔV and the characteristic length is of order the viscous boundary layer δ:

$$S \approx \Delta V/\delta = (\Delta\rho/\rho_c)(a\omega/\delta). \quad (15)$$

The effect of a periodic shear flow ($S= 600$ s^{-1}) has been already investigated in Ref. [14] in a binary liquid (nitrobenzene and n-hexane) near its liquid-liquid critical point. However, there is a number of important differences with the present situation: the shear was uniform in the sample (it shows here the period L_m) and the frequency (0.03 – 0.3 s^{-1}) was much smaller. A steady state of partial phase separation was obtained that corresponded to the balance between the homogenization effect of shear and phase separation. The domains were highly anisotropic.

In the present situation, the shear is periodic in time and space and remains confined in the viscous boundary layer between the domains. The domains are stretched as St in the velocity or vibration direction, modifying the growth in Eq.3 as

$$(dLm/dt) = USt \sin\omega t. \quad (16)$$

During a period of vibration $\theta = 2\pi/\omega$, the domains are then periodically stretched if the condition $S\theta > 1$ is fulfilled. (This is the case in Fig.5 a-b where $\theta \approx 50$ ms, $\tau \approx 0.6$ mK and $S \approx 100$ s^{-1} according to Eqs. 4, 10). The stretching is averaged on a time larger than θ but smaller than the inverse growth rate $(L_m/(dLm/dt) = t)$:

$$L_m = b(\sigma/\eta)(1+S\theta/4\sqrt{2})\,t. \quad (17)$$

It is interesting to note that the time average of the angle of deformation of the domains is zero because the excitation is periodic: the pattern remains in average isotropic.

From Eqs. 6, 17, one finally gets a growth velocity increase

$$\Delta U_S = gU(\Delta\rho/\rho_c)(a/\delta). \quad (18)$$

Here g is a numerical factor of order unity. As above, this mechanism corresponds to stage (b) of Figs. 5-6. The anisotropic growth (c) is not accounted for.

(3) Another mechanism similar to a Bernoulli pressure can be invoked. The velocity difference between the domains induces an average Bernoulli pressure difference ΔP, directed perpendicularly to the vibration direction, which scales as

$$\Delta P \sim (\Delta\rho/\rho_c)(a\omega)^2. \quad (19)$$

This pressure adds to the capillary pressure (σ/L_m). A dimensional analysis of the Navier-Stokes Equations as in Ref. [8] leads to $\nabla P = \eta\Delta V$, or

$$(\sigma/L_m^2) + (g/\delta)(\Delta\rho/\rho_c)(a\omega)^2 = (\eta/L_m^2)(dLm/dt). \quad (20)$$

Here the Bernoulli gradient has been estimated by the ratio of the pressure difference across the boundary layer δ. After some algebra, and for small times, it comes eventually

$$\Delta U_B \approx h(\Delta\rho/\rho_c)(a/\delta)(a\omega). \quad (21)$$

where h is a numerical factor of order unity.

In contrast to the two former models, this mechanism gives rise to a preferential growth rate perpendicularly to the vibration direction. It looks therefore more appropriate to account for stage (c) that stage (b).

In order to test the three proposed behaviors, we note that Eqs. 14, 18, 21 can be written in terms of the experimental quantities (ΔU, U, a, ω) as:

$$\Delta U = a^x \omega^y U^z, \quad (22)$$

with exponent x, y, z varying according to the above assumptions (1) $x = 1$, $y = 1$, $z = 0.26$, (2) $x = 1$, $y = 0.5$, $z = 1.26$, (3) $x = 2$, $y = 1.5$, $z = 0.26$. We define the dimensionless parameter

$$C = \Delta U a^{-x} \omega^{-y} U^{-z}, \quad (23)$$

with unknown exponents x, y, z. The exponents most probable values can be determined by calculating the set (x, y, z) that makes C as constant as possible with respect to ΔU for the various a, ω, U combinations. The test is the minimization of the relative deviation $s = <\Delta C>/<C>$. In Table 2 are reported the results. When x, y, z are all free within [-2, 2], the minimum is found for $s_0 = 0.6$ and the set $x = 0.35 \pm 0.5$, $y = 0.35 \pm 0.5$, $z = 1.3 \pm 0.4$. The values for y and z are close to Eq. 18 (shear influence), but x is too small. The sensitivity of the method, however, is not very high, as one can see in Table 2 where are reported also the s – values for the exponents of Eqs. 14, 18, 21. Due to the shallow minimum of s_0 in the x, y, z space, this method cannot give a definitive answer - although there is some tendency to highlight the shear values.

TABLE 2. Result of the minimization of the relative deviations of reduced parameter C (see text). 14 useful experiments have been considered with vibration amplitude ranging from 53 to 410 μm, angular frequency ω between 127 and 314 s^{-1}, temperature between 0.17 and 1.1 mK.

	$\frac{<\Delta C>}{<C>}$	x	y	z
minimum	0.60	0.35 ± 0.5	0.35 ± 0.5	1.3 ± 0.4
ΔU_G	0.76	1	1	0.26
ΔU_S	0.75	1	0.5	1.26
ΔU_B	1.08	2	1.5	0.26

Phase transition at $\phi < 0.3$

Let us now investigate the case where the pattern is not interconnected. For obvious reasons of wetting by the liquid phase, phase separation is performed at liquid density such as only vapor bubbles are allowed to nucleate and grow. The measurement of temperature does not matter in this case, as noted above in the introduction. In Figs. 8-9 are reported for $a = 0.47$ mm and $f = 20$ s^{-1} the different observed regimes. We report the bubble diameter D (the direct experimental determination), which can be related to L_m by Eq.2. As the dependence in ϕ is weak (in $\phi^{1/3}$), in the range $\phi = 0.2 - 0.3$, $L_m \approx 1.4\, D$.

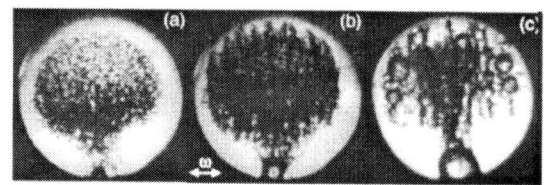

FIGURE 8. Phase transition in H_2 under vibration ($a = 0.47$ mm, $f = 20$ s^{-1}) for low volume fraction (gas bubble pattern (a): time 0.9 s after quench; (b): time 2.1 s after quench; (c): time 3.3 s after quench. The drop at the sample bottom is attached to the filling capillary. (No bubbles nucleate near the walls due to the Piston Effect thermalization process that induce a density rise in a boundary layer near the wall [1, 9]).

Figure 9 shows an evolution similar to the interconnected case of Figure 6. The effect of vibration remains invisible (case "a" in Figure 8) when D does not exceed the viscous boundary layer. In Figure 9, the effect of vibration is seen for $D > L'_0$ ($L'_0 \approx 160$ μm for $f = 20$ s^{-1}). From Eq. 11, one indeed verifies that L_0 ($\omega = 125$ rd/s) = 150 μm, i. e. a value close to L'_0. In this domain, the bubbles begin to align in periodic, parallel layers, a phenomenon which increases the coalescence rate (cases "b" – "c" in Figure 8). A new population of bubbles that grow faster due to enhanced coalescences thus appears, with an average drop radius evolution that can be approximated by a linear law (see Fig. 9).

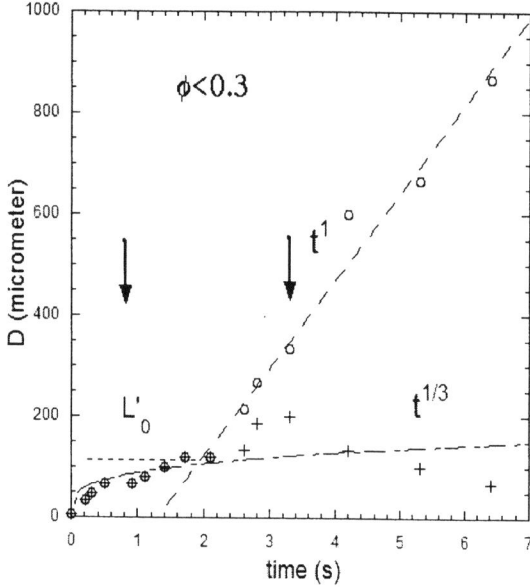

FIGURE 9. Typical evolution of the vapor bubble diameter during phase separation under vibrations ($a = 0.47$ mm, $f = 20$ s^{-1}). Crosses: bubble population not affected by vibrations. Circles: bubble coalescing under the influence of vibration when $D > L'_0$. The interrupted curves are fit to $t^{1/3}$ and t^1 growth law. The arrows correspond to Figs. 8 a -.b.

A number of drops are not affected by this process, however, and their growth can be successfully fitted to the classical $t^{1/3}$ growth law, Eq. 5. Indeed, by fitting the data to $D = d_0 t^w$, we find $w = 0.3 \pm 0.1$.

These findings can be compared to the effect of a shear flow between bubbles and liquid. However, for a population of drops or bubbles, the effect of a uniform shear flow is found insignificant [15], mainly because the rotational part of a simple shear makes the droplets rotate on each other, without coalescing.

Another effect comes from the hydrodynamic forces that arise between droplets submitted to high frequency vibrations. According to Ref. [16], two bubbles whose line between the centers is parallel to the vibration direction repulse each other and those whose line between the centers is perpendicular to the vibration direction attract each other. The forces are in $(D/d)^4$, with d the distance between the drops. (In addition, the drops are (slightly) flattened in the direction perpendicular to the vibration direction). At least qualitatively, the pattern evolution in Figs. 8 b-c can be well understood by the presence of these hydrodynamical forces. However, the periodic layering is not explained.

Simple arguments can explain the linear growth law. A time dt is needed for one drop to move close its neighbor and coalesce with it. In Eq. 5, dt is of the order of the Brownian diffusion time over the mean distance L_m between bubbles, which is of order D, leading to $dt \sim (6\pi\eta/kT)D^3$. Under vibrations, the bubbles attract each other. The time of approach becomes negligible compared to the time coming from the viscous flow during coalescence, Eq. 6, thus leading to a constant growth velocity $U \sim \sigma/\eta$. As σ is much temperature dependent, such a model could be tested by varying the temperature quench.

CONCLUSION

These experiments, designed to study the influence of vibrations on phase transition in fluids, are very difficult to handle because of a necessary close approach to the critical point and the need to compensate gravity forces. Temperature measurements are particularly difficult in the magnetic set-up used to compensate gravity.

It appears that the effect of vibrations is primarily governed by inertial effects that initiate velocity differences in the gas – liquid growing domains. A major result is that the growth is unaffected as long as the domain size is smaller than the viscous boundary layer.

Growth of interconnected gas-liquid separating domains or well identified bubbles is always accelerated. The velocity difference not only induce shear flows, but also more complex behavior where the domains can order in periodic planes perpendicular to the vibration direction. In any cases, coalescence process is enhanced.

These phenomena deserve further studies. In particular, the origin of the spatial periodicity of domains in the direction perpendicular to vibrations is unclear. This phenomenon is also observed with solid inhomogeneities immersed in liquids under vibrations [17] – where it is not explained.

ACKNOWLEDGMENTS

This work has been partially supported by the Centre National d'Etudes Spatiales.

REFERENCES

1. Onuki, A., *Phase transition dynamics*, Cambridge University Press, Cambridge, 2002.
2. Moldover, M. R., Sengers, J. V., Gammon, R. W., and Hocken, R. J., *Rev. Mod. Phys.* **51**, 79-99 (1979).
3. Beysens, D., *Microgravity Q.* **5**, 35-43 (1995).
4. Wunenburger, R., Chatain, D., Garrabos Y., and Beysens, D., *Phys. Rev.E* **62**, 460-476 (2000).
5. Beysens, D., and Garrabos, Y., *Physica A* **281**, 361-380 (2000) and Refs. therein.
6. Nikolayev, V. S., Beysens, D., and Guenoun, P., *Phys. Rev. Lett.* **76**, 3144-3147 (1996); Nikolayev, V. S., and Beysens, D., *Phys. of Fluids* **9**, 3227-3234 (1997).
7. Stanley, H. E., *Introduction to phase transitions and critical point phenomena*, Oxford University Press, Oxford, 1971.
8. Siggia, E. D., *Phys. Rev. A* **20**, 595-605 (1979).
9. Zappoli, B., Bailly, D., Garrabos, Y., Le Neindre, B., Guenoun, P., and Beysens, D., *Phys. Rev. A* **41**, 2264-2267 (1990).
10. Binder, K., in *Material Science and Technology: phase transformations in materials*, vol. 5, edited by P. Haasen, VCH Verlagsgesellschaft Weinheim, Germany, 1991, pp 405-471 and references therein.
11. Moldover, M.R., *Phys. Rev. A* **31**, 1022-1033 (1985).
12. Diller, D.E., *J. Chem. Phys.* **42**, 2089-2098 (1965).
13. Landau, L.D., and Lifshitz, E.M., *Fluid Mechanics*, Pergamon Press, Elmsford, New York, 1975.
14. Beysens, D., and Perrot, F., *J. Physique. Lett.* **45**, 31-38, (1984).
15. Baumberger, T., *Ph. D. thesis*, University Paris XI, 1992, p.129
16. Lyubimov, D.V., Lyubimova, T.P., and Shklyaev, S.V., "Behaviour of a Drop (Bubble) in a Pulsating Flow near Vibrating Rigid Surface" in *1st International Symposium on Microgravity Research & Applications in Physical Sciences and Biotechnology*, Sorrento (Italy), Sept. 10-15 2000, edited by ESA-SP, Conference Proceedings 454, 2001, pp. 879-886.
17. Wunenburger, R., Carrier, V., and Garrabos, Y., *Phys. of Fluids* **14**, 2350-2359, (2002).

The Flow and Adsorption of DNA Polymers Near Surfaces

Ronald G. Larson, Lei Li[1], Manish Chopra[2], and Mark A. Burns

Department of Chemical Engineering, University of Michigan, Ann Arbor, MI 48109-2136, U.S.A.

[1]*Current address: Department of Pharmaceutical Sciences, University of Michigan, Ann Arbor, MI.*
[2]*Current address: Mckinsey & Company, Detroit, MI.*

Abstract. We discuss how agreement between single-molecule imaging methods applied to DNA molecules in flow [1,2] and Brownian dynamics simulations using bead-spring or bead-rod course-grained models [3,4] can be extended to study the interactions of flowing DNA polymers with surfaces, which are of importance in the development of microfluidic devices for processing of DNA and other large molecules for genomics, bio-assays, combinatorial polymer science, etc. Using single-molecule experiments and Brownian dynamics simulations we review work on isolated DNA molecules near adsorbing and non-adsorbing walls in the presence of a simple shearing flow and in an evaporating droplet. The former flow is predicted to produce highly stretched adsorbed molecules due to the prevalence of end-sticking, following by regular unraveling from one end to the other and laying down of the molecule onto the surface. In the drying-droplet flow, this process is inhibited by the downward convection, which drives the molecule towards the surface, resulting in complete adhesion before unraveling is complete. Experimental studies using surfaces treated with APTES (3-aminopropyltriethoxysilane) to produce strong sticking of DNA confirm the Brownian dynamics predictions for the drying flow containing DNA. In simple shearing flow, an unusual, and unexplained, interaction of DNA with the surface inhibits stretching, at distances as great as 20 microns from the surface.

INTRODUCTION

Recent advances in single-molecule imaging methods applied to DNA molecules in flow [1,2] and advances in computer speed have allowed detailed comparisons to be made between observed and predicted behavior of dilute polymeric DNA molecules in simple flows. These have shown that the conformations and rheology of DNA molecules in bulk solution can be predicted with high accuracy by Brownian dynamics simulations using bead-spring or bead-rod course-grained models that include hydrodynamic interactions only in the choice of the effective bead drag coefficient [3,4]. We have shown that more complex effects of *deformation-dependent* hydrodynamic interactions are expected to be important in synthetic polymers such as high-molecular-weight polystyrene in dilute solution, but that these effects are negligible in DNA due to its open configuration at rest, owing to its large persistence length [5].

A logical next step is to extend these methods to complex flows, and to the interactions of flowing DNA polymers with surfaces, which are of importance in the application of microfluidic devices to the processing of DNA and other large molecules for genomics, bio-assays, combinatorial polymer science, etc. Using single-molecule experiments and Brownian dynamics simulations we have considered isolated DNA molecules near adsorbing and non-adsorbing walls in the presence of a simple shearing flow and in an evaporating droplet. We review this work here.

FLOW FIELDS

In simple shearing flow, Brownian dynamics simulations in the absence of hydrodynamic interactions predict that the molecules will become highly stretched as they become adsorbed irreversibly onto a surface [6]. This occurs becomes of end-sticking, i.e., the tendency of the molecule to adsorb first at its end, and the to unravel onto the surface in the shearing flow. The end sticking prevents the molecule from undergoing continual tumbling and collapse, as well as stretching, in bulk shearing flow. The result is that the molecule, once anchored to the surface, can become highly stretched in the shearing flow, before it becomes irreversibly adsorbed onto the surface.

To extend these results to a complex flow, we need to solve for the flow field of a complex flow, and use this flow in a Brownian dynamics simulation to predict molecular stretch during DNA deposition. As a test problem, we have chosen the flow in a drying water droplet resting on a substrate. Because of the pinned contact line, the droplet does not shrink its radius until the very last stages of drying, but instead shrinks its height. As a result, fluid that evaporates from the edge of the droplet must be replaced by fluid flowing to the edge from the droplet center [7]; see Fig. 1.

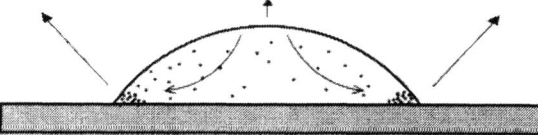

Figure 1. The flow in a drying droplet with pinned contact line.

This flow convects solute towards the droplet edge, where it deposits it in a ring [7], or "water spot," frequently seen on dishware that has been left to dry.

Besides being an annoyance, the "coffee ring" effect can be used to advantage in genomics applications. Schwartz and co-workers [8] have shown that the flow in a drying droplet can be used to stretch out and deposit DNA molecules, whereupon they can be subjected to a restriction enzyme digestion, and the length and relative positions of the fragments measured by simple fluorescence optical microscopy, used DNA stained by intercalating dyes.

The droplet drying flow, then, is an attractive one in which to carry out Brownian dynamics analyses of DNA stretch and deposition. Since the flow is not only multi-dimensional, but also time dependent, to carry out Brownian dynamics simulations, one would need either an analytic solution to the droplet-drying flow, or a numerical solution for this flow that is carried out simultaneously with the Brownian dynamics simulations. We have pursued the former approach, but used a numerical solution to check the accuracy of our analytic solution.

To obtain an analytic solution, we start with the approximate solution of Deegan et al. [7] for the height-averaged radial velocity field, which is

$$\bar{u}_r = \frac{1}{4}\frac{1}{t_f - t}\frac{R}{\tilde{r}}\left[(1-\tilde{r}^2)^{-\lambda(\theta)} - (1-\tilde{r}^2)\right] \quad (1)$$

Here, λ is a parameter reflecting the uniformity of evaporation, which is 0.5 for a flat droplet. The accuracy of this expression has been thoroughly explored in a recent experimental and finite-element study of droplet evaporation by Hu and Larson [9], who found that for small droplets whose shape is that of a spherical cap resting on a surface with contact angle θ, the above expression for evaporative flux is accurate, with $\lambda = \frac{1}{2} - \frac{\theta}{\pi}$, where θ is the contact angle in radians. From the approximate evaporative flux expression, the entire axisymmetric flow field inside the droplet, including both radial and axial components, can be obtained by using the lubrication approximation and the zero-shear-stress boundary condition on the free air-liquid surface and the no-slip boundary condition on the glass substrate. The zero-shear-stress boundary condition neglects Marangoni flow produced by temperature gradients that arise due to evaporative cooling. However, for water, thermal Marangoni flows are generally not observed, and so these surface-tension-gradient effects can be neglected and the no-shear-stress boundary condition on the free surface can be used. With these boundary conditions and the continuity equation, the radial and axial velocity components in dimensionless form can be given as [6]:

$$\tilde{u}_r = \frac{3}{8}\frac{1}{1-\tilde{t}}\frac{1}{\tilde{r}}\left[(1-\tilde{r}^2) - (1-\tilde{r}^2)^{-\lambda(\theta)}\right]\left\{\frac{\tilde{z}^2}{\tilde{h}^2} - 2\frac{\tilde{z}}{\tilde{h}}\right\} \quad (2)$$

$$\tilde{u}_z = \frac{3}{4}\frac{1}{1-\tilde{t}}\left[1 + \lambda(\theta)(1-\tilde{r}^2)^{-\lambda(\theta)-1}\right]\left\{\frac{\tilde{z}^3}{3\tilde{h}^2} - \frac{\tilde{z}^2}{\tilde{h}}\right\} +$$

$$\frac{3}{2}\frac{1}{1-\tilde{t}}\left[(1-\tilde{r}^2) - (1-\tilde{r}^2)^{-\lambda(\theta)}\right]\left\{\frac{\tilde{z}^2}{2\tilde{h}^2} - \frac{\tilde{z}^3}{3\tilde{h}^3}\right\}\dot{\tilde{h}}(0,\tilde{t}) \quad (3)$$

Here the dimensionless variables are defined as follows:

$$\tilde{u}_r = \frac{u_r t_f}{R}, \quad \tilde{u}_z = \frac{u_z t_f}{h_0}, \quad \tilde{t} = \frac{t}{t_f}, \quad \tilde{r} = \frac{r}{R},$$

$$\tilde{z} = \frac{z}{h_0}, \quad \tilde{h} = \frac{h}{h_0} \quad (4)$$

and u_r and u_z are the dimensional radial and axial velocity components respectively. R is the radius of the droplet, which is assumed to be constant with time and r is the radial position. t_f is the total drying time, t is the time, h_0 is the initial height of the droplet, h is the instantaneous height at time t at a radial position r, θ is the instantaneous contact angle and is a linearly decreasing function of time. A key feature of this solution is that the velocities become singular at the end of drying.

This flow field was tested by using a particle-tracking velocimetry method, described elsewhere and found to be reasonably accurate [10]. We therefore combined this flow field with Brownian dynamics simulations of lambda-phage DNA molecules, to predict their deformation and adhesion to a sticky glass substrate during the droplet-drying flow [11]. Irreversible DNA adhesion was incorporated into the Brownian dynamics simulations by freezing onto the glass substrate any bead of the bead-spring chain that contacted the surface during the simulations. Effects of "lubrication forces" or local electrostatic interactions with the substrate were neglected.

BROWNIAN DYNAMICS SIMULATIONS

Brownian dynamics simulations of a lambda-phage DNA molecule were carried out using the well-known bead-spring model, using methods described elsewhere [6,11]. The parameters of the bead-spring model, including bead drag coefficient and elastic spring constants, are chosen a priori using the known molecular properties of lambda-phage DNA [3]. Interactions between one molecule and another are neglected, since we are focusing on dilute solutions with low surface coverage. We ignore hydrodynamic lubrication forces of the thin fluid layers between the polymer and the substrate, and hydrodynamic interactions between the DNA and the wall. We also neglect monomer-monomer excluded-volume interactions within a single molecule. We analyze 2,000 chains at each evaporation rate (drying time), distributed initially uniformly throughout the volume of the droplet. We store the maximum stretch length (defined as the distance between two farthest-separated segments) of the molecule at the moment of "touch-down" of each bead until the molecule adsorbs fully (i.e., all the beads have been bound to the surface) [11]. We validated the accuracy of the simulations by varying the number of beads and the time step, and found the simulations to be quantitatively robust to variations in these parameters.

The simulations show that, in the drying-droplet flow, the process by which the chain unravels onto the surface, seen in simulations of the simple shear flow, is inhibited by the downward convection, which drives the molecule towards the surface, resulting in complete adhesion before unraveling is complete [11]. Thus, in the simulations, the molecular stretch is highly heterogeneous, and on average, rather weak.

EXPERIMENTS

As described in more detail elsewhere [11], we carried out experimental studies of DNA deposition from a drying droplet using glass surfaces treated with APTES (3-aminopropyltriethoxysilane) to produce a positively charged surface, and therefore strong sticking of negatively charged DNA molecules. Specifically, 0.5µl droplets of 50pg/ml DNA solutions were deposited onto the APTES-coated glass coverslips. The droplet was shielded from air currents using a cylindrical cap (6 mm in diameter and 6 mm in height) with an open top to permit evaporation. The evaporation rate could be controlled using a Millipore membrane on top of the droplet chamber. We used 10mM Tris-HCl and 1mM EDTA at pH 8.0 as buffer solutions, with 50pg/ml 48.5-kbp λ-phage DNA (New England BioLab) stained with YOYO-1 fluorescent dye (Molecular Probe Inc.) at a dye-to-base-pair molar ratio of 1:8, and 10% β-mercaptoethanol (Sigma). We note that we use a higher concentration of β-mercaptoethanol than the 4% used in Smith and Chu's experiments (1999) because the former yields greater delaying of photo-bleaching. The effect of this on the DNA molecular persistence length is here neglected. After the droplet had dried out, 10ml of 10mM Tris-HCl and 1mM EDTA buffer solution with 10% β-mercaptoethanol were deposited onto the dried spot to re-wet the surface for observation of the DNA. Stained DNA molecules, either in bulk solution or on the coated glass slips, were visualized using a Nikon TE200 fluorescent microscope with a 100x objective, with a digital interline CCD camera MicroMax 1300YHS (Princeton Instruments, distributed by Fryer Co.) to capture both still and dynamic images at a resolution of 1300x1030 using full-chip acquisition. The image acquisition software MetaView/MetaMorph version 4.5 (Universal Image, distributed by Fryer Co.) was used to control the camera, the XYZ stage motor (Prior Inc.), and the electronic shutter (Uniblitz VMM-D1, Vincent Associates), and to run the image-analysis software.

These experiments confirm the Brownian dynamics predictions for the drying flow containing DNA. We showed that the combination of a continuum fluid dynamical solution for the flow in an evaporating droplet with Brownian dynamics simulations of DNA molecules in that flow, are able accurately to predict the stretching and deposition of DNA molecules onto a substrate during droplet drying; see Fig. 2. The average stretch of the deposited lambda DNA molecules is only around 15% of full stretch, even under rather rapid evaporation conditions in which a 1-mm radius droplet dries out in three minutes. The distribution of stretch is well predicted, as is the distribution of orientation angles with respect to the flow direction [11].

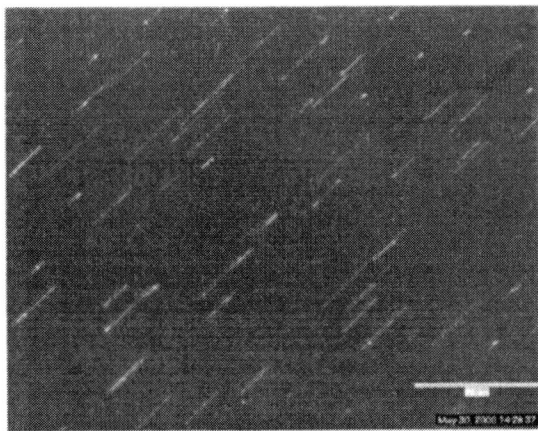
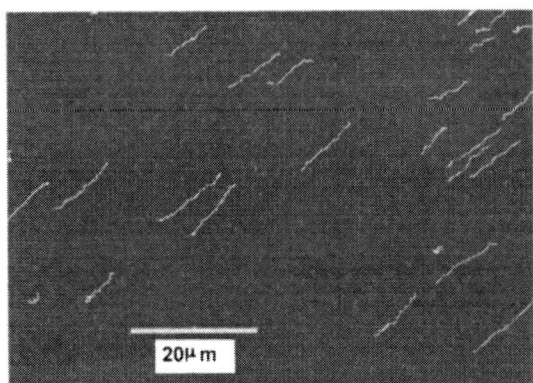

Figure 2. DNA images at the edge of a droplet (radial position > 90% of the droplet radius) at 3 mins. drying time, 20 μm from the edge of the droplet. The top is an experimental image and the bottom is the simulated one, under conditions identical to those in the experiments. Histograms showing distribution of stretch and orientation from the simulations are in good agreement with those measured. The scale bar represents 20μm in length. (The figure is taken from ref. 11, used with permission, from the Journal of Rheology).

We also carried out experiments in simple shearing flows, using a torsional shearing apparatus, consisting of a pair of parallel glass disks, the upper of which could be rotated using a motor, and the lower of which was the cover-slip for an inverted fluorescence microscope. In this flow, we also observed DNA stretch, not only after deposition onto an APTES-treated glass coverslip, but also in the fluid undergoing shear, above the cover slip. We observed that, both after adsorption on the coverslip, and in the fluid near the coverslip, the DNA was much less stretched than predicted by theory [12]. At this point, we do not have an explanation for this unusual behavior.

REFERENCES

1. Perkins, T.T., Smith, D.E., and Chu, S., *Science*, **276**, 2016-2021 (1997).
2. Smith, D.E., and Chu, S. *Science* **281**, 1335-1340 (1998).
3. Larson, R.G., Hu, H. Smith, D.E., and Chu, S., *J. Rheol.*, **43** 267-304 (1999).
4. Hur, J.S., Shaqfeh, E.S.G.S., and Larson, R.G., *J. Rheol.* **44** 713-742 (2000).
5. Hsieh, C.-C., and R.G. Larson, J.Non-Newt. Fluid Mech.,in press (2003).
6. Chopra, M. and Larson, R.G., *J. Rheol.*, **46**, 831-862, (2002).
7. Deegan, R. D., Bakajin, O., Dupont, T. F., Huber, G., Nagel, S. R., Witten, T. A., *Nature* **389**, 827-829 (1997).
8. Jing J., Reed, J., Huang, J., Hu, X., Clarke, C. V., Edington, J., Housman, D., Anantharaman, T. S., Huff, E.J., Mishra, B., Porter, B., Shenker, A., Wolfson, E., Hiort, C., Kantor, R., Aston, C., Schwartz, D. C., *Proc. Natl. Acad. Sci. USA* **95**, 8046-8051 (1998).
9. Hu, H. and Larson, R.G., *J. Phys. Chem.* **106**, 1334-1344 (2002).
10. Hu, H. and Larson, R.G., Micro-Fluid Dynamics in an Evaporating Sessile Droplet, to be published.
11. Chopra, M., Li, L., Burns, M.A., and Larson, R.G., *J. Rheol.*, in press. (2003).
12. Lei, L., Hu, H., and Larson, R.G., *Rheol Acta*, submitted (2003).

Experimental study of self organized criticality on a three dimensional pile of rice

C. M. Aegerter, K. A. Lőrincz, M. S. Welling, and R. J. Wijngaarden

Department of Physics, Faculty of Sciences, Vrije Universiteit, 1081HV Amsterdam, The Netherlands

Abstract. The dynamics of a driven, three dimensional rice-pile is studied. When the pile is fully grown, its activity takes place in power-law distributed avalanches. The observation of finite-size scaling in the observed cut-off size indicates that the pile is in a true critical state, as is demanded by self-organized criticality. Before the stationary state is reached, the maximum slope of the pile is increasing towards a critical value, where the critical state is reached asymptotically. The exponent governing this approach to the critical state is related to the exponents determined in the critical state for the avalanche dimension and distribution. This is in good accord with an analytical theory of self-organized criticality, based on extremal dynamics.

INTRODUCTION

The ubiquitous appearance of power-laws in nature has lead Bak *et al.* in 1987 to propose that slowly driven non-equilibrium systems self-organize into a critical (SOC) state, which naturally leads to power-law behavior [1]. While in the past 15 years, much progress has been made in the theoretical foundations of a process that naturally leads a system to its critical state [2, 3], controlled experiments are a rarity in the field. There are less than a handful of experiments actually demonstrating the existence of a true critical state in a slowly driven system, showing finite size scaling of the cut-off size in addition to a power-law distribution. Two such experiments were carried out in two dimensional piles of granular material (confined between two glass plates), with in one case rice [4] and in the other case steel balls [5] as the granular material. An extension to three dimensional piles has only been published this year, where again two independent experiments were carried out, one on a pile of rice [6] and one on a pile of beads [7]. Here we describe the experiment on the three dimensional pile of rice, however going beyond the mere existence of a critical state.

From the theoretical investigations it has been shown that in order to get a proper understanding of SOC, the way to the critical state is of utmost importance, as this allows the self-organization process to be studied [8]. Therefore, we here also study the behavior of the rice pile in the transient regime in the approach to the critical state and concentrate in particular on the maximum slope of the pile, as the critical state is reached. Using the theory that extremal dynamics is what underlies SOC, the approach to the critical state can be described analytically by what is called the Gap-Equation [2]. This results in the fact that the exponent governing the maximum slope to its critical value, δ, is related to the exponents characterizing the critical state, i.e. the avalanche distribution exponent, τ, the avalanche dimension, D and the fractal dimension of the active sites, d_B. Combining the results of the experiments on the critical state with those from the transient behavior, can therefore lead to a confirmation that extremal dynamics indeed lies at the heart of SOC.

EXPERIMENTAL DETAILS

The experiments were carried out on a pile of rice with the surface area of $\sim 1 \times 1\ m^2$ [6]. Long grained rice of dimensions $\sim 2 \times 2 \times 7\ mm^3$ is fed continuously at one edge of the pile in a uniformly distributed line and an image of the pile surface is taken every 30s with a high resolution charge-coupled device (CCD) camera (1596x2048 pixels). The driving rate corresponds to the dropping of ~ 1500 grains between two images. This does however qualify as slow driving, as at each point along the line of growth, this only corresponds to about 2 grains being added each time step. Furthermore, one has to compare the number of added grains with that already in the pile, which is of the order of $10^7 - 10^8$ grains. In order to reconstruct the surface coordinates of the pile, a set of colored lines (red-green-blue) is projected onto the pile perpendicularly. The images are taken at an angle of $45°$ to the projection direction leading to a distortion of the lines corresponding to the surface properties, with and accuracy and precision of $\sim 1-2\ mm$ [9]. After identification of the different lines from the image, a simple calculation leads to the surface coordinates as shown in Fig. 1.

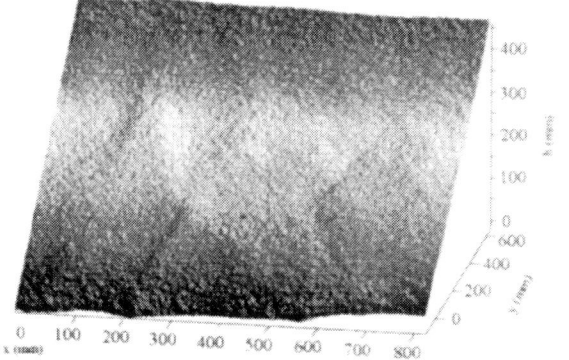

FIGURE 1. A typical image of the reconstruction of the rice-pile surface. Due to the number of lines projected, and the high resolution of the CCD camera, the surface can be reconstructed with an accuracy of $\sim 1-2mm$, which is comparable to the size of a grain.

In the first part we will discuss experiments in a statistically stationary state. In this case, the pile was fully grown when the experiment was started and the average removal of material at the bottom of the pile corresponded to the average material added. Here, each experiment consisted of about 400 images and four separate experiments were analyzed. The volume of rice displaced by an avalanche, ΔV, is then determined from the height-difference between two consecutive time steps

$$\Delta V = 1/2 \int |h(x,y,t) - h(x,y,t+\Delta t)| dx dy, \quad (1)$$

where $h(x,y,t)$ is the height of the pile at position x,y and at time t. The height difference between two time steps also presents a measure for the distribution of active sites, which shows self-similar behavior. While these results are used below, they are not shown here in detail, see Ref. [6].

For the experiments on the build-up towards the critical state, we first created a flat pile surface, at an angle far below the critical angle ($\phi_0 \simeq 0.55$ compared to $\phi_c \simeq 0.8$). Each experiment consisted of about 500 images and 9 separate experiments were analyzed [10]. Here, we want to determine a measure for the distance from the critical state (the Gap of Ref. [2]). We therefore determine the maximum local slope of the pile, $f(t)$, as a function of time (see inset of Fig. 5). As the pile gets closer to the critical state, the maximum local angle will approach a critical value, f_c. The Gap is then given by the difference $G(t) = f_c - f(t)$ of the maximum local slope to its critical value.

RESULTS IN THE STATIONARY STATE

Determining the size of an avalanche as described above, the evolution of the rice pile in the stationary state is intermittent, as can be seen in Fig. 2. There the avalanche

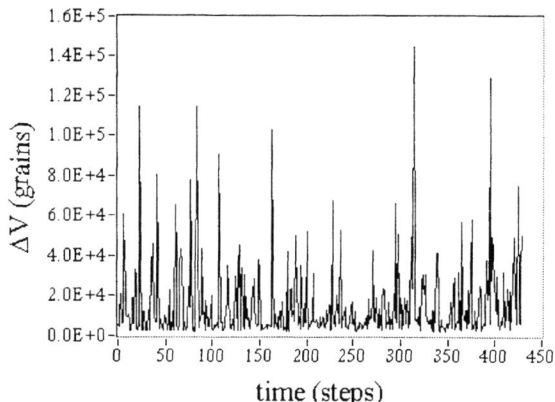

FIGURE 2. The temporal evolution of the avalanche sizes ΔV for one experiment. The activity is clearly intermittent, which already indicates SOC behavior. A proper test however consists of studying the size distribution as well as its finite size scaling. This is done in Figs. 3 and 4.

size as a function of time is given for one experiment. A histogram of this time evolution leads to the avalanche size distribution, which is a central issue in SOC physics. By studying subsets of the whole surface separately, it

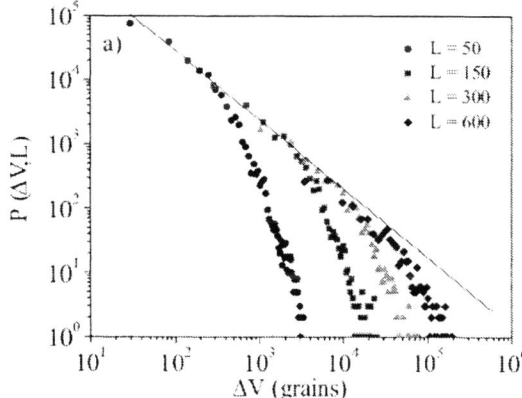

FIGURE 3. The avalanche size distributions for different size subsets of the pile on a double logarithmic plot. The size distributions are obtained from a histogram of the temporal evolution as shown in Fig. 2. As system sizes have a scaling region, where power-law behavior is observed with an exponent of $\tau \simeq 1.2$. The cut-off size, where the power-law breaks down, increases with the linear size of the system. This indicates the presence finite size scaling, which is shown explicitly in Fig. 4.

is also possible to check the data for finite size scaling, which in a true critical state should be observed. The histograms corresponding to different system sizes (with a linear extent of 50, 150, 300, and 600mm respectively) are shown in Fig. 3. As can be seen the avalanche size distribution is a power-law in all data sets, where the exponent can be estimated from a direct fit to be $\tau = 1.20(5)$. The different data sets do however show a cut-off increasing with the system size L. This is a strong indication of finite size scaling, which can be checked by performing a curve collapse. On scaling the avalanche size with L^{-D} and the avalanche probability with ΔV^{τ}, we obtain a curve collapse for these data, as can be seen Fig. 4. Here, $D = 1.99(2)$ is the avalanche dimension and $\tau = 1.21(2)$ is the avalanche size distribution exponent. These values and their errors correspond to the best curve collapse, as shown in Fig. 4. This indicates the presence of finite size scaling and hence the fact that the rice pile is indeed in a critical state. On the other hand, the fractal

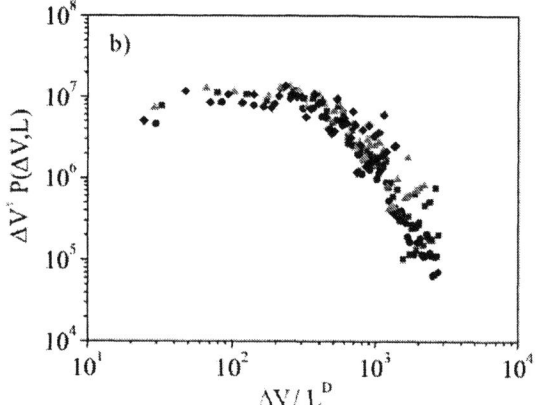

FIGURE 4. The same data as in Fig. 3, where the avalanche sizes are scaled with the system size L^{-D} and the probabilities are scaled with the avalanche size ΔV^{τ}. The data sets for the different subsets collapse onto the same curve, which is a strong indication of finite size scaling and hence the presence of a critical state in the dynamics of the rice-pile.

dimension of the active sites was determined using a box counting method to be $d_B = 1.58(3)$ [6] from the spatial distribution of the height differences.

RESULTS IN THE TRANSIENT STATE

When the rice pile approaches the critical state, the maximum local slope increases towards its critical value. In the context of extremal dynamics, this process can be described by the Gap-equation, which predicts that $G(t) \propto t^{-\delta}$. Furthermore, the value of the exponent characterizing the approach to the critical state, δ, is directly related to the avalanche size distribution exponent, τ, the avalanche dimension D and the fractal dimension of the active sites d_B in the stationary state via

$$\delta = 1 - \frac{1 - d_B/D}{2 - \tau}. \quad (2)$$

The time dependence of the Gap is shown in Fig. 5, where clearly the approach to the critical state follows a power law over two decades. Experimentally, δ is determined to be $0.8(1)$, where the biggest contribution of the error comes from the determination of f_c, which was obtained independently from a direct experiment on the maximum possible slope of a small part of the pile to be $f_c = 0.92(2)$. The value for δ obtained in the direct experiment on the transient slopes is in good agreement with the expectation from Eq. 2 of $\delta = 0.75(3)$ using the values of τ, D and d_B determined above.

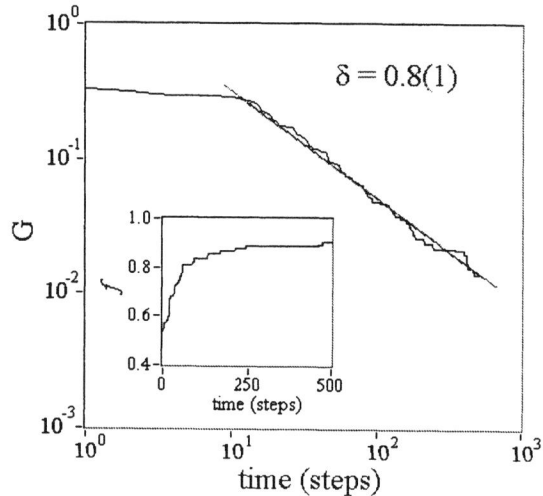

FIGURE 5. The maximum local slope as the critical state is approached is shown directly in the inset. The main figure shows the difference of the maximum slope with its critical value. This Gap asymptotically reaches zero as a power-law, where scaling is observed over two decades. The exponent is found to be $\delta = 0.8(1)$, which is in good agreement with the expectation for an extremal system using the avalanche distribution exponent in the critical state (see above).

CONCLUSIONS

We have shown that a three dimensional pile of rice in the stationary state does show the characteristic properties of a critical state, namely power-law distributed avalanches and more notably finite-size scaling. This implies that indeed the system has been driven to a critical state. Experimentally however, there was no tuning of parameters necessary to reach this critical state, which implies that the rice-pile is in a SOC state. However, in order to understand the nature of the critical state itself, we have further studied the approach of the rice-pile to the critical state. Here, the maximum local slope approaches its critical value according to a power law with an exponent predicted from the theory of systems with extremal dynamics. Thus interpreting the maximum of the local slopes as the Gap leading to critical behavior, it can be seen that the three dimensional rice-pile studied here does self-organize into a critical state according to the law of systems with extremal dynamics. This shows that the critical state does not just fall out of the air also in an experimental system, but can be studied in great detail. Theories of extremal dynamics are intimately connected to those of interface roughening in random media. Indeed there have recently been several mappings [11, 12, 13] of SOC models on interface models similar to the Kardar-Parisi-Zhang equation [14]. This connection between avalanche dynamics and surface roughening can also be seen in our experiments. Here the roughening exponents, which describe the self-affine structure [15] of the rice-pile surface are connected to the avalanche distribution exponents and the avalanche dimension [6]. Furthermore, Paczuski has pointed out that a formal indication for SOC can be obtained directly from the critical state without studying the transient behaviors using the multi-scaling properties of the pile surface in time [8]. Due to the presence of the transient timescale in the dynamics of the pile, there should accordingly be no generic scaling in the temporal behavior. The experimentally determined dependence of the dependence of the growth exponent on the multi-scaling moment q is in fact in good agreement with the theoretical prediction [16] as well. This indicates for instance that indeed, Paczuski's criterion can be used to distinguish generic critical behavior from SOC, in systems where the transient stat is unavailable to the experiment.

ACKNOWLEDGMENTS

This work was supported by FOM (Stichting voor Fundamenteel Onderzoek der Materie), which is financially supported by NWO (Nederlandse Organisatie voor Wetenschappelijk Onderzoek).

REFERENCES

1. Bak P., Tang C., and Wiesenfeld K., *Phys. Rev. Lett.* **59**, 381-384 (1987) and *Phys. Rev. A* **38**, 364-374 (1988).
2. Paczuski M., Maslov S., and Bak P., *Phys. Rev. E* **53**, 414-443 (1996).
3. Dickman R., Muñoz M. A., Vespignani A., and Zapperi S., *Braz. J. Phys.* **30**, 27-41 (2000); *cond-mat* 9910454.
4. Frette V., Christensen K., Malthe-Srenssen A., Feder J., Jssang T., and Meakin P., *Nature (London)* **379**, 49-52 (1996).
5. Altshuler E., Ramos O., Martínez C., Flores L. E., and Noda C. *Phys. Rev. Lett.* **86**, 5490-5493 (2001).
6. Aegerter C. M., Günther R., and Wijngaarden R. J., *Phys. Rev. E* **67**, 051306/1-6 (2003).
7. Costello R. M., Cruz K. L., Egnatuk C., Jacobs D. T., Krivos M. C., Louis T. S., Urban R. J., and Wagner H., *Phys. Rev. E* **67**, 041304/1-9 (2003).
8. Paczuski M., *Phys. Rev. E* **52**, R2137-2140 (1995).
9. Günther R., *Master's Thesis, Vrije Universiteit* unpublished (2002).
10. Aegerter C. M., Lőrincz K. A., and Wijngaarden R. J., submitted to *Phys. Rev. Lett.* (2003).
11. Alava M. J., and Lauritsen K. B., *Europhys. Lett.* **53**, 563-569 (2001).
12. Pruessner G., *Phys. Rev. E* **67**, 030301/1-4 (2003).
13. Vespignani A., Dickman R., Muñoz M. A., and Zapperi S., *Phys. Rev. Lett.* **81**, 5676-5679 (1998); Muñoz M. A., Dickman R., Vespignani A., and Zapperi S., *Phys. Rev. E* **59** 6175-6179 (1999).
14. Kardar M., Parisi G., and Zhang Y.-C., *Phys. Rev. Lett.* **56**, 889-892 (1986).
15. Barabasi A. L. and Stanley H. E. *Fractal Concepts in Surface Growth*, Cambridge University Press, 1995.
16. Aegerter C. M., Lőrincz K. A., and Wijngaarden R. J., submitted to *Europhys. Lett.* (2003).

Sorption Dynamics of Cr(VI) on Used Black Tea Leaves

Mohammad Abul Hossain*, Mikio Kumita and Shigeru Mori

Graduate School of Natural Science and Technology, Kanazawa University
2-40-20, Kodatsuno, Kanazawa 920-8667, Japan

Abstract. Sorption efficiency of Cr(VI) on used black tea leaves from aqueous solutions was evaluated. Kinetic studies were conducted using a batch process, and the effects of Cr(VI) concentration, solution pH and temperature on the adsorption and reduction performance were investigated. The adsorption kinetics follows pseudo-second order rate equation better than pseudo-first order one. The rate constant of pseudo-second order adsorption decreases with increasing an initial concentration of Cr(VI), up to a certain limit, then becomes steady. The maximum value of the rate constant was observed at an initial solution pH = 1.3. The rate constant was found to linearly increase with an increase in temperature, showing that the process is endothermic. The activation energy of adsorption calculated from Arrhenius plot is 16.3 kJ/mol, indicating that the adsorption occurred easily.

INTRODUCTION

Hexavalent chromium, Cr(VI) is a toxic heavy element typically found in the various industrial effluents. Depending on the solution pH and concentration, Cr(VI) can exist mainly as $Cr_2O_7^{2-}$, $HCrO_4^-$, CrO_4^{2-} and H_2CrO_4 [1]. Therefore, the removal of Cr(VI) from aqueous solutions is more complicated than that of other heavy metal ions. Currently, Cr(VI) removal technologies such as the precipitation, the reverse osmosis, the adsorption on activated carbon and the ion exchange may not be applicable to small industries especially in under-developing countries because of their high capital investment and running costs. Sorption of metal ions from aqueous solutions plays an important role in the wastewater treatment strategy. In recent years, there has been an interest in the use of low-cost sorbents for the removal of metal ions from aqueous system. Used black tea leaves (UBTLs) as a complex material are one of the attractive sorbents for low-cost removal process of Cr(VI) because of their high sorption capacity for it [2]. Moreover, the earlier investigations were mainly focused on the study of the conformational aspects of the sorption capacity and significant attempts were scarcely made to follow the kinetic path of the adsorption process. The dynamics is an essential information of sorption process especially for practical applications. Since the sorption dynamics of Cr(VI) on used black tea leaves also is not well understood, the present investigation has been undertaken for studying this dynamic behavior through batch experiments performed under different conditions: initial concentration of Cr(VI), solution pH and processing temperature.

EXPERIMENTAL

Materials

Used black tea leaves were obtained after extracting tea liquor from fresh black tea leaves (CTC Manufacturing Process, Bangladesh Tea Research Institute) by boiling with distilled water for 8 hours. Extracted leaves were dried at 105 °C for 24 hours and then sieved. Characteristic features of the UBTLs are given in Table 1. Elemental analysis using an Energy-dispersive X-ray Microanalyzer (EMAX-5770W, HORIBA) shows that UBTLs contain 65.3 % of C, 34.2 % of O, 0.1 % of Ca and less than 0.1 % S and P.

Analytical grade reagents were used in all cases. A stock solution of synthetic wastewater containing 1000 mg/L of Cr(VI) was prepared by dissolving $K_2Cr_2O_7$ (Wako Pure Chem. Ind., Japan) into pure water. All working solutions of various concentrations of Cr(VI) were obtained by diluting the stock solution with

TABLE 1. Physical properties of UBTL

Mean particle diameter	[mm]	0.38
Bulk density	[g/cm^3]	0.34
BET surface area/Kr	[m^2/g]	1.34

distilled water. Nitric acid was used to adjust the initial solution pH to minimize the reduction of Cr(VI).

Methods

Batch sorption experiments were carried out in capped conical flasks (100mL, Taplon) at a specified temperature, by suspending 0.1g/L of the UBTLs in 50mL of Cr(VI) solution. The solution pH was adjusted in the range of 1.00-2.00. The suspensions were mixed on a shaker with a constant speed of 125 rpm. The solutions were withdrawn at certain time intervals and separated solutions were analyzed with a reversed-phase HPLC-UV system for simultaneous determination of Cr(VI) and Cr(III) in the solution. The amounts adsorbed and reduced were calculated based on the analysis of remaining solutions. Similarly, the batch sorption experiments were conducted with various initial concentrations of Cr(VI), initial solution pH values and processing temperatures.

RESULTS AND DISCUSSION

The experimental results showed that both adsorption and reduction were involved in this sorption of Cr(VI) on the UBTLs. According to the previous study, in the low solution pH region, the adsorption is predominant over the reduction [3]. Since the adsorption is more important than the reduction for the treatment process, the kinetics was studied in the low solution pH region. Both of the adsorption and the reduction are affected by the initial concentration of Cr(VI), the solution pH and the processing temperature. These effects were investigated from the dynamic aspects of this sorption process.

Effect of Concentration

The concentration decays of Cr(VI) by both of adsorption and reduction for different initial concentrations are shown in Fig. 1, for the specified amount of UBTLs, W_s = 0.1 g/L. At low concentration of Cr(VI), the sorption is purely by the adsorption. As the initial concentration of Cr(VI) increases from about 150 mg/L, the reduction increases gradually with time. This effect of concentration might be due to the unavailability of the free adsorption site of the UBTLs surface after specific amount was adsorbed, as discussed later. Within the first one day, the amount of Cr(VI) adsorbed on the UBTLs rapidly reached 50-70 % of the equilibrium value, q_e, and then gradually increased for 10-15 days. First, the pseudo-first order rate equation, Eq. (1) given by Lagergren [4], was tried to analyze the kinetic behavior of this adsorption process.

$$q_e - q_t = q_e e^{-k_{ad} t} \quad (1)$$

For the first several days, the change in the amount adsorbed with time was found to fit Eq. (1) and thereafter deviate from it for low initial concentration of Cr(VI) to shift to the second step.

The change in the amount of the adsorbed Cr(VI) with time was found to fit the pseudo-second order rate equation, Eq. (2), as shown in Fig. 2 [5].

$$\frac{t}{q_t} = \frac{1}{kq_e^2} + \frac{1}{q_e}t \quad (2)$$

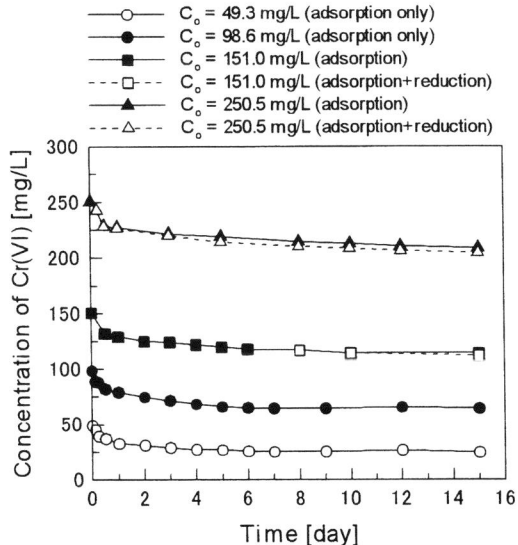

FIGURE 1. Plot of concentration decay of Cr(VI) with time for different initial concentrations. (W_s = 0.1g/L, pH = 1.54 and T = 25 °C).

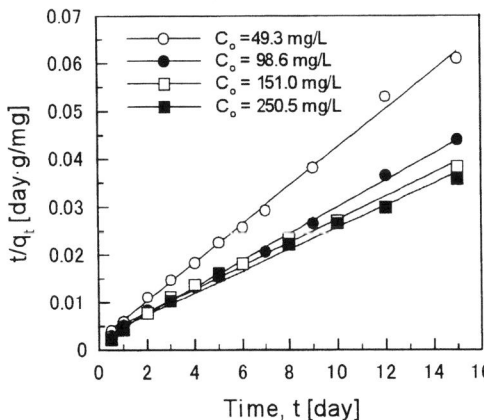

FIGURE 2. Plot of the pseudo-second order kinetics for adsorption of Cr(VI) on UBTLs (pH = 1.54, W_s = 0.1 g/L and T = 25 °C)

where, q_t = the amount adsorbed at time t and q_e = the equilibrium amount adsorbed, calculated from t/q_t vs t plot. The pseudo-second order rate constant, k, was calculated by using Fig. 2 and plotted in Fig. 3 as a function of the initial concentration of Cr(VI). Figure 3 shows that the rate constant decreases with an increase in initial concentration of Cr (VI), C_o, up to a certain limit, then, becomes steady.

Equation (2) was derived for divalent ion, by assuming that two surface sites could be occupied by one adsorbate ion. Thus Eq. (2) would be expected to be applicable for the adsorption of $Cr_2O_7^{2-}$ ions only. But the present experimental results are against this expectation. In the test solution case (pH = 1.54), in the range of $K_2Cr_2O_7$ concentrations below 152 mg/L, $HCrO_4^-$ ions predominantly exist in the solution over $Cr_2O_7^{2-}$ ions [6]. In addition, the ionic size of $HCrO_4^-$ is smaller than that of $Cr_2O_7^{2-}$ [7]. At the lower concentration of $K_2Cr_2O_7$, thus, these mono-valent $HCrO_4^-$ ions should be more easily transferred and adsorbed onto the surface of the UBTL, compared to $Cr_2O_7^{2-}$ ions. Our obtained values of pseudo-second order rate constant decreases with an increase in initial concentration of $K_2Cr_2O_7$ (i.e., decreasing of $HCrO_4^-$ ions) up to 151.0 mg/L, and also support the predominant adsorption of $HCrO_4^-$ ions over that of $Cr_2O_7^{2-}$ at low concentration (<151 mg/L). Despite of this fact, our experimental results show that the adsorption dynamics well follows Eq. (2), better than pseudo-first order one. On the other hand, Davis and Leckie (1980) also found that each divalent ion (CrO_4^{2-}) covers 3 to 4 hydroxyl (protonated) surface sites (C_xOH) in the adsorption process [8]. Similarly, our experimental results and the electro-affinity of surface functional-group concept, weaker than that of the normal groups in the solution, suggest that one $HCrO_4^-$ ion be adsorbed with two surface functional-groups.

Furthermore, at high concentration of $K_2Cr_2O_7$, predominant species $Cr_2O_7^{2-}$ ions might produce Cr(III) and $HCrO_4^-$ ions as the following Eq. (3), which indicates the reduction of Cr(VI) at high concentration [6].

$$3C_xOH + Cr_2O_7^{2-} + 4H^+ \rightleftharpoons$$
$$3C_xOH + HCrO_4^- + Cr^{3+} + 3H_2O \quad (3)$$

Effect of Solution pH

The effect of the initial solution pH on the sorption was studied in the range of 1.0 to 2.0 with 100 mg/L of Cr(VI) and 0.1 g/L of UBTLs at 25 °C. Figure 4 shows that at the initial pH = 2.00, the sorption is purely by the adsorption process and at the initial pH = 1.54, the reduction is slightly involved in the sorption process. As the initial pH decreases from 1.54, the amount reduced increases and also the reduction starts earlier as the initial pH decreases. Even at the different initial pH's, the adsorption of Cr(VI) on UBTLs was found to follow Eq. (2). The rate constant strongly depends on the initial solution pH and increases rapidly with increasing solution pH in the pH range of 1.0 to 1.3, and then decreases. This behavior can be interpreted as follows: the pH dependence of metal adsorption can largely be related to the type and the ionic state of surface functional groups and also on the metal chemistry in the solution. Previous article reported that the zero point charge pH (pH_{zpc}) of UBTLs is 3.6, indicating that the surface of UBTLs is positive at solution pH less than this pH_{zpc} value [3]. On the other hand, the distribution of the Cr(VI) species in aqueous solution depends on pH and Cr(VI) concentration

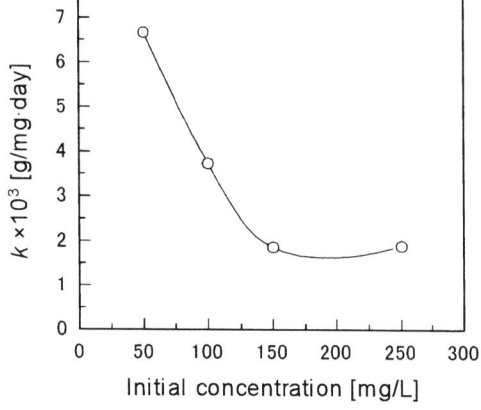

FIGURE 3. Variation of the pseudo-second order rate constant with initial concentration of Cr(VI) (pH = 1.54, W_s = 0.1 g/L and T = 25 °C)

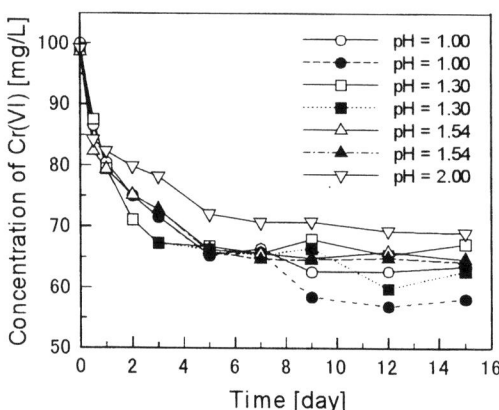

FIGURE 4. Plot of concentration decay of Cr(VI) with time for different initial pHs (C_o = 98.9 mg/L, W_s = 0.1 g/L and T = 25 °C) (——: Adsorption, - - -: Adsorption+ Reduction)

according to equilibrium equation, Eq. (4) [6]. Therefore, the neutral H_2CrO_4, formed predominantly at pH<1.0, is not rapidly adsorbed on the positively charged UBTLs surface. Experimentally it was also observed that at pH<1.3, the reduction of Cr(VI) to Cr(III) become predominant, accompanying the decomposition of UBTLs. That is, the maximum protonation of UBTLs surface occurs at pH = 1.3. Thus the adsorption rate constant is high at this pH and decreases with increasing pH. Furthermore, in the higher pH region, larger $Cr_2O_7^{2-}$ ions become predominant over $HCrO_4^-$, and this fact also supports such type of decreasing of rate constant.

$$H_2CrO_4 \underset{LogK=0.38}{\overset{H^+}{\longleftrightarrow}} HCrO_4^- \underset{LogK=6.14}{\overset{H^+}{\longleftrightarrow}} CrO_4^{2-} \quad (4)$$

$$\text{pH}<1 \quad LogK=1.71 \updownarrow H^+ \quad \text{pH}>6$$

$$Cr_2O_7^{2-}$$
$$\text{pH}=2\text{-}6$$

Effect of Temperature

The effect of temperature on the sorption of Cr(VI) on UBTLs was also investigated in the range of 10 to 65 °C at pH = 2.0. The influence of temperature on the sorption kinetics is presented in Fig. 5. The reduction was observed in this sorption process when the processing temperature is 50 °C or above. This observation might be due to the enhancement of the reaction given in Eq. (3). Pure adsorption kinetics in the range of 10 to 65 °C was shown to follow the pseudo-second order rate equation. The obtained rate constant increases linearly with an increase in temperature. The equilibrium amount absorbed also increases with an increase in processing temperature (q_e = 312.5 and 476.2 mg-Cr(VI)/g-UBTLs at 10 and 65 °C, respectively), indicating that this adsorption system might be controlled by chemisorption. The activation energy of adsorption was calculated from the Arrhenius plot as 16.3 kJ/mol. This low value indicates that the adsorption occurs easily.

CONCLUSION

Both of the adsorption and the reduction, involved in the sorption of Cr(VI) on used black tea leaves, are affected by the initial concentration of Cr(VI), the solution pH, and the processing temperature. The pure adsorption kinetics follows the pseudo–second order rate equation better than the pseudo-first order one. The rate constant of the adsorption increases with increasing an initial concentration of Cr(VI) up to a certain limit, where the reduction starts, then becomes steady. The solution pH has a profound effect on the adsorption rate and the maximum rate constant was observed at solution pH = 1.3. The rate constant increases linearly with an increase in temperature, showing the endothermic process. The Arrhenius plot of the rate constants gives the low activation energy value, indicating that Cr(VI) is easily adsorbed on UBTLs.

ACKNOWLEDGMENTS

The authors would like to thank to Mr. Y. Michigami, Environmental Protection Engineering Center, Kanazawa University, for his support in RP-HPLC-UV for Cr analysis. We are also grateful to the Bangladesh Tea Research Institute for providing black tea leaves.

REFERENCES

1. Cotton, F.A., and Wilkinson, G.W., *Advanced Inorganic Chemistry*, 3rd Ed., Wiley, New York, 1972, pp. 828-833.
2. Hossain, M.A., *Treatment of Wastewater Containing Toxic Heavy Metal [Cr(VI)] with Used Tea Leaves*, MS. thesis, Kanazawa University, Kanazawa, Japan, (2003).
3. Hossain, M.A., and Tajmeri, S.A.I., *J. Bangla-desh Academy Sciences*, **2**, 91-99 (1998).
4. Ho, Y.S., and McKay, G., *Wat. Res.*, **33**(2), 578-584 (1999).
5. Ho, Y.S. and McKay, G., *Wat. Res.*, **34**(3), 735-742 (2000).
6. Cimino, G., Passerini, A., and Toscano, G., *Wat. Res.*, **34**(11), 2955-2962 (2000).
7. Brito, F., Ascanio, J., Mateo, S., Hernandez, C., and Mederos, A., *Polyhedron*, **16**, 3835-3846 (1997).
8. Davis, J.A., and Leckie J.O., *J. Colloid Interface Sci.*, **74**, 32-43 (1980).

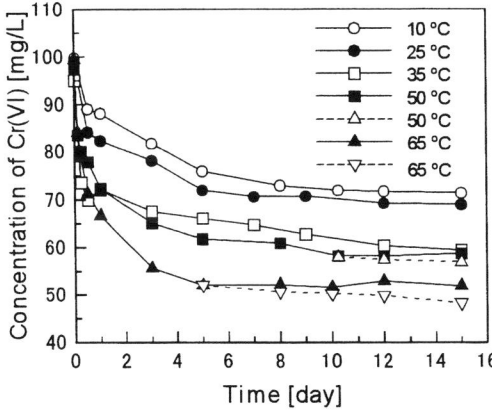

FIGURE 5. Plot of concentration decay of Cr(VI) with time for different temperatures. (C_o = 98.9 mg/L, W_s = 0.1 g/L and pH = 2.00) (—— : Adsorption, -- : Adsorption+ Reduction)

Kikuchi-Kossel diffraction line analysis on crystallization in salt-free aqueous colloidal suspensions

I. S. Sogami*, T. Shinohara*, M. Okuno*, T. Kurokawa*, M. M. Arishiro*, T. Itoh*, M. Tanigawa* and T. Yoshiyama*

*Department of Physics, Kyoto Sangyo University, Kamigamo, Kita-Ku, Kyoto 603-8555, Japan

Abstract. The Kikuchi-Kossel laser diffraction analysis proved that the crystallization in salt-free aqueous suspensions of highly-charged latex particles proceeds by way of the following multi-phase transitions: two-dimensional hcp structure → random layer structure → layer structure with one sliding degree of freedom → stacking disorder structure → stacking structure with multi-variant periodicity → fcc structure with (111) twin → normal fcc structure. For less concentrated suspensions (<2 vol%), the phase transition progresses further from the normal fcc structure to the normal bcc structure via the bcc twin structure. In this note, we report results of the Kikuchi-Kossel line analysis on the newly discovered phase transitions at an intermediate stage from the stacking structure with multi-variant periodicity to the fcc (twin) structures.

INTRODUCTION

Crystallization in salt-free aqueous suspensions of highly-charged colloidal particles was observed by laser diffraction method for dilute and semi-dilute specimens of concentration 0.1–5.0 vol%. Owing to long time scales of state changes in colloidal suspensions, it is possible to observe and record ordering processes in suspensions in real time with the Kikuchi-Kossel laser diffraction patterns [1, 2] which represent faithfully accurate information on lattice symmetries [3, 4, 5, 6] and lattice constants [7] of colloidal crystals. The results show that the crystallization proceeds by way of the following multi-stage phase transitions:

	two dimensional hcp structure
→ ($\ell 1$)	random layer structure
→ ($\ell 2$)	layer structure with one sliding degree of freedom
→ ($\ell 3$)	stacking disorder structure
→ (ℓc)	stacking structure with multi-variant periodicity
→ (c1)	fcc twin structure with twin plane (111)
→ (c2)	normal fcc structure
→ (c3)	bcc twin structure with twin plane $(1\bar{1}2)$ or $(\bar{1}12)$
→ (c4)	normal bcc structure

The first three stages of three-dimensional ordering from ($\ell 1$) to ($\ell 3$) are characterized by strong anisotropy originating in the wall effect which initiates the formation of a two-dimensional hcp arrangement of the colloidal particles. We call this stage an era of layer structure. After the intermediate stage of stacking structure with multi-variant periodicity (ℓc), the ordering process passes slowly into the era of cubic structure from (c1) to (c4). Thermal agitation and interactions among particles rectify gradually the anisotropy and complete the formation of crystals with cubic symmetry. While the crystallization terminates with the fcc structure in the semi-dilute suspensions, it develops further to the bcc structure in the dilute suspensions. Therefore, the fcc (bcc) structure is more stable thermodynamically than the bcc (fcc) structure in the semi-dilute (dilute) suspensions.

The intermediate stage (ℓc) has not yet been well investigated. In this note, we make a report on the new states discovered at the (ℓc) stage by the Kikuchi-Kossel line analysis.

EXPERIMENTAL

In this experiment, we used colloidal suspensions consisting of polystyrene-based latex N-150 (Sekisui Chemical Co., Osaka, Japan). The average diameter of the spherical colloidal particles was $2R = 150$ nm and the standard deviation of the particle size distribution was 0.0035. Sample latices were dialyzed with pure water and purified by anion-cation exchange resin. Suspensions of well-deionized latices which began to show iridescence were introduced into rectangular quartz cu-

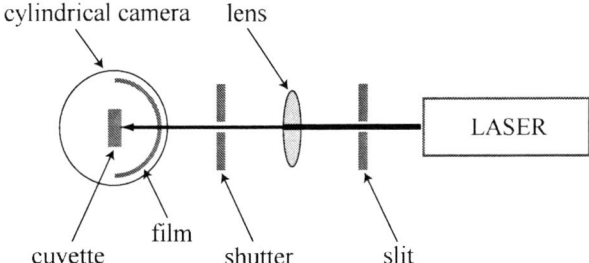
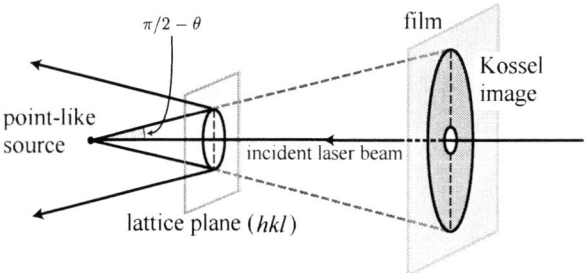

FIGURE 1. Apparatus for photography of Kikuchi-Kossel images of colloidal crystals. Incident laser beams collimated by a slit and focused by a lens pass through a hole punched in a film and illuminate the colloidal crystals. The specimen cuvette is placed on the head of a goniometer at the center of a cylindrical camera.

FIGURE 2. Mechanism for generation of a Kikuchi-Kossel diffraction image. Secondary waves diverging from a point-like source in the colloidal crystal are incident onto the (hkl) lattice planes. The component waves satisfying the Bragg condition are reflected and other components pass through the lattice planes. As a result, the cross-sections of the Kossel diffraction cones with the film surface are recorded as conical sections.

vettes of dimension 1mm×10mm×4mm. It took several hours (minutes) for dilute (concentrated) suspensions to be in iridescent states again after being introduced into the cuvettes. We used the suspensions with the volume fraction of $\phi=0.0204$ for laser diffraction analyses on crystallization when they began to show iridescence in the cuvettes.

The specimen cuvette was mounted upright on a goniometer head and the incident laser beams (wavelength 488 nm for Ar laser, 532 nm for green laser and 632.8 nm for HeNe laser) were incident normally to the wide surface of the cuvette (see Fig. 1). Backward Kikuchi-Kossel images were photographed on Fuji FG films with a modified rotating crystal camera consisting of a cylindrical film holder with diameter $2R_{cam} = 57.3$ mm. Time of exposure was about ten seconds for Ar laser (3 mW). The Miller indices and spacings of lattice planes were obtained for each Kikuchi-Kossel line and the crystal structure was determined from symmetry of diffraction pattern. The lattice constant of the colloidal crystal can be determined with a high precision of 2 % [7] in the Kikuchi-Kossel diffraction method.

Measurements of the coordinates of three points on a Kikuchi-Kossel line determine the directions of beams on the Kossel cone relative to the incident laser beams and the semi-apex angle $\alpha(=\pi/2 - \theta)$ of the Kossel cone (Fig. 2). Photographed Kikuchi-Kossel lines are deformed owing to the curvature of a cylindrical film holder and by the effects of refraction at the air-quartz and quartz-suspension interfaces. The deformation due to the geometry of camera is corrected with high accuracy. The refractive effect of the quartz cuvette plate is merely a parallel displacement of beams by the thickness of the plate. To correct the refractive effect by using Snell's law, we adopted the formula [8]:

$$n(\phi) = n_{\text{water}}(1-\phi) + n_{\text{particle}}\phi, \quad (1)$$

for the index of refraction of colloidal suspension $n(\phi)$, where n_{water} and n_{particle} are indices of refraction of water and polystyrene particle ($n_{\text{water}} = 1.33$ and $n_{\text{particle}} = 1.60$). ϕ is the volume fraction of the suspension.

RESULTS AND DISCUSSION

Figures 3-6 are photographs of the Kikuchi-Kossel images from a single crystal grain by Ar laser taken on the 16th day after preparation. The grain with the size of 1.5mm was at the height of 10 mm from the bottom of the container. The computer simulation is made for the black diffraction lines and the Miller indices are given in Figures 4 and 6. The lapse of time corresponds to Figures: 3→4→5→6.

In Figure 3, we can identify the Kikuchi-Kossel black lines with indices (222), ($3\bar{1}1$) and ($31\bar{1}$) and complex patterns of white lines. In addition to the black lines with indices (222), ($3\bar{1}1$), ($31\bar{1}$), ($\bar{1}31$) and ($\bar{1}13$), Figure 4 has the black lines with indices (220), (202), (022) and their twin components, and complex patterns of white lines. Figure 5 has the black lines with indices (222), ($3\bar{1}1$), ($31\bar{1}$), ($\bar{1}31$), ($\bar{1}13$), (220), (202), (022), (311), (131) and (113) and, complex patterns of white lines.

In the computer simulation of Figure 4, the solid lines with integral indices and the dashed lines with fractional indices represent the main Kikuchi-Kossel lines for the fcc twin structure and the stacking structure with 6 period, respectively. In contrast to Figures 3-5, the Kikuchi-Kossel image in Figure 6 with the three-fold symmetry proves that the crystal has the normal fcc structure. It should also be noticed that all of the Miller indices assigned in Figure 6 are either even or odd integers. This is one of the characteristic of the fcc crystal structure.

It is remarkable that all of the Kikuchi-Kossel black lines in Figures 3-6 lead to the same value $a = 650\pm 1$ nm for the lattice constant. From these results, we are able

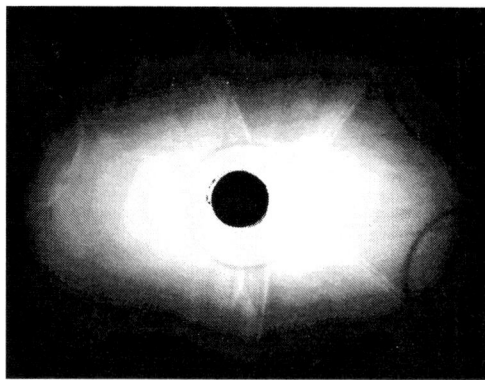

FIGURE 3. The Kikuchi-Kossel image for colloidal crystal observed using Ar laser at an early intermediate stage for the specimen of latex N-150 with a volume fraction $\phi=0.0204$.

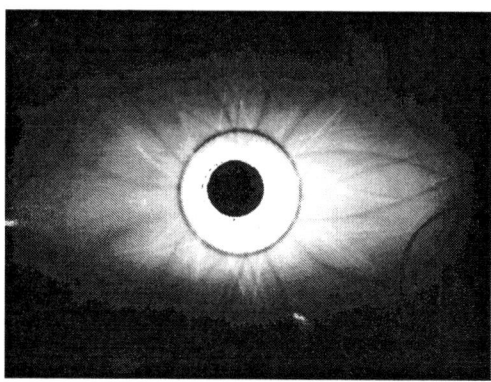

FIGURE 5. The Kikuchi-Kossel image at a later intermediate stage showing that the crystal has the normal fcc structure accompanying extra stacking structure with multi-variant periodicity.

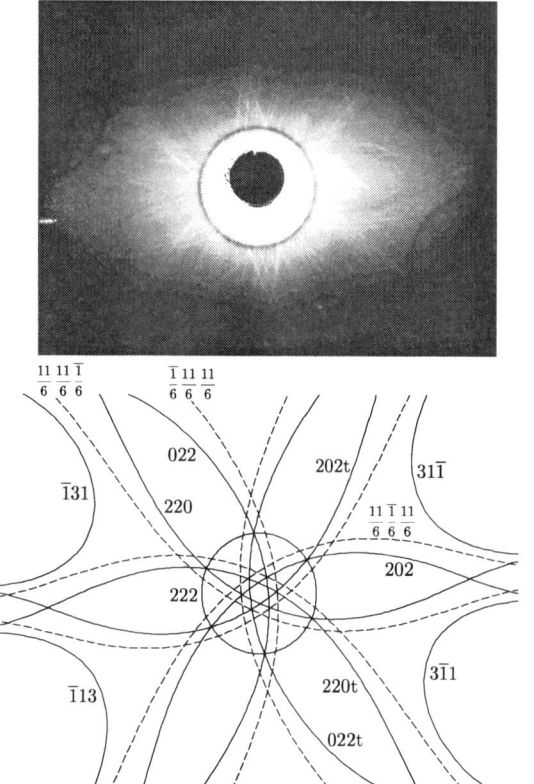

FIGURE 4. The Kikuchi-Kossel image for colloidal crystal observed using Ar laser and the computer simulation. The crystal is in an intermediate state. Solid lines and dashed lines represent the main Kikuchi-Kossel lines for fcc twin structure and stacking structure with 6 period.

FIGURE 6. The Kikuchi-Kossel image for colloidal crystal by Ar laser and the computer simulation. The three-fold symmetry proves that colloidal crystal has the normal fcc structure.

to interpret that the ordering process evolved in the observed crystal grain through, at least, three states of stacking structure with multi-variant periodicity in Figures 3-5 to the normal fcc structure recorded in Figure 6.

Figures 7 and 8 show the Kikuchi-Kossel images taken at the same hour with Figure 4 using, respectively, the Green and HeNe lasers. We can visibly identify the stacking structure with 12 period in Figure 7. Note that only the 6 period is observed in Figure 6. The fcc twin

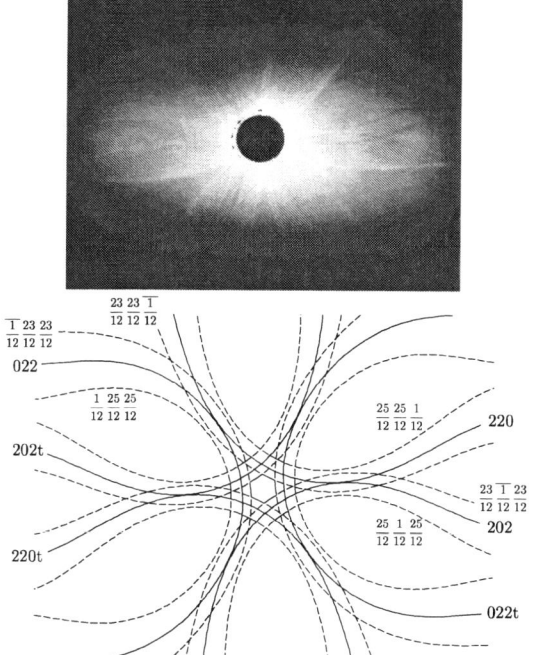

FIGURE 7. The Kikuchi-Kossel image for colloidal crystal by Green laser taken at the same hour with Figure 4 and the computer simulation. Solid lines and dashed lines represent, respectively, the main Kikuchi-Kossel lines for the fcc twin structure and the stacking structure with 12 period.

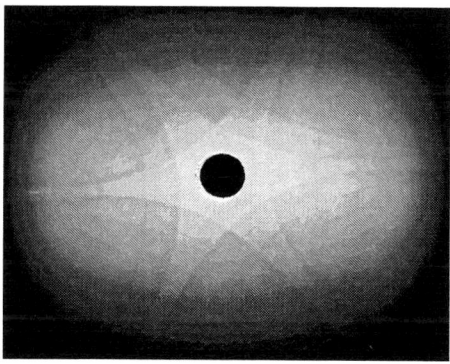

FIGURE 8. The Kikuchi-Kossel image for colloidal crystal by HeNe laser (At the same hour with Figure 4). The six-fold symmetry proves the crystal to have the fcc twin structure.

structure can clearly seen in Figure 8.

CONCLUSIONS

In this way, we observed new intermediate states of stacking structure with multi-variant periodicity which evolve into the state with normal fcc structure. This result shows that the intermediate stage (ℓc) has a rich variety of transient phase transitions.

In colloidal chemistry, charged particles are usually assumed to interact with the screened repulsive Coulomb potential. Such theory with medium- and long-range repulsion, however, seems to be difficult in explaining the observed multi-stage phase transitions through which the crystallization proceeds in the colloidal suspension.

In this connection it is essential to note that, if the whole suspension is filled uniformly with a single crystal of the normal fcc structure, the lattice constant satisfies the relation

$$a_0 = \left(\frac{16\pi}{3\phi}\right)^{\frac{1}{3}} R \qquad (2)$$

where R is the radius of particle. This relation leads to the lattice constant $a_0 = 702$ nm for the specimen used in this experiment. The reduction of the observed lattice constant $a = 650 \pm 1$ nm from this value necessarily requires coexistence of the ordered (concentrated) and disordered (dilute) states in the suspension. To explain this phenomena of the reduction of lattice constant and the occurrence of multi-stage phase transitions, it seems natural to assume that the interaction of colloidal particles in suspensions has also a long-range attractive component [9].

ACKNOWLEDGMENTS

This study is partially funded by "Ground Research for Space Utilization" promoted by NASDA and Japan Space Forum.

REFERENCES

1. Kikuchi, S., *Jpn. J. Phys.* **5**, 83-96 (1928).
2. Kossel, W., Loeck, V., and Voges, H., *Zeit. für Physik* **94**, 139-145 (1935).
3. Yoshiyama, T., Sogami, I., and Ise, N., *Phys. Rev. Lett.* **53**, 2153-2156 (1984).
4. Yoshiyama, T., and Sogami, I. S., *Phys. Rev. Lett.* **56**, 1609-1612 (1986); "Crystal Growth in Colloidal Suspensions: Kossel Line Analysis," in *Ordering and Phase Transitions in Charged Colloids*, edited by A. K. Arora and B. V. R. Tata, VHC Publishers, New York, 1996, pp. 41-68.
5. Sogami, I. S., and Yoshiyama, T., *Phase Transition* **21**, 171-182 (1990).
6. Dosho, S., et al., *Langmuir* **9**, 394-411 (1993).
7. Shinohara, T., et al., *Langmuir* **17**, 8010-8015 (2001); *Langmuir* (2003), submitted.
8. Hiltner, P. A., and Krieger, I. M., *J. Chem. Phys.* **73**, 2386-2389 (1969).
9. Sogami, I., *Phys. Lett.* **96A**, 199-203 (1983); Sogami, I., and Ise, N., *J. Chem. Phys.* **81**, 6320-6332 (1984).

Phase Separation of Polymer Mixtures Driven by Temporally and Spatially Periodic Forcing

Satonori Komatsu, Shinsuke Nishigami, Shinsuke Yoshida,
Tomohisa Norisuye and Qui Tran-Cong-Miyata*

Department of Polymer Science and Engineering
Kyoto Institute of Technology
Matsugasaki, Kyoto 606-8585 JAPAN
(* Corresponding author : qui@ipc.kit.ac.jp)

Abstract. Effects of temporally and spatially forcing on phase separation of polymer mixtures were investigated by using anthracene-labeled polystyrene/poly(vinyl methyl ether) (PSA/PVME) and stilbene-labeled polystyrene/poly(vinyl methyl ether) (PSS/PVME) mixtures. Phase separation was induced by either photodimerization of anthracene or *trans* → *cis* photoisomerization of stilbene. Specific mode-selection processes were found in these experiments.

INTRODUCTION

Phase separation of chemically reacting mixtures is not only a fundamental physico-chemical problem related to the mode selection process in critical phenomena, but it is also one of the important research topics in polymer materials because of the relation to morphology control, a practical problem in polymer industries [1]. We have experimentally demonstrated that the length scale as well as the spatial symmetry of the morphology of polymer mixtures can be manipulated and controlled by using appropriate chemical reactions [2]. Here, we report the temporal and spatial modulation effects on the morphology and kinetics of reaction-induced phase separation of polymer mixtures.

EXPERIMENTAL SECTION

Principle of Light-Induced Phase Separation

In this work, phase separation of polymer mixtures was induced by photochemical reactions of a specific molecules, i.e. the "*phase-separation driver*", chemically attached to polymer chains of one component in a binary mixture. Upon irradiation with ultraviolet light, the driver molecule undergoes appropriate chemical transformation, leading to phase separation because of a change in Gibbs free-energy of the mixture triggered by the photochemical reactions. The mechanism of this thermodynamic instability varies with the reaction mechanism. For photodimerization of anthracene, the PSA networks are generated and phase separation is induced by a decrease in entropy of the mixture. On the other hand, photoisomerization of stilbene modifies the segmental volumes of polymer as well as the enthalpic interactions between the two components. Both the variation has been experimentally verified by using small-angle neutron scattering [3]. In our work, the mixture is initially at the miscible state and phase separation can only occur in the presence of light. Taking advantage of this feature of light-induced phase separation, temporally and spatially forcing of phase separation were carried out by irradiation.

Samples Synthesis and Characterization

Anthracene-labeled polystyrene (PSA) and *trans*-stilbene-labeled polystyrene (PSS) were prepared according to the procedure reported previously [2, 3]. Their molecular parameters are as follows: PSA (M_w = 2.8 x 10^5, M_w / M_n = 1.8). The label contents are 1anthracene/55 repeat monomer units for PSA and 1 stilbene/23 monomer units for PSS. Poly (vinyl methyl ether) (PVME, $M_w = 1.0 \times 10^5$, $M_w/M_n = 2.5$) was obtained by purifying a commercially available sample. The light intensity was kept at 3.0 mW/cm^2. The thickness of the blends was adjusted at ca. 30 μm for all the experiments. It is worth noting that both PSA/PVME and PSS/PVME blends possess a lower critical solution temperature (LSCT), undergoing phase separation upon increasing temperature. All the details of the phase behavior of these mixtures were reported previously [1-3].

Phase Separation Driven by Temporally Forcing

Temporally periodic forcing of phase separation was performed by irradiating a PSA/ PVME blend with a 365 nm light beam chopped at various frequencies ranging from 200 to 0 Hz (corresponding to continuous irradiation). Figure 1 shows the effects

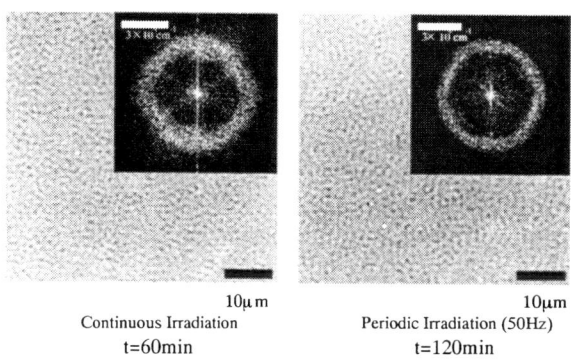

FIGURE 1. Morphology and the corresponding 2D-Fourier intensity distribution (power spectra) of a PSA/PVME (20/80) blend irradiated with continuous and periodic irradiation at 100°C.

of temporally periodic forcing on phase separation of a PSA/PVME (20/80) blend observed at 100°C which is located in the miscible region at 20°C below the phase boundary of the blend. To compare the phase separation induced by periodic irradiation with the case of continuous irradiation, we observed the morphology induced by irradiation with the same amount of photons in these two cases. This criterion allows us to compare the unstable modes developing in the irradiated blend under the similar cross-link densities. As seen in Figure 1, it is obvious that the width of the 2D power spectra in the case of periodic irradiation is much narrower than the case of continuous irradiation. Furthermore, as illustrated in Figure 2, there clearly exists a strong dependence of the morphological length scale distribution,

represented by a width of the 1D power spectra, on the external forcing frequency. To quantify this characteristic length scale distribution, the 2D Fourier

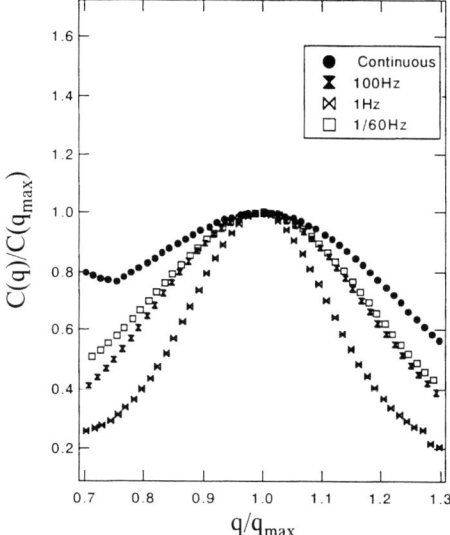

FIGURE 2. Dependence of the 1D power spectra observed for a PSA/PVME (20/80) blend irradiated at various irradiation frequencies at 110 °C.

power spectra of the blend obtained for several irradiation frequencies were circularly averaged and converted into 1D data. Subsequently, the width of the 1D power spectra, i.e. the distribution of the characteristic length scales of the morphology, was calculated by fitting the 1D data to the following empirical formula:

$$C(q) = \frac{a}{\sqrt{2\pi\sigma^2}} \exp[-\frac{(q-q_{max})^2}{2\sigma^2}] + bq^{-c} + d$$

where:
$C(q)$: the 1D Fourier intensity distribution of the optical image obtained by averaging the 2D-FFT data.

σ: the distribution of the structural period of the morphology.

a, b, c, d: the constants characterizing the optical image.

q_{max}: the frequency corresponding to the maximal Fourier intensity.

The length scale distribution Γ of the morphology obtained under different forcing frequencies was calculated from the standard deviation σ in Eq. (1)

using $\sqrt{2\ln 2}\,\sigma = \Gamma$. Figure 3 shows the results obtained for a PSA/PVME (15/85) blend cross-linked

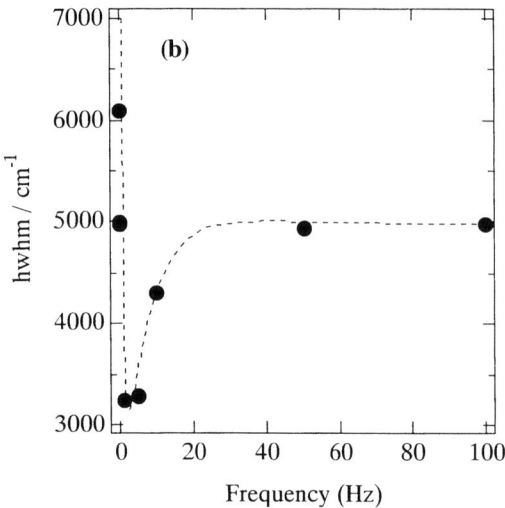

FIGURE 3. Dependence of the characteristic length scale distribution on the forcing frequency observed for a PSA/PVME (15/85) at 110°C.

by irradiation at 110°C. As the forcing frequency f increases, the distribution Γ drops very quickly in the region of low frequency and becomes almost constant in the range of high frequency f after passing a minimum around 1Hz. On the other hand, the Fourier intensity $C(q_{max})$, an indication for the morphological regularity, changes with the forcing frequency in a way contrary to the behavior depicted in Figure 3 for Γ, verifying the adequateness of the analysis.

The experimental results described above suggest the existence of a specific response of a polymer system undergoing critical phenomena induced by periodic forcing.

Phase Separation Driven by Spatially Forcing

To perform spatially periodic forcing of phase separation, a PSS/PVME (20/80) blend was irradiated with uv light through photomasks with various spacing (Λ) in the range 100 - 20 μm. These experimental data are compared to those obtained with Λ = 8 mm which corresponds to the maskless case. It was found that the blend irradiated without a photomask exhibits an isotropic spinodal structure. As illustrated in Figure 4a, this conventional spinodal structure remains unchanged until Λ is reduced beyond 50 μm. Beyond this particular length scale, the morphology induced by irradiation starts showing anisotropy. The most significant phase separated structure obtained by this spatial confinement was obtained with Λ = 20 μm as

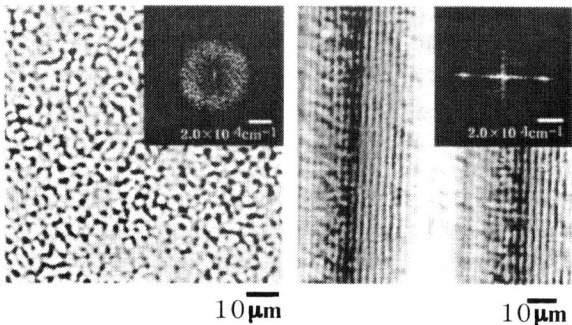

FIGURE 4. Morphology of a PSS/PVME(20/80) irradiated using a He-Cd laser (325 nm, 30 mW/cm^2) through photomasks with different spacings: 8 mm (left) and 20 μm (right). Both data were obtained at 95°C after 30 min of irradiation.

shown in Figure 4. Here was observed a superlattice structure composing of two co-existing lattices with different spacing. One is the *macro-lattice* with the period 20 μm corresponding to the refractive-index grating generated by the *trans-cis* photoisomerization of the stibene moieties chemically labeled on the PSS chains in the blend. The other lattice, i.e. the *micro-lattice*, corresponds to the morphology resulting from phase separation of the blend confined in the irradiated area of the macro-lattice. The micro-lattice induced by the experimental conditions described above did not exhibit significant evolution with time. To confirm the existence of this superlattice structure, two-dimensional (2D) light scattering from the irradiated blend was monitored. Shown in Figure 5 is the 2D scattering pattern of the PSS/PVME (20/80) blend

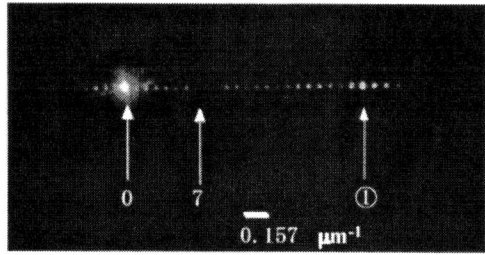

FIGURE 5. Diffraction patterns of a PSS/PVME (20/80) blend irradiated with a He-Cd laser through a 20 μm photomask. The number in the circle indicates the order of diffraction from the micro-lattice.

with the morphology illustrated in Figure 4. The diffraction spots with short spacing originate from the macro-lattice of refractive-index generated in the PSS/PVME blend by irradiation through the photomask. These high-order diffraction spots show upto 40th order. The extinction observed at the 7th order suggests that the volume ratio of the refractive-index macro-lattice is not exactly 50:50. On the other hand, the diffraction intensity becomes very strong around the 23th order which corresponds to a spacing of 1.7 µm. This length scale is in very good agreement with the spacing of the micro-lattices observed under a phase-contrast microscope as illustrated in Fig.4. From the structural analysis using both phase-contrast microscopy and light scattering, we conclude that the superlattice structure shown in Fig. 4 is the morphology of the blend under the spatial confinement induced by light. For experiments at the length scales smaller than 20 µm, interference patterns generated by a He-Cd uv laser (325nm, 30 mW/cm^2), was used for irradiation. No phase separation was observed under a phase-contrast optical microscope when an interference pattern with a spacing of Λ = 4.6 µm was used for irradiation. Furthermore, this particular blend only shows diffraction patterns corresponding to the refractive-index grating (the macro-lattice). This particular result would suggest that the unstable modes of concentration fluctuations were suppressed under this particular confinement.

Effects of Elastic Stress Induced by Chemical Reactions on Morphology under Spatial Confinement

So far we have described the effects of spatial confinement on phase separation of *trans*-stilbene labeled polystyrene/poly(vinyl methyl ether) (PSS/PVME) blends. Upon irradiation, the blend was thermodynamically destabilized by the changes in both segmental volumes and the enthalpic interactions between PSS segments bearing the stilbene isomers and the PVME segments. It was found that the elastic stress generated by the inhomogeneity of the chemical reaction in the bulk state of polymer blends plays an important role in the resulting morphology. Here, these particular effects were examined by comparing the morphology obtained by photoisomerization of stilbene, a non-crosslinking reaction to those induced by photodimerization of anthracene, a cross-linking reaction. For this purpose, these spatial confinement experiments were repeated with polystyrene labeled with anthracene. The resulting morphology was illustrated in Figure 6 for these two kinds of reaction. Obviously, there exists a distinct difference between the two phase-separated structures. For the case of non-cross-linking reaction, the micro-lamellar structure is continuous, whereas these lamellae are disrupted upon cross-linking. The difference in morphology between the two cases was confirmed by using both light scattering and laser confocal scanning

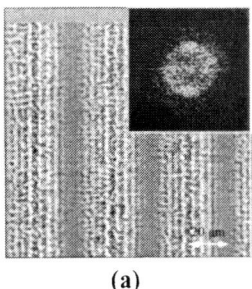 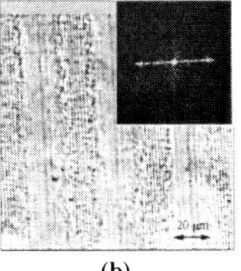

(a) (b)

FIGURE 6. Morphology of polymer blends induced by : (a) photo-cross-linking reaction for a PSA/PVME (20/80) blend; (b) photoisomerization for a PSS/PVME (20/80) blend.

miscroscope (LCSM). Furthermore, Mach-Zehnder interferograms of the irradiated blends show that a large elastic stress was generated inside the cross-linked blend, whereas this elastic stress field is almost negligible for the case of photoisomerization [4]. From these results, we conclude that the disconnecting lamellar structures shown in Fig. 6-a come from the effects of cross-link induced elastic stress on phase separation. Further quantitative examination is in progress and will be reported later.

ACKNOWLEDGMENTS

This work is supported by the Ministry of Education, Japan (MONKASHO) through a grant-in-aid No. 13031054 on the Priority-Area-Research *"Dynamic Control of Strongly Correlated Soft Materials"*.

REFERENCES

1. Tran-Cong, Q. , "Phase Separation and Morphology of Chemically Reacting Polymer Blends", in *Structure and Properties of Multiphase Polymeric Materials,* edited by T. Araki, Q. Tran-Cong and M. Shibayama, New York: Marcel Dekker, 1998, pp. 155-194.
2. Tran-Cong-Miyata, Q. *Macromolecular Symposia* **160**, 91-97 (2000).
3. Urakawa, O., Yano, O.; Q. Tran-Cong, A. I. Nakatani, C. C. Han *Macromolecules* **31**, 7962-7965 (1998)
4. Komatsu, S.; Inoue, K., Norisuye, T. and Q. Tran-Cong-Miyata, to be published.

Statistical Mechanics of Phase Unwrapping Problem by the Q-Ising Model

Y. Saika[*] and H. Nishimori[†]

[*]Department of Electrical Engineering, Wakayama National College of Technology, 77 Noshima, Nada, Gobo, Wakayama 644-0023, Japan
[†]Department of Physics, Tokyo Institute of Technology, 2-12-1 Oh-okayama, Meguro, Tokyo 152-8551, Japan

Abstract. We construct the statistical mechanical formulation for the problem of phase unwrapping, appearing in adaptive optics. We estimate the performance of our method using the replica theory and the time-dependent Ginzburg Landau theory for the infinite-range model. The replica theory clarifies that our method works well if we select appropriate model of the prior. Then, the time-dependent Ginzburg Landau theory estimates the dynamical property of the simulated annealing. These results are qualitatively confirmed by the Monte Carlo simulations for the realistic model.

INTRODUCTION

In resent years, a lot of researchers in statistical physics have been working on problem related to information processing [1-5], such as image analysis, statistics for spatial data, and Markov random fields. Various methods of statistical mechanics, such as the mean-field theory [6], have been applied to problems of image restoration and error-correcting codes.

In adaptive optics [7-9], phases often carry information through noisy transmission. Retrieving phases is therefore necessary using the measured phase differences in the principle interval $[-\pi, \pi]$. Especially this problem becomes difficult due to the under-sampling, even if no noise is introduced into measured variables.

In the present research, we formulate the problem of phase reconstruction on the basis of the statistical mechanics of the Q-Ising model. Further, we estimate the performance of our method using the replica theory for the infinite-range model and the Monte Carlo simulation for the realistic model. The replica theory derives the result that our method due to the MPM estimate works well irrespective of the initial condition when the Nyquist condition holds. This result is confirmed by Monte Carlo simulation for a typical phase pattern in adaptive optics. If the Nyquist condition does not completely hold at every sampling point, the Monte Carlo simulation derives the result that our method using the simulated annealing works well when we start from appropriate initial phase pattern.

GENERAL FORMULATION

We first show the statistical mechanical formulation for phase reconstruction using the Q-Ising model. Here we use the language of adaptive optics.

First, we consider a set of original phase pattern $\{\xi_{(x,y)}\}$, which is generated with the probability $P(\{\xi\})$. Here $\xi_{(x,y)} = -R/2 + kR/Q, k = 1, \ldots, Q, x, y = 1, 2, \ldots$. Then, the complex phase difference expressed as $\hat{\xi}_{(x,y)}\hat{\xi}^*_{(x',y')} = \exp(-j(\xi_{(x,y)} - \xi_{(x',y')}))$ is transmitted through the noisy channel and is then rewritten into $\hat{J}_{(x,y|x',y')} = |J_{(x,y|x',y')}|\exp(-jA_{(x,y|x',y')})$ with the conditional probability:

$$P(\{J\}|\{\xi\}) = \frac{1}{(2\pi J^2)^{N_B}} \exp\left(-\frac{1}{2J^2}\sum_{n.n}|\hat{J}_{(x,y|x',y')} - J_0\hat{\xi}_{(x,y)}\hat{\xi}^*_{(x',y')}|^2\right) \quad (1)$$

where $|A_{(x,y|x',y')}| < \pi$. Then J_0 and J^2 are the mean and variance of the Gaussian noise with each other.

Next, we retrieve phases by the MPM estimate by using the system $\{z_{(x,y)}\}$ where $z_{(x,y)} = -R/2 + k/RQ$, $k = 1,...,Q$, if observed data are considered to be sufficiently sampled. Then, the retrieved information is given by

$$\hat{z}_{(x,y)} = \arg\max_{z_{(x,y)}} \mathrm{Tr}_{\{z\}/z_{(x,y)}} P(\{z\}|\{J\}). \quad (2)$$

Here this posterior probability is given by the Bayes formula:

$$P(\{z\}|\{J\}) \propto P(\{z\})P(\{J\}|\{z\}) \quad (3)$$

where we assume the models of the true prior and the noise probability as

$$P(\{z\}) \propto \exp\left(-\frac{1}{T_m}\sum\left((z_{(x+1,y)} - z_{(x,y)})^2 + (z_{(x-1,y)} - z_{(x,y)})^2\right)\right), \quad (4)$$

$$P(\{J\}|\{z\}) \propto \exp\left(-\frac{\beta_J T_s}{T_m}\sum_{nn}\cos(A_{(x,y|x',y')} - z_{(x,y)} + z_{(x',y')})\right). \quad (5)$$

On the other hand, when measured data are considered not to be sufficiently sampled, we use the initial phase pattern constructed by the following procedure:

$$z'_{(x,y)} = \frac{1}{2}(u_{(x,y)} + v_{(x,y)}). \quad (6)$$

$$u(x,y) = A_{(x-1,y|x,y)} + 2\pi m_{(x,y)} - z'_{(x-1,y)} + z'_{(x-2,y)}, \quad (7)$$

$$v(x,y) = A_{(x,y-1|x,y)} + 2\pi n_{(x,y)} - z'_{(x,y-1)} + z'_{(x,y-2)}, \quad (8)$$

where $m_{(x,y)}$ and $n_{(x,y)}$ are determined so as to minimize $(\partial_x z_{(x,y)} - \partial_x z_{(x-1,y)})^2$ and $(\partial_y z_{(x,y)} - \partial_y z_{(x-1,y)})^2$. Here $\partial_x z_{(x,y)} = z_{(x,y)} - z_{(x-1,y)}$ and $\partial_y z_{(x,y)} = z_{(x,y)} - z_{(x,y-1)}$. Using this initial pattern, we retrieve phases by the conventional simulated annealing. Here the retrieved phase is given by

$$\hat{z}_{(x,y)} = \arg\max_{\{z\}} P(\{z\}|\{J\}) \quad (9)$$

When we estimate the performance of our method, we use the sample average of the mean square error between original and the retrieved phase patterns as

$$\hat{\sigma} = \left[\sum_{(x,y)}\frac{1}{(2\pi L)^2}\left(\xi_{(x,y)} - \hat{z}_{(x,y)}\right)^2\right]_{\xi,J} \quad (10)$$

INFINITE-RANGE MODEL

Replica Symmetric Theory

The replica theory for the infinite-range model is useful to give a guide to the qualitative understanding of system properties. Here we evaluate how the mean square error depends on the parameter T_m. We first note three assumptions should be introduced when we use the replica theory: (1) the large number limit of lattice/sampling points; (2) the infinite-range versions of the true prior and the noise probability as

$$P(\{\xi\}) \propto \exp\left(-\frac{1}{NT_s}\sum_{i<j}(\xi_i - \xi_j)^2 - h\sum_i \xi_i\right), \quad (11)$$

$$P(\{J\}|\{\xi\}) = \frac{N}{2\pi J^2}\exp\left(-\frac{1}{2J^2}\left|J_{ij} - \frac{J_0}{N}\hat{\xi}_i\hat{\xi}_j\right|^2\right) \quad (12)$$

where i and j are the sites on the grid and N is the number of sites; (3) the replica theory within the replica symmetric assumption. Next, under these assumptions, we derive the self-consistent equations for m_0, m_x, m_y, q and m due to the saddle-point conditions of the free energy. As is shown in Fig. 1, with the use of the solutions of these equations, we obtain the result that the mean square error $\hat{\sigma}$ takes

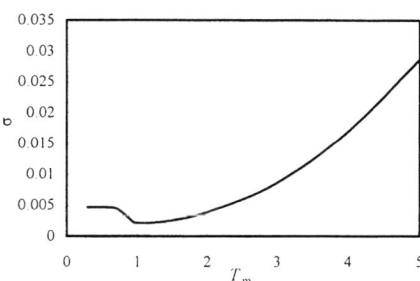

FIGURE 1. The mean square error σ as a function of T_m using the replica theory. Other parameters are set as $T_s = 1.0, J_0 = 1.0, J = 0.3, \beta_J = 1.0, h = 0.06$.

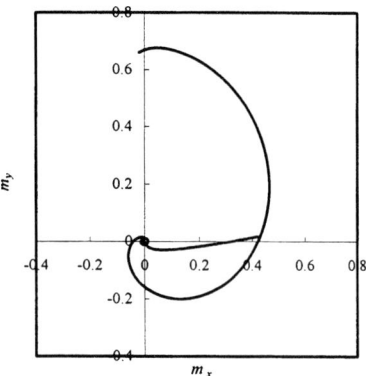

FIGURE 2. The time evolution of the macroscopic state on the $m_x - m_y$ plane due to the TDGL theory. Here $T_s = 1.0, J_0 = 1.0, J = 0.3, \beta_J = 1.0, h = 0.06$.

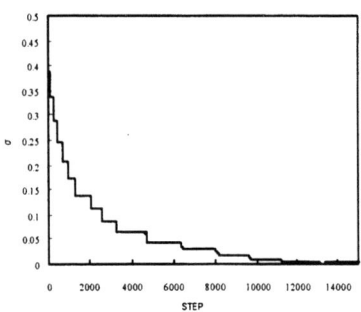

FIGURE 3. The time evolution of the mean square error σ obtained by the TDGL theory where $T_s = 1, h_0 = 0.05, J_0 = 1, J = 0.3, \beta_J = 0.5, h = 1$

its minimum when we appropriately tune parameter T_m, where other parameters are set as $J_0 = 1, J = 0.3, \beta_J = 0.5, T_s = 1$. In this case, the observed data is considered not to be sampled sufficiently. This result also means that it is not necessary to lower temperature T_m than $T_m < 1.0$ when we retrieve phases using the simulated annealing.

Time-Dependent Ginzburg Landau Theory for Infinite-Range Model

Next, we qualitatively estimate the dynamical property of our method due to the simulated annealing using the time-dependent Ginzburg Landau theory for the infinite-range model. Here, we evaluate the time evolution of macroscopic states m_0, m_x, m_y, q, m and the mean square error $\hat{\sigma}$.

As is shown in Fig. 2 and Fig. 3, the TDGL theory clarifies the result that the macroscopic state closes to the optimal state by going around the origin on the $m_x - m_y$ plane through the annealing procedure.

MONTE CARLO SIMULATION

MPM Estimate

In order to confirm the above results of the infinite-range model, we carry out here Monte Carlo simulations for the two-dimensional realistic model.

Here, we estimate the performance by our method for the typical phase pattern:

$$\theta^z_{(x,y)} = A_s \exp\left(-\frac{1}{2J_s^2}\left\{\left(x-\frac{L}{2}\right)^2 + \left(y-\frac{L}{2}\right)^2\right\}\right) \quad (13)$$

in adaptive optics. This pattern is corrupted by the Gaussian noise with a mean $\hat{\xi}(x,y)\hat{\xi}^*(x',y')$ and a variance $J = 0.3$. To estimate the performance of our method, we use the mean square error $\hat{\sigma}$ which is averaged over 10 patterns corrupted from the original pattern.

We evaluate how the mean square error $\hat{\sigma}$ depends on T_m when the Nyquist condition holds. Shown in Fig. 4, we clarify that the mean square error σ gradually closes to its minimum as T_m decreases and that it takes its minimum value $T_m < 2.0$ where $L = 64, A_s = 10.0, J_s = 6.0$.

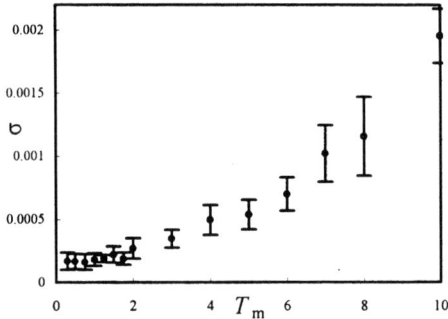

FIGURE 4. The mean square error σ as a function of T_m using the Monte Carlo simulation for typical pattern when $T_s = 1.0, J_0 = 1.0, J = 0.3, \beta_J = 1.0, h = 0.06$.

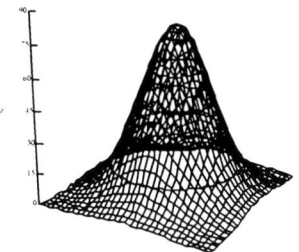

FIGURE 5. The retrieved pattern by the Monte Carlo simulation for a typical one in adaptive optics where $T_s = 1.0, J_0 = 1.0, J = 0.3, \beta_J = 1.0, h = 0.06$.

Simulated Annealing

Next, when the Nyquist condition does not hold at every sampling point, that is, the sample pattern we use is given as $L = 32, A_s = 90, J_s = 6.0$ in (13), we estimate the performance of our method by the Monte Carlo simulation. Figure 5 shows the retrieved phases by our method.

SUMMARY AND DISCUSSION

In the previous chapters, we construct the statistical mechanical formulation for the problem of phase reconstruction by the Q-Ising model. We then estimate the performance of our method using the replica theory and the time-dependent Ginzburg Landau theory for the infinite-range model. We show that our method works well when we tune appropriate parameters and that the phases are gradually retrieved through the annealing procedure irrespective of the initial condition. We estimate the performance of our method for the two-dimensional realistic model using the Monte Carlo simulations. When the Nyquist condition holds, phases are retrieved irrespective of the initial condition. On the other hand, when the Nyquist condition does not hold completely at every sampling point, our method works well when we start from the appropriate initial condition.

ACKNOWLEDGMENTS

We would like to thank Prof. Y. Iba for his fruitful discussions and comments. One of the authors (YS) would like to thank Dr. M. Okada, Dr. T. Aonishi and Dr. M. Inoue for fruitful discussions.

REFERENCES

1. Nishimori, H., *Statistical Physics of Spin Glasses and Information Processing: An Introduction*, Oxford University Press, Oxford, 2001, pp74-213.
2. Geman, S., and Geman, D., *IEEE Trans. PAMI* **6**, 721-741 (1984).
3. Sourlas, N., *Nature* **339**, 693-695 (1989).
4. Rujan, P., *Phys. Rev. Lett.* **70**, 2968-2971 (1993).
5. Pryce, J. M., and Bruce, D. B., *J. Phys. A* **28**, 511-532 (1995).
6. Tanaka, K., and Morita, T., *Phys. Lett.* **203A**, 122-128 (1995).
7. Fried, D. L., *J. Opt. Soc. Am.* **67**, 370-375 (1976).
8. Takajo, H., and Takahashi, T., *J. Opt. Soc. Am. A* **5**, 416-425 (1986).
9. Nico, G., Palubinskas, G., and Datcu, M., *IEEE Trans. Signal Processing* **48**, 2545-2555 (2000).

Feedback Coupling and Chemical Reactions

Steffen Trimper* and Knud Zabrocki*

Department of Physics, Martin-Luther-University, D-06099 Halle, Germany

Abstract. When the entities undergoing a chemical reaction are not available simultaneously, the classical rate equation in the reaction-limited regime, should be extended by including non-Markovian memory effects. We consider the two cases of an external feedback, realized by fixed functions and an internal feedback originated in a self-organized manner by the relevant concentration itself. Whereas in the first case the fixed points are not changed, although the dynamical process is altered, the second case offers a complete new behavior, characterized by the existence of a time persistent solution. As an example we consider a single-species pair annihilation $A + A \to \emptyset$ process combined with a spontaneous creation of particles $\emptyset \to A$.

INTRODUCTION

Many physical phenomena in complex systems are governed by nonlinear rate equations. Applying such kind of equations for chemical reactions, to determine the time evolution of the concentration of an entity $c(t)$, the forward and the backward reactions has to be balanced by the product of the concentrations [1]. Apparently this approach is based on the assumption that the reactants are available simultaneously. In addition to competing interaction forces the transport and reactive properties of strongly disordered systems should be also influenced by spatio-temporal feedback-couplings. The present paper is focused on such delayed-couplings, modelling a non-Markovian behavior as a further unifying feature of systems far from equilibrium. Our model can be grouped into the increasing interest of incoperating delay and feedback mechanism in diverse systems like physical [2, 3] as well as biological systems [4, 5, 6, 7]. In particular, a Fokker-Planck equation with a nonlinear memory term can be used to discuss anomalous diffusion in disordered systems [8], where the analytical results could be confirmed by numerical simulations including diffusion on fractals [9, 10, 11]. As a limiting case it had been demonstrated [12] that mobile particles remain localized due to the feedback-coupling. Even the non-Gaussian fluctuations of the asset price can be traced back to memory effects [13], too. An additional cumulative feedback coupling within the Lotka-Volterra model, which may stem from mutations of the species or a climate change, leads to an entirely different behavior [14] compared to the conventional model. If the Ginzburg-Landau model for the time evolution of an order parameter is supplemented by a competing memory term, the asymptotic behavior and the phase diagram may be completely dominated by such a term [15].

Generally, evolution equations with memory kernels can be derived following the well-established projector formalism due to [16], see also [17], which had been successfully applied for density-density correlation functions in studying processes in undercooled liquids [18, 19]. Following that line let us study the subsequent rate equation throughout the paper:

$$\partial_t c(t) = \mathcal{M}[c(t)] - \int_0^t \mathcal{K}(t,t';c)\,dt' \qquad (1)$$

Here the first term $\mathcal{M}[c(t)]$ is typically a nonlinear and time local quantity. As a special realization we discuss $\mathcal{M} = c(t) - c^2(t)$, which displays the rate equation for the combination of the single-species pair annihilation $A + A \to \emptyset$ and the spontaneous creation of particles $\emptyset \to A$. The memory kernel $\mathcal{K}$ is determined by the accumulation of the reacting entities in the time interval $t - t'$. Moreover, we are interested in a coupling of the rates leading to the following form of the kernel $\mathcal{K}$:

$$\mathcal{K}(t,t';c) = K(t,t';c)\,\partial_{t'} c(t') \qquad (2)$$

Because $\mathcal{K}$ represents a weighted coupling between the initial time $t = 0$ and the observation time t, the long-time behavior is significantly modified. As a new ingredient we treat also the case, that the memory term K depends on the concentration $c(t)$ in a self-organized manner. To be specific we study two different realizations, namely the memory effect is given by an external feedback, where $K(t,t')$ is a concentration independent quantity, or by an internal feedback, where the memory kernel depends on the concentration $c(t)$ itself and is in so far state-dependent.

EXTERNAL FEEDBACK

In this section we specify the model, Eq. (1), by fixing the memory kernel K in terms of deterministic functions. We consider four different cases.

(i) Firstly we introduce a discrete time-delay in accordance with $K(t,t';\tau) = \delta(t-t'-\tau)$. In case of a short ranged time-delay $\tau \ll t$ the stability is determined by the exponents

$$\kappa_\pm = \frac{1}{2\mu\tau}\left(1+\mu \pm \sqrt{(1+\mu)^2 - 4\mu\tau}\right). \quad (3)$$

For the memory strength $\mu > 0$ both exponents are positive leading to instablities. Likewise for negative $\mu < 0$ one of the exponents is always negative and the stability is lost.

(ii) An usual choice of a kernel in problems dealing with continuously distributed time-delay is the exponential kernel $K(t,t') = \exp[-\lambda(t-t')]$ ($\lambda > 0$), which satisfies some properties like the boundedness and positivity of the kernel. Whereas the fixed points $c_s = 0$ and $c_s = 1$ are independent of the memory parameter μ, their stability is actually controlled by the memory strength. To gain information on the stability we make a linear analysis. Using Laplace transformation it results

$$c(z) = c_0 \frac{z+\lambda+\mu}{(z-1)(z+\lambda)+\mu z}. \quad (4)$$

The zeros of the denominator of the latter formula are given by $z_\pm = 1/2[-(\lambda+\mu-1)\pm\sqrt{D}]$ with $D = (\lambda+\mu-1)^2 + 4\lambda$. It is obvious that $D > 0$, because of the assumption $\lambda > 0$. Further one observes that $\sqrt{D} > |\lambda+\mu-1|$ and so z_+ is positive and z_- is negative. Since the solution can be written as $c(t) = A\exp(z_+ t) + B\exp(z_- t)$, we find only unstable or unbounded solutions (for $t \to \infty$), respectively. That means that the exponential growth, which is observed in case of $\mu = 0$, could not be restricted through the additional feedback term. Alternatively, the evolution equation can be analyzed by a repeated differentiating with respect to t and a renewed use of the first order equation to eliminate the integral. The procedure results in the second order equation

$$\frac{d^2c}{dt^2} + [2c+\lambda+\mu-1]\frac{dc}{dt} + \lambda c[c-1] = 0 \quad (5)$$

whose linear equivalent exhibits the same behavior found above. The nonlinear Eq. (5) gives rise to a more complex stability behavior. To get insight into the stability of both fixed points we take the ansatz $c(t) = c_s + \varphi(t)$. Whereas the trivial fixed point $c_s = 0$ is always unstable, the second fixed point offers a broader behavior elucidated via the Laplace transformed function $\varphi(z)$. It results

$$\varphi(z) = \varphi_0 \frac{z+\lambda+\mu}{(z+1)(z+\lambda)+\mu z}. \quad (6)$$

In this case the zeros of the denominator are given by $1/2[-(1+\lambda+\mu)\pm\sqrt{D}]$, where the discriminate is $D = (1+\lambda+\mu)^2 - 4\lambda$. The solution of $\varphi(t)$ (for $D \neq 0$) can be written as $\varphi(t) = \varphi_0[A_+\exp(z_+ t) + A_-\exp(z_- t)]$ with the coefficients

$$A_\pm = \frac{\pm(\lambda+\mu-1)+\sqrt{D}}{2\sqrt{D}}. \quad (7)$$

The last formula implies that for $D \geq 0$ the solution is real and hence physically relevant. The limiting case $D = 0$ delivers critical values either for μ or for λ. Choosing $\mu = \mu(\lambda)$ the critical values for μ are $\mu_\pm^c(\lambda) = -(\lambda+1)\pm 2\sqrt{\lambda}$. If μ is situated in the intervals $\mu > \mu_+^c$ or $\mu < \mu_-^c$, $D > 0$ is fulfilled and the resulting solution is real. The solution for φ offers only a stable behavior, if $\mu > -(1+\lambda)$. All together the parameter μ has to fulfil $\mu > \mu_+^c$ to get stable, physically relevant solutions. In the limiting case, $\mu = \mu_+^c$, one specifies

$$\varphi(t) = \varphi_0 \exp\left(-\sqrt{\lambda}\,t\right)\left[1+(\sqrt{\lambda}-1)t\right], \quad (8)$$

which is a bounded solution for $t \to \infty$. In case of $\lambda < 1$, $\varphi(t)$ changes its sign at $t^* = (1-\sqrt{\lambda})^{-1}$. Altogether we can determine the stability domain for the fixed point $c_s = 1$ to the area in the (λ, μ)-plane, where $\mu \geq \mu_+^c(\lambda)$ is fulfilled. Eq. (5) is an evolution equation, which is well-known as equation for a damped oscillator. Interpreting the equation as an equation of motion for a particle in a potential $U(c) = -c^2/2 + c^3/3$, then the factor in front of the first derivative has the meaning of the damping constant. Due to the nonlinearity of the underlying evolution equation the damping parameter $\gamma(c) = 2c(t)+\lambda+\mu-1$ is driven by the time dependent concentration. If one regognizes that only a positive damping parameter $\gamma(c) \geq 0$ is reasonable, it is obvious, that the stability criteria is additionally dependent on the initial value c_0. That is indeed a feature of systems including feedback or memory effects. With the following constraints

$$\begin{aligned}\mu+\lambda+1-2\sqrt{\lambda} &\geq 0 \\ 2c_0+\mu+\lambda-1 &\geq 0\end{aligned} \quad (9)$$

one gets the stability domain depicted in Fig. 1, where the (μ,λ,c_0)-plane is displayed. The area, limited by the curved area and the plane area at the forefront, is the region where the solution is supposed to be stable.

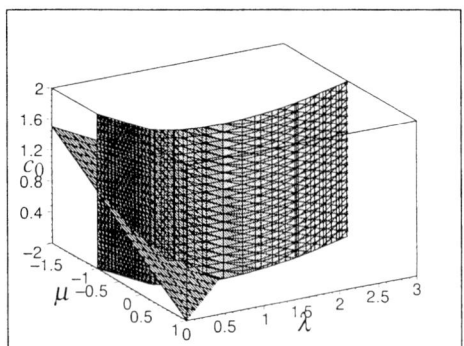

FIGURE 1. Stability domain for an exponential kernel in the (μ, λ, c_0) plane.

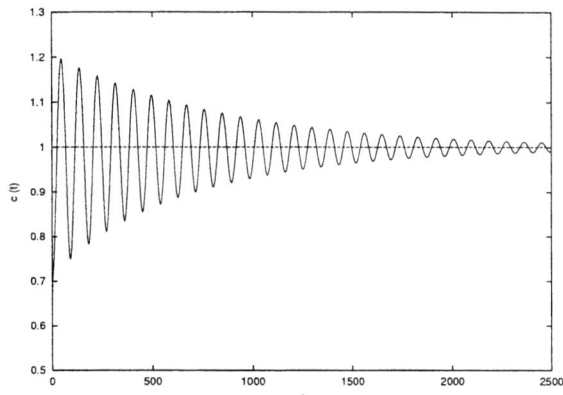

FIGURE 2. Time evolution of $c(t)$ with a periodic kernel $\lambda = 0, \omega = 0.1, \mu = 1$ and $c_0 = 0.5$.

(iii) Since the dynamical behavior is affected by the sign of the memory parameter μ, we consider a periodic kernel

$$K(t,t') = \exp\left[-\lambda(t-t')\right]\cos\left[\omega(t-t')\right].$$

Again the two stationary values are determined to $c_s = 0$ and $c_s = 1$. Using Laplace transformation for the corresponding linear evolution equation the quantity $c(z)$ is a rational function with a cubic polynominal denominator. At least one of the three solutions has a positive real part in case of $\lambda \neq 0$ and so there are only unbounded solutions for the linear equivalent of evolution equation for long times, i.e. the trivial fixed point $c_s = 0$ becomes unstable. Like in the case of the exponential kernel it can be written as a conventional differential equation which is in the present case of third order. The fixed point $c_s = 1$ is stable. An example is presented in Fig. 2 for the pure periodic case, i. e. for $\lambda = 0$. The trajectory reveals a damped oscillatory behavior which is apparently caused by the memory. In that case the memory generates an additional dissipative mechanism. This attenuation is enforced by the exponential part of the kernel. Obviously, an increasing value of the freuqency ω leads to an overdamped behavior.

(iv) The special case of a constant memory kernel $K(t,t') = 1$ can be treated exactly leading to

$$c(t) = \frac{1-\mu}{2} + \frac{\hat{D}}{2}\tanh\left\{\frac{1}{2}\left(\hat{D}t + \ln\left[\frac{\hat{D}+\zeta}{\hat{D}-\zeta}\right]\right)\right\} \quad (10)$$

with $\hat{D} = \sqrt{(1-\mu)^2 + 4\mu c_0}$ and $\zeta = 2c_0 + \mu - 1$. The decay of the concentration decay depends directly on the initial value c_0 making the memory apparently. Notice that the solution is only reasonable within the interval $0 < c_0 < 1$. The special case $\mu = 1$ must be treated seperately leading to

$$c(t) = \sqrt{c_0}\tanh\left[\sqrt{c_0}t + \frac{1}{2}\ln\left(\frac{1+\sqrt{c_0}}{1-\sqrt{c_0}}\right)\right]. \quad , \quad (11)$$

The stationary fixed points are

$$c_s(\mu) = \frac{1}{2}[1-\mu \pm \hat{D}]$$

different from $c_s = 0$ or $c_s = 1$, respectively. The last result can be generalized to an arbitrary kernel $K(t)$. Namely, the fixed points remain unchanged, whenever the Laplace-transformed memory kernel satisfies the relation

$$\lim_{z\to 0} zK(z) = 0 \quad . \quad (12)$$

INTERNAL FEEDBACK

Now let us discuss the case that the time scale of the kernel $K(t)$ in Eqs. (1,2) is determined by the concentration $c(t)$ itself with the simplest realization

$$K(t,t';c) \equiv K\left(c(t-t')\right) \quad . \quad (13)$$

In that case the memory term is a competitive one to the conventional quadratic reaction term and violates the condition of Eq. (12). Furthermore, the phase diagram is strongly affected by the sign of the memory strength μ. Using Laplace transformation we find a non-trivial, memory controlled solution

$$c_s(\mu) = \frac{1+\mu c_0}{1+\mu} \quad . \quad (14)$$

The complete phase diagram, depicted in Fig. 3, is obtained employing a linear stability analysis. The non-trivial stationary solution is stable for $\mu > -1$ in the interval $0 < c_0 \leq 1$ and for $c_0 \geq 1$ in case of $\mu \geq \mu_1(c_0) = -1 + \sqrt{1-c_0^{-1}}$. There exists a second domain where the non-trivial stationary solution should be

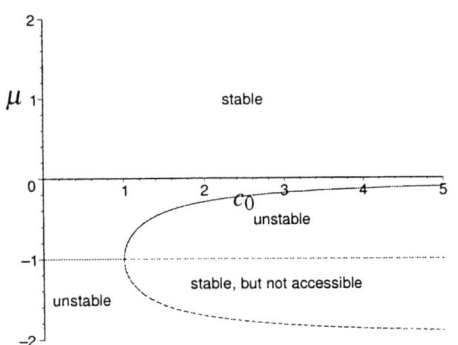

FIGURE 3. Stability of the internal memory controlled stationary solution in the $c_0 - \mu$-plane.

stable. This area is limited by the lines $\mu = -1$ and $\mu_2(c_0) = -1 - \sqrt{1 - c_0^{-1}}$ (dashed lines in Fig. 3). However the system can not achieve this region, because the initial value $c_0 > 1$ leads to $c_s(\mu) > c_0$ in contradiction to the monotone decrease of the function $c(t)$ as indicated by the self-organized evolution equation. We omit the discussion of the other cases with different signs of the parameter r and u in Eq. (1). The inclusion of an internal memory leads to a non-trivial stationary point which is controlled strongly by the memory parameter μ. Because the feedback is characterized by the concentration, relation Eq. (12) is violated leading to a new fixed point in comparison to the non-Markovian situation.

CONCLUSIONS

In the present paper we have generalized the conventional rate equations in the reaction-limited case by including memory effects. Once, the memory can be originated by external constraints which hints the particle to encounter simultaneously. As a consequence the rate of the concentration at the present time may be coupled to the rate at a previous time, where between both processes an additional delay-time τ appears. Although the stationary points are unchanged by such kind of external delay, the dynamical behavior to reach those fixed points is altered. In case of an internal feedback the stationary solution is changed and depends on the memory coupling and the initial value. The reason for that consists of the self-organized coupling to the concentration inherent in the memory. Such a situation may be realized whenever the reacting entities are not available simultaneously. Thus, the reaction is determined by the accumulation of the species at previous times. To capture the influence of that delay process, a memory term is included into the evolution equation. The non-Markovian part is generally found adopting projection methods of statistical mechanics. As further steps we are interested in inclusion of spatial fluctuations in the diffusion-limited regime.

ACKNOWLEDGMENTS

We thank Michael and Beatrix M. Schulz as well as Marian Brandau for fruitful discussions. This work was supported by DFG (SFB 418).

REFERENCES

1. ben Avraham, D., and Havlin, S., *Diffusion and Reactions in Fractals and Disordered*, Cambridge University Press, Cambridge, UK, 2000, pp. 170–191.
2. Tsimring, L. S., and Pikovsky, A., *Phys.Rev.Lett*, **87**, 250602-1–250602-4 (2001).
3. Masoller, C., *Phys.Rev.Lett*, **88**, 034102-1–034102-4 (2002).
4. Murray, J. D., *Mathematical Biology*, Springer Verlag, Berlin, Germany, 2000, pp. 13–17.
5. Banks, R. B., *Growth and Diffusion Phenomena*, Springer Verlag, Berlin, Germany, 1994.
6. Freeman, M., *Nature*, **408**, 313–319 (2000).
7. Kuang, Y., *Delay Differential Equations with Applications in Population Dynamics*, Academic Press, London, UK, 1993.
8. Schulz, M., and Stepanow, S., *Phys. Rev. B*, **59**, 13528–13530 (1999).
9. Schulz, B. M., and Trimper, S., *Phys. Lett. A*, **256**, 266–271 (1999).
10. Schulz, B. M., Trimper, S., and Schulz, M., *Eur. Phys. J. B*, **15**, 499–505 (2000).
11. Schulz, B. M., Schulz, M., and Trimper, S., *Phys. Rev. E*, **66**, 031106-1–031106-6 (2002).
12. Schulz, M., and Trimper, S., *Phys. Rev. B*, **64**, 233101-1–233101-4 (2001).
13. Schulz, M., Trimper, S., and Schulz, B. M., *Phys. Rev. E*, **64**, 026104-1–026104-5 (2001).
14. Trimper, S., Zabrocki, K., and Schulz, M., *Phys.Rev.E*, **65**, 056106-1–056106-6 (2002).
15. Trimper, S., Zabrocki, K., and Schulz, M., *Phys.Rev.E*, **66**, 026114-1–026114-7 (2002).
16. Mori, H., *Prog. theor. Phys.*, **34**, 399–416 (1965).
17. Fick, G., and Sauermann, E., *Quantenstatistik dynamischer Prozesse*, vol. 1, Akademische Verlagsgesellschaft Geest & Portig K.G., Leipzig, 1983.
18. Leutheusser, E., *Phys. Rev. A*, **29**, 2765–2773 (1984).
19. Götze, W., *J. Phys.: Condens. Matter*, **11**, A1–A45 (1999).

Phase transitions of nematic rubbers

Ko Okumura* and Pierre-Gilles de Gennes[†]

*Department of Physics, Graduate School of Humanity and Sciences, Ochanomizu University, 3-1-1, Otsuka,
Bunkyo-ku, Tokyo 112-8610, Japan
[†]Physique de la Matière Condensée, Collège de France, 75231 Paris Cedex 05, France

Abstract. Single crystal nematic elastomers undergo a transition from a strongly ordered phase N to an "isotropic" phase I. We show that: (a) samples produced under tension by the Finkelmann procedure are intrinsically anisotropic and should show a small (temperature-dependent) birefringence in the high temperature I phase. (b) even without such an anisotropy, there is no instability at the standard spinodal temperature for the the transition from N to I phase via heating, although there remains a spinodal limit for the opposite transition via cooling. (c) the transition from N to I is reminiscent of a martensitic transformation: nucleation of the I phase should occur in the form of platelets, making a well defined angle with the director.

INTRODUCTION

When mesogenic groups are incorporated in flexible chain polymers and the solution or the melt is cross-liked, a nematic crystalline elastomer is obtained. Such nematic elastomer or rubbers were first constructed by H. Finkelmann and coworkers [1]. Subsequently the single-domain samples came to be available [2] and they show a spectacular change in shape when they are switched from low temperatures ($T < T_c$) to high temperatures ($T > T_c$) undergoing a transition from a strongly ordered nematic phase N to an "isotropic" phase I. Many properties resulting from the coupling between nematic order and elastic deformations [3] have been analyzed by M. Warner, E. M. Terentjev and coworkers [4]. We are concerned here mainly by the N $\rightarrow$ I transition.

Two striking facts should be mentioned here: (a) the transition is expected to be first order, but the plots of birefringence versus temperature T are continuous [5]. (b) the transition is very slow (minutes) [6].

Our aim here is to discuss some effects of the nematic / elastic coupling on the phase transition [7].

ANISOTROPY IN MONO-DOMAIN SAMPLES

We discuss the anisotropy induced by the preparation method for mono-domain samples. In the Finkelmann scheme, a very weak network is prepared first, and is put under a prescribed deformation ε_0 along one axis [2]. Then a second reaction is started and the final nematic rubber is generated. The free energy of the system can be written as

$$F = F_0(Q) + \mu \text{Tr}(\varepsilon - \varepsilon_0)^2 - \Lambda \text{Tr}(Q\varepsilon) \quad (1)$$

where $F_0(Q)$ is the standard Landau free energy of the nematic order and the second term is the shear elastic energy (we consider only incompressible systems: Q and ε are *symmetric traceless tensors*). The last term describes the coupling (Λ) between deformation and order. The crucial point is that the elastic energy (with a strong coefficient μ) is minimal *in the original state* ($\varepsilon = \varepsilon_0$) while the nematic order is directly coupled to the deformation ε itself. With the shift in the definition of the deformation, $e \equiv \varepsilon - \varepsilon_0$, there appears a fixed *external field* $\sigma_0 = \Lambda \varepsilon_0$ coupled to the nematic order Q. Via minimization of this energy with respect to e, we find that *the deformation is always proportional to the order* ($e \sim Q$) and the deformation e contained in F can be eliminated:

$$F = -\text{Tr}(\sigma_0 Q) + a\text{Tr}Q^2 - b\text{Tr}Q^3 + \cdots \quad (2)$$

Under the field σ_0, the nematic order Q is nonzero even above the transition temperature: the high temperature "I phase" shows a small (temperature-dependent) birefringence, and the discontinuity jump in the plots of order versus temperature tends to be small.

ABSENCE OF THE N $\rightarrow$ I SPINODAL POINT

Can we achieve the transition by soft phonon modes? Since our materials are essentially incompressible, we must investigate the transverse phonons. We assume rapid equilibration for the elastic degrees of freedom: we always use $e \sim Q$.

We analyze T-jumps in the nematic rubbers based on the energy in *an ideal case* where the order disappears above the transition point:

$$F = a\mathrm{Tr}Q^2 - b\mathrm{Tr}Q^3 + c\left(\mathrm{Tr}Q^2\right)^2 + \cdots \quad (3)$$

which is already minimized for the elasticity. The realistic case with the anisotropy effect will be discussed later.

We find a traditional spinodal transition for I → N transition. The coefficient a vanishes at the spinodal temperature for a cooling jump. Amplitudes of our transverse phonons are proportional to Q and the energy required for the phonon generation vanishes at $T = T^{**}$: the phonons are soft at this temperature.

But for a heating jump we show that the elastic couplings tend to suppress the spinodal instability, although there are two soft modes at the standard spinodal temperature. In this N → I transition, we start with a uniaxial nematic phase described by a diagonal tensor, $(Q_{xx}, Q_{yy}, Q_{zz}) = (-S, -S, 2S)$; we can find a candidate for the spinodal temperature T^* for a special value $S = S^*$, satisfying the conditions

$$\frac{\partial F}{\partial S} = \frac{\partial^2 F}{\partial S^2} = 0 \quad \text{(at } S = S^* \text{ and } T = T^*) \quad (4)$$

We expand the above free energy to second order around the spinodal point to find:

$$F(Q) - F(Q^*) = \frac{9b^2}{64c}\left[(Q_{xx} - Q_{yy})^2 + 4Q_{xy}^2\right] \quad (5)$$

This form may be understood generally from the Nambu-Goldstone theorem. $F(Q)$ is invariant under SO(3) (with three generators T_i). However, *the ground state*, Q^*, does not have the full SO(3) symmetry and is invariant only for a subgroup generated by T_3 (rotations around the z axis). Considering the invariance of $F(Q^*)$ under the subgroups generated by T_1 and T_2, we find two Goldstone modes: terms proportional to Q_{zx}^2 and Q_{zy}^2 are absent. In addition, the Q_{zz}^2 term vanishes under the spinodal condition.

We can now investigate the two possible transverse phonon modes in Fig. 1 whose wave vector and displacement are $\mathbf{q}$ and $\mathbf{u}$, respectively ($e_{\alpha\beta} = (q_\alpha u_\beta + q_\beta u_\alpha)/2$). We can show that both transverse modes are not soft for all nonzero angles ($\theta \neq 0$) at $T = T^*$ because of eq. (5): the first mode makes $Q_{xx} - Q_{yy}$ nonzero while the second makes Q_{xy} nonzero [7]. The two $\theta = 0$ modes ($e_{xz} \sim Q_{xz} \neq 0$ and $e_{yz} \sim Q_{yz} \neq 0$) are soft at all temperatures, if $F(Q)$ has full rotational invariance. But excitation of these modes corresponds to rotations of a uniaxial Q tensor around x or y axis and cannot cause a change in the magnitude of the order parameter (see Appendix); these modes do not catalyze the transition.

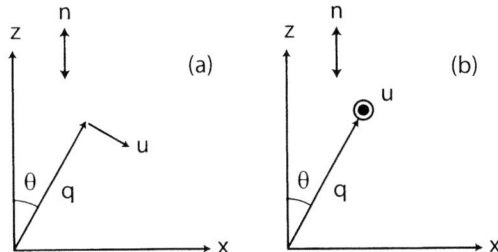

FIGURE 1. (a) First transverse mode $\mathbf{u} = (u\cos\theta, 0, -u\sin\theta)$. (b) Second transverse mode $\mathbf{u} = (0, u, 0)$. Note the difference from Fig. 2: the director vector is parallel to the z-axis.

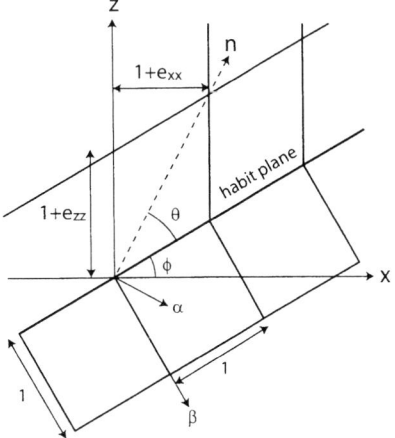

FIGURE 2. Habit plane separating nematic and isotropic regions. In two dimensions the "isotropic" network is represented by a square array, while the nematic network is represented by parallelograms. Along the habit plane the two networks match.

NUCLEATION OF I PHASE IN A NEMATIC CRYSTAL

We discuss the nucleation of an I phase inside a nematic single crystal in the ideal case (the I phase is literally isotropic). This N → I transition is reminiscent of a *martensitic transformation* [8]: nucleation of the I phase should occur in the form of platelets, making a well defined angle with the director. As a result, the homogeneous nucleation is prohibited in nematic rubbers – just as it is in martensites.

A) Choice of a habit plane

We now investigate a possible plane boundary (*habit plane*) between I and N phase. Here, we consider a T-jump from a temperature just below the thermodynamic transition point T_c towards a higher temperature T, in the region where spinodal instabilities are ruled out.

The two-dimensional example (plane strain in the $x - z$ plane) is shown on Fig. 2, where *we impose compati-*

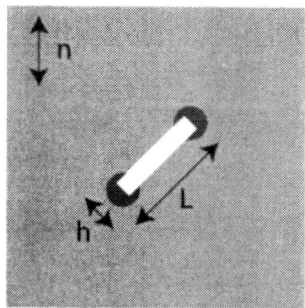

FIGURE 3. Platelet nucleus in the nematic phase accompanied by a large strain around the periphery.

bility between an I phase (represented symbolically by a square unit cell) and a uniaxial N phase (represented by parallelograms). The squares have been rotated by an angle ϕ from the x-axis. The N phase has an elongation e_{zz} along the z axis. Note here that the director is *not* parallel to the z axis. The parallelograms are sheared with a strain

$$e_{xz} = \frac{1}{2}(\partial_x u_z + \partial_z u_x) = \frac{1}{2}\tan\phi \quad (6)$$

There is also a deformation $e_{xx} = -e_{zz}$ in the N phase (we assume incompressibility) and *matching* imposes $1 + e_{xx} = \cos\phi$. Thus from Fig. 2 we can show the angle θ between the director axis and the habit plane is given by

$$\tan(\theta + \phi) = \frac{2 - \cos\phi + \sin\phi}{\cos\phi} \quad (7)$$

Note that, for small ϕ, the parallelogram is a lozenge, and the director is parallel to the long axis of the lozenge – at $\pi/4$ from the habit plane. The magnitude of ϕ is fixed by the condition that the elastic energy of the N state be equal to the equilibrium value $f(T_i)$ in the initial nematic phase.

For this two-dimensional case, we achieve a perfectly isotropic state I on one side and the standard N state of the other side. However, this perfect matching is not possible in 3D (exactly as in similar martensitic transformations [8]). [7]

B) Homogeneous nucleation

Again we think of a T-jump from an initial temperature T_i just below T_c (we take $T_i = T_c$ in practice for simplicity) up to a final temperature T lying a bit above T_c. Our assumed shape for the nucleus is a platelet corresponding to the habit plane orientation, with a thickness h and dimension L (Fig. 3).

Due to distortions at the *periphery* of our platelet where there is no matching at all (the matching was achieved only at the habit plane boundary), we have *large deformations* taking place in a toroidal region near the periphery (Fig. 3). As a result, we expect a large barrier for the formation of a critical droplet and the nucleation rate can be thus dramatically suppressed. [7]

CONCLUSION

Clearly, we need heterogeneities to nucleate, and we do expect them in networks: we can expect a slow transition. The heterogeneous distribution of the transition temperatures may disguise the transition as a "continuous" transition (The anisotropy effect, which reduces the discontinuity jump, may also help this). Even when dealing with heterogeneous nucleation, it is reasonable to think that platelets (or needles) close to the habit angle will be preferentially generated. This would possibly be observed by electron microscopy or by force microscopy on the outer surface. From a practical point of view, it should be beneficial to accelerate the commutation process – we can think of at least two ways: (a) adding colloidal platelets (clay ?) at the correct angle (possibly orienting them by fields during the synthesis). (b) shearing the sample with a shear plane near the habit angle.

DISCUSSION

If the anisotropy effect is so strong that the transition becomes continuous, we do not need the nucleation (after T-jumps there is only a single local minimum) and we may not expect the habit angle. However, this may not correspond to real situations: 1) In the Finkelmann procedure, the network at stage 1 is very weak. 2) In the current experiments of P. Keller, [9] the alignment is provided by contact with an oriented solid surface: a very weak drive.

We examined instabilities and nucleation only in the ideal case. In real mono-domain sample, however, we expect a finite order for the high T phase (although the external field is not too strong). In such cases, both the N $\rightarrow$ I and I $\rightarrow$ N spinodal points may be stabilized due to the phonon. As for the N $\rightarrow$ I transition, we expect as in the above the heterogeneous nucleation; the transition looks continuous and is rather slow. As for the I $\rightarrow$ N transition, we also need the nucleation and expect a slow transition in the realistic case. In both N $\rightarrow$ I and I $\rightarrow$ N transitions, the nucleus may be still in the form of platelet but the situation would be a bit different (due to the nonzero order in the "I phase," the matching condition shall be different). In the case of martensite, it is experimentally known that the transition times of the N $\rightarrow$ I and I $\rightarrow$ N counterparts are always of the same order of magnitude (if N $\rightarrow$ I is slow, I $\rightarrow$ N is slow, too). This also suggests the above scenario. It is worth studying these points elsewhere.

ACKNOWLEDGMENTS

We have benefited from exchanges with Y. Quéré and E. Raphaël. K. O. thanks M. Warner, E. Terentjev, and T. Kakeshita for informative communications. K. O. appreciates H. Nakanishi, T. Ohta, and, S. Uchida for useful comments given in a conference held at Yukawa Institute for Theoretical Physics.

APPENDIX

The tensorial order is transformed as $Q \to RQR^{-1}$ under rotation where $R = \exp(i\theta_j T_j)$. The generators can be represented by

$$T_1 = \begin{pmatrix} 0 & 0 & 0 \\ 0 & 0 & -i \\ 0 & i & 0 \end{pmatrix}, \tag{8}$$

$$T_2 = \begin{pmatrix} 0 & 0 & i \\ 0 & 0 & 0 \\ -i & 0 & 0 \end{pmatrix}, \tag{9}$$

$$T_3 = \begin{pmatrix} 0 & -i & 0 \\ i & 0 & 0 \\ 0 & 0 & 0 \end{pmatrix} \tag{10}$$

The uniaxial order $(Q_{xx}, Q_{yy}, Q_{zz}) = (-S, -S, 2S)$ is invariant under infinitesimal rotations around the z axis ($R_3 \simeq i\delta\theta_3 T_3$). For rotations around x and y axes, the variation in Q under the transformation is respectively given by

$$\delta Q = \delta\theta_1 \begin{pmatrix} 0 & 0 & 0 \\ 0 & 0 & 3S \\ 0 & 3S & 0 \end{pmatrix}, \delta\theta_2 \begin{pmatrix} 0 & 0 & -3S \\ 0 & 0 & 0 \\ -3S & 0 & 0 \end{pmatrix} \tag{11}$$

Thus, Q_{yz} and Q_{xz} are the elements which appear after rotations around x and y axes, respectively. In turn, these elements can be removed by appropriate rotations to return to the uniaxial tensor.

REFERENCES

1. Finkelmann, H., Kock, H.-J., and Rehage, G., *Macromol. Rapid Comm.* **2**, 317-322 (1981).
2. Küpfer, J., and Finkelmann, H., *Macromol. Chem. Rap. Commun.* **12**, 717-726 (1991).
3. De Gennes, P.-G., *C. R. Acad. Sci. (Paris) B*, **281**, 101-103 (1975).
4. Warner, M., and Terentjev, E.M., *Liquid Crystal Elastomers*, Oxford Univ. Press press, Oxford, 2003.
5. Selinger, J.V., Jeon, H.G., and Ratna, B.R., *Phys. Rev. Lett.* **89**, 225701-1-225701-4 (2002).
6. Warner, M., and Terentjev, E.M., private communication.
7. De Gennes, P.-G., and Okumura, K., *Europhys. Lett.* **63**, 76-82 (2003).
8. Wayman, C.M., "Phase Transformations, Nondiffusive," in *Physical Metallurgy II*, edited by R. W. Cahn and P. Haasen, North-Holland Phys. Pub., Amsterdam, 1983, pp. 1031-1074.
9. Keller, P., private communication.

Driven Motion of Extended Defects Wetted by a New Phase

Alexander L. Korzhenevskii[*], Richard Bausch[†] and Rudi Schmitz[**]

[*]*Institute for Problems of Mechanical Engineering, RAS, Bol'shoi prosp. V. O., 61, St Petersburg, 199178, Russia*
[†]*Institut für Theoretische Physik IV, Heinrich-Heine-Universität Düsseldorf, Universitätsstrasse 1, D-40225 Düsseldorf, Germany*
[**]*Institut für Theoretische Physik A, RWTH Aachen, Templergraben 55, D-52056 Aachen, Germany*

Abstract. Moving extended defects can close to a bulk phase transition be coated by nuclei of a new phase. Depending on the order of the transition and on the type of the defects, the co-moving nuclei display various types of non-equilibrium phase transitions and instabilities. These are expected to affect significantly the plastic behavior of real materials.

INTRODUCTION

Plastic flow of solid materials is driven by the motion of dislocations, twin boundaries, and of other extended defects [1]. At temperatures, sufficiently close to a bulk phase transition of the material, defects often give rise to heterogeneous nucleation of a new phase [2]. Since the nuclei need some time to build up, their formation only occurs in some finite region of the temperature-velocity diagram [3].

Nuclei, attached to static dislocations have cylindrical shapes whereas at twin boundaries they form a kind of wetting layer. In stationary motion the nuclei develop a trail which grows with increasing velocity [4]. In case of a first-order bulk transition this surprisingly can lead to a kinetic complete-wetting transition of the nucleation layer, attached to a planar defect [5]. This transition marks a boundary between regimes of reversible and irreversible plastic behavior [1].

Energy dissipation in co-moving nuclei gives rise to a viscous-friction force, acting on the defects, and, in this respect, dominates all formerly known mechanisms, at least in the case of dislocations [6]. In the case of a planar defect the most remarkable observation is the emergence of a characteristic shift of the drag coefficient at the kinetic complete-wetting transition which allows to observe this transition in real materials [5].

Another interesting implication of the nuclei-induced friction is the appearance of various types of morphological instabilities of the defects. Gliding dislocations, coated by a meta-stable nucleus, can e.g. accomplish a kinetic roughening instability [7]. In case of a twin boundary, covered by a wetting layer, a quite different instability occurs which is reminiscent of a de-wetting process by spinodal decomposition [3].

THE MODEL

All features, described so far, can be established for a class of model systems, described by the equations of motion

$$\partial_t \varphi = -\Lambda \, \delta H/\delta \varphi + \theta , \quad (1)$$

$$\mathbf{N} \cdot \partial_t \mathbf{R} = -\frac{1}{B} \frac{1}{|\partial \mathbf{R}|} \mathbf{N} \cdot \frac{\delta H}{\delta \mathbf{R}} + \mathbf{N} \cdot \mathbf{F} + \mathbf{N} \cdot \mathbf{f}. \quad (2)$$

Here, $\varphi(\mathbf{r},t)$ is a scalar order parameter at position $\mathbf{r}$ and time t, and $\mathbf{r} = \mathbf{R}(\xi,\mathbf{t})$ describes the location of a single defect, depending on a set of internal parameters ξ. $\mathbf{N}(\xi,t)$ is a normal vector of a twin boundary, or of a dislocation in the glide plane. Covariance under reparametrizations leads to the appearance of projections onto $\mathbf{N}$, and to the Jacobian $|\partial \mathbf{R}|$. For a dislocation in the glide plane $y = 0$ the Monge representation $\mathbf{R}(x,t) = [x,0,Z(x,t)]$ leads to a Kardar-Parisi-Zhang - like form [8] for eqn. (2).

In case of pure relaxation of φ the operator Λ is a simple relaxation constant λ whereas for a conserved order parameter $\Lambda = -D\nabla^2$. Similarly, the coefficient $1/B$ measures the mobility of the defect. Thermal noise or disorder, e.g. by frozen-in point defects, are described by the random forces θ and $\mathbf{f}$.

The effective Hamiltonian $H[\varphi, \mathbf{R}]$ of the system is given by

$$H = \int d^d r \left[\frac{1}{2}(\nabla\varphi)^2 + f(\varphi) + \frac{1}{2} U[\mathbf{R}]\varphi^2 \right] + \int d^D \xi \, \frac{\sigma}{2} (\partial \mathbf{R})^2 . \quad (3)$$

Here, $f(\varphi)$ is a free-energy density which for a second-order transition is of the form

$$f = \frac{\tau}{2}\varphi^2 + \frac{u}{4}\varphi^4 \qquad (4)$$

where $u > 0$, and $\tau \equiv \alpha(T - T_c)$ measures the distance from the critical point. In case of a first-order transition $u < 0$ which requires a stabilizing term of order φ^6 Moreover, σ is a stiffness coefficient of the defect which accounts for the elastic energy, generated in the medium.

The coupling of φ to a dislocation is, in the simplest case, described by the term $U(\mathbf{r},\mathbf{R}) = \kappa\varepsilon$ where κ is a coupling constant, and ε is the dilatation field, caused by the defect. For an edge dislocation with Burgers vector $\mathbf{b} = (0,0,b)$, extending along $z = Z(x,t)$ in the glide plane $y = 0$ of an isotropic medium with Poisson ratio v, the explicit form of $U(\mathbf{r},\mathbf{R})$ is

$$U = \kappa \frac{b_z}{2\pi} \frac{1-2v}{1-v} \frac{y}{[z-Z(x,t)]^2 + y^2}. \qquad (5)$$

The simplest model for a twin boundary, moving in z-direction, is

$$U = -\kappa \delta[z - Z(x,y,t)]. \qquad (6)$$

Generally, the coupling term $\propto U$ can be combined with the term $\propto \tau$ in (4), producing a local shift of the transition temperature. This roughly explains why a nucleus can exist above the global transition point of the host material.

A more detailed source of information can be obtained for the case where the nucleus appears via a second-order transition. Slightly below the corresponding nucleation threshold eqn. (1) can be linearized in φ, and then, in a mean-field picture, looks like a Schroedinger equation where U enters as a potential. The lowest bound state of this potential determines the shape of nucleus close to the nucleation threshold. Furthermore, in case of a stationary motion of the defect, its velocity V appears as an imaginary vector potential, from which the trail of the nucleus emerges.

If the bulk phase transition in the host material is of first order, an analytic treatment of the shape of moving nuclei can be achieved by choosing for $f(\varphi)$ instead of a polynomial double-well potential a piece-wise parabolic form [5].

KINETIC WETTING

In this case, for a nucleus joining a planar defect which is moving in normal direction, the following order-parameter profiles are found with increasing velocity V.

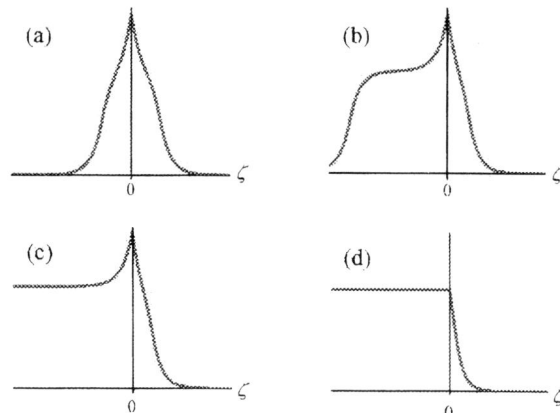

FIGURE 1. Order-parameter profiles for moving nuclei

The profile (a) in Fig.1 is that of a static nucleus. At non-zero velocity a shoulder develops in the profile, as illustrated in (b). This shoulder is captured by a metastable minimum in the free-energy density and limited by a kink from behind. The kink tries to destroy the metastable phase by moving towards the defect. Above some critical velocity $V = V_c(T)$ the kink cannot follow the defect any more, so that the trail of the profile extends to minus infinity, as shown in (c). Eventually, the stage (d) is reached where the order-parameter amplitude at the defect has reached the hight of the shoulder.

In the T,V - "phase diagram", the solid lines enclose the nucleation region, outside of which a stable nucleus does not exist. The broken line is defined by the function $V = V_c(T)$ and, due to the above discussion, represents a line of kinetic complete-wetting transitions of the nucleation layer [5]. Along the dotted line the points (a) - (d) correspond to the nucleus profiles of Fig.1.

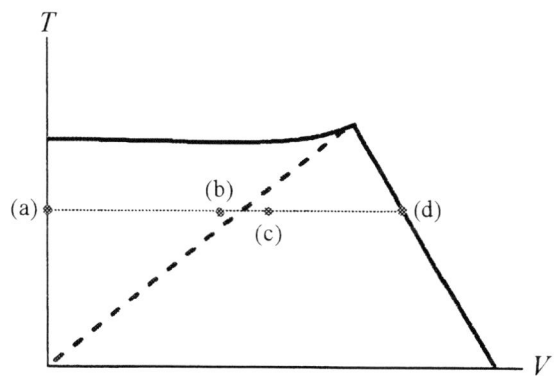

FIGURE 2. Temperature-velocity - phase diagram

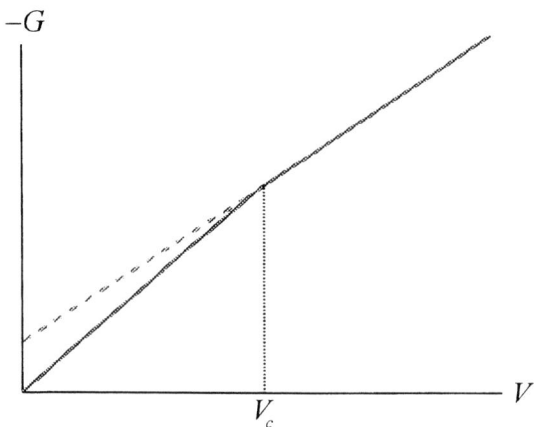

FIGURE 3. Nucleus-induced friction versus velocity

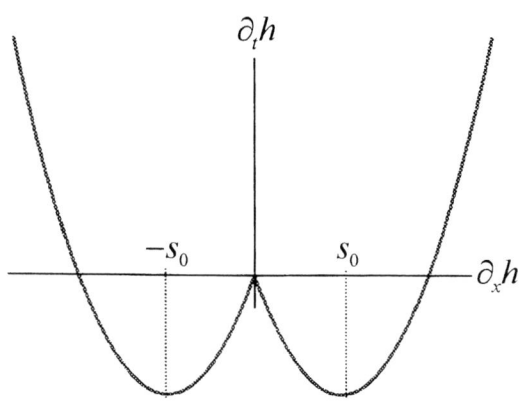

FIGURE 4. Plot of the reduced evolution equation

Fig.3 shows the nucleus-induced friction force G as a function of the velocity. The singular behavior at $V = V_c(T)$ corresponds to a jump of the drag coefficient which should allow to observe the kinetic complete-wetting transition in real materials [5].

It should also be mentioned that the thickness of the nucleation layer is found [5] to display a logarithmic singularity at $V = V_c(T)$, just as in a complete-wetting transition in thermal equilibrium.

ROUGHENING INSTABILITY

Due to (5) the size of a nucleus, attached to a dislocation, is proportional to the component of the Burgers vector in normal direction of the defect line. Accordingly, a screw dislocation is unable to create nuclei, except via slope fluctuations within some glide plane. In this way local friction forces are generated which in turn enhance the slope fluctuations.

The friction, arising in this way in a complete-wetting regime, is linear in the local slope of the dislocation line [7]. In the evolution equation of the line profile this term competes with a Kardar-Parisi-Zhang nonlinearity [8]. Ignoring in a first run smoothing effects due to the line-tension term, the evolution equation reduces for $h(x,t) \equiv Z(x,t) - Vt$ to the form

$$\partial_t h = (\partial_x h)^2/2 - s_0|\partial_x h|, \quad (7)$$

plotted graphically in Fig.4.

A stability analysis for screw dislocation, disturbed by a triangular section as in Fig.5, shows that all slopes of this section in the interval $-s_0 < \partial_x h < s_0$ are unstable. As a result, zigzag-like configurations of the dislocation line with random position of slopes $\pm s_0$ are expected to arise [7].

FIGURE 5. Variational form of a dislocation line

SPINODAL INSTABILITY

The mechanism behind the spinodal-like instability of a two-dimensional defect is based on the fact that some time is needed for the nucleation layer to build up. Its formation, accordingly, is impeded by the motion of the defect which, eventually, leads to a decreasing order-parameter amplitude of the layer with increasing defect velocity.

A fluctuation-induced local excursion of the defect plane in forward direction, therefore, is connected with a local reduction of this amplitude, and, consequently, with a weakening of the friction force. This in turn leads to an enhancement of the initial excursion, initiating a self-amplifying process. A stabilizing counter-effect is provided by the surface tension of the defect. This effect dominates below some velocity threshold $V_s(T)$ whereas above this threshold the described instability mechanism wins [3].

The situation is visualized in Fig.6 where the driving force is plotted versus the defect velocity. The solid line represents an isotherm, showing the anomalous behavior $\partial F/\partial V < 0$ for velocities V inside the interval $V_s(T) < V < V_c(T)$. This interval shrinks to zero at some critical isotherm $T = T_1$, represented by the broken line, which at a kind of critical point $T_1, V_1 \equiv V_s(T_1)$ has a horizontal slope.

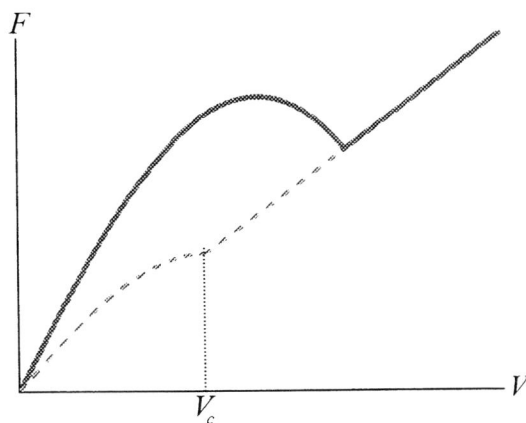

FIGURE 6. Subcritical and critical force-velocity isotherm

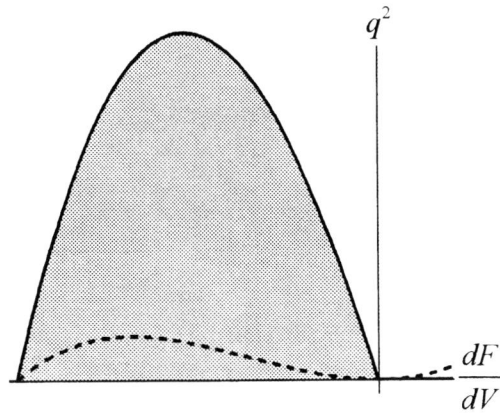

FIGURE 8. Instability region in the $q^2, \partial F/\partial V$ - plane

It should be mentioned that the scenario of a spinodal-like instability of a moving planar defect can also be established for the case of a conserved order parameter of the nucleus [9]. Moreover, such an instability can also be shown to occur in the glide motion of dislocations [10]. We, finally, expect that a similar instability also will arise near the surface of a growing crystal, if surface reconstruction is taken into account.

ACKNOWLEDGMENTS

This work has been supported by the Deutsche Forschungsgemeinschaft under BA 944/21

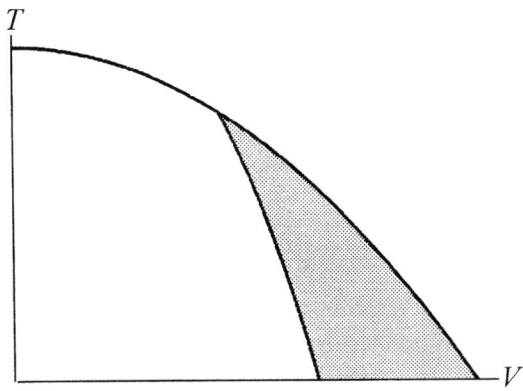

FIGURE 7. Kinetic spinodal line in the T, V - phase diagram

REFERENCES

1. Boyko V. S., Garber R. I., and Kosevich A. M., *Reversible Crystal Plasticity*, Amer. Inst. Phys., New York, 1994.
2. Boulbitch A. A., and Toledano P., *Phys. Rev. Lett.* **81**, 838-841 (1998).
3. Korzhenevskii A. L., Bausch R., and Schmitz R., *Europhys. Lett.* **59**, 533-539 (2002).
4. Bulbich A. A., and Pumpyan P. E., *Sov. Phys. Crystallogr.* **35**, 156-158 (1990).
5. Korzhenevskii A. L., Bausch R., and Schmitz R., *Phys. Rev. Lett.* **91**, 236101-1-4 (2003).
6. Korzhenevskii A. L., Bausch R., and Schmitz R., *Phys. Rev. B* **67**, 100103(R)-1-3 (2003).
7. Korzhenevskii A. L., Bausch R., and Schmitz R., *Phys. Rev. Lett.* **83**, 4578-4581 (1999).
8. Kardar M., Parisi G., and Y. C. Zhang Y. C., *Phys. Rev. Lett.* **56**, 889-892 (1986).
9. Korzhenevskii A. L., Bausch R., and Schmitz R., *J. Phys. A* **35**, 4565-4569 (2002).
10. Korzhenevskii A. L., Bausch R., and Schmitz R., *Phys. Rev. E* **63**, 056105-1-4 (2001).

Fig.7 shows the nucleation region for the case of a second-order bulk transition [3]. The instability occurs in the shaded subregion which is bounded from below by the spinodal line $V = V_s(T)$.

The instability shows up most naturally in the response function $R(x,t)$ of the defect position $Z(x,t)$ to a space-time-dependent driving force $F(x,t)$. According to the presence of two modes in the model (1) - (5), $R = R_1 + R_2$ where, in Fourier representation,

$$R_i(q,t) = \Theta(t) R_i(q) \exp\left[-\Omega_i(q)t\right], \quad (8)$$

involving the Heaviside step function $\Theta(t)$.

An instability only occurs, if the real part of any of the rates Ω_i becomes negative. Within linear response theory these rates have been determined in [3] as a function of q^2 and of $\partial F/\partial V$. The shaded region in Fig.8 refers to the resulting unstable regime where below the broken line both rates Ω_i are real.

Effects of temperature chaos on rejuvenation and memory in Migdal-Kadanoff spin glasses

M. Sasaki[*] and O.C. Martin[†]

[*]*Institute for Solid State Physics, University of Tokyo, Kashiwa-no-ha 5-1-5, Kashiwa, 277-8581, Japan.*
[†]*Laboratoire de Physique Théorique et Modèles Statistiques, bât. 100, Université Paris-Sud, F–91405 Orsay, France.*

Abstract. We study aging phenomena of Migdal-Kadanoff spin glasses in order to clarify relevancy of temperature chaos to rejuvenation and memory. By exploiting renormalization, we do efficient dynamical simulations in very wide time/length scales including the chaos length. As a consequence, we find that temperature chaos and temperature dependence of speed of equilibration cause two significantly different effects against temperature variations, i.e., rejuvenation for *positive* temperature variation and memory for *negative* temperature variation, as are observed experimentally in spin glasses.

It is well known that in spin glasses, dynamical effects strongly depend on the history of the system after quench from above the transition temperature T_c. These phenomena are called aging and have been studied using various experimental protocols [1, 2, 3]. Measurement of ac-susceptibility during T-cycle [4], which is employed in this work, is one of them. This experiment consists of the following three stages. In the first stage, the system is quenched from above T_c and it is kept at a temperature T ($< T_c$) during a time t_1. In the second stage the temperature is temporarily changed to $T \pm \Delta T$ ($< T_c$) during a time t_2, and then it is set back to T in the third stage. Ac-susceptibility χ is measured during all the three stages. In the case $-\Delta T$, χ in the third stage resumes its relaxation from the value at the end of the period t_1 as if the system remembers how far the relaxation at T had proceeded before the perturbation (*memory effect*). On the other hand, the system is *rejuvenated* by positive T-cycle, and we observe strong relaxation of χ in the third stage.

From a theoretical point of view, "temperature chaos" [5], decorrelation of the equilibrium states at two temperatures beyond the so-called chaos length $\ell(T, T')$, has been one of the most conceivable causes of rejuvenation. However, temperature chaos seems to be incompatible with memory if one naively thinks. Therefore, relation among temperature chaos, rejuvenation and memory has been eagerly studied [6, 3] as a key to understand aging phenomena. In this work, we also address this issue by studying Migdal-Kadanoff (MK) spin glasses. There are mainly two merits in working on MK spin glasses. First the existence of temperature chaos is shown in this model, and whose exact renormalization [7] allows one to measure the chaos length $\ell(T, T')$. Second we can do efficient dynamical simulations at very long *time* scales by exploiting renormalization [8]. (See also [9] as a similar approach.) In the previous work [8], we have investigated dynamics of MK spin glasses by utilizing these advantages, and found that temperature chaos causes rejuvenation, but it also destroys most of the memory if the length scale equilibrated during the second stage is much larger than $\ell(T, T')$. The main purpose of this work is to show that memory is preserved if we take temperature dependence of speed of equilibration into account.

The model —. We consider MK lattices following the standard real space renormalization group approximation [7] to the Edwards Anderson (EA) model [10]. The recursive construction of such hierarchical lattices is described in Fig. 1; edges are replaced by $2b$ edges so the "length" of the lattice is multiplied by 2. We call generation "level" the order of the recursion and G the total number of these. Then the lattice length L is 2^G and the number of bonds is $(2b)^G$ (which is also roughly the number of sites); one can thus identify $1 + \ln b / \ln 2$ with the dimension of space on such a lattice. When all the edges are constructed, each is assigned a random coupling J_{ij}^0. The superscript "0" implies that J_{ij}^0 are bare couplings. Similarly, on each site i we put an Ising spin $S_i = \pm 1$. The Hamiltonian is

$$H_J(\{S_i\}) = -\sum_{<ij>} J_{ij}^0 S_i S_j, \quad (1)$$

where the sum is over all the nearest neighbor spins of the lattice. All of the work presented here will be for three dimensions ($b = 4$) with couplings J_{ij}^0 taken from

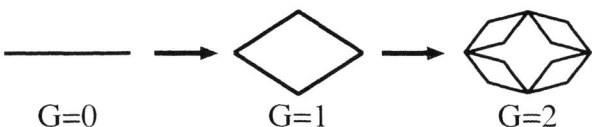

FIGURE 1. Construction of a MK lattice ($b = 2$).

TABLE 1. Size dependence of $C(L, T = 0.7, T' = 0.65)$.

L	2^5	2^7	2^9	2^{11}	2^{13}	2^{15}
C	1.00	0.99	0.92	0.62	0.10	0.00

a Gaussian of mean 0 and width 1. The model then undergoes a spin glass transition at $T_c \approx 0.896$ [11].

Temperature chaos in MK spin glasses —. A great advantage of MK lattices is that it allows us to do renormalization procedure exactly [7]. Now let us denote a set of spins of level n as $\{S_n\}$. By tracing over spins of lower levels, we explicitly obtain

$$P_k(\{S_0\},\cdots,\{S_{G-k}\}) \equiv \frac{\text{Tr}_{\{S_{G-k+1}\}\cdots\{S_G\}}\exp(-\mathscr{H}_J^0)}{\text{Tr}_{\{S_0\}\cdots\{S_G\}}\exp(-\mathscr{H}_J^0)}$$
$$\propto \exp[-\mathscr{H}_J^k(\{S_0\},\cdots,\{S_{G-k}\})], (2)$$

$$\mathscr{H}_J^k(\{S_0\},\cdots,\{S_{G-k}\}) = -\sum_{<ij>_{G-k}} \tilde{J}_{ij}^k S_i S_j, \quad (3)$$

where the sum $\sum_{<ij>_n}$ is over all the nearest neighbor spins of a MK lattice with generation n. Effective couplings $\tilde{J}_{ij}^k$ are J_{ij}^0/T for $k=0$. Otherwise, they are calculated by the recursion formula

$$\tilde{J}_{ij}^{k+1} = \sum_{l=1}^{4}\text{arctanh}[\tanh\tilde{J}_{il}^k \tanh\tilde{J}_{lj}^k], \quad (4)$$

where $\tilde{J}_{il}^k$ and $\tilde{J}_{lj}^k$ lie on the l-th path connecting i and j.

Now let us turn to temperature chaos in MK spin glasses. If we start from the same set of bare couplings, $\{\tilde{J}_{ij}^0(T)\}$ and $\{\tilde{J}_{ij}^0(T+\Delta T)\}$ are completely correlated for any T and ΔT. Does the correlation still survive after renormalization is repeated again and again? This question was first addressed by Banavar and Bray [12], and they have found that for arbitrarily small ΔT $\{\tilde{J}_{ij}^k(T)\}$ and $\{\tilde{J}_{ij}^k(T+\Delta T)\}$ become completely decorrelated when k is large enough, indicating that spin polarizations at two temperatures are different at all. In table 1, we show the linear correlation coefficient

$$C(L,T,T') = \frac{\overline{\tilde{J}^k(T)\tilde{J}^k(T')}}{\sigma(T)\ \sigma(T')}, \quad (5)$$

for $T = 0.7$ and $T = 0.65$. In this definition, $L = 2^k$, $\overline{\cdots}$ is the disorder average, σ is the standard deviation of $\tilde{J}^k$, and we have used the fact that $\overline{\tilde{J}^k} = 0$. We see $C(L,T,T')$ rapidly drops from 1 to 0 around $L \approx 2^{11}$. We hereafter utilize $C(L,T,T')$ as an indicator of temperature chaos, and define the chaos length $\ell(T,T')$ as the value of L where $C = 1/e$. (In this case, $\ell(T,T') \approx 2^{12}$.)

Exploiting renormalization for dynamics —. Since our purpose is to examine relevancy of temperature chaos on dynamics, we should do dynamical simulations at very long time so that the length scale equilibrated during simulation is comparable with the chaos length. However, this condition is hardly satisfied by usual Monte-Carlo simulations because frustration and randomness inherent in spin glasses make their dynamics extremely slow. In fact, measurements of $L_{eq}(t)$, the equilibrated length during t, in 3d-EA spin glass model have shown that $L_{eq}(t)$ for $t = 10^6$ Monte Carlo Sweeps (MCS) is less than 10 at any temperatures below T_c [13], while it is almost impossible to go beyond the time scale by the present computer resource. This length scale seems to be hopelessly shorter than the chaos length. (Recall that $\ell(T,T') \approx 2^{12}$ in the previous case though the temperature difference is not so small.)

In order to overcome the difficulty, we exploit renormalization for dynamics [8]. The basic idea is as follows. Suppose we focus on a time window $t_{\min} \leq t \leq t_{\max}$. Between $t = 0$ and $t = t_{\min}$ the system has had time to equilibrate up to the length scale $l(t_{\min})$; essentially all out of equilibrium physics comes from larger length scales. On MK lattices, this means that the spins whose generation "level" is larger than $G_{\min}$ (with $2^{NRG} = l(t_{\min})$ and $NRG \equiv G - G_{\min}$) are in local equilibrium; the other spins have dynamics that is well described by the effective Hamiltonian $\mathscr{H}_J^{NRG}(\{S_0\},\cdots,\{S_{G_{\min}}\})$. In practice, we adopt the following procedure to implement this idea with taking temperature dependence of speed of equilibration into account.

1. Calculate the effective couplings at T_H (the higher temperature used in T-cycle protocol) and those at T_L (the lower one). We first generate a large number of bare couplings from a Gaussian of mean 0 and width 1. Then, we do renormalization procedure to produce an ensemble of effective couplings. This process is iterated NRG times. The final effective couplings are then randomly assigned to the edges of a MK lattice of size $2^{G_{\min}}$.

2. The direction of each spin at $t = 0$ is chosen randomly with equal probability, corresponding to a quench from an infinitely high temperature.

3. At T_L, we simply do standard Monte Carlo by using $\mathscr{H}_J^{NRG}(\{S_0\},\cdots,\{S_{G_{\min}}\})$ prepared at step 1.

4. At T_H, dynamics is further accelerated by

the following procedure. We first calculate $\mathcal{H}_J^{NRG'}(\{S_0\},\cdots,\{S_{G'_{min}}\})$, where NRG$'$ = $G - G'_{min}$ and NRG$'$ > NRG. Then we do Monte Carlo by using $\mathcal{H}_J^{NRG'}$ to update the spins whose level is smaller (or equal to) G'_{min}. After each MCS, the lower spins $\{S_k\}$ ($G'_{min} < k \leq G_{min}$) are locally equilibrated with fixing the spins of smaller levels.

Note that one MCS at T_L (T_H) on the renormalized lattice corresponds to a huge number of sweeps on the non-renormalized lattice, in fact to the number needed to equilibrate on the length scale 2^{NRG} ($2^{NRG'}$).

Results —. We use the standard temperature cycling protocol and measure a quantity similar to the ac-susceptibility defined as [14]

$$\chi(\omega,t) = \frac{1 - Q(t + \frac{2\pi}{\omega},t)}{T}, \quad (6)$$

where $Q(t,t') \equiv \sum_i \langle S_i(t)S_i(t') \rangle / N$ and N is the number of spins. Every MCS updates all the spins once. In the both positive and negative T-cycle simulations, we use $T_H = 0.7$ and $T_L = 0.65$. Note that these two temperatures are the same as those in Table 1. The period $2\pi/\omega$ of ac-field is 16 MCS. All the simulations are done on MK lattices with five generations ($G_{min} = 5$) using $0 \leq$ NRG ≤ 15. Since we calculate renormalized couplings at T_L and T_H from the *same* set of bare couplings, they are highly correlated for small NRG. However, their correlation vanishes for large NRG due to temperature chaos. The difference NRG$'$ − NRG is 1 for all the simulations. We hereafter denote χ with a negative (positive) T-cycle as χ_{Ncycle} (χ_{Pcycle}) and the isothermal χ at T as $\chi_{iso}(T)$.

In Fig. 2, we show three typical behaviors observed in positive T-cycle simulations. In the main frame, we omit t_2 part of data and connect t_1 and t_3 parts to compare with the isothermal data drawn by line. For small NRG, χ_{Pcycle} is remarkably below $\chi_{iso}(T_L)$ in the third stage, as illustrated in the main frame of Fig. 2(a). This means that equilibration at T_L is sharply accelerated in the second stage because renormalized couplings at T_L and T_H are strongly correlated and equilibration is accelerated at T_H. This trend (acceleration of equilibration) arises until NRG ≈ 8. Fig. 2(b) shows that χ_{Pcycle} begins to have a strong curvature in the third stage as a sign of rejuvenation, while renormalized couplings at T_L and T_H are still highly correlated. (As shown in Table 1, $C = 0.92$ for NRG = 9.) However, χ_{Pcycle} is still below $\chi_{iso}(T_L)$ at later times of the third stage. Finally, Fig. 2(c) is the result for NRG = 13. Renormalized couplings are very decorrelated now ($C = 0.10$). We see strong rejuvenation in the third stage, as is found in experiments. We have checked that rejuvenation is *perfect* in the sense that χ_{Pcycle} in the first stage and that in the third stage com-

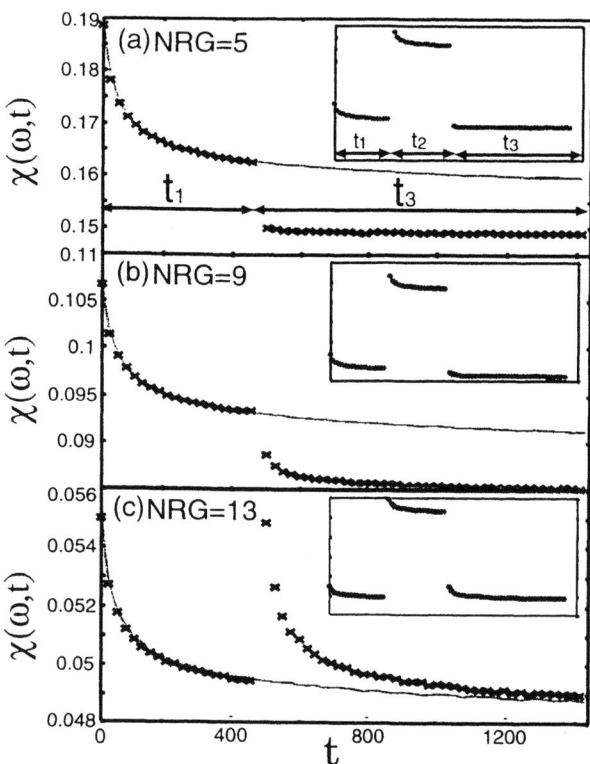

FIGURE 2. $\chi(\omega,t)$ with a positive T-cycle (crosses) for NRG = 5, 9 and 13 (inset). Temperature is temporarily increased from $T_L = 0.65$ to $T_H = 0.7$ for t_2. In the main frame, t_1 and t_3 parts are connected after omitting t_2 part. The line is the isothermal data at T_L. The average is from 6×10^3 samples.

pletely overlap. We have also found that χ_{Pcycle} in the second stage and $\chi_{iso}(T_H)$ collapse into a single curve for any NRGs, meaning that aging at lower temperature is not helpful in equilibration at higher temperature.

For negative T-cycle case, we only show two extreme cases in Fig. 3, i.e., a highly correlated case and a highly decorrelated one. Now a surprising fact is that *perfect* memory is observed in the both cases. Especially, in the decorrelated case (NRG = 13), memory appears in the third stage though rejuvenation is observed for positive T-cycle (Fig. 2(c)). This result is very contrast with [8] which has shown that negative T-cycle leads to strong (but not perfect) rejuvenation when NRG = NRG$'$. These findings tell us that temperature dependence of speed of equilibration is crucial for memory effect. Lastly, we have compared χ_{Ncycle} in the second stage with χ_{Pcycle} in the third stage, and found that they perfectly overlap for any NRGs. Since their difference is whether previous aging at T_L exists or not, this result suggests that equilibration at higher temperature makes previous aging at lower temperature insignificant.

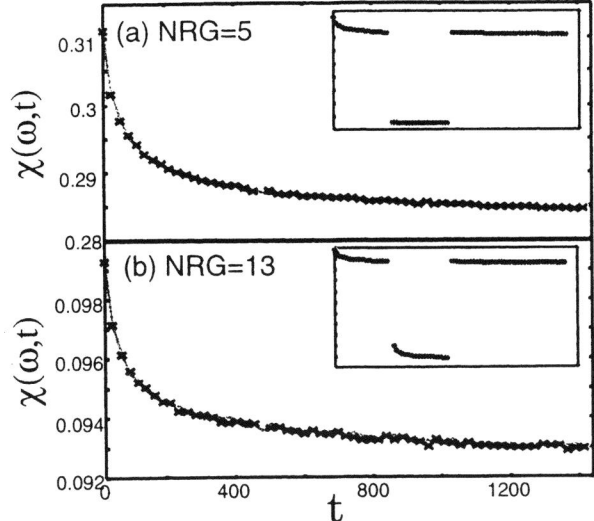

FIGURE 3. $\chi(\omega,t)$ with a negative T-cycle for NRG = 5 and 13. Temperature is temporarily decreased from $T_H = 0.7$ to $T_L = 0.65$ for t_2. In the main frame, t_1 and t_3 parts are connected after omitting t_2 part. The line is the isothermal data at T_H. The average is from 6×10^3 samples.

Discussion and conclusions —. In this work, we have found that positive T-cycle and negative one cause quite different effect, i.e., rejuvenation in the former and memory effect in the latter, if we take temperature dependence of speed of equilibration into account. In fact, the both lead to strong rejuvenation when renormalized couplings are decorrelated and NRG = NRG$'$ [8]. These results are interpreted as follows. Concerning negative T-cycle, spin polarizations are reconstructed only at shorter length scales in the second stage. Since the structure at larger length scales (polarizations of spins with smaller generations in our model) created in the first stage is not destroyed at the time, we see memory in the following third stage. On the other hand, for positive T-cycle, the length scale reconstructed in the second stage is larger than the equilibrated length scale in the first stage. As a result, the order created in the first stage is completely destroyed, and we see rejuvenation. This scenario is very similar to the picture of [15] where separation of the relevant length scale at each temperature plays a crucial role in memory and rejuvenation. However, the only difference is that temperature chaos exists for sure in MK spin glasses, while their picture relies on not temperature chaos but reweighing of Boltzmann factor by temperature variations as the cause of rejuvenation.

Lastly, we comment on acceleration of equilibration observed in positive T-cycle for small NRGs. In experiments, this behavior is observed in many glassy systems like polymer glasses [16] and orientational glasses [17]. Furthermore, the same behavior is also observed in spin glasses if the equilibrated length scale during the second stage is much smaller than $\ell(T,T')$. (See Fig. 14 of [3].) These results may suggest that temperature chaos is absent, or the equilibrated length scale in experimental time scale is much smaller than $\ell(T,T')$ for these systems.

ACKNOWLEDGMENTS

M. S. was partially supported by the Japan Society for the Promotion of Science for Japanese Junior Scientists. The present simulations have been performed on SGI 2800/384 at the Supercomputer Center, Institute for Solid State Physics, the University of Tokyo. This work was also supported in part by the European Community under contract HPRN-CT-2002-000307 (Dyglagemem). M. S. acknowledges support from the MENRT while he was in France. The LPTMS is an Unité de Recherche de l'Université Paris XI associée au CNRS.

REFERENCES

1. Vincent, E. et al., "Slow Dynamics and Aging in Spin-Glasses," in *Complex Behaviour of Glassy Systems*, edited by E. Rubi, Springer-Verlag, 1997, pp. 476.
2. Nordblad, P., and Svedlindh, P., "Experiments on Spin Glasses," in *Spin Glasses and Random Fields*, edited by A. P. Young, World Scientific, Singapore, 1998, pp. 1-27.
3. Jönsson, P. E. et al., cond-mat/0307640 (2003).
4. Vincent, E., Bouchaud, J.-P., Hammann, J., and Lefloch, F., *Phil. Mag. B* **71**, 489-500 (1995).
5. Bray, A. J., and Moore, M. A., *Phys. Rev. Lett.* **58**, 57-60 (1987).
6. Yoshino, H., Lemaître, A., and Bouchaud, J.-P., *Eur. Phys. J. B* **20**, 367-395 (2001).
7. Southern, B. W., and Young, A. P., *J. Phys. C* **10**, 2179-2195 (1977).
8. Sasaki, M., and Martin, O. C., *Phys. Rev. Lett.* **91**, 097201 (2003).
9. Scheffler, F., Yoshino, H., and Maass, P., *Phys. Rev. B* **68**, 060404(R) (2003).
10. Edwards, S. F., and Anderson, P. W., *J. Phys. F* **5**, 965-974 (1975).
11. Ney-Nifle, M., and Hilhorst, H., *Phys. Rev. Lett.* **68**, 2992-2995 (1992).
12. Banavar, J. R., and Bray, A. J., *Phys. Rev. B* **35**, 8888-8890 (1987).
13. Komori, T., Yoshino, H., and Takayama, H., *J. Phys. Soc. Jpn.* **68**, 3387-3393 (1999).
14. Komori, T., Yoshino, H., and Takayama, H., *J. Phys. Soc. Jpn. Suppl. A* **69**, 228-237 (2000).
15. Bouchaud, J.-P., Dupuis, V., Hammann, J., and Vincent, E., *Phys. Rev. B* **65**, 024439 (2001).
16. Bellon, L., Ciliberto, S., and Laroche, C., cond-mat/9905160 (1999).
17. Alberich, F., Doussineau, P., and Levelut, A., *J. Phys. I (France)* **7**, 329-348 (1997).

Searching for Backbones – An Efficient Parallel Algorithm for Finding Groundstates in Spin Glass Models

Johannes J. Schneider

Department of Physics, Johannes Gutenberg University of Mainz, Staudinger Weg 7, 55099 Mainz, Germany

Abstract. Comparing different good solutions for an optimization problem, one usually finds structures which are common to all solutions. Thus, one can assume that these "backbones" are part of any good solution and also of the global optimum of the considered problem. For spin glass models like the Sherrington-Kirkpatrick-model, these backbones consist of spins which are parallel and antiparallel, rsp., to each other in all solutions. The Searching for Backbones algorithm produces a number of good solutions in a first iteration, finds these backbones, and holds them constant in the next iteration in which better solutions are found which have more structures in common. This approach is iterated until all optimization runs end at the same solution.

INTRODUCTION

Spin glasses are unordered magnetic materials which are dominated by frustration effects due to the competing interactions between the single spins. A widely used model for spin glasses is the Sherrington-Kirkpatrick-model (short: SK-model) [1]: the Hamiltonian of this model is given by

$$\mathcal{H} = -\sum_{i,j} J_{ij} S_i S_j \qquad (1)$$

with N Ising spins $S_i = \pm 1$ and interactions J_{ij} which are Gaussian distributed around zero. Thus, all spins interact with each other, the interaction between two spins can be either positive or negative.

This model can be considered as an optimization problem: the task is to find a setting of the spins S_i for a proposed interaction matrix J such that the energy is minimum. This optimization problem is NP-complete, i.e., there is no exact algorithm solving this problem in polynomial time. As the number of configurations explodes exponentially with the system size N, only relatively small configurations can be solved exactly with mathematical algorithms such as Branch&Cut. For larger instances, heuristics have to be used for finding optimum or at least very good configurations.

In computational physics, Simulated Annealing [2] has become the main heuristics for finding quasi optimum configurations of complex problems: starting out from a random configuration σ_0, a series of moves $\sigma_i \to \sigma_{i+1}$ is applied to the system. Usually, these moves are accepted or rejected according to the Metropolis criterion with the transition probability [3]

$$W(\sigma_i \to \sigma_j) = \begin{cases} 1 & \text{if } \Delta\mathcal{H} \leq 0 \\ \exp\left(-\frac{\Delta\mathcal{H}}{kT}\right) & \text{otherwise} \end{cases} \qquad (2)$$

with the energy difference $\Delta\mathcal{H} = \mathcal{H}(\sigma_j) - \mathcal{H}(\sigma_i)$.

There are also other optimization algorithms which are closely related to Simulated Annealing and which differ only in the choice of the acceptance criterion: Threshold Accepting [4] is a deterministic variant [5] of Simulated Annealing and accepts a move with the acceptance criterion

$$W(\sigma_i \to \sigma_j) = \begin{cases} 1 & \text{if } \Delta\mathcal{H} \leq T \\ 0 & \text{otherwise} \end{cases} . \qquad (3)$$

Thus, the outline of Simulated Annealing and related algorithms is as follows:

- Simulated Annealing then starts out at a randomly created configuration σ_0 at some large initial temperature.
- Then a loop over several temperature steps is performed:
  - In each temperature step, the system performs a series of moves:
    * Starting from the current configuration σ_i, the system creates a tentative new configuration σ_{i+1}.
    * This move is either accepted or rejected according to the acceptance criterion of the algorithm:
      · In case of acception, the system moves to the new configuration σ_{i+1},

such that σ_{i+1} becomes the current configuration.
- In case of rejection, the system stays in σ_i. One sets then $\sigma_{i+1} := \sigma_i$.
- After a certain number of moves, the temperature is decreased to some extent.
• The final configuration is printed as the result of the optimization algorithm.

There are various ways how to change a configuration with a move. Using the Local Search approach, the configuration is only changed slightly, e.g., by a spinflip $S_i \to -S_i$. But one can also change the system to a larger extent by changing several spins at the same time [6].

MOTIVATION FOR SEARCHING FOR BACKBONES

Simulated Annealing is a heuristic optimization algorithm which leads to different quasi optimum configurations if different seeds for the random number generator are used and if the problem is very difficult to solve. However, if comparing these solutions, which were independently generated, one finds that these solutions have many structures in common: the nature of these structures depends on the considered optimization problem. For example, in sequencing problems like the Traveling Salesman Problem, in which optimum sequences of nodes have to be found, these structures are partial sequences of a few nodes which are common to all solutions [7, 8, 9].

In a spin glass model like the SK-model, each spin S_i can take the two values $S_i = +1$ and $S_i = -1$. Here such a structure is given by a group of spins which are parallel or antiparallel to each other in all solutions, as shown schematically in Fig. 1, in which groups of such spins can be identified.

Usually, after already having a "quite good" solution for the proposed optimization problem, one performs further optimization runs in order to achieve an "even better" solution. But it would be a waste of calculation time if these structures, which were found in all previous solutions, have to be determined again and again in further optimization runs. As they were already found by several independently performed optimization runs, they can be assumed to be part of any very good configuration and also of the optimum solution.

Thus, a way has to be found to tell further optimization processes about the existence of these backbones which shall be kept constant during the whole optimization runs. Thus, the computing time can be concentrated on parts of the system which are obviously more difficult to solve in an optimum way. Then the new solutions are on average better than the previous solutions and have more structures in common than those. Thus, a new set of backbones has to be determined by comparing the new solutions. It might be that two or more backbones can be connected to one backbone, such that the number of backbones decreases and the backbones grow in size. Then again new optimization runs are performed in which these new backbones are kept constant. This approach is iterated until all optimization runs end up at the same solution.

OUTLINE OF THE ALGORITHM

Summarizing, the outline of this algorithm if applied to a spin glass model with Ising spins is as follows:

- M independent optimization runs for the proposed spin glass instance are performed, leading to M solutions $\sigma_\mu = (S_1^\mu, S_2^\mu, \ldots)$, $\mu = 1, \ldots, M$. These optimization runs can be most efficiently performed in parallel.
- Then the solutions are compared for common structures:
  - An overlap matrix $\eta(i,j)$ between the single spins is created:

  $$\eta(i,j) = \text{abs}\left(\sum_{\mu=1}^{M} S_i^\mu S_j^\mu\right) \quad (4)$$

  If $\eta(i,j) = M$, then either $S_i = S_j$ in all solutions or $S_i = -S_j$ in all solutions.

  - If $\eta(i,j) = M$ then the spins i and j can be connected to one backbone. This backbone can be expressed as a block spin α. The Hamiltonian of the SK-model can then be rewritten with these block spins as

  $$\mathcal{H} = -\sum_{\alpha,\beta} \mathcal{J}_{\alpha\beta} \mathcal{S}_\alpha \mathcal{S}_\beta \quad (5)$$

  with

  $$\mathcal{J}_{\alpha\beta} \mathcal{S}_\alpha \mathcal{S}_\beta = \sum_{\substack{i \in \alpha \\ j \in \beta}} J_{ij} S_i^\nu S_j^\nu \quad (6)$$

  with some chosen spin configuration $\sigma^\nu = (S_1^\nu, S_2^\nu, \ldots)$.
  There remains some arbitrariness in the new interaction matrix $\mathcal{J}$ due to the explicit choice of the signs of the backbone spins $\mathcal{S}_\alpha$. Here we set $\mathcal{S}_\alpha = 1$ by which we decide also about the signs of the $\mathcal{J}_{\alpha\beta}$.

- The spins S_i are replaced by the backbone spins $\mathcal{S}_\alpha$, the interaction matrix J is replaced by the matrix $\mathcal{J}$.

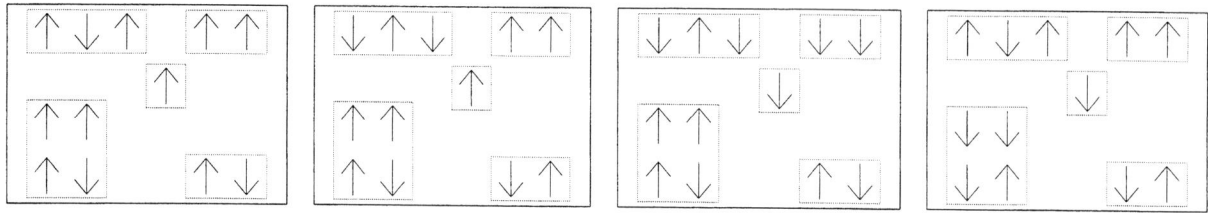

FIGURE 1. Four different solutions of a spin glass instance with 12 spins: five groups of spins can be identified which are common to all solutions.

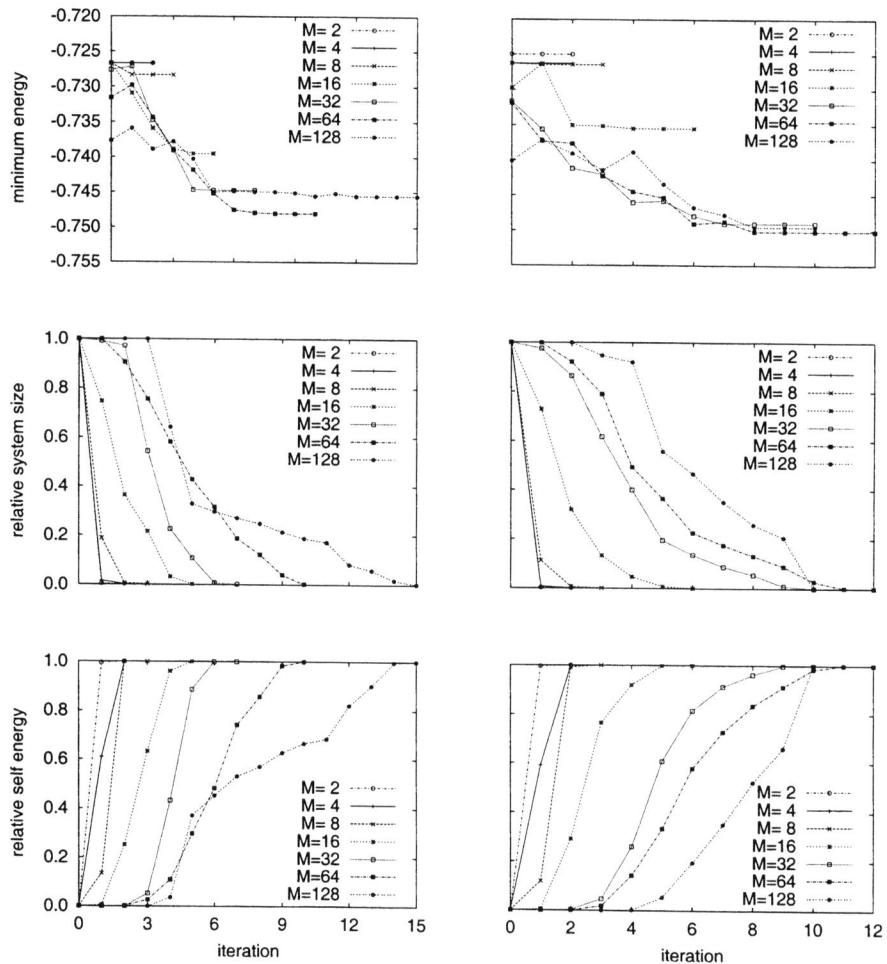

FIGURE 2. Solution of two instances of the SK-model (left: instance with 500 spins, right: instance with 1000 spins) with the Searching for Backbones algorithm: at the top, the decrease of the minimum energy found in each iteration is shown. In the middle, the relative system size, i.e., the number of spins left divided through the original number of spins, is shown. At the bottom, the relative self energy according to formula (7) is shown.

- The algorithm ends if only one spin is left and $\mathcal{H}(\sigma^\nu) = -\mathcal{J}_{11}$, i.e., if the whole energy is stored in the self interaction of the only remaining spin. Otherwise, the algorithm jumps back to step 1.

Please note that the interaction matrix $\mathcal{J}$ between the block spins is still symmetric, but self energies of the block spins are stored as the diagonal elements $\mathcal{J}_{\alpha\alpha}$.

COMPUTATIONAL RESULTS

Figure 2 shows computational results for two spin glass instances, one with 500 spins and one with 1000 spins. The underlying optimization algorithm used was Threshold Accepting with large moves [6] with which M solutions were generated in optimization runs which were performed independently of each other on a parallel computer. We see at first sight that the Searching for Backbones algorithm converges rather rapidly within only a few iterations, in contrast to the convergence results found for the Traveling Salesman Problem [7, 9] where the algorithm did not converge for large M. Furthermore, we find that the algorithm takes more iterations to converge for larger numbers of compared solutions but that also the quality of solutions found is better for larger M. Thus, we find that the Searching for Backbones algorithm is really able to improve the results found with the underlying heuristics.

The convergence can best be seen by looking at the system size, which is shown in the middle of Fig. 2: as the single spins are replaced by fewer backbone spins, the system size decreases gradually. The relative system size is given by the number of backbone spins in each iteration divided through the original number of spins. A further parameter measuring the convergence of the algorithm is the relative self energy

$$r = \frac{\sum_\alpha |\mathcal{J}_{\alpha\alpha}|}{\sum_{\alpha,\beta} |\mathcal{J}_{\alpha\beta}|}, \quad (7)$$

which is a measure for the energy fraction which is already part of the self energies of the backbone spins. This is a measure how much the algorithm can still optimize.

SUMMARY AND REMARK

The Searching for Backbones algorithm compares different solutions of a proposed optimization problem for common parts called backbones and holds them constant in further optimization runs. Thus, the system size and also the complexity of the problem instance are reduced. This approach is iterated until all solutions are identical at the end.

Please note that the insight of Albert Einstein that "Common sense is the collection of prejudices acquired by age eighteen." does not apply for our computational results. In the real world, there are no real independencies, everybody is in contact with other persons such that prejudices can spread and are even given from one generation to the next one. Furthermore, the human mind is selective and weighs experiences stronger which fit in already existing schemes, such that prejudices keep rather stable. Here however we have no such existing schemes and we have no information from outside, such that we can really trust in that these structures are something special of the problem instance, if no construction heuristics or other intelligence was used and if the optimization runs were really performed independently of each other.

ACKNOWLEDGMENTS

I kindly acknowledge large grants of computing time on the Intel paragons at the John von Neumann-Institute of Computing (former Höchstleistungsrechenzentrum) at the Forschungszentrum Jülich and the Swiss Federal Institute of Technology Zurich (ETHZ) 1995-1999. Furthermore, I want to thank K. Binder, S. Kirkpatrick, S. Kobe, U. Krey, I. Morgenstern, J. M. Singer, and E. P. Stoll for fruitful discussions.

REFERENCES

1. Sherrington, D., and Kirkpatrick, S., *Phys. Rev. Lett.* **35**, 1792-1796 (1975).
2. Kirkpatrick, S., Gelatt Jr., C.D., and Vecchi, M.P., *Science* **220**, 671-680 (1983).
3. Metropolis, N., Rosenbluth, A.W., Rosenbluth, M.N., Teller, A.H., and Teller, E., *J. Chem. Phys.* **21**, 1087-1092 (1953).
4. Dueck, G., and Scheuer, T., *J. Comp. Phys.* **90**, 161-175 (1990).
5. Moscato, P., and Fontanari, J.F., *Phys. Lett. A* **146**, 204-208 (1990).
6. Schneider, J., Stoll, E.P., Meier, P.F., Schrimpf, G., Stamm-Wilbrandt, H., Dueck, G., Morgenstern, I., and Kobe, S., "Optimization of Spin Glasses with Ruin & Recreate," preprint, 2001.
7. Schneider, J., Froschhammer, Ch., Morgenstern, I., Husslein, Th., and Singer, J.M., *Comput. Phys. Comm.* **96**, 173-188 (1996).
8. Schneider, J., Britze, J., Ebersbach, A., Morgenstern, I., and Puchta, M., *Int. J. Mod. Phys. C* **11**, 949-972 (2000).
9. Schneider, J., *Future Generation Computer Systems* **19**, 121-131 (2003).

Convective flow driven by chemical reaction

Hiroyuki Kitahata*, Nobuyuki Magome* and Kenichi Yoshikawa*

Department of Physics, Graduate School of Science, Kyoto University

Abstract. The generation of convective flow by a chemical wave was studied experimentally on Belousov-Zhabotinsky (BZ) reaction medium. A propagating chemical wave causes a transient increase in interfacial tension, which induces convection in the bulk phase. The observed flow profile was reproduced with a numerical simulation by coupled equation of a Navier-Stokes equation including interfacial tension with the chemical kinetics, named Oregonator. The present experimental system serves as a model toward the deeper understanding on the mechanism of direct coupling of chemical energy with mechanical motion.

Molecular machineries in the living organisms convert chemical energy into mechanical work under isothermal conditions. This mechanism is quite different from that of the thermal engines, which are often discussed under the framework of the Carnot cycle on thermodynamics. In order to elucidate the mechanism of chemo-mechanical energy conversion, many researchers have been studying both experimentally and theoretically. However, in spite of numerous past effects, it remains to be an open problem. We are approaching this problem by building and analyzing a model experimental system in which chemical energy is converted into mechanical work or vectorial motion. In the present study, we adopted Belousov-Zhabotinsky (BZ) reaction, an oscillatory chemical reaction, which is often used as a experimental model of the nonequilibrium open systems (dissipative systems) [1].

It was reported that the interfacial tension of BZ medium changes synchronized with the chemical oscillation, that is, BZ medium in the oxidized state has higher interfacial tension than that in the reduced state [2]. Driven by the repetitive change in interfacial tension, it was found that Marangoni convection is induced at the free surface of BZ medium [3, 4]. However, the convection is induced on the perpendicular plane, so the gravitational effects are unavoidable. Here, we adopted a system made of two bulk phases of BZ reaction medium and oleic acid (oil), and observed the flow near the interface of the two phase by adding small particles for visualization. Due to the change in interfacial tension, Marangoni convection was induced synchronizing with the chemical wave propagation (Fig. 1). Near the interface between the both phases, the strong convective flow was observed toward the oxidized area, and two rolls were formed in the bulk phases. In addition, the chemical wave propagated slower than that in the bulk phase. This convection was induced in the horizontal plane, so there were no gravitational effects on this phenomenon. The details of these experiments are in Ref.[5].

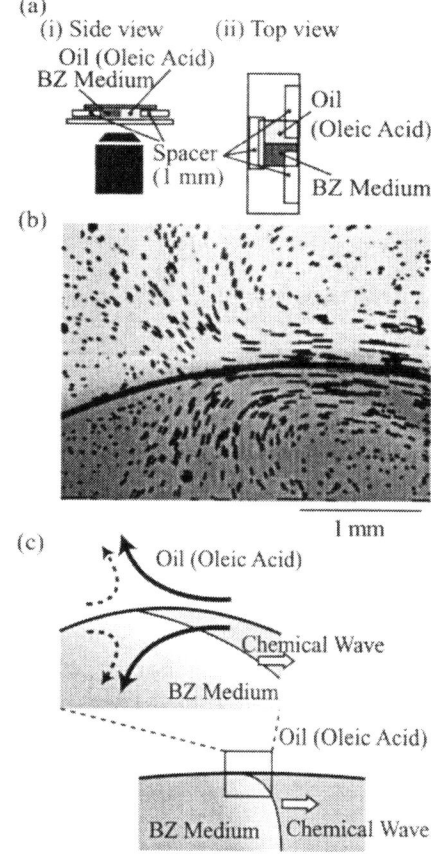

FIGURE 1. Experimental results of convection induced by BZ reaction. (a) Experimental system. (b) Streamline for 1 s and (c) the schematic representation. Convective flow is induced toward the position where the chemical wave touched the interface [5].

For a numerical simulation, we adopted an Oregonator, which has often been used for reproducing BZ reaction. In many past studies, the pattern in the settled BZ reaction medium is described as the reaction-diffusion equation using Oregonator as a reaction term. However, in this system, the medium can move. Therefore, we have to consider the dynamics of the medium, or field, itself. Consequently, we adopted Navier-Stokes equation for the movement of the medium. The medium is affected by the interfacial tension due to the change in the chemical components coupled with BZ reaction. Thus, we can write as follows[3]:

$$\rho \left(\frac{\partial}{\partial t} + \mathbf{v} \cdot \nabla \right) \mathbf{v} = \eta \nabla^2 \mathbf{v} - \nabla p + F_i, \quad (1)$$

$$\nabla \cdot \mathbf{v} = 0, \quad (2)$$

$$\left(\frac{\partial}{\partial t} + \mathbf{v} \cdot \nabla \right) U = F(U,V) + D_U \nabla^2 U, \quad (3)$$

$$\left(\frac{\partial}{\partial t} + \mathbf{v} \cdot \nabla \right) V = G(U,V) + D_V \nabla^2 V, \quad (4)$$

$$F(U,V) = \frac{1}{\varepsilon} \left\{ U(1-U) - fV \frac{U-q}{U+q} \right\}, \quad (5)$$

$$G(U,V) = U - V, \quad (6)$$

where ρ is the density, $\mathbf{v}$ is the velocity of the fluid, p is the pressure, η is the viscosity, U and V are the concentrations of represented chemical reagents, and f, ε, and q is the parameters corresponding to the nature of BZ medium. Here, interfacial tension is assumed to be as a volume force, which has non-zero value in the area within the distance, δ, from the interface. The interfacial tension is proportional to the gradient of the concentration of oxidized catalyst, V:

$$F_i = \begin{cases} k \frac{\partial V}{\partial x} \mathbf{e}_x & (0 < y < \delta), \\ 0 & (y > \delta), \end{cases} \quad (7)$$

where k is a constant and $\mathbf{e}_x$ is the unit vector in the x direction.

In fig. 2, the profiles in the bulk phase calculated using the above equations are shown. Near the interface, toward the point where chemical wave touches the interface. The chemical wave near the interface propagate slower than that in the bulk phase. These characteristics are similar to those of the experimental results.

The generation of convection as shown in the present study serves as a simple representation on the transduction from the chemical energy to vectorial motion. To extend such an idea further, we developed another system, in which a small droplet of BZ medium was floated on an

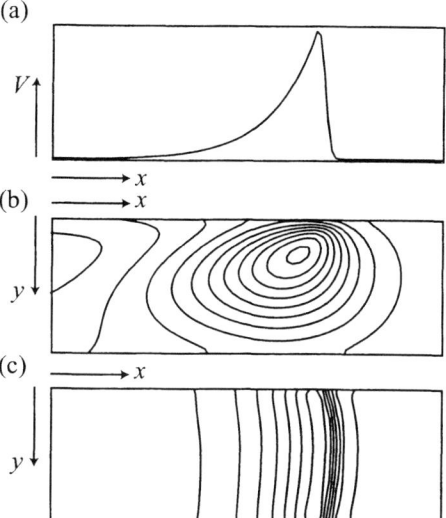

FIGURE 2. Results of numerical simulation. (a) Profiles of a traveling chemical wave represented as the concentration of the inhibitor (catalyst in the oxidized state), V, at the interface, (b) The streamline in the bulk phase, and (c) the chemical wave in the bulk phase are shown. The experimental trends have been thus represented in a qualitative manner. The parameters used in this simulation are $\rho = 1.0$, $\eta = 3.0$, $D_U = D_V = 0.5$, $\varepsilon = 0.04$, $f = 3.0$, $q = 0.0008$, $k = 0.01$, and $\delta = 0.5$.

oil phase. We observed the droplet spontaneously oscillated synchronized with chemical reaction in it (data not shown) [5]. BZ reaction is known to generate little heat, and the system can be thought as an isothermal system. In other words, we can build up a system in which chemical energy is directly converted into mechanical work. Analyzing this experimental model system may help further understanding of biological molecular machineries.

The authors thank Professor Masaharu Nagayama (Kyoto university, Japan), Dr. Ryoichi Aihara (RIKEN, Japan) and Dr. Takatoshi Ichino (Kyoto university, Japan) for their helpful discussion on this subject. This study is partly supported by a Grant-in-Aid for JPSJ fellows from the Ministry of Education, Culture, Sports, Science and Technology of Japan (No. 5490).

REFERENCES

1. Kapral, R., and Showalter, K., *Chemical Waves and Patterns*, Kluwer Academic, Dordrecht, 1995.
2. Yoshikawa, K., Kusumi, T., Ukitsu, M., and Nakata, S., *Chem. Phys. Lett.*, **211**, 211-213 (1993).
3. Miike, H., Müller, S. C., and Hess, B., *Phys. Rev. Lett.*, **61**, 2109-2112 (1988).
4. Matthiessen, K., Wilke, H., and Müller, S. C., *Phys. Rev. E*, **53**, 6056-6060 (1996).
5. Kitahata, H., Aihara, R., Magome, N., and Yoshikawa, K., *J. Chem. Phys.*, **116**, 5666-5672 (2002).

Crack Patterns in Drying Process Show Memories Contained Inside Granular Networks

A. Nakahara and Y. Matsuo

Laboratory of Physics, College of Science and Technology, Nihon University, Funabashi 274-8501, Japan

Abstract. We perform experiments in which we dry the mixture of powder and water. When the proportion of water to powder in the mixture is high and the mixture is Newtonian viscous fluid, crack patterns that will appear after the drying process are cellular. On the other hand, when the proportion of water to powder is low and thus the mixture becomes visco-plastic with a non-zero yield stress, the morphology of crack patterns can change from homogeneous cellular patterns to inhomogeneous patterns. For example, if we add initial vibration to the container of the visco-plastic mixture, we get a laminar crack pattern, the direction of which is perpendicular to the initial external vibration. That is, when the mixture is visco-plastic, the granular network inside the mixture remembers the direction of the initial external vibration.

INTRODUCTION

When we mix powder with water, pour the mixture into a container and keep them in an air-conditioned room of fixed temperature and humidity, the crack pattern appears as the mixture is dried. In many cases, the crack patterns are homogeneous cellular patterns [1].

However, when we apply an external vibration to the container as soon as we pour the mixture into the container, the morphology of crack patterns changes drastically. In this manuscript, we show the morphological change of the crack pattern as we change both the direction of the initial external vibration and the proportion of water to powder in the mixture, and investigate the mechanism of how granular network inside the mixture keeps the memory of the initial external vibration.

MORPHOLOGICAL CHANGE

We perform experiments using powder of Calcium Carbonate ($CaCO_3$). We mix powder with water, pour the mixture into acryl dishes or boxes and keep them at the condition of 23 °C in temperature and 23% in humidity.

First, we prepare the mixture by setting the value of the proportion of water to powder as 2, so that the mixture becomes visco-plastic non-Newtonian fluid with non-zero yield stress. In Figure 1, we apply the initial external vibration to the container in the horizontal direction and get the laminar crack pattern. When we apply the external vibration in the circular direction, a radial crack pattern emerges, as is shown in Fig. 2.

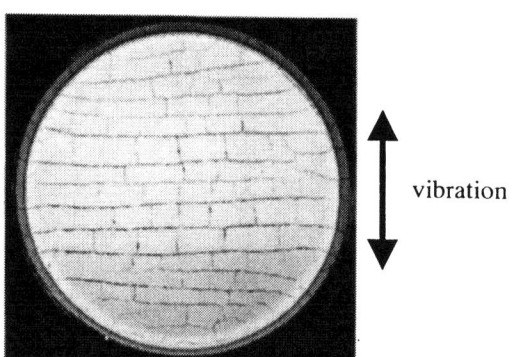

FIGURE 1. Laminar Crack Pattern [2].

Notice that the direction of the laminar crack pattern is always perpendicular to the direction of the initial external vibration [2]. That is, the granular network of powder inside the mixture remembers the direction of the initial external vibration, and, as a result, the morphology of the crack pattern changes.

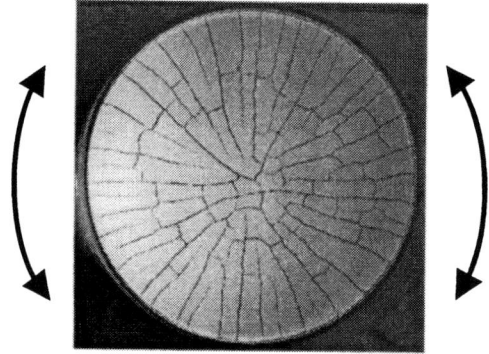

FIGURE 2. Radial Crack Pattern [2].

CHANGING THE RHEOLOGY OF THE MIXTURE

Next, we change the proportion of water to powder in the mixture. In following experiments, we fix the amount of powder per unit base of the container so that the height of the bulk of powder after it is dried can be kept constant. As a control parameter, we change the proportion of water to powder in the mixture, because the rheology of the mixture changes drastically as the proportion of water to powder changes. Decreasing the value of the proportion of water to powder, the fluidity of the mixture changes from the water-rich Newtonian viscous fluid to the water-poor non-Newtonian visco-plastic fluid with a non-zero yield stress.

In Fig. 3, we prepare the mixtures with different values of the proportion of water to powder, and these values are listed in Table 1.

FIGURE 3. Crack Pattern of Calcium Carbonate. The value of the proportion of water to powder is 1/2 (top-left), 2/3 (top-right), 1 (bottom-right), 2 (bottom-left), respectively [3].

TABLE 1. Value of the water fraction and the yield stress of each mixture in Fig.3.

Box in Fig. 3	Proportion of Water to Powder	Yield Stress (Pa)	Crack Pattern
Top-left	1/2	2.2	Laminar
Top-right	2/3	1.8	Laminar
Bottom-right	1	0.2	Laminar
Bottom-left	2	0.0	Cellular

We apply the initial external vibration in horizontal direction. We find that, when the proportion of water to powder is high and the mixture is the Newtonian viscous fluid with no yield stress, the crack pattern is cellular pattern with no memory effect. On the other hand, when the proportion of water to powder is low and the mixture is visco-plastic with a non-zero yield stress, crack patterns become laminar, showing that they remember the direction of the initial external vibration.

CONCLUDING REMARKS

We have performed the experiments of drying the mixture of powder and water, and found that the rheology of the mixture plays an important role in the memory effects.

ACKNOWLEDGMENTS

We would like to thank H. Uematsu, S. Sasa, T. S. Komatsu, and M. Ohtsuki for valuable discussions.

REFERENCES

1. Groisman, G., and Kaplan, E., *Europhys. Lett.* **26**, 415-420 (1994).

2. Nakahara, A., and Matsuo, Y., *Bussei Kenkyu (Kyoto)* **74**, 650-653 (2000).

3. Nakahara, A., and Matsuo, Y., *Bussei Kenkyu (Kyoto)* **81**, 184-185 (2003).

Diffusing-Wave Spectroscopy of Gelling Dairy Systems

Y. Hemar[1,*], P. Hebraud[2], R. Sarcia[2] and D. N. Pinder[3]

1. Institute of Food, Nutrition and Human Health, Massey University, Palmerston North, New Zealand
2. ESPCI, 10 Rue Vauquelin, F-75005 Paris, France
3. Institute of Fundamental sciences Massey University, Palmerston North, New Zealand

Abstract. Conventional diffusing-wave spectroscopy (DWS) was combined with a CCD-camera based diffusing-wave spectroscopy (MSDWS) to study acid induced skim milk gels. This allowed the measurement of intensity autocorrelation functions from 10^{-6} to 3×10^3 s, giving access to both fast and slow dynamic modes of these gels.

INTRODUCTION

Milk is a colloidal dispersion containing casein micelles with diameters in the range 50-300 nm. These micelles are stabilised by κ-casein, located at their surfaces, through an electro-steric stabilizing mechanism. The removal of electrostatic charges by acidification induces destabilisation and the aggregation of the casein micelles, this route is used in yogurt manufacture.

In the past, several laboratory techniques have been used to study acid-induced aggregation of casein micelles, particularly rheology and light scattering. However, rheology is intrusive as it requires the application of a mechanical stress or strain to characterise the milk gel and conventional light scattering techniques demand extensive sample dilution.

Diffusing-wave spectroscopy (DWS) [1] is a non-intrusive light scattering technique not requiring sample dilution. This technique has proved to be very useful in the investigation of casein micelle aggregation [2]. In the present work we further extend the use of DWS by combining conventional DWS and a CCD-camera based DWS [3] to investigate acidified milk gels.

MATERIALS AND METHODS

Reconstituted skim milk solutions with total solids were made by dissolving low heat skim milk powder (SMP) in distilled water using a magnetic stirrer. To ensure full hydration, the dispersions were stored overnight at 4°C. Acidified milk gels were prepared by the addition of 1.3 wt% of Gluco-δ-lactone (GDL) (Sigma Chemical Co, St. Louis, MO), stirring for 2 minutes and incubating at 30°C for 16 hours.

Confocal laser microscopy (CSLM) was performed using a Leica TCS 4D confocal microscope (Leica Lasertechnik GmbH, Heidelberg, Germany) with a 100 mm oil immersion objective lens. Excitation was with the 488 nm line of an air-cooled Ar/Kr laser. The sample was stained with Fast Green (Sigma Chemical Co, St. Louis, MO), placed in a glass slide cavity, covered with a coverslip and incubated at 30 °C for 16 hours before observation under the microscope.

A conventional DWS set-up used in this study consisted of a high-power Melles Griot (Carlsbad, USA) HeNE Laser, operating at 632.8 nm and delivering 35 mW as a light source. The laser beam was expanded to approximately 8 mm diameter at the sample cell. The transmitted light was collected by a single-mode fiber (P1-3224-PC-5, Thorlabs Inc. Germany) fitted with GRIN lenses (F230FC-B FC, Thorlabs Inc. Germany). The optic fibre was connected to a photomultiplier tube (Hamamatsu HC120-08 module from correlator.com) and a Malvern 7132 correlator was used to obtain the correlation function.

The CCD-camera based DWS (MSDWS) system [2] uses a Leutron LV-75 monochrome CCD camera which is run at 25 frames per second. The camera is placed 10 cm behind the diaphragm. The images are transferred to a computer running at 2.2 GHz using a National Instruments data acquisition board IMAQ-1409.

Both conventional DWS and MSDWS measurement were performed simultaneously in the backscattering geometry.

RESULTS AND DISCUSSION

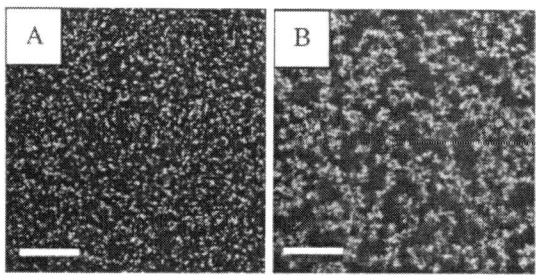

Figure 1. Confocal micrographs of GDL acidified milk gels containing (A) 5 wt%, (B) 10 wt% skim milk. Scale bare = 20 μm.

The milk proteins were stained so they appear white in the confocal micrographs. It is clear from the micrographs that the casein micelles have aggregated, causing the formation of an inhomogeneous system with voids (see Fig. 1).

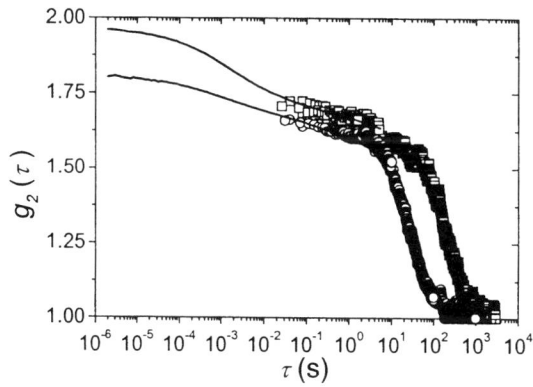

Figure 2. Intensity auto-correlation function of GDL acidified milk gels containing 5 wt% skim milk (— ○) and 10 wt% skim milk (— □). Conventional DWS (line) and CCD-based MSDWS (symbols).

Fig. 2 shows the DWS measurements obtained by combining both the conventional DWS (line) and the MSDWS (open symbols). It can be seen that the values of intensity auto-correlation function $g2(t)$ at zero time are not close to 2. Nonetheless information about the dynamics of these gels can still be derived.

At both concentrations, two relaxation modes were clearly observed. A fast mode of $6 \cdot 10^{-4}$ s for both samples, and a slow mode at 15s and 153 s for 5 wt% and 10 wt% skim milk gels respectively. We are tempted to advance that one mode could be associated with the dynamic of the casein micelles within the aggregates, while the other mode is associated with the gel network.

CONCLUSION

Both the fast and slow modes of a milk system gelled by acidification were observed. Overall, this work demonstrated clearly the usefulness of combining DWS and MSDWS to study the gelation of dairy systems.

ACKNOWLEDGMENTS

This work is partly funded by the Macdiarmid Centre for Advanced Materials and Nanotechnology under grant PR57123.

REFERENCES

1. Weitz, D.A. and Pine D.J., "Diffusing-wave spectroscopy," in *Dynamic light scattering, the method and some applications*, edited by W. Brown, Oxford Science, Clarendon Press, Oxford, 1993, pp 652-720.

2. Horne, D.S., and Davidson, C.M., *Milchwissenschaft* **45**, 712-715 (1990).

2. Viasnoff, V., Lequeux, F., and Pine, D.J., *Rev. Sci. Instruments* **73**, 2336-2344 (2002).

Dynamic structure of liquid Se, Te and Se-Te mixtures by neutron scattering measurements

Ayano Chiba*, Makoto Yao*, Yoshinori Ohmasa*, Jon Taylor† and Stephen M. Bennington†

*Department of Physics, Graduate School of Science, Kyoto University, Kyoto 606-8502, Japan
†Rutherford Appleton Laboratory, Chilton, Didcot OX11 0QX, UK

Abstract. Inelastic neutron scattering measurements are performed to investigate the dynamic structure of liquid Se, Te, and $Te_{50}Se_{50}$. The bond-stretching modes for liquid Se and $Te_{50}Se_{50}$ (both are in the semiconducting phase) are clearly observed at higher-energy regions than that for their trigonal phase. This shift is a reflection of their pronounced molecular-like properties without prominent inter-molecular interactions, whereas the vibrational modes for liquid Te (in the metallic phase) show its metallic-like bonding character with remarkable inter-molecular interactions. We thus observed a change in dynamic structure that accompanies the semiconductor-to-metal transition; the change that would be related to the slow dynamics induced by the transition.

INTRODUCTION

Metal-nonmetal (M-NM) transitions have been intensively studied in various liquid systems, but still very little is known about their dynamic aspects. Recently, we have found, on the basis of observations of anomalous sound attenuation, that there occurs a slow dynamics with a nanosecond time scale in the M-NM transition range for liquid Hg [1, 2]. We have also suggested that the Semiconductor-to-Metal (S-M) transition in liquid Se-Te systems is also accompanied by a slow dynamics [2, 3] in the time scale of nanoseconds.

In this paper we will consider how the dynamic properties are influenced by the S-M transition in liquid Se-Te systems. In the solid state, both Se and Te are semiconductors that are composed of helical chains in which each atom is bonded to two adjacent atoms by covalent bonds. When Te is melted by heating, it exhibits metallic properties in contrast to Se in which the helical chain structure is more or less preserved even in the liquid state. It is well known that in liquid Te-Se mixtures a S-M transition occurs by raising temperature and that the transition temperature increases with the Se-concentration [4].

EXPERIMENTAL

The neutron scattering experiment was performed on the MARI spectrometer installed at ISIS, Rutherford Appleton Laboratory in the United Kingdom. The E-resolution at the elastic position was 0.8 meV in half-width at half maximum (HWHM), estimated from the vanadium run.

RESULTS AND DISCUSSION

Figure 1 shows $S(Q,E)$ at $E \sim 4$ Å^{-1} for liquid Se, $Te_{50}Se_{50}$ and Te at slightly above their melting points. As to liquid Se, the bond-stretching modes around ± 31 meV are clearly observed, and the bond-bending modes around ± 15 meV can be also seen as weak features. These peak positions agreed with former works for supercooled liquid Se [5]. It may be worth pointing out that similar vibrational density of states (VDOS) are also observed for amorphous Se [6]. The point about the similarity is the position of the stretching mode; the stretching modes for liquid and amorphous Se show remarkable shifts to higher energies (higher-$|E|$) than that for trigonal Se where the mode lies below 30 meV [7]. The reason for this shift would be the reduced inter-molecular correlations in the liquid and amorphous phase; it is well known for this system that the reduced inter-molecular coupling leads to increased intra-molecular bonding [8], i.e., a blue shift of the stretching mode. The stretching modes for liquid $Te_{50}Se_{50}$ are also clearly observed around ± 25 meV, as shown in Fig. 1, and also show shifts to higher-$|E|$, compared to the stretching mode for trigonal $Te_{50}Se_{50}$ that lies around 22 meV [9].

These clear stretching modes and their blue shifts for liquid Se and $Te_{50}Se_{50}$ show their pronounced molecular-like behavior without prominent inter-molecular interactions. We suppose that these liquids

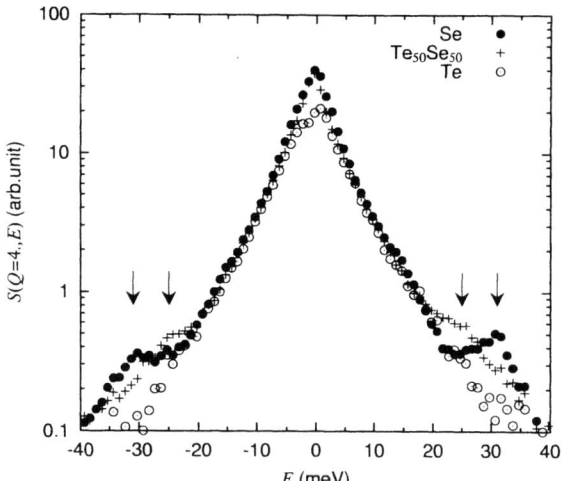

FIGURE 1. $S(Q \sim 4\text{Å}^{-1}, E)$ for liquid Se (semiconductor, denoted by closed circles), liquid $Te_{50}Se_{50}$ (semiconductor, crosses) and liquid Te (metal, open circles) at slightly above their melting points; at 237°C for Se, at 360°C for $Te_{50}Se_{50}$ and at 470°C for Te.

would consist of isolated chain-like molecules. Note that liquid Se and $Te_{50}Se_{50}$ are semiconductors whereas liquid Te is metal. For the metallic phase, it is suggested that the bond breaking and rearrangement of chains occur frequently [10]. Thus the structure in the metallic phase could not be regarded as that composed of simple quasi-one-dimensional chains; the inter-molecular correlations are thought to play an important role in the metallic nature. Related to these views is the fact that the spectrum for liquid Te do not show clear vibrational modes, as shown in Fig. 1. This result supports the weak covalent bonds and frequent bond switching in the liquid Te, and suggests that interactions between atoms for liquid Te probably develop a metallic-like bonding character.

CONCLUSION

We have presented the inelastic part of neutron scattering spectra for liquid Se, $Te_{50}Se_{50}$, and Te at slightly above their melting points, where the former two liquids are in the semiconducting phase, and the last one is in the metallic phase. Most importantly, the vibrational modes for liquid Se and $Te_{50}Se_{50}$ are clearly observed, and their peak positions reflect their molecular-like properties, whereas that for liquid Te could be regarded as a reflection of somewhat metallic-like bonding character. Specifically, the vibrational spectrum for liquid Se is observed to be similar to that of supercooled and amorphous states, that is, the bond-stretching modes show a blue shift compared to that for trigonal phase. The same applies to the stretching mode for liquid $Te_{50}Se_{50}$. We thus supposed that liquid Se and $Te_{50}Se_{50}$ consist of isolated chain-like molecules without prominent inter-molecular interactions. Since Se is much investigated from the aspect of the glass transition, the dynamic structure for its liquid state reported here would, it is hoped, give clues to understand glassy Se.

ACKNOWLEDGMENTS

We would like to thank O. Petrenko, Y. Kawakita and K. Nagaya for their collaboration on the experiments. We are grateful to M. Arai and J.-B. Suck for encouragement. This work was partially supported by Grant-in-Aid for Scientific Research from the MEXT, Japan. The experiments were carried out under the UK-Japan collaboration on neutron scattering. One of the authors (A. C.) has been supported by Hayashi Memorial Foundation for Female Natural Scientists and JSPS Research Fellowships for Young Scientist.

REFERENCES

1. Kohno, H., and Yao, M., *J. Phys.: Condens. Matter* **11**, 5399-5413 (1999); *ibid.*, **13**, 10293-10305 (2001).
2. Yao, M., Kohno, H, and Kajikawa, H, *Z. Phys. Chem.* **217**, 803-816 (2003).
3. Yao, M., Hirano, I., Kajikawa, H., Kohno, H., Itokawa, N., Kajihara, Y., and Hiejima, Y., *J. Non-Cryst. Solids* **312-314**, 361-365 (2001).; Yao, M., Itokawa, N., Kohno, H., Kajihara, Y., and Hiejima, Y., *J. Phys.: Condens. Matter* **12**, 7323-7339 (2000); See also Kajihara, H., Yao, M., Kohno, H., and Kobayashi, K., in this Symposium.
4. Endo, H., Tamura, K., Yao, M., *Can. J. Phys.* **65**, 266-285 (1987).
5. Phillips, W. A., Buchenau, U., Nücker, N., Dianoux, A. J., and Petry, W., *Phys. Rev. Lett.* **63**, 2381-2384 (1989).
6. Zhang, X., and Drabold, D. A., *Phys. Rev. Lett.* **83**, 5042-5045 (1999) [See Fig. 1 (c) measured by Kamitakahara, W. A.].
7. See, e.g., Needham, L. M., Cutroni, M., Dianoux, A. J., and Rosenberg, H. M., *J. Phys.: Condens. Matter* **5**, 637-646 (1993); See also Ref. [5].
8. See, e.g., Hohl. D., and Jones, R. O., *Phys. Rev. B* **43**, 3856-3870 (1991).
9. Chiba, A., Ohmasa, Y., and Yao, M., *J. Chem. Phys.* **119** 9047-9062 (2003).
10. Shimojo, F., Hoshino, K., Watabe, Y., and Zempo, Y., *J. Phys.: Condens. Matter* **10**, 1199-1210 (1998); Hoshino, K., and Shimojo, F., *J. Phys.: Condens. Matter* **10**, 11429-11438 (1998); Chiba, A., Ohmasa, Y., Yao, M., Petrenko, O., and Kawakita, Y., *J. Phys. Soc. Jpn.* **71**, 504-508 (2002).

Slow dynamics induced by the metal-nonmetal transition in liquids

H. Kajikawa, M. Yao, H. Kohno and K. Kobayashi

Department of Physics, Graduate School of Science, Kyoto University, Kyoto 606-8502, Japan

Abstract. We have measured the sound attenuations of liquid Hg and liquid Se-Te mixtures at some tens of MHz by means of an ultrasonic pulse transmission/echo method, and observed anomalous attenuations in the metal-nonmetal (M-NM) transition region in both systems. The anomalous attenuations are attributed to the increase of bulk viscosity. Assuming a simple Debye-type relaxation, we have estimated the relaxation time to be of the order of nanoseconds, which is very long compared with the typical time scale of atomic motion in the liquid state. It is expected from such a long relaxation time that a global rearrangement of atoms should take place in the M-NM transition region.

INTRODUCTION

It is well known that the liquid dynamics slows down on approaching the liquid-gas critical point or cooling to the glass transition temperature.

Recently, we have found from the sound attenuation measurements of liquid metals that slow dynamics is induced by radical changes of electronic structure in the M-NM transition region. In this paper, we present the experimental results of sound attenuation measurements for liquid Hg [1, 2, 3] and liquid Se-Te mixtures [1, 4, 5], which despite large difference in atomic structure show similar acoustical properties in the M-NM transition region.

Liquid Hg is a monatomic liquid and is transformed from a metal to a semiconductor around 8 to 9 g/cm^3, which is intermediate between the critical density (5.8g/cm^3) and the triple point density (13.7g/cm^3). The M-NM transition in liquid Hg is mainly due to the lack of overlapping between the 6s and 6p bands.

Se and Te are semiconductors and form twofold coordinated structure in the solid state. Te is changed to a liquid metal on melting, while Se remains to a semiconductor. In liquid Se-Te mixtures, the semiconductor-to-metal (S-M) transition occurs at a moderate temperature by heating up. The transition temperature range shifts higher with increasing Se concentration. In the semiconducting state, Se-Te mixtures consist of twofold coordinated polymeric chains like liquid Se. Chain length becomes shorter with increasing temperature. The S-M transition is accompanied by a structural change with volume contraction. In the metallic state, Se-Te mixtures are suggested to consist of densely stacked planar zigzag chains like liquid Te.

EXPERIMENTAL

Ultrasonic measurements were performed at some tens of MHz by means of an ultrasonic pulse transmission/echo method [4], which is useful under high temperature and high pressure conditions. The liquid samples were located in the gap between the buffer rods. Two Z-cut Pb(Zr·Ti)O$_3$ transducers were bonded to the cold ends of the rods. Sound absorption coefficient, α, was deduced from the ratio between the amplitudes of transmitted signals and those of reflected signals.

RESULTS

Figure 1 shows the sound absorption coefficient, α, of expanded liquid Hg at 20, 32, and 44 MHz [3]. In addition to the critical attenuation α^{CP} around 6g/cm^3, which is commonly observed in various systems, there appears anomalous attenuation around 9g/cm^3, α^{M-NM}. When the frequency f increases, the magnitude of α^{M-NM} tends to be smaller than that expected from the normal f^2-dependence. This behavior suggests that some kind of relaxation process should take place in the M-NM transition region. The anomalous sound attenuation α^{M-NM} may be attributed to the increase of bulk viscosity [2], which is proportional to the imaginary part of the inverse of the frequency dependent adiabatic compressibility, $\beta(\omega)$. Assuming a Debye-type relaxation for $\beta(\omega)$, we have estimated the relaxation time τ to be 2.2ns [3].

In liquid Se-Te mixtures, we have observed more complicated behavior in the temperature dependence of sound attenuation [4, 5]. Figure 2 shows the temperature dependence of α/f^2 for Se$_{50}$Te$_{50}$ mixture at 150MPa.

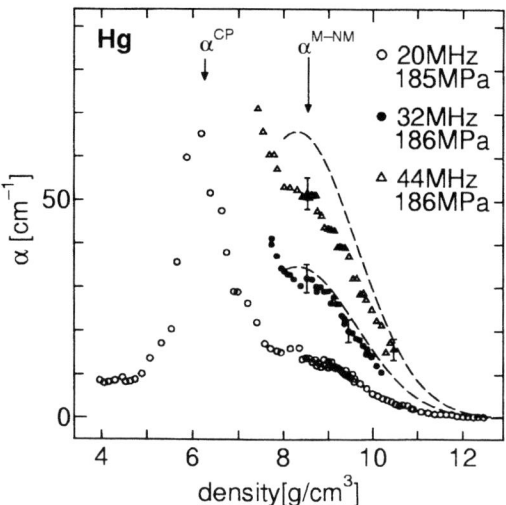

FIGURE 1. The sound absorption coefficient, α, at 20(○), 32(●) and 44MHz(△) around 186MPa versus density for liquid Hg. The dashed lines denote the variations of α^{M-NM} expected for 32 and 44MHz from the f^2-dependence when α^{M-NM} at 20MHz is taken as a standard.

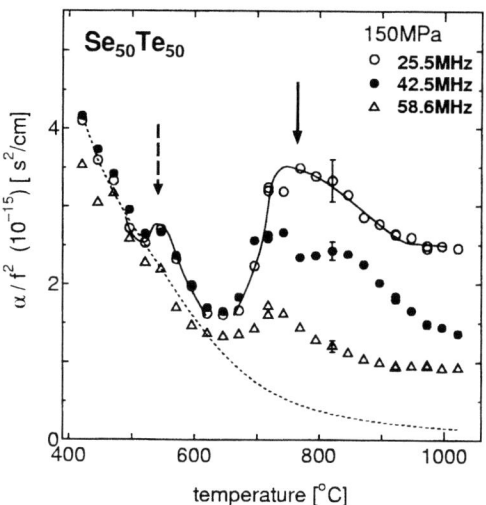

FIGURE 2. The sound absorption coefficient α/f^2 at 25.5 (○), 42.5 (●) and 58.6MHz (△) at 150MPa versus temperature for liquid $Se_{50}Te_{50}$ mixture. The dashed line indicates the normal attenuation expected from the temperature variation of α/f^2 near the melting temperature. The dashed and solid arrows indicate IT and HT peaks, respectively. The thin lines in the figure are guides for the eyes.

Near the melting point, α/f^2 decreases with increasing temperature, which we call 'normal attenuation'. The normal attenuation is expected from the temperature dependence of shear viscosity, and may be assigned to conformational changes within the chain molecules. In addition, there are two anomalous peaks in the S-M transition region. These peaks are attributed to the increase of bulk viscosity because there is no anomalous behavior of shear viscosity in the S-M transition region. One peak exists near 550°C, which we call an 'intermediate - temperature (IT) peak', and the other exists above 700°C, which we call a 'high-temperature(HT) peak'. HT peaks show strong frequency dependence. Assuming a Debye-type relaxation as in the case of liquid Hg, we have estimated the relaxation time to be about 5 ns.

DISCUSSION AND SUMMARY

The relaxation time of the order of nanoseconds is very long compared with the typical time scale of atomic motion in the liquid state, which is of the order of picoseconds. It is expected from such a long relaxation time that a global rearrangement of atoms should take place in the M-NM transition region. In this region, two local structures coexist under large density fluctuations, one is metallic structure, whose density is high due to metallic cohesion, and the other non-metallic. Applying sound pressure causes radical changes of local electronic structures followed by a global atomic rearrangement. This may be a reason for such a long relaxation time.

In conclusion, we have observed anomalous sound attenuations in the M-NM transition region of liquid Hg and liquid Se-Te mixtures. The anomalous attenuations are attributed to the increase of bulk viscosity. Assuming a Debye-type relaxation, we have estimated the relaxation time to be of the order of nanoseconds. Changes of electronic structure in the M-NM transition region may cause a global atomic rearrangement and slow dynamics.

ACKNOWLEDGMENTS

The author's presentation in the symposium is supported by a Grant-in-Aid for the 21st Century COE 'Center for Diversity and Universality in Physics'.

REFERENCES

1. Yao, M., Kohno, H., and Kajikawa, H., *Z. Phys. Chem.* **217**, 803-816 (2003)
2. Kohno, H., and Yao, M., *J. Phys.: Condens. Matter* **11**, 5399-5413 (1999)
3. Kohno, H., and Yao, M., *J. Phys.: Condens. Matter* **13**, 10293-10305 (2001)
4. Yao, M., Itokawa, N., Kohno, H., Kajihara, Y., and Hiejima, Y., *J. Phys.: Condens. Matter* **12**, 7323-7339 (2000)
5. Yao, M., Hirano, I., Kajikawa, H., Kohno, H., Itokawa, N., Kajihara, Y., Hiejima, Y., *J. Non-Cryst. Solids* **312-314**, 361-365 (2002)

Experimental Study of One-Dimensional Spinodal Decomposition in Liquid Crystals

Tomoyuki Nagaya* and Jean-Marc Gilli[†]

*Department of Electrical and Electronic Engineering, Faculty of Engineering,
Okayama University, Okayama, 700-8530, Japan*
[†]*Institue Non-Linéaire de Nice, UMR6618, CNRS-UNSA, Valbonne, 06560, France*

Abstract. The coarsening dynamics of a zigzag wall formed in nematic liquid crystals under external fields is investigated experimentally. The vertexes of zigzag can be considered as kinks in a one-dimensional order parameter system and the geometrical constrain associated to the necessary equal length sum of zig and zag segments impose a conserved quantity in this Cahn-Hilliard type problem. We consequently use this experiment to investigate the spinodal decomposition in a one dimentional conserved order parameter system. It is found that, as expected by the theories for one-dimensional kink dynamics, the characteristic length of the system $L(t)$ increases logarithmically in time and the dynamical scaling law holds in the coarsening process.

INTRODUCTION

Phase ordering dynamics has been investigated intensively in the last two decades. A number of experimental studies that utilize the properties of liquid crystals have been undertaken for various kinds of non-conserved order parameter systems. However, probably due to experimental difficulties, there have not been any detailed experimental studies for the one-dimensional situation.

Recently, two similar experimental systems for studying an Ising conserved order parameter system in one-dimensional space have been proposed[1, 2]. In the first case, 4-cyano-4'-pentyl biphenyl (5CB), whose dielectric anisotropy is positive $\delta\varepsilon > 0$, was used[1]. In the second one, p-methoxybenzilidene-p-n-butylaniiliinee (MBBA), $\delta\varepsilon < 0$, was used [2]. In both systems, a straight splay-bend wall, or so-called "Ising wall", formed under a magnetic field is spontaneously transformed into a zigzag wall by application of a suitable electric voltage. Though the zigzag deformation increases the length of the wall and consequently tends to increase free energy, the twist deformation besides the wall compensates for the increase of wall energy and reduces the total deformation energy of the system. This occurs as the twist deformation is more favorable than the other elastic deformations. The neighboring zigzag vertexes attract each other and disappear by coalescence while the zigzag angle remains constant. As a result, the number of zigzags decreases and the average width of the zigzag increases with time. In the present study, we have investigated the coarsening process of the zigzag wall in detail from the viewpoint of spinodal decomposition.

EXPERIMENTS

The liquid crystal used is MBBA which is sandwiched between two glass plates coated with a homeotropic anchoring agent (Nissan Chemical Industries, SUNEVER SE-1211). The thickness and the size of the cell are $d=50\mu$m and $2.0\text{cm} \times 1.5\text{cm}$, respectively. The sandwiched cell is placed above two permanent Nd-Fe-B magnets that produced a slightly inhomogeneous magnetic field inside the cell. Below the magnets stage a polarizer is placed. The distance between the cell and the magnets is fixed. Though the inhomogeneous magnetic field exerts a restoring force on the Ising wall if the wall is moved from its equilibrium position, the strength of the restoring force is considered as negligible, the magnetic field being almost uniform and fixed 315mT in the vicinity of the wall. The straight wall in the absence of an electric field is perpendicular to the magnetic field. For a narrow range of low frequency ac electric field values applied perpendicular to the glass plates, a zigzag instability of this wall appears. The voltage and frequency are fixed to 8.0Vr.m.s. and 20Hz, respectively. The threshold voltage for the zigzag instability is 6.9Vr.m.s. A sequence of 12 snapshots from $t=0$s to 2042s with exponential time step based on 2s is captured and positions of the vertexes of the zigzag wall are estimated. The statistical quantities are evaluated from 10 runs of the same experimental condition.

RESULTS AND DISCUSSIONS

Typical coarsening patterns are shown in Fig.1. After $t \sim 16$s, the undulation of the wall becomes visible and transforms into a zigzag. The zigzag angle approaches an equilibrium value $\Psi(r,t)$ at about $t \sim 64$s and the coarsening of the zigzag wall proceeds, where r is a coordinate along the straight wall at $t=0$.

Since each tilted segment moves along the wall while keeping constant their tilt angle $\Psi(r,t)$, the respective total lengths for positive and negative slope segments remain constant. Unlike normal phase separation phenomena in binary systems, in which a conservative property of order parameter comes from a conservation of components, simple geometrical constrains ((i) the continuity of wall and (ii) the preservation of zigzag angle) of the zigzag wall are here at the origin of the order parameter conservation in this experiment.

The time evolution of the mean width of zigzag, $W(t)$, are shown in Fig.2. It is remarkable that $W(t)$, i.e. the characteristic length of this system, increases logarithmically in time, which agrees with the theories for the kink dynamics under an attractive exponential interaction in the conserved order parameter system [3, 4, 5].

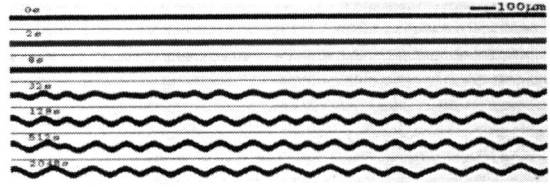

FIGURE 1. Time evolution of zigzag wall.

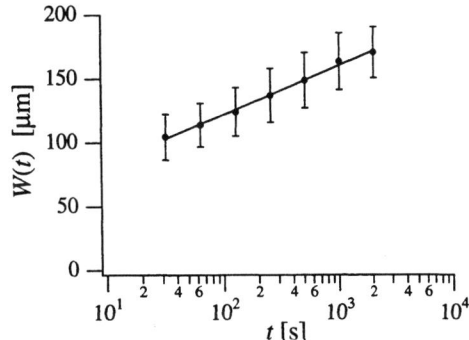

FIGURE 2. Logarithmic growth of width of zigzag.

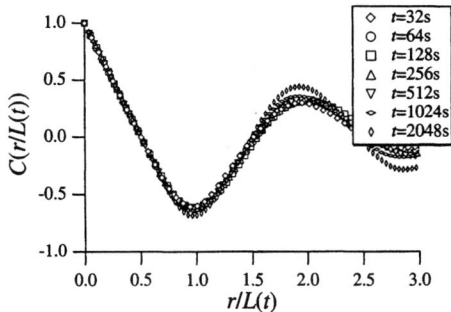

FIGURE 3. Scaled spatial correlation function.

Now let us introduce an Ising order parameter $\phi(r,t)$ defined as

$$\phi(r,t) = \begin{cases} 1 & \text{if } \Psi(r,t) > 0, \\ -1 & \text{if } \Psi(r,t) < 0. \end{cases} \quad (1)$$

Then, the statistical features of the coarsening is analyzed by measuring the spatial autocorrelation function, $C(r,t) = <\phi(r,t)\phi(0,t)>$. The presence of dynamical scaling laws is observed on curves $C(r,t)$ vs. $r/L(t)$ in Fig.3 where $L(t)$ is the characteristic length of the system is defined by $L(t) = W(t)/2 \sim \log(t)$.

In conclusion, we have measured important statistical quantities in this new 1D liquid crystal experiment. The logarithmic growth of the characteristic length and the dynamical scaling law predicted by theories are confirmed in our experiment. Unfortunately, there is actually an obvious lack of theoretical prediction for the correlation function and other important quantities. We hope this work will stimulate further developments of the theoretical aspects concerning this problem.

ACKNOWLEDGMENTS

This work was supported by a Grant-in-Aid for Scientific Research from the Japan Society for the Promotion of Science (KAKENHI No.15540371).

REFERENCES

1. Chevallard C., Clerc M., Coullet P. and Gilli J. M., *Eur. Phys. J. E* **1**, 179-188 (2000); Chevallard C., Nobili M., and Gilli J. M., *Liq. Cryst.* **28**, 179-189 (2001).
2. Nagaya T. and Gilli J. M., *Phys. Rev. E* **65**, 51708-1-51708-8 (2002).
3. Rutenberg A. D. and Bray A. J., *Phys. Rev. E* **51** 5499-5514 (1995).
4. Kawakatsu T. and Munakata T., *Prog. Theoret. Phys.* **74** 11-19 (1985).
5. Majumdar S. N. and Huse D. A., *Phys. Rev. E* **52** 270-284 (1995).

Sonoluminescing bubbles from a diffusing glycerol droplet

Shigeo Hayashi[*] and Naser Harba[*†]

[*]*Department of Applied Physics and Chemistry, University of Elecrtro-Communiations, Chofu, Tokyo 182-8585, Japan*
[†]*National Research and Calibration Laboratory, Damascus, Syria*

Abstract. Several air-filled drops of glycerol, injected into and trapped in the standing acoustic field within water, were found to serve as a stage for the transformation from multibubble to single-bubble sonoluminescence during ultrasound-aided diffusion of glycerol. Initially, a cluster of bubbles were formed. Its diameter was of the order of 1 mm, but the number of bubbles varied. The cluster eventually disapeared, leaving a single sonoluminescing bubble. The transformation process was observed using a stroboscopic telemicroscope.

INTRODUCTION

The diffusion of ordinary liquids in water is essentially a slow process, since the diffusion constants are of the order of 10^{-9} m^{2}s^{-1}. Mixing behavior, as exemplified by the dropping of a drop onto water, may differ considerably, depending on the viscosity η, as can be seen from a comparison of glycerol ($\eta = 1500$ cP) and ethanol ($\eta = 1.06$ cP). The mixing process is sped up dramatically if the water is excited by ultrasounds (usually of the order of 10^4 Hz), because the interface of the liquids is agitated violently, which is in effect a kind of stirring. This ultrasound-aided diffusion works provided that the liquids have different values of the product ρc, where ρ and c are density and sound velocity, respectively.

If a drop of glycerol is injected into water, while an acoustic standing field exists within it, the diffusion also proceeds rapidly. In this case, however, a peculiar phenomenon may occur; since the drop contains air, it gives rise to a seed for a bubble rather than dissolved into water. The bubble oscillates violently, generating light called sonoluminescence.

Thus, ultrasounds can generate sonoluminescence [1] when they are applied to liquids, usually water. The energy conversion is indirect since it is mediated by cavitation [2], i.e., violent collapse of a bubble wherein tremendous heat is generated. There are two classes of sonoluminescence, multibubble sonoluminescence (MBSL) [3] and single-bubble sonoluminescence (SBSL) [4]. The former is vague light originating from many bubbles distributed within water, and the phase of the bubbles is essentially random. On the other hand, the latter is easier to discern because it is caused by a single bubble trapped at a fixed position in water and because it is emitted from the bubble collapsing at a higher temperature, say 20,000 K, leading to brighter sonoluminescence with the emission spectrum peaked in the ultraviolet region.

In preceding papers [5, 6] we devised a procedure for creating SBSL; a glycerol droplet was injected into water while it was irradiated with ultrasounds at a resonant frequency. The glycerol diffused into water quickly leaving behind a sonoluminescing bubble with the help of the acoustic field.

In the present paper, we report on a new aspect of the ultrasound-aided diffusion when many drops of water are injected into water. The process is the same as the previous one in that a single sonoluminescing bubble (SBSL bubble) is finally left behind. During the course of the diffusion, however, it is possible to observe many bubbles in the form of a cluster of bubbles because the diffusion takes a longer time to sustain the cluster. Those bubbles may be identified as the so-called MBSL bubbles.

Even though the ultrasound-aided diffusion, taking less than a minute, is fast in comparison with ordinary diffusion, it is still a slow process considering characteristic time scales involved in sonoluminescence: the life time of SBSL is of the order of $10^{-11} - 10^{-10}$ s [4], the lifetime of MBSL, essentially energy conversion within a molecule, is expected to be of the order of 10^{-9} s and the period of bubble oscillation is of the order of $10^{-5} - 10^{-4}$ s.

EXPERIMENTAL

The experimental setup was detailed in [6]. The acoustic cell, 40 mm × 40 mm × 53 mm, was driven by a

Langevin-type transducer oscillating at the resonant frequency of the cell, 29.5 kHz. The wavelength of ultrasound was thus 5.1 cm. The driving voltage was controlled manually by the output of the function generator. A simple stroboscope was employed: the object in question was illuminated by a light emitting diode which was lit for 100 ns. Its phase was kept fixed during each acoustic cycle of the period 34 μs and advanced by $\Delta\varphi$ in the end, where $\Delta\varphi$ was of the order of 1°.

The method for injecting glycerol drops was as follows: glycerol was bubbled using a syringe so that as much air could be contained as possible, and several glycerol drops, i.e., a more amount of glycerol, were injected using a syringe. The driving voltage for the ultrasonic transducer was then adjusted until the glycerol drops, moving around in water, were finally trapped at the antinode of the acoustic field.

RESULTS AND DISCUSSION

Most striking images are shown in a sequence of Fig. 1. The drop contains many air bubbles, and white tailing objects that emanate outward are probably such bubbles. In Frame (b), a considerable amount of glycerol has dissolved into water but there still remain air bubbles. The image also reveals the diffusion of glycerol. In Frame (c), a single bubble begins to emerge within the region rich in glycerol. In Frame (d), a single bubble is still surrounded by glycerol, but by this time the bubble oscillates in accordance with the usual SBSL-bubble dynamics, which explains why the dark spot is bigger than that in the preceding frame. This bubble is emitting light, as confirmed by the naked eye.

It should be noted that the figure shows just an example; the ultrasound-aided diffusion of air-filled glycerol is so complex a process that one will observe different images every time one does an experiment. For example, we observed a cluster which contained less number of bubbles that showed no eruption. Thus, the behavior of the glycerol drop is diversified, depending on many factors such as the size of the glycerol drop, the amount of air therein, the acoustic field strength, etc.

The erupting behavior of Frame (a) is difficult to account for; The bubbles should probably be experiencing attractive secondary Bjerknes force since they must oscillate in phase [2] at an interbubble distance which was much less than the wavelength of acoustic field. More elaborate study is thus needed.

In conclusion, we have successfully observed the transformation of a cluster of multiple bubbles into a single sonoluminescing bubble within the same experimental setup, which is a kind of slow and complex process. Our approach should provide with a new methodology

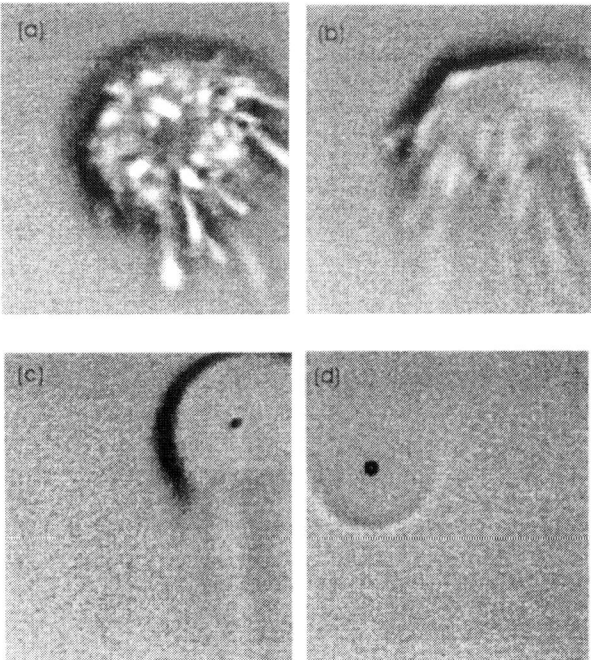

FIGURE 1. Stroboscopic images of a glycerol drop diffusing in the acoustic field. A cluster of many bubbles, (a), transforms into a single, sonoluminescing bubble, (d). The width is 0.93 mm for Frame (a). The elapsed time is (a) 0 s (starting frame), (b) 13/30 s, (c) 20/30 s, (d) 153/30 s.

for the experimental study of sonoluminescence from a wider perspective.

ACKNOWLEDGMENTS

The authors are grateful to Dr. S. Hatanaka for helpful discussion.

REFERENCES

1. Puttermann, S. J, *Scientific American*, **272**, 32-37 (1995).
2. Young, F. R, *Cavitation*, Imperial College, London, 1999.
3. Verral, R. E. and Sehgal, C. M., *Ultrasound: Its Chemical, Physical, and Biological Effects*, edited by K. S. Suslick, VCH, New York, 1988, pp. 227- 286.
4. Brenner M. P., Hilgenfeld S., and Lohse D., *Rev. Mod. Phys.* **74**, 425-484 (2002).
5. Harba N., and Hayashi S., *Jpn. J. Appl. Phys.* **42**, 716-720 (2003).
6. Harba N., and Hayashi S., *Jpn. J. Appl. Phys.* **42**, 2971-2974 (2003).

Shear-Induced Structure and Velocity Fluctuations in Particulate Suspensions Probed by Ultrasonic Correlation Spectroscopy and Rheology

Anatoliy L. Strybulevych, Del M. Leary and John H. Page.

Ultrasonics Research Laboratory, Department of Physics and Astronomy, University of Manitoba, Winnipeg R3T 2N2, Canada

Abstract. We report measurements of the steady-state shear viscosity and shear-induced structure of neutrally buoyant non-Brownian particulate suspensions. Four distinct types of behavior were found for the viscosity of the suspensions as the shear rate and concentration were varied. Possible interpretations of the data in term of the shear-induced structure of the suspensions are discussed. A new ultrasonic technique was used to determine the relaxation time required for the particles to come to rest after the shear rate is set to zero.

The flow properties and dynamics of particulate suspensions are of considerable current interest both scientifically and practically. Since the pioneering work of Gadala-Maria and Acrivos [1], much of this interest has focused on how the flow properties are influenced by the microstructure, the hydrodynamic inter-particle interactions and the redistribution of particles in the suspension under shear. For concentrated suspensions of non-Brownian, neutrally buoyant solid particles, they first reported that the apparent shear viscosity in a Couette viscometer decreased with time during prolonged shear. Leighton and Acrivos [2] attributed this long-term viscosity decrease to shear-induced particle migration to low shear stress regions. The application of an NMR imaging technique [3] and a novel correlation method [4] have shed light on the evolution of particle concentration profiles and shear-induced self-diffusion in Couette flow. To further investigate this behaviour, we report measurements of the steady-state shear viscosity and particle velocities in neutrally buoyant suspensions of non-Brownian particles using rheology and Dynamic Sound Scattering [5].

The neutrally buoyant suspensions were assembled by first repeatedly sieving spherical borosilicate glass beads to obtain a nearly monodisperse size distribution of uniform-density particles (diameter $d = 127 \pm 22$ μm, density $\rho = 2,220$ kg/m^3). The glass beads were then immersed in a density-matched liquid consisting of a mixture of LST heavy liquid (a low viscosity aqueous solution of lithium hetero-polytungstates) and water. Suspensions with particle volume fractions ϕ of 0, 0.1, 0.2, 0.3, 0.4, 0.5, 0.55, and 0.6 were prepared.

The viscosity measurements were carried out in a concentric cylindrical cell at a temperature of 27° C using a TA instruments AR2000 rheometer. The viscosity of the suspensions was investigated under steady shear flow at constant angular velocity from 0.1 to 20 rad/s for both increasing and decreasing angular velocity, in a series of steps or cycles that allowed long-term evolution of the suspension microstructure to be investigated.

Figure 1 illustrates general trends of the viscosity during Couette flow for the suspension at different volume fractions. Four distinct types of behaviour were found for the viscosity of the suspensions as the shear rate and concentration were varied. For dilute suspensions as the (ϕ up to 10%), normal Newtonian viscosity was found, as expected. For ϕ between 20 and 30%, reversible shear thinning was observed, suggesting that the particles rearrange in a layered structure to facilitate the flow. At higher ϕ ($\geq 40\%$), considerable hysteresis was found at increasing and decreasing shear rate, reflecting a slow dynamic evolution of the suspension microstructure. At even higher concentration ($\phi \geq 55\%$), the viscosity was found to increase with shear rate, as particle interactions and jamming tend to break up the layered structure that was favored at lower ϕ.

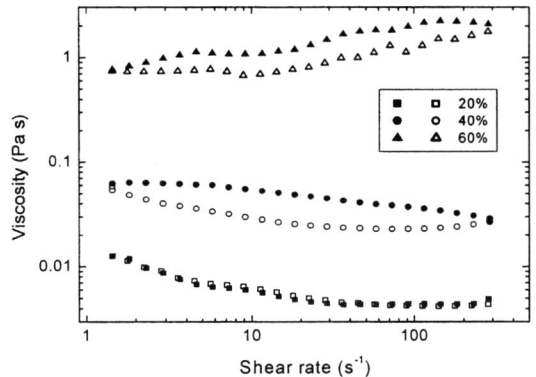

FIGURE 1. Measured viscosity as a function of shear rate for suspensions of neutrally buoyant glass beads at volume fractions of 0.2, 0.4, and 0.6. Solid and open symbols represent data at increasing and decreasing shear rates.

To measure the dynamics of the particles directly, we use Dynamic Sound Scattering (DSS), a new technique in ultrasonic correlation spectroscopy that allows the temporal evolution of the mean square displacement of the particles $(\Delta r^2(\tau))$ to be determined from the fluctuations of singly scattered ultrasound [5]. The DSS measurements were performed in a cylindrical shear cell, with a stationary outer wall and a rotating inner wall, separated by a 4 mm gap. Figure 2 shows typical results taken in reflection using a focusing transducer with the beam axis perpendicular to the cell wall. In this orientation, the technique is still sensitive to the dominant motion of the particles in the direction of the shear flow because of the angular distribution of incident and scattered waves about the beam axis, allowing the average velocity of the particles V_{rms} parallel to the applied shear to be probed.

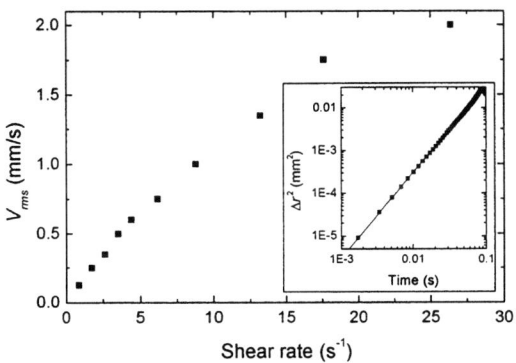

FIGURE 2. Particle velocity as a function of shear rate. The inset shows the expected quadratic time dependence of Δr^2 at one shear rate: $\Delta r^2(\tau) = V_{rms}^2 \tau^2$.

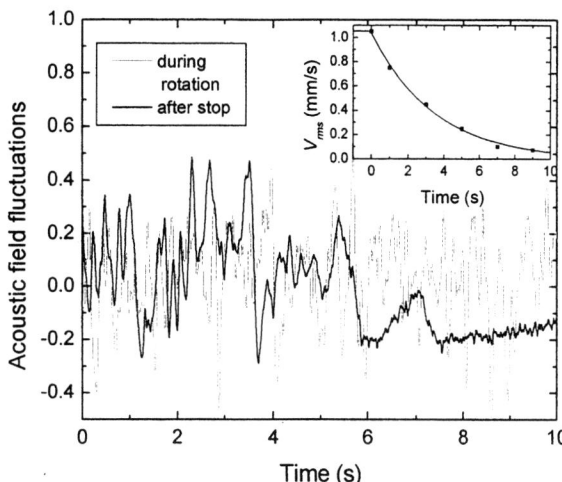

FIGURE 3. Field fluctuations during steady shear (grey line) and after the rotor was stopped (black line). The inset shows the decay of the particle motion after the shear rate was set to zero.

Figure 3 shows the field fluctuations both during shear and immediately after the rotor was stopped. It can be seen that the field fluctuations gradually slow down after the shear rate is set to zero, allowing the relaxation time of the particle dynamics, including the time required for shear-induced structures to break up, to be determined. From the exponential decay of V_{rms} shown in the inset, this relaxation time was found to be 3.4 s for this system at $\phi = 0.30$.

ACKNOWLEDGMENTS

Research support from the Canadian Space Agency is gratefully acknowledged.

REFERENCES

1. Gadala-Maria, F., and Acrivos, A., *J. Rheol.* **24**, 799-814 (1980).

2. Leighton, D., and Acrivos, A., *J. Fluid Mech.* **181**, 415- (1987).

3. Chow A.W., Sinton, S.W., and Iwamiya, J.H., *Phys. Fluids* **6**, 2561-2576 (1994).

4. Breedveld, V., van den Ende, D., Bosscher, M., Jongschaap, R.J.J., and Mellema, J., *Phys. Rev. E* **63**, 021403 (2001).

5. Cowan, M.L., Page, J.H., and Weitz, D.A., *Phys. Rev. Lett.* **85**, 453-456 (2000).

Generalization of the Ohta-Kawasaki Theory for Microphase Separation of Block Copolymer Melts

Takashi Uneyama*, Yuichi Masubuchi*, Jun-ichi Takimoto* and Masao Doi*

Department of Computational Science and Engineering, Graduate School of Engineering, Nagoya University, Furou-cho, Chikusa-ku, Nagoya, 464-8603, JAPAN

Abstract. The Ohta-Kawasaki theory for the microphase separation of the diblock copolymer melts has been generalized to study arbitrary block copolymer melts and blends. The results of the simulation with the generalized theory were compared with those of the self consistent field (SCF) theory and it has been shown that they give qualitatively the same result.

INTRODUCTION

The microphase separation of the diblock copolymer melts has been theoretically studied by many authors with the density functional theory of Leibler [1] or Ohta-Kawasaki [2]. Although their theories agree with the experiments well, they cannot be applied to the other systems such as the star polymer or comb polymer systems.

The self consistent field (SCF) theory is a widely used theory to study the microphase separation of the block copolymer systems. The SCF theory can handle arbitrary block copolymers and the blends of them, and the simulation with the SCF theory gives quantitatively good results. The simulation with the SCF theory, however, needs high computational costs compared with that with the density functional theory.

In this paper, we propose a generalization of the Ohta-Kawasaki theory to arbitrary block copolymer blends and melts, and apply it to some systems such as the ABC triblock copolymer melts. It has been shown that our generalized thoery gives qualitatively similar results to the SCF theory.

FREE ENERGY FUNCTIONAL FOR BLOCK COPOLYMER BLENDS

We consider the system consisting of one or more polymer species. The pth polymer species consists of contains some subchains one of which is represented as the ith subchain.

The scattering function (the correlation function of the second order) for the system is defined as follows.

$$S^{(2)}_{pi,qj}(\bm{r}-\bm{r}') = \left\langle \delta\phi_{pi}(r)\delta\phi_{qj}(r') \right\rangle \quad (1)$$

where $\delta\phi_{pi}(\bm{r}) = \phi_{pi}(\bm{r}) - \bar{\phi}_{pi}$, $\phi_{pi}(\bm{r})$ is the volume fraction of the ith subchain of the pth polymer, $\bar{\phi}_{pi}$ is the spacial average of $\phi_{pi}(\bm{r})$, and $\langle\ \rangle$ means the ensemble average.

The vertex function of the second order $\Gamma^{(2)}$ is defined by

$$\sum_{qj} S^{(2)}_{pi,qj}(\bm{q}) \Gamma^{(2)}_{qj,rk}(\bm{q}) = \delta_{pi,rk} \quad (2)$$

in the wavenumber space.

Ohta and Kawasaki approximated the second order vertex function $\Gamma^{(2)}$ for diblock copolymer melts of the ideal chains by using the interpolation between $\bm{q}^2 \to 0$ and $\bm{q}^2 \to \infty$ limits with the extreme forms of $\Gamma^{(2)}$ for $\bm{q}^2 \to 0, \infty$. $\Gamma^{(2)}$ can be reduced to a scalar in the case of diblock copolymer melts, while it cannot be reduced to a scalar in the case of arbitrary block copolymer blends. We also approximate $\Gamma^{(2)}$ for the general systems by the similar interpolation, and obtain the following expression;

$$\Gamma^{(2)}_{pi,qj}(\bm{q}) \approx \frac{\delta_{pq}}{\bar{\phi}_p}\left[G^{(-1)}_{p,ij}\bm{q}^{-2} + G^{(0)}_{p,ij} + \frac{\delta_{ij}b_{pi}^2}{12 f_{pi}}\bm{q}^2 \right] + \chi_{pi,qj} \quad (3)$$

Here $G^{(-1)}_{p,ij}$ and $G^{(0)}_{p,ij}$ are the matrices which are determined from the scattering function $S^{(2)}_{pi,qj}$, b_{pi} is the segment size, and f_{pi} is the volume fraction of the ith segment in the pth polymer, and $\bar{\phi}_p$ is the volume fraction of the pth polymer. The interaction among the segments is taken into account by using the random phase approximation (RPA) with the parameters $\chi_{pi,qj}$.

We can calculate the free energy functional from eq (3).

$$F\left[\{\phi_{pi}(\boldsymbol{r})\}\right] = \\ \sum_{p,ij} \frac{G_{p,ij}^{(-1)}}{2\bar{\phi}_p} \int d\boldsymbol{r} d\boldsymbol{r}' \mathcal{G}(\boldsymbol{r}-\boldsymbol{r}') \delta\phi_{pi}(\boldsymbol{r}) \delta\phi_{pj}(\boldsymbol{r}') \\ + \sum_{pi} f_{pi} G_{p,ii}^{(0)} \int d\boldsymbol{r}\, \phi_{pi}(\boldsymbol{r}) \ln \phi_{pi}(\boldsymbol{r}) \\ + \sum_{pi} \frac{b_{pi}^2}{24\bar{\phi}_{pi}} \int d\boldsymbol{r} \left|\nabla \delta\phi_{pi}(\boldsymbol{r})\right|^2 \\ + \sum_{pi,qj} \left[\frac{1}{2}\chi_{pi,qj} + \frac{\delta_{pq} G_{p,ij}^{(0)}}{2\bar{\phi}_p}\right] \int d\boldsymbol{r}\, \phi_{pi}(\boldsymbol{r}) \phi_{qj}(\boldsymbol{r}) \quad (4)$$

where $\mathcal{G}(\boldsymbol{r}-\boldsymbol{r}')$ is the Green function that satisfies $-\nabla^2 \mathcal{G}(\boldsymbol{r}-\boldsymbol{r}') = \delta(\boldsymbol{r}-\boldsymbol{r}')$. We used some approximations so that eq (4) reproduces the Flory-Huggins free energy for the homogeneous polymer blend systems.

RESULTS OF SIMULATIONS

Homopolymer Blends

The equilibrium structure simulations for A,B homopolymer blends have been done to study the interface profiles given by the generalized Ohta-Kawasaki theory and the SCF theory. The SCF simulation has been done by SUSHI [3].

The condition for the simulation is as follows: $N_A = N_B = 40, b_A = b_B = 1, \bar{\phi}_A = \bar{\phi}_B = 0.5, \chi_{AB} = 0.5$, one dimensional periodic system (the length is $l_x = 32$, divided into 256 lattice points).

Figure 1 is the result of the simulation by the generalized Ohta-Kawasaki theory and the SCF theory. It can be observed that two profiles agree well around the average density $\bar{\phi}_A = 0.5$ while they differ where the fluctuation $\delta\phi_A(\boldsymbol{r})$ is large.

Triblock Copolymer Melts

The equilibrium structure for triblock copolymer melts has been studied by the simulation using eq (4). The simulation has been done for the ABC triblock star copolymer melts ($N = 30, f_A = f_B = f_C = 1/3, b_A = b_B = b_C = 1, \bar{\phi} = 1, \chi_{AB} = \chi_{BC} = \chi_{CA} = 1$) in the two dimensional periodic system ($l_x = l_y = 32$, divided into 64×64 lattice points).

Figure 2 is the result of the simulation. The SCF simulation gives the similar morphology.

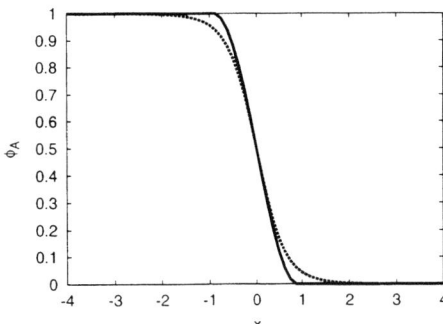

FIGURE 1. The profile of ϕ_A near the interface of the A,B homopolymer blend. The solid line is the result of the generalized Ohta-Kawasaki theory and the dashed line is the result of the SCF.

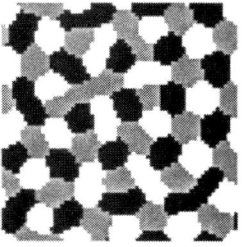

FIGURE 2. The equilibrium structure for the ABC triblock star copolymer. A, B and C segment densities are shown as black, gray and white, respectively.

DISCUSSION

It has been shown that the results of our generalized Ohta-Kawasaki theory qualitatively agree with those of the SCF theory, although there remains some difference which may be due to the incomplete treatment of the higher order terms of $\delta\phi_{pi}(\boldsymbol{r})$.

The simulation by using our generalized theory is faster than that using the SCF theory by factor of ten or more.

REFERENCES

1. Leibler, L., *Macromolecules*, **13**, 1602–1617 (1980).
2. Ohta, T., and Kawasaki, K., *Macromolecules*, **19**, 2621–2632 (1986).
3. http://octa.jp

Dynamics in a Bistable-Element-Network with Delayed Coupling and Local Noise

Daniel Huber and Lev Tsimring

Institute for Nonlinear Science, University of California, San Diego, La Jolla, CA 92093-0402

Abstract. The dynamics of an ensemble of bistable elements under the influence of noise and with global time-delayed coupling is studied numerically by using a Langevin description and analytically by using 1) a Gaussian approximation and 2) a dichotomic model. We find that for a strong enough positive feedback the system undergoes a phase transition and adopts a non-zero stationary mean field. A variety of coexisting oscillatory mean field states are found for positive and negative couplings. The magnitude of the oscillatory states is maximal for a certain noise temperature, i.e., the system demonstrates the phenomenon of coherence resonance. While away form the transition points the system dynamics is well described by the Gaussian approximation, near the bifurcations it is more adequately described by the dichotomic model.

INTRODUCTION

Stochastic rate processes in bi- or multi-stable systems lead to many interesting phenomena observed in various scientific areas ranging from physics to social science, and have thus been studied for a long time.

Here we consider a network of stochastically driven bistable elements whose distinct feature is a time-delayed coupling. The time delays are considered as uniform and the network elements are assumed to be highly interconnected, so that the connectivity can be approximated by a global all to all coupling.

The dynamics of the network is numerically explored by using a Langevin model and analytically by using a Gaussian approximation and a dichotomous model, derived from the corresponding Fokker-Planck equations and Master equation, respectively.

LANGEVIN MODEL

The Langevin model consists of N equations, each describing the overdamped motion of a particle in a bistable potential in the presence of noise and global coupling to a time-delayed mean field $X(t-\tau) = N^{-1} \sum_{i=1}^{N} x_i(t-\tau)$,

$$\dot{x}_i(t) = x_i(t) - x_i(t)^3 + \varepsilon X(t-\tau) + \sqrt{2D}\xi(t), \quad (1)$$

where τ is the time delay, ε is the coupling strength of the feedback and D denotes the variance of the Gaussian fluctuations $\xi(t)$.

GAUSSIAN APPROXIMATION

In order to theoretically study the dynamical properties of a globally coupled set of noisy bistable elements (with no time delay), Desai and Zwanzig [1] derived a hierarchy of equations for the cumulant moments of the distribution function from the multi-dimensional Fokker-Planck equation for the joint probability distribution for all elements. For large noise intensities, when the statistics of individual elements are approximately Gaussian, this hierarchy can be truncated. Applying this approach to our system yields the following set of equations for the mean field X and the variance $M = N^{-1} \sum (x_i - X)^2$,

$$\begin{aligned}
\dot{X} &= X - X^3 - 3XM + \varepsilon X(t-\tau), \\
\frac{1}{2}\dot{M} &= M - 3X^2M - 3M^2 + D.
\end{aligned} \quad (2)$$

DICHOTOMOUS MODEL

To study the dynamics of a single bistable element with time-delayed feedback Tsimring and Pikovsky [2] used a dichotomous approximation. In the dichotomous approximation intra-well fluctuations of x_i are neglected, so that a bistable element can be replaced by a discrete two-state system. Applying this complementary approach to our system, allows us to express the mean field dynamics in terms of the hopping rates $p_{12,21}$ which denote the probability of a bistable element to change its state form -1 to $+1$ and vice versa, respectively. The equation for the

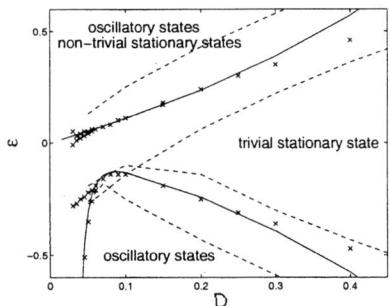

FIGURE 1. Phase diagram of the Langevin model (crosses), the Gaussian approximation (dashed lines) and the dichotomous theory (solid lines). Double lines indicate hysteretic transitions among phases.

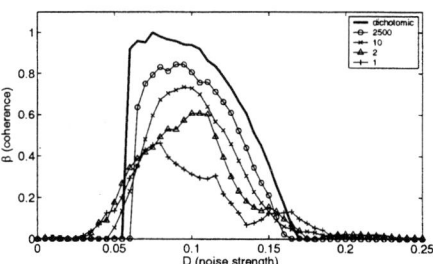

FIGURE 2. The normalized coherence β of the oscillatory state at $\varepsilon = -0.2$. as a function of the noise strength D, for networks of different size N as well as for the dichotomic theory. The coherence is given through $\beta = H\omega_p/\delta\omega$, where H is the hight of the dominant spectral peak at ω_p and $\delta\omega$ is the half-width of the peak.

mean field reads

$$\dot{X} = p_{12} - p_{21} - (p_{21} + p_{12})X, \quad (3)$$

where the hopping probabilities are given by the Kramers transition rate, which in the limit of small noise and small coupling strength reads (cf. [2])

$$p_{12,21} = \frac{\sqrt{2 \mp 3\varepsilon X(t-\tau)}}{2\pi} \exp\left(-\frac{1 \mp 4\varepsilon X(t-\tau)}{4D}\right). \quad (4)$$

PHASE DIAGRAM

A numerical study of the Langevin model (Eq. 1) shows that the system undergoes ordering transitions and demonstrates multistability. That is, for a strong enough positive coupling the system exhibits a non-zero stationary mean field and a variety of stable oscillatory states are accessible for positive and negative feedback. While the transition to the non-zero stationary mean field is second order (continuous), the type of the oscillatory transitions depends on the system parameters and can be first order (discontinuous), associated with hysteretic behavior, or second order.

A linear stability analysis of Eq. (3) yields the critical coupling strengths ε and the frequencies ω of the accessible oscillatory states.

A comparison of the Langevin model with the Gaussian approximation and the dichotomous theory shows that while near the transition points the system dynamics is strongly non-Gaussian, it is in this regime well described by a two-state model (see Fig. 1), which allows for a complete analysis of the bifurcations of the trivial equilibrium.

COHERENCE RESONANCE

The considered system exhibits the phenomena of coherence resonance and array-enhanced resonance (see Fig. 2).

Both, Kramers random switching frequency p (see 4) and the frequency of the oscillatory states ω, resulting from the coupling with the time-delayed mean field, depend on the noise strength, i.e., $p = p(D)$ and $\omega = \omega(D)$. Thus, the noise can be tuned so that the random hoping between the potential wells of the bistable oscillators is synchronized with the periodic modulation of the meanfield. This statistical synchronization takes place when $\omega = \pi p$, where the coherence of the oscillatory states becomes maximal.

For the $N = 1$ case the coherence resonance was observed by Tsimring and Pikovsky [2]. We observe that the resonance phenomenon not only persists in globally coupled networks with large N, but is enhanced, a property which was found in other systems and is sometimes referred to as array-enhanced resonance [3].

ACKNOWLEDGMENTS

This work was supported by the Swiss National Science Foundation (D.H.) and by the U.S. Department of Energy, Office of Basic Energy Sciences under grant DE-FG-03-96ER14592 (L.T.).

REFERENCES

1. Desai, R. C., and Zwanzig, R., *J. Stat. Phys.* **19**, 1-24 (1978).
2. Tsimring, L. S., and Pikovsky, A., *Phys. Rev. Lett.* **87**, 2506021-4 (2001).
3. Zhou, C., Kurths, J., and Hu, B., *Phys. Rev. Lett.* **87**, 981011-4 (2001).

Photo-induced wave propagation in Langmuir monolayers

Tohru Okuzono[*], Yuka Tabe[*†] and Hiroshi Yokoyama[*†]

[*] *Yokoyama Nano-structured Liquid Crystal Project, ERATO, Japan Science and Technology Agency, 5-9-9 Tokodai, Tsukuba, Ibaraki 300-2635, Japan*
[†] *Nanothechnology Research Institute, National Institute of Advanced Industrial Science and Technology, 1-1-1 Umezono, Tsukuba, Ibaraki 305-8568, Japan*

Abstract. A phenomenological model of photo-induced wave propagation in liquid-crystalline Langmuir monolayers is constructed. The effects of the spontaneous splay deformation of liquid crystal order and the anisotropic photo-excitation of molecules are taken into consideration in this model. Most of qualitative features of the phenomenon are successfully explained by numerical simulations and theoretical analyses of the model.

Spatiotemporal structures in illuminated Langmuir monolayers [1] provide a good example of dynamic structures far from equilibrium observed in soft condensed matter where a simple reaction-diffusion model cannot be applied [2]. In this paper, we construct a phenomenological model of photo-induced wave propagation in liquid-crystalline Langmuir monolayers that well explains a recent experimental observation [3].

We consider a monolayer system that consists of rod-like molecules whose directions are tilted from the layer normal. The local orientation of the molecules can be described by a vector field $\boldsymbol{c}(\boldsymbol{r},t)$ defined as a projection of the director $\boldsymbol{n}$ of the molecules onto the two-dimensional space $\boldsymbol{r} = (x,y)$, namely, the monolayer at time t. When the system is illuminated with linearly polarized light, the constituent molecules undergo trans-cis photo-isomerization, that is, a molecule changes its shape from a rod (trans) to a bent (cis) shape, and vice versa. Introducing a scalar field $\psi(\boldsymbol{r},t)$ defined as the local concentration difference between trans- and cis-isomers, we may write the free energy F in the dimensionless units as [4, 5, 6],

$$F = \int d\boldsymbol{r} \left[\frac{1}{2}\sum_i |\nabla c_i|^2 - \frac{\tau}{2}|\boldsymbol{c}|^2 + \frac{u}{4}|\boldsymbol{c}|^4 - \lambda \psi \nabla \cdot \boldsymbol{c} + \frac{D}{2}|\nabla \psi|^2 + \frac{\chi}{2}\psi^2 \right], \quad (1)$$

where the sum is taken over the component c_i ($i = x, y$) of $\boldsymbol{c}$ and τ, u, D, and χ are positive constants. The second and third terms of the integrand express the smectic-A to smectic-C transition. Since we are concerned with the smectic-C phase only, we assume $\tau > 0$. The coupling term with a coupling constant λ is allowed to enter F because there is no inversion symmetry about $\boldsymbol{c}$ in this system [7]. This term causes the spontaneous splay deformation of liquid crystal order which plays a crucial role in the wave propagation phenomenon. In this paper we only consider the case $\chi > 0$, that is, the concentration field does not tend to phase-separate in itself.

The kinetic equations for $\boldsymbol{c}$ and ψ are written as

$$\frac{\partial \boldsymbol{c}}{\partial t} = -\frac{\delta F}{\delta \boldsymbol{c}} + \boldsymbol{f}(\boldsymbol{c}, \psi), \quad (2)$$

$$\frac{\partial \psi}{\partial t} = M\nabla^2 \frac{\delta F}{\delta \psi} + g(\boldsymbol{c}, \psi), \quad (3)$$

where M is the mobility and $\boldsymbol{f}(\boldsymbol{c}, \psi)$ and $g(\boldsymbol{c}, \psi)$ are the reactive terms due to the photo-isomerization that are given by

$$\boldsymbol{f}(\boldsymbol{c}, \psi) = -\gamma_2 \frac{1-\psi}{1+\psi}\boldsymbol{c}, \quad (4)$$

$$g(\boldsymbol{c}, \psi) = -(\gamma_1 + \gamma_2)\psi - (\gamma_1 - \gamma_2), \quad (5)$$

where γ_1 and γ_2 are the trans-to-cis and cis-to-trans reaction rates, respectively. The rate equations of this type have been used by Reigada et al. [8] with constant reaction rates. However, the trans-to-cis reaction rate should depend on the relative orientation of $\boldsymbol{c}$ to the polarization of excitation light. Here we take the anisotropy of the reaction rate into account by using the expression

$$\gamma_1 = g_0 |\boldsymbol{c}|^2 + g_1 (\hat{\boldsymbol{E}} \cdot \boldsymbol{c})^2, \quad (6)$$

where g_0 and g_1 are positive constants and $\hat{\boldsymbol{E}} \equiv (\cos\theta, \sin\theta)$ with the orientation θ of the polarization of light. Henceforth, we assume $g_0 = 0$, since we are concerned with a strongly ordered state of liquid-crystalline monolayer system where the isotropic part of

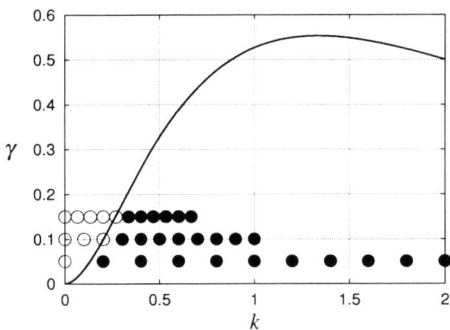

FIGURE 1. The linear stability diagram for the uniform equilibrium solution of Eqs. (7) and (8). Below solid line and $\gamma > 0$ the equilibrium solution is unstable, otherwise stable. For the parameters shown by closed or open circles, traveling waves or no patterns are observed, respectively, in the numerical simulations of Eqs. (2)–(6) with the uniform initial condition.

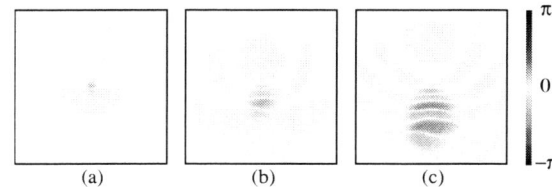

FIGURE 2. Spatial distributions of azimuth $\phi(\mathbf{r},t)$ at $t = 10$ (a), 100 (b), and 200 (c) are displayed using the gray scale shown in the right.

γ_1 is negligible. Furthermore, we put $M = 0$ taking the slow diffusion limit of ψ.

In order to carry out a theoretical analysis of Eqs. (2)–(6) we simplify the model equations assuming that the tilt order $|\mathbf{c}|^2$ is constant. For simplicity we take $|\mathbf{c}| = 1$ so that $\mathbf{c} = (\cos\phi, \sin\phi)$. Then Eqs. (2) and (3) become

$$\frac{\partial \phi}{\partial t} = \nabla^2 \phi + \lambda \left(\sin\phi \frac{\partial \psi}{\partial x} - \cos\phi \frac{\partial \psi}{\partial y} \right), \quad (7)$$

$$\frac{1}{\gamma}\frac{\partial \psi}{\partial t} = 1 - \psi - k\cos^2(\theta - \phi)(1 + \psi), \quad (8)$$

where $k \equiv g_1/\gamma_2$ and $\gamma \equiv \gamma_2$. The parameter k and γ are related to the property of excitation light. Actually, k depends on the wavelength of excitation light and γ is proportional to the intensity of illumination.

Equations (7) and (8) have a uniform equilibrium solution: $\phi = \phi_0$, $\psi = [1 - k\cos^2(\theta - \phi_0)]/[1 + k\cos^2(\theta - \phi_0)]$, where ϕ_0 is an arbitrary constant. Performing the linear stability analysis around this uniform solution with $\phi_0 = 0$ to the perturbation $\exp(iqy + \sigma_q t)$ with wave number q, we find that an oscillatory instability at finite wave number occurs (Re $\sigma_q > 0$ and Im $\sigma_q \neq 0$ for $q \neq 0$) for lower value of γ. The linear stability line in the parameter space (k,γ) is shown in Fig. 1 for $\theta = \pi/4$. Below this line and $\gamma > 0$ the uniform state is unstable and traveling waves are expected to emerge. Indeed, traveling waves associated with ϕ are observed in numerical simulations of Eqs. (2)–(6) for the parameters shown by closed circles while no patterns are observed for the parameters shown by open circles in Fig. 1. Snapshots of $\phi(\mathbf{r},t)$ are shown in Fig. 2 obtained by the numerical simulation for the uniform initial state ($\phi_0 = 0$) with a localized perturbation at the center ($k = 1$, $\gamma = 0.05$, $\theta = \pi/4$, $\lambda = 1$ and $\tau = u = 2$). We observe in these figures that the wave associated with the azimuth ϕ propagates downward.

In the experiment [3] the propagation directions of waves are reversed when the polarization of illumination is switched by $\pi/2$. This phenomenon can be reproduced by the numerical simulations of our model. The result from the linear stability analysis that the eigenvalue $\sigma_q(\theta)$ as a function of θ has the opposit sign of $\sigma_q(\theta + \pi/2)$ also implies the switching phenomenon. Note that the equilibrium solution always stable for $\theta = n\pi/2$ (n is integer) and the above argument holds for $\theta \neq n\pi/2$.

In this paper we have constructed a model for nonequilibrium structure formation in illuminated liquid-crystalline monolayers. We have demonstrated numerically and theoretically that our model well describes most of qualitative features of the phenomenon observed in the experiments. We believe that our model forms a new class of pattern-forming system in soft condensed matter far from equilibrium.

ACKNOWLEDGMENTS

We would like to thank Professor A. S. Mikhailov for valuable discussions.

REFERENCES

1. Tabe, Y. and Yokoyama, H., *Langmuir* **11**, 4609–4613 (1995).
2. Mikhailov, A. S. and Ertl, G., *Science* **272**, 1596–1597 (1996).
3. Tabe, Y., Yamamoto, T., and Yokoyama, H., *New J. Phys.* **5**, 65.1–65.11 (2003).
4. Selinger, J. V., Wang, Z.-G., Bruinsma, R. F., and Knobler, C. M., *Phys. Rev. Lett.* **70**, 1139–1142 (1993).
5. Tabe, Y. and Yokoyama, H., *J. Phys. Soc. Jpn.* **63**, 2472–2476 (1994).
6. Tabe, Y., Shen, N., Mazur, E., and Yokoyama, H., *Phys. Rev. Lett.* **82**, 759–762 (1999).
7. Meyer, R. B. and Pershan, P. S., *Solid State Commun.* **13**, 989–992 (1973).
8. Reigada, R., Sagués, F., and Mikhailov, A. S., *Phys. Rev. Lett.* **89**, 038301-1–038301-4 (2002).

Molecular Dynamics Simulation of Liquid Crystalline Polymer Networks and Flexible Polymer Network in Liquid Crystal Solution

A. Zarembo [1], A. Darinskii[2], I. Neelov [1,2,3], N. Balabaev[4], F. Sundholm[1]

[1] *Laboratory of Polymer Chemistry, University of Helsinki, P.O.Box 55, FIN-00014, Helsinki, Finland*
[2] *Institute of Macromolecular Compounds, St.Petersburg, Bolshoi pr.31, 199004, Russia*
[3] *IRC in Polymer Science&Technology, University of Leeds, LS2 9JT, Leeds, UK*
[4] *Institute of Mathematical Problems of Biology, Pushchino, 142290, Moscow region, Russia*

Abstract. Molecular dynamics have been carried out for a flexible regular tetrafunctional network swollen in a low molecular liquid-crystal (LC) solvent. The LC solvent comprises of anisotropic rod-like semiflexible linear molecules (mesogens) composed of particles bonded into the chain by FENE potential. Rigidity of LC molecules was induced by a bending potential, proportional to the cosine of the angle between adjoining bonds. All interactions between nonbonded particles are described by a repulsive Lennard-Jones potential. The size of the network elementary cell was chosen to be close to the length of the mesogen. Simulated systems differ from each other by the volume fraction of LC solvent. For comparison the simulation of network swollen in monomer solvent was carried out. The static and dynamic characteristics of the systems were studied. Both nematic and smectic phases are observed for LC in the network but their density regions are shifted to higher densities in comparison to the pure LC. The presence of the LC solvent results in the anisotropy of translational diffusion of the network. Anisotropy of diffusion increases with increase of the fraction of LC solvent. Computer code was modified to simulate the transformation of these systems into liquid crystalline polymer networks. For such a system the smectic phase is observed at smaller densities as compare to the nonpolymerized LC+network system. The effect of polymerization on the translational mobility of polymer network was studied.

INTRODUCTION

Polymer-low molecular liquid crystal composites are of the great interest due to many applications in electrooptics [1]. There are a lot of experimental works devoted to such systems, but much less attention is paid to the modeling of polymer-LC composites. Recently we applied the method of molecular dynamics to simulate polymer-dispersed liquid crystal using the simple model of flexible regular polymer network swollen in low molecular LC solvent [2]. As a model of low molecular LC the system of semiflexible rod-like molecules consisting of linear sequence of 7 beads was used. A network was described using the model of the regular tetrafunctional diamond-like network consisting of flexible chains composed of 31 beads each. The network beads were similar to those of the solvent. The beads are bonded by the FENE potential. To induce rigidity into LC solvent the bending potential, proportional to the cosine of the angle between adjoining bonds was used. All the beads in the system interact by repulsive Lennard-Jons potential. The network fraction was varied in rather narrow range from to 0,26 to 0,33. The procedure of preparation, simulation and equilibration of the systems at different densities is described in detail in our recent work [2]. The systems obtained after sufficiently long trajectory simulation were transformed into liquid crystalline polymer network with side chain mesogens. Molecules are polymerized by introduction a new FENE potential between a bead belongs to network and a bead belongs to LC molecule. Polymerization is carried out using NVT ensemble, degree of polymerization is about 99%.

For the non-polymerized system it was shown that by increase of the density the low molecular solvent immersed into the polymer network goes through the same LC phase states as the pure solvent does: the smectic phase (S), the nematic (N) and the isotropic (I) phase with the density decrease. The presence of the network only shifts the region of the isotopic-nematic (I-N) transition to larger total densities but does not influence the position of the nematic-smectic (N-S) transition. The periodicity of the LC phase determines the morphology of the network.

The polymerization of the system shifts the position of the N-S transition to the lower densities and decrease the order parameter of the system. Here we will present the effect of polymerization on the translational mobility of polymer network. The result are presented for the system with the initial polymer network fraction equal to 0,288.

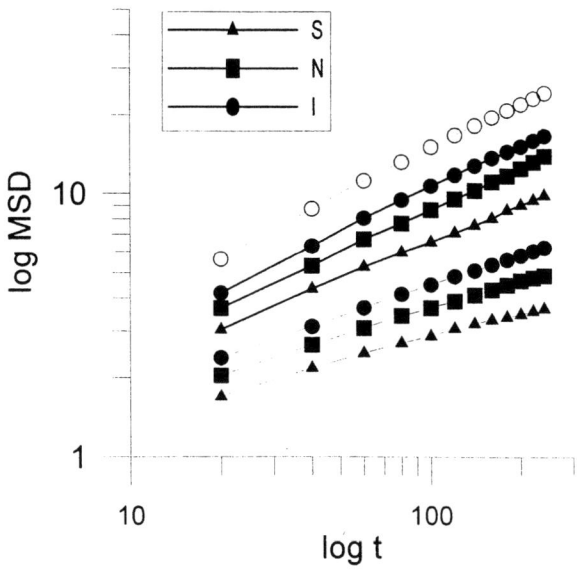

FIGURE 1. Time dependences of MSD of middle chain monomers of network in non-polymerized (solid lines) and polymerized (dashed lines) system. Open circles- network in monomer solvent.

RESULTS AND DISCUSSION

To investigate the translational mobility of the network the mean square displacements (MSD) for middle monomers of network chain and crosslinks in different LC phases were calculated. For the non-polymerized system the simulated time dependence is close to linear in the log-log scales (see solid lines in Fig.1). The slope is about 0.53 for all systems except of the smectic phase, where slope is less – 0.46. The translational mobility of chain monomers in the monomer solvent is larger than that in isotropic phase of LC solvent but the slope of is about the same. The difference between these two systems can be prescribed to the difference of viscosities of solvents consisting of short and long particles.

There is no anisotropy of diffusion for the network crosslinks and middle monomers in the isotropic state. However when the solvent is in LC states the anisotropy of the diffusion of middle chain monomer as well as crosslinks is also observed. In the smectic state the anisotropy of diffusion is practically the same for MSD of crosslinks and monomers. In the nematic state the diffusion anisotropy of chain beads is larger than that for crosslinks.

After the polymerization the translation mobility of the network decrease, the time dependence of MSD remains close to the linear one and the slope of middle monomer MSD become in the range from 0,31 to 0,39 (see dashed lines in Fig.1). The anisotropy of diffusion has the same character as for non-polymerized system, but becomes more pronounced in the smectic phase.

ACKNOWLEDGMENTS

We are very grateful to Professor K.Kremer for helpful discussions. This paper is a part of the SUPERNET network research of European Science Foundation and INTAS project 99-1114. The financial support from the Academy of Finland and the Magnus Ehrnrooth foundation is gratefully acknowledged. Russian participants are grateful to the Russian Foundation of Basic Research (grants 02-03-33135).

REFERENCES

1. Crawford, G.P., and Zumer, S., "Historical perspective of Liquid Crystals Confined to Curved Geometries", in *Liquid Crystals: Applications and Uses*, edited by B. Bahadur, World Scienific, Singapore, 1990, pp.1-20.

2. Darinskii, A., Zarembo, A., Balabaev, N., Neelov, I., Sundholm, F., *Polymer* (2003), submitted.

Irreversible Sequential Adsorption of Line Segments with Diffusional Relaxation on a One-Dimensional Lattice

Kyung Eun Lee*, Heung Sik Park* and Jae Woo Lee*

Department of Physics, Inha University, Incheon 402-751 Korea

Abstract. We consider a random sequential adsorption of line segments(k-mer) with diffusional relaxation. The line segments of a lenght k depsosit with a probability p or diffuse up to a hopping length $l(l \leq k)$ with a probability $1-p$ on a one-dimensional lattice. For the dimer, the empty area fraction decays according to $1 - \theta(t) \sim [(1-p)pt]^{-1/2}$, regardless of the diffusion length and the adsorption probability. For $k \geq 3$, the empty area fraction decays according to the power law as $1 - \theta(t) = A(k,l)[(1-p)pt]^{-\alpha(k,l)}$. The decaying exponents depend on the length of the line segment and the depositing probability p. The kinetics of the empty area fraction of the dimers is equivalent to the diffusion-limited reaction, $A + A \to 0$, at the long time limits. However, for $k \geq 3$, the model with $l > 1$ stepping corresponds to reactions where the particles (gaps of size l) hop in a correlated way. We found that new power law behavior for l-group-diffusion limited k-particle reactions and the exponents of the power law depend on the hopping length l.

INTRODUCTION

The random sequential adsorption(RSA) is a typical model for the irreversible and sequential adsorption of macromolecules. In the lattice model of RSA, the coverage fraction $\theta(t)$ of the monolayer surface shows the exponential behavior, such as $\theta(t) = \theta(\infty) - B\exp(-Ct)$ where $\theta(\infty)$ is the jamming limits at the long time and B, C are the constants. When the deposited particles are subject to diffusion, the coverage fraction of the surface approaches the close-packing limit on the one dimensional lattice[1]. Privman and Nielaba reported that the coverage fraction approaches the jamming limit according to a power-law $\theta(t) = 1 - Bt^{-1/2}$ for the dimer on a one-dimensional lattice with diffusional relaxation[2]. Fusco et. al reported the RSA and diffusion of dimers and k-mers on a square lattice. They observed that the coverage fraction decays with $t^{-1/2}$ as a leading term[3].

We consider the random sequential adsorption of line segments with diffusional relaxation on a one-dimensional lattice by using Monte Carlo method. The line segments with a length k deposit with a probability p or diffuse up to a hopping length $l(l \leq k)$ with a probability $1 - p$.

MONTE CARLO METHOD

Consider an initially empty one-dimensional lattice of size $L = 10^4$ with a periodic boundary condition. A line segment of length k is deposited with a deposition probability p or diffuses with the probability $1 - p$. The diffusion length is greater than $l \geq 1$. Let's choose a site randomly. If a chosen site is occupied by a deposited object, the diffusion is tried with the probability $1 - p$ when the l-consecutive neighbors are empty. The diffusion direction is selected randomly. When the dimer diffuses, they choose equally the left or right direction. When there are l-consecutive empty sites, the dimer moves to l-lattice sites to the chosen direction. If a chosen site is empty, we try the deposition attempts with a probability p. In the deposition trial, we check the chosen site and a nearest neighbor site. If all checked sites are empty, we deposit the chosen line segment. Otherwise, we reject the deposition trial. One Monte Carlo time is defined by the total number of attempts to select a site divided by the lattice size. The data are averaged over 100-independent runs for each choice of parameters.

RESULTS AND DISCUSSION

For the dimer $k = 2$, the empty area fraction decays according to a power-law $1 - \theta(t) = Bp^{0.68}(1-p)^{0.10}[(1-p)pt]^{-0.50}$, regardless of the hopping length and the adsorption probability, where B is a constant. These observations are consistent with the previous results of Privman and Nielaba for $l = 1$[2]. The kinetics of the empty area fraction of the dimers is equivalent to the diffusion-limited reaction(DRL), $A + A \to 0$, at the asymptotic limit, where A is a chemical reactant[4, 5]. In this case a single empty site at the RSA corresponds to the reactant A at the DRL.

For $k \geq 3$, the empty area fraction also decays according to the power-law, such as $1 - \theta(t) = B(k,l)[(1-p)pt]^{-\alpha(k,l)}$, where $B(k,l)$ is a constant depending on k and l. For $k = 3$, there is a logarithmic correction for $l = 1$. The decaying exponents $\alpha(k,l)$ depend on the length of the line segment k and the hopping length l. In Fig. 1 we gave the log-log plot of the empty area fraction $1 - \theta(t)$ as a function of pt for $k = 4, l = 1$ with the adsorption probabilities $p = 0.1, 0.5, 0.9$. In the inset figure of Fig. 1, we presented the collapses of the data $[1 - \theta(t)][p(1-p)]^{\alpha(k,l)}$ agagint the time t with $\alpha(k,l) = 0.213(7)$.

For $k \geq 3$, the kinetics of the empty area fraction is not interpreted by the kinetics of the ordinary diffusion-limited reaction $kA \to 0$. For $k \geq 3$, the model with $k > 1$ stepping corresponds to the reactions where the particles(gaps of size l) hop in *a correlated way*. Thus, our model of l-group diffusion-limited k-particles reactions is different from those of the ordinary reaction $kA \to 0$. We found the new power-law behavior for l-group diffusion-limited k-particles reactions and the exponents of the power-law depends on the hopping length l and the length of the line segment[6].

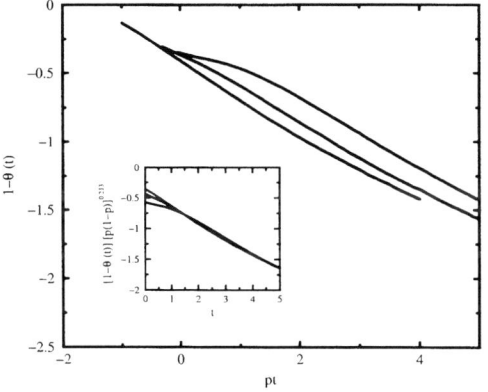

FIGURE 1. Log-log plots of the empty area fraction $1 - \theta(t)$ as a function of pt for $k = 4, l = 1$ with the adsorption probabilities $p = 0.1, 0.5$, and 0.9 from bottom to top. Inset: the data collapses $[1 - \theta(t)][p(1-p)]^{\alpha(k,l)}$ against the time t for $k = 4, l = 1$ with $\alpha = 0.213(7)$.

We consider the vacancy dynamics of 4-mers. We calculated the total number of gaps such as the number of an isolated empty sites $N_1(t)$, of the isolated two contiguous empty sites $N_2(t)$, of the isolated three contiguous empty sites $N_3(t)$, of the isolated four contiguous empty sites $N_4(t)$, of the isolated contiguous empty sites greater than four contiguous empty sites $N_>(t)$, and the total number of empty sites $N_t(t)$. For $k = 4, p = 0.9$ and $l = 1$, we observed that the gap greater than $g \geq 3$ decays rapidly at the late stage and the dynamics is controlled by the $g = 1, 2$. The density of gaps also follows the power law such as $N_1(t) \sim t^{-0.18(4)}$, and $N_2(t) \sim t^{-0.30(5)}$. For $k = 4$ and $p = 0.9$ the $g = l$ vacancies control the dynamics of the gaps. We obtained the power law as $N_2(t) \sim t^{-0.30(4)}$, $N_1(t) \sim t^{-0.43(5)}$ for $l = 2$ as shown in Fig. 8 (b), $N_3(t) \sim t^{-0.42(5)}$, $N_1(t) \sim t^{-0.52(3)}$ for $l = 3$, $N_4(t) \sim t^{-0.55(5)}$, $N_1(t) \sim t^{-0.57(6)}$ for $l = 4$. The gap dynamics is controlled dominantly by the gap of the diffusion length, i.e., $g = l$, and the isolated one-site vacancies $g = 1$ at the late stage. The dynamics of empty area fraction is also determined by the gap creations and annihilations by the diffusion process of the line segments.

CONCLUSION

We consider the random sequential adsorption of the line segment with diffusional relaxation on a one-dimensional lattice. For the adsorption and diffusion of the dimer, the kinetics of RSA is equivalent to the diffusion-limited reaction, $A + A \to 0$. For the trimer the empty area fraction has a logarithmic correction, but the exponent of the empty area fraction depends on the diffusion length of the trimer. For $k \geq 4$, the empty area fraction follows a power-law as $1 - \theta(t) = A(k,l)[p(1-p)t]^{-\alpha(k,l)}$ where $A(k,l)$ is a constant and the exponent $\alpha(k,l)$ depends only on the length of the line segment k and the diffusion length l. We found that the our model is equivalent to l-group-diffusion limited k-particle reactions. For $k \geq 4$, there is a mixed dynamics of the gap creations, splitting, and annihilations at the long time.

ACKNOWLEDGMENTS

This work has been supported by Inha University Research Grant(INHA-30205).(2003)

REFERENCES

1. Evans J. W., *Rev. Mod. Phys.* **65**, 1281-1329 (1993).
2. Privman V., and Nielaba N., *Europhys. Lett.* **18**, 673-678 (1992).
3. Fusco C., Gallo P., Petri A., and Rovere M., *cond-mat/0103582*.
4. Lee J. W., *J. Chem phys.* **113**, 9702-9705 (2000).
5. Lee J. W., *Phys. Rev. E* **62**, 2959-2962 (2000).
6. Hong B. H., and Lee J. W., *J. Chem. Phys.* **119**, 533-537 (2003).

Complex History-Dependent Freeze-Slip Transition in the Population of Phase Oscillators

Akinori Awazu

Department of Pure and Applied Sciences University of Tokyo
Komaba 3-8-1, Meguro-ku, Tokyo 153-8902, Japan

Abstract. History-dependent transition between the slipping and freezing states is investigated by use of a dynamical system with repulsive particles in the external force. In this system, we found that the freeze-slip transition point is multi-valued and strongly influenced by their histories. We show non-trivial relations between transition points in some cases with seven or eight particles system; I) The threshold-type response of the system against change of external force is observed that obeys a rule 'if A, then B'. II) The system realizes not only the freeze-slip transition obeying a rule but also the change of the transition rule itself.

We study a 1-dimensional system containing N particles on a circle whose length is L. The particles have repulsive interactions and are under a uniform external force. As heterogeneity by particles, we introduce the dispersion of the sensitivity to the external force in each particle. The motion of each particle is given by the following equation,

$$\dot{x}\{i\} = \sum \theta(1/2 - |x\{i\} - x\{j\}|) \sin 2\pi(x\{i\} - x\{j\}) + c\{i\}F$$

where $x\{i\}$, $a\{i\}$, and $c\{i\}$ are the i th particle's position ($x\{i\}$ ... mod L), interaction strength, and sensitivity, and F is the external force ($\theta()$ is the step function.). In this model, the maximum value of repulsive forces working between particles is finite and two particles can exchange their relative positions.

Figure 1 gives a typical temporal evolution of the position and velocity of each particle. Here, we set $c\{i\} = (N-2i-1)/N$. This system has basically two steady states; I) The freezing state in which all particles stop as in the upper and the lower parts of Fig. 1, II) the slipping state in which at least one particle continues to rotate and changes their relative positions as in the middle of Fig. 1. In slipping state, the motion of each particle is always periodic. In general, the freezing state is realized if |F| is small, while the slipping state is realized for large |F|.

Figure 2 shows the value of the force $F\{c\}$ at which the system shows a transition from freezing to slipping by the increase of |F| [2]. It is plotted as a function of N/L for N=4~9. For the case with N/L > 3, $F\{c\}$ has more than three or four values as expected previously while $F\{c\}$ has only one or two values when N/L < 3.

Then, we focused on the correlations between each transition point for cases with multi-valued $F\{c\}$. In order to clarify the correlations of each $F\{c\}$, we plot transition diagrams for each condition [3]. By using these diagrams, one can select an aimed state by operating F as such. When N is larger with N/L>3, we can obtain complicated diagrams.

In some cases, these complex diagrams also contain some structures. For example, for cases with N=7 and L=1.8 (Fig.3), we find a bifurcation as follows. We start from the state described by a solid box [F<0.8 (F>-0.8)]. Through the operations, 1) decrease F, 2) increase F until system start to slip, 3) decrease F, we have found following rules of freezing-slipping transition. If F is decreased to the range (i) 0 ~ -0.49, (ii) -0.5 ~ -0.6, (iii) -0.6 ~ -0.78, (iv)-0.79 ~ -0.84 by the operation 1), and then $F\{s\}$ takes the value (i) 0.8, (ii) 0.85, (iii) 0.8, (iv) 0.81 by the operation 2). By the operation 3), the system comes back to the starting box.

As an example of most complex cases, we show the diagram for the case with N=8 and L=1.2 (Fig. 4), where hierarchical transitions are observed. In this case, the system obeys simple transition rules described in the three colored areas, respectively, if F changes smoothly in a given amplitude. However, if F changes specifically, for example to take to large value, transitions between these three areas occur. This means the freeze-slip transition rule itself is changed.

Here, the sign * over boxes in Fig. 4 mean that the transition indicated arrows from these boxes sometimes do not occur and the system switches to any other states. The occurrence of these transitions depends on the timing of the change from $F\{before\}$ to $F\{after\}$. Such timing dependency appears when the phase space orbit (limit cycle) of the system with F{before} crosses to two or more attractors of the system with F{after}. In this case, the occurrence of the transition from area I or area III to area II depends on the timing of the change of F, while the transition of the opposite direction is independent of the timing. Hence, the transition between each area can be asymmetric.

History-dependent phenomena are observed in many situations in physics and biology. Although the model is simple and abstract, the result here may be relevant to understand adaptations, memory, and functional response in a biological system.

REFERENCE

1. Awazu. A., Physica D **178**, 19-25 (2003).
2. In this paper, F varies always quasi-statically; i.e., F is incremented by 0.01 after waiting long enough for the system to settle down to an attracter.
3. (Manual for diagrams.) In the diagram, each arrow represents path of operations. The signs + and − mean quasi-static increase and decrease of F. 'Slip' means the occurence of the transition from a freezing to a slipping state, and 'slip stop' means the opposite transition. Solid boxes and solid arrows indicate the operations in freezing state, and dotted boxes and arrows indicate the operations during slipping.

In order to simplify the transition diagrams, we prepare two types of operations, Regular operation and Opposite operation. Regular operation is written by $+$ and $-$, and Opposite is written by (+) and (−). For Regular operations, the value written out of ( ) in each box gives a threshold of F, while for Opposite operations, the value written in ( ) gives a threshold of F. At an arrow by the crossing arrows sandwiched in between +, −, (+) and (−), the type of operations changes from Regular to Opposite, or from Opposite to Regular. As an initial condition, we set F at large enough value, and decrease to 0 quasi-statically.

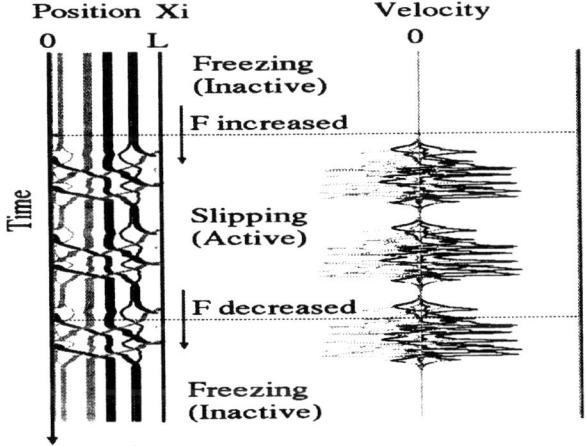

FIGURE 1. Temporal evolution of each particle.

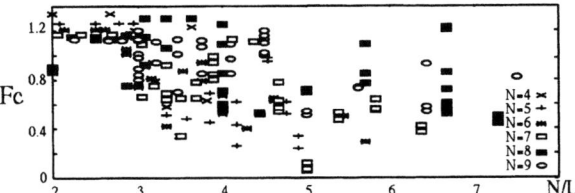

FIGURE 2. F{c} as a function of N/L for N=4 ~9

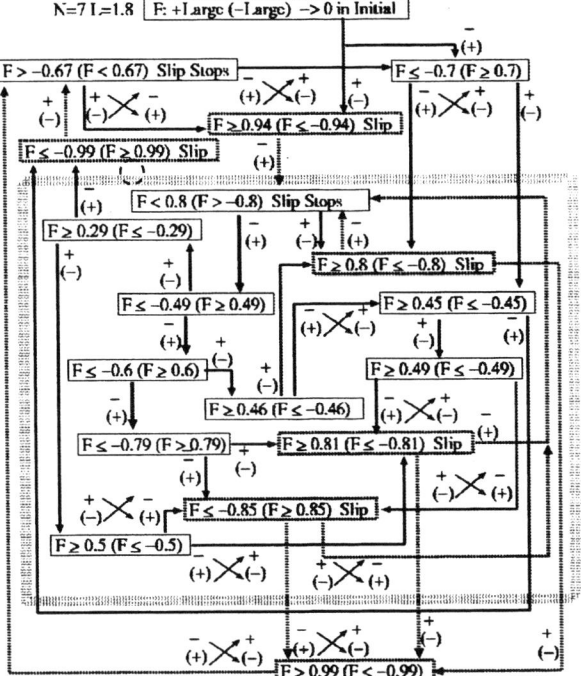

FIGURE 3. Transition diagram for N=7 and L=1.8

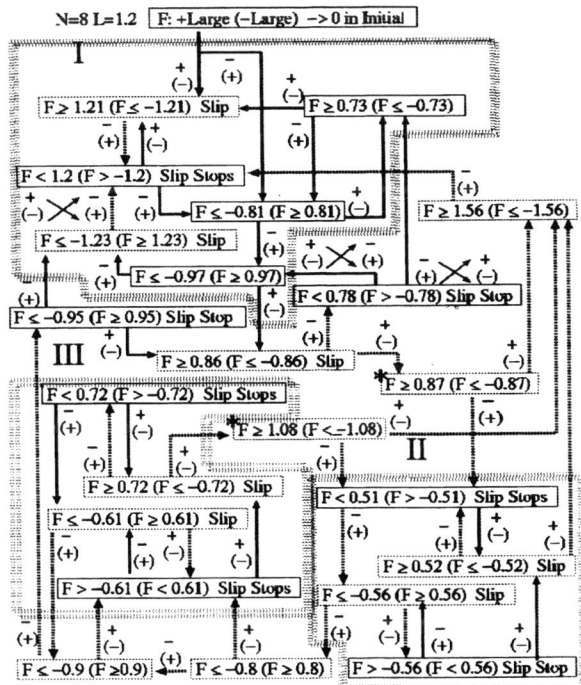

FIGURE 4. Transition diagram for N=8 and L=1.2

One Idea of Portfolio Risk Control Focusing on States of Correlation

Noboru Nishiyama

Asahi Life Asset Management Co,. Ltd.
34TH FLOOR, SHINJUKU CENTER BLDG.
1-25-1, NISHI-SHINJUKU, SHINJUKU-KU,
TOKYO, 163-0657 JAPAN

Abstract. In the modern portfolio theory there are 2 major risk parameters that mean and variance. Correlations should be playing important role as well but variance is thought to be most important risk parameter for risk control in the theory. I focused on states of correlation to calculate eigen values as risk control parameter.

INTRODUCTION

In this paper I try to bring eigen value into portfolio risk control. I analyzed nonlinear type of fluctuation in the market to capture complicated market movement using threshold point and risk adjustments by scenario correlation as implicit signals. Threshold becomes control parameter of risk exposure to set downside floor and forecast extreme nonlinear type of fluctuation under a certain probability

CORRELATION AND RISK IN THE MODERN PORTFOLIO THEORY

In the modern portfolio theory, total risk can be divided into two parts. First part is systematic risk, which is not able to reduce completely and second part is unsystematic risk, which is able to reduce by increasing number of assets.

Here I have alternative idea that total risk can be decomposed into three parts and they are market risk, group risk and specific risk.

$$\sigma^2_{INDEX} = \sum_i a_i^2 \sigma_i^2 + \rho \sum_i \sum_{i \neq j} a_i a_j \sigma_i \sigma_j \quad (1)$$

In formula (1), idea of two decomposition that first term of right hand side is unsystematic part and second term of right hand side is systematic part which consists of ρ defined as group risk which is introduced in this paper and product sum of individual standard deviations. So far in practice, in order to control specific risk, we focus on first term of right hand side and increasing number of assets

Group risk is variable parameter between systematic and unsystematic, which was dominated by correlation structure. Portfolio risk monopolized by whether systematic risk or unsystematic risk. If correlation were 'one neighborhood' then group risk would become almost systematic risk and impossible to control. If correlation were 'zero neighborhood' then group risk would become unsystematic risk and possible to control portfolio risk by increasing number of assets.

CORRELATION AND EIGEN VALUE

In order to capture group risk, I conduct singular decomposition of the correlation and consider discrete distribution of Eigen Value. Below two graphs show simulation results. At Figure1 vertical axis is eigen value and horizontal axis is eigen number of two hypothetical correlation matrix $\rho = 0.1$ and $\rho = 0.9$. In the case of special correlation, the higher ρ is, the more shrinkage eigen value No.1 as F1. At Figure2 vertical axis is eigen value and horizontal axis are nine hypothetical various correlation matrix increase from ρ =0.1 as left to ρ =0.9 as right by 0.1. The higher ρ are in the matrix, the wider difference between maximum eigen value and minimum eigen value are calculated.

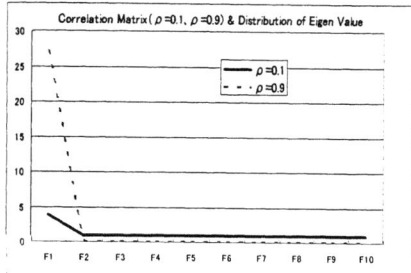

FIGURE 1. Distribution Eigen Value descending Eigen No.

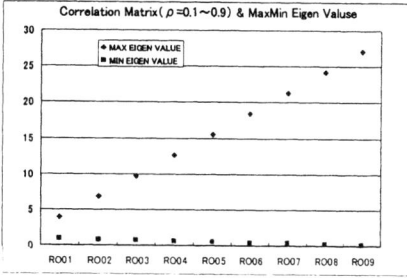

FIGURE 2. Distribution Eigen Value ascending hypothetical correlation.

I looked into states of correlation and assumed its behavior would be ex-ante signals of market crisis. Almost all people in the financial world have already known that correlation is important and employ Markowitz's Mean-Variance Approach. They accept minimizing Variance could be sufficient methods.

In this research I emphasized importance of correlation rather than variance itself. Studies in terms of correlation from Econophysics field could appeal to the practical financial phenomena.

The calculation method of scenario correlation is as follows.

R : stress scenario correlation matrix

R* : scenario correlation matrix

R0 : real correlation matrix

$$R^* = (1-\alpha)R + \alpha I \quad (2)$$

If $\alpha = 1$ then R*=I $\to$ No correlation among components

If $\alpha = 0$ then R*=R $\to$ Identical with stress scenario correlation

α is a coefficient for calculation of scenario correlation. λ are eigen value which generated from correlation matrix.

$$\lambda_{MAX} = \lambda_1 \geq \lambda_2 \geq \cdots \geq \lambda_N = \lambda_{MIN}$$

$$\lambda_{MAX\,\lambda\geq1}(R) = \underset{x\neq0}{Max}\frac{x'Rx}{x'x} = \left[\underset{x'x=1}{Max\, x'Rx}\right]$$

$$\lambda_{MIN\,\lambda\geq1}(R) = \underset{x\neq0}{Min}\frac{x'Rx}{x'x} = \left[\underset{x'x=1}{Min\, x'Rx}\right]$$

$$\lambda_{MAX\lambda\geq1}(R^*) - \lambda_{MIN\lambda\geq1}(R^*)$$
$$= (1-\alpha)[\underset{max\,\lambda\geq1}{\lambda}(R) - \underset{min\,\lambda\geq1}{\lambda}(R)] \quad (3)$$

I summarize relationship between α and risk exposure.

If α approach to one, then group risk would be lower and higher risk exposure can be taken at portfolio construction.

If α approach to zero, then group risk would be higher and lower risk exposure can be taken at portfolio construction.

REFERENCES

1. Nishiyama, N., *Physica A* **301**, 457-472 (2001)

2. Stanley, H. E. and Mantegna R. N., *Japanese translation of An Introduction to Econophysics: Correlation and Complexity in Finance*, Economist Co,. Ltd., Tokyo, 2000, pp134-135

Dynamics in a Complex-Fracture-Subterranean-System with Application to HDR Geothermal Reservoirs

Kei Yoshida, Sergei Fomin, Zhenzi Jing and Toshiyuki Hashida

Fracture Research Institute, Tohoku University, Sendai 980-8579, Japan

Abstract. Hydraulic fracturing or hydraulic stimulation is one of the most effective methods of enhancing hot dry rock (HDR) geothermal system productivity. In the present study, network models of "fractal geometry" approximate a 3D structure of fractured rocks. The fracture network models are generated by distributing fractures randomly in space and assuming the fractal equation correlating the number of fractures and fracture lengths. Based on this approach, a mathematical model of hydraulic rock fracturing is proposed. The model incorporates approximations of the fracture mechanical behavior drawn from the rock mechanics literature, a very simplified analysis of the operative physical processes, and mapping of the connectivity of fracture network to a cubic regular grid. It is assumed that the flow properties of a stochastic fracture network depend on the fluid pressure. Along with the fractal-type distribution of the fracture lengths, the fracture surfaces are also assumed to follow fractal geometry. The latter allows numerical simulation of the natural rock fracture dilation caused by fracture shear offset. On of the problems that can be resolved by fracture surface modeling is the apparent limitation on the number of fractures that can be analyzed experimentally. In this respect, the suggested mathematical model can be used to simulate fractal surfaces identical to fractures found in the natural rocks. Taken together, these approaches permit the approximate engineering resolution of the multi-parametric, highly complex mechanical problem.

The numerical model FRACSIM-3D developed by the research group in Tohoku University is proved to be an appropriate approximate model capable to address the problems associated with hydraulic stimulation and can quantitatively predict the 3D reservoir growth behaviour. One of the aspects of the general problem of reservoir modelling is the proper approximation of the fracture distribution within the rock. In fractured rocks, groundwater flow occurs predominantly through the connected network of discrete fractures. A series of geophysical investigations has confirmed that subsurface fracture networks can commonly be described by fractal geometry. The model presented here is based on the relationship between the fracture length and a number of fractures, as suggested in [1]. It incorporates the elements of the approximate model discussed in [2], where the fracture shear displacements and openings, variation of the shape of the stimulating rock volume, pressure compliant fracture apertures are taken into account. The natural rock fracture surfaces, which are proved to be of fractal geometry as well [3], are modelled on the basis of the spectral synthesis method [4], which accounts for the fractal nature of the fracture surfaces. This model is used for numerical simulation of the fracture dilation due to the offset of the fracture surfaces and accounts for the experimentally acquired data of the mating rough fracture surfaces profiles.

The models of fracture networks are generated by distributing penny-shaped subsurface fractures randomly in space and assuming the fractal correlation $N_r = Cr^{-D}$ that incorporates the fracture length r, number of fractures N_r whose characteristic length is greater than r, fractal dimension D, and fracture density within the rock mass C. Hence, the number of fractures between the specified upper and lower limits is given by $N_{r_{min}}^{r_{max}} = C(r_{min}^{-D} - r_{max}^{-D})$, where r_{min} and r_{max} are the lower and upper fracture radius limits, respectively. Consider some fraction, α, of this total number counting from r_{min} upward and the corresponding size r_α of the largest object in that fraction, then $N_{r_{min}}^{r_\alpha} = \alpha C(r_{min}^{-D} - r_{max}^{-D}) = C(r_{min}^{-D} - r_\alpha^{-D})$. The latter yields the fractal fracture length distribution $r_\alpha = [(1-\alpha)r_{min}^{-D} + \alpha r_{max}^{-D}]^{-1/D}$, where α is a random parameter in the interval [0,1]. Using this equation, fracture r_α can be generated by simply changing the α value. For any generated fracture, the initial (i.e. undisturbed) fracture aperture, a_0, when evaluated at zero effective stress, is assumed to be proportional to the fracture radius, $a_{i0} = \beta \cdot r_i$, where $i=1, 2,\ldots, N_f$, and r_i is the radius of each fracture from the whole set of fractures N_f; β is a constant of proportionality, which is chosen to allow the undisturbed fracture network to match (at least approximately) the *in situ* measured permeability. Fracture apertures are affected by the effective normal stress at the fracture surface and by shear displacement that determines the fit of the opposing rough surfaces. Shear stability is expressed by a simple friction law, when slip is taking place if $\tau > (\sigma_n - P)\tan(\phi_i + \phi_{dil}^{eff})$, where τ is a shear stress, σ_n is the rock stress normal to the fracture surface, P is a fluid pressure in fracture, and ϕ_{dil}^{eff} is an effective shear dilation angle at a given effective normal stress. The basic friction angle ϕ_i is a material property of the fracture walls. The effective shear dilation angle ϕ_{dil}^{eff} is a property of both the fracture wall asperities and effective normal stress

$$\tan(\phi_{dil}^{eff}) = \tan(\phi_{dil})/[1 + 9(\sigma_n - P)/\sigma_{nref}], \quad (1)$$

where σ_{nref} is the effective normal stress applied to cause a 90% reduction in the aperture and ϕ_{dil} is the shear dilation angle at very low effective stress. The amount of shear displacement depends on the fracture shear stiffness and on the amount of "excess" shear stress available. Referring to the theory of elasticity, the shear displacement of a fracture U can be expressed as:

$$U = [\tau - (\sigma_n - P)\tan(\phi_i + \phi_{dil}^{eff})]/H_s, \quad (2)$$

where H_s is a shear stiffness of the fracture. The change in aperture is a product of the displacement and the tangent of the effective shear dilation angle, $a_s = U \tan(\phi_{dil}^{eff})$. An expression for the aperture a of a sheared fracture is following

$$a = (a_0 + U \tan(\phi_{dil}))/[1 + 9(\sigma_n - P)/\sigma_{nref}]. \quad (3)$$

For the typical conditions of the geothermal reservoir exploitation, flow is laminar and the linear Darcy momentum equations can be employed, namely, $u_m = -(K_m/\mu)(\partial P/\partial x_m)$, where $m=1, 2, 3$ and K_1, K_2, K_3 are the diagonal components of the permeability tensor. Accounting for the mass conservation equation, the Darcy flow model leads to the following equation for pressure distribution

$$\sum_{m=1}^{3} \partial[K_m(\partial P/\partial x_m)]/\partial x_m = 0, \quad (4)$$

The above model contains the unknown value of shear dilation angles ϕ_{dil}, which can be predicted for the specified rock by analyzing the effect of shears displacement of the syntactic fracture on the fracture dilation. This can be made with analytic-experimental method (it couples the analytical algorithm for fracture mathematical modeling and experimental data for the natural fracture topography). The algorithm of natural fracture surface simulation involves: (i) measuring the fracture surface topography; (ii) computing the power spectral density for the fracture surface as a function of spatial frequency; (iii) plotting this function in logarithmic coordinates and computing the plot's slope β ($1 \leq \beta \leq 3$) and surface fractal dimension $D_s = (5-\beta)/2$; (iv) applying these parameters to constructing synthetic fractal surfaces that share the same physical properties as the natural fracture surfaces. A detailed description of the spectral synthesis method in application to fractal surface modeling and corresponding computer codes can be found in [3] and [4]. To model the fracture dilation induced by shear offset, two synthetic fractal surfaces that make up the fracture can be assumed to match each other perfectly—the upper surface is been taken to be an exact but inverted duplicate of the lower surface. If these surfaces are laterally offset, they maintain contact at least at one point. If fracture asperities are assumed to be unabrading, then this schematic geometric approach for modeling the shear induced growth of the fracture aperture is quite reasonable [3]. If there is no interpenetration, increasing the lateral displacement of the synthetic surface causes the aperture to increase, which can be readily computed in geometrical terms. Our computations show that fracture dilation is more pronounced for the fractal surfaces with lower fractal dimension (greater β). After the variation of fracture dilation *vs* shear displacement is computed, the shear dilation angle ϕ_{dil} can be obtained (and subsequently substituted into (3)) as the mean angle between the tangent to the shear dilation plot and the horizontal axis. A series of computations was carried out to estimate the effect of the fracture fractal dimension D_s and fracture length on of the shear dilation angle and of the fractal dimension of fracture distribution D and fracture density C on the reservoir volume.

REFERENCES

1. Watanabe, K. and Takahashi, T., *J. Geophys. Res.* **100**, 521-528 (1995).
2. Jing, Z., Willis-Richards, J., Watanabe, K. and Hashida, T., *J. Geophys. Res.* **105**, 23663-23679 (2000).
3. Brown, S.R., *J. Geophys. Res.* **100**, 5941-5952 (1995).
4. Peitgen, H.O. and Saupe, D., *The science of Fractal Images*, Springer-Verlag, New York, 1988.

The stability of the critical scaling against the time-dependent perturbation

Heungsik Park[*] and Hyunggyu Park[†]

[*]Department of Physics, Inha University, Inchon 402-751, Korea.
[†]School of Physics, Korea Institute for Advanced Study, Seoul 130-722, Korea.

Abstract. We study the stability of critical scaling against the time-dependent perturbation in the contact process(CP) model. The critical probability of the particle varies as $p = p_0 + ct^{-\alpha}$. we perform the static Monte Carlo simulation using the finite size scaling theory in the steady state. For the $\alpha > 1/v_{\|}$, the time dependent perturbation is irrelevant. therefore, the critical exponents $\beta/v_{\|}, \beta/v_{\perp}$ have the DP value. For the $\alpha = 1/v_{\|}$, $\beta/v_{\|}$ is DP value but $\beta/v_{\perp}$ is varied with perturbation strength c. For the $\alpha < 1/v_{\|}$, the particle density is decayed with $\rho \sim t^{\alpha\beta}$ in thermodynamic limit. However, for the all case, z have DP value. To study the stability of critical scaling, we introduce the time-dependent perturbation and know that critical scaling function is satisfied in all cases. Numerical simulations confirm our predictions.

Various kinds of non-equilibrium lattice models exhibiting absorbing phase transitions have been studied extensively during last few decades[1].
Two distinct types of absorbing phase transitions have been identified in one dimension: the directed percolation (DP) and directed Ising (DI) universality class. Most models have been found to belong to the DP class, which involves typically a single absorbing state or multiple absorbing states without any symmetry.
DI-type critical behavior appears in models with two equivalent absorbing states or two equivalent group of absorbing states [2, 3, 4].
In this paper, we study the stability of the critical scaling against the time-dependent perturbation in the contact process. The ordinary contact process is composed of two sub-process: a catalytic creation and a spontaneous annihilation of particles. Randomly selecting particles are annihilated spontaneously with the probability $1-p$ and create the particle on the nearest neighbor empty sites with the probability p.

$$\begin{aligned} A &\to \emptyset \text{ with } 1-p \\ A &\to 2A \text{ with } p \end{aligned} \quad (1)$$

The absorbing configuration is without particle. This model has one absorbing state. Thus, this model belongs to the DP universal class. To study the stability of critical scaling, we introduce the time dependent perturbation. The probability of creation varies as $p = p_c + Ct^{-\alpha}$. Under rescaling, with $\Delta = b\Delta'$ and $t = b^{-v_{\|}}t'$, where $\Delta \equiv p - p_c$, The constants C transforms according to

$$\frac{1}{C'} = b^{1-\alpha v_{\|}}\frac{1}{C} \quad (2)$$

When $\alpha = 1/v_{\|}$, the time-dependent perturbation becomes marginal and the dynamic critical exponents, δ and η vary continuously with the perturbation strength C. However, the sum $\delta + \eta$ remains unchanged. To identify the scaling behaviors near the critical point, we analyze the finite-size effects on the steady-state particle density (ρ_s). Using the finite-size scaling theory on $\rho(\Delta, t, L)$[5],

$$\rho_s(t) \simeq t^{\beta/v_{\|}}, \bar{\rho}(L) \simeq L^{\beta/v_{\perp}}, \tau \simeq L^{v_{\|}/v_{\perp}} \quad (3)$$

$$\rho(\Delta, t, L) = t^{-\beta/v_{\|}} f(\Delta t^{1/v_{\|}}, t^{1/z}/L) \quad (4)$$

where $\Delta \equiv ct^{-\alpha}$.
For the irrelevant case, $\alpha = 1 > 1/v_{\|}$, we perform the static Monte Carlo simulation in $L = 32 \sim 2048$, at the critical point $p_c = 0.767325(5)$ varying c from 0.01 to 0.3.
In small size system, we know that only $\beta/v_{\|}$ is varied with c. We assume that system size is vary large,

$$\begin{aligned} \rho(t, \Delta, L) &= t^{-\beta/v_{\|}} f(ct^b, t^{1/z}/L) \quad (5) \\ &= t^{-\beta/v_{\|}} f(ct^b, 0) \quad (L \to \infty) \quad (6) \end{aligned}$$

where $b \equiv 1/v_{\|} - \alpha$

$$\rho(t, \Delta, L) \sim \begin{cases} t^{-\beta/v_{\|}} & \text{for } t \gg \tau^* \\ t^{-\beta/v_{\|}} f'(ct^b) & \text{for } t \ll \tau^* \end{cases} \quad (7)$$

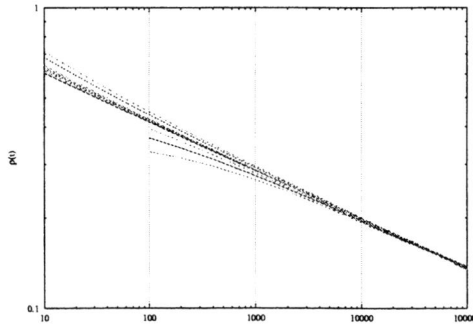

FIGURE 1. $\rho(t)$ versus t from $c = -0.3$ to $c = 0.3$ for irrelevant case($\alpha = 1$

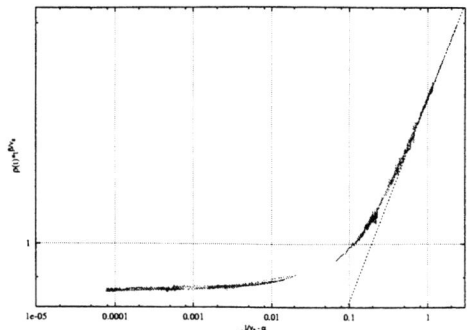

FIGURE 3. scaling plot for relevant case ($\alpha = 0.4$)

particle density function in thermodynamic limit scale as

$$\rho(t,\Delta,L \to \infty) \sim \begin{cases} c^\beta t^{-\alpha\beta} & \text{for } t \gg \tau_{rel} \\ t^{-\beta/\nu_\|} f(ct^b, 0) & \text{for } t \ll \tau_{rel} \end{cases} \quad (11)$$

$$f(\Delta t^{1/\nu_\|}, t^{1/z}/L) \sim \Delta^\beta \text{ for } (t \to \infty, L \to \infty) \quad (12)$$

$$\tau_{rel} \sim c^{-1/b} \quad (13)$$

Using the scaling plot, we measured the order parameter exponent $\beta = 0.2578(5) \simeq \beta_{DP}$.

To study the stability of critical scaling, we introduce the time-dependent perturbation and know that critical scaling function is satisfied in all cases. For $\alpha > 1/\nu_\|$, The time dependent perturbation is irrelevant. Therefore, the critical exponents $\beta/\nu_\|, \beta/\nu_\perp$ have the DP value. For $\alpha = 1/\nu_\|$, $\beta/\nu_\|$ is DP value but $\beta/\nu_\perp$ is varied with perturbation strength c. For $\alpha < 1/\nu_\|$, the particle density decayed with $\rho \sim t^{-\alpha\beta}$ in thermodynamic limit. However, for the all case, z have DP value. Numerical simulations confirm our predictions.

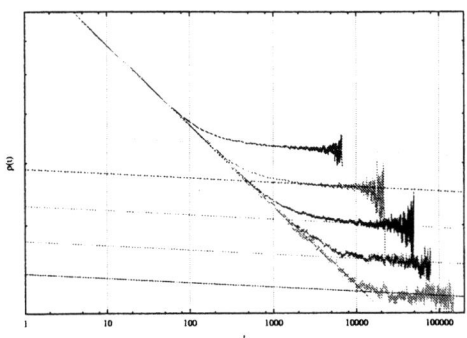

FIGURE 2. $\rho(t)$ versus t at $c = 0.1$ from $L = 64$ to $L = 2048$ for marginal case($\alpha = 1/\nu_\|$

where is $\tau^* \sim c^{-1/b}$ After the characteristic time τ^*, $\beta/\nu_\|(c) = \beta/\nu_\|(DP)$.

Because of $\tau^* \gg \tau_{saturate}$, in small size system, there is seemed that $\beta/\nu_\|$ is varied by c. For the marginal case, $\alpha = 1/\nu_\|$, the particle density function in thermodynamic limit scale as

$$\rho(t,\Delta,l) = t^{-\beta/\nu_\|} f(c, t^{1/z}/L), x \equiv t^{1/z}/L \quad (8)$$

$$\rho(t,\Delta,L) = \begin{cases} f'(c) t^{-\beta/\nu_\|} & \text{for } t \ll \tau_{mar} \\ At^{\frac{-\gamma c}{z}} & \text{for } t \gg \tau_{mar} \end{cases} \quad (9)$$

where is $A \equiv L^{-\beta/\nu_\perp + \gamma c}$
The scaling function $f(c, t^{1/z}/L)$ scale as

$$f(c, t^{1/z}/L) = \begin{cases} f'(c) & \text{for } t \to 0 \\ x^{\alpha'} & \text{for } t \to \infty \end{cases} \quad (10)$$

where is $\alpha' \equiv \beta/\nu_\perp - \gamma c$. We estimate $\gamma = 0.139$ by static M.C simulation. The $\beta/\nu_\|, z$ is unchanged and the α' vary in dependent on c. For the $\alpha = 0.4 < 1/\nu_\|$, the

REFERENCES

1. Marro, J., and Dickman, R., *Nonequilibrium Phase Transition in Lattice Models*, Cambridge University Press, London, 1995, pp100-228
2. Park, H., and Park, H., *Physica A* **221**, 97-103 (1995); M. H. Kim and H. Park, *Phys. Rev. Lett* **73**, 2579-2582 (1994).
3. Hwang, W., Kwon, S., Park, H., and Park, H., *Phys. Rev. E* **57**, 6438-6450 (1998).
4. Hwang, H., and Park, H., *Phys. Rev. E* **59**, 4683-4686 (1999).
5. Aukrust, T., Browne, D.A., and Webman, I., *Phys. Rev. A* **41**, 5294-5301 (1990).

Internal Motion of Confined Molecules in Fullerene

Yasuteru Shigeta

Department of Basic Science, Graduate School of Arts and Sciences, University of Tokyo, Komaba 153-8902, Japan.e

Abstract. We performed quantum mechanical and molecular mechanics molecular dynamics simulations of hydrogen-absorbed Be endohedral metallofulleren (Be+nH$_2$)@C$_{60}$ (n=1,2). By estimating the correlation dimension of trajectories, it is found that dynamical behaviors of the internal rotation for n=1 and n=2 cases are regular and weakly chaotic, respectively.

Be endohedral fullerene is generated by a recoil process of nuclear reactions rather than by the conventional ion implantation technique [1]. Its generation process was also investigated in detail by the first principle calculation. Türker carried out calculations of the formation energy and an effective charge on Be atom using the semi-empirical calculation [2]. The author found that the effective charge on Be atom in C$_{60}$ fullerene is slightly negative and becomes positive when two more H$_2$ molecules are encapsulated. However, it is expected that the structure and effective charges on atoms remarkably vary with a change of the environment such as interactions with other molecules, temperature and so on. We have recently carried out quantum mechanical and molecular mechanics molecular dynamics (QM/MM MD) simulations of (Be+nH$_2$)Be@C$_{60}$ (n=1 and 2) in order to investigate dynamical changes due to hydrogen absorption. We found dynamically induced effective charge fluctuation on Be atom for n=2 case [3]. In the work, we analyze an internal motion of the encopshlated Be atom in the cage by using the GP method [4].

The simulations are carried out only at AM1/MM2 level using a modified version of GAMESS package programs [5]. The simulation conditions are dt=0.20 (fs) and T=200 (K) by employing a Nosé-Hoover's thermostat [6]. After 1 ps equilibration has been carried out, the subsequent 9 ps simulations have been performed to calculate ensemble averages. One (Be+nH2)@C$_{60}$ is placed in a simple cubic cell (a=14.520 Å), where we set a cut-off length at 30 (Å) instead of imposing the periodic boundary condition.

To see relation between internal motions and a position of Be atom or an effective charge on Be, Z_{Be}, we first define the angle between initial position of Be and that at time t as follows:

$$\cos\theta(t) = \frac{\mathbf{r}_{Be-COM}(t) \cdot \mathbf{r}_{Be-COM}(0)}{r_{Be-COM}(t) r_{Be-COM}(0)} \quad (1)$$

where $\mathbf{r}_{Be-COM}(t)$ means a position of Be atom from the center of mass (COM) of C$_{60}$ at time t. This angle reflects the internal rotation (revolution-like motion) of the Be atom in the cage. We plot correlations of short time averages of distance $r_{Be-COM}(t)$ and effective charge Z_{Be} with that of θ in Figs. 1-(a) and (b). It is evident from Fig. 1-(a) that the Be atom moves around a sphere inside of the cage whose effective radius may be regarded as the averaged distance $<r_{Be-COM}(t)>$ of 0.82 (Å) for n=1 and 1.23 (Å) for n=2. It is expected that a period of the internal rotation for n=2 is longer than that for n=1 because of the length of the radius. In other words, an angular velocity of n=1 is faster than that of n=2. Therefore, electrostatic properties such as a direction of the dipole moment for n=2 case changes more slower than that for n=1 case. According to Fig. 1-(a), it is also evident that there is no characteristic

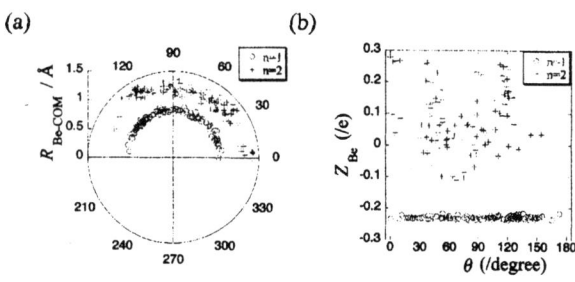

FIGURE 1. Correlations of short time average of (a) $r_{Be-COM}(t)$ and (b) Z_{Be} with that of θ at T=200 (K).

angle, because the averaged values are distributed almost uniformly for both cases. Both Figs. 1-(a) and (b) show that a deviation from the ensemble average for n=2 is much shaper than that for n=1. These results indicate that an effective potential for Be atom around the bottom is evenly spherical for n=1 and those is unevenness for n=2. It is concluded here that the magnitude of Z_{Be} depends mainly on $r_{Be-COM}(t)$ for n=2 case, while that is stable with respect to applied temperature for n=1 case.

According to the obtained results it seems reasonable to suppose that a turbulence of the potential is due to the cooperative motions of the hydrogen molecules and Be atoms. Since the given energy is too low not to break a chemical bond in the hydrogen molecule, the hydrogen molecule is regarded as one particle. Therefore, n=1 case is considered as a two-body problem and n=2 as a three-body problem with small perturbations from the cage and surrounding molecules. In Figs. 2 we depict trajectories of Be position from COM of cage onto X-Y and X-Z planes. From Fig. 2-(a) we found the internal rotation of the Be atom. On the other hand, it is observed in Fig. 2-(b) that Be atom is intermittently trapped at a local minima and then moves. Although the trajectory for n=2 looks like random, this results in the cooperative motions with hydrogen molecules and cage. In order to show differences in dynamical behaviors between two cases more clearly, we here compare the correlation dimension of the obtained time series of $r_{Be-COM}(t)$. We here employ the GP method to estimate the correlation dimension of the time series. In the GP method, the correlation integral of the time series of $s(t)$ is defined as

$$C^m(s) = \lim_{N \to \infty} \frac{1}{N^2} \sum_{i>j}^{N} h(s - |\mathbf{s}_i - \mathbf{s}_j|) \quad (2)$$

where m means the embedding dimension, N is a number of sampling points, and $h(s)$ is the heviside function. $\mathbf{s}_i = (r(t_i), r(t_i+\tau), \ldots, r(t_i+(m-1)\tau))$ is a m-dimensional vector of time series given with the delay time τ. A correlation dimension corresponds to the asymptotic value of a correlation exponent, $v(m)$, that is defined through $\log C^m(s) \approx v(m) \log(s)$. The correlation exponent of $r_{Be-COM}(t)$ for both cases becomes a constant with increase of m. The correlation dimensions of $r_{Be-COM}(t)$ are about $d=1.0$ for n=1 and $d=1.2$ for n=2, respectively. These results indicate that the dynamics of $r_{Be-COM}(t)$ for n=1 is almost regular and that for n=2 is weakly chaotic. That is to say, this system may be a good example for a molecular chaos in the low-energy region that usually appears rather in the high-energy region of molecular vibrations.

In summary, we have shown great effective charge fluctuation of confined Be atom in C_{60} by using

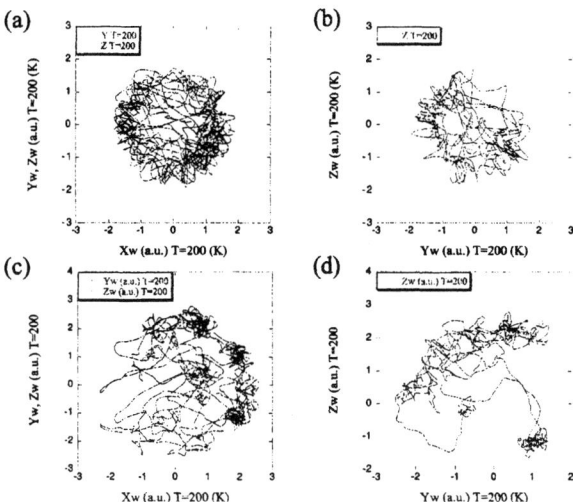

FIGURE 2. Trajectories onto X-Y and X-Z planes at T=200 (K) (a) for n=1 and (b) for n=2, respectively.

QM/MM MD method. Final results told us that the trajectory of the Be position for n=2 case shows chaotic behavior rather than regular one. It is interested to understand how the chaotic behavior is observed in actual experiments. Moreover, in contrary to the metallofullerene with heavy elements, the present systems consist of light elements. Within the present condition, all confined elements are trapped locally at the bottom of potentials for a while. It is expected that temperature becomes lower, quantum tunneling of these light elements in the chaotic effective potential affect the dynamics of the Be atom. It seems that the path integral MD [7] and a wave function based [8] methods are useful for investigating the quantum effects. These analyses will be performed in the future works.

YS has been assisted by a Research Fellowship of the Japan Society for the Promotion of Science for Young Scientists.

REFERENCES

1. Ohtsuki T., et al., *Phys. Rev. Lett.* **77**, 3522-3524 (1996).
2. Türker L., *J. Mol. Struct. (Theochem)* **577**, 205-210 (2002).
3. Shigeta Y., Saito H., *Synth. Met.* **135-136**, 765-766 (2003).
4. Grassberger P., Pricaccia I., *Phys. Rev. Lett* **50**, 346-349 (1983).
5. Schmidt M. W. et al. *J. Comput. Chem.* **14**, 1347-1363 (1993).
6. Nosé S., *Prog. Theo. Phys. Suppl.* **103**, 46 (1991).
7. Shiga M., Tachikawa M., Miura S., *J. Chem. Phys.* **115**, 9149-9159 (2001).
8. Shigeta Y., Nagao H., Nishikawa K., Yamaguchi K., *J. Chem. Phys.* **111**, 6171-6179 (1999).

Granular Flow Simulation by Granular Element Method

Y. Kishino

Civil Eng. Dept., Tohoku University, Aoba 06, Sendai 980-8579, Japan

Abstract. Granular media behave solid-likely when they are put in quasi-static environment and fluid-likely when once they lose their static equilibrium. The granular element method (GEM), proposed originally as a discrete element method for quasi-static problems, has recently been extended to a dynamic code. After explaining its algorithm briefly, this paper demonstrates granular flow experiments conducted by this method for a wide range of strain rate.

INTRODUCTION

As is observed in debris flows or any other granular flows, granular media behave solid-likely when they are put in quasi-static environment and fluid-likely when once they lose their static equilibrium. Using a discrete particle computer simulation of two-dimensional Couette flows, Zhang et al.[1] studied the phase changes from fluid-like to solid-like behavior, and found that the transition from solid to fluid is a yield-like phenomenon. Recently, the granular element method (GEM) has been extended from original quasi-static code[2] to three-dimensional dynamic one. Unlike the conventional distinct element method[3], the granular element method utilizes an iterative algorithm in getting equilibrium states accurately, and it may be a useful tool in the research of transitional behaviors of granular media.

GRANULAR ELEMENT METHOD

In GEM, as in other soft particle methods, material properties at each contact point are characterized by the spring constants in normal and tangential directions (k_n, k_t), the damping coefficients in normal and tangential directions (η_n, η_t) and the friction angle (φ). The motion of granular assembly is characterized by the configuration vector **x** whose components are the coordinates and rotations with respect to the gravity centers of particles that constitute the assembly. GEM adopts an iterative process to solve the dynamic equilibrium equation. In incremental loading from an equilibrated configuration $\mathbf{x}^0$, the Nth approximation of configuration is expressed as

$$\mathbf{x}^N = \mathbf{x}^0 + \sum_{n=1}^{N} \mathbf{u}^n \quad (1)$$

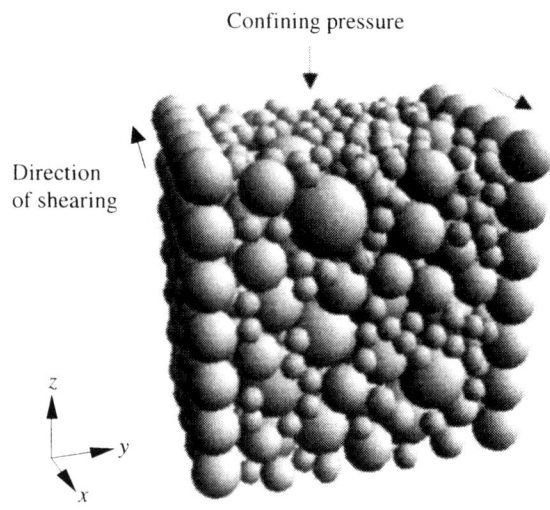

FIGURE 1. Initial configuration of particles

When we assume that the acceleration changes linearly during a time step Δt, we can derive the following iterative equation:

$$\left[\frac{6}{\Delta t^2}\mathbf{M} + \frac{3}{\Delta t}\mathbf{C} + \mathbf{K}\right]\mathbf{u}^{N+1} = -\mathbf{M}\ddot{\mathbf{x}}^N - \mathbf{C}\dot{\mathbf{x}}^N + \mathbf{F}_S^N + \mathbf{F}_B \quad (2)$$

In the above equation, **M**, **C** and **K** are the mass, damping and stiffness matrices, and $\mathbf{F}_S^N$ and $\mathbf{F}_B$ are the spring-force and body-force vectors. As components of **K** are tangential stiffnesses that varies with the configuration, $\mathbf{F}_S^N$ can not generally be represented in terms of **K**. In the GEM algorithm, the inversion of the whole matrix in the left hand side of Equation (2) does not take place. Instead, the algorithm utilizes a relaxation procedure for a set of particle matrices with 6 × 6 components that we define letting neighboring particles fixed.

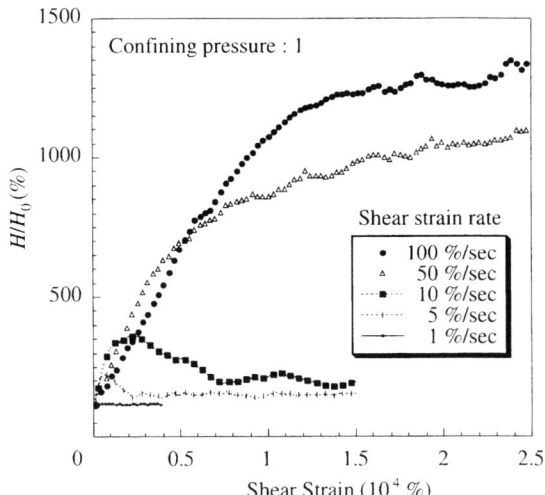

FIGURE 2. Change in hight of assembly

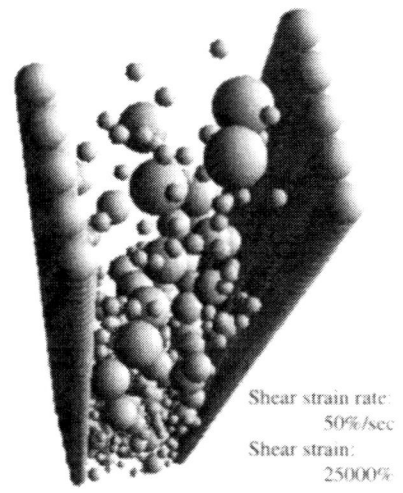

FIGURE 3. Typical inertia-dominant flow

SHEAR FLOW SIMULATIONS

The new dynamic code of granular element method was utilized to simulate experiments of shear flows under the non-gravitational field. **FIGURE 1.** shows the initial packing of granular assembly used in the numerical experiments. A pair of shear walls are located in parallel with $y = const.$ plane, and each shear wall consists of regular packing of uniform particles with radius of 1. The number of inner particles, randomly arranged between two shear walls, is 439, and the radii range from 0.5 to 1.5. The initial particle concentration is 0.61. The assumed values of density of particle (ρ) and other material constants are as follows: $\rho = 1000, k_n = 20000, k_t = 14000, \eta_n = 200, \eta_t = 140, \varphi = 15°$.

We assume that the upper and lower boundaries are smooth planes, so that no damping and friction forces act between each of these planes and contacting particles. The lower plane is assumed to be fixed and the upper plane is controlled such that it moves vertically without rotation and that a constant confining stress is maintained. The magnitude of confining stress assumed in the following simulation is 1. In x direction, we assume periodic boundaries with a constant length of periodicity. These boundary conditions simulate a shear cell that has two co-axial cylinders with different radii that can rotate in opposite directions. This type of shear cell has been proposed by a group of Prof. J. T. Jenkins[4].

The shear walls should prepare for the hight change of granular assembly. However, in **FIGURE 1.**, it was omitted to draw the upper parts of shear walls for convenience. Shear flows are activated in the granular assembly by the opposite movements of shear walls as shown in the figure. The shear-strain rates adopted in the following simulations range from 1%/sec to 100%/sec. The periods of time step adopted are 10^{-3} sec for the slower two cases and 10^{-4} sec for the faster three cases.

FIGURE 2. shows the changes in the height of granular assembly (H) normalized by the initial value (H_0). In the slowest shearing, the hight change is very limited and the shear flow is quasi-static. The maximum height becomes higher as the strain rate increases. In the fastest shearing, the maximum height is more than 13 times higher than the initial height. In some cases, marked hight change is observed only in the initial transient stage. However, in such cases, the initial fabric may not be expected to be preserved.

FIGURE 3. is a snapshot at the end of the shear flow whose shear-strain rate is 50%/sec. In this case, the flow is inertia-dominant and the confining stress at the upper boundary is equilibrated by the momentum transfer from colliding particles. Between quasi-static and inertia-dominant flows, various types of transitional granular flows are anticipated. To establish general mechanics of granular media, such transitional behaviors should be fully investigated.

REFERENCES

1. Zhang, Y., and Campbell, C. S., *J. Fluid Mech.*, **237**, 541-568 (1992).
2. Kishino, Y., "Disc Model Analysis of Granular Media," in *Micromechanics of Granular Materials*, edited by M. Satake et al., Elsevier, Amsterdam, 1988, pp.143-152.
3. Cundall, P. A. and Strack, O. D. L., *Gèotechnique*, **29**, 47-65 (1979).
4. Louge, M. Y., Xu, H. and Jenkins, J. T., "Studies of Gas-Particle Interactions in a Microgravity Flow Cell", in *Powders & Grains 2001*, edited by Y. Kishino, Swets & Zeitlinger, Lisse, 2001, pp.557-560.

Molecular Hydrodynamics: From Kubo To Smoluchowski Example of a Montmorillonite Clay

J.-F. Dufrêche[1], N. Malikova[1], V. Marry[1], F. Grün[2], M. Jardat[1], E. Dubois[1], P. Turq[1]

[1]*Laboratoire Liquides Ioniques et Interfaces Chargées, LI2C UMR 7612 CC51
Université Pierre et Marie Curie, 4 Place Jussieu, 75252 Paris Cedex 05, FRANCE*
[2]*UMR Pasteur, Ecole Normale Supérieure, 24 rue Lhomond, 75005 Paris, FRANCE*

Abstract. Electro-osmosis is an electro-kinetic phenomenon corresponding to the motion of a liquid phase adjacent to a charged surface under an applied electric field. The classical theory of electro-osmosis was initially proposed by Smoluchowski. In this mesoscopic approach, the macroscopic law of Navier-Stokes is solved taking into account the geometry of the pores. The electro-osmotic flow is induced by the force of the external electric field on the ions in the liquid phase of the system and the flow is controlled by the concentration profile of the ions. Ionic concentration profiles in the interlayer of a montmorillonite clay have been obtained using microscopic and mesoscopic models. Further, the profile of electro-osmotic velocities was calculated from the microscopic model (Kubo relation) and compared to the solution of the Poisson-Boltzmann and Navier-Stokes equations. It was concluded that Smoluchowski theory can be applied only if slipping hydrodynamic boundary conditions are introduced.

INTRODUCTION

Charged porous media, of which montmorillonite clay is an example, are described as pores with overall non-zero electric charge with compensating unbound ions. Porosity, closely linked to the water content in the case of clays, is the key factor controlling the dynamics of unbound species in these systems, above all their self-diffusion coefficients and electrical conductivity. Thus, as a function of water content, electro-kinetic phenomena are readily studied in the case of clays.

Clays consist of alumino-silicate sheets with an overall negative charge stemming from substitution of Al and Si atoms by lower valence atoms, such as Mg. Compensating ions reside between the sheets in so-called interlayers. The extremes of clay hydration are well-defined, at the lower limit of a few water molecules per compensating ion, water phase forms hydration shells around ions, at the upper limit colloidal suspensions of clay sheets are observed and dynamical properties of interlayer species tend towards properties of a bulk solution. In this study, hydration states corresponding to interlayer spacing significantly greater than the atomic size of the ions and water molecules were considered.

CONCENTRATION PROFILES OF INTERLAYER IONS

Classical theory of Smoluchowski predicts that electro-osmotic flow in this system is ruled by the concentration profile of the interlayer cations as a function of perpendicular distance from the clay sheets (z). These profiles were modeled in order of increasing scale by Molecular Dynamics, Brownian Dynamics and analytical solution of the Poisson-Boltzmann equation.

Molecular Dynamics (MD) simulations consider the atomic details of both the clay sheets and interlayer ion and water. Electrostatic and van der Waals interaction between all atoms were considered and the rigid SPC/E model of water used, as it closely predicts the dielectric constant of bulk water (81.5 ± 5). Starting from an equilibrium configuration of Monte Carlo simulations in the NPT ensemble, concentration profiles were obtained from Molecular Dynamics simulations in the NVT ensemble, where motion of interlayer species evolves according to Newton's equations of motion. Under Brownian Dynamics (BD) simulation the detailed structure of the clay sheets is neglected, water phase is treated as a continuum characterized by its dielectric constant and ions are described by a sphere of radius corresponding to hydrated ions. Interactions considered are electrostatic and short-range repulsion between the ions and the clay sheets ($1/r^{12}$). Under Poisson Boltzmann (PB) model, the charge distribution is obtained by solving the PB equation for a uniformly charged layer.

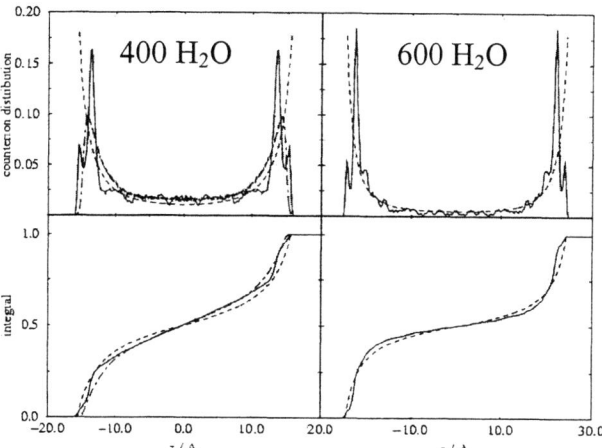

FIGURE 1. Cation concentration profiles and their integrals for two hydration states as indicated. Solid line: MD simulation, dashed: PB solution, dot-dashed: BD simulation.

Agreement was found in the ionic distributions obtained by the three above methods for high hydration states (interlayer spacing greater than 20 Å) (Figure 1). We note that the PB and BD approaches cannot reproduce the oscillations in the profiles seen in the MD simulations arising from the finite molecular size of the solvent.

ELECTRO-OSMOTIC VELOCITIES

Two approaches were used to obtain the values of electro-osmotic velocities as a function of the distance from the clay sheet. Initially [1], following the Smoluchowski treatment, the Navier-Stokes (NS) equation was solved for the given geometry, with no-slip boundary conditions ($\mathbf{v} = \mathbf{0}$). Thereafter, following the microscopic description, electro-osmotic velocities were obtained from equilibrium molecular dynamics simulations [2]. Within the linear response theory, electro-osmotic velocity can be evaluated from the correlation function of charge flow and mass flow under equilibrium conditions (Kubo relation).

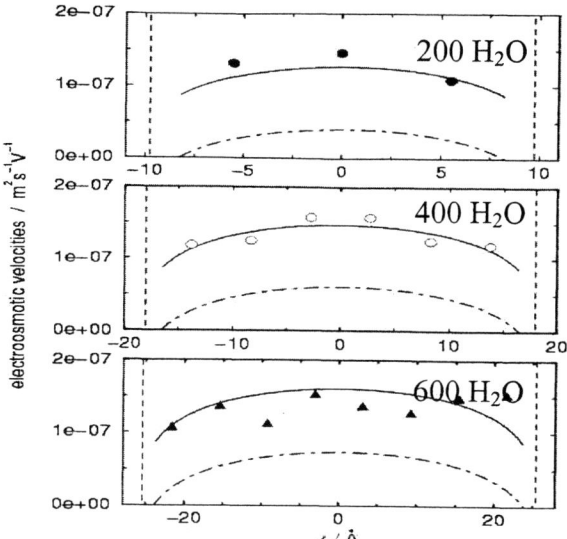

FIGURE 2. Comparison of electro-osmotic profiles. Point data: Kubo evaluation (MD), dot-dashed line: PB/NS no-slip b.c., solid line: PB/NS slip b.c. with slipping length 6 Å.

From the comparison of these two descriptions, Smoluchowski theory was found not to hold even for large interlayer spacings, where the ionic concentration profiles of the microscopic and mesoscopic methods superimposed. After revision of the boundary conditions for the solution of the Navier-Stokes equation [3], microscopic data was explained by introducing slip boundary conditions into the mesoscopic treatment. A slipping length of approximately 6 Å was found (Figure 2).

ACKNOWLEDGMENTS

This research was supported by ANDRA (Agence Nationale pour la Gestion des Déchets Radioactifs). We thank Eric Giffaut (ANDRA) for his helpful comments.

REFERENCES

1. Dufrêche J.-F., Marry V., Bernard O., Turq P., *Coll. and Surf. A* **195**, 171-180 (2001).

2. Marry V., Dufrêche J.-F., Jardat M., Turq P., *Mol. Phys.*, in press (2003).

3. Bocquet L., Barrat J.-L., *Phys. Rev. Lett.* **70**, 2726-2729 (1993), Bocquet L., Barrat J.-L., *Phys. Rev. E* **49**, 3079-3092 (1994).

V. GLASS TRANSITION

Vibrational dynamics and thermodynamics, ideal glass transitions and folding transitions, in liquids and biopolymers

C. Austen Angell*, Li-Min Wang*, Stefano Mossa**, Yuanzheng Yue[#], and John R. D. Copley[@].

*Dept. of Chemistry and Biochemistry, Arizona State University, Tempe, AZ 85287
[#]Department of Chemistry, Aalborg University, 9220 Aalborg, Denmark
[@]National Institute of Standards and Technology, Gaithersburg, MD 20899-8562
[A]Scottsdale Community College, Scottsdale, AZ 85256-2626
**Center for Statistical Mechanics and Complexity
**Universita di Roma "La Sapienza", Piazzale Aldo Moro 2, I-00185, Roma, Italy

Abstract. We use recent studies on hyperquenched glasses, both laboratory and computer simulated, to demonstrate that boson peak vibrations become more intense with increasing fictive temperature, and that this forces a revision of the standard textbook rendering of glass transition thermodynamics. The correct depiction depends on the thermodynamic condition, constant volume or pressure. The absence of a boson peak in glassy water, along with other dynamic and thermodynamic data, is used to argue that water yields the most ordered (near-ideal) glass, due to cooperativity. The similarity of events in this transition, to the folding of proteins into the native form, is emphasized by "funnel" diagrams, in which diversion to fibril states of proteins is seen as the analog of cubic ice formation from deeply supercooled water. A method of studying the energetic details of protein folding, using a special solvent to suppress ice formation and aggregation, is described.

INTRODUCTION

In laboratory studies of glassformers, the natural thermodynamic condition is that of constant pressure, usually that of the atmosphere. By contrast, the "default" condition for theoretical and computer simulation studies, is that of constant volume. This is primarily because of the fact that systems at constant volume provide a simpler target for analysis. For instance, the potential energy landscape, that is so frequently invoked in discussions of complex systems, is only uniquely defined by the potential of interaction of the particles if the volume remains constant [1-3]. If the volume changes, the landscape changes. Also for computer simulations, the calculations are simpler and less time-consuming if the periodic box within which the particles interact, remains constant in volume throughout the calculation. In this paper we first examine some problems in the science of "glasses" that can arise from this cultural distinction.

The data we will use for this purpose, taken from a recent computer simulation of the single component fragile glassformer, orthoterphenyl (OTP) [4,5], will show us that a diagram widely used in the glass science literature to illustrate changes in degrees of freedom on passage through the glass transition, is incorrectly interpreted in most discussions unless the liquid in question has no change in thermal expansion coefficient at T_g (which is not the case in the vast majority of glassforming substances). The same data will serve to illustrate the behavior of the much discussed "boson" peak in this type of system.

This will provide the motivation for examining the corresponding behavior in laboratory systems, which can only be obtained by employing great variations in the rate at which samples of glass for study are cooled into the glassy state. We will compare the properties of glasses formed by "hyperquenching" (cooling at a rate of 10^6 K/s) [6-9] with the properties of those formed by cooling at the "standard rate" of 20K/min. This will allow us to identify the boson

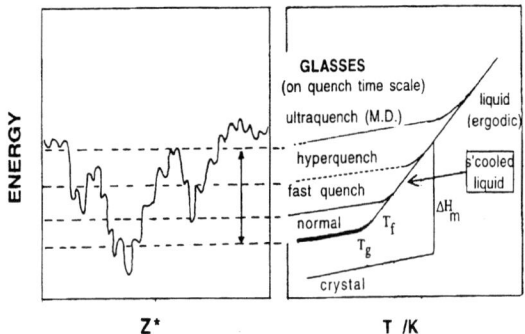

FIGURE 1. Depiction of the relation between the energy of glasses and the rate of quenching. LHS shows the trapped system energy in relation to its energy landscape, represented in the common (but highly over-simplified) two dimensional form appropriate to constant volume systems.

peak intensity with the level of disorder in the system, introduced by change of "fictive" temperature. The fictive temperature of a constant pressure system is usually taken as the temperature at which the system departed from equilibrium during cooling. This is best chosen as the temperature at which the property which changes through the glass transition (e.g. heat capacity) is half way between its ergodic and final glassy values. This choice is made so that, for systems cooled and heated at the same rate, the fictive temperature is the same as the common "onset heating" glass temperature [10,11].

In landscape terms (Fig. 1), the change of fictive temperature from standard glass to hyperquenched glass values is that associated with moving the system point from a low level on its energy landscape to a higher level on a *different* energy landscape, namely that characteristic of the system at the volume that it had when arrested during the hyperquench. For economy, we show only a single landscape in Fig. 1, which must therefore be thought of as representing a system with a negligible expansion coefficient. Once we have established the height of the boson peak as a measure of disorder in the system, we will look for glassy systems of very low disorder, judged by magnitude of the boson peak obtained from low energy neutron scattering studies. We find such a case in water vitrified by pressure amorphization, followed by relaxation to the low density amorphous (LDA) state [12]. This glassy state of water is similar to, though not identical with [13], the state of water produced by (a) hyperquenching and annealing tiny droplets of water, and (b) by vapor deposition of water onto a cryosurface, and then "sintering" the deposit. Following additional physical evidence that this state of water is a glassy state of unusually low levels of disorder [14-16], we confirm the physical evidence by reference to independent measurements of the entropy of LDA which establish it to be of extraordinarily small "excess" entropy.

The process of forming such a nearly ideal glass [14], can be represented, in energy landscape terms, by descent within an energy "funnel", very similar to that invoked for the description of the mesoscopic complex systems represented by proteins [17]. (Proteins of the small globular type are frequently found to fit the description "two-state folder" [18]). It is even more like the "folding funnel" of the typical two-state protein that must be invoked subsequent to the discovery by Dobson and co-workers [19] that there is generally available a further low energy state for proteins. This is the "fibril" state which involves an organized aggregate of identically folded molecules which may not be in the lowest energy state for the individual molecule, but in which the average free energy is lower than that of same number of optimally folded protein molecules in solution. The "fibril state" [19] is the equivalent, for proteins, of the crystalline state of water molecules.

Excluding the fibril state, we will then show how our exploitation of the cooling rate variable (quenching strategies) can be applied to obtain extra information of a useful type on the energetics of protein folding, using the case of lysozyme.

THE BOSON PEAK, AND THE GLASS TRANSITIONS AT CONSTANT PRESSURE VS. CONSTANT VOLUME

In Fig. 2 we show the results of some computer simulation studies of systems that have a common

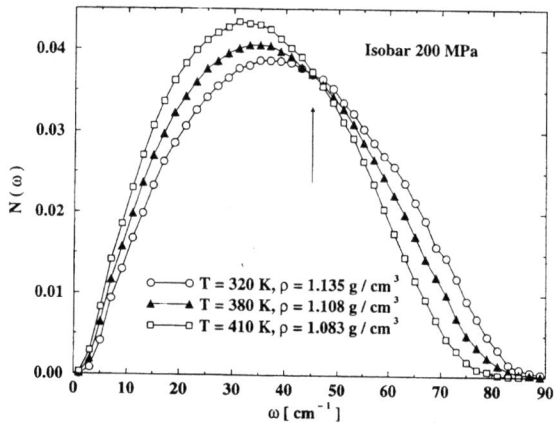

FIGURE 2. The vibrational density of states VDOS for orthoterphenyl OTP at 2000 mPa, in the Lewis-Wahnstrøm model, showing the manner in which, at constant pressure, the increase of temperature causes an increase in the density of states at low frequency at the expense of modes of high frequency (from ref. 9 by permission).

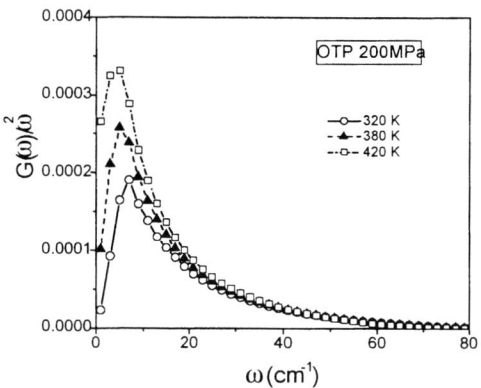

FIGURE 3. Data of Fig. 2 shown in the Boson peak representation $(G(\omega)/\omega^2)$. The boson peak is seen to generation of structures with a specific size. This increase in intensity and move to lower frequencies as fictive temperature is increased.

pressure, 200 MPa. The particular property represented in Fig. 2 is the density of vibrational states $G(\omega)$ for the inherent structure at each of three different temperatures. For inherent structures the temperature corresponds to the fictive temperature. Since the inherent structure for a given temperature is not unique but rather has a statistical probability over a narrow band of energies, the three different $G(\omega)$ values should not be thought of as unique to each temperature, but rather as representative. We note that $G(\omega)$ for the glass of highest fictive temperature is considerably richer in low frequency modes than are the other two and the difference is systematic with temperature. The boson peak, that has been identified from light scattering studies [20-22] is related to the excess density of states as $G(\omega)/\omega^2$ [23] We show the boson peak implications of the Fig. 2 data in Fig. 3 where the boson peak is seen to gain in intensity and decrease in peak frequency with increasing fictive temperature.

While experimental studies of glasses of different fictive temperature have been made [24], the range of fictive temperatures has been small, and the effects observed also small. We recently showed that much larger effects could be obtained by studying the DOS of hyperquenched glasses. Results are shown in Fig. 4 for a mineral glass of basalt composition of which hyperquenched samples were readily available [8]. In this case, the DOS changes were also studied for a restricted range of Q vectors, corresponding to the distances 6-11Å in real space. The data, shown in lower part of Fig. 4, imply that some sort of point defect with a narrow distribution of topologies may serve as the elementary excitation in these systems [9].

Even more striking effects in the boson peak behavior have been obtained in recent studies of the fragile CaO-SiO_2 system, which will be published separately [25].

Thermodynamic Consequence of Increased Low Frequency DOS at High Fictive Temperature

In assessing the "configurational" heat capacity of liquids, it has rather generally been the practice to simply extrapolate the glassy state heat capacity to temperatures above T_g, using the behavior of the glass up to T_g, and the crystal above T_g, as a guide [26,27]. In making this construction, it is assumed

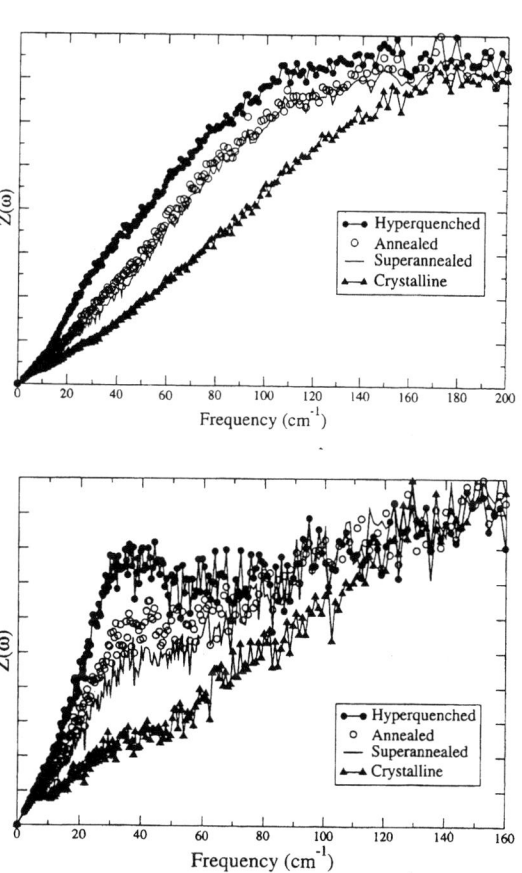

FIGURE 4. Low frequency part of the $Z(\omega)$ (~VDOS) for a laboratory system, showing the effect of temperature on the intensity of low frequency modes qualitatively like Fig. 2. Fig. 4, lower part, shows the same data from the same study now restricted in Q values to those corresponding to the distance range $2\pi/Q$ of 6-11Å, suggesting that some sort of spatially confined defect is involved in the excitations. (From ref. 9, by permission)

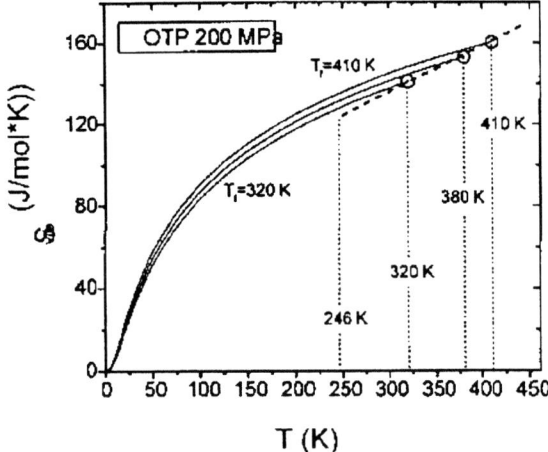

FIGURE 5. Vibrational entropies of the three glasses whose VDOSs are shown in Fig. 2. The lines each stop at the fictive temperature for the glass since at higher temperatures the system would move to a higher energy level of basins and the VDOS will change. The difference in total entropy between glasses, caused by the different populations of low frequency modes is not great. However, the *slope* of the vibrational entropy vs temperature of the equilibrated liquid at constant pressure ($T = T_f$,) obtained by joining the end-points of each curve – see dashed line) is very different from that of the single glass. The single glass C_p is similar to that of the crystal, whose structure also remains fixed during heating. In Fig. 5, the ergodic liquid C_p(vib) has been extrapolated to the experimental T_g, 246K.

that the vibrational heat capacity behaves like that of the crystal (fixed structure) regardless of how high the fictive temperature is. This amounts to assuming that the shapes of the basins on the energy landscape (crudely represented on the LHS of Fig. 1) are not changing with height on the landscape. The data of Figs. 2 and 4 show that this is not the case at all, for at least one simulated single component molecular system and one multicomponent ionic system. Since these two systems are almost totally unrelated, it is reasonable to assume that their behavior represents the general case. (However, it would be surprising and provocative if the simple manner in which the vibrational spectrum can be divided up into three Gaussian components, the central member of which does not change with changing fictive temperature [9], were found to be general).

If the latter is indeed the general case then we need to examine the consequences of the changing DOS quantitatively, to see how greatly the general picture must be modified.

To assess the thermodynamic consequences of the generation of low frequency modes of Fig. 2 at the expense of the high frequency modes with increasing fictive temperature (seen in Fig. 2) we simply apply the harmonic oscillator heat capacity equations to each of the three DOS, to obtain the harmonic vibrational entropy from 0K to the fictive temperature. The harmonic vibrational entropy of each glass (each inherent structure) is shown in Fig. 5. The vibrational entropy of the supercooled liquid at different temperatures can be assessed by joining the values of the vibrational entropy at each fictive temperature together, since the states giving the three DOS of Fig. 2 are the ergodic states at the respective fictive temperatures. It is shown in Fig. 5 as a dashed line. It is clear from the dashed line in Fig. 5, that the rate of *vibrational* entropy increase in the (ergodic) liquid state of this system is very different from any individual fixed structure glass. Since the crystal has fixed structure, the vibrational entropy behavior of the liquid is therefore also very different from that of the crystal which has served as a guide for the vibrational properties of the supercooled liquid in most previous representations of glassy system thermodynamics (represented by Fig. 6a). How different, will be assessed after making the following important point.

The deviation of actual behavior from previously supposed behavior we have just described is only for the *constant pressure* system. When the same analysis is applied to the results obtained at constant volume [4], the opposite situation applies. In the constant volume case, high temperatures favor high frequency modes [4,28], and so the rate of entropy increase above T_g will be *smaller* than for the crystal. When the temperature dependence of the vibrational contribution to the total entropy is extrapolated to the lower fictive temperature of T_g (inaccessible to simulation), we can obtain an idea of what the total excess vibrational entropy (i.e. excess over crystal of fixed structure) should be like relative to the glass value (dashed line in Fig. 5). Indeed, it is very like that sketched in Fig. 4 of Ref. 29.

The continuous change in slope at T_g in Fig. 5 means that there is an almost discontinuous change in vibrational heat capacity due to the unfreezing of the structure. The change must be as abrupt as the change in total heat capacity registered in a differential scanning calorimetry scan through T_g. However, to the best of our knowledge, this change in vibrational heat capacity has never been rendered in any graphical representation of the heat capacity behavior. It is therefore represented explicitly in Fig. 6 part (b). The corresponding breakdown of the contributions to the total heat capacity that would be measured at constant volume, is shown in Fig. 6 part (c).

This analysis warns us that the configurational heat capacity of liquids, which is the total measured heat capacity less the vibrational heat capacity, is much lower than is normally supposed, at least for fragile liquids of the OTP type. Although this effect has been known since Goldstein's analysis in 1972 [30], it has

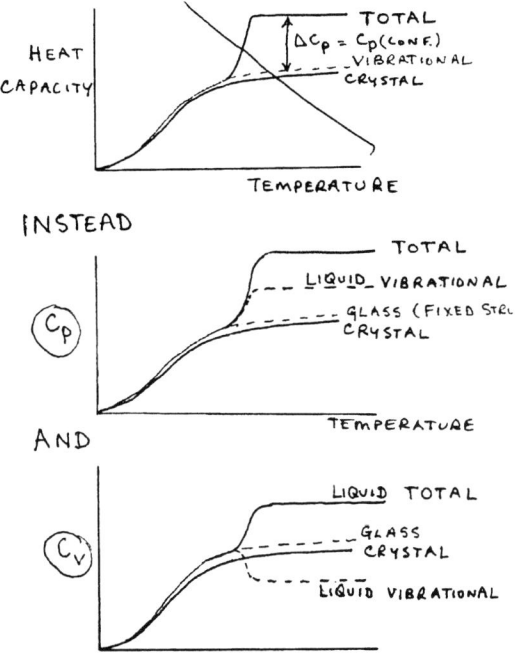

FIGURE 6. (a) Usual representation of the C_p behavior of liquid, crystal and glass states of a substance showing the vibrational component of the heat capacity following the crystal behavior (though somewhat higher).

(b) Correct representation of the same system taking into account the behavior observed for the Lewis-Wahnstrom model, at constant pressure.

(c) Correct representation of the total heat capacity at constant volume, showing the decrease in vibrational heat capacity at T_g required by the oppositely behaving VDOS.

generally been disregarded as a small effect, which we see here is by no means correct. It will therefore be necessary to take a more careful look at the vibrational heat capacity of glasses as a function of fictive temperature before deciding what the *configurational* entropy of a given liquid actually is. It will considerably change at least two types of analyses of glassy system thermodynamics [27,31] in which the behavior of the vibrational heat capacity was not properly taken into account.

Firstly it will complicate the application of the Adam-Gibbs equation to the analysis of such quantities as the minimum size cooperative group as a function of temperature [27], and will in fact cause the revision of all such numbers *upward*. This will be a welcome change since the numbers obtained from the careful analyses of Yamamuro Matsuo, and coworkers [27] have seemed to be unphysically small. Secondly it will considerably change the estimate of the temperature at which a system will reach the top of its energy landscape. In earlier assessments it was supposed that the vibrational excess heat capacity would not amount to more that some 10% of the excess C_p, and was neglected. The effect will shift the estimate of the $T_{ToL}1.69\ T_K$ [31] to $2.5 - 3T_K$, as simulation studies would suggest.

BOSON PEAKS, DISORDER, AND THE NEARLY IDEAL GLASSY STATE OF LDA WATER

In the previous section we showed how the generation of vibrational modes of low frequency during the configurational excitation of glasses, had important thermodynamic consequences, and indeed required us to considerably modify some basic conceptions in glass science (Fig. 6). An important implication of Figure 5 is that, at sufficiently low temperatures, the Boson peak and the contribution of its modes to the entropy of the system will disappear. In Fig. 4 of ref. 29 this was depicted as happening at the Kauzmann temperature. Thus the boson peak serves as a spectroscopically accessible measure of structural disorder. We now utilize this aspect of the boson peak as the starting point of an analysis of a particular glass that would appear to be capable of existing in a state of exceptionally low structural disorder, indeed to represent an almost ideal glass state.

The disconcerting aspect of this analysis is that the substance in question, namely, water in its low density amorphous form (LDAW, (or annealed amorphous solid water ASW, or annealed hyperquenched glassy water HQGW), is not normally thought of as a glassforming substance. Our conclusions may well apply to a series of substances that can only be obtained in the glassy state by roundabout routes, but which, when finally vitrified, find themselves in states of very low disorder, as assessed from both dynamic (boson peak etc) and thermodynamic (excess entropy over crystal) criteria.

In a recent review on the subject of Amorphous Water [14], we drew attention to three surprising aspects of the behavior of water in both high and low density vitreous states, HDA and LDA, obtained by the pressure amorphization route. The latter is structurally close to the amorphous states of water that are obtained by vapor deposition, and by liquid hyperquenching, when each process is followed by some suitable annealing (to remove most of the frozen in disorder) has been carried out. There are minor differences [13] which we will not concern ourselves with here. The advantage of the pressure amorphization route is that large quantities of material can be prepared, with which accurate data are more easily acquired.

The three observations that were notable were, firstly, the absence of a significant excess of low frequency vibrational modes over those of the crystal

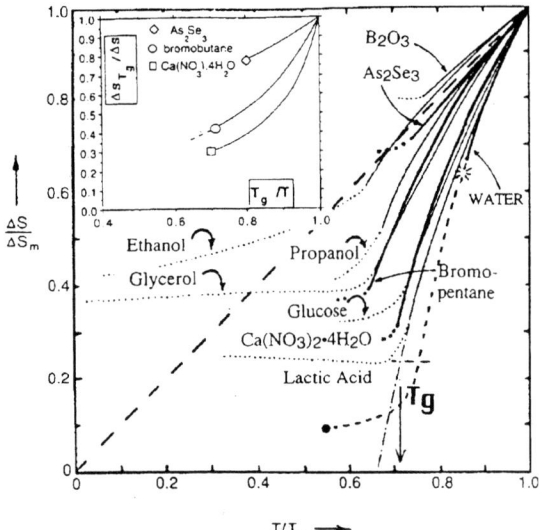

FIGURE 7. Kauzmann diagram, showing excess entropy ΔS vs temperature both scaled by the property at T_m. Added to previous plot [36] is the value for water measured at low temperature, and the connection to high temperature data deduced by Starr et al [39], in order to emphasize the near ideality of glassy water [14]. The starburst on the water plot is the homogeneous nucleation temperature for supercooled water. Here $\Delta S = S_{ex}$. (adapted from Fig. 1 of ref. 36)

[15a], i.e. to good approximation the boson peak was missing (from both polyamorphs). Secondly, the LDA sample showed a dispersion relation (obtained from inelastic X-ray scattering) that was almost as well defined as for ice I_c crystals [15b]. Thirdly, the sample of LDA (but not HDA) had a thermal conductivity that increased with decreasing temperature like a crystal [16], meaning the disorder is so small that heat-carrying phonons are not scattered at low temperatures.

Such striking indications of dynamical order are supported by the thermodynamic measurement of order represented by total entropy relative to crystal entropy. The entropy of the LDA phase [32-34] estimated from the measured free energy (from vapor pressure data) and enthalpy data [32] and by other methods [33,34], showed that the excess entropy of LDA over that of ice I_h, is very small. It is only one tenth of the entropy of fusion, compared with 1/3rd for the most fragile liquids in Kauzmann's original comparisons [35]. The analysis of ref. 33 also yielded a value for HDA which, though larger than for LDA, was still much smaller than for other glasses. Thus it seems that, although water is not easily obtained in the vitreous state, once it is obtained in that state it provides an almost ideal, defect-free, example of the glassy state.

This conclusion is consistent with the pattern of behavior of other liquids of varying fragility originally represented by Kauzmann in terms of entropy and temperature variables, scaled by their values at the melting point. We show this diagram, embellished by three other cases of different fragilities [36], in Fig. 7.

We include the vitreous water data in Fig. 7, showing it at the temperature 150 K where it was measured [32]. Just how the connection is to be made between the data acquired above 236 K, where the system is ergodic, and the temperature 150 K where the system is non-ergodic (according to several recent arguments [10,11,14,37,38], is a matter for discussion. This problem was addressed by Starr et al [39], and their thermodynamically reasoned construction is used in Fig 7. It shows water as the most fragile liquid of all, and the one that, in consequence, reaches the lowest entropy state (relative to the entropy of fusion) before the structure becomes fixed by the glass transition. The assignment of relative ideality on the latter basis deserves some discussion. It must be asked if judging the relative ideality of a glass by the fraction of the entropy of fusion that is residual at T_g is appropriate, since it implies something absolute about the entropy of fusion [40]. This is particularly dubious when the system is not glassforming by the usual liquid cooling methods. Substances that are not glassforming by normal cooling methods (usually meaning $T_m/T_g > 1.5$), will have larger entropies of fusion than the same substance would have if it satisfied the $T_m/T_g = 1.5$ rule that applies to most natural glassformers. However, even if water were assigned an entropy of fusion of only half the measured value, it would still have the smallest known relative excess entropy, $S_{ex}(T_g)/\Delta S_{fusion}$ at T_g, so its assignment as thermodynamically the most ideal glass on record would seem secure.

ENERGY FUNNELS AND THE RELATION BETWEEN FORMING THE IDEAL GLASS OF WATER AND THE FOLDING OF PROTEINS

It has become a popular concept to view the progress of a protein from the high temperature unfolded state to the low energy folded state as the progressive descent within an energy "funnel" that guides the system to a final low energy minimum. We believe this is an equally appropriate description of the way in which the hydrogen bond formation, starting at very high temperatures, guides liquid water into the ideal glass state, even including a cooperative "rush" towards the ground state that sets in at sub-zero temperatures. The cooperative rush is best seen in the heat capacity behavior of supercooled water as deduced for the non-crystallizing case by Starr et al [39], using thermodynamic constraint arguments in combination with available data for glassy and supercooled water. This is shown in Fig. 8 (taken from ref. 39). Fig. 8 is the analog of the heat capacity maximum seen at the changeover from unfolded to folded state that occurs Water is distinguished from the usual representation of the folding protein energy funnel [17] by the presence of an additional narrow (low entropy) energy well representing the crystalline form, see Fig. 9a, the (nucleated) transition to which is unfortunately highly probable. This distinction has, however, recently been diminished by the discovery that proteins in general have an analog lower energy state that can be reached if the system is held long enough close to but below the folding temperature [19]. At this temperature frequent fold-unfold transitions open the possibility of aggregation of non-native folded states into the fibril state - which is the analog of ice I_c. In the fibril state, the overall energy per mole of protein molecules is lower, even if the individual molecules are not in their lowest possible energy nucleation event, which requires passage over an energy barrier. The entropy loss implied by the change in basin width is compensated by an energy gain which, at equilibrium, is given by the relation $\Delta E = T\Delta S$, and under metastable conditions is given by the inequality $T\Delta S > \Delta E$. Since we believe the evidence that two-state folding is a nucleated process [43-45], the bottom of the folding funnel would, we suggest, be better represented by the depiction in Fig. 9c. Fig 9c is a symmetrical version of the "two megabasin" representation given originally in ref. 46. configurations. This is because the fibril state has the equivalent of the lattice energy of the usual crystalline state. Fig. 9a,b shows the comparison. In each case the transition from the liquid (molten globule) state to the crystal (fibril) state requires a Insight into the folding of proteins from experimental analogs of water hyper-quenching studies

To enquire further into the energetics of the protein folding process, we report briefly on some studies, detailed elsewhere [47], in which the folding was suppressed by a fast quenching process equivalent to that which avoids crystallization in the case of water. By use of a novel solvent [48] in which the protein lysozyme can unfold and fold repeatedly without aggregating, and in which no ice forms during cool/heat cycles, we have been able to study the folding of a protein at low temperatures. In cold refolding, as this process may be called, the different energetic steps in the process may be seen. The resolution of the refolding process into distinct stages can be seen by comparison of the heating scan of a quenched sample in which the protein has never left the folded state, with that of a sample quenched from just above the unfolding temperature, 67°C. The whole scan starting from below the glass transition temperature of the solvent (-70 °C) is shown in Fig.

10a. The interesting part, that from the sudden onset of the cold refolding process at about 0°C up to the completion of the folding (which overlaps the restart of the unfolding), is shown in Fig. 10b. Note how closely the second unfolding endotherm overlaps the original unfolding endotherm, proving that essentially no protein was lost to aggregation in the cycle.

There are three aspects of our cold refolding exotherm that are of special interest. The first is that the total energy evolved is the same as that absorbed in the "remelting" at higher temperature, which provides a consistency check for the interpretation of the observed exotherm. We note, with interest and satisfaction, that the enthalpy of work. The dashed plot is for a solution *before* any unfolding has occurred. The solid plot is the same solution quenched from 80K (i.e., after denaturing). This scan shows an exotherm equal in area to the unfolding *endotherm*, suggesting folding was completely suppressed during the quench. Unfolding is the same

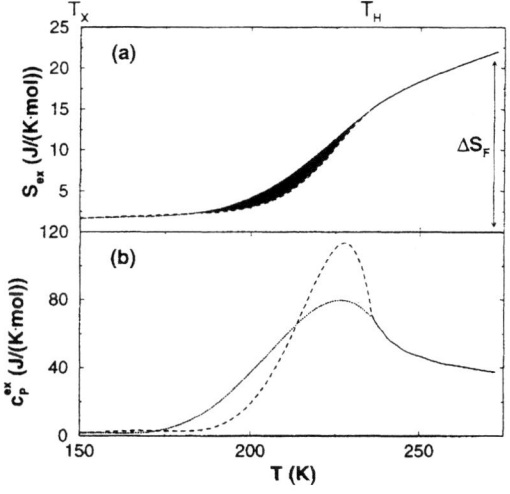

FIGURE 8. The excess entropy (a) and heat capacity (b) of supercooled water in the absence of the renaturation temperature during cooling of a protein. However, for the case of protein folding, the individual molecules fold by fast two-state "on-off" events, and the continuous appearance of the cooling exotherm is due to small system effects. The analogy would be more complete if water were to enter the low temperature, nearly ideal glass, state through a first order liquid-liquid transition such as is known to occur in the topologically similar case of liquid silicon [41,42]. crystallization, deduced by thermodynamic constraint arguments [39] from available data on glassy and supercooled water. The two curves represent the limiting forms permitted by the data uncertainties, excluding the existence of a first order liquid-liquid transition like that in liquid silicon. T_X is near the LDA/ASW crystallization temperature and T_H is the homogeneous nucleation temperature of supercooled water. (From Ref. [39], ©2003, reprinted with permission from Elsevier).

479

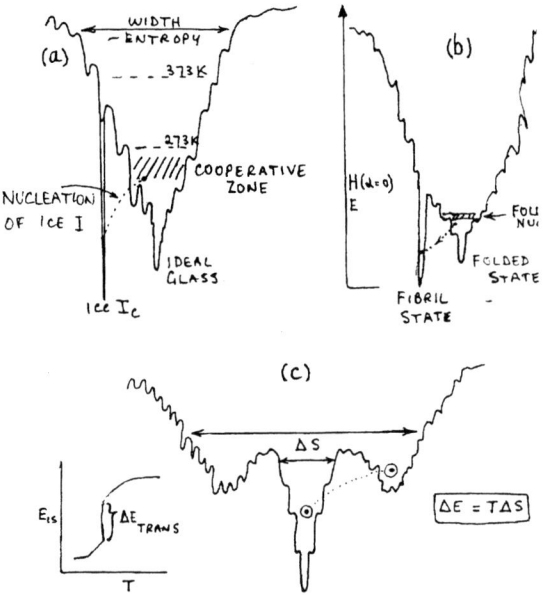

FIGURE 9 (a) Funnel representation of the energy landscape guiding the system water towards the fully hydrogen bonded amorphous ground state. Entropy is represented by the width of the funnel [17]. The narrow and deep well to the left represents the low entropy state of ice Ic, into which the system can transform by a nucleation and growth process.

(b) Funnel representation (Wolynes and co-workers [17]) of the folding of a protein into its low energy (but still conformationally labile) tertiary structure, modified by the addition of a fibril state [19] that is the analog of the crystalline state of normal molecular glassformers. Since the native state is non-periodic, the analogy could be better made with the transformation of liquid silicon in its 4.5 coordinated state, to high temperature

(c) Energy funnel for system with nucleated transitions (two state folding hetero-polymer or liquid-liquid phase transition).

in our solution as it is in normal buffer, notwithstanding the great decrease in water activity that we have affected.

The second aspect of interest is the sudden start to the process at 0°C. At first it was thought that this was an artifact due to ice condensation on the sample pan during transition from the liquid nitrogen quench bath to the DSC sample compartment (since the drybox in which the process is carried out is simple). Indeed, by subjecting the reference pan to the same quench and mount procedure, it is possible to remove or invert the endothermic spike seen in Fig. 10. However, the sudden start to the refolding exotherm seems to be a robust feature. In this case it requires interpretation. Our tentative interpretation is that this is the temperature at which the system explores its configuration space sufficiently rapidly to nucleate the process by achieving the critical number of native contacts (all of them low energy) to nucleate the final folding into the native state [17b]. This nucleation step is the bottleneck to the process, which can then proceed continuously if not immediately to the final state. Holding the sample at any temperature at 5°C or higher for five minutes is sufficient for the process to go to completion.

Unfortunately we cannot make satisfactory observations of the whole process by isothermal scans at 5°C because of baseline uncertainties. The way to see the kinetic details of the folding process after its initiation at 0°C is to continuously scan so that the slower parts of the final assembly are encouraged by the higher temperatures, and this constitutes the third item of interest in Fig. 10b. The deceleration (but not arrest) of the folding process seen at 25°C corresponds to the intermediate stage in the lysozyme folding reported by Dobson and coworkers [49-51] using the concentration jump method. In this latter method a solution of protein, denatured by guanidinium chloride, is suddenly diluted to concentrations where the folded state is stable. This alternative method of performing cold refolding studies allows the process to be studied under more natural circumstances than does ours, but does not allow the requenching to trap intermediate states with the same efficiency. It would be very good if a microcalorimetry study of the refolding energetics, following concentration jump, could be made for comparison with our isocompositional result.

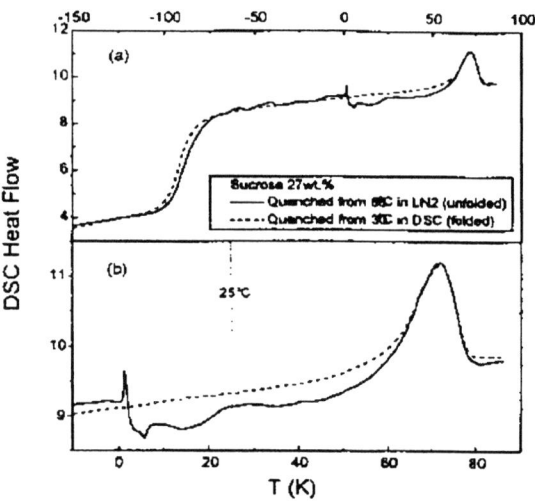

FIGURE 10. Differential scanning calorimetry up scans of LN2-quenched solutions of lysozyme, in the special non-crystallizing non-aggregating solvent developed in this Panel (b) shows a blow-up of this part: the energetic structure of the cold refolding process is revealed. Note the bump at 25°C. The sharp initial exotherm at about 0°C is discussed in text in relation to evidence for a possible nucleation step in the refolding process.

The conclusion from Fig. 10 is that the folding of lysozyme, rather than being a smooth descent of a folding funnel as depicted in Fig. 9b, and ref. 17, is much more consistent with Fig. 9c (and Fig. 8 of ref. 43), but with a shoulder structure in the lower energy megabasin. Water vitrification on the other hand, in view of Fig. 8, is probably intermediate between the two, while liquid silicon vitrification during an appropriate hyperquench, is unambiguously Fig. 9c in type, since its first order character has recently been clearly established [42].

It is hoped that further work along these lines, with more detailed thermal studies (including the use of temperature steps such as those used in ref. 45 for the study of nucleation and growth of crystals) might result in an improved understanding of both the thermodynamics of water and the folding of proteins. It is expected that there will be a number of other processes of biological interest in which the use of non-crystallizing solvents such as those used to obtain the Fig. 9 data, will lead to improved understanding. of the dynamics and thermodynamics of the processes.

ACKNOWLEDGEMENTS

This work was supported by the NSF-DMR Solid State Chemistry program, under grant no. DMR-0082535. The measurements at NIST utilized facilities supported in part by the National Science Foundation under Agreement No. DMR-0086210.

REFERENCES

1. Goldstein, M., *J. Chem. Phys.* **51**, 3728 (1969).
2. (a) Stillinger, F. H., *Science* **267**, 1935 (1995).
   (b) Stillinger, F. H., and Weber, T. A. *Science* **228**, 983 (1984).
3. Stillinger, F. H., and Debenedetti, P. G., *Nature* **410**, 259-267 (2001)
4. Mossa, S., La Nave, E., Stanley, H. E., Donati, C., Sciortino, F., and Tartaglia, P., *Phys. Rev.* **E65**, 041205 [1-8] (2002).
5. La Nave, E., Sciortino, F., Tartaglia, P., De Michele, C., and Mossa, S., *J. Phys. Condens. Matter* **15**, S1085-S1094 Sp. Iss. SI (2003)
6. Chen, H. S., and Inoue, A., Sub-T_g enthalpy relaxation in PdNiSi alloy glasses. *J. Non-Cryst. Sol.* **805**, 61-62 (1984).
7. Huang, J. and Gupta, P., *J. Non-Crystalline Solids*, **151**, 175, (1992).
8. Yue, Y. Z., Christiansen, J. deC., Jensen, and S. L., *Appl. Phys. Lett.* **81**, 2983-2985 (2002).
9. Angell, C. A., Yue, Y. Z., Wang, L-M., Copley, J. R. D., Borick, S., and Mossa, S., *J. Phys. Cond. Mat.* **15**, S1051-S1068 (2003).
10. Velikov, V., Borick, S., and Angell, C. A., *Science*, **294**, 2335-2338 (2001).
11. Angell, C. A., *Chem. Rev.* **102**, 2627-2649 (2002).
12. (a) Mishima, D., Calvert, L. D., and Whalley, E., *Nature* **310**, 393-395. 1984,
    (b) Mishima, O. J., *Chem. Phys.*, **100**, 5910 (1994).
13. Johari G. P., Hallbrucker, E., and Mayer, E.. *Science* **273**, 90-92 (1996).
14. Angell, C. A., *Annu. Rev. Phys. Chem.* (2004, in press).
15. (a) Schober, H., Koza, M., Tölle, A., Fujara, F., Angell, C. A., Bohmer, R., *Physica B.* **241-243**, 897-902 (1998).
    (b) Schober, H., Koza, M. M., Tölle, A., Masciovecchio, C., Sette, F., Ans, A., Fujara, F., *Phys. Rev. Lett.* **85**, 4100 (2000).
16. Andersson, O., Suga, H., *Phys. Rev. B.* **65**:140201(R) (2002).
17. (a) Wolynes, P. G., Onuchic, J. N., and Thirumalai, D., *Science* **267**, 1619-1620 (1995).
    (b) Onuchic, J. N., Luthey Schulten, Z., and Wolynes, P. G., *Annu. Rev. Phys. Chem.* **48**, 545-600 (1997).
18. (a) Privalov, P. I., *Adv. Protein Chem.* **13**, 167 (1979).
    (b) Shakhnovich, E. I., and Finkelstein, A., *Biopolymers* **28**, 1667 (1989).
19. (a) MacPhee, C. E., Dobson, C. M., *J. Am.Chem. Soc.* **122**, 12707-12713, (2000).
    (b) Morozova-Roche, L. A., Zurdo, J., Spencer, A., et al. *J. Struct. Biol.*, **130**, 339, (2000).
20. Duval, E., Boukenter, A., and Achibat, T., *J. Phys. Condensed Matt. Phys.* **2**, 10227 (1990).
21. (a) Malinovsky, V. K., and Sokolov, A. P., *Solid State Communications* **67**, 757-761, (1986).
    (b) Sokolov, A. P., Roessler, E., Kisliuk A., and Quitmann, D., *Phys. Rev. Lett.* **71**, 2062-2065, (1993).
22. Wischnewski, U. Buchenau, A. J. Dianoux, W. A. Kamitakahara and J. L. Zarestky, *Phys. Rev. B* **57**, 2663-2666 (1998).
23. Engberg, D., Wischnewski, A., Buchenau, U., Borjesson, L., Dianoux, A. J., Sokolov, A. P., and Torell, L. M., Phys. Rev. B **58** 14 (1998).
24. Suck, J.-B., in *Dynamics of Disordered Materials*, ed. D. Richter, A. J. Dianoux, W. Petry, and J. Teixera: Springer, Berlin, 1989, p. 182. A related observation is that of Sokolov that the quasi-elastic scattering of an annealed glass is less than that of a quenched glass of the same material at the same temperature.
25. Schober, H., Yue, J., et al, new $CaO-SiO_2$ results (to be published).
26. Ediger, M. D., Angell, C. A., and Nagel, S. R., *J. Phys. Chem.* **100**, 13200, (1996).
27. a) Takahara, S., Yamamuro, O., and Matsuo, T., *J. Phys. Chem.* **99**, 9589, (1995).
    b) Yamamuro, O, Tsukushi, I, Lindqvist, A, Takahara S, Ishikawa M, Matsuo T *J. Phys. Chem. B*, **102**, 1605, (1998).
28. Sastry, S., *Nature*, **409**, 164. (2001).
29. Martinez, L.-M., and Angell, C. A., *Nature* **410**, 663-667 (2001).
30. Goldstein, M., *J. Chem. Phys.* **64**, 4767 (1976).
31. C. A. Angell, in "Complex Behavior of Glassy Systems" Ed. M. Rubi, Springer, 1997, p. 1.

32. Speedy, R. J., Debenedetti, P. G., Smith, R. S., Huang, C., and Kay, B. D., *J. Chem Phys.*, **105** (1), 240 (1996).
33. Whalley, E., Klug D. D., and Handa, Y.P., *Nature*, **342,** 89 (1989).
34. Kouchi, A., *Nature*, **330**, 550 (1987).
35. Kauzmann, W., *Chem. Rev.*, **43**, 218 (1948).
36. Ito K., Moynihan C. T., and Angell, C. A. *Nature* **398**, 492 (1999).
37. Yue, Y-Z., and Angell, C. A., *Nature* 2003, (in press).
38. Minoguchi, A., Richert, R., and Angell, C. A., (to be published).
39. Starr, F. W., Angell, C. A., and Stanley, H. E., *Physica, A* **3223**, 51-66 (2003).
40. More appropriate would be the residual entropy per re-arrangeable sub-unit of the substance under consideration, but unfortunately there is no definitive method of assigning the number of such sub-units, and simple rigid molecules are in general not glass-forming.
41. Angell, C. A., Borick S., and Grabow, M., *J. Non-Cryst. Solids*, **205-207**, 463-471 (1996).
42. (a) Sastry, S., and Angell, C. A. *Nature Materials*, **2**, 739-743 (2003).
    (b) Angell, C. A., *Physica D* **107**, 122-142 (1997).
43. Shakhnovich, E., Abkevich, V., and Pitsyn, O., *Nature* **379**, 96-98, (1996).
44. Dokholyan, N. V., Buldyrev, S. V., Stanley, H. E., and Shakhnovich, E. I., *Folding and Design* **3**, 577-587 (1998).
45. Dokholyan, N. V., Buldyrev, S. V., Stanley, H. E., and Shakhnovich, E. I., *J. Mol. Biol.* **296**, 1183-1188 (2000).
46. Angell, C. A., *Physica D*, **107**, 122-142 (1997).
47. Angell C. A., and Wang, L.-M., *Biophys. Chem.* **105**, 621-637, (2003).
48. The solvent consists of approximately equal parts of water, ionic liquid (ethyl ammonium nitrate) and sugar (sucrose or glucose, but not fructose or trehalose. The latter exclusions are because of the presence, in the two excluded sugars of an anomeric equilibrium in the temperature range of the denaturation, the endothermic character of which confuses the thermogram.
49. Matagne, A., Jamin, M. Chung, E. W., Robinson, C. V., Radford, S. E., and Dobson, C. M., *J. Mol. Biol.* **297**, 193-210, (2000).
50. Radford, S. E., Dobson, C. M., and Evans, P. A., *Nature* **358**, 302-307, (1992).
51. Ptitsyn O. B., Finkelstein A.V., and Dobson C.M., *Mol. Biol.* **33**, 893-896, (1999).

Dynamic Heterogeneities in Liquid Water

Nicolas Giovambattista*, S. V. Buldyrev*, Francis W. Starr[†] and H. E. Stanley*

*Center for Polymer Studies and Department of Physics, Boston University, Boston, MA 02215 USA
[†]Department of Physics, Wesleyan University, Middletown, CT 06459 USA

Abstract. We investigate the dynamic heterogeneities of liquid water by performing molecular dynamics simulations using the SPC/E model. We identify clusters of mobile molecules consistent with spatially heterogeneous dynamics. We study the temperature and time dependence of the cluster size and find that clusters grow as temperature decreases and reach their maximum size at the time scale corresponding to the escape of the molecules from the cages formed by neighboring molecules. We relate the average mass n^* of mobile particle clusters to the diffusion constant, D, and the configurational entropy, S_{conf}. We find that n^* can be interpreted as the mass of the "cooperatively rearranging regions" that form the basis of the Adam-Gibbs theory of the dynamics of supercooled liquids.

INTRODUCTION

Supercooled liquids are characterized by the non-exponential decay of ensemble-averaged time correlation functions [1, 2, 3, 4]. According to the mode coupling theory (MCT) [5, 6], this decay can be expressed in terms of a stretched exponential function, $\exp\left[-(t/\tau)^\beta\right]$ with $\tau \sim (T - T_{MCT})^{-\gamma}$, where T_{MCT} is the mode coupling temperature which is slightly above the glass transition temperature, T_g.

Two microscopic scenarios have been proposed to explain this behavior; they are schematically shown in Fig. 1. In the spatially "homogeneous dynamics" scenario, correlation functions for different molecules decay in the same way, e.g., by a unique stretched exponential function with a characteristic relaxation time τ and exponent β. As shown in Fig. 1, in the "homogeneous" scenario, all molecules are equivalent. As temperature is lowered, the locally averaged molecular displacement is the same at every point in the system. The homogeneous scenario is inconsistent with experiments [7, 8, 9, 10, 11, 12] and simulations [13, 14, 15, 16], which identify dynamical heterogeneities in supercooled liquids and spin glasses [17].

In the spatially "heterogeneous dynamics" (SHD) scenario, correlation functions for different molecules also decay exponentially, but with a distribution of relaxation times[18]. The superposition of these individual exponential contributions produces a non-exponential decay of the ensemble-averaged time correlation function, and the exponent β is a measure of the width of the distribution of relaxation times. In the heterogeneous scenario, the locally averaged molecular displacements are different depending on the part of the system box we are looking at. One finds groups of molecules that are more mobile and groups that are less mobile than the average molecule in the system. As the temperature is lowered patches formed by mobile molecules increase in size. These patches of mobile molecules have a short lifetime; they appear and disappear constantly in different parts of the system. In the next section we show that SHD describe the dynamics in supercooled water.

The number of different configurations available to the system has been related to the configurational entropy S_{conf}, a concept first introduced by Adam-Gibbs (AG) [29]. The AG theory predicts that the diffusion coefficient can be expressed as a function of temperature and S_{conf}. The AG prediction was confirmed in computer simulations using the SPC/E model [30]. However, the AG theory is based on the concept of cooperative rearranging regions (CRR), which are not precisely defined. In the subsequent section we relate the clusters formed by mobile molecules found in simulations [31] with the CRR from AG theory.

SPATIALLY HETEROGENEOUS DYNAMICS

Clusters composed of particles with high mobility have been found in numerical simulations of simple systems, e.g., Lennard-Jones (LJ) mixtures, indicating the presence of spatially heterogeneous dynamics (SHD) [13, 16, 32, 33, 34, 7, 35, 36, 37, 38, 39, 40]. Hence, the SHD scenario for the dynamics of cold liquids was confirmed. In this section we show that SHD are also present in computer simulations of the SPC/E [41] wa-

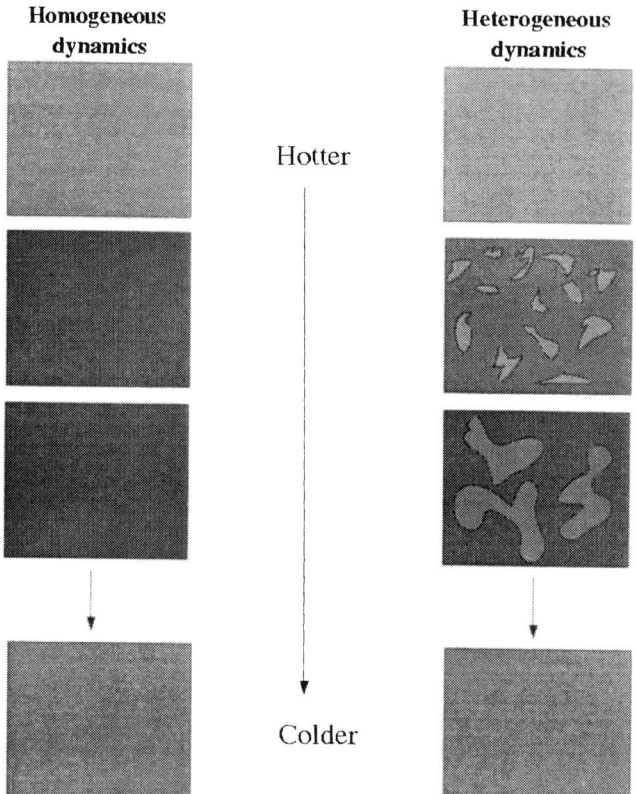

FIGURE 1. Two possible scenarios proposed to describe diffusion in cold liquids. In the spatially homogeneous dynamics scenario molecules relax in the same way, while in the SHD scenario, sets of more mobile molecules (in comparison to the average motion of the molecules in the system) form patches or clusters. The size of these clusters increases upon cooling.

ter model. We study a system with $N = 1728$ molecules at fixed density $\rho = 1.0$ g/cm^3 varying the temperature T from 200 K to 260 K in steps of 10 K. In order to increase statistics, we performed two independent simulations for every temperature. We find that the T-dependence of the diffusion constant can be expressed by:

$$D \sim (T - T_{MCT})^\gamma, \quad (1)$$

where the mode coupling temperature $T_{MCT} = 193$ K and the diffusivity exponent $\gamma = 2.80$.

We use the approach to define SHD clusters which was introduced in a study of a LJ mixture [33] and in experiments on colloids [40]. We calculate the self part of the time-dependent van Hove correlation function [42] $G_s(r,t)$,

$$G_s(r,t) \equiv \frac{1}{N} \sum_{i=1}^{N} \langle \delta(|\vec{r}_i(t) - \vec{r}_i(0)| - r) \rangle, \quad (2)$$

where $\langle \cdots \rangle$ represents average over configurations and $\vec{r}_i(t)$ are the coordinates of the oxygen atom of the i-th molecule. The probability of finding an oxygen atom at a distance r at time t from its position at $t = 0$ is given by $4\pi r^2 G_s(r,t) dr$.

For both short times (when particles move ballistically) and long times (when particle motion can be described by the diffusion equation), $G_s(r,t)$ can be fitted by a Gaussian approximation

$$G_0(r,t) = \left[\frac{3}{2\pi \langle r^2(t) \rangle}\right]^{3/2} \exp\left[-3r^2/2\langle r^2(t) \rangle\right], \quad (3)$$

where $\langle r^2(t) \rangle$ is the mean square displacements of the oxygen atoms. However, deviations of $G_s(r,t)$ from $G_0(r,t)$ are well pronounced at intermediate times, corresponding to the vibrations of the particle within the cage formed by neighbor molecules. We define t^* as the value of time at which the deviation of $G_s(r,t)$ from $G_0(r,t)$ is maximum. In order to do this, we find the maximum of the non-Gaussian parameter [1, 2]

$$\alpha_2(t) \equiv \frac{3}{5} \langle r^4(t) \rangle / \langle r^2(t) \rangle^2 - 1 \quad (4)$$

In Fig. 2, we see that $G_s(r,t^*)$ and $G_0(r,t^*)$ intersect for large r at r^*, and that $G_s(r,t^*)$ develops a tail

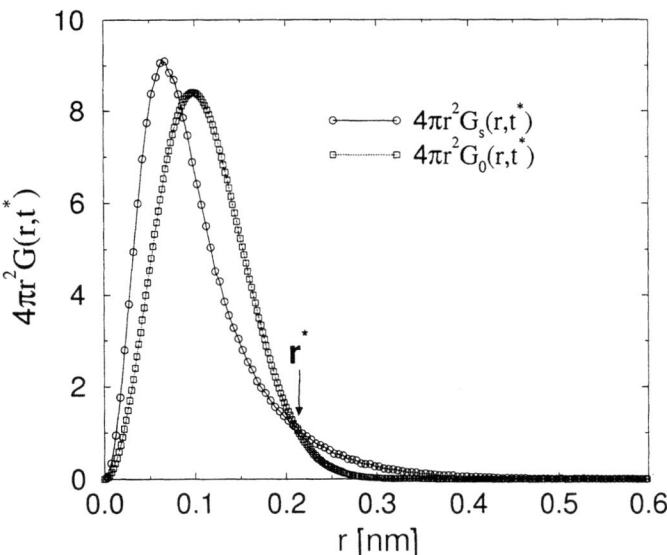

FIGURE 2. Van Hove correlation function $G_s(r,t^*)$ and its Gaussian approximation $G_0(r,t^*)$ obtained using $\langle r^2(t^*)\rangle$, for $T = 220$ K. The tails of the distributions cross at $r^* \approx 0.225$ for all temperature.

for large r falling outside the Gaussian distribution. Molecules with displacements $r > r^*$ can be considered as molecules that move more than expected (in comparison to $G_0(r,t^*)$). We find r^* is in the range 0.20–0.25 nm for all T (the oxygen-hydrogen distance in a molecule for SPC/E is 0.1 nm). If there are SHD, one would expect molecules with displacements $r > r^*$ to get involved in the SHD clusters. The fraction of molecules with $r > r^*$ at $t = t^*$ is given by $\phi \equiv \int_{r^*}^{\infty} 4\pi r^2 G_s(r,t^*)dr$. Depending on T, we find $6\% < \phi < 8\%$. For simplicity we fix $\phi = 7\%$ for all T. Similar values of ϕ were found in atomic systems [32, 33, 40] and in polymer melts [43].

Following Refs. [33] and [40], we define the mobility of molecule i at a given time t_0 as the maximum displacement of the oxygen atom in the interval $[t_0, t_0 + \Delta t]$,

$$\mu_i(t_0, \Delta t) = max\{|\vec{r}_i(t_0) - \vec{r}_i(t + t_0)|, t_0 \leq t \leq t_0 + \Delta t\}. \quad (5)$$

We will be interested in the "mobile" molecules defined as the fraction ϕ of molecules with larger μ_i. Finally, we define a SHD cluster for an observation time Δt as those mobile molecules whose nearest-neighbor oxygen-oxygen distance at time t_0 is less than 0.315 nm, the first minimum of the oxygen-oxygen radial distribution function.[1] We find in water that SHD-clusters are similar to those in models of simpler liquids. Fig. 3 shows two snapshots of mobile particle clusters at $T = 260$ K for $\Delta t = t^*$. In LJ systems [32], monatomic liquids [44], and polymers [45], complex clusters are composed of more elementary "strings" in which particles are arranged in a roughly linear fashion. This is not so clear in simulations of water because the hydrogen bond network constrains the geometry of the clusters.

Dependence of Cluster Size on t and Δt

We first address the issue of the dependence of SHD clusters on the observation time Δt. The quantities we study are average cluster size, $\langle n(\Delta t)\rangle$, and the weight average cluster size,

$$\langle n(\Delta t)\rangle_w \equiv \frac{\langle n^2(\Delta t)\rangle}{\langle n(\Delta t)\rangle}, \quad (6)$$

which is the average size of a cluster to which a randomly chosen molecule belongs. Figure 4(a) shows $\langle n(\Delta t)\rangle$ and $\langle n(\Delta t)\rangle_w$ for $T = 210$ K. It is known from percolation theory [46] that clusters of randomly chosen molecules have non-trivial dependence of their sizes on the fraction of chosen molecules, ϕ. In order to define how the cooperativity affects the cluster sizes, we use a normalized quantity, $\langle n(\Delta t)\rangle_w / \langle n_r\rangle_w$, where $\langle n_r\rangle_w$ is the weight average cluster size when choosing randomly ϕN molecules. For comparison, we also show in Fig. 4(b) the non-Gaussian parameter $\alpha_2(\Delta t)$ and the mean-squared displacement $\langle r^2(\Delta t)\rangle$. The three characteristic time regimes: ballistic, cage and diffusive are indicated.

---

[1] Alternatively, we also consider using a separation of 0.35 nm, the distance criterion commonly used by hydrogen bond definitions [F. Sciortino and S. L. Fornili, J. Chem. Phys. **90**, 2786 (1989)]. Preliminary calculations indicated this alternative choice does not qualitatively affect our results.

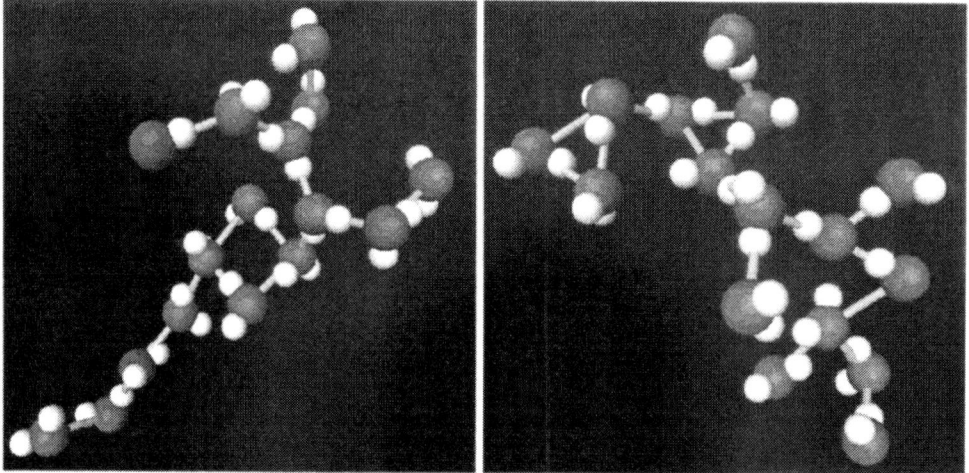

FIGURE 3. Two of the larger clusters of mobile molecules found at $T = 260$ K defined with an observation time $\Delta t = t^* \approx 3$ ps. Tubes connect neighboring molecules whose oxygen-oxygen distance is less than 0.315 nm, the first minimum in the oxygen-oxygen radial distribution function.

We find that $\langle n(\Delta t)\rangle_w/\langle n_r\rangle_w$ behaves in a way similar to polymer systems [43] but differs in that there is a clear increase in $\langle n(\Delta t)\rangle_w/\langle n_r\rangle_w$ at the time scale at which molecules go from a ballistic to a cage regime. We attribute this to strong correlations in the vibrational motion of the first-neighbor molecules, owing to the presence of hydrogen bonds.

In Fig. 4(c) we show $\langle n(\Delta t)\rangle_w/\langle n_r\rangle_w$ for all considered T. For $T \leq 240$ K, the maximum in $\langle n(\Delta t)\rangle_w/\langle n_r\rangle_w$ increases in magnitude and shifts to larger time scales with decreasing T. The plateau at the crossover from the ballistic regime is nearly T-independent, as expected since the mean collision time is nearly T-independent. For $T \geq 250$ K, the maximum and the plateau merge, so it is not possible to separately distinguish these features.

In the SHD scenario, clusters of mobile molecules appear and disappear continually in time. In Fig. 5, we show the time evolution of a cluster defined using $\Delta t = t^* \approx 65$ ps, at $T = 210$ K. We set $t_0 = 0$ as the time at which the cluster is defined. The corresponding snapshot is shown in panel (a). It shows the molecules that will have the largest displacements during the time interval $[t_0, t_0 + \Delta t]$. Their positions at the end of this interval are shown on panel (d). The times of each snapshot are indicated schematically by arrows at the bottom of the figure. Times separating the three regimes (ballistic, cage and diffusive) are also identified. During the ballistic regime and until the beginning of the cage regime, we observe no change in the cluster structure. During the cage regime, the collisions of the molecules with the neighbor molecules produce little effect. At $t \approx 16.4$ ps only one molecule of the cluster (at the right end) changes hydrogen bonds. These cumulative small effects produce a split of the cluster as time reaches the end of the cage regime.[2] At $t = t^*$ the cluster splits into several subclusters. This fact shows that the cluster does not behave any more as a whole entity but its members possess certain degree of independence from each other. As expected, molecules for longer times diffuse farther and farther from each other, and no memory of the starting structure of the cluster remains. During this process, other clusters appear and disappear in the system.

Dependence of Clusters on Temperature

We focus now on the temperature dependence of the clusters obtained with $\Delta t = t^*$. Because we use the same definition of clusters as in Ref.[33] we can compare our results with those found there for a LJ mixture. Fig. 6 shows the probability distribution $P(n,T)$ to find a cluster with n molecules for different temperatures T using a fraction $\phi = 0.07$. We note that for this fraction there are no percolating clusters for this system size.

We fit the distributions with the *Ansatz* defined in percolation theory [46]

$$P(n,T) \sim n^{-\tau(T)} \exp\left[\frac{-n}{n_0(T)}\right], \quad (7)$$

where $n_0(T)$ is a characteristic cluster size at T and $\tau(T)$ is the Fisher exponent, which may also depend on T. A similar expression for $P(n,T)$ was found in [43] for a polymer melt. The parameters in this expression are tabulated in Table I. Changing the value of ϕ does not affect

---

[2] For our particular case, at the beginning of the diffusive regime molecules are able to recombine forming two residual subclusters.

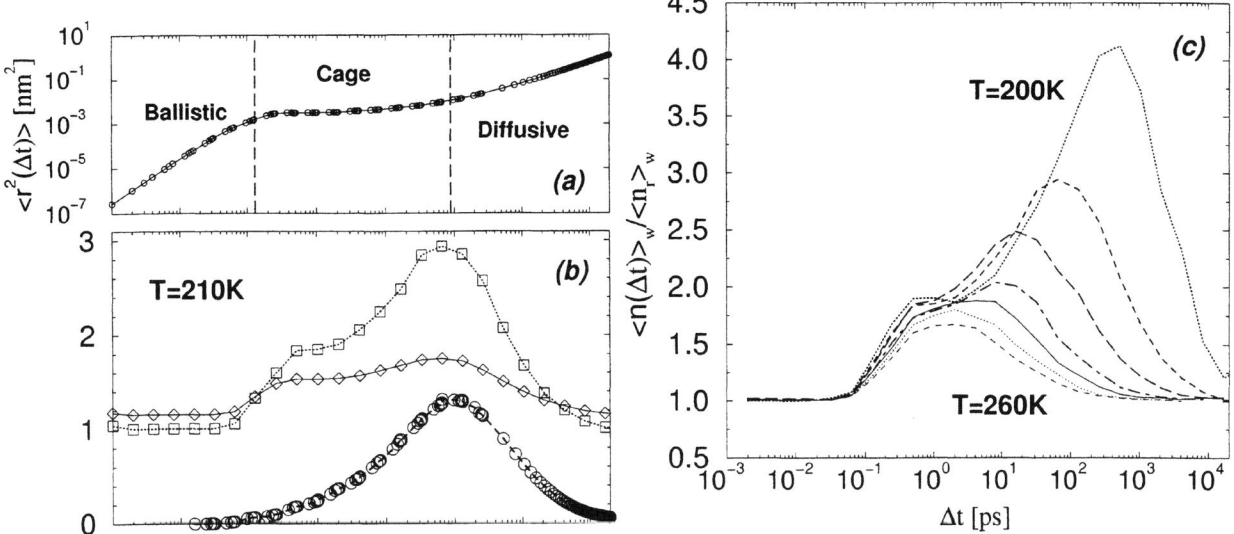

FIGURE 4. (a) Mean square displacement $\langle r^2(\Delta t)\rangle$ at $T = 210$ K showing the ballistic, cage and diffusive regimes. (b) Average number of molecules $\langle n(\Delta t)\rangle$ ($\diamond$) and normalized weight cluster size $\langle n(\Delta t)\rangle_w$ ($\square$). The behavior of these quantities correlate with $\langle r^2(\Delta t)\rangle$. The maxima of $\langle n(\Delta t)\rangle_w$ and $\langle n(\Delta t)\rangle$ occur at times slightly smaller than the time for the maximum in $\alpha_2(\Delta t)$ ($\bigcirc$), the non-Gaussian parameter. (c) Weight average cluster size $\langle n(\Delta t)\rangle_w/\langle n_r\rangle_w$ for temperatures ranging from 200 K to 260 K in intervals of 10 K. Note the T-independent plateau at the crossover from ballistic motion to cage behavior.

TABLE 1. Fitting parameters for $P(n,T) \sim n^{-\tau}\exp(-n/n_0(t))$.

T	$\tau(T)$	$n_0(T)$
200	2.17	33.44
210	1.99	7.43
220	2.09	7.98
230	1.96	4.28
240	2.04	4.89
250	2.01	4.05
260	1.96	3.58

the functional dependence of $P(n,T)$ although it changes the values of $n_0(T)$ and $\tau(T)$. The values of $n_0(T)$ do not follow a simple dependence on T. However, the behavior of $n_0(T)$ indicates that clusters grow upon cooling. From Table I, $\tau(T) \approx 2$ for all T. This value coincides with the value obtained for LJ particles [$\tau(T) \approx 1.9$] [33] and for colloids [$\tau(T) \approx 2.2 \pm 0.2$] [40].

Clusters found in water seem to be smaller than those found in LJ systems. In Ref.[33] it was found that $P(n,T)$ for $T = 1.07\ T_{MCT}$ and $\phi = 0.05$ is non-zero up to $n \approx 80$ while from Fig. 6 we see that even at $T = 200$ K (i.e., $1.03\ T_{MCT}$) there are no clusters containing more than 50 molecules. Moreover the system studied in [33] contains 4 times more molecules than our system. This may explain the difference in the cluster size distribution.

SPATIALLY HETEROGENEOUS DYNAMICS AND THE COOPERATIVELY REARRANGING REGIONS OF THE ADAM-GIBBS THEORY

Many years ago Adam and Gibbs (AG) proposed a theory to describe the dynamics of supercooled liquids[29, 47, 48]. They introduced a concept of "cooperatively rearranging regions" (CRR) to describe the diffusion in a low temperature liquid. The AG theory assumes that the SHD scenario is correct.

The theory predicts the empirical Williams-Landel-Ferry equation which determines the evolution of the relaxation time with temperature. Another important result is the relation between the diffusion constant D, the temperature T and the configurational entropy of the system S_{conf},

$$D \propto \exp\left(\frac{-A}{TS_{conf}}\right). \quad (8)$$

S_{conf} is interpreted, in the thermodynamic limit, as $k_B \log W_c$, where W_c is the number of configurations accessible to the system and k_B is the Boltzmann constant. More recently, W_c has been identified as the number of basins in the potential energy landscape (PEL) accessible to the system in equilibrium, and this allows an easier di-

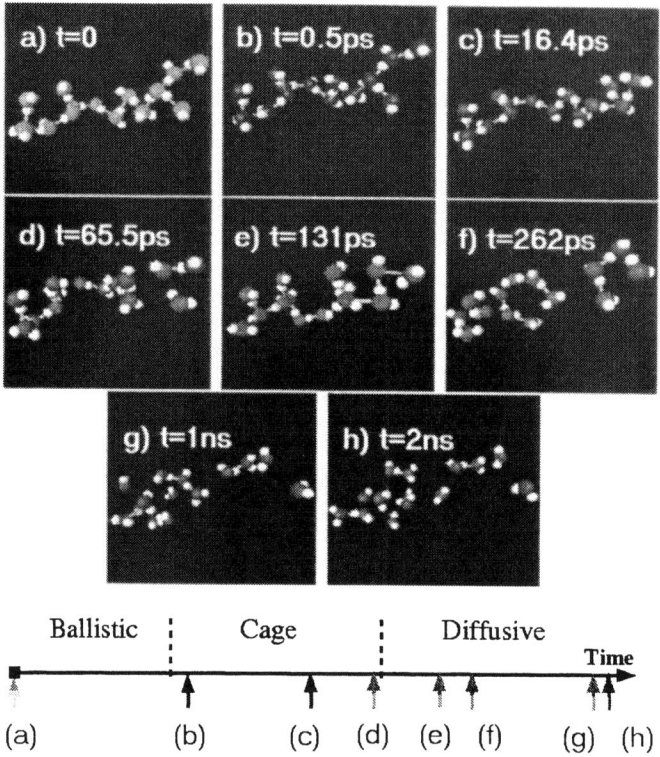

FIGURE 5. Time evolution of a cluster of mobile molecules identified with $\Delta t = t^* \approx 65$ ps at $T = 210$ K. Time increases from left-to-right and top-to-bottom and ais indicated along the time axis at the bottom of the figure. The three regimes (ballistic, cage and diffusive) defined in Fig. 4(a) are also indicated along the time axis.

rect calculation of S_{conf} by computer simulations[49, 50].

Equation (8) has been tested and appears to be valid across a wide spectrum of liquids [30, 51, 52]. However it is based on a somewhat imprecise definition of CRR. In their work, AG define a CRR as "...a subsystem of the sample which, upon a sufficient fluctuation in energy (or, more correctly, enthalpy), can rearrange into another configuration independently of its environment." There is no quantitative definition of CRR. AG predict that the characteristic mass z of the CRR is related to the configurational entropy of the CRR $s_{\text{conf}}(z)$ and the total configurational entropy S_{conf} by,

$$z = \frac{N s_{\text{conf}}(z)}{S_{\text{conf}}} \quad (9)$$

where N is the number of molecules in the liquid.

Based on Eq. 9, we will develop a quantitative definition of CRR in the context of the SHD analysis described in the section above. In order to find a relation between SHD and AG predictions, we calculate the average cluster mass $\langle n(\Delta t) \rangle$ for each T for $\Delta t = t^*$. Motivated by the recent results that the average instantaneous cluster mass scales inversely with the entropy in a model of living polymers [53] and based on Eq. (9), we use $n^* \equiv \langle n(t^*) \rangle$ as a measure of z, since at t^* correlations are very pronounced and $\langle n(t) \rangle$ is nearly maximal.[3] Using the values of S_{conf} from Ref. [30], we find a linear relationship between n^* and $1/S_{\text{conf}}$ (Fig. 7(a)),

$$n^* - 1 \propto \frac{1}{S_{\text{conf}}}. \quad (10)$$

This finding is consistent with the possibility that $n^* - 1$ can be regarded as a measure of z and provides a quantitative connection between SHD clusters and the AG approach.[4] It is necessary to subtract one from n^* to obtain direct proportionality, implying that a cluster of unit size does not correspond to a CRR[33]. Equation (10) provides a clear link between a cluster property, n^*, and a property of the PEL, S_{conf}. Since S_{conf} and the diffusion constant D are related [30], we expect to find

$$D \sim e^{-A(n^*-1)/T}. \quad (11)$$

---

[3] The maximum of $\langle n(\Delta t) \rangle$ occurs at time slightly before t^*. Our conclusions are unaffected by choosing n^* or the maximum of $\langle n(\Delta t) \rangle$.
[4] This connection relies on the assumption that the T dependence of $s_{\text{conf}}(z)$ is weak in comparison to that of S_{conf}, as can be expected since $z \ll N$ and the configurational entropy is an extensive property.

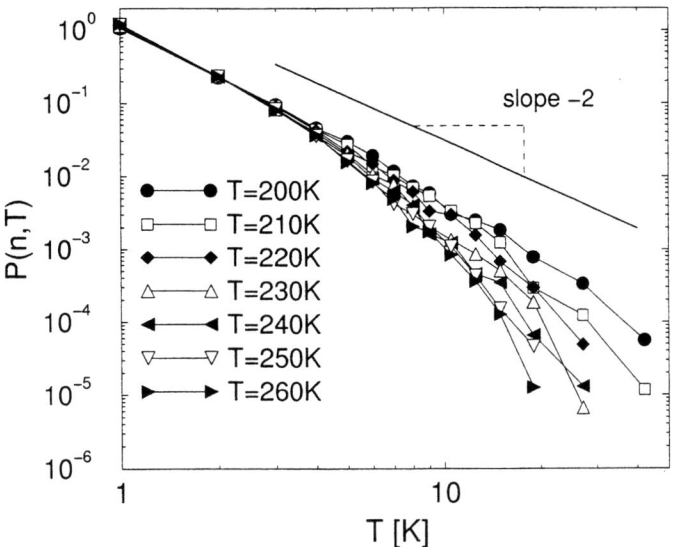

FIGURE 6. Probability distribution $P(n,T)$ to find a cluster with n molecules at temperature T. Data can be well fitted by $P(n,T) \sim n^{-\tau(T)} \exp(-n/n_0(T))$. Values of the parameters $n_0(T)$ and $\tau(T)$ are in Table I.

Indeed, Fig. 7(b) confirms this expectation.

REFERENCES

1. F. Sciortino, P. Gallo, P. Tartaglia, and S. H. Chen, Phys. Rev. E **54**, 6331 (1996).
2. F. Sciortino, L. Fabbian, S.-H. Chen, and P. Tartaglia, Phys. Rev. E **56**, 5397 (1997).
3. W. Kob and H. C. Anderson, Phys. Rev. Lett. **73**, 1376 (1994).
4. F. W. Starr, F. Sciortino, and H. E. Stanley, Phys. Rev. E **60**, 6757 (1999).
5. W. Götze, J. Phys.: Condens. Matter **11**, A1 (1999).
6. W. Götze, in *Liquids, Freezing and Glass Transition*, edited by J. P. Hansen, D. Levesque, and J. Zinn-Justin (North-Holland, Amsterdam, 1991).
7. K. Schmidt-Rohr and H. W. Spiess, Phys. Rev. Lett. **66**, 3020 (1991).
8. J. Liesen, K. Schmidt-Rohr, and H. W. Spiess, J. Non-Cryst. Solids **172-174**, 737 (1994).
9. A. Heuer, M. Wilhelm, H. Zimmermann, and H. W. Spiess, Phys. Rev. Lett. **75**, 2851 (1995).
10. M. T. Cicerone and M. D. Ediger, J. Chem. Phys. **103**, 5684 (1995).
11. F. Fujara, B. Geil, H. Sillescu, and G. Fleischer, Z. Phys. B **88**, 195 (1992).
12. M. T. Cicerone, F. R. Blackburn, and M. D. Ediger, J. Chem. Phys. **102**, 471 (1995).
13. M. Hurley and P. Harrowell, Phys. Rev. E **52**, 1694 (1995).
14. A. I. Mel'cuk, R. A. Ramos, H. Gould, W. Klein, and R. D. Mountain, Phys. Rev. Lett. **75**, 2522 (1995).
15. T. Muranaka and Y. Hiwatari, Phys. Rev. E **51**, R2735 (1995).
16. W. Kob, C. Donati, S. J. Plimpton, P. H. Poole, and S. C. Glotzer, Phys. Rev. Lett. **79**, 2827 (1997).
17. P. H. Poole, S. C. Glotzer, A. Coniglio, and N. Jan, Phys. Rev. Lett. **78**, 3394 (1997).
18. C.A. Angell, K.L. Ngai, G.B. McKenna, P.F. McMillan, and S.W. Martin, J. Appl. Phys. **88**, 3113 (2000).
19. M. Goldstein, J. Chem. Phys. **51**, 3278 (1969).
20. F. H. Stillinger and T. A. Weber, Phys. Rev. A **28**, 2408 (1983).
21. F. H. Stillinger, Science **267**, 1935 (1995).
22. C. A. Angell, Science **267**, 1924 (1995).
23. S. Sastry, P. G. Debenedetti, and F. H. Stillinger, Nature (London) **393**, 554 (1998).
24. P. G. Debenedetti, and F. H. Stillinger, Nature (London) **410**, 259 (2001).
25. T. B. Schrøder, S. Sastri, J.C. Dyre, and S.C Glotzer, J. Chem. Phys. **112**, 9834 (2000).
26. A. Heuer, Phys. Rev. Lett. **78**, 4051 (1997); S. Buchner and A. Heuer, Phys. Rev. E **60**, 6507 (1999).
27. L. Angelani, G. Parisi, G. Ruocco, and G. Villani, Phys. Rev. Lett. **81**, 4648 (1998).
28. F. Sciortino, W. Kob, and P. Tartaglia, Phys. Rev. Lett. **83**, 3214 (1999).
29. G. Adam and J. H. Gibbs, J. Chem. Phys. **43**, 139 (1965).
30. A. Scala, F. W. Starr, E. La Nave, F. Sciortino and H. E. Stanley, Nature (London) **406**, 166 (2000).
31. N. Giovambattista, F. W. Starr, F. Sciortino, S. V. Buldyrev, and H. E. Stanley, Phys. Rev. E **65**, 041502 (2002).
32. C. Donati J. F. Douglas, W. Kob, S. J. Plimpton, P. H. Poole, and S. C. Glotzer, Phys. Rev. Lett. **80**, 2338 (1998).
33. S. C. Glotzer, P. H. Poole, W. Kob, S. J. Plimpton, Phys. Rev. E **60**, 3107 (1999).
34. B. Doliwa and A. Heuer, Phys. Rev. Lett. **80**, 4915 (1998).
35. R. Böhmer et al., Europhys. Lett. **36**, 55 (1996).
36. B. Schiener et al., Science **274**, 752 (1996).
37. W. K. Kegel, and A. van Blaaderen, Science **287**, 290 (2000).
38. H. Sillescu, J. Non-Cryst. Solids **243**, 81 (1999).

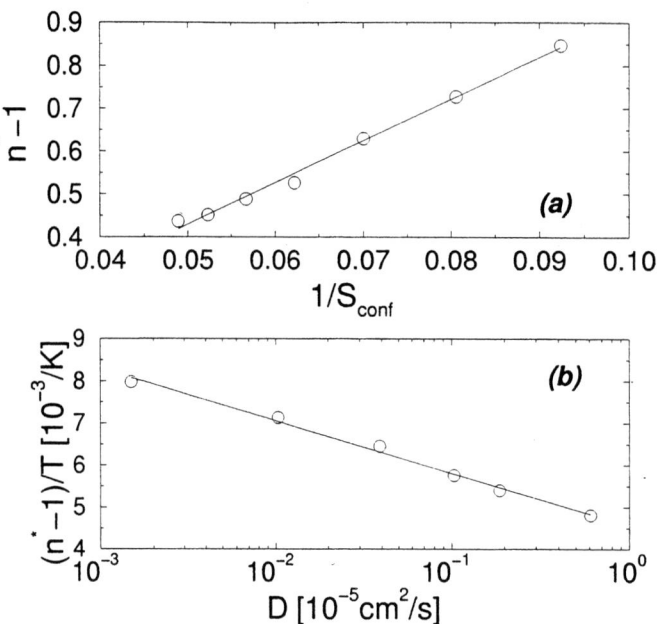

FIGURE 7. (a) The average cluster size n^* is proportional to the inverse of the configurational entropy S_{conf}, suggesting that $n^* - 1$ can be used as a measure of the size of the cooperatively rearranging regions hypothesized by Adam and Gibbs. (b) Log-linear plot of $(n^* - 1)/T$ as a function of the diffusion constant D. The AG prediction $D \sim \exp(A/TS_{conf})$ implies that $\log D \sim (n^* - 1)/T$. This relationship holds for almost three decades in D.

39. M. D. Ediger, Ann. Rev. Phys. Chem. **51**, 99 (2000).
40. E. R. Weeks et al., Science **287**, 627 (2000).
41. H. J. Berendsen et al. J. Phys. Chem. **91**, 6269 (1987).
42. J. P. Hansen and I. R. McDonald, *Theory of Simple Liquids* (Academic Press, London, 1986).
43. Y. Gebremichael T. B. Schrøder, F. W. Starr, and S. C. Glotzer, Phys. Rev. E **64**, 051503 (2001).
44. C. Bennemann, J. Baschnagel, and S. C. Glotzer, Nature **399**, 246 (1999).
45. Y. Gebremichael, T. B. Schrøder, and S. C. Glotzer, Abstr. Am. Chem. Soc. **220**, 412 (2000).
46. S. Stauffer and A. Aharony, *Introduction to Percolation Theory* (Taylor and Francis, London, 1998).
47. For a clear description of the physical basis of AG theory, see P. G. Debenedetti, *Metastable Liquids* (Princeton University Press, Princeton, 1996).
48. P. G. Debenedetti, J. Phys.: Cond. Mat. **15**, R1669–R1726 (2003).
49. F. H. Stillinger, and T. A. Weber, J. Phys. Chem. **87**, 2833 (1983).
50. B. Coluzzi, G. Parisi, and P. Verrocchio, Phys. Rev. Lett. **84**, 306 (1999).
51. S. Sastry, Nature (London) **409**, 164 (2001).
52. S. Mossa et al., Phys. Rev. E **65**, 041205 (2002).
53. J. Dudowicz, K. F. Freed, and J. F. Douglas, J. Chem. Phys. **111**, 7116 (1999).
54. B. Doliwa and A. Heuer, Phys. Rev. E **67**, 030501 (2003).
55. B. Doliwa and A. Heuer, Phys. Rev. E **67**, 031506 (2003).
56. R. A. Denny, D. R. Reichman, and J.-P. Bouchaud, Phys. Rev. Lett. **90**, 025503 (2003).
57. I. Ohmine, and H. Tanaka, Chem. Rev. **93**, 2545 (1993) and references therein.
58. *Hydrogen Bonded Liquids*, edited by J. Dore and J. Texeira (Kluwer, Dordrecht, 1991), pp.171-183.
59. E. Grunwald, J. Am. Chem. Soc. **108**, 5719 (1986).
60. A. H. Narten and H. A. Levy, Science **165**, 447 (1969).
61. P. A. Giguere, J. Chem. Phys. **87**, 4835 (1987).
62. G. E. Walrafen, M. S. Hokmabadi, W. H. Yang, Y. C. Chu, and B. Monosmith, J. Phys. Chem. **93**, 2909 (1989).

Self-diffusion and Spatially Heterogeneous Dynamics in Supercooled Liquids Near T_g

Stephen F. Swallen[a], Osamu Urakawa[b], Marie Mapes[a], and M.D. Ediger[a]

[a]*Department of Chemistry, University of Wisconsin-Madison, Madison, WI 53706 USA*
[b]*Department of Macromolecular Science, Graduate School of Science, Osaka University, Toyonaka, Osaka, 560-0043, JAPAN*

Abstract. In spite of the fundamental and practical importance of the self-diffusion coefficient, values of this quantity are not generally available near the glass transition temperature T_g. Here we describe experimental results for three different single component glass-forming systems. In *tris*-naphthylbenzene, the self-diffusion coefficient at T_g is 400 times larger than the value anticipated by the Stokes-Einstein equation. Preliminary results on *ortho*-terphenyl indicate similarly enhanced translational diffusion near T_g. For oligomers of polystyrene (20-mers), the temperature dependence of the terminal relaxation times completely tracks that of the self-diffusion coefficient.

INTRODUCTION

In spite of significant progress in the last decade, glasses and glass formation remain challenging problems in condensed phase physical science[1-3]. In a temperature span of 50 K above the glass transition temperature T_g, a typical supercooled liquid or amorphous polymer increases its viscosity by 8 orders of magnitude. In contrast to the first order solidification transition (crystallization), the transition to an amorphous solid occurs without any significant structural change. As observed in the lab, the glass transition is entirely kinetic, occurring when molecular motion becomes so slow that the liquid can no longer achieve equilibrium on the time scale of the experiment.

As described here, our recent work has focused on measuring self-diffusion near T_g in single component glass forming liquids. This work is motivated partly by the intrinsic importance of the self-diffusion coefficient as a parameter for characterizing molecular motion in supercooled liquids and by practical applications where knowledge of diffusion is critically important. For example, the pharmaceutical industry currently produces almost all consumer drugs in a crystalline form. For many otherwise useful drugs, the crystalline forms are so insoluble that bioavailability is negligible. Bioavailability of the amorphous drug can be 1000 times higher but inhibiting crystallization for the required multi-year shelf-life is a considerable challenge that requires, among other things, accurate information about molecular transport near T_g. A second source of motivation for our study of self-diffusion near T_g is work in the last decade that gives one reasons to suspect that self-diffusion near T_g will be "interesting". This recent work will be summarized in the remainder of this introduction.

Previous Work on Diffusion in Supercooled Liquids

Sillescu's group first reported the remarkable result that translational and rotational motions have different temperature dependences as T_g is approached[4]. Translational motion at low temperature is faster than predicted by the Stokes-Einstein equation:

$$D_T = kT/6\pi\eta r_s \quad (1)$$

This enhancement of translational motion is as large as a factor of 4 in the range in which Sillescu and coworkers could obtain self-diffusion data ($\geq 1.2\ T_g$). Much larger enhancements have been observed at lower temperatures for *probe* diffusion. An example of this data is shown in Figure 1, for two different probes in o-terphenyl (OTP). Each panel compares the rotational correlation times and the translational diffusion coefficients for a probe to the temperature dependence of η/T. The larger probe (rubrene), shown in the upper panel, has rotational and translation motions which exactly follow the hydrodynamic predictions (Stokes-Einstein and Debye-Stokes-Einstein) right down to T_g. The smaller probe (tetracene) shows more complex behavior. While the rotational correlation times track η/T, the translational diffusion coefficients show a substantially weaker temperature dependence than T/η. This deviation amounts to about two decades at temperatures near T_g. Which probe might be indicating what the OTP molecules are doing? Since OTP and tetracene are similar in size, one would guess that self-diffusion of OTP would be similar to that shown in Figure 1b.

Are the results shown in Figure 1b some weird probe effect? Would an enhancement of translational diffusion also be observed for the host molecules, *i.e.*, in the self-diffusion coefficient? It is surprising that so little information is available about self-diffusion near T_g in glass forming liquids. Here are the most relevant aspects of the literature: 1) Amorphous metals. Single component systems cannot be studied near T_g due to crystallization. The diffusion coefficients of some components in amorphous alloys have been measured near T_g.[5] Since different components may well have different diffusion coefficients with different temperature dependences, it is difficult to make a comparison to the temperature dependence of the viscosity. 2) Inorganic glasses. Diffusion has been measured in many systems; often these systems have many components.[6] In these studies, the different components can have substantially different diffusion coefficients, again making the comparison to the viscosity difficult. 3) Amorphous polymers. While many measurements of self-diffusion have been made, very few measurements of diffusion near T_g have been performed. 4) Single component, small

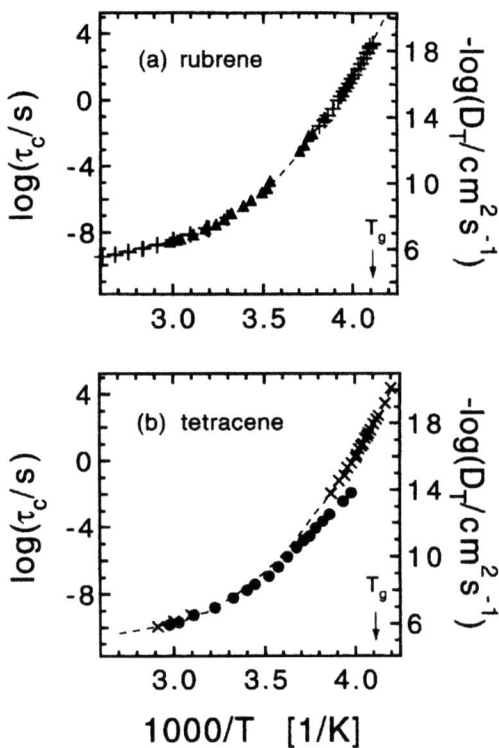

FIGURE 1. Temperature dependence of translational diffusion coefficients (filled symbols) and rotational correlation times (+,x) for rubrene and tetracene in OTP. In both panels the dashed lines show the temperature dependence of η/T. Reproduced from reference[7].

molecule glass formers. To our knowledge, prior to our recent work, the only diffusion coefficients available near T_g were for water.[8] While these results are very interesting, one cannot know if translation is enhanced since no viscosity values are available near T_g.

The possibility that self-diffusion shows a strong enhancement near T_g is striking because, if one insists on a homogeneous view of the liquid, it implies that individual molecules translate further and further before reorienting as the temperature is lowered. In contrast, in a heterogeneous scenario, this unphysical conclusion can be avoided by attributing enhanced translation to the different ways in which rotational and translational motion average over heterogeneity. Thus our current efforts to measure self-diffusion near T_g connects with recent progress in understanding dynamic heterogeneity in supercooled liquids.

Spatially heterogeneous dynamics

For more than fifty years, it has been known than relaxation processes in supercooled liquids are typically not exponential. Often these are characterized by fitting to the Kohlrausch-Williams-Watts (KWW) function:

$$CF(t) = e^{-(t/\tau)^\beta} \quad (2)$$

Typically, ß values of 0.3-0.6 are observed near T_g. This is the case for dielectric relaxation, NMR, and depolarized light scattering measurements, all of which are sensitive to reorientational motions. Such ß values have often been taken as an indication of heterogeneity, i.e., that different regions of the sample are relaxing at different rates. Only in the last ten years have experiments been able to show definitively that this is the case. Solid-state NMR[9], dielectric holeburning[10], and near-surface dielectric relaxation[11] have all given definitive evidence for spatially heterogeneous dynamics in neat glass formers[12, 13]. Probe measurements[12], including single molecule experiments[14], also support this view. The picture[12] that emerges from these measurements is that regions of different dynamics are 1-4 nm in size and retain their dynamic identities at least as long as τ_α, the average reorientation time in a molecular liquid. Slow and fast regions differ by up to four orders of magnitude in local relaxation times and relaxation within a given region can be nearly exponential. While the experimental evidence indicating heterogeneity comes almost exclusively from experiments very near T_g, simulations done at higher temperature also indicate significant heterogeneity[15-18]. These simulations are broadly in accord with experimental observations but do not indicate as pronounced heterogeneity, presumably because of the higher temperature. There is great interest in understanding what variation in local structure gives rise to the variation in local relaxation times. Candidates include density[19], local orientational order[20], local configurational entropy[21], local potential energy[22], and structures related to a frustrated phase transition[23].

SELF-DIFFUSION IN *TRIS*-NAPHTHYLBENZENE (TNB) NEAR T_g

We have measured, for the first time, self-diffusion at T_g in a single component glass former, *tris*-naphthylbenzene (TNB)[24]. Vapor deposition was used to create a bilayer of TNB (200 nm) and TNB-d_{14} (50 nm) well below T_g. Forward Recoil Spectrometry[25,26] (FReS) was used to measure concentration profiles of deuterio and protio TNB before and after annealing above T_g. Figure 2 shows the concentration profiles for TNB-d_{14} as deposited (open symbols) and after annealing (closed symbols) for 1100 s at T_g + 14 K. Both profiles are blurred due to the 35 nm resolution of these experiments. After accounting for this broadening, D_T can be calculated assuming Fickian diffusion; this assumption has been tested and found adequate by obtaining the same D_T for different annealing times.

Figure 3 shows the diffusion coefficients measured with FReS along with higher temperature measurements by static field gradient NMR[27]; both are compared to T/η, the temperature dependence of the SE equation (equation 1).

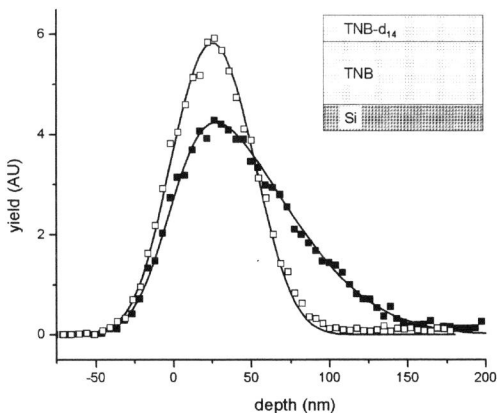

FIGURE 2. Deuterium FReS data, converted to yield versus depth from the air/TNB-d_{14} interface. Hollow symbols are an unannealed sample while the filled symbols are for a sample annealed for 1140 sec at 356 K (T_g + 14 K). The solid lines are the best fits, giving a self-diffusion coefficient $D_T = 1.2 \times 10^{-14}$ cm^2/sec. The top inset is a representation of the deuterio/protio bilayer of TNB as initially deposited from the vapor phase. Reprinted with permission from S.F. Swallen, P.A. Bonvallet, R.J. McMahon, and M.D.Ediger, Phys. Rev. Lett. 90, 015901-2 (2003). Copyright 2003 by the American Physical Society.

Clearly diffusion has a weaker temperature dependence than is predicted by the SE equation. At T_g, D_T is enhanced by a factor of 400 relative to the prediction of SE. While enhancements of this magnitude have been reported previously for probes in supercooled liquids, the observed enhancement depends critically upon probe size and the self-diffusion coefficient could not be accurately predicted from the probe diffusion.

The enhanced translational diffusion observed in TNB has an important impact on crystallization rates. The crystal growth rate of TNB changes less rapidly with temperature than does the viscosity[28]. Recently, this data (which extends down to T_g+30 K) has been used together with a model of crystal growth to make a quantitative prediction of the temperature dependence of D_T[29], assuming that D_T is the relevant transport coefficient. Reference 29 predicts that $D_T \propto \eta^{-0.74}$ from 373-413 K, in reasonable agreement with the direct measurements of $D_T \propto \eta^{-0.77}$. The non-trivial implication of this agreement is that even in a spatially heterogeneous system, crystal growth depends on the long time diffusion coefficient, as opposed to some other average of the local diffusivities.

Strictly speaking, Figure 3 shows inter-diffusion coefficients for TNB and TNB-d_{14} rather than self-diffusion. Given that ΔT_g for these two samples is less than 2 K,[31] and that diffusion is measured in essentially a 1:1 mixture, we estimate that the measured D_T values differ from true self-diffusion by about 0.15 decades. This correction has been applied to the data so that the reported diffusion coefficients represent self-diffusion of protio-TNB to an excellent approximation.

Self-diffusion in Other Single Component Glass Formers

We have also employed FReS to measure self-diffusion in oligo-styrene with 20 repeat units[32]. Again these measurements were combined with NMR field gradient measurements at high temperature. Here the temperature dependence of the SE equation is nearly perfectly obeyed. The small deviations which remain at low temperature can be quantitatively explained by detailed viscoelastic measurements which show that the product $D_T\tau_R$, where τ_R is the longest relaxation time of the chain, shows no temperature dependence. This result is consistent with the view that regions of enhanced mobility are extremely small, thus even a 20-mer is long enough that different parts of the molecule experience different dynamics. (Also, note the large difference observed in Figures 1a and 1b when the probe size increases by only a factor of two.) This precludes the enhanced translation possible for a smaller glass-former.

Preliminary measurements have been made on the self-diffusion of OTP using a temperature programmed desorption technique similar to that used by Kay and coworkers to study diffusion in water. Data acquired at $T_g + 9$ K indicate an enhancement of translational diffusion of about 2 decades relative to the viscosity. Thus it appears that OTP self-diffusion is similar to the tetracene probe diffusion coefficients shown in Figure 1b.

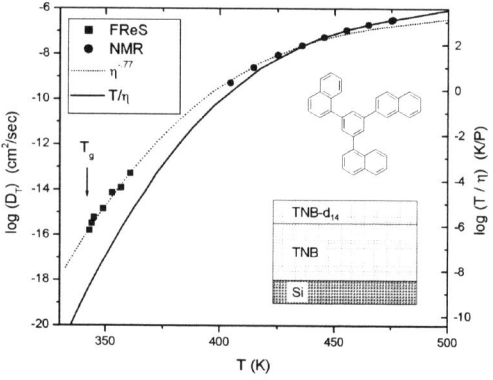

FIGURE 3. Comparison of the temperature dependence of translational diffusion with the viscosity. D_T for TNB determined from FReS and NMR[27] is given on the left axis. Temperature over viscosity[30] is given by the solid line, with scale on the right hand axis, shifted in order to permit overlap of the high temperature values of the viscosity and D_T values. The dashed line is $\eta^{-0.77}$, vertically shifted. Reprinted with permission from S.F. Swallen, P.A. Bonvallet, R.J. McMahon, and M.D.Ediger, Phys. Rev. Lett. 90, 015901-2 (2003). Copyright 2003 by the American Physical Society.

SUMMARY

The translational self-diffusion coefficient is arguably the most fundamental molecular parameter for describing motion in liquids. For single component supercooled liquids, this quantity is known at T_g for only a few materials. Almost all of the known information is summarized in this paper. The information available at this time indicates that for fragile glass-formers the self-diffusion coefficient and the viscosity can have substantially different temperature dependences and that diffusion at T_g can be enhanced by more than a factor of 400 relative to the prediction of the Stokes-Einstein equation. There may be a correlation between the extent of enhanced translation and the fragility, but this remains to be further explored. A second critical question is how small a length scale must be studied in order to find evidence of non-Fickian diffusion. At high temperatures, it is known that the mean-square-displacement becomes linear in time after the average molecule has moved only two diameters. If the large enhancements observed for self-diffusion near T_g are due to spatially heterogeneous dynamics, then non-Fickian diffusion should be observable out to much larger length scales. Information about this length scale and the manner in which diffusion becomes Fickian as the length scale is increased will provide critical tests for molecular descriptions of dynamics near the laboratory glass transition temperature.

ACKNOWLEDGMENTS

We thank the National Science Foundation (CHE-0245674) for supporting this work.

REFERENCES

[1] C. A. Angell, K. L. Ngai, G. B. McKenna, et al., J. Appl. Phys. **88**, 3113 (2000).
[2] M. D. Ediger, C. A. Angell, and S. R. Nagel, J. Phys. Chem. **100**, 13200 (1996).
[3] P. G. Debenedetti and F. H. Stillinger, Nature **410**, 259 (2001).
[4] F. Fujara, B. Geil, H. Sillescu, et al., Z. Phys. B Con. Mat. **88**, 195 (1992).
[5] H. Ehmler, A. Heesemann, K. Ratzke, et al., Phys. Rev. Lett. **80**, 4919 (1998).
[6] G. Brebec, R. Seguin, C. Sella, et al., Acta Metallurgica **28**, 327 (1980).
[7] M. T. Cicerone and M. D. Ediger, J. Chem. Phys. **104**, 7210 (1996).
[8] R. S. Smith and B. D. Kay, Nature **398**, 788 (1999).
[9] K. Schmidt-Rohr and H. Spiess, Phys. Rev. Lett. **66**, 3020 (1991).
[10] B. Schiener, R. V. Chamberlin, G. Diezemann, et al., J. Chem. Phys. **107**, 7746 (1997).
[11] E. V. Russell and N. E. Israeloff, Nature **408**, 695 (2000).
[12] M. D. Ediger, Annu. Rev. Phys. Chem. **51**, 99 (2000).
[13] H. Sillescu, J. Non-Cryst. Solids **243**, 81 (1999).
[14] L. A. Deschenes and D. A. Vanden Bout, Science **292**, 255 (2001).
[15] M. M. Hurley and P. Harrowell, Phys. Rev. E **52**, 1694 (1995).
[16] C. Donati, J. F. Douglas, W. Kob, et al., Phys. Rev. Lett. **80**, 2338 (1998).
[17] G. Johnson, A. I. Mel'cuk, H. Gould, et al., Phys. Rev. E **57**, 5707 (1998).
[18] R. Yamamoto and A. Onuki, Phys. Rev. Lett. **81**, 4915 (1998).
[19] R. E. Robertson, J. Polym. Sci.: Polym. Symp. **63**, 173 (1978).
[20] R. M. Ernst, S. R. Nagel, and G. S. Grest, Phys. Rev. B **43**, 8070 (1991).
[21] M. D. Ediger, J. Non-Cryst. Solids **235**, 10 (1998).
[22] C. Donati, S. C. Glotzer, P. H. Poole, et al., Phys. Rev. E **60**, 3107 (1999).
[23] D. Kivelson and G. Tarjus, Philos. Mag. B **77**, 245 (1998).
[24] S. F. Swallen, P. A. Bonvallet, R. J. McMahon, et al., Phys. Rev. Lett. **90**, 015901 (2003).
[25] P. F. Green, P. J. Mills, and E. J. Kramer, Polymer **27**, 1063 (1986).
[26] B. L. Doyle and D. K. Brice, Nucl. Instrum. Methods Phys. Res. Sect. B-Beam Interact. Mater. Atoms **35**, 301 (1988).
[27] I. Chang and H. Sillescu, J. Phys. Chem. B **101**, 8794 (1997).
[28] J. H. Magill and D. J. Plazek, J. Chem. Phys. **46**, 3757 (1967).
[29] K. L. Ngai, J. H. Magill, and D. J. Plazek, J. Chem. Phys. **112**, 1887 (2000).
[30] D. J. Plazek and J. H. Magill, J. Chem. Phys. **49**, 3678 (1968).
[31] R. Richert, K. Duvvuri, and L.-T. Duong, J. Chem. Phys. **118**, 1828 (2003).
[32] O. Urakawa, S. F. Swallen, M. D. Ediger, et al., Macromolecules, to be submitted (2002).

Crystallisation and Local Order in Glass-Forming Binary Mixtures

Julián R. Fernández[*] and Peter Harrowell

School of Chemistry, University of Sydney, New South Wales, 2006, Australia
and
[*]Comisión Nacional de Energía Atómica, Av. Libertador 8250
Capital Federal, Buenos Aires, Argentina

Abstract. The local organisation of a simulated glass-forming mixture due to Kob and Andersen is analysed. Evidence is presented for a structural transition from triangulated coordination polyhedra to cubic as the number fraction of the smaller species B increases towards equimolar. The impact of the change on the partial radial distribution function $g_{BB}(r)$ is established. The dependence of the crystallisation rate on the composition is determined and the related to the observed structural changes.

INTRODUCTION

Mechanical stability confers a special status on the associated configuration. Such special structures, no matter how apparently 'disordered', are subject to configurational restrictions by virtue of their stability. The 'deeper' the stability, the greater the configurational constraint, with long-range order (crystalline or orientational only) representing possible endpoints of this reduction in configuration space. The question of amorphous structure corresponds to establishing just how these constraints are manifested for the given stable configurations and whether these constraints provide a useful means of quantifying a type of order.

In this paper we present a preliminary study of the structure of the amorphous states of a popular model of a glass-forming alloy based on a binary mixture of particles interacting via Lennard-Jones (LJ) potentials. This work extends our recent study [1,2] of the stable and metastable crystalline phases of LJ mixtures.

The literature on amorphous structures predates the development of modern liquid theory and, therefore, may not be familiar to all readers. For this reason we have included a brief and, unavoidably, subjective summary of the history of this topic. We then introduce the model mixture and algorithms. Our results are divided into four sections. First, we examine the nature of coordination geometry about the minority component. We then present a simple argument which establishes a relationship between the composition and the average number of coordination polyhedra that must be packed about each majority component. This constraint is then used to rationalise the anomalous behaviour observed in the radial distribution function and the crystallisation rates.

A VERY BRIEF HISTORY OF AMORPHOUS STRUCTURE

A persistent dichotomy between *homogeneous* and *heterogeneous* pictures of glass structure can be traced back at least as far as the 1930's. In 1932 Zacharaisen [3] described a random network some 30 years before Bernal [4] and Finney [5] gave it substance in their packing studies of hard spheres. It was Bernal, in particular, who vigorously asserted the homogeneous character of such random packing. The random close packed model was extended to amorphous metal-metalloid alloys by Polk [6] but failed to account for the observed variation in metal-metal scattering with the choice of the metalloid [7]. The idea of the homogeneous random network has proved most useful in the low coordinated glasses such as silica and the chalcogenide mixtures. The constraint theory of Phillips [8] and Thorpe [9] provides a powerful set of limits on the stability of homogeneous random networks.

The heterogeneous picture of glass structure found an early proponent in Tammann [10]. In contrast with Zacharaisen's *structural* prescription, Tammann pictured the glass forming *process* as analogous to that of clays in which rigid clusters gradually come into contact as the intervening liquid is removed. Most of the efforts to 'flesh out' this idea begin with the suggestion of Frank [11] in 1952 that the stability of liquids to crystallisation might be due to the stability of icosahedral clusters. This idea has since been considerably extended in both application and sophistication [12]. Hoare [13] has presented one of the most lucid accounts of the program to demonstrate that Tammann's clusters could be constructed out of clusters characterised by 5- and 10-fold symmetry. Hoare, in particular, has emphasised the importance of explaining how the growth of these 'rigid clusters' comes to be self-limiting. An alternative, non-structural, perspective on glass formation has developed over the last 10 years that presents a picture very much in the spirit of Tammann's idea. This approach is based on the measurement of the growth of dynamic heterogeneities with supercooling.

Given the perennial popularity of the idea of icosahedral organisation in the context of amorphous structure it is worth making three points. First, the low energy of isolated icosahedral clusters of a single spherical species arises from their low surface energy and is not reproduced in condensed phases except in the case of oscillatory interactions such as those examined by Dzugutov [14]. Second, while the icosahedron cannot uniformly fill space it certainly forms stable crystals and, therefore, its presence alone is insufficient to explain the absence of crystallisation. Finally, of the 110 convex polyhedra with regular faces (excluding the prisms and antiprisms) only 3 can uniformly fill space. The icosahedron, therefore, must merely take its place among the many local coordination geometries that might frustrate crystallisation. Gaskell [15], for example, has provided strong evidence for the important role of a 9-coordinated polyhedra, the tricapped trigonal prism, in the amorphous Ni-P alloys. The successful packing criteria developed by Egami [16] to identify glass-forming alloys also allows for a wide range of coordination polyhedra.

This, then, is the point from which we shall start. Our initial perspective on amorphous structure is local (i.e nearest neighbour), as determined by the short range of the particle interactions. We shall, however, be led to consider the intermediate range, i.e. those lengths that cover the packing of adjacent coordination polyhedra. We shall argue that is unlikely that structure over any longer length scales is necessary to stabilise an amorphous state over the limited run times and system sizes accessible computationally.

MODEL AND ALGORITHM

The Lennard-Jones (LJ) potential for a mixture has the form

$$V_{ij}(r) = 4\varepsilon_{ij}\left[\left(\frac{\sigma_{ij}}{r}\right)^{12} - \left(\frac{\sigma_{ij}}{r}\right)^{6}\right] \quad (1)$$

where the sub-indices i and j could take the values A or B. We truncate the potential at a distance $2.5\sigma_{ij}$ and shift the potential so that it equals zero at the cut-off. (Here we shall set the masses of both components equal to m.) We shall work in the following reduced units throughout this paper: the unit of length is σ_{AA}, the unit of energy ε_{AA}, and the unit of time $\tau = \sigma_{AA}(m/\varepsilon_{AA})^{1/2}$. We shall follow Kob and Andersen (KA) [17] and set $\sigma_{AB} = 0.8$, $\sigma_{BB} = 0.88$, $\varepsilon_{AB} = 1.5$ and $\varepsilon_{BB} = 0.5$. This mixture at a composition of $x_B = N_B/N = 0.2$ has been studied extensively as a model glass former. The choice of parameters was originally made to model the Ni-P system. Based on the stable crystal phases [1], we have suggested that it better represents the Ni-Be mixture.

Molecular dynamics simulations have been carried out at constant NPT using a Nosé-Poincaré-Andersen Hamiltonian and a generalised leapfrog algorithm [18]. All calculations were performed at zero pressure. Enthalpy minimizations were carried out using a conjugate gradient scheme which ensures a fixed pressure.

RESULTS

The Stable and Metastable Crystal Structures

To establish a well-defined point of departure, we shall begin with the stable and metastable crystal structures of the KA mixture. We have recently completed a study of the lattice energies of a range of LJ mixtures [2]. In Figure 1 we plot the lattice energies per particle of a number of crystal structures as a function of σ_{AB} for A_3B mixtures (i.e. $x_B = 0.25$).

As reported previously, the most stable crystal state of the KA mixture over the composition range $0.0 \leq x_B \leq 0.5$ consists of coexisting face centered cubic (fcc) of pure A and the CsCl structure with composition AB. In order of ascending lattice energies we have the $PuBr_3$ structure, coexisting Pd_2Zr structure and fcc, the cementite Fe_3C structure and then the Ni_3P structure. For reference, the lowest 'lattice energy' per particle obtained for the amorphous state following a similar enthalpy minimization is -7.92, significantly higher than the lattice energy of any of the crystalline states identified.

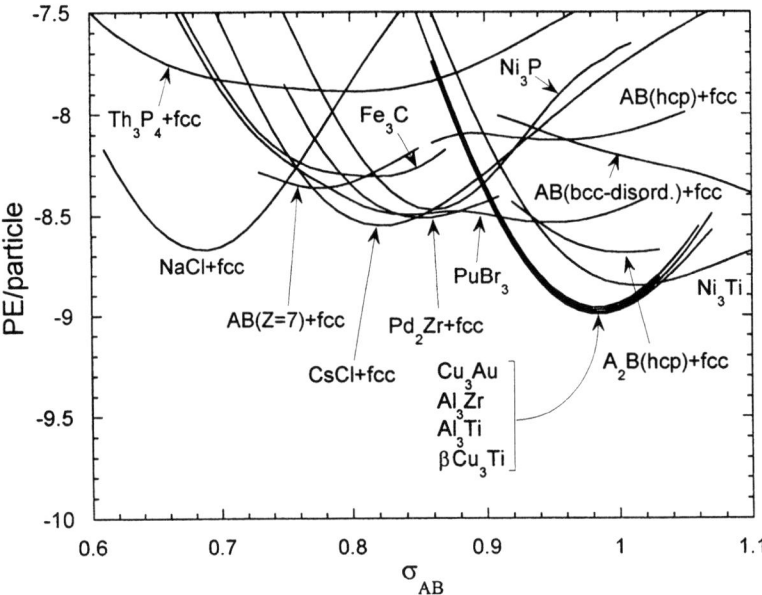

FIGURE 1. Lattice enthalpies per particle vs σ_{AB} of a number of binary crystal structures at a composition $x_B = 0.25$. Note that the KA mixture corresponds to $\sigma_{AB} = 0.8$.

In the CsCl structure each B particle lies in the center of a cube of eight A particles. In the $PuBr_3$, Fe_3C and Ni_3P structures, each B particle lies in the center of a tricapped trigonal prism (see Figure 4) consisting of nine A particles.

The Coordination of B Particles in the Amorphous State

In the KA mixture, the A-B interaction is strongly favoured over the B-B interaction. As a result, for $x_B < 0.5$ each B particle has only A nearest neighbours at low temperatures. In Figure 2 we plot the fraction of B particles with 7, 8 or 9 A neighbours for the amorphous mixture with $x_B = 0.25$ as a function of change in the local coordination of the B particles on cooling with 8- and 9-fold clusters dominating at low temperatures. The low temperature limit of the B coordination has not yet been established.

As each B particle represents the centre of a polyhedron of A particles, then it follows that any two B particles that share four A neighbours represent two polyhedra sharing a 4-fold face. Similarly, two B particles that share three A neighbours correspond to adjacent polyhedra sharing a triangular face, and so on. We shall refer to such B particles pairs as "B^nA bonds" where n is 4, 3, 2 or 1 depending on the number of A neighbours shared by the two B particles.

temperature. Note that there is a significant systematic dominated by the triangular dodecahedron and the square antiprism.

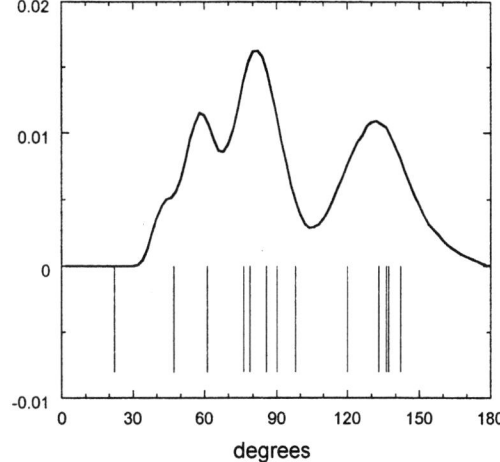

FIGURE 3. The distribution of angles (curve) between pairs of B^3A bonds about individual B particles with 9-fold coordination in an $x_B = 0.25$ mixture at $T = 0.4$. The vertical lines represent the angles between surface normals of the tricapped trigonal prism.

Illustrations of the triangular dodecahedron (TD) and the tricapped trigonal prism (TTP) are provided in Figure 4. For the discussion that follows it is worth emphasising that both of these polyhedra have only triangular faces, in contrast to the square faces associated with B coordination in the stable CsCl crystal. It is also worth noting that while neither of these polyhedra can uniformly fill space both do occur in stable crystals; Th_3P_4 in the case of the TD polyhedron and, in the case of the TTP polyhedron, a number of crystals including Fe_3C, as already mentioned.

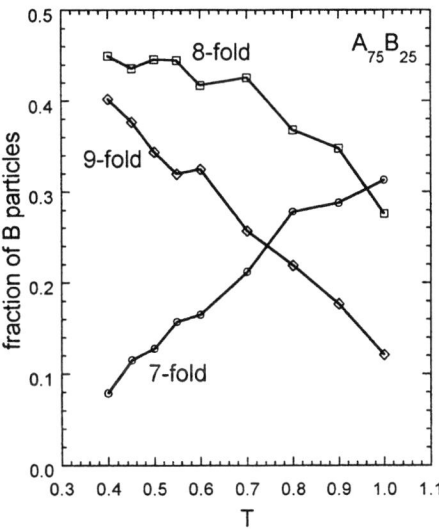

FIGURE 2. The fraction of B particles with 7, 8 or 9 A neighbours as a function of temperature in a $x_B = 0.25$ mixture.

This identification of B^4A and B^3A bonds with shared faces of the coordination polyhedra allows us to establish a picture of the geometry of the B coordination. By way of example, we plot in Figure 3 the distribution of angles between pairs of B^3A bonds about individual B particles with 9-fold coordination in the $x_B = 0.25$ mixture at $T = 0.4$. We find that the peaks in this distribution correspond closely with the angles between surface normals of the perfect tricapped trigonal prism and conclude that that this geometry provides a reasonable description of the 9-fold coordination. The 8-fold coordination in the amorphous state is more ambiguous and it appears to be best described as a mixture of geometries

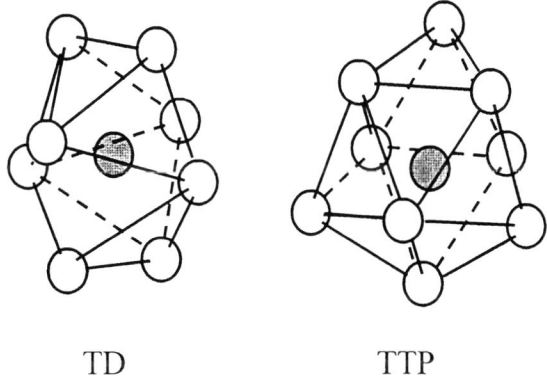

FIGURE 4. Illustrations of the triangular dodecahedron (TD) and tricapped trigonal prism (TTP).

Packing Constraints for Polyhedra

Having characterised the coordination polyhedra, we need to consider how they pack in space. In spite of the considerable interest in the sphere packing problem, there is surprisingly little known about the packing of polyhedra. Here we present a simple argument that relates the composition and average number of B-centered polyhedra that share a vertex (i.e. an A particle). The argument assumes homogeneity of composition.

Let the average coordination of a B particle by A particles only by η_B. Then the average number of AB bonds is $N_B\eta_B = N_A\eta_A$, where η_A is the average number of B neighbours about each A particle. We thus have

$$\eta_A = \eta_B N_B/N_A = \eta_B x_B/(1-x_B) \qquad (2)$$

In the previous section we established that at low temperatures η_B is either 8 or 9 and, according to Eq. 1, η_A must be less than that as long as $x_B < 0.5$. As each B particle lies at the centre of a polyhedron of A's, η_A is equal to the number of polyhedra that share the vertex occupied by that A particle. There must be a geometrical limit as to how many polyhedra can physically share a vertex. For some guidance, lets look at this number in some crystals. In Fe_3C with TTP coordination polyhedra, each A particle has only 3 B neighbours. In Al_2Cu with the square antiprism coordination, four of these polyhedra can meet at any vertex. In Th_3P_4, six TD coordination polyhedra meet about any A particle. The maximum possible packing of polyhedra goes to the cubic coordination found in the CsCl structure in which 8 B particles surround each A particle. At this stage these values represent our best estimates of the limit of packing for each type of polyhedron in an amorphous phase. We have, of course, not considered the case of mixed polyhedra.

As η_A increases with composition, this geometrical constraint (whatever it turns out to be) must be met, resulting, ultimately, in the exclusion of all coordination polyhedra except cubic. We find that the properties associated with the amorphous states with composition x_B close to equimolar include rapid crystallisation and relatively low diffusion constants.

The Anomalous Behaviour of $g_{BB}(r)$

The change in the geometry of B coordination with changing composition predicted in the previous section should be evident in the partial radial distribution functions; $g_{AA}(r)$, $g_{AB}(r)$ and $g_{BB}(r)$. While g_{AA} and g_{AB} are dominated by the internal structure of the coordination polyhedra, the correlation function g_{BB} corresponds to the correlation between polyhedra. In Figure 5 we resolve g_{BB} for a $x_B = 0.25$ mixture at $T = 0.4$ into the contributions from B^nA bonds with $1 \leq n \leq 5$. We find that individual peaks in $g_{BB}(r)$ can be quite cleanly associated with particular types of B_nA bonds. The small first peak in the A_3B mixture can thus be directly attributed with the relatively small number of polyhedra sharing square faces. We can also understand the anomalous temperature dependence of $g_{BB}(r)$ in which the height of the first peak *decreases* on cooling. If the triangulated coordination polyhedra such as TD and TTP are more stable at this composition, then we would expect the number of B^4A bonds to decrease with the temperature.

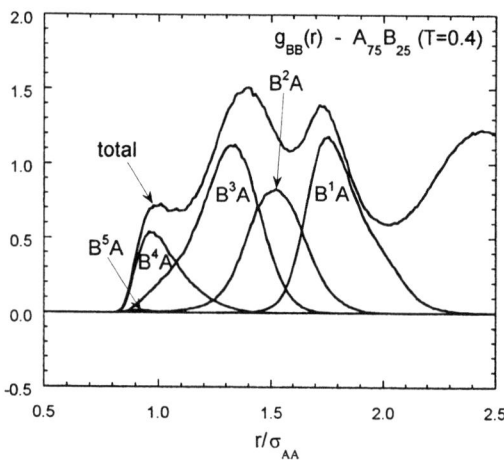

FIGURE 5. The contributions to the B-B radial distribution function $g_{BB}(r)$ from the B^nA bonds as described in the text with $n = 1-5$ for the $x_B = 0.25$ mixture at $T = 0.4$. Note that the first peak is due almost exclusively to the B^4A bonds and that the second peak is largely due to the B^3A bonds which, at this composition, significantly exceed the B^4A bonds in number.

These arguments lead us to expect a significant change in g_{BB} as we increase x_B. We have plotted g_{BB} for a range of compositions at $T = 0.6$ in Figure 6. Note that the relative heights of the first and second peaks undergo an inversion as we go from $x_B = 0.25$ towards the equimolar mixture. One possible explanation of this inversion is that a change in the intermediate structure as occurred, from one based on polyhedra sharing triangular faces to a structure characterised by shared square faces.

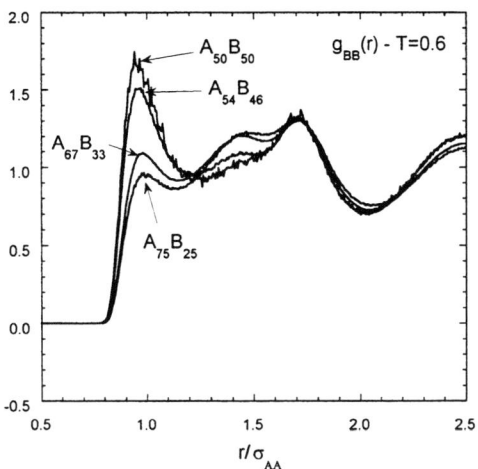

FIGURE 6. The B-B radial distribution function $g_{BB}(r)$ at T = 0.6 for the following compositions: x_B = 0.25, 0.33, 0.46 and 0.5. Note the significant increase in the height of the first peak as x_B is increased from 0.33 to 0.46 and the accompanying decrease in the height of the second peak.

Crystallisation

In the previous section we noted that at compositions $x_B < 0.4$ the liquid structure, as seen through g_{BB}, exhibited correlations incommensurate with those in the CsCl crystal structure. Such a difference in the structures of the crystal and liquid phases would be expected to increase the interfacial free energy between the two phases. Through its inclusion in the free energy of the critical crystal nucleus in the classical theory of homogeneous nucleation, this increase in surface free energy would translate as a significant slow down in the time required for crystal growth to occur.

To explore this point, we have made a rough estimate of the minimum crystallisation time as a function of composition. The crystallisation time at a given composition and temperature is defined here as the time required after the quench for the potential energy to drop to the value midway between that of the initial disordered state and the final crystalline state. Locating the minimum such time at a given composition involves quenching the mixture to a number of different temperatures. The time required for these runs has meant that we can only perform single run at any given temperature and composition with the result that there is considerable statistical uncertainty in the estimate of the crystallisation time.

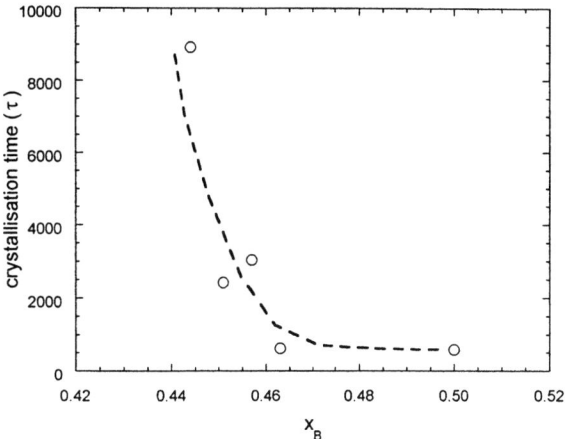

FIGURE 7. The minimum crystallisation time (see text for the definition) as a function of the composition x_B. The dashed line is included as a guide to the eye.

These minimum crystallisation times (our estimate of the 'nose' of the time-temperature transformation curves) are plotted against composition in Figure 7. Note the rapid increase in this time as the x_B decreases from 0.5. For $x_B < 0.42$, we can no longer observe any sign of crystallisation over runs of $10^4 \tau$. We note that this composition lies within the range of composition identified in the previous section over which we see the inversion in the relative heights of the first and second peaks of g_{BB}.

CONCLUSIONS

In this paper we have presented a preliminary account of the structure of the amorphous binary mixture introduced by Kob and Andersen. Our main results can be summarised as follows. At low temperatures we find almost all the B particles in either 8- or 9-fold coordination polyhedra made up of A particles. These polyhedra are largely triangular-faced with the result that the first peak of $g_{BB}(r)$, which reflects the number of shared square faces, is considerably smaller than the second peak. The proposal that it is the stability of these triangular-faced polyhedra that suppresses crystallisation of the CsCl phase gains support from the coincidence of the increase in magnitude of the first peak of $g_{BB}(r)$ and the onset of crystallisation as the composition x_B is increased.

What remains to be explained is the extended structure and stability of the triangular-faced polyhedra for $0.2 < x_B < 0.4$ and the nature of the structural transition in the amorphous state as the composition is varied. In terms of face-sharing between adjacent polyhedra, there would seem to be some incompatibility between polyhedra with triangular faces and those with square faces which could result in an 'all-or-nothing' collective selection of one or other type of coordination. This argument neglects, however, the role of edge sharing and the stability of crystals like Al_2Cu in which the local coordination exhibits both square and triangular faces. Clearly we have more work to do to understand the collective character of packing of 'soft' polyhedra.

With respect to the stability of the structure we note that recent studies of 2D glass-forming mixtures [19] have demonstrated that domains as small as a particle and its nearest neighbour shell can play the role of Tammann's 'rigid clusters' over the time scale accessible to simulations. This would suggest that fragments rather than extended structures are probably sufficient to account for the stability of simulated glasses. This does not mean that we do not need to look for more extended forms of structure but simply that we may exhaust the information that current molecular dynamics simulation methods can access.

ACKNOWLEDGMENTS

We would like to acknowledge the support of the Australian Research Council and the Comisión Nacional de Energía Atómica of Argentina.

REFERENCES

1. Fernández, J. R. and Harrowell, P., *Phys. Rev. E* **67**, 011403/1-7 (2003).

2. Fernández, J. R. and Harrowell, P., to be published in *J. Chem. Phys.*

3. Zachariasen, W. H., *J. Am. Chem. Soc.* **54**, 3841-3851 (1932).

4. Bernal, J. D., *Nature* **185**, 68-70 (1960); *Proc. Roy. Soc.* **A280**, 289-322 (1964).

5. Finney, J. L., *Proc. Roy. Soc. Lond. A* **319**, 479-493 (1970).

6. Polk, D. E., *Acta Metall.* **20**, 485-491 (1972).

7. Gaskell, P. H. *Acta Metall.* **29**, 1203-1211 (1981).

8. Phillips, J. C., *J. Non-Cryst. Solids* **34**, 153-181 (1979).

9. Thorpe, M. F., *J. Non-Cryst. Solids* **57**, 355-370 (1983).

10. Tamman, G., *Der Glasszustand*, Leopold Voss, Leipzig, 1933.

11. Frank, F. C., *Proc. Roy. Soc.* **A215**, 43-46 (1952).

12. Sadoc, J-F and Mosseri, R., *Geometrical Frustration*, Cambridge University Press, Cambridge, 1999.

13. Hoare, M. R., *Ann. New York Acad. Sci.* **279**, 186-194 (1976); Hoare, M. R. and Barker, J. A., "Tammann revisited: cluster theories of the glass transition with special reference to soft packings" in *The Structure of Non-Crystalline Materials*, ed. Gaskell, P. H., Taylor and Francis, London, 1977, pp.175-180.

14. Dzugutov, M., *Phys. Rev. A* **46**, R2984-R2987 (1992).

15. Gaskell, P. H., *Nature* **289**, 474-476 (1981).

16. Egami, T. and Waseda, Y., *J. Non-Cryst. Solids* **64**, 113-134 (1984).

17. Kob, W. and Andersen, H. C., *Phys. Rev. E* **51**, 4626-4641 (1995).

18. Sturgeon, J. B. and Laird, B., *J. Chem. Phys.* **112**, 3474-3482 (2000).

19. Widmer-Cooper, A. and Harrowell, P., in preparation.

The Structure of Energy Landscape and the Non-Arrhenius behaviour of supercooled Liquids

M. Schulz

Abteilung für Theoretische Physik, Universität Ulm, 89069 Ulm, Germany

Abstract. A theoretical explanation of the non-Arrhenius behavior that can be observed in supercooled liquids at sufficiently low temperatures is given. The starting point of the investigations is the determination of the density of minimum points of the energy landscape. The knowledge of this density allows the determination of characteristic length scales in the multidimensional configurational space and their corresponding scales in the real space. From this point of view it is possible to study the dynamics of the glass transition. Four regimes are obtained from this analysis. While the non-activated regime is mainly described in terms of the mode-coupling theory and the frozen low temperature regime is related to the spectrum of vibrations and possible local rearrangements, the weak and the strong activated regime becomes important close to the glass transition temperature.

The weak activated regime requires considers a diffusion like motion of the system in the configurational space. Characteristic properties of a supercooled liquid, like relaxation time and viscosity, can be expressed in terms of material constants which are well known from or which are measurable in the thermodynamic equilibrium. The obtained relations contain no free parameters (with the exception of a simple gauge point) in contrast to usual heuristic equations with some free fit parameters.

In case of a strong activation the density of minimum points allows the explanation of the transition from ergodicity to nonergodicity in the thermodynamic limit. Furthermore, experimental results of the glass transition in finite regimes may be at least qualitatively predicted by the application of this theory on a finite number of degrees of freedom.

obtains at least qualitatively These results are supported by several molecular dynamics simulations demonstrating complicated structure of the energy landscape and the distribution of the minimum points.

INTRODUCTION

A large number of liquids has been observed to develop very slow dynamics without the evolution of any observable long range correlated or ordered structure. The most apparent feature of this glass transition process is the rapid increase of the characteristic relaxation time. Usually, the relaxation time increases faster than Arrhenius dependence. The curved trajectory at low temperatures may be fitted by the popular Williams-Landel-Ferry curve, $\ln \tau \sim (T - T_0)^{-1}$, with a finite Vogel temperature T_0 or by another laws, e.g. $\ln \tau \sim T^{-\mu}$. Because τ becomes inaccessible large in course of the cooling procedure, it is impossible to decide if the relaxation time actually diverges at a finite temperature T_0 or not. A possible description of this phenomena starts from an analysis of the so called energy landscape, firstly introduced by Goldstein [1]. In such a theory, the supercooled liquid is defined by their position $\mathbf{R} = \{\mathbf{r}_1, \mathbf{r}_2, ..., \mathbf{r}_N\}$ in the configurational space, spanned by a reference frame of $3N$ axes corresponding to the positions $\mathbf{r}_k$ ($k = 1...N$) of the N microscopic particles. as the spatial subspace of the phase space. The landscape itself corresponds to a suitable energy function. In the following we use the mechanical potential $U(\mathbf{R})$. The merit of this approach is that the landscape is independent from the temperature in opposition to the application of a microscopic defined free energy landscape.

The point is wether a universal relation exists which describes the behavior of the relaxation time of supercooled liquids close to the glass transition. This relation should be independent of free fit parameters, i.e. all open parameters of such a theory should be determinable from such material constants which are obtainable from thermodynamic equilibrium data.

DISTRIBUTION OF MINIMUM POINTS

For the sake of simplicity, we restrict the following investigations on a homogeneous liquid. The more general case of liquids consisting of different components requires in principle the same approach presented below [2]. Hence the classical partition function Z_{liq} is

$$Z_{\text{liq}} = \int \frac{DrDp}{N!} \exp\left\{-\frac{1}{k_B T}\left[\sum_{k=1}^{N} \frac{\mathbf{p}_k^2}{2m} + U\right]\right\} \quad (1)$$

The functional structure of the potential $U(\mathbf{R})$ shows many valleys, hills and saddles. The system occupies mainly the environment of the minimum points at sufficiently low temperatures. Each minimum point corresponds to a configuration $\mathbf{R}_0^{(j)}$ (the label j indicates the minimum). It can be expected that the global minimum of the system corresponds to the crystalline structure, i.e. the potential can be gauged by $U(\mathbf{R}_0^{(\text{crystal})}) = 0$. All other existing minimum points have an energy level $\varepsilon^{(j)} = U(\mathbf{R}_0^{(j)}) > U(\mathbf{R}_0^{(\text{crystal})}) = 0$. The total number of minimum points increases dramatically with increasing particle number N. Hence, the distribution of the local minimum points with respect to their energy becomes a quasi continuous representation in the case of the thermodynamic limit. This situation suggests the following question: It is possible to calculate a density $\rho(\varepsilon)$ which determines the number of minimum points $dv = \rho(\varepsilon)d\varepsilon$ within an energy interval $[\varepsilon, \varepsilon + d\varepsilon]$? A positive answer can be given at least for low energy levels [2]. For example, an ideal simple glass is characterized by the following distribution functions of minimum points

$$\rho(\varepsilon) = \frac{N!}{\Gamma\left(\frac{N\Delta c_p}{k_B}\right)} \left(\frac{\varepsilon}{k_B T_m}\right)^{\frac{N\Delta c_p}{k_B}} \frac{\exp\left\{\frac{N(q - \Delta c_p T_m)}{k_B T_m}\right\}}{\varepsilon} \quad (2)$$

Here, T_m and q are the melting temperature and the melting heat per particle, respectively. Note, that an ideal simple glass is defined by a constant specific heat capacity $c_p^{\text{liq}} = \Delta c_p + c_p^{\text{sol}}$ well above the glass transition temperature. However, it should be remarked that this apparently very special law (2) is a good approach to the thermodynamics of many supercooled liquids [2].

LANDSCAPE GEOMETRY

It is recognizable that the minimum points of the above introduced energy landscape have various heights. How many minimum points can be occupied with a finite probability by a system of temperature T in course of it's motion through the configuration space? We get from equation (2) that the number of these relevant minimum points increases with increasing temperature. The mean square distance $\langle \xi^2(T) \rangle$ between neighbored relevant minimum points is given by

$$\langle \xi^2(T) \rangle \sim \left(\frac{N}{V}\right)^{-\frac{2}{3}} \left(\frac{(\Delta c_p + 3k_B)T}{\Delta c_p T_m}\right)^{-\frac{2\Delta c_p}{3k_B}} \quad (3)$$

The physical interpretation of this length is obviously. ξ^2 is the sum over the mean square displacements of all particles in the real space which are necessary for a displacement of the total system from one relevant minimum point to another neighbored relevant minimum point in the configurational space. Let us assume for the moment that the total system is trapped in the environment of a relevant minimum point. Obviously this trapped system visits now only a limited volume of the configuration space. The averaged radius of this vibrational sphere can be also determined from equation (2) and the assumption, that the potential close to the minimum point can be approached by a square form of the particle coordinates. We get

$$\frac{R_0(T)}{\sqrt{N}} \sim x_m \left(\frac{T}{T_m}\right)^{1/2} \left(\frac{2\Delta c_p}{2\Delta c_p + 3k_B}\right)^{\frac{\Delta c_p}{3k_B}} \quad (4)$$

Here, x_m is the well known Lindemann melting parameter. Note, that contributions of higher (non-harmonic) terms to the dynamics of the vibrations are not considered in this formula. These terms may lead to a correction of (4) but not to a change of the general idea.

HIGH TEMPERATURE REGIME

In principle, one can distinguish between a high temperature regime corresponding to the supercooled liquid between the melting temperature and a certain transition temperature T^* and the low temperature regime which existing below T^*.

Let us start with an investigation of a supercooled liquid as sufficiently high temperatures. From the above geometrical studies we arrive at the following scenario. The configurational space consists of a set of minimum points. A minimum is a relevant point if its energy level is smaller than the total energy of the system. In other words, there is a finite probability that the supercooled liquid reaches in course of it's dynamical evolution the basin of attraction of a given relevant minimum point. Each relevant minimum point is the center of a vibrational shell corresponding to a region which is relatively often visited by the total system. The vibrational spheres of neighbored minimum points are connected by mutual overlaps. At a given time the system may be localized in the vibrational sphere of a certain relevant minimum point for a finite time interval. The motion of the system inside the sphere is mainly determined by a superposition of $3N$ oscillators. The averaged oscillation time is almost independent of the temperature. Note that the temperature determines the radius of the vibrational sphere and consequently the amplitude of the oscillations, but the temperature has no influence on the oscillation times in the harmonic approximation.

Definitely the system is not fixed in the initial sphere. The system changes to a neighbored vibrational sphere after a characteristic time scale which may be estimated from the averaged transition rate between neighbored relevant minimum points. This rate is defined by a microscopic time scale which is of the order of magnitude of the averaged oscillation time and the overlap between both the new and the old vibrational sphere. The knowledge of the transition rate allows us to determine the diffusion coefficient of the whole system in the configurational space and therefore also the diffusion coefficients of the particles inside the supercooled liquid. Following [2] one gets the result

$$D(T) = D_0 \left(\frac{T_m}{T}\right)^{(\gamma-2)/2} \exp\left\{-B\left(\frac{T_m}{T}\right)^\gamma\right\} \quad (5)$$

with

$$B = \frac{(36\pi)^{2/3}}{8x_m^2} \left(\frac{\gamma}{\gamma+1}\right)^\gamma \exp\left\{-\frac{2q}{3k_B T_m}\right\} \quad (6)$$

and the exponent

$$\gamma = 1 + \frac{2\Delta c_p}{3k_B} \quad . \quad (7)$$

It should be remarked that these results are typical mean field solutions. The distribution functions (2) was exclusively used for the determination of some averaged geometrical quantities from which we have obtained finally the diffusion coefficient. Therefore, this regime may be interpreted also as a non-activated ore weak activated diffusion of the supercooled liquid in the configurational space. However, (5) allows the determination of the relaxation time of the structural changes above the glass transition temperature. Since the non-activated regime should be equivalent to the regime of an apparently coupling of various kinetic observables, we may assume that this characteristic relaxation time τ can be obtained from the simple relation $D(T)\tau \sim 1$ and the simple measurable static viscosity is determined by an analogous relation, $\eta \sim \tau$. Hence

$$\ln\frac{\tau}{\tau_0} = \ln\frac{\eta}{\eta_0} = \frac{\gamma-2}{2}\ln\left(\frac{T}{T_m}\right) + B\left(\frac{T_m}{T}\right)^\gamma \quad . \quad (8)$$

We remark that similar equations are used as an empirical description of the non–Arrhenius behavior of the α-process for a long time [3, 4]. But in contrast to these empirical equations, Equation (8) contains no fit parameter, except that an irrelevant calibration (given by τ_0 or η_0) is necessary. All other parameters (B, γ) are defined by quantities of the thermodynamical equilibrium.

Equation (8) contains only one calibration. Therefore, one must define the viscosity (or the relaxation time) at a given temperature. To this aim we use here the gauge point $\{T_g, \eta_g = \eta(T_g)\}$ with the empirically defined glass transition temperature T_g. Although this definition follows some standard procedures, this choice emphasizes no general physical meaning. However, one obtains now

$$\ln\frac{\eta}{\eta_g} = \frac{2-\gamma}{2}\ln x + B\left(\frac{T_m}{T_g}\right)^\gamma [x^\gamma - 1] \quad (9)$$

with $x = T_g/T$. Furthermore, B is related to the well known fragility

$$F = \left[\frac{\partial \log_{10}\eta}{\partial T_g/T}\right]_{T=T_g} = \frac{2-\gamma}{2\ln 10} + \frac{B\gamma}{\ln 10}\left(\frac{T_m}{T_g}\right)^\gamma . \quad (10)$$

In order to check the validity of (7) we replace B by the fragility F. It remains a general law, which is exclusively controlled by the extended gauge point $\{T_g, \eta_g, F\}$ and by the exponent γ and therefore by the gap in the specific heat capacity Δc_p between the liquid and the amorphous solid. Fig.1 represents the relation for various materials. It is obviously that the experimental data show a satis-

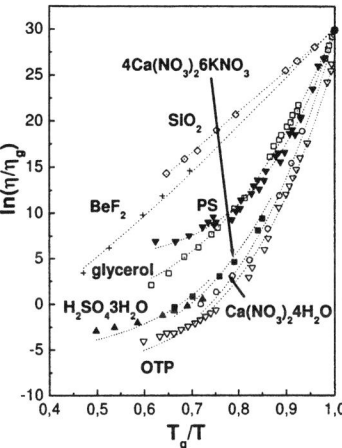

FIGURE 1. The rescaled temperature-dependence of the viscosity. The dotted lines are obtained from eq.9 while the experimental curves come from [5]

factory agreement with the theoretical estimations. The fragility F is no independent material constant since it may be determined directly from (6). For a check of this relation we use the fragility as input parameter and determine via the equations (10) and (6) the Lindemann melting parameter x_m. The expected value of x_m should be in a range of $0.15 \leq x_m \leq 0.25$. Fig. 2 shows the calculated melting parameters for various glass formers. It is obviously that the experimental data yield also a sufficiently precise agreement with our prediction. It is obviously that the experimental data yield also a sufficiently precise agreement with our prediction.

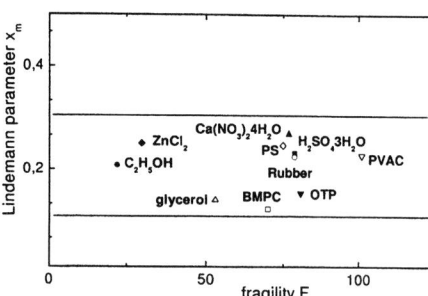

FIGURE 2. Calculated Lindemann parameter x_m for various glass formers versus the experimentally obtained fragility F. The stripe between the two lines corresponds to physical reasonable values of x_m.

LOW TEMPERATURE REGIME

The low temperature regime or the thermodynamically activated regime appears if the energetic barrier between neighbored minimum points becomes of an order of magnitude of $k_B T$. On the other hand, the assumption of a simple harmonic structure around each minimum point allows us to calculate an approximation of the averaged height of a barrier as $E_{\text{barr}}/E_{\text{total}} = (\xi/2R)^2$. From here, we may estimate the temperature T^* at which the crossover from the high temperature regime to the low temperature regime occurs. We obtain with (3) and (4)

$$\frac{T^*}{T_m} = [2B(1+\gamma)]^{1/\gamma} \quad (11)$$

Although the accuracy of the available data is not very high, we find as a rough approximation $T_m > T^* \sim T_g$. Below the crossover temperature T^*, the rate for a transition from a minimum point to a neighbored minimum is mainly determined by the high δE of the barrier between both points. For the following investigations let us assume a simple trap model with a constant absolute energy level E_0 for all barriers, but different depths $E_0 - \varepsilon$ of the local minima. Then, the transition rate for a jump from a minimum point of the energy level ε into an arbitrary other minimum is simply given by $p(\varepsilon) \sim \exp\{-(E_0 - \varepsilon)/k_B T\}$. The typical value of the energy scale E_0 may be estimated by the application of the distribution (2). For this purpose, we first determine the volume $\Omega(\varepsilon)$ of a single basin of attraction around a minimum of the energy ε. This is in the framework of the assumed harmonic approximation around a minimum always possible. We get $\Omega(\varepsilon) \sim (E_0 - \varepsilon)^{3N/2}$, for the detailed result see [2]. number to this. Subsequently, we may estimate the total volume of all basins of attraction via

$$U = \int_0^{E_0} \Omega(\varepsilon)\rho(\varepsilon)d\varepsilon \quad (12)$$

Finally, the requirement that this volume is equivalent to the total volume V^N of the system in the configurational space leads the wanted approximation of E_0. However the present estimation is very rough, we find at least that E_0/Nk_B has an order of magnitude of T_g. In order to obtain the averaged occupation time τ_{jump} of the basin of attraction of an arbitrary minimum point and therefore the characteristic relaxation time of the supercooled liquid, we have to calculate

$$\tau_{\text{jump}} \sim \frac{\int_0^{E_0} p(\varepsilon)^{-1}\rho(\varepsilon)d\varepsilon}{\int_0^{E_0} \rho(\varepsilon)d\varepsilon} \quad (13)$$

This expression can be written as

$$\tau_{\text{jump}} \sim \mu^{-\frac{N\Delta c_p}{k_B}} \gamma\left(\frac{N\Delta c_p}{k_B}, \mu\right)\exp\{\mu\} \quad (14)$$

with $\mu = E_0/k_B T$ and the incomplete Γ-function $\gamma(a,x) = \int_0^x t^{a-1}e^{-t}dt$. The thermodynamic limit $N \to \infty$ leads finally to

$$\ln \tau_{\text{jump}} \sim -\ln(T - T_0) \quad (15)$$

with the Vogel-temperature $T_0 = E_0/N\Delta c_p$. Although this divergence is too weak in comparison to the Williams-Landel-Ferry behavior ($\ln(T - T_0)$ instead $(T - T_0)^{-1}$), we may conclude that the power law (8) of the non-activated regime must be replaced by another, possible divergent relation for the activated diffusion in the configurational space.

NUMERICAL INVESTIGATIONS

As discussed in the previous sections, the system will be trapped in the environment of a local minimum during a characteristic time interval. Small deviations from the local minimum lead to oscillations around the minimum point. With regard to liquids near the glass transition such a picture means the supercooled liquid behaves like an amorphous solid for a short time interval where the structure corresponds to that one obtained by a rapid cooling down of the liquid to $T \to 0$. After a certain waiting time the system should be able to leave the basin of attraction around the minimum and should move towards another neighbored basin of attraction overcoming the potential barrier between both regions. Here, it will be localized again temporarily during a short time interval. This behavior allows the construction of a so-called pseudo-trajectory which means a piecewise straight line in the configurational space. All the local minima in configurational space are connected step by step by the pseudo-trajectory meanwhile the real trajectory touches the corresponding basins of attraction. Such pseudo-trajectories are obtainable from molecular dynamic simulations [6]. The concept of pseudo-trajectories allows a clear separation between pure oscillations and structural changes. Oscillations correspond to simple shifts inside a given basin of attraction. Whereas the conventional trajectory shows a continuous change with respect to the evolution time, the pseudo-trajectory stays at the corresponding minimum point. On the other hand, if the system crosses the border between two basins of attraction, the real trajectory shows no essential change whereas the pseudo-trajectory realizes a sufficiently long jump from the previous minimum point to the actual minimum point. The new minimum point corresponds to another structural configuration, so that each jump of the pseudo-trajectory indicates a change of the structure of the supercooled liquid. The jump lengths $l(t)$ of the pseudo-trajectory form a random time series which suggests the existence of both, relatively long and short jumps, simultaneously, see fig.3. The concept of pseudo-trajectories is comparable

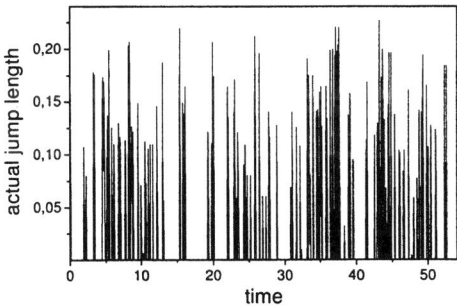

FIGURE 3. Representation of the time series of jump lengths of a pseudo-trajectory for $T/T_0 \simeq 1.2$

with the analysis of the energy landscape elaborated in a series of other publications [7, 8, 9, 10]. Both approaches in common are the analysis of multidimensional potential minima and their basins of attraction.

We define an elementary jump as a jump connecting two neighbored minimum points. Thus, it seems to be possible that long jumps may be composed of a set of correlated elementary jumps. The decision, whether a jump is an elementary jump or it is a combined object can be obtained from a change of the numerical time steps. Refining the time scale we conclude that long jumps detected for a sufficiently small time steps are not superpositions of shorter jumps, i.e., these long jumps are predominantly elementary jumps. The probability that a jump of the pseudo-trajectory has a length between l and $l + dl$ is $P(l)dl$. For the further discussion it seems reasonable to introduce another probability

$$H(l) = \frac{lP(l)}{\int_0^\infty \zeta P(\zeta)d\zeta} \quad (16)$$

which defines the contribution of the elementary jumps of length to the total length of the pseudo-trajectory. Fig. 4 shows the distribution function $H(l)$ for various temperatures. The position of the maximum of the dis-

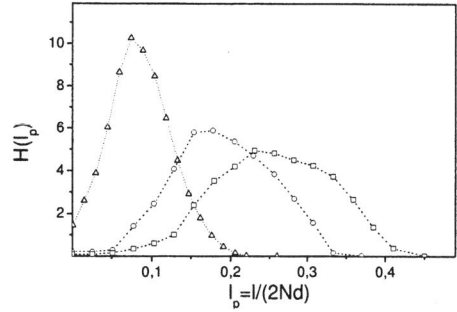

FIGURE 4. Normalized distribution function $H(l_p)$ of the relative jump length per particle $l_p = l/(2Nd)$ where d is the averaged diameter of the particles. The curves correspond to the temperatures $T/T_0 \simeq 1.2$ (triangles), $T/T_0 \simeq 1.6$ (circles) and $T/T_0 \simeq 2.0$ (squares)

tributions $H(l)$ decreases with decreasing temperature. In other words, with decreasing temperature the systems occupy mainly such regions of the configurational space with a high density of the minimum points so that the jump length between neighbored minimum points is short. It is still an open numerical problem, whether these regions becomes isolated areas below the above introduced Vogel temperature T_0 and for the thermodynamic limit $N \to \infty$, so that a real ergodicity–nonergodicity transition occurs. It should be remarked, that these results can be obtained qualitatively also from an analysis of the real trajectories of the system in the configurational space [11]. The length l of a jump of the system between two neighbored minimum points corresponds to an effective jump l/N of a single particle of the system. On the other hand, the components of the jump vector of the pseudo trajectory, $\mathbf{R}(t+\Delta t) - \mathbf{R}(t) = \{\mathbf{r}_1(t+\Delta t) - \mathbf{r}_1(t),...,\mathbf{r}_N(t+\Delta t) - \mathbf{r}_N(t)\}$ define the jump length of a particle $l_i = |\mathbf{r}_i(t+\Delta t) - \mathbf{r}_i(t)|$. We define, that the particle i contributes essentially to an elementary jump of the system if its actual jump length l_i is larger than the actual averaged jump length l/N. Such a particle is denoted as an actual active particle. The set of all particles with $l_i > l/N$ may be interpreted as the actual cooperatively rearranging region[12]. This definition needs neither information about the topological structure of the regions not of possible correlations between single particles. The size of these regions is simply characterized by the number of active particles. Fig. 5 shows the probability distribution $G(n)$, that n active particles contribute to a jump of the pseudo-trajectory essentially. These numerical find-

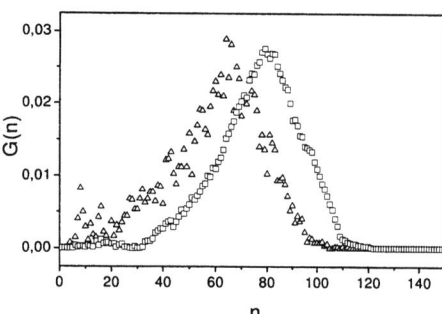

FIGURE 5. Normalized distribution function $G(n)$ of the size of cooperatively rearranging regions for the temperatures $T/T_0 \simeq 1.2$ (triangles) and $T/T_0 \simeq 2.0$ (squares)

ings suggest the surprising result that the cooperatively rearranging regions decreases with decreasing temperature. Obviously, the number of particles contributing to a jump between two neighbored minimum points in the configurational space and consequently leading to a change of the structure of the glass becomes sufficiently large at high temperatures because jumps over high energy barriers are favored at rather high temperatures. If the number of particles joined in such jumps is rather large there appear likewise sufficiently large regions which are able to perform a rearrangement in a cooperative manner. In the opposite case of low temperatures an elementary jump over a high barrier is a rather rare event and hence, the motion between neighbored minimum points can be realized only by using other paths consisting of elementary jumps over lower barriers with a smaller numbers of active particles. Thus, the cooperatively rearranging regions are also shrunk in their size. At very low temperatures, the size of that region may be eventually of the order necessary for a simple exchange of particles. In so far the case is similar to the situation in a crystalline solid. The remaining hopping processes maintain the diffusion of single particles and therefore of the total system in the configurational space. An elementary hopping process is induced by the nearest environment of the jumping particles, i.e., the energy barrier between the initial and the final state is finite. Note that the structure of the glass is not changed by such particle exchange processes. Thus our numerical simulations suggest that if there is a temperature T_0 below which only particle exchange processes remain active, the system undergoes a transition from an ergodic to a nonergodic state. The ergodicity breaking implies the existence of at least two stable solids with different structures separated by an infinite large energy barrier. Such a situation can be realized obviously only in case of an infinite large system, $N \to \infty$, but precursor effects of ergodicity breaking seems to be observable already in the present finite systems.

CONCLUSIONS

The presented results supports the hope that a deeper theoretical understanding of the non–Arrhenius behavior of supercooled liquids may be obtained by an analysis of the dynamics of large system in the multidimensional configurational space. It was pointed out that with only some few assumptions a estimation of characteristic quantities of glasses can be reproduced without further free fit parameters. Consequently, the extremely slow dynamics of glasses is mainly characterized by the information obtainable from the thermodynamic dynamic equilibrium. The presented analytical results are supported by several numerical simulations which demonstrate the diffusion like behavior of the total system in the relatively complex energetic landscape of the configurational space.

ACKNOWLEDGMENTS

The author thanks B.M. Schulz and S. Trimper for many helpful discussions and suggestions. This work has been supported by the Deutsche Forschungsgemeinschaft DFG.

REFERENCES

1. Goldstein, M., *J. Chem. Phys.*, **51**, 3728-3735 (1969).
2. Schulz, M., *Phys. Rev. B*, **57**, 11319-11333 (1998).
3. Bässler, H., *Phys. Rev. Lett.*, **58**, 767-770 (1987).
4. Arkhipov, V., and Bässler, H., *J. Phys. Chem.*, **98**, 662-682 (1994).
5. Böhmer, R., Ngai, K., Angell, C., and Plazek, D., *J. Chem. Phys.*, **99**, 4201-4209 (1993).
6. Schulz, B. M., Schulz, M., and Trimper, S., *J. Chem. Phys.*, **114**, 10402-10410 (2001).
7. Stillinger, F. H., *Phys. Rev. B*, **41**, 2409-2416 (1990).
8. Stillinger, F. H., *Science*, **267**, 1953-1955 (1995).
9. Sastry, S., Debenedetti, P. G., and Stillinger, F., *Nature (London)*, **393**, 554-557 (1998).
10. Büchner, S., and Heuer, A., *Phys. Rev. Lett.*, **84**, 2168-2171 (2000).
11. Gaukel, C., and Schober, H. R., *Solid State Commun.*, **107**, 1-17 (1998).
12. Adam, G., and Gibbs, J., *J. Chem. Phys.*, **43**, 139-145 (1965).

Computer Simulation of the Glass Transition in Thin Films

K. Binder*, F. Varnik†, J. Baschnagel**, P. Scheidler* and W. Kob‡

*Institut für Physik, Johannes Gutenberg Universität Mainz, Staudinger Weg 7, D-55099 Mainz, Germany
†LPMCN, Bat. Louis Brillouin, Université Lyon I, 69622 Villeurbanne Cedex, France
**Institut Charles Sadron, 6 rue Boussingault, 67083 Strasbourg Cedex, France
‡Laboratoire des Verres, Université Montpellier II, 34095 Montpellier, France

Abstract. Molecular Dynamics studies of simple models (short non-entangled polymer chains and a binary Lennard-Jones mixture) of glassforming fluids confined between various types of walls are discussed, with an emphasis on the understanding of the nature of the glass transition from the analysis of confinement effects. Particular attention is paid to the possible conclusions on a characteristic length that grows, as the glass transition is approached. Particularly useful is the analysis of the "local relaxation time" $\tau(z)$, z being the distance from the closest confining wall. Both smooth and rough walls are considered (the latter are constructed such that there is almost no layering effect). In all cases it is found that the characteristic lengths do increase with decreasing temperature, albeit rather slowly; the relevance of these lengths for the slowing down near the glass transition remains doubtful.

INTRODUCTION

The nature of the glass transition is still a big puzzle [1, 2, 3, 4, 5, 6]: although there is such a dramatic slowing down in undercooled fluids as the glass transition temperature T_g is approached, one does not see significant changes in the structure (as observed by diffuse scattering of X-rays or neutrons). The structural relaxation time $\tau(T)$ of the fluid increases from very small values at high temperatures ($\tau \approx 10^{-13}$ s) up to macroscopic times at the glass transition $\tau(T=T_g) = 100$ s. While the first few decades of this slowing down are often rather well accounted for by mode coupling theory (MCT), which in its idealized form implies [2] $\tau(T) \approx (T-T_c)^{-\gamma}$ with (typically) $\gamma \approx 2$, near the MCT critical temperature T_c, one actually finds a crossover to a different relaxation law, sometimes modeled by an Arrhenius relation $\tau(T) \propto \exp(E_{\text{act}}/k_B T)$ or a Vogel-Fulcher-Tammann (VFT) relation [1] $\tau(T) \propto \exp[E_{\text{act}}/k_B(T-T_{\text{VFT}})]$, E_{act} being some activation energy. The VFT relation implies that $\tau(T)$ diverges at the temperature T_{VFT} ($<T_g$).

One idea that often is invoked (e.g. [3]) to explain the dramatic increase of $\tau(T)$ is that there is an underlying correlation length $\xi(T)$ that grows to large values as T_g is approached. This length could either characterize static correlations of a subtle character so that they do not show up in scattering experiments (this situation is known to occur for spin glasses [7]), or the length can be seen in suitable dynamic quantities only (e.g. the size of cooperatively rearranging regions [8], etc.)

Since it is unclear what quantity of a system in the bulk should be measured in order to gain information on this length, it is tempting to invoke the idea to perturb the system by a surface and probe this correlation length from the range over which the surface effects lead to deviations from bulk behavior. Note that neither the magnitude nor the sign of the surface effect is important, but only the range: using the local mobility as a probe, one expects an increase at a free surface, but a decrease at a wall with strong attraction to the molecules of the glassforming fluid. Therefore the glass transition in a very thin film may either increase or decrease with decreasing thickness, depending on these precise boundary conditions of the surface [9]. Here we shall not focus on this question, but rather on the correlation length which is a bulk property (to be extracted from thick enough films where bulk behavior is indeed reached in the center of the film). This concept of estimating ξ from the range over which a quantity locally deviates from bulk behavior near a surface is well established in the study of ordinary phase transitions [10].

A COARSE GRAINED MODEL FOR GLASSFORMING POLYMERS

For computer simulation, polymers are a particular challenge [11]: macromolecules containing from 100 to 1000 monomers exhibit structure from the scale of chemical bonds, the Ångstrøm scale, over the persistence length (≈ 1 nm) to the coil size (≈ 10 nm) or even larger collective length scales. There is an even more dramatic spread of time scales [11]. Thus one has to restrict the study to very short chains and coarse-grained models. The idea of coarse-graining means to integrate a few chemical monomers along the backbone of the chain into an effective segment. Remember that many chemically different polymers show qualitatively a rather similar freezing in towards the amorphous structure [4, 5]. So one has good reasons to believe that chemical details are not so important here.

These considerations justify the use of a bead-spring model, where the effective monomers interact with a Lennard-Jones (LJ) potential, while the springs contain in addition the finitely extensible nonlinear elastic (FENE) potential [12],

$$U_{LJ}(r) = 4\varepsilon\left[(\sigma/r)^{12} - (\sigma/r)^6\right] + C, \ r \leq r_c, \quad (1)$$
$$U_{LJ}(r \geq r_c) = 0,$$

$$U_{FENE}(r) = -\frac{k}{2}R_0^2 \ln\left[1 - (r/R_0)^2\right]. \quad (2)$$

Here the constant C is chosen such that $U_{LJ}(r)$ is continuous at the cutoff distance $r_c = 2 \times 2^{1/6}\sigma$. Note that $\sigma = 1$ is the length unit in all what follows, $\varepsilon = 1$ sets the scale for the temperature. Choosing $k = 30$ and the maximum extension of the FENE potential $R_0 = 1.5$, one obtains the minimum of the bond potential at $r_{min}^b \approx 0.96$, while the LJ-minimum is at $r_{min}^{LJ} \approx 1.13$: The ratio of these distances is incompatible with a simple crystal structure for the monomers (such as a fcc or bcc lattice). Of course, the system can form a crystal when all chains are stretched out linearly like rigid rods, but this structure does not form easily from the melt. If one creates the crystal artificially as an initial condition [13], one finds that it melts at about $T_m \approx 0.75$. For this model, one hence can obtain deeply supercooled melts (metastable equilibrium can be obtained on the time scales accessible for Molecular Dynamics simulations for T down to $T = 0.46$ at a pressure $p = 1$ [12]), and the structure factor $S(q)$ looks qualitatively very similar [12] to corresponding experiments [14]. The key feature why this model is a very good glassformer is the competition between length scales — monomers alone would easily crystallize.

Now it remains to clarify where in this model the glass transition actually occurs. Experimentally a convenient procedure is to cool down the system at constant cooling rate and constant pressure and monitor the volume $V(T)$ [1]. The glass transition then is located by the kink point in the straight line extrapolations of the $V(T)$ curves for the liquid and glass phases, respectively. Of course, this estimate for T_g will depend on the cooling rate somewhat, since the kink signifies where the system falls out of equilibrium. One can do the same in the simulation [5] and one finds for the model of Eqs. (1,2) that $T_g \approx 0.41$ in Lennard-Jones units [5]. Note, however, that the natural time scale for Molecular Dynamics (MD) is 10^{-12} s, and the cooling rate used [12] is about 10^9 K/s, many orders of magnitude larger than in the experiment.

From these simulations, one can then extract information on dynamic properties, such as the mean square displacement $g_1(t)$ of inner monomers as function of time t, and the mean square displacement of the center of mass, $g_3(t)$. In MD simulations, time is measured in units of $\tau_{MD} = \sigma(m/\varepsilon)^{1/2}$, $m (\equiv 1)$ being the mass of the effective monomers. Simulations were carried out at constant pressure p (e.g. for $p = 1$ [12]). At high temperatures (such as $T = 1$) and for $t \ll 1$ one then encounters a ballistic regime $g_1(t) = 3Tt^2$, $g_3(t) = 3Tt^2/N$, $N = 10$ being the chain length [12], since then the displacements are small in comparison with the size of the cage ($\sim r_{sc}$) formed by the nearest neighbors of each monomer. At $t \approx 1$ a crossover to Rouse-like relaxation sets in (ideally one should have $g_1(t) \propto t^{1/2}$ [15], while in practice rather $g_1(t) \propto t^{0.63}$ is observed [12, 16]). The Einstein relation $g_1(t) = g_3(t) = 6D_s t$, D_s being the self-diffusion constant of the chains, is only observed for late times and large enough displacements ($g_1(t) > \langle R^2 \rangle$, $\langle R^2 \rangle$ being the mean square end-to-end distance of a chain). Therefore an accurate estimation of $D_s(T)$ requires considerable computational effort, particularly at lower temperatures, where $D(T)$ is small, and where between the regime of ballistic motion and Rouse-like motion there enters another regime in which $g_1(t)$ is almost constant (near $6r_{sc}^2$), i.e. a plateau-like region occurs due to the cage effect [2]. One finds that for $T \leq T_m$ the diffusion constant $D_s(T) = D_s(\infty)\exp[-E/(T - T_{VFT})]$, with $T_{VFT} \approx 0.34$ [12, 16, 17]. The incoherent intermediate scattering function, defined by

$$F_s(q,t) = \frac{1}{M}\sum_{i=1}^{M}\left\langle \exp\left\{i\vec{q}\cdot\left[\vec{r}_i(t) - \vec{r}_i(0)\right]\right\}\right\rangle, \quad (3)$$

where the sum is extended over all M monomers in the system ($M = 1200$ [12], and it makes sense to choose q of the order of the peak position q_{max} of the static structure factor $S(q)$, $q_{max} = 6.9$ [12]), develops a clear two-step decay as $T \to T_c$ with $T_c \approx 0.45$. The second step obeys the time-temperature superposition principle, $F_q(t) \propto \exp[-(t/\tau_q)]^{\beta_q}$, with $0.62 \leq \beta_q \leq 0.78$ dependent on q [12]. Such a stretched exponential decay is typical of glass-forming fluids [1]. The relaxation time τ_q is compatible with a power law, $\tau_q \propto (T - T_c)^{-\gamma_q}$, $\gamma_{q=6.9} \approx 2.09$ [12, 16]. Thus it can be concluded that the model reproduces well the qualitative behavior of glassforming polymers.

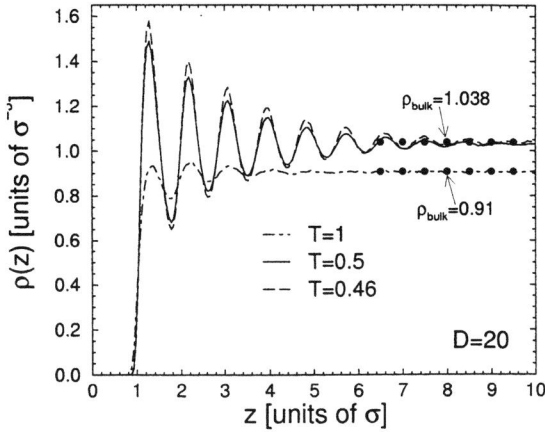

FIGURE 1. Density $\rho(z)$ plotted vs. z for a film of thickness $D = 20$, choosing here the origin of z at the left wall (since $\rho(z)$ is symmetric around $z = D/2$, only the regime $z \leq 10$ is shown). The temperatures are characteristic of the high-temperature state ($T = 1$) and supercooled states ($T = 0.5, 0.46$). The filled dots indicate the bulk densities $\rho_{bulk} = 0.91$ and $\rho_{bulk} = 1.038$ at $T = 1$ and $T = 0.46$ for $p = 1$, respectively. From Varnik et al. [21].

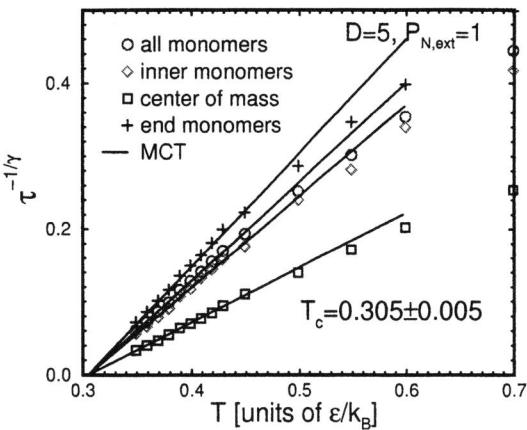

FIGURE 2. Plot of $\tau^{-1/\gamma}$ versus T for films of thickness $D = 5$. Relaxation times τ_i were measured from the condition $g_i(t = \tau) = 1$, using the mean square displacements of inner, end, and all monomers, and of the center of mass. Straight lines show the MCT fits, using $\gamma = 2.5$. From the intersection of all these lines with the abscissa the critical temperature $T_c(D = 5) = 0.305 \pm 0.005$ is determined. From Varnik et al. [21].

PROPERTIES OF THE POLYMER MELT CONFINED BETWEEN FLAT STRUCTURELESS WALLS

Experimentally [18] the glass transition temperature $T_g(D)$ of thin films was recorded as function of the film thickness D. Assuming that the shifted glass transition occurs when $\xi \approx D/2$, one can try to estimate the glass correlation length $\xi(T)$ from such observations [18]. However, the advantage of simulations is that one cannot just reproduce experimental behavior in terms of simple models, but one can go beyond experiment: For the present model system, one can obtain arbitrarily detailed information on both statics and dynamics, not just only of the thin film as a whole, but spatially resolved on an atomistic scale.

The model that is discussed here assumes flat structureless purely repulsive walls at $z_{wall} = \pm D/2$, with a potential $U_{wall}(z) = (\sigma/z)^9$, $z = |z_{monomer} - z_{wall}|$ acting on the monomers. In x, y directions, periodic boundary conditions are used, and the linear dimension L of the box in x, y directions fluctuates, to strictly maintain a normal pressure of $p_N = 1$ [19, 20].

Using $D = 20$, a chain length $N = 10$ and $M = 2000$ monomers, one finds that for $T = 1$ the density is almost everywhere constant across the film. Only below T_m pronounced density oscillations near the wall ("layering") sets in (Fig. 1) [21]. From the range of these oscillations one can extract a static length, which will be discussed below. It should also be stressed that this regular structure in z-direction near the walls does not imply that there is any tendency towards surface-induced crystallization: studying the radial pair distribution function $g(z, r)$ between two points a distance z from the walls and a distance $|\vec{r}|$ parallel to the walls apart reveals at most minor differences from bulk behavior [22]. Also the total static structure factor $S(q)$ with $\vec{q}$ oriented parallel to the film, looks qualitatively just as the bulk $S(q)$.

When one compares thin film and bulk data on a quantitative level, one finds that the dynamics is sped up near flat repulsive walls [21, 22], and hence also the critical temperatures $T_c(D)$ and VFT-temperatures $T_{VFT}(D)$ get reduced with decreasing film thickness. Thus, while $T_c = 0.45$ in the bulk, one finds $T_c(D = 20) \approx 0.415$, $T_c(D = 10) \approx 0.39$ and $T_c(D = 5) \approx 0.31$ [21]. An analysis of various relaxation times implies that $\tau \propto [(T/T_c(D) - 1)]^{-\gamma(D)}$ with $\gamma(D = 5) = 2.5$, $\gamma(10) = \gamma(20) = 2.1$, while $\gamma = 1.95 \pm 0.10$ in the bulk [21] (see Fig.2).

Many more data corroborate and complement this analysis [21, 22, 23]. Here we only focus on the results that allow to conclude on the glass correlation length. Figure 3 shows the layer-resolved mean-square displacement of the innermost monomers versus time. While in the bulk there is a clear cut plateau near $g_1(z,t) = 6r_{sc}^2 \approx 0.054$, particles close to the wall move much faster, and for $0 < z \leq 2$ no plateau at all is developed. From the condition $g_1(t = \tau(z), z) = 1$ one can define a local relaxation time $\tau(z)$, see Fig. 4 [21]. From this plot of $\tau(z)$ vs. z, one can define a correlation length ξ_τ from the distance over which $\tau(z)$ deviates from its bulk value. To this end, we have used the following (empirical) formula

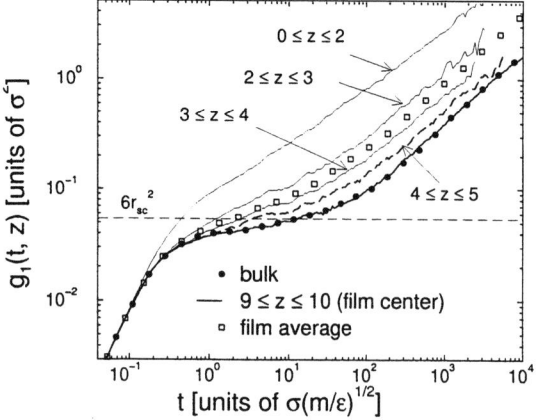

FIGURE 3. Layer-resolved mean square displacements of the innermost monomer $g_1(z,t)$ plotted versus time for a film of thickness $D = 20$ at $T = 0.46$. Here z denotes the distance from the left wall (cf. Fig. 1). The displacements are calculated parallel to the walls and multiplied by $3/2$ to put them on the same scale as the bulk data (full dots). Only those trajectories are included in the average for which the z-coordinate stays in the specified intervals throughout the time-span shown. From Varnik et al. [21].

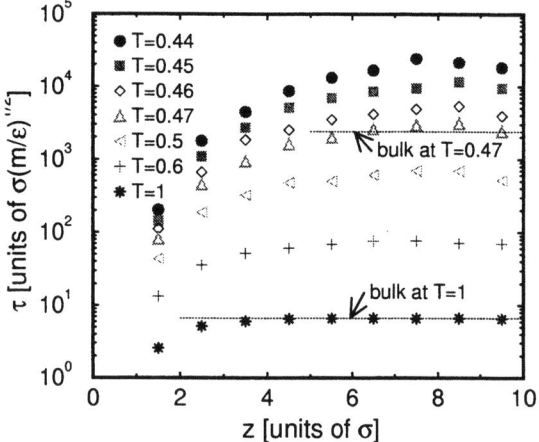

FIGURE 4. Relaxation time $\tau(z)$ plotted vs distance z from the wall for $D = 20$ and several temperatures as indicated. From Varnik et al. [21].

to describe the dependence of $\tau(z)$ on z:

$$\ln(\tau(z)/\tau_{\text{bulk}}) = -A\exp(-z/\xi_\tau). \quad (4)$$

Here, A is a constant, z is the distance from the wall and $\tau_{\text{bulk}}(=\tau(z\to\infty))$ the relaxation time in the bulk. Figure 5 compares the length ξ_τ with ξ_ρ and ξ_g extracted from exponential fits to the decay of the amplitude of the density oscillations [see Fig. 1] and of the pair correlation function respectively [$g(r)$ not shown here]. We see that ξ_τ is the length that increases fastest. However, all these

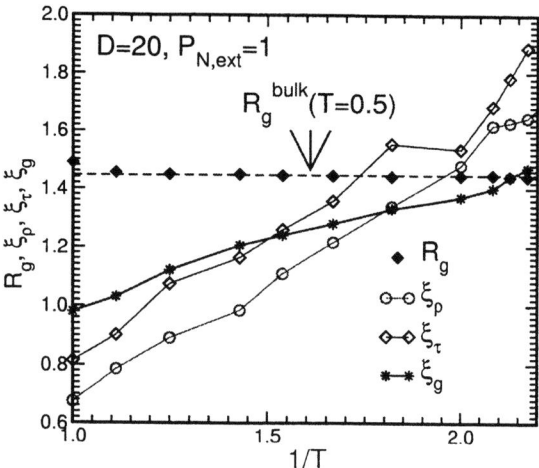

FIGURE 5. Plot of characteristic lengths of polymers in thin films ($D = 20$, normal pressure $p_N = 1$) vs. inverse temperature: gyration radius $R_g(N = 10)$, and correlation lengths extracted from $\rho(z)$, $\tau(z)$, and $g(r)$ (circles, diamonds, and asterisks, respectively). From Varnik et al. [24].

lengths are not larger than two Lennard-Jones diameters, in the temperature range where they can be observed. In this range all lengths increase by at most a factor of three, while the relaxation time increases many orders of magnitude. Thus it is not clear whether the increase of the lengths with $1/T$ shown in Fig. 5 is fundamental or accidental!

THE BINARY LENNARD JONES MIXTURE: CONFINEMENT BY ROUGH VS SMOOTH WALLS

One very successful model for simulation studies of the glass transition is the binary (A,B) mixture with 80% A-particles interacting with a LJ potential V_{AA}, 20% B-particles interacting with V_{BB}, and the potential V_{AB} is chosen such as to provide the deepest minimum of the closest distance [25], $\sigma_{AB} = 0.8$, $\varepsilon_{AB} = 1.5$, while $\sigma_{AA} = 1$, $\varepsilon_{AA} = 1$, $\sigma_{BB} = 0.88$, $\varepsilon_{BB} = 0.5$. These potentials are chosen such that neither phase separation nor crystallization occurs.

One now creates confined systems as follows [22, 26, 27, 28]: First a bulk system in a cubic box with periodic boundary conditions is equilibrated. Then inside the box virtual cylindrical or planar boundaries are introduced, freezing all particles outside the cylindrical tube or planar slit pore thus created, in their instantaneous positions. Of course, the LJ interactions of these frozen particles with the mobile particles inside the cylinder pore or the thin film are still effective. The system is then equilibrated again, and static and dynamic properties of the

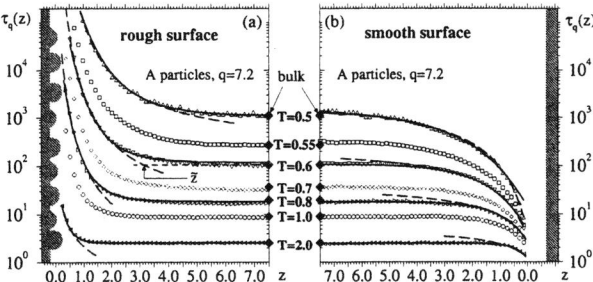

FIGURE 6. Structural relaxation time $\tau_q(z)$ at $q = 7.2$ as a function of particle distance z from the wall for (a) rough and (b) smooth surfaces at different temperatures. The large diamonds are the bulk values. Curves describe fits to phenomenological expressions discussed in [28]. From Scheidler et al. [28].

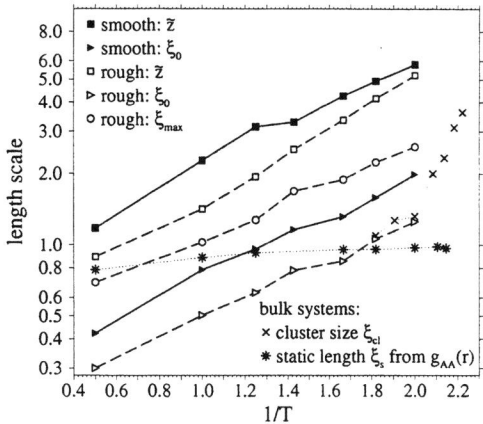

FIGURE 7. Different dynamical length scales (see Ref. [28] for definitions) in confined binary LJ systems as a function of inverse temperature in a logarithmic plot. Note that the asterisks refer to a static length scale [range of oscillations in the pair distribution function $g_{AA}(r)$]. From Scheidler et al. [28].

model can be studied. In this way, the fluid is confined by its own glass! This method has the merit that it creates a "most natural" rough wall boundary condition which provides almost no perturbation of the static structure. When one studies the density profiles across the tube or thin film, respectively, one finds [26, 27, 28] that there is almost no layering whatsoever, and the lateral correlations (parallel to the tube axis or planar walls, respectively) are completely identical to the bulk correlations at all temperatures.

A local correlation time $\tau(z,T)$ is defined using the intermediate incoherent structure function $F_s(q,z,t)$ where $\vec{q}$ in Eq. (3) is oriented parallel to the walls of the thin film, and the summation over particles i is constrained to particles that have their z-coordinate z_i, at time $t=0$ in an interval $\Delta z = 0.5$ centered at a distance z from the wall, requiring $F_s(q,z,t=\tau) = 1/e$.

One sees that from these natural rough walls the relaxation time near the wall is always higher than in the bulk (Fig. 6), while also in this model for smooth walls $\tau(z,T)$ near the wall is reduced (in this case a special potential to suppress layering was added [28]). Nevertheless, the characteristic lengths that one can extract from the range over which $\tau(z,T)$ differs from the bulk are very similar for both boundary conditions (Fig. 7). They increase like those for the polymer model (Fig. 5), compatible with an Arrhenius law, and do not exceed a few LJ diameters near T_c. Since in this model layering is absent, we can at least conclude that the static lengths extracted from the pair distribution or from $\rho(z)$ (Fig. 5) must be irrelevant for the glass transition. Whether the modest increase of the various dynamic length scales in Fig. 7 should be made responsible for the dramatic slowing down near the glass transition remains again doubtful, however.

Acknowledgements: We gratefully acknowledge the financial support of the Deutsche Forschungsgemeinschaft (DFG), (projects SFB 262/D1,D2 and VA 205/101), the Bundesministerium für Bildung und Forschung (BMBF) under the project No 03N6015, the European Community's Human Potential Program under contract HPRN-CT-2002-00307, DYGLAGEMEM, and the European Science Foundation (ESF) under the SUPERNET program. We are grateful to the NIC Jülich for generous grants of computer time at the CRAY-T3E.

REFERENCES

1. J. Jäckle, Rep. Progr. Phys. **49**, 171-291 (1986).
2. W. Götze, J. Phys.: Condens. Matter **10**, A1-A45, (1999).
3. E. Donth, *The Glass transition. Relaxation Dynamics in Liquids and Disordered Materials* (Springer, Berlin–Heidelberg, 2001).
4. K. Ngai (ed.) *Proceedings of the Fourth International Discussion Meeting on Relaxation in Complex Systems, Chersonissos, Crete, June 2001*, J. Non-Cryst. Solids **307-310** (2002).
5. K. Binder, J. Baschnagel, and W. Paul, Progr. Polym. Sci. **28**, 115-172 (2003).
6. W. Kob, preprint.
7. K. Binder and A. P. Young, Rev. Mod. Phys. **58**, 801-976 (1986).
8. G. Adam and H. J. Gibbs, J. Chem. Phys. **43**, 139-146 (1965).
9. J. Torres, P. Nealey, and J. de Pablo, Phys. Rev. Lett. **85**, 3221-3224 (2000).
10. K. Binder *"Critical behavior at surfaces"*, in *Phase Transitions and Critical Phenomena, Vol. 8* (C. Domb and J. L. Lebowitz, eds.) p. 1-144 (Academic, New York, 1983).
11. K. Binder (ed.) *Monte Carlo and Molecular Dynamics Simulations in Polymer Science* (Oxford Univ. Press, New

York 1995).
12. C. Bennemann, W. Paul, K. Binder, and B. Dünweg, Phys. Rev. E **57**, 843-851 (1998).
13. J. Buchholz, W. Paul, F. Varnik, and K. Binder, J. Chem. Phys. **117**, 7364-7372 (2002).
14. A. Arbe, D. Richter, J. Colmenero, and B. Farago, Phys. Rev. E **54**, 3853-3869 (1996).
15. M. Doi and S. F. Edwards, *The Theory of Polymer Dynamics* (Clarendon Press, Oxford, 1986).
16. C. Bennemann, J. Baschnagel, W. Paul, and K. Binder, Comput. Theor. Polym. Sci. **9**, 217-226 (1999).
17. J. Baschnagel, C. Bennemann, W. Paul, and K. Binder, J. Phys.: Condens. Matter **12**, 6365-6374 (2000).
18. J. Keddie, R. Jones, and R. Cory, Europhys. Lett. **27**, 59-64 (1994).
19. F. Varnik, J. Baschnagel, and K. Binder, J. Chem. Phys. **113**, 4444-4453 (2000).
20. F. Varnik, Computer Phys. Commun. **149**, 61-70 (2002).
21. F. Varnik, J. Baschnagel, and K. Binder, Phys. Rev. E **65**, 021507, 1-14, (2002).
22. F. Varnik, P. Scheidler, J. Baschnagel, W. Kob, and K. Binder, Mat. Res. Proc **651**, T3.1-T3.6 (2001).
23. F. Varnik, J. Baschnagel, and K. Binder, Eur. Phys. J. E **8**, 175-192 (2002).
24. F. Varnik *et al.*, in preparation.
25. W. Kob and H. C. Andersen, Phys. Rev. E **51**, 4626-4641, (1995); *ibid.* **52**, 4134-4153 (1995).
26. P. Scheidler, W. Kob, and K. Binder, J. Phys. IV (France) **10**, Pr. 7-33-Pr. 7-36 (2000).
27. P. Scheidler, W. Kob, K. Binder, and G. Parisi, Philos. Mag. B **82**, 283-290 (2002).
28. P. Scheidler, W. Kob, and K. Binder, Europhys. Lett. **52**, 277-283 (2000); *ibid.* **59**, 701-707 (2002).

Johari-Goldstein Or Primitive Relaxation: Terminator Of Caged Dynamics And Precursor Of α-Relaxation

K.L. Ngai

Naval Research Laboratory, Washington, DC 20375-5320, USA

Abstract. Advance in understanding glass transition dynamics is made by identifying (1) the "universal" Johari-Goldstein (JG) relaxation with the primitive relaxation of the coupling model and (2) the nearly constant loss (NCL) as the short time caged dynamics, all with support from experimental data. Dynamics proceeds with time in the order of vibration (boson peak), caged motions (NCL), primitive or JG relaxation, and finally the fully cooperative α-relaxation. The primitive or JG relaxation plays two pivotal roles, the terminator of the caged dynamics (NCL) and the precursor of the α-relaxation. The dependence of molecular mobility on thermodynamic variables such as volume and entropy enters already at early times in caged dynamics (NCL) and primitive or JG relaxation, as borne out by experimental data, and is magnified in the α-relaxation by many-molecule heterogeneous dynamics caused by intermolecular coupling.

INTRODUCTION

Secondary or β- relaxations occur at shorter times or higher frequencies and they are often considered to play no important role in determining the glass transition. This attitude is justified for some β-relaxations that involve isolated motion of a part of flexible molecules. However, Johari and Goldstein (JG) showed that not all β-relaxations have such trivial intramolecular origin by finding β-relaxations even in glass-formers composed of completely rigid molecules [1]. This intriguing finding of JG implies that some β-relaxation does not involve intramolecular degree of freedom, but instead certain motion involving essentially all parts of the molecule, and its existence is universal [1-5]. This is definitely the case of rigid molecular glass-formers, and possibly even in all other kinds of glass-formers as shown by dielectric, mechanical and nuclear magnetic resonance measurements in the last thirty years. To honor their important discovery that has fundamental implication for glass transition, in my view, only this class of β-relaxations should be called the Johari-Goldstein (JG) β-relaxations. However, in the current literature, there is an unfortunate tendency that any observed β-relaxation is referred to as the JG β-relaxation without qualification. We have recently classified observed β-relaxations by their dynamic properties and established several criteria to distinguish genuine JG β-relaxation (according to our definition) from non-JG relaxations [6]. If more than one secondary relaxation are present, the faster one is likely non-JG because the genuine JG has to be the slowest of all. Genuine JG relaxation, though exists, may not be resolved as a peak or shoulder in the loss spectrum if it is located not far from the usually more intense α-relaxation loss peak [7,8]. Non-JG β-relaxation has little impact on glass transition, and they are not discussed in this work.

An important property of JG β-relaxations often overlooked is that it also senses the glass transition. The relaxation strength, $\Delta\varepsilon_\beta$, of the JG relaxation [9] is found to change on heating through the glass transition temperature in a similar manner as the changes observed in the enthalpy H, entropy S, and volume V. The derivative of $\Delta\varepsilon_\beta$ with respect to temperature, $d\Delta\varepsilon_\beta/dT$, is raised from lower values at temperatures below T_g to higher values at temperatures above T_g, mimicking the same behavior of the specific heat C_p and the expansion coefficient, which are the derivatives dH/dT and dV/dT respectively. The angle of rotation of the JG relaxation, and hence $\Delta\varepsilon_\beta$, likely is dependent on the specific volume and entropy. Thus $d\Delta\varepsilon_\beta/dT$ is similar to dH/dT and dV/dT, which define what is called "glass transition". The temperature dependence of the JG relaxation time τ_β is invariably Arrhenius below T_g, but whether the Arrhenius dependence continues above T_g is not clear. Above T_g, τ_β cannot be determined accurately due to proximity of the β-relaxation to the primary α-relaxation. Some arbitrary fitting procedure has to be used to resolve the JG relaxation and the result is never certain. Recently this difficulty is circumvented by application of high pressure to clearly resolve the JG relaxation in sorbitol and xylitol above T_g [10]. No fitting is needed and the data show that, above T_g, τ_β decreases more rapidly with increasing temperature than the Arrhenius dependence established below T_g and extrapolated to above T_g. Hence the temperature dependence of τ_β changes across T_g, just like the α-relaxation time τ, to reflect that the liquid is falling out of equilibrium to form the non-equilibrium glassy state on cooling.

JG relaxation in the glass was found to depend on the thermal history including the cooling rate used to vitrify the liquid and the aging time [1]. The angle of rotation or $\Delta\varepsilon_\beta$ is expected to be dependent on the thermal history of the glass. A denser glass has a smaller $\Delta\varepsilon_\beta$. More recent work confirms this by aging studies of $\Delta\varepsilon_\beta$ in the glassy state [9], although PMMA may be an exception. Another recent detailed study of the JG β-relaxation in dipropylglycol dibenzoate (DPGDB) [11] demonstrates several effects. Thermal history of the glass has a strong influence on the JG β-relaxation. On aging the glass, τ_β of the JG relaxation increases significantly, mimicking the well-known increase of τ_α with aging.

From the above, we have seen that the properties of τ_β and $\Delta\varepsilon_\beta$ of JG relaxation in the equilibrium liquid and the glassy state are different, undergoing a change when crossing T_g. These are characteristics of the primary α-relaxation that we usually associate with glass transition. Therefore we can infer that there is some fundamental connection of the JG β-relaxation to the α-relaxation and glass transition.

THE JG RELAXATION IDENTIFIED WITH THE PRIMITIVE RELAXATION

The importance of the JG relaxation to glass transition ascends to a higher level through the recent remarkable findings from experimental data of many glass-formers, at temperatures above and below T_g and various pressures P, that [6-8]

$$\tau_\beta(T,P) \approx \tau_0(T,P). \qquad (1)$$

Here τ_0 is the primitive (independent) relaxation time of the coupling model (CM) [12-15] that can be calculated from the α-relaxation time τ_α by

$$\tau_0 = (t_c)^n (\tau_\alpha)^{1-n}, \qquad (2)$$

In Eq.2, $(1-n)$ is the stretch exponent in the Kohlrausch-Williams-Watts (KWW) function,

$$\phi(t) = \exp[-(t/\tau_\alpha)^{1-n}], \qquad (3)$$

that fits the time dependence of the α-relaxation. t_c is the crossover time from independent relaxation to cooperative relaxation and has the approximate value of 2×10^{-12} s [15] for molecular glass-formers. The findings that Eq.(1) holds for many glass-formers is remarkable because it is a cross relation between τ_β and τ_0. The former is the JG β-relaxation time, while the latter is determined entirely by parameters of the α-relaxation, e.g. τ_α and n in Eq.3.

Note that from Eq.(2), the separation between τ_α and τ_0 is given by $(\log\tau_\alpha - \log\tau_0) = n(\log\tau_\alpha - \log t_c)$. Hence with Eq.(1) the separation between τ_α and τ_β is approximately given by

$$(\log\tau_\alpha - \log\tau_\beta) = n(\log\tau_\alpha - \log t_c). \qquad (4)$$

For the same τ_α, the separation decreases as n becomes smaller, and vanishes when n is exactly equal to zero. This has been verified for many glass-formers, which have resolved JG β-relaxation peak or shoulder [6-8] In hindsight, a few of the β-relaxations considered in the first work of this kind [7a] are not genuine JG β-relaxation [6] and should not have been included. By eliminating them, the correlation becomes even stronger. The dependence of the separation ($\log\tau_\alpha - \log\tau_\beta$) on n is best demonstrated for glass-formers of the same family [16,17], and it also explains why the JG relaxation is not resolved in glass-formers with smaller n because of smaller separation, and instead appears as an excess wing on the high frequency flank of the α-loss peak [6,7]. However, it can seen from the right-hand-side of Eq.4 that the separation can be increased for a better chance of resolving the JG β-relaxation by increasing $\log\tau_\alpha$ through aging even without change in n. This goal was achieved in glycerol, propylene carbonate and propylene glycol, which all have smaller n, by aging these glass-formers for a long period of time [7b]. Again from Eq.4, another method of increasing the separation in the glass-former such as picoline that show only an excess wing is by enhancing n. This can be done by mixing it with a much less mobile glass-former (i,e. with a higher T_g) such as ortho-terphenyl and tri-styrene as carried out by Blochowicz and coworkers and discussed in Ref.[6]. Indeed in the mixtures, the JG β-relaxation of picoline has been resolved.

Identification of τ_β with τ_0 is reasonable because, like JG β-relaxation, the primitive relaxation of the coupling model (CM) entails the local motion of the entire molecule and is not cooperative. Thus, it bears a strong resemblance to the JG relaxation, and it is expected that the primitive relaxation time, τ_0, is approximately located near the most probably relaxation time τ_β of the JG relaxation at all temperatures and pressures. Note that the interpretation of the dielectric spectrum according to the coupling model [8] differs from the conventional view that the JG relaxation is a broad distribution of relaxation times, and there is no contradiction of identifying τ_0 with τ_β. This point will become clearer later, after the evolution of dynamics from the short times caged regime through the intermediate primitive or JG relaxation to the fully cooperative α-relaxation regime has been described.

The identification is mutually beneficial for the JG relaxation and the CM. The primitive relaxation time τ_0, a key quantity in the CM, becomes even more real because it is exemplified experimentally by the JG relaxation time τ_β. On the other hand, the JG relaxation acquires an important role in glass transition through τ_0, because the primitive relaxation is the precursor of the cooperative (i.e., intermolecularly coupled) α-relaxation. Eqs.(2) and (3) basically have taken care of only the effects of intermolecular coupling and many-molecule-dynamics. To consider glass transition, which involves changes of temperature or pressure, the dependence of molecular mobility on the thermodynamic variables such as specific volume V and entropy S (or alternatively free volume v_f and configurational entropy S_c) has to be taken into account by any theory. In the CM, the intrinsic mobility of molecules is determined by τ_0, and hence τ_0 and its relaxation strength depend on V and S. Taking into account explicitly the dependence of τ_0 on T and P, we have $\tau_0(T,P,V,S)$ or $\tau_0(T,P,v_f,S_c)$. These properties of the primitive relaxation are passed on to the JG relaxation because of Eq.(1). Conversely, if we start with the JG relaxation and the observed properties showing that it is sensitive to volume and/or entropy (e.g., the similarity between $d\Delta\varepsilon_\beta/dT$ and dH/dT or dV/dT across T_g [9]), then the same is true for the primitive relaxation because of Eq.(1).

Only in the glassy state where structural relaxation is arrested will τ_0 and τ_β has the Arrhenius T-dependence, $\tau_0 \approx \tau_\beta = \tau_\infty \exp(E_\beta/RT)$. In the equilibrium liquid state, the dependence of τ_0 and τ_β on V and S renders the temperature dependence of τ_0 and τ_β non-Arrhenuius. Hence the temperature dependences of τ_β and $\Delta\varepsilon_\beta$ change when crossing T_g, as observed experimentally [9,10].

At $T=T_g$, a corollary of Eqs.(1) and (2) combined is the expression $[(1-n)\ln\tau_\alpha(T_g) + n\ln t_c - \ln\tau_\infty]$ for the ratio E_β/RT_g. The value of E_β/RT_g computed by this expression is in agreement with the experimental value for many glass-formers [18]. In fact for many small molecular glass-formers, the computed values of E_β/RT_g are about 24 as found empirically [19].

Precursor Of The α-Relaxation

τ_β is the reciprocal of the angular frequency, $\omega_\beta \equiv 2\pi\nu_\beta$, of the JG β-relaxation loss peak. Traditionally, the peak is modeled by a Cole-Cole distribution. In the extended coupling model [8] τ_β is interpreted as merely an indicator of the primitive relaxation time τ_0. With further increase in time beyond τ_0, there is increasing independent relaxations and many-body relaxation dynamics becomes increasingly important (or increasing degree of "cooperativity"). This evolution of dynamics accounts for the dispersion at frequencies below ν_β (at times longer than τ_β) and continues until the onset of full many-body cooperative dynamics, describable by the KWW function, Eq.(3), is reached. After the onset, the relaxation becomes the KWW α-relaxation and is amenable to the CM description to be discussed in the next section. Because of the increasing development of many-body dynamics with increasing time in the JG relaxation range as conventionally defined, the JG β-relaxation is thus heterogeneous with respect to relaxation time constants as found by dielectric hole burning experiments [3,20].

THE α-RELAXATION

Numerous experimental data and molecular dynamics simulations have shown at long times the observed correlation functions are invariably well approximated by the Kohlrausch function (Eq.3), one of the components of the CM. Thus, we do not need any more theory just to assure us that the correlation function is approximately a Kohlrausch function, or to wait for a theory that gives us the exact coupling parameter n for a real glass-former. The latter is an impossible task for any theory at the present time. We just shrewdly take what the experiment gives us for the value of n, at any temperature or pressure [21]. t_c has already been determined to be approximately 2 ps for molecular liquids and polymers [15]. The dependences on T, P, V and S have already entered into τ_0 or τ. Then, via Eqs.(1) and (2), τ_α is given by

$$\tau_\alpha(T,P,V,S) = [t_c^{-n}\tau_0(T,P,V,S)]^{1/(1-n)}$$
$$\approx [t_c^{-n}\tau_\beta(T,P,V,S)]^{1/(1-n)} \quad (5)$$

All the determining factors of τ_α have been accounted for in this expression. They are: (1) the dependences on T, P, V and S; and (2) many-molecules relaxation dynamics that arise from intermolecular coupling. Note that the dependences of τ_α on volume and entropy are mere consequences of the dependence of τ_0 or τ_β on these thermodynamic factors, and not vice versa. This view is supported by the fact that the JG β-relaxation and the primitive relaxation have transpired at times long before the α-relaxation. The effect of intermolecular coupling on the α-relaxation, obtained by Eq.(5), magnifies the dependences of τ_α on T, P, V and S from that of τ_0 or τ_β by the superlinear power of $1/(1-n)$. The resultant τ_α has both the dependence on T, P, V and S, and the effect of intermolecular coupling. It is certainly responsible for the kinetics of the glass transition. Nevertheless, the dependences of τ_0 or τ_β on V and S are the root cause.

Below T_g and in the absence of structural relaxation (i.e., iso-structural state or constant fictive temperature condition), τ_α has the Arrhenius temperature dependence, $\tau_\alpha = \tau_{\alpha\infty}\exp(E_\alpha/RT)$, because of Eqs.(1) and (2) and $\tau_0 \approx \tau_\beta = \tau_\infty\exp(E_\beta/RT)$. Furthermore, we have the relation, $(1-n)E_\alpha = E_\beta$, between the two activation enthalpies in the glassy state. We have verified this relation in a number of small molecular glass-formers [22].

As demonstrated before in glass-formers and other coupled systems [14,15,23,24], when used judiciously Eq.(5) can explained other observed properties of τ_α and the associated transport coefficients such as viscosity, diffusion coefficient and d.c. conductivity, especially the anomalous ones. The origin of several found correlations between n (or anticorrelation between $1-n$) and the properties of τ_α [25] are easily traced to the dependence of τ_α on n in Eq.(5) [14,15]. An example is the anomalous dependence, $\tau_\alpha \propto Q^{-2/(1-n)}$, on the scattering wave-vector Q found in neutron scattering [26] and light scattering [27]. This is immediately explained by Eq.2 from the normal Q^{-2}-dependence of τ_0 or τ_β.

SHORT TIME CAGED DYNAMICS

For glass-formers that have a resolved JG β-relaxation and/or a non-JG secondary relaxation, the combined dispersion of the α- and β-relaxation of the equilibrium liquid occupies many decades of frequency, making difficult the characterization of the dynamics by experiment at times shorter than the β-relaxation. At higher temperatures, the α- and β-relaxation merge together but the relaxation times becomes short, and again does not allow the dynamics to be seen over a broad frequency range. Therefore, the best glass-formers for optimal observation of caged dynamics within the equilibrium liquid state are those that have small n so that the separation between τ_α and τ_β of the JG relaxation is short (Eq.4), and no other non-JG β-relaxation is present. There is no such problem in the glassy state of any glass-former because, by lowering the temperature sufficiently, all secondary relaxations move to lower frequencies out of the experimental window, enabling the faster caged dynamics to be observed. Dielectric data of such glass-formers [7b,8,28-30] show the loss at high frequencies can be approximately described by the v^λ-dependence with a small positive exponent λ, or a logarithmic function of v, or some other slowly varying function of frequency that increases with decreasing v. Similar dispersion is found in glassy, molten and even crystalline ionic conductors at temperatures where the ions are caged and not able to move to another site in the experimental frequency/time range [8a,31]. The variation of this dispersion with frequency is very slow and hence it is called the nearly constant loss (NCL) in the field of ionic conductors. A good example is the glass-forming molten salt, $0.4Ca(NO_3)_2$-$0.6KNO_3$ (CKN), which is both a glass-former and an ionic conductor [8a,31]. The observation of the NCL in glass-formers is not restricted to dielectric relaxation measurement. Dynamic light scattering measurements on polyisobutylene [32], poly(methyl methacrylate) [33] and glycerol [33] have found the NCL in the imaginary part of the depolarized and polarized susceptibility, $\chi''(v)$, at high frequencies. The optical Kerr effect studies of several organic glass-forming liquids [34] have found presence of the near constant from the measured time derivative of the correlation function at temperatures below the critical temperature T_c of the mode coupling theory [35].

JG Relaxation Signals The Termination Of Caging

For glass-formers, the NCL regime is found [8a, 36] to start at frequencies (times) higher (shorter) than some v_{x1} (t_{x1}) which is roughly a decade or more higher (shorter) than the calculated independent relaxation frequency $v_0 \equiv 1/2\pi\tau_0$ (time τ_0). The same applies to the ionic conductors, where now τ_0 is the independent ion hopping relaxation time [24,37]. From its very definition, τ_0 is the shortest possible relaxation time of the entire molecule and the same applies to the corresponding JG relaxation time τ_β. Thus, whatever the nature of the NCL, the responsible dynamics proceed when essentially all molecules remain caged. The motion within the cage has very limited length scale and the integrated total loss from the onset frequency to ν_{x1} has to be small, and thus the slow increase of the loss with decreasing frequency, i.e., the NCL behavior. Using $1/\nu_{x1}$ or τ_0 as the upper bound of the time regime of the NCL, we have $<u^2(T)>_{NCL} \propto 1/\log\tau_0$ [31,37]. From this result we have derived the observed weak temperature dependence of the NCL approximately described by $\exp(T/T_0)$.

Caged Dynamics: The NCL Or The β-Process Of Mode Coupling Theory?

We believe the NCL is the true caged dynamics for any temperature, at least for molecular glass-formers and ionic conductors. The cage is not permanent and it decays with time due to the onset of the primitive relaxation process. The cages decay time is about τ_0 or τ_β, thus justifying calling the primitive relaxation or the JG relaxation the terminator of the caged dynamics. One would like to use the mode coupling theory (MCT) [35], which is the best-known and respectable theory for caged dynamics, to describe the experimental data. Unfortunately, the standard mode coupling theory does not predict the NCL except under very special potential and condition [38], but NCL is found in glass-formers of many chemical forms with different potentials and unrestricted in the temperature range. NCL as a background loss can give rise to a susceptibility minimum having ν^b and ν^β-dependences on the low and high frequency sides respectively. This happens at high temperatures (such as above T_c of the mode coupling theory), when τ_α becomes short or the α-loss peak frequency ν_α sufficiently high. Then, the extent of the NCL is shrunk by the encroaching high frequency flank of the α-loss peak having the $\nu^{-(1-n)}$-dependence, and on the high frequency side by the low frequency flank of the vibrational or Boson peak contribution. The result is a susceptibility minimum as observed by dielectric relaxation, neutron and light scattering experiments at higher temperatures where τ_α becomes short, typically of the order of nanoseconds. This also explains at T_c, obtained by using the scaling laws of various quantities related to the susceptibility minimum of MCT, invariably τ_α is in the neighborhood of nanoseconds. On the other hand, it is obvious that a susceptibility minimum predicted by the standard mode coupling theory cannot account for the NCL at temperatures below T_c. One certainly expects at temperatures lower than T_c that the molecules or ions are still caged and even for a longer period of time. Thus the caged dynamics persist below T_c, and even below T_g. This is the case if caged dynamics are exemplified by the NCL, which has been observed at lower temperatures and even in the glassy state, indicating that NCL is the true caged dynamics. This is also consistent with the CM description of caging, because the termination time of caged dynamics, τ_0 or τ_β, becomes longer at lower temperatures [8,31,36,37]. In fact the NCL, caged dynamics in the CM description, is observed experimentally to extend to longer times or lower frequencies on decreasing temperature and this trend persists down to T_g and beyond [28-31]. This property is to be contrasted with the caged dynamics of the standard mode coupling theory, which are limited to $T>T_c$.

Study of 1.4-polybutadiene (PBD) by neutron scattering were able to find different temperature and pressure (T,P) combinations at which the density is constant and the measured static structure factor $S(Q)$ is the same [39], but the α-relaxation and the fast relaxation both change. For two (T,P) combinations having the same density, it was observed that the one with a higher T has shorter α-relaxation time and higher intensity of the fast relaxation. The fast relaxation shows up like a susceptibility minimum, but the minimum is so flat, particularly for the (T,P) combination with a higher temperature, that it resembles the PIB data at 260 K in Ref.[33]. Thus the flat minimum suggests it may actually be the NCL, as in PIB which is definitely the case because of availability of solid evidence at lower temperatures when the α-relaxation time, τ_α, becomes longer. A shorter τ_α engenders shorter τ_0 (Eq.2). From this and the relation, $<u^2(T)>_{NCL} \propto 1/\log\tau_0$ [31,37] we draw the conclusion that the combination having a shorter τ_α corresponds to a larger NCL. The two combinations (T=314 K, P=200 MPa) and (T=220 K, P=0.1 MPa) have the same density [39]. But the combination (T=314 K, P=200 MPa) has a shorter α-relaxation time than the (T=220 K, P=0.1 MPa) combination. Hence we expect a higher NCL intensity for the (T=314 K, P=200 MPa) combination. The higher NCL intensity relative to the boson peak makes susceptibility minimum more flat for (T=314 K, P=200 MPa) to reveal the fast relaxation observed has the true color of the NCL. Thus by interpreting the fast relaxation as the NCL, the change from one (T,P) combination to another can be predicted and is in accord with the findings of Ref.[39].

High-frequency susceptibility of ortho-terphenyl measured by light scattering for different (T,P) combinations [40] also shows the susceptibility "minimum" gets very broad at high pressures where the α-relaxation has been moved to low frequencies out of the experimental frequency range. For example at 600 bar and 290 K, the flat "minimum" seen may as well be identified as the NCL. The existence of NCL in polyisoprene at low temperatures has been confirmed by precision dielectric measurements [41a]. These measurements have found the NCL in the 10 Hz to 10^5 Hz range at temperatures below T_g. At higher temperatures, the NCL in polyisoprene is observed in the 1-400 GHz range by light scattering [41b]. These measurements combined show that NCL can be observed over a wide temperature range.

Usually when a susceptibility minimum is observed, it is profitable to fit the fast relaxation with the MCT predictions. However, for PBD there are serious deviations from the MCT relation between the critical exponents a and b determined from the frequency dependence of the susceptibility obtained by incoherent neutron scattering experiment [42]. This deviation had long since suggested the fast relaxation in polybutadiene may not be the kind predicted by the MCT. Furthermore, in the idealized version of the MCT, temperature and pressure determine the dynamics through the static structure factor $S(Q)$. Hence, for the same static structure factor, similar fast relaxation and α-relaxation should be observed. This prediction of the MCT contradicts the experimental findings of Ref.[39]. As discussed above, both the fast relaxation and the α-relaxation were found to change for different (T,P) combinations that have the same $S(Q)$. Altogether the data of Refs.[39] and [42] indicate that the fast relaxation observed in PBD is not the fast β-process of MCT. Similar conclusion was drawn from a high-frequency light scattering study of orth-terphenyl under pressure discussed above [40]. The work found that the exponents a and b are temperature and pressure dependent. The exponent b strongly decreases with pressure, reaching values lower than 0.2 at lower temperatures, which is indicative of the emergence of the NCL.

Origin Of The NCL

Molecular dynamics simulations of Li metasilicate glass have been carried out recently to investigate the origin of the NCL coming from caged Li ions [43]. By analysis of all the trajectories of the ions obtained up a fixed long time but essentially all ions are still confined within cages, it was found that the length scales of the caged Li ion motions are distributed according to a Lévy distribution that has a long tail. This fractal nature of the length scales of the trajectories of caged ions is an important discovery and its origin may be traced to the dynamic anharmonic potential experienced by the caged ions. A nonlinear Hamiltonian dynamics treatment is ultimately needed for a fundamental understanding of the source of the Lévy distribution of the amplitudes of the ion motions contributing to the NCL. Nevertheless the results suggest that the intensity of the NCL increases with anharmonicity and strength of the cage potential. In molecular liquids a molecule is caged by neighboring molecules, and hence the cage potential is related to the intermolecular potential. In the CM slowing down of the cooperative α-relaxation is by intermolecular coupling, and hence the coupling parameter n also increases with anharmonicity and strength of the intermolecular potential [13]. Therefore, the intensity of the NCL should correlate with n, as found empirically [44]. From the implications that anharmoncity is involved, a fundamental solution of caged dynamics and the NCL has to be based on nonlinear Hamiltonian dynamics (e.g., chaos) in phase space.

Since the intermolecular potential in the Hamiltonian governs all dynamics and thermodynamics, all properties of the NCL, the JG relaxation, the α-relaxation (i.e., nonexponentiality, dynamic heterogeneity, fragility, etc.), and even the volume and entropy are ultimately determined by it. Thus correlations between the NCL, the α-relaxation, and thermodynamic quantities can be expected. Previously found examples include: (*i*) the correlations between the T_g-scaled temperature dependences of the NCL, τ_α, and S_c (the so-called thermodynamic fragility) [25]; and (*ii*) all three T_g-scaled temperature dependences correlate with n, with a few exceptions. More examples are given in the following sections after neutron and light scattering data are considered.

Fast Caged Dynamics (NCL) Senses T_g

The NCL constitute a major part of the fast relaxation observed from picosecond to nanosecond time range by quasielastic neutron and light scattering [25,44]. From the incoherent scattering function, $S(Q,\omega,T)$, the elastic part of the scattering, $S_{el}(Q,\Delta\omega,T)$, is operationally defined by the integral of $S(Q,\omega,T)$ over ω within $-\Delta\omega<\omega<\Delta\omega$, where Q is the momentum transfer and $\Delta\omega$ is the resolution frequency width. After normalizing $S_{el}(Q,\Delta\omega,T)$ measured at temperature T by its value at $T=0$, it is expressed as $\exp(-Q^2<u^2(T)>/3)$, in terms of the mean square displacement $<u^2(T)>$. Comparing $<u^2(T)>$ at the respective T_g of various glass-formers, a correlation between $<u^2(T_g)>$ and n was found [25,44].

The fast relaxation intensity from neutron scattering data expressed as a mean square displacement, $<u^2(T)>$, shows a change in slope around T_g, similar to that of $<u^2(T)>_{NCL}$ and of the volume, enthalpy and entropy. This was first found by neutron scattering in selenium [45] and the same property was found later on in other polymeric and small molecular glass-formers [25,44,46]. Like τ_β and $\Delta\varepsilon_\beta$ of JG relaxation, the intensity of the NCL [25,44] and the fast relaxation also senses the glass transition. These behavior of the relaxation processes occurring at much shorter time than τ_α, in particular the fast relaxation transpiring in the pico- to nano- seconds range, is puzzling to those associating glass transition with nothing else but the α-relaxation by its relaxation time becomes long, causing the structure of the liquid to fall out of equilibrium at some temperature known as T_g.

Comparison of the intensity of the fast process or $<u^2(T)>$ seen by quasielastic neutron scattering with the hole volume measured by the positronium annihilation lifetime spectroscopy (PALS) has been made in cis-1,4 PB, PIB and aPP [46] and in small molecular glass-formers, OTP, propylene carbonate, glycerol and propylene glycol [47]. In PALS, the hole volume is determined by the positronium life-time, τ_3. The data show that the temperature dependence of $<u^2>$ and the hole volume are remarkably similar. In particular, like $<u^2(T)>$, the hole volume as a function of temperature exhibits a change in slope at T_g. Duval et al. [48] subtracted from the measured $<u^2(T)>$ of PMMA a linear temperature dependent term contributed by harmonic vibrations. The result, $\Delta<u^2(T)>$, has nearly the same temperature dependence as the dynamic hole volume fraction, $\Delta F_h(T)$, obtained from PALS. These findings that links the hole volume, V_h, to $<u^2(T)>$ of fast relaxation have an explanation from the dependence of τ_0 on hole volume. Indeed let us assume that τ_0 depends on the hole volume according to $\tau_0 \propto \exp[V_0/V_h(T)]$, where V_0 is a constant. Originally, such dependence on V_0 was suggested by Doolittle for viscosity [49], but we shall adopt the same form for τ_0 here. Then it, together with the relation, $<u^2(T)>_{NCL} \propto 1/\log(\tau_0)$ discussed before, lead us to $<u^2(T)>_{NCL} \propto V_h(T)$. The result explains the relation between free volume and the fast relaxation found from experimental studies of Bartos et al [46b], Duval et al. [48], and Ngai et al. [47]. It can also be considered as another evidence that dependence on hole volume has entered first and foremost into the NCL and τ_0, and the dependence of τ_α on hole volume comes merely as a consequence of Eq.(5).

EVOLUTION OF DYNAMICS: CAGED MOTION (NCL) / PRIMITIVE OR JG β-RELAXATION / α-RELAXATION

The data of the fast relaxation and NCL discussed above indicate that at early times when molecules are still caged, their motion already sense the change of volume and entropy as the liquid is cooled. The data of the JG relaxation occurring at much shorter times than τ_α show that the local non-cooperative relaxation (τ_β and $\Delta\varepsilon_\beta$) of individual molecules also sense the changes in volume and entropy long before many-molecule cooperative dynamics take hold. Therefore the seed of the dependence of τ_α on V and S was sowed earlier in the JG relaxation or the primitive relaxation of the CM. The dependence of τ_α on V and S is much stronger than that of τ_β due to amplification by raising the latter to the superlinear power, $1/(1-n)$, in Eq.5, and this has led to the common belief that for glass transition phenomenon one needs only to consider the α-relaxation. The evidences presented here indicate otherwise and it is τ_0 or τ_β that is originally endowed with the dependence on V and S. The many-molecule cooperative dynamics, which accounts for the slowing down of the τ_0 or τ_β to become the cooperative τ_α by heterogeneous many-molecule dynamics, magnify the dependence of τ_α on V and S (Eq.5) and cause the liquid to fall off from equilibrium at some temperature T_g. The temperature dependence of V and S are different in the equilibrium liquid state than in the glassy state. The changes of in slope of V and S on crossing T_g are fed back to τ_0 or τ_β and even the fast dynamics or the NCL, thus explaining the observations of changes of temperature dependence in τ_β, $\Delta\varepsilon_\beta$ and $<u^2>$ when T_g is crossed.

The description of the dynamics here is the evolution with time from caged motion to primitive relaxation, then continuous development of intermolecular coupling and finally the onset of fully cooperative α-relaxation describable of the KWW correlation function. Thus the NCL of caged motion is not an additive contribution, a point which we have proven by comparing dynamic light scattering and dielectric relaxation data of 0.4Ca(NO$_3$)$_2$-0.6KNO$_3$ [36,50,51]. The NCL is the *precursor* of the primitive or JG relaxation, or conversely the primitive or JG relaxation is the *terminator* of the NCL. The primitive or JG relaxation is not an additive contribution either, but is a *precursor* of the α-relaxation. The primitive or JG relaxation plays the pivotal role in bridging the short time caged dynamics to the long time fully cooperative KWW α-relaxation. It is interesting to contrast this view of the evolution of dynamics with that of MCT, which considers caged dynamics (in terms of the fast β-process of MCT) to be the precursor of the α-relaxation, bypassing the JG β-relaxation.

ACKNOWLEDGMENTS

The work was supported by ONR. I thank all my collaborators for their important contributions.

REFERENCES

1. Johari, G.P., and Goldstein, M. *J. Chem. Phys.* **53**, 2372-2381 (1970). Johari, G.P *Annals New York Acad.Sci.* **279**, 117-130 (1976). G. P. Johari, J. Non-Cryst. Solids, **307-310**, 317-325 (2002).
2. Vogel, et al. *J. Non-Cryst. Solids* **307-310**, 326-335 (2002).
3. Richert, R. *Europhys.Lett.* **54**, 767 (2001).
4. (a) Nozaki R, et al. *J.Non-Crys.Solids* **235-237**, 393 (1998). (b) Pisignano, D. et al., *J.Phys.:Condens.Matter* **13**, 4405-4414 (2001). (c) Olsen N.B., *J.Non-Cryst.Solids* **235-237**, 399-407 (1998). (d) Casalini, R., and Roland, C.M., *Phys.Rev.Lett.* **91**, 015702 (2003). (e) Kahle, S. et al. *Macromolecules* **30**, 7214-7223 (1997).
5. Rault, J., *J. Non-Cryst. Solids* **271**, 177-193 (2000).
6. Ngai, K.L. and Paluch, M. *J.Chem.Phys.* in press.
7. (a) Ngai, K.L. *J.Chem.Phys.* **109**, 6982 (1998). (b) Ngai, K.L., Lunkenheimer, P., León, C., Schneider, U., Brand, R., and Loidl, A. *J.Chem.Phys.***115**, 1405-1418 (2001).
8. (a) Ngai, K.L. *J.Phys.:Condens.Matter* **15**, S1107-1120 (2003). (b) Ngai, K.L., and Paluch, M. *J.Phys.Chem.B* **107**, 6865-6871 (2003).
9. Johari, G.P., Power, G., and Vij, J.K. *J.Chem.Phys.* **116**, 5908-5911 (2002); **117**, 1714-1719 (2002).
10. Paluch, M., Roland, C.M., Pawlus, S., Zio_o, J., and Ngai, K.L. Phys.Rev.Lett. 91, 115701 (2003).
11. Prevosto, D., Capaccioli, S., Lucchesi, M., and Rolla, P.A., to be published.
12. Ngai, K.L. Comment Solid State Phys. 9, 127-140 (1979); 9, 141-149 (1979).
13. Tsang K.Y. and Ngai, K.L. Phys. Rev. E 54, 3067-3071. (1997). Ngai, K.L. and Tsang, K.Y. *Phys.Rev.E* **60**, 4511-4517 (1999).
14. Ngai, K.L. *IEEE Transactions in Dielectrics and Electrical Insulation* **8**, 329-347 (2001).
15. Ngai, K.L. and Rendell, R.W. "Coupling model explanation of salient dynamic properties of glass-forming substances", in *Supercooled Liquids, Advances and Novel Applications*, edited by J.T. Fourkas, et al. ACS Symposium Series Vol. **676**, Washington, DC.: Am.Chem.Soc., 1997, pp. 45-63.
16. Döß, A., Paluch, M., Sillescu, H. and Hinze, G. *J.Chem.Phys.* **117**, 6582-6589 (2002).
17. Mattsson, J., Bergman, R., Jacobsson, P., and Börjesson, L. *Phys.Rev.Lett.* **90**, 0757021-0757024 (2003).
18. Ngai, K.L., and Capaccioli, S. *Phys.Rev.E* (submitted).
19. Kudlik, A., Tschirwitz, C., Benkhof, S., Blochowicz T., and E. Rössler, *Europhys.Lett.* **40**, 649-654 (1997).
20. Duvvuri, K. and Richert, R. *J.Chem.Phys.* **118**, 1356-1363 (2003).
21. Only for a homogenous liquid comprised of identical molecules. In mixtures, concentration fluctuations give rise to a distribution of coupling parameters.
22. Capaccioli, S., and Ngai, K.L., to be submitted.
23. Ngai, K.L., Plazek, D.J., and Rendell, R.W. *Rheol. Acta*, **36**, 307-319 (1997).
24. Ngai, K.L., and León, C., *J.Non-Cryst.Solids* **315**, 124-133 (2003).
25. Ngai, K.L. *J.Non-Cryst.Solids* **275**, 7-31 (2000).
26. Arbe, A., Colmenero, J., Richter, D., Gomez, J., and Farago, B. *Phys.Rev.Lett.* **60**, 1103-1107 (1999).
27. Segre, P.N., and Pusey, P.N. *Phys.Rev.Letters*, **77**, 771-775 (1996).
28. Hofmann, A., Kremer, F., Fischer, E.W., and Schönhals, A. in *Disorder Effects on Relaxational Processes*, edited by R. Richert and A. Blumen, Springer, Berlin, 1994, p.309-325.
29. León, C., and Ngai, K.L. *J.Phys.Chem.B*, **103**, 4045-4053 (1999).
30. Kudlik, A., Benkhof, S., Blochowicz, T., Tschirwitz, C., and Rössler, E. *Mol.Struture.* **479**, 210-225 (1999).
31. Ngai, K.L. and León, C. *Phys.Rev.B*, **66**, 0643081-0643088 (2002).
32. Sokolov, A.P., Kisluik, A., Novikov, V.N. and Ngai K.L. *Phys. Rev. B* **63**, 172204 (2001).
33. Caliskan, G., Kisluik, A., Sokolov, A.P., and Novikov, V.N. *J.Chem.Phys.*, **114**, 10189-11195 (2001).
34. Cang, H., Novikov, V.N., and Fayer, M.D. *J.Chem.Phys.* **118**, 2800-2807 (2003).
35. Götze, W., *J.Phys.:Condens.Matter* **11**, A1-A45 (1999).
36. Ngai, K.L., and Casalini, R. *Phys.Rev.B* **66**, 1322051-1322054 (2002).
37. Habasaki, J., Ngai, K.L., and Y. Hiwatari, *Phys.Rev.E* **66**, 0212051-0212059 (2002).
38. Fabbian, L., Götze, W., Sciortino, F., and Tartaglia, P. *Phys.Rev.E*, **60**, 2430-2442 (2002).
39. Frick, B., Alba-Simionesco, C., Anderson, K.H., and Willner, L. *Phys.Rev.E* **67**, 0518011-0518017 (2003).
40. Patkowski, A., Matos Lopes, M., and Fischer, E.W. *J.Chem.Phys.* **119**, 1579-1585 (2003).
41. (a) Schroder, M., Ngai, K.L., and Roland, C.M., to be submitted (2003).(b) Sokolov, A.P. et al. to be published.
42. Zorn, R, Richter, D. Frick, B. and Farago, B. *Physica A* **201**, 52-59 (1993).
43. Habasaki, J., Hiwatari, Y., and Ngai, K.L. to be published (2003).
44. Casalini, R., and Ngai, K.L. *J.Non-Cryst.Solids* **293**, 318-327 (2001).
45. Buchenau, U. and Zorn, R. *Europhys.Lett.*, **18**, 523-527 (1992).
46. (a) Kanaya, T., Tsukushi, T., Kaji, K., Bartos, J. and Kristiak, J., *Phys.Rev.E* **60**, 1906-1915 (1999). (b) Bartos, J., Kristiak, J. and Kanaya, T. *Physica B*, **234-236**, 435-438 (1997).
47. Ngai, K.L., Bao, L.-R., Yee, A.F. and Soles, C.L., *Phys.Rev.Lett.*, **87**, 2159011-2159014 (2001).
48. Duval E., Mermet, A., Surovtsev, N., Jal J.F., and Dianoux A.J. *Phil.Mag.B*, **77**, 457-462 (1997); *J.Non-Cryst.Solids* **235-237**, 203-208 (1998).
49. Doolittle, A.K. *J.Appl.Phys.* **22**, 1031-1039 (1951).
50. Li, G, Du, W.M., Chen, X.K., Cummins, H.Z., and Tao, N.J. *Phys.Rev.A* **45**, 3867-3878 (1992).
51. Lunkenheimer, P., Pimenov, A., and Loidl, A. *Phys.Rev.Lett.* **78**, 2995-2999 (1997). Lunkenheimer, P. *Dielectric Spectroscopy of Glassy Dynamics*, Shaker, Aachen, 1999, pp.1-188.

Johari-Goldstein Relaxations during Physical Aging of Propylene Glycol Oligomers under High Pressure

C.M. Roland and R. Casalini

Naval Research Laboratory, Chemistry Division, Code 6120, Washington, DC 20375-5342 USA

Abstract. Dielectric loss spectra at elevated pressure of the dimer and trimer of propylene glycol reveal the existence of an excess wing, which may evolve into a distinct peak upon physical aging. This relaxation process occurs simultaneously with the higher frequency secondary relaxation observed at low pressure. From the properties of the excess wing, we conclude that it has an intermolecular origin; that is, for these liquids the excess wing is a Johari-Goldstein relaxation, which serves as a precursor to the α-relaxation.

INTRODUCTION

The rich phenomenology of the glass transition remains to be completely understood, with various aspects not reconcilable within a single theoretical framework. Among the processes transpiring near the glass transition temperature, T_g, especially intriguing are the secondary β-relaxations. These fall at higher frequencies and usually have weaker amplitudes than the primary α-relaxation. When the molecular motion underlying the secondary relaxation involves only intermolecular degrees of freedom, it is referred to as a Johari-Goldstein (JG) process.[1] The JG process is related to the highly cooperative α-relaxation, with the details of the relationship the focus of much current research.[2,3,4,5,6,7,8,9,10,11]

The properties of the JG relaxation (Arrhenius behavior below T_g, with an extrapolated merging[12] with the α-relaxation at T ~ 1.2 to 1.5 × T_g, and a dielectric strength that increases with temperature, showing a change in slope at T_g[13]) appear to be universal, although in some supercooled liquids, a distinct β-peak is absent from the dielectric loss spectrum. Instead, an excess intensity (excess wing, "EW") is observed, toward the high frequency side of the α-relaxation peak. Various experimental evidences suggest that the EW is a distinct process, submerged by the close-lying α-peak. For example, materials having very similar α relaxations may not have the same EW.[5] The α-peak and EW can also be separated by physical aging, as shown for propylene carbonate and glycerol[2], and by high pressure, as observed for the xyliltol[10] and propylene glycol trimer.[14]

We describe herein dielectric measurements on the dimer (DPG) and trimer (TPG) of propylene glycol, at elevated pressure. These liquids exhibit a secondary relaxation, commonly regarded as a JG process. Previously, we reported ambient pressure dielectric measurements of the α-relaxation and this secondary process, for polypropylene glycols having molecular weights varying from the monomer to 4000 g/mol.[15] As shown herein, however, under high pressure, an EW arises in both DPG and TPG, which may evolve into a distinct peak during the course of physical aging under high pressure. From these results, we conclude that the secondary peaks observed[15] at ambient pressure in PG oligomers are not JG processes. Rather, this designation belongs to the relaxation (EW or distinct peak) occurring at frequencies much closer to the α-peak. Only by slowing the α-process down, using pressure and physical aging, can the JG-process be resolved from the close-lying, intense α-peak.

This conclusion can be drawn from recently published dielectric measurements on TPG at elevated pressure.[14] We extend that work herein with further measurements, which show clearly the effect of physical aging on the prominence of the JG relaxation. We also present new results on DPG, which exhibits similar behavior, although the closer proximity to the α-peak makes resolution of the JG more difficult.

Propylene glycol materials are especially interesting, because the hydrogen bonds are affected by both temperature and pressure. In the absence of H-bonding, the EW has an activation volume much closer to that of the α-relaxation,[4] and therefore the two processes are more difficult to resolve. The degree of H-bonding varies inversely with the number of PG units in the molecule, since only the terminal hydroxyl groups form H-bonds. This results in an increasing contribution of volume to the temperature-dependence of the α-relaxation with increasing molecular weight, although thermal energy remains the more dominant control variable.[16] This is generally the case for associated glass-formers, but not for van der Waals liquids.[17]

EXPERIMENTAL

DPG and TPG, having the formula H-$(C_3H_6O)_n$-OH where $n = 2$ and 3 respectively, were obtained from Aldrich Chemical Co. The liquids were placed over molecular sieves and, immediately prior to measurements, dried for one hour at 125C in a nitrogen atmosphere. Dielectric spectra were obtained with a parallel plate geometry using an IMASS time domain dielectric analyzer (10^{-4} to 10^3 Hz) and a Novocontrol Alpha Analyzer (10^{-2} to 10^6 Hz). For measurements at elevated pressure, the sample was contained in a Manganin cell (Harwood Engineering), with pressure applied using a hydraulic pump (Enerpac) in combination with a pressure intensifier (Harwood Engineering). After equilibration, constant pressures could be retained for up to 6 days, with a pressure loss of less than 1 %. Pressures were measured with a Sensotec tensometric transducer.

RESULTS

Propylene glycol dimer

Displayed in Fig. 1 is the dielectric loss for DPG. At low pressure, only the dc conductivity and the α-relaxation peak are observed. However, with increasing pressure, the α-peak moves toward lower frequencies, revealing more clearly the excess wing. The EW is obscured at lower pressures by the secondary relaxation at *ca.* 10^4 Hz. This peak's position is essentially insensitive to pressure. Since the JG process involves virtually the entire molecule, the slowest secondary relaxation must be the JG. For this reason we identify the EW in Fig. 1 as the JG relaxation. With the available range of T and P, we cannot separate the EW from the overlapping α-peak.

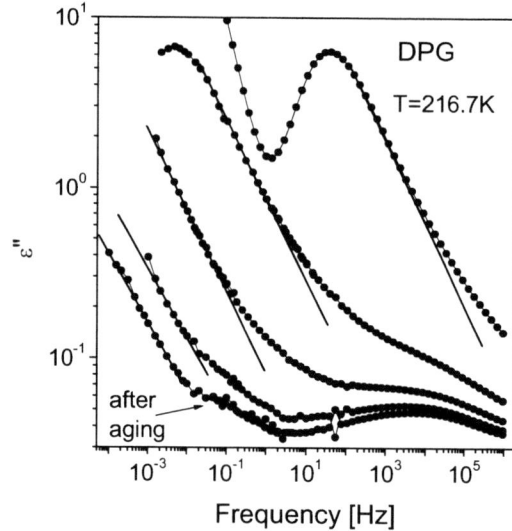

FIGURE 1. Dielectric loss for PPG dimer at pressures (from right to left) of 67.6, 248.7, 335.7, 520 and 510 MPa. The latter was measured after 12 hours aging. There is a pressure-independent secondary peak at ~10^4 Hz, which exhibits a negligible response to aging.

Propylene glycol trimer

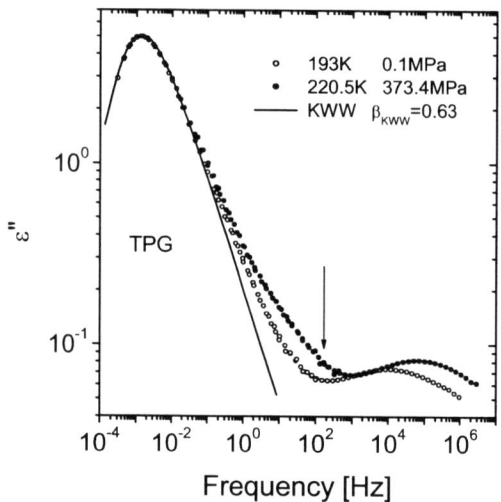

FIGURE 2. Dielectric loss spectra for TPG at atmospheric and high pressure. The data at high pressure where shifted on the abscissa by a factor 2.2 toward higher frequency, with ε" divided by 0.78, to superimpose the maxima. The arrow represents the location of the JG peak calculated by eq. 1.

524

In Fig. 2 the dielectric loss is shown for TPG at 0.1 MPa (193K) and 373.4 MPa (220.5K). These conditions are such that the τ_α are almost the same.

After scaling the high pressure spectra to superimpose the α-peaks, the β-peak at high pressure is seen to be further separated from the α-relaxation, and the EW has become more prominent. We fit the primary peak to the KWW function, which yields β_{KWW}=0.63. From this value of the stretch exponent, we can calculate the JG relaxation time, using[15,18]

$$\tau_{JG} = t_c^{1-\beta_{KWW}} \tau_\alpha^{\beta_{KWW}} \quad (1)$$

where t_c is an temperature and pressure insensitive constant equal to 2 ps. The calculated peak frequency, = $1/2\pi\tau_{JG}$ is denoted by the arrow in Fig. 2. It is at least two decades slower than the β-peak, affirming the idea that the latter is not a JG relaxation.

In Fig. 3 the dielectric loss spectrum for TPG is shown for various pressures at fixed temperature. The systematic slowing down of the α-relaxation reveals the frequency-invariant secondary peak at *ca.* 20 kHz, along with a prominent excess wing.

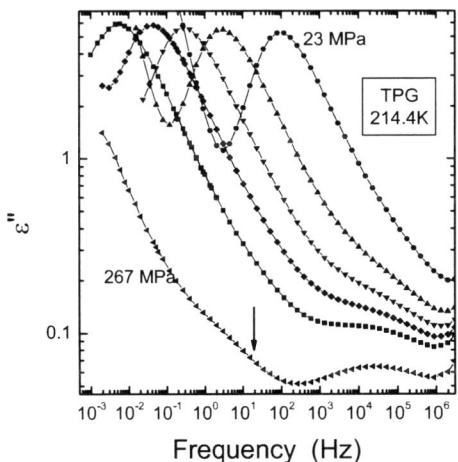

FIGURE 3. Dielectric loss spectra for TPG at increasing pressures (23, 51, 92, 130, 160, 190 and 267 MPa). All data except for the latter are for equilibrium.

At the highest pressure in Fig. 3, τ_α has reached a sufficiently large value that it falls out of the experimentally accessible range. However, we can estimate it by determining the shift factor required to superimpose the high frequency side of the -peak (note that there is negligible change in the shape of the α-peak with pressure); this yields $\tau_\alpha = 3.6\times10^3$ s for P = 267 MPa. Since equilibrium has not necessarily been attained, this represents a lower bound on the α-

relaxation time. Using this in eq. 1, we calculate τ_{JG} = 8.2 ms, which corresponds to the frequency indicated by the arrow in Fig. 3. Note that at the highest pressure, the EW is developing a shoulder; moreover, the calculated τ_{JG} is in the vicinity of the EW, not the peak at higher frequency. The conclusion is that this EW or nascent shoulder, observed only at high pressure, is the JG relaxation, not the higher frequency peak.

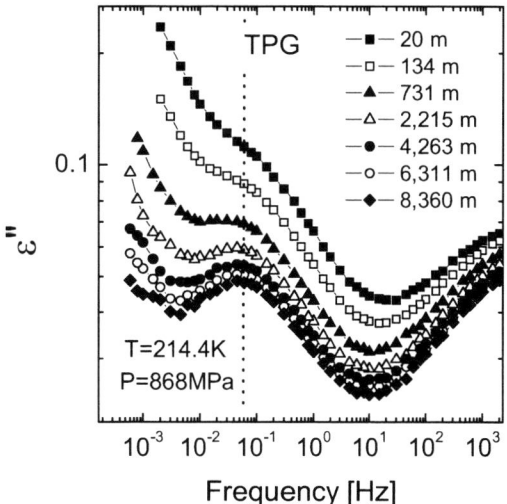

FIGURE 4. Dielectric loss for PPG trimer at the indicated time periods (in minutes) after application of 868 MPa pressure at 214.3K. The longest aging time almost 6 days.

In Fig. 4 are the dielectric loss spectra of TPG at fixed values of T = 214.3K and P = 868 MPa. The large pressure causes a substantial shift of the α-peak, beyond our experimental window. With T and P held constant, the system moves toward equilibrium, giving rise to further slowing down of τ_α with time. This physical aging enables a distinct peak to be clearly observed, at the frequency associated with the EW at shorter annealing times. This new peak, which is completely distinct from the higher frequency relaxation at > 1 kHz, is the JG relaxation emerging at lower pressure in Fig. 3.

SUMMARY

The dielectric spectra presented herein show that the EW in propylene glycol dimer and trimer is actually a distinct peak, submerged at low pressures by the α−process. This EW is completely separated from the secondary relaxation observed at higher frequencies. There are various reasons to designate this EW as a JG process. Since motion of the entire molecule is responsible for the JG relaxation (it involves intermolecular barriers), τ_{JG} is expected to be longer than all other secondary relaxation processes. Moreover, the frequency of the EW in both DPG and TPG value is consistent with eq. 1. We also note that the EW in the propylene glycol oligomers has been shown to exhibit an activation volume that is of the same order of magnitude as that of the α-relaxation, but much larger than the activation volume smaller of the higher frequency peak.[20] Such a result is consistent with an intermolecular origin.

Since the terminal hydroxyls in PPG can form hydrogen bonds, the extent number of H-bonding varies inversely with molecular weight. More H-bonds gives rise to a weaker sensitivity of the α-relaxation time to pressure.[19] This is seen in the pressure coefficient of T_g, which equals (in units of K/GPa) 37 for propylene glycol,[16] 80 for the dimer,[16] and 109 for the trimer[20]. The displacement towards lower frequency of the α–peak with pressure is expected to allow resolution of the JG-process, which has a somewhat weaker pressure sensitivity. This is reflected in the results in Figs. 2 - 4. For the more strongly H-bonded dimer, we cannot resolve a distinct JG peak, whereas the stronger effect of pressure on τ_α for the trimer enables its JG process to be uncovered.

The evolution of the EW in TPG into a distinct peak during physical aging at elevated pressure is similar to results from physical aging at ambient pressure of propylene carbonate and glycerol.[2] Likewise, the EW in sorbitol can be transformed into a peak using pressure (or lower temperature). On the other hand, non-associated, van der Waals liquids appear to behave differently. Under pressure, the shape of the α-peak and EW remains invariant, when comparisons are made for conditions of constant. τ_α.[4] A better understanding of these differences between the EW and JG properties of various glass-formers remains an important objective towards obtaining a complete understanding of the vitrification process.

ACKNOWLEDGMENTS

This work was supported by the Office of Naval Research. Stimulating discussions with M. Paluch and K.L. Ngai are gratefully acknowledged.

REFERENCES

1. Johari, G.P., and Goldstein.M, *J. Chem. Phys.* **53**, 2372-2388 (1970); Goldstein, M. *J. Chem. Phys.* **51**, 3728-3735 (1969).

2. Schneider, U., Brand, R., Lunkenheimer, P., Loidl, A., *Phys. Rev. Lett.* **84**, 5560-5563 (2000).

3. Paluch, M., Casalini, R., Hensel-Bielowka, S., Roland, C.M., *J. Chemi. Phys.* **116**, 9839-9844 (2002).

4. Roland, C.M., Casalini, R., Paluch, M., *Chem. Phys. Lett.* **367**, 259-264 (2003).

5. Casalini, R, Roland, C.M. *Phys. Rev. B* **66**, Art. No. 180201 (2002).

6. Brand, R., Lunkenheimer, P., Loidl, A. *J.Chem.Phys.* **116**, 10386-10401 (2002).

7. Hansen, C., Stickel, F., Berger, T., Richert, R., Fischer, E.W., *J.Chem.Phys.* **107**, 1086-1093 (1997).

8. Hensel-Bielowka, S., Paluch, M., Ziolo, J., Roland, C.M. *J.Phys.Chem.B* **106**, 12459-12463 (2002).

9. Köplinger, J., Kasper, G., Hunklinger, S., *J.Chem.Phys.* **113**, 4701-4706 (2000).

10. Paluch, M., Casalini, R., Hensel-Bielowka, S., Roland, C.M., *J. Chem. Phys.* **116**, 9839-9844 (2002).

11. Hensel-Bielowka, S., Paluch, M., *Phys. Rev. Lett.* **89**, Art. No. 25704 (2002).

12. Such an extrapolation is incorrect, as shown in Paluch, M., Roland, C.M., Pawlus, S., Zioło, J., Ngai, K.L, *Phys. Rev. Lett.* **91**, Art. No. 115701 (2003).

13. Johari, G.P., Powers, G. and Vij, J.K. *J. Chem. Phys.* **116**, 5908-5909 (2002).

14. Casalini, R., Roland, C.M., *Phys. Rev. Lett.* **91**, Art. No. 15702 (2003).

15. León, C.; Ngai, K.L., Roland, C.M. *J. Chem. Phys.* **110**, 11585-11591 (1999).

16. Casalini, R., Roland, C.M., *J. Chem. Phys.*, in press.

17. Roland, C.M., Casalini, R., *Macromolecules* **36**, 1361-1367 (2003).

18. Ngai, K.L., *J. Phys. Cond. Mat.*, **15**, S1107-S1125 (2003).

19. Roland, C.M., Casalini, R., Paluch, M. *Chem. Phys. Lett.* **367**, 259-264 (2003)

20. Casalini, R., Roland, C.M. *Phys. Rev. B*, in press.

Non-Debye Dielectric Response and non-Arrhenius Kinetics in Complex Systems at Mesoscale

Yu. Feldman[*], A. Puzenko[*], Ya. Ryabov[**] and A. Gutina[*]

[*]Department of Applied Physics, The Hebrew University of Jerusalem, Givat-Ram, 91904, Jerusalem, Israel
[**]Department of Chemistry & Biochemistry, University of Maryland, 1109 Agriculture/Life Sciences Surge Building College Park, MD 20742-3360, USA

Abstract. The paper considers several examples of non-Debye dielectric response in complex heterogeneous media. The percolation phenomenon and Cole-Cole relaxation in disordered matter are discussed in detail. The models enable us to establish the relationship between the parameters of dielectric relaxation broadening, structural properties of the media and transport features of charge carriers in the considered systems. In addition, the origins of "strange kinetic" phenomena and the specific features of relaxation kinetics in systems with different kinds of confinements are discussed in the paper. In contrast to the usual Arrhenius or Vogel-Fulcher-Tammann patterns, a quite unusual non-monotonic dependence of relaxation time versus temperature is observed in such systems. Based on the free volume concept, a model for this type of kinetics was illustrated by several particular examples: water confined in porous glasses and doped ferroelectric crystal.

INTRODUCTION

The general interest of this contribution is centered on the area of "soft" condensed matter science for investigation of the structure, dynamics and macroscopic behavior of complex systems. Complex systems (CS) are a very broad and general class of materials that are typically non-crystalline. Glass forming liquids, polymers, biopolymers, colloid systems (emulsions and microemulsions), cells, porous materials and liquid crystals can be considered as complex systems. All of these materials involve a common feature such as the appearance of a new ("mesoscopic") length scale, intermediate between molecular and macroscopic. A simple exponential relaxation law and the classical model of Brownian diffusion cannot describe the relaxation phenomena and kinetics in such materials. This kind of non-exponential relaxation behavior and anomalous diffusion phenomena is today called "strange kinetics" [1, 2].

Generally, the complete characterization of these relaxation behaviours requires the use of a variety of techniques in order to span the relevant ranges in frequency. In this approach, Dielectric Spectroscopy (DS) has its own advantages. Modern DS techniques may overlap extremely wide frequency (10^{-6} to 10^{11} Hz) and temperature (- 170 °C to +500 °C) ranges. DS is especially sensitive to intermolecular interactions and is able to monitor cooperative processes at the molecular level. Therefore, this method is more appropriate than any other to monitor such different scales of molecular motion. It provides a link between the investigation - via molecular spectroscopy - of the properties of the individual constituents of the complex material and the characterization of its bulk properties.

In most cases of non-Debye dielectric spectrums the data have been described by the so-called Havriliak-Negami (HN) relationship

$$\varepsilon^*(\omega) = \varepsilon_\infty + \frac{\varepsilon_s - \varepsilon_\infty}{\left[1 + (i\omega\tau_m)^\alpha\right]^\beta}, \quad 0 < \alpha, \beta \le 1. \quad (1)$$

Here α and β are empirical exponents. The specific case $\alpha=1$, $\beta=1$ gives the Debye relaxation law, $\beta=1$, $\alpha\ne1$ corresponds to the so-called Cole-Cole (CC) equation, whereas the case $\alpha=1$, $\beta\ne1$ corresponds to the Cole-Davidson (CD) formula.

An example of a phenomenological decay function that has different short- and long-time asymptotic forms (with the different characteristic times) can be presented as follows [1, 2]

$$\phi(t) = A\left(\frac{t}{\tau_1}\right)^{-\mu} \exp\left\{-\left(\frac{t}{\tau_m}\right)^{\nu}\right\}, \quad (2)$$

where A is an amplitude, exponent $\mu > 0$ and a characteristic time τ_1 is associated with the effective relaxation time of the microscopic structural unit, τ_m is a characteristic relaxation time and empirical exponent $0 < \nu \leq 1$. This function is the product of KWW and power-law dependencies. The relaxation law (2) in time domain and the HN law (1) in frequency domain are rather generalized representations that lead to the known dielectric relaxation laws. The fact that these functions have the power-law asymptotic has inspired numerous attempts to establish a relationship between their various parameters. In this regard, the exact relationship between the parameters of Eq. (2) and the HN law should be a consequence of the Laplace transform. However, there is currently no concrete proof that this is indeed so. Thus, the relationship between the parameters and the HN law seems to be valid only asymptotically.

In this work, we will review several particular examples of DS application to CS study. The percolation phenomenon and Cole-Cole relaxation in disordered matter will be discussed in details. The proposed models were discussed in relation with several examples of different nature [3-10] and in this paper we will review some of these examples in order to show the general ideas underlying these models.

To demonstrate versatility of DS technique we will also discuss another feature inherent to some CS i.e. non-Arrhenius relaxation kinetics [11], which for a wide variety of systems can be treated in the framework of the model

$$\ln\left(\frac{\tau}{\tau_0}\right) = \frac{E_a}{k_B T} + C \exp\left(-\frac{E_b}{k_B T}\right), \quad (3)$$

where τ relaxation time, T is absolute temperature, E_a is activation energy of relaxation process, E_b is the energy required to make an inert particle be able to participate in relaxation, k_B is Boltzmann constant and C is dimensionless factor reflecting free volume in the system under consideration [11].

SYMMETRIC DIELECTRIC SPECTRUM BROADENING IN DISORDERED MATERIALS

In this section we would like to review several particular examples illustrating the model [12] that establishes the relationship between dynamic and structural parameters of empirical Cole-Cole expressions in the form

$$\alpha = \frac{d_G}{2} \frac{\ln(\tau \omega_s)}{\ln(\tau/\tau_0)}, \quad (4)$$

where $\omega_s = 2d_E G^{2/d_G} D / R_0^2$ is the characteristic frequency of the diffusion process. This equation establishes the relationship between the CC exponent α, the relaxation time τ, the geometrical properties (fractal dimension d_G), and the diffusion coefficient (through ω_s).

The first mention of the $\alpha(\tau)$ dependencies was in the experimental work [13]. The dielectric relaxation data of the water in the mixtures of seven water-soluble polymers was presented there. It was found that in all these solutions relaxation of water obeys the CC law, while the bulk water exhibits the well-known Debye-like pattern [14, 15]. Another observation was that α is dependent not only on the concentration of solute but also on the hydrophilic (or hydrophobic) properties of the polymer [13].

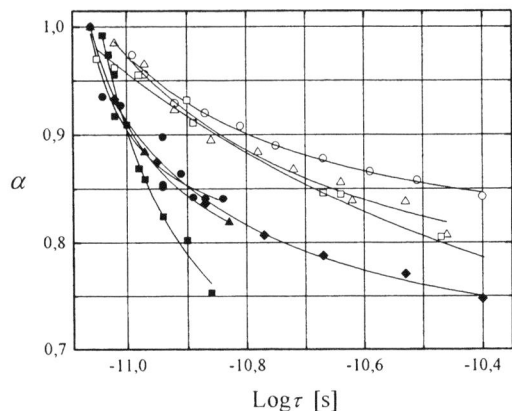

FIGURE 1. Cole-Cole exponent α versus relaxation time τ for PVA (●), PAlA (▲), PAA (■), PEI (◆), PEG (○), PVME (△) and PVP (□) samples. The curves correspond to the model (4). The full symbols correspond to the hydrophilic polymers and the open symbols correspond to the hydrophobic samples. (Reproduced from [12]. Copyright 2002 American Institute of Physics).

In Fig. 1 one can see an example of such dependencies for seven water soluble polymers studied in [12] together with the theoretical curves that obey Eq. (4). In this case the temperature of all samples was kept constant and $\alpha(\tau)$ dependencies were obtained for different compositions of polymer-water mixtures.

In the framework of the model (4) the microscopic relaxation time of water molecule is equal to the cutoff time of the scaling in time domain τ_0. The results of paper [12] suggest that for hydrophilic polymers the strong interaction between the polymer and the water molecule results in larger values of the microscopic relaxation time τ_0, while for hydrophobic polymers this time constant became shorter. This fact signifies that weakening the hydrophilic properties (or intensifying the hydrophobic properties) results in a decreasing of interaction between the water and the polymer and consequently in the decrease of τ_0.

According to the model (4) the interaction between the water and the polymer occurs in the vicinity of the polymer chains [4]. Only the water molecules situated in this interface are affected by the interaction. The space fractal dimension d_G in this case is the dimension of the macromolecule chain. If a polymer chain is stretched as a line, then its dimension is 1. In any other conformation, the wrinkled polymer chain has a larger space fractal dimension, which falls into the interval $1 < d_G < 2$. Thus, it is possible to argue that the value of the fractal dimension is a measure of polymer chain meandering. Straighter (probably more rigid) polymer chains have d_G values close to 1. More wrinkled polymer (probably more flexible) chains have d_G values close to 2 [12].

Another example of similar behavior is a micro-composite polymer material (nylon 66 reinforced with Kevlar fibers) [10]. Fig. 2 presents $\alpha(\tau)$ dependencies for quenched nylon (*QN*), crystalline nylon (*CN*) and microcomposite crystalline nylon (*MCN*) samples, together with the fitting curves. In contrast to the previous example the composition of samples was kept constant and $\alpha(\tau)$ dependencies were obtained for different sample temperature. Nevertheless, even in this case model (4) enables us to extract information about the nylon 66 chain structure and dynamics, as an example the characteristic fractal dimension d_G, the characteristic time scales τ_0 and $1/\omega_s$ [10, 16, 17].

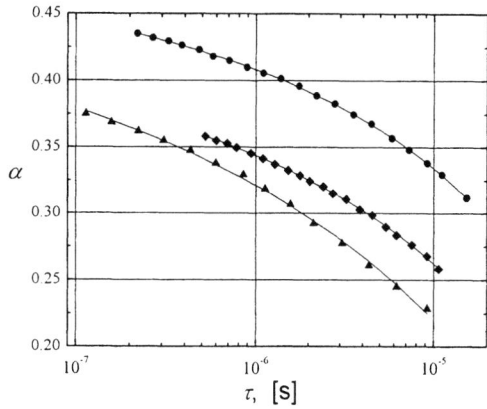

FIGURE 2. The dependence of α versus relaxation time τ for QN (●), CN (▲) and MCN (◆) samples. The curves correspond to the model (4). (Reproduced with permission from [2]. Copyright 2002 Elsevier Science B.V.)

NON-DEBYE RELAXATION CAUSED BY PERCOLATION

Non-Debye dielectric relaxation in porous silica glasses is another example of the dynamic properties of complex systems on a mesoscale. The porous silica glasses obtained from sodium borosilicate glasses are defined as bicontinuous random structures of two interpenetrating percolating phases, the solid and the pore networks. The pores in the glasses are connected to each other and the pore size distribution is narrow. The characteristic pore spacing depends on the method of preparation, and can be between 2 and 500 nm [18]. A rigid SiO_2 matrix represents the irregular structure of porous glasses. Water can be easily adsorbed on the surface of this matrix. The dielectric response is found to be very sensitive to the geometrical nano- and mesostructural features of the porous media and water molecules in the adsorptive layer on the pore.

The dielectric relaxation properties of silica glasses over broad frequency and temperature ranges have been investigated recently [5, 6, 19, 20]. The complex dielectric behavior of them can be described in terms of several distributed relaxation processes [6, 19, 20].

However, in this section we are going to discuss only the process that is caused by the percolation of charge carries trough the open-pore network of such samples (see Fig. 3). The movement of charge carriers results in a transfer of the electric excitation within the channels along random paths.

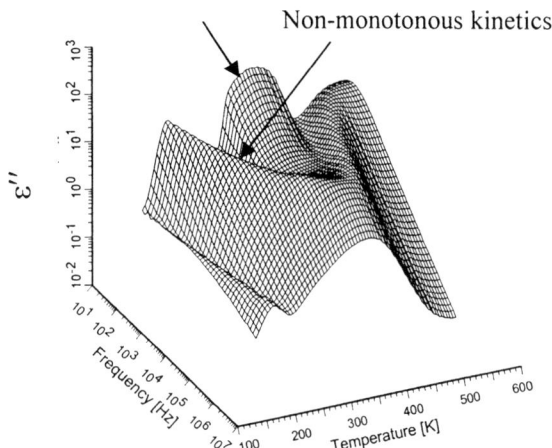

FIGURE 3. The typical three-dimensional plot of the complex dielectric permittivity real imaginary part ε" versus frequency and temperature for a porous glass sample. (Reproduced with permission from [2]. Copyright 2002 Elsevier Science B.V.)

A detailed description of the relaxation mechanism associated with an excitation transfer based on regular and statistical fractal models is introduced in [2-4], where it was applied to the cooperative relaxation of ionic microemulsions at percolation.

The dielectric relaxation at percolation in porous glasses is analyzed in terms of the dipole correlation function DCF $\Psi(t)$. For static percolation that takes place in porous glasses the correlation function $\Psi(t)$ may be written in the asymptotic form (2) with $\mu \to 0$ and where the exponent ν, at the percolation threshold, is related to the fractal dimension of pore network surface D_f [4] as follows:

$$\nu = D_f / 3 \qquad (5)$$

The averaged porosity $\langle \Phi_p \rangle$ of the porous glass can be calculated by the simple formula [5]:

$$\langle \Phi_p \rangle \approx \frac{1}{4 - D_f} \qquad (6)$$

The results of the porosity calculation together with the fractal dimension determined from dielectric measurements are in good agreement with the porosities values obtained by the relative decrement values method [20] for the samples with comparatively large pore sizes. For samples with ultra small porous structures the porosity calculated from approximation (6) is bigger than porosity values obtained by other methods. This result reflects the fact that the percolation model expressed by Eq. (2) takes into account only the open network structure where charge carriers can penetrate through the macroscopically relevant volumes. Thus, in the case of a net of super small pores, the dielectric response is more sensitive and accurate in the determination of real open porosity than any other conventional method.

Similar ideas were used for the treatment of percolation processes in other porous materials like porous silicon, where model (6) also enables us to calculate samples porosity [7].

NON-ARRHENIUS KINETICS IN COMPLEX SYSTEMS

The non-monotonous relaxation kinetics (3) is also one of the features of complex systems [11]. For example this kind of relaxation kinetics is inherent to dielectric relaxation of water absorbed on inner surface of porous glasses (see Figs. 3 and 4).

The key heuristic idea of model (3) is a confinement [11]. Impact of such confinement is regulated by the constant $C = v_0 n_0 / V$ in Eq. (3), where v_0 is a volume of single particular relaxing unit, n_0 their concentration and V is the total volume of the system. Thus, if the total volume of the system V is sufficiently large and the maximum possible concentration of relaxing particles is sufficiently small $n_0/V \ll 1/v_0$, then the free volume arguments become irrelevant and relaxation kinetics obtains an Arrhenius form. However, in the case of a constraint, when the volume of a system is small and $n_0/V \approx 1/v_0$, an increase of temperature leads to the significant decrease of free volume and slows down the relaxation.

As we are going to show further this situation usually occurs for "small" systems where relaxing particles become able to participate in relaxation due to the formation of some "defects" in the ordered structure. In this case n_0 could be regarded as the maximum possible defect concentration. Therefore, confinement provides comparatively large concentration of defects n_0/V since for such a system the confining geometry affects comparatively larger amount of system constituents.

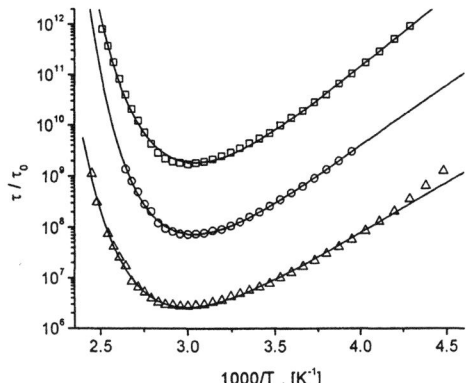

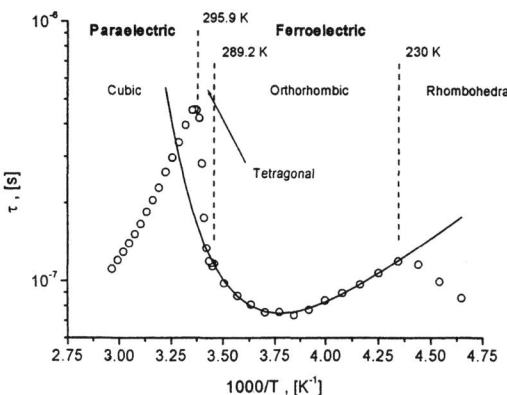

FIGURE 4. Temperature dependency of the dielectric relaxation time of water confined in porous glasses. Symbols represent experimental data. Full lines correspond to the best fit according to Eq. (3). (Reproduced with permission from [11]. Copyright 2004 The American Physical Society).

FIGURE 5. Temperature dependency of the dielectric relaxation time for the process attributed to the Cu doping in KTN. Open cycles represent experimental data from [11, 21]. The full line corresponds to the best fit according to Eq. (3) he dashed lines mark transitions between the phases with rhombohedral, orthorhombic, tetragonal and cubic unit cells. (Reproduced with permission from [11]. Copyright 2004 The American Physical Society).

Just such a mechanism for confined water relaxation was proposed recently [6], where E_a can be regarded to the activation energy of reorientation of a water molecule and E_b as the defect formation energy.

In the case of dielectric relaxation presented in Figs. 3 and 4 such a relation units are water molecules Thus, the reorientation of a water molecule, leading to the dielectric relaxation, may occur only in the vicinity of a defect in the hydrogen bonds network structure. A similar mechanism for confined water relaxation was proposed recently [6]. In this case, E_a can be regarded as the activation energy of reorientation of a water molecule and E_b as the defect formation energy.

Similar non-monotonous relaxation kinetics was also found in the variety of other experimental situations [11]. For example, this kind of kinetics was observed for KTN ($KTa_{0.65}Nb_{0.35}O_3$) ferroelectric crystal with a perovskite structure doped, with Cu in a concentration of about one copper ion per thousand unit cells of KTN [21].

Under temperature variations, KTN undergoes three transitions between the phases with rhombohedral, orthorhombic, tetragonal and cubic unit cells as outlined in Fig. 5 [21, 22]. In the cubic phase above 295.9 K the crystal exhibits paraelectric properties. Below this temperature, the crystal demonstrates ferroelectric behavior and both subsequent transitions at 289.2 K and at 230 K are transitions between the different unit cells in the ferroelectric phase.

It was observed [21, 23] that the copper doping in KTN generates a specific relaxation process that does not exist in non-doped crystals. This process demonstrates non-monotonous temperature dependence as presented in Fig. 5. It was assumed [21] that this process in the paraelectric phase is due to the reorientation of virtual dipoles provided by the Cu ions hopping between different states of local equilibrium in a multi-well potential created by local fields in the KTN unit cell.

It was found that the non-monotonic relaxation kinetics in the ferroelectric orthorhombic phase of KTN is phenomenon dependent on two processes. The first process is the tilt of oxygen octahedron, which changes the orientation of the elementary dipole moments. This orientation is associated with the unit cell and described by the Arrhenius term with the activation energy E_a in Eq. (3). The second process is the Cu ion jump between the several local minima positions with the activation energy E_b. For the latter process, when a Cu ion reaches a certain place in the multi-well potential, it creates a "defect" in the unit cell structure and blocks the possibility for oxygen octahedron to be tilted. Thus, the increase of such "defects" with temperature growth slows down the relaxation process, as is described by the exponential term in the right-hand side of Eq. (3). In this case, the necessary free volume needed for the oxygen octahedron tilt is provided by the difference in the ionic radii of the Cu^{+2} and K^+ ions, while the transition from paraelectric to ferroelectric phase provides necessary confinement [21].

In conclusion we would like to recapitulate the main goal of our paper: that cooperativity and confinement can be responsible for non-Debye and non-monotonous relaxation kinetics.

REFERENCES

1. Shlesinger, M.F., Zaslavsky, G.M., and Klafter J., *Nature* **363**, 31-37 (1993).
2. Feldman, Yu., Puzenko, A., and Ryabov, Ya., *Chem. Physics* **284**, 139-168 (2002).
3. Feldman, Yu., Kozlovich, N., Nir, I., and Garti, N., *Phys. Rev. E.* **51**, 478-491 (1995).
4. Feldman, Yu., Kozlovich, N., Alexandrov, Yu., Nigmatullin, R., and Ryabov, Ya., *Phys. Rev. E.* **54**, 20-28 (1996).
5. Puzenko, A., Kozlovich, N., Gutina, A., and Feldman, Yu., *Phys. Rev. B* **60**, 14348-14359 (1999).
6. Ryabov, Ya., Gutina, A., Arkhipov, V., and. Feldman, Yu., *J. Phys. Chem. B* **105**, 1845-1850 (2001).
7. Axelrod, E., Givant, A., Shappir, J., Feldman, Y., and Sa'ar, A., *Phys. Rev. B* **65**, 1654291-7 (2002).
8. Axelrod, E., Givant, A., Shappir, J., Feldman, Yu., and Sa'ar, A., *J. Non-Cryst. Solids* **305**, 235-242 (2002).
9. Ryabov, Ya., and Feldman, Yu., *Physica A* **314**, 370-378 (2002).
10. Ryabov, Ya., Nuriel, H., Marom, G., and Feldman, Yu., *J. Polym. Sci. Part B: Polym. Phys.* **41**, 217-223 (2003).
11. Ryabov, Ya., Puzenko, A., and Feldman, Yu., *Phys. Rev. B* **69**, (in press) (2004).
12. Ryabov, Ya., Shinyashki, N., Yagihara, S.,and Feldman, Yu., *J. Chem. Phys.* **116**, 8610-8615 (2002).
13. Shinyashiki, N., Yagihara, S., Arita I., and Mashimo S., *J. Phys. Chem. B* **102**, 3249-3251 (1998).
14. Hasted, J.B., "Liquid Water: Dielectric Properties", in *Water: a comprehensive treatise*, edited by F. Franks, Plenum Press, New York, 1972, pp. 255-309.
15. Kaatze, U., *J. Chem. Eng. Data* **34**, 371-374 (1989).
16. Korbakov, N., Feldman, Yu., and Marom, G., *Macromolecular Chemistry and Physics* **203**, 2267-2272 (2002).
17. Ryabov, Ya., Marom, G., and Feldman, Yu., *Journal of Thermoplastic Composite Materials*, (in press) (2003).
18. Rysiakiewicz-Pasek, E., and Marczuk, K., *J. Porous Materials* **3**, 17-22 (1996).
19. Gutina A., Axelrod, E., Puzenko, A., Rysiakiewicz-Pasek, E., Kozlovich, N., and Feldman, Yu., *J. Non-Cryst. Solids* **235**, 302-307 (1998).
20. Gutina, A., Antropova, T., Rysiakiewicz-Pasek, E., Virnik, K., and Feldman, Yu., *Microporous and Mesoporous Materials* **58**, 237-254 (2003).
21. Ben Ishai, P., Ryabov Ya., Feldman, Yu., and Agranat, A.J., *Phys. Rev. Lett,.*, (submitted) (2003).
22. Triebwasser, S., *Phys. Rev.* **114**, 63-70 (1959).
23. Bitton, G., Feldman, Yu., and Agranat, A.J., *J. Non-Cryst. Solids* **305**, 362-367 (2002).

Dynamic Crossover in Complex Systems: From a "Simple" Liquid to a Protein

Alexei Sokolov and Vladimir Novikov

Department of Polymer Science, The University of Akron, Akron, OH 44325-3909 USA

Abstract. In this contribution we present an overview of experimental data on dynamic crossover in various molecular and polymeric liquids, van-der-Waals, ionic, covalent and hydrogen bonded systems. We demonstrate that the structural relaxation time τ measured at T_D has nearly the same value, $\tau(T_D) \sim 10^{-7\pm1}$ sec, for different glass-forming systems. We speculate that decoupling of various relaxation processes observed at T_D might be the reason for failure of time-temperature superposition known for polymers. We demonstrate that the dynamic transition in biological macromolecules follows the scenario known for the dynamic crossover in glass forming liquids. Moreover, our analysis reveals that T_D observed in dynamics of proteins is close to T_D of pure solvents. We speculate that the solvent's dynamic crossover controls the dynamic transition of proteins. In other words, proteins and DNA are "slaves" of a solvent and this might open ways for a control of their biochemical activity.

INTRODUCTION

Dynamics of complex systems show a behavior general for vast variety of materials, including small molecules and polymers, van-der-Waals and ionic liquids, hydrogen and covalently bonded systems and even biological macromolecules. Dynamic structure factor $S(Q,E)$ of these systems usually exhibits collective low-frequency vibrations, the so-called boson peak (overdamped at high temperatures), fast conformational fluctuations (picosecond relaxation) and a slow relaxation process, the so-called α-relaxation (frozen at low temperatures). One of the interesting findings of the last two decades was an observation of a dynamic crossover (or dynamic arrest) in many glass forming systems [1,2] and also in proteins and DNA [3-5].

Looking back to earlier literature one finds that predictions of qualitative changes in dynamics of glass forming liquids at temperatures much above the conventional glass transition temperature T_g were proposed already in sixties [6,7]. Boyer and co-workers proposed the existence of a third order thermodynamic liquid-liquid transition in polymer melts at $T \sim 1.2 T_g$ [6]. The proposed idea has been based in part on non-correct measurements (see, for example [8]) and up to now we don't have clear experimental evidences for the thermodynamic transition. In 1969, Martin Goldstein proposed the existence of a purely dynamic crossover from liquid-like dynamics to a viscous flow driven by over-barrier relaxation in glass forming liquids at temperatures much above T_g [7]. He argues that at a particular relaxation time τ_α (in his estimates $\sim 10^{-9}$ sec) molecular motion in a liquid slows down significantly so that an over-barrier relaxation becomes a dominating mechanism of the viscous flow.

However, only developments of mode-coupling theory (MCT) of the glass transition in mid eighties [1] attract significant attention of scientific community to the problem of the dynamic crossover. Evidences of qualitative changes in dynamics at some temperature T_c much above T_g have been observed in various experimental and theoretical studies, and in computer simulations [2,9-13]. Nevertheless, the microscopic nature of the dynamic crossover remains a subject of intensive debates [14,15].

In this contribution we present an overview of experimental data collected to date on the dynamic crossover in various molecular and polymeric liquids, hydrogen and covalently bonded systems. We focus on estimates of the characteristic relaxation time τ_α at T_C that appears to be universal, $\tau_\alpha(T_c) \sim 10^{-7\pm1}$ sec [16]. Possible reasons for such a universal relaxation time that appears to be independent on system density, chemistry and intermolecular interactions remain unclear. Implication of the dynamic crossover on polymer dynamics, in particular, on break down of time-temperature superposition, is discussed in the second part of the paper. Dynamic transition in biological macromolecules (proteins and DNA) and its relationship to a dynamic crossover in solvents are discussed in the final part of the paper.

"MAGIC" RELAXATION TIME

There are a few experimental evidences that dynamics of glass forming systems experience qualitative changes at some temperature much above T_g [17]. They include among others decoupling of rotational and translational motions [9], decoupling of primary (α-) and secondary (β-) relaxation processes [18], change in temperature variations of the α-process identified through a derivative analysis (Stickel plot [2]) and in temperature behavior of the stretching exponent [19], model dependent MCT analysis of dynamic structure factor or intermediate scattering function [20]. In most cases all these changes appear at the same temperature range. This suggests their relationship to qualitative changes in the underlying dynamics that occurs in the crossover region. The crossover marks a transition from liquid like to a solid like dynamics on a molecular time and length scales. The crossover expands over a substantial temperature range that makes it difficult (and may be useless) to identify an exact value of T_c. Glycerol presents a nice example where estimates of T_c using MCT analysis scatters from ~225 K [21] up to ~300 K [22].

Table 1 presents a collection of experimental data obtained to date on estimates of T_c using MCT analysis and estimates of T_B using the so-called Stickel plot [2]. In most cases $T_B \sim T_c$ (Table 1). The ratio of T_c/T_g seems to correlate with fragility of the systems and varies from ~1.1 in extremely fragile CKN up to ~1.5-1.7 in relatively strong glass former B_2O_3 (Table 1). The most interesting point is that the relaxation time at T_c or T_B, τ_c or τ_B, appears to have rather universal value, $\tau \sim 10^{-7\pm1}$ sec, for so different materials presented in the Table 1. The list includes small molecules and polymers, van-der-Waals and ionic liquids, hydrogen and covalently bonded system, and even orientationally disordered crystal.

One significant deviation, $\tau_\alpha \sim 10^{-4.4}$ sec, is observed for the case of BMPC. However, the universal value, $\tau_\alpha \sim 10^{-6.5}$ sec, appears in a very similar molecular system BMMPC (Table 1). So, it remains unclear whether there might be a few systems with strong deviation from the universal τ_α, or the deviation may depend on accuracy of T_B estimates.

The proposed recently by Rössler et al. [23] universal scaling of viscosities between T_g and T_x is another evidence of the observed universality in τ_c. Cassalini et al. have analyzed influence of pressure on T_B in phenolphthalein-dimethyl-ether and in polychlorinated biphenyls [24] and reported universal value of τ_B regardless of the temperature or density at which the crossover occurs. We need to emphasize that $\tau_B \sim 10^{-4}$ sec has been obtained in [24]. The difference with reported above "magic" time might be related to different ways used for estimates of T_B and T_c. The authors did not use the analysis proposed by Stickel et al [2]. Instead, they use their own definition that might underestimate T_B. This result emphasizes that the dynamic crossover occurs over rather broad temperature range and estimates of T_c (or T_B) depend on the way of the data extrapolation.

The collection of the data presented in the Table 1 and in earlier publication [16] suggests that the dynamic crossover from liquid like to solid like behavior is observed in various systems at a particular "magic" relaxation time, regardless of details of their chemical structure, intermolecular interaction, density and temperature. Collection of the presented data, including pressure data from [24], suggests that neither density, nor temperature or intermolecular interactions are crucial for the dynamic crossover. It seems that the structural relaxation time, τ_α, appears to be the most important parameter for the dynamic crossover. This result supports the idea proposed more than 30 years ago by Martin Goldstein [25], although the characteristic relaxation time τ_c appears to be ~100 times slower than the predicted one. This unusual universality remains unexplained. It suggests that some general, may be a geometric, parameter or some particular property of a cage effect, controls the dynamic crossover. It is known that the characteristic molecular time scale τ_0 (attempt rate) is $\sim 10^{-12}$-10^{-13} sec, characteristic structural relaxation time at T_g (human time scale) $\sim 10^1$-10^3 sec. However, why the time range $t_\alpha \sim 10^{-8}$-10^{-6} sec appears to be critical for the dynamic crossover remains to be explained. Some

TABLE 1. The values of glass transition temperature T_g, critical temperature T_c, crossover temperature T_B, ratio T_c/T_g (T_B/T_g in the cases of BMPC and BMMPC), and fragility m. In the last column, $-\log \tau_c$ (sec) ($-\log \tau_B$ in the cases of BMPC and BMMPC) is shown. In cases when the reference is not shown, the data are from Ref. [16].

System	Tg [K]	T_C [K]	T_B [K]	T_C/Tg	m	-log[τ_C)]
Small molecules						
propylene glycol	167	251	280	1.50	52	7.5
propylene carbonate	158	176-196	189	1.11-1.24	104	5.2-7.7
orthoterphenyl	243	285-293	290	1.17-1.19	81	7.0-7.8
Salol	218	256-275	265	1.17-1.26	73	6.6-8.1
glycerol	186	225-300	285	1.18-1.61	53	4.4-9.0
sorbitol	264	309		1.17	93	6.5
toluene	118	143-153		1.21-1.30	107	7-8
m-fluoroaniline	173	212		1.22	109	7.3-7.8
picoline	133	162		1.22		7.5
n-butylbenzene	128	150-160		1.17-1.25		6.0-7.2
m-tricresyl phosphate	210	260	250	1.24	63	7
dibuthylphthalate	170	227		1.34	69	6.8
isopropylenbenzene	125	150		1.20		5.8
2,4,6-trimethylheptane	125	150		1.20		6.3
aab-tris-naphtylbenzene	345	407-415		1.18-1.19	86	6.6-7.0
BMPC [2]	275		275		63 [50]	4.4
BMMPC [2]	325		325		63 [50]	6.5
Polymers						
PB	180	216		1.20	59	7.1
PPG	200	250		1.25	117	7.1
PIB	200	270		1.35	46	6.5
PS	370	420		1.14	139	6.0
Ionic systems						
CKN	333	368-388		1.10-1.14	93	6.3-7.5
CRN	333	365-378		1.10-1.13	97	6.6-7.5
$ZnCl_2$	375	563		1.50	30	7.9
$Na_{0.5}Li_{0.5}PO_3$	515	620		1.20		6.5
Covalent systems						
B_2O_3	526	800-900		1.52-1.71	32	6.2-6.7
Orientationally disordered crystall						
($NPA_{0.7}NPG_{0.3}$)	156	227		1.45	30	6

qualitative ideas proposed for an explanation of the "magic" relaxation time have been presented in Ref. [16].

BREAKDOWN OF TIME-TEMPERATURE SUPERPOSITION IN POLYMERS

Time-temperature superposition (TTS) is a traditional approximation used in polymer science [26]. It assumes that all viscoelastic properties, reptation, Rouse and segmental relaxation, have the same temperature variations. It has been demonstrated [27,28], however, that the temperature dependence of chain relaxation (Rouse, terminal) differs from the temperature dependence of segmental relaxation and the difference increases with approaching T_g. This breakdown of TTS appears, in particular, in shrinking of rubbery plateau in mechanical relaxation [27] and has been clearly observed in dielectric relaxation in polymers where both chain (normal) and segmental modes are observed [29]. Detailed review on the breakdown of TTS is presented in [28].

On the other hand, analysis of polymer dynamics at higher temperatures shows reasonable agreement between temperature variations of chain and segmental relaxation (shift factors) [30]. Thus, apparently, TTS works at higher T but does not work with temperature approaching T_g.

The question appears: Does the TTS breakdown appear at the dynamic crossover? This could be a reasonable expectation, because it is known that decoupling of various processes and broadening of the segmental relaxation spectrum are usually observed at temperatures below T_c [9,17,31]. Schönhals [29] presented the ratio of frequencies of the maximum of normal modes f_n to f_α of segmental relaxation measured simultaneously in dielectric spectra of PPG. The ratio f_α/f_n (Fig.1) remains rather constant at high temperatures (T>240K) and drops significantly below this temperature. The behavior presented in Fig.1 is very similar to the well known decoupling of rotational and translational motions [9]. Moreover, the temperature where the ratio f_α/f_n starts to drop coincides well with the value of T_c~236K reported for PPG in [32].

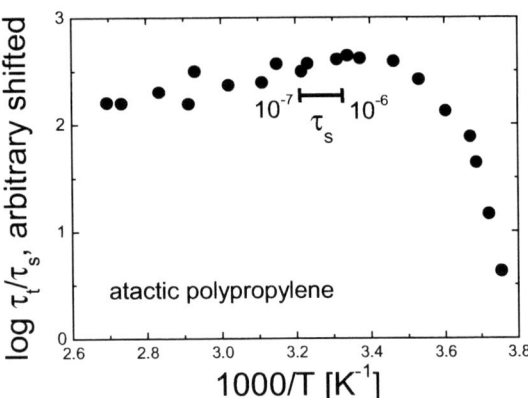

FIGURE 2. Ratio of terminal relaxation time τ_t to segmental relaxation time τ_S for atactic polypropylene (data from [33]). T_c of the polymer is not known. The bar indicates the temperature range where τ_S~10^{-7}-10^{-6} sec, i.e. where T_c can be expected.

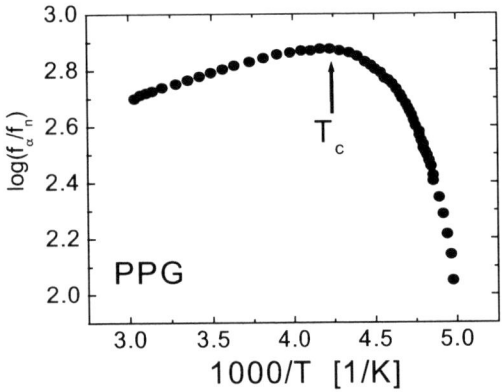

FIGURE 1. Ratio of the frequency of the maximum of segmental relaxation f_α to that of the chain relaxation f_n in dielectric spectra of PPG (data from [29]). The arrow indicates T_c=236K estimated for PPG in [32] using MCT analysis.

Figure 2 presents similar analysis of large collection of data presented in [33] for terminal and segmental relaxation in another polymer, atactic polypropylene. Once again, a constant ratio is observed at higher temperature and significant drops occurs at T approaching T_g. T_c (or T_B) of atactic polypropylene is not known for us. But using the idea of "magic" relaxation time one expects T_c~300-310 K. This value is indicated in the Fig.2. It coincides with the temperature at which the ratio of two shift factors starts to drop (Fig.2).

Ediger and co-workers [31] demonstrated decoupling of rotational and translational motions of guest molecules in polymer melts. Analysis of their data show (Fig.3) that the decoupling in polyisobutylene (PIB) and polystyrene (PS) appears around T_c estimated for these polymers using traditional MCT scenario [34].

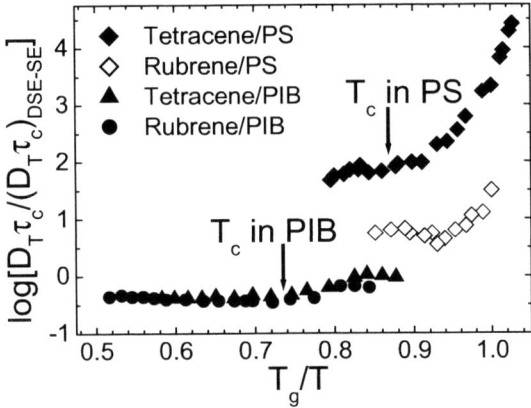

FIGURE 3. Ratio of rotation to translation time scales for guest molecules in polymer melt (data from [31]). The arrows indicate T_c estimated using MCT approach for these polymers [34].

The obvious assumption is that similar decoupling might happen between translational, rotational and other motions of monomers in a polymer chain. That might be the reason for the observed breakdown of TTS. Recent results by Ediger, et al. [35] demonstrate that diffusion of PS chains follows terminal shift factor while viscosity has stronger temperature variations and follows segmental relaxation. The question of breakdown of TTS is very important for correct description of polymer dynamics. Its understanding will not only explain the effect itself, but might bring important insight into long standing problem of microscopic mechanism behind the segmental friction coefficient in polymer melts.

DYNAMIC TRANSITION IN PROTEINS AND DNA

Dynamics of biological macromolecules is very similar to dynamics of glass forming liquids. It includes low-frequency collective vibrations (the boson peak), fast picosecond relaxation and a slow process (Fig.4) [36,37]. Proteins and DNA also exhibit a dynamic transition. It is usually observed as a sharp change in temperature variations of mean-squared atomic displacements, $<x^2>$, from nearly harmonic behavior, $<x^2> \propto kT$, at low T to much stronger anharmonic behavior at higher T [38-42]. Temperature of the dynamic transition, T_D, appears to be essentially the same ~200K-230K in different hydrated proteins and DNA [37-42]. It has been shown that the dynamic transition does not appear in dry proteins and DNA, at least, up to T~320-350K [37-42]. It is also suppressed in proteins placed in trehalose [43] and appears to be shifted to higher T in protein placed in glycerol [42]. The interest to the dynamic transition was stimulated by the observation that below T_D biochemical activity of proteins decreases essentially to zero. Thus crossing the dynamic transition in some sense is a disabling the protein's function. Understanding the mechanism of the dynamic transition might help in understanding how proteins function.

Analysis of DNA spectra presented in Fig.4 shows that the slow process disappears from the accessible frequency range at temperatures below T_D~220K. The slow process does not appear in the spectra of dry DNA even at highest temperature of the measurements ~320K. These results clearly suggest that the dynamic transition might be related to an activation of the slow process [37,44]. The same conclusion has been achieved from analysis of light scattering spectra of protein lysozyme [46]. Thus enabling the slow process apparently enables protein functions.

Most of the authors tried to relate T_D to some kind of a glass transition temperature, T_g, in proteins. Doster et al. proposed an analogy between the dynamic transition in proteins and the dynamic crossover in glass forming systems and demonstrated that one can get a reasonable fit of neutron scattering spectra of myoglobin using MCT scenario [36]. Similar approach has been applied to analysis of neutron scattering spectra of wet DNA [37,44]. The authors were able to fit the DNA spectra using only one free fit parameter - τ_S of the slow process (Fig.4).

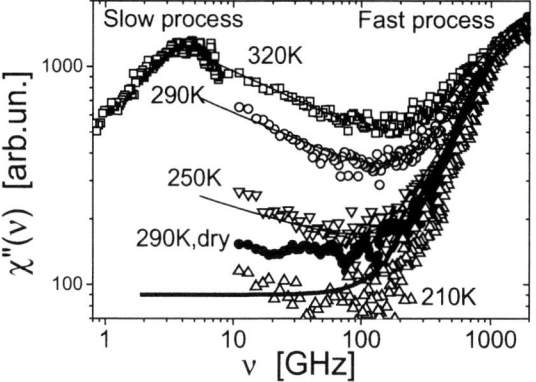

FIGURE 4. Neutron scattering (time-off-flight) data of wet (open symbols) and dry (,) DNA at different temperatures (data from [44,45]). The spectra at T=320K are extended to lower frequency using back-scattering data [45]. The slow process is strongly suppressed in wet DNA at T=210 K and is not observed in the dry sample even at T=290K. The thin solid lines show the fits using MCT approximation with the fast process shown by the thick solid line. The spectrum at T=210K appears to be below the fast spectrum at ν>100 GHz indicating variations of the fast relaxation at these temperatures. The maximum at ν~4 GHz observed at T=320K gives an estimate of τ_s~4×10^{-11} sec.

Later the same authors were able to extend their data down to lower energy (Fig.4) that allows us to estimate temperature variations of τ_S in absolute values. The characteristic relaxation time of the slow process shows strong temperature variations (Fig.5).

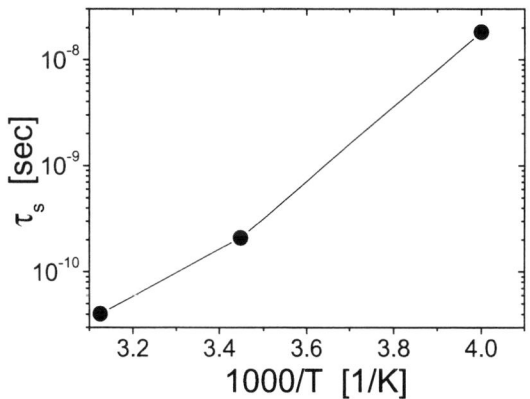

Figure 5. Temperature variation of the characteristic relaxation time of the slow process in wet DNA estimated from the fit of the data presented in Fig. 4.

It is know that MCT predicts a power-law dependence for variation of τ:

$$\tau_S \propto (T - T_C)^{1/\gamma} \quad (1)$$

where the exponent γ is directly related to the stretching exponent b of the slow process [1]. In particular, the exponent γ is ~3.9 when the value of b is ~0.3 (the slope of the high frequency tail of the slow process in DNA, Fig.4 [44]). Fig.6 shows that τ_S follows the predicted power law dependence and provides an estimate of T_D~230K. Thus MCT can provide reasonable description of the dynamics of proteins and DNA and the dynamic transition in biological macromolecules might be interpreted as a dynamic crossover from liquid-like to solid-like behavior on a molecular length and time scales.

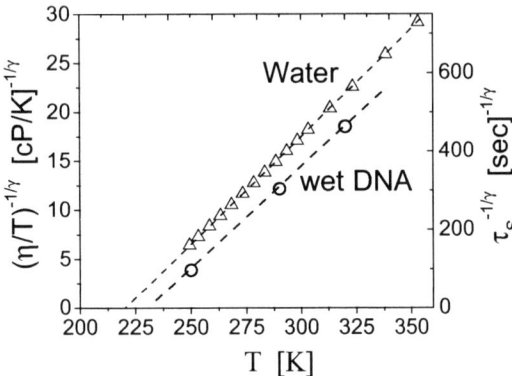

Figure 6. Critical temperature behavior of characteristic relaxation time τ_S in DNA. T_c appears to be close to T_c of bulk water (data from [44]).

The most interesting observation, however, is a coincidence of the dynamic transition temperature in wet proteins and DNA, T_D~200K-230K, with the dynamic crossover temperature in bulk water, T_C~225K [47]. This observation leads the authors of [37,44] to speculate that the dynamic crossover in a solvent controls the dynamic transition in biological macromolecules.

Analysis of a protein lysozyme in glycerol provides strong support to this idea (Fig.7) [46]. First of all, no slow process is observed in dry lysozyme and the slow process is strongly suppressed in lysozyme/glycerol solution at low temperatures (Fig.7). Much simpler analysis has been applied to the spectra of lysozyme/glycerol samples. The tail of the slow process is clearly visible in the spectra of lysozyme at high T. It can be approximated by a power law:

$$\chi''_{slow}(\nu) \propto (\nu \tau_S)^{-b} \quad (2)$$

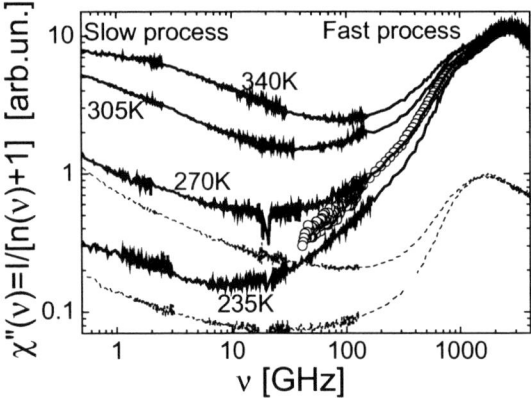

Figure 7. Light scattering susceptibility spectra of lysozyme/glycerol sample (solid lines) at different temperatures, dry lysozyme (symbols) at T=295K and spectra of glycerol (dashed lines) at T=270K and 235K. The spectra of glycerol are scaled to the spectra of lysozyme/glycerol sample at high frequency vibrational modes (~10-15 THz, for details, see [46]). No slow process is observed in the spectra of dry lysozyme.

Because of a critical temperature dependence of τ_S (eq.1) the amplitude of the power-law tail should also have a critical behavior:

$$\chi''_{slow}(\nu) \propto \nu^{-b}(T - T_C)^{-1/b\gamma} \quad (3)$$

At rather high stretching that is observed in spectra of proteins and DNA (b~0.25-0.35), the exponent 1/bγ is ~1-0.8. Thus nearly linear change of the intensity of the slow process with temperature should be observed. Figure 8 shows the temperature dependence of the slow process estimated using the spectra presented in the Fig.7. Indeed nearly linear temperature dependence is observed. This dependence gives an estimate of T_C~270K in agreement with analysis of mean-squared displacement measured by neutron scattering [42]. This temperature is close to the crossover temperature of bulk glycerol T_c~270-290 K [16].

These results (Figures 6 and 8) clearly demonstrate that the dynamic crossover in a solvent controls the dynamic transition in proteins and DNA. Moreover, the analysis suggests that the dynamic transition might be considered as a dynamic arrest of the slow process. This arrest leads to suppression of biochemical activities of proteins. What is the microscopic picture behind the correlation between T_D of a protein and T_c of a solvent? Results of recent computer simulations [48] might provide an additional insight. The authors constrained diffusion of water molecules by placing oxygen atoms in very deep energy minima. The water molecules could rotate, lifetime of hydrogen bonds remains similar to a normal case. However, analysis of intermediate scattering function I(Q,t) of the wet protein in the environment with restricted water diffusion became similar to I(Q,t) in a dry protein [48]. The slow process appears to be suppressed [48]. It means that restriction of the water diffusion puts wet protein even at room T below its dynamic transition. According to MCT ideas, the dynamic crossover is essentially a dynamic arrest on a molecular time and length scale. In that case, crossing T_c of a solvent is equivalent to restricting a diffusion of the solvent molecules. This restriction brings the protein below its dynamic transition.

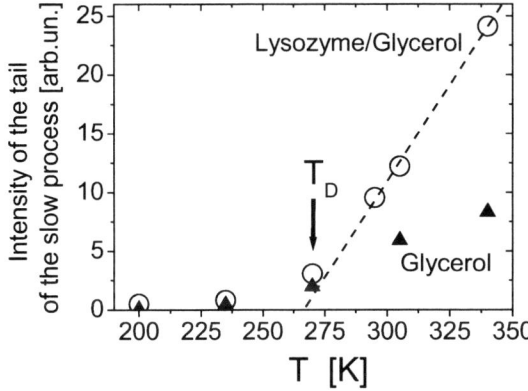

Figure 8. Temperature dependence of the intensity of the tail of the slow process in the frequency range 1-5 GHz for lysozyme/glycerol sample (open symbols) and bulk glycerol (7). The arrow shows T_D estimated for lysozyme/glycerol in [42] from $<x^2>$.

This picture provides a microscopic explanation for the observed importance of T_c of the solvents for the dynamic transition in biological macromolecules. It agrees with the idea proposed long time ago [49] that proteins are "slaves" of solvents. Of course, the presented picture is a strong oversimplification. It is known that dynamics of water bonded to protein surface differ from the dynamics of bulk water. That might affect T_c of solvent molecules. This effect, however, might be of secondary importance.

Thus it appears that T_c of a solvent is crucial for the dynamic transition of biological macromolecules. This observation might open a way for control of biochemical activities of proteins by changing T_c of a solvent. As a result, it might be possible to design a solvent where enzymes would work at temperature range at which they are not working in native environment, or, reverse, their biochemical activity might be disabled at the temperature range where they are active in native environment.

CONCLUSIONS

We present an overview of a dynamic crossover (transition) in various molecular systems, including small molecules with simple (van-der-Waals) interactions, polymers and biological macromolecules. Our analysis shows existence of a universal relaxation time at which dynamic crossover happens in glass forming systems regardless of their chemistry, density and characteristic temperature. This result is close to the naïve idea proposed by Martin Goldstein [7] and suggests that neither temperature, or density or intermolecular interaction is important for the dynamic crossover. It seems that the time scale of the structural relaxation appears as a crucial parameter.

Implications of the dynamic crossover on dynamics of synthetic polymers and biological macromolecules are discussed. In particular, we speculate that the decoupling of various relaxation processes observed below T_c might be the reason for the breakdown of time-temperature superposition in polymer melts at temperatures close to T_g. Detailed analysis of this phenomenon might provide additional insight into the microscopic picture behind the segmental friction coefficient. We also demonstrate that the dynamic transition observed in proteins and DNA might be related to the dynamic crossover in a solvent. This idea explains the observation that all wet proteins and even DNA demonstrate the dynamic transition in the same temperature range. Thus the dynamic transition is not an intrinsic property of biological macromolecules but is a transition imposed by a solvent due to restriction of its diffusion below T_c. We speculate that these ideas might open a way to manipulate by biochemical activities of proteins.

ACKNOWLEDGMENTS

This work has been supported by NSF (DMR-031538). We also thank OBR and NIST for the financial support.

REFERENCES

1. Götze, W., Sjogren, L., *Rep. Prog. Phys.* **55**, 241-376 (1992).
2. F. Stickel, E.W. Fischer, and R. Richert, *J. Chem. Phys.* **104**, 2043-2055 (1996).
3. Parak F., Frolov, E. N., Moessbauer, R. L., Goldanskii, V. I., *J. Molec. Biol.* **15**, 825-833 (1981).
4. Rasmussen, B.F., Stock, A. M., Ringe, D., Petsko, G. A., *Nature* **357**, 423-424 (1992).
5. Caliskan, G., Kisliuk, A., Sokolov, A. P., *J. Non-Cryst. Sol.* **307–310**, 868–873 (2002).
6. Boyer, R. F., *Polym. Eng. Sci.* **19**, 732-742 (1979).
7. Goldstein, M., *J. Chem. Phys.* **51**, 3728–3739 (1969).
8. Plazek, D. J., *J. Polym. Sci., Polym. Phys.* **20**, 1533–1541 (1982).
9. Rössler, E., *Phys. Rev. Lett.* **65**, 1595-1598 (1990).
10. H. Maekawa, H., Inagaki, Y., Shimokawa, S., and Yokokawa, T., *J. Chem. Phys.* **103**, 371-376 (1995).
11. Adichtchev, S. V., Benkhof, St., Blochowicz, Th., Novikov, V.N., Rössler, E., Tschirwitz, Ch., and Wiedersich, J., *Phys. Rev. Lett.* **88**, 055703-4 (2002).
12. Ngai, K. L., Bao, L.-R., Yee, A. F., and Soles, C. L., *Phys. Rev. Lett.* **87**, 215901-4 (2001).
13. Schönhals, A., *Europhys. Lett.* **56**, 815-821 (2001).
14. Hinze, G., Brace, David D., Gottke, S. D., and Fayer, M. D. *J. Chem. Phys.* **113**, 3723-3733 (2000).
15. Casalini, R., Ngai, K. L., and Roland, C. M., *J. Chem. Phys.* **112**, 5181-5189 (2000).
16. Novikov, V.N., and Sokolov, A.P., *Phys. Rev. E* **67**, 031507-6 (2003).
17. Sokolov, A. P., *Science*, **273**, 1675-1676 (1996).
18. Rössler, E., Novikov, V. N., Sokolov, A. P., *Phase Transitions* **63**, 201-225 (1997).
19. León, C., Ngai., K. L., *J. Phys. Chem. B* **103**, 4045-4051 (1999).
20. Götze, W., *J. Phys.: Cond. Matter* **11**, A1-A45 (1999).
21. Wuttke, J., Hernamdez, J., Coddens, G., Cummins, H. Z., Fujara, F., Petry, W., and Sillescu, H., *Phys. Rev. Lett.* **72**, 3052-3055 (1994).
22. Rössler, E., Sokolov, A. P., Kisliuk, A., and Quitmann, D., *Phys. Rev. B* **49**, 14967-14978 (1994).
23. Rössler, E., Hess, K.-U., and Novikov, V. N., *J. Non-Cryst. Solids* **223**, 207-222 (1998).
24. Casalini, R., Paluch, M., Roland, C. M., *J. Chem. Phys.* **118**, 5701-5703 (2003).
25. Goldstein, M., *J. Chem. Phys.* **51**, 3728-3738 (1969).
26. Flory, P. J., *Principles of Polymer Chemistry*, New York: Cornell University Press, 1953.
27. Plazek, D. J., *Polym. J.* **12**, 43-53 (1980).
28. Ngai, K. L., and Plazek, D. J., *Rubber Chemistry and Technology* **68**, 376-434 (1995).
29. Schönhals, A., "Dielectric Properties of Amorphous Polymers" in *Dielectric Spectroscopy of Polymeric Materials,* edited by J.P. Rant and J.J. Fitgerald, Washington, DC: American Chemical Society, 1997, pp. 81-105.
30. Arbe A., Richter D., Colmenero J., Farago B., *Phys. Rev. E* **54**, 3853-3869 (1996).
31. Bainbridge, D., Ediger, M. D., *Rheol.Acta* **36**, 209-216 (1997).
32. Bergman, R., Borjesson, L., Torell, L. M., and Fontana, A., *Phys. Rev. B* **56**, 11619–11628 (1997).
33. Roland, C. M., Ngai, K. L., Santangelo, P. G., Qiu, XH., Ediger, M. D., and Plazek, D. J., *Macromolecules* **34**, 6159-6160 (2001).
34. Kisliuk, A., Mathers, R. T., and Sokolov, A., *J. Polym. Sci., Part B: Polym. Phys.* **38**, 2785-2790 (2000).
35. Urakawa, Osamu, Swallen, Stephen F., Ediger, M. D., von Meerval, Ernst D., *Macromolecules,* in print.
36. Doster, W., Cusack, S., Petry, W., *Phys.Rev.Lett.* **65**, 1080-1083 (1990).
37. Sokolov, A. P., Grimm, H., Kahn, R., *J. Chem. Phys.* **110**, 7053-7057 (1999).
38. Doster, W., Cusack, S., Petry, W., *Nature* **337**, 754–756 (1989).
39. Iben, I. E., *et al.*, *Phys. Rev. Lett.* **62**, 1916-1919 (1989).
40. Ferrand, M., Dianoux, A.J., Petry, W., Zaccai, G., *Proc. Natl. Ac. Sci. USA* **90**, 9668-9678 (1993).
41. Lichtenegger, H., Doster, W., Kleinert, T., Birk, A., Sepiol, B., Vogl, G. *Biophys. J.* **76**, 414-422 (1999).
42. Tsai, A.M., Neumann, D.A., and Bell, L.N., *Biophys. J.* **79**, 2728-2732 (2000).
43. Cordone, L., Ferrand, M., Vitrano, E., Zaccai, G., *Biophys. J.* **76**, 1043-1047 (1999).
44. Sokolov, A. P., Grimm, H., Kisliuk, A., Dianoux, A.J., *J.Biological Physics* **27**, 313-320 (2001).
45. Grimm, H., Sokolov, A. P., Dianoux, A. J., *Applied Physics A* **74**, S1248-S1250 (2002).
46. Caliskan, G., Kisliuk, A., Sokolov, A. P., *J. Non-Cryst. Sol.* **307–310**, 868–873 (2002).
47. Sokolov, A.P., Hurst, J., Quitmann, D., *Phys. Rev. B* **51**, 12865-12868 (1995).
48. Tarek, M., Tobias, D.J., *Phys.Rev.Lett.* **88**, 138101-4 (2002).
49. Beece, D., Eisenstein, L., Frauenfelder, H., Good, D., Marden, M. C., Reinisch, L., Reynolds, A. H., Sorensen, L. B., Yue, K. T., *Biochemistry* **19**, 5147-5157 (1980).
50. Beiner, M., Huth, H., Schröter, K., *J. Non-Cryst. Sol.* **279**, 126-310 (2001).

Two-order-parameter model of liquid: Water-like thermodynamic anomaly, liquid-liquid transition, and liquid-glass transition

Hajime Tanaka

*Institute of Industrial Science, University of Tokyo,
Komaba 4-6-1, Meguro-ku, Tokyo 153-8505, Japan.*

Abstract. Recently we have proposed a two-order-parameter model of liquid to understand water-like thermodynamic and kinetic anomalies of liquid, liquid-liquid phase transition in a single-component liquid, and liquid-glass transition. Here we present a general framework of the two-order-parameter model of liquid to describe all these phenomena in a unified manner and discuss how these phenomena, which are apparently independent of each other, can be closely related.

INTRODUCTION

Contrary to the common-sense view that a liquid is in a completely disordered state, a real liquid generally has a tendency to form short-range bond order and thus the structure of a liquid becomes locally more ordered with decreasing temperature. A liquid is in a disordered state in the long range, but it can locally possess short-range bond order. This short-range order is due to specific interactions between liquid atoms or molecules that have the symmetry-selective nature. They may stem from van der Waals interactions, hydrogen bonding, covalent bonding, or electrostatic interactions. Most typical examples of short-range bond order is a tetrahedral structure for water, silicon, silica, and germania and an icosahedral structure for metallic glass formers [1, 2].

On the basis of a physical picture that the formation of locally favored structure, or the short-range bond ordering, is intrinsic to a liquid state in general, we express a liquid state by a simple two-state model with cooperativity of bond ordering. Namely, we introduce the bond order parameter in addition to the density order parameter and proposed the two-order-parameter (TOP) model of liquid. We applied this TOP model to (i) water-like thermodynamic and kinetic anomaly of liquid [3, 4], (ii) liquid-liquid phase transition in a single-component liquid [5], and (iii) liquid-glass transition [6, 7, 8].

According to our model, (i) water-like thermodynamic anomaly of liquids is a result of the local ordering of bond order parameter, (ii) liquid-liquid transition is a result of the gas-liquid-like cooperative ordering of the bond order parameter (while a gas-liquid transition is that of the density order parameter), and (iii) vitrification is a result of the competition (frustration in the preferred symmetry) between the two order parameters, namely, between long-range density ordering (crystallization) and short-range bond ordering. We point out that a liquid having a strong tendency of short-range bond ordering may even achieve long-range bond ordering. In our view, such phenomena can be seen in (a) water and water-like tetrahedral liquids [4] and (b) metallic liquids. For case (a), a crystal having a larger specific volume than a liquid is formed, while for case (b) quasicrystal is formed. Both ordered states can be viewed as the ordered state of the bond order parameter with translational order.

In this paper we present a general framework of TOP model of liquid to describe all these phenomena (i)-(iii) in a unified manner, and discuss how these phenomena, which are apparently independent of each other, can be closely related.

TWO-ORDER-PARAMETER MODEL OF LIQUID

Our TOP model of liquid [3, 4, 5, 6] relies on the physical picture that (i) there exist unique locally favored structures in any liquids and (ii) such structures are formed in a sea of normal-liquid structures and its number density increases upon cooling since they are energetically more favorable by ΔE than normal-liquid structures. The specific volume and the entropy are larger and smaller for the former than the latter, respectively, by Δv and $\Delta \sigma$. We identify locally favored structures as a minimum structural unit [symmetry element]. For metallic glass formers, it should be icosahedron, as first suggested by Frank [2]. To express such short-range bond ordering in liquids, we introduce the so-called bond-orientational order parameter Q_{lm} (see Refs. [9] for its definition). We take the normalized average of Q_{lm} over a small volume located at $\vec{r}$, which we express by $\bar{Q}_{lm}(\vec{r})$. Then, its rotationally invariant combination can be defined as $Q_l(\vec{r}) = [\frac{4\pi}{2l+1}\Sigma_{m=-l}^{l}|\bar{Q}_{lm}(\vec{r})|^2]^{1/2}$. We can use this $Q_l(\vec{r})$ to define the local bond order parameter, which is the local fraction of locally favored structures: $S(\vec{r}) \equiv Q_l(\vec{r})$. Note that $l = 6$ for icosahedron [9]. Then the liquid-state free energy functional associated with locally favored structures is given by [6, 3, 4, 5]

$$f(S) = \int d\vec{r}[-\Delta G S(\vec{r}) + JS(\vec{r})(1-S(\vec{r}))$$
$$+ k_BT(S(\vec{r})\ln S(\vec{r}) + (1-S(\vec{r}))\ln(1-S(\vec{r})))],$$

where $\Delta G = \Delta E - T\Delta\sigma - \Delta vP$, J represents the cooperativity, k_B is the Boltzmann constant, T is the temperature, and P is the pressure.

Next we consider density ordering, which describes crystallization [6]. Density fluctuations $\delta\rho$, which we express by ρ for a while, in the liquid phase indicating the instability toward the solid phase have a maximum at nonzero wavenumber q_0. This ordering is described by the following free energy functional:

$$f(\rho) = k_BT \int d\vec{r}\rho(\vec{r})(\ln\rho(\vec{r}) - 1)$$
$$- \frac{k_BT}{2}\int d\vec{r}d\vec{r}'\rho(\vec{r})c(|\vec{r}-\vec{r}'|)\rho(\vec{r}'),$$

where $c(r)$ is the direct correlation function. This density ordering takes place near the melting point T_m, which is the liquidus temperature for metallic glass formers. Finally we include effects of coupling between ρ and S and those of possible long-range crystal or quasicrystal ordering. Considering only the lowest-order coupling, we obtain the total free energy functional of the system as

$$f(\rho,S) = f(\rho) + \int d\vec{r}\, c\, \rho(\vec{r})S(\vec{r}) + f(S) + f_{LOS}. \quad (1)$$

Here f_{LOS} represents a free energy functional describing the long-range ordering of S (long-range translational and orientational (Q_{lm}) ordering). It can be diamond-like crystal for tetrahedral order, and quasicrystal for icosahedral order (see, e.g., Ref. [9] for its possible form).

In the above free energy, there are new important effects of short-range bond ordering, which have so far not been considered to describe liquid-glass transition: (1) random field effects of $S(\vec{r})$ on density ordering, (2) thermodynamic effects of short-range bond ordering stemming from $f(S)$, and (3) long-range crystalline (or quasicrystal) ordering consistent with the symmetry of S

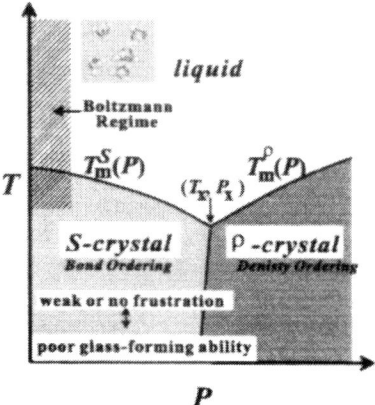

FIGURE 1. P-T phase diagram of water-type liquids including water itself and water-type atomic liquids (Si, Ge, Bi, Sb, and Ga).

(f_{LOS}). Hereafter, we consider problems of thermodynamic and kinetic anomaly of liquid, liquid-liquid transition, and liquid-glass transition, focusing on these three effects (1)-(3).

THERMODYNAMIC AND KINETIC ANOMALY OF WATER-LIKE LIQUIDS

What makes water so different from ordinary liquids?

First we consider the unusual behavior of water [1]. In our view [3], water is the only exceptional molecular liquid for the above statement that bond ordering is inconsistent with any crystallographic symmetry: The locally favored structure of water due to hydrogen bonding is consistent with the crystallographic symmetry of ice Ih. We argue that all thermodynamic anomalies of water originate from (i) this dominance of bond ordering below a crossover pressure P_x (~ 2 kbar), where the melting point of ice crystals has a minimum, and (ii) an unusually large Δv. Below P_x, the crystallization is due to bond ordering, while above P_x it is due to density ordering as in ordinary liquids (see Fig. 1). This gives a natural explanation for the unusual pressure dependence of the melting point of ice crystals, including its minimum around 2 kbar. We propose that ice Ih is S-crystal, long-range ordering of S driven by $f_{LOS}(S)$, while high-pressure ices are ρ-crystals. Our model also provides us with simple analytical predictions for the thermodynamic and dynamic anomaly of water-type liquids, as described below.

Analytical prediction for thermodynamic anomaly of water-like liquids

Here we consider a simple two-state model of liquid, which corresponds to the weak-coupling limit of our TOP model [3]. We first estimate how the average fraction of locally favored structures, $\bar{S}$, increases with a decrease in T. From the condition $\partial f(S)/\partial S = 0$, $\bar{S}$ can be obtained as

$$\bar{S} = \frac{\sigma_S \exp(-\beta E_S)}{\sigma_\rho \exp(-\beta E_\rho) + \sigma_S \exp(-\beta E_S)}, \quad (2)$$

where $\beta = 1/k_B T$ and E_i and σ_i are the energy level and the number of degenerate states of i-type structure, respectively. Here we assume $J \cong 0$ for simplicity, but J may play an important role in the presence of liquid-liquid transition [3, 5]. $i = \rho$ corresponds to normal liquid structures of water, while $i = S$ to locally favored structures. The uniqueness of a locally favored structure and the existence of many possible configurations for normal-liquid structures lead to the conclusion $\sigma_\rho \gg \sigma_S$. Then, $\bar{S}$ is estimated as

$$\bar{S} \sim S_0 \exp[\beta(\Delta E - \Delta v_S P)], \quad (3)$$

We stress that this relation should hold even for a nonzero J if $\bar{S} \ll 1$ [3]. In the above, the relation $E_\rho - E_S \cong \Delta E - \Delta v_S P$ is used, where ΔE and Δv_S are the energy gain and the volume increase upon the formation of a locally favored structure, respectively, and P is the pressure.

According to the above picture, the unusual decrease in ρ upon cooling below 4 °C in water can simply be explained by an increase in the number density of locally favored structures, $\bar{S}$, upon cooling. Thus, the specific volume v_{sp} and the density ρ are, respectively, given by

$$v_{sp}(T,P) = v_{sp}^B(T,P) + \Delta v_S \bar{S}, \quad (4)$$

$$\rho(T,P) \sim \rho_B(T,P) - \rho_B(T,P) \frac{\Delta v_S}{v_{sp}} \bar{S}, \quad (5)$$

where $\rho_B(T,P) = M/v_{sp}^B(T,P)$ (M: molar mass). Note that v_{sp}^B and ρ_B depend almost linearly on T as those of ordinary liquids. Then, $K_T = -\frac{1}{v_{sp}}(\frac{\partial v_{sp}}{\partial P})_T$ can straightforwardly be calculated from Eq. (4) as

$$K_T = -\frac{1}{v_{sp}}(\frac{\partial v_{sp}^B}{\partial P})_T + \frac{1}{v_{sp}}[-(\frac{\partial \Delta v_S}{\partial P})_T + \beta \Delta v_S^2]\bar{S}. \quad (6)$$

The anomalous increase in C_P upon cooling can also be explained as follows. The locally favored structure has a rather unique configuration and the associated degree of freedom is much smaller for it than for the normal structure of water. Thus, entropy s increases upon heating, reflecting a decrease in $\bar{S}$: $s = s_B(T,P) - \Delta s \bar{S}$. Thus,

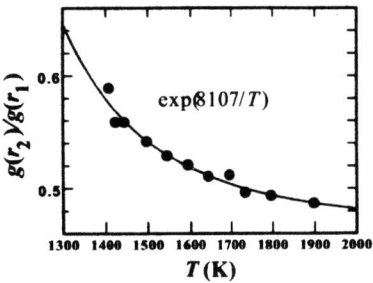

FIGURE 2. Temperature dependence of $g(r_2)/g(r_1)$ of Si calculated from the experimentally measured $g(r)$ [10]. The solid curve is our prediction: $g(r_2)/g(r_1) = a + b\bar{S}$, where a and b are positive constants [4]. Its anomalous increase upon cooling is very well described by the Boltzmann factor, $\exp(8107/T)$ (solid line).

$C_P = T(\partial s/\partial T)_P$ should increase upon cooling as

$$C_P = T(\frac{\partial s_B}{\partial T})_P + [-T(\frac{\partial \Delta s}{\partial T})_P + \beta \Delta s(\Delta E - \Delta v_S P)]\bar{S}. \quad (7)$$

The relevance of these predictions was confirmed for water [3] and water-like atomic liquids [4].

Structural change upon cooling in liquid Si: Evidence of short-range ordering

For water, liquid Si, Ge, and Ga, the existence of short-range bond order with tetrahedral symmetry is evidenced by the shoulder in the high wave number (q) side of the first peak of the structure factor $F(q)$, or the second peak of the radial distribution function $g(r)$, For Si, for example, the first peak of $g(r)$ is located around $r_1 = 2.4$ Å, while the second one is around $r_2 = 3.5$ Å [10]. The ratio of 3.5/2.4 = 1.46 is compatible with that of the two characteristic interatomic distances of the tetrahedral unit, $2\sqrt{6}/3 = 1.63$. For Si, the temperature dependence of the ratio of the height of the second peak to that of the first one of $g(r)$, $g(r_2)/g(r_1)$, which is a direct measure of the population of tetrahedral units, is found to be well described by $\bar{S}$ with $\Delta E = 8107$ K (see Fig. 2). This suggests the validity of our argument. We also found that the anomalies of ρ and C_P of liquid Si can also be well explained by our predictions [Eqs. (5) and (7)] with the same ΔE. Thus, the anomalous thermodynamic behavior can be well explained by the surprisingly simple scenario. We believe that critical phenomena do not play a primary role in the anomaly [3, 4], contrary to the currently popular scenario of a second critical point associated with liquid-liquid transition (see Ref. [1]). Our model can also explain the non-Arrhenius behavior of water viscosity without invoking either dynamic critical phenomena or glassy slow dynamics [3].

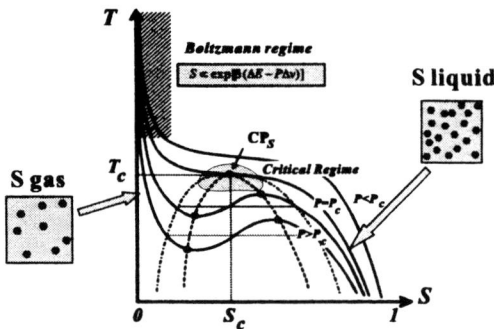

FIGURE 3. Schematic phase diagram of liquid-liquid transition in T-S plane [5].

LIQUID-LIQUID TRANSITION

Thermodynamics

We now consider a possible liquid-liquid phase transition, or *cooperative medium-range bond ordering*, on the basis of the above free energy f (see Fig. 3) [5]. The equilibrium value of S is determined by the condition $\partial f/\partial S = 0$, or

$$\beta[-\Delta E + \Delta v P + J(1-2S)] + \ln \frac{g_\rho S}{g_S(1-S)} = 0, \quad (8)$$

where $\Delta E = E_\rho - E_S > 0$, $\Delta v = v_S - v_\rho$, and $\beta = 1/k_B T$. It is worth noting that the degeneracy of each state, or the entropy difference between the two states, strongly affects the phase behavior. A critical point is determined by the conditions, $f'_S(S_c) = 0$, $f''_S(S_c) = 0$, $f^{(3)}_S(S_c) = 0$, and $f^{(4)}_S(S_c) > 0$, as

$$S_c = 1/2, \quad (9)$$
$$T_c = J/(2k_B), \quad (10)$$
$$P_c = [\Delta E - T_c \Delta \sigma]/\Delta v. \quad (11)$$

A first-order phase-transition temperature T_t is obtained as

$$T_t = (\Delta E - P\Delta v)/\Delta \sigma. \quad (12)$$

Note that a first-order transition occurs only if $T_t < T_c$. Δv may be positive in most cases (e.g., water), but it can also be negative in principle. The sign of Δv determines the slope of $T_t(P)$.

Dynamic coupling between density and bond order parameters

The Hamiltonian of ideal liquids associated with density fluctuations is approximately given by

$$\beta H_\rho = \int d\vec{r}\, \frac{\tau}{2} \delta \rho^2 = \int d\vec{r}\, f(\delta \rho),$$

where $\tau = \beta(\bar{\rho}^2 K_T)^{-2}$ and it is positive. Here $\bar{\rho}$ is the average density and a decreasing function of T (note that $\rho = \bar{\rho} + \delta\rho$). In a real liquid, however, the bond order parameter plays essential roles, as explained above. Using $\delta S = S - \bar{S}$, we introduce the following minimal Landau-type Hamiltonian by expanding $f(S)$ in terms of δS, which governs S fluctuations near a gas-liquid-like critical point or mean-field spinodal lines of bond ordering:

$$\beta H_S = \int d\vec{r}\, [\frac{\kappa}{2}\delta S^2 + \frac{b_4}{4}\delta S^4] = \int d\vec{r}\, g(\delta S),$$

where $\kappa = b_2(T - T_S^*)$ (T_S^*: a critical or spinodal temperature of bond ordering without the coupling to ρ) and b_2 and b_4 are positive constants. By further including the gradient terms and the lowest-order (bilinear) couplings between $\delta\rho$ and δS, we obtain the following Hamiltonian that we believe is relevant to the physical description of liquid near a gas-liquid-like transition of locally favored structures [5]:

$$\beta H_{\rho S} = \int d\vec{r}\, [h(\delta\rho, \delta S) + \frac{K_\rho}{2}|\nabla\delta\rho|^2 + \frac{K_S}{2}|\nabla\delta S|^2]. \quad (13)$$

$$h(\delta\rho, \delta S) = f(\delta\rho) + g(\delta S) - c_{1\rho}\delta\rho(\bar{S} + \delta S) - c_{1S}(\bar{\rho} + \delta\rho)\delta S. \quad (14)$$

Note that f, g, and h are dimensionless free-energy densities. For $\Delta v > 0$, which is a usual case, an increase in S leads to a decrease in ρ and an increase in τ, while an increase in ρ leads to a decrease in S and T_S^*. Hence, all the coupling constants c_i in Eq. (14) may be negative for usual cases, although there exist some liquids, for which c_i is positive.

Next we consider how the dynamics of $\delta\rho$ and δS should be described near the mean-field spinodal line of bond ordering. Here we treat $\delta\rho$ as a conserved order parameter obeying a diffusion-type equation, while δS as a non-conserved order parameter. Thus, we have the following dynamic equations [5]:

$$\frac{\partial \delta\rho}{\partial t} = L_\rho \nabla^2 \left[-K_\rho \nabla^2 \delta\rho + \frac{\partial h(\delta\rho, \delta S)}{\partial \delta\rho(\vec{r},t)} \right], \quad (15)$$

$$\frac{\partial \delta S}{\partial t} = -L_S \left[-K_S \nabla^2 \delta S + \frac{\partial h(\delta\rho, \delta S)}{\partial \delta S(\vec{r},t)} \right], \quad (16)$$

where L_ρ and L_S are kinetic coefficients. Here the Gaussian noise terms are not written explicitly. We propose that Eqs. (13)-(16) are the fundamental equations universally describing ultraslow critical-like dynamics of supercooled liquids. These dynamic equations are basically the same as those of the so-called "model C".

We argue that dynamic critical phenomena associated with the mean-field spinodal (Fischer clusters) and the kinetics of liquid-liquid transition can be described by the above coupled equations [5]. The details will be described elsewhere.

LIQUID-GLASS TRANSITION

In the above, we demonstrate the importance of locally favored structures in the problem of liquid-state thermodynamic anomaly and liquid-liquid transition. We argue that the same physical picture can also be applied to any liquids [6] and Eq. (3) holds for any liquids. For example, spherical particles are known to form icosahedral structures locally, whose energy is even lower than the corresponding fcc or bcc crystals. Differently from water, the symmetry of a locally favored structure is not consistent with any crystallographic symmetry. This energetic frustration hidden in the interaction potential causes the frustration effects on density ordering (crystallization).

Hamiltonian relevant to the problem

On the basis of the above picture, we here make a simple physical model. By including the lowest-order coupling between $\delta\rho$, which we express by ρ below, and S into the standard theory of a liquid-solid transition, we obtain the following Hamiltonian that we believe is relevant to glass transition:

$$f_G(\rho, S) = f(\rho) + \int d\vec{r}\, c\rho(\vec{r})S(\vec{r}). \quad (17)$$

Strictly speaking, we need to describe the tensorial character of short-range bond order for treating energetic frustration due to symmetric mismatch, we treat it as a scalar variable for simplicity. In the above, the coupling between ρ and S is introduced through the coupling constants c_i. A case of the "negative" coupling ($c_i < 0$) corresponds to the situation that the formation of active bonds, or locally favored structures, leads to the decrease of local density. For the "positive" coupling ($c_i > 0$), on the other hand, the formation of active bonds leads to the increase in the local density. Irrespective of the sign of coupling between ρ and S, random fluctuating parts of S produce the random disorder effects on the density ordering an lead to vitrification.

Our model tells us that $S(\vec{r})$ acts as random fields, which disturb crystallization and help vitrification [6], reflecting the frustration between the two order parameters. It should be noted [6] that the part of the free energy, $f(\rho) + \int d\vec{r}\, c\rho(\vec{r})S(\vec{r})$, which is responsible for vitrification, is equivalent to that of Kirkpatrick and Thirumalai [11] if we regard $S(\vec{r})$ as a quenched random field and take the limit that the amplitude of $S(\vec{r})$ goes to zero. We note that this limit corresponds to the fragile limit in our model. It was also shown [11] that their theory is equivalent to the schematic mode coupling theory.

In our picture, disorder effects on density ordering sets in only below $\sim T_m$. This reflects the change in the free-energy landscape from a simple to a multi-valley structure

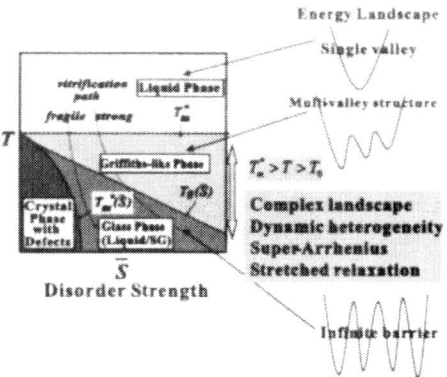

FIGURE 4. Schematic phase diagram of liquid under frustration in symmetry [6]. S represents the degree of energetic frustration due to symmetric mismatch in the interacting potential.

[6] (see Fig. 4). Note that $\delta f(\rho, S)/\delta\rho = 0$ starts to have multiple solutions below T_m due to random-field effects of $S(\vec{r})$. Thus, a transition from the Arrhenius to the non-Arrhenius behavior (the onset of cooperativity) should occur around T_m. Upon further cooling, the system eventually becomes spin-glass-like, non-ergodic state at the Vogel-Fulcher temperature T_0.

Next we consider how the local bond ordering affects glass formability. Conventional models of glass transition cannot answer the question of what controls the glass formability, since they do not put focus on crystallization itself. Our model, on the other hand, focuses on crystallization itself and explains the glass formability in terms of the strength of frustration between long-range crystal ordering and short-range bond ordering [7, 8]. By modifying the classical theory [12] with including the effect of translational-rotational decoupling, the nucleation frequency I is given by $I = k_n D_T \exp[-\Delta F^c/k_B T]$, where k_n is a constant and D_T is the translational diffusion constant [13]. ΔF^c is the free-energy barrier for nucleation of a critical nucleus, which is estimated as $\Delta F^c = 16\pi\gamma_{l-c}^3/(3\delta\mu)$ ($\delta\mu$: the Gibbs free energy of a supercooled liquid over the crystal per unit volume, γ_{l-c}: interface tension between liquid and crystal). Usually, it is assumed that $\delta\mu = \Delta H_f(1 - T/T_m)$, where ΔH_f is the enthalpy of fusion. According to our model, however, it should be modified due to the existence of locally favored structures as follows:

$$\delta\mu \cong \Delta H_f(1 - T/T_m) + \Delta G(T_m)\bar{S}(T_m) - \Delta G(T)\bar{S}(T).$$

The downward deviation of $\delta\mu$ from the linear temperature dependence is indeed observed for various metallic glass formers [14, 15]. Furthermore, this deviation is larger for a stronger (better) glass former [14, 15]. According to our model, a stronger glass former should have larger $\bar{S}$. Thus, the above observation is quite consistent with our model. We also note that γ_{l-c} should be larger for larger $\bar{S}$. Thus, we conclude that the better glass formability is due to smaller $\delta\mu$ and larger γ_{l-c}, which are induced by *a stronger tendency of short-range bond ordering (larger $\bar{S}$) of a stronger liquid with larger D*. Thus, our model suggests that a better glass former should have a stronger tendency of short-range bond ordering and be a stronger liquid. Glass formability is often characterized by the critical cooling rate R_c, which is the slowest cooling rate to form a glassy state from a supercooled liquid without crystallization. We note that the positive correlation between R_c and D can naturally be explained by our model [8].

We also show that this short-range bond ordering affects the conventional picture even qualitatively. For example, the excess entropy of liquid over the crystal should be modified, reflecting the entropy decrease due to the short-range bond ordering. This leads to the violation of the well-known relation $T_0 = T_K$ (T_0: Vogel-Fulcher temperature; T_K: Kauzmann temperature). The deviation of T_K from T_0 should be larger for a stronger liquid, which has recently been confirmed for a wide variety of liquids [7].

Here we consider an interesting problem of the relationship among local icosahedral ordering, glass formability, and quasicrystal formation in bulk metallic glass formers [16]. Chen et al. [17] recently reported the structural similarity between a supercooled liquid and an icosahedral phase in $Zr_{65}Al_{7.5}Ni_{10}Cu_{12.5}Ag_5$. They found that (i) the effective activation energy of transition from a supercooled liquid to an icosahedral quasicrystalline phase is much lower than that from a supercooled liquid to eutectic crystalline phases and (ii) the activation energy of transition from an icosahedral to a crystalline phase is almost the same as that from a supercooled liquid to a crystalline phase. These facts strongly suggest the similarity of the local atomic structure between the supercooled and the icosahedral phase. Our model provides us with a natural scenario for the close relationship among the degree of local icosahedral ordering in liquid, glass formability, and quasicrystal formability.

Finally, we mention the origin of the so-called Boson peak. We recently proposed that the boson peak is due to the localized vibrational modes characteristic of clusters of locally favored structures (icosahedral structures for metallic glass formers) [18]. Our model is consistent with the fact that the boson peak exists even in an equilibrium liquid state (above T_m) of some strong liquids, which seems to be difficult to be explained by conventional models.

SUMMARY

In summary, we present the two-order-parameter model of liquid, which can describe thermodynamic anomalies of liquids, liquid-liquid transition, and liquid glass transition, in a unified manner. Our model also naturally explain the close relationship among icosahedral short-range ordering, fragility, glass formability, and quasicrystal formation in metallic glass formers. Further experimental and theoretical studies are highly desirable to check the validity of this physical view.

REFERENCES

1. Debenedetti, P. G., *Metastable Liquids*, Princeton Univ. Press, Princeton, 1997.
2. Frank, F. C., *Proc. R. Soc. A* **215**, 43-46 (1952).
3. Tanaka, H. *Phys. Rev. Lett.* **80**, 5750-5753 (1998); *J. Chem. Phys.* **112**, 799-809 (2000); *Europhys. Lett.* **50**, 340-346 (2000); *J. Phys.: Condens. Matter* **15**, L703-L711 (2003).
4. Tanaka, H., *Phys. Rev. B* **66**, 064202 (2002).
5. Tanaka, H., *J. Phys.: Condens. Matter* **11**, L159-L168 (1999); *Phys. Rev. E* **62**, 6968-6976 (2000).
6. Tanaka, H., *J. Phys.: Condens. Matter* **10**, L207-L214 (1998); *J. Chem. Phys.* **111**, 3163-3174 (1999); *J. Chem. Phys.* **111**, 3175-3182 (1999).
7. Tanaka, H., *Phys. Rev. Lett.* **90**, 055701 (2003).
8. Tanaka, H., *J. Phys.: Condens. Matter* **15**, L491-L498 (2003).
9. Steinhardt, P. J., Nelson, D. R., and Ronchetti, M., *Phys. Rev. B* **28**, 784-805 (1983).
10. Kimura, H., Watanabe, M., Izumi, K., Hibiya, T., Holland-Moritz, D., Schenk, T., Bauchspiess, K. R., Schneider, S., Egry, I., Funakoshi, K., and Hanfland, M., *Appl. Phys. Lett.* **78**, 604-606 (2001).
11. Kirkpatrick, T. R. and Thirumalai, D., *J. Phys.: Math. Gen.* **22**, L149-L155 (1989).
12. Turnbull, D., *Contemp. Phys.* **10**, 473-488 (1969).
13. Tanaka, H., *Phys. Rev. E* **68**, 011505 (2003).
14. Glade, S. G., Busch, R., Lee, D. S., Johnson, W. L., Wunderlich, R. K., and Frecht, H. J., *J. Appl. Phys.* **87**, 7242-7248 (2000).
15. Lu, Z. P., Li, Y., and Liu, C. T., *J. Appl. Phys.* **93** 286-290 (2003).
16. Inoue, A., Zhang, T., and Masumoto, T., *Mater. Trans. JIM* **31**, 425-428 (1991).
17. Chen, M. W., Dutta, I., Zhang, T., Inoue, A., and Sakurai, T., *Appl. Phys. Lett.* **79**, 42-44 (2001).
18. Tanaka, H., *J. Phys. Soc. Jpn.* **70**, 1178-1181 (2001).

Bulk Nonequilibrium Alloys by Stabilization of Supercooled Liquid: Fabrication and Functional Properties

Akihisa Inoue and Akira Takeuchi

Institute for Materials Research, Tohoku University, Sendai 980-8577, Japan

Abstract. Since the recognition of stabilization phenomenon of supercooled liquid in multi-component metallic alloys without metalloid elements in 1988, a number of nonequilibrium alloys in a bulk form have been fabricated with the aim of clarifying the origin for its novel phenomenon and finding useful characteristics. This review deals with alloy systems in which the stabilization can be achieved, stabilization mechanism, and physical, chemical, mechanical and magnetic properties of the resulting glassy and nonequilibrium crystalline bulk alloys.

INTRODUCTION

Metallic alloys without metalloid elements are ordinary in a metallic bonding state and the diffusion rate of the constituent elements is very fast in the temperature range just below their melting temperature. The diffusivity is further enhanced in the liquid state and hence the transition from supercooled liquid to a crystalline state is completed within an extremely short time. Based on the instantaneous solidification phenomenon to crystalline phases, various kinds of crystalline alloys have been developed and used, accompanied by the progress of physical, chemical and processing metallurgies [1]. Recently, such an instantaneous solidification has been suppressed through an increase in the stability of supercooled liquid against crystallization for a number of special multi-component metallic alloys, leading to the formation of glassy alloys in a bulk form even in the slow cooling condition from liquid state at cooling rates of less than 100 K/s [2-6]. The developments of a number of glassy and metastable crystalline alloys in a bulk form as well as the stabilization of supercooled metallic liquid have opened up a new basic science and engineering field of "supercooled liquid metallurgy". This paper reports the fundamental concept, present state and future prospect of supercooled liquid metallurgy which has been developed mainly by our group.

STABILIZATION PHENOMENON OF METALLIC SUPERCOOLED LIQUID

Figure 1 shows a schematic illustration of continuous cooling transformation curves of some typical metallic liquids. The incubation time for the transition of supercooled liquid to crystalline phase is as short as less than 10^{-5} s for conventional crystalline alloys and of the order of 10^{-4} s for ordinary amorphous alloys which require high cooing rates of about 10^6 K/s. These incubation times are too short to control the supercooled liquid state. However, since 1988, the incubation time increases dramatically to several hundreds seconds and reaches the longest time of about 4000 s and the lowest critical cooling rate is as low as 0.007 K/s [7]. Thus, the stability of metallic supercooled liquid increases by hundred million times for the last 15 years. Such a high stability of supercooled liquid has enabled us to fabricate various nonequilibrium alloys exhibiting functional properties in a bulk form. As a result, we can obtain bulk glassy alloys with various outer shapes such as massive form of 40 to 80 mm in diameter, cylindrical rods of 25 mm in diameter and 250 to 400 mm in length, and long tubes of 10 mm in inner diameter, 12 mm in outer diameter and 1.5 to 1.8 m in length [8]. The glassy alloy tubes have been produced in Zr-, La- and Pd-based alloy systems. Thus, bulk metallic alloys used for human life after recorded history had been limited to a crystalline structure for long period before 1990, but since then we can also utilize another type of bulk metallic alloys consisting of a glassy structure.

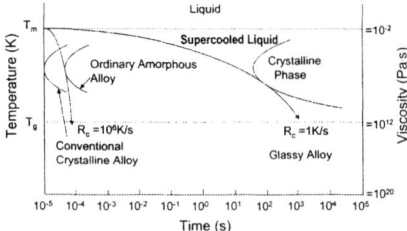

FIGURE 1. Schematic illustration of continuous cooling transformation curves of metallic liquids for different alloys in conventional crystalline, ordinary amorphous and bulk glassy types.

FEATURES OF ALLOY COMPONENTS

Table 1 summarizes typical bulk glassy alloy systems reported up to date together with the calendar years when the first paper or patent of each alloy system was published. The alloy systems can be divided into tow types, i.e., nonferrous and ferrous alloy types. As the nonferrous alloy type, one can see Mg-, lanthanide (Ln)-, Zr-, Ti-, Pd- and Cu-based alloys, while the ferrous alloy type is composed of a variety of Fe-, Co- and Ni-based alloys. It is also noticed that more than 50 % in the numbers of the alloy systems were developed for the last five years, indicating that this research field has been significantly extended even at present. Furthermore, we can recognize that bulk glassy alloys consisting only of metallic components are formed in a number of alloy systems except Fe- and Co-based alloys. Thus, most of the bulk glassy alloys found after 1988 do not include metalloid elements, in good contrast to Pd-Ni-P [9] and Pt-Ni-P [10] systems which are previously known as typical bulk glassy alloy systems. When we look at the features of alloy components for more than one thousand kinds of alloy systems shown in Table 1, one can notice the existence of simple three component rules [2-5], i.e., (1) multi-component consisting of more than three elements, (2) significant atomic size mismatches above 12 % among the three main elements, and (3) negative heats of mixing among the main three elements. As described later, the total solute contents in the bulk glassy alloys with the three component rules are in the range of 25 to 45 at% where eutectic points with low liquidus temperatures are always included. The solute content factor can be regarded as a complementary parameter which is necessary for optimization of an alloy composition for stabilization of supercooled liquid.

TABLE 1. Typical bulk glassy alloy systems reported to date together with the calendar years when the first paper or patent of each alloy system was published.

1. Nonferrous alloy systems	Year
Mg-Ln-M (Ln=lanthanide metal, M=Ni,Cu,Zn)	1988
Ln-Al-TM (TM=VI ~ VIII group transi. metal)	1989
Ln-Ga-TM	1989
Zr-Al-TM	1990
Ti-Zr-TM	1993
Zr-Ti-TM-Be	1993
Zr-(Ti,Nb,Pd)-Al-TM	1995
Pd-Cu-Ni-P	1996
Pd-Ni-Fe-P	1996
Ti-Ni-Cu-Sn	1998
Cu-(Zr,Hf)-Ti	2001
Cu-Zr, Cu-Hf	2001
Cu-(Zr,Hf)-Ti-(Y,Be)	2001
Cu-(Zr,Hf)-Ti-(Fe,Co,Ni)	2002
Ca-Mg-Ag-Cu	2002
Cu-(Zr,Hf)-Al	2003
Cu-(Zr,Hf)-Al-(Ag,Pd)	2003
2. Ferrous alloy systems	
Fe-(Al,Ga)-(P,C,B,Si,Ge)	1995
Fe-(Nb,Mo)-(Al,Ga)-(P,B,Si)	1995
Co-(Al,Ga)-(P,B,Si)	1996
Fe-(Zr,Hf,Nb)-B	1996
Co-(Zr,Hf,Nb)-B	1996
Fe-Co-Ln-B	1998
Fe-Ga-(Cr,Mo)-(P,C,B)	1999
Fe-(Nb,Cr,Mo)-(C,B)	1999
Ni-(Nb,Cr,Mo)-(P,B)	1999
Co-Ta-B	1999
Fe-Ga-(P,B)	2000
Ni-Zr-Ti-Sn-Si	2001
Ni-(Nb,Ta)-Zr-Ti	2002
Fe-B-Si-Nb	2002
Co-Fe-B-Si-Nb	2002
Co-Fe-Ta-B	2003
Ni-Nb-Sn	2003

STABILIZATION MECHANISM

It is important to clarify the reason why the alloys with the three component rules can have high glass-forming ability through stabilization of supercooled liquid. We paid attention to local atomic configurations in their multi-component glassy alloys. Therefore, we tried to clarify their glassy structure by using various advanced analytical techniques such as anomalous X-ray scattering, neutron scattering, small angle scattering and high-resolution transmission electron microscopy. As a result, it has been reported that the bulk glassy alloys have novel local atomic configurations with the following three features [2-5,8], i.e., (1) high degree of dense packed atomic configurations, (2) new local atomic configurations which are different from those for the corresponding equilibrium phases, and (3) long-range homogeneity with attractive interaction. In addition, it is important to point out the difference in the local atomic configurations between the metal-metal and the metal-metalloid type alloys. Figure 2 shows a schematic illustration of the local atomic configurations among the three types of bulk glassy alloys. The metal-metal type glassy alloys as exemplified for Zr-Al-Ni-Cu and Hf-Al-Ni-Cu systems are mainly composed of an icosahedral atomic configuration, while the metal-metalloid type glassy alloys such as Fe-Ln-B and Fe-M-B (M=Zr, Hf, Nb, Ta)-B systems have a network-like structure consisting of trigonal prisms through glue atoms of Ln or M element. The Pd-Cu-Ni-P glassy alloys are mainly composed of transformed tetragonal dodecahedrons consisting of Pd, Cu and P elements and triacontrahedrons consisting of Pd, Ni and P elements. Any types of these local atomic configurations can be regarded as more homogenized, more densely packed and novel structures. Although the bulk glassy alloys include such ordered atomic configurations on a short-range scale, their structures on a much longer range scale have been analyzed to consist of random atomic configurations without any periodic or quasi-periodic atomic arrays.

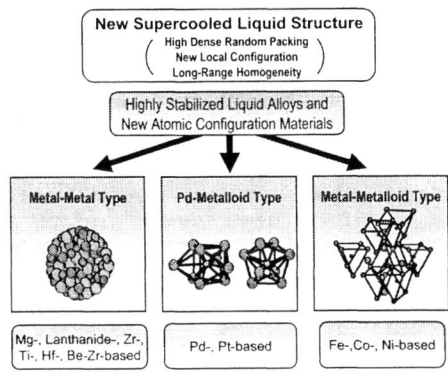

FIGURE 2. Schematic illustration of new supercooled liquid structure in metal-metal, metal-metalloid and Pd-based bulk glassy alloys.

Here some experimental data on the local atomic configurations are presented for Zr- and Fe-based bulk glassy alloys. In the high-resolution TEM image of the as-cast $Zr_{70}Al_{7.5}Ni_{10}Cu_{12.5}$ alloy shown in Fig. 3, one can see a number of unique contrast regions with a size of about 0.5 nm which look like a transverse cross section of vegetable onions [11]. This unique contrast image and its size agree well with the computer simulated contrast image [12] which is obtained in the assumption of the existence of icosahedral clusters with a size of about 1 nm. Based on the coordination numbers and atomic distances obtained by the anomalous X-ray scattering technique, the local atomic structure models of Fe-Ln-B and Fe-TM-B glassy alloys are shown in Fig. 3 [13,14]. These glassy alloys are composed of network-like structures in which trigonal prisms consisting of Fe and B atoms are connected with each other through glue atoms of Ln or TM element.

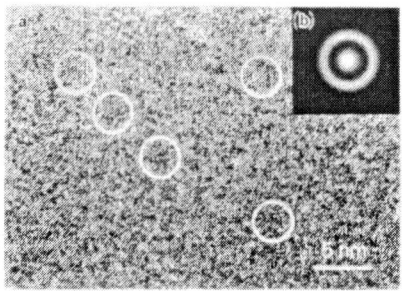

FIGURE 3. High-resolution TEM image and selected-area electron diffraction pattern of $Zr_{70}Al_{7.5}Ni_{10}Cu_{12.5}$ glassy alloy.

The special liquids with the unique local atomic configurations which can be obtained in the alloys with the three component rules can have high liquid/solid interfacial energy leading to the suppression of nucleation of a crystalline phase. Atomic rearrangements are also suppressed in the liquids, resulting in an increase of glass transition temperature (T_g). The liquid also has the necessity of atomic rearrangements on a long-range scale for the progress of crystallization, resulting in the suppression of growth reaction of a crystalline phase. Owing to the combination of these three factors, the special multi-component alloys can have higher reduced glass transition temperatures (T_g/T_l) through a decrease of liquidus temperature (T_l) and an increase of T_g.

In addition, we examined a primary crystallization phase from supercooled liquid because the phase is expected to have a close relation to the local atomic configurations of the supercooled liquid. Table 2 summarizes primary crystallization phases in the alloy systems examined in the present study [8]. When much attention is paid to metal-metal type glassy alloys, the primary crystallization phase can be limited to big-cube and icosahedral phases in spite of a variety of alloy components. In the Zr-Al-Ni-Cu and Hf-Al-Ni-Cu glassy alloys with the largest supercooled liquid region, the primary crystallization phase is a big cube Zr_2Ni-type phase with a large lattice parameter of 1.12 nm including 96 atoms. In addition, the big cube phase always includes local icosahedral atomic configurations. When special elements with nearly zero or positive heats of mixing against the other constituent elements are added, the primary crystallization phase changes to an icosahedral phase. It is thus said that the addition of the elements leading to the deviation from the three component rules causes the precipitation of the icosahedral phase. The precipitation of the icosahedral phase from the glassy matrix takes place very homogeneously. The particle size is as small as 10 to 20 nm and the volume fraction of the precipitates is above 70 %. We have also clarified that the critical size for the transition from the glassy phase to the icosahedral phase lies in the vicinity of 1 nm from the high-resolution TEM image and nano-beam electron diffraction patterns shown in Fig. 4 [11]. These results indicate that the metal-metal type bulk glassy alloys in the Zr-, Hf and Ti-based alloy systems are composed mainly of icosahedral atomic configurations with a size less than 0.5 nm.

TABLE 2. The summary of the primary phase of the crystallization process of a number of bulk glassy alloys.

*P	System	Typical alloy composition (at%)
B	Zr-Al-Ni-Cu	$Zr_{65}Al_{7.5}Ni_{10}Cu_{17.5}$
	Zr-Al-Ni-Cu-(Cr,Mo)	$Zr_{65}Al_{7.5}Ni_{10}Cu_{12.5}(Cr,Mo)_5$
	Zr-Ni-(Fe,Co)	$Zr_{70}Ni_{10}(Fe,Co)_{20}$
	Hf-Al-Ni-Cu	$Hf_{65}Al_{7.5}Ni_{10}Cu_{17.5}$
I	Zr-Al-Cu-O*	$(Zr_{65}Al_{7.5}Cu_{27.5})+O$
	Zr-Al-Ni-Cu-O*	$(Zr_{65}Al_{7.5}Ni_{10}Cu_{17.5})+O$
	Zr-Ti-Al-Ni-Cu-O*	$(Zr_{59}Ti_3Al_{10}Ni_8Cu_{20})+O$
	Zr-Al-Ni-Cu-(Ag,Pd,Au,Pt)	$Zr_{65}Al_{7.5}Ni_{10}Cu_{7.5}(Ag,Pd,Au,Pt)_{10}$
	Zr-Al-Ni-Cu-(Nb,Ta,V)	$Zr_{65}Al_{7.5}Ni_{10}Cu_{12.5}(Nb,Ta,V)_5$
	Zr-Al-Ni-(Ag,Pd,Au,Pt)	$Zr_{65}Al_{7.5}Ni_{10}(Ag,Pd,Au,Pt)_{17.5}$
	Zr-Al-Cu-Pd	$Zr_{65}Al_{7.5}Cu_{17.5}Pd_{10}$
	Zr-Ni-(Pd,Au,Pt)	$Zr_{70}Ni_{10}(Pd,Au,Pt)_{10}$
	Zr-Cu-Pd	$Zr_{70}Cu_{10}Pd_{20}$
	Zr-Pd	$Zr_{70}Pd_{30}$
	Zr-Pt	$Zr_{80}Pt_{20}$
	Hf-Al-Ni-Cu-Pd	$Hf_{65}Al_{7.5}Ni_{10}Cu_{12.5}Pd_5$
	Hf-Al-Ni-Ag	$Hf_{65}Al_{7.5}Ni_{10}Ag_{17.5}$
	Hf-Cu-(Pd,Pt)	$Hf_{70}Cu_{20}(Pd,Pt)_{10}$
	Ti-Zr-Ni	$Ti_{45}Zr_{38}Ni_{17}$
	Ti-Zr-Ni-Cu	$Ti_{60}Zr_{15}Ni_{15}Cu_{10}$
C	Fe-(Nb,W,Ta)-B	$Fe_{70}(Nb,Ta,W)_{10}B_{20}$

*P: Phase,, B: Big cube, I: Icosahedral, C: Complex fcc

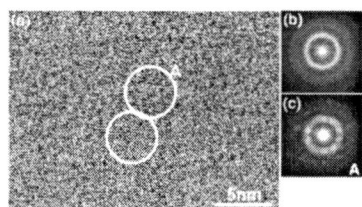

FIGURE 4. TEM images of $Zr_{70}Pd_{30}$ alloy annealed at 690 K for 120 s. (a) High resolution TEM image, (b) selected-area electron diffraction pattern taken from an area of 1 mm in diameter, and (c) nano-beam electron diffraction pattern taken from the area marked with A in (a).

ESTIMATIONS OF GLASS-FORMING ABILITY AND GLASS COMPOSITION RANGE

By using the three dominant factors for stabilization of supercooled liquid, the glass-forming ability has been estimated [15]. In the estimation, the three factors were changed to thermodynamical parameters of mismatch entropy for the atomic size mismatch in the assumption of hard sphere model and mixing enthalpy for negative heats of mixing in the assumption of regular solution model. The relation including the two thermodynamical factors to estimate the critical cooling rate has been proposed. It has subsequently been reported that the critical cooling rate obtained from the new relation can predict rather well the order of the critical cooling rate for a number of bulk glassy alloys. Furthermore, the glass-forming composition ranges and the best alloy compositions with the largest supercooled liquid region prior to crystallization have also been predicted by using Miedema's semi-empirical model [16] and quasi-chemical model [17] including interaction parameters among constituent elements. In the criterion of ΔH (Solid Solution) $- \Delta H$ (Glass) > 0 for glass formation, it has been reported that the glass-forming composition ranges in La-Al-Ni and Zr-Al-Ni systems agree rather well with the experimental data, as exemplified for La-Al-Ni system in Fig. 5 [18]. It has also been recognized that the alloy composition with Cowley's order parameter [19] of the largest negative value agrees well with that with the largest supercooled liquid region prior to crystallization, as shown for La-Al-Ni alloys in Fig. 5 [20]. The good agreement between Cowley's order parameter and the highest stability of supercooled liquid indicates that the high stability of supercooled liquid may originate from the development of short-range ordered atomic configurations such as icosahedral or network-like atomic configuration, being consistent with the above-described interpretation.

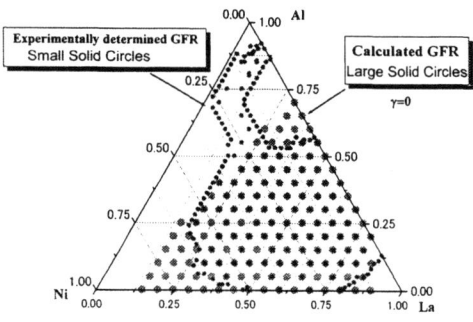

FIGURE 5. Predicted and experimentally determined composition ranges for glass formation in La-Al-Ni system.

EXPERIMENTAL DETERMINATION OF THERMODYNAMICAL PROPERTIES

The use of bulk glassy alloys has enabled us to measure the temperature dependence of specific heat over the whole supercooled liquid region. Based on the experimental data, the temperature dependence of enthalpy, entropy and Gibbs free energy has been determined quantitatively as shown for $Pd_{40}Cu_{30}Ni_{10}P_{20}$ alloy in Fig. 6 [21]. The extremely high glass-forming ability for the Pd-based alloy is partly due to the low free energy for the liquid/crystal transition. In addition, the fictive glass transition temperature can be also determined from the temperature dependence of entropy.

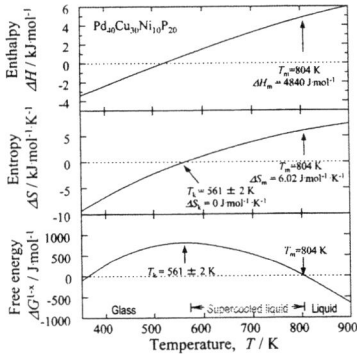

FIGURE 6. Temperature dependence of enthalpy (ΔH), entropy (ΔS) and Gibbs free energy (ΔG) for $Pd_{40}Cu_{30}Ni_{10}P_{20}$ glassy alloy.

MECHANICAL PROPERTIES

The combination of new alloy components, new local atomic configurations and three dimensional (bulk) material form has resulted in unique mechanical properties for bulk glassy alloys. Figure 7 summarizes the relation between tensile strength and Young's modulus for typical bulk glassy alloys, together with the data of conventional crystalline alloys [2-5]. In comparison with the tensile strength at the same Young's modulus, the strength of the bulk glassy alloys is about three times higher than those for conventional crystalline alloys. On the other hand, the Young's modulus of the bulk glassy alloys at the same tensile strength is about one-third as high as those for crystalline alloys. In addition, the elastic strain limit represented by the slope in the relation is 0.019 for bulk glassy alloys, being also about three times larger than that (0.65 %) for conventional crystalline alloys. Figure 7 also shows that the high strength exceeding 2000 MPa which cannot be obtained for conventional crystalline alloys is achieved by choosing Ni- and Cu-based bulk glassy alloys. We have also examined torsional rupture strength and elastic torsional strain limit of bulk glassy alloys by using the rod form specimens of 10 mm in diameter which have the size and dimension in the JIS criterion, in comparison with those for conventional plain carbon (0.4 mass%C) steel. As shown in Fig. 8, the elastic twist angle is about 18 degrees for $Pd_{40}Cu_{30}Ni_{10}P_{20}$ glassy alloy rod and 4 degrees for the carbon steel [22]. In addition, the elastic shearing stress is about 900 MPa which is about three times higher than that (300 MPa) for the plain carbon steel. This difference is the same as that for the uniaxial tensile strength and Young's modulus data.

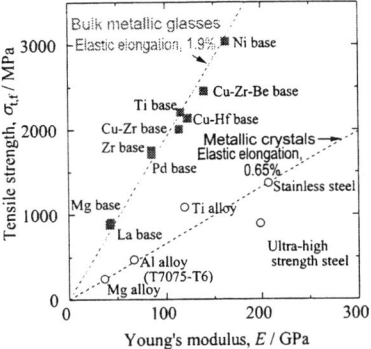

FIGURE 7. Relation between tensile strength and Young's modulus for bulk glassy alloys. The data of crystalline metallic alloys are also shown for comparison.

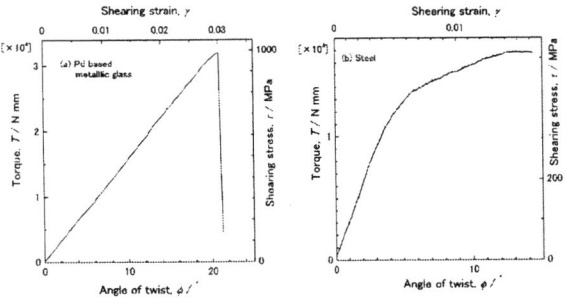

FIGURE 8. Relation between torque and twist angle for Pd-based bulk glassy alloys and 0.65 mass% carbon steel. The shearing stress and strain calculated from the torque and twist angle relation are also shown for comparison.

We have further examined fatigue strength and fatigue crack propagation behavior for the bulk glassy alloys by using three kinds of specimens in smooth and notch forms. Figure 9 shows the relation between the applied cycled stress ratio (cycled applied stress/tensile strength) and the number of cycles up to failure for Zr- and Pd-based bulk alloys in glassy single phase and nanocrystalline phase states [23].

The fatigue limit defined by the maximum applied cycle ratio at which no final fracture is observed after 10^7 cycles is about 0.04 for two kinds of Zr-based bulk glassy alloys and 0.2 for the Pd-based glassy alloy. It is further noticed that the fatigue limit increases significantly to 0.13 by the dispersion of nanocrystals into the glassy phase. It has been reported that the crack initiation of the Zr-based bulk glassy alloys in glassy single and nanocrystalline phases occurs at a very early stage of about 5 % to the total number of cycles and most of the cycles are spent for propagation of fatigue crack [23]. Figure 10 shows the relation between fatigue crack propagation rate (da/dn) and normalized stress intensity factor range ($\Delta K/E$) for $Zr_{55}Al_{10}Cu_{30}Ni_5$ glassy alloy [23]. The fatigue crack propagation rate is in proportional to the stress intensity factor range in spite of different fatigue test conditions and can be measured as da/dn mm/cycle. In addition, the amplitude of displacement at the fatigue crack tip as a function of cycle remains unchanged for the un-ruptured specimen (at low applied stress level) and decreases gradually for the ruptured specimen (at high applied stress level). The gradual decrease suggests that structural relaxation-induced hardening occurs in the vicinity of fatigue crack tip.

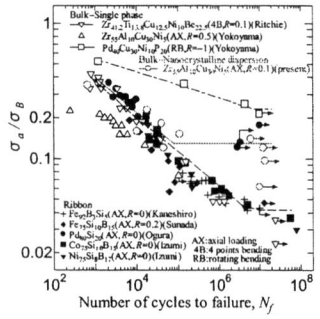

FIGURE 9. S-N curve of the Zr-based nanocrystalline bulk metallic glass.

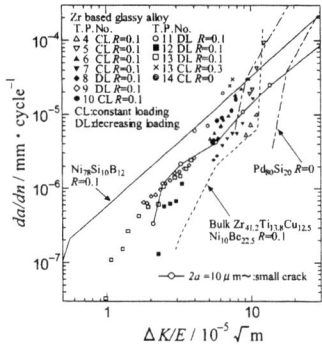

FIGURE 10. The relation between fatigue crack propagation rate (da/dn) and normalized stress intensity factor range ($\Delta K/E$) for $Zr_{55}Al_{10}Cu_{30}Ni_5$ glassy alloy.

CORROSION RESISTANCE

Considering the future application field of bulk glassy alloys exhibiting good mechanical properties, it is important for the glassy alloys to exhibit simultaneously high corrosion resistance. The newly developed Ni-based bulk glassy alloys with high tensile strength and good ductility [24] also exhibited high corrosion resistance in various chemical solutions such as 1 N hydrochloric acid, 6N hydrochloric acid, 1N sulfuric acid and 3% sodium chloride [25]. Figure 11 shows the cathodic/anodic polarization curves of $Ni_{55}Co_5Nb_{20}Zr_{10}Ti_{10}$ glassy alloy rod with a diameter of 3 mm which was measured in 1 N hydrochloric acid at 298 K in air. The Ni-based bulk glassy alloy has higher pitting corrosion potential and much lower anodic current density as compared with pure metals such as Ni, Co, Ti, Zr and Nb. The addition of Ta to the Ni-based glassy alloy causes a further increase in the corrosion resistance [26]. It has also recognized that the addition of Nb or Ta to $(Cu_{0.6}Hf_{0.25}Ti_{0.15})_{100-x}M_x$ (M=Nb or Ta) bulk glassy alloys increases drastically the corrosion resistance through the increase in the pitting corrosion potential and the decrease in the anodic current density. We could not detect any weight loss for the Cu-based glassy alloys containing 8 %Nb in 1 N hydrochloric acid and 4 %Nb in 3% sodium chloride [27]. The simultaneous achievement of high strength, high ductility, high glass-forming ability leading to the formation of bulk glassy alloy and high corrosion resistance has not been obtained for any kinds of amorphous-type alloys and is encouraging for future development of the Ni- and Cu-based bulk glassy alloys as a new type of functional materials utilizing their chemical properties.

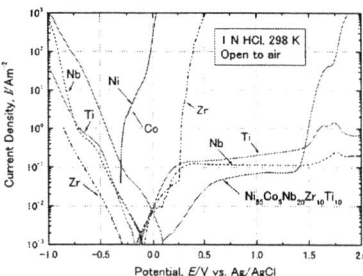

FIGURE 11. The cathodic/anodic polarization curves of $Ni_{55}Co_5Nb_{20}Zr_{10}Ti_{10}$ glassy alloy in 1N HCl solution open to air at 298 K.

MAGNETIC PROPERTIES

Novel bulk glassy alloys in Fe- and Co-based systems also exhibit good soft magnetic properties.

However, their characteristics were expected to be the same as those for amorphous type alloys which required high cooling rates for their formation. Recently, it has been clarified [28] that the soft magnetic properties of Fe-based glassy alloys are much superior to those [29] for Fe-based amorphous alloys. Figure 12 summarizes the relation between coercivity and magnetostriction to magnetization ratio for Fe-based glassy alloys, together with the data of Fe-based amorphous alloys [28]. A rather good linear relation is recognized, indicating that the relation can be expressed by the following relation in Eq. (1);

$$H_c \propto \Delta V \sqrt{\rho_d} \frac{\lambda_s}{J_s}, \qquad (1)$$

where ΔV is the volume of defects, ρ_d the density of defects, λ_s the saturation magnetostriction and J_s the saturated magnetization.

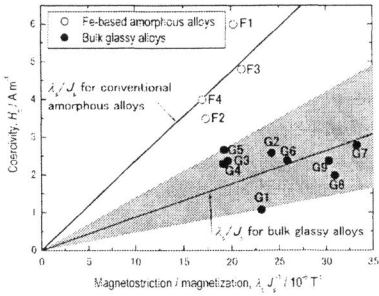

FIGURE 12. The relation between coercivity and magnetostriction to magnetization ratio for Fe-based glassy alloys (G1 to G9), together with the data of Fe-based amorphous alloys (F1 to F4).

The slope is dependent on the volume and density of internal defects, i.e., structural inhomogeneity in their alloys [29]. It is seen that the slope of Fe-based glassy alloys is much smaller than that of Fe-based amorphous alloys, indicating that the Fe-based glassy alloys have more homogenized atomic configurations as compared with ordinary Fe-based amorphous alloys. It is therefore concluded that such a more homogenized structure is the origin for the lower coercivity of Fe-based glassy alloys. Reflecting the lower coercivity, Fe-based glassy alloys exhibit much higher permeability as compared with Fe-based amorphous alloys in conjunction with rather high saturation magnetization of 1.3 to 1.5 T [30]. By use of the good soft magnetic properties, Fe-based glassy alloys have been used as consolidated magnetic cores of common mode choke coils and noise filters for electrical power supplies through the development of inexpensive mass-production process consisting of water atomization, followed by mixing with insulating and lubricating materials, drying, crushing, classification, consolidation and then heat treatment [31]. Figure 13 shows some characteristics of $Fe_{74.43}Cr_{1.96}P_{9.04}C_{2.16}B_{7.54}Si_{4.87}$ magnetic cores thus produced [31]. The magnetic core exhibits a nearly constant relative permeability in a wide frequency range up to several MHz, good linearity in the relation between permeability and DC bias field, and much lower core losses as compared with Fe-Al-Si sendust and Ni-Fe-Mo supermalloy.

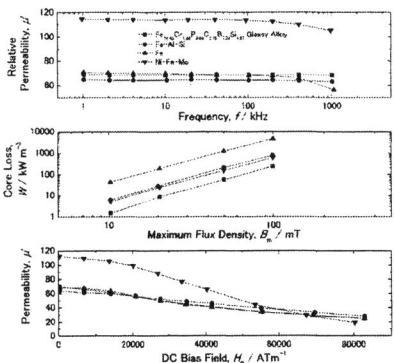

FIGURE 13. Relative permeability (μ), core loss (W) and peameability (μ), of $Fe_{74.43}Cr_{1.96}P_{9.04}C_{2.16}B_{7.54}Si_{4.87}$ magnetic core. The data of Fe, Fe-Al-Si sendust and Ni-Fe-Mo supermalloy are also shown for comparison.

Bulk glassy alloy rods with diameters up to 2 mm can be produced for $Fe_{67}Co_{9.5}Nd_3Dy_{0.5}B_{20}$ alloy by copper mold casting. When the glassy alloy is annealed for 420 s at 903 K, the annealed alloy exhibits rather good hard magnetic properties, i.e., maximum energy product $(BH)_{max}$ of 110 kJ/m^3 and remanence of 1.35 T [32]. The annealed alloy has a nanostructure consisting of Fe_3B and $Nd_2Fe_{14}B$ phases surrounded by the remaining glassy phase. The particle size is about 20 nm for Fe_3B and 15 to 20 nm for $Nd_2Fe_{14}B_{1.35}$ and the width of the remaining glassy phase is measured as 2 to 5 nm. It is believed that the achievement of the rather good hard magnetic properties in the nanostructure state including the remaining glassy phase is the first evidence for the Fe-Ln-B type alloys with high B concentrations.

VISCOUS FLOW WORKING

Bulk glassy alloys exhibit a unique transformation sequence of glass transition, followed by a supercooled liquid region and then crystallization which is in contrast to the direct transformation from amorphous to crystalline phase for amorphous alloys. Figure 14 shows the viscosity and true stress as a function of strain rate in a wide temperature rage including the supercooled liquid region for $Zr_{65}Al_{10}Ni_{10}Cu_{15}$ alloy [2-5]. It is noticed that the viscosity in the supercooled liquid region remains constant in a wide strain rate range, indicating the achievement of the Newtonian viscous flow. In the Newtonian flow region, a good linear relation between true stress (σ) and strain rate ($\dot{\varepsilon}$) is recognized and can be expressed by the relation of $\sigma = k\dot{\varepsilon}^m$. The slope corresponding to the strain rate sensitivity exponent (m-value) shows 1.0, indicating the achievement of the ideal superplasticity in the supercooled liquid region. By use of the ideal superplasticity, it has been confirmed that the La-Al-Ni-Co bulk glassy alloy rod exhibits extremely large elongations reaching 1.8×10^6 % in the supercooled liquid region. In addition, the bulk glassy alloys have good transcription ability. In a one cycle die pressing treatment in the supercooled liquid region, the Zr-Al-Ni-Cu bulk glassy alloys exhibit very smooth surface on a nanometer scale as well as surface steps with a minimum size of 0.5 μm reflecting the shape and size of V-grooved silicon die [33]. By the same pressing treatment using silicon die, nanoscale imprinted patterns with a size of about 200 nm have been produced for the Zr-based bulk glassy alloys [34]. The imprinted pattern is expected to be used as the original disc plates for information-storage medium and hence the further refinement of the imprinted pattern to about 20 nm has been tried.

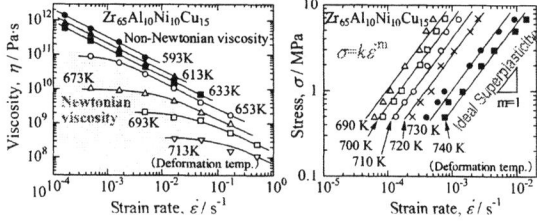

FIGURE 14. The viscosity and true stress as a function of strain rate in a wide temperature rage including the supercooled liquid region for $Zr_{65}Al_{10}Ni_{10}Cu_{15}$ alloy.

NANOCRYSTALLINE AND NANOQUAICRYSTALLINE BULK ALLOYS

All the bulk glassy alloys described above are obtained in the alloy systems where the three component rules are satisfied. When the three component rules are deviated intentionally by adding special elements with strongly positive heats of mixing or nearly zero and positive heats of mixing against the other constituent elements, the annealing of their glassy phase as well as the control of cooling rate from the liquid state causes the formation of nanocrystalline and nanoquasicrystalline bulk glassy alloys, as illustrated in Fig. 15 [2-5]. Furthermore, the alloys consisting of three elements with nearly zero and positive heats of mixing can have supersaturated bcc and fcc solid solutions with nanograin sizes. Figure 16 shows bright-field TEM images and selected-area electron diffraction patterns of $Zr_{60}Cu_{20}Pd_{10}Al_{10}$ glassy alloy in as-quenched and annealed states. Nanoscale big cube type compounds with a size below 20 nm precipitate homogeneously in the glassy matrix. The nanostructure alloys exhibit improved mechanical properties, i.e., from 1800 to 1920 MPa for tensile strength, from 88 to 97 GPa for Young's modulus and from 2.2 to 2.3 % for elongation in the volume fraction range of about 30 %. When the precipitation size is controlled to be about 5 nm, the ductility is significantly improved as exemplified in Fig. 17. The nanoscale mixed phase alloy does not fracture during three point bending deformation, in good contrast to the fracture behavior for the glassy single phase alloy. In addition, the $Ti_{50}Zr_{30}Nb_{10}Ta_{10}$ bcc solid solution alloy including solute elements with nearly zero and positive heats of mixing against Ti which was subjected to cold rolling and then annealing consists of fine grains with a size of about 5 nm and exhibits unique mechanical properties, i.e., high tensile strength of 1070 MPa, low Young's modulus of 42 GPa and large elastic elongation limit of 1.7 %. It is noticed that the features of these mechanical properties resemble to those for bulk glassy alloys.

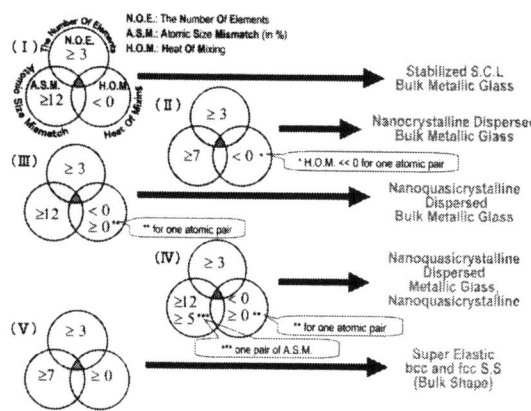

FIGURE 15. The relations between the factors described in the three component rules and the resultant alloys.

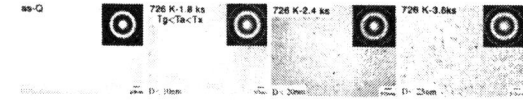

FIGURE 16. Bright-field TEM images and selected-area electron diffraction patterns of $Zr_{60}Cu_{20}Pd_{10}Al_{10}$ glassy alloy in as-quenched and annealed states.

FIGURE 17. Bright-field TEM image of $Ti_{50}Zr_{10}Nb_{10}Ta_{30}$ alloy subjected to cold rolling to 99 % reduction in thickness and then annealing for 0.36 ks at 873 K.

SUPERCOOLED LIQUID METALLURGY AND ITS MATERIAL SCIENCE MEANING

Figure 18 summarizes new material science and technological fields achieved by stabilization of supercooled metallic liquid [2-5,8]. The use of stabilized supercooled liquid with unique local atomic configurations has opened up new basic science fields leading to the clarification of structure, properties and phase transformation of supercooled liquid and liquid/solid interface structure. We can also produce bulk metallic glassy and bulk nonequilibrium crystalline alloy groups with various useful properties by various conventional casting processes. In addition, we have developed the novel forming processes which can be named as "supercooled liquid forming" as exemplified for near-net shape casting and forging, precision forming and fusion welding by use of Newtonian flow.

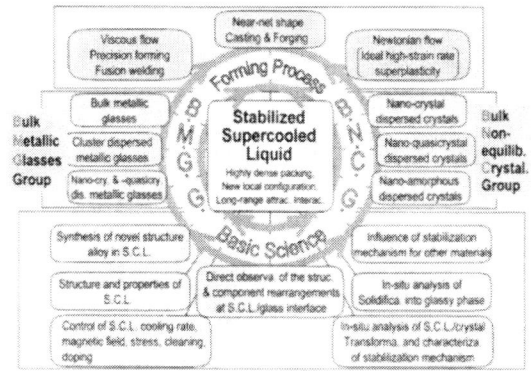

FIGURE 18. New material science and technological fields achieved by stabilization of supercooled metallic liquid.

FABRICATION AND PROPERTIES OF NONEQUILIBRIUM BULK ALLOYS BY

USE OF STABILIZATION EFFECT OF SUPERCOOLED LIQUID

When we choose the alloy systems with the three component rules, bulk glassy alloy groups can be obtained in the higher solute concentration range of 25 to 45 at%. Even in the lower solute concentration range, the alloys in which the three component rules are satisfied can keep the stabilization effect of supercooled liquid and have various nonequilibrium phases as summarized in Fig. 19, i.e., amorphous, nanocrystalline dispersed amorphous and nanoscale composite crystalline phases in the medium solute concentration range of 7 to 25 at%, nanocrystalline and nanoquasicrystalline dispersed alloys in the solute concentration range of 5 to 7 at%, and nonequilibrimum solid solution phase with nanograin size in the low solute concentration range below 5 at%. These nonequilibrium alloys exhibit various engineering properties depending on alloy components and solute contents.

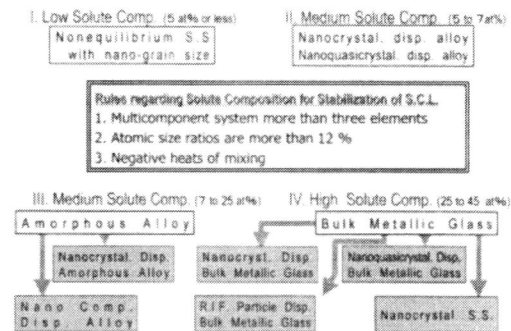

FIGURE 19. Fabrication and industrialization of high function nonequilibrium materials by use of stabilization phenomena of supercooled liquid.

Multi-component alloys in the solute concentration range of 7 to 25 at%

As a typical alloy system, it is meaning to present some data on Fe-TM-B (TM=Zr, Hf, Nb) alloys with the three component rules. Figure 20 shows a bright-field TEM image and selected-area electron diffraction patterns of $Fe_{88}Hf_{10}B_2$ amorphous alloy annealed for 3.6 ks at 873 K [35]. Fine bcc grains with a size of about 10 nm precipitate at a high volume fraction of about 70 % and are surrounded by the remaining glassy phase. The selected-area diffraction patterns taken from the regions 1 and 2 can be identified as bcc and glassy phases, respectively. The nanostructure bcc alloys exhibit excellent soft magnetic properties of high permeability of 20000 to 80000 at 1 kHz and rather high saturation magnetization of 1.5 to 1.7 T as well as very low core losses of 10^{-2} to 10^{-1} W/kg at 50 Hz in the maximum magnetic flux density of 0.4 to 1.5 T [35]. Owing to the excellent soft magnetic properties which cannot be obtained for any kinds of other soft magnetic alloys, the nanostructure bcc alloys have been used in some application fields of power transformers for switched power supplies, common mode choke coils and magnetic shielding etc [35]. In the application field to soft magnetic alloys, the use as core materials in pole transformers is thought to be the most important and challenging subject. With the aim of applying the nanostructure bcc alloys to pole transformers, we constructed a large scale melt-spinning equipment to produce a wide amorphous alloy sheet of 100 to 150 mm in width. Figure 21 shows outer shape and surface morphology of the melt-spun $Fe_{86}Nb_6B_8$ alloy sheet with a width of 100 mm produced by the large scale melt-spinning equipment [36]. By using the wide nanostructure bcc alloy sheets, we constructed a 20kVA pole-transformer as shown in Fig. 22 [36]. The pole-transformer exhibits core losses ranging from 0.15 to 0.18 W/kg at 50 Hz and 1.33 T, being much lower than those (0.28 to 0.45 W/kg) for conventional Fe-Si-B amorphous alloy sheet and oriented Fe-Si steel sheet. The development of mass-production technique of the wide nanocrystalline alloy sheet is important for the progress to the next stage for the achievement of real application to pole-transformer.

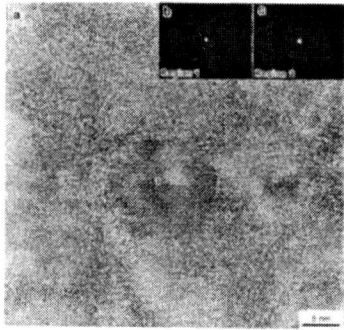

FIGURE 20. Bright-field TEM image and selected-area electron diffraction patterns of $Fe_{88}Hf_{10}B_2$ amorphous alloy annealed for 3.6 ks at 873 K.

FIGURE 21. Outer shape and surface morphology of the melt-spun $Fe_{86}Nb_6B_8$ alloy sheet with a width of 100 mm produced by the large scale melt-spinning equipment.

FIGURE 22. 20kVA pole-transformer conducted by using the wide nanostructure bcc alloy sheets.

Nanostructure Al- and Mg-based Alloys with Solute Concentrations of 3 to 15 at%

Figure 23 shows schematic illustrations of nonequilibrium structures in atomized Al- and Mg-based alloy powders and microstructures and mechanical properties of bulk alloys produced by warm extrusion of their atomized powders [37]. Even for Al- and Mg-based alloys with low solute concentrations of 3 to 15 at%, various kinds of nonequilibrium phases such as amorphous, amorphous plus fcc-Al, icosahedral quasicrystal (Q) plus fcc-Al and nanogranular hcp-Mg solid solution can be obtained in Al-Ni-Ln, Al-M-Ln-TM (M=V, Cr, Mn) and Mg-Zn-Y systems with the three component rules. By warm extrusion of their nonequilibrium alloy powders, we can produce various nanocomposite bulk alloys exhibiting good mechanical properties. The nanocomposite alloys consisting of Al + Al_3TM + $Al_{11}Ln_2$ and Al + Q produced by warm extrusion of amorphous (or amorphous + fcc-Al) and Al + Q phases, respectively, exhibit high tensile strength of 500 to 1000 MPa and elongation of 1 to 25 % which are superior to those for JIS Al-based alloys. These Al-based alloys have been used in wide application fields such as robot parts, machine parts, die cast molds for plastics, sporting goods materials, light weight tools, fishing reel, gears in bicycle and wheelchair parts.

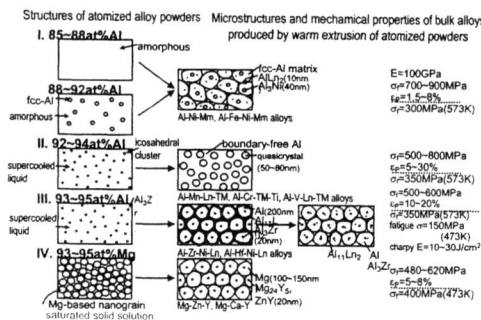

FIGURE 23. Schematic illustrations of nonequilibrium structures in atomized Al- and Mg-based alloy powders and microstructures and mechanical properties of bulk alloys produced by warm extrusion of their atomized powders.

The $Mg_{97}Zn_1Y_2$ alloy produced by warm extrusion of rapid solidification powders exhibits the highest specific tensile yield strength of 0.33 $MPa.kg^{-1}.m^3$ [38] as compared with conventional metallic alloys. Figure 24 shows bright-field TEM image and selected-area electron diffraction pattern of the Mg-Zn-Y alloy. The alloy consists of hcp-Mg phase with fine grain sizes of 100 to 200 nm which have not been obtained for any kinds of Mg-based alloys. In addition, one can observe a high density of plane faults in almost all hcp-Mg grains. This is believed to be the first evidence for the formation of plane faults in the hcp-Mg phase. The nanogranular Mg-based alloy exhibits high tensile yield strength of 450 to 620 MPa and large elongation of 5 to 8 % depending on the extrusion temperature. The strength values are about three times higher than those for conventional Mg-based alloys. The high strength values obey the Hall-Petch relation which can be extrapolated on the basis of the previous data on Mg-based alloys and hence the high strength may be concluded to result from the formation of the fine grain structure. As the origin for the formation of such a fine grain structure, we paid attention to the plane fault structure. When the specimen is tilted to an appropriate Bragg reflection condition, we can always observe a high density of plane faults as exemplified in Fig. 25 [39,40]. The selected-area electron diffraction pattern includes extra reflection spots at the positions of n/3 to (0001)hcp reflection spot, in addition to the ordinary hcp reflection spots, indicating the formation of a long-periodic hexagonal structure in which the periodicity is three times longer than that for ordinary hcp-Mg phase. From the high-resolution TEM image shown in Fig. 26, the atomic array in the faulted region takes place in a configuration mode of ABACAB which is different from the ACACAC configuration mode for hcp-Mg phase. In the Mg-Zn-Y alloys produced at higher cooling rates, we also detected the formation of another type of long-periodic hexagonal structure with 14 layered periodicity. Figure 27 shows the high-resolution TEM image revealing the atomic arrays of the constituent atoms in the 14-layered hexagonal phase [41]. The atoms array in the configuration mode of ACACACABCBCBCB and one misfit array per seven atomic arrays can be recognized as marked with arrows. In the misfit regions of atomic arrays, one can see the periodic enrichment of Y and Zn elements in the high-angle annular dark-field (HAADF) image. The periodic enrichment arrays of Y and Zn elements have been interpreted to originate from the preferential precipitation of short-range ordered Y-Zn atomic pairs with more strongly negative heat of mixing as compared with the other atomic pairs in the misfit atomic arrays per seven stacking atomic layers. The enrichment of Y and Zn elements on an atomic scale at the special misfit sites seems to increase thermal stability of faulted atomic configurations which can play a dominant role in the formation of nanogranular hcp Mg phase.

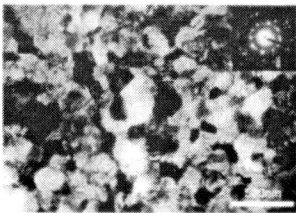

FIGURE 24. Bright-filed TEM image and selected-area electron diffraction pattern of $Mg_{97}Zn_1Y_2$ alloy produced by warm extrusion of atomized powder at 573 K.

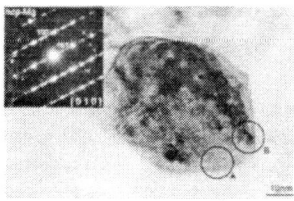

FIGURE 25. High-resolution TEM image and selected-area electron diffraction pattern of the RS/PM $Mg_{97}Zn_1Y_2$ alloy.

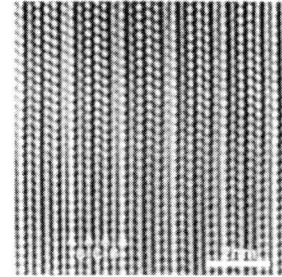

FIGURE 26. High-resolution TEM image of the RS/PM $Mg_{97}Zn_1Y_2$ alloy.

CONCLUSIONS

By use of spontaneous formation of novel short-range ordered atomic configurations in the supercooled liquid of the special component alloys with the three component rules, we have developed various kinds of nonequilibrium alloys with functional properties in a bulk form. The utilization of such local atomic configurations can be also regarded as a kind of subnanoscale structure control method. It is therefore believed that the development of various nonequilibrium bulk alloys with useful engineering properties by use of the subnanoscale structure control method becomes more and more significant in the near future.

REFERENCES

1. Cahn, R. W., *Rapidly Solidified Alloys*, edited by H. H. Liebermann Marcel Dekker, New York, 1993, pp.1-788.

2. Inoue, A., *Acta Mater.* **48** 279-306 (2000).

3. Inoue, A., *Mater. Sci. Eng.* **A304-306** 1-10 (2001).

4. Inoue, A., and Takeuchi, A., *J. Metastable and Nanocrystalline Mater.* **14** 1-12 (2002).

5. Inoue, A., and Takeuchi, A., *Mater. Trans.* **43** 1892-1906 (2002).

6. Inoue, A., Kimura, H. M., and Takeuchi, A., Mater. Sci. Forum. **426-432** 3-10 (2003).

7. Nishiyama, N. and Inoue, A., *Intermetallics* **10** 1141-1147 (2002).

8. Inoue, A., and Takeuchi, A., *Mater. Sci. Eng.* (2004) in press.

9. Bagley, B. G., and DiSalvo F. J., Amorphous Magnetism, edited by H.O. Hooper and A.M. deGraaf, New York, Plenum, 1973, pp. 143-148.

10. Chen, H. S., *Acta Metall.* **22** 1505-1511 (1974).

11. Saida, J., Matsushita, M., and Inoue, A., *Appl. Phys. Lett.* **79** 412-414 (2001).

12. Takagi, T., Ohkubo, T., Hirotsu, Y., Murty, B. S., Hono, K., and Shindo, D., *Appl. Phys. Lett.* **79** 485-487 (2001).

13. Imafuku, M., Sato, S., Koshiba, H., Matsubara, E., and Inoue, A., *Mater. Trans., JIM* **41** 1526-1529 (2000).

14. Imafuku, M., Yaota, K., Sato, S., Zhang, W., and Inoue, A., *Mater. Trans., JIM* **40** 1144-1148 (1999).

15. Takeuchi, A., and Inoue, A., *Mater. Trans., JIM* **41** 1372-1378 (2000).

16. *Cohesion in Metals*: edited by de Boer, F.R.; Boom, R.; et al. North-Holland Publishing Co., 1988, pp. 1-758.

17. Desre, P. J., *Mat. Res. Soc. Symp.* **554** 51-62 (1999).

18. Shindo, T., Waseda, Y., and Inoue, A., Mater. Trans. **43** 2502-2508 (2002).

19. Cowley, J. M., *Phys. Rev.* **77** 669-675 (1950).

20. Shindo, T., Waseda, Y., and Inoue, A., Mater. Trans. **43** 351-357 (2003).

21. Nishiyama, N, Horino, M., Haruyama, O., and Inoue, A., *Appl. Phys. Lett.*, 76 3914-3916. (2000).

22. Fujita, K., Inoue, A., Zhang, T., and Nishiyama: N., Mater. Trans. **43** 1957-1960 (2002).

23. Fujita, K., Inoue, A., and Zhang, T., Mater. Trans., **42**, 1502-1508 (2001).

24. Inoue, A., Zhang, W., Zhang, T., Kurosaka, K., and Louzguine, D. V., *J. Metas. Nanocryst. Mater.* **15-16** 3-10 (2003).

25. Pang, S., Zhang, T., Asami, K., and Inoue, A., Mater. Trans. **43** 1771-1773 (2002).

26. Pang, S., Zhang, T., Asami, K., and Inoue, A., Mater. Trans. **44** (2002), in press.

27. Qin C., Asami, K., Zhang, T., Zhang, W., and Inoue, A., Mater. Trans. **44** 749-753 (2003).

28. Bitoh, T., Makino, A., and Inoue, A., Mater. Trans. **44** 2020-2024 (2003).

29. O'Handley, R. C., Narasimhan, M. C., and Sullivan, M. O., *J. Appl. Phys.* **50** 1633–1635 (1979).

30. Inoue, A., and Shen, B. L., *Mater. Sci. Eng.* (2004) in press.

31. Yoshida, S., Mizushima, T., Hatanai, T., and Inoue, A., IEEE Trans. Mag. **36** 3424-3429 (2000).

32. Zhang, W., Long, Y., Imafuku, M., and Inoue, A., Mater. Trans. **43** 1974-1978 (2002).

33. A. Inoue, private communication.

34. Saotome, Y., Ito, K., Zhang, T., and Inoue, A., *Scripta Mater.* **44** 1541-1545 (2001).

35. Inoue, A., and Makino, A., *Magnetic Properties of Nanocrystalline Materials* in *Nanoctructured Materials* edited by C. C., Koch, Data Corp (Noyes Publications), New York, 2002), pp. 355-395.

36. NEDO repert " Research and Development of materials for ultra-low-loss transformer" (2003) (in Japanese).

37. Inoue, A. *Hand book on the Physics and Chemistry of Rare Earths* **24** 83-219 (1997).

38. Kawamura, Y., Hayashi, K., Inoue, A., and Masumoto, T., *Mater. Trans.* **42** 1172-1176 (2001).

39. Abe, E., Kawamura, Y., Hayashi, K., and Inoue, A., *Acta Mater.* **50** 3845-3857 (2002).

40. Ping, D. H., Hono, K., Kawamura, Y., and Inoue, A., *Phil. Mag. Lett.* **82** 543-551 (2002).

41. Amiya, K., Ohsuna, T., and Inoue, A., *Mater. Trans.* **44** 2151-2156 (2003).

Logarithmic decay in a two-component model

Matthias Sperl

Physik-Department, Technische Universität München, 85747 Garching, Germany

Abstract. The correlation functions near higher-order glass-transition singularities are discussed for a schematic two-component model within the mode-coupling theory for ideal glass-transitions. The correlators decay in leading order like $-\ln(t/\tau)$ and the leading correction introduces characteristic convex and concave patterns in the decay curves. The time scale τ follows a Vogel-Fulcher type law close to the higher-order singularities.

INTRODUCTION

Mode-coupling theory for ideal glass-transitions (MCT) describes the transition from a liquid to a glass as a bifurcation in the equation for the long-time limit of the density-autocorrelation function $\phi_q(t) = \langle \rho_q^*(t)\rho_q \rangle / \langle |\rho_q|^2 \rangle$ for density fluctuations of wave-number modulus $q = |\vec{q}|$. If the long-time limit $f_q = \lim_{t\to\infty} \phi_q(t)$ reaches zero, $f_q = 0$, the system is in the liquid state. For $f_q > 0$, the system is in a glass state [1]. The theory has been worked out in detail for the hard-sphere system (HSS) [1, 2] where the results of experiments [3] and computer simulations [4] support the validity of MCT. In addition to liquid-glass transitions, MCT also allows for bifurcations of higher order [5]. It can be shown that only bifurcations of a certain hierarchy A_ℓ can arise from MCT [6] which are equivalent to the singularities in the real roots of polynomials of order ℓ upon variation of parameters [7]. The simplest singularity A_2 or *fold* generically occurs when only a single control parameter is changed and the liquid-glass transition is identified with this bifurcation. At the critical value of the control parameter the long-time limit f_q jumps from zero to f_q^c. Close to the singularity and for small $|\phi_q(t) - f_q|$, the correlation functions can be expanded in asymptotic series yielding power laws. A detailed analysis including leading and next-to-leading order results was given for the HSS [2, 8, 9].

An A_3-singularity or *cusp* can appear when two control parameters are varied. The A_3-singularity is the endpoint of a line of A_2-singularities which are identified as glass-glass-transition points where a first glass state, characterized by $f_q^1 > 0$, transforms discontinuously into a second glass state with $f_q^c > f_q^1$. At the A_3, this discontinuity vanishes and the glass-glass-transition line ends. Such cusp was predicted for particles interacting with a hard core and a short-ranged attraction [10, 11] where density and strength of the attraction are the control parameters. The glass-glass transition takes place between a state dominated by repulsion as in the HSS and a state dominated by attraction, the latter arrested state was proposed to be related to a gel [11].

Tuning a third control-parameter, the extension of the glass-glass-transition line can be varied and once the latter vanishes, the A_3-singularity merges with the liquid-glass-transition line and gives birth to a *swallowtail* or A_4-singularity. If the range of the attraction is of the order of 5% of the hard-core diameter such singularity is predicted for a square-well system with a strength of the attraction of several $k_\mathrm{B}T$ and a density which is slightly larger than the one for the glassy arrest in the HSS [12].

In contrast to the power-laws at the A_2-singularity, the dynamics close to higher-order singularities is ruled by the logarithm of the time [5]. The wave-vector dependent asymptotic solutions along with the corrections have been calculated in full generality [13]. The asymptotic laws were demonstrated for the A_3-singularity of a two-component model [13], and for the higher-order singularities of microscopic models with short-ranged attraction [14]. Logarithmic decays compatible with the predicted scenarios were found in recent computer simulation studies [15, 16].

In the vicinity of an A_3-singularity, different liquid-glass-transition lines cross, and the dynamics in the liquid regime close to this crossing is influenced by three different singularities. Dynamical scenarios predicted by theory in that region [10, 12, 17] were found in experiments [18, 19, 20, 21, 22, 23] and computer-simulation studies [19, 24]. These findings indicate that such rich dynamics is indeed relevant for these colloidal systems and that asymptotic expansions at the MCT-singularities can interpret the scenario qualitatively and explain the data even quantitatively [17].

In the following, the two-component schematic model

used already in Ref. [13] shall be reconsidered with respect to the A_4-singularity and the evolution of the time scales. The advantage of the schematic model is that all transition points can be calculated analytically. Even in this simple model typical features of the full microscopic theory are identified. First, the asymptotic laws for the higher-order singularities and the schematic model shall be introduced briefly, further details are found in Ref. [13]. Second, the asymptotic approximation is shown to work well for the correlation functions at the A_4-singularity. Third, the time scales for the logarithmic decays at A_3- and A_4-singularity are discussed for specific paths in control-parameter space.

ASYMPTOTIC LAWS

The A_2-singularities are characterized by a two-step relaxation around a plateau value, and the asymptotic decay laws are given by scaling functions [2]. For short times the decay onto the plateau is given by a critical decay $\phi_q(t) - f_q(t) = h_q(t_0/t)^a$, while for long times one gets the von Schweidler law $\phi_q(t) - f_q(t) = -h_q(t/t'_\sigma)^b$. The exponents are determined by a single exponent parameter $\lambda = \Gamma(1-a)^2/\Gamma(1-2a) = \Gamma(1+b)^2/\Gamma(1+2b) < 1$. For $\lambda = 1$, the asymptotic solution by power laws becomes invalid and a higher-order singularity occurs. The dynamics at higher-order singularities A_l, $l > 2$ is described up to next-to-leading order by [13]

$$\phi_q(t) = (f_q^c + \hat{f}_q) + h_q[(-B + B_1)\ln(t/\tau) \\ + (B_2 + K_q B^2)\ln^2(t/\tau) \\ + B_3 \ln^3(t/\tau) + B_4 \ln^4(t/\tau)]. \quad (1)$$

The wave-vector-dependent numbers f_q^c, h_q, and K_q are characteristic for a specific singularity, while B, B_i, and $\hat{f}_q$ are in addition functions of the separation parameters ε_1 and ε_2 quantifying the distance of the actual state from the higher-order singularity [13]. The leading order in Eq. (1) is given by the first line with $B_1 = \hat{f}_q = 0$ and the leading order prefactor for the logarithm is $B \propto \sqrt{|\varepsilon_1|}$. The time scale τ is fixed by matching the asymptotic approximation with the numerical solution for $\phi_q(t)$ at the plateau $(f_q^c + \hat{f}_q)$.

THE MODEL

To mimic the q-dependence, which is an important feature of the asymptotic expansions above, we use a two-component model that was introduced for the description of a symmetric molten salt [25]. The model has three control parameters which are combined to the vector $\mathbf{V} = (v_1, v_2, v_3)$. We will use Brownian dynamics, so the equations of motion for the correlators $\phi_q(t)$, $q = 1, 2$, read

$$\tau_q \partial_t \phi_q(t) + \phi_q(t) + \int_0^t m_q(t-t')\partial_{t'}\phi_q(t')dt' = 0, \quad (2a)$$
$$m_1(t) = v_1 \phi_1^2(t) + v_2 \phi_2^2(t), \quad (2b)$$
$$m_2(t) = v_3 \phi_1(t)\phi_2(t). \quad (2c)$$

For the long-time limit of Eq (2), one gets a parameterized representation of the transition surface [13],

$$v_3^c = x, \quad f_1^c = y. \quad (3a)$$

$$v_1^c = \frac{3 - (2+x)y}{2(1-y)^2 y(2-xy)}, \quad (3b)$$

$$v_2^c = \frac{x^2 y(y^2 - 2y^3)}{2(1-y)^2(x^2 y^2 - 3xy + 2)}. \quad (3c)$$

The variables x and y with $x > 4$ and $1/2 \le y \le 3/(2+x)$ serve as surface parameters. The exponent parameter $\lambda = 1 - \mu_2$ is determined by

$$\mu_2 = \frac{(3x^2 + 6x)y^3 - (x^2 + 18x + 8)y^2 + (6x+18)y - 6}{(2x^2 + 4x)y^3 - 12xy^2 + (2x+4)y}. \quad (3d)$$

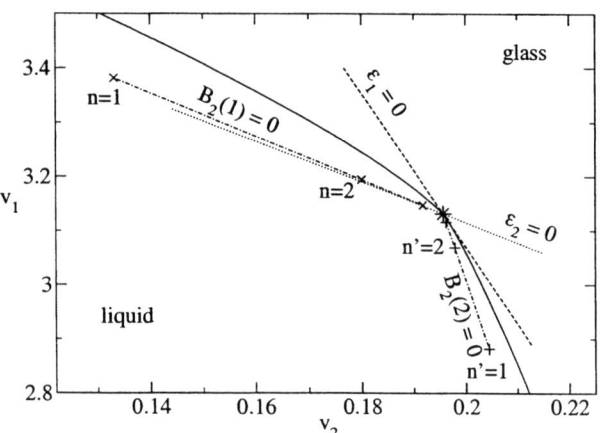

FIGURE 1. Glass-transition diagram for the two-component model for the cut $v_3 = v_3^*$. Liquid-glass transition points are shown by the full line, the A_4-singularity is shown by a star. The vanishing separation parameters ε_1 and ε_2 are shown by dashed and dotted straight lines, Chain lines mark the location of vanishing correction $B_2(q) = B_2 + K_q B^2$ for $q = 1, 2$ with paths indicated by crosses and pluses, respectively.

The A_3-singularities are given by $\lambda = 1$ or equivalently $\mu_2 = 0$ in Eq. (3d). For x large enough, such a solution exists with two roots, $y_1(x) < y_2(x)$, where only $y_2(x)$ is relevant. For small x, no such solution exists. Varying x, the two cusp values $y_1(x)$ and $y_2(x)$ coalesce at x^* with $y_1(x^*) = y_2(x^*) = y^*$, defining the A_4-singularity, $x^* = 24.779392\ldots$, $y^* = 0.24266325\ldots$.

The cut through the transition surface for $v_3 = x^*$ is shown in Fig. 1 as pair of light full lines joining at the A_4-singularity which is indicated by a star, $(v_1^*, v_2^*, v_3^*) = (3.132, 0.1957, 24.78)$. Attached to the A_4-singularity we find the lines of vanishing separation parameters, $\varepsilon_1(\mathbf{V}) = 0$ (dashed) and $\varepsilon_2(\mathbf{V}) = 0$ (dotted), which represent a local coordinate system. The correction amplitudes at the A_4-singularity are $K_1 = 0.3244$ and $K_2 = -2.109$; these yield two lines of vanishing quadratic correction shown by the chain lines in Fig. 1.

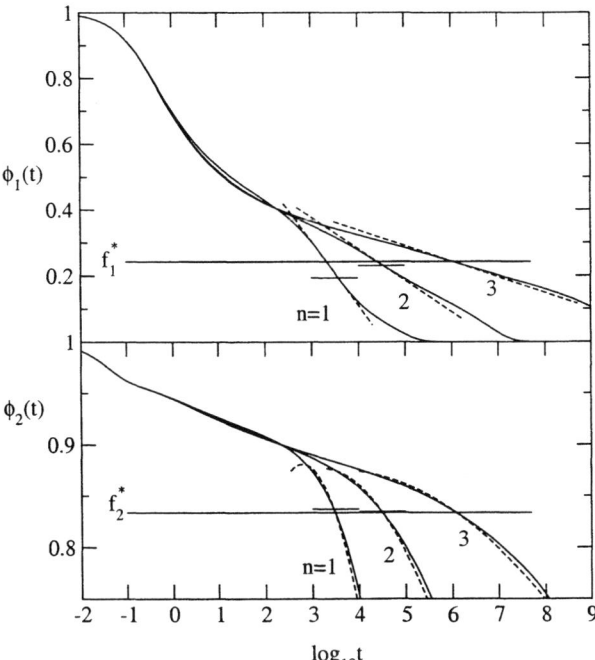

FIGURE 2. Logarithmic decay at the A_4-singularity on the path ($\times$) with $n = 1, 2, 3$ in Fig. 1 with $v_1 - v_1^* = 1/2^{n+1}$, $v_2^* - v_2 = 0.25009/2^{n+1}$. The solutions of Eq. (2) are shown as full lines, the approximation (1) as dashed lines. Long horizontal lines exhibit the critical plateau values f_q^*, short lines the corrected plateau values $f_q^* + \hat{f}_q$.

ASYMPTOTIC APPROXIMATIONS

We first analyze the path labeled n ($\times$) in Fig. 1. The solutions of Eq. (2) are shown as full curves in Fig. 2. The approximations (1) for the correlators are displayed as dashed lines. The time scale τ was matched for $\phi_1(t)$ and $\phi_2(t)$ independently at the corrected plateau values. The approximation describes the decay around the plateaus f_q^* reasonably for both correlators. Since $B_2 + K_1 B^2$ is set to zero on the present path and $B_3 = B_4 = 0$ for an A_4-singularity [13], only the first line in Eq. (1) is relevant for the upper panel. This approximation describes the decay of $\phi_1(t)$ from 0.4 down to 0.1 and for a window increasing window in time with increasing n. Since $K_2 < K_1$, the prefactor for the quadratic correction in Eq. (1) for $\phi_2(t)$ is negative and hence the decay around the plateau f_2^* is concave which is described well by the asymptotic approximation. However, for both q the prefactor $(B - B_1)$ slightly overestimates the absolute slope of the solutions. This is due to a relatively large positive next-to-leading-order correction C_1 that renormalizes the prefactor to $(B - B_1 - C_1)$ [26].

Another feature of the curves in Fig. 2 is noteworthy. There appears a window in time outside the transient dynamics between $t \approx 1$ and $t \approx 10^3$ where the description by Eq. (1) is not applicable. This is caused by the close-by A_2-singularities on the chosen path, cf. Fig. 1, where $f_q^c > f_q^*$, $q = 1, 2$ which introduces an additional slowing down of the dynamics before the logarithmic decay is encountered. In addition, we observe that $\phi_2(t)$ varies almost linearly in $\ln t$ between $t \approx 1$ and $t \approx 250$. It is, however, easy to distinguish that effective logarithmic variation, that stays the same upon further approaching the singularity, from the characteristic behavior of the correlators around the plateau where the prefactor of the logarithmic decay vanishes with the square root of the distance. The almost linear relaxation in $\log t$ seen for $\phi_2(t)$ is related to the β-peak phenomenon [27].

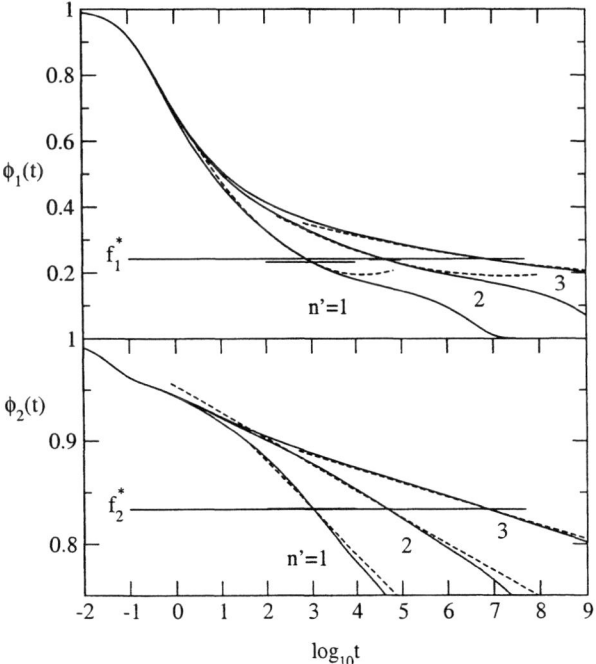

FIGURE 3. Logarithmic decay on the path (+) for $n' = 1, 2, 3$ in Fig. 1. $v_1^* - v_1 = 1/4^n$, $v_2 - v_2^* = 0.3541/4^n$. Line styles and notation are the same as in Fig. 2.

In Fig. 3 the quadratic corrections in Eq. (1) for the correlator $\phi_2(t)$ are zero up to higher orders. The quadratic corrections to the first correlator are positive, $B_2(1) > 0$, and $\phi_1(t)$ is convex in $\ln t$ around f_1^*. There are close-by A_2-singularities on the chosen path with $f_q^c < f_q^*$, $q = 1, 2$, causing a decay which is encountered after the logarithmic decay, which is seen best for curves $n' = 1$ and $n' = 2$ in the upper panel of Fig. 3 for $\phi_1(t) < 0.15$. While in Fig. 2 the validity of the asymptotic approximation around f_q^* was limited by a higher plateau from above, in Fig. 3 the lower plateaus present a boundary for the application of Eq. (1) from below. Because higher plateaus do not interfere, the asymptotic description applies three decades earlier in Fig. 3 than in Fig. 2 for comparable distances from the A_4-singularity.

The estimate for the sign of C_1 given above mainly depends on the relative distance of the path chosen in control-parameter space to the lines $\varepsilon_1 = 0$ and $\varepsilon_2 = 0$. As the ordering of the latter two lines and the path is reversed for the case $B_2(2) = 0$ in Fig. 1, C_1 is expected to be negative there. The path labeled n' in Fig. 1 is now closer to the line $\varepsilon_1 = 0$ than to $\varepsilon_2 = 0$. In Fig. 3 we find indeed, that a comparison of solutions and approximations (1) indicates a negative value for C_1 to account for a steeper slope $(B - B_1 - C_1)$. Despite those small deviation, the correlators are described well by the asymptotic laws.

The comparison of Figs. 2 and 3 is summarized as follows. For each value of q there exists a line with vanishing quadratic correction $B_2(q)$ for the specified q in the approximation of Eq. (1). On this line, the logarithmic decay is displayed best for the correlator specified by q. Moving to control parameter values above this line, $B_2(q) < 0$, introduces concave decay of the correlator $\phi_q(t)$ in $\ln t$. Going below the line, $B_2(q) > 0$, yields a convex decay in the correlator $\phi_q(t)$. For increasing the value of K_q, the curves specified by $B_2(q) = 0$ rotate clockwise around the A_4-singularity.

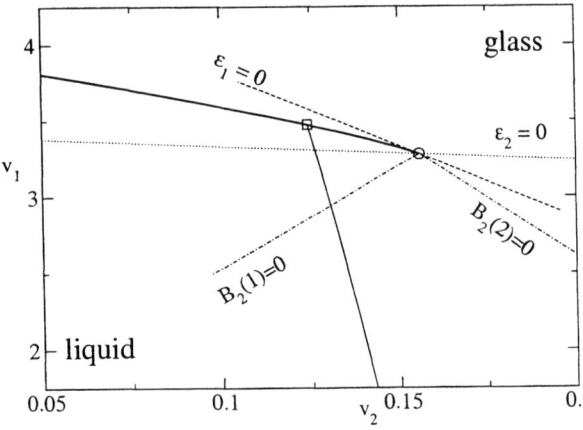

FIGURE 4. Glass-transition diagram for the two-component model for $v_3 = 45$. Notation and line styles are the same as in Fig. 1. The circle marks the A_3-singularity. Heavy and light full lines denote A_2-transition points with their respective f_q^c higher and lower than the value f_q° at the A_3-singularity. The crossing point of the two lines is indicated by a square.

For control-parameter values $v_3 > v_3^*$, the A_4-singularity is replaced by an A_3-singularity which terminates a glass-glass-transition line that extends further into the arrested region as v_3 increases. The lines $\varepsilon_1 = 0$, $\varepsilon_1 = 0$, $B_2(1) = 0$, and $B_2(2) = 0$ for the A_3-singularities emerge from smooth transformations of the ones at the A_4-singularity. Figure 4 shows a cut through the transition diagram for $v_3 = 45$. Different from the situation in Fig. 1, the line $B_2(1) = 0$ is now well separated from liquid-glass-transition lines while the line $B_2(2) = 0$ is located completely in the arrested regime. The dynamics on the line $B_2(1) = 0$ for this A_3-singularity and the related crossing scenario on the same line is discussed in Ref. [13] in great detail.

TIME SCALE

It is obvious from Figs. 2 and 3 that the time scales τ for a given state determined by matching, e.g., $\phi_1(t)$ at f_1^* or $f_1^* + \hat{f}_1$ can deviate considerably. The resulting scale is also different if τ is fixed for different correlators. However, asymptotically close to the higher-order singularity, $\hat{f}_q$ approaches zero and also the time scales determined by using different correlators converge towards each other as explained earlier [13]. Therefore, only the scale fixed by $\phi_1(\tau) = f_1^*$ shall be considered in the following.

The time scale τ when considering only the leading order result in Eq. (1) is given by [28]

$$\log_{10} \tau \propto 1/|\varepsilon_1|^{1/6} \quad (4a)$$

for the A_3-singularity and from a similar argument [29] for the A_4-singularity as

$$\log_{10} \tau \propto 1/|\varepsilon_1|^{1/4}. \quad (4b)$$

Thus, the logarithm of the time scale is the relevant quantity to be discussed further on, and $\log_{10} \tau$ is expected to diverge like a power law when the separation from the singularity vanishes. Figure 5 shows that this divergence is indeed stronger for the A_4-singularity than for the A_3-singularity. For $|\varepsilon_1| < 10^{-3}$ the time scale τ is described qualitatively by the asymptotic laws. The laws in Eq. (4) were verified quantitatively by extending the analysis to times as large as $\log_{10} \tau \approx 100$. With one exception, the time scales deviate to lower values for τ for larger separations $|\varepsilon_1|$. Hence, the slope of the $\log_{10} \tau$-versus-$\log_{10}(-\varepsilon_1)$ curve is even larger in that regime than given by the asymptotic laws in Eq. (4). In the lower panel for $|\varepsilon_1| > 10^{-3}$, the time scales for the A_3-singularity are described better by the law for the close-by A_4-singularity.

It is seen for the case of the A_4-singularity that the proximity of a line of liquid-glass transitions (scales τ

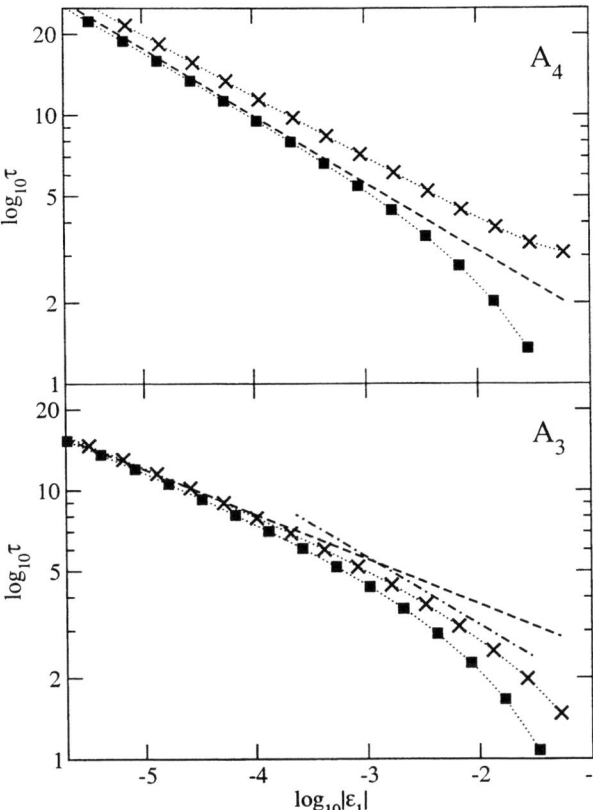

FIGURE 5. Time scale τ determined by matching $\phi_1(t)$ at f_1^* or f_1° for the paths $B_2(1) = 0$ (crosses) and $B_2(2) = 0$ (squares) from Figs. 1 and 4, respectively. The dotted lines are guides to the eye. The corresponding asymptotic laws (dashed lines), $\tau \propto \exp[1/|\varepsilon_1|^{1/4}]$ for the A_4 and $\tau \propto \exp[1/|\varepsilon_1|^{1/6}]$ for the A_3 are fitted to the data for the paths where $B_2(1) = 0$ around $\tau \approx 10^{100}$. The asymptotic law for the A_4-singularity is redrawn in the lower panel as chain line.

indicated by crosses) influences also the validity of the asymptotic law for τ: Only for extremely large times, τ follows the law (4b). The time scales are also much larger on that path as discussed already in connection with Fig. 2.

CONCLUSION

It was shown that the asymptotic expansion in Eq. (1) provides a way to divide the control-parameter space into distinct regions by setting the dominant correction to the logarithmic decay laws to zero for different correlators. If plotted in the form of Figs. 1 and 4, the glass-transition diagrams of the two-component model are easily seen to be topologically equivalent to the ones shown for the square-well system in Refs. [14, 17]. For increasing K_q in Eq. (1), the lines of vanishing quadratic corrections rotate clockwise around the higher-order singularity in the two-component model, which mimics the behavior in the microscopic model for increasing wave vector.

Using the asymptotic expansion to describe the solutions of the equations of motion (2), the approximations fit rather accurately already in a regime that should be accessible to experiments and molecular-dynamics simulations. However, the presence of other glass-transition singularities in the vicinity of higher-order singularities can have drastic influence on the window in time where the logarithmic laws are applicable, cf. Figs. 2 and 3. The validity of Eq. (1) for $\phi_q(t)$ around f_q^* can be bound either from above at earlier times, cf. Fig. 2, or from below to later times, Fig. 3, depending on the specific location of the chosen path in the transition diagram.

The time scale τ asymptotically follows the Vogel-Fulcher-type law of Eq. (4). That the relaxation processes close to higher-order glass-transition singularities are extremely slow is manifested by the fact, that instead of τ itself, the logarithm of the time scale is the relevant quantity that diverges in some power law. The different behavior of τ for A_4- and A_3-singularities is clearly seen in the numerical solution of the MCT equations, cf. Fig. 5. However, these asymptotic laws only show up for extremely long times and small distances from the singularity. In addition, other close-by glass-transition singularities can significantly modify the validity of the asymptotic law especially for further separation from the higher-order singularity under consideration.

ACKNOWLEDGMENTS

Fruitful discussion with W. Götze is gratefully acknowledged. This work was supported by the Deutsche Forschungsgemeinschaft Grant No. Go154/13-2.

REFERENCES

1. Bengtzelius, U., Götze, W., and Sjölander, A., *J. Phys. C*, **17**, 5915–5934 (1984).
2. Franosch, T., Fuchs, M., Götze, W., Mayr, M. R., and Singh, A. P., *Phys. Rev. E*, **55**, 7153–7176 (1997).
3. van Megen, W., *Transp. Theory Stat. Phys.*, **24**, 1017–1051 (1995).
4. Kob, W., "Supercooled liquids, the glass transition, and computer simulations," in *Slow Relaxations and Nonequilibrium Dynamics in Condensed Matter*, edited by J.-L. Barrat, M. Feigelman, J. Kurchan, and J. Dalibard, Springer, Berlin, 2003, vol. Session LXXVII (2002) of *Les Houches Summer Schools of Theoretical Physics*, pp. 199–269.
5. Götze, W., and Haussmann, R., *Z. Phys. B*, **72**, 403–412 (1988).
6. Götze, W., and Sjögren, L., *J. Math. Analysis and Appl.*, **195**, 230–250 (1995).

7. Arnol'd, V. I., *Catastrophe Theory*, Springer, Berlin, 1992, 3rd edn, pp. 29–39.
8. Fuchs, M., Götze, W., and Mayr, M. R., *Phys. Rev. E*, **58**, 3384–3399 (1998).
9. Götze, W., and Mayr, M. R., "The dynamics of a hard sphere moving in a hard-sphere system near the liquid-glass transition point," in *Slow Dynamics in Complex Systems*, edited by M. Tokuyama and I. Oppenheim, AIP, New York, 1999, vol. 469 of *AIP Conference Proceedings*, pp. 358–378.
10. Fabbian, L., Götze, W., Sciortino, F., Tartaglia, P., and Thiery, F., *Phys. Rev. E*, **59**, R1347–R1350 (1999); **60**, 2430 (1999).
11. Bergenholtz, J., and Fuchs, M., *Phys. Rev. E*, **59**, 5706–5715 (1999).
12. Dawson, K., Foffi, G., Fuchs, M., Götze, W., Sciortino, F., Sperl, M., Tartaglia, P., Voigtmann, T., and Zaccarelli, E., *Phys. Rev. E*, **63**, 011401 (2001).
13. Götze, W., and Sperl, M., *Phys. Rev. E*, **66**, 011405 (2002).
14. Sperl, M., *Phys. Rev. E*, **68**, 031405 (2003).
15. Puertas, A. M., Fuchs, M., and Cates, M. E., *Phys. Rev. Lett.*, **88**, 098301 (2002).
16. Sciortino, F., Tartaglia, P., and Zaccarelli, E., cond-mat/0304192.
17. Sperl, M., *Phys. Rev. E*, in press (2003), cond-mat/0308425.
18. Mallamace, F., Gambadauro, P., Micali, N., Tartaglia, P., Liao, C., and Chen, S.-H., *Phys. Rev. Lett.*, **84**, 5431–5434 (2000).
19. Pham, K. N., Puertas, A. M., Bergenholtz, J., Egelhaaf, S. U., Moussaïd, A., Pusey, P. N., Schofield, A. B., Cates, M. E., Fuchs, M., and Poon, W. C. K., *Science*, **296**, 104–106 (2002).
20. Chen, W.-R., Chen, S.-H., and Mallamace, F., *Phys. Rev. E*, **66**, 021403 (2002).
21. Poon, W. C. K., Pham, K. N., Egelhaaf, S. U., and Pusey, P. N., *J. Phys.: Condens. Matter*, **16**, S269–S275 (2003).
22. Chen, S.-H., Chen, W.-R., and Mallamace, F., *Science*, **300**, 619–622 (2003).
23. Pham, K. N., Egelhaaf, S. U., Pusey, P. N., and Poon, W. C. K., *Phys. Rev. E*, in press (2003), cond-mat/0308250.
24. Zaccarelli, E., Foffi, G., Dawson, K. A., Buldyrev, S. V., Sciortino, F., and Tartaglia, P., *Phys. Rev. E*, **66**, 041402 (2002).
25. Bosse, J., and Krieger, U., *J. Phys. C*, **19**, L609–L613 (1987).
26. Sperl, M., *Asymptotic Laws near Higher-Order Glass-Transition Singularities*, Ph.D. thesis, TU München (2003), pp. 131–134.
27. Götze, W., and Sperl, M., in preparation.
28. Götze, W., and Sjögren, L., *J. Phys.: Condens. Matter*, **1**, 4203–4222 (1989).
29. Flach, S., Götze, W., and Sjögren, L., *Z. Phys. B*, **87**, 29–42 (1992).

The Boson peak and the phonons in glasses

S. Ciliberti*, T. S. Grigera†, V. Martín-Mayor**, G. Parisi* and P. Verrocchio**

*INFM UdR Roma1, Universitá di Roma "La Sapienza", and Center for Statistical Mechanics and Complexity (SMC), P.le A. Moro 2, I-00185 Roma, Italy
†Centro di Studi e Ricerche "Enrico Fermi", via Panisperna 89/A, I-00184 Roma, Italy
**Departamento de Fisica Teorica I, Universidad Complutense de Madrid, Madrid 28040, Spain; Instituto de Biocomputación y Física de Sistemas Complejos (BIFI). Universidad de Zaragoza, 50009 Zaragoza, Spain.

Abstract. Despite the presence of topological disorder, phonons seem to exist also in glasses at very high frequencies (THz) and they remarkably persist into the supercooled liquid. A universal feature of such a systems is the Boson peak, an excess of states over the standard Debye contribution at the vibrational density of states. Exploiting the euclidean random matrix theory of vibrations in amorphous systems. we show that this peak is the signature of a phase transition in the space of the stationary points of the energy, from a minima-dominated phase (with phonons) at low energy to a saddle-point dominated phase (without phonons). The theoretical predictions are checked by means of numeric simulations.

INTRODUCTION

X-ray and neutron scattering techniques allow to obtain very detailed physical insight into the high-frequency (0.1–10 THz) vibrational dynamics of supercooled liquids and glasses. Within this range of frequencies their spectra reveal several universal properties [1], related with the presence of sound-like excitations even for momenta p of the same order of magnitude of p_0, the first maximum of the static structure factor (typically corresponding to wave numbers of a few nm^{-1}). This *high-frequency sound* is revealed as Brillouin-like peaks in the Thz region of the dynamic structure factor. An accessible quantity to experiments is the vibrational density of states (VDOS), $g(\omega)$, whose most striking feature is the presence of an excess of states over the Debye ω^2 law in the "low" frequency region, (i.e. where the dispersion relation is linear, but still in the Thz region)[2]. This excess of states is seen as a peak when plotting $g(\omega)/\omega^2$ and has been named *Boson peak* (BP) [1]. The peak position ω_{BP} usually shifts to lower frequency on heating [3], except for the case of silica [4]. In this material the shift is seen on lowering the density [5].

Due to its universality, the relevant physics underlying the Boson peak can be hopefully captured by some simple model. Furthermore, several recent numerical simulations have shown that a model of harmonic vibrations is wholly adequate to describe this frequency range [6] and that anharmonicity need not be invoked. Given the presence of well formed local structures (SiO_2 tetrahedra, for instance) a natural approximation is to consider that the oscillation centers form a crystalline structure, the disorder in the atomic positions being mimicked by randomness in their interaction potential [7, 8] (disordered lattice models [9]). The main drawback of such a models is that they dramatically underestimate the scattering of sound waves [10]. A different approach studies vibrations around a topologically disordered [9] (liquid like) structure. It is followed by two different theories: modified mode-coupling theory [11] (which is not limited to harmonic excitations), and euclidean random matrix theory (ERMT) [12, 13, 14]. ERMT owes its name to the fact that it formulates the vibrational problem as random matrix problem [15]. The matrices involved are called Euclidean random matrices [16], and their study has required the development of new analytical tools. Both MCT and ERMT predict an enhanced scattering of sound waves as compared to disordered crystals.

On the other hand, even within the harmonic framework the nature of the extra low frequency modes giving rise to the BP is still an open point. At a qualitative level, the frequency ω_{BP} is close to the Ioffe-Regel [17] frequency ω_{IR}, suggesting the possibility that the excess BP modes are localized [18]. However, numerical simulations have shown that the localization edge is at frequencies greater than ω_{BP} and ω_{IR} [19]. The Ioffe-Regel criterion signals rather a crossover to a region where the harmonic excitations are not longer propagating, due to the

---

[1] There exist alternative ways of defining the boson peak from experiments, for example as a peak in Raman scattering data or as a peak in the difference between the observed VDOS of the glass and that of the corresponding crystal

strong interaction with the disorder. We call these modes *glassons* (since they do not propagate but "diffuse", they have also been called *diffusons* [19]). A large bump of glassons is generally found around the Ioffe-Regel frequency, due to the flattening of the dispersion relation. This can be considered as the glass counterpart of the van Hove singularity of crystals [8, 20]. All the recently proposed theoretical frameworks predict that this peak of glassons should move to lower frequencies when approaching an instability transition, where negative eigenvalues (imaginary frequencies) appear. The aim of the paper is showing that the ERM theory makes sharp predictions about the values of universal critical exponents describing the approach to this singularity and comparing them with numeric results. The emerging scenario describes the BP modes as given by the hybridization between the phonons and the low-energy tail of the glasson peak which softens when the system approaches the instability transition [21, 22].

THE EUCLIDEAN RANDOM MATRIX THEORY

The starting approximation is that particles can only oscillate around fixed random positions, so that the position of particle i at time t is $\mathbf{x}_i(t) = \mathbf{x}_i^{eq} + \varphi_i(t)$; the $\mathbf{x}_i^{eq}$ are quenched equilibrium positions (whose distribution must be specified) and $\varphi_i(t)$ are the displacements. Hence the Hamiltonian is

$$H[\mathbf{x}] = \sum_{i,j}^{1,N} V(\mathbf{x}_i - \mathbf{x}_j) \simeq \frac{1}{2}\sum_{i,j}^{1,N}\sum_{\mu,\nu}^{1,3} M_{i\mu,j\nu}[\mathbf{x}^{eq}]\varphi_i^\mu \varphi_j^\nu \quad (1)$$

where the dynamical matrix M is an Euclidean Random Matrix:

$$M_{i\mu,j\nu}[\mathbf{x}^{eq}] \equiv -f_{\mu\nu}(\mathbf{x}_i^{eq} - \mathbf{x}_j^{eq}) + \delta_{ij}\sum_{k=1}^{N} f_{\mu\nu}(\mathbf{x}_i^{eq} - \mathbf{x}_k^{eq}), \quad (2)$$

with $f_{\mu\nu}(\mathbf{x}) \equiv \partial_{\mu\nu}V(\mathbf{x})$.

In the one-excitation approximation the dynamic structure factor is

$$S^{(1)}(\mathbf{p},\omega) = \overline{\frac{k_B T}{m\omega^2}\sum_n \left|\sum_i \mathbf{p}\cdot\mathbf{e}_{n,i}e^{i\mathbf{p}\cdot\mathbf{x}_i^{eq}}\right|^2 \delta(\omega - \omega_n)}, \quad (3)$$

where $\mathbf{e}_n$ are the eigenvectors of the dynamical matrix and ω_n its eigenfrequencies (= square root of eigenvalues). The overline means average over the disordered quenched positions, whose distribution $P[\mathbf{x}^{eq}]$ has to be specified. The density of states (VDOS) is obtained in the limit of large momenta:

$$g(\omega) = \lim_{p\to\infty}\frac{m\omega^2}{k_B T p^2}S^{(1)}(p,\omega). \quad (4)$$

We can obtain $S^{(1)}(\mathbf{p},\omega)$ through the resolvent $G(\mathbf{p},z)$:

$$G_{\mu\nu}(\mathbf{p},z) \equiv \frac{1}{N}\overline{\sum_{jk}e^{i\mathbf{p}\cdot(\mathbf{x}_j^{eq}-\mathbf{x}_k^{eq})}\left[\frac{1}{z-M}\right]_{j\mu,k\nu}}$$

$$\equiv G_L(p,z)\frac{p_\mu p_\nu}{p^2} + G_T(p,z)\left(\delta_{\mu\nu} - \frac{p_\mu p_\nu}{p^2}\right) \quad (5)$$

separating the axial tensor in a longitudinal term and a transversal one. The dynamic structure factor is then:

$$S^{(1)}(\mathbf{p},\omega) = -\frac{2k_B T p^2}{\omega\pi}\mathrm{Im}\, G_L(\mathbf{p},\omega^2 + i0^+). \quad (6)$$

A transverse dynamic structure factor (not measurable in experiments) can be defined in an analogous way. However, a most important and general result is that for $p\to\infty$ the resolvent becomes isotropic:

$$G_{\mu\nu}^\infty(z) = \frac{1}{N}\overline{\sum_j\left[\frac{1}{z-M}\right]_{j\mu,j\nu}} = \delta_{\mu\nu}\frac{1}{N}\overline{\mathrm{Tr}\,[z-M]^{-1}}. \quad (7)$$

So both longitudinal and transverse structure factors tend to a common limit (the VDOS, see eq. 4) at infinite momentum.[2] Leaving the potential $V(r)$ unspecified and taking the simplified case $P[\mathbf{x}^{eq}] = 1/V^N$ (V being the volume), one finds that:

$$G_{\mu\nu}(\mathbf{p},z) = \left[\frac{1}{z - \rho\hat{f}(0) + \rho\hat{f}(\mathbf{p}) - \Sigma(\mathbf{p},z)}\right]_{\mu\nu}, \quad (8)$$

The self-energy $\Sigma(\mathbf{p},z)$ is a matrix with the standard form

$$\Sigma_{\mu\nu}(\mathbf{p},z) = \Sigma_L(p)\frac{p_\mu p_\nu}{p^2} + \Sigma_T(p)\left(\delta_{\mu\nu} - \frac{p_\mu p_\nu}{p^2}\right). \quad (9)$$

which vanishes at $\rho = \infty$ and that can be computed in a series expansion in $1/\rho$. The main point is that the sum of all the infinite diagrams obtained recursively starting from this next-to-leading order result gives a self-consistent integral equation[22]:

$$\Sigma_{\mu\nu}(\mathbf{p},z) = \frac{1}{\rho}\int\frac{d^3 q}{(2\pi)^3}V_{\mu\lambda}(\mathbf{q},\mathbf{p})G_{\lambda\sigma}(\mathbf{q},z)V_{\sigma\nu}(\mathbf{q},\mathbf{p}). \quad (10)$$

where the vertices have the form $V_{\mu\nu}(\mathbf{q},\mathbf{p}) = \rho(\hat{f}_{\mu\nu}(\mathbf{q}) - \hat{f}_{\mu\nu}(\mathbf{p}-\mathbf{q}))$. Let us remark that the self energy renormalizes the dispersion relations and gives a finite width to the Brillouin peaks:

$$\omega_{L,T}^2(p) = (\omega_{L,T}^0)^2(p) + \mathrm{Re}\,\Sigma_{L,T}(p,\omega_{L,T}(p)),$$
$$\Gamma_{L,T}(p) = \mathrm{Im}\,\Sigma_{L,T}(p,\omega_{L,T}(p))/\omega_{L,T}(p). \quad (11)$$

---

[2] Consequently, both the dispersion relations saturate at the same value. However due to the broadening of the line, they are rather ill-defined when $\omega \sim \omega_{IR}$

The correlations between the equilibrium positions of the particles can be taken into account quite easily at the level of the *superposition approximation* in the above approach. The results derived above for the case without correlations are translated to the correlated case by replacing the functions $f(\mathbf{x})$ by $g^{(2)}(\mathbf{x})f(\mathbf{x})$. In this way the usual power law divergence of the pair potential for $|\mathbf{x}| \to 0$ is balanced by the exponential behaviour of the pair distribution function, and this ensures the existence of the Fourier transform of the product $f(\mathbf{x})g^{(2)}(\mathbf{x})$.

The phase transition

From equation (10) it is possible to derive a few analytic model-independent results about the arising of the Boson Peak. These results are expressed in form of scaling laws, whose exponents are predicted in this approximation. Simulations (see below) and experiments will allow to clarify the dependence of the exponents on the approximation. The VDOS can be obtained from

$$g(\omega) = -\frac{2\omega}{\pi} \operatorname{Im} G^\infty(\omega^2 + i0^+), \quad (12)$$

where $z = \omega^2 + i0^+$ and $G^\infty(z) \equiv \lim_{p \to \infty} G(\mathbf{p}, z)$.

Hence one have to solve the integral equation (10) in the $p \to \infty$ limit:

$$\frac{1}{G^\infty(z)} = z - \rho \hat{f}(0) - \rho A G^\infty(z) - \rho \int \frac{d^3q}{(2\pi)^3} \hat{f}^2(\mathbf{q}) G(\mathbf{q}, z) \quad (13)$$

where $G^\infty(z)$, $A \equiv (2\pi)^{-3} \int d^3q \hat{f}^2(\mathbf{q})$ and the last term are matrices proportional to the identity.

The solution of the above integral equation yields a VDOS which contains both the phonons, since $g(\omega) \propto \omega^2$ at $\omega \to 0$, and the extended but not propagating glassons, described by a semicircle with center at $\omega = \rho \hat{f}(0)$ and radius $2\sqrt{\rho A}$ [20]. If we limit (for pedagogical purposes) to the case where the VDOS changes because of changes in the density, the key point is the existence of a phase transition in the space of the eigenvalues of the Hessian matrix. In fact $G^\infty(0)$ develops an imaginary part when $\rho < \rho_c$ (ρ_c being a critical density), then the transition separates the stable phase (all positive eigenvalues) and the unstable phase (negative and positive eigenvalues). The glassons are the modes which move towards the negative zone of the spectrum (hybridizing the phonons) when approaching the transition. The order parameter is $\varphi = -\operatorname{Im} G^\infty(i0^+)$ which vanishes as $\varphi \sim |\Delta|^\beta$, with $\beta = 1/2$ and $\Delta \equiv \rho - \rho_c$

The relation with the Boson Peak becomes quite clear when one writes down the VDOS in the stable phase arising from the theory without any reference to the control parameter, which then does not need to be ρ. In fact, one has

$$g(\omega, \Delta) = \omega^\gamma h(\omega \Delta^{-\rho}), \quad h(x) \sim \begin{cases} x^{2-\gamma} & x \ll 1 \\ \text{const.} & x \gg 1 \end{cases}, \quad (14)$$

with Δ defined in terms of an arbitrary control parameter. The ERMT (in the cactus approximation) predicts $\rho = 1$, $\gamma = 3/2$. Hence it exist a crossover frequency (in the region where the dispertion relation is still linear) between a ω^2 and a ω^γ region. We shall identify that with the BP frequency ω_{BP}. This implies that $\omega_{BP} \sim \Delta^\rho$ and $g(\omega_{BP}, \Delta)/\omega_{BP}^2 \sim \Delta^{-\eta}$, with

$$\eta = \rho(2-\gamma). \quad (15)$$

Let us note that the BP is indicated from a peak in the function $g(\omega)/\omega^2$m not in $g(\omega)$. Summarizing, according to ERMT the BP frequency moves linearly toward 0 when approaching the transition (from the stable side) and its height diverges as a power law whose exponent is $\eta = 1/2$. Eq. (4) shows that at the level of the one-phonon approximation it can also be detected in the large p limit of the dynamic structure factor $S(p, \omega)$.

BOSON PEAK IN A GAUSSIAN MODEL

In order to confirm that the saddle-phonon transition described by the Euclidean Random Matrix theory is not an artifact of the approximation involved (cactus resummation), we solved numerically the cactus equation for the case where $f(p)$ has a Gaussian form and compare with direct numerical results for the same model[22]. The model is described by

$$\hat{f}_{\mu\nu}(\mathbf{p}) = \hat{f}_L(p)\frac{p_\mu p_\nu}{p^2} + \hat{f}_T(p)\left(\delta_{\mu\nu} - \frac{p_\mu p_\nu}{p^2}\right),$$

$$\hat{f}_{L,T}(p) = \left(\frac{2\pi}{\sigma_0^2}\right)^{3/2} \exp(-p^2/2\sigma_{L,T}^2). \quad (16)$$

This choice for $\hat{f}(p)$ is mainly due to its simplicity. However the behaviour of the Boson peak close to the saddle-phonon transition have to be independent of the details of the model. Moreover the superposition approximation takes $\hat{f}_{\mu\nu}(p) = \mathscr{F}[g(r)v_{\mu\nu}(r)]$ yielding a $\hat{f}_{L,T}(p)$ finite at $p = 0$, like in the Gaussian model. We shall consider various values of the density, which is here the control parameter, comparing the analytical (cactus) results with the numerical spectra and dynamic structure factor obtained from the method of moments [23]. In the high density regime the approximations used in deriving the integral equation (13) are quite good since the analytic solution reproduces the numerical spectrum (and in particular the Debye behaviour) rather accurately (fig.1).

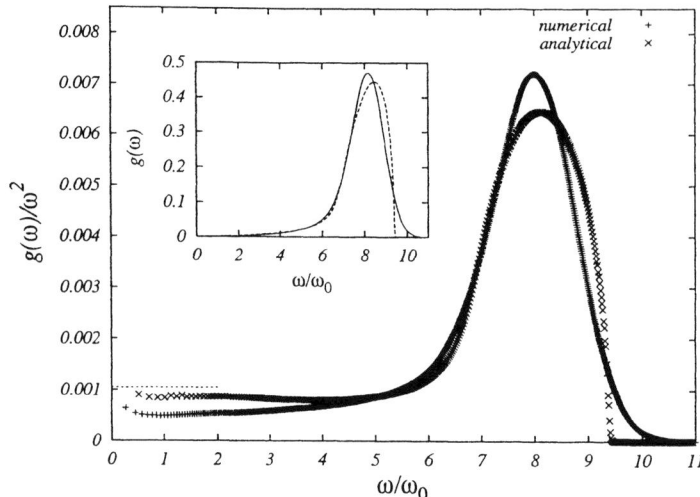

FIGURE 1. The VDOS $g(\omega)$ as a function of eigenfrequencies divided by the Debye behaviour ω^2 for $\rho = 4 > \rho_c$, both numerical (obtained via the method of moments) and analytical. In the inset, we show $g(\omega)$ vs. ω.

However, the crucial check regards the exponents of the transition. Figs. 2a and 2c show that the position of the BP is linear with respect to $\Delta \equiv (\rho - \rho_c)$ and that the height of BP diverges as $\Delta^{-1/2}$. This confirms the theoretical predictions $\nu = 1$ and $\eta = 1/2$. In Fig. 2b we determine the value of γ by studying the fraction of unstable modes. In fact, in the region of parameters where $\rho < \rho_c$ the fraction of unstable modes, defined as $f_u = \int_{-\infty}^{0} g_\lambda(\lambda) d\lambda$, is given by

$$f_u(\Delta) = \int_0^\infty d\omega\, \omega^\gamma \tilde{g}(\omega/|\Delta|) \sim |\Delta|^{1+\gamma}. \quad (17)$$

We find numerically (Figs. 3b) that $f_u \sim (\rho_c - \rho)^{5/2}$, i.e. $\gamma = 3/2$. Finally, the order parameter φ vanishes as $(\rho_c - \rho)^\beta$ with $\beta = 1/2$ (Fig. 2d).

Hence our analytic treatment based on euclidean random matrix theory describe quite well the vibrational features of simple topologically disordered systems[22]. The following step is understanding to what degree of accurateness ERMT could describe the high frequency properties of more realistic systems[21].

BOSON PEAK IN A FRAGILE GLASS

Starting from the hypothesis that the Thz region of supercooled liquids and glasses can be described in terms of purely harmonic excitations, the origin of the Boson peak in glasses can be understood if we consider the ensemble of generalized inherent structures (GIS). For each equilibrium configuration the associated GIS is the nearest stationary point of the Hamiltonian. If we start from an equilibrium configuration at low temperature, the GIS is a local minimum, and it coincides with the more frequently used inherent structures (IS) [24](i.e. the nearest minimum of the Hamiltonian). On the contrary, if we start from high temperature, the GISs are saddle points. In the GIS ensemble there is a sharp phase transition separating these two regimes. It takes place in glass-forming liquids [25] at the Mode Coupling temperature [26] (T_{MC}), above which liquid diffusion is no longer ruled by rare "activated" jumps between ISs but by the motion along the unstable directions of saddles. Phonons are present in the spectrum of the VDOS in the low temperature phase (IS dominated) but are absent in the saddle phase. The key point is that the minima obtained starting from configurations below T_{MC} and the saddles obtained starting above T_{MC} join smoothly at T_{MC}. Thus we can study GIS as a single ensemble parametrized by their energy [25].

Since this transition separates a phase where all the eigenvalues are positive from another one where even negative eigenvalues exist, we expect that ERMT is able to describe correctly this phenomenun. Hence we measured numerically the values of some exponents predicted by the theory in a simple model of a fragile glass [21]. We simulated a soft-spheres binary mixture [27] in the stable (phonon) phase with the Swap Monte Carlo algorithm [28], and computed the VDOS of the ISs obtained starting from equilibrium configurations at temperatures below T_{MC}[3].

In Fig.3 we show that the theoretical predictions agree with the numerical data. Taking the IS's energy as the

---

[3] At very low T, where equilibrium is not achieved, runs were followed until e_{IS} got very close to its asymptotic value

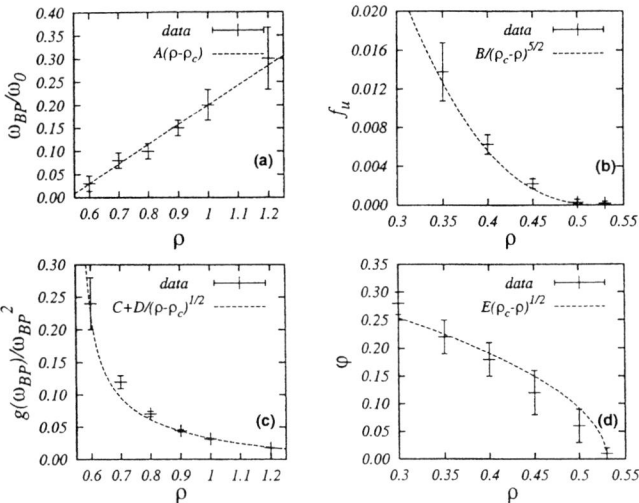

FIGURE 2. Numeric results. The critical density in the following fits has been fixed to $\rho_c = 0.54$ and capital letters are the fitting parameters. **(a)**: The position of Boson peak as a function of the density near the critical point. ω_{BP} vanishes linearly in $\Delta = \rho - \rho_c$. **(b)**: The fraction of unstable modes vanishes as $(\rho_c - \rho)^{2.5}$, thus yielding $\gamma = 3/2$. **(c)**: The height of the BP, defined by $g(\omega_{BP})/\omega_{BP}^2$, diverges as $\Delta^{-\eta}$, with $\eta \sim 1/2$. **(d)**: The order parameter $\varphi \equiv -\mathrm{Im}\,G^\infty(0)$ vanishes as $(\rho_c - \rho)^\beta$, with $\beta \sim 1/2$.

relevant parameter for describing the spectral properties, one has $\Delta = e_c - e_{IS}$, e_{IS} being the energy of the ISs and e_c the critical value. In fact, plotting $g(\omega)/\omega^2$ a peak is clearly identified, which is seen to grow in height and shift to lower frequency on rising the IS's energy. Using all the spectra for which the peak position can be clearly identified, we find that the relationship between ω_{BP} and the energy of the IS is linear (Fig. 3a). The energy at which ω_{BP} becomes zero, e_c, is found from a linear fit as $e_c = 1.74\,\varepsilon$ (ε is the energy scale), quite close to the value where the GIS stop to be minima (IS) and become saddles [25]. As for the height of the peak (Fig. 3b), the results are compatible with a power-law divergence. Fixing e_c at the value $1.74\,\varepsilon$ arising from the linear fit of ω_{BP} vs. e_{IS}, a power-law fit yields an exponent $\beta = 0.40(15)$, while if one fixes the exponent at $\beta = 1/2$, then the critical value is $e_c = 1.752(2)\,\varepsilon$. Thus the numerical data are compatible with the theoretically predicted scaling, although we have not been able to work close to the critical point, and thus cannot get a great accuracy on the critical exponents or the critical point.

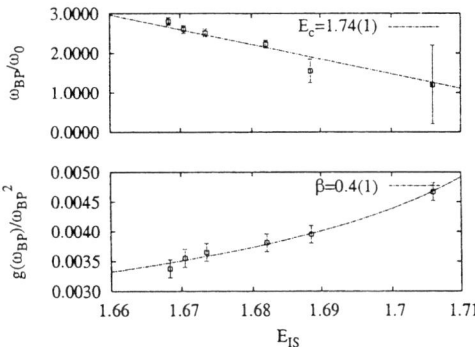

FIGURE 3. Scaling of the position, ω_{BP}, and height of the Boson peak near the saddle-phonon transition (energies and frequencies in units of ε and ω_0 respectively). **Top** ω_{BP} is linear in the control parameter e_{IS} and vanishes at $e_{IS} = e_c = 1.74(1)\,\varepsilon$. **Bottom** The height of the Boson peak diverges as a power law with exponent $\beta \sim 0.4$. Height and position of the BP were obtained by fitting a parabola to the peak of $g(\omega)/\omega^2$. *Reprinted with permission from Nature [21], Copyright (2003) Macmillan Magazines Limited*

CONCLUSIONS

In summary, we have shown that the saddle (negative eigenvalues)-phonon (no negative eigenvalues) transition, a common feauture of vibrating topologically disordered systems, is well described by the euclidean random matrix theory. It provides a coherent scenario for the arising of a Boson Peak, which results from the hybridization of acoustic modes with high-energy modes that soften upon approaching the transition. Hence we applied the theory to describe the saddle-phonon transition and the BP in supecooled liquids, comparing the predicted scaling laws with the numeric results obtained for a simple fragile glass former. The agreement found is quite encouraging. The present discussion applies to experiments as long as one is in the regime where the inverse frequency is much larger than the structural relaxation time, when the harmonic approximation makes sense. We expect that the saddle-phonon transition point of view will be able to bridge the realms of experiment and numerical studies of the energy landscape. As a matter of fact, a recent experiment on the poly(methyl methacrylate) (PMMA) glass gave the first experimental confirmation of the ERMT predictions[29].

ACKNOWLEDGMENTS

V.M.-M. is a *Ramón y Cajal* research fellow (MCyT, Spain). P.V. was supported through the European Community's Human Potential Programme under contract HPRN-CT-2002-00307, DYGLAGEMEM.

REFERENCES

1. Sette, F., Krisch, M., Masciovecchio, C., Ruocco, G. and Monaco, G. *Science* **280**, 1550-1555 (1998); Ruocco, G. and Sette, F. *J. Phys.: Cond. Matt.* **13**, 9141-9164 (2001); Courtens, E.; Foret, M., Hehlen, B., Vacher, R., *Solid State Commun.* **117**, 187-200 (2001).
2. Foret, M., Courtens, E., Vacher, R., Suck, J.-B., *Phys, Rev. Lett.* **77**, 3831-3834 (1996); Benassi, P., Krisch, M., Masciovecchio, C., Mazzacurati, V., Monaco, G., Ruocco, G., Sette, F. and Verbeni, R., *Phys, Rev. Lett.* **77**, 3835-3838 (1996); Pilla, O., Cunsolo, A., Fontana, A., Masciovecchio, C., Monaco, G., Montagna, M., Ruocco, G., Scopigno, T., and Sette, F. *Phys, Rev. Lett.* **85**, 2136-2139 (2000); Rufflé, B., Foret, M., Courtens, E., Vacher, R. and Monaco, G., *Phys. Rev. Lett.* **90**, 095502-1 - 095502-4 (2003).
3. Sokolov, A. P., Buchenau, U., Steffen, W., Frick, B., Wischnewski, A., *Phys. Rev. B* **52**, 9815-9818 (1995); Tao, N. J., Li, G., Chen, X., Du, W. M., Cummins, H. Z., *Phys. Rev. A* **44**, 6665-6676 (1991); Engberg, D., Wischnewski, A., Buchenau, U., Börjesson, L., Dianoux, A. J., Sokolov, A. P., Torell, L. M., *Phys. Rev. B* **59**, 4053-4057 (1999).
4. Wischnewski, A., Buchenau, U., Dianoux, A. J., Kamitakahara, W. A., Zarestky, J. L., *Phys. Rev. B* **57**, 2663-2666 (1998).
5. Sugai, S. and Onodera, A., *Phys. Rev. Lett.* **77** 4210-4213 (1996); Inamura, Y., Arai, M., Kitamura, N., Bennington, S. M., Hannon, A. C., *Physica B* **241-243**, 903-905 (1998); Inamura, Y., Arai, M., Yamamuro, O., Inaba, A., Kitamura, N., Otomo, T., Matsuo, T., Bennington, S. M., Hannon, A. C., *Physica B* **263-264**, 299-302 (1999); Jund, P. and Jullien, R., *J. Chem. Phys.* **113**, 2768-2771 (2000).
6. Horbach, J., Kob, W. and Binder, K., *J. Phys. Chem. B* **103**, 4104-4108 (1999); Ruocco, G., Sette, F., Di Leonardo, R., Monaco, G., Sampoli, M., Scopigno, T. and Viliani, G. *Phys. Rev. Lett.* **84**, 5788-5791 (2000).
7. Schirmacher, W., Diezemann, G., Ganter, C., *Phys. Rev. Lett.* **81**, 136-139 (1998).
8. Taraskin, S. N., Loh, Y. L., Natarajan, G. and Elliott, S. R., *Phys. Rev. Lett.* **86**, 1255-1258 (2001); Taraskin S. N. and Elliott, S. R., *J. Phys.: Cond. Matt.* **14**, 3143-3166 (2002); Simdyankin, S. I., Taraskin, S. N., Elenius, M., Elliott, S. R. and Dzugutov, M. *Phys. Rev. B* **65**, 104302-1 - 104302-7 (2002).
9. Elliott, S. R., *Physics of Amorphous Materials,* Longman, New York (1990).
10. Martín-Mayor, V., Parisi, G. and P. Verrocchio, *Phys. Rev. E,* **62** 2373-2379 (2000).
11. Götze, W. and Mayr, M. R., *Phys. Rev. E* **61**, 587-606 (2000).
12. Mézard, M., Parisi G. and Zee, A., *Nucl. Phys. B* **559**, 689-701 (1999).
13. Martín-Mayor, V., Mézard, M., Parisi, G. and Verrocchio, P., *J. Chem. Phys.* **114**, 8068-8081 (2001).
14. Grigera, T. S., Martín-Mayor, V., Parisi, G. and Verrocchio, P., *Phys. Rev. Lett.* **87**, 085502-1 - 085502-4 (2001).
15. Mehta, M. L., *Random Matrices,* Academic Press, London (1991).
16. Wu T. M., and Loring, R. F., *J. Chem. Phys.* **97**, 8568-8575 (1992); Wan Y. and Stratt, R., *J. Chem. Phys.* **100**, 5123-5138 (1994); Cavagna, A., Giardina I. and Parisi, G., *Phys. Rev. Lett.* **83**, 108-111 (1999).
17. Ioffe, A. F. and Regel, A. R., *Prog. Semicond.* **4**, 237-291 (1960); Taraskin S. N. and Elliott, S. R., *Phys. Rev. B* **61**, 12017-12030 (2000).
18. Alexander, S., *Phys. Rev. B* **40**, 7953-7965 (1989).
19. Fabian, J. and Allen, P. B., *Phys. Rev. Lett.* **77**, 3839-3842 (1996); Feldman, J. L., Allen, P. B. and Bickham, S. R., *Phys. Rev. B* **59**, 3551-3559 (1999).
20. Grigera, T. S., Martin-Mayor, V., Parisi, G. and Verrocchio, P., *J. Phys.: Cond. Matt.* **14**, 2167-2179 (2002).
21. Grigera, T. S., Martín-Mayor, V., Parisi, G. and Verrocchio, P, *Nature* **422**, 289-292 (2003).
22. Ciliberti, S., Grigera, T. S., Martín-Mayor, V., Parisi, G. and Verrocchio, P., *J. Chem. Phys.* **119**, 8577-8591 (2003).
23. Benoit, C., Royer E. and Poussigue, G., *J. Phys.: Condens. Matter* **4**, 3125-3152 (1992).
24. Rahman, A., Mandell, M. and McTague, J. P. *J. Chem. Phys.* **64**, 1564-1568 (1976); Seeley, G., and Keyes, T., *J. Chem. Phys.* **91**, 5581-5586 (1989); Madan, B. and Keyes, T., *J. Chem. Phys.* **98**, 3342-3350 (1992); Keyes, T., *J. Chem. Phys.* **101**, 5081-5092 (1994); Cho, M., Fleming, G. R., Saito, S., Ohmine, I. and Stratt, R. M., *J. Chem. Phys.* **100**, 6672-6683 (1994); Bembenek, S. and Laird, B., *Phys. Rev. Lett.* **74**, 936-939 (1995); Bembenek, S., and Laird, B., *J. Chem. Phys.* **104**, 5199-5208 (1996).
25. Angelani, L., Di Leonardo, R., Ruocco, G., Scala, A. and Sciortino, F., *Phys. Rev. Lett.* **85**, 5356-5359 (2000); Broderix, K., Bhattacharya, K. K., Cavagna, A., Zippelius, A. and Giardina, I *Phys. Rev. Lett.* **85**, 5360-5363 (2000); Grigera, T. S., Cavagna, A., Giardina, I. and Parisi, G., *Phys. Rev. Lett.* **88**, 055502-1 - 055502-4 (2002);
26. Götze, W. and Sjorgen, L., *Rep. Prog. Phys.* **55**, 241-376 (1992); Kob, W. and Andersen, H. C., *Phys. Rev. E* **51**, 4626-4641 (1995).
27. Bernu, B., Hansen, J.-P., Hiwatari, Y., and Pastore, G., *Phys. Rev. A* **36**, 4891-4903 (1987).
28. Grigera, T. S. and Parisi, G., *Phys. Rev. E* **63**, 045102-1 - 045102-4 (2001).
29. Duval, E., Saviot, L., David, L., Etienne, S., Jal, J.F., *Europhys. Lett.* **63**, 778-784 (2003).

Structural Relaxations in Silica Glass

Akira J. Ikushima, Kazuya Saito, and Hiroshi Kakiuchida

Research Center for advanced Photon Technology, Toyota Technological Institute
Hisakata, Tempaku-ku, Nagoya-city, Aichi-prefecture 468-8511, Japan

Abstract. Structural relaxation is a key factor to control structure and physical properties of silica glass. We investigated structural disorder during structural relaxation processes, and have found that the disorder reduces with decreasing the fictive temperature. Consequently, various optical properties, e.g., Rayleigh light scattering, Urbach edge, and formation of photo-induced defects are straightforwardly correlated to the fictive temperature. The structural relaxation has been tentatively analyzed assuming two processes: main-and sub-relaxations. Both processes seem to be encouraged tremendously by the halogen doping, especially with fluorine and chlorine. Both main- and sub-relaxations can be shortened by the order of 5-6 orders of magnitude by doping of several percents of F. The result implies that a great possibility of breakthrough that silica glass can be more transparent, as the density fluctuation in silica glass that is frozen at the fictive temperature can be reduced through the relaxations. This in turn means that the fiber drawing process under the best-optimized temperature condition should yield much more lucent fibers. We have also found that UV absorption edge is very much affected by the structural relaxation. The absorption edge shifts to shorter wavelengths with decreasing the fictive temperature.

INTRODUCTION

Silica glass is undoubtedly a key material that has been exclusively used in photonic industry. Optical telecommunication network has so far been constructed worldwide by silica fibers because of its outstanding advantages such as transparency at telecommunication wavelengths, its strength in mechanical, thermal and chemical senses, and even more.

However, transparency, or the optical loss, in silica fibers had not been improved over 18-19 year since in 1985 Sumitomo Electric Co. ever established a champion data of the loss to be 0.154 dB/km [1].

Then, in 2002 the same company again gave an improved data, 0.151 dB/cm [2]. The reason why the progress or improvement has been so slow might be that the observed optical loss, 0.15 dB/km had been thought to be the theoretical limit on the assumption that the glass transition temperature of silica glass is around 1400-1500K, and that the invention of optical amplifier discouraged efforts to be paid to reduce the loss.

However, the tremendously fast increase of information to be transferred by the optical telecommunication has lead to WDM. This implies that quite a number of optical signals should be on a single fiber, leading to a situation that high light intensity should be sustained by a single fiber and the self-focusing effect due to the third-order optical nonlinearity may cause damage of the fiber. Therefore, the intensity of individual optical signal should be made as low as possible. Moreover, further increase of information exchange should it necessary to expand the wavelength width to both sides of 1.55 μ m, the wavelength now employed for the present long-haul telecommunication because that here the optical loss is minimum. This also implies that we should use other wavelengths even if the loss is not at the minimal point, and the characteristic behavior of the optical loss *vs.* wavelength should be lowered as a whole.

In these at least two senses, optical loss in fibers should still be strongly desired to be reduced to as low a value as possible.

The optical loss consists of mainly two components; the Rayleigh scattering due to the density fluctuation in glass, and the multi-phonon absorption due to phonons in silica glass. Although the latter could not be changed unless we change silica glass to any another glass with different phonon structure, the former could be reduced if the density fluctuation is made smaller. The density fluctuation, seen by the laser light passing through the glass, is formed by the freezing of thermal fluctuation at the temperature where a supercooled liquid becomes to a glass state as the viscosity reaches a value of approximately 10^{13} Poise and impede the liquid to flow during the reasonable period of time that human being can wait. This in turn means that any glass at room temperature, say, has the same amount of density fluctuation or the disorder at the glass transition temperature.

More precisely speaking, the glassy state at room temperature is not necessarily the frozen-in state at the glass transition temperature, but the state at the *fictive temperature*, T_f. Figure 1 clearly shows the situation, where the Rayleigh light scattering in two kinds of silica glass are shown [3]. In the sample with just a very small amount of OH shows a clear kink at a certain temperature, whereas another sample containing 1,200 wt ppm OH shows a peculiar behavior around the glass-forming temperature, suggesting that the glass-forming process is more complicated. In the latter case, the room temperature state is corresponding to a temperature that is different from the point where glassy state starts to be formed. The temperature where the room temperature state is frozen-in is called the fictive temperature. T_f is lower if the cooling rate of the supercooled liquid state is slower, meaning that degree of structural disorder in glasses is getting less with decreasing T_f.

Structural disorder in glasses affects various optical and other properties. UV (ultra-violet) absorption edge of silica glass seems also to be relevant to the structural disorder. Silica glass is also an important material in UV range, as its UV absorption edge is located around 150-160 nm. Absorption edge should be, as is with any other material in general, of a fundamental interest correlated to the electronic structure. Moreover, silica glass should be a strong candidate to be utilized in microlithography application in the very near future, as it is supposed to be carried out at 157nm by using F_2 excimer laser. Therefore, correlation between transparency at 157nm and structural disorder of the glass should be the very key issue.

We have devoted to elucidate the structural disorder in silica glass as functions of temperature and species and concentration of dopant. This paper is a apart of report we have carried out over last 8 years [4].

STRUCTURAL RELAXATIONS IN SILICA GLASS

We have a conventional way to deduce the fictive temperature. The group in Rensselaer Polytchnic Institute established an empirical relationship between the fictive temperature and IR absorption peak near 2,260 cm^{-1} [5]. We employed this method with an FTIR measurement (Perkin Elmer) of a high precision. We could determine T_f within the accuracy of ± 5K, and determined such relationships in pure and silica glasses doped with OH, F, Cl etc. as well [6].

We carried out experiments to know structural relaxation times as follows: first, we put a sample in a furnace of T_0 for a sufficiently long time so that the structure is in equilibrium at T_0. Then, we put the sample to another furnace of temperature T, and started to observe drift of IR absorption peak position around 2,260 cm^{-1}. In this way, we could see the change of T_f

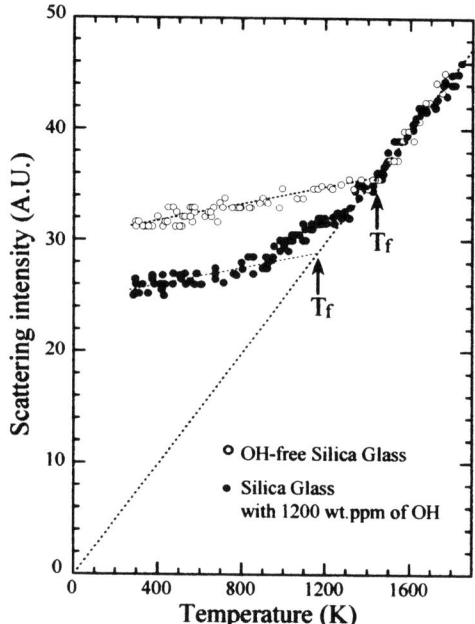

FIGURE 1. Temperature dependences of the Rayleigh light scattering

as a function of time at temperature equal to T. We always took $|T - T_0| = 100$K.

We could not analyze the data of T_f vs. time, t, by a single-relaxation function, and therefore used conventionally a double-exponential function,

$$T_f = A_1 \exp(-t/\tau_1) + A_2 \exp(-t/\tau_2) \quad (1)$$

We tentatively called the first and the second terms the main- and the sub-relaxations.

Figure 2 is a typical result with F-containing silica glass expressed as usual Arrhenius plots of τ_1 and τ_2 [7]. As F concentration increases, both τ_1 and τ_2 get shorter remarkably, or, in other words, F-doping

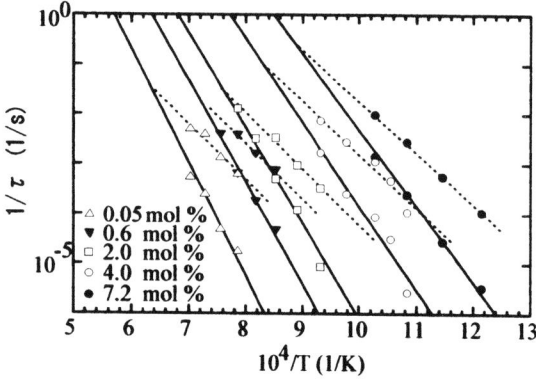

FIGURE 2. Structural relaxation time in F-doped silica

glass.

remarkably accelerates both main- and sub-relaxation processes by 5-6 orders of magnitude.

Figure 3 then shows plots of the activation energies deduced from the former figure as functions of F concentration, together with the result from the viscosity [8]. This indicates that the main-relaxation is the viscous flow process of the glass. The sub-relaxation is, however, the process that we have not yet been able to understand its mechanism. A recent study [9] has shown that the relaxation time seem to have a distribution which depends on the temperature and even temperature difference, $|T - T_0|$.

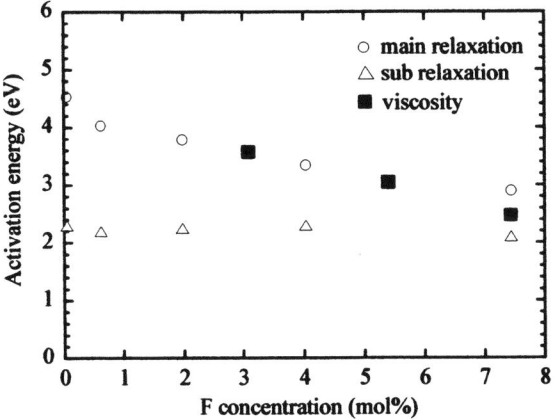

FIGURE 3. Activation energy of main- and subrelaxations and viscosity.

From the standpoint of optical fiber manufacturing, the dramatic acceleration of relaxations strongly indicates a possibility of great breakthrough that silica glass doped by a proper dopant at a proper concentration can relax even during a very short period of time, a few tenths of a second, say, under an optimized thermal condition. The condition during fiber drawing process can be simulated being based on the present result [10]. And, this implies that much more lucent optical fibers could be realized if the drawing is made under the optimized condition.

UV ABSORPTION EDGE

The UV absorption edge was measured by a self-designed system. See Ref. [11] for details. The system had a capability to measure over the wavelength ranging from 130 to 310 nm, and temperature from 10 to 2200 K.

We first measured the absorption edge of pure silica glass. The result is shown in Fig. 4 [12]. We then analyzed variation of the Urbach energy as a function of temperature, assuming the Einstein model to estimate effects due to the thermal fluctuation coming from the structural vibration of silica glass, alike the

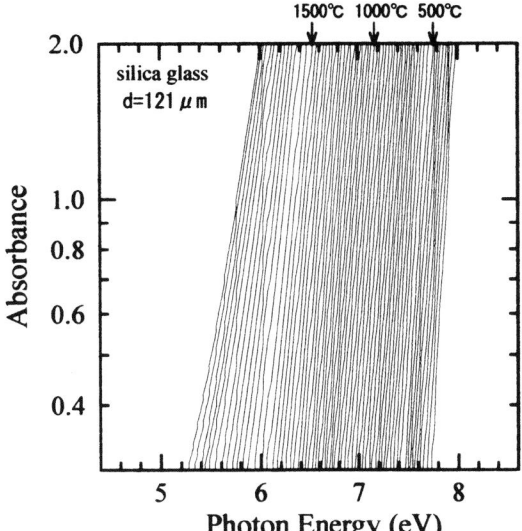

FIGURE 4. Absorption edge of pure silica glass at various temperatures.

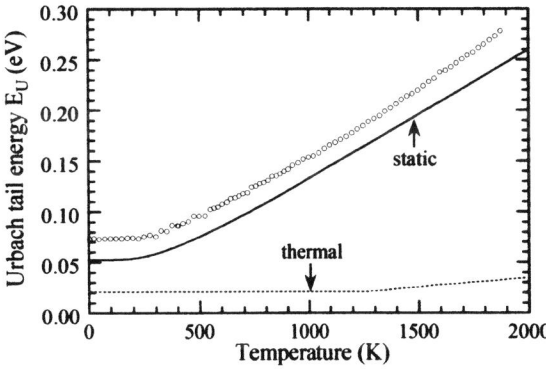

FIGURE 5. Temperature dependence of the Urbach tail energy in silica glass.

lattice vibration in crystals. Figure 5 is the result, where the Urbach energy values deduced from the experiment, and behaviors expected from the thermal and static fluctuations are shown. Surprisingly enough, the static effect due to the disordered structure of glass is less than the thermal effect, only approximately 30%.

We then extended the same measurement to silica glasses, with various F concentrations. Employing the same way of data analysis, and deduced the Einstein temperatures. The absorption edge shifts remarkably as we change F concentration, and the fictive temperature, too. Figure 6 is a summary of the behavior, where both quantities, F concentration and the fictive temperature T_f, are important [7]. The result is a very valuable indication to materialize silica glass that is sufficiently transparent for the microlithography at 157 nm.

Figure 7 then shows the Einstein temperature as a function of F concentration [13]. The Einstein temperature does not seem to depend on F concentration up to even 7.2 mole%.

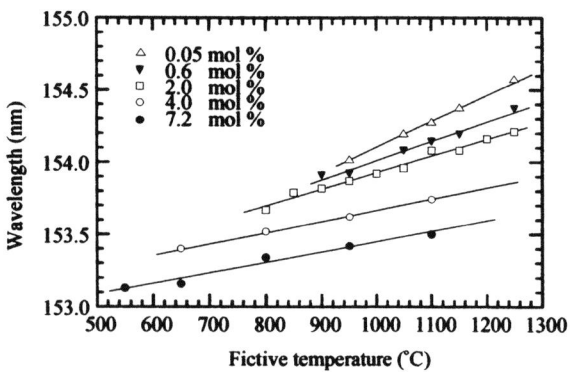

FIGURE 6. Shift of the absorption edge in F-doped silica glasses. The wave length where the transparency is 40 % in samples with thickness of 2 mm.

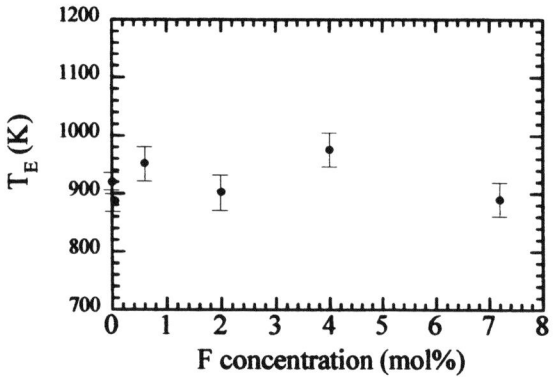

FIGURE 7. Einstein temperatures in F-doped silica glasses.

CONCLUSIVE REMARKS
--STRUCTURE OF SILICA GLASS MODIFIED BY DOPANT--

The results given by the present study have to be interpreted from the standpoint of silica glass structure modified by dopant. The following discussion will mostly focus on F-doped silica glass, as in this case we could change the concentration up to more than 7 mol%.

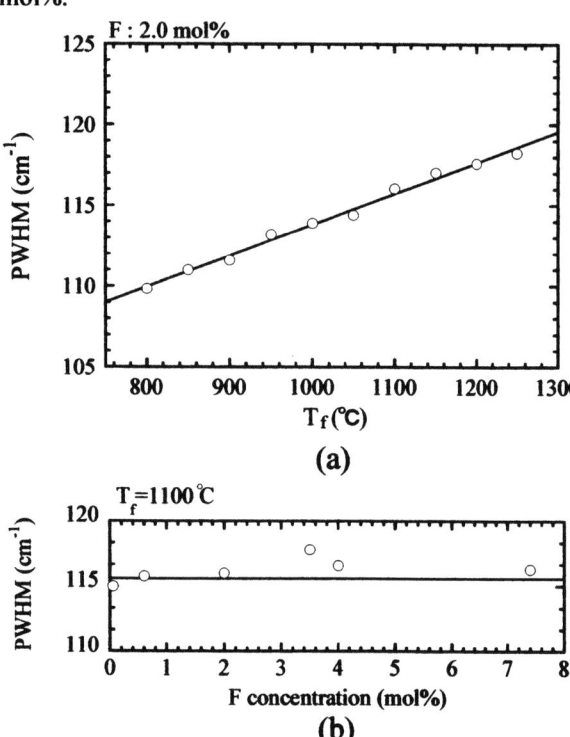

FIGURE 8. PWHM of the 2260 cm-1 IR absorption band. (a) fictive temperature dependence, and (b) F contents dependence.

First of all, we wonder why the Einstein temperature is so insensitive to F concentration. Moreover, PWHM (peak width at half maximum) of the IR absorption around 2,260 cm^{-1}, as shown in Fig. 8 [7], seems to mean that doping of F does not modify the main structure of silica glass, while Si-F bonds are formed since that the bonding energy between Si and F is larger than Si and O.

We also made measurements of other two macroscopic quantities, density and refractive index [14], and they are both quite dependent on F concentration. The density increases linearly against F concentration, which is accompanied by a linear increase of the refractive index.

These results clearly indicate an existence of a region around F that is more vacant than the "normal" region unaffected by F-doping. The "normal" region sustains structures of silica glass in both static and thermal senses, while F's as dopants tend to expand surrounding volume and affect the refractive index, and so on.

In many cases, more often in high-polymer science, a concept called *free volume* has been employed to interpret experimental results, and the concept has also been introduced in glass science in some cases. However, to the present authors' opinion, the free volume has not been clearly verified to exist in glasses so far, while it would be a convenient of thinking in some cases. The present study clearly indicates the existence of "free volume" or "micro void" at least in F-doped silica glass.

ACKNOWLEDGMENTS

The authors thank This work was partially supported by Grant-in Aid for Scientific Research from the Ministry of Education, Culture, Sports, Science, and Technology (Nos. 09355021, 09640408, 11750595, 11694179, 12355025).

REFERENCES

1. Kanamori, H., Yokota, H., Tanaka, G., Watanabe, M., Ishiguro, Y., Yoshida, I., Kakii, T., Itoh, S., Asano, Y., and Tanaka, S., *IEEE J. Lightwave Technol.* **LT-4**, 1144-1150 (1986).
2. Nagayama, K., Saitoh, T., Kakui, M., Kawasaki, K., Matsui, M.,Takamizawa, H., Miyaki, H., Ooga, Y., Tsuchiya, I., and Chigusa, Y., *OFC2002 Postdeadline papers* **FA10**, 1-2 (2002).
3. Saito, K., and Ikushima, A. J., *Appl. Phys. Lett.* **70**, 3504-3506 (1997).
4. Ikushima, A. J., Fujiwara, T., and Saito, K., *J. Appl. Phys.* **88**, 1201-1213 (2000).
5. Agarwal, A., Davis, K. M., and Tomozawa, M., *J. Non-Cryst. Solids* **185**, 191-198 (1995).
6. Saito, K., and Ikushima, A. J., *J. Appl. Phys.*, **91**, 4886-4890 (2002).
7. Kakiuchida, H., Saito, K., and Ikushima, A. J., *J. Appl. Phys.*, **93**, 777-779 (2003).
8. Kyoto, Ohba, Y., Ishikawa, S., and Ishiguro, Y., *J. Mater. Sci.*, **28**, 2738-2744 (1993).
9. Kakiuchida, H., Saito, K., and Ikushima, A. J., *J. Appl. Phys.* **94**, 1705-1708 (2003).
10. Kakiuchida, H., Saito, K., and Ikushima, A. J., *Proc of SPIE*, **4905**, 355-362 (2002).
11. Saito, K., and Ikushima, A. J., *J. Non-Cryst. Solids*, **259**, 81-86 (1999).
12. Saito, K., and Ikushima, A. J., *Phys. Rev.*, **62**, 8584 –8587 (2000).
13. Saito, K., and Ikushima, A. J., *Phys. Rev.* submitted.
14. Kakiuchida, H., Sekiya, E., Saito, K., and Ikushima, A. J., *Jpn. J. Appl. Phys.*, (2003) to be published.

Dynamics of a rod in a homogeneous/inhomogeneous frozen disordered medium: Correlation functions and non-Gaussian effects

Angel J. Moreno*† and Walter Kob*

*Laboratoire des Verres. Université de Montpellier II. Place E. Bataillon. CC 069. F-34095 Montpellier, France.
†Present address: Dipartimento di Fisica and INFM Udr and Center for Statistical Mechanics and Complexity, Università di Roma "La Sapienza". Piazzale Aldo Moro 2. I-00185 Roma, Italy.

Abstract. We present molecular dynamics simulations of the motion of a single rigid rod in a disordered static 2d-array of disk-like obstacles. Two different configurations have been used for the latter: A completely random one, and which thus has an inhomogeneous structure, and an homogeneous "glassy" one, obtained from freezing a liquid of soft disks in equilibrium. Small differences are observed between both structures for the translational dynamics of the rod center-of-mass. In contrast to this, the rotational dynamics in the glassy host medium is strongly slowed down in comparison with the random one. We calculate angular correlation functions for a wide range of rod length L and density of obstacles ρ as control parameters. A two-step decay is observed for large values of L and ρ, in analogy with supercooled liquids at temperature close to the glass transition. In agreement with the prediction of the Mode Coupling Theory, a time-length and time-density scaling is obtained. In order to get insight on the relation between the heterogeneity of the dynamics and the structure of the host medium, we determine the deviations from Gaussianity at different length scales. Strong deviations are obtained even at spatial scales much larger than the rod length. The magnitude of these deviations are independent of the nature of the host medium. This result suggests that the large scale translational dynamics of the rod is affected only weakly by the presence of inhomogeneities in the host medium.

INTRODUCTION

Since it was initially introduced by Lorentz as a model for the electrical conductivity in metals [1], the problem of the Lorentz gas has given rise to a substantial theoretical effort aimed to understand its properties [2, 3, 4, 5, 6, 7, 8]. In this model, a *single* classical particle moves through a disordered array of *static* obstacles. It can thus be used as a simplified picture of the motion of a light atom in a disordered environment of heavy particles having a much slower dynamics. In the simplest case, where the diffusing particle and the obstacles are modeled as hard spheres, an exact solution for the diffusion constant exists in the limit of low densities of obstacles [9]. However, the problem becomes highly non-trivial with increasing density, where dynamic correlations and memory effects start to become important for the motion of the diffusing particle, and the system shows the typical features of the dynamics of supercooled liquids or dense colloidal systems, such as a transition to a non-ergodic phase of zero diffusivity [2, 3, 4, 6]. In particular, diffusion constants and correlation functions are non-analytical functions of the density [2, 4, 7, 8]. Moreover, correlation functions show non-exponential long-time decays.

Diffusing particles and obstacles are generally modeled as disks or spheres in two and three dimensions, respectively. Much less attention has been paid to systems that have orientational degrees of freedom. Motivated by this latter question, we have recently started an investigation, at low and moderate densities of obstacles, on a generalization of the Lorentz gas, namely a model in which the diffusing particle is a rigid rod [10]. An array of randomly distributed disks has been used for the host medium. For simplicity simulations have been done in two dimensions, reducing the degrees of freedom to the center-of-mass position and the orientation of the rod axis. As in the case of supercooled liquids [11, 12, 13, 14, 15] or dense colloidal systems [16, 17, 18], one observes at intermediate times a caging regime for the rod center-of-mass motion, which is due to the steric hindrance produced by the presence of the neighboring obstacles.

More interestingly, strong deviations from Gaussianity have been obtained for the incoherent intermediate scattering function at wavelengths much longer than the rod length, giving evidence for a strongly heterogeneous character of the long-time dynamics at such length scales. The inhomogeneous structure of the model used for the host medium has been pointed out as a possible origin of such non-Gaussian effects since a random configuration features large holes on one side but on the other side also dense clusters of obstacles. The presence of holes might lead to a finite probability of jumps that are much longer than the average "jump length" and thus to a heterogeneous dynamics.

In order to shed new light on this question, we have repeated the simulations by taking a disordered but *homogeneous* configuration of the obstacles instead of a random one. We will see, however, that large scale non-Gaussian effects are not significantly affected by the particular choice of the configuration of the obstacles. Some information on angular correlation functions is also presented. The article is organized as follows: In Section II we present the model and give details of the simulation. Translational and angular mean-squared displacements are presented in Section III. The behavior of angular correlation functions is shown and discussed in Section IV, in terms of the Mode Coupling Theory. Non-Gaussian effects in the random and glassy models of the host medium are investigated in Section V. Conclusions are given in Section VI.

MODEL AND DETAILS OF THE SIMULATION

The rigid rod, of mass M, was modeled as N aligned point particles of equal mass $m = M/N$, with a bond length 2σ. The rod length is therefore given by $L = (2N-1)\sigma$. The positions of the obstacles in the random configuration were generated by a standard Poisson process. In order to obtain the glassy host medium, we equilibrated at a reduced particle density $\rho = 0.77$ and at temperature $T = \varepsilon/k_B$ a two-dimensional array of point particles interacting via a soft-disk potential $V(r) = \varepsilon(\sigma/r)^{12}$. This procedure produced an homogeneous liquid-like configuration. The latter was then permanently frozen and was expanded or shrunk to obtain the desired density of obstacles, defined as $\rho = n_{obs}/l_{box}^2$, with n_{obs} the number of obstacles and l_{box} the length of the square simulation box used for periodic boundary conditions.

The same soft-disk potential $V(r)$ was used for the interaction between the particles forming the rod and the obstacles. For computational efficiency, $V(r)$ was truncated and shifted at a cutoff distance of 2.5σ. In the following, space and time will be measured in the reduced units σ and $(\sigma^2 m/\varepsilon)^{1/2}$, respectively. Typically 600-1000 realizations of the ensemble rod-obstacles were considered. The set of rods was equilibrated at $T = \varepsilon/k_B$. After the equilibration, a production run was done at constant energy and results were averaged over the different realizations. These runs covered 10^6 time units, corresponding to $(1-5) \cdot 10^8$ time steps, depending on the step size used for the different rod lengths and densities. Run times were significantly longer than the relaxation times of the system.

MEAN-SQUARED DISPLACEMENTS

Figs. 1a and 1b show respectively for $\rho = 6 \cdot 10^{-3}$ a comparison between the random and the glassy configuration for the mean-squared displacement of the rod center-of-mass $\langle(\Delta r(t))^2\rangle$ and the mean-squared angular displacement $\langle(\Delta\Phi(t))^2\rangle$. Brackets denote ensemble average. In order to facilitate the observation of the different dynamic regimes, data have been divided by the time t. The comparison covers a wide range of rod lengths from $L \sim 0.1 d_{nn}$ to $L \sim 10 d_{nn}$, where $d_{nn} = \rho^{-1/2} \approx 13$, is the average distance between obstacles for the mentioned density. At short times the rod does not feel the presence of the neighboring obstacles and $\langle(\Delta r(t))^2\rangle$ and $\langle(\Delta\Phi(t))^2\rangle$ show the quadratic time-dependence characteristic of a ballistic motion. For the shortest rods a sharp transition to the long-time linear regime is observed. In contrast to this, for long rods a crossover regime between both limits, showing a weaker time dependence that the ballistic motion, is present over 1-2 decades of intermediate times. Such a crossover corresponds to the well-known caging regime [11, 12, 13, 14, 15, 16, 17, 18] observed in supercooled liquids or dense colloidal systems. Due to the presence of the neighboring obstacles, the particle is trapped within an "effective cage" for some time until it escapes from it and begins to show a diffusive behavior.

No significant differences are observed between the values of $\langle(\Delta r(t))^2\rangle$ for a same rod length in the two different, random and glassy, configurations. This is more clearly seen by calculating for each L the ratio $D_{CM}^{ran}/D_{CM}^{gla}$ between the center-of-mass translational diffusion constant D_{CM} in both configurations. The latter is calculated as the long-time limit of $\langle(\Delta r(t))^2\rangle/4t$. As shown in Fig. 2, differences in D_{CM} between the random and the glassy host medium are less than a factor 1.5. The values of D_{CM} are systematically lower for the glassy host medium, evidencing a stronger backscattering for the diffusion, as intuitively expected from the homogeneous character of this latter configuration in contrast to the random one (see Fig. 3), where the presence of holes facilitates diffusion. It is noteworthy that the maximum difference between the values of D_{CM} in both configurations takes place for $L \sim 10-20$, i.e., when the rod becomes longer than d_{nn}. Thus, while in the homogeneous glassy host medium the rod will not be able to

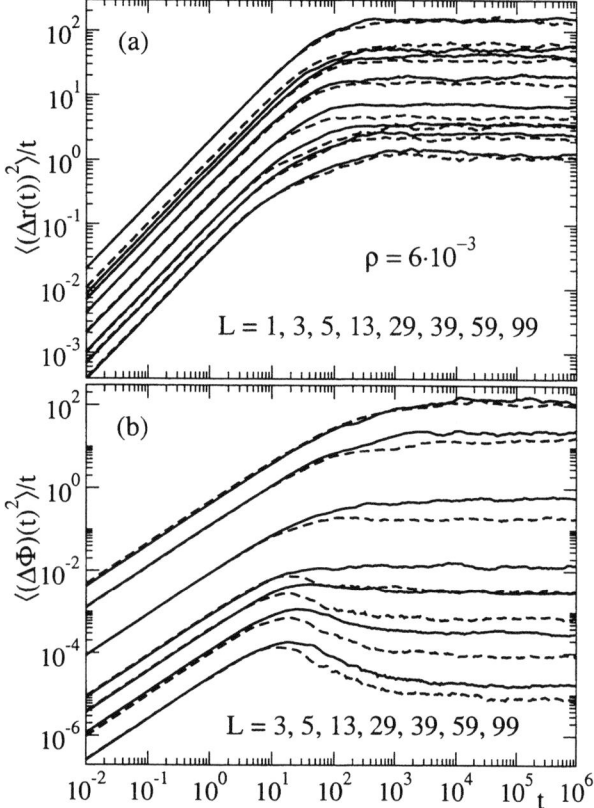

FIGURE 1. Mean-squared displacement of the center-of-mass (a) and mean-squared angular displacement (b), both divided by the time t, for $\rho = 6 \cdot 10^{-3}$ and different rod lengths, for the random (solid lines) and glassy (dashed lines) configuration of the obstacles.

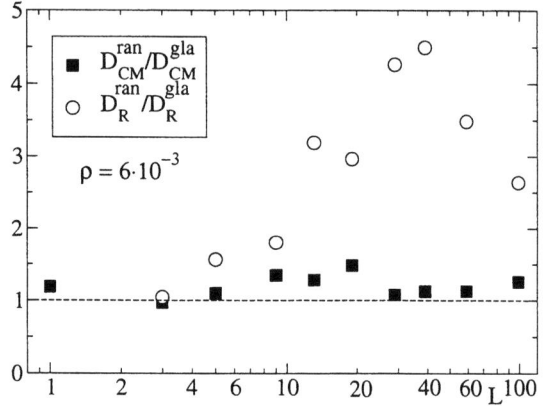

FIGURE 2. Ratio between the rotational and center-of-mass translational diffusion constants for the random and the glassy host medium at $\rho = 6 \cdot 10^{-3}$ and different rod lengths.

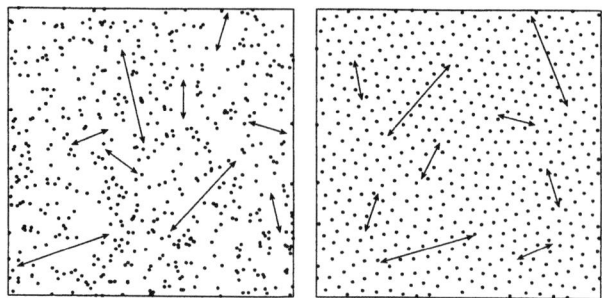

FIGURE 3. Two configurations of the obstacles for a same density $\rho = 6 \cdot 10^{-3}$. Left: A random configuration. Right: A homogeneous glassy configuration. Short and long arrows correspond, respectively, to distances of 40 and 100.

pass transversally between two neighboring obstacles, in the inhomogeneous random configuration this diffusion channel will still be present due to the presence of holes and will only be suppressed for very long rods.

Strong differences are observed between both models of the host medium in the case of the rotational dynamics, as shown in Fig. 1b for $\langle (\Delta\Phi(t))^2 \rangle$ and in Fig. 2 for the ratio of the rotational diffusion constants, $D_R^{\text{ran}}/D_R^{\text{gla}}$. ($D_R$ is calculated as the long-time limit of $\langle (\Delta\Phi(t))^2 \rangle/2t$.) For rod lengths smaller than d_{nn} the rotational dynamics is only weakly sensitive to the configuration of the medium and the ratio $D_R^{\text{ran}}/D_R^{\text{gla}}$ remains close to unity. However, for rods longer than d_{nn} the latter ratio strongly increases up to a maximum of ~ 4.5 for $L \sim 40$. The position of the maximum can be again rationalized from the inhomogeneous structure of the random host medium. Short arrows in the configurations of Fig. 3 for $\rho = 6 \cdot 10^{-3}$ indicate distances of 40. From the comparison between both configurations it is clear that while in the glassy medium rods of $L \sim 40$ can perform only small rotations within the tubes formed by the neighboring obstacles, in the random medium they can go into the holes and thus rotate freely over a larger angle. Therefore, the long-time angular displacement will grow up more quickly than for the motion between narrow tubes in the glassy configuration. It must also be mentioned that even for the largest investigated rod length, $L = 99$, where motion between tubes also dominates the diffusion in the random medium, the ratio $D_R^{\text{ran}}/D_R^{\text{gla}}$ is still significantly different from unity. Thus, while in the glassy medium the walls of the tube are formed by uniformly distributed obstacles, in the random configuration long distances are allowed between some of the neighboring obstacles forming the tube, leading to additional escaping channels (see long arrows in Fig. 3). In principle, only for extremely long rods the ratio $D_R^{\text{ran}}/D_R^{\text{gla}}$ is expected to approach unity.

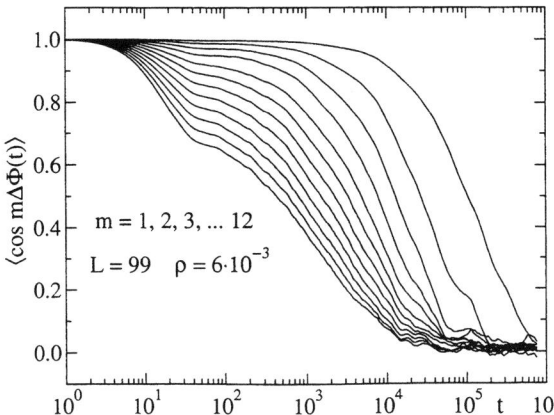

FIGURE 4. Correlation function $\langle \cos m\Delta\Phi(t)\rangle$ for $L = 99$ and $\rho = 6 \cdot 10^{-3}$, with $m = 1,2,3...12$ (from top to bottom). Data correspond to the random host medium.

CORRELATION FUNCTIONS

At present, the only theory that is able to make precise quantitative predictions on the dynamics of supercooled liquids or dense colloidal systems is the Mode Coupling Theory (MCT) [19, 20, 21, 22]. In its idealized version, it predicts a dynamic transition from an ergodic to a non-ergodic phase at some critical value of the control parameters. These are usually the temperature or the density, though in principle the MCT formalism can be generalized to other control parameters. Moreover, it has been recently tested in a kind of systems very different from liquids or colloids, such as plastic crystals [23], suggesting a more universal character for this theory. Motivated by this possibilities, we test some predictions of MCT for the rotational dynamics of the rod with L and ρ as control parameters.

According to MCT, upon approaching the critical point, the correlation function shows a two-step decay. The initial decay corresponds to the dynamic transition from the ballistic to the caging regime. As a consequence of the temporary trapping in the cage formed by the neighboring obstacles, the particle does not lose the memory of its initial position and the correlation functions decay very slowly, giving rise to a plateau at intermediate times between the ballistic and the diffusive regime. The dynamics close to this plateau is usually referred as the β-relaxation. The closer the control parameters are to the critical values, the slower is the mobility of the particles and the longer is the lifetime of the cage, leading to a longer plateau in the correlation function. Finally, at much longer times, the particle escapes from the cage and loses memory of its initial environment, leading to a second long-time decay of the correlation function to zero, known as the α-relaxation.

Fig. 4 shows the behavior of the angular correlators

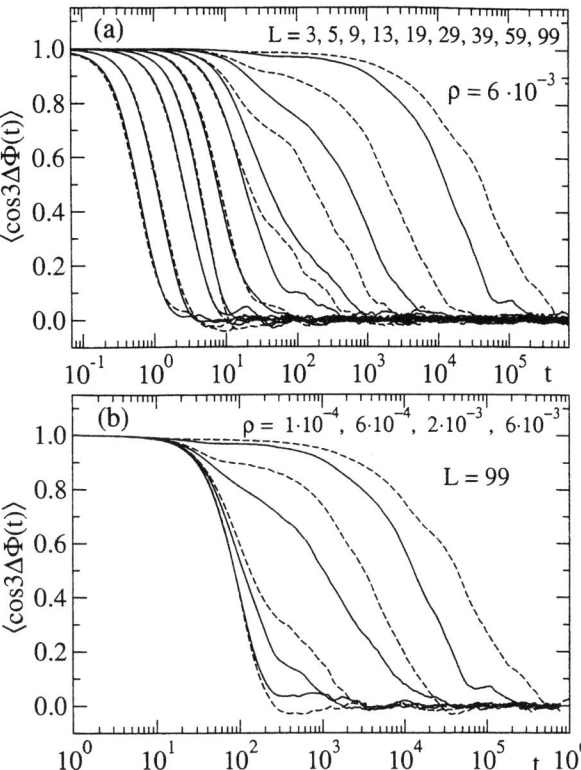

FIGURE 5. (a) Correlation function $\langle \cos 3\Delta\Phi(t)\rangle$ for $\rho = 6 \cdot 10^{-3}$ and different rod lengths; (b) The same function for $L = 99$ and different densities. Solid and dashed lines correspond, respectively, to the random and the glassy configurations of the host medium.

$\langle \cos m\Delta\Phi(t)\rangle$ for $m = 1,2,3,...12$ for a density of obstacles $\rho = 6 \cdot 10^{-3}$ and a rod length $L = 99$ much longer than d_{nn}. Data are shown for the random host medium, though for the glassy one they show the same qualitative behavior. Though it is difficult to see it for the smallest values of m, the plateau is clearly visible for $m \geq 4$. Thus, long rods will only perform small rotations before hitting the walls of the tube. As a consequence, the correlators for small m will decrease only very weakly during the ballistic regime, and hence it will be difficult to see the subsequent plateau. In contrast to this, for higher-order correlators, the angle rotated during the ballistic regime will be amplified by a factor m, leading to a stronger decay before the caging regime and facilitating the observation of the plateau. The latter begins to develop around $t \sim 30$. As can be seen in Fig. 1b for the curve for $L = 99$, this time corresponds to the beginning of the caging regime for the mean-squared angular displacement, in agreement with the MCT prediction.

Fig. 5 shows for $m = 3$, and for the two models of host medium, how the plateau develops with increasing rod length and density of obstacles. While for values of L below d_{nn}, the correlators show a simple decay

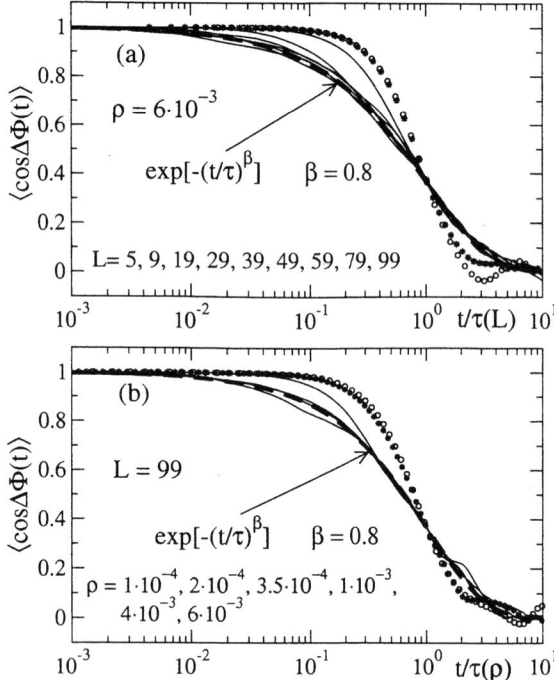

FIGURE 6. (a) Time-length superposition for $\langle\cos\Delta\Phi(t)\rangle$ at density $\rho = 6\cdot 10^{-3}$; (b) Time-density scaling for the same function for length $L = 99$. For the sake of clarity the lowest values of ρ and L have been represented with points. Dashed lines are fits to KWW functions. The stretching exponents are indicated in the figure. Data correspond to the random host medium.

and within the error bar show no difference between both models of the host medium, a clear difference is observed for longer rods. Thus, for the glassy medium rotational relaxation times become about a decade larger and the plateau is significantly higher than in the random configuration, in agreement with the results presented in Section III for the mean-squared angular displacement.

Another important prediction of the MCT is the so called "second universality": given a correlator $g(t,\zeta)$, with ζ a control parameter, then one has that, in the time scale of the α-regime, the correlator follows a scaling law $g(t,\zeta) = \tilde{g}(t/\tau(\zeta))$, where $\tau(\zeta)$ is the ζ-dependence of the relaxation time τ of the α-regime for such a correlator, and $\tilde{g}$ is a master function. The relaxation time is in practice defined as the time where the correlator decays to an arbitrary but small fraction of its initial value, or is obtained from fitting the α-decay of the correlator to a Kohlrausch-Williams-Watts (KWW) function, $\exp(-(t/\tau)^\beta)$, widely used in the analysis of relaxations in complex systems [24]. In most of the experimental situations the relevant control parameter is the temperature and for that reason, the second universality is often referred as the "time-temperature superposition principle". Now we test the existence of a time-length and a time-density superposition principle for the angular correlators of the rod. Fig. 6 shows the correlators $\langle\cos\Delta\Phi(t)\rangle$ as a function of the scaled times $t/\tau(L)$ for constant density $\rho = 6\cdot 10^{-3}$, and $t/\tau(\rho)$ for constant rod length $L = 99$. Data are shown for the random host medium. The relaxation times τ have been defined as $\langle\cos\Delta\Phi(\tau)\rangle = 1/e$. Apart from the trivial scaling in the limit of short rods and low densities, a good superposition to a master curve is obtained for larger values of L and ρ, confirming the second universality of the MCT for this system. The master curve can be well reproduced by a KWW function with a stretching exponent $\beta = 0.8$, as shown in Fig. 6. Analogous results, with a very similar stretching exponent, are obtained for the glassy host medium.

NON-GAUSSIAN EFFECTS

In Einstein's random walk model for diffusion, particles move under the effect of collisions with the others. When a particle undergoes a collision, it changes its direction randomly, and completely loses the memory of its previous history. When the time and spatial observational scales are much larger than the characteristic time and the mean free path between collisions, the van Hove self-correlation function, i.e., the time and spatial probability distribution $G_s(r,t)$ of a particle initially located at the origin, is given by a Gaussian function [25]. This functional form is exact for an ideal gas and for an harmonic crystal. It is also valid in the limit of short times, where atoms behave as free particles. When the system is observed at time and length scales comparable to those characteristic of the collisions, the possible intrinsic dynamic heterogeneities of the system will be reflected by strong deviations from Gaussianity in $G_s(r,t)$. Such deviations are usually quantified by the so-called second-order non-Gaussian parameter $\alpha_2(t)$, which in two-dimensions is defined as $\alpha_2(t) = [\langle(\Delta r(t))^4\rangle/2\langle(\Delta r(t))^2\rangle^2] - 1$. For a Gaussian function in two dimensions, $\alpha_2(t) = 0$, while deviations from Gaussianity result in finite values of $\alpha_2(t)$.

Our previous investigation [10] on the non-Gaussian parameter in the random host medium revealed some features similar to those observed in supercooled liquids or dense colloids, such as the development of a peak (see also Fig. 7), which grows with increasing rod length - in analogy to decreasing temperature in supercooled liquids or increasing density in colloids. As also observed in these latter systems [11, 12, 13, 14, 15, 16, 17, 18], the region around the maximum of the peak corresponds to the time interval corresponding to the end of the caging and the beginning of the long-time diffusive regime. Thus, the breaking of the cage leads to a finite probability of

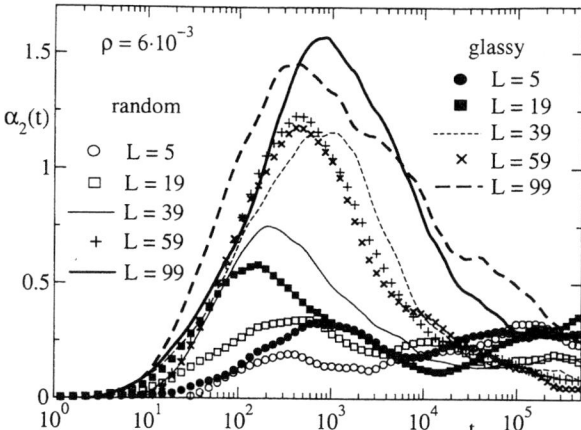
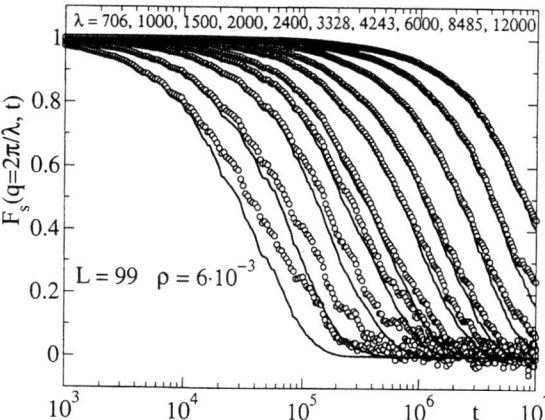

FIGURE 7. Time-dependence of the non-Gaussian parameter $\alpha_2(t)$ for different values of L and for the random and the glassy host medium at density $\rho = 6 \cdot 10^{-3}$.

FIGURE 8. Intermediate incoherent scattering function $F_s(q = 2\pi/\lambda, t)$ at different wavelengths λ for $\rho = 6 \cdot 10^{-3}$ and $L = 99$. Points correspond to simulation data in the random host medium. Lines are the curves in Gaussian approximation as calculated from the center-of-mass mean-squared displacement obtained from the simulation (see text).

jumps much longer that the size of the cage, resulting in a strongly heterogeneous dynamics at that intermediate time and spatial scale, which is reflected by a peak in the non-Gaussian parameter.

Another interesting result was the observation that, in particular for long rods, in the time span of the simulation, i.e. for time scales that are much longer than the typical relaxation time of the system, $\alpha_2(t)$ did not decay to zero but remained finite [10]. Thus, the long-time dynamics is still significantly non-Gaussian, i.e. heterogeneous, at the spatial scale -much larger than the rod length- covered by the simulation. In order to investigate the possibility of a relation of this large-scale non-Gaussian dynamics with the presence of holes in the random structure of obstacles, that might lead to a finite probability of jumps much longer than the average, as pointed out in Ref.[10], we have calculated $\alpha_2(t)$ for the dynamics of the rod in the glassy host medium. Fig. 7 shows a comparison between the non-Gaussian parameter in both structures for several rod lengths and density $\rho = 6 \cdot 10^{-3}$. While in the limit of short and long rods $\alpha_2(t)$ takes similar values for both configurations in all the time window investigated, a much more pronounced peak is observed at intermediate rod lengths for the glassy configuration. This result can be rationalized by the absence of holes in the homogeneous glassy host medium. Thus in the latter, jumps that lead to an escape from the cage formed by the neighboring obstacles will be necessarily long, since long rods will be confined in tubes which they will be able to leave only by making a long longitudinal motion. In contrast to this, the above mentioned inhomogeneous nature of the tube walls in the random structure will allow the rod to escape the cage also by shorter jumps, resulting in a less heterogenous caging regime. Only rods much longer than the hole size will need long jumps to escape from the tubes in the random host medium and the caging regime will become as heterogeneous as in the glassy host medium, as evidenced by the similar peak heights in Fig. 7 for $L = 59$ and 99.

Concerning the long-time dynamics, the non-Gaussian parameter also remains finite for the glassy host medium, and no important differences with the random configuration are observed. Therefore, contrary to what was previously pointed out [10], the non-Gaussianity observed at large length scales for the long-time dynamics in the inhomogeneous random host medium is also present in the homogeneous glassy one, and is not related to the presence of holes in the configuration of obstacles, the latter having effects only in the intermediate caging regime.

Another way of visualize the non-Gaussian effects at large length scales is obtained by representing the intermediate incoherent scattering function $F_s(q,t) = \langle \exp(-i\mathbf{q} \cdot \Delta \mathbf{r}(t)) \rangle$, at very long wavelengths $\lambda = 2\pi/q$. Brackets denote ensemble and angular average. Fig. 8 shows this latter function in the random host medium for $L = 99$ and $\rho = 6 \cdot 10^{-3}$, at different (long) wavelengths. (Note that the simulations have been extended one order of magnitude respect to those presented in Ref. [10].) The presence of non-Gaussian effects are made clear by comparing the different curves with those corresponding to the Gaussian case in two-dimensions [25], $\exp[-\langle(\Delta r(t))^2\rangle q^2/4]$, where we use the center-of-mass mean-squared displacement $\langle(\Delta r(t))^2\rangle$ calculated from the simulations. From this comparison it is clear that significant non-Gaussian effects are still present at length scales of at least ~ 9000, i.e. two orders of magnitude larger than the rod length. Whether this result indicates that the diffusion of a rod in a disordered array of obstacles is actually a Poisson process at any length scale is an open question.

CONCLUSIONS

By means of molecular dynamics simulations, we have compared the dynamics of a rigid rod in two models of a 2d-disordered static host medium: a random configuration of soft disks and another "glassy" one obtained from freezing a liquid of soft disks in equilibrium. While the former is characterized by the presence of big holes and clusters of close obstacles, the latter presents an homogeneous structure. No significant differences have been observed in the translational dynamics of the rod center-of-mass. However, rotations are much more hindered in the glassy host medium.

Angular correlation functions have been calculated for a wide range of rod length and density of obstacles. In agreement with the predictions of the Mode Coupling Theory, these functions show a plateau at the time scale of the caging regime, and follow a time-length and a time-density scaling for the long-time dynamics.

Strong non-Gaussian behavior has been observed at large length scales, though no significant differences are evidenced between the random and the glassy configuration of the obstacles. This result suggests that the long-time translational dynamics is not controlled by the presence of inhomogeneities in the host medium.

ACKNOWLEDGMENTS

We thank E. Frey for useful discussions. A.J.M acknowledges a postdoctoral grant from the Basque Government. Part of this work was supported by the European Community's Human Potential Program under contract HPRN-CT-2002-00307, DYGLAGEMEM.

REFERENCES

1. Lorentz, H.A., *Arch. Neerl.* **10**, 336 (1905).
2. Bruin, C., *Phys. Rev. Lett.* **29**, 1670-1674 (1972).
3. Alder, B.J., and Alley, W.E., *J. Stat. Phys.* **19**, 341-349 (1978).
4. Götze, W., Leutheusser, E., and Yip, S., *Phys. Rev. A* **23**, 2634-2643 (1981); *ibid.* **24**, 1008-1015 (1981); *ibid.* **25**, 533-539 (1982).
5. Masters, A., and Keyes, T., *Phys. Rev. A* **26**, 2129-2139 (1982).
6. Keyes, T., *Phys. Rev. A* **28**, 2584-2587 (1983).
7. Machta, J., and Moore, S.N., *Phys. Rev. A* **32**, 3164-3167 (1985).
8. Binder, P.M., and Frenkel, D., *Phys. Rev. A* **42**, R2463-R2466 (1990).
9. McQuarrie, D.A., *Statistical Physics*, University Science Books, Sausalito, CA, USA, 2000.
10. Moreno, A.J., and Kob, W., *Philos. Magaz. B* (to be published); cond-mat/0303510.
11. Kob, W., and Andersen, H.C., *Phys. Rev. E* **51**, 4626-4641 (1995).
12. Sciortino, F., Gallo, P., Tartaglia, P., and Chen, S.H., *Phys. Rev. E* **54**, 6331-6343 (1996).
13. Mossa, S., Di Leonardo, R., Ruocco, G., and Sampoli, M., *Phys. Rev. E* **62**, 612-630 (2000).
14. Caprion, D., and Schober, H.R., *Phys. Rev. B* **62**, 3709-3716 (2000).
15. Colmenero, J., Alvarez, F., and Arbe, A., *Phys. Rev. E* **65**, 041804 (2002).
16. Van Megen, W., *J. Phys.: Condens. Matter* **14**, 7699-7717 (2002).
17. Weeks, E.R., and Weitz, D.A., *Chem. Phys.* **284**, 361-367 (2002).
18. Puertas, A.M., Fuchs, M., and Cates, M.E., *Phys. Rev. E* **67**, 031406 (2003).
19. Bengtzelius, U., Götze, W., and Sjölander, A., *J. Phys. C* **17**, 5915-5934 (1984).
20. Leutheusser, E., *Phys. Rev. A* **29**, 2765-2773 (1984).
21. Götze, W., and Sjögren, L., *Rep. Prog. Phys.* **55**, 241-376 (1992).
22. Götze, W., *J. Phys.: Condens. Matter* **11**, A1-A45 (1999).
23. Affouard, F., and Descamps, M., *Phys. Rev. Lett.* **87**, 035501 (2001).
24. Phillips, J.C., *Rep. Prog. Phys.* **59**, 1133-1207 (1996).
25. Hansen, J.P., and McDonald. I.R., *Theory of Simple Liquids*, Academic Press London, 1986.

Microscopic Dynamics in Non-Simple Liquid Metals

S. Hosokawa and W.-C. Pilgrim

Institut für Physikalische-, Kern-, und Makromolekulare Chemie, Philipps Universität Marburg, D-35032 Marburg, Germany

Abstract. Our recent inelastic X-ray scattering (IXS) experiments on several non-simple liquid metals have revealed characteristic common features in the collective dynamics. 1) Clear indications for propagating phonon modes were found as in simple liquid metals as liquid alkalis. 2) They exhibit a *positive* dispersion of about 20 % (except in liquid Ge) again as in liquid simple metals. 3) A very short time (sub-picosecond) retaining of the nearest-neighbor correlation is visualized by a Gaussian component in the quasielastic line shape, which may be related to short living covalent species. In this paper, we review the IXS experiment on liquid Si as a typical non-simple liquid metal, and discuss its dynamic properties in connection with results of an *ab initio* molecular dynamics simulation.

INTRODUCTION

Our recent inelastic X-ray scattering (IXS) experiments on several non-simple liquid metals, such as Si [1], Ge [2], Sn [3], and Ga [4], have revealed characteristic common features in the collective dynamics. 1) Clear indications for propagating phonon modes were found as in simple liquid metals as liquid alkalis [5], whereas their lifetimes are much shorter. 2) They exhibit a *positive* dispersion of about 15-25 % (except in liquid Ge) again as in liquid simple metals. 3) A very short time (sub-picosecond) retaining of the nearest-neighbor correlation is visualized by a Gaussian component in the quasielastic line shape, which may be related to short living covalent species. In this paper, we review our recent findings by focusing on liquid Si [1] as a typical non-simple liquid metal.

In the crystalline phase, Si is a typical semiconductor with a diamond structure. Upon melting, it undergoes a semiconductor-metal transition [6] where the density increases by about 10% [6]. This transition is accompanied by significant changes in the local structure; e.g., the coordination number increases from four in the solid to ~6.5 in the liquid [7]. Despite the metallic nature, its structural properties are considerably more complicated than in simple liquid metals such as liquid alkalis. In addition to the low coordination number, the structure factor, $S(Q)$, of liquid Si is characterized by a distinct shoulder at the high-Q side of the first maximum, a feature that cannot be reproduced by a simple hard-sphere approach. *Ab initio* molecular dynamics (MD) simulation studies [8,9] were carried out on this system, and many experimental results could well be reproduced. In Ref. [8], evolution of electron charges around moving atoms could be visualized, indicating chemical bonds with a very short lifetime (< 30 fs), whereas the valence-band electron density of states calculated is remarkably free-electron-like, which is in good agreement with photoemission results [10]. In contrast to the intensive theoretical efforts, experimental investigations on the collective microscopic motion have been hindered by the fact the collective modes in liquid Si are out of reach for thermal neutrons due to the fast sound velocity (~4000 ms^{-1} [11]) and the kinematical restrictions of this technique. On the contrary, IXS is a technique that allows the study of the Q dependence of the excitation in the meV range, but in contrast to neutron scattering, it has no kinematical restrictions. In the energy range of interest, the scattered radiation is entirely coherent. Hence, this technique is ideally suited for the investigation of the collective dynamics in liquids and disordered solids.

We have investigated the microscopic particle dynamics in liquid Si using high-resolution IXS, and for the first time obtained the dynamic scattering law, $S(Q,\omega)$, of this metallic liquid. In this paper, we will discuss its dynamic properties in connection with results of an *ab initio* MD simulation.

EXPERIMENTAL PROCEDURE

The experiments were carried out at the beamline BL-35XU of the SPring-8 using a horizontal IXS spectrometer [12]. A monochromatized beam of 3×10^9 photons s^{-1} was obtained from a cryogenically cooled Si(111) double crystal followed by a Si(11 11 11) monochromator operating in backscattering geometry (89.975°, 21.75 keV). The same backscattering geometry of four two-dimensionally curved Si analyzers was used for the energy analysis of the scattered photons. The energy scans were performed through a thermal variation of the lattice parameter of the monochromer crystal. The energy resolution was determined from a Plexiglas sample and values of 1.5-1.9 meV (FWHM) were found for all four detecting systems. The Q resolution was about ± 0.48 nm^{-1}. The sample was located in a single-crystal sapphire cell, which was a slight modification of the so-called Tamura-type cell [13]. It was placed in a vessel equipped with continuous Be windows [14] capable of covering scattering angles between 0° and 25°. The high temperature of 1733 K was achieved using a W resistance heater, and monitored with two W-5%Re/W-26%Re thermocouples. The IXS experiments were carried out at 26 Q-values between 2.0 and 29.6 nm^{-1} covering an energy transfer range from -50 to +50 meV. Empty cell measurements were separately performed for background corrections.

RESULTS AND DISCUSSION

Fig. 1 shows selected spectra normalized to the integral intensity which is nearly identical to $S(Q,\omega)/S(Q)$. Also given is a typical example of the resolution function (dashed curve). At low Q values, distinct inelastic excitations are visible, which are superimposed by a sharp quasielastic line. With increasing Q, the energy of the excitations rises, indicating that the particle dynamics in liquid Si is dominated by propagating modes, as in simple liquid metals [5].

For analyzing the obtained $S(Q,\omega)$ spectra, we used a generalized Langevin formalism [15] using a model memory function for the density fluctuations. Since $S(Q,\omega)$ is the frequency spectrum of the intermediate scattering function, $F(Q,t)$, it is possible to obtain the latter from the experimental data of sufficient quality. Within the generalized Langevin formalism [15], $F(Q,t)$ is determined by

$$\ddot{F}(Q,t) + \omega_0^2(Q) F(Q,t)$$
$$+ \int_0^t M(Q,t') F(Q,t-t') dt' = 0 \quad (1)$$

where ω_0^2 is the second normalized frequency moment of $S(Q,\omega)$, and $M(Q,t)$ is the first order memory function of $F(Q,t)$. For $M(Q,t)$, we used a well-known approximation containing two exponential decay channels for viscous relaxation and one exponential for thermal relaxation. This approach has proven to be useful in describing results of computer simulation studies on simple liquids [16], and also more recently of experimental IXS data on liquid alkali metals [17]. For each-Q value, the real part of the Laplace-Fourier transform of $F(Q,t)$ convoluted with the experimental resolution was fitted to the present $S(Q,\omega)$. Solid curves in Fig. 1 show the best fits of this analysis.

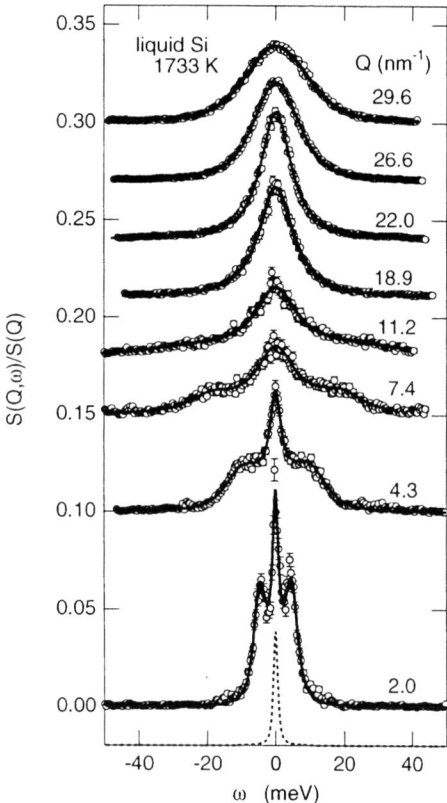

FIGURE 1. Selected $S(Q,\omega)$ spectra of liquid Si at 1733 K normalized to $S(Q)$. Circles are experimental data and solid curves are best fits of a model $S(Q,\omega)$ within the generalized Langevin formalism convoluted with the resolution function (dashed curve). A memory function was employed containing two exponential decay channels for viscous relaxations. See text for details.

From the resolution-deconvoluted $S(Q,\omega)$ data obtained using the above memory function analysis, the longitudinal current correlation spectra, $J_l(Q,\omega) = (\omega^2/Q^2)S(Q,\omega)$ were calculated, and the dispersion relation of the collective modes was determined from the maxima ω_l of these functions. The result is shown in Fig. 2. The dashed line represents the dispersion of hydrodynamic sound. Its slope is given by the bulk adiabatic sound velocity $c_s = 3952$ ms^{-1} at 1733 K [11]. As clearly seen in the figure, the frequencies of the collective excitations ω_l increase noticeably faster with Q than predicted by classical hydrodynamics. This is the so-called *positive* dispersion, which was already found in liquid alkali metals [5,17] and also in some other non-simple liquid metals [3,4]. It can be understood within the framework of generalized hydrodynamics [18], and is related to the onset of shear viscosity on microscopic scales at high frequencies. It should be noted that the obtained magnitude of the *positive* dispersion in liquid Si, about 18 %, is very similar to that in liquid alkali metals (typically 20%).

At Q-values near the $S(Q)$ maximum, beyond 20 nm^{-1}, the quasielastic line shape cannot be approximated by a Lorentzian, but a Gaussian component is additionally needed [1]. The fraction of the Gaussian component increases with increasing Q, and eventually reaches about 50% at ~30 nm^{-1}. This finding is generally observed in the non-simple liquid metals [3,4] which we have recently studied using IXS. In order to unravel the physical origin of this unusual line shape behavior, we have searched for traces of this effect directly in time domain.

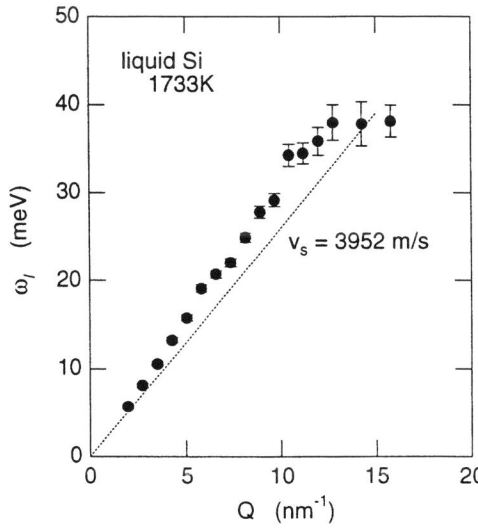

FIGURE 2. Dispersion relation of the collective modes in liquid Si at 1733 K. The dashed line represents the dispersion of hydrodynamic sound, and its slope is given by the adiabatic velocity of sound [11].

Fig. 3 shows selected $F(Q,t)$ spectra obtained from Fourier-transforming the resulting $S(Q,\omega)$. They are normalized to their initial values, $F(Q,0) = S(Q)$. At low Q, the spectra exhibit oscillatory behavior, which is the time-domain analogue of the inelastic excitations in $S(Q,\omega)$. At 18.9 nm^{-1}, the decay of $F(Q,t)$ is nearly exponential, which reflects the Lorentzian shape of the quasielastic line in $S(Q,\omega)$ in this Q-range. Close to the $S(Q)$ maximum, however, it is evident that the decays are no longer exponential. The dashed curves represent the Gaussian components in $F(Q,t)$. Below $t = 0.1$ ps, $F(Q,t)$ is rather Gaussian, which is in accord with the observation that a Gaussian contribution is needed to model the quasielastic line. We interpret the associated slower initial decay of $F(Q,t)$ as an indication for an additional enhancement of the correlation time between neighboring particles on the sub-picosecond level. We can estimate the timescale of the corresponding correlation from the width of the Gaussian contribution in $F(Q,t)$ at the $S(Q)$ maximum to be ~ 0.09 ps.

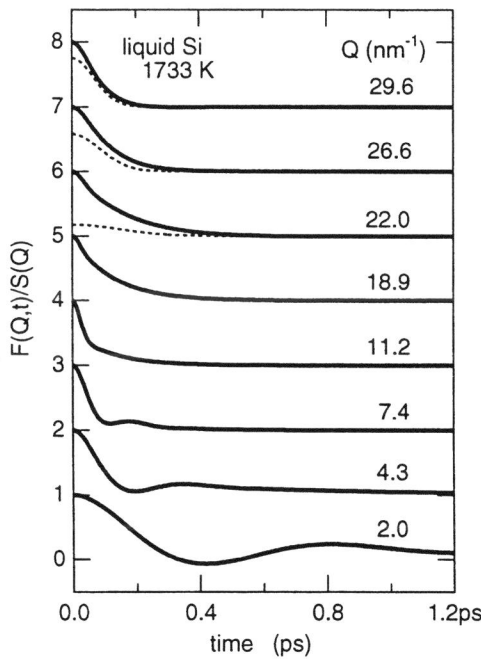

FIGURE 3. $F(Q,\omega)$ normalized to the initial value $S(Q)$ at selected Q values. The dashed curves represent the Gaussian components.

It should be noted that this contribution sets in at Q-values in the vicinity of the $S(Q)$ maximum where the structural correlations to next neighboring particles dominate. Such short time correlations between neighbors have already been observed in an *ab initio* MD simulation [8] on liquid Si under similar conditions, and interpreted as *striking evidence for the persistence of covalent bonds in the liquid*. In this MD work, it was found that on a timescale of some ten femtoseconds, a substantial amount of charge piles up between atoms approaching closer than a critical distance. Interestingly, it was furthermore shown that the bond angle distribution function of these short-lived species peaked at an angle close to tetrahedral, similar to that in amorphous Si although there, the bond angle distribution is considerably narrower. In a later work [9], the MD findings were confirmed, and it could be shown that localizing spins play an important role in the attractive interactions between the Si atoms. These results are in accord with the line shape variation found in our experiment close to the $S(Q)$ maximum. Also, the estimated correlation period lies in the same range as the time scale observed in the MD simulations. Our observation is therefore the first direct experimental evidence for these sub-picosecond density fluctuations.

It is furthermore tempting to speculate that these short time fluctuations of forming and breaking bonds may be interpreted as a precursor of the electron localization that takes place either on freezing but more likely during the bond formation process when the supercooled metallic Si-melt transforms into a semiconducting amorphous fourfold coordinated state at lower temperature.

ACKNOWLEDGMENTS

This IXS work on liquid Si was performed at the SPring-8 with the approval of the Japan Synchrotron Radiation Research Institute (JASRI) in collaboration with Prof. S. Takeda and his group from Kyushu University, and with Dr. A. Q. R. Baron and the BL35XU/SPring-8 beamline staff. The IXS studies were supported by the Deutsche Forschungsgemeinschaft (DFG) and the Fonds der Chemischen Industrie.

REFERENCES

1. Hosokawa, S., Greif, J., Demmel, F., and Pilgrim, W.-C., *Nucl. Instrum. Method B* **199**, 161-164 (2003); Hosokawa, S., Pilgrim, W.-C., Kawakita, Y., Ohshima, K., Takeda, S., Ishikawa, D., Tsutsui, S., Tanaka, Y., and Baron, A. Q. R., *J. Phys: Condens. Matter*, in press.

2. Hosokawa, S., Kawakita, Y., Pilgrim, W.-C., and Sinn, H., *Phys. Rev. B* **63**, 134205-1-5 (2001); *Phisica B* **316-317**, 610-612 (2002).

3. Hosokawa, S., Greif, J., Demmel, F., and Pilgrim, W.-C., *Chem. Phys.* **292**, 253-261 (2003).

4. Hosokawa, S., Pilgrim. W.-C., Sinn, H., and Alp, E. E., *Proc. 3rd European Conf. on Neutron Scattering (Montpellier, 3-6 Sept. 2003)*, submitted.

5. Pilgrim, W.-C., Hosokawa, S., Saggau, H., Sinn, H., and Burkel, E., *J. Non-Cryst. Solids* **250-252**, 96-101 (1999), and references therein.

6. Glazov, V. M., Chizhevskaya, S. N., and Glagoleva, N. N., *Liquid Semiconductors*, Plenum, New York, 1969, Chap. 3.

7. Waseda, Y. and Suzuki, K., *Z. Phys. B* **20**, 339-343 (1975).

8. Stich, I., Car, R., and Parrinello, M., *Phys. Rev. B* **44**, 4262-4274 (1991).

9. Stich, I., Parrinello, M., and Holender, J. M., *Phys. Rev. Lett.* **76**, 2077-2080 (1996).

10. Gantner, G., Boyer, H.-G., and Oelhafen, P., *Europhys. Lett.* **31**, 163-168 (1995).

11. Yoshimoto, N., Ikeda, M., Yoshizawa, M., and Kimura, S., *Physica B* **219&220**, 623-625 (1996).

12. Baron, A. Q. R., Tanaka, Y., Goto, S., Takeshita, K., Matsushita, T., and Ishikawa, T., *J. Phys. Chem. Solids* **61**, 461-465 (2000).

13. Tamura, K., Inui, M., and Hosokawa, S., *Rev. Sci. Instrum.* **70**, 144-152 (1999).

14. Hosokawa, S. and Pilgrim, W.-C., *Rev. Sci. Instrum.* **72**, 1721-1728 (2001).

15. Mori, H., *Prog. Theor. Phys.* **33**, 423-455 (1965); Zwanzig, R., *J. Chem. Phys.* **39**, 1714-1721 (1963).

16. Levesque, D., Vervet, L., and Kürkijarvi, J., *Phys. Rev. A* **7**, 1690-1700 (1973).

17. Scopigno, T., Balucani, U., Ruocco, G., and Sette, F., *J. Phys.: Condens. Matter* **12**, 8009-8034 (2000); *Phys. Rev. E* **65**, 031205-1-7 (2002).

18. Boon, J. P., and Yip, S., *Molecular Hydrodynamics*, McGraw-Hill, New York, 1980.

Volume Effects on the Molecular Rearrangements in Vicinity of Glass Transition

M. Paluch[*], K. Grzybowska[*], A. Grzybowski[*], and C. M. Roland[†]

[*]Institute of Physics, Silesian University, Universytecka 4, 40-007 Katowice Poland
[†]Naval Research Laboratory, Chemistry Division, Code 6120, Washington, DC 20375-5342 USA

Abstract. Analysis of the temperature and pressure dependence of the relaxation times in polymethylphenylsiloxane indicate that volume and thermal energy exert a similar influence. Accordingly, we employ the Dynamic Liquid Lattice model of Pakula to describe the measured relaxation times. In this model, the activation energy for local motion is explicitly considered to depend on volume. From the analysis, the change of the activation energy with volume is calculated.

INTRODUCTION

One of the most fascinating feature of the dynamics of supercooled liquids and polymer melts is a continuous increase of the viscosity and structural (α-) relaxation time over many decades as temperature is lowered or pressure increased toward the glass transition. A simple Arrhenius form, $\tau = \tau_A \exp[E_A/kT]$, satisfactorily describes experimental data only in the temperature range well above the melting point, whereas it fails in the vicinity of the glass transition. A primary aim of many researchers is finding an adequate description of the $\tau(T)$ dependence of glass-formers in this regime.

As temperature is reduced, the molecular motions of glass-forming liquids become more restricted, due in part to a reduction of their kinetic energy. Consequently, molecules cannot surmount potential barriers, and they can become trapped within minima on the potential energy surface. The lowering of temperature also increases molecular crowding (increase of density), which likewise constrains molecular motions. Thus, it is reasonable to expect that the molecular rearrangements in dense media would be controlled by fluctuations in both thermal energy and density. Intuitively, molecular rearrangements near the glass transition are comprised of molecular surmounting of potential barriers (thermal energy fluctuation), into new sites provided that latter are unoccupied (density fluctuations).

In recent years, much effort has been expended examining the relative contribution of thermal and volume effects to the non-Arrhenius behavior of the α-relaxation times [1]. Initially, it was postulated that the observed behavior was due primarily to thermal energy fluctuations. This view was formulated on the basis of analysis of the temperature and density dependence of viscosity data for glycerol and triphenyl phosphite [1]. However, more recently, experimental results for a wide range of glass formers indicate unambiguously that fluctuations in both thermal energy and free volume contribute to the dynamics of supercooled liquids [2] These findings are at odds with the conclusion of Ferrer and co-workers [1] that temperature is the dominant variable controlling the supercooled dynamics.

With few exceptions [3], theoretical efforts to understand the molecular dynamics in supercooled liquids generally concentrate solely on either thermal [4] or density effects [5]. This makes them somewhat inadequate for describing the $\tau(T)$ dependences of glass-formers. Recently, in the framework of the dynamic lattice liquid (DLL) model, Pakula [6] proposed a very general approach for analyzing the dynamics in glass-forming liquids and polymers. Especially noteworthy is the fact that the DLL model takes into account both temperature and volume effects. The general assumption of the model is to regard relaxation as encompassing thermally activated processes, with activation energies that depend on local volume.

In this work we focus on the relative role of volume and activation energy in the dynamics of polymethylphenylsiloxane (PMPS). We demonstrate that the DLL model provides an adequate description of the isothermal and isobaric volume-dependences of structural relaxation for this material.

DESCRIPTION OF THE DLL MODEL

The DLL model of Pakula [6] for describing molecular rearrangements in liquids employs a lattice structure, with each molecule assigned to a lattice site. Since all lattice sites are occupied, molecular displacements to another site require cooperative motion of neighboring molecules. Such cooperative rearrangement can be visualized as a collective displacement, involving more than two molecules, along the trajectory to form a closed loop. Consequently, the sum of the displacements of all molecules involved in the process is zero. The probability that a given molecule participates in the collective displacement determines the structural (or dielectric α-relaxation) time. This probability can be calculated by simply counting the number of self-avoiding loops including n molecules (or segments in the case of a polymer).

One of the main outcomes of the Pakula approach is the following formula for the relaxation time:

$$\tau = \left[B \sum_{n=3}^{\infty} n^{-h+1} (\mu_0 p_s)^n \right]^{-1} \quad (1)$$

where B is a lattice–dependent constant, h is positive and dependent on the dimensionality of the lattice, and μ_0 is the connectivity of the lattice in the athermal case. The dependence of the relaxation time on the thermodynamic state parameters (P, V, T) can be introduced into the above equation via the term p_s that modifies the connectivity of the lattice. In order to obtain an explicit temperature dependence of the relaxation times, Pakula assumed:

1. A local volume v is assigned to each molecule. This volume can fluctuate, assuming values not smaller than a minimum volume v_0. The excess volume has an exponential distribution:

$$\phi(v) = \frac{1}{\overline{v} - v_0} \exp\left(-\frac{v - v_0}{\overline{v} - v_0}\right) \quad (2)$$

2. The system expands thermally with an expansion coefficient α, so that

$$\overline{v} = v_0 [1 + \alpha(T - T_0)] \quad (3)$$

3. Molecular transport is driven by a thermally activated process with potential energy barriers $E(v)$ dependent on the local density of the system. The probability for a molecule to take part in a local rearrangement is given by the Boltzman factor:

$$p(v, T) = \exp\left(-E(v)/kT\right) \quad (4)$$

Since the higher terms in equation 1 can usually be neglected, taking into account the above assumption, the expression for relaxation time can be written as:

$$\tau \propto \langle \phi(v) p(v, T) \rangle^{-1} = \left\{ \int_{v_0}^{\infty} \phi(v) p(v, T) dv \right\}^{-1} \quad (5)$$

In order to obtain an explicit analytical expression for the temperature dependence of the relaxation time, a dependence of E on v has to be assumed. Herein we consider a linear decrease of the activation energy from E_{a1} to E_{a2} in the range between v_0 and v_c, as depicted schematically in Figure 1.

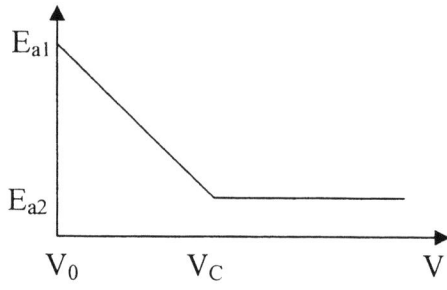

FIGURE 1. The assumed dependence of the activation energy for local rearrangement on local volume.

This leads to the following dependence of the relaxation times on temperature and volume:

$$\tau = \tau_0 \left\{ \frac{1}{1 + \frac{(E_1 - E_2)}{kTw}} \left[e^{-\frac{E_1}{kT}} - e^{-\frac{E_2}{kT}} \exp(w) \right] + e^{-\frac{E_2}{kT}} \exp(w) \right\}^{-1} \quad (6)$$

in which $w = (v_0 - v_c)/(\bar{v} - v_0)$ and v_c is a constant having the units of volume.

Below we present the results of an analysis of the isothermal and isobaric dependence of segmental relaxation time in PMPS [7] using equation 6.

RESULT AND DISCUSSION

Figure 2 illustrates the temperature dependence of the dielectric relaxation times measured for PMPS at ambient pressure. From this data, it can be seen that the α-relaxation times increase with decreasing of temperature in a non-Arrhenius fashion. Materials that exhibit strong deviations from Arrhenius behavior are commonly termed fragile. The steepness index is often used as a quantitative measure of this departure from Arrhenius behavior. Glass formers having values of m around 100 or more are considered as fragile. In the case of PMPS, $m = 98 \pm 3$; accordingly, this polymer can be classified as a fragile glass-former.

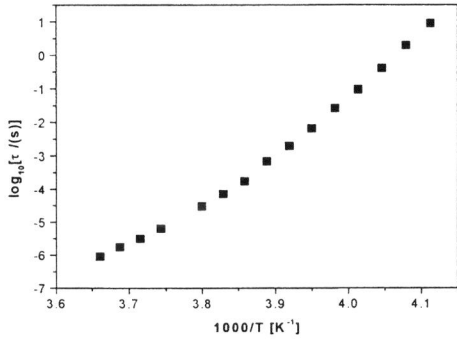

FIGURE 2. Temperature dependence of relaxation times of PMPS at ambient pressure.

In Figure 3 are shown the isothermal and isobaric dependences of segmental relaxation times on specific volume for PMPS [7]. A cursory inspection of the data reveal that volume effects are not negligible, exerting along with thermal energy a significant role in determining the fragile character of the temperature dependence of the relaxation times. From Figures 2 and 3, we note that the isothermal pressure-induced changes in volume are equivalent to ca. a 20° temperature change (i.e., from -10°C to -30°C), causing τ to vary by almost two and one-half decades. This same temperature change, on the other hand, leads to a change of τ of about 5.5 decades, whereby we can conclude roughly half of this increase is due to the volume change alone. Thus, the contributions from volume and thermal energy to the segmental relaxation times in PMPS are comparable.

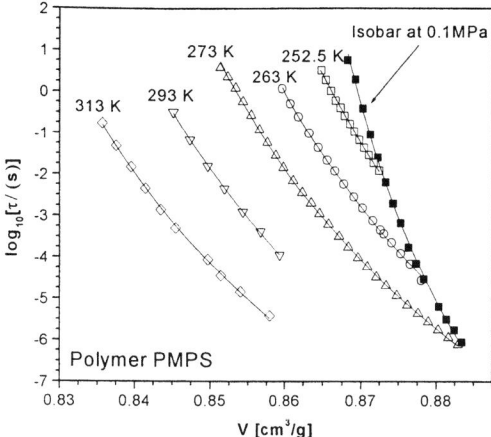

FIGURE 3. Isothermal and isobaric dependences of the segmental relaxation times of PMPS as a function of specific volume at the indicated temperatures and pressure. Solid lines represent fits to the DLL model.

The relative contribution of the two effects can be quantified from the ratio of the activation energy at constant volume [8], $E_V = \left(\partial \ln \tau / \partial T^{-1}\right)_V$, to the activation energy at constant pressure, $E_P = \left(\partial \ln \tau / \partial T^{-1}\right)_P$ (Since the liquid structure changes with temperature and pressure, strictly speaking, these represent apparent slopes of Arrhenius plots, rather than activation energies.) In the case of volume-governed relaxation, this ratio is close to zero, whereas it is equal to unity for simple thermally activated processes. As pointed in ref. 7, the average value of this ratio lies around 0.5 for PMPS. This value confirms that the temperature dependence of τ_α is due almost as much to the volume changes accompanying a change in temperature, as to the change in thermal energy.

As discussed in the previous section, in the DLL model, the barrier heights are quantitatively related to the local density via the volume dependence of the activation energy. From the analysis of the relaxation times by eq. 6, it is possible to determine to what extent activation energy changes with specific volume. To this end, we numerically fit eq. 6 to the relaxation data in Figure 3. The solid lines obtained from the fitting procedure describe fairly well the experimental points. The values of the five adjustable parameters are collected in Table 1. In the DLL model, v_0 and v_c are defined analogously to their definition in free volume models, and they may be dependent on

TABLE 1. DLL parameters (eq. 6) for the data shown in Fig. 3.

Parameters	Isobar P = 0.1MPa	Isotherm T = 252.5 K	Isotherm T = 263 K	Isotherm T = 273 K	Isotherm T = 293 K	Isotherm T = 313 K
$\log\tau_0$	-18.297	-18.297	-18.297	-18.297	-18.297	-18.297
E_1 [K]	44600	44600	44600	44600	44600	44600
E_2 [K]	265	265	265	265	265	265
V_0 [cm^3/g]	0.843	0.814	0.805	0.797	0.788	0.777
V_c [cm^3/g]	1.951	2.959	3.061	3.131	3.112	3.124

temperature and particle density. For this reason, these parameters were allowed to vary when eq. 6 was simultaneously fit to all the experimental curves in Figure 3.

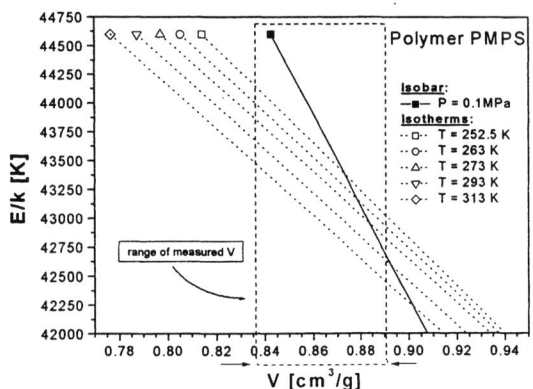

FIGURE 4. The change of activation energy with specific volume, as determined from the values of the fitting parameters in the DLL model.

From the parameters collected in Table 1, the dependence of activation energy on volume is calculated (Figure 4). In the range of our experimental data, the linear decrease in activation energy is not large. For instant, for the isotherm at 0.1 MPA, E increases from 42,980 K to 43,578 K for a specific volume change from 0.883 to 0.868. Since the change of activation energy in this measured range of volume is surprisingly small, it may just indicate that the slowing down of the dynamics in PMPS reflects to a large extent the reduction of free volume. In this context, it is important to note that PMPS has an extraordinarily large value of dT_g/dP, equal to about 300 K/GPa. This very strong dependence of the glass transition temperature on pressure is a reflection of the substantial role that molecular crowding has in the arrest of the structural dynamics.

As mentioned previously, the assumption of a linear decrease of activation energy with volume leads to an analytical expression for the relaxation time (eq. 6), which provides a satisfactory fit to the experimental data. However, this is only a first approximation, since $E(v)$ certainly may possess a more complicated nature. Obviously, further studies are needed to clarify this important problem in future. Indeed, the determination of $E(v)$ represents an essential aspect of understanding the dynamical properties of liquids in the vicinity of their glass transition.

ACKNOWLEDGMENTS

Financial support of the Committee for Scientific research, Poland (KBN, Grant No 5PO3B 022 20) is gratefully acknowledged. Work at NRL was supported by the Office of Naval Research.

REFERENCES

1. Ferrer M. L., Lawrence C., Demirjian B. G., Kivelson D., Alba-Simionesco C., and Tarjus G., *J. Chem. Phys.* **109**, 8010-8015 (1998).

2. Paluch M., Casalini R., and Roland C. M., *Phys. Rev. B.* **66**, 092202-3 (2002).

3. Macedo P. B., Litovitz T. A., *J. Chem. Phys.* **42**, 245-256 (1965)

4. Glasstone S., Laidler K., and Eyring H., *The theory of Rate Processes*, McGraw-Hill, New York, 1941.

5. Ferry J. D., *Viscoelastic Properties of Polymers*, Wiley, New York, 1980.

6. Pakula T., *J. Mol. Liq.* **86**, 109-121(2000).

7. Paluch M., Casalini R., Patkowski A., Pakula T., and Roland C. M, *Phys. Rev. E* **68**, 031802-5 (2003).

8. Naoki M., Endou H., and Matsumoto K., *J. Phys. Chem.* **91**, 4169-4174 (1987).

Glassy crystals: above and below Tg

M. Descamps and J.F. Willart

Laboratoire de Dynamique et Structure des Matériaux Moléculaires. UMR. CNRS 8024.
University LILLE1. Bat P5 - 59655 Villeneuve d'Ascq CEDEX, France

Abstract. Results clarifying the entrance of a glassy crystal cyanoadamantane into a non-ergodic situation are shortly reviewed. Special attention has been paid to the short range order which develops on approaching the glass transition. The specific ability of the system to relax more deeply than a liquid below Tg is also shortly presented.

INTRODUCTION

Glass formation in plastic crystals – giving the so-called Glassy Crystal (GC) state [1] – offers an exciting opportunity to apply the experimental methods used to understand the structure and dynamics of crystals, which are mainly useless in studying conventional glasses. In contrast to the usually named "orientational glasses" [2], GC are not frustrated by a quenched disorder extrinsically imposed by a dilution effect somewhat similar to a spin-glass situation. GC share many of their properties with conventional glasses. These include –(1) Vitrification, which needs to bypass a first order transition. –(2) The thermodynamic signature of the glass transition at Tg. –(3) The dynamic decoupling occurring in the high temperature ns-ps regime, and (4) the plateau of thermal conductivity and an increase of the specific heat at helium temperature faster than in normal crystals. To be definitive experiments must be performed on single crystals of the plastic phase, which must also be safely undercooled. These very stringent experimental requirements have been the main causes, which so far limited the exploitation of the GC structural periodicity in full.

We have found in some adamantane derivatives the possibility to obtain rich structural and dynamic informations on single crystals of extremely high quality [3-9]. The present paper summarizes the experimental situation for adamantane derivatives which have allowed an unprecedented investigation in which calorimetry is couple to single crystal time resolved X-Ray investigation in the plastic state both above and below Tg. The experimental results give a clear image of the local structure changes above Tg and specificities of the sub-Tg relaxations [6].

CYANOADAMANTANE ABOVE Tg

At room temperature (RT), cyanoadamantane (Can) is in an equilibrium rotationally disordered cubic phase, called R in the following [10]. The main feature of the structure is a dynamical disorder of the molecular dipoles, which mimics that of a Potts model. The molecular dipoles can flip randomly among six possible orientations defined by the fourfold axis of the cubic cell. At low temperature (LT), the stable phase in monoclinic (called m) and its transition toward the rotator phase occurs at $T_{m-R} \cong 283K$. A transient metastable state (called t) which forms, with phase m, a monotropic set of phases, was also recently detected [11,12]. From x-ray investigations and numerical simulations, the structure of phase t was inferred to be pseudotetragonal with an antiferroelectric order of the dipoles. The transition temperature T_{t-R} between phase t and phase R takes place around 208K. Upon a rapid cooling, phase R can, however, be easily undercooled, so that the transitions $t \rightarrow R$ and $m \rightarrow R$ can only be detected upon heating, and after appropriate annealing at LT. Moreover, for a deep enough cooling, differential scanning calorimetry (DSC) scans show the C_p jump signature of conventional glasses. The onset of this Cp jump at Tg is about 170 K [6].

We have performed calorimetric and x-ray diffraction experiments to fully characterize the entrance of the rotator phase R of CNa into the non-ergodic glassy state. The calorimetric results show that the CNa glassy crystal has thermodynamic properties fully similar to those of structural glasses. In particular, the approach of the glass transition is characterized by a continuous decrease of the configurational entropy Sc. The structural interpretations of the observed Cp and Sc evolutions are made easy in this specific glass forming system. Very perfect single crystals can indeed be grown in the phase R which can also be quenched without being destroyed. A full investigation of the structure factore S(**q**) can thus be performed at room temperature both in the temperature domain of moderated undercooling and in the temperature range of Tg. The quenched state remains perfectly cubic at least in the time immediately following the quenching. However a systematic investigation of the reciprocal lattice reveals that broad diffuse scattering spots appear at the X boundary point of the Brillouin zone which are forbidden reflections of the fcc lattice of phase R. This proves that anti-parallel inter-molecular correlations of a short range nature are set up in the system during the cooling process.

At room temperature traces of local dipolar ordering can already be detected through a very broad bump. This scattering develops considerably and sharpens in the quenched state. From the width of the peak we can estimate the correlation length of the short range order as well as its temperature evolution. This correlation length increases by a factor of 3/2 between room temperature ant Tg. Just after quenching at Tg a correlation length of about 25 Å is calculated. Beginning of diverging behaviour suggested by the heat capacity evolution in the metastable phase R indicates that most of the increase in the correlation length occurs between 225 and 178K. At lower temperatures this evolution is arrested.

Since the decay of the metastable state is not observed during the course of the above measurements, this local ordering cannot be confused with the formation and growth of heterophase nuclei. The evolution of this local order is likely to be responsible for the evolution of the configurational entropy. In CNa, we have thus obtained evidence that the **q**-dependence is a characteristic feature of the evolution of the molecular organization of this glassy crystal. This must certainly results in a **q**-dependence of the dynamics. One could thus expect a critical slowing down at the specific **q** position driving the structural fluctuations.

SUB-Tg RELAXATIONS AND FRUSTRATION

Enthalpy relaxation: DSC systematic study of the isothermal aging of CNa and also mixed compounds of the latter with chloroadamantane have revealed signatures of the aging fully similar to that observed with conventional glasses. Glassy crystals annealed for a long time can show extremely large Cp overshoot which end the Cp jump occurring at Tg [13]. The data were analysed in term of the fictive temperatures Tf derived by the usual equal area construction [14]. In the course of isothermal aging experiments it was observed that Tf rapidly becomes increasingly lower than the aging temperature itself. This totally unexpected dependence is incompatible with the hypothesis – usually made for liquid glasses – that the undercooled crystal relaxes towards the equilibrium metastable plastic phase in internal equilibrium. It shows rather that, after some annealing period, the system has relaxed towards a more stable state of lower enthalpy, which is different from the equilibrium metastable plastic phase. The endothermic overshoot is however the reversion signature of the state which takes place during the relaxation. The enthalpy release which is reflected in the evolution of Tf shows that the system is far from equilibrium after 5 h of aging at a temperature so close to the calorimetric glass transition Tg for which relaxation times of a few hundred seconds are expected for a normal rate of 10°C/min. The corresponding kinetics also reveal a change of regime. An early fast relaxation can correspond to the classical equilibration of the metastable state. It then crosses over to a much slower relaxation to the state of lower enthalpy.

Structural relaxation below Tg: Isothermal X-ray experiments performed at temperatures slightly below Tg revealed the structural features correlated to the aging observed above Tg in calorimetry [7,15,16]. This appears in an amplification and sharpening of the X boundary peaks whose intensity already underwent strong increases in intensity upon quenching. This evolution of the X-ray diagram corresponds to an increase in orientational order. The process is extremely slow and saturation is far from being reached even after eight days at a temperature of Tg – 12K. It appears also that this slow process is also very weakly time dependant. The inverse peak width reflects the emerging characteristic length. On a log-log scale of the time evolution of this length the slope never exceeds 0.1; it even tends to a lower value towards the late stage of the process. For a non conserved order parameter system as we have, we expect the temporal evolution of the characteristic size to obey a power law with a growth exponent of ½ in the coarsening late stage. Here it was demonstrated that even the late stage is not still reached because the volume fraction of sample transformed in a more ordered phase continue to increases with time. Below Tg we thus face a aging situation which is not that of the approach of an ideal glassy state. We clearly face an orientational ordering process. This ordering is however likely very frustrated as indicated by the very weak time dependence of the growth law of the ordered zones, and the small characteristic size (~20Å) reached after long sub-Tg annealing. The structural analysis of the systems reveals several possible sources of frustration. It is, for instance, the strong steric hindrance between the molecules which makes some molecular configurations unreachable [16,17]. Strong elastic interactions required to make coherent the interface between the highly degenerated ordered domains and the orientationnally disordered matrix can also be invoked [18].

ACKNOWLEDGMENTS

This work was supported by the FEDER in the frame of an interreg III program (Nord Pas de Calais – Haute Normandie – Kent)

REFERENCES

1. Suga, H., and Seki, S., *Journal of Non Crystalline Solids* **16** (2), 171-194 (1974).

2. Höchli, U., Knorr, K., and Loïdl, A., *Advence in Physics* **39**, 405-615 (1990).

3. Willart, J. F., and Descamps, M., *Progress of Theoretical Physics* **126**, 239-243 (1997).

4. Willart, J. F., Descamps, M., and Benzakour, N., *Journal of Chemical Physics* **104** (7), 2508-2517 (1996).

5. Willart, J. F., Descamps, M., Bertault, M., et al., *Journal of physics Condensed matter* **4** (48), 9509-9516 (1992).

6. Willart, J. F., Descamps, M., and van Miltenburg, J. C., *Journal of Chemical Physics* **112** (24), 10992-10997 (2000).

7. Willart, J. F., Descamps, M., and van Miltenburg, J. C., *Journal of thermal analysis and calorimetry* **51** (3), 943-949 (1998).

8. Descamps, M., and Caucheteux, C., *Journal of physics C Solid state physics* **20** (31), 5073-5095 (1987).

9. Descamps, M., Caucheteux, C., Odou, G., et al., *Journal de Physique (Lettres)* **45** (14), L719-L727 (1984).

10. Amoureux, J. P., Sauvajol, J. L., and Bee, M., *Acta Crystallographica, Section A* **37** (1), 97-104 (1981).

11. Descamps, M., Willart, J. F., Kuchta, B., et al., *Journal of Non Crystalline Solids* **235** (237), 559-566 (1998).

12. Willart, J. F., Descamps, M., and van Miltenburg, J. C., *Phase Transitions* **76** (3), 239-246 (2003).

13. Delcourt, O., Descamps, M., Even, J., et al., *Chemical Physics* **215** (1), 51-57 (1997).

14. Descamps, M., Willart, J. F., and Delcourt, O., *Physica A* **201** (1-3), 346-362 (1993).

15. Descamps, M., and Willart, J. F., *Journal of non crystalline solids* **172** (1), 510-519 (1994).

16. Willart, J. F., Descamps, M., and Naudts, J., *Phase Transitions B* **31** (1-4, pt.4-6), 261-270 (1991).

17. Willart, J. F., Mouritsen, O. G., Naudts, J., et al., *Physical Review B* **46** (13), 8089-8098 (1992).

18. Descamps, M., Willart, J. F., Odou, G., et al., *Journal de physique I* **2** (6), 813 (1992).

Neutron Scattering and Dielectric Study on the Structural and Dynamical Peculiar Properties of Poly(vinyl chloride)

A. Arbe*, A. Moral[†], A. Alegría*[†], J. Colmenero*[†], W. Pyckhout-Hintzen**, D. Richter**, B. Farago[‡] and B. Frick[‡]

*Unidad de Física de Materiales CSIC-UPV/EHU Apartado 1072, 20080 San Sebastián, SPAIN
[†]Departamento de Física de Materiales UPV/EHU, Apartado 1072, 20080 San Sebastián, SPAIN
**Institut für Festkörperforschung, Forschungszentrum Jülich GmbH, D-52425 Jülich, GERMANY
[‡]Institut Laue–Langevin, BP 156, 38042 Grenoble Cedex 9, FRANCE

Abstract. In this work we have studied the anomalous dynamical behavior of poly(vinyl chloride) (PVC) searching its origin in the dynamical heterogeneities arising from the structural peculiarities of this polymer. For this purpose we have combined dielectric spectroscopy, coherent and incoherent neutron scattering for the dynamics investigation, and SANS for resolving the heterogeneous structure of PVC. The SANS experiments indicate the existence of structural modulations that persist in the temperature range $T < 430$ K. We show that a distribution of glass transition temperatures due to these density modulations causes the broadening of the response from the structural relaxation and the anomalous momentum transfer dependence of the incoherent scattering function.

INTRODUCTION

The nature of poly(vinyl chloride) (PVC) has revealed to be so particular that even books have been devoted to it (see, e.g. [1]). Despite the big effort made, some of its peculiarities – structural as well as dynamical – are not yet understood. Its "crystalline" nature is subject of strong controversy [1, 2, 3]. On the other hand, the anomalous broadening of the α-relaxation, which rapidly increases with decreasing temperature approaching the glass transition temperature ($T_g \approx 353$ K), is known since 1941 [4]. Though this behavior was associated to some heterogeneity characteristic for semicrystalline polymers, such a hypothesis has never been quantitatively checked. In order to contribute to these questions, in this work we present a thorough investigation of the thermal evolution of the structure and the dynamics of PVC in a wide T-range around and above T_g [5].

EXPERIMENTS

The experiments related to the structure were carried out by means of the Small Angle Neutron Scattering (SANS) diffractometer KWS1 at the FRJ-2 research reactor in Jülich, Germany. The static structure factor $S(Q)$ was determined from a fully deuterated sample (PVCd) and the chain form factor from a blend consisting of 10% deuterated chains in a protonated matrix: PVCd/PVCh 10%/90%. Dielectric spectroscopy (DS) measurements were performed in the frequency domain from 10^{-2} Hz to 10^9 Hz. The Neutron Spin Echo (NSE) experiments were performed by the IN11 spectrometer (Institute Laue-Langevin (ILL) in Grenoble, France) on PVCd at the first static structure factor peak (Q_{max}=1.2 A^{-1}). These measurements reveal thus the time-decay of the intermolecular correlations $S(Q_{max},t)/S(Q_{max})$. Finally, by means of the backscattering (BS) instrument IN10 also at the ILL, the self-motion of the protons was investigated on PVCh through the incoherent scattering function $S_{inc}(Q,\omega)$. Further experimental details can be found elsewhere [5].

RESULTS

SANS revealed the presence of structural heterogeneities which manifest in a peak of $S(Q)$ centered at about 0.05 A^{-1} [see Fig. 1(a)]. This peak evidences the existence of a long period within the PVCd sample reflecting some periodic modulation of the density. For $T > 430$ K the maximum position and the intensity of the peaks are strongly T-dependent, indicating that above 430 K the system becomes gradually homogeneous with increasing T. On the other hand, the single chain form factor measured also by this technique reflects random-walk statistics, as can be realized from Fig.1(b): the slope of the measured intensity at high Q clearly follows a Q^{-2} de-

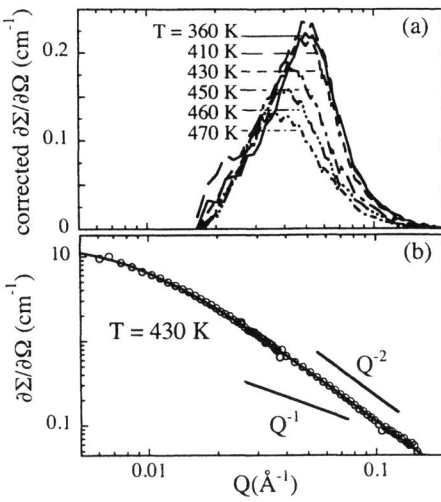

FIGURE 1. SANS structural results on PVC. (a): $S(Q)$ peak. (b): Cross section measured for the blend. The solid line is the description obtained assuming Gaussian chains.

of polymers can be well described by an analogous expression to Eq. (1) (or equivalently its Fourier transform in the frequency domain, as it is the case for BS data) with a Q-dependent characteristic relaxation time. This time displays a $\tau_w(Q) \propto Q^{-2/\beta}$ dependence [7]. Such behavior is due to the Gaussian form of the incoherent scattering function and the motion of the protons can be interpreted in terms of a sublinear diffusion (their mean squared displacement grows as t^β) during the α-relaxation process. For the case of PVC [see some examples of the $S_{inc}(Q,\omega)$ results in Figs.2(c) and (d)] we observe a much weaker Q-dependence of the associated characteristic time: $\tau_w(Q) \propto Q^{-1.35/\beta}$ [5].

MODEL OF HETEROGENEITIES

The SANS results on $S(Q)$ show clear evidence for the existence of a density modulation in the sample. For $T \leq 430$ K the weak T-dependence of the maximum of the peak, revealing the average domain size, simply follows thermal expansion. Furthermore, the intensity of the SANS peaks in this T-range is constant, indicating that the number of regions or the density contrast does not significantly change, i.e., the structural modulation remains unaltered. However, above 430 K a qualitatively different behavior sets on. The peak shifts very rapidly and the intensity strongly decreases with increasing T. This observation might relate to the melting of the "crystalline" parts. A description of the heterogeneities in terms of micro-crystallites distributed according to a Gaussian function [5] yields a steady average size of such micro-crystallites around 32 A in the whole T-range investigated. On the contrary, the separation between these crystallites increases with T above 430 K: from its value of about 11 A for $T \leq 430$ K to 146 A at 470 K. This implies a decrease of the crystallinity from 4% to 1.6%. These results support the nodular structure assumed for the interpretation of SAXS data in Ref. [2]. Moreover, our results on the chain form factor of PVC also support this scenario against the lamellar structure proposed by other authors [3]. As pointed out before, the chains display a Gaussian conformation. The Q-dependence observed for the scattering from the PVCd/PVCh blends at high Q-values excludes considerable sections of parallel stems as would be expected in long nodules. Figure 2(b) clearly shows that our results are not compatible with the Q^{-1} dependence expected in such case.

Qualitatively, the dynamical results here obtained for PVC by DS and incoherent NS remind those found for polymer blends (see, e.g. [8]). In such systems, the peculiar dynamical features find a natural explanation in the existence of different environments for the different regions in the material, i.e., in the presence of hetero-

pendence, underlining the Gaussian chain conformation.

For the dynamics study, DS facilitated in a wide T-range a precise determination of the dynamic response related to the segmental relaxation [see Fig. 2(a)]. The DS spectra above T_g show three contributions that are displayed in this figure for the case of 380 K: that of the conductivity and interfacial polarization, the β-relaxation and the α-relaxation. At first sight, it becomes evident the increase of the broadening of the α-relaxation with decreasing T. Moreover, in this figure can also be appreciated that close to T_g the lineshape strongly deviates from the usual Kohlrausch-Williams-Watts (KWW) functional form found for common glass-forming systems (which at 380 K would give rise to the curve shown by the dotted line). It is found that only at high T ($T > 410$ K) a KWW-like function

$$\Phi(t) \propto exp\left[-\left(\frac{t}{\tau_w}\right)^\beta\right], \quad (1)$$

provides a good description for the relaxation function of the α-relaxation in the time domain. The value of the shape parameter β found then is $\beta \approx 0.5$.

Turning to Neutron Scattering (NS), the NSE results [Fig.2(b)] confirm such kind of functional form for the α-process at high temperature. It is noteworthy that the decay of $S(Q_{max},t)$ is due to this relaxation without contributions of secondary processes [6]. Fits of KWW functions to the NSE data above 420 K deliver values close 0.5 for the β-parameter. Finally, concerning the incoherent NS results, again PVC shows peculiarities with respect to conventional glass forming polymers. The incoherent scattering function in the α-relaxation regime

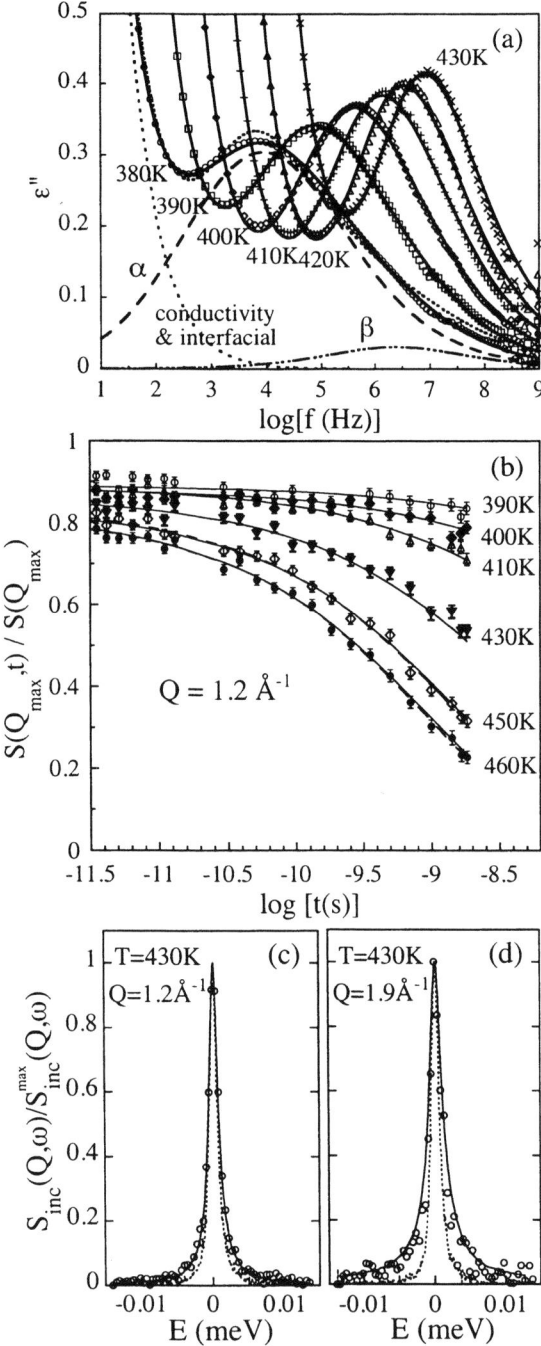

FIGURE 2. Dynamical results on PVC. (a): DS spectra. For 380 K the different contributions are shown as well as the description obtained when a single KWW is assumed for the α-relaxation contribution (dotted line). (b): NSE-results on the dynamic structure factor. (c) and (d): Incoherent scattering functions. The dotted lines show the instrumental resolution. In all cases the solid lines are the fitting curves obtained from the heterogeneous model.

geneities in the system. This analogy can now be taken as a starting point for our interpretation of PVC behavior. We have invoked a distribution of sizes for explaining the heterogeneities evidenced by SANS. There are also indications in the literature that the non-crystalline portions may exist in various degrees of order with different densities. Density fluctuations associated to different packing degrees in our homopolymer would be the counterpart for concentration fluctuations in blends. It can be expected that regions with different density would show different dynamical behavior. Thus we propose a model that considers the coexistence of regions with different dynamical properties leading to a distribution of characteristic relaxation times. We assume that within each region the α-relaxation takes place in a similar way as observed for many polymers. Thus, the decay of $S(Q_{max},t)$ due to this relaxation follows the typical KWW behavior [Eq.(1)], and the self-motion of the protons occurs according to a sublinear diffusion showing a Gaussian scattering function. For all regions the same functional form for the α-relaxation is assumed. This can be univocally determined from the NSE data at Q_{max}, where, as above commented, the α-relaxation is observed without contamination by secondary processes. Furthermore, at high enough T, the nearly T-independent spectral shape was described with $\beta = 0.5$, a reasonable value for a polymer. Also DS measurements in the same T-range result in a similar β value. We now allow for different time scales in the different sample regions. These are assumed to be distributed by a log normal distribution, leading to the same kind of distribution for the characteristic time:

$$g(log\tau_w) = \frac{1}{\sigma\sqrt{2\pi}}exp\left[-\frac{1}{2}\left(\frac{log\tau_w - \langle log\tau_w \rangle}{\sigma}\right)^2\right] \quad (2)$$

where $\langle log\tau_w \rangle$ corresponds to the average time and σ is the width of the distribution function. Both parameters can be T-dependent. The global relaxation observed by different experimental methods can be considered as a weighted integral over the local relaxations.

Since the DS measurements cover the widest frequency/temperature range, the model was first applied to the permittivity data. As can be appreciated from Fig. 2(a), the achieved description of the experimental data is very satisfactory. The distribution functions $g(log\tau_w)$ derived from these fits are depicted in Fig. 3(a). With increasing T a clear narrowing of the distribution function is observed, and at high temperatures the timescales of the different regions become very similar. $\langle log\tau_w \rangle$ follows the usual Vogel Fulcher (VF) dependence

$$\langle log\tau_w \rangle = \langle log\tau_w \rangle_o \exp\left(\frac{B}{T-T_o}\right) \quad (3)$$

shown in Fig. 3(a) with $\langle log\tau_w \rangle_o = -11.66\ s$, $B = 1005.6$ K and the VF-temperature $T_o = 317$ K. Fixing the distribution functions and the T-dependence for the average time deduced from the DS study, an almost perfect description is obtained for the NS results: both, the NSE as well as the incoherent data are fairly reproduced by the model proposed (see Fig. 2). We emphasize the fact that the distribution functions used implicitly impose the Q-dependence of the incoherent scattering functions – the observed deviations from Gaussian behavior of the experimental data are thus fully taken into account by the proposed scenario [5].

Finally we may ask about the origin of the distributions of relaxation times found. Again we invoke the knowledge on blend dynamics. For blends, these distributions could be related to a distribution of VF-temperatures $h(T_o)$ [8]. If this holds, $g(log\tau_w)$ and $h(T_o)$ are related through $g(\tau_w)\ d(log\tau_w) = h(T_o)\ d(T_o)$. Taking into account the VF equation for the timescale in each region, $h(T_o)$ can easily be obtained from $g(log\tau_w)$. As in the case of blends, for PVC a unique function $h(T_o)$ is obtained starting from distribution functions $g(log\tau_w)$ corresponding to different temperatures, independently of the temperature [see Fig. 3(b)]. This result corroborates the hypothesis that the behavior found is a consequence of the distribution of only the VF-temperature, or, equivalently, the T_g. The resulting $h(T_o)$ shows a maximum around 320 K, is asymmetric and can be considered as T-independent at least in the range $380\ K \leq T \leq 430\ K$. The features found could in fact be expected from the structural study: SANS revealed a T-independent distribution of density domains in the same T-range where $g(log\tau_w)$ – and consequently $h(T_o)$ – has been determined. The T-independence of the density domains demands T-independent distributions $h(T_o)$. Let us note that, though the density modulations hardly vary below ≈ 430 K, the timescales of the different domains approach each other with increasing T. This is due to their VF-like T-dependencies, that lead to the observed narrowing of $g(log\tau_w)$. Thus, a kind of "dynamical homogenization" occurs in the system, that is not related to the structural changes. Above 430 K, an additional homogeneization of the dynamics related to the vanishing structural heterogeneities is expected. However, this effect cannot be resolved experimentally for the dynamics because the values for the variances of $g(log\tau_w)$ are already very small at such high temperatures. Nevertheless, we can conclude that the proposed scenario enables us to finally understand the dynamical anomalies of PVC in terms of its structural properties.

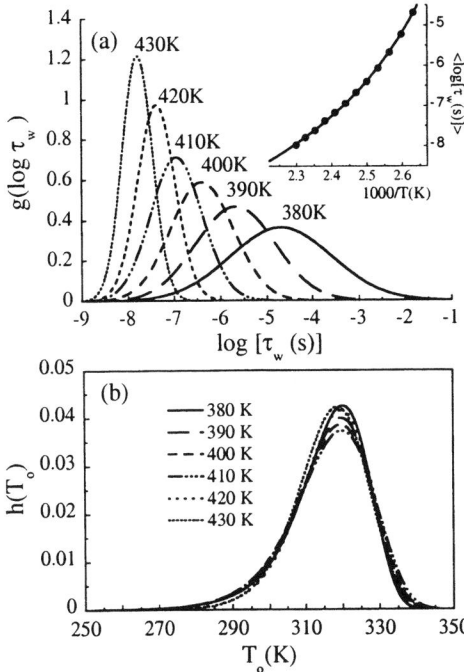

FIGURE 3. (a): T-dependence of the distributions of characteristic times and the average time obtained from the DS study. (b): Corresponding distributions of the VF-temperature.

ACKNOWLEDGMENTS

A. A., A. A., and J. C. acknowledge support from the following projects: DGICYT, PB97-0638; GV, EX 1998-23; UPV/EHU, 206.215-G20/98; 9/UPV 00206.215-13568/2001 and from "Donostia International Physics Center".

REFERENCES

1. Gilbert, M., *Journal of Macromolecular Science Reviews in Macromolecular Chemistry and Physics*, **C34**, 77–135 (1994).
2. Blundell, D. J., *Polymer*, **20**, 934–938 (1979).
3. Wenig, W., *J. Polym. Sci., Polym. Phys. Ed.*, **16**, 1635–1649 (1978).
4. Fuoss, R. M., and Kirkwood, J. G., *J. Am. Chem. Soc.*, **63**, 385–394 (1941).
5. Arbe, A., Moral, A., Alegría, A., Colmenero, J., Pyckhout-Hintzen, W., Richter, D., Farago, B., and Frick, B., *J. Chem. Phys.*, **117**, 1336–1350 (2002).
6. Arbe, A., Richter, D., Colmenero, J., and Farago, B., *Phys. Rev. E*, **54**, 3853–3869 (1996).
7. Colmenero, J., Alegría, A., Arbe, A., and Frick, B., *Phys. Rev. Lett.*, **69**, 478–481 (1992).
8. Cendoya, I., Alegría, A., Alberdi, J. M., Colmenero, J., Grimm, H., Richter, D., and Frick, B., *Macromolecules*, **32**, 4065–4078 (1999).

Effect of Polymer-Substrate Interactions on the Glass Transition of Polymer Thin Films

Ophelia K. C. Tsui,[1] T. P. Russell,[2] C. J. Hawker[3]

[1]*Physics Department, Hong Kong University of Science and Technology, Clear Water Bay, Hong Kong.*
[2]*Polymer Science and Engineering Department, University of Massachusetts, Amherst, MA 01003, USA.*
[3]*IBM Almaden Research Center, 650 Harry Road, San Jose, CA, USA.*

Abstract. It has been suggested that when the polymer-substrate interaction, γ_s, in a polymer film supported by substrate is strongly favorable, the T_g of the polymer film may increase with decreasing film thickness; but the opposite prevails if the interaction is less than weakly favorable. We present a quantitative study of the glass transition temperature, T_g, in thin films of polystyrene (PS) as a function of γ_s by measuring the change in the thermal expansion using x-ray reflectivity. Using random copolymer of styrene and methylmethacryalte anchored to the substrate, γ_s could be varied by varying the styrene content, f. With a fixed PS film thickness of 33 nm, the T_g was depressed by ~20 °C as f was decreased from 1 to 0.7. An analysis analogous to the Gibbs-Thompson model indicated that the surface energy was not a suitable parameter to use to describe the effect of interfacial interactions on the T_g of polymer thin films. Instead, an associated local fractional change in the polymer mass density at the substrate interface was introduced to describe the observed change in T_g with different γ_s.

INTRODUCTION

Previous studies [1-3] had demonstrated that the glass transition temperature, T_g, of polymer thin films supported by substrates is strongly dependent on the interaction, γ_s, between the polymer and the underlying substrate. In particularly, it was suggested that when γ_s is strongly favorable, the T_g of a polymer film increases with decreasing film thickness, but the opposite prevails if it is less than weakly favorable. In this study, we determine this alleged effect of γ_s quantitatively by measuring the T_g of polystyrene (PS) thin films with thickness fixed at 33 nm as a function of γ_s. Variation of γ_s was realized by using random copolymers of styrene and methyl methacrylate P(S-r-MMA) with different styrene fraction, f, anchored to silicon. Result of previous contact angle measurements showed that γ_s varies continuously with f according to $\gamma_f = 0.8 - 1.1f$ (in ergs cm^{-2}) [4]. With f varied between 0.7 and 1 in this experiment, γ_s was correspondingly varied by ~0.33 erg/cm^2.

EXPERIMENT

End-functionalized P(S-r-MMA) (M_w ~ 10,000, M_w/M_n ~ 1.1 – 1.8) was synthesized by a "living" free-radical polymerization technique. The copolymers were end-grafted onto cleaned Si(111) surfaces by annealing the copolymer coated substrates at 170 °C for 2 days allowing the terminal OH group of the copolymer to react with the native oxide layer on the silicon. The thickness of the P(S-r-MMA) brush layer thus produced is ~ 3 nm. To prepare the PS homopolymer films with thickness = 33 nm, solutions of 2 wt% PS (M_w = 96,000, M_w/M_n = 1.04) in toluene was spin coated on the P(S-r-MMA) end-grafted silicon substrates at a spinning speed of 3000 rpm. X-ray reflectivity was used to measure the thickness of the PS film as a function of temperature. T_g was identified as the temperature at which the slope in a plot of thickness versus temperature changed. Measurements were performed from 150 – 180 °C to 30 °C (cooling only) in 10 °C decrements with 45 min. being allowed for the sample to thermally equilibrate.

Throughout the measurement, the sample was kept under a 10^{-2} torr vacuum with the temperature controlled within ± 0.5 °C.

RESULTS AND DISCUSSIONS

Figure 1 shows the T_g as a function of f for the 33 nm thickness PS films (solid circles). As f was decreased from one, the T_g decreased systematically from a value consistent with the bulk T_g (=100 °C within error bars) to ~78 °C at f =0.7. This result is in keeping with the conventional notion that the T_g of a polymer film decreases with weakening in the polymer-substrate interactions.

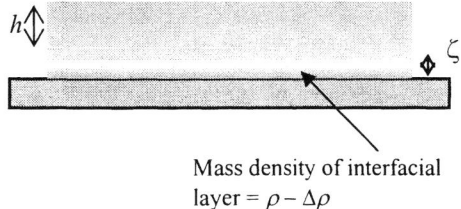

FIGURE 2. Diagram illustrating the model involving a local fractional change in the mass density at the substrate induced by the polymer-substrate interactions.

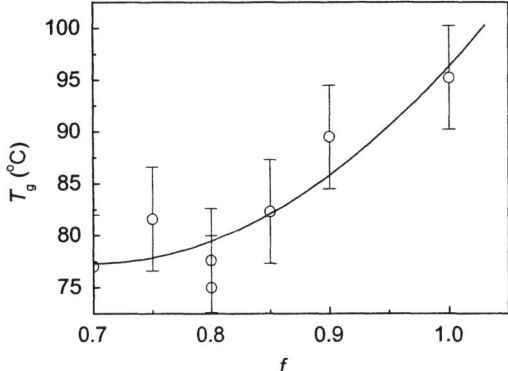

FIGURE 1. T_g of 33 nm thick PS films spin coated onto P(S-r-MMA) brushes as a function of the styrene content f in the copolymer (open symbols). The solid line is a guide to the eye.

As mentioned in the introduction, the amount of change in γ_s resulting from variation of f from 1 to 0.7 is ~0.33 erg/cm². We analyze if this amount of change in γ_s should lead to a change of 20 °C in T_g in analogy to the Gibbs-Thompson model for the melting temperature. From the literature, the activation energy for segmental motion near T_g is ~130 Jcm⁻³ [5]. Assuming the size of cooperativity at glass transition to be ~3 nm - a typical value, the activation energy per unit area, ε_a, is ~130 Jcm⁻³ × 3 nm = 390 erg/cm². The fact that ε_a is much larger than 0.33 erg/cm² suggests that the interfacial interactions cannot have made the noted influence on T_g by direct modification of the activation energy. Instead, they must have acted through their effects upon parameters that T_g is very sensitive to. One possibility for such quantity is the mass density, ρ. It has been estimated that for each g/cm³ change in ρ, T_g changes by 1350 °C (= $\delta T_g/\delta\rho$) [6]. Thus, only a small decrease of 1.4% in the average density can produce a 20°C depression in T_g found in this experiment. However, such a large change in the average density of a thin film polymer has not been observed in either the twin neutron reflectivity measurements of Wallace et al. [7] or the Brillouin light scattering studies of Forrest et al. [8] where, to within ~ 0.5%, the densities of thin PS films were found to be equal to that of the bulk. Thus, a reduced average film density cannot be used to explain the observations here.

We consider a bilayer model in which there is a local change in the mass density, $\Delta\rho$, of the polymer within a boundary layer of thickness ζ at the substrate interface due to perturbations by γ_s (Figure 2). The T_g of the film as whole is thus given by T_g^{bulk} + $(\delta T_g/\delta\rho)(\zeta/h)(\Delta\rho)$. For a 10 nm thick film, a 20 °C reduction in T_g (as noted for PS films supported by silicon [1,2]) would be produced when $\zeta\Delta\rho$ ~ 1.48 x 10^{-8} g/cm². If we take ζ = 3 nm, then $\Delta\rho$ = -0.05, a 5% density decrease at the interface. It should be remarked that $\Delta\rho\zeta/h$ gives the fractional density change due to the interfacial layer, and it is not possible to determine ζ and $\Delta\rho$ separately using the present simple model. As the film thickness is increased, this effect will diminish accordingly to $\Delta\rho\zeta/h$. Comparing the previous result [1] that the thickness dependence of changes in T_g follows ~ $1/h^{1.8}$, it thus requires $\zeta\Delta\rho$ ~ $1/d^{0.8}$. Confinement effect may give rise to strong perturbations to chain conformations [9]. However, the thickness at which changes in T_g start to occur (~ 400Å for PS on SiO_2 as previous experiments showed [1,6]), being bigger than the radius of gyration of the polymer molecules, R_G (~15 nm, here), is too large to produce much foreseeable effect. Computer simulation studies [10,11] show that perturbations to the segment density due to the interface persist over a distance $\sim R_G$. Insofar, these studies have been performed on systems with sizes sufficiently large that the density always returns to the bulk value at the center of the confined space. It will be interesting to repeat these studies on smaller systems where this does not hold and see if a reduction in the system dimension will modify the density perturbations coming from the interface. In particular, one would like to see whether it will reproduce the $1/d^{1.8}$ dependence of the fractional density change deduced from the present simple argument.

We recognize that a three-layer model - one that includes also a higher mobility surface layer at the polymer/air interface [1,2,12] - will probably offer a more realistic picture. However, most measurements of T_g in thin films are based on an average characteristic of the film, for example the film thickness, so are not going be able to distinguish between influences of the two surfaces. The bilayer model elaborated here can be considered describing the *combined* effects of the air- and the substrate-interfaces, parametrized by the product ($\Delta\rho\,\zeta$), which provides a quantitative measure for the consequential perturbations on the overall segment density. The foregoing discussion has been focused on cases where T_g decreases with decreasing film thickness. By considering favorable interactions that should make $\Delta\rho$ positive, the present model will also describe systems with enhanced T_g with decreasing film thickness. Under the special condition where $\Delta\rho = 0$, the T_g of a film will remain the same as the bulk. This is expected to occur when the interface produces no perturbation to the chain density, which is likely to be realized by a boundary coated with brushes of the same composition as the polymer itself. Indeed, no change in T_g is found with the $f = 1$ brush (see Figure 1). On the other hand, such an interface will be attractive (not neutral) to the polymer. Our result highlights the inappropriateness to use the interfacial interaction to characterize the effect of an interface on the T_g of polymer thin films. The density parameter ($\Delta\rho\,\zeta$), on the other hand, evidently offers a much better choice.

Finally, it is worth mentioning that existing arguments for the reduction in T_g found in polymer films either based on a reduction in the coupling parameter for chains confined in thickness less than R_G [13] or existence of a novel sliding motion in free-standing films [14] do not address the issue concerning the effect of polymer-substrate interactions. It is hoped that the present study could stimulate theoretical investigations along this line.

REFERENCES

1. Keddie, J. L., Jones, R. A. L., and Cory, R. A., *Europhys. Lett.* **27**, 59-64 (1994).

2. Keddie, J. L., Jones, R. A. L., and Cory, R. A., *Faraday Diss.* **98**, 219-230 (1994).

3. Wallace, W. E., van Zanten, J. H., and Wu, W. L., *Phys. Rev.* **E52**, R3329-R3332 (1995).

4. Mansky, P., Liu, Y., Huang, E., Russell, T. P., and Hawker, C. J., *Science* **275**, 1458-1460 (1997).

5. The data is obtained by using the data of Santangelo, P. G., and Roland, C. M., *Macromolecules* **31**, 4581-4585 (1998), and assume the density of PS to be 1g/cm^3.

6. Tsui, O. K. C., Russell, T. P., and Hawker, C. J., *Macromolecules* **34**, 5535-5539 (2001).

7. Wallace, W. E., Beck-Tan, N. C., and Wu, W. L., *J. Chem. Phys.* **108**, 3798-3804 (1998).

8. Forrest, J. A., Kalnoki, K., and Dutcher, J. R. *Phys. Rev. E* **58**, 6109-6114 (1998).

9. De Gennes, P. -G. *Scaling Concepts in Polymer Physics*, Ithaca and London: Cornell University Press, 1996.

10. Baschnagel, J., and Binder, K. *Macromolecules* **28**, 6808-6818 (1995).

11. Bitsanis, I., and Hadziioannou, G. *J. Chem. Phys.* **92**, 3827-3847 (1990).

12. Kajiyama, T., Tanaka, K., and Takahara, A. *Macromolecules* **30**, 280-285 (1997).

13. Ngai, K. L. *Eur. Phys. J. E* **8**, 225-235 (2002).

14. Dalnoki-Veress, K., Forrest, J. A., de Gennes, P. G., and Dutcher, J. R. *J. Phys. IV France* **10**, Pr7-221-232 (2000).

Universal Reference Temperature for Melt Viscosity Temperature Relationship

Norimasa Okui

*Department of Organic and Polymeric Materials, International Research Center of Macromolecular Science
Tokyo Institute of Technology, Ookayama, Meguroku, Tokyo, JAPAN*

Abstract. There are many characteristic temperatures commonly observed on a mechanical relaxation spectrum, DSC and the other methods, such as T_m, T_g, T_β, T_{α_c}, and T_{cmax}. Among these characteristics, there are several empirical rules. The reference temperature (T_r) in WLF equation was attributed to many characteristic temperatures such as T_{cmax}, T_{α_c}, T_{ll}, T_{cross} and T_{mode} with value of ca.$1.25T_g$ for polymeric materials. According to these phenomenological relationships, the universal curve of crystal growth rate was obtained based on the maximum crystal growth rate and the universal plot of melt viscosity was observed based on the reference temperature (T_r) for polymeric materials, inorganic substances and organic compounds.

RELATIONSHIP AMONG THERMODYNAMIC TEMPERATURES

There are many characteristic temperatures commonly observed on a mechanical relaxation spectrum, DSC and the other methods. Such characteristics are melting temperature (T_m), glass transition temperature (T_g), β-relaxation temperature (T_β) at which polymer segmental motion is ceased, liquid-liquid transition temperature (T_{ll}), crystalline dispersion temperature ($T_{\alpha c}$) and maximum crystallization temperature (T_{cmax}) at which crystal growth rate shows a maximum. Also, there are second order transition (T_2) and the temperature at which a free volume is zero (T_0), although hypothetical temperature. T_0, however, is identical with T_2. Among these characteristics, there are several empirical rules. Constancy of the ratio of T_g/T_m is widely known for many materials. The ratio of T_g/T_m is found to be about 2/3 not only for polymers but also for inorganic substances and organic compounds. According to WLF relationship, $T_0/T_g = 1 - f_g/\Delta\alpha T_g$ where f_g is a free volume at T_g and $\Delta\alpha$ is the difference of the thermal expansivity between a glass and a super-cooled liquid. The mean values of $\Delta\alpha T_g$ and f_g are approximated to be 0.1 and 0.025 [1], respectively. Thus, the ratio of T_0/T_g yields to be 3/4. In other words, the free volume f_g is expressed as $\Delta\alpha T_g/4$. We can obtain the values for $\Delta\alpha$ and T_g in many reference data. The histogram of $\Delta\alpha T_g$ shows that the distribution curve is considerably wide with the maximum value of 0.1 for polymer (the most probable value). Such wide distribution of the free volume gives rise to the wide distribution of the ratio of T_g/T_m.

The relationship between T_{cmax} and T_m is formulated by equating to zero the derivative of equation (1) with respect to the temperature.

$$G = G_0 \exp\{-\Delta E/RT - KT_m/RT\Delta T\} \quad ...(1)$$

Thus obtained the ratio of T_{cmax}/T_m is expressed as $C/(1+C)$ where $C=(1+\Delta E/K)^{0.5}$ [2]. ΔE and K are the activation energy of migration through the nucleus melt interface and the nucleation parameter associated with the mean surface energy (σ) and the heat of fusion (ΔH_m). Here it is interesting to note that the ratio of $\Delta E/K$ can be divided into two factors such as the ratios of $\sigma/\Delta H_m$ and $\Delta E/\Delta H_m$. These two factors show the constant values yielding the constant value of the ratio of T_{cmax}/T_m. In fact, most of polymers show that the ratio of T_{cmax}/T_m is about 0.83. However the ratio of T_{cmax}/T_m is strongly dependent on a type of materials [3].

Moreover, similar relationship between T_{cmax} and T_g is found as $T_{cmax}/T_g = 1.25$ which value almost equals to that of $T_{\alpha c}/T_g$ [4]. In addition, the ratio of T_{ll}/T_g is also found to be about 1.2, which value coincides with the ratio of T_{mode}/T_g, where T_{mode} is a critical temperature in the mode coupling theory of the liquid to glass transition. This might be suggested that there is some correlation between the molecular motion in amorphous state and that in crystalline state. The origin of these characteristic temperatures may be based on a similar mechanism associated with large molecular motions in the amorphous and crystalline states.

In addition to these characteristic ratio, it is interested to note that the reference temperature (T_r) for WLF equation expressed as $T_r/T_g = 1 + f_g/\Delta\alpha T_g$, ~5/4. On the temperature dependence of melt viscosity, the melt viscosity increases as a decrease of temperature. In the higher temperature regions, the viscosity obeys Arrhenius law. On the other hand, the viscosity obeys WLF law or VFT law in the temperature regions between T_g and T_r. A crossover temperature (T_{cross}) from Arrhenius to WLF appears at about $1.25T_g$ for polymer and $0.95T_m$ for fragile organic compounds. It is interested to note that these crossover temperatures coincide with T_{cmax}. In other words, the reference temperature in WLF equation is attributed to many characteristic temperatures such as T_{cmax}, $T\alpha_c$, T_{ll}, T_{cross} and T_{mode} with value of ca. $1.25T_g$ for polymeric materials.

Based on these phenomenological relationships, crystal growth behavior based on T_{cmax} and temperature dependence of melt viscosity based on T_{cross} are discussed in this report.

UNIVERSAL CURVE OF CRYSTAL GROWTH RATE

Maximum growth rate (G_{max}) can be observed by equating to zero the derivative of equation (1) with respect to the temperature. The relation thus obtained is expressed as follows [5].

$$Ln(G/G_{max}) = \{Ln(G_{max}/G_o)\}\{(1-X)^2/X(A-X)\} \quad ...(2)$$

X is the reduced crystallization temperature of T/T_{cmax} and A is the ratios of T_m/T_{cmax}. In polymer crystallization data obtained over a wide range of temperature through a maximum crystal growth rate, data sets of G_{max}, T_{cmax} and T_m for many polymers are available in the referenced literature. G_{max}, T_{cmax}, and T_m show remarkable molecular weight dependence. The G_{max} decreases with the molecular weight (M) as expressed by power low of $G_{max} \propto M^{0.5}$ for chain folding crystallization [6]. T_{cmax} increases with molecular weight similar to the molecular weight dependence of T_m. The ratio of T_{cmax}/T_m is almost independent of the molecular weight, yielding the constant value of 0.83. When the crystal growth data are plotted according to equation-2, the reduced growth rate (G/G_{max}) shows a linear relationship with the second term in the right hand side of equation-2. The linear relationships indicate that ratios of G_{max}/G_o are independent of molecular weight. In other words, the molecular weight dependence of G_{max} is the mainly a consequence of the molecular weight dependence of G_o. However, each polymer gives a different value to the ratio of G_{max}/G_o indicating a material constant. Therefore, the reduced growth rate (G/G_{max}) in each polymer should be normalized by the value of G_{max}/G_o. All experimental crystal growth data (total numbers of data points are over 400) are drawn in to a single universal curve as seen in figure-1 [7]. The universal curve is true for other materials such as inorganic substances as shown in figure-2.

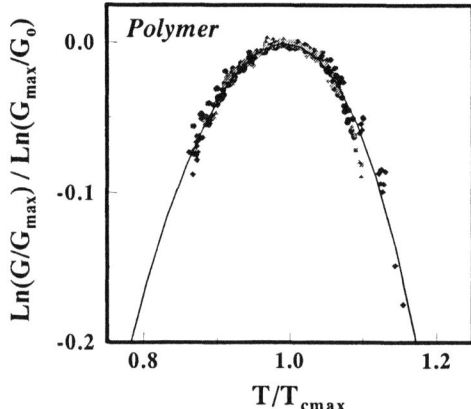

Figure-1 Universal curve of crystal growth rate for various polymers (i-PP, i-PS, PTMPS, PESU, PET, PLLA, cis-PIP and Nylon-6).

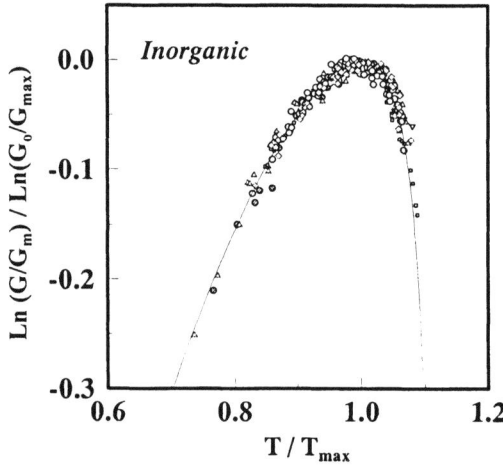

Figure-2 Universal curve of crystal growth rate for various inorganic substances (Li_2O-$2SiO_2$, K_2O-B_8O_{13}, Na_2O-$2SiO_2$, Na_2O-BaO-$4SiO_2$, PBO-$2B_2O_3$, and GeO_2)

UNIVERSAL PLOTS OF MELT VISCOSITY

According to the Doolittle equation, the temperature dependence of melt viscosity can be express as,

$$Ln(\eta/\eta_r) = b(1/f - 1/f_r) \quad ...(3)$$

where f and f_r are free volume at a given temperature (T) and the reference temperature (T_r) and f is commonly expressed as $f=\Delta\alpha(T-T_o)$. Here, T_r is employed as $1.25T_g$ for polymer. The temperature-viscosity relationship is divided into three temperature regions such as ①: $T>T_m$, ②: $T_m<T<T_r$, ③: $T_r<T<T_g$. In the temperature region higher than T_m, melt viscosity can be expressed by Arrhenius expression and then the following equation is obtained.

$$Ln(\eta/\eta_r) = K_m(T_g/T - 1) \quad ...(4)$$

In the temperature region between T_g and T_r, viscosity of super cooled melt can be expressed by WLF expression yielding the following equation.

$$Ln(\eta/\eta_r) = K_g\{(T_g-T_o)(T-T_o) - (T_r-T_o)(T_r-T_o)\} \quad ...(5)$$

K_m and K_g is a function of free volume at T_r.

According to various reference data for temperature-viscosity relationship, figure-3 shows the relationship between the reduced viscosity at T_r and the ratio of T_g/T for polymeric materials [8,9]. Most of data in the temperature region between T_r and T_g obey a single WLF curve. On the other hand, in the temperature region above T_m, various linear relations with merging into a single point are obverted. These slopes give the activation energies for viscous flow which can be related to the free volume at T_r. These viscosity relationships are true for inorganic substances and organic compounds as shown in figures 3 and 4, respectively [10].

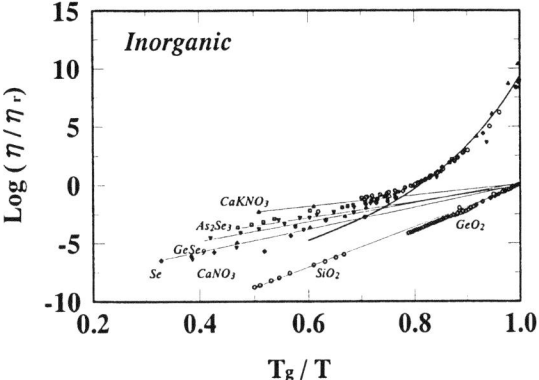

Figure-4 Universal plots of melt viscosity for inorganic materials. Data are based on Angell who kindly provided the original data.

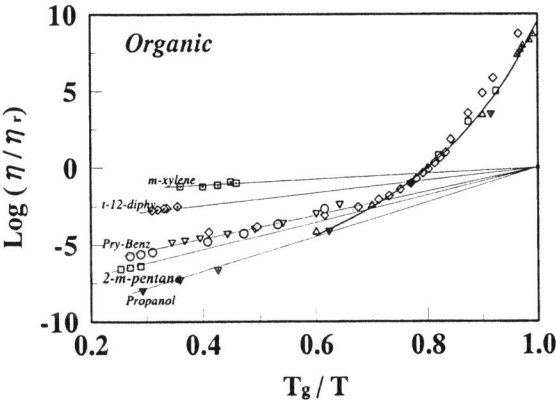

Figure-4 Universal plots of melt viscosity for organic materials. Data are based on Angell who kindly provided the original data.

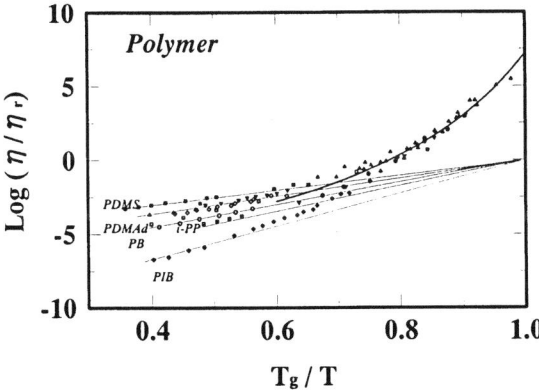

Figure-3 Universal plots of melt viscosity for polymeric materials. Data are based on Magill et.al [8] and van Krevelm et.al [9].

REFERENCES

1. van Krevelen, D.W., *Properties of Polymers*, Elsevier, London (1976)
2. Okui, N., *Polymer J.*, **19**, 1309-1315 (1987)
3. Okui, N., *J. Materials Sci.*, **25**, 1623-1631 (1990)
4. Okui, N., *Polymer*, **31**, 92-94 (1990)
5. Okui, N., *Polym.Bull.*, **23**, 111-118 (1990)
6. Umemoto, S., and Okui, N., *J.Macromol.Sci.*, **B41**, 923-938 (2002)
7. Umemoto, S., and Okui, N., *Polymer*, **43**, 1423-1427 (2002)
8. Breitling, S.M., and Magill, J.H., *J.Appl.Phys.* **45**, 4167-4171 (1974)
9. van Krevelen, D.W., and Hoftyzer, P.J., *Angew. Makromol. Chem.*, **52** 101-109 (1976)
10. Angel, C.A, *J.Non-Cryst.Solid*, **73**, 1-17 (1985)

The Role Of Configurational Entropy In Chemical Vitrification

S. Corezzi

INFM and Dipartimento di Fisica, Università di Perugia, I-06123, Perugia, Italy

Abstract. Glasses can be formed in numerous ways, involving very different microscopic processes. This article reviews some recent results on chemical vitrification, a process where the slowdown in the dynamics of a liquid is controlled by the irreversible formation of chemical bonds. Making a connection between the reduction in configurational entropy and the number of chemical bonds, the dynamics of vitrification in chemically reactive systems is explained in terms of their configurational restrictions in the same manner as in stable glass-forming liquids under cooling or compression.

INTRODUCTION

Glass formation entails slowing down continuously the particles in a liquid, till the system fails to rearrange its structure on the experimental timescale and then acts effectively like a solid. The formation of glasses on cooling or compression (*physical vitrification*) is normal for most liquids. However, in chemically reactive systems the glassy state may also result from the progressive polymerization of the constituent molecules via the formation of irreversible chemical bonds (*chemical vitrification*). The hardening of natural and synthetic resins and the formation of most of the materials used in engineering plastics and high-performance composites are based on chemical vitrification. Explaining the surprising similarities in the relaxational slowdown following physical and chemical changes may reveal a common origin of vitrification processes and improve general understanding of the liquid-glass transition.

It is a widespread belief that molecules rearrange cooperatively as a liquid nears its glass transition. In particular, the notion of cooperatively rearranging region (CRR) is central in the theory of Adam and Gibbs (AG) [1]. Here the CRR's size growth reflects a configurational entropy (S_c) decrease, which controls the structural relaxation time τ according to

$$\tau = \tau_0 \exp(C/TS_c) \quad (1)$$

with τ_0 and C nearly constant. The entropy $S_c \equiv S^{liquid} - S^{nonstruct} = k_B \ln W_c$ (k_B is the Boltzmann's constant) quantifies the number W_c of structurally distinct configurational states available to the liquid. The non-structural contribution, $S^{nonstruct}$, to the liquid's entropy, S^{liquid}, arises from vibrational and secondary relaxational degrees of freedom.

Direct experimental tests of equation (1) are hampered because S_c is not accessible to experiments. Notwithstanding, results of computer simulations on model glass formers [2, 3] and experiments performed as a function of both temperature and pressure [4, 5], provide a support to equation (1) and suggest configurational entropy as a tenable basis for a theory of the glass transition. Here we investigate application of the entropy model to polymerization reactions and provide evidence that chemical vitrification shares with physical vitrification the same fundamental interpretation.

An entropy-based picture would qualitatively apply to reactions of polymerization if the system's evolution could be seen as a succession of quasi-equilibrium states. As reaction proceeds, the formation of covalent bonds in place of weaker interactions is expected to impose configurational restrictions and force the molecules to move cooperatively over an increasing length scale. A correlation of molecular slowdown with loss of configurations is therefore naturally believed, though it cannot be easily proved. Indeed, a direct proof needs parallel information on how τ and S_c change throughout the reaction, but the character of the process prevents from doing an experimental determination of S_c. An alternative method can be followed in the presence of a step reaction mechanism.

In step polymerizations [6], it is not unusual to realize conditions where a connection can be made between a decrease in configurational entropy and the chemical conversion $p(t)$, which measures the extent of reaction at a time t through the fraction p of functional groups that have reacted till then.

Owing to the random nature of the step-growth mechanism, at any fixed temperature, one can expect monomers linked with each other approximately to behave as a single cooperative unit. As they are interlocked segments, a change in configuration of any one of them causes a change in configuration of the others, and it is realistic to expect that the average number of different configurations accessible to one cooperative region remains the same. Accordingly, one has $W_c(p) = W_c(0)^{1/x_n}$, where x_n is the number-average degree of polymerization (i.e., the average number of monomers per molecule in the system), $W_c(p)$ is the number of configurations available to the macroscopic system at conversion p, and $W_c(0)$ the number of configurations at $p = 0$. Consequently,

$$S_c(p) = S_c(0)/x_n \qquad (2)$$

Equation (2) establishes how configurational entropy reduces with respect to its initial value due to an advancement of reaction. It is important to note that in the AG formalism, where the size of a CRR relates inversely to S_c, this equation asserts that the CRR's size grows in proportion to the average size of the molecules in the system.

On the basis of simple considerations [7], for a general mixture in which no intramolecular reactions occur, the dependence of x_n on p can be written as $x_n(p) = 1/(1 - \bar{f}p)$, where $\bar{f}$ is the average number per monomer of functional groups whose conversion is p. Thereby, equation (2) gives $S_c(p) = S_c(0)(1 - \bar{f}p)$, and equation (1) becomes

$$\tau = \tau_0 \exp\left(\frac{B(T)}{1 - \bar{f}p}\right) \qquad (3)$$

with τ_0 approximately independent of both T and p, and $B(T) = C/TS_c(0)$ only dependent on temperature, on the assumption that the variation of C with p is negligible. Very strong predictions follow from equation (3), which can be tested experimentally. (*i*) τ tends to diverge where p equals the theoretical value $1/\bar{f} \equiv p_0^{theor}$; (*ii*) because the average functionality $\bar{f}$ may be controlled by varying the molar ratio of reagents, p_0^{theor} may also be varied in this way irrespective of any reaction details; (*iii*) for a given reaction, the same p_0^{theor} is expected at different T.

EXPERIMENTS

Relaxation and conversion data have been collected and re-examined for several isothermal polymerizations of epoxy-amine systems (see Table 1), where reaction proceeds by polyaddition. Three epoxy prepolymers (diglycidyl ether of bisphenol A) are involved, with functionality $f=2$ and different epoxy equivalent weight: EPON828 (e.e.w.=190), EPONX22 (e.e.w.=172), DGEBA348 (e.e.w.=174). EPON828

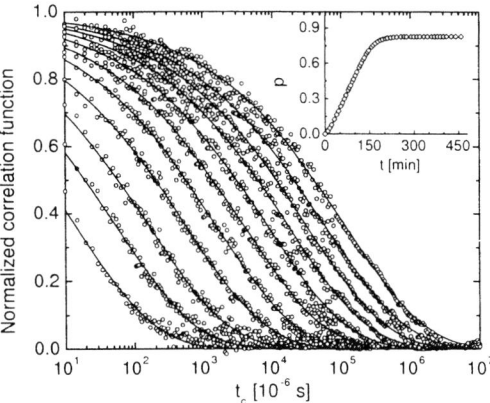

FIGURE 1. Normalized correlation functions vs. the correlation time t_c, measured by depolarized PCS at different reaction times during the isothermal polymerization of DGEBA348/DETA 4:3 at 30 °C. The reaction time increases from the left to the right spectra. Solid lines are the best fit with a stretched exponential function. In the inset: For the same reaction, conversion p of epoxy groups vs. the reaction time t.

and DGEBA348 were cured with aliphatic amines: *n*-butylamine (BAM; $f=2$), cyclohexylamine (CHA; $f=2$), ethylenediamine (EDA; $f=4$), and diethylenetriamine (DETA; $f=5$); EPONX22 was cured with an aromatic amine: 4,4'- diaminodiphenylmethane (DDM; $f=4$). Reactions with difunctional amines (BAM and CHA) produce the growth of linear-chain polymers; reactions with multifunctional amines (EDA, DDM, and DETA) produce network polymers.

As reaction proceeds, the extent of reaction at a time t can be measured through the fraction p of epoxy groups that have reacted with amino groups at elapsed time t since the beginning. Except for the EPONX22/DDM 2:1 system, where Fourier-transform infrared spectroscopy was used [8], the conversion of all of the systems was measured by calorimetry as the ratio between the partial heat $\Delta H(t)$ released up to the time t and the total heat of reaction ΔH_{tot}. Typical uncertainty in the calorimetric measure of p is 2-3%, mainly due to errors in the determination of ΔH_{tot} and the samples weight.

Measurements of structural relaxation times suitable for a test of equation (3) require the reaction to be slow enough for having no appreciable reaction during a time interval of duration τ (quasi-equilibrium condition) and during the acquisition of any spectrum. Measurements of τ under these conditions have been performed by means of dielectric (DS), depolarized photon-correlation (PCS), and heat capacity spectroscopy (HCS) by several research groups, cited in

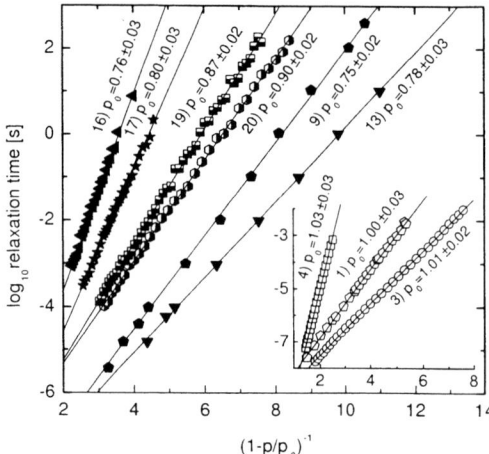

FIGURE 2. Linearized plot of the dependence of the relaxation time on conversion, for some isothermal step polymerizations (the reactions are numbered as in Table 1). The straight lines represent the best fit with equation (3). For each reaction, the value p_0 obtained from the fit must be compared with the value predicted theoretically, which is: $p_0^{theor}=1.00$ (reactions in the inset), $p_0^{theor}=0.88$ (semiclosed-symbols reactions), $p_0^{theor}=0.75$ (closed-symbols reactions).

Table 1. We have reduced all of the reported values of relaxation time to $\tau_{max} = (2\pi\nu_{max})^{-1}$ values, with ν_{max} the frequency at which the imaginary part of the measured susceptibility has a maximum, in order to remove the differences among τ produced by different fit equations of the spectra. The data of τ_{max} as a function of the reaction time t have been combined with the data of chemical conversion p as a function of t, in order to reconstruct $\tau_{max}(p)$ curves. Fig. 1 illustrates the evolution of PCS spectra for the mixture DGEBA348/DETA 4:3 during reaction at 30 °C. The inset shows the corresponding behavior of the chemical conversion.

RESULTS AND DISCUSSION

Depending on the functionalities of the reagents and the molar ratio between them, the reactions studied can be divided into three groups characterized by a different average epoxy-functionality, $\bar{f}$, and accordingly different $p_0^{theor} = 1/\bar{f}$ as follows: $p_0^{theor}=1.00$ in 1:1 mixtures, $p_0^{theor}=0.88$ in 4:3 mixtures, and $p_0^{theor}= 0.75$ in 2:1 mixtures. In testing equation (3) experimentally, it should be noted that it is based on some ideal assumptions which limit its validity. Especially high temperature tends to remove quasi-equilibrium conditions and to favor the occurrence of intramolecular reactions. Polymerizations in Table 1, however, are expected to depart negligibly from ideal conditions.

For each isothermal reaction, experimental data of τ_{max} as a function of conversion have been fitted by the form $\tau_{max} = \tau_0 \exp[B/(1-p/p_0)]$, which reproduces at constant T the model equation (3), with τ_0, B, and p_0 as free parameters. The values p_0 obtained experimentally are listed in Table 1. When plotted against $(1-p/p_0)^{-1}$, $\log_{10}\tau_{max}$ should display linear behavior. This data representation is shown in Fig. 2, which emphasizes the remarkable agreement (over the whole range investigated) with the functional dependence on p expected from equation (3). Interestingly, such a dependence is formally similar to a Vogel-Fulcher dependence, which was originally proposed for epoxy-amine systems on a phenomenological basis [9, 12] or a semi-empirical T_g-based approach [16]. More importantly, the results in Table 1 meet our expectations for the values of the p_0 parameter. In all cases examined, the divergence of τ_{max} and therefore the graphic linearization in Fig. 2 occurs, within quite acceptable experimental error, strictly around the ideal values $p_0^{theor}=1.00$, 0.88, and 0.75 that are expected for our mixtures solely on the basis of their functionality and molar ratios, independently of T. This astonishing result constitutes strong evidence in favor of the idea behind equation (3).

Further observations support the entropy-based equation. First, experimental results are consistent with the expectation of a negligible T dependence of the pre-exponential factor τ_0, and a variation of B according to $B(T) = D/(T - T_0)$, with D and T_0 constant parameters [10]. Second, using this form of $B(T)$ in equation (3), an expression can be obtained for the variation with p of the 'kinetic fragility', a parameter quantified by the index K_{VF} in the Vogel-Fulcher equation written as $\tau_{max} = \tau_0 \exp(1/[K_{VF}(T/T_0 - 1)])$. It is $K_{VF} = (T_0/D)[1-\bar{f}p]$. We note that a decreasing behavior of fragility in step polymerizations is consistent with the presence of an increasing fraction of highly directional interactions, as typical of 'strong' systems. Although an isothermal determination of K_{VF} is difficult to be done, a tendency of network-forming mixtures to become less fragile seems confirmed by experiments [17, 18].

In step polymerizations leading to network polymers, in addition to vitrification, manifested by structural arrest, another critical phenomenon, gelation, takes place, manifested by shear viscosity η tending to infinity [6]. Different from the behavior in supercooled liquids, the divergence of η and τ in network-forming systems is not at once, gelation occurring well before vitrification. It can be noted that equations (1)-(3), by establishing a connection of the relaxation time with x_n and then with the number-average molecular weight $M_n = M_0 x_n$ (M_0

TABLE 1. Isothermal step-polymerization reactions: best-fit parameter p_0

Reaction	Refs.	p_0	Reaction	Refs.	p_0
1) EPON828/EDA 1:1 - 25 °C *	[9, 10]	1.00±0.03	11) EPONX22/DDM 2:1 - 68 °C *	[8]	0.79±0.03
2) EPON828/EDA 1:1 - 25 °C †	[11]	1.00±0.03	12) EPONX22/DDM 2:1 - 75 °C *	[8]	0.77±0.03
3) EPON828/BAM 1:1 - 25 °C *	[12, 10]	1.01±0.02	13) EPONX22/DDM 2:1 - 82 °C *	[8]	0.78±0.03
4) EPON828/CHA 1:1 - 27 °C *	[13]	1.03±0.03	14) EPONX22/DDM 2:1 - 90 °C *	[8]	0.80±0.03
5) EPON828/CHA 1:1 - 41 °C *	[14]	1.05±0.03	15) EPONX22/DDM 2:1 - 98 °C *	[8]	0.79±0.03
6) EPON828/CHA 1:1 - 27 °C **	[15]	0.98±0.07	16) EPON828/EDA 2:1 - 23 °C †	[11]	0.76±0.03
7) EPON828/CHA 1:1 - 40 °C **	[15]	1.00±0.02	17) EPON828/EDA 2:1 - 25 °C †	[11]	0.80±0.03
8) EPON828/CHA 1:1 - 50 °C **	[15]	1.04±0.02	18) EPON828/EDA 2:1 - 32.1 °C †	[11]	0.75±0.03
9) EPONX22/DDM 2:1 - 55 °C *	[8]	0.75±0.02	19) DGEBA348/DETA 4:3 - 25 °C †	[11]	0.87±0.02
10) EPONX22/DDM 2:1 - 60 °C *	[8]	0.77±0.03	20) DGEBA348/DETA 4:3 - 30 °C †	[11]	0.90±0.02

* Relaxation times measured by DS
† Relaxation times measured by PCS
** Relaxation times measured by HCS

is the average molecular weight of the monomers), properly reflect this behavior. At the gel-point, an infinite network begins to form, and the weight-average molecular weight M_w tends to diverge. However, relatively low values of x_n are reached and M_n remains finite.

CONCLUSIONS

We have reported a study of the role of configurational entropy in vitrification processes induced by spontaneous growth of macromolecules in reactive systems. A successful comparison is made between experimental data and the dynamic evolution predicted within the entropy model for step polymerizations (equation (3)). Our test spans (i) linear and network polymerizations; (ii) different compositions of the same reagents (e.g., both stoichiometric and non-stoichiometric balance of mutually reactive functional groups is realized in the EPON828/EDA mixture); (iii) different curing temperatures, for the same reactive mixture (e.g., EPONX22/DDM 2:1 mixture). The validity of equation (3) has two implications. First, it validates the connection we make between the amount of chemical bonds created (a quantity experimentally accessible) and the associated reduction in configurational entropy; and second, it provides evidence that the dynamics-to-thermodynamics correlation, $\tau(S_c)$, appearing in the form of equation (3) along physical vitrification paths (by varying T or P), is preserved in the same functional form along chemical vitrification paths (by varying p). Our findings state that configurational entropy plays a crucial role in controlling slow dynamics in chemically vitrifying systems, just like it does in glass forming liquids under cooling or compression.

ACKNOWLEDGMENTS

The author thanks D. Fioretto, P. A. Rolla, J. M. Kenny, and D. Puglia for assistance and discussions.

REFERENCES

1. Adam, G., and Gibbs, J. H., *J. Chem. Phys.*, **43**, 139–146 (1965).
2. Scala, A., Starr, F. W., La Nave, E., Sciortino, F., and Stanley, H. E., *Nature*, **406**, 166–169 (2000).
3. Sastry, S., *Nature*, **409**, 164–167 (2001).
4. Casalini, R., Capaccioli, S., Lucchesi, M., Rolla, P. A., and Corezzi, S., *Phys. Rev. E*, **63**, 031207 (2001).
5. Prevosto, D., Lucchesi, M., Capaccioli, S., Casalini, R., and Rolla, P. A., *Phys. Rev. B*, **67**, 174202 (2003).
6. Young, R. J., and Lovell, P. A., *Introduction to polymers*, Chapman and Hall, New York, 1991, pp. 15–43.
7. Stockmayer, W. H., *J. Polym. Sci.*, **9**, 69–71 (1952).
8. Deng, Y., and Martin, G. C., *J. Polym. Sci.: Part B:Polym. Phys.*, **32**, 2115–2125 (1994).
9. Casalini, R., Corezzi, S., Fioretto, D., Livi, A., and Rolla, P. A., *Chem. Phys. Lett.*, **258**, 470–476 (1996).
10. Corezzi, S., Fioretto, D., and Rolla, P., *Nature*, **420**, 653–656 (2002).
11. Corezzi, S., Fioretto, D., Puglia, D., and Kenny, J. M., *Macromolecules*, **36**, 5271–5278 (2003).
12. Gallone, G., Capaccioli, S., Levita, G., Rolla, P. A., and Corezzi, S., *Polym. Int.*, **50**, 545–551 (2001).
13. Johari, G. P., Ferrari, C., Salvetti, G., and Tombari, E., *Phys. Chem. Chem. Phys.*, **1**, 2997–3005 (1999).
14. Tombari, E., Ferrari, C., Salvetti, G., and Johari, G. P., *J. Phys.: Condens. Matter*, **9**, 7017–7037 (1997).
15. Presto, S., Tombari, E., Salvetti, G., and Johari, G. P., *Phys. Chem. Chem. Phys.*, **4**, 3415–3421 (2002).
16. Enns, J. B., and Gillham, J. K., *J. Appl. Polym. Sci.*, **28**, 2567–2591 (1983).
17. Parthun, M. G., and Johari, G. P., *J. Chem. Phys.*, **103**, 440–450 (1995).
18. Fitz, B., Andjelic, S., and Mijovic, J., *Macromolecules*, **30**, 5227–5238 (1997).

A crossover region in the glassy freezing of deuteron dipole glass

Y.-S. Choi and J.-J. Kim [1]

Physics Department, Korea Advanced Institute of Science and Technology - Taejon 305-701, Korea

Abstract. Dipole glass phase of $Rb_{1-x}(ND_4)_xD_2PO_4$ (x=0.47) mixed crystal was studied by low frequency dielectric measurements. A crossover behavior of dielectric relaxation from a power law decay to a stretched exponential decay was observed in the region of glass transition, which was explained on the basis of the dynamically correlated domains model form Chamberlin *et al.*. There is a narrow region of coexistence for the two relaxations crossing over with glass freezing of deuterons.

INTRODUCTION

Slow relaxation dynamics of nonexponential decay in glass is described by either Kohlrausch-Williams-Watts (KWW) stretched-exponential decay function $\phi(t) = \phi_0 \exp(-(t/\tau)^\beta)$ ($0 < \beta < 1$) or Curie-von Schweidler (CvS) power law decay function $\phi(t) = \phi_0 (t/\tau)^{-\alpha}$ ($0 < \alpha < 2$) [1, 2, 3, 4, 5]. The KWW stretched-exponential decay function gives a long tail in the high frequency side of the relaxation frequency spectrum while the CvS power law decay function gives a tail extension in the low frequency side. This long tail extension asymmetry becomes more apparent when the frequency dependent dielectric data is analyzed in terms of relaxation time distribution functions [6]. The temperature dependence of the relaxation time distribution function was studied in details for a DRADP-x (x=0.40) dipole glass approaching the glassy freezing [6].

DYNAMICALLY CORRELATED DOMAINS

Chamberlin *et al.* proposed the dynamically correlated domains (DCDs) model [7, 8, 9, 10, 11, 12, 13] to explain the universal relaxation functions of the glassy freezing. In the DCDs model the decay function is given by

$$\Phi(t) = \Phi_0 \int_D ds\, s\, n_s \exp(-t\omega_s) \quad (1)$$

where s represents domain size, the size dependent relaxation frequency ω_s of domains is given by $\omega_s = \omega_\infty \exp(-C/s)$, and the domain size distribution function n_s is given for quenched random systems of spin glass or dipole glass as $n_s = s^{1/9} \exp(-(s/\sigma)^{2/3})$. This derivation is based on the assumption of an infinite percolation cluster formation at the freezing temperature with randomly competing bonds in the nearest and next nearest neighbor couplings [14, 15]. From the Fourier transform of the above decay function $\Phi(t)$ to the frequency domain we can obtain the corresponding dielectric loss function $\varepsilon''(\omega)$ as follows:

$$\varepsilon''(\omega) = \Delta\varepsilon \int_D ds\, s^{10/9} \exp(-s^{2/3}) \frac{\omega/\omega_s}{1+(\omega/\omega_s)^2} \quad (2)$$

This result of the DCDs model exhibits both CvS power law relaxation for $C > 0$ and KWW stretched-exponential relaxation for $C < 0$, where C of $\omega_s = \omega_\infty \exp(-C/s)$ represents a dynamic correlation parameter.

EXPERIMENTAL

We want to report our experimental observation of the two universal modes of glassy relaxation in coexistence in one and the same system of dipole glass DRADP-x (x= 0.47), that is, the CvS power law relaxation and the KWW stretched-exponential relaxation together simultaneously in a narrow temperature range from 42K to 40K.

Sample crystals DRADP-x were grown from saturated mixed solutions between RbD_2PO_4 (DRDP) and $ND_4D_2PO_4$ (DADP). Deuteration ratio was confirmed by a proton NMR analysis to be about 93% and the cationic mixing concentration of x= 0.47 was confirmed by the inductive coupling plasma (ICP) mass spectrometry. Dielectric susceptibilities were obtained by use of

---

[1] jjkim@kaist.ac.kr

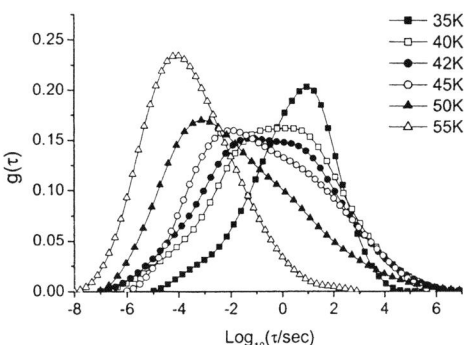

FIGURE 1. Relaxation time distribution functions $g(\tau)$ of a dipole glass DRADP-x (x=0.47) at selected temperatures across the crossover region of glassy freezing.

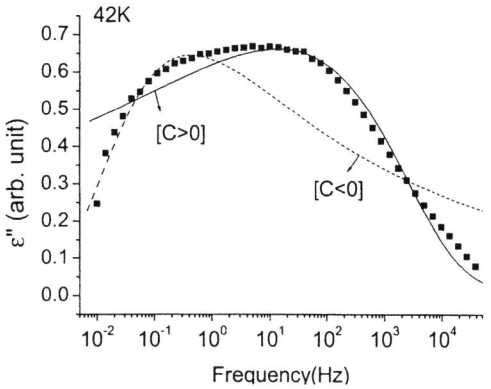

FIGURE 2. As temperature enters the crossover region (42K~40K) we can not fit the observed data by use of only one (either $C > 0$ or $C < 0$) relaxation function of the DCDs model.

a lock-in amplifier (EG&G, DSP-7260) in the low frequency measurements and by an impedance analyzer (HP-4192A) in the high frequency measurements. Gain controls of the respective amplifier systems are needed to match the low frequency data with the high frequency data of separate measurements.

RESULTS AND DISCUSSION

In the region of x≃0.5 frustration effects [16] and anisotropic cluster correlations [17] may be significant to show x-dependent variations of anisotropy strength and freezing temperature [18]. In Fig. 1 we have collected our dielectric relaxation data obtained at various temperatures from 55K to 35K, where we can clearly see an apparent crossover change from the long-time (low-frequency) tailing (T> 42K) to the short-time (high frequency) tailing (T< 40K). Between the two regions of asymmetric tailing extensions to the opposite sides, there is a narrow region of temperature between 42K and 40K where the relaxation spectrum loses the one-sided tailing asymmetry and becomes rather symmetric with a larger broadening.

In the DCDs theoretical model the peak position of the relaxation bands is always shifted to the tailing side as the system enters the glassy freezing, that is, for the $C > 0$ systems the peak position moves to the low frequency side with increasing $|C|$ but, for the $C < 0$ systems the peak position moves to the high frequency side with increasing $|C|$. In the DCDs model a crossover change without discontinuity from $C > 0$ to $C < 0$ in the same system may be possible only when the system is crossing $C = 0$, that is, $|C|$ should decrease to zero as the system approaches the crossover at $C = 0$. In Fig. 1 of our experimental data [6, 19], however, we can observe that this crossover change is accompanied by a monotonic shift of the peak position always to lower frequencies as we pass the crossover region to low temperatures. Furthermore, the band width of distribution is observed to increase as we enter the crossover region of our experimental data, whereas in the DCDs model the band width would diminish as we enter the region of $C = 0$ in the course of a crossover change from $C > 0$ to $C < 0$.

The Arrhenius-type relaxations ($C > 0$) can be derived from a 3-level scheme [5, 7, 10] of hopping transitions over activation barriers at $T \geq T_f$, where T_f represents a freezing temperature, whereas the inverse Arrhenius-type relaxations ($C < 0$) from a 2-level scheme of detailed balance transitions inside the deep freezing valleys at $T < T_f$. A global change in the free energy landscape associated with the nonergodic glass transition as in the spin glass system [5, 14] can bring about this microscopic change in the transition kinetics between energy levels, and thus a change from $C > 0$ to $C < 0$ in the DCDs model. In Fig. 2 we illustrate our fits at 42K, where we can see the deviation is so large that the best fits by use of one single ($C > 0$ or $C < 0$) response function of the DCDs model can no longer work.

Our experimental data of $\varepsilon''(\omega)$ observed at both 42K and 40K exhibit very broad symmetric bands without the one-sided long tail asymmetry, and can not be fitted well by one single relaxation function (either $C > 0$ or $C < 0$) of the DCDs model although this single relaxation function ($C > 0$ at $T > T_f$ or $C < 0$ at $T < T_f$) fits well the data outside this narrow crossover region [6, 19]. Instead, we could best fit the experimental data in this crossover region by use of two (both $C > 0$ and $C < 0$) relaxation functions of the DCDs model as depicted in Fig. 3(a,b). This implies coexistence of two competing relaxation modes corresponding to $C > 0$ and $C < 0$.

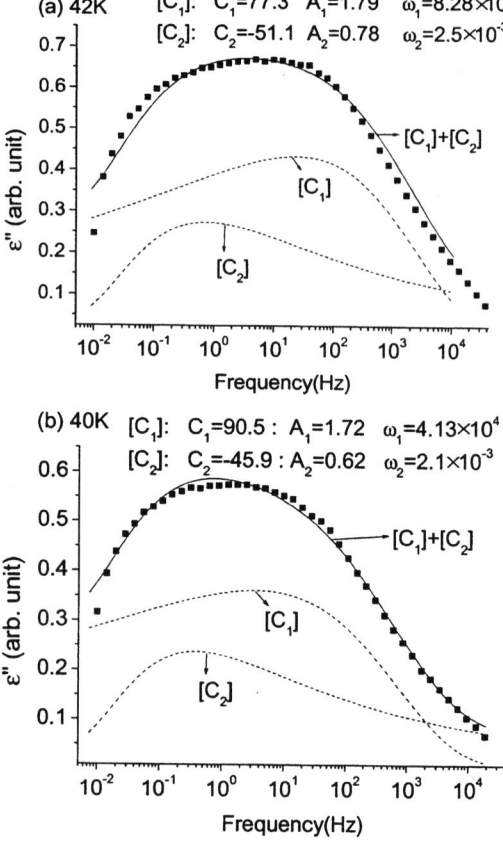

FIGURE 3. DCDs-model best fits to the experimental data in the crossover region by use of two relaxation functions (both $C > 0$ and $C < 0$) at (a)T=42K and (b)T=40K.

In Fig. 3 we show our results of the best fits to the experimental data in the crossover region by use of two relaxation functions ($C > 0$ and $C < 0$), where we can see an increasing contribution from the $C < 0$ mode in addition to a larger contribution of the $C > 0$ mode. Even at higher or lower temperatures outside the crossover region we may well improve the best fits when we use two relaxation functions instead of only one single relaxation function of either $C > 0$ or $C < 0$. This deviation grows as we enter the crossover region.

This observation of a crossover change in the bandshape of $\varepsilon''(\omega, T)$ may be closely related to the neutron scattering observation of a diffuse-incommensurate freezing [20] prior to the zone center dielectric freezing of deuterons. Furthermore, the dielectric data from the complex systems such as a ferroelectric relaxor and a dipole glass were also shown to require a sign change from $C > 0$ to $C < 0$ in the response function [12].

CONCLUSION

In the present work we have observed a narrow coexistence region, where we have an overlap between $C > 0$ and $C < 0$ responses, before an apparent sign change from $C > 0$ to $C < 0$. This crossover region is shown first to require two separate relaxation functions instead of a simple sign change of C in one single relaxation function.

ACKNOWLEDGMENTS

This work was supported in part by the Korea Research Foundation (KRF-2003).

REFERENCES

1. Höchli U. T., *Phys. Rev. B* **32**, 1985, pp. 4546–4550.
2. Nordblad P., *Phys. Rev. B* **33**, 1986, pp. 645–648.
3. Stillinger F. H., *Phys. Rev. B* **41**, 1990, pp. 2409–2416.
4. Kenning G. G., Joh Y. G., Chu D., and Orbach R., *Phys. Rev. B* **52**, 1995, pp. 3479–3483.
5. Chamberlin R. V., *Phase. Transition* **65**, 1998, pp. 169–209.
6. Kim B.-G. and Kim J.-J., *Phys. Rev. B* **55**, 1998, pp. 5558–5561.
7. Chamberlin R. V. and Haines D. N., *Phys. Rev. Lett.* **65**, 1990, pp. 2197–2220.
8. Chamberlin R. V., *Phys. Rev. B* **48**, 1993, pp. 15638–15645.
9. Chamberlin R. V., *J. Appl. Phys.* **76**, 1994, pp. 6401–6406; Chamberlin R. V. and Kingsbury D. W., *J. Non-Cryst. Solids*, **172-174**, 1994, pp. 318–395.
10. Chamberlin R. V., *EuroPhys. Lett.* **33** (7), 1996, pp. 545–550.
11. Kleemann W. and Klossner A., *Europhys. Lett.* **35** (5), 1996, pp. 391-395.
12. Kleemann W., Albertini A., Chamberlin R. V., and Bednorz J. G., *Europhys. Lett* **37** (2), 1997, pp. 145–150.
13. Kleemann W. and Lindner R., *Ferroelectrics* **199**, 1997, pp. 1–10.
14. Mydosh J. A., *Spin glasses : an experimental introduction* (Taylor and Francis, London, 1993) pp. 234–237.
15. Malozemoff A. P. and Barbara B., *J. Appl. Phys.* **57**, 1985, pp. 3410–3412.
16. Selke W. and Courtens E., *Ferroelectrics Lett.* **5**, 1986, pp. 173–183.
17. Matsushita M. and Matsubara T., *J. Phys. Soc. Jpn.* **55**, 1986, pp. 666–671.
18. Ko J.-H., Kim B.-G., Kim J.-J., Fujimori H., and Miyajima S., *J. Phys.: Condens. Matt.* **9**, 1997, pp. 4403–4410; Ko J.-H., Choi Y.-S., and Kim J.-J., *Ferroelectrics* **240**, 2000, pp. 257–264.
19. Kim B.-G., Kim J.-J., and Jang H.-M., *Phys. Rev. B* **60**, 1999, pp. 7170–7177.
20. Moussa F., Courtens E., and Launois P., *Europhys. Lett.* **33** (2), 1996, pp. 129–134.

Confined and bulk dynamics of a simple glass-former

C. Svanberg, R. Bergman, P. Jacobsson and L. Börjesson

Department of Applied Physics, Chalmers University of Technology, SE-412 96 Göteborg

Abstract. We have measured the dielectric response of propylene carbonate (PC) subject to different surrounding environments. The dynamics was studied in the supercooled regime in bulk, in 2D inter-platelet spaces of vermiculite clays and in polymer solutions. We show that the main α-relaxation in PC is influenced by both the type and the degree of confinement. In the polymer solutions the main peak becomes systematically slower, broader and weaker with increasing polymer concentration. The primary influence of the clay layers on the other hand is a smaller relaxation strength. Furthermore, bulk PC lacks a well-resolved β-peak but instead a second power law regime is detected at high frequencies. Our observations of the dynamics under spatial restrictions implies that the excess wing in PC is due to a secondary relaxation process.

INTRODUCTION

When the temperature of a glass former is lowered towards the glass transition temperature, T_g, the dynamics of the main (α-) relaxation process is rapidly slowing down. The α-process is generally believed to be cooperative, i.e. a many-body process, resulting in a relaxation peak much broader than a Debye process and with a non-Arrhenius temperature dependence. Characterizing the dynamical length scale of the glass transition has inspired many investigations of glass-formers in various types of confinements. However, the results of these studies are far from conclusive and no generally accepted theoretical model exists. One of the primary reasons for the sometimes conflicting results is the differences in the interaction with the confining surfaces.

Most glass formers also exhibits a secondary β-relaxation that is less temperature dependent than the α-relaxation. However, already in the seminal paper by Johari and Goldstein [1] some exceptions from the normal α-β scheme were noted. Instead of a well resolved secondary relaxation some glass formers reveal a transition to another power law behavior as compared with the high frequency extrapolation of the α-relaxation. The physical origin of this so called excess wing has been extensively debated in the literature, and two opposing scenarios has been proposed: *i)* the excess wing is an inherent feature of the α-relaxation or *ii)* the excess wing is a secondary process partly hidden by the α-relaxation.

Annealing experiments on some glass-formers have shown that at temperatures slightly below T_g the excess wing gradually transforms towards a secondary peak [2–4]. This implies that the excess wing is not an intrinsic feature of the α-relaxation. Furthermore, a systematic study on a series of oligomers show that the secondary loss peak tends to merge with the α-relaxation with decreasing chain-length [5]. This clearly shows that in simple glass formers the secondary peak may be observable only as an excess contribution to the α-relaxation.

Another approach to investigate the properties of the excess wing is to use confinement. Secondary processes are generally believed to be more localized than the cooperative α-relaxation and is therefore expected to be less influenced by spatial restriction. We have recently reported dielectric studies on simple glass-formers in various types of confinement [6–8]. Here we will focus on the dielectric response of propylene carbonate, which is a classical glass-former with an excess wing, in three types of experimental configurations: in bulk, dissolved in a polymer and in 2D confinement.

EXPERIMENTAL

We have measured the dielectric response of propylene carbonate (PC) in three different configurations: *i)* bulk, *ii)* dissolved in poly(methyl methacrylate) (PMMA), and *iii)* confined to the inter-platelet galleries of a Na-vermiculate clay. Anhydrous PC was purchased from Fluka and used as received. The solutions of PC in PMMA were prepared over the concentration range from 10 % to 50 % PMMA as described in Refs. [6, 9]. The clay samples were prepared by submersion of macroscopic (10×10 mm^2) clay pieces into the PC. The structure of the vermiculite clays is a stack of essentially two dimensional silicate platelets with very thin galleries (3-5 Å) in between[10]. The clay pieces were placed with the platelets parallel to the electrodes. The experiments of all the configurations were performed with a high resolution dielectric spectrometer (Novocontrol Alpha) from

10^{-2} to 10^7 Hz over the temperature range from 156 K to 200 K.

To describe our data we performed curve-fitting using two different approaches: classical superposition and using Williams ansatz. For the α-relaxation we used the Havriliak-Negami (HN) expression [11, 12],

$$\varepsilon^*(f) = \frac{\Delta\varepsilon}{[1 + (\iota 2\pi f \tau)^\alpha]^\gamma}, \quad (1)$$

where $\Delta\varepsilon$ is the relaxation strength, α and γ the shape parameters, and τ the relaxation time. For the β-process we used the symmetric Cole-Cole (CC) function [13], i.e. $\gamma = 1$ in Eq. 1. Of specific interest for this study is the low and high frequency power law exponents, which we denote a and b, respectively. These are directly obtained from the HN-equation as $a = \alpha$ and $b = \alpha\gamma$.

In Williams ansatz[14] the merging of the two processes in the time domain is described by

$$\varphi(t) = f\varphi_\alpha(t) + (1-f)\varphi_\alpha(t)\varphi_\beta(t). \quad (2)$$

In Eq. 2 the second term of the right hand side is the effective process β_{eff}. For the frequency dependent permittivity the multiplication of the processes in time domain corresponds to a complex convolution, for which we have previously published a calculation scheme [7]. Similarly as for superposition analysis we use the HN function and the CC function for the α- and β-relaxation, respectively. The difference is that we assume that the underlying β-relaxation is a CC function, which results in an effective β-relaxation distinctly different from the simple CC expression [7].

RESULTS

In Fig. 1 the dielectric response at 160 K of bulk PC, PC/PMMA mixtures and PC in clays are shown. The main feature for bulk PC is the α-relaxation. However, at high frequencies there is a transition to another power law regime as compared with an extrapolation of the α loss peak (see dotted line in Fig. 1). This extra contribution is the so called excess wing. For PC dissolved in PMMA the basic features remain similar with a main α-relaxation and an extra contribution at high frequencies. However, the α-relaxation becomes systematically slower and broader with increasing polymer concentration. As a consequence the main loss peak moves towards lower frequencies and the extra contribution becomes easier to characterize. With increasing polymer concentration the extra contribution systematically transforms from a wing to a shoulder. The obtained shape parameters have been extensively discussed in previous publications [6, 7]. One important finding is that the amplitude of the α-relaxation *decreases* while the amplitude of the

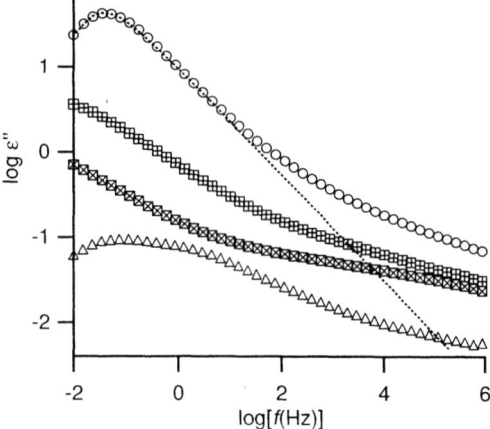

FIGURE 1. Imaginary part of the dielectric response of bulk PC (circles), PC dissolved in PMMA (squares) and PC in clay (triangles). All data are obtained at 160 K. Two concentrations of PC/PMMA samples are shown; squares with + corresponds to 70% PC and squares with × corresponds to 50 % PC. The dotted line is an extrapolation of the high frequency behavior of the main loss peak of bulk PC.

β process *increases* with increasing temperature. This in agreement with general observations of the glass formers with a well resolved β process. We also observe that both shape parameters of the α-relaxation, a and b, decreases with decreasing temperature and increasing polymer concentration.

For PC confined between 2D clay layers the main loss peak appears at the same frequency as for bulk PC, but it has a bimodal feature. We note that the shape of this main loss peak is almost completely independent of temperature. Since the slowest peak frequency of PC between clay layers perfectly match that of bulk PC for all investigated temperatures we identify this with the α-relaxation of PC. There is also a third contribution visible at high frequencies.

To further examine the shape of the relaxation pattern we have also calculated the logarithmic derivative, which is shown in Fig. 2. First we note that the α-peak becomes slower and broader with increasing polymer concentration, while the derivative for the clay sample reflects the bimodal features of the main loss peak. We also observe that the derivatives for all the samples tend towards the same value, ~ -0.2, at high frequencies.

The difficulty with a superposition analysis of bulk PC is to extract the features of the secondary peak, which is submerged into the main loss peak. In Fig. 3 an Arrhenius plot of the obtained inverse of the peak frequencies is shown. In agreement with previous observations we observe a very strong temperature dependence of the

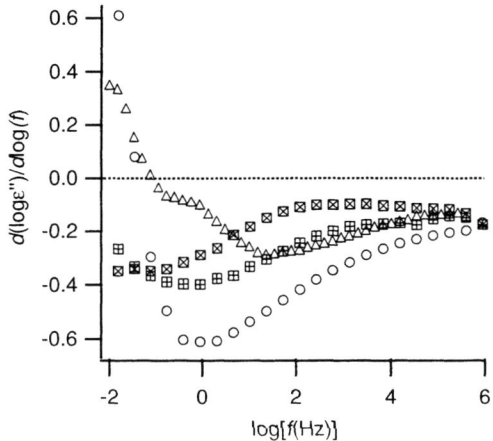

FIGURE 2. Logarithmic derivative of the imaginary part of the dielectric response for the data presented in Fig. 1. The symbols and the temperature are the same as in Fig. 1

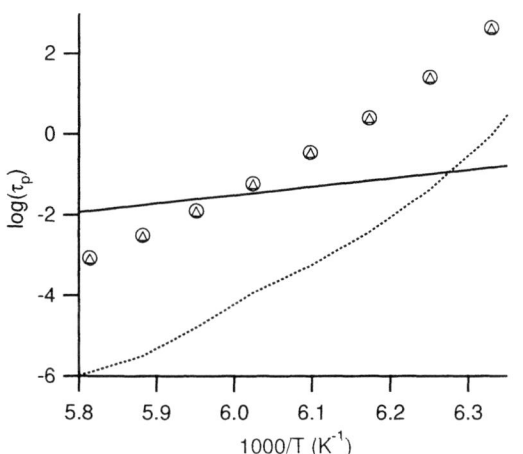

FIGURE 3. Arrhenius plot for the relaxation times obtained for bulk PC. The dotted and the full lines are the relaxation times for the β-process obtained from superposition analysis and with Williams ansatz, respectively. Similarly, the triangles and the circles are the relaxation times for the α-process using superposition and Williams ansatz, respectively.

secondary peak [4, 8]. In Fig. 3 we have included the results of an analysis following Williams ansatz for bulk PC. It is reassuring to note that the α-relaxation times are identical for the two procedures. In the same way as with the superposition analysis it is difficult to extract the β-relaxation parameters since it is submerged into the α-peak. However, a systematic investigation of the PC/PMMA mixtures [7] implies that the β-process has the Arrhenius behavior shown in Fig. 3. This behavior could then be used to describe bulk PC provided Williams ansatz is used. We would like to emphasize that the number of free parameters is equal in the two procedures and that the quality of the fits are equally good.

DISCUSSION

One interesting observation is that PC crystallize in all the investigated geometries, except for samples with high polymer concentrations. The crystallization is observed as a sharp decrease in the loss peak and in all systems it occurs just around 180 K, in agreement with previous observations for bulk PC [3]. Thus there are distinct similarities between the bulk behavior and the confined systems also concerning the tendency to crystallize.

Concerning the α-relaxation of PC in the polymer solutions the main influence on the dynamics is an increasingly slower and broader process for higher polymer concentrations[6]. We attribute this slowing down to a dynamical coupling between the solvent and the slower polymer matrix. This also implies more heterogenous surroundings of the PC molecules, which results in broader loss peaks in consistency with our experimental observations. Another interesting observation for the α-relaxation is that the high frequency parameter, b, extrapolates towards unity at the same temperature, $\sim$220 K, for almost all investigated samples. Since b equal to unity corresponds to a Debye-like process, at least at high frequencies, this can be interpreted as the onset of cooperativity. Our experimental data therefore suggest that the onset of cooperativity occurs around 220 K, which, within experimental uncertainty, is the melting temperature of bulk PC.

The shape of the main peak of PC in clay layers is more complex. It has a bimodal signature, with one of the peaks being practically identical to the α-relaxation of bulk PC. This implies that there exists a bulk-like relaxation also in the clay samples. However, the relaxation strength is much smaller than in bulk implying that the confined α-process is less efficient. The additional feature of the main peak is the extra shoulder roughly one decade above the peak frequency for all temperatures. This shoulder can either be due to a second peak or it may be an intrinsic feature of PC in the clay confinement. We have previously shown that this secondary peak has many features in common with a submerged secondary peak of bulk PC obtained via a superposition analysis [8]. However, the shape of the total loss peak remains remarkably similar for all temperatures, which could imply that these peaks has a common origin. An alternative interpretation is therefore that origin of the whole main loss peak is due to a modified α-relaxation. The excess wing could then be connected to another contribution visible at the lowest temperatures and the highest frequencies. This idea is supported by Fig. 2 where we show that the power law exponent of the clay sample at high frequencies is similar to that of the excess wing of bulk PC.

One very important conclusion is that we can obtain completely different β-relaxation times depending on the curve-fitting procedure as shown in Fig. 3. A superposition analysis yields a very strong temperature dependence of the secondary process, which is not the case for normal Johari-Goldstein β-relaxations. However, a simple addition of the two loss peak is valid only if the two processes are completely separated, either in time or in space. For bulk PC the two loss peaks occurs at the same time scales and it is generally believed that the same molecules participate in the α- and β-relaxations. Therefore a simple addition of the two processes is questionable. An attempt to include the basic feature of the correlation between the processes has been given by Williams [14]. In a simplified approach the processes are assumed to be independent, which certainly is a oversimplification for the α- and β-processes. However, this approach is still more correct than a pure superposition and it also has been argued that the effects of the neglected cross terms is quite small [14]. It is therefore our opinion that it is more physically sound to use the Williams ansatz than a simple superposition. In this context it is interesting to note that using Williams ansatz we can restore an Arrhenius behavior similar to that of normal slow β-relaxations. This implies that the excess wing in PC can be explained as a normal Johari-Goldstein relaxation submerged into the main α peak.

CONCLUSIONS

We have investigated the dielectric response of propylene carbonate in bulk and various confinement. We show that the cooperative α-relaxation is influenced by spatial restrictions. Generally, the relaxations becomes broader, which we attribute to a more heterogenous dynamical surrounding of the PC molecules. The bulk and solution data were analyzed using two different approaches, superposition and Williams ansatz, yielding conflicting results. The analysis based on Williams ansatz shows that the excess wing in PC can in a consistent manner be described by a secondary relaxation submerged in the α-relaxation.

ACKNOWLEDGMENTS

Financial support from the Swedish Research Council is gratefully acknowledged.

REFERENCES

1. Johari, G. P., and Goldstein, M., *J. Chem. Phys.*, **53**, 2372–2388 (1970).
2. Schneider, U., Brand, R., Lunkenheimer, P., and Loidl, A., *Phys. Rev. Lett.*, **84**, 5560–5563 (2000).
3. Schneider, U., Lunkenheimer, P., Brand, R., and Loidl, A., *Phys. Rev. E*, **59**, 6924–6936 (1999).
4. Ngai, K. L., Lunkenheimer, P., Leon, C., Schneider, U., Brand, R., and Loidl, A., *J. Chem. Phys.*, **115**, 1405–1413 (2001).
5. Mattsson, J., Bergman, R., Jacobsson, P., and Börjesson, L., *Phys. Rev. Lett.*, **90**, 075702 (2003).
6. Svanberg, C., Bergman, R., Jacobsson, P., and Börjesson, L., *Phys. Rev. B*, **66**, 054304 (2002).
7. Svanberg, C., Bergman, R., and Jacobsson, P., *Europhys. Lett.*, **64**, 358-363 (2003).
8. Bergman, R., Mattsson, J., Svanberg, C., Schwartz, G. A., and Swenson, J., *Europhys. Lett.* **64**, 675-681 (2003).
9. Svanberg, C., Bergman, R., Börjesson, L., and Jacobsson, P., *J. Phys. IV France*, **10 (7)**, 313–316 (2000).
10. Skipper, N. T., Soper, A. K., and McConnell, J. D. C., *J. Chem. Phys.*, **94**, 5751–5760 (1991).
11. Havriliak, S., and Negami, S., *Polymer*, **8**, 161–210 (1967).
12. Havriliak, S., and Havriliak, S. J., *J. Non-Cryst. Sol.*, **172-174**, 297–310 (1994).
13. Cole, K. S., and Cole, R. H., *J. Chem. Phys.*, **9**, 341–351 (1941).
14. Williams, G., *Adv. Polymer Sci.*, **33**, 60–92 (1979).

Different routes to an understanding of the excess wing in the dielectric loss of glass-formers

R. Bergman[*], J. Mattsson[*,†] and C. Svanberg[*]

[*]*Department of Applied Physics, Chalmers University of Technology, SE-41296 Göteborg, Sweden*
[†]*Department of Physics and DEAS, Harvard University, Cambridge, MA 02138, USA*

Abstract.
Results are presented from three separate studies aimed at elucidating the nature of the much debated excess wing in the dielectric response of glass-formers. The work includes studies of *i)* a range of glass-formers subject to geometrical confinement, *ii)* an archetypal wing-glass-former within a polymer matrix and *iii)* an oligomer system with systematic variations of the chain-length. The results suggest that the excess wing should be attributed to a relaxation process separate from, and thus not a feature of, the α relaxation. We also found that this underlying secondary relaxation has an Arrhenius temperature dependence of its relaxation time.

INTRODUCTION

The main feature of glassy dynamics is the rapid slowing down of the main (α-) relaxation process when temperature is lowered towards the glass transition temperature, T_g. Explaining the non-Arrhenius and non-Debye nature of this process remains an outstanding challenge in condensed matter physics. In addition to the main α-relaxation, the dielectric response of most glass-formers includes a faster secondary β-relaxation. This so called Johari-Goldstein β-relaxation has been suggested to be a general feature of the glass transition [1]. Some glass-formers, however, lack a clear β-relaxation loss-peak and instead show a change of slope at high frequencies in a double logarithmic plot of the dielectric loss. Thus a crossover to a second power-law behaviour is observed. The origin of this so called "excess wing" is of large importance for an understanding of the glass transition in general. Two opposing scenarios have been suggested as explanations for the excess wing: *i)* it is an intrinsic feature of the α-relaxation loss peak or *ii)* it is due to a separate secondary relaxation that is partly hidden beneath the high-frequency power-law contribution of the α loss.

We will here present results from dielectric relaxation studies aimed specifically at providing an increased understanding of the origin of the excess wing. These studies are reported separately and in more detail elsewhere [2–4].

EXPERIMENTAL

The glass formers studied were propylene carbonate (PC), propylene glycol (PG), dipropylene glycol (2-PG), 3-fluoroaniline (3-FA) and a homologous series of propylene glycol based dimethyl ethers, $CH_3 - O - [CH_2 - CH(CH_3) - O]_n - CH_3$. The effects of confinement on the dielectric response of the PC, PG, 2-PG and 3-FA were studied by confining the liquids within the inter-platelet quasi-2D-galleries of a Vermiculite clay. The distance between the platelets are $\sim$ 3-5 Å. PC was also studied in mixtures with polymethylmethacrylate (PMMA). The studied concentrations were in the range 50 - 100 % PC. Finally the effect of chain length was studied in the above mentioned series of oligomers with n=1, 2, 3 and 7. The sample preparation is described in the respective original publications [2–4]. In all three studies the samples were placed between the gold plated electrodes (20 mm in diameter) of a high resolution and broadband (10^{-2}- 10^7 Hz) dielectric spectrometer (Novocontrol Alpha). The complex dielectric permittivity, $\varepsilon^*(f) = \varepsilon'(f) - i\varepsilon''(f)$, was recorded isothermally every second or third degree in the relevant temperature ranges with a temperature stability better than 0.1 K.

RESULTS AND DISCUSSION

In the present studies the $\varepsilon''(f)$ data were analyzed using both a simple superposition ansatz and the more sophisticated so called Williams ansatz [5]. The β-relaxations were described with the symmetric Cole-Cole expression

[6]. For the α-relaxations we used either the Havriliak-Negami equation [7] or a recently proposed equation [8] that is able to parametrize the asymmetric broadening and also gives a good frequency representation of the Kohlrausch-Williams-Watts (KWW) equation [9, 10]. The analyses used in the different studies are described in more detail in the respective recent publications [2–4]. In the following we briefly present and discuss the main results from the analyses.

From the study on a number of molecular glass-formers, including systems displaying both β relaxations (2-PG and 3-FA) and excess wings (PC and PG) respectively, we found that the α and β processes are differently affected by geometrical confinement [3]. When the molecular movements of glass-formers are restricted to the thin quasi-2D galleries of a Na-vermiculite clay, there is a relative enhencement of the magnitude of the β-relaxation loss peak compared with the one for the α relaxation. For glass-formers exhibiting excess wings, the dielectric loss becomes clearly bimodal when subject to the same confinement, see inset of fig.1. It was demonstrated that the dielectric loss spectra of confined and bulk samples can be described within the same framework, the only major differerence being the amplitudes of the different processes. From the performed simple superposition analysis the obtained relaxation time for the secondary (wing) process was found to be non-Arrhenius, see fig. 1. However, for the bulk samples exhibiting an excess wing the data could also be described using the Williams' ansatz and an Arrhenius temperature dependence for the secondary process responsible for the wing. This issue will be discussed further below.

The confinement induced relative enhancement of the secondary process thus seems to be a general effect. From the result that the wing contribution is affected in the same way as the β process it would be natural to conclude that the excess wing has to be due to the β process or at least a process separate from the α relaxation. However, the response of the confined liquid is rather weak and since the two processes are so close in time a quantitative analysis becomes difficult. It is hence not clear from this study if the high frequency part of the bimodal response is due to a "normal" β relaxation.

In order to facilitate a thorough investigation of the α and β relaxations, it is thus desirable to separate the timescales of the two relaxations. One way to do this is to slow down the primary relaxation of the glass-former by mixing it with a slower relaxing component, such as a polymer. By varying the concentration of PC in PMMA we found that the position of the PC α loss peak systematically moves relative to the corresponding β-relaxation peak, which stays largely unchanged, see fig. 2. The shift of the α loss towards lower frequencies is directly related to the increasing constraint formed by the surrounding PMMA matrix for higher polymer concentrations. It was

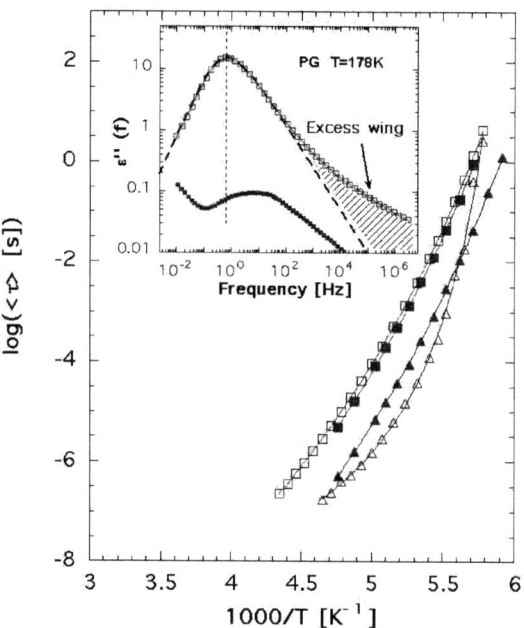

FIGURE 1. Relaxation times of PG in bulk (open symbols) and when confined (filled symbols) in Na-vermiculite clay. The α- and secondary-relaxation times are marked with squares and triangles, respectively. The inset shows the dielectric loss of PG in bulk and when confined at the same temperature, T = 178 K. For the bulk sample the excess wing is clearly visible as an extra contribution compared to the KWW-fit (dashed line) at the high frequency part of the spectra (shaded). For PG in the clay, apart from the general decrease of the signal, a slight shift to higher frequencies and a bimodal shape of the peak is observed.

shown [4] that for increasing PMMA concentration the excess wing clearly observed in pure PC gradually develops into a shoulder as the two relaxation processes become more separated in time, see inset of fig. 2. It is reassuring to note that a similar separation of time scales for the α and β relaxations were observed also for toluene, which have a well separated β relaxation, when mixed with polystyrene [11]. In general, this result, together with the studies by Schneider et al. [12] where the time scales of the α and β relaxations were separated through annealing, provide strong evidence for the excess wing originating in a partly hidden secondary relaxation.

Further evidence concerning the origin of the excess wing comes from chain-length studies. Results from two series of molecular glass-formers [13, 14] have shown that the wing develops into a shoulder when the chain-length increases. However, the analysis in these studies is complicated by the fact that a chain-length dependent hydrogen-bonding strongly affects the dynamics. By the use of the dimethyl ethers of propylene glycol oligomers,

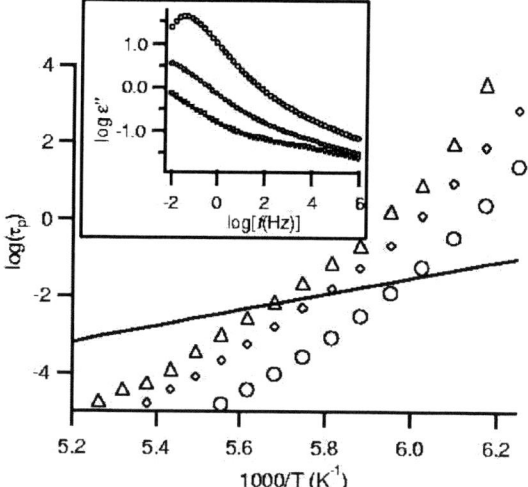

FIGURE 2. α and β relaxation times of PC/PMMA mixtures. The concentration of PC is 100 % (circles), 70 % (diamonds) and 50 % (triangles). The inset shows the normalized dielectric loss at T = 160 K for the same concentrations. Pure PC exhibit a clear excess wing while for the sample with most PMMA the excess contribution shows a clear shoulder. For all samples the dielectric response can be described using Williams ansatz, see text, and an underlying β-relaxation with an Arrhenius temperature dependence of its relaxation time (full line).

we were able to strongly reduce intermolecular interactions normally caused by the hydroxyl end-groups of the glycols [2]. This choice of system thus facilitated detailed studies of the relaxational behaviour as a function of chain length. In particular, we found that when the effect of hydrogen-bonding was removed the fragility becomes practically independent of chain-length so that all α-relaxations coincide in a T_g-scaled Arrhenius plot (fig. 3). We also found, as shown in fig. 3, that for decreasing oligomer chain-lengths, the merging of the α and β relaxations systematically move to lower temperatures and concurrently to longer relaxation times [2]. The result is a relaxation scenario gradually changing from one with clearly separated α and β loss peaks for the longer oligomers to one lacking a separate β peak, but instead displaying an excess wing for the monomer, see inset of fig. 3. In conclusion, the results from all three studies strongly suggest that the excess wing, at least in the investigated systems, is due to a β relaxation.

An important finding from these studies concerns the apparent temperature dependence of the relaxation process responsible for the wing. The usual way to analyze dielectric data of the α- and β-relaxations in glass-formers has been to fit the total spectrum to an addition of the two contributions. However, according to Williams [5] the total normalized relaxation function in the time domain is given by;

$$\phi(t) = A\phi_\alpha(t) + (1-A)\phi_\alpha(t)\phi_\beta(t) \quad (1)$$

where $\phi_\alpha(t)$ and $\phi_\beta(t)$ are the normalized relaxation functions for the α- and β-relaxation processes, respectively. The parameter A is the relative strength of the α-process and the product $\phi_\alpha(t)\phi_\beta(t)$ constitutes what can be called an effective β-relaxation, $\phi_{\beta,eff}(t)$. According to this picture, it is this effective β-relaxation that is seen in the data and in the merging region it is not equal to the underlying or pure β-relaxation. In particular, the main difference in the outcome of analyses based on the two approaches is that when analyzed according to the superposition ansatz one typically obtains a β-relaxation time with a non-Arrhenius temperature dependence, parallelling the α-relaxation time, see fig. 1. This, for β-relaxations unexpected behviour, is removed when the data is analyzed according to Williams' ansatz. It is for instance possible to describe the full relaxation scenario for PC mixed with PMMA for all investigated concentrations with Williams ansatz and an Arrhenius temperature dependence of the β-relaxation time [4]. In the chain-length study it was possible to determine the relaxation time of the "hidden" secondary process by the use of a scaling approach [2]. Also here evidence for an Arrhenius temperature dependence of its relaxation time was found, see dotted line in fig. 3. Thus it seems as, in general, the secondary process responsible for the wing can be described with an Arrhenius temperature dependence of its relaxation time.

Based on the studies described above, it would be natural to conclude that the secondary process responsible for the wing should be identified with a normal Johari-Goldstein β-relaxation. However, recent studies of tri(propylene glycol) under pressure show the coexistence of a wing and a β-relaxation at intermediate pressures [15]. Also this pressure induced wing is undoubtedly due to a secondary relaxation as it develops into a separate peak for increasing pressure. This shows that in relatively simple glass-formers a very complex relaxation scenario with multiple secondary relaxations can be found. The possible universality and relationship with the α-relaxation of these relaxations are questions that should be further investigated in future. Clearly, much work remains before a full picture is obtained of the glass-transition related dynamics.

CONCLUSION

The results from the presented studies clearly show that the excess wing is not a part of the α-relaxation. Rather, it is due to a partly hidden secondary relaxation which has many features in common with normal β-relaxations

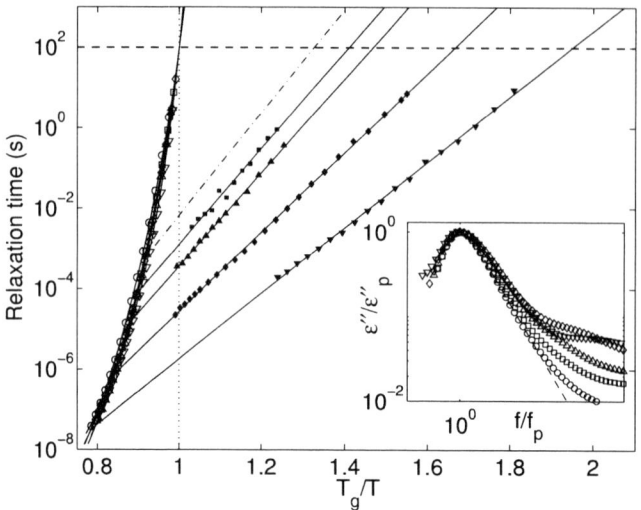

FIGURE 3. α and β relaxation times of the oligomers. The temperature axis is scaled with the T_g of the samples; $T_g(n=1) = 113$ K, $T_g(n=2) = 136$ K, $T_g(n=3) = 149$ K, $T_g(n=7) = 170$ K. The inset shows the normalized dielectric loss at temperatures corresponding to the same peak frequency. The data include the monomer (circles), dimer (squares), trimer (triangles-up), heptamer (diamonds) and polymer (tringles-down). The polymer data is from PPG4000 for which the OH-endgroups have a marginal effect on the dynamics. The dash-dotted line indicates the β relaxation times of the monomer obtained from a scaling analysis [2].

such as an Arrhenius temperature dependence of its relaxation time. The Arrhenius temperature dependence of the relaxation time of the underlying secondary relaxation follws from two different approaches; a scaling analysis for the chain-length study and an analysis according to Williams ansatz for the PC/PMMA study.

ACKNOWLEDGMENTS

Thanks are due to L. Börjesson, P. Jacobsson, J. Swenson and G.A. Schwartz for fruitful discussions and collaboration in the original studies. This work was supported by the Swedish Research Council. JM also acknowledges support from the Wenner-Gren and Hans Werthén foundations.

REFERENCES

1. Johari, G. P., and Goldstein, M., *J.Chem. Phys.*, **53**, 2372–2388 (1970).
2. Mattsson, J., Bergman, R., Jacobsson, P., and Börjesson, L., *Phys. Rev. Lett.*, **90**, 075702 (2003).
3. Bergman, R., Mattsson, J., Svanberg, C., Schwartz, G. A., and Swenson, J., *Europhys. Lett.*, **64**, 675–681 (2003).
4. Svanberg, C., Bergman, R., and Jacobsson, P., *Europhys. Lett.*, **64**, 358–363 (2003).
5. Williams, G., *Adv. Polymer Sci.*, **33**, 60–92 (1979).
6. Cole, K. S., and Cole, R. H., *J. Chem. Phys.*, **9**, 341–351 (1941).
7. Havriliak, S., and Negami, S., *Polymer*, **8**, 161–210 (1967).
8. Bergman, R., *J. Appl. Phys.*, **88**, 1356–1365 (2000).
9. Kohlrausch, R., *Pogg. Ann. Phys.*, **91**, 179–214 (1854).
10. Williams, G., and Watts, D. C., *Trans. Faraday Soc.*, **66**, 80–85 (1970).
11. Matic, A., Mattsson, J., and Bergman, R., *these proceedings* (2003).
12. Schneider, U., Brand, R., Lunkenheimer, P., and Loidl, A., *Phys. Rev. Lett.*, **84**, 5560–5563 (2000).
13. Leon, C., Ngai, K. L., and Roland, C. M., *J. Chem. Phys.*, **110**, 11585–11591 (1999).
14. Paluch, A. D. M., Sillescu, H., and Hinze, G., *Phys. Rev. Lett.*, **88**, 095701 (2002).
15. Casalini, R., and Roland, C. M., *Phys. Rev. Lett.*, **91**, 015702 (2003).

Observation on Surface Change of Fragile Glass: Temperature - Time Dependence Studied by X-Ray Reflectivity

Hiroyuki Kikkawa, Amane Kitahara and Isao Takahashi

Advanced Research Center of Science, School of Science and Technology,
Kwansei Gakuin University (ARCS - KGU), Gakuen 2-1, Sanda, Hyogo 669-1337, Japan

Abstract. The structural change of a fragile glass surface close to the glass transition temperature T_g is studied by using X-ray reflectivity. Measurements were performed on surfaces of maltitol, which is a typical polyalcohol fragile glass with T_g = 320K. Upon both heating and cooling, we find the following features which are also noticed in silicate glass surfaces: (i) On heating, the surface morphology indicates a variation at temperatures below T_g; (ii) A drastic increase in surface roughness occurs at a temperature about 333K on heating, which is 13K higher than T_g; (iii) During the cooling of the sample, formation of a low-density surface layer (3nm at 293K) is observed. Prior to the crystallization, nm - μm sized domains were grown at the surface, which might not be reported for other glasses.

INTRODUCTION

Every substance has the surface. The surface is the region of crossing different phases, materials and symmetries. Thereby, many surfaces are intrinsically unstable and processes occurring at surfaces are extremely significant for enormous physical phenomena in general. For materials forming a glassy state, molecules on the surface should also exhibit intriguing behaviors different from those of bulk, i.e., glass transition, relaxation and crystallization are expected to be strikingly affected at these surfaces. Therefore, an understanding of phenomena related to the glass surfaces helps comprehend the nature of glassy state in a deeper level.

When viscosity (or relaxation time) vs. $1/T$ is plotted, strong glass would show a steady, linear increase on $1/T$, whereas fragile glass shows a much steeper variation on $1/T$ [1,2]. Around the glass transition temperature T_g, fragile glasses vary its characteristics much more drastically than strong glasses. Thus, properties peculiar to the fragility would also be expected on fragile glass surfaces. From the Kauzmann temperature arguments, fragile glasses belong to a more suitable system for clarifying the unsolved problems of glass and glass transition [3,4].

For small angle incidence, X-rays are able to afford surface information, and an optical treatment is possible [5,6]. X-ray reflectivity (XR) is an invaluable tool to study the surface structure quantitatively in a nondestructive manner, e.g., enormous soft matter surfaces of polymers, delicate organic multilayers, and the surfaces of liquids have been investigated [7-9].

In this report, we present XR measurements on the surface of maltitol forming a fragile glass of polyalcohol close to its glass transition temperature T_g = 320K. Upon both heating and cooling, we find the following features some of which are common in silicate glasses: (i) On heating process, the surface morphology indicates a variation at temperatures below T_g under nitrogen atmosphere; (ii) A drastic increase in surface roughness occurs at a temperature about 333K on heating, which is 13K above T_g; (iii) During the cooling of the sample, we observe a low-density surface layer of which thickness is 3nm at 293K. Moreover, nm - μm sized domains are found to form at the surface prior to the crystallization. As far as the authors knowledge, it has not been reported for other glass surfaces.

EXPERIMENTAL DETAILS

Maltitol is produced by hydrogenation of the disaccharide maltose. The molecular formula and molecular weights of maltitol are $C_{12}H_{24}O_{11}$ and 344.3 and its glass transition temperature T_g is 320K. Like as sorbitol, xylitol and erythritol, maltitol is one of the fragile glasses classified as polyalcohol. Since maltitol has an excellent stability against moisture absorption and its glass transition temperature is slightly above the room temperature, it is suitable for the study of fragile glass surface. To obtain flat maltitol glassy surfaces, crystalline powders (Tokyo Kasei Kogyo Co., Ltd.) on a metallic plate (50mmϕ) were heated sufficiently above the melting temperature T_m (= 423K) in a low vacuum, and then quenched to 293K. The root mean square (rms) roughness of these surfaces was revealed to be about 0.4nm by analyzing the specular XR. The value is small enough to study the surface quantitatively.

As the index of refraction in X-ray regime is slightly smaller than unity, total external reflection occurs when the angle of incidence is sufficiently small (< 0.2 degrees for maltitol). Specular XR usually measured by longitudinal scans indicates variation in electron density normal to the sample surface: Film thickness, density, surface and interface roughness can be obtained. Peak profile of specular XR measured by transverse scans generally contains information on lateral surface structures. Wavevector transfer q is defined as $q = k_{out} - k_{in}$, where k_{in} and k_{out} are the wavevectors of the incident and scattered X-rays, respectively. The magnitude of q is equal to $4\pi\sin\theta/\lambda$, where 2θ is the scattering angle and λ is the X-ray wavelength. In the present study, q is decomposed as $q = q_x + q_z$, where q_x is the horizontal component and q_z represents the vertical component [5,6]. Measurements of XR were conducted by using a diffractometer with a rotating-anode X-ray generator, designed for studying free liquid surfaces [10-12]. As an incident beam, CuKα_1 (λ = 0.15405nm) radiation was monochromated and collimated by using a two-bounced Ge(220) crystal and several slits. During the experiments, the maltitol surfaces were kept under a nitrogen gas atmosphere to exclude the moisture. Temperatures were controlled by a heater and temperature controller, ranging from 293K to 373K with a stability better than 0.1K. The heating rate and the cooling rate were 0.6K/min.

RESULTS

Figure 1 shows specular XR peak profiles obtained by transverse scans as a function of q_x. As the temperature raises, the profile shows a distinct broadening at temperatures below T_g (See the XR profile at 313K). Since the profile of specular XR contains information on lateral surface structure, this broadening indicates that large deformation of a lateral surface structure occurs below T_g. Figure 2 displays the XR as a function of q_z obtained by longitudinal scans. Open symbols are observations and solid lines are fit curves. The slope in XR shows a sudden increase at 333K, which is 13K above T_g. This is due to a drastic increase in surface roughness at this temperature. The inset represents the rms roughness of the surface, indicating the noticeable roughening at 333K in heating. Since XR obtained by longitudinal scans includes information on vertical structure near the surface, structural change at the surface along the normal direction starts at above T_g. It contrasts with the surface deformation in lateral direction, which begins below T_g.

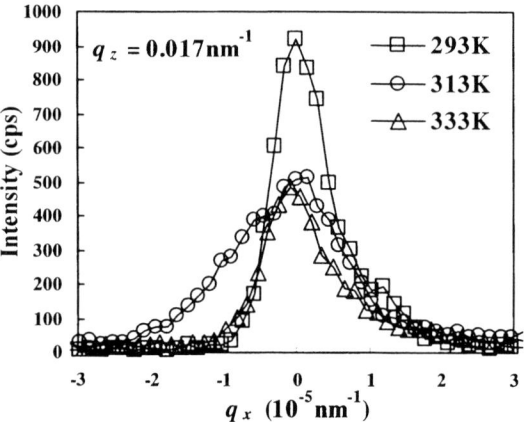

FIGURE 1. Distinct broadening due to lateral deformation is observed below T_g. The lines are guides to eyes. At 333K, sharpening of XR is recognized.

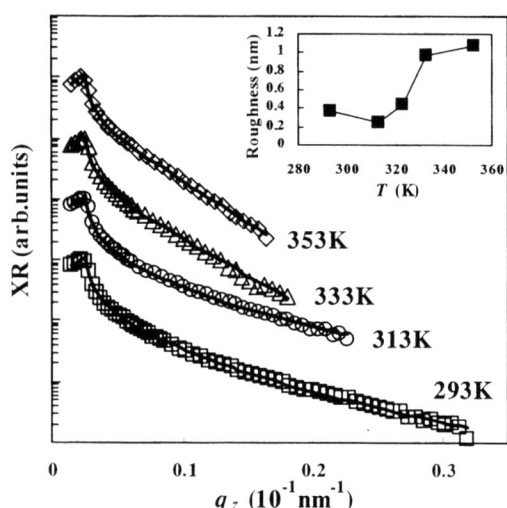

FIGURE 2. Several XRs by longitudinal scans. From the bottom: 293K, 313K, 333K and 353K. Solid lines are fit curves. The XRs are shifted vertically for clarity on the intensity scale. Inset shows the fitted rms surface roughness.

Figure 3 shows $R(q_z)/R_F(q_z)\exp(-q_z^2\sigma^2)$ at 353K in the heating and at 293K in the cooling, where $R(q_z)$ is the observed XR, $R_F(q_z)$ is the Fresnel Reflectivity from an ideally flat surface and σ denotes the rms roughness. A one-layer model almost reproduces the $R(q_z)/R_F(q_z)\exp(-q_z^2\sigma^2)$ at 353K: The dotted line in Fig. 3 is the calculation of the one-layer model. However, the one-layer model failed to reproduce the 293K data; a two-layer model showed a better fit, giving the solid curve in Fig. 3. Such a surface layer is also confirmed at 353K, 333K and 313K in the coolings. Therefore, the present study shows that the surface layer forms in the cooling process. Average thickness of the surface layer was obtained as 3nm, and average density of the surface layer was 94.4 (±0.2) % of the bulk density, indicating that the surface layer has a lower density.

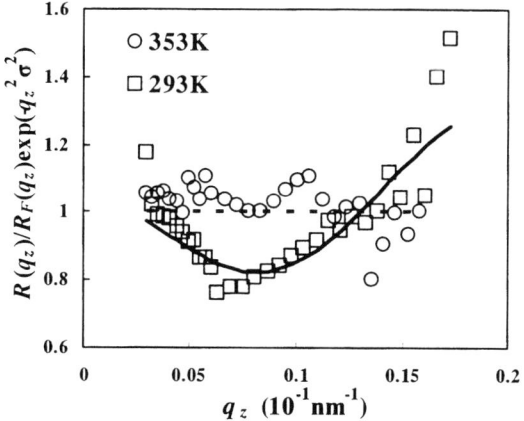

FIGURE 3. The XRs normalized by the Fresnel reflectivity, in which an effect of the rms roughness is included. Open symbols are observation and curves represent the calculation. The XR of 353K is fitted by a one-layer model, whereas the 293K data is fitted by a two-layer model.

In heating process, crystallization happened to be observed for several samples. At first stage of the crystallization, subsidary peaks are seen in the XR profile of transverse scans (Fig. 4). These subsidary peaks can originate from diffraction scattered in various directions due to surface corrugations; the corrugations can be caused by nm - μm sized domains in the surface region. In general, crystallization occurs from the surface as a result of heterogeneous nucleation. For maltitol, crystallization also occurs from the surface. In order to create nucleus for crystallization, molecules in the surface region have to be moved and reoriented. We consider that the multi-domain surface giving the subsidary peaks is the result of this heterogeneous nucleation [12].

We also performed the measurement of XR in air. Here, the broadening in the XR profile was much larger above T_g than that in nitrogen atmosphere. Moreover, a temporal variation was observed at 353K.

Within 5 hours, the width of XR profile varied from 0.96 to 0.074 degrees. The latter is the initial width of XR. Since any temporal variation was not observed for the moisture-free surface (in the nitrogen ambient). We conclude that water significantly affects the structural relaxation (molecular mobility) on the maltitol surface.

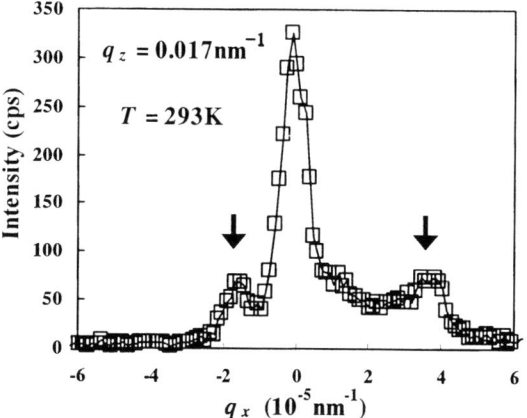

FIGURE 4. Specular XR profile and subsidary peaks. The subsidary peaks indicated by arrows are observed in the early stage of crystallization. The lines are guides to eyes.

DISCUSSION

Broadening of XR profile at 313K (Fig. 1) shows that the surface morphology significantly changes along the lateral direction below T_g, although bulk structure can not vary at this temperature due to the large bulk viscosity. Such an XR profile broadening is explained if low viscosity regions (LVRs) appear in the surface below T_g. However, the LVRs should not be a layer covering the whole surface, because the longitudinal scan of XR (= specular XR), sensitive to the existence of overlayers, does not indicate any noticeable change below T_g (See 313K data in Fig. 2). Thus, it is likely that high viscosity regions (HVRs) of which surface is still flat are surrounded by the LVRs below T_g. When the mean distance between the LVRs is larger than the coherence length of X-ray (in this case it is less than μm), the LVRs would not be recognized as a layer even if they tilt the flat surface of HVRs from the horizontal plane and reduce the area of each HVR. Number of LVRs becomes larger as the temperature becomes higher. At a temperature near T_g, the LVRs would almost cover the maltitol surface. Then, morphological flat surface larger than $\sim\mu$m^2 in area can recover, but the microscopic roughness detected as the rms roughness should significantly increase due to the atomic thermal fluctuations in the thin layer with low viscosity. We consider that the temperature around 333K, 13K higher than T_g, is the one at which the LVRs spread all over the surface: Sharpening of XR profile shown in Fig. 1 (= recovery of the morphological flatness) and sudden decrease in specular XR shown in Fig. 2 (= increase of roughness), both of them can be the evidence of this phenomena. One may call the surface layer as a supercooled 2D quasi-liquid, although we do not give any structural description except for the maximum layer thickness and thermal vibration of outermost atoms: They should be less than 4nm and 1nm, respectively. Such a surface roughening above T_g is also observed for glycerol [13], pyrex glass and soda-lime glass. Figure 5 shows $\log_{10}$(viscosity) vs. T_g/T for these glasses [1,14]. The roughening temperatures above T_g are also indicated by symbols. As far as the data plotted in Fig. 5, the surface roughening temperature seems to lie in the range of $0.7 < T_g/T < 0.95$, corresponding to $5 < \log_{10}$(viscosity) < 11. Measuring the surface roughening temperature of other glasses is important for the sake of confirming the hypothesis that the surface roughening occurs in such a restricted region in the Arrhenius plot.

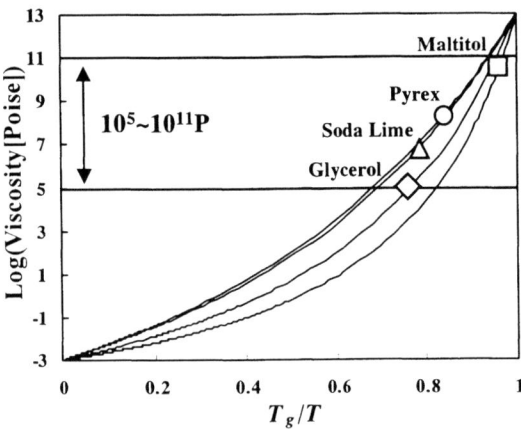

FIGURE 5. Log(viscosity) vs. T_g/T. The temperature and viscosity in which the surface roughening occurs are indicated as symbols.

Since molecules in the LVR have a large mobility above the roughening temperature, relaxation can easily be realized at these surfaces. We consider that this is the reason why the surface layer up to 3nm thickness was observed in the cooling. The relaxed region should be solid-like thus, it has a higher density than that of the glassy state at the same temperature. We think that the relaxed high density region is formed and distributed on the surface as lots of islands with mean height of 3nm.

In conclusions, we observed the atomically flat maltitol surface by XR. Morphological change below T_g and thermal roughening above T_g are related with the growth of LVRs in the surface. Moreover, nucleation of crystalline phase in the heating, relaxed layer formed in the cooling, and temporal variation in air are also observed on the maltitol surface.

We thank professor H. Terauchi for insightful discussions. Part of this study was financially supported by Kwansei Gakuin University (Special Grant 2003).

REFERENCES

1. Böhmer, R., Ngai, K. L., Angell, C. A., and Plazek, D. J., *J. Chem. Phys.* **99**, 4201-4209 (1993).
2. Ediger, M. D., Angell, C. A., and Nagel, S. R., *J. Phys. Chem.* **100**, 13200-13212 (1996).
3. Kauzmann W., *Chem. Rev.* **43**, 219-256 (1948).
4. Angell, C. A., *J. Res. Natl. Inst. Stand. Technol.* **102**, 171-185 (1997).
5. Als-Nielsen, J., and McMorrow, D., *Elements of Modern X-Ray Physics*, New York: Wiley, 2000, pp.61-106.
6. Tolan, M., *X-Ray Scattering from Soft-Matter Thin Films*, Berlin: Springer, 1999, pp. 5-43.
7. Weber, R., Zimmermann, K.-M., Tolan, M., Setettner, J., and Press, W., *Phys. Rev. E* **64**, 061508 (2001).
8. Jensen, T. R., Jensen, M. Ø., Reitzel, N., Balashev, K., Peters, G. H., Kjaer, K., and Bjørnholm, T., *Phys. Rev. Lett.* **28**, 086101 (2003).
9. Braslau, A., Pershan, P. S., Swislow, G., Ocko, B. M., and Als-Nielsen, J., *Phys. Rev. A* **38**, 2457-2470 (1988).
10. Takahashi, I., Ueda, K., Tsukahara, Y., Ichimiya, A., and Harada, J., *J. Phys.: Condens. Matter* **10**, 4489-4497 (1998).
11. Takahashi, I., Tanaka, N., and Doi, S., *J. Appl. Cryst.* **36**, 244-248 (2003).
12. Doi, S., and Takahashi, I., *Philosophical Magazine A* **80**, 1889-1899 (1999).
13. Seydel, T., Tolan, M., Ocko, B. M., Weber, R., DiMasi, E., and Press, W., *Phy. Rev. B* **65**, 1842071-1842077, (2002).
14. Bustin, O., and Descamps, M., *J. Chem. Phys.* **110**, 10982-10992 (1999).

Competition Between Crystallization and Glass-Transition Processes in Binary Amorphous Molecular Systems

Kikujiro Ishii, Miki Murai, Masatsugu Yamamoto, Masaki Takei, and Hideyuki Nakayama

Department of Chemistry, Gakushuin University, 1-5-1 Mejiro, Toshimaku, Tokyo, 171-8588 Japan

Abstract. Crystallization and glass transition phenomena of binary amorphous molecular systems were studied as the function of composition. For chlorobenzene/toluene and chlorobenzene/ethylbenzene systems, it was found that the temperatures of direct crystallization and glass transition from the amorphous state and the temperature of crystallization from supercooled liquid state come across each other in a certain region of composition, where the three processes are considered to compete with each other. Molecular motions related to these processes are discussed on the basis of the structure of crystals that appear as the results of the annealing of the samples.

INTRODUCTION

Amorphous samples prepared by vapor deposition of simple organic compounds on cold substrates crystallize in two types of the manner when the temperature is raised [1-3]. One is the direct crystallization (DC) from the amorphous state, and in the other type of the process, the sample undergoes first the glass transition (GT) and crystallizes finally through the supercooled liquid state (the crystallization of supercooled liquids, hereafter CL). DC is considered to take place when an amorphous sample reaches a temperature where short-range diffusion of individual molecules is possible. On the other hand, GT is considered to take place if the cooperative molecular motion becomes easy at a low temperature before the diffusion of individual molecules occurs.

To clarify the competition between the molecular motions related to DC and GT processes, we have studied the relaxation process of binary amorphous systems consisting of a compound that shows DC and a compound that shows GT respectively from their neat amorphous states. We found for chlorobenzene(CB)/toluene(TL) and CB/ethylbenzene(EB) systems that the characteristic temperatures of DC, GT, and CL come across each other in a certain region of the composition. Thus molecular motions related to these processes compete with each other in the same region. In other words, there is a crossover of the relaxation mechanisms of amorphous molecular systems. In this paper, we summarize these data, and discuss the molecular motions related to the three processes on the basis of the X-ray diffraction data of crystals that appear as the result of the annealing of the samples.

EXPERIMENTAL

To study the behavior of amorphous molecular systems made of simple organic molecules, we employed as the sample-preparation method the vacuum deposition of the vapor onto cold substrate. The apparatus has been described previously [3]. The sample vapor was introduced into the vacuum chamber with the base pressure of about 10^{-7} Pa, and deposited onto a gold-plated copper substrate at 78 K. Since the composition of the deposited sample deviates from that of the source material that is liquid mixture at the room temperature, we determined the composition by comparing the Raman-band intensities of the component molecules in the spectra of the deposited sample. The sample thickness was made to be about 10 μm by monitoring the interference fringe of laser light reflected at the sample surface. Actually, the reflected light is the superposition of the light reflected at the top surface of the sample film and the light reflected at the interface between the organic film and metal substrate. However, we call the monitored light the transmission light hereafter for convenience.

We determined the characteristic temperatures of DC, GT, and CL by employing three methods. Crystallizations were recognized when an abrupt decay of the transmission-light intensity occurred. This is due to the appearance of small crystalline grains in the sample. We also recognized the crystallization by analyzing the change of Raman-band width. That is, Raman bands of the component molecules narrow their width when the sample undergoes crystallization. We found that the crystallization temperature T_c determined by the transmission-light intensity was a little lower (by about 1 K) than that determined by the

Raman-band width. Light scattering is caused by rather a small amount of structural inhomogeneity, while Raman spectra reflect overall ratio between the amounts of molecules in different circumstances. We thus use below the T_c values determined by Raman spectra.

As to the determination of the glass-transition temperature T_g, we first encountered serious problem, since almost no change was observed in X-ray diffraction patterns or Raman spectra. However, we found that the interference pattern of the transmission light recorded during the temperature elevation of the sample is reproducible if the rate of temperature elevation was kept at fixed value. We further found that a turning of the phase change on the interference pattern occur at the glass transition [3]. This enabled us to determine T_g.

In addition to the optical methods described so far, we performed X-ray diffraction measurements on similarly prepared samples during the annealing by temperature elevation. The apparatus was the same as in our previous study [4], but the substrate in this time was the (1 0 0) face of a silicon crystal. A Cu-Kα X-ray source was used along with a graphite monochromator. For monitoring the thickness and state of the sample, a diode laser and a photodiode were equipped on the vacuum chamber.

RESULTS AND DISCUSSION

Diagram Obtained by Optical Methods

For the CB/TL binary systems, we have reported the results of the characteristic temperatures of DC, TG, and CL [5]. We reproduce these in Fig. 1 to compare with the corresponding results for CB/EB systems in Fig. 2. All these experiments were performed with the same rate of temperature elevation, 0.28 K/min. We employ hereafter the notation T'_c for the temperature of CL to distinguish this from T_c for DC. Amorphous neat CB undergoes DC at 109 K, and T_c increases as TL or EB is added in the sample (full circles in Figs. 1 and 2). On the other hand, amorphous neat TL undergoes TG at 117 K and crystallizes from the supercooled-liquid state around 128 K (Fig. 1). Amorphous neat EB undergoes similar processes, T_g being 117 K and T'_c being about 130 K (Fig. 2).

T_g of both TL- and EB-rich systems increases as CB is added (open squares in Figs. 1 and 2). The behavior observed here for CB/EB systems are essentially the same as that observed in the earlier study by Angell et al. [6] in determining the fictive T_g of CB, but the results plotted in Fig. 2 show the change in detail in the most EB-rich region. As to T'_c (open circles in Figs. 1 and 2), the CB/TL and CB/EB systems show behaviors distinct from each other. Although T'_c of CB/TL systems decreases as CB is added to TL (Fig. 1), T'_c of CB/EB systems seems to increase as small amount of CB is added to EB but the inclination turns to decrease if CB concentration is increased further (Fig. 2). This behavior will be discussed below in relation to the structure of crystals appearing by CL.

It is remarkable in Figs.1 and 2 that the three kinds of the characteristic temperatures merge with each other at a certain composition region; around x(TL) = 0.75 for CB/TL systems and around x(EB) = 0.2 for CB/EB systems. These facts imply first that DC and TG compete with each other in this composition region. In other words, molecular motions related to the DC and TG processes show a crossover in the above composition regions. As already mentioned, DC is related to the short-range diffusion of individual molecules, while GT is related cooperative molecular

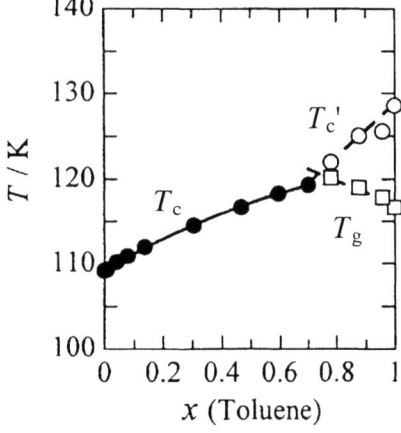

Fig. 1 The direct crystallization temperature T_c (●), glass-transition temperature T_g (□), and crystallization temperature from supercooled liquids T'_c (○) of chlorobenzene(CB)/toluene(TL) systems as the function of toluene mole fraction.

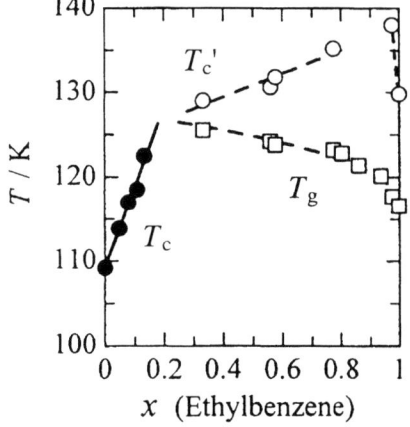

Fig. 2 The crystallization and glass-transition temperatures of chlorobenzene(CB)/ethylbenzene(EB) systems as the function of ethylbenzene mole fraction. The same symbols are used as in Fig. 1.

motions in scales much larger than the size of the component molecules. What does donate the different natures to the molecules with similar structures, CB and TL, and CB and EB? This will be discussed a little later in connection with the X-ray results.

The second remarkable point in Figs. 1 and 2 is that the temperatures of the two types of crystallization processes, T_c and T'_c, merge with each other at the same composition regions as mentioned above. DC and CL have been considered to involve different types of microscopic processes. CL has been studied well, being considered to occur through the seed-formation and crystal-growth processes in the melt. On the other hand, DC is considered to occur by short-range diffusion of molecules in the solid state. However, the mechanism of DC has not been studied extensively. This is emphasized especially for amorphous molecular systems, although some proposal has been given recently [7]. The present results shown in Figs. 1 and 2 imply that there must be some continuity between the mechanisms of DC and CL processes.

As to the overall features of Figs. 1 and 2, it is worth mentioning that these figures have a substantial meaning as phase diagrams, although they are not for phases in thermodynamically stable states. In other words, one can predict on the basis of these diagrams the sample state at particular composition and temperature if the original amorphous sample was prepared with the same condition and if the sample temperature was elevated with the same rate. Since DC, GT, and CL processes are irreversible processes, they are often said to depend greatly on the thermal history of the sample. However, the present results indicate that these processes can be reproduced well if we carefully control the sample conditions.

X-Ray Diffraction

To study further the DC and CL processes of CB/TL and CB/EB systems, we measured X-ray diffraction on binary samples with selected compositions. The measurements were performed raising the sample temperature with a constant rate as before, and X-ray apparatus scanned repeatedly the region of the diffraction pattern where significant Bragg peaks were expected to appear. For all the samples, only a broad diffraction hump was observed initially indicating the amorphousness of the structure. By raising the temperature to a point determined by the composition, several Bragg peaks came out and grew in a narrow temperature range. T_c or T'_c estimated by the middle point of the Bragg peaks' growth were almost consistent with those estimated previously from Raman-band widths. Since the ambiguity in the composition of the deposited samples is considered also to accompany the X-ray experiment, we estimated the sample composition from the change of the trans-

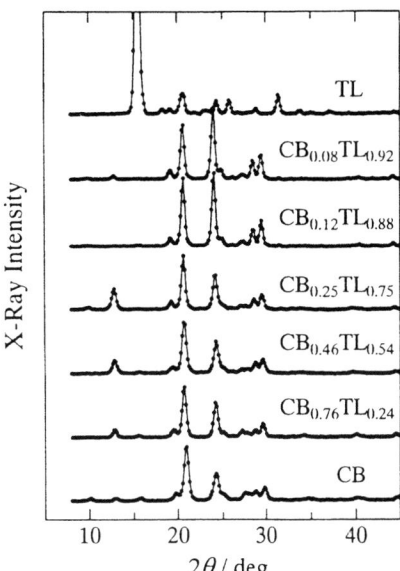

Fig. 3 X-Ray diffraction patterns of CB/TL binary systems after the temperature elevation. The bottom and top patterns are those of neat CB and neat TL, respectively, similarly measured after the temperature elevation. All the patterns, except neat TL (133 K), were measured at 130 K.

mission-light intensity on the basis of T_c, T_g, or T'_c data in Figs. 1 and 2.

Figure 3 displays the selected diffraction patterns measured for CB/TL samples for which the compositions are indicated in the figure. The patterns for the neat CB and TL are also shown in the same figure. All the data in this figure were obtained at 130 K except the data for the neat TL (133 K). At these temperatures, the samples were considered to have completed the crystallization. The patterns for neat CB and TL are almost reproduced using the literature data [8,9], although the temperatures are different from those in the literatures.

One may notice that many diffraction patterns in Fig. 3 are very similar to that of neat CB. Let us check the patterns in increasing order of TL mole fraction. The intensity of the peak around $2\theta = 13$ degrees shows irregular increase in the region of medium TL content. This may reflect some irregular orientation of the crystal grains in the samples. Since the system undergoes DC for the TL mole fraction up to 0.75, the resemblance of the pattern to that of CB implies that TL molecules in the samples in this composition region are considered incorporated into the CB crystal by short-range diffusion in the amorphous state. In addition, it is surprising that even for the samples beyond $x(\text{TL}) = 0.75$, the resultant patterns look like that of the CB crystal. This implies that, in the CL process from a TL-rich melt, the seeds are those of the CB crystal and TL molecules then join to grow the stranger crystals further.

What is the reason that CB and TL make uniform

crystals in a so wide range of TL concentration? Obviously, it is a fact that the volumes of the CB and TL molecules are very similar to each other. By analyzing further the diffraction patterns, we found that the unit-cell volume of the crystal increases a little by the increase of TL mole fraction. This is in harmony with the slightly larger molecular volume of TL than CB [10].

In Fig. 4, we summarize the X-ray diffraction patterns for CB/EB systems. In this case, the displayed patterns are those at 150 K, except for the neat CB sample (130 K). It is the same as in CB/TL systems that the structures of the crystals appeared in the CB/EB systems are similar to that of neat CB, if the EB mole fraction is less than 0.7. Thus the seeds appearing in the EB-rich supercooled liquid are that of the CB crystal as in CB/TL systems. In the pattern for the sample of x(EB) = 0.76, a weak hump around 2θ = 20 degree is seen implying the poor crystallinity of the sample. Further more, for the sample of x(EB) = 0.88, we could not detect Bragg peaks. These facts indicate that EB molecules are not incorporated easily in the CB crystal as compared with TL molecules.

The diffraction pattern for the sample of x(EB) = 0.99 does coincide neither with that of CB nor with that of EB. This reflect the fact that T'_c of CB/EB systems shows a curious change in the most EB rich region (Fig. 2). Further studies are required to clarify the crystallization phenomena of supercooled liquid states of CB/EB systems in the EB-rich region.

Rigid Molecules Like Crystallization

From the experimental results described so far, it is evident that CB molecules have strong tendency to make crystals, while TL and EB molecules have opposite nature and thus show glass transitions if the temperature of the neat amorphous sample is raised. Obviously, the difference comes from the flexibility (degree of freedom for deformation) of the substituent on the phenyl rings in the molecules. Thus EB molecules have stronger nature to undergo glass transition than TL. It is remarkable that the tendency of CB molecules to make crystals is effective even in solutions that are rich with molecules of opposite nature. Such a viewpoint may be important to understand why three lines representing the DC, TG, and CL processes merge with each other in a certain region in the diagram such as Figs. 1 and 2. Namely, the crystallizing tendency of CB molecules seems to dominate the relaxation process in both sides of the crossing point of the three kinds of processes. It is noted finally that the strong resemblance in molecular volume between CB and TL made us to perform a systematic work on complicated phenomena in which crystallization and glass transition processes compete with each other.

REFERENCES

1. Ishii, K., Yoshida, M., Suzuki, K., Sakurai, H., Shimayama, T., and Nakayama, H., *Bull. Chem. Soc. Jpn.* **74**, 435-440 (2001).
2. Ishii, K., Okamura, T., Ishikawa, N., and Nakayama, H., *Chem. Lett.* 52-53 (2001).
3. Ishii, K., Nakayama, H., Okamura, T., Yamamoto, M., and Hosokawa, T., *J. Phys. Chem. B* **107**, 876-881 (2003).
4. Ishii, K., Nakayama, H., Yoshida, T., Usui, H., and Koyama, K., *Bull. Chem. Soc. Jpn.* **69**, 2831-2838 (1996).
5. Murai, M., Nakayama, H., and Ishii, K. *J. Therm. Anal. Cal.* **69**, 953-959 (2002).
6. Angell. C. A., Sare, J. M., and Sare, E. J. *J. Phys. Chem.* **82**, 2622-2629 (1978).
7. Okamoto, N., Oguni, M., and Sagawa, Y. *J. Phys. C* **9**, 9187-9198 (1997).
8. Andre, D., Fourme, R., and Renaud, M. *Acta Cryst. B* **27**, 2371-2380 (1971).
9. Anderson, M., Bosio, L., Bruneaux-Poulle, J., and Fourme, R. *J. Chim. Phys.* **74**, 68-73 (1977).
10. Kitaigorodsky, A. I., *Molecular Crystals and Molecules*, Academic Press, New York, 1973, pp.18-21.

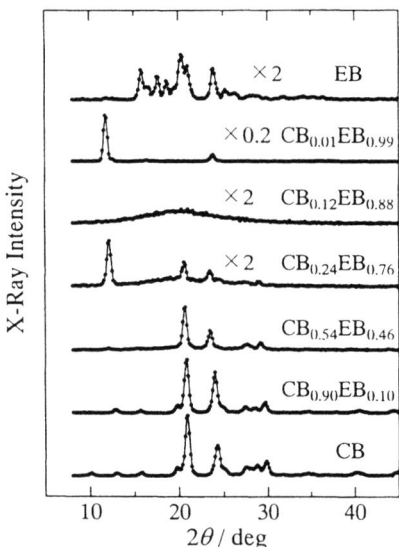

Fig. 4 X-Ray diffraction patterns of CB/EB binary systems after the temperature elevation. The bottom and top patterns are those of neat CB and neat EB, respectively, similarly measured after the temperature elevation. All the patterns, except neat CB (130 K), were measured at 150 K.

Structural Relaxation and Low-energy Excitation in Amorphous Ice and Related Glasses

Osamu Yamamuro

Department of Chemistry, Graduate School of Science, Osaka University, 1-1 Machikaneyama-cho, Toyonaka, Osaka 560-0043, Japan

Abstract. The elastic and inelastic neutron scattering from amorphous ice were measured in a wide momentum transfer ($0.006 < Q < 50 Å^{-1}$) and energy transfer ($E < 100$ meV) ranges. Similar experiments were performed also for the amorphous aqueous solution of methanol (CD_3OH), sulfur hexafluoride (SF_6), and xenon (Xe). All of the amorphous samples were prepared by vapor-deposition at 8 K in a cryostat designed for this experiment. We have obtained several interesting results on (1) the structural relaxation which occurs below the glass transition temperature (135 K), (2) the low-energy excitation below 6 meV, and (3) the effects of the solutes on the above two phenomena. Based on these results, we discuss the relation between the extent of hydrogen-bond formation and the low energy excitations characteristic to amorphous states.

INTRODUCTION

Amorphous ice [1] has been one of the most attractive objects of research for many scientists. Amorphous ice is categorized into two types according to their mass density; one is low density amorphous ice (LDA) with density of ca. 0.94 g cm^{-3} and the other is high density amorphous ice (HDA) with density of ca. 1.17 g cm^{-3}. HDA is prepared by compressing ice I_h up to ca. 1 GPa at 77 K [2]. LDA is formed mainly by three different methods: (1) transition from HDA at 120 K [2], (2) condensation of water droplet of μm size at temperatures below 100 K [3], (3) deposition of water vapor to a substrate at a temperature below 100 K [4]. LDA exhibits a glass transition at 130—135 K and immediately transform to cubic ice I_c [2,3,5]. On further heating, cubic ice irreversibly transformed to hexagonal ice I_h around 200 K [6,7]. In the present study, we prepared LDA by a vapor-deposition method and measured elastic and inelastic neutron scattering from LDA. Part of the present data have been published elsewhere [8,9].

The first purpose of the present study is to investigate the "structural relaxation" of amorphous ice below T_g, which is equivalent to "annealing effect" from the experimental point of view. Up to this work, there has been no systematic structural study on the annealing effect. The second purpose is to investigate the "low energy excitation" which is observed in most of amorphous solids and sometimes called a "boson peak". The density of states of amorphous ice has been measured by neutron scattering works [10,11] but no clear evidence for the excess excitation has been obtained. The third purpose is to investigate the effect of solute on the low-energy excitation in amorphous aqueous solutions. We have chosen methanol (CD_3OH), sulfur hexafluoride (SF_6), and xenon (Xe) as solutes. It is expected that methanol produces defects in the hydrogen-bond network of water while SF_6 and Xe produce local cage-like structure and enhance hydrogen-bond formation as in the case of clathrate hydrate crystals [12]. The formation of the amorphous aqueous solutions has been confirmed by the X-ray [13] and NMR [14] studies. Our final goal is to clarify the relation between the low-energy excitation and the local and intermediate-range structures of the amorphous ice.

EXPERIMENTAL

The sample preparation and the neutron scattering experiments were all performed by using a novel cryostat described elsewhere [8,9,15]. Vapor of D_2O

and H_2O was deposited for the elastic and inelastic scattering experiments, respectively. The temperature of the cold substrate was kept at ca. 8 K, the deposition rate was ca. 10 μm h^{-1}, and the total mass of the deposited sample was ca. 0.5 g corresponding to a thickness of ca. 0.1 mm assuming a uniform thickness. The amorphous aqueous solutions of methanol CD_3OH, SF_6 and Xe were similarly prepared by depositing mixed vapor of the solute and water. The use of CD_3OH was to observe only hydrogen-bonding hydrogen atoms. The mole fraction of CD_3OH was 5 % and 10 %, and those of SF_6 and Xe were both 5.6 %, corresponding to the stoichiometric composition of the type II clathrate hydrate crystal ($M \cdot 17H_2O$).

The total scattering of amorphous ice was measured with a high intensity neutron diffractometer HIT covering the momentum transfer (Q) region of 0.5—50 Å^{-1}. The small-angle scattering data were collected with a small- and wide-angle diffractomter SWAN [16] with a Q range covering over 0.006—20 Å^{-1}. The inelastic neutron scattering (INS) experiment was performed using an inverted geometry spectrometer LAM-D [17] with energy resolution of 0.35 meV and maximum energy transfer of 100 meV. The scattering angles 2θ of this spectrometer are 35° and 85°, corresponding to Q values (at the elastic position) of 0.9 and 2.0 Å^{-1}, respectively. All of the above three spectrometers are of time-of-flight (TOF) type and installed at the pulsed spallation neutron source in the High Energy Accelerator Research Organization (KEK), Tsukuba, Japan.

For all of the experiments, as-deposited amorphous sample was measured first at 50 K where no structural relaxation was observed. The sample was then annealed at 73 K, 96 K, 120 K, 150 K, and 240 K each for 2 h. After each annealing, the sample was cooled back (quenched) to 50 K and the data were collected to investigate the effect of annealing. The duration of the measurements was 5—10 h for the elastic scattering and 20—30 h for the inelastic scattering.

RESULTS AND DISCUSSION

Figure 1 shows the neutron diffraction data (only low-Q region) measured by HIT for the as-deposited D_2O and for the sample annealed at 50, 73, 96, 120 K. The halo patterns indicate that amorphous state was successfully obtained by our vapor-deposition technique. The Bragg peak of the sample annealed at 150 K indicates that the sample crystallized into ice I_c at 150 K. These data show that the first sharp diffraction peak (FSDP) of the amorphous ice shifted

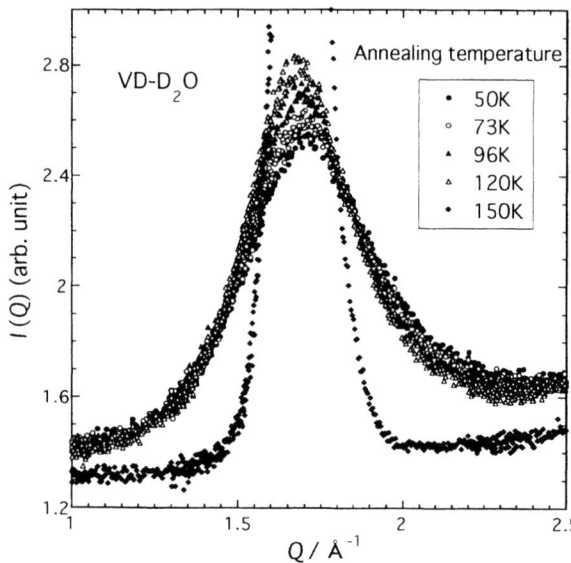

FIGURE 1. The diffraction intensity around FSDP of the vapor-deposited amorphous ice annealed at various temperatures for 2 h.

to the low-Q side with peak sharpening as the annealing temperature became higher. It is of interest that the annealing effect appeared from 73 K which is as low as the half of the T_g of amorphous ice (135 K). The data around the FSDP were fitted well by a Lorentzian $I(Q) = A/[(Q-Q_0)^2+\Delta Q^2]+B$ where Q_0 is a peak position, ΔQ is a HWHM, and A and B are fitting parameters.

The parameters thus determined are plotted together in Fig. 2. For molecular glasses, FSDP position corresponds to the mass density and HWHM to the inverse of the correlation length ($R_c = 2\pi/\Delta Q$). Therefore, these results indicate that volume expansion and ordering of local structure take place together as the structural relaxation proceeds on annealing.

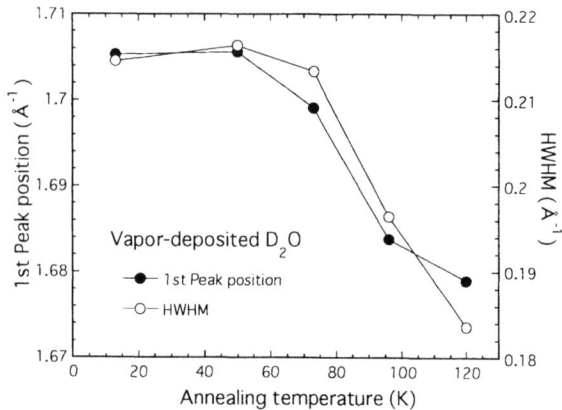

FIGURE 2. The position (left axis) and the HWHM (right axis) of the FSDP as a function of annealing temperature.

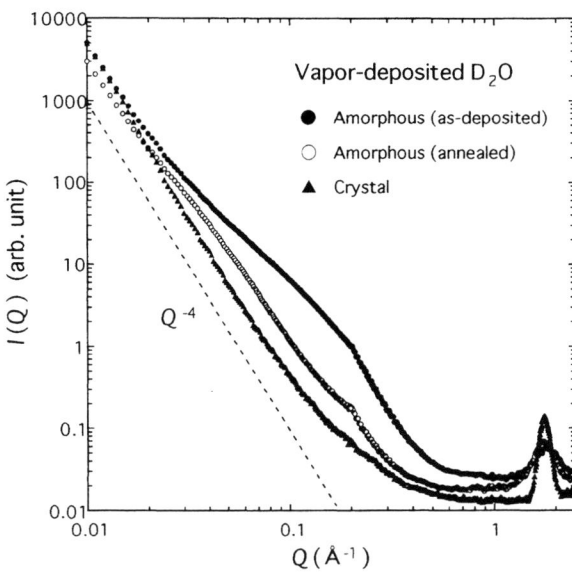

FIGURE 3. Small-angle neutron scattering intensities of the as-deposited and annealed amorphous ices and ice I_c.

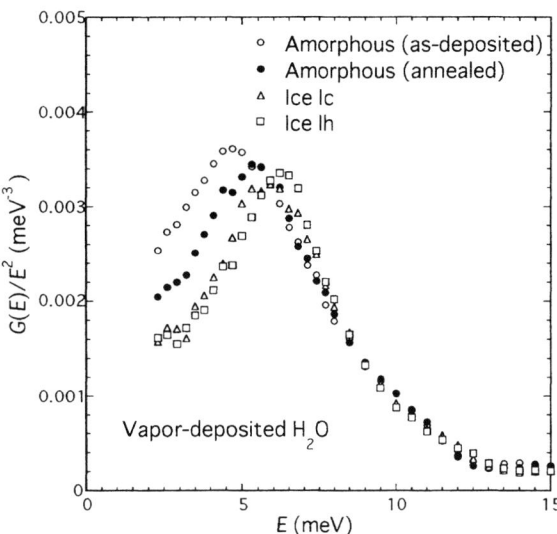

FIGURE 4. Vibrational density of states divided by squared energy. The data are normalized to a water molecule.

Figure 3 shows the SWAN data of the as-deposited amorphous ice, the amorphous ice annealed at 120 K, and ice I_c obtained by annealing the sample at 150 K. A large small-angle scattering was observed in every curve. The intensity at lower-Q region is proportional to Q^{-4}, indicating that the samples had 3-dimensional smooth surfaces of the order of more than 1000 Å. This type of small-angle scattering has been observed also in vapor-deposited CCl_4 and CS_2 [15,18]. It is likely that this is not due to the intrinsic property of the glasses but some macroscopic cracks or voids which appeared in the sample during the vapor-deposition. The intensity data of the as-deposited sample was drastically reduced by the annealing in the Q range between 0.05 and 1 Å^{-1}. The maximum annealing effect appeared at around 0.2 Å^{-1} corresponding to the real space scale of 30 Å. This indicates that some density fluctuation with average size of 30 Å was produced by the vapor-deposition and reduced by the annealing.

The $S(2\theta,E)$ spectra measured by LAM-D are essentially equal to the incoherent dynamic structure factor $S_{inc}(2\theta,E)$ since the observed neutron cross section was dominated by incoherent scattering from hydrogen atoms. Therefore, we calculated the vibrational density of states $G(E)$ from $S_{inc}(2\theta,E)$ by assuming one phonon scattering process. A scale factor to give the density of states per meV and per water molecule was determined from the experimental heat capacity of ice I_h [19]. Further details of this analysis is described elsewhere [8]. Figure 4 shows $G(E)$ thus obtained for the as-deposited amorphous ice, the amorphous ice annealed at 120 K, ice I_c, prepared by annealing at 150 K and ice I_h prepared by annealing at 240 K. $G(E)$ was divided by E^2 to magnify the low energy region and to show the deviation from the Debye model. The $G(E)$ of amorphous ices is larger than that of crystalline ices below 6 meV and they do not obey the Debye model even at 2 meV. The excess contribution below 6 meV was reduced by annealing. All of the data above 6 meV coincide with each other. The excess densities of states $\Delta G(E)$ for the as-deposited and annealed amorphous ices (relative to $G(E)$ of ice) were integrated to give 0.060 and 0.039 degrees of freedom per water molecule, respectively. $\Delta G(E)$ similarly calculated for LDA obtained by Schober et al [11] was 0.019. The previous ambiguous results on the low-energy excitation [10,11] may be due to the deference in preparation and annealing condition of amorphous ice. The magnitude of $\Delta G(E)$ of amorphous ice is much smaller than those of van der Waals glasses (1—2) and comparable with those of covalent-bond network glasses (0.01—0.1).

Figure 5 shows the $S(2\theta,E)$ spectra of the amorphous aqueous solution of $(H_2O)_{1-x}(CD_3OH)_x$ (x = 0.051, 0.106], $(H_2O)_{0.94}(SF_6)_{0.056}$, and $(H_2O)_{0.94}(Xe)_{0.056}$, and pure amorphous ice all measured at 50 K. All of the samples were annealed at 120 K for 2 h before the measurements. The low-energy-excitation ($E < 6$ meV) depends strongly on the solutes; i.e., the excitation was enhanced by methanol but drastically reduced by SF_6 and Xe. This may be because the hydrogen-bond network was broken by methanol due to the substitution of CD_3 groups for H atoms but rather constructed by SF_6 and Xe due to the formation of local cage-like structure stabilized by hydrophobic interaction as in clathrate hydrate crystals.

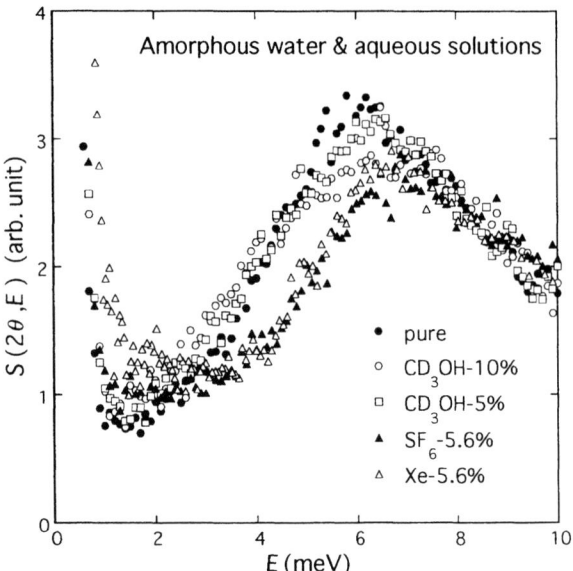

FIGURE 5. Dynamic structure factors of the amorphous ice and amorphous aqueous solutions of CD_3OH, SF_6 and Xe.

CONCLUSION

The present neutron diffraction and small-angle scattering studies have revealed that the annealing below T_g enhances the hydrogen-bond formation and reduces density fluctuation in amorphous ice. The low energy excitation of amorphous ice is reduced by annealing below T_g or adding SF_6 and Xe (forming hydrogen bonds) but increased by adding methanol (breaking hydrogen bonds). This is consistent with the results found in alcohol glasses [20] and LiCl aqueous solution glasses [21]; the low-energy excitation was enhanced as the hydrogen-bond formation is reduced due to the alkyl-group of alcohol molecules and the strong coordination of water molecules to Li ions. From all of the above experimental results, we conclude that the low-energy excitation of amorphous ice may be produced by the disorder, distortion, or defects of hydrogen-bond network.

ACKNOWLEDGMENTS

The author sincerely thanks Prof. T. Matsuo (Osaka University), Mr. N. Tanaka (Osaka University), Dr. N. Onoda-Yamamuro (Tokyo Denki University), Dr. I. Tsukushi (Chiba Institute of Technology), Dr. K. Takeda (Naruto University of Education), Dr. T. Otomo (KEK), Prof. M. Misawa (Niigata University), Dr. K. Ito (Kyoto University), Prof. T. Fukunaga and Dr. K. Shibata (Japan Atomic Energy Research Institute) for their experimental assistance and helpful discussion throughout the present study.

REFERENCES

1. Petrenko, V. F., and Whitworth, R. W., The Other Phases of Ice in *Physics of Ice*, Oxford Univ., Oxford, 1999, pp. 252-286.
2. Mishima, O., Calvert, L. D., and Whalley, E., *Nature* **310**, 393-395 (1984).
3. Mayer, E., *J. Appl. Phys.* **58**, 663-667 (1985).
4. Narten, A. H., Venkatesh, C. G., and Rice, S. A., *J. Chem. Phys.* **64**, 1106-1121 (1976).
5. Sugisaki, M., Suga, H., and Seki, S., *Bull. Chem. Soc. Jpn.* **41**, 2591-2599 (1968).
6. Handa, Y. P., Klug, D. D., and Whalley, E., *J. Chem. Phys.* **84**, 7009-7010 (1986).
7. Yamamuro, O., Oguni, M., Matsuo, T., and Suga, H., *J. Phys. Chem. Solids* **48**, 935-942 (1987).
8. Yamamuro, O., Madokoro, Y., Yamasaki, H., Matsuo, T., Tsukushi, I., and Takeda, K., *J. Chem. Phys.*, **115**, 9808-9814 (2001).
9. Yamamuro, O., Matsuo, T., Tsukushi, I., and Onoda-Yamamuro, N., *Can. J. Phys.*, **81**, 107-114 (2003).
10. Kolesnikov, A. I., Li, J. -C., Parker, S. F., Eccleston, R. S., Hahn, W., and Loong, C. -K., *Phys. Rev. B* **59**, 3569-3578 (1999).
11. Schober, H., Koza, M., Tölle, A., Fujara, F., Angell, C. A., and Böhmer, R., *Physica B* **241-243**, 897-902 (1998).
12. Davidson, D. W., Clathrate Hydrates in *Water: A comprehensive treatise Vol. 2*, edited by F. Franks, Plenum, New York, 1982, pp. 115-234.
13. Hallbrucker, A., and Mayer, E., *J. Chem. Soc. Faraday Trans. 1* **86**, 3785-3792 (1990).
14. Nakayama, H., Omi, H., Eguchi, T., Klug, D. D., Tse, J. S., Ratcliffe, C. I. and Ripmeester, J. A., *The Review of High Pressure Science and Technology* **12**, 10-15 (2002).
15. Yamamuro, O., Matsuo, T., Onoda-Yamamuro, N., Takeda, K., Munemura, H., Tanaka, S., and Misawa, M., *Europhys. Lett.*, **63**, 368-373 (2003).
16. Otomo, T., Furusaka, M., Satoh, S., Ito, S., Adachi, T., Simizu, S., and Takeda, M., *J. Phys. Chem. Solids*, **60**, 1579-1582 (1999).
17. Inoue, K., Kanaya, T., Kiyanagi, Y., Shibata, K., Kaji, K., Ikeda, S., Iwasa, H., and Izumi, Y., *Nucl. Instrum. Methods. Phys. Res.*, A **327**, 433-440 (1993).
18. Yamamuro, O., Yamasaki, H., Madokoro, Y., Matsuo, T., Tsukushi, I., and Otomo, T., *KENS Report-XIII*, 132-133 (2001).
19. Flubacher, P., Leadbetter, A. J., and Morrison, J. A., *J. Chem. Phys.* **33**, 1751-1755 (1960).
20. Yamamuro, O., Harabe, K., Matsuo, T., Takeda, K., Tsukushi, I., and Kanaya, T., *J. Phys.: Cond. Matter*, **12**, 5143-5154 (2000).
21. Madokoro, Y., Yamamuro, O., Yamasaki, H., Tsukushi, I., Kamiyama, T. and Ikeda, S., *J. Chem. Phys.*, **116**, 5673-5679 (2002).

A unified theory of the liquid-glass transition

Toyoyuki Kitamura

Nagasaki Instite of Applied Science, Nagasaki 851-0193, Japan

Abstract. A unified theory of the liquid-glass transition based on the two band model in the harmonic potential approximation is presented. The Kauzmann paradox on the Kauzmann entropy crisis and the Vogel-Tamman-Fulcher (VTF) law on the relaxation times and the transport coefficients are elucidated. The hopping of the particles relates to the Kauzmann entropy crisis and VTF law, the α-relaxation, and the randomness of the harmonic frequencies to the β-relaxation. The Kauzmann entropy yields the gap of the specific heat at the glass transition temperature. The phonon frequency in the short wavelength regime corresponds to the boson peak.

INTRODUCTION

Here we present a unified theory of the liquid-glass transition in the framework of the two-band model proposed for sound and phonons to elucidate the origin of the Kauzmann paradox and the VTF law. We develop the theory within the mean field approximation which is enough to explain the characteristic features in the liquid-glass transition. We are concerned with a system composed of one kind of particles. A particle can be an atom, a molecule, a polymer and even a living cell. In the condensed state such as a liquid and a solid, the particles constitute the spatial structure contrary to the gassy state. A crystal has a periodic structure. In a liquid and a glass, the particles are randomly distributed in space, but the liquid and the glass also have the structure. The structure is represented by the pair distribution function, which can be observed experimentally. The Hamiltonian of the system has the spatially translational and rotational invariance, but in the crystal the symmetry spontaneously breaks down. In the liquid the averaged pair distribution over time and space is formed, but the hopping of particles changes a pair distribution to the other distribution; ergodicity holds. In the glass a pair distribution is freezed at T_g and the symmetry of time and space spontaneously breaks down; ergodicity breaks down. The pair distribution function is a remnant of the structure of the crystal. Here in the liquid state we assume the spontaneous break down of the translational and rotational symmetry.

MODEL

The Ward-Takahashi (WT) relations at finite temperatures associated with the spontaneous breakdown of the spatially translational and rotational symmetry play an essential role in the condensed state. The WT relation requires the existance of the Nambu-Goldstone (NG) bosons, phonons. The phonons determine the thermodynamical character of the new phase. The WT relation also requires that the phonons are the particle-hole pairs composed of the ground state and the first excited states of a particle in the potential; the wave function of a first excited state is also proprtional to the spatial derivative of the ground state. This fact verifies the two band model and the harmonic potential approximation.

The two band model is composed of the two parts:

(i) Random harmonic frequencies.

Particles are randomly distributed. A particle at R_m is in a harmonic potential. We consider the two levels; the ground state $\tilde{\omega}_{m0} = \frac{3}{2}\omega_m$ and the first excited state $\tilde{\omega}_{mi} = \frac{5}{2}\omega_m$, where the first excited state $\tilde{\omega}_{mi}$ consists of the first excited state in an x_i-component, $\omega_{mi} = \frac{3}{2}\omega_m$ and the ground states in the other components. The tilde on $\omega_{m0,i}$ means the three dimensional eigenfrequencies and ω_m is the harmonic frequency. The pair distribution function $g(R)$ breaks the spatially translational invariance. The propagations of up and down transitions between the two levels are phonons.

(ii) Random hopping amplitudes.

A particle at a site R_m feels a harmonic potential made up by the surrounding particles. A surrounding particle at R_n makes a negative potential at R_m the particle site and a potential wall to the particle at R_m. If there is a vacancy at R_n around a particle at R_m, since the particle misses the negative potential at R_m and the wall which should be made by a particle at R_n, the potential of the particle at R_m is lifted and has a saddle point upwards convex in the direction to the vacancy x_i and downwards convex in the direction perpendicular to x_i. The particle

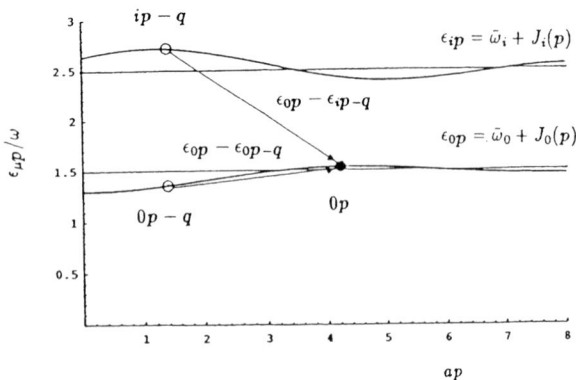

FIGURE 1. The energy dispersion curves of the two bands and, an intra-band and an inter-band elementary excitations. Reprinted from [1] © 2003, with permission from Elsevier.

can hop to the vacancy.

Thus a model Hamiltonian consists of random harmonic frequencies and random hopping amplitudes in the harmonic potential approximation:

$$H = \sum_{m\mu} \hbar \bar{\omega}_{m\mu} b_{m\mu}^\dagger b_{m\mu} + \sum_{mn\mu} \hbar J_{m\mu n\mu} b_{m\mu}^\dagger b_{n\mu}, \quad (1)$$

where $J_{m\mu n\mu}$ is a hopping amplitude and $b_{m\mu}$ an annihilation operator. Taking configurationally averaged harmonic eigenfrequencies and hopping amplitudes, we construct two band picture as shown in Fig.1: the lower and upper bands, $\varepsilon_\mu(p)$ are constructed by the ground state and the first excited state eigenfrequencies, $\bar{\omega}_\mu$ with the band widths corresponding to the Fourier transformed hopping amplitudes $J_\mu(p)$, respectively. $\mu = 0$ for the ground state, $\bar{\omega}_0 = 3\omega/2$ and $\mu = i = 1,2,3$ for the first excited state, $\bar{\omega}_i = 5\omega/2$, where ω is the configurationally averaged harmonic frequency.

$$\varepsilon_{\mu p} = \bar{\omega}_\mu + J_\mu(p). \quad (2)$$

Near the liquid-glass transition, particles move little so that the bandwidths are very narrow. Since the band gap is nearly equal to $\hbar\omega$ which is of the order of the boiling temperature, the majority of particles stay in the lower band. Thus here, we consider the lower temperature approximation: $\beta\hbar\omega \gg 1$, $\beta\hbar|J_\mu| \ll 1$, $\omega\tau \gg 1$, where $\beta = (k_B T)^{-1}$ and τ is a relaxation time.

THE KAUZMANN PARADOX, THE VTF LAW AND BOSON PEAK

The density fluctuations consist of intra-band and inter-band density fluctuations. The modes of intra-band density fluctuations ρ_{0q} and inter-band density fluctuations ρ_{iq}, which correspond to density and current fluctuations in the classical theory, yield sound and phonons, respectively. Hereafter, when we consider the sound, we confine ourselves to the lower band. In the liquid state, the hopping is essential. The hopping causes the intra-band elementary excitations, $\omega_q^0 = (1/N)\sum_p(\varepsilon_{0p} - \varepsilon_{0p-q})$. The correlation function of the intra-band density fluctuations ρ_{0q} yields the dispersion curve of the sound:

$$\omega_{sq}^2 \cong \{1 + \beta V_{00}(q)\}(\omega_q^0)^2 \equiv v_p^2 q^2/(3S(q)), \quad (3)$$

where $\omega_q^0 \cong v_p q/\sqrt{3}$ in the long wavelength regime, $S(q)$ is the static structure factor and v_p is the mean velocity of particles, $v_p = \frac{1}{N}\sum_p \frac{\partial \varepsilon_{0p}}{\partial p}$. The sound velocity is $v_s = v_p/\sqrt{3S(0)}$. $V_{00}(q)$ is an interaction potential between intra-band elementary excitations and the magnitude of $V_{00}(q)$ has the negative minimum value at a reciprocal particle distance $\bar{K}$. The sound instability occurs at a temperature T_0 very close to the Kauzmann temperature T_K at the reciprocal particle distance $\bar{K}$.

$$1 + \beta_0 V_{00}(\bar{K}) = 0. \quad (4)$$

The magnitude of T_0 indicates the fragility.

The entropy of the intra-band density fluctuations ρ_{0q} consists of three parts: the sound, the intra-band fluctuation and the dissipative (diffusion) entropies. The fluctuation entropy has a negative value. The dissipative entropy compensates the fluctuation entropy and yields a local equilibrium. The sound and the fluctuation entropy near the reciprocal particle distance $\bar{K}$ yields the entropy crisis at T_0:

$$S_K \cong -\frac{N_0 k_B}{2} \frac{T_0}{T - T_0}, \quad (5)$$

where N_0 is the number of states near $q = \bar{K}$. We call S_K the Kauzmann entropy. S_K explains the experimental results.

A hopping of a particle from a site to a vacancy corresponds to a jump from a deep valley to another deep valley in the multi-dimensional configuration space in the energy landscape model (ELM) as the α-relaxation process. The succesive hoppings constitute a configuration space. The hopping probability is proportional to the configuration number, which is e^{S_K/k_B} from the Einstein relation. The hopping probability of a particle is proportional to the hopping amplitude:

$$J = e^{\frac{zS_K}{Nk_B}} = e^{-\frac{E}{T-T_0}} \quad (6)$$

where $E = zN_0T_0/2N$ and z is of the order of the number of the surrounding particles. This equation is just the VTF law. The Adam-Gibbs formula incorporating to ELM includes the inverse form of the configurational entropy corresponding to S_K in the exponent. The present

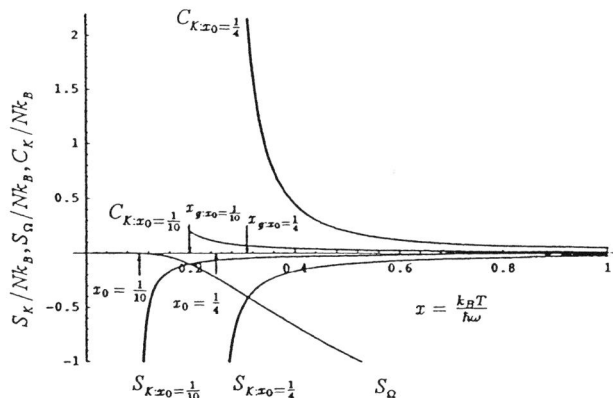

FIGURE 2. The Kauzmann entropy, the inter-band fluctuation entropy and the specific heat due to the Kauzmann entropy. Reprinted from [1] © 2003, with permission from Elsevier.

theory can resolve some conflicting points in the Adam-Gibbs formula.

In the glassy state, the propagation of the up and down transition, phonon is essential. The correlation functions of the inter-band density fluctuations ρ_{iq} yield the phonon dispersion curves:

$$\omega_{\lambda q} \cong \frac{1}{M} \int d^3 R g(R)(1 - e^{i q \cdot R}) V_\lambda(R), \quad (7)$$

where V_λ is the interaction potential between inter-band elementary excitations, $\omega_{iq}^0 \cong (1/N) \sum_p (\varepsilon_{ip} - \varepsilon_{0p-q}) \cong \omega$ and M is the mass of a particle. If we take the z-Cartesian coordinate as the direction of q, the z-component of the phonon corresponds to a logitudinal mode; $\lambda = l$ and, x and y-components correspond to transverse modes; $\lambda = t$. In the short wavelength limit,

$$\lim_{q \to \infty} \omega_{\lambda q} = \omega. \quad (8)$$

This ω corresponds to the boson peak.

In the liquid state, the hopping yields the bands to $\varepsilon_{\mu p}$ so that the elementary excitations $\omega_{\lambda q}^0$ constitutes a continuum. When phonons merge into the continuum, the phonons have the life times, which smear the structure of phonons. Thus the life time of phonons is intrinsic in the liquid state.

The entropy of the inter-band density fluctuations ρ_{iq} also consists of three parts: the phonon, the inter-band fluctuation and the dissipative (viscosity) entropies. The fluctuation entropy yields $S_\Omega = -\frac{1}{T} \frac{3Nh\omega}{e^{\beta h\omega}-1}$. The dissipative entropy compensates the fluctuation entropy and yields a local equilibrium.

A propagation of an up and down transition from a site to a surrounding site corresponds to a jump from a shallow valley to another shallow valley in the multi-dimensional configuration space in ELM as the β-relaxation process. The succesive propagations constitute another configuration space different from hopping. The probability of the magnitude of the randomness of harmonic frequencies is proportional to the configuration number, e^{S_Ω/k_B}. The magnitude of randomness of harmonic frequencies of a particle is proportional to

$$\Omega = e^{\frac{iS_\Omega}{Nk_B}} = \exp\{-\frac{3z\beta\hbar\omega}{e^{\beta\hbar\omega}-1}\}. \quad (9)$$

The intra-band fluctuation entropy S_Ω crossovers the Kauzmann entropy S_K above the temperature T_0. The intra-band fluctuation entropy S_Ω defends from the entropy crisis. Thus the crossover temperature is just the glass transition temperature T_g:

$$S_K = S_\Omega|_{T=T_g}. \quad (10)$$

Sound is a collision wave essential in a fluid, while phonons are elastic waves essential in a solid. The glass transion is a sort of dynamical transition.

The Kauzmann entropy reflects the specific heat. The magnitude of T_0 indicates the fragility of the liquid. Since a fragile liquid releases more excess entropy than a strong one, the specific heat of the fragile is larger than that of the strong one. This fact explains the experimental results shown in Fig.2. The specific heat due to the remaining entopy of sound yields $k_B(N - N_0)$ around the glass transition temperature. This latent heat is released near 0 K as the linear dependent specific heat. This phenomenon is well-known as the specific heat anormaly, which comes from the band width of the lower particle energy, $\hbar\tau_0^{-1} \sim 1K$ due to the uncertainty of the life time τ_0.

THE RELAXATION TIMES AND THE TRANSPORT COEFFICIENTS

The dissipative processes are constructed by elementary scattering processes due to simultaneously scattering processes of two particles in μ- and ν-bands by the same random potentials; the interaction Hamiltonian is given by

$$H_I = \sum_{m\mu} \hbar(\bar{\omega}_{m\mu} - \bar{\omega}_\mu) b_{m\mu}^\dagger b_{m\mu}$$
$$+ \sum_{mn\mu} \hbar(J_{m\mu n\mu} - J_\mu(R_m - R_n)) b_{m\mu}^\dagger b_{n\mu}, \quad (11)$$

One elementary scattering process comes from the scattering process due to the random harmonic frequencies at the same site: $\bar{\omega}_{\mu m} - \bar{\omega}_\mu$. The sum of configurationally averaged elementary scattering processes over all the

TABLE 1. The Vogel-Fulcher law of transport coefficients, relaxation times and velocity of modes. $J = \exp\{-\frac{E}{T-T_0}\}$, $\Omega = \exp\{-\frac{3z\beta\hbar\omega}{e^{\beta\hbar\omega}-1}\}$.[a]

	Trasport coefficient				Relaxation time				Velocity			
	D	v_λ	κ_s	κ_λ	τ_0	τ_M	τ_{es}	$\tau_{e\lambda}$	v_s	v_λ	u_s	u_λ
$T_g < T$	J	J^{-1}	-	J^{-1}	J^{-1}	J^{-1}	J^{-1}	J^{-2}	J^{-1}	J	-	J	-
$T < T_g$	J^2	Ω^{-1}	J	Ω^{-1}	Ω^{-1}	Ω^{-1}	Ω^{-1}	J^{-1}	Ω^{-1}	J	-	J	-

[a] Reprinted from [1] © 2003, with permission from Elsevier.

sites yields a correlation function of random harmonic frequencies:

$$U_\omega^{\mu\nu} \cong \frac{1}{N^2} \sum_m \{<\bar{\omega}_{\mu m}\bar{\omega}_{\nu m}>_c - \bar{\omega}_\mu \bar{\omega}_\nu\}, \quad (12)$$

where $<\cdots>_c$ means the configurational average. $U_\omega^{\mu\nu} \propto \Omega^2$ which determines the β-relaxation. Another elementary scattering process comes from the scattering process due to the random hopping magnitudes at the same sites: $J_{m\mu n\mu} - J_\mu(R_m - R_n)$. The sum of configurationally averaged random scattering processes over all the sites yields a correlation function of the random hopping magnitudes:

$$U_J^{\mu\nu} \cong \frac{1}{N} \int d^3 R g(R) e^{-iq\cdot R}$$
$$\{<J_{m\mu n\mu} J_{m\nu n\nu}>_c - J_\mu(R)J_\nu(R)\}_{R=R_m-R_n}, \quad (13)$$

where $U_J^{\mu\nu} \propto J^2$, which determines the α-relaxation. The relaxation time of sound τ_0 is given by

$$\frac{1}{\tau_0} \cong \sqrt{N(U_\omega^{00} + U_J^{00})}. \quad (14)$$

and the relaxation time of phonon is given by the Maxwell relaxation time τ_M, $1/\tau_M \propto 1/\tau_0$.

The dynamical processes including dissipation processes yield:

$$q_0 - \frac{\omega_{\mu q}^2}{q_0 + \frac{i}{\tau_\mu}} = 0, \quad (15)$$

where $\boldsymbol{q} = q\boldsymbol{e}_z$ and in the long wavelength regime, for sound $\mu = 0$: $\omega_{0q}^2 \equiv \omega_{sq}^2 \cong \frac{v_s^2 q^2}{3S(q)} = v_s^2 q^2$ and for phonons $\mu = \lambda$: $\omega_{\lambda q}^2 \cong v_\lambda^2 q^2$, and $\tau_\lambda = \tau_M$. Eq.(15) is just the dynamical equation for the mode coupling theory (MCT). The terms $\omega_{\mu q}^2$ and i/τ_μ in Eq.(15) are replaced by a local frequency of a particle in a cage, $\Omega_q^2 = \frac{q^2 k_B T}{MS(q)}$ and the memory function $M(q_0, \boldsymbol{q})$ in MCT, respectively. In our theory the terms $\omega_{\mu q}^2$ and i/τ_μ are determined self-consistently, v_s, $\tau_0^{-1} \propto J$ and $v_p \cong$ constant, while in MCT Ω_q^2 is not determined self-consistently, but only $M(q_0, \boldsymbol{q})$ is determined self-consistently as a nonlinear term. Thus Eq.(15) correctly leads to the diffusivity, $D = v_s^2 \tau_0 \propto J$ and the viscosity, $v_\lambda/M\rho = \tau_M v_\lambda^2 \propto J^{-1}$.

Including the multi-phonon processes, we can investigate thermal conductivity. The thermal conductivity is calculated by the correlation functions of entropy density fluctuations. The entropy density fluctuations consist of sound density fluctuations and phonon density fluctuations. The random scattering processes due to the random harmonic frequencies and random hopping amplitudes are included by the elementary scattering amplitudes through the bubble diagrams constructing the sound and the phonons. Taking into account the effective interaction Hamiltonians and the scattering processes in the correlation functions of sound and phonon density fluctuations in a similar manner to intra-band and inter-band density fluctuations, we obtain respective entropy fluctuation modes $\bar{\omega}_{sq}$, $\bar{\omega}_{\lambda q}$ at high frequencies and respective thermal conductivities $\kappa_s = C_s \tau_{es} u_s^2$, $\kappa_\lambda = C_\lambda \tau_{e\lambda} u_\lambda^2$ at lower frequencies, where $C_{s,\lambda}$, $\tau_{es,e\lambda}$ and $u_{s,\lambda}$ are respective specific heats, relaxation times and velocities of entropy fluctuation modes.

The VTF law governs the transport coefficients, the relaxation times and the velocities of modes through the hopping rate J. The summary of the results are shown in Table 1.

CONCLUDING REMARKS

The two band model is involved with ELM in Eqs.(6, 9), and MCT in Eq.(15). However, the Adam-Gibbs formula incorporating to ELM includes the inverse form of the configurational entropy corresponding to S_K in the exponent. Eq.(6) resolves this conflicting points. In the dynamical equation of MCT corresponding to Eq.(15), MCT should take into account the local frequency Ω_q^2 self-consistently including the hopping process and the glassy state.

REFERENCES

1. T. Kitamura, *Phys. Rep.* **383**, 1-94 (2003).

Supercooled liquids under shear: A mode-coupling theory approach

Kunimasa Miyazaki*, Ryoichi Yamamoto[†][**] and David R. Reichman*

*Department of Chemistry and Chemical Biology, Harvard University, 12 Oxford Street, Cambridge, MA 02138, U.S.A
[†]Department of Physics, Kyoto University, Kyoto 606-8502, Japan
[**]PRESTO, Japan Science and Technology Agency, 4-1-8 Honcho Kawaguchi, Saitama, Japan.

Abstract. We generalize the mode-coupling theory of supercooled fluids to systems under stationary shear flow. Our starting point is the generalized fluctuating hydrodynamic equations with a convection term. The method is applied to a two dimensional colloidal suspension. The shear rate dependence of the intermediate scattering function and shear viscosity is analyzed. The results show a drastic reduction of the structural relaxation time due to shear and strong shear thinning behavior of the viscosity which are in qualitative agreement with recent simulations. The microscopic theory with minimal assumptions can explain the behavior far beyond the linear response regime.

INTRODUCTION

Many complex fluids such as suspensions, polymer solutions, and granular fluids exhibit very diverse rheological behavior. Shear thinning is among the most-known phenomena. Recently, it was found by experiments[1] and simulations[2] that supercooled liquids near the glass-transition also show strong shear thinning behavior. They have observed that near the transition temperature, the structural relaxation time and the shear viscosity both decrease as $\dot{\gamma}^{-\nu}$, where $\dot{\gamma}$ is the shear rate and ν is an exponent which is less than but close to 1. For such systems driven far from equilibrium, the nonequilibrium parameter $\dot{\gamma}$ is not a small perturbation parameter but plays a role more like an intensive parameter which characterizes the "thermodynamic state" of the system[3]. Such rheological behavior is interesting in its own right, but understanding the dynamics of supercooled liquids in a nonequilibrium state is more important because it has possibility to shed light on an another typical and perhaps more important nonequilibrium problem, non-stationary aging. Aging is the slow relaxation after a sudden quench of temperature below the glass transition temperature. In this case, the waiting time plays a similar role to (the inverse of) the shear rate. Aging behavior has been extensively studied for spin glasses (see Ref.[4] and references therein). The relationship between aging and a system driven far away from the equilibrium was considered using a schematic model based on the exactly solvable p-spin spin glass by Berthier, Barrat and Kurchan[5] and its validity was tested numerically for supercooled liquids[6]. There are attempts to study the aging of structural glasses theoretically[7] but it has not been analyzed and compared with the simulation results[8].

In this paper, we investigate the dynamics of supercooled liquids under shear theoretically, by extending the standard mode-coupling theory (MCT). We start with generalized fluctuating hydrodynamic equations with a convection term. Using several approximations, we obtain a closed nonlinear equation for the intermediate scattering function for the sheared system. The theory is applicable to both normal liquids and colloidal suspensions in the absence of hydrodynamic interactions. Numerical results will be presented only for the colloidal suspensions, but generalization to liquids are straightforward. Some of the preliminary results have already been published in Ref.[9].

THEORY

We shall consider a two dimensional colloidal suspension under a stationary simple shear flow given by

$$\mathbf{v}_0(\mathbf{r}) = \Gamma \cdot \mathbf{r} = (\dot{\gamma}y, 0), \quad (1)$$

where $(\Gamma)_{\alpha\beta} = \dot{\gamma}\delta_{\alpha x}\delta_{\beta y}$ is the velocity gradient matrix. The hydrodynamic fluctuations for density $\rho(\mathbf{r},t)$ and the velocity field $\mathbf{v}(\mathbf{r},t)$ obey the following set of

Langevin equations[10].

$$\frac{\partial \rho}{\partial t} = -\nabla \cdot (\rho \mathbf{v}),$$
$$m\frac{\partial (\rho \mathbf{v})}{\partial t} + m\nabla \cdot (\rho \mathbf{v}\mathbf{v}) = -\rho \nabla \frac{\delta \mathscr{F}}{\delta \rho} - \zeta_0 \rho (\mathbf{v} - \mathbf{v}_0) + \mathbf{f}_R,$$
(2)

where ζ_0 is the collective friction coefficient for colloidal particles. $\mathbf{f}_R(\mathbf{r},t)$ is the random force which satisfies the fluctuation-dissipation theorem of the second kind (2nd FDT); $\langle \mathbf{f}_R(\mathbf{r},t)\mathbf{f}_R(\mathbf{r}',t')\rangle_0 = 2k_B T \rho(\mathbf{r})\zeta_0 \delta(\mathbf{r}-\mathbf{r}')\delta(t-t')$ for $t \geq t'$, where $\langle \cdots \rangle_0$ is an average over the conditional probability for a fixed value of $\rho(\mathbf{r})$ at $t = t'$. Note that the random force depends on the density and thus the noise is multiplicative. We assumed that the 2nd FDT holds even in nonequilibrium state since the correlation of the random forces are short-ranged and short-lived, and thus the effect of the shear is expected to be negligible. The friction term is specific for the colloidal case. In the case of liquids, it should be replaced by a stress term which is proportional to the gradient of the velocity field multiplied by the shear viscosity. Both cases, however, lead to the same dynamical behavior on long time scales. The first term in the right hand side of the equation for the momentum is the pressure term and $\mathscr{F}$ is the total free energy in a stationary state. Here we assume that the free energy is well approximated by that of the equilibrium form and is given by a well-known expression;

$$\beta \mathscr{F} \simeq \int d\mathbf{r}\, \rho(\mathbf{r})\{\ln \rho(\mathbf{r})/\rho_0 - 1\} - \frac{1}{2}\int d\mathbf{r}_1 \int d\mathbf{r}_2\, \delta\rho(\mathbf{r}_1)c(r_{12})\delta\rho(\mathbf{r}_2),$$
(3)

where $\beta = 1/k_B T$ and $c(r)$ is the direct correlation function. Under shear, it is expected that $c(r)$ will be distorted and should be replaced by a nonequilibrium, steady state form $c_{NE}(\mathbf{r})$, which is an anisotropic function of $\mathbf{r}$. It is, however, natural to expect that this distortion is very small in the molecular length scale, which plays the most important role in the slowing down of the structural relaxation near the glass transition. We confirmed this by numerical simulation[11]. By linearizing eq.(2) around the stationary state as $\rho = \rho_0 + \delta\rho$ and $\mathbf{v} = \mathbf{v}_0 + \delta\mathbf{v}$, where ρ_0 is the average density, we obtain the following equations,

$$\left(\frac{\partial}{\partial t} - \mathbf{k}\cdot\Gamma\cdot\frac{\partial}{\partial \mathbf{k}}\right)\delta\rho_{\mathbf{k}}(t) = -ikJ_{\mathbf{k}}(t),$$

$$\left(\frac{\partial}{\partial t} - \mathbf{k}\cdot\Gamma\cdot\frac{\partial}{\partial \mathbf{k}} + \hat{\mathbf{k}}\cdot\Gamma\cdot\hat{\mathbf{k}}\right)J_{\mathbf{k}}(t)$$
$$= -\frac{ik}{m\beta S(k)}\delta\rho_{\mathbf{k}}(t)$$
(4)
$$-\frac{1}{m\beta}\int_{\mathbf{q}} i\hat{\mathbf{k}}\cdot\mathbf{q}c(q)\delta\rho_{\mathbf{k}-\mathbf{q}}(t)\delta\rho_{\mathbf{q}}(t) - \frac{\zeta_0}{m}J_{\mathbf{k}}(t) + f_{R\mathbf{k}}(t),$$

where $c(q)$ is the Fourier transform of $c(r)$, $\hat{\mathbf{k}} \equiv \mathbf{k}/|\mathbf{k}|$, $J_{\mathbf{k}}(t) = \rho_0 \hat{\mathbf{k}} \cdot \delta\mathbf{v}_{\mathbf{k}}(t)$ is the longitudinal momentum fluctuation, and $\int_{\mathbf{q}} \equiv \int d\mathbf{q}/(2\pi)^2$. Note that our approximate equation does not contain coupling to transverse momentum fluctuations even in the presence of shear.

In order to construct equations for the appropriate correlations from the above expressions, an approximate symmetry is necessary. In the presence of shear, translational invariance is violated. In other words, correlations of arbitrary fluctuations, $f(\mathbf{r},t)$ and $g(\mathbf{r},t)$, do not satisfy $\langle f(\mathbf{r},t)g(\mathbf{r}',0)\rangle \neq \langle f(\mathbf{r}-\mathbf{r}',t)g(\mathbf{0},0)\rangle$. Instead, it has the following symmetry[12];

$$\langle f(\mathbf{r},t)g(\mathbf{r}',0)\rangle = \langle f(\mathbf{r}-\mathbf{r}'(t),t)g(\mathbf{0},0)\rangle,$$
(5)

or in wavevector space $\langle f_{\mathbf{k}}(t)g^*_{\mathbf{k}'}(0)\rangle = \langle f_{\mathbf{k}}(t)g^*_{\mathbf{k}(t)}(0)\rangle \times \delta_{\mathbf{k}(t),\mathbf{k}'}$, where we defined the time-dependent position and wave vector by $\mathbf{r}(t) \equiv \exp[\Gamma t]\cdot\mathbf{r} = \mathbf{r} + \dot{\gamma}ty\hat{\mathbf{e}}_x$, where $\hat{\mathbf{e}}_x$ is an unit vector oriented along the x-axis and $\mathbf{k}(t) = \exp[{}^t\Gamma t]\cdot\mathbf{k} = \mathbf{k} + \dot{\gamma}tk_x\hat{\mathbf{e}}_y$, where ${}^t\Gamma$ denotes the transpose of Γ and $\delta_{\mathbf{k},\mathbf{k}'} \equiv (2\pi)^2 V^{-1}\delta(\mathbf{k}-\mathbf{k}')$ for a system of volume V. Using this approximation, it is straightforward to construct the mode-coupling equations for the appropriate correlation functions. We shall derive the equation for the intermediate scattering function defined by $F(\mathbf{k},t) \equiv N^{-1}\langle \delta\rho_{\mathbf{k}(-t)}(t)\delta\rho^*_{\mathbf{k}}(0)\rangle$, where N is the total number of the particles in the system. Note that the wave vector in $\delta\rho_{\mathbf{k}}(t)$ is now replaced by a time-dependent one $\mathbf{k}(-t)$.

Eq.(4) has a quadratic nonlinear term in $\delta\rho_{\mathbf{k}}(t)$. This term can be renormalized to give a generalized friction coefficient or the memory kernel following the standard procedure of derivation of the mode-coupling equation[13]. To the lowest order in the loop expansions, we obtains the equation for the velocity-density correlation $C(\mathbf{k},t) \equiv N^{-1}\langle J_{\mathbf{k}(-t)}(t)n^*_{\mathbf{k}}(0)\rangle$;

$$\frac{dC(\mathbf{k},t)}{dt} - \hat{\mathbf{k}}(-t)\cdot\Gamma\cdot\hat{\mathbf{k}}(-t)C(\mathbf{k},t) = -\frac{ik(-t)F(\mathbf{k},t)}{m\beta S(k(-t))}$$
$$-\frac{1}{m}\int_0^t dt' \int d\mathbf{k}'\, \zeta(\mathbf{k}(-t),\mathbf{k}',t-t')C(\mathbf{k}',t').$$
(6)

Note that in the above equation, the differential operator $\mathbf{k}\cdot\Gamma\cdot\partial/\partial\mathbf{k}$ disappears and $\mathbf{k}$ is replaced by $\mathbf{k}(t)$. $\zeta(\mathbf{k},\mathbf{k}',t)$ is the generalized friction coefficient. $\zeta(\mathbf{k},\mathbf{k}',t)$ is given by the sum of the bare friction coefficient and the mode-coupling term as $\zeta(\mathbf{k},\mathbf{k}',t) = \zeta_0 \times 2\delta(t) + \delta\zeta(\mathbf{k},t)\delta_{\mathbf{k}(t),\mathbf{k}'}$ with the mode-coupling contribution given by

$$\delta\zeta(\mathbf{k},t) = \frac{\rho_0}{2\beta}\int_{\mathbf{q}} \mathscr{V}(\mathbf{k},\mathbf{q})\mathscr{V}(\mathbf{k}(t),\mathbf{q}(t)) \times F(\mathbf{k}(t)-\mathbf{q}(t),t)F(\mathbf{q}(t),t),$$
(7)

where $\mathscr{V}(\mathbf{k},\mathbf{q})$ is the vertex function given by $\mathscr{V}(\mathbf{k},\mathbf{q}) = \hat{\mathbf{k}} \cdot \{\mathbf{q}c(\mathbf{q}) + (\mathbf{k}-\mathbf{q})c(\mathbf{k}-\mathbf{q})\}$. Note that, in the derivation of eq.(7), we have assumed the fluctuation-dissipation theorem of the first kind (1st FDT), which relates the response function to the correlation function. In order to derive the renormalized friction coefficient, propagators (the response function to the random forces) as well as correlation functions[13] are naturally introduced. The 1st FDT makes it possible to eliminate the propagators in favour of the correlation functions. In the overdamped limit, we may neglect the time-derivative of $C(\mathbf{k},t)$. The second term on the left hand side is also neglected if Péclet number Pe= $\dot{\gamma}\sigma^2/D_0$ (σ is the diameter of the particle and $D_0 = k_B T/\zeta_0$ is the diffusion coefficient) is small. Therefore, combining the first equation of eq.(4), one may eliminate $C(\mathbf{k},t)$ from the above equations and arrive at the closed equation for $F(\mathbf{k},t)$;

$$\frac{dF(\mathbf{k},t)}{dt} = -\frac{D_0 k(-t)^2}{S(k(-t))} F(\mathbf{k},t) - \int_0^t dt' \, M(\mathbf{k}(-t), t-t') \frac{dF(\mathbf{k},t')}{dt'}, \quad (8)$$

where

$$M(\mathbf{k},t) = \frac{\rho_0 D_0}{2} \frac{k}{k(t)} \int_\mathbf{q} \mathscr{V}(\mathbf{k},\mathbf{q}) \mathscr{V}(\mathbf{k}(t),\mathbf{q}(t)) \times F(\mathbf{k}(t)-\mathbf{q}(t),t) F(\mathbf{q}(t),t). \quad (9)$$

In the absence of the shear, they reduce to the conventional mode-coupling equation[14].

RESULTS

Solving eqs.(8) and (9) numerically is more demanding than the equations in the equilibrium state because the wavevectors are distorted by shear and the system is not isotropic. We have considered a two dimensional colloidal suspension consisting of hard disks. For the static correlation function $c(k)$ and $S(k)$, the analytic expressions derived by Baus et al.[15] were used. We have divided the two dimensional wavevector space into N_k grids for each direction. The cut-off wavevector was chosen to be $k_c\sigma = 10\pi$. For the self-consistent calculation of the mode-coupling equation, we have used the algorithm developed by Fuchs et al.[16]. In the following results, we have used the grid number $N_k = 55$, for which the ergodic-nonergodic transition occurs at the volume fraction $\phi_c = \pi\sigma^2\rho_0/4 = 0.7665$. Apparently, N_k is not large enough to give the right transition density which is 10% smaller. However, the qualitative behaviors does not change by increasing the grid number.

In Figure 1, we show the behavior of $F(\mathbf{k},t)$ for $\phi = \phi_c \times (1-10^{-4})$ for various shear rate Pe=10^{-10} to 10. The wavevectors were chosen to be $\mathbf{k}\sigma = (0,3)$ and

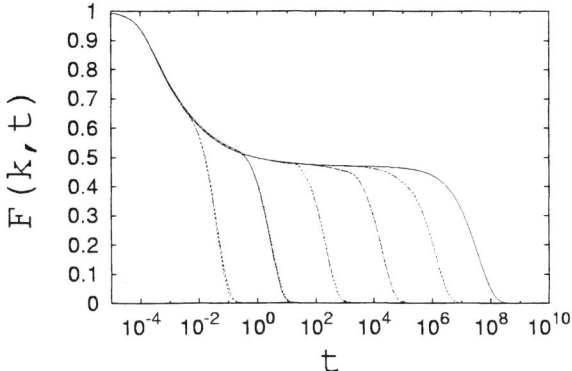

FIGURE 1. $F(\mathbf{k},t)/S(k)$ for $\mathbf{k}\sigma = (0,3), (3,0)$ for various shear rates $\dot{\gamma}$. From the right to the left, Pe = $10^{-10}, 10^{-7}, 10^{-5}, 10^{-3}, 10^{-1}$, and 10. The time t is scaled by σ^2/D_0.

$(3,0)$, parallel and perpendicular to the shear flow, respectively. For the smaller shear rate, Pe $< 10^{-10}$, we did not see any shear effect. For higher shear rates, we have observed the drastic reduction of the relaxation time due to shear. This is similar to the behavior reported in recent molecular dynamics simulations for a binary soft-core liquid[2].

The two lines for a fixed shear magnitude but for wavevectors along different directions collapsed onto each other and we do not see a noticeable difference. It is surprising that, though the perturbation of shear flow is highly anisotropic, the dynamics of fluctuations are almost isotropic. This fact is also observed in the simulation for a binary liquid[11]. The reason for the isotropic nature is understood as follows. The shear flow perturbs and randomizes the phase of coupling between different modes. This perturbation dissipates the cage that transiently immobilizes particles. Mathematically, this is reflected through the time dependence of the vertex. This "phase randomization" occurs irrespective of the direction of the wavevector, which results in the isotropic behavior of relaxation. This mechanism is very different from that of many complex fluids and dynamic critical phenomena under shear, in which the faster relaxation occurs mainly due to the distortion of the structures at small wavevectors which are stretched out by the shear flow and pushed to larger wavevectors where faster relaxation occurs.

The shear viscosity, η, is easily evaluated by modifying the Green-Kubo formula for the sheared system[17];

$$\eta(\dot{\gamma}) = \eta_0 + \frac{1}{2\beta} \int_0^\infty dt \int_\mathbf{k} \frac{k_x k_x(t)}{S^2(k) S^2(k(t))} \frac{\partial S(k)}{\partial k_y} \frac{\partial S(k(t))}{\partial k_y(t)} F^2(\mathbf{k}(t),t), \quad (10)$$

where η_0 is the viscosity of the solvent alone. The integral can be implemented for the set of $F(k,t)$ evaluated using eq.(8). Shear rate dependence of the reduced viscosity $\eta_R(\dot{\gamma}) \equiv \{\eta(\dot{\gamma}) - \eta_0\}/\eta_0$ is plotted in Figure 2 for various densities around ϕ_c. The strong non-Newtonian

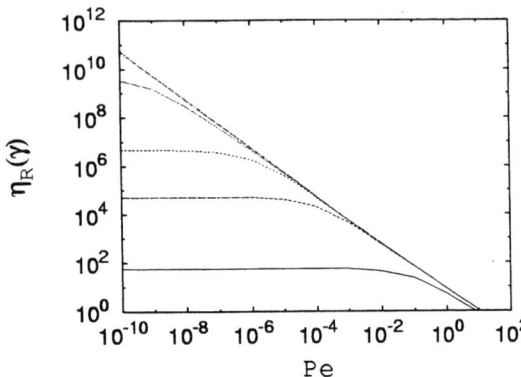

FIGURE 2. The reduced viscosity is plotted for $Pe = \dot{\gamma}\sigma^2/D_0$ for various densities. From above to bottom; $\phi = 0.766549, 0.7665, 0.766453, 0.7664, 0.766$, and 0.756. The highest density is 4×10^{-5} % larger than ϕ_c.

behavior is observed at high shear rate and large densities, which is again in qualitative agreement with the simulation results for liquids. Slightly above ϕ_c, the plastic behavior which implies the presence of the yield stress is also observed. The shear thinning exponent is extracted from this plot between $10^{-10} < Pe < 1$ and we obtained $\eta_R(\dot{\gamma}) \propto \dot{\gamma}^{-\nu}$ with $\nu \simeq 0.99$. For the larger shear rate, $Pe > 1$, the exponent becomes smaller. In this regime, it is expected that other mechanism such as the distortion of structure $c(k)$ and $S(k)$ by shear becomes important.

The mode-coupling theory developed in this paper is far from complete. The most crucial approximation is the use of the 1st FDT, which was employed when we close the equation in terms of the correlation functions alone. It is already known that the 1st FDT is violated for supercooled systems under shear as well as during aging[6]. Without the 1st FDT, one has to solve simultaneously the set of mode-coupling equations for the propagator and correlation function, which couple each other through the memory kernels. Research in this direction is under way. Another important approximation was to neglect the small distortion of the structure, $c(k)$ and $S(k)$, due to shear. The construction of the equation for such an equal-time correlation functions might be more subtle and should be considered in future. It is surprising, however, that despite of these approximations, the theory reproduces the major features which was seen in simulations; the drastic reduction of the relaxation time, the isotropic nature of the dynamics, and plastic-like strong shear thinning.

An analogous effort has been made by Fuchs and Cates[18]. They have derived a mode-coupling expression for $F(k,t)$ using a projection operator for the Smoluchowski equation for N-particle colloidal suspensions. They have observed similar shear thinning behavior for approximated expressions, for an isotropic model, where the anisotropy hidden in the equations are neglected.

The details of analytical and numerical calculations for results given in the present paper are given elsewhere[11].

ACKNOWLEDGMENTS

The authors acknowledge support from NSF grant #0134969. The authors would like to express their gratitude to Prof. Matthias Fuchs for for useful discussions and helpful instruction about the program codes to solve the mode-coupling equation.

REFERENCES

1. Simmons, J. H., Mohr, R. K., and Montrose, C. J., *J. Appl. Phys.*, **53**, 4075-4080 (1982).
2. Yamamoto, R., and Onuki, A., *Phys. Rev. E*, **58**, 3515-3529 (1998).
3. Liu, A. J., and Nagel, S. R., *Nature*, **396**, 21-22 (1998).
4. Cugliandolo, L. F., "Dynamics of Glassy Systems", in *Slow relaxations and nonequilibrium dynamics in condensed matter*, ed. J.-L. Barrat, M. Feigelman, and J. Kurchan, Springer-Verlag, New York, (2003), pp. 371-521.
5. Berthier, L., Barrat, J.-L., and Kurchan, J., *Phys. Rev. E*, **61**, 5464-5472 (2000).
6. Berthier, L., and Barrat, J.-L., *Phys. Rev. Lett.*, **89**, 095702 (2002).
7. Latz, A., *J. Stat. Phys.*, **109**, 607-622 (2002).
8. Kob, W., Barrat, J.-L., Sciortino, F., and Tartaglia, P., *J. of Phys.: Condens. Matter*, **12**, 6385-6394 (2000).
9. Miyazaki, K., and Reichman, D. R., *Phys. Rev. E*, **66**, 050501(R) (2002).
10. Kirkpatrick, T. R., and Nieuwoudt, J. C., *Phys. Rev. A*, **33**, 2651-2657 (1986).
11. Miyazaki, K., Yamamoto, R., and Reichman, D. R., (unpublished).
12. Onuki, A., *J. of Phys.: Condens. Matter*, **9**, 6119-6157 (1997).
13. Martin, P. C., Siggia, E. D., and Rose, H. A., *Phys. Rev. A*, **8**, 423-437 (1973).
14. Götze, W., and Sjögren, L., *Rep. Prog. Phys.*, **55**, 241-376 (1992).
15. Baus, M., and Colot, J. L., *Phys. Rev. A*, **36**, 3912-3925 (1987).
16. Fuchs, M., Götze, W., and Latz, A., *J. of Phys: Condens. Matter*, **3**, 5047-5071 (1991).
17. Kirkpatrick, T. R., *J. of Non-Crystalline Solids*, **75**, 437-442 (1985).
18. Fuchs, M., and Cates, M. E., *Phys. Rev. Lett.*, **89**, 248304 (2002).

Fluctuation-dissipation relations in ageing and driven non-mean field glass models

Suzanne M. Fielding[*] and Peter Sollich[†]

[*]Polymer IRC and Department of Physics & Astronomy, University of Leeds, Leeds LS2 9JT, United Kingdom
[†]Department of Mathematics, King's College London, Strand, London, WC2R 2LS, United Kingdom

Abstract. We study the fluctuation-dissipation theorem (FDT) in the glass phase of (1) Bouchaud's trap model and (2) its driven counterpart, the "soft glassy rheology" model. We incorporate into the models an arbitrary observable m and obtain its correlation and response functions in closed form. A limiting non-equilibrium FDT plot (of correlator *vs.* response) is approached at long times for most choices of m. In contrast to standard mean field models, however, the plot, in general, (i) depends non trivially on the observable, (ii) has a continuously varying slope (even though there is a single scaling of relaxation times with age) and (iii) differs in the ageing and driven regimes. Despite this, all plots share the same limiting slope for well separated times, suggesting that a meaningful non-equilibrium effective temperature could apply in this limit. Beyond the trap model, we discuss more generally the status of FD temperatures in such non-mean field systems.

INTRODUCTION

Glasses relax very slowly at low temperatures. They thus stay far from equilibrium long after preparation, and show ageing [1]: the time scale for response to perturbations (or decay of correlations) increases with the time t_w since the temperature quench, eventually exceeding any experimental time scale. Time translational invariance is lost. Because of this sluggishness, glasses are highly susceptible to external driving, which typically stabilises a non-equilibrium steady (TTI) state of apparent age $O(1/\dot{\gamma})$, for drive-rate $\dot{\gamma}$ [2].

Let $C(t,t_w) = \langle m(t)m(t_w)\rangle - \langle m(t)\rangle\langle m(t_w)\rangle$ be the autocorrelation function for an observable m, $R(t,t_w) = \delta\langle m(t)\rangle/\delta h(t_w)|_{h=0}$ the linear response of $m(t)$ to a small impulse in its conjugate field h at time t_w, and $\chi(t,t_w) = \int_{t_w}^{t} dt' R(t,t')$ the response to a field step $h(t) = h\Theta(t-t_w)$. In *equilibrium*, $C(t,t_w) = C(t-t_w)$ by TTI (similarly for R and χ), and the FDT reads $-\frac{\partial}{\partial t_w}\chi(t-t_w) = R(t,t_w) = \frac{1}{T}\frac{\partial}{\partial t_w}C(t-t_w)$, with T the thermodynamic temperature. (We set $k_B = 1$.) A parametric FDT plot of χ *vs.* C is thus a straight line of slope $-1/T$.

Out of equilibrium, FDT violation is measured by a fluctuation-dissipation ratio (FDR), X, defined by [3]

$$-\frac{\partial}{\partial t_w}\chi(t,t_w) = R(t,t_w) = \frac{X(t,t_w)}{T}\frac{\partial}{\partial t_w}C(t,t_w). \quad (1)$$

In ageing systems, violation ($X \neq 1$) can persist even at long times $t_w \to \infty$, indicating far from equilibrium behaviour even when one-time quantities, *e.g.* entropy have settled to stationary values. Similarly, driven glasses can violate FDT even for weak driving, $\dot{\gamma} \to 0$.

Remarkably, however, the FDR for several mean field models [3] assumes a special form at long times (ageing case). Taking $t_w \to \infty$ at constant $C = C(t,t_w)$, $X(t,t_w) \to X(C)$ becomes a (nontrivial) function of the single argument C. If the equal-time correlator $C(t,t)$ also approaches a constant C_0 for $t \to \infty$, it follows that

$$\chi(t,t_w) = \int_{C(t,t_w)}^{C_0} dC\, X(C)/T. \quad (2)$$

A limiting non-equilibrium FDT plot is then obtained by plotting χ *vs.* C for increasingly large times; from its slope $-X(C)/T$, an *effective temperature* [4] can be defined as $T_{\text{eff}}(C) = T/X(C)$. An equivalent FD relation has been suggested to hold in slowly *driven* glasses [4] with $T_{\text{eff}}(C, \dot{\gamma} \to 0) = T_{\text{eff}}(C, t_w \to \infty)$.

In the most general ageing scenario, a system evolves on several characteristic time scales, each with its own functional dependence on t_w. If these become infinitely separated as $t_w \to \infty$, they form distinct 'time sectors'. In mean field, $T_{\text{eff}}(C)$ is *constant* in each sector [3]. It has thus been interpreted as a time scale dependent non equilibrium temperature, and shown to have many of the properties of a thermodynamic temperature (*e.g.* in controlling of heat flow) [4]. Of crucial importance to its interpretation as a temperature, it is *independent* of the observable m used to construct the FD plot.

While this picture is well established in mean field, evidence beyond mean field is limited. Limiting FD plots were found in, *e.g.*, Refs. [5, 6]. Observable independence is largely unestablished, but see Ref. [5] for encouraging results. Evidence for equivalent ageing and driven FD plots is limited, but has been found in models of supercooled liquids [7]. In this work, therefore, we study a simple non-mean field model for which FD plots can be calculated for arbitrary observables, allowing detailed study of whether the mean field picture applies [8].

TRAP MODEL

The (undriven) trap model [9] comprises an ensemble of uncoupled particles exploring a landscape of energy traps by thermal activation. The traps descend from a common level, with depths E chosen from a 'prior' distribution $\rho(E)$ ($E > 0$). A particle in a trap of depth E escapes on a time scale $\tau(E) = \tau_0 \exp(E/T)$ and hops into another trap, with a depth chosen randomly from $\rho(E)$. The probability, $P(E,t)$, of finding a randomly chosen particle in a trap of depth E at time t thus obeys

$$\partial_t P(E,t) = -\tau^{-1}(E) P(E,t) + Y(t) \rho(E) \quad (3)$$

in which the first (second) term on the RHS represents hops out of (into) traps of depth E, and $Y(t) = \langle \tau^{-1}(E) \rangle_{P(E,t)}$ is the average hop rate. For a prior distribution $\rho(E) \sim \exp(-E/T_g)$ the model shows a glass transition at a temperature T_g, because for $T \leq T_g$ the equilibrium state $P_{eq}(E) \propto \tau(E)\rho(E) \propto \exp(E/T)\exp(-E/T_g)$ is unnormalizable and the average lifetime $\langle \tau \rangle_\rho$ is infinite. Following a quench to $T \leq T_g$, the system cannot equilibrate and instead ages. At large times $t_w \to \infty$ a scaling limit is reached with $P(\tau, t_w) = [T/\tau(E)]P(E, t_w)$ concentrated on traps of lifetime $\tau = O(t_w)$. The model thus has just one characteristic time scale, growing linearly with age. We set $T_g = 1$, $\tau_0 = 1$.

To study FDT we assign to each trap a generic observable m. The landscape is then characterized by the joint prior distribution $\sigma(m|E)\rho(E)$, where $\sigma(m|E)$ is the distribution of m across traps of a fixed energy E. We consider non-equilibrium dynamics after a quench at $t = 0$ from $T = \infty$ to $T < 1$. The initial condition is thus $P_0(E,m) = \sigma(m|E)\rho(E)$, with subsequent evolution

$$\partial_t P(E,m,t) = -\frac{P(E,m,t)}{\tau(E,m)} + Y(t)\rho(E)\sigma(m|E) \quad (4)$$

where the activation times are modified by a small field h as $\tau(E,m) = \tau(E)\exp(mh/T)$. (Other choices of $\tau(E,m)$ that maintain detailed balance are also possible [10, 11].)

To find the autocorrelation function C for m at $h = 0$ we need the probability that a particle with m_w and energy E_w at time t_w subsequently has m and E at time t:

$$P(E,m,t|E_w,m_w,t_w) =$$

$$\delta(m - m_w)\delta(E - E_w)e^{-(t-t')/\tau(E_w)}P(E_w, m_w, t_w)$$
$$+ \int_{t_w}^{t} dt' \frac{e^{-(t'-t_w)/\tau(E_w)}}{\tau(E_w)} P(E, t-t')\sigma(m|E). \quad (5)$$

The first (second) term on the RHS corresponds to a particle not having hopped since t_w (first having hopped at t'). After hopping the particle evolves as if "reset" to time zero since it selects its new trap from the prior distribution, which also describes the initial state. From Eqn. 5, an exact expression for $C(t, t_w)$ can be found.

To find the response function, we proved

$$T\frac{\partial}{\partial t_w}\chi(t,t_w) = \frac{\partial}{\partial t}C(t,t_w) + \frac{\partial}{\partial t}\langle m(t) \rangle \langle m(t_w) \rangle \quad (6)$$

in which $\langle m(t) \rangle = \langle \overline{m}(E) \rangle_{P(E,t)}$ is the global mean of m. This generalizes the results of [11, 12] to non-zero means. We rescale the field $h \to Th$, absorbing a factor $1/T$ into the response function. The slope of the FDT plot is then $-X = -T/T_{\text{eff}}$ ($= -1$ in equilibrium).

Our expressions for C and χ each comprise two additive components, depending separately on the mean $\overline{m}(E)$ and variance $\Delta^2(E)$ of $\sigma(m|E)$. Using them, we numerically calculated C and χ for several different distributions $\sigma(m|E)$, each specified by given functional forms of $\overline{m}(E)$ and $\Delta^2(E)$. For simplicity, we considered only distributions of zero mean (but non-zero variance); or of zero variance (but non-zero mean).

As expected, the decay of C in general depends on the waiting time t_w: TTI is lost. For any observable m that is correlated with E, the equal-time correlator $C(t,t)$ can also depend on t (either decaying or diverging). While Eqn. 2 implies that an FD plot can be produced either with t as the curve parameter (at fixed t_w), or vice versa, Eqn. 1 in general only ensures a slope of $-X(t,t_w)/T$ with t_w as the parameter. This issue is important if, as here, $C(t,t)$ is time dependent, requiring pre-normalisation of χ and C to ensure a limiting FD plot of time independent size.[1] To preserve the connection between X and the slope, the normalisation factor must be the same for χ and C, and independent of t_w. We therefore use $C(t,t)$, rather than $C(t_w,t_w)$, denoting the normalized quantities by $\tilde{\chi}$ and $\tilde{C}$. A limiting plot *may* then be approached at long times. If so, either t_w or t could be used as the curve parameter. We choose t_w, because this ensures that FD plots constructed with the switch-on or switch-off response functions are trivially related, as discussed more fully in Ref. [13].

For zero mean observables, $\overline{m}(E) = 0$, we consider a variance $\Delta^2(E) = \exp(En/T)$. For different values of n, the correlator probes different moments of the prior

---

[1] This issue has not arisen in mean field studies, where variables are usually sufficiently "neutral" that $C(t,t) \to$ const.

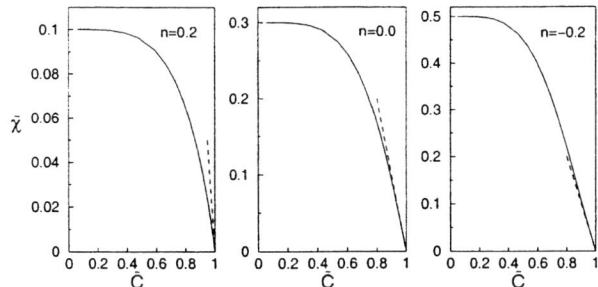

FIGURE 1. FDT plots of $\tilde{\chi}$ vs $\tilde{C}$ for a distribution $\sigma(m|E)$ of variance $\exp(nE/T)$ (but zero mean) for $n = 0.2, 0.0, -0.2$; $T = 0.3$. For each n data are shown for times $t = 10^6, 10^7$; these are indistinguishable, confirming that a limit FDT plot has been attained. Dashed: asymptote $\tilde{\chi} = 1 - \tilde{C}$ for $t \to \infty$ and $\tilde{C} \to 1$.

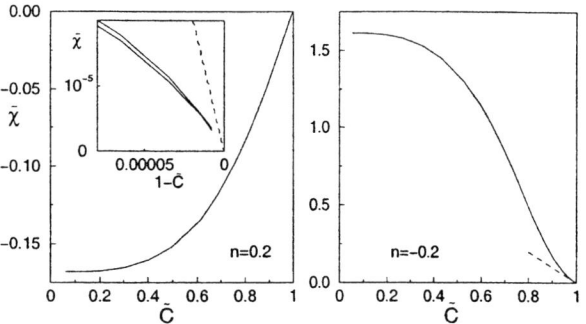

FIGURE 2. FDT plots of $\tilde{\chi}$ vs $\tilde{C}$ for a distribution with mean $\exp(nE/2T)$ (but zero variance), for $n = 0.2, -0.2$; $T = 0.3$. Curves are shown for times $t = 10^6, 10^7$, but are indistinguishable except for the zoom-inset on the left (upper: $t = 10^7$). Dashed: predicted asymptote $\tilde{\chi} = 1 - \tilde{C}$ for $t \to \infty$, $\tilde{C} \to 1$.

distribution $\rho(E)$. For $n < T - 1$, it is sensitive only to shallow traps and decays on time scales $t - t_w = O(1)$, probing only quasi-equilibrium behaviour and so yielding an FD plot that is a straight line of slope -1 as $t \to \infty$. In contrast [2], for $T - 1 < n < T$ the correlator is dominated by traps with $\tau(E) = O(t)$, and decays on ageing time scales $t - t_w = O(t_w)$. Equilibrium FDT is then violated. A limiting non-equilibrium FD plot is nevertheless approached at long times (Fig. 1) since $\tilde{C}$ and $\tilde{\chi}$ then share the same scaling variable $(t - t_w)/t_w$. The slope of each plot varies continuously with $\tilde{C}$. In contrast to mean field, this is not due to an infinite hierarchy of time sectors: the variation occurs across the single time sector $t - t_w = O(t_w)$. More seriously, different observables give different plots: at a fixed value of $\tilde{C}$ the slopes $-X$ depend on n. For variables with zero variance and mean $\overline{m}(E) = \exp(En/2T)$ we again find FDT violation for $T - 1 < n < T$ (Fig. 2) with a limiting non-equilibrium plot that depends (now obviously) on m.

The concept of a non equilibrium FD temperature is therefore not straightforward in the trap model. Can it nonetheless be rescued? One difficulty is the non-uniqueness of the FD plots. Observable dependent FD plots have also been found in the zero-temperature Glauber-Ising chain (ZTGIC) [13, 14], for different correlation lengths of the applied field, h. One could argue that to probe an inherent T_{eff}, the properties of the observable must not change much across the phase space regions visited during ageing. Applying this to the trap model, where the typical trap depth E increases without bound for $t \to \infty$, a "neutral" observable requires $\Delta^2(E), \overline{m}(E) \to$ const. as $E \to \infty$. With this restriction, we do indeed get a unique FD plot. The same is true for the ZTGIC, for neutral (random) fields.

---

[2] The regime $n > T$ is meaningless: It gives $C(t,t) = \infty \forall t$.

Even with a judicious choice of neutral observable, however, X still varies continuously across the single time sector in the trap model. Two thermometers probing time scales that differ only by a factor of order unity would thus measure different (and so meaningless) effective temperatures. Similarly rounded plots are seen in the ZTGIC [13, 14]. There is, however, the possibility that the limit of X obtained at large time separation $X_\infty = \lim_{t_w \to \infty} \lim_{t \to \infty} X(t, t_w)$ may still give a meaningful T_{eff}. Indeed, in the trap model $X_\infty = 0$ is the same for all variables considered; the ZTGIC also has $X_\infty = 0$ for domain wall variables (but $X_\infty = 1/2$ for spin variables) [14].

We now turn to the driven model, as first defined to study "soft glassy rheology". Each particle is assigned a local elastic "strain" l and "stress" kl. (We set $k = 1$.) After any hop, l resets to zero. Between hops, $\dot{l} = \dot{\gamma}$, the external strain rate. A particle in a trap of depth E strained by l sees a reduced energy barrier $E - \frac{1}{2}l^2$, so

$$[\partial_t + \dot{\gamma}\partial_l]P(E, l, t) = -\tau^{-1}(E)e^{l^2/2T}P + Y(t)\rho(E)\delta(l). \tag{7}$$

In the glass phase, steady driving ($\dot{\gamma} = $ const.) interrupts ageing and restores a steady state. In the limit $\dot{\gamma} \to 0$, the steady state distribution $P_\infty(E) = \int dl\, P_\infty(E, l)$ approaches a scaling state with all relaxation times $O(1/\dot{\gamma})$.

In our study of driven FDT, we focus on neutral observables, $\overline{m}(E) = 0$ and $\Delta^2(E) = $ const. For these, the autocorrelation function

$$C(t, \dot{\gamma}) = \int_{\dot{\gamma}t}^\infty dl \int_0^\infty dE\, P_\infty(E, l) \tag{8}$$

where $P_\infty(E, l) = \lim_{t \to \infty} P(E, l, t)$, which can be calculated exactly. (Because TTI is restored, C and χ do not depend explicitly upon the waiting time t_w, so we have set $t_w = 0$.) Eqn. 8 can be understood by noting that

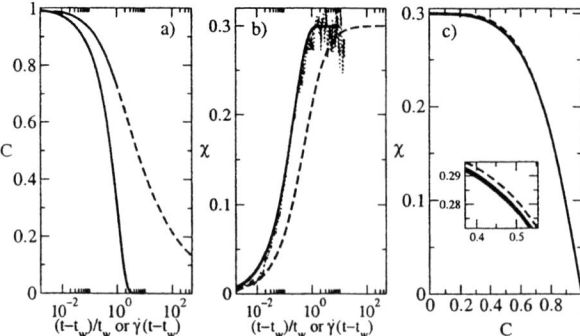

FIGURE 3. a) Correlator and b) response *vs.* scaled time for the neutral observable $\Delta^2(E) = 1$: driven (solid lines); ageing (dashed). Waiting times (or inverse driving rates) 10^3, 10^4, 10^5, 10^6 are shown (but indistinguishable). c) FD plots: driven (solid); ageing (dashed). For the driven case, driving rate decreases downwards at fixed C. Temperature $T = 0.3$.

CONCLUSIONS

We have shown that the mean field concept of a non equilibrium FD temperature is not straightforward in the trap model. FD plots, in general, (i) depend on observable, (ii) have a slope varying continuously across a single time sector, (iii) differ in the ageing and driven cases. Nonetheless, *neutral* observables do all share a unique (observable independent) FD plot, which is the same (to within logs) for ageing and driven systems. Although the slope of the plot varies continuously, there remains the intriguing possibility that the limit of X obtained for large time separations $X_\infty = \lim_{t_w \to \infty} \lim_{t \to \infty} X(t,t_w)$ may still corresponds to a meaningful T_{eff}. Indeed, for the trap model, X_∞ is the same (zero) for all variables considered; this value even holds in trap models with slow dynamics arising from entropy rather than energy barriers [15].

only particles that have not hopped since time $t = 0$ (*i.e.* those with strains $l \geq \dot{\gamma}t$) contribute to the correlator. The switch-on response function can be shown to be

$$\chi(t,\dot{\gamma}) = \int_0^\infty dE \int_0^\infty dl \frac{[P_\infty(E,l) - P_\infty(E,l+\dot{\gamma}t)]}{\dot{\gamma}\tau(E)} I(l,T) \quad (9)$$

in which $I(l,T) = \int_0^l ds \exp\left(\frac{s^2}{2T}\right)$.

Using these expressions, we evaluated $C(t,\dot{\gamma})$ and $\chi(t,\dot{\gamma})$ numerically. The results are shown in Fig. 3a,b (with simulation results as a check). In the limit $\dot{\gamma} \to 0$, $t \to 0$ at fixed $\dot{\gamma}t$, $C(t,\dot{\gamma})$ and $\chi(t,\dot{\gamma})$ depend on t and $\dot{\gamma}$ only through the scaling variable $\dot{\gamma}t$, as expected.

Although the scaling functions $C(\dot{\gamma}t)$ and $\chi(\dot{\gamma}t)$ both differ strongly from their ageing counterparts $C(\frac{t-t_w}{t_w})$, $\chi(\frac{t-t_w}{t_w})$ (compare solid and dashed lines in Fig. 3) the ageing and driven FD relations are remarkably similar (Fig. 3c). Both start with a quasi-equilibrium slope $-X(C=1) \equiv \chi'(C=1) = -1$ and finish with slope $\chi'(C=0) = 0$ at intercept $\chi(C=0) = T$. Between these limits, these is little discernible difference between the ageing and driven plots. This non-trivial result is consistent with the predictions of Cugliandolo *et al.*, that the relationship between correlation and response should be the same in ageing and weakly driven glasses. Despite this, the inset of Fig. 3c, does reveal a small discrepancy. To investigate this further, we examined the behaviour of $X \equiv -\chi'(C)$ in the limit $C \to 0$, finding $X \sim C^{1/T}$ in the ageing case, but $C \sim X^T[\log(1/X)]^{(T-1)/2}$ in the driven case. Therefore, the driven and ageing FD plots are only equivalent to within minor logarithmic corrections. Finally, this similarity of ageing and driven FD relations is not robust with respect to non-neutrality of observable, or to driving mechanisms that respect detailed balance [8].

REFERENCES

1. Bouchaud, J. P., Cugliandolo, L. F., Kurchan, J., and Mézard, M., "Out of equilibrium dynamics in spin-glasses and other glassy systems," in A. P. Young, editor, *Spin glasses and random fields*, pages 161-223, Singapore, 1998. World Scientific.
2. Cugliandolo, L. F., Kurchan, J., LeDoussal, P., and Peliti, L., *Phys. Rev. Lett.*, **78**, 350–353 (1997).
3. Cugliandolo, L. F., and Kurchan, J., *Phys. Rev. Lett.*, **71**, 173–176 (1993). Cugliandolo, L. F., and Kurchan, J., *J. Phys. A*, **27**, 5749–5772 (1994). A recent review is Crisanti, A., and Ritort, F., *J. Phys. A*, **36**, R181–R290 (2003).
4. Cugliandolo, L. F., Kurchan, J., and Peliti, L., *Phys. Rev. E*, **55**, 3898–3914 (1997).
5. Kob, W., and Barrat, J. L., *Eur. Phys. J. B*, **13**, 319–333 (2000). Arenzon, J. J., Ricci-Tersenghi, F., and Stariolo, D. A., *Phys. Rev. E*, **62**, 5978–5985 (2000).
6. Marinari, E., Parisi, G., Ricci-Tersenghi, F., and Ruiz-Lorenzo, J. J., *J. Phys. A*, **31**, 2611–2620 (1998). Barrat, A., *Phys. Rev. E*, **57**, 3629–3632 (1998).
7. Barrat, J. L., and Berthier, L., *Phys. Rev. E*, **63**, 012503 (2001). Berthier, L., and Barrat, J. L., *Phys. Rev. Lett.*, **89**, 095702 (2002). Berthier, L., and Barrat, J. L., *J. Chem. Phys.*, **116**, 6228–6242 (2002).
8. Fielding, S. M., and Sollich, P., *Phys. Rev. Lett.*, **88**, 050603 (2002). Fielding, S. M., and Sollich, P., *Phys. Rev. E*, **67**, 011101 (2003).
9. Bouchaud, J. P., *J. Phys. (France) I*, **2**, 1705–1713 (1992).
10. Rinn, B., Maass, P., and Bouchaud, J.-P., *Phys. Rev. Lett.*, **84**, 5403–5406 (2000).
11. Bouchaud, J. P., and Dean, D. S., *J. Phys. (France) I*, **5**, 265–286 (1995).
12. Sasaki, M., and Nemoto, K., *J. Phys. Soc. Jpn.*, **68**, 1148–1161 (1999).
13. Sollich, P., Fielding, S., and Mayer, P., *J. Phys.-Condens. Matter*, **14**, 1683–1696 (2002).
14. Mayer, P., Berthier, L., Garrahan, J. P., and Sollich, P., *Phys. Rev. E*, **68**, 016116 (2003).
15. Sollich, P., *e-print cond-mat/0303637*.

Stochastic Approach to Glass Transition

Tadao Ishii

Faculty of Engineering, Okayama University, Tsushimanaka 3-1-1, Okayama 700-8530, Japan

Abstract. On the basis of the ion hopping model which causes the extended relaxation modes of glass forming liquid, the relaxation mode approach to a glass transition under an external perturbation is examined where a nonequilibrium disorder parameter is explicitly taken into account. It is shown that a cooling rate dependence of the glass transition temperature T_g reproduces experimental findings quite well, and also predicted therein that two characteristic transition temperatures T_0 and T_∞ exist in the zero and infinite limits of the cooling rate, respectively, leading to a possible glass transition regime, approximately, of $T_0 < T_g < T_\infty$.

INTRODUCTION

The structural glass transition has theoretically been studied in two ways, dynamical mode coupling [1,2] and stochastic equation methods [3,4]. However, since both methods do not explicitly take into account an external disturbance accompanied by cooling process, one can not understand a cooling rate dependence of the glass transition temperature [5,6]. In this work, we try to investigate a glass transition with cooling in a simplified model supercooled liquid, on the basis of the relaxation mode theory [7,8]. Since a relaxation mode is characterized by an inherent random density distribution in a nonequilibrium state or by a glassy state [9], we can define "glass transition" by its freezing. The relaxation mode theory has made clear, so far [9,10], mechanisms of non-Debye conductivities [11] and diffusions [12] in glassy ionic conductors.

The supercooled liquid is now modeled by the lattice liquid (interacting lattice gas ions) [13], consisting of cations and anions confined, respectively, on sublattices A and B where anions play a role only to provide a distribution of random energies. In the mean field approximation, we have the conventional master equation [13-15] for the probability $p_n(t)$ of finding a cation at site n on A sublattice and time t as

$$\frac{\partial p_n(t)}{\partial t} = \sum_d \left[(1-p_n(t))\Gamma_{n,n+d}(t) p_{n+d}(t) - (1-p_{n+d}(t))\Gamma_{n+d,n}(t) p_n(t) \right] \quad (1)$$

where $\Gamma_{m,n}(t)$ is the transition probability from site n to its nearest neighbor site m. Further we consider a symmetric hopping $\Gamma^0_{n,m} = \Gamma^0_{m,n}$ for 0 implying a thermal equilibrium state.

RELAXATION MODES EXCITATION

In the present symmetric hopping case, we have the probability $p^0 = c$ per site in thermal equilibrium. While in the presence of a small external perturbation Δg, the probability is written as $p_n(t) = p_n^0 + \Delta p_n(t)$. Then linearization of eq.(1) in $\Delta p_n(t)$ leads to [8]

$$\frac{\partial |\psi(t)\rangle}{\partial t} = -H|\psi(t)\rangle - H'|\Phi\rangle, \quad H' = H\Delta g$$
$$|\psi(t)\rangle = |\Phi^{-1}\Delta p(t)\rangle, \quad \Phi^2 = p^0(1-p^0) \quad (2)$$

where $|\psi(t)\rangle = \sum_n \psi_n(t)|n\rangle$, and $H_{nm} = \sum_{m' \neq n} \Gamma^0_{m'n}$ for $n = m$ and $-(\Gamma^0_{nm}\Gamma^0_{mn})^{1/2}$ for $n \neq m$, respectively.

In the limit $H' \to 0$, we have the eigenvalue equation $H|\psi_{q\varepsilon}\rangle = E_{q\varepsilon}|\psi_{q\varepsilon}\rangle$: in thermal equilibrium, the solution follows the detailed balance condition as $E_{q=0,\varepsilon=0} = 0$ for $|\psi_{q=0,\varepsilon=0}\rangle \equiv |\phi\rangle$ where $|\phi\rangle$ satisfies the normalization $\langle\phi|\phi\rangle = 1$ and is given by $|\Phi\rangle = (N(1-c))^{1/2}|\phi\rangle$ with the number of particles N. The eigenvalues are labeled by the wave number q and band index $\varepsilon = 0, 1, 2, 3, ...$ corresponding to a complete set of the eigenfunctions concerned. It is also noted that the eigenvalue defines the inverse of the relaxation time $\tau_{q\varepsilon}$ by $E_{q\varepsilon} = \tau_{q\varepsilon}^{-1}$, and so $\tau_{q\varepsilon}$ does not mean the relaxation of a single hopping event but of a mode in the system. As has intensively been studied so far [8,9], we have the diffusive modes ($\varepsilon = 0$) and non-diffusive modes ($\varepsilon \neq 0$) in that the former determines the dc conductivity and the latter does the ac conductivity, respectively, at $q = 0$. It is also important to declare that the diffusive mode at $q = 0$ implies the thermal equilibrium state and all other modes signify non-equilibrium states. Since the q-fluctuation of the relaxation modes does not work in the present aperiodic random system, only $E_{q=0,\varepsilon} \equiv E_\varepsilon$ is considered for later discussion.

In the presence of the perturbation from outside, one has the solution from eq.(2) in the form:

$$|\psi(t)\rangle = -\sqrt{N(1-c)} e^{-\int_0^t H(t')dt'} \times \int_0^t dt' e^{\int_0^{t'} H(t'')dt''} H(t') \Delta g(t') |\phi\rangle \quad (3)$$

since the temperature depends on time and ϕ is a constant determined only by N and the number of lattice sites.

Let us consider the cooling process starting at a temperature T_h, which we call the hopping-initiation temperature higher than the melting temperature T_M. The important process for the perturbation stems from an effective change of the binding energy V, which depends on the volume variation with time. With an introduction of the energy V^0 and transition rate Γ^0 at T_h, the perturbation is given by $\Delta\Gamma_{m,n} = \Gamma_{m,n}(t) - \Gamma_{m,n}^0$:

$$\Delta\Gamma_{m,n} = \omega_s e^{-\beta V_n^0}\left[\exp\left(-\beta\int_0^t (dV_n(t)/dt)\,dt\right) - 1\right] \\ \sim \omega_s e^{-\beta V_n^0}[\exp(-\eta_n t) - 1], \quad \eta = \beta V^0 \alpha_L v \quad (4)$$

where α_L is the linear thermal expansion coefficient given by $\alpha_L = (d\Omega/dT)/3\Omega$ for Ω a local volume and the temperature variation follows as $T - T_h = -vt$ in the cooling process. Consequently the perturbation term becomes $\Delta g = e^{-\eta t} - 1$. Thus we obtain

$$\langle \psi_\varepsilon | \psi(t) \rangle = \delta\sqrt{N(1-c)}\, e^{-\int_0^t E_\varepsilon(t')dt'} \times \int_0^t dt' e^{\int_0^{t'} E_\varepsilon(t'')dt''} E_\varepsilon(t')\left(1 - e^{-\lambda t'}\right) \quad (5)$$

to the first-order cumulant which is given by

$$\langle \psi_\varepsilon | e^{-\eta t} - 1 | \phi \rangle \sim \delta\left(e^{-\lambda t} - 1\right), \quad \lambda = \beta \Delta_\varepsilon \alpha v \quad (6)$$

with $\beta\langle\psi_\varepsilon | V^0 \alpha_L |\phi\rangle v \sim \delta\lambda$ where Δ_ε and α imply the average energy for the relaxation modes ε and linear thermal expansion coefficient of the lattice liquid at T_h, respectively. The non-dimensional smallness parameter δ is also introduced.

GLASS TRANSITION

Now we will discuss the glass transition when the perturbation (6) is introduced. From eq.(5), we obtain a disorder parameter ξ_ε at a nonequilibrium glassy state $|\psi_\varepsilon\rangle$ at time t via $W_\varepsilon = |\langle\psi_\varepsilon|\psi(t)\rangle|^2/\langle\Phi|\Phi\rangle$ as

$$W_\varepsilon = \delta^2 I_\varepsilon^2 \equiv \delta^2 \xi_\varepsilon \\ I_\varepsilon = e^{-\int_0^t E_\varepsilon(t')dt'}\int_0^t dt' e^{\int_0^{t'} E_\varepsilon(t'')dt''} E_\varepsilon(t')(1 - e^{-\lambda t'}) \quad (7)$$

and its time-derivative by

$$w_\varepsilon = \frac{dW_\varepsilon}{dt} = 2E_t I_\varepsilon\left(1 - e^{-\lambda t} - I_\varepsilon\right)\delta^2. \quad (8)$$

The excess heat capacity can be obtained as proportional to eq.(8) from the entropy, and thus the glass transition temperature is determined by the temperature when eq.(8) takes the maximum value. It should be noted that $W_\varepsilon \sim |\langle\psi_\varepsilon|\psi(t)\rangle|^2 \sim \langle\psi_\varepsilon|\Delta p\rangle^2$ implies a square of the density fluctuation for the mode ε, being a nonequilibrium disorder parameter characterizing a glass. On the other hand, $\langle\phi|\psi(t)\rangle = 0$ means that the average of the density fluctuation gives zero such that $\langle\Delta p\rangle_\phi = \langle\phi|\Delta p\rangle = 0$.

A smallness parameter δ gives no principal effect on the glass transition as we see from eqs.(7) and (8), and thus we study $\widehat{W}_\varepsilon = W_\varepsilon/(2\delta)^2 = \xi_\varepsilon/4$. Although all nondiffusive modes should be considered for the transition, we take account of the most extended relaxation mode $\varepsilon = 1$, having the largest relaxation time $\tau = \tau_1$ in them, for simplicity, since in the cooling process, the longest-lived relaxation mode controls the glass transition. The relaxation time corresponding to this mode is confirmed previously given in the form $E(t) = \tau(t)^{-1} \sim \omega_s e^{-\beta(t)\Delta}$ [9]. On these simplifications, we numerically calculate eqs.(7) and (8).

Figure 1(a) shows the temperature dependences of $\widehat{W}_\varepsilon$ for the cooling rates $v = 10^{-4} - 10\,[K/s]$, which saturate at smaller temperatures. This saturation means that the system gets frozen in the nonequilibrium glassy state. A set of used parameters, with which we will discuss the glass transition hereinafter, are such that $\alpha = 2.32 \times 10^{-4}\,[\deg^{-1}]$, $T_h = 740\,[K]$, $\omega_s = 1 \times 10^{12}$ [rad/s] and $\Delta = 1.555\,[eV]$. The glass transition temperatures are also obtained by the maximum value of eq.(8) [Fig.1(b)] where the normalized time-derivative by the cooling velocity $w_{nor} = w/\{(2\delta)^2 v\}$ is shown. It is readily recognized that the maximum temperatures decrease and gradually approach to a lower-limit temperature with decreasing cooling rate.

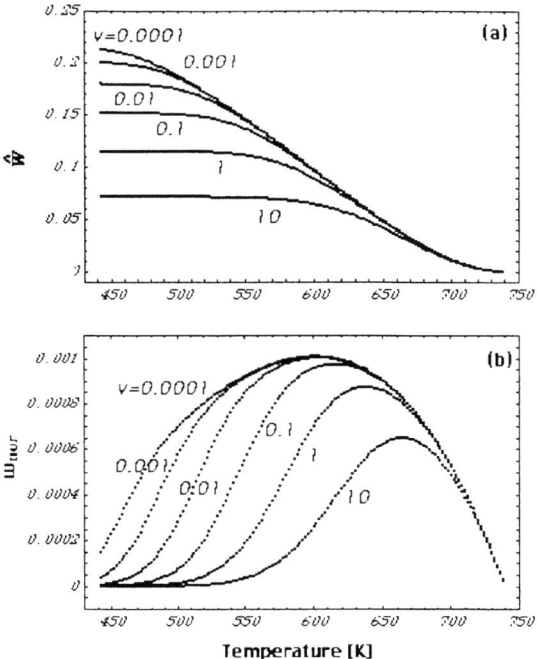

FIGURE 1. Temperature dependences of (a) the disorder parameter $\hat{W}$ and (b) its time-derivative normalized by the cooling rate w_{nor} in the cooling process.

In order to discuss the glass transition quantitatively, the cooling rate dependence of the transition temperature T_g is obtained in Fig.2, together with the experimental results of a Johnson alloy [16] using the same set of parameters. As has been mentioned, T_g saturates as $v \to 0$ to a lower-limit temperature $T_{g,v\to 0} = T_0$ which may correspond to the Kauzmann temperature T_K [17]. Analytically, T_0 is obtained from $dw/dt = 0$ as

$$\exp\left(\frac{1}{\hat{T}} - \frac{1}{\hat{T}_h}\right) - 1 = \frac{1}{1 - 2\hat{T}} \quad (9)$$

which is universal for $\hat{T}_h$, where $\hat{T} = T/(\alpha \Delta T_h)$ and $\Delta = \Delta/k_B$ in units of $k_B = 1$. For the Johnson alloy in Figs.1 and 2, $T_0 = 601[\text{K}]$. On the other hand, in the limit $v \to \infty$, we have the glass transition temperature $T_{g,v\to \infty} = T_\infty$ by

$$\frac{\Delta}{T^2} e^{-(\lambda/v)x}\left(e^{(\lambda/v)x} - 1 - \alpha T_h\right)\int_0^x E\left(1 - e^{-(\lambda/v)y}\right)dy$$
$$= E\left(e^{(\lambda/v)x} - 1\right)^2, \quad x = T_h - T \quad (10)$$

which gives the value of $T_\infty \sim 690[\text{K}]$. As can be seen in Fig.2, the peak transition temperature $T_p \sim 691[\text{K}]$ at $v \sim 10^{2.6}[\text{K/s}]$ exists which is quite close to T_∞. Thus for the present physical parameters, the possible transition temperatures exist in $T_0 < T_g < T_\infty$.

But in general, we have the upper limit temperature T_U in the limit $v \to \infty$. From $e^{(\lambda/v)x} - 1 - \alpha T_h > 0$ in eq.(10), we have

$$\hat{T}_\infty < \hat{T}_U, \quad \hat{T}_U = \frac{\hat{T}_h}{1 + \hat{T}_h \ln(1 + \alpha T_h)}. \quad (11)$$

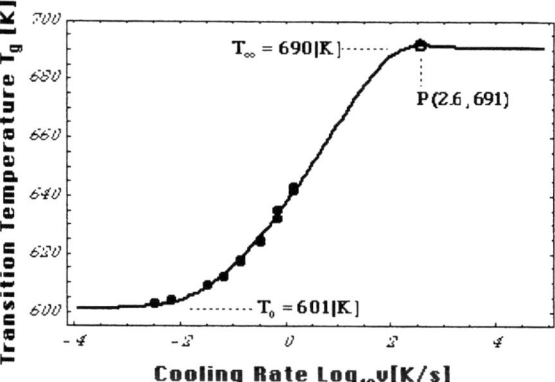

FIGURE 2. Cooling rate dependence of the glass transition temperature: the solid circles are experimental obtained in a Johnson alloy [16] and solid line is theoretical. Also are shown the two transition temperatures, $T_0 \sim 600.916[K]$ and $T_\infty \sim 690.337[K]$.

Consequently, the glass transition temperature exists in the region $T_0 < T_g < T_U$. This upper limit temperature T_U may correspond to the melting temperature T_M, whose experimental value for the Johnson alloy is $T_M \sim 710[\text{K}]$ [18], while the theoretical estimate gives $T_U \sim 713[\text{K}]$.

The dependence of T_g on the cooling rate has sometimes been discussed by the Vogel-Fulcher-Tammann [VFT] relation $T_g = T_{g0} + A/\ln(B/v)$ [5,6]. Utilizing the experimental data, we have $T_{g0} = 582[\text{K}]$, $A = 193[\text{K}]$ and $B = 33.4[\text{K/s}]$ (Fig.3(b)). From the present theoretical prediction of Fig.3(a), which is the part of Fig.2, the lower-limit transition temperature, corresponding to T_{g0}, is $T_0 = 601[\text{K}]$ which is larger than T_{g0} by about $20[\text{K}]$. The experimental results are, however, well fitted by both within the experimental cooling rate range.

In the present numerical calculation, the irreducible minimum parameters have been chosen, and thus, are considered to leave some arbitrariness in their values. Nevertheless, we have reasonable physical parameters. One more important parameter is the participation of the attempt frequency ω_s. This frequency comes from the vibration of ions sitting on localized potential wells, in the present model, which is now taken to be constant at any site. This suggests the Einstein-like optic mode which almost has no dispersion, and more specifically, means the low lying mode whose energy amounts to be around $1[\text{meV}] \sim 1.52 \times 10^{12}[\text{rad/s}]$.

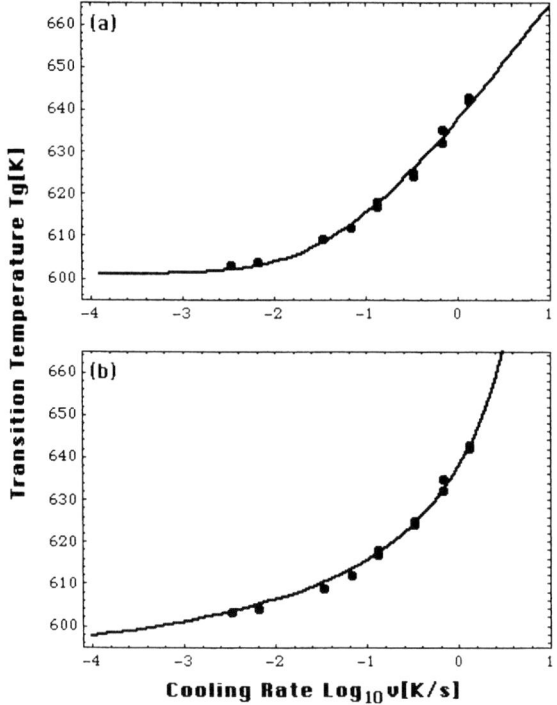

FIGURE 3. (a) Theoretical fit and (b) VFT fit to the experimental data where the latter is approximately given by $T_g = 582 + 83.8/(1.52 - \log_{10} v)$ [see text].

In this paper, the glass transition is proposed where it is charcterized by (1) the growth and freeze of the disorder parameter W_ε, especially a freeze of the non-equilibrium glass state, and (2) the peak of the w_ε which represents the peak of the specific heat, although the vibrational part is not considered [19]. The result explains quite well the cooling rate dependence of the transition temperature, and in addition, gives the upper and lower transition temperatures. The freezing of the disorder parameter means that of the random density distribution inherent to the relaxation mode concerned. The distribution map of relaxations in supercooled liquid has been visualized in the MD simulation [20]. This corresponds to the density map of the relaxation modes. Thus from the above facts, the present freezing of the relaxation modes are the glass transition.

The hopping-initiation temperature T_h is considered to be the ideal crossover from normal to lattice liquid, which means that the region $T > T_h$ is the normal liquid, $T_h > T > T_g$ is the lattice liquid and thus $T_g > T$ is the glass region. The glass transition occurs within $T_0 < T_g < T_U$.

ACKNOWLEDGMENTS

Helpful discussion and permission for use of the experimental data by Dr. H. Takahashi are gratefully acknowledged.

REFERENCES

1. Leutheusser, E., *Phys. Rev.* **A29**, 2765-2773 (1984).
2. Gotze, W., "Liquids, Freezing and Glass Transition" edited by J. P. Hansen et al., North-Holland, Amsterdam, 1991, pp. 287-503.
3. Dyre, J.C., *Phys. Rev. Lett.* **58**, 792-795 (1987).
4. Odagaki, T., and Hiwatari, Y., *Phys. Rev.* **A41**, 929-937 (1990).
5. Bruning, R., and Samwer, K., *Phys. Rev.* **B46**, 11318-11322 (1992).
6. Vollmayr, K., Kob, W., and Binder, K., *J. Chem.Phys.* **105**, 4714-4728 (1996).
7. Ishii, T., *Prog. Theor. Phys.* **73**, 1084-1097 (1985).
8. Ishii, T., *Prog. Theor. Phys.* **77**, 1364-1375 (1987).
9. Ishii, T., and Abe, T., *J. Phys. Soc. Jpn.* **69**, 2549-2558 (2000).
10. Ishii, T., *Solid State Commun.* **116**, 327-331 (2000).
11. Nowick, A.S., Lim, B.S., and Vaysleyb, A.V., *J. Non-Cryst. Solids* **172-174**, 1243-1251 (1994).
12. Colmenero, J., Arbe, A., and Alegria, A., *Phys. Rev. Lett.* **71**, 2603-2606 (1993).
13. Ishii, T., *J. Phys. Soc. Jpn.* **69**, 139-148 (2000).
14. Ishii, T., *Recent Res. Devel. Physics*, Transworld, Res. Network, India, **3**, 2002, pp.613-640.
15. Van Kampen, N.G., *Stochastic Processes in Physics and Chemistry*, North-Holland, Amsterdam, 1981, Chaps.V and XII.
16. Hiki, Y., and Takahashi, H., *14th Symposium thermophysical properties*, Boulder, 2000 (The results obtained is the heating rate dependence of the glass transition temperature).
17. Kauzmann, W., *Chem. Rev.* **43**, 219-256 (1948).
18. Takahashi, H., private communication.
19. Tao, T., Yoshimori, A., and Odagaki, T., *Phys. Rev.* **E64**, 46112-46116 (2001).
20. Perera, D., and Harrowell, P., *J. Chem. Phys.* **111**, 5441-5454 (1999).

Molecular Dynamics simulations and Neutron Spin Echo experiments of difluorotetrachloroethane glassy crystal

F. Affouard, E. Cochin, R. Decressain, M. Descamps* and W. Haeussler[†]

*Laboratoire de Dynamique et Structure des Matériaux Moléculaires
CNRS UMR 8024, Universite Lille 1, 59655 Villeneuve d'Ascq Cedex, France
[†]Institut Laue-Langevin, 38042 Grenoble, France

Abstract. Neutron Spin Echo experiments and Molecular Dynamics simulations (MD) have been performed on difluorotetrachloroethane (CFCl2-CFCl2) glassy crystal. This complementary investigation shows that systems whose dynamics are almost completely controlled by orientational degrees of freedom share some common dynamical features with glass-forming liquids relatively well described by the idealized version of the Mode Coupling Theory (MCT). It also reveals the existence of two remarkable dynamical crossover temperatures in the pico-nanosecond regime very much in keeping with recent views proposed for supercooled liquids. The highest one, $T_A \simeq 190$ K, marks the onset of slow dynamics i.e. non exponential and non-Arrhenian relaxations. The lowest one, $T_c \simeq 125$ K, corresponds to the critical temperature predicted by MCT.

INTRODUCTION

Some molecular crystals generally made of globular molecules exhibit a partially disordered phase, called *plastic*, in which the average position of the centers of mass are ordered on a lattice while the orientations are dynamically disordered. Some of them, called *glassy crystals* [Suga and Seki(1974)], such as cyanoadamantane (CNa) [Affouard et al.(2002)], ethanol [Criado et al.(2000), Benkhof et al.(1998)], cyclooctanol [Brand et al.(1997)] or difluorotetrachloroethane [Krüger et al.(1994)] can be deeply supercooled and present many properties characteristic of the conventional molecular liquid glasses such as a step in the specific heat at the glass transition temperature T_g or a non-Arrhenius behavior of the relaxation times. In opposition to the so called "orientational glasses" [Höchli et al.(1990)], glass formation is not induced by quenched diluted disorder in glassy crystals, but it occurs similarly to conventional glasses. Glassy crystals offer valuable possibilities to focus mainly on the role of the orientational degrees of freedom whose the importance during the glass formation remains a matter of debate at present.

Plastic crystals have also been recently found relevant to be investigated in the pico-nanosecond regime. In this regime, the existence of two remarkable temperatures has been particularly demonstrated from molecular dynamics (MD) simulation for glass-forming liquids [Schroeder et al.(2000)]: (i) T_c, the critical temperature predicted by the mode coupling theory MCT [Götze and Sjögren(1992)] where a dynamical decoupling is expected. It is now well accepted that T_c also marks the crossover to the *landscape-dominated* regime, as advocated long time ago by Goldstein [Goldstein(1969)], (ii) T_A, the temperature associated with the onset of non-exponential and non-Arrhenius relaxations where the dynamics of the system has been identified as *landscape-influenced* [Sastry et al.(1998)]. In [Affouard et al.(2001a), Affouard et al.(2001b), Affouard and Descamps(2001)], we particularly showed from NMR and Raman experiments, and molecular dynamics (MD) computer simulations that both crossover temperatures T_A and T_c where rotational dynamics change in nature can be extracted in different plastic crystals.

A fundamental question concerns the microscopic description of the cooperative mechanisms which develop over a temperature range in which dynamics start being both relatively well described by the predictions of MCT and influenced by the potential energy landscape. We have found glassy crystal difluorotetrachloroethane (DFTCE) to be a very favorable and relevant system to investigate dynamics in this pico-nanosecond regime by means of Neutron Spin Echo (NSE) experiments and MD simulations. This compound is composed of simple molecules $CFCl_2 - CFCl_2$ (see figure 1) close to dumbbells extensively used in MD calculations as prototype of molecular glass-forming liquids [Kämmerer et al.(1997), Michele and Leporini(2001), Chong and Götze(2002)]. Furthermore, DFTCE has been experimentally widely studied and presents a rich variety of interesting properties such as a glass transition of the overall rotation of the molecules at $T_g = 86$ K [Kishimoto et al.(1978)] and two other specific heat anomalies found at 60 and 130 K and associated respectively to a sub-T_g β process and the freezing of the transformation between *trans* and *gauche* conformations of the molecule. Changes concerning the nature of dynamics in this system have been reported from NMR experiments [Stokes et al.(1979)], Brillouin and dielectric spectroscopy [Krüger et al.(1994)] where it was suggested that the freezing process could be described on the basis of MCT.

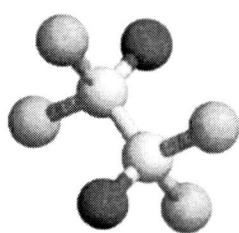

FIGURE 1. *Trans*-conformer of the difluorotetrachloroethane (CFCl2-CFCl2) molecule.

EXPERIMENTS AND DETAILS OF THE SIMULATION

The NSE experiments were performed on the IN11 spectrometer at the Institut Laue-Langevin (ILL), Grenoble, France. We used the multidetector version of IN11 which allows simultaneous measurements at different scattering angles. A setup with an incident wavelength of $\lambda = 5.5$ Å ($\Delta\lambda/\lambda \simeq 16$ %) gave access to wave numbers approximately between 1.28 and 1.72 Å^{-1} and we covered the time range 8 ps - 1 ns. The sample consists of commercial 1,2-difluorotetrachloroethane purchased from Sigma-Aldrich and purified by melting zone. We made measurements of the coherent signal of a polycrystalline system filled into a flat Al container from $T = 150$ to 220 K in steps of 10 K. For normalization, a resolution scan of the elastic scattering was realized at $T = 5$ K in the glassy phase and all data were divided by the polarization at this temperature.

MD calculations were performed on a system of $N = 686$ ($7 \times 7 \times 7$ bcc crystalline cells) molecules. Each DFTCE molecule $CFCl_2 - CFCl_2$ is described by its 8 atoms and considered as a rigid unit. They interact through a Buckingham short range atom-atom potential (see parameters in table 1) and the electrostatic contributions have been neglected since the DFTCE molecule possesses a weak dipolar moment. No structural or dynamical change has been found for a system where electrostatic interactions are taken into account. It should be noted that both structure and dynamics were found in good agreement with experimental results (see figure 2). Newton's equations of motion were solved with a time step of $\Delta t = 5$ fs. We worked in the NPT statistical ensemble with periodic boundaries conditions where the simulation box is allowed to change in size and shape. MD simulations were done for a sample corresponding to the DFTCE orientationally disordered phase at 14 different temperatures from $T = 130$ to 260 K in steps of 10 K. It should be mentioned that our very simple DFTCE model allows us to perform very long MD runs of about 50 ns.

TABLE 1. Simulation coefficients taken from literature for Buckingham potential $\phi(r) = A\exp(-\rho.r) - C/r^6$.

Site-Site	A (kJmol^{-1})	ρ (Å^{-1})	C (kJmol^{-1}Å^6)
C - C	226307	0.288	2420
C - F	196747	0.260	1168
C - Cl	390940	0.284	3864
F - F	171038	0.237	565
F - Cl	320883	0.258	1808
Cl - Cl	586389	0.284	5798

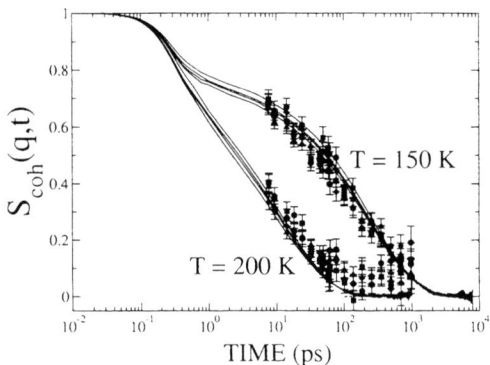

FIGURE 2. Coherent intermediate scattering function $S_{coh}(q,t)$ at $q = 1.54$ Å^{-1} (full circle), 1.59 Å^{-1} (square), 1.64 Å^{-1} (diamond) and 1.70 Å^{-1} (triangle) obtained from NSE experiments at $T = 150$ and 200 K. $S_{coh}(q,t)$ obtained from computer simulations are also displayed at same temperatures and wave vectors (solid lines).

RESULTS AND DISCUSSION

The reorientational motions can be described by the coherent intermediate scattering function which has been calculated as time correlation function of the density operator $S_{coh}(q,t) = \langle \rho_{\vec{q}}(t)\rho_{\vec{q}}(0)\rangle$. The density operator is

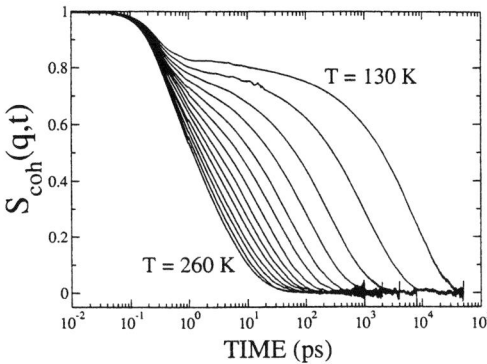

FIGURE 3. Coherent intermediate scattering function $S_{coh}(q,t)$ at $q = 1.54$ $^{-1}$ obtained from MD computer simulations at different temperatures from $T = 130$ to 260 K in steps of 10 K.

defined as $\rho_{\vec{q}}(t) = \sum_{\alpha} b_{\alpha} exp(i\vec{q}.\vec{r}_{\alpha})$ where the sum is over all the atoms of the system, b_{α} and $\vec{r}_{\alpha}$ are respectively the scattering length and the position of the α atom. An average over isotropically distributed q-vectors having the same modulus q is performed in order to obtain $S_{coh}(q,t)$. Good agreement between experimental and numerical $S_{coh}(q,t)$ is found (see figure 2). However, it should be noted that some discrepancies are particularly observed at high temperatures at the longest Fourier times for which the NSE signal does not decay completely to zero. Since no elastic contribution is expected outside the Bragg Peaks, this feature is most likely due to uncertainties in the background substraction.

Fig. 3 (see also figure 2) show the coherent intermediate scattering function $S_{coh}(q,t)$, for the wave vector $q = 1.54$ $^{-1}$ at different temperatures. At high temperatures, it decays rapidly to zero and can be described by a simple exponential shape. With lowering the temperature, the rotator phase of DFTCE exhibits a two-step relaxation as already observed in chloroadamantane plastic crystals [Affouard et al.(2001a)] and classically seen in supercooled liquids [Kämmerer et al.(1997)]. At short time, correlation functions decay to a plateau-like region which reveals the existence of an orientational cage effect *i.e* the rotational analogue of the translational cage effect observed in liquids [Bordat et al.(2003)]. The long-time decay can only be well fitted with a stretched exponential. This regime can be associated in our system, with large tumbling motion between directions where molecules are preferably localized as usually observed in plastic crystals [Affouard et al.(2001a)]. This long process requires collective orientational motions of neighboring molecules and corresponds to the breaking of the cages.

A long relaxation time τ_q can be defined as the time it takes for the correlator $S_{coh}(q,t)$ ($q = 1.54$ $^{-1}$) to decay from 1 to $1/e$. At high temperatures, τ_q can be well fitted with an Arrhenius law $\tau_q(T) = \tau_0 \exp(E_0/T)$ where the parameters (τ_0, E_0) allow us to calculate the temperature dependent activation energy $E(T)/E_0 = T.log(\tau_q/\tau_0)$. Figure 4a displays $E(T)/E_0$ as function of the temperature. Clearly, close to $T_A \simeq 190$ K, the activation energy start diverging from unity associated with the Arrhenian high temperature behavior. T_A indicates the onset of slow dynamics and suggests the existence of a crossover from free rotational diffusion to activated geared tumbling motions corresponding to the *landscape-influenced* regime. Similarly, this feature has been demonstrated in [Sastry et al.(1998)] for modeled glass-forming liquids.

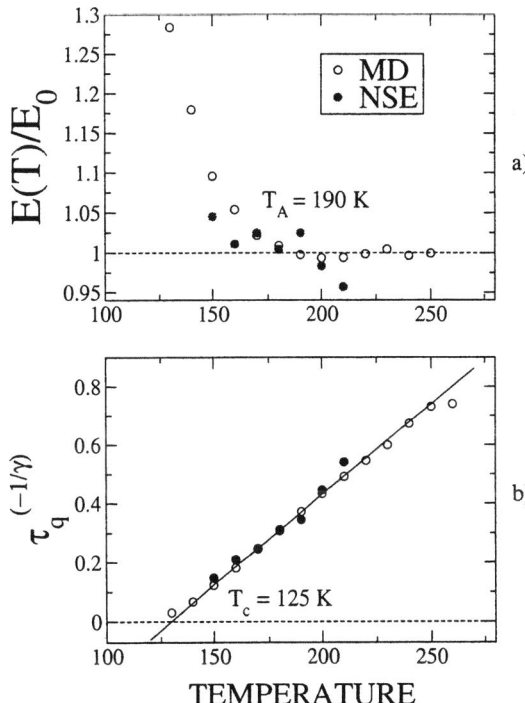

FIGURE 4. a) Temperature dependent activation energy $E(T)/E_0 = T.\ln(\tau_q/\tau_0)$. E_0 and τ_0 parameters are calculated from a fitting procedure of τ_q obtained from MD simulations at high temperatures with an Arrhenius law. b) α relaxation time $\tau_q^{-1/\gamma}$ as function of temperature obtained from neutron spin echo experiments and MD simulation. The exponent $\gamma = 2.41$ has been extracted from the MCT analysis in the β regime. Solid lines indicate MCT-fit using the linear law $\tau_q^{-1/\gamma} \sim (T - T_c)$.

In the MCT framework, a two-step relaxation scenario (fast β, slow α) of all the time dependent correlators is predicted. At short times, MCT particularly predicts that all correlators decay to a plateau value, the so-called non-ergodicity parameter f_q^c. Above T_c, the late β regime or the early α relaxation can be described with the classical von Schweidler law $S_{coh}(q,t) = f_q^c - h_q^{(1)}.(t/\tau)^b$ where $\tau \propto (T - T_c)^{-\gamma}$. The parameters γ and b are temperature and q independent and related via $\gamma = 1/2a + 1/2b$ and $\Gamma^2(1-a)/\Gamma(1-2a) = \Gamma^2(1+b)/\Gamma(1+2b)$ where $\Gamma(x)$ is the gamma function. In order to verify the validity of those scaling laws, we first fixed the exponent b using the factorization theorem in the β regime as described in [Affouard and Descamps(2001)]. Using a fitting procedure performed at several different temperatures for different time correlators such as $S_{coh}(q,t)$ we obtained the best results for $b = 0.60$ which corresponds to $a = 0.32$ and $\gamma = 2.41$. Then, fixing these values, we performed individual fits of $S_{coh}(q,t)$ in the late β regime using the above power law including a second order correction. The critical temperature $T_c \simeq 125$ K was found for the different correlators. At long time, MCT also predicts that the previous fast regime is followed by a slow relaxation, called α, with the characteristic time τ. It is assumed that any characteristic times belonging to the α regime show asymptotically the same temperature dependence $\tau_q \sim \tau$. According to MCT, the α relaxation times $\tau^{-1/\gamma}(T)$ should yield straight lines intersecting the abscissa at $T = T_c$. Fig. 4 shows that for the correlator $S_{coh}(q,t)$ at $q = 1.54$ this prediction holds well over a relatively large temperature range both for MD and NSE data. Extrapolation of the temperature dependence gives a critical temperature of $T_c \simeq 125$ K consistent with the temperature found in the β regime.

CONCLUSION

Convincing experimental and numerical evidence for two dynamical crossovers ($T_A = 190$ K, $T_c = 125$ K) in glassy crystal DFTCE in the ps-ns regime have been obtained, very much in keeping with recent views on structural glass-formers. Our work also revealed that dynamics of orientationally disordered crystal can be well described by some MCT prediction in the temperature range $[T_c, T_A]$. This is a not trivial result since MCT was originally developed for simple Van Der Waals atomic liquids. Results of the present study suggest new interesting possibilities for testing the different theoretical approaches of the glass formation. Promising results concerning the Potential Energy Landscape description of glassy crystals have recently been obtained.

ACKNOWLEDGMENTS

The authors wish to acknowledge the use of the facilities of the IDRIS (Orsay, France) and the CRI (Villeneuve d'Ascq, France) where calculations were carried out.

This work was supported by the INTERREG III program (Nord Pas de Calais/Kent).

REFERENCES

[Suga and Seki(1974)] Suga, H., and Seki, S., *J. of Non-Cryst. Solids*, **16**, 171–194 (1974).
[Affouard et al.(2002)] Affouard, F., Willart, J.-F., and Descamps, M., *J. of Non-Cryst. Solids*, **307-310**, 9–15 (2002).
[Criado et al.(2000)] Criado, A., Jimenez-Ruiz, M., Cabrillo, C., Bermejo, F. J., Fernandez-Perea, R., Fischer, H. E., and Trouw, F. R., *Phys. Rev. B*, **61**, 12082–12093 (2000).
[Benkhof et al.(1998)] Benkhof, S., Kudlik, A., Blochowicz, T., and Rössler, E., *J. Phys.: Condens. Matter*, **10**, 8155–8171 (1998).
[Brand et al.(1997)] Brand, R., Lunkenheimer, P., and Loidl, A., *Phys. Rev. B*, **56**, 5713–5716 (1997).
[Krüger et al.(1994)] Krüger, J. K., Schreiber, J., Jimenez, R., and Bohn, K.-P., *J. Phys.: Condens. Matter*, **6**, 6947–6964 (1994).
[Höchli et al.(1990)] Höchli, U. T., Knorr, K., and Loidl, A., *Adv. Phys.*, **39**, 405–615 (1990).
[Schroeder et al.(2000)] Schroeder, T. B., Sastry, S., Dyre, J. C., and Glotzer, S. C., *J. Chem. Phys.*, **112**, 9834–9840 (2000).
[Götze and Sjögren(1992)] Götze, W., and Sjögren, L., *Rep. Prog. Phys.*, **55**, 241–376 (1992).
[Goldstein(1969)] Goldstein, M., *J. Chem. Phys.*, **51**, 3728–3739 (1969).
[Sastry et al.(1998)] Sastry, S., Debenedetti, P. G., and Stillinger, F. H., *Nature*, **393**, 554–557 (1998).
[Affouard et al.(2001a)] Affouard, F., Cochin, E., Decressain, R., and Descamps, M., *Europhys. Lett.*, **53**, 611–617 (2001a).
[Affouard et al.(2001b)] Affouard, F., Hédoux, A., Guinet, Y., Denicourt, T., and Descamps, M., *J. Phys.: Cond. Matter*, **13**, 7237–7248 (2001b).
[Affouard and Descamps(2001)] Affouard, F., and Descamps, M., *Phys. Rev. Lett.*, **87**, 035501/1–4 (2001).
[Kämmerer et al.(1997)] Kämmerer, S., Kob, W., and Schilling, R., *Phys. Rev. E*, **56**, 5450–5461 (1997).
[Michele and Leporini(2001)] Michele, C. D., and Leporini, D., *Phys. Rev. E*, **63**, 036702/1–10 (2001).
[Chong and Götze(2002)] Chong, S.-H., and Götze, W., *Phys. Rev. E*, **65**, 041503/1–17 (2002).
[Kishimoto et al.(1978)] Kishimoto, K., Suga, H., and Seki, S., *Bull. Chem. Soc. Japan*, **51**, 1691–1696 (1978).
[Stokes et al.(1979)] Stokes, H., Case, T., Ailion, D., and Wang, C., *J. Chem. Phys.*, **70**, 3563–3571 (1979).
[Bordat et al.(2003)] Bordat, P., Affouard, F., Descamps, M., and Müller-Plathe, F., *J. Phys.: Cond. Matter*, **15**, 5397–5407 (2003).

Single Particle Jumps in a Glass: A Computer Simulation

K. Vollmayr-Lee

Department of Physics, Bucknell University, Lewisburg, PA 17837, USA

Abstract. We study a binary Lennard-Jones system below the glass transition via molecular dynamics simulations. To investigate the dynamics of the system we define single particle jumps via their single particle trajectories. We find two kinds of jumps: "reversible jumps" where a particle jumps back and forth between two or more states and "irreversible jumps" where a particle does not return to any of its former states. For both the irreversible and reversible jumps we present as a function of temperature the number of jumping particles, jump sizes in position and energy, and times during the jump and between successive jumps.

INTRODUCTION

If a liquid is cooled rapidly enough, so that crystallization is avoided, one obtains a glass [1]. During the transition from liquid to glass drastic changes occur in the dynamics. One commonly used picture for the dynamics of glasses is that the molecules forming the glass are caged in, i.e. trapped by their neighbors, and, after long enough waiting time, escape their cage. We focus here on this escape out of the cage ("jump"). Using molecular dynamics simulations we define a jump via single particle trajectories. We find two kinds of jumps, reversible and irreversible jumps, which we analyze separately.

In the following we present the model and simulation details. We then define jump-occurrence and jump-type, followed by a presentation of the number of jumping particles, time scales of the jumps and jump sizes in position and energy and their dependence on temperature. We finish with concluding remarks.

MODEL AND SIMULATION

We use a binary Lennard-Jones (LJ) mixture of 800 A and 200 B particles with the same mass. The interaction potential for particles i and j at positions $\mathbf{r}_i$ and $\mathbf{r}_j$ and of type $\alpha, \beta \in \{A,B\}$ is

$$V_{\alpha\beta}(r) = 4\varepsilon_{\alpha\beta}\left(\left(\frac{\sigma_{\alpha\beta}}{r}\right)^{12} - \left(\frac{\sigma_{\alpha\beta}}{r}\right)^{6}\right), \quad (1)$$

where $r = |\mathbf{r}_i - \mathbf{r}_j|$ and $\varepsilon_{AA} = 1.0$, $\varepsilon_{AB} = 1.5$, $\varepsilon_{BB} = 0.5$, $\sigma_{AA} = 1.0$, $\sigma_{AB} = 0.8$ and $\sigma_{BB} = 0.88$. We truncate and shift the potential at $r = 2.5\sigma_{\alpha\beta}$ [2]. In the following we will use reduced units where the unit of length is σ_{AA}, the unit of energy is ε_{AA} and the unit of time is $\sqrt{m\sigma_{AA}^2/(48\varepsilon_{AA})}$.

We carry out molecular dynamics (MD) simulations using the velocity Verlet algorithm with a time step of 0.02. The volume is kept constant at $V = 9.4^3 = 831$ and we use periodic boundary conditions. To study microscopic dynamics below the glass transition, which is according to previous simulations around 0.435 [2, 3], we analyze here simulations at $T = 0.15$, 0.2, 0.25, 0.30, 0.35, 0.38, 0.40, 0.41, and 0.43, as they have been described in [4]. For each temperature we use 10 independent initial configurations and run NVE simulations with $5 \cdot 10^6$ MD steps. During each production run the positions of all particles (configurations) are stored every 2000 MD steps which are then used for analysis.

The relaxation times for these temperatures are significantly larger than the waiting times before the production runs. We therefore study relaxation processes out of equilibrium and accordingly find signs of aging effects.

DEFINITION OF JUMP-OCCURRENCE AND JUMP-TYPE

In this paper we focus on events where a particle escapes its cage, using single particle trajectories $\mathbf{r}_i(t)$ given by the periodically stored configurations.

As sketched in Fig. 1a, we identify jumps for each particle i by comparing changes in its time averaged positions (160000 MD steps apart) $|\Delta \bar{\mathbf{r}}_i|$ with an estimate for its fluctuations in position $\sigma_{i,\text{est}}$. A jump is defined to occur whenever $|\Delta \bar{\mathbf{r}}_i|^2 > 20\sigma_{i,\text{est}}^2$.

When we apply this jump definition, we find two types of jumps, which we call irreversible and reversible

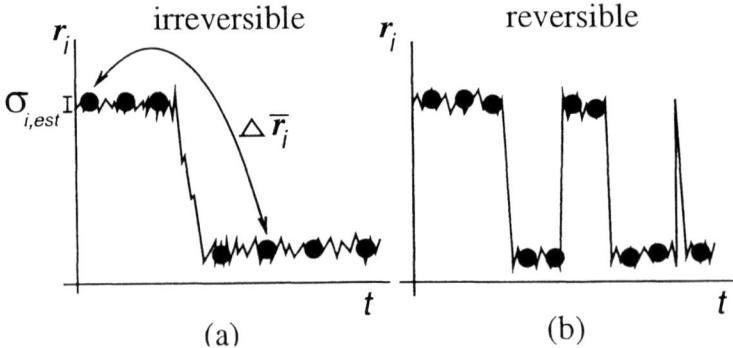

FIGURE 1. Sketch of single particle jumps to illustrate the definitions of jump-occurrence and the jump-types: (a) irreversible and (b) reversible.

jumps. As sketched in Fig. 1a,b, a particle which undergoes an irreversible jump succeeds in escaping its cage (for the time window of the simulation) whereas a particle undergoing a reversible jump returns back to one of its previous average positions. To increase our time resolution we use both the time averaged positions $\overline{\mathbf{r}}_i(t)$ and the stored configurations $\mathbf{r}_i(t)$ for the check of a return to a previous position (see Fig. 1b). We distinguish irreversible and reversible jumps in all following analysis. To distinguish not only irreversible and reversible *jumps* but also *particles*, we define a jumping particle to be irreversible if it undergoes only irreversible jumps and otherwise to be a reversible jumping particle. For further details of the definition of jump-occurrence and jump-type, and for a comparison with previous work on jumps, see [5].

NUMBER OF JUMPING PARTICLES

Applying the above definitions of jump-occurrence and jump-type we obtain the number of jumping particles as a function of temperature T (see Fig. 2). As one might expect, the number of jumping particles increases with increasing temperature consistent with an increasing number of relaxation processes. More surprisingly, we find that not only the smaller B-particles but also A-particles are jumping. However, relative to the number of particles in the system ($N_A = 800, N_B = 200$) a larger fraction of B-particles jump due to their smaller size. Both irreversible and reversible jumps occur at all temperatures.

Fig. 3 shows how the number of irreversible jumping particles, normalized by the number of jumping particles, depends on temperature. We find that this ratio is increasing with increasing temperature for temperatures $T > 0.35$, and interpret this increase as sketched in Fig. 4. Both irreversible and reversible jumping particles begin with a jump out of their cage of neigh-

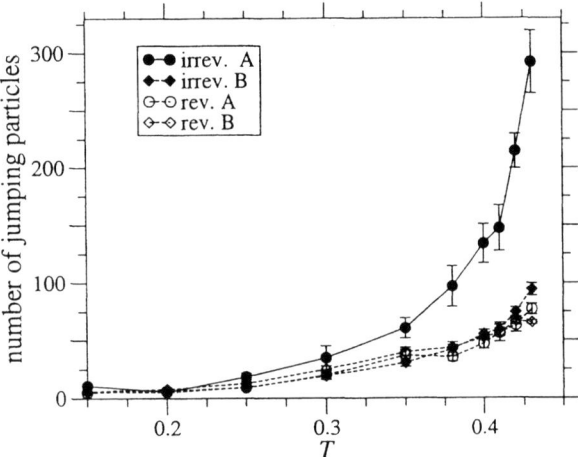

FIGURE 2. Number of jumping particles as a function of temperatures for irreversible and reversible jumpers.

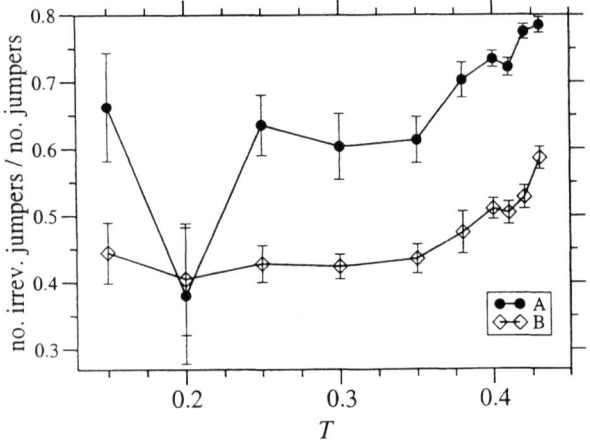

FIGURE 3. Number of irreversible jumping particles divided by the number of both reversible and irreversible jumping particles.

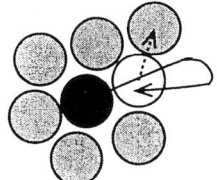
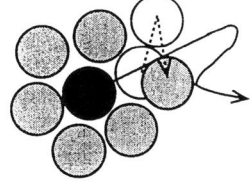

FIGURE 4. Illustration for our interpretation of Fig. 3.

boring particles possibly through an "opening" in the cage. After this jump, however, the reversible jumper returns back to its cage, whereas the irreversible jumper can no longer return because of rearrangements of the cage forming neighboring particles which block the "entrance" back into the cage. The fraction of irreversible jumps increases with increasing temperature because the mobility of the cage forming particles and therefore their rearrangements become more likely at larger temperatures.

TIMES

We investigate next the time duration of a jump Δt_d, the time between two successive jumps Δt_b, the time before the first jump of a particle Δt_{head}, and the time after the last jump of a particle Δt_{tail}.[1] For precise definitions of these times and the approximate correction of the finite time window of the simulation see [5]. Fig. 5 shows that $\Delta t_d \ll \Delta t_b$, which explains why the above definition of jump-occurrence allowed us to identify jumps. Furthermore, Δt_b is temperature independent, which might seem contrary to increased particle mobility with increasing temperature. We interpret this temperature independence of Δt_b as an aging effect for two reasons: (1) $\Delta t_{tail} > \Delta t_{head}$ (see Fig. 5) because jumps are more likely to occur at the beginning of the production run, and (2) because Doliwa et al. [6], who investigated very long simulation runs, found initially a temperature independence of waiting times Δt_b but a temperature dependence of Δt_b at later times.

---

[1] Please notice that Δt_b is only defined in the case of multiple jumps of a single particle, whereas Δt_d, Δt_{tail} and Δt_{head} are also defined in the case of single jumps.

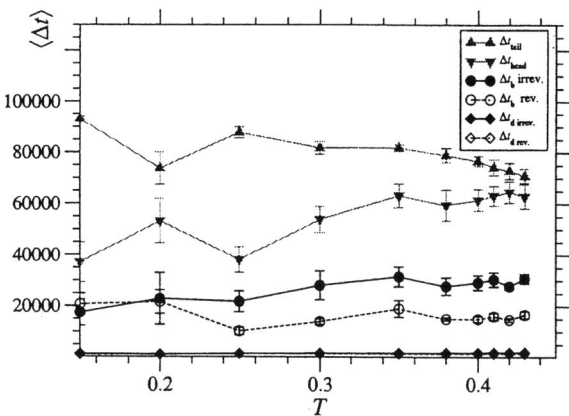

FIGURE 5. Times Δt_{tail}, Δt_{head} for both reversible and irreversible jumps and Δt_b, Δt_d separately for reversible and irreversible jumps of both A and B particles.

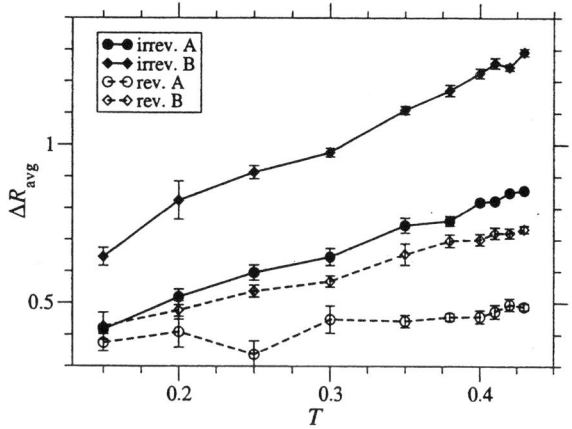

FIGURE 6. Jump size in position ΔR_{avg} as a function of temperature separately for irreversible and reversible jumps of A and B particles.

JUMP SIZES IN POSITION AND ENERGY

To investigate how far particles jump we define ΔR_{avg} as the distance between average positions before and after the jump (for details see [5]). Fig. 6 shows that the smaller B particles jump farther than the A particles, and the reversible jumps are longer (i.e. further away from the original cage) than the irreversible jumps. With increasing temperature relaxation processes are enhanced by larger jumps in position.

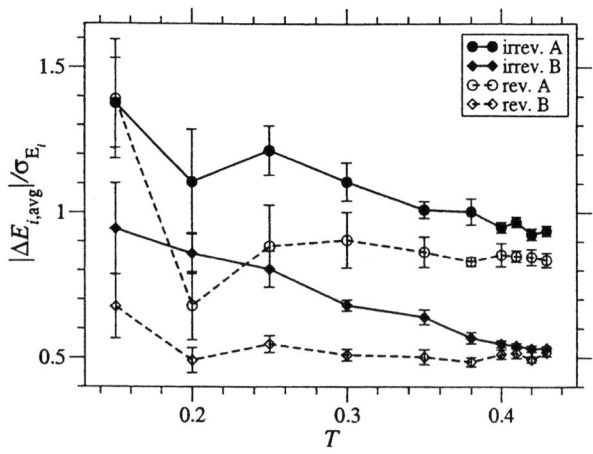

FIGURE 7. Absolute value of the change in single particle potential energy divided by single particle fluctuation in single particle potential energy.

To study these relaxation processes energetically, we use the single particle potential energy $E_i(t) = \sum_{j \neq i} V_{\alpha\beta}(r_{ij}(t))$ and investigate the change in the time averages of $E_i(t)$ before and after the jump $\Delta E_{i,\text{avg}}$. Fig. 7 shows the absolute value of $\Delta E_{i,\text{avg}}$ normalized by the fluctuations in single particle potential energy σ_{E_i}. We find that this ratio $|\Delta E_{i,\text{avg}}|/\sigma_{E_i}$ decreases for irreversible jumps with increasing temperature, which we interpret such that relaxation processes are increasingly more driven by thermal fluctuations.

CONCLUSIONS

We study a binary Lennard-Jones system below the glass transition. Using single particle trajectories we identify jumps of particles escaping their cage. At all investigated temperatures two types of jumps occur: reversible and irreversible jumps. With increasing temperature relaxation processes are enhanced by an increasing number of jumping particles and larger jumps in position and energy. The waiting times between successive jumps seem temperature independent, which we interpret to be due to aging in our system. Further investigations thereof are in progress. With increasing temperature the jumps are increasingly driven by thermal fluctuations.

ACKNOWLEDGMENTS

I would like to thank K. Binder and A. Zippelius for hospitality, financial support and for their helpful discussions. I also thank J. Horbach for fruitful discussions and B. Vollmayr-Lee for a reading of this manuscript. Financial support from SFB 262 and DFG Grant No. Zi 209/6-1 is gratefully acknowledged.

REFERENCES

1. For a review see C. A. Angell, *Science* **267**, 1924 – 1935 (1995).
2. W. Kob and H.C. Andersen, *Phys. Rev. E* **51**, 4626 – 4641 (1995); *Phys. Rev. Lett.* **73**, 1376 – 1379 (1994); *Phys. Rev. E* **52**, 4134 – 4153 (1995).
3. J. -L. Barrat and W. Kob, *Europhys. Lett.* **46**, 637 – 642 (1999).
4. K. Vollmayr-Lee, W. Kob, K. Binder and A. Zippelius, *J. Chem. Phys.* **116**, 5158 – 5166 (2002).
5. K. Vollmayr-Lee, submitted to *J. Chem. Phys.*.
6. B. Doliwa and A. Heuer, *Phys. Rev. E* **67**, 031506 (2003); e-print cond-mat/0306343 (2003); *Phys. Rev. E* **67**, 030501(R) (2003).

Self-Atomic Motions in Glass-Forming Polymers: Neutron Scattering and Molecular Dynamics Simulations Results

J. Colmenero*†, F. Alvarez*†, A. Narros†, A. Arbe*, M. Monkenbusch**, D. Richter** and B. Farago‡

*Unidad de Física de Materiales CSIC-UPV/EHU Apartado 1072, 20080 San Sebastián, SPAIN
†Departamento de Física de Materiales UPV/EHU, Apartado 1072, 20080 San Sebastián, SPAIN
**Institut für Festkörperforschung, Forschungszentrum Jülich GmbH, D–52425 Jülich, GERMANY
‡Institut Laue–Langevin, BP 156, 38042 Grenoble Cedex 9, FRANCE

Abstract. By a combined effort of neutron scattering and molecular dynamics simulation, we have recently shown the existence of a crossover from Gaussian to non-Gaussian character of the self-correlation function of hydrogen atoms in the α-relaxation regime of a glass forming polymer, polyisoprene (PI). Following these previous results, here we present new data displaying different features of the above mentioned crossover. In particular, neutron scattering and molecular dynamics simulations corresponding to another glass-forming polymer, poly(vinyl ethylene) (PVE), and simulation results about the effect of temperature and density on the crossover in PI. We found that PVE displays a similar behavior to PI, showing the generality of the crossover picture. Moreover, we have also found that for PI the main features of the crossover do not depend on temperature and density. These results can be rationalized in the framework of the anomalous jump diffusion model which was previously introduced by us for self atomic motions in the α-regime of glass-forming polymers.

INTRODUCTION

Valuable information on the dynamics of the α-relaxation can be obtained by quasielastic neutron scattering (NS). For instance, NS on protonated samples is directly related with the self part of the van Hove correlation function $G_s(\vec{r},t)$ corresponding to the hydrogens in the system. $G_s(\vec{r},t)$ is the probability to find an atom at time t at $\vec{r}$ if it was at $\vec{r}=0$ for $t=0$. Neutron spin echo (NSE) accesses its Fourier transform $F_s(Q,t)$, and e.g. time of flight or backscattering techniques its counterpart $S_{inc}(Q,\omega)$ ($\hbar Q, \hbar\omega$: momentum- and energy-transfer). As NS delivers the spatial information only in the reciprocal space, the interpretation of experimental results is sometimes not straightforward. Magnitudes as $G_s(\vec{r},t)$ and its moments are not directly accessed. In this direction, computer simulations on fully atomistic polymer models have proven to be an useful complementary tool for unraveling NS data, providing one has realistic enough models [1, 2, 3].

In a previous work [3, 4] we have investigated the Q-dependence for the H-self motion in the α-relaxation regime of a glass forming polymer, polyisoprene (PI), by a combined effort involving fully atomistic molecular dynamic (MD) simulations and NS measurements. As it is usually found, we obtained that the slow decay of $F_s(Q,t)$, which is due to the α-process, can be well described by a Kohlrausch-Williams-Watts (KWW) function

$$F_s(Q,t) = A\exp\left[-\left(\frac{t}{\tau_w}\right)^\beta\right], \quad (1)$$

where β is the shape parameter ($0<\beta \leq 1$) and with a Q-dependent characteristic time $\tau_w(Q)$. A is a Lamb-Mössbauer factor giving account for the fast first step of $F_s(Q,t)$. A depends on Q as

$$A = \exp\left[-Q^2\frac{\langle u^2 \rangle}{3}\right] \quad (2)$$

where $\langle u^2 \rangle$ is the associated mean squared displacement. Moreover, by studying the combined NS and simulation results we have established the existence of a crossover in the resulting Q-dependence of $F_s(Q,t)$. In the low Q-regime $F_s(Q,t)$ follows the Gaussian behavior

$$F_s(Q,t) = \exp\left[-Q^2\frac{\langle r^2(t) \rangle}{6}\right] \quad (3)$$

corresponding to a sublinear diffusion [$\langle r^2(t) \rangle \propto t^\beta$]. At larger Q's (shorter distances), $F_s(Q,t)$ strongly deviates from the Gaussian behavior. This crossover mani-

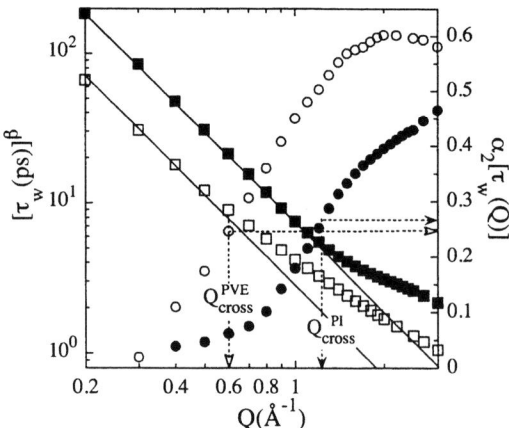

FIGURE 1. Q-dependence of τ_w^β (squares) and $\alpha_2[t = \tau_w(Q)]$ (circles) for PI at 363 K (full) and PVE at 418 K (empty). Solid lines show the Gaussian behavior. PVE times have been shifted (divided by 2) for clarity.

fests in a change in the Q-dependence of τ_w, as can be seen in Fig. 1. The Gaussian behavior of $F_s(Q,t)$ corresponds to a Q^{-2} dependence of τ_w^β while deviations from this dependence are a clear signature of non-Gaussian behavior. As can be seen in Fig. 1, the crossover takes place in the case of PI at $Q_{cross} \approx 1.2 \text{ Å}^{-1}$. The deviations from the Gaussian behavior of $F_s(Q,t)$ are usually quantified in terms of the so called non-Gaussian parameter $\alpha_2(t)$ which is defined as

$$\alpha_2(t) = \frac{3}{5} \frac{\langle r^4(t) \rangle}{\langle r^2(t) \rangle^2} - 1. \quad (4)$$

The mean squared displacement $\langle r^2(t) \rangle$ and $\langle r^4(t) \rangle$ are moments of $G_s(\vec{r},t)$ and can be easily calculated from the atomic trajectories in a MD-simulation. Figure 1 shows that the deviations of $\tau_w(Q)$ from the Gaussian behavior begin to be noticed when α_2 evaluated at $t = \tau_w(Q)$ takes values of the order and above 0.25. Following these previous results, here we present new data displaying different features of the above described crossover. In particular, NS and MD-simulations corresponding to another glass-forming polymer, poly(vinyl ethylene) (PVE), and simulation results about the effect of temperature and density on the crossover in PI.

RESULTS

Figure 2(a) shows the $F_s(Q,t)$ for PVE measured at 418 K by NSE at different Q-values in comparison with those calculated from MD-simulations at the same Q-values and temperature. As can be seen, the agreement is almost perfect indicating that our simulations of PVE provide a very convincing mimic of reality. The second decay of $F_s(Q,t)$ is due to the α-relaxation. Also for PVE, it can be well described by a KWW function with a constant value of the shape parameter ($\beta \approx 0.55$) and a Q-dependent τ_w. Figure 2(b) shows the mean squared displacement $\langle r^2(t) \rangle$ as well as the non-Gaussian parameter $\alpha_2(t)$ calculated from the simulation results. As in the case of PI [1], $\alpha_2(t)$ shows two peaks. It was shown in Ref. [1] that the first peak, which occurs at rather short times corresponding to the first decay of $F_s(Q,t)$, is related with the librational motions of C-H bonds. The second peak of $\alpha_2(t)$ is centered at about 4 ps as in PI at 363 K. This time range corresponds to the "decaging" regime of $\langle r^2(t) \rangle$ after the microscopic behavior. Figure 1 shows that also for PVE, the second peak of $\alpha_2(t)$ is directly connected with the crossover of $\tau_w(Q)$. In this figure, we have also represented $\tau_w^\beta(Q)$ for PVE. We realize that for this polymer the deviation from the Q-dependence which characterizes the Gaussian behavior of $F_s(Q,t)$ takes place at rather low Q-values of Q ($\approx 0.6 \text{ Å}^{-1}$). However, as for PI, this deviation also corresponds to values of about 0.25 of $\alpha_2(t)$ calculated at $t = \tau_w(Q)$. In conclusion, we can say that the crossover from Gaussian to non-Gaussian behavior found for PI definitively is not a particularity of this system but a universal feature of the α-relaxation dynamics in glass forming polymers. The fact that the crossover occurs in PVE at lower Q-values can be understood taking into account that PVE is a polymer with a large side group in the monomer. This introduces an additional heterogeneity in the dynamics.

Now we may also ask whether the crossover from Gaussian to non-Gaussian behavior of the self-correlation function depends on temperature. To do this we have carried out MD-simulations on our PI-model but at different temperatures ranging from 314 K to 513 K. Details of the simulations, as well as a direct comparison with NSE experimental data, can be found in references [1, 4]. We may should only mention that the β-value obtained results to be temperature dependent. This tendency was also experimentally found. Here we will focus on the question of the temperature dependence of the crossover. For this purpose, using appropriate temperature shift factors, we have displayed in Fig. 3(a) the $\tau_w^\beta(Q)$ values obtained from the simulation at different temperatures in form of a master plot. As may be seen, the data from all temperatures collapse to a single master curve. Thus the crossover to non-Gaussianity seems to be not affected by temperature, but appears to be an intrinsic property of the sublinear diffusion process.

The simulations carried out at different temperatures also involve different densities, because at each temperature the density of the simulated cell is adjusted to the experimental density corresponding to that temperature. Now, trying to unravel the density and temperature ef-

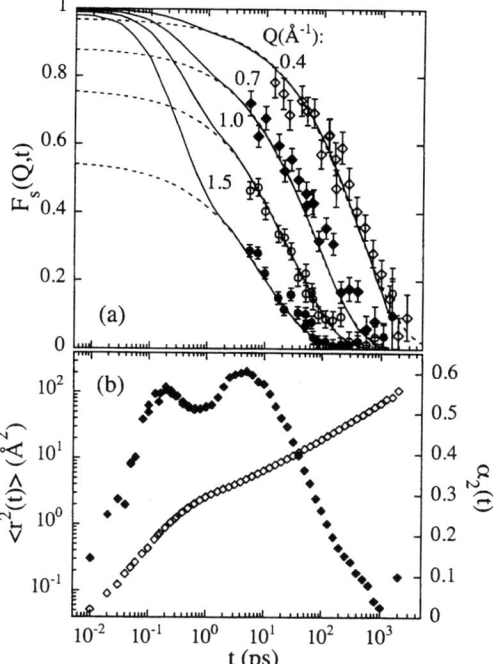
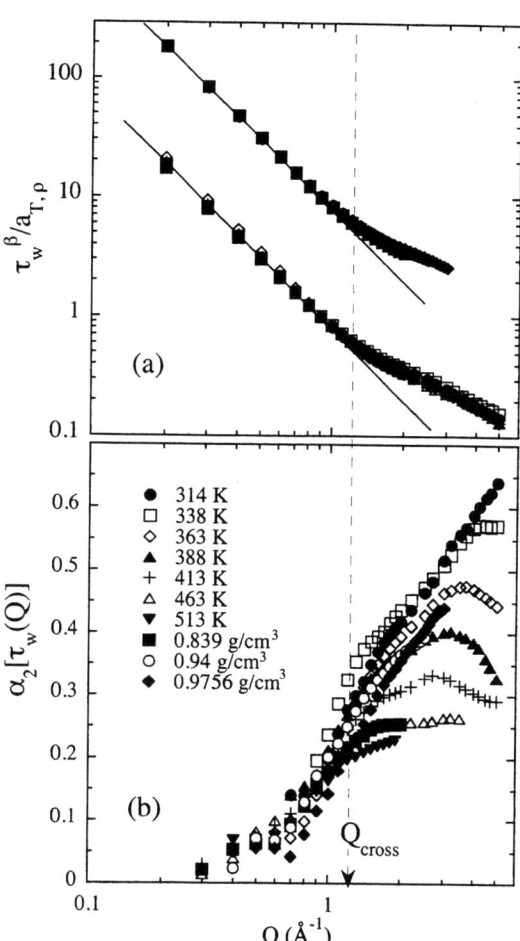

FIGURE 2. Results corresponding to all Hydrogens in PVE at 418 K: (a) $F_s(Q,t)$ from simulation (lines) and NSE results (symbols) at the Q-values indicated. Dotted lines: KWW fits of the second decay. (b): Time evolution of the mean squared displacement $\langle r^2 \rangle$ (empty) and the non-Gaussian parameter α_2 (full) calculated from the MD-simulations.

FIGURE 3. (a): Q-dependence of the master curves obtained for τ_w^β from different starting temperatures (lower curve) and from different starting densities at 513 K (upper curve). Solid lines show the Gaussian behavior. (b): $\alpha_2[t = \tau_w(Q)]$ for all the different starting conditions considered (listed in the figure). The vertical dotted line shows the approximate position of Q_{cross}.

fects, we have also carried out MD-simulations at three different densities of the simulated cell, but keeping constant the temperature (513 K). The results obtained are shown in Fig. 3(a) again in form of a master plot. As we can see, also in this case all data collapse to a single master curve indicating that density does not affect the crossover, at least in the range of densities explored. These results can be rationalized taking into account the value of the non-Gaussian parameter $\alpha_2(t)$ calculated at $t = \tau_w(Q)$ for the different $\tau_w(Q)$ included in Fig. 3(a). The results obtained are displayed in Fig. 3(b). We can see that the data corresponding to different temperatures and densities and those corresponding to different densities at constant temperature, all seem to show the same low-Q asymptotic behavior. Thereby, the crossover for all temperatures and densities always takes place when $\alpha_2[t = \tau_w(Q)]$ takes values of the order of 0.25.

DISCUSSION

As it was discussed in Ref. [4], different theoretical approaches can be invoked to understand the crossover form Gaussian to non-Gaussian behavior of the proton self-correlation function in the α-regime of glass-forming polymers. The results described in this paper and, in particular, the fact that the crossover hardly depends on both density and temperature, are a real challenge for those models. In Ref. [4], we also developed a simple picture which, in principle, seems to be compatible with the framework of the Mode Coupling Theory. This simple interpretation is based on the existence of a distribution of discrete jumps underlying the atomic motions in the α-process. In this framework, at low Q corresponding to large distances, we deal with a Gaussian sublinear diffusion process. At higher Q (shorter distances) the Gaussianity breaks down due to the discrete character of the sublinear diffusion. A distribution of jump lengths ℓ, $f(\ell) = (\ell/\ell_o^2) exp(-\ell/\ell_o)$ results from the disorder characterizing glass-forming polymers. ℓ_o is the preferred jump distance. The results described here indicate that this distribution hardly depends on temperature and density. This very simple model was able to capture the main features of $\tau_w(Q)$ and $\alpha_2(t)$ for PI at 363 K [4]. Now we can see whether this model also reproduces the "universal" features displayed by $\alpha_2[t = \tau_w(Q)]$ in Fig. 3(b). In the framework of such a model (see Ref. [4])

$$\tau_w = \tau_o \left[1 + \frac{1}{Q^2 \ell_o^2}\right]^{\frac{1}{\beta}} \quad (5)$$

where τ_o is a microscopic residence time. On the other hand, the non-Gaussian parameter approximately reads:

$$\alpha_2(t) = \frac{72 \ell_o^4 \left(\frac{t}{\tau_o}\right)^\beta}{\left[2 \langle u^2 \rangle + 6 \ell_o^2 \left(\frac{t}{\tau_o}\right)^\beta\right]^2} . \quad (6)$$

$\alpha_2[t = \tau_w(Q)]$ is then straightforwardly calculated:

$$\alpha_2[\tau_w(Q)] = \frac{72 \ell_o^4 \left(1 + \frac{1}{Q^2 \ell_o^2}\right)}{\left[2 \langle u^2 \rangle + 6 \ell_o^2 \left(1 + \frac{1}{Q^2 \ell_o^2}\right)\right]^2} . \quad (7)$$

It is worthy of remark that the expression found does not depend on the β-value, which depends on temperature [4] and density. This expression only depends on ℓ_o and $\langle u^2 \rangle$. Taking into account the values of these two parameters deduced from simulations ($\ell_o \approx 0.42$ A) and NS data ($\langle u^2[A^2] \rangle \approx -0.489 + 0.036T$ [K] in the T-range investigated) in PI, we have calculated $\alpha_2[t = \tau_w(Q)]$ at different temperatures. The results are shown in Fig. 4. As we can see, they nicely capture the main qualitative features displayed in Fig. 3(b) and in particular the low-Q asymptotic behavior which depends on Q as Q^2. The agreement with the simulated data of Fig. 3(b) is even semiquantitative, in particular in the Q-range until about the crossover. In the high Q-regime, the agreement cannot obviously be quantitative because the simulated data are affected by the first peak of $\alpha_2(t)$ which is not considered in the simple anomalous jump diffusion model here invoked.

CONCLUSIONS

By combining NS and MD simulations we have established the general character of the crossover from Gaussian to non-Gaussian behavior of the self-atomic motions in the α-regime of glass-forming polymers. The main features of this crossover do not depend significantly on temperature and density. These new results can also be

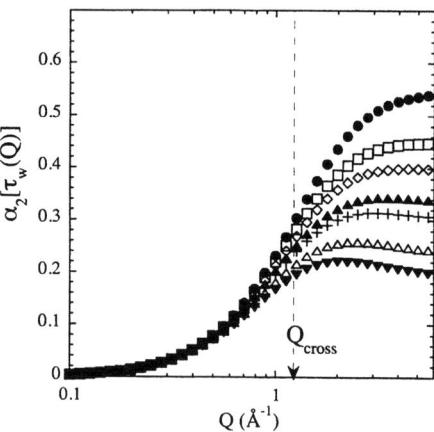

FIGURE 4. Q-dependence of $\alpha_2[t = \tau_w(Q)]$ calculated from the anomalous jump diffusion model for the same temperatures shown in Fig. 3. Symbols as in Fig. 3.

understood in the framework of the anomalous jump diffusion model previously introduced by us.

ACKNOWLEDGMENTS

We acknowledge support from the following projects: DGICYT, PB97-0638; GV, EX 1999-11; UPV/EHU, 206.215-G20/98, and A. Narros the FPI grant of the Spanish Ministry of Science and Technology. Support from "Donostia International Physics Center" is also acknowledged.

REFERENCES

1. Colmenero, J., Alvarez, F., and Arbe, A., *Phys. Rev. E*, **65**, 041804 (2002).
2. Alvarez, F., Colmenero, J., Zorn, R., Willner, L., and Richter, D., *Macromolecules*, **36**, 238–248 (2003).
3. Arbe, A., Colmenero, J., Alvarez, F., Monkenbusch, M., Richter, D., Farago, B., and Frick, B., *Phys. Rev. Lett.*, **89**, 245701 (2002).
4. Arbe, A., Colmenero, J., Alvarez, F., Monkenbusch, M., Richter, D., Farago, B., and Frick, B., *Phys. Rev. E*, **67**, 051802 (2003).

Broadband Dielectric Study on Alpha- and Beta-Process for Poly(Ethylene Glycol)-Water Mixtures

Naoki Shinyashiki, Seiichi Sudo, Mayumi Shimomura, and Shin Yagihara

Department of Physics, Tokai University, Hiratsuka, Kanagawa, 259-1292, Japan

Abstract. Broadband dielectric measurements of poly(ethylene glycol)(PEG)-water mixtures with various molecular weights of PEG were carried out in the frequency range of 1μHz to 30GHz and in the temperature range of 130K to 298K. The shape of the loss peak in the liquid state at high temperature is strongly related to the types of the separation of the α and β processes in low temperature. The characteristic length of the molecular motion concerned with the relaxation processes were estimated by the PEG molecular size dependence of the relaxation time.

INTRODUCTION

We have studied dielectric properties of various water mixtures in liquid state. One of the interesting topics for the aqueous binary systems is the solute molecular size dependence of the shape of the relaxation curve of water. For the polymer-water mixtures, a relaxation process of water was observed and the loss curve broadened symmetrically about the logalism of frequency [1,2]. On the other hand, for small organic compound-water mixtures, one relaxation peak due to both the water and solute is observed and the high frequency side of the loss peak is broader than the lower frequency side [2-4]. The shape of relaxation curve is attributed to the degree of cooperativity of water and solute molecules. The symmetric loss curve observed for the polymer-water mixtures is a result of variation of local structure of large molecules. The large molecules cannot move cooperatively with water molecules and behave as a geometrical constraint to the water molecules. On the other hand, the cooperative motion of water and small organic molecules causes the asymmetric loss curve observed for the small organic molecular liquid-water mixture.

Broadband dielectric measurements for various glass-forming materials from liquid to glassy state have revealed the common futures. When the material takes liquid state in higher temperature range, a single loss peak, i.e., a process, is observed in the GHz range. The a process shifts to lower frequency with decreasing temperature. When the relaxation process is located in the MHz range, the process spread to two processes at the crossover temperature, T_C. Below T_C, the loss peak at higher frequency is the β process with symmetric shape and that at the lower frequency is the α process with asymmetric shape.

We expected that the shape of relaxation curve observed in the aqueous solution should reflect the α-β separation. In addition, it is also expected that changing molecular size of solute, we can estimate the space scale of the molecular motion contributing to the relaxation processes.

EXPERIMENT

The numbers of repeat unit of PEG, n_{PEG}, used in these experiments were 1 – 9. Water mixtures with 65 wt% of PEG were prepared to avoid the crystallization of water in the mixture. Broadband dielectric measurement for the PEG-water mixtures with the various number of repeat unit of PEG were carried out in the temperature range from 298K down to 130K in the frequency range of 20GHz to 1μHz.

RESULTS AND DISCUSSION

Figure 1 shows dielectric constant and loss for

65 wt% of tetraethylene glycol-water mixture at various temperatures. According to the dielectric constant and losses for PEG-water mixtures with various PEG molecular weights, a single relaxation process was observed for all the PEG-water mixtures in the higher temperature range, and two relaxation processes were observed in the lower temperature range in analogy with the various single-component materials. The symmetric loss curve observed for the larger PEG-water mixture ($n_{PEG} \geq 3$) in higher temperature range continues to the symmetric β process at higher frequency below T_C and additional α process appeared at lower frequency. On the other hand, the asymmetric loss curve observed for the smaller PEG-water mixtures ($n_{PEG} \leq 2$) above T_C continues to the asymmetric α process at lower frequency below T_C and additional β process appeared at higher frequency. These results imply that the shape of the loss peak in the liquid state is strongly related to the types of the separation of the α and β processes.

The molecular motion has its own characteristic length. If the characteristic length is smaller than PEG molecule, the relaxation time is independent of n_{PEG} because of only the local motion of PEG molecules concerned with the relaxation process. On the other hand, if the characteristic length is larger than PEG molecule, the over all motion of PEG should reflect the relaxation time. We can estimate the characteristic

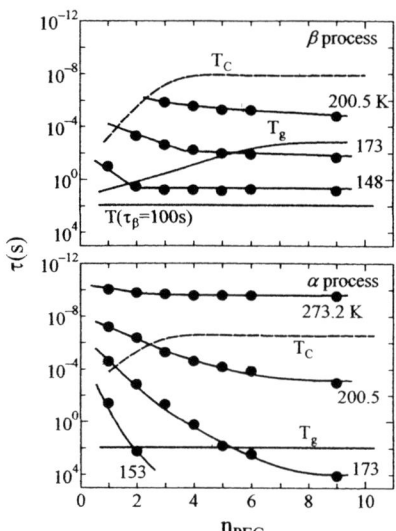

FIGURE 2. n_{PEG} dependence of τ at various temperatures. Upper: β process. Lower: α process.

length of the molecular motion related with the each relaxation process from n_{PEG} dependence of the relaxation time. Figure 2 shows the n_{PEG} dependence of relaxation time. The temperatures T_g and $T(\tau_\beta=100s)$ were defined as the relaxation time of the α and β process are 100 seconds, respectively. In the temperature range above T_C, the relaxation time is independent of n_{PEG}. Below T_C, n_{PEG} dependence of the relaxation time of the α process increases with decreasing temperature and implies that the cooperativity of the molecular motion of the alpha process increases. On the other hand, n_{PEG} dependence of the relaxation time of the β process is small in comparison with that of the α process, and is negligibly small below T_g. This result indicates that the local motion with a small characteristic length should cause the β process. These results indicate that the molecular size dependence of the relaxation time of PEG-water mixtures is a very effective analysis to estimate the characteristic length of relaxation process.

REFERENCES

1. Shinyashiki N., Yagihara, S., Arita, I., and Mashimo, S., *J. Phys. Chem. B* **102**, 3249-3251 (1998).
2. Shinyashiki, N., Sudo, S., Abe, W., and Yagihara, S., *J. Chem. Phys.* **109**, 9843-9847(1998).
3. Shinyashiki, N. and Yagihara, S., *J. Phys. Chem.* **103** 4481-4484 (1999).
4. Sudo, S., Shinyashiki, N., Kitsuki, Y., and Yagihara, S., *J. Phys. Chem. A* **106**, 458-464 (2002).

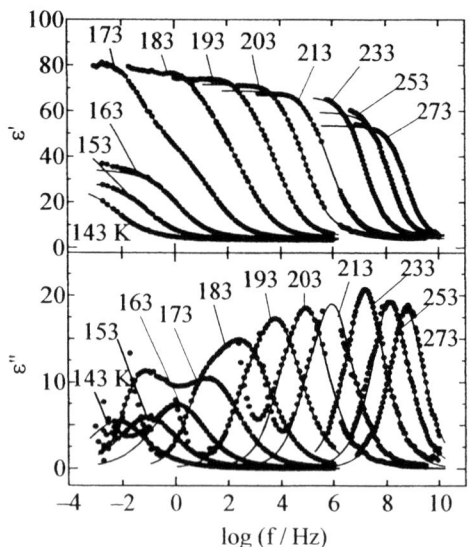

FIGURE 1. Dielectric constant and loss for 65 wt% tetraethylene glycol-water mixture at various temperatures.

Mechanical Relaxations in Metallic Glasses at Higher Temperatures

Y. Hiki[a], T. Aida[b], and S. Takeuchi[b]

a *Faculty of Science, Tokyo Institute of Technology, 39-3-303 Motoyoyogi, Shibuya-ku, Tokyo 151-0062, Japan*
b *Department of Materials Science and Technology, Tokyo University of Science, Noda, Chiba 278-8510, Japan*

Abstract. Low frequency internal friction (IF) of metallic glasses was measured from the room temperature through the glass transition up to the crystallization. Materials adopted were six kinds of metallic glasses with various compositions and largely different glass-forming abilities (GFA). Several IF peaks were observed overlapping to the background IF which increased with temperature. The temperature dependence of background IF was analyzed and the result was related to GFAs of materials.

INTRODUCTION

Studies of metallic glasses, especially the bulk metallic glasses (BMG), are meaningful because of both scientific and technological importance [1,2]. On the one hand, one of the most important problems in glass science is the study of mechanical relaxations in glass-forming materials. Internal friction (IF) is one of the representative quantities related to the relaxation. We are carrying on IF measurement of BMGs, where various kinds of glasses with largely different glass forming ability (GFA) have been selected to study whether a correlation is seen between GFA and IF [3,4].

METHOD

The internal friction measurement is performed using an inverted torsion pendulum with the free decay method. A ribbon sample of the metallic glass with a gauge length of 10-20 mm is used. The amplitude of the damping oscillation is optically detected, and three cycles of oscillation are used for evaluating the IF value Q^{-1}. The frequency of the oscillation is $f \sim 1$ Hz. The IF measurement is carried out from the room temperature, through the glass transition temperature T_g, up to the crystallization temperature T_x, and the measurement is made with a heating rate of 1 K/min, and the data are taken with a temperature interval of 3 K.

The method of specimen preparation is as follows. A mother alloy ingot is produced by melting together appropriate amounts of constituent elements in an arc-melting furnace under argon atmosphere. The mother alloy is melt-quenched with a single-roller melt-spinning apparatus in an argon atmosphere to produce a glassy ribbon sample of about 30 μm thick, 1 mm in width, and several m in length. The range of the melt-spinning speed is 2600-5200 m/s. Binary, ternary, quaternary, and quitary metallic glasses were prepared.

RESULTS AND DISCUSSION

An example of the experimental data of the temperature dependences of Q^{-1} and f^2, which is proportional to the shear modulus of the specimen, is shown in Fig. 1 for the case of well-annealed (40 h at 40 K below T_g) Zr-Ti-Cu-Ni-Be (Johnson alloy). Q^{-1} increases with increasing temperature, and a large peak

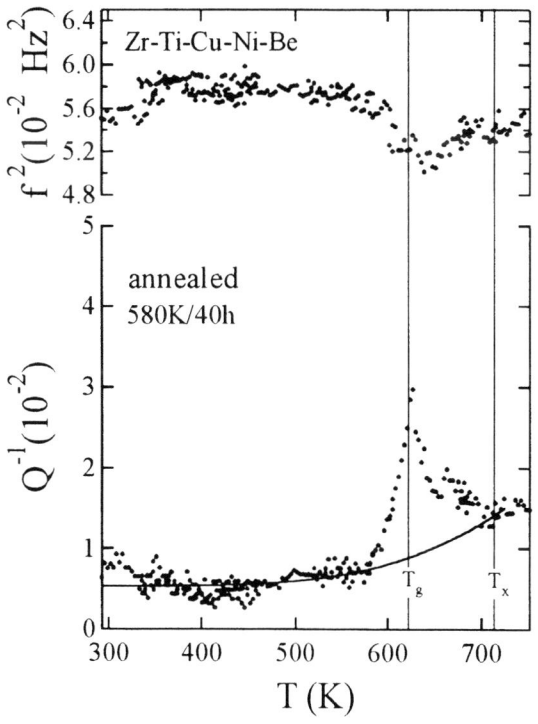

FIGURE 1. Temperature dependences of internal friction Q^{-1} and squared frequency f^2 in well-annealed Johnson alloy.

TABLE 1. Specification of metallic glasses and data of activation energy.

No.	Composition	T_g (K)	T_x (K)	ΔT (K)	R_c (K/s)	E (eV)
1	$Pd_{82}Si_{18}$	628.9	642.8	13.9	10^3-10^4	0.16
2	$Pd_{77}Cu_6Si_{17}$	630.7	674.3	43.6	10^3	0.28
3	$Pd_{40}Cu_{40}P_{20}$	472.9	520.8	47.9	10^2-10^3	0.15
4	$Zr_{60}Al_{15}Ni_{25}$	699.1	760.4	61.3	10^1-10^2	0.50
5	$Zr_{65}Al_{10}Ni_{10}Cu_{15}$	609.7	659.0	49.3	10^0-10^1	0.49
6	$Zr_{41.2}Ti_{13.8}Cu_{12.5}Ni_{10.0}Be_{22.5}$	621.8	712.9	91.1	10^0	0.38

T_g: glass transition temperature, T_x: crystallization temperature, $\Delta T = T_x - T_g$, R_c: critical cooling rate, E: activation energy.

anomalies are also seen in both quantities near T_x. Thus, the observed Q^{-1} value is considered to be composed of a background IF and IF peaks.

The temperature dependence of the background IF can well be expressed by a formula

$$Q^{-1} = A + B\exp(-E/k_BT), \quad (1)$$

where E represents a kind of activation energy, A and B are constants, and other quantities have their usual meanings. The parameter-fitted curve with choosing E, A, and B as parameters is shown in the figure. The ranges of temperature where the IF peaks appear are omitted for the fitting. The result of the fitting seems to be acceptable. Values of the activation energy have been determined for six kinds of metallic glasses in the same manner. All specimens were heat treated in the same way: for 40 h at the temperature 40 K below T_g. The obtained values of the activation energy E are shown in Table 1, where other important quantities specifying the alloy materials are also included.

In the table, the values of T_g and T_x are those measured for the specimens used for the present IF measurements. Note that GFA is very different among the six metallic glasses. The most convenient measure of GFA is the critical cooling rate R_c. A material in liquid state can be transferred to the glassy state when the liquid is cooled faster than the critical rate. GFA is said to be great when the value of R_c is small. The values of R_c in the table were reproduced from existing data [2]. The value of ΔT can also be a measure of GFA: the larger this difference the greater the GFA [2].

It is interesting to see that there is a correlation between the R_c value and the value of activation energy E. Namely, when the R_c value is smaller, or the GFA higher, the E value tends to be greater. This result tentatively interpreted as follows. Consider a glass with high GFA. The material is more easily transferred to the glassy state when the liquid state is cooled down. Meanwhile, the glassy state is not so easily transferred to the liquid state when the glass is heated up. The apparent activation energy observed during the course of the heating process in the present experiment is considered to represent the energy barrier for the glass-to-liquid transition. Accordingly, the observed activation energy is great in such material of high GFA. This is a qualitative explanation for the results.

Finally the IF peaks observed above the background IF will be considered. The small peak or peaks are almost always observed near T_x. The peak is apparently due to the mechanical loss originating from the superlcooled-to-liquid phase transition. The origin of the peak is understandable even if accurate discussion about this phenomenon cannot be made so easily. Meanwhile, experimental situation concerning the peak observed around T_g is very complicated. Here only typical examples of the complexity will be shown:

No.1 Alloy; in the as-prepared specimen a peak appears at a temperature very close to T_g, no peak can be seen in annealed specimens.

No.3 Alloy; in the as-prepared specimen a peak appears at a temperature 44 K below T_g, in the specimen annealed for 40 h at the temperature 40 K below T_g the peak shifts to 27 K below T_g.

No.6 Alloy; in the as-prepared specimen a peak appears at a temperature 30 K above T_g, in the specimen annealed for 20 h at the temperature 20 K below T_g the peak shifts to 50 K above T_g, in the specimen annealed for 40 h at the temperature 40 K below T_g the peak further shifts to a temperature very close to T_g (as shown in Fig. 1).

At present we have no idea for explaining these complex behaviors. These are probably due to the complexity of the glass transition phenomenon itself.

ACKNOWLEDGMENTS

The authors thank Dr. T. Yagi for his contribution to the present study in the early stage of the project.

This work was supported in part by a grant from High-Damping Material Project of 'Research for the Future' of Japan Society for the Promotion of Science (JSPS).

REFERENCES

1. Johnson, W. L., *MRS Bulletin/October*, 42-56 (1992).
2. Inoue, A., *Acta mater.* **48**, 279-306 (2000).
3. Hiki, Y., Yagi, T., Aida, T., and Takeuchi, S., *Mat. Sci. Eng.* A (2003), in press.
4. Hiki, Y., Yagi, T., Aida, T., and Takeuchi, S., *J. Alloys Comp.* **355**, 42-46 (2003).

Concentration Dependence of the α- and β-Processes for Alcohol-Water Mixtures

S. Sudo, M. Shimomura, S. Tsubotani, N. Shinyashiki, and S. Yagihara

Department of Physics, Tokai University, Hiratsuka, Kanagawa 259-1292, Japan.

Abstract. We performed broadband dielectric measurements for triethylenelgycol-water mixtures with various water contents in the frequency range between 1 μHz and 30 GHz in the temperature range between 130 K and 300 K. For the lower temperature range, the α- and β-processes were observed for each mixture. The β-process depends on the water content. The β-process for lower water content relates to the local motion of solute molecules and for higher water content mixtures relates to the motion of the water molecules.

INTRODUCTION

In order to clarify the mechanism of the α-β separation for aqueous solutions, we had performed broadband dielectric measurements in wide temperature range for some aqueous solutions, systematically changing water content and molecular structure of the solute molecules. Water mixtures investigated were ethyleneglycol- and glycerol-water mixtures with various concentrations [1,2], and 65wt% ethyleneglycol oligomer (EGO)-water mixtures with various number of repeat units [3]. These mixtures show the α-β separation in the lower temperature range. At higher temperature range, for the alcohol- and smaller EGO-water mixtures, the motion of the solute molecules cooperatively occurs with water molecules. In contrast, for the larger EGO-water mixtures, the solute molecule does not exhibit the cooperativity with the water molecules, and then the EGO molecules behave as the geometrical constraint to the motion of the water molecules. Below the α-β separation temperature, the cooperative motion of part of the water and solute molecules leads to the α-process, and the motion of other water molecules leads to the β-process for water mixtures with any solute molecular size. The mechanism of the α-β separation for alcohol- and smaller EGO-water mixtures is explained by heterogeneous local structure [2,4]. We believe that the mechanism of the α-β separation for larger solute molecule-water mixtures can be discussed in details by the water content dependence of the dielectric behavior. In this work, we performed broadband dielectric measurements for triethyleneglycol-water mixtures with various water contents.

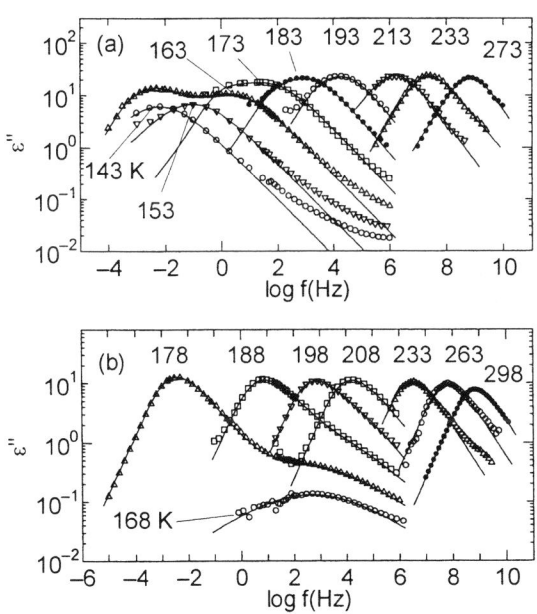

FIGURE 1. Frequency dependence of the dielectric loss for (a) 65wt% triethyleneglycol-water mixture and (b) 100wt% triethyleneglycol at various temperatures.

EXPERIMENT

Samples used in these works are 65-100wt% triethyleneglycol (3EG)-water mixtures. Dielectric complex permittivity of these mixtures was measured in the frequency range between 1 μHz and 30 GHz in the temperature range between 130 K and 300 K.

RESULTS AND DISCUSSION

Figure 1 shows the frequency dependence of the dielectric loss for 65wt% 3EG-water mixture and 100wt% 3EG. At higher temperature range, one relaxation peak is observed for each mixture. The shape of the relaxation curve is symmetric for 65-70wt% and asymmetric for 80-100wt%. The α-β separation occurs for each mixture at lower temperature range. In the frequency range 2 decades higher than the peak frequency of the β-process, the existence of small relaxation process was recognized for each mixture.

Figures 2 (a)-(c) show plots of the relaxation strength against the reciprocal of temperature for 65, 80, and 100wt% 3EG-water mixtures, respectively. For 65wt% 3EG-water mixture, the relaxation strength at higher temperatures increases with decreasing temperature, and it continues to the β-process at lower temperatures. For 80-100wt% 3EG-water mixtures, the relaxation strength at higher temperatures increases with decreasing temperature, and it continues to the α-process at lower temperatures. At lower temperature range, the relaxation strength of the β-process for high water content is much larger than that for lower water content.

In generally, the relaxation curve for the polymer-water mixtures and molecular liquids confined in porous systems shows a symmetric shape. The common feature of the molecules contributing to the relaxation process in these systems is to be spatially confined and to exist under geometrical constraint. For 3EG-water with higher water content mixtures, the symmetric shape of the relaxation curve at higher temperatures is mainly contributed by the motion of water molecules. In lower temperature range, the relaxation strength of the β-process for the pure 3EG still remains in the mixtures, though it is very small. In contrast, the relaxation strength of the β-process for higher water content the mixture is large. These results suggest that the β-process for lower water content relates to the local motion of solute molecules, and for higher water content relates to the motion of water molecules. Therefore, for the higher water content 3EG-water mixtures, continuing from the process at higher temperature range to the β-process is dominantly due to the motion of water molecules. For the lower water content 3EG-water mixtures, the local motion of solute molecules leads to the β-process.

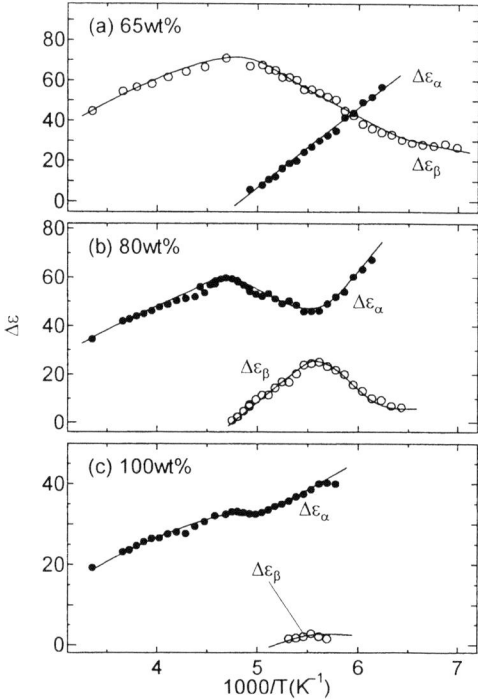

FIGURE 2. Plots of the relaxation strength against reciprocal temperature for (a) 65, (b), 80, and (c) 100wt% triethyleneglycol-water mixture.

REFERENCES

1. Sudo, S., Shinyashiki, N., and Yagihara, S., *J. Mol. Liq.* **90**, 113-117 (2001).
2. Sudo, S., Shimomura, M., Shinyashiki, N., and Yagihara, S., *J. Non-Cryst. Solids* **307-310**, 356-363 (2002).
3. Sudo, S., Shimomura, M., Saito, T., Kasiwagi, T., Shinyashiki, N., and Yagihara, S., *J. Non-Cryst. Solids* **305**, 197-203 (2002).
4. Sudo, S., Shinyashiki, N., Kitsuki, Y., and Yagihara, S., *J. Phys. Chem. A* **106**, 458-464 (2002).

Low-Temperature Specific Heat and Brillouin Scattering Measurements on Hydrogen-Bonded Glasses

Miguel A. Ramos, César Talón, Rafael J. Jiménez-Riobóo*, and Sebastián Vieira

Laboratorio de Bajas Temperaturas, Depto. de Física de la Materia Condensada, C-III, Instituto "Nicolás Cabrera", Universidad Autónoma de Madrid, Cantoblanco, 28049 Madrid, Spain
*Instituto de Ciencia de Materiales de Madrid, CSIC, Cantoblanco, 28049 Madrid, Spain

Abstract. We review specific-heat and Brillouin-scattering experiments, that have been conducted as a function of temperature on different hydrogen-bonded glasses. Specifically, we have measured the low-temperature specific heat for a set of glassy alcohols: normal and fully-deuterated ethanol, 1- and 2- propanol, and glycerol. In addition, we are currently performing Brillouin-scattering experiments on the same glasses in order to obtain their sound velocities and hence their Debye coefficients. First acoustic measurements conducted on 1- and 2- propanol are presented.

INTRODUCTION

Glasses are known to exhibit universal properties at low temperatures very different from those of crystalline solids. At very low temperatures, $T < 1$ K, the specific heat depends approximately linear on temperature, $C_p \propto T$, in contrast to the expected Debye's behavior $C_p \propto T^3$ found in crystals. At $T > 1$ K, C_p still deviates from that cubic dependence, presenting a broad maximum in C_p/T^3, what is originated from a difference or excess in the vibrational density of states $g(\nu)$ at low frequencies over the crystalline Debye behavior, leading to a ubiquitous maximum in $g(\nu)/\nu^2$, which is known as the "boson peak". It seems therefore interesting to study these issues in molecular glasses such as simple alcohols, which also allow a direct comparison with the corresponding properties in crystalline phases of the same substance.

EXPERIMENTAL RESULTS

Ethanol exhibits an interesting polymorphism presenting three different solid phases [1] at low temperature: (i) a fully-ordered (monoclinic) crystal; (ii) an orientationally-disordered (cubic) crystal or "glassy crystal"; (iii) the ordinary structural (amorphous) glass. We have measured their specific heats at low temperature, both for normal (H-) and fully-deuterated (D-) ethanol [2-4]. Fig. 1 shows the obtained C_p/T^3 data for D-ethanol. As can be seen, the glassy crystal phase exhibits the same *glassy behavior* as the structural glass, in contrast to the the Debye behavior of the stable crystal at low temperatures. Very similar curves [5] were obtained for the two isomeric phases of propanol and for glycerol.

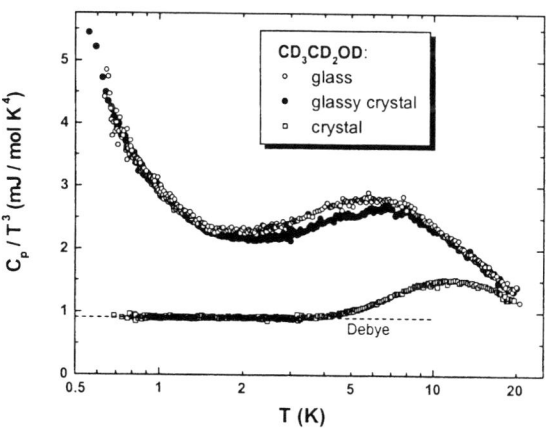

FIGURE 1. Low-temperature specific heat C_p/T^3 data of (amorphous) glass, "glassy crystal", and stable (monoclinic) crystal phases of fully-deuterated ethanol.

We have shown [5] that a consistent analysis of the low-temperature specific heat of glasses can be done based upon the Soft-Potential Model (SPM), by fitting the data in a C_p/T vs T^2 representation to a quadratic polynomial

$$C_p = C_{TLS}T + C_D T^3 + C_{sm} T^5 \quad (1)$$

with C_{TLS} being the contribution from two-level systems, C_{sm} that from additional "soft modes" within the SPM, and C_D the expected Debye coefficient. To be consistent, the fits must be performed in the temperature range $0 < T < 3/2\, T_{min}$, or up to a close temperature which lies in the middle between T_{min} and T_{max}, the temperatures where the minimum and maximum of C_p/T^3 occur, respectively. In Table 1, we present the results of these fits for all the glassy alcohols that we have measured.

In order to check the validity of this method, it is very interesting to obtain independently the Debye coefficient C_{Debye} from elastic measurements. From published data of sound velocity and density of glassy glycerol, we estimated [5] C_{Debye} = 0.835 mJ mol^{-1} K^{-4}, in good agreement with the fitted C_D.

With such an aim, we plan to measure the longitudinal v_L and transverse v_T sound velocities at low temperatures for all these glassy alcohols, by means of right-angle Brillouin scattering experiments. By extrapolating preliminary results to zero kelvin, we would obtain v_L (v_T) = 3125 (1562) m/s and 3008 (1508) m/s, and hence C_{Debye} = 1.45 mJ·mol^{-1}·K^{-4} and 1.7 mJ·mol^{-1}·K^{-4}, for 1-propanol and 2-propanol, respectively, only in rough agreement with the fitted C_D coefficients of Table 1. Further Brillouin measurements are currently in progress.

TABLE 1. Specific-heat data of glassy alcohols. Fit parameters and coefficents are defined in the text.

| | H-ethanol | | D-ethanol | | 1-propanol | 2-propanol | Glycerol |
	Glass	Glassy crystal	Glass	Glassy crystal	Glass	Glass	Glass
P_{mol} (g mol^{-1})	46.1		52.1		60.1	60.1	92.1
T_{min} (K)	2.3	2.6	2.1	2.3	1.8	1.6	2.0
T_{max} (K)	6.1	6.8	6.0	6.4	6.7	5.0	8.7
C_{TLS} (mJ mol^{-1} K^{-2})	1.2	1.27	1.05	1.13	0.424	0.516	0.157
C_D (mJ mol^{-1} K^{-4})	1.55	1.45	1.80	1.72	1.77	2.54	0.855
C_{sm} (mJ mol^{-1} K^{-6})	0.0432	0.0288	0.0572	0.0419	0.0367	0.0845	0.0139

CONCLUSION

We have measured the low-temperature specific heat of several well-known alcohols (H- and D-ethanol, 1- and 2-propanol, and glycerol) both in glassy and crystalline states. In contrast to the Debye behavior found in the latter, all glassy phases (including the "glassy crystal" of ethanol, i.e. a crystal with mere orientational disorder) strongly exhibit the universal low-temperature behavior typical of all glasses, namely a linear contribution to C_p at the lowest temperatures due to two-level systems and a broad peak in C_p/T^3 or "boson peak", in excess of the Debye contribution from lattice vibrations.

Furthermore, the very similar qualitative and *quantitative* behavior observed [1-5] in the "glassy crystal" phase of ethanol in comparison with that of the true (amorphous) glass, provides a strong evidence for the irrelevance of the lack of long-range crystalline order characteristic of amorphous solids to account for the universal properties of glasses.

ACKNOWLEDGMENTS

This work has been supported by MCyT (Spain) within project BFM2000-0035-C02.

REFERENCES

1. Haida, O., Suga, H., and Seki, S., *J. Chem. Thermodyn.* **9** 1133-1148 (1977).

2. Ramos, M. A. et al., *Phys. Rev. Lett.* **78**, 82-85 (1997).

3. Talón, C. et al., *Phys. Rev. B* **58**, 745-755 (1997).

4. Talón, C., Ramos, M. A., and Vieira, S., *Phys. Rev. B* **66**, 012201 (2002).

5. Ramos, M. A., Talón, C., Jiménez-Riobóo, R. J., and Vieira, S., *J. Phys.: Condens. Matter* **15**, S1007-S1018 (2003).

Ultra-Slow Dielectric Relaxation Process in Polyols

Yoshiki Yomogida, Ayumi Minoguchi, and Ryusuke Nozaki

Division of Physics, Graduate School of Science, Hokkaido University
Sapporo 060-0810, Japan

Abstract. Dielectric relaxation processes with relaxation times larger than that for the structural α process are reported for glycerol, xylitol, sorbitol and their mixtures for the first time. Appearance of this ultra-slow process depends on cooling rate. More rapid cooling gives larger dielectric relaxation strength. However, relaxation time is not affected by cooling rate and shows non-Arrhenius temperature dependence with correlation to the α process. It can be considered that non-equilibrium dynamic structure causes the ultra-slow process. Scale of such structure would be much larger than that of the region for the cooperative molecular orientations for the α process.

INTRODUCTION

One of the interesting features of supercooled liquids is coexistence of many dynamic processes with different characteristic times. Resent studies by means of dielectric spectroscopy have revealed that there are at least four processes commonly recognized at a temperature near the glass transition temperature T_g [1]. These are the α process, the slow β process, the fast β process and the boson peak. The α process has the longest relaxation time and is considered to be the structural relaxation process directly connected to the glass transition phenomena. The slow β process shows up at a temperature where the α process starts to follow a non-Arrhenius manner with decreasing temperature. Recent investigations on polyols (sorbitol, xylitol and so on) by means of a broadband dielectric spectroscopy have suggested that the β process is concerned with local fluctuations among molecules, which are also correlated with molecules [2-5]. The relaxation times of these two processes have strong temperature dependence, especially for the non-Arrhenius α process. On the other hand, the fast β process and the boson peak show essentially no temperature dependence and the relaxation times are very small. It is considered that there are connected to local librations of the molecules and/or atomic groups.

In this paper, we show another dielectric relaxation process exists below the α process. Relaxation time of this ultra-slow process is much larger than that of the α process. It can be considered that a non-equilibrium dynamic structure causes the ultra-slow process because of its strong cooling rate dependence. Scale of this structure would be much larger than that for the origin of the α process.

EXPERIMENTAL

Crystal powder of xylitol and sorbitol, and liquid glycerol were purchased from Kishida Chemical.

We have measured the complex permittivity of sorbitol, xylitol and their mixtures in the frequency range between 10mHz and 10GHz at temperatures between 250K and 346K. In these measurements, we have paid attention to lower frequency side of the α relaxation to see what is going on below the α process. Usually, this kind of dielectric measurement is very difficult due to the effect of conductivity. We solved such problem by careful dehydration and evaluation of conductive parts in the complex permittivity data..

RESULTS AND DISCUSSION

As an example, temperature dependence of relaxation frequencies of the α, the slow β and the ultra-slow processes is indicated for xylitol-sorbitol (7:3) mixture in Fig.1. The ultra-slow processes, which were newly found for all systems measured in this work, are always observed below the α processes. The temperature dependence of the ultra-slow process is not of the Arrhenius type like the α process, however, the dependence is less strong than that for the α process. As temperature increases, dielectric strength of the ultra-slow process decreases and the process disappears at about 330 K, corresponding to symbol (○) at highest temperature in Fig.1.

which we have observed the ultra-slow process is $T_g < T < T_g+30$ K, which is significantly lower than their range T_g+50 K $< T$ for the Fischer's process, we can't compare our process directly with their process at this moment. However, relation between the two processes must be of great interest.

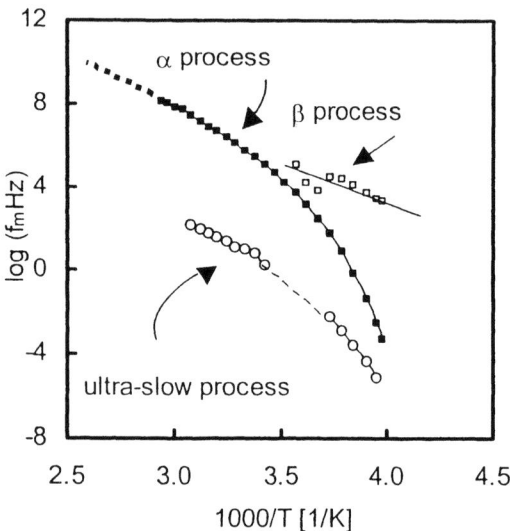

FIGURE 1. Arrhenius diagram for xylitol-sorbitol (7:3) mixture.

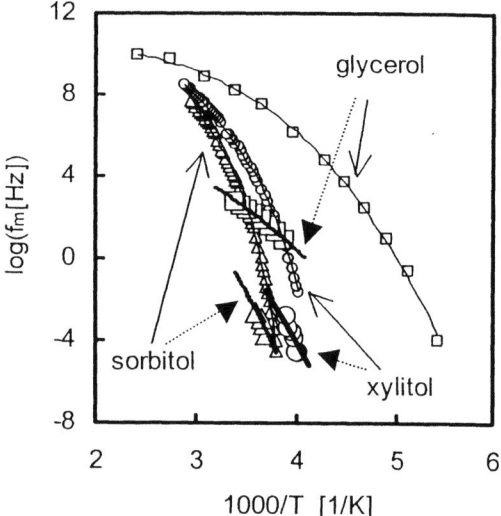

FIGURE 2. Arrhenius diagram for glycerol (□), xylitol (○) and solbitol (△). Allows with solid line indicate α process and those with dotted line indicate ultra-slow process of each materials. Data for α process of glycerol are from Ref. [6]. Note that β processes are omitted in this figure to avoid complication.

Fig.2 shows dielectric relaxation map for glycerol, xylitol and sorbitol together. The ultra-slow process depends on the molecular size. Larger size brings longer relaxation time. This situation is similar to that for the α process while the β process is not affected by the molecular size [5].

The ultra-slow process strongly depends on cooling rate. In the case of glycerol, for example, measurement at 258 K after rapid cooling with liquid nitrogen indicates large dielectric strength while direct cooling to the same temperature brings small strength. It should be pointed out that relaxation time of the ultra-slow process is not affected by cooling rate.

It can be considered that the ultra-slow process is due to kinds of non-equilibrium dynamic structures. Scale of such structure would be considerably larger than that for the molecular region corresponding to the origin of the α process.

Fischer and his coworkers have reported similar relaxation process from their static light scattering measurements of polymers and supercooled liquids [7,8]. Here, we call that as "Fischer's process". They claimed that the relaxation times of the Fischer's process are 8-10 orders of magnitudes longer than the α relaxation time. Since the temperature range at

REFERENCES

1. for example: Kremer, F., and Schönhals, A., "The Scaling of the Dynamics of Glasses and Supercooled Liquids" in *Broadband Dielectric Spectroscopy*, edited by F. Kremer and A. Schönhals, Berlin: Springer, 2003, pp. 99-129.
2. Nozaki, R., Suzuki, D., Ozawa, S., and Shiozaki, Y., *J. of Non-Cryst. Solids* **235-237**, 393-398 (1998).
3. Minoguchi, A., and Nozaki, R., *J. of Non-Cryst. Solids* **307-310**, 246-251 (2002).
4. Nozaki, R., Zenitani, K., Minoguchi, A., and Kitai, K., *J. of Non-Cryst. Solids* **307-310**, 349-355 (2002).
5. Minoguchi, A., Kitai, K., and Nozaki, R., *Phy. Rev. E* **68**, 031501 (2003).
6. Schneider, U., Lunkenheimer, P., Brand, R., and Loidl, A., *J. of Non-Cryst. Solids* **235-237**, 173-179 (1998).
7. Fischer, E.W., *Physica A* **201**, 183-206 (1993).
8. Kanaya, T., Patkowski, A., Fischer, E.W., Seils, J., Gläser, H., and Kaji, K., *Acta Polymer.* **45**, 137-142 (1994).

Terahertz Time Domain Spectroscopy of Boson Peak

S. Kojima, H. Kitahara[a], S. Nishizawa[b] and M. Wada Takeda[a]

Institute of Materials Science, University of Tsukuba, Tsukuba, Ibaraki 305-8573, Japan
[a]*Department of Physics, Faculty of Science, Shinshu University, Matsumoto, Nagano 390-8621, Japan*
[b]*Research Center for Development of Far-Infrared Region, Fukui University, Fukui 910-8507, Japan*

Abstract. Using terahertz time domain spectroscopy (THz-TDS) we have observed both the real and the imaginary parts of complex dielectric constants in the far-infrared region for a silica glass. The intensity and phase of transmittance were accurately measured in the frequency range between 10 cm^{-1} and 80 cm^{-1}. The absorption appeared around a boson peak (BP) frequency. The BP frequency is much lower than that of Raman scattering but is approximately equal to those of neutron inelastic scattering and hyper-Raman scattering and far-infrared Fourier transform spectroscopy.

The dynamical properties of glass-forming liquids have been extensively studied on the basis of the mode coupling theory [1]. In experimental studies, the broadband dielectric spectroscopy (BDS) has become one of the important techniques, and the conventional measurements have been applied for various liquids below 1 cm^{-1} (= 30 GHz). The detailed behaviors of α and slow β-relaxation processes have been well studied. While, the understanding is still unclear for fast relaxation processes and boson peaks. The experimental study requires higher frequency dynamics above 1 cm^{-1}. The boson peaks have been observed by the low-frequency Raman scattering technique [2,3], and the imaginary part of Raman susceptibility $\chi''(\omega)$ was discussed. However, the physical meaning of $\chi''(\omega)$ is different from that of $\varepsilon''(\omega)$, the combination of $\varepsilon''(\omega)$ determined by BDS up to higher frequency is impossible by light scattering. Therefore, the extension of $\varepsilon(\omega)$ to the terahertz (THz) region is strongly hoped for the ultra-broadband analysis of dielectric spectra.

The recent progress in the generation of coherent THz radiation by a femto-second pulse laser enables the very promising THz-TDS. It has better signal to noise ratio at frequency below 100 cm^{-1} in comparison with the far-infrared Fourier transform spectroscopy. This time-gated coherent nature makes possible to determine not only transmission intensity but also its phase delay accurately. It newly enables unique applications for the determination of complex dielectric constants in the THz region [4-6]. Moreover, in THz-TDS we can avoid the uncertainty caused by Kramers-Kronig analysis in the traditional far-infrared spectroscopy. These unique features of THz-TDS are very important to investigate glassy materials. As a typical organic glass, the boson peak of PMMA has

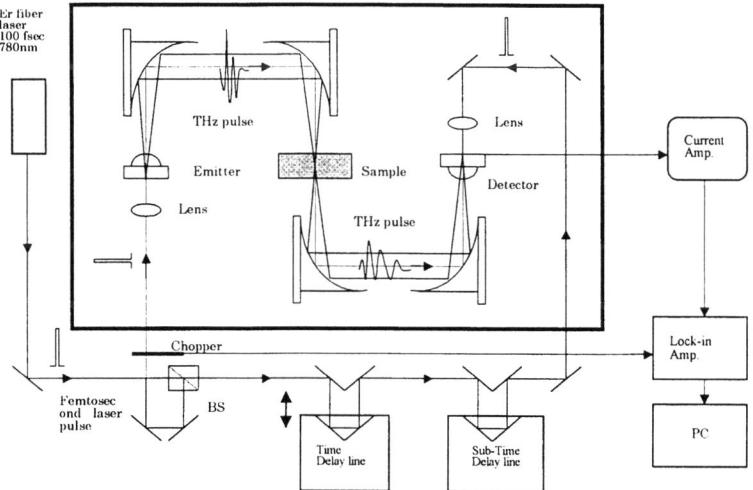

FIGURE 1. Experimental setup of THz time domain spectrometer.

been studied [6]. In this paper, the complex dielectric constants of a boson peak of a silica glass have been investigated by THz time domain spectrometer as shown in Fig.1. Both the real and the imaginary parts of dielectric constants were successfully reproduced in the frequency range between 10 cm^{-1} and 80 cm^{-1}.

As a typical inorganic glass we studied a silica glass, because its boson peak has been well studied in both theoretical and experimental points of view. The various physical properties are available in literature [7-9]. The transmission spectra of a silica glass were observed at room temperature using THz-TDS. First, the time-domain THz signals with and without a sample were recorded. Secondly, these signals were converted into the frequency-domain transmission power T(ω) and phase shift $\phi(\omega)$. Thirdly the transmission power was normalized by the reference signal without a sample to be measured. The transmittance shows the broad absorption in the THz region, which corresponds to the boson peak frequency.

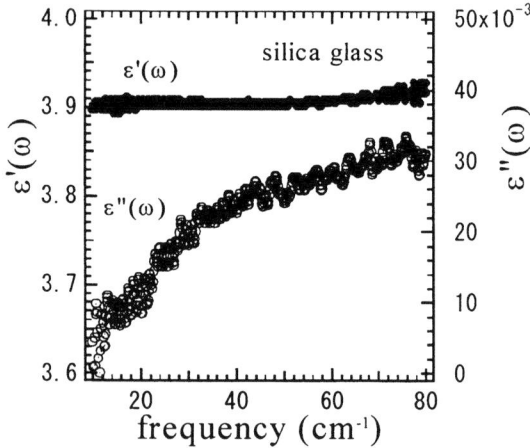

FIGURE 2. Real and imaginary parts of complex dielectric constant of a silica glass.

Figure 2 shows the real and imaginary parts of dielectric constant of a silica glass calculated from transmission power and phase shift. According to the hyper Raman scattering study of a silica glass [8], the boson peak is inactive in infrared and Raman. In contrast, $\varepsilon''(\omega)$ obtained by THz-TDS shows the broad absorption around the boson peak frequency. In the damped harmonic oscillator model, the quantity $\varepsilon''(\omega)/\omega$ indicates the peak at the mode frequency. Figure 3 shows the frequency dependence of $\varepsilon''(\omega)/\omega$. It is found that the peak frequency is about 30cm^{-1}, which is in a good agreement with the boson peak frequency observed by inelastic neutron scattering and hyper-Raman scattering. The frequency of the boson peak around 30cm^{-1} is much lower than that of Raman scattering about 60cm^{-1}. However, according to the traditional far-infrared measurements, the absorption corresponds to the boson peak is also about 30cm^{-1} [9]. Since the boson peak absorption and its frequency depend on the content of OH [9], further study on such dependences is now in progress.

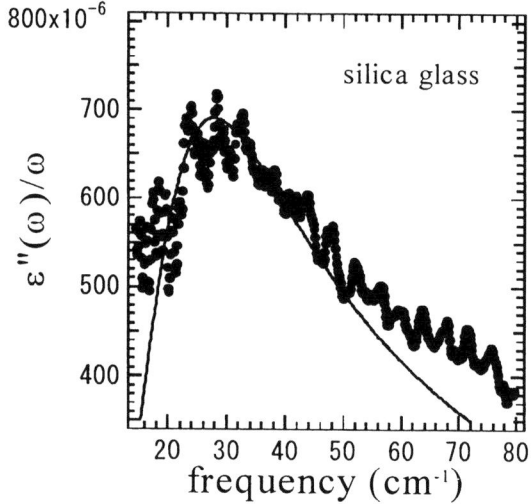

FIGURE 3. The frequency dependence of imaginary parts of dielectric constants divided by frequency in a silica glass. The solid line shows the fitted curve by the empirical function of a boson peak.

In summary, the complex dielectric constant of a silica glass has been studied using THz-TDS. The frequency dependence of $\varepsilon''(\omega)/\omega$ shows the peak about 30cm^{-1}, which is in agreement with the boson peak frequency obtained by far-infrared spectroscopy, neutron inelastic and hyper-Raman scattering.

REFERENCES

1. Gotze W. and Sojogren L., *Rep. Prog. Phys.* **55**, 241-376 (1992).

2. Kojima S., *Phys. Rev.* **B47**, 2924-2928 (1993).

3. Kojima S., Novikov V. and Kodama M., *J. Chem. Phys.* **113**, 6344-6350 (2000).

4. Nishizawa S. et al., Proc.1999 IEEE 7th Int. Conf. on *Terahertz Electronics (THz'99)*, Nara, 1999, pp.50-53.

5. Kojima S. et al, *Phys. Rev.* **B 67**, 035102-1-5 (2003).

6. Kojima S. et al, *J. Mol. Struc.* **651**, 285-287 (2003).

7. Winterling G., *Phys. Rev.* **B 12**, 2432-2440 (1975).

8. Hehlen B. et al., *Phys. Rev. Lett.* **84**, 5355-5358 (2000).

9. T. Ohsaka and S. Oshikawa, *Phys. Rev.* **B 57**, 4995 (1998).

Slow Dynamics and Dielectric Relaxation in Water / Glycerol Mixtures

Yoshihito Hayashi, Yaroslav E Ryabov, Anna Gutina and Yuri Feldman[*]

Department of Applied Physics, The Hebrew University of Jerusalem, Givat Ram, 91904 Jerusalem, Israel

Abstract. Broadband dielectric spectroscopy (BDS) has been applied for studies of glass forming liquids, such as anhydrous glycerol and its mixtures with water. Usually, glycerol does not undergo crystallization. However, we showed that the BDS with a special protection of water absorption from air to the sample and a special temperature controlling can observe the crystallization while heating of anhydrous glycerol around 263 K and the melting at 293 K. In the case of a sample without the special protection of water absorption, the crystallization and the melting were not observed, and the sample remained in liquid, supercooled liquid or glassy states. These results indicate that very small amount of water absorption can change the dynamic structure of glycerol significantly and disables the crystallization. We also found non-Arrhenius behaviors of dc-conductivity and so called "excess wing" as well as the main relaxation process, and discussed changes of dielectric properties of water / glycerol mixtures with increase of water content.

INTRODUCTION

Although glycerol exists only in liquid, supercooled liquid or glassy states in general, anhydrous glycerol can be crystallized by cooling down below the glass transition temperature (T_g = 190 K) and following slow heating [1]. However, the crystallization of glycerol is a very unusual and unstable process that depends on the temperature history and impurities of the sample, therefore almost no experimental studies of crystallized glycerol have been performed.

Glycerol and its water mixtures are widely used to study dielectric properties of glass forming and hydrogen bonded liquids [e.g., 2]. In these systems, the main dielectric relaxation process so called "α-process" is observed together with dc conductivity and "excess wing", and some cases (i.e., low temperature and high water content) additional relaxation process can be observed [2]. These dielectric behaviors relate to the glass-forming and cooperative dynamics, and are current scientific interests.

In this paper, we have studied crystallization of anhydrous glycerol and cooperative dynamics of water/glycerol mixtures by broadband dielectric spectroscopy (BDS).

MATERIALS AND METHODS

Anhydrous glycerol was obtained from Fluka (Buchs, Switzerland). Glycerol samples for BDS were prepared by two different protocols: (1) all preparation was done under a dry nitrogen atmosphere to protect water absorption from air to the sample; (2) the sample cells were filled and sealed without the special protection of the water absorption. Water/glycerol mixtures (25 to 95 mol% of glycerol with interval of 5%) were prepared from the anhydrous glycerol and double distillated water. BDS measurements of anhydrous glycerol with and without special protection of water absorption were done through wide ranges of temperature (133 to 325 K with interval of 3 K) and frequency (10 m to 3 MHz). The samples were cooled down from room temperature to 133 K quickly, and then BDS measurements were started. BDS measurements of water / glycerol mixtures were done over the ranges of temperate (173 to 323 K), and frequency (1 Hz to 1 MHz).

---

[*] Author to whom correspondence should be addressed; email: yurif@vms.huji.ac.il

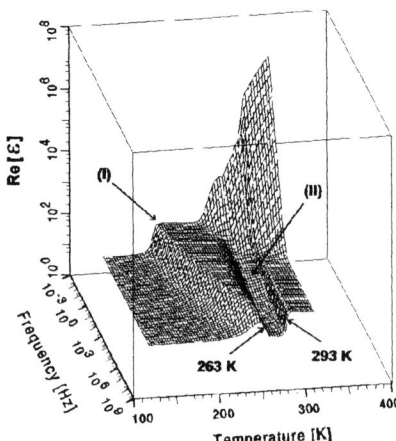

FIGURE 1. The real part of the complex permittivity ε of anhydrous glycerol (protocol #1) versus temperature and frequency. The arrows mark temperatures of crystallization (263 K), melting point (293 K), and the main relaxation process before (I) and after (II) the crystallization. (Reproduced with permission from [1], Copyright 2003 the American Physical Society.)

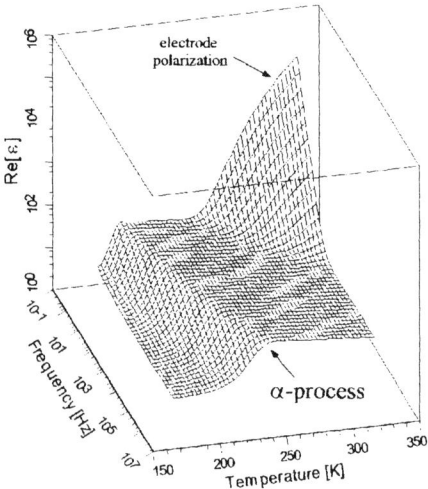

FIGURE 2. The real part of the complex permittivity of water/glycerol mixture (50 mol%) versus temperature and frequency.

RESULTS AND DISCUSSION

Dielectric permittivity of anhydrous glycerol with the special protection of water absorption displayed significant changes at 263 and 293 K as shown in Fig. 1. The temperature 293 K is known as the melting point of glycerol [1]. It is considered that mobility of glycerol molecules was frozen below the glass transition temperature (T_g = 190 K) therefore it was impossible to form the crystal structure; even above T_g, the mobility was still not enough and remained a supercooled state; at the specific temperature of 263 K, it became possible to form the crystal structure. In the crystal phase, observed main relaxation process (α-process) indicated by (I) in Fig. 1 disappeared and other process (II) appeared in lower frequency region. In contrast to this, no unusual behaviors were observed in the case of the sample without the special protection of water absorption. Although there is not numerical estimation of water absorption yet, it would be very small, and such small water absorption disabled the crystallization; it may be due to changes of the hydrogen-bonds networks and the dynamic structure of glycerol molecules.

Figure 2 shows typical results of BDS measurements of water/glycerol mixture in glycerol rich region from 50 up to 100 mol%. The relaxation time showed non-Arrhenius behavior, and it can be described by Vogel-Fulcher-Tammann (VFT) law [2,3]. Non-Arrhenius behaviors were also found on the excess wing and dc-conductivity. These three non-Arrhenius behaviors may be based on the same origin that could be related to so-called orientational and ionic defects of the hydrogen-bonds networks similar to defect mobility in ice [3]. At the same time, dielectric properties were changed below 40 mol% of glycerol, and the secondary relaxation process can be observed. The main and secondary processes can be ascribed to the dynamics of the cooperative domains of glycerol and water, respectively. The number of hydrogen bonds of ice is known as n_w = 4, and that of glycerol can be $n_g \approx 6$ [3], that is estimated by computer simulation [4,5]. Thus, the critical concentration for the hydrogen-bonds networks can be estimated as $x = 100!% \, n_w /(n_w + n_g) \approx 40$ mol% that is good agreement with the present study.

REFERENCES

1. Ryabov, Ya. E., Hayashi, Y., Gutina, A., and Feldman, Y., *Phys.Rev. B* **67**, 132202 (2003).

2. Sudo, S., Shimomura, M., Shinyashiki, N., and Yagihara, S., *J. Non-Cryst. Solids* **307-310**, 356-363 (2002).

3. Ryabov, Ya. E., Hayashi, Y., Balin, I., Puzenko, A., Feldman, Y., Kaatze, U., and Behrends, R., *submitted for publication*.

4. Padró, J. A., Saiz, L., Guàrdia, E., *J. Mol. Struct.* **416**, 243-248 (1997).

5. Chelli, R., Procacci, P., Gardini G., and Califano, S., *Phys. Chem. Chem. Phys.* **1**, 879-885 (1999).

Relaxation dynamics in supercooled liquids studied by time-resolved spectroscopy

Mika Kobayashi*†, Yuhji Tsujimi* and Toshirou Yagi*

*Research Institute for Electronic Science, Hokkaido University, Sapporo 060-0812, JAPAN
†Present address: Institute of Industrial Science, University of Tokyo, Tokyo 153-8505, JAPAN

Abstract. We have carried out impulsive stimulated thermal scattering (ISTS) on supercooled liquids to investigate q-dependence of relaxation dynamics in the liquid-glass transition. It is found that relaxation time τ_R of density fluctuation and thermal relaxation time τ_H become their maximum values at a temperature much higher than glass transition temperature T_g, and show strong q-dependence. These results indicate the crossover between some scales.

Experiments at finite wave vector magnitude may provide information on spatial correlation in supercooled liquids toward liquid-glass transition.

Impulsive stimulated thermal scattering (ISTS) is a time-resolved spectroscopy to observe density response of sample to the given transient heat having the selected wave vector magnitude q [1]. Density fluctuation with q of about $10^2 \sim 10^4 \mathrm{cm}^{-1}$ can be excited selectively through the response and its time dependence is detected in real time above 1 ns. We have carried out ISTS on supercooled liquids to investigate q-dependence of relaxation dynamics in the liquid-glass transition.

In ISTS experiment, the density fluctuation with the selected wave vector q is excited in the sample, and its time-dependent intensity is observed. The excitation light source was a infrared pulse laser (Quantronix 4217, wavelength λ_E=1053 nm, pulse width 200 ps). The laser was mode-locked and Q-switched to produce the energy of 50 μJ per one pulse. Two output pulses were crossed in the sample at angle θ_E to make the density fluctuation $\delta\rho(t)$ with wave vector magnitude,

$$q = (2\pi/\lambda_E) \cdot 2\sin(\theta_E/2). \quad (1)$$

Another continuous wave laser (Spectra Physics Beam-Lok 2060, wavelength 514.5 nm) was used for the probe light to detect $\delta\rho(t)$ through Bragg diffraction from the excited density fluctuation. The ISTS signal, the diffraction intensity $I(t)$, is given by

$$I(t) \sim |\delta\rho(t)|^2 = [A\exp(-t/\tau_H) - B\exp(-(t/\tau_R)^\beta)]^2. \quad (2)$$

This formula has been proved by hydrodynamic approach in the time range after the acoustic phonon attenuates [1]. The first term in Eq. 2 means the contribution to the thermal relaxation due to the thermal diffusion and τ_H is the thermal relaxation time. The second term expresses the relaxation of the density fluctuation which is described with stretched exponential function. The parameter τ_R is relaxation time of the density fluctuation and β is stretching parameter. A and B are the coefficients dependent on temperature.

It has been found that the density response observed by ISTS does not become slower monotonously with decrease in temperature, but is the slowest at temperature much higher temperature of about 30 K $\sim$ 60 K than the calorimetric glass transition temperature T_g [2, 3]. Figure 1 shows the typical ISTS signal intensity obtained from a glass-forming liquid glycerol (melting point T_m = 291 K, T_g = 185 K). The intensity represents the time-dependent magnitude of excited density fluctuation after

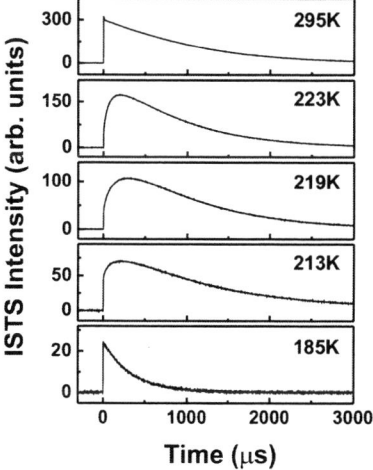

FIGURE 1. Typical ISTS intensity $I(t)$ on glycerol at excitation wave vector magnitude $q = 8.3 \times 10^2 \mathrm{cm}^{-1}$.

the heat is given at t=0. In the data at 295 K (liquid state), the intensity increases and decreases quickly since sample responds fast. The gradual rise and decay appear with decreasing temperature down to the temperature around 219 K (supercooled liquid state). However, the rise and decay become faster again below 219 K.

The relaxation times τ_R and τ_H were obtained by the successful fit with eq. 2 to ISTS signals. The temperature dependence of τ_R and τ_H are plotted in Fig. 2, reflecting the anomalous temperature dependence of density response shown in Fig. 1. The relaxation times τ_R and τ_H reach their maximum values at a temperature much higher than T_g.

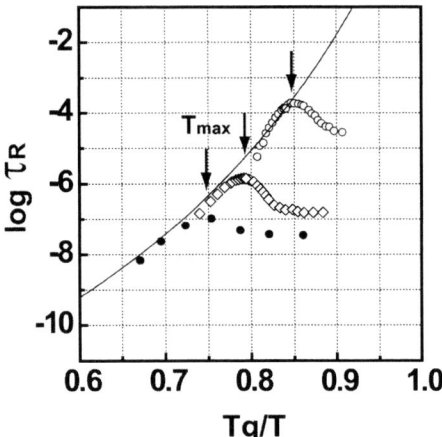

FIGURE 3. Arrhenius representation of τ_R for glycerol ((○) $q = 8.3 \times 10^2 \text{cm}^{-1}$, (◊) $q = 8.6 \times 10^3 \text{cm}^{-1}$, (●) $q = 3.4 \times 10^4 \text{cm}^{-1}$;) and the structural relaxation time τ_α determined from dielectric measurements (curved line: $q \sim 0$ [4]).

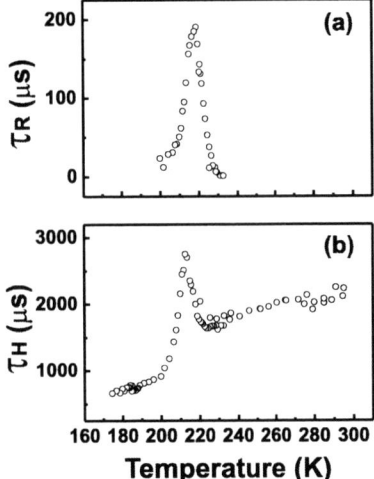

FIGURE 2. Temperature dependence of the relaxation times for glycerol at $q = 8.3 \times 10^2 \text{cm}^{-1}$, (a)relaxation time τ_R of density fluctuation, (b)thermal relaxation time τ_H.

Arrhenius representation of τ_R is shown in Fig. 3 at three wave vector magnitudes. The temperature dependence of the structural relaxation time τ_α [4] is also added to Fig. 3 with a curved full line. It is shown from this figure that the temperature dependence of τ_R is in good agreement with that of τ_α above T_{max}. On the other hand, the apparent difference is recognized between the temperature dependencoes of τ_R and τ_α, and strong q-dependence exists below T_{max}.

The wave vector q-dependence indicates the crossover at T_{max} between the observation length scale $\Lambda = 2\pi/q$ and a length scale such as the characteristic length scale ξ that grows with decrease in temperature. At high temperature $T > T_{max}$, ξ is supposed to be smaller than Λ. The whole motion characterized by ξ can be detected by ISTS in this case. Since the time scale of the whole motion is τ_α, τ_R must be equal to τ_α at $T > T_{max}$. Conversely, at low temperature $T < T_{max}$, ξ is supposed to be larger than Λ. The whole motion characterized by ξ cannot be detected by ISTS anymore, but only the partial motion inside the whole motion can be detected. However, the observation length Λ in our experiments (1.9 ~ 75 μm) is very large and it is not consistent with the characteristic length that has been estimated to be a few nm in other reports [5].

Another possibility is the crossover at T_{max} between the detectable timescale limit for ISTS given by τ_H and the distribution of relaxation time. ISTS may not be able to detect slower components of relaxation time than τ_H if non-exponential relaxation corresponds the existence of the distribution. However, this interpretation cannot explain the temperature dependence of τ_R below T_{max}.

In summary, anomalous temperature dependence of relaxation time observed by ISTS and its q-dependence indicate a crossover between some scales, however, further research is needed to specify.

The present work has been supported in part by Grants-in-Aid for Scientific Research from the Ministry of Education, Culture, Sports, Science and Technology of Japan, No. 13010720 and No. 12640365.

REFERENCES

1. Yang, Y., and Nelson, K.A., *J. Chem. Phys.* **103**, 7722-7731 (1995).
2. Kobayashi, M., Tsujimi, Y., and Yagi, T., *J. Korean Phys. Soc.* **35**, S1331-S1334 (1999).
3. Kobayashi, M., Nakanishi, M., Tsujimi, Y., and Yagi, T., *J. Non-Cryst. Solids* **307-310**, 252-256 (2002).
4. Menon, N., O'Brien, K.P., Dixon, P.K., Wu, L., Nagel, S.R., Williams, B.D., and Carini, J.P., *J. Non-Cryst. Solids* **141**, 61-65 (1992).
5. Sillescu, H., *J. Non-Cryst. Solids* **243**, 81-108 (1999).

Thermal Properties of Supercooled Water Confined within Silica Gel Pores

Satoshi Maruyama, Kenji Wakabayashi and Masaharu Oguni

Department of Chemistry, Graduate School of Science and Engineering, Tokyo Institute of Technology
2-12-2, Ookayama, Meguro-ku, Tokyo, 152-8551, Japan

Abstract. Adiabatic calorimetry of water confined within nano-pores of silica gel was carried out. The water within pores was classified into two parts, namely interfacial water and internal water. Water within 3 nm pores was well prevented from crystallization, and showed a heat capacity peak at 227 K and two glass transitions at 115 K and 160 K. Meanwhile, in the cases of 6 nm and 50 nm, water showed only one glass transition at 125 K and 132 K, respectively. The glass transition at 160 K and the broad heat capacity peak at 227 K are attributed to the internal water. The interfacial water shows a glass transition at lower temperature than internal water and no heat capacity anomaly at around 227 K. The heat capacity curve of internal water seemed to be connected smoothly with the other data obtained so far, indicating that the properties of the internal water is close to those of bulk water.

INTRODUCTION

Thermodynamic properties of supercooled water are quite strange and have been one of the most attractive subjects to be clarified. The thermal behaviors of vapor-deposited amorphous solid water (ASW), reported by Sugisaki et al.[1], suggest that the glass transition takes place at 136 K with a big heat capacity jump. On the other hand, Velikov et al. insist on the glass transition occurring at around 165 K, based on the temperature dependence of enthalpy relaxation rate of hyperquenched water[2]. Meanwhile, the heat capacity of supercooled water is known to increase, with decreasing temperature below 0 °C, apparently according to a critical-point equation with the critical temperature T_c = 228 K. Sastry et al., however, explain this anomaly without critical phenomenon[3]. The experimental determination of those issues has been prevented from by the crystallizations of supercooled water at around 236 K and of ASW at around 136 K. Here we determine the thermal behaviors by using the supercooled water confined within pores of silica gel.

EXPERIMENTAL

Here we use the silica gel with pore diameter of 3, 6 and 50 nm. The samples were prepared by loading the distilled water into the pores of silica gel on a vacuum line by distillation. The thermal measurements were carried out with the adiabatic calorimeter equipped in our laboratory. The heat capacities were measured by an intermittent heating method. During the heat capacity measurement, spontaneous heat absorption or evolution effects were also measured. In this study, these measurements were performed in the temperature range of 80 – 300 K.

RESULTS AND DISCUSSION

Most of water crystallized in the cases of 6 nm and 50 nm pores. In the case of 3 nm pores, because of the presence of a little distribution in pore sizes, a small part of water crystallized but the majority of water remained in the liquid state down to the lowest temperature measured.

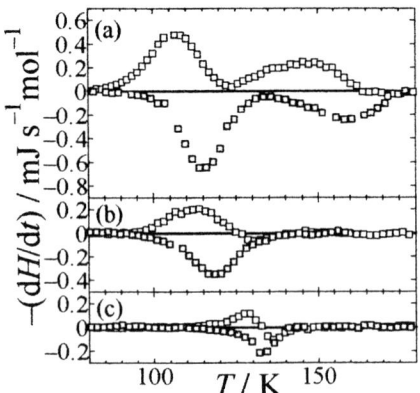

FIGURE 1. Rates of spontaneous heat absorption or evolution due to water confined within silica gel pores:(a) pore size of 3 nm; (b) 6 nm; (c) 50 nm.

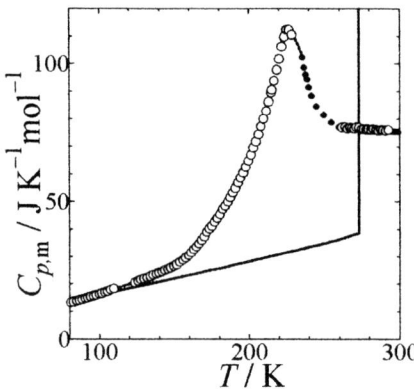

FIGURE 2. Heat capacity of internal water: ○, internal water; ●, emulsified water; —, ice.

In the heat capacity measurements, big heat capacity peaks appeared at around 260 K and 270 K in the cases of 6 nm and 50 nm pores, respectively, due to fusion of ice. For 3 nm pore sample, there appeared a small peak at 227 K and another peak, due to fusion, at 240 K. During the measurements, each sample exhibited spontaneous heat absorption or evolution effects depending on whether the sample was prepared by cooling rapidly or slowly. This thermal hysteresis is characteristic of a glass transition. Fig.1 shows these effects. For 6 nm and 50 nm samples, only one glass transition was observed at 125 K and 132 K, respectively. For 3 nm sample, two glass transitions were observed at 115 K and 160 K.

The water confined within pores are classified into two parts according to the regions in which it is located; internal water and interfacial water. Since the interfacial water is known to undergo no crystallization into ice, it is expected in the cases of 6 nm and 50 nm that the interfacial water exhibited the glass transition while that the internal water crystallized rather completely and the glass transition did not appear. In the case of 3 nm pores, the majority of internal water is interpreted as being in the liquid state even at low temperatures and showed the glass transition in a different temperature region from that of interfacial water. The glass transition which occurred at 115 K in 3 nm pores is rather close to that in 6 nm and 50 nm pores in temperature, and these glass transitions are attributed to interfacial water. Another glass transition at 160 K is then concluded to originate from the freezing in the configuration of the internal water.

The heat capacities of the interfacial water were derived by subtracting the contribution of the ice in the cases of 6 nm and 50 nm pores. The values in the 6 nm and 50 nm pore cases are in complete agreement with each other in the range 140 – 190 K, indicating the reasonableness of the data processing. The disagreement between the two below 140 K is caused by the presence of glass transition, and above 190 K is caused by premelting effect that is not removed in the calculation. If these two contributions are removed, these heat capacities are expected to agree with each other in more large temperature region. The heat capacities of the internal water within the 3 nm pores were derived by subtracting the contribution of interfacial water from the total of interfacial and internal water. As shown in Fig.2, the heat capacities show a broad peak at 227 K. That indicates that no phase transition is associated with the peak but some kind of cooperative effect functions around there. Gradual jump around 160 K is also shown in the heat capacities, which stems from the glass transition. The heat capacities of internal water seem to be connected smoothly with other data, and it indicates that the thermodynamic, and possibly kinetic, properties of the internal water are close to those of bulk water, even if the molecular aggregation of the internal water was not really bulk water.

ACKNOWLEDGMENTS

We would like to thank Fuji Silysia Chemicals Co Ltd., for providing us kindly with silica gel utilized in the present work.

REFERENCES

1. Sugisaki, M., Suga, H., Seki, S., *Bull. Chem. Soc. Jap.*, **41**, 2591-2599(1968).

2. Velikov, V., Borick, S., Angell, C.A., *Science*, **294**, 2335-2338(2001)

3. Sastry, S., Debenedetti, P.G., Sciortino, F., Stanley, H.E., *Phys. Rev. E*, **53**, 6, 6144-6154(1996)

Intramolecular Rotational Diffusion Crossover in Supercooled Liquids

Nobuaki Yonekura

Department of Chemistry, Biology and Marine Sciences, University of the Ryukyus, Okinawa 903-0213, Japan

Abstract. Intramolecular rotational diffusion processes of molecular rotors and phenylmethanes in supercooled van der Waals liquids and associated liquids have been studied by measuring temperature dependences of fluorescence intensity of the molecules. Crossovers have been observed in the dependences for all the liquids, of which the temperatures are close to MCT critical temperatures.

INTRODUCTION

Studies on the rotational diffusion of molecular probes provide detail information on microscopic dynamics of supercooled liquids [1]. Most of them are based on measurements of the rotation of the whole molecular frame. It has been known that some sorts of molecules having freely rotating groups exhibit strong couplings between the fluorescence properties and the intramolecular rotational diffusion; when the molecule is set into a viscous medium, the intramolecular rotation is inhibited and, as a result, the fluorescence intensity increases and the lifetime becomes longer. Such fluorescence properties of the probes have been utilized in studies on microscopic viscosities of various fluids or conformational changes of biomocromolecules. Thus, it should be expected that studies on intramolecular rotational diffusion in supercooled liquid reveal dynamical aspects of liquid differing from those obtained in studies on rotational diffusion of molecular frame. However, only few studies have been reported to date [2]. In the present work, we have studied the intramolecular rotational diffusions of probes in supercooled van der Waals liquids and associated liquids by measuring the temperature dependences of fluorescence intensity. Intermolecular rotational probes we utilized are neutral molecular rotors (CCVJ and DCQ) and cationic phenylmethanes (auramine O, malachite green, crystal violet and methyl green), of which the structures are shown in Fig.1.

EXPERIMENTAL

Propylene carbonate or glycerol containing the intramolecular rotational probes was sealed in a pyrex cell attached with a thermocouple and cooled down in a quartz dewar to 77K at a rate of about 15K/sec. The fluorescence intensity of the probe was monitored continuously using a fluorescence spectrometer from T_g of the liquid to room temperature with a rate of about 10K/min.

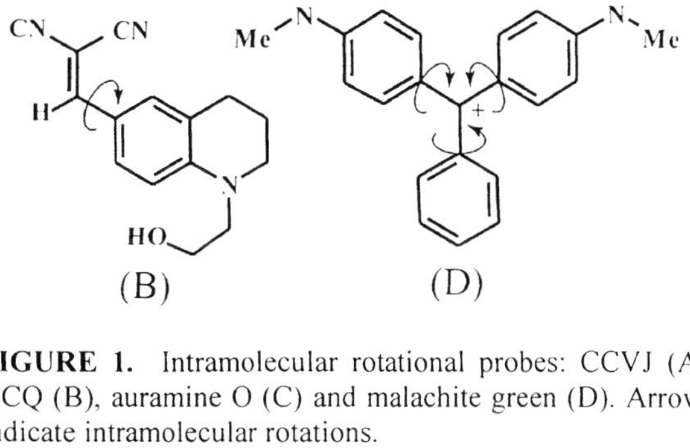

FIGURE 1. Intramolecular rotational probes: CCVJ (A), DCQ (B), auramine O (C) and malachite green (D). Arrows indicate intramolecular rotations.

CP708, *Slow Dynamics in Complex Systems: 3rd International Symposium*, edited by M. Tokuyama and I. Oppenheim

RESULTS AND DISCUSSION

Figure 2 shows temperature dependences of fluorescence intensity of the intramolecular rotational probes in supercooled propylene carbonate. For the all probes, crossovers of the dependences can be seen at several tens Kelvin above the glass transition temperatures; at higher temperatures than the crossover temperatures, the dependences are quite strong, but, at lower ones, they are weak. Such difference in the dependence is rather remarkable for molecular rotors. Similarly, for supercooled glycerol, such crossovers have been observed. The crossover temperatures of glycerol and propylene carbonate, listed in Table 1, are close to the MCT critical temperatures, which have been reported in many experimental studies stimulated by the theory [3].

To elucidate whether the temperature dependence of fluorescence intensity at higher temperatures than the crossover temperature can be attributed to coupling between the macroscopic viscosity of the supercooled liquid and the intramolecular rotational diffusion, we have carried out a simulation based on an excited state relaxation model [4]; the fluorescence quantum yield Φ_F can be expressed using radiative rate k_0, intersystem crossing rate k_T and internal conversion rate k_N which correlates with the intramolecular rotational diffusion as follows:

$$\frac{1}{\Phi_F} = 1 + \frac{k_T}{k_0} + \frac{k_N}{k_0} \quad . \qquad (1)$$

TABLE 1. Crossover Temperature T_x

Probe	T_x / K	
	Propylene carbonate	Glycerol
CCVJ	171	252
DCQ	168	252
Auramine O	167	242
Malachite green	161	248
Crystal Violet	161	248
Methyl Green	163	248

Assuming that k_N can be described by the DSE law, we have simulated the temperature dependence of fluorescence intensity. As a result, it was found that, for propylene carbonate, the intramolecular rotational diffusion couples with the viscosity over a wide temperature range, but dose not for glycerol.

For auramine O in supercooled o-terphenyl, two crossovers are clearly observed in the temperature dependence of fluorescence intensity. Interestingly, the similar crossovers have been found in the temperature dependence of rotational diffusion time of a spin probe in o-terphenyl [5]. The similarity between the two probes could be attributed to that the size of rotating group of auramine O and the molecular size of the spin probe are almost same and thus the time scales of rotation at the crossover temperatures are not so different.

REFERENCES

1. Ediger, M. D., *Annu. Rev. Phys. Chem.* **51**, 99-128 (2000).

2. Ye, J. Y., Hattori, T., Nakatsuka, H., Maruyama, Y., Ishikawa, M., *Phys. Rev.* **B56**, 5286-5296 (1997).

3. Götze, W., *J. Phys. Cond. Matt.* **11**, A1-A45 (1999).

4. Oster, G., and Nishijima, Y., *J. Am. Chem. Soc.* **78**, 1581-1585 (1956).

5. Andreozzi, L., Faetti, M., Giordano, M., Leporini, D., *J. Phys. Chem.* **103**, 4097-4103 (1999).

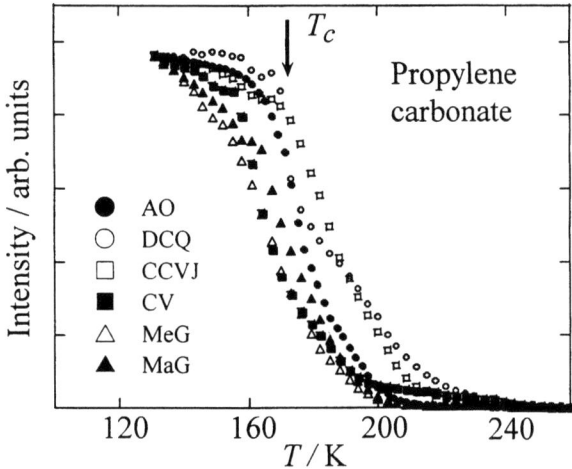

FIGURE 2. Temperature dependences of fluorescence intensities of intramolecular rotational probes, auramine O (AO), DCQ, CCVJ, crystal violet (CV), malahite green (MaG) and methyl green (MeG), in supercooled propylene carbonate. T_c refers to the MCT critical temperature.

History memorized on glass transition

Yoshihisa Miyamoto*, Hiromi Yamao* and Ken Sekimoto[†]

*Graduate School of Human and Environmental Studies, Kyoto University, Kyoto 606-8501 Japan
[†]Institue de Physique, Université Louis Pasteur, 67084 Strasbourg Cedex, France

Abstract. The memory effect is studied in the glass to rubber transition of polyisoprene rubber with the strain as a controlling parameter. A phenomenological model is proposed taking the history of the temperature and the strain into account, by which the experimental results are interpreted. It is shown both experimentally and theoretically that the glassy state memorizes the time-course of strain upon glassification, not as a single parameter but as the history itself.

INTRODUCTION

The glass transition is one of the phenomena in which the exchange of timescales of the system's relaxation time and the observation time plays a crucial role. These phenomena can be also characterized by the existence of a frozen structure which depends on the history, i.e., the memory effect. In this study, we have chosen the rubber as a material and controlled the relaxation time by imposing uniaxial strains. The glass to rubber transition under strain is measured for the samples glassified under the different conditions. A phenomenological model equation is proposed as a natural extension of the linear viscoelastic model, and the memory effect on the glass transition is discussed[1].

RESULTS AND DISCUSSION

The material used is made of synthetic cis-1,4 polyisoprene kindly supplied by Toyo Tire & Rubber Co., Ltd. The stretched glassy rubbers are prepared by the following protocols: *(HT)* The samples are stretched at room temperature (the rubbery region), and then cooled to $-100°C$. *(GT)* The samples are quenched to $-61 \pm 1°C$ (the glass transition region), stretched, kept at a prescribed length for a given time, and then cooled to $-100°C$. The inset of Fig.1 shows the aging of stress at $-61°C$ in the protocol-*GT*, just after the stretching up to an elongation ratio $\lambda = 4$. After a given time of aging (indicated by the vertical arrows), the sample is cooled down to $-100°C$. The stress of the samples prepared by protocols-*HT* and -*GT*, is released at $-100°C$ and then the length of the sample is fixed. After that the tension is measured on heating at 1K/min with the sample length being kept fixed.

In the protocol-*HT*, no stress emerges in the glassy state and the stress recovers the rubbery values corresponding to a given strain at the glass-rubber transition as shown by the label 'HT' in the main part of Fig. 1. In

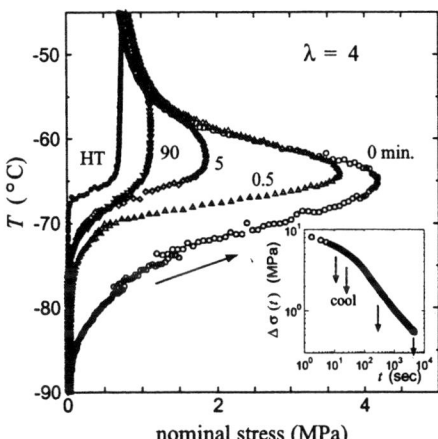

FIGURE 1. Change in stress on heating from the glassy state. ●, protocol-*HT*. Protocol-*GT*: aging time, ○, 10 s, △, 0.5 min, ◇, 5 min, ▽, 90 min. Inset: The aging of stress at $T = -61°C$.

the protocol-*GT*, the results of stress recovery on heating are shown in Fig.1 (the numbers indicate the aging time at $-61°C$): the tension shows a prominent maximum near the T_g of the protocol-*HT* with $\lambda_{HT} = 4$. The effect of aging is evident: the temperature at which the tension departs from zero increases as a function of the aging time.

We propose the following phenomenological equation for the stress $\sigma(t)$ at time t as a functional of the strain $\gamma(t') = \lambda(t') - 1$ and the temperature $T(t')$ at $t'(<t)$:

$$\sigma(t) = \sigma_R(T(t), \gamma(t))$$

$$+ \ G_\infty \int_{-\infty}^{t} [\gamma(t) - \gamma(t')] \frac{\partial \mathscr{G}(\tilde{t}(t,t'))}{\partial t'} dt', \quad (1)$$

where σ_R represents a rubberlike response, G_∞ is the modulus of the glass well below T_g, $\tilde{t}(t,t')$ is defined by $\tilde{t}(t,t') \equiv \int_{t'}^{t} [\tau(T(u), \gamma(u))]^{-1} du$, giving the *intrinsic time lapse* between t' and $t(>t')$, measured with the instantaneous relaxation time $\tau(T, \gamma)$ as a function of T and γ, and $\mathscr{G}(z)$ is the scaled relaxation function which decreases from $\mathscr{G}(0) = 1$ to $\mathscr{G}(\infty) = 0$. The second term on the right hand side of Eq. (1) indicates that the fraction $\frac{\partial \mathscr{G}(\tilde{t}(t,t'))}{\partial t'} dt'$ of the mechanical elements is effective from the time slice $[t', t' + dt']$, and each engaged element contributes to the stress by $G_\infty[\gamma(t) - \gamma(t')]$ at time t.

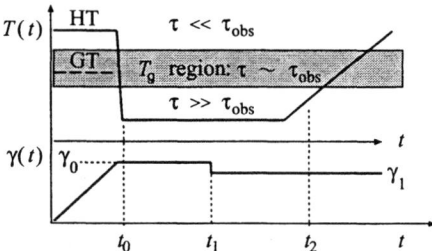

FIGURE 2. Schematic diagram of sample history for protocols-*HT* and -*GT* (The aging times are not shown.).

Figure 2 shows the history of temperature and strain. After the samples are stretched to the strain γ_0 in the rubbery region (protocol-*HT*, $T \gg T_g$) or in the glass transition region (protocol-*GT*, $T \sim T_g$), they are quenched to $T \ll T_g$ at a time t_0. The stress is released in the glassy state at t_1, by which the strain is reduced to γ_1, then the samples are reheated keeping γ_1 being fixed. On reheating, t_2 is the time at which T approaches $T_g(\gamma_1)$.

We assume $\tau(T,\gamma) \ll \tau_{obs}$ for $T \gg T_g(\gamma)$, $\tau(T,\gamma) \sim \tau_{obs}$ for $T \sim T_g(\gamma)$, and $\tau(T,\gamma) \gg \tau_{obs}$ for $T \ll T_g(\gamma)$, where τ_{obs} is the experimental time scale. Since the intrinsic time lapse $\tilde{t}$ does not increase when $T \ll T_g$, i.e., $\tilde{t}(t,t') = \tilde{t}(t_0, t')$ for $t_0 < t < t_2$, upon releasing the stress at $t = t_1$, $\gamma_1 \equiv \gamma(t_1)$ satisfies

$$0 = \sigma_R(T,\gamma_1) + G_\infty \int_{-\infty}^{t} [\gamma_1 - \gamma(t')] \frac{\partial \mathscr{G}(\tilde{t}(t_0,t'))}{\partial t'} dt', \quad (2)$$

which implies that the rubbery stress and the stress by the mechanical elements balance each other.

If we stretch the sample at $T \gg T_g(\gamma)$ (protocol-*HT*), the integral of Eq. (1) is negligible and the stress is rubberlike since τ is of the microscopic order and $\tilde{t} \sim \infty$. Then on releasing the stress in the glassy state at $t = t_1$, Eq. (2) gives $0 = \sigma_R(T, \gamma_1) + G_\infty(\gamma_1 - \gamma_0)$. Note that $\gamma_1 - \gamma_0 < 0$. If we fix $\gamma(t)$ at γ_1 on reheating, $\sigma(t)$ in Eq. (1) deviates from zero when $T(t)$ approaches T_g: $\tilde{t}(t,t')$ abruptly increases and the integral tends towards 0 rapidly, leaving only the rubberlike elasticity for $\sigma(t)$ (Fig. 1, *HT*).

If we stretch the sample at $T \sim T_g(\gamma)$ (protocol-*GT*), the integrand $[\gamma_1 - \gamma(t')] \frac{\partial \mathscr{G}}{\partial t'}$ in Eq.(2) is non-negligible for a substantial range of t' up to t_0, since τ is comparable to the experimental time scale in this case. When we start reheating, we have $\tilde{t}(t,t') = \tilde{t}(t_0, t') + \tilde{t}(t, t_2)$. Since $\tilde{t}(t, t_2)$ is independent of t', this process leads to a common "shift" of the function $\mathscr{G}$ by α, $\mathscr{G}(z) \mapsto \mathscr{G}(z + \alpha)$, in the integral of Eq.(1), with $\alpha = \tilde{t}(t, t_2)$. Except for the special case of $\mathscr{G}(z + \alpha) \propto \mathscr{G}(z)$ for all z (i.e. $\mathscr{G} = e^{-z}$), the equality (2) will be broken under the above shift of $\mathscr{G}$ at $t > t_2$, implying that the nonzero stress reappears upon reheating (Fig.1, *GT*). This is the origin of the memory effect. A numerical check has been done with a simpli-

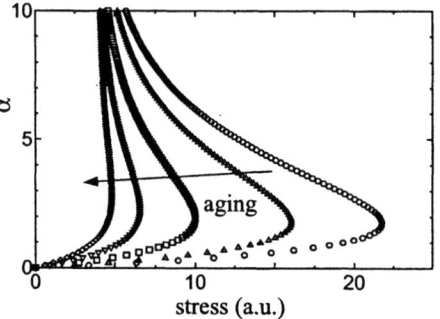

FIGURE 3. Reproduction of the memory effect.

fied case where $\mathscr{G}(z) = \theta e^{-\frac{z}{z_1}} + (1-\theta)e^{-\frac{z}{z_2}}$ and $\tau(T, \gamma_2)$ at the glass to rubber transition region being a constant independent of $T(t)$. Figure 3 shows the result of σ as the function of the shift $\alpha \equiv \tilde{t}(t, t_2)$, which reproduces the memory effect. Its peak amplitude decreases with the duration of the aging, while the peak position is found at virtually the same value. More extensive results will be discussed elsewhere.

From the experimental results shown in Fig.1 and the phenomenological equation (1), we have shown that the time-course of the strain and the temperature is memorized as a history in the glassy state, more precisely, if $\mathscr{G}(t)$ is characterized by $N (\leq \infty)$ relaxation times, then the system can memorize the *history* of $\gamma(t')$ and $T(t')$ with N order parameters.

Acknowledgements. The authors thank Dr. R. Ohara of Toyo Tire & Rubber Co. Ltd. for supplying the material. This work is partly supported by a Grant-in-Aid for Scientific Research from Japan Society for the Promotion of Science and from the Ministry of Education, Culture, Sports, Science and Technology of Japan.

REFERENCES

1. Y. Miyamoto, K. Fukao, H. Yamao and K. Sekimoto, Phys. Rev. Lett. **88**, 225504-1–225504-4 (2002).

Dielectric Study on Poly(vinyl pyrrolidone)-Alcohol Mixtures

D. Imoto, S. Sudo, N. Shinyashiki, and S. Yagihara

Department of Physics, Tokai University, Hiratsuka, Kanagawa 259-1292, Japan

Abstract. Using broadband dielectric spectroscopy, we observed dielectric relaxation phenomena for alcohol mixtures of poly(vinyl prrolidone) (PVP) with various alcohols and concentrations in the frequency range from 20Hz up to 20GHz at 25°C. PVP concentration dependence of the relaxation strength for each relaxation process suggest that, the high frequency process is caused by the rotational motion of alcohol molecules, the middle frequency process is caused by the micro Brownian motion of PVP chain, and the low frequency process is caused by the electrode polarization. For the PVP-monohydric alcohol mixtures, the shift factor obtained linearly depends on the number of OH groups per unit volume of pure alcohol, ρ_{OH}. On the other hand, for the PVP-polyhydric alcohol mixtures, the shift factor is independent of ρ_{OH}. The relaxation time for PVP-alcohol mixtures with polyhydric alcohol and small monohydric alcohol is larger than that of the pure alcohol. On the other hand, it is noted that the relaxation time of large monohydric alcohol in the mixture is smaller than that of the pure alcohol.

INTRODUCTION

We have extensively performed dielectric measurements on various binary systems, in which intermolecular interaction was systematically changed with the composition. Relaxation curves observed for water in seven kinds of synthetic polymer-water mixtures could be well described by the Cole-Cole equation and were symmetrically broadened with decreasing water content. The correspondence between the relaxation time and the symmetric broadening could be classified into two groups of polymer-water mixtures. One group contains nonelectrolyte polymers-water mixtures, and another group contains water mixtures of electrolyte polymers and poly (vinylalcohol). This result suggests that water structures in the former group are more uniform than those in the latter group. [1,2]

Chemical structures of polymers determine the local structure of water in polymer-water mixtures. Considering the change in the relaxation time and the shape of relaxation curve shown by the previous results, we can evaluate the effect of the polymer structure on the local water structure. The polymer chain should also bring upon a variety of local structures of alcohols, which are small hydrogen-bonding molecules like water. The localstructure of alcohol affected by the polymer chain offers information of the intermolecular interaction between small alcohol molecules and polymer and also explanations about these dielectric properties from the local structure of hydrogen-bonding liquids. [3]

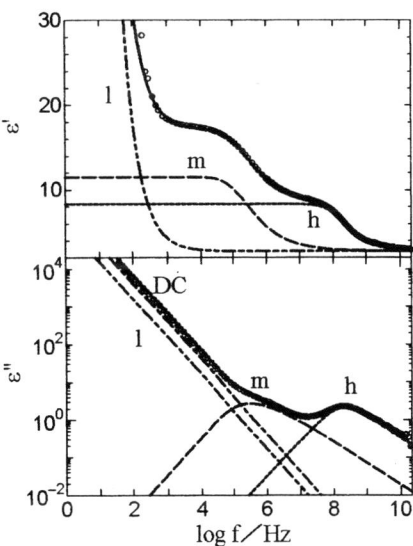

FIGURE 1 Dielectric constant and loss for 30wt% PVP-1-hexanol mixture at 298K. The dotted line: h process; the dashed line: m process; the dashed double-dotted line: ℓ process; dashed dotted line: DC conductivity; and the solid line: sum of these processes.

EXPERIMENT

In order to clarify the effect of polymer to the local structure of alcohol, dielectric measurements were carried out for mixtures of PVP (weight average molecular weight M_W=40,000) and various kinds of alcohols in the frequency range of 20Hz-20GHz. The alcohols used in this study were listed in Table 1. PVP-alcohol mixtures with concentrations of PVP from 0 to 60wt% were prepared.

RESULTS AND DISCUSSION

Figure 1 shows dielectric constant and loss for 30wt% PVP-1-hexanol mixture at 298K with various concentrations as example. Three relaxation processes were observed. PVP concentration dependence of the relaxation strength suggest that the high frequency process (h process) is caused by the rotational motion of alcohol molecules. The middle frequency process (m process) is caused by the micro Brownian motion of PVP chain. the low frequency process (ℓ process) is caused by the electrode polarization.

Figure 2 shows PVP concentration dependence of the relaxation time for some PVP solutions. In order to evaluate the effect of PVP to the dynamics of alcohols, the relaxation time of alcohol in PVP-alcohol mixtures were normalized by the relaxation time of each pure alcohol as a shift factor of the relaxation time, τ/τ_{pure}. The shift factor thus obtained indicates characteristic behaviors with the relaxation time in Figure 2.

Figure 3 provides plots of the shift factor of the mixture with 30wt% PVP against the number of OH groups per unit volume of pure alcohol, ρ_{OH}. For the PVP-polyhydric alcohol mixtures, the shift factor is independent of ρ_{OH}. On the other hand, for the PVP-monohydric alcohol mixtures, the shift factor increases linearly with increasing ρ_{OH}. The relaxation time of PVP-alcohol mixtures with polyhydric alcohol and small monohydric alcohol is larger than that of the pure alcohol. However, it is noted that the shift factor of the PVP-monohydric alcohol mixtures with small ρ_{OH} is smaller than unity, i.e., the relaxation time of large monohydric alcohol in the mixture is smaller than that of the pure alcohol. The local structure of alcohol molecules is dominated by the hydrogen-bonding network. The difference in the hydrogen-bonding networks between the monohydric and polyhydric alcohols should bring about the different ρ_{OH} dependence of the shift factor.

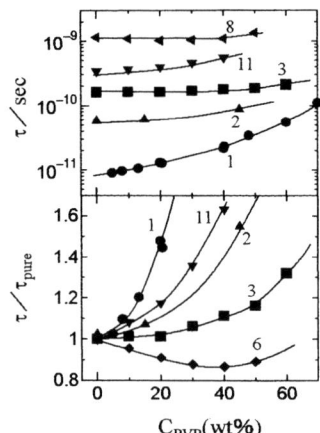

FIGURE 2 PVP concentration dependences of the relaxation time, τ, and the shift factor, τ/τ_{pure}.

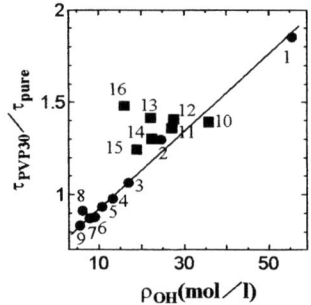

FIGURE 3 Plots of the shift factor of 30wt% PVP-alcohol mixture against the number of OH groups per unit volume of pure alcohol, ρ_{OH}.

TABLE 1. Corresponding samples to numbers indicated for plots in figures.

No.	Sample	No.	Sample
1	Water	9	1-Nonanol
2	Methanol	10	EthylenGlycol
3	Ethanol	11	1,2-Propanediol
4	1-Propanol	12	1,3-Propanediol
5	1-Butanol	13	1,2-Butanediol
6	1-Pentanol	14	1,4-Butanediol
7	1-Hexanol	15	1,5-Pentanediol
8	1-Octanol	16	1,2-Hexanediol

REFERENCES

1. Shinyashiki, N., Yagihara, S., Arita, I., and Mashimo, S., *J. Phys. Chem. B* **102**,3249-3251 (1998).

2. Ryabov, Y. E., Feldman, Y., Shinyashiki, N., and Yagihara, S., *J. Chem. Phys.* **116**, 8610-8615 (2002).

3. Asaka, N., Shinyashiki, N., Umehara, T., and Mashimo, S., *J. Chem. Phys.* **93**, 8273-8275 (1990).

Broadband Dielectric Study on Dynamics of Poly(vinyl pyrrolidone)-Poly(ethylene glycol) Blend

S. Tsubotani, S. Sudo, H. Nakamura, N. Shinyashiki, S. Yagihara, and R.J. Sengwa*

Department of Physics, Tokai University, Hiratsuka, Kanagawa, 259-1292, Japan.
*Department of Physics, JNV University, Jodhpur-342 001, India.

Abstract. Dielectric measurements for blends of poly(vinyl pyrrolidone) (PVP) (M_w=40,000) and poly(ethylene glycol) (PEG) (M_w=400) with various compositions were carried out in the frequency range of 1μHz to 10GHz and temperatures range between 298 and 173K. Three relaxation processes were observed above 298K. The high frequency process (h1 process) is caused by the chain motion of PEG, the middle frequency process (m process) is caused by the segmental motion of PVP chains, and the low frequency process is caused by ionic impurities in the mixture. The relaxation time of h1 process increased with decreasing temperature and separated into two processes at 253K. Moreover h1 process was separated again at 223K. The relaxation time of h1 process was 100s at 208K. The glass transition is attributed to the motion of unfrozen PEG molecules.

INTRODUCTION

We have studied dielectric relaxation phenomena observed for various two-component systems in broad frequency and temperature range. Two-component systems are useful, as the degree of intermolecular interaction and cooperativity between guest and host molecules can be controlled. Especially, the molecular size dependence of dielectric properties for two-component systems shows various characteristic properties [1,2].

DSC studies on blend of poly(vinyl pyrrolidone) (PVP) with liquid poly(ethylene glycol) (PEG) were carried out by Feldstein et.al. [3-5] Heating process for PVP (M_w=1,000,000) – PEG (M_w=400) blend containing 31wt% of PVP exhibits glass transition, cold crystallization of PEG, melting of PEG, and vaporization of absorbed water. Though they interpreted that the glass transition is caused by the PVP-PEG network, a lack of the dynamical aspect still remains. Thus, in this work, dielectric measurements of PVP-PEG blend were carried out in order to clarify the mechanism of the glass transition and the endothermal peak observed by DSC in low temperature range.

EXPERIMENT

Polymers used for sample preparations were PVP with the weight-average molecular weight of 40,000 and PEG with the weight-average molecular weight of 400.

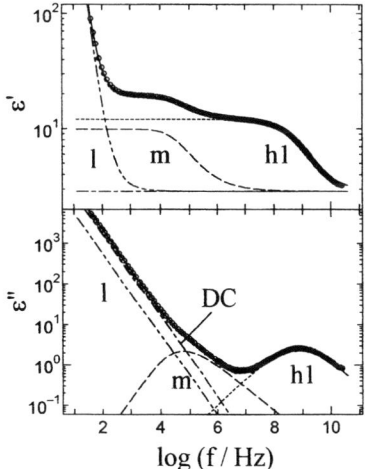

FIGURE 1. Dielectric dispersion and absorption for 20wt% PVP-PEG blend at 298K. The dotted line: h1 process; the dashed line: m process; the dashed double-dotted line: l process; the dashed dotted line: DC conductivity; the solid line: sum of these processes.

Dielectric measurements were performed in a frequency range of 1µHz to 10GHz by time domain reflectometry (TDR) (100MHz-10GHz), Impedance/Material Analyzer (IMA) (1MHz-1.8GHz, Hewlett Packard), LCR meter (20Hz-1MHz, Hewlett Packard), AC phase analysis (ACPA) (1mHz-100Hz), and DC trangent current (DCTC) (1µHz-100mHz) in the temperature range between 298 and 173K.

RESULTS AND DISCUSSION

Figure 1 shows dielectric dispersion and absorption for 20wt% PVP-PEG blend at 298K. Three relaxation processes are shown in Fig.1. According to the composition dependence of each relaxation process, the high frequency process (h1 process) is caused by the chain motion of PEG, the middle frequency process (m process) is caused by the segmental motion of PVP chains, and the low frequency process is caused by dynamical behaviors of ionic impurities. The m process was observed in five-decade lower frequency range than the h1 process.

Figures 2 and 3 show temperature dependences of the relaxation strength and the relaxation time of each process. The dotted lines in the figures indicate temperatures of glass transition, T_g=209.4K, PEG cold crystallization, T_c=231.0K, and PEG melting, T_m=276.9K determined by DSC heating thermo grams by Feldstein et.al. [3-5]

Below 278K, the relaxation strength of h1 process drastically decreased because of the crystallization of PEG. This temperature agreed with T_m=276.9K reported by Feldstein et.al. [3-5] However, the relaxation time of h1 process observed in the liquid state continued to that in the frozen state as shown in Fig.3. The relaxation time increased with decreasing temperature and h1 process separated into two processes at 253K. Moreover h1 process was separated again at 223K.

In general, according to the dynamical viewpoint, the glass transition temperature, T_g, is defined as a temperature at which the relaxation time is 100s. Figure 3 suggests that T_g determined by the relaxation time of h1 process is 208K. This temperature agrees well with T_g=209.4K reported by Feldstein et.al. [3-5] The present results indicate that the glass transition is attributed to the motion of PEG molecules, and segmental motion of PVP does not participate in the glass transition.

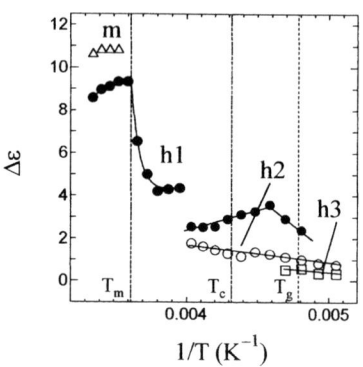

FIGURE 2. Temperature dependence of the relaxation strength of h1, h2, h3 and m processes. ●: h1 process; ○: h2 process; □: h3 process; △: m process.

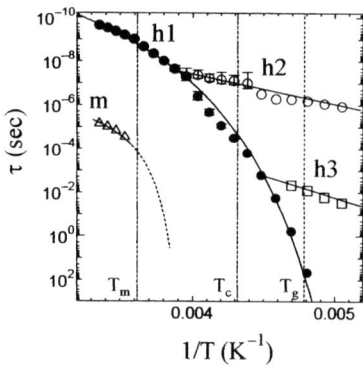

FIGURE 3. Temperature dependence of the relaxation time of h1, h2, h3 and m processes. ●: h1 process; ○: h2 process; □: h3 process; △: m process.

REFERENCES

1. Shinyashiki N., Sudo S., Abe W., and Yagihara S., *J. Chem. Phys.* **109**, 9843-9847 (1998).

2. Shinyashiki N. and Yagihara S., *J. Phys. Chem.* **103**, 4481-4484(1999).

3. Feldstein M. M., and co-workers, *Polymer Science, Ser.A*, **41**, 854-866(1999).

4. Feldstein M. M., and co-workers, *Polymer Science, Ser.A*, **41**, 867-875(1999).

5. Feldstein M. M., and co-workers, *Polymer*, **41**, 5327-5338(2000).

Glass Transition in the Stable Crystalline State of 3-Chlorothiophene

H. Fujimori[1], A. Todoroki[1], T. Asaji[1], and M. Oguni[2]

[1] Department of Chemistry, College of Humanities and Sciences, Nihon University
Sakurajosui, Setagaya-ku, Tokyo 156-8550, Japan
[2] Department of Chemistry, Graduate School of Science and Engineering, Tokyo Institute of Technology
Ookayama, Meguro-ku, Tokyo 152-8551, Japan

Abstract. Heat capacities of 3-chlorothiophene were measured between 13 and 300 K using an adiabatic calorimeter. Fusion temperature at the triple point was found to be 214.16 K and the enthalpy and entropy of fusion were determined to be 9.389 kJmol^{-1} and 43.84 JK^{-1}mol^{-1}, respectively. The glass transition temperature and the associated heat capacity jump in the stable crystalline phase were determined to be 125 K and 0.35 JK^{-1}mol^{-1}, respectively.

INTRODUCTION

A glass transition is a freezing-in phenomenon of a rearrangement motion of molecules. It is observed not only in a liquid state but also in a crystalline state. In the former case, the both positions and orientations of whole molecules are in the disordered state, and frozen in simultaneously. In the latter case, the positions of the center of molecules are fixed and only their orientations remain disordered to be frozen in. The relaxation phenomena observed around the glass transition temperature in the crystalline state are thus simpler in the understanding than those in the liquid state.

Chlorothiophenes are expected to be a planar molecule. It has been shown by an adiabatic calorimetry that 2-chlorothiophene, which has its fusion temperature of 201.3 K at the triple point, exhibits two glass transitions at 164 K and 186 K in the stable crystalline state[1]. In general, the correlation time of the molecular motion shows an Arrhenius behavior with about 10^{-14} s at infinite temperature and about 10^3 s at the glass transition temperature[2]. Therefore, the activation energies for the reorientational motion attributed to the glass transition can be roughly estimated to be about 53 kJmol^{-1} and 60 kJmol^{-1}, respectively. These values were in agreement with those estimated from T_1 measurement of ^{35}Cl NQR[3].

In the present study, Heat capacities of 3-chlorothiophene were measured between 13 and 300 K using an adiabatic calorimeter. A glass transition in the stable crystalline phase was discovered at around 125 K.

EXPERIMENTAL

3-Chlorothiophene, purchased from Wako Pure Chemical Industries, Ltd., was distilled fractionally at reduced pressure by use of a home-made rectifier with about 17 theoretical plates. Heat capacities were measured between 13 and 300 K using an adiabatic calorimeter[4]. The sample was loaded into a calorimeter cell under an atmosphere of helium gas at p = 100 kPa and T = 297 K. The mass of the sample was 21.210 g (corresponding to 0.17855 mol).

RESULTS AND DISCUSSION

Figure 1 shows the molar heat capacities obtained. A fractional-melting experiment was carried out to determine the temperature of fusion and purity of the sample. Fusion temperature at the triple point was determined to be 214.16 K and the mole-fraction purity of the sample to be 0.9937. The molar enthalpy and entropy of fusion were determined to be 9.389 kJmol^{-1} and 43.84 JK^{-1}mol^{-1}, respectively.

Figure 2 shows the temperature dependence of spontaneous temperature drift rates in two different series of heat-capacity measurements on heating in the stable crystalline state precooled at rates of 3 Kmin^{-1} and 30 mKmin^{-1}, respectively. The systematic change in the drift was accompanied with a heat capacity jump as shown in Fig. 3. The glass transition temperature and the associated heat-capacity jump were determined to be 125 K and 0.35 JK^{-1}mol^{-1}, respectively. Only one glass transition existed in the stable crystalline state of 3-chlorothiophene. The activation energy of the reorientational molecular motion related with the glass transition was estimated to be 41 kJmol^{-1}. The heat-capacity jumps in 2-chlorothiophene were 0.20 JK^{-1}mol^{-1} and 4.22 JK^{-1}mol^{-1} at 164 K and 186 K, respectively. The activation energy and the heat-capacity jump in 3-chlorothiophene are similar to those of the lower-temperature glass-transition in 2-chlorothiophene. One possible motion responsible to the glass transition in 3-chlorothiophene is the out-of-plane reorientation about the in-plane axis through the chlorine atom and the molecular center of mass[1].

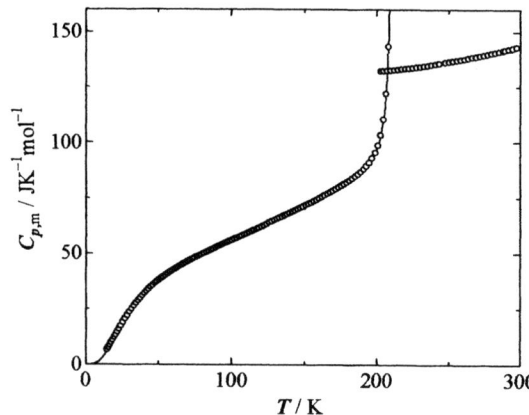

FIGURE 1. Molar heat capacities of 3-chlorothiophene.

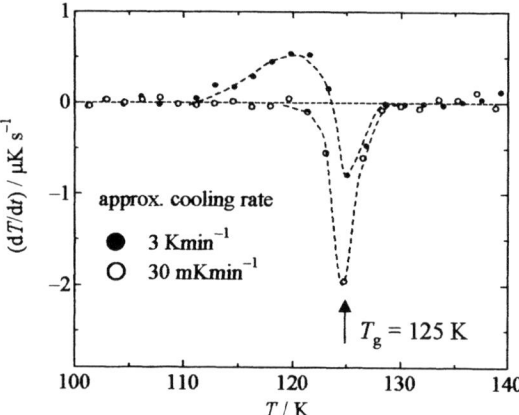

FIGURE 2. Temperature dependence of spontaneous temperature drift rates observed in two different series of heat capacity measurements on heating in the stable crystalline state.

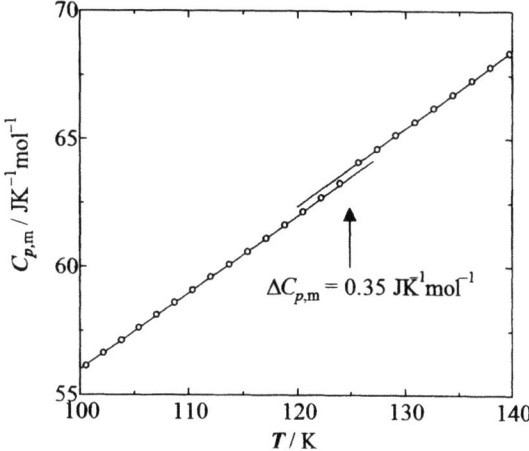

FIGURE 3. Heat capacities in the glass transition region on an enlarged scale.

REFERENCES

1. Fujimori, H., and Oguni, M., *J. Phys. Chem. Solids* **54**, 607-612 (1993).

2. Fujimori, H., and Oguni, M., *Solid State Commun.* **94**, 157-162 (1995).

3. Fujimori, H., and Asaji, T., *Z. Naturforsch.* **55a**, 183-185 (2000).

4. Fujimori, H., and Oguni, M., *J. Phys. Chem. Solids* **54**, 271-280 (1993); Kobashi, K., Kyomen, T., and Oguni, M., *J. Phys. Chem. Solids* **59**, 667-677 (1998).

Origin of the Exceptional Behaviors of Lower Alcohols in the Supercooled Liquid State

Yusuke Hiejima and Makoto Yao

Department of Physics, Graduate School of Science, Kyoto University, 606-8502 Kyoto, Japan

Abstract. The dielectric relaxation processes of lower alcohols are reasonably explained by our dielectric relaxation model for hydrogen bonding fluids, which is based on a single-molecular picture. The second slowest process, whose importance was suggested recently, is interpreted as the hydrogen bond breaking process. The intermolecular stretching modes, which are the most frequent hydrogen bond vibrations, are found to be important to the bond-breaking mechanism. The difference of the slowest and the second slowest relaxation processes may be assigned to a waiting time for the reorientation of dipolar molecular.

INTRODUCTION

Monohydric alcohols have been known as exceptional glass-forming materials [1]. For 1-propanol dielectric relaxation time is by far longer than the mechanical relaxation time, unlike van der Waals glass-forming liquids, and the relaxation function is nearly exponential with time in spite of the intermediate fragility. Hansen et al. [2] suggested that these exceptional behaviors of 1-propanol could be rationalized if the second slowest relaxation process (process II) is taken to be relevant to the structural relaxation. Since they gave no clear physical picture to the slowest process (process I), which accounts for ~96% of the total relaxation strength, the interpretation remains still controversial up to now.

In this paper, the dielectric relaxation processes of lower alcohols are reasonably explained by our model. Our model also provides a microscopic basis for the glass transition scenario by Hansen et al. [2] by considering the breaking of hydrogen bond.

OUR MODEL FOR ALCOHOLS

We have measured the Debye-type dielectric relaxation time τ_D of water [3] and lower alcohols [4, 5] up to the supercritical state. The Debye-type relaxation time τ_D (open circles in Figure 1), which corresponds to the relaxation time τ_1 for process I, is obtained in a wide range of temperature and pressure, and the relaxation time τ_2 for process II (closed circles) is also obtained near room temperature. By utilizing the experimental τ_D, we have obtained a simple model model of dielectric relaxation processes for water [3] and lower alcohols [4, 5] in the whole fluid phase. The interpretation begins with simple gas-dynamics and proceeds to complex liquid-dynamics.

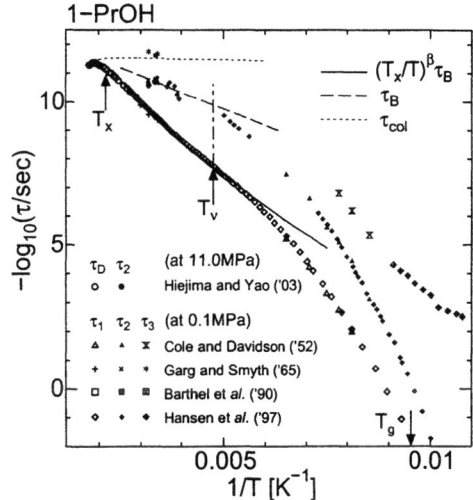

FIGURE 1. Temperature dependence of the dielectric relaxation times for 1-propanol. The binary collision time τ_{col}, and the lifetime of hydrogen bond τ_B are given by Equation (1) and (2), respectively.

At high temperatures, dielectric relaxation of lower alcohols is represented by a Debye function [4, 5]. In our model [3, 4, 5], the Debye-type relaxation time τ_D is simply described by a binary collision time τ_{col} and a lifetime of hydrogen bond τ_B.

The binary collision time τ_{col} is given by [4, 5]

$$\tau_{col} = \frac{1}{4\rho\sigma_{eff}}\sqrt{\frac{m\pi}{k_B T}} \qquad (1)$$

where ρ is number density, m is the mass of a molecule, and σ_{eff} is the effective cross section of binary collision. At low densities, where the relevant parameter is density, the dielectric relaxation is dominated by the binary collisions of the free molecules.

The lifetime of hydrogen bond τ_B is given by [4, 5]

$$\tau_B = v_{\text{stretch}}^{-1} \exp[\Delta H/k_B T]. \quad (2)$$

Here v_{stretch} is the frequency of the intermolecular stretching modes, which are the most frequent intermolecular modes, and ΔH is the enthalpy of hydrogen bonding. v_{stretch} and ΔH are estimated from infrared absorption and Raman spectroscopy measurements, respectively. In the liquid state, where τ_D strongly depends on temperature, the contribution of the molecules bound to the hydrogen bond (HB) network is characterized by τ_B.

The parameters for our interpretation of the dielectric relaxation processes are given in Table 1.

TABLE 1. Parameters for the interpretation of dielectric relaxation of lower alcohols. The detailed meaning of these parameters are described in the text and Ref [5].

	MeOH	EtOH	1-PrOH
σ_{eff} [Å^2]	7 ($\pm$1)	7 ($\pm$1)	9 ($\pm$1)
v_{stretch}^{-1} [ps]	0.28	0.30	0.23
ΔH [kJ/mol]	11.3 ($\pm$0.1)	10.5 ($\pm$0.1)	$\sim$11
T_x [K]	400 ($\pm$10)	440 ($\pm$10)	460 ($\pm$10)
T_v [K]	173	158	209
β	2.0	4.4	6.4

RELEVANCE OF THE BREAKING OF HYDROGEN BOND

For 1-propanol, in which the dielectric relaxation is represented by three distinct processes, the relaxation times (τ_1, τ_2 and τ_3) are plotted against the inverse of temperature in Figure 1. Near room temperature, τ_2, which was considered to be relevant [2], is well reproduced by τ_B (dashed line in Figure 1),

$$\tau_2 \simeq \tau_B. \quad (3)$$

Then, τ_2 corresponds to the lifetime of hydrogen bond, where the bond-breaking is promoted by thermal excitation of the intermolecular stretching modes. Moreover, a crossover from the Arrhenius law to the Vogel-Fulcher-Tammann's (VFT) law is predicted by the temperature, T_v, where the intermolecular stretching modes are no longer thermally excited.

$$T_v = hv_{\text{stretch}}/k_B. \quad (4)$$

The crossover at T_v strongly supports our picture of the mechanism of the breaking of hydrogen bonds.

According to our interpretation in the low-temperature liquid state [4, 5], $\tau_D(=\tau_1)$ is empirically described by

$$\tau_D \simeq \left(\frac{T_x}{T}\right)^\beta \tau_B. \quad (5)$$

The empirical power law (solid line in Figure 1) reproduces the non-Arrhenius temperature dependence of τ_D in a wide range of temperature, though the physical meaning is not clear at this state. τ_3 is close to τ_{col} (dotted line in Figure 1),

$$\tau_3 \sim \tau_{\text{col}}. \quad (6)$$

Equations (3), (5) and (6) suggest that τ_3 is assigned as a characteristic time for microscopic interaction such as binary collision, τ_2 as the HB breaking time and τ_D as an enhanced escape time from the HB network.

The difference between τ_D and τ_2 may be assigned to a waiting time for the reorientation of dipolar molecular. In contrast to the process II, where the molecule follows the applied oscillating field immediately, the molecular reorientation takes place after a number of stochastic processes, such as recapturing by the HB network, in the process I.

The crossover of τ_D and τ_2 at T_v suggests a change of the mechanism of the HB breaking. Above T_v, the breaking of hydrogen bond is promoted by thermal excitation of the intermolecular stretching modes, while a more cooperative mechanism may be required below T_v.

ACKNOWLEDGMENTS

This presentation is supported by a Grant-in-Aid for the 21st Century COE "Center for Diversity and Universality in Physics"

REFERENCES

1. Böhmer, R., and Angell, C.A., "Local and Global Relaxations in Glass Forming Materials", in *Disorder Effects on Relaxational Processes*, edited by Richert, R., and Blumen, A., Springer-Verlag, Berlin, Heidelberg, 1994, pp.11-54.
2. Hansen, C., Stickel, F., Berger, T., Richert, R., and Fischer, E.W., *J. Chem. Phys.* **107**, 1086-1093 (1997).
3. Okada, K., Yao, M., Hiejima, Y., Kohno, H., and Kajihara, Y., *J. Chem. Phys.* **110**, 3026-3036 (1999).
4. Hiejima, Y., Kajihara, Y., Kohno, H., and Yao, M., *J. Phys. Condens. Matter* **13**, 10307-10320 (2001).
5. Hiejima, Y., and Yao, M., *J. Chem. Phys.* **119**, 7931-7942 (2003).

Study of dielectric relaxations in glucose-water mixtures

Ji young Oh, Jeong-Ah Seo, Hyung Kook Kim, Yoon-Hwae Hwang*

RCDAMP and Department of Physics, Pusan National University

Abstract. We have studied two types of relaxation phenomena, the · and the secondary relaxations, in the glucose-water mixtures by frequency-dependent dielectric constant measurement. Dielectric measurements of 5 % and 10 % glucose-water mixtures were performed in the frequency range of 10^{-1}-10^7 Hz and in the temperature range of 260-320 K. The addition of water into glucose is believed to enhance the contribution of the secondary relaxation.

INTRODUCTION

In the dielectric-loss spectra of molecular glass materials, supercooled liquids usually exhibit at least two relaxation processes, α and β relaxations. The α-relaxation is related to a long-time scale and corresponds to the overall structure rearrangement of a system. The β-relaxation is related to a short-time scale and corresponds to local dynamics[1]. In addition to the α-relaxation, the secondary relaxation is often observed in the intermediate timescales in between the α-relaxation and the β-relaxation[2]. Secondary relaxation classify into two groups according to contribution of peak and the temperature dependence of the relaxation time[3]. When the dielectric loss spectrum exhibits an excess wing on the high frequency side of the α-relaxation peak, we call these glass formers type A systems. The temperature dependence of the secondary relaxation time obeys a Vogel-Fulcher-Tamman equation[4, 5, 6]. In contrast, the temperature dependence of the secondary relaxation time obeys an Arrhenius law in type B glass formers, which exhibit a well distinguished slow β-relaxation peak.

We studied dielectric relaxations of the liquid and glassy states of glucose-water mixtures. We approach toward better understanding of the secondary relaxation is to compare the properties of 5 % and 10 % glucose-water mixtures which exhibit the secondary relaxation.

EXPERIMENTS

The sample used in this study was α-D glucose(Aldrich Chemical Company, Inc. USA) and was mixed with purified water. The glucose-water mixtures with appropriate concentration were heated up to ~373 K and quenched into room temperature. Dielectric measurements were carried out over 8 decades of frequency (10^{-1}-10^7 Hz), using a dual phase lock-in amplifier and a HP4194A impedance analyzer. Sample thermalization was achieved by using a He closed-cycle refrigerator. It was controlled by a temperature controller(Lakeshore 330, USA) with 0.01 K precision with a Platinum sensor and a Silicon diode sensor.

RESULTS

Figures 1(a) and 1(b) show the dielectric constant(the imaginary part of the relative permittivity) of 5 and 10 % glucose-water mixtures as a function of frequency. As the temperature lowered, the α-relaxation and secondary relaxation shifted to lower frequencies. At high temperatures two relaxation superposed but secondary relaxation is successively separated from the α-relaxation with decreasing temperature. The α peaks can be described by the empirical Havriliak-Negami[6] function. In materials with well-pronounced β peaks their spectral shape can be described by the Cole-Cole[7] function.

As the water weight percent increased, T_g of the glucose-water mixtures lowered because of a plasitization. The effect of water content on the contribution of the α and the secondary relaxation of glucose-water mixtures can be seen in Fig. 1. For the secondary relaxation of 10 % glucose-water mixture is seen to be more contributed than the 5 % glucose-water mixture.

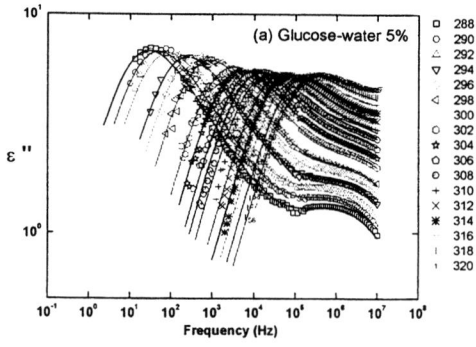

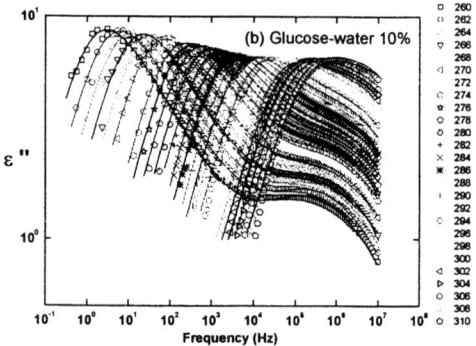

FIGURE 1. Frequency dependence of the imaginary parts of the dielectric constant of (a) 5 %, (b) 10 % glucose-water mixtures at various temperatures. All curves were taken in thermodynamics equilibrium and two-relaxation fittings with the use of Havriliak-Negami and Cole-Cole equations are shown by solid lines.

The α- and secondary relaxation times are plotted logarithmically against the reciprocal temperature in Fig. 2. For materials at $T > T_g$, as in most glass formers, $\tau_\alpha(T)$ can be parameterized using the VFT equation[4, 5, 8]. In addition to the VFT analysis, we can determine $\tau_\alpha(T)$ by using different fitting functions such as the Mode coupling theory[1] and Souletie[9] form. The quality of the fits was as good as that using VFT. That is, all three different fitting functions can describe $\tau_\alpha(T)$ well. And the temperature dependence of the secondary relaxation time $\tau_\beta(T)$ of glucose-water mixtures showed the Arrhenius behavior.

CONCLUSIONS

In the present work, dielectric spectroscopy of the two glucose-water mixtures have been studied in order to investigate the dynamics in the supercooled and glassy phase. Both · and secondary relaxation show systematic variations in the dynamics depending on water content. The glucose-water mixtures are a candidate for a research about the origin of secondary

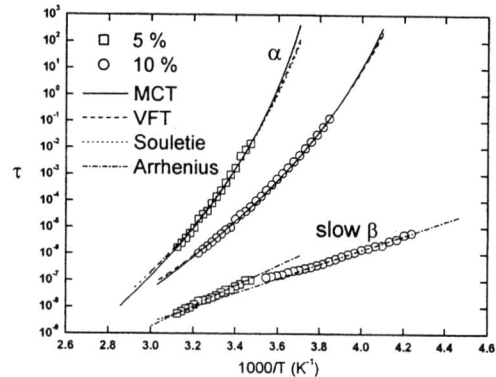

FIGURE 2. Temperature dependence of the α- and secondary relaxation times of 5 % and 10 % glucose-water mixture as determined from fits as shown in Fig. 1. The solid, dash, dot and dash dot lines are fit to MCT, VFT, Souletie and Arrhenius behavior for $\tau_\alpha(T)$, respectively.

relaxation because its secondary relaxation depend on water content. Fully understanding the secondary relaxation of the glucose-water mixtures remains a future challenge.

ACKNOWLEDGMENTS

This work was supported by Grant No. R01-2002-000-00038-0 from Basic research program of the KOSEF. We would like to thank Prof. S. Kojima (University of Tsukuba) for helpful discussion.

REFERENCES

[1] Gotze, and L. Sjogren, Rep. Prog. Phys. **55**, 241-376 (1992)
[2] G. P. Johari and M. Goldstein, J. Chem. Phys. **53**, 2372-2388 (1970)
[3] A. Kudlki, S. Bemkhof, T. Blochowicz, C. Tschirwitz and E. Rossler, J. Mol. Struct. **479**, 201-218 (1999)
[4] H. Vogel, Phys. Z. **22**, 645-646 (1921)
[5] G. S. Fulcher, J. Am. Ceram. Soc. **8**, 339-355 (1923)
[6] S. Havriliak and Jr, S. Negami, Polymer **8**, 161-210 (1967)
[7] K. S. Cole, R. H. Cole, J. Chem. Phys. **9**, 341-351 (1941)
[8] G. Tamman and W. Hesse, Z. Anorg. Allg. Chem. **156**, 245-247 (1926)
[9] J. Souletic, D. Bertrand, J. Phys. I **1**, 1627-1637 (1991)

Ultra-slow Dynamics in Glass-forming Polybutadiene

R. Inoue, N. Takahashi, K. Nishida and T. Kanaya

Institute for Chemical Research, Kyoto University, Uji, Kyoto fu 611-0011

Abstract. We studied the long-range density fluctuations (Fischer cluster) of polybutadiene using static and dynamic light scattering. We employed low molecular weight polybutadiene to neglect the effects of chain entanglement. The observed features of Fischer cluster of polybutadiene are very similar to those reported for other glass-forming materials, indicating that the Fischer cluster is a general feature of glass-forming materials.

INTRODUCTION

It was reported [1], [2], [3], [4] that some glass-forming liquids exhibit long-range density fluctuations besides the density fluctuations described by isothermal compressibility. These long-range fluctuations are termed Fischer cluster which show unusual but common features, i.e. (i) excess isotropic Rayleigh ratio, and (ii) high Landau-Placzek ratio, (iii) ultra-slow process that is extraordinary slower than the so-called α process.

In order to clarify the nature of the Fischer cluster, we studied the low molecular polybutadiene that is free from entanglement effects by means of static and dynamic light scattering.

EXPERIMENTAL

Glass-forming polybutadiene (PB) ($M_w = 3.4 \times 10^4$, $M_w/M_n = 3.9$, cis=56%, trans=42%, and vinyl=2%, T_g=172K [5]) was used in this experiments. Static light scattering measurements were carried out using System 4700, Malvern Instruments Ltd (Ar$^+$ laser, λ=488nm, 75mW). The Rayleigh ratio was calculated using the value of toluene to be R_{tol}=39.6 $\times$ 10^{-6} cm^{-1} for λ=488nm at 298K. Time correlation functions of the scattered light intensity were measured using a digital correlator ALV-5000.

RESULTS AND DISCUSSION

Static Light Scattering

In the static scattering measurements speckle pattern was observed when it was not rotated, indicating that the time average intensity $<I>_T$ is not equal to the ensemble average one $<I>_E$ in the experimental timescale. In order to reduce the effects of the speckle, the sample cell was rotated in the static light scattering measurements. The static scattered intensity $I(Q)$ was measured after equilibration at 323K, 333K, 338K, 343K, 353K, and 388K and indicated in Fig.1 as a function of the scattering vector Q. As the temperature increases, the excess scattering intensity decreases and was not observed at and above 388K. In order to observe the melting process of the cluster directly, we monitored the scattering intensity from the sample equilibrated at 298K during the temperature jump from 298K to 388K. The excess scattering intensity completely disappeared immediately after the temperature jump.

We also conducted the temperature jump experiment from 353K to 298K and from 393K to 298K. The data was not reproducible in the former but reproducible in the latter. These results were strongly affected by thermal history. At 393K the cluster melts but the cluster is

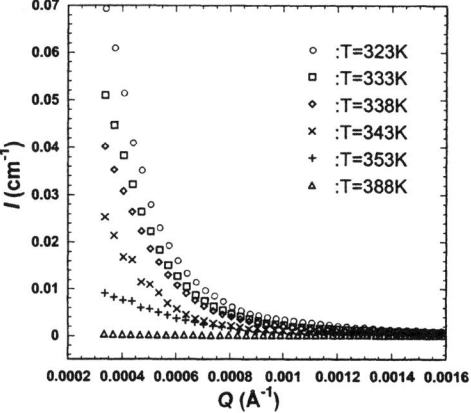

FIGURE 1. The scattering intensity $I(Q)$ from the cluster equilibrated at 323K, 333K, 338K, 343K, 353K and 388K.

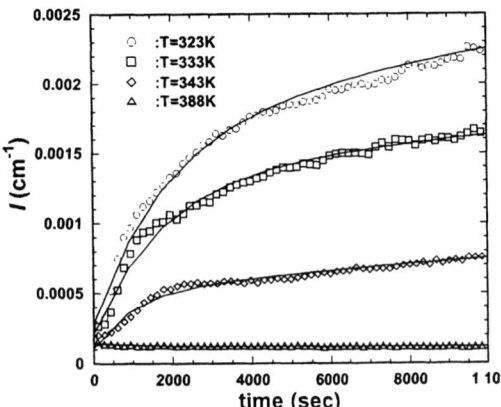

FIGURE 2. Time evolution of the scattering intensity from the cluster after the temperature jump from 393K.

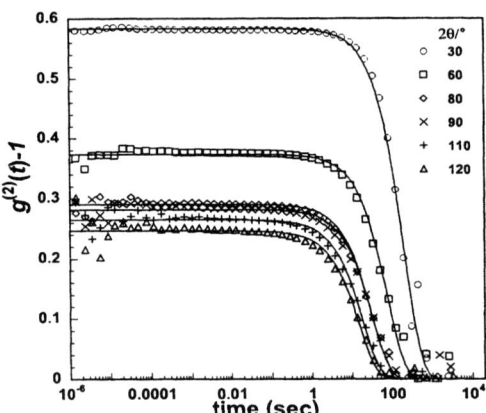

FIGURE 3. The intensity correlation function $g^{(2)}(t) - 1$ equilibrated at 333K.

still observed at 353K as seen in Fig.1. This suggests that annealing at the temperatures above the melting is necessary to remove the effect of thermal history.

We studied the time evolution of the scattering intensity of the cluster after the temperature jump from above the melting. It was suggested that the time evolution of the intensity during the equilibrium process of the cluster was well described by equation (1) [2], [3].

$$I(2\theta = 30°) = I_{\text{ini}} + \Delta I(1 - \exp(-t/\tau)) \quad (1)$$

where I_{ini} is the intensity immediately after the quench, ΔI is the change of the intensity, and τ is the equilibration time. It was found that equation (1) did not describe the time evolution in this experiment well, but the equation (2) did.

$$\begin{aligned} I(2\theta = 30°) &= I_{\text{ini}} + \Delta I_1(1 - \exp(-t/\tau_1)) \\ &+ \Delta I_2(1 - \exp(-t/\tau_2)) \end{aligned} \quad (2)$$

This indicates that two relaxation processes exist to the equilibrium process of the cluster. In Fig.2 the time evolution of the scattering intensity from the cluster and the result of the fitting by equation (2) are indicated.

Dynamic Light Scattering

The intensity correlation functions $g^{(2)}(t)$-1 at various scattering angles were measured for this sample equilibrated at 333K, which are indicated in Fig. 3. The ultra-slow process about 10^{13} times slower than the α-process [6] was observed in a time range from hundred to thousand seconds. We evaluated the relaxation time τ_{ultra} by fitting to equation (3) and found that τ_{ultra} is 1 or 2 orders of magnitude shorter than the equilibration time of the cluster.

$$g_2(t) - 1 = c|\exp(-t/\tau_{\text{ultra}})^\beta|^2 \quad (3)$$

where c is a constant. The experimental data were well described by assuming $\beta=1$ and the Q-dependence of the relaxation rate ($\Gamma = \tau^{-1}$) is similar to the simple diffusion.

CONCLUSIONS

The general features of the Fischer cluster were observed for glass-forming polybutadiene that has no special interactions like polyethylene oxide [4], suggesting that special interactions like hydrogen bonding are not necessary for Fischer cluster.

REFERENCES

1. Fischer, E. W., Meier, G., Rabenau, T., Patkowski, A., Steffen, W., and Thonnes, W., *J. Non-Cryst. Solids*, **131-133**, 134–138 (1991).
2. Kanaya, T., Patkowski, A., Fischer, E. W., Seils, J., and Glaser, H., *Acta Polym*, **45**, 137–142 (1994).
3. Patkowski, A., Fischer, E. W., Steffen, W., Glaser, H., Baumann, M., Ruths, T., and Meier, G., *Phys. Rev. E*, **63**, 061503-1-12 (2001).
4. Walkenhorst, R., Selser, J. C., and Piet, G., *J. Chem. Phys*, **109**, 11043–11050 (1998).
5. Ferry, J. D., *Viscoelastic Properties of Polymers*, Jhon Wiley and Sons Inc, New York, 1980, pp. 277–279.
6. Kanaya, T., Kawaguchi, T., and Kaji, K., *Macromolecules*, **32**, 1672–1678 (1999).

High-Resolution Brillouin Scattering Study of Intermediate Glass-Forming Materials

Yuji Ike, Seiji Kojima

Institute of Materials Science, University of Tsukuba, Tsukuba Ibaraki, 305-8573, Japan

Abstract. An angular dispersion-type Fabry-Perot interferometer (ADFPI) was used to measure the Brillouin spectra of glass-forming poly(propylene glycol)-diglycidylether (PPGDE) compounds with different average number of monomeric units in the temperature range from 400K to 160K. A high reflectivity solid etalon and a highly sensitive charge-coupled device (CCD) detector were employed to obtain a high finesse above 100 in the short acquisition time. The non-Debye relaxation process was discussed.

INTRODUCTION

Although the liquid-glass transition has been studied for many decades, the exact nature of this phenomenon have not been yet clearly understood. Brillouin scattering measurements have been widely used to study the viscoeloastic properties of various liquids and glassy systems. Therefore the Brillouin scattering spectroscopy has become a powerful tool in examining dynamical properties of glass-forming materials. In the present study, the Brillouin scattering of poly(propylene glycol) diglycidylether with two different average molecular weight was investigated. They are good glass-forming compounds, which undergo a glass transition without crystallization even in slow cooling. A high reflectivity solid etalon employed angular dispersion-type Fabry-Perot interferometer with a relatively small sized pixel of highly sensitive CCD detector was used.

EXPERIMENT

The samples of PPGDE with two different average number of monomeric units (molecular weight M=380 and 640, by Aldrich) were prepared with filtration by passing through 0.2μm Millipore membrane filters (Whatman). The experimental setup of the ADFPI is shown in Fig. 1. The Brillouin scattering spectra were measured at right angle scattering geometry using a diode-pumped solid-state laser (DPSS532), operating in a single mode at 532nm with 150mW. A solid etalon is made of a single fused silica plate for mechanical stability and the degree of its flatness and parallelism is better than $\lambda/100$. Its free spectral range and reflectivity are 30 GHz and 99 %, respectively. The transmitted light through the etalon was focused onto a CCD detector (AP32E-2, Apogee) by the lens L3 (f=800mm). It has a full frame resolution of 2184×1472 with a pixel size of 6.8μm×6.8μm. The CCD detector operated at −25C° by a thermoelectric cooler to reduce thermal noise. The temperature of sample was changed from 400K to 160K with heating or cooling rate about 2K/min. Details of apparatus can be found in the previous work. [1]

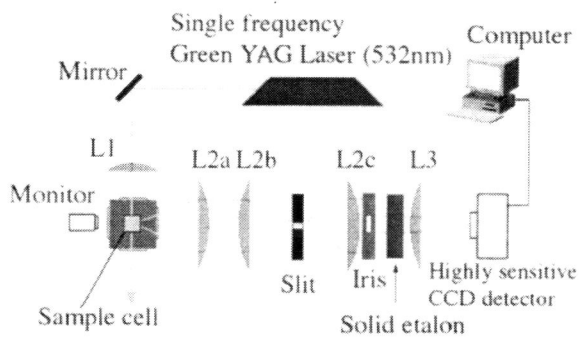

FIGURE1. Schematic diagram of the experimental setup.

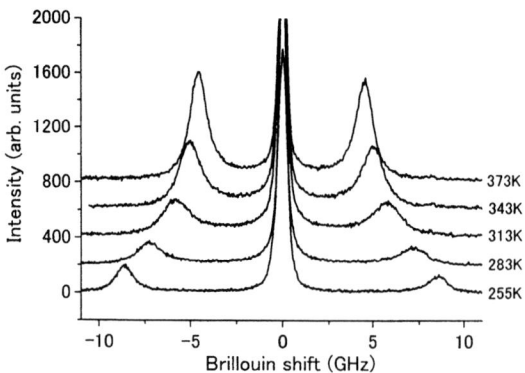

FIGURE 2. Brillouin spectra of PPGDE (M=640) at selected temperatures.

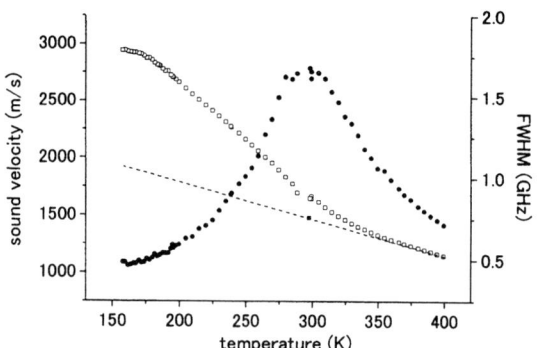

FIGURE 3. Temperature dependences of sound velocity and FWHM of PPGDE (M=380). Solid square denotes the sound velocity measured by the ultrasonic method at room temperature. The break line represents the low-frequency limit of sound velocity.

RESULTS AND DISCUSSION

In Fig. 2 the Brillouin spectra of PPGDE (M=640) at several temperatures are shown. As the temperature decreases, the Brillouin shift increases while the full width at half maximum (FWHM) show the maximum about 300K. Due to the high flatness and reflectivity of the etalon with the iris, the finesse is relatively high above 100 even its system is single-pass Fabry-Perot interferometer and the finesse in overall experiment was about 100 with an iris of a diameter of 10mm. A simplified Mountain theory [2] was examined to fit the Brillouin doublet for the spectral analysis. The longitudinal sound velocity and the FWHM obtained from Brillouin spectra are shown in Fig. 3. To derive the relaxation time and frequency, the Cole-Cole function was used. This equation for the complex sound velocity $V^*(\omega)$ was given by

$$V^*(\omega) = V_0 + (V_\infty - V_0)\frac{(i\omega\tau)^\beta}{1+(i\omega\tau)^\beta}. \quad (1)$$

where V_0, V_∞, ω, τ, β are the low-frequency limit of sound velocity, high-frequency limit of sound velocity, angular frequency, relaxation time and shape parameter, respectively. The shape parameter of (1) was assumed to $\beta=0.2$. In the case of Debye approximation, $\beta=1$. Figure 4 illustrates the relaxation frequency derived from Cole-Cole function and the results of α and slow-β dielectric relaxation fitting lines by S.Capaccioli et al. [3]

Two kinds of PPGDE of Different weight were investigated by ADFPI with a high finesse of more than 100. Brillouin spectra were analyzed using eq. (1), and the temperature dependence of relaxation frequency was determined as shown in Fig. 4.

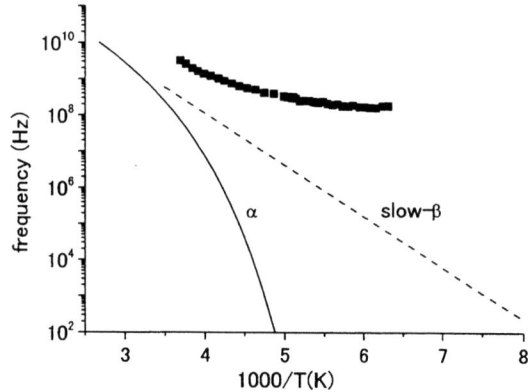

FIGURE 4. The Cole-Cole relaxation frequencies of PPGDE (M=380) and dielectric relaxation frequencies. [3]

REFERENCES

1. Ko, J.-H., and Kojima, S., *Rev. Sci. Instrum.* **73**, 4390-4392 (2002).

2. Mountain, R. D., *J. Res. Natl. Bur. Stand.* **70A**, 207-220 (1966).

3. Capaccioli, S., Casalini, R., Lucchesi, M., Lovicu, G., Prevosto, D., Pisignano, D., Romano, G., and Rolla, P. A., *J. Non-Cryst. Solids* **307-310**, 238-245 (2002).

SAXS analysis of rapidly solidified $Al_{92}V_3Fe_3Zr_2$ ribbon

Tomoaki Kamiyama, Hisamichi M. Kimura and Akihisa Inoue

Institute for Materials Research, Tohoku University, Sendai 980-8577, Japan

Abstract. Deformed scattering of Al-based ribbons prepared by the rapid solidification showed isotropic isointensity curves in the q region higher than 0.035 $Å^{-1}$. The size distribution of the precipitates was calculated from the isotropic parts of the scattering intensity and compared with the result of the Lifshitz-Slyozov-Wagner theory.

A nonequilibrium structure consisting of nanoscale particles of amorphous or icosahedral phase surrounded by fcc-Al phase was formed for Al-based ribbons through rapid solidification by single roller melt-spinning. Sizes of the precipitates may be kept within nano-meter length scales during nucleation and growth, probably due to the relatively low precipitation temperature of Al and low growth rate of precipitates resulting from the low diffusivity of the solute elements V and Fe [1]. Anisotropic scattering patterns were detected for an $Al_{92}V_3Fe_3Zr_2$ ribbon (Fig. 1), which presents the scattering intensities calculated by the radial averages of the two-dimensional small-angle X-ray scattering (SAXS) data with an azimuthal angle of 30 degrees over the sectors for parallel (_X) and perpendicular (_Y) directions of the ribbons, respectively, together with the circularly averaged (_cir) scattering intensity [2]. The deformed SAXS patterns become highly anisotropic, with increasing the circumferential velocity V_c of the melt-spinning roller from 20 to 50 m/s.

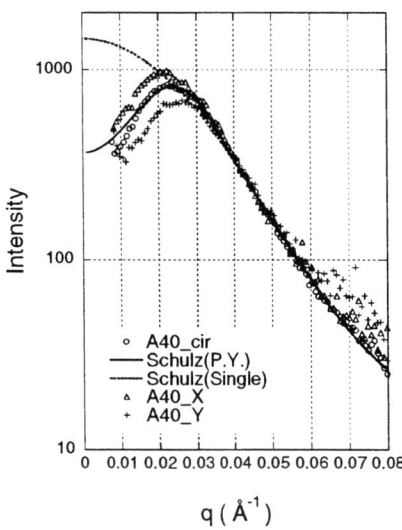

FIGURE 1. Scattering intensity for $Al_{92}V_3Fe_3Zr_2$ ribbons prepared at circumferential velocity of 40 m/s, where $q = 4\pi \sin\theta/\lambda$, 2θ is the scattering angle, and λ is the radiation wavelength $\lambda = 1.54$ for Cu-Kα radiation.

REDUCTION TO SHEAR PROBLEMS

In the single roller spinning for metallic alloys, a solid boundary layer will form adjacent to the chill surface and propagate into the melt puddle to form the ribbon. From a relation $\delta_T/\delta_M \sim (Pr)^{-1/2}$, where Pr is the Prandtl number of the fluid, δ_T is the thermal boundary layer thickness, and δ_M is the momentum boundary layer thickness; value 8.9 for the ratio δ_T/δ_M was given for pure Al melt at $1000°C$ [3]. Taking the ribbon thickness about $20\mu m$ as δ_T approximately leads to a value about $2.2\mu m$ for δ_M. And, the shear rate S at the boundary layer reaches about $1.8 \times 10^7 s^{-1}$ by melt-spinning at 40 m/s. As the SAXS results [2] showed that the microstructure of $Al_{92}V_3Fe_3Zr_2$ ribbons was deformed during the melt-spinning with shear rates of the order of $10^7 s^{-1}$, the Deborah number $D(= S\tau)$ must become comparable to one, where τ is the longest characteristic structural relaxation time. Generally large R_g values may imply large τ.

The Ostwald ripening, where larger precipitates grow at the expense of smaller ones, is expected to be affected strongly by the shear, compared to the nucleation and growth. The size distribution of a dilute system of particles that evolved through Ostwald ripening was predicted by the Lifshitz-Slyozov-Wagner (LSW) theory of domain coarsening [5], [6].

SAXS ANALYSIS

The scattering intentities show radially symmetric patterns in the q region higher than about 0.035^{-1} and decay following to the Porod law, as illustrated in Fig. 1. Transmission electron micrographs for the $Al_{92}V_3Fe_3Zr_2$ ribbons showed that the microstructure consisted of approximately spherical particles being dispersed in nearly close contact with each other [2]. This leads to the interpretation of the scattering intensities in terms of an equation

$$I(\boldsymbol{q}) = nP(\boldsymbol{q})S(\boldsymbol{q}) \qquad (1)$$

which holds in the system which consists of a population of spherical scatterers, where n is the number density of spheres. $P(\boldsymbol{q})$ is the particle form factor and $S(\boldsymbol{q})$ is the structure factor for the correlation between particles. The scattering intensities in the larger q range where they show radially symmetric patterns can be attributed largely to the form factor $P(q)$, because the form factor reflects the shape of the particle.

For the scattering intensity from polydisperse hardsphere fluids whose size distribution is represented by a Γ (Schulz) distribution

$$f(\sigma) = (\sigma/b)^{c-1} e^{-\sigma/b}/[b\Gamma(c)], \qquad (2)$$

Griffith et al. drove an analytic solution for equation (1) within the Percus-Yevick approximation [4].

The scattering intensities for $Al_{92}V_3Fe_3Zr_2$ ribbons prepared at V_c of 30, 40 and 50 m/s could be fit to a rather good approximation as shown in Fig.1, where the solid line is the result of a nonlinear least squares fit using a modified Marquardt method, and the dotted line is the single particle term obtained concomitantly. The scattering intensities for the ribbons prepared at 10 and 20 m/s could not be fit. Size distributions for $Al_{92}V_3Fe_3Zr_2$ ribbons prepared at 30, 40 and 50 m/s were calculated from eq. (2) using the evaluated parameters b and c (Fig. 2).

The radii of gyration of 75 Å 55 Å and 44 Å were calculated using $R_g^2 = 3\langle\sigma^8\rangle/20\langle\sigma^6\rangle$, where symbol $\langle...\rangle$ indicates the average over $f(\sigma)$ of spheres, for $Al_{92}V_3Fe_3Zr_2$ ribbons prepared at 30, 40 and 50 m/s, respectively. The radius of gyration can be also evaluated from the slope of the Guinier plot of scattering intensity. Values evaluated from the Guinier approximation were 68.4 ± 1.6, 53.1 ± 1.3 and 42.1 ± 1.5 Å for the ribbons, correspondingly.

DISCUSSION AND CONCLUSIONS

The values of the radius of gyration evaluated from the size distributions were comparable to that evaluated from the Guinier approximation, and Fig. 2 shows that similar shapes of $f(\sigma)$ with the mean sizes are gradually reduced with increasing the circumferential velocities from 30 to 50 m/s. Goldburg and Min observed that the shear ruptures all droplets exceeding a certain size, $R = R_{burst}$.

Fig. 2 shows that the maxima of the size distributions are located at lower size than the mean sizes in the $f(\sigma)$ curves of $Al_{92}V_3Fe_3Zr_2$ ribbons prepared at 30, 40 and 50 m/s, and the mean size decreases with increasing the shear rate S, as well as the radius of gyration. Taking into account the LSW result that the maxima of the size distributions are located at higher size than the mean size, we conclude that the shear tends to alter the shape of the size distribution of the precipitates and shift the maximum position towards lower size, as well as aligns the precipitates in the ribbon direction, for $Al_{92}V_3Fe_3Zr_2$ ribbons prepared at 30, 40 and 50 m/s.

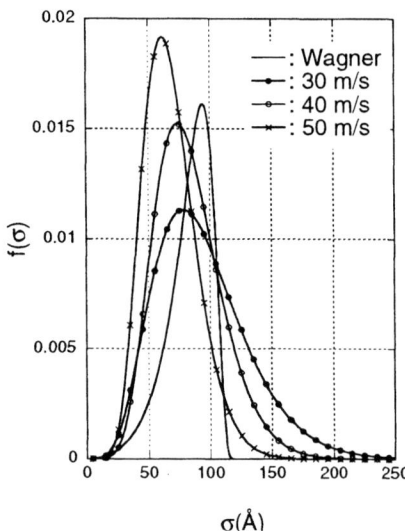

FIGURE 2. Particle size distributions of $Al_{92}V_3Fe_3Zr_2$ ribbons prepared at V_c of 30, 40 and 50 m/s, and the LSW result with the mean diameter 83 Å evaluated for the ribbon prepared at 40 m/s.

REFERENCES

1. Inoue, A., Kimura H.M., Sasamori, K., and Masumoto, T., *Nanostruct. Mater.* **7**, 363-382 (1996).
2. Kamiyama, T., Kimura, H.M., Sasamori, K., and Inoue, A., *Mater. Trans.* **42**, 1552-1560 (2001).
3. Kavesh, S., *in Metallic Glasses*, edited by J.J. Gilman and H.J. Leamy, American Society for Metals, pp. 36-73, (1977).
4. Griffith, W.L., Triolo, R., and Compere, A.L., *Phys. Rev.* **A 35**, 2200-2206 (1987).
5. Lifshitz, I.M., and Slyozov, V.V., *J. Phys. Chem. Solids* **19**, 35-50 (1961).
6. Wagner, C., *Z. Elektrochemie* **65**, 581-591 (1961).
7. Goldburg, W.I., and Min, K.Y., *Physica* **A 204**, 246-260 (1994).

Study of Glass Transition Temperatures in Sugar Mixtures by DSC

Jeong-Ah Seo, Su Jae Kim, Jiyoung Oh, Y.S.Yang[a],
Hyung Kook Kim, Yoon-Hwae Hwang*

RCDAMP and Department of Physics, Pusan National University
[a] *Institute of Nanoscience and Technology, Pusan National University*

Abstract. We studied the glass transition temperatures of mono-monosaccharide and mono-disaccharide mixtures by differential scanning calorimetry. We found that glass transition temperatures of mono-monosaccharide mixtures can be described by the Gordon-Taylor equation. However the glass transition temperatures of mono-disaccharide showed a deviation form the Gordon-Taylor equation, indicating that the molecule size has an effect on the glass transition temperature of sugar mixtures.

INTRODUCTION

We were interested in sugar glasses mainly by two reasons. There have been no published studies of the glass transition dynamics in sugars designed to reveal the full structural relaxation process and sugars are also useful materials for studying aging phenomena. Aging phenomena is a special property of non-equilibrium state of a glass phase and the study of aging phenomena including two-time scaling, fluctuation-dissipation violations, and rejuvenation effects is a relatively unknown area. The motivation for selecting sugars as the materials for this study was the glass transition temperature of several sugars lies above room temperature. This characteristic is particularly attractive for physical aging experiments which require a rapid temperature quench in order to study the evolution of relaxation dynamics with time. In this study, we measured the glass transition temperature of sugar mixtures, mono-monosaccharide and mono-disaccharide mixtures by differential scanning Calorimetry (DSC).

EXPERIMENTS

It is very difficult to make a glass phase of a sugar because sugars are very sensitive to heat. When a sugar was heated, it changed into brown colored material called caramel due to a dehydration [1]. To prevent the caramelization, we have to heat the sugar uniformly and quickly. For melting sugar without caramelization we used Thermogravimetry – Differential Thermal Analysis (TG-DTA : MAC science, DTA 2000S, Japan) because TG-DTA was effective method to heat the sample uniformly and quickly. The melting temperatures of sorbitol, glucose and sucrose were 99℃, 155℃ and 186℃, respectively. The difference of melting temperatures, ΔT, between glucose and sorbitol was about 60℃ and between glucose and sucrose was about 30℃. But we need the higher heating rate for glucose-sucrose mixture than sorbitol-glucose mixture because the caramelization phenomenon was holdback in sorbitol. IR(infrared) furnace was used for glucose-sucrose mixtures with 600℃/min heating rate and electric furnace was used for sorbitol-glucose mixtures with maximum heating rate of 40℃/min.

The glass transition temperatures of mono-mono saccharide and mono-disaccharide mixtures were measured by using Differential Scanning Calorimetry (DSC ; MAC science, DSC3100, Japan). The heating rate used in this measurement was 4℃/min and the glass transition temperatures were taken as midpoint between the onset and end point. We used a cylindrical shape aluminum cell and the size of sample cell was

5.2 mm in diameter and 5.1 mm in height, and Al_2O_3 was used for a reference material. All sugar chemicals (D-glucose, sorbitol and sucrose) were purchased from Sigma Chemical Co. and were used without further purification.

RESULTS & DISCUSSIONS

The glass transition temperature of sugar mixtures has been expressed by the Gordon-Taylor equation.

$$T_g = \frac{w_1 T_{g1} + k w_2 T_{g2}}{w_1 + k w_2}$$

Here, T_g is glass transition temperature of the mixture, w_1 and w_2 are weight fractions of components 1 and 2, T_{g1} and T_{g2} are glass transition temperatures of component 1 and 2, and k is a constant [2].

We found that glass transition temperatures of mono-monosaccharide mixtures can be described by the Gordon-Taylor equation as shown in Fig. 1. The value of constant k was 0.46. But we observed different result in the mono-disaccharide mixtures. Figure 2 shows a glass transition temperatures of glucose-sucrose mixtures (mono-disaccharide mixtures). The glass transition temperatures showed a deviation form the Gordon-Taylor equation. The value of constant k was 0.43. A careful inspection revealed that the curve was crossed at the point where the weight fraction of monosaccharide was 0.35. In the mono-mono saccharide mixture the weight fraction W and the number fraction N are the same but in the mono-disaccharide mixture W and N are different because the molecular weight of glucose, sorbitol and sucrose are 180.16, 182.2 and 342.3, respectively. The point where $W = 0.35$ represents when the number of sugar units of monosaccharide and disaccharide are the same. From this result, we found that the molecule size has an effect on the glass transition temperature of sugar mixtures.

In summary, we measured the glass transition temperature of mono-monosaccharide and mono-di saccharide mixture, respectively. The results of mono-mono saccharide obeyed the Gordon-Taylor equation but mono-disaccharide mixture showed a deviation form Gordon-Taylor equation. Mono-di saccharide mixture data were crossed with the Gordon- Taylor equation at the point where the number of monosaccharide and disaccharide were the same. From this result, we found that the molecule size has an effect on the glass transition temperature of sugar mixtures.

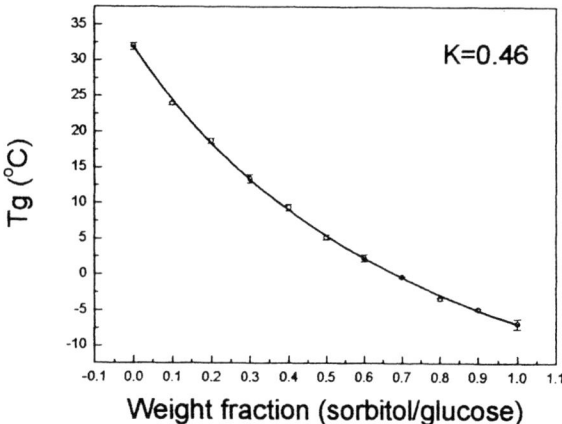

FIGURE 1. Glass transition temperatures of sorbitol-glucose mixtures. (Mono-monosaccharide mixtures)

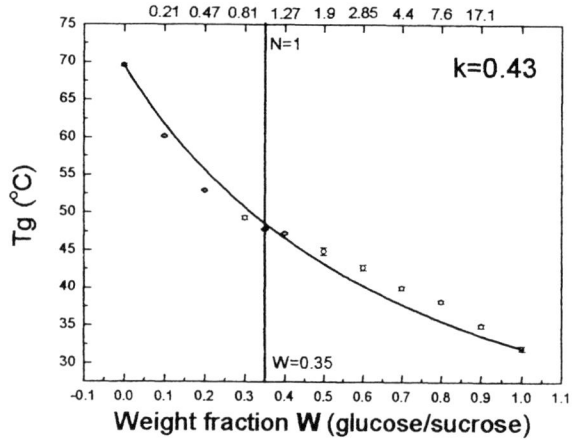

FIGURE 2. Glass transition temperatures of glucose-sucrose mixtures. (Mono-disaccharide mixtures)

ACKNOWLEDGMENTS

We thank H.Z.Cummins for suggesting sugars for a glass transition study. This work was supported by Grant No. R01-2002-000-00038-0 from Basic Research Program of the KOSEF.

REFERENCES

1. Owen R. Fennema, *Food Chemistry*, Marcel Dekker, Inc., New York, 1985, p. 69-137.
2. Gordon, M.; Taylor, J.S. *J.Appl.Chem.* **2**, 493-500 (1952).

NMR Hole-Burning Experiments on Superionic Conductor Glasses

J. Kawamura, N. Kuwata and T. Hattori,

*Institute of Multidisciplenary Research for Advanced Materials,
Tohoku University, Sendai, 2-1-1, Japan*

Abstract. Inhomogeneity is an inherent nature of glass, which is the density and concentration fluctuation frozen at glass transition temperature. The inhomogeneity of the glass plays significant role in so called superionic conductor glasses (SIG), since the mobile ions seek to move through energetically favorable paths. The localization of mobile ions in SIG near the 2nd glass transition is a remaining issue, where the trapping, percolation and many-body interactions are playing the roles. In order to investigate the trapping process in SIG, the authors have applied 109Ag NMR Hole-Burning technique to AgI containing SIG glasses. By using this technique, the slowing down process of the site-exchange rates between different sites were evaluated.

GLASS TRANSITION OF SIG

Through the glass transition from a supercooled liquid to a glass, the density and the concentration fluctuations are frozen to remain as a spatial inhomogeneity in the glassy state. Such an inhomogeneity causes some peculiar properties of the glass.

Superionic conductivity is a good example of them. Some molten salts containing silver halides, lithium chalcogenides etc. can be vitrified through normal glass transition (**Tg1**) by quenching techniques. However, even in the glassy state, they show very high ionic conductivity like molten salts, which are called as Superionic Conductor Glasses (**SIG**) [1,2]. Thus, the mobile ions in the SIG can be regarded as in sub-liquid state. During this glass transition Tg1, the density and concentration fluctuations in the molten salt are frozen as a spatial inhomogeneity in nano-meter or sub-nano-meter scale. The mobile ions seek to move in the preferable region of the inhomogeneous structure. The motion of mobile ions in SIG can be considered as diffusion in a restricted space. Thus, the ionic transport is a good probe of the inhomogeneous structure of the glass.

The mobile ions sub-liquid in the SIG is frozen at very low temperature (ex. ~80K for AgI containing glass), which can be regarded as the second glass transition (**Tg2**) [3]. Approaching to this temperature, some anomolies are observed in dynamical responses such as frequency dependence of ac conductivity[2], Non-Arrhenius behavior of dc conductivity[6], NMR relaxation [7,8], specific heat etc. Those anomalies are owing to the localization of mobile ions in the inhomogeneous SIG structure relating to the 2nd glass transition. The localization of mobile ions in the glass matrix is found to be a complex process, where the trapping, percolation and many-body interactions are playing the roles.

NMR OF MOBILE IONS IN GLASS

Nuclear magnetic resonance (NMR) is a powerful tool to investigate the structure and the dynamics of the ions in glasses. Actually, some NMR relaxation experiments revealed non-BPP universal behavior in SIG, which have been attributed to the inhomogeneity and/or many-body interactions[7,8]. However, the conventional NMR technique is limited to clarify the localization process of ions in the glass, since it deals with the whole spectrum from ions in different environments. Particularly, the migration of mobile ions in the inhomogeneous structure of SIG results in a complex temperature dependence of the NMR spectra [5,8].

In order to make further advance from the phenomenological approach based on the overall spectrum to more precise understanding of the localization mechanism, it is necessary to develop new experimental techniques, which are sensitive to the local environment as the site-selective excitation techniques. In this paper is demonstrated our recent application of site-selective NMR hole-burning technique to some superionic conductor glasses [4,5].

HOLE-BURNING SPECTRA OF SIG

^{109}Ag NMR spectra were measured by Bruker MSL-300 spectrometer whose resonance frequency for ^{109}Ag nucleus was 13.9MHz. Its rather long T_1 value (~ 100ms~10s) allows us to apply the hole burning NMR. Hole burning spectrum was observed after selective long pulse of 5 ms, which saturates the ^{109}Ag nuclear spins only in selected sites. Recovery process of the spectrum was observed after various waiting times τ.

At room temperature just below Tg1, only a narrow Lorentian peak was observed, which was broadened to be a Gaussian like spectrum as the temperature was lowered close to Tg2. This broadening is mainly due to the chemical shift distribution of the mobile ions frozen at different sites, which is schematically shown in figure 1. The temperature dependence of the spectrum could be well expressed by considering a modified Bloch equation with site-exchange term based on the multi-site exchange model [4,5]

If the inhomogeneously broadened spectrum is irradiated by a narrow frequency pulse, only the nuclear spin at the selected site is excited. If the excited spin moves to other site during the spin is in excited state, one can observe the frequency diffusion; see fig. 1.

An example of the NMR hole burning spectra is shown in figure 2 for 20AgI-80AgPO$_3$ glasses at -20 C. In this figure, the recovery process after the hole burning saturation pulse at 400, 300 and 200 ppm are shown. The hole at 400 ppm diffused quickly (~10ms) to the other sites, since this hole is created on the Ag$^+$ ions surrounded mainly by I-, which diffuse fast and quickly transfer to the other sites. On the other hand, the hole created at 200 ppm does not diffuse to the other sites, since this hole is on the site of less mobile Ag$^+$ surrounded mainly by oxygens.

From these experiments, we can conclude that the Ag$^+$ ions distributed randomly among differently coordinated sites in the glasses, which exchange their positions very fast at high temperature. When the temperature is lowered the exchange rates among them decrease to zero near the 2nd glass transition Tg2, where the mobile ion sub-liquid is also frozen.

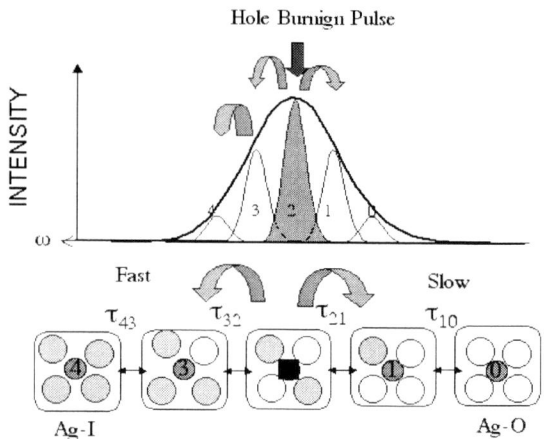

Fig.1 Principle of NMR Hole-Burning to detect site exchange of mobile ions (Ag$^+$) in SIG.

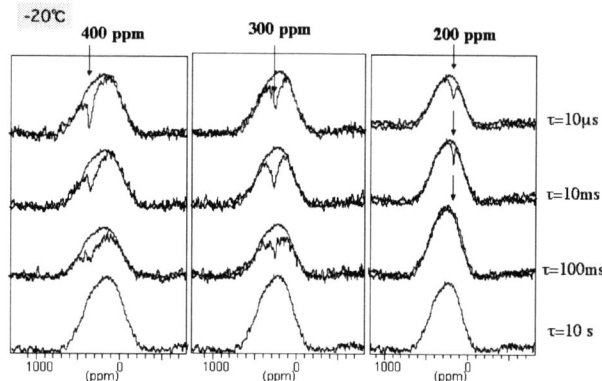

Fig.2 NMR Hole-Burning spectra of AgI-AgPO$_3$ glass excited at different sites.

REFERENCES

1. C. A. Angell, Ann.Rev.Phys.Chem. **43**, 693-717 (1992).
2. J. Kawamura and M. Shimoji, Materials Chem.Phys. **23**, 99-120 (1989).
3. M. Nakayama, M. Hanaya, and M. Oguni, Solid State Commun **89**, 403 (1994).
4. J. Kawamura, N. Kuwata, Y. Nakamura, T. Erata, and T. Hattori, Solid State Ionics ***154-455***, 183-188 (2002).
5. J. Kawamura, N. Kuwata, T. Hattori, and Y. Nakamura, Proceedings of 8th Asian Conference on Solid State Ionics , ***751-762*** (2002)
6. N. Kuwata, T. Saito, M. Tatsumisago, T. Minami and J. Kawamura, Ext.Abs.14th Int.Nat.Conf.Solid State Ionics, 336 (2003)/ to be published in Solid State Ionics.
7. S. H. Chung, K. R. Jeffrey, J. R. Stevens, and L. Borjesson, Phys.Rev.B ***41***, 6154-6164 (1990).
8. N. Kuwata, T. Saito, M. Tatsumisago, T. Minami, and J. Kawamura, J.Non-Cryst.Solids, ***324***, 79-91 (2003).

Ground-State Memories Survive Strenuous Thermal Fluctuations: Dynamics of Dimerized Spin Chains

Isao Sawada

Department of General Education
Ishikawa National College of Technology, Tsubata, Ishikawa 929-0392, Japan

Abstract. We present a new feature of the $T = \infty$ dynamics: the dynamical memory of the ground state surviving extremely strong thermal fluctuations in the aligned spin dimers. The eigenfrequencies are determined by the topology of interaction and dominantly written by Ω for the perpendicular chain, 2Ω for the two-leg ladder, and both Ω and 2Ω for the alternating chain, where $\Omega = 2|J|\sqrt{S(S+1)/3}$ with the exchange integral J in a dimer with arbitrary spin S. These modes are short-ranged owing to strenuous thermal disturbance and thus stable for weak disorder with less than 10 % concentration.

INTRODUCTION

The ground-state time correlations would die out as the temperatures rise up. Corresponding to the thermodynamics proportional to $1/T$, one expect that the dynamics also become featureless due to strong thermal fluctuations. In the high temperature limit the density operator change into the indentity operator; however, the specific frequency modes may survive for the time correlations to show the intrinsic oscillations over the monotonical decay. Researchers of the $T = \infty$ dynamics have much interest in the long-time tails, e.g., especially in the spin diffusion [3]. In this work the author presents another interesting feature of the $T = \infty$ dynamics.

We study the dynamics of systems composed of the spin dimers in a line and investigate the topological effects of weak dimer interaction on the high frequency mode. Very short-ranged spin time correlation as one of the ground-state memories survives strenuous thermal fluctuations. The correlation in the systems with 10 % or less impurities doped maintains this behavior.

SYSTEMS AND FORMALISM

Aligned dimers

The Heisenberg dimerized spin chains are given below:

$$H_\pi = J\sum_i (\hat{S}_{4i} \cdot \hat{S}_{4i+1} + \hat{S}_{4i+2} \cdot \hat{S}_{4i+3})$$
$$+ \pi \sum_i \Big((\hat{S}_{4i-2} + \hat{S}_{4i-1}) \cdot \hat{S}_{4i}$$
$$+ \hat{S}_{4i+1} \cdot (\hat{S}_{4i+2} + \hat{S}_{4i+3})\Big), \quad (1)$$

$$H_\lambda = J\sum_i \hat{S}_{2i} \cdot \hat{S}_{2i+1} + \lambda \sum_i \hat{S}_i \cdot \hat{S}_{i+2}, \quad (2)$$

$$H_\alpha = J\sum_i \hat{S}_{2i} \cdot \hat{S}_{2i+1} + \alpha \sum_i \hat{S}_{2i-1} \cdot \hat{S}_{2i}. \quad (3)$$

The aligned dimers in the ground state given by Eqs. (1)-(3) are the perpendicular chains, the two-leg ladders, and the alternating chains, respectively. The spin operator with arbitarary magnitude is conventionally denoted on a cite i. The exchange integral in a dimer is J, and the much smaller dimer interactions are written in Greek lower cases, i.e., $\pi^2, \lambda^2, \alpha^2 \ll J^2$.

Recursion method of dynamics

A dynamical variable of interest is chosen with respect to a dimer and then written as f_0 for later use

$$A = S_j^z + S_{j+1}^z = f_0. \quad (4)$$

At $T = \infty$, i.e., $\beta = 1/k_B T \to 0$, the canonical correlation converges to the ensemble average: $(A(t), A) \equiv \beta^{-1}\int_0^\beta \langle A(t - \hbar\lambda)A^\dagger\rangle d\lambda \to \langle A(t)A\rangle$ with $\langle O \rangle = \text{Tr}[O]/\text{Tr}[1]$. The Laplace transform of $a_0(t) \equiv \langle A(t)A\rangle/\langle A^2\rangle = (1/2\pi i)\oint dz e^{zt}\bar{a}_0(z)$ is written in a continued fraction [1] with a recurrence relation [2],

$$\bar{a}_0(z) = 1/z + \Delta_1/z + \Delta_2/z + \cdots \quad (5)$$

where

$$\Delta_n = (f_n, f_n^\dagger)(f_{n-1}, f_{n-1}^\dagger)^{-1} \quad (6)$$
$$f_{n+1} = (i/\hbar)[H, f_n]_- + \Delta_n f_{n-1} \quad (7)$$

with the boundary conditions $\Delta_0 = 1$ and $f_{-1} = 0$. In fact, e.g., $(f_0, f_0) = \langle A^2 \rangle = 2a$ with $a = S(S+1)/3$ and $\hbar = 1$. We note that the memory of a system is in the continued fraction coefficients $\{\Delta_n\}$ through the evaluation of the comutator in Eq. (7).

SURVIVING MEMORIES

When dimers are perfectly isolated, Eq. (4) is a constant of motion leading to $f_1 = \Delta_1 = 0$ and Re $\bar{a}_0(-i\omega^+) = \pi\delta(\omega)$. While dimer interactions are weakly introduced as given in Eqs. (1)-(3), $\{\Delta_n\}$ of Eq. (6) are

$$\{4a\pi^2, 2aJ^2, 2aJ^2, O(\pi^2), O(J^2)_{n\geq 5}\}, \quad (8)$$

$$\{4a\lambda^2, 4aJ^2, 12aJ^2, O(\lambda^2), O(J^2)_{n\geq 5}\}, \quad (9)$$

$$\{2a\alpha^2, 4aJ^2, 6aJ^2, 6aJ^2, 4aJ^2, O(\alpha^2), O(J^2)_{n\geq 7}\}, \quad (10)$$

respectively. Since $\pi^2, \lambda^2, \alpha^2 \ll J^2$, we find a dip in the space $\{\Delta_n\}$ at $n = 4$ for Eq. (8) and also Eq. (9) [5], while at $n = 6$ for Eq. (10) [4].

After some manipurations using the approximation of $\Delta_{n\geq 6} = \Delta_5$ for (8) and (9) and $\Delta_{n\geq 8} = \Delta_7$ for (10), well defined poles exist in the corresponding Eq. (5). The dynamical modes at high energies appear in Re $\bar{a}_0(-i\omega^+)$ as

$$\omega_\pi = \Omega \quad (11)$$

$$\omega_\lambda = 2\Omega, \quad (12)$$

$$\omega_\alpha = \Omega, 2\Omega \quad (13)$$

with

$$\Omega = 2|J|\sqrt{S(S+1)/3}. \quad (14)$$

Specifically, we have $\Omega/|J| \simeq (1, 1.6, 2.2, 2.8, 3.4, \cdots)$ for the case $S = (1/2, 1, 3/2, 2, 5/2, \cdots)$, respectively. The above approximation gives the life time for the modes of Eqs. (11)-(13).

Each residue is never vanishing even at $T = \infty$ and proportional to interaction squared, $\pi^2, \lambda^2, \alpha^2 \ll J^2$. These modes are strongly localized around the dimer considered by Eq. (4), and furthermore, stable also in the systems with 10 % or less impurities doped [6].

Here we consider the diamond chains as a reference system for the aligned dimers Eqs. (1)-(3),

$$H_\delta = J\sum_i \hat{S}_{3i+1} \cdot \hat{S}_{3i+2}$$
$$+ \delta\sum_i \left((\hat{S}_{3i-2} + \hat{S}_{3i-1}) \cdot \hat{S}_{3i} \right.$$
$$\left. + \hat{S}_{3i} \cdot (\hat{S}_{3i+1} + \hat{S}_{3i+2})\right) \quad (15)$$

with $\delta^2 \ll J^2$. This system contains a monomer between the adjacent dimers. High-energy mode like Eqs. (11)-(13) in terms of Eq. (14) does not appear in a reference system of Eq. (15) since $\{\Delta_n\} = \{4a\delta^2, 14a\delta^2, (37/7)a\delta^2, O(J^2)_{n\geq 4}\}$ without a noticeable dip. This is how we elucidate that the high-energy dynamics specified by Eqs. (11), (12), and (13) is inherent to the systems composed of connected dimers alone.

SUMMARY

The $T = \infty$ dynamics possesses well-defined high-energy poles and they are determined by the topological array of dimer interactions. The eigenfrequencies are dominantly written by Ω for the perpendicular chain, 2Ω for the two-leg ladder, and both Ω and 2Ω for the alternating chain, where $\Omega = 2|J|\sqrt{S(S+1)/3}$ with the exchange integral J in a dimer with arbitrary spin S. The ground-state memories survives strenuous thermal fluctuations.

REFERENCES

1. Mori, H., *Prog. Theor. Phys.* **34**, 399-416 (1965).
2. Lee, M. H., *Phys. Rev. B* **26**, 2547-2551 (1982).
3. Viswanath, V. S., and Müller, G., *The Recursion Method*, Springer, Berlin, 1994, ch. 10, pp. 136-182.
4. Sawada, I., *Phys. Rev. Letters* **83**, 1668-1671 (1999).
5. Sawada, I., *J. Phys. Chem. Solids* **62**, 373-376 (2001).
6. Sawada, I., *Physica B* **329-333**, 998-999 (2003).

Exact Non-Equilibrium Fluctuation Dissipation Relations for Multi-Spin Observables in the Glauber-Ising Spin Chain

P. Mayer* and P. Sollich*

*Department of Mathematics, King's College London, Strand, London, WC2R 2LS, United Kingdom

Abstract. We investigate the relation between two-time, multi-spin, correlation and response functions in the non-equilibrium dynamics of the Glauber-Ising chain quenched to zero temperature. We find fluctuation-dissipation relations qualitatively similar to those reported in various glassy materials, but quantitatively dependent on the chosen observable. Our results can be understood by considering separately the contributions from large wavevectors, which are at quasi-equilibrium and obey the fluctuation dissipation theorem (FDT), and from small wavevectors where a generalized FDT holds with a non-trivial fluctuation-dissipation ratio X^∞. For spin observables, reflecting critical aspects of the $T=0$ quench, we get $X^\infty = \frac{1}{2}$ while defect observables produce $X^\infty = 0$, revealing the underlying ordered phase. Our results suggest that the definition of an effective temperature $T_{\text{eff}} = T/X^\infty$ for non-equilibrated large length scales is generically possible in non-equilibrium critical dynamics.

INTRODUCTION

The fluctuation-dissipation theorem (FDT) may be used to define the thermodynamic temperature T of a system in equilibrium. An accumulating body of analytical and numerical results suggest a possibility to extend FDT – and thus the concept of temperature – to non-equilibrium systems. It amounts to the introduction of the fluctuation-dissipation ratio (FDR) $X(t, t_w)$ via

$$R(t, t_w) = \frac{X(t, t_w)}{T} \frac{\partial}{\partial t_w} C(t, t_w). \quad (1)$$

Here $C(t, t_w) = \langle O(t)O(t_w) \rangle - \langle O(t) \rangle \langle O(t_w) \rangle$ denotes the two-time connected correlation function of the observable O. The complementary two-time response function is $R(t, t_w) = \delta \langle O(t) \rangle / \delta h(t_w)|_{h=0}$, with h the thermodynamically conjugate field to O. While in equilibrium one has $X \equiv 1$ according to FDT, it has been found that for non-equilibrium system, such as glassy, coarsening or critical systems, a non-trivial limit form $X(t, t_w) \to X(C(t, t_w))$ emerges at large times.

In the context of p-spin models, $T_{\text{eff}} = T/X$ has been shown to play the role of a time-scale dependent effective temperature [1]. Beyond mean-field models, however, the relevance of $X(t, t_w)$ remains unclear. In particular the minimum requirement of *observable independence* of the FDR defined by (1) remains an open issue. To address this question we consider the dynamics of multi-spin observables in the one-dimensional Glauber-Ising chain and derive exact results for the FDR.

MULTI-SPIN FUNCTIONS

The basis of our work is a novel, specifically tailored approach to solve the hierarchy of differential equations for correlation functions [2]. It essentially consists in first solving the inhomogeneous sub-systems of evolution equations for k-spin correlations. From that general expressions for arbitrary correlation functions are then constructed in a recursive manner. We also show in [2] that multi-spin two-time correlations

$$C^{(k,l)}_{\mathbf{i},\mathbf{j}}(t, t_w) = \langle \sigma_{i_1}(t) \cdots \sigma_{i_k}(t) \sigma_{j_1}(t_w) \cdots \sigma_{j_l}(t_w) \rangle \quad (2)$$

and response functions

$$R^{(k,l)}_{\mathbf{i},\mathbf{j}}(t, t_w) = \left. \frac{\delta \langle \sigma_{i_1}(t) \cdots \sigma_{i_k}(t) \rangle}{\delta h_{\mathbf{j}}(t_w)} \right|_{h_{\mathbf{j}}=0}, \quad (3)$$

where $h_{\mathbf{j}}$ is the conjugate field to $\sigma_{j_1} \cdots \sigma_{j_l}$, satisfy formally equivalent evolution equations as one-time correlations. This allows us to obtain closed exact expressions for multi-spin two-time correlation and response functions after quenching the system from a random initial configuration to zero temperature.

RESULTS

We focus on the FDR associated with the family of spin observables $O_s = \sum_i \varepsilon_i \sigma_i$ and defect observables $O_d = \sum_i \varepsilon_i \sigma_i \sigma_{i+1}$. The ε_i are quenched random variables with zero mean $[\varepsilon_i] = 0$ and translation invariant covariances $[\varepsilon_i \varepsilon_j] = q_{i-j}$; $[\cdot]$ denotes the average over the distribution of ε. By tuning the covariances between $q_n = \delta_{n,0}$ and $q_n = 1$ the two-time connected correlation and response functions associated to O_s, O_d as defined below (1) effectively interpolate between those for local, incoherent observables (individual spins and defects) and the ones for global, coherent observables (magnetisation and energy), respectively. Between the two extremes, we distinguish between short range $\sum_n |q_n| < \infty$ and infinite range $\sum_n |q_n| = \infty$ correlated fields.

Our results [3] show that varying q_n has the same *qualitative* effect on the FDR for both spin and defect observables in the limit of large times: Incoherent observables produce an FDR that continuously crosses over from quasi-equilibrium $X = 1$ at $t - t_w \ll t_w$ to its asymptotic value

$$X^\infty = \lim_{t_w \to \infty} \lim_{t \to \infty} X(t, t_w), \quad (4)$$

whereas the FDR for coherent observables is constant at large times and coincides with X^∞ (see Fig. 1). Any short range correlated field – apart from pathological cases – eventually leads to the same FDR as the incoherent observable. Infinite range correlated fields yield new intermediate limit forms for X but again cross over from $X = 1$ to the very same X^∞.

We explain this robustness of X^∞ by introducing a generic, length scale dependent FDT. This consists in spatially Fourier transforming the correlation and response functions (2) and (3) with respect to $i - j$ and linking them via (1) by the FDR $X(k; t, t_w)$ for modes with wave-vector k. In this notation the FDR for the observables O_s, O_d may be written as

$$X(t, t_w) = \frac{\int dk\, X(k; t, t_w)\, q(k)\, \frac{\partial}{\partial t_w} C(k; t, t_w)}{\int dk\, q(k)\, \frac{\partial}{\partial t_w} C(k; t, t_w)}. \quad (5)$$

Based on this representation we show in [3] that independently of the covariances q_n the limit (4) is always dominated by the FDR $X(k=0; t, t_w)$. Therefore the definition of an effective temperature $T_{\text{eff}} = T/X^\infty$ is generically possible for large, non-equilibrated length scales.

Interestingly it turns out that the *quantitative* values of X^∞ are different for O_s and O_d. For spin observables one finds $X_s^\infty = 1/2$ whereas defect observables produce $X_d^\infty = 0$. This ambiguity may be related to the fact that spin observables reflect the critical aspects of the $T = 0$ quench while defect observables reveal the underlying ferromagnetic phase of the Glauber-Ising chain.

SUMMARY

We have investigated the robustness of fluctuation-dissipation relations in the Glauber-Ising spin chain at $T = 0$

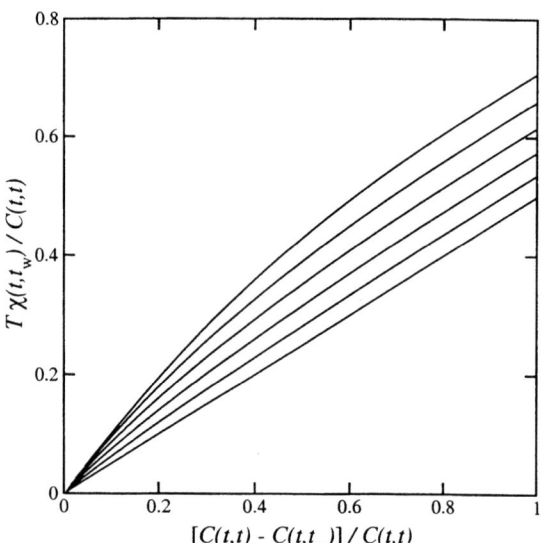

FIGURE 1. Long time fluctuation-dissipation limit plots for various spin observables O_s normalised by $C(t,t)$; here $\chi(t, t_w) = \int_{t_w}^t d\tau R(t, \tau)$. The slopes of the curves are given by the FDR $X(t, t_w)$. The top curve is obtained for a local spin, or any observable O_s with short range correlated fields, while the bottom curve corresponds to the magnetisation. Long range correlated fields that behave asymptotically like $q_n \sim |n|^{-\alpha}$ produce the intermediate curves for $\alpha = 0.8, 0.6, 0.4, 0.2$ from top to bottom, respectively. (Figure taken from [3])

for a wide range of observables. While generally observable dependent, the FDT violations fall into well defined classes. All spin observables share the same $X_s^\infty = \frac{1}{2}$ while $X_d^\infty = 0$ for defect observables. Our results suggest that the definition of an effective temperature $T_{\text{eff}} = T/X^\infty$ is generically possible for non-equilibrated large length scales in critical systems.

ACKNOWLEDGMENTS

We acknowledge financial support from the Austrian Academy of Sciences, the Wilhelm Macke Foundation and EPSRC Grant No. 00800822 (PM), and the Nuffield Foundation Grant No. NAL/00361/G (PS).

REFERENCES

1. Cugliandolo, L. F., Kurchan, J., and Peliti, L., *Phys. Rev. E*, **55**, 3898–3914 (1997).
2. Mayer, P., and Sollich, P., e-print cond-mat/0307214 (2003).
3. Mayer, P., Berthier, L., Garrahan, J. P., and Sollich, P., *Phys. Rev. E*, **68**, 016116 (2003).

Duality Symmetry, the Disorder Parameter, and the Glass Transition

I. Kanazawa

Department of Physics, Tokyo Gakugei University, Koganei-shi, Tokyo 184-8501, Japan

Abstract. We introduce generalized views of the physical origin of the Boson peak, the viscosity of the supercooled metallic liquids, and the relationship between the duality symmetry and the melting in the gauge-invariant formula.

INTRODUCTION

Incompatibility in gauge theory is a generalization of the theory of defects in condensed matter. If an internal parameter (matter field) is given everywhere within a continuum, this defines a mapping between the base space and the manifold, M, of states of the internal parameter (the fiber of the bundle) [1]. Extending the Sethna-Sachdev-Nelson formula [2,3], the present author [4,5] has introduced the effective Lagrangian in the gauge-invariant formula with spontaneous symmetry breaking for two dimensional and three dimensional metallic glasses. In this study, we will discuss the origin of the Boson peak, the viscosity of the supercooled metallic liquids, and the relationship between the duality symmetry and the melting, in the gauge-invariant formula.

THE PHYSICAL ORIGIN OF THE BOSON PEAK

In glasses and amorphous materials, one observes a thermal conductivity plateau at $\sim 10K$, and a low energy broad peak is observed in Raman and neutron scattering, the boson peak. It is thought that the vibrational states responsible for the boson peak contribute also to the thermal conductivity plateau, because the energy range spanned by the plateau covers that of the boson peak spectra, indicating that acoustic excitations must cease to propagate when their wavelength λ reaches the nm range. That is, acoustic modes may become strongly localized modes, satisfying the Ioffe-Regel condition. By a computer simulation of a soft sphere glass, it is found that there are (quasi-) localized modes with effective masses ranging from 10 atomic masses upwards, which are related to the boson peak. In the present theoretical formulation, the effective Lagrangian represents three massive vector fields A_μ^1, A_μ^2, and A_μ^3, which are localized within a radius, $\sim 1/|m_1|$, around the hedgehog-like clusters [5]. Thus, it is suggested that the localized gauge fields A_μ^1, A_μ^2, and A_μ^3 around the hedgehog-like clusters (solitions) are related to the (quasi-) localized modes of the boson peak. That is, the localized gauge fields A_μ^1, A_μ^2, and A_μ^3 introduce the localized strain tensor $u_{k\mu} \sim C_{ijk\mu}^{-1}\varepsilon_{ik\mu}\varepsilon_{jm\alpha}\partial_k\partial_m A_\mu^\alpha$ ($\alpha = 1, 2, 3$) around the hedgehog-like cluster (soliton), where $C_{ijk\mu}$ is the elastic tensor. It should be noted that localized modes around the hedgehog-like clusters (solitions) are required naturally through the gauge invariant condition in the present theory. Expanding the present formula, we can introduce a more generalized view of the origin of the boson peak. We adopt the generalized parameter,

$$\rho(r,u) \equiv \rho^a \quad (a = 1, 2, 3, 4, 5, \cdots N).$$

When the locally favored cluster is created, we set the symmetry breaking of $\langle 0|\rho^b|0\rangle \neq 0$, in which b represents components within the components from $a = m$ to N. As a result, $m - 1$ massive gauge modes (the localized modes) are introduced around the locally favored cluster through the Higgs mechanism.

THE VISCOSITY OF SUPERCOOLED METALLIC LIQUIDS

We shall consider how is configuration of aggregation of the icosahedral clusters in the supercooled metallic liquid phase. Using the formulation of the Ising-like cluster model by Coniglio-Klein [6], it is shown that by throwing bouds between nearest-neighbour pairs of the icosahedral clusters with a probability, $P = 1 - e^{-\frac{J}{2k_BT}}$. J is the binding energy between icosahedral clusters. Then we find the percolation-like line of the icosahedral clusters below $P = 1$. This line ends at the Ising critical point. At the Ising critical point, we find the connected length $\xi \sim (\frac{J}{k_B T_c} - \frac{J}{k_B T})^{-\nu} \sim (\frac{k_B}{J}\frac{T_c T}{(T - T_c)})^{\nu}$, where $\nu \sim 0.63$ [7] in three dimensions. From the relaxation mechanism in star entangled icosahedral cluster lines showing arm retraction and tube reconfiguration [8], the viscosity $\eta \sim \exp[u\xi] \sim \exp[u(\frac{k_B}{J}\frac{T_c T}{(T - T_c)})^{0.63}]$ with u a constant is introduced approximatelly. In this system, icosahedral cluster branched-lines may correspond to the cooperatively rearranging regions (CRR). It is suggested that the critical temperature T_c is similar to Vogel-Fulcher temperature T_0.

THE DISORDER PARAMETER $B(C)$ AND THE DUALITY SYMMETRY

In this simple model, the hedgehog-like solitons are invariant under the $Z(2)$ symmetry in $3 + 1$ dimensions. We introduce the order parameter in the usual way as follows,

$$A(C') = T_r P \exp \int_{C'} g_1 \cdot A_k(x) dx^k.$$

We can introduce the disorder parameter $B(C)$ from the beautiful commutation relation, $A(C')B(C) = B(C)A(C')\exp(2\pi i n/2)$, in $3 + 1$ dimensions, the disorder parameter, $B(C)$, creates a Rivier line, which is analogous to the Nielsen-Olesen voltex, where n is the number of times the closed curve C winds around the closed curve C'. It is illustrative to compare the qualitative behaviour of $\langle A(C) \rangle$ and $\langle B(C) \rangle$ for large C in a theory with a complete Higgs effect, and a small coupling constant g_1 (low temperature). Because the gauge fields are short-range, their propagators drop exponentially at large distance, so the main contribution to $\langle A(C) \rangle$ comes from the gauge field propagators that connect only closely separated points on C. For a large, smooth, C, we get the following result,

$$\langle A(C) \rangle \propto \exp(-\alpha L(C)).$$

Here $L(C)$ is the total length of C and α is a fixed constant. In this condition, a stretch of a Rivier string of length l, whose ends are attached to monopole-like solitons, is created, evolves for a Euclidean time of length τ and is then annihilated. In Euclidean space the amplitude for such a process approaches $e^{-Ml\tau}$ for large τ, where M is the mass of the hedgehog(mono-pole)-like soliton. Thus we find $\langle B(C) \rangle \sim \exp(-Ml\tau) = \exp(-M\Sigma(C))$, where $\Sigma(C)$ is the area enclosed by C. In this phase, the hedgehog(mono-pole)-like solitons are confined. For the case of a strong coupling constant g_1 (high temperature), we have the opposite to the previous phase,

$$A(C) \sim \exp(-\alpha \sum(C))$$
$$B(C) \sim \exp(-ML(C)).$$

The self-duality symmetry under $g_1 \to g_1^{-1}$ implies that this point has to be $g_1 = 1$. This point might correspond to the melting temperature.

REFERENCES

1. River, N., "GAUGE THEORY AND GEOMETRY OF CONDENSED MATTER" in *GEOMETRY IN CONDENSED MATTER PHYSICS*, edited by J. F. Sadoc- world Science, Singapore, 1990, pp.1-88.

2. Sethna, J.P., *Phys. Rev. Letters* **51**, 2198-2201(1983).

3. Sachdev, S., and Nelson, D. R., *Phys. Rev. Letters* **53**,1947-1950(1984).

4. Kanazawa, I., *Prog. Theor. Phys. Suppl.* **126**,393-397(1997).

5. Kanazawa, I., *J. Non-Cryst. Solids* **293**,615-619(2001).

6. Conglio, A., and Klein, W., *J. Phys.* **A13**, 2775-2780(1980).

7. Berber, M.N., *Phys. Reports* **C29**, 1-84(1977).

8. McLeish, T.C.B., and Milner, S.T., *Adv. Polym. Sci.* **143**, 195-256(1999).

9. 't Hooft, G., *Acta Phys. Austr. Suppl.* **22**, 531-586(1980).

Glass transition of hard sphere systems–Molecular dynamics and density functional approaches

Kang Kim [*,†] and Toyonori Munakata [**]

[*]*Department of Physics, Kyoto University, Kyoto 606-8502, Japan*
[†]*PRESTO, Japan Science and Technology Agency, 4-1-8 Honcho Kawaguchi, Saitama, Japan*
[**]*Department of Applied Mathematics and Physics, Graduate School of Informatics,
Kyoto University, Kyoto 606-8501, Japan*

Abstract. The glass transition of a hard sphere system is investigated within the framework of the density functional theory (DFT). Molecular dynamics (MD) simulations are performed to study dynamical behavior of the system on the one hand and to provide the data to produce the density field for the DFT on the other hand. Energy landscape analysis based on the DFT shows that there appears a metastable (local) free energy minimum representing an amorphous state as the density is increased. This state turns out to become stable, compared with the uniform liquid, at some density, around which we also observe sharp slowing down in MD simulations.

In order to understand the universal mechanism of various phase transitions, the density functional theory (DFT) is recently gathering much attention. The glass transition has been investigated also based on the DFT by some workers [1, 2]. In the earlier works, the random close packing (RCP) has been produced by the Bennett's algorithm and the free energy from the DFT has been calculated. It is remarked here that since the RCP configurations were produced by a kind of aggregation method, we can not study dynamical aspects of the glass transition found by the energetics based on DFT. Our purpose is first to produce a glassy state for a hard sphere system, relying on molecular dynamics (MD) method and then to study both dynamical and static properties of the state [3]. Especially from the particle configuration data, we can discuss free energy within the DFT framework.

We first try to obtain hard sphere glasses by MD simulations without recourse to the Bennett algorithm. Our system consists of $N = 1372$ identical hard spheres with the mass m and the diameter σ in a cubic box of volume V with periodic boundary conditions. Hereafter, we take to be the nondimensional number density $\tilde{n} = n\sigma^3$ ($n \equiv N/V$). As is well known, the system freezes at $\tilde{n} \simeq 0.94$. To avoid the crystallization and obtain amorphous glassy states, a compressing procedure has been introduced [4]. Following Lubachevsky et al., we could obtain various high density states, $0.86 \leq \tilde{n} \leq 1.21$ without crystallization.

We next consider energetics of the system based on the DFT and the particle configuration generated by the MD simulations. For a practical calculation of the DFT, we employ the Ramakrishnan and Yussouff free energy functional:

$$F[n(\mathbf{r})] = F_{id} + F_{int}^{(2)}, \quad (1)$$

where

$$F_{id} = k_B T \int n(\mathbf{r}) \ln\left[\frac{n(\mathbf{r})}{n}\right] d\mathbf{r}, \quad (2)$$

$$F_{int}^{(2)} = -\frac{k_B T}{2} \int \int [n(\mathbf{r}) - n] C(|\mathbf{r} - \mathbf{r}'|)[n(\mathbf{r}') - n] d\mathbf{r} d\mathbf{r}'. \quad (3)$$

Here, F_{id} and $F_{int}^{(2)}$ represent the ideal gas term and the second order interaction term in the expansion around the uniform liquid state, respectively. We note that $C(\mathbf{r})$ is the direct correlation function of the uniform liquid with the density n. In order to evaluate the free energy of the system, we need the trial density field $n(\mathbf{r})$, for which we employ a conventional Gaussian superposition[1, 2],

$$n(\mathbf{r}) = \sum_{i=1}^{N} \left(\frac{\alpha}{\pi}\right)^{3/2} \exp[-\alpha(\mathbf{r} - \mathbf{r_i})^2], \quad (4)$$

where α is a variational parameter for the calculation of the free energy. It is noted that the particle sites $\{\mathbf{r_i}\}$ are given by our MD simulations. The total free energy per particle relative to uniform state,

$$\Delta f(\alpha) = \frac{F_{id}(\alpha) + F_{int}^{(2)}(\alpha)}{Nk_B T}, \quad (5)$$

is calculated as a function of the localization parameter α (see Figs. 1). While only the uniform liquid state is stable at low density, the free energy local minimum at finite α,

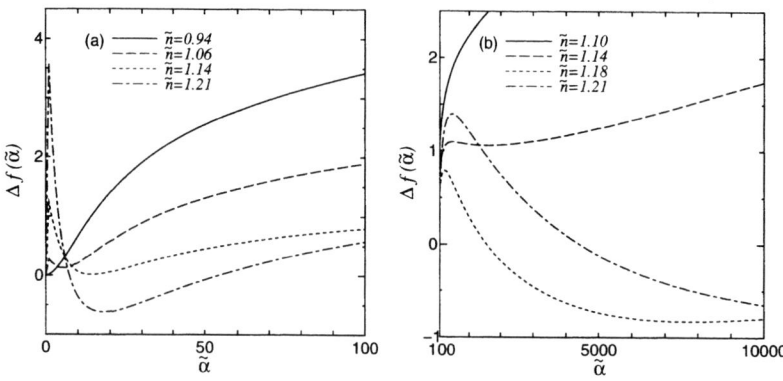

FIGURE 1. Total free energy per particle relative to uniform liquid $\Delta f(\tilde{\alpha})$ for $\tilde{\alpha} \leq 100$ (a) and $\tilde{\alpha} \geq 100$ (b). The units of $\tilde{n}$ and $\tilde{\alpha}$ are σ^{-3} and σ^{-2}.

which represents an amorphous state, begins to appear at high density $\tilde{n} \simeq 1.06$. In addition, Figs 1 show that two local minima are located at $\tilde{\alpha}(\equiv \alpha\sigma^2) \simeq 13$ and 1600 for $\tilde{n} = 1.14$. Das et al. also observe two local minima of $\Delta f(\alpha)$, which are called the weakly localized state for small α and the highly localized state for large α, respectively [2].

In Fig. 2, we plotted the free energy differences Δf of the weakly and highly localized states as a function of the density $\tilde{n}$. From Fig. 2, we notice that the weakly localized state appears for $\tilde{n} \geq 1.06$ and the highly localized state appears $\tilde{n} \geq 1.14$. For higher densities $\tilde{n} \geq 1.15$, it is seen that the highly localized state is more stable than the weakly localized state. Moreover, it is found in Fig. 2 that both the weakly and highly localized states become more stable than the uniform state at around $\tilde{n}_{g,DFT} \simeq 1.15$, which is the liquid-glass transition density from the energetics based on the DFT.

Finally, we compare our results from the energetics above with dynamical informations supplied by our MD [3]. We calculate the intermediate scattering function $F_s(q,t)$ and find that $F_s(q,t)$ begins to exhibit the two-step relaxation at the density $\tilde{n} \simeq 1.06$, which corresponds precisely to the density where the free energy local minimum begins to appear in our DFT (see Figs. 1). As to the structural relaxation time τ, we notice that the density dependence could be described by the power-law $\tau \propto (\tilde{n}_{g,MD} - \tilde{n})^{-\gamma}$ with $\tilde{n}_{g,MD} \simeq 1.15$. This density happened to coincide with the density $n_{g,DFT}$ beyond which the localized state is more stable than the uniform liquid in the present DFT. From these results, we consider that the DFT based energetics and dynamical behaviors related to slow dynamics are well correlated with each other.

In summary, we reconsidered the DFT approach to the glass transition in the hard sphere system, which was first undertaken by Singh et al. We obtained hard sphere glasses by MD simulations without recourse to the Bennett algorithm and the information on particle configurations produced by the MD simulations are used as input data when the free energy is calculated based on the DFT. While only the uniform liquid state is stable at low density, the free energy local minimum begins to appear at high density $\tilde{n} \simeq 1.06$, where our MD shows that two-step relaxation begins to appear. This metastable glassy state becomes stable relative to uniform liquid at $\tilde{n}_{g,MD} = 1.15$.

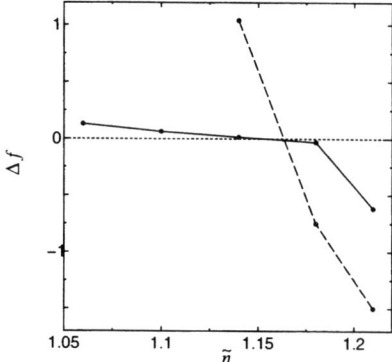

FIGURE 2. Free energy differences Δf of the weakly localized state (solid line) and the highly localized state (dashed line) as a function of density $\tilde{n}$.

REFERENCES

1. Singh, Y., Stoessel, J.P., and Wolynes, P.G., *Phys. Rev. Lett.* **54**, 1059-1062 (1985).
2. Kaur, C., and Das, S.P., *Phys. Rev. Lett* **86**, 2062-2065 (2001).
3. Kim, K., and Munakata, T., *Phys. Rev. E* **68**, 021502 (2003).
4. Lubachevsky, B.D, and Stillinger, F.H., *J. Stat. Phys.* **60**, 561-583 (1990).

Transition from Annealed to Quenched Dynamics

F. Tagawa and T. Odagaki,

Department of Physics, Kyushu University, Hakozaki 6-10-1, Higashiku, Fukuoka, 812-8581, Japan

Abstract. A framework to calculate the specific heat near the glass transition point is presented within the energy landscape picture. This framework gives the characteristics of the glass transition that the glass transiton a transition from annealed to quenched system and the glass transition point is decreased with increasing observation time. We apply this framework to classic model and one constructed with Debye oscillators, and they shows the characteristics of the glass transition.

INTRODUCTION

Near the glass transition temperature, it is considered that the energy landscape plays dominant role in determining dynamic and thermodynamic properties. In this scheme, the system is explained with the motion of a representative point in the energy landscape. It has been shown that the dynamical characteristic near the glass transition can be understood in a unifying manner by the single particle motion in the energy landscape.[1] The thermodynamics of a energy landscape composed of Einstein oscillators with various frequencies and it is shown that the tharacteristics of the glass transition that the glass transition is a thermodynamic transition from annealed to quenched dynamics and the glass transition temperature becomes lower as the observation time is increased.[2][3][4] It is an open and important problem to show if the results obtained in the landscape picture depend on the model of the landscape and the dynamics within basins of the landscape.

We apply a formalism proposed in Ref[2] to a classic oscillator model and Debye oscillator model (hereinafter reffered to as Clasic model and Debye model) in the energy landscapes and we show that the glass transition characteristics do not depend on the detail of the landscape.

LANDSCAPE DESCRIPTION

In the system in the super cooled state, the potential energy has basins. In high temperature region, the motion of the system is hardly influenced by basins, however at low temperature, which is near the glass transition point, the exisistence of basins influences the motion of the system strongly. In this regime, there is two types of the motion; vibration motion in one basin and jumping motion between basins. Since the long time is needed to observe thermodynamic properties like specific heat, it is important to include jump motion to the theory. Tao et al.[2] proposed the framework to calculate the phycical property for the system with landscapes. In the energy-landscape picture, the energy of the system at time t is given by

$$\langle E(t) \rangle = \sum_i E_i P_i(t,T) \quad (1)$$

where E_i is the energy of basin i and $P_i(t,T)$ is the probability that the system is in basin i at time t at tempearute T. Then the specific heat can be expressed as

$$C_v = \sum_i \left[\frac{dE_i}{dT} P_i(t,T) + E_i \frac{\partial P_i(t,T)}{\partial T} \right]. \quad (2)$$

The energys of the Classic model is

$$E_i = 3Nk_B T + V_i \quad (3)$$

and the one of the Debye model is

$$E_i = 9Nk_B T \left(\frac{T}{T_D}\right)^3 \int_0^{T_D/T} \frac{x^3}{e^x - 1} dx + V_i. \quad (4)$$

Here, V_i is the potential local minimum of basin i.

The probability P_i obeys the master equation

$$\frac{\partial P_i(t,T)}{\partial t} = \sum_j \left(W_{j,i}(T) P_j - W_{i,j}(T) P_i \right) \quad (5)$$

where $W_{i,j}$ is the transition rate from basin i to basin j.

RESULTS

We assume the system is in equilibrium at temperature T for $t < 0$ and change the temperaute to $T + \Delta T$ at $t = 0$.

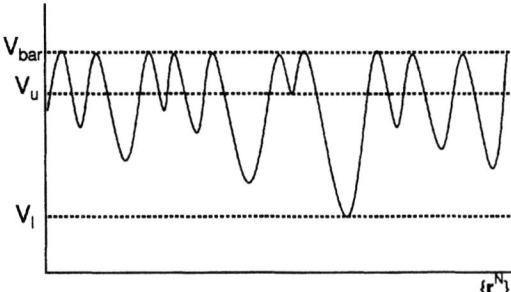

FIGURE 1. The schematic view of the potential energy landscape; Assumptions we set are the potential minimums are distributed uniformly between V_u and V_l and the curvature at local minimum are same at all basins.

At $t_{obs} = 0$, where t_{obs} is observation time, the specific heat at T is represented as

$$C_v = \sum_i \frac{\partial E_i}{\partial T} P_i^{eq}(T), \qquad (6)$$

where $P_i^{eq}(T)$ is the equilibrium value of the probability $P_i(t,T)$, which is the stationary solution of equation (5). This corresponds to the quenched system. At $t_{obs} = \infty$, when we take $\Delta T \to 0$ limit, the specific heat at T is represented as

$$C_v = \sum_i \left(\frac{\partial E_i}{\partial T} P_i^{eq}(T) + E_i \frac{\partial P_i^{eq}(T)}{\partial T} \right) \qquad (7)$$

This coreesponds to the annealed system.

The potential energy landscape is shown in fig 1. The potential minimum is distributed uniformly between V_u and V_l, where V_u is the upper limit and V_l is the lower limit of the potential minimum. The potential energy of barrier is set as $(V_{bar} - V_u)/(V_u - V_l) = 1/10$. Assuming all basins are mutually connected and curvatures of all basins are same, the transition rate is represented as

$$W_{i,j} = C \exp\{-\beta(V_{wall} - V_i)\}, \qquad (8)$$

where C is a constant.

We calculate the specific heat at $Ct = 0$(quenched system),$1.0 \times 10, 1.0 \times 10^2, 1.0 \times 10^4, \infty$ (annealed system). Fig 2, 3 show the behavior of the specific heat of the Classic model and the Debye model. We can see that as the tempeareture decreases, a transition from annealed to quenched system occur, and the transition point is decreases with increasing observation time.

CONCLUSION

We have calculated the specific heat of the Classic model and the Debye model having many potential minimums.

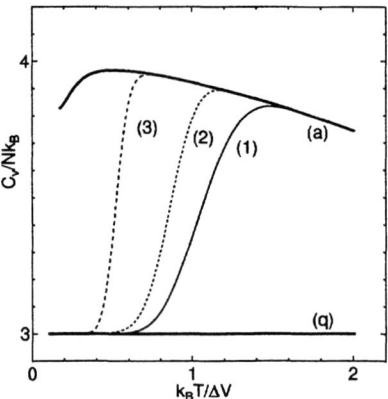

FIGURE 2. Temperature dependence of a specific heat for the classic model; where $\Delta V = (V_{wall} - V_u)$, the lines represent (q) the quenched system $Ct = 0$, (a) the annealed system $Ct = \infty$, (1) $Ct = 1.0 \times 10$, (2) $Ct = 1.0 \times 10^2$, (3) $Ct = 1.0 \times 10^4$ respectively.

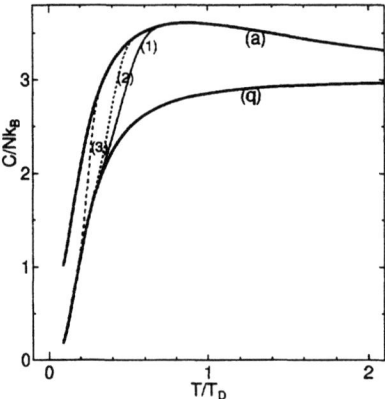

FIGURE 3. Temperature dependence of a specific heat for the debye model; The lines represent (q) the quenched system $Ct = 0$, (a) the annealed system $Ct = \infty$, (1) $Ct = 1.0 \times 10$, (2) $Ct = 1.0 \times 10^2$, (3) $Ct = 1.0 \times 10^4$ respectively.

We have shown the behavior of the specific heat doesn't depend strongly on the detail of the characteristic and dynamics of landscape basins and the glass transition can be understood as annealed to quenched transition phenomenologically.

REFERENCES

1. Odagaki,T., Hiwatari,Y., *Phys.Rev.A* **41**, 929-937 (1991).
2. Tao,T., Yoshimori,A., and Odagaki,T., *Phys.Rev.E* **66** 041103 (5pages) (2002).
3. Odagaki,T., Tao,T., and Yoshimori,A., *J.Non-Cryst.Solids* **307**, 407-411 (2002).
4. Odagaki,T.,Yoshidome,T.,Tao,T.,and Yoshimori,A., *J.Chem.Phys* **117**, 10151-10155 (2002)

How reproducible is the structure of dynamic heterogeneity in glass forming liquids?

Asaph Widmer-Cooper and Peter Harrowell

School of Chemistry, University of Sydney, Sydney, NSW 2006, Australia

Abstract. Near T_g the dynamics in glass-formers vary by orders of magnitude from one region to another. The existence of such spatially heterogenous dynamics is now well established, both through simulation and experiment. However, it remains unclear to what extent this dynamic heterogeneity is a reflection of some underlying structural properties of the glass-former. Here we provide a general theoretical tool for establishing the connection between a particle configuration and the spatial heterogeneity of the dynamics that originates from that configuration, and apply this to a model soft-disk glass. We find that the dynamic heterogeneity in a trajectory is not completely determined by the initial configuration and establish a rigorous causal connection between a given configuration and the propensity of particles to move. Finally, we discuss a number of generalisations of this tool.

Dynamic heterogeneities, both temporal and spatial, are now an established aspect of glass phenomenology. The existence of long-lived kinetic subpopulations has resolved a number of puzzles associated with supercooled liquids. These include the anomalous time scaling of crystal nucleation [1] and the breakdown of the Stokes-Einstein relation between diffusion and viscosity [2]. How much more can we learn from dynamic heterogeneity? The aim of this paper is to establish the useful information content of these heterogeneities by considering their reproducibility.

The spatial distribution of dynamic heterogeneities has been characterised in simulations [3, 4] and, more recently, experiments [2]. While this spatial variation in kinetics points strongly to an underlying structural origin, there has been little progress in establishing an explicit connection between structure and dynamics. In this paper we present a general theoretical tool for establishing the connection between a particle configuration and the spatial heterogeneity of the dynamics that originates from that configuration, and apply this to a model glass.

A number of workers have looked for correlations between a particle's mobility and energy [3], local topology [4], or local free volume [5]. However, no correlation has been established strong enough to indicate a causal link. In this paper instead of trying to directly answer the question "What aspect of the structure gives rise to the observed dynamic heterogeneity?" we first address the question, "What aspect of the dynamic heterogeneity is the result of the structure?"

Our approach is based on an analysis of the correlations among the set of N-particle trajectories that pass through a specific particle configuration. We shall refer to this collection of trajectories as an *iso-configurational ensemble*. In practice, the trajectories are generated by running many MD simulations using the same initial configuration but with each run having a different random assignment of initial particle momenta, taken from the appropriate Boltzmann distribution. Thus the only constant in phase space is the original configuration.

We consider a 2D glass-forming liquid consisting of a binary mixture of particles interacting via purely repulsive softcore potentials of the form $u_{ab}(r) = \varepsilon[\frac{\sigma_{ab}}{r}]^{12}$. For further details of the model see [6, 4]. A total of 1024 particles were enclosed in a square box with periodic boundary conditions. Simulations were run in the isothermal-isobaric ensemble, using the Nosé-Poincaré-Andersen Hamiltonian and algorithm of Sturgeon and Laird.[7] All units refer to reduced units.

In Figure 1, we present the plots of the particle displacement vectors following two different runs starting from the same configuration. Each run was carried out at a temperature $T = 0.4$ and a pressure $P = 13.5$. This represents a temperature below that at which a two-step character first appears in the decay of the intermediate scattering function.[6] A time interval of 1000τ was used, chosen as one and a half times the structural relaxation time, τ_e.

The plots in Figure 1 exhibit the now familiar features of dynamics in deeply supercooled liquids: large (non-Gaussian) variations in the particle displacements, clear clustering of the 'slow' particles, and aggregation of the more mobile particles, sometimes in 'string-like' features. What is just as striking is that each plot dif-

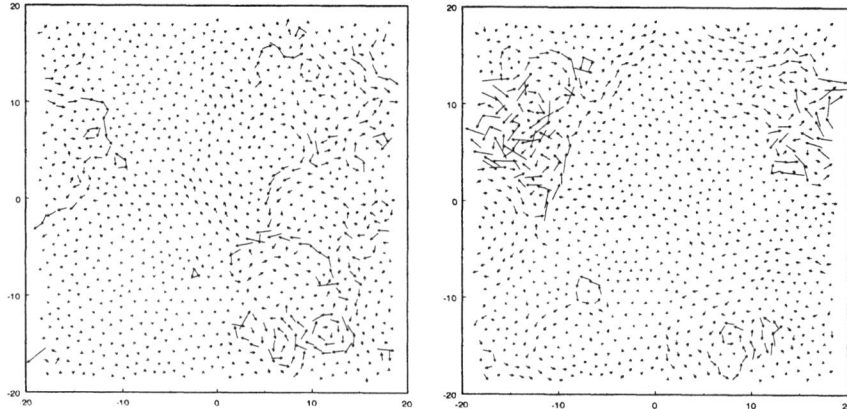

FIGURE 1. Displacement vectors for two different runs of $1.5\tau_e$ at $T = 0.4$. Both were started from the same initial configuration and differ only in the distribution of initial momenta.

fers markedly from each other. While some particles exhibit a mobility that is reproducible from run to run, the dynamics of other particles can vary substantially. The variation from run to run of the distribution of mobilities in magnitude and space demonstrates that significant aspects of the dynamic heterogeneities observed in an individual trajectory cannot be attributed to the structure of the initial particle configuration.

Given the large variation in the spatial arrangement of fast and slow particles from run to run, it is nontrivial to ask what, if any, dynamic heterogeneity will survive averaging over many runs? In Figure 2 are plotted the displacement vectors for 50 runs of $1.5\tau_e$ at $T = 0.4$. These constitute a sampling of our iso-configurational ensemble. It is clear that some spatial variation in kinetics persists after our averaging. Given that the different trajectories have nothing in common except the initial configuration, we can conclude that the spatial structure of average mobility in Figure 2 is determined by the initial configuration. This ensemble average mobility represents the likelihood that a particle will move during a trajectory through this configuration, i.e. the *propensity* for motion of a particle in a given configuration. It is this aspect of the dynamics that is directly related to structure. Finally, we note that the observed differences in propensity are not an artefact of the large variance in single-run mobilities. Increased sampling of the ensemble reduces the error bars on the individual propensities, but the spatial variations remain.

The iso-configurational ensemble technique that we have demonstrated here represents a general tool for determining that part of the dynamics that is related to structure. It is straightforward to extend this, for example, to explore the change in the propensity map due to a temperature change, a change in initial configuration, or both. It is also directly extendable to the study of glass-formation in three dimensions.

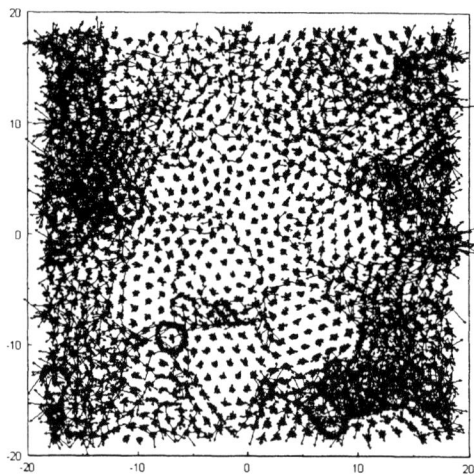

FIGURE 2. Displacement vectors for 50 runs of $1.5\tau_e$ at $T = 0.4$. All were started from the same initial configuration used in Figure 1 and differ only in the distribution of initial momenta.

REFERENCES

1. Harrowell, P., and Oxtoby, P. W., *Ceramic Trans.*, **30**, 35–44 (1993).
2. Ediger, M. D., *Ann. Rev. Phys. Chem.*, **51**, 99–128 (2000).
3. Donati, C., Glotzer, S. C., Poole, P. H., Kob, W., and Plimpton, S. J., *Phys. Rev. E*, **60**, 3107–3119 (1999).
4. Perera, D. N., and Harrowell, P., *J. Chem. Phys.*, **11**, 5441–5454 (1999).
5. Starr, F. W., Sastry, S., Douglas, J. F., and Glotzer, S. C., *Phys. Rev. Lett.*, **89**, 125501/1–125501/4 (2002).
6. Perera, D. N., and Harrowell, P., *Phys. Rev. E*, **59**, 5721–5743 (1999).
7. Sturgeon, J. B., and Laird, B. B., *J. Chem. Phys.*, **112**, 3474–3482 (2000).

Computer simulations of a model glass with the internal structures

T. Muranaka

Aichi Institute of Technology, 1247 Yachigusa, Yagusa-cho, Toyota, 470-0392, Japan

Abstract. I have two configurations constracted by the model short polymers. That polymers have the internal structure by means of it's potentials and the systems also have the internal structures. One system has some small crystalline domains and the long time relaxation. The another system has no long time relaxation. I think that the long displacement of polymers in short time distinguishes the two system's dynamics.

INTRODUCTION

I have calculated some moleculer dynamics simulations of the model glasses. They were the two-dimensional (2-D) fluid model of soft-sphere(disk) mixtures[1,2], the three-dimensional (3-D) model[3], the various system sizes of each model[4], and the model polymer with only bond between the soft-spheres in 3-D[5]. I have found the correlated motions of atoms[1], the coupling between jump motions and correlated motions in 2-D[2,6], and that the dynamics have the system size dependency[4]. Then I wanted to know the dynamic properties of more realistic model glasses.

SYSTEM

The molecules have the bonds between atoms nearby, a bond have the bends to the next bond, and I adopt the torsional potential as usual. In short, the DREIDING[7] potential parameters have been used in this research. The model molecule has 20 united atoms, the system has 2,000 molecules. The volume and density(set to 0.9 g/cm^3) are constant for the all simulations. I have created five initial configurations, but I will show one which is called "ini5,000". This was created in the temperature at 700K. The fluid sample was created to anneal it for some time. One sample which is called "straight" was made by cooling straightway to 300K from the fluid. Another sample which is called "stepwise2" was made by cooling stepwise down 100K from the fluid, the enough annealing time were accomplished after each cooling process, and four cooling processes are necessary to 300K. I calculated for 6.8 nsec below NVE ensemble for each sample.

RESULTS

I have two configurations of a model glass with the internal structures. The density auto-correlation function is shown in Fig.1.

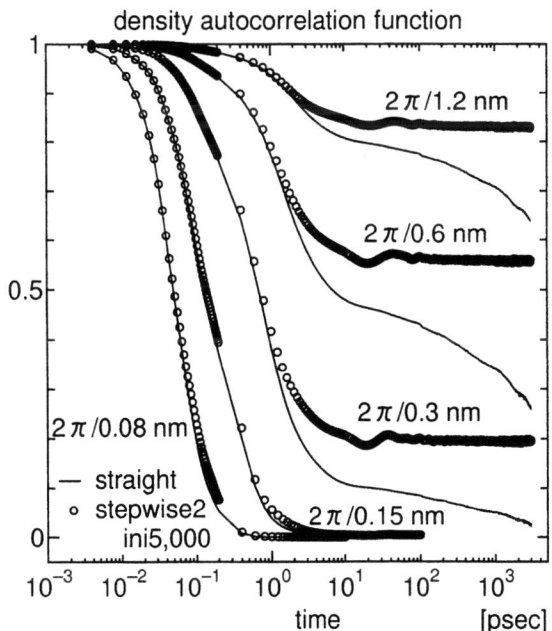

FIGURE 1. The density auto-correlation function. The solid line shows straight, the circle shows stepwise2. Five wave number's curves are shown.

The straight sample has some domains like a crystalline, but the long time relaxation exists. The stepwise2 sample has the orientational order in hole system, and the long time relaxation disappears. There are two-step relaxations in the straight sample. Before several psec, the relaxations are same for each sample. The velocity autocorrelation becomes zero first at 0.04 psec, the united atoms move like the thermal vibration. The longest time region of relaxation is laid in several nsec. That seems to exist in the same time region for every wave numbers smaller than $q=2\pi/0.3$ nm. What motion of atoms bring about the longest relaxation?

The global orientational order parameter in the long range (5 nm) becomes about 0.8 for the stepwise2 and 0.0 for the straight. This is most clear difference in the structure. But where is the dynamics difference. The displacement between 10 psec of the first united atom and the first molecule(which includes the first united atom) and the first molecule from the surrounding some molecules($\sim$10 molecules) are shown in Fig.2.

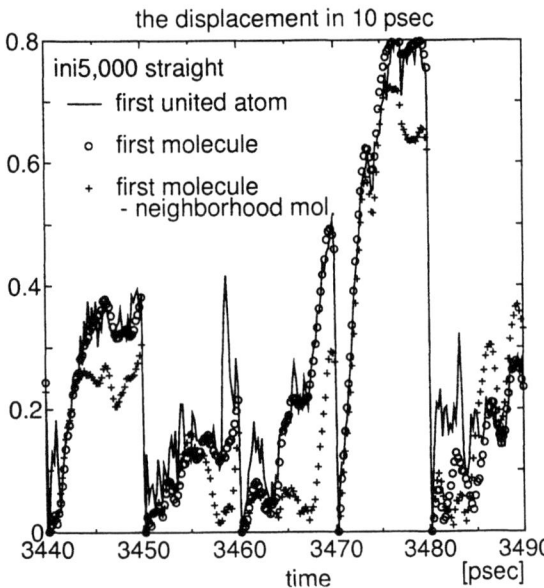

FIGURE 2. The displacement between 10 psec for ini 5,000. The solid line shows the displacement of the first united atom, the circle shows the displacement of the first molecule, the cross shows the displacement of the first molecule from the surrounding some molecules.

I have analized for each sample with many time regions and molecules. The motion of molecules are almost single motion of each molecule like between 3,440 and 3,450 psec, but some motion of molecules are cooperative motion with the surrounding molecules like between 3,460 and 3,470 psec. And a part of molecule happens to travel very long distance for short time. This long displacement reaches over 1 nm for 6 psec. The end to end distance of molecule is about 2.35 nm, it reveales that the displacement of such molecules is extraordinary long distance. The opportunity of this motion is very rare, but the probability of this motion between the two systems is clearly different. For example, 300 molecules in stepwise2 and 933 molecules in straight that the molecule happens to travel over 0.8 nm in 6 psec for 6.8 nsec.

CONCLUSION

I obtained two samples like glass (straight) and crystalline (stepwise2). The dynamics in the short time of these are almost same. But the long time properties of these are different, the glass continues to relax and the crystalline stops to relax. The structural difference of these are clear, the snapshots of these configurations and the global orientational order parameters lead the same conclusion. The dynamical difference of these are not clear, but I have found that the long displacement of molecules distinguishes the dynamics of glass and crystalline.

REFERENCES

1. Muranaka, T., and Hiwatari, Y., *Phys. Rev. E* **51**, R2735–R2738 (1995).
2. Muranaka, T., and Hiwatari, Y., "Coupling between Jump Motions and Correlated Motions", in *Dynamics of Glass Transition and Related Topics*, edited by T. Odagaki et al., PROGRESS OF THEORETICAL PHYSICS SUPPLEMENT, Kyoto, 1997, pp.403–406.
3. Hiwatari, Y., Matsui, J., Muranaka, T., and Odagaki, T., *Journal of Molecular Liquids* **65/66**, 123–130 (1995).
4. Muranaka, T., "Self-similarity in the regions of the cooperative displacement", in *JOURNAL DE PHYSIQUE IV PROCEEDINGS INTERNATIONAL WORKSHOP ON DYNAMICS IN CONFINEMENT*, edited by B. Frick et al., EDP Sciences, Les Ulis, 2000, pp.47–50.
5. Muranaka, T., and Hiwatari, Y., "Temperature Dependence of the Susceptibility on the Glassy Systems", in *SLOW DYNAMICS IN COMPLEX SYSTEMS*, edited by M. Tokuyama and I. Oppenheim, AIP CONFERENCE PROCEEDINGS 469, New York, 1998, pp.569–570.
6. Hiwatari, Y., and Muranaka, T., *Journal of NON-Crystalline Solids* **235-237**, 19–26, (1998).
7. Mayo, S. L., Olafson, B. D., and Goddard III, W. A., *J. Phys. Chem.* **94**, 8897–8909 (1990).

Simulation of the Effect of Interstitials to Shear Modulus in Aluminum and an Ionic Crystal

Ouyang Lizhi and Da-Ming Zhu

Department of Physics
University of Missouri-Kansas City
5110 Rockhill Road, Kansas City, Missouri 64110, USA

Abstract. In this study, we conducted an *ab initio* calculation of Al which contains self-interstitials with concentration up to 6% and an empirical molecular dynamics simulation of an ionic crystal $Ca_{1-x}La_xF_{2+x}$ with the concentration of charge balanced interstitials up to 50%. For Al, we find that the shear modulus decreases with the concentration of self-interstitials and becomes unstable when interstitialcy concentration reaches about 6%, a trend agrees with that predicted by an interstitialcy model. For the ionic crystal, the shear modulus changes very little with the interstitialcy concentration and shows a dip as the concentration is about 13%.

INTRODUCTION

A recent proposed interstitialcy theory considers that basic units that are responsible for all the fundamental properties of liquids and glasses are intrinsic interstitial-like defects.[1] The theory is based upon earlier theoretical calculations[2] and information about the configuration and dynamics of interstitials in face-centered-cubic (fcc) simple metals.[3-5] The key element of the theory is the assumption that the shear modulus, G, of a crystal is a function of the interstitial concentration of the form[1]

$$G(c) = G(0)e^{-a_1\beta c} \qquad (1)$$

where a_1 is on the order of 1, β is the shear susceptibility and $a_1\beta \cong 40$, c is the interstitial concentration. Experimentally testing this dependence is a challenging task, primarily because vacancies are always present in large numbers and tend to combine with interstitials in simple metals; the maximum self-interstitial concentration that can be experimentally realized is very limited.[3,6] It is more challenging to determine what structural entity in a glass has the properties that resemble the interstitials in simple metals.

In this work, we carried out a computer *ab initio* calculation of elastic constant C_{44} of Al as a function of concentration of self-interstitials over a wide range, to verify the exponential dependence of Eq (1). We have also carried out an empirical molecular dynamics (MD) simulation of elastic constants of $Ca_{1-x}La_xF_{2+x}$ as a function of x, to investigate the roles of charge balance interstitials play in fluorite mixed crystals of the form $M_{1-x}R_xF_{1-x}$, where M can be Ba, Ca, Cd, Pb, or Sr and R is La, Y or any rare earth element.[7] In these crystals, dopant R atoms randomly replace M atoms in a cubic structure. The dominant defect is an interstitial F^{-1} ion neighboring a R^{+3} ion which substitutes for a M^{+2} ion to balance the charge,[7] so the defect is termed as charge balance interstitial. These crystals do display various glass-like behaviors, observed in their low temperature thermal conductivity, specific heat, and internal friction studies.[7]

RESULTS AND DISCUSSIONS

Interstitial effect to shear modulus in Al is *ab initio* calculated using Vienna *ab-initio* Simulation Package (VASP) with ultrasoft pseudopotential.[8] All the calculations were carried out using a local density approximation for exchange-correlation function. We find that the interstitial atoms in Al indeed form a dumb-bell configuration with an extended displacement in surrounding atoms along the closed packed directions.[1] The elastic constant C_{44} is obtained from the numerically fitted total energy and strain equations. The concentration dependence of C_{44} shows an exponential dependence; agree with key element of the interstitialcy theory.[1]

Fig. 1 displays normalized elastic constant $C_{44}/C_{44}(0)$ versus the concentration of self-interstitial, where

C(0) is the elastic constant for a perfect crystal. The calculated C_{44} agrees with that predicted by the interstitialcy model, showing an exponential dependence on the interstitial concentration and a large diaelastic softening. The dashed line is the exponential fit, with dC_{44}/dc to be −29, a value very close to that found for Cu.[3] This result is consistent with what is predicted by the interstitialcy theory $\beta = 4\pi^2 \approx 40$.[1] A linear extrapolation of the changes of C_{44} for Cu at low interstitial concentration implies the lattice becomes unstable for about 3% of interstitials.[1] We find that the Al lattice structure becomes unstable when the concentration increases to above 6%, a trend entirely consistent with the prediction of interstitialcy theory.[1]

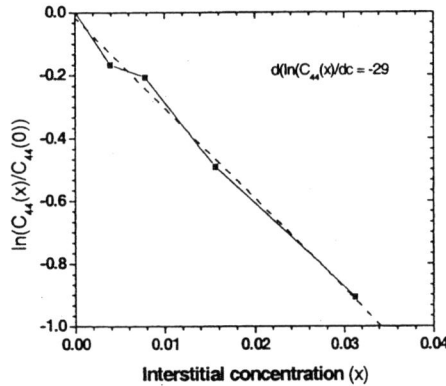

FIGURE 1. *Ab Initio* calculated C_{44} versus interstitial concentration in fcc Al.

In the MD simulation, we employed General Lattice Utility Program (GULP) with the atomic pair potentials from the package.[9] Fig. 2 plots the results of the shear modulus of $Ca_{1-x}La_xF_{2+x}$. The results differ distinctly from that found for Al. The bulk modulus shows a very small decrease with the charge balance interstitial concentration, while C_{44} rises lightly with the concentration. Both the bulk modulus and C_{44} show a small dip at around 13% of the interstitial concentration. The glassy properties of the crystal become most prominent when $x \approx 30\%$, as observed experimentally.[7] Our preliminary results also show excess low energy excitations in the lattice vibration of $Ca_{1-x}La_xF_{2+x}$, which are consistent with the findings of the low temperature studies.[7]

An important characteristics of an interstitial in simple fcc metals is the dumb-bell configuration aligned in (100) direction.[1] The charge balance interstitials in $Ca_{1-x}La_xF_{2+x}$ do not possess this feature. If some kind of interstitialcy-like defects in $Ca_{1-x}La_xF_{2+x}$ are responsible for its glassy properties they would involve a structural entity that are more complex than a single charge balance interstitial. Identifying this entity and that in glasses in general remains to be an interesting and challenging task in testing the merit of the interstitialcy theory.[1]

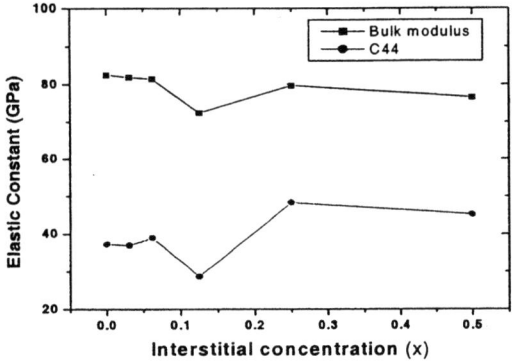

FIGURE 2. The results of simulated elastic constant of $Ca_{1-x}La_xF_{2+x}$. The lines are the guides to the eye.

REFERENCES

1. Granato, A. V., *Phys. Rev. Lett.*, **68**, 974-977 (1992); *J. Phys. Chem. Solids,* **55**, 931-939 (1994);
2. Huntington, H. B., and Seitz, F., *Phys. Rev.*, **61**, 315-325 (1942).
3. Holder, J. T., Granato, A. V., and Rehn, L. E., *Phys. Rev. Lett.*, **32**, 1054-1057 (1974); *Phys. Rev. B*, **10**, 363-375 (1974); Rehn, L. E., Holder, J. T., Granato, A. V., Coltman, R. P., and Young, F. W. Jr., *Phys. Rev. B,* **10**, 349-362 (1974).
4. Ehrhadt, P., Haubold, A. G., and Schilling, W. *Adv. Solid State Phys.*, **14**, 87-110 (1974).
5. Dederichs, P. H., Lehmann, C., Schober, H. R., Scholz, A., and Zeller, R., *J. Nucl. Mater.*, **69**, 176-199 (1978).
6. Robrock, K. H., Spiric, V., Rehn, L. E., *Radiation Effects*, **27**, 189-190 (1976); Rehn, L. E., And Robrock, K. H., *J. Phys. F: Metal Phys.*, **7**, 1107-1113 (1977).
7. Pohl, R. O., Liu, X., and Crandall, R. S., *Current Opinion in Solid State and Materials Sciences*, **4**, 281-287 (1999); Cahill, D. G., and Pohl, R. O., *Phys. Rev. B*, **39**, 10477-10480 (1989).
8. Kresse, G., and Hafner, J., *Phys. Rev. B*, **47**, 558-561 (1993); Kresse, G., and Furthmuller, J., *Comput. Mat. Sci.*, **6**, 15-50 (1996); Kresse, G., and Furthmuller, J., *Phys. Rev. B*, **54**, 11169-11186 (1996); Vanderbilt, D., *Phys. Rev. B*, **41** 7892-7895 (1990).
9. Gale, J. D., *JCS Faraday Trans.*, **93**, 629-637 (1997); *Phil. Mag. B*, **73**, 3-19 (1996).

Supercooled Liquids under Shear: Computational Approach

R. Yamamoto[*†], K. Miyazaki[**] and D.R. Reichman[**]

[*]*Department of Physics, Kyoto University, Kyoto 606-8502, Japan.*
[†]*PRESTO, Japan Science and Technology Agency, 4-1-8 Honcho Kawaguchi, Saitama, Japan.*
[**]*Department of Chemistry and Chemical Biology, Harvard University, MA 02138, USA.*

Abstract. We have performed molecular dynamics simulations for a model two-dimensional soft-core mixture in a supercooled state. The mixture exhibits a slow structural relaxation in a quiescent state, however, the relaxation is much enhanced in sheared states. There observed surprisingly small anisotropy both in the coherent and incoherent density correlation functions even under extremely strong shear which is 10^3 times faster than the structural relaxation rate. The present simulation results agree well with predictions of the recently developed mode-coupling theory in shear.

As liquids are cooled toward the glass transition, the dynamics is drastically slowed down, while only small changes can be detected in the static properties. One of the main targets in theoretical investigations on the glass transition is to identify the mechanism of the drastic slowing-down. Beside this fundamental problem, a striking example occurs when one brings glassy materials away from equilibrium, for instance, by changing temperature rapidly or applying shear flow to them. There appears a variety of unique phenomena such as aging or shear thinning. Although these phenomena are not only conceptually but also practically important, physical properties of glassy materials in nonequilibrium conditions has not yet been understood well. A couple of years ago, we performed extensive molecular dynamics (MD) simulations in two dimensions (2D) and three dimensions (3D) on binary soft core mixtures with and without shear flow. It was found that the dynamical properties of the mixtures under shear can be fairly mapped onto those at quiescent states at higher temperatures [1, 2]. We found also the surprising isotropy in the tagged particle motions even under extremely high shear, which may justify the simple mapping idea. In the present study, we calculate intermediate scattering functions by using the method proposed by Onuki [3, 4] to investigate microscopic dynamics of glassy materials in shear flow. Simulations have been done in 2D to compare the present computational results directry with the theory developed recently in 2D [5].

Our model system is composed of two different particle spices 1 and 2, which interact via the soft-core potential

$$v_{\alpha\beta}(r) = \varepsilon(\sigma_{\alpha\beta}/r)^{12}, \quad \sigma_{\alpha\beta} = (\sigma_\alpha + \sigma_\beta)/2, \quad (1)$$

where r is the distance between two particles, and $\alpha, \beta \in$ 1, 2. We take the mass ratio $m_2/m_1 = 2$, the size ratio $\sigma_2/\sigma_1 = 1.4$, and the numbers of particles $N_1 = N_2 = 5000$. Simulations were performed in the absence and presence of shear flow with being particle density and temperature fixed at $n = (N_1 + N_2)/V = 0.8/\sigma_1^2$ and $k_B T = 0.526\varepsilon$, respectively. In the sheared case, we kept the temperature at a constant using the Gaussian constraint thermostat to eliminate the viscous heating effect. Our method of applying shear is as follows: The system was at rest for $t < 0$ for a very long equilibration time and was then sheared for $t > 0$. Here we added the average velocity $\dot{\gamma} y$ to the velocities of all the particles in the x direction at $t = 0$ and afterwards maintained the shear flow by using the Lee-Edwards boundary condition. Simulation data have been taken and accumulated in steady states which can be realized after transient states.

Figure 1 (a) shows the geometry of shear flow in the present simulation. As shown in Figure 1 (b), shear flow with the rate $\dot{\gamma}$ advect a positional vector $\mathbf{r}$ as

$$\mathbf{r}(t) = \mathbf{r} + \dot{\gamma} t r_y \mathbf{e}_x \quad (2)$$

in the time duration t, where $\mathbf{e}_\alpha$ is a unit vector in $\alpha \in x, y$ axis. A similar advection can be defined in Fourier space for a wave vector $\mathbf{k}$ as

$$\mathbf{k}(t) = \mathbf{k} - \dot{\gamma} t k_x \mathbf{e}_y. \quad (3)$$

The above definition enable us to calculate Fourier component $\mathbf{k}$ of the time correlation function

$$C(\mathbf{k}, t) \equiv \langle A_{-\mathbf{k}}(0) A_{\mathbf{k}(t)}(t) \rangle \quad (4)$$

in shear flow. We calculated the incoherent and coherent scattering functions by using the definitions

$$Fs(\mathbf{k}, t) = \frac{1}{N}\left\langle \sum_{i=1}^{N} e^{-i\{\mathbf{k}(t)\cdot\mathbf{r}_i(t) - \mathbf{k}\cdot\mathbf{r}_i(0)\}} \right\rangle \quad (5)$$

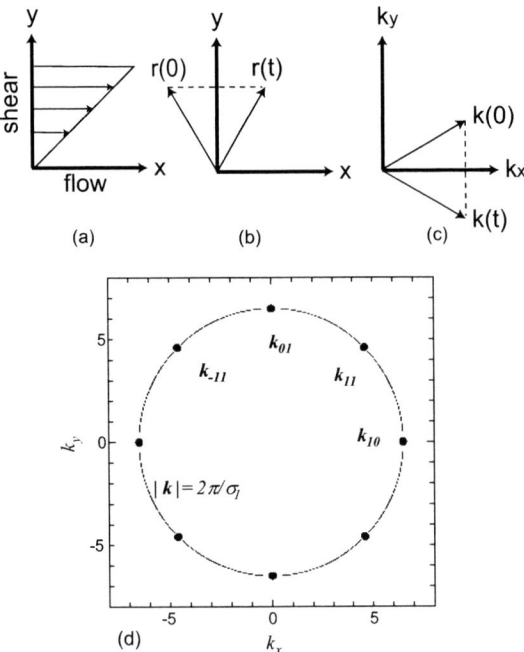

FIGURE 1. (a) Geometry of shear flow. (b) Shear advection in real space. (c) Shear advection in Fourier space. (d) Sampled wave vectors (t=0).

and

$$F(\mathbf{k},t) = \frac{1}{N}\left\langle \sum_{i=1}^{N} e^{[-i\mathbf{k}(t)\cdot\mathbf{r}_i(t)]} \sum_{j=1}^{N} e^{[i\mathbf{k}\cdot\mathbf{r}_j(0)]} \right\rangle, \quad (6)$$

respectively. To examine anisotropy in the dynamics, the wave vector is taken in four different directions $\mathbf{k}_{10}$, $\mathbf{k}_{01}$, $\mathbf{k}_{11}$, and $\mathbf{k}_{-11}$, where

$$\mathbf{k}_{\alpha\beta} = \frac{k_0}{\sqrt{\alpha^2+\beta^2}}(\alpha,\beta), \quad (7)$$

$k_0 = 2\pi/\sigma_1$, and $\alpha,\beta \in 0,1$ as shown in Figure 1 (d).

Figure 2 (a) and (b) show $Fs_1(\mathbf{k},t)$ and $F_{11}(\mathbf{k},t)$, respectively, at $T = 0.526$ with and without shear flow. Here the subscript 1 denotes the smaller particle component. The so-called α relaxation time τ_α of the present mixture is defined by

$$F_{11}(k_0,\tau_\alpha) \simeq Fs_1(k_0,\tau_\alpha) = e^{-1} \quad (8)$$

in the quiescent state. The followings have been found; i) $Fs_1(\mathbf{k},t)$ and $F_{11}(\mathbf{k},t)$ behave quite similarly both in quiescent and sheared states. ii) Shear accelerates drastically the microscopic structural relaxation in the supercooled state. The structural relaxation time τ_α decreases strongly with increasing shear rate as $\tau_\alpha \sim \dot{\gamma}^{-\nu}$ with $\nu \sim 1$. iii) The acceleration in the dynamics due to shear occurs almost isotropically. There observed surprisingly

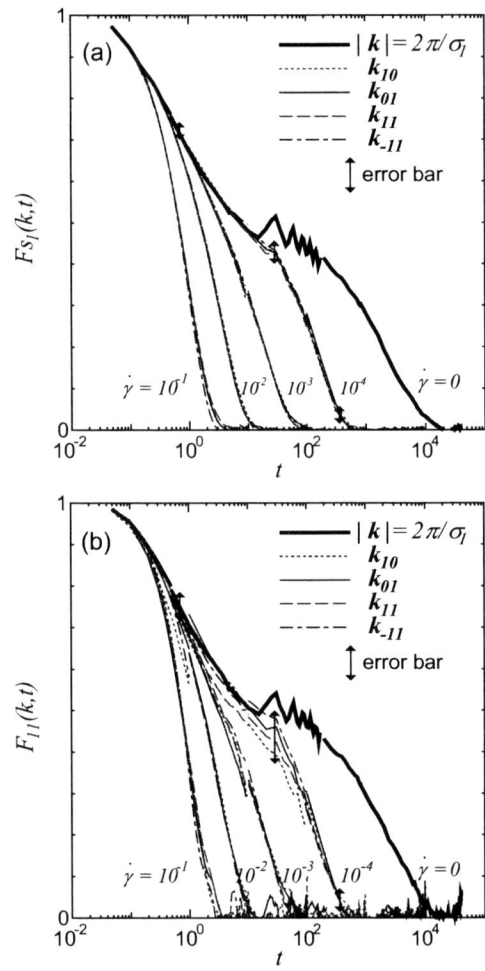

FIGURE 2. Intermediate scattering functions under shear; (a) incoherent part and (b) coherent part.

small anisotropy in the correlation functions even in extremely strong shear $\dot{\gamma}\tau_\alpha \simeq 10^3$. This simplicity in the dynamics is quite different from behaviors of other complex fluids such as critical fluids or polymers in shear. Finally we note that the recent mode-coupling theory in shear flow [5] predicts almost the same behaviors.

REFERENCES

1. Yamamoto, R., and Onuki, A., *Phys. Rev. E* **58**, 3515-3529 (1998).
2. Yamamoto, R., and Onuki, A., *J. Chem. Phys.* **117**, 2359-2367 (2002).
3. Onuki, A., and Kawasaki, K., *Ann. Phys. (N.Y.)* **121**, 456-528 (1979).
4. Onuki, A., *J. Phys.: Condens. Matter* **9**, 6119-6157 (1997).
5. Miyazaki, K., and Reichman, D.R., *Phys. Rev. E* **66**, 050501 (2002).

VI. OTHER RELATED TOPICS

Atomic dynamics of a bulk and hyperquenched metallic glass

Jens-Boie Suck

Materials Research and Liquids, Institute of Physics, Technical University Chemnitz, D-09107 Chemnitz, Germany

Abstract. The atomic dynamics of hyperquenched glasses proceeds in basins in a different region of the 3N-dimensional energy surface than are the basins of slow cooled glasses. As has been shown by computer simulations and experiments very recently, the atomic dynamics can be different in these basins in different regions of the energy surface. Using neutron inelastic scattering this question is here investigated on the basis of the total dynamic structure factor of a bulk metallic glass and the same glass after melt spinning, leading to a difference in quenchrate of approximately three orders of magnitude. The investigation of the wavelength dependence of the atomic dynamics demonstrates that the additional modes found in the dynamic structure factor of the hyperquenched glass show no momentum transfer dependence and are therefore most likely due to non- propagating modes.

INTRODUCTION

Most of the glass production processes will very likely produce hyperquenched glasses, staying on the "save side" of the necessary quench rate for glass formation. However, only quench rate differences of several orders of magnitude will produce glasses in appreciably different regions of the 3N-dimensional energy surface [1] and with this will lead to measurable different physical properties of two glass samples with the same chemical composition. While for SiO_2-based glasses this has always been possible and was used in glass-fiber and rock-wool production and has been investigated recently [2], for metallic glasses large differences in quench rates have only become possible with the advent of bulk metallic glasses. In the context of hyperquenching the latter glasses are of special interest, as nearly all computer simulations of the shape of the energy surface were done using a model potential designed to describe the metallic glass $Ni_{80}P_{20}$ [3, 4]. As $Ni_{80}P_{20}$ does not form an amorphous alloy on slow cooling, half of the Ni was replaced by Pd atoms, allowing then cooling rates below 100K/s. Even with this replacement the samples remain simple glasses with mainly spherical interactions. Thus the assumption that the results of the simulations will still apply to this bulk metallic glass seems very reasonable.

SAMPLE PRODUCTION AND EXPERIMENTS

The rapidly quenched $Pd_{40}Ni_{40}P_{20}$ samples were produced with a quench rate above $10^5 K/s$ by melt spinning ingots of the prealloy, which were produced from 99,9% pure elements in an arc furnace. Both processes were guided under Ar gas of high purity. Part of the sample material was subsequently relaxed at 520K for 1h. The bulk glass was produced from the same kind of prealloy by quenching the filled quartz-tubes of 2 mm inner diameter in water at ambient temperature. This way fully amorphous sticks of 2 mm diameter and approximately 60 mm length were obtained. The structure of the complete sample material was investigated by neutron diffraction, proving that all three samples were fully amorphous.

The experiments were performed at the thermal neutron time-of-flight spectrometer MARI at the neutron spallation source ISIS, UK. The incident energy was 75.4 meV and scattering angles between 3 and 138 degrees were covered. For elastic scattering at zero energy transfer $\hbar\omega$ this corresponds to momentum transfers $\hbar Q$ with Q-values between 4 and 110 nm^{-1}. Of these, however, only data measured above approximately 80 nm^{-1} were used for the investigation of the atomic dynamics because of the paramagnetic scattering at lower Q-values, where the magnetic form factor still has a non-negligible weight. As the contribution of multiphonon processes to the scattered intensity increases rapidly (Q^{2n} for an n-phonon process) with increasing Q-values, the experiments were done at 200K to lower the multi-phonon scattering probability. All necessary corrections have been

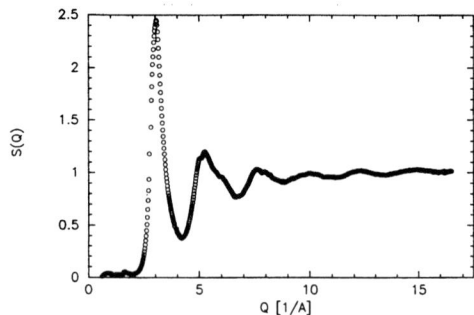

FIGURE 1. Static structure factor S(Q) of glassy $Ni_{40}Pd_{40}P_{20}$

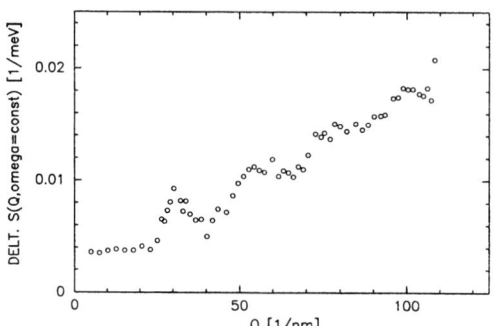

FIGURE 2. Cut through the dynamic structure factor $S(Q,\omega)$ of hyperquenched $Ni_{40}Pd_{40}P_{20}$ at 4.76 meV

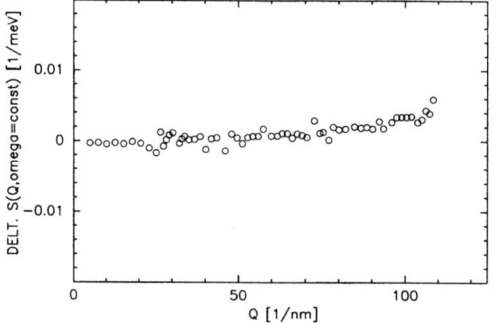

FIGURE 3. Cut through the difference of the dynamic structure factors of the hyperquenched - bulk metallic glass at 4.76 meV

applied to the raw data except those for multiple scattering of the neutrons in the sample and for the resolution function of the spectrometer. The lack of both corrections is not severe for the results presented here, because their effect is strongly reduced in the differential quantities investigated here.

RESULTS AND DISCUSSION

The total dynamic structure factor, as determined *experimentally* is composed of the weighted sum of the $n(n+1)/2$ *partial* dynamic structure factors, $S_{ij}(Q,\omega)$, where n is the number of different elements in the sample, plus the weighted sum of the three (for a ternary system in our case) self parts of the dynamic structure factor, $S_s(Q,\omega)$ representing the dynamical autocorrelation of the elements and with this the incoherent scattering:

$$\sigma^{sc} S(Q,\omega) = 4\pi \sum_{i}^{n}\sum_{j}^{n} b_i b_j c_i c_j S_{ij}(Q,\omega) + \sum_{i}^{n} c_i \sigma_i^{inc} S_i^{self}(Q,\omega)$$

where:

c_i = atomic concentration of atomic species i
b_i = coherent scatt. length of atomic species i
σ^{sc} = $4\pi \sum_i \sum_j b_i b_j c_i c_j + \sum_i c_i \sigma_i^{inc}$
n = number of atomic species in the sample
ω = $\frac{\hbar}{2m}(k_o^2 - k^2)$
Q = $k_o^2 + k^2 - 2k_0 k \cdot \cos\Theta$ (isotropic sample)
Θ = scattering angle
k_0, k = the modulus of the wavevector of the neutron

In the case studied here, the weighting factors of the Ni correlations are dominant, followed by those of the heavier Pd atoms.

The static structure factor S(Q) in Fig. 1 demonstrates that our samples were completely amorphous: no Debye-Scherrer peaks are seen in the diagram. The small substructure seen is due to some incomplete normalization of the 8 detector sections of the D4 diffractometer at the HFR in Grenoble, where the data were measured.

In the following we analyse the total dynamic structure factor $S(Q,\omega)$ at fixed energy transfer $\hbar\omega$ as a function of the momentum transfer $\hbar Q$, As we know from the analysis of the generalized vibrational density of states [5] that the main difference between the atomic dynamics of the hyperquenched and the bulk glass is to be expected at energies below 20 meV with a maximal difference at 10 meV, we concentrate the analysis of $S(Q,\omega)$ on the region of low energy transfers. Fig. 2 shows a cut through the neutron energy loss part of the dynamic structure factor of the hyperquenched glass at 4.67 meV.

The cut clearly demonstrates the coherent scattering

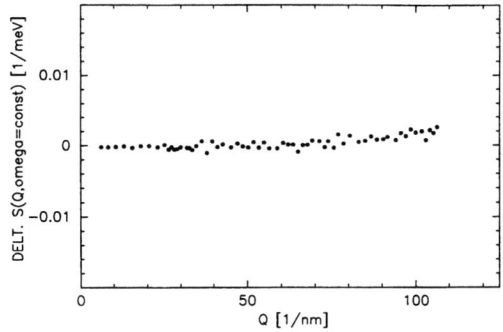

FIGURE 4. Cut through the difference of the dynamic structure factors of the hyperquenched - bulk metallic glass at 10.06 meV

from in-phase vibrating atoms with a dominant excitation of transverse modes at the feet of the broad maxima of the structure factor in a similar way as one finds this next to the Debye- Scherrer lines in the case of a polycrystal. In contrast to this, the cut through the *difference* of the two dynamic structure factors $\Delta S(Q,\omega)$, which shows the intensity of the modes *additionally* found in the dynamics of the hyperquenched glass at the same energy, does not show any appreciable Q-dependence in the whole (large!) range of reciprocal space covered in this experiment (see Fig.3).

As is demonstrated in Fig. 4 the same is true at other energies, especially also at 10 meV, where the largest difference between the generalized vibrational density-of- states of the hyperquenched and the slow cooled glass was found in an earlier investigation [5] . Even when one takes into account that the error bars in Fig. 3 and 4 are definitely larger than in Fig. 2, this conclusion is still valid.

CONCLUSION

From the wavelength dependence of the atomic dynamics as investigated via the total dynamic structure factor, which clearly reflects the difference in the excitation probability of longitudinal and transverse modes with varying momentum transfer, we conclude that the *additional* intensity found in $S(Q,\omega)$ of the hyperquenched glass does not show any appreciable variation on momentum transfer apart from a steady increase with increasing Q. This property strongly suggest that these modes are non-propagating but are more likely localized modes.

ACKNOWLEDGMENTS

I am indepted to H. Teichmann for making my samples. It is a pleasure to acknowledge the help obtained at the start up of the experiments by O. Petrenko on MARI and G. Cuello on D4. Financial support of this project by the BMBF under grant no. 03SUE8C1 is gratefully acknowledged.

REFERENCES

Goldstein, M., *J. Chem. Phys.* **51** 3728-3739 (1969)

Angell, C. A.,, et al. *J. Phys.: Condensed Matter* **15** S1051-S1068 (2003)

Weber, T. A., and Stillinger, F. H.*Phys. Rev.* **B31** 1954-1963 (1985)

Kob, W., and Anderson, H. C.,*Phys. Rev. E* **51** 4626-4641 (1995)

Suck, J.-B.,"Low energy dynymics in glasses" in *Adv. in Solid State Phys.* **42** edited by B. Kramer, Springer, Berlin, Heidelberg 2002 pp.393-403

New Aspect of the Spontaneous Formation of a Bilayer Lipid Membrane

Hisashi Fujiwara, Masayuki Fujihara, Takahiro Koyama, and Takashi Ishiwata

Faculty of Information Sciences, Hiroshima City University, Hiroshima 731-3194, Japan

Abstract. An artificial lipid bilayer in planar form, well known as bilayer lipid membrane (BLM), spontaneously forms from a lipid droplet (diphytanoyl phosphatidylcholine in n-decane and chloroform in this work) in an aperture of a thin partition in aqueous solution. The thinning dynamics of the lipid droplet or membrane has been studied by simultaneous capacitance and image recording. The simultaneous measurements have revealed the two-step thinning of the lipid membrane from its specific capacitance value: first, the initial droplet thins to yield a membrane of 60 nm thickness (0.03 μF/cm^2), and second, within this thin lipid membrane, a lipid bilayer of 4 nm thickness (0.45 μF/cm^2) suddenly emerges and grows with keeping a bilayer structure. The revealed dynamics provides a quantitative support for a "zipper" mechanism proposed by Tien and Dawidowicz; in the mechanism, the first thinning results in a sandwich consisting of the organic solvent between two adsorbed lipid monolayers whose distance is the order of 100 nm, and then a chance contact of both monolayers initiates the formation and growth of a lipid bilayer in a zipper-like manner. However, because of the existence of the two solvent-water interfaces containing surface-active molecules, phospholipids, this work claims that the zipper mechanism should be modified in view of the Marangoni effect. The present formation and growth of a lipid bilayer can be explained by the classic nucleation theory of two-dimensional crystallization. BLM systems with the simultaneous measurements can be considered as a useful environment for the study of soft-matter chemical physics.

INTRODUCTION

An artificial lipid bilayer in planar form, well known as bilayer or black lipid membrane (BLM), has been used in view of biochemistry, biophysics, bioelectronics, and physiology. The preparation of a BLM is quite simple [1]. A minute droplet of lipid solution is introduced to a small aperture in a thin partition, and under favorable conditions the lipid droplet or membrane thins spontaneously to a BLM. This formation process is interesting in view of the chemical physics of self-assembly, and has been investigated quantitatively by simultaneous capacitance and image recording. First we chose L-α-phosphatidylcholine as a bilayer-forming lipid [2], because this phospholipid has been frequently used as a representative one for more than 40 years of BLM experiments. Second, in this work, we have chosen diphytanoyl phosphatidylcholine, because, as a single constituent, this phospholipid can form a lipid bilayer at room temperature, and thus is suitable for the research of fundamental chemical physics.

EXPERIMENT

The lipid bilayer was made in an aperture (of diameter ≈ 1 mm, but slightly out of round) in a partition of a Teflon cell having two compartments filled with 1 M KCl aqueous solutions. This BLM cell was equipped with two platinized platinum electrodes and two glass windows, and enclosed in a Faraday cage to avoid electromagnetic interferences. The membrane-forming solution was prepared by dissolving diphytanoyl phosphatidylcholine (Avanti Polar Lipids) 20 mg in n-decane 1 ml and chloroform 0.3 ml. This solution was introduced onto the aperture by using a microliter syringe, and the resulting lipid droplet or membrane spontaneously thinned to yield a lipid bilayer.

The details of the simultaneous electrical and optical measurements were described elsewhere [2]. In summary, the membrane capacitance was measured with an impedance analyzer (HP4192A, Hewlett Packard) at $f = 10$ kHz, and transferred to a personal computer. Simultaneously with the electrical measurement, the lipid membrane was observed vertically on a long distance microscope with transmitted-light. The microscope, whose working-distance is 12 cm, was equipped with a monochrome charge-coupled device (CCD) camera of 2.1 megapixels (Flovel). The light source was a combination of a 500W high-pressure mercury lamp, an interference filter (450 nm), and neutral density filters. The membrane images were recorded as a bitmap file on a hard disk drive of another personal computer. The contrast of the images was enhanced by using a commercial software after the measurements. All the measurements were done at room temperature.

RESULTS AND DISCUSSION

Figure 1 shows the temporal change of the capacitance and image of the lipid membrane introduced in the cell aperture by applying the membrane-forming solution at 0 s. The membrane capacitance gradually increased until 152 s. In the same period, the following two processes were optically observed: the formation of the peripheral region of the Plateau-Gibbs border and the emergence of a shadow region from a lower part inside the Plateau-Gibbs border [2]. At 152 s, on the other hand, the capacitance suddenly started to increase steeply until ca. 230 s. In the same period, according to the optical observation, a new region emerged suddenly in the bottom of the shadow region and spread like a blast wave (e.g., images 2 and 3).

From both capacitance and image, we can derive the specific capacitance of the lipid membrane [2]. The specific capacitance C_s is related to the membrane thickness d by $d = \varepsilon\varepsilon_0 / C_s$ where ε is the relative permittivity of the membrane and ε_0 is the permittivity of vacuum; we assume $\varepsilon = 2$. The specific capacitance of the shadow region is 0.03 µF/cm² at 152 s, which value corresponds to the thickness of 60 nm. After 152 s, the specific capacitance of the new region shows significant characteristics (Fig. 2). Namely, its value is clearly independent of the area, and the averaged value is 0.45 µF/cm². This value corresponds to the bilayer thickness of 4 nm.

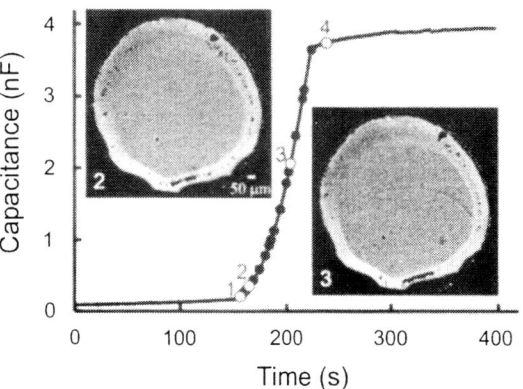

FIGURE 1. Simultaneous measurements of both capacitance and image of the thinning lipid membrane. The closed circles and the numbered open circles represent the capacitance data corresponding to the specific capacitance data in Fig. 2. The white line in image 2 corresponds to 50 µm.

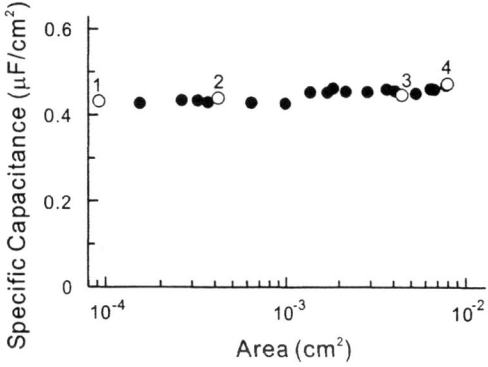

FIGURE 2. Specific capacitance of the emerged region. Numbered data correspond to those of the capacitance data shown in Fig. 1.

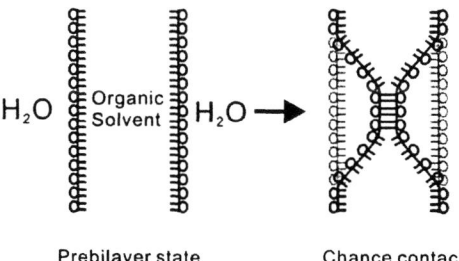

FIGURE 3. Zipper mechanism of the bilayer formation.

These data indicate that a lipid bilayer emerges within the lipid membrane of 60 nm thickness, and grows with keeping a bilayer structure. This picture provides a quantitative support for a "zipper" mechanism proposed by Tien and Dawidowicz [1-3]. In the mechanism (Fig. 3), the first thinning results in the prebilayer state, a sandwich consisting of the organic solvent between two adsorbed lipid monolayers whose distance is the order of 100 nm.

Then a chance contact of both monolayers initiates the formation and growth of a lipid bilayer in a zipper-like manner.

However, because of the existence of the two solvent-water interfaces containing surface-active molecules, phospholipids, we claim that the zipper mechanism should be modified in view of the Marangoni effect [2, 4-6]. The fluid interfaces adsorbed by surface-active molecules have a resistance to deformation, which is an aspect of what is known as the Marangoni effect, namely, the carrying of bulk material through motions energized by surface (or interfacial) tension gradients [6].

Imagine a small bilayer is tentatively formed, by a chance contact, within the lipid membrane in the prebilayer state. It accompanies the local deformation and expansion of the interfaces around the bilayer, and the interfacial concentration of the lipid decreases in the deformed region. The resulting interfacial tension gradients, produced by the gradients in lipid concentration, causes a shear stress at the solvent-water interfaces (in the direction of increasing interfacial tension) which induces motion in the interfaces and the adjoining liquid layers [7]. This motion induces a tentative increase of the liquid pressure adjacent to the tentative bilayer and thus works to separate the bilayer into the original monolayers. Namely, the Marangoni effect works as repulsive interaction between the two phospholipid layers of the tentative bilayer.

This repulsive interaction should be proportional to the peripheral length of the bilayer, whereas to the bilayer area the net attractive interaction based on both van der Waals attraction and Born repulsion [1]. This qualitative discussion allows us to consider the existence of the critical nucleus size of an emerged bilayer where the two counter interactions are in balance; once a bilayer larger than this size emerges, it can exist stably and serves as a nucleus for the further bilayer growth. The existence of the critical nucleus size in the bilayer formation can be treated qualitatively by the classic nucleation theory of two-dimensional crystallization [2, 6, 8]. The concept of two-dimensional crystallization in BLM experiments has been proposed intuitively [1], and is addressed, for the first time, by the simultaneous measurements [2].

As shown above, BLM systems with the simultaneous measurements provide a useful environment for studying spontaneous formation of a phospholipid bilayer, in other words, self-assembly of phospholipids. Furthermore they can address both two-dimensional crystallization and dynamics of an interface containing surface active molecules (e.g., the Marangoni effect). These two topics and the first topic, self-assembly of phospholipids, correspond to important subjects of soft-matter chemical physics, namely, phase transition, interface activity, and structural organization, respectively [9, 10]. The simultaneous electrical and optical measurements open new possibilities of treating these subjects in BLM systems, which is a new aspect of BLM experiments.

ACKNOWLEDGMENTS

We thank A. Tamura and K. Tatibana for technical assistance. H.F. was supported by a Grant-in-Aid for Encouragement of Young Scientists (No. 15750013) from the Ministry of Education, Science, Sports and Culture, Japan. H.F. was also supported by a Hiroshima City University Grant for Special Academic Research (General Studies).

REFERENCES

1. Tien, H. T., *Bilayer Lipid Membranes (BLM): Theory and Practice*, New York : Dekker, 1974.

2. Fujiwara, H., Fujihara, M., and Ishiwata, T., *J. Chem. Phys.* **119**, 6768-6775 (2003).

3. Tien, H. T., and Dawidowicz, E. A., *J. Colloid Interface Sci.* **22**, 438-453 (1966).

4. Sternling, C. V., and Scriven, L. E., *AIChE J.* **5**, 514-523 (1959).

5. Scriven, L. E., and Sternling, C. V., *Nature* **187**, 186-188 (1960).

6. Adamson, A. W., and Gast, A. P., *Physical Chemistry of Surfaces, 6th ed.*, New York: Wiley, 1997.

7. Troian, S. M., Wu, X. L., and Safran, S. A., *Phys. Rev. Lett.* **62**, 1496-1499 (1989).

8. Saito, Y., *Statistical Physics of Crystal Growth*, Singapore: World Scientific, 1996.

9. Safran, S. A., *Statistical Thermodynamics of Surfaces, Interfaces, and Membranes*, Reading: Perseus Books, 1994.

10. Hamley, Ian W., *Introduction to Soft Matter – Polymers, Colloids, Amphiphiles and Liquid Crystals*, New York: Wiley, 2000.

Nonperturbative anharmonic phenomena in crystal lattice dynamics

M. I. Katsnelson* and A. V. Trefilov[†]

Department of Physics, Uppsala University, Box 530, SE-751 21 Uppsala, Sweden
[†]*Russian Science Center "Kurchatov Institute", 123182, Moscow, Russia*

Abstract. Slow dynamics of energy transfer between different phonon modes under the resonance conditions is considered. It may result in new effects in the inelastic and quasielastic neutron scattering spectra.

INTRODUCTION

Crystal lattice dynamics [1, 2] is a prototype theory for many-body physics as a whole. In this case all principal approximations and concepts can be justified accurately on the basis of adiabatic (Born-Oppenheimer) approximation of quantum mechanics. Due to the smallness of typical atomic displacements in crystals $\bar{u}$ in comparison with interatomic distances d one can pass rigorously from a problem of strongly interacting *particles* (atoms, ions, or molecules) to a problem of *weakly* interacting *quasiparticles* (phonons). In the leading order in the smallness parameter $\eta = \bar{u}/d$ the crystal lattice dynamics and thermodynamics can be described in terms of an *ideal* phonon gas (harmonic approximation) [1, 2]. With the temperature T increase the parameter η increases as well, however, due to a semiempirical Lindemann criterion (see, e.g., [2, 3] and Refs. therein) $\eta \simeq 0.1$ at the melting point $T = T_m$ so higher-order (*anharmonic*) contributions to thermodynamic properties are usually small up to the melting temperature [4, 5]. This statement is true for the most of *average* characteristics. At the same time, for some *peculiar* modes anharmonicity can be crucially strong, especially in the vicinity of some structural transformations (see, e.g., computational results for Ba [6] and Zr [7, 8], as well as inelastic neutron scattering data for A15 structure compounds [9] and for high-temperature bcc phases of Ti, Zr, and Hf [10]). For these cases a standard phonon picture is not adequate, for instance, additional non-phonon peaks in the dynamical structural factor may appear [11]. A "slow" (low-frequency) lattice dynamics in the form of a "central peak" in quasielastic neutron scattering is also typical for these strongly anharmonic modes [9, 12, 13].

Phonon picture can be broken for these modes since the corresponding effective potential $V(u)$ turns out to be essentially nonparabolic [6, 7, 11, 13, 14]. In the vicinity of structural phase transitions (e.g., ferroelectric transitions [13] or martensitic transformations in metals [6, 14]) this potential normally has several minima corresponding to several competing phases, the height of the barrier being much smaller than a typical cohesion energy. It is not surprising that the harmonic approximation is an extremally poor starting point to describe this situation. Here we discuss the following issue: is the phonon picture always adequate in a *generic* case when $\eta \ll 1$ for all phonon modes? To answer this question it is instructive to consider a crystal from the point of view of a contemporary theory of dynamic systems rather than from that of statistical thermodynamics [14]. According to the KAM (Kolmogorov-Arnold-Moser) theory [15] a system of noninteracting oscillators (that is, an ideal phonon gas) is stable with respect to small interphonon interactions (anharmonicity) except the cases of the resonance where the ratio of phonon frequencies is close to the ratio of small enough integer numbers. One can expect that this exceptional case is a situation where even small anharmonicities can lead to essentially nonperturbative effects in the lattice dynamics [14, 16, 17].

RESONANCE PHENOMENA IN PHONON SUBSYSTEM

The well-known Fermi resonance in the Raman molecular spectra under the conditions of integer ratio of phonon frequencies [18, 2] is a prototype example of these nonperturbative resonance phenomena. It was first discovered for CO_2 molecule where the ratio of two molecular vibration frequencies ω_1 and ω_2 is close, by accident, to 1:2. This leads to a quasidegeneracy of the states with energies $\hbar\omega_1$ and $2\hbar\omega_2$. As a result the first harmonic of the first mode is strongly coupled with the second harmonic of the second mode which leads to the violation of symmetry-induced selection rules and appearance of additional lines in the Raman spectra. Fermi resonance is typical for many organic compounds, especially containing CH group, and is well investigated by molecular spectroscopists [19]. The theory of the Fermi resonance is rather simple since the ultraquantum limit $T \ll \hbar\bar{\omega}$ (where $\bar{\omega}$ is a mean phonon frequency) is typical for all these cases and therefore there are only few states which are involved in the resonance.

When we consider Fermi-resonance-like phenomena for crystals we immediately face several difficulties which make the problem much less trivial [14, 16, 17]. First, the classical limit $T > \hbar\bar{\omega}$ is interesting now and it turns out to be much more complicated than the ultraquantum one since here we have an infinite number of relevant states. Second, now we have to satisfy the resonance conditions not only for the frequencies but also for the phonon wave vectors $\mathbf{q}$. Then, there is a continuum of nonresonant phonons which is nevertheless also important working as a thermal bath and leading to the phonon damping and "Brownian motion" fluctuations. The most important point is the occurrence of essentially new phenomenon in the classical case, namely, a low-frequency dynamics due to energy transfer between the modes participating in the resonance; this effect is exponentially small at low temperatures (ultraquantum regime) when the corresponding modes are not excited. This "slow" dynamics corresponds to transitions between the states split by the anharmonic interaction under the Fermi resonance conditions. In this sense it relates to the Fermi resonance like the electron paramagnetic resonance relates to the Zeeman effect. Experimentally it may result in the appearance of low-energy peaks in the quasielastic neutron scattering spectra [14, 16, 17].

These new phenomena result from a correlated character of atomic displacements corresponding to different phonon branches under the Fermi resonance conditions. In a generic case the smallness of adiabatic parameter justifies the applicability of the perturbation theory for the consideration of anharmonic effects [20, 2]. From the technical point of view, it means the decoupling of the higher-order phonon correlators according to the Wick's theorem, or, in classical terms, we suppose that the initial phases of the interacting phonons are random. This is a standard approximation which is used to calculate the temperature dependence of phonon frequencies and dampings [21, 22, 23]. However, when the phonon frequency ratio turns out to be integer (and provided that some symmetry conditions are satisfied) a well-known nonlinear effect, the "phase locking", appears [24] and the perturbation theory fails. In particular, it leads to an essential enhancement of multiphonon contributions to the dynamical structure factor in comparison with the perturbative treatment, and to the appearance of quasistatic atomic displacements. This is a consequence of subtle phase coherence phenomena which are completely ignored by a standard perturbative treatment. The situation is similar to the Anderson localization in disordered systems which also cannot be described by any consideration of the *average* Green function [25].

To illustrate general mechanisms consider first the simplest case of the resonance, namely, phonons with the wave vector $\mathbf{q}_0 = (\frac{2}{3}, \frac{2}{3}, \frac{2}{3})$ and the frequency ratio 1:2 [16, 17]. It corresponds to BCC phases of alkali and alkaline earth metals (for instance, for potassium the frequencies of longitudinal and transverse phonons in this point at $T = 4.2K$ are 0.27 ± 0.01 and 0.55 ± 0.01 of ionic plasma frequency, respectively). Transverse phonons are double degenerate, however, microscopic estimations of the corresponding anharmonic coupling constants [16] show that only one transverse branch participates in the resonance since the coupling constant for another one is four orders of magnitude smaller. So we can consider a simplified model with the interaction of one longitudinal branch (with the displacement field $u(\mathbf{r},t)$) and one transverse one (with the displacement field $v(\mathbf{r},t)$). Taking into accoun only resonant interaction term $V = \lambda \int d\mathbf{r} u v^2$ one can write a set of equations of motion,

$$\begin{cases} \ddot{u} + \omega^2(-i\nabla)u + 2\gamma(-i\nabla)\dot{u} + \lambda v^2 = 0 \\ \ddot{v} + \Omega^2(-i\nabla)v + 2\Gamma(-i\nabla)\dot{v} + 2\lambda uv = 0 \end{cases} \quad (1)$$

where γ, Γ are corresponding phonon damping parameters. We try the solutions of Eq.(1) in the form

$$\begin{cases} u = A(\mathbf{r},t)\exp i[\mathbf{q}_0\mathbf{r} - \omega(\mathbf{q}_0)t] + \\ \quad B(\mathbf{r},t)\exp i[\mathbf{q}_0\mathbf{r} + \omega(\mathbf{q}_0)t] + c.c. \\ v = C(\mathbf{r},t)\exp i[\mathbf{q}_0\mathbf{r} + \omega(\mathbf{q}_0)t/2] + \\ \quad D(\mathbf{r},t)\exp i[\mathbf{q}_0\mathbf{r} - \omega(\mathbf{q}_0)t/2] + c.c. \end{cases} \quad (2)$$

Amplitudes A, B, C, D are slowly varying in space and time functions. It is important that the wave vector $\mathbf{q}_0$ is equivalent to $-2\mathbf{q}_0$ since $\exp(3i\mathbf{q}_0\mathbf{r}) = 1$ for the crystal lattice sites. Substituting Eq.(2) into Eq.(1), taking into account only resonant terms and only leading approximations in the anharmonic smallness parameter (for more details, see Ref. [17])

$$\begin{cases} \frac{\partial A}{\partial t} + \left(\frac{\partial \omega}{\partial \mathbf{q}}\right)_0 \nabla A - \frac{i}{4\omega_0}\left(\frac{\partial^2 \omega^2}{\partial q_\alpha \partial q_\beta}\right)_0 \frac{\partial^2 A}{\partial x_\alpha \partial x_\beta} \\ \quad + \gamma_0 A + i\Lambda C^{*2} = 0 \\ \frac{\partial C}{\partial t} - \frac{1}{2}\left(\frac{\partial \omega}{\partial \mathbf{q}}\right)_0 \nabla C + \frac{i}{8\omega_0}\left(\frac{\partial^2 \omega^2}{\partial q_\alpha \partial q_\beta}\right)_0 \frac{\partial^2 C}{\partial x_\alpha \partial x_\beta} \\ \quad + \left(\Gamma_0 + \frac{i\nu_0}{2}\right)C - 4i\Lambda A^*C^* = 0 \end{cases} \quad (3)$$

where $\omega = 2\Omega + \nu$ ($\nu \ll \omega$), $\Lambda = \lambda/2\omega_0$ and subscript "0" means $\mathbf{q} = \mathbf{q}_0$. Equations for B and D differ from Eq.(3) by the replacement $\nu \to -\nu, \Lambda \to -\Lambda$.

One can demonstrate [17] that at small enough ν (realistic estimations for alkali metals show that "small" means 5-7% of ω) purely sine waves $A, B, C, D = const$ turn out to be unstable with respect to a self-modulation and the solitons of the envelopes can form. This modulation is connected with a slow (in comparison with a characteristic phonon times) dynamics of energy transfer between longitudinal and transverse phonons. Numerical simulations of the thermal noise effects [26, 27] show that the latter do not suppress this slow dynamics. It appeared that there are *two* limit circles in this system (which correspond to two phase locking regimes with different relative phases for A and C waves) and this energy transfer dynamics can be described as a stochastic resonance between these two limit circles [27].

Similar phenomena can be also considered for acoustic (long-wavelength) phonons as an energy transfer between two ultrasound waves with different polarization vectors and integer ratio of sound velocities [28]. Alkali metals near the melting point (1:3 ratio for different transverse sound waves propagating into <110> direction) or $W_{1-x}Re_x$, $Mo_{1-x}Re_x$ alloys (1:1 ratio for the same sound waves) might be an interesting examples. Computer simulations [28] demonstrate that, depending on the initial phases of the waves, this energy transfer may be both chaotic and quasiperiodic. It would be interesting to check this prediction experimentally.

It is worthwhile to note that for the case of 1:2 frequency ratio considered above the resonance conditions for phonon wave vectors can be satisfied only in some peculiar points of the Brillouin zone. For the case of 1:3 resonance they can be satisfied in a generic case; for high-symmetry directions the number of waves participating in the resonance coupling can be very large. FCC La is an interesting example of such resonance (below we follow our work [29]). The lattice dynamics of FCC La is characterized by a drastic nonmonotonicity of the dispersion curves $\omega(\mathbf{q})$ for transverse phonons in $\mathbf{q} \parallel \langle 111 \rangle$ direction and by a significant temperature dependence of their frequencies, i.e. by strong anharmonic interactions [30]. As it was discussed in Ref. [30] this behavior can be accounted for by the electron-phonon interaction (Kohn anomalies). One can see from these experimental data that the ratio of longitudinal to transverse phonon frequencies in $\langle 111 \rangle$ direction varies around the integer ratio $\frac{\omega_2(\mathbf{q}_0)}{\omega_1(\mathbf{q}_0)} = 3$, the value of wave vector $\mathbf{q}_0$ varying with the temperature within sufficiently wide limits. This makes it possible to shift the resonance watching simultaneously the change in the lattice dynamics behavior.

Let us set up the simplest model to describe the resonance effects under consideration in the lattice dynamics of FCC La. We shall proceed with the equations of motion for the amplitudes of longitudinal (u) and transverse (v) phonons allowing only for the "resonance" anharmonic interaction $V = \lambda \int d\mathbf{r} u v^3$:

$$\begin{cases} \ddot{u} + \omega^2 u + \lambda v^3 = 0 \\ \ddot{v} + \Omega^2 v + 3\lambda u v^2 = 0 \end{cases} \quad (4)$$

Because of the FCC lattice symmetry we should consider phonons propagating in four equivalent directions $\langle 111 \rangle, \langle 1\bar{1}\bar{1} \rangle, \langle \bar{1}1\bar{1} \rangle, \langle \bar{1}\bar{1}1 \rangle$; the relevant wave vectors meeting condition $\omega(\mathbf{q}) = 3\Omega(\mathbf{q})$ will be denoted as $\mathbf{q}_j$ ($j = 0, 1, 2, 3$). Similarly to Ref. [17] we try the solutions of the equations (4) as

$$\begin{cases} u = \sum_{j=0}^{3} (A_j \exp i(\mathbf{q}_j \mathbf{r} - \omega t) + B_j \exp i(\mathbf{q}_j \mathbf{r} + \omega t)) + c.c. \\ v = \sum_{j=0}^{3} (C_j \exp i(\mathbf{q}_j \mathbf{r} - \Omega t) + D_j \exp i(\mathbf{q}_j \mathbf{r} + \Omega t)) + c.c. \end{cases} \quad (5)$$

where A_j, B_j, C_j, D_j are slowly varying (due to λ smallness) functions of $\mathbf{r}$ and t, i.e. the envelopes of the phonons under consideration. Substituting Eqs. (5) into Eqs. (4) and neglecting the second time derivatives of the envelopes as well as the nonresonance terms we obtain the following set of equations

$$\begin{cases} \dot{A}_0 + i\Lambda(6D_1^* D_2^* D_3^* + 3C_0^2 D_0^* + 6C_0 \sum_{j=1}^{3} C_j D_j^*) = 0 \\ \dot{B}_0 - i\Lambda(6C_1^* C_2^* C_3^* + 3D_0^2 C_0^* + 6D_0 \sum_{j=1}^{3} D_j C_j^*) = 0 \\ \dot{C}_0 + 9i\Lambda(2B_1^* C_2^* C_3^* + 2B_2^* C_1^* C_3^* + 2B_3^* C_1^* C_2^* + \\ \qquad 2A_0 \sum_{j=0}^{3} C_j^* D_j + 2D_0 \sum_{j=1}^{3} A_j C_j^* + \\ \qquad D_0^2 B_0^* + 2D_0 \sum_{j=1}^{3} D_j B_j^*) = 0 \\ \dot{D}_0 - 9i\Lambda(2A_1^* D_2^* D_3^* + 2A_2^* D_1^* D_3^* + 2A_3^* D_1^* D_2^* + \\ \qquad 2B_0 \sum_{j=0}^{3} D_j^* C_j + 2C_0 \sum_{j=1}^{3} B_j D_j^* + \\ \qquad C_0^2 A_0^* + 2C_0 \sum_{j=1}^{3} C_j A_j^*) = 0; \end{cases} \quad (6)$$

the rest twelve equations are obtained from (6) by cyclic permutations of the indices. Here $\Lambda = \frac{\lambda}{2\omega}$.

The dynamics of the energy transfer in this system has been investigated in Ref. [29] by numerical simulations. It was shown that depending on the initial phases of the involved phonons this energy transfer can be either regular or chaotic. One can expect from the dimension considerations that a typical frequency of this slow dynamics will be of order of

$$\omega^* \simeq \bar{\omega} \left(\frac{\bar{u}^2}{a^2} \right) \frac{\lambda}{M \bar{\omega}^2 a^6}, \quad (7)$$

Here M is the ion mass, a is the lattice constant, $\bar{u}^2$ is the average square of atomic displacements. For the room temperature $\omega^* \approx 10^{-3} \bar{\omega}$ can be taken for the estimation. However, it was shown that in reality it is much larger,

$\triangle\omega \simeq 10^2 \omega^*$, due to a large number of interacting waves (32 real fields).

Similarly to Ref. [17] (see also above) it can be shown that the account for intermode or non-resonant intramode anharmonicities may result in the appearance of the peaks in quasielastic neutron scattering spectra with a frequency width about $\triangle\omega$. The largest contributions result from the lowest-order (three-phonon) anharmonic processes. For example, the term $V = \mu \int d\mathbf{r} v^3$ in the potential energy leads to the appearance of the "low-frequency" contribution to the transverse phonon field, $\delta v = -3\mu v^2/\Omega^2$. Assuming all the amplitudes A_j and C_j in the equation (5) to be constant we would obtain static contributions to v with the wave vectors $\mathbf{Q}_a = \mathbf{q}_i \pm \mathbf{q}_j + \mathbf{g}$ where $\mathbf{g}$ is a reciprocal lattice vector (which can be, in particular, equal to zero). One can expect therefore the peaks in the quasielastic neutron scattering spectra near the wavevectors $\mathbf{Q}_a$ which reflect the dynamics of the envelopes. In a special case considered above for the frequency ratio 2:1 one of these vectors coincides with the vector $\mathbf{q}_i$ but for the ratio 3:1 it is impossible. The vectors $\mathbf{Q}_a$ for $i = j$ are just the reciprocal lattice vectors but the rest of them should be temperature dependent due to the temperature dependence of the phonon frequencies shifting the Fermi resonance point. It can be important for the experimental verification of the effects under consideration. A similar contribution of order of u^2 appears in the longitudinal phonon field because of the potential energy term $V = v \int d\mathbf{r} u^3$.

CONCLUSIONS

To conclude, we predict a new class of nonperturbative anharmonic phenomena in lattice dynamics under resonance conditions in the phonon system. This resonance results in the instability of phonons and complicated picture of energy transfer between the modes in the resonance. It can be experimentally investigated by inelastic and quasielastic neutron scattering as well as by acoustic methods. There is an interesting open issue what is the role of these phenomena in the development of lattice instabilities of the crystals, in particular, in the melting [14]. This question is motivated by two general remarks: (i) anharmonic effects in a generic case are small up to the melting point and (ii) instability of a generic weakly anharmonic dynamical system, according to the KAM theory, is connected with the resonances. Therefore it might appear that the phenomenon under consideration is not a kind of exotic but important for any crystal.

REFERENCES

1. Born, M., and Huang, K., *Dynamical Theory of Crystal Lattices*, Oxford University Press, Oxford, 1998.
2. Katsnelson, M. I., and Trefilov, A. V., *Crystal Lattice Dynamics and Thermodynamics*, Atomizdat, Moscow, 2002.
3. Bratkovskii, A. M., Vaks, V. G., and Trefilov, A. V., *J. Exper. Theor. Phys.* **59**, 1245-1255 (1984).
4. Vaks V. G., Kravchuk S. P., and Trefilov A. V., *J. Phys. F* **10**, 2325-2344 (1980).
5. Katsnelson, M. I., Maksytov A. F., and Trefilov, A. V., *Phys. Lett. A* **295**, 50-54 (2003).
6. Chen, Y., Ho, K. M., and Harmon, B. N., *Phys. Rev. B* **37**, 283-288 (1988).
7. Ye, Y. Y., Chen, Y., Ho, K. M., Harmon, B. N., and Lindgard, P. A., *Phys. Rev. Lett.* **58**, 1769-1772 (1987).
8. Morris, J. R., and Ho, K. M., *Phys. Rev. B* **63**, 224116 (2001).
9. Axe, J. D., and Shirane, G., *Phys. Rev. B* **8**, 1965-1977 (1973).
10. Petry, W., *Phase Transit.* **31**, 119-136 (1991).
11. Gornostyrev, Yu. N., Katsnelson, M. I., Trefilov, A. V., and Tret'jakov, S. V., *Phys. Rev. B* **54**, 3286-3294 (1996).
12. Axe, J. D., Keating, D. T., and Moss, S. C., *Phys. Rev. Lett.* **35**, 530-533 (1975).
13. Aksenov, V. L., Plakida, N. M., and Stamenkovic, S., *Neutron Scattering by Ferroelectrics*, World Scientific, Singapore, 1989.
14. Katsnelson, M. I., and Trefilov, A. V., *Fiz. Metal. Metalloved.* **64**, 629-642 (1987).
15. Arnold, V. I., *Mathematical Methods of Classical Mechanics*, Springer, Berlin, 1989.
16. Katsnelson, M. I., and Trefilov, A. V., *JETP Lett.* **45**, 634-638 (1987).
17. Katsnelson, M. I., and Trefilov, A. V., *J. Exper. Theor. Phys.* **70**, 1067-1071 (1990).
18. Fermi, E., *Molecules, Crystals, and Quantum Statistics*, Benjamin, New York, 1966.
19. Lisitsa, M. P., and Yaremko, A. M., *The Fermi Resonance*, Naukova Dumka, Kiev, 1984.
20. Cowley, R. A., *Adv. Phys.* **12**, 421-480 (1963).
21. Vaks V. G., Kravchuk S. P., and Trefilov A. V., *J. Phys. F* **10**, 2105-2124 (1980).
22. Katsnelson, M. I., Trefilov, A. V., and Khromov, K. Y., *JETP Lett.* **69**, 688-693 (1999).
23. Katsnelson, M. I., Trefilov, A. V., Khlopkin, M. N., and Khromov, K. Y., *Phil. Mag. B* **81**, 1893-1913 (2001).
24. Haken, H., *Advanced Synergetics. Instability Hierarchies of Self-Organizing Systems and Devices*, Springer, Berlin, 1987.
25. Anderson, P. W., *Phys. Rev.* **109**, 1492-1505 (1958).
26. Gornostyrev, Yu. N., Katsnelson, M. I., Platonov, A. P., and Trefilov, A. V., *Phys. Rev. B* **51**, 12817-12820 (1995); *J. Exper. Theor. Phys.* **80**, 525-530 (1995).
27. Gornostyrev, Yu. N., Zhdakhin, D. I., Katsnelson, M. I., and Trefilov, A. V., *JETP Lett.* **69**, 630-635 (1999).
28. Katsnelson, M. I., Platonov, A. P., and Trefilov, A. V., *JETP Lett.* **69**, 453-458 (1999).
29. Katsnelson, M. I., Stroev, A. Y., and Trefilov, A. V., *Phys. Rev. B* **66**, 092303 (2002).
30. Stassis, C., Smith, G. D. Harmon, B. N. Ho, K. M. and Chen, Y., *Phys. Rev. B* **31**, 6298-6304 (1985).

Dynamics and its stability of Boltzmann-machine learning algorithm for gray scale image restoration

Jun-ichi Inoue* and Kazuyuki Tanaka[†]

*Complex Systems Engineering, Graduate School of Engineering
Hokkaido University N13-W8, Kita-ku, Sapporo 060-8628, Japan
[†]Department of Applied Information Science, Graduate School of Information Sciences
Tohoku University, Aramaki-aza-aoba 09, Aoba-ku, Sendai 090-8579, Japan

Abstract. Dynamic behavior and its stability of Boltzmann-machine learning algorithm for Bayesian gray scale image restoration are investigated. We derive the differential equations by which we attempt to maximize the marginal likelihood function with respect to hyper-parameters. Average-case performance and linear stability of the algorithm are evaluated exactly at the mean-field level. We conclude that the solution of the Boltzmann-machine learning equation is *asymptotically stable* as long as the solution is identical to the correct value of the hyper-parameters.

INTRODUCTION

In the context of Bayesian image restoration, estimation of *hyper-parameters* which specify the posterior distribution is the most important procedure to improve the quality of the restored image. For this problem, the present authors have investigated the average case performance of the *maximum marginal likelihood method* to infer the hyper-parameters [1]. If we consider the image restoration as probabilistic systems, hyper-parameter estimation can be regarded as problem to obtain the maximum likelihood estimate for a given probabilistic model. However, in image restoration, one cannot use any information about the original image, and as the result, it is impossible for us to use the complete data log-likelihood function as the cost to determine the appropriate hyper-parameters and we should use the incomplete data log-likelihood function (*marginal likelihood*). In order to maximize the marginal likelihood, *Boltzmann-machine learning* [1] and *EM algorithm* (Expectation Maximum algorithm) [2] are applicable to this problem. In this paper, we show the dynamics and its stability of the Boltzmann-machine learning for gray scale image restoration by using *exact analysis of the mean-field model*.

THE DEFINITION OF THE MODEL

In this section, we introduce the model system for Bayesian gray scale image restoration [3]. As the original images, we use snapshots from the following Boltzmann-Gibbs distribution :

$$P_{\beta_s}(\{\xi\}) = \frac{e^{-(\beta_s/2N)\sum_{ij}(\xi_i-\xi_j)^2}}{\text{tr}_\xi e^{-(\beta_s/2N)\sum_{ij}(\xi_i-\xi_j)^2}} \quad (1)$$

where each pixel in the original image $\{\xi\} \equiv (\xi_1,\cdots,\xi_N)$ takes Q values $\xi_i = 0,1,2,\cdots,Q-1$, and $\beta_s = T_s^{-1}$ is inverse temperature and specifies the macroscopic property of the original image. This original image $\{\xi\}$ is degraded by the *Gaussian channel* :

$$P_{a_\tau}(\{\tau\}|\{\xi\}) = \frac{e^{-(1/2a_\tau^2)\sum_i(\tau_i-\xi_i)^2}}{\sqrt{2\pi}a_\tau} \quad (2)$$

and we receive the degraded image $\{\tau\} \equiv (\tau_1,\cdots,\tau_N)$. For this model, we construct the posterior, that is, the probability of the estimate $\{\sigma\} \equiv (\sigma_1,\cdots,\sigma_N)$ for the original image provided that one receives the degraded image $\{\tau\}$, according to the Bayes rule :

$$P(\{\sigma\}|\{\tau\}) = \frac{e^{-(\beta/2N)\sum_{ij}(\sigma_i-\sigma_j)^2-h\sum_i(\tau_i-\sigma_i)^2}}{\text{tr}_\sigma e^{-(\beta/2N)\sum_{ij}(\sigma_i-\sigma_j)^2-h\sum_i(\tau_i-\sigma_i)^2}} \quad (3)$$

where the hyper-parameters to be determined are β and h. In both *Maximum A Posteriori (MAP)* and *Maximizer of Posterior Marginal (MPM)* estimations, the quality of restoration, namely, the mean-square error $D \equiv (1/N)\sum_i(\xi_i-\hat{\xi}_i^{(\text{MAP,MPM})})^2$ between the original image ξ_i and its estimate $\hat{\xi}_i$ depends on (β,h), and takes the minimum at $(\beta_s,1/2a_\tau^2)$.

In the context of the Bayesian gray scale image restoration, the estimates $\hat{\xi}_i^{(\text{MAP,MPM})}$ are represented by $\{\xi\}^{(\text{MAP})} \equiv \text{argmax}_{\{\sigma\}} P(\{\sigma\}|\{\tau\})$ and

$\hat{\xi}_i^{(\text{MPM})} \equiv \Omega(\arg\max_{\sigma_i} \text{tr}_{\{\sigma\}\neq\sigma_i} P(\{\sigma\}|\{\tau\}))$ with $\Omega(x) \equiv \sum_{k=0}^{Q-1} k[\Theta(x-(2k-1)/2) - \Theta(x-(2k+1)/2)]$, where $\Theta(x)$ means Heviside step function. In the next section, we describe the learning equations that maximize the *marginal likelihood* to determine the hyper-parameters correctly.

BOLTZMANN-MACHINE LEARNING

We determine these hyper-parameters so as to maximize the following marginal likelihood $-K_{\beta,h}(\{\xi,\tau\})$:

$$e^{-K_{\beta,h}(\{\xi,\tau\})} \equiv \text{tr}_{\{\sigma\}} \frac{e^{-(\beta/2N)\sum_{ij}(\sigma_i-\sigma_j)^2 - h\sum_i(\sigma_i-\tau_i)^2}}{Z_\Pi Z_L} \quad (4)$$

where we defined the following normalization constants: $Z_\Pi \equiv \text{tr}_{\{\sigma\}} e^{-(\beta/2N)\sum_{ij}(\sigma_i-\sigma_j)^2}$ and $Z_L \equiv \text{tr}_{\{\tau\}} e^{-h\sum_i(\sigma_i-\tau_i)^2}$. We attempt to maximize this cost function directly *via* the following gradient descent

$$\Gamma_\beta \dot{\beta} = -\partial_\beta K_{\beta,h}(\{\xi,\tau\}) \quad (5)$$
$$\Gamma_h \dot{h} = -\partial_h K_{\beta,h}(\{\xi,\tau\}) \quad (6)$$

where we defined $\dot{X} \equiv dX/dt$, $\partial_\beta X \equiv \partial X/\partial \beta$, and Γ_β, Γ_h are relaxation times. These equations (5)(6) are referred to as *Boltzmann-machine learning equations* in the context of *neural networks*. By taking the derivatives in (5)(6), we obtain

$$\Gamma_\beta \dot{\beta} = -(1/2N)\langle \sum_{ij}(\sigma_i-\sigma_j)^2 \rangle_{\text{pos.}}$$
$$+ (1/2N)\langle \sum_{ij}(\sigma_i-\sigma_j)^2 \rangle_{\text{pri.}} \quad (7)$$
$$\Gamma_h \dot{h} = -\langle \sum_i (\sigma_i-\tau_i)^2 \rangle_{\text{pos.}} + N/2h. \quad (8)$$

These expressions tell us that we need expectations $\langle \cdots \rangle_{\text{pos.}}, \langle \cdots \rangle_{\text{pri.}}$ which mean the averages over the posterior and the prior at each time step. It may cost us a lot of computational time to evaluate these expectations *via* Markov Chain Monte Carlo sampling due to the *quenched disorders* $\{\xi,\tau\}$. As the result, the system exhibits *slow relaxation* to thermodynamically equilibrium states.

AVERAGE-CASE PERFORMANCE

Obviously, the equations derived in the previous section depend on the data $\{\xi,\tau\}$. In order to investigate the average case performance of the algorithm, we should evaluate the averages $[\partial_\beta K_{\beta,h}(\{\xi,\tau\})]_{\{\xi,\tau\}}$ and $[\partial_h K_{\beta,h}(\{\xi,\tau\})]_{\{\xi,\tau\}}$ for each time step.

For the full-connected mean-field model, the quantities appearing in the right hand sides of each equation (5) and (6) become *self-average* in the thermodynamic limit and converge to $[\partial_\beta K_{\beta,h}(\{\xi,\tau\})]_{\{\xi,\tau\}}$ and $[\partial_h K_{\beta,h}(\{\xi,\tau\})]_{\{\xi,\tau\}}$, respectively. Moreover, in our model system, the average of the derivative, that is,

$$[\partial_\beta K_{\beta,h}(\{\xi,\tau\})]_{\{\xi,\tau\}}$$
$$= \partial_\beta \text{tr}_{\{\xi\}} P_{\beta_s}(\{\xi\}) \int_{-\infty}^{\infty} d\{\tau\} P_{a_\tau}(\{\tau\}|\{\xi\}) K_{\beta,h}(\{\xi,\tau\})$$

with

$$\text{tr}_{\{\xi\}} \equiv \sum_{\xi_1=0}^{Q-1} \sum_{\xi_2=0}^{Q-1} \cdots \sum_{\xi_N=0}^{Q-1}, \quad d\{\tau\} \equiv d\tau_1 d\tau_2 \cdots d\tau_N$$

can be evaluated *without the replica method*. Thus, we obtain the averaged Boltzmann-machine learning equations as follows.

$$\Gamma_\beta \dot{\beta} = -m^2 + \ll \int_{-\infty}^{\infty} Dx \langle 2m\sigma - \sigma^2 \rangle_{**}^{(\beta,h:a_\tau)} \gg^{(\beta_s)}$$
$$- \langle 2m_1\sigma - \sigma^2 \rangle_*^{(\beta)} + m_1^2 \quad (9)$$
$$\Gamma_h \dot{h} = 1/2h$$
$$- \ll \int_{-\infty}^{\infty} Dx \langle (\sigma - a_\tau x - \xi)^2 \rangle_{**}^{(\beta,h:a_\tau)} \gg^{(\beta_s)} \quad (10)$$
$$m_1 = \langle \sigma \rangle_*^{(\beta)} \quad (11)$$
$$m = \ll \int_{-\infty}^{\infty} Dx \langle \sigma \rangle_{**}^{(\beta,h:a_\tau)} \gg^{(\beta_s)} \quad (12)$$

where $Dx \equiv dx e^{-x^2/2}/\sqrt{2\pi}$ and we defined three kinds of averages

$$\ll \cdots \gg^{(\beta_s)} \equiv \frac{\text{tr}_\xi(\cdots) e^{2\beta_s m_0 \xi - \beta_s \xi^2}}{\text{tr}_\xi e^{2\beta_s m_0 \xi - \beta_s \xi^2}} \quad (13)$$

$$\langle \cdots \rangle_*^{(\beta)} \equiv \frac{\text{tr}_\sigma(\cdots) e^{2\beta m_1 \sigma - \beta \sigma^2}}{\text{tr}_\sigma e^{2\beta m_1 \sigma - \beta \sigma^2}} \quad (14)$$

$$\langle \cdots \rangle_{**}^{(\beta,h:a_\tau)} \equiv \frac{\text{tr}_\sigma(\cdots) e^{-h(\sigma - a_\tau x - \xi)^2 + 2\beta m\sigma - \beta \sigma^2}}{\text{tr}_\sigma e^{-h(\sigma - a_\tau x - \xi)^2 + 2\beta m\sigma - \beta \sigma^2}} \quad (15)$$

It is helpful for the reader to keep in mind that the data-averaged Boltzmann-machine learning algorithm depends on the hyper-parameters through these averages.

Before we proceed to the analysis of the equations (9)-(12), it should be mentioned that the mean-square error also becomes self-average as

$$D \equiv \lim_{N\to\infty} \frac{1}{N} \sum_i (\xi_i - \hat{\xi}_i^{(\text{MAP,MPM})})^2 = [D]_{\{\xi,\tau\}}. \quad (16)$$

For the MPM estimation, D is written explicitly by

$$D = \ll \int_{-\infty}^{\infty} Dx [\xi - \Omega(\langle \sigma \rangle_{**}^{(\beta,h:a_\tau)})]^2 \gg^{(\beta_s)}. \quad (17)$$

In [3], the mean-square error D was shown as a function of $T = \beta^{-1}$ keeping the ratio $h/\beta = T_s/2a_\tau^2$. They found that D takes its minimum at $T = T_s$ and the difference between the MAP and the *best possible* MPM estimations $\Delta D \equiv D_{T=T_s} - D_{T=0}$ is always negative [3].

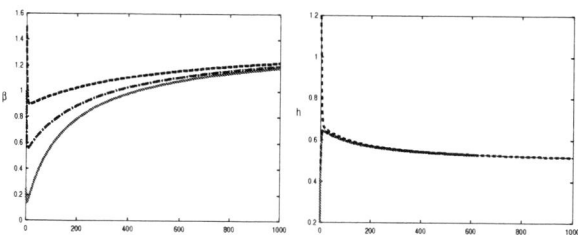

FIGURE 1. Behavior of the averaged Boltzmann-machine learning equations (5)(6) for the case of $Q = 3$. Both β and h converge to the correct solutions $\beta_s^{-1} = 0.75$ and $1/2a_\tau^2 = 0.5$, respectively.

In FIGURE 1, we plot the time evolutions of β and h for the case of $Q = 3$. We find that these hyperparameters converge to their correct solutions. In the next section, we investigate the stability around the solution.

ANALYSIS OF THE STABILITY

In this section, we investigate the stability around the solution of the equations (9)(10). This section is a main part of this paper.

In order to check the stability of the equations around the solutions $s_* \equiv (\beta_*, h_*)$, we set $s = s_* + \delta s$ and substitute s into the equations (9)(10), and expand them with respect to δs up to $\mathcal{O}(\delta s)$. Then, we easily obtain $d(\delta s)/dt = A\delta s + \mathcal{O}(\delta s^2)$. Hessian matrix A is defined as

$$A = \begin{pmatrix} \frac{[\partial_\beta^2 K_{\beta,h}(\{\xi,\tau\})]_{\{\xi,\tau\}}}{-\Gamma_\beta} & \frac{[\partial_h \partial_\beta K_{\beta,h}(\{\xi,\tau\})]_{\{\xi,\tau\}}}{-\Gamma_\beta} \\ \frac{[\partial_\beta \partial_h K_{\beta,h}(\{\xi,\tau\})]_{\{\xi,\tau\}}}{-\Gamma_h} & \frac{[\partial_h^2 K_{\beta,h}(\{\xi,\tau\})]_{\{\xi,\tau\}}}{-\Gamma_h} \end{pmatrix}$$

$$\equiv \begin{pmatrix} A_{11} & A_{12} \\ A_{21} & A_{22} \end{pmatrix} \quad (18)$$

and each matrix element is given by

$$\Gamma_\beta A_{11} \equiv \ll \int_{-\infty}^{\infty} Dx \langle (2m\sigma - \sigma^2)^2 \rangle_{**}^{(\beta_*,h_*:a_\tau)} \gg^{(\beta_s)}$$
$$- \ll \int_{-\infty}^{\infty} Dx \left\{ \langle 2m\sigma - \sigma^2 \rangle_{**}^{(\beta_*,h_*:a_\tau)} \right\}^2 \gg^{(\beta_s)}$$
$$- \langle (2m_1\sigma - \sigma^2)^2 \rangle_*^{(\beta_*)} + \left\{ \langle 2m_1\sigma - \sigma^2 \rangle_*^{(\beta_*)} \right\}^2 \quad (19)$$

$$\Gamma_\beta A_{12} \equiv \ll \int_{-\infty}^{\infty} Dx \langle 2m\sigma - \sigma^2 \rangle_{**}^{(\beta_*,h_*:a_\tau)}$$
$$\times \langle (\sigma - a_\tau x - \xi)^2 \rangle_{**}^{(\beta_*,h_*:a_\tau)} \gg^{(\beta_s)}$$
$$- \ll \int_{-\infty}^{\infty} Dx \langle (2m\sigma - \sigma^2)$$
$$\times (\sigma - a_\tau x - \xi)^2 \rangle_{**}^{(\beta_*,h_*:a_\tau)} \gg^{(\beta_s)} = \Gamma_h A_{21} \quad (20)$$

$$\Gamma_h A_{22} \equiv \ll \int_{-\infty}^{\infty} Dx \langle (\sigma - a_\tau x - \xi)^4 \rangle_{**}^{(\beta_*,h_*:a_\tau)} \gg^{(\beta_s)}$$
$$- \ll \int_{-\infty}^{\infty} Dx \left\{ \langle (\sigma - a_\tau x - \xi)^2 \rangle_{**}^{(\beta_*,h_*:a_\tau)} \right\}^2 \gg^{(\beta_s)}$$
$$- 1/2h_*^2. \quad (21)$$

Therefore, we should check the sign of eigenvalues λ_1 and λ_2 of the matrix A. In FIGURE 2, we plot the

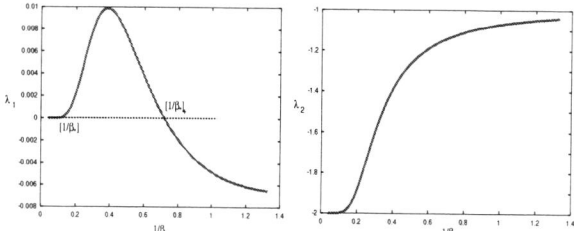

FIGURE 2. Eigenvalues $\lambda_1 > \lambda_2$ of the Hessian matrix, as a function of $1/\beta_*$ for the case of $\beta_s^{-1} = 0.75$ and $1/2a_\tau^2 = h_s = h_* = 0.5$.

eigenvalues λ_1, λ_2 as a function of $1/\beta_*$ for the case of $\beta_s^{-1} = 0.75$ and $1/2a_\tau^2 = h_s = 0.5 = h_*$. We find that in the region $[1/\beta_*]_- < 1/\beta_* < [1/\beta_*]_+$, the equations (9)(10) around (β_*, h_*) become asymptotically unstable. In FIGURE 3, the h_*-β_*^{-1} phase diagram for the case

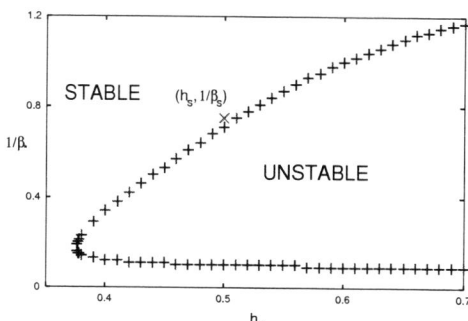

FIGURE 3. The h_*-β_*^{-1} phase diagram for the case of $\beta_s^{-1} = 0.75$ and $h_s = 0.5$. The solution which is identical to (β_s^{-1}, h_s) is asymptotically stable.

of $(h_s, \beta_s^{-1}) = (0.5, 0.75)$ is displayed. In this figure, the solution (h_*, β_*^{-1}) located in the label **STABLE** is asymptotically stable, while the region labeled by **UNSTABLE** means asymptotically unstable. We find that if the fixed point of the equations (9)(10) is identical to the correct noise level and the source temperature, namely, (h_s, β_s^{-1}), the solution is asymptotically stable. Apparently, the region $[1/\beta_*]_- < 1/\beta_* < [1/\beta_*]_+$ in

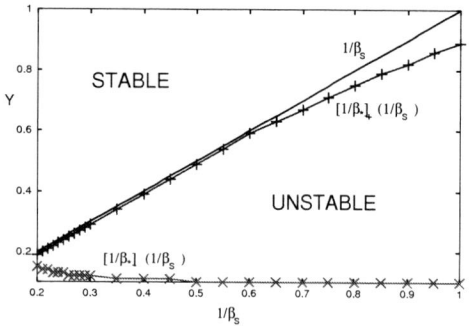

FIGURE 4. $Y = [1/\beta_*]_+, [1/\beta_*]_-$ and $Y = 1/\beta_s$ as a function of $1/\beta_s$ for the case of $h_* = h_s = 0.5$.

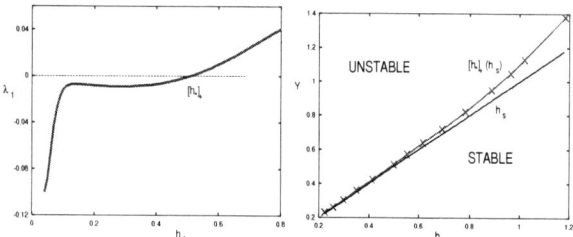

FIGURE 5. The largest eigenvalue of the Hessian matrix (left). The right panel shows $Y = [h_*]_+$ and $Y = h_s$ as a function of h_s for the case of $\beta_* = 0.75$.

which the solution becomes asymptotically unstable depends on the source temperature β_s^{-1} and the noise level $h_s = 1/2a_\tau^2$. In FIGURE 4, we plot $Y = [1/\beta_*]_+, [1/\beta_*]_-$ and $Y = 1/\beta_s$ as a function of $1/\beta_s$ for the case of $h_* = h_s = 0.5$. From this figure, we find that for any choice of the source temperature, the solution of Boltzmann-machine learning algorithm (9)(10) is asymptotically stable.

On the other hand, In FIGURE 5 (left), we plot the eigenvalue $\lambda_1 (> \lambda_2)$ of the matrix A as a function of h_* for the case of the source temperature $\beta_s^{-1} = 0.75 = \beta_*^{-1}$ and the noise revel $h_s = 1/2a_\tau^2 = 0.5$. The solution falls in the domain $h > [h_*]_+$ becomes asymptotically unstable.

In order to investigate the noise revel h_s-dependence of the domain $h > [h_*]_+$ for a fixed source temperature β_s^{-1}, we plot $Y = [h_*]_+$ and $Y = h_s$ as a function of h_s in FIGURE 5 (right). We find that for any case of the noise revel h_s, the solution of the equations (9)(10) becomes asymptotically stable as long as the solution h_* is identical to h_s.

CONCLUSIONS AND DISCUSSION

In this paper, we investigated the dynamical properties of the Boltzmann-machine learning algorithm for the Bayesian gray scale image restoration. We derived the differential equations with respect to the hyper-parameters and checked the stability around the solution. We conclude that the solution is asymptotically stable as long as the solution is identical to the true value of the hyper-parameters.

Although we focused on the linear stability of the solutions, it is easy for us to derive the relation between $\delta\beta$ and δh near the solution (h_*, β_*). The relation is simply given by

$$\alpha_{11}^{(\beta_*,h_*)} \delta\beta + \alpha_{12}^{(\beta_*,h_*)} \delta h = C^{\lambda_1,\lambda_2}(\beta_*, h_*)(\alpha_{21}^{(\beta_*,h_*)} \delta\beta + \alpha_{22}^{(\beta_*,h_*)} \delta h)^{\frac{\lambda_1}{\lambda_2}} \quad (22)$$

where a constant $C^{\lambda_1,\lambda_2}(\beta_*, h_*)$ is independent of $(\delta\beta, \delta h)$, and $\alpha_{11}^{\beta_*,h_*}$, etc. are related to the eigenvectors of the Hessian matrix A. It is important to bear in mind that around the solution (β_s, h_s), the ratio λ_1/λ_2 is small (for example, $\lambda_1/\lambda_2 = (-0.000797/-1.1124304) = 0.000708$ for the case of $\beta_s^{-1} = 0.75, h_s = 0.5$. See also FIGURE 2.) and equation (22) approximately leads to

$$\alpha_{11}^{(\beta_s,h_s)} \delta\beta + \alpha_{12}^{(\beta_s,h_s)} \delta h \simeq C^{\lambda_1,\lambda_2}(\beta_s, h_s). \quad (23)$$

In FIGURE 6, we plot the h-β flows of the Boltzmann-

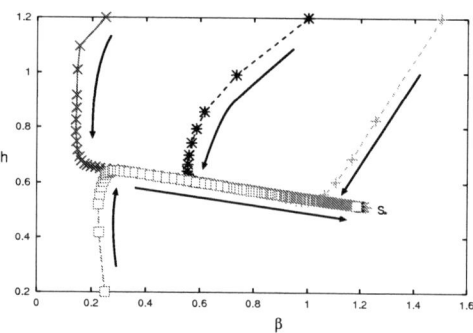

FIGURE 6. β-h flows of the Boltzmann-machine learning equations (9)(10). All flows converge to the correct value (β_s, h_s).

machine learning equations (9)(10). We find that around the solution, the linear relation between $\delta\beta$ and δh, namely, equation (23) is actually observed.

REFERENCES

1. Inoue, J. and Tanaka, K., Phys. Rev. E **65** 016125-1-016125-11 (2002).
2. Inoue, J. and Tanaka, K., J. Phys. A : Math. Gen. **36** 10997-11010(2003).
3. Inoue, J and Carlucci, D. M., Phys. Rev. E **64** 036121-036121-18 (2001).

Simple models of unusual elastic properties

K. W. Wojciechowski

Institute of Molecular Physics, Polish Academy of Sciences, Smoluchowskiego 17/19, 60-179 Poznań, Poland

Abstract. Elastic properties of a class of two-dimensional model systems, consisted of hard cyclic multimers, are discussed. Each multimer is composed of $m = 3k$ (where k is a positive integer) hard discs of diameter σ and centers forming a perfect polygon of m-sides, where the side length is l. Close packed structures of such systems, which are isotropic from the point of view of elastic properties, were solved exactly in the close packing limit at zero temperature. It was shown that the Poisson ratio, v_p, of the multimers is negative when their roughness parameter, defined as $\alpha \equiv l/(2\sigma)$, is large. In the limit $m \to \infty$ one obtains hard disc-like particles, which in contrast to the standard hard discs are *rough*. It is conjectured that the formula obtained for the Poisson ratio of the $3k$-multimers, $v_p = (1 - 2\alpha^2)/(3 - 2\alpha^2)$, is valid also for $m \neq 3k$ in the limit $m \to \infty$.

INTRODUCTION

A system of negative Poisson ratio [1] (NPR) increases its transverse dimensions when expanded longitudinally. This is in contrast to common systems which decrease the transverse dimensions at such a deformation [1], see Fig.1. This unusual property is not only of interest from the point of view of fundamental research but it is also useful for various practical applications. Hence, since manufacturing NPR structures by Lakes [2] and Evans [3], an increasing research activity is observed in this field [4, 5].

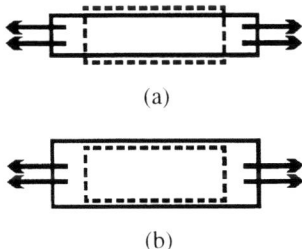

FIGURE 1. Deformation accompanying stretching of a sample of (a) a common material showing positive Poisson ratio and (b) a sample of material exhibiting negative Poisson ratio.

Structures on macro-, mezo- and microscopic level have been found which exhibit $v_p < 0$. The structures on the microscopic level are of particular interest. One of the possibilities to obtain such structures is by using molecules which can form NPR *phases*. The first step in this direction can be done by constructing various model particles of such a property. To simplify the analysis it is meaningful to start with two-dimensional models.

Analysis of various molecules indicates that one of the crucial molecular parameters characterizing the Poisson ratio is the molecular *shape* [6]. The simplest interactions which can characterize the molecular shape are the hard-body interactions, infinite when the bodies overlap and zero otherwise. The hard-body systems have been used to model various structures of real matter and phenomena occurring in them, see e.g. [7, 8] and references therein. Being athermal, the hard-body systems are convenient reference systems for fluids, liquid crystals and plastic crystals. As the hard potential is non-analytic they also constitute demanding test models for various theoretical approximations and simulation methods. In the present note a class of two-dimensional particles, interacting through the hard potential is considered in the aspect of searching for systems of negative Poisson ratio.

HARD CYCLIC MULTIMERS IN TWO DIMENSIONS

The examples of even and odd cyclic multimers, respectively consisting of $m = 9, 12$ hard discs, are shown in Fig.2. The disc centers are assumed to form perfect polygons which are rigid (i.e cannot change their shape); the internal stability of the multimers is not discussed in this work.

It can be rigorously proven that, when $m = 3k$ and the positive integers k are even, the multimer centers form the triangular lattice at close packing [6]. Computer simulations indicate [9] that the triangular lattice of the multimer centers corresponds also to the close packed structure of the multimers when k is any odd integer. Thus, when $m = 3k$, the close packed structures of the cyclic multimers must show the 3-fold symmetry axis,

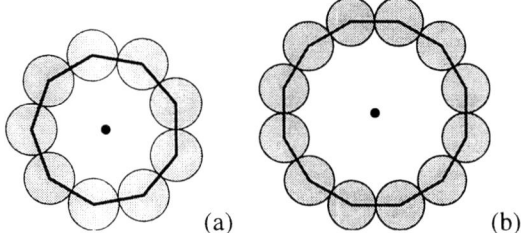

FIGURE 2. Geometry of the hard cyclic multimers for (a) $k = 3$ and (b) $k = 4$. In both cases the roughness parameter α is equal to 1/2. It is worth to note that the multimer (a) and other multimers with odd k do not show the center of symmetry. This is in contrast to the multimer (b) and other multimers of even k which show such a symmetry.

see Fig.3.

The 3-fold symmetry axis implies that for small deformations the elastic properties of the system do not depend on the direction [10], i.e. the system is elastically isotropic. As the system is isotropic, one needs only two elastic constants (and pressure) to describe its elastic properties [1]. This can be seen by expanding the system elastic energy per unit volume in powers of the (Lagrange) strain components ε_{ij}

$$E_{elast}/V_{ref} = -p(\varepsilon_{xx} + \varepsilon_{yy}) + 2\lambda_{\xi\eta\xi\eta}(\varepsilon_{xx} + \varepsilon_{yy})^2$$
$$+ \lambda_{\xi\xi\eta\eta}\left[(\varepsilon_{xx} - \varepsilon_{yy})^2 + 4\varepsilon_{xy}^2\right], \quad (1)$$

where V_{ref} is the two-dimensional 'volume' (area) of the reference state corresponding to the equilibrium state at the pressure p; the linear terms in the strain components come from the fact that the pressure can be different from zero, in general. The bulk modulus, B, and the shear modulus, μ, are related to the above defined elastic constants as follows [11]

$$B = 4\lambda_{\xi\eta\xi\eta}, \quad \mu = 2\lambda_{\xi\xi\eta\eta} - p. \quad (2)$$

As the stability of the system requires positive values of the bulk modulus and the shear modulus [1]

$$B > 0, \mu > 0, \quad (3)$$

the Poisson ratio v_P must fulfill the relation [12]

$$-1 \leq v_P \equiv \frac{B - \mu}{B + \mu} \leq 1. \quad (4)$$

The simplest way to determine the elastic properties of the static structures of the hard cyclic multimers is to replace the (non-analytic) hard potential through which the hard discs of different multimers interact, $u(r > \sigma) = 0, u(r < \sigma) = \infty$, by the limit of an analytic, n-inverse power interactions between the disc centers

$$u(r) = \lim_{n \to \infty} \left(\frac{\sigma}{r}\right)^n. \quad (5)$$

Within this approach one can restrict the interactions to the nearest neighboring discs of different multimers only

$$E_{tot} \equiv E_{elast} + E_{ref} = \sum_{1 \leq i < j \leq N} \sum_{k,l=n.n.} u(r_{k_i l_j}), \quad (6)$$

where $N \to \infty$ is the number of the multimers in the system, E_{ref} is the energy of the reference state, and the second summation on the right hand side concerns only the nearest-neighboring atoms k_i, l_j of the neighboring molecules i, j which are taken into account in the first summation.

The pressure, the bulk modulus and the shear modulus can be determined by differentiating the total energy per unit volume E_{tot}/V_{ref} with respect to the components of the strain tensor. Differentiation with respect to ε_{xx} (or ε_{yy}) at the reference state, gives the pressure

$$p = -\frac{1}{v_{ref}} \frac{\partial E}{\partial \varepsilon_{xx}}\Big|_{\varepsilon=0}, \quad (7)$$

where $\varepsilon = 0$ indicates that after the differentiation all the strain tensor components should be replaced by zero.

By double differentiation of the energy one obtains the elastic constants:

$$\frac{1}{v_{ref}} \frac{\partial^2 E}{\partial \varepsilon_{xx}^2}\Big|_{\varepsilon=0} = 4\lambda_{\xi\eta\xi\eta} + 2\lambda_{\xi\xi\eta\eta}, \quad (8)$$

$$\frac{1}{v_{ref}} \frac{\partial^2 E}{\partial \varepsilon_{xx} \partial \varepsilon_{yy}}\Big|_{\varepsilon=0} = 4\lambda_{\xi\eta\xi\eta} - 2\lambda_{\xi\xi\eta\eta}. \quad (9)$$

Combining (2), (4) with (7)-(9) one gets the Poisson ratio for the systems with even and odd k. The obtained dependences show that the Poisson ratio is negative when the roughness parameter, $\alpha = l/(2\sigma)$, is large [6].

Taking the limits $n, m \to \infty$ and using the definition of the roughness parameter one obtains the same dependence in both the cases considered above

$$v_P = \frac{1 - 2\alpha^2}{3 - 2\alpha^2}. \quad (10)$$

The obtained Poisson ratio dependence on the roughness parameter is shown in Fig. 4.

It can be seen that when the roughness parameter is zero the Poisson ratio is equal to 1/3 which is the value of the Poisson ratio of the static (i. e. zero temperature), close packed structure of hard discs. It can be also seen that the Poisson ratio decreases with increasing α, and for $\alpha > 2^{-1/2}$ the Poisson ratio is negative, reaching its minimum value -1 at the stability limit when $\alpha = 1$.

ROUGH DISC SYSTEM

In the above considerations the multimers which were constituted of $m = 3k + 1 \equiv m^{(1)}$ and $m = 3k + 2 \equiv m^{(2)}$

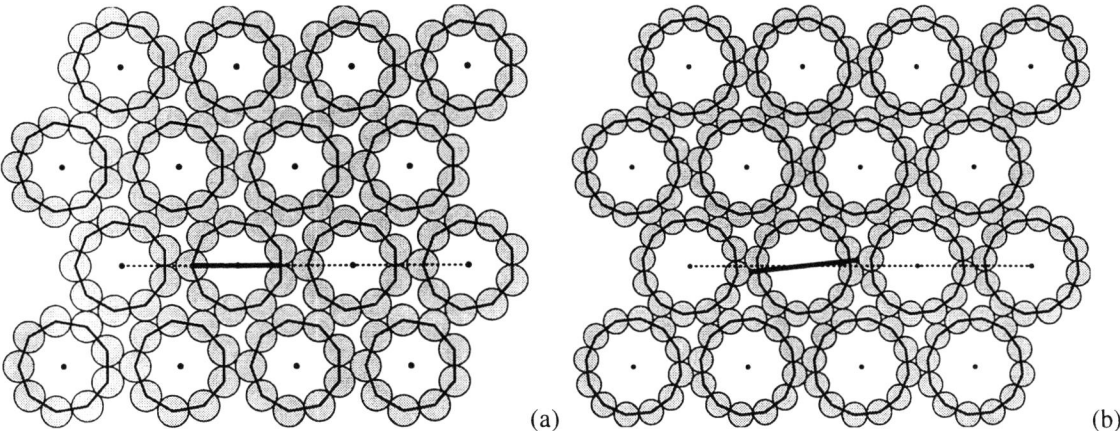

FIGURE 3. Geometry of the close packed structures of the hard cyclic multimers shown in Fig.1. It is worth to note that the structure shown in (a) exhibits the mirror symmetry with respect to the crystalline axes; the same is true for each structure with odd k. This is in contrast to structures formed by multimers with even k, including the structure illustrated in (b), which do not show such a symmetry, i.e. they are chiral.

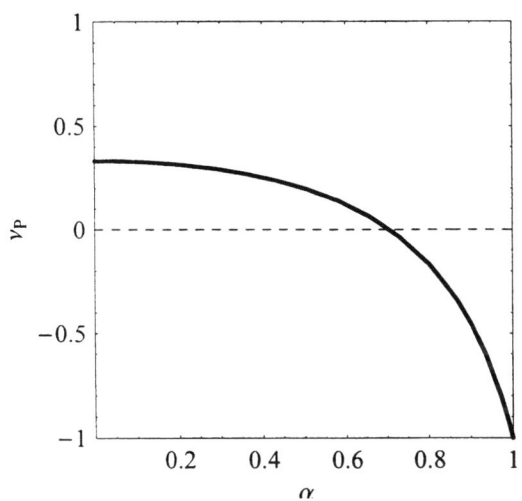

FIGURE 4. The Poisson ratio ν_P of the close packed rough disc system as the function of the roughness parameter α.

hard discs have not been taken into account. The reason is that the structure of close packed configurations formed by such multimers is not known and these structures are not elastically isotropic for finite m, in general (see e.g. the heptamers considered in the reference [8]). In the limit of $m \to \infty$ the multimer shape tends to the hard disc shape. In contrast to the standard hard discs, which are smooth, the limiting particles are *rough*. As the ratio of the thickness of the rough surface to the rough disc diameter tends to zero, the structure of their centers at close packing must tend to the triangular lattice.

We conjecture that the formula (10), which has been proved above for $m = 3k$, is valid also for static structures with $m^{(i)} \to \infty$, where $i = 1, 2$. In other words, we conjecture that the Poisson ratio of close packed structure of rough discs at the zero temperature depends on the roughness parameter in the way given by (10).

SIMULATIONS IN THE GENERALIZED CONSTANT PRESSURE ENSEMBLE

No analytic methods are known to the present Author which allow one to study quantitatively the elastic properties of the defined above hard cyclic multimers at positive temperatures. These systems can be studied, however, by computer simulations. One might add here that determination of the elastic properties of hard-body systems is not a trivial task, in general. There are several reasons why it is so. (a) No zero-temperature harmonic approximation can be used for systems of hard bodies below close packing density. (b) Microscopic formulae for the elastic constants in the constant strain ensembles include second derivatives of the hard-body interaction potential which are difficult to compute. (c) Standard computations in the constant stress (or thermodynamic tension) ensembles require precise knowledge of the equilibrium state which is used as the reference state in the strain-fluctuation methods. When the symmetry of the structure is not known, determination of the latter state can be, however, almost as time consuming as the computation of the elastic constants itself.

The mentioned difficulties can be overcome within the Parrinello-Rahman (variable-shape-box) generalization [13, 14] of the constant pressure ensemble (NpT) [15]. Computations performed in that ensemble do neither require any harmonic approximation nor microscopic formula for the elastic constants. Moreover, the reference state can be determined during the same run in which the elastic properties of the system are simulated [16, 8]. Although the NpT method is rather slowly convergent [8,17-19], it has been shown that for hard spheres [18] and hard discs [19] it gives results which are in a good agreement with those obtained by other, more elaborate methods.

Simplicity of the NpT method encouraged us to study elastic properties of the hard multimer systems. The simulations have proven the existence of thermodynamically stable *phases* that exhibit negative Poisson ratio. Rough molecules have been found for which such phases exist in the whole stability range of the solid, from the close packing up to melting. The results of the simulations will be published elsewhere [9].

CONCLUSIONS

The hard cyclic multimer system discussed in this paper shows that very simple molecular interactions can be used to model rather unusual effects like negative Poisson ratio. The hard cyclic multimer model not only increases the amount of known mechanisms [2-5,10-12,20-26] which can lead to $v_p < 0$ but also offers very simple microscopic examples of chiral and non-chiral structures which properties can be studied by generalizations of the elasticity theory which include the orientational degrees of freedom of the elastic medium [27-29].

In the limit of infinitely many atoms forming the hard cyclic multimers, the latter molecules can be thought of as the rough discs. The analysis presented in this paper indicates that the Poisson ratio of static structures of the rough discs can be uniquely characterized by the roughness parameter.

Computer simulations performed for the hard cyclic multimers prove that it is possible to construct very simple model molecules which form thermodynamically stable solid phases exhibiting the negative Poisson ratio in a broad range of the thermodynamic parameters [9].

In three dimensions, analogous considerations can be done for globular multimers, leading to the notion of rough spheres. This will be the subject of separate works.

ACKNOWLEDGMENTS

The Author thanks Dr. A. Alderson for careful reading of the manuscript. Part of this work was performed at the Poznań Computer and Networking Center (PCSS) in the framework of the Polish Committee for Scientific Research grant 4T11F 010 23. This work was also partially supported by the Centre of Excellence "Magnetic and Molecular Materials for Future Electronics" within the European Commission contract No. G5MA-CT-2002-04049.

REFERENCES

1. Landau, L. D., Lifshits, E. M., Kosevich, A. M., and Pitaevskii, I. P., *Theory of Elasticity*, Pergamon Press, London, 1986 pp. 1-37.
2. Lakes, R., Science **235**, 1038-1040 (1987).
3. Caddock, B. D., and Evans, K. E., J. Phys. D: Appl Phys. **22**, 1877-1882 (1989).
4. Lakes R., Advanced Materials **5**, 293-296 (1993).
5. Evans, K. E., and Alderson, A., Advanced Materials **12**, 617-628 (2000).
6. Wojciechowski, K. W., unpublished.
7. Allen, M. P., Evans, G. T., Frenkel, D., and B. M. Mulder, Adv. Chem. Phys. LXXXVI, 1-166 (1993).
8. Wojciechowski, K. W., Tretiakov, K. V., and Kowalik, M., Phys. Rev. E**67**, 036121 1-14 (2003).
9. Wojciechowski, K. W., and Tretiakov, K. V., unpublished.
10. Wojciechowski, K. W., J. Phys. Soc. Japan **76** 1819-1820 (2003).
11. Wojciechowski, K. W., J. Phys. A: Math. Gen. **36**, 11765-11778 (2003).
12. Wojciechowski, K. W., Phys. Lett. A**137**, 60-64 (1989).
13. Parrinello, M., and Rahman, A., J. Appl. Phys. **52**, 7182-7190 (1981).
14. Parrinello, M., and Rahman, A., J. Chem. Phys. ?**76**, 2662-2668 (1982).
15. Wojciechowski, K. W., and Tretiakov, K. V., Computer Phys. Commun. **121-122**, 528-530 (1999).
16. Wojciechowski, K. W., Computational Methods in Science and Technology **8**, 77-83 (2002).
17. Tretiakov, K. V., and Wojciechowski, K. W., J. Phys.: Cond. Matter **14**, 1261-1273 (2002).
18. Wojciechowski, K. W., and Tretiakov, K. V., Computational Methods in Science and Technology **8**, 84-92 (2002).
19. Wojciechowski, K. W., Tretiakov, K. V., Brańka, A. C., and Kowalik, M., J. Chem. Phys. **119**, 939-946 (2003).
20. Almgren, R. F., J. Elasticity **15**, 427-430 (1985).
21. Kolpakov, A. G., Prikl. Matem. Mekh., **49**, 969-977 (1985).
22. Milton, G. W., J. Mech. Phys. Solids, **40**, 1105-1137 (1992).
23. Wojciechowski, K. W., Mol. Phys. Reports **10**, 129-136 (1995).
24. Novikov, V. V., and Wojciechowski, K. W., Phys. Solid State **41**, 1970-1975 (1999).
25. Grima, J. N., and Evans, K. E., Journal of Material Science Letters **19**, 1563-1565 (2000).
26. Ishibashi, Y., and Iwata, M., J. Phys. Soc. Jpn. **69**, 2702-2703, (2000).
27. Eringen, E. A. C., *Polar and Nonlocal Field Theories, vol. IV of Continuum Physics*, Academic Press, New York, 1976.
28. Vasiliev, A. A., and Dmitriev, S. V., and Ishibashi, Y., and Shigenari, T., Phys. Rev. B**65**, 094101 1-7 (2002).
29. Dmitriev, S.V., Vasiliev, A.A., Miroshnichenko, A.E., Shigenari, T., Liu, Y., Kagawa, Y., and Ishibashi, Y., arXiv: cond-mat/0209386.

Temperature Dependence of the Ultrafast Solvation of a Dye Molecule in Alcohol

Hiroshi Murakami and Momoko Tanaka

Japan Atomic Energy Research Institute, Umemidai 8-1, Kizucho, Sorakugun, Kyoto 619-0215, Japan

Abstract. We have performed femtosecond time-resolved fluorescence spectroscopy up to 10 ns for a dye molecule in a alcohol mixture at 170 K and 296 K. It has been found that the energy relaxation occurs in the electronic excited state of the molecule owing to the structural dynamics of the surroundings of the molecule with time constants of the order of 0.1 ps, 1 ps and a slower one, and the temperature dependence of it implies that there is some connection among the relaxation components with different time constants.

INTRODUCTION

The structural dynamics of complex systems such as liquid and polymers have been an issue of the condensed matter physics for several decades [1]. The structural dynamics (relaxation) on an ultrafast time scale can be probed in real time by femtosecond fluorescence time-resolved (FTRF) spectroscopy [2]. As shown for a molecule in complex systems schematically in Fig.1, the surroundings of the molecule have different configurations, corresponding to the energy minima displaced, according to the electronic state of the molecule, and so the energy relaxation occurs owing to the configurational change of the surroundings in the electronic excited state by light irradiation. Since the molecule in the electronic excited state emits fluorescence whose lifetime is of the order of ns for dye molecules, the energy relaxation is investigated from fs to ns through the peak-energy shift of the FTRF spectrum. Hence, low-frequency vibrational relaxation, the so-called fast beta-process and the alpha-relaxation are expected to be investigated from the FTRF spectrum. Most of the measurements, however, were made only at room temperature with a liquid-flow optical-cell system in order to prevent the molecule from damage due to dense light excitation. One method to study the mechanism of the relaxation in detail is to measure the FTRF spectrum as a function of temperature. We have recently developed a liquid nitrogen flow-type cryostat system which is free from such damage at low temperatures [3]. Thus, in the present study, we have performed FTRF spectroscopy for coumarin 540A in an ethanol/methanol mixture (glass-transition temperature:130 K) with a voluminal ratio of 4:1 (C540A/EtMt), under excitation at 410 nm around the absorption peak of the sample, at 170 K and 296 K.

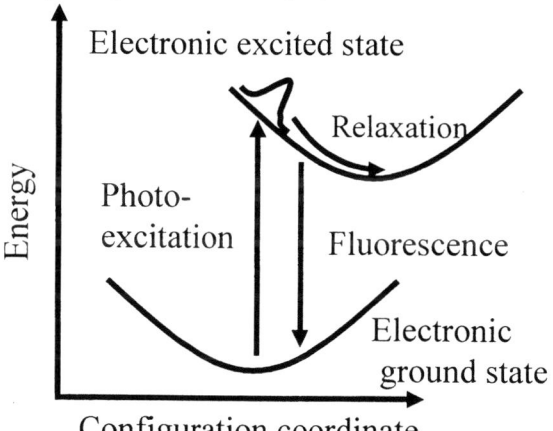

FIGURE 1. A schematic energy diagram of a molecule in complex systems.

The time profiles of the peak-energy of the FTRF spectrum of C540A/EtMt at 296 K and 170 K are depicted as circles and triangles, respectively, in Fig.2, where the peak energy shift is normalized by $(E_p(t)-E_p(t=\infty))/(E_p(t=0)-E_p(t=\infty))$ (E_p: the peak energy of the FTRF spectrum). The FTRF spectrum shifts to the low-energy side with time at the two temperatures, while the time profile at 296 K is different from that at 170 K. The time profile at 296 K is fitted using a sum of three exponentials with time constants of the order of 0.1 ps, 1 ps and 10 ps, while done not by a sum of three exponentials but rather by a sum of two exponentials with time constants of the order of 0.1 ps and 1 ps and a stretched exponential with a time constant of 1ns and a stretching exponent of 0.4 at 170 K. Hereafter, we refer to the three relaxation components in order of increasing time constant as I, II, III, where component III whose time constant depends on temperature is considered to be attributed to the diffusion-like motion of the solvent molecules, as shown by a sub-ns fluorescence spectroscopy [4], and the time profile of the stretched-exponential will come from that the system is in the supercooled state at 170 K. On the other hand, components I and II, from the viewpoint of the time scales, correspond to the vibrational motions of the matrix and fast beta process, respectively. Further, the value of the pre-exponential factor of the exponentials of components I and II has been found to decrease considerably, especially for II, at 170 K, while that of component III increases. Therefore, the difference between the time profiles at the two temperatures is due to the changes in the magnitude of the relaxation components and in the time profile of component III. The decrease in the magnitude of components I and II is compensated by the increase in that of component III at 170 K. If the relaxation components with different time scales are independent of each other, the magnitude of them should not change irrespective of temperature. Therefore, these relaxation components are not regarded to be independent of each other.

Hydrogen-bonded networks are formed in the solvent of C540A/EtMt, and the strength of such a network will become large considerably in the supercooled state, compared with that in the liquid state. The motions responsible for components I and II will be more confined under the strong hydrogen-bonded network at 170 K, while the motion for component III will become cooperative owing to the network and show nonexponential relaxation. In fact, the peak-energy shift of the FTRF spectrum obtained from the preliminary result at 230 K was able to be fitted using three exponentials, where the values of the pre-exponential factors of components I, II, and III hardly change from those at 296 K, respectively, and the time constant of component III becomes larger at 230 K.

A model of hierarchically constrained dynamics (HCD) has been proposed in order to explain the nonexponential relaxation observed widely in complex systems [5]. It is assumed in the HCD that the relaxation of a certain group of atoms in a system becomes possible only when some groups of atoms with a smaller size attain the right configuration and such a process occurs hierarchically for groups of atoms with various sizes. Actually, reverse constraints will also appear, that is, the degrees of freedom with a large spatial scale constrain those with a small one. Accordingly, it seems to be natural that the fast motions (I, II) are constrained by the slow motion (III) in C540A/EtMt. We have discussed the temperature dependence of the HCD to explain the energy relaxation of the electronic excited state of a dye molecule in a polymer and protein [6] and considered that an abrupt descent of the barrier height for a given constraint occurs when the constraint is released at a certain temperature. According to this idea, the fast relaxations are hampered by an energy barrier which is responsible for the slow one, on the way at 170 K, while the height of the energy barrier is too low to hamper the fast relaxations at 296 K.

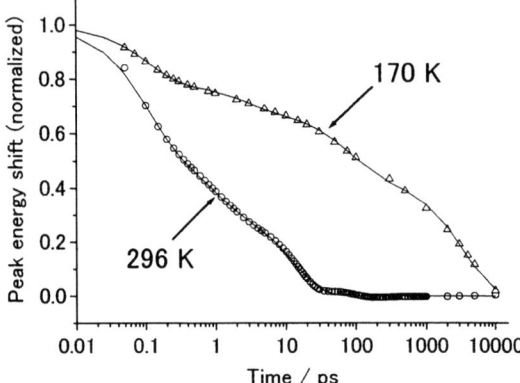

FIGURE 2. The time profile of the peak-energy shift of the FTRF spectrum of C540A/EtMt. (see the text for the details of the notation).

REFERENCES

1. See, for example, Donth, E., *The Glass Transition: Relaxation Dynamics in Liquids and Disordered materials*, Springer, Berlin, 2001.

2. Horng, M. L., Gardecki, J. A., Papazyan, A., and Maroncelli, M., *J. Phys. Chem.* **99**, 17311-17337 (1995); Barbara, P. F., and Jarzeba, W., *Adv. Photochem.* **15**, 1-68 (1990).

3. Murakami, H., *J. Lumin.* **102-103**, 295-300 (2003); Murakami, H., *J. Mol. Liq.* **89**, 33-45 (2000).

4. Kinoshita, S., and Nishi, N., *J. Chem. Phys.* **89**, 6612-6622 (1988).

5. Palmer, R. G., Stein, D. L., Abrahams, E., and Anderson, P. W., *Phys. Rev. Lett.* **53**, 958-961 (1984).

6. Murakami, H., Kushida, T., and Tashiro, H., *J. Chem. Phys.* **108**, 10309-10318 (1998); Murakami, H., and Kushida, T., *Phys. Rev.* **B54**, 978-989 (1996).

Dynamical Susceptibility close to a critical point in $Sr_3Ru_2O_7$

R. A. Borzi*, S. A. Grigera* and A. P. Mackenzie*

*School of Physics and Astronomy, University of St. Andrews, KY16 9SS St. Andrews, UK.

Abstract. In this paper we explore the possibilities given by the ac-susceptibility $\chi(\omega)$, to gain information about the dynamical response of a system in the critical region. $Sr_3Ru_2O_7$ is known to be a strongly correlated metal with an itinerant metamagnetic first order transition, that can also be associated with a new type of quantum critical point. We study the $\chi(\omega)$ of this system and show that the (tuneable) finite characteristic time associated with the measurement allows us, in addition to determining the experimental H,T phase diagram, to get information about the dynamics of this system in the critical region.

Even though first order phase transitions are a commonplace occurrence, the dynamics of such transitions remains an issue which is still not fully resolved[1]. Itinerant metamagnetism provides a very interesting test-case where many of these issues can be studied in a controlled way. The presence of a symmetry breaking field conjugated to the order parameter makes the metamagnetic transition formally equivalent to the liquid to vapour transition in water. An important advantage from the experimental point of view is that magnetic systems have easily controllable external parameters, such as the magnetic field, and dynamic and thermodynamic quantities that can be measured easily.

It is in this context, then, that $Sr_3Ru_2O_7$ is a particularly good system for studying. We have shown[2, 3, 4] that it has an anisotropic metamagnetic transition consisting of a line of first order transition ending at a critical point at a temperature T^*. We also showed[3, 4] that this last can be tuned down to $T = 0$ using the angle between the field and the crystallographic axis, creating in this way a quantum critical point. The determination of the thermodynamic phase diagram of this system was done using a combination of results from magnetisation and, in a greater proportion, from measurements of the ac-susceptibility at very low temperatures ($T<1K$).

The measurement of ac-susceptibility has several advantages over the static techniques. From a practical point, it is very easy to implement, even when measuring in the mK range of temperatures. Its main advantage however lies in the fact that the complex susceptibility contains a considerable amount of information about the dynamics the system. From the out-of-phase component, for example, it is possible to learn about dissipative processes occurring at the transition. The magnitude of each of the components of the ac-susceptibility depends on the characteristic time of response of the system as compared

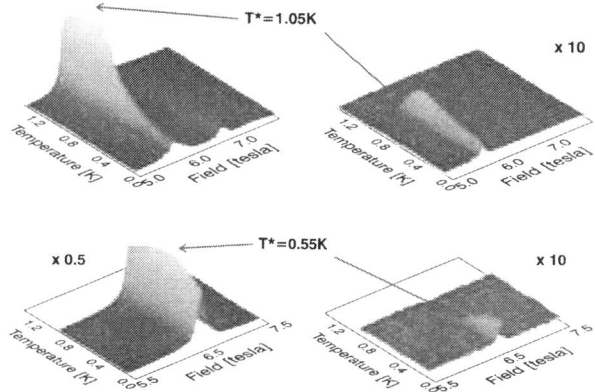

FIGURE 1. Real (left) and imaginary (right) components of the ac-susceptibility for two orientations of the magnetic field with respect to the *ab*-plane: 40° (top) and 60° (bottom)

to the characteristic time of the measurement, which in this case is inversely proportional to the excitation frequency. In this way, by changing the frequency of the excitation field, it is possible to extract information about the characteristic times of the system. In this paper we give a brief example by performing such a study in the vicinity of the metamagnetic critical point in $Sr_3Ru_2O_7$.

Figure 1 shows the real and imaginary components of the susceptibility at two different angles of the magnetic field (40° (top) and 60° from the *ab*-plane). We can see that the in-phase component displays an abrupt maximum as a function of field, related to the metamagnetic transition. The maximum peak height, in turn, displays a non monotonous behaviour, forming itself a wide peak as a function of temperature, with an absolute maximum at a temperature T^* which depends on the angle. As noted by the arrows in the figure, it is precisely below T^* that a signal appears in the out-of-phase component: a ridge

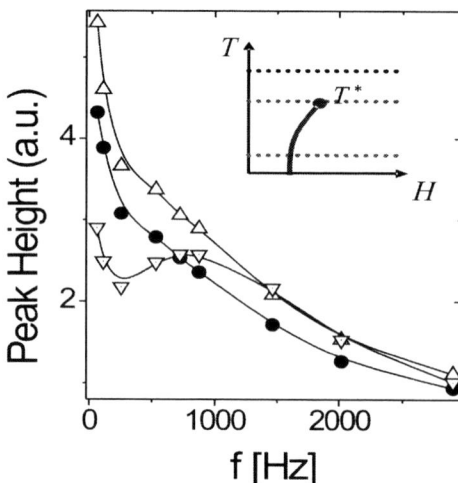

FIGURE 2. Value of the absolute maximum of χ' as a function of the excitation frequency at three different temperatures (above, below and at T^*): $T = 100mK$ inverted triangles, $T = T^* = 700mK$ open triangles, and $T = 1000mK$ black circles. The field is at 56° from the ab-plane. Inset: schematical phase diagram showing the phase diagram. The dotted lines correspond to the temperatures where each of the measurements has been done.

associated with dissipation. This behaviour can be understood in terms of a simple phase diagram: a line of first order transitions ending at a critical point at T^* (see the inset of Fig. 2). The location of the critical point is determined by the position of the absolute maximum in the real part, and the ridge of dissipation corresponds to hysteresis at the line of first order transitions. The role of the angle is simply to move T^* towards lower temperatures as the field is turned from the ab-plane to the c-axis.

When dealing with a metamagnetic transition, we expect the static susceptibility (i.e. dM/dH) to diverge at the metamagnetic field. In a real situation, demagnetisation effects would affect the measured susceptibility, which tend to saturate at the inverse of the demagnetisation value, while a mixed phase state is stabilized. The crossover range of temperatures would be characterised by the absence of this coexistence region. If these statements are true, we are then left with the problem of explaining the absence of a divergence in the data.

The existence of a *maximum* rather than a divergence in χ' can be understood taking into account the finite characteristic time of our measurements. The response of the system increases when approaching T^* from above, as the correlation lengths diverges. However, at the same time the characteristic time diverges, and the system is not able to fully respond during the time window of the measurement, which cuts down the signal of the susceptibility. At temperatures lower than T^*, the transition becomes first order. In this regime a different –and slower– dynamics drives the system, which may involve domain wall motion. This last phenomenon may also be made even slower due to the very low temperatures involved and the presence of some disorder in our samples, which may act as pinning centres for domain walls.

Figure 2 shows the frequency dependence of the height of the absolute maximum of the in-phase part of the susceptibility as a function of frequency for a particular angle, for temperatures above and below T^*. It can be seen that, irrespective of the temperature, there is a steep increase on this curve at low frequencies –including the 80 Hz at which the curves on Fig. 1 have been measured. Again, there is a clear distinction in the curves behaviour above and below T^*. The monotonous decrease in response with frequency for temperatures above T^* reflects the fact that the system is less able to follow the driving field when the measuring time decreases. At temperatures lower than T^*, where a first order occurs, we find a different behaviour, characterised by the presence of a local maximum at finite frequencies. This non-monotonous behaviour may be explained in terms of a resonant coupling of the driving ac-field with pinned domain walls.

In summary we have shown that the ac-susceptibility can be used as a valuable tool to gain insight into the dynamical behaviour of a magnetic system close to a critical point. We have shown that in such circumstances, relaxation times can be so long that even at frequencies which are usually considered to be in the static limit the system shows a strong frequency dependence. In the case of first order phase transitions, a careful analysis of this frequency dependence can give information about the formation and dynamics of domain walls.

ACKNOWLEDGMENTS

We would like to acknowlege discussions with S. R. Julian, G. G. Lonzarich, Y. Maeno, A. J. Millis, A. J. Schofield, The crystals where this study was performed were provided by R. S. Perry and Y. Maeno. This work was partially supported by the Engineering and Physics Research Council (UK) and the Royal Society (UK).

REFERENCES

1. Onuki, A., *Phase transition dynamics*, Cambridge University Press, Cambridge, 2002, pp. 371-700.
2. Perry R. S., *et al.*, *Phys. Rev. Lett.* **86**, 2661-2664 (2001).
3. Grigera S. A., *et al.*, *Science* **294**, 329-332 (2001).
4. Grigera S. A., *et al.*, *Phys. Rev. B* **67**, 214427-01-08 (2003).

Deformations of adhering elastic tubes

K. Tamura*, S. Komura* and T. Kato*

*Department of Chemistry, Tokyo Metropolitan University, Tokyo 192-0397, Japan

Abstract. Deformation of an elastic tube adhering onto a substrate due to van der Waals attractive interaction is investigated by means of computer simulation and scaling theory. The sum of the stretching, bending, and van der Waals energies of the tube is numerically minimized using the conjugate gradient method. The onset of the deformation and the total energy can be scaled with a variable C_b/N^2, where C_b is the bending constant and N the size of the tube. For a significantly deformed tube, the scaling relation between the bending energy and the bending constant is explained within the shell theory.

INTRODUCTION

Carbon nanotubes have attracted great interests due not only to their peculiar structure, but also to the electrical, chemical, and mechanical properties associated with these structures. Examples of the possible applications are such as nanowires or electronic devices. The electric transport through nanotubes is studied after their deposition on a substrate with which they interact. It is known, however, that the resistivity of the nanotube is affected by their elastic deformations. Since there is little control over the alignment and the shape of adsorbed nanotubes, it is crucial to know how they deform on the substrate. There are several works which report on the deformations of nanotubes due to the van der Waals (vdW) interaction. Multiwalled nanotubes can even fully collapse along their length [1, 2].

In this paper, we theoretically investigate the deformation of an elastic nanotube adhering onto a rigid substrate due to the vdW attractive interaction.

MODEL AND RESULTS

Consider a cross section of an elastic tube interacting with a rigid substrate. We assume that the axial deformation is uniform along the tube. The elastic tube is modeled by a circular network of N beads connected by harmonic springs. The adhesion energy is taken into account through the vdW interaction between each of the bead and the substrate.

The discretized stretching energy is given by the sum over Hooke's law of each spring:

$$E_s = \sum_i \frac{1}{2} C_s \left(\frac{L_i - L_0}{L_0} \right)^2. \qquad (1)$$

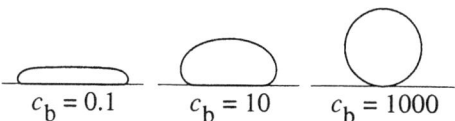

FIGURE 1. Equilibrium configurations of the deformed tubes for various values of the scaled bending constant $c_b = C_b/\varepsilon$. The tube globally flattens for $c_b = 0.1$, but hardly deforms for $c_b = 1000$.

Here, C_s is the spring constant, L_i is the length of the i-th spring, and L_0 is the natural length of the spring taken here as a constant. The discretized bending energy, on the other hand, is calculated by

$$E_b = \sum_{\langle ij \rangle} \frac{1}{2} C_b |\hat{n}_i - \hat{n}_j|^2, \qquad (2)$$

where C_b is the bending constant, $\hat{n}_i$ is the unit normal vector of the i-th spring, and the sum is taken over each pair of springs which share a common bead. The adhesion energy of the tube is taken into account through the vdW interaction between each of the bead and the substrate:

$$W = \sum_i \frac{2^{8/3}}{3} \varepsilon \left[\left(\frac{\sigma}{z_i} \right)^{12} - \left(\frac{\sigma}{z_i} \right)^3 \right], \qquad (3)$$

where z_i is the height of the i-th bead from the substrate. When the adhesion energy per bead is plotted against z_i, ε corresponds to the depth of the energy minimum.

The total energy $E_{tot} = E_s + E_b + W$ is numerically minimized in the computer using the conjugate gradient method. As for the initial condition of the simulation,

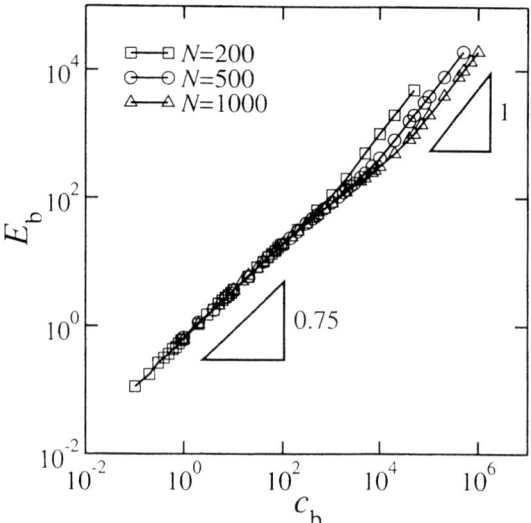

FIGURE 2. The bending energy E_b as a function of c_b for size $N = 200, 500, 1000$. We see two scaling behaviors: $E_b \sim c_b/N$ and $E_b \sim c_b^{0.75}$ for large and small c_b, respectively.

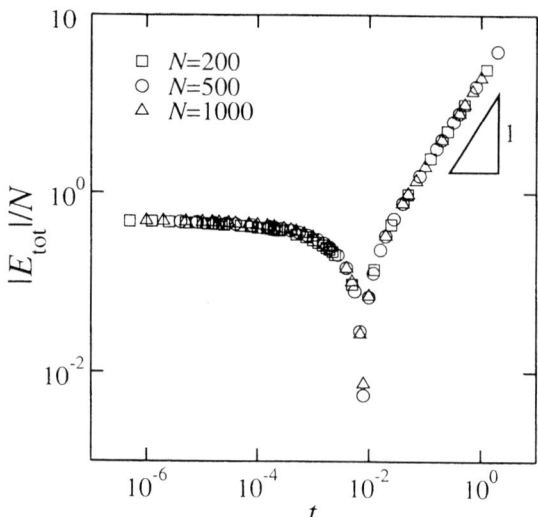

FIGURE 3. The absolute value of the total energy per bead $|E_{tot}|/N$ as a function of the scaling variable $t = c_b/N^2$. The scaling $E_{tot}/N \sim t$ holds in the "bending regime", whereas E_{tot}/N asymptotically approaches to -1 in the "adhesion regime".

each bead is located on a circle with a distance being equal to the natural length of the spring L_0. Since there is no spontaneous curvature in our model, even the undeformed tubes cost certain curvature energy.

In equilibrium, each spring relaxes almost at its natural length. For a large bending constant such as $c_b = 1000$, the tube hardly deforms and keeps its circular shape in spite of the adhesion. As c_b is reduced to $c_b = 10$, a considerable deformation occurs and the contact area (line) increases significantly. Further decrease of c_b results in a configuration such as $c_b = 0.1$ in Fig. 1. Here a flattening of the tube is observed, and the curvature is localized at the regions close to the contact line.

For almost undeformed tubes on which the curvature is uniformly distributed, the bending energy can be readily estimated. Since the radius of curvature is proportional to the number of beads N in such a case, the bending energy becomes $E_b \sim N(C_b/N^2) \sim C_b/N$. This linear dependence of E_b on C_b in the undeformed region can be checked in Fig. 2 where we have plotted E_b against C_b. For smaller c_b, on the other hand, all the data collapses on a single straight line regardless of the tube size N. In this strongly deformed region, we observe a nontrivial scaling behavior, i.e., $E_b \sim c_b^{0.75}$. This nontrivial scaling can be understood within the shell theory.

When a tube of radius R is largely deformed on a substrate, a contact area develops. For such a deformation, the elastic energy is localized in the two parallel straight strips near the edge of the bulge. We apply the argument of Ref. [3] for a tube, and the total elastic energy becomes

$$E_b \sim \left(\frac{C_s}{L_0^2}\right)^{1/4} C_b^{3/4} \frac{H}{R^{3/2}}, \quad (4)$$

where H is the depth of the bulge being fixed and given. The scaling relation Eq. (4) accounts for the dependence of E_b on C_b as seen in Fig. 2. Notice that, in our simulation results, the contribution of the equilibrated stretching energy E_s is negligibly small compared to that of other energies, i.e., $E_s + E_b \approx E_b$.

In Fig. 3, we have plotted the absolute value of the total energy per bead $|E_{tot}|/N$ as a function of $t = c_b/N^2$ for three different tube sizes. Notice that E_{tot} can take negative value due to the vdW interaction when the tube is strongly adsorbed. It is remarkable that all the data collapse onto a single curve irrespective of the tube size N. Hence the total energy can be scaled with c_b/N^2.

REFERENCES

1. Yu, M.-F., Kowalewski, T., and Ruoff, R. S., *Phys. Rev. Lett.* **86**, 87-90 (2000).
2. Chopra, N. G., Benedict, L. X., Crespi, V. H., Cohen, M. L., Louie, S. G., and Zettl, A., *Nature* **377**, 135-138 (1995).
3. Landau, E. D., and Lifshitz, E. M., *Theory of Elasticity*, Pergamon Press, Oxford, 1986, pp. 54-57.

Exact Renormalization Group for the Brazovskii Model

Y. Shiwa

Statistical Mechanics Laboratory, Kyoto Institute of Technology,
Matsugasaki, Sakyo-ku, Kyoto 606-8585, Japan

Abstract. We consider the coarse graining of the generalized Brazovskii free energy functional for striped patterns. The technique developed by Shankar for the Fermi liquids is combined with the irreducible version of the exact renormalization group to calculate the recursion relations for interaction vertices. We perform the one-loop calculations from this method taking the eight-point vertex into account.

The sytems that undergo transitions from an isotropic, disordered phase to a nonuniform, spatially periodic phase are commonly referred to as Brazovskii systems [1]. An essential feature of these systems is that the ordered phase is described by the spatial period $2\pi/k_0$ and the fluctuation spectrum has a minimum at nonzero wave vectors $\bm{k}$ with $|\bm{k}| = k_0$, represented by a hypersphere in reciprocal space. Because of the large phase space for one-dimensional fluctuations in the direction transverse to the hypersphere, the Brazovskii systems are quite distinctive in comparison with the usual systems, where the periodic structure is determined by isolated points in reciprocal space and consequently the phase volume of fluctuations is small.

In order to deal with the large fluctuations in the vicinity of a shell of nonzero wave vectors, the momentum-shell renormalization-group (RG) theory has been worked out by Shankar [2]. It turns out [3] that all interaction parameters are relevant, and that, within a $(\phi^2)^3$ theory, the recursion relations cannot be integrated to obtain the bulk (thermodynamic) quantities in certain parameters region. Thus the often successful RG techniques simply fail for the Brazovskii class.

Confronted with the above problems, we here present an exact RG equation for the generating functional (Γ) of the one-particle-irreducible correlation functions. Since the equation describes the scale dependence of the free energy functional, the physical properties at a given length or momentum scale can be computed.

The Brazovskii model is a scalar field theory with a bare action

$$S\{\phi\} = S_0\{\phi\} + S_{int}\{\phi\}, \quad (1)$$

where the free part is given by

$$S_0\{\phi\} = \frac{1}{2} \int_{|k-k_0|<\Lambda_0} \frac{d\bm{k}}{(2\pi)^d} \phi_{\bm{k}} G_0^{-1}(\bm{k}) \phi_{-\bm{k}}, \quad (2)$$

$\phi_{\bm{k}} = \int d\bm{r} e^{-i\bm{k}\cdot\bm{r}} \phi(\bm{r})$ being the Fourier transform of an order parameter $\phi(\bm{r})$ in d dimensions, and

$$G_0^{-1}(\bm{k}) = r_0 + \xi_0^2 (k - k_0)^2, \quad (3)$$

with $k \equiv |\bm{k}|$. We assume that initially the order-parameter fields $\phi_{\bm{k}}$ with large momenta $|k - k_0| \geq \Lambda_0$ have been integrated out. The interaction part $S_{int}\{\phi\}$ is a local function of the fields that is invariant under $\phi \to -\phi$.

The first step of the exact RG method [4, 5] employs the coarse graining procedure whereby the fluctuating degrees of freedom with wave vectors in the range $\Lambda < |k - k_0| < \Lambda_0$ are averaged over to obtain the effective Γ, denoted as $\Gamma^{\Lambda,\Lambda_0}\{\varphi\}$, for modes with $|k - k_0| \leq \Lambda$. (The field $\varphi(\bm{r})$ is the expectation value of $\phi(\bm{r})$ in the ground state of the source dependent action.) The exact flow equation can be derived as formal identity from the functional integral which defines this coarse graining, and it is cast in the form of functional differential equations [6]. In this way we obtain

$$\partial_\Lambda \Gamma^{\Lambda,\Lambda_0}$$
$$= -\frac{1}{2} \int_{\bm{k}} \frac{\delta(\Omega_k - \Lambda)}{r_0 + \xi_0^2 \Lambda^2 + \Sigma^{\Lambda,\Lambda_0}(\bm{k})}$$
$$\times \left[\mathcal{U}^{\Lambda,\Lambda_0} \left(1 + G^{\Lambda,\Lambda_0} \mathcal{U}^{\Lambda,\Lambda_0}\right)^{-1} \right]_{\bm{k},-\bm{k}}$$
$$- \frac{V}{2} \int_{\bm{k}} \delta(\Omega_k - \Lambda) \ln\left[\frac{r_0 + \xi_0^2 \Lambda^2 + \Sigma^{\Lambda,\Lambda_0}(\bm{k})}{r_0 + \xi_0^2 \Lambda^2}\right], \quad (4)$$

where $\int_{\bm{k}} \equiv (2\pi)^{-d} \int d\bm{k}$, V being the volume of the system. Here $[\mathcal{O}]_{\bm{k},\bm{k}'}$ denotes the matrix element of $\mathcal{O}$ in $\bm{k}$-space, and $\Sigma^{\Lambda,\Lambda_0}$ is defined as the field independent part of the second functional derivative of $\Gamma^{\Lambda,\Lambda_0}$:

$$\delta^2 \Gamma^{\Lambda,\Lambda_0}\{\varphi\}/\delta\varphi_{\bm{k}}\delta\varphi_{\bm{k}'}$$
$$= (2\pi)^d \delta(\bm{k}+\bm{k}') \Sigma^{\Lambda,\Lambda_0}(\bm{k}) + \mathcal{U}^{\Lambda,\Lambda_0}_{\bm{k},\bm{k}'}\{\varphi\}.$$

The interacting propagator G^{Λ,Λ_0} is related to the non-interacting one via $(G^{\Lambda,\Lambda_0})^{-1} = (G_0^{\Lambda,\Lambda_0})^{-1} + \Sigma^{\Lambda,\Lambda_0}$, where

$$[G_0^{\Lambda,\Lambda_0}]_{\boldsymbol{k},\boldsymbol{k}'} = (2\pi)^d \delta(\boldsymbol{k}+\boldsymbol{k}') \frac{\theta(\Omega_k - \Lambda) - \theta(\Omega_k - \Lambda_0)}{r_0 + \xi_0^2 (k-k_0)^2},$$

with $\Omega_k \equiv |k - k_0|$, $\theta(x)$ being the Heaviside function.

We expand the generating functional in powers of φ,

$$\Gamma^{\Lambda,\Lambda_0}\{\varphi\} = \Gamma^{(0)}_{\Lambda,\Lambda_0}$$
$$+ \sum_{n=1}^{\infty} \frac{1}{n!} \int_{\boldsymbol{k}_1} \cdots \int_{\boldsymbol{k}_n} (2\pi)^d \delta(\boldsymbol{k}_1 + \cdots + \boldsymbol{k}_n)$$
$$\times \Gamma^{(n)}_{\Lambda,\Lambda_0}(\boldsymbol{k}_1,\cdots,\boldsymbol{k}_n) \varphi_{\boldsymbol{k}_1} \cdots \varphi_{\boldsymbol{k}_n}, \quad (5)$$

which defines the irreducible n-point vertex functions $\Gamma^{(n)}(\boldsymbol{k}_1,\cdots,\boldsymbol{k}_n)$. Note that only the even-point vertices are non-zero. Identifying the terms with the same powers of φ's on both sides of Eq. (4), we obtain an infinite hierarchy of flow equations for $\Gamma^{(n)}$s.

After the elimination of degrees of freedom, we must put the coarse-grained generating functional $\Gamma^{\Lambda,\Lambda_0}$ into the same form as the original Γ to complete the RG transformation. This can be achieved by rescaling the momenta and fields:

$$k - k_0 = \Lambda q, \quad \varphi_{\boldsymbol{k}} = \Lambda^{-3/2} \bar{\varphi}_{\boldsymbol{Q}}. \quad (6)$$

We have labeled the momentum $\boldsymbol{k}$ by its direction $\hat{\boldsymbol{n}} = \boldsymbol{k}/k$ and by the dimensionless variable $q \equiv (k-k_0)/\Lambda$, so that we now use the notation $\boldsymbol{Q} \equiv (\hat{\boldsymbol{n}}, q)$ instead of $\boldsymbol{k}$. Substituting the above rescaled variables into Eq. (4) and using Eq. (5), we obtain the flow equation for the rescaled vertices.

Unfortunately, hindered by the complexity of the functional differential equations thus obtained, the practical application of the exact RG method has not yet been successful. In order to proceed further in this paper, we resort to truncation of the hierarchical set of flow equations. Let us set $\Gamma^{(n)} = 0$ for $n \geq 10$, and perform the one-loop calculation (Fig. 1).

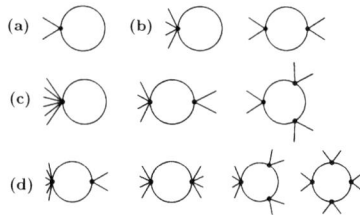

FIGURE 1. One-loop diagrams contributing to irreducible vertex functions; diagrams contributing to $\Gamma^{(2)}$ (a), $\Gamma^{(4)}$ (b), $\Gamma^{(6)}$ (c), and to $\Gamma^{(8)}$ (d).

Due to the argument given in Ref. [3], we know that only a part of the n-point vertex functions ($n > 2$) whose wave vectors are equal and opposite in pairs with their magnitude equal to k_0 survives the iterated application of RG transformations. Furthermore, we replace the vertex functions in the flow equations by the angle-averaged quantities. The recursion relation for these vertex "constants" now follows. The recursion equations are rather lengthy, and will be given elsewhere. We remark that if we set $\Gamma^{(8)} = 0$, the resulting recursion equations are exactly of the same form as obtained in Ref. [3].

We have solved the recursion equations numerically with the initial conditions

$$\Gamma_R^{(2)} = \tau, \ \Gamma_R^{(4)} = 1 \text{ and } \Gamma_R^{(m)} = 0 \text{ for } m = 6, 8 \quad (7)$$

where the subscript R stands for the rescaled quantity. However, we find that the solutions are not well-defined for all τ and Λ (see Figs. 2). Nevertheless, we note that the unphysical parameter region is rendered shrunk by including the eight-vertex interactions.

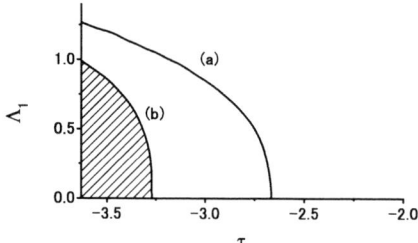

FIGURE 2. The recursion relations have a singularity in the shaded region (b) for the Brazovskii model with the eight vertex function. In the case $\Gamma^{(8)} = 0$, the solutions of the recursion relations are not defined in the underlying region of the curve (a). The ordinate Λ_1 is the coarse-graining scale Λ in the scaled units.

Thus, even with φ^8-interactions taken into consideration, the RG recursion relations contain singularities for $\tau < 0$, and the difficulty to use RG method for Brazovskii model remains unresolved.

REFERENCES

1. Brazovskii, S.A., *Zh. Eksp. Theor. Fiz.* **68**, 175-185 (1975) [*Sov. Phys. JETP* **41**, 85-89 (1975)].
2. Shankar, R., *Rev. Mod. Phys.* **66**, 129-192 (1994).
3. Hohenberg, P.C., and Swift, J.B., *Phys. Rev. E* **52**, 1828-1845 (1995).
4. Wegner, F.J., and Houghton, A., *Phys. Rev. A* **8**, 401-412 (1973).
5. For a review of recent development, see Bagnuls, C., and Bervillier, C., *Phys. Rep.* **348**, 91-157 (2001); Berges, J., Tetradis, N., and Wetterich, C., *ibid.* **363**, 223-386 (2002).
6. Kopietz, P., *Nucl. Phys. B* **595**, 493-518 (2001); Kopietz, P., and Busche, T., *Phys. Rev. B* **64**, 155101(14 pages) (2001).

A Concept of Effectively Global Search in Optimization by Local Search Heuristics

M. Hasegawa

Institute of Engineering Mechanics and Systems, University of Tsukuba, Tsukuba 305-8573, Japan

Abstract. A method for evaluating the effectively global exploration in the cost space is introduced and the result is used for an adaptive stopping criterion in optimization by local search heuristics. As an example, an adaptive cooling schedule in simulated annealing (SA) is properly designed and its performance is empirically examined on the Euclidean traveling salesman problem. The present adaptive SA outperforms a conventional non-adaptive one both in quality and in stability. The performance seems to be affected by the total number of search steps and by the final temperature; the parameters used for an conventional standard design of cooling schedule seem to influence the performance indirectly, only through the former two factors.

INTRODUCTION

Local search is one of powerful tools for solving combinatorial optimization problems [1]. In its advanced search strategies, a local optimization performance is supplemented by global exploration mechanisms to avoid poor local optima. This enhancement of global exploration offers us hope of obtaining much better results. In a usual implementation of the algorithm, the search process is stopped if a chosen number of search steps elapsed either in total or since the latest improvement of the solution. Because the solution space is vast and complex, we often wonder if we should stop the search. In fact, however, the thoroughly global exploration is not always necessary even if no helpful technique for limiting the search space exists. The quality of a solution is evaluated from its cost (the value of the objective function), not by the solution itself. If we know a global exploration in the cost space, we may use this information for the termination of the search process.

Here we introduce a method for evaluating the effectively global exploration in the cost space and use the results for an adaptive termination of the search process. In the present study, we consider simulated annealing (SA), which was motivated by an analogy to the behavior of physical systems in the presence of a heat bath [2]. In its implementation, the design of a cooling schedule of the heat bath is important. The above idea can be applied to a proper design of an adaptive scheduling, which has been considered to be a potential improvement on a conventional standard SA [3]. For a full evaluation of this approach, more information on its performance has been expected [3]. Here we empirically examined the performance of the adaptive algorithm through numerical experiments on the Euclidean traveling salesman problem (TSP).

ADAPTIVE SCHEDULING IN SIMULATED ANNEALING

The effectively global exploration in the cost space is evaluated in the following way. First we combine a conventional standard SA with random multi-start strategy and monitor cumulative cost distributions in each search process independently. Then we evaluate the degree of agreement among the distributions at each temperature; here, it is measured by variance of the average costs for simplicity. We denote the value of variance evaluated at a search step t by $\Omega(t)$. A quasi-equilibrium condition, $\Omega(t)/\Omega(0) < \varepsilon$, is used as a criterion for the termination of the search process at each temperature. We call a simple combination of a conventional standard SA and random multi-start strategy a Random Multi-start Simulated Annealing (RMSA) and call the algorithm using the above quasi-equilibrium condition an Adaptive Random Multi-start Simulated Annealing (ARMSA).

NUMERICAL EXPERIMENTS

Numerical experiments were performed on various instances of TSP. In the present paper, we summarize typical results observed in random instances. These are prepared by sampling N points from uniform distribution in a square, whose size is adjusted such that the cost of the best solution obtained in a preliminary experiment equals N. Discussions about another instances will be shown elsewhere. In all through the present experiments, 2-opt neighborhood [3] is used.

To compare the performance between RMSA and ARMSA, optimization by each method was done under the same total number, t_{total}, of search steps. First RMSA is performed with a fixed value of I (the number of simultaneous searches), T_s (the initial temperature), γ (the cooling ratio), and L (the temperature length, that is the fixed number of steps for search at each temperature in RMSA). Each search process is stopped when five consecutive temperatures have passed without an improvement and without the acceptance percentage going above 2%, which is the condition used in a baseline implementation of SA in Ref. 3. Then ARMSA is performed with the same value of I, T_s, and γ as used in RMSA and with the various values of ε. The total number of steps is equalized with the average number of steps employed in the corresponding RMSA.

Moreover, to examine the best choice of parameter values in ARMSA, we performed the experiment with the various combinations of the values of I, T_s, γ, and ε and with some fixed values of t_{total}.

In all of the present experiments, I solutions, the best ones found in each search process, are modified by a simple local optimization algorithm.

RESULTS AND DISCUSSION

In the present adaptive cooling schedule, the number of steps necessary for satisfying a quasi-equilibrium condition increases with a decrease of temperature: according to our observation for the case of $N=100$ with $I=50$, $\gamma=0.95$, and $\varepsilon=0.3$ and with high enough T_s, an adaptive length seems to be inversely proportional to the power of temperature. This was observed in a wide range of temperature including the point giving the maximum specific heat.

In an extensive comparative study between RMSA and ARMSA, we observed that the latter always outperformed the former if the value of ε was chosen appropriately. On the same parameter condition, standard deviation of the costs of I solutions was also minimized. This was observed irrespective of the combination of the values of N, I, T_s, γ, and L. Tuned ARMSA always provides higher quality solutions more stably in comparison to RMSA.

Fig. 1 shows the results for our parametric study on ARMSA: the cost of the median of I solution values are shown as a function of the final temperature T_f. The results for the case of $N=100$ with all combinations of two Is (10 and 100), three T_ss (acceptance percentage is 90, 60, and 30%), four εs (0.1, 0.3, 0.5, and 0.7), and three γs (0.95, 0.75, and 0.55), that is 72 results are plotted for each t_{total} using the same mark. The average results in 10 runs on each combination are plotted. There appears three clusters classified by the value of t_{total}: 72 results are roughly on a curve, which has minimum point at almost the same T_f irrespective of the value of t_{total}. Its performance seems to be affected by t_{total} and by T_f; another parameters used for a standard design of scheduling seem to influence the performance indirectly, only through the former two factors.

In a finite time search in SA, local optimization performance cannot be expected at high temperatures, on the other hand global exploration cannot at low temperatures. As a result, the optimal performance can be obtained at an intermediate temperature. In the case of random instance, tuned ARMSA improves its performance with the enhancement of the optimum temperature search.

REFERENCES

1. Aarts, E.H.L., and Lenstra, J.K., "Introduction," in *Local Search in Combinatorial Optimization*, edited by E.H.L.Aarts and J.K.Lenstra, John Wiley & Sons, Chichester 1997, pp.1-17.
2. Kirkpatrick, S., Gelatt, C.D., and Vecchi, M.P., *Science*, **220**, 671-680 (1983); Černý, V., *Journal of Optimization Theory and Applications*, **45**, 41-51 (1985).
3. Johnson, D.S., and McGeoch, L.A., "The traveling salesman problem: a case study," in *Local Search in Combinatorial Optimization*, edited by E.H.L.Aarts and J.K.Lenstra, John Wiley & Sons, Chichester 1997, pp.215-310.

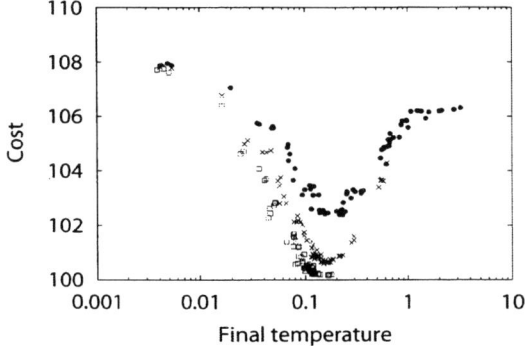

FIGURE 1. Results of the average cost of solutions in ARMSA as a function of the final temperature ($N=100$).
● : $t_{total} = 10^5$, × : $t_{total} = 10^6$, □ : $t_{total} = 10^7$.

Behaviors of thermodynamic quantities of a noise-driven nonlinear oscillator

M. Akimoto* and A. Suzuki

*Center for Solid-State Physics and Department of Physics, Science University of Tokyo
1-3 Kagurazaka, Shinjuku-ku, Tokyo 162-8601, Japan.

Abstract. We obtain an exact stationary distribution of a noise-driven modified van der Pol oscillator and discuss thermodynamic behaviors of the system.

INTRODUCTION

The self-sustained nonlinear oscillators have been studied extensively in nonlinear dynamics and can be found in different fields of general sciences such as physics, electronics, chemistry, biology, and other disciplines. The appearance of a limit cycle attractor via bifurcation, that is a kind of temporal order, is of special interest and it is important to elucidate the influence of noise on the onset of such a dissipative structure.

In this study, we investigate a certain class of self-sustained oscillators driven by noise, namely, a *noise-driven modified van der Pol oscillator* expressed by,

$$\dot{v} = -x - [u_1 + u_2(x^2 + v^2)]v + \sigma g(x,v)\xi(t), \quad (1)$$

where $u_1, u_2 \in \mathbb{R}$ ($u_2 > 0$) are nonfluctuating parameters assumed to be controllable, $g(x,v)$ is some smooth function of x and $v (= \dot{x})$, which makes the noise term multiplicative, and $\xi(t)$ is Gaussian white noise with an intensity σ so as to satisfy $\langle \xi(t) \rangle = 0, \langle \xi(t)\xi(t') \rangle = \delta(t-t')$. For a noise-free case ($\sigma = 0$), Eq. (1) exhibits the deterministic Andronov-Hopf bifurcation when a bifurcation parameter u_1 takes the critical value, $u_1 = 0$ while the parameter u_1 takes the value $u_1 \in [-\infty, 0)$, a limit cycle oscillation emerges with radius $\sqrt{-u_1/u_2}$.

We are interested here in the stationary distribution function in order to study the influence of noise on the system. Once one obtain an exact stationary distribution, thermodynamic quantities such as entropy and free energies could be determined from it. Then it could be feasible to understand how the thermodynamic quantities behave as the parameters are varied.

METHOD

The modified van der Pol oscillator (1) is classified to the following two-dimensional stochastic system of Ito-Langevin type,

$$\begin{aligned}\dot{x} &= \frac{\partial H}{\partial y} + \alpha^2 f(H;u)\frac{\partial H}{\partial x} + \alpha\sigma g(H;u)\xi_1(t), \\ \dot{y} &= -\frac{\partial H}{\partial x} + \beta^2 f(H;u)\frac{\partial H}{\partial y} + \beta\sigma g(H;u)\xi_2(t),\end{aligned} \quad (2)$$

which express a kind of system obeying the so-called *mixed canonical-dissipative dynamics* (MCDD) first formulated by Enz [1] in critical dynamics and later extended by Hongler and Ryter [2, 3] to apply to the problems of nonlinear dynamical systems with noise. In Eq. (2), $H = H(x,y)$ plays the role of Hamiltonian, $f(H;u)$ and $g(H;u)$ are some functions of H and parametrized by $u = \{u_1, u_2\}$, α and β are real parameters, and $\xi_1(t), \xi_2(t)$ stand for Gaussian white noises: $\langle \xi_i(t) \rangle = 0, \langle \xi_i(t)\xi_j(t') \rangle = \delta_{ij}\delta(t-t'), (i,j = 1,2)$. Here, the modified van der Pol oscillator corresponds to the case, where $H = \frac{1}{2}(x^2 + y^2)$, $\alpha = 0$, $\beta = 1$, $f(H) = -(u_1 + 2u_2 H)$ and g is arbitrary but restricted to a function of H [3].

By following the procedure of the methods of MCDD, the corresponding Fokker-Planck equation is established and from which we can exactly obtain the stationary solution for a stationary distribution as

$$P_s(x,y) = P_s(H) = \frac{C}{g(H;u)}\exp\left(\frac{2}{\sigma^2}\int \frac{f(H;u)}{g^2(H;u)}dH\right), \quad (3)$$

where C represents the normalization constant. One of the most important results of the methods of MCDD is the fact that the stationary distribution of Eq. (2) depends only on the Hamiltonian H instead of original variables x, y.

STATIONARY DISTRIBUTION

We treat here the more specialized case by introducing three parameters $b, q, r \in \mathbb{R}$, which are redefined from u_1, u_2: $u_1 = 1 - \frac{1}{2}(1-q), u_2 = \frac{rb}{2}(q-1), (q > 1, r > 0)$. In this case the deterministic bifurcation occurs at $q = 3$ and a deterministic stable limit cycle has a radius $R = \sqrt{x^2 + y^2} = \sqrt{\frac{q-3}{rb(q-1)}}$. In the limit $q \to \infty$, the radius R saturates at $1/\sqrt{rb}$ although the radius of a limit cycle usually increases infinitely. By specifying simultaneously the function of multiplicative noise as $g(H; b, q) = \sqrt{1 + b(q-1)H}$, the stationary distribution of the system considered here is obtained from Eq. (3) as

$$P_s(H) = C[1 - b(1-q)H]^{\frac{a}{b} \cdot \frac{1-r}{1-q} + \frac{a}{2b} - \frac{1}{2}} e^{-arH}, \quad (4)$$

where we put a new parameter a instead of $2/\sigma^2$ for brevity. The limit $a \to \infty$ implies the weak noise limit (deterministic limit). We note that this distribution function is characterized by four *independent* parameters, a, b, q, r. Because Eq. (4) is a function of H only, the stationary distribution (4) is rotationally invariant in phase space spanned by (x, y). It should be noted that Eq. (4) leads to the extended Tsallis distribution [4] as $a/b \to 1$,

$$P_s(H) \to C[1 - a(1-q)H]^{\frac{1-r}{1-q}} e^{-raH}. \quad (5)$$

We note that this limit induces striking phenomena which have no analogy with deterministic systems. The most important feature of the stationary distribution (4) is the modality (number of maxima). Maxima occurs at $R^2 = \frac{q-3}{rb(q-1)} - \frac{1}{ra}$ and the bifurcation of stochastic system is delayed by multiplicative noise owing to the correction term $1/(ra)$. Moreover it can be shown that a condition $R^2 > 0$ is equivalent to $q > 3 + \frac{2b}{a-b}$ if $a > b$. From this inequality the bifurcation on the distribution (change to bimodality) is suppressed *perfectly* in the limit $a/b \to 1$ and it coincides with the extended Tsallis distribution (5). This is a new kind of noise induced phenomena in the noise-driven modified van der Pol oscillator.

Fig. 1 shows the profiles of the stationary distributions as a function of the radius R, where noise intensity a and other parameters b, r are fixed. The stationary distribution has one maximum after the bifurcation of underlying mechanics occurs. Fig. 2 shows the profiles of distributions but in this case the bifurcation parameter q is fixed in the limit cycle region. As intensity of noise is weakened, the maximum, that corresponds to the presence of limit cycle, approaches to the radius of deterministic case and the shape of distribution becomes sharpen like a delta function. Here, strong noise suppresses the shape of distribution, and in the limit $a/b \to 1$ the maximum is vanished while underlying deterministic system has even a stable limit cycle. In consequence, the modality of dis-

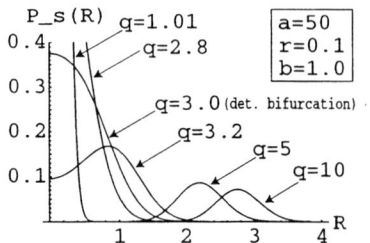

FIGURE 1. P_s vs. R. (a, b, r: fixed.)

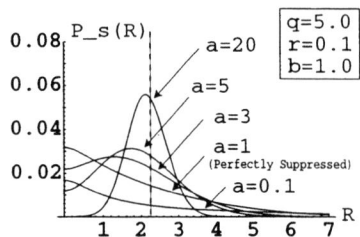

FIGURE 2. P_s vs. R. (q: fixed.)

tributions changes according to the bifurcation of underlying deterministic systems and it is affected strongly by multiplicative noise.

THERMODYNAMIC QUANTITIES

Using Eq. (4), we can investigate the behaviors of thermodynamic quantities such as conventional entropy: $S = -\iint dxdy P_s(H) \log P_s(H)$ and Tsallis entropy indexed by q': $S_{q'} = (\iint dxdy (P_s(H))^{q'} - 1)/(1 - q')$. We found that: (i) The entropy increases monotonically as bifurcation parameter q grows. This fact is also applied for Tsallis entropies. (ii) In the weak noise limit the occurrence of deterministic bifurcation could be regarded as a kind of phase transition. By using Tsallis entropy with large q', one can estimate the transition for moderately strong noise case. These facts suggest that thermodynamical characterization of nonlinear dissipative systems may be worked out more nicely with the help of the framework of the Tsallis statistics. The detailed discussions will be reported elsewhere.

REFERENCES

1. Enz, C. P., *Physica*, **89A**, 1-36 (1977).
2. Hongler, M.-O., and Ryter, D. M., *Z. Phys. B*, **31**, 333-337 (1978).
3. Ebeling, W., and Engel-Herbert, H., *Physica*, **104A**, 378-396 (1980).
4. Kaniadakis, G., and Quarati, P., *Physica A*, **237**, 229-239 (1997).

Thermal Contact in Quantum Systems

A. Iwaya* and A. Suzuki

*Center for Solid-State Physics and Department of Physics, Science University of Tokyo
1-3 Kagurazaka, Shinjuku-ku, Tokyo 162-8601, Japan.

Abstract. We propose a theory which establishes a consistency between thermodynamics and quantum pure states. In the theory, thermodynamical entropy is defined in terms of parameter that characterizes quantum pure states.

INTRODUCTION

Thermodynamical studies on quantum systems have been done by many authors [1, 2, 3]. In those studies a change of entropy is a central issue since it is not widely recognized whether there exist the laws of thermodynamics in quantum systems. The issue lies mainly in the fact that entropy is a constant of motion for a closed system which evolves unitarily according to a Hamiltonian [4]. Actually, there exists a situation where thermodynamics is required: let us consider two isolated quantum systems A and A' interacting each other. The interaction induces an energy level shift caused by external parameters and a change of probability of finding states of these systems. The latter is caused by the thermal contact between these systems in which energy is transferred from one system to the other with no energy level shift. In a case where external parameters are fixed, energy levels of each system are not affected and the change of mean energy for each system is solely caused by the thermal contact (i.e., heat transfer between them) [5, 6]. In such a case, thermodynamics should be introduced since an unitary evolution generated by a Hamiltonian could not describe a change of probability of finding states. In this paper, we shall formulate a theory which parameterizes quantum pure states, so that thermodynamics can be applied to quantum pure states.

In order to formulate the problem of thermal contact, let us clarify a physical situation considered here: at first two quantum systems are prepared independently, where a system A is in a state $|\psi_\alpha\rangle$ and A' in $|\psi_\beta\rangle$, respectively. In this paper, Greek letters, $\alpha, \beta, \gamma, \cdots$ will be used to label the quantum states. Those state vectors describe pure states. Second, let two systems be in thermal contact as they are totally isolated (no external perturbation exists, i.e., $\Delta W = 0$) in which the total energy remains constant. For such thermodynamical processes, the first law of thermodynamics should be applied to each system. In other words, we treat the quantum systems, where the change of their energies are due only to heat transfer $\Delta E = \Delta Q$ (note that $\Delta W = 0$). After thermal contact, two systems change their states. To analyze the situation, let us prepare a set of all state vectors which possibly represent the states of these systems. The preparation can be set up in the following way: we expand a state vector by a complete orthonormal system (CONS). From the fact that the expansion coefficients of state vectors are satisfied with probabilistic conditions [2], we can describe all state vectors in terms of a set of probability functions:

$$\{p(\theta_\alpha), p(\theta_\beta), p(\theta_\gamma), \cdots\}. \quad (1)$$

A set (1) is introduced to maintain correspondence between a state vector $|\psi_\alpha\rangle$ and a probability function $p(\theta_\alpha)$, $|\psi_\beta\rangle$ and $p(\theta_\beta)$ and so on, where $\theta_\alpha, \theta_\beta, \cdots$, denote parameters. We will show that the zeroth law of thermodynamics can be represented in terms of parameter θ for quantum systems. For example, in the case, $\theta_\alpha = \theta_\beta$, the states $|\psi_\alpha\rangle, |\psi_\beta\rangle$ are in an equilibrium. Thus θ could be regarded as a thermodynamical variable. To justify the statement of the zeroth law of thermodynamics, however, we have to introduce a restriction: all state vectors $\{|\psi_\alpha\rangle, |\psi_\beta\rangle, \cdots\}$ are expanded by the same CONS to obtain a set of probability functions (1). We will show that a parameter θ introduced in the theory enables us to define a thermodynamical entropy and so that the second law of thermodynamics can be applied to quantum pure states.

THEORY

The theory of thermal contacts presented here is constructed in the following way: first, we represent all possible states in terms of a set of probability functions, whereby the zeroth law of thermodynamics is introduced. Next, a thermodynamical entropy is defined so as to apply the second law of thermodynamics to quantum pure states. We shall show a one-to-one correspondence between a probability function p and a state vector $|\psi\rangle$ by recalling a proposition of quantum theory: *a quantum state is determined by preparations and tests*. A pure state is determined by maximal tests. Let us prepare a pure state, say, $|\psi_\alpha\rangle$ (which is normalized to unity). Then, $|\psi_\alpha\rangle$ can be expanded by CONS:

$$|\psi_\alpha\rangle = a_1^\alpha |1\rangle + a_2^\alpha |2\rangle + \cdots + a_i^\alpha |i\rangle + \cdots + a_n^\alpha |n\rangle, \quad (2)$$

where a ket $|i\rangle$ $(i = 1, 2, \cdots, n)$ is a basis vector in CONS. Greek letter α indicates that the expansion coefficients belong to the same state vector $|\psi_\alpha\rangle$. Those expansion coefficients as in Eq. (2) satisfy the conditions:

$$|a_i^\alpha|^2 \geq 0, \quad \sum_i^n |a_i^\alpha|^2 = 1. \quad (3)$$

Physically, $|a_i^\alpha|^2$ represents a probability of finding the state i. Let us introduce a probability function $p(\theta_\alpha; x)$, where x denotes a random variable and θ_α is a parameter. Clearly, $p(\theta_\alpha; x = i)$ represents the *probability* of finding the state i. Therefore, there exists correspondence between the expansion coefficient a_i^α and the probability function $p(\theta_\alpha; x)$:

$$|a_i^\alpha|^2 \to p(\theta_\alpha; x = i) \text{ for } \forall i. \quad (4)$$

Conversely, the expansion coefficient a_i^α is derived from a probability function $p(\theta_\alpha; x)$ by taking a square root of the probability function, viz., $\sqrt{p(\theta_\alpha; x = i)} = a_i^\alpha$. Though a_i^α is a complex number, as far as taking maximal tests, it is not a problem that the phase factor of an expansion coefficient is indefinite. Thus one can obtain the expansion coefficient from the probability function:

$$p(\theta_\alpha; x = i) \to |a_i^\alpha|^2 \text{ for } \forall i. \quad (5)$$

From Eqs. (4) and (5), the probability function is, therefore, proved to be well defined so as to represent a pure state.

Now let us consider a set of state vectors. The correspondence between a state vector and a probability function can be applied to other state vectors as well. Accordingly, a correspondence between a set of state vectors and a set of probability functions $\mathcal{M}$ can be established:

$$\{|\psi_\alpha\rangle, |\psi_\beta\rangle, \cdots\} \to \{p(\theta_\alpha), p(\theta_\beta), \cdots\} =: \mathcal{M}. \quad (6)$$

This relation states that a set of probability functions corresponds to a set of state vectors and *there exists a correspondence between a parameter θ and a probability function*, viz., $\theta \to p(\theta)$. Therefore, the set $\mathcal{M}$ can be replaced by a set $\mathcal{M}'$:

$$\mathcal{M}' := \{\theta_\alpha, \theta_\beta, \cdots\}, \quad (7)$$

and pure states are thus parametrized by θ. Accordingly, the zeroth law of thermodynamics could be represented in terms of θ and a thermodynamical entropy also introduced naturally into the theory. Indeed, the zeroth law of thermodynamics (viz., an equivalence law) can be represented by $\theta_\alpha = \theta_\beta$. In such a case two states $|\psi_\alpha\rangle$ and $|\psi_\beta\rangle$ are said to be comparable (or in equilibrium) in a sense that two pure states are not distinguished any more.

An entropy S is defined as a map from a set $\mathcal{M}'$ to a real number $\boldsymbol{R}$:

$$S : \mathcal{M}' \to \boldsymbol{R}. \quad (8)$$

The entropy defined in (8) is a quantity of states since a parameter θ_α ($\in \mathcal{M}'$) refers to a state $|\psi_\alpha\rangle$ via the relation (6). Our main result so far is that *by parametrizing pure states, any quantum state can be represented by an entropy as a function of θ_α*. Our ultimate goal is to find an entropy function that can express the second law of thermodynamics in terms of θ_α which characterizes a probability function $p(\theta_\alpha)$. A remaining task is, however, that we must clarify whether the entropy defined in (8) describes the second law of thermodynamics in quantum system. To see this, let the states of two systems A and A' be prepared by θ_α and θ_β, respectively. After thermal contact, the states of these systems possibly change to respective states characterized by $\theta_{\alpha'}$ and $\theta_{\beta'}$. By the second law of thermodynamics, the possible new states, $|\psi_{\alpha'}\rangle$ and $|\psi_{\beta'}\rangle$, should satisfy the conditions $\theta_\alpha \leq \theta_{\alpha'} \leq \theta_{\beta'} \leq \theta_\beta$ and $0 \leq [S(E'_A + E'_{A'}) - \{S(E_A) + S(E_{A'})\}]$ in case of $\theta_\alpha \leq \theta_\beta$ being satisfied. If this is true, it can be proved that the entropy defined in (8) is a thermodynamical entropy and a parameter θ can be regarded as a temperature in classical thermodynamics. It should be noted that the energy of respective systems is calculated from $\langle E \rangle = \sum_x p(\theta; x) E_x$.

We have established the consistency between pure states representing quantum systems and thermodynamics. Therefore, we can study the effect of thermal contact in quantum systems in terms of thermodynamics if we know the entropy function along with the probability function $p(\theta; x)$. The detail of thermal contact in quantum systems will be published in a separate paper.

REFERENCES

1. von Neumann, J., *Mathematische Grundlagen der Quantenmechanik*, Springer, Berlin, 1932, chap. 5.
2. Peres, A., *Quantum Theory: Concepts and Methods*, Kluwer-Academic, Boston, 1993, chaps. 2, 3 and 9.
3. Partovi, M. H., *Phys. Lett. A*, **137**, 440-444 (1989).
4. Wehrl, A., *Rev. Mod. Phys.*, **50**, 221-260 (1978).
5. Lieb, E. H., and Yngvason, J., *Phys. Rep.*, **310**,1-96 (1999).
6. Leif, F., *Fundamentals of Statistical and Thermal Physics*, McGraw-Hill, New York, 1965, chap. 2.

Theoretical Study On Photoexcited States Of Strongly Correlated Electron Systems

K. Sugiyama, T. Yamaguchi, I. Sakamoto, T. Yoshimoto, H. Nagao and K. Nishikawa

Department of Computational Science, Faculty of Science, Kanazawa University, Kanazawa 920-1192, Japan

Abstract. We investigate the oxygen rearrangement model by using a cluster model of the copper oxide. The potential energy surface is estimated by the semiempirical molecular orbital calculation (AM1). We discuss the jumping of oxygen ions in the Cu-O chain in relation to the possibility of the laser control of the oxygen ion.

INTRODUCTION

It is well known that copper oxide systems yield high-Tc superconductivity under the hole or electron doping. The hole or electron doping is usually made either by chemical substitution of element as in $La_{2-x}(Ba, Sr)_xCuO_{4-y}$, or by varying the oxygen content as in $YBa_2Cu_3O_{6+x}$.

Kudinov et al. [1] have reported that the illumination with visible laser causes an increase of the superconducting Tc of $YBa_2Cu_3O_{6+x}$ thin film. To explain this phenomenon, so called photoinduced superconductivity, three models have been proposed; (A) the charge transfer model, (B) oxygen rearrangement model and (C) the charge transfer and oxygen rearrangement model. In the first model, electrons transfer from the CuO_2 plane to the CuO chain by the photoinducing. The increase of the hole concentration of the CuO_2 plane leads to the change of superconducting properties. In the second model, oxygen atoms in the CuO chain are rearranged that the CuO chain segments elongate, by the photoinducing. This oxygen rearrangement causes the change of the hole concentration of the CuO_2 plane. In the third model, the hole concentration changes by the charge transfer and the oxygen rearrangement.

Although the above models can explain some properties of the photoinduced superconductivity, the mechanism of the photoinduced superconductivity is still under debate. Recently, we have investigated the charge transfer model by using a transistor model [2]. The dependence of the superconducting Tc on the applied field agrees with that of experimental results. We have pointed out that the change of the carrier concentration is an important role for the photoinduced superconductivity.

In this study, we investigate the oxygen rearrangement model by using a cluster model. The potential energy surface is estimated by semiempirical molecular orbital calculation (AM1). We discuss the jumping of oxygen ions in the Cu-O chain in relation to the possibility of the laser control of the oxygen ion by strong field.

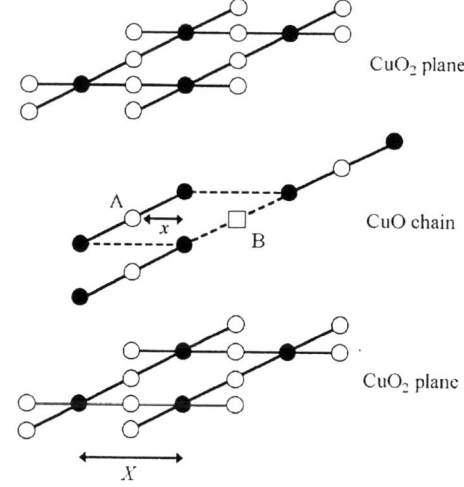

FIGURE 1. The cupper oxide model cluster. Open (filled) circle refers O (Cu) atom. Open square refers lattice defect.

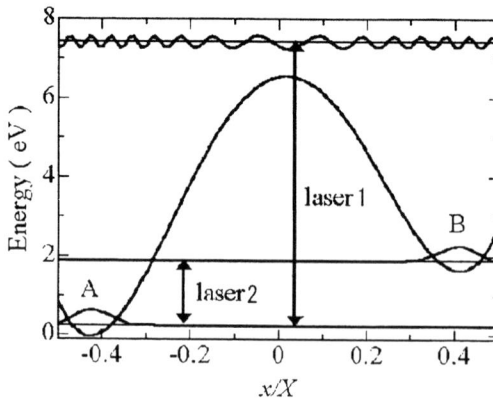

FIGURE 2. The potential energy surface for the O atom rearrangement and its eigenstates used to STIRAP simulation.

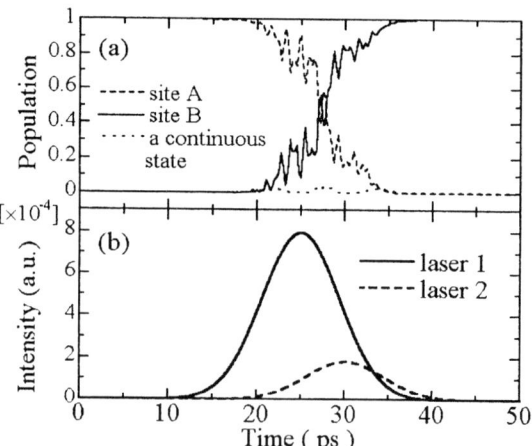

FIGURE 3. The time variation of (a) population of eigenstates and (b) pulse sequence.

RESULTS AND DISCUSSION

The model cluster used in this study is shown in figure 1. This model consists of two CuO_2 planes and one CuO chain. We assumed the charge of Cu ion and O ion as +2 and –2, respectively. AM1 calculations were performed by MOPAC 2002 [3].

We estimated the potential energy surface (PES) of which the O atom of the CuO chain moves from the site A ($x/X = -0.50$) to the site B ($x/X = 0.50$). We found that the PES has two local minima in the vicinity of the site A and the site B, and the former is lower than latter. We fitted this PES by the following equation,

$$V(x/X) = \frac{\delta}{2q}(x/X + q) + \frac{(\Delta - \delta/2)}{q^4}(x/X - q)^2(x/X + q)^2, \quad (1)$$

where δ refers the energy difference between two local minima, Δ refers the barrier height, and q refers the position of local minimum. The fitted values of δ, Δ and q are 1.66 eV, 6.55 eV and 0.44. The fitting PES is shown in figure 2. We also found that electron transfers from the CuO_2 plane to the CuO chain at $x/X = -0.03$, and about 0.6 electron transfers to CuO chain at $x/X = 0.50$. This result means that hole concentration of the CuO_2 plane increases by the rearrangement of the O atom of the CuO chain, and then the superconducting Tc may increases or the superconducting state may appears by inducing photon.

To investigate the reality of the laser control of the O atom rearrangement, we performed the stimulated Raman adiabatic passage (STIRAP) [4] simulation with fitting PES and its three eigenstates. The STIRAP is well known as very effective technique for the population transfer by using counterintuitive pulse sequences. The time variation of population of each eigenstates and pulse sequence are shown in figure 3. We found that the O atom of the CuO chain moves from site A to site B by inducing the laser pulse sequence as shown in figure 3 (b). This result suggests that there is the possibility of the appearance of the photoinduced superconductivity by the oxygen rearrangement model.

In summary, we have investigated the oxygen rearrangement model by using cupper oxide cluster model with AM1 calculation. We found that hole concentration of CuO_2 plane increases by rearrangement of the O atom of the CuO chain. We also suggested the possibility of the photoinduced superconductivity by the oxygen rearrangement model by the STIRAP simulation.

Finally, we propose a new model; (D) electron photo-radiation model. In this model, electrons in the CuO_2 plane are directly radiated from the cupper oxide crystal by a high-intensity laser ($\sim 10^{16}$ W/cm^2) that is recently developed. In the future work, we will investigate the electron photo-radiation model by the STIRAP and so on.

REFERENCES

1. Kudinov, V.I., Chaplygin, I.L., Kinlyuk, A.I., and Kreines, M.N., *Phys. Rev. B* **47**, 9017-9028 (1993).
2. Nagao, H., Mitani, M., Nishino, M., Shigeta, Y., Yoshioka. Y., and Yamaguchi, K., *Int.J.Quantum Chem.* **75**, 549-561 (1999).
3. Stewart, J. J. P., "MOPAC 2002", Fujitsu Limited. Tokyo, Japan (2001).
4. Ohta, Y., Yoshimoto, T., Saito, H., Nagao, H., and Nishikawa, K., *Recent Res. Devil. Quantum Chem.* **3**, 211-244 (2002).

Theoretical study on the effect of solvent and intermolecular fluctuations in proton transfer reactions: General theory

Nobuhiko Kato, Tomonori Ida and Kazunaka Endo

Department of Chemistry, Faculty of Science, Kanazawa University, Kakuma, Kanazawa, 920-1192, Japan

Abstract. We present a theory of proton transfer reactions which incorporate the modulation of the proton's potential surface by intermolecular vibrations and the effect of coupling to solvent degree of freedom. The proton tunnels between states corresponding to it being localized in the wells of a double minimum potential. The resulting tunnel splitting depends on the intermolecular separation. The solvent response to the proton's charge is modeled as that of a continuous distribution of harmonic oscillators and the intermolecular stretching mode is also damped because of the interaction with solvent degree of freedom. The transition rate is given by the Fermi Gorlden Rule expression.

INTRODUCTION

Proton tunneling transitions are an important class of chemical ractions. A characteristic feature of these reactions may appear in an activation at high temperature and fall to an almost constant value at low temperature. The constant value is much greater than that predicted by extrapolating the Arrhenius value to low temperature. Cukier et.al.[1]. have stressed that the activation energy of tunneling reactions does not arise from the barrier along the proton's reaction coordinate but rather arises from the reorganization energy of the solvent required approximately to equalize the proton's energy in the initial and final state. While, Borgis et.al[2]. have discussed the inter- and intramolecular contribution by use of a time-dependent point of view, in a regime where the solvent motion can be treated classically and Morillo et.al[3]. have dealt with the low temperature regime in proton transfer reaction which incorporate the modulation of the proton's potential surface by intermolecular vibration and the effect of coupling to solvent degrees of freedom. Our approach is closest in spirit to that of Morillo et.al. in the sense of methodology, but we also incorporate intermolecular mode's dampimg because of the interaction with the solvent degree of freedom. The physical model we adopt is the similar one as Morillo et.al have used. Then a protonic potential surface can be defined for each solvent configuration and intermolecular distance. A plot of potential surfaces for solvent motion is shown in Figure 1. The eigenvalues of the proton Schrodinger equation will then be denoted by $E_n^L(q)$, $E_n^R(q)$ where L and R refer to the left and right wells. We assume that the spacings of the energy of these states are large compared with thermal energies and therefore we need only consider the two lowest state.(n=0,m=0). On the other hand, the effect of intermolecular mode is the typical dependence of splitting. The dependence of splitting on x, as long as the splitting is small, is exponential.

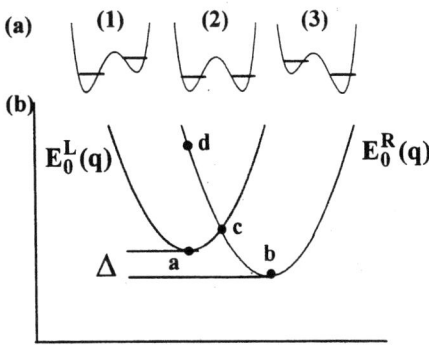

Figure 1. (a) proton potential surface for three different solvent configurations. (b) Plot of potential surfaces for solvent motion. The solvent coordinates at points a, b, c correspond to the proton potential surfaces in (1), (2) and (3), respectively, of (a).

FORMALISM

A Hamiltonian that incorporates the modulation of the proton's potential by the solvent and intermolecular mode is

$$\hat{H} = V(x_v)\hat{\sigma}_x + \frac{\Delta}{2}\hat{\sigma}_z + (\frac{p_v^2}{2} + \frac{\omega_v^2}{2}x_v^2)$$
$$+ \sum_k \{\frac{P_k^2}{2} + \frac{1}{2}\Omega_k^2(q_k - \frac{\gamma_k}{\Omega_k^2}\hat{\sigma}_z)^2\}$$
$$+ \sum_l \{\frac{P_{vl}'^2}{2} + \frac{1}{2}\Omega_{vl}'^2(q_{vl}' + \frac{C_{vl}}{\Omega_{vl}'^2}x_v)^2\}$$
$$\equiv \hat{H}_0 + V\hat{\sigma}_x \qquad (1)$$

with

$$V(x_v) = V_0 \exp(-\alpha x_v). \qquad (2)$$

The x_v represents the intermolecular coordinate which depend on the tunnel splitting; they are measured from their equilibrium distances. The harmonic motion of this indirect contribution is given by the $\{p_v, x_v\}$ term in Eq. (1). The $\{P_k, q_k\}$ are the momenta and positions of the solvent coordinates, respectively. $\{\Omega_k\}$ and $\{\gamma_k\}$ indicate their frequencies the coupling constants. The $\{P_{vl}', q_{vl}'\}$ term is the solvent mode which interacts with intermolecular stretching mode x_v. $\{\Omega_{vl}\}$ is also their frequencies and $\{C_{vl}\}$ the coupling constants. We have set the oscillator masses to unity for convenience. The parameter Δ is defined in Figure 2. The $\hat{\sigma}_x, \hat{\sigma}_y$ and $\hat{\sigma}_z$ denote the Pauli spin matrices. We assume that there is a rate constant describing proton transfer between localized state basis; $|1\rangle$, $|2\rangle$. H_0 defined by Eq. (1) is diagonal in the basis, while $V\hat{\sigma}_x$ is responsible for the transition. We assume that V is so small that the transition rate is given by the Fermi Golden Rule expression.

$$\Gamma_{1\to 2} = \frac{1}{\hbar^2}\int_{-\infty}^{+\infty} dt\, e^{-i\Delta t/\hbar}\langle V(x)V(x(t))\rangle$$
$$\times \langle e^{iH_{2q}t/\hbar}e^{-iH_{1q}t/\hbar}\rangle \qquad (3)$$

The evolution of the $\{q_k\}$ modes is specified by the Hamiltonians

$$H_{\substack{1q\\(2q)}} = \sum_k \{\frac{P_k^2}{2} + \frac{1}{2}\Omega_k^2(q_k \mp \frac{\gamma_k}{\Omega_k^2})^2\}. \qquad (4)$$

While the Hamiltonian of the x_v mode is specified by

$$H_{1x} = (\frac{p_v^2}{2} + \frac{\omega_v^2}{2}x_v^2)$$
$$+ \sum_l \{\frac{P_{vl}'^2}{2} + \frac{1}{2}\Omega_{vl}'^2(q_{vl}' + \frac{C_{vl}}{\Omega_{vl}'^2}x_v)^2\} \qquad (5)$$

The $\langle\cdots\rangle$ are defined by equilibrium averages with their corresponding Hamiltonian H_{1q} and H_{1x} and

$$V(x(t)) = e^{iH_{1x}t/\hbar}V(x)e^{-iH_{1x}t/\hbar}. \qquad (6)$$

In the model which Morillo et.al. have discussed, the correlation function of the x_v mode does not incorporate the effect of damping by the interaction between solvent mode and intermolecular stretching mode. But in our model, the damping can be treated through Eq. (5). The solvent mode correlation function has been evaluated by using Eq.(4).(see in Ref.3).

CONCLUSION

We formulate the transition rate of proton transfer reactions which incorporate the effect of the intermolecular and solvent mode. Intermolecular mode is damped by solvent mode through the second term in Eq.(5). We will present the interpretation and application of Eq.(3) elsewhere.

REFERENCES

1. Cukier.R.I. and Morillo.M., *J.Chem.Phys.* **91**, 857-863 (1989)

2. Borgis.D, Lee.S, and Hynes.J.T, *Chem. Phys. Lett.* **162**, 19-26 (1989)

3. Morillo.M and Cukier.R.I., *J.Chem.Phys.* **92**, 4833-4838 (1990)

Bottleneck in Energy Relaxation and its Self-Organization

Hidetoshi Morita and Kunihiko Kaneko

Department of Pure and Applied Sciences, University of Tokyo, Komaba 3-8-1, Meguro-ku, Tokyo 153-8902, Japan

Abstract. We study an energy relaxation process after many degrees of freedom are excited in a Hamiltonian system with a large number of degrees of freedom. Bottlenecks of relaxation, where the relaxation of the excited elements is drastically slowed down, are discovered. By defining an internal state for the excited degrees of freedom, it is shown that the drastic slowing down occurs when the internal state is in a critical state. The relaxation dynamics brings the internal state into the critical state, and the critical bottleneck of relaxation is self-organized.

Relaxation process to equilibrium has gathered much attention of physicists over several decades. When a system is excited weakly, it is described by a superposition of elementary excitations with weak interactions. When the excitation is strong, on the other hand, it is inevitable to take into account strong interactions of the excited modes, which may behave cooperatively and cause an internal state. The relaxation depends on the internal state, and, in turn, the internal state dynamically changes with the relaxation. We consider this interplay, by setting up such a situation in a simple model.

As a specific model, we adopt a Hamiltonian system [2,3];

$$H = \sum_{i=1}^{N} \frac{p_i^2}{2} + \frac{K}{2(2\pi)^2 N} \sum_{i=1}^{N} \sum_{j=1}^{N} [1 - \cos 2\pi(\theta_i - \theta_j)], \quad (1)$$

where $K=1$ throughout this paper. This system is known to exhibit continuous phase transition in equilibrium at the critical temperature $T_c = 0.5/(2\pi)^2$.

$N^{(K)}$ ($<N$) elements are highly excited instantaneously with the same momentum $P^{(K)}$. Meanwhile, we consider the case that the rest system is huge enough. Instead of carrying out numerical simulations of such a huge system, we add a heat bath with the temperature T to the rest ($N - N^{(K)}$) elements (in numerical simulations, we use Langevin equation.)

Subsequently, the excited elements ($N^{(E)}$) relax one by one from $N^{(K)}$ to zero. Its time series (FIGURE 1-(a)) shows several plateaus, where the relaxation is drastically slowed down. We call these plateaus as *bottlenecks* of the relaxation.

Since the interaction of two elements is very week if their momenta are much different [4], our system is represented by two partial thermodynamic systems which are weakly coupled. One is composed of not-excited elements, the state of which is basically determined by the heat bath. The other is composed of the excited elements, the state of which, on the other hand, changes as the relaxation progresses. Thus it is relevant to regard a thermodynamic state of the excited part as an *internal state* of the whole system. To quantify the internal state, we introduce *effective temperature of the excited part* $T^{(E)} \equiv \sum_i^{(E)} (p_i - P^{(E)})^2 / N^{(E)}$, by taking an inertial frame with the center-of-mass momentum of the excited part $P^{(E)} \equiv \sum_i^{(E)} p_i / N^{(E)}$, where $\Sigma^{(E)}$ denotes the summation over the excited elements.

To see how the relaxation depends on the internal state, we calculate the time (τ) when the first one of $N^{(K)}$ excited elements relaxes to the rest system, whose dependence on $T^{(E)}$ is FIGURE 2. A peak is discernible. This peak corresponds to the effective *critical* temperature of the excited part $T_c^{(E)} \equiv T_c N^{(E)} / N$ [5], when the excited part is regarded as in equilibrium. Actually, in the time series, the plateau of $N^{(E)}$ corresponds to the time when the effective temperature of the excited part is around the critical temperature (FIGURE 1-(b),(c)). Thus it is concluded that the relaxation bottleneck occurs when the internal state is in the critical state.

Note that this phenomenon is different from the so-called critical slowing down in two points. First, the whole system is not in the critical state. In fact we verified that the coincidence of the peak position to the effective critical point is qualitatively independent of the heat bath temperature T. Secondly, such a long relaxation bottleneck is not found when disregarding the rest system. In fact we estimated the first escape time of an element in only a single system, by calculating the time until one of the elements first takes a momentum smaller than a threshold. This escape time monotonously decreases with the increase of the temperature, without showing any peak. Hence the present slow relaxation is *inter*-system relaxation, not *intra*-system relaxation to which the so-called critical slowing down belongs.

Next, we consider how the internal state changes between before and after one of the excited elements relaxes, by introducing *effective temperature change* $\Delta T^{(E)} \equiv T^{(E)}_{after} N^{(E)} / (N^{(E)} - 1) - T^{(E)}_{before}$, where $T^{(E)}_{before}$ and $T^{(E)}_{after}$ are the effective temperature before and after τ, respectively, while the factor $N^{(E)} / (N^{(E)} - 1)$ comes from the contribution of scaling [5]. Its dependence on $T^{(E)}_{before}$ (FIGURE 3) shows that $T^{(E)}$ tends to increase when $T^{(E)}_{before} < T^{(E)}_c$, and decrease when $T^{(E)}_{before} > T^{(E)}_c$. Thus the internal state tends to be attracted into the critical state. Note that this dependence can be roughly evaluated analytically (not shown here [1]). We conclude, therefore, that the bottleneck is self-organized.

Although we have demonstrated the self-organized bottleneck at a critical state by using a simple Hamiltonian system, the mechanism is expected to be general, as long as the system can exhibit continuous phase transition, and one part of the system is highly excited. Hence this phenomenon can be observed in condensed matters, and in molecules with sufficiently large degrees of freedom, when a part of them is highly excited with light or something like that.

ACKNOWLEDGMENTS

The authors thank S. Sasa, N. Nakagawa, and A. Shimizu for discussion. This work was supported by a Grant-in-Aid for Scientific Research from the Ministry of Education, Science, and Culture of Japan.

REFERENCES

1. Morita, H. and Kaneko, K., cond-mat/0304649.
2. Konishi, T. and Kaneko, K., *J. Phys. A*, **25**, 6283-6296 (1992).
3. Antoni, M. and Ruffo, S., *Phys. Rev. E* **52**, 2361-2374 (1995); Yamaguchi, Y. Y., *Prog. Theor. Phys.* **95**, 717-731 (1996); Latora, V., Rapisarda, A., and Ruffo, S., *Phys. Rev. Lett.* **83**, 2104-2107 (1999).
4. Nakagawa, N. and Kaneko, K., *J. Phys. Soc. Jpn.* **69**, 1255-1258 (2000); *Phys. Rev. E* **64**, 055205(R) (2001).
5. Hamiltonian (1) is scaled with K, by $(\theta, p, t, H) \rightarrow (\theta, \sqrt{K} p, t/\sqrt{K}, KH)$. Each element evolves by the equation of motion with factor K/N. Hence the excited part as a thermodynamic system is scaled with the effective coupling constant $K^{(E)}$ defined as $K/N = K^{(E)}/N^{(E)}$. Since the temperature is also scaled as $T \rightarrow KT$, the effective critical temperature $T^{(E)}_c = T_c K^{(E)}/K = T_c N^{(E)}/N$.

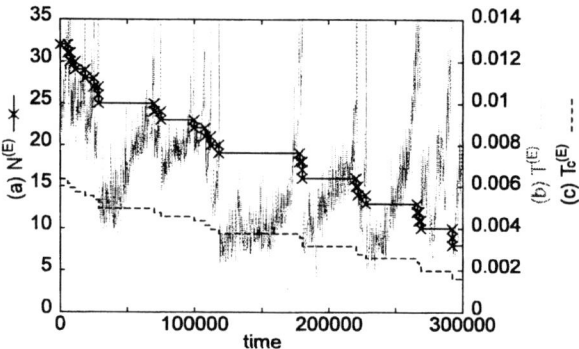

FIGURE 1. A typical time series of (a) the population of the excited elements $N^{(E)}$, (b) the effective temperature of the excited part $T^{(E)}$, and (c) the effective critical temperature $T^{(E)}_c$. $N = 64$, $T = 0.01$, $\gamma = 0.01$, $N^{(K)} = 32$, $P^{(K)} = 0.6$.

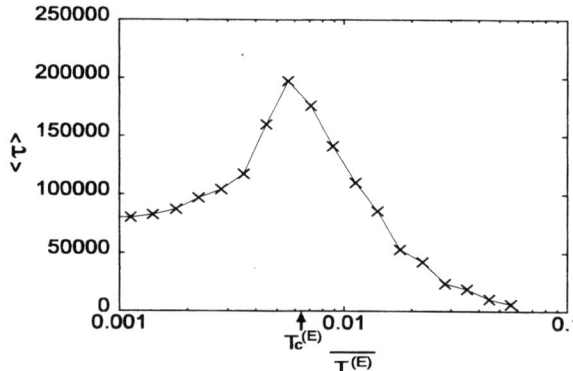

FIGURE 2. The first relaxation time τ versus the effective temperature $T^{(E)}$. $T^{(E)}$ is time-averaged over a period of 1000. 2000 samples are computed in all, and τ is averaged for the samples within the bin size of 0.1 for log scale. $N = 32$, $T = 0.01$, $\gamma = 0.01$, $N^{(K)} = 16$, $P^{(K)} = 0.8$.

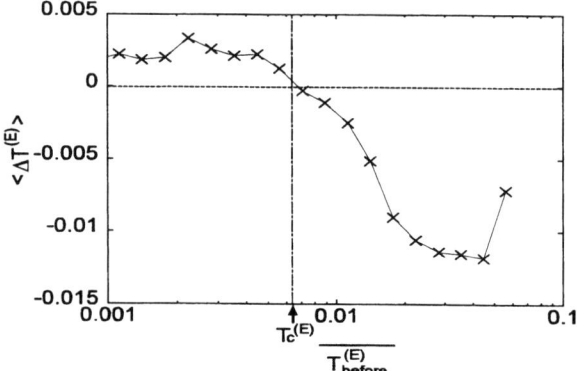

FIGURE 3. The effective temperature change $\Delta T^{(E)}$ versus the effective temperature before relaxation $T^{(E)}_{before}$, computed from the simulations of FIGURE 2.

Exact Relations for Quantum Many-Body Correlation Functions

T. Toyoda[(1)], H. Koizumi[(2)], K. Ito[(3)] and K. Takiuchi[(4)]

[(1)]Dept. of Physics, Tokai Univ., Kitakaname 1117, Hiratsuka, Kanagawa, Japan 259-1292
[(2)]Nippon Gear Co., Ltd., Kirihara-cho 7, Fujisawa, Kanagawa, Japan 252-0811
[(3)]NCUEE, Komaba 2-19-23, Meguro, Tokyo, Japan 153-8501
[(4)]Japan Electronics College, Hyakunin-cho 1-25-4, Shinjuku, Tokyo, Japan 169-8522

Abstract. Basic theoretical mechanism of the nonperturbative canonical formulation of Matsubara temperature Green's functions, including the finite temperature generalized Ward-Takahashi relations, is examined.

Almost half a century ago Matsubara [1] initiated the quantum field theoretical method of statistical physics by introducing the Matsubara temperature Green's functions. Immediately after Matsubara's work, Martin and Schwinger [2] thoroughly examined the analytical properties of the Matsubara Green's functions and showed the way to apply them to time-dependent problems by analytic continuation of the frequency. Matsubara temperature Green's functions have been further elaborated by Abrikosov, Gor'kov and Dyaloshinskii [3], who developed a systematic expansion technique on the basis of the finite temperature version of Wick's theorem found by Ezawa, Tomozawa and Umezawa [4]. These works contributed enormously to the subsequent developments in condensed matter theory in general. A great number of important theoretical contributions in condensed matter physics have been made by using this perturbative method of the Matsubara finite temperature Green's functions.

However, there is an essential fundamental difficulty in such a perturbative method. Important results such as theories of plasmons, superconductivity, superfluidity, etc., have been obtained by going beyond the rigorous scheme of the perturbational expansion. In the languages of Feynman diagrams, only certain types of diagrams were considered and other diagrams were neglected to obtain these results. There were no satisfactory theoretical formulation that could be used to justify or to correct such seemingly arbitrary approximations. The difficulty was finally settled by the nonperturbative canonical formulation of Matsubara temperature Green's function developed by Toyoda [5]. The formulation yields the finite temperature generalized Ward-Takahashi relations (FTGWTR), which are exact relations between Matsubara temperature Green's functions that reflect symmetry properties of hamiltonians. The FTGWTR can be used to enforce relevant conservation laws in perturbative approximations. The FTGWTR can be also used to relate different physical quantities without approximations.

In this paper we shall illustrate the basic theoretical mechanism of the nonperturbative canonical formulation of Matsubara temperature Green's function. We shall consider the quantized Schroedinger field. One of the most straightforward results of the nonperturbative canonical formulation is the FTGWTR

$$\Sigma(p+q) - \Sigma(p) = \frac{1}{\hbar} \int_k \{G(k+q) - G(k)\} \tilde{\Gamma}(k, p+q; p, k+q) \quad (1)$$

which relates the proper self-energy and the proper vertex part. This relation is a direct consequence of the equation of continuity of the particle number density and plays a crucial role in the quantum transport theory. The peturbative approach to the non-equilibrium properties of a Fermi liquid was developed by Eliashberg [3]. The essential features of the Eliashberg theory are the Bethe-Salpeter equation for the two-particle Green's function and the Fermi liquid Ansatz for the one-particle Green's function. From the two-particle Green's function one can calculate various response functions such as the retarded current response function. A perturbative approximation for calculating the two-particle Green's function can be characterized by a model for the proper vertex part appeared as an integral kernel in the Bethe-Salpeter equation. The above FTGWTR shows such a model proper vertex part is not arbitrary but related to the proper self-energy that appears in the Fermi liquid Ansatz. For example, it has been clearly shown that in the calculation of the electrical conductivity the use of the FTGWTR is essential to obtain the correct result [6]. The use of the FTGWTR is equivalent to the conserving approximation formulated by Baym and Kadanoff [7].

Another important consequence of the nonperturbative canonical formulation is an exact generalization of the linear response formula. We consider a model Hamiltonian for an interacting electron gas

$$H = H_A + H_S + H_{int} + H_{imp} \tag{2}$$

where the first term is

$$H_A = \int d^3x \, \psi_\alpha^\dagger(x) \xi_A(\nabla) \psi_\alpha(x) \tag{3}$$

with the differential operator

$$\xi_A(\nabla) = \frac{-\hbar^2}{2m} \sum_k \left(\partial_k + \frac{ie}{\hbar} A_k(\mathbf{x}) \right)^2 - \mu \tag{4}$$

and the second term is the coupling between the magnetic field and the electron spin

$$H_S = \frac{g\mu_B}{2} \int d^3x \sum_i B_i(\mathbf{x}) S_i(\mathbf{x}) \tag{5}$$

where the electron spin operator

$$S_i(x) \equiv \psi_\alpha^\dagger(x) \sigma_{\alpha\beta}^i \psi_\beta(x) \tag{6}$$

is used. Using the non-perturbative canonical formulation of the Matsubara Green's function the following result

$$\langle S_i(x) \rangle = \frac{g\mu_B}{4\hbar} \int d^3\mathbf{x} \int_0^{\beta\hbar} d\tau' \times$$

$$\times \left[-B_i(\mathbf{x}') \sum_k \langle T_\tau \{ S_k(x') S_k(x) \} \rangle \right.$$

$$\left. + \sum_k B_k(\mathbf{x}') \langle T_\tau \{ S_i(x') S_k(x) \} \rangle \right] \tag{7}$$

has been derived [8]. This results looks like a linear response formula. However, the result is derived without any approximations. The quantum statistical average <...> contains the external field. Many similar useful relations for the Green's functions have been derived on the basis of the nonperturbative canonical formulation [9,10,11,12].

REFERENCES

1. Matsubara, T., *Progr. Theor. Phys.*, **14**, 351-367 (1955).

2. Martin, P. C., and Schwinger, J., *Phys. Rev.*, **115**, 1342-1373 (1959).

3. Abrikosov, A. A., Gor'kov, L. P., and Dzyaloshinskii, L. P. Ye., *Quantum Field Theoretical Methods in Statistical Physics*, Pergamon, Oxford, 1965.

4. Ezawa, E., Tomozawa, Y., and Umezawa, H., *Nuovo Cimento*, **10**, 810-841 (1957).

5. Toyoda, T., *Ann. Phys. (NY)*, **173**, 226-245 (1987).

6. Toyoda, T., *Phys. Rev. A*, **39**, 2659-2671 (1989).

7. Baym, G., and Kadanoff, L. P., *Phys. Rev.*, **124**, 287-299 (1961).

8. Toyoda, T., and Okada, M., *Phys. Rev. B*, **58**, 1210-1217 (1998).

9. Toyoda, T., and Takiuchi, K., *Physica A*, **261**, 471-481 (1998)

10. Takiuchi, K., and Toyoda, T., *Phys. Lett. A*, **262**, 40-43 (1999).

11. Ito, K., Fukuda, M., Toyoda, T., *Phys. Rev. B*, **56**, 10161-10164 (1997).

12. Toyoda, T., and Koizumi, H., *J. Phys. A: Math. Gen.*, **31**, 6209-6218 (1998).

New Canonical Transformations to Eliminate External Fields in Quantum Many-Body Problems

M. Fujita[1], D. Anma[1], T. Fukuda[1], H. Koizumi[2], T. Toyoda[1]

[1] *Department of Physics, Tokai University*
Kitakaname 1117, Hiratsuka, Kanagawa, Japan 259-1292
[2] *Nippon Gear Co., Ltd.*
Kirihara-cho 7, Fujisawa, Kanagawa, Japan 252-0811

Abstract. The wave function of an electron in a time varying electric field, a static magnetic field and a impurity potential can be transformed to the wave function of an electron without the electric field. By a unitary transformation, the electric field can be rigorously eliminated.

In the conventional quantum many-body theory, the effects of an external field are treated within the framework of the linear response approximation. However, if the external field is not weak enough, it is necessary to go beyond the linear response approximation. Recently two seemingly different new approaches to overcome the deficiency of the linear response approximation have been proposed, particularly in the problem of an external electric field applied to a many-electron problem. Toyoda and Koizumi extended the Avron-Herbst transformation to an interacting many-electron system and showed that an external electric field can be removed by a canonical transformation [1]. Their external electric field is static and spatially uniform. On the other hand Chao [2, 3] showed a time-dependent external electric field acting on an electron wave function can be eliminated by a unitary transformation. The unitary transformation derived by Chao has an advantage that one can consider an oscillating external electric field such as laser. In this work we extend Chao's transformation to the electron system with a uniform and static magnetic field as well as a potential describing the effects of impurity atoms.

We consider a two dimensional electron gas under a uniform static magnetic field and an intense laser radiation. We choose the static magnetic field to be along the z-direction and the laser electric field to be along the y-direction,, where E and ω are the amplitude and frequency of the laser field. We use the following vector potentials to describe the static magnetic field and the time-dependent laser field

$$A_y = \frac{Ec}{\omega}\cos(\omega t) + Bx \tag{1}$$

where B is the magnetic flux density. Then the Hamiltonian is given as

$$H(t) = \frac{1}{2m^*}\hat{p}^2_x + \frac{1}{2m^*}\left(\hat{p}_y - \frac{e}{c}B\hat{x} - \frac{Ec}{\omega}\cos(\omega t)\right)^2 + V(\hat{x},\hat{y}) \tag{2}$$

where m^* is the electron effective mass, e is the electron. The last term $V(x, y)$ is the potential due to impurity atoms. We define the unitary transformation

$$|\Psi(t)\rangle = U(t)|\Phi(t)\rangle \tag{3}$$

with the unitary operator

$$U \equiv \exp\left(\frac{i}{\hbar}u_1(t)\right)\exp\left(\frac{i}{\hbar}u_2(t)\hat{x}\right) \times \exp\left(\frac{i}{\hbar}u_3(t)\hat{p}_x\right)\exp\left(\frac{i}{\hbar}u_5(t)\hat{p}_y\right) \tag{4}$$

Now we require that the Schrödinger equation for the state does not contain the $A(t)$ term, i.e.,

$$i\hbar \frac{\partial}{\partial t}|\Phi(t)\rangle = \{H_0 + V[\hat{x} - u_3(t), \hat{y} - u_5(t)]\}|\Phi(t)\rangle \quad (5)$$

where H_0 is

$$H(t) = \frac{1}{2m^*}\hat{p}^2_x + \frac{1}{2m^*}\left(\hat{p}_y - \frac{e}{c}B\hat{x}\right)^2 + V(\hat{x}, \hat{y}) \quad (6)$$

This requirement leads to a set of differential equations for the u_i's:

$$-\dot{u}_1 + \dot{u}_2 u_3 = \frac{1}{2m}\left(\frac{eB}{c}u_3 - \frac{e}{c}A(t)\right)^2 + \frac{1}{2m}u_2^2 \quad (7)$$

$$-\dot{u}_2 = \frac{-eB}{m}\left(\frac{eB}{c}u_3 - \frac{e}{c}A(t)\right) \quad (8)$$

$$-\dot{u}_3 = \frac{1}{m}u_2 \quad (9)$$

$$-\dot{u}_5 = \frac{1}{m}\left(\frac{eB}{c}u_3 - \frac{e}{c}A(t)\right) \quad (10)$$

These differential equations can be solved analytically. We have obtained the following results[4]

$$u_1 = \frac{e^2 E^2}{2m^*}\left(\frac{1}{\omega_c^2 - \omega^2}\right)^2 \times$$

$$\times \left\{\frac{\omega_c^3}{2\omega^2}\sin(2\omega_c t) + \frac{\omega}{4}\left(\frac{3\omega_c^2}{\omega^2} - 1\right)\sin(2\omega t) + \quad (11)\right.$$

$$\left. \left(\frac{\omega_c^2 - \omega^2}{2}\right)t - \frac{\omega_c^3}{\omega^2}\{\sin[(\omega_c - \omega)t] + \sin[(\omega_c + \omega)t]\}\right\}$$

$$u_2(t) = eE\left(\frac{\omega_c}{\omega_c^2 - \omega^2}\right)\left[\frac{-\omega_2}{\omega}\sin(\omega_c t) + \sin(\omega t)\right] \quad (12)$$

$$u_3(t) = \frac{eE}{m^*}\frac{\omega_c}{\omega}\left(\frac{1}{\omega_c^2 - \omega^2}\right)[-\cos(\omega_c t) + \cos(\omega t)] \quad (13)$$

$$u_5(t) = \frac{eE}{m^*}\left(\frac{1}{\omega_c^2 - \omega^2}\right)\left[\frac{\omega_c}{\omega}\sin(\omega_c t) - \sin(\omega t)\right] \quad (14)$$

Substituting these functions into the unitary operator given by (4), we can transform the wave function for the electron in the time varying electric field to that without the electric field. The transformed wave function satisfies the Schrödinger equation (5), where the impurity potential now has an explicit time dependence. The effects of the time dependent impurity potential can be treated as a time-dependent perturbation.

The present formulation opens a new way to treat the effects of a time dependent strong electric field acting on the electrons. No approximations, such as linear approximation, is required to obtain the result. Therefore, various interesting phenomena in strongly coupled electron-laser systems such as resonant absorption [5], THz switching effect in resonant tunneling [6], THz cyclotron resonance [7], etc., can be theoretically examined within the framework of the present theory.

REFERENCES

1. Toyoda, T., and Koizumi, H., *J. Phys. A: Math. Gen.*, **31**, 6209-6218 (1998).

2. Zhang, C., and Xu, W., *Physica B*, **298**, 333-338 (2001).

3. Zhang, C., *Phys. Rev. B*, **65**, 153107 (2002).

4. Fujita, M., Toyoda, T., Cao, J. C., Zhang, C., *Phys. Rev. B*, **67**, 075105 (2003).

5. Koehler, R., *Nature*, **417**, 156-159 (2002).

6. Asmar, N. G., Makelz, A. G., Gwinn, E. G., Cerne, J., Sherwin, M. S., Campman, K. L., Hopkins, P. F., Gossard, A. C., *Phys. Rev. B*, **51**, 18041-18044 (1995).

7. Orellana, P., Claro, F., Anda, E., *Phys. Rev. B*, **62**, 9959-9961 (2000).

Quantum Many-Body Virial Theorem And Matsubara Green's Function

D. Anma[a], T. Fukuda[a], M. Fujita[a], K. Takiuchi[b], T. Toyoda[a]

[a] *Department of Physics, Tokai University*
Kitakaname 1117, Hiratsuka, Kanagawa, Japan 259-1292
[b] *Japan Electronics College*
Hyakunin-cho 1-25-4, Shinzyu-ku, Tokyo, Japan 169-8522

Abstract. We discuss the quantum field theoretical formulation of the virial theorem on the basis of the canonical field theory of the generalized coordinate transformation and show the equation of motion of a charged Fermion system coupled to an electromagnetic field. Possible application to Fermion-Boson mixtures is also discussed.

The general coordinate transformation has been one of the most basic problems in classical field theory. However its extension to the theory of quantized field had not been developed for decades. An infinitesimal general coordinate transformation is given as

$$\mathbf{r} \to \mathbf{r}' = \mathbf{r} + \eta \mathbf{f}(\mathbf{r}) \qquad (1)$$

In order to deal with this general coordinate transformation in quantum field theory, the best approach is to use the canonical formulation. The first step is to examine the transformation property of the field operator and to find the Lie derivatives. In 1989 Toyoda [1] studied the quantized Schrödinger field and found the following Lie derivatives

$$\delta^L \psi(\mathbf{r}) = -\left\{\frac{1}{2}(\nabla \cdot \mathbf{f}(\mathbf{r})) + \mathbf{f}(\mathbf{r}) \cdot \nabla\right\} \psi(\mathbf{r}) \qquad (2)$$

$$\delta^L \psi^\dagger(\mathbf{r}) = -\left\{\frac{1}{2}(\nabla \cdot \mathbf{f}(\mathbf{r})) + \mathbf{f}(\mathbf{r}) \cdot \nabla\right\} \psi^\dagger(\mathbf{r}) \qquad (3)$$

The canonical generator of these Lie derivatives is given as

$$G = \frac{1}{2}\int d^3\mathbf{r} \, \mathbf{f}(\mathbf{r}) \cdot \left\{\psi^\dagger(\mathbf{r})(\nabla - \overleftarrow{\nabla})\psi(\mathbf{r})\right\} \qquad (4)$$

which was obtained for the first time by Toyoda [1]. The general coordinate transformation includes the conformal transformation and the scale transformation. This general formulation has been applied to a scale transformation of the spatial coordinates to derive a quantum field theoretical version of the virial theorem by Toyoda and Takiuchi [2]. An infinitesimal scale transformation of the spatial coordinates is given by

$$\mathbf{r} \to \mathbf{r}' = (1 + \eta)\mathbf{r} \qquad (5)$$

Then, the corresponding Lie derivatives of the field operators are

$$\delta^L \psi(\mathbf{r}) = -\left(\frac{3}{2} + \mathbf{r} \cdot \nabla\right)\psi(\mathbf{r}) \qquad (6)$$

$$\delta^L \psi^\dagger(\mathbf{r}) = -\left(\frac{3}{2} + \mathbf{r} \cdot \nabla\right)\psi^\dagger(\mathbf{r}) \qquad (7)$$

Then the canonical generator that yields these Lie derivatives can be obtained by straightforwardly obtained from (4),

$$G = \frac{1}{2}\int d^3\mathbf{r} \, \mathbf{r} \cdot \left\{\psi^\dagger(\mathbf{r})(\nabla - \overleftarrow{\nabla})\psi(\mathbf{r})\right\} \qquad (8)$$

By making use of this canonical generator of the scale transformation, the virial theorem can be extended to the quantum many-body problem and then the equation of states

$$PV = \frac{2}{3}\int d^3\mathbf{r}\left\langle \psi^\dagger(\mathbf{r})\left(\frac{-\hbar^2}{2m}\nabla^2\right)\psi(\mathbf{r})\right\rangle \\ -\frac{1}{6}\int d^3\mathbf{r}\int d^3\mathbf{r}'\{(\mathbf{r}-\mathbf{r}')\cdot\nabla V(\mathbf{r}-\mathbf{r}')\}\langle\rho(\mathbf{r})\rho(\mathbf{r}')\rangle \quad (9)$$

can be derived rigorously. Subsequently, Takiuchi and Toyoda succeeded to include the coupling between the Schrödinger field and an electromagnetic field [3]. The canonical generator of the scale transformation for the Schrödinger field that coupled to the electromagnetic gauge field is found to be

$$G_A = G + \frac{ie}{\hbar c}\sum_\alpha \int d^3\mathbf{r}\,\mathbf{r}\cdot\mathbf{A}(\mathbf{r})\psi_\alpha^\dagger(\mathbf{r})\psi_\alpha(\mathbf{r}) \quad (10)$$

Using this new canonical generator Takiuchi and Toyoda derived the equation of state

$$PV \\ = \frac{2}{3}\sum_\alpha \int d^3\mathbf{r}\left\langle \psi_\alpha^\dagger(\mathbf{r})\frac{-\hbar^2}{2m}\left(\nabla + \frac{ie}{\hbar c}\mathbf{A}(\mathbf{r})\right)^2\psi_\alpha(\mathbf{r})\right\rangle \\ -\frac{1}{3c}\int d^3\mathbf{r}\,\mathbf{r}\cdot\{\mathbf{B}(\mathbf{r})\times\langle\mathbf{J}(\mathbf{r})\rangle\} \\ -\frac{g\mu_B}{6}\int d^3\mathbf{r}\{(\mathbf{r}\cdot\nabla)\mathbf{B}(\mathbf{r})\}\cdot\langle\mathbf{S}(\mathbf{r})\rangle \\ -\frac{1}{6}\int d^3\mathbf{r}\int d^3\mathbf{r}'\{(\mathbf{r}-\mathbf{r}')\cdot\nabla V_0(\mathbf{r}-\mathbf{r}')\}\langle\rho(\mathbf{r})\rho(\mathbf{r}')\rangle \\ -\frac{1}{6}\int d^3\mathbf{r}\int d^3\mathbf{r}'\{(\mathbf{r}-\mathbf{r}')\cdot\nabla V_1(\mathbf{r}-\mathbf{r}')\} \\ \times \sum_{i=1}^{3}\frac{1}{4}\langle S^{(i)}(\mathbf{r})S^{(i)}(\mathbf{r}')\rangle \quad (11)$$

which includes the electron spin correlation function.

In 1957 Ezawa, Tomozawa and Umezawa tried to develop a canonical quantum theory of the scale transformation [4] but failed to derive the quantum field theoretical version of the virial theorem.

From both experimental and theoretical point of views a quantum many-body system of Fermion-Boson mixture is a fascinating object. In 2002 we started to extend the above described formulation of the quantum field theoretical virial theorem to a Fermion-Boson mixture. Experimental studies of Fermion-Boson mixtures are mainly performed on the ^{3}He-^{4}He mixtures.

In 1965 Edwards et al. [5] discovered that there is no phase separation in the mixtures of liquid ^{3}He and ^{4}He containing less than 6 % of ^{3}He at very low temperatures. Their finding was confirmed next year by Anderson et al. [6]. In 1966 Emery [7] proposed a phenomenological effective model Hamiltonian. In the same year, Bardeen, Baym, and Pines [8] derived the approximate form of the effective interaction between ^{3}He atoms in liquid ^{4}He from the experimental data on spin diffusion and phase separation in dilute mixtures of ^{3}He in ^{4}He. Although a number of theoretical investigations have been carried out since then, it seems there have been no serious attempts to derive the equation of state that describes the liquid ^{3}He-^{4}He mixtures. The canonical formulation of the field theoretical virial theorem described in this paper can be applied to such a Fermion-Boson mixture. The canonical generator for the mixture can be written as

$$G_{mixture} = G_{Fermion} + G_{Boson} \quad (12)$$

where the first term is the Fermion generator and the second term is the Boson generator. Because these two generators commute with each other, the entire formulation can be straightforwardly applied [9].

REFERENCES

1. Toyoda, T., *Phys. Rev. A*, **39**, 2659-2671 (1989).

2. Toyoda, T., and Takiuchi, K., *Physica A*, **65**, 471-481 (1998).

3. Takiuchi, K., and Toyoda, T., *Phys. Letters A*, **262**, 40-43 (1999).

4. Ezawa, H., Tomozawa, Y., and Umezawa, H., *Nuovo Cimento*, **5**, 810-814 (1957).

5. Edwards, D. O., Brewer, D. F., Seligman, P., Skertic, M., and Yaqub, M., *Phys. Rev. Lett.*, **15**, 773-775 (1965).

6. Anderson, A. C., Roach, W. R., Sarwinski, R. E., and Wheatley, J. C., *Phys. Rev. Lett.*, **16**, 263-264 (1966).

7. Emery, V. J., *Phys. Rev.*, **148**, 138-145 (1966).

8. Bardeen, J., Baym, G., and Pines, D., *Phys. Rev.*, **156**, 207-221 (1967).

9. Anma, D., Takiuchi, K., and Toyoda, T., in preparation.

Quasi-Particle Lifetime of Quantum Coulomb Systems

T. Fukuda[1], M. Fujita[1], D. Anma[1], K. Ito[2], and T. Toyoda[1]

[1] *Department of Physics, Tokai University*
Kitakaname 1117, Hiratsuka, Kanagawa, Japan 259-1292
[2] *The National Center for University Entrance Examinations*
Komaba 2-19-23, Meguro-ku, Tokyo, Japan 153-8501

Abstract. A formula that relates the imaginary part of the proper self-energy and the proper vertex part can be derived from the finite temperature generalized Ward-Takahashi relation. Using the formula, the effects of the electron-electron Coulomb interaction on the lifetime of the electron quasi-particles can be calculated within the well-established RPA scheme.

The concept of quasi-particles is one of the most useful tools in understanding a wide range of phenomena in condensed matter physics, particularly to describe and to understand the properties of electrons in metals or semiconductors. In the quantum field theoretical formulation of quantum many-body problem, the concept of quasi-particles is directly related to the field operators in the Heisenberg picture. If we consider a system of an interacting d-dimensional electron gas (d=2 or 3), a model Hamiltonian may be written as

$$H = \int d^d x \, \psi_\alpha^\dagger(\mathbf{x}) \left[\frac{-\hbar^2}{2m} \nabla^2 - \mu \right] \psi_\alpha(\mathbf{x}) + H_{\text{int}} \qquad (1)$$

where m and μ are the mass and the chemical potential of the electrons, respectively. Here we adopt Einstein's convention with respect to the Greek subscripts denoting the electron spin variables unless otherwise stated. If the field operator is acted on the ground state, we have a quasi-particle state. Because of the interaction term H_{int} in the Hamiltonian, such a quasi-particle state cannot be an energy eigenstate. Hence the corresponding wave function of the quasi-particle will change during the time evolution. This can be described in terms of a lifetime and an effective mass. Then an important theoretical problem is to calculate the effects of the electron-electron interaction on these quantities.

A number of authors used perturbative expansion of Matsubara Green's function to calculate the effects of the electron-electron interaction on the lifetime of the quasi-particles. Because the lifetime appears in the poles of the one-particle Green's function, a finite perturbation series cannot give any shift of the lifetime. A shift of the pole of the Green's function cannot be calculated by a finite perturbation series. Instead, one has to consider an infinite number of perturbation terms to obtain such a shift of the lifetime. Hence an approximation scheme to calculate the effects of the electron-electron interaction on the lifetime must be accompanied by a summation of a particular set of infinite number of terms. The choice of such a particular set of the diagrams characterizes the approximation. It has turned out that there are no well established reliable criteria that can be used to justify the choice of such diagrams in various approximations. In view of the fact that the lifetime of the electrons quasi-particles is one of the most important physical quantities to link various experiments and theories, the lack of solid basis for the approximation schemes is quite disturbing.

Meanwhile a powerful method in quantum many-body theory has been developed since 1987 by Toyoda and his collaborators [1], i.e., the nonperturbative canonical formulation of finite temperature quantum many-body field theory, which is a rigorous extension of Matsubara's temperature Green's function theory on the basis of the canonical commutation relation of the field operators and current operators. One of the most useful features obtained in this theory is the set of the finite temperature generalized Ward-Takahashi relations (FTGWTR). Among the number of useful FTGWTR derived, one of the most powerful FTGWTR is

$$\Sigma(p+q) - \Sigma(p) = \frac{1}{\eta} \int_k \{G(k+q) - G(k)\} \Gamma(k, p+q; p, k+q) \quad (2)$$

which gives a rigorous relation between the proper self-energy Σ and the proper vertex part Γ of the electron Green's functions. The proper self-energy gives the effective mass of the electron field and the proper vertex part describes the electron-electron interaction. To choose a particular model of the proper vertex part corresponds to select a particular type of diagrams for summation in the diagrammatic perturbative expansion. The well known approximation such as the RPA or the ladder approximation can be obtained by selecting relevant model for the proper vertex part. The formula (2) is extremely important in the theoretical study of the lifetime of the quasi-particles. If we take a limit on the left-hand side, it gives the imaginary part of the proper self-energy, which is directly related to the lifetime. Hence this formula can be used to calculate the lifetime of the quasi-particles directly from the proper vertex part.

This FTGWTR can be further rewritten in the following form [2]

$$\mathrm{Im}\Sigma^R(\mathbf{q}, \omega) = \frac{1}{\pi\eta} \int \frac{d^d p}{(2\pi)^d} \int_{-\infty}^{\infty} dx [f(x) + \bar{f}(x-\omega)] \quad (3)$$
$$\times \mathrm{Im} G^R(\mathbf{p}, x) \mathrm{Im} \Pi_2(\mathbf{p}-\mathbf{q}, \omega-x)$$

where the functions f and $\bar{f}$ are defined as

$$f(x) \equiv \frac{1}{\exp(\beta\eta x) + 1}, \quad \bar{f}(x) \equiv \frac{1}{\exp(\beta\eta x) - 1}. \quad (4)$$

and Π_2 is an exchange part of the vertex part [3]. The expression (3) has a very important physical implication. In the perturbative calculations of the lifetime of the electron quasi-particles, it has been known that the ladder type diagrams are relevant. The ladder diagrams correspond to the boson exchange type proper vertex function in the B-S equation for the two-particle Green's function. From the expression (3) one can immediately conclude that the exchange type proper vertex part contributes to the non-vanishing imaginary part of the proper self-energy that corresponds the finite lifetime of the electron quasi-particles. The formula (3) also has a practical importance in the calculation of the lifetime of the electron quasi-particles. If we apply the formula to a two-dimensional electron gas, we obtain

$$\mathrm{Im}\Sigma^R(\mathbf{p}, \omega) =$$
$$\frac{-am^*}{2\pi^2 \eta^2} \int_{\Omega_1(\omega)}^{\Omega_2(p,\omega)} d\omega' \int_{k_-}^{k_+} dk Z(p, k, \omega, \omega')$$
$$+ \frac{-am^*}{2\pi^2 \eta^2} \int_{\Omega_2(p,\omega)}^{\infty} d\omega' \int_{-k_-}^{k_+} dk Z(p, k, \omega, \omega') \quad (5)$$

where we have defined

$$Z = [f(\omega' + \omega) + \bar{f}(\omega')] F \mathrm{Im} \Pi_2^R(k, \omega') \quad (6)$$

and

$$F \equiv \left[(2pk)^2 - \left(\frac{2m^*}{\eta}(\omega' + \omega) - p^2 - k^2 + p_F^2 \right)^2 \right]^{-1/2} \quad (7)$$

The frequency variables and the wave-numbers are defined as

$$\Omega_1 \equiv \frac{-\eta p_F^2}{2m^*} - \omega, \quad \Omega_2 \equiv \frac{-\eta(p^2 - p_F^2)}{2m^*} - \omega \quad (8)$$

and

$$k_\pm \equiv p \pm \sqrt{p_F^2 + \frac{2m^*}{\eta^2}(\omega' + \omega)} \quad (9)$$

The proper vertex part can be chosen as the RPA dynamical potential, i.e., the screened Coulomb potential with the frequency dependence. Thus the calculation of the quasi-particle lifetime can be directly linked the RPA scheme [4].

REFERENCES

1. Toyoda, T., *Ann. Phys. (N.Y.)*, **173**, 226-245 (1987).
2. Toyoda, T., *Phys. Rev. A*, **48**, 3492-3498 (1993).
3. Toyoda, T., and Ito, K., *Phys. Rev. B*, **64**, 073104 (2001).
4. Fukuda, T., and Toyoda, T., in preparation.

Behavior of observables in nonequilibrium state

H. Majima* and A. Suzuki

*Center for Solid-State Physics and Department of Physics, Science University of Tokyo
1-3 Kagurazaka, Shinjuku-ku, Tokyo 162-8601, Japan.

Abstract. One possible way to treat the change in thermal states in thermal nonequilibrium processes is presented without any ad hoc assumption on the system Hamiltonian on the basis of thermo field dynamics.

INTRODUCTION

Many attempts have been made to understand thermal nonequilibrium processes. However, no satisfactory microscopic theory has yet been developed since those phenomena like heat conduction and diffusion can not be expressed in terms of Hamiltonian. This makes difficulties in developing a theory for such thermal irreversible processes [1, 2].

One way to avoid such difficulties is that we should pay our attention to the change in the state of physical system instead of making any *ad hoc* assumptions into Hamiltonian. Here we show one possible way to treat such a problem with the help of the theory of thermo field dynamics (TFD) originally developed by Takahashi and Umezawa [3].

In TFD, the statistical averaging of an observable A in ordinary quantum-statistical mechanics is expressed in terms of the thermal vacuum $|\beta\rangle$:

$$\langle A \rangle = \mathrm{Tr}(\rho A) \equiv \langle \beta | A | \beta \rangle, \quad (1)$$

where ρ is a density operator and $|\beta\rangle$ denotes the thermal vacuum state characterized by an inverse temperature β ($\equiv 1/k_B T$). Thermal states are generated by the heat-up operator

$$B[\beta] = \sum_{\mathbf{k}} \theta_k(\beta) a_{\mathbf{k}}^{a\dagger} \tau_2^{ab} a_{\mathbf{k}}^{b} \quad (2)$$

that operates on the zero temperature vacuum $|0\rangle \otimes |0\rangle$ as

$$|\beta\rangle = e^{iB[\beta]} |0\rangle \otimes |0\rangle. \quad (3)$$

In Eq. (2), τ_2^{ab} is the Pauli matrix

$$\tau_2^{ab} = \begin{pmatrix} 0 & -i \\ i & 0 \end{pmatrix} \quad (4)$$

and the thermal doublet notation

$$a_{\mathbf{k}}^{a} = \begin{pmatrix} a_{\mathbf{k}} \\ \tilde{a}_{\mathbf{k}}^{\dagger} \end{pmatrix} \quad (5)$$

is used for the annihilation ($a_{\mathbf{k}}$) and creation ($\tilde{a}_{\mathbf{k}}^{\dagger}$) operators in TFD and will be used hereafter.

In this paper we will show how the particle distribution of a system including the effect of irreversible process can be constructed from constituting thermal state of the system on the basis of a thermal Heisenberg operator (THO) formulation of TFD and derive the formulas for the particle number density and the propagator on the basis of TFD mentioned above. In this paper, we work out in Thermal Heisenberg Picture (THP), recently established by the present authors [4]

THERMO FIELD DYNAMICS OF IRREVERSIBLE PROCESS

We consider a system where a local temperature is extending spatiotemporally. Here we construct the theory for such a nonequilibrium phenomenon. In TFD, states of the system at time t can be defined by introducing a local temperature since once a local temperature of the system is defined, subsequent thermal states can be specifically definded in TFD due to the fact that when one temperature β^{-1} is decided, one state will become settled. Therefore, to formulate the problem, we assume a system initially in local equilibrium with inhomogeneous temperature by introducing a local temperature. By this assumption, we can construct a theory within the flamework of TFD without any modification or ad hoc assumption on the Hamiltonian of a system.

Let us consider a system initially in local equilibrium with inhomogeneous temperature by introducing an inverse local temperature $\beta = \beta_0 + \beta'(\mathbf{x},t)$ instead of $\beta = \beta_0$. Here time (t) dependence is entered through β'. Since the macroscopic change of temperature usually takes place in high-temperature regime [5], we shall consider the case $|\beta'| \ll |\beta_0|$. In this case, the heat-up operator $B[\beta]$ can be expanded with respect to β' and approxi-

mated as
$$B[\beta] \approx B[\beta_0] + \Gamma[\beta'], \quad (6)$$

where $\Gamma[\beta]$ is given by

$$\Gamma[\beta'] = i\sum_{\mathbf{k}} a_{\mathbf{k}}^{a\dagger} \tau_2^{ab} a_{\mathbf{k}}^b g_k[\beta_0]\beta', \quad g_k[\beta_0] = \left.\frac{\partial \theta_k}{\partial \beta'}\right|_{\beta=\beta_0}. \quad (7)$$

For bosons, θ is determined from [4]

$$\tanh \theta_k = e^{-\beta \omega_k}, \quad \omega_k = \mathbf{k}^2/(2m) - \mu, \quad (8)$$

where $\mathbf{k}$ is a wave vector and μ denotes chemical potential, and thus $g[\beta_0]$ for bosons is calculated as

$$\left.\frac{\partial \theta_k}{\partial \beta'}\right|_{\beta=\beta_0} = -\frac{\omega_k}{2}\frac{e^{-\beta_0 \omega_k}}{1 - e^{-\beta_0 \omega_k}}. \quad (9)$$

Now let us work out the particle number $\langle a_{\mathbf{k}}^\dagger a_{\mathbf{k}} \rangle$ in THP [4]: THO and a state vector are respectively defined as

$$A_H(\beta) = e^{-iB[\beta]} A e^{iB[\beta]}, \quad (10)$$

$$|\beta\rangle_H = e^{-iB[\beta]}|\beta\rangle = |0\rangle \otimes |0\rangle =: |0\rangle. \quad (11)$$

Hereafter we shall drop the subscript H. Noting that we can write $e^{iB[\beta]}$ as $e^{iB[\beta_0]}e^{i\Gamma[\beta']}$ since $[B[\beta_0], \Gamma[\beta']]_- = 0$, the annihilation and creation operators in TFD can be thus expressed in terms of THO and they are related to the corresponding THO by the following Bogoliubov transformation:

$$\begin{pmatrix} a_{\mathbf{k}}(\beta_0, \beta') \\ \tilde{a}_{\mathbf{k}}^\dagger(\beta_0, \beta') \end{pmatrix} = U_\beta \begin{pmatrix} a_{\mathbf{k}} \\ \tilde{a}_{\mathbf{k}}^\dagger \end{pmatrix}, \quad U_\beta = \begin{pmatrix} u_k & v_k \\ v_k & u_k \end{pmatrix}, \quad (12)$$

where the matrix elements, u_k and v_k, are respectively given by

$$u_k = \cosh(g_k[\beta_0]\beta')\cosh\theta_k + (\cosh \Rightarrow \sinh), \quad (13)$$
$$v_k = \sinh(g_k[\beta_0]\beta')\cosh\theta_k + (\cosh \Leftrightarrow \sinh). \quad (14)$$

A particle number operator in THP is thus expressed in terms of THO by

$$n_{\mathbf{k}}(\beta_0, \beta') = a_{\mathbf{k}}^\dagger(\beta_0, \beta') a_{\mathbf{k}}(\beta_0, \beta'). \quad (15)$$

Using Eq. (11), the expectation value of boson number at $\beta [= \beta_0 + \beta'(\mathbf{x}, t)]$ is given by

$$\begin{aligned}\langle n_{\mathbf{k}}(\beta_0, \beta')\rangle &= \langle 0|a_{\mathbf{k}}^\dagger(\beta_0, \beta')a_{\mathbf{k}}(\beta_0, \beta')|0\rangle \\ &= \langle n_{\mathbf{k}}(\beta_0)\rangle\left(1 - \frac{\beta'\omega_k}{1 - e^{-\beta_0 \omega_k}}\right), \quad (16)\end{aligned}$$

where $\langle n_{\mathbf{k}}(\beta_0)\rangle$ denotes the bose distribution function at temperature β_0^{-1}. It should be noted that the factor of $\langle n_{\mathbf{k}}(\beta_0)\rangle$ comes from the inhomogeneity of local temperature and contains the information of thermal effects like temperature gradient.

Let us obtain a thermal propagator. Assuming that the field operator

$$\phi^a(x) = (2\pi)^{3/2}\int d^3k \begin{pmatrix} a_{\mathbf{k}} \\ \tilde{a}_{\mathbf{k}}^\dagger \end{pmatrix} e^{ikx}, \quad (17)$$

obeys the Schrödinger equation [6]

$$\left[i\frac{\partial}{\partial t} - \omega(-i\nabla)\right]\phi(x) = 0, \quad (18)$$

we can easily obtain the thermal propagator $G(x-y)$ as

$$\begin{aligned}G(x-y) &= \langle 0|T\phi^a(x;\beta)\phi^{b\dagger}(y;\beta)|0\rangle \\ &= \langle 0|TU_\beta \phi^a(x)\phi^{b\dagger}(y)U_\beta|0\rangle \\ &= U_\beta\langle 0|T\phi^a(x)\phi^{b\dagger}(y)|0\rangle U_\beta \\ &= U_\beta G_0(x-y)U_\beta, \quad (19)\end{aligned}$$

where the symbol T denotes a time-ordering operator and the relation $U_\beta^\dagger = U_\beta$ was used. The $G_0(x-y)$ is a zero temperature propagator and the Fourier transform of Eq. (20) gives $G_0(k)$ as

$$G_0(k) = \frac{i\tau_3}{k_0 - \omega_k + i\varepsilon\tau_3}, \quad (20)$$

where τ_3 denotes the Pauli matrix:

$$\tau_3 = \begin{pmatrix} 1 & 0 \\ 0 & -1 \end{pmatrix}. \quad (21)$$

In our theory including the effect of local temperature extended spatiotemporally, the propagator (20) along with the Lagrangian or the Hamiltonian of the system enables us to perform a perturbation expansion with respect to the interaction and to analyze those vertex terms based on the Feynman diagram technique as in the usual TFD.

REFERENCES

1. Kubo, R., Toda, M., and Hashitume, N., *Statistical Physics II: Nonequilibrium Statistical Mechanics* (Springer Series in Solid-State Sciences, Vol. 3, 1985).
2. Zubarev, D. N., *Nonequilibrium Statistical Thermodynamics*, Consultants Bureau, New York (1974).
3. Takahashi, Y., and Umezawa, H., *Collect. Phenom.*, **2**, 55-80 (1975).
4. Majima, H., and Suzuki, A., in *Similarity in Diversity*, ed. by Morabito, D. L. and Okamura, Y., Nova Science, New York (2003).
5. Ezawa, H., in *Progress in Quantum Field Theory*, ed. by Ezawa, H., and Kamefuchi, S., pp. 305-324, North Holland, Amsterdam (1991).
6. Non-relativistic case is considered. Thus, in Eq. (18), $kx \equiv \mathbf{k}\cdot\mathbf{x} - \omega_k t$.

Resonant Charge-Exchange Processes in Nonideal Plasmas

Mi-Young Song and Young-Dae Jung

Department of Physics, Hanyang University, Ansan, Kyunggi-Do 425-791, South Korea

Abstract. The resonant charge-exchange by ion-ion collisions in nonideal plasmas are investigated. The interaction potential in nonideal plasmas is given by an pseudopotential model taking into account the plasma screening effect as well as the nonideal collective effect. The screened atomic states in nonideal plasmas are obtained by the variational principle and the perturbational analysis. The semiclassical straight-line trajectory method is used to describe the behavior of the projectile ion. The maximum position of the resonant charge-exchange probability, i.e., the position which takes place the resonant change-exchange process, is found to be receded from the target ion with increasing nonideal effect.

The resonant charge-exchange process[1,2] in atom-charged particle collisions has been of great interest since this process has many applications in atomic physics, plasma physics, and astrophysics. Recently, ion-ion collision processes[3,4] in plasma have received many attentions since theses processes can be used for plasma diagnostics. Recently, an integro-differential equation for the effective potential of the particle interaction taking into account the simultaneous correlations of many particles was obtained by Baimbetov[5] and his collaborators. This interaction potential can be described by the pseudopotential model. Using the pseudopotential taking into account the plasmas screening and the collective effects, the interaction potential[6,7] $V_{NI}(r)$ between the bound electron and the ion with charge Z in nonideal plasmas can be represented by

$$V_{NI}(r) = -\frac{Ze^2}{r}\exp(-\frac{r}{\Lambda})\frac{1+\gamma f(r/\Lambda)/2}{1+c(\gamma)}, \quad (1)$$

where $f(r/\Lambda) = \left[e^{(-\sqrt{\gamma}\, r/\Lambda)} - 1\right]\left[1 - e^{(-2r/\Lambda)}\right]/5$,
$c(\gamma) \cong -0.0086 + 0.4559\gamma - 0.1084\gamma^2 + 0.0094\gamma^3$,
$\Lambda\ (\equiv \sqrt{k_B T_e/4\pi e^2 n_e})$ is Debye length, $\gamma\ (\equiv e^2/\Lambda k_B T_e)$ is the nonideality plasma parameter, $c(\gamma)$ is the correction coefficient, k_B is the Boltzmann constant, T_e is the electron temperature, and n_e is the electron density. When $\gamma \to 0$, the pseudopotential [Eq. (1)] goes over into the classical Debye-Hückel potential. When a hydrogenic ion is embedded in the nonideal plasma, the radial Schrödinger equation for a hydrogenic ion with nuclear charge Z in nonideal plasmas would be given by

$$\left[-\frac{\hbar^2}{2m}\left(\frac{d^2}{dr^2} - \frac{l(l+1)}{r^2}\right) + V_{NI}(r)\right]P_{nl}(r) = E_{nl}P_{nl}(r), \quad (2)$$

where $P_{nl}(r)$ and E_{nl} are the radial wave functions and energy eigenvalues of the bound electron in the nl-th shell, respectively. Here, we assume the following normalized variational $1s$ wave function:

$$P_{1s}(r) = 2\alpha_{1s}^{-3/2} r \exp\left(-\frac{r}{\alpha_{1s}}\right), \quad (3)$$

The solution for the variation parameter $\alpha_{1s}(\gamma, \Lambda)$ can be obtained by the minimization[8] $\delta\langle E_{1s}(\alpha_{1s})\rangle = 0$. Using the perturbational analysis, then the approximate analytic solution for $\alpha_{1s}(\gamma, \Lambda)$ becomes

$$\alpha_{1s}(\gamma, \Lambda) \cong a_Z A(\gamma)\left[1 - \frac{3}{4}\left(\frac{a_Z A(\gamma)}{\Lambda}\right)^2\left(1 - \frac{2}{5}\gamma^{3/2}\right) \right.$$
$$\left. + \left(\frac{a_Z A(\gamma)}{\Lambda}\right)^3\left(1 - \frac{12}{5}\gamma^{3/2} - \frac{3}{5}\gamma^2\right)\right]^{-1}, \quad (4)$$

where $A(\gamma) \equiv 1 + c(\gamma)$. This correction to the Bohr radius a_Z is quite reliable for our interesting domain ($0 \leq \gamma \leq 0.5$). Then, the ground state energy is obtained as

$$\langle E_{1s}(\gamma, \Lambda)\rangle = -Z^2 Ry A(\gamma)^{-2}\left[1 - \Delta_{1s}(\gamma, \Lambda)\right], \quad (5)$$

where $Ry\ (\equiv me^4/2\hbar^2 \approx 13.6\ \text{eV})$ is the Rydberg constant and $\Delta_{1s}(\gamma, \Lambda)$ is the correction effect on the $1s$ ground state:

$$\Delta_{1s}(\gamma, \Lambda) = 2\left(\frac{a_Z A(\gamma)}{\Lambda}\right) - \frac{3}{2}\left(\frac{a_Z A(\gamma)}{\Lambda}\right)^2\left(1 - \frac{2}{5}\gamma^{3/2}\right)$$
$$+ \left(\frac{a_Z A(\gamma)}{\Lambda}\right)^3\left(1 - \frac{12}{5}\gamma^{3/2} - \frac{3}{5}\gamma^2\right). \quad (6)$$

Using the semiclassical straight-line trajectory analysis[9] the cross section for the resonant change-exchange from the ground state of the target ion (Z) to the ground state of the projectile ion (Z) can be given by

$$\sigma_{fi} = \int d^2\mathbf{b}\ |T_{fi}(b)|^2, \quad (7)$$

TABLE 1. The maximum values of the scaled resonant charge-exchange probability

γ	$\bar{b}\|T_{1s,1s}(\bar{b},\gamma,\varepsilon,a_\Lambda)\|^{2*}$	$\bar{b}\|T_{1s,1s}(\bar{b},\gamma,\varepsilon,a_\Lambda)\|^{2\dagger}$	$\bar{b}\|T_{1s,1s}(\bar{b},\gamma,\varepsilon,a_\Lambda)\|^{2**}$	$\bar{b}\|T_{1s,1s}(\bar{b},\gamma,\varepsilon,a_\Lambda)\|^{2\ddagger}$
$0.0(\bar{b}_{max}=0.19)$	237.497	79.166	225.074	75.025
$0.1(\bar{b}_{max}=0.19)$	226.331	75.443	214.283	71.428
$0.4(\bar{b}_{max}=0.24)$	201.453	67.151	189.238	63.079

* The resonant charge-exchange probabilities at $\varepsilon=100$, $a_\Lambda=0.05$.
† The resonant charge-exchange probabilities at $\varepsilon=300$, $a_\Lambda=0.05$.
** The resonant charge-exchange probabilities at $\varepsilon=100$, $a_\Lambda=0.1$.
‡ The resonant charge-exchange probabilities at $\varepsilon=300$, $a_\Lambda=0.1$.

where **b** is the impact parameter. From the first-order time-dependent perturbation theory, the transition amplitude T_{fi} is represented as

$$T_{fi} = -\frac{i}{\hbar}\int_{-\infty}^{\infty} dt\, e^{i(E_f-E_i)t/\hbar}\langle f|V_{NI}(\mathbf{r},\mathbf{R})|i\rangle, \quad (8)$$

where $V_{NI}(\mathbf{r},\mathbf{R})$ is the interaction Hamiltonian, **r** and $\mathbf{R}(t)$ [$\equiv b\hat{y}+vt\hat{z} \equiv b\hat{y}+\tau\hat{z}$] are, respectively, the position vector of the bound electron and the position vector of the projectile ion from the ion core of the target system, v is the collision velocity, $\tau(=vt)$ is the scaled time, E_i and E_f are the energies of ground state of the target ion and the projectile ion. After some algebra using the semiclassical straight-line trajectory analysis with the total initial and final eigenfunctions of the collision system, the transition probability for the $1s \to 1s$ resonant charge-exchange process becomes

$$\bar{b}|T_{1s,1s}(\bar{b},\gamma,Z,M,\varepsilon,a_\Lambda)|^2 = \frac{2^{10}\eta_{1s}^8}{\pi^2}\frac{1}{[1+c(\gamma)]^2}\frac{M}{\varepsilon}$$

$$\times \left\{ 2Z\eta_{1s}\int dQ \frac{Q}{(Q^2+\eta_{1s}^2)^4}\int_{-\infty}^{\infty} d\bar{\tau}\frac{\sin(Q\sqrt{\bar{b}^2+\bar{\tau}^2})}{\bar{b}^2+\bar{\tau}^2}\right.$$

$$\left[e^{-\sqrt{\bar{b}^2+\bar{\tau}^2}a_\Lambda} + \frac{\gamma}{10}\left(e^{-(\sqrt{\gamma}+1)\sqrt{\bar{b}^2+\bar{\tau}^2}a_\Lambda} - e^{-\sqrt{\bar{b}^2+\bar{\tau}^2}a_\Lambda}\right.\right.$$

$$\left.\left. + e^{-3\sqrt{\bar{b}^2+\bar{\tau}^2}a_\Lambda} - e^{-(\sqrt{\gamma}+3)\sqrt{\bar{b}^2+\bar{\tau}^2}a_\Lambda}\right)\right]$$

$$- \int dQ \frac{Q}{(Q^2+\eta_{1s}^2)^2}\int_{-\infty}^{\infty} d\bar{\tau}\frac{\sin(Q\sqrt{\bar{b}^2+\bar{\tau}^2})}{\sqrt{\bar{b}^2+\bar{\tau}^2}}$$

$$\left[(\bar{\beta}^2+Q^2)^{-1} + \frac{\gamma}{10}\left((\bar{\beta}_1^2+Q^2)^{-1} - (\bar{\beta}^2+Q^2)^{-1}\right.\right.$$

$$\left.\left.\left. + (\bar{\beta}_2^2+Q^2)^{-1} - (\bar{\beta}_3^2+Q^2)^{-1}\right)\right]\right\}^2, \quad (9)$$

where $\bar{b}(\equiv b/a_Z)$ is the scaled impact parameter, $a_\Lambda(\equiv a_Z/\Lambda)$ is the scaled reciprocal Debye length, $\varepsilon(\equiv \mu v^2/2Z^2Ry)$ is the scaled collision energy, $M \equiv \mu/m$, μ is the reduced mass of the collision system, $Q(\equiv qa_Z)$ is the scaled momentum transfer, $\bar{\tau} \equiv \tau/a_Z$, $\eta_{1s} \equiv a_0/\alpha_{1s}$, $\bar{\beta} \equiv \eta_{1s}+a_\Lambda$, $\bar{\beta}_1 \equiv \eta_{1s}+(\sqrt{\gamma}+1)a_\Lambda$, $\bar{\beta}_2 \equiv \eta_{1s}+3a_\Lambda$, and $\bar{\beta}_3 \equiv \eta_{1s}+(\sqrt{\gamma}+3)a_\Lambda$. In order to investigate the plasma screening and collective effects on the resonant charge-exchange probability we consider three cases of the nonideality parameter: $\gamma = 0$, 0.1, and 0.4. The maximum values of the resonant charge-exchange probabilities are given in Table 1. The positions of the maximum resonant change-exchange probabilities are also given in parenthesis. As we see in this table, we found that the resonant charge-exchange probabilities is considerably reduced with increasing plasma screening and collective effects. It is important note that the position of the maximum resonant change-exchange probability is almost unchanged with changing collision energy.

ACKNOWLEDGMENTS

This work was supported by the Korea Basic Science Institute through the HANBIT User Development Program (FY2003).

REFERENCES

1. Miyamoto, K., *Plasma Physics for Nuclear Fusion*, revised ed., Cambridge: MIT Press, 1989, pp. 509.
2. Hutchinson, I. H., *Principles of Plasma Diagnostics*, 2nd ed., Cambridge: Cambridge University Press, 2002, pp. 344-354
3. Shevelko, V. P., and Tawara, H., *Atomic Multielectron Processes*, Berlin: Springer-Verlag, 1998, pp. 54-72
4. Smirnov, B. M., *Physics of Ionized Gases*, New York: Wiley, 2001, pp. 45-62.
5. Baimbetov, F. B., Nurekenov, Kh. T., and Ramazanov, T. S., *Phys. Lett. A***202**, 211-214(1995).
6. Arkhipov, Yu. V., and Davletov, A. E., *Phys. Lett. A***247**, 339-342(1998).
7. Arkhipov, Yu. V., Baimbetov, F. B., and Davletov, A. E., *Eur. Phys. J. D***8**, 299-304(2000).
8. Salzmann, D., *Atomic Physics in Hot Plasma*, New York: Oxford, 1998, pp. 77-121.
9. McGuire, J. H., *Electron Correlation Dynamics in Atomic Collisions*, Cambridge: Cambridge University Press, 1997, PP. 12-17.

Light Transmittance of Solid Polymeric Film including Spherulites

Takashi Taniguchi*, N. Kobayashi[†], M. Doi**, M. Sugimoto* and K. Koyama*

*Dept. of Polymer Sci. and Eng., Yamagata University, 4-3-16, Yonezawa, Yamagata, 992-8510, Japan
[†]Material Science Laboratory, Mitsui Chemicals, Inc. Ichihara, 299-0108, Japan
**Department of Comp. Sci. and Eng., Nagoya University, Chikusaku, Nagoya, 464-8603, Japan

Abstract. We formulate a method to calculate a turbidity of a polymeric material having a spatial fluctuation of refractive index coming from spherulites of comparable size with the wave length of visible light. We carry out computer simulations of spherulites growth satisfying Avrami equation. Using the obtained spherulites structures, we estimate a turbidity of polyethylene and compare the turbidity with actual measurements of Polyethylene haze.

When we view an object through a polymer film, a degradation of image of the object takes place by a reduction of an intensity of transmitted light through the film. The reason of the reduction of intensity of the visible light mainly comes from a diffuse reflection at the rough film surface and a light scattering at the bulk region. The light scattering at the bulk region originates from a spatial inhomogeneity of medium, i.e., a spatial fluctuation of a refractive index in a film. Since the spatial distribution of refractive index is determined by a solidification process from a melt state, the optical property of solid polymeric material can be connected with the physical properties in the bulk region. In the case of a crystalline polymer, since the bulk is occupied by spherulites that consist of crystal and amorphous regions, the spatial distribution of refractive index depends on that of spherulites and an internal structure of spherulite.

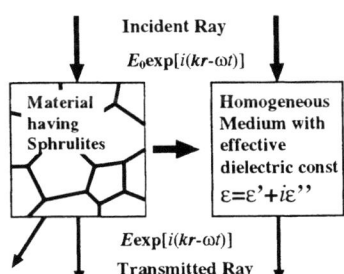

FIGURE 1. Schematic views of a light passing through a solid of polyethylene (left) and a corresponding homogeneous medium (right) having an effective dielectric constant $\bar{\varepsilon}_{\alpha\beta}$.

Depending on a spatial and size distribution of spherulite, the material which is originally transparent becomes muddy. Since most of commercially mass-produced polymers are crystalline ones, we focus on the turbidity of the crystalline polymer, especially, that of polyethylene solid which has a structure occupied by spherulites. The aim of this paper is to construct a method that can predict the turbidity of a polymeric material having an internal spherulite structure whose characteristic size is comparable to the wavelength of the visible light.

The decrease of intensity of the light after propagating by a distance L in the material can be described by the following empirical Lambert's equation as $I_{\text{bulk}}(L) = I_0 \exp(-\tau L)$ where the value τ is called the turbidity of the material, and it is an important material constant. In order to connect the spatial inhomogeneity inside the material with the turbidity τ, here we consider a crystalline polymer, say polyethylene, with a spatially non-uniform structure whose characteristic size of spherulite is comparable to the wave length of the incident beam. In crystalline polymers, the heterogeneity of the structure originates from the nucleus formation at the time of creation of the solid phase. The whole system is occupied by spherulites as a result of the nucleus formation of solid phase and the interior of spherulite consists of the crystal and amorphous part. The statistical property of the spatial and size distribution of spherulites are determined by the process of the solidification.

Now we consider a situation that the plane waves with a wave number vector **k** goes into the solid crystalline polymer and is scattered at the inside of the material occupied by spherulites as shown in Figure 1 (left). In order to estimate the decrease of the intensity of light by the scattering, usually we have to solve numerically the Maxwell equation for propagating waves in the medium with a spatial distribution of permittivity $\varepsilon_{\alpha\beta}(\mathbf{r})$. But, instead of it, we estimate an effective complex dielectric constant of a uniform system (Fig.1:right) which gives an equivalent decrease of the light intensity to that in the system on the left of Fig.1. For the present, we assume that the value of a dielectric tensor $\varepsilon_{\alpha\beta}(\mathbf{r})$ is known at an arbitrary positions $\mathbf{r}$ in the material. If deviations of eigen values of the dielectric tensor $\varepsilon_{\alpha\beta}(\mathbf{r})$ from the spatial average ε^o are sufficiently smaller than ε^o, we can apply a perturbative method to solve the Maxwell equation. By a perturbative calculation [1], the turbidity can be expressed by the following equation

$$\tau(k) \simeq \frac{k^4}{4\varepsilon^o(2\pi)^2}(\delta_{\alpha\beta} - \hat{k}_\alpha\hat{k}_\beta)\int(\delta_{\mu\nu} - \hat{q}_\mu\hat{q}_\nu)$$
$$\langle \tilde{\varepsilon}_{\alpha\mu}(\mathbf{k} - k\hat{\mathbf{q}})\tilde{\varepsilon}_{\nu\beta}(-\mathbf{k} + k\hat{\mathbf{q}})\rangle d\Omega_q \quad (1)$$
$$\hat{k}_\alpha \equiv k_\alpha/|\mathbf{k}|, \quad \hat{q}_\mu \equiv q_\mu/|\mathbf{q}|$$

where $\langle X \rangle$ means statistical average of X, $\tilde{\varepsilon}_{\alpha\beta}(\mathbf{k})$ is the Fourier transformation of $\varepsilon_{\alpha\beta}(\mathbf{r})$, $\mathbf{k}$ the wavelength of incident light and $d\Omega_q$ the solid angle in the q-space. The equation (1) means that it is necessary to know the spatial distribution and the correlation function of dielectric tensor in order to calculate the turbidity. Hence, it is important to establish the way to estimate a spatial distribution of dielectric tensor from a structure obtained by a spherulite growth simulation. However, since the detail of the structure inside the polyethylene spherulite is still not clear experimentally and theoretically, the structure used in the calculation of turbidity was estimated by the experimentally known evidences. It is well known experimentally for the polyethylene spherulite that the polymer crystal lamellae grow radially from a nucleus and each lamella is twisting with a constant pitch. The

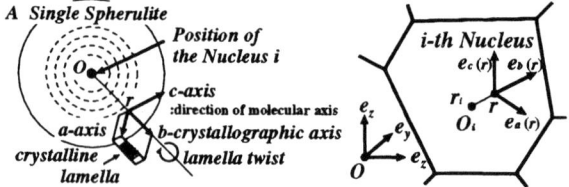

FIGURE 2. Schematic view of a single spherulite.

local dielectric constant $\varepsilon_{\alpha\beta}(\mathbf{r})$ is assumed to be given by a superposition of contributions from the crystal part with a dielectric tensor and from the amorphous one as $\varepsilon_{\alpha\beta}(\mathbf{r}) = \phi_c \varepsilon_{\alpha\beta}^{(cryst)}(\mathbf{r}) + (1 - \phi_c)\varepsilon^{(amor)}\delta_{\alpha\beta}$ where ϕ_c is the constant crystallinity of the system as a whole. When the position $\mathbf{r}$ is in the interior of the spherulite i as shown in Fig.2 the relative unit vector from the position $\mathbf{r}_i$ of the nucleus of spherulite i to the position $\mathbf{r}$ is give by $\mathbf{e}_b = (\mathbf{r} - \mathbf{r}_i)/|\mathbf{r} - \mathbf{r}_i|$. This direction of $\mathbf{e}_b$ is in agreement with that of the b-axis of polyethylene crystal. Since the a-axis and the c-axis are perpendicular to the b-axis and the lamella is twisted, the a- and c-axis are rotating around the b-axis. From the above discussion, we can obtain the following expression for the unit vector $\mathbf{e}_a(\mathbf{r})$ along the a-axis as $\mathbf{e}_a(\mathbf{r}) =$ $\cos(|\mathbf{r} - \mathbf{r}_i|/L_t)\mathbf{e}_a(\mathbf{0}) - \sin(|\mathbf{r} - \mathbf{r}_i|/L_t)\mathbf{e}_c(\mathbf{0})$ where L_t is a constant pitch. The unit vector along the c-axis automatically determined by $\mathbf{e}_c(\mathbf{r}) = \mathbf{e}_a(\mathbf{r}) \times \mathbf{e}_b(\mathbf{r})$. It is assumed that the unit vector $\mathbf{e}_a(\mathbf{0})$ is given randomly for each spherulite. Since the crystallographic axes are much the same as the optic axes, if we take the directions of crystallographic axes as the bases, the dielectric tensor is diagonalized and expressed as $(\varepsilon_a, \varepsilon_b, \varepsilon_c) = (n_a^2, n_b^2, n_c^2)$, where n_a, n_b, n_c are refractive indices for each optic axis of polyethylene crystal. Since directions of the crystallographic axes (base directions of three axis $\{\mathbf{e}_a(\mathbf{r}), \mathbf{e}_b(\mathbf{r}), \mathbf{e}_c(\mathbf{r})\}$) at any point $\mathbf{r}$ in the system can be estimated by the above equations, the transformation of the bases to the laboratory coordinate system $\{\mathbf{b}_x, \mathbf{b}_y, \mathbf{b}_z\}$ is given by the tensor, $\mathcal{R}_{\alpha\alpha'} = \mathbf{b}_\alpha \cdot \mathbf{e}_{\alpha'}(\mathbf{r})$, $(\alpha' \in \{a,b,c\})$. Using this transformation tensor, the dielectric tensor field in a laboratory coordinate system is given as $\varepsilon_{\alpha\beta}(\mathbf{r}) = \mathcal{R}_{\alpha\alpha'}(\mathbf{r})\varepsilon_{\alpha'\beta'}(\mathbf{r})({}^t\mathcal{R}(\mathbf{r}))_{\beta'\beta}$. where ${}^t\mathcal{R}$ denotes the transpose of $\mathcal{R}$.

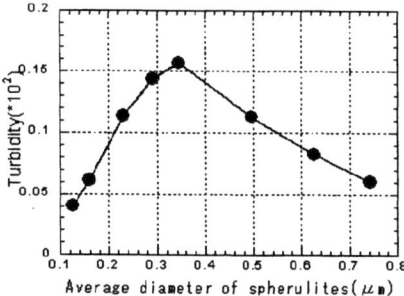

FIGURE 3. Integrated turbidity over the range of wave length of visible light is plotted versus the averaged diameter of spherulites.

In the evaluation of turbidity, first we obtained a Voronoi division which corresponds to a material occupied by spherulites. Using the above method to estimate the distribution of $\varepsilon_{\alpha\beta}(\mathbf{r})$ from the Voronoi division and refractive indices (1.514, 1.519, 1.575) along the three optical axes of the crystal part, and 1.510 for the isotropic amorphous part, the turbidity for the visible light is calculated by the integration of the wavelength dependent turbidity obtained by Eq.(1). In Fig.3, We show the calculated turbidity as a function of the averaged size of spherulite. This behavior of turbidity is good agreement with the Haze obtained in an experiment.

ACKNOWLEDGMENTS

This work is supported by Grant (No.14045247) of the Ministry of Education, Culture, Sports, Science and Technology, Japan.

REFERENCES

1. Onuki, A., and Doi, M., *J. Chem. Phys.* **85**, 1190-1197 (2002).

Localization in Disordered Two-Chain System with Long-Range Correlation

Hiroaki Yamada

Department of Material Science and Technology, Niigata University, Ikarashi 2-nochou 8050, Niigata 950-2181, Japan. [1]

Abstract. Localization property in disordered two-chain system with long-range correlation is numerically investigated. We apply the chain system with correlated disorder in the interchain hopping for a simple model of the double strand DNA. Numerical result for the Lyapunov exponent (inverse localization length) of the wave function is given. It is found that the correlation effect enhances the localization length.

INTRODUCTION

The localization phenomena in one-dimensional disordered system have been extensively studied. It is well known that almost all the eigenstates are exponentially localized and the system has pure point energy spectrum under the presence of any disorder. Recently, the effect of the long-range correlated disorder in the potential field on the localization have been reported by some groups [1, 2]. Indeed, it have been found that the base (nucleotide) sequence of the various genes such as human chromosome has strong correlation characterized by power spectrum $S(f) \sim f^{-\alpha}$ ($0.2 < \alpha < 0.8$) [2, 3].

On the other hand, the recent development of the nanoscale fabrication enable us to measure direct DNA transport phenomena [4, 5, 6]. However, transport property though DNA are still controversial mainly due to the complexity of the experimental environment and the molecule itself. Although the theoretical explanations for the phenomena have been tried by some standard pictures in the solid state physics such as polarons, solitons, electrons or holes [7, 8, 9], the situation is still far from unifying theoretical scheme.

In the present paper we investigate the localization property of the electron in the disordered two-chain system (ladder model) with long-range correlation as a simple model for electronic property in DNA. We will present some numerical results for the Lyapunov exponent (inverse localization length) of the wave function.

---

[1] Present address: Aoyama 5-7-14-205, Niigata 950-2002, Japan.

MODEL

Consider a one-electron system described by following tightly binding Hamiltonian $\hat{H}$ consisting of the two chains. This model for the DNA was first considered by Iguchi [7] as a model for considering the electronic properties of a double strand of DNA.

$$\hat{H} = \sum_{c=A,B} H_c + H_{AB},$$

$$H_c = \sum_n \{c_{n,n}|c:n\rangle\langle c:n| - (c_{n+1,n}|c:n+1\rangle\langle c:n| + h.c.)\},$$

$$H_{AB} = -\sum_n V_n(|A:n\rangle\langle B:n| + |B:n\rangle\langle A:n|),$$

where the $\{|A:n>, |B:n>\}$ denotes an orthonormalized set and $A_{n+1,n}$ ($B_{n+1,n}$) means the hopping integral between the nth and $(n+1)$th sites and $A_{n,n}$ ($B_{n,n}$) the on-site energy at site n in chain A (B), and V_n is the hopping integral from chain $A(B)$ to chain $B(A)$ at site n, respectively. The Schrödinger equation $\hat{H}|\Psi\rangle = E|\Psi\rangle$ becomes,

$$A_{n+1,n}\phi_{n+1}^A + A_{n,n-1}\phi_{n-1}^A + A_{n,n}\phi_n^A + V_n\phi_n^B = E\phi_n^A,$$
$$B_{n+1,n}\phi_{n+1}^B + B_{n,n-1}\phi_{n-1}^B + B_{n,n}\phi_n^B + V_n\phi_n^A = E\phi_n^B,$$

where $\phi_n^A \equiv \langle A:n|\Psi\rangle$ and $\phi_n^B \equiv \langle B:n|\Psi\rangle$. According to the parameter sets given by Iguchi [7], we set $A_{n+1,n} = B_{n+1,n} = a(= b)$ at odd (even) site n respectively and $V_n = 0$ at even sites (phosphate sites) for simplicity. The chain A and B are constructed by repetition of the sugar-phosphate sites, and the inter-chain hopping V_n at the sugar sites come from nucleotide base-pairs, i.e., $A - T$ or $G - C$ pairs. (See Fig.1.)

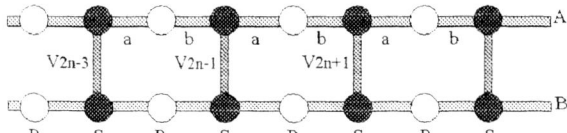

FIGURE 1. The model of double strand DNA.

The correlated sequence $\{V_n\}$ of the interchain hopping at the odd sites can be generated by a modified Bernoulli map [1].

$$X_{n+1} = \begin{pmatrix} X_n + 2^{B-1} X_n^B & (0 \leq X_n < 1/2) \\ X_n - 2^{B-1}(1-X_n)^B & (1/2 \leq X_n \leq 1) \end{pmatrix},$$

where B is a bifurcation parameter which controls the correlation of the sequence. We use the symbolized sequence $\{V_n\}$ by the following rule:

$$\begin{pmatrix} 0 \leq X_n < 1/2 & \to & V_{2n-1} = W \\ 1/2 \leq X_n < 1 & \to & V_{2n-1} = W' \end{pmatrix}.$$

We set $W' = W/2$ in this paper. Then the binary sequence can be roughly regarded as the base-pair sequence as observed in λ−DNA or human chromosome 22. The correlation function $C(n) (\equiv <V_{n_0} V_{n_0+n}>)$ decays by inverse power-law depending on the value B as $C(n) \sim n^{-\frac{2-B}{B-1}}$ for large n ($3/2 < B < 2$). The power spectrum becomes $S(f) \sim f^{-\frac{2B-3}{B-1}}$ for small f. We focus on the Gaussian ($1 < B < 3/2$) and non-Gaussian stationary region $2/3 \leq B < 2$, corresponding to the DNA sequence.

RESULTS AND DISCUSSION

We give a preliminary numerical result of the energy dependence of the Lyapunov exponents. We set the non-negative Lyapunov exponents by $\gamma_1 \geq \gamma_2 \geq 0$. First, it is easily seen that if the hopping integrals of the intra-chain are constant ($a = b$), the equations can be decoupled to one-dimensional Anderson model with diagonal randomness by following Unitary transform;

$$\begin{pmatrix} \psi_n^A \\ \psi_n^B \end{pmatrix} = U \begin{pmatrix} \phi_n^A \\ \phi_n^B \end{pmatrix}, U = \frac{1}{\sqrt{2}} \begin{pmatrix} 1 & 1 \\ 1 & -1 \end{pmatrix}.$$

Figure 2 shows the energy dependence of the γ_1 and γ_2 in some cases. The value of the minimum Lyapunov exponent (γ_2) effects on the transport phenomena in the system. It found that the correlation enhance the localization length (γ_2^{-1}) around $|E| < 2$, although the γ_1 is almost remained.

It must be noted that if we take into account the $\pi - \pi$ interaction between the stacked base pairs, the

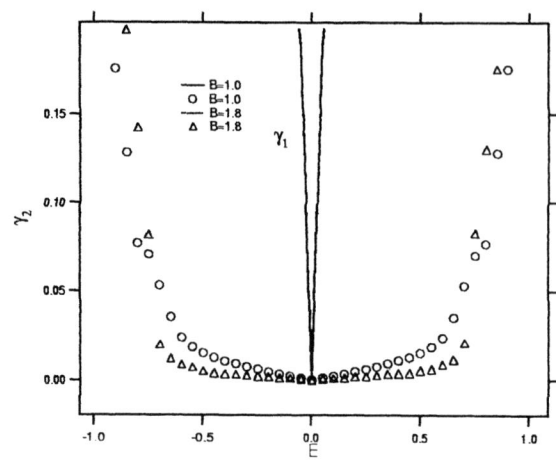

FIGURE 2. Lyapunov exponents as a function of energy. The parameters $W = 2.0, a = 1.0, b = 0.5$. The on-site energy is set at $A_{n,n} = B_{n,n} = 0$.

model system becomes three chains model [10]. The details of the energy dependence, correlation dependence and potential strength dependence of the DOS and the Lyapunov exponents, will be given elsewhere.

ACKNOWLEDGMENTS

The author would like to thank Dr. Kazumoto Iguchi for stimulating and useful discussions and providing his papers and preprints.

REFERENCES

1. Yamada H., Goda M. and Aizawa Y., *J. Phys.:Condens. Matter* **3**, 10043-10055 (1991); Yamada H. and Okabe T., *Phys. Rev.* **E63**, 26203-26217 (2001); Yamada H., *J. Phys. Soc. Jpn. Suppl. A*, **72**, 123-124 (2003); preprint; and references therein.
2. Carpena P., Galvan P.B., Ivanov P.Ch. and Stanley H.E., *Nature* **418**, 955-959 (2002); ibid, **84**, 764(2003).
3. Voss R.F., *Phys.Rev.Lett.* **68**, 3805-3808 (1992); Isohata Y. and Hayashi M., *J. Phys. Soc. Jpn.* **72**, 735-742(2003).
4. Tran P., Alavi B., and Gruner G., *Phys. Rev. Lett.* **85**, 1564-1567 (2000).
5. Porath D., Bezryadin A., de Varies S. and Dekker C., *Nature* **403**, 635-638 (2000).
6. *Technical Proceedings of the 2003 Nanotechnology Conference and Trade Show*, vol.2.
7. Iguchi K., *J. Phys. Soc. Jpn.* **70**, 593-597 (2001); *Int. J. Mod. Phys.* **17**, 2565-2578 (2003); preprint (2003); and references therein.
8. Hennig D., Archilla J.F.R. and Agarwal J., *Physica D* **180**, 256-272 (2003); and references therein.
9. Ladik J.J., *Phys. Rep.* **313**, 171-235(1999).
10. Yamada, H., in preparation.

Molecular Dynamics Simulations of "The Cooperativity Blockage Effect" in Alkali Metasilicate

J. Habasaki, K. L. Ngai* and Y. Hiwatari**

Tokyo Institute of Technology, 4259 Nagatsuta-cho, Yokohama, Kanagawa, 226-8502, Japan
*Naval Research Laboratory, Washington, DC 20375-5320, USA
** Kanazawa University, Kakuma, kanazawa 920-1192, Japan

Abstract. Molecular dynamics simulations were carried out at 700 K in single alkali and mixed alkali metasilicate glasses to reproduce the dramatic mixed alkali effect occurring in the dilute foreign alkali region. Both small numbers of frozen ions and another kind of alkali appears to immobilize a large number of host ions as previously found in experiment. This large number can now be rationalized as being due to blockage by the less mobile foreign ions of the cooperative ion dynamics that originate from ion-ion interaction and correlation.

INTRODUCTION

The relaxation dynamics of a complex interacting system can be drastically changed with the introduction of another component with different dynamics by mixing. Examples include the segmental dynamics in miscible polymer blends and the conductivity relaxation of mixed alkali ions in oxide glasses. Replacement of host alkali ions in a glass with guest alkali ions of a different kind lowers the d.c. conductivity σ. The derivative, $(\partial \ln\sigma/\partial x)$, is a measure of how rapidly the conductivity changes with replacement of the majority ions. Here x is the fraction of the majority host ions and $(1-x)$ the fraction of the minority guest ions. The most rapid change, measured by the largest value of $(\partial \log\sigma/\partial x)$, occurs at $x=1$, i.e., the dilute limit of the foreign alkali region [1]. In this work, we elucidate the effect of the less mobile component on the dynamics of the more mobile component by molecular dynamics simulations of lithium ions motion in lithium metasilicate glass by freezing some randomly chosen lithium ions (5 % and 10 %) at their initial locations at 700 K.

MOLECULAR DYNAMICS SIMULATIONS

MD simulations were performed in the same way as in previous studies [2,3]. Contained in the basic cell were 144 M (M=Li or Li+K), 72 Si and 216 O for Li_2SiO_3. MD runs of the confined Li_2SiO_3 systems were started from the same equilibrated configurations and the results are compared with the standard run (Run I) without modification of the system. This procedure enables us to reduce the effect of fluctuation of the dynamics due to exchange between slow and fast dynamics observed in these systems [2]. Randomly chosen several percents ((a) 5, (b) 10, (c) 25) of Li ions of Li_2SiO_3 system are frozen in Run II. Randomly chosen 10 % of oxygen atoms of Li_2SiO_3 system were frozen in Run III. The mixed alkali systems ($Li_{1-x}K_x)_2SiO_3$ with x= (a) 10, (b) 25, (c) 50 % and (d) 75 % were examined in Run IV. The dynamics of Li ions in the Li metasilicate glass confined by walls made of the same glass except all Li ions therein are frozen were examined in Run V.

MD Simulations in the present work were performed up to 2-8 ns (500,000-2000,000 steps). Temperatures of all systems were kept at 700 K during the run by simple scaling. The volume V was fixed as that derived by NPT (constant pressure and temperature) ensemble simulation at atmospheric pressure. Pair potential functions of Gilbert-Ida type and r^{-6} terms were used. The parameters of the potentials used were previously derived on the basis of *ab initio* molecular orbital calculations, and their validity was checked in the liquid, glassy and crystal states under constant pressure conditions.

RESULTS AND DISCUSSIONS

In Figure 1, the self-part of the density-density correlation functions, $F_s(k,t)$, of the majority of the Li ions in the run II, where 10% of Li ions are frozen, are compared with those in run I. A remarkable slowing down of the dynamics of the mobile Li ions was observed both in the $F_s(k,t)$ and in the mean squared displacements, MSD. On the other hand, there is no significant change in the structure. The motion of the Li ions in the unadulterated Li metasilicate glass has previously been shown to be dynamically heterogeneous [2] and the fast and slow ions have been divided into two groups. Similar classifications are made in the present case when 5 or 10 % of the Li ions are frozen. The particles showing a squared displacement less than the square of the distance equal to the first minimum of the Li-Li pair correlation function g(r) for the time period t_m=920 ps are defined as type A. These ions are located within neighboring sites during t_m. Li ions having squared displacements greater than the squared of the distance equal to the first minimum of g(r) are defined as type B, and they contribute to the diffusion at longer times. The plot of squared displacement against the number of jumps of each Li ion within the time period t_m reveals that the number of type B ions, which shows faster dynamics (Lévy flight) facilitated by cooperative jumps, decreases considerably. It follows that there is a large reduction of the mobility of the Li ions, much more than expected from freezing 5 or 10 % of the mobile Li ions. The situation also is quite similar to that observed in molecular dynamics simulations of the mixed alkali system, $LiKSiO_3$, where the suppression of the cooperative jumps occurs due to the jump paths being intercepted by other kind of alkali metal ions [3]. Large reduction of the mobility of Li ions is also observed in the mixed alkali systems with small K contents. These results are in accordance with the experimental finding in mixed alkali silicate glasses that the most dramatic reduction of ionic conductivity occurs in the dilute foreign alkali limit [1]. Naturally we describe the effect found here by simulation as "cooperativity blockage". Besides the effect by direct blockage of the jump path, there is indirect blockage effect by the suppression of the pre-jump motions by matrix atoms. In the case of freezing 10 % of oxygen atoms chosen at random, the slowing down of the motion of Li ions was observed, although the motion of Li ions in Li_2SiO_3 system is fairly de-coupled with that of framework.

We also examined the "cooperativity blockage" by confining the Li metasilicate glass by parallel walls, where all Li ions therein are frozen. The self-part of the density-density correlation function, $F_s(k,t;z)$ obtained at any distance z from the wall has the Kohlrausch stretched exponential time dependence. Stretching as well as slowing down of the Li ion dynamics are observed. The effect is largest near the wall and decreases monotonically with distance from the wall. It is noteworthy that the wall even suppresses the jumps to the opposite direction of the wall, where the jump path is open. These results rule out the mechanism of ion conduction based on the percolation with the single jumps or the fixed random energy barriers. Quite similar behaviors were found in the confined Lennard-Jones system [4]. This indicates that the same physics governs the Li ion in glasses and particles in Lennard-Jones liquid.

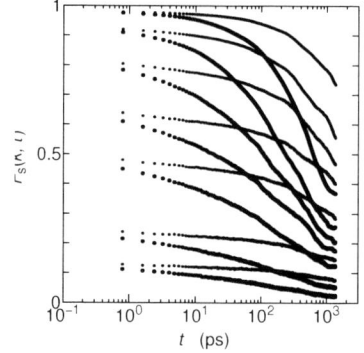

FIGURE 1. Self-part of the density-density correlation function, Fs(k, t), of mobile Li ions in Li_2SiO_3, where randomly chosen 10 % of Li ions are fixed (small dot) and that in Li_2SiO_3 as a standard (large dot) for k=2π/n (Å^{-1}) (n=10, 5, 3, 2, 1.5, 1.0, 0.8 from top to bottom) both at 700 K.

CONCLUSION

Using molecular dynamic simulations, the large mixed alkali effect in dilute foreign alkali limit is reproduced. Similar effect is found by confining a single alkali system by freezing the ions. This large blockage effect rules out single particle models for dynamics of ions, because one foreign ion can at most immobilize one or at most a few host ion in these models. Since there is no significant structural change by freezing of the motion of Li ions, the main controlling factor of the dynamics is the changes in dynamical and geometrical correlations (*i.e.,* cooperativity) among the mobile ions.

REFERENCES

1. Moynihan, C.T. , Saad, N.S. , Tran, D.C., and Lesikar, A.V., *J. Am..Ceram. Soc.* **63,** 458-464 (1980); Ngai, K.L., Wang Y. and Moynihan, C.T. , *J. Non-Cryst. Solids* **307-310,** 999-1011 (2002).
2. Habasaki, J. Okada, I. and Hiwatari, Y., *Phys. Rev.* **E52** 2681-2687 (1995). Habasaki, J. and Hiwatari, Y., *Phys. Rev.* **B55,** 6309-6315 (1997); *Phys. Rev.* **E65,** 021604 1-8 (2002).
3. Habasaki, J. and Hiwatari, Y., *J. Non-Cryst. Solids,* **320,** 281-290 (2003).
4. Scheidler, P., Kob, W., Binder, W. and Parisi, G., *Philos. Mag.* **B82,** 283 –290(2002).

Kinetic Monte Carlo Simulation of Via Filling

Y. Kaneko, Y. Hiwatari [a], K. Ohara [b] and T. Murakami [b]

*Department of Applied Analysis and Complex Dynamical Systems, Graduate School of Informatics,
Kyoto University, Sakyo-ku, Kyoto 606-8501, Japan*
[a] *Department of Computational Science, Faculty of Science, Kanazawa University,
Kakuma-machi, Kanazawa 920-1192, Japan*
[b] *Central Research Laboratory, C. Uyemura & Co., Ltd., 1-5-1 Deguchi, Hirakata,
Osaka 573-0065, Japan*

Abstract. In this paper, we study the influence of additives in the filling process of via holes for LSI Cu interconnections by using the kinetic Monte Carlo method. As a model for electroplating, we extended the Solid-by-Solid model for crystal growth to include additives which inhibit the adsorption of new atoms. This enables us to control the local surface growth rate to find out the optimal deposition condition for void-free filling. The distribution of additives on the surface and their influence on the surface and void structures are carefully examined.

INTRODUCTION

The copper damascene electroplating has been studied extensively as a new technology for LSI interconnects since the announcement of IBM to replace the vapor deposition of aluminum with electroplating copper. [1] The damascene process is a technique to fill the lines and via holes for electronic circuits by copper electroplating. The important requirement for the success of this process is to fill the lines and holes completely without voids or seams. A lot of works have been done to find out the optimal condition of electrolytic solutions for void-free filling by putting additives in solution to control the local overpotential. Two types of additives, inhibitors and accelerators, are commonly used in experiments. Recently, we have developed a new model for crystal growth, which includes void formation during the thin film growth. [2] The new model, which we call Solid-by-Solid (SBS) model, has been applied to investigate the void formation mechanism during the filling of V-shaped and flat-bottomed grooves. [3] In the present work, we include additives in the SBS model to control the local growth rate to suppress the void formation within the hole. We pay attention to inhibitors, that is, the additives which prevent the adsotption of new atoms on the surface to reduce the local growth rate. The distribution of additives and their influence on the surface and void structures are studied in detail.

THE SBS MODEL WITH ADDITIVES

The outline of the SBS model is the following. [2] The system is a square lattice each site of which is occupied by a liquid atom or a solid atom, otherwise, vacant. We assume three events, that is, adsorption of new atoms on the surface, desorption of surface atoms and diffusion of surface atoms. The adsorption rate k_n^+, the desorption rate k_n and the rate of surface diffusion k_{nm}, depend upon the number of bonds (n, m) at each site. The following relations are assumed for the rate constants. [2]

$$\frac{k_n}{k_n^+} = \exp\left[\frac{(2-n)\phi}{k_B T} - \frac{\mu}{k_B T}\right] \quad (1)$$

$$k_{nm} = \frac{k_n k_m^+}{k_0^+} \exp\left[\frac{\phi - E_d}{k_B T}\right] \quad (2)$$

Here ϕ is the binding energy between atoms, μ is the electrochemical potential and E_d is the activation energy for surface diffusion. In the SBS model, the adsorption occurs on any one of the nearest neighbor vacant sites of the surface solid atoms. The empty sites, which are surrounded by solid atoms and have no connection to the solution part, are regarded as vacancies. Following these rates, the state of each site is changed sequentially using the algorithm of kinetic Monte Carlo method of Bortz et.al. [4]

In this work, we introduce additives in the SBS model. The additives occupy lattice sites in the solution part and move to the surface by random walk. The distribution of additives in solution is described by the Fick's law. The random work of additives and the surface growth of the SBS model are simulated simultaneously. Additives are assumed to be inhibitors. When they reach the metal surface, they stick to the surface with the probability k_{ad} and have the effect of preventing the adsorption on the surface sites within the range l_{inh}. That is, when the adsorption is selected in the simulation, it is rejected if the adsorption site is within the range l_{inh} from any one of the inhibitors. The range l_{inh} is regarded as the size of the inhibitors. The parameters which characterize the inhibitors are the mobility, the concentration (the ratio of the number of additives to that of lattice sites), l_{inh} and k_{ad}.

RESULTS

Figure 1 shows the result of the simulation of via filling without additives. The parameters for the SBS model are $\phi/k_B T = 9.44$, $\mu/k_B T = 10.0$ and $E_d = \phi/2$. The width of the hole is 100 lattice sites (approximately 30 nm) and the aspect ratio is 2. The final surface is almost flat after the filling. In the film, however, large voids which are elongated in the growth direction remain in the middle of the hole. Such voids are also observed in experiments using the solution without additives. Figure 2 shows the results of the simulation with inhibitors. The concentration of inhibitors is 0.125%, the range l_{inh} is 2 lattice spacing (~ 0.6 nm) and $k_{ad} = 0.5$. Figure 2 (a) shows the distribution of inhibitors after the filling. It is observed that inhibitors are embedded in the film and distributed around the upper part of the hole. It is expected that the inhibitors reduce the growth rate of the upper surface, preventing the upper parts of the side walls from being connected to each other. As a result, the void size is found to be smaller than the case without additives. These results show that the void size can be reduced by the effect of inhibitors.

In Fig.2 (b), however, voids still remain in the hole, which are created when the surfaces from both sides of the hole meet during the growth. Although we performed a series of simulations by changing the parameters characterizing the inhibitors, it is difficult to remove the voids completely by using only inhibitors. This is because inhibitors control only the growth of the upper part of the hole. Therefore, although the inhibitors can be used to reduce the void size, they are not enough to realize the complete void-free filling. In order to fill the hole without voids, it is necessary to increase the growth rate of the bottom part of the hole, which would be realized by the effect of accelerators, the additives which increase the local growth rate. The next step of this work is to include accelerators in the SBS model to investigate the interplay of two kinds of additives for the bottom-up process.

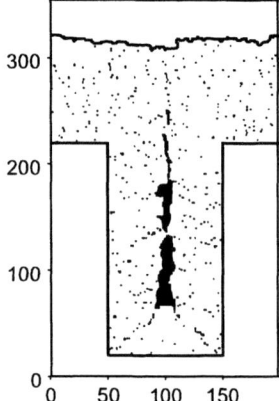

FIGURE 1. Result of the simulation of via filling without additives. The lines show the initial and the final surfaces. Dots show vacancies and the labels denote the number of lattice sites.

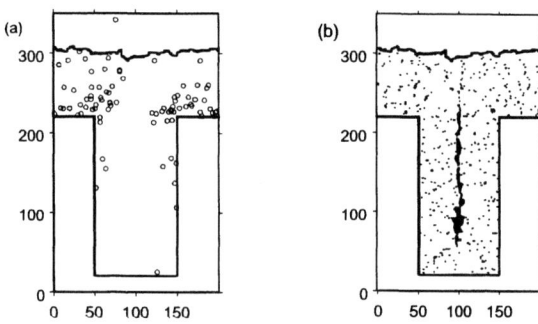

FIGURE 2. (a) Distribution of inhibitors after the filling. (denoted by circles) (b) Film structure generated by the simulation with inhibitors.

REFERENCES

1. Ducovic, J. O., *IBM Res. Develop.* **37**, 125-141, 1993.

2. Kaneko, Y., Hiwatari, Y., Ohara, K. and Murakami, T., *J. Phys. Soc. Japan* **69** (2000) 3607-3613.

3. Kaneko, Y., Hiwatari, Y., Ohara, K. and Murakami, T., *Technical Proceedings of the fifth International Conference on Modeling and Simulation of Microsystems*, 2002, pp430-433.

4. Bortz, A. B., Kalos, M. H. and Lebowitz, J. L., *J. Comput. Phys.* **17** (1975) 10-18.

Simulation Study on Strain-Mediated Coarsening of Quantum Dots

Y. Enomoto, H. Itamoto, and M. Okada

Department of Environmental Technology, Nagoya Institute of Technology, Nagoya 466-8555, Japan

Abstract. We propose a physical model which describes the post-deposition coarsening of coherently strained three-dimensional islands on a flat substrate. In this model, formulated in terms of a set equations of motion for the island volume and position, both Ostwald ripening mechanism and misfit strain induced elastic effects are taken into account. Large scale computer simulations demonstrate that the repulsive inter-island elastic interaction causes the island motion, leading to the self-organized formation of a regular array of islands with both uniform size and spacing.

INTRODUCTION

Clustering on surfaces by nucleation and growth during atom deposition is of great interest in basic and applied science [1, 2]. Especially, nanometer-scale clusters or quantum dots (QDs) have attracted much attention, since semiconductor QDs are potentially valuable for optical, electronic, and magnetic device applications.

In this work, as one of important open problems regarding the regular array formation of three dimensional (3D) QDs, we study the coarsening of 3D strained islands in heteroepitaxial systems after the deposition process of materials has been stopped (the post-deposition QDs coarsening). In these cases there exist two important factors regarding the mechanism of the self-assembled QDs formation, that is, the surface diffusion induced coarsening (the Ostwald ripening mechanism) and the coherent misfit strain induced elastic effects.

THE MODEL

We consider an array of coherently strained 3D islands on a flat substrate, formed at the early stage of heteroepitaxial growth, and discuss their coarsening after the deposition has been stopped.

The total energy, E, of the system is a sum of the surface energy and strain energy (self relaxation energy and the elastic interaction energy).

An explicit form of E has been discussed in Ref.[3, 4] Due to the surface energy, the islands, without strain effects, would tend to ripen to reduce the overall surface energy. On the other hand, the inter-island interaction generates the repulsive inter-island force.

In a standard coarsening process, in which mass transport between islands is limited by attachment/detachment of adatoms to/from island perimeters, the rate of change of the island volume can be expressed as [3, 4]

$$\frac{dV_i}{dt} = \xi_V V_i^{1/3} [\bar{\mu} - \mu_i] \quad (1)$$

where ξ_V is a kinetic coefficient related to the adatom attachment/detachment rate, the chemical potential of the i-th island is defined as $\mu_i \equiv v \partial E / \partial V_i$ with the atomic volume v, and $\bar{\mu}$ is the mean chemical potential averaged over all the islands. The mean chemical potential, $\bar{\mu}$, is determined by mass conservation, $\sum_i dV_i/dt = 0$.

The strain induced inter-island interaction not only changes the average chemical potential of the island, but also introduces a chemical potential gradient. Such chemical potential gradient induces the mass transport within an island, leading to the strain-directed island motion. The speed of the island motion is thought to be proportional to the magnitude of the chemical potential gradient at the center of each island and to be inversely proportional to its volume. Thus, the island migration is simply modeled as [3,4]

$$\frac{d}{dt}\vec{X}_i = -\xi_X V_i^{-1} \nabla \mu_i \qquad (2)$$

where ξ_X is the kinetic coefficient associated with the island motion.

SIMULATION RESULTS

We perform the computer simulations of coarsening of strained 3D islands.

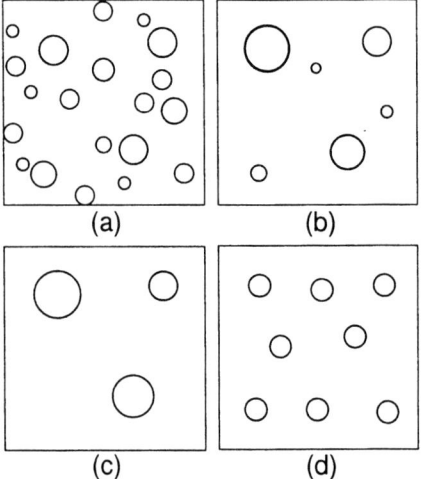

FIGURE 1. (a) Initial island configuration for surface coverage $q = 0.1$ and initial average island volume $V_{init} = 100V_0$, and simulated configurations at $t/t_0 = 20$ for three different cases: (b) SE, (c) SR, and (d) EI. For abbreviations, see the text.

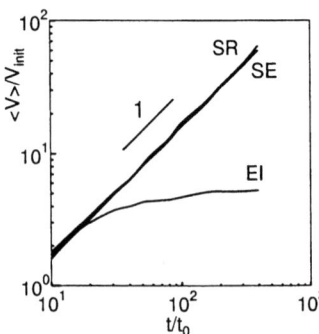

FIGURE 2. Time evolution of the average island volume $<V>$ for $q = 0.1$ and $V_{init} = 100V_0$. The slope of a line indicates the growth law exponent for $<V>$. For abbreviations, see the text.

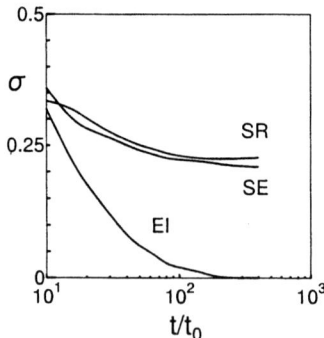

FIGURE 3. Time evolution of the standard deviation of the island volume distribution σ for $q = 0.1$ and $V_{init} = 100V_0$. For abbreviations, see the text.

From large scale simulations we have obtained the following results: In the case SE, the well-known Ostwald ripening with the growth law of the average island volume $<V> \sim t^1$ occurs, since the island chemical potential is inversely proportional to its size with the positive island surface energy.

Even for the case SR, the Ostwald ripening also takes place. It is because the island chemical potential associated with the self-relaxation energy is a constant value, and thus the system situation is the same as the case SE. On the other hands, in the case EI, the inter-island interaction directs the island motion, leading to the self-organized formation of a regular island lattice with uniform size, even for the case of positive surface energy.

REFERENCE

1. Shchukin, V, A., and Bimberg, D., *Rev.Mod.Phys.* **71**, 1125-1130(1999).
2. Politi, P., Grenet, G., Marty, A., Ponchet, A., and Villain, J., *Phys.Rep.* **324**, 271-284(2000).
3. Liu, F., Li, A, H., and Lagally, M, G., *Phys.Rev.Lett.* **87** 125103-125108(2001).
4. Enomoto, Y., Sawa, M., and Itamoto, H., *phys.stat.sol(a)* (2003), in press.

Simulation Study on Slow Dynamics in Magnetic Fluids

M. Okada and Y. Enomoto

Department of Environmental Technology, Nagoya Institute of Technology, Nagoya 466-8555, Japan

Abstract. We present our computer simulation results on the slow dynamics in magnetic fluids. Magnetic fluids are modeled as an ensemble of interacting ferromagnetic nanoparticles suspended in a viscous fluid. From the Brownian dynamics simulations of the model, it is found that the origin of slow dynamics in magnetic fluids is related to the structure formation of magnetic particles.

INTRODUCTION

Magnetic fluids are oil- or water-based colloidal suspensions of ferromagnetic nanoparticles such as Co–ferrite, magnetite, and Ba–ferrite [1,2]. Since the properties and location of these fluids can be easily influenced by an external field, they have recently attracted much scientific and technological interest [1,2].

Recently, we have proposed the Langevin–type microscopic equations of motion for magnetic fluids [3]. Magnetic fluids are modeled as an ensemble of interacting ferromagnetic nanoparticles suspended in a viscous fluid. The model is described in terms of position vectors of magnetic particles and orientation vectors of their magnetic moments. In this model, forces and torques arising from the magnetic origin and the surrounding fluid are included.

In this paper, based on the Brownian dynamics simulations of the above model, we study the slow dynamics in magnetic fluids. Especially, we discuss the relation between the origin of slow dynamics and the microstructure formation of magnetic particles.

THE MODEL

A model magnetic fluid system studied here is supposed to consist of N interacting, uniformly sized particles of a point magnetic dipole moment, suspended in an incompressible Newtonian fluid of viscosity η. Magnetic particles are assumed to have an identical spherical shape of the radius a with the center of mass position $\mathbf{r}_i$ ($1 < i < N$) and the magnetic dipole moment $m_0 \mathbf{n}_i$ centered at $\mathbf{r}_i$. It is assumed that the unit vector along the dipole moment, $\mathbf{n}_i$, is frozen to the particle, and the moment magnitude, m_0, has an identical constant value [3,4].

For the case of low Reynolds number at low particle volume fractions, the equations of motion of $\mathbf{r}_i$ (the translational motion) and $\mathbf{n}_i$ (the rotational motion) for interacting spherical particles can be described as [3]

$$\frac{d\mathbf{r}_i}{dt} = \frac{1}{\zeta_t} \left[\sum_{j(\neq i)}^{N} (\mathbf{F}_{ij}^1 + \mathbf{F}_{ij}^2) + \mathbf{R}_i(t) \right] \quad (1)$$

$$\frac{d\mathbf{n}_i}{dt} = \frac{1}{\zeta_r} \left(\sum_{j(\neq i)}^{N} \mathbf{T}_{ij}^1 + \mathbf{T}_i^2 + \mathbf{N}_i(t) \right) \times \mathbf{n}_i \quad (2)$$

where $\mathbf{F}_{ij}^1$ is the force acting on the i-th particle due to the dipolar interaction between two particles i and j, $\mathbf{F}_{ij}^2$ the short range repulsive force due to surface surfactants, $\mathbf{T}_{ij}^1$ the torque due to the dipolar interaction, and $\mathbf{T}_{ij}^2$ the torque due to the external magnetic field, with the translational drag force coefficient $\zeta_t \equiv 6\pi a \eta$ and the rotational drag force coefficient $\zeta_r \equiv 8\pi a^3 \eta$. Explicit forms of these quantities have been discussed in Ref.[3], as well as the random force $\mathbf{R}_i$ and the random torque $\mathbf{N}_i \equiv \mathbf{n}_i \times \mathbf{R}_i^r$.

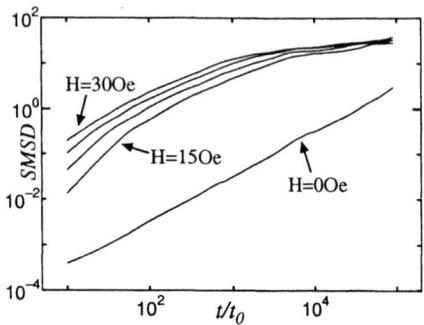

FIGURE 1. The SMSD as a function of the time for several values of the external field.

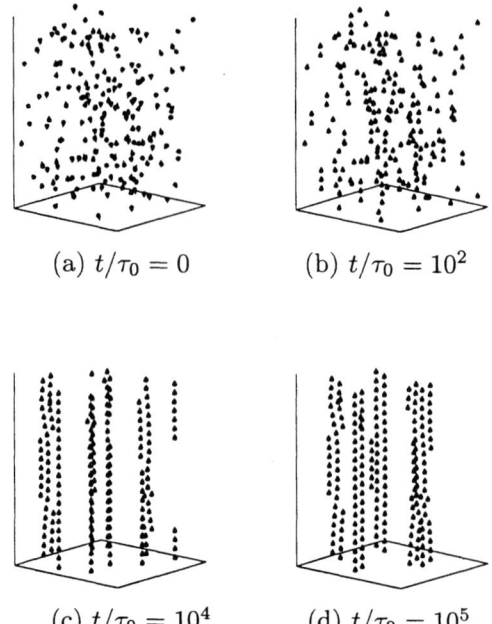

FIGURE 2. Structure formation of magnetic particles for $H = 15\ Oe$.

SIMULATION RESULTS

We carry out the Brownian dynamics simulations of the above equations to study microstructure formation processes of interacting magnetic particles. Detailed simulation methods have been discussed in [3,4]. From the simulation data, we calculate the self mean–square displacement (SMSD) of the particles, defined by

$$SMSD = \frac{1}{N}\sum_{i=1}^{N}(\vec{r}_i(t) - \vec{r}_i(0))^2 \quad (3)$$

In Fig.1 we show the SMSDs as a function of the time for several cases. We can see that in the presence of the external field the SMSDs deviate from the well–known t linear behavior to the slowing down behavior. Of course, in the absence of the external field only the diffusive behavior can be seen. In Fig.2 we show the structure formation of interacting magnetic particles for $15\ Oe$. In the absence of the external field we can see random patterns at long times. On the other hand, for non–zero magnetic field cases we can see the crossover from random to chain like patterns in the presence of the external fields. From these results the chain formations might result in the slowing down of the SMSD. To confirm this, the large scale computer simulations are needed and now under the way, as well as changing various conditions.

REFERENCE

1. Rosensweig, R.E., *Ferrohydrodynamics*, Cambridge University Press, New York, 1987.
2. Odenbach, S., *Magnetoviscous Effects in Ferrofluids*, Springer, Berlin, 2002.
3. Enomoto, Y., Oba, K., and Okada, M., *Physica A*, **331**, 189-199(2004)..
4. Enomoto, Y., and Oba, K., *Physica A* **309**, 15-20(2002).

Vortex Dynamics in Superconducting Films with Twin Boundary Networks

H. Itamoto and Y. Enomoto

Department of Environmental Technology, Nagoya Institute of Technology, Nagoya 466-8555, Japan

Abstract. We present two–dimensional computer simulation results of magnetic vortex systems driven by Lorentz force through polycrystalline superconductors with random network structures of twin boundaries. Langevin dynamics simulations of a point vortex model demonstrate that (1) the threshold behavior of I–V curves is observed, and (2) the threshold current is a decreasing function of an average superconducting grain size.

INTRODUCTION

Study on pinning properties and its dynamics of magnetic vortex systems is of great interest for applications of type–II superconductors [1]. Various defects structures have been suggested to act as pinning centers of vortex motion in cuprate superconductors. Among them, twin boundaries (TBs) are the most important anisotropic pinning defects in the orthorhombic compound YBCO samples.

So far, computer simulations have been done only for cases of a single TB and unidirectional or parallel TBs [2–6]. In this work, we simulate the vortex dynamics on random TB networks, which model polycrystalline materials. Especially, we study effects of the superconducting grain size on vortex flow patterns and I-V curves.

THE MODEL

Our simulation geometry is that of an infinite slab of superconductors under a magnetic field perpendicular to the slab surface. We also treat the vortices as stiff pillars, so that we need to model only a two–dimensional (2D) slice (x–y plane) of the 3D slab.

The equation of motion for the i–th vortex can be described by the overdamped equation as [2–6]

$$\eta \frac{d}{dt}\mathbf{r}_i = \mathbf{F}_L + \mathbf{F}_{vv}(\mathbf{r}_i) + \mathbf{F}_{tb}(\mathbf{r}_i) + \mathbf{F}_{th}(t) \quad (1)$$

where $\mathbf{r}_i$ denotes the i–th vortex position vector and η is viscous coefficient. $\mathbf{F}_L$ is the Lorentz force, due to an applied current density j is assumed to be along the x direction. $\mathbf{F}_{vv}$ is the repulsive intervortex interaction force, $\mathbf{F}_{tb}$ is the TBs pinning force, and $\mathbf{F}_{th}(t)$ is the random force. Explicit forms of these quantities have been discussed in Ref.[2-6].

SIMULATION RESULTS

We simulate how vortices flow through the sample with random TB network structures by changing the Lorentz force strength f_L. We consider two different random TB network structures: one has a small average grain size and the other has a large one. The numerical method to create TB network structures have been discussed in Ref. [7, 8].

Typical flow patterns for temperature $T = 50K$, vortex number $N_v = 100$ and the large grain size case are shown in Fig.1(a)–(c).

The x-component of the mean vortex velocity, V_x, is plotted in Fig.1(d) as a function of the Lorentz force f_L. We can see the depinning behavior or threshold behavior at a certain threshold

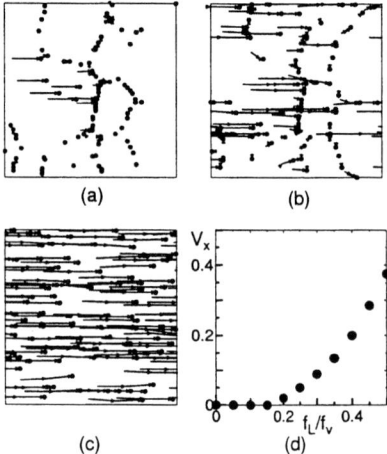

FIGURE 1. Typical flow patterns for $T = 50$K, $N_v = 100$ and the large grain size at $f_L/f_v = 0.1$ (a), 0.2 (b), and 0.5 (c).
The mean vortex velocity V_x is plotted as a function of f_L in (d).

current ($\sim 0.17 f_v$ in this case).

Threshold behaviors of I–V curves are summarized in Fig.2 for several cases.

It is found that the threshold current is a decreasing function of T, N_v, and the average grain size. These behaviors are understood as results of the thermal depinning from TBs (▲), the escape from TBs by the enhanced repulsive force due the large number of vortices (×), and the increase of the number of TBs pinning centers (○), compared with the result of Fig.1(d).

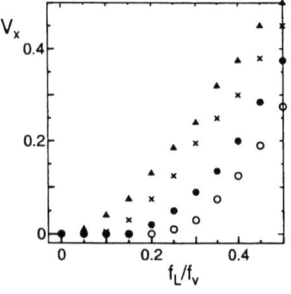

FIGURE 2. The mean vortex velocity V_x as a function of the Lorentz force f_L for several cases. Conditions are the same as those of Fig.1(d), except for $T = 100$K (▲), $N_v = 400$ (×), and the small grain size (○). For comparison, the result of Fig.1(d) is also shown by ●.

CONCLUSION

We have studied vortex flow patterns driven by Lorentz force through the polycrystalline superconductors with random TB networks. Langevin dynamics simulations of the point vortex model have demonstrated the threshold behavior of the I–V curves.

REFERENCE

1. Blatter, G., Feigel'man, M, F., Geshkenbein, V, B., Larkin, A, I., and Vinokur, V, M., *Rev.Mod.Phys.* **66**, 1125-1130 (1994).
2. Groth, J,. Reichharrdt, C., Olson, C, J., Field, S, B., and Nori, F., *Phys.Rev.Lett.* **77**, 3625-3630 (1996).
3. Zhu, B, Y., Dong, J., Xing, D, Y., and Wang, Z, S., *Phys.Rev.B* **57**, 5063-5068 (1998).
4. Reichharrdt, C., Olson, C, J., and Nori, F., *Phys.Rev.B* **61**, 3665-3670 (2001).
5. Enomoto, Y., and Mitsuda, T., *Physica C* **367**, 60-64 (2002).
6. Enomoto, Y., and Mitsuda, T., *Sing.J.Phys.* **18**, 105-108(2002).
7. Enomoto, Y., and Kato, R., *Phys.Lett.A* **142**, 256-259 (1989).
8. Chen, L, Q., *Script Metall.* **32**, 15-30 (1995).

Numerical simulation for collisions of a rigid disk on fluid surface

Shin-ichiro Nagahiro and Yoshinori Hayakawa

Department of Physics, Tohoku University, Aoba-ku, Sendai, 980-8578, Japan

Abstract. We studied collision between a fluid surface and a rigid disk using smoothed particle hydrodynamics (SPH) technique. Analytical treatment of the problem is extremely difficult because the free surface of the fluid largely deforms. SPH is an effective method to solve such problems which involve time-dependent boundary condition. In our model, a collision between the disk and the fluid surface is characterized by Reynolds number, Froude number, angle of incidence of the colliding disk and the ratio of disk density to fluid density. For oblique impact, the disk will go down into fluid or rebound. We numerically investigated the conditions for the rebounds.

INTRODUCTION

There are many phenomena in which impacts between a solid body and a fluid-surface play an important role. Some examples are estimation for the action of waves on ship-hull or breakwaters, control of spacecrafts reentering the atmosphere and so on. Basilik lizards inhabiting Central and South American rainforests are famous for their ability running on the water surface [1]. A familiar example is so called "stone skipping", *i.e.* a stone thrown to water surface rebouds against it [2].

It is difficult to simulate these problems because boundary conditions are defined on a moving surface. In recent years, some numerical methods have been developed to treat moving bolundary [3]. For our simulation, we used SPH technique [4, 5, 6]. SPH is a lagrangian particle method which does not require a fixed grid and can be used to simulate fluids with moving boundaries. In this method, the pressure which acts on a solid objects is replaced by a force between the particles and solid surface. Thus, it is suitable for problems of fluid impact.

In this paper, we mainly focus on conditions for the rebound of the disk against fluid surface through the 2-dimensional simulation using SPH technique. Other interests are pressure force acting on the disk or energy dissipation during the impact. The impact is characterized by Reynolds number R, Froude number F, angle of incidence θ and ratio of fluid density to disk density. We examined the conditions for the disk to rebound by changing these parameters.

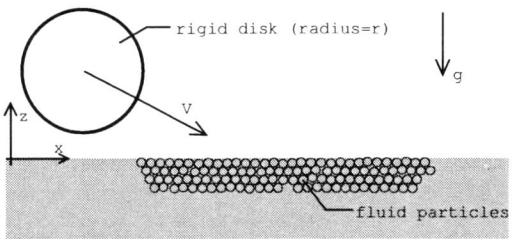

FIGURE 1. A schematic view of the model.

MODEL

In Fig.1, we show a schematic view of our model for fluid impact. A disk with radius r collides with a fluid-surface with angle θ under the action of gravity g. We assume that the disk is rigid, and therefore only translational and rotational degree of freedom are taken into account.

Motion of the fluid is calculated with SPH. The method interpolates pressure, density or other thermodynamic variables with a Gaussian karnel. The half width of the gaussian karnel h must be sufficiently smaller than the radius of the disk. We set $h/r = 0.05$ for our simulations. The sound velocity c of the fluid is decided with artificial equation of states (see Ref.[7]) which relates pressure of particles to density. In general case, the Mach number $M(= v_0/c)$, where v_0 is impact velocity of the disk, is much smaller than unity. We keep $M = 0.04$ fixed.

Reynolds number and Froude number are defined as follows,

$$R = \frac{v_0 \rho d}{\mu}, \quad F = \frac{v_0^2}{gd}, \qquad (1)$$

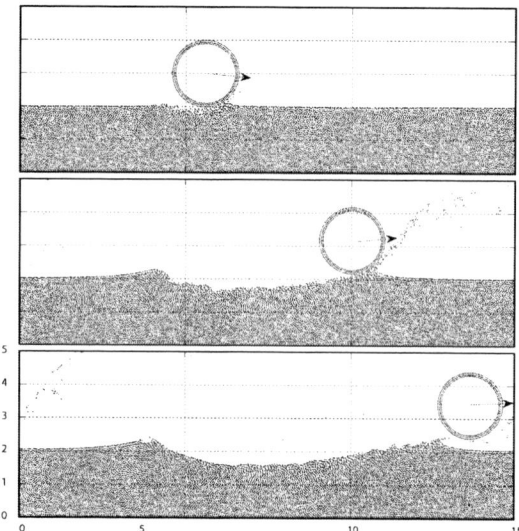

FIGURE 2. Snap shots of a disk and fluids undergoing collision for $F = 55$ and $R = 10$. From the top to bottom, time is 4.5, 9.5 and 13.0 respectively.

where ρ is the density of fluid and μ is the dynamic viscosity, respectively.

To calculate the solid-fluid interaction, we assume a thin layer of fluid particles which is fixed on the edge of the disk. The pressure acting on the disk surface is obtained by calculating the force between these fixed particles and fluid particles.

RESULTS AND DISCUSSION

We show the snapshots of the collision on a condition $F = 55, R = 10, \theta = 0$ in Fig. 2. The simulation was performed with 1.2×10^4 fluid particles. Our model practically reproduces the rebound against fluid surface. As the first step, we investigate the condition for the rebound by changing R and F, and fixing $\theta = 0$ and $\rho/\rho_{Disk} = 1$. The result is shown in Fig. 3. In the figure, the open circles indicate that the disk rebounds after impact and black circles indicate that the disk goes down into the fluid. In the present case, density of the disk is equal to that of fluid and the buoyancy which acts on the disk are vanishing. Hence, we judge that the disk has "rebounded" when momentum of z-component has changed its sign.

The rebound occurs when Froude number exceeds $F^* = 25$ and it does not strongly depend on the value of R. This result leads to following inequaruty

$$v_0 > \sqrt{F^* g r}. \quad (2)$$

In the case $R < 1$, large part of the energy of the disk is transformed into the fluid. When the angle of incident is

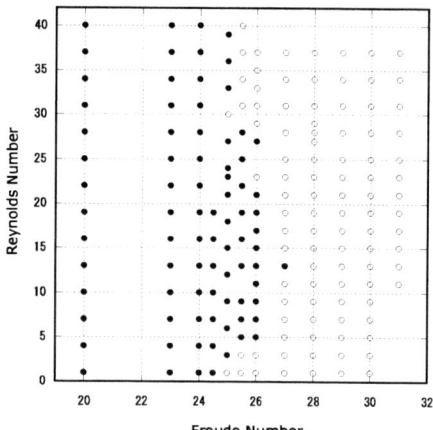

FIGURE 3. The open circles indicate that the disk rebound after the impact. The black circles indicate that the disk goes down into the fluid.

sufficiently small, steady flow takes place in x-direction because we take periodic boundary condition. Hence we cannot confirm the validity of eq.(2) in the range $R < 1$.

SUMMARY

We present results of our simulation for impact between a disk and a fluid surface. Conditions for rebound are investigated by changing Reynolds number and Froude number.

ACKNOWLEDGMENTS

One of the authors (S.N) acknowledges helpful discussions with Dr. H. N. Kono.

REFERENCES

1. Glasheen, J. W. and McMahon, T. A. *Nature*, **380**, 28, 340-342 (1996)
2. Bocquet, L. *Am. J. Phys.* **71**, 2, 150-155 (2003)
3. Takewaki, H., Nishiguchi, A. and Yabe, T. *J. Comput. Phys.* **61**, 261-268 (1987)
4. Lucy, L. B. *Astron. J.* **82**, 12, 1013-1024 (1977)
5. Gingold, R. A. and Monaghan, J. J. *Mon. Not. R. astr. Soc.* **181**, 375-389 (1977)
6. Takeda, H., Miyata, S. M. and Sekiya, M. *Prog. Theo. Phys*, **92**, 5, 939-960 (1994)
7. Cleary, P. W. *Appl. Math. Mod.* **22**, 981-993 (1998)

Laser control of non-stationary proton state in hydrogen-bonded system

T. Yamaguchi*, I. Sakamoto*, Y. Ohta†, H. Nagao* and K. Nishikawa*

*Department of Computational Science, Faculty of Science, Kanazawa University,Kakuma, Kanazawa 920-1192, Japan
†PRI, Natinoal Institute of Advanced Industrial Science and Technology, Kansai center,Osaka 563-8577, Japan

Abstract. Stimulated Raman adiabatic passage method (STIRAP) is well known to be a very effective method for the efficient and selective transition between two molecular eigenstates in molecule. In this study, we generalize the STIRAP to be able to treat the transition to the non-stationary state in the hydrogen bonded model system.

INTRODUCTION

Recently, various methods have been proposed to control the chemical reaction by means of laser pulses. The purpose of these works is to control the chemical reaction and develop the optical molecular device. One promising approach is the optimal control theory (OCT) [1, 2, 3], which makes it possible to design the pulse shape that could transfer the initial state of molecule to a desired state. However, the pulse shape produced by the OCT is usually quite complicated to be made experimentally. One the other hand, the STIRAP [4, 5] is also very interesting for the efficient and selective transition in molecule. This method is very flexible with respect to the change of laser parameters such as intensity, width, delay time, detuning and frequency.

Hydrogen bond plays fundamentally a very important role in a variety of fields, and determines the physical and chemical properties of molecule and solid. Malonaldehyde have an intramolecular hydrogen bond, so that many studies have been performed theoretically and experimentally to investigate the proton tunneling in molecules. A substituted malonaldehyde could have two geometrical isomers as shown in Fig.1 due to the position of proton related to the hydrogen bond.

The STIRAP is significantly effective for the complete population transfer between two eigenstates in the three-level system. However, contray to the OCT, it is not aplicable to the transition to the arbitrary non-stationary state. In this study, we generalize the STIRAP to be able to treat the transition from the ground state to the non-stationary state.

MODEL SYSTEM

In order to simulate the control of proton motion, we need to calculate the potential curve of proton motion along the reaction coordinate. The ab-initio calculation of the potential curve is performed with Gaussian98 software package, where we employed B3LYP/6-31G** for geometry optimization and IRC calculation. Fig.2 shows the resultant potential curve of proton in 1-methyl malonaldehyde molecule. The ground state and 1st excited state are almost localized to the left and right potential minimum, respectively, while the transition state corresponds to the potential maximum.

In the ordinary STIRAP, the pulse sequence is composed of two laser pulses, and the external electric field $E(t)$ is given by

$$E(t) = \sum_{i=1}^{2} A_i \cos(\omega_i t) g(t_i, \sigma_i), \qquad (1)$$

where $A_i (i = 1, 2)$ is the maximum amplitude of i-th laser pulses. $g(t_i, \sigma_i)$ is the Gaussian shape function with time center t_i, width σ_i, and frequency ω_i.

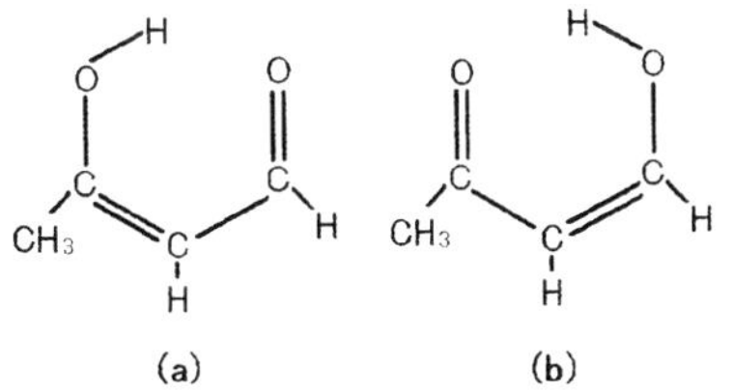

FIGURE 1. Two tautomers of 1-methyl malonaldehyde.

CP708, *Slow Dynamics in Complex Systems: 3rd International Symposium*, edited by M. Tokuyama and I. Oppenheim

$$g(t_i, \sigma_i) = \exp\left[-\left(\frac{t-t_i}{\sigma_i}\right)^2\right] \quad (2)$$

We have already simulated the complete transition from the ground state $|\phi_0\rangle$ to the 1st excited state $|\phi_3\rangle$ by the ordinary STIRAP [6]. In order to make a transition to a non-stationary state being a superposed state of some eigenstates, we have to treat many states within the STIRAP by changing some external laser parameters.

RESULTS AND DISCUSSION

In our simulation, we consider the model system with five levels, and realize a non-stationary state being composed of three eigenstates as target state, so that we take into account for 4 laser pulses. Namely, by changing the time t_i and the Rabi frequency Ω_i, we simulate numerically the coherence transfer from the ground state $|\phi_0\rangle$ to the superposed state $C_1|\phi_1\rangle + C_2|\phi_2\rangle + C_3|\phi_3\rangle$. Here we try to obtain two non-stationary states, where the ratio of the expansion coefficient is given by $C_1 : C_2 : C_3 = 1 : 1 : 1$ and $1 : 2 : 3$. In order to find such situation by changing the laser parameters, we adopted a GA-like algorism, which searches effectively the target state with the required laser parameters.

Fig.3 represent the population change with respect to time t (changing from 0 to 50 ps.), while fixing the other parameters as $t_1 = 30, t_2 = 24.5, t_3 = 30.5, t_4 = 18$ and Rabi frequencies $\Omega_1 = 4.5*10^{-5}, \Omega_2 = 6.9*10^{-5}, \Omega_3 = 6.7*10^{-5}, \Omega_4 = 9.2*10^{-5}$ au. From this figure, we easily find out that we could obtain the desired target state and the required laser pulse sequence. In a similar way, it is easily shown that the population change in Fig.4 realize the superposed state with the ratio of expansion coefficient such as $C_1 : C_2 : C_3 = 1 : 2 : 3$.

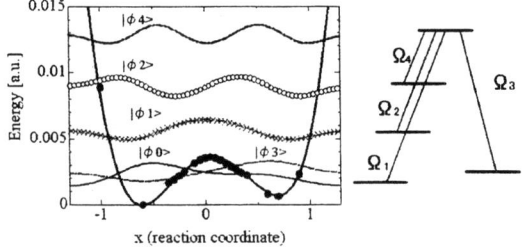

FIGURE 2. Model potential and some eigenstates. Dot points mean the value of IRC calculation, and are used to fit the potential curve. Schematic diagram shows the relation between energy levels and Rabi frequencies.

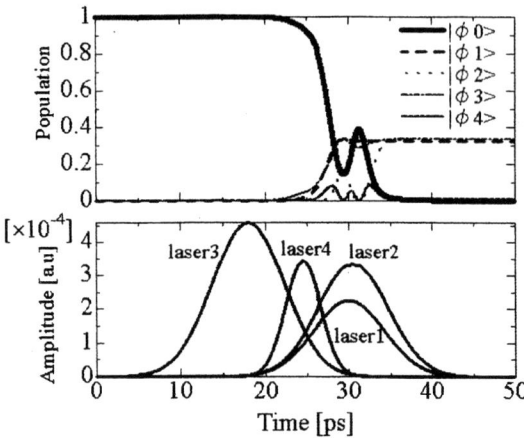

FIGURE 3. Time-development of population and laser field (the ratio of expansion coefficient is $C_1 : C_2 : C_3 = 1 : 1 : 1$)

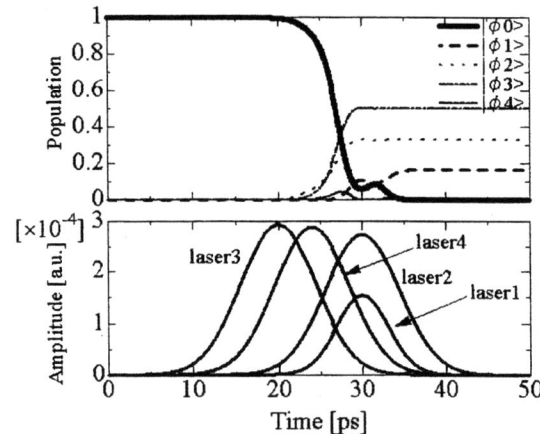

FIGURE 4. Time-development of population and laser field (the ratio of expansion coefficient is $C_1 : C_2 : C_3 = 1 : 2 : 3$)

REFERENCES

1. Zhu, W., Botina, J., and Rabitz, H., *J.Chem.Phys*, **108**, 1953–1963 (1998).
2. Došlić, N., Kühn, O., Manz, J., and Sundermann, K., *J.Phys.Chem.A*, **102**, 9645–9650 (1998).
3. Fujimura, Y., Gonzalez, L., Hoki, K., Manz, J., and Ohtsuki, Y., *Chem.Phys.Lett.*, **306**, 1–8 (1999).
4. Gaubantz, U., Rudecki, P., Schiemann, S., and Bergmann, K., *J.Phys.Chem.*, **92**, 5363–5376 (1990).
5. Bergmann, K., Theuer, H., and Shore, B. W., *Rev.Modern Phys.*, **70**, 1003–1025 (1998).
6. Ohta, Y., Yoshimoto, T., and Nishikawa, K., *Chem.Phys.Lett.*, **316**, 551–557 (2000).

Simulation of Vortex Creep in Type-II Superconductors

Ryuzo Kato* and Yoshihisa Enomoto[†]

Department of Bioengineering, Kagoshima University, 1-21-40 Koorimoto, Kagoshima 890-0065, Japan
†Department of Environmental Technology, Nagoya Institute of Technology, Gokiso, Nagoya 466-8555, Japan

Abstract. The experimental study pointed out that voltage noise power spectrum has two types. One is broad-band noise (BBN), which results from bulk pinning. The other is narrow-band noise (NBN), which is observed near critical current. Therefore, it is expected to occur in the region of vortex creep. NBN is sensitive to even direction of external current, which is supposed to result from surface condition of superconductors. To find out origin of NBN, we perform simulations of the vortex dynamics in the 2D system with pinning array. Simulations show that the periodic pinning potential generates NBN and the first peak frequency is proportional to voltage. We find that the initial vortex arrangement has effect on the first peak frequency.

INTRODUCTION

There has been growing interest in vortex dynamics of high-T_c superconductors in the mixed state as it was found the mixed state has some phases and is complicated because of large fluctuations due to the 2D layer structure and high temperature.

Moreover pinning plays a great role in transport properties in real superconductors. Noise measurement is a powerful tool for investigation of dynamics properties of vortex under the influence of pinning. The measurements for $YBa_2Cu_3O_{7-\delta}$[1] and $Bi_2Sr_2CaCu_2O_y$[2] were performed. In the both cases the power spectrum of voltage noise consists of broad-band noise (BBN) and narrow-band noise (NBN).

BBN is a Lorentzian form and results from bulk pinning. While NBN is sensitive to even direction of external current and only observed in the region of vortex creep near critical current.

We simulate the vortex creep state in type-II superconducting films with pinning array to investigate the origin of NBN.

MODEL

A type-II superconducting film is in the x–y plane in the presence of an external magnetic field applied along the z–axis. Transport current flows in the y-direction. Our precedent simulations for the system with 2D vortex-vortex interaction show that the random media potential and the periodic potential of pinning generate respectively BNB and NBN. To investigate origin of NBN, the pinning potential is only periodic pinning array.

The equation of motion for the i–th two dimensional vortex position $\mathbf{r}_i$ can be described by

$$\Gamma^{-1}\frac{\partial \mathbf{r}_i}{\partial t} = -\sum_{j\neq i}^{N_v} \nabla_i V_v(\mathbf{r}_i - \mathbf{r}_j) - \sum_{k=1}^{N_p} \nabla_i V_p(\mathbf{r}_i - \mathbf{R}_k) + \mathbf{j} \times \frac{\Phi_0}{c}, \quad (1)$$

where Γ is the friction coefficient of a single vortex, N_v is the total number of vortices and N_p is the total number of pinning center. Here $\mathbf{R}_k$ denotes the position of k–th pinning center.

The vortex-vortex interaction is repulsive and long range force[3]. On the other hand, the pinning-vortex interaction is attractive and Gaussian potential, given by $V_p(\mathbf{r}) \propto -\exp(-|\mathbf{r}|^2/2a^2)$, where $a = 2\sqrt{2}\,\xi$ and ξ is coherence length. In our simulations, the pinning array form tetragonal lattice and the periodic boundary condition is applied.

RESULTS

In the case of 4 × 4 pinning array, current-voltage characteristic with $N_v = 16$ is shown in Fig.1, which is obtained by 10 simulations where each initial arrangement of vortices is random. Current-voltage curve is convex, which is consistent with experimental results where NBN appears in the noise measurement. There are many variety in Fig. 1 in the same system except for the initial condition, which indicate vortex arrangement is dominant in the current-voltage characteristic.

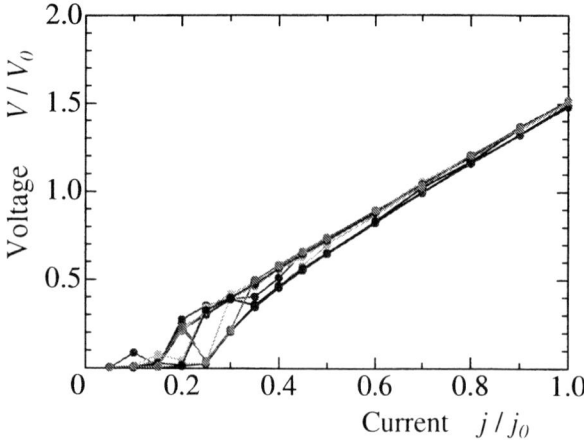

FIGURE 1. Current-voltage characteristics of the system with the pinning array of 4 × 4 tetragonal lattice for 10 samples with different initial conditions.

Shown in Fig. 2 is a simulation of the voltage noise power spectrum at $j/j_0 = 0.5$, which is above critical current. It represents only NBN with some peaks, where the first peak frequency f_{p1} is fundamental frequency and other peaks are generated by higher harmonics. The frequency of the first peak f_{p1} has various value at the same condition expect for initial arrangements of vortices. NBN is obtained by the simulations of vortex motion in the region of vortex creep.

The experimental results show that f_{p1} is proportional to logarithm of external current[1]. In our simulations there is no relationship beween external current and f_{p1}, while all f_{p1} are on an alternative line in Fig. 3. Since the average velocity of vortices $\bar{v}$ is proportional to voltage,

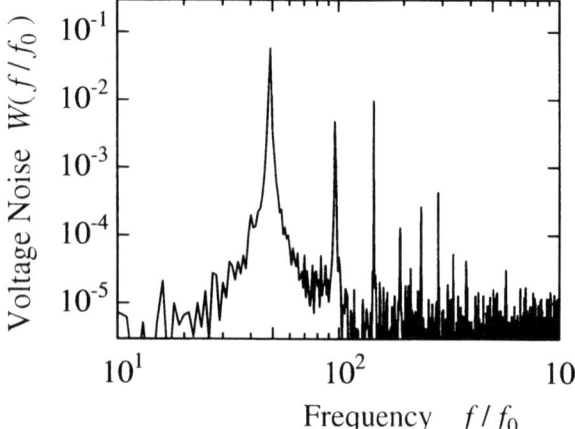

FIGURE 2. Voltage noise power spectrum at $j/j_0 = 0.5$. Only NBN appears and there are some peaks. The frequency of the i-th peak is n times of that of the first peak.

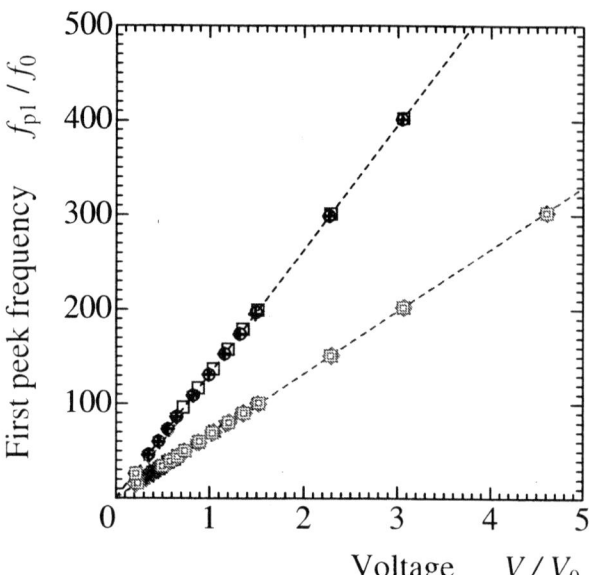

FIGURE 3. Plot of the first peak frequency vs voltage for 10 samples with different initial conditions. All points are on an alternative line. Steeper gradient is twice of the other.

it results in

$$f_{p1} \propto \bar{v} / \ell, \tag{2}$$

where ℓ is pinning lattice spacing.

SUMMARY

Our simulations show that vortex creep in the system with periodic pinning potential generates NBN. The first peak frequency f_{p1} is equal to either $\bar{v}/\ell$ or $2\bar{v}/\ell$. The peak frequency of NBN is dependent on the initial arrangement of vortices but the transport current. Difference of factor 2 indicates that matching between the vortex lattice and pinning array has effect on NBN. It coincides with the experimental results where the surface effect is major in appearance of NBN.

REFERENCES

1. D'Ann, G. *et al.*, *Phys. Rev. Lett.*, **75**, 3521–3524 (1995).
2. Tsuboi, T., Hanaguri, T., and Maeda, A., *Phys. Rev. Lett.*, **80**, 4550–4553 (1998).
3. Fetter, A. L., and Hohenberg, P. C., *Phys. Rev.*, **159**, 330–343 (1967).

Proton Dynamics Simulation of *p*-Chloro and *p*-Bromobenzyl Alcohol Crystals

T. Ida[1], D. Matsumoto[1], M. Hamada[1], M. Mizuno[1], K. Endo[1], M. Hashimoto[2]

[1] *Department of Chemistry, Faculty of Science, Kanazawa University, Kanazawa, 920-1192, Japan,*
[2] *Department of Chemistry, Faculty of Science, Kobe University, Nada-ku, Kobe, 657-8501, Japan*

Abstract. The structure of *para*-chlorobenzyl alcohol (*p*CBA) and *para*-bromobenzyl alcohol (*p*BBA) crystals are characterized by the O–H⋯O hydrogen bonded chains along c axis. The direction of the hydrogen bond in the low temperature phase (LTP) is opposite to that in the room temperature phase (RTP). These transitions are related to the hydrogen bonds. We performed the molecular dynamics simulation of *p*CBA and *p*BBA crystals in the LTP and RTP using forces determined by gradient of the energy within semiempirical molecular orbital calculation (PM3). From the distribution of the hydrogen atom of –OH group, we found that the hydrogen atoms in the LTP vibrate at the equilibrium position within one site or asymmetric potential. In the RTP, the hydrogen atoms of *p*BBA are jumping between the two sites in the symmetric double minimum potential.

INTRODUCTION

para-Chlorobenzyl alcohol (*p*CBA) and *para*-bromobenzyl alcohol (*p*BBA) have been reported to undergo a first-order phase transition at T_{c1}=236 and 217K, respectively [1,2]. Crystals of *p*CBA and *p*BBA are isomorphous (monoclinic, space group $P2_1$, Z=2). The structure is characterized by the O–H⋯O hydrogen bonded chains along *c* axis [3]. The results of ^{35}Cl, ^{85}Br NQR and dielectric measurements predicted another high-order phase transition at T_{c2}=218 and 195K for *p*CBA and *p*BBA, respectively [1,2]. The direction of the hydrogen bond in the low temperature phase (LTP) is opposite to that in the room temperature phase (RTP) [3]. These transitions should be closely related to local structure of hydrogen bonds.

Recently we investigated the local dynamic behavior for the phase transition from ^{2}H NMR spectra, the relaxation time of the spin-lattice (T_1) and the quadrupole order (T_{1Q}) of the –OH group [4]. Then we concluded that, in the LTP of both crystals, the hydrogen atoms of –OH groups are jumping between the two sites and the jump of hydrogen atoms occurs in asymmetric potential wells. We also founded that the high-order phase transitions of both crystals are closely related to the change of these asymmetry potential wells to symmetric ones.

Molecular dynamics (MD) simulation is a powerful tool to find out the microscopic structural and dynamical properties of molecule. Classical MD simulation cannot circumvent the limitations stemming from the uncertainties in pair potentials between arbitrary molecules. Thus, we think that MD simulation including the quantum mechanical calculations are necessary for consistency and accuracy of chemically complex system. In the simulation, the inter- and intra-molecular forces are computed directly from the electronic structure within the molecular orbital calculation.

In the present work, we perform a molecular dynamics simulation for *p*CBA and *p*BBA crystals using forces determined by gradient of the energy within parameter modified 3 (PM3) method [5], which is a semi-empirical quantum molecular orbital calculation. Our aims of this investigation are to examine the local structure and kinetics associated with rotational and vibrational motion of –OH group in the *p*CBA and *p*BBA crystals and to study the relation between symmetry and asymmetry potential wells and the motions of –OH group in the crystals.

COMPUTATIONAL DETAILS

We performed the MD simulation of *p*CBA and *p*BBA crystals in the LTP (150K) and the RTP (300K).

The equations of nuclear motion were integrated using 5-value Gear method [6], as a predictor-corrector algorithm, with a time step of 0.5 fs, and each simulation run was performed for 10 ps (2×10^4 steps). In order to determine properties averaged within the NVT ensemble, the temperature was controlled with Nose-Hoover thermostats [7] and sampling position data were carried out after 2 ps (4×10^3 steps). The simulations were performed using a cluster model with reflecting the hydrogen bonded chain consists of two unit cells of the crystal and are surrounded by $3 \times 3 \times 3$ image cells with point charges on each atom.

RESULTS AND DISCUSSION

Fig. 1 shows distributions of dihedral angle between the C-O-H plane and C-O-c axis plane in the pCBA and pBBA at the low temperature phase (a) and the room temperature phase (b). These distributions correspond to the direction of the hydrogen bonds of –OH group with respect to c axis, i.e. 0 degree means that the hydrogen bond orients to up direction along the c axis and 180 degree does to down direction.

In the LTP, the hydrogen atoms in both pCBA and pBBA vibrate at the vicinity of the equilibrium position (at 26 degree). The thermal vibrations are in the region of about ±5 degree. On the other hand, the distributions in the RTP show different behavior of proton motions between the pCBA and pBBA. It is clear that the motion in the pCBA fluctuates only at the equilibrium position (at 165 degree) in the vibrational region of ±15 degree. The hydrogen bonds in the pBBA take three directions at 25, 120 and 160 degrees. Then, we define each direction as peak-A, B and C, respectively, as shown in Fig.1 (b). In the figure, peak-C is obviously the equilibrium position in the RTP, and peak-A corresponds to the equilibrium direction in the LTP. This result indicates that the hydrogen atoms of –OH groups are jumping between the two sites.

From analysis of the local structure of –OH group at time of 5 ps, the peak-B is due to lattice defect. The defect means the directions of the nearest hydrogen bonds toward the up and the down directions along c axis. Furthermore we obtained the almost equal populations in the peak-A and peak-B+C from the results of population analysis. This result shows that the hydrogen atoms of pBBA in the RTP occur in the double minimum potential between stable sites in the RTP and LTP. These results are in good agreement with the results of our ^{2}H NMR study [4]. Thus we concluded that the hydrogen atoms of both crystals in the LTP vibrate at the equilibrium position within one site or asymmetric potential. In the RTP, the hydrogen atoms of pBBA are jumping between the two sites in the symmetric double minimum potential.

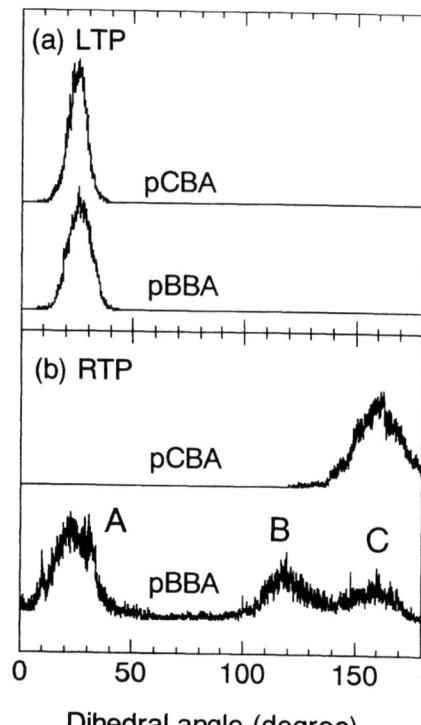

FIGURE 1. Distributions of dihedral angle between the C-O-H plane and C-O-c axis plane in the pCBA and pBBA crystal. (a) and (b) are corresponding to the low temperature phase and the room temperature phase, respectively.

REFERENCES

1. Niki H., Kano K., and Hashimoto M., *Z. Naturforsch.* **51a**, 731-735 (1996).

2. Hashimoto M., Monobe Y., Terao H., Niki H. and Mano K., *Z. Naturforsch.* **53a**, 436-441 (1998).

3. Hashimoto M. and Nakamura Y., *Acta. Crystallogr.* **C44**, 482-484 (1988).

4. (a) Hashimoto M., Harada M., Mizuno M., Hamada M., Ida T. and Suhara M., *Z. Naturforsch.* **57a**, 381-387 (2002). (b) Mizuno M., Hamada M., Ida T., Suhara M. and Hashimoto M., *Z. Naturforsch.* **57a**, 388-394 (2002).

5. Stewart J.J.P., *J. Comp. Chem* **10**, 209-264 (1989).

6. Gear C.W., "The Numerical Integration of Ordinary Differential Equations of Various Orders", *Report ANL 7126*, Argonne National Laboratry (1985).

7. (a) Nose S., *J. Chem. Phys.* **81**, 511-519 (1984). (b) Hoover W.G., *Phys. Rev. A*, **31**, 1695-1697 (1985).

Control of cis-trans isomerization by stimulated Raman adiabatic passage method

T. Yamaguchi[*], Y. Ohta[†], K. Sugiyama[*], H. Nagao[*], and K. Nishikawa[*]

[*]Department of Computational Science, Faculty of Science, Kanazawa University,
Kakuma, Kanazawa 920-1192, Japan
[†]PRI, Natinoal Institute of Advanced Industrial Science and Technology, Kansai center,
Osaka 563-8577, Japan

Abstract. Stimulated Raman adiabatic passage method (STIRAP) is well known to be a very effective method for the efficient and selective transition between two molecular eigenstates in molecule. In this study, we generalize the STIRAP to be able to complete transfer between all-trans and 11-cis in the retinal.

INTRODUCTION

Due to the great development of laser technique, several theoretical methods using the laser pulses have been proposed to control the chemical reaction and to develop the optical molecular device. For example, the π-pulse method, the chirped pulse method, the stimulated Raman adiabatic passage (STIRAP) method, and the optimal control theory have been applied to the highly selective transition from the ground state to the specified excited state. We have noticed that the STIRAP proposed by Bergman et al. [1,2] realizes the complete population transfer between two molecular states. The STIRAP method has a counterintuitive pulse sequence, a rather large overlap between two pulses, and a larger pulse width compared to the π-pulse width. This method has very flexible conditions for the laser parameters such as amplitude, frequency, pulse area, delay time or pulse width, and is well known as very effective method for population transfer in three-level systems. Due to the robustness with respect to the laser parameters, we have applied the STIRAP to the control of the proton motion in hydrogen bond system [3], and have found the effective pulse sequence to yield the complete proton transfer.

The cis-trans isomerization in a molecule is one of the most fundamental unimolecular chemical reactions, and plays also a very important role in a biochemical reaction. Especially, retinal and phytochrome have been attracting much interest in relation to their biochemical importance in photoreceptor proteins. For example, the chromophore in rhodopsin is a protonated Schiff base of 11-cis retinal bound to the protein opsin. After the retinyl chromophores absorb light, the cis-trans isomerization of this chromophore rapidly triggers the following vision process, where the surrounding protein plays also a important role as well as the chromophore molecule [4], but the first cis-trans isomerization is a crucial process in the photochemical process in vision. Thus the photoisomerization of free retinal molecule has been studied extensively by various methods. The retinal molecule has some double bonds, where the isomerization might occur through the rotation around this bond. Thus the mechanism of the isomerization has been analyzed in detail in relation to the excited state involved and the relaxation process, and it is clarified that the photoisomerization from mono-cis isomers is more efficient than that from all-trans ones [5].In this study, we investigate the control of cis-trans isomerization of retinal by the STIRAP method.

MODEL AND SIMULATION

The retinal could become a candidate of the optical molecular device, so the purpose of this work is to control the cis-trans isomerization of retinal molecule itself using the laser pulses. First we calculate the adiabatic potential around the specified double bond, which induce the cis-trans isomerization as shown in Fig.1. We used Gaussian98 software package, where we employed B3LYP/6-31G** for energy calculation. Then we calculate numerically the eigenstates of the system required for the STIRAP simulation. Finally we perform the STIRAP simulation to control the geometrical isomerization of retinal.

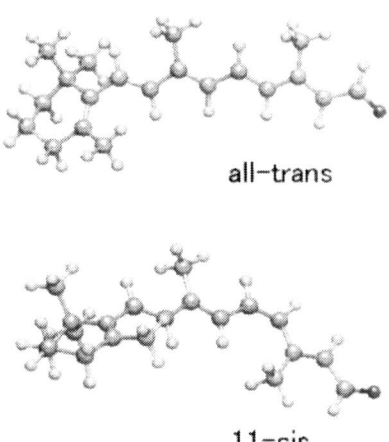

FIGURE 1. Photoisomerization of 11-cis and all-trans retinal molecule

RESULTS AND DISCUSSION

In this simulation, we consider the model system of retinal with three levels, and try to realize a complete population transfer from trans-form to cis form. Fig.2 shows the model potential curve of the isomerization obtained from the ab-initio calculation, and three eigenstates involved in the required STIRAP simulation.

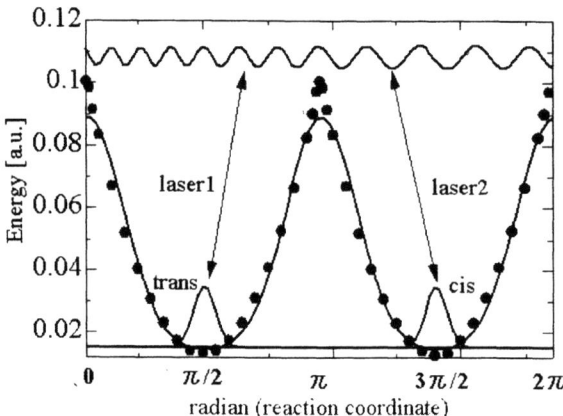

FIGURE 2. The potential energy curve and some eigenstates The values of the dot points is from the Gaussian calculation, and are used to fit the potential curve.

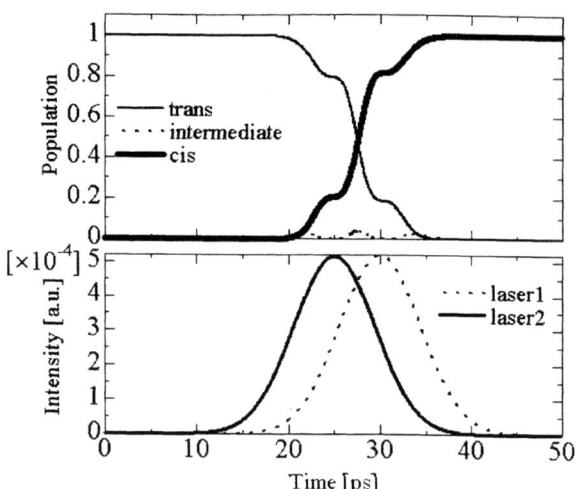

FIGURE 3. Time-development of population and laser field.

Fig.2 shows the model potential curve of the isomerizarion obtained from the ab-inito calculation, and three eigenstates involved in the required STIRAP simulation. Fig.3 shows the results of our simulation, and we could realize the complete population transfer from trans-form to cis-form, where we changed the time between two laser pulses by 5.0 ps. The Rabi frequencies are assumed to be $\Omega_1 = 4.8*10^{-5}$ and $\Omega_2 = 4.9*10^{-5}$.

The retinal molecule has some double bonds, and could have many geometrical isomers with respect to the rotation around the double bond. Therefore, in future work, we will try to consider the formation of the specified isomer from the possible configuration by means of the laser control, namely to change the molecular shape at will by the STIRAP method.

REFERENCES

1. Kuklinski, J. R., Gaubatz, U., Hioe, F. T., and Bergmnn, K., *Phys. Rev. A*, **40**, 6741-6744 (1989).

2. Bergmann, K., Theuer, H., and Shore, B. W., *Rev. Modern Phys.*, **70**, 1003-1025 (1998).

3. Ohta, Y., Yoshimoto, T., and Nishikawa, K., *Chem. Phys. Lett.*, **316**, 551-557 (2000).

4. Grigoerieff, N., Ceska, T. A., Downing, K. H. Baldwin, J. M., and Henderson, R., *J. Mol. Biol.*, **259**, 393-421 (1996).

5. Tahara, T., Toleutaev, B. N., and Hamaguchi, H., *J. Chem. Phys.* **100**, 786-796 (1994).

Quantum algorithm in quantum network systems

I. Sakamoto*, T. Yamaguchi*, H. Nagao* and K. Nishikawa*

Department of Computational Science, Faculty of Science, Kanazawa University, Kakuma, Kanazawa, 920-1192, Japan

Abstract. Recently, the quantum computer (QC) using the nano-devices have significantly attracted attention, because a large-scale extention of the qubits could be easily realized in the nano-devices. However, some problems for the realization of the QC with nano-devices arise from the short decoherence time and the interaction of qubits only between nearest-neighbor qubits. Therefore, we try to design the optimal quantum circuit of the quantum Fourier transform in various network system by means of the genetic algorithm (GA).

INTRODUCTION

The QC has attracted many theorists and experimentists of all over the world since Shor discovered the factorization algorithm in 1994 [1], where the factorization of a large integer could be solved in polynomial time. Since then, many experimental efforts have actively been made to realize quantum hardwares, and many candidates for the quantum device have been proposed to realize the QC by using nuclear magnetic resonance(NMR) [2], electronic spins [3], quantum dots [4], Josephson Junction [5] and so on. Among them, Chuang realized a NMR QC with 7 qubits. However, it is very difficult to extent the QC with more than 7 qubits. In order to realize the practical QC, we need to install many qubits effectively.

The advantage of the nano-devices is that a large-scale extention of the qubits could be easily realized. However, we have still a few problems for the realization of QC with nano-devices. In the QC, it is significant that all qubits interact with each others. Although it is relatively easy to acumulate qubits with nano-devices, it is difficult to create C-NOT gate due to the loss of the long-range interactions. Moreover, the decoherence time of nano-devices is usually very short. If the decoherence is shorter than the quantum computational time, the result of the computation will be meaningless.

In this study, we consider some quantum network models for the nano-devices QC as shown in Fig.1. We optimize quantum circuts of each quantum network to implement effectively the quantum Fourier transform (QFT) using the GA. The optimization of one-dimentional 5 qubits chain system shown in Fig.1 (a) has already been presented [6]. However we try to find out more efficient circuit in the network system, and investigate the dependence of the operational number of the quantum computation on the number of interactions between qubits in order to realize the practical QC.

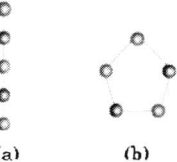

FIGURE 1. The model system of quantum network. (a) 1-dimensional chain. (b) 1-dimensional ring.

MODEL AND METHOD

The QFT is the factorization alogrithm proposed by Shor. The QFT means the following operation,

$$|x\rangle \longrightarrow \frac{1}{\sqrt{N}} \sum_{y=0}^{N-1} e^{\frac{2\pi i xy}{N}} |y\rangle \qquad (1)$$

The rotation of j-th qubit around X and Y axes, and the interactions between j- and k-th qubit can be expressed by, respectively,

$$X_j(\theta) = e^{-i\frac{\theta}{2}\sigma_x} \qquad (2)$$

$$Y_j(\theta) = e^{-i\frac{\theta}{2}\sigma_y} \qquad (3)$$

$$T_{jk}(\theta) = e^{i\frac{\theta}{2}\sigma_{jz}\sigma_{kz}} \qquad (4)$$

The Hadamard tranform H_j and the gate B_{jk} acting on two qubits can be represented as follows,

$$H_j = Y_j(-\pi/2)X_j(\pi)$$
$$= \frac{1}{\sqrt{2}} \begin{pmatrix} 1 & 1 \\ 1 & -1 \end{pmatrix} \qquad (5)$$

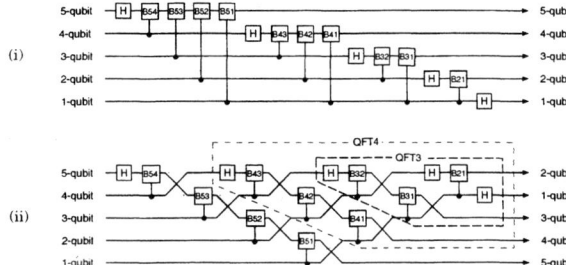

FIGURE 2. (i) The general QFT circuit, which involves the interactions between all qubits. (ii) Quantum circuit of the general QFT using swap gate [6]. The QFT_5 circuit contains the QFT_4 circuit.

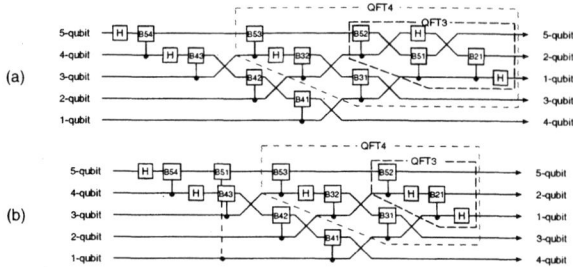

FIGURE 3. The optimal QFT circuit of 5 qubits in the nano-devices, where (a) and (b) represent 1-dimensional chain and ring system. The QFT_5 circuit in each network system contains the parts of QFT_4 circuit.

$$B_{jk} = e^{i\theta_{jk+2}}X_j(\pi/2)Y_j(\theta_{jk+1})X_j(-\pi/2)$$
$$X_k(\pi/2)Y_k(\theta_{jk+1})X_k(-\pi/2)T_{jk}(\theta_{jk+1})$$
$$= \begin{pmatrix} 1 & 0 & 0 & 0 \\ 0 & 1 & 0 & 0 \\ 0 & 0 & 1 & 0 \\ 0 & 0 & 0 & e^{i\theta_{jk}} \end{pmatrix} (\theta_{jk} = \frac{\pi}{2^{k-j}}) \quad (6)$$

The Hadarmard transform H_j is composed of 2 elementary gates, and this operation requires 2 steps. On the other hand, B_{jk} has 7 elementary gates, and it requires clearly 4 steps. The QFT can be constructed by using H_j and B_{jk} as follows,

$$H_n B_{n,n-1} B_{n,n-2} B_{n,n-3} \cdots B_{3,2} B_{2,1} H_{n-1}$$
$$B_{n-1,n-2} B_{n-2,n-3} \cdots H_3 B_{3,2} B_{3,1} H_2 B_{2,1} H_1 \quad (7)$$

For example, Fig.2 (i) shows 5 quibts QFT circuit representing equation (6), where all qubits interact with each other. Now, we consider the interaction only between nearest-neighbor qubits in the nano-devices. To realize the QFT in the nano-devices, we introduce the swap gate S_{jk}, which exchange 2 qubits with C-NOT gate C_{jk}.

$$C_{jk} = Y_j(-\pi/2)X_j(\pi/2)Y_j(-\pi/2)X_j(-\pi/2)X_k(\pi/2)$$
$$Y_k(\pi/2)X_k(-\pi/2)T_{jk}(\pi/2)Y_j(\pi/2) \quad (8)$$
$$S_{jk} = C_{jk}C_{kj}C_{jk} \quad (9)$$

The swap gate is constructed of 27 elementary gates, and the number of steps are 16. However, we could reduce the steps of the swap gate using the GA. The refined swap gate we obtained is as follows

$$S_{jk} = e^{-i3\pi/4}Y_k(-\pi/2)X_j(\pi/2)T_{jk}(-\pi/2)Y_j(-\pi/2)$$
$$X_k(-\pi/2)T_{jk}(\pi/2)Y_k(-\pi/2)X_j(-\pi/2)Y_j(-\pi)$$
$$T_{jk}(\pi/2) \quad (10)$$

The above swap gate can be implemented by 7 steps.

RESULTS AND DISCUSSION

We could obtain the optimal QFT as shown in Fig.3. We find out that the QFT_n circuit contains the parts of QFT_{n-1} circuit in each network system. Therefore, we could design the optimal circuit of n qubits. It is important to compare the operational steps of the optimal circuit and nonoptimal circuit. The nonoptimal circuit is a simple circuit obtained using the swap gate in Fig.2 (ii), and this circuit of n qubits requires $22n - 36$ steps. Contrary to this, the optimal QFT circuit in Fig.3 (a) and (b) can be implemented with $22n - 41$ and $22n - 52$ ($n > 3$) steps, respectively.

We considered the optimal quantum circuit of the QFT in the nano-devices. As a conclusion, the circuits (a) and (b) in Fig.3 could be implemented $O(n)$ steps, so that we could create the QFT circuit more efficient than circuit in Fig.2 (ii). Moreover, we could succeed in creating the optimal QFT circuit of n qubits. As a future subject, we will consider more practical network systems (2-dimensional lattice, star type, and so on), and perform the other quantum algorithm (grover algorithm, Deutcsh-Jozsa algorithm, and so on).

REFERENCES

1. Shor, P. W., *SIAM J. Comput.*, **26**, 1484–1509 (1997).
2. Vandersypen, L. M. K., Steffen, M., Breyta, G., Yannoni, C. S., sherwood, M. H., and Chuang, I. L., *Nature*, **414**, 883–887 (2001).
3. Kikkawa, J. M., Smorchkova, I. P., Samarth, N., and Awschalom, D. D., *Science*, **277**, 1284–1287 (1997).
4. Kamada, H., Gotoh, H., Temmyo, J., Takagahara, T., and Ando, H., *Phys. Rev. Lett.*, **87**, 246401 1–4 (2001).
5. Nakamura, Y., Pashkin, Y. A., and Tsai, J. S., *Nature*, **398**, 786–788 (1999).
6. Blais, A., *Phys. Rev. A*, **64**, 022312 1–5 (2001).

Quantum Molecular Dynamics Simulation of Guest Molecules in Gas Hydrate

D. Matsumoto, T. Ida, N. Kato, M. Mizuno, K. Endo

Department of Chemistry, Faculty of Science, Kanazawa University, Kanazawa, 920-1192, Japan

Abstract. We performed molecular dynamics simulation on methane, ethane and carbon dioxide hydrate. It was found that in the small cage the methane and ethane are moving near the center of the cavity, while carbon dioxide exists apart from the center of the cavity. We also evaluated the stretching vibrational spectra of the methane in both cages. The calculated spectra are in considerably good agreement with experimental results.

INTRODUCTION

Gas hydrates are well known to be the typical clathrate hydrate, and one of the new energy resources in the near future. Gas hydrates are mainly classified into the sI and sII crystal structures, which include a small guest molecule (methane, ethane, carbon dioxide), and larger molecules (propane, iso-pentane), respectively from X-ray diffraction and other analytical methods [1,2]. The small and large cavities in the sI structure in Figure 1, where we deal with this in the present work, are composed of 12 pentagonal faces (5^{12}) of 20 water molecules with equal edge lengths and equal angles, and 12 pentagonal and two hexagonal faces of 24 water molecules as labeled $5^{12}6^2$. We are interested in investigating the local static and dynamic behavior of gas molecules in the small and large hydrogen-bonded cluster cavities.

Already the molecular dynamics (MD) simulation, which is based on the use of empirical interatomic potential functions parametrized to experimental data has been used to investigate these behaviors [3]. The MD simulation has some limitations and it can not consider polarization effects because it uses an empirical interaction potential as input. In order to avoid using empirical interaction potentials as input, recently quantum mechanical MD simulation has been developed [4]. In this simulation interatomic forces are computed directly from the electronic structure.

In this study we perform a quantum mechanical molecular dynamics simulation based on semi-empirical molecular orbital method. As a model of the gas hydrate, we use cluster molecules, which consist of small and large cavities, including 2 guest molecules and sI hydrate surrounding water molecules connected to two cavities (a part of sI). Our aim of this investigation is to examine the differences of the local structures and dynamics in the water cavity between several kinds of guest molecules.

COMPUTATONAL DETAILS

Interatomic potential. To estimate interatomic potentials, we used semi-empirical PM3 method, based on NDDO (Neglect of Diatomic Differential Overlap) approximation.

Method of calculation. We performed molecular dynamics simulation on a cluster molecule, which consists of small and large cavities, including two guest molecules and sI hydrate surrounding water molecules connected to two cavities. Oxygen of the surrounding water molecules are fixed as the time independence and considered only in estimating interatomic potentials. The initial positions of water molecules were determined by X-ray diffraction of methane hydrate crystals and those of guest molecules were positioned at the center of the cage. As guest molecules, we chose methane, ethane and carbon dioxide in our simulation. The equations of motions were solved using the velocity Verlet algorithm with time step of 0.5 fs and each simulation was performed for 5 ps (10000 steps).

RESULTS AND DISCUSSION

Trajectories of guest molecules. Trajectories of the methane (C-atom) in small and large cages were projected on the pentagonal and hexagonal planes of the cages, respectively. In the small cavity, the methane moves around very close to the center of the cavity, while in the large cavity the methane moves apart from the center of cavity. The ethane also moves like the methane, but the carbon dioxide moves apart from the center of cavity even in small cage. We found that the carbon dioxide could not be stable in the cavities.

Vibrational spectra. We calculated stretching vibrational frequency of C-H bonds of methane encaged in the cages and isolated methane. In each case the calculated values were much greater than experimental values. In order to correct the calculated values, we estimated scaling factor dividing experimental frequency of isolated methane by calculated one. Then the scaled frequencies are in good agreement with experimental ones. In the two cavities only one peak was found as the C-H bond of the methane in small cage and in large cage. The C-H bonds seem to interact with the cages isotropically.

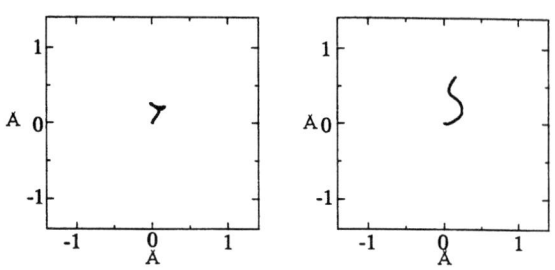

FIGURE 2. Trajectries of (a) methane and (b) carbon dioxide in small cage.

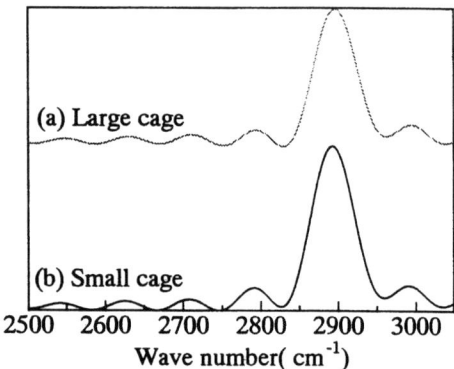

FIGURE 3. Stretching vibrational spectra of methane in (a) Large cage and (b) Small cage.

REFERENCES

1. Sloan, E. D. Jr., *Clathrate Hydrates of Natural Gases*, 2nd ed., New York: Marcel Dekker, Inc, 1998.
2. McMullan, R. K., Jeffrey, G. A., *J. Chem. Phys* **42** 2725-2737 (1965).
3. Tse, J. S., Klein, M. L., McDonald, I. R., *J. Phys. Chem* **87** 4198-4203 (1983).
4. Tuckerman, M. E., Ungar, P. J., von Rosenvinge, T., Klein, M. L., *J. Phys. Chem* **100** 12878-12887 (1996).

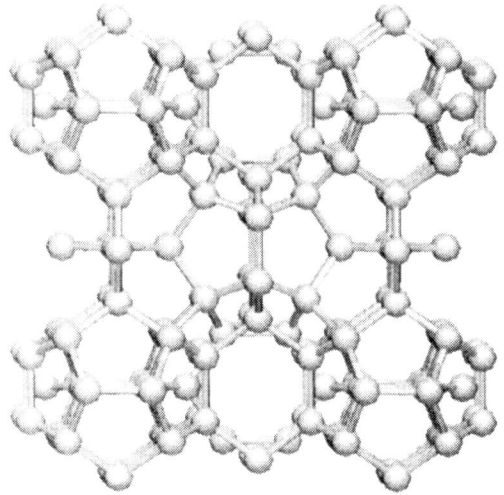

FIGURE 1. Structure of sI hydrate

TABLE 1. Vibrational spectra for methane.

	Calculation(cm^{-1})	Corrected frequencies	Experiment(cm^{-1})
Small cage	4690	2893	2915
Large cage	4695	2896	2905
Isolated	4731	2918	2918

*scaling factor = 0.6186

Dendritic Side Branching Structure of CML model

Masako Ohtaki*, Haruo Honjo† and Hidetsugu Sakaguchi†

*Advanced Materials Institute, Fukuoka University, Nanakuma 8-19-1, Jyonan-ku, Fukuoka, Japan
m-ohtaki@cis.fukuoka-u.ac.jp
†Faculty of Engineering Science, Department of Applied Science and Materials, Kyushu University, Kasugakouen
6-1, Kasuga, Fukuoka, Japan

Abstract. We suppose the dendrite has three regions along the crystal surface from the tip. (1) The stable region, (2) The competition region, (3) The regular interval region . We divided (2) The competition region into three more with Coupled Map Lattice (CML) model without external noise.

INTRODUCTION AND MODEL

Growing interfaces make a pattern in the diffusion field. There are various patterns such as dendrite, DLA (Diffusion-Limited Aggregation), DBM (Dense-Branching Morphology) and Dendrite which have been studied as a non-linear and non-equilibrium problem [1]. In addition, there exists a symmetry-broken dendrite, called Doublon. We have studied these patterns using a discrete model that is a Coupled Map Lattice (CML) model for melt growth. CML model has succeeded in generating DLA, and DBM, Dendrite patterns[2, 3, 4] and also Doublon pattern[5].

Dendrite has a main stem with parabolic tip and many side branches. The stable tip grows with constant velocity. We show a typical pattern of dendrite using CML model in FIGURE.1. The length that is between side branches is determined by competition of those. The surviving dynamics , however, does not clearly understand yet. The experiment shows that side branches grow faster than the tip. We suppose the dendrite has three regions along the crystal surface from the tip.

(1) The stable region(tip region)
(2) The competition region
(3) The regular interval region

In the stable region, the tip remains stable parabola. In the competition region, the surface is no more stable and side branches appears. The competition appears within diffusion length. In the regular interval region, it seems there is the regular interval between branches. The intelval is lager than diffusion length and the global structure of the side branches makes a straight line.
We divided the competition region into three furthermore with CML model without external noise. The coupled map lattice model is composed of two steps. The first step is a diffusion process and another is a crystallization process. The diffusion process for non-dimensional temperature field is written as,

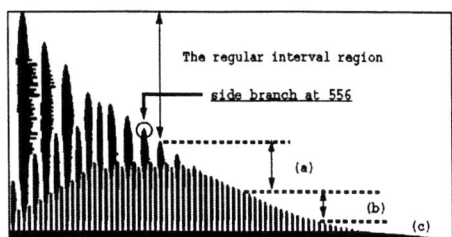

FIGURE 1. A typical pattern of dendrite using CML model on 800×1200 lattices. The branch ois screened in the reagion (a), A branch grows in acceleration in the reagion (b) and The unstable mode appears in the reagion (c) .

$$u_{n+1}(i,j) = u_n(i,j) - D(u_n(i+1,j) + u_n(i-1,j) + u_n(i,j+1) + u_n(i,j-1) - 4u_n(i,j)) \quad (1)$$

where i and j denote the lattice point and D denotes the diffusion constant. The second step is crystallization process, which occurs only at the crystal interface. We consider a variable $x(i,j)$ at every site as an order parameter. If $x(i,j)$ is 0, the (i,j) site is the unoccupied site (melt site) and if $x(i,j)$ is over 1, the (i,j) site is the occupied site (crystallized site). The variable $x(i,j)$ changes from 0 to 1 in sequence only at the crystal interface. At the interface the temperature changes due to latent heat, which is released in proportion to the crystallization. We assume the following process for $x(i,j)$ and $u(i,j)$ at the interface,

$$u_{n+2}(i,j) = u_{n+1}(i,j) + c_1(u_{int} - u_{n+1}(i,j))$$
$$x_{n+2}(i,j) = x_{n+1}(i,j) + c_2(u_{int} - u_{n+1}(i,j)) \quad (2)$$

If the (i,j) site is not the interface, the crystallization process is expressed by $x_{n+2}(i,j) = x_{n+1}(i,j), u_{n+2}(i,j) =$

$u_{n+1}(i,j)$. The u_{int} is the normalized interface temperature, and the value depends on the curvature at the interface. In this model we have assumed the boundary condition as

$$u_{int} = u_M(i,j) + d[N-3] \quad (3)$$

Here u_M is melting temperature and is set as unity. This equation corresponds to Gibbs-Thomson condition [2, 3]. The second term in the right hand side denotes the effect of the surface tension, which is proportional to the curvature. The parameter d corresponds to capillary length. N is the number of the nearest occupied sites and number of the next-nearest occupied sites around the (i,j) site. We have subtracted 3 from N, because $N = 3$ for the flat interface, and u_{int} should be equal to u_M for the flat interface.

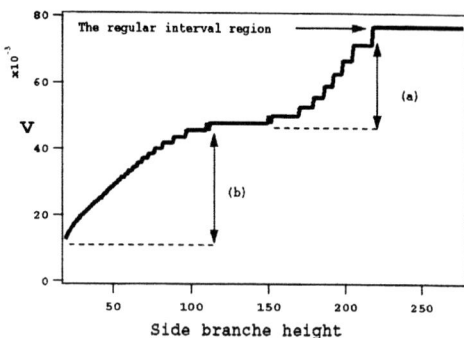

FIGURE 2. The growth velocity V of side branch at 556 in FIGURE1. In the region (b), all side branches grows in acceleration together. In the region (a), only the side branch survived as a result of competition goes up growth velocity again.

THE COMPETITION REGION

In this model, we divided the competition region into three furthermore, show in FIGURE1. We pay attention to the region (a) and (b) in FIGURE.1 where length between side branches can be defined. FIGURE.2 shows the growth velocity of the branch at 556 in FIGURE1. The height of side branch at 556 in FIGURE1 is 278. At that time, the growth velocity of side branch at 556 is same as that of the tip. The growth velocity is increase accelerative in the region (b) and is constant when a side branch grows in the region (a). The constant growth velocity is, however, not stable. If the side branch survived, the growth velocity is increased again shown in FIGURE2. Since there are competitions between side branches, the length between side branches in the region (a) and (b) is smaller than the diffusion length. When the growth velocity is small, the interaction between side branches is weak. Because there are few latent heat and that is transported quickly away from a crystal interface. From the region (a), in a back interface, a side branch interval is more than diffusion length. In that region the growth velocity of side branch is same as the growth velocity at a tip.

SUMMARY

Dendritic side branching structure of CML model can be divided into the region which can define a side branch interval, and the region which is not made. Furthermore, as for the side branch, competition showed that a side branch interval survived above diffusion length, and grows steadily. The dependency of the degree of non-equilibrium of growth velocity of the side branch in the region (a) and (b) and the mode selection in the region (c) are further work.

ACKNOWLEDGMENTS

This work has been supported by Advanced Materials Institute Fukuoka University.

REFERENCES

1. J. S. Langer, *Rev. Mod. Phys.* **52**, (1980) 1-28.
2. H. Sakaguchi, *J. Phys. Soc. Jpn.* **67**, (1998) 96-98.
3. H. Sakaguchi and M. Ohtaki, *Physica A* **272**, (1999) 300-313.
4. M. Ohtaki, H. Honjo and H. Sakaguchi, *J. Crys. Growth* **237**, (2002) 159-163.
5. M. Ohtaki, H. Honjo and H. Sakaguchi, *IJMPB*, (Inpress).

PROGRAM

November 3, 2003 (Monday)

9:30-10:00	Opening Address
10:10-11:00	J. Nishizawa (Semiconductor Research Inst./Photodynamics Research Center, RIKEN)
	Pioneering Work of THz Wave and Its Application for Molecular Sciences
11:00-11:20	Coffee Break
11:20-12:20	A. J. Heeger (U.C.S.B.)
	Ultrafast Photoinduced Electron Transfer: "Superquenching" as a Route to Biosensors Using Luminescent Conjugated Polymers
12:20-13:30	Lunch
13:30-14:20	C. A. Angell (Arizona State Univ.)
	Slow Processes in Liquids and Biomolecules: Ergodic versus 'Quench-then-Look' Strategies
14:20-14:40	Coffee Break
14:40-15:15	D. A. Weitz (Harvard Univ.)
	Jamming Phase Diagram for Colloidal Particles
15:15-15:50	H. Löwen (Heinrich-Heine-Univ. Düsseldorf)
	Colloidal Suspensions Driven by External Fields
15:50-16:10	M. Tokuyama (Tohoku Univ.)
	Universal Features of Collective Interactions in Hard-Sphere Systems at Higher Volume Fractions
16:10-16:35	Coffee Break
16:35-17:10	S.-H. Chen (M.I.T.)
	Neutron and Light Scattering Studies of the Liquid-to-Glass and Glass-to-Glass Transitions in a Copolymer Micellar Systems
17:10-17:30	E. Del Gado (Univ. Montpellier II)
	Glassy Dynamics in Gelling Systems: From Chemical Gels to Colloidal Glasses
17:30-18:10	Long Discussion
	Discussion Leader: D. A. Weitz
19:00-20:30	Welcome Party at Sendai Hotel

November 4, 2003 (Tuesday)

9:30-10:05	P. Pincus (U.C.S.B.)
	Polyelectrolyte Animals and Slow Modes
10:05-10:40	K. Kremer (Max Planck Inst. for Polymer Research)
	Polymer Melts and Networks:
	The Polymer Entanglement Concept Revisited
10:40-11:00	J. Rault (Univ. Paris Sud)
	Slow Dynamics and Mechanical Properties of Polymer Glass:
	Ageing Properties
11:00-11:20	Coffee Break
11:20-12:20	S. Chu (Stanford Univ.)
	Watching Molecular Systems Work, One at a Time
12:20-14:00	Lunch
14:00-14:20	M. Tanaka (National Inst. for Fusion Science)
	Charge Inversion of a Macroion in Electrolyte Solvent under
	Monovalent and Polyelectrolyte Salts
14:20-14:45	Coffee Break
14:45-15:20	B. J. Alder (Lawrence Livermore National Lab.)
	Molecular Dynamics Compared to Hydrodynamics for
	Rayleigh-Taylor Instability
15:20-15:55	D. Beysens (ESEME)
	Dynamics of Liquid-Vapor Phase Transition under
	High Frequency Vibrations
15:55-16:30	R. Larson (Univ. of Michigan)
	The Flow and Adsorption of DNA Polymers Near Surfaces
16:30-17:00	Coffee Break
17:00-17:35	D. Richter (Forschungszentrum Jülich)
	Intermediate Length Scale Dynamics in Polymer Melts
17:35-18:10	M. Rubinstein (Univ. of North Carolina)
	"Gelling" Transition of Hydrophobic Polyelectrolytes
18:10-18:30	T. Kanaya (Kyoto Univ.)
	Glass Transition and Dynamics of Polymer Thin Films
18:30-19:10	Long Discussion
	Discussion Leader: P. Pincus

November 5, 2003 (Wednesday)

9:30-10:05 M. E. Cates (Univ. of Edinburgh)
  Glass Transitions, Jamming and Colloid Rheology

10:05-10:40 G. Maret (Univ. of Konstanz)
  Glass Transition in a Two-Dimensional System of Magnetic Colloids

10:40-11:00 H. Pleiner (Max Planck Inst. for Polymer Research)
  General Nonlinear 2-Fluid Hydrodynamics of Complex Fluids
  and Soft Matter

11:00-11:20 Coffee Break

11:20-12:20 I. Giaever (Rensselaer Polytechnic Inst. and Applied Biophysics, Inc.)
  Electrical Impedance Analysis of Mammalian Cells

12:20-13:20 Lunch

13:20-15:20 Poster Session

15:20-15:55 H. E. Stanley (Boston Univ.)
  Slow Dynamics of the Complex Fluid, Liquid Water

15:55-16:30 M. D. Ediger (Univ. of Wisconsin-Madison)
  Self-Diffusion near the Glass Transition Temperature:
  Influence of Spatially Heterogeneous Dynamics

16:30-17:05 P. Harrowell (Univ. of Sydney)
  Crystallisation and Local Order in Glass-Forming Binary Mixtures

17:05-17:30 Coffee Break

17:30-18:05 G. Tarjus (Univ. Pierre et Marie Curie)
  The Glass Transition of Liquids: A Theoretical Approach
  in Terms of Frustration

18:05-18:40 S. Sastry (JNCASR)
  Dynamics and the Glass Transition in Liquids

18:40-19:00 M. Schulz (Univ. Ulm)
  The Structure of the Energy Landscape and the Non-Arrhenius
  Behavior of Supercooled Liquids and Glasses

19:00-19:40 Long Discussion
  Discussion Leader: P. Harrowell

November 6, 2003 (Thursday)

9:30-10:05 K. Binder (Johannes Gutenberg Univ.)
  Computer Simulation of the Glass Transition in Thin Films

10:05-10:40　K. L. Ngai (Naval Research Lab.)
　　　　　　From Fast Caged Dynamics to Slow Many- Body Coupled Dynamics
　　　　　　in Complex Systems
10:40-11:00　C. M. Roland (Naval Research Lab.)
　　　　　　The Pressure Dependence of the Johari-Goldstein
　　　　　　Relaxation and the Excess Wing in Supercooled Liquids
11:00-11:20　　　　Coffee Break
11:20-12:20　R. B. Laughlin (Stanford Univ.)
　　　　　　Configurational Memory of RNA Polymerase
　　　　　　in Transcription Regulation
12:40-16:00　Excursion to Johgi
19:00-21:00　Banquet at Ichinoboh Inn

November 7, 2003 (Friday)

10:30-11:45　Contributed Talks 1-1
10:30-10:45　U. A. Handge (ETH Zürich)
　　　　　　Interplay between Morphology and Rheology in Recovery of Blends of
　　　　　　Immiscible Polymers after Melt Elongation
10:45-11:00　J. Yamamoto (Yokoyama NanoLC Project ERATO)
　　　　　　Photonic Liquid Crystals in Chiral Twin/Monomer Commensurate Mixture
11:00-11:15　H. Tanaka (Univ. of Tokyo)
　　　　　　Aging and Shear Rejuvenation: Avalanches in Soft Glassy Materials
11:15-11:30　F. Scheffold (Univ. of Fribourg)
　　　　　　Light Scattering Probes of Complex Fluids and Solids
11:30-11:45　L. Cipelletti (Univ. Montpellier II and CNRS)
　　　　　　Temporal Heterogeneity in the Slow Dynamics of Jammed Soft Matter
10:30-11:45　Contributed Talks 1-2
10:30-10:45　H. Taub (Univ. of Missouri-Columbia)
　　　　　　Intramolecular Diffusive Motion in Alkane Monolayers Probed by
　　　　　　High-Resolution Quasielastic Neutron Scattering
10:45-11:00　T. Uematsu (Chalmers Univ. of Technology)
　　　　　　Power Laws in the Dynamics of Polymer Solutions
11:00-11:15　D. E. Dunstan (Univ. of Melbourne)
　　　　　　Direct Measurement of Polymer Segment Orientation and Distortion
　　　　　　in Shear: Semi- Dilute Solution Behaviour of a Conjugated System

11:15-11:30	H. Kim (Sogang Univ.)	

11:15-11:30   H. Kim (Sogang Univ.)
Polymer Film Dynamics with Coherent X-Ray Scattering

11:30-11:45   A. V. Zvelindovsky(Leiden Univ.)
Defect Dynamics in Block Copolymer Mesophases

10:30-12:00   Contributed Talks 1-3

10:30-10:45   A. Ikai (Tokyo Inst. of Technology)
Dynamics of Protein Extraction and Extension by Force Spectroscopy and Molecular Dynamics Simulation

10:45-11:00   M. Irisa (Kyushu Inst. of Technology)
A Brownian Ratchet Model of Actin Polymerization Motor by Using Extended Scaled Particle Theory

11:00-11:15   G. J. A. Sevink (Leiden Inst. of Chemistry)
Modelling Polymersomes: a Prototype for Complex Cellular Structures

11:15-11:30   T. Komatsuzaki (Kobe Univ.)
Hierarchical Regularity in Multi-Basin Dynamics on Protein Landscapes

11:30-11:45   M. Osaka (Nippon Medical School)
Slow Oscillations of Sympathetic Nerve Activity Cause 1/f Fluctuation of Heartbeat Intervals

11:45-12:00   M. Pinak (Japan Atomic Energy Research Inst.)
Molecular Dynamics of 8-Oxoguanine Lesioned B-DNA Molecule- Structure and Energy Analysis

10:30-12:00   Contributed Talks 1-4

10:30-10:45   C. M. Aegerter (Vrije Univ.)
Experimental Study of Self Organized Criticality on a Three Dimensional Pile of Rice

10:45-11:00   M. A. Hossain (Kanazawa Univ.)
Sorption Dynamics of Cr(VI) on Used Black Tea Leaves

11:00-11:15   I. S. Sogami (Kyoto Sangyo Univ.)
Kikuchi-Kossel Diffraction Line Analysis on Crystallization in Salt-Free Aqueous Colloidal Suspensions

11:15-11:30   Q. Tran-Cong-Miyata (Kyoto Inst. of Technology)
Phase Separation of Polymer Mixtures Driven by Temporally and Spatially Periodic Forcing

11:30-11:45   Y. Saika (Wakayama National College of Technology)
Statistical Mechanics of Phase Unwrapping Problem by the Q-Ising Model

11:45-12:00 S. Trimper (Martin-Luther-Univ.)
  Feedback Coupling and Chemical Reactions

10:30-12:00 Contributed Talks 1-5

10:30-10:45 S. Hosokawa (Philipps Univ. Marburg)
  Microscopic Dynamics in Non-Simple Liquid Metals

10:45-11:00 M. Paluch (Silesian Univ.)
  Volume Effects on the Molecular Rearrangements in Vicinity of Glass Transition

11:00-11:15 M. Descamps (Univ. Lille1)
  Glassy Crystals: a System that Removes the Difficulty of Linking Glass Transition with Stability Limit

11:15-11:30 A. Arbe (CSIC-UPV/EHU)
  Heterogeneous Structure of Poly(Vinyl Chloride) as the Origin of Anomalous Dynamical Behavior

11:30-11:45 O. K. C. Tsui (Hong Kong Univ. of Science and Technology)
  Effect of Interfacial Interactions on the Glass Transition of Polymer Thin Films

11:45-12:00 N. Okui (Tokyo Inst. of Technology)
  Universal Reference Temperature for Melt Viscosity-Temperature Relationship

10:30-12:00 Contributed Talks 1-6

10:30-10:45 S. Corezzi (Univ. di Perugia)
  The Role of Configurational Entropy in Physical and Chemical Vitrification

10:45-11:00 J.-J. Kim (Korea Advanced Inst. of Science and Technology)
  A Crossover Change in Relaxation Dynamics of Glass Freezing in a Dipole Glass DRADP-x

11:00-11:15 C. Svanberg (Chalmers Univ. of Technology)
  Confined and Bulk Dynamics of a Simple Glass-Former

11:15-11:30 R. Bergman (Chalmers Univ. of Technology)
  Different Routes to an Understanding of the Excess Wing in the Dielectric Loss of Glass-Formers

11:30-11:45 R. Casalini (George Mason Univ./Naval Research Lab.)
  Dielectric Spectroscopy in Supercooled Liquids under High Pressure: Evidence of a Dynamic Crossover

11:45-12:00 T. Scopigno (Univ. di Roma)
  Connection between the Low Temperature Vibrational Motion in a Glass and the Fragility of Its Supercooled Liquid

12:00-13:30	Lunch
13:30-15:30	Tour of Katahira Campus of Tohoku Univ.
15:30-17:00	Poster Session
17:00-18:30	Contributed Talks 2-1
17:00-17:15	R. Di Leonardo (Univ. di Roma)
	Dynamic Light Scattering from a Sheared Colloidal Glass
17:15-17:30	A. Matic (Chalmers Univ. of Technology)
	Dynamics around the Sol-Gel Transition in Thermoreversible Atactic-Polystyrene Gels
17:30-17:45	H. König (Univ. of Konstanz)
	Investigation of the Local Particle Rearrangements in a Two-Dimensional Colloidal Glass Former
17:45-18:00	M. W. Dreischor (Univ. of Amsterdam)
	What Is the Excluded Volume for an Excluded Volume Chain?
18:00-18:15	A. Aradian (Univ. of Edinburgh)
	Modelling Rheological Chaos in the Flow of Complex Fluids: Non-Linear Rheology and Slow Relaxations
18:15-18:30	T. Terao (Hokkaido Univ.)
	Molecular Dynamics Simulation of Dendrimers: Structural Formation and Internal Charge Distribution
17:00-18:30	Contributed Talks 2-2
17:00-17:15	F. Tanaka (Kyoto Univ.)
	Thermoreversible Gelation Driven by Coil-to-Helix Transition of Polymers
17:15-17:30	T. Indei (Kyoto Univ.)
	Analysis of Shear-Thickening in Physical Gel by Transient Network Theory
17:30-17:45	C.-H. Cheng (National Central Univ.)
	Adsorption of a Single Charged Polymer on a Charged Substrate
17:45-18:00	F. Y. Hansen (Technical Univ. of Denmark)
	Analysis of the Center of Mass-, Rotational- and Intramolecular Diffusive Motions in a Monolayer Film of Intermediate-Length Alkane Molecules Adsorbed on a Solid Surface
18:00-18:15	C. C. Liew (A.I.S.T.)
	Nanostructure Formation in Polymer Thin Films: Dissipative Particle Dynamics Simulation Using a Space-Time Coarse-Grained Model

18:15-18:30	C. P. Lowe (Univ. of Amsterdam)
	Long Polymer Dynamics from Short Model Polymers
17:00-18:00	Contributed Talks 2-3
17:00-17:15	K. Okumura (Ochanomizu Univ.)
	Phase Transitions of Nematic Rubbers
17:15-17:30	R. Bausch (Heinrich-Heine-Univ. Düesseldorf)
	Driven Motion of Extended Defects, Wetted by a New Phase
17:30-17:45	M. Sasaki (Univ. of Tokyo)
	Effects of Temperature Chaos on Rejuvenation and Memory in Migdal- Kadanoff Spin Glasses
17:45-18:00	J. J. Schneider (Johannes Gutenberg Univ. of Mainz)
	Searching for Backbones - An Efficient Parallel Algorithm for Finding Groundstates in Spin Glass Models
17:00-18:45	Contributed Talks 2-4
17:00-17:15	P. Habdas (Emory Univ.)
	Local Perturbations of Colloidal Glasses
17:15-17:30	H. Kikkawa (Kwansei Gakuin Univ.)
	Observation on Surface Change of Fragile Glass: Temperature - Time Dependence Studied by X-Ray Reflectivity
17:30-17:45	K. Ishii (Gakushuin Univ.)
	Competition between Crystallization and Glass-Transition Processes in Binary Amorphous Molecular Systems
17:45-18:00	O. Yamamuro (Osaka Univ.)
	Structural Relaxation and Low-Energy Excitation in Amorphous Ice and Related Glasses
18:00-18:15	F. Affouard (Univ. Lille 1)
	Onset of Slow Dynamics in Orientationally Disordered Crystals: Molecular Dynamics Simulation and Neutron Spin Echo
18:15-18:30	K. Vollmayr-Lee (Bucknell Univ.)
	Single Particle Jumps in a Glass: A Computer Simulation
18:30-18:45	J. Colmenero (CSIC-UPV/EHU)
	Self-Atomic Motions in Glass-Forming Polymers: Neutron Scattering and Molecular Dynamics Simulations Results
17:00-18:00	Contributed Talks 2-5
17:00-17:15	T. Kitamura (Nagasaki Inst. of Applied Science)
	A Unified Theory of the Liquid-Glass Transition
17:15-17:30	K. Miyazaki (Harvard Univ.)
	Supercooled Liquids under Shear: A Mode-Coupling Theory Approach

17:30-17:45 S. M. Fielding (Univ. of Leeds)
    Fluctuation-Dissipation Relations in Ageing
    and Driven Non-Mean Field Glass Models
17:45-18:00 T. Ishii (Okayama Univ.)
    Stochastic Approach to Glass Transition
17:00-18:15 Contributed Talks 2-6
17:00-17:15 J.-B. Suck (Technical Univ. Chemnitz)
    Dynamic Structure Factor and Generalized Vibrational Density of States
    of a Simple Glass after Hyperquenching, Structural Relaxation and
    Slow Cooling
17:15-17:30 H. Fujiwara (Hiroshima City Univ.)
    New Aspect of the Spontaneous Formation of a Bilayer Lipid Membrane
17:30-17:45 M. I. Katsnelson (Uppsala Univ.)
    Nonperturbative Anharmonic Phenomena in Crystal Lattice Dynamics
17:45-18:00 J. Inoue (Hokkaido Univ.)
    Dynamical Properties of EM Algorithm for Gray Scale Image Restoration
18:00-18:15 K. W. Wojciechowski (Polish Academy of Sciences)
    Simple Systems of Unusual Elastic Properties

November 8, 2003 (Saturday)

 9:30-10:05 Y. Feldman (The Hebrew Univ. of Jerusalem)
    Non-Debye Dielectric Response in Complex Systems at Mesoscale
10:05-10:40 A. Sokolov (The Univ. of Akron)
    Dynamics of Complex Systems: from a "Simple Liquid" to a Protein
10:40-11:00 H. Tanaka (Univ. of Tokyo)
    Two-Order-Parameter Model of Liquid: Water-Like Thermodynamic
    Anomaly, Liquid-Liquid Transition, and Liquid-Glass Transition
11:00-11:20   Coffee Break
11:20-11:55 A. Inoue (Tohoku Univ.)
    Stabilization of Supercooled Metallic Liquid and Bulk Glassy Alloys
11:55-12:30 M. Sperl (Technische Univ. München)
    Dynamics near a Higher-Order Glass-Transition Singularity
12:30-13:30   Lunch
13:30-15:30 Poster Session
15:30-15:50   Coffee Break

15:50-16:10   P. Verrocchio (Univ. Complutense de Madrid)
               Phonon Interpretation of the Boson Peak in Supercooled Liquids
16:10-16:30   K. Saito (Toyota Technological Inst.)
               Structural Relaxation in Silica Glass
16:30-17:05   W. Kob (Univ. de Montpellier II)
               On the Relaxation Dynamics of a Rigid Rod in a Disordered Environment
17:05-17:45   Long Discussion
               Discussion Leader: K. L. Ngai
17:45-18:15   Closing Address
               Commentator: C. A. Angell

PARTICIPANTS

Aegerter, Christof Markus (Vrije Universiteit)

Affouard, Frederic (Universite Lille 1)

Akimoto, Mitsuhiro (Tokyo University of Science)

Akiyama, Jotaro (Tohoku University)

Alder, Berni J. (Lawrence Livermore National Laboratory)

Ando, Masahiko (Hitachi, Ltd.)

Angell, C. Austen (Arizona State University)

Anma, Daisuke (Tokai University)

Aoki, Atsunori (Toyo University)

Aoki, Yasuyoshi (Tohoku Pole Co., Ltd.)

Aradian, Achod Andre (University of Edinburgh)

Araki, Takeaki (The University of Tokyo)

Arbe, Arantxa (CSIC-UPV/EHU)

Asano, Megumi (Tokai University)

Awazu, Akinori (The University of Tokyo)

Bausch, Richard (Heinrich-Heine-University Düesseldorf)

Ben Ishai, Paul (Hebrew University)

Bergman, Rikard (Chalmers University of Technology)

Berkenbos, Arjen (University of Amsterdam)

Beysens, Daniel Andre-Marie (Commissariat à l'Energie Atomique)

Binder, Kurt (Johannes Gutenberg Universität Mainz)

Bonn, Daniel (Ecole Normale Supérieure)

Borzi, Rodolfo A. (University of St. Andrews)

Casalini, Riccardo (George Mason University, Naval Research Laboratory)

Cates, Michael E. (University of Edinburgh)

Chen, Sow-Hsin (MIT)

Cheng, Chi-Ho (National Central University)

Chiba, Ayano (Kyoto University)

Chida, Tatsuya (Tohoku Univercity)

Choi, Yoon Seok (Korea Advanced Institute of Science and Technology)

Chu, Steven (Stanford University)

Cipelletti, Luca (Université Montpellier II and CNRS)

Colmenero, Juan (CSIC-UPV/EHU, UPV/EHU)

Corezzi, Silvia (Università di Perugia)

Deguchi, Tetsuo (Ochanomizu University)

Del Gado, Emanuela (Université Montpellier II)

Descamps, Marc (Universite Lille 1)

Di Leonardo, Roberto (Universitá di Roma)

Doi, Masao (Nagoya University)

Dotera, Tomonari (Kyoto University)

Dreischor, Menno Wilko (University of Amsterdam)

Dubois, Emmanuelle (University of Pierre and Marie Curie)

Dunstan, Dave E. (The University of Melbourne)

Dupuis, Vincent (Universite Paris 6)

Ediger, Mark D. (University of Wisconsin-Madison)

Elias, Florence (University Paris 6)

Enomoto, Yoshihisa (Nagoya Institute of Technology)

Feldman, Yuri (The Hebrew University of Jerusalem)

Fielding, Suzanne Margaret (University of Leeds)

Fomin, Sergei A. (Tohoku University)

Frusawa, Hiroshi (Kochi University of Technology)

Fujima, Takuya (The Institute of Physical and Chemical Research)

Fujimori, Hiroki (Nihon University)

Fujishima, Musashi (University of Tsukuba)

Fujita, Maho (Tokai University)

Fujiwara, Hisashi (Hiroshima City University)

Fukao, Koji (Kyoto Institute of Technology)

Fukuda, Ikuo (National Institute of Advanced Industrial Science and Technology)

Fukuda, Jun-ichi (Yokoyama Nano-Structured Liquid Crystal Project)

Fukuda, Tatsuro (Tokai University)

Fukunaga, Kenji (Ube Industries, Ltd.)

Fukunishi, Yu (Tohoku University)

Furukawa, Akira (Kyoto University)

Giaever, Ivar (Rensselaer Polytechnic Institute and Applied Biophysics, Inc.)

Grigera, Santiago A. (University of St. Andrews)

Habasaki, Junko (Tokyo Institute of Technology)

Habdas, Piotr (Emory University)

Hagita, Katsumi (Keio University)

Handge, Ulrich Alexander (ETH Zürich)

Hansen, Flemming Yssing (Technical University of Denmark)

Harrowell, Peter (University of Sydney)

Hasegawa, Manabu (University of Tsukuba)

Hashi, Yasuo (Tokyo University of Science)

Hashimoto, Hideo (Tohoku Electric Power Co., Inc.)

Hayase, Toshiyuki (Tohoku University)
Hayashi, Shigeo (University of Electro-Communications)
Hayashi, Yoshihito (The Hebrew University of Jerusalem)
Hayashi, Yoshikatsu (Lund University)
Head, David Andrew (Vrije Universiteit)
Heeger, Alan J. (U.C.S.B.)
Hemar, Yacine (Massey University)
Henn, Francois Edouard (Université de Montpellier II)
Hibino, Masahiro (Muroran Institute of Technology)
Hidaka, Kuniaki (Tohoku University)
Hideshima, Taketoshi (Chiba University)
Hiejima, Yusuke (Kyoto University)
Hiki, Yosio (Tokyo Institute of Technology)
Hiwatari, Yasuaki (Kanazawa University)
Hoshino, Kyoko (Kobe University)
Hosokawa, Shinya (Philipps Universität Marburg)
Hossain, Mohammad Abul (Kanazawa University)
Huber, Daniel (University of California)
Hwang, Yoon-Hwae (Pusan National University)
Ichikawa, Masatoshi (Kyoto University and CREST)
Ida, Tomonori (Kanazawa University)
Idaka, Nao (Kansai Paint Co., Ltd.)
Ikai, Atsushi (Tokyo Institute of Technology)
Ike, Yuji (University of Tsukuba)
Ikohagi, Toshiaki (Tohoku University)
Ikushima, Akira J. (Toyota Technological Institute)
Imai, Masayuki (Ochanomizu University)
Imoto, Daizo (Tokai University)
Indei, Tsutomu (Kyoto University)
Inoue, Akihisa (Tohoku University)
Inoue, Jun-ichi (Hokkaido University)
Inoue, Osamu (Tohoku University)
Inoue, Rintaro (Kyoto University)
Inoue, Tadashi (Kyoto Univeristy)
Irisa, Masayuki (Kyushu Institute of Technology)
Ishii, Kikujiro (Gakushuin University)
Ishii, Koji (Osaka Prefecture University)
Ishii, Tadao (Okayama University)

Ishizuka, Tomoyuki (Tohoku University)
Isobe, Masaharu (Nagoya Institute of Technology)
Itamoto, Hidenori (Nagoya Institute of Technology)
Ito, Kei (The National Center for University Entrance Examinations)
Iwashita, Yasutaka (The University of Tokyo)
Iwaya, Akiyuki (Tokyo University of Science)
Jacobsson, Per (Chalmers University of Technology)
Joti, Yasumasa (Japan Atomic Energy Research Institute)
Kajikawa, Hiroaki (Kyoto University)
Kamijo, Kenjiro (Tohoku University)
Kamiyama, Shinichi (Akita Prefectural University)
Kamiyama, Tomoaki (Tohoku University)
Kanaya, Toshiji (Kyoto University)
Kanazawa, Ikuzo (Tokyo Gakugei University)
Kaneko, Yutaka (Kyoto University)
Kato, Nobuhiko (Kanazawa University)
Kato, Ryuzo (Kagoshima University)
Kato, Tadashi (Tokyo Metropolitan University)
Kato, Takuma (Tohoku University)
Katsnelson, Mikhail I. (Uppsala University)
Kawabata, Youhei (Tokyo Metropolitan University)
Kawai, Kiyoshi (Tokyo University of Fisheries)
Kawakatsu, Toshihiro (Tohoku University)
Kawamura, Junichi (Tohoku University)
Kawasaki, Masashi (Tohoku University)
Kikkawa, Hiroyuki (Kwansei Gakuin University)
Kim, Hyeon-Deuk (Kyoto University)
Kim, Hyunjung (Sogang University)
Kim, Jong-Jean (Korea Advanced Institute of Science and Technology)
Kim, Jung Eun (University of Tsukuba)
Kim, Kang (Kyoto University)
Kimura, Yasuyuki (The University of Tokyo)
Kinouchi, Sumie (Tokyo Metropolitan University)
Kishino, Yuji (Tohoku University)
Kitahara, Amane (Kwansei Gakuin University)
Kitahata, Hiroyuki (Kyoto University)
Kitamura, Toyoyuki (Nagasaki Institute of Applied Science)
Kitsunezaki, So (Nara Women's University)

Kob, Walter (Université de Montpellier II)

Kobayashi, Hideaki (Tohoku University)

Kobayashi, Mika (Hokkaido University)

Kobayashi, Tsunehiro (Tsukuba College of Technology)

Koda, Tomonori (Yamagata University)

Kohama, Yasuaki (Tohoku University)

Koibuchi, Hiroshi (Ibaraki College of Technology)

Koizumi, Hideki (Nippon Gear Co., Ltd.)

Kojima, Seiji (University of Tsukuba)

Komatsu, Satonori (Kyoto Institute of Technology)

Komatsuzaki, Tamiki (Kobe University)

Komura, Shigeyuki (Tokyo Metropolitan University)

König, Hans (University of Konstanz)

Koopman, Evert Adriaan (University of Amsterdam)

Koyama, Akira (Kyoto University)

Koyama, Takehito (The University of Tokyo)

Kremer, Kurt (Max Planck Institute for Polymer Research)

Kumemura, Momoko (Tokyo Metropolitan University)

Kuninaka, Hiroto (Kyoto University)

Kurita, Rei (The University of Tokyo)

Kuroda, Akiyoshi (Yamagata University)

Lai, Pik-Yin (National Central University)

Lairez, Didier (CEA Saclay)

Larson, Ronald G. (University of Michigan)

Laughlin, Robert B. (Stanford University)

Lee, Jae Woo (Inha University)

Liew, Chee Chin (National Institute of Advanced Industrial Science and Technology)

Lin, Tsang-Lang (National Tsinghua University)

Long, Didier (Université de Paris-Sud)

Longeville, Stephane (CEA-CNRS)

Lowe, Christopher Paul (University of Amsterdam)

Löwen, Hartmut (Heinrich-Heine-Universität Düsseldorf)

Lu, David (National Central University)

Maekawa, Toru (Toyo University)

Majima, Hiroki (Tokyo University of Science)

Maret, Georg (University of Konstanz)

Maruta, Kaoru (Tohoku University)

Maruyama, Satoshi (Tokyo Institute of Technology)

Maruyama, Shigenao (Tohoku University)
Maruyama, Youhei (Tohoku University)
Masubuchi, Yuichi (Nagoya University)
Masui, Tomomi (Ochanomizu University)
Matic, Aleksandar (Chalmers University of Technology)
Matsuda, Hiromitsu (Yamagata University)
Matsumoto, Daisuke (Kanazawa University)
Matsunaga, Yasuhiro (Kobe University)
Matsuura, Hiroyuki (National Graduate Institute for Policy Studies)
Matsuyama, Akihiko (Kyusyu Institute of Technology)
Mayer, Peter (King's College London)
Meyer, Andreas (Technische Universität München)
Mikami, Masuhiro (National Institute of Advanced Industrial Science and Technology)
Mitsutake, Ayori (Keio University)
Miyamoto, Yoshihisa (Kyoto University)
Miyazaki, Kunimasa (Harvard University)
Miyazaki, Yuji (Osaka University)
Mori, Noriyasu (Osaka University)
Mori, Shigeru (Kanazawa University)
Moriguchi, Naoki (Osaka University)
Morikawa, Ryota (Tokyo University of Pharmacy and Life Science)
Morimoto, Hisao (The University of Tokyo)
Morita, Hidetoshi (The University of Tokyo)
Muguruma, Chizuru (Chukyo University)
Murakami, Hiroshi (Japan Atomic Energy Research Institute)
Muranaka, Tadashi (Aichi Institute of Technology)
Murata, Katsumi (The Graduate University for Advanced Studies)
Murayama, Yoshihiro (The University of Tokyo)
Muromoto, Takayuki (Osaka University)
Nagahiro, Shin-ichiro (Tohoku University)
Nagao, Hidemi (Kanazawa University)
Nagao, Michihiro (The University of Tokyo)
Nagaoka, Yutaka (Toyo University)
Nagasawa, Yutaka (Osaka University)
Nagaya, Tomoyuki (Okayama University)
Nakahara, Akio (Nihon University)
Nakaya, Kaori (Ochanomizu University)
Nakayama, Hideyuki (Gakushuin University)

Nakayama, Yasuya (Japan Science and Technology Agency)
Nanbu, Kenichi (Tohoku University)
Narikiyo, Osamu (Kyushu University)
Ngai, Kia Ling (Naval Research Laboratory)
Niioka, Takashi (Tohoku University)
Nishikawa, Kiyoshi (Kanazawa University)
Nishikawa, Yuya (The University of Tokyo)
Nishiyama, Hideya (Tohoku University)
Nishiyama, Noboru (Asahi Life Asset Management Co., Ltd.)
Nishizawa, Junichi (Semiconductor Research Inst., Photodynamics Research Center)
Nomata, Atsushi (Tokyo University of Science)
Norizoe, Yuki (Tohoku University)
Nozaki, Ryusuke (Hokkaido University)
Obayashi, Shigeru (Tohoku University)
Oguni, Masaharu (Tokyo Institute of Technology)
Oh, Ji Young (Pusan National University)
Ohm, Vincent (University of Amsterdam)
Ohtaki, Masako (Fukuoka University)
Okada, Masafumi (Nagoya Institute of Technology)
Okui, Norimasa (Tokyo Institute of Technology)
Okumura, Ko (Ochanomizu University)
Okuzono, Tohru (Yokoyama Nano-Structured Liquid Crystal Project)
Onoda-Yamamuro, Noriko (Tokyo Denki University)
Onuki, Akira (Kyoto University)
Oppenheim, Irwin (MIT)
Orihara, Hiroshi (Nagoya University)
Osaka, Motohisa (Nippon Medical School)
Paluch, Marian (Silesian University)
Park, Heungsik (Inha University)
Paul, Alok K. R. (Kyoto University)
Perzynski, Regine (UMR CNRS)
Pinak, Miroslav (Japan Atomic Energy Research Institute)
Pincus, Philip (UCSB)
Pleiner, Harald (Max Planck Institute for Polymer Research)
Qiu, Jinhao (Tohoku University)
Ramos, Miguel Angel (Universidad Autónoma de Madrid)
Rault, Jacques (Université de Paris-Sud)
Reichman, David R (Harvard University)

Richter, Dieter Oswald (Forschungszentrum Jülich)
Roland, C Michael (Naval Research Laboratory)
Rubinstein, Michael (University of North Carolina)
Saika, Yohei (Wakayama National College of Technology)
Saito, Tsutomu (Tohoku University)
Sakamoto, Isao (Kanazawa University)
Sakamoto, Naoto (University of Tsukuba)
Sakaue, Takahiro (Kyoto University)
Samukawa, Seiji (Tohoku University)
Sandkühler, Peter (Swiss Federal Institute of Technology)
Sandre, Olivier (Universite Pierre et Marie Curie)
Sano, Ryoko (Shizuoka University)
Santangelo, Patrick Gerard (Naval Research Laboratory)
Sasai, Masaki (Nagoya University)
Sasaki, Munetaka (The University of Tokyo)
Sasaki, Shigeo (Kyushu University)
Sasoh, Akihiro (Tohoku University)
Sastry, Srikanth (Jawaharlal Nehru Centre for Advanced Scientific Research)
Sawada, Isao (Ishikawa National College of Technology)
Scheffold, Frank (University of Fribourg)
Schneider, Johannes Josef (Johannes Gutenberg University of Mainz)
Schulz, Beatrix Mercedes (Martin Luther Universität Halle-Wittenberg)
Schulz, Michael (Universität Ulm)
Scopigno, Tullio (Universitá di Roma)
Seo, Jeong-Ah (Pusan National University)
Seto, Hideki (Kyoto University)
Sevink, Agur Gj (Leiden Institute of Chemistry)
Sferrazza, Michele (University of Surrey)
Sheu, Sheh-Yi (National Yang-Ming University)
Shibayama, Mitsuhiro (The University of Tokyo)
Shida, Kazuhito (Tohoku University)
Shigeta, Masaya (Tohoku University)
Shigeta, Yasuteru (The University of Tokyo)
Shimura, Tsutomu (Tohoku University)
Shinoda, Wataru (National Institute of Advanced Industrial Science and Technology)
Shinyashiki, Naoki (Tokai University)
Shirotori, Hisashi (Tokyo Metropolitan University)
Shito, Akifumi (Toyo University)

Shiwa, Yasuhiro (Kyoto Institute of Technology)
Sogami, Ikuo S. (Kyoto Sangyo University)
Sokolov, Alexei (The University of Akron)
Song, Mi-Young (Hanyang University)
Sperl, Matthias (Technische Universität München)
Stanley, H. Eugene (Boston University)
Strybulevych, Anatoliy Leonidovych (University of Manitoba)
Suck, Jens-Boie (Technical University Chemnitz)
Sudo, Seiichi (Tokai University)
Suezaki, Yukio (Saga Medical School)
Suga, Takanori (Osaka University)
Sugiyama, Ayumu (Kanazawa University)
Sugiyama, Kiichiro (Kanazawa University)
Suzuki, Akira (Tokyo University of Science)
Suzuki, Yasuo Y. (Takushoku University)
Svanberg, Christer (Chalmers University of Technology)
Tadokoro, Makoto (Osaka City University)
Tagawa, Fumitaka (Kyushu University)
Takada, Akira (Asahi Glass Co. Ltd.)
Takagi, Shinsaku (The University of Tokyo)
Takagi, Toshiyuki (Tohoku University)
Takagi, Yasunari (The University of Electro-Communications)
Takahashi, Isao (Kwansei Gakuin University)
Takano, Hiroshi (Kochi National College of Technology)
Takasu, Masako (Kanazawa University)
Takeda, Takayoshi (Hiroshima University)
Takekida, Hideto (Tohoku University)
Takiuchi, Ken-ichi (Japan Electronics College)
Tamura, Keizo (Tokyo Metropolitan University)
Tanaka, Fumihiko (Kyoto University)
Tanaka, Hajime (The University of Tokyo)
Tanaka, Motohiko (National Institute for Fusion Science)
Tanaka, Tomoki (Shizuoka University)
Tani, Junji (Tohoku University)
Taniguchi, Takashi (Yamagata University)
Tarjus, Gilles (Université Pierre et Marie Curie)
Taub, Haskell (University of Missouri-Columbia)
Terada, Yayoi (Tohoku University)

Terao, Takamichi (Hokkaido University)
To, Kiwing (Academia Sinica)
Togashi, Hiroyuki (Tohoku Electric Power Co., Inc.)
Tokuyama, Michio (Tohoku University)
Toyoda, Tadashi (Tokai University)
Toyokawa, Seiko (Tokyo University of Science)
Tran-Cong-Miyata, Qui (Kyoto Institute of Technology)
Trimper, Steffen (Martin-Luther-University)
Tsubotani, Sousuke (Tokai University)
Tsui, Ophelia K. C. (Hong Kong University of Science and Technology)
Uchida, Nariya (Tohoku University)
Uchiyama, Masaru (Tohoku University)
Uematsu, Takashi (Chalmers University of Technology)
Ueno, Yoshihiro (Tokyo University of Pharmacy and Life Science)
Ukai, Tomofumi (Toyo University)
Uneyama, Takashi (Nagoya University)
Verrocchio, Paolo (Universidad Complutense de Madrid)
Vollmayr-Lee, Katharina (Bucknell University)
Wada, Hirofumi (The University of Tokyo, Hiroshima University)
Weitz, David A. (Harvard University)
Widmer-Cooper, Asaph (University of Sydney)
Wojciechowski, Krzysztof Witold (Polish Academy of Sciences)
Wu, Ten-Ming (National Chiao-Tung University)
Yada, Makoto (Yokoyama Nano-Structured Liquid Crystal Project)
Yamada, Hiroaki (Niigata University)
Yamaguchi, Tomoya (Kanazawa University)
Yamamoto, Jun (Yokoyama Nano-Structured Liquid Crystal Project)
Yamamoto, Ryoichi (Kyoto University)
Yamamoto, Takao (Gunma University)
Yamamoto, Takehiro (Osaka University)
Yamamuro, Osamu (Osaka University)
Yamazaki, Hiroyuki (Fuji Photo Film Co., Ltd.)
Yamazaki, Masahito (Shizuoka University)
Yang, Dah-Yen (Academia Sinica)
Yokojima, Yasunori (Toyota Techno Service Corp., Kyoto Institute of Technology)
Yonekura, Nobuaki (University of the Ryukyus)
Yoshida, Kei (Tohoku University)
Yoshinaga, Natsuhiko (Kyoto University)

Zarembo, Anna (University of Helsinki)
Zhou, Xin (Tokyo Institute of Technology)
Zhu, Da-Ming (University of Missouri-Kansas City)
Zvelindovsky, Andrei V. (Leiden University)

Author Index

A

Aegerter, C. M., 390
Affouard, F., 647
Afrin, R., 291
Aida, T., 661
Akimoto, M., 749
Akiyama, J., 168
Alder, B. J., 376
Alegría, A., 594
Alvarez, F., 655
Andelman, D., 130
Angell, C. A., 473
Anma, D., 761, 763, 765
Aradian, A., 84
Arbe, A., 189, 594, 655
Arimitsu, T., 225
Arishiro, M. M., 398
Asaji, T., 685
Auer, S., 3
Awazu, A., 456
Azuma, C., 267

B

Baba, T., 352
Bakker, A. F., 241
Balabaev, N., 452
Ballesta, P., 68
Baschnagel, J., 509
Basu, J. K., 213
Bausch, R., 418
Bennington, S. M., 436
Bergman, R., 72, 611, 615
Berkenbos, A., 273
Beysens, D. A., 379
Binder, K., 509
Blaak, R., 3
Bonn, D., 60
Börjesson, L., 611
Borzi, R. A., 741
Buldyrev, S. V., 483
Burns, M. A., 386

C

Casalini, R., 523
Cates, M. E., 33, 84
Chang, S. L., 162
Chen, H., 170
Chen, S.-H., 16

Chen, W.-R., 16
Cheng, C.-H., 229
Chiba, A., 436
Choi, Y.-S., 608
Chon, K. H., 306
Chopra, M., 386
Ciliberti, S., 565
Cipelletti, L., 68
Cochin, E., 647
Colmenero, J., 189, 594, 655
Coniglio, A., 28
Copley, J. R. D., 473
Corezzi, S., 604
Coussot, P., 60
Criswell, L., 201

D

Darinskii, A., 452
de Arcangelis, L., 28
Decressain, R., 647
de Gennes, P.-G., 414
Del Gado, E., 28
Descamps, M., 591, 647
Diama, A., 201
Dimeo, R. M., 201
Dimonte, G., 376
Doi, M., 261, 446, 771
Doi, S., 255
Dotera, T., 257, 267
Dreischor, M. W., 80, 241
Dubois, E., 124, 468
Dufrêche, J.-F., 468
Dunstan, D. E., 209
Dupuis, V., 124
Dzubiella, J., 3

E

Ediger, M. D., 491
El kharrat, D., 122
Endo, K., 755, 791, 797
Enomoto, Y., 779, 781, 783, 789

F

Farago, B., 594, 655
Feldman, Y., 527, 671
Fernández, J. R., 496
Fielding, S. M., 639
Fierro, A., 28

Fomin, S., 275, 460
Fraaije, J. G. E. M., 298
Frick, B., 594
Froltsov, V., 3
Frusawa, H., 102
Fuhrmann, D., 201
Fujihara, M., 724
Fujima, T., 102
Fujimori, H., 685
Fujimoto, H., 310
Fujita, M., 761, 763, 765
Fujiwara, H., 724
Fukao, K., 251, 277
Fukuda, I., 356
Fukuda, J., 146
Fukuda, T., 761, 763, 765
Fukunaga, K., 245
Furuike, S., 320
Furukawa, A., 140
Furuta, M., 100

G

Germann, T. C., 376
Gilli, J.-M., 440
Giovambattista, N., 483
Go, N., 358
Greco, F., 261
Grigera, S. A., 741
Grigera, T. S., 565
Grün, F., 468
Grzybowska, K., 587
Grzybowski, A., 587
Gutina, A., 527, 671

H

Habasaki, J., 775
Hadjiconstantinou, N. G., 376
Haeussler, W., 647
Hagita, K., 279
Hamada, M., 791
Handge, U. A., 52
Hansen, F. Y., 201, 233
Harba, N., 442
Harden, J. L., 46
Harreis, H. M., 3
Harrowell, P., 496, 711
Hasegawa, M., 747
Hashida, T., 275, 460
Hashimoto, M., 791
Hashimoto, T., 245
Hato, M., 352
Hattori, T., 699

Hawker, C. J., 598
Hayakawa, H., 132, 156
Hayakawa, Y., 785
Hayashi, M., 354
Hayashi, S., 442
Hayashi, Y., 671
Head, D. A., 364
Hebraud, P., 434
Hemar, Y., 434
Herminghaus, S., 126
Hertadi, R., 291
Herwig, K. W., 201
Hibino, M., 326
Hideshima, T., 322
Hiejima, Y., 687
Hiki, Y., 661
Hill, E. K., 209
Hirabayashi, M., 257, 267
Hiwatari, Y., 775, 777
Holian, B. L., 376
Honjo, H., 799
Hoshino, K., 344
Hosokawa, S., 583
Hossain, M. A., 394
Hu, Y., 328
Huber, D., 448
Hwang, Y.-H., 689, 697

I

Ianniruberto, G., 261
Ichikawa, M., 110
Ida, T., 755, 791, 797
Ikai, A., 291
Ike, Y., 693
Ikeda, S., 134, 152
Ikushima, A. J., 571
Imai, M., 94, 106, 108, 114
Imoto, D., 681
Indei, T., 225
Inoue, A., 547, 695
Inoue, J.-i., 731
Inoue, R., 197, 691
Inoue, T., 253
Irisa, M., 294
Ishii, K., 112, 623
Ishii, T., 643
Ishikawa, K., 112
Ishikawa, T., 100
Ishiwata, T., 724
Isobe, M., 158
Itamoto, H., 779, 783
Ito, K., 759, 765
Ito, T., 320
Itoh, T., 398

Iwamoto, M., 170
Iwamoto, Y., 138
Iwaya, A., 751

J

Jacobsson, P., 205, 611
Jardat, M., 468
Jeng, U.-S., 328
Jiménez-Riobóo, R. J., 665
Jing, Z., 460
Joti, Y., 358
Jung, Y.-D., 769

K

Kadau, K., 376
Kajikawa, H., 438
Kakiuchida, H., 571
Kamiyama, T., 695
Kanaya, T., 197, 691
Kanazawa, I., 705
Kaneko, K., 757
Kaneko, Y., 777
Kato, N., 755, 797
Kato, R., 789
Kato, T., 94, 334, 338, 743
Katsnelson, M. I., 727
Kawabata, Y., 92, 94, 118, 120
Kawakatsu, T., 114, 265
Kawamura, J., 699
Kikkawa, H., 619
Kim, H., 213
Kim, H.-D., 132
Kim, H. K., 689, 697
Kim, J.-J., 608
Kim, K., 707
Kim, S. J., 697
Kimura, H. M., 695
Kimura, Y., 98
Kinouchi, S., 338
Kishino, Y., 466
Kitahara, A., 255, 619
Kitahara, H., 669
Kitahata, H., 430
Kitamura, T., 631
Kitao, A., 358
Kitsunezaki, S., 154
Kob, W., 509, 576
Kobayashi, K., 438
Kobayashi, M., 326, 673
Kobayashi, N., 771
Kobayashi, T., 142
Koda, T., 134, 152

Kohlbrecher, J., 64
Kohno, H., 438
Koibuchi, H., 144
Koizumi, H., 759, 761
Kojima, S., 669, 693
Komatsu, S., 402
Komatsuzaki, T., 302, 342, 344
Komura, S., 106, 130, 334, 338, 743
König, H., 40, 76
Koopman, E. A., 271
Korenaga, T., 104
Korzhenevskii, A. L., 418
Kostov, K. S., 302
Koyama, A., 277
Koyama, K., 269, 771
Koyama, T., 724
Koyama, Y., 110
Kremer, K., 179
Kumagai, H., 306
Kumemura, M., 104
Kumita, M., 394
Kuninaka, H., 156
Kuroda, A., 269
Kurokawa, T., 398
Kusano, N., 144
Kuwata, N., 699

L

Lai, P.-Y., 229, 281
Lal, J., 213
Larson, R. G., 386
Leary, D. M., 444
Lee, J. W., 454
Lee, K. E., 454
Leon, S., 179
Levine, A. J., 364
Li, L., 386
Li, S. J., 314
Licinio, P., 122
Liew, C. C., 237
Ligoure, C., 68
Likos, C. N., 3
Lin, T.-L., 328
Lizhi, O., 715
Lomdahl, P. S., 376
Lörincz, K. A., 390
Lowe, C. P., 80, 241, 271, 273
Löwen, H., 3
Lurio, L. B., 213
Lyakhova, K. S., 217

M

Mackenzie, A. P., 741
MacKintosh, F. C., 364
Magome, N., 430
Majima, H., 767
Malikova, N., 468
Mallamace, F., 16
Mapes, M., 491
Maret, G., 40
Marrucci, G., 261
Marry, V., 468
Martin, O. C., 422
Martín-Mayor, V., 565
Maruyama, S., 675
Masubuchi, Y., 261, 446
Masui, T., 108
Masum, S. M., 314
Matic, A., 72
Matsuda, H., 134
Matsumoto, D., 791, 797
Matsunaga, Y., 302, 342, 344
Matsuo, Y., 432
Matsuura, H., 336
Matsuyama, A., 128
Matsuzawa, Y., 110
Mattsson, J., 72, 615
Mayer, P., 703
Mériguet, G., 124
Meunier, J., 60
Mikami, M., 237, 352
Miller, M., 344
Minewaki, K., 94
Minoguchi, A., 667
Mitsutake, A., 350
Miyamoto, Y., 277, 679
Miyasaka, H., 249
Miyazaki, K., 94, 635, 717
Miyazaki, T., 197
Mizuno, D., 98
Mizuno, M., 791, 797
Mo, H., 201
Mochrie, S. G. J., 213
Monkenbusch, M., 189, 655
Moral, A., 594
Moreno, A. J., 576
Mori, N., 96, 148
Mori, S., 394
Mori, Y., 249, 318
Moriguchi, N., 96
Morikawa, R., 354
Morita, H., 757
Mossa, S., 473
Munakata, T., 707
Murai, M., 623
Murakami, H., 739

Murakami, T., 777
Muranaka, T., 713
Murata, K., 332
Murayama, Y., 324
Muromoto, T., 249, 318

N

Nagahiro, S., 785
Nagao, H., 360, 362, 753, 787, 793, 795
Nagao, M., 92, 118, 120
Nagasawa, Y., 249, 318
Nagaya, T., 440
Nakagawa, Y., 318
Nakahara, A., 432
Nakamura, H., 356, 683
Nakano, M., 336
Nakaya, K., 106, 108, 114
Nakayama, H., 112, 623
Nakayama, T., 88
Nanbu, K., 172
Narikiyo, O., 150
Narros, A., 655
Neelov, I., 452
Nemoto, T., 310, 336
Neumann, D. A., 201
Ngai, K. L., 515, 775
Nidaira, A., 144
Nishida, K., 197, 691
Nishigami, S., 402
Nishikawa, K., 360, 362, 753, 787, 793, 795
Nishimori, H., 406
Nishiyama, H., 174
Nishiyama, I., 56
Nishiyama, N., 458
Nishizawa, J.-i., 369
Nishizawa, S., 669
Norisuye, T., 402
Novikov, V., 533
Nozaki, R., 667
Nydén, M., 205

O

Odagaki, T., 709
Oguni, M., 675, 685
Oh, J. Y., 689, 697
Ohara, K., 777
Ohashi, K., 320
Ohmasa, Y., 436
Ohta, S., 291
Ohta, Y., 787, 793
Ohtaki, M., 799
Okada, M., 779, 781

Okada, T., 249, 318
Okamoto, Y., 332, 350
Okui, N., 601
Okumura, K., 414
Okuno, M., 398
Okuzono, T., 450
Olmsted, P. D., 334
Onami, T., 306
O'Neill, P., 310
Oppenheim, I., 8
Orihara, H., 116
Osaka, M., 306

P

Page, J. H., 444
Paluch, M., 587
Parisi, G., 565
Park, H. S., 454
Park, He., 462
Park, Hy., 462
Perzynski, R., 122, 124
Pilgrim, W.-C., 583
Pinak, M., 310
Pinder, D. N., 434
Pleiner, H., 46
Puzenko, A., 527
Pyckhout-Hintzen, W., 594

R

Ramos, M. A., 665
Rault, J., 183
Reichman, D. R., 635, 717
Richter, D., 189, 594, 655
Roland, C. M., 523, 587
Rühm, A., 213
Russell, T. P., 598
Ryabov, Y., 527
Ryabov, Y. E., 671

S

Saika, Y., 406
Saito, H., 360, 362
Saito, K., 571
Sakaguchi, H., 799
Sakamoto, I., 753, 787, 795
Sakata, K., 306
Sakaue, T., 340
Sakikawa, W., 150
Sakurai, T., 360, 362
Sandre, O., 122

Sano, M., 324
Sano, R., 316, 320
Sarcia, R., 434
Saruta, T., 306
Sasaki, M., 422
Sawada, I., 701
Scheffold, F., 64
Scheidler, P., 509
Schlag, E. W., 346
Schmitz, R., 418
Schneider, J. J., 426
Schulz, B. M., 126
Schulz, M., 126, 503
Schurtenberger, P., 64
Sekimoto, K., 679
Selzle, H. L., 346
Sengwa, R. J., 683
Seo, J.-A., 689, 697
Seto, H., 92, 118, 120
Sevink, G. J. A., 217, 298
Sheu, S.-Y., 330, 346
Shibata, K., 197
Shibayama, M., 92
Shibuya, T., 116
Shigeta, M., 174
Shigeta, Y., 464
Shih, M.-C., 328
Shimomura, M., 659, 663
Shimura, T., 166
Shinoda, W., 352
Shinohara, T., 398
Shinyashiki, N., 659, 663, 681, 683
Shirotori, H., 334
Shiwa, Y., 259, 745
Sinha, S. K., 213
Sogami, I. S., 398
Sokolov, A., 533
Sollich, P., 639, 703
Song, M.-Y., 769
Sperl, M., 559
Stanley, H. E., 483
Stark, H., 146
Starr, F. W., 483
Stradner, A., 64
Strybulevych, A. L., 444
Suck, J.-B., 721
Sudo, S., 659, 663, 681, 683
Suga, T., 148
Sugimoto, M., 771
Sugita, Y., 332, 350
Sugiyama, A., 360, 362
Sugiyama, K., 753, 793
Sundholm, F., 452
Suzuki, A., 749, 751, 767
Suzuki, K., 144
Suzuki, Y. Y., 257, 338

Svanberg, C., 205, 611, 615
Swallen, S. F., 491

T

Tabe, Y., 450
Tagawa, F., 709
Takahashi, I., 619
Takahashi, N., 691
Takano, H., 100, 279
Takasu, M., 263
Takeda, T., 118, 120
Takei, M., 623
Takekida, H., 172
Takeuchi, A., 547
Takeuchi, S., 661
Takimoto, J., 261, 446
Takiuchi, K., 759, 763
Talón, C., 665
Tamba, Y., 314
Tamura, K., 338, 743
Tamura, Y., 221
Tanaka, F., 221
Tanaka, H., 60, 541
Tanaka, K., 731
Tanaka, M., 285, 739
Tanaka, T., 316
Tanigawa, M., 398
Taniguchi, T., 771
Tanizawa, T., 100
Taub, H., 201, 233
Taylor, J., 436
Terada, Y., 8, 164, 166, 168
Terao, T., 88
Thakahashi, I., 255
To, K., 247
Todoroki, A., 685
Tokita, N., 267
Tokuyama, M., 8, 164, 166, 168
Tomita, J., 263
Toyoda, T., 759, 761, 763, 765
Tran-Cong-Miyata, Q., 402
Trefilov, A. V., 727
Trimper, S., 410
Tsimring, L., 448
Tsubotani, S., 663, 683
Tsui, O. K. C., 598
Tsujimi, Y., 673
Tsukushi, I., 197
Turq, P., 468

U

Uchida, N., 160
Uematsu, T., 205
Ueno, Y., 354

Umeyama, H., 112
Uneyama, T., 446
Urakami, N., 106
Urakawa, O., 491

V

Varnik, F., 509
Vavrin, R., 64
Verrocchio, P., 565
Vieira, S., 665
Volkmann, U. G., 201
Vollmayr-Lee, K., 651

W

Wada, H., 136
Wada Takeda, M., 669
Wakabayashi, K., 675
Wales, D. J., 344
Wang, L.-M., 473
Watanabe, M. A., 306
Wei, Y., 209
Welling, M. S., 390
Widmer-Cooper, A., 711
Wijngaarden, R. J., 390
Willart, J. F., 591
Wojciechowski, K. W., 735
Wu, J.-C., 328
Wu, T.-M., 162
Wysocki, A., 3

Y

Yagi, T., 673
Yagihara, S., 659, 663, 681, 683
Yamada, H., 773
Yamada, N. L., 118
Yamagami, A., 316
Yamaguchi, T., 753, 787, 793, 795
Yamamoto, J., 56
Yamamoto, M., 623
Yamamoto, R., 635, 717
Yamamoto, T., 96, 138, 148, 277
Yamamuro, O., 627
Yamano, H., 197
Yamao, H., 679
Yamashita, Y., 314
Yamazaki, H., 8, 166
Yamazaki, M., 314, 316, 320
Yang, C.-P., 328
Yang, D.-Y., 330, 346
Yang, Y. S., 697

Yao, M., 436, 438, 687
Yasuda, K., 96
Yokojima, Y., 259
Yokoyama, H., 56, 146, 450
Yomogida, Y., 667
Yonekura, N., 677
Yoneya, M., 146
Yoshida, K., 460
Yoshida, S., 402
Yoshikawa, K., 110, 348, 430
Yoshimoto, T., 360, 362, 753
Yoshinaga, N., 348

Yoshiyama, T., 398
Yue, Y., 473

Z

Zabrocki, K., 410
Zahn, K., 40
Zarembo, A., 452
Zhou, X., 170
Zhu, D.-M., 715
Zvelindovsky, A. V., 217